INTRODUCTION TO ASTRONOMY
THE SOLAR SYSTEM &
THE SEARCH FOR LIFE BEYOND EARTH

SECOND CUSTOM EDITION FOR UNIVERSITY OF MICHIGAN

Taken from:

The Cosmic Perspective: The Solar System, Fourth Edition
by Jeffrey Bennett, Megan Donahue, Nicholas Schneider, and Mark Voit

Life in the Universe, Second Edition
by Jeffrey Bennett and Seth Shostak

PEARSON Custom Publishing

PEARSON Addison Wesley

Taken from:

The Cosmic Perspective: The Solar System, Fourth Edition
by Jeffrey Bennett, Megan Donahue, Nicholas Schneider, and Mark Voit

Published by Addison-Wesley
Boston, Massachusetts 02116

Life in the Universe, Second Edition
by Jeffrey Bennett and Seth Shostak

Published by Addison-Wesley

This special edition is published in cooperation with Pearson Custom Publishing.

Printed in the United States of America

10 9 8 7 6 5 4 3 2 1

ISBN 0-536-46996-2

2007400076

TB

Please visit our web site at *www.pearsoncustom.com*

PEARSON Custom Publishing

PEARSON CUSTOM PUBLISHING
501 Boylston Street, Suite 900, Boston, MA 02116
A Pearson Education Company

Brief Contents

From Bennett, Donahue, Schneider, and Voit, *The Cosmic Perspective: The Solar System* Fourth Edition:

From Bennett and Shostak, *Life in the Universe* Second Edition:

Detailed Contents

From Bennett, Donahue, Schneider, and Voit, *The Cosmic Perspective: The Solar System* Fourth Edition:

PART II
KEY CONCEPTS FOR ASTRONOMY

PART III
LEARNING FROM OTHER WORLDS

APPENDIXES

Detailed Contents

From Bennett and Shostak, *Life in the Universe*
Second Edition:

PART III

LIFE IN THE SOLAR SYSTEM

PART IV

LIFE AMONG THE STARS

THE COSMIC PERSPECTIVE

Fourth Edition

The Solar System

1

Our Place in the Universe

LEARNING GOALS

1.1 Our Modern View of the Universe
- What is our place in the universe?
- How did we come to be?
- How can we know what the universe was like in the past?
- Can we see the entire universe?

1.2 The Scale of the Universe
- How big is Earth compared to our solar system?
- How far away are the stars?
- How big is the Milky Way Galaxy?
- How big is the universe?
- How do our lifetimes compare to the age of the universe?

1.3 Spaceship Earth
- How is Earth moving in our solar system?
- How is our solar system moving in the Milky Way Galaxy?
- How do galaxies move within the universe?
- Are we ever sitting still?

1.4 The Human Adventure of Astronomy
- How has the study of astronomy affected human history?

We shall not cease from exploration
And the end of all our exploring
Will be to arrive where we started
And know the place for the first time.

—T.S. Eliot

Far from city lights on a clear night, you can gaze upward at a sky filled with stars. Lie back and watch for a few hours, and you will observe the stars marching steadily across the sky. Confronted by the seemingly infinite heavens, you might wonder how Earth and the universe came to be. If you do, you will be sharing an experience common to humans around the world and in thousands of generations past.

Modern science offers answers to many of our fundamental questions about the universe and our place within it. We now know the basic content and scale of the universe. We know the age of Earth and the approximate age of the universe. And, although much remains to be discovered, we are rapidly learning how the simple ingredients of the early universe developed into the incredible diversity of life on Earth.

In this first chapter, we will survey the content and history of the universe, the scale of the universe, and the motions of Earth. We'll develop a "big picture" perspective on our place in the universe that will provide a base on which we can build a deeper understanding in the rest of the book.

1.1 Our Modern View of the Universe

If you observe the sky carefully, you can see why most of our ancestors believed that the heavens revolved about a stationary Earth. The Sun, Moon, planets, and stars appear to circle around our sky each day, and we cannot feel the constant motion of Earth as it rotates on its axis and orbits the Sun. Thus, it seems quite natural to assume that we live in an Earth-centered, or *geocentric,* universe.

Nevertheless, we now know that Earth is a planet orbiting a rather average star in a vast universe. The historical path to this knowledge was long and complex. In later chapters, we'll see that many ancient beliefs seemed to make a lot of sense in their day and changed only when people were confronted by strong evidence to the contrary. We'll also see how the process of science has enabled us to acquire this evidence and thereby discover that we are connected to the stars in ways our ancestors never imagined. First, however, it's useful to have at least a general picture of the universe as we know it today.

• What is our place in the universe?

We can describe our place in the universe with what we might call our "cosmic address," illustrated in Figure 1.1. Earth is a planet in our **solar system**, which consists of the Sun and all the objects that orbit it: the planets and their moons, the chunks of rock we call *asteroids,* the balls of ice we call *comets,* and countless tiny particles of interplanetary dust.

Our Sun is a star, just like the stars we see in our night sky. The Sun and all the stars we can see with the naked eye make up only a small part of a huge, disk-shaped collection of stars called the **Milky Way Galaxy**. A **galaxy** is a great island of stars in space, containing from a few hundred million to a trillion or more stars. The Milky Way Galaxy is relatively large, containing more than 100 billion stars. Our solar system is located a little over halfway from the galactic center to the edge of the galactic disk.

Billions of other galaxies are scattered throughout space. Some galaxies are fairly isolated, but many others are found in groups. Our Milky Way, for example, is one of the two largest among about 40 galaxies in the **Local Group**. Groups of galaxies with more than a few dozen members are often called **galaxy clusters**.

On a very large scale, the universe appears weblike, with galaxies and galaxy clusters loosely arranged in giant chains and sheets. The galaxies and galaxy clusters are more tightly packed in some places than in others, forming giant structures called **superclusters**. The supercluster to which our Local Group belongs is called, not surprisingly, the **Local Supercluster**. Between the vast groupings of galaxies lie huge voids containing few, if any, galaxies. Finally, the **universe** is the sum total of all matter and energy, encompassing the superclusters and voids and everything within them.

THINK ABOUT IT

Some people think that our tiny physical size in the vast universe makes us insignificant. Others think that our ability to learn about the wonders of the universe gives us significance despite our size. What do *you* think?

• How did we come to be?

According to modern science, we humans are newcomers in an old universe. We'll devote much of the rest of this textbook to studying the scientific evidence that backs up this idea. To help prepare you for this study, let's look at a quick overview of the scientific history of the universe, as summarized in Figure 1.2 (page 5).

The Big Bang and the Expanding Universe As we'll discuss in Section 1.3, telescopic observations of distant galaxies show that the entire universe is *expanding.* That is, average distances between galaxies are increasing with time. If the universe is expanding, everything must have been closer together in the past. From the observed rate of expansion, astronomers estimate that the expansion started about 14 billion years ago. We call this beginning the **Big Bang**. The three cubes in the upper left corner of Figure 1.2 represent the expansion through time of a small piece of the entire universe.

The universe as a whole has continued to expand ever since the Big Bang, but in many smaller regions, the force of gravity has drawn matter together. Structures such as galaxies and clusters of galaxies occupy regions where gravity

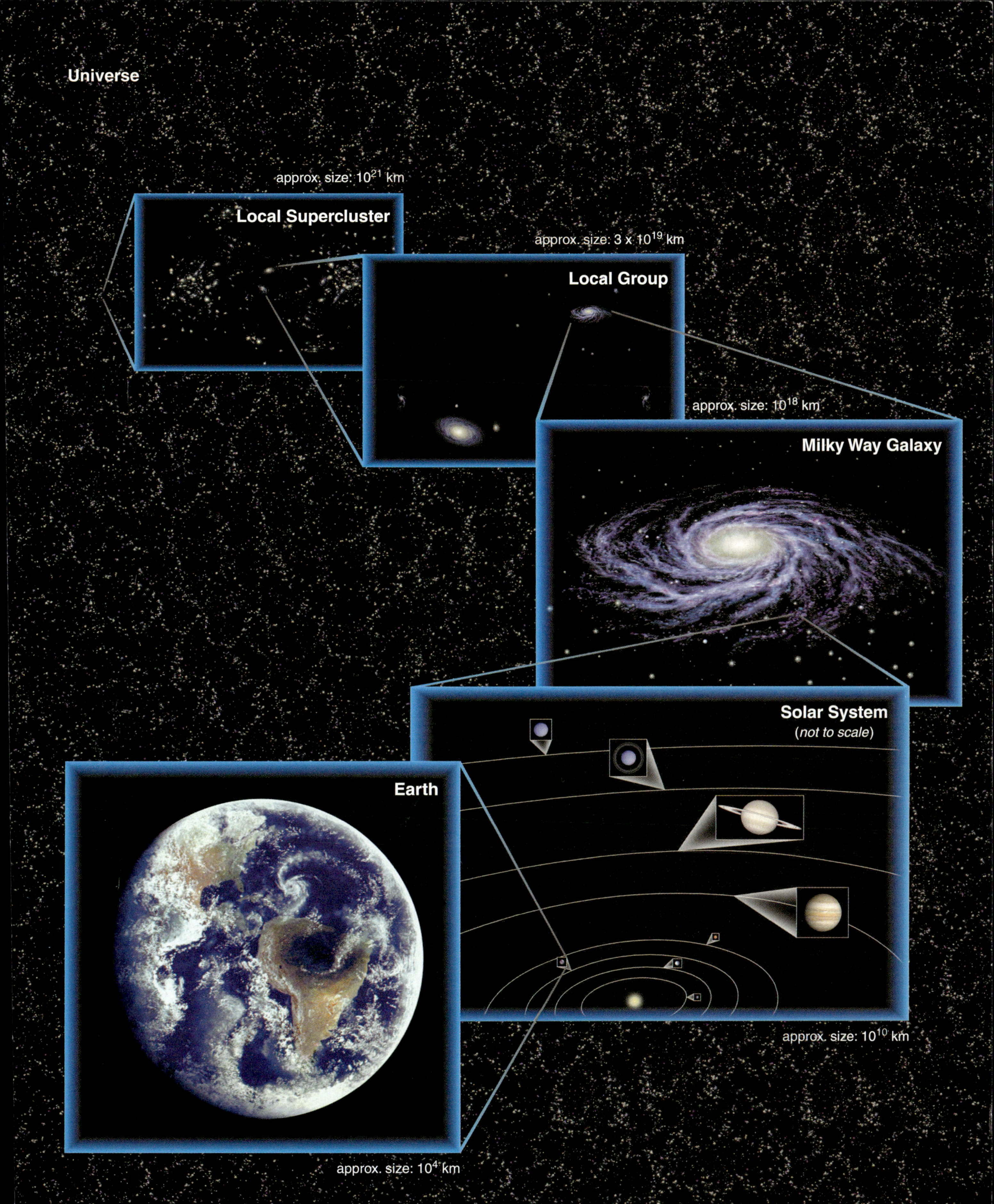

Figure 1.1 Our cosmic address.

has won out against the overall expansion. That is, while the universe as a whole continues to expand, individual galaxies and their contents do *not* expand. This idea is illustrated by the expanding cube in Figure 1.2. Notice that as the cube as a whole grows larger, the matter within it tends to clump into galaxies and galaxy clusters. Most galaxies, including our own Milky Way, probably formed within a few billion years after the Big Bang.

Stellar Lives and Galactic Recycling Within galaxies, gravity drives the collapse of clouds of gas and dust to form stars and planets. Stars are not living organisms, but they nonetheless go through "life cycles." After their births in giant clouds of gas and dust, stars shine for millions or billions of years. The energy that makes them shine comes from **nuclear fusion**, the process in which lightweight atomic nuclei smash together and stick (or fuse) to make heavier nuclei. Nuclear fusion occurs deep in a star's core throughout its life. A star "dies" when it finally exhausts all its usable fuel for fusion.

In its final death throes, a star blows much of its content back out into space. In particular, massive stars die in titanic explosions called *supernovae.* The returned matter mixes with other matter floating between the stars in the galaxy, eventually becoming part of new clouds of gas and dust from which new generations of stars can be born. Thus, galaxies function as cosmic recycling plants, recycling material expelled from dying stars into new generations of stars and planets. This cycle is illustrated in the lower right of Figure 1.2. Our own solar system is a product of many generations of such recycling.

Basic Astronomical Objects, Units, and Motions

This box summarizes a few key astronomical definitions introduced in this chapter and used throughout the book.

Basic Astronomical Objects

star Our Sun and other ordinary stars are large, glowing balls of gas that generate heat and light through nuclear fusion in their cores.

planet A moderately large object that orbits a star and shines primarily by reflecting light from its star. More precisely, according to a definition approved in 2006, an object can be considered a planet only if it: (1) orbits a star; (2) is large enough for its own gravity to make it round; and (3) has cleared most other objects from its orbital path. An object that meets the first two criteria but has not cleared its orbital path, like Pluto, is designated a *dwarf planet.*

moon (or satellite) An object that orbits a planet. The term *satellite* is also used more generally to refer to any object orbiting another object.

asteroid A relatively small and rocky object that orbits a star. Asteroids are officially considered part of a category known as "small solar system bodies."

comet A relatively small and ice-rich object that orbits a star. Like asteroids, comets are considered "small solar system bodies."

Collections of Astronomical Objects

solar system Our solar system consists of the Sun and all the material that orbits it, including the planets, dwarf planets, and small solar system bodies. The term *solar system* technically refers only to our own star system (because *solar* means "of the Sun"), but it is sometimes applied to other star systems.

star system A star (sometimes more than one star) and any planets and other materials that orbit it. (Roughly half of all star systems contain two or more stars.)

galaxy A great island of stars in space, containing from a few hundred million to a trillion or more stars, all held together by gravity and orbiting a common center.

cluster (or *group*) of galaxies A collection of galaxies bound together by gravity. Small collections (up to a few dozen galaxies) are generally called *groups,* while larger collections are called *clusters.*

supercluster A gigantic region of space where many individual galaxies and many groups and clusters of galaxies are packed more closely together than elsewhere in the universe.

universe (or cosmos) The sum total of all matter and energy—that is, all galaxies and everything between them.

observable universe The portion of the entire universe that can be seen from Earth, at least in principle. The observable universe is probably only a tiny portion of the entire universe.

Astronomical Distance Units

astronomical unit (AU) The average distance between Earth and the Sun, which is about 150 million kilometers. More technically, 1 AU is the length of the semimajor axis of Earth's orbit.

light-year The distance that light can travel in 1 year, which is about 9.46 trillion kilometers.

Terms Relating to Motion

rotation The spinning of an object around its axis. For example, Earth rotates once each day around its axis, which is an imaginary line connecting the North Pole to the South Pole through the center of Earth.

orbit (revolution) The orbital motion of one object around another. For example, Earth orbits around the Sun once each year.

expansion (of the universe) We say that the universe is expanding because the average distance between galaxies is increasing with time. Note that while the universe as a whole is expanding, individual galaxies and their contents (as well as groups and clusters of galaxies) are *not* expanding.

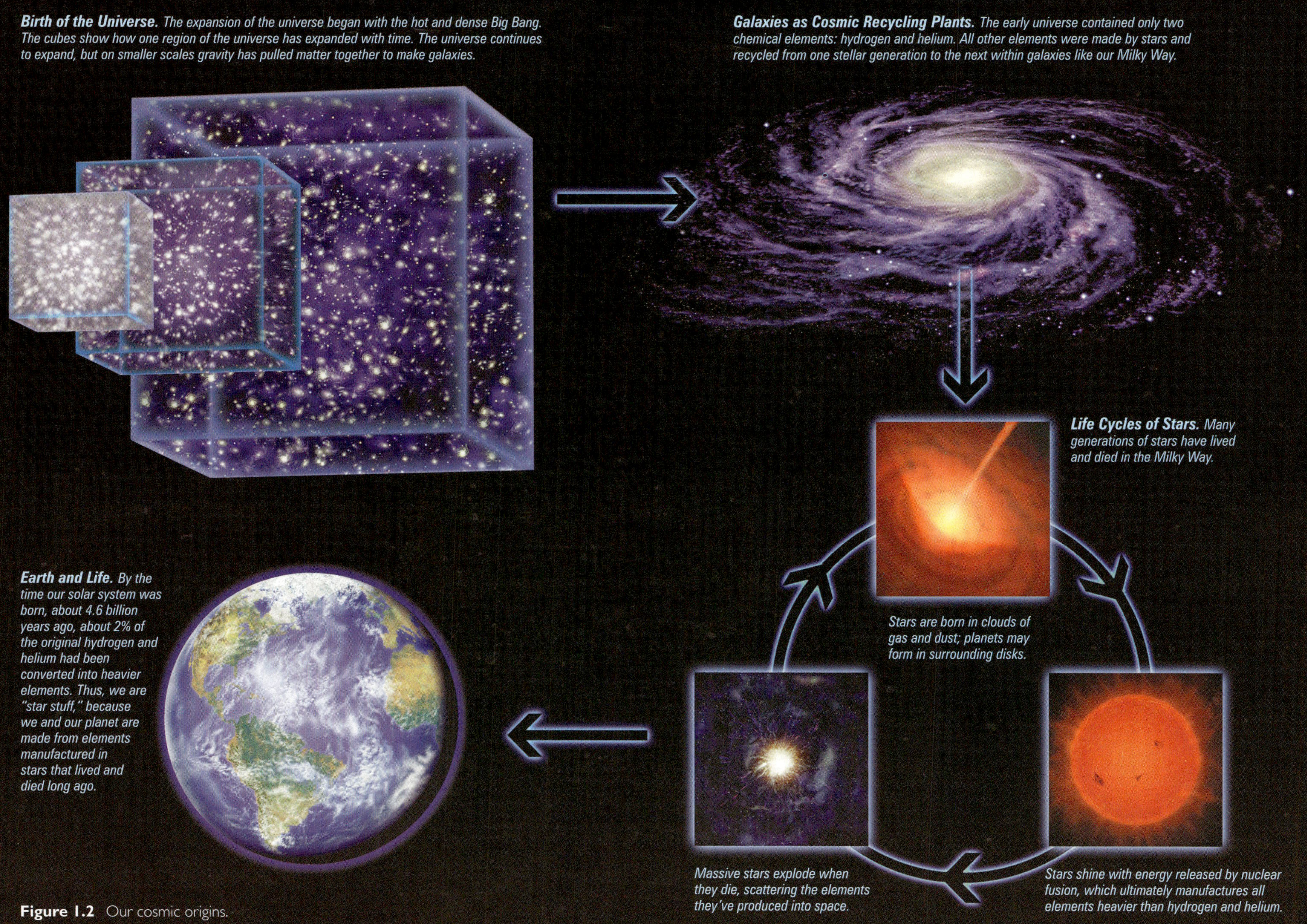

Figure 1.2 Our cosmic origins.

Stars Manufacture the Elements of Earth and Life The recycling of stellar material is connected to our existence in an even deeper way. By studying stars of different ages, we have learned that the early universe contained only the simplest chemical elements: hydrogen and helium (and a trace of lithium). We and Earth are made primarily of other elements, such as carbon, nitrogen, oxygen, and iron. Where did these other elements come from? Apparently, all the elements besides hydrogen and helium were manufactured by stars, either through the nuclear fusion that makes stars shine or through nuclear reactions accompanying the explosions that end stellar lives.

The recycling of matter in the Milky Way Galaxy had already been taking place for several billion years by the time our solar system formed, about 4.6 billion years ago. By that time, stars had converted about 2% of the original hydrogen and helium into heavier elements. Thus, the cloud that gave birth to our solar system was made of about 98% hydrogen and helium and 2% of everything else. This 2% may sound small, but it was more than enough to make the small rocky planets of our solar system, including Earth. On Earth, some of these elements became the raw ingredients of life, ultimately blossoming into the great diversity of life on Earth today.

In summary, most of the material from which we and our planet are made was created inside stars that died before the birth of our Sun. As astronomer Carl Sagan (1934–1996) used to say, we are "star stuff."

• How can we know what the universe was like in the past?

You may wonder how we can claim to know anything about what the universe was like in the distant past. It's possible because we can actually see into the past by studying light from distant stars and galaxies.

Light travels extremely fast by earthly standards. The speed of light is 300,000 kilometers per second, a speed at which it would be possible to circle Earth nearly eight times in just 1 second. Nevertheless, even light takes a substantial amount of time to travel the vast distances in space. For example, light takes about 1 second to reach Earth from the Moon, and about 8 minutes to reach Earth from the Sun. Light from the stars takes many years to reach us, so we measure distances to the stars in units called **light-years**. One light-year is the distance that light can travel in 1 year—about 10 trillion kilometers, or 6 trillion miles. (You can calculate this distance by multiplying the speed of light by the number of seconds in 1 year; see Mathematical Insight 1.1.) Note that a light-year is a unit of *distance,* not time.

COMMON MISCONCEPTIONS

The Meaning of a Light-Year

You'll often hear advertisements say things like "It will be light-years before anyone builds a better product." But this advertisement doesn't really make sense, because light-years are a unit of *distance* and the statement makes them sound like a unit of time. If you are unsure whether the term *light-years* is being used correctly, try testing the statement by remembering that 1 light-year is approximately 10 trillion kilometers, or 6 trillion miles. The advertisement then reads "It will be 6 trillion miles before anyone builds a better product," which clearly does not make sense.

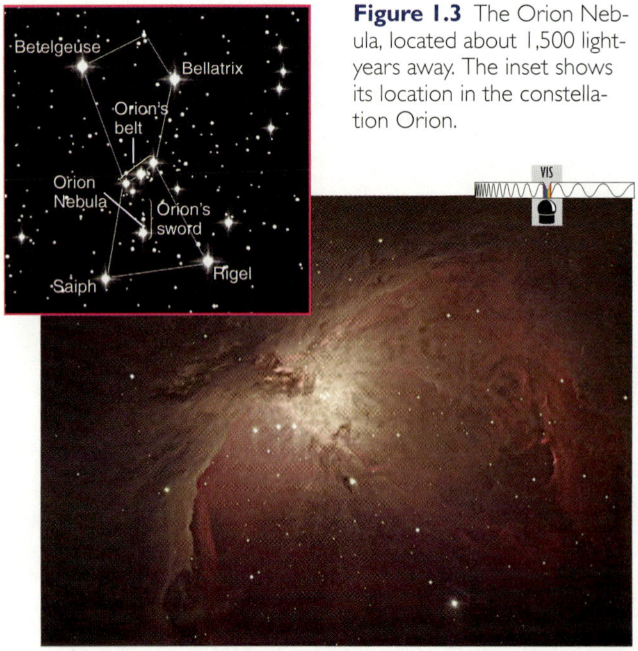

Figure 1.3 The Orion Nebula, located about 1,500 light-years away. The inset shows its location in the constellation Orion.

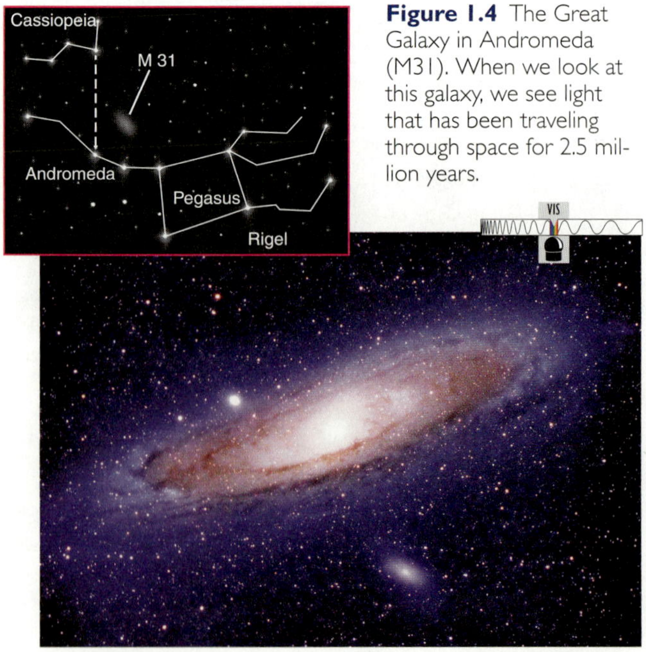

Figure 1.4 The Great Galaxy in Andromeda (M31). When we look at this galaxy, we see light that has been traveling through space for 2.5 million years.

Because light takes time to travel through space, we are led to a remarkable fact:

The farther away we look in distance, the further back we look in time.

For example, the brightest star in the night sky, Sirius, is about 8 light-years away, which means its light takes about 8 years to reach us. Thus, when we look at Sirius, we are seeing it as it was about 8 years ago.

The effect is more dramatic at greater distances. The Orion Nebula, a giant cloud in which stars and planets are forming, is located about 1,500 light-years from Earth (Figure 1.3). Thus, we see the Orion Nebula as it looked about 1,500 years ago—about the time of the fall of the Roman Empire. If any major events have occurred in the Orion Nebula since that time, we cannot yet know about them because the light from these events has not yet reached us. Figure 1.4 shows an even more distant object: the Great Galaxy in Andromeda (also known as the Andromeda Galaxy and M 31), which lies about 2.5 million light-years from Earth. Its photograph is therefore a picture of this galaxy as it looked about 2.5 million years ago, when early humans were first walking on Earth.

It's also amazing to realize that any "snapshot" of a distant galaxy is a picture of both space and time. For example, because the Andromeda Galaxy is about 100,000 light-years in diameter, the light we see from the far side of the galaxy must have left on its journey to us some 100,000 years before the light from the near side. Thus, the single photograph in Figure 1.4 not only shows us how the Andromeda Galaxy looked more than 2 million years ago, but also shows different parts of the galaxy spread over a time period of 100,000 years. When we study the universe, it is impossible to separate space and time.

MATHEMATICAL INSIGHT 1.1 How Far Is a Light-Year? An Introduction to Astronomical Problem Solving

The text states that a light-year is about 10 trillion kilometers, but how do we know? After all, this distance is far greater than any person or spacecraft has ever traveled, so we cannot possibly have measured it. Instead, we know the distance that light travels in a year through mathematics, which allows us to answer many questions that we cannot answer through direct measurement.

Although people often expect the mathematics needed in astronomy to be quite difficult (and it's true that professional astronomers use advanced mathematics in their work), many astronomical problems can be solved with little more than arithmetic or simple algebra; as we'll see below, answering the question "How far is a light-year?" requires only multiplication and division. Thus, the "trick" to solving this and other astronomical problems lies in finding an appropriate problem-solving process. Problem solving is more of an art than a science, but the following three-step process should help you whenever you have difficulty with a quantitative problem.

Step 1 Understand the problem: Ask yourself what the solution will look like (for example, What units will it have? Will it be big or small?) and what information you need to solve the problem. Draw a diagram or think of a simpler analogous problem to help you decide how to solve it.

Step 2 Solve the problem: Carry out the necessary calculations.

Step 3 Explain your result: Be sure that your answer makes sense, and consider what you've learned by solving the problem.

You can remember this process by thinking of the first letters of each step: USE for Understand, Solve, and Explain. As you gain problem-solving experience, you may not always need to write out the three steps explicitly. However, they may help you whenever you get "stuck" while trying to solve a quantitative problem.

Example: How far is a light-year?

Solution: Let's go through the three-step process.

Step 1 Understand the problem: The question asks how *far*, so we are looking for a *distance*. By remembering the definition of a light-year, we know we are looking for the *distance that light can travel in 1 year*. A simpler but analogous problem might therefore ask, say, "How far can your car travel in 2 hours?" By thinking about this simpler problem, you'll realize that all you need to know is the car's speed; for example, if you travel at a speed of 50 kilometers per hour, then you'll travel 100 kilometers in 2 hours. From there, you can realize that finding a distance from a speed just requires multiplying the speed by the time, that is: distance = speed × time. Since this problem asks how far light travels in a year, the speed is the speed of light and the time is 1 year.

Step 2 Solve the problem: From Step (1), we know that we need to multiply the speed of light by 1 year. The text states that the speed of light is 300,000 km/s. Because this speed is given *per second* and the time is 1 *year*, we'll need to convert a year to seconds. We can do this by knowing that there are 60 seconds in 1 minute, 60 minutes in 1 hour, 24 hours in 1 day, and 365 days in 1 year. We now carry out the calculations:

$$\begin{aligned} 1 \text{ light-year} &= (\text{speed of light}) \times (1 \text{ yr}) \\ &= \left(300{,}000\frac{\text{km}}{\cancel{\text{s}}}\right) \times \left(1\,\cancel{\text{yr}} \times \frac{365\,\cancel{\text{days}}}{1\,\cancel{\text{yr}}} \times \frac{24\,\cancel{\text{hr}}}{1\,\cancel{\text{day}}} \times \frac{60\,\cancel{\text{min}}}{1\,\cancel{\text{hr}}} \times \frac{60\,\cancel{\text{s}}}{1\,\cancel{\text{min}}}\right) \\ &= 9{,}460{,}000{,}000{,}000 \text{ km (or 9.46 trillion km)} \end{aligned}$$

Notice that the unit conversions are really just clever ways of multiplying by 1; see Appendix C.3 for a more detailed review of proper ways of working with units.

Step 3 Explain your result: In complete sentence form, our answer is: "One light-year is about 9.46 trillion kilometers." This answer makes sense: It has the expected units of distance (kilometers) and it is a long way, which we expect for the distance that light can travel in a year. We say "about" in the answer because we know it is not exact: For example, we know that a year is not exactly 365 days long. In fact, for most purposes, it makes more sense to approximate the answer further as: "One light-year is about 10 trillion kilometers."

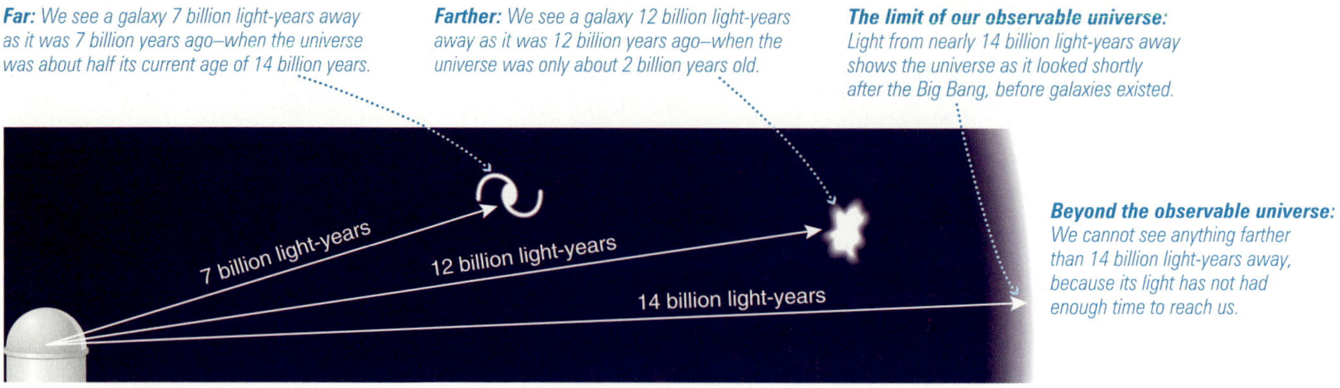

Figure 1.5 The farther away we look in space, the further back we look in time. Thus, the age of the universe puts a limit on the size of the *observable* universe—the portion of the entire universe that we can observe, at least in principle.

THINK ABOUT IT

Suppose that, at this very moment, students are studying astronomy on a planet somewhere in the Andromeda Galaxy. What would they see as they look from afar at our Milky Way? Could they know that we exist here on Earth? Explain.

• Can we see the entire universe?

We see more distant galaxies as they were even further back into the past, a fact that allows us to observe the ways in which the universe has changed through time (Figure 1.5). Remember that we think the universe is about 14 billion years old. If we look at a galaxy that is 7 billion light-years away, we see it as it looked 7 billion years ago*—which means we see it as it was when the universe was only half its current age. If we look at a galaxy that is 12 billion light-years away, we see it as it was 12 billion years ago, when the universe was only 2 billion years old. Thus, by looking to great distances, we can see what the universe looked like when it was younger. The remarkable photo that opens this chapter (p. 1), taken by the Hubble Space Telescope, shows some galaxies that are more than 12 billion light-years away.

Even with the most powerful telescopes imaginable, there's a limit to how far we can see into space. This limit is imposed by the age of the universe. In a universe that is 14 billion years old, we cannot possibly see anything more than 14 billion light-years away. Beyond that—say, at a distance of 15 billion light-years—we'd be looking back to a time before the universe existed, which means there's nothing to see. Thus, our **observable universe**—the portion of the entire universe that we can potentially observe—consists only of objects that lie within 14 billion light-years of Earth. This fact does not put any limit on the size of the *entire* universe, which may be far larger than our observable universe. We simply have no hope of seeing or studying anything beyond the bounds of our observable universe. Anything that exists beyond our observable universe is so far away that its light has not yet had time to reach us.

*We will find in Chapter 20 that distances to very faraway galaxies are rather tricky to define in an expanding universe. In these examples we will simply define distances to galaxies in terms of how far their light has travelled.

Mastering ASTRONOMY Scale of the Universe Tutorial, Lessons 1–3

1.2 The Scale of the Universe

The numbers in our description of the size and age of the universe probably have little meaning for you—after all, they are literally astronomical. In this section, we will try to give those figures some meaning.

Mastering ASTRONOMY Virtual Tour of the Solar System

• How big is Earth compared to our solar system?

One of the best ways to develop perspective on cosmic sizes and distances is to imagine our solar system shrunk down to a scale on which you could walk through it. The Voyage scale model solar system in Washington, D.C.,

Figure 1.6 This photo shows the pedestals housing the Sun (the gold sphere on the nearest pedestal) and the inner planets in the Voyage scale model solar system (Washington, D.C.). The building at the left is the National Air and Space Museum. (The model planets are encased in the sidewalk-facing disks that are visible at about eye level on the planet pedestals.)

makes such a walk possible (Figure 1.6). The Voyage model shows the Sun and the planets, and the distances between them, at *one ten-billionth* their actual sizes and distances. Figure 1.7a shows the planets at their scaled sizes, and 1.7b shows a map of the locations of the planets in the Voyage model. If you study the figures carefully, you'll notice several key features of our solar system:

- The Sun and planets are tiny compared to the distances between them. On the Voyage scale, the Sun is about the size of a large grapefruit, while the planets range in size from barely visible dots to marble-size Jupiter. Earth is about the size of the ball point in a pen and is located about 15 meters (16.5 yards) from the Sun.
- The planets fall into two clear groups by distance. The four inner planets—Mercury, Venus, Earth, and Mars—all lie within just a few steps of the Sun on this scale. The four large outer planets—Jupiter, Saturn, Uranus, and Neptune—are much more widely separated. For example, Jupiter is more than three times as far as Mars from the Sun.
- One of the most striking features of the solar system is its emptiness (that's why we call it *space!*). To show the complete orbits of the planets around the Sun, the Voyage model would require an area measuring over a kilometer (0.6 mile) on a side, which is larger than most college campuses. Spread over this large area, only the grapefruit-size Sun, the planets, and a few moons would be big enough to see with the naked eye.

THINK ABOUT IT

Earth is the only place in our solar system—and the only place we yet know of in the universe—with conditions suitable for human life. How does visualizing the Earth to scale affect your perspective on human existence? How does it affect your perspective on our planet? Explain.

Seeing our solar system to scale also helps us put space exploration into perspective. Although robotic spacecraft have visited all the planets through Neptune, humans have

SPECIAL TOPIC How Many Planets Are There In Our Solar System?

Everyone in college today grew up in an era when we were taught that there are nine planets in our solar system. However, as you've probably heard, astronomers recently voted to demote Pluto to being a "dwarf planet," leaving our solar system with only eight official planets. Why the change, and is this really the end for Pluto as a planet?

When it was discovered in 1930, Pluto was assumed to be similar to other planets. But as we'll discuss in Chapter 12, we've known for some time that Pluto is much smaller than any of the first eight planets, and that it shares the outer solar system (a region called the *Kuiper Belt*) with thousands of similarly icy objects. Still, as long as Pluto was the largest known of these objects, most astronomers were content to leave the planetary status quo. Change was forced by discovery of the object now named Eris, identified in 2005 by Caltech astronomer Mike Brown. Because Eris is slightly larger than Pluto, astronomers could no longer avoid the question of what objects should count as planets.

Official decisions on astronomical names and definitions rest with the International Astronomical Union (IAU), an organization made up of professional astronomers from around the world. At a contentious meeting in August 2006, members of the IAU considered various options, ultimately voting to define a "planet" as an object that: (1) orbits a star (but is itself neither a star nor a moon); (2) is massive enough for its own gravity to give it a nearly round shape; and (3) has cleared the neighborhood around its orbit. Objects that meet the first two criteria but that have not cleared their orbital neighborhoods—including Pluto, Eris, and the asteroid Ceres—are now designated *dwarf planets.* Dozens of other objects probably also qualify as dwarf planets, but an object can join the offical list only after observations confirm that it is round. The myriad smaller objects that orbit the Sun, including asteroids and comets, become members of a class called *small solar system bodies.*

The important thing to remember is that this new definition was established by a vote, making it politics, not science. The politics may yet change again (some astronomers have already voiced objections to the new definition), but Pluto will remain the same. Moreover, many people are likely to keep thinking of Pluto as a planet regardless of what the professioinal astronomers say, much as many people still think of Europe and Asia as separate continents even though both belong to the single land mass of Eurasia. So if you're a Pluto fan, don't despair: It's good to know the official definitions, but it's better to understand the science behind them.

Notes left at the Voyage Scale Model Solar System Pluto plaque upon Pluto's demotion to "dwarf planet."

a This painting shows the Sun and planets at one ten-billionth of their actual sizes. (Distances are not to scale in this painting.)

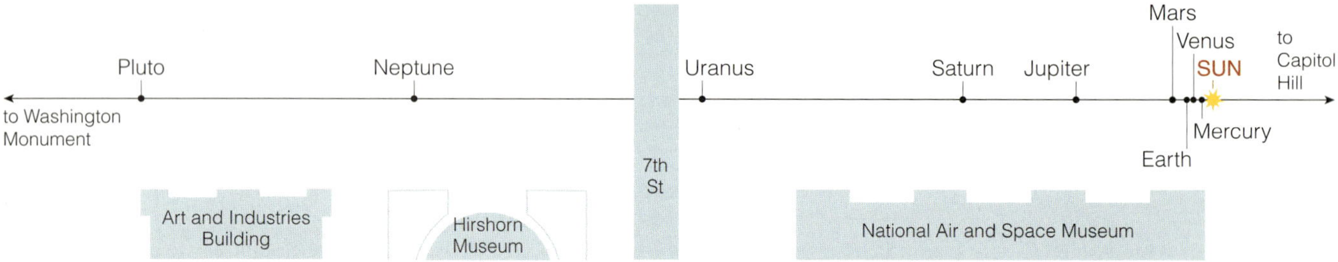

b This map shows the locations of the Sun and planets in the Voyage model. Planets are lined up in the model, but in reality each planet orbits the Sun independently and a perfect alignment never occurs.

Figure 1.7 The Voyage scale model represents the solar system at *one ten-billionth* of its actual size. Pluto is included in the Voyage model, which was built before the International Astronomical Union decided that Pluto would be considered a "dwarf planet" rather than a true planet.

traveled only as far as the Moon. Twelve astronauts walked on the Moon between 1969 and 1972 as part of the Apollo program (Figure 1.8). On the 1-to-10-billion Voyage scale, the Moon is a barely visible speck about two thumb-widths (4 centimeters) from Earth (see Figure 2.21). Thus, if you place the model Earth in the center of your palm, the farthest place humans have ever reached also lies within your palm. Sending humans to Mars will require a journey some 200 times as far.

Figure 1.8 The Moon is the most distant place ever visited by humans, yet on the 1-to-10-billion scale of the Voyage model, it is only about 4 centimeters (1.5 inches) from Earth. This famous photograph from the first Moon landing (*Apollo 11* in July 1969) shows astronaut Buzz Aldrin, with Neil Armstrong reflected in his visor. Armstrong was the first to step onto the Moon's surface, saying, "That's one small step for [a] man, one giant leap for mankind."

• How far away are the stars?

Imagine that you start at the Voyage model Sun in Washington, D.C. You walk the roughly 600-meter distance to Pluto (just over $\frac{1}{3}$ mile) and then decide to keep going to find the nearest star besides the Sun. How far would you have to go?

Amazingly, you would need to walk to California. That is, on the same scale that allows you to walk from the Sun to Pluto in minutes, even the nearest stars would be more

MATHEMATICAL INSIGHT 1.2 The Scale of Space and Time

Scaling is a great way to put space and time into perspective, and it usually requires nothing more than division. For example, in a 1-to-20 architectural scale model, a building that is actually 60 feet tall will be only $60 \div 20 = 3$ feet tall in the model. The idea is the same for astronomical scaling, except that we must divide by much larger numbers. As a result, it's much easier to work with numbers expressed in *scientific notation;* that is, with the aid of powers of 10. The following examples show the idea. If you are unfamiliar with powers of 10 and scientific notation, or if you need a review, please see Appendices C.1 and C.2.

Example 1: How big is the Sun on the 1-to-10-billion scale described in the text?

Solution:

Step 1 Understand: We are looking for the *size* of the Sun on the scale, which means we need to know its actual size. From Appendix E.1, we find that the Sun's radius is 695,000 km, which we express in scientific notation as 6.95×10^5 km. To find the Sun's radius on the 1-to-10-billion scale, we need to divide its actual radius by 10 billion, or 10^{10}.

Step 2 Solve: Now we carry out the division:

$$\begin{aligned}\text{scaled radius of Sun} &= \frac{\text{actual radius}}{10^{10}} \\ &= \frac{6.95 \times 10^5 \text{ km}}{10^{10}} \\ &= 6.95 \times 10^{(5-10)} \text{ km} \\ &= 6.95 \times 10^{-5} \text{ km}\end{aligned}$$

Notice that we used the rule that dividing powers of 10 means subtracting their exponents [Appendix C.1].

Step 3 Explain: We have found an answer, but it's hard to explain in its original form because most of us don't have a good sense of what 10^{-5} kilometer looks like. We need to go further and convert the answer into units that will be easier to interpret. We know there are 1,000 (or 10^3) meters in a kilometer and 100 (or 10^2) centimeters in a meter, so we can convert our answer to centimeters:

$$6.95 \times 10^{-5} \,\cancel{\text{km}} \times \frac{10^3 \,\cancel{\text{m}}}{1 \,\cancel{\text{km}}} \times \frac{10^2 \text{ cm}}{1 \,\cancel{\text{m}}} = 6.95 \text{ cm}$$

We can also convert this answer to inches by using data from Appendix Table C.3, which tell us that 1 cm = 0.3937 in.:

$$6.95 \text{ cm} \times 0.3937 \tfrac{\text{in.}}{\text{cm}} = 2.74 \text{ in.}$$

We've found that the scaled radius of the Sun on the 1-to-10-billion scale is just under 7 centimeters or $2\frac{3}{4}$ inches, which means a diameter of just under 14 centimeters or $5\frac{1}{2}$ inches. In other words, the Sun on a 1-to-10-billion scale is about the size of a large grapefruit.

Example 2: What scale do we need to use if we want the 100,000-light-year diameter of the Milky Way Galaxy to fit on a 100-meter-long football field?

Solution:

Step 1 Understand: This time we are looking to find *how many times larger* is the actual diameter of the galaxy than 100 meters, which means we need to divide the larger size by the smaller size. We can't do the division until we've put both sizes in the same units. We already know that a light-year is about 10^{13} kilometers (see Mathematical Insight 1.1), so we can use this fact and the fact that there are 10^3 meters in a kilometer to put both numbers in terms of meters.

Step 2 Solve: We now put the numbers in the same units and carry out the division; to make things easier to read, we write all the numbers as powers of 10, including the galaxy's diameter of 100,000 light-years, or 10^5 ly:

$$\begin{aligned}\frac{\text{galaxy diameter}}{\text{football field diameter}} &= \frac{10^5 \,\cancel{\text{ly}} \times \frac{10^{13} \,\cancel{\text{km}}}{1 \,\cancel{\text{ly}}} \times \frac{10^3 \text{ m}}{1 \,\cancel{\text{km}}}}{10^2 \text{ m}} \\ &= 10^{(5+13+3-2)} = 10^{19}\end{aligned}$$

Notice how we've worked with powers of 10 in the calculation, and that the answer has no units because it simply tells us how many times larger one thing is than the other.

Step 3 Explain: We've found that we need a scale of 1 to 10^{19} to make the galaxy fit on a football field.

Example 3: The cosmic calendar in Figure 1.11 represents 14 billion years of real time. How much time does each second represent on the cosmic calendar?

Solution:

Step 1 Understand: We are looking for a real time that corresponds to the scaled time on the calendar.

Step 2 Solve: There are several ways to solve this problem, but it's particularly easy once we realize that the calendar scale is 1 to 14 billion, since 1 calendar year represents 14 billion real years. Thus, 1 second on the cosmic calendar represents 14 billion seconds of real time.

Step 3 Explain: Our initial answer is difficult to explain without further work because most people don't have a good sense of "14 billion seconds." We can make better sense of the answer by converting it to years:

$$1.4 \times 10^{10} \,\cancel{\text{s}} \times \frac{1 \,\cancel{\text{min}}}{60 \,\cancel{\text{s}}} \times \frac{1 \,\cancel{\text{hr}}}{60 \,\cancel{\text{min}}} \times \frac{1 \,\cancel{\text{day}}}{24 \,\cancel{\text{hr}}} \times \frac{1 \text{ yr}}{365 \,\cancel{\text{day}}} = 443.9 \text{ yr}$$

Recognizing that we've approximated the number of days in the year (and that we've used an approximate value of 14 billion years for the age of the universe), we'll say: "One second on the cosmic calendar represents about 440 years of real time." Thus, a tenth of a second represents about 44 years and a twentieth of a second (0.05 s) represents about 22 years—roughly the age of the average college student.

than 4,000 kilometers (2,500 miles) away. If this answer seems hard to believe, you can check it for yourself. A light-year is about 10 trillion kilometers, which becomes 1,000 kilometers on the 1-to-10-billion scale (because 10 trillion ÷ 10 billion = 1,000). The nearest star system to our own, a three-star system called Alpha Centauri (Figure 1.9), is about 4.4 light-years away. That distance becomes about 4,400 kilometers (2,700 miles) on the 1-to-10-billion scale, or roughly equivalent to the distance across the United States.

The tremendous distances to the stars give us some perspective on the technological challenge of astronomy. For example, because the largest star of the Alpha Centauri system is roughly the same size and brightness as our Sun, viewing it in the night sky is somewhat like being in Washington, D.C., and seeing a very bright grapefruit in San Francisco (neglecting the problems introduced by the curvature of the Earth). It may seem remarkable that we can see this star at all, but the blackness of the night sky allows the naked eye to see it as a faint dot of light. It looks much brighter through powerful telescopes, but we still cannot see any features of the star's surface.

Now, consider the difficulty of seeing *planets* orbiting nearby stars. It is equivalent to looking from Washington, D.C., and trying to see ball points or marbles orbiting grapefruits in California or beyond. You probably won't be surprised to learn that we cannot yet see any surface details on such planets. Indeed, the bigger surprise may be that we *have* discovered more than 150 planets around other stars, most through indirect techniques [Section 13.1].

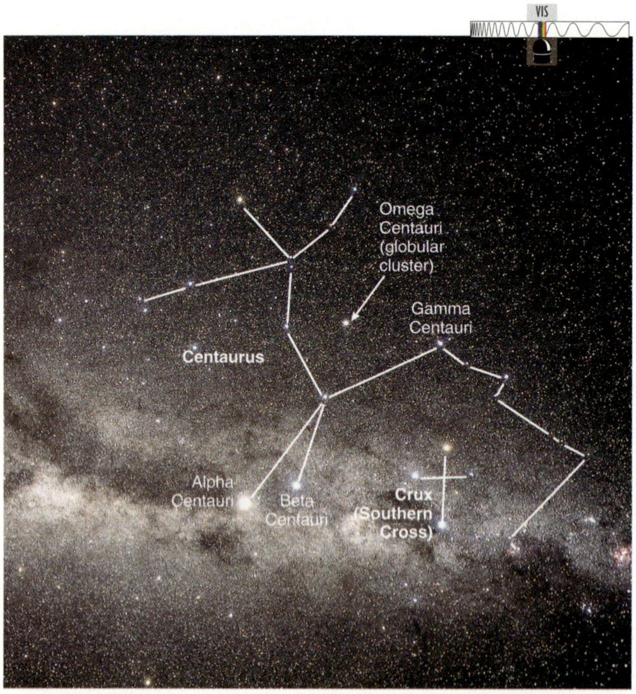

Figure 1.9 This photograph and diagram show the constellation Centaurus, which is visible from tropical and southern latitudes. Alpha Centauri's real distance is about 4.4 light-years, which becomes 4,400 kilometers (2,700 miles) on the 1-to-10-billion Voyage scale.

Our examination of stellar distances also offers a sobering lesson about the possibility of travel to the stars. Although science fiction shows like *Star Trek* and *Star Wars* make interstellar travel seem easy, the reality is far different. Consider the *Voyager 2* spacecraft. Launched in 1977, *Voyager 2* flew by Jupiter in 1979, Saturn in 1981, Uranus in 1986, and Neptune in 1989. (Its trajectory did not take it near Pluto.) *Voyager 2* is now bound for the stars at a speed of close to 50,000 kilometers per hour—about 100 times as fast as a speeding bullet. Even at this speed, *Voyager 2* would take about 100,000 years to reach Alpha Centauri if it were headed in that direction (which it's not). Convenient interstellar travel remains well beyond our present technology.

• How big is the Milky Way Galaxy?

The vast separation between our solar system and Alpha Centauri is typical of the separations among star systems here in the outskirts of the Milky Way Galaxy. Thus, the 1-to-10-billion scale is useless for thinking about distances beyond the nearest stars, because more distant stars would not fit on Earth with this scale. Visualizing the entire galaxy requires a new scale.

Let's further reduce our solar system scale by a factor of 1 billion (making it a scale of 1 to 10^{19}). On this new scale, each light-year becomes 1 millimeter, and the 100,000-light-year diameter of the Milky Way Galaxy becomes 100 meters, or about the length of a football field. Visualize a football field with a scale model of our galaxy centered over midfield. Our entire solar system is a microscopic dot located around the 20-yard line. The 4.4-light-year separation between our solar system and Alpha Centauri becomes just 4.4 millimeters on this scale—smaller than the width of your little finger. If you stood at the position of our solar system in this model, millions of star systems would lie within reach of your arms.

Another way to put the galaxy into perspective is to consider its number of stars—more than 100 billion. Imagine that tonight you are having difficulty falling asleep (perhaps

COMMON MISCONCEPTIONS

Confusing Very Different Things

Most people are familiar with the terms *solar system* and *galaxy*, but people sometimes mix them up. Notice how incredibly different our solar system is from our galaxy. Our solar system is a single star system consisting of our Sun and the various objects that orbit it, including Earth and the other planets. Our galaxy is a collection of more than 100 billion star systems—so many that it would take thousands of years just to count them. Thus, confusing the terms *solar system* and *galaxy* means making a mistake by a factor of 100 billion—a fairly big mistake!

because you are contemplating the scale of the universe). Instead of counting sheep, you decide to count stars. If you are able to count about one star each second, on average, how long would it take you to count 100 billion stars in the Milky Way? Clearly, the answer is 100 billion (10^{11}) seconds, but how long is that? Amazingly, 100 billion seconds turns out to be more than 3,000 years. (You can confirm this by dividing 100 billion by the number of seconds in 1 year.) Thus, you would need thousands of years just to *count* the stars in the Milky Way Galaxy, and this assumes you never take a break—no sleeping, no eating, and absolutely no dying!

• How big is the universe?

As incredible as the scale of our galaxy may seem, the Milky Way is only one of at least 100 billion galaxies in the observable universe. Just as it would take thousands of years

MATHEMATICAL INSIGHT 1.3 Order of Magnitude Estimation

We can often get a good feel for an idea by making an estimate rather than an exact calculation. In astronomy, the numbers are often so large that an estimate is useful even if it's good only to about the nearest power of 10. For example, the text states that there are about 10^{22} stars in the observable universe, a number found by estimating 100 billion galaxies with an average of 100 billion stars each. Because of the "ballpark" nature of the numbers we used, the real number of stars could easily be 10 times as large as our estimate (or 10^{23} stars), or 10 times smaller (10^{21} stars.) Nevertheless, the estimate of 10^{22} still gives us a good sense of the incredible number of stars in our universe. Estimates good to about the nearest power of 10 are usually called **order of magnitude estimates.**

Example 1: The text claims that "roughly speaking, the number of grains of sand [on all the beaches on Earth] is comparable to the number of stars in the observable universe." Check to see if this claim makes sense.

Solution:

Step 1 Understand: To verify the claim, we need to estimate the total number of grains of sand and see whether it is indeed close to the 10^{22} stars we've estimated to be in the observable universe. We find the total number of sand grains by dividing the total volume of sand on Earth's beaches by the average volume of an individual sand grain. The latter number is relatively easy to estimate: Find some sand and measure out a particular volume (such as 0.1 cm^3), then count the number of grains in this volume; dividing the volume by the number of grains will tell you the average volume of each grain. Some sand is more fine or coarse than other sand, but you'll typically find that the volume of individual sand grains is between about 0.1 and 1 cubic millimeter. For our order of magnitude estimate, let's call it 0.1 mm^3, which you should confirm to be equivalent to 10^{-10} m^3.

Now let's consider the total volume of sand on beaches. We can find a volume by multiplying a width, depth, and length. In this case, the depth and width are the average depth and width of dry sand on beaches that have sand, while the length is the total length of such sandy beaches along all the coastlines on Earth. The width and depth are not too hard to estimate. If you've ever dug down to dirt or wet sand on a beach, you'll know that the dry sand is certainly deeper than 0.1 meter on average but less deep than 10 meters. We can therefore use the intermediate value of 1 meter as a good order of magnitude estimate for beach depth. Similarly, the width of sandy beaches varies widely but the average is certainly more than 1 meter and less than 100 meters, so we can choose the intermediate value of 10 meters as our order of magnitude estimate for beach width. The total length of sandy beach on Earth is a much more difficult number to estimate. A good strategy is to "solve" the problem without this number, then see what number it would have to be for the claim about the total number of grains to be true.

Step 2 Solve: As we've discussed, we have the following simple formula for the total number of grains of dry sand:

$$\text{total number of grains} = \frac{\text{beach depth} \times \text{beach width} \times \text{beach length}}{\text{average volume of 1 sand grain}}$$

Now, let's put in the estimates we've already made—1 m for beach depth, 10 m for beach width, and 10^{-10} m^3 for the volume of a sand grain:

$$\text{total number of grains} = \frac{1\cancel{m} \times 10\cancel{m} \times \text{beachlength}}{10^{-10}\cancel{m^3}} = \frac{10^{11}}{\text{m}} \times \text{beach length}$$

The claim we are trying to verify is that the total number of grains is roughly 10^{22}. For this to be the case, the beach length would have to be 10^{11} meters, since that would give us $10^{11} \times 10^{11} = 10^{22}$ on the right side of the equation above. (Notice how the units work out correctly.)

Step 3 Explain: We've found that the claim of 10^{22} grains of sand is reasonable *if* the total length of a sandy beach is close to 10^{11} meters, or 10^{8} kilometers (which is 100 million kilometers). It's not easy to estimate the total length of a sandy beach, since beaches do not follow straight lines and it's hard to account for all the islands and inlets on Earth. Nevertheless, if you spend some time studying a globe or atlas, you can probably convince yourself that 100 million kilometers is an overestimate of the total length of beach. In fact, it's probably an overestimate by at least a factor of 10. Thus, the actual number of grains of sand may be closer to 10^{20} or 10^{21} than to 10^{22}. However, this does not mean the claim was wrong. When you consider all the order of magnitude estimates we've made—which include the number of galaxies, the number of stars per galaxy, the volume of beaches, and the volume of individual sand grains—getting to within a factor of 10 or even 100 is pretty good. Thus, it's reasonable to say that the number of stars in the observable universe is comparable to the number of grains of sand on all the beaches on Earth.

to count the stars in the Milky Way, it would take thousands of years to count all the galaxies.

Think for a moment about the total number of stars in all these galaxies. If we assume 100 billion stars per galaxy, the total number of stars in the observable universe is roughly 100 billion × 100 billion, or 10,000,000,000,000,000,000,000 (10^{22}).

How big is this number? Visit a beach. Run your hands through the fine-grained sand. Imagine counting each tiny grain of sand as it slips through your fingers. Then imagine counting every grain of sand on the beach and continuing to count *every* grain of dry sand on *every* beach on Earth. If you could actually complete this task, you would find that, roughly speaking, the number of grains of sand is comparable to the number of stars in the observable universe (Figure 1.10).

Figure 1.10 The number of stars in the observable universe is comparable to the number of grains of dry sand on all the beaches on Earth.

THINK ABOUT IT

Contemplate the fact that there may be as many stars in the observable universe as grains of sand on all the beaches on Earth and that each star is a potential sun for a system of planets. With so many possible homes for life, do you think it is conceivable that life exists only on Earth? Why or why not?

• How do our lifetimes compare to the age of the universe?

Now that we have developed some perspective on the scale of space, we can do the same for the scale of time. Imagine the entire history of the universe, from the Big Bang to the present, compressed into a single year. We can represent this history with a *cosmic calendar,* on which the Big Bang takes place at the first instant of January 1 and the present day is just before the stroke of midnight on December 31 (Figure 1.11). For a universe that is about 14 billion years old, each month on the cosmic calendar represents a little more than 1 billion years.

On this time scale, the Milky Way Galaxy probably formed sometime in February. Many generations of stars lived and died in the subsequent cosmic months, enriching the galaxy with the "star stuff" from which we and our planet are made.

Our solar system and our planet did not form until early September on this scale, or 4.6 billion years ago in

Figure 1.11 The cosmic calendar compresses the 14-billion-year history of the universe into 1 year, so that each month represents a little more than 1 billion years (more precisely, 1.17 billion years). This cosmic calendar is adapted from a version created by Carl Sagan.

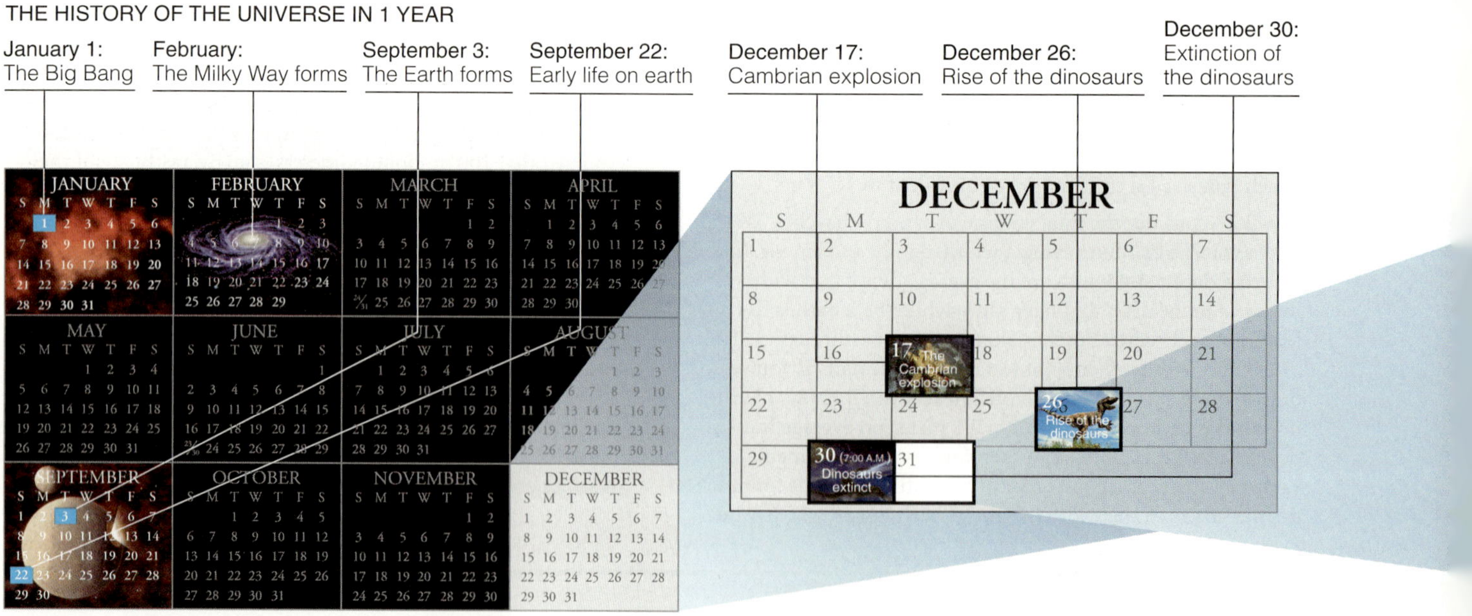

real time. By late September, life on Earth was flourishing. However, for most of Earth's history, living organisms remained relatively primitive and microscopic in size. On the scale of the cosmic calendar, recognizable animals became prominent only in mid-December. Early dinosaurs appeared on the day after Christmas. Then, in a cosmic instant, the dinosaurs disappeared forever—probably due to the impact of an asteroid or a comet [Section 12.4]. In real time, the death of the dinosaurs occurred some 65 million years ago, but on the cosmic calendar it was only yesterday. With the dinosaurs gone, small furry mammals inherited Earth. Some 60 million years later, or around 9 P.M. on December 31 of the cosmic calendar, early hominids (human ancestors) began to walk upright.

Perhaps the most astonishing thing about the cosmic calendar is that the entire history of human civilization falls into just the last half-minute. The ancient Egyptians built the pyramids only about 11 seconds ago on this scale. About 1 second ago, Kepler and Galileo proved that Earth orbits the Sun rather than vice versa. The average college student was born about 0.05 second ago, around 11:59:59.95 P.M. on the cosmic calendar. On the scale of cosmic time, the human species is the youngest of infants, and a human lifetime is a mere blink of an eye.

1.3 Spaceship Earth

Wherever you are as you read this book, you probably have the feeling that you're "just sitting here." Nothing could be further from the truth. In fact, you are being spun in circles as Earth rotates, you are racing around the Sun in Earth's orbit, and you are careening through the cosmos in the Milky Way Galaxy. In the words of noted inventor and philosopher R. Buckminster Fuller (1895–1983), you are a traveler on *spaceship Earth.* In this section, we'll take a brief look at the motion of spaceship Earth through the universe.

• How is Earth moving in our solar system?

The most basic motions of Earth are its **rotation** (spin) and its **orbit** (sometimes called *revolution*) around the Sun.

Earth rotates once each day around its axis, which is the imaginary line connecting the North Pole to the South Pole through the center of Earth. Earth rotates from west to east—counterclockwise as viewed from above the North Pole—which is why the Sun and stars appear to rise in the east and set in the west each day.

Although we do not feel any obvious effects from Earth's rotation, the speed of rotation is substantial (Figure 1.12). Unless you live very near the North or South Poles, you

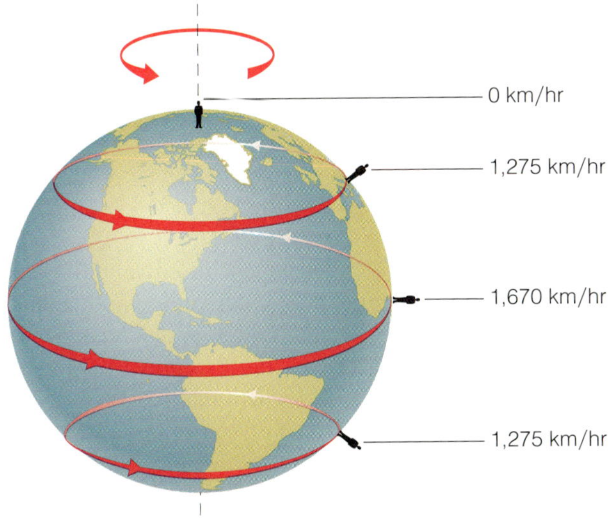

Figure 1.12 As Earth rotates, your speed around Earth's axis depends on your location: The closer you are to the equator, the faster you are traveling with rotation. Notice that Earth rotates from west to east—which is why the Sun appears to rise in the east and set in the west.

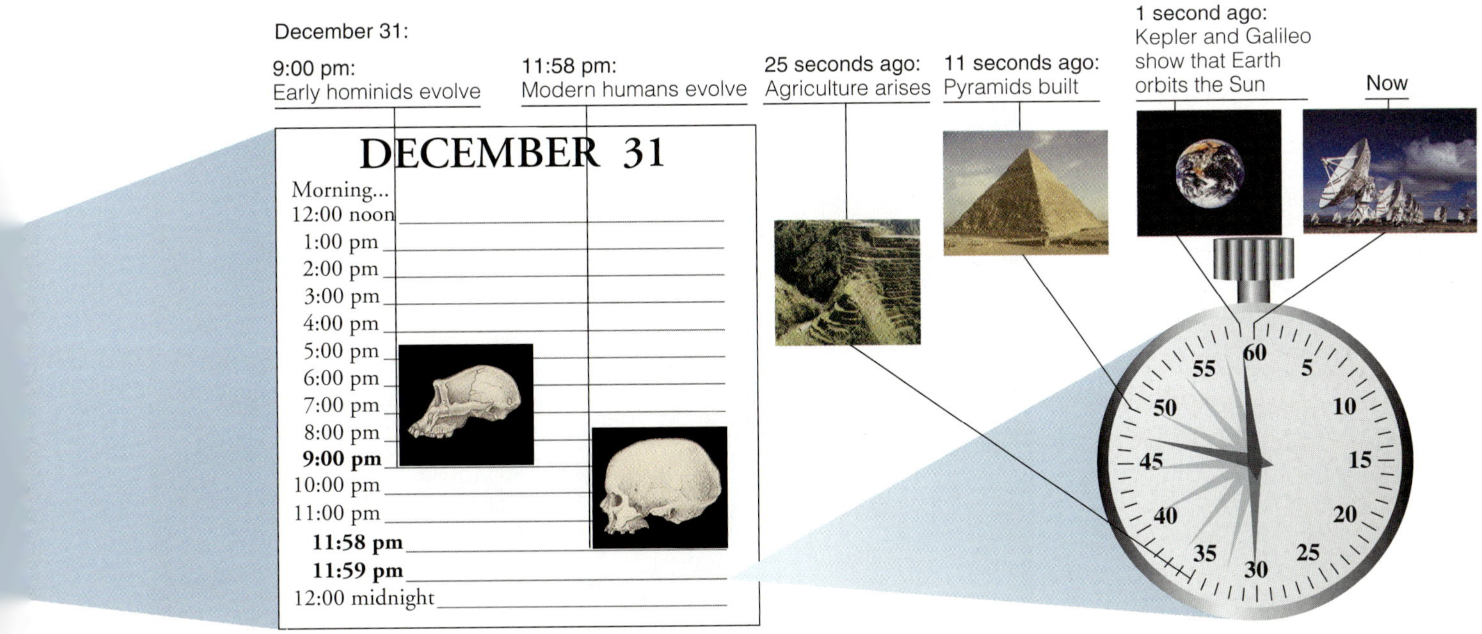

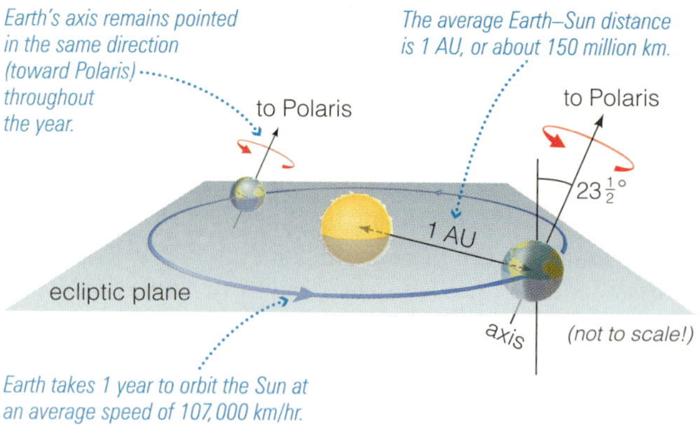

Figure 1.13 Earth takes a year to complete an orbit of the Sun, but its orbital speed is still surprisingly high. Notice that Earth both rotates and orbits counterclockwise as viewed from above the North Pole.

are whirling around Earth's axis at a speed of more than 1,000 kilometers per hour (600 miles per hour)—faster than most airplanes travel.

At the same time Earth is rotating, it is also orbiting the Sun, completing one orbit each year (Figure 1.13). Earth's average orbital distance is called an **astronomical unit**, or **AU**, which is equivalent to about 150 million kilometers (93 million miles). Again, even though we don't feel the effects of this motion, the speed is quite impressive. The orbit is not quite a perfect circle, which makes Earth's orbital speed vary slightly [Section 3.3], but at all times we are racing around the Sun at a speed in excess of 100,000 kilometers per hour (60,000 miles per hour). This is about 100 times as fast as a speeding bullet and faster than the launch speed of any spacecraft.

As you study Figure 1.13, notice that Earth's orbital path defines a flat plane that we call the **ecliptic plane**. Earth's axis is tilted by $23\frac{1}{2}°$ from a line *perpendicular* to the ecliptic plane. This **axis tilt** happens to be oriented so that it points almost directly at a star called *Polaris*, or the *North Star*. Keep in mind that the idea of axis tilt makes sense only in relation to the ecliptic plane. That is, the idea of "tilt" by itself has no meaning in space, where there is no absolute up or down. In space, "up" and "down" mean only "away from the center of Earth (or another planet)" and "toward the center of Earth," respectively.

THINK ABOUT IT

If there is no up or down in space, why do you think that nearly all globes have the North Pole on top and the South Pole on the bottom? Would it be equally correct to have the South Pole on top or to turn the globe sideways? Explain.

Notice also that Earth orbits the Sun in the same direction that it rotates on its axis: counterclockwise as viewed from above the North Pole. This is not a coincidence but a consequence of the way our planet was born. As we'll discuss in Chapter 8, Earth and the other planets were born in a spinning disk of gas that surrounded our Sun when it was young, and Earth rotates and orbits in the same direction as the disk was spinning. Notice also that we give a special name to the average distance between the Sun and Earth.

• How is our solar system moving in the Milky Way Galaxy?

Rotation and orbit are only a small part of the travels of spaceship Earth. In fact, our entire solar system is on a great journey within the Milky Way Galaxy (Figure 1.14).

Figure 1.14 This painting illustrates the motion of our solar system within the Milky Way Galaxy. The "zoom in" box shows that stars in our local solar neighborhood move essentially at random relative to one another. But the entire galaxy is rotating at the same time, so that all these stars orbit around the center of the galaxy.

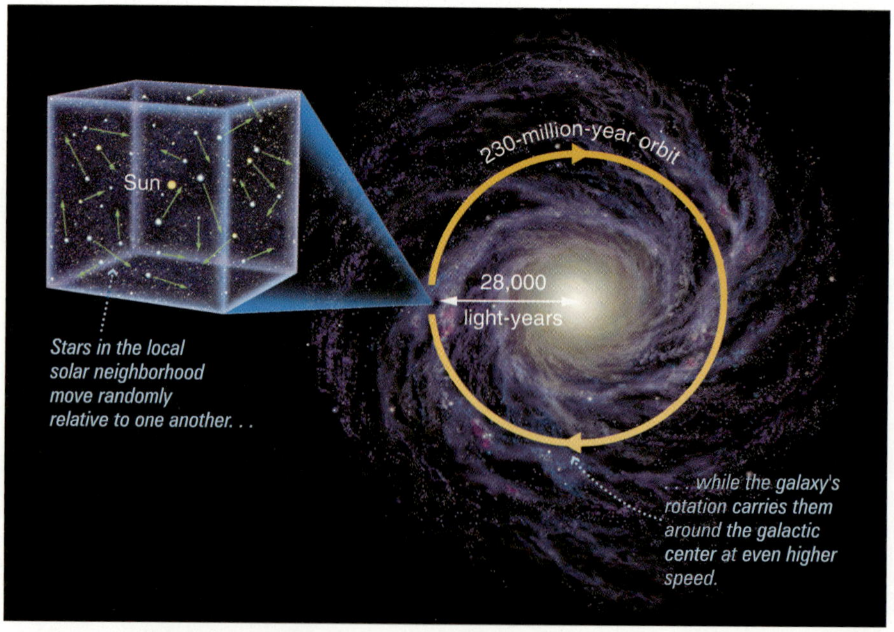

Our Local Solar Neighborhood Let's begin with the motion of our solar system relative to nearby stars in what we call our *local solar neighborhood,* the region of the Sun and nearby stars. As you can see in Figure 1.14, this neighborhood is tiny compared to the entire galaxy. It also offers an important reminder of the incredible scale of the galaxy. Imagine drawing the tiniest dot that you can make on this painting of the galaxy. Your dot will probably be about 10,000 times smaller than the picture of the galaxy as a whole, which means the dot will cover a region representing more than 10 million stars! (The entire galaxy contains more than 100 billion stars, and 100 billion ÷ 10,000 = 10 million.) We usually think of our local solar neighborhood as an even smaller region of the galaxy, containing just a few thousand to a few million of the nearest stars.

The box in Figure 1.14 shows that stars within the local solar neighborhood move essentially at random relative to one another. These stars (or the stars of any other small region of the galaxy) also generally move quite fast relative to one another. For example, we are moving relative to nearby stars at an average speed of about 70,000 kilometers per hour (40,000 miles per hour), about three times as fast as the Space Station orbits Earth. Given these high speeds, why don't we see nearby stars racing around our sky?

The answer lies in their vast distances from us. You've probably noticed that a distant airplane appears to move through your sky more slowly than one flying close overhead. If we extend this idea to the stars, we find that even at speeds of 70,000 kilometers per hour, stellar motions would be noticeable to the naked eye only if we watched them for thousands of years. That is why the patterns in the constellations seem to remain fixed. Nevertheless, in 10,000 years the constellations will be noticeably different from those we see today. In 500,000 years they will be unrecognizable. If you

MATHEMATICAL INSIGHT 1.4 Speeds of Rotation and Orbit

In the first three Mathematical Insights, we investigated general problem-solving techniques, the use of units, the use of scientific notation, and estimation. Now we will see how simple formulas can expand the range of astronomical problems we can solve. In this case, we'll use the simple formula for the circumference of a circle, $C = 2 \times \pi \times \text{radius}$ (or $C = 2\pi r$), to confirm the speeds of rotation and orbit given in the text.

Example 1: How fast is a person on Earth's equator moving with Earth's rotation?

Solution:

Step 1 Understand: The question *how fast* tells us we are looking for a *speed* as the answer. If you remember that highway speeds are posted in miles per hour, you'll always remember that we find speed by dividing a distance (such as miles) by a time (such as hours). In this case, the distance is the circumference of Earth around the equator, because that is how far a person travels with each rotation of our planet (see Figure 1.12). The time is 24 hours, because that is how long each rotation takes.

Step 2 Solve: From Appendix E.1, Earth's equatorial radius is 6,378 km, so its circumference is $2 \times \pi \times 6{,}378\ \text{km} = 40{,}074\ \text{km}$. We divide this distance by the time of 24 hours:

$$\begin{aligned}\text{rotation speed at equator} &= \frac{\text{equatorial circumference}}{\text{length of day}}\\ &= \frac{40{,}074\ \text{km}}{24\ \text{hr}}\\ &= 1{,}670\ \text{km/hr}\end{aligned}$$

We can convert the speed to miles per hour by using the conversion from kilometers to miles given in Appendix Table C.3: $1{,}670\ \text{km} \times 0.6214\ \text{mi/km} = 1{,}038\ \text{mi/hr}$.

Step 3 Explain: A person at Earth's equator is being carried around with Earth's rotation at a speed of about 1,670 kilometers per hour, or a little over 1,000 miles per hour. If you know that commercial jets typically travel at about 500 miles per hour, you'll see that Earth's equatorial rotation speed is twice as fast. (See Problem 52 at the end of the chapter to learn how to find the speed at other latitudes.)

Example 2: How fast is Earth orbiting the Sun?

Solution:

Step 1 Understand: We are again looking for a speed, so we need to divide a distance by a time. In this case the distance is the circumference of Earth's orbit, because that is how far Earth travels each time it goes around the Sun. (Earth's orbit is not a perfect circle, but it's close enough for a circle to be a good approximation.) The time is 1 year, because that is how long each orbit takes.

Step 2 Solve: The text tells us that Earth's average distance from the Sun (1 AU) is about 149.6 million (1.496×10^8) km, so the orbit circumference is about $2 \times \pi \times 1.496 \times 10^8\ \text{km} \approx 9.40 \times 10^8\ \text{km}$. We find the orbital speed by dividing this orbital distance by 1 year, which we convert to hours so that our speed will come out in kilometers per hour rather than kilometers per year:

$$\begin{aligned}\text{orbital speed} &= \frac{\text{orbital circumference}}{1\ \text{yr}}\\ &= \frac{9.40 \times 10^8\ \text{km}}{1\ \cancel{\text{yr}} \times \frac{365\ \cancel{\text{day}}}{\cancel{\text{yr}}} \times \frac{24\ \text{hr}}{\cancel{\text{day}}}}\\ &\approx 107{,}000\ \text{km/hr}\end{aligned}$$

Step 3 Explain: Earth's average speed as it orbits the Sun is about 107,000 km/hr. (You should confirm that this is about 66,000 mi/hr.) On the Web, you can look up the fact that a typical speeding bullet travels between about 800 and 1,800 km/hr. Thus, roughly speaking, Earth's orbital speed is about 100 times as fast as a speeding bullet.

could watch a time-lapse movie made over millions of years, you *would* see stars racing across our sky.

THINK ABOUT IT

Despite the chaos of motion in the local solar neighborhood over millions and billions of years, collisions between star systems are extremely rare. Explain why. (*Hint:* Consider the sizes of star systems, such as the solar system, relative to the distances between them.)

Galactic Rotation If you look closely at leaves floating in a stream, their motions relative to one another might appear random, just like the motions of stars in the local solar neighborhood. As you widen your view, you see that all the leaves are being carried in the same general direction by the downstream current. In the same way, as we widen our view beyond the local solar neighborhood, the seemingly random motions of its stars give way to a simpler and even faster motion: The entire Milky Way Galaxy is rotating. Our solar system, located about 28,000 light-years from the galactic center, completes one orbit of the galaxy in about 230 million years. Even if you could watch from outside our galaxy, this motion would be unnoticeable to your naked eye. However, if you calculate the speed of our solar system as we orbit the center of the galaxy, you will find that it is close to 800,000 kilometers (500,000 miles) per hour.

Careful study of the galaxy's rotation reveals one of the greatest mysteries in science—one that we will study in depth in Chapter 22. Stars at different distances from the galactic center orbit at different speeds, and we can learn how mass is distributed in the galaxy by measuring these speeds. Such studies have turned up an enormous surprise: It seems that the stars in the disk of the galaxy represent only the "tip of the iceberg" compared to the mass of the entire galaxy (Figure 1.15). That is, most of the mass of the galaxy seems to be located outside the visible disk, in what we call the galaxy's *halo*. We don't know the nature of this mass. Because we have not detected any light coming from it, we call it **dark matter**. Studies of other galaxies suggest that they also are made mostly of dark matter. In fact, most of the mass in the universe seems to be made of this mysterious dark matter, but we do not yet know what it is.

• How do galaxies move within the universe?

The billions of galaxies in the universe also move relative to one another. Within the Local Group (see Figure 1.1), some of the galaxies move toward us, some move away from us, and at least two small galaxies (known as the Large and Small Magellanic Clouds) apparently orbit our Milky Way Galaxy. Again, the speeds are enormous by earthly standards. For example, the Milky Way is moving toward the Great Galaxy in Andromeda at about 300,000 kilometers per hour (180,000 miles per hour). Despite this high speed, we needn't worry about a collision anytime soon. Even if the

Figure 1.15 This painting shows an edge-on view of the Milky Way Galaxy. Most visible stars reside within the galaxy's central bulge and thin *disk*, which run horizontally across the page in this figure. However, careful study of galactic rotation suggests that most of the mass lies in the galactic *halo*—a large, spherical region that surrounds and encompasses the disk. Because this mass emits no light that we have detected, we call it *dark matter*.

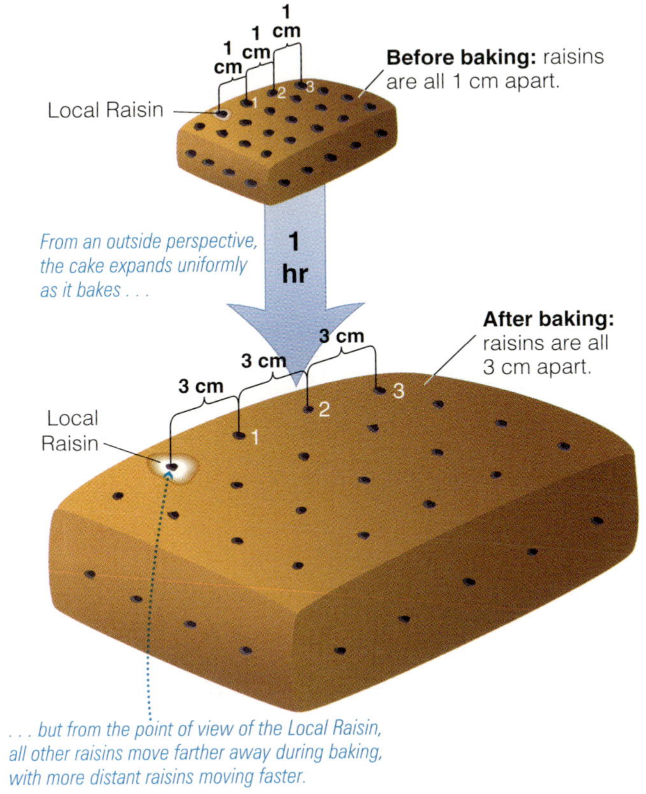

Distances and Speeds as Seen from the Local Raisin

Raisin Number	Distance Before Baking	Distance After Baking (1 hour later)	Speed
1	1 cm	3 cm	2 cm/hr
2	2 cm	6 cm	4 cm/hr
3	3 cm	9 cm	6 cm/hr
⋮	⋮	⋮	⋮

Figure 1.16 Interactive Figure An expanding raisin cake offers an analogy to the expanding universe. Someone living in one of the raisins inside the cake could figure out that the cake is expanding by noticing that all other raisins are moving away, with more distant raisins moving away faster. In the same way, we know that we live in an expanding universe because all galaxies outside our Local Group are moving away from us, with more distant ones moving faster.

Milky Way and Andromeda Galaxies are approaching each other head-on (which they might not be), it will be billions of years before any collision begins.

When we look outside the Local Group, however, we find two astonishing facts recognized in the 1920s by Edwin Hubble, for whom the Hubble Space Telescope was named:

1. Virtually every galaxy outside the Local Group is moving *away* from us.
2. The more distant the galaxy, the faster it appears to be racing away.

These facts might make it sound like we suffer from a cosmic case of chicken pox, but there is a much more natural explanation: *The entire universe is expanding.* We'll save details about this expansion and how Hubble discovered it for Chapter 20, but you can understand the basic idea by thinking about a raisin cake baking in an oven.

Imagine that you make a raisin cake in which the distance between adjacent raisins is 1 centimeter. You place the cake into the oven, where it expands as it bakes. After 1 hour, you remove the cake, which has expanded so that the distance between adjacent raisins has increased to 3 centimeters (Figure 1.16). The expansion of the cake seems fairly obvious. But what would you see if you lived *in* the cake, as we live in the universe?

Pick any raisin (it doesn't matter which one), call it the Local Raisin, and identify it in the pictures of the cake both before and after baking. Figure 1.16 shows one possible choice for the Local Raisin, with three nearby raisins labeled. The accompanying table summarizes what you would see if you lived within the Local Raisin. Notice, for example, that Raisin 1 starts out at a distance of 1 centimeter before baking and ends up at a distance of 3 centimeters after baking, which means it moves a distance of 2 centimeters away from the Local Raisin during the hour of baking. Hence, its speed as seen from the Local Raisin is 2 centimeters per hour. Raisin 2 moves from a distance of 2 centimeters before baking to a distance of 6 centimeters after baking, which means it moves a distance of 4 centimeters away from the Local Raisin during the hour. Hence, its speed is 4 centimeters per hour, or twice as fast as the speed of Raisin 1. Generalizing, the fact that the cake is expanding means that all raisins are moving away from the Local Raisin, with more distant raisins moving away faster.

Hubble's discovery that galaxies are moving in much the same way as the raisins in the cake, with most moving away from us and more distant ones moving away faster, implies that the universe in which we live is expanding much like the raisin cake. If you now imagine the Local Raisin as representing our Local Group of galaxies and the other raisins as representing more distant galaxies or clusters of galaxies, you have a basic picture of the expansion of the universe. Like the expanding dough between the raisins in the cake, *space* itself is growing between galaxies. More distant galaxies move away from us faster because they are carried along with this expansion like the raisins in the expanding cake. Many billions of light-years away, we see galaxies moving away from us at speeds approaching the speed of light.

There's one important distinction between the raisin cake and the universe: A cake has a center and edges, but we do not think the same is true of the entire universe. Anyone living in any galaxy in an expanding universe sees just what we see—other galaxies moving away, with more distant ones moving away faster. Because the view from each point in the universe is about the same, no place can claim to be any more "central" than any other place. The idea that the universe has no center or edges may sound strange, and it is not something that we can easily visualize. However, as we'll discuss in Chapters S2 and S3, space and time don't necessarily obey our commonsense notions about them. Einstein showed that the three dimensions of space—length, width, and depth—are intertwined with a fourth dimension that we perceive as time. Once you learn how to think about the universe in four dimensions, the idea that the universe has no center or edges won't seem so strange.

It's also important to realize that, unlike the case with a raisin cake, Hubble couldn't actually *see* galaxies moving apart with time—the distances are far too vast for any motion to be noticeable on the time scale of a human life. Instead, he measured the speeds of galaxies by spreading their light into spectra and observing what we call *Doppler shifts* [Section 5.5]. We still use the same technique today, which illustrates how modern astronomy depends both on making careful observations and on using current understanding of the laws of nature to explain what we see.

• Are we ever sitting still?

As we have seen, we are never truly sitting still. Figure 1.17 summarizes the motions we have discussed. We spin around Earth's axis at more than 1,000 km/hr, while our planet orbits the Sun at more than 100,000 km/hr. Our solar system moves among the stars of the local solar neighborhood at typical speeds of 70,000 km/hr, while also orbiting the center of the Milky Way Galaxy at a speed of about 800,000 km/hr. Our galaxy moves among the other galaxies of the Local Group, while all other galaxies move away from us at speeds that grow greater with distance in our expanding universe. Spaceship Earth is carrying us on a remarkable journey!

1.4 The Human Adventure of Astronomy

In relatively few pages, we've laid out a fairly complete overview of our modern scientific ideas about the universe. But our goal in this book is not for you simply to be able to recite these ideas. Rather, it is to help you understand the evidence supporting them and the extraordinary story of how they developed.

• How has the study of astronomy affected human history?

Astronomy is a human adventure in the sense that it affects virtually everyone—even those who have never looked at the sky—because the development of astronomy has been so deeply intertwined with the development of civilization as a whole. Revolutions in astronomy have gone hand in hand with the revolutions in science and technology that have shaped modern life.

Witness the repercussions of the Copernican revolution, which showed us that Earth is not the center of the universe but rather just one planet orbiting the Sun. This revolution, which we will discuss further in Chapter 3, began when Copernicus published his idea of a Sun-centered solar system in 1543. Three later figures—Tycho Brahe, Johannes Kepler, and Galileo—provided the key evidence that eventually led to wide acceptance of the Copernican idea. The revolution culminated with Isaac Newton's uncovering of the laws of motion and gravity. Newton's work, in turn,

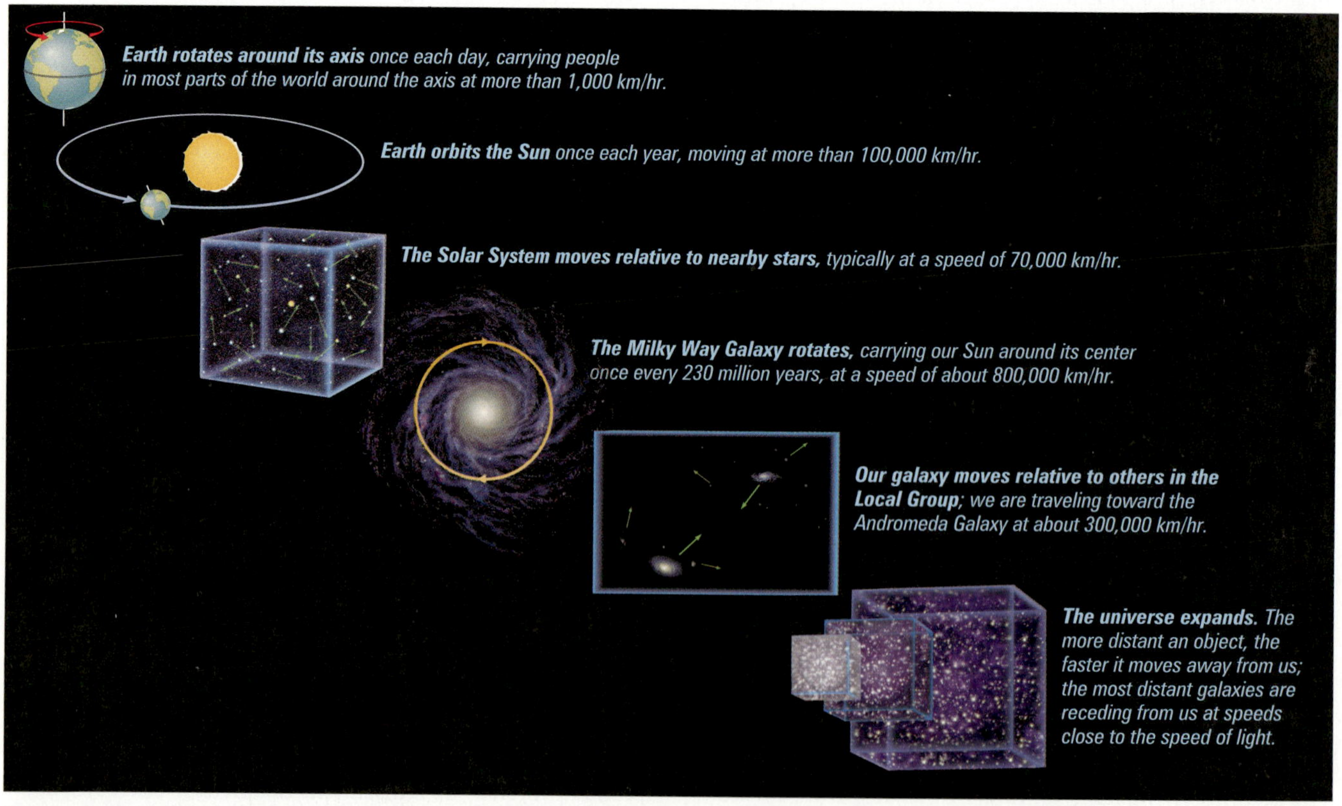

Figure 1.17 This figure summarizes the basic motions of Earth in the universe, along with their associated speeds.

became the foundation of physics that helped fuel the industrial revolution.

More recently, the development of space travel and the computer revolution have helped fuel tremendous progress in astronomy. We've learned a lot about our solar system by sending probes to the planets, and many of our most powerful observatories, including the Hubble Space Telescope, reside in space. On the ground, computer design and control have led to tremendous growth in the size and power of telescopes, particularly in the past decade.

Many of these efforts, and the achievements they spawned, led to profound social change. The most famous example is the fate of Galileo, whom the Vatican put under house arrest in 1633 for his claims that Earth orbits the Sun. Although the Church soon recognized that Galileo was right, he was formally vindicated only in 1992 with a statement by Pope John Paul II. In the meantime, his case spurred great debate in religious circles and profoundly influenced both theological and scientific thinking.

As you progress through this book and learn about astronomical discovery, try to keep in mind the context of the human adventure. You will then be learning not just about a science, but also about one of the great forces that has shaped our modern world. This context will also lead you to think about how the many astronomical mysteries that remain—such as the makeup of dark matter, the events in the first instant of the Big Bang, and the possibility of life beyond Earth—may influence our future.

What would it mean to us if we were ever to learn the complete story of our cosmic origins? How would our view of Earth be changed if we were to learn that Earth-like planets are common, or exceedingly rare? Only time may answer these questions, but the chapters ahead will give you the foundation you need to understand how we changed from a primitive people looking at patterns in the night sky to a civilization capable of asking deep questions about our existence.

THE BIG PICTURE

Putting Chapter 1 into Context

In this first chapter, we developed a broad overview of our place in the universe. As we consider the universe in more depth in the rest of the book, remember the following "big picture" ideas:

- Earth is not the center of the universe but is instead a planet orbiting a rather ordinary star in the Milky Way Galaxy. The Milky Way Galaxy, in turn, is one of billions of galaxies in our observable universe.
- We are "star stuff." The atoms from which we are made began as hydrogen and helium in the Big Bang and were later fused into heavier elements by massive stars. When these stars died, they released their atoms into space, where our galaxy recycled them into new stars and planets. Our solar system formed from such recycled matter some 4.6 billion years ago.
- Cosmic distances are literally astronomical, but we can put them into perspective with the aid of scale models and other scaling techniques. When you think about these enormous scales, don't forget that every star is a sun and that every planet is a unique world.
- We are latecomers on the scale of cosmic time. The universe was already more than half its current age when our solar system formed, and several billion more years passed before humans arrived on the scene.
- All of us are being carried through the cosmos on spaceship Earth. Although we cannot feel this motion in our everyday lives, the associated speeds are surprisingly high. Learning about the motions of spaceship Earth gives us a new perspective on the cosmos and helps us understand its nature and history.

SUMMARY OF KEY CONCEPTS

1.1 Our Modern View of the Universe

- **What is our place in the universe?** Earth is a planet orbiting the Sun. Our Sun is one of more than 100 billion stars in the **Milky Way Galaxy.** Our galaxy is one of about 40 galaxies in the **Local Group.** The Local Group is one small part of the **Local Supercluster,** which is one small part of the **universe.**

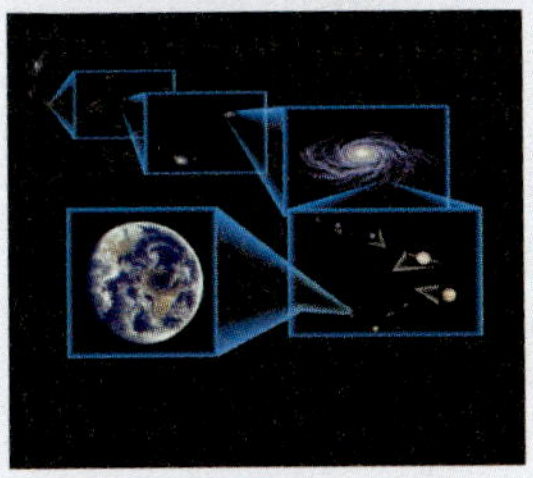

- **How did we come to be?** The universe began in the **Big Bang** and has been expanding ever since, except in localized regions where gravity has caused matter to collapse into galaxies and stars. The Big Bang essentially produced only two chemical elements: hydrogen and helium. The rest have been produced by stars and recycled within galaxies from one generation of stars to the next, which is why we are "star stuff."

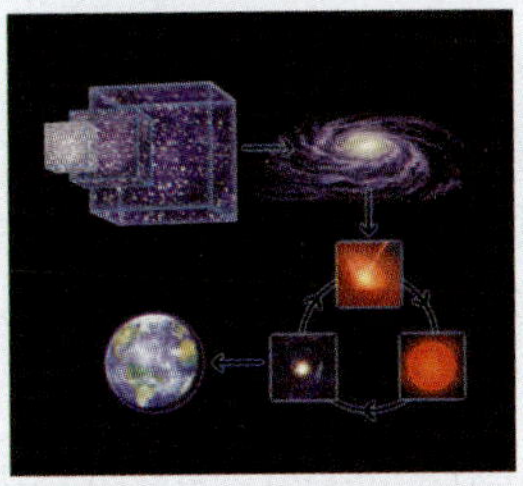

- **How can we know what the universe was like in the past?** We can actually see into the past by studying light from distant stars and galaxies. Light takes time to travel through space, so the farther away we look in distance, the further back we look in time. Thus, when we look billions of **light-years** away, we see pieces of the universe as they were billions of years ago.

- **Can we see the entire universe?** No. The age of the universe limits the extent of our **observable universe.** For a universe that is 14 billion years old, our observable universe extends to a distance of 14 billion light-years. If we tried to look beyond that distance, we'd be trying to look to a time before the universe existed.

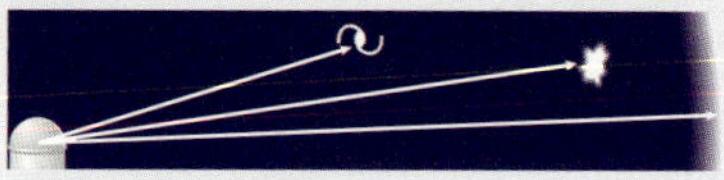

1.2 The Scale of the Universe

- **How big is Earth compared to our solar system?** On a scale of 1 to 10 billion, the Sun is about the size of a grapefruit. Planets are much smaller, with Earth the size of a ball point and Jupiter the size of a marble on this scale. The distances between planets are huge compared to their sizes, with Earth orbiting 15 meters from the Sun on this scale.

- **How far away are the stars?** On the 1-to-10-billion scale, it is possible to walk from the Sun to Pluto in just a few minutes. On the same scale, the nearest stars besides the Sun are thousands of kilometers away.
- **How big is the Milky Way Galaxy?** Using a scale on which the Milky Way galaxy is the size of a football field, the distance to the nearest star would be only about 4 millimeters. There are so many stars in our galaxy that it would take thousands of years just to count them.
- **How big is the universe?** The observable universe contains at least 100 billion galaxies. The total number of stars in the observable universe is comparable to the number of grains of dry sand on all the beaches on Earth.
- **How do our lifetimes compare to the age of the universe?** On a cosmic calendar that compresses the history of the universe into 1 year, human civilization is just a few seconds old, and a human lifetime lasts only a fraction of a second.

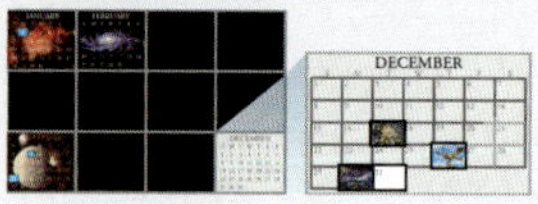

1.3 Spaceship Earth

- **How is Earth moving in our solar system?** Earth **rotates** on its axis once each day and **orbits** the Sun once each year. Earth orbits at an average distance from the Sun of 1 **AU** and with an **axis tilt** of $23\frac{1}{2}°$ to a line perpendicular to the **ecliptic plane.**

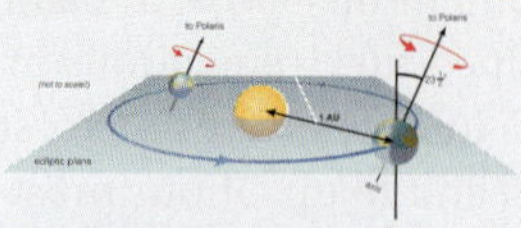

- **How is our solar system moving in the Milky Way Galaxy?** We move seemingly randomly relative to other stars in our local solar neighborhood. The speeds are substantial by earthly standards, but the stars are so far away that their motions are undetectable to the naked eye. Our Sun and other stars in our neighborhood orbit the center of the galaxy every 230 million years, because the entire galaxy is rotating.

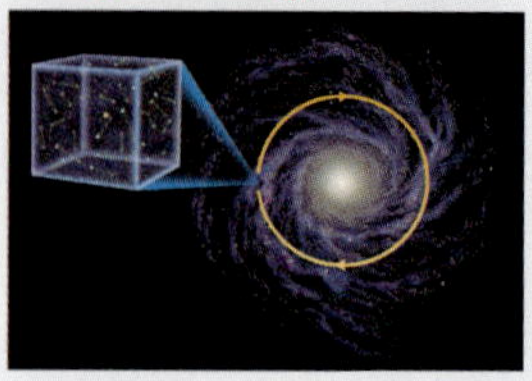

- **How do galaxies move within the universe?** Galaxies move essentially at random within the Local Group, but all galaxies beyond the Local Group are moving away from us. More distant galaxies are moving faster, which tells us that we live in an expanding universe.
- **Are we ever sitting still?** We are never truly sitting still. We spin around Earth's axis and orbit the Sun. Our solar system moves among the stars of the local solar neighborhood while orbiting the center of the Milky Way Galaxy. Our galaxy moves among the other galaxies of the Local Group, while all other galaxies move away from us in our expanding universe.

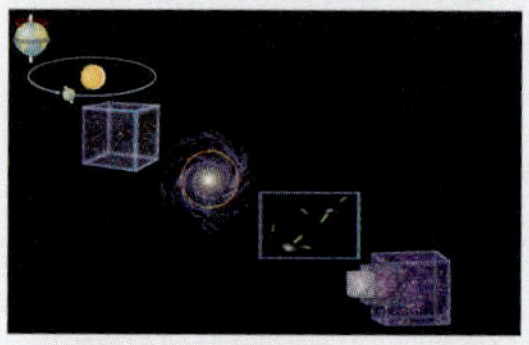

1.4 The Human Adventure of Astronomy

- **How has the study of astronomy affected human history?** Throughout history, astronomy has developed hand in hand with social and technological development. Astronomy thereby touches each one of us and is a human adventure that all can enjoy.

EXERCISES AND PROBLEMS

For instructor-assigned homework go to MasteringASTRONOMY www.masteringastronomy.com

Review Questions

Short-Answer Questions Based on the Reading

1. What do we mean by a *geocentric* universe? In broad terms, contrast a geocentric view of the universe with our modern view of the universe.
2. Briefly describe the major levels of structure (such as planet, star, galaxy) in the universe.
3. What do we mean when we say that the universe is *expanding?* How does expansion lead to the idea of the *Big Bang?*
4. What did Carl Sagan mean when he said that we are "star stuff"?
5. How fast does light travel? What is a *light-year?*
6. Explain the statement *The farther away we look in distance, the further back we look in time.*
7. What do we mean by the *observable universe?* Is it the same thing as the entire universe?
8. Describe the solar system as it looks on the 1-to-10-billion scale used in the text. How far away are other stars on this same scale?

9. Describe at least one way to put the scale of the Milky Way Galaxy into perspective and at least one way to put the size of the observable universe into perspective.
10. Imagine describing the cosmic calendar to a friend. In your own words, give your friend a feel for how the human race fits into the scale of time.
11. Define *astronomical unit, ecliptic plane,* and *axis tilt.* Explain how each is related to Earth's rotation and/or orbit.
12. What is the shape of the Milky Way Galaxy? Where is our solar system located within the galaxy? How does our solar system move within the galaxy?
13. Distinguish between our galaxy's *disk* and *halo.* Where do most visible stars reside? Where does the mysterious *dark matter* seem to reside?
14. What key observations by Edwin Hubble lead us to conclude that the universe is expanding? Use the raisin cake model to explain how these observations imply expansion.

Test Your Understanding

Does It Make Sense?

Decide whether the statement makes sense (or is clearly true) or does not make sense (or is clearly false). Explain your reasoning.

Example: I walked east from our base camp at the North Pole.

Solution: The statement does not make sense because *east* has no meaning at the North Pole—all directions are south from the North Pole.

15. Our solar system is bigger than some galaxies.
16. The universe is billions of light-years in age.
17. It will take me light-years to complete this homework assignment!
18. Someday we may build spaceships capable of traveling a light-year in only a decade.
19. Astronomers recently discovered a moon that does not orbit a planet.
20. NASA plans soon to launch a spaceship that will photograph our Milky Way Galaxy from beyond its halo.
21. The observable universe is the same size today as it was a few billion years ago.
22. Photographs of distant galaxies show them as they were when they were much younger than they are today.
23. At a nearby park, I built a scale model of our solar system in which I used a basketball to represent Earth.
24. Because nearly all galaxies are moving away from us, we must be located at the center of the universe.

Quick Quiz

Choose the best answer to each of the following. Explain your reasoning with one or more complete sentences.

25. Which of the following correctly lists our "cosmic address" from small to large? (a) Earth, solar system, Milky Way Galaxy, Local Group, Local Supercluster, universe; (b) Earth, solar system, Local Group, Local Supercluster, Milky Way Galaxy, universe; (c) Earth, Milky Way Galaxy, solar system, Local Group, Local Supercluster, universe.
26. When we say the universe is *expanding,* we mean that: (a) Everything in the universe is growing in size. (b) The average distance between galaxies is growing with time. (c) The universe is getting older.
27. If stars existed but galaxies did not: (a) We would probably still exist anyway. (b) We would not exist because life on Earth depends on the light produced by galaxies. (c) We would not exist because we are made of material that required recycling in galaxies.
28. Could we see a galaxy that is 20 billion light-years away? (a) Yes, if we had a big enough telescope. (b) No, because it would be beyond the bounds of our observable universe. (c) No, because a galaxy could not possibly be that far away.
29. The star Betelgeuse is about 425 light-years away. If it explodes tonight: (a) We'll know because it will be brighter than the full Moon in the sky. (b) We'll know because debris from the explosion will rain down on us from space. (c) We won't know about it until 425 years from now.
30. If we represent the solar system on a scale that allows us to walk from the Sun to Pluto in a few minutes, then: (a) The planets would be the size of basketballs and the nearest stars would be a few miles away. (b) The planets would all be marble size or smaller and the nearest stars would be thousands of miles away. (c) The planets would be microscopic and the stars would be light-years away.
31. The total number of stars in the observable universe is roughly equivalent to: (a) the number of grains of sand on all the beaches on Earth. (b) the number of grains of sand on Miami Beach. (c) infinity.
32. The age of our solar system is about: (a) one-third of the age of the universe. (b) three-fourths of the age of the universe. (c) two billion years less than the age of the universe.
33. An astronomical unit is: (a) any planet's average distance from the Sun. (b) Earth's average distance from the Sun. (c) any large astronomical distance.
34. The fact that nearly all galaxies are moving away from us, with more distant ones moving faster, helped us to conclude that: (a) The universe is expanding. (b) Galaxies repel each other like magnets. (c) Our galaxy lies near the center of the universe.

Investigate Further

In-Depth Questions to Increase Your Understanding

Short-Answer/Essay Questions

35. *Our Cosmic Origins.* Write one to three paragraphs summarizing why we could not be here if the universe did not contain both stars and galaxies.
36. *Alien Technology.* Some people believe that Earth is regularly visited by aliens who travel here from other star systems. For this to be true, how much more advanced than our own technology would the space travel technology of the aliens have to be? Write one to two paragraphs to give a sense of the technological difference. (*Hint:* The ideas of scale in this chapter can help you contrast the distance the aliens would have to travel easily with the distances we are now capable of traveling.)
37. *Stellar Collisions.* Is there any danger that another star will come crashing through our solar system in the near future? Explain.
38. *Raisin Cake Universe.* Suppose that all the raisins in a cake are 1 centimeter apart before baking and 4 centimeters apart after baking.
    a. Draw diagrams to represent the cake before and after baking.
    b. Identify one raisin as the Local Raisin on your diagrams. Construct a table showing the distances and speeds of other raisins as seen from the Local Raisin.
    c. Briefly explain how your expanding cake is similar to the expansion of the universe.

39. *Scaling the Local Group of Galaxies.* Both the Milky Way Galaxy and the Great Galaxy in Andromeda (M 31) have a diameter of about 100,000 light-years. The distance between the two galaxies is about 2.5 million light-years.
    a. Using a scale on which 1 centimeter represents 100,000 light-years, draw a sketch showing both galaxies and the distance between them to scale.
    b. How does the separation between galaxies compare to the separation between stars? Based on your answer, discuss the likelihood of galactic collisions in comparison to the likelihood of stellar collisions.
40. *The Cosmic Perspective.* Write a short essay describing how the ideas presented in this chapter affect your perspectives on your own life and on human civilization.

Quantitative Problems

Be sure to show all calculations clearly and state your final answers in complete sentences.

41. *Distances by Light.* Just as a light-year is the distance that light can travel in 1 year, we define a light-second as the distance that light can travel in 1 second, a light-minute as the distance that light can travel in 1 minute, and so on. Calculate the distance in both kilometers and miles represented by each of the following:
    a. 1 light-second;
    b. 1 light-minute;
    c. 1 light-hour;
    d. 1 light-day.
42. *Moonlight and Sunlight.* How long does it take light to travel from:
    a. the Moon to Earth?
    b. the Sun to Earth?
43. *Communication with Mars.* We use radio waves, which travel at the speed of light, to communicate with robotic spacecraft. How long does it take a message to travel from Earth to a spacecraft on Mars when:
    a. Mars it at its closest to Earth?
    b. Mars is at its farthest from Earth? (*Data:* The distance from Earth to Mars ranges between about 56 and 400 million kilometers.)
44. *Communication with Pluto.* About how long does it take light to travel from Earth to Pluto? You can assume that both Earth and Pluto are at their average distances from the Sun.
45. *Moon to Stars.* How many times greater is the distance from Earth to Alpha Centauri (4.4 light-years) than the distance from Earth to the Moon?
46. *Saturn vs. the Milky Way.* Photos of Saturn and photos of galaxies can look so similar that children often think the photos show similar objects. (After all, like Saturn, at least some galaxies have a large, round center surrounded by a disk that looks like rings.) In reality, a galaxy is far larger than any planet. About how many times larger is the diameter of the Milky Way Galaxy than the diameter of Saturn's rings? (*Data:* Saturn's rings are about 270,000 km in diameter; the Milky Way is 100,000 light-years in diameter.)
47. *Galaxy Scale.* Consider the 1-to-10^{19} scale on which the disk of the Milky Way Galaxy fits on a football field. On this scale, how far is it from the Sun to Alpha Centauri (real distance: 4.4 light-years)? How big is the Sun itself on this scale? Compare the Sun's size on this scale to the size of a typical atom (real diameter: about 10^{-10} meter).
48. *Universal Scale.* Suppose we wanted to make a scale model of the Local Group of Galaxies, in which the Milky Way Galaxy was the size of a marble (about 1 cm in diameter).
    a. How far from the Milky Way Galaxy would the Andromeda Galaxy be on this scale?
    b. How far would the Sun be from Alpha Centauri on this scale?
    c. How far would it be on this scale from the Milky Way Galaxy to the most distant galaxies in the observable universe?
49. *Driving Trips.* Imagine that you could drive your car at a constant speed of 100 km/hr (62 mi/hr), even across oceans and in space. (In reality, the law of gravity would make driving through space at a constant speed all but impossible.) How long would it take to drive:
    a. around Earth's equator?
    b. from the Sun to Earth?
    c. from the Sun to Pluto?
    d. to Alpha Centauri (4.4 light-years away)?
50. *Faster Trip.* Suppose you wanted to reach Alpha Centauri in 100 years.
    a. How fast would you have to go, in km/hr?
    b. How many times faster is the speed you found in (a) than the speeds of our fastest current spacecraft (around 50,000 km/hr)?
51. *Galactic Rotation Speed.* We are located about 28,000 light-years from the galactic center and we orbit the center once every 230 million years. How fast are we traveling around the galaxy? Give your answer in both km/hr and mi/hr.
52. *Earth Rotation Speed.* Mathematical Insight 1.3 shows how to find Earth's equatorial rotation speed. To find the rotation speed at any other latitude, you need the following fact: The radial distance from Earth's axis at any latitude is equal to the equatorial radius times the cosine of the latitude. Use this fact to find the rotation speed for a person at latitude:
    a. 30°N;
    b. 60°N;
    c. your latitude. (*Hint:* When using the cosine (cos) function on your calculator, be sure your calculator is set to recognize angles in degree mode, as opposed to radian or gradient mode.)

Discussion Questions

53. *Eliot Quote.* Think carefully about the chapter-opening quotation from T.S. Eliot. What do you think he means? Explain clearly.
54. *Infant Species.* In the last few tenths of a second before midnight on December 31 of the cosmic calendar, we have developed an incredible civilization and learned a great deal about the universe, but we also have developed technology with which we could destroy ourselves. The midnight bell is striking, and the choice for the future is ours. How far into the next cosmic year do you think our civilization will survive? Defend your opinion.
55. *A Human Adventure.* Astronomical discoveries clearly are important to science, but are they also important to our personal lives? Defend your opinion.

MEDIA EXPLORATIONS

For self-study activities go to MasteringASTRONOMY www.masteringastronomy.com

Interactive Tutorials

Tutorial Review of Key Concepts

Use the interactive **Tutorial** at **www.masteringastronomy.com** to review key concepts from this chapter.

Scale of the Universe Tutorial

Lesson 1 Distance Scales: The Solar System
Lesson 2 Distance Scales: Stars and Galaxies
Lesson 3 Powers of 10

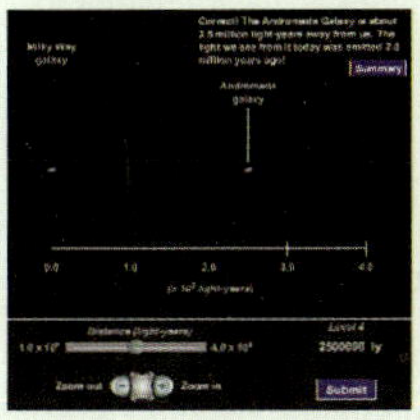

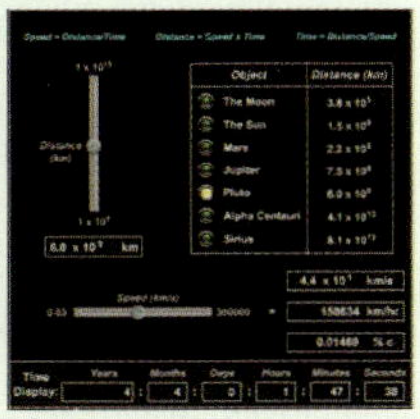

Supplementary Tutorial Exercises

Use the interactive **Tutorial Lessons** to explore the following questions.

Scale of the Universe Tutorial, Lesson 2

1. Could we use spaceships like those that we use to explore our solar system to explore planets around other stars? Why or why not?
2. How does the distance between galaxies in the Local Group compare to the sizes of the galaxies?
3. Why are collisions between galaxies more likely than collisions between stars within galaxies?

Scale of the Universe Tutorial, Lesson 3

1. Why are powers of 10 useful for describing and comparing distances in the universe?
2. If you begin from the scale of the observable universe, by how many powers of 10 must you zoom in to see the orbits of the planets in our solar system?
3. How many powers of 10 separate the scale on which you can see all the planetary orbits from the scale on which you can see individual people?

Exploring the Sky and Solar System

Of the many activities available on the ***Voyager: SkyGazer* CD-ROM** accompanying your book, use the following files to observe key phenomena covered in this chapter.

Go to the **File: Basics** folder for the following demonstrations.

1. Chicago 10000AD
2. Dragging the Sky

Go to the **Explore** menu for the following demonstrations.

1. Solar Neighborhood
2. Paths of the Planets

Movies

Check out the following narrated and animated short documentaries available on **www.masteringastronomy.com** for a helpful review of key ideas covered in this chapter.

From the Big Bang to Galaxies Movie

Web Projects

1. *Astronomy on the Web.* The Web contains a vast amount of astronomical information. Starting from the links on the textbook Web site, spend at least an hour exploring astronomy on the Web. Write two or three paragraphs summarizing what you learned from your Web surfing. What was your favorite astronomical Web site, and why?
2. *Tour Report.* Take the virtual tour of the Voyage scale model solar system on **www.masteringastronomy.com**. After completing it, imagine that a friend asks you the following questions. Answer each question in one paragraph.
   a. Is the Sun really much bigger than Earth?
   b. Is it true that the Sun uses nuclear energy?
   c. Would it be much harder to send humans to Mars than to the Moon?
   d. In elementary school, I heard that Neptune is farther from the Sun than Pluto. Is this true?
   e. I read that Pluto is not really a planet. What's the story?
   f. Why didn't they have any stars besides the Sun in the scale model?
   g. What was the most interesting thing you learned during your tour?
3. *NASA Missions.* Visit the NASA Web site to learn about upcoming astronomy missions. Write a one-page summary of the mission you feel is most likely to give us new astronomical information during the time you are enrolled in your astronomy course.
4. *The Hubble Ultra Deep Field.* The photo that opens this chapter is called the Hubble Ultra Deep Field. Find the photo on the Hubble Space Telescope Web site, and learn how it was taken, what it shows, and what we've learned from it. Write a short report on your findings.

2

Discovering the Universe for Yourself

LEARNING GOALS

2.1 Patterns in the Night Sky
- What does the universe look like from Earth?
- Why do stars rise and set?
- Why do the constellations we see depend on latitude and time of year?

2.2 The Reason for Seasons
- What causes the seasons?
- How do we mark the progression of the seasons?
- How does the orientation of Earth's axis change with time?

2.3 The Moon, Our Constant Companion
- Why do we see phases of the Moon?
- What causes eclipses?

2.4 The Ancient Mystery of the Planets
- What was once so mysterious about planetary motion in our sky?
- Why did the ancient Greeks reject the real explanation for planetary motion?

We had the sky, up there, all speckled with stars, and we used to lay on our backs and look up at them, and discuss about whether they was made, or only just happened.

Mark Twain, *Huckleberry Finn*

This is an exciting time in the history of astronomy. A new generation of telescopes is probing the depths of the universe. Increasingly sophisticated space probes are collecting new data about the planets and other objects in our solar system. Rapid advances in computing technology are allowing scientists to analyze the vast amount of new data and to model the processes that occur in planets, stars, galaxies, and the universe.

A goal of this book is to help you share in the ongoing adventure of astronomical discovery. One of the best ways to become a part of this adventure is to do what other humans have done for thousands of generations: Go outside, observe the sky around you, and contemplate the awe-inspiring universe of which you are a part. In this chapter, we'll discuss a few key ideas that will help you understand what you see in the sky.

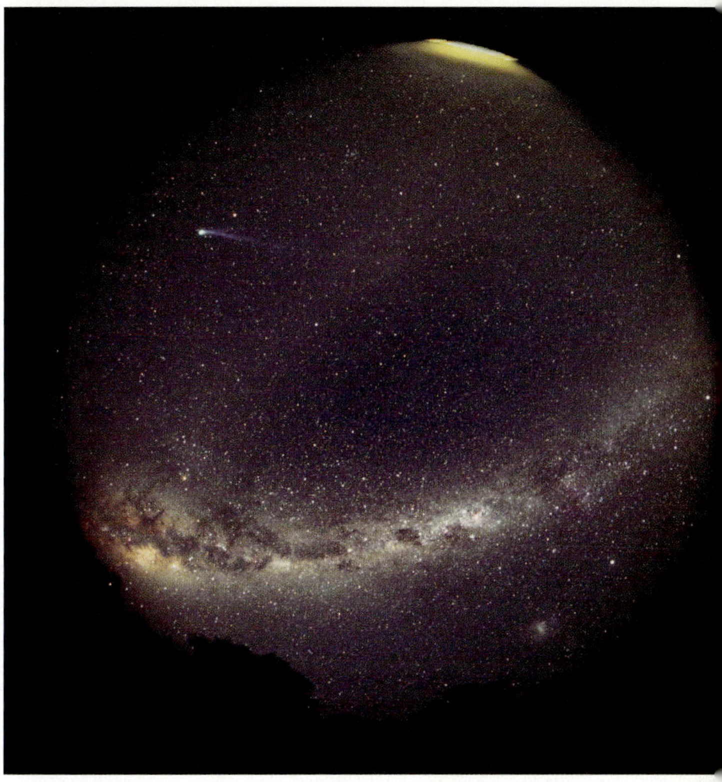

Figure 2.1 A "fish-eye" photograph of the Australian night sky. The Milky Way is the prominent whitish band with dark lanes running through it. A comet is also visible near the upper left of this photo. (It is Comet Hyakutake, which shined in Earth's sky for a few months in 1996.)

2.1 Patterns in the Night Sky

Today we take for granted that we live on a small planet orbiting an ordinary star in one of many galaxies in the universe. But this fact is not obvious from a casual glance at the night sky, and we've learned about our place in the cosmos only through a long history of careful observations of patterns in the sky. In this section, we'll discuss the major features of the night sky, considering both how they appear to our eyes and how we understand them in light of our current knowledge of the universe.

• What does the universe look like from Earth?

Shortly after sunset, as daylight fades to darkness, the sky appears to fill slowly with stars. On clear, moonless nights far from city lights, more than 2,000 stars may be visible to your naked eye, along with the whitish band of light that we call the *Milky Way* (Figure 2.1). As you look at the stars, your mind may group them into patterns that look like familiar shapes or objects. If you observe the sky night after night or year after year, you will recognize the same patterns of stars. Indeed, the patterns we see today are virtually the same ones that generations of our ancestors saw over the last several thousand years.

Constellations People of nearly every culture gave names to patterns they saw in the sky. The pattern that the Greeks named Orion, the hunter (see Figure 1.3), was seen by the ancient Chinese as a supreme warrior called *Shen.* Hindus in ancient India also saw a warrior, called *Skanda,* who rode a peacock and was the general of a great celestial army. Aborigines of northern Australia saw the three stars of Orion's belt as three fishermen in a canoe. In southern California, these three stars climb almost straight up into the sky as they rise in the east, which may explain why the Chemehuevi Indians of the California desert saw them as a line of three sure-footed mountain sheep.* These are but a few of the many names, each accompanied by a rich folklore, that have been given to the pattern of stars we call Orion.

We usually refer to Orion and other patterns of stars as constellations, but to astronomers the term has a more precise meaning: A **constellation** is a *region* of the sky with well-defined borders. Just as every spot of land in the United States is part of some state, every point in the sky belongs to some constellation. For example, Figure 2.2 shows the borders of the constellation Orion and several of its neighbors. The familiar patterns of stars merely help us locate these constellations.

The official borders of the constellations were set in 1928 by members of the International Astronomical Union (IAU), an association of astronomers from around the world. The IAU divided the sky into 88 constellations (see Appendix I), choosing constellation names that were famil-

*These and other constellation stories are found in E. C. Krupp, *Beyond the Blue Horizon* (Oxford University Press, 1991).

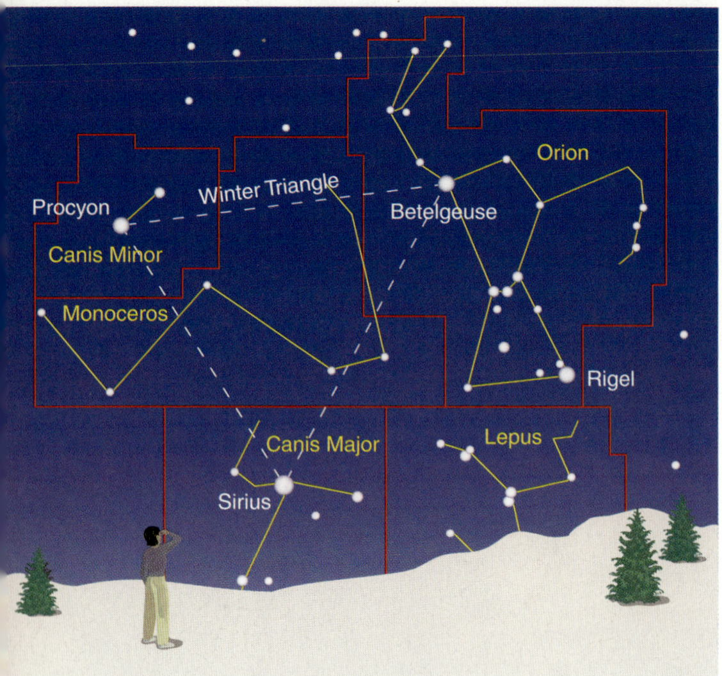

Figure 2.2 Red lines mark official borders of several constellations near Orion. Yellow lines connect recognizable patterns of stars within constellations. Sirius, Procyon, and Betelgeuse form a pattern that spans several constellations and is called the *Winter Triangle*. It is easy to find on clear winter evenings.

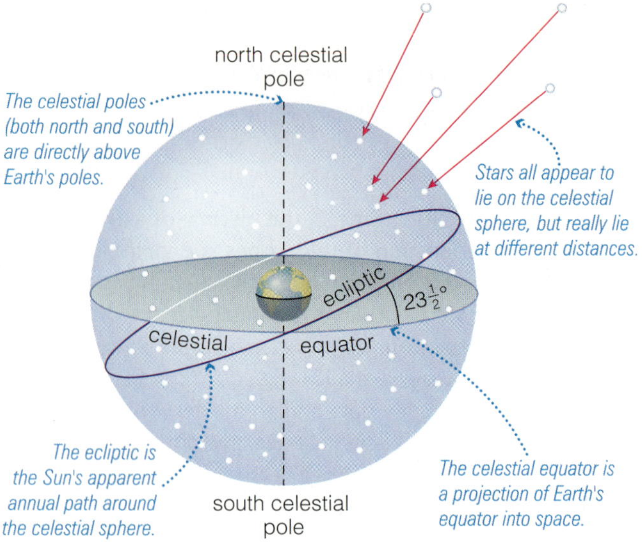

Figure 2.3 Stars appear to lie on a great celestial sphere that surrounds the Earth.

iar to European and American astronomers. Most constellations visible in the Northern Hemisphere have names that can be traced back to civilizations of the ancient Middle East. Many Southern Hemisphere constellations carry official names that originated with seventeenth-century European explorers.

Learning your way around the constellations is no more difficult than learning your way around your own neighborhood, and recognizing the patterns of just 20–40 constellations is enough to make the entire sky seem familiar. The best way to learn the constellations is to go out and view them, guided by a few visits to a planetarium and star charts like the ones in the back of this book (Appendix J). The *SkyGazer* software that comes with this book can also help you learn constellations.

The Celestial Sphere The stars in a particular constellation appear to lie close to one another but may be quite far apart in reality, because they may lie at very different distances from Earth. This illusion occurs because we lack depth perception when we look into space, a consequence of the fact that the stars are so far away [Section 1.2]. The ancient Greeks mistook this illusion for reality, imagining the stars to lie on a great **celestial sphere** that surrounds the Earth (Figure 2.3).

We now know that Earth does not really lie in the center of a giant ball of stars, but we can still use the idea of a celestial sphere to help us understand and map the sky. For reference, we identify four special points and circles.

- The **north celestial pole** is the point directly over Earth's North Pole.
- The **south celestial pole** is the point directly over Earth's South Pole.
- The **celestial equator**, which is a projection of Earth's equator into space, makes a complete circle around the celestial sphere.
- The **ecliptic** is the path the Sun follows as it appears to circle around the celestial sphere once each year.

Notice that the ecliptic crosses the celestial equator at a $23\frac{1}{2}°$ angle, because that is the tilt of Earth's axis. In fact, the ecliptic on the celestial sphere is the projection into space of Earth's orbital plane (called the *ecliptic plane* [Section 1.3]).

Just as a globe shows us the layout of Earth's surface, a model of the celestial sphere shows how the stars appear to be arranged in the sky. A typical model of the celestial sphere (Figure 2.4) shows patterns of bright stars, borders of the 88 official constellations, and the celestial poles, celestial equator, and ecliptic.

The Milky Way As your eyes adapt to darkness at a dark site, you'll begin to see the whitish band of light called the *Milky Way* (see Figure 2.1). Our Milky Way Galaxy gets its name from this band of light. You can see only part of the Milky Way at any particular time, but it circles all the way around the celestial sphere, passing through more than a dozen constellations. The widest and brightest parts of the Milky Way are most easily seen from the Southern Hemisphere, which probably explains why the Aborigines of Australia gave names to the patterns they saw within the Milky Way in the same way other cultures named patterns of stars.

The band of light called the Milky Way bears an important relationship to the Milky Way Galaxy: *It traces our galaxy's disk of stars—the galactic plane—as it appears from our location in the outskirts of the galaxy.* Figure 2.5 shows the idea. The Milky Way Galaxy is shaped like a thin pan-

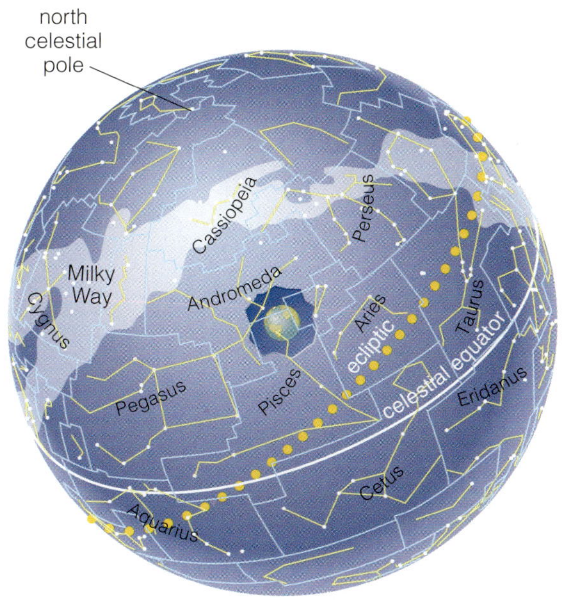

Figure 2.4 A model of the celestial sphere showing bright stars and the borders of constellations. Remember that the celestial sphere represents the view from Earth, so we imagine Earth to reside in the center of the sphere.

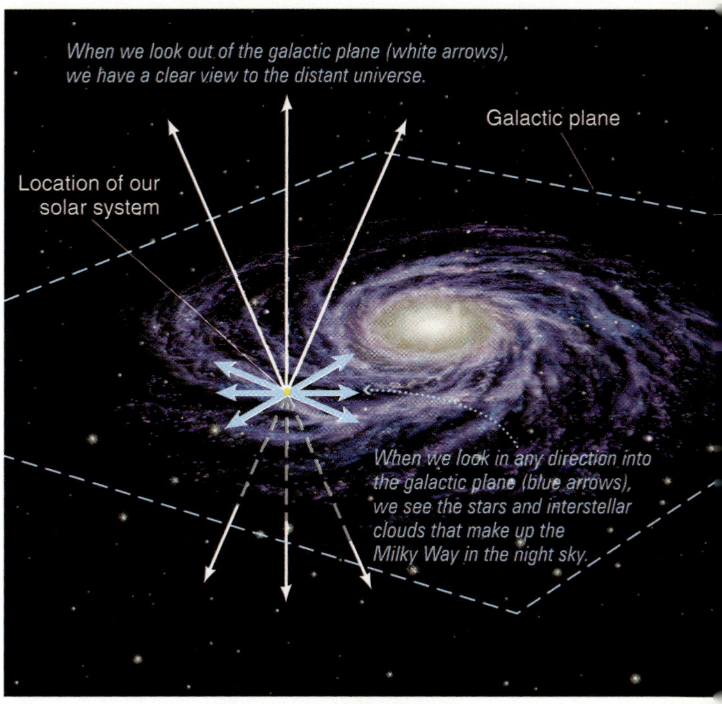

Figure 2.5 This painting shows how our galaxy's structure affects our view from Earth. Remember that all the stars visible to the naked eye lie quite close to us compared to the size of the galaxy.

cake with a bulge in the middle. We view the universe from our location a little more than halfway out from the center of this "pancake." No matter what direction we look *within* the disk of the galaxy, our view takes in countless stars and vast clouds of interstellar gas and dust. These stars and clouds form the Milky Way in our night sky. We see far fewer stars and relatively little gas and dust in our own galaxy when we look in directions pointing *away* from the galactic plane. Thus, as shown by the white arrows in Figure 2.5, these directions offer a clear view to the far reaches of the universe, as long as we look with sufficiently powerful telescopes.

If you examine the Milky Way closely, you'll notice that it varies in width and has dark lanes running through it. The width depends on the direction we are looking into the galactic disk: It is widest in the direction of the constellation Sagittarius, because that is the direction in which we are looking toward the galaxy's wide, central bulge. The dark lanes appear in regions where particularly dense interstellar clouds obscure our view of stars behind them. The Milky Way's abundant gas and dust prevents us from seeing more than a few thousand light-years into our galaxy's disk.* This limitation kept much of our own galaxy hidden from view until just a few decades ago, when new technologies allowed us to peer through the clouds by observing forms of light that are invisible to our eyes (such as radio waves and X-rays [Section 5.2]).

THINK ABOUT IT

Consider a distant galaxy located in the same direction from Earth as the center of our own galaxy (but much farther away). Could we see it with our eyes? Explain.

*There are a few small "windows" through which we can see much farther; see Figure 15.4 for an example.

The Local Sky The celestial sphere, banded by the Milky Way, is a useful way of thinking about the appearance of the universe from Earth. But it is not what we actually see when we go outside. Picture yourself standing in a flat, open field. The sky appears to take the shape of a dome, making it easy to understand why people of many ancient cultures imagined that we lived on a flat Earth under a great dome encompassing the world. We see only half of the celestial sphere at any particular moment from any particular location, while the other half is blocked from view by the ground. The half of the celestial sphere that you see at any time represents what we call your **local sky**—the sky as seen from wherever you happen to be standing.

We use several key features of the local sky to serve as references for locating objects (Figure 2.6). The boundary between Earth and sky defines the **horizon**. The point directly overhead is the **zenith**. The **meridian** is an imaginary half-circle stretching from the horizon due south, through the zenith, to the horizon due north.

We can pinpoint the position of any object in the local sky by stating its **direction** along the horizon and its **altitude** above the horizon. For example, Figure 2.6 shows a person pointing to a star located in the direction of southeast at an altitude of 60°. (The zenith has an altitude of 90° but no direction, because it is straight overhead.)

In addition to stating an object's location in our sky, we often want to describe its size or how far it is from other objects in the sky. For example, to describe the pattern of stars in a particular constellation, we might wish to say how large the pattern appears in the sky and where individual stars are located relative to others. It's easy to do this, as long as we remember that we can describe only *angular*

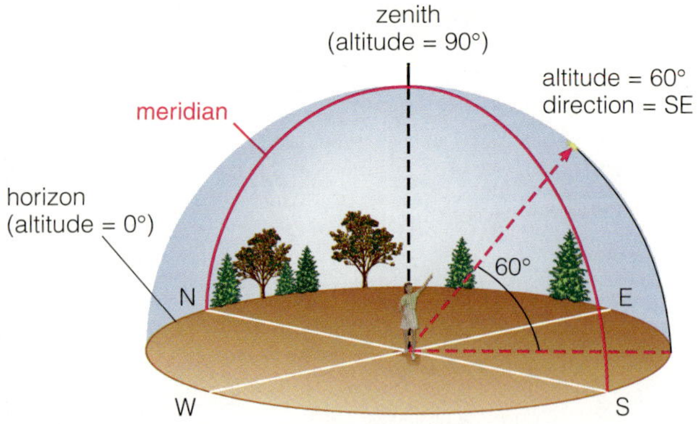

Figure 2.6 From any place on Earth, the local sky looks like a dome (hemisphere). This diagram shows key reference points in the local sky. It also shows how we can describe any position in the local sky by its altitude and direction.

sizes and distances in the sky. Our lack of depth perception on the celestial sphere means that we have no way to judge the true sizes or separations of the objects we see.

The **angular size** of an object is the angle it appears to span in your field of view. For example, the angular sizes of the Sun and Moon are each about $\frac{1}{2}°$ (Figure 2.7a). Notice that angular size does not by itself tell us much about an object's true size, because angular size also depends on distance: The farther away an object is, the smaller it will look in angular size. For example, the Sun is about 400 times larger in diameter than the Moon, but it has the same angular size in our sky because it is also about 400 times farther away.

The **angular distance** between a pair of objects in the sky is simply the angle that appears to separate them. For example, the angular distance between the "pointer stars" at the end of the Big Dipper's bowl is about 5° (Figure 2.7b). You can use your outstretched hand to make rough estimates of angles in the sky (Figure 2.7c).

For more precise astronomical measurements, we subdivide each degree into 60 **arcminutes** and subdivide each arcminute into 60 **arcseconds** (Figure 2.8). We abbreviate arcminutes with the symbol ′ and arcseconds with the symbol ″. For example, we read 35°27′15″ as "35 degrees, 27 arcminutes, 15 arcseconds."

COMMON MISCONCEPTIONS

The Moon Illusion

You've probably noticed that the full moon appears to be larger when it is near the horizon than when it is high in your sky. However, this apparent size change is an illusion. If you measure the angular size of the full moon on a particular night, you'll find that it is about the same whether the Moon is near the horizon or high in the sky. The Moon's angular size in the sky depends only on its true size and its distance from Earth. Although this distance varies over the course of the Moon's monthly orbit, it does not change enough to cause a noticeable effect on a single night. You can confirm that the Moon's angular size remains the same by measuring it. You may also be able to make the illusion go away by viewing the Moon upside down between your legs when it is on the horizon.

THINK ABOUT IT

Children often try to describe the sizes of objects in the sky (such as the Moon or an airplane) in inches or miles, or by holding their fingers apart and saying "it was THIS big." Can we really describe objects in the sky in this way? Why or why not?

• Why do stars rise and set?

If you spend a few hours out under a starry sky, you'll soon notice that the universe seems to be circling around us. Stars move gradually across the sky from east to west, and

Figure 2.7 We measure *angular sizes* or *angular distances*, rather than actual sizes or distances, when we look at objects in the sky.

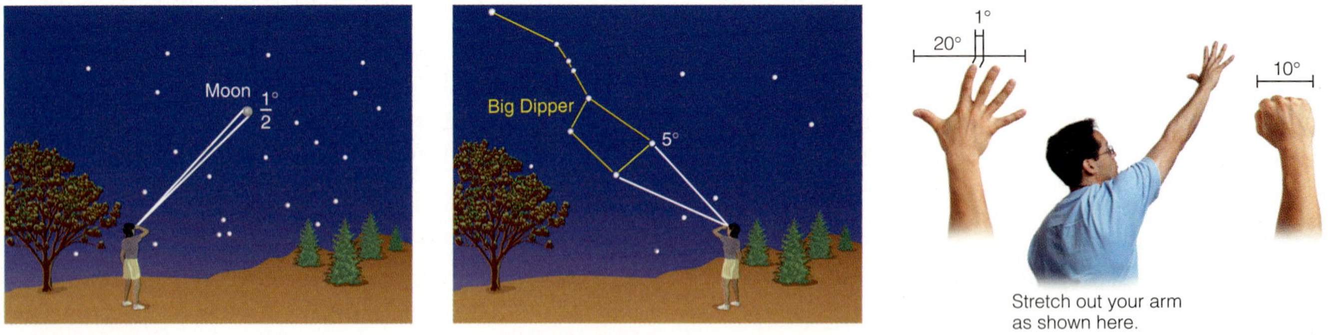

a The angular size of the Moon is about 1/2° (which is also the angular size of the Sun).

b The angular distance between the two "pointer stars" of the Big Dipper (which point to the North Star, Polaris; see Figure 2.13a) is about 5°.

c You can estimate angular sizes or distances with your outstretched hand.

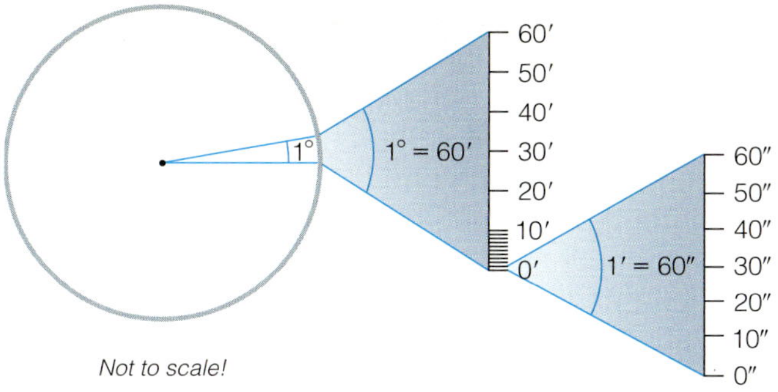

Figure 2.8 We subdivide each degree into 60 arcminutes and each arcminute into 60 arcseconds. (The figure is not drawn to scale.)

you'll always see some stars rising in the east while others are setting in the west. Many ancient people took this appearance of movement at face value, concluding that we lie in the center of a universe that rotates around us each day. Today we know that the ancients had it backward: It is Earth that rotates each day, not the rest of the universe. The stars, along with the Sun, Moon, and planets, all appear to rise in the east and set in the west because Earth rotates in the opposite direction, from west to east.

THINK ABOUT IT

Do distant galaxies also rise and set like the stars in our sky? Why or why not?

Because the stars all appear to reside on a celestial sphere, we can picture the movement of the sky by showing the celestial sphere rotating around Earth (Figure 2.9). From this perspective it is very easy to see how the universe seems to turn around us: Every object on the celestial sphere appears to make a simple daily circle around Earth. However, the motion can look a little more complex in the local sky.

In the local sky, the horizon cuts the celestial sphere in half. Figure 2.10 shows the idea for a typical location in the Northern Hemisphere. As Earth rotates—making the celestial sphere appear to circle around us—we see stars and other objects on the celestial sphere move across our local sky. Different stars follow different paths through the sky, depending on where they are located on the celestial sphere. If you study Figure 2.10 carefully, you'll notice the following key facts about the paths of stars (and other celestial objects) through the sky:

- Stars relatively near the north celestial pole remain perpetually above the horizon. They never rise or set but instead make daily counterclockwise circles around the north celestial pole. We say that such stars are **circumpolar**.
- Stars relatively near the south celestial pole never rise above the horizon at all.
- All other stars have daily circles that are partly above the horizon and partly below it. Because Earth rotates from west to east (counterclockwise as viewed from above the North Pole), the stars appear to rise in the east and set in the west.

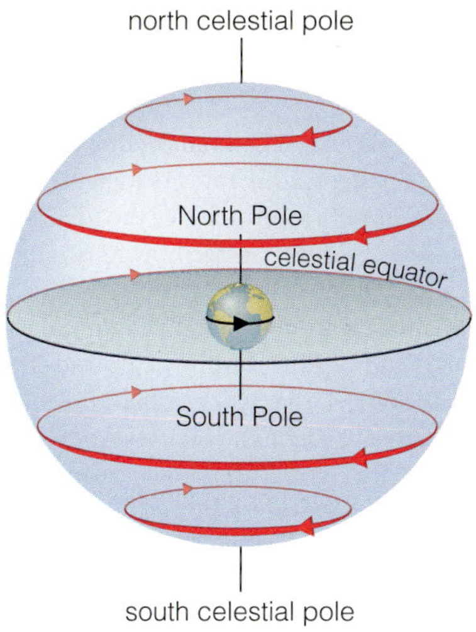

Figure 2.9 Earth rotates from west to east (black arrow), making the celestial sphere *appear* to rotate around us from east to west (red arrows).

The photo that opens this chapter (p. 26) shows a beautiful time-exposure photograph that helps us see the daily paths of stars. It was taken at Arches National Park in Utah. Paths of circumpolar stars are visible within the arch; notice that the complete daily circles for these stars are above the horizon (although the photo shows only a portion of each circle). The north celestial pole lies at the center of these circles. The circles grow larger for stars farther from the north celestial pole. If they are large enough, the circles cross the horizon, so that the stars rise in the east and set in the west. The same ideas apply in the Southern Hemisphere, except that circumpolar stars are those near the south celestial pole.

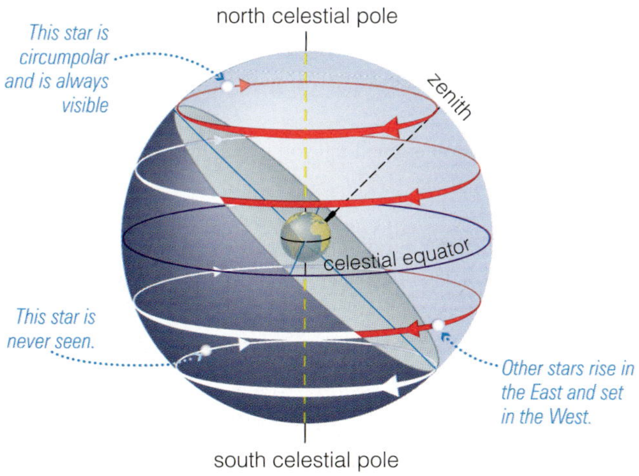

Figure 2.10 This diagram shows the local sky for a typical location in the Northern Hemisphere. The local horizon slices through half the celestial sphere, which is why the local sky looks like a dome. Because the slice is at an angle to the equator, the simple daily circles of stars are tilted in the local sky. Note: It may be easier to follow the star paths in the local sky if you rotate the page so that the zenith points up.

 Angular Size, Physical Size, and Distance

If you hold a quarter in front of your eye, it can block your entire field of view. But as you move it farther way, it appears to get smaller and it blocks less of your view. Figure (a) summarizes the idea by showing the quarter in cross section, so we can see how its angular diameter decreases with distance.

It's useful to have a formula telling us how an object's angular size depends on its physical size and distance, and we can find the formula with a little mathematical trick that works when the angular size is small. In Figure (b), we've made the quarter from Figure (a) look like a tiny piece of a circle going all the way around your eye. The radius of the circle is the *distance* from your eye to the quarter, the angle from your eye is the quarter's *angular size*, and we've labeled the quarter's actual diameter as its *physical size*. Now, notice that as long as the angular size is relatively small (less than a few degrees), we can pretend that the quarter's physical size (diameter) is a small piece of the circle we've drawn. This is the trick we needed: The quarter's angular size is now the *same fraction* of the full 360° circle as its physical size is of the circle's physical circumference. Since the circumference of this circle is $2\pi \times (distance)$, we can write what we've found as:

$$\frac{\text{angular size}}{360^\circ} = \frac{\text{physical size}}{2\pi \times \text{distance}}$$

Multiplying both sides by 360° and rearranging a bit, we have a formula that allows us to determine angular size when we know physical size and distance:

$$\text{angular size} = \text{physical size} \times \frac{360^\circ}{2\pi \times \text{distance}}$$

(This formula is sometimes called the *small-angle formula*, since it is valid only when the angular size is small.) In astronomy, we generally measure an object's angular size and often have a way of determining its distance (we'll discuss distance measurement techniques in later chapters). Thus, it's useful to rearrange the formula so we can calculate physical size when we know angular size and distance. You should confirm that a little algebra tells us that:

$$\text{physical size} = \text{angular size} \times \frac{2\pi \times \text{distance}}{360^\circ}$$

The context of a problem will tell you which form of the formula to choose.

Example 1: The angular diameter of the Moon is about 0.5° and the Moon is about 380,000 km away. Estimate the Moon's actual diameter.

Solution:

Step 1 Understand: We are asked to find the Moon's actual diameter given its angular diameter and distance. The formula below tells us how to find physical *size* from angular size and distance; we can use this formula once we realize that, in this case, the "size" is a diameter.

Step 2 Solve: We now use the formula to calculate the Moon's physical size (diameter) from the given values of its angular size and distance:

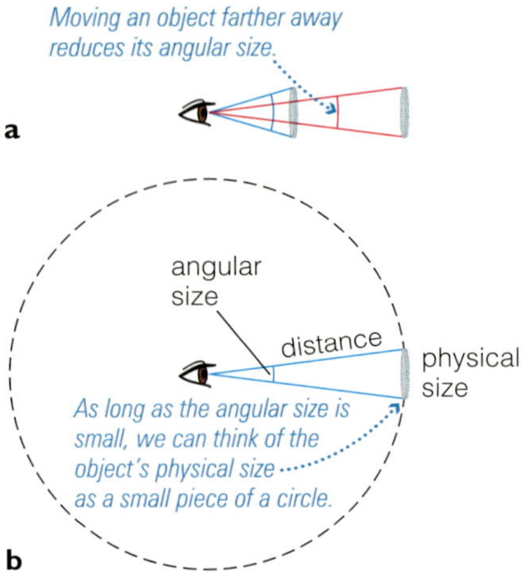

Angular size depends on physical size and distance.

$$\begin{aligned}\text{physical size} &= \text{angular size} \times \frac{2\pi \times \text{distance}}{360^\circ}\\ &= 0.5^\circ \times \frac{2\pi(380{,}000\ \text{km})}{360^\circ}\\ &\approx 3{,}300\ \text{km}\end{aligned}$$

Step 3 Explain: We have found that the Moon's diameter is about 3,300 kilometers. We can check that our answer makes sense by comparing it to the value for the Moon's diameter given in Appendix E. Our estimate of 3,300 kilometers is fairly close to the Moon's actual diameter of 3,476 kilometers, which we could find by using more precise values for the Moon's angular diameter and distance.

Example 2: Suppose the two headlights on a car are separated by 1.5 meters and you are looking at the car from a distance of 500 meters. What is the angular separation of the headlights?

Solution:

Step 1 Understand: In this case we are asked about "separation" between two lights rather than size, but the idea is the same. We simply replace size with separation in the below formulas and we have all the information we need to solve the problem.

Step 2 Solve: Because we know the physical separation and distance in this case, we use the formula in the first form that we found above.

$$\begin{aligned}\text{angular separation} &= \text{physical separation} \times \frac{360^\circ}{2\pi \times \text{distance}}\\ &= 1.5\ \text{m} \times \frac{360^\circ}{2\pi(500\ \text{m})} = 0.17^\circ\end{aligned}$$

Step 3 Explain: We have found that the angular separation of the two headlights is 0.17°. However, remember that it is more common to express fractions of a degree in arcminutes or arcseconds. Remembering that there are 60 arcminutes in 1°, our answer of 0.17° is equivalent to $0.17^\circ \times 60\ \text{arcmin}/1^\circ =$ 10.2 arcminutes. In other words, the angular separation of the headlights is about 10 arcminutes, which is about $\frac{1}{3}$ of the 30 arcminute (0.5°) angular diameter of the full moon.

• Why do the constellations we see depend on latitude and time of year?

If you stay in one place, the basic patterns of motion in the sky will stay the same from one night to the next. However, you'll see different constellations at different times of year. For example, the constellation Orion is prominent in the February evening sky but visible only shortly before dawn in September. In addition, you'll see a different set of constellations from those you see at home if you travel far to the north or south. For example, you'll never see the constellation Centaurus (see Figure 1.9) from the northern United States, but it will be circumpolar in your sky if you travel to southern Chile. Let's explore why the constellations depend on location and time of year.

Variation with Latitude To understand why the visible constellations vary with northward or southward travel, we must first review how we locate points on Earth (Figure 2.11a). **Latitude** measures north-south position on Earth and is defined to be 0° at the equator and increases northward and southward. The North Pole and South Pole have latitude 90°N and 90°S, respectively. Note that "lines of latitude" are imaginary *circles* running parallel to the equator. **Longitude** measures east-west position, so "lines of longitude" are semicircles (half-circles) extending from the North Pole to the South Pole. By international treaty, the line of longitude passing through Greenwich, England, is defined to be longitude 0°. This line is called the **prime meridian** (Figure 2.11b). Stating a latitude and a longitude pinpoints a location on Earth. For example, Rome lies at about 42°N latitude and 12°E longitude and Miami lies at about 26°N latitude and 80°W longitude.

You can see why visible constellations vary with latitude by looking again at Figure 2.10. If we move the person standing on Earth to a different latitude, it shifts the position of her horizon and zenith on the celestial sphere. For example, Figure 2.12 shows how the local sky appears relative to the celestial sphere for the latitudes of the North Pole (90°N) and Sydney, Australia (34°S).

Note that although the sky varies with latitude, it does *not* vary with longitude. For example, Charleston (South Carolina) and San Diego are at about the same latitude, so people in both cities see the same set of constellations at night. (However, because they are in different time zones,

COMMON MISCONCEPTIONS

Stars in the Daytime

Because we don't see stars in the daytime, some people believe that the stars vanish in the daytime and "come out" at night. In fact, the stars are always present. The reason your eyes cannot see stars in the daytime is that their dim light is overwhelmed by the bright daytime sky. You *can* see bright stars in the daytime with the aid of a telescope, and you may see stars in the daytime if you are fortunate enough to observe a total eclipse of the Sun. Astronauts can also see stars in the daytime. Above Earth's atmosphere, where no air is present to scatter sunlight through the sky, the Sun is a bright disk against a dark sky filled with stars. (However, because the Sun is so bright, astronauts must block its light and allow their eyes to adapt to darkness if they wish to see the stars.)

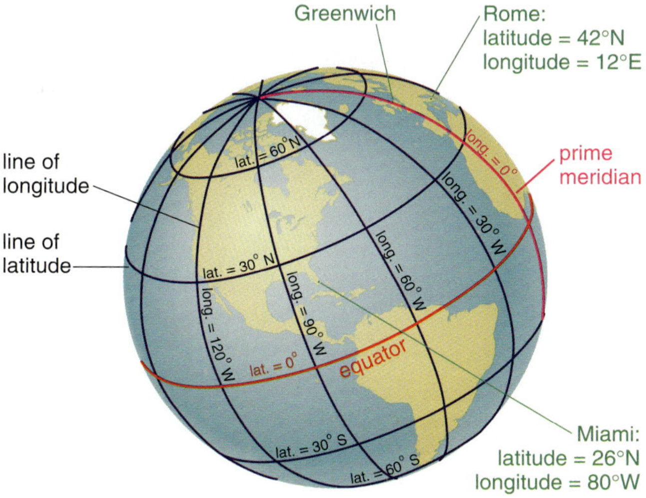

a Latitude measures angular distance north or south of the equator. Longitude measures angular distance east or west of the prime meridian, which passes through Greenwich, England.

b The entrance to the Old Royal Greenwich Observatory, near London. The line emerging from the door marks the prime meridian.

Figure 2.11 We can locate any place on Earth's surface by its latitude and longitude.

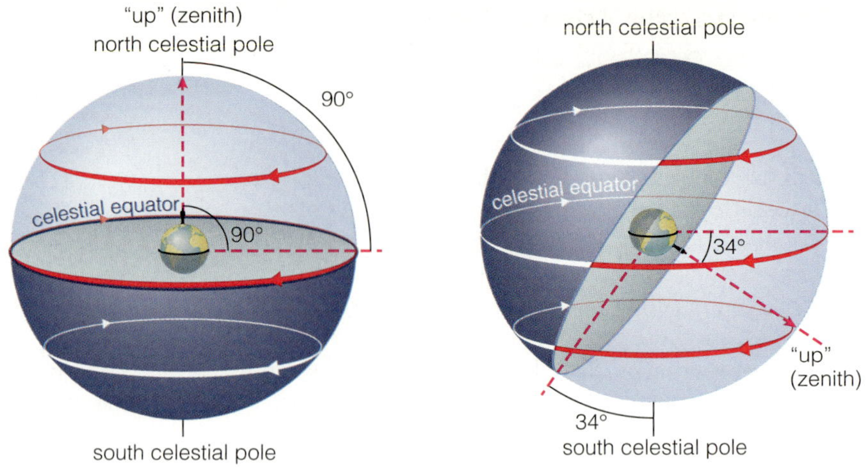

Figure 2.12 Interactive Figure The sky varies with latitude. Notice that the altitude of the celestial pole that is visible in your sky is always equal to your latitude.

a The local sky at the North Pole (latitude 90°N).

b The local sky at latitude 34°S.

night comes to Charleston about 3 hours before it comes to San Diego.)

If you study diagrams like those in Figures 2.10 and 2.12 carefully, you can discover many interesting facts about the sky. For example, notice that all stars visible at the North Pole are circumpolar, and you can see only the stars that lie on the northern half of the celestial sphere. These facts explain why the Sun remains above the horizon for 6 months at the North Pole: The Sun lies on the northern half of the celestial sphere for half of each year (see the yellow dots representing the Sun in Figure 2.4), so during these 6 months it must circle the North polar sky daily just like a circumpolar star.

The diagrams also show a fact that is very important to navigation:

> **The altitude of the celestial pole in your sky is equal to your latitude.**

For example, if the north celestial pole appears in your sky at an altitude of 40° above your north horizon, your latitude is 40°N. Similarly, if the south celestial pole appears in your sky at an altitude of 34° above your south horizon, your latitude is 34°S. Thus, you can determine your latitude just by finding the celestial pole in your sky. Finding the north celestial pole is fairly easy, because it lies very close to the star Polaris, also known as the North Star (Figure 2.13a).

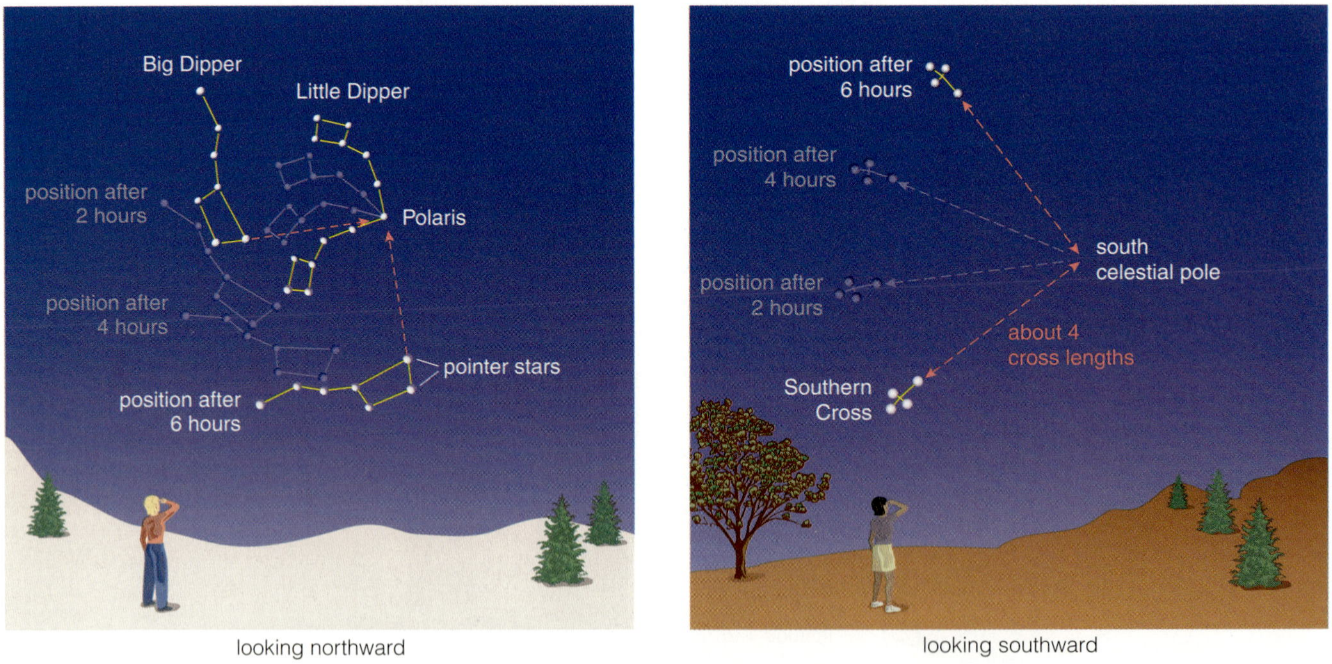

a In the Northern Hemisphere, the pointer stars of the Big Dipper point to the North Star, Polaris, which lies within 1° of the north celestial pole. Note that the sky appears to turn counterclockwise around the north celestial pole.

b In the Southern Hemisphere, the Southern Cross points to the south celestial pole, which is not marked by any bright star. The sky appears to turn clockwise around the south celestial pole.

Figure 2.13 Interactive Figure You can determine your latitude by measuring the altitude of the celestial pole in your sky.

In the Southern Hemisphere, you can find the south celestial pole with the aid of the Southern Cross (Figure 2.13b). We'll discuss celestial navigation and how the sky varies with latitude in more detail in Chapter S1.

THINK ABOUT IT

What is your latitude? Based on your latitude, where is the north (or south) celestial pole in your sky? Where should you look to see circumpolar stars? What portion of the celestial sphere is never visible in your sky?

Variation with Time of Year The night sky changes throughout the year because of Earth's changing position in its orbit around the Sun. Figure 2.14 shows how this works. As we orbit the Sun over the course of a year, the Sun *appears* to move against the background of the distant stars in the constellations. We don't see the Sun and the stars at the same time, but if we could we'd notice the Sun gradually moving eastward along the ecliptic, completing one circuit each year. The constellations along the ecliptic are called the constellations of the **zodiac**. (Tradition places 12 constellations along the zodiac, but the official borders include a wide swath of a thirteenth constellation, Ophiuchus.)

The Sun's apparent location along the ecliptic determines which constellations we see at night. For example, Figure 2.14 shows that the Sun appears to be in Leo in late August. We therefore cannot see Leo in late August, because it moves with the Sun through the daytime sky. However, we can see Aquarius all night long, since it is opposite Leo on the celestial sphere. Six months later, in February, we see Leo at night while Aquarius is above the horizon only in the daytime.

COMMON MISCONCEPTIONS

What Makes the North Star Special?

Most people are aware that the North Star, Polaris, is a special star. Contrary to a relatively common belief, however, it is *not* the brightest star in the sky. More than 50 other stars are either considerably brighter or comparable in brightness. Polaris is special because it is so close to the north celestial pole. This position makes it very useful in navigation, because it closely marks the direction of due north and because its altitude in your sky is nearly equal to your latitude.

THINK ABOUT IT

Based on Figure 2.14 and today's date, in what constellation does the Sun currently appear? What constellation of the zodiac will be on your meridian at midnight? What constellation of the zodiac will you see in the west shortly after sunset? Go outside at night to confirm your answers.

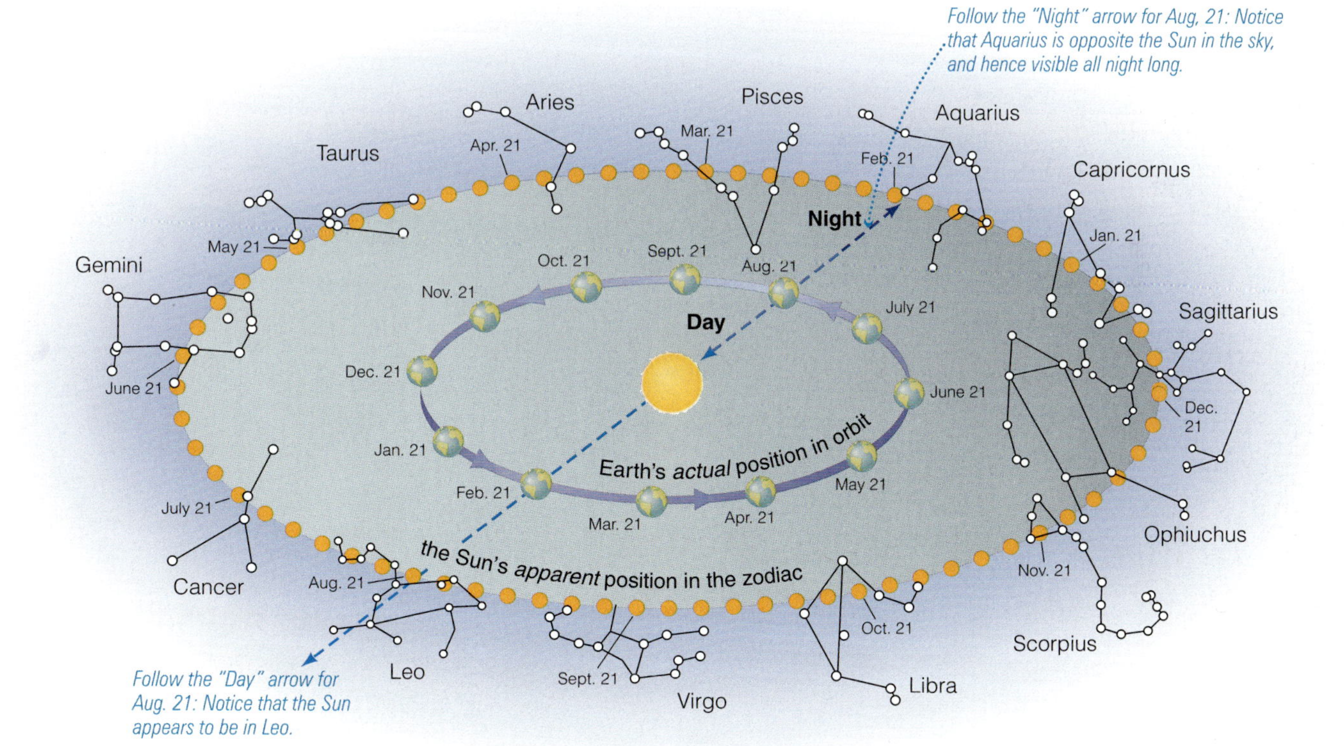

Figure 2.14 Interactive Figure This diagram shows why the Sun appears to move steadily eastward along the ecliptic, through the constellations of the zodiac. As Earth orbits the Sun, we see the Sun against the background of different zodiac constellations at different times of year. For example, on August 21 the Sun appears to be in Leo, because it is between us and the much more distant stars that make up Leo.

Seasons Tutorial, Lessons 1–3

2.2 The Reason for Seasons

We have seen how Earth's rotation makes the sky appear to circle us daily and how the night sky changes as Earth orbits the Sun each year. The combination of Earth's rotation and its orbit also leads to the progression of the seasons. In this section, we'll explore the reason for seasons.

• What causes the seasons?

You know that we have seasonal changes, such as longer and warmer days in summer and shorter and cooler days in winter. But why do the seasons occur? The answer is that the tilt of Earth's axis causes sunlight to fall differently on Earth at different times of year. Figure 2.15 shows the idea.

Notice that Earth's axis remains pointed in the same direction in space (toward Polaris) throughout the year [Section 1.3]. However, because Earth orbits the Sun, the orientation of the axis *relative to the Sun* changes over the course of a year. For example, look at the left side of Figure 2.15. The Northern Hemisphere is tipped toward the Sun, making it summer there, while the Southern Hemisphere is tipped away from the Sun, making it winter there. Half an orbit later (the right side of Figure 2.15), the situation is reversed. The Northern Hemisphere is in winter because it is tipped away from the Sun, while the Southern Hemisphere is in summer because it is tipped toward the Sun. Thus, the two hemispheres experience opposite seasons.

We can also see why the days are longer and warmer in summer and shorter and cooler in winter. Sunlight strikes the summer hemisphere at a steeper angle than it strikes the winter hemisphere. The steeper angle means sunlight is more concentrated, which is why summer tends to be warmer than winter. The steeper angle also means that the Sun follows a longer and higher path through the summer sky, which is why the days are long and midday shadows are short. You can see the annual change in how high the Sun rises in your sky by observing the Sun's position at the same time each day (Figure 2.16).

Notice that the seasons on Earth are caused only by the axis tilt and *not* by any change in Earth's distance from the Sun. Although Earth's orbital distance varies over the course of each year, the variation is fairly small: Earth is only about 3% farther from the Sun at its farthest point than at

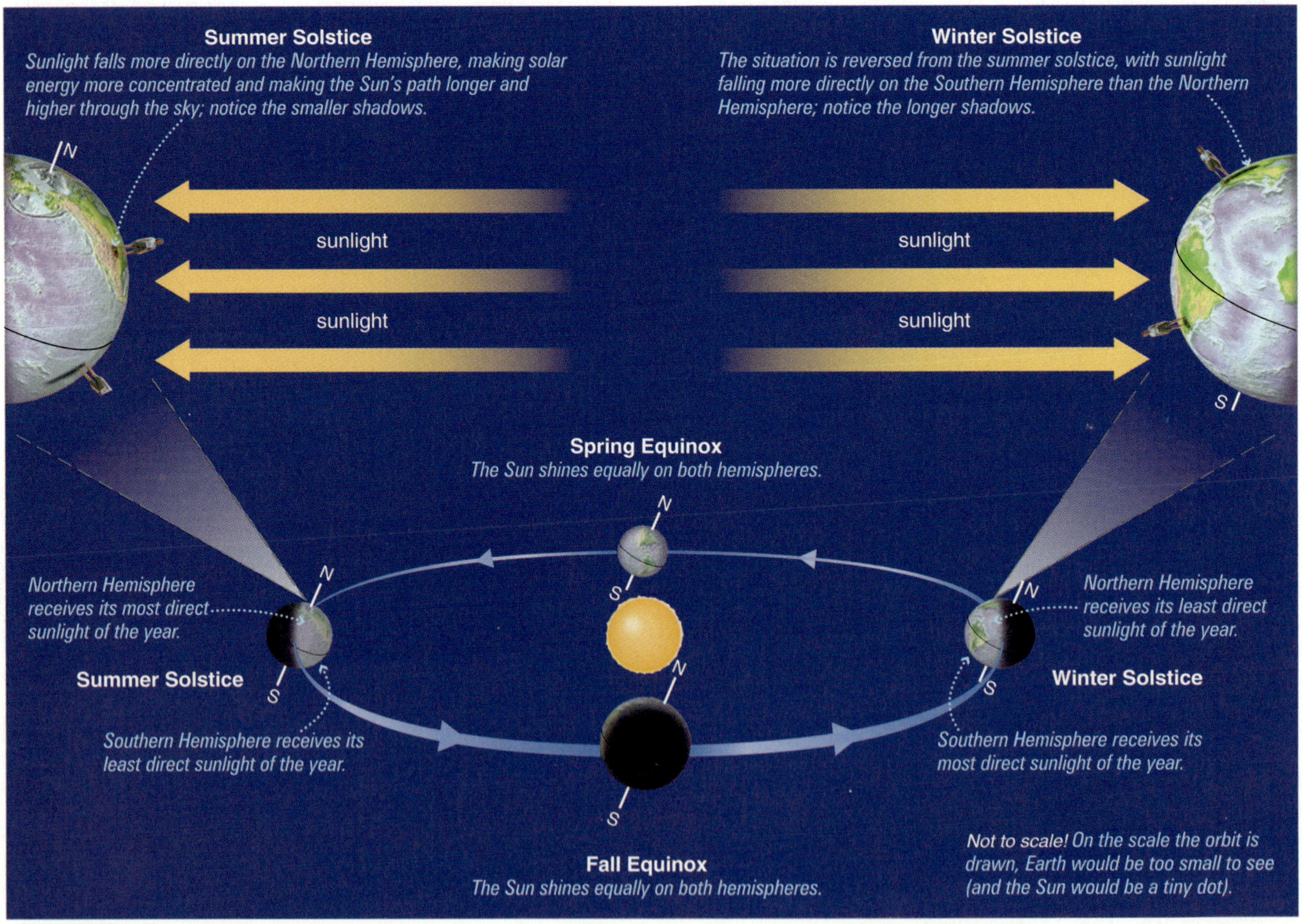

Figure 2.15 Interactive Figure Earth's seasons are caused by the tilt of its axis. Notice that the axis points in the same direction (toward Polaris) throughout the year, which means the Northern Hemisphere is tipped toward the Sun on one side of the orbit and away from the Sun on the other side of the orbit. The same is true for the Southern Hemisphere, but on opposite sides of the orbit.

Figure 2.16 This composite photograph shows midday images of the Sun at 7- to 11-day intervals over the course of a year, always from the same spot (the Parthenon in Athens, Greece) and at the same time of day (technically, at the same "mean solar time"). Notice the dramatic change in the Sun's midday altitude over the course of the year. We'll discuss the reasons for the "figure 8" (called an *analemma*) in Chapter S1.

its nearest. The difference in the strength of sunlight due to this small change in distance is easily overwhelmed by the effects caused by the $23\frac{1}{2}°$ axis tilt. (*Note:* Some people also misinterpret diagrams like Figure 2.15, getting the mistaken impression that orbital position can make one hemisphere closer to the Sun. The mistake happens because figures like this are not drawn to scale. If the figure were drawn to scale, Earth would be microscopic, showing that there can be no practical difference in the distances of the two hemispheres from the Sun.)

THINK ABOUT IT

Jupiter has an axis tilt of about 3°, small enough to be insignificant. Saturn has an axis tilt of about 27°, or slightly greater than that of Earth. Both planets have nearly circular orbits around the Sun. Do you expect Jupiter to have seasons? Do you expect Saturn to have seasons? Explain.

• How do we mark the progression of the seasons?

Today, we use a calendar to mark the progression of the seasons. In ancient times, however, people tracked the seasons by observing the Sun's changing position in the sky. You can see these changes for yourself. For example, in summer the Sun traces a long, high arc across the sky, while in winter its path is short and low. Let's explore how the Sun's changing path allows us to mark the changing of the seasons.

COMMON MISCONCEPTIONS

The Cause of Seasons

When asked what causes the seasons, many people mistakenly answer that the seasons are caused by variations in Earth's distance from the Sun. By recognizing that the Northern and Southern Hemispheres experience opposite seasons, you'll see that Earth's varying distance from the Sun *cannot* be the cause of the seasons. If that were the case, then both hemispheres would have summer at the same time. The real cause of the seasons is Earth's axis tilt, which causes the Northern and Southern Hemispheres to alternately receive more or less direct sunlight.

Solstices and Equinoxes To help us mark the changing seasons, we define four special moments in the year, each of which corresponds to one of the four special positions in Earth's orbit shown in Figure 2.15.

- The **summer solstice**, which occurs around June 21 each year, is the moment when the Northern Hemisphere receives its most direct sunlight (and the Southern Hemisphere gets its least direct sunlight).
- The **winter solstice**, which occurs around December 21, is the moment when the Northern Hemisphere receives its least direct sunlight.
- The **spring equinox** (or *vernal equinox*) occurs around March 21, at the moment when the Northern Hemisphere goes from being tipped slightly away from the Sun to being tipped slightly toward the Sun.
- The **fall equinox** (or *autumnal equinox*) occurs around September 22 and marks the opposite change, when the Northern Hemisphere first starts to be tipped away from the Sun.

The exact dates and times of the solstices and equinoxes vary slightly from year to year, but stay within a couple of days of the dates given here. In fact, our modern calendar was designed specifically to ensure that the solstices and equinoxes occur around the same dates each year. As we'll discuss in Chapter S1, the calendar keeps this correspondence between dates and seasons by including a leap year in almost but not quite every four-year period. The pattern, set in the late 1500s, makes the average length of the calendar year almost precisely equal to the time between one spring equinox and the next.

Identifying the Solstices and Equinoxes Ancient people recognized the days on which the solstices and equinoxes occurred by observing the Sun in the sky. Many ancient structures were used for this purpose, including Stonehenge in England and the Sun Dagger in New Mexico [Section 3.1].

The equinoxes occur on the only two days of the year on which the Sun rises precisely due east and sets precisely due west (Figure 2.17). These are also the only two days when sunlight falls equally on both hemispheres. The summer solstice occurs on the day that the Sun follows its longest and highest path through the Northern Hemisphere sky (and its shortest and lowest path through the Southern Hemisphere sky). It is therefore the day that the Sun rises and sets farther to the north than on any other day of the year, and on which the noon Sun reaches its highest point in the Northern Hemisphere sky. The opposite is true on the day of the winter solstice, when the Sun rises and sets farthest to the south and the noon Sun is lower in the Northern Hemisphere sky than on any other day of the year.

First Days of Seasons We usually say that each equinox and solstice marks the first day of a season. For example, the day of the summer solstice is usually said to be the "first day of summer." Notice, however, that the summer solstice occurs when the Northern Hemisphere has its *maximum* tilt toward the Sun. Thus, you might wonder why we consider the summer solstice to be the beginning rather than the midpoint of summer.

In part, the choice of the summer solstice as the "first" day of summer is somewhat arbitrary. However, the choice makes sense in at least two ways. First, it was much easier for ancient people to identify the days on which the Sun reached extreme positions in the sky—such as when it reached its highest point on the summer solstice—than other days in between. Second, we usually think of the seasons in terms of weather, and the solstices and equinoxes correspond quite well with the beginnings of seasonal weather patterns. For example, although the Sun's path through the Northern Hemisphere sky is longest and highest around the time of the summer solstice, the warmest days tend to come one to two months later. To understand why, think about what happens when you heat a pot of cold soup. Even though you may have the stove turned on high from the start, it takes a while for the soup to warm up. In the same way, it takes some time for sunlight to heat the ground and oceans from the cold of winter to the warmth of summer. Thus, "midsummer" in terms of weather comes in late July and early August, which makes the summer solstice a pretty good choice for the "first day of summer." For similar reasons, the winter solstice is a good choice for the first day of winter, and the spring and fall equinoxes are good choices for the first days of those seasons.

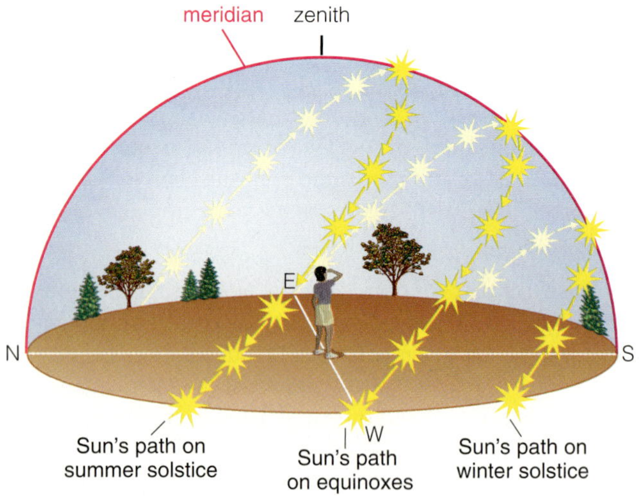

Figure 2.17 This diagram shows the Sun's path on the solstices and equinoxes for the Northern Hemisphere sky (latitude 40°N). Notice that the Sun rises exactly due east and sets exactly due west only on the equinoxes. The summer solstice occurs on the day that the Sun rises and sets farthest to the north and reaches its highest point in the sky. The winter solstice occurs on the day that the Sun rises and sets farthest to the south and traces its lowest path through the sky.

Seasons Around the World The names of the solstices and equinoxes reflect the northern seasons, which can make things sound strange when we talk about seasons in the Southern Hemisphere. For example, on the *summer* solstice it is *winter* in the Southern Hemisphere. This apparent injustice to people in the Southern Hemisphere arose because the solstices and equinoxes were named long ago by people living in the Northern Hemisphere. A similar injustice is inflicted on people living in equatorial regions. If you study Figure 2.14 carefully, you'll see that Earth's equator gets its most direct sunlight on the two equinoxes and its least direct sunlight on the solstices. Thus, people living near the equator don't experience four seasons in the same way as people living at mid-latitudes. Instead, equatorial regions generally have rainy and dry seasons, with the rainy seasons coming when the Sun is higher in the sky.

In addition, seasonal variations around the times of the solstices are more extreme at high latitudes. For example, Alaska has much longer summer days and much longer winter nights than Florida. In fact, the Sun becomes circumpolar at very high latitudes (within the *Arctic* and *Antarctic Circles*) in the summer. The Sun never sets during these summer days in what we call the *land of the midnight Sun*

COMMON MISCONCEPTIONS

High Noon

When is the Sun directly overhead in your sky? Many people answer "at noon." It's true that the Sun reaches its *highest* point each day when it crosses the meridian, giving us the term "high noon" (though the meridian crossing is rarely at precisely 12:00; see Section S1.2). However, unless you live in the Tropics (between latitudes 23.5°S and 23.5°N), the Sun is *never* directly overhead. In fact, any time you can see the Sun as you walk around, you can be sure it is *not* at your zenith. Unless you are lying down, seeing an object at the zenith requires tilting your head back into a very uncomfortable position.

Figure 2.18 This sequence of photos shows the progression of the Sun all the way around the horizon on the summer solstice at the Arctic Circle. Notice that the Sun does not set but instead skims the northern horizon at midnight. It then gradually rises higher, reaching its highest point at noon, when it appears due south.

(Figure 2.18). Of course, the name "land of noon darkness" would be more appropriate in the winter, when the Sun never rises above the horizon at these high latitudes.

Why Orbital Distance Doesn't Affect Our Seasons As we've discussed, Earth's distance from the Sun varies slightly over the course of each year, but not enough to be a factor in causing the seasons. Still, we might expect the varying orbital distance to make the seasons slightly more extreme in one or the other hemisphere. For example, Earth is closest to the Sun (perihelion) in early January and farthest from the Sun (aphelion) in early July, which means that the Northern Hemisphere receives slightly more intense winter sunlight and less intense summer sunlight than does the Southern Hemisphere. This "extra" winter sunlight should tend to make the Northern Hemisphere winter less cold, while the reduced summer sunlight should tend to make the Northern Hemisphere summer less hot. Thus, we would expect the Northern Hemisphere to have slightly more moderate seasons than the Southern Hemisphere. However, weather records show that the opposite is true: Northern Hemisphere seasons are slightly more extreme than those of the Southern Hemisphere.

The main reason for this surprising fact becomes obvious when you look at a map of Earth (Figure 2.19). Most of Earth's land lies in the Northern Hemisphere, with far more ocean in the Southern Hemisphere. As you can see at any beach, lake, or pool, water takes longer to heat or cool than soil or rock (largely because sunlight heats bodies of water to a depth of many meters while heating only the very top layer of land). The water temperature therefore remains fairly steady both day and night, while the ground can heat up and cool down dramatically. The Southern Hemisphere's larger amount of ocean moderates its climate. The Northern Hemisphere, with more land and less ocean, heats up and cools down more easily, explaining why it has the more extreme seasons.

A secondary reason for the Northern Hemisphere's more extreme seasons is that Earth moves slightly faster in its orbit when it is closer to the Sun and slightly slower when it is farther from the Sun (a fact embodied in Kepler's second law [Section 3.3]). Thus, Earth moves slightly slower during the Northern Hemisphere summer (when Earth is farther from the Sun), which makes the northern summer last about 2–3 days longer than the southern summer. A couple of extra days of more direct sunlight means more time to heat up.

In summary, Earth's variation in distance from the Sun makes virtually no difference at all to the seasons, because other effects are far more important. However, the same is not necessarily true for other planets, especially if they

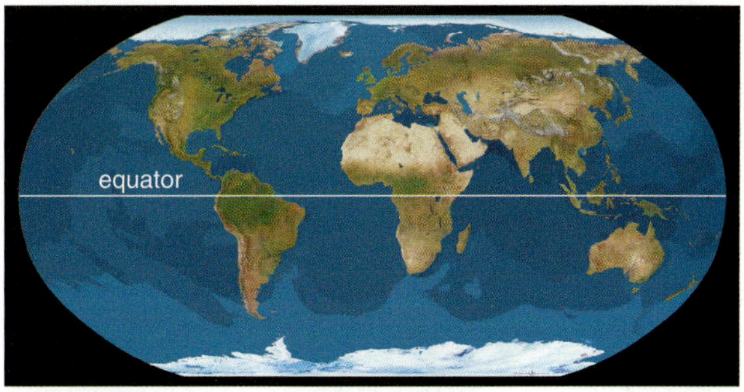

Figure 2.19 A map of Earth shows that most land lies in the Northern Hemisphere while most ocean lies in the Southern Hemisphere. The climate-moderating effects of water therefore make Southern Hemisphere seasons less extreme than Northern Hemisphere seasons.

have significantly greater distance variations. For example, Mars varies in distance from the Sun much more than does Earth. As a result, Mars has significantly more extreme seasons in one hemisphere than the other, even though it has nearly the same axis tilt as Earth [Section 10.4].

• How does the orientation of Earth's axis change with time?

We have now discussed both daily and seasonal changes in the sky. These changes are all you are likely to notice in your daily life. However, if you could observe the sky over many centuries, you would begin to notice other changes in its appearance. One of the most important of these slow changes is called **precession**, which gradually changes the orientation of Earth's axis in space.

Precession occurs with many rotating objects. You can see it easily by spinning a top (Figure 2.20a). As the top spins rapidly, you'll notice that its axis sweeps out a circle at a somewhat slower rate. We say that the top's axis *precesses.* Earth's axis precesses in much the same way, but far more slowly (Figure 2.20b). Each cycle of Earth's precession takes about 26,000 years. Today, the axis points toward Polaris, making it our North Star. However, because of precession, Polaris has been a good North Star for only a few centuries, and a few centuries from now it will no longer be a good marker for due north. Some 13,000 years from now, the axis will point toward the star Vega (within a few degrees), making Vega the North Star at that time. In 26,000 years, the axis will again point the way it does now. At most other times, the axis will not point near any bright star.

Notice that while precession affects the orientation of Earth's axis, it does not affect the *amount* of the axis tilt and therefore does not affect the pattern of the seasons. However, because the solstices and equinoxes correspond to points in Earth's orbit that depend on the direction the axis points in space, their positions in the orbit gradually shift with the cycle of precession. As a result, the constellations associated with the solstices and equinoxes gradually change. For example, a couple thousand years ago the Sun appeared in the constellation Cancer on the day of the summer solstice, but it now appears in Gemini on that day. That explains something you can see on any world map: The latitude at which the Sun is directly overhead on the summer solstice ($23\frac{1}{2}°$ N) is called the *Tropic of Cancer,* telling us that it got its name back when the Sun used to appear in Cancer on the summer solstice.

THINK ABOUT IT

What constellation will the Sun be in on the summer solstice about 2,000 years from now? (*Hint:* Figure 2.14 shows the names and order of the zodiac constellations.)

Why does precession occur? It is caused by gravity's effect on a tilted, rotating object that is *not* a perfect sphere. You have probably seen how gravity affects a top. If you try to balance a nonspinning top on its point, it will fall over almost immediately. This happens because a top that is not spherical will inevitably be leaning a little to one side. No matter how slight this lean, gravity will quickly tip the nonspinning top over. However, if you spin the top rapidly, it does not fall over so easily. The spinning top stays upright because rotating objects tend to keep spinning around the same rotation axis (a consequence of something known as the *law of conservation of angular momentum* [Section 4.3]). This tendency prevents gravity from immediately pulling the spinning top over, since falling over would mean a change

Figure 2.20 Interactive Figure Precession affects the orientation of a spinning object's axis but not the amount of its tilt.

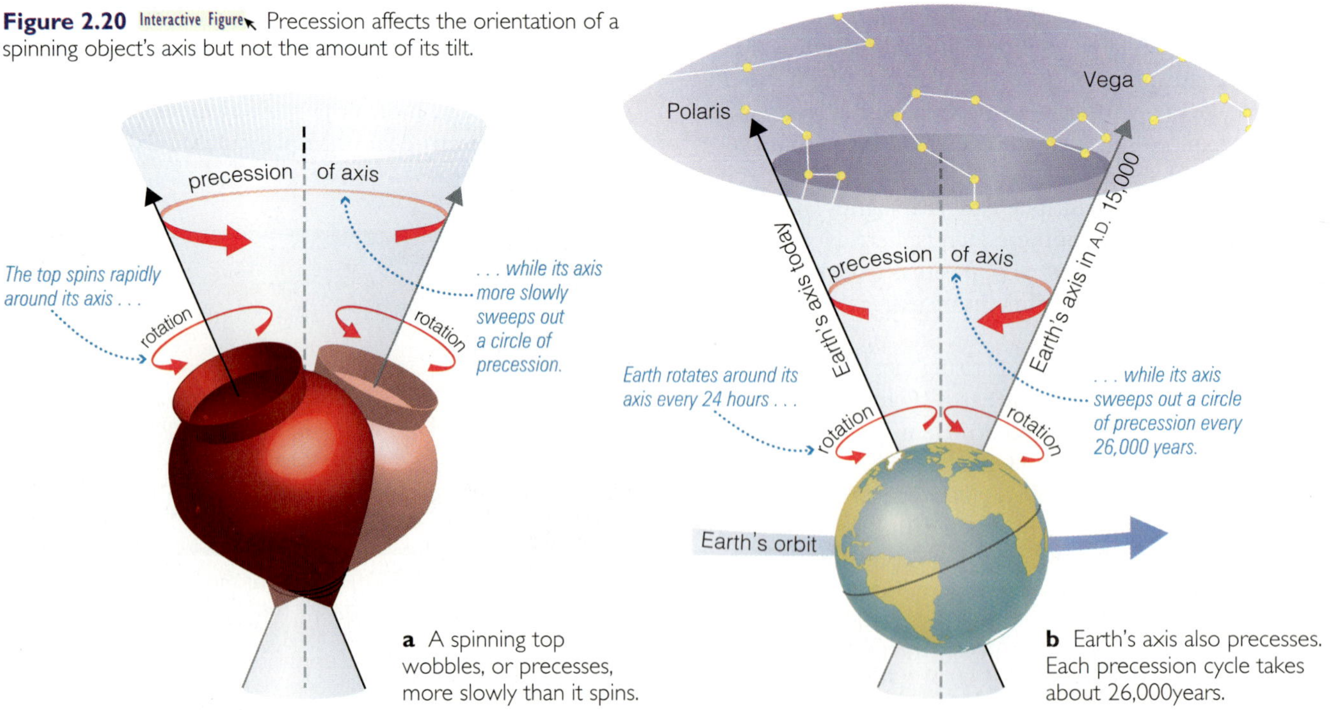

a A spinning top wobbles, or precesses, more slowly than it spins.

b Earth's axis also precesses. Each precession cycle takes about 26,000years.

COMMON MISCONCEPTIONS

Sun Signs

You probably know your astrological "Sun sign." When astrology began a few thousand years ago, your Sun sign was supposed to represent the constellation in which the Sun appeared on your birth date. However, because of precession, this is no longer the case for most people. For example, if your birthday is the spring equinox, March 21, a newspaper horoscope will show that your Sun sign is Aries but the Sun actually appears in Pisces on that date. The astrological Sun signs are based on the positions of the Sun among the stars described by the Greek scientist Ptolemy in his book *Tetrabiblios* [Section 3.5], which was written in about A.D. 150. Because almost 2,000 years have passed since that time, Earth's axis has moved about $\frac{1}{13}$ of the way through its 26,000-year precession cycle since the Sun signs got their names and dates. As a result, the astrological Sun signs are off by nearly a month from the actual positions of the Sun among the constellations today.

in the spin axis from near-vertical to horizontal. Instead, gravity succeeds only in making the axis trace circles of precession. As friction slows the top's spin, the circles of precession get wider and wider, and ultimately the top falls over. If there were no friction to slow its spin, the top would spin and precess forever.

The rotating Earth precesses because of gravitational tugs from the Sun and Moon. Earth is not quite a perfect sphere, instead bulging slightly at its equator. Because the equator is tilted $23\frac{1}{2}°$ to the ecliptic plane, the gravitational attractions of the Sun and Moon try to pull the equatorial bulge into the ecliptic plane, effectively trying to "straighten out" Earth's axis tilt. However, like the spinning top, Earth tends to keep rotating around the same axis. Gravity therefore does not succeed in changing Earth's axis tilt and instead only makes the axis precess. If you want to gain a better understanding of precession and how it works, you might wish to purchase a simple toy *gyroscope.* Gyroscopes are essentially rotating wheels mounted in a way that allows them to move freely, which makes it easy to see how their spin rate affects their motion. (The fact that gyroscopes tend to keep the same rotation axis makes them very useful in aircraft and spacecraft navigation.)

MasteringASTRONOMY Phases of the Moon Tutorial, Lessons 1–3

2.3 The Moon, Our Constant Companion

Aside from the Sun, the Moon is the brightest and most noticeable object in our sky. But the Moon behaves in some ways that may at first seem rather strange. Sometimes it is visible in the daytime and sometimes only at night. Its appearance also changes, so that sometimes we see less of it and sometimes we see more of it. Occasionally, the Moon's position is just right to cause a lunar or solar eclipse. The Moon behaves in these special ways because it is our constant companion in space, orbiting Earth as we orbit the Sun. In this section, we'll discuss the how and why of the Moon's appearance in our sky.

Before we get into the details, it's helpful to know a few basic facts about the Moon. The Moon orbits Earth about once every $27\frac{1}{3}$ days. Its average distance from Earth is about 380,000 kilometers (240,000 miles), but the precise distance varies because the orbit is not quite a perfect circle. Figure 2.21 shows the Moon's orbit on the same scale we used for the model solar system in Section 1.2. Remember that on this scale, the Sun is about the size of a large grapefruit and is located about 15 meters from Earth. Thus, the entire orbit of the Moon would fit easily inside the Sun, and for practical purposes we can consider the Earth and Moon to share the same orbit around the Sun.

Like all objects in space, the Moon appears to reside on the celestial sphere. Earth's daily rotation makes the Moon appear to rise in the east and set in the west each day. In addition, because of its orbit around Earth, the Moon appears to move eastward from night to night through the constellations of the zodiac. Each circuit through the constellations takes the same $27\frac{1}{3}$ days that the Moon takes to orbit Earth. If you do the math, you'll see that this means the Moon moves relative to the stars by about $\frac{1}{2}°$—its own

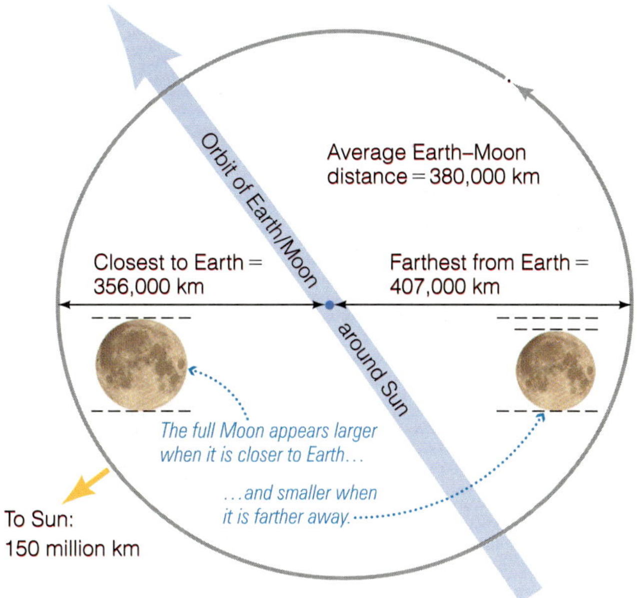

Figure 2.21 The Moon's orbit around Earth, shown on the 1-to-10-billion scale used in Section 1.2 (see Figure 1.6). The segment shown of our orbit around the Sun looks nearly straight because the distance to the Sun is so great in comparison to the size of the Moon's orbit. The inset photos contrast the relative angular size of the full moon in our sky when the Moon is at the near and far points of its orbit; of course, full moon occurs only when the Moon is opposite the Sun as seen from Earth.

angular size—each hour. You can notice this gradual motion in just a few hours by checking the Moon's position in comparison to bright stars near it in the sky.

• Why do we see phases of the Moon?

As the Moon moves through the sky, both its appearance and the times at which it rises and sets change with the cycle of **lunar phases**. The phase of the Moon on any given day depends on its position relative to the Sun as it orbits Earth.

The easiest way to understand the lunar phases is with the simple demonstration illustrated in Figure 2.22. Take a ball outside on a sunny day. (If it's dark or cloudy, you can use a flashlight instead of the Sun; put the flashlight on a table a few meters away and shine it toward you.) Hold the ball at arm's length to represent the Moon while your head represents Earth. Slowly spin around (counterclockwise), so that the ball goes around you just like the Moon orbits Earth. As you turn, you'll see the ball go through phases just like the Moon. If you think about what's happening, you'll realize that the phases of the ball result from just two basic facts:

1. Half the ball always faces the Sun (or flashlight) and therefore is bright, while the other half faces away from the Sun and therefore is dark.
2. As you look at the ball at different positions in its "orbit" around your head, you see different combinations of its bright and dark faces.

For example, when you hold the ball directly opposite the Sun, you see only the bright portion of the ball, which represents the "full" phase. When you hold the ball at its "first-quarter" position, half the face you see is dark and the other half is bright.

We see lunar phases for the same reason. Half the Moon is always illuminated by the Sun, but the amount of this illuminated half that we see from Earth depends on the Moon's position in its orbit. The photographs in Figure 2.22 show how the phases look. Each complete cycle of phases, from one new moon to the next, takes about $29\frac{1}{2}$ days—hence the origin of the word *month* (think "moonth"). This is slightly longer than the Moon's actual orbital period because of Earth's motion around the Sun during the time the Moon is orbiting around Earth [see Figure S1.3].

The different phases not only look different but also rise, reach their highest points, and set at different times. For example, because full moon occurs when the Moon is

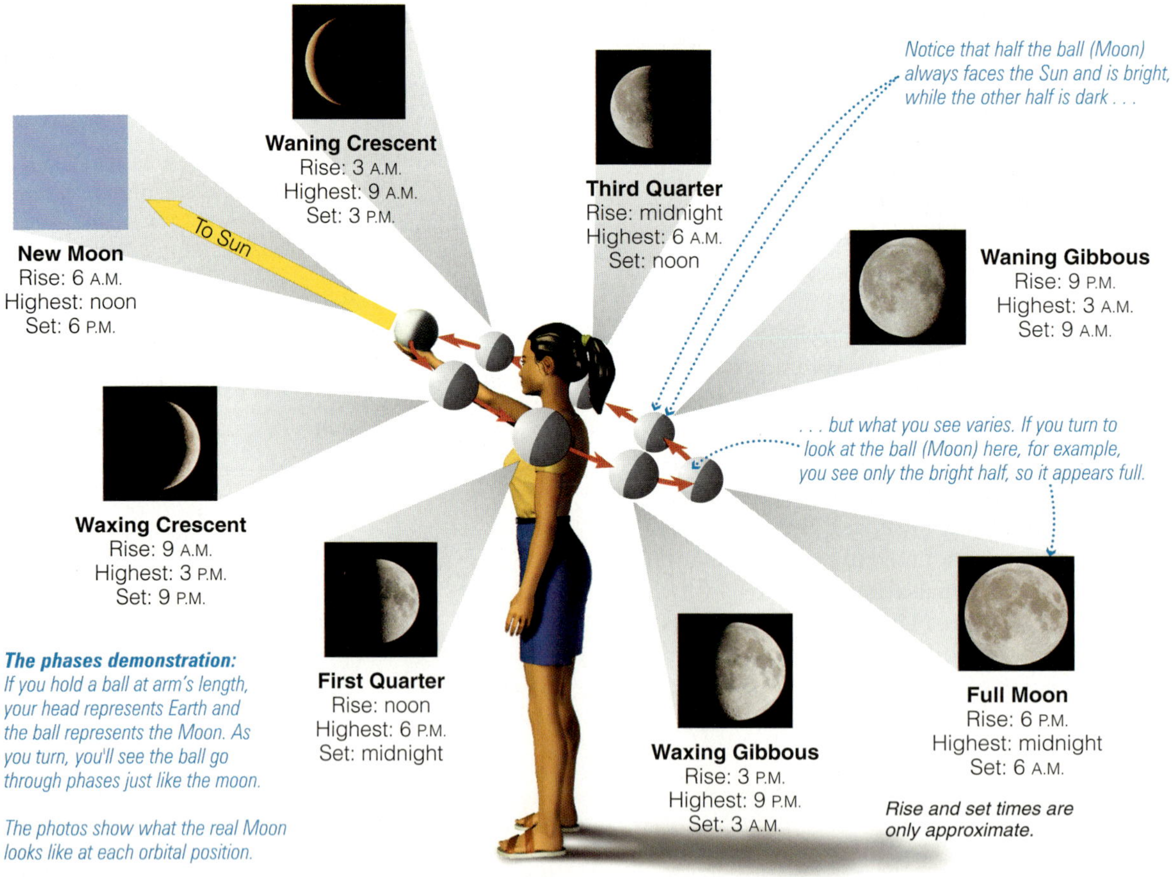

Figure 2.22 Interactive Figure A simple demonstration illustrates the phases of the Moon. The half of the ball (Moon) facing the Sun is always illuminated while the half facing away is always dark, but you see the ball go through phases as it orbits around your head (Earth). (The new moon photo shows blue sky, because a new moon is always close to the Sun in the sky and hence hidden from view by the bright light of the Sun.)

opposite the Sun in the sky, the full moon must rise around sunset, reach its highest point in the sky at midnight, and set around sunrise. Similarly, because first-quarter moon occurs when the Moon is about 90° east of the Sun in our sky, it must rise around noon, reach its highest point around 6 P.M., and set around midnight. Figure 2.22 lists the approximate rise, highest point, and set times for each phase. (The exact times vary with latitude, time of year, and other factors.)

THINK ABOUT IT

Suppose you go outside in the morning and notice that the visible face of the Moon is half-light and half-dark. Is this a first-quarter or third-quarter moon? How do you know?

As you study the phases in Figure 2.22, it's worth noting at least three interesting things about their names. First, notice that the phases from new to full are said to be *waxing*, which means "increasing"; the phases from full to new are said to be *waning*, which means "decreasing." Second, notice that there's no phase called a "half moon." Instead, what most people think of as a "half moon" is either a first-quarter or third-quarter phase. These phases get their names because the cycle of phases is usually taken to begin with new moon. Thus, a first-quarter moon occurs when the Moon is one-quarter of the way through its cycle, and a third-quarter moon occurs at the three-quarters point of the cycle. Third, while nearly everyone is familiar with the term *crescent* for the waxing and waning crescent phases, few people know the name *gibbous* (pronounced with a hard *g* as in "gift") for the waxing and waning gibbous phases. Notice that a gibbous moon is essentially the opposite of a crescent moon—a crescent moon has a small sliver of light while a gibbous moon has a small sliver of dark. The term *gibbous* literally means "hump-backed," so you can probably see how the gibbous moon gets its name. We see the two gibbous phases just as often as we see the crescent phases, so it's worth knowing their names.

The Moon's Synchronous Rotation Although we see many *phases* of the Moon, we do not see many *faces*. In fact, from Earth we always see the same face of the Moon.* This fact tells us that the Moon must rotate once on its axis in the same time that it makes a single orbit of Earth, a trait that we call **synchronous rotation** (because the periods of rotation and orbit are the same, or synchronized).

To see why the Moon must rotate synchronously to keep the same face toward Earth, try the following demonstration. Place a ball on a table to represent Earth while you represent the Moon. Start by facing the ball. If you do not rotate as you walk around the ball, you'll be looking away from it by the time you are halfway around your orbit (Figure 2.23a). The only way you can face the ball at all times is by completing exactly one rotation while you complete one orbit (Figure 2.23b). (We'll learn in Chapter 5 *why* the Moon's periods of rotation and orbit are the same.)

The View from the Moon A good way to solidify your understanding of the lunar phases is to imagine that you live on the side of the Moon that faces Earth. For example, ask yourself what you would see as you looked at Earth at a time when people on Earth saw a new moon. By remembering that a new moon occurs when the Moon is between the Sun and Earth, you'll realize that from the Moon you'd be looking at Earth's daytime side and hence would see a *full earth*. Similarly, at full moon you would be facing the night side of Earth and would see a *new earth*. In general, you'd always see Earth in a phase opposite the phase of the Moon being seen by people on Earth at the same time. Moreover, because the Moon always shows nearly the same face to Earth, Earth would appear to hang nearly stationary in your sky as it went through its cycle of phases.

THINK ABOUT IT

How long would each day and night last if you lived on the Moon? Explain.

*Because the Moon's orbital speed varies while its rotation rate is steady, the visible face appears to wobble slightly back and forth as it orbits Earth. This effect, called *libration*, allows us to see a total of about 59% of the Moon's surface over the course of a month, even though we see only 50% of the Moon at any single time.

a If you do not rotate while walking around the model, you will not always face it.

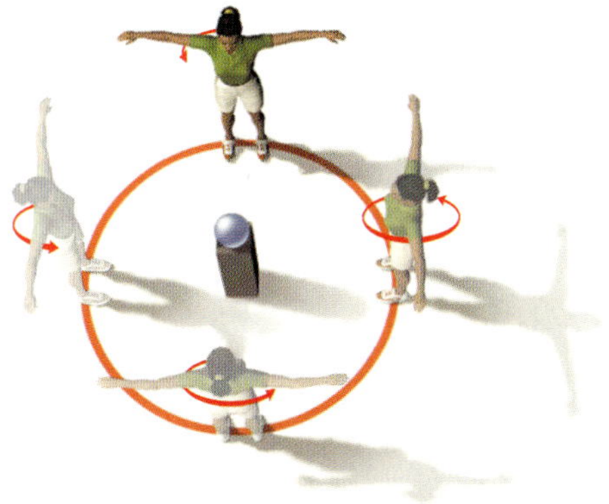

b You will face the model at all times only if you rotate exactly once during each orbit.

Figure 2.23 The fact that we always see the same face of the Moon means that the Moon must rotate once in the same amount of time that it takes to orbit Earth once. You can see why by walking around a model of Earth while imagining that you are the Moon.

COMMON MISCONCEPTIONS

Moon in the Daytime and Stars on the Moon

In traditions and stories, night is so closely associated with the Moon that many people mistakenly believe that the Moon is visible only in the nighttime sky. In fact, the Moon is above the horizon as often in the daytime as at night, though it is easily visible only when its light is not drowned out by sunlight. For example, a first-quarter moon is easy to spot in the late afternoon as it rises through the eastern sky, and a third-quarter moon is visible in the morning as it heads toward the western horizon.

Another misconception appears in illustrations that show a star in the dark portion of the crescent moon. The star in the dark portion appears to be in front of the Moon, which is impossible because the Moon is much closer to us than is any star.

COMMON MISCONCEPTIONS

The "Dark Side" of the Moon

The term *dark side of the Moon* really should be used to mean the night side—that is, the side facing away from the Sun. Unfortunately, *dark side* traditionally meant what would better be called the *far side*—the face that never can be seen from Earth. Many people still refer to the far side as the "dark side," even though this side is not necessarily dark. For example, during new moon the far side faces the Sun and hence is completely sunlit (see Figure 2.22). The only time the far side is completely dark is at full moon, when it faces away from both the Sun and Earth.

Thinking about the view from the Moon clarifies another interesting feature of the lunar phases: The dark portion of the lunar face is not *totally* dark. Imagine that you are standing on the Moon when it is in a crescent phase. Because it's nearly new moon as seen from Earth, you would see nearly full earth in your sky. Just as we can see at night by the light of the Moon, the light of Earth would illuminate your night moonscape. (In fact, because Earth is much larger than the Moon, the full earth is much bigger and brighter in the lunar sky than the full moon is in Earth's sky.) This faint light illuminating the "dark" portion of the Moon's face is often called the *ashen light* or *earthshine.* This light enables us to see the outline of the full face of the Moon even when the Moon is not full.

Eclipses Tutorial, Lessons 1–3

• What causes eclipses?

The phases of the Moon are the most obvious effect of the Moon's orbit around Earth. Occasionally, however, we see something much more dramatic. The Moon and Earth cast shadows in sunlight, and these shadows can create **eclipses** when the Sun, Earth, and Moon fall into a straight line. There are two basic types of eclipse:

- A **lunar eclipse** occurs when Earth lies directly between the Sun and Moon, so that Earth's shadow falls on the Moon.
- A **solar eclipse** occurs when the Moon lies directly between the Sun and Earth, so that the Moon's shadow falls on the Earth. People living within the area covered by the Moon's shadow will see the Sun blocked or partially blocked from view.

To understand when and why eclipses occur, we must investigate the Moon's orbit of Earth a little more deeply.

Conditions for Eclipses Look once more at Figure 2.22. The figure makes it look like the Sun, Earth, and Moon line up with every new and full moon. Thus, if this figure told the whole story of the Moon's orbit, we would have both a lunar and a solar eclipse every month—but we don't.

The missing piece of the story in Figure 2.22 is that the Moon's orbit is slightly inclined (by about 5°) to the ecliptic plane (the plane of Earth's orbit around the Sun). To visualize this inclination, imagine the ecliptic plane as the surface of a pond, as shown in Figure 2.24. Because of the inclination of its orbit, the Moon spends most of its time either above or below this surface. It crosses *through* this surface only twice during each orbit: once coming out and once going back in. The two points in each orbit at which the Moon crosses the surface are called the **nodes** of the Moon's orbit.

As you study Figure 2.24, notice that the nodes are aligned approximately the same way (diagonally on the page) throughout the year. As a result, the nodes lie in a straight line with the Sun and Earth only about twice each year. (For reasons we'll discuss shortly, it is not *exactly* twice each year.) Thus, we find the following two conditions that must be met for an eclipse to occur:

1. The phase of the Moon must be full (for a lunar eclipse) or new (for a solar eclipse), since those are the only phases at which the Sun, Earth, and Moon can lie in a straight line.
2. The new or full moon must occur during one of the periods when the nodes of the Moon's orbit are aligned with the Sun and Earth, since those are the only times when the new and full moons lie in the ecliptic plane.

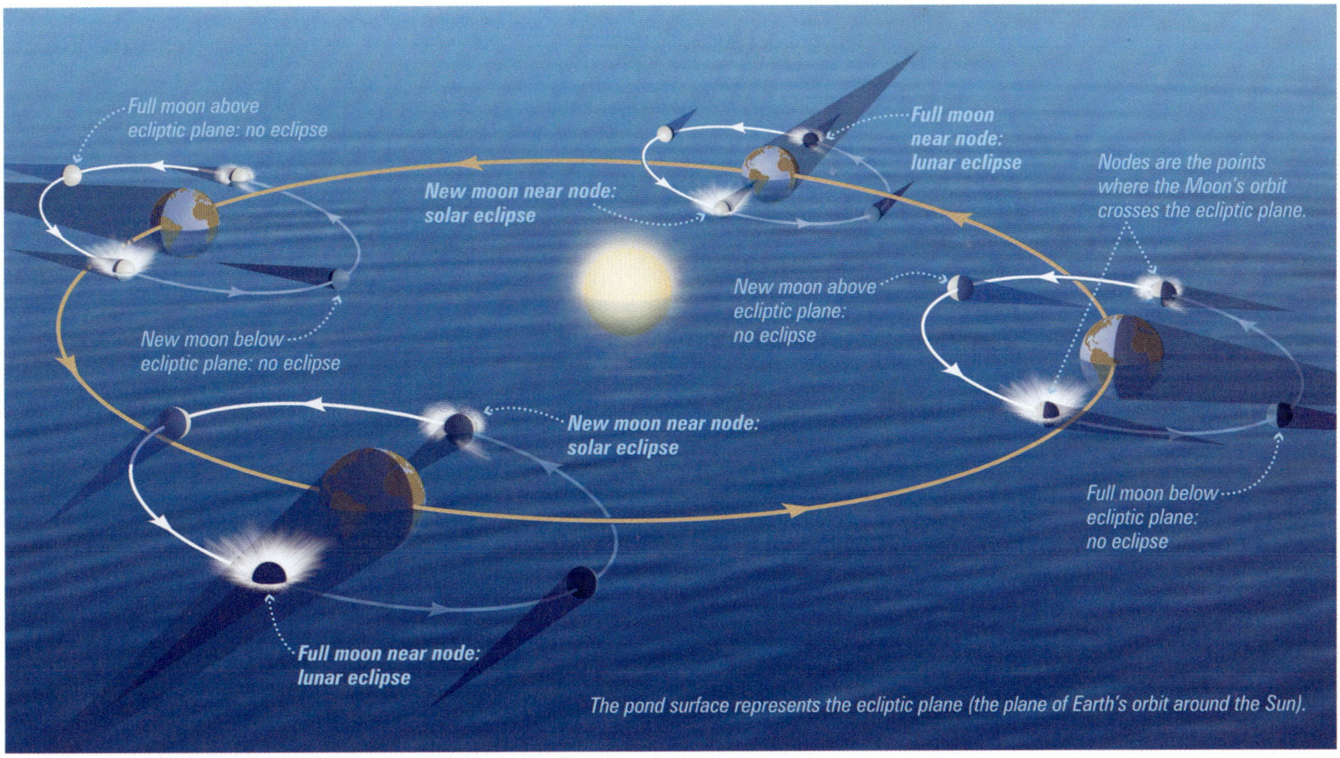

Figure 2.24 Interactive Figure This illustration represents the ecliptic plane as the surface of a pond. The Moon's orbit is tilted by about 5° to the ecliptic plane, so the Moon spends half of each orbit above the plane (the pond surface) and half below it. Eclipses occur only when the Moon is at both a node (passing through the pond surface) *and* a phase of either new moon (for a solar eclipse) or full moon (for a lunar eclipse)—as is the case with the lower left and top right orbits shown. (Not to scale.)

Although there are two basic types of eclipse—lunar and solar—each of these types can look different depending on precisely how the shadows fall. The shadow of the Moon or Earth consists of two distinct regions: a central **umbra**, where sunlight is completely blocked, and a surrounding **penumbra**, where sunlight is only partially blocked (Figure 2.25). Thus, an umbral shadow is totally dark, while a penumbral shadow is only slightly darker than no shadow. Let's see how this affects eclipses.

Lunar Eclipses A lunar eclipse begins at the moment when the Moon's orbit first carries it into Earth's penumbra. After that, we will see one of three types of lunar eclipse (Figure 2.26). If the Sun, Earth, and Moon are nearly perfectly aligned, the Moon will pass through Earth's umbra, and we will see a **total lunar eclipse**. If the alignment is somewhat less perfect, only part of the full moon will pass through the umbra (with the rest in the penumbra), and we will see a **partial lunar eclipse**. If the Moon passes through Earth's penumbra only, we will see a **penumbral lunar eclipse**.

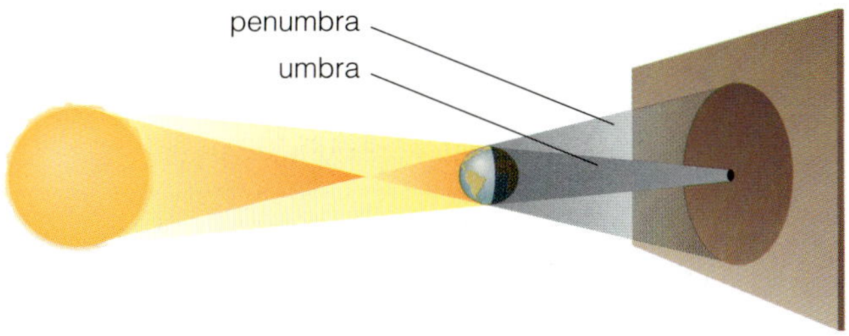

Figure 2.25 The shadow cast by an object in sunlight. Sunlight is fully blocked in the umbra and partially blocked in the penumbra.

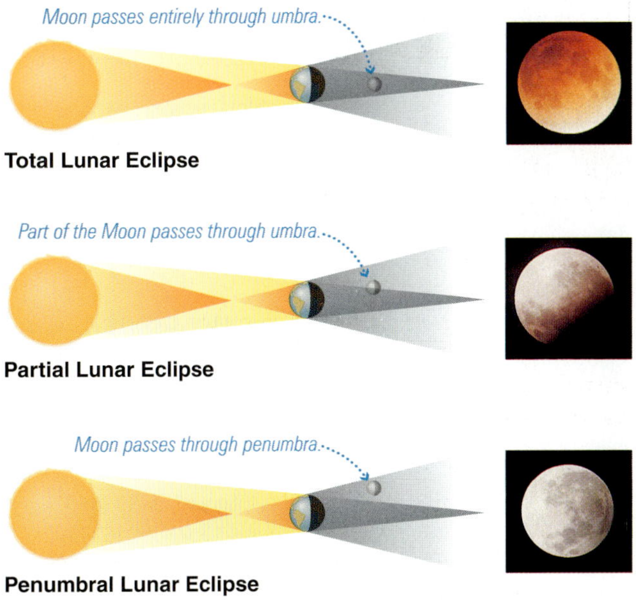

Figure 2.26 Interactive Figure The three types of lunar eclipse.

Figure 2.27 This sequence of photos, taken at 5-minute intervals, shows the progression of a total lunar eclipse; the central photos show more detail during totality. Notice Earth's curved shadow advancing across the Moon during the partial phases, and the redness of the full moon during totality. The arc-shaped arrangement was chosen for effect and has no inherent meaning.

Penumbral eclipses are the most common type of lunar eclipse, but they are the least visually impressive because the full moon darkens only slightly. Earth's umbral shadow clearly darkens part of the Moon's face during a partial lunar eclipse, and the curvature of this shadow demonstrates that Earth is round. A total lunar eclipse is particularly spectacular because the Moon becomes dark and eerily red during **totality**—the time during which the Moon is entirely engulfed in the umbra (Figure 2.27). Totality typically lasts about an hour. The Moon becomes dark because it is in shadow, and red because Earth's atmosphere scatters some of the red light from the Sun enough so that it is directed toward the Moon.

Solar Eclipses We can also see three types of solar eclipse (Figure 2.28). If a solar eclipse occurs when the Moon is relatively close to Earth in its orbit, the Moon's umbra touches a small area of Earth's surface (no more than about 270 kilometers in diameter). Anyone within this area will see a **total solar eclipse**. Surrounding the region of totality is a much larger area (typically about 7,000 kilometers in diameter) that falls within the Moon's penumbral shadow.

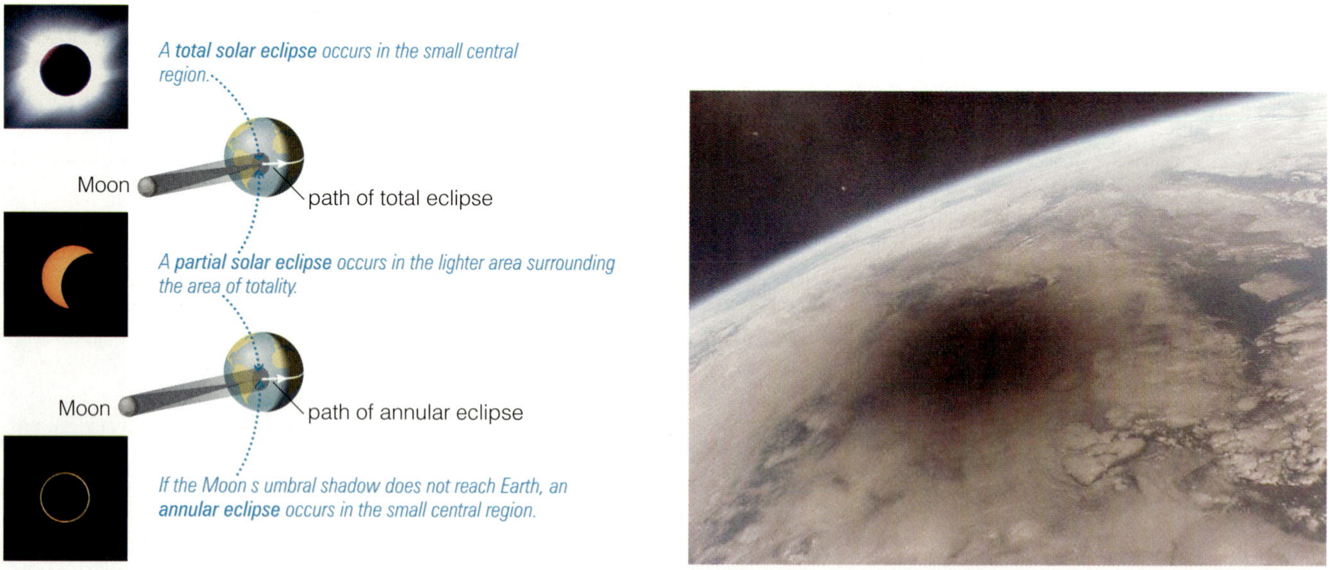

a The three types of solar eclipse. The diagrams show the Moon's shadow falling on Earth; note the dark central umbra surrounded by the much lighter penumbra.

b This photo from Earth orbit shows the Moon's shadow (umbra) on Earth during a total solar eclipse. Notice that only a small region of Earth experiences totality at any one time.

Figure 2.28 Interactive Figure During a solar eclipse, the Moon's small shadow moves rapidly across the face of the Earth.

Figure 2.29 This multiple-exposure photograph shows the progression of a total solar eclipse. Totality (central image) lasts only a few minutes, during which time we can see the faint corona around the outline of the Sun. This photo was taken July 22, 1990, in La Paz, Mexico. The foreground church was photographed at a different time of day.

Anyone within this region will see a **partial solar eclipse**, in which only part of the Sun is blocked from view.

If the eclipse occurs when the Moon is relatively far from Earth in its orbit, the Moon's slightly smaller angular size means its umbral shadow may not quite reach Earth. In that case, anyone in the small region of Earth directly behind the umbra will see an **annular eclipse**, in which a ring of sunlight surrounds the disk of the Moon. Again, anyone in the surrounding penumbral shadow will see a partial solar eclipse.

The combination of Earth's rotation and the orbital motion of the Moon causes the Moon's umbral and penumbral shadows to race across the face of Earth at a typical speed of about 1,700 kilometers per hour. As a result, the umbral shadow traces a narrow path across Earth, and totality never lasts more than a few minutes in any particular place.

A total solar eclipse is a spectacular sight. It begins when the disk of the Moon first appears to touch the Sun. Over the next couple of hours, the Moon appears to take a larger and larger "bite" out of the Sun. As totality approaches, the sky darkens and temperatures fall. Birds head back to their nests, and crickets begin their nighttime chirping. During the few minutes of totality, the Moon completely blocks the normally visible disk of the Sun, allowing the faint *corona* to be seen (Figure 2.29). The surrounding sky takes on a twilight glow, and planets and bright stars become visible in the daytime. As totality ends, the Sun slowly emerges from behind the Moon over the next couple of hours. However, because your eyes have adapted to the darkness, totality appears to end far more abruptly than it began.

Predicting Eclipses Few phenomena have so inspired and humbled humans throughout the ages as eclipses. For many cultures, eclipses were mystical events associated with fate or the gods, and countless stories and legends surround

SPECIAL TOPIC Does the Moon Influence Human Behavior?

From myths of werewolves to stories of romance under the full moon, human culture is filled with claims that our behavior is influenced by the phase of the Moon. Can we say anything scientific about such claims?

The Moon clearly has important influences on Earth. For example, the Moon is primarily responsible for the tides [Section 4.5]. However, the Moon's tidal force cannot directly affect objects as small as people.

If a physical force from the Moon cannot affect human behavior, could we be influenced in other ways? Certainly, anyone who lives near the oceans is influenced by the rising and falling of the tides. For example, fishermen and boaters must follow the tides. Thus, although the Moon does not influence their behavior directly, it does so indirectly through its effect on the oceans.

Physiological patterns in many species appear to follow the lunar phases; for example, some crabs and turtles lay eggs only at full moon. No human trait is so closely linked to lunar phases, but the average human menstrual cycle is so close in length to a lunar month that it is difficult to believe the similarity is mere coincidence. Nevertheless, aside from the physiological cycles and the influence of tides on people who live near the oceans, claims that the lunar phase affects human behavior are difficult to verify scientifically. For example, although it is possible that the full moon brings out certain behaviors, it may also simply be that some behaviors are easier to exhibit when the sky is bright. A beautiful full moon may bring out your desire to walk on the beach under the moonlight, but there is no scientific evidence to suggest that the full moon would affect you the same way if you were confined to a deep cave.

them. One legend holds that the Greek philosopher Thales (c. 624–546 B.C.) successfully predicted the year (but presumably not the precise time) that a total eclipse of the Sun would be visible in the area where he lived, which is now part of Turkey. Coincidentally, the eclipse occurred as two opposing armies (the Medes and the Lydians) were massing for battle. The eclipse so frightened the armies that they put down their weapons, signed a treaty, and returned home. Because modern research shows that the only eclipse visible in that part of the world at about that time occurred on May 28, 585 B.C., we know the precise date on which the treaty was signed—the earliest historical event that can be dated precisely.

Much of the mystery of eclipses probably stems from the relative difficulty of predicting them. Look again at Figure 2.24. The two periods each year when the nodes of the Moon's orbit are nearly aligned with the Sun are called **eclipse seasons**. Each eclipse season lasts a few weeks, so some type of lunar eclipse occurs during each eclipse season's full moon, and some type of solar eclipse occurs during its new moon.

If Figure 2.24 told the whole story, eclipse seasons would occur every 6 months and predicting eclipses would be easy. For example, if eclipse seasons always occurred in January and July, eclipses would always occur on the dates of new and full moons in those months. But the figure does not show one important thing about the Moon's orbit: The nodes slowly move around the orbit. As a result, eclipse seasons occur slightly less than 6 months apart (about 173 days apart) and do not recur in the same months year after year.

The combination of the changing dates of eclipse seasons and the $29\frac{1}{2}$-day cycle of lunar phases makes eclipses recur in a cycle of about 18 years $11\frac{1}{3}$ days. This cycle is called the **saros cycle**. Astronomers in many ancient cultures identified the saros cycle and thus could predict *when* eclipses would occur. However, the saros cycle does not account for all the complications involved in predicting eclipses. If a solar eclipse occurred today, the one that would occur 18 years $11\frac{1}{3}$ days from now would not be visible from the same places on Earth and might not be of the same type. For example, one might be total and the other only partial. No ancient culture achieved the ability to predict eclipses in every detail.

Today, we can predict eclipses because we know the precise details of the orbits of Earth and the Moon. Many astronomical software packages, including the *SkyGazer* package that comes with this book, do the necessary calculations automatically. Table 2.1 lists upcoming lunar eclipses of all types; notice that, as we expect, eclipses generally come a little less than 6 months apart. Figure 2.30 shows paths of totality for total solar eclipses (but not for partial or annular eclipses) over the next couple of decades, using color coding to show eclipses that repeat with the saros cycle.

THINK ABOUT IT

In Table 2.1, notice that there's one exception to the "rule" of eclipses coming a little less than 6 months apart: the 2009 lunar eclipses of July 7 and August 6. How can eclipses occur just a month apart like this? Should you be surprised that both these lunar eclipses are penumbral? Explain.

2.4 The Ancient Mystery of the Planets

We've now covered the appearance and motion of the stars, Sun, and Moon in the sky. That leaves us with the planets yet to discuss. As we'll soon see, planetary motion posed an ancient mystery that played a critical role in the development of modern civilization.

Table 2.1 Lunar Eclipses 2006–2010*

Date	*Type*	*Where You Can See It*
Mar. 14, 2006	penumbral	Americas, Europe, Africa, Asia
Sep. 7, 2006	partial	Europe, Africa, Asia, Australia
Mar. 3, 2007	total	Americas, Europe, Africa, Asia
Aug. 28, 2007	total	eastern Asia, Australia, Pacific, Americas
Feb. 21, 2008	total	central Pacific, Americas, Europe, Africa
Aug. 16, 2008	partial	South America, Europe, Africa, Asia, Australia
Feb. 9, 2009	penumbral	eastern Europe, Asia, Australia, Pacific, western North America
Jul. 7, 2009	penumbral	Australia, Pacific, Americas
Aug. 6, 2009	penumbral	Americas, Europe, Africa, western Asia
Dec. 31, 2009	partial	Europe, Africa, Asia, Australia
Jun. 26, 2010	partial	eastern Asia, Australia, Pacific, western Americas
Dec. 21, 2010	total	eastern Asia, Australia, Pacific, Americas, Europe

*Dates are based on Universal Time and hence are those in Greenwich, England, at the time of the eclipse; to see an eclipse, check a news or Web source for the exact local time and date. Data from NASA's Eclipse Home Page maintained by Fred Espanak.

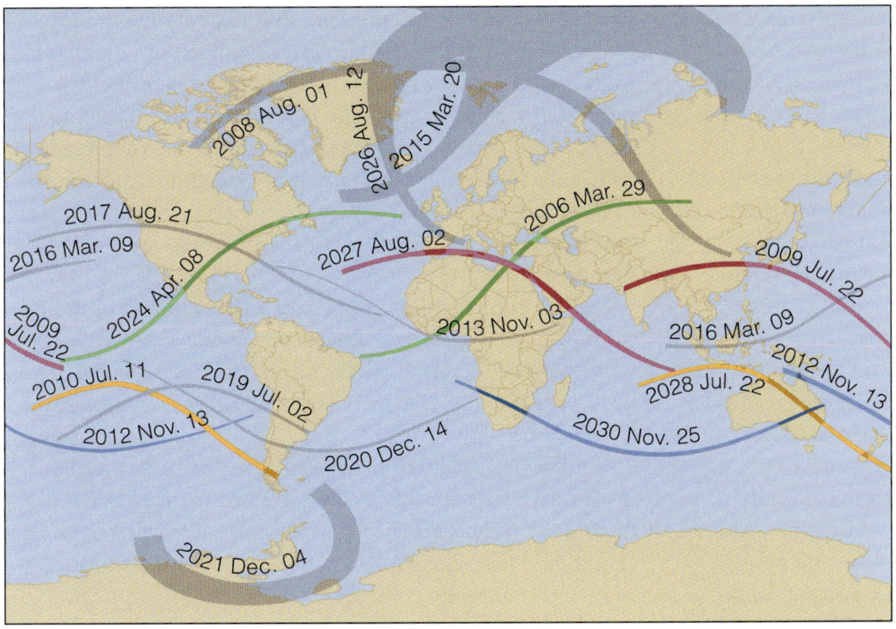

Figure 2.30 This map shows the paths of totality for solar eclipses from 2006 to 2030. Paths of the same color represent eclipses occurring in successive saros cycles, separated by 18 years 11 days. For example, the 2024 eclipse occurs 18 years 11 days after the 2006 eclipse (both shown in green). (Data from NASA's Eclipse Home Page maintained by Fred Espanak.)

Five planets are easy to find with the naked eye: Mercury, Venus, Mars, Jupiter, and Saturn. Mercury is visible only infrequently, and then only just after sunset or just before sunrise because it is so close to the Sun. Venus often shines brightly in the early evening in the west or before dawn in the east. If you see a very bright "star" in the early evening or early morning, it is probably Venus. Jupiter, when it is visible at night, is the brightest object in the sky besides the Moon and Venus. Mars is often recognizable by its reddish color, though you should check a star chart to make sure you aren't looking at a bright red star. Saturn is also easy to see with the naked eye, but because many stars are just as bright as Saturn, it helps to know where to look. (It also helps to know that planets tend not to twinkle as much as stars.) Sometimes several planets may appear close together in the sky, offering a particularly beautiful sight (Figure 2.31).

Figure 2.31 This photograph shows a rare planetary grouping in which all five planets that are easily visible to the naked eye appeared close together in the sky. It was taken near Chatsworth, New Jersey, just after sunset on April 23, 2002. The next such close grouping of these five planets in our sky will not occur until September 2040.

• What was once so mysterious about planetary motion in our sky?

Ancient people carefully observed the movements of the planets among the stars, and you can observe the same motions for yourself. On any particular night, the planets move through the sky just like stars, rising in the east and setting in the west. However, if you observe planetary positions among the stars over a period of weeks or months, you'll notice that the planets wander slowly through the constellations of the zodiac. (The word *planet* comes from the Greek for "wandering star.") By itself, this wandering might not be too surprising. After all, the Sun and Moon also move among the constellations. But the planets move in what seems to be a very strange way.

Unlike the Sun and Moon, which move steadily eastward relative to the stars, the planets vary substantially in both speed and brightness as they move among the stars. Moreover, while the planets *usually* move eastward relative to the stars, they occasionally reverse course completely, moving westward rather than eastward through the zodiac. A period during which a planet appears to move westward relative to the stars is called a period of **apparent retrograde motion** (*retrograde* means "backward"). These periods may last from a few weeks to a few months, depending on the planet (Figure 2.32).

Ancient astronomers could easily "explain" the daily paths of the stars through the sky by imagining that the celestial sphere was real and that it really rotated around Earth each day. But the apparent retrograde motion of the planets posed a far greater mystery: What could cause the planets sometimes to go backward? As we'll discuss in Chapter 3, the ancient Greeks came up with some very clever ways to explain the occasional backward motion of the planets, despite being wedded to the incorrect idea of an Earth-centered universe. However, the Greek explanation was quite complex, and ultimately wrong.

In contrast, apparent retrograde motion has a simple explanation in a Sun-centered solar system. You can demonstrate it for yourself with the help of a friend (Figure 2.33a). Pick a spot in an open field to represent the Sun. You can represent Earth, walking counterclockwise around the Sun, while your friend represents a more distant planet (such as Mars, Jupiter, or Saturn) by walking counterclockwise around the Sun at a greater distance. Your friend should walk more slowly than you, because more distant planets orbit the Sun more slowly. As you walk, watch how your friend appears to move relative to buildings or trees in the distance. Although both of you always walk the same way around the Sun, your friend will appear to move backward against the background during the part of your "orbit" at which you catch up to and pass him or her. To understand the apparent retrograde motions of Mercury and Venus, which are closer to the Sun than is Earth, simply switch places with your friend and repeat the demonstration.

This demonstration applies directly to the planets. For example, because Mars takes about 2 years to orbit the Sun (actually 1.88 years), it covers about half its orbit during the 1 year in which Earth makes a complete orbit. If you trace lines of sight from Earth to Mars from different points in their orbits, you will see that the line of sight usually moves eastward relative to the stars but moves westward during the time when Earth is passing Mars in its orbit (Figure 2.33b). Like your friend in the demonstration, Mars never actually changes direction. It only *appears* to change direction from our perspective on Earth.

• Why did the ancient Greeks reject the real explanation for planetary motion?

If the apparent retrograde motion of the planets is so readily explained by recognizing that Earth orbits the Sun, why wasn't this idea accepted in ancient times? In fact, the idea that Earth goes around the Sun was suggested as early as 260 B.C. by the Greek astronomer Aristarchus. No one knows why Aristarchus proposed a Sun-centered solar system, but the fact that it explains planetary motion so naturally

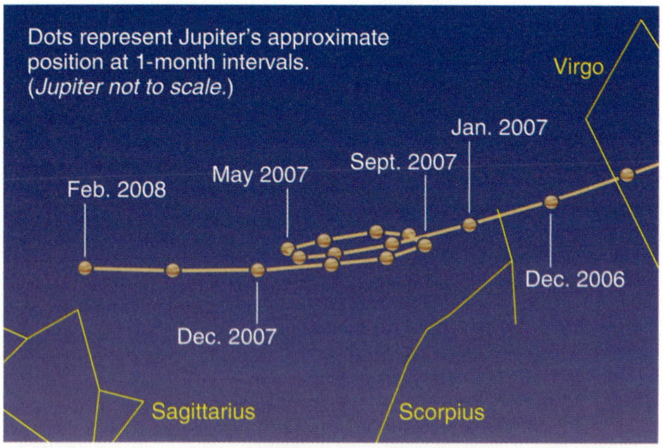

a This diagram shows Jupiter s approximate position among the stars in our sky from late 2006 to early 2008. Notice the backward path between May and September 2007.

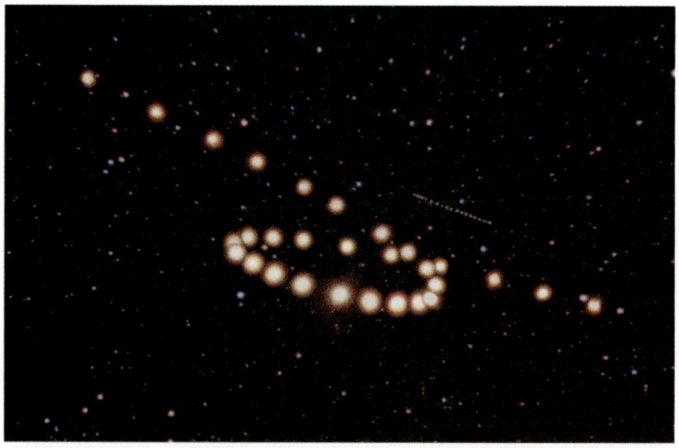

b This composite photograph shows Mars between early June and late November 2003. The progression goes from right to left, with apparent retrograde motion in the middle. Notice that Mars is biggest and brightest in the middle of the retrograde loop, because that is where it is closest to Earth in its orbit. The 29 individual photos were taken at 5- to 8-day intervals. (The white dots in a line just right of center are the planet Uranus, which by coincidence was in the same part of the sky.)

Figure 2.32 Apparent retrograde motion.

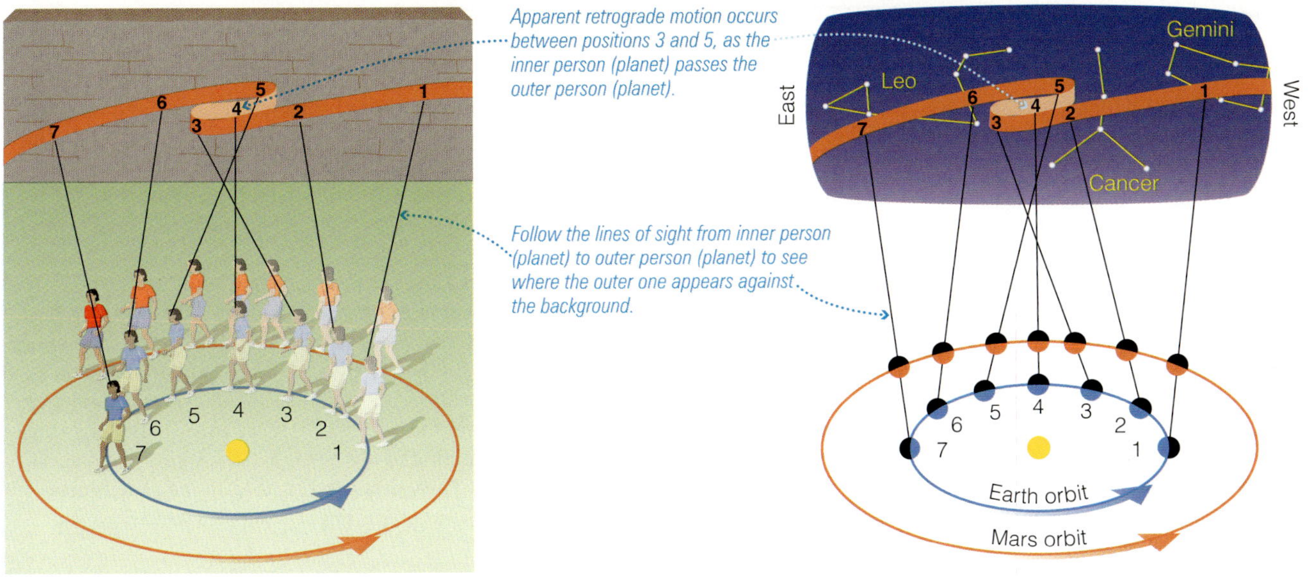

a The retrograde motion demonstration: Watch how your friend (in red) usually appears to you (in blue) to move forward against the background of the building in the distance but appears to move backward as you catch up to and pass him or her in your "orbit."

b This diagram shows how the idea from the demonstration applies to a planet. Follow the lines of sight from Earth to Mars in numerical order. Notice that Mars appears to move westward relative to the distant stars as Earth passes it by in its orbit (roughly from points 3 to 5 in the diagram).

Figure 2.33 Interactive Figure Apparent retrograde motion—the occasional "backward" motion of the planets relative to the stars—has a simple explanation in a Sun-centered solar system.

probably played a role (see box, p. 52). Nevertheless, Aristarchus's contemporaries rejected his idea, and the Sun-centered solar system did not gain wide acceptance until almost 2,000 years later.

Although there were many reasons why the Greeks were reluctant to abandon the idea of an Earth-centered universe, one of the most important was their inability to detect something called **stellar parallax.** Extend your arm and hold up one finger. If you keep your finger still and alternately close your left eye and right eye, your finger will appear to jump back and forth against the background. This apparent shifting, called *parallax,* occurs because your two eyes view your finger from opposite sides of your nose. If you move your finger closer to your face, the parallax increases. If you look at a distant tree or flagpole instead of your finger, you may not notice any parallax at all. Thus, parallax depends on distance, with nearer objects exhibiting greater parallax than more distant objects.

If you now imagine that your two eyes represent Earth at opposite sides of its orbit around the Sun and that your finger represents a relatively nearby star, you have the idea of stellar parallax. That is, because we view the stars from different places in our orbit at different times of year, nearby stars should *appear* to shift back and forth against the background of more distant stars (Figure 2.34).

Because the Greeks believed that all stars lie on the same celestial sphere, they actually expected to see stellar parallax in a slightly different way. If Earth orbited the Sun, they reasoned that at different times of year we would be closer to different parts of the celestial sphere and thus would notice changes in the angular separations of stars. However, no matter how hard they searched, they could find no sign

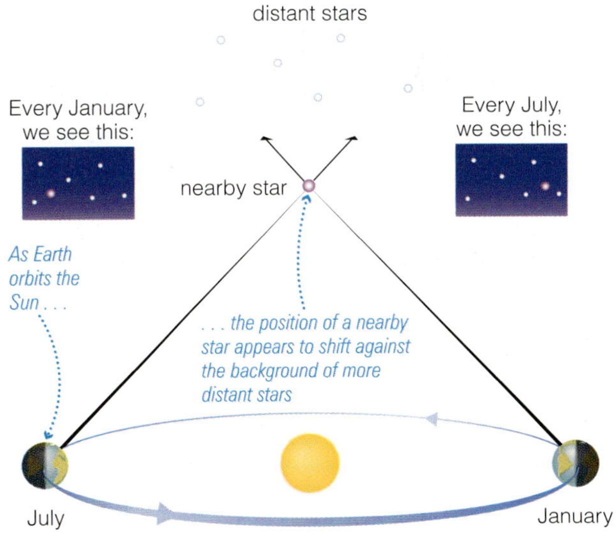

Figure 2.34 Stellar parallax is an apparent shift in the position of a nearby star as we look at it from different places in Earth's orbit. This figure is greatly exaggerated; in reality, the amount of shift is far too small to detect with the naked eye.

SPECIAL TOPIC **Who First Proposed a Sun-Centered Solar System?**

You've probably heard of Copernicus, whose work in the mid-1500s started the revolution that ultimately overturned the ancient belief in an Earth-centered universe [Section 3.3]. However, Copernicus was *not* the first person to propose the idea that Earth goes around the Sun. By the time it was adopted by Copernicus, the Sun-centered idea had already been around for nearly 2,000 years.

As far as we know, the Greek scientist Aristarchus (c. 310–230 B.C.) was the first person to suggest that Earth goes around the Sun rather than vice versa. He proposed the idea in about 260 B.C. Little of Aristarchus's work survives to the present day, so we cannot know what motivated him to suggest an idea so contrary to the prevailing wisdom of an Earth-centered universe. However, it's quite likely that he was motivated by the fact that a Sun-centered system offers a much more natural explanation for the apparent retrograde motion of the planets. To account for the lack of detectable stellar parallax, Aristarchus suggested that the stars were extremely far away.

Aristarchus further strengthened his argument by estimating the sizes of the Moon and the Sun. By observing the shadow of Earth on the Moon during a lunar eclipse, he estimated the Moon's diameter to be about one-third of Earth's diameter—only slightly higher than the actual value. He then used a geometric argument, based on measuring the angle between the Moon and the Sun at first- and third-quarter phases, to conclude that the Sun must be larger than Earth. (Aristarchus's measurements were imprecise, so he estimated the Sun's diameter to be about seven times Earth's rather than the correct value of about 100 times.) His conclusion that the Sun is larger than Earth may have been another reason why he felt that Earth should orbit the Sun, rather than vice versa.

Although he was probably the first to suggest that Earth orbits the Sun, Aristarchus's ideas were in large part built upon the work of earlier scholars. For example, Heracleides (c. 388–315 B.C.) had previously suggested that Earth rotates. Aristarchus may have drawn on this idea to explain the apparent daily rotation of the stars in our sky. Heracleides also suggested that not all heavenly bodies circle Earth: Based on the fact that Mercury and Venus always stay fairly close to the Sun in the sky, he argued that these two planets must orbit the Sun. Thus, in suggesting that *all* the planets orbit the Sun, Aristarchus was extending the ideas of Heracleides and others before him.

Aristarchus gained little support among his contemporaries, but his ideas never died. They were kept alive by scholars throughout history, and Copernicus was well aware of Aristarchus's claims when he proposed his own version of the Sun-centered system. Thus, our modern understanding of the universe owes at least some debt to the remarkable vision of a man who lived more than 2,200 years ago.

---

of stellar parallax. They concluded that one of the following must be true:

1. Earth orbits the Sun but the stars are so far away that stellar parallax is undetectable to the naked eye.
2. There is no stellar parallax because Earth remains stationary at the center of the universe.

Unfortunately, with notable exceptions such as Aristarchus, the Greeks rejected the correct answer (the first one) because they could not imagine that the stars could be *that* far away. Today, we can detect stellar parallax with the aid of telescopes, providing direct proof that Earth really does orbit the Sun. Careful measurements of stellar parallax also provide the most reliable means of measuring distances to nearby stars [Section 15.1].

THINK ABOUT IT

How far apart are opposite sides of Earth's orbit? How far away are the nearest stars? Describe the challenge of detecting stellar parallax. It may help to visualize Earth's orbit and the distance to the stars on the 1-to-10-billion scale used in Chapter 1.

The ancient mystery of the planets drove much of the historical debate over Earth's place in the universe. In many ways, the modern technological society we take for granted today can be traced directly to the scientific revolution that began in the quest to explain the strange wandering of the planets among the stars in our sky. We will turn our attention to this revolution in the next chapter.

THE BIG PICTURE

Putting Chapter 2 into Context

In this chapter, we surveyed the phenomena of our sky. Keep the following "big picture" ideas in mind as you continue your study of astronomy:

- You can enhance your enjoyment of learning astronomy by observing the sky. The more you learn about the appearance and apparent motions of the sky, the more you will appreciate what you can see in the universe.
- From our vantage point on Earth, it is convenient to imagine that we are at the center of a great celestial sphere—even though we really are on a planet orbiting a star in a vast universe. We can then understand what we see in the local sky by thinking about how the celestial sphere appears from our latitude.
- Most of the phenomena of the sky are relatively easy to observe and understand. The more complex phenomena—particularly eclipses and apparent retrograde motion of the planets—challenged our ancestors for thousands of years. The desire to understand these phenomena helped drive the development of science and technology.

SUMMARY OF KEY CONCEPTS

2.1 Patterns in the Night Sky

- **What does the universe look like from Earth?** Stars and other celestial objects appear to lie on a great **celestial sphere** surrounding Earth. We divide the celestial sphere into **constellations** with well-defined borders. From any location on Earth, we see half the celestial sphere at any one time as the dome of our **local sky**, in which the **horizon** is the boundary between Earth and sky, the **zenith** is the point directly overhead, and the **meridian** runs from due south to due north through the zenith.

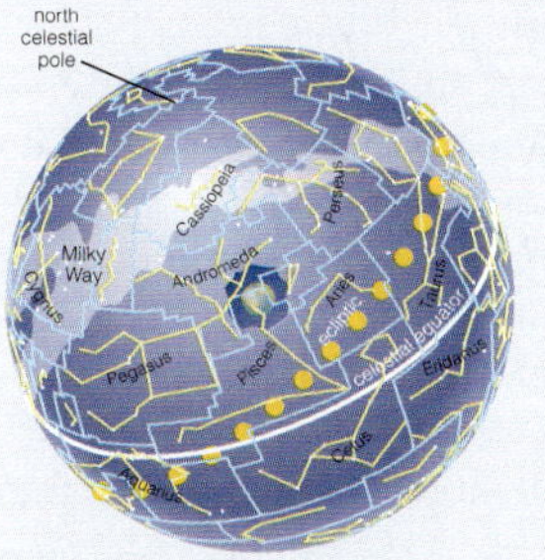

- **Why do stars rise and set?** Earth's rotation makes stars appear to circle around Earth each day. A star whose complete circle lies above our horizon is said to be **circumpolar.** Other stars have circles that cross the horizon, making them rise in the east and set in the west each day.

- **Why do the constellations we see depend on latitude and time of year?** The visible constellations vary with time of year because our night sky lies in different directions in space as we orbit the Sun. The constellations vary with **latitude** because your latitude determines the orientation of your horizon relative to the celestial sphere. The sky does not vary with **longitude.**

2.2 The Reason for Seasons

- **What causes the seasons?** The tilt of Earth's axis causes the seasons. The axis points in the same direction (toward Polaris) throughout the year. Thus, as Earth orbits the Sun, sunlight hits different parts of Earth more directly at different times of the year.

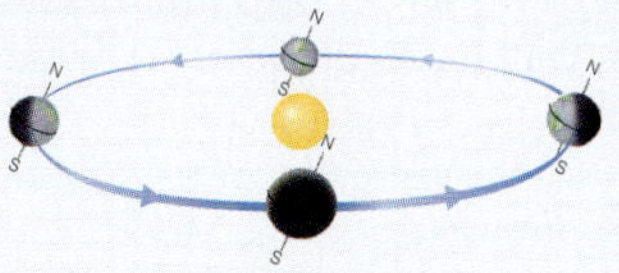

- **How do we mark the progression of the seasons?** The **summer** and **winter solstices** are the times during the year when the Northern Hemisphere gets its most and least direct sunlight, respectively. The **spring** and **fall equinoxes** are the two times when both hemispheres get equally direct sunlight.

- **How does the orientation of Earth's axis change with time?** Earth's 26,000-year cycle of **precession** changes the orientation of the axis in space, although the tilt remains about $23\frac{1}{2}°$. The changing orientation of the axis does not affect the pattern of seasons, but it changes the identity of any "north star" and shifts the locations of the solstices and equinoxes in Earth's orbit.

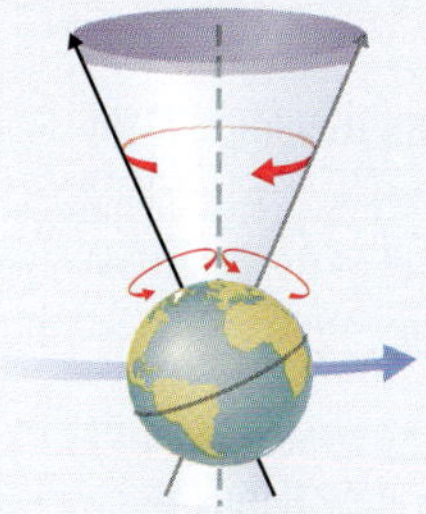

2.3 The Moon, Our Constant Companion

- **Why do we see phases of the Moon?** The **phase** of the Moon depends on its position relative to the Sun as it orbits Earth. The half of the Moon facing the Sun is always illuminated while the other half is dark, but from Earth we see varying combinations of the illuminated and dark faces.

- **What causes eclipses?** We see a **lunar eclipse** when Earth's shadow falls on the Moon, and a **solar eclipse** when the Moon blocks our view of the Sun. We do not see an eclipse at every new and full moon because the Moon's orbit is slightly inclined to the ecliptic plane. Eclipses come in different types, depending on where the dark **umbral** and lighter **penumbral** shadows fall.

2.4 The Ancient Mystery of the Planets

- **What was once so mysterious about planetary motion in our sky?** Planets generally drift eastward relative to the stars over the course of the year, but for weeks or months they reverse course in their periods of **apparent retrograde motion.** This motion occurs when Earth passes by (or is passed by) another planet in its orbit, but it posed a major mystery to ancient people who assumed Earth to be at the center of the universe.

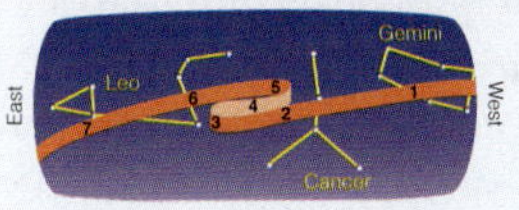

- **Why did the ancient Greeks reject the real explanation for planetary motion?** The Greeks rejected the idea that Earth goes around the Sun in part because they could not detect **stellar parallax**—slight apparent shifts in stellar positions over the course of the year. To most Greeks, it seemed unlikely that the stars could be so away as to make parallax undetectable to the naked eye, even though that is, in fact, the case.

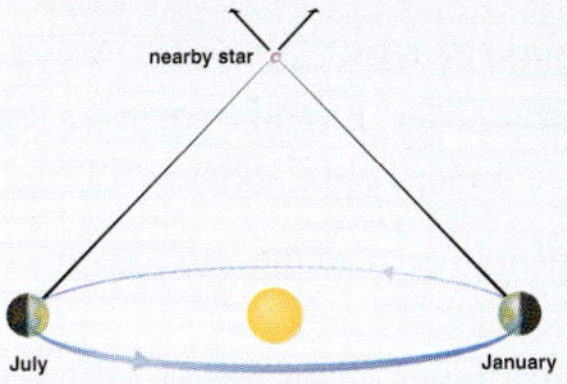

EXERCISES AND PROBLEMS

For instructor-assigned homework go to MasteringASTRONOMY www.masteringastronomy.com

Review Questions

Short-Answer Questions Based on the Reading

1. What are *constellations?* How did they get their names?
2. Suppose you were making a model of the celestial sphere with a ball. Briefly describe all the things you would need to mark on your celestial sphere.
3. On a clear, dark night, the sky may appear to be "full" of stars. Does this appearance accurately reflect the way stars are distributed in space? Explain.
4. Why does the *local sky* look like a dome? Define *horizon, zenith,* and *meridian.* How do we describe the location of an object in the local sky?
5. Explain why we can measure only *angular sizes* and *angular distances* for objects in the sky. What are *arcminutes* and *arcseconds?*
6. What are *circumpolar stars?* Are more stars circumpolar at the North Pole or in the United States? Explain.
7. What are *latitude* and *longitude?* Does the sky vary with latitude? Does it vary with longitude? Explain.
8. What is the *zodiac,* and why do we see different parts of it at different times of year?
9. Suppose Earth's axis had no tilt. Would we still have seasons? Why or why not?
10. Briefly describe what is special about the summer and winter solstices and the spring and fall equinoxes.
11. What is *precession,* and how does it affect the sky that we see from Earth?
12. Briefly describe the Moon's cycle of *phases.* Can you ever see a full moon at noon? Explain.
13. What do we mean when we say that the Moon exhibits *synchronous rotation?* What does this tell us about the Moon's periods of rotation and orbit?
14. Why don't we see an *eclipse* at every new and full moon? Describe the conditions that must be met for us to see a *solar* or *lunar eclipse.*
15. What do we mean by the *apparent retrograde motion* of the planets? Why was it difficult for ancient astronomers to explain but easy for us to explain?
16. What is *stellar parallax?* Briefly describe the role it played in making ancient astronomers believe in an Earth-centered universe.

Test Your Understanding

Does It Make Sense?

Decide whether the statement makes sense (or is clearly true) or does not make sense (or is clearly false). Explain your reasoning. (For an example, see Chapter 1, "Does It Make Sense?")

17. The constellation Orion didn't exist when my grandfather was a child.
18. When I looked into the dark lanes of the Milky Way with my binoculars, I saw what must have been a cluster of distant galaxies.
19. Last night the Moon was so big that it stretched for a mile across the sky.
20. I live in the United States, and during my first trip to Argentina I saw many constellations that I'd never seen before.
21. Last night I saw Jupiter right in the middle of the Big Dipper. (*Hint:* Is the Big Dipper part of the zodiac?)
22. Last night I saw Mars move westward through the sky in its apparent retrograde motion.
23. Although all the known stars appear to rise in the east and set in the west, we might someday discover a star that will appear to rise in the west and set in the east.
24. If Earth's orbit were a perfect circle, we would not have seasons.
25. Because of precession, someday it will be summer everywhere on Earth at the same time.
26. This morning I saw the full moon setting at about the same time the Sun was rising.

Quick Quiz

Choose the best answer to each of the following. Explain your reasoning with one or more complete sentences.

27. Two stars that are in the same constellation: (a) must both be part of the same cluster of stars in space. (b) must both have been discovered at about the same time. (c) may actually be very far away from each other.
28. The North Celestial Pole is 35° above your northern horizon. This tells you that: (a) you are at latitude 35°N. (b) you are at longitude 35°E. (c) it is the winter solstice.
29. Beijing and Philadelphia have about the same latitude but very different longitudes. Therefore, tonight's night sky in these two places: (a) will look about the same. (b) will have completely different sets of constellations. (c) will have partially different sets of constellations.
30. In winter, Earth's axis points toward the star Polaris. In spring: (a) the axis also points toward Polaris. (b) the axis points toward Vega. (c) the axis points toward the Sun.
31. When it is summer in Australia, it is: (a) winter in the United States. (b) summer in the United States. (c) spring in the United States.
32. If the Sun rises precisely due east: (a) you must be located at Earth's equator. (b) it must be the day of either the spring or fall equinox. (c) it must be the day of the summer solstice.
33. A week after full moon, the Moon's phase is: (a) first quarter. (b) third quarter. (c) new.
34. The fact that we always see the same face of the Moon tell us that: (a) the Moon does not rotate. (b) the Moon's rotation period is the same as its orbital period. (c) the Moon looks the same on both sides.
35. If there is going to be a total lunar eclipse tonight, then you know that: (a) the Moon's phase is full. (b) the Moon's phase is new. (c) the Moon is unusually close to Earth.
36. When we see Saturn going through a period of apparent retrograde motion, it means: (a) Saturn is temporarily moving backward in its orbit of the Sun. (b) Earth is passing Saturn in its orbit, with both planets on the same side of the Sun. (c) Saturn and Earth must be on opposite sides of the Sun.

Investigate Further

In-Depth Questions to Increase Your Understanding

Short-Answer/Essay Questions

37. *New Planet.* Suppose we discover a planet in another solar system that has a circular orbit and an axis tilt of 35°. Would

you expect this planet to have seasons? If so, would you expect them to be more extreme than the seasons on Earth? If not, why not?

38. *Your View.*
    a. Find your latitude and longitude, and state the source of your information.
    b. Describe the altitude and direction in your sky at which the north or south celestial pole appears.
    c. Is Polaris a circumpolar star in your sky? Explain.
39. *View from the Moon.* Assume you live on the Moon near the center of the face that looks toward Earth.
    a. Suppose you see a full Earth in your sky. What phase of the Moon would people on Earth see? Explain.
    b. Suppose people on Earth see a full moon. What phase would you see for Earth? Explain.
    c. Suppose people on Earth see a waxing gibbous moon. What phase would you see for Earth? Explain.
    d. Suppose people on Earth are viewing a total lunar eclipse. What would you see from your home on the Moon? Explain.
40. *View from the Sun.* Suppose you lived on the Sun (and could ignore the heat). Would you still see the Moon go through phases as it orbits Earth? Why or why not?
41. *A Farther Moon.* Suppose the distance to the Moon were twice its actual value. Would it still be possible to have a total solar eclipse? Why or why not?
42. *A Smaller Earth.* Suppose Earth were smaller. Would solar eclipses be any different? If so, how? What about lunar eclipses? Explain.
43. *Observing Planetary Motion.* Find out what planets are currently visible in your evening sky. At least once a week, observe the planets and draw a diagram showing the position of each visible planet relative to stars in a zodiac constellation. From week to week, note how the planets are moving relative to the stars. Can you see any of the apparently wandering features of planetary motion? Explain.
44. *A Connecticut Yankee.* Find the book *A Connecticut Yankee in King Arthur's Court* by Mark Twain. Read the portion that deals with the Connecticut Yankee's prediction of an eclipse (or read the entire book). In a one- to two-page essay, summarize the episode and explain how it helped the Connecticut Yankee gain power.

Quantitative Problems

Be sure to show all calculations clearly and state your final answers in complete sentences.

45. *Arcminutes and Arcseconds.* There are 360° in a full circle.
    a. How many arcminutes are in a full circle?
    b. How many arcseconds are in a full circle?
    c. The Moon's angular size is about $\frac{1}{2}$°. What is this in arcminutes? In arcseconds?
46. *Latitude Distance.* Earth's radius is approximately 6,370 km.
    a. What is Earth's circumference?
    b. What distance is represented by each degree of latitude?
    c. What distance is represented by each arcminute of latitude?
    d. Can you give similar answers for the distances represented by a degree or arcminute of longitude? Why or why not?
47. *Angular Conversions I.* The following angles are given in degrees and fractions of degrees. Rewrite them in degrees, arcminutes, and arcseconds.
    a. 24.3° d. 0.01°
    b. 1.59° e. 0.001°
    c. 0.1°
48. *Angular Conversions II.* The following angles are given in degrees, arcminutes, and arcseconds. Rewrite them in degrees and fractions of degrees.
    a. 7°38′42″ d. 1′
    b. 12′54″ e. 1″
    c. 1°59′59″
49. *Moon Speed.* The Moon takes about $27\frac{1}{3}$ days to complete each orbit of Earth. About how fast is the Moon going as it orbits Earth? Give your answer in km/hr.
50. *Scale of the Moon.* The Moon's diameter is about 3,500 km and its average distance from Earth is about 380,000 km. How big and how far from Earth is the Moon on the 1-to-10-billion scale used in Chapter 1? Compare the size of the Moon's orbit to the size of the Sun on this scale.
51. *Angular Size of Your Finger.* Measure the width of your index finger and the length of your arm. Based on your measurements, calculate the angular width of your index finger at arm's length. Does your result agree with the approximations shown in Figure 2.7c? Explain.
52. *Find the Sun's Diameter.* The Sun has an angular diameter of about 0.5° and an average distance of about 150 million km. What is the Sun's approximate physical diameter? Compare your answer to the actual value of 1,390,000 km.
53. *Find a Star's Diameter.* The supergiant star Betelgeuse (in the constellation Orion) has a measured angular diameter of 0.044 arcsecond. Its distance has been measured to be 427 light-years. What is the actual diameter of Betelgeuse? Compare your answer to the size of our Sun and the Earth-Sun distance.
54. *Eclipse Conditions.* The Moon's precise equatorial diameter is 3,476 km, and its orbital distance from Earth varies between 356,400 km and 406,700 km. The Sun's diameter is 1,390,000 km and its distance from Earth ranges between 147.5 and 152.6 million km.
    a. Find the Moon's angular size at its minimum and maximum distances from Earth.
    b. Find the Sun's angular size at its minimum and maximum distances from Earth.
    c. Based on your answers to (a) and (b), is it possible to have a total solar eclipse when the Moon and Sun are both at their maximum distances? Explain.

Discussion Questions

55. *Earth-Centered Language.* Many common phrases reflect the ancient Earth-centered view of our universe. For example, the phrase "the Sun rises each day" implies that the Sun is really moving over Earth. We know that the Sun only *appears* to rise as the rotation of Earth carries us to a place where we can see the Sun in our sky. Identify other common phrases that imply an Earth-centered viewpoint.
56. *Flat Earth Society.* Believe it or not, there is an organization called the Flat Earth Society. Its members hold that Earth is flat and that all indications to the contrary (such as pictures of Earth from space) are fabrications made as part of a conspiracy to hide the truth from the public. Discuss the evidence for a round Earth and how you can check it for yourself. In light of the evidence, is it possible that the Flat Earth Society is correct? Defend your opinion.

For self-study activities go to MasteringASTRONOMY www.masteringastronomy.com

Interactive Tutorials

Tutorial Review of Key Concepts

Use the interactive **Tutorials** at **www.masteringastronomy.com** to review key concepts from this chapter.

Seasons Tutorial

Lesson 1 Factors Affecting Seasonal Changes
Lesson 2 The Solstices and Equinoxes
Lesson 3 The Sun's Position in the Sky

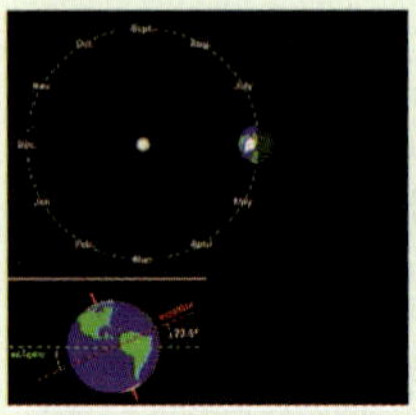
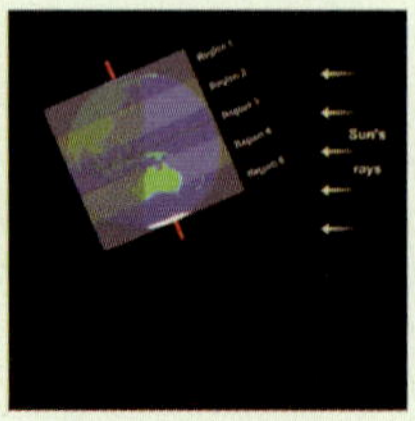
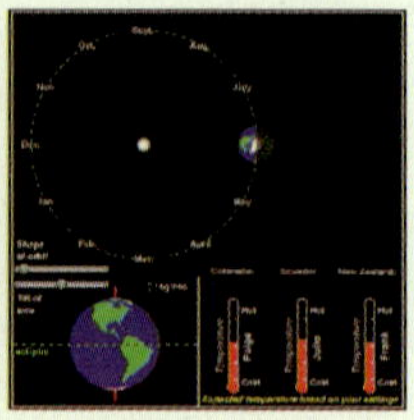

Phases of the Moon Tutorial

Lesson 1 The Causes of Lunar Phases
Lesson 2 Time of Day and Horizons
Lesson 3 When the Moon Rises and Sets

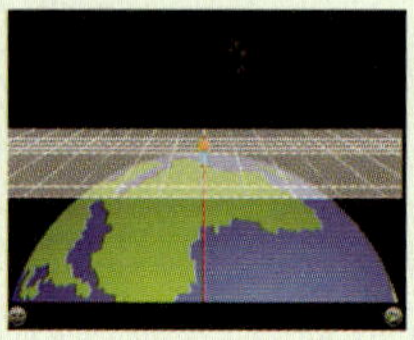
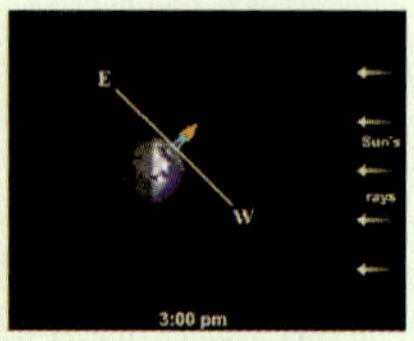

Eclipses Tutorial

Lesson 1 Why and When Do Eclipses Occur?
Lesson 2 Types of Solar Eclipses
Lesson 3 Lunar Eclipses

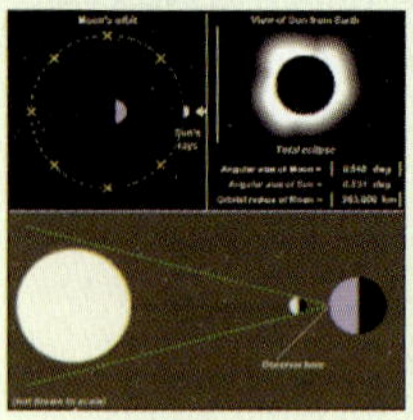
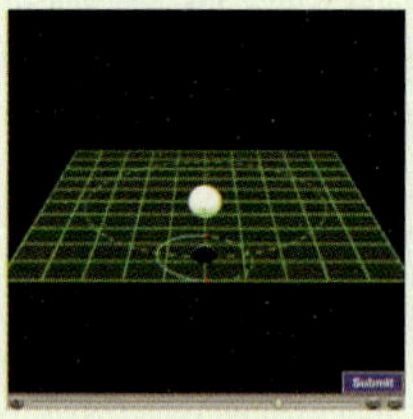
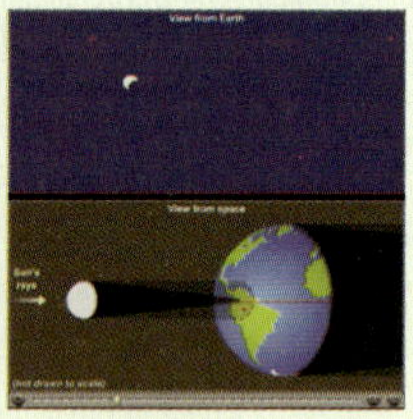

Supplementary Tutorial Exercises

Use the interactive **Tutorial Lessons** to explore the following questions.

Seasons Tutorial, Lesson 1

1. What factors affect a planet's surface temperature?
2. Why does axis tilt have a greater effect than orbital distance on Earth's seasons?
3. Why is the equator always hot and why are the poles always cold despite the passage of the seasons?

Phases of the Moon Tutorial, Lesson 3

1. When does a new moon rise and set?
2. In which phase does the Moon set just after the Sun?
3. What factors influence our view of the Moon from Earth?

Eclipses Tutorial, Lesson 1

1. What happens during a solar eclipse?
2. What happens during a lunar eclipse?
3. What role does the Moon's orbit play in the appearance of a solar eclipse?

Exploring the Sky and Solar System

Of the many activities available on the ***Voyager: SkyGazer*** **CD-ROM** accompanying your book, use the following files to observe key phenomena covered in this chapter.

Go to the **File: Basics** folder for the following demonstrations.

1. Wide Field Milky Way
2. Eclipse 1991–1992 Views
3. Winter Sky

Go to the **File: Demo** folder for the following demonstrations.

1. Russian Midnight Sun
2. Earth Orbiting the Moon
3. Mars in Retrograde

Go to the **Explore** menu for the following demonstrations.

1. Shadows on Earth
2. Phases of the Planets

Web Projects

1. Sky Information. Search the Web for sources of daily information about sky phenomena (such as lunar phases, times of sunrise and sunset, or dates of equinoxes and solstices). Identify and briefly describe your favorite source.
2. Constellations. Search the Web for information about the constellations and their mythology. Write a short report about one or more constellations.
3. Upcoming Eclipse. Find information about an upcoming solar or lunar eclipse that you might have a chance to witness. Write a short report about how you could best witness the eclipse, including any necessary travel to a viewing site, and what you can expect to see. Bonus: Describe how you could photograph the eclipse.

3

The Science of Astronomy

LEARNING GOALS

3.1 The Ancient Roots of Science
- In what ways do all humans employ scientific thinking?
- How did astronomical observations benefit ancient societies?
- What did ancient civilizations achieve in astronomy?

3.2 Ancient Greek Science
- Why does modern science trace its roots to the Greeks?
- How did the Greeks explain planetary motion?
- How was Greek knowledge preserved through history?

3.3 The Copernican Revolution
- How did Copernicus, Tycho, and Kepler challenge the Earth-centered idea?
- What are Kepler's three laws of planetary motion?
- How did Galileo solidify the Copernican revolution?

3.4 The Nature of Science
- How can we distinguish science from nonscience?
- What is a scientific theory?

3.5 Astrology
- How is astrology different from astronomy?
- Does astrology have any scientific validity?

We especially need imagination in science. It is not all mathematics, nor all logic, but is somewhat beauty and poetry.

Maria Mitchell (1818–1889), astronomer and the first woman elected to the American Academy of Arts and Sciences

Today we know that Earth is a planet orbiting a rather ordinary star, in a galaxy of more than a hundred billion stars, in an incredibly vast universe. We know that Earth, along with the entire cosmos, is in constant motion. We know that, on the scale of cosmic time, human civilization has existed for only the briefest moment. How did we manage to learn these things?

It wasn't easy. Astronomy is the oldest of the sciences, with roots extending as far back as recorded history allows us to see. But the most impressive advances in knowledge have occurred in just the past few centuries.

In this chapter, we will trace how modern astronomy grew from its roots in ancient observations, including those of the Greeks. We'll pay special attention to the unfolding of the Copernican revolution, which overturned the ancient belief in an Earth-centered universe and laid the foundation for nearly all of modern science. Finally, we'll explore the nature of modern science and the scientific method.

3.1 The Ancient Roots of Science

A common stereotype holds that scientists walk around in white lab coats and somehow think differently than other people. In reality, scientific thinking is a fundamental part of human nature. In this section, we will trace the roots of science to experiences common to nearly all people and nearly all cultures.

• In what ways do all humans employ scientific thinking?

Scientific thinking comes naturally to us. By about a year of age, a baby notices that objects fall to the ground when she drops them. She lets go of a ball—it falls. She pushes a plate of food from her high chair—it falls too. She continues to drop all kinds of objects, and they all plummet to Earth. Through powers of observation, the baby learns about the physical world, finding that things fall when they are unsupported. Eventually, she becomes so certain of this fact that, to her parents' delight, she no longer needs to test it continually.

One day somebody gives the baby a helium balloon. She releases it, and to her surprise it rises to the ceiling! Her understanding of nature must be revised. She now knows that the principle "all things fall" does not represent the whole truth, although it still serves her quite well in most situations. It will be years before she learns enough about the atmosphere, the force of gravity, and the concept of density to understand *why* the balloon rises when most other objects fall. For now, she is delighted to observe something new and unexpected.

The baby's experiences with falling objects and rising balloons exemplify scientific thinking. In essence, it is a way of learning about nature through careful observation and trial-and-error experiments. Rather than thinking differently than other people, modern scientists simply are trained to organize everyday thinking in a way that makes it easier for them to share their discoveries and employ their collective wisdom. This type of clear and organized thinking is at the heart of science.

THINK ABOUT IT

Describe a few cases where you have learned by trial and error while cooking, participating in sports, fixing something, on the job, or in any other situation.

Just as learning to communicate through language, art, or music is a gradual process for a child, the development of science has been a gradual process for humanity. Science in its modern form requires painstaking attention to detail, relentless testing of each piece of information to ensure its reliability, and a willingness to give up old beliefs that are not consistent with observed facts about the physical world. For professional scientists, these demands are the "hard work" part of the job. At heart, professional scientists are like the baby with the balloon, delighted by the unexpected and motivated by those rare moments when they—and all of us—learn something new about the universe.

• How did astronomical observations benefit ancient societies?

We will discuss modern science shortly, but first we will explore how it arose from the observations of ancient peoples. Our exploration begins in central Africa, where people of many indigenous societies learned to predict the weather with reasonable accuracy by making careful observations of the Moon. Remember that the Moon begins its monthly cycle as a crescent in the western sky just after sunset. Through long traditions of sky watching, central African societies discovered that the orientation of the crescent "horns" relative to the horizon is closely tied to local rainfall patterns (Figure 3.1).

No one knows when central Africans first developed the ability to predict weather using the lunar crescent, but we do know that people began paying attention to the sky long ago. The earliest known written astronomical record comes from central Africa near the border between modern-day Congo and Uganda. It consists of an animal bone (known as the *Ishango* bone) etched with patterns that appear to be part of a lunar calendar, probably carved around 6500 B.C.

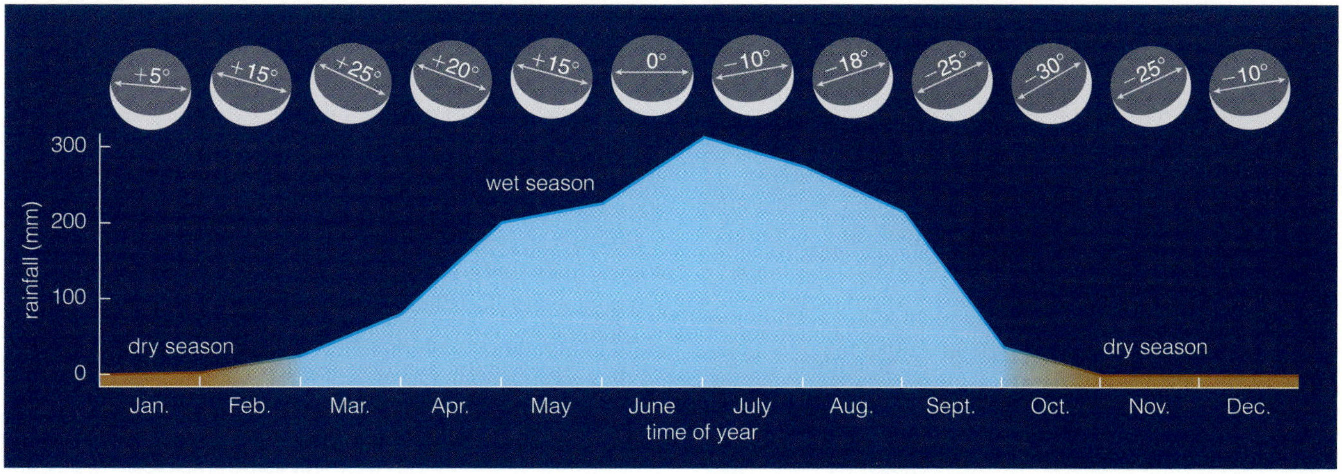

Figure 3.1 The roots of science lie in careful observation of the world around us. This diagram shows how central Africans used the orientation of the waxing crescent moon to predict the rainfall. The graph depicts the annual rainfall pattern in central Nigeria, and the Moon diagrams show the varying angle of the "horns" of a waxing crescent moon relative to the western horizon. The technique works because the orientation depends on the relative positions of the Sun and Moon along the ecliptic, which varies with the time of year, and because the equatorial climate features a rainy season and a dry season rather than the four seasons we experience at temperate latitudes. (Adapted from *Ancient Astronomers* by Anthony F. Aveni.)

Why did ancient people make such careful and detailed observations of the sky? In part, it was probably to satisfy their inherent curiosity. In the daytime, they surely recognized the importance of the Sun to their lives. At night, without electric light, they were much more aware of the starry sky than we are today. Thus, it's not surprising that they paid attention to patterns of motion in the sky and developed ideas and stories to explain what they saw.

Astronomy also played a practical role in ancient times by enabling people to keep track of time and seasons, a crucial skill for anyone who depends on agriculture for survival. This ability may seem quaint today, when digital watches tell us the precise time and date, but it required considerable knowledge and skill in ancient times, when the only clocks and calendars were in the sky.

Modern measures of time come directly from ancient observations of motion in the sky. The length of our day is the time it takes the Sun to make one full circuit of the sky. The length of a month comes from the Moon's cycle of phases [Section 2.3], and our year is based on the cycle of the seasons [Section 2.2]. The seven days of the week were named after the seven naked-eye objects that appear to move among the constellations: the Sun, the Moon, and the five planets* recognized in ancient times (Table 3.1).

• What did ancient civilizations achieve in astronomy?

Nearly all ancient civilizations practiced astronomy at some level. Many built remarkable structures for observing the sky. Let's explore a few of the many ways that ancient societies studied the sky.

Determining the Time of Day In the daytime, ancient peoples could tell time by observing the Sun's path through the sky. Many cultures probably used the shadows cast by sticks as simple sundials [Section S1.2]. The ancient Egyptians built huge obelisks, often inscribed or decorated in homage to the Sun, that probably also served as clocks (Figure 3.2).

At night, ancient people could estimate the time from the position and phase of the Moon by using the ideas we discussed in Chapter 2 (see Figure 2.22). For example, recall that a first-quarter moon sets around midnight, so it is not yet midnight if the first-quarter moon is still above your western horizon. The positions of the stars can also

Table 3.1 The Seven Days of the Week and the Astronomical Objects They Honor

In English, the correspondence between celestial objects and days is clear only if we look at the names of the objects used by the Teutonic tribes that lived in the region of modern-day Germany. French and Spanish are included because they make some of the correspondences easier to see.

Object	*Teutonic Name*	*English*	*French*	*Spanish*
Sun	Sun	Sunday	dimanche	domingo
Moon	Moon	Monday	lundi	lunes
Mars	Tiw	Tuesday	mardi	martes
Mercury	Woden	Wednesday	mercredi	miércoles
Jupiter	Thor	Thursday	jeudi	jueves
Venus	Fria	Friday	vendredi	viernes
Saturn	Saturn	Saturday	samedi	sábado

*The word *planet,* which means "wanderer," originally referred to the Sun and the Moon as well as the five visible planets. Earth was not considered a planet in ancient times, since it was assumed to be stationary at the center of the universe.

Figure 3.2 This ancient Egyptian obelisk, which stands 83 feet tall and weighs 331 tons, resides in St. Peter's Square at the Vatican in Rome. It is one of 21 surviving obelisks from ancient Egypt, most of which are now scattered around the world. Shadows cast by the obelisks may have been used to tell time.

indicate the time, as long as you know the approximate date. For example, in December the constellation Orion rises around sunset, reaches the meridian around midnight, and sets around sunrise. Hence, if it is winter and Orion is setting, dawn must be approaching. Most ancient peoples probably were adept at estimating the time of night, although written evidence is sparse.

We can trace the origins of our modern clock to ancient Egypt, some 4,000 years ago. The Egyptians divided the daylight into 12 equal parts, and we still break the 24-hour day into 12 hours each of A.M. and P.M. (The abbreviations *A.M.* and *P.M.* stand for the Latin terms *ante meridiem* and *post meridiem*, respectively, which mean "before the middle of the day" and "after the middle of the day.") However, unlike our modern hours, the Egyptian "hours" varied in length because the amount of daylight varies during the year. For example, the daylight "summer hours" were longer than the daylight "winter hours," because one-twelfth of the daylight lasts longer in summer than in winter. Only much later in history did the hour become a fixed amount of time, subdivided into 60 equal minutes each consisting of 60 equal seconds.

The Egyptians also divided the night into 12 equal parts. Egyptian *star clocks*, often found painted on the coffin lids of Egyptian pharaohs, essentially cataloged where particular stars appear in the sky at particular times of night and particular times of year. By knowing the date from their calendar and observing the positions of particular stars in the sky, the Egyptians could use the star clocks to estimate the time of night.

By about 1500 B.C., Egyptians had abandoned star clocks in favor of clocks that measure time by the flow of water through an opening of a particular size, just as hourglasses measure time by the flow of sand through a narrow neck.* These *water clocks* had the advantage of being useful even when the sky was cloudy. They eventually became the primary timekeeping instruments for many cultures, including the Greeks, Romans, and Chinese. Water clocks, in turn, were replaced by mechanical clocks in the 1600s and by electronic clocks in the twentieth century. Despite the availability of other types of clocks, sundials were common throughout ancient times and remain popular today both for their decorative value and as reminders that the Sun and stars once were our only guides to time.

Marking the Seasons Many ancient cultures built structures to help them mark the seasons. One of the oldest standing human-made structures served such a purpose: Stonehenge in southern England, which was constructed in stages from about 3100 B.C. to about 1550 B.C. (Figure 3.3). Stonehenge also served as a social gathering place and probably as a religious site. No one knows whether its original purpose was social or astronomical. Perhaps it was built for both—in ancient times, social rituals and practical astronomy probably were deeply intertwined.

Among the most spectacular structures used to mark the seasons was the Templo Mayor in the Aztec city of Tenochtitlán, located on the site of modern-day Mexico City (Figure 3.4). Twin temples stood on a flat-topped, 150-foot-high pyramid. From the vantage point of a royal observer watching from the opposite side of the plaza, the Sun rose through the notch between the temples on the equinoxes. Like Stonehenge, the Templo Mayor served important social and religious functions in addition to its astronomical role. Before the Conquistadors destroyed it, other Spanish visitors reported elaborate rituals there, sometimes including human sacrifice, at times determined by astronomical observations. After its destruction, stones from the Templo Mayor were used to build the nearby cathedral in the great plaza of Mexico City.

Ancient buildings and cities were often aligned with the cardinal directions (north, south, east, and west), enabling people to work and live in an environment that essentially represented a miniature version of the heavens. This type of alignment is common to ancient architecture around the world, and is found at such diverse sites as the Egyptian pyramids, the Forbidden City in China, and ceremonial kivas built by the Anasazi people of the American southwest

*Hourglasses using sand were not invented until about the eighth century A.D., long after the advent of water clocks. Natural sand grains vary in size, so making accurate hourglasses required technology for making uniform grains of sand.

a The remains of Stonehenge today.

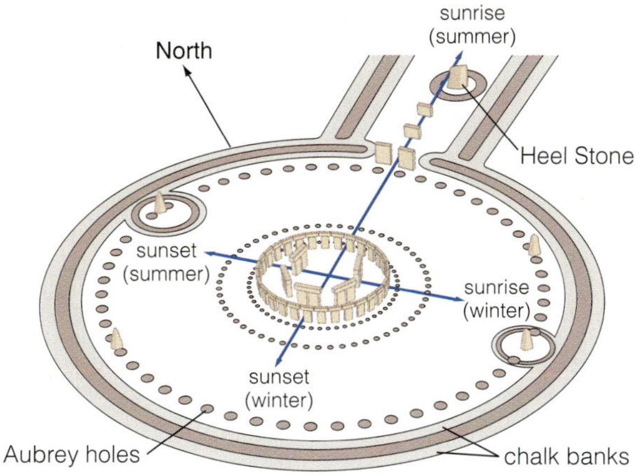

b This sketch shows how archaeologists believe Stonehenge looked upon its completion in about 1550 B.C. Several astronomical alignments are shown as they appear from the center. For example, the Sun rises directly over the Heel Stone on the summer solstice.

Figure 3.3 Stonehenge, in southern England.

(Figure 3.5). Many modern cities retain this layout, which is why you'll find so many streets that run directly north-south or east-west.

Other structures were used to mark the Sun's position on special dates such as the winter or summer solstice. For example, someone among the ancient Anasazi people carved a 19-turn spiral—often called the *Sun dagger*—on a vertical cliff face near the top of Fajada butte in Chaco Canyon, New Mexico (Figure 3.6). Three large slabs of rock in front of the spiral produce different patterns of light and shadow throughout the year. The slabs form a dagger of sunlight that pierces the center of the carved spiral only once each year—at noon on the summer solstice. Two daggers of light bracket the spiral at winter solstice, and a single dagger cuts through the center of a smaller spiral on each equinox.

Lunar Calendars Many ancient civilizations paid particular attention to lunar phases, often using them as the basis for calendars. The months on a lunar calendar generally have either 29 or 30 days, chosen to make the average agree with the approximately $29\frac{1}{2}$-day lunar cycle. Thus, a 12-month lunar calendar has only 354 or 355 days, or about 11 days fewer than a calendar based on the Sun. Such a calendar is still used in the Muslim religion. That is why the month-long fast of Ramadan (the ninth month) begins about 11 days earlier with each subsequent year.

Some lunar calendars remain roughly synchronized with solar calendars by taking advantage of an interesting coincidence: On a solar calendar, the lunar phases repeat on the same dates about every 19 years. For example, there was a full moon on October 17, 2005, and there will be a full moon 19 years later, on October 17, 2024. This 19-year cycle on which the dates of lunar phases repeat is called the *Metonic cycle*, because the Greek astronomer Meton recognized it in 432 B.C. (However, Babylonian astronomers almost certainly knew of the cycle centuries earlier.)

Lunar phases repeat with the 19-year Metonic cycle because 19 years on a solar calendar is almost precisely

Figure 3.4 This scale model shows the Templo Mayor and the surrounding plaza as they are thought to have looked before the Spanish Conquistadors destroyed Aztec civilization. The structure was used to help mark the seasons.

Figure 3.5 This large structure, more than 20 meters in diameter, is an Anasazi kiva in Chaco Canyon, New Mexico. It was built approximately 1,000 years ago. Its main axis is aligned almost precisely north-south.

a A single dagger of sunlight pierces the center of the carved spiral only at noon on the summer solstice. Other special patterns occur on the winter solstice and the equinoxes, and the Sun Dagger can also be used to observe an 18.6-year cycle of the moon.

b This photo shows Fajada butte in Chaco Canyon, New Mexico, where the Sun dagger is located on a vertical cliff near the top.

Figure 3.6 The Sun Dagger. Three large slabs of rock lie in front of a 19-turn carved spiral, creating patterns of light and shadow that vary throughout the year.

235 months on a lunar calendar. (See Problem 47 at the end of this chapter to verify this fact.) Thus, a lunar calendar will remain roughly in sync with a solar calendar if it has 235 months in each 19-year period. Because 19 years with 12 months each would mean only $19 \times 12 = 228$ months, or 7 short of the needed 235 months, lunar calendars that follow the Metonic cycle add a thirteenth month to 7 of every 19 years. For example, the Jewish calendar has a thirteenth month in the third, sixth, eighth, eleventh, fourteenth, seventeenth, and nineteenth years of each 19-year cycle. This scheme keeps the dates of Jewish holidays within about a 1-month range on a solar calendar, with precise dates repeating every 19 years. It also explains why the date of Easter changes from year to year: The New Testament ties the date of Easter to the Jewish festival of Passover. In a slight modification of the original scheme, most Western Christians now celebrate Easter on *the first Sunday after the first full moon after March 21.* If the full moon falls on Sunday, Easter is the following Sunday. (Eastern Orthodox churches calculate the date of Easter differently, because they base the date on the Julian rather than the Gregorian calendar [Section S1.1].)

In addition to following the lunar phases, some ancient cultures discovered other lunar cycles. In the Middle East more than 2,500 years ago, the ancient Babylonians achieved remarkable success in predicting eclipses, thanks to their recognition of the approximately 18-year saros cycle [Section 2.3]. The Mayans of Central America also appear to have been experts at eclipse prediction. The Mayan calendar featured a sacred cycle that almost certainly was related to eclipses. This Mayan cycle, called the *sacred round,* lasted 260 days—almost exactly $1\frac{1}{2}$ times the 173.32 days between successive eclipse seasons. Unfortunately, we know little more about the extent of Mayan knowledge because the Spanish Conquistadors burned most Mayan writings.

The Anasazi people also appear to have understood lunar cycles quite well. The rise and set positions of the full moon vary in an 18.6-year cycle (due to precession of the Moon's orbit), so that the full moon rises at its most southerly point along the eastern horizon only once every 18.6 years. At this time, known as a "major lunar standstill," the shadow of the full moon passes through the slabs of rock to lie tangent to the edge of the spiral in the Sun dagger (see Figure 3.6); 9.3 years later the lunar shadow cuts through the center of the spiral. The major lunar standstill was apparently also observed by the Anasazi at nearby Chimney Rock and at Mesa Verde in Colorado.

Ancient Structures and Archaeoastronomy Many ancient cultures apparently also made careful observations of planets and stars. The Chinese, for example, began keeping remarkably detailed records of astronomical observations at least 5,000 years ago. Their careful record keeping allowed Chinese astronomers to make many important discoveries. By the 1400s, the Chinese had built a great observatory in Beijing, which still stands today (Figure 3.7).

It's easy to establish the astronomical intentions of ancient cultures that left extensive written records, such as the Chinese and the Egyptians. In other cases, however, claims that ancient structures served astronomical purposes can be much more difficult to evaluate.

Figure 3.7 This photo shows a model of the celestial sphere and other instruments on the roof of the ancient astronomical observatory in Beijing. The observatory was built in the 1400s; the instruments shown here were built later and show a European influence brought by Jesuit missionaries.

Figure 3.8 The Big Horn Medicine Wheel in Wyoming. A study once claimed that its "spokes" have astronomically significant alignments, but later research showed the claim was in error—making this a good example of how science adapts as new data come to light.

Figure 3.9 Hundreds of lines and patterns are etched in the sand of the Nazca desert in Peru. This aerial photo shows a large etched figure of a hummingbird.

The study of ancient structures in search of astronomical connections is called *archaeoastronomy*, a word that combines archaeology and astronomy. Scientists engaged in archaeoastronomy usually start by evaluating an ancient structure to see whether it shows any particular astronomical alignments. For example, they may check to see whether an observer in a central location would see particular stars rise above specially marked stones, or whether sunlight enters through a window only on special days like the solstices or equinoxes. However, the mere existence of astronomical alignments is not enough to establish that a structure had an astronomical purpose; the alignments may be coincidental. Moreover, many claimed alignments have later proven to be in error.

Native American Medicine Wheels offer a good example of the pitfalls that can occur in trying to establish the intentions of ancient builders. Medicine wheels are stone circles found throughout the northern plains of the United States. In the 1970s, a study of the Big Horn Medicine Wheel in Wyoming (Figure 3.8) seemed to indicate that its 28 "spokes" were aligned with the rise and set of particular stars. However, later research showed that the original study had failed to take into account the motion of stars as they rise above the horizon and the way the atmosphere affects the visibility of stars at the latitude of Big Horn. In reality the spokes do *not* show any special alignments with bright stars. Moreover, if Medicine Wheels really did serve an astronomical purpose, we'd expect all of them to have been built with consistent alignments—but that is not the case. Nevertheless, many popular books and articles still report the original claims regarding astronomical purposes for Medicine Wheels, even though there is no evidence to support these claims.*

Other cases are more ambiguous. For example, ancient people in what is now Peru etched hundreds of lines and patterns in the sand of the Nazca desert. Many of the lines point to places where bright stars or the Sun rise at particular times of year, but that doesn't prove anything: With hundreds of lines, random chance ensures that many will have astronomical alignments no matter how or why they were made. The patterns, many of which are large figures of animals (Figure 3.9), have evoked even more debate. Some people think they may be representations of constellations recognized by the people who lived in the region, but we really do not know for sure.

THINK ABOUT IT

Animal figures like that in Figure 3.9 show up clearly only when seen from above. As a result, some UFO enthusiasts argue that the patterns must have been created by aliens. What do you think of this argument? Defend your opinion.

In some cases, scientists can use other clues to establish the intentions of ancient builders. For example, lodges built by the Pawnee people in Kansas feature strategically placed holes for observing the passage of constellations that figure prominently in Pawnee folklore. In this case, the correspondence between the folklore and the structural features provides a strong case for deliberate intent rather than coincidence. Similar evidence suggests that people of the Inca Empire of South America built cities and structures with deliberate astronomical alignments. Inca traditions held that its rulers were descendents of the Sun and therefore demanded that movements of the Sun be watched closely. Many of the cities and ceremonial centers of the Inca, such as the World Heritage Site of Machu Picchu, appear to have been designed so that particular arrangements of roads, temples, and pillars on the horizon marked the positions where the Sun rose or set on special dates like the solstices (Figure 3.10). Other Inca cities and ceremonial centers show similar alignments for the Sun, as well as alignments with the rise and set positions of bright stars such as the Pleiades.

A different type of evidence makes a convincing case for the astronomical sophistication of ancient Polynesians.

*Textbooks are not immune to this problem; early editions of this textbook also reported the original research without the later corrections.

Figure 3.10 The World Heritage Site of Machu Picchu has structures aligned with sunrise at the winter and summer solstices.

The Polynesians colonized and traveled among many islands of the mid- and South Pacific, even though the next island in a journey usually was too distant to be seen. Their success clearly depended upon superb navigational skills, including skills of celestial navigation. Indeed, anthropological studies have shown that the most esteemed position in Polynesian culture was that of the Navigator, a person who had acquired the knowledge necessary to navigate great distances among the islands. The Navigators employed a combination of detailed knowledge of astronomy and equally impressive knowledge of the patterns of waves and swells around different islands (Figure 3.11). The stars provided the broad navigational sense, pointing the Navigators in the general direction of their destination. As they neared a destination, the wave and swell patterns guided them to their precise landing point. A Navigator memorized all his skills and passed them to the next generation through a well-developed program for training future Navigators. Unfortunately, with the advent of modern navigational technology, many of the skills of the Navigators have been lost.

Figure 3.11 A Micronesian stick chart, an instrument used by Polynesian Navigators to represent swell patterns around islands.

From Observations to Science Before a structure such as Stonehenge or the Templo Mayor could be built, careful observations had to be made and repeated over and over to ensure their validity. Some long-term cycles became clear only when generations of observers recorded and preserved their discoveries. Carefully recorded, repeatable observations also underlie modern science. Thus, at least some elements of modern science were present in many early human cultures.

The extent to which scientific ideas developed in different societies depended on practical needs, social and political customs, and interactions with other cultures. For example, despite their "head start" in astronomical record keeping, Chinese science and technology had clearly fallen behind that of Europe by the end of the 1600s. Some historians argue that a primary reason for this decline was that the Chinese tended to regard their science and technology as state secrets. This secrecy may have slowed Chinese scientific development by preventing the broad-based collaborative science that fueled the European advance beginning in the Renaissance.

In Central America, the ancient Mayans also were ahead of their time in many ways. For example, their system of numbers and mathematics looks distinctly modern. They invented the concept of zero some 500 years before its introduction in the Eurasian world (by Hindu mathematicians, around A.D. 600). The Aztecs, Incas, Anasazi, and other ancient peoples of the Americas may have been quite advanced in many other areas as well, but few written records survive to tell the tale.

If the circumstances of history had been different, any one of these or many other cultures might have been the first to develop what we consider to be modern science. In the end, however, history takes only one of countless possible paths. The path that led to modern science emerged from the ancient civilizations of the Mediterranean and the Middle East—and especially from ancient Greece.

3.2 Ancient Greek Science

By 3000 B.C., civilization was well established in two major regions of the Middle East: Egypt and Mesopotamia (Figure 3.12). Their geographical locations placed these civilizations at a crossroads for travelers, merchants, and armies of Europe, Asia, and Africa. This mixing of cultures fostered creativity, and the broad interactions among peoples ensured that new ideas spread throughout the region.

Over the next 2,500 years, numerous great cultures arose. The ancient Egyptians built the Great Pyramids between 2700 and 2100 B.C., using their astronomical knowledge to orient the Pyramids with the cardinal directions. They also invented papyrus scrolls and ink-based writing. The Babylonians invented a method of writing on clay tablets and developed arithmetic to serve in commerce and later in astronomical calculations. Many more of our mod-

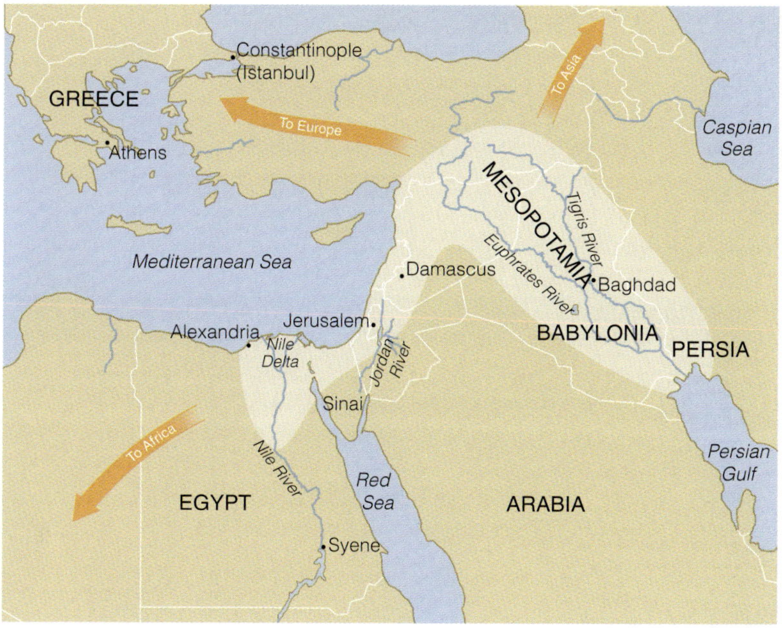

Figure 3.12 The highlighted region on this map shows the approximate extent of the ancient civilizations of Egypt and Mesopotamia. The map also identifies a few other places discussed in this text. The overlying white borders show modern country borders.

ern principles of commerce, law, and religion originated with the cultures of Egypt and Mesopotamia. But much of what we now call *science* came from ancient Greece, which rose as a power in the Middle East around 500 B.C.

• Why does modern science trace its roots to the Greeks?

The ancient Greeks stand out as the first people who tried to explain the motions of astronomical objects by using logic and geometry, without resorting to the supernatural. In this sense, they were the first people to seek knowledge in a way that looked something like what we now call science.

Greek philosophers sought to explain the architecture of the universe by constructing **models** of nature. More than two thousand years later, this idea of modeling nature is central to modern science.

Scientific models differ somewhat from the models you may be familiar with in everyday life. In our daily lives, we tend to think of models as miniature physical representations, such as model cars or airplanes. In contrast, a scientific model is a conceptual representation whose purpose is to explain and predict observed phenomena. For example, a model of Earth's climate uses logic and mathematics to represent what we know about how climate works. Its purpose is to explain and predict climate changes, such as the changes that may occur with global warming. Just as a model airplane does not faithfully represent every aspect of a real airplane, a scientific model may not fully explain all our observations of nature. Nevertheless, even the failings of a scientific model can be useful, because they often point the way toward building a better model.

The Greek models of nature sought to explain things such as the properties of matter and the motions of the stars, Sun, Moon, and planets. Many of the scientific ideas discussed in this book originated with the Greeks. (Figure 3.13 shows a time line of the most influential Greek thinkers about astronomy.) Although the Greek models may seem primitive from our modern perspective, they were an enormous step forward in scientific thinking. As an example of the modeling process, we'll look next at how the Greeks developed models to explain celestial motion.

• How did the Greeks explain planetary motion?

As we discussed in Section 2.4, most ancient Greek philosophers assumed that Earth must reside at the center of the universe. The Greeks therefore developed a sophisticated **geocentric model** of the cosmos—so named because it placed a spherical Earth at the center of the universe—that they used to explain the motions of the Sun, Moon, planets, and stars.

Early Development of the Geocentric Model We generally trace the origin of Greek science to the philosopher Thales (c. 624–546 B.C.; pronounced *thay-lees*). We encountered Thales earlier because of his legendary prediction of a solar eclipse [Section 2.3]. Thales was the first person known to have addressed the question, "What is the universe made of?" without resorting to supernatural explanations. His own guess—that the universe fundamentally consists of water and that Earth is a flat disk floating in an infinite ocean—was not widely accepted even in his own time. Nevertheless, just by asking the question he suggested that the world is inherently understandable and thus inspired others to come up with better models for the structure of the universe.

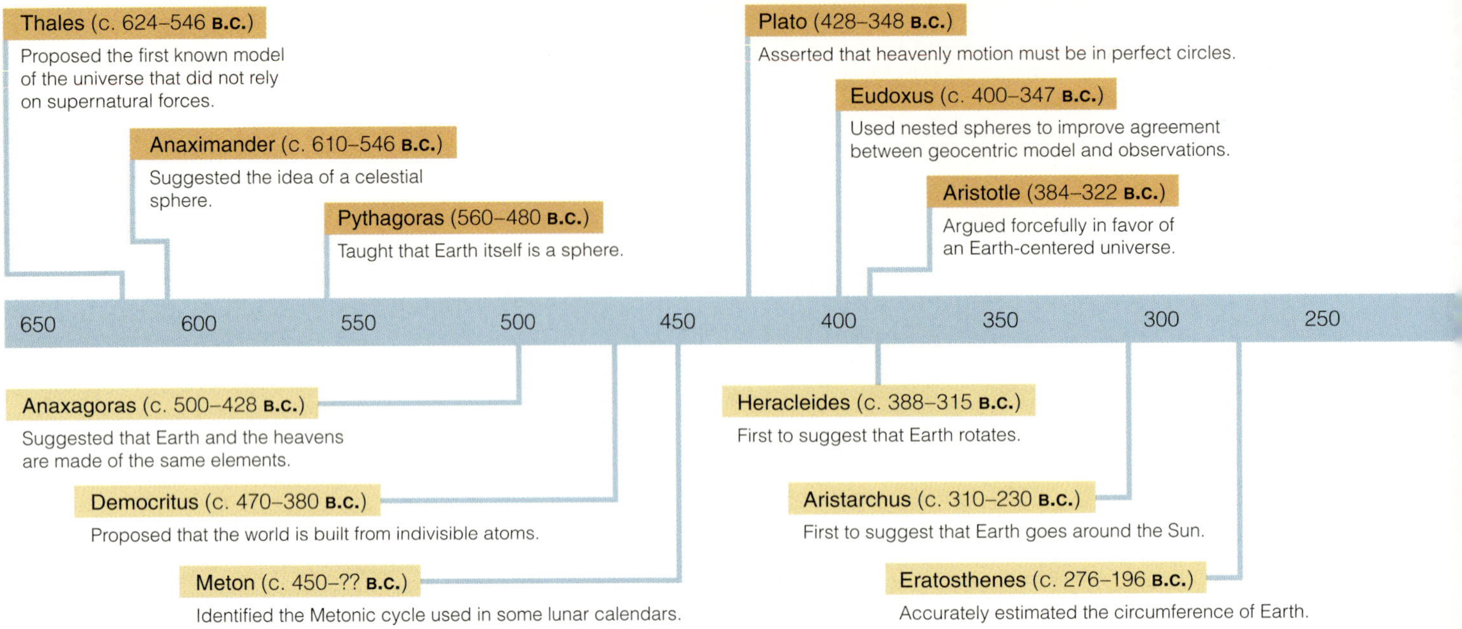

Figure 3.13 Time line for major Greek figures in the development of astronomy. (All these individuals are discussed in this book but not necessarily in this chapter.)

A more sophisticated idea followed soon after, proposed by a student of Thales named Anaximander (c. 610–546 B.C.). Anaximander suggested that Earth floats in empty space surrounded by a sphere of stars and two separate rings along which the Sun and Moon travel. Thus, we credit him with inventing the idea of a celestial sphere [Section 2.1]. Interestingly, Anaximander imagined Earth itself to be cylindrical rather than spherical in shape. He probably chose this shape because he knew Earth had to be curved in a north-south direction to explain changes in the constellations with latitude. Because the visible constellations do not change with longitude, he apparently saw no need for curvature in the east-west direction.

We do not know precisely when the Greeks first thought that Earth is round, but this idea was already being taught by about 500 B.C. by the famous mathematician Pythagoras (c. 560–480 B.C.). He and his followers imagined Earth as a sphere floating at the center of the celestial sphere. Much of their motivation for adopting a spherical Earth probably was philosophical: The Pythagoreans had a mystical interest in mathematical perfection, and they considered a sphere to be geometrically perfect. More than a century later, Aristotle cited observations of Earth's curved shadow on the Moon during lunar eclipses as evidence for a spherical Earth.

The supposed heavenly perfection of spheres influenced models of the cosmos for many centuries. Plato (428–348 B.C.), whose philosophy was based much more on pure thought than on observations, asserted that all heavenly objects move in perfect circles at constant speeds and therefore must reside on huge spheres encircling Earth (Figure 3.14). Greeks who took observations more seriously found this model problematic: The apparent retrograde motion of the planets [Section 2.4], already well known by that time, clearly showed that planets do not move at constant speeds around Earth.

An ingenious solution came from one of Plato's colleagues, Eudoxus (c. 400–347 B.C.). He created a model in which the Sun, the Moon, and the planets each had its own sphere, which in turn was nested within several other spheres. Individually, the nested spheres turned in perfect circles. By carefully choosing the sizes, rotation axes, and rotation

Figure 3.14 This model represents the Greek idea of the heavenly spheres (c. 400 B.C.). Earth is a sphere that rests in the center. The Moon, the Sun, and the planets each have their own spheres. The outermost sphere holds the stars.

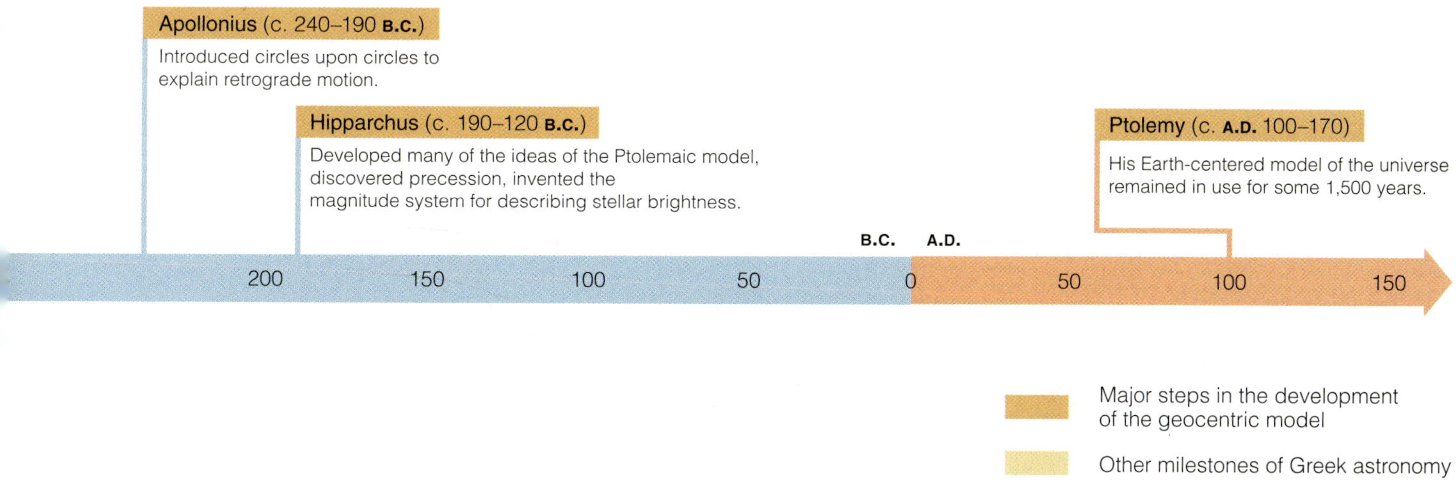

speeds for the invisible spheres, Eudoxus was able to make them work together in a way that reproduced many of the observed motions of the Sun, Moon, and planets in our sky. Other Greeks refined the model by comparing its predictions to observations and adding more spheres to improve the agreement.

This is how things stood when Aristotle (384–322 B.C.) arrived on the scene. Whether Eudoxus and his followers thought of the nested spheres as real physical objects is not clear, but Aristotle certainly did. In the cosmic model of Aristotle, all the spheres responsible for celestial motion were transparent and interconnected like the gears of a giant machine. Furthermore, Earth's position at the center of everything—as well as its composition and shape—was explained as a natural consequence of gravity. Aristotle argued that gravity pulled heavy things toward the center of the universe and allowed lighter things to float toward the heavens. Thus, all the dirt, rock, and water of the universe collected at the center, forming a spherical Earth. We now know that Aristotle was wrong about both gravity and Earth's location. However, largely because of his persuasive arguments for an Earth-centered universe, the geocentric view dominated Western thought for almost 2,000 years.

SPECIAL TOPIC Eratosthenes Measures Earth

In a remarkable feat, the Greek astronomer and geographer Eratosthenes estimated the size of Earth in about 240 B.C. He did it by comparing the altitude of the Sun on the summer solstice in the Egyptian cities of Syene (modern-day Aswan) and Alexandria.

Eratosthenes knew that the Sun passed directly overhead in Syene on the summer solstice. He also knew that in the city of Alexandria to the north, the Sun came within only 7° of the zenith on the summer solstice. He therefore reasoned that Alexandria must be 7° of latitude to the north of Syene (see figure). Because 7° is $\frac{7}{360}$ of a circle, he concluded that the north-south distance between Alexandria and Syene must be $\frac{7}{360}$ of the circumference of Earth.

Eratosthenes estimated the north-south distance between Syene and Alexandria to be 5,000 stadia (the *stadium* was a Greek unit of distance). Thus, he concluded that:

$$\frac{7}{360} \times \text{circumference of Earth} = 5000 \text{ stadia}$$

From this he found Earth's circumference to be about 250,000 stadia.

Today, we don't know exactly what distance a stadium meant to Eratosthenes. Based on the actual sizes of Greek stadiums, it must have been about $\frac{1}{6}$ kilometer. Thus, Eratosthenes estimated the circumference of the Earth to be about $\frac{250{,}000}{6} = 42{,}000$ kilometers—remarkably close to the real value of just over 40,000 kilometers.

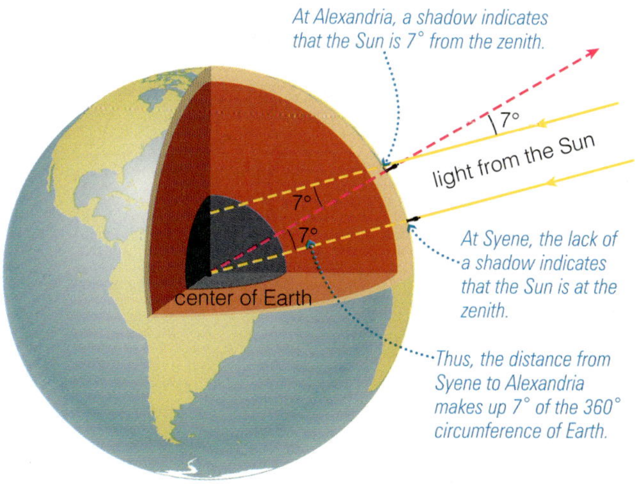

At noon on the summer solstice, the Sun appears at the zenith in Syene but 7° shy of the zenith in Alexandria. Thus, 7° of latitude, which corresponds to a distance of $\frac{7}{360}$ of Earth's circumference, must separate the two cities.

Ptolemy's Synthesis of the Geocentric Model Greek modeling of the cosmos became much more quantitative after Aristotle, culminating in the work of Claudius Ptolemy (c. A.D. 100–170; pronounced *TOL-e-mee*). Ptolemy's model still placed Earth at the center of the universe, but it differed in significant ways from the nested spheres of Eudoxus and Aristotle. We refer to Ptolemy's model as the **Ptolemaic model** to distinguish it from earlier geocentric models.

To explain the apparent retrograde motion of the planets, the Ptolemaic model applied an idea first suggested by Apollonius (c. 240–190 B.C.). This idea held that each planet moves around Earth on a small circle that turns upon a larger circle (Figure 3.15). (The small circle is sometimes called an *epicycle,* and the larger circle is called a *deferent.*) A planet following this circle-upon-circle motion traces a loop as seen from Earth, with the backward portion of the loop mimicking apparent retrograde motion.

Ptolemy also relied heavily on the work of Hipparchus (c. 190–120 B.C.), considered one of the greatest of the Greek astronomers. Among his many accomplishments, Hipparchus developed the circle-upon-circle idea of Apollonius into a model that could predict planetary positions. To do this, Hipparchus had to add several features to the basic idea; for example, he included even smaller circles that moved upon the original set of small circles, and he positioned the large circles slightly off-center from Earth.

Ptolemy's great accomplishment was to adapt and synthesize these earlier ideas into a single system that agreed quite well with the astronomical observations available at the time. In the end, he created and published a model that could correctly forecast future planetary positions to within a few degrees of arc—roughly equivalent to the size of your hand viewed at arm's length against the sky. His model generally worked so well that it remained in use for the next 1,500 years. When his book describing the model was translated by Arabic scholars around A.D. 800, they gave it the title *Almagest,* derived from words meaning "the greatest compilation."

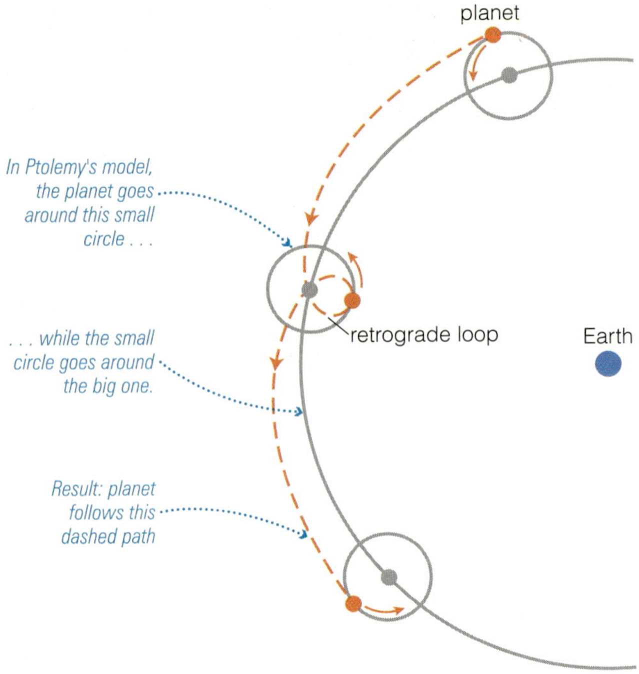

Figure 3.15 Interactive Figure This diagram shows how the Ptolemaic model accounted for apparent retrograde motion. Each planet is assumed to move around a small circle that turns upon a larger circle. The resulting path (dashed) includes a loop in which the planet goes backward as seen from Earth.

COMMON MISCONCEPTIONS

Columbus and a Flat Earth

A widespread myth holds that Columbus proved Earth to be round rather than flat. In fact, knowledge of the round Earth predated Columbus by nearly 2,000 years. However, it is probably true that most people in Columbus's day believed Earth to be flat, largely because of the poor state of education. The vast majority of the public was illiterate and unaware of scholarly evidence for a spherical Earth. Interestingly, Columbus's primary argument with other scholars concerned the *distance* from Europe to Asia going westward—and it was Columbus who was wrong. He underestimated the true distance and as a result was woefully unprepared for the voyage to Asia he thought he was undertaking. Indeed, his voyages would almost certainly have ended in disaster had it not been for the presence of the Americas, which offered a safe landing well to the east of Asia.

• How was Greek knowledge preserved through history?

One reason Greek thought gained such broad influence was that the Greeks proved as adept at politics and war as they were at philosophy. In about 330 B.C., Alexander the Great (356–323 B.C.) began a series of conquests that expanded the Greek Empire throughout the Middle East, absorbing all the former empires of Egypt and Mesopotamia. Alexander was more than just a military leader. He was also keenly interested in science and education. As a teenager, he had been tutored by none other than Aristotle.

Alexander encouraged the pursuit of knowledge and respect for foreign cultures. On the Nile delta in Egypt, he founded the city of Alexandria, which soon became a center of world culture. Not long after his death, the city of Alexandria commenced work on a great library and research center: the Library of Alexandria, which opened in about 300 B.C. (Figure 3.16). It remained the world's preeminent center of research for some 700 years, surviving well past the fall of the classical Greek Empire and into the heyday of the Roman Empire.

a This rendering shows an artist s reconstruction, based on scholarly research, of how the Great Hall may have looked in the ancient Library of Alexandria.

b A similar rendering to (a), this time showing a scroll room in the ancient library.

c The New Library of Alexandria, in Egypt, which opened in 2003.

Figure 3.16 The ancient Library of Alexandria thrived for some 700 years, starting in about 300 B.C.

Destruction of the Library of Alexandria Much of the Library of Alexandria's long history remains unknown today, in part because the books that recorded this history were deliberately destroyed along with the Library. Nevertheless, historians are confident that the Library's demise was intertwined with the life and death of one of the few prominent female scholars of the ancient world: a woman named Hypatia (A.D. 370–415). Hypatia was one of the last resident scholars of the Library, as well as the director of the observatory in Alexandria and one of the leading mathematicians and astronomers of her time. Unfortunately, she became a scapegoat during a time of rising sentiment against free inquiry and was murdered by an anti-intellectual mob in A.D. 415. The final destruction of the Library took place not long after her death.

At its peak, the Library of Alexandria may have held more than a half million books, handwritten on papyrus scrolls. Most of these scrolls were ultimately burned, their contents lost forever. In some cases, we were left with small fragments or titles or descriptions from other works, teasing us with intriguing subject matter about which we will never read. In commemoration of the ancient Library, the country of Egypt recently built a New Library of Alexandria (more commonly called the *Bibliotheca Alexandrina*). Egypt hopes that this new library, which opened in 2003, will once again make Alexandria a global center for research and scientific excellence.

THINK ABOUT IT

Estimate the number of books you're likely to read in your lifetime, and compare this number to the half million books once housed in the Library of Alexandria. Can you think of other ways to put into perspective the loss of ancient wisdom resulting from the destruction of the Library of Alexandria?

Islamic Preservation of Science Much of Greek knowledge was lost with the destruction of the Library of Alexandria. That which survived was preserved primarily thanks to the rise of a new center of intellectual inquiry in Baghdad (in present-day Iraq). While European civilization fell into the period of intellectual decline known as the Dark Ages, scholars of the new religion of Islam sought knowledge of mathematics and astronomy in hopes of better understanding the wisdom of Allah. During the eighth and ninth centuries A.D., scholars working in the Muslim empire centered in Baghdad translated and thereby saved many ancient Greek works.

Around A.D. 800, the Islamic leader Al-Mamun (A.D. 786–833) established a "House of Wisdom" in Baghdad with a mission much like that of the destroyed Library of Alexandria. Founded in a spirit of openness and tolerance, the House of Wisdom employed Jews, Christians, and Muslims, all working together in scholarly pursuits. Using the translated Greek scientific manuscripts as building blocks, these scholars developed the mathematics of algebra and many new instruments and techniques for astronomical observation. Most of the official names of constellations and stars come from Arabic because of the work of the scholars at Baghdad. If you look at a star chart, you will see that the names of many bright stars begin with *al* (for example, Aldebaran, Algol), which simply means "the" in Arabic.

Toward a Scientific Renaissance The Islamic world of the Middle Ages was in frequent contact with Hindu scholars from India, who in turn brought knowledge of ideas and discoveries from China. Hence, the intellectual center in Baghdad achieved a synthesis of the surviving work of the ancient Greeks and that of the Indians and the Chinese. The accumulated knowledge of the Arabs spread throughout the Byzantine Empire (the eastern part of the former Roman Empire). When the Byzantine capital of Constantinople (modern-day Istanbul) fell to the Turks in 1453, many Eastern scholars headed west to Europe, carrying with them the knowledge that helped ignite the European Renaissance. It is to this story that we turn next in our quest to understand the origins of modern science.

3.3 The Copernican Revolution

The Greeks and other ancient peoples developed many important ideas of science, but what we now think of as science arose during the European Renaissance. Within a half century after the fall of Constantinople, Polish scientist Nicholas Copernicus began the work that ultimately overturned the Earth-centered Ptolemaic model. Over the next century and a half, philosophers and scientists (who were often one and the same) debated and tested his radical view of the cosmos. Ultimately, the new ideas introduced by Copernicus fundamentally changed the way we perceive our place in the universe. This dramatic change, known as the **Copernican revolution**, spurred the development of virtually all modern science and technology.

• How did Copernicus, Tycho, and Kepler challenge the Earth-centered idea?

Recall that Aristarchus suggested the idea of a Sun-centered solar system more than 1,700 years before Copernicus was born [Section 2.4]. However, before the time of the Copernican revolution, few people took the idea seriously . Let's examine how three key individuals—Nicholas Copernicus, Tycho Brahe, and Johannes Kepler—challenged the prevailing dogma that our planet must be the center of the universe.

Copernicus Nicholas Copernicus was born in Toruń, Poland, on February 19, 1473. His family was wealthy and he received a first-class education in mathematics, medicine, and law. He began studying astronomy in his late teens. By that time, tables of planetary motion based on the Ptolemaic model were noticeably inaccurate. But few people were willing to undertake the difficult calculations required to revise the tables. Indeed, the best tables available had been compiled some two centuries earlier under the guidance of Spanish monarch Alphonso X (1221–1284). Commenting on the tedious nature of the work required to make these *Alfonsine Tables,* the monarch is said to have complained, "If I had been present at the creation, I would have recommended a simpler design for the universe."

Copernicus (1473–1543)

In his quest for a better way to predict planetary positions, Copernicus adopted Aristarchus's Sun-centered idea. He was probably motivated in large part by the much simpler explanation for apparent retrograde motion offered by a Sun-centered system (see Figure 2.33). As he worked out the mathematical details of his model, Copernicus also discovered simple geometric relationships that allowed him to calculate each planet's orbital period around the Sun and its relative distance from the Sun in terms of the Earth-Sun distance (see Mathematical Insight S1.1, p. 94). The success of his model in providing a geometric layout for the solar system further convinced him that the Sun-centered idea must be correct.

Copernicus was hesitant to publish his work, fearing that his suggestion that Earth moved would be considered absurd. Nevertheless, he discussed his system with other scholars, generating great interest. At the urging of some of these scholars, including some high-ranking officials of the Catholic Church, he finally agreed to publish his work. Copernicus saw the first printed copy of his book, *De Revolutionibus Orbium Coelestium* ("Concerning the Revolutions of the Heavenly Spheres"), on the day he died—May 24, 1543.

Early supporters of Copernicus were drawn to the aesthetic advantages of his model. However, it did not make substantially better predictions than Ptolemy's model, largely because Copernicus still believed that heavenly

SPECIAL TOPIC Aristotle

Aristotle (384–322 B.C.) is among the best-known philosophers of the ancient world. Both his parents died when he was a child, and he was raised by a family friend. In his 20s and 30s, he studied under Plato (428–348 B.C.) at Plato's Academy. He later founded his own school, called the Lyceum, where he studied and lectured on virtually every subject. Historical records tell us that his lectures were collected and published in 150 volumes. About 50 of these volumes survive to the present day.

Many of Aristotle's scientific discoveries were about the nature of plants and animals. He studied more than 500 animal species in detail, dissecting specimens of nearly 50 species, and came up with a strikingly modern classification system. For example, he was the first person to recognize that dolphins should be classified with land mammals rather than with fish. In mathematics, he is known for laying the foundations of mathematical logic. Unfortunately, he was far less successful in physics and astronomy, areas in which many of his claims turned out to be wrong.

Despite his wide-ranging discoveries and writings, Aristotle's philosophies were not particularly influential until many centuries after his death. His books were preserved and valued by Islamic scholars but were unknown in Europe until they were translated into Latin in the twelfth and thirteenth centuries. Aristotle achieved his near-reverential status only after St. Thomas Aquinas (1225–1274) integrated Aristotle's philosophy into Christian theology. In the ancient world, Aristotle's greatest influence came indirectly, through his role as the tutor of Alexander the Great.

motion must occur in perfect circles. Because the true orbits of the planets are *not* circles, Copernicus found it necessary to add circles upon circles to his system, just as Ptolemy had done. As a result, his complete model was no more accurate and no less complex than the Ptolemaic model, and the Sun-centered idea won relatively few converts in the 50 years after it was published. After all, why trade thousands of years of tradition for a new model that worked equally poorly?

Tycho Part of the difficulty faced by astronomers who sought to improve either the Ptolemaic or the Copernican system was a lack of quality data. The telescope had not yet been invented, and existing naked-eye observations were not very accurate. In the late 1500s, Danish nobleman Tycho Brahe (1546–1601), usually known simply as Tycho (pronounced *tie-koe*), set about correcting this problem.

When Tycho was a young boy, his family discouraged his interest in astronomy. He therefore followed his passion in secret, learning the constellations from a miniature model of a celestial sphere that he kept hidden. As he grew older, Tycho was often arrogant about both his noble birth and his intellectual abilities. At age 20, he fought a duel with another student over which of them was the better mathematician. Part of Tycho's nose was cut off, and he designed a replacement piece made of silver and gold.

In 1563, Tycho decided to observe a widely anticipated alignment of Jupiter and Saturn. To his surprise, the alignment occurred nearly two days later than Copernicus had predicted. Resolving to improve the state of astronomical prediction, he set about compiling careful observations of stellar and planetary positions in the sky.

Tycho's fame grew after he observed what he called a *nova,* meaning "new star," in 1572. By measuring its parallax and comparing it to the parallax of the Moon, he proved that the nova was much farther away than the Moon. (Today, we know that Tycho saw a *supernova*—the explosion of a distant star [Section 17.3].) In 1577, Tycho made similar observations of a comet and proved that it too lay in the realm of the heavens. Others, including Aristotle, had argued that comets were phenomena of Earth's atmosphere. King Frederick II of Denmark decided to sponsor Tycho's ongoing work, providing him with money to build an unparalleled observatory for naked-eye observations (Figure 3.17). After Frederick II died in 1588, Tycho moved to Prague, where German emperor Rudolf II supported his work.

Tycho Brahe (1546–1601)

Over a period of three decades, Tycho and his assistants compiled naked-eye observations accurate to better than 1 arcminute—less than the thickness of a fingernail viewed at arm's length. Because the telescope was invented shortly after his death, Tycho's data remain the best set of naked-eye observations ever made. Despite the quality of his observations, Tycho never succeeded in coming up with a satisfying explanation for planetary motion. He was convinced that the *planets* must orbit the Sun, but his inability to detect stellar parallax [Section 2.4] led him to conclude that Earth must remain stationary. Thus, he advocated a model in which the Sun orbits Earth while all other planets orbit the Sun. Few people took this model seriously.

Kepler Tycho failed to explain the motions of the planets satisfactorily, but he succeeded in finding someone who could: In 1600, he hired the young German astronomer Johannes Kepler (1571–1630). Kepler and Tycho had a strained relationship, but Tycho recognized the talent of his young apprentice. In 1601, as he lay on his deathbed, Tycho begged Kepler to find a system that would make sense

Figure 3.17 Tycho Brahe in his naked-eye observatory, which worked much like a giant protractor. He could sit and observe a planet through the rectangular hole in the wall as an assistant used a sliding marker to measure the angle on the protractor.

Johannes Kepler (1571–1630)

of the observations so "that it may not appear I have lived in vain."*

Kepler was deeply religious and believed that understanding the geometry of the heavens would bring him closer to God. Like Copernicus, he believed that Earth and the other planets traveled around the Sun in circular orbits. He worked diligently to match circular motions to Tycho's data.

Kepler labored with particular intensity to find an orbit for Mars, which posed the greatest difficulties in matching the data to a circular orbit. After years of calculation, Kepler found a circular orbit that matched all of Tycho's observations of Mars's position along the ecliptic (east-west) to within 2 arcminutes. However, the model did not correctly predict Mars's positions north or south of the ecliptic. Because Kepler sought a physically realistic orbit for Mars, he could not (as Ptolemy and Copernicus had done) tolerate one model for the east-west positions and another for the north-south positions. He attempted to find a unified model with a circular orbit. In doing so, he found that some of his predictions differed from Tycho's observations by as much as 8 arcminutes.

Kepler surely was tempted to ignore these discrepancies and attribute them to errors by Tycho. After all, 8 arcminutes is barely one-fourth the angular diameter of the full moon. But Kepler trusted Tycho's careful work. The small discrepancies finally led Kepler to abandon the idea of circular orbits—and to find the correct solution to the ancient riddle of planetary motion. About this event, Kepler wrote:

> *If I had believed that we could ignore these eight minutes [of arc], I would have patched up my hypothesis accordingly. But, since it was not permissible to ignore, those eight minutes pointed the road to a complete reformation in astronomy.*

Kepler's key discovery was that planetary orbits are not circles but instead are a special type of oval called an **ellipse**. You probably know how to draw a circle by putting a pencil on the end of a string, tacking the string to a board, and pulling the pencil around (Figure 3.18a). Drawing an ellipse is similar, except that you must stretch the string around *two* tacks (Figure 3.18b). The locations of the two tacks are called the **foci** (singular, **focus**) of the ellipse. The long axis of the ellipse is called its *major axis*, each half of which is called a **semimajor axis**; as we'll see shortly, the length of the semimajor axis is particularly important in astronomy. The short axis is called the *minor axis*. By altering the distance between the two foci while keeping the length of string the same, you can draw ellipses of varying **eccentricity**, a quantity that describes how much an ellipse deviates from a perfect circle (Figure 3.18c). A circle is an ellipse with zero eccentricity, and greater eccentricity means a more elongated ellipse.

Kepler's decision to trust the data over his preconceived beliefs marked an important transition point in the history of science. Once he abandoned perfect circles in favor of ellipses, Kepler soon came up with a model that could predict planetary positions with far greater accuracy than

*For a particularly moving version of the story of Tycho and Kepler, see Episode 3 of Carl Sagan's *Cosmos* video series.

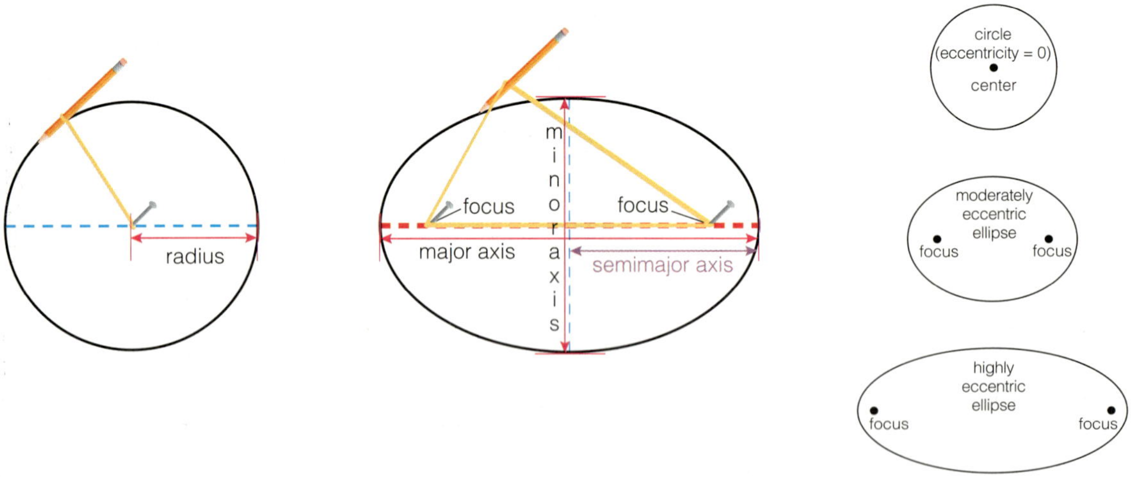

a Drawing a circle with a string of fixed length.

b Drawing an ellipse with a string of fixed length.

c *Eccentricity* describes how much an ellipse deviates from a perfect circle.

Figure 3.18 Interactive Figure An ellipse is a special type of oval. These diagrams show how an ellipse differs from a circle and how different ellipses vary in their eccentricity.

Ptolemy's Earth-centered model. Kepler's model withstood the test of time and became accepted not only as a model of nature but as a deep, underlying truth about planetary motion.

Orbits and Kepler's Laws Tutorial, Lessons 2–4

• What are Kepler's three laws of planetary motion?

Kepler summarized his discoveries in three simple laws that we now call **Kepler's laws of planetary motion**. He published the first two laws in 1610 and the third in 1618.

Kepler's First Law Kepler's first law tells us that *the orbit of each planet around the Sun is an ellipse with the Sun at one focus* (Figure 3.19). (Nothing is at the other focus.) In essence, this law tells us that a planet's distance from the Sun varies during its orbit. It is closest at the point called **perihelion** (from the Greek for "near the Sun") and farthest at the point called **aphelion** (from the Greek for "away from the Sun"). The *average* of a planet's perihelion and aphelion distances is the length of its *semimajor axis.* We will refer to this simply as the planet's average distance from the Sun.

Kepler's Second Law Kepler's second law states that *as a planet moves around its orbit, it sweeps out equal areas in equal times.* As shown in Figure 3.20, this means the planet moves a greater distance when it is near perihelion than it does in the same amount of time near aphelion. That is, the planet travels faster when it is nearer to the Sun and slower when it is farther from the Sun.

Kepler's Third Law Kepler's third law tells us that *more distant planets orbit the Sun at slower average speeds, obeying*

MATHEMATICAL INSIGHT 3.1 Eccentricity and Planetary Orbits

We describe how much a planet's orbit differs from a perfect circle by stating its orbital *eccentricity.* There are several equivalent ways to define the eccentricity of an ellipse, but the simplest is shown in Figure 1. We define c to be the distance from each focus to the center of the ellipse and a to be the length of the semimajor axis. The eccentricity is defined to be

$$e = \frac{c}{a}$$

Notice that $c = 0$ in a perfect circle, because a circle is an ellipse with both foci *in* the center, so this formula gives an eccentricity of 0 for a perfect circle, just as we expect.

You can find the eccentricity of planetary orbits in many tables of planetary data, such as Table E.2 in Appendix E of this book. Once you know the eccentricity, the following formulas allow you to calculate the perihelion and aphelion distances of the planet (Figure 2):

$$\text{perihelion distance} = a(1 - e)$$
$$\text{aphelion distance} = a(1 + e)$$

Example: What are Earth's perihelion and aphelion distances?

Solution:

Step 1 Understand: From the given formulas, we see that calculating Earth's perihelion and aphelion distances requires knowing Earth's orbital eccentricity and the length of its semimajor axis, which is its average distance from the Sun. From Table E.2, Earth's orbital eccentricity is $e = 0.017$ and its average distance is $a = 149.6$ million km.

Step 2 Solve: We use the values we've found for e and a in the above equations:

$$\begin{aligned} \text{Earth's perihelion distance} &= a(1 - e) \\ &= (149.6 \times 10^6 \text{ km})(1 - 0.017) \\ &= 147.1 \times 10^6 \text{ km} \\ \text{Earth's aphelion distance} &= a(1 + e) \\ &= (149.6 \times 10^6 \text{ km})(1 + 0.017) \\ &= 152.1 \times 10^6 \text{ km} \end{aligned}$$

Step 3 Explain: We have found that Earth's perihelion distance is 147.1 million kilometers and its aphelion distance is 152.1 million kilometers. Because perihelion is Earth's closest point to the Sun and aphelion is its farthest point, we've learned that Earth's distance from the Sun varies between 147.1 and 152.1 million kilometers over the course of each orbit.

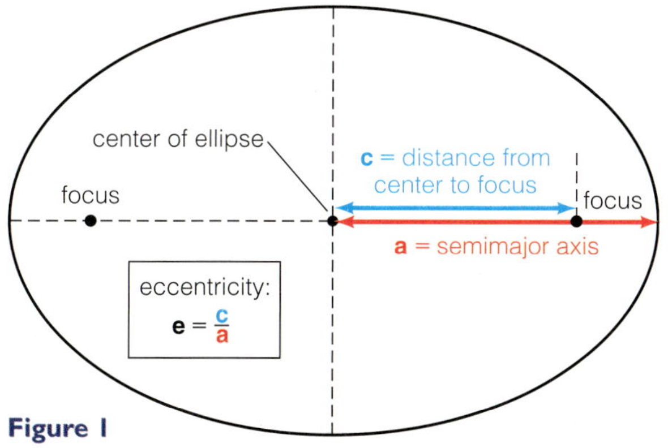

Figure 1

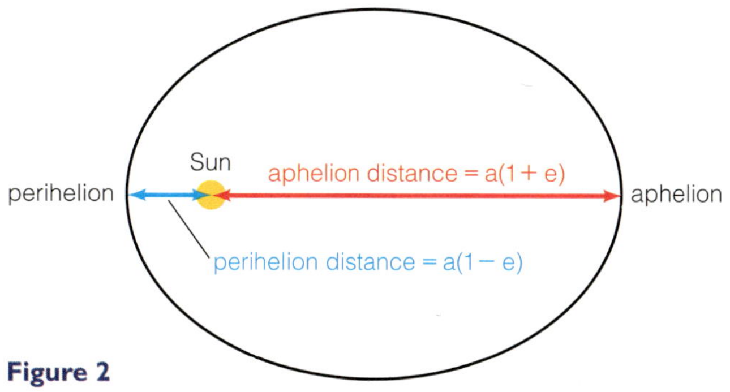

Figure 2

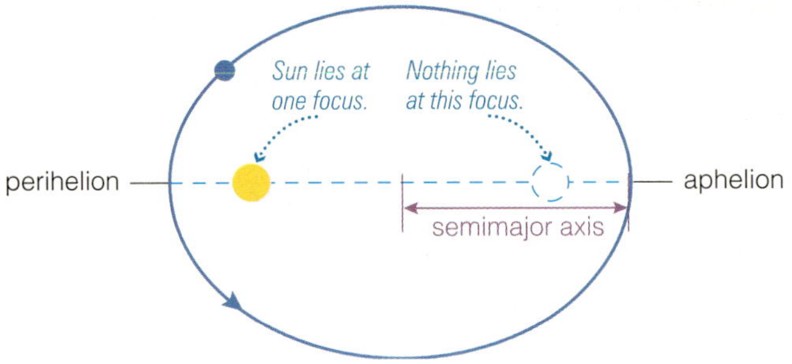

Figure 3.19 Interactive Figure Kepler's first law: The orbit of each planet about the Sun is an ellipse with the Sun at one focus. (The eccentricity shown here is exaggerated compared to the actual eccentricities of the planets.)

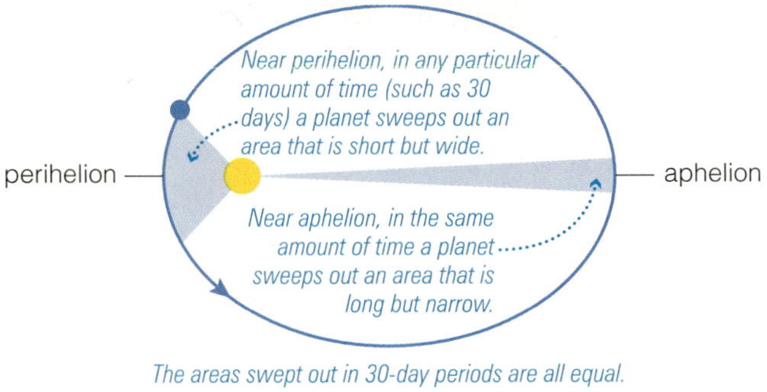

Figure 3.20 Interactive Figure Kepler's second law: As a planet moves around its orbit, it sweeps out equal areas (the shaded regions) in equal times.

a precise mathematical relationship. The relationship is written

$$p^2 = a^3$$

where p is the planet's orbital period in years and a is its average distance from the Sun in astronomical units. Figure 3.21a shows the $p^2 = a^3$ law graphically. Notice that the square of each planet's orbital period (p^2) is indeed equal to the cube of its average distance from the Sun (a^3). Because Kepler's third law relates a planet's orbital distance to its orbital time (period), we can use the law to calculate a planet's average orbital speed.* Figure 3.21b shows the result, confirming that more distant planets orbit the Sun more slowly.

*To calculate orbital speed from Kepler's third law: Remember that speed = distance/time. For a complete planetary orbit, the distance is orbital circumference, which is $2\pi a$ (where a is the semimajor axis, roughly the "radius" of the orbit); the time is the orbital period, p. Thus, the orbital speed is $(2\pi a)/p$. From Kepler's third law, $p = a^{3/2}$. Plugging this value for p into the orbital speed equation, we find that a planet's orbital speed is $2\pi/\sqrt{a}$; the graph of this equation is the curve in Figure 3.21b.

The fact that more distant planets move more slowly led Kepler to suggest that planetary motion might be the result of a force from the Sun. He even speculated about the nature of this force, guessing that it might be related to magnetism. (This idea, shared by Galileo, was first suggested by William Gilbert [1544–1603], an early believer in the Copernican system.) Kepler was right about the existence of a force, but wrong in his guess of magnetism. More than a half century later, Isaac Newton explained planetary motion as a consequence of the gravitational force attracting planets to the Sun [Section 4.4].

THINK ABOUT IT

Suppose a comet has an orbit that brings it quite close to the Sun at its perihelion and beyond Mars at its aphelion, but with an average distance (semimajor axis) of 1 AU. According to Kepler's laws, how long would the comet take to complete each orbit of the Sun? Would it spend most of its time close to the Sun, far from the Sun, or somewhere in between? Explain.

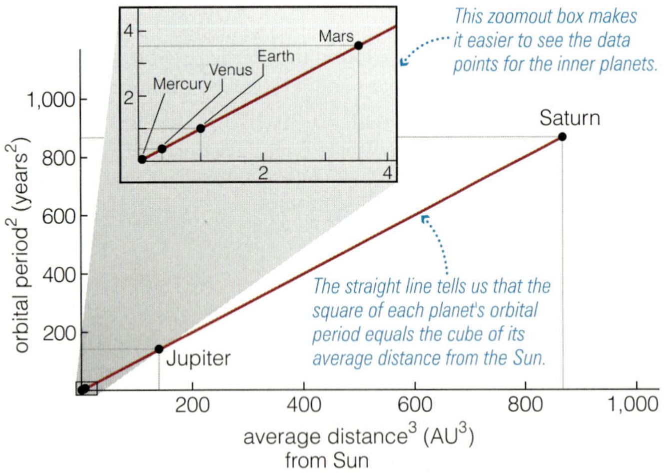

a This graph shows that Kepler s third law ($p^2 = a^3$) does indeed hold true; for simplicity, the graph shows only the planets known in Kepler s time.

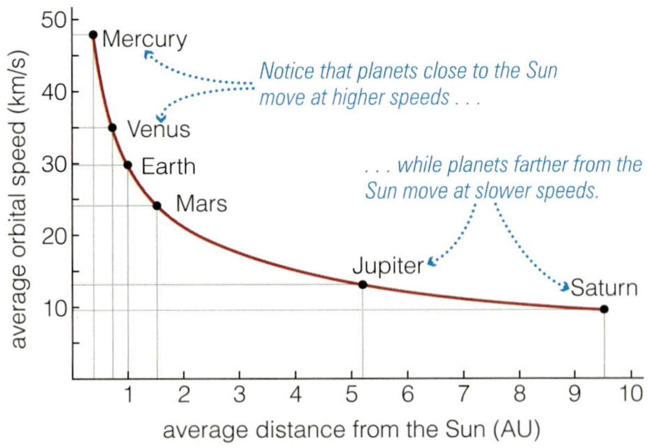

b This graph shows how the orbital speeds of the planets depend on their distances from the Sun: More distant planets orbit the Sun more slowly. (Kepler knew the form of the relationship but could not determine speeds in km/s because the numerical value of the astronomical unit was not yet known.)

Figure 3.21 Graphs based on Kepler's third law.

• How did Galileo solidify the Copernican revolution?

The success of Kepler's laws in matching Tycho's data provided strong evidence in favor of Copernicus's placement of the Sun, rather than Earth, at the center of the solar system. Nevertheless, many scientists still voiced reasonable objections to the Copernican view. There were three basic objections, all rooted in the 2,000-year-old beliefs of Aristotle and other ancient Greeks.

- First, Aristotle had held that Earth could not be moving because, if it were, objects such as birds, falling stones, and clouds would be left behind as Earth moved along its way.
- Second, the idea of noncircular orbits contradicted Aristotle's claim that the heavens—the realm of the Sun, Moon, planets, and stars—must be perfect and unchanging.
- Third, as the ancients had argued, stellar parallax ought to be detectable if Earth orbits the Sun [Section 2.4].

Galileo (1564–1642)

Galileo Galilei (1564–1642), a contemporary and correspondent of Kepler, answered all three objections.

Galileo (nearly always known by only his first name) defused the first objection with experiments that almost single-handedly overturned the Aristotelian view of physics. In particular, he used experiments with rolling balls to demonstrate that a moving object remains in motion *unless* a force acts to stop it (an idea now codified in Newton's first law of motion [Section 4.2]). This contradicted Aristotle's claim that the natural tendency of any

MATHEMATICAL INSIGHT 3.2 Kepler's Third Law

When Kepler discovered his third law ($p^2 = a^3$), he knew only that it applied to the orbits of planets around the Sun. However, it turns out to apply much more generally; for example, it works for asteroids and comets as well as planets. We'll see its most general form in Mathematical Insight 4.3, but even in its original form we can use Kepler's third law under the following conditions:

1. In its original form $p^2 = a^3$, Kepler's third law applies only to objects *orbiting the Sun* or another star of precisely the same mass as the Sun.
2. We can use Kepler's third law in its original form only if we measure orbital *periods in years* and orbital *distances in AU.*

Example 1: The largest asteroid, Ceres, orbits the Sun at an average distance (semimajor axis) of 2.77 AU. What is its orbital period?

Solution:

Step 1 Understand: We know that Kepler's third law will allow us to find an orbital period (p) from an average distance (a), but only if the problem meets the two conditions described above. The first condition is met because the asteroid orbits the Sun. The second is met because we are given the orbital distance in AU. Thus, we can use Kepler's third law in its original form.

Step 2 Solve: Notice that Kepler's law is written in a way that gives us p^2, but we are looking for the orbital period, p. Thus, we first solve Kepler's third law for p and then substitute the given value of $a = 2.77$ AU.

$$p^2 = a^3 \Rightarrow p = \sqrt{a^3} = \sqrt{2.77^3} = 4.6$$

Notice that because of the special conditions attached to the use of Kepler's third law in its original form, we do *not* include units when working with it; we know we'll get a period in years as long as we start with a distance in AU.

Step 3 Explain: We have found that Ceres has an orbital period of 4.6 years, meaning that it takes 4.6 years to complete each orbit around the Sun.

Example 2: A new planet is discovered to be orbiting a star with the same mass as our Sun. The planet orbits the star every 3 months. What is its average distance from its star?

Solution:

Step 1 Understand: This time we are given an orbital period and want to find the orbital distance. Again, we can use Kepler's third law in its original form only if the problem meets the two conditions above. The first condition is met because the planet is orbiting a star with the same mass as our Sun. The second is not met, because the orbital period is given in months rather than years. However, we can easily convert it to years: Three months is $\frac{1}{4}$ year, so we have $p = 0.25$ year.

Step 2 Solve: Kepler's law is written in a way that gives us a^2, but we are looking for the average distance, a. We therefore solve Kepler's third law for a and then substitute the orbital period of $p = 0.25$ year:

$$p^2 = a^3 \Rightarrow a = \sqrt[3]{p^2} = \sqrt[3]{0.25^2} = 0.40$$

Step 3 Explain: We have found that the planet orbits its star at an average distance of 0.4 AU. By comparing this result to the distances of planets in our own solar system given in Table E.2, we find that this planet's average orbital distance is just slightly larger than that of the planet Mercury in our own solar system.

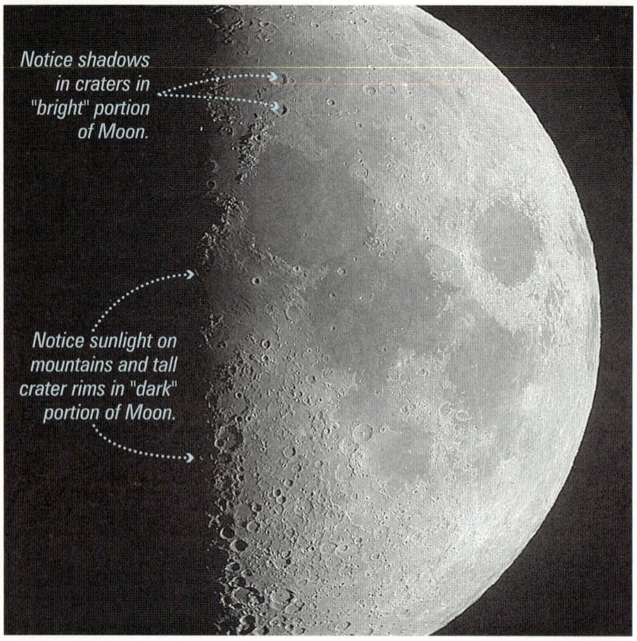

Figure 3.22 Interactive Photo The shadows cast by mountains and crater rims near the dividing line between the light and dark portions of the lunar face prove that the Moon's surface is not perfectly smooth.

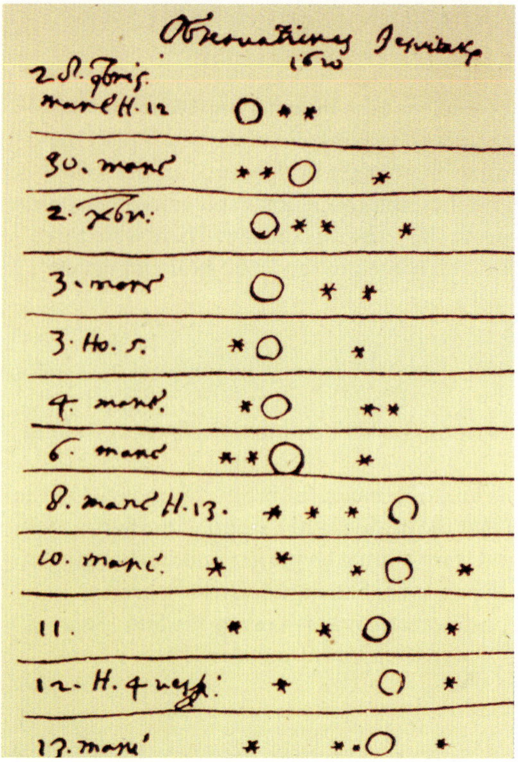

Figure 3.23 A page from Galileo's notebook written in 1610. His sketches show four "stars" near Jupiter (the circle) but in different positions at different times (and sometimes hidden from view). Galileo soon realized that the "stars" were actually moons orbiting Juptier.

moving object is to come to rest. Galileo concluded that objects such as birds, falling stones, and clouds that are moving with Earth should stay with Earth unless some force knocks them away. This same idea explains why passengers in an airplane keep moving with the airplane even when they leave their seats.

Tycho's supernova and comet observations already had challenged the validity of the second objection by showing that the heavens could change. Galileo shattered the idea of heavenly perfection after he built a telescope in late 1609. (Galileo did *not* invent the telescope; Hans Lippershey invented it in 1608. However, Galileo took what was little more than a toy and turned it into a scientific instrument.) Through his telescope, Galileo saw sunspots on the Sun, which were considered "imperfections" at the time. By noticing the shadows cast near the dividing line between the light and dark portions of the lunar face, he was also able to use his telescope to prove that the Moon has mountains and valleys like the "imperfect" Earth (Figure 3.22). If the heavens were in fact not perfect, then the idea of elliptical orbits (as opposed to "perfect" circles) was not so objectionable.

The third objection—the absence of observable stellar parallax—had been of particular concern to Tycho. Based on his estimates of the distances of stars, Tycho believed that his naked-eye observations were sufficiently precise to detect stellar parallax if Earth did in fact orbit the Sun. Refuting Tycho's argument required showing that the stars were more distant than Tycho had thought and therefore too distant for him to have observed stellar parallax. Although Galileo didn't actually prove this fact, he provided strong evidence in its favor. For example, his telescope allowed him to discover that the Milky Way is made up of countless individual stars. This discovery helped him argue that the stars were far more numerous and more distant than Tycho had imagined.

In hindsight, the final nails in the coffin of the Earth-centered universe came with two of Galileo's earliest discoveries through the telescope. First, he observed four moons clearly orbiting Jupiter, *not* Earth (Figure 3.23). (By itself, this observation still did not rule out a stationary, central Earth. However, it showed that moons can orbit a moving planet like Jupiter, which overcame some critics' complaints that the Moon could not stay with a moving Earth.) Soon thereafter, he observed that Venus goes through phases in a way that makes sense only if it orbits the Sun and not Earth (Figure 3.24).

With Earth clearly removed from its position at the center of the universe, the scientific debate turned to the question of whether Kepler's laws were the correct model for our solar system. The most convincing evidence came in 1631, when astronomers observed a transit of Mercury across the Sun's face. Kepler's laws had predicted the transit with overwhelmingly better success than any competing model.

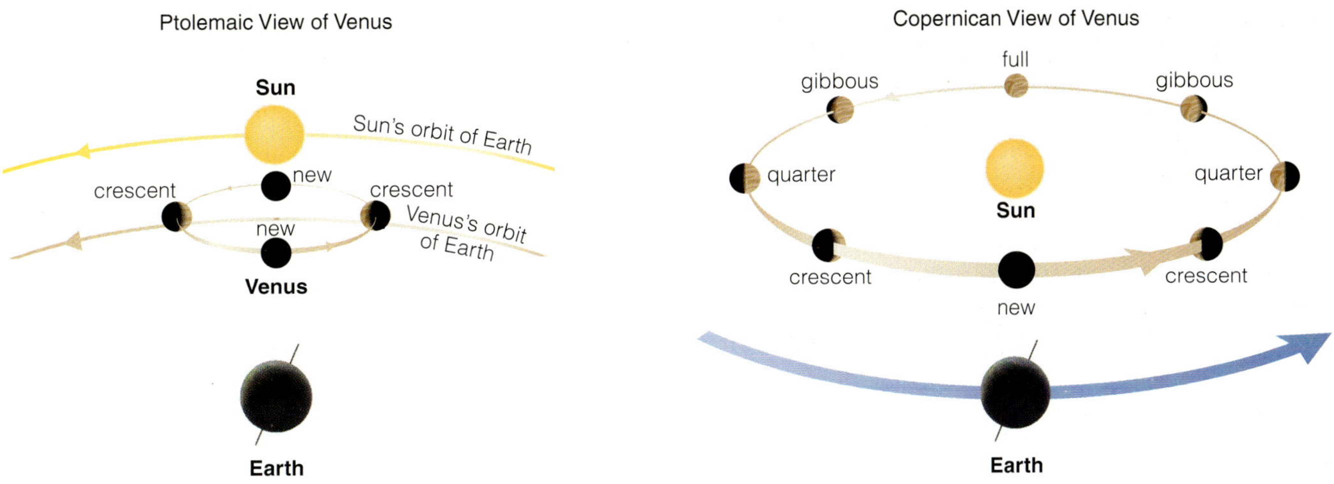

a In the Ptolemaic system, Venus orbits Earth, moving around a small circle on its larger orbital circle; the center of the small circle lies on the Earth-Sun line. Thus, if this view were correct, Venus's phases would range only from new to crescent.

b In reality, Venus orbits the Sun, so from Earth we can see it in many different phases. This is just what Galileo observed, allowing him to prove that Venus really does orbit the Sun.

Figure 3.24 Interactive Figure Galileo's telescopic observations of Venus proved that it orbits the Sun rather than Earth.

Although we now recognize that Galileo won the day, the story was more complex in his own time, when Catholic Church doctrine still held Earth to be the center of the universe. On June 22, 1633, Galileo was brought before a Church inquisition in Rome and ordered to recant his claim that Earth orbits the Sun. Nearly 70 years old and fearing for his life, Galileo did as ordered and his life was spared. However, legend has it that as he rose from his knees, he whispered under his breath, *Eppur si muove*—Italian for "And yet it moves." (Given the likely consequences if Church officials had heard him say this, most historians doubt the legend.)

The Church did not formally vindicate Galileo until 1992, but Church officials gave up the argument long before that: In 1757, all works backing the idea of a Sun-centered solar system were removed from the Church's Index of banned books. Today, Catholic scientists are at the forefront of much astronomical research, and official Church teachings are compatible not only with Earth's planetary status but also with the theories of the Big Bang and the subsequent evolution of the cosmos and of life.

3.4 The Nature of Science

The story of how our ancestors gradually figured out the basic architecture of the cosmos exhibits many features of what we now consider "good science." For example, we have seen how models were formulated and tested against observations, and modified or replaced when they failed those tests. The story also illustrates some classic mistakes, such as the apparent failure of anyone before Kepler to question the belief that orbits must be circles. The ultimate success of the Copernican revolution led scientists, philosophers, and theologians to reassess the various modes of thinking that played a role in the 2,000-year process of discovering Earth's place in the universe. Let's examine how the principles of modern science emerged from the lessons learned in the Copernican revolution.

• How can we distinguish science from nonscience?

Perhaps surprisingly, it turns out to be quite difficult to define the term *science* precisely. The word comes from the Latin *scientia*, meaning "knowledge," but not all knowledge is science. For example, you may know what music you like best, but your musical taste is not a result of scientific study.

The Idealized Scientific Method One reason science is difficult to define is that not all science works in the same way. It is often said that science follows the "scientific method," which means we propose an idea and then perform experiments or observations to test it. As an example of the scientific method, consider what you would do if your flashlight suddenly stopped working. You might wonder why it has stopped working, and you might *hypothesize* that the batteries have died. In other words, you've created a tentative explanation, or **hypothesis**, for the flashlight's failure. A hypothesis is sometimes called an *educated guess*—in this case, it is "educated" because you already know that flashlights need batteries. Your hypothesis then allows you to make a simple prediction: If you replace the batteries with new ones, the flashlight should work. You can test this prediction by replacing the batteries. If the flashlight now works, you've confirmed your hypothesis. If it doesn't, you must revise or discard your hypothesis, usually in favor of some other one that you can also test (such as that the bulb is dead). Figure 3.25 illustrates the basic flow of this process.

However, while the scientific method can be a useful idealization, science rarely progresses in such an orderly way. Science can also progress when someone goes out and looks at nature in a general way in hopes of learning something new and unexpected, rather than conducting a careful set of experiments. For example, Galileo wasn't looking for anything in particular when he pointed his telescope at the sky and made his first startling discoveries. Furthermore, scientists are human beings, and their intuition and personal beliefs inevitably influence their work. Copernicus, for example, adopted the idea that Earth orbits the Sun not because he had carefully tested it but because he believed it made more sense than the prevailing view of an Earth-centered universe. While his intuition guided him to the right general idea, he erred in the specifics because he still clung to Plato's ancient belief that heavenly motion must be in perfect circles.

Given that the idealized scientific method is an overly simplistic characterization of science, how can we tell what is science and what is not? To answer this question, we must look a little deeper at the distinguishing characteristics of scientific thinking.

SPECIAL TOPIC And Yet It Moves

The case of Galileo is often portrayed as a simple example of conflict between science and religion, but the reality was much more complex, with deep divisions inside the Church hierarchy. Perhaps the clearest evidence for a more open-minded Church comes from the case of Copernicus, whose revolutionary work was strongly supported by many Church officials. A less-well-known and even earlier example concerns Nicholas of Cusa (1401–1464), who published a book arguing for a Sun-centered solar system in 1440, more than a century before Copernicus. Nicholas was ordained a priest in the same year that his book was published, and he was later elevated to Cardinal. Clearly, his views caused no problems for Church officials of the time. (Copernicus probably was not aware of this earlier work by Nicholas of Cusa.)

Many other scientists received similar support from within the Church. Indeed, for most of his life, Galileo counted Cardinals (and even the Pope who later excommunicated him) among his friends. Some historians suspect that Galileo got into trouble less for his views than for the way in which he portrayed them. For example, in 1632—just a year before his famous trial—he published a book in which two fictional characters debated the geocentric and Sun-centered views. He named the character taking the geocentric position Simplicio—essentially "simple-minded"—and someone apparently convinced the Pope that the character was meant to represent him. If it was personality rather than belief that got Galileo into trouble, he was not the only one. Another early supporter of Copernicus, Giordano Bruno (1548–1600), drew the wrath of the Church after essentially writing that no rational person could disagree with him—even though the evidence for the Copernican system was still quite weak at the time, as Kepler had not yet discovered his laws of planetary motion. Bruno was branded a heretic and burned at the stake.

The evidence supporting the idea that Earth rotates and orbits the Sun was quite strong by the time of Galileo's trial in 1633, but it was still indirect. Thus, conservatives within the Church could at least make some case against Galileo. Today, we have much more direct proof that Galileo was correct when he supposedly whispered of Earth *Eppur si muove*—"And yet it moves."

French physicist Jean Foucault provided the first direct proof of *rotation* in 1851. Foucault built a large pendulum that he carefully started swinging. Any pendulum tends to swing always in the same plane, but Earth's rotation made Foucault's pendulum appear to twist slowly in a circle. Today, *Foucault pendulums* are a popular attraction at many science centers and museums. A second direct proof that Earth rotates is provided by the *Coriolis effect*, first described by French physicist Gustave Coriolis (1792–1843). The Coriolis effect [Section 10.2], which would not occur if Earth were not rotating, is responsible for things such as the swirling of hurricanes and the fact that missiles that travel great distances on Earth deviate from straight-line paths.

The first direct proof that Earth orbits the Sun came from English astronomer James Bradley (1693–1762). To understand Bradley's proof, imagine that starlight is like rain, falling straight down. If you are standing still you should hold your umbrella straight over your head, but if you are walking through the rain, you should tilt your umbrella forward, because your motion makes the rain appear to be coming down at an angle. Bradley discovered that observing light from stars requires that telescopes be tilted slightly in the direction of Earth's motion—just like the umbrella. This effect is called the *aberration of starlight.* Stellar parallax also provides direct proof that Earth orbits the Sun, and it was first measured in 1838 by German astronomer Friedrich Bessel.

Foucault pendulums are popular attractions at many science centers, museums, and other public places.

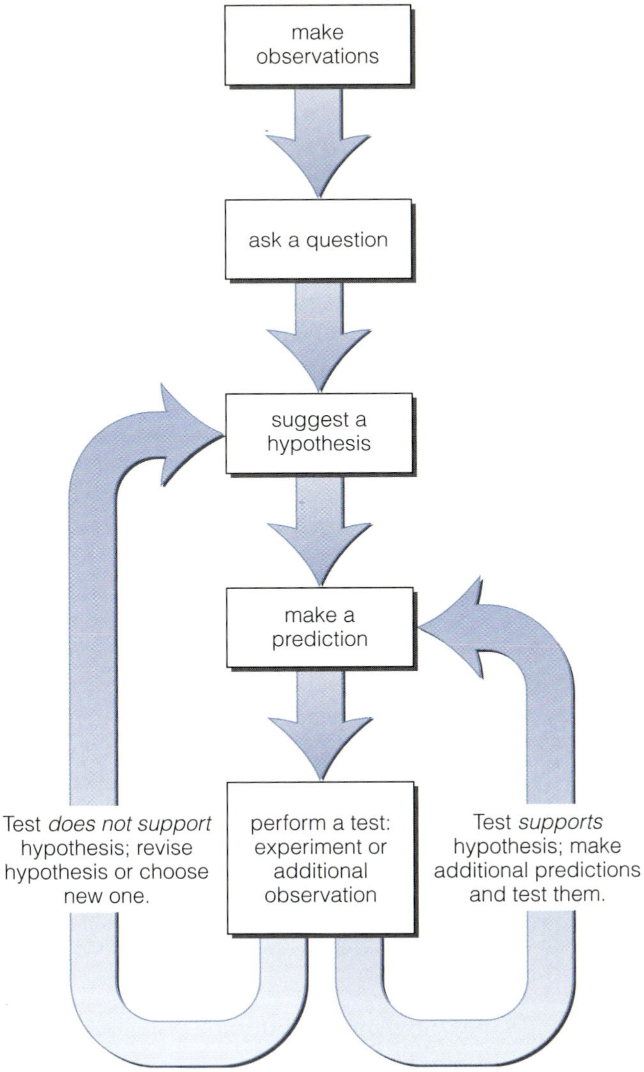

Figure 3.25 This diagram illustrates what we often call the *scientific method*.

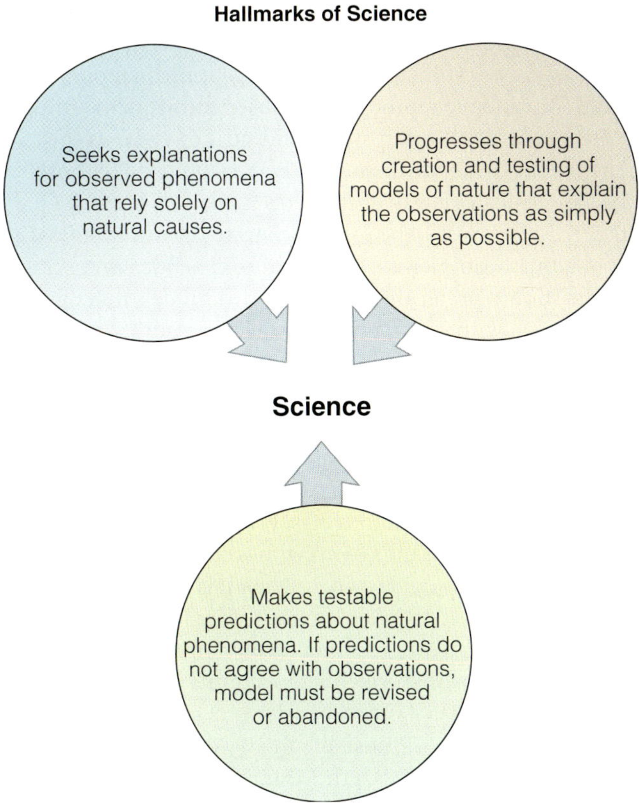

Figure 3.26 Hallmarks of science.

Hallmarks of Science One way to define scientific thinking is to list the criteria that scientists use when they judge competing models of nature. Historians and philosophers of science have examined (and continue to examine) this issue in great depth, and different experts express somewhat different viewpoints on the details. Nevertheless, everything we now consider to be science shares the following three basic characteristics, which we will refer to as the "hallmarks" of science (Figure 3.26):

- Modern science seeks explanations for observed phenomena that rely solely on natural causes.
- Science progresses through the creation and testing of models of nature that explain the observations as simply as possible.
- A scientific model must make testable predictions about natural phenomena that would force us to revise or abandon the model if the predictions do not agree with observations.

Each of these hallmarks is evident in the story of the Copernican revolution. The first shows up in the way Tycho's exceptionally careful measurements of planetary motion motivated Kepler to come up with a better explanation for those motions. The second is evident in the way several competing models were compared and tested, most notably those of Ptolemy, Copernicus, and Kepler. We see the third in the fact that each model could make precise predictions about the future motions of the Sun, Moon, planets, and stars in our sky. When a model's predictions failed, the model was modified or ultimately discarded. Kepler's model gained acceptance in large part because its predictions matched Tycho's observations much better than Ptolemy's model.

Occam's Razor The criterion of simplicity in the second hallmark deserves further explanation. Remember that the original model of Copernicus did *not* match the data noticeably better than Ptolemy's model. Thus, if scientists had judged Copernicus's model solely on the accuracy of its predictions, they might have rejected it immediately. However, many scientists found elements of the Copernican model appealing, such as the simplicity of its explanation for apparent retrograde motion. They therefore kept the model alive until Kepler found a way to make it work.

In fact, if agreement with data were the sole criterion for judgment, we could imagine a modern-day Ptolemy adding millions or billions of additional circles to the geocentric model in an effort to improve its agreement with observations. A sufficiently complex geocentric model could in principle reproduce the observations with almost perfect accuracy—but it still would not convince us that Earth is the center of the universe. We would still choose the Copernican view over the geocentric view because its predictions would be just as accurate yet would follow from a much simpler model of nature. The idea that scientists should prefer the simpler of two models that agree equally well with observations is called *Occam's razor*, after the medieval scholar William of Occam (1285–1349).

Verifiable Observations The third hallmark of science asks the question of what counts as an "observation" against which a prediction can be tested. Consider the claim that aliens are visiting Earth in UFOs. Proponents of this claim say that many thousands of eyewitness observations of UFO encounters provide evidence that it is true. But should these personal testimonials count as *scientific* evidence? On the surface, the answer may not be obvious, because all scientific studies involve eyewitness accounts on some level. For example, only a handful of scientists have personally made detailed tests of Einstein's theory of relativity, and it is their personal reports of the results that have convinced other scientists of the theory's validity. However, there's a very important difference between personal testimony about a scientific test and an observation of a UFO: The first is at least in principle verifiable by anyone, while the second is not.

Understanding this difference is crucial to understanding what counts as science and what does not. Even though you may never have conducted a test of Einstein's theory of relativity yourself, there's nothing stopping you from doing so. It might require several years of study before you have the necessary background to conduct the test, but you could then confirm the results reported by other scientists. In other words, while you may currently be trusting the eyewitness testimony of scientists, you always have the option of verifying their testimony for yourself.

In contrast, there is no way for you to verify someone's eyewitness account of a UFO. Without hard evidence such as photographs or pieces of the UFO, there is nothing that you could evaluate for yourself, even in principle. (And in those cases where "hard evidence" for UFO sightings has been presented, scientific study has never yet found the evidence to be strong enough to support the claim of alien spacecraft [Section 24.5].) Moreover, scientific studies of eyewitness testimony show it to be notoriously unreliable. For example, different eyewitnesses often disagree on what they saw even immediately after an event has occurred. As time passes, memories of the event may change further. In some cases in which memory has been checked against reality, people have reported vivid memories of events that never happened at all. This explains something that virtually all of us have experienced: disagreements with a friend about who did what and when. Since both people cannot be right in such cases, at least one person must have a memory that differs from reality.

Because of its demonstrated unreliability, eyewitness testimony alone should *never* be used as evidence in science, no matter who reports it or how many people offer similar testimony. It can be used in support of a scientific model only when it is backed up by independently verifiable evidence that anyone could in principle check for him- or herself. (For the same reason, eyewitness testimony alone is usually insufficient for a conviction in criminal court; at least some other evidence, such as motive, is required.)

Science and Pseudoscience It's important to realize that science is not the only valid way of seeking knowledge. For example, suppose you are shopping for a car, learning to play drums, or pondering the meaning of life. In each case, you might make observations, exercise logic, and test hypotheses. Yet these pursuits clearly are not science, because they are not directed at developing testable explanations for observed natural phenomena. As long as nonscientific searches for knowledge make no claims about how the natural world works, they do not conflict with science.

However, you will often hear claims about the natural world that seem to be based on observational evidence but do not treat evidence in a truly scientific way. Such claims are often called **pseudoscience**, which literally means "false science." To distinguish real science from pseudoscience,

COMMON MISCONCEPTIONS

Eggs on the Equinox

One of the hallmarks of science holds that you needn't take scientific claims on faith. In principle, at least, you can always test them for yourself. Consider the claim, repeated in news reports every year, that the spring equinox is the only day on which you can balance an egg on its end. Many people believe this claim, but you'll be immediately skeptical if you think about the nature of the spring equinox. The equinox is merely a point in time at which sunlight strikes both hemispheres equally (see Figure 2.15). It's difficult to see how sunlight could affect an attempt to balance eggs (especially if the eggs are indoors), and there is no difference in the strength of either Earth's gravity or the Sun's gravity on that day compared to any other day.

More important, you can test this claim directly. It's not easy to balance an egg on its end, but with practice you'll find that you can do it on any day of the year, not just on the spring equinox. Not all scientific claims are so easy to test for yourself, but the basic lesson should be clear: Before you accept any scientific claim, you should demand at least a reasonable explanation of the evidence that backs it up.

a good first step is to check whether a particular claim exhibits all three hallmarks of science. Consider the example of people who claim a psychic ability to "see" the future. If they truly can see the future, psychics should be able to make specific, testable predictions. In this sense, "seeing" the future sounds scientific, since we can test it. However, numerous studies have tested the predictions of "seers" and have found that their predictions come true no more often than would be expected by pure chance. If the "seers" were scientific, they would admit that this evidence undercuts their claim of psychic abilities. Instead, they generally make excuses. For example, they might say that the predictions didn't come true because of some type of "psychic interference." Making testable claims but then ignoring the results of the tests marks as pseudoscience the claimed ability to see the future.

Objectivity in Science The idea that science is objective, meaning that all people should be able to find the same results, is very important to the validity of science as a means of seeking knowledge. However, there is a difference between the overall objectivity of science and the objectivity of individual scientists. In particular, because science is practiced by human beings, individual scientists bring their personal biases and beliefs to their scientific work.

Personal bias can influence the way a scientist proposes or tests a model. For example, most scientists choose their research projects based on personal interests rather than on some objective formula. In some extreme cases, scientists have even been known to cheat—either deliberately or subconsciously—to obtain a result they desire. For example, in the late nineteenth and early twentieth centuries, some astronomers (including, most notably, Percival Lowell) claimed to see artificial canals in their blurry telescopic images of Mars [Section 9.4]. They hypothesized that Mars was home to a dying civilization that used the canals to transport water to thirsty cities (leading H. G. Wells to write his novel *The War of the Worlds*). However, no such canals actually exist. These astronomers apparently allowed their beliefs about extraterrestrial life to influence the way they interpreted blurry images—in essence, a form of cheating, though certainly not intentional.

Bias can sometimes show up even in the thinking of the scientific community as a whole. Some valid ideas may not be considered by any scientist because the ideas fall too far outside the general patterns of thought, or **paradigm**, of the time. Einstein's theory of relativity provides an example. Many scientists in the decades before Einstein had gleaned hints of the theory but did not investigate them, at least in part because the ideas seemed too outlandish. (Chapters S2 and S3 discuss Einstein's special and general theories of relativity.)

The beauty of science is that it encourages continued testing by many people. Even if personal biases affect

SPECIAL TOPIC Logic and Science

In science, we attempt to acquire new knowledge through logical reasoning. We carry out the process of reasoning by constructing arguments in which we begin with a set of premises and try to draw appropriate conclusions. This concept of a logical argument differs somewhat from the definition of argument in everyday life. In particular, a logical argument need not imply any animosity or dissension.

There are two basic types of argument: *deductive* and *inductive.* Both are important in science.

The following is a simple example of a deductive argument:

PREMISE: All planets orbit the Sun in ellipses with the Sun at one focus.

PREMISE: Earth is a planet.

CONCLUSION: Earth orbits the Sun in an ellipse with the Sun at one focus.

As long as the two premises are true, the conclusion must also be true—this is the essence of deduction. Note the construction of a deductive argument: The first premise is a general statement that applies to all planets, and the conclusion is a specific statement that applies only to Earth. As this example suggests, we often use deduction to *deduce* a specific prediction from a more general theory. If the specific prediction proves to be false, then something must be wrong with the premises from which it was deduced. If it proves true, then we've acquired a piece of evidence in support of the premises.

Contrast the deductive argument above with the following example of an *inductive argument*:

PREMISE: Birds fly up into the air but eventually come back down.

PREMISE: People who jump into the air fall back down.

PREMISE: Rocks thrown into the air come back down.

PREMISE: Balls thrown into the air come back down.

CONCLUSION: What goes up must come down.

Because each premise supports the conclusion, you might at first agree that this is a strong inductive argument and that its conclusion probably is true. However, no matter how many more examples you consider of objects that go up and come down, you could never *prove* that the conclusion is true—only that it seems likely to be true. Moreover, a single counterexample can prove the conclusion of an inductive argument to be false. For example, in this case we know that the conclusion is false because the spacecraft *Voyager 2* was launched from Earth and never fell back down.

Inductive arguments *generalize* from specific facts to a broader model or theory. Thus, inductive arguments are the arguments used to build scientific theories, because we use them to *infer* general principles from observations and experiments. That is why theories can never be proved true beyond all doubt—they can only be shown to be consistent with ever larger bodies of evidence. Theories *can* be proved false, however, if they fail to account for observed or experimental facts.

some results, tests by others will eventually uncover the mistakes. Similarly, if a new idea is correct but falls outside the accepted paradigm, sufficient testing and verification of the idea will eventually force a change in the paradigm. Thus, although individual scientists rarely follow the idealized scientific method, the collective action of many scientists over many years ensures that science as a whole remains objective.

• What is a scientific theory?

The most successful scientific models explain a wide variety of observations in terms of just a few general principles. When a powerful yet simple model makes predictions that survive repeated and varied testing, scientists elevate its status and call it a **theory**. Some famous examples are Isaac Newton's theory of gravity, Charles Darwin's theory of evolution, and Albert Einstein's theory of relativity.

Note that the scientific use of the word *theory* is quite different from its everyday use, in which we equate a theory more closely with speculation or a hypothesis. In everyday life, someone might get a new idea and say, for example, "I have a new theory about why people enjoy the beach." Without the support of a broad range of evidence that others have tested and confirmed, this idea is really only a hypothesis. In contrast, Newton's theory of gravity qualifies as a scientific theory because it uses simple physical principles to explain a great many observations and experiments.

Despite its success in explaining observed phenomena, a scientific theory can never be proved true beyond all doubt, because ever more sophisticated observations may eventually disagree with its predictions. However, anything that qualifies as a scientific theory must be supported by a large, compelling body of evidence.

In this sense, a scientific theory is not at all like a hypothesis or any other type of guess. We are free to change a hypothesis at any time, because it has not yet been carefully tested. In contrast, we can discard or replace a scientific theory only if we have an alternate way of explaining the evidence that supports it.

Again, the theories of Newton and Einstein offer good examples. A vast body of evidence supports Newton's theory of gravity, but by the late 1800s scientists had begun to discover cases where its predictions did not perfectly match observations. These discrepancies were explained only when Einstein developed his general theory of relativity in the early twentieth century, which was able to match the observations. Still, the many successes of Newton's theory could not be ignored, and Einstein's theory would not have gained acceptance if it had not been able to explain these successes equally well. It did, and that is why we now view Einstein's theory as a broader theory of gravity than Newton's theory. Some scientists today are seeking a theory of gravity that will go beyond Einstein's. If any new theory ever gains acceptance, it will have to match all the successes of Einstein's theory as well as work in new realms where Einstein's theory does not.

THINK ABOUT IT

When people claim that something is "only a theory," what do you think they mean? Does this meaning of "theory" agree with the definition of a theory in science? Do scientists always use the word *theory* in its "scientific" sense? Explain.

3.5 Astrology

We have discussed the development of astronomy and the nature of science in some depth. Now let's talk a little about a subject often confused with the science of astronomy: the similarly named subject of *astrology*. Today, astronomy and astrology are very different practices. In ancient times, however, astrology and astronomy often went hand in hand, and astrology played an important role in the historical development of astronomy.

• How is astrology different from astronomy?

The basic tenet of astrology is that the apparent positions of the Sun, Moon, and planets among the stars in our sky influence human events. The origins of this idea are easy to understand. After all, the position of the Sun in the sky certainly influences our lives, since it determines the seasons and the times of daylight and darkness. Similarly, the Moon directly affects us by creating the tides and sometimes providing moonlight to see by at night. Because the planets also appear to move among the stars, it seemed reasonable to ancient people that planets also influence our lives, even if these influences were much more subtle and difficult to discover.

Ancient astrologers hoped that they might learn *how* the positions of the Sun, Moon, and planets influence our lives. They charted the skies, seeking correlations with events on Earth. For example, if an earthquake occurred when Saturn was entering the constellation Leo, might Saturn's position have caused the earthquake? If the king became ill when Mars appeared in the constellation Gemini and the first-quarter moon appeared in Scorpio, might it mean another tragedy for the king when this particular alignment of the Moon and Mars next recurred? Surely, the ancient astrologers thought, the patterns of influence would eventually become clear. The astrologers hoped that they might someday learn to forecast human events with the same reliability with which astronomical observations of the Sun could forecast the coming of spring.

Because forecasts of the seasons and forecasts of human events were imagined to be closely related, astrologers and astronomers usually were one and the same in the ancient world. For example, in addition to his books on astronomy, Ptolemy published a treatise on astrology called *Tetrabiblios* that remains the foundation for much of astrology today. But Ptolemy himself recognized that astrology stood upon a far shakier foundation than astronomy. In the introduction to *Tetrabiblios*, Ptolemy compared astronomical and astrological predictions:

> *[Astronomy], which is first both in order and effectiveness, is that whereby we apprehend the aspects of the movements of sun, moon, and stars in relation to each other and to the earth. . . . I shall now give an account of the second and less sufficient method [of prediction (astrology)] in a proper philosophical way, so that one whose aim is the truth might never compare its perceptions with the sureness of the first, unvarying science. . . .*

Other ancient scientists probably also recognized that their astrological predictions were far less reliable than their astronomical ones. Nevertheless, if there were even a slight possibility that astrologers could forecast the future, no king or political leader would dare to be without one. Astrologers held esteemed positions as political advisers in the ancient world and were provided with the resources they needed to continue charting the heavens and human history. Indeed, wealthy political leaders' support of astrology made possible much of the development of ancient astronomy.

Throughout the Middle Ages and into the Renaissance, many astronomers continued to practice astrology. For example, Kepler cast numerous *horoscopes*—the predictive charts of astrology (Figure 3.27)—even as he was discovering the laws of planetary motion. However, given Kepler's later description of astrology as "the foolish stepdaughter of astronomy" and "a dreadful superstition," he may have cast the horoscopes solely as a source of much-needed income. Modern-day astrologers also claim Galileo as one of their own, in part for his having cast a horoscope for the Grand Duke of Tuscany. However, Galileo's astronomical discoveries changed human history, while the horoscope he cast was just plain wrong: The Duke died just a few weeks after Galileo's horoscope predicted that he would have a long and fruitful life.

The scientific triumph of Kepler and Galileo in showing Earth to be a planet orbiting the Sun heralded the end of the linkage between astronomy and astrology. Astronomy has since gained status as a successful science that helps us understand our universe, while astrology no longer has any connection to the modern science of astronomy.

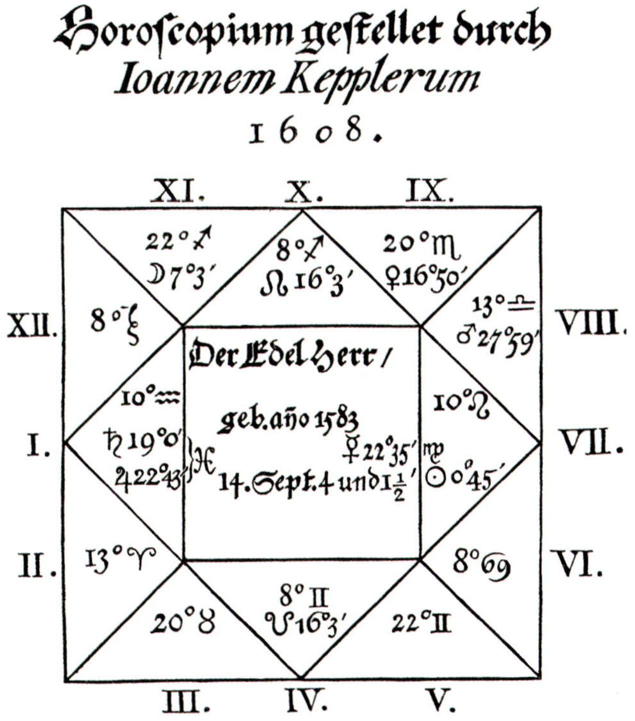

Figure 3.27 This chart is an example of a horoscope, in this case one cast by Kepler, who called astrology "a dreadful superstition."

• Does astrology have any scientific validity?

Although astronomers gave up on it centuries ago, astrology remains popular with the general public. Many people read their daily horoscopes in newspapers, and some pay significant fees to have personal horoscopes cast by "professional" astrologers. In fact, more people earn incomes by casting horoscopes than through astronomical research, and books and articles on astrology often outsell all but the most popular books on astronomy. With so many people giving credence to astrology, is it possible that it has some scientific validity after all?

Testing Astrology The question of scientific validity can be more difficult to answer than it sounds, because there's no general agreement among astrologers even on such basic things as what astrology is or what it can predict. For example, "Western astrology" is quite different in nature from the astrology practiced in India and China. Some astrologers no longer make testable predictions at all; rather, they give vague guidance about how to live one's life. Most newspaper horoscopes fall into this category. Although your horoscope may seem to ring true at first, a careful read will usually show it to be so vague as to be untestable. For example, a horoscope that says, "It is a good day to spend time with your friends" may be good advice but doesn't offer much to test. If you read the horoscopes for all 12 astrological signs, you'll probably find that several of them apply equally well to you. When astrology offers only vague advice rather than testable predictions, the question of scientific validity does not apply.

THINK ABOUT IT

Look in a local newspaper for today's weather forecast and for your horoscope. Contrast the nature of their predictions. By the end of the day, you will know if the weather forecast was accurate. Will you know whether your horoscope was accurate? Explain.

Nevertheless, most professional astrologers still earn their livings by casting horoscopes that either predict future events in an individual's life or describe characteristics of the person's personality and life. If the horoscope predicts future events, we can check to see whether the predictions come true. If it describes a person's personality and life, the description can be checked for accuracy. A scientific test of astrology requires evaluating many horoscopes and comparing their accuracies to what would be expected by pure chance. For example, suppose a horoscope states that a person's best friend is female. Because roughly half the population of the United States is female, an astrologer who casts 100 such horoscopes would be expected by pure chance to be right about 50 times. Thus, we would be impressed with the predictive ability of the astrologer only if he or she were right much more often than 50 times out of 100.

In hundreds of scientific tests, astrological predictions have never proved to be accurate by a substantially greater margin than expected from pure chance. Similarly, in tests in which astrologers are asked to cast horoscopes for people they have never met, the horoscopes fail to match actual personality profiles more often than expected by chance. The verdict is clear: The methods of astrology are useless for predicting the past, the present, or the future.

Examining the Underpinnings of Astrology In science, observations and experiments are the ultimate judge of any idea. No matter how outlandish an idea might appear, it cannot be dismissed if it successfully meets observational or experimental tests. The idea that Earth rotates and orbits the Sun at one time seemed outlandish, yet today it is so strongly supported by the evidence that we consider it a fact. The idea that the positions of the Sun, Moon, and planets among the stars influence our lives might sound outlandish today, but if astrology were to make predictions that came true, adherence to the principles of science would force us to take astrology seriously. However, given that scientific tests of astrology have never found any evidence that its predictive methods work, it is worth looking at its premises to see whether they make sense. Might there be a few kernels of wisdom buried within the lore of astrology?

Let's begin with one of the key premises of astrology: There is special meaning in the patterns of the stars in the constellations. This idea may have seemed quite reasonable in ancient times, when the stars were assumed to be fixed on an unchanging celestial sphere. Today we know that the patterns of the stars in the constellations are accidents of the moment. Long ago the constellations did not look the same, and they will look still different far in the future [Section 1.3]. Moreover, the stars in a constellation don't necessarily have any *physical* association (see Figure 2.3). Because stars vary in distance, two stars that appear on opposite sides of our sky might well be closer together than two stars in the same constellation. Constellations are only *apparent* associations of stars, with no more physical reality than the water in a desert mirage.

Astrology also places great importance on the positions of the planets among the constellations. Again, this idea might have seemed quite reasonable in ancient times, when it was thought that the planets truly wandered among the stars. Today we know that the planets only *appear* to wander among the stars. In reality, the planets are in our own solar system, while the stars are vastly farther away. It is difficult to see how mere appearances could have profound effects on our lives.

Many other ideas at the heart of astrology are equally suspect. For example, most astrologers claim that a proper horoscope must account for the positions of *all* the planets. Does that mean that all horoscopes cast before the discoveries of Neptune in 1846 and Pluto in 1930 were invalid? If so, why didn't astrologers notice that something was wrong with their horoscopes and predict the existence of these planets? (Astronomers *did* predict the existence of Neptune by using Newton's universal law of gravitation; see p. 130.) In recent years, astronomers have discovered many Pluto-like objects orbiting in the outer region of the solar system [Section 12.3]. Given their similarity to Pluto, shouldn't these objects also be included when casting horoscopes, and does that mean that horoscopes cannot be accurate until we've discovered every single one of these objects? What about planets orbiting other stars, which far outnumber the planets of our own solar system? Given seemingly unanswerable questions like these, there seems little hope that astrology will ever meet its ancient goal of forecasting human events.

THE BIG PICTURE

Putting Chapter 3 into Context

In this chapter, we focused on the scientific principles through which we have learned so much about the universe. Key "big picture" concepts from this chapter include the following:

- The basic ingredients of scientific thinking—careful observation and trial-and-error testing—are a part of everyone's experience. Modern science simply provides a way of organizing this everyday thinking to facilitate the learning and sharing of new knowledge.
- Although our understanding of the universe is growing rapidly today, each new piece of knowledge rests on ideas that came before.
- The Copernican revolution, which overthrew the ancient Greek belief in an Earth-centered universe, did not occur instantaneously. It unfolded over a period of more than a century, during which many of the characteristics of modern science first appeared.
- Several key hallmarks distinguish science from other ways of gathering knowledge: Science seeks natural explanations for observed phenomena, it progresses through the creation and testing of models that explain the observations as simply as possible, and it revises or abandons models whose predictions disagree with observations.
- Astronomy and astrology once developed hand in hand, but today they represent very different things.

3.1 The Ancient Roots of Science

- **In what ways do all humans employ scientific thinking?** Scientific thinking relies on the same type of trial-and-error thinking that we use in our everyday lives, but in a carefully organized way.

- **How did astronomical observations benefit ancient societies?** Ancient cultures used astronomical observations to help them keep track of time and the seasons, crucial skills for people who depended on agriculture for survival.

- **What did ancient civilizations achieve in astronomy?** Ancient astronomers were accomplished observers who learned to tell the time of day and the time of year, to track cycles of the Moon, and to observe planets and stars. Many ancient structures aided in astronomical observations.

3.2 Ancient Greek Science

- **Why does modern science trace its roots to the Greeks?** The Greeks developed **models** of nature and emphasized the importance of having the predictions of those models agree with observations of nature.

- **How did the Greeks explain planetary motion?** The Greek **geocentric model** reached its culmination with the model of Ptolemy. The **Ptolemaic model** explained apparent retrograde motion by having each planet move on a small circle whose center moves around Earth on a larger circle.

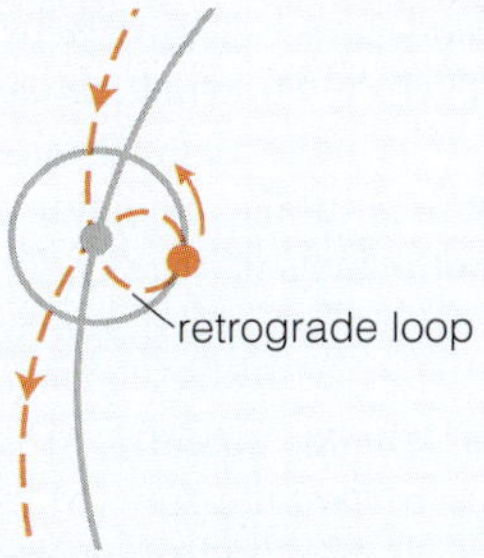

- **How was Greek knowledge preserved through history?** While Europe was in its Dark Ages, Islamic scholars in the Middle East preserved and extended ancient Greek knowledge. After the fall of Constantinople, scholars moved west to Europe, where their knowledge helped ignite the European Renaissance.

3.3 The Copernican Revolution

- **How did Copernicus, Tycho, and Kepler challenge the Earth-centered idea?** Copernicus created a Sun-centered model of the solar system designed to replace the Ptolemaic model, but it was no more accurate than Ptolemy's because Copernicus still used perfect circles. Tycho's accurate, naked-eye observations provided the data needed to improve on Copernicus's model. Kepler found a model of planetary motion that fit Tycho's data.

- **What are Kepler's three laws of planetary motion?** (1) The orbit of each planet is an ellipse with the Sun at one focus. (2) As a planet moves around its orbit, it sweeps out equal areas in equal times. (3) More distant planets orbit the Sun at slower average speeds, obeying the precise mathematical relationship $p^2 = a^3$.

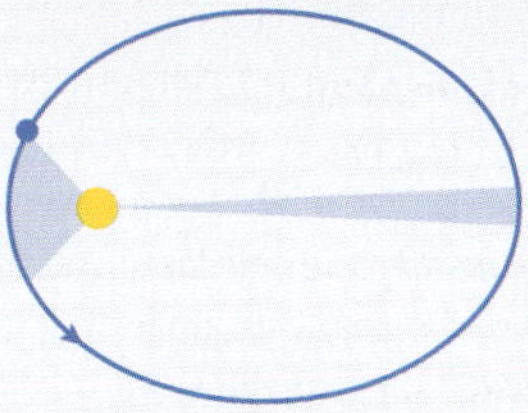

- **How did Galileo solidify the Copernican revolution?** Galileo's experiments and telescopic observations overcame remaining objections to the Copernican idea of Earth as a planet orbiting the Sun. Although not everyone accepted his results immediately, in hindsight we see that Galileo sealed the case for the Sun-centered solar system.

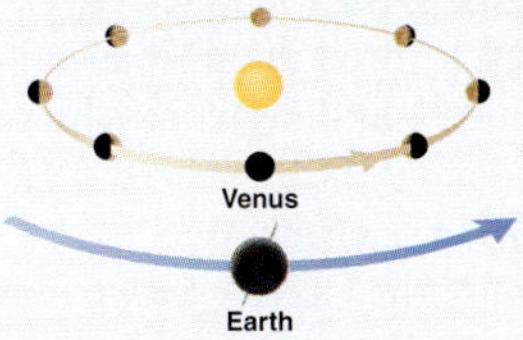

3.4 The Nature of Science

- **How can we distinguish science from nonscience?** Science generally exhibits these three hallmarks: (1) Modern science seeks explanations for observed phenomena that rely solely on natural causes. (2) Science progresses through the creation and testing of models of nature that explain the observations as simply as possible. (3) A scientific model must make testable predictions about natural phenomena that would force us to revise or abandon the model if the predictions do not agree with observations.

- **What is a scientific theory?** A scientific **theory** is a simple yet powerful model that explains a wide variety of observations in terms of just a few general principles, and has attained the status of a theory by surviving repeated and varied testing.

3.5 Astrology

- **How is astrology different from astronomy?** Astronomy and astrology share common roots in ancient observations of the sky but they are very different today. Astronomy is a modern science that has taught us much about the universe. Astrology is a search for hidden influences on human lives based on the positions of planets and stars in the sky.

- **Does astrology have any scientific validity?** Scientific tests have shown that astrological predictions do not prove to be accurate more than we can expect by pure chance, showing that the predictions have no scientific validity.

EXERCISES AND PROBLEMS

For instructor-assigned homework go to MasteringASTRONOMY www.masteringastronomy.com

Review Questions

Short-Answer Questions Based on the Reading

1. In what way is scientific thinking natural to all of us? How does modern science differ from this everyday type of thinking?
2. Why did ancient peoples study astronomy? Describe the astronomical achievements of at least four ancient cultures.
3. How are the names of the seven days of the week related to astronomical objects?
4. Describe at least three ways that ancient people determined either the time of day or the time of year.
5. What is a lunar calendar? What is the Metonic cycle? Explain why the dates of Ramadan cycle through our solar calendar while the dates of Jewish holidays and Easter remain within about a 1-month period.
6. What do we mean by a *model* in science?
7. Summarize the development of the Greek *geocentric model.*
8. Who was Ptolemy? How did the Ptolemaic model account for the apparent retrograde motion of planets in our sky?
9. What was the *Copernican revolution*, and how did it change the human view of the universe?
10. Why wasn't the Copernican model immediately accepted? Describe the roles of Tycho, Kepler, and Galileo in the eventual triumph of the Sun-centered model.
11. What is an *ellipse*? Define the *focus* and the *eccentricity* of an ellipse. Why are ellipses important in astronomy?
12. Clearly state each of *Kepler's laws of planetary motion.* For each law, describe in your own words what it means in a way that almost anyone could understand.
13. What is the difference between a *hypothesis* and a *theory* in science?
14. Describe each of the three hallmarks of science and give an example of how we can see each one in the unfolding of the Copernican revolution. What is *Occam's razor*? Why doesn't science accept personal testimony as evidence?
15. What do we mean by *pseudoscience*? How is it different from other types of nonscience?
16. What is the basic idea behind *astrology*? Explain why this idea seemed reasonable in ancient times but is no longer given credence by scientists.

Test Your Understanding

Does It Make Sense?

Decide whether the statement makes sense (or is clearly true) or does not make sense (or is clearly false). Explain your reasoning. (For an example, see Chapter 1, "Does It Make Sense?")

17. Ancient astronomers failed to realize that Earth goes around the Sun because they just weren't as smart as people today.
18. In ancient Egypt, children whose parents gave them "1 hour" to play got to play longer in the summer than in the winter.
19. If the planet Uranus had been identified as a planet in ancient times, we'd probably have eight days in a week.
20. The date of Christmas (December 25) is set each year according to a lunar calendar.
21. When navigating in the South Pacific, the Polynesians found their latitude with the aid of the pointer stars of the Big Dipper.
22. The Ptolemaic model reproduced apparent retrograde motion by having planets move sometimes counterclockwise and sometimes clockwise in their circles.
23. According to Kepler's laws, Earth would take longer to orbit the Sun if it had a larger mass.
24. In science, saying that something is a theory means that it is really just a guess.
25. A scientific theory should never gain acceptance until it has been proved true beyond all doubt.
26. Ancient astronomers were convinced of the validity of astrology as a tool for predicting the future.

Quick Quiz

Choose the best answer to each of the following. Explain your reasoning with one or more complete sentences.

27. Stonehenge was useful for: (a) telling the time of day; (b) determining the season; (c) predicting lunar eclipses.
28. With each 19-year Metonic cycle: (a) the lunar phases repeat on the same dates of the year; (b) solar eclipses repeat at the same times and places; (c) Ramadan occurs on the same dates of the year.
29. In the Greek geocentric model, the retrograde motion of a planet occurs when: (a) Earth is about to pass the planet in its orbit around the Sun; (b) The planet actually goes backward in its orbit around Earth; (c) The planet is aligned with the Moon in our sky.
30. Which of the following was *not* a major advantage of Copernicus's Sun-centered model over the Ptolemaic model? (a) It made significantly better predictions of planetary positions in our sky. (b) It offered a more natural explanation for the apparent retrograde motion of planets in our sky. (c) It allowed calculation of the orbital periods and distances of the planets.
31. When we say that a planet has a highly eccentric orbit, we mean that: (a) it is spiraling in toward the Sun; (b) its orbit is an ellipse with the Sun at one focus; (c) in some parts of its orbit it is much closer to the Sun than in other parts.
32. Earth is closer to the Sun in January than in July. Therefore: (a) Earth travels faster in its orbit around the Sun in July than in January. (b) Earth travels faster in its orbit around the Sun in January than in July. (c) It is summer in January and winter in July.
33. According to Kepler's third law: (a) Mercury travels fastest in the part of its orbit in which it is closest to the Sun. (b) Jupiter orbits the Sun at a faster speed than Saturn. (c) Pluto has a highly eccentric orbit.
34. Tycho Brahe's contribution to astronomy included: (a) inventing the telescope; (b) proving that Earth orbits the Sun; (c) collecting data that enabled Kepler to discover the laws of planetary motion.
35. Galileo's contribution to astronomy included: (a) discovering the laws of planetary motion; (b) discovering the law of

gravity; (c) making observations and conducting experiments that dispelled scientific objections to the Sun-centered model.

36. Which of the following is not true about scientific progress? (a) Science progresses through the creation and testing of models of nature. (b) Science advances only through the scientific method. (c) Science avoids explanations that invoke the supernatural.

Investigate Further

In-Depth Questions to Increase Your Understanding

Short-Answer/Essay Questions

37. *Earth's Shape.* It took thousands of years for humans to deduce that Earth is spherical. For each of the following alternative models of Earth's shape, identify one or more observations that you could make for yourself and that would invalidate the model.
    a. A flat Earth.
    b. A cylindrical Earth (shaped like a fallen log), like that proposed by Anaximander.
    c. A football-shaped Earth.
38. *Lunar Calendars.*
    a. Find the dates of the Jewish festival of Chanukah (also spelled "Hanukah") for this year and the next three years. Based on what you have learned in this chapter, explain why the dates change as they do.
    b. Find the dates of the Muslim fast for Ramadan for this year and the next three years. Based on what you have learned in this chapter, explain why the dates change as they do.
39. *Copernican Players.* Using a bulleted list format, make a one-page "executive summary" of the major roles that Copernicus, Tycho, Kepler, and Galileo played in overturning the ancient belief in an Earth-centered universe.
40. *What Makes It Science?* Choose a single idea in the modern view of the cosmos as discussed in Chapter 1, such as "The universe is expanding," "The universe began with a Big Bang," "We are made from elements manufactured by stars," or "The Sun orbits the center of the Milky Way Galaxy once every 230 million years."
    a. Briefly describe how the idea you have chosen is rooted in each of the three hallmarks of science discussed in this chapter. (That is, explain how it is based on observations, how our understanding of it depends on a model, and how the model is testable.)
    b. No matter how strongly the evidence may support a scientific idea, we can never be certain beyond all doubt that the idea is true. For the idea you have chosen, describe an observation that might cause us to call the idea into question. Then briefly discuss whether you think that, overall, the idea is likely or unlikely to hold up to future observations. Defend your opinion.
41. *Influence on History.* Based on what you have learned about the Copernican revolution, write a one- to two-page essay about how you believe it altered the course of human history.
42. *Scientific Test of Astrology.* Find out about at least one scientific test that has been conducted to test the validity of astrology. Write a short summary of how the test was conducted and what conclusions were reached.
43. *Your Own Astrological Test.* Devise your own scientific test of astrology. Clearly define the methods you will use in your test and how you will evaluate the results. Then carry out the test and write a report on your methods and results.
44. *Cultural Astronomy.* Choose a particular culture of interest to you, and research the astronomical knowledge and accomplishments of that culture. Write a two- to three-page summary of your findings.
45. *Astronomical Structures.* Choose an ancient astronomical structure of interest to you (e.g., Stonehenge, Templo Mayor, Pawnee lodges) and research its history. Write a two- to three-page summary of your findings. If possible, also build a scale model of the structure or create detailed diagrams to illustrate how the structure was used.
46. *Venus and the Mayans.* The planet Venus apparently played a particularly important role in Mayan society. Research the evidence and write a one- to two-page summary of current knowledge about the role of Venus in Mayan society.

Quantitative Problems

Be sure to show all calculations clearly and state your final answers in complete sentences.

47. *The Metonic Cycle.* The length of our calendar year is 365.2422 days, and the Moon's monthly cycle of phases averages 29.5306 days in length. By calculating the number of days in each, confirm that 19 solar years is almost precisely equal to 235 cycles of the lunar phases. Show your work clearly, then write a few sentences explaining how this fact can be used to keep a lunar calendar roughly synchronized with a solar calendar.
48. *Chinese Calendar.* The traditional Chinese lunar calendar has 12 months in most years but adds a thirteenth month to 22 of every 60 years. How many days does this give the Chinese calendar in each 60-year period? How does this compare to the number of days in 60 years on a solar calendar? Based on your answers, explain how this scheme is similar to the scheme used by lunar calendars that follow the Metonic cycle. (*Hint:* You'll need the data given in Problem 47.)
49. *Method of Eratosthenes I.* You are an astronomer on planet Nearth, which orbits a distant star. It has recently been accepted that Nearth is spherical in shape, though no one knows its size. One day, while studying in the library of Alectown, you learn that on the equinox your sun is directly overhead in the city of Nyene, located 1,000 kilometers due north of you. On the equinox, you go outside in Alectown and observe that the altitude of your sun is 80°. What is the circumference of Nearth? (*Hint:* Apply the technique used by Eratosthenes to measure Earth's circumference.)
50. *Method of Eratosthenes II.* You are an astronomer on planet Tirth, which orbits a distant star. It has recently been accepted that Tirth is spherical in shape, though no one knows its size. One day, you learn that on the equinox your sun is directly overhead in the city of Tyene, located 400 kilometers due north of you. On the equinox, you go outside in Alectown and observe that the altitude of your sun is 86°. What is the circumference of Tirth? (*Hint:* Apply the technique used by Eratosthenes to measure Earth's circumference.)
51. *Mars Orbit.* Find the perihelion and aphelion distances of Mars. (*Hint:* You'll need data from Appendix E.)

52. *Most Eccentric Orbit.* Which planet in our solar system has the most eccentric orbit? Find the planet's perihelion and aphelion distances.
53. *Least Eccentric Orbit.* Which planet in our solar system has the least eccentric orbit? Find the planet's perihelion and aphelion distances.
54. *Sedna Orbit.* The recently discovered object Sedna orbits our Sun beyond Pluto, at an average distance (semimajor axis) of 509 AU. What is its orbital period?
55. *New Planet Orbit.* A newly discovered planet orbits a distant star with the same mass as the Sun at an average distance of 112 million kilometers. Its orbital eccentricity is 0.3. Find the planet's orbital period and its nearest and farthest orbital distances from its star.
56. *Halley Orbit.* Halley's comet orbits the Sun every 76.0 years and has an orbital eccentricity of 0.97.
    a. Find its semimajor axis distance.
    b. Find its perihelion and aphelion distances.
    c. Does Halley's comet spend most of its time near its perihelion distance, its aphelion distance, or halfway in between? Explain.

Discussion Questions

57. *The Impact of Science.* The modern world is filled with ideas, knowledge, and technology that developed through science and application of the scientific method. Discuss some of these things and how they affect our lives. Which of these impacts do you think are positive? Which are negative? Overall, do you think science has benefited the human race? Defend your opinion.
58. *The Importance of Ancient Astronomy.* Why was astronomy important to people in ancient times? Discuss both the practical importance of astronomy and the importance it may have had for religious or other traditions. Which do you think was more important in the development of ancient astronomy, its practical or its philosophical role? Defend your opinion.
59. *Astronomy and Astrology.* Why do you think astrology remains so popular around the world even though it has failed all scientific tests of its validity? Do you think the popularity of astrology has any positive or negative social consequences? Defend your opinions.

MEDIA EXPLORATIONS

For self-study activities go to MasteringASTRONOMY www.masteringastronomy.com

Interactive Tutorials

Tutorial Review of Key Concepts

Use the interactive **Tutorial** at **www.masteringastronomy.com** to review key concepts from this chapter.

Orbits and Kepler's Laws Tutorial
Lesson 2 Kepler's First Law
Lesson 3 Kepler's Second Law
Lesson 4 Kepler's Third Law

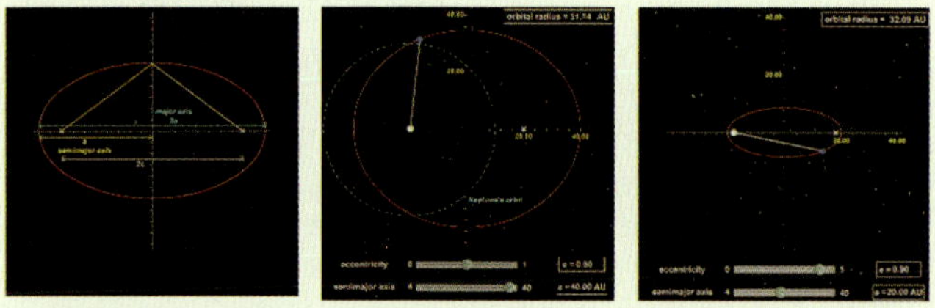

Supplementary Tutorial Exercises

Use the interactive **Tutorial Lesson** to explore the following questions.

Orbits and Kepler's Laws Tutorial, Lesson 2

1. When is an ellipse a circle?
2. Use the ellipse tool in Lesson 2 to show that the semimajor axis of an ellipse is a planet's average orbital radius:
    a. Orbital radius when the planet is closest to the Sun: perihelion = ___ AU
    b. Orbital radius when the planet is farthest from the Sun: aphelion = ___ AU
    c. Average of perihelion and aphelion = ___ AU
    d. Length of the major axis of the ellipse = ___ AU
    e. Length of the semimajor axis of the ellipse = ___ AU

    Are the average orbital radius (c) and the semimajor axis (e) equal?

Exploring the Sky and Solar System

Of the many activities available on the ***Voyager: SkyGazer* CD-ROM** accompanying your book, use the following files to observe key phenomena covered in this chapter.

Go to the **File: Basics** folder for the following demonstrations.

1. Ptolemy on Venus
2. Pluto's Orbit

Go to the **File: Demo** folder for the following demonstrations.

1. Hale-Bopp Path
2. Venus-Earth-Moon

Go to the **Explore** menu for the following demonstrations.

1. Paths of the Planets

Web Projects

1. *Easter.* Find out when different sects of Christianity celebrate Easter. Then research how and why different sects set different dates for Easter. Summarize your findings in a one- to two-page report.
2. *Greek Astronomers.* Many ancient Greek scientists had ideas that, in retrospect, seem well ahead of their time. Choose one or more of the following ancient Greek scientists, and

learn enough about their work in science and astronomy to write a one- to two-page "scientific biography."

Thales	Anaximander	Pythagoras
Anaxagoras	Empedocles	Democritus
Meton	Plato	Eudoxus
Aristotle	Callipus	Aristarchus
Archimedes	Eratosthenes	Apollonius
Hipparchus	Seleucus	Ptolemy
Hypatia		

3. *The Ptolemaic Model.* This chapter gives only a very brief description of Ptolemy's model of the universe. Investigate the model in greater depth. Using diagrams and text as needed, give a two- to three-page description of the model.
4. *The Galileo Affair.* In recent years, the Roman Catholic Church has devoted a lot of resources to learning more about the trial of Galileo and to understanding past actions of the Church in the Galilean case. Learn more about such studies, and write a short report about the current Vatican view of the case.
5. *Science or Pseudoscience.* Choose some pseudoscientific claim that has been in the news recently, and learn more about it and how scientists have "debunked" it. Write a short summary of your findings.

SI

Celestial Timekeeping and Navigation

Supplementary Chapter

LEARNING GOALS

SI.I Astronomical Time Periods

- How do we define the day, month, year, and planetary periods?
- How do we tell the time of day?
- When and why do we have leap years?

SI.2 Celestial Coordinates and Motion in the Sky

- How do we locate objects on the celestial sphere?
- How do stars move through the local sky?
- How does the Sun move through the local sky?

SI.3 Principles of Celestial Navigation

- How can you determine your latitude?
- How can you determine your longitude?

Socrates: *Shall we make astronomy the next study? What do you say?*
Glaucon: *Certainly. A working knowledge of the seasons, months, and years is beneficial to everyone, to commanders as well as to farmers and sailors.*
Socrates: *You make me smile, Glaucon. You are so afraid that the public will accuse you of recommending unprofitable studies.*

Plato, *Republic*

In ancient times, the practical need for timekeeping and navigation was one of the primary reasons for the study of astronomy. The celestial origins of timekeeping and navigation are still evident. The time of day comes from the location of the Sun in the local sky, the month comes from the Moon's cycle of phases, and the year comes from the Sun's annual path along the ecliptic. The very name of the "North Star" tells us how it can be an aid to navigation.

Today, we can tell the time by glancing at an inexpensive electronic watch and navigate with handheld devices that receive signals from satellites of the global positioning system (GPS). But knowing the celestial basis of timekeeping and navigation can still be useful, particularly for understanding the rich history of astronomical discovery. In this chapter, we will explore the apparent motions of the Sun, Moon, and planets in greater detail, which will allow us to study the principles of celestial timekeeping and navigation.

S1.1 Astronomical Time Periods

Today, our clocks and calendars are beautifully synchronized to the rhythms of the heavens. Precision measurements allow us to ensure that our clocks keep pace with the Sun's daily trek across our sky, while our calendar holds the dates of the equinoxes and solstices as steady as possible. In earlier chapters, we saw how this synchronicity took root in ancient observations of the sky. However, it is only much more recently that we have come to understand timekeeping in all its details. In this section, we will look in more detail at basic measures of time and our modern, international system of timekeeping.

• How do we define the day, month, year, and planetary periods?

By now you know that the length of the day corresponds to Earth's rotation, the length of the month to the cycle of lunar phases, and the length of the year to our orbit around the Sun. However, when we look carefully at each case, we find that the correspondence is not quite as simple as we might at first guess. Instead, we are forced to define two different types of day, month, and year. We also define planetary periods in two different ways. Let's take a look at how and why we make these distinctions.

The Length of the Day We usually think of a day as the time it takes for Earth to rotate once, but if you measure this time period you'll find that it is *not* exactly 24 hours. Instead, Earth's rotation period is about 4 minutes short of 24 hours. What's going on?

We can understand the answer by thinking about the movement of the stars and Sun across our sky. Remember that the daily circling of the stars in our sky is an illusion created by Earth's rotation (see Figure 2.9). Thus, you can measure Earth's rotation period by measuring how long it takes for any star to go from its highest point in the sky one day to its highest point the next day (Figure S1.1a). This time period, which we call a **sidereal day**, is about 23 hours 56 minutes (more precisely, $23^h56^m4.09^s$). *Sidereal* (pronounced *sy-DEAR-ee-al*) means "related to the stars"; note that you'll measure the same time no matter what star you choose. For practical purposes, the sidereal day is Earth's precise rotation period.

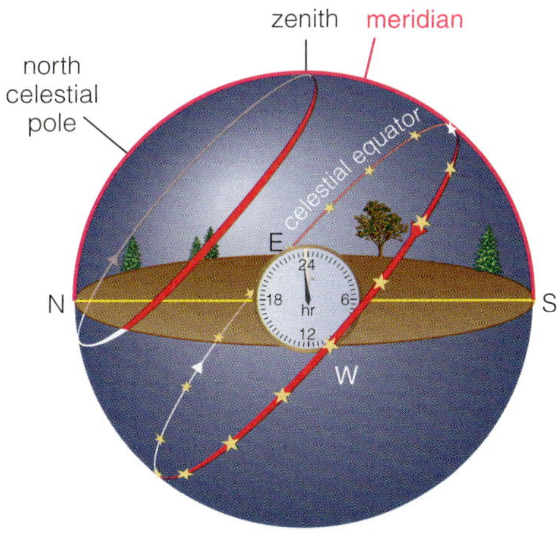

a A sidereal day is the time it takes any star to make a circuit of the local sky. It is about 23 hours 56 minutes.

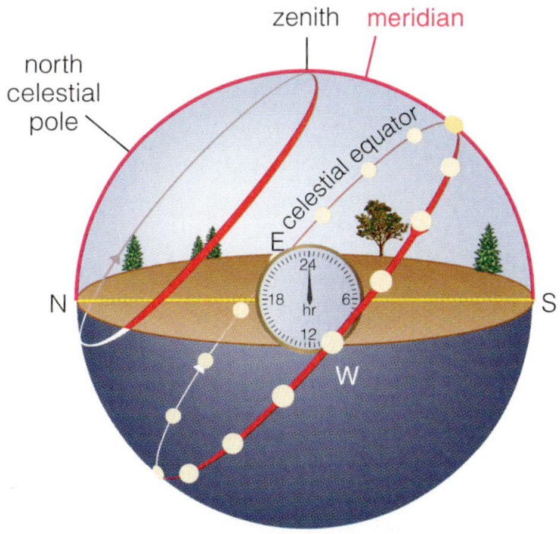

b A solar day is measured similarly, but by timing the Sun rather than a distant star. The length of the solar day varies over the course of the year but averages 24 hours.

Figure S1.1 Using the sky to measure the length of a day.

Our 24-hour day, which we call a **solar day**, is based on the time it takes for the *Sun* to make one circuit around the local sky. You can measure this time period by measuring how long it takes the Sun to go from its highest point in the sky one day to its highest point the next day (Figure S1.1b). The solar day is indeed 24 hours on average, although it varies slightly (up to 25 seconds longer or shorter than 24 hours) over the course of a year.

A simple demonstration shows why the solar day is about 4 minutes longer than the sidereal day. Set an object to represent the Sun on a table, and stand a few steps away to represent Earth. Point at the Sun and imagine that you also happen to be pointing toward a distant star that lies in the same direction. If you rotate (counterclockwise) while standing in place, you'll again be pointing at both the Sun and the star after one full rotation (Figure S1.2a). However, because Earth rotates at the same time it orbits the Sun, you can make the demonstration more realistic by taking a couple of steps around the Sun (counterclockwise) while you rotate (Figure S1.2b). After one full rotation, you will again be pointing in the direction of the distant star, so this represents a sidereal day. But it does not represent a solar day, because you will not yet be pointing back at the Sun. If you wish to point again at the Sun, you need to make up for your orbital motion by making slightly more than one full rotation. This "extra" bit of rotation makes a solar day longer than a sidereal day.

The only problem with this demonstration is that it exaggerates Earth's daily orbital motion. Because Earth takes about 365 days (1 year) to make a full 360° orbit around the Sun, Earth actually moves only about 1° per day around its orbit. Thus, a solar day represents about 361° of rotation, rather than the 360° for a sidereal day (Figure S1.2c). The extra 1° rotation takes about $\frac{1}{360}$ of Earth's rotation period, which is about 4 minutes.

The Length of the Month As we discussed in Chapter 2, our month comes from the Moon's $29\frac{1}{2}$-day cycle of phases (think "moonth"). More technically, the $29\frac{1}{2}$-day period required for each cycle of phases is called a **synodic month**. The word *synodic* comes from the Latin *synod*, which means "meeting." A synodic month gets its name because the Sun and the Moon "meet" in the sky with every new moon.

Just as a solar day is not Earth's true rotation period, a synodic month is not the Moon's true orbital period. Earth's motion around the Sun means that the Moon must complete more than one full orbit of Earth from one new moon to the next (Figure S1.3). The Moon's true orbital period, or **sidereal month**, is only about $27\frac{1}{3}$ days. Like the sidereal day, the sidereal month gets its name because it describes how long it takes the Moon to complete an orbit relative to the positions of distant stars.

The Length of the Year A year is related to Earth's orbital period, but again there are two slightly different definitions for the length of the year. The time it takes for Earth to complete one orbit relative to the stars is called a **sidereal year**. But our calendar is based on the cycle of the seasons, which we measure as the time from the spring equinox one year to the spring equinox the next year. This time period, called a **tropical year**, is about 20 minutes shorter than the sidereal year. A 20-minute difference might not seem like much, but it would make a calendar based on the sidereal year get out of sync with the seasons by 1 day every 72 years—a difference that would add up over centuries.

The difference between the sidereal year and the tropical year arises from Earth's 26,000-year cycle of axis precession [Section 2.3]. Precession not only changes the orientation of the axis in space but also changes the locations in Earth's orbit at which the seasons occur. Each year, the location of the equinoxes and solstices among the stars shifts about

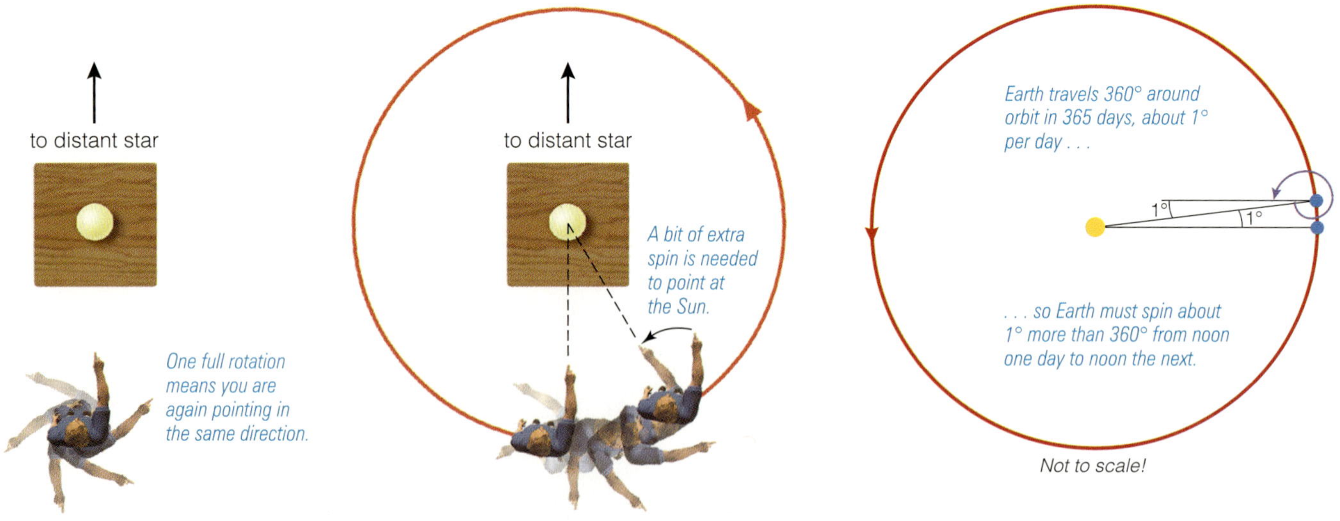

a One full rotation represents a sidereal day and means that you are again pointing to the same distant star.

b While you are "orbiting" the Sun, one rotation still returns you to pointing at a distant star, but you need slightly more than one full rotation to return to pointing at the Sun.

c Earth travels about 1° per day around its orbit, so a solar day requires about 361° of rotation.

Figure S1.2 A demonstration showing why a solar day is slightly longer than a sidereal day.

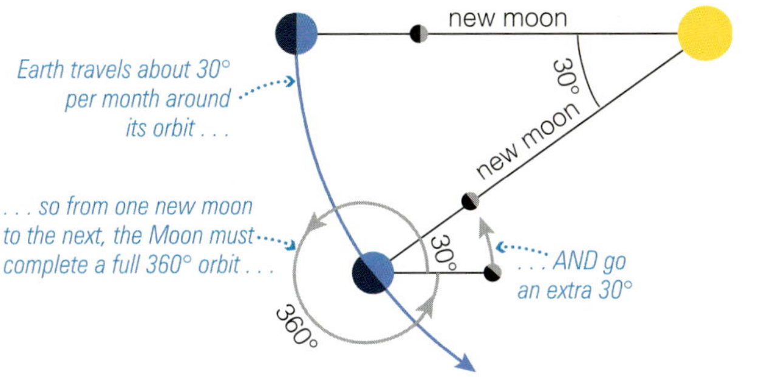

Figure S1.3 Interactive Figure The Moon completes one 360° orbit in about $27\frac{1}{3}$ days (a sidereal month), but the time from new moon to new moon is about $29\frac{1}{2}$ days (a synodic month).

$\frac{1}{26{,}000}$ of the way around the orbit. If you do the math, you'll find that $\frac{1}{26{,}000}$ of a year is about 20 minutes, which explains the 20-minute difference between the tropical year and the sidereal year.

Planetary Periods Although planetary periods are not used in our modern timekeeping, they were important to many ancient cultures. For example, the Mayan calendar was based in part on the apparent motions of Venus. In addition, Copernicus's ability to determine orbital periods of planets with his Sun-centered model played an important role in keeping the model alive long enough for its ultimate acceptance [Section 3.3].

A planet's **sidereal period** is the time the planet takes to orbit the Sun. (As usual, it has the name *sidereal* because it is measured relative to distant stars.) For example, Jupiter's sidereal period is 11.86 years, so it takes about 12 years for Jupiter to make a complete circuit around the constellations of the zodiac. Thus, Jupiter appears to move through roughly one zodiac constellation each year. If Jupiter is currently in Libra (as it is for much of 2006), it will be in Scorpius at this time next year and Sagittarius the following year, returning to Libra in 12 years.

A planet's **synodic period** is the time between when it is lined up with the Sun in our sky one time and the next similar alignment. (As with the Moon, the term *synodic* refers to the planet's "meeting" the Sun in the sky.) Figure S1.4 shows that the situation is somewhat different for planets nearer the Sun than Earth (that is, Mercury and Venus) and planets farther away (all the rest of the planets).

Look first at the situation for the more distant planet in Figure S1.4. As seen from Earth, this planet will sometimes line up with the Sun in what we call a **conjunction.** At other special times, it will appear exactly opposite the Sun in our sky, or at **opposition.** We cannot see the planet during conjunction with the Sun because it is hidden by the Sun's glare and rises and sets with the Sun in our sky. At opposition, the planet moves through the sky like the full moon, rising at sunset, reaching the meridian at midnight, and setting at dawn. Note that the planet is closest to Earth at opposition and hence appears brightest in our sky at this time.

Now look at the planet that is *nearer* than Earth to the Sun in Figure S1.4. This planet never has an opposition but instead has two conjunctions—an "inferior conjunction" between Earth and the Sun and a "superior conjunction" when the planet appears behind the Sun as seen from Earth. Two other points are important for the inner planets: their points of **greatest elongation,** when they appear farthest from the Sun in our sky. At its greatest eastern elongation, Venus appears about 46° east of the Sun in our sky, which means it shines brightly in the evening sky. Similarly, at its greatest western elongation, Venus appears about 46° west of the Sun and shines brightly in the predawn sky. In between the times when Venus appears in the morning sky and the times when it appears in the evening sky, Venus disappears from view for a few weeks with each conjunction. Mercury's pattern is similar, but because it is closer to the Sun, it never appears more than about 28° from the Sun in our sky. That makes Mercury difficult to see, because it is almost always obscured by the glare of the Sun.

THINK ABOUT IT

Do we ever see Mercury or Venus at midnight? Explain.

As you study Figure S1.4, you might wonder whether Mercury or Venus ever falls directly in front of the Sun at inferior conjunction, creating mini-eclipses as they block a little of the Sun's light. They do, but only rarely, because

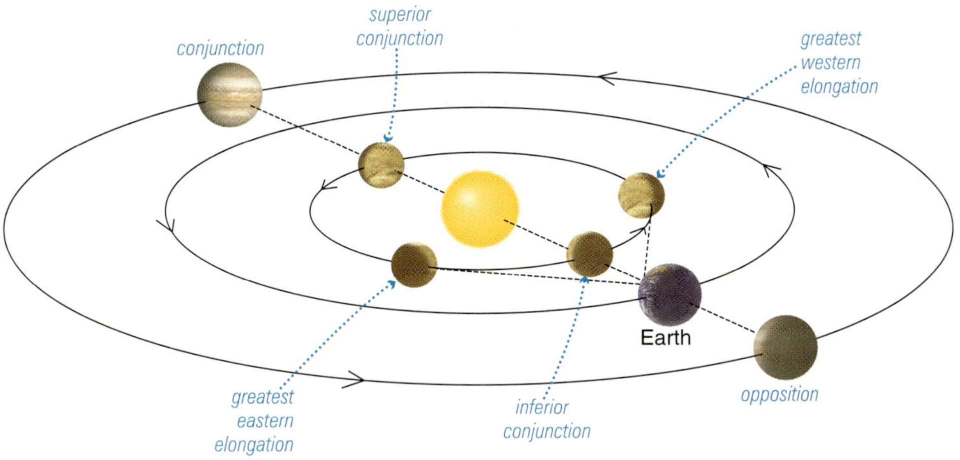

Figure S1.4 This diagram shows important positions of planets relative to Earth and the Sun. For a planet farther from the Sun than Earth (such as Jupiter), conjunction occurs when the planet appears aligned with the Sun in the sky, and opposition occurs when the planet appears on our meridian at midnight. Planets nearer the Sun (such as Venus) have two conjunctions and never get farther from the Sun in our sky than at their greatest elongations. (Adapted from *Advanced Skywatching,* by Burnham et al.)

their orbital planes are slightly tilted compared to Earth's orbital plane (the ecliptic plane). As a result, Mercury and Venus usually appear slightly above or below the Sun at inferior conjunction. But on rare occasions, we do indeed see Mercury or Venus appear to pass directly across the face of the Sun during inferior conjunction (Figure S1.5). Such events are called **transits**. Mercury transits occur an average of a dozen times per century, with the first two of this century on November 8, 2006, and May 9, 2016. Venus transits come in pairs 8 years apart, with more than a century between the second of one pair and the first of the next. We are currently between the two transits of a pair: The first occurred on June 8, 2004, and the second will occur on June 6, 2012. After that, it will be 105 years until the first of the next pair of Venus transits, which will occur in 2117 and 2125.

• How do we tell the time of day?

Telling time seems simple, but in fact there are several different ways to define the time of day, even after we agree that time should be based on the 24-hour solar day. Let's explore some of the ways of telling time, and see how they ultimately led to our modern system in which we can synchronize clocks anywhere in the world.

Apparent Solar Time If we base time on the Sun's *actual* position in the local sky, as is the case when we use a sundial (Figure S1.6), we are measuring **apparent solar time.** Noon is the precise moment when the Sun is highest in the sky (on the meridian) and the sundial casts its shortest shadow. Before noon, when the Sun is rising upward through the sky, the apparent solar time is *ante meridiem* ("before the

MATHEMATICAL INSIGHT S1.1 The Copernican Layout of the Solar System

As discussed in Chapter 3, Copernicus favored the Sun-centered model partly because it allowed him to calculate orbital periods and distances for the planets. Let's see how the Copernican system allows us to determine orbital (sidereal) periods of the planets.

We cannot directly measure orbital periods, because our own movement around the Sun means that we look at the planets from different points in our orbit at different times. However, we can measure synodic periods simply by seeing how much time passes between one particular alignment (such as opposition or inferior conjunction) and the next. Figure 1 shows the geometry for a planet *farther from the Sun than Earth* (such as Jupiter), under the assumption of circular orbits (which is what Copernicus assumed). Study the figure carefully to notice the following key facts:

- The dashed brown curve shows the planet's complete orbit. The time the planet requires for one complete orbit is its orbital (sidereal) period, P_{orb}.
- The solid brown arrow shows how far the planet travels along its orbit from one opposition to the next. The time between oppositions is defined as its synodic period, P_{syn}.
- The dashed blue curve shows Earth's complete orbit; Earth takes $P_{Earth} = 1$ yr to complete an orbit.
- The solid red curve (and red arrow) shows how far Earth goes during the planet's synodic period; it is *more* than one complete orbit because Earth must travel a little "extra" to catch back up with the planet. The time it takes Earth to travel the "extra" distance (the thick part of the red curve) must be the planet's synodic period minus 1 year, or $P_{syn} - 1$ yr.
- The angle that the planet sweeps out during its synodic period is equal to the angle that Earth sweeps out as it travels the "extra" distance. Thus, the *ratio* of the planet's complete orbital period (P_{orb}) to its synodic period (P_{syn}) must be equal to the *ratio* of Earth's orbital period (1 yr) to the time required for the "extra" distance (see Appendix C.5 for a review of ratios). Since we already found that the time required for this extra distance is $P_{syn} - 1$ yr, we write:

$$\frac{P_{orb}}{P_{syn}} = \frac{1\text{ yr}}{(P_{syn} - 1\text{ yr})}$$

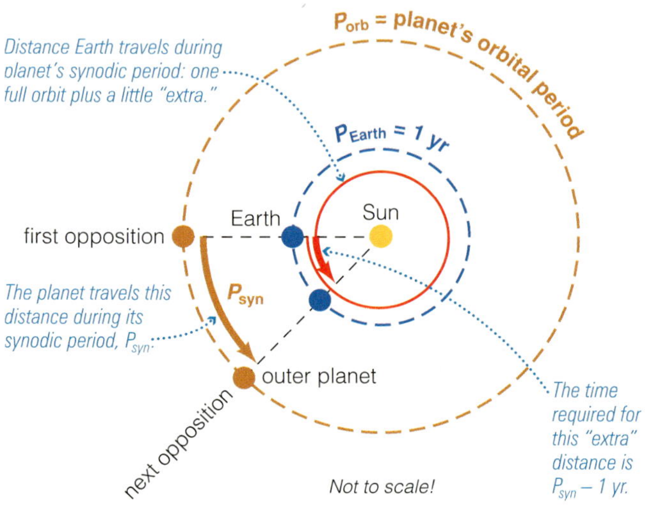

Figure 1

Multiplying both sides by P_{syn} gives us the final equation:

$$P_{orb} = P_{syn} \times \frac{1\text{ yr}}{(P_{syn} - 1\text{ yr})}$$

[for planets farther from the Sun than Earth]

The geometry is slightly different when Earth is the outer planet, as shown in Figure 2. You should convince yourself that in this case the two equal ratios are $1\text{ yr}/P_{syn} = P_{orb}/(P_{syn} - P_{orb})$. With a little algebra, you can solve this equation for P_{orb}:

$$P_{orb} = P_{syn} \times \frac{1\text{ yr}}{(P_{syn} + 1\text{ yr})}$$

[for planets closer to the Sun than Earth]

Copernicus knew the synodic periods of the planets and was therefore able to use the above equations (in a slightly different form) to calculate their true orbital periods. From there, he was able to use the geometry of planetary alignments to compute the

Figure S1.5 This photo was taken in Florida during the transit of Venus on June 8, 2004. Venus is the small black dot near the right edge of the Sun's face.

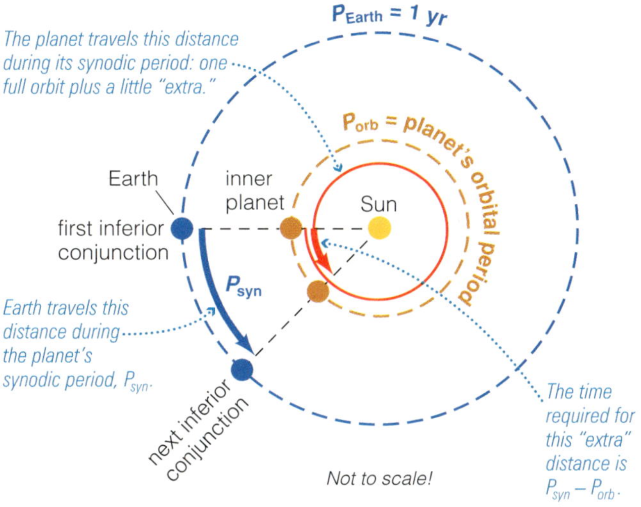

Figure 2

distances to the planets in terms of the Earth-Sun distance. (That is, he calculated distances in AU, but he did not have a method for determining how far the Earth is from the Sun.) His results, which were quite close to modern values, made so much sense to him that he felt he must have uncovered some deep truth about nature.

Example 1: Jupiter's synodic period is 398.9 days, or 1.092 years. What is its actual orbital period?

Solution:

Step 1 Understand: We are given Jupiter's synodic period (P_{syn} in the above equations) and are asked to find its orbital period (P_{orb}). Because Jupiter is farther from the Sun than Earth, we can use the first equation.

Step 2 Solve: We use the equation for a planet farther from the Sun than Earth with P_{syn} = 1.092 yr (Jupiter's synodic period):

$$\begin{aligned} P_{orb} &= P_{syn} \times \frac{1\text{ yr}}{(P_{syn} - 1\text{ yr})} \\ &= 1.092\text{ yr} \times \frac{1\text{ yr}}{(1.092\text{ yr} - 1\text{ yr})} \\ &= 11.87\text{ yr} \end{aligned}$$

Step 3 Explain: We have found that Jupiter's orbital period is 11.87 years. In other words, simply by measuring the time that passes from when Jupiter is opposite the Sun in one year to when it is again opposite the next year (which is Jupiter's synodic period), we have learned that Jupiter takes a little less than 12 years to orbit the Sun. Notice that the answer makes sense in that it is longer than Earth's orbital period of 1 year, just as we expect for a planet that is farther than Earth from the Sun.

Example 2: Venus's synodic period is 583.9 days. What is its actual orbital period?

Solution:

Step 1 Understand: We are given Venus's synodic period, P_{syn} = 583.9 days, and are asked to find its orbital period. Because Venus is closer to the Sun than Earth, we need the second equation. Also, because we are given Venus's synodic period in days, for unit consistency we need to convert it to years; you should confirm that 583.9 days = 1.599 yr.

Step 2 Solve: We simply plug the value of P_{syn} into the equation for a planet closer to the Sun than Earth:

$$\begin{aligned} P_{orb} &= P_{syn} \times \frac{1\text{ yr}}{(P_{syn} + 1\text{ yr})} \\ &= 1.599\text{ yr} \times \frac{1\text{ yr}}{(1.599\text{ yr} + 1\text{ yr})} \\ &= 0.6152\text{ yr} \end{aligned}$$

Setp 3 Explain: We have found that Venus takes 0.6152 year to orbit the Sun. This number is easier to interpret if we convert it to days or months; you should confirm that it is equivalent to 224.7 days, or about $7\frac{1}{2}$ months. Notice that the answer makes sense in that it is shorter than Earth's orbital period of 1 year, just as we expect for a planet that is closer than Earth to the Sun.

Figure S1.6 A basic sundial consists of a dial marked by numerals, and a stick, or *gnomon*, that casts a shadow. Here, the shadow is on the Roman numeral III, indicating that the apparent solar time is 3:00 P.M. (The portion of the dial without numerals represents nighttime hours.) Because the Sun's path across the local sky depends on latitude, a particular sundial will be accurate only for a particular latitude.

middle of the day"), or A.M. For example, if the Sun will reach the meridian 2 hours from now, the apparent solar time is 10 A.M. After noon, the apparent solar time is *post meridiem* ("after the middle of the day"), or P.M. If the Sun crossed the meridian 3 hours ago, the apparent solar time is 3 P.M. Note that, technically, noon and midnight are *neither* A.M. nor P.M. However, by convention we usually say that noon is 12 P.M. and midnight is 12 A.M.

THINK ABOUT IT

Is it daytime or nighttime at 12:01 A.M.? 12:01 P.M.? Explain.

Mean Solar Time Suppose you set a clock to precisely 12:00 when a sundial shows noon today. If every solar day were precisely 24 hours, your clock would always remain synchronized with the sundial. However, while 24 hours is the *average* length of the solar day, the actual length of the solar day varies throughout the year. As a result, your clock will not remain perfectly synchronized with the sundial. For example, your clock is likely to read a few seconds before or after 12:00 when the sundial reads noon tomorrow, and within a few weeks your clock time may differ from the apparent solar time by several minutes. Your clock (assuming it is accurate) will again be synchronized with the Sun on the same date next year, since it keeps track of the average length of the solar day.

If we average the differences between the time a clock would read and the time a sundial would read, we can define **mean solar time** (*mean* is another word for *average*). A clock set to mean solar time reads 12:00 each day at the time that the Sun crosses the meridian *on average*. The actual mean solar time at which the Sun crosses the meridian varies over the course of the year in a fairly complex way (see "Solar Days and the Analemma," p. 98). The result is that, on any given day, a clock set to mean solar time may read anywhere from about 17 minutes before noon to 15 minutes after noon (that is, from 11:43 A.M. to 12:15 P.M.) when a sundial indicates noon.

Although the lack of perfect synchronization with the Sun might at first sound like a drawback, mean solar time is actually more convenient than apparent solar time (the sundial time)—as long as you have access to a mechanical or electronic clock. Once set, a reliable mechanical or electronic clock can always tell you the mean solar time. In contrast, precisely measuring apparent solar time requires a sundial, which is useless at night or when it is cloudy.

Like apparent solar time, mean solar time is a *local* measure of time. That is, it varies with longitude because of Earth's west-to-east rotation. For example, clocks in New York are set 3 hours ahead of clocks in Los Angeles. If clocks were set precisely to local mean solar time, they would vary even over relatively short east-west distances. For example, mean solar clocks in central Los Angeles would be about 2 minutes behind mean solar clocks in Pasadena, because Pasadena is slightly farther east.

Standard, Daylight, and Universal Time Clocks displaying mean solar time were common during the early history of the United States. However, by the late 1800s, the growth of railroad travel made the use of mean solar time increasingly problematic. Some states had dozens of different "official" times, usually corresponding to mean solar time in dozens of different cities, and each railroad company made schedules according to its own "railroad time." The many time systems made it difficult for passengers to follow the scheduling of trains.

On November 18, 1883, the railroad companies agreed to a new system that divided the United States into four time zones, setting all clocks within each zone to the same time. That was the birth of **standard time**, which today divides the world into time zones (Figure S1.7). Depending on where you live within a time zone, your standard time may vary somewhat from your mean solar time. In principle, the standard time in a particular time zone is the mean solar time in the *center* of the time zone, so that local mean solar time within a 1-hour-wide time zone would never differ by more than a half-hour from standard time. However, time zones often have unusual shapes to conform to social, economic, and political realities, so larger variations between standard time and mean solar time sometimes occur.

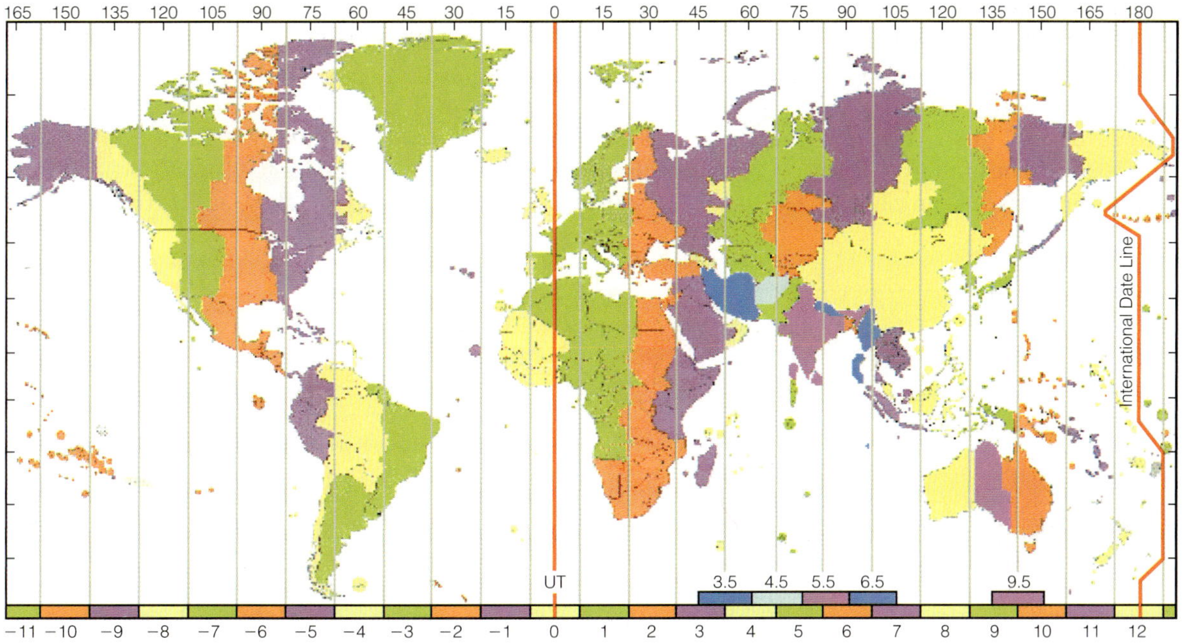

Figure S1.7 Time zones around the world. The numerical scale at the bottom shows hours ahead of (positive numbers) or behind (negative numbers) the time in Greenwich, England; the scale at the top is longitude. The vertical lines show standard time zones as they would be in the absence of political considerations. The color-coded regions show the actual time zones. Note, for example, that all of China uses the same standard time, even though the country is wide enough to span several time zones, and that the time zones of Russia and other countries of the former USSR are all shifted by about an hour. Note also that a few countries use time zones centered on a half-hour (the upper set of four colored bars), rather than an hour, relative to Greenwich time.

In most parts of the United States, clocks are set to standard time for only part of the year. Between the first Sunday in April and the last Sunday in October,* most of the United States changes to **daylight saving time**, which is 1 hour ahead of standard time. Because of the 1-hour advance with daylight saving time, clocks read around 1 P.M. (rather than around noon) when the Sun is on the meridian.

For purposes of navigation and astronomy, it is useful to have a single time for the entire Earth. For historical reasons, this "world" time was chosen to be the mean solar time in Greenwich, England—the place that also defines longitude 0° (see Figure 2.11). Today, this *Greenwich mean time* (*GMT*) is often called **universal time (UT)**. (Outside astronomy, it is more commonly called *universal coordinated time* [UTC]. Many airlines and weather services call it "Zulu time," because Greenwich's time zone is designated "Z" and "zulu" is a common way of phonetically identifying the letter *Z*.)

Seasons Tutorial, Lesson 2

• When and why do we have leap years?

Our modern calendar is based on the length of the tropical year, which is the amount of time from one spring equinox to the next. Thus, the calendar should stay synchronized with the seasons. Getting this synchronization just right was a long process in human history.

The origins of our calendar go back to ancient Egypt. By 4200 B.C., the Egyptians were using a calendar that counted 365 days in a year. However, because the length of a year is about $365\frac{1}{4}$ days (rather than exactly 365 days), the Egyptian calendar drifted out of phase with the seasons by about 1 day every 4 years. For example, if the spring equinox occurred on March 21 one year, 4 years later it occurred on March 22, 4 years after that on March 23, and so on. Over many centuries, the spring equinox moved through many different months. To keep the seasons and the calendar synchronized, Julius Caesar decreed the adoption of a new calendar in 46 B.C. This *Julian calendar* introduced the concept of **leap year**: Every fourth year has 366 days, rather than 365, so that the average length of the calendar year is $365\frac{1}{4}$ days.

The Julian calendar originally had the spring equinox falling around March 24. If it had been perfectly synchronized with the tropical year, this calendar would have ensured that the spring equinox occurred on the same date every 4 years (that is, every leap-year cycle). It didn't work perfectly, however, because a tropical year is actually about 11 minutes short of $365\frac{1}{4}$ days. Thus, the moment of the spring equinox slowly advanced by an average of 11 minutes per year. By the late 1500s, the spring equinox was occurring on March 11.

Concerned by this drift in the date of the spring equinox, Pope Gregory XIII introduced a new calendar in 1582. This *Gregorian calendar* was much like the Julian calendar,

*Under legislation passed in 2005, beginning in 2007 daylight saving time will extend four weeks longer in the United States: It will start on the second Sunday in March and end on the first Sunday in November.

with two important adjustments. First, Pope Gregory decreed that the day in 1582 following October 4 would be October 15. By eliminating the 10 dates from October 5 through October 14, 1582, he pushed the date of the spring equinox in 1583 from March 11 to March 21. (He chose March 21 because it was the date of the spring equinox in A.D. 325, which was the time of the Council of Nicaea, the first ecumenical council of the Christian church.) Second, the Gregorian calendar added an exception to the rule of having leap year every 4 years: Leap year is skipped when a century changes (for example, in years 1700, 1800, 1900) *unless* the century year is divisible by 400. Thus, 2000 was a leap year because it is divisible by 400 ($2{,}000 \div 400 = 5$), but 2100 will *not* be a leap year. These adjustments make

SPECIAL TOPIC Solar Days and the Analemma

The average length of a solar day is 24 hours, but the precise length varies over the course of the year. Two effects contribute to this variation.

The first effect arises from Earth's varying orbital speed. Recall that, in accord with Kepler's second law, Earth moves slightly faster when it is closer to the Sun in its orbit and slightly slower when it is farther from the Sun. Thus, Earth moves slightly farther along its orbit each day when it is closer to the Sun. This means that the solar day requires more than the average amount of "extra" rotation (see Figure S1.2) during these periods—making these solar days longer than average. Similarly, the solar day requires less than the average amount of "extra" rotation when it is in the portion of its orbit farther from the Sun—making these solar days shorter than average.

The second effect arises from the tilt of Earth's axis, which causes the ecliptic to be inclined by $23\frac{1}{2}°$ to the celestial equator on the celestial sphere. Because the length of a solar day depends on the Sun's apparent *eastward* motion along the ecliptic, the inclination would cause solar days to vary in length even if Earth's orbit were perfectly circular. To see why, suppose the Sun appeared to move exactly 1° per day along the ecliptic. Around the times of the solstices, this motion would be entirely eastward, making the solar day slightly longer than average. Around the times of the equinoxes, when the motion along the ecliptic has a significant northward or southward component, the solar day would be slightly shorter than average.

Together, the two effects make the actual length of solar days vary by up to about 25 seconds (either way) from the 24-hour average. Because the effects accumulate at particular times of year, the apparent solar time can differ by as much as 17 minutes from the mean solar time. The net result is often depicted visually by an **analemma** (Figure 1), which looks much like a figure 8. You'll find an analemma printed on many globes (Figure 2.16 shows a photographic version).

By using the horizontal scale on the analemma (Figure 1), you can convert between mean and apparent solar time for any date. (The vertical scale shows the declination of the Sun, which is discussed in Section S1.2.) For example, the dashed line shows that on November 10, a mean solar clock is about 17 minutes "behind the Sun," or behind apparent solar time. Thus, if the apparent solar time is 6:00 P.M. on November 10, the mean solar time is only 5:43 P.M. The discrepancy between mean and apparent solar times is called the **equation of time**. It is often plotted as a graph (Figure 2), which gives the same results as reading from the analemma.

The discrepancy between mean and apparent solar time also explains why the times of sunrise and sunset don't follow seasonal patterns perfectly. For example, the winter solstice around December 21 has the shortest daylight hours (in the Northern Hemisphere), but the earliest sunset occurs around December 7, when the Sun is still well "behind" mean solar time.

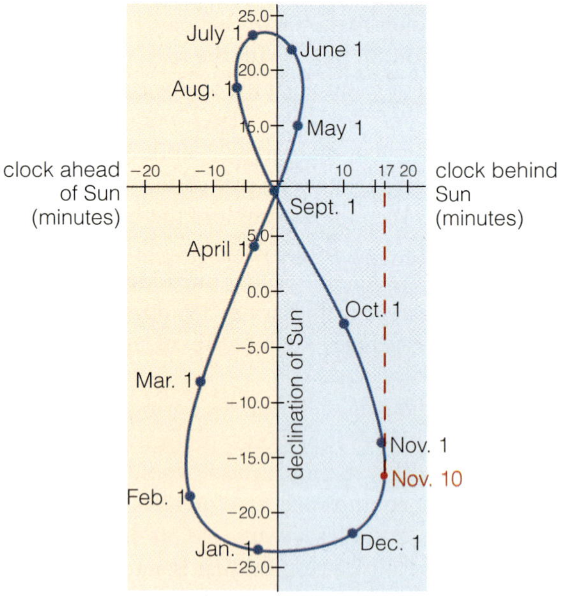

Figure 1 The analemma shows the annual pattern of discrepancies between apparent and mean solar time. For example, the dashed red line shows that on November 10, a mean solar clock reads 17 minutes behind (earlier than) apparent solar time.

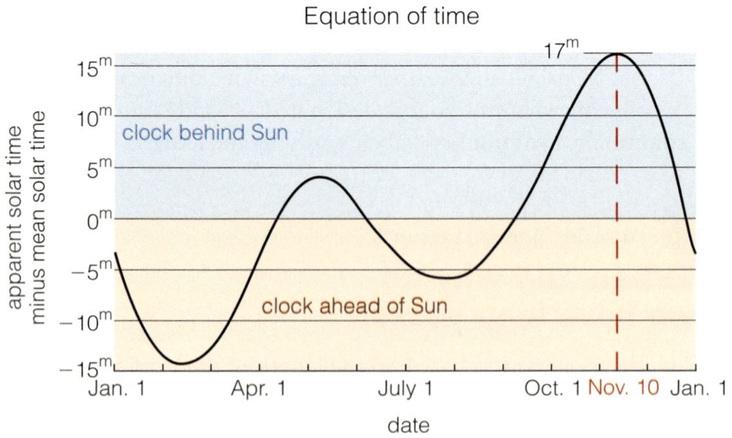

Figure 2 The discrepancies can also be plotted on a graph as the equation of time.

the average length of the Gregorian calendar year almost exactly the same as the actual length of a tropical year, which ensures that the spring equinox will occur on March 21 every fourth year for thousands of years to come.

Today, the Gregorian calendar is used worldwide for international communication and commerce. (Many countries still use traditional calendars, such as the Chinese, Islamic, and Jewish calendars, for cultural purposes.) However, as you might guess, the Pope's decree was not immediately accepted in regions not bound to the Catholic Church. For example, the Gregorian calendar was not adopted in England or in the American colonies until 1752, and it was not adopted in China until 1912 or in Russia until 1919.

S1.2 Celestial Coordinates and Motion in the Sky

We are now ready to turn our attention from timekeeping to navigation. The goal of celestial navigation is to use the Sun and the stars to find our position on Earth. Before we can do that, we need to understand the apparent motions of the sky in more detail than we covered in Chapter 2. We'll begin in this section by discussing how we locate objects on the celestial sphere, which will then allow us to explore how positions on the celestial sphere determine motion in the local sky. With this background, we'll be ready to explore the principles of celestial navigation in the final section of this chapter.

• How do we locate objects on the celestial sphere?

Recall from Chapter 2 that the celestial sphere is an illusion, but one that is quite useful when looking at the sky. We can make the celestial sphere even more useful by giving it a set of **celestial coordinates** that function much like the coordinates of latitude and longitude on Earth. Just as we can locate a city on Earth by its latitude and longitude, we will use an object's celestial coordinates to describe its precise location on the celestial sphere.

We have already discussed the basic features of the celestial sphere that will serve as starting points for our coordinate system: the north and south celestial poles, the celestial equator, and the ecliptic. Figure S1.8 shows these locations on a schematic diagram. The arrow along the ecliptic indicates the direction in which the Sun appears to move over the course of each year. It is much easier to visualize the celestial sphere if you make a model with a simple plastic ball. Use a felt-tip pen to mark the north and south celestial poles on your ball, and then add the celestial equator and the ecliptic. Note that the ecliptic crosses the celestial equator on opposite sides of the celestial sphere at an angle of $23\frac{1}{2}°$ (because of the tilt of Earth's axis).

Equinoxes and Solstices Remember that the equinoxes and solstices are special moments in the year that help define

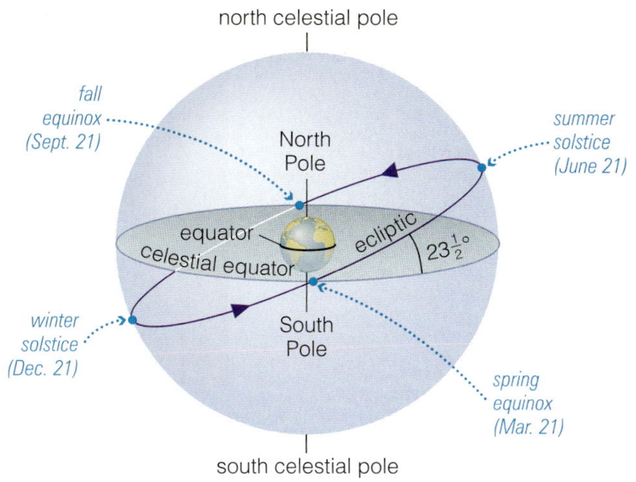

Figure S1.8 Schematic diagram of the celestial sphere, shown without stars. As a study aid, you should use a plastic ball as a model of the celestial sphere, marking it with the same special locations.

the seasons [Section 2.2]. For example, the *spring equinox,* which occurs around March 21 each year, is the moment when spring begins for the Northern Hemisphere and fall begins for the Southern Hemisphere. These moments correspond to positions in Earth's orbit (see Figure 2.15) and hence to apparent locations of the Sun along the ecliptic. As shown in Figure S1.8, the spring equinox occurs when the Sun is on the ecliptic at the point where it crosses from south of the celestial equator to north of the celestial equator. This point is also called the spring equinox. Thus, the term *spring equinox* has a dual meaning: It is the *moment* when spring begins and also the *point* on the ecliptic at which the Sun appears to be located at that moment.

Figure S1.8 also shows the points marking the summer solstice, fall equinox, and winter solstice, with the dates on which the Sun appears to be located at each point. Remember that the dates are approximate because of the leap-year cycle and because a tropical year is not exactly $365\frac{1}{4}$ days. (For example, the spring equinox may occur anytime between about March 20 and March 23.)

Although no bright stars mark the locations of the equinoxes or solstices among the constellations, you can find them with the aid of nearby bright stars (Figure S1.9). For example, the spring equinox is located in the constellation Pisces and can be found with the aid of the four bright stars in the Great Square of Pegasus. Of course, when the Sun is located at this point around March 21, we cannot see Pisces or Pegasus because they are close to the Sun in our daytime sky.

THINK ABOUT IT

Using your plastic ball as a model of the celestial sphere (which you have already marked with the celestial poles, equator, and ecliptic), mark the locations and approximate dates of the equinoxes and solstices. Based on the dates for these points, approximately where along the ecliptic is the Sun on April 21? On November 21? How do you know?

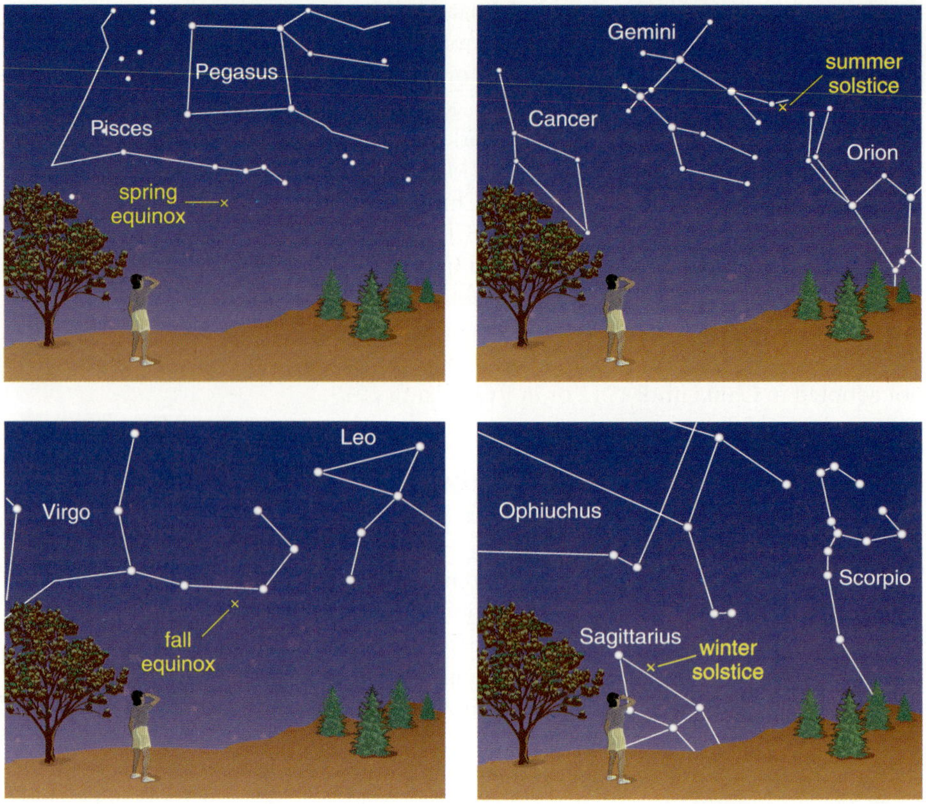

Figure S1.9 These diagrams show the locations among the constellations of the equinoxes and solstices. No bright stars mark any of these points, so you must find them by studying their positions relative to recognizable patterns. The time of day and night at which each point is above the horizon depends on the time of year.

Celestial Coordinates We can now add our system of celestial coordinates to the celestial sphere. Because this will be the third coordinate system we've used in this book, it's easier to understand if we first review the other two. Figure S1.10a shows the coordinates of *altitude* and *direction* (or *azimuth*[*]) we use in the local sky. Figure S1.10b shows the coordinates of *latitude* and *longitude* we use on Earth's surface. Finally, Figure S1.10c shows our new system of celestial coordinates. As you can see in the figure, these coordinates are called **declination (dec)** and **right ascension (RA)**.

[*]Azimuth is usually measured clockwise around the horizon from due north. By this definition, the azimuth of due north is 0°, of due east is 90°, of due south is 180°, and of due west is 270°.

If you compare Figures S1.10b and c, you'll see that declination on the celestial sphere is very similar to latitude on Earth, while right ascension is very similar to longitude. Let's start by looking at declination; notice the following key points:

- Just as lines of latitude are parallel to Earth's equator, lines of declination are parallel to the celestial equator.
- Just as Earth's equator has lat = 0°, the celestial equator has dec = 0°.
- Latitude is labeled *north* or *south* relative to the equator, while declination is labeled *positive* or *negative*. For example, the North Pole has lat = 90°N, while the north celestial pole has dec = +90°; the South Pole has lat = 90°S, while the south celestial pole has dec = −90°.

Figure S1.10 Celestial coordinate systems.

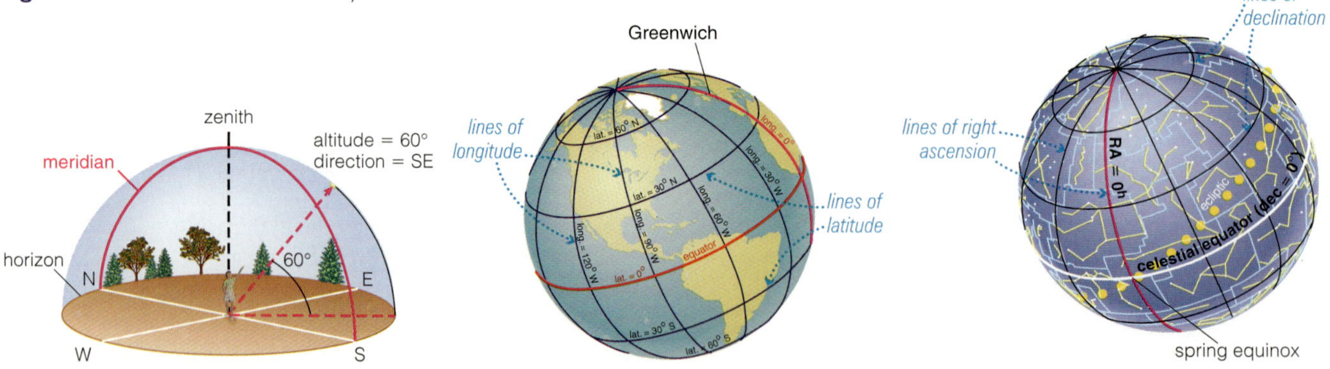

a We use altitude and direction to pinpoint locations in the local sky.

b We use latitude and longitude to pinpoint locations on Earth.

c We use declination and right ascension to pinpoint locations on the celestial sphere.

Next, notice the close correspondence between right ascension and longitude:

- Just as lines of longitude extend from the North Pole to the South Pole, lines of right ascension extend from the north celestial pole to the south celestial pole.
- Just as there is no natural starting point for longitude, there is no natural starting point for right ascension. By international treaty, longitude zero (the prime meridian) is the line of longitude that runs through Greenwich, England. By convention, right ascension zero is the line of right ascension that runs through the spring equinox.
- Longitude is measured in *degrees* east or west of Greenwich, while right ascension is measured in *hours* (and minutes and seconds) east of the spring equinox. A full 360° circle around the celestial equator goes through 24 hours of right ascension, so each hour of right ascension represents an angle of $360° \div 24 = 15°$.

As an example of how we use celestial coordinates to locate objects on the celestial sphere, consider the bright star Vega. Its coordinates are dec $= +38°44'$ and RA $= 18^h35^m$ (Figure S1.11). The positive declination tells us that Vega is 38°44′ *north* of the celestial equator. The right ascension tells us that Vega is 18 hours 35 minutes east of the spring equinox. Translating the right ascension from hours to angular degrees, we find that Vega is about 279° east of the spring equinox (because 18 hours represents $18 \times 15° = 270°$ and 35 minutes represents $\frac{35}{60} \times 15° \approx 9°$).

THINK ABOUT IT

On your plastic ball model of the celestial sphere, add a scale for right ascension along the celestial equator and also add a few circles of declination, such as declination 0°, ±30°, ±60°, and ±90°. Where is Vega on your model?

The Vega example also shows why right ascension is measured in units of time: It is because time units make it easier to track the daily motions of objects through the local sky. All objects with a particular right ascension cross the meridian at the same time, which means that all stars with RA $= 0^h$ cross the meridian at the same time that the spring equinox crosses the meridian. Thus, when we use time units, the right ascension of any object tells us how long *after* the spring equinox the object crosses the meridian. For example, Vega's right ascension, 18^h35^m, tells us that on any particular day, Vega crosses the meridian about 18 hours 35 minutes after the spring equinox. (This is 18 hours 35 minutes of *sidereal time* later, which is not exactly the same as 18 hours 35 minutes of solar time; see Mathematical Insight S1.2.)

Stars are so far away that they take thousands of years or more to move noticeably on the celestial sphere. Nevertheless, the celestial coordinates of stars are not quite constant, because they are tied to the celestial equator and the celestial equator gradually moves relative to the constellations with Earth's 26,000-year cycle of axis precession [Section 2.2]. (Axis precession does not affect Earth's orbit, so it does not affect the location of the ecliptic among the constellations.) Even over just a few decades, the coordinate changes are significant enough to make a difference in precise astronomical work—such as aiming a telescope at a particular object. As a result, careful observations require almost constant updating of celestial coordinates. Star catalogs therefore always state the year for which coordinates are given (for example, "epoch 2000"). Astronomical software can automatically calculate day-to-day celestial coordinates for the Sun, Moon, and planets as they wander among the constellations.

Celestial Coordinates of the Sun Unlike stars, which remain fixed in the patterns of the constellations on the celestial sphere, the Sun moves gradually along the ecliptic. It takes a year for the Sun to make a full circuit of the ecliptic, which means it moves through all 24 hours of right ascension over the course of the year. Each month, the Sun moves approximately one-twelfth of the way around the ecliptic, meaning that its right ascension changes by about $24 \div 12 = 2$ hours per month. Figure S1.12 shows the ecliptic marked with

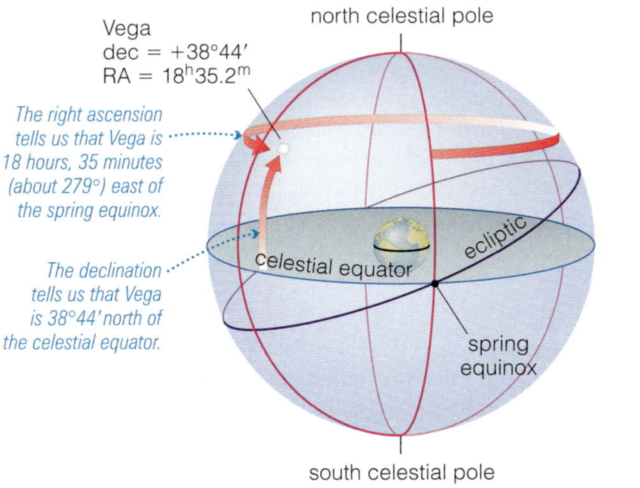

Figure S1.11 This diagram shows how we interpret the celestial coordinates of Vega.

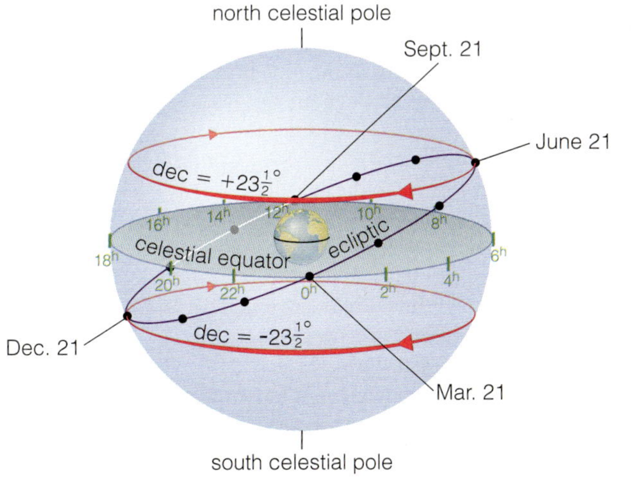

Figure S1.12 We can use this diagram of the celestial sphere to determine the Sun's right ascension and declination at monthly intervals.

the Sun's monthly position and a scale of celestial coordinates. From this figure, we can create a table of the Sun's month-by-month celestial coordinates.

Table S1.1 starts from the spring equinox, when the Sun has declination 0° and right ascension 0^h. You can see in the shaded areas of the table that while RA advances steadily through the year, the Sun's declination changes much more rapidly around the equinoxes than around the solstices. For example, the Sun's declination changes from $-12°$ on February 21 to 12° on April 21, a difference of 24° in just 2 months. In contrast, during the 2 months around the summer solstice (that is, between May 21 and July 21), the declination varies only between $+20°$ and its maximum of $+23\frac{1}{2}°$. This behavior explains why the number of daylight

MATHEMATICAL INSIGHT S1.2 Time by the Stars

The clocks we use in daily life are set to solar time, ticking through 24 hours for each day of mean solar time. In astronomy, it is also useful to have clocks that tell time by the stars, or **sidereal time**. Just as we define *solar time* according to the Sun's position relative to the meridian, *sidereal time* is based on the positions of stars relative to the meridian. We define the **hour angle (HA)** of any object on the celestial sphere to be the time since it last crossed the meridian. (For a circumpolar star, hour angle is measured from the *higher* of the two points at which it crosses the meridian each day.) For example:

- If a star is crossing the meridian now, its hour angle is 0^h.
- If a star crossed the meridian 3 hours ago, its hour angle is 3^h.
- If a star will cross the meridian 1 hour from now, its hour angle is -1^h or, equivalently, 23^h.

By convention, time by the stars is based on the hour angle of the spring equinox. That is, the **local sidereal time (LST)** is

$$LST = HA_{\text{spring equinox}}$$

For example, the local sidereal time is 00:00 when the spring equinox is *on* the meridian. Three hours later, when the spring equinox is 3 hours west of the meridian, the local sidereal time is 03:00.

Note that, because right ascension tells us how long after the spring equinox an object reaches the meridian, the local sidereal time is also equal to the right ascension (RA) of objects currently crossing your meridian. For example, if your local sidereal time is 04:30, stars with RA $= 4^h30^m$ are currently crossing your meridian. This idea leads to an important relationship among any object's current hour angle, the current local sidereal time, and the object's right ascension:

$$HA_{\text{object}} = LST - RA_{\text{object}}$$

This formula will make sense to you if you recognize that an object's right ascension tells us the time by which it trails the spring equinox on its daily trek through the sky. Because the local sidereal time tells us how long it has been since the spring equinox was on the meridian, the difference $LST - RA_{\text{object}}$ must tell us the position of the object relative to the meridian.

Sidereal time has one important subtlety: Because the stars (and the celestial sphere) appear to rotate around us in one sidereal day (23^h56^m), sidereal clocks must tick through 24 hours of sidereal time in 23 hours 56 minutes of solar time. That is, a sidereal clock gains about 4 minutes per day over a solar clock. As a result, you cannot immediately infer the local sidereal time from the local solar time, or vice versa, without either doing some calculations or consulting an astronomical table. Of course, the easiest way to determine the local sidereal time is with a clock that ticks at the sidereal rate. Astronomical observatories always have sidereal clocks, and you can buy moderately priced telescopes that come with sidereal clocks.

Example 1: Suppose the local apparent solar time is 9:00 P.M. on the spring equinox (March 21). What is the local sidereal time?

Solution:

Step 1 Understand: The trick to this problem is understanding exactly what we are looking for. We are asked to find the local sidereal time, which is defined as the hour angle of the spring equinox in the local sky. Thus, we need to know where the spring equinox is located in the local sky. We are given the key clue: The date is the day of the spring equinox, the one day of the year on which the Sun is located in the same position as the spring equinox in the sky.

Step 2 Solve: We can now find the hour angle of the spring equinox from the hour angle of the Sun. We are told that the local apparent solar time is 9:00 P.M., which means that the Sun is 9 hours past the meridian and thus has an hour angle of 9 hours. Because the spring equinox and the Sun are located in the same place on this one date of the year, the hour angle of the spring equinox is also 9 hours.

Step 3 Explain: The hour angle of the spring equinox is 9 hours, which means the local sidereal time is LST = 09:00.

Example 2: Suppose the local sidereal time is LST = 04:00. When will Vega cross the meridian?

Solution:

Step 1 Understand: We are given the local sidereal time, which tells us the hour angle of the spring equinox in the local sky. To determine when Vega will cross the meridian, we need to know its hour angle, which we can calculate from its right ascension and the formula given above. Figure S1.11 shows us that Vega has RA $= 18^h35^m$, so we have all the information we need. (Appendix F also gives RA and dec for Vega and other stars.)

Step 2 Solve: We now use the formula to find Vega's hour angle from the local sidereal time and Vega's right ascension:

$$HA_{\text{Vega}} = LST - RA_{\text{Vega}} = 4{:}00 - 18{:}35 = -14{:}35$$

Step 3 Explain: Vega's hour angle is -14 hours 35 minutes, which means it will cross your meridian 14 hours and 35 minutes from now. This also means that Vega crossed your meridian 9 hours and 25 minutes ago (because $14^h35^m + 9^h25^m = 24^h$). (Note that these are intervals of sidereal time.)

Table S1.1 The Sun's Approximate Celestial Coordinates at 1-Month Intervals

Approximate Date	*RA*	*Dec*
Mar. 21 (spring equinox)	0 hr	0°
Apr. 21	2 hr	+12°
May 21	4 hr	+20°
June 21 (summer solstice)	6 hr	$+23\frac{1}{2}°$
July 21	8 hr	+20°
Aug. 21	10 hr	+12°
Sept. 21 (fall equinox)	12 hr	0°
Oct. 21	14 hr	−12°
Nov. 21	16 hr	−20°
Dec. 21 (winter solstice)	18 hr	$-23\frac{1}{2}°$
Jan. 21	20 hr	−20°
Feb. 21	22 hr	−12°

hours increases rapidly in spring and decreases rapidly in fall, while the number of daylight hours remains long and nearly constant for a couple of months around the summer solstice and short and nearly constant for a couple of months around the winter solstice.

THINK ABOUT IT

On your plastic ball model of the celestial sphere, add dots along the ecliptic to show the Sun's monthly positions. Based on your model, what are the Sun's approximate celestial coordinates on your birthday?

Seasons Tutorial, Lesson 3

• How do stars move through the local sky?

We can use our system of celestial coordinates to gain a better understanding of the way stars move through the local sky. Earth's rotation makes all celestial objects appear to circle around Earth each day (see Figure 2.9), but what we actually see in the local sky is more complex because we see only half the celestial sphere at one time (the ground blocks our view of the other half). Let's explore the appearance of the local sky. As we'll see, the path of any star through your local sky depends on only two things: (1) your latitude; and (2) the declination of the star whose path you want to know.

The Sky at the North Pole The daily paths of stars are easiest to understand for the local sky at the North Pole, so let's begin there before moving on to other latitudes. Figure S1.13a shows the rotating celestial sphere and your orientation relative to it when you are standing at the North Pole. Your "up" points toward the north celestial pole, which therefore marks your zenith. Earth blocks your view of anything south of the celestial equator, which therefore runs along your horizon. To make it easier for you to visualize the local sky, Figure S1.13b shows your horizon extending to the celestial sphere. The horizon is marked with directions, but remember that all directions are south from the North Pole. We therefore cannot define a meridian for the North Pole, since a meridian would have to run from the north to the south points on the horizon and there are no such unique points at the North Pole.

Notice that the daily circles of the stars keep them at constant altitudes above or below the North Polar horizon.

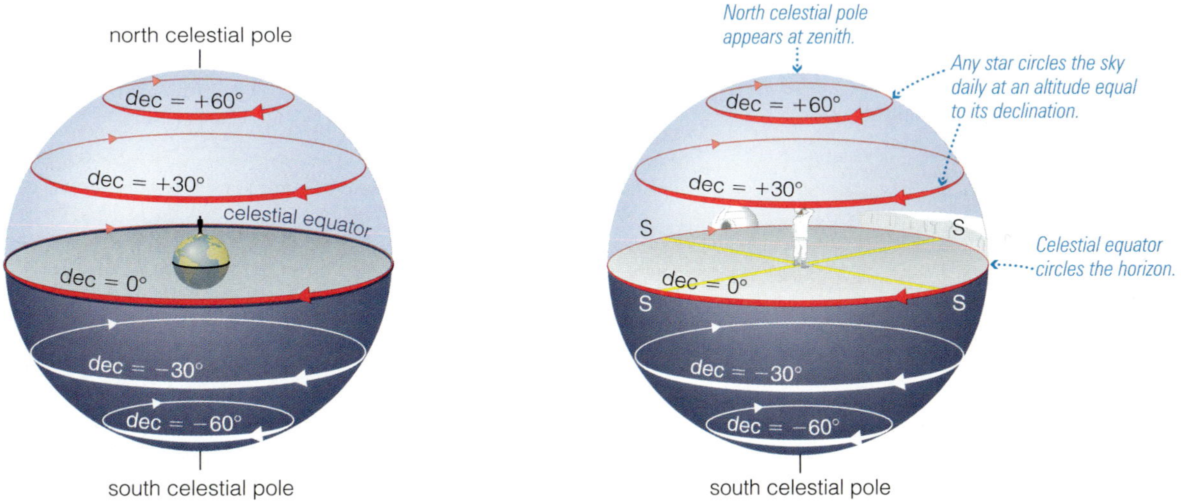

a The orientation of the local sky, relative to the celestial sphere, for an observer at the North Pole.

b Extending the horizon to the celestial sphere makes it easier to visualize the local sky at the North Pole. (Note: To understand the extension, think of Earth in part (a) as being extremely small compared to the celestial sphere.)

Figure S1.13 Interactive Figure The sky at the North Pole.

Moreover, the altitude of any star is equal to its declination. For example, a star with declination +60° circles the sky at an altitude of 60°, and a star with declination −30° remains 30° below your horizon at all times. As a result, all stars north of the celestial equator are circumpolar at the North Pole, meaning that they never fall below the horizon. Stars south of the celestial equator can never be seen at the North Pole. If you are having difficulty visualizing the star paths, it may help you to watch star paths as you rotate your plastic ball model of the celestial sphere.

You should also notice that right ascension does not affect a star's path at all: The path depends only on declination. As we'll see shortly, this rule holds for all latitudes. Right ascension affects only the *time* of day and year at which a star is found in a particular position in your sky.

The Sky at the Equator After the Poles, the equatorial sky is the next easiest case to understand. Imagine that you are standing somewhere on Earth's equator (lat = 0°), such as in Ecuador, in Kenya, or on the island of Borneo. Figure S1.14a shows that "up" points directly away from (perpendicular to) Earth's rotation axis. Figure S1.14b shows the local sky more clearly by extending the horizon to the celestial sphere and rotating the diagram so the zenith is up. As everywhere except at the poles, the meridian extends from the horizon due south, through the zenith, to the horizon due north.

Look carefully at how the celestial sphere appears to rotate in the local sky. The north celestial pole remains stationary on your horizon due north. As we should expect, its altitude of 0° is equal to the equator's latitude [Section 2.1]. Similarly, the south celestial pole remains stationary on your horizon due south. At any particular time, half the celestial equator is visible, extending from the horizon due east, through the zenith, to the horizon due west. The other half lies below the horizon. As the equatorial sky appears to turn, all star paths rise straight out of the eastern horizon and set straight into the western horizon, with the following features:

- **Stars with dec = 0°** lie *on* the celestial equator and therefore rise due east, cross the meridian at the zenith, and set due west.
- **Stars with dec > 0°** rise north of due east, reach their highest point on the meridian in the north, and set north of due west. Their rise, set, and highest point depend on their declination. For example, a star with dec = +30° rises 30° north of due east, crosses the meridian 30° to the north of the zenith—that is, at an *altitude* of 90° − 30° = 60° in the north—and sets 30° north of due west.
- **Stars with dec < 0°** rise south of due east, reach their highest point on the meridian in the south, and set south of due west. For example, a star with dec = −50° rises 50° south of due east, crosses the meridian 50° to the south of the zenith—that is, at an *altitude* of 90° − 50° = 40° in the south—and sets 50° south of due west.

Notice that exactly half of any star's daily circle lies above the horizon, which means that every star at the equator is above the horizon for exactly half of each sidereal day, or just under 12 hours, and below the horizon for the other half of the sidereal day. Also, notice again that right ascension does not affect star paths, although it does affect the time of day and year at which a star will be in a particular place along its path.

THINK ABOUT IT

Are any stars circumpolar at the equator? Are there stars that never rise above the horizon at the equator? Explain.

Skies at Other Latitudes Star tracks may at first seem more complex at other latitudes, with their mixtures of circum-

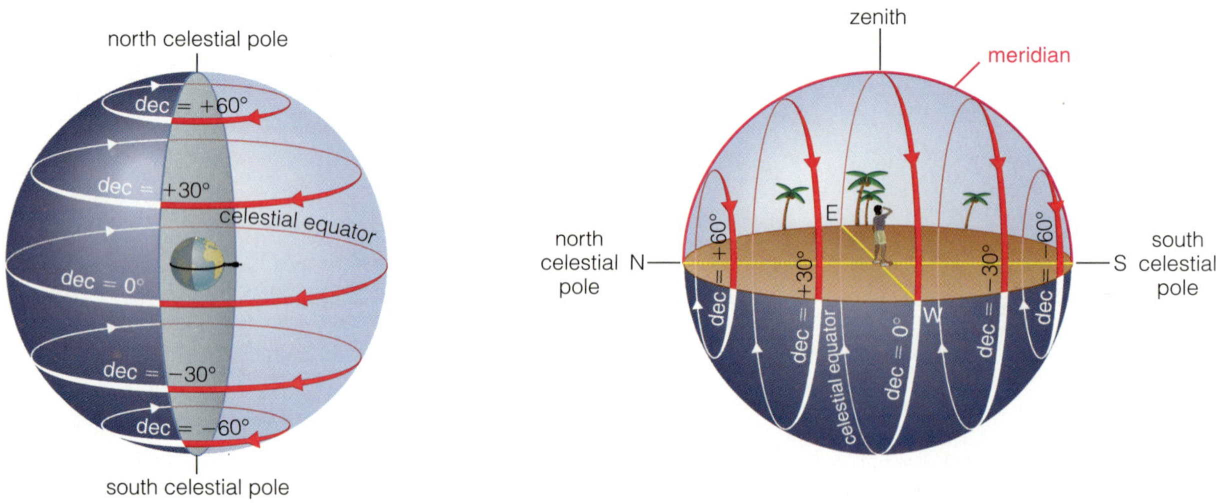

a The orientation of the local sky, relative to the celestial sphere, for an observer at Earth's equator.

b Extending the horizon and rotating the diagram make it easier to visualize the local sky at the equator.

Figure S1.14 Interactive Figure The sky at the equator.

polar stars and stars that rise and set. However, they are easy to understand if we apply the same basic strategy we've used for the North Pole and equator. Let's consider latitude 40°N, such as in Denver, Indianapolis, Philadelphia, or Beijing. First, as shown in Figure S1.15a, imagine standing at this latitude on a basic diagram of the rotating celestial sphere. Note that "up" points to a location on the celestial sphere with declination +40°. To make it easier to visualize the local sky, we next extend the horizon and rotate the diagram so the zenith is up (Figure S1.15b).

As we would expect, the north celestial pole appears 40° above the horizon due north, since its altitude in the local sky is always equal to the latitude. Half the celestial equator is visible. It extends from the horizon due east, to the meridian at an altitude of 50° in the south, to the horizon due west. By comparing this diagram to that of the local sky for the equator, you can probably notice a general rule for the celestial equator at any latitude except the poles:

The celestial equator always extends from due east on your horizon to due west on your horizon, crossing the meridian at an altitude of 90° *minus* your latitude.

The celestial equator crosses the meridian south of the zenith for locations in the Northern Hemisphere and north of the zenith for locations in the Southern Hemisphere.

If you study Figure S1.15b carefully (or, better yet, rotate your plastic ball model of the celestial sphere with the same orientation), you'll notice the following features of the sky for latitude 40°N:

- **Stars with dec = 0°** lie *on* the celestial equator and therefore follow the path of the celestial equator through the local sky. That is, for latitude 40°N, these stars rise due east, cross the meridian at altitude 90° − 40° = 50° in the south, and set due west.
- **Stars with dec > (90° − lat)** are circumpolar. Thus, for latitude 40°N, stars with declination greater than 90° − 40° = 50° are circumpolar, because they lie *within* 40° of the north celestial pole.
- **Stars with dec > 0° but which are not circumpolar** follow paths parallel to but north of the celestial equator: They rise north of due east and set north of due west, and cross the meridian to the north of the place where the celestial equator crosses it by an amount equal to their declination. For example, because the celestial equator at latitude 40° crosses the meridian at altitude 50° in the south, a star with dec = +30° crosses the meridian at altitude 50° + 30° = 80° in the south. Similarly, a star with dec = +60° crosses the meridian 60° farther north than the celestial equator, which means at altitude 70° in the north. (To calculate this result, note that the sum 50° + 60° = 110° goes 20° past the zenith altitude of 90°, making it equivalent to 90° − 20° = 70°.)
- **Stars with dec < (−90° + lat)** never rise above the horizon. Thus, for latitude 40°N, stars with declination less than −90° + 40° = −50° never rise above the horizon, because they lie within 40° of the south celestial pole.
- **Stars with dec < 0° but which are sometimes visible** follow paths parallel to but south of the celestial equator: They rise south of due east and set south of due west, and cross the meridian south of the place where the celestial equator crosses it by an amount equal to their declination. For example, a star with dec = −30° crosses the meridian at altitude 50° − 30° = 20° in the south.

You should also notice that the fraction of any star's daily circle that is above the horizon—and hence the amount of time it is above the horizon each day—depends on its declination. Because exactly half the celestial equator is above the horizon, stars on the celestial equator (dec = 0°) are

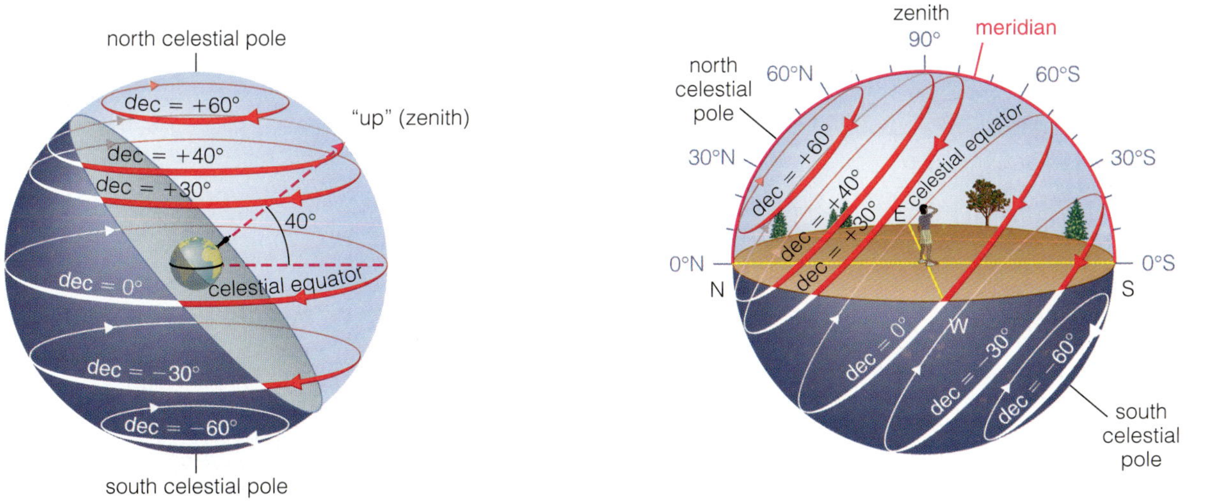

a The orientation of the local sky, relative to the celestial sphere, for an observer at latitude 40°N. Because latitude is the angle to Earth's equator, "up" points to the circle on the celestial sphere with declination +40°.

b Extending the horizon and rotating the diagram so the zenith is up make it easier to visualize the local sky. The blue scale along the meridian shows altitudes and directions in the local sky.

Figure S1.15 Interactive Figure The sky at 40°N latitude.

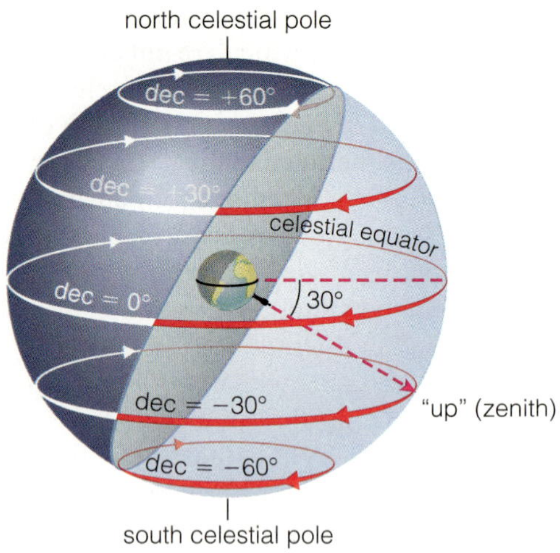

a The orientation of the local sky for an observer at latitude 30°S, relative to the celestial sphere. "Up" points to the circle on the celestial sphere with declination −30°.

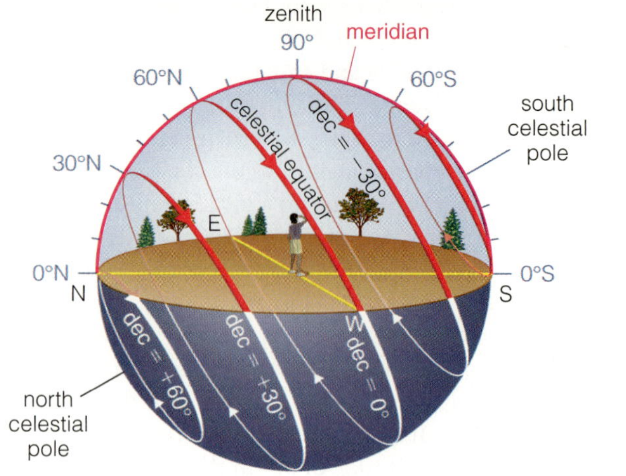

b Extending the horizon and rotating the diagram so the zenith is up make it easier to visualize the local sky. Note that the south celestial pole is visible at altitude 30° in the south, while the celestial equator stretches across the northern half of the sky.

Figure S1.16 Interactive Figure The sky at 30°S latitude.

above the horizon for about 12 hours per day. For northern latitudes like 40°N, stars with positive declinations have more than half their daily circles above the horizon and hence are above the horizon for more than 12 hours each day (with the range extending to 24 hours a day for the circumpolar stars). Stars with negative declinations have less than half their daily circles above the horizon and hence are above the horizon for less than 12 hours each day (with the range going to zero for stars that are never above the horizon).

We can apply the same strategy we used in Figure S1.15 to find star paths for other latitudes. Figure S1.16 shows the process for latitude 30°S. Note that the south celestial pole is visible to the south and that the celestial equator passes through the northern half of the sky. If you study the diagram carefully, you can see how star tracks depend on declination.

THINK ABOUT IT

Study Figure S1.16 for latitude 30°S. Describe the path of the celestial equator. Does it obey the 90° − lat rule given earlier? Describe how star tracks differ for stars with positive and negative declinations. What declination must a star have to be circumpolar at this latitude?

• How does the Sun move through the local sky?

Like the stars and other objects on the celestial sphere, the Sun's path depends only on its declination and your latitude. However, because the Sun's declination changes over the course of the year, the Sun's path also changes.

Figure S1.17 shows the Sun's path on the equinoxes and solstices for latitude 40°N. On the equinoxes, the Sun is on the celestial equator (dec = 0°) and therefore follows the celestial equator's path: It rises due east, crosses the meridian at altitude 50° in the south, and sets due west. Like any object on the celestial equator, it is above the horizon for 12 hours. On the summer solstice, when the Sun has dec = $+23\frac{1}{2}°$ (see Table S1.1), the Sun rises well north of due east,* reaches an altitude of $50° + 23\frac{1}{2}° = 73\frac{1}{2}°$ when it crosses the meridian in the south, and sets well north of due west. The daylight hours are long because much more than half the Sun's path is above the horizon. On the winter solstice, when the Sun has dec = $-23\frac{1}{2}°$, the Sun rises well south of due east, reaches an altitude of only $50° - 23\frac{1}{2}° = 26\frac{1}{2}°$ when it crosses the meridian in the south, and sets well

*Calculating exactly how far north of due east the Sun rises is beyond the scope of this book, but *SkyGazer, Starry Night,* and other astronomical software packages can tell you exactly where (and at what time) the Sun rises and sets each day.

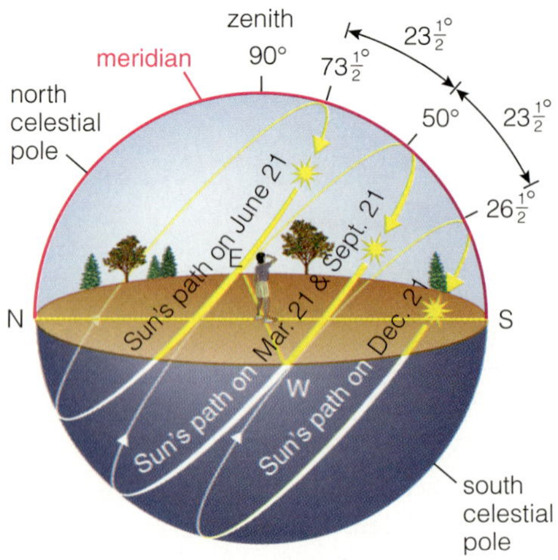

Figure S1.17 Interactive Figure The Sun's daily paths for the equinoxes and solstices at latitude 40°N.

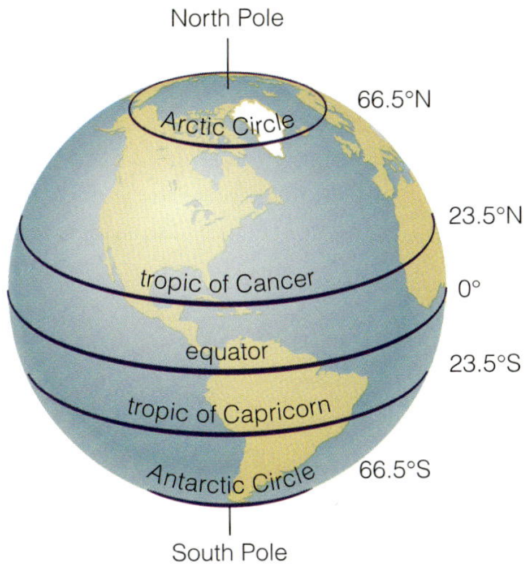

Figure S1.18 Special latitudes defined by the Sun's path through the sky.

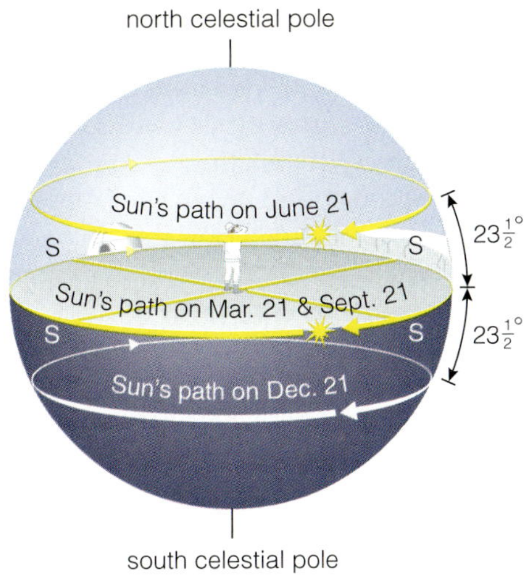

Figure S1.19 Interactive Figure Daily paths of the Sun for the equinoxes and solstices at the North Pole.

south of due west. The daylight hours are short because much less than half the Sun's path is above the horizon.

We could make a similar diagram to show the Sun's path on various dates for any latitude. However, the $23\frac{1}{2}°$ tilt of Earth's axis makes the Sun's path particularly interesting at the special latitudes shown in Figure S1.18. Let's investigate.

The Sun at the North and South Poles Remember that the celestial equator circles the horizon at the North Pole. Figure S1.19 shows how we use this fact to find the Sun's path in the North Polar sky. Because the Sun appears *on* the celestial equator on the day of the spring equinox, the Sun circles the North Polar sky *on the horizon* on March 21, completing a full circle in 24 hours (1 solar day). Over the next 3 months, the Sun continues to circle the horizon each day, circling at gradually higher altitudes as its declination increases. It reaches its highest point on the summer solstice, when its declination of $+23\frac{1}{2}°$ means that it circles the North Polar sky at an altitude of $23\frac{1}{2}°$. After the summer solstice, the daily circles gradually fall lower over the next 3 months, reaching the horizon on the fall equinox. Then, because the Sun's declination is negative for the next 6 months (until the following spring equinox), it remains below the North Polar horizon. Thus, the North Pole essentially has 6 months of daylight and 6 months of darkness, with an extended twilight that lasts a few weeks beyond the fall equinox and an extended dawn that begins a few weeks before the spring equinox.

The situation is the opposite at the South Pole. Here the Sun's daily circle first reaches the horizon on the fall equinox. The daily circles then rise gradually higher, reaching a maximum altitude of $23\frac{1}{2}°$ on the *winter* solstice, and then slowly fall back to the horizon on the spring equinox. Thus, the South Pole has the Sun above the horizon during the 6 months it is below the north polar horizon.

Although we've correctly described the Sun's true position in the polar skies over the course of the year, two effects complicate what we actually see at the poles around the times of the equinoxes. First, the atmosphere bends light enough so that the Sun *appears* to be slightly above the horizon even when it is actually slightly below it. Near the horizon, this bending makes the Sun appear about 1° higher than it would in the absence of an atmosphere. Second, the Sun's angular size of about $\frac{1}{2}°$ means that it does not fall below the horizon at a single moment but instead sets gradually. Together, these effects mean that the Sun appears above each polar horizon for slightly longer (by several days) than 6 months each year.

The Sun at the Equator At the equator, the celestial equator extends from the horizon due east, through the zenith, to the horizon due west. The Sun therefore follows this path on each equinox, reaching the zenith at local noon (Figure S1.20). Following the spring equinox, the Sun's increasing declination means that it follows a daily track that takes it gradually northward in the sky. It is farthest north on the summer solstice, when it rises $23\frac{1}{2}°$ north of due east,

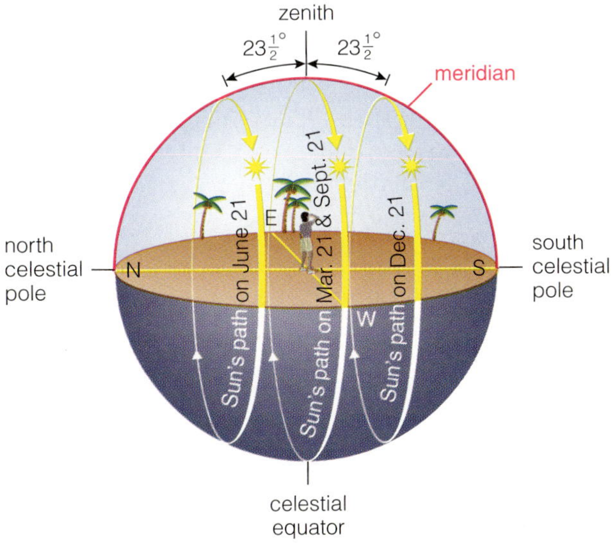

Figure S1.20 Interactive Figure Daily paths of the Sun for the equinoxes and solstices at the equator.

crosses the meridian at altitude $90^\circ - 23\frac{1}{2}^\circ = 66\frac{1}{2}^\circ$ in the north, and sets $23\frac{1}{2}^\circ$ north of due west. Over the next 6 months, it gradually tracks southward until the winter solstice, when its path is the mirror image (across the celestial equator) of its summer solstice path.

Like all objects in the equatorial sky, the Sun is always above the horizon for half a day and below it for half a day. Moreover, the Sun's track is highest in the sky on the equinoxes and lowest on the summer and winter solstices. That is why equatorial regions do not have four seasons like temperate regions [Section 2.2]. The Sun's path in the equatorial sky also makes it rise and set perpendicular to the horizon every day of the year, making for a more rapid dawn and a briefer twilight than at other latitudes.

The Sun at the Tropics The circles of latitude 23.5°N and 23.5°S are called the **tropic of Cancer** and the **tropic of Capricorn**, respectively (see Figure S1.18). The region between these two circles, generally called the *tropics*, represents the parts of Earth where the Sun can sometimes reach the zenith at noon.

Figure S1.21 shows why the tropic of Cancer is special. The celestial equator extends from due east on the horizon to due west on the horizon, crossing the meridian in the south at an altitude of $90^\circ - 23\frac{1}{2}^\circ$ (the latitude) $= 66\frac{1}{2}^\circ$. The Sun follows this path on the equinoxes (March 21 and September 21). As a result, the Sun's path on the summer solstice, when it crosses the meridian $23\frac{1}{2}^\circ$ northward of the celestial equator, takes it to the zenith at local noon. Because the Sun has its maximum declination on the summer solstice, the tropic of Cancer marks the northernmost latitude at which the Sun ever reaches the zenith. Similarly, at the tropic of Capricorn the Sun reaches the zenith at local noon on the winter solstice, making this the southernmost latitude at which the Sun ever reaches the zenith. Between the two tropic circles, the Sun passes through the zenith twice a year; the precise dates vary with latitude.

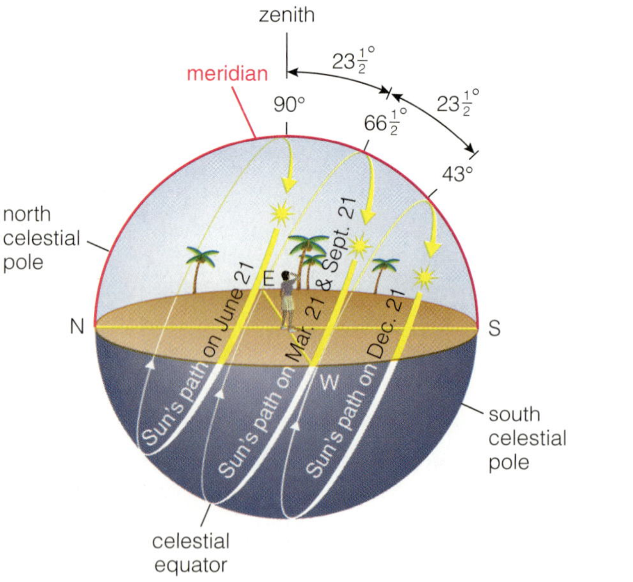

Figure S1.21 Interactive Figure Daily paths of the Sun for the equinoxes and solstices at the tropic of Cancer.

The Sun at the Arctic and Antarctic Circles At the equator, the Sun is above the horizon for 12 hours each day year-round. At latitudes progressively farther from the equator, the daily time that the Sun is above the horizon varies progressively more with the seasons. The special latitudes at which the Sun remains continuously above the horizon for a full day each year are the polar circles: the **Arctic Circle** at latitude 66.5°N and the **Antarctic Circle** at latitude 66.5°S (see Figure S1.18). Poleward of these circles, the length of continuous daylight (or darkness) increases beyond 24 hours, reaching the extreme of 6 months at the North and South Poles.

Figure S1.22 shows why the Arctic Circle is special. The celestial equator extends from due east on the horizon to due west on the horizon, crossing the meridian in the south at an altitude of $90^\circ - 66\frac{1}{2}^\circ$ (the latitude) $= 23\frac{1}{2}^\circ$. As a result, the Sun's path is circumpolar on the summer solstice: It skims the northern horizon at midnight, rises through the eastern sky to a noon maximum altitude of 47° in the south (which is the celestial equator's maximum altitude of $23\frac{1}{2}^\circ$ plus the Sun's summer solstice declination of $23\frac{1}{2}^\circ$), and then gradually falls through the western sky until it is back on the horizon at midnight (see the photograph of this path in Figure 2.18). At the Antarctic Circle, the Sun follows the same basic pattern on the winter solstice, except that it skims the horizon in the south and rises to a noon maximum altitude of 47° in the north.

However, as at the North and South Poles, what we actually see at the polar circles is slightly different from this idealization. Again, the bending of light by Earth's atmosphere and the Sun's angular size of about $\frac{1}{2}^\circ$ make the Sun *appear* to be slightly above the horizon even when it is slightly below it. Thus, the Sun seems not to set for several days, rather than for a single day, around the summer solstice at the Arctic Circle (the winter solstice at the Antarctic Circle). Similarly, the Sun appears to peek above the horizon momentarily, rather than not at all, around the winter solstice at the Arctic Circle (the summer solstice at the Antarctic Circle).

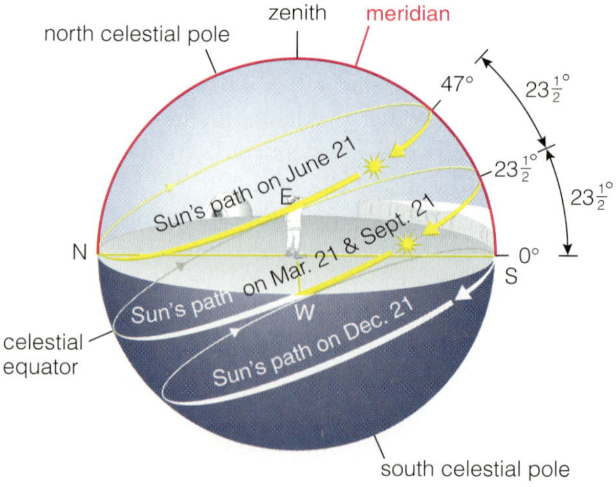

Figure S1.22 Interactive Figure Daily paths of the Sun for the equinoxes and solstices at the Arctic Circle.

S1.3 Principles of Celestial Navigation

We now have all the background we need to cover the basic principles of celestial navigation. Imagine that you're on a ship at sea, far from any landmarks. How can you figure out where you are? It's easy if you understand the apparent motions of the sky that we have already discussed in this chapter.

• How can you determine your latitude?

Determining latitude is particularly easy if you can find the north or south celestial pole: Your latitude is equal to the altitude of the celestial pole in your sky. In the Northern Hemisphere at night, you can determine your approximate latitude by measuring the altitude of Polaris. Because Polaris has a declination within 1° of the north celestial pole, its altitude is within 1° of your latitude. For example, if Polaris has altitude 17°, your latitude is between 16°N and 18°N.

If you want to be more precise, you can determine your latitude from the altitude of *any* star as it crosses your meridian. For example, suppose Vega happens to be crossing your meridian right now and it appears in your southern sky at altitude 78°44′. Because Vega has dec = +38°44′ (see Figure S1.11), it crosses your meridian 38°44′ north of the celestial equator. As shown in Figure S1.23a, you can conclude that the celestial equator crosses your meridian at an altitude of precisely 40° in the south. Your latitude must therefore be 50°N because the celestial equator always crosses the meridian at an altitude of 90° minus the latitude. You know you are in the Northern Hemisphere because the celestial equator crosses the meridian in the south.

In the daytime, you can find your latitude from the Sun's altitude on your meridian if you know the date and have a table that tells you the Sun's declination on that date. For example, suppose the date is March 21 and the Sun crosses your meridian at altitude 70° in the north (Figure S1.23b). Because the Sun has dec = 0° on March 21, you can conclude that the celestial equator also crosses your meridian in the north at altitude 70°. You must be in the Southern Hemisphere, because the celestial equator crosses the meridian in the north. From the rule that the celestial equator crosses the meridian at an altitude of 90° minus the latitude, you can conclude that you are at latitude 20°S.

• How can you determine your longitude?

You can determine your longitude by comparing the current position of an object in your sky with its position as seen from some known longitude. As a simple example, suppose you use a sundial to determine that the apparent solar time is 1:00 P.M., which means the Sun passed the meridian 1 hour ago. You immediately call a friend in England and learn that it is 3:00 P.M. in Greenwich (or you carry a clock that keeps Greenwich time). You now know that your local time is 2 hours earlier than the local time in Greenwich, which means you are 2 hours west of Greenwich. (An earlier time means that you are *west* of Greenwich, because Earth rotates from west to east.) Each hour corresponds to 15° of longitude, so "2 hours west of Greenwich" means longitude 30°W.

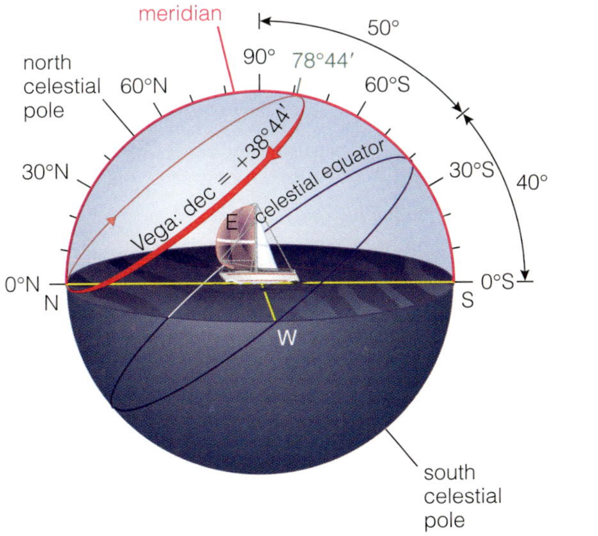

a Because Vega has dec = +38°44′, it crosses the meridian 38°44′ north of the celestial equator. Because Vega crosses the meridian at altitude 78°44′ in the south, the celestial equator must cross the meridian at altitude 40° in the south. Thus, the latitude must be 50°N.

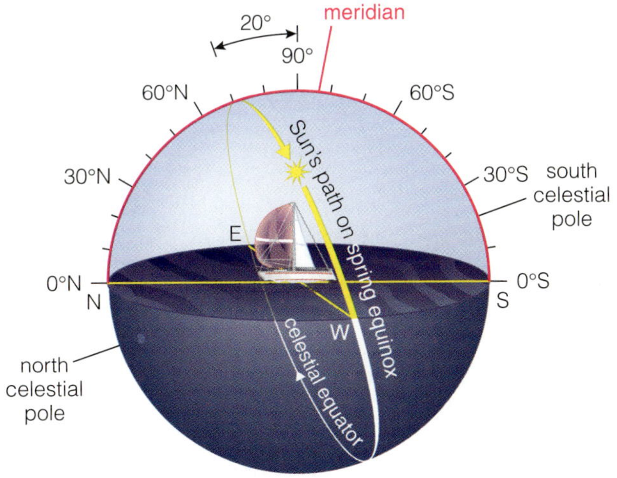

b To determine latitude from the Sun's meridian crossing, you must know the Sun's declination, which you can determine from the date. The case shown is for the spring equinox, when the Sun's declination is 0° and hence follows the path of the celestial equator through the local sky. Because the celestial equator crosses the meridian at 70° in the north, the latitude must be 20°S.

Figure S1.23 Determining latitude from a star and from the Sun.

At night, you can find your longitude by comparing the positions of stars in your local sky and at some known longitude. For example, suppose Vega is on your meridian and a call to your friend reveals that it won't cross the meridian in Greenwich until 6 hours from now. In this case, your local time is 6 hours later than the local time in Greenwich. Thus, you are 6 hours east of Greenwich, or at longitude 90°E (because $6 \times 15° = 90°$).

Celestial Navigation in Practice Although celestial navigation is easy in principle, at least three practical considerations make it more difficult in practice. First, finding either latitude or longitude requires a tool for measuring angles in the sky. One such device, called an *astrolabe,* was invented by the ancient Greeks and significantly improved by Islamic scholars during the Middle Ages. The astrolabe's faceplate (Figure S1.24a) could be used to tell time, because it consisted of a rotating star map and horizon plates for specific latitudes. Today you can buy similar rotatable star maps, called *planispheres.* Most astrolabes contained a sighting stick on the back that allowed users to measure the altitudes of bright stars in the sky. These measurements could then be correlated against special markings under the faceplate (Figure S1.24b). Astrolabes were effective but difficult and expensive to make. As a result, medieval sailors often measured angles with a simple pair of calibrated perpendicular sticks, called a *cross-staff* or *Jacob's staff* (Figure S1.24c). A more modern device called a *sextant* allows much more precise angle determinations by incorporating a small telescope for sightings (Figure S1.24d). Sextants are still used for celestial navigation on many ships. If you want to practice celestial navigation yourself, you can buy an inexpensive plastic sextant at many science-oriented stores.

A second practical consideration is knowing the celestial coordinates of stars and the Sun so that you can determine their paths through the local sky. At night, you can use a table listing the celestial coordinates of bright stars. In addition to knowing the celestial coordinates, you must either know the constellations and bright stars extremely well or

Figure S1.24 Navigational instruments.

a The faceplate of an astrolabe; many astrolabes had sighting sticks on the back for measuring positions of bright stars.

b A copper engraving of Italian explorer Amerigo Vespucci (for whom America was named) using an astrolabe to sight the Southern Cross. The engraving by Philip Galle, from the book *Nova Reperta,* was based on an original by Joannes Stradanus in the early 1580s.

c A woodcutting of Ptolemy holding a cross-staff (artist unknown).

d A sextant.

carry star charts to help you identify them. For navigating by the Sun in the daytime, you'll need a table listing the Sun's celestial coordinates on each day of the year.

The third practical consideration applies to determining longitude: You need to know the current position of the Sun (or a particular star) in a known location, such as Greenwich, England. Although you could find this out by calling a friend who lives there, it's more practical to carry a clock set to universal time (that is, Greenwich mean time). In the daytime, the clock makes it easy to determine your longitude. If apparent solar time is 1:00 P.M. in your location and the clock tells you that it is 3:00 P.M. in Greenwich, then you are 2 hours west of Greenwich, or at longitude 30°W. The task is more difficult at night, because you must compare the position of a *star* in your sky to its current position in Greenwich. You can do this with the aid of detailed astronomical tables that allow you to determine the current position of any star in the Greenwich sky from the date and the universal time.

Historically, this third consideration created enormous problems for navigation. Before the invention of accurate clocks, sailors could easily determine their latitude but not their longitude. Indeed, most of the European voyages of discovery beginning in the 1400s relied on little more than guesswork about longitude, although some sailors learned complex mathematical techniques for estimating longitude through observations of the lunar phases. More accurate longitude determination, upon which the development of extensive ocean commerce and travel depended, required the invention of a clock that would remain accurate on a ship rocking in the ocean swells. By the early 1700s, solving this problem was considered so important that the British government offered a substantial monetary prize for the solution. John Harrison claimed the prize in 1761, with a clock that lost only 5 seconds during a 9-week voyage to Jamaica.*

The Global Positioning System In the past couple of decades, a new type of celestial navigation has supplanted traditional methods. It finds positions relative to a set of satellites in Earth orbit. These satellites of the **global positioning system (GPS)** in essence function like artificial stars. The satellites' positions at any moment are known precisely from their orbital characteristics. The GPS currently uses about two dozen satellites orbiting Earth at an altitude of 20,000 kilometers. Each satellite transmits a radio signal that a small radio receiver can receive—rain or shine, day or night. Each GPS receiver has a built-in computer that calculates your precise position on Earth by comparing the signals received from several GPS satellites.

The United States originally built the GPS in the late 1970s for military use. Today, the many applications of the GPS include automobile navigation systems as well as systems for helping airplanes land safely, guiding the blind around town, and helping lost hikers find their way. Geologists have used the GPS to measure *millimeter*-scale changes in Earth's crust.

With rapid growth in the use of GPS navigation, the ancient practice of celestial navigation is in danger of becoming a lost art. Fortunately, many amateur clubs and societies are keeping the skills of celestial navigation alive.

*The story of the difficulties surrounding the measurement of longitude at sea and how Harrison finally solved the problem is chronicled in *Longitude*, by Dava Sobel (Walker and Company, 1995).

COMMON MISCONCEPTIONS

Compass Directions

Most people determine direction with the aid of a compass rather than the stars. However, a compass needle doesn't actually point to true geographic north. Instead, the compass needle responds to Earth's magnetic field and points to *magnetic* north, which can be substantially different from true north. If you want to navigate precisely with a compass, you need a special map that takes into account local variations in Earth's magnetic field. Such maps are available at most camping stores. They are not perfectly reliable, however, because the magnetic field also varies with time. In general, celestial navigation is much more reliable for determining direction than using a compass.

THE BIG PICTURE

Putting Chapter S1 into Context

In this chapter, we built upon concepts from the first three chapters to form a more detailed understanding of celestial timekeeping and navigation. We also learned how to determine paths for the Sun and the stars in the local sky. As you look back at what you've learned, keep in mind the following "big picture" ideas:

- Our modern systems of timekeeping are rooted in the apparent motions of the Sun through the sky. Although it's easy to forget these roots when you look at a clock or a calendar, the sky was the only guide to time for most of human history.
- The term *celestial navigation* sounds a bit mysterious, but it refers to simple principles that allow you to determine your location on Earth. Even if you're never lost at sea, you may find the basic techniques of celestial navigation useful to orient yourself at night (for example, on your next camping trip).
- If you understand the apparent motions of the sky discussed in this chapter and also learn the constellations and bright stars, you'll feel very much "at home" under the stars at night.

SUMMARY OF KEY CONCEPTS

S1.1 Astronomical Time Periods

- **How do we define the day, month, year, and planetary periods?** Each of these is defined in two ways. A **sidereal day** is Earth's rotation period, which is about 4 minutes shorter than the 24-hour **solar day** from noon one day to noon the next day. A **sidereal month** is the Moon's orbital period of about $27\frac{1}{3}$ days; a **synodic month** is the $29\frac{1}{2}$ days required for the Moon's cycle of phases. A **sidereal year** is Earth's orbital period, which is about 20 minutes longer than the **tropical year** from one spring equinox to the next. A planet's **sidereal period** is its orbital period, and its **synodic period** is the time from one opposition or conjunction to the next.

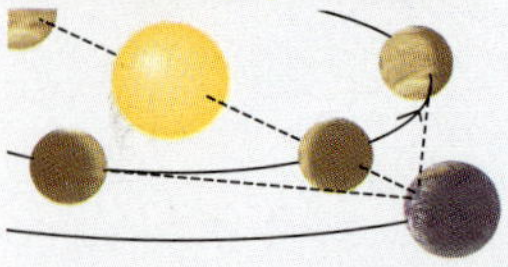

- **How do we tell the time of day?** There are several time measurement systems. **Apparent solar time** is based on the Sun's position in the local sky. **Mean solar time** is also local, but it averages the changes in the Sun's rate of motion over the year. **Standard time** and **daylight saving time** divide the world into time zones. **Universal time** is the mean solar time in Greenwich, England.

- **When and why do we have leap years?** We usually have a **leap year** every 4 years because the length of the year is $365\frac{1}{4}$ days. However, it is not exactly $365\frac{1}{4}$ days, so our calendar skips a leap year in century years not divisible by 400.

S1.2 Celestial Coordinates and Motion in the Sky

- **How do we locate objects on the celestial sphere?** **Declination** is given as an angle describing an object's position north or south of the celestial equator. **Right ascension**, usually measured in hours (and minutes and seconds), tells us how far east an object is located relative to the spring equinox.

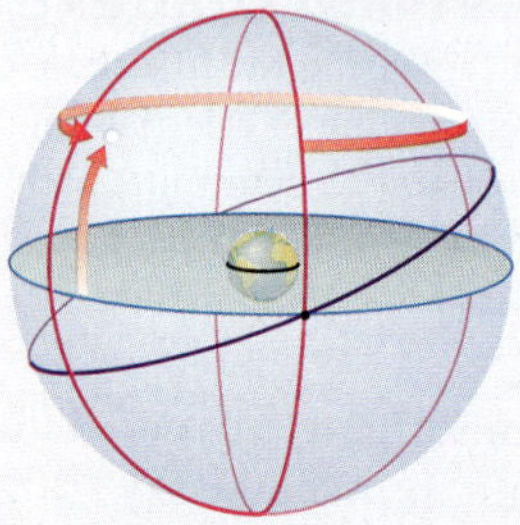

- **How do stars move through the local sky?** A star's path through the local sky depends on its declination and your latitude. Latitude tells you the orientation of your sky relative to the celestial sphere, while declination tells you how a particular star's path compares to the path of the celestial equator through your sky.

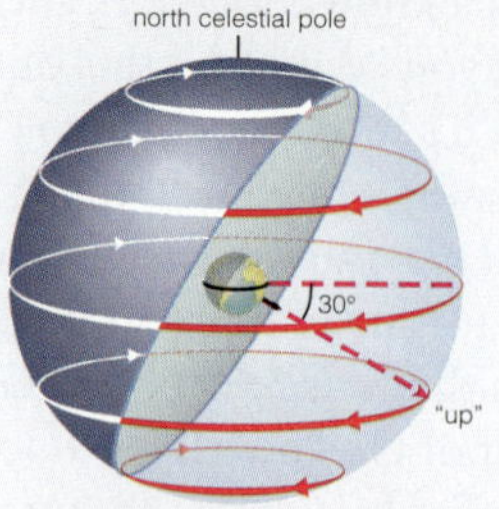

- **How does the Sun move through the local sky?** The Sun's path also depends on its declination and your latitude, but it varies throughout the year because its declination varies. The Sun's varying path helps define special latitudes, including the **tropics of Cancer and Capricorn** and the **Arctic and Antarctic Circles**.

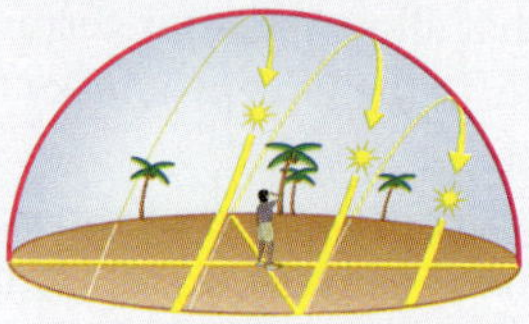

S1.3 Principles of Celestial Navigation

- **How can you determine your latitude?** You can determine your latitude from the altitude of the celestial pole in your sky or by measuring the altitude and knowing the declination of a star (or the Sun) as it crosses your meridian.

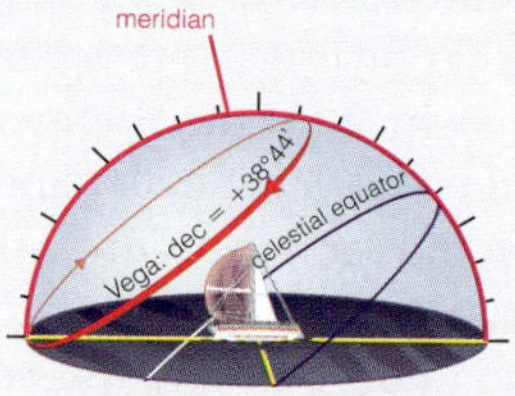

- **How can you determine your longitude?** To determine longitude you must know the position of the Sun or a star in your sky and its position at the same time in the sky of Greenwich, England (or some other specific location). This is most easily done if you have a clock that tells universal time.

EXERCISES AND PROBLEMS

Review Questions

Short-Answer Questions Based on the Reading

1. Why is a *sidereal day* shorter than a *solar day*?
2. What is the difference between a *sidereal month* and a *synodic month*? Between a *sidereal year* and a *tropical year*? Between a planet's *sidereal period* and its *synodic period*?
3. What do we mean by *opposition*, *conjunction*, and *greatest elongation* for planets? Explain both for planets closer than Earth to the Sun and for planets farther than Eartth from the Sun.
4. Under what circumstances do we see a *transit* of a planet across the Sun?

5. What is *apparent solar time*? Why is it different from *mean solar time*? How are *standard time, daylight saving time*, and *universal time* related to mean solar time?
6. Describe the origins of the Julian and Gregorian calendars. Which one do we use today?
7. What do we mean when we describe the equinoxes and solstices as points on the celestial sphere? How are these points related to the times of year that we call the equinoxes and solstices?
8. What are *declination* and *right ascension*? How are these celestial coordinates similar to latitude and longitude on Earth? How are they different?
9. How and why do the Sun's celestial coordinates change over the course of each year?
10. Suppose you are standing at the North Pole. Where is the celestial equator in your sky? Where is the north celestial pole? Describe the daily motion of the sky. Do the same for the sky at the equator and at latitude 40°N.
11. Describe the Sun's paths through the local sky on the equinoxes and on the solstices for latitude 40°N. Do the same for the North Pole, South Pole, and equator.
12. What is special about the tropics of Cancer and Capricorn? Describe the Sun's path on the solstices at these latitudes. Do the same for the Arctic and Antarctic Circles.
13. Briefly describe how you can use the Sun or stars to determine your latitude and longitude.
14. What is the global positioning system?

Test Your Understanding

Does It Make Sense?

Decide whether the statement makes sense (or is clearly true) or does not make sense (or is clearly false). Explain your reasoning. (For an example, see Chapter 1, "Does It Make Sense?")

(*Hint:* For statements that involve coordinates—such as altitude, longitude, or declination—check whether the correct coordinates are used for the situation. For example, it does not make sense to describe a location on Earth by an altitude since altitude makes sense only for positions of objects in the local sky.)

15. Last night I saw Venus shining brightly on the meridian at midnight.
16. The apparent solar time was noon, but the Sun was just setting.
17. My mean solar clock said it was 2:00 P.M., but a friend who lives east of here had a mean solar clock that said it was 2:11 P.M.
18. When the standard time is 3:00 P.M. in Baltimore, it is 3:15 P.M. in Washington, D.C.
19. Last night around 8:00 P.M. I saw Jupiter at an altitude of 45° in the south.
20. The latitude of the stars in Orion's belt is about 5°N.
21. Today the Sun is at an altitude of 10° on the celestial sphere.
22. Los Angeles is west of New York by about 3 hours of right ascension.
23. The summer solstice is east of the vernal equinox by 6 hours of right ascension.
24. Even though my UT clock had stopped, I was able to find my longitude by measuring the altitudes of 14 different stars in my local sky.

Quick Quiz

Choose the best answer to each of the following. Explain your reasoning with one or more complete sentences.

25. The time from one spring equinox to the next is the: (a) sidereal day; (b) tropical year; (c) synodic month.
26. Jupiter is brightest when it is: (a) at opposition; (b) at conjunction; (c) closest to the Sun in its orbit.
27. Venus is easiest to see in the evening when it is: (a) at superior conjunction; (b) at inferior conjunction; (c) at greatest eastern elongation.
28. In the winter, your wristwatch tells: (a) apparent solar time; (b) standard time; (c) universal time.
29. A star that is located 30° north of the celestial equator has: (a) declination = 30°; (b) right ascension = 30°; (c) latitude = 30°.
30. A star's path through your sky depends on your latitude and the star's: (a) declination; (b) right ascension; (c) both declination and right ascension.
31. At latitude 50°N, the celestial equator crosses the meridian at altitude: (a) 50° in the south; (b) 50° in the north; (c) 40° in the south.
32. At the North Pole on the summer solstice, the Sun: (a) remains stationary in the sky; (b) reaches the zenith at noon; (c) circles the horizon at altitude $23\frac{1}{2}°$.
33. If you know a star's declination, you can determine your latitude if you also: (a) measure its altitude when it crosses the meridian; (b) measure its right ascension; (c) know the universal time.
34. If you measure the Sun's position in your local sky, you can determine your longitude if you also: (a) measure its altitude when it crosses the meridian; (b) know its right ascension and declination; (c) know the universal time.

Investigate Further

In-Depth Questions to Increase Your Understanding

Short-Answer/Essay Questions

35. *Opposite Rotation.* Suppose Earth rotated in the opposite direction from its revolution; that is, suppose it rotated clockwise (as seen from above the North Pole) while still orbiting counterclockwise around the Sun each year. Would the solar day still be longer than the sidereal day? Explain.
36. *No Precession.* Suppose Earth's axis did *not* precess. Would the sidereal year still be different from the tropical year? Explain.
37. *Fundamentals of Your Local Sky.* Answer each of the following for *your* latitude.
    a. Where is the north (or south) celestial pole in your sky?
    b. Describe the location of the meridian in your sky. Specify its shape and at least three distinct points along it (such as the points at which it meets your horizon and its highest point).
    c. Describe the location of the celestial equator in your sky. Specify its shape and at least three distinct points along it (such as the points at which it meets your horizon and crosses your meridian).
    d. Does the Sun ever appear at your zenith? If so, when? If not, why not?
    e. What range of declinations makes a star circumpolar in your sky? Explain.
    f. What is the range of declinations for stars that you can never see in your sky? Explain.

38. *Sydney Sky.* Repeat Problem 37 for the local sky in Sydney, Australia (latitude 34°S).
39. *Path of the Sun in Your Sky.* Describe the path of the Sun through your local sky for each of the following days.
    a. The spring and fall equinoxes.
    b. The summer solstice.
    c. The winter solstice.
    d. Today. (*Hint:* Estimate the right ascension and declination of the Sun for today's date by using the data in Table S1.1).
40. *Sydney Sun.* Repeat Problem 39 for the local sky in Sydney, Australia (latitude 34°S).
41. *Lost at Sea I.* During an upcoming vacation, you decide to take a solo boat trip. While contemplating the universe, you lose track of your location. Fortunately, you have some astronomical tables and instruments, as well as a UT clock. You thereby put together the following description of your situation:
    - It is the spring equinox.
    - The Sun is on your meridian at altitude 75° in the south.
    - The UT clock reads 22:00.

    a. What is your latitude? How do you know?
    b. What is your longitude? How do you know?
    c. Consult a map. Based on your position, where is the nearest land? Which way should you sail to reach it?
42. *Lost at Sea II.* Repeat Problem 41, based on the following description of your situation:
    - It is the day of the summer solstice.
    - The Sun is on your meridian at altitude $67\frac{1}{2}°$ in the north.
    - The UT clock reads 06:00.
43. *Lost at Sea III.* Repeat Problem 41, based on the following description of your situation:
    - Your local time is midnight.
    - Polaris appears at altitude 67° in the north.
    - The UT clock reads 01:00.
44. *Lost at Sea IV.* Repeat Problem 41, based on the following description of your situation:
    - Your local time is 6 A.M.
    - From the position of the Southern Cross, you estimate that the south celestial pole is at altitude 33° in the south.
    - The UT clock reads 11:00.
45. *Powering Spirit.* Suppose that it is currently northern summer on Mars, and that the Mars Exploration Rover *Spirit* is in Gusev Crater near 15° north latitude. Spirit's operators have discovered that its solar panels need to receive just a few more minutes of sunlight each day to power the rover through the Martian night. What should they do? Explain.
46. *The Sun from Mars.* Mars has an axis tilt of 25.2°, only slightly larger than that of Earth. Compared to Earth, is the range of latitudes on Mars for which the Sun can reach the zenith larger or smaller? Is the range of latitudes for which the Sun is circumpolar larger or smaller? Make a sketch of Mars similar to the one for Earth in Figure S1.18.

Quantitative Problems

Be sure to show all calculations clearly and state your final answers in complete sentences.

47. *Solar and Sidereal Days.* Suppose Earth orbited the Sun in 6 months rather than 1 year but had the same rotation period. How much longer would a solar day be than a sidereal day? Explain.
48. *Saturn's Orbital Period.* Saturn's synodic period is 378.1 days. What is its actual orbital period?
49. *Mercury's Orbital Period.* Mercury's synodic period is 115.9 days. What is its actual orbital period?
50. *New Asteroid.* You discover an asteroid with a synodic period of 429 days. What is its actual orbital period?
51. *Using the Analemma I.* It's February 15 and your sundial tells you the apparent solar time is 18 minutes until noon. What is the mean solar time?
52. *Using the Analemma II.* It's July 1 and your sundial tells you that the apparent solar time is 3:30 P.M. What is the mean solar time?
53. *Find the Sidereal Time.* It is 4 P.M. on the spring equinox. What is the local sidereal time?
54. *Where's Vega?* The local sidereal time is 19:30. When will Vega cross your meridian?
55. *Find Right Ascension.* You observe a star that has an hour angle of 13 hours (13^h) when the local sidereal time is 8:15. What is the star's right ascension?
56. *Where's Orion?* The Orion Nebula has declination of about $-5.5°$ and right ascension of 5^h25^m. If you are at latitude 40°N and the local sidereal time is 7:00, approximately where does the Orion Nebula appear in your sky?
57. *Meridian Crossings of the Moon and Phobos.* Estimate the time between meridian crossings of the Moon for a person standing on Earth. Repeat your calculation for meridian crossings of the Martian moon Phobos. Use the Appendices in the back of the book if necessary.
58. *Mercury's Rotation Period.* Mercury's sidereal day is approximately $\frac{2}{3}$ of its orbital period, or about 58.6 days. Estimate the length of Mercury's solar day. Compare to Mercury's orbital period of about 88 days.

Discussion Questions

59. *Northern Chauvinism.* Why is the solstice in June called the *summer solstice,* when it marks winter for places like Australia, New Zealand, and South Africa? Why is the writing on maps and globes usually oriented so that the Northern Hemisphere is at the top, even though there is no up or down in space? Discuss.
60. *Celestial Navigation.* Briefly discuss how you think the benefits and problems of celestial navigation might have affected ancient sailors. For example, how did they benefit from using the north celestial pole to tell directions, and what problems did they experience because of the difficulty in determining longitude? Can you explain why ancient sailors generally hugged coastlines as much as possible on their voyages? What dangers did this type of sailing pose? Why did the Polynesians become the best navigators of their time?

MEDIA EXPLORATIONS

For self-study activities go to MasteringASTRONOMY www.masteringastronomy.com

Interactive Tutorials

Tutorial Review of Key Concepts

Use the interactive **Tutorials** at **www.masteringastronomy.com** to review key concepts from this chapter.

Seasons Tutorial

Lesson 2 The Solstices and Equinoxes
Lesson 3 The Sun's Position in the Sky

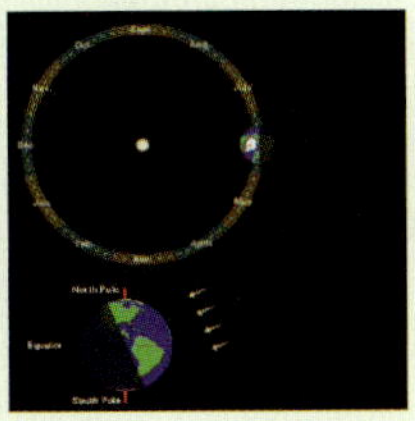

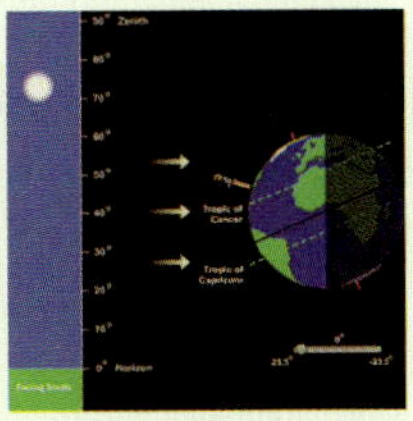

Supplementary Tutorial Exercises

Use the interactive **Tutorial Lessons** to explore the following questions.

Seasons Tutorial, Lesson 2

1. How are the solstices and equinoxes related to Earth's orbital position around the Sun?
2. What are day and night like at the North Pole?
3. How does the length of the day vary with the seasons at the Antarctic Circle?

Seasons Tutorial, Lesson 3

1. When can you see the Sun directly over the equator?
2. Where can you see the Sun directly overhead on the winter solstice?
3. When is the Sun seen directly overhead at your latitude?

Exploring the Sky and Solar System

Of the many activities available on the ***Voyager: SkyGazer*** **CD-ROM** accompanying your book, use the following files to observe key phenomena covered in this chapter.

Go to the **File: Basics** folder for the following demonstrations.

1. Analemma
2. Rubber Horizon
3. Three Cities

Go to the **File: Demo** folder for the following demonstrations.

1. Venus Transit of 1769
2. Celestial Poles
3. Russian Midnight Sun

Movies

Check out the following narrated and animated short documentaries available on **www.masteringastronomy.com** for a helpful review of key ideas covered in this chapter.

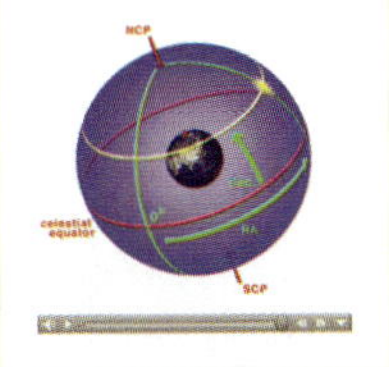

The Celestial Sphere Movie

Time and Seasons Movie

Web Projects

1. *Sundials.* Although they are no longer necessary for timekeeping, sundials remain popular for their cultural and artistic value. Search the Web for pictures and information about interesting sundials around the world. Write a short report about at least three sundials that you find particularly interesting.
2. *The Analemma.* Learn more about the analemma and its uses from information available on the Web. Write a short report on your findings.
3. *Calendar History.* Investigate the history of the Julian or Gregorian calendar in greater detail. Write a short summary of some interesting aspect of the history you learn from your Web research. (For example, why did Julius Caesar allow one year to have 445 days? How did our months end up with 28, 30, or 31 days?)
4. *Global Positioning System.* Learn more about the global positioning system and its uses. Write a short report summarizing how you think the growing availability of GPS will affect our lives over the next 10 years.

4

Making Sense of the Universe

Understanding Motion, Energy, and Gravity

LEARNING GOALS

4.1 Describing Motion: Examples from Daily Life
- How do we describe motion?
- How is mass different from weight?

4.2 Newton's Laws of Motion
- How did Newton change our view of the universe?
- What are Newton's three laws of motion?

4.3 Conservation Laws in Astronomy
- Why do objects move at constant velocity if no force acts on them?
- What keeps a planet rotating and orbiting the Sun?
- Where do objects get their energy?

4.4 The Universal Law of Gravitation
- What determines the strength of gravity?
- How does Newton's law of gravity extend Kepler's laws?

4.5 Orbits, Tides, and the Acceleration of Gravity
- How do gravity and energy together allow us to understand orbits?
- How does gravity cause tides?
- Why do all objects fall at the same rate?

If I have seen farther than others, it is because I have stood on the shoulders of giants.

Isaac Newton

The history of the universe is essentially a story about the interplay between matter and energy since the beginning of time. Interactions between matter and energy began in the Big Bang and continue today in everything from the microscopic jiggling of atoms to gargantuan collisions of galaxies. Understanding the universe therefore depends on becoming familiar with how matter responds to the ebb and flow of energy.

You might guess that it would be difficult to understand the many interactions that shape the universe, because they occur on so many different size scales. However, we now know that just a few physical laws govern the movements of everything from atoms to galaxies. The Copernican revolution spurred the discovery of these laws, and Galileo deduced some of them from his experiments. But it was Sir Isaac Newton, one of the most influential human beings of all time, who put all of the pieces together into a simple system of laws describing both motion and gravity.

In this chapter, we'll discuss the laws that describe motion and energy in the universe, including Newton's laws of motion, the laws of conservation of angular momentum and of energy, and the universal law of gravitation. By understanding these laws, you will be able to make sense of the wide-ranging phenomena you will encounter as you continue your study of astronomy.

4.1 Describing Motion: Examples from Daily Life

Think about what happens when you throw a ball to a dog. The ball leaves your hand, traveling in some particular direction at some particular speed. During its flight, the ball is pulled toward Earth by gravity, slowed by air resistance, and pushed by gusts of wind. Despite the complexity of the ball's motion, the dog still catches it.

We humans can perform an even better trick: We have learned how to figure out where the ball will land even before throwing it. In fact, we can use the same basic trick to predict the motions of objects throughout the universe, and we can perform it with such extraordinary precision that we can land a spaceship on target on Mars after sending it on a journey of hundreds of millions of kilometers.

Our primary goal in this chapter is to understand how humans have learned to make sense of motion in the universe. We all have a great deal of experience with motion and a natural intuition about how motion works, but in science we need to define our ideas and terms precisely. In this section, we'll use examples from everyday life to explore some of the fundamental ideas that underlie our understanding of motion. After this introduction to the basic ideas, we'll go on to explore the laws that govern all motion.

• How do we describe motion?

You are probably familiar with the terms used to describe motion in science—terms such as *velocity, acceleration,* and *momentum.* However, their scientific definitions may differ subtly from those you use in casual conversation. Let's investigate the precise meanings of these terms.

Speed, Velocity, and Acceleration We use three basic terms to describe how an object is moving. These terms are all familiar to you from driving a car:

- The **speed** of the car tells us how far it will go in a certain amount of time. For example, 100 km/hr (about 60 mi/hr) is a speed, and it tells us that the car will cover a distance of 100 kilometers if it is driven at this speed for an hour.
- The **velocity** of the car tells us both its speed and direction. For example, "100 km/hr going due north" describes a velocity.
- The car has an **acceleration** if its velocity is changing in any way, whether in speed or direction or both.

You are undoubtedly familiar with the term *acceleration* as it applies to increasing speed. In science, we also say that you are accelerating when you slow down or turn (Figure 4.1). Slowing occurs when acceleration is in a direction opposite to the motion. In this case, we say that your acceleration is negative, causing your velocity to decrease. Turning changes your velocity because it changes the direction in which you are moving. Thus, turning is a form of acceleration even if your speed remains constant.

You can often feel the effects of acceleration. For example, as you speed up in a car you feel yourself being pushed back into your seat. As you slow down you feel yourself being pulled forward from the seat. As you drive around a curve you feel yourself being pushed away from the

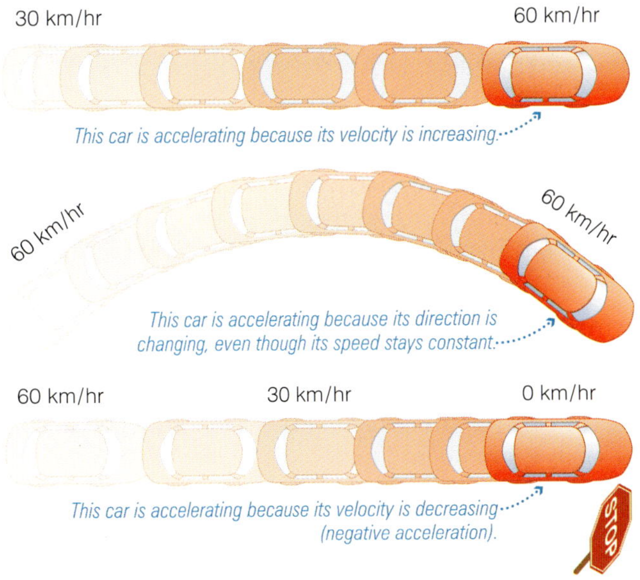

Figure 4.1 Speeding up, turning, and slowing down are all examples of acceleration.

direction of your turn. In contrast, you don't feel such effects when moving at *constant velocity.* That is why you don't feel any sensation of motion when you're traveling in an airplane on a smooth flight.

The Acceleration of Gravity One of the most important types of acceleration is that caused by gravity, which makes objects accelerate as they fall. In a legendary experiment in which he supposedly dropped weights from the Leaning Tower of Pisa, Galileo demonstrated that gravity accelerates all objects by the same amount, regardless of their mass. This fact may be surprising because it seems to contradict everyday experience: A feather floats gently to the ground, while a rock plummets. However, air resistance causes this difference in acceleration. If you dropped a feather and a rock on the Moon, where there is no air, both would fall at exactly the same rate.

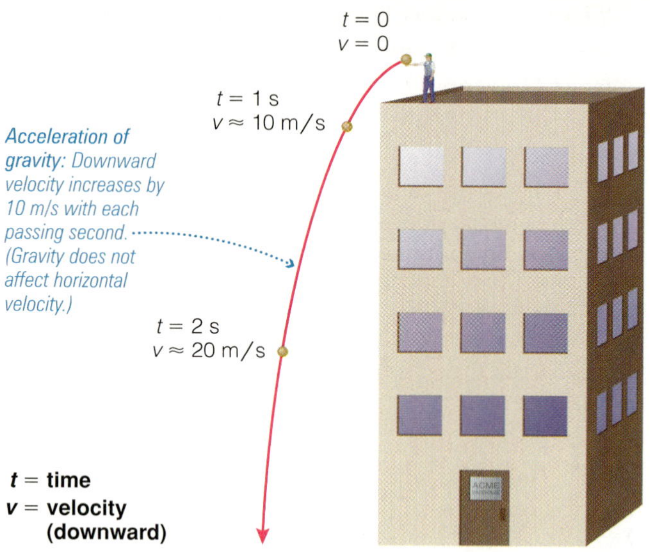

Figure 4.2 On Earth, the acceleration of gravity is about 10 m/s^2 (more precisely, 9.8 m/s^2), which means an unsupported object's downward velocity increases by about 10 m/s with each passing second (assuming we neglect air resistance).

THINK ABOUT IT

Find a piece of paper and a small rock. Hold both at the same height, one in each hand, and let them go at the same instant. The rock, of course, hits the ground first. Next, crumple the paper into a small ball and repeat the experiment. What happens? Explain how this experiment suggests that gravity accelerates all objects by the same amount.

The acceleration of a falling object is called the **acceleration of gravity**, abbreviated g. On Earth, the acceleration of gravity causes falling objects to fall faster by 9.8 meters per second (m/s), or about 10 m/s, with each passing second. For example, suppose you drop a rock from a tall building. At the moment you let it go, its speed is 0 m/s. After 1 second, the rock will be falling downward at about 10 m/s. After 2 seconds, it will be falling at about 20 m/s. In the absence of air resistance, its speed would continue to increase by about 10 m/s each second until it hits the ground (Figure 4.2). We therefore say that the acceleration of gravity is about 10 *meters per second per second,* or 10 *meters per second squared,* which we write as 10 m/s^2. (More precisely, $g = 9.8$ m/s^2.)

Momentum and Force The concepts of speed, velocity, and acceleration describe how an individual object moves, but most of the interesting phenomena we see in the universe result from interactions between objects. We need two additional concepts to describe these interactions:

- An object's **momentum** is its combination of mass and velocity. (Mathematically, we express this idea as momentum = mass × velocity.)
- The only way to change an object's momentum is to apply a **force** to it.

We can understand these concepts by considering the effects of collisions. Imagine that you're innocently stopped at a red light in your car when a bug flying at a velocity of 30 km/hr due south slams into your windshield. What will happen to your car? Not much, except perhaps a bit of a mess on your windshield. Next imagine that a 2-ton truck runs the red light and hits you head-on with the same velocity as the bug. Clearly, the truck will cause far more damage. To see why, let's consider the momentum and force in each collision.

Before the collisions, the truck's much greater mass means that it has far more momentum than the bug, even though both the truck and the bug are moving with the same velocity. During the collisions, the bug and the truck each transfer some of their momentum to your car. The bug has very little momentum to give to your car, so it does not exert much of a force. In contrast, the truck imparts enough of its momentum to cause a dramatic and sudden change in your car's momentum. You feel this sudden change in momentum as a force, and it can do great damage to you and your car.

The mere presence of a force does not always cause a change in momentum. For example, a moving car is always affected by forces of air resistance and friction with the road—forces that will slow your car if you take your foot off the gas pedal. However, you can maintain a constant velocity, and hence constant momentum, if you step on the gas pedal hard enough to overcome the slowing effects of these forces.

In fact, forces of some kind are always present. For example, on Earth we always feel the force of gravity and the electromagnetic forces acting between atoms. The **net force** (or *overall force*) acting on an object represents the combined effect of all the individual forces put together. There is no net force on your car when you are driving at constant velocity, because the force generated by the engine to turn the wheels precisely offsets the forces of air resistance and road friction. A change in momentum occurs only when the net force is not zero.

Changing an object's momentum means changing its velocity, as long as its mass remains constant. Thus, the presence of a net force causes an object to accelerate. Moreover, anytime an object accelerates, a net force must be causing the acceleration. That is why you feel forces (pushing you forward, backward, or to the side) when you accelerate in your car. We can use the same ideas to understand many astronomical processes. For example, planets are always accelerating as they orbit the Sun, because their direction of travel constantly changes as they go around their orbits. We can therefore conclude that some force must be causing this acceleration. As we'll discuss shortly, Isaac Newton identified this force as gravity.

Moving in Circles Think about an ice skater spinning in place. She isn't going anywhere, so she has no overall velocity and hence no overall momentum. Nevertheless, every part of her body is moving in a circle as she spins, so these parts have momentum even though her overall momentum is zero. Is there a way to describe the total momentum from each part of her body as she spins? Yes—we say that her spin gives her **angular momentum**, which you can also think of as "circling momentum" or "turning momentum." (The term *angular* arises because a complete circle turns through an *angle* of 360°.)

Any object that is either spinning or moving along a curved path has angular momentum, which makes angular momentum very important in astronomy. For example, Earth has angular momentum due both to its rotation (its *rotational angular momentum*) and to its orbit around the Sun (its *orbital angular momentum*). In Section 4.3, we'll use the idea of angular momentum to explain why Earth and other planets don't need any fuel to keep rotating and orbiting the Sun.

Because angular momentum is a special type of momentum, an object's angular momentum can change only when a special type of force is applied to it. To see why, consider what happens when you try to open a swinging door. Opening the door means making it rotate on its hinges, which means giving the door some angular momentum. If you push directly on the hinges, it will have no effect on the door, even if you push with a very strong force. However, even a light force can make the door rotate if you push on the part of the door that is farthest from the hinges. The type of force that can change an object's angular momentum is called a **torque**, which you can think of as a "twisting force." As the door example shows, the amount of torque depends not only on how much force you use to push on the door, but also on where you push.

Changing a tire offers another familiar example of torque. Turning the bolts on a tire means making them rotate, which requires giving them some angular momentum. A longer wrench means you can push from much farther out than you can with a short wrench, so you can turn the bolts with less strength. We will see many more applications of angular momentum in astronomy throughout the rest of the book.

• How is mass different from weight?

In discussing momentum and force, we've referred several times to the idea of *mass.* In daily life, we usually think of mass as something you can measure with a bathroom scale, but technically the scale measures your weight, not your mass. The distinction between mass and weight rarely matters when we are talking about objects on Earth, but it is very important in astronomy:

- Your **mass** is the amount of matter in your body.
- Your **weight** describes a *force* that is acting on your mass. In other words, it is what a scale measures when you stand on it.*

To understand the difference between mass and weight, imagine standing on a scale in an elevator (Figure 4.3). Your mass will be the same no matter how the elevator is moving, but your weight can vary. When the elevator is stationary or moving at constant velocity, the scale reads your "normal" weight. When the elevator accelerates upward, the floor exerts a greater force than it does when you are at rest. You feel heavier, and the scale verifies your greater weight. When the elevator accelerates downward, the floor and the scale exert a weaker force on you, so the scale registers less weight. Be sure to note that the scale shows a weight different from your "normal" weight only when the elevator is *accelerating,* not when it is going up or down at constant speed.

THINK ABOUT IT

Find a small bathroom scale and take it with you on an elevator ride. How does your weight change when the elevator accelerates upward or downward? Does it change when the elevator is moving at constant speed? Explain your observations.

Thus, your mass depends only on the amount of matter in your body and is the same anywhere, but your weight can vary because the forces acting on you can vary. For example, your mass would be the same on the Moon as on Earth, but you would weigh less on the Moon because of its weaker gravity.

Free-Fall and Weightlessness Now consider what happens if the elevator cable breaks (see the last frame in Figure 4.3). The elevator and you are suddenly both in **free-fall**—falling without any resistance to slow you down. The floor drops away at the same rate that you fall, allowing you to "float" freely above it, and the scale reads zero because you are no longer held to it. In other words, your free-fall has made you **weightless**.

In fact, you are in free-fall whenever there's nothing to *prevent* you from falling. For example, you are in free-fall

*Some physics texts distinguish between your "true weight" measured at rest and your "apparent weight" measured when you stand on an accelerating scale (as in an elevator). We will not make this distinction in this book.

Figure 4.3
Interactive Figure

Mass is not the same as weight. The man's mass never changes, but his weight is different when the elevator accelerates.

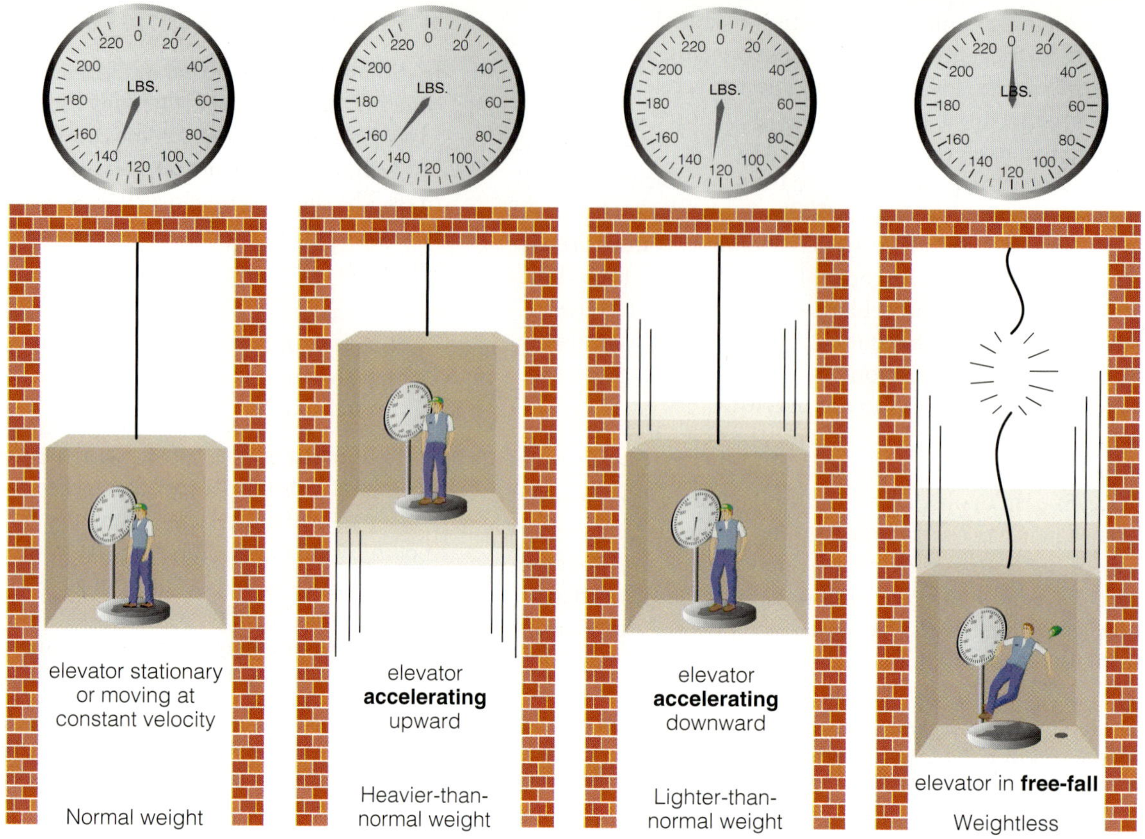

when you jump off a chair or spring from a diving board or trampoline. Surprising as it may seem, you have therefore experienced weightlessness many times in your life. You can experience it right now simply by jumping off your chair. Of course, your weightlessness lasts for only the very short time before you hit the ground.

Weightlessness in Space You've probably seen video of astronauts floating weightlessly in the Space Shuttle or the Space Station. But why are they weightless? Many people guess that there's no gravity in space, but that's not true. After all, it is gravity that makes the Space Shuttle and Space Station orbit Earth. Astronauts are weightless for the same reason you are weightless when you jump off a chair: They are in free-fall.

More specifically, astronauts are weightless the entire time they are orbiting Earth because they are in a *constant state of free-fall.* The idea that an orbiting object is in a "constant state of free-fall" may sound a bit strange, but it's easy to understand. Imagine a tower that goes all the way to the Space Station's orbit, about 350 kilometers above Earth. Figure 4.4 shows the idea. If you stepped off the tower, you would fall downward, remaining weightless until you hit the ground (or until air resistance had a noticeable effect on you). Now, imagine that instead of stepping off the tower, you ran and jumped out of the tower. You'd still fall to the ground, but because of your forward motion you wouldn't fall straight down. You'd land a short distance away from the base of the tower.

The faster you ran out of the tower, the farther you'd go before landing. If you could somehow use a rocket to run fast enough—about 28,000 km/hr (17,000 mi/hr) at the orbital altitude of the Space Station—a very interesting thing would happen: By the time gravity had pulled you downward as far as the length of the tower, you'd already have moved far enough around Earth that you'd no longer be going down at all. Instead, you'd be just as high above Earth's surface as you'd been all along, but a good portion of the way around the world. In other words, you'd be orbiting Earth.

Thus, the Space Shuttle, the Space Station, and all other orbiting objects stay in orbit because they are constantly "falling around" the Earth. Their constant state of free-fall makes these spacecraft and everything in them weightless.

THINK ABOUT IT

In the *Hitchhiker's Guide to the Galaxy* books, author Douglas Adams says that the trick to flying is to throw yourself at the ground and miss. Although this phrase does not really explain flying, which involves lift from air, it does describe an orbit fairly well. Explain.

Mastering Astronomy — Motion and Gravity Tutorial, Lesson 1

4.2 Newton's Laws of Motion

The human "trick" of being able to figure out where a ball will land before it is thrown or to predict how other

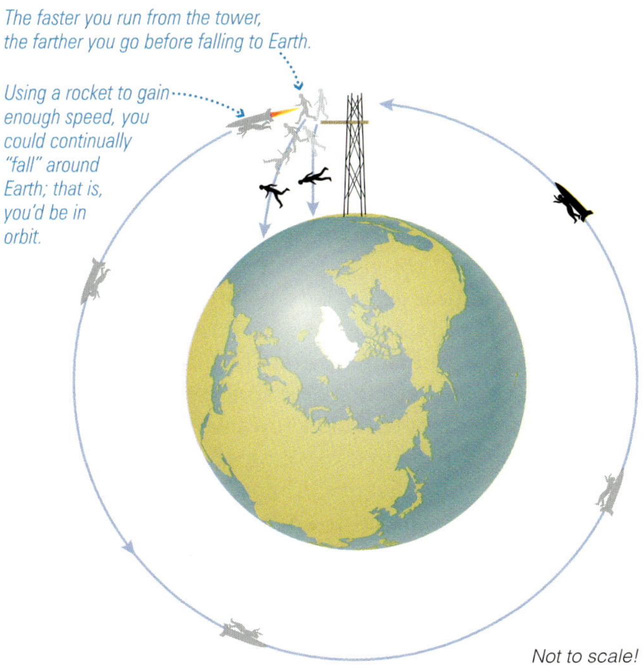

Figure 4.4 Interactive Figure This figure explains why astronauts are weightless and float freely in space. It shows that if you could leap from a tall tower with enough speed (perhaps with the aid of a rocket) you could travel forward so fast that you'd orbit Earth. You'd then be in a constant state of free-fall, which means you'd be weightless. *Note:* On the scale shown here, the tower extends far higher than the Space Station's orbit; the rocket orientation assumes that it rotates once with each orbit, as is the case for the Space Shuttle. (Adapted from *Space Station Science* by Marianne Dyson.)

COMMON MISCONCEPTIONS

No Gravity in Space?

If you ask people why astronauts are weightless in space, one of the most common answers is "There is no gravity in space." But you can usually convince people that this answer must be wrong by following with another simple question: Why does the Moon orbit Earth? Most people know that the Moon orbits Earth because of gravity, proving that there is gravity in space. In fact, at the altitude of the Space Station's orbit, the acceleration of gravity is scarcely less than it is on Earth's surface.

The real reason astronauts are weightless is that they are in a constant state of free-fall. Imagine being an astronaut. You'd have the sensation of free-fall—just as when you are falling from a diving board—the entire time you were in orbit. This constant falling sensation makes most astronauts sick to their stomachs when they first experience weightlessness. Fortunately, they quickly get used to the sensation, which allows them to work hard and enjoy the view.

motions will unfold requires understanding precisely how forces affect objects in motion. The complexity of motion in daily life might lead you to guess that the laws governing motion would also be complex. For example, if you watch a falling piece of paper waft lazily to the ground, you'll see it rock irregularly back and forth in a seemingly unpredictable pattern. However, the complexity of this motion arises because the paper is affected by a variety of forces, including gravity and the changing forces caused by air currents. If you could analyze the forces individually, you'd find that each force affects the paper's motion in a simple, predictable way. Sir Isaac Newton (1642–1727) discovered the remarkably simple system of laws that govern all motion.

• How did Newton change our view of the universe?

Newton was born in Lincolnshire, England, on Christmas Day in 1642. His father, a farmer who never learned to read or write, died 3 months before his premature birth. Newton had a difficult childhood and showed few signs of unusual talent. He attended Trinity College at Cambridge, where he earned his keep by performing menial labor, such as cleaning the boots and bathrooms of wealthier students and waiting on their tables.

The plague hit Cambridge shortly after Newton graduated, and he returned home. By his own account, he experienced a moment of inspiration in 1666 when he saw an apple fall to the ground. He suddenly realized that the gravity making the apple fall was the same force that held the Moon in orbit around Earth. In that moment, Newton shattered the remaining vestiges of the Aristotelian view of the world, which for centuries in Europe had been taken as unquestioned truth.

Aristotle (384–322 B.C.) had made many claims about the physics of motion, using his ideas to support his belief in an Earth-centered cosmos. He had also maintained that the heavens were totally distinct from Earth, so that physical laws on Earth did not apply to heavenly motion. By the time Newton saw the apple fall, the Copernican revolution had displaced Earth from a central position, and Galileo's experiments had shown that the laws of physics were not what Aristotle had believed.

Sir Isaac Newton (1642–1727)

Newton's sudden insight delivered the final blow to Aristotle's physics. When Newton realized that gravity operated in the heavens as well as on Earth, he eliminated Aristotle's distinction between the two realms. For the first time in history, the heavens and Earth were brought together as one *universe.* Newton's insight also heralded the birth of the modern science of *astrophysics* (although the term wasn't coined until much later). Astrophysics applies physical laws discovered on Earth to phenomena throughout the cosmos.

Over the next 20 years, Newton's work completely revolutionized mathematics and science. He quantified the laws of motion and gravity, conducted crucial experiments regarding the nature of light, built the first reflecting telescopes, and invented the mathematics of calculus. The compendium of Newton's discoveries is so tremendous that it would take a complete book just to describe them, and many more books to describe their influence on civilization. When Newton died in 1727, at age 84, English poet Alexander Pope composed the following epitaph:

> *Nature, and Nature's laws lay hid in the Night.*
> *God said,* Let Newton be! and *all was Light.*

We'll discuss Newton's laws of motion in the rest of this section, and later in the chapter we'll turn our attention to Newton's discoveries about gravity.

• What are Newton's three laws of motion?

Newton published the laws of motion and gravity in 1687, in a book usually called *Principia,* short for *Philosophiae Naturalis Principia Mathematica* ("Mathematical Principles of Natural Philosophy"). He enumerated three laws that apply to all motion, so we now call them **Newton's laws of motion.** These laws govern the motion of everything from our daily movements here on Earth to the movements of planets, stars, and galaxies throughout the universe. Figure 4.5 summarizes the three laws, which we now discuss in a little more detail.

Newton's First Law Newton's first law of motion essentially restates Galileo's discovery that objects will remain in motion unless a force acts to stop them. It can be stated as follows:

> *Newton's first law: An object moves at constant velocity if there is no net force acting upon it.*

Thus, objects at rest (velocity = 0) tend to remain at rest, and objects in motion tend to remain in motion with no change in either their speed or their direction.

The idea that an object at rest should remain at rest is rather obvious: A car parked on a flat street won't suddenly start moving for no reason. But what if the car is traveling along a flat, straight road? Newton's first law says that the car should keep going at the same speed forever *unless* a force acts to slow it down. You know that the car eventually will come to a stop if you take your foot off the gas pedal, so we must conclude that one or more forces are stopping the car—in this case forces arising from friction and air resistance. If the car were in space, and therefore unaffected by friction or air, it would keep moving forever (though gravity would eventually alter its speed and direction). That is why interplanetary spacecraft need no fuel to keep going after they are launched into space, and why astronomical objects don't need fuel to travel through the universe.

Newton's first law also explains why you don't feel any sensation of motion when you're traveling in an airplane on a smooth flight. As long as the plane is traveling at constant velocity, no net force is acting on it or on you. Therefore, you feel no different from the way you would feel at rest. You can walk around the cabin, play catch with a person a few rows forward, or relax and go to sleep just as though you were "at rest" on the ground.

Newton's Second Law Newton's second law of motion tells us what happens to an object when a net force *is* present. We have already seen that a net force will change an object's momentum, accelerating it in the direction of the force. Newton's second law quantifies this relationship, telling us that the amount of the acceleration depends on the object's

Newton's first law of motion:
An object moves at constant velocity unless a net force acts to change its speed or direction.

Example: A spaceship needs no fuel to keep moving in space.

Newton's second law of motion:
Force = mass × acceleration

Example: A baseball accelerates as the pitcher applies a force by moving his arm. (Once the ball is released, the force from the pitcher's arm ceases, and the ball's path changes only because of the forces of gravity and air resistance.)

Newton's third law of motion:
For any force, there is always an equal and opposite reaction force.

Example: A rocket is propelled upward by a force equal and opposite to the force with which gas is expelled out its back.

Figure 4.5 Newton's three laws of motion.

mass and the strength of the net force. We usually write this law as an equation:

Newton's second law: **Force = mass × acceleration** (*or* $F = ma$).

(The law is sometimes instead written as force = rate of change in momentum.)

This law explains why you can throw a baseball farther than you can throw a shot-put. The force your arm delivers to both the baseball and the shot-put equals the product of mass and acceleration. Because the mass of the shot-put is greater than that of the baseball, the same force from your arm gives the shot-put a smaller acceleration. Because of its smaller acceleration, the shot-put leaves your hand with less speed than the baseball and thus travels a shorter distance before hitting the ground. Astronomically, Newton's second law explains why a large planet such as Jupiter has a greater effect on asteroids and comets than a small planet such as Earth [Section 12.4]. Jupiter exerts a stronger gravitational force on passing asteroids than does Earth, so Newton's second law tells us that Jupiter sends the asteroids scattering with a greater acceleration.

We can also use Newton's second law of motion to understand acceleration around curves. Suppose you swing a ball on a string around your head (Figure 4.6a). The ball is accelerating even if it has a steady speed, because it is constantly changing direction and therefore has a changing velocity. What makes it accelerate? According to Newton's second law, the taut string must be applying a force to the ball. We can understand this force by considering what happens when the string breaks and the force disappears (Figure 4.6b). In that case, the ball simply flies off in a straight line. Thus, when the string is intact, the force must be pulling the ball *inward* to keep it from flying off. Because acceleration must be in the same direction as the force, we conclude that the ball has an inward acceleration as it moves around the circle.

The same idea helps us understand the force on a car moving around a curve or a planet orbiting around the Sun. In the case of the car, the force comes from friction between the tires and the road. The tighter the curve or the faster the car is going, the greater the force needed to keep the car moving around it. If the force is not great enough, the car skids outward. For a planet moving around the Sun, the force pulling inward is gravity, which we'll discuss shortly. Thus, an orbiting planet is always accelerating toward the Sun. Indeed, it was Newton's discovery of the precise nature of this acceleration that helped him deduce the law of gravity.

Newton's Third Law Think for a moment about standing still on the ground. Your weight exerts a downward force, so if this force were acting alone, Newton's second law would demand that you be accelerating downward. The fact that you are not falling means that the ground must be pushing back up on you with exactly the right amount of force to offset the force you exert downward on the floor. This fact is embodied in Newton's third law of motion, which tells us that any force is always opposed by an equal and opposite reaction force.

Newton's third law: For any force, there is always an equal and opposite reaction force.

This law is very important in astronomy, because it tells us that objects always attract *each other* through gravity. For example, your body always exerts a gravitational force on the Earth identical to the force that Earth exerts on you, except that it acts in the opposite direction. Of course, the same force means a much greater acceleration for you than for Earth (because your mass is so much smaller than Earth's), which is why we see you falling toward Earth when you jump off a chair, rather than Earth falling toward you.

Newton's third law also explains how a rocket works: A rocket engine generates a force that drives hot gas out the back, which creates an equal and opposite force that propels the rocket forward.

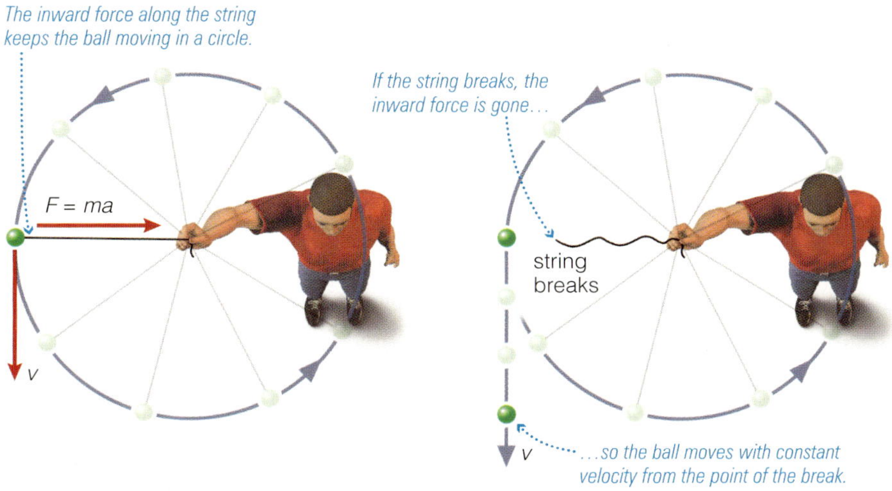

a When you swing a ball on a string, the string exerts a force that pulls the ball inward.

b If the string breaks, the ball flies off in a straight line at constant velocity.

Figure 4.6 Newton's second law of motion tells us that an object going around a curve has an acceleration pointing toward the inside of the curve.

COMMON MISCONCEPTIONS

What Makes a Rocket Launch?

If you've ever watched a rocket launch, it's easy to see why many people believe that the rocket "pushes off" the ground. In fact, the ground has nothing to do with the rocket launch. The rocket's launch is explained by Newton's third law of motion. To balance the force driving gas out the back of the rocket, an equal and opposite force must propel the rocket forward. Thus, rockets can be launched horizontally as well as vertically, and a rocket can be "launched" in space (for example, from a space station) with no need for any nearby solid ground.

4.3 Conservation Laws in Astronomy

Newton's laws of motion are easy to state, but they may seem a bit arbitrary. Why, for example, should every force be opposed by an equal and opposite reaction force? In the centuries since Newton first stated his laws, we have learned that they are not arbitrary at all, but instead reflect deeper aspects of nature known as *conservation laws.* In this section, we'll explore three of the most important conservation laws for astronomy: *conservation of momentum, conservation of angular momentum,* and *conservation of energy.* We'll see some immediate examples of how they apply to astronomy, and then use these laws over and over throughout the rest of the book.

• Why do objects move at constant velocity if no force acts on them?

The first of our conservation laws, the law of **conservation of momentum**, states that the total momentum of interacting objects cannot change as long as no external force is acting on them; that is, their total momentum is *conserved.* An individual object can gain or lose momentum only if some other object's momentum changes by a precisely opposite amount.

The law of conservation of momentum is implicit in Newton's laws. To see why, watch a game of pool. Newton's second law tells us that when one pool ball strikes another, it exerts a force that changes the momentum of the second ball. At the same time, Newton's third law tells us that the second ball exerts an equal and opposite force on the first one—which means that the first ball's momentum changes by precisely the same amount as the second ball's momentum, but in the opposite direction. Thus, the total combined momentum of the two balls remains the same both before and after the collision (Figure 4.7). Note that no *external* forces are accelerating the balls.

Rockets offer another good example of conservation of momentum in action. When you fire a rocket engine, the total momentum of the rocket and the hot gases it shoots out the back must stay the same. In other words, the amount of forward momentum the rocket gains is equal to the amount of backward momentum in the gas that shoots out the back. That is why forces between the rocket and the gases are always equal and opposite.

From the perspective of conservation of momentum, Newton's first law makes perfect sense. When no net force acts on an object, there is no way for the object to transfer any momentum to or from any other object. Thus, in the absence of a net force, an object's momentum must remain unchanged—which means the object must continue to move exactly as it has been moving.

According to current understanding of the universe, conservation of momentum is an absolute law that always holds true. For example, it holds even when you jump up into the air. So you may wonder, Where do I get the momentum that carries me upward? The answer is that

MATHEMATICAL INSIGHT 4.1 **Units of Force, Mass, and Weight**

Newton's second law, $F = ma$, shows that the unit of force is equal to a unit of mass multiplied by a unit of acceleration. For example, if a mass of 1 kilogram accelerates at 10 m/s^2, the magnitude of the responsible force is:

$$\begin{aligned}\text{force} &= \text{mass} \times \text{acceleration}\\ &= 1\text{ kg} \times 10\frac{\text{m}}{\text{s}^2} = 10\frac{\text{kg} \times \text{m}}{\text{s}^2}\\ &= 10\text{ newtons}\end{aligned}$$

The standard unit of force is the *kilogram-meter per second squared,* called the **newton** for short.

We can now further clarify the difference between mass and weight. When you stand on a scale, it records the downward force that you exert on it, which is equal and opposite to the upward force it exerts on you. If the scale were suddenly pulled out from under your feet, you would begin accelerating downward with the acceleration of gravity. Thus, when you are standing still, the scale must support you with a force equal to your mass times the acceleration of gravity. Your weight must also equal this force (but in an opposite direction):

$$\text{weight} = \text{mass} \times \text{acceleration of gravity}$$

Your apparent weight may differ from this value if forces besides gravity are affecting you.

Like any force, weight has units of mass times acceleration. Thus, although we commonly speak of weights in *kilograms,* this usage is not technically correct: Kilograms are a unit of mass, not of force. You may safely ignore this technicality as long as you are dealing with objects on Earth that are not accelerating. In space or on other planets, the distinction between mass and weight is important and cannot be ignored.

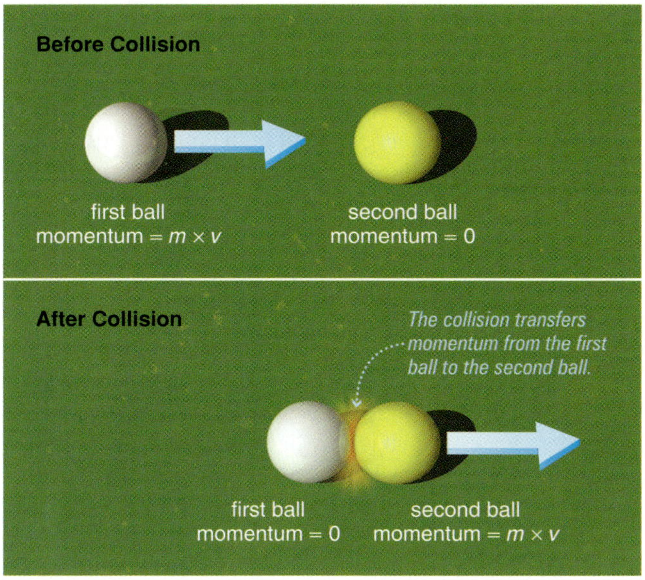

Figure 4.7 Conservation of momentum demonstrated on a pool table.

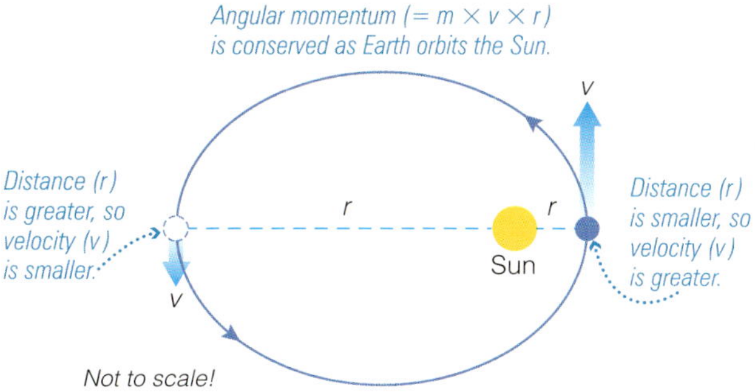

Figure 4.8 Earth's angular momentum always stays the same as it orbits the Sun, so it moves faster when it is closer to the Sun and slower when it is farther from the Sun. It needs no fuel to keep orbiting because no forces are acting in a way that could change its angular momentum.

as your legs propel you skyward, they are actually pushing Earth in the other direction, giving Earth's momentum an equal and opposite kick. However, Earth's huge mass renders its acceleration undetectable. During your brief flight, the gravitational force between you and Earth pulls you back down, transferring your momentum back to Earth. The total momentum of you and the rest of Earth remains the same at all times.

• What keeps a planet rotating and orbiting the Sun?

Perhaps you've wondered how Earth manages to keep rotating and going around the Sun day after day and year after year. The answer comes from our second conservation law: the law of **conservation of angular momentum**. Recall that rotating or orbiting objects have angular momentum because they are moving in circles or going around curves, and that angular momentum can be changed only by a "twisting force," or *torque*. The law of conservation of angular momentum states that as long as there is no external torque, the total angular momentum of a set of interacting objects cannot change. An individual object can change its angular momentum only by transferring some angular momentum to or from another object. Because astronomical objects can have angular momentum due to both their rotation and orbit, let's consider both cases.

Orbital Angular Momentum Consider Earth's orbit around the Sun. A simple formula tells us Earth's angular momentum at any point in its orbit:

$$\text{angular momentum} = m \times v \times r$$

where m is Earth's mass, v is its speed (or velocity) around the orbit, and r is the "radius" of the orbit, by which we mean its distance from the Sun (Figure 4.8). Because there are no objects around to give or take angular momentum from Earth as it orbits the Sun, Earth's orbital angular momentum must always stay the same. This explains two key facts about Earth's orbit:

1. Earth needs no fuel or push of any kind to keep orbiting the Sun—it will keep orbiting as long as nothing comes along to take angular momentum away.
2. Because Earth's angular momentum at any point in its orbit depends on the product of its speed and orbital radius (distance from the Sun), Earth's orbital speed must be faster when it is nearer to the Sun (and the radius is shorter) and slower when it is farther from the Sun (and the radius is longer).

The second fact is just what Kepler's second law of planetary motion tells us [Section 3.3]. Thus, the law of conservation of angular momentum tells us *why* Kepler's law is true.

Rotational Angular Momentum The same idea explains why Earth keeps rotating. As long as Earth isn't transferring any of the angular momentum of its rotation to another object, it keeps rotating at the same rate. (In fact, Earth is very gradually transferring some of its rotational angular momentum to the Moon, and as a result, Earth's rotation is gradually slowing down; see Section 4.5.)

Conservation of angular momentum also explains why we see so many spinning disks in the universe, such as the disks of galaxies like the Milky Way and disks of material orbiting young stars. The idea is easy to illustrate with an ice skater spinning in place (Figure 4.9). Because there is so little friction on ice, the angular momentum of the ice skater remains essentially constant. When she pulls in her extended arms, she decreases her radius—which means her velocity of rotation must increase. Stars and galaxies are both born from clouds of gas that start out much larger in size. These clouds almost inevitably have some small net rotation, though it may be imperceptible. However, like the spinning skater as she pulls in her arms, they must spin faster as gravity makes them shrink in size. (We'll discuss why the clouds also flatten into disks in Chapter 8.)

Figure 4.9 A spinning skater conserves angular momentum.

THINK ABOUT IT

How does conservation of angular momentum explain the spiraling of water going down a drain?

Mastering ASTRONOMY Energy Tutorial, Lesson 1

• Where do objects get their energy?

Our third crucial conservation law for astronomy is the law of **conservation of energy**. This law tells us that, like momentum and angular momentum, energy cannot appear out of nowhere or disappear into nothingness. Thus, objects can gain or lose energy only by exchanging energy with other objects. Because of this law, the story of the universe is a story of the interplay of energy and matter: All actions in the universe involve exchanges of energy or the conversion of energy from one form to another.

Throughout the rest of this book, we'll see numerous cases in which we can understand astronomical processes simply by tracing the ways in which energy is transformed and exchanged. For example, we'll see that planetary interiors cool with time only because they radiate energy into space, and that the Sun became hot because of energy released by the gas that formed it. It's no exaggeration to say that, by applying the laws of conservation of momentum, conservation of angular momentum, and conservation of energy, we can understand almost every major process that occurs in the universe.

Basic Types of Energy Before we can fully understand the law of conservation of energy, we need to know what energy is. In essence, energy is what makes matter move. Because this statement is so broad, we often distinguish between many different types of energy. For example, we talk about the energy we get from the food we eat, the energy that makes our cars go, and the energy a light bulb emits. Fortunately, we can classify all these various types of energy into just three major categories (Figure 4.10):

- Energy of motion, or **kinetic energy** (*kinetic* comes from a Greek word meaning "motion"). Falling rocks, orbiting planets, and the molecules moving in the air around us are all examples of objects with kinetic energy.
- Energy carried by light, or **radiative energy** (the word *radiation* is often used as a synonym for *light*). All light carries energy, which is why light can cause changes in matter. For example, light can alter molecules in our eyes—thereby allowing us to see—or warm the surface of a planet.
- Stored energy, or **potential energy**, which might later be converted into kinetic or radiative energy. For example, a rock perched on a ledge has *gravitational* potential energy because it will fall if it slips off the edge, and gasoline contains *chemical* potential energy that can be converted into the kinetic energy of a moving car.

Regardless of which type of energy we are dealing with, we can measure the amount of energy with the same standard units. For Americans, the most familiar units of energy are *Calories*, which are shown on food labels to tell us how much energy our bodies can draw from the food. A

Figure 4.10 The three basic categories of energy. Energy can be converted from one form to another, but it can never be created or destroyed, an idea embodied in the law of conservation of energy.

typical adult needs about 2,500 Calories of energy from food each day. In science, the standard unit of energy is the **joule**. One food Calorie is equivalent to about 4,184 joules, so the 2,500 Calories used daily by a typical adult is equivalent to about 10 million joules. Table 4.1 compares various energies in joules.

Thermal Energy—The Kinetic Energy of Many Particles Although there are only three major categories of energy, we sometimes divide them into various subcategories. In astronomy, the most important subcategory of kinetic energy is **thermal energy**, which represents the collective kinetic energy of the many individual particles (atoms and molecules) moving randomly within a substance like a rock or the air or the gas within a distant star. In such cases, it is much easier to talk about the thermal energy of the object rather than about the kinetic energies of its billions upon billions of individual particles. Note that all objects contain thermal energy even when they are sitting still, because the particles within them are always jiggling about randomly. These random motions can contain substantial energy—the thermal energy of a parked car due to the random motion of its atoms is much greater than the kinetic energy of the car moving at highway speed.

Thermal energy gets its name because it is related to temperature, but temperature and thermal energy are not quite the same thing. Thermal energy measures the *total* kinetic energy of all the randomly moving particles in a substance, while **temperature** measures the *average* kinetic energy of the particles. Thus, for a particular object, a higher temperature simply means that the particles on average have more kinetic energy and hence are moving faster (Figure 4.11). You're probably familiar with temperatures measured in *Fahrenheit* or *Celsius,* but in science we often use the **Kelvin** temperature scale (Figure 4.12). The Kelvin scale never has negative temperatures, because it starts from the coldest possible temperature, known as *absolute zero* (0 K).

Thermal energy depends on temperature, because a higher average kinetic energy for the particles in a substance must also lead to a higher total energy. However, thermal energy also depends on the number or density of the particles. You can see why by considering what would happen if you quickly thrust your arm into and out of a hot oven and a pot of boiling water.

The air in a hot oven is much hotter in temperature than the water boiling in a pot (typically 400°F for the oven versus 212°F for boiling water). However, the boiling water would scald your arm almost instantly, while you can safely put your arm into the oven air for a few seconds. The reason for this difference is density (Figure 4.13). If air or water is hotter than your body, molecules striking your skin transfer thermal energy to molecules in your arm. The higher temperature in the oven means that the air molecules strike

Table 4.1 Energy Comparisons

Item	*Energy (joules)*
Energy of sunlight at Earth (per m^2 per second)	1.3×10^3
Energy from metabolism of a candy bar	1×10^6
Energy needed to walk for 1 hour	1×10^6
Kinetic energy of a car going 60 mi/hr	1×10^6
Daily food energy need of average adult	1×10^7
Energy released by burning 1 liter of oil	1.2×10^7
Thermal energy of parked car	1×10^8
Energy released by fission of 1 kilogram of uranium-235	5.6×10^{13}
Energy released by fusion of hydrogen in 1 liter of water	7×10^{13}
Energy released by 1-megaton H-bomb	5×10^{15}
Energy released by major earthquake (magnitude 8.0)	2.5×10^{16}
Annual U.S. energy consumption	10^{20}
Annual energy generation of Sun	10^{34}
Energy released by a supernova	10^{44}–10^{46}

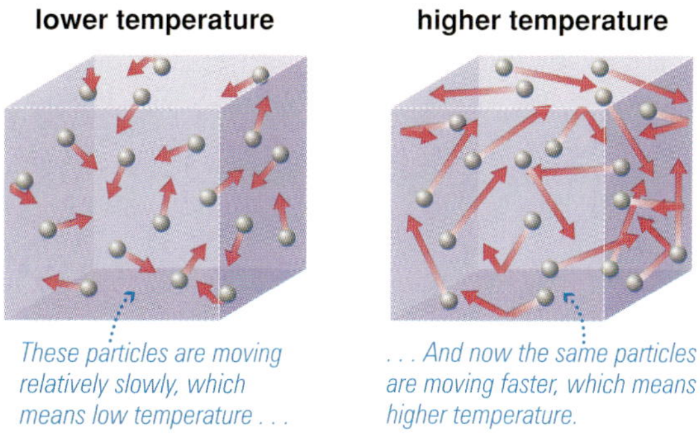

Figure 4.11 Temperature is a measure of the average kinetic energy of the particles (atoms and molecules) in a substance. (Longer arrows represent higher speeds.)

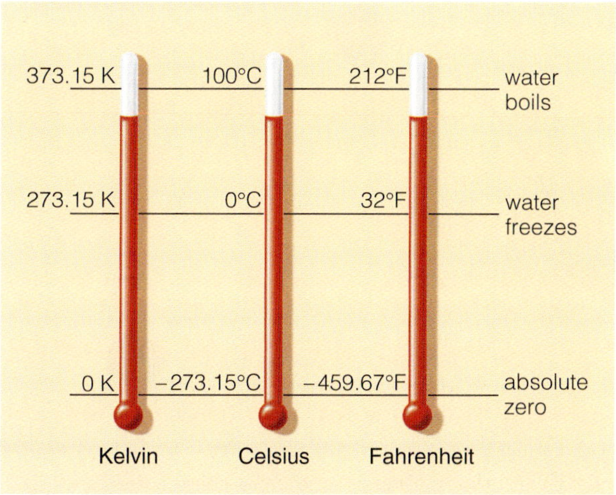

Figure 4.12 Three common temperature scales: Kelvin, Celsius, and Fahrenheit. Scientists generally prefer the Kelvin scale. (The degree symbol ° is not usually used with the Kelvin scale.)

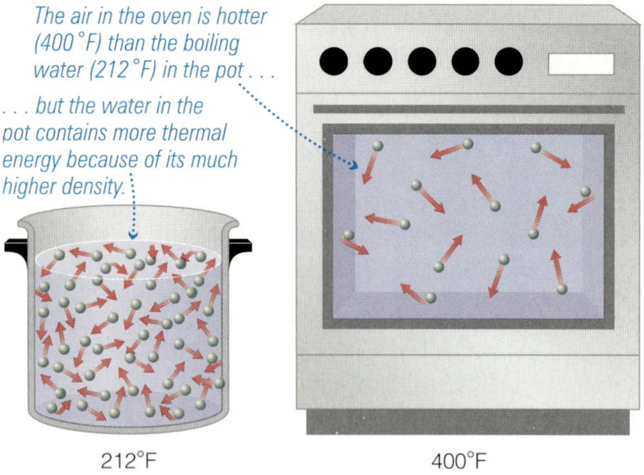

Figure 4.13 Thermal energy depends on both the temperature and density of particles in a substance.

your skin harder, on average, than the molecules in the boiling water. However, because the *density* of water is so much higher than the density of air (meaning water has far more molecules in the same amount of space), many more molecules strike your skin each second in the water. Thus, while each individual molecule that strikes your skin transfers a little less energy in the boiling water than in the oven, the sheer number of molecules hitting you in the water means that more thermal energy is transferred to your arm. That is why the water causes a rapid burn.

THINK ABOUT IT

In air or water that is colder than your body temperature, thermal energy is transferred from you to the surrounding cold air or water. Use this fact to explain why falling into a 32°F (0°C) lake is much more dangerous than standing naked outside on a 32°F day.

The environment in space provides another example of the difference between temperature and heat. Surprisingly, the temperature in low Earth orbit can be several thousand degrees. However, astronauts working in Earth orbit (outside the Space Shuttle or Station) can get very cold and sometimes need heated space suits and gloves. The astronauts can get cold despite the high temperature, because the extremely low density of space means that relatively few particles are available to transfer thermal energy to them. (You may wonder how the astronauts become cold given that the low density also means the astronauts cannot transfer much of their own thermal energy to the particles in space. It turns out that they lose their body heat by emitting *thermal radiation*, which we will discuss in Section 5.3.)

Potential Energy in Astronomy Many types of potential energy are important in astronomy, but two are particularly important: gravitational potential energy and the potential energy of mass itself, or mass-energy. Let's examine each of these two types of potential energy in a little more depth.

An object's **gravitational potential energy** depend on its mass and how far it can fall as a result of gravity. An object has more gravitational potential energy when it is higher and less when it is lower. For example, if you throw a ball up into the air, it has more potential energy when it is high up than it does near the ground. Because energy must be conserved during the ball's flight, the ball's kinetic energy must increase when its gravitational potential energy decreases, and vice versa (Figure 4.14a). That is why the ball travels fastest (has the most kinetic energy) when it is closest to the ground, where it has the least gravitational potential energy. The higher it is, the more gravitational potential energy it has and thus the slower the ball travels (less kinetic energy).

The same general idea explains how stars become hot (Figure 4.14b). Before a star forms, its matter is spread out

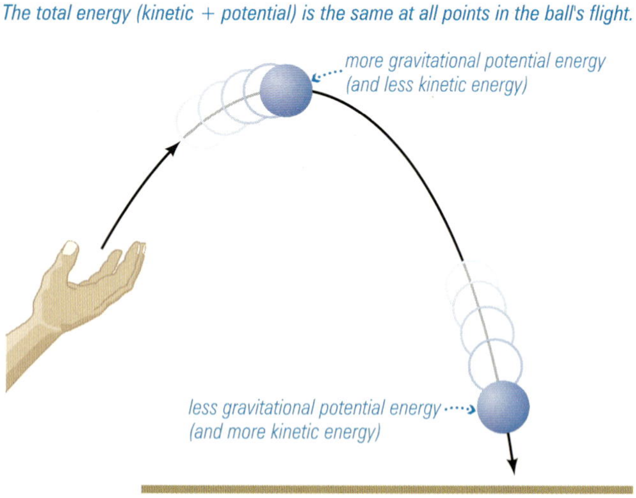

a The ball has more gravitational potential energy when it is high up than when it is near the ground.

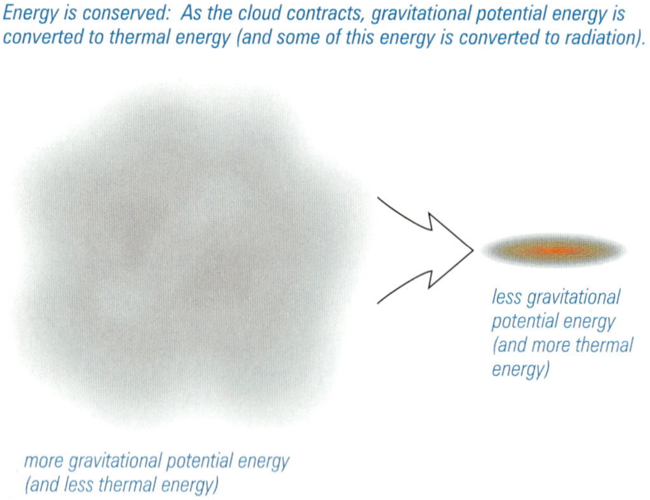

b A cloud of interstellar gas can contract due to its own gravity. It has more gravitational potential energy when it is spread out than when it shrinks in size.

Figure 4.14 Two examples of gravitational potential energy.

in a large, cold cloud of gas. Most of the individual gas particles are far from the center of this large cloud and therefore have a lot of gravitational potential energy. The particles lose gravitational potential energy as the cloud contracts under its own gravity, and this "lost" potential energy ultimately gets converted into thermal energy, making the center of the cloud hot.

The second particularly important form of potential energy in astronomy is the energy contained in mass itself, often called **mass-energy**. Einstein discovered that mass is a form of potential energy, and he expressed the idea that mass can be converted to other forms of energy with his famous formula

$$E = mc^2$$

where E is the amount of potential energy, m is the mass of the object, and c is the speed of light.

A small amount of mass contains a huge amount of energy. For example, the energy released by a 1-megaton H-bomb—enough to destroy a major city—comes from converting only about 0.1 kilogram of mass (about 3 ounces—a quarter of a can of soda) into energy (Figure 4.15). The Sun generates energy by converting a tiny fraction of its mass into energy through a similar process of nuclear fusion [Section 14.2].

Figure 4.15 The energy released by this H-bomb comes from the conversion of only about 0.1 kilogram of mass into energy in accordance with the formula $E = mc^2$.

Just as Einstein's formula tells us that mass can be converted into other forms of energy, it also tells us that energy can be transformed into mass. This process is especially important to understanding what we think happened during the early moments in the history of the universe, when some of the energy of the Big Bang turned into the mass from which we are now made [Section 23.1]. Scientists also use this idea to search for undiscovered particles of matter,

MATHEMATICAL INSIGHT 4.2 Mass-Energy

It's easy to calculate mass-energies with Einstein's formula $E = mc^2$. Once we calculate an energy, we can compare it to other known energies.

Example: Suppose a 1-kilogram rock were completely converted to energy. How much energy would it release? Compare this to the energy released by burning 1 liter of oil.

Solution:

Step 1 Understand: We are asked how much energy would be released by converting all the mass of a 1-kilogram rock to energy. Thus, we need to know the total mass-energy of the rock, which we can then compare to the energy released by burning a liter of oil.

Step 2 Solve: The total mass-energy of the rock is given by $E = mc^2$, where m is the 1-kilogram mass and $c = 3 \times 10^8$ m/s:

$$\begin{aligned} E = mc^2 &= 1\text{ kg} \times \left(3 \times 10^8 \frac{\text{m}}{\text{s}}\right)^2 \\ &= 1\text{ kg} \times \left(9 \times 10^{16} \frac{\text{m}^2}{\text{s}^2}\right) \\ &= 9 \times 10^{16} \frac{\text{kg} \times \text{m}^2}{\text{s}^2} \\ &= 9 \times 10^{16}\text{ joules} \end{aligned}$$

We now compare this energy to the energy released by burning 1 liter of oil, which is 12 million joules (see Table 4.1). Dividing the mass-energy of the rock by the energy released by burning 1 liter of oil, we find:

$$\frac{9 \times 10^{16}\text{ joules}}{1.2 \times 10^7\text{ joules}} = 7.5 \times 10^9$$

Step 3 Explain: We have found that converting a 1-kilogram rock completely to energy would release 9×10^{16} joules of energy, which is about 7.5 billion times as much energy as we get from burning 1 liter of oil. Moreover, by looking it up, we can find that 7.5 billion liters of oil is roughly the amount of oil that *all* cars in the United States use in a week. Thus, complete conversion of the mass of a single 1-kilogram rock to energy could yield enough energy to power all the cars in the United States for a week. Unfortunately, no technology available now or in the foreseeable future can release all the mass-energy of a rock.

by using large machines called *particle accelerators* to create subatomic particles from energy.

Conservation of Energy We have seen that energy comes in three basic categories—kinetic, radiative, and potential—and explored several subcategories that are especially important in astronomy: thermal energy, gravitational potential energy, and mass-energy. Now we are ready to return to the question of "Where do objects get their energy?" Because energy cannot be created or destroyed, objects always get their energy from other objects. Ultimately, we can always trace an object's energy back to the Big Bang [Section 1.1], the beginning of the universe in which all matter and energy is thought to have come into existence.

For example, imagine that you've thrown a baseball. It is moving, so it has kinetic energy. Where did this kinetic energy come from? The baseball got its kinetic energy from the motion of your arm as you threw it. Your arm, in turn, got its kinetic energy from the release of chemical potential energy stored in your muscle tissues. Your muscles got this energy from the chemical potential energy stored in the foods you ate. The energy stored in the foods came from sunlight, which plants convert into chemical potential energy through photosynthesis. The radiative energy of the Sun was generated through the process of nuclear fusion, which releases some of the mass-energy stored in the Sun's supply of hydrogen. The mass-energy stored in the hydrogen came from the birth of the universe in the Big Bang. After you throw the ball, its kinetic energy will ultimately end up as thermal energy of the air or ground. According to present understanding, the total energy content of the universe was determined in the Big Bang. It remains the same today and will stay the same forever into the future.

4.4 The Universal Law of Gravitation

Newton's three laws of motion describe how objects in the universe move in response to forces. The laws of conservation of momentum, angular momentum, and energy offer an alternative and often simpler way of thinking about what happens when a force causes some change in the motion of one or more objects. Either way, however, we cannot fully understand motion unless we also understand the forces that lead to changes in motion. In astronomy, the most important force is gravity, which governs virtually all large-scale motion in the universe.

Motion and Gravity Tutorial, Lesson 2

• What determines the strength of gravity?

Isaac Newton discovered the basic law that describes how gravity works. Newton expressed the force of gravity mathematically with his **universal law of gravitation**. Three simple statements summarize this law:

- Every mass attracts every other mass through the force called *gravity*.
- The strength of the gravitational force attracting any two objects is *directly proportional* to the product of their masses. For example, doubling the mass of *one* object doubles the force of gravity between the two objects.
- The strength of gravity between two objects decreases with the *square* of the distance between their centers. That is, the gravitational force follows an **inverse square law** with distance. For example, doubling the distance between two objects weakens the force of gravity by a factor of 2^2, or 4.

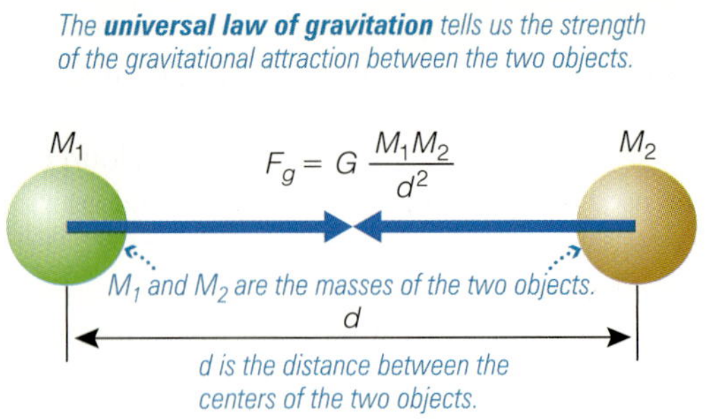

Figure 4.16 The universal law of gravitation is an *inverse square law*, which means that the force of gravity declines with the *square* of the distance d between two objects.

These three statements tell us everything we need to know about Newton's universal law of gravitation. Mathematically, all three statements can be combined into a single equation, usually written like this:

$$F_g = G\frac{M_1M_2}{d^2}$$

where F_g is the force of gravitational attraction, M_1 and M_2 are the masses of the two objects, and d is the distance between their centers (Figure 4.16). The symbol G is a constant called the **gravitational constant**, and its numerical value has been measured to be $G = 6.67 \times 10^{-11}\ m^3/(kg \times s^2)$.

THINK ABOUT IT

How does the gravitational force between two objects change if the distance between them triples? If the distance between them drops by half?

Orbits and Kepler's Law Tutorial, Lessons 1–4

• How does Newton's law of gravity extend Kepler's laws?

By the time Newton published *Principia* in 1687, Kepler's three laws of planetary motion [Section 3.3] had already been known and tested for some 70 years. Kepler's laws had proven so successful at predicting planetary positions that there was little doubt about their validity. However, there was great debate among scientists about *why* Kepler's laws hold true.

Newton solved the mystery by showing that Kepler's laws are consequences of the laws of motion and the universal law of gravitation. For example, we've already seen how we can understand Kepler's second law, that a planet moves faster when it is closer to the Sun, by thinking about conservation of angular momentum. Kepler's third law, that average orbital speed is higher for planets with smaller average orbital distance, comes directly from the fact that planets closer to the Sun have a stronger gravitational attraction to the Sun. In essence, Newton used the mathematical expressions of his laws of gravity and motion to show that a planet orbiting the Sun should automatically have an elliptical orbit that obeys Kepler's laws.

Newton's explanation of Kepler's laws sealed the triumph of the Copernican revolution. Prior to Newton, it was still possible to see Kepler's model of planetary motion as "just" another model, albeit one that fit the observational data far better than any previous model. By explaining Kepler's laws in terms of basic laws of physics, Newton removed virtually all remaining doubt about the legitimacy of the Sun-centered solar system.

In addition, Newton found that Kepler's laws were only part of the story of how objects move in response to gravity. Remember that Kepler had discovered his laws by analyzing the orbits of planets around the Sun, and he therefore had no reason to think that his laws would apply to any other cases, such as moons orbiting planets or comets orbiting the Sun. However, when Newton analyzed his equations for gravity and motion, he discovered that Kepler's laws are just one special case of a more general set of rules about orbiting objects. These more general rules explain the motion of objects throughout the universe; as we will see throughout this book, they are a crucial part of the foundations of modern astronomy. Let's explore four ways in which Newton extended Kepler's laws.

Planets Are Not the Only Objects with Elliptical Orbits Kepler wrote his first two laws for planets orbiting the Sun, but Newton showed that any object going around another object will obey these two laws. For example, the orbits of a satellite around Earth, of a moon around a planet, and of an asteroid around the Sun are all ellipses in which the orbiting object moves faster at the nearer points in its orbit and slower at the farther points.

Ellipses Are Not the Only Possible Orbital Paths Kepler was right when he found that ellipses (which include circles) are the only possible shapes for **bound orbits**—orbits in which an object goes around another object over and over again. (The term *bound orbit* comes from the idea that gravity creates a *bond* that holds the objects together.) However, Newton discovered that objects can also follow **unbound orbits**—paths that bring an object close to another object just once. For example, some comets that enter the inner solar system follow unbound orbits. They come in from afar just once, loop around the Sun, and never return.

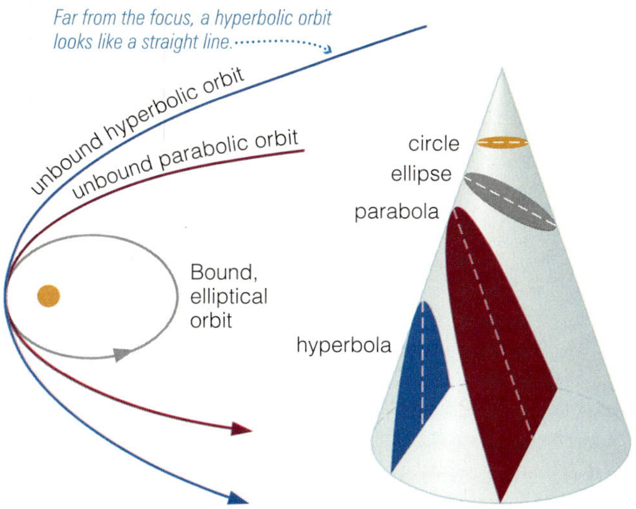

a Orbits allowed by the law of gravity.

b Ellipses (which include circles), parabolas, and hyperbolas are conic sections, made by slicing a cone at different angles.

Figure 4.17

More specifically, Newton showed that the allowed orbital paths are ellipses, parabolas, and hyperbolas (Figure 4.17a). Bound orbits are ellipses, while unbound orbits can be either parabolas or hyperbolas. Together, these shapes are known in mathematics as the *conic sections*, because they can be made by slicing through a cone at different angles (Figure 4.17b). Note that objects on unbound orbits still obey the basic principle of Kepler's second law: They move faster when they are closer to the object they are orbiting, and slower when they are farther away.

Objects Orbit Their Center of Mass We usually think of one object orbiting another object, like a planet orbiting the Sun or the Moon orbiting Earth. However, Newton showed that two objects attracted by gravity actually *both* orbit around their common **center of mass**—the point at which the two objects would balance if they were somehow connected (Figure 4.18). For example, in a binary star system in which both stars have the same mass, we would see both stars tracing ellipses around a point halfway between them. When one object is more massive than the other, the center of mass lies closer to the more massive object.

The idea that objects orbit their common center of mass holds even for the Sun and planets. However, the Sun is so much more massive than the planets that the center of mass between the Sun and any planet actually lies either inside or nearly inside the Sun, making it difficult for us to notice the Sun's motion about this center. Nevertheless, with precision measurements we can indeed detect the Sun's slight motion around this center of mass. As we will see in Chapter 13, similar measurements of other stars are

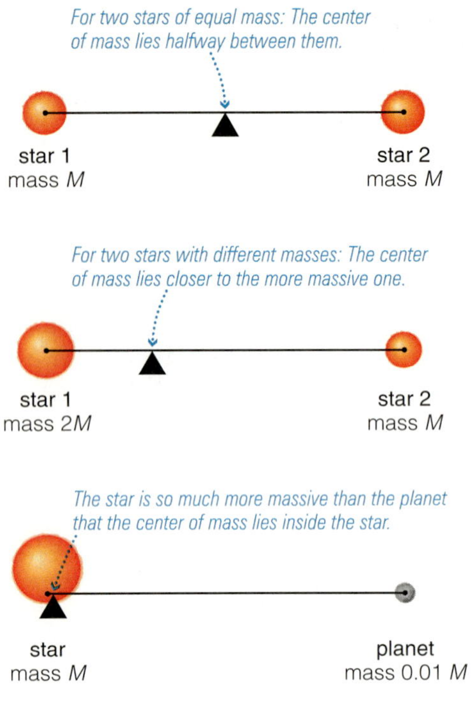

Figure 4.18 Interactive Figure Two objects attracted by gravity orbit their common *center of mass*—the point at which they would balance if they were somehow connected.

Figure 4.19 Newton's version of Kepler's third law enables us to measure masses throughout the universe. In addition, it shows that when one object orbits a much more massive object, the orbital period depends only on its average orbital distance. Thus, the astronaut and the Space Shuttle share the same orbit and stay together—even as both orbit Earth at a speed of some 25,000 km/hr.

the primary method astronomers have used to discover planets around those stars.

Orbital Characteristics Tell Us the Masses of Distant Objects Recall that the precise statement of Kepler's third law is $p^2 = a^3$, where p is a planet's orbital period in years and a is the planet's average distance from the Sun in AU. Newton found that this statement is actually a special case of a more general equation that we call **Newton's version of Kepler's third law** (see Mathematical Insight 4.3). This more general equation allows us to measure orbital period and distance in any units we wish (rather than only in years and AU) and also shows that the relationship between the orbital period and the average distance depends on the masses of the orbiting objects. In particular, when an object is much less massive than the object it orbits, we can calculate the mass of the central object from the orbital period and average distance of the orbiting object.

Newton's version of Kepler's third law is the primary means by which we determine masses throughout the universe. For example, it allows us to calculate the mass of the Sun from Earth's orbital period (1 year) and its average distance from the Sun. Similarly, measuring the orbital period and average distance of one of Jupiter's moons allows us to calculate Jupiter's mass, and measuring the orbital periods and average distances of stars in binary star systems allows us to determine their masses.

Newton's version of Kepler's third law also explains another important characteristic of orbital motion. It shows that the orbital period of a *small* object orbiting a much more massive object depends only on its orbital distance, not on its mass. That is why an astronaut does not need a tether to stay close to the Space Shuttle or the Space Station during a space walk (Figure 4.19). The spacecraft and the astronaut are both much smaller than Earth and thus stay together because they have the same orbital distance and hence the same orbital period.

4.5 Orbits, Tides, and the Acceleration of Gravity

Newton's universal law of gravitation has applications that go far beyond explaining Kepler's laws. In this final section, we'll explore three important concepts that we can understand with the help of the universal law of gravitation: how orbits can change, how gravity produces tides, and the acceleration of gravity.

• How do gravity and energy together allow us to understand orbits?

We've seen that Newton's law of universal gravitation explains Kepler's laws of planetary motion, which describe the simple and stable orbits of the planets. By extending Kepler's laws, Newton also explained many other stable orbits, such as the orbit of a satellite around Earth or of a moon around a planet. But orbits do not always stay the same. For example, you've probably heard of satellites crashing to Earth from orbit, proving that orbits can sometimes change dramatically. To understand how and why orbits sometimes change, we need to consider the role of energy in orbits.

Orbital Energy Consider the orbit of a planet around the Sun. An orbiting planet has both kinetic energy (because it is moving around the Sun) and gravitational potential energy (because it would fall toward the Sun if it stopped orbiting). The planet's kinetic energy depends on its orbital speed, and its gravitational potential energy depends on its distance from the Sun. Because the planet's distance and speed both vary as it orbits the Sun, the amounts of

MATHEMATICAL INSIGHT 4.3 Newton's Version of Kepler's Third Law

Newton's version of Kepler's third law relates the orbital periods, distances, and masses of any pair of orbiting objects, such as the Sun and a planet, a planet and a moon, or two stars in a binary star system. Mathematically, we write it as follows:

$$p^2 = \frac{4\pi^2}{G(M_1 + M_2)}a^3$$

where M_1 and M_2 are the masses of the two objects, p is their orbital period, and a is the distance between their centers. The term $4\pi^2$ is simply a number ($4\pi^2 \approx 4 \times 3.14^2 = 39.44$), and G is the gravitational constant. Any units for p and a can be used in this equation, as long as the units are consistent with the units used for the gravitational constant G.

Newton's version of Kepler's third law gives us the power to measure the masses of distant objects. Any time we measure the orbital period and distance of an orbiting object, we can use Newton's equation to calculate the sum $M_1 + M_2$ of the two objects involved in the orbit. If one object is much more massive than the other, we essentially learn the mass of the massive object. For example, in the case of a planet orbiting the Sun, the sum $M_{Sun} + M_{planet}$ is pretty much just M_{Sun} because the Sun is so much more massive than any of the planets. Thus, knowing any planet's orbital period and distance from the Sun allows us to calculate the mass of the Sun.

Historically, the value of G wasn't determined experimentally until more than 100 years after Newton published *Principia*. Only then did it become possible to use Newton's version of Kepler's third law to measure masses in absolute units. The following examples show some of the remarkable power of Newton's version of Kepler's third law.

Example 1: Use the fact that Earth orbits the Sun in 1 year at an average distance of 150 million kilometers (1 AU) to calculate the mass of the Sun.

Solution:

Step 1 Understand: We know that we can calculate the mass with Newton's version of Kepler's third law. We start by realizing that for Earth orbiting the Sun, this law takes the form:

$$(p_{Earth})^2 = \frac{4\pi^2}{G(M_{Sun} + M_{Earth})}(a_{Earth})^3$$

Because the Sun is so much more massive than Earth, the sum of their masses is approximately the mass of the Sun alone: $M_{Sun} + M_{Earth} \approx M_{Sun}$. Using this approximation, we can write the equation as:

$$(p_{Earth})^2 \approx \frac{4\pi^2}{G \times M_{Sun}}(a_{Earth})^3$$

Step 2 Solve: Because we already know Earth's orbital period (p_{Earth}) and distance (a_{Earth}), the above equation contains only one unknown: the Sun's mass (M_{Sun}). We can therefore find the Sun's mass by solving the equation for this one unknown. We multiply both sides by M_{Sun} and divide both sides by $(p_{Earth})^2$:

$$M_{Sun} \approx \frac{4\pi^2}{G}\frac{(a_{Earth})^3}{(p_{Earth})^2}$$

We now plug in the known values. Earth's orbital period is $p_{Earth} = 1$ yr, which is the same as 3.15×10^7 seconds. Its average distance from the Sun is $a_{Earth} \approx 150$ million km, or 1.5×10^{11} m. Using these values and the experimentally measured value $G = 6.67 \times 10^{-11}\ m^3/(kg \times s^2)$, we find:

$$M_{Sun} \approx \frac{4\pi^2}{\left(6.67 \times 10^{-11}\frac{m^3}{kg \times s^2}\right)}\frac{(1.5 \times 10^{11}\ m)^3}{(3.15 \times 10^7\ s)^2}$$
$$= 2 \times 10^{30}\ kg$$

Step 3 Explain: Simply by knowing Earth's orbital period and distance from the Sun and the gravitational constant G, we have used Newton's version of Kepler's third law to find that the Sun's mass is about 2×10^{30} kilograms.

Example 2: A *geosynchronous satellite* orbits Earth in the same amount of time that Earth rotates: 1 sidereal day. If a geosynchronous satellite is also in an equatorial orbit, it is said to be *geostationary* because it remains fixed in the sky (it maintains a constant altitude and direction) as seen from the ground (see Problem 47). Calculate the orbital distance of a geosynchronous satellite.

Solution:

Step 1 Understand: Again, we start by putting Newton's version of Kepler's third law into a form we can apply to this problem. A satellite is much less massive than Earth, so $M_{Earth} + M_{satellite} \approx M_{Earth}$, so we can use the law in this form:

$$(p_{satellite})^2 \approx \frac{4\pi^2}{G \times M_{Earth}}(a_{satellite})^3$$

Step 2 Solve: Because we want to know the satellite's distance, we solve for $a_{satellite}$ by multiplying both sides of the equation by $(G \times M_{Earth})$, dividing both sides by $4\pi^2$, and then taking the cube root of both sides:

$$a_{satellite} \approx \sqrt[3]{\frac{G \times M_{Earth}}{4\pi^2}(p_{satellite})^2}$$

We know that $p_{satellite} = 1$ sidereal day $\approx 86{,}164$ seconds. You should confirm that substituting this value and the mass of Earth yields $a_{satellite} \approx 42{,}000$ km.

Step 3 Explain: We have found that a geosynchronous satellite orbits at a distance of 42,000 kilometers above the *center* of the Earth, which is about 35,600 kilometers above Earth's surface.

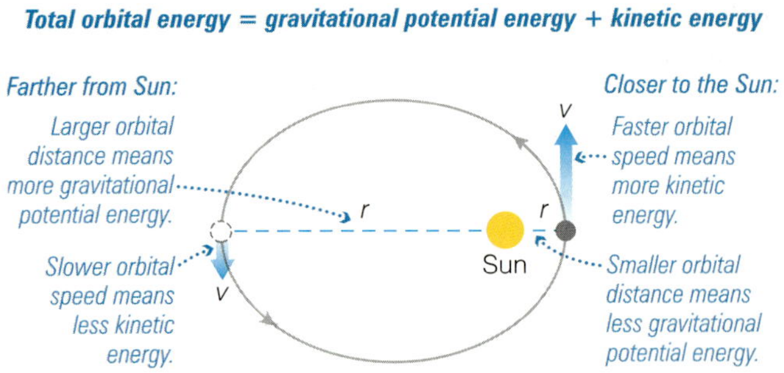

Figure 4.20 The total orbital energy of a planet stays constant throughout its orbit, because its gravitational potential increases when its kinetic energy decreases, and vice versa.

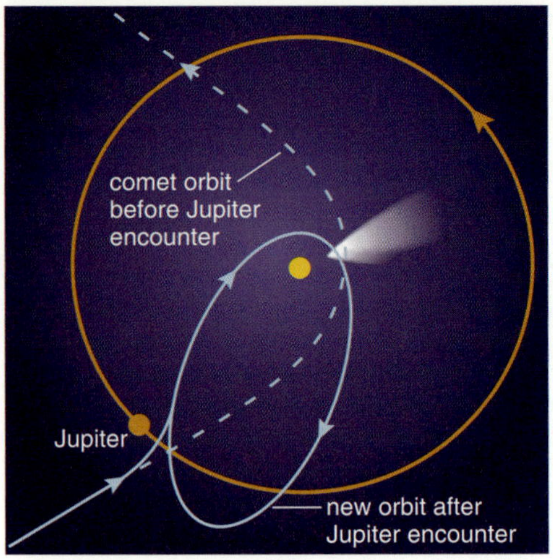

Figure 4.21 Depiction of a comet in an unbound orbit of the Sun that happens to pass near Jupiter. The comet loses orbital energy to Jupiter, changing its unbound orbit to a bound orbit around the Sun.

its gravitational potential energy and kinetic energy also vary (Figure 4.20). However, the planet's total **orbital energy**—the sum of its kinetic and gravitational potential energies—always stays the same. This fact is a consequence of the law of conservation of energy. As long as no other object causes the planet to gain or lose orbital energy, its orbital energy cannot change and its orbit must remain the same.

Generalizing from planets to other objects leads us to a very important idea about motion throughout the cosmos: *Orbits cannot change spontaneously.* Left undisturbed, planets would forever keep the same orbits around the Sun, moons would keep the same orbits around their planets, and stars would keep the same orbits around their galaxies.

Gravitational Encounters Although orbits cannot change spontaneously, they can change through exchanges of energy. One way that two objects can exchange orbital energy is through a **gravitational encounter**, in which they pass near enough so that each can feel the effects of the other's gravity. For example, in the rare cases in which a comet happens to pass near a planet, the comet's orbit can change dramatically. Figure 4.21 shows a comet headed toward the Sun on an unbound orbit. The comet's close passage by Jupiter allows the comet and Jupiter to exchange energy. In this case, the comet loses so much orbital energy that its orbit changes from unbound to a bound, elliptical orbit. Jupiter gains exactly as much energy as the comet loses, but the effect on Jupiter is unnoticeable because of its much greater mass.

Spacecraft engineers can use the same basic idea in reverse. The two *Voyager* spacecraft that visited the outer planets in the 1980s and the *Cassini* spacecraft now orbiting Saturn were all deliberately sent past Jupiter on paths that caused them to gain orbital energy at Jupiter's expense. This extra orbital energy helped the spacecraft reach more distant planets. Of course, the effect of the tiny spacecraft on Jupiter was unnoticeable.

A similar dynamic sometimes occurs naturally and may explain why most comets orbit so far from the Sun. Astronomers think that most comets once orbited in the same region of the solar system as the large outer planets [Section 12.2]. Gravitational encounters with Jupiter or the other large planets then caused some of these comets to be "kicked" into much more distant orbits around the Sun. In some cases, the comets may have been ejected from the solar system completely.

Atmospheric Drag Friction can cause objects to lose orbital energy. For example, consider a satellite orbiting Earth. If the orbit is fairly low—say, just a few hundred kilometers above Earth's surface—the satellite experiences a bit of drag from Earth's thin upper atmosphere. This drag gradually causes the satellite to lose orbital energy until it finally plummets to Earth. The satellite's lost orbital energy is converted to thermal energy in the atmosphere, which is why a falling satellite usually burns up.

Friction may also have played a role in shaping the current orbits of some moons and planets. For example, some of the small moons of Jupiter and the other outer planets may once have orbited the Sun independently. Their orbits could not have changed spontaneously. However, the outer planets probably once were surrounded by clouds of gas [Section 8.4], and friction would have slowed objects passing through this gas. Some of these small objects probably lost just enough energy to friction to allow them to be "captured" as moons.

Escape Velocity An object that gains orbital energy moves into an orbit with a higher average altitude. For example, if we want to boost the orbital altitude of a spacecraft, we can give it more orbital energy by firing a rocket. The chemical potential energy released by the rocket fuel is converted to orbital energy for the spacecraft. The Hubble Space Telescope is still in orbit only because it has gotten periodic boosts during visits by the Space Shuttle. Without these boosts, atmospheric drag would by now have caused the telescope to fall to Earth.

If we give a spacecraft enough orbital energy, it may end up in an unbound orbit that allows it to *escape* Earth completely (Figure 4.22). For example, if we want to send a

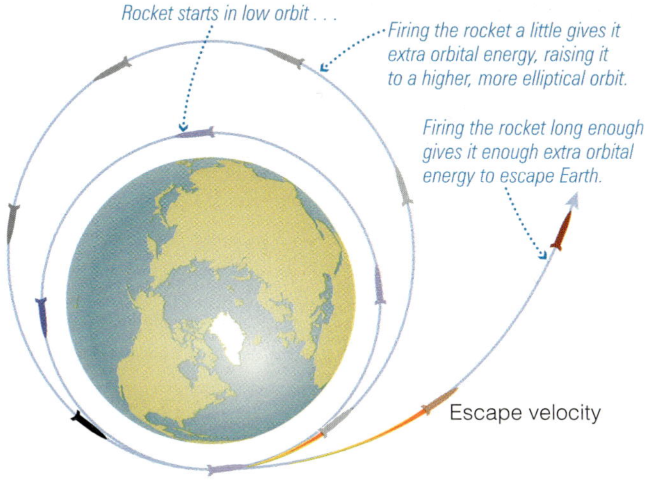

Figure 4.22 Interactive Figure An object with escape velocity has enough orbital energy to escape Earth completely.

space probe to Mars, we must use a large rocket that gives the probe enough energy to leave Earth orbit. Although it would probably make more sense to say that the probe achieves "escape energy," we instead say that it achieves **escape velocity**. The escape velocity from Earth's surface is about 40,000 km/hr, or 11 km/s, meaning that this is the minimum velocity required to escape Earth's gravity if a spacecraft starts near the surface.

Notice that the escape velocity does not depend on the mass of the escaping object—*any* object must travel at a velocity of 11 km/s to escape from Earth, whether it is an individual atom or molecule escaping from the atmosphere, a spacecraft being launched into deep space, or a rock blasted into the sky by a large impact. Escape velocity *does* depend on whether the object starts from the surface or from someplace high above the surface. Because gravity weakens with distance, it takes less energy—and hence a lower escape velocity—to escape from a point high above Earth than from Earth's surface.

• How does gravity cause tides?

If you've spent time near an ocean, you're probably aware of the rising and falling of the tides. In most places, tides rise and fall twice each day. We can understand the basic

MATHEMATICAL INSIGHT 4.4 Escape Velocity

A simple formula allows us to calculate the escape velocity from any planet, moon, or star:

$$v_{escape} = \sqrt{2 \times G \times \frac{M}{R}}$$

where M is the object's mass, R is the starting distance above the object's center, and G is the gravitational constant. If you use this formula to calculate the escape velocity from an object's surface, replace R with the object's radius.

Example 1: Calculate the escape velocity from the Moon. Compare it to that from Earth.

Solution:

Step 1 Understand: The text has already told us that the escape velocity from Earth's surface is 11 km/s. We can find the Moon's escape velocity with the formula given above. We can find the Moon's mass and radius in Appendix E; they are, respectively, $M = 7.4 \times 10^{22}$ kg and $R = 1.7 \times 10^6$ m.

Step 2 Solve: We substitute the Moon's mass and radius into the escape velocity formula:

$$v_{escape} = \sqrt{2 \times G \times \frac{M_{Moon}}{R_{Moon}}}$$

$$= \sqrt{\frac{2 \times \left(6.67 \times 10^{-11} \frac{m^3}{kg \times s^2}\right) \times (7.4 \times 10^{22}\ kg)}{1.7 \times 10^6\ m}}$$

$$\approx 2{,}400\ m/s = 2.4\ km/s$$

Step 3 Explain: We have found that the escape velocity from the Moon is 2.4 km/s, which is less than one-fourth the 11-km/s escape velocity from Earth.

Example 2: Suppose a future space station orbits Earth in geosynchronous orbit, which is 42,000 kilometers above the center of Earth (see Mathematical Insight 4.3). At what velocity must a spacecraft be launched from the station to escape Earth? Is there any advantage to launching from the space station instead of from Earth's surface?

Solution:

Step 1 Understand: In this case, we want to find the escape velocity from a satellite orbiting 42,000 kilometers above the center of Earth. We therefore will use the escape velocity formula with the mass of Earth ($M_{Earth} = 6.0 \times 10^{24}$ kg) and the satellite's distance above the center of Earth ($R = 42{,}000\ km = 4.2 \times 10^7\ m$).

Step 2 Solve: We substitute the needed values into the escape velocity formula:

$$v_{escape} = \sqrt{2 \times G \times \frac{M_{Earth}}{R_{orbit}}}$$

$$= \sqrt{\frac{2 \times \left(6.67 \times 10^{-11} \frac{m^3}{kg \times s^2}\right) \times (6.0 \times 10^{24}\ kg)}{4.2 \times 10^7\ m}}$$

$$= 4{,}400\ m/s = 4.4\ km/s$$

Step 3 Explain: The escape velocity from geosynchronous orbit is 4.4 km/s—considerably lower than the 11-km/s escape velocity from Earth's surface. Thus, it requires substantially less fuel to launch the spacecraft from the space station than from Earth. (In addition, the spacecraft would already have the orbital velocity of the space station.) Of course, this assumes that the space station is already in place and that the spacecraft is assembled at the space station.

cause of tides by examining the gravitational attraction between Earth and the Moon. We'll then see how the same ideas explain many other phenomena that we can observe throughout the universe, including the synchronous rotation of our own Moon and many other worlds.

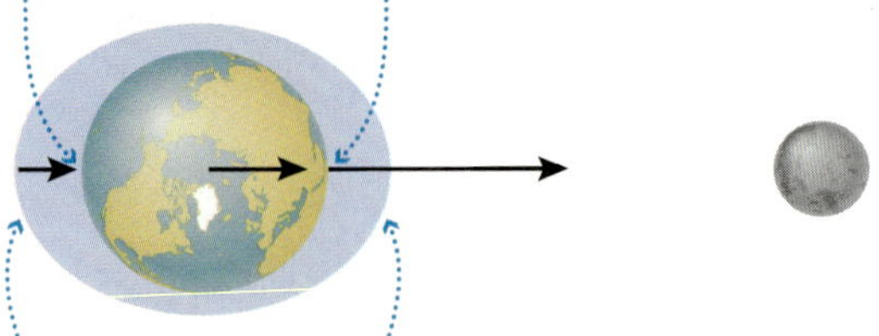

Figure 4.23 Tides are created by the difference in the force of attraction between different parts of Earth and the Moon. There are two daily high tides as any location on Earth rotates through the two tidal bulges. (The diagram highly exaggerates the tidal bulges, which raise the oceans only about 2 meters and the land only about a centimeter.)

The Moon's Tidal Force Gravity attracts Earth and the Moon toward each other (with the Moon staying in orbit as it "falls around" Earth), but it affects different parts of Earth slightly differently: Because the strength of gravity declines with distance, the gravitational attraction of each part of Earth to the Moon becomes weaker as we go from the side of Earth facing the Moon to the side facing away from the Moon. This difference in attraction creates a "stretching force," or **tidal force**, that stretches the entire Earth to create two tidal bulges—one facing the Moon and one opposite the Moon (Figure 4.23). If you are still unclear about why there are *two* tidal bulges, think about a rubber band: If you pull on a rubber band it will stretch in both directions relative to its center, even if you pull on only one side. In the same way, Earth stretches on both sides even though the Moon is tugging harder on only one side.

Tides affect both land and ocean, but we generally notice only the ocean tides because water flows much more readily than land. Earth's rotation carries any location through each of the two bulges each day, creating two high tides. Low tides occur when the location is at the points halfway between the two tidal bulges. The two "daily" high tides actually come slightly more than 12 hours apart. Because of its orbital motion around Earth, the Moon reaches its highest point in the sky at any location about every 24 hours 50 minutes, rather than every 24 hours. Thus, the tidal cycle of two high tides and two low tides actually takes about 24 hours 50 minutes, with each high tide occurring about 12 hours 25 minutes after the previous one.

The height and timing of tides can vary considerably from place to place around the Earth, depending on factors such as latitude, the orientation of the coastline (such as whether it is north-facing or west-facing), and the depth and shape of any channel through which the rising tide must flow. For example, while the tide rises gradually in most locations, the incoming tide near the famous abbey on Mont-Saint-Michel, France, moves much faster than a person can swim (Figure 4.24). In centuries past, the Mont was an island twice a day at high tide but was connected

Figure 4.24 Photographs of high and low tide at the abbey at Mont-Saint-Michel, France, one of the world's most popular tourist destinations. Here the tide rushes in much faster than a person can swim. Before a causeway was built (visible to the left, with cars on it), the Mont was accessible by land only at low tide. At high tide, it became an island.

COMMON MISCONCEPTIONS

The Origin of Tides

Many people believe that tides arise because the Moon pulls Earth's oceans toward it. But if that were the whole story, there would be a bulge only on the side of Earth facing the Moon, and hence only one high tide each day. The correct explanation for tides must account for why Earth has *two* tidal bulges.

Only one explanation works: Earth must be stretching from its center in both directions (toward and away from the Moon). The stretching force, or tidal force, arises from the *difference* between the force of gravity attracting different parts of Earth to the Moon. In fact, stretching due to tides affects many objects, not just Earth. Many moons are stretched into slightly oblong shapes by tidal forces caused by their parent planets, and mutual tidal forces stretch close binary stars into teardrop shapes. In regions where gravity is extremely strong, such as near a black hole, tides could even stretch spaceships or people.

to the mainland at low tide. Many pilgrims drowned when they were caught unprepared by the tide rushing in. (Today, a human-made land bridge keeps the island connected to the mainland.) Another unusual tidal pattern occurs in coastal states along the northern shore of the Gulf of Mexico, where topography and other factors combine to make only one noticeable high tide and low tide each day.

The Tidal Effect of the Sun The Sun also exerts a tidal force on Earth, causing Earth to stretch along the Sun-Earth line. You might at first guess that the Sun's tidal force would be more than the Moon's, since the Sun's mass is more than a million times that of the Moon. Indeed, the *gravitational* force between Earth and the Sun is much greater than that between Earth and the Moon, which is why Earth orbits the Sun. However, the much greater distance to the Sun (than to the Moon) means that the *difference* in the Sun's pull on the near and far sides of Earth is relatively small.

The overall tidal force caused by the Sun is about one-third that caused by the Moon (Figure 4.25). When the tidal forces of the Sun and the Moon work together, as is the case at both new moon and full moon, we get the especially pronounced *spring tides* (so named because the water tends to "spring up" from the Earth). When the tidal forces of the Sun and the Moon counteract each other, as is the case at first- and third-quarter moon, we get the relatively small tides known as *neap tides.*

THINK ABOUT IT

Explain why any tidal effects on Earth caused by the other planets would be extremely small (in fact, so small as to be unnoticeable).

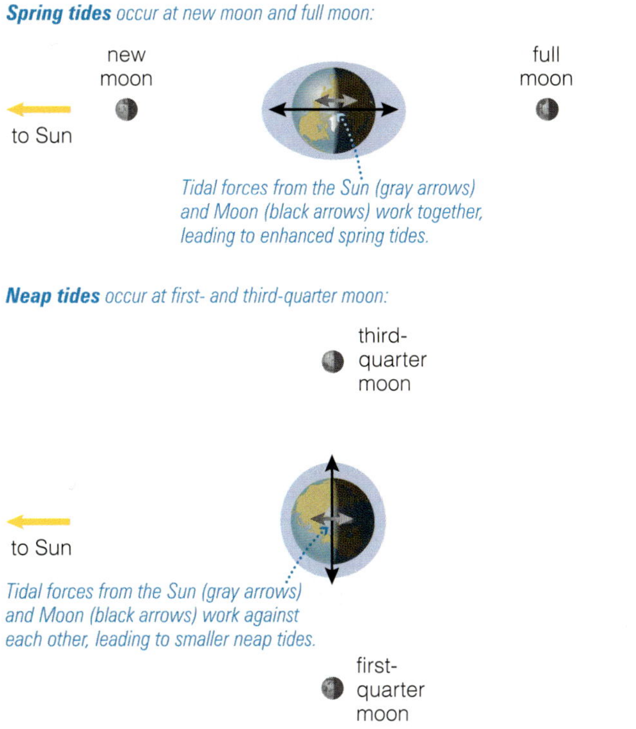

Figure 4.25 Interactive Figure The Sun exerts a tidal force on Earth about one-third as strong as that from the Moon. When the tidal forces from the Sun and Moon work together at new moon and full moon, we get enhanced *spring tides.* When they work against each other, at first- and third-quarter moons, we get smaller *neap tides.*

Tidal Friction So far, we have talked as if Earth slides smoothly through the tidal bulges as it rotates. But because tidal forces stretch the Earth itself, the process involves some friction, called **tidal friction.** Figure 4.26 shows the effects of this friction. In essence, the Moon's gravity tries to keep the tidal bulges on the Earth-Moon line, while Earth's rotation tries to pull the bulges around with it. The resulting "compromise" keeps the bulges just ahead of the Earth-Moon line at all times.

The slight misalignment of the tidal bulges with the Earth-Moon line causes two important effects. First, because the Moon's gravity is always pulling back on the bulges, Earth's rotation slows very gradually. Second, the gravity of the bulges pulls the Moon slightly ahead in its orbit, thereby adding to the Moon's orbital energy and causing the Moon to move gradually farther from Earth. These effects are barely noticeable on human time scales. For example, tidal friction increases the length of a day by only about 1 second every 50,000 years.* But the effects add up over billions of years. Early in Earth's history, a day may have been only 5 or 6 hours long and the Moon may have been one-tenth or less its current distance from Earth. These changes also provide a great example of conservation of angular momentum: The Moon's growing orbit gains the angular momentum that Earth loses as its rotation slows.

*This effect is overwhelmed on short time scales by effects due to slight changes in Earth's internal mass distribution; these changes can alter Earth's rotation period by a second or more per year, which is why "leap seconds" are occasionally added to or subtracted from the year.

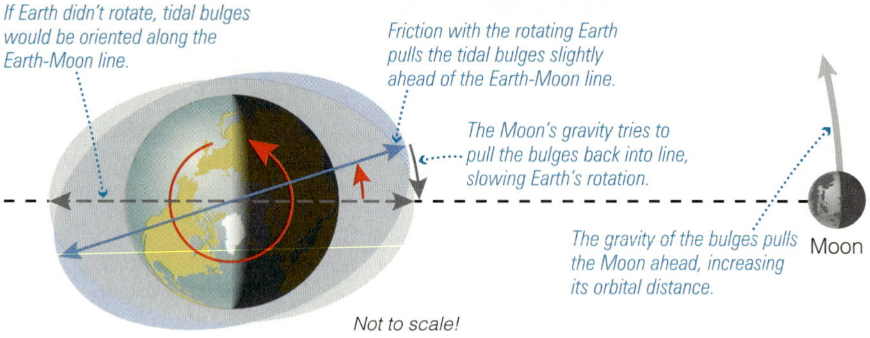

Figure 4.26 Earth's rotation pulls its tidal bulges slightly ahead of the Earth-Moon line, leading to gravitational effects that gradually slow Earth's rotation and increase the Moon's orbital energy and distance.

The Moon's Synchronous Rotation Recall that the Moon always shows (nearly) the same face to Earth, a trait called **synchronous rotation** (see Figure 2.23). Synchronous rotation may seem like an extraordinary coincidence, but it is a natural consequence of tidal friction.

Because Earth is more massive than the Moon, Earth exerts a greater tidal force on the Moon than the Moon does on Earth. This tidal force gives the Moon two tidal bulges along the Earth-Moon line, much like the two tidal bulges that the Moon creates on Earth. (The Moon does not have visible tidal bulges, but it does indeed have excess mass along the Earth-Moon line.) Thus, if the Moon were rotating through its tidal bulges in the same way that Earth rotates through its tidal bulges, the resulting friction would cause the Moon's rotation to slow down. This is exactly what we think happened long ago.

The Moon probably once rotated much faster than it does today. As a result, it *did* rotate through its tidal bulges, and its rotation gradually slowed. Once the Moon's rotation slowed to the point at which the Moon and its bulges rotated at the same rate—that is, synchronously with the orbital period—there was no further source for tidal friction. Thus, the Moon's synchronous rotation was a natural outcome of Earth's tidal effects on the Moon. In fact, if the Moon and Earth stay together long enough (for another few hundred billion years), the gradual slowing of Earth's rotation will eventually make Earth keep the same face to the Moon as well.

Tidal Effects on Other Worlds Tidal forces affect not only Earth and the Moon, but also many other objects. Synchronous rotation caused by tidal friction is especially common. For example, Jupiter's four large moons (Io, Europa, Ganymede, and Callisto) keep nearly the same face toward Jupiter at all times, as do many other moons. Pluto and its moon Charon *both* rotate synchronously: Like two dancers, they always keep the same face toward each other. Many binary star systems also rotate in this way. Some moons and planets exhibit variations on synchronous rotation. For example, Mercury rotates exactly three times for every two orbits of the Sun. This pattern ensures that Mercury's tidal bulge always aligns with the Sun at perihelion, where the Sun exerts its strongest tidal force.

Tidal forces play other roles in the cosmos as well. They can alter the shapes of objects by stretching them along the line of tidal bulges. More dramatically, in Chapter 11 we'll see how tidal forces have led to the astonishing volcanic activity of Jupiter's moon Io and the possibility of a subsurface ocean on its moon Europa. As you study astronomy, you'll encounter many more cases where tides and tidal friction play important roles.

• Why do all objects fall at the same rate?

Throughout the remainder of the text, we will see many more applications of the universal law of gravitation. For now, let's look at just one more: Galileo's discovery that the acceleration of a falling object is independent of its mass.

If you drop a rock, the force acting on the rock is the force of gravity. The two masses involved are the mass of Earth and the mass of the rock, which we'll denote M_{Earth} and M_{rock}, respectively. The distance between their *centers* is the distance from the *center of Earth* to the center of the rock. If the rock isn't too far above Earth's surface, this distance is approximately the radius of Earth, R_{Earth} (about 6,400 kilometers). That is, $d \approx R_{\text{Earth}}$. Thus, the force of gravity acting on the rock is:

$$F_g = G\frac{M_{\text{Earth}}M_{\text{rock}}}{d^2} \approx G\frac{M_{\text{Earth}}M_{\text{rock}}}{(R_{\text{Earth}})^2}$$

According to Newton's second law of motion ($F = ma$), this force is equal to the product of the mass and the acceleration of the rock. That is:

$$G\frac{M_{\text{Earth}}\cancel{M_{\text{rock}}}}{(R_{\text{Earth}})^2} = \cancel{M_{\text{rock}}}a_{\text{rock}}$$

Note that M_{rock} "cancels" because it appears on both sides of the equation (as a multiplier), giving Galileo's result that the acceleration of the rock—or of any falling object—does not depend on the object's mass.

The fact that objects of different masses fall with the same acceleration struck Newton as an astounding coincidence, even though his own equations showed it to be so. For the next 240 years, this seemingly odd coincidence remained just that—a coincidence—in the minds of scientists. However, in 1915 Einstein discovered that it is not a coincidence at all. Rather, it reveals something deeper about the nature of gravity and of the universe. Einstein described the new insights in his *general theory of relativity* (the topic of Chapter S3).

MATHEMATICAL INSIGHT 4.5 The Acceleration of Gravity

The text shows that the acceleration of a falling rock near the surface of Earth is:

$$a_{\text{rock}} = G \times \frac{M_{\text{Earth}}}{(R_{\text{Earth}})^2}$$

Because this formula applies to *any* falling object on Earth, it is the *acceleration of gravity, g*. Calculating g is easy. Simply look up Earth's mass (6.0×10^{24} kg) and radius (6.4×10^6 m) in Appendix E, and then "plug in":

$$\begin{aligned} g &= G \times \frac{M_{\text{Earth}}}{(R_{\text{Earth}})^2} \\ &= \left(6.67 \times 10^{-11} \frac{\text{m}^3}{\text{kg} \times \text{s}^2}\right) \times \frac{6.0 \times 10^{24}\ \text{kg}}{(6.4 \times 10^6\ \text{m})^2} \\ &= 9.8 \frac{\text{m}}{\text{s}^2} \end{aligned}$$

We can find the acceleration of gravity on the surface of any other world by using the same formula, using the other world's mass and radius instead of Earth's.

Example 1: What is the acceleration of gravity on the surface of the Moon?

Solution:

Step 1 Understand: We want to find the acceleration of gravity on the Moon's surface, so we can simply use the formula given above. We look up the values for the Moon's mass (7.4×10^{22} kg) and radius (1.7×10^6 m).

Step 2 Solve: We substitute the values of the Moon's mass and radius into the formula for the acceleration of gravity:

$$\begin{aligned} g_{\text{Moon}} &= G \times \frac{M_{\text{Moon}}}{(R_{\text{Moon}})^2} \\ &= \left(6.67 \times 10^{-11} \frac{\text{m}^3}{\text{kg} \times \text{s}^2}\right) \times \frac{7.4 \times 10^{22}\ \text{kg}}{(1.7 \times 10^6\ \text{m})^2} \\ &= 1.7 \frac{\text{m}}{\text{s}^2} \end{aligned}$$

Step 3 Explain: The acceleration of gravity on the Moon is $1.7\ \text{m/s}^2$, or about one-sixth that on Earth. Thus, objects on the Moon weigh about one-sixth of what they would weigh on Earth. If you can lift a 50-kilogram barbell on Earth, you'll be able to lift a 300-kilogram barbell on the Moon.

Example 2: The Space Station orbits at an altitude roughly 300 kilometers above Earth's surface. What is the acceleration of gravity at this altitude?

Solution:

Step 1 Understand: Because the Space Station is significantly above Earth's surface, we cannot use the approximation $d \approx R_{\text{Earth}}$ that we used in the text. Instead, we must go back to Newton's second law and set the gravitational force on the Space Station equal to its mass times acceleration. The acceleration in this equation will be the acceleration of gravity at the Space Station's altitude.

Step 2 Solve: We write Newton's second law with the force being the force of gravity acting between Earth and the Space Station, which we set equal to the Space Station's mass times its acceleration:

$$G \times \frac{M_{\text{Earth}} \cancel{M_{\text{station}}}}{d^2} = \cancel{M_{\text{station}}} \times a_{\text{station}}$$

You should confirm that when we solve this equation for the acceleration of gravity, we find:

$$a_{\text{station}} = G \times \frac{M_{\text{Earth}}}{d^2}$$

In this case, the distance d is the 6,400-kilometer radius of Earth *plus* the 300-kilometer altitude of the Station, or d = 6,700 km = 6.7×10^6 m. Thus, the gravitational acceleration of the Space Station when orbiting Earth is:

$$\begin{aligned} a_{\text{station}} &= G \times \frac{M_{\text{Earth}}}{d^2} \\ &= \left(6.67 \times 10^{-11} \frac{\text{m}^3}{\text{kg} \times \text{s}^2}\right) \times \frac{6.0 \times 10^{24}\ \text{kg}}{(6.7 \times 10^6\ \text{m})^2} \\ &= 8.9 \frac{\text{m}}{\text{s}^2} \end{aligned}$$

Step 3 Explain: The acceleration of gravity in low-Earth orbit is $8.9\ \text{m/s}^2$. Notice that this is only about 10% less than the $9.8\ \text{m/s}^2$ acceleration of gravity at Earth's surface. Thus, we see again that lack of gravity cannot be the reason why astronauts are weightless in orbit; rather, they are weightless because they are in free-fall.

THE BIG PICTURE

Putting Chapter 4 into Context

We've covered a lot of ground in this chapter, from the scientific terminology of motion to the story of how Newton explained universal motion. Be sure you grasp the following "big picture" ideas:

- Understanding the universe requires understanding motion. Motion may seem complex, but it can be understood simply through Newton's three laws of motion.
- Today, we comprehend Newton's laws of motion through deeper principles, including the laws of conservation of angular momentum and of energy. We can understand many processes in the universe by following how energy is exchanged between different objects.
- Newton also discovered the universal law of gravitation, which explains how gravity holds planets in their orbits and much more—including how satellites can reach and stay in orbit, the nature of tides, and why the Moon rotates synchronously with Earth.
- Perhaps most important, Newton's discoveries showed that the same physical laws we observe on Earth apply throughout the universe. The universality of physics opens up the entire cosmos as a possible realm of human study.

4.1 Describing Motion: Examples from Daily Life

- **How do we describe motion?** **Speed** is the rate at which an object is moving. **Velocity** is speed in a certain direction. **Acceleration** is a change in velocity, meaning a change in either speed or direction. **Momentum** is mass × velocity. A **force** can change an object's momentum, causing it to accelerate.

- **How is mass different from weight?** An object's mass is the same no matter where it is located, but its weight varies with the strength of gravity or other forces acting on the object. An object becomes **weightless** when it is in **free-fall**, even though its mass is unchanged.

4.2 Newton's Laws of Motion

- **How did Newton change our view of the universe?** Newton showed that the same physical laws that operate on Earth also operate in the heavens, making it possible for us to learn about the universe by studying physical laws here on Earth.

- **What are Newton's three laws of motion?** (1) An object moves at constant velocity if there is no net force acting upon it. (2) Force = mass × acceleration ($F = ma$). (3) For any force, there is always an equal and opposite reaction force.

4.3 Conservation Laws in Astronomy

- **Why do objects move at constant velocity if no force acts on them?** **Conservation of momentum** means that an object's momentum cannot change unless it transfers momentum to or from other objects. When no force is present, no momentum can be transferred so an object must maintain its speed and direction.

- **What keeps a planet rotating and orbiting the Sun?** **Conservation of angular momentum** means

that a planet's rotation and orbit cannot change unless it transfers angular momentum to another object. The planets in our solar system do not exchange substantial angular momentum with each other or anything else, so their orbits and rotation rates remain quite steady.

- **Where do objects get their energy?** Energy is

always conserved—it can be neither created nor destroyed. Thus, objects must have gotten whatever energy they now have from exchanges of energy with other objects. Energy comes in three basic categories—**kinetic**, **radiative**, and **potential**—though sometimes we use subcategories for convenience.

4.4 The Universal Law of Gravitation

- **What determines the strength of gravity?**

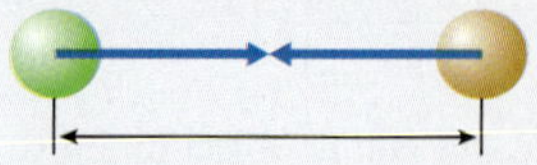

According to the **universal law of gravitation,** every object attracts every other object with a gravitational force that is directly proportional to the product of the objects' masses and declines with the square of the distance between their centers:

$$F_g = G\frac{M_1 M_2}{d^2}$$

- **How does Newton's law of gravity extend Kepler's laws?** Newton extended Kepler's laws in

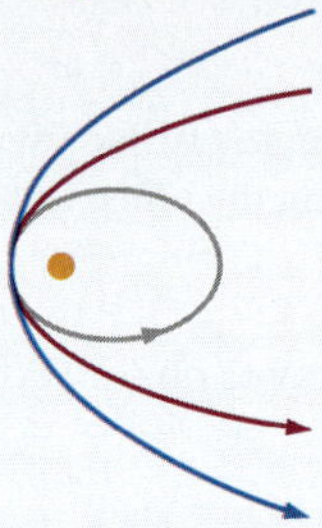

four main ways: (1) He showed that any object going around another object will obey Kepler's first two laws. (2) He showed that **bound orbits** in ellipses (which include circles) are not the only possible orbital shape—orbits can also be **unbound** parabolas or hyperbolas. (3) He showed that two objects actually orbit their common **center of mass.** (4) **Newton's version of Kepler's third law** allows us to calculate the masses of orbiting objects from their orbital periods and distances.

4.5 Orbits, Tides, and the Acceleration of Gravity

- **How do gravity and energy together allow us to understand orbits?** Gravity determines orbits,

and an object cannot change its orbit unless it gains or loses **orbital energy**—the sum of its kinetic and gravitational potential energies—through energy transfer with other objects. If an object gains enough orbital energy, it may achieve **escape velocity** and leave the gravitational influence of the object it was orbiting.

- **How does gravity cause tides?** The Moon's

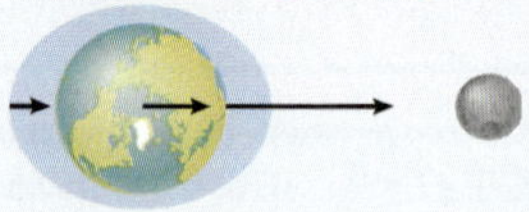

gravity creates a **tidal force** that stretches Earth along the Earth-Moon line, causing the tidal bulges responsible for our daily tides. Tidal forces can also lead to **tidal friction**, which is gradually slowing Earth's rotation and explains the synchronous rotation of the Moon.

- **Why do all objects fall at the same rate?** Newton's equations show that the acceleration of gravity is independent of the mass of a falling object, so all objects fall at the same rate.

EXERCISES AND PROBLEMS

For instructor-assigned homework go to MasteringASTRONOMY www.masteringastronomy.com

Review Questions

Short-Answer Questions Based on the Reading

1. How does *speed* differ from *velocity?* Give an example in which you can be traveling at constant speed but not at constant velocity.
2. What do we mean by *acceleration?* What is the *acceleration of gravity?* Explain what we mean when we state an acceleration in units of m/s^2.
3. What is *momentum?* How can momentum be affected by a *force?* What do we mean when we say that momentum will be changed only by a *net force?*
4. What is *free-fall*, and why does it make you *weightless?* Briefly describe why astronauts are weightless in the Space Station.
5. State each of *Newton's three laws of motion*. For each law, give an example of its application.
6. What are the laws of *conservation of momentum, conservation of angular momentum*, and *conservation of energy?* For each, give an example of how it is important in astronomy.
7. Define *kinetic energy, radiative energy*, and *potential energy*. For each type of energy, give at least two examples of objects that either have it or use it.
8. Define *temperature* and *thermal energy*. How are they related? How are they different?
9. Which has more gravitational potential energy: a rock on the ground or a rock that you hold out the window of a 10-story building? Explain.
10. What do we mean by *mass-energy?* Is it a form of kinetic, radiative, or potential energy? How is the idea of mass-energy related to the formula $E = mc^2$?
11. Summarize the *universal law of gravitation* in words. Then state the law mathematically, explaining the meaning of each symbol in the equation.
12. What is the difference between bound and unbound orbit? What orbital shapes are possible?
13. What do we need to know if we want to measure an object's mass with *Newton's version of Kepler's third law?* Explain.
14. Explain why orbits cannot change spontaneously. How can atmospheric drag affect an orbit? How can a *gravitational encounter* cause an orbit to change? How can an object achieve *escape velocity?*
15. Explain how the Moon creates tides on Earth. Why do we have two high and low tides each day?
16. How do the tides vary with the phase of the Moon? Why?
17. What is *tidal friction*? What effects does it have on Earth? How does it explain the Moon's synchronous rotation?
18. Would you fall at the same rate on the Moon as on Earth? Explain.

Test Your Understanding

Does It Make Sense?

Decide whether the statement makes sense (or is clearly true) or does not make sense (or is clearly false). Explain your reasoning. (For an example, see Chapter 1, "Does It Make Sense?")

19. If you could go shopping on the Moon to buy a pound of chocolate, you'd get a lot more chocolate than if you bought a pound on Earth. (*Hint:* Pounds are a unit of weight, not mass.)
20. Suppose you could enter a vacuum chamber (on Earth), that is, a chamber with no air in it. Inside this chamber, if you dropped a hammer and a feather from the same height at the same time, both would hit the bottom at the same time.
21. When an astronaut goes on a space walk outside the Space Station, she will quickly float away from the station unless she has a tether holding her to the station or constantly fires thrusters on her space suit.
22. Newton's version of Kepler's third law allows us to calculate the mass of Saturn from orbital characteristics of its moon Titan.
23. If we could somehow replace the Sun with a giant rock that has precisely the same mass, Earth's orbit would not change.
24. The fact that the Moon rotates once in precisely the time it takes to orbit Earth once is such an astonishing coincidence that scientists probably never will be able to explain it.
25. Venus has no oceans, so it could not have tides even if it had a moon (which it doesn't).
26. If an asteroid passed by Earth at just the right distance, it would be captured by Earth's gravity and become our second moon.
27. When I drive my car at 30 miles per hour, it has more kinetic energy than it does at 10 miles per hour.
28. Someday soon, scientists are likely to build an engine that produces more energy than it consumes.

Quick Quiz

Choose the best answer to each of the following. Explain your reasoning with one or more complete sentences.

29. Which one of the following describes an object that is accelerating? (a) A car traveling on a straight, flat road at 50 miles per hour. (b) A car traveling on a straight uphill road at 30 miles per hour. (c) A car going around a circular track at a steady 100 miles per hour.
30. Suppose you visit another planet: (a) Your mass and weight would be the same as they are on Earth. (b) Your mass would be the same as on Earth, but your weight would be different. (c) Your weight would be the same as on Earth, but your mass would be different.
31. Which person is weightless? (a) A child in the air as she plays on a trampoline. (b) A scuba diver exploring a deep-sea wreck. (c) An astronaut on the Moon.
32. Consider the statement "There's no gravity in space." This statement is: (a) Completely false. (b) False if you are close to a planet or moon, but true in between the planets. (c) Completely true.
33. If you want to make a rocket turn left, you need to: (a) Fire an engine that shoots out gas to the left. (b) Fire an engine that shoots out gas to the right. (c) Spin the rocket counterclockwise.
34. Earth's angular momentum when it is at perihelion (nearest to the Sun) in its orbit is: (a) Greater than its angular momentum at aphelion. (b) Less than its angular momentum at aphelion. (c) Exactly the same as its angular momentum at aphelion.
35. As an interstellar gas cloud shrinks in size, its gravitational potential energy: (a) stays the same at all times. (b) gradually is transformed into other forms of energy. (c) gradually grows larger.

36. If Earth were twice as far from the Sun, the force of gravity attracting Earth to the Sun would be: (a) twice as strong. (b) half as strong. (c) one-quarter as strong.
37. According to the law of universal gravitation, what would happen to Earth if the Sun were somehow replaced by a black hole of the same mass? (a) Earth would be quickly sucked into the black hole. (b) Earth would slowly spiral in to the black hole. (c) Earth's orbit would not change.
38. If the Moon were closer to Earth, high tides would: (a) be higher than they are now. (b) be lower than they are now. (c) occur three or more times a day rather than twice a day.

Investigate Further

In-Depth Questions to Increase Your Understanding

Short-Answer/Essay Questions

39. *Units of Acceleration.*
    a. If you drop a rock from a very tall building, how fast will it be going after 4 seconds? Explain.
    b. As you sled down a steep, slick street, you accelerate at a rate of 4 m/s^2. How fast will you be going after 5 seconds? Explain.
    c. You are driving along the highway at a speed of 60 miles per hour when you slam on the brakes. If your acceleration is at an average rate of –20 miles per hour per second, how long will it take to come to a stop?
40. *Weightlessness.* Astronauts are weightless when in orbit in the Space Shuttle. Are they also weightless during the Shuttle's launch? How about during its return to Earth? Explain.
41. *Gravitational Potential Energy.*
    a. Why does a bowling ball perched on a cliff ledge have more gravitational potential energy than a baseball perched on the same ledge?
    b. Why does a diver on a 10-meter platform have more gravitational potential energy than a diver on a 3-meter diving board?
    c. Why does a 100-kilogram satellite orbiting Jupiter have more gravitational potential energy than a 100-kilogram satellite orbiting Earth, assuming both satellites orbit at the same distance from their planets' centers?
42. *Einstein's Famous Formula.*
    a. What is the meaning of the formula $E = mc^2$? Be sure to define each variable.
    b. How does this formula explain the generation of energy by the Sun?
    c. How does this formula explain the destructive power of nuclear bombs?
43. *The Gravitational Law.*
    a. How does quadrupling the distance between two objects affect the gravitational force between them?
    b. Suppose the Sun were somehow replaced by a star with twice as much mass. What would happen to the gravitational force between Earth and the Sun?
    c. Suppose Earth were moved to one-third of its current distance from the Sun. What would happen to the gravitational force between Earth and the Sun?
44. *Allowable Orbits?*
    a. Suppose the Sun were replaced by a star with twice as much mass. Could Earth's orbit stay the same? Why or why not?
    b. Suppose Earth doubled in mass (but the Sun stayed the same as it is now). Could Earth's orbit stay the same? Why or why not?
45. *Head-to-Foot Tides.* You and Earth attract each other gravitationally, so you should also be subject to a tidal force resulting from the difference between the gravitational attraction felt by your feet and that felt by your head (at least when you are standing). Explain why you can't feel this tidal force.
46. *Synchronous Rotation.* Suppose the Moon had rotated *more slowly* when it formed than it does now. Would it still have ended up in synchronous rotation? Why or why not?
47. *Geostationary Orbit.* A satellite in geostationary orbit appears to remain stationary in the sky as seen from any particular location on Earth.
    a. Briefly explain why a geostationary satellite must orbit Earth in 1 *sidereal* day, rather than 1 solar day.
    b. Explain why a geostationary satellite must orbit around Earth's equator, rather than in some other orbit (such as around the poles).
    c. Home satellite dishes (such as those used for television) receive signals from communications satellites. Explain why these satellites must be in geostationary orbit.
48. *Elevator to Orbit.* Some people have proposed building a giant elevator from Earth's surface to geosynchronous orbit. The top of the elevator would then have the same orbital distance and period as any satellite in geosynchronous orbit.
    a. Suppose you were to let go of an object at the top of the elevator. Would the object fall? Would it orbit Earth? Explain.
    b. Briefly explain why (not counting the huge costs for construction) the elevator would make it much cheaper and easier to put satellites in orbit or to launch spacecraft into deep space.

Quantitative Problems

Be sure to show all calculations clearly and state your final answers in complete sentences.

49. *Energy Comparisons.* Use the data in Table 4.1 to answer each of the following questions.
    a. Compare the energy of a 1-megaton hydrogen bomb to the energy released by a major earthquake.
    b. If the United States obtained all its energy from oil, how much oil would be needed each year?
    c. Compare the Sun's annual energy output to the energy released by a supernova.
50. *Moving Candy Bar.* We can calculate the kinetic energy of any moving object with a very simple formula: kinetic energy $= \frac{1}{2}mv^2$, where m is the object's mass and v is its velocity or speed. Table 4.1 shows that metabolizing a candy bar releases about 10^6 joules. How fast must the candy bar travel to have the same 10^6 joules in the form of kinetic energy? (Assume the candy bar's mass is 0.2 kilogram.) Is your answer faster or slower than you expected?
51. *Spontaneous Human Combustion.* Suppose that all the mass in your body were suddenly converted into energy according to the formula $E = mc^2$. How much energy would be released? Compare this to the energy released by a 1-megaton hydrogen bomb (see Table 4.1). What effect would your disappearance have on your surroundings?
52. *Fusion Power.* No one has yet succeeded in creating a commercially viable way to produce energy through nuclear fusion. However, suppose we could build fusion power plants

using the hydrogen in water as a fuel. Based on the data in Table 4.1, how much water would we need each minute in order to meet U.S. energy needs? Could such a reactor power the entire United States with the water flowing from your kitchen sink? Explain. (*Hint:* Use the annual U.S. energy consumption to find the energy consumption per minute, and then divide by the energy yield from fusing 1 liter of water to figure out how many liters would be needed each minute.)

53. *Understanding Newton's Version of Kepler's Third Law I.* Imagine another solar system, with a star of the same mass as the Sun. Suppose there is a planet in that solar system with a mass twice that of Earth orbiting at a distance of 1 AU from the star. What is the orbital period of this planet? Explain. (*Hint:* The calculations for this problem are so simple that you will not need a calculator.)
54. *Understanding Newton's Version of Kepler's Third Law II.* Suppose a solar system has a star that is four times as massive as our Sun. If that solar system has a planet the same size as Earth orbiting at a distance of 1 AU, what is the orbital period of the planet? Explain. (*Hint:* The calculations for this problem are so simple that you will not need a calculator.)
55. *Using Newton's Version of Kepler's Third Law I.*
    a. The Moon orbits Earth in an average time of 27.3 days at an average distance of 384,000 kilometers. Use these facts to determine the mass of Earth. (*Hint:* You may neglect the mass of the Moon, since its mass is only about $\frac{1}{80}$ of Earth's.)
    b. Jupiter's moon Io orbits Jupiter every 42.5 hours at an average distance of 422,000 kilometers from the center of Jupiter. Calculate the mass of Jupiter. (*Hint:* Io's mass is very small compared to Jupiter's.)
    c. You discover a planet orbiting a distant star that has about the same mass as the Sun. Your observations show that the planet orbits the star every 63 days. What is its orbital distance?
56. *Using Newton's Version of Kepler's Third Law II.*
    a. Pluto's moon Charon orbits Pluto every 6.4 days with a semimajor axis of 19,700 kilometers. Calculate the *combined* mass of Pluto and Charon. Compare this combined mass to the mass of Earth, which is about 6×10^{24} kilograms.
    b. Calculate the orbital period of the Space Shuttle in an orbit 300 kilometers above Earth's surface.
    c. The Sun orbits the center of the Milky Way Galaxy every 230 million years at a distance of 28,000 light-years. Use these facts to determine the mass of the galaxy. (As we'll discuss in Chapter 22, this calculation actually tells us only the mass of the galaxy *within* the Sun's orbit.)
57. *Escape Velocity.* Calculate the escape velocity from each of the following.
    a. The surface of Mars (mass $= 0.11M_{\text{Earth}}$, radius $= 0.53R_{\text{Earth}}$).
    b. The surface of Mars's moon Phobos (mass $= 1.1 \times 10^{16}$ kg, radius = 12 km).
    c. The cloud tops of Jupiter (mass $= 317.8M_{\text{Earth}}$, radius $= 11.2R_{\text{Earth}}$).
    d. Our solar system, starting from Earth's orbit. (*Hint:* Most of the mass of our solar system is in the Sun; $M_{\text{Sun}} = 2.0 \times 10^{30}$ kg.)
    e. Our solar system, starting from Saturn's orbit.
58. *Weights on Other Worlds.* Calculate the acceleration of gravity on the surface of each of the following worlds. How much would *you* weigh, in pounds, on each of these worlds?
    a. Mars (mass $= 0.11M_{\text{Earth}}$, radius $= 0.53R_{\text{Earth}}$).
    b. Venus (mass $= 0.82M_{\text{Earth}}$, radius $= 0.95R_{\text{Earth}}$).
    c. Jupiter (mass $= 317.8M_{\text{Earth}}$, radius $= 11.2R_{\text{Earth}}$). Bonus: Given that Jupiter has no solid surface, how could you weigh yourself on Jupiter?
    d. Jupiter's moon Europa (mass $= 0.008M_{\text{Earth}}$, radius $= 0.25R_{\text{Earth}}$).
    e. Mars's moon Phobos (mass $= 1.1 \times 10^{16}$ kg, radius = 12 km).
59. *Gees.* Acceleration is sometimes measured in *gees*, or multiples of the acceleration of gravity: 1 gee ($1g$) means $1 \times g$, or 9.8 m/s^2; 2 gees ($2g$) means $2 \times g$, or 2×9.8 m/s^2 = 19.6 m/s^2; and so on. Suppose you experience 6 gees of acceleration in a rocket.
    a. What is your acceleration in meters per second squared?
    b. You will feel a compression force from the acceleration. How does this force compare to your normal weight?
    c. Do you think you could survive this acceleration for long? Explain.
60. *Earth's 2nd Moon.* Suppose Earth had a second moon, called Swisscheese, with an average orbital distance double the Moon's and a mass about the same as the Moon's.
    a. Is Swisscheese's orbital period longer or shorter than the Moon's? Explain.
    b. The Moon's orbital period is about one month. Apply Kepler's 3rd law to find the approximate orbital period of Swisscheese. (*Hint:* If you form the ratio of the orbital distances of Swisscheese and the Moon, you can solve this problem with Kepler's original version of his third law rather than looking up all the numbers you'd need to apply Newton's version of Kepler's third law.)
    c. In words, describe how tides would differ due to the presence of this second moon. Consider the cases when the two moons are on the same side of Earth, on opposite sides of Earth, and 90° apart in their orbits.

Discussion Questions

61. *Knowledge of Mass-Energy.* Einstein's discovery that energy and mass are equivalent has led to technological developments that are both beneficial and dangerous. Discuss some of these developments. Overall, do you think the human race would be better or worse off if we had never discovered that mass is a form of energy? Defend your opinion.
62. *Perpetual Motion Machines.* Every so often, someone claims to have built a machine that can generate energy perpetually from nothing. Why isn't this possible according to the known laws of nature? Why do you think claims of perpetual motion machines sometimes receive substantial media attention?
63. *Tidal Complications.* The ocean tides on Earth are much more complicated than they might at first seem from the simple physics that underlies tides. Discuss some of the factors that make the real tides so complicated and how these factors affect the tides. Consider the following factors: the distribution of land and oceans; the Moon's varying distance from Earth in its orbit; and the fact that the Moon's orbital plane is not perfectly aligned with the ecliptic and that neither the Moon's orbit nor the ecliptic is aligned with Earth's equator.

MEDIA EXPLORATIONS

For self-study activities go to MasteringASTRONOMY www.masteringastronomy.com

Interactive Tutorials

Tutorial Review of Key Concepts

Use the interactive **Tutorial** at **www.masteringastronomy.com** to review key concepts from this chapter.

Motion and Gravity Tutorial

Lesson 1, Lesson 2

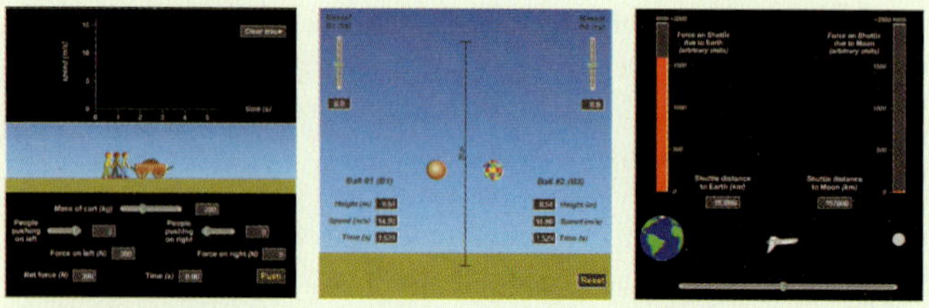

Energy Tutorial

Lesson 1 Kinetic Energy and Gravitational Potential Energy

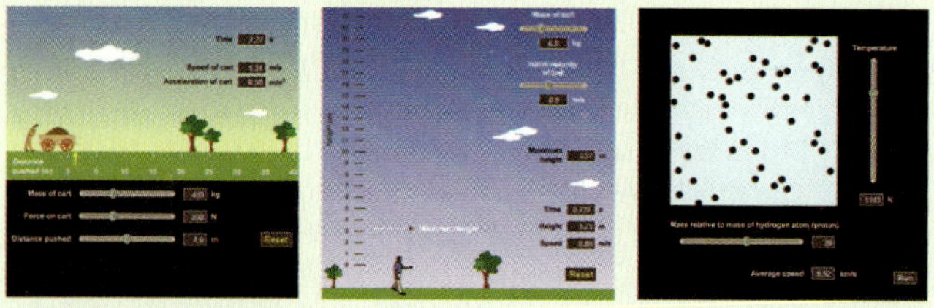

Orbits and Kepler's Laws Tutorial

Lessons 1–4

Supplementary Tutorial Exercises

Use the interactive **Tutorial Lessons** to explore the following questions.

Motion and Gravity, Lesson 2

1. Suppose Earth were twice as large in radius but still had the same mass. How much would you weigh? Explain.
2. Does the force of gravity between Earth and a spaceship launched from Earth ever become zero? Explain.
3. If a spaceship somehow came to a stop precisely $\frac{3}{4}$ of the way from Earth to the Moon, would it fall toward Earth, toward the Moon, or stay where it is? Explain.

Energy, Lesson 1

1. What properties of an object determine its kinetic energy? Its gravitational potential energy?
2. What condition must be met for an object launched from Earth to escape Earth's gravity?

Orbits and Kepler's Laws Tutorial, Lesson 4

1. Use the tool in Lesson 4 to predict the orbital period of an asteroid that has an orbital radius of 4 AU ($p =$ ____).
2. Use Kepler's third law, $p^2 = a^3$, to check your prediction ($a^3 =$ ____). Now take the square root of the result ($p =$ ____). Is your orbital period the same as the answer you calculated for question 1?
3. Use the tool to predict the orbital radius of an asteroid that has a period of 2 years. Does your answer agree with what you would expect from Kepler's third law?
4. Does the eccentricity of the orbit affect your answers to the last three questions?

Exploring the Sky and Solar System

Of the many activities available on the ***Voyager: SkyGazer* CD-ROM** accompanying your book, use the following files to observe key phenomena covered in this chapter.

Go to the **File: Basics** folder for the following demonstrations.

1. Planet Paths 2. Planet Orrery 3. Follow a Planet

Go to the **File: Demo** folder for the following demonstrations.

1. Earth and Venus 2. Hyakutake at Perihelion
3. Pluto's Orbit

Go to the **Explore** menu for the following demonstrations.

1. Solar System 2. Paths of the Planets

Movies

Check out the following narrated and animated short documentary available on **www.masteringastronomy.com** for a helpful review of key ideas covered in this chapter.

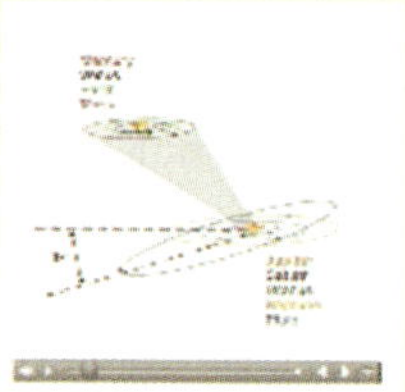

Orbits in the Solar System

Web Projects

1. *Space Station.* Visit a NASA site with pictures from the Space Station. Choose two photos that illustrate some facet of Newton's laws of motion or gravity. Explain how what is going on is related to Newton's laws.
2. *Energy Comparisons.* Using information from the Energy Information Administration Web site, choose some aspect of U.S. or world energy use that interests you. Write a short report on this issue.
3. *Tide Tables.* Find a tide table or tide chart for a beach town that you'd like to visit. Explain how to read the table and discuss any differences between the actual tidal pattern and the idealized tidal pattern described in this chapter.
4. *Space Elevator.* Read more about space elevators (see Problem 48) and how they might make it easier and cheaper to get to Earth orbit or beyond. Write a short report about the feasibility of building a space elevator, and briefly discuss the pros and cons of such a project.

5

Light and Matter

Reading Messages from the Cosmos

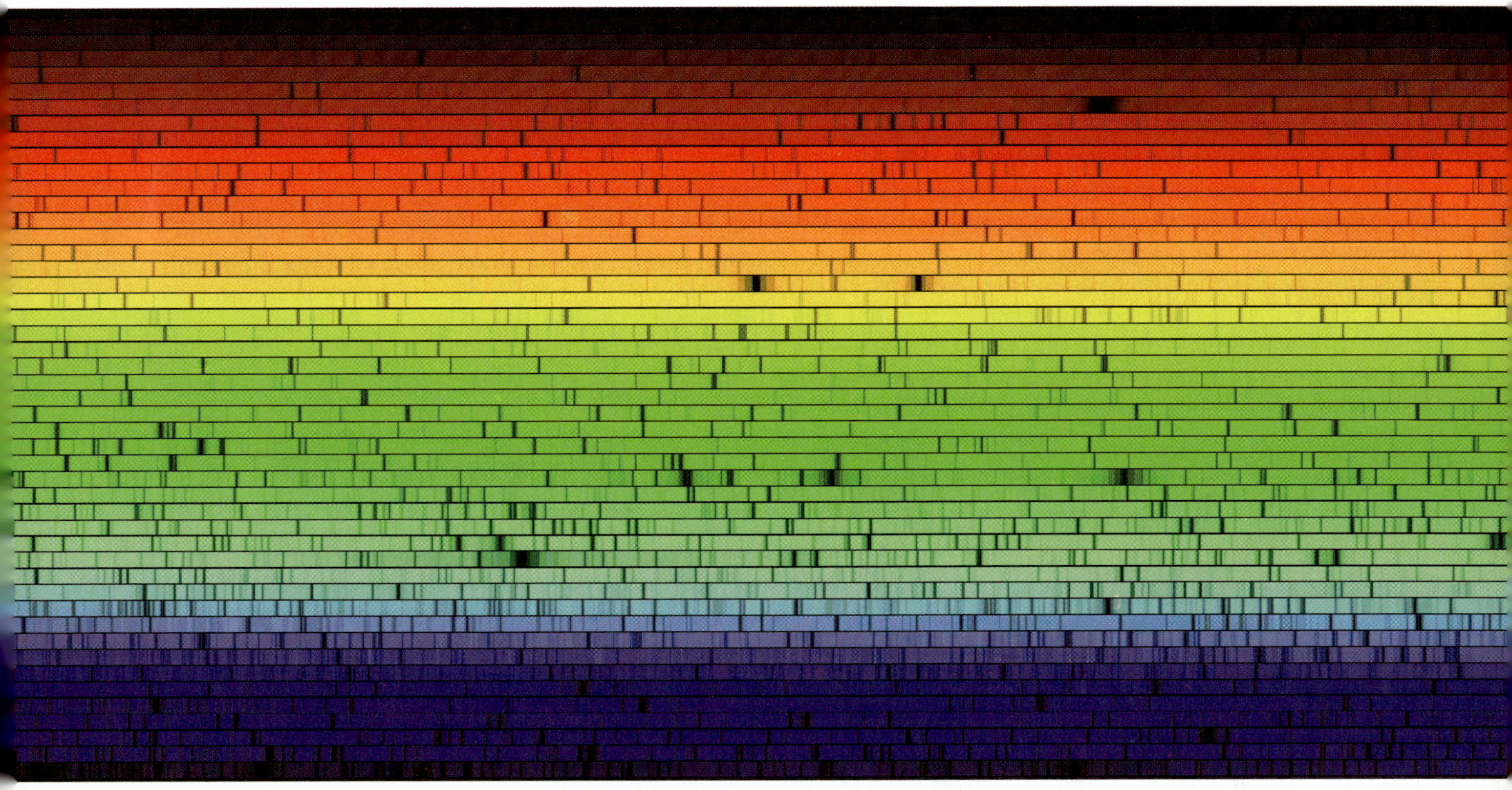

LEARNING GOALS

5.1 Light in Everyday Life
- How do we experience light?
- How do light and matter interact?

5.2 Properties of Light
- What is light?
- What is the electromagnetic spectrum?

5.3 Properties of Matter
- What is the structure of matter?
- What are the phases of matter?
- How is energy stored in atoms?

5.4 Learning from Light
- What are the three basic types of spectra?
- How does light tell us what things are made of?
- How does light tell us the temperatures of planets and stars?
- How do we interpret an actual spectrum?

5.5 The Doppler Shift
- How does light tell us the speed of a distant object?
- How does light tell us the rotation rate of a distant object?

May the warp be the white light of morning,
May the weft be the red light of evening,
May the fringes be the falling rain,
May the border be the standing rainbow.
Thus weave for us a garment of brightness.

Song of the Sky Loom (Native American)

Ancient observers could discern only the most basic features of the light that they saw—features such as color and brightness. Over the past several hundred years, we have discovered that light carries far more information. Today, we can analyze the light of distant objects to learn what they are made of, how hot they are, how fast they are moving, and much more.

We are lucky that light carries so much information. Our present spacecraft can reach only objects within our own solar system, and except for an occasional meteorite falling from the sky, the cosmos does not come to us. In contrast to the limited reach of spacecraft, light travels throughout the universe, carrying its treasury of information wherever it goes. Light is truly the cosmic messenger, bringing the stories of distant objects to our home here on Earth.

In this chapter, we will focus our attention on learning how to read the messages carried to us by light from the cosmos. We'll begin with a brief look at the basic interactions of light and matter that create those messages, and then study the properties of light and matter individually and in some detail. With that background, we'll be ready to explore the formation of light spectra, so that we can understand how light can encode so much information about distant objects.

Light and Spectroscopy Tutorial, Lesson 1

5.1 Light in Everyday Life

Look around you right now. What do you see? You may be tempted to list nearby objects, but all you're really seeing is light that has interacted with those objects. Through intuition and experience, you're able to interpret the colors and patterns of the light and turn them into information about the objects and substances that surround you.

Astronomers do much the same thing when studying the universe with telescopes. Telescopes collect the light of distant objects, and astronomers then extract information from the light. Knowing how to extract the maximum amount of information from light requires a deeper understanding of what light is and how it interacts with matter. We'll start developing that understanding by taking a closer look at our everyday experience with light.

• How do we experience light?

Even without opening your eyes, you can deduce that light is a form of energy. Outside on a hot, sunny day you can feel your skin warm as it absorbs sunlight. If you remember that greater warmth means more molecular motion, you'll realize that sunlight must be transferring its energy to the molecules in your skin. The energy that light carries is called *radiative energy*; recall that it is one of the three basic categories of energy, along with kinetic and potential energy (see Figure 4.10). But how do we measure the amount of energy that light carries?

Energy and Power We measure energy in units of joules [Section 4.3]. However, in astronomy we are often more interested in the *rate* at which light carries energy toward or away from us than in the total amount of energy it carries. After all, because light always travels through space at the speed of light, we cannot hold light in our hands in the same way that we can hold a hot potato, which has thermal energy, or a rock, which has gravitational potential energy. In science, the rate of energy flow is called **power**. The standard units of power are called **watts**. A power of 1 watt means an energy flow of 1 joule per second:

$$1 \text{ watt} = 1 \text{ joule/s}$$

Thus, for every second that you leave a 100-watt light bulb turned on, you will have to pay the utility company for 100 joules of energy. Interestingly, the power requirement of an average human—about 10 million joules per day—is about the same as that of a 100-watt light bulb.

Light and Color Everyday experience tells us that light comes in different forms that we call *colors*. You've probably seen a prism split light into the rainbow of light called a **spectrum** (Figure 5.1). The basic colors in a rainbow-like spectrum are red, orange, yellow, green, blue, and violet. We see *white* when the basic colors are mixed in roughly equal proportions. Light from the Sun or a light bulb is often called *white light*, because it contains all the colors of

Figure 5.1 When we pass white light through a prism, it disperses into a rainbow of color that we call a *spectrum*.

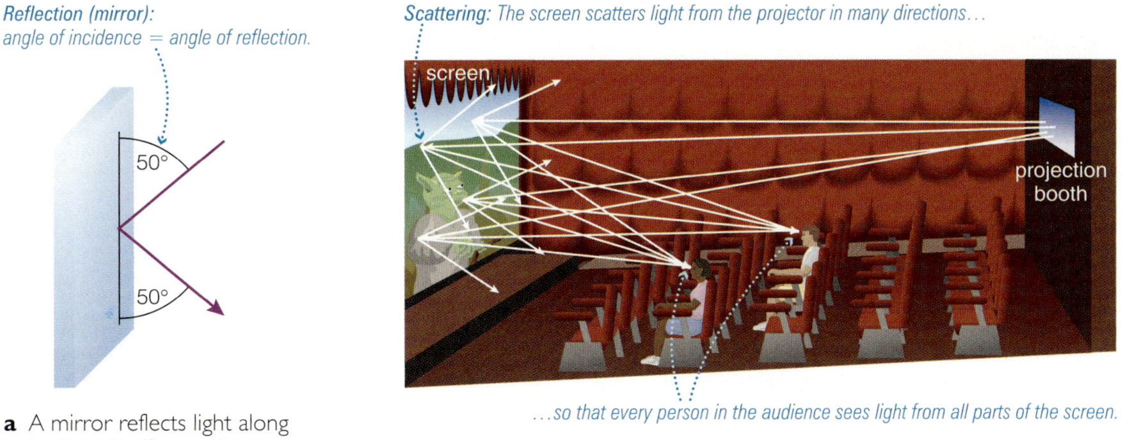

a A mirror reflects light along a simple path: The angle at which the light strikes the mirror is the same angle at which it is reflected.

b A movie screen scatters light in many different directions, so that each member of the audience can watch the movie. The pages in a book do the same thing, which is why you can read them from different angles and distances.

Figure 5.2 Reflection and scattering.

the rainbow. *Black* is what we perceive when there is no light and hence no color.

The wide variety of all possible colors comes from mixtures of just a few colors in varying proportions. Your television takes advantage of this fact to simulate a huge range of colors by combining only red, green, and blue light; these three colors are often called the primary colors of vision, because they are the colors directly detected by cells in your eyes. Colors tend to look different on paper, so artists generally work with an alternate set of primary colors: red, yellow, and blue. If you do any graphic design work, you may be familiar with the "CMYK" process that mixes the four colors cyan, magenta, yellow, and black to produce a great variety of color; the four-color CMYK process was used to print this book.

THINK ABOUT IT

If you have a magnifying glass handy, hold it close to your TV set to see the individual red, blue, and green dots. If you don't have a magnifying glass, try splashing a few droplets of water onto your TV screen (carefully!). What do you see? What are the drops of water doing?

You can produce a spectrum with either a prism or a **diffraction grating**, which is a piece of plastic or glass etched with many closely spaced lines. If you have a compact disc or DVD handy, you can make a spectrum for yourself. The bottom of a CD or DVD is etched with many closely spaced circles and therefore acts like diffraction grating. That's why you see rainbows of color on the bottom of the disc when you hold it up to the light.

• How do light and matter interact?

If you think about the interactions between light and matter that you see in everyday life, you'll realize that light can interact with matter in four basic ways:

- **Emission**: When you turn on a lamp, electric current heats the filament of the light bulb to a point at which it *emits* visible light.
- **Absorption**: If you place your hand near a lit light bulb, your hand *absorbs* some of the light, and this absorbed energy warms your hand.
- **Transmission**: Some forms of matter, such as glass or air, *transmit* light. That is, they allow light to pass through.
- **Reflection/scattering**: Light can bounce off matter, leading to what we call *reflection* (when the bouncing is all in the same general direction) or *scattering* (when the bouncing is more random). Figure 5.2 contrasts reflection and scattering.

Materials that transmit light are said to be *transparent*, and materials that absorb light are called *opaque*. Many materials are neither perfectly transparent nor perfectly opaque. For example, dark sunglasses and clear eyeglasses are both at least partially transparent, but the dark glasses absorb more and transmit less light. In addition, many materials affect different colors of light differently. For example, red glass transmits red light but absorbs other colors, while a green lawn reflects (scatters) green light but absorbs all other colors.

Let's put these ideas together to understand what happens when you walk into a room and turn on the light switch (Figure 5.3). The light bulb begins to emit white light, which is a mix of all the colors in the spectrum. Some of this light exits the room, transmitted through the windows. The rest of the light strikes the surfaces of objects inside the room, and the material properties of each object determine the colors it absorbs or reflects. The light coming from each object therefore carries an enormous amount of information about the object's location, shape and structure, and composition. You acquire this information when light enters your eyes, where special cells in your retina absorb it and send signals to your brain. Your brain interprets the messages that light carries, recognizing materials and objects in the process we call vision.

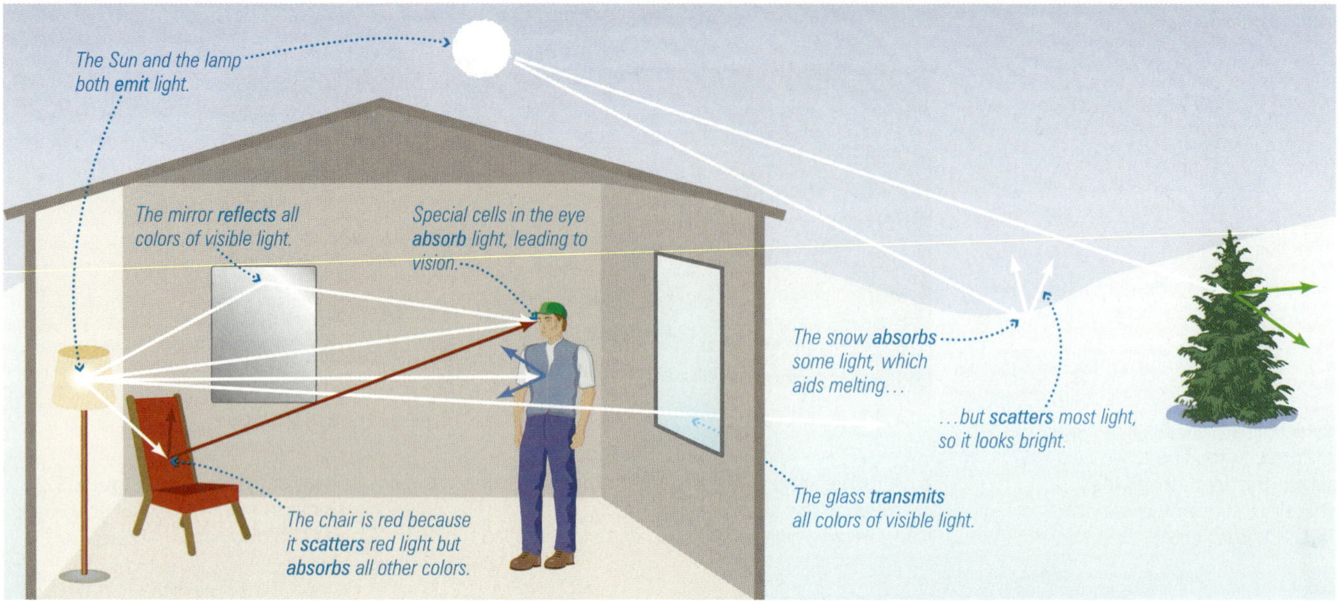

Figure 5.3 When light strikes any piece of matter in the universe, that matter reacts in one or a combination of four ways: emission, absorption, transmission, and reflection (or scattering).

All the information that light brings to Earth from the universe was encoded by one of the four basic interactions between light and matter common to our everyday experience. However, our eyes perceive only a tiny fraction of all the information contained in light. Modern instruments can break light into a much wider variety of colors and can analyze those colors in far greater detail. In order to understand how to decode that information, we need to examine the nature of light and matter more closely.

5.2 Properties of Light

Light has always been familiar to people, but its actual nature remained a mystery for most of human history. The first real insights into the nature of light came with experiments that Isaac Newton performed in the 1660s. It was already well known that passing light through a prism produced a rainbow of colors, but most people assumed that the colors were a property of the prism rather than of the light itself. Newton dispelled this belief by placing a second prism in front of the light of just one color, such as red, from the first prism. If the rainbow of color came from the prism itself, the second prism would have produced a rainbow just like the first. But it did not: When only red light entered the second prism, only red light emerged, proving that the color was a property of the light and not of the prism.

Newton's work tells us something about the nature of color, but it still does not tell us exactly what light *is*. In this section, we'll investigate modern understanding of the nature of light.

• What is light?

Newton himself guessed that light is made up of countless tiny particles. However, other scientists soon conducted experiments that demonstrated that light behaves like waves. Thus began one of the most important debates in scientific history: Is light a wave or a particle? To understand this question, and our modern answer to it, we must first understand the differences between particles and waves.

Particles and Waves in Everyday Life Marbles, baseballs, and individual atoms and molecules are all examples of *particles.* A particle of matter can sit still or it can move from one place to another. If you throw a baseball at a wall, it moves from your hand to the wall.

Now think about what happens when you toss a pebble into a pond (Figure 5.4). The ripples moving out from the place where the pebble lands are *waves* consisting of *peaks*, where the water is higher than average, and *troughs*, where the water is lower than average. If you watch as the waves pass by a floating leaf, you'll see the leaf rise up with the peak and drop down with the trough, but the leaf itself does *not* travel across the pond's surface with the wave. This observation tells us that the particles (molecules) of water are moving up and down but are not flowing outward. That is, the waves carry *energy* outward from the place where the pebble landed but do not carry matter along with them. In essence, a particle is a *thing*, while a wave is a *pattern* revealed by its interaction with particles.

For our purposes in this book, we can characterize the waves on the pond by three basic properties: wavelength, frequency, and speed.* Their **wavelength** is the distance from one peak to the next (or one trough to the next). Their **frequency** is the number of peaks passing by any point each second. For example, if we see the leaf bobbing up and down three times each second, we know that three peaks

*A fourth characteristic of a wave is its amplitude, defined as half the height from trough to peak. Amplitude is related to the brightness of light, but we will not use it in our study of light in this book.

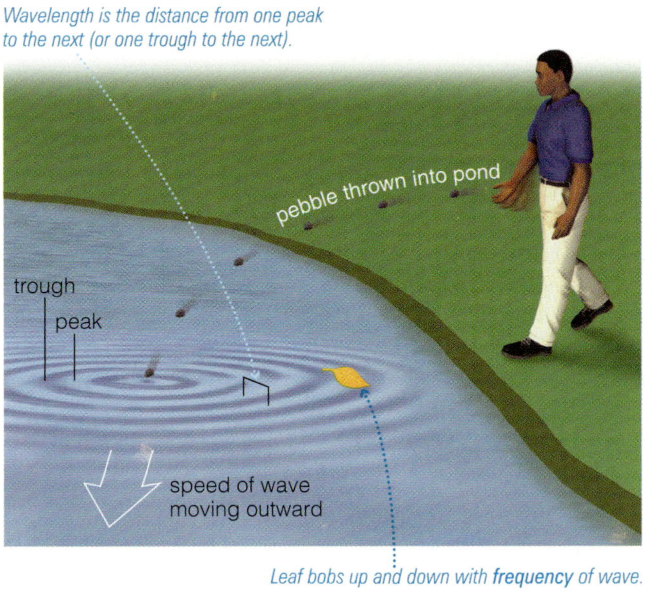

Figure 5.4 Interactive Figure Tossing a pebble into a pond generates waves. The waves carry energy outward, but matter does *not* travel outward. Matter, such as a floating leaf or the water itself, bobs up and down and sloshes back and forth as the waves pass by.

must be passing by the leaf each second. We then say that the waves have a frequency of three **cycles per second** (referring to the up-and-down "cycles" of the passing waves). Cycles per second often are called **hertz (Hz)**, so we can also describe this frequency as 3 Hz. The **speed** of the waves tells us how fast their peaks travel across the pond. Because the waves carry energy, the speed essentially tells us how fast the energy travels from one place to another.

A simple formula relates the wavelength, frequency, and speed of any wave. Suppose a wave has a wavelength of 1 centimeter and a frequency of 3 hertz. The wavelength tells us that each time a peak passes by, the wave peak has traveled 1 centimeter. The frequency tells us that three peaks pass by each second. Thus, the speed of the wave must be 3 centimeters per second. If you try a few more similar examples, you'll find the general rule:

$$\text{wavelength} \times \text{frequency} = \text{speed}$$

Light as an Electromagnetic Wave You've probably heard that light is a wave, but it isn't quite like the waves we see in everyday life. More familiar waves always move through some form of matter. For example, the waves on the pond move through the water, causing particles (molecules) of water to vibrate up and down and slosh back and forth, while sound waves move through air and cause air molecules to vibrate back and forth. The vibrations of matter allow the waves to transmit energy from one place to another, even though particles of matter do not travel along with the waves. In contrast to these everyday examples of waves, we do not see anything move up and down when light travels through space. So what, exactly, is "waving" when a light wave passes by?

The answer is what scientists call electric and magnetic fields. The concept of a **field** is a bit abstract. Fields associated with forces, such as electric and magnetic fields, describe the strength of the force that any particle would experience at any point in space. For example, recall that the force Earth exerts on an orbiting satellite depends on the mass of Earth, the mass of the satellite, and the distance of the satellite from Earth's center [Section 4.4]. We can therefore characterize the force that *any* object would experience at the satellite's distance by saying that Earth creates a *gravitational field* equal to the force on the satellite divided by the satellite's mass. In other words, Earth's gravitational field at any point in space depends only on Earth's mass and the distance of that point from Earth's center. Likewise, we can describe the forces that charged particles exert on one another in terms of *electric fields* and *magnetic fields.*

Light waves are vibrations of both electric and magnetic fields caused by the motions of charged particles. We therefore say that light is an **electromagnetic wave**. Just as the ripples on a pond will cause a leaf to bob up and down, the vibrations of the electric field in an electromagnetic wave will cause an electron to bob up and down. If you could set up electrons in a row, they would wriggle like a snake as light passed by (Figure 5.5a). The distance between peaks in this row of electrons would tell us the wavelength of the light wave, while the number of times each electron bobs up and down would tell us the frequency (Figure 5.5b).

All light travels through empty space at the same speed—the **speed of light** (often represented by the letter *c*), which is about 300,000 kilometers per second. Because the speed of any wave is its wavelength times its frequency, we find a very important relationship between wavelength and frequency for light: *The longer the wavelength, the lower the frequency, and vice versa.* For example, light waves with a wavelength of 1 centimeter must have half the frequency

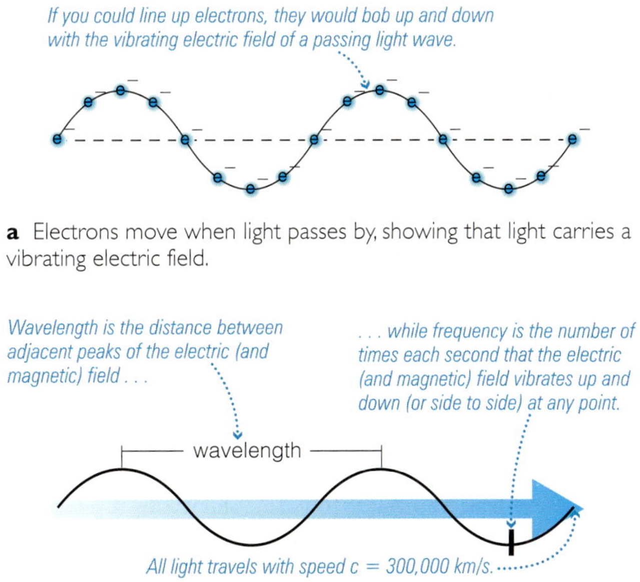

a Electrons move when light passes by, showing that light carries a vibrating electric field.

b The vibrations of the electric field determine the wavelength and frequency of a light wave. Light also carries a magnetic field (not shown) that vibrates perpendicular to the direction of the electric field vibrations.

Figure 5.5 Interactive Figure Light is an electromagnetic wave.

SPECIAL TOPIC What Do Polarized Sunglasses Have to Do with Astronomy?

If you go to the store to buy a pair of sunglasses, you'll face a dizzying array of choices. Sunglasses come in different styles, different tints, and with different efficiencies in blocking ultraviolet and infrared light. Most of these choices should at least make sense to you (well, perhaps not all of the styles), but one option may be less familiar: The labels on some sunglasses say that they are "polarized." What does this mean? The term comes from a property of light, called *polarization*, that has to do with the direction in which a light wave vibrates and how those vibrations change when light bounces off or passes through matter. Polarization turns out to be important not only to your sunglasses but also to astronomy.

It's easiest to understand the idea of polarization if you think about how waves move on a string when you shake one end of it. The string vibrates either up and down or back and forth while the wave itself moves along the string, perpendicular to the direction of vibration. Light waves move in a similar way, with the electric and magnetic fields vibrating either up and down or side to side compared with the direction of travel. For example, the wave shown in Figure 5.5b is moving to the right on the page while its electric field vibrates up and down on the page.

The direction of vibration affects the way light interacts with matter. As Figure 5.5a indicates, an electric field that vibrates up and down will make electrons move up and down as the wave passes by. Thus, the direction in which the electric field vibrates determines the direction in which charged particles will vibrate as the wave passes by. Because the direction of wave vibration matters, we give it a name: the *polarization* of the wave. In other words, an individual wave moving toward you can be polarized either with its vibrations up and down or side to side, or some direction in between those two. (In all cases, the polarization is perpendicular to the direction of travel.)

We rarely notice polarization in everyday life because our eyes are not equipped to detect this property of light. Each of the light waves entering our eyes (or, more technically, each individual photon) has a certain direction of polarization. If all the waves taken together have no preferential direction of vibration, we say that the light is *unpolarized*. However, some physical processes can produce waves with a particular direction of polarization—which is where your sunglasses and astronomy come in.

When light reflects off a flat, horizontal surface like the ground or a lake, the surface tends to reflect light in which the electric field is vibrating along (parallel to) the surface and to transmit or absorb light with electric fields vibrating in other directions. As a result, the reflected light becomes polarized, with its electric field vibrating horizontally. Polarized sunglasses are designed to block out horizontally polarized light, which means they block the light reflected from horizontal surfaces—which is otherwise known as "glare." Of course, the polarized glasses work only if you are wearing them horizontally as well; if you turn a pair of polarized sunglasses so that the two lenses are no longer horizontal to the ground, they will not block glare effectively.

In astronomy, we aren't worried about glare from distant objects, but if we learn that a light source is producing polarized light, it tells us something about the nature of the source. For example, light that passes through clouds of interstellar dust tends to be polarized, telling us that the dust grains in the cloud must be preferentially absorbing light with electric fields vibrating in a particular direction. More detailed analysis has taught us that the polarization arises because the microscopic dust grains have an elongated shape and all tend to be aligned in the same way due to magnetic fields within the clouds. Polarization arises in many other astronomical contexts as well, including the study of the leftover radiation from the Big Bang. However, while polarization has provided important insights into many astronomical processes, its analysis can be fairly technical, and we will not discuss it further in this book.

---

of light waves with a wavelength of $\frac{1}{2}$ centimeter, and one-fourth the frequency of light waves with a wavelength of $\frac{1}{4}$ centimeter (Figure 5.6).

Photons: "Particles" of Light In our everyday lives, waves and particles appear distinctly different: No one would confuse the ripples on a pond with a baseball. However, experiments have shown that light behaves as *both* a wave and a particle. We therefore say that light comes in individual "pieces," called **photons**, that have properties of both particles and waves. Like baseballs, photons of light can be counted individually and can hit a wall one at a time. Like waves, each photon is characterized by a wavelength and a frequency. The idea that light can be both a wave and a particle may seem quite strange, but it is fundamental to our modern understanding of physics. (We will discuss some of the implications of this "wave-particle duality" in Chapter S4.)

Just as a moving baseball carries a specific amount of kinetic energy, each photon of light carries a specific amount of radiative energy. The shorter the wavelength of the light (or, equivalently, the higher its frequency), the higher the energy of the photons.

To sum up, our modern understanding maintains that light is both a particle and a wave, an idea we describe by saying that light consists of individual photons character-

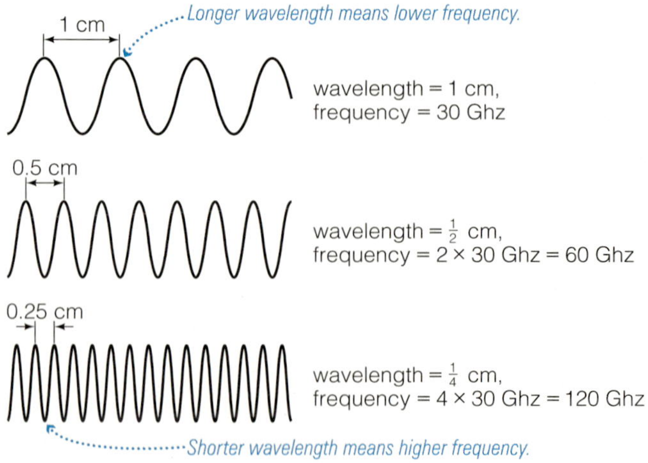

Figure 5.6 Because all light travels through space at the same speed, light of longer wavelength must have lower frequency, and vice versa. (GHz stands for gigahertz, or 10^9 hertz.)

ized by wavelength, frequency, and energy. The wavelength, frequency, and energy are related in a very specific way so that all photons of light travel through space at the same speed—the speed of light.

THINK ABOUT IT

If you assume that each of the three waves shown in Figure 5.6 represents a photon of light, which one has the most energy? Which one has the least energy? Explain.

• What is the electromagnetic spectrum?

Newton's experiments proved that white light is actually a mix of all the colors in the rainbow. Later scientists found that there is light "beyond the rainbow" as well. Just as there are sounds that our ears cannot hear (such as the sound of a dog whistle), there is light that our eyes cannot see. In fact, the spectrum of **visible light** that splits into the rainbow of color is only a tiny part of the complete range of light's wavelengths. Visible light differs from other forms of light only in the wavelength and frequency of the photons.

Because we sometimes describe light as an electromagnetic wave, the complete spectrum of light is usually called the **electromagnetic spectrum**. Light itself is often called **electromagnetic radiation**. Figure 5.7 shows the complete electromagnetic spectrum.

Notice that we give names to various portions of the electromagnetic spectrum. Visible light has wavelengths ranging from about 400 nanometers at the blue or violet end of the rainbow to about 700 nanometers at the red end. (A nanometer [nm] is a billionth of a meter.) Light with wavelengths somewhat longer than red light is called **infrared**, because it lies beyond the red end of the rainbow. **Radio waves** are the longest-wavelength light. Thus, radio waves are a form of light, *not* a form of sound. The region near the border between infrared and radio waves, where

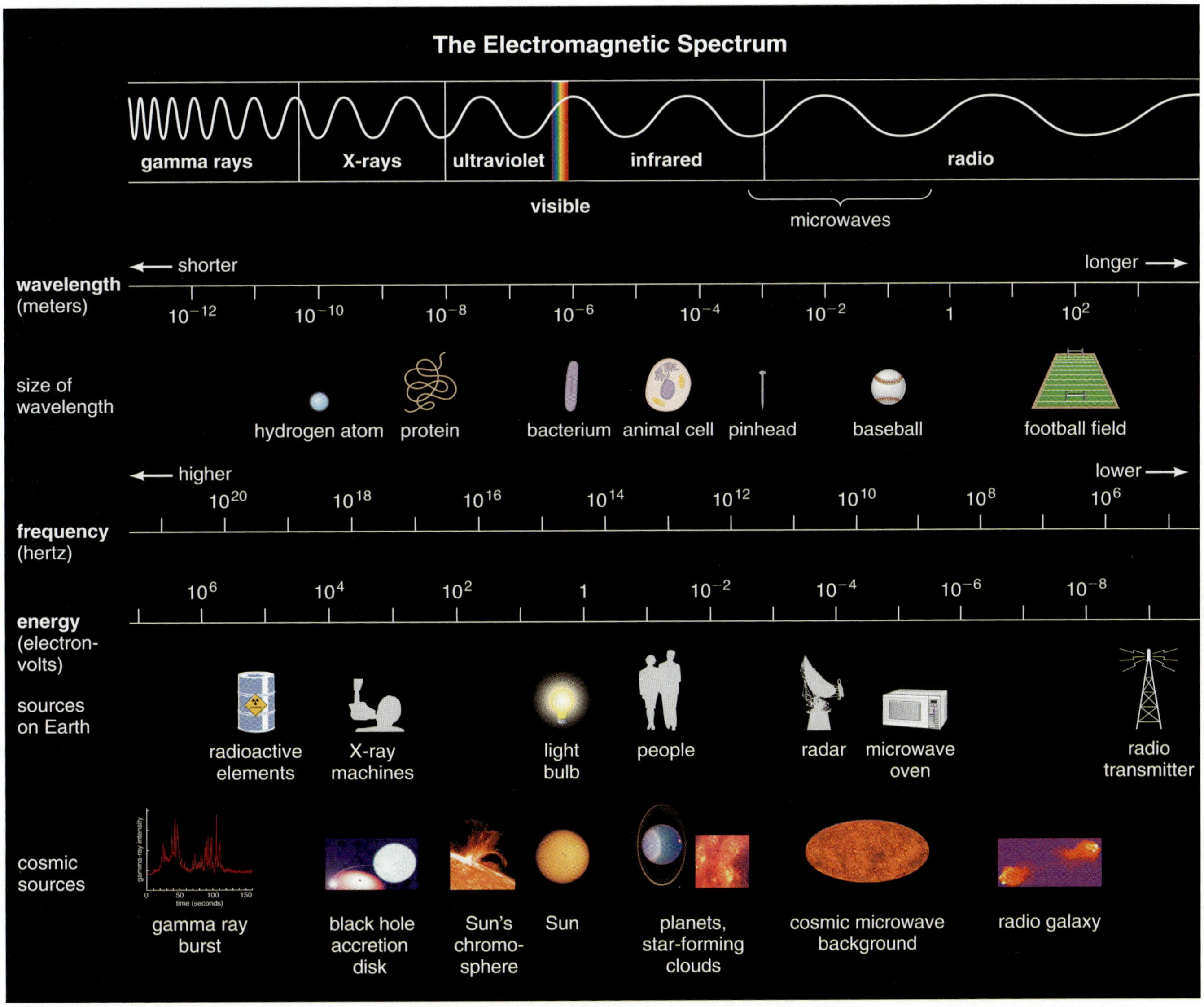

Figure 5.7 The electromagnetic spectrum. Notice that wavelength increases as we go from gamma rays to radio waves, while frequency and energy increase in the opposite direction.

wavelengths range from micrometers to millimeters, is sometimes given the name **microwaves.**

On the other side of the spectrum, light with wavelengths somewhat shorter than blue light is called **ultraviolet**, because it lies beyond the blue (or violet) end of the rainbow. Light with even shorter wavelengths is called **X rays**, and the shortest-wavelength light is called **gamma rays.** Notice that visible light is an extremely small part of the entire electromagnetic spectrum: The reddest red that our eyes can see has only about twice the wavelength of the bluest blue, but the radio waves from your favorite radio station are a billion times longer than the X rays used in a doctor's office.

MATHEMATICAL INSIGHT 5.1 Wavelength, Frequency, and Energy

The speed of any wave is the product of its wavelength and its frequency. Because all forms of light travel at the same speed (in a vacuum) of $c = 3 \times 10^8$ m/s, we can write:

$$\lambda \times f = c$$

where λ (the Greek letter *lambda*) stands for wavelength and f stands for frequency. This formula is simple but revealing: Because the speed c is constant, the formula tells us that frequency must go up when wavelength goes down, and vice versa.

The formula for the radiative energy (E) carried by a photon of light is:

$$E = h \times f$$

where h is *Planck's constant* ($h = 6.626 \times 10^{-34}$ joule $\times$ s). Thus, energy increases in proportion to the frequency of the photon.

Example 1: The numbers on a radio dial for FM radio stations are the frequencies of radio waves in megahertz (MHz), or millions of hertz. If your favorite radio station is "93.3 on your dial," it broadcasts radio waves with a frequency of 93.3 million cycles per second. What is the wavelength of these radio waves?

Solution:

Step 1 Understand: Radio waves are a form of light, so we know they obey the relationship

$$\lambda \times f = c$$

The speed of light is always the same and, in this case, we are given the frequency ($f = 93.3$ MHz). Thus we simply solve for the wavelength (λ).

Step 2 Solve: We solve the above equation for wavelength by dividing both sides by the frequency (f), which gives us:

$$\lambda = \frac{c}{f}$$

Now we plug in the speed of light (3×10^8 m/s) and the frequency. Remember that 1 MHz $= 10^6$ Hz, so the frequency in this case is 93.3×10^6 Hz. Also remember that hertz (Hz) is an abbreviation for "cycles per second"; the "cycles" have no units, so the actual units of hertz are simply "per second," equivalent to 1/s. Thus, we find:

$$\lambda = \frac{c}{f} = \frac{3 \times 10^8 \frac{\text{m}}{\text{s}}}{93.3 \times 10^6 \frac{1}{\text{s}}} = 3.2 \text{ m}$$

Step 3 Explain: We have found that radio waves with a frequency of 93.3 MHz have a wavelength of 3.2 meters. That is why radio towers are so large—they need to be longer than the size of the waves they are transmitting.

Example 2: The wavelength of green visible light (in the middle of the visible spectrum) is about 550 nanometers (1 nm $= 10^{-9}$ m). What is the frequency of this light?

Solution:

Step 1 Understand: Again, we know that visible light obeys the relation $\lambda \times f = c$. This time we are given the wavelength ($\lambda = 550$ nm $= 550 \times 10^{-9}$ m), so we need to solve the equation for the frequency.

Step 2 Solve: We solve the equation ($\lambda \times f = c$) for frequency by dividing both sides by the wavelength (λ), which gives us:

$$f = \frac{c}{\lambda}$$

Now we plug in the speed of light (3×10^8 m/s) and the wavelength ($\lambda = 550 \times 10^{-9}$ m) to find:

$$f = \frac{c}{\lambda} = \frac{3 \times 10^8 \frac{\text{m}}{\text{s}}}{550 \times 10^{-9} \text{ m}} = 5.45 \times 10^{14} \frac{1}{\text{s}}$$

Step 3 Explain: Our answer is about 5.5×10^{14} 1/s; remembering that the units 1/s are called hertz, this answer is equivalent to 5.5×10^{14} Hz, or 550 trillion Hz. Thus, green visible light has a frequency of about 550 trillion Hz. Notice that this frequency is extremely high, which explains in part why it took so long for humans to realize that light behaves like a wave.

Example 3: What is the energy of a visible-light photon with wavelength 550 nanometers?

Solution:

Step 1 Understand: The energy of a photon is given by the formula $E = h \times f$. From Example 2, we also know that $f = c/\lambda$. Thus, we can write the energy formula as:

$$E = h \times f = h \times \frac{c}{\lambda}$$

Step 2 Solve: We plug in the wavelength ($\lambda = 550 \times 10^{-9}$ m) and Planck's constant ($h = 6.626 \times 10^{-34}$ joule $\times$ s) to find:

$$E = h \times \frac{c}{\lambda} = (6.626 \times 10^{-34} \text{ joule} \times \text{s}) \times \frac{3 \times 10^8 \frac{\text{m}}{\text{s}}}{550 \times 10^{-9} \text{ m}}$$
$$= 3.6 \times 10^{-19} \text{ joule}$$

Step 3 Explain: The energy of a single visible-light photon is about 3.6×10^{-19} joule. Note that this energy for a single photon is extremely small compared to, say, the energy of 100 joules used each second by a 100-watt light bulb.

COMMON MISCONCEPTIONS

Can You Hear Radio Waves?

Most people associate the term *radio* with sound, but radio waves are a form of *light* with long wavelengths—too long for our eyes to see. Radio stations encode sounds (such as voices and music) as electrical signals and broadcast the signals as radio waves. What we call "a radio" in daily life is an electronic device that receives these radio waves and decodes them to re-create the sounds played at the radio station. Televisions, cell phones, and other wireless devices also work by encoding and decoding information in the form of light called radio waves. Moreover, there's nothing special about encoding sounds with radio waves—any form of light could work in principle. Indeed, fiber-optic cables transmit sounds and data by encoding them into infrared or visible light. Thus, we cannot actually hear radio waves; we hear only the decoded signals that they carry.

The different energies of different forms of light explain many familiar effects in everyday life. Radio waves carry so little energy that they have no noticeable effect on our bodies. However, radio waves can make electrons move up and down in an antenna, which is how the antenna of your car radio receives the radio waves coming from a radio station. Molecules moving around in a warm object emit infrared light, which is why we sometimes associate infrared light with heat. Receptors in our eyes respond to visible-light photons, making vision possible. Ultraviolet photons carry enough energy to harm cells in our skin, causing sunburn or skin cancer. X-ray photons have enough energy to penetrate through skin and muscle but can be blocked by bones or teeth. That is why doctors and dentists can see our underlying bone structures on photographs taken with X-ray light.

Just as different colors of visible light may be absorbed or reflected differently by the objects we see (see Figure 5.3), the various portions of the electromagnetic spectrum may interact in very different ways with matter. For example, a brick wall is opaque to visible light but transmits radio waves, which is why we can listen to radios inside buildings. Similarly, glass that is transparent to visible light may be opaque to ultraviolet light. In general, certain types of matter tend to interact more strongly with certain types of light, so each type of light carries different information about distant objects in the universe. Astronomers therefore seek to observe light of all wavelengths, using telescopes adapted to detecting each form of light, from radio to gamma rays [Section 6.4].

COMMON MISCONCEPTIONS

What Could You See with X-Ray Vision?

The idea of having "X-ray vision" that would allow you to see through walls is part of our Hollywood culture, and dates back to old Superman comics. But is it really possible? At first, the idea might seem reasonable; after all, you've probably seen a doctor or dentist holding up "X rays" that allow you to see into your bones or teeth. However, the "X rays" that the doctor or dentist holds up are not a form of light; they are just pieces of film. The images of your bones and teeth are made with the help of a special machine that works somewhat like the flash on an ordinary camera but emits X rays instead of visible light. This machine flashes the X rays at you, and a piece of photographic film (or an electronic detector) records the X rays that are transmitted through your body. Thus, what you see is the image left by these X rays; you cannot see the X rays themselves, because they have wavelengths far too short for human eyes to see. If you think about this process, you'll realize that X-ray vision would do you no good, even if it were possible. People, walls, and other ordinary objects do not emit any X rays of their own, so there'd be nothing to see with your X-ray vision.

5.3 Properties of Matter

Light carries information about matter across the universe, but we are usually more interested in the matter with which the light interacts than we are in the light itself. Planets, stars, and galaxies are made of matter, and we must understand the nature of matter if we are to decode the messages we receive in light.

• What is the structure of matter?

Like the nature of light, the nature of matter remained mysterious for most of human history. Nevertheless, ancient philosophers came up with some ideas that are still with us today.

The ancient Greek philosopher Democritus (c. 470–380 B.C.) wondered what would happen if we broke a piece of matter, such as a rock, into ever smaller pieces. Democritus believed that the rock would eventually break into particles so small that nothing smaller could be possible. He called these particles *atoms*, a Greek term meaning "indivisible." Building upon the beliefs of earlier Greek philosophers, Democritus thought that all materials were composed of just four basic *elements:* fire, water, earth, and air. He proposed that the different properties of the elements could be explained by the physical characteristics of their atoms. For example, Democritus suggested that atoms of water were smooth and round, so water flowed and had no fixed shape, while burns were painful because atoms of fire were thorny. He also imagined atoms of earth to be rough and jagged, so that they could fit together like pieces of a three-dimensional jigsaw puzzle, and he used this idea to suggest that the universe began as a chaotic mix of atoms that slowly clumped together to form Earth.

COMMON MISCONCEPTIONS

Is Radiation Dangerous?

Many people associate the word *radiation* with danger. However, the word *radiate* simply means "to spread out from a center" (note the similarity between *radiation* and *radius* [of a circle]). Radiation is energy being carried through space. If energy is being carried by particles of matter, such as protons or neutrons, we call it *particle radiation*. If energy is being carried by light, we call it *electromagnetic radiation*.

High-energy forms of radiation, such as particles from radioactive materials or X rays, are dangerous because they can penetrate body tissues and cause cell damage. Low-energy forms of radiation, such as radio waves, are usually harmless. The visible-light radiation from the Sun is necessary to life on Earth. Thus, while some forms of radiation are dangerous, others are harmless or beneficial.

Although Democritus was wrong about there being only four types of atoms and about their specific properties, he was on the right track. All ordinary matter is indeed composed of **atoms**, and the properties of ordinary matter depend on the physical characteristics of their atoms. However, by modern definition, atoms are *not* indivisible because they are composed of even smaller particles.

Atoms come in different types, and each type corresponds to a different chemical **element**. Today, we have identified more than 100 chemical elements, and fire, water, earth, and air are *not* among them. Some of the most familiar chemical elements are hydrogen, helium, carbon, oxygen, silicon, iron, gold, silver, lead, and uranium. Appendix D gives the periodic table of all the elements.

Atomic Structure Each chemical element consists of a different type of atom, and atoms are in turn made of particles that we call **protons, neutrons**, and **electrons** (Figure 5.8). Protons and neutrons are found in the tiny **nucleus** at the center of the atom. The rest of the atom's volume contains electrons, which surround the nucleus. Although the nucleus is very small compared to the atom as a whole, it contains most of the atom's mass, because protons and neutrons are each about 2,000 times as massive as an electron. Note that atoms are incredibly small: Millions could fit end to end across the period at the end of this sentence. The number of atoms in a single drop of water (typically, 10^{22} to 10^{23} atoms) may exceed the number of stars in the observable universe.

The properties of an atom depend mainly on the **electrical charge** in its nucleus; an object's electric charge is a measure of how strongly it will interact with electromagnetic fields. We define the electrical charge of a proton as the basic unit of positive charge, which we write as $+1$. The electron

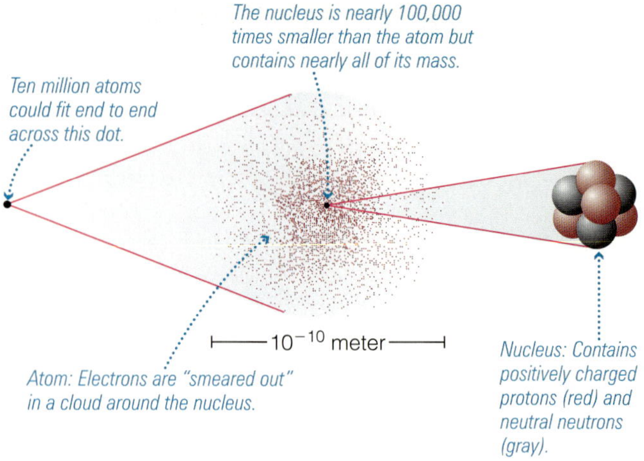

Figure 5.8 The structure of a typical atom. Notice that atoms are extremely tiny: The atom shown in the middle is magnified to about 1 billion times its actual size, and the nucleus on the right is magnified to about 100 trillion times its actual size.

has an electrical charge that is precisely opposite that of a proton, so we say it has negative charge (-1). Neutrons are electrically neutral, meaning that they have no charge.

Oppositely charged particles attract one another, and similarly charged particles repel. An atom is held together by the attraction between the positively charged protons in its nucleus and the negatively charged electrons that surround the nucleus. Ordinary atoms have identical numbers of electrons and protons, making them electrically neutral overall.*

Although we can think of electrons as tiny particles, they are not quite like tiny grains of sand and they don't orbit the nucleus the way planets orbit the Sun. Instead, the electrons in an atom form a kind of "smeared out" cloud that surrounds the nucleus and gives the atom its apparent size. The electrons aren't really cloudy, but it is impossible to pinpoint their positions in the atom.

In Figure 5.8, you can see that the electrons give the atom a size far larger than its nucleus even though they represent only a tiny portion of the atom's mass. If you imagine an atom on a scale that makes its nucleus the size of your fist, its electron cloud would be many miles wide.

Atomic Terminology Astronomical objects contain many different types of atoms, and it's important to have a way to describe the atoms. You've probably learned the basic terminology of atoms in past science classes, but let's review it just to be sure.

Each different chemical element contains a different number of protons in its nucleus. This number is its **atomic number**. For example, a hydrogen nucleus contains just one proton, so its atomic number is 1. A helium nucleus contains two protons, so its atomic number is 2.

*You may wonder why electrical repulsion doesn't cause the positively charged protons in a nucleus to fly apart from one another. The answer is that an even stronger force, called the *strong force*, overcomes electrical repulsion and holds the nucleus together [Section S4.2].

COMMON MISCONCEPTIONS

The Illusion of Solidity

Bang your hand on a table. Although the table feels solid, it is made almost entirely of empty space! Nearly all the mass of the table is contained in the nuclei of its atoms. But the volume of an atom is more than a trillion times the volume of its nucleus, so relatively speaking, the nuclei of adjacent atoms are nowhere near to touching one another. The solidity of the table comes from a combination of electrical interactions between the charged particles in its atoms and the strange quantum laws governing the behavior of electrons. If we could somehow pack all the table's nuclei together, the table's mass would fit into a microscopic speck. Although we cannot pack matter together in this way, nature can and does—in *neutron stars*, which we will study in Chapter 18.

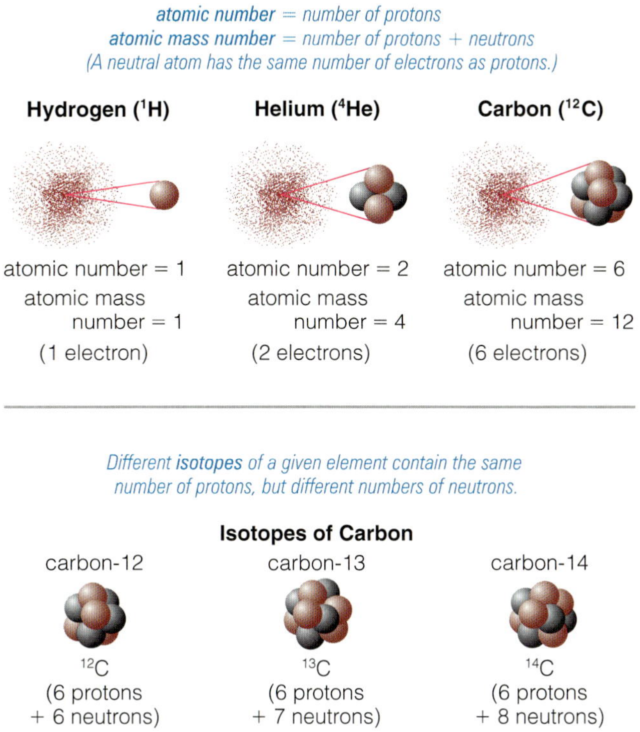

Figure 5.9 Terminology of atoms.

The *combined* number of protons and neutrons in an atom is called its **atomic mass number**. The atomic mass number of ordinary hydrogen is 1 because its nucleus is just a single proton. Helium usually has two neutrons in addition to its two protons, giving it an atomic mass number of 4. Carbon usually has six protons and six neutrons, giving it an atomic mass number of 12.

Every atom of a given element contains exactly the same number of protons, but the number of neutrons can vary. For example, all carbon atoms have six protons, but they may have six, seven, or eight neutrons. Versions of an element with different numbers of neutrons are called **isotopes** of the element. Isotopes are named by listing their element name and atomic mass number. For example, the most common isotope of carbon has 6 protons and 6 neutrons, giving it atomic mass number $6 + 6 = 12$, so we call it carbon-12. The other isotopes of carbon are carbon-13 (six protons and seven neutrons give it atomic mass number 13) and carbon-14 (six protons and eight neutrons give it atomic mass number 14). We can also write the atomic mass number of an isotope as a superscript to the left of the element symbol: ^{12}C, ^{13}C, ^{14}C. We read ^{12}C as "carbon-12." Figure 5.9 summarizes basic atomic terminology.

THINK ABOUT IT

The symbol ^{4}He represents helium with an atomic mass number of 4. ^{4}He is the most common form of helium, containing two protons and two neutrons. What does the symbol ^{3}He represent?

Molecules The number of different material substances is far greater than the number of chemical elements because atoms can combine to form **molecules**. Some molecules consist of two or more atoms of the same element. For example, we breathe O_2, oxygen molecules made of two oxygen atoms. Other molecules, such as water, are made up of atoms of two or more different elements. The symbol H_2O tells us that a water molecule contains two hydrogen atoms and one oxygen atom. Substances composed of molecules with two or more different types of atoms are called **compounds**. Thus, water is a compound.

The chemical properties of a molecule are different from those of its individual atoms. For example, molecular oxygen (O_2) behaves very differently from atomic oxygen (O), and water behaves very differently from pure hydrogen or pure oxygen.

• What are the phases of matter?

Interactions between light and matter also depend on the physical state of the matter. Everyday experience tells us that a substance can behave dramatically differently in different *phases*, even though it is still made of the same atoms or molecules. For example, molecules of H_2O can exist in three familiar phases: as **solid** ice, as **liquid** water, and as the **gas** we call water vapor. How can the same molecules look and act so differently in these different phases?

You are probably familiar with the idea of a **chemical bond**, the name we give to the interactions between electrons that hold the atoms in a molecule together. For example, we say that chemical bonds hold the hydrogen and oxygen atoms together in a molecule of H_2O. Similar but much weaker interactions among electrons hold together the many water molecules in a block of ice or a pool of water. Thus, we can think of the interactions that keep neighboring atoms or molecules close together as other types of bonds.

If we think in terms of bonds, the phases of solid, liquid, and gas differ in the strength of the bonds between neighboring atoms and molecules. Phase changes occur when one type of bond is broken and replaced by another. Changes in either pressure or temperature (or both) can cause phase changes, but it's easier to think first about temperature: As a substance is heated, the average kinetic energy of its particles increases, enabling the particles to break the bonds holding them to their neighbors.

Phase Changes in Water Water is the only familiar substance that we see in all three phases (solid, liquid, gas) in our everyday lives. Let's therefore take water as an example of what happens as we heat a substance and observe the phase changes.

At low temperatures (below 0°C at sea level on Earth), water molecules have a relatively low average kinetic energy. Each molecule is therefore bound tightly to its neighbors, making the *solid* structure of ice. As long as the temperature remains below freezing, the water molecules in ice remain rigidly held together. However, the molecules within this rigid structure are always vibrating, and higher temperature means greater vibrations. Thus, if we start with ice at a very low temperature, the molecular vibrations will grow gradually stronger as the temperature rises toward the melting point of 0°C.

The melting point (0°C at ordinary pressure) is the temperature at which the molecules finally have enough energy to break the solid bonds of ice. The molecules can then move much more freely among one another, allowing the water to flow as a *liquid.* However, the molecules in liquid water are not completely free of one another, as we can tell from the fact that droplets of water can stay intact. Thus, adjacent molecules in liquid water are still held together by a type of bond, though a much looser bond than the one that holds them together in solid ice.

If we continue to heat the water, the increasing kinetic energy of the molecules will ultimately break the bonds between neighboring molecules altogether. The molecules will then be able to move freely, and freely moving particles constitute what we call a *gas.* Above the boiling point (100°C at sea level), all the bonds between adjacent molecules are broken so that the water can exist only as a gas.

We see ice melting into liquid water and liquid water boiling into gas so often that it's tempting to think that's the end of the story. However, a little thought should convince you that the reality has to be more complex. For example, you know that Earth's atmosphere contains water vapor that condenses to form clouds and rain. But Earth's temperature is well below the boiling point of water (luckily for us!), so how is it that our atmosphere can contain water in the gas phase?

You'll understand the answer if you remember that temperature is a measure of the *average* kinetic energy of the particles in a substance [Section 4.3]; individual particles may have substantially lower or higher energies than the average. Thus, even at the low temperatures at which most water molecules are bound together as ice or liquid, a few molecules will always have enough energy to break free of their neighbors and enter the gas phase. In other words, some gas (water vapor) is always present along with solid ice or liquid water. The process by which molecules escape from a solid is called **sublimation**, and the process by which molecules escape from a liquid is called **evaporation**. Higher temperatures lead to higher rates of sublimation or evaporation.

THINK ABOUT IT

Global warming is expected to increase Earth's average temperature by up to a few degrees over the coming century. Based on what you've learned about phase changes, how would you expect global warming to affect the total amount of cloud cover on Earth? Explain.

Molecular Dissociation and Ionization The state of matter changes even further at higher temperatures. To see what happens, let's return to the example of heating the water. Above the boiling point, all the water will have entered the gas phase. What happens if we continue to raise the temperature?

The molecules in a gas move freely, but they often collide with one another. As the temperature rises, the molecules move faster and the collisions become more violent. At high enough temperatures, the collisions will be so violent that they can break the chemical bonds holding individual water molecules together. The molecules then split into pieces, a process we call **molecular dissociation**. (In the case of water, molecular dissociation usually breaks one hydrogen atom away from a remaining molecule that

COMMON MISCONCEPTIONS

One Phase at a Time?

In daily life, we usually think of H_2O as being in the phase of either solid ice, liquid water, or water vapor, with the phase depending on the temperature. In reality, two or even all three phases can exist at the same time. In particular, some sublimation *always* occurs over solid ice, and some evaporation *always* occurs over liquid water. You can tell that evaporation always occurs, because an uncovered glass of water will gradually empty as the liquid evaporates into gas. You can see that sublimation occurs by watching the snow pack after a winter storm: Even if the snow doesn't melt into liquid, it will gradually disappear as some of the ice sublimates into water vapor.

consists of one hydrogen atom and one oxygen atom [OH]; at even higher temperatures, the OH dissociates into individual atoms.)

At still higher temperatures, collisions can break the bonds holding electrons around the nuclei of individual atoms, allowing the electrons to go free. The loss of one or more negatively charged electrons leaves the remaining atom with a net positive charge. Such charged atoms are called **ions**. The process of stripping electrons from atoms is called **ionization**.

Thus, at temperatures of several thousand degrees, what once was water becomes a hot gas consisting of freely moving electrons and positively charged ions of hydrogen and oxygen. This type of hot gas, in which atoms have become ionized, is called a **plasma**. Because a plasma contains many charged particles, its interactions with light are different from those of a gas consisting of neutral atoms, which is one reason that plasma is sometimes referred to as "the fourth phase of matter." However, because the electrons and ions are not bound to one another, it is also legitimate to call plasma a gas. That is why we sometimes say that the Sun is made of hot gas and sometimes say that it is made of plasma; both statements are correct.

The degree of ionization in a plasma depends on its temperature and composition. A neutral hydrogen atom contains only one electron, so hydrogen can be ionized only once; the remaining hydrogen ion, designated H^+, is simply a proton. Oxygen, with atomic number 8, has eight electrons when it is neutral, so it can be ionized multiple times. *Singly ionized* oxygen is missing one electron, so it has a charge of +1 and is designated O^+. *Doubly ionized* oxygen, or O^{++}, is missing two electrons; *triply ionized* oxygen, or O^{+3}, is missing three electrons; and so on. At temperatures of several million degrees, oxygen can be *fully ionized,* in which case all eight electrons are stripped away and the remaining ion has a charge of +8.

Figure 5.10 summarizes the changes that occur as we heat water from ice to a fully ionized plasma. Other chemical substances go through similar phase changes, but the changes generally occur at different temperatures for different substances.

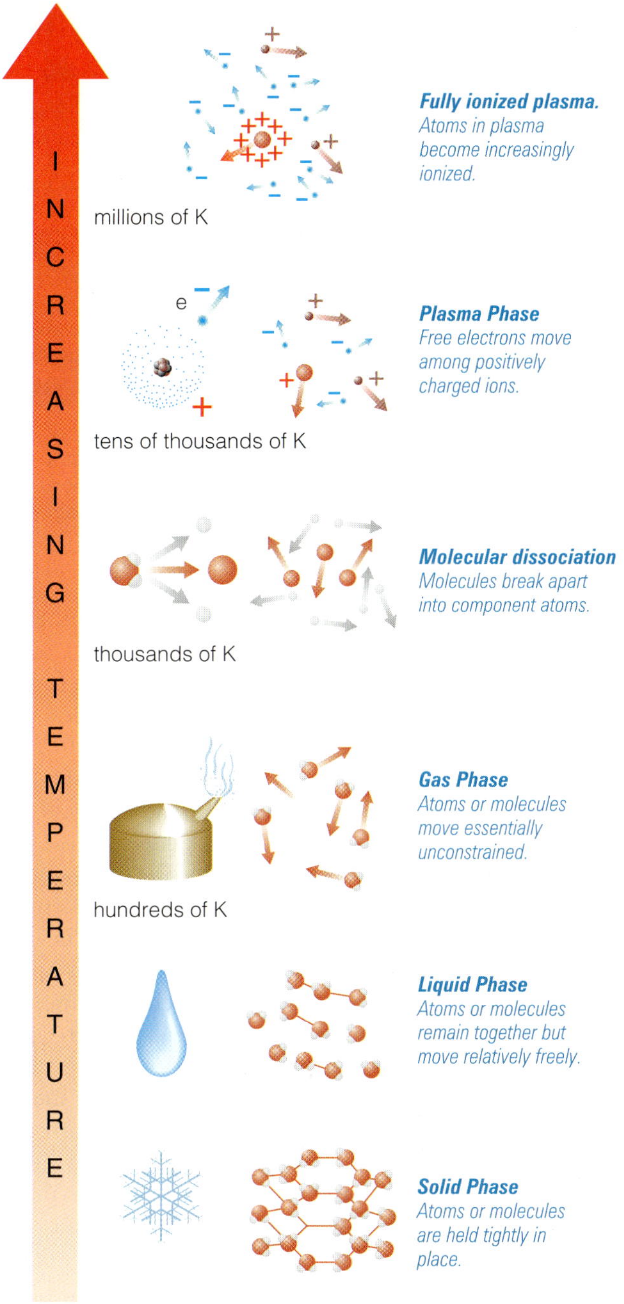

Figure 5.10 Interactive Figure The general progression of phase changes in water.

Phases and Pressure Temperature is the primary factor determining the phase of a substance and the ways in which light interacts with it, but pressure also plays a role. You're undoubtedly familiar with the idea of pressure in an everyday sense: For example, you can put more pressure on your arm by squeezing it. In science, we use a more precise definition: **pressure** is the *force per unit area* pushing on an object's surface. You feel more pressure when you squeeze your arm because squeezing increases the force on each square centimeter of your arm's surface. Similarly, piling rocks on a table increases the weight (force) on the table, which therefore increases the pressure on the surface of the table; if the pressure becomes too great, the table breaks. The gas in an atmosphere also creates pressure, because the weight of the gas bears down on everything beneath it. For example, at sea level on Earth, the weight of the atmosphere creates a pressure of about 14.7 pounds per square inch. That is, the total weight of all the air above each square inch of Earth's surface is about 14.7 pounds [Section 10.1].

Pressure can affect phases in a variety of ways. For example, deep inside Earth, the pressure is so high that Earth's inner metal core remains solid, even though the temperature is high enough that the metal would melt into liquid under less extreme pressure conditions [Section 9.1]. On

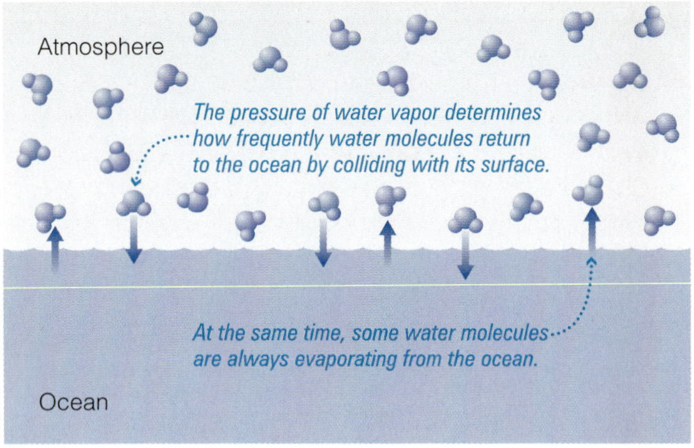

Figure 5.11 Evaporation of water molecules from the ocean is balanced in part by molecules of water vapor in Earth's atmosphere returning to the ocean. The rate at which these molecules return is directly related to the pressure created by water vapor in the atmosphere.

a planetary surface, atmospheric pressure can determine whether water is stable in liquid form.

Remember that liquid water is always evaporating (or ice sublimating) at a low level, because a few molecules randomly get enough energy to break the bonds holding them to their neighbors. On Earth, enough liquid water has evaporated from the oceans to make water vapor an important ingredient of our atmosphere. Some of these atmospheric water vapor molecules collide with the ocean surface, where they can "stick" and rejoin the ocean—essentially the opposite of evaporation (Figure 5.11). The greater the pressure created by water vapor molecules in our atmosphere,* the higher the rate at which water molecules return to the ocean. This direct return of water vapor molecules from the atmosphere helps keep the total amount of water in Earth's oceans fairly stable.** On the Moon, where the lack of atmosphere means no pressure from water vapor at all, liquid water would evaporate quite quickly (as long as the temperature were high enough that it did not freeze first). The same is true on Mars, because the atmosphere lacks enough water vapor to balance the rate of evaporation.

High pressure can also cause gases to dissolve in liquid water. For example, sodas are made by putting water in contact with high-pressure carbon dioxide gas. Because of the high pressure, many more carbon dioxide molecules enter the water than are released, so the water becomes "carbonated"—that is, it has a lot of dissolved carbon dioxide. When you open a bottle of carbonated water, exposing it to air with ordinary pressure, the dissolved carbon dioxide quickly bubbles up and escapes.

*Technically, this is known as the *vapor pressure* of water in the atmosphere. We can also measure vapor pressure for other atmospheric constituents, and the total gas pressure is the sum of all the individual vapor pressures.
**Rain and snow also contribute, of course; however, even if Earth's temperature rose enough so that raindrops and snowflakes could no longer form, only a small fraction of Earth's ocean water would evaporate before the return rate of water vapor molecules balanced the evaporation rate.

• How is energy stored in atoms?

Now that we have reviewed the structure of matter and how its phase depends on temperature and pressure, it is time to return to the primary goal of this chapter: understanding how we learn about distant objects by studying their light. To produce light, these objects must somehow transform energy contained in matter into the vibrations of electric and magnetic fields that we call light. We therefore need to focus on the charged particles within atoms, particularly the electrons, because only particles that have charge can interact with light.

Atoms actually contain energy in three different ways. First, by virtue of their mass, they possess mass-energy in the amount mc^2. Second, they possess kinetic energy by virtue of their motion. Third, and most important to reading the messages encoded in light, atoms contain *electrical potential energy* that depends on the arrangement of their electrons around their nuclei. To interpret the messages carried by light, we must understand how electrons store and release their electrical potential energy.

Energy Levels in Atoms The energy stored by electrons in atoms has a very strange but important property: The electrons can have only particular amounts of energy, and not other energies in between. As an analogy, suppose you're washing windows on a building. If you use an adjustable platform to reach high windows, you can stop the platform at any height above the ground. But if you use a ladder, you can stand only at *particular* heights—the heights of the rungs of the ladder—and not at any height in between. The possible energies of electrons in atoms are like the possible heights on a ladder. Only a few particular energies are possible, and energies between these special few are not possible. The possible energies are known as the **energy levels** of an atom.

Let's consider the energy levels of hydrogen, shown in Figure 5.12. Hydrogen has the simplest set of energy levels of any atom, because it has only one electron. Notice that, on the far right of the diagram, the various energy levels are labeled with energies in units of **electron-volts**, or **eV** for short. ($1 \text{ eV} = 1.60 \times 10^{-19}$ joule.)

The lowest possible energy level is called level 1 or the **ground state**. It represents the minimum amount of energy that the electron can have when it is part of a hydrogen atom, so its energy is labeled 0 eV. Each of the other possible energy levels represents an **excited state**, so named because an electron in any of these levels is "excited" and will soon (usually in a tiny fraction of a second) "calm down" and return to the ground state. Each excited state, from level 2 on upward, is also labeled with an energy in electron-volts; this energy represents the "extra" energy of an electron in that level compared to the ground state.

Energy Level Transitions An electron can rise from a low energy level to a higher one or fall from a high level to a lower one; such changes are usually called **energy level**

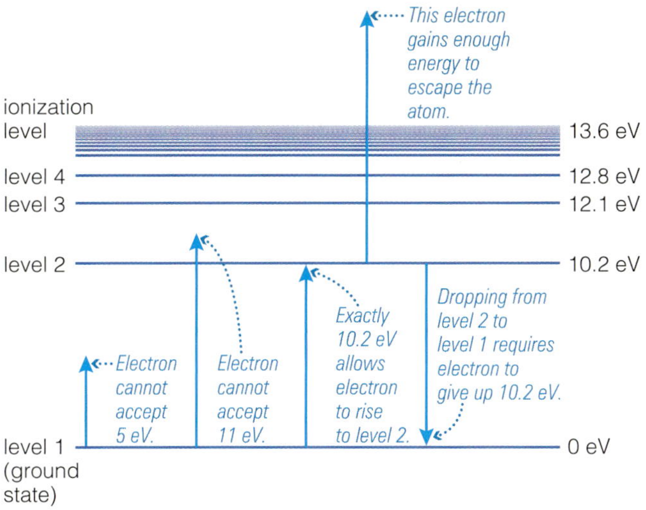

Figure 5.12 Energy levels for the electron in a hydrogen atom. The electron can change energy levels only if it gains or loses the amount of energy separating the levels. If the electron gains enough energy to reach the ionization level, it can escape from the atom, leaving behind a positively charged ion. (The many levels between level 4 and the ionization level are not labeled.)

transitions. Because energy must be conserved, energy level transitions can occur only when an electron gains or loses the specific amount of energy separating two levels. For example, an electron in level 1 can rise to level 2 only if it gains 10.2 eV of energy. If you try to give the electron 5 eV of energy, it won't accept it because that is not enough energy to reach level 2. Similarly, if you try to give it 11 eV, it won't accept it because it is too much for level 2 but not enough to reach level 3. Once in level 2, the electron can return to level 1 by giving up 10.2 eV of energy. Figure 5.12 shows several examples of allowed and disallowed energy level transitions.

Notice that the amount of energy separating the various levels gets smaller near the top. For example, it takes more energy to raise the electron from level 1 to level 2 than from level 2 to level 3, which in turn takes more energy than the transition from level 3 to level 4. If the electron gains enough energy to reach the *ionization level,* it escapes the atom completely, thereby ionizing the atom. Any excess energy beyond that needed for ionization becomes kinetic energy of the free-moving electron.

THINK ABOUT IT

Are there any circumstances under which an electron in a hydrogen atom can lose 2.6 eV of energy? Explain.

Quantum Physics If you think about it, the idea that electrons in atoms are restricted to particular energy levels is quite bizarre. It is as if you had a car that could go around a track only at particular speeds and not at speeds in between. How strange it would seem if your car suddenly changed its speed from 5 miles per hour to 20 miles per hour without first passing through a speed of 10 miles per hour! In scientific terminology, the electron's energy levels in an atom are said to be *quantized,* and the study of the energy levels of electrons (and other particles) is called *quantum physics* (or *quantum mechanics*). We will explore more of the astonishing implications of quantum physics in Chapter S4.

Electrons have quantized energy levels in all atoms, not just in hydrogen. Moreover, the allowed energy levels differ from element to element and from one ion of an element to another ion of the same element. Even molecules have quantized energy levels. As we will see shortly, the different energy levels of different atoms and molecules allow light to carry "fingerprints" that can tell us the chemical composition of distant objects.

5.4 Learning from Light

Matter leaves its fingerprints whenever it interacts with light. Examining the color of an object is a crude way of studying the clues left by the matter it contains. For example, a red shirt absorbs all visible photons except those in the red part of the spectrum, so we know that it must contain a dye with these special light-absorbing characteristics. If we take light and disperse it into a spectrum, we can see the spectral fingerprints more clearly. For example, the photograph that opens this chapter (p. 145) shows the Sun's visible-light spectrum in great detail, with the rainbow of color stretching in horizontal rows from the upper left to the lower right of the photograph. We see similar dark or bright lines when we look at almost any spectrum, whether it is the spectrum of the flame from the gas grill in someone's backyard or the spectrum of a distant galaxy whose light we collect with a gigantic telescope. As long as we collect enough light to see details in the spectrum, we can learn many fundamental properties of the object we are viewing, no matter how far away the object is located.

The process of obtaining a spectrum and reading the information it contains is called **spectroscopy**. Spectra often look much like rainbows, at least if they are made only from visible light. However, for the purposes of examining spectra in greater detail, it's often more useful to display them as graphs that show the amount of radiation, or **intensity**, at different wavelengths. The graphs of intensity in this book simply show the relative amount of energy received from an object in each wavelength of light. In other words, the intensity at a given wavelength is proportional to the number of photons observed at that wavelength times the energy of those photons.

For example, Figure 5.13 shows a schematic spectrum of light from a celestial body such as a planet. Notice that this spectrum extends all the way from the ultraviolet on the left to the infrared on the right, which means the graph is showing us details that would be invisible to our eyes in a photograph. At wavelengths where a lot of light is coming from the celestial body, the intensity is high, while at wavelengths where there is little light, the intensity is low.

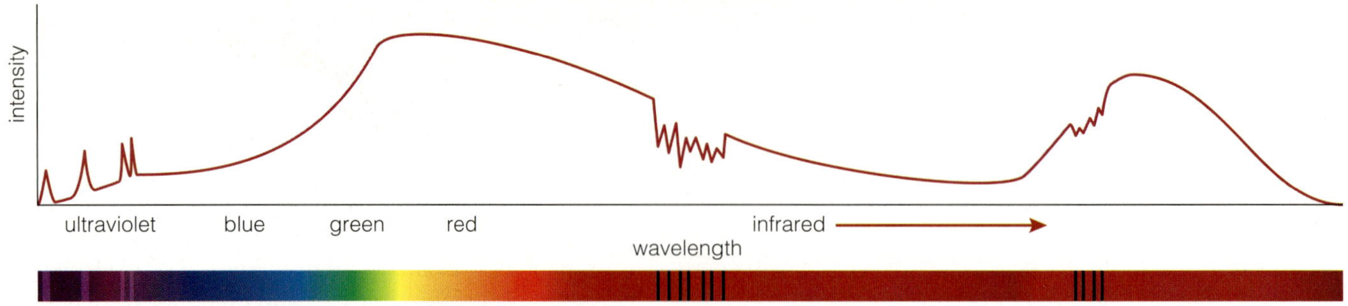

Figure 5.13 A schematic spectrum obtained from the light of a distant object. The "rainbow" at bottom shows how the light would appear when viewed through a prism or diffraction grating; of course, our eyes cannot see the ultraviolet or infrared light. The graph shows the corresponding intensity of the light at each wavelength. The intensity is high where the rainbow is bright and low where it is dim (such as in places where the rainbow shows dark lines).

Our goal in this section is to learn how to interpret astronomical spectra. The bumps and wiggles in Figure 5.13 arise from several different processes, making it a good case study for spectral interpretation. We'll consider these processes one at a time, then we'll return to interpret Figure 5.13 at the end of this section.

Light and Spectroscopy Tutorial, Lessons 2–4

• What are the three basic types of spectra?

Let's begin our study of spectra by learning how to classify them. Laboratory studies show that spectra come in three basic types: continuous, emission line, and absorption line. Figure 5.14 shows an example of each type, along with the conditions that produce it. (The rules that specify these conditions are often called *Kirchhoff's laws.*) Study the figure carefully to notice the following key ideas:

1. The spectrum of a common (incandescent) light bulb is a rainbow of color. Because the rainbow spans a broad range of wavelengths without interruption, we call it a **continuous spectrum**.
2. A thin or low-density cloud of gas does not produce a continuous spectrum. Instead, it emits light only at specific wavelengths that depend on its composition and temperature. The spectrum therefore consists of bright **emission lines** that we see against a black background, making what we call an **emission line spectrum**.
3. If the cloud of gas lies between us and a light bulb, we still see most of the continuous light emitted by the light bulb. However, the cloud absorbs light of specific wavelengths, so that the spectrum shows dark **absorption lines** over the background rainbow from the light bulb.* We call this an **absorption line spectrum**.

*More technically, we'll see an absorption line spectrum as long as the cloud is cooler in temperature than the source of illumination, which in this case means the light bulb filament.

Each of the three spectra is shown both as a band of light (essentially as it would be recorded on a photograph) and as a graph of intensity versus wavelength. Notice that absorption lines appear as dips on a background of relatively high-intensity light, while emission lines look like spikes on a background with little or no intensity.

We can apply some of these ideas to the solar spectrum that opens this chapter. Notice that it shows numerous absorption lines over a background rainbow of color. This tells us that we are essentially looking at a hot light source through gas that is absorbing some of the colors, much as we see when looking through the cloud of gas to the light bulb in Figure 5.14c. For the solar spectrum, the hot light source is the hot interior of the Sun, while the "cloud" is the relatively cool and low-density layer of gas that makes up the Sun's visible surface.

• How does light tell us what things are made of?

We have just seen *how* different viewing conditions lead us to see different types of spectra, so we are now ready to discuss *why.* Let's start with emission and absorption line spectra. Emission and absorption lines form as a direct consequence of the fact that each type of atom, ion, or molecule possesses a unique set of energy levels. This fact allows us to learn the compositions of distant objects in the universe, as we can see by considering what happens in a cloud of gas consisting solely of hydrogen atoms.

Emission Line Spectra The atoms in any cloud of gas are constantly colliding with one another, exchanging energy in each collision. Most of the collisions simply send the atoms careening off in new directions. However, a few of the collisions transfer the right amount of energy to bump an electron from a low energy level to a higher energy level.

Electrons can't stay in higher energy levels for long. They always fall back down to the ground state, level 1, usually in a tiny fraction of a second. The energy the electron loses when it falls to the lower energy level must go somewhere,

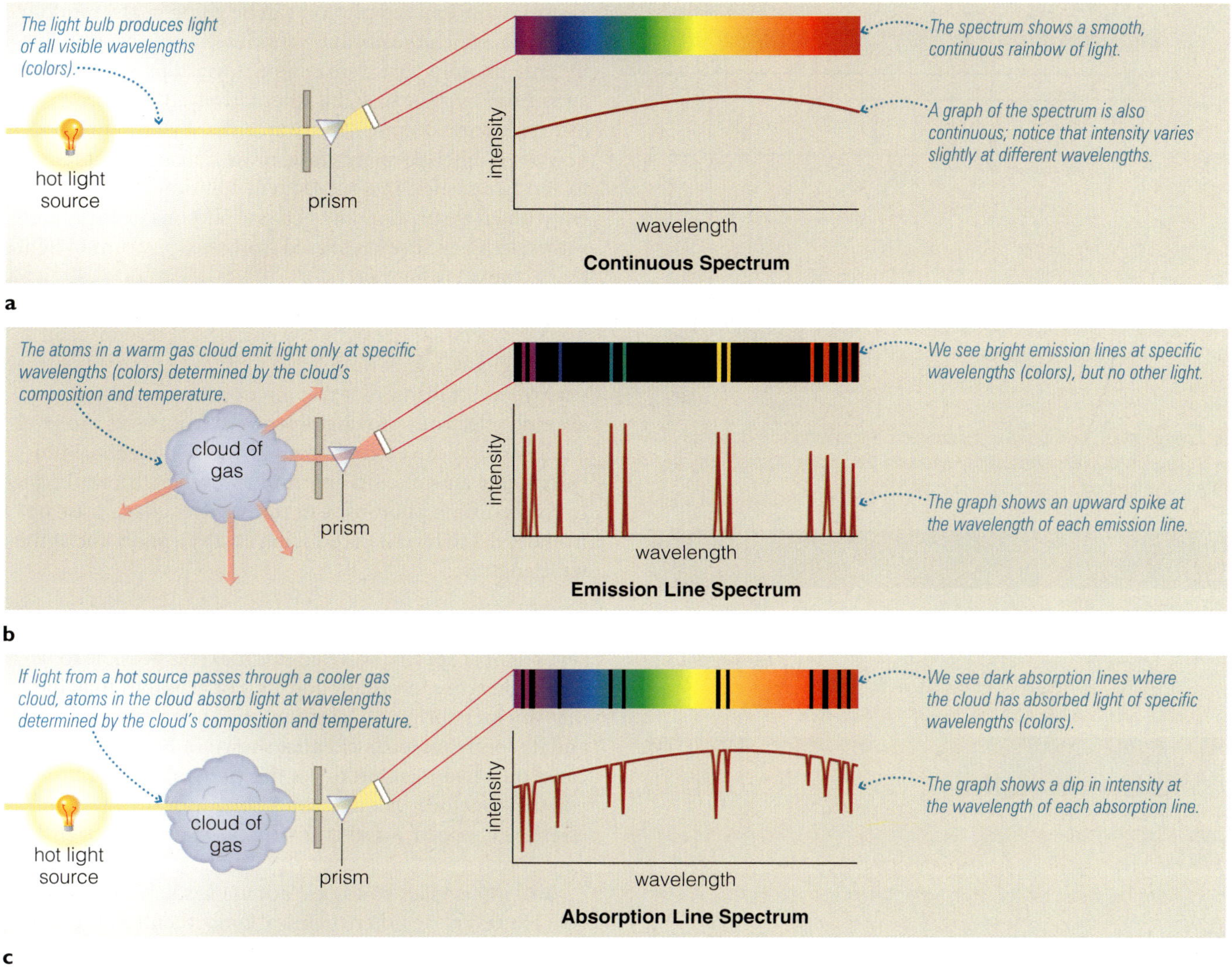

Figure 5.14 Interactive Figure These diagrams show examples of the conditions under which we see the three basic types of spectra.

and often it goes to *emitting* a photon of light. The emitted photon must have the same amount of energy that the electron loses, which means that it has a specific wavelength (and frequency). Figure 5.15a shows the energy levels in hydrogen that we saw in Figure 5.12. This time, however, it is also labeled with the wavelengths of the photons emitted by various downward transitions of an electron from a higher energy level to a lower one. For example, the transition from level 2 to level 1 emits an ultraviolet photon of wavelength 121.6 nm, and the transition from level 3 to level 2 emits a red visible-light photon of wavelength 656.3 nm.*

Although electrons that rise to higher energy levels in a gas quickly return to level 1, new collisions can raise other electrons into the higher levels. Thus, as long as the gas remains even moderately warm, collisions are always bumping some electrons into levels from which they fall back down and emit photons with some of the wavelengths shown in Figure 5.15a. The gas therefore emits light with these specific wavelengths. That is why a warm gas cloud produces an emission line spectrum, as shown in Figure 5.15b. The bright emission lines appear at the wavelengths that correspond to downward transitions of electrons, and the rest of the spectrum is dark (black). The specific set of lines that we see depends on the cloud's temperature as well as its composition: At higher temperatures, electrons are more likely to be bumped to higher energy levels.

*Astronomers call transitions between level 1 and other levels the *Lyman* series of transitions. The transition between level 1 and level 2 is Lyman α, between level 1 and level 3 Lyman β, and so on. Similarly, transitions between level 2 and higher levels are called *Balmer* transitions. Other sets of transitions also have names.

THINK ABOUT IT

If nothing continues to heat the hydrogen gas, all the electrons eventually will end up in the lowest energy level (the ground state, or level 1). Use this fact to explain why we should *not* expect to see an emission line spectrum from a very cold cloud of hydrogen gas.

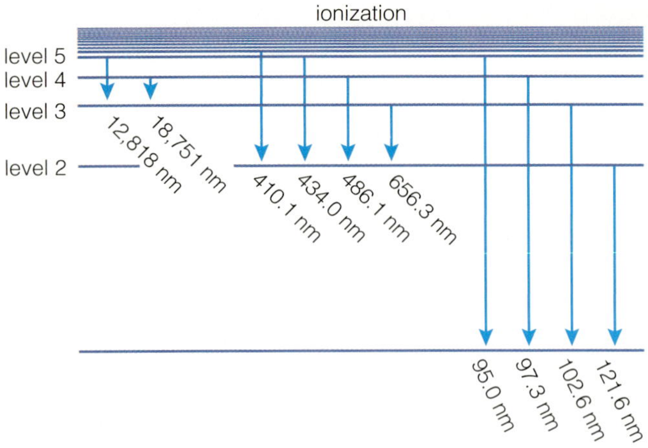

a Energy level transitions in hydrogen correspond to photons with specific wavelengths. Only a few of the many possible transitions are labeled.

b This spectrum shows emission lines produced by downward transitions between higher levels and level 2 in hydrogen.

c This spectrum shows absorption lines produced by upward transitions between level 2 and higher levels in hydrogen.

Figure 5.15 Interactive Figure An atom emits or absorbs light only at specific wavelengths that correspond to changes in the atom's energy as an electron undergoes transitions between its allowed energy levels.

Absorption Line Spectra Now, suppose a light bulb illuminates the hydrogen gas from behind (as in Figure 5.14c). The light bulb emits light of all wavelengths, producing a spectrum that looks like a rainbow of color. However, the hydrogen atoms can absorb those photons that have the right amount of energy needed to raise an electron from a low energy level to a higher one. Figure 5.15c shows the result. It is an absorption line spectrum, because the light bulb produces a continuous rainbow of color while the hydrogen atoms remove (absorb) light at specific wavelengths.

Given that electrons in high energy levels quickly return to lower levels, you might wonder why photons emitted in downward transitions don't cancel out the effects of those absorbed in upward transitions. Finding the answer requires looking more deeply at what happens to the absorbed photons. Two things can happen after an electron absorbs a photon and rises to a higher energy level. The first is that the electron quickly returns to its original level, emitting a photon of the same energy that it absorbed. However, the emitted photon can be going in any random direction, which means that we still see an absorption line because photons that were coming toward us are redirected away from our line of sight. Alternatively, the electron can lose its energy in some other way, either by dropping back down to a different energy level or by transferring its energy to another particle in a subsequent collision. Again, we are left with an absorption line because photons of a specific wavelength have been removed from the spectrum of light that's coming toward us.

You can now see why the dark absorption lines in Figure 5.15c occur at the same wavelengths as the emission lines in Figure 5.15b: Both types of lines represent the same energy level transitions, except in opposite directions. For example, electrons moving downward from level 3 to level 2 in hydrogen can emit photons of wavelength 656.3 nm (producing an emission line at this wavelength), while electrons absorbing photons with this wavelength can rise up from level 2 to level 3 (producing an absorption line at this wavelength).

Chemical Fingerprints The fact that hydrogen emits and absorbs light at specific wavelengths makes it possible to detect its presence in distant objects. For example, imagine that you look through a telescope at an interstellar gas cloud, and its spectrum looks like that shown in Figure 5.15b. Because only hydrogen produces this particular set of lines, you can conclude that the cloud is made of hydrogen. In essence, the spectrum contains a "fingerprint" left by hydrogen atoms.

Real interstellar clouds are not made solely of hydrogen. However, the other chemical constituents in the cloud leave fingerprints on the spectrum in much the same way. Every type of atom has its own unique spectral fingerprint, because it has its own unique set of energy levels. For example, Figure 5.16 shows emission line spectra for helium, sodium, and neon.

Not only does each chemical element produce a unique spectral fingerprint, but *ions* of a particular element (atoms that are missing one or more electrons) produce fingerprints different from those of neutral atoms (Figure 5.17). For example, the spectrum of doubly ionized neon (Ne^{++}) is different from that of singly ionized neon (Ne^{+}), which in turn is different from that of neutral neon (Ne). These differences can help us determine the temperature of a hot

Figure 5.16 Visible-light emission line spectra for helium, sodium, and neon. The patterns and wavelengths of lines are different for each element, giving each a unique spectral fingerprint.

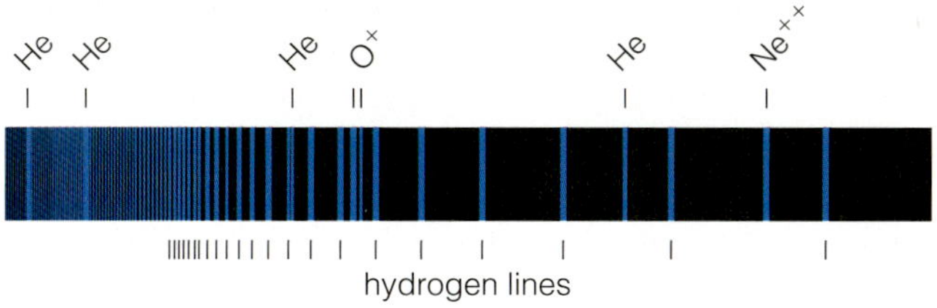

Figure 5.17 The emission line spectrum of the Orion Nebula in a portion of the ultraviolet (about 350–400 nm). The lines are labeled with the chemical elements or ions that produce them (He = helium; O = oxygen; Ne = neon). The many hydrogen lines are all transitions from high levels to level 2.

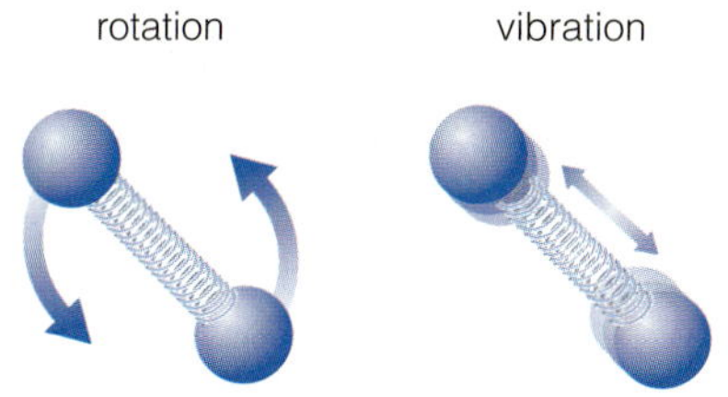

a We can think of a two-atom molecule as two balls connected by a spring. Although this model is simplistic, it illustrates how molecules can rotate and vibrate. The rotations and vibrations can have only particular amounts of energy and therefore produce unique spectral fingerprints.

b This spectrum of molecular hydrogen (H_2) shows that molecular spectra consist of lines bunched into broad molecular bands.

Figure 5.18 Like atoms and ions, molecules also emit or absorb light at specific wavelengths.

gas or plasma, and they come in handy when we're trying to measure the surface temperatures of stars [Section 15.1]. At higher temperatures, more highly charged ions will be present, so we can estimate the temperature by identifying the ions that are creating spectral lines.

Molecules also produce spectral fingerprints. Like atoms, molecules can produce spectral lines when their electrons change energy levels. But molecules can also produce spectral lines in two other ways. Because they are made of two or more atoms bound together, molecules can vibrate and rotate (Figure 5.18a). Vibration and rotation also require energy, and the possible energies of rotation and vibration in molecules are quantized much like electron energy levels in atoms. Thus, a molecule can absorb or emit a photon when it changes its rate of vibration or rotation. The energy changes in molecules are usually smaller than those in atoms and therefore produce lower-energy photons, and the energy levels also tend to be bunched more closely together than in atoms. Molecules therefore produce spectra with many sets of tightly bunched lines, called **molecular bands** (Figure 5.18b), that are usually found in the infrared portion of the electromagnetic spectrum. That is one reason why infrared telescopes and instruments are so important to astronomers.

Over the past century, scientists have conducted laboratory experiments to identify the spectral lines of every chemical element and of many ions and molecules. Thus, when we see lines in the spectrum of a distant object, we can usually determine what chemicals produced them. For example, if we see spectral lines of hydrogen, helium, and carbon in the spectrum of a distant star, we know that all three elements are present in the star. Moreover, with detailed analysis, we can determine the relative proportions of the various elements. That is how we have learned the chemical compositions of objects throughout the universe.

• How does light tell us the temperatures of planets and stars?

We have seen how emission and absorption line spectra form, and how we can use them to determine the composition of a cloud of gas. Now we are ready to turn our attention to continuous spectra. Although continuous spectra can be produced in more than one way, light bulbs, planets, and stars produce them in a way that can help us determine their temperatures.

Thermal Radiation: Every Body Does It In a cloud of gas that produces a simple emission or absorption line spectrum, the individual atoms or molecules are essentially independent of one another. Most photons pass easily through such a gas, except those that cause energy level transitions in the atoms or molecules of the gas. However, the atoms and molecules within most of the objects we encounter in everyday life, such as rocks, light bulb filaments, and people, cannot be considered independent. Thus, their energy levels are far more complex. They tend to absorb light across a broad range of wavelengths, meaning that any light striking them cannot easily pass through and that any light emitted inside them cannot easily escape. In fact, the same is true of almost any large or dense object, including planets and stars.

In order to understand the spectra of such objects, let's consider an idealized case, in which an object absorbs all the photons that strike it and does not allow photons inside it to escape easily. Photons trying to escape are quickly absorbed by an atom or molecule, which quickly reemits the photon—but often with a slightly different wavelength and in a different direction.* In effect, the emitted photons bounce randomly around inside the object, constantly exchanging energy with the object's atoms or molecules. By the time the photons finally escape the object and fly off

*One of the reasons that photons are reemitted with a slightly different wavelength has to do with the Doppler effect (see Section 5.5). If the absorbing atom (or molecule) is moving, there will be a shift in wavelength that depends on the atom's velocity and the direction in which the photon escapes.

into space, their radiative energies have become randomized so that they are spread over a wide range of wavelengths. The wide wavelength range of the photons explains why the spectrum of light from such an object is smooth or *continuous,* like a pure rainbow without any absorption or emission lines.

Most important, the spectrum from such an object depends on only one thing: the object's *temperature.* To understand why, remember that temperature represents the average kinetic energy of the atoms or molecules in an object [Section 4.3]. Because the randomly bouncing photons interact so many times with those atoms or molecules, they end up with energies that match the kinetic energies of the object's atoms or molecules—which means the photon energies depend only on the object's temperature, regardless of what the object is made of. The temperature dependence of this light explains why we call it **thermal radiation** (sometimes known as *blackbody* radiation), and why its spectrum is called a **thermal radiation spectrum**. Thermal radiation spectra are the most common type of continuous spectra.

No real object emits a perfect thermal radiation spectrum, but almost all familiar objects—including the Sun, the planets, rocks, and even you—emit light that approximates thermal radiation. Figure 5.19 shows graphs of the idealized thermal radiation spectra of three stars and a human, each with its temperature given on the Kelvin scale (see Figure 4.12). Be sure to notice that these spectra show the intensity of light emitted from each square meter of an object's surface, not the total amount of light emitted by the object; that is, they tell us the emitted intensity *per unit area.* Thus, for example, a very large 3,000 K star can emit more total light than a small 15,000 K star, even though the hotter star emits much more light per unit area of its surface.

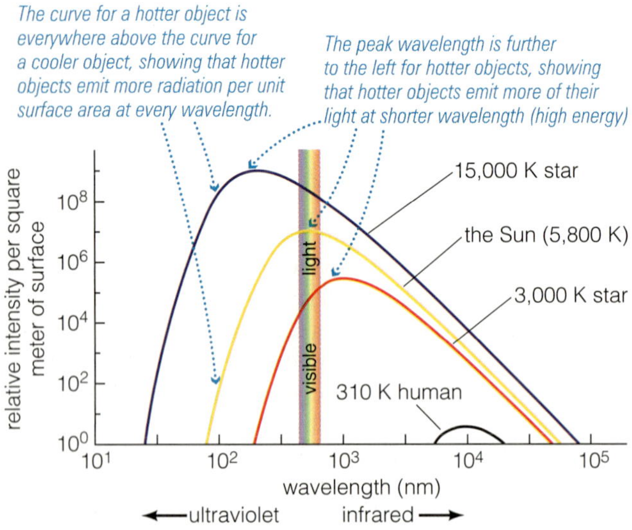

Figure 5.19 Interactive Figure Graphs of idealized thermal radiation spectra demonstrate the two laws of thermal radiation: (1) Each square meter of a *hotter object's surface emits more light at all wavelengths;* (2) hotter objects emit photons with a higher average energy. Notice that the graph uses power-of-10 scales on both axes, so that we can see all the curves even though the differences between them are quite large.

The Two Laws of Thermal Radiation If you compare the spectra in Figure 5.19, you'll see that temperature has two effects on the spectrum; the quantitative descriptions of these effects (see Mathematical Insight 5.2) are often called the *laws* of thermal radiation:

- Law 1 (Stefan-Boltzmann law): *Each square meter of a hotter object's surface emits more light at all wavelengths.* For example, each square meter on the surface of the 15,000 K star emits a lot more light at every wavelength than each square meter of the 3,000 K star, and the hotter star emits light at some ultraviolet wavelengths that the cooler star does not emit at all.
- Law 2 (Wien's law ["Wien" is pronounced *veen*]): *Hotter objects emit photons with a higher average energy,* which means a shorter average wavelength. That is why the "humps" of the spectra are at shorter wavelengths for hotter objects. For example, the hump for the 15,000 K star is in ultraviolet light, the hump for the 5,800 K Sun is in visible light, and the hump for the 3,000 K star is in the infrared.

You can see these properties in action by playing with a light that has a dimmer switch. The first property is apparent as you turn the dimmer switch up: The filament in the light bulb gets hotter and the light brightens. When you turn the switch back down, the filament gets cooler and the light dims. (You can verify the changing temperature by placing your hand near the bulb.) The color of the light illustrates the second property. Before you turn on the switch, the filament is too cool to emit visible light at all, instead emitting only infrared radiation that your eyes cannot see. When you first turn the switch on low, the filament begins to glow red or orange—the longest wavelength of visible light. As you turn the switch higher and the filament gets hotter, the average wavelength of the emitted photons moves toward the blue end of the visible spectrum. The mix of colors emitted at this higher temperature makes the filament and its light look white to your eyes.

Because thermal radiation spectra depend only on temperature, we can use them to measure the temperatures of distant objects. In many cases we can estimate temperatures simply from the object's colors. Notice that while hotter objects emit more light at *all* wavelengths, the biggest difference appears at the shortest wavelengths. An object with a temperature of 310 K, which is about human body temperature, emits mostly in the infrared and emits no visible light at all—which explains why we don't glow in the dark! A relatively cool star, with a 3,000 K surface temperature, emits mostly red light. That is why some bright stars in our sky, such as Betelgeuse (in Orion) and Antares (in Scorpius), appear reddish in color. The Sun's 5,800 K surface emits most strongly in green light (around 500 nm), but the Sun looks yellow or white to our eyes because it also emits other colors throughout the visible spectrum. Hotter stars emit mostly in the ultraviolet, but because our eyes cannot see ultraviolet, they appear blue-white in color. If an object were heated to a temperature of millions of degrees, it would radiate mostly X rays. Some astronomical objects are indeed hot enough to emit X rays, such as disks of gas encircling exotic objects like neutron stars and black holes (see Chapter 18).

• How do we interpret an actual spectrum?

We've covered the processes that form the three idealized types of spectra (continuous or thermal, absorption line, and emission line). The spectra of real astronomical objects are usually combinations of these three idealized types. In addition, real spectra often show features produced by one other process familiar from everyday life: reflection (or scattering). Let's consider how reflection affects a spectrum, and then we'll be ready to return to the interpretation of the spectrum we began with in Figure 5.13.

Reflected Light Spectra Most of the light we see in daily life is either reflected or scattered. The source of the light is thermal radiation from the Sun or a lamp. After this light strikes the ground, clouds, people, or other objects, we see only the wavelengths of light that are reflected (or scattered). For example, a red sweatshirt absorbs blue light and reflects red light, so its visible spectrum looks like the thermal radiation spectrum of its light source—the Sun—but with blue light missing.

In the same way that we can distinguish lemons from limes, we can use color information in reflected light to learn about celestial objects. Different fruits, different rocks, and even different atmospheric gases reflect and absorb light at different wavelengths. Although the absorption features that show up in spectra of reflected light are not as distinct as the emission and absorption lines for thin gases, they still provide useful information. For example, the surface materials of a planet determine how much light of different colors is reflected or absorbed. The reflected light gives the planet its color, while the absorbed light heats the surface and helps determine its temperature.

Putting It All Together Figure 5.20 shows the same spectrum we began with in Figure 5.13, but this time with labels indicating the processes responsible for its various features. What can we say about this object from its spectrum? The hump of thermal emission peaking in the infrared shows that this object has a surface temperature of about 225 K, well below the 273 K freezing point of water. The absorption bands in the infrared come mainly from carbon dioxide, which tells us that the object has a carbon dioxide atmosphere. The emission lines in the ultraviolet come from hot gas in a high, thin layer of the object's atmosphere. The reflected light looks like the Sun's 5,800 K thermal

MATHEMATICAL INSIGHT 5.2 Laws of Thermal Radiation

The two rules of thermal radiation have simple mathematical formulas. Law 1, called the *Stefan-Boltzmann law* (named after its discoverers), is expressed as:

$$\text{emitted power (per square meter of surface)} = \sigma T^4$$

where σ (Greek letter *sigma*) is a constant with a measured value $\sigma = 5.7 \times 10^{-8}$ watt/($m^2 \times$ Kelvin^4). Note that this equation is for power *per square meter of surface*. Finding the total power that an object radiates requires multiplying this power by the object's surface area.

Law 2, called *Wien's law*, is expressed approximately as:

$$\lambda_{max} \approx \frac{2{,}900{,}000}{T\,(\text{Kelvin})}\text{ nm}$$

where λ_{max} (read as *lambda-max*) is the wavelength (in nanometers) of maximum intensity, which is the peak of the hump in a thermal radiation spectrum.

Example: Consider a 15,000 K object that emits thermal radiation. How much power does it emit per square meter? What is its wavelength of peak intensity?

Solution:

Step 1 Understand: We can calculate the total emitted power per square meter from the Stefan-Boltzmann law and the wavelength of maximum intensity from Wien's law.

Step 2 Solve: First we use the Stefan-Boltzmann law to calculate the emitted power per square meter. We plug in the object's temperature of T = 15,000 K:

$$\sigma T^4 = 5.7 \times 10^{-8}\frac{\text{watt}}{\text{m}^2 \times \text{K}^4} \times (15{,}000\text{ K})^4$$
$$= 2.9 \times 10^9\text{ watt/m}^2$$

We find the wavelength of maximum intensity with Wien's law:

$$\lambda_{max} \approx \frac{2{,}900{,}000}{15{,}000\text{ (Kelvin)}}\text{ nm} \approx 190\text{ nm}$$

Step 3 Explain: A 15,000 K object emits a total power of 2.9 billion watts per square meter. Its wavelength of maximum intensity is 190 nm, which is in the ultraviolet portion of the electromagnetic spectrum. By using these facts in reverse, we can learn about astronomical objects. For example, if an object has a thermal radiation spectrum that peaks at a wavelength of 190 nm, we know that its surface temperature is about 15,000 K. Furthermore, knowing the temperature of the object tells us how much power it emits per square meter of surface, allowing us to calculate its total size from its total radiated power.

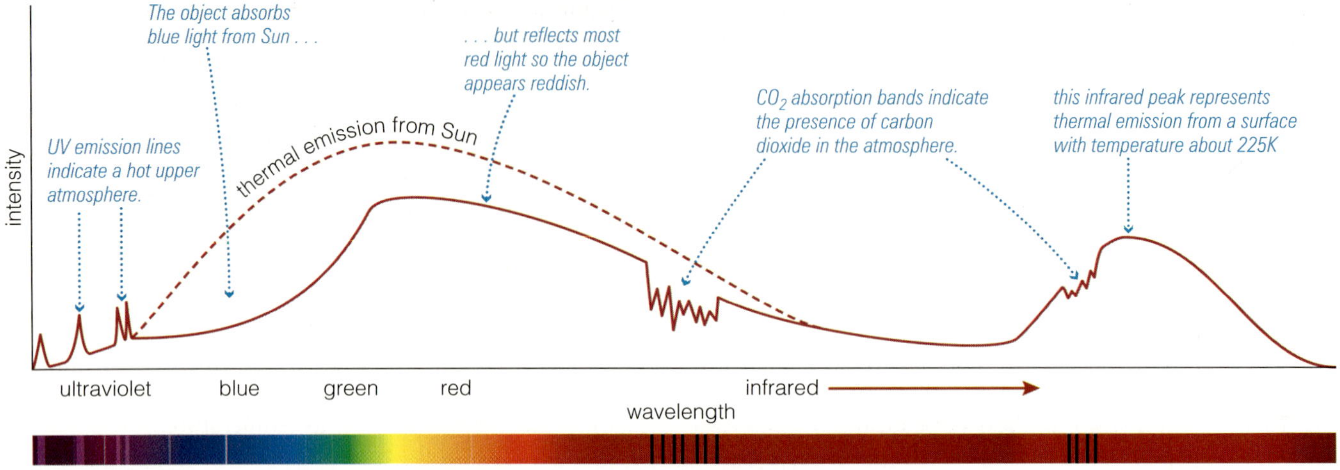

Figure 5.20 The spectrum of Figure 5.13, with interpretation. We conclude that the object looks red in color because it absorbs more blue light than red light from the Sun. The absorption lines tell us that the object has a carbon dioxide atmosphere, and the emission lines tell us that its upper atmosphere is hot. The hump in the infrared (near the right of the diagram) tells us that the object has a surface temperature of about 225 K. It is an idealized spectrum of the planet Mars.

radiation except that much of the blue light is missing, so the object must be reflecting sunlight and must look red in color. Perhaps by now you have guessed that this figure represents the spectrum of the planet Mars.

Mastering Astronomy The Doppler Effect Tutorial, Lessons 1–2

5.5 The Doppler Shift

You're probably already amazed at the volume of information contained in light, but there is still more. In particular, we can learn about the motion of a distant object (relative to us) from changes in its spectrum caused by the **Doppler effect**.

• How does light tell us the speed of a distant object?

You've probably noticed the Doppler effect on the *sound* of a train whistle near train tracks. If the train is stationary, the pitch of its whistle sounds the same no matter where you stand (Figure 5.21a). But if the train is moving, the pitch will sound higher when the train is coming toward you and lower when it's moving away from you. Just as the train passes by, you can hear the dramatic change from high to low pitch—a sort of "weeeeeeee–ooooooooh" sound. To understand why, we have to think about what happens to the sound waves coming from the train (Figure 5.21b). When the train is moving toward you, each pulse of a

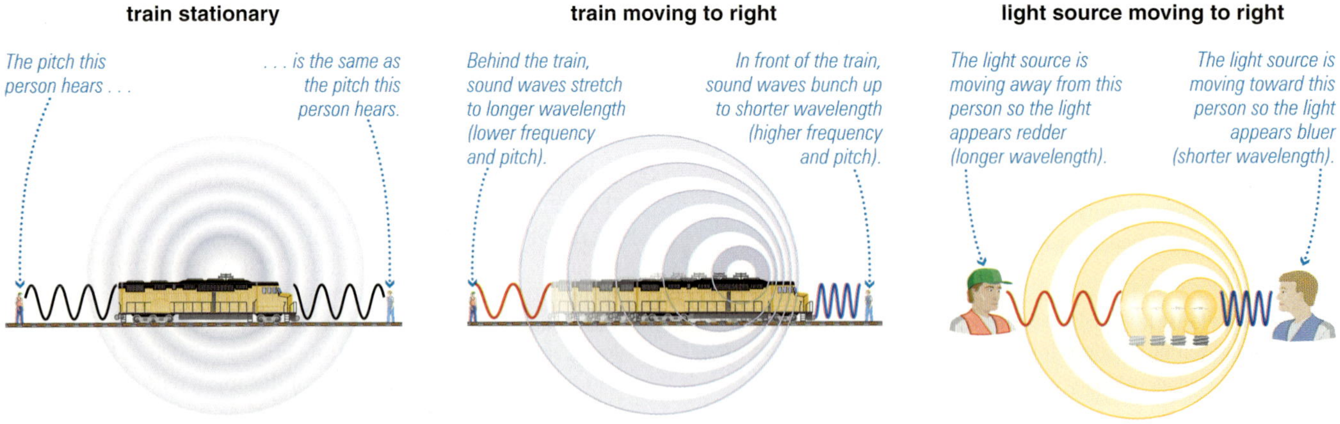

a The whistle sounds the same no matter where we stand near a stationary train.

b For a moving train, the sound you hear depends on whether the train is moving toward you or away from you.

c We get the same basic effect from a moving light source (although the shifts are usually too small to notice by eye).

Figure 5.21 The Doppler effect. Each circle represents the crests of sound (or light) waves going in all directions from the source. For example, the circles from the train might represent waves emitted 0.001 second apart.

sound wave is emitted a little closer to you. The result is that waves are bunched up between you and the train, giving them a shorter wavelength and higher frequency (pitch). After the train passes you by, each pulse comes from farther away, stretching out the wavelengths and giving the sound a lower frequency.

The Doppler effect causes similar shifts in the wavelengths of light (Figure 5.21c). If an object is moving toward us, the light waves bunch up between us and the object, so that its entire spectrum is shifted to shorter wavelengths. Because shorter wavelengths of visible light are bluer, the Doppler shift of an object coming toward us is called a **blueshift**. If an object is moving away from us, its light is shifted to longer wavelengths. We call this a **redshift** because longer wavelengths of visible light are redder. For convenience, astronomers use the terms *blueshift* and *redshift* even when they aren't talking about visible light.

Spectral lines provide the reference points we use to identify and measure Doppler shifts (Figure 5.22). For example, suppose we recognize the pattern of hydrogen lines in the spectrum of a distant object. We know the **rest wavelengths** of the hydrogen lines—that is, their wavelengths in stationary clouds of hydrogen gas—from laboratory experiments in which a tube of hydrogen gas is heated so that the wavelengths of the spectral lines can be measured. If the hydrogen lines from the object appear at longer wavelengths, then we know they are redshifted and the object is moving away from us. The larger the shift, the faster the object is moving. If the lines appear at shorter wavelengths, then we know they are blueshifted and the object is moving toward us.

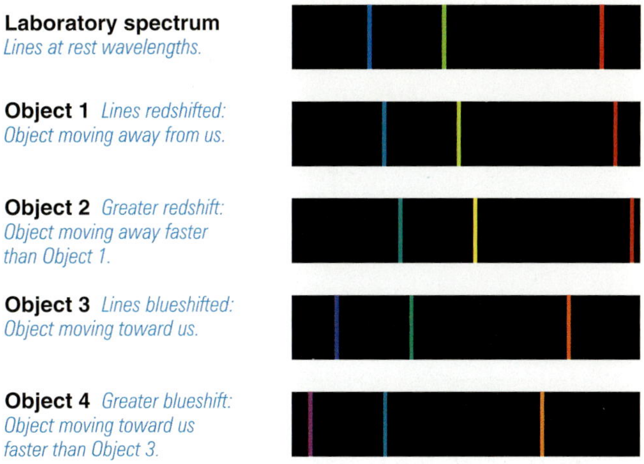

Figure 5.22 Interactive Figure Spectral lines provide the crucial reference points for measuring Doppler shifts.

THINK ABOUT IT

Suppose the hydrogen emission line with a rest wavelength of 121.6 nm (the transition from level 2 to level 1) appears at a wavelength of 120.5 nm in the spectrum of a particular star. Given that these wavelengths are in the ultraviolet, is the shifted wavelength closer to or farther from blue visible light? Why, then, do we say that this spectral line is *blueshifted?*

Notice that the Doppler shift tells us only the part of an object's full motion that is directed toward or away from us (the object's *radial* component of motion). Doppler shifts do not give us any information about how fast an object is

MATHEMATICAL INSIGHT 5.3 The Doppler Shift

As long as an object's radial velocity is small compared to the speed of light (that is, less than a few percent of c), we can use a simple formula to calculate the radial velocity (toward or away from us) of an object from its Doppler shift:

$$\frac{v_{rad}}{c} = \frac{\lambda_{shift} - \lambda_{rest}}{\lambda_{rest}}$$

where v_{rad} is the radial velocity of the object, λ_{rest} is the rest wavelength of a particular spectral line, and λ_{shift} is the shifted wavelength of the same line. (And, as always, c is the speed of light.)

If the result is positive, the object has a redshift and is moving away from us. A negative result means that the object has a blueshift and is moving toward us.

Example: The rest wavelength of one of the visible lines of hydrogen is 656.285 nm. This line is easily identifiable in the spectrum of the bright star Vega, but it appears at a wavelength of 656.255 nm. What is the radial velocity of Vega?

Solution:

Step 1 Understand: We notice that the line's wavelength in Vega's spectrum is slightly shorter than its rest wavelength. Thus, the line is blueshifted and Vega's radial motion is *toward* us. We can calculate the radial velocity from the given formula.

Step 2 Solve: In the radial velocity formula, the rest wavelength is $\lambda_{rest} = 656.285$ nm. The shifted wavelength is the wavelength in Vega's spectrum, $\lambda_{shift} = 656.255$ nm. We find:

$$\frac{v_{rad}}{c} = \frac{\lambda_{shift} - \lambda_{rest}}{\lambda_{rest}}$$

$$= \frac{656.255 \text{ nm} - 656.285 \text{ nm}}{656.285 \text{ nm}}$$

$$= -4.5712 \times 10^{-5}$$

Step 3 Explain: The answer we have found tells us Vega's radial velocity as a fraction of the speed of light; it is negative because Vega is moving toward us. To convert our answer to a velocity in km/s, we multiply by the speed of light:

$$v_{rad} = -4.5712 \times 10^{-5} \times c$$

$$= -4.5712 \times 10^{-5} \times (3 \times 10^{5} \text{ km/s})$$

$$= -13.7 \text{ km/s}$$

Vega is moving *toward* us at 13.7 km/s. This velocity is typical of stars in our neighborhood of the galaxy.

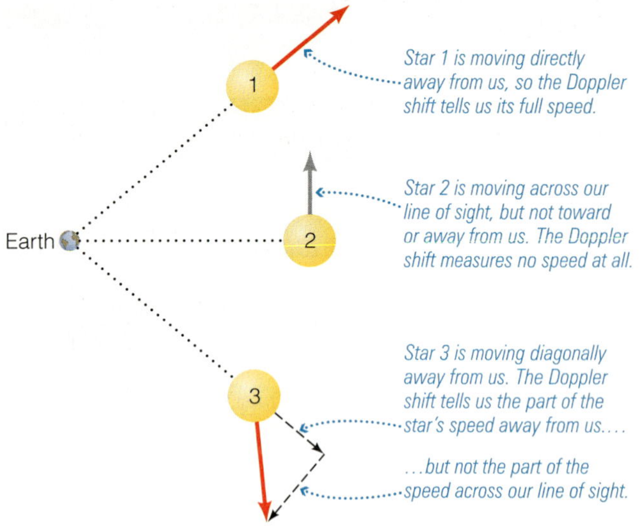

Figure 5.23 Interactive Figure The Doppler shift tells us only the portion of an object's speed that is directed toward or away from us. It does not give us any information about how fast an object is moving across our line of sight.

moving across our line of sight (the object's *tangential* component of motion). For example, consider three stars all moving at the same speed, with one moving directly away from us, one moving across our line of sight, and one moving diagonally away from us (Figure 5.23). The Doppler shift will tell us the full speed only of the first star. The Doppler shift will not measure any speed for the second star, because none of its motion is directed toward or away from us. For the third star, the Doppler shift will tell us only the part of the star's speed that is directed away from us. To measure how fast an object is moving across our line of sight, we must observe it long enough to notice how its position gradually shifts across our sky.

• How does light tell us the rotation rate of a distant object?

The Doppler effect not only tells us how fast a distant object is moving toward or away from us but also can reveal information about motion *within* the object. For example, suppose we look at spectral lines of a planet or star that happens to be rotating (Figure 5.24).

As the object rotates, light from the part of the object rotating toward us will be blueshifted, light from the part rotating away from us will be redshifted, and light from the center of the object won't be shifted at all. The net effect, if we look at the whole object at once, is to make each spectral line appear *wider* than it would if the object were not rotating. The faster the object is rotating, the broader in wavelength the spectral lines become. Thus, we can determine the rotation rate of distant objects by measuring the width of their spectral lines. We will see later that the Doppler effect on the spectra of celestial objects reveals even more information.

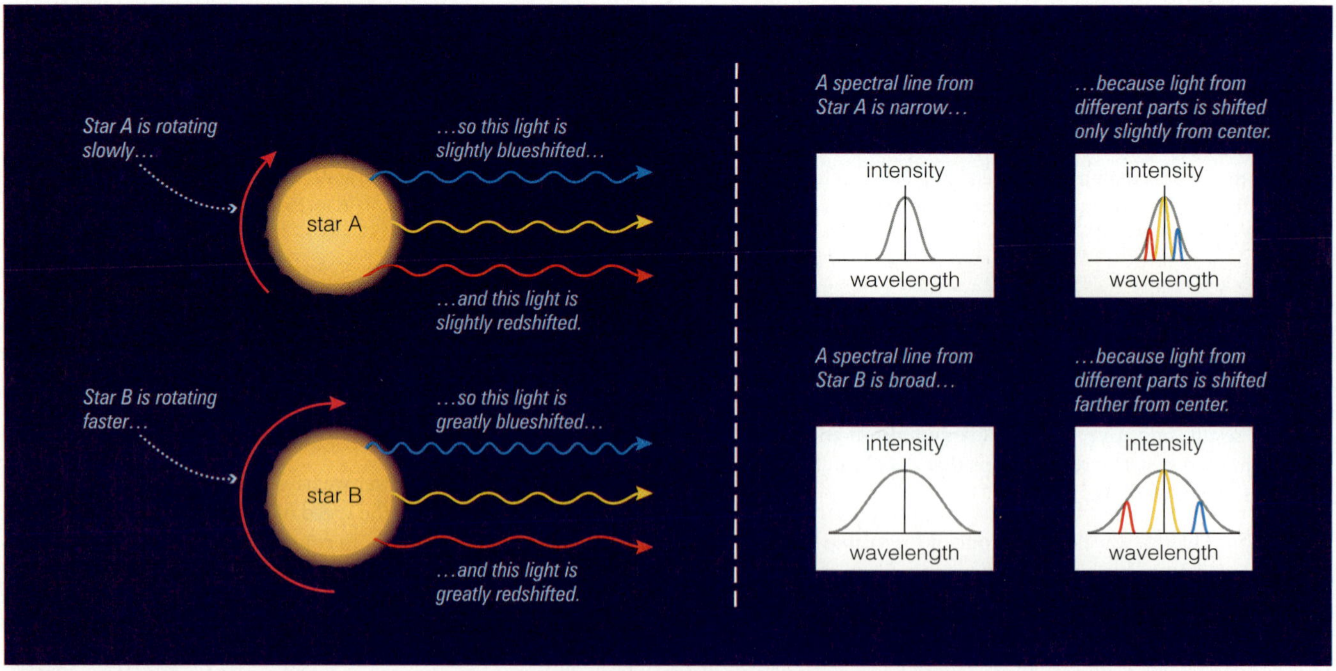

Figure 5.24 The Doppler effect broadens the widths of the spectral lines of rotating objects. One side of a rotating object is moving toward us, creating a blueshift, while the other side is rotating away from us, creating a redshift. The faster the rotation, the greater the spread in wavelength between the light from the two sides. When we look at a spectral line for the object as a whole, we see light from all parts of the object—that is, the parts with blueshifts as well as the parts with redshifts—at the same time. Thus, the greater Doppler shifts in a fast-rotating object make the overall spectral line wider because the light is spread over a greater range of wavelengths.

THE BIG PICTURE

Putting Chapter 5 into Context

This chapter was devoted to one essential purpose: understanding how we learn about the universe by observing the light of distant objects. "Big picture" ideas that will help you keep your understanding in perspective include the following:

- Light and matter interact in ways that allow matter to leave "fingerprints" on light. Thus, we can learn a great deal about the objects we observe by carefully analyzing their light. Most of what we know about the universe comes from information that we receive from light.
- The visible light that our eyes can see is only a small portion of the complete electromagnetic spectrum. Different portions of the spectrum may contain different pieces of the story of a distant object, so it is important to study all forms of light.
- There is far more to light than meets the eye. By dispersing the light of a distant object into a spectrum, we can determine the object's composition, surface temperature, motion toward or away from us, rotation rate, and more.

SUMMARY OF KEY CONCEPTS

5.1 Light in Everyday Life

- **How do we experience light?** Light carries radiative energy that it can exchange with matter. **Power** is the *rate* of energy transfer, measured in **watts**: 1 watt = 1 joule/s. The colors of light contain a great deal of information about the matter with which it has interacted.
- **How do light and matter interact?** Matter can emit light, absorb light, transmit light, or reflect (or scatter) light.

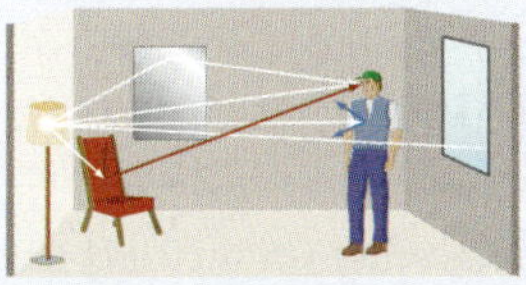

5.2 Properties of Light

- **What is light?** Light is an **electromagnetic wave**, but also comes in individual "pieces" called **photons**. Each photon has a precise wavelength, frequency, and energy: The shorter the wavelength, the higher the frequency and energy.

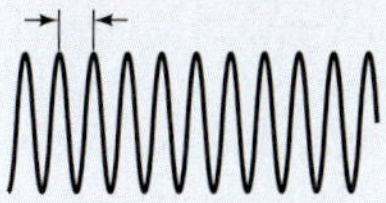

- **What is the electromagnetic spectrum?** In order of decreasing wavelength (increasing frequency and energy), the forms of light are **radio waves**, **microwaves**, **infrared**, **visible light**, **ultraviolet**, **X rays**, and **gamma rays**.

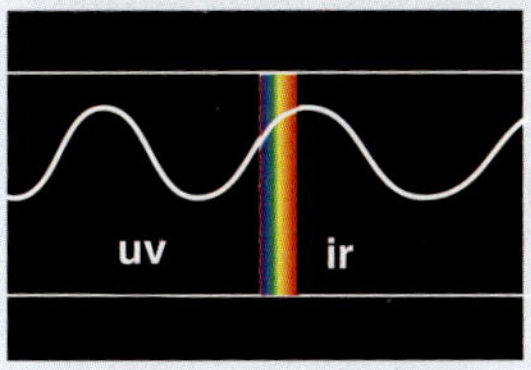

5.3 Properties of Matter

- **What is the structure of matter?** Ordinary matter is made of **atoms**, which are made of **protons**, **neutrons**, and **electrons**. Atoms of different **chemical elements** have different numbers of protons. **Isotopes** of a particular chemical element all have the same number of protons but different numbers of neutrons. **Molecules** are made from two or more atoms.

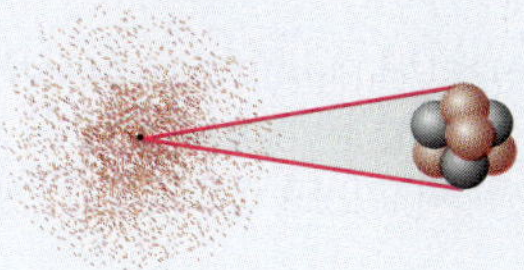

- **What are the phases of matter?** The appearance of matter depends on its phase: **solid**, **liquid**, or **gas**. Gas is always present along with solid or liquid phases; solids **sublimate** into gas and liquids **evaporate** into gas. At very high temperatures, molecular dissociation breaks up molecules and **ionization** strips electrons from atoms; an ionized gas is called a **plasma**.
- **How is energy stored in atoms?** Electrons can exist only in particular **energy levels** within an atom. **Energy level transitions**, in which an electron moves from one energy level to another, can occur only when the electron gains or loses just the right amount of energy.

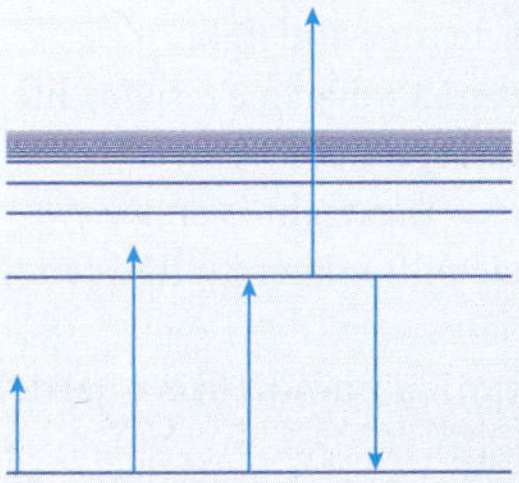

5.4 Learning from Light

- **What are the three basic types of spectra?** There are three basic types of spectra: a **continuous** spectrum, which looks like a rainbow of light; an **absorption line** spectrum, in which specific colors are missing from the rainbow; and an **emission line** spectrum, in which we see light only with specific colors against a black background.

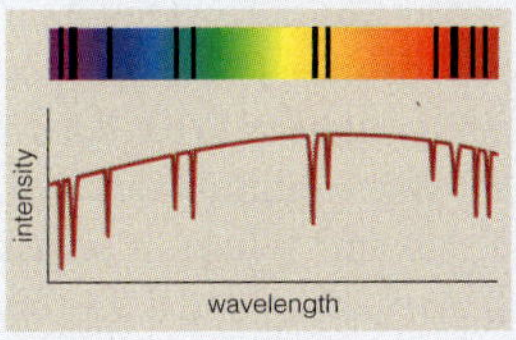

- **How does light tell us what things are made of?** Emission or absorption lines occur only at specific wavelengths that correspond to particular energy level transitions in atoms or molecules. Every kind of atom, ion, and molecule produces a unique set of spectral lines, so we can determine composition by identifying these lines.
- **How does light tell us the temperatures of planets and stars?** Objects such as planets and stars produce **thermal radiation** spectra, the most common type of continuous spectra. We can determine temperature from these spectra because hotter objects emit more total radiation per unit area and emit photons with a higher average energy.

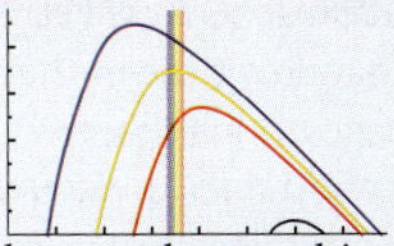

- **How do we interpret an actual spectrum?** The spectrum of a real object generally shows a combination of features of emission line, absorption line, and thermal radiation spectra, as well as features produced by reflection. By carefully studying the spectral features, we can learn a great deal about the object that produced them.

5.5 The Doppler Shift

- **How does light tell us the speed of a distant object?** The **Doppler effect** tells us how fast an object is moving toward or away from us. Spectral lines are shifted to shorter wavelengths (a **blueshift**) in objects moving toward us and to longer wavelengths (a **redshift**) in objects moving away from us.

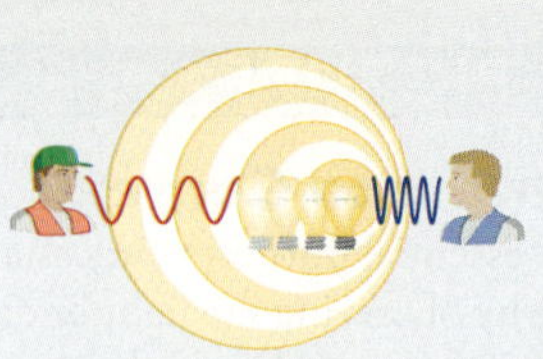

- **How does light tell us the rotation rate of a distant object?** The Doppler effect broadens spectral lines of rotating objects. The faster the rotation, the wider the spectral line.

EXERCISES AND PROBLEMS

For instructor-assigned homework go to MasteringASTRONOMY www.masteringastronomy.com

Review Questions

Short-Answer Questions Based on the Reading

1. What is the difference between *energy* and *power*? What units do we use to measure power?
2. What is a *spectrum*, and how do we see one?
3. Give an example from everyday life of each of the four major types of interaction between matter and light.
4. How does a particle differ from a wave in everyday life? Define *wavelength, frequency,* and *speed* for a wave.
5. What do we mean when we say that light is an *electromagnetic wave*? How is wavelength related to frequency for electromagnetic waves? Explain.
6. What is a *photon*? In what way is a photon like a particle? In what way is it like a wave?
7. List the different forms of light in order from lowest to highest energy. Would the list be different if you went in order from lowest to highest frequency? From shortest to longest wavelength? Explain.
8. Briefly describe the structure of an atom. How big is an atom? How big is the *nucleus* in comparison to the entire atom?
9. What determines an atom's *atomic number*? What determines its *atomic mass number*? Under what conditions are two atoms different *isotopes* of the same element? What is a *molecule*?
10. What is *electrical charge*? Will an electron and a proton attract or repel one another? Will two electrons attract or repel one another? Explain.
11. Describe the phase changes of water as you heat it starting from its solid phase, ice. What happens at very high temperatures? What is a *plasma*?
12. What do we mean when we say that *energy levels* are quantized in atoms? Under what circumstances can *energy level transitions* occur?
13. How do we convert a spectrum shown as a band of light (like a rainbow) into a graph of the spectrum?
14. Describe the conditions that would cause us to see each of the three basic types of spectra. What do we see in the Sun's spectrum shown on the opening page of this chapter?
15. How can we use emission or absorption lines to determine the chemical composition of a distant object?
16. Describe two ways in which the thermal radiation spectrum of an 8,000 K star would differ from that of a 4,000 K star.
17. Describe each of the key features of the spectrum in Figure 5.20 and explain what it tells us about the object.
18. Describe the *Doppler effect* for light and what we can learn from it. What does it mean to say that radio waves are *blueshifted*? Why does the Doppler effect widen the spectral lines of rotating objects?

Test Your Understanding

Does It Make Sense?

Decide whether the statement makes sense (or is clearly true) or does not make sense (or is clearly false). Explain your reasoning. (For an example, see Chapter 1, "Does It Make Sense?")

19. The walls of my room are transparent to radio waves.
20. Because of their higher frequencies, X rays must travel through space faster than radio waves.
21. If you could see infrared light, you would see a glow from the backs of your eyelids when you closed your eyes.
22. If you had X-ray vision, you could read this entire book without turning any pages.
23. Two isotopes of the element rubidium differ not only in their number of neutrons, but also in their number of protons.
24. A "white hot" object is hotter than a "red hot" object.
25. If the Sun's surface became much hotter (while the Sun's size remained the same), the Sun would emit more ultraviolet light but less visible light than it currently emits.
26. If you could view a spectrum of light reflecting off a blue sweatshirt, you'd find the entire rainbow of color (looking the same as a spectrum of white light).
27. Nearly all galaxies show redshifts, so they must all be red in color.
28. If a distant galaxy has a substantial redshift (as viewed from our galaxy), then anyone living in that galaxy would see a substantial redshift in a spectrum of the Milky Way Galaxy.

Quick Quiz

Choose the best answer to each of the following. Explain your reasoning with one or more complete sentences.

29. Why is a sunflower yellow? (a) It emits yellow light. (b) It absorbs yellow light. (c) It reflects yellow light.

30. Blue light has higher frequency than red light. Thus, blue light has: (a) higher energy and shorter wavelength than red light; (b) higher energy and longer wavelength than red light; (c) lower energy and shorter wavelength than red light.
31. Radio waves are: (a) a form of sound; (b) a form of light; (c) a type of spectrum.
32. Compared to an atom as a whole, an atomic nucleus: (a) is very tiny but has most of the mass; (b) is quite large and has most of the mass; (c) is very tiny and has very little mass.
33. Some nitrogen atoms have seven neutrons and some have eight neutrons. These two forms of nitrogen are: (a) ions of each other; (b) phases of each other; (c) isotopes of each other.
34. Sublimation is the process by which: (a) solid material enters the gas phase; (b) liquid material enters the gas phase; (c) solid material becomes a liquid.
35. If you heat a rock until it glows, its spectrum will be: (a) a thermal radiation spectrum; (b) an absorption line spectrum; (c) an emission line spectrum.
36. The set of spectral lines that we see in a star's spectrum depends on the star's: (a) interior temperature; (b) chemical composition; (c) rotation rate.
37. A star whose spectrum peaks in the infrared is: (a) cooler than our Sun; (b) hotter than our Sun; (c) larger than our Sun.
38. A spectral line that appears at a wavelength of 321 nm in the laboratory appears at a wavelength of 328 nm in the spectrum of a distant object. We say that the object's spectrum is: (a) redshifted; (b) blueshifted; (c) skewed.

Investigate Further

In-Depth Questions to Increase Your Understanding

Short-Answer/Essay Questions

39. *Atomic Terminology Practice I.*
    a. The most common form of iron has 26 protons and 30 neutrons in its nucleus. State its atomic number, atomic mass number, and number of electrons if it is electrically neutral.
    b. Consider the following three atoms: Atom 1 has 7 protons and 8 neutrons; atom 2 has 8 protons and 7 neutrons; atom 3 has 8 protons and 8 neutrons. Which two are *isotopes* of the same element?
    c. Oxygen has atomic number 8. How many times must an oxygen atom be ionized to create an O^{+5} ion? How many electrons are in an O^{+5} ion?
40. *Atomic Terminology Practice II.*
    a. Consider fluorine atoms with nine protons and 10 neutrons. What are the atomic number and atomic mass number of this fluorine? Suppose we could add a proton to this fluorine nucleus. Would the result still be fluorine? Explain. What if we added a neutron to the fluorine nucleus?
    b. The most common isotope of gold has atomic number 79 and atomic mass number 197. How many protons and neutrons does the gold nucleus contain? If it is electrically neutral, how many electrons does it have? If it is triply ionized, how many electrons does it have?
    c. The most common isotope of uranium is ^{238}U, but the form used in nuclear bombs and nuclear power plants is ^{235}U. Given that uranium has atomic number 92, how many neutrons are in each of these two isotopes of uranium?
41. *The Fourth Phase of Matter.*
    a. Explain why nearly all the matter in the Sun is in the plasma phase.
    b. Based on your answer to part (a), explain why plasma is the most common phase of matter in the universe.
    c. If plasma is the most common phase of matter in the universe, why is it so rare on Earth?
42. *Energy Level Transitions.* The following labeled transitions represent an electron moving between energy levels in hydrogen. Answer each of the following questions and explain your answers.

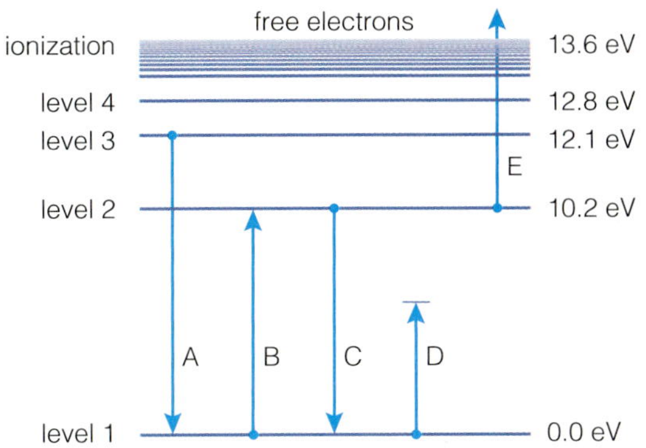

    a. Which transition could represent an atom that *absorbs* a photon with 10.2 eV of energy?
    b. Which transition could represent an atom that *emits* a photon with 10.2 eV of energy?
    c. Which transition represents an electron that is breaking free of the atom?
    d. Which transition, as shown, is *not* possible?
    e. Would transition A represent emission or absorption of light? How would the wavelength of the emitted or absorbed photon compare to that of the photon involved in transition C? Explain.
43. *Spectral Summary.* Clearly explain how studying an object's spectrum can allow us to determine each of the following properties of the object.
    a. The object's surface chemical composition.
    b. The object's surface temperature.
    c. Whether the object is a low-density cloud of gas or something more substantial.
    d. Whether the object has a hot upper atmosphere.
    e. Whether the object is reflecting blue light from a star.
    f. The speed at which the object is moving toward or away from us.
    g. The object's rotation rate.
44. *Orion Nebula.* To the eye (through a telescope), much of the Orion Nebula looks like a glowing cloud of gas. What type of spectrum would you expect to see from the glowing parts of the nebula? Why?
45. *Neptune's Spectrum.* The planet Neptune is colder than Mars and it appears blue in color. (a) Make a sketch similar to Figure 5.20 for Mars, but instead showing the spectrum you'd expect to see from Neptune. Label the axes clearly, and briefly describe each of the features shown in your spectrum in much the same way that Figure 5.20 describes the features in Mars's spectrum. (b) Suppose a very large asteroid crashed into Neptune, causing its atmosphere to become 10K warmer for a short time. List two ways in which the spectrum you drew in part (a) would differ when the atmosphere became

warmer. (c) Suppose Neptune rotated much faster. How would you expect its spectral lines to change?

46. *The Doppler Effect.* In hydrogen, the transition from level 2 to level 1 has a rest wavelength of 121.6 nm. Suppose you see this line at a wavelength of 120.5 nm in Star A, at 121.2 nm in Star B, at 121.9 nm in Star C, and at 122.9 nm in Star D. Which stars are coming toward us? Which are moving away? Which star is moving fastest relative to us (either toward or away from)? Explain your answers without doing any calculations.

Quantitative Problems

Be sure to show all calculations clearly and state your final answers in complete sentences.

47. *Human Wattage.* A typical adult uses about 2,500 Calories of energy each day. Use this fact to calculate the typical adult's average *power* requirement, in watts. (*Hint:* 1 Calorie = 4,184 joules.)
48. *Electric Bill.* Your electric utility bill probably shows your energy use for the month in units of *kilowatt-hours.* A kilowatt-hour is defined as the energy used in 1 hour at a rate of 1 kilowatt (1,000 watts); that is, 1 kilowatt-hour = 1 kilowatt $\times$ 1 hour. Use this fact to convert 1 kilowatt-hour into joules. If your bill says you used 900 kilowatt-hours, how much energy did you use in joules?
49. *Radio Station.* What is the wavelength of a radio photon from an "AM" radio station that broadcasts at 1,120 kilohertz? What is its energy?
50. *UV Photon.* What is the energy (in joules) of an ultraviolet photon with wavelength 120 nm? What is its frequency?
51. *X-Ray Photon.* What is the wavelength of an X-ray photon with energy 10 keV (10,000 eV)? What is its frequency? (*Hint:* 1 eV = 1.60×10^{-19} joule.)
52. *How Many Photons?* Suppose that all the energy from a 100-watt light bulb came in the form of photons with wavelength 600 nm. (This is not quite realistic; see Problem 59.)
    a. Calculate the energy of a *single* photon with wavelength 600 nm.
    b. How many 600-nm photons must be emitted each second to account for all the light from this 100-watt light bulb?
    c. Based on your answer to part (b), explain why we don't notice the particle nature of light in our everyday lives.
53. *Thermal Radiation Laws I.* Consider a 3,000 K object that emits thermal radiation. How much power does it emit per square meter? What is its wavelength of peak intensity?
54. *Thermal Radiation Laws II.* Consider a 50,000 K object that emits thermal radiation. How much power does it emit per square meter? What is its wavelength of peak intensity?
55. *Hotter Sun.* Suppose the surface temperature of the Sun were about 12,000 K, rather than 6,000 K.
    a. How much more thermal radiation would the Sun emit?
    b. What would happen to the Sun's wavelength of peak emission?
    c. Do you think it would still be possible to have life on Earth? Explain.
56. *Taking the Sun's Temperature.* The Sun radiates a total power of about 4×10^{26} watts into space. The Sun's radius is about 7×10^{8} meters.
    a. Calculate the average power radiated by each square meter of the Sun's surface. (*Hint:* The formula for the surface area of a sphere is $A = 4\pi r^2$.)
    b. Using your answer from part (a) and the Stefan-Boltzmann law, calculate the average surface temperature of the Sun. (*Note:* The temperature calculated this way is called the Sun's *effective temperature.*)
57. *Doppler Calculations I.* In hydrogen, the transition from level 2 to level 1 has a rest wavelength of 121.6 nm. Suppose you see this line at a wavelength of 120.5 nm in Star A and at 121.2 nm in Star B. Calculate each star's speed, and be sure to state whether it is moving toward or away from us.
58. *Doppler Calculations II.* In hydrogen, the transition from level 2 to level 1 has a rest wavelength of 121.6 nm. Suppose you see this line at a wavelength of 121.9 nm in Star C and at 122.9 nm in Star D. Calculate each star's speed, and be sure to state whether it is moving toward or away from us.
59. *Understanding Light Bulbs.* A standard (incandescent) light bulb uses a hot tungsten coil to produce a thermal radiation spectrum. The temperature of this coil is typically about 3,000 K.
    a. What is the wavelength of maximum intensity for a standard light bulb? Compare this to the 500-nm wavelength of maximum intensity for the Sun.
    b. Overall, do you expect the light from a standard bulb to be the same as, redder than, or bluer than light from the Sun? Why? Use your answer to explain why professional photographers use a different type of film for indoor photography than for outdoor photography.
    c. Do standard light bulbs emit all their energy as visible light? Use your answer to explain why light bulbs are usually hot to touch.
    d. *Fluorescent* light bulbs primarily produce emission line spectra rather than thermal radiation spectra. Explain why, if the emission lines are in the visible part of the spectrum, a fluorescent bulb can emit more visible light than a standard bulb of the same wattage.
    e. *Compact fluorescent* light bulbs are designed to produce so many emission lines in the visible part of the spectrum that their light looks very similar to the light of standard bulbs. However, they are much more energy efficient: A 15-watt compact fluorescent bulb typically emits as much visible light as a standard 75-watt bulb. Although compact fluorescent bulbs generally cost more than standard bulbs, is it possible that they could save you money? Besides initial cost and energy efficiency, what other factors must be considered?

Discussion Questions

60. *The Changing Limitations of Science.* In 1835, French philosopher Auguste Comte stated that science would never allow us to learn the composition of stars. Although spectral lines had been seen in the Sun's spectrum by that time, not until the mid-1800s did scientists recognize that spectral lines give clear information about chemical composition (primarily through the work of Foucault and Kirchhoff). Why might our present knowledge have seemed unattainable in 1835? Discuss how new discoveries can change the apparent limitations of science. Today, other questions seem beyond the reach of science, such as the question of how life began on Earth. Do you think such questions will ever be answerable through science? Defend your opinion.
61. *Your Microwave Oven.* A *microwave oven* emits microwaves that have just the right wavelength needed to cause energy level changes in water molecules. Use this fact to explain how

a microwave oven cooks your food. Why doesn't a microwave oven make a plastic dish get hot? Why do some clay dishes get hot in the microwave? Why do dishes that aren't themselves heated by the microwave oven sometimes still get hot when you heat food on them?

62. *Democritus and the Path of History.* Besides his belief in atoms, Democritus held several other strikingly modern notions. For example, he maintained that the Moon was a world with mountains and valleys and that the Milky Way was composed of countless individual stars—ideas that weren't generally accepted until the time of Galileo, more than 2,000 years later. Unfortunately, we know of Democritus's work only secondhand because none of the 72 books he is said to have written survived the destruction of the Library of Alexandria. Do you think history might have been different if the work of Democritus had not been lost? Defend your opinion.

MEDIA EXPLORATIONS

For self-study activities go to MasteringASTRONOMY www.masteringastronomy.com

Interactive Tutorials

Tutorial Review of Key Concepts

Use the interactive **Tutorials** at **www.masteringastronomy.com** to review key concepts from this chapter.

Light and Spectroscopy Tutorial

Lesson 1 Radiation, Light, and Waves
Lesson 2 Spectroscopy
Lesson 3 Atomic Spectra—Emission and Absorption Lines
Lesson 4 Thermal Radiation

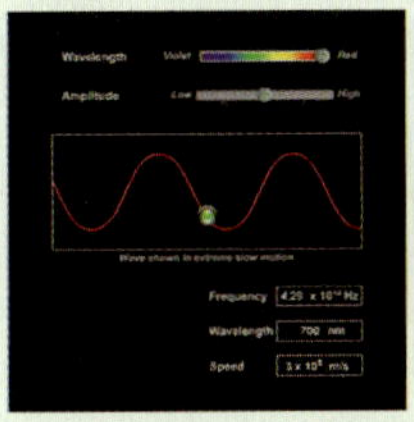
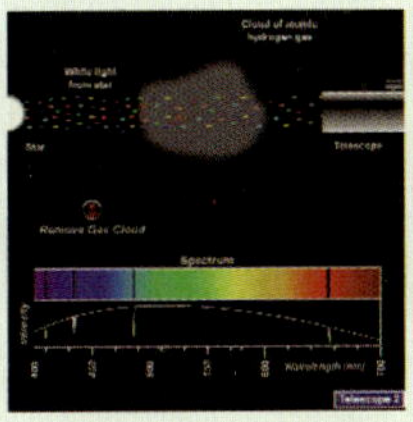
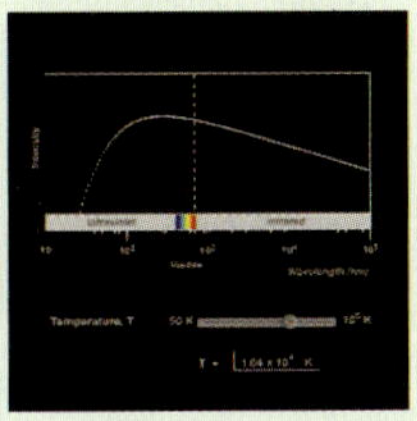

Doppler Effect Tutorial

Lesson 1 Understanding the Doppler Shift
Lesson 2 Using Emission and Absorption Lines to Measure the Doppler Shift

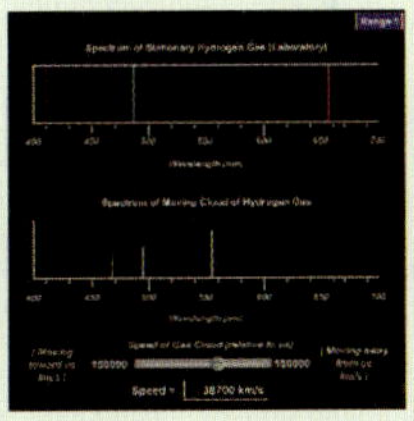
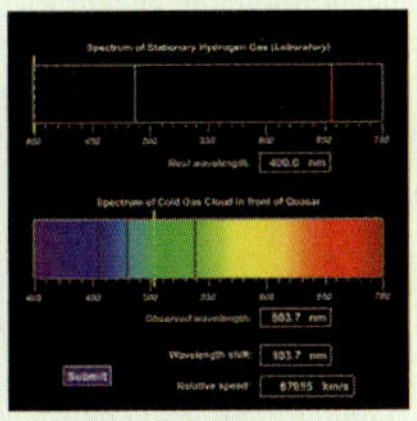

Supplementary Tutorial Exercises

Use the interactive **Tutorial Lessons** to explore the following questions.

Light and Spectroscopy Tutorial, Lesson 3

1. Why do neon lights come in so many different colors?
2. How can we tell the difference between lithium and carbon in the spectrum of a distant object?
3. How will an interstellar cloud affect the spectrum of light that we see from a star that lies behind it?

Light and Spectroscopy Tutorial, Lesson 4

1. How would an object's spectrum influence the choice of telescope used to observe it?
2. Consider the spectra of hot stars as compared to those of cooler stars. Which stars would be best observed with an ultraviolet telescope? Explain.

Doppler Effect Tutorial, Lesson 1

1. As the source of a sound passes you, which two properties will you hear change? Which change is due to the Doppler effect?
2. Which property of a sound wave is responsible for the pitch that your ears hear? How does the Doppler effect change the sound you hear?
3. How can the Doppler effect for light be used to measure the speed of a car?

Doppler Effect Tutorial, Lesson 2

1. How can we use precise wavelengths of spectral lines to measure the speed of a distant object?
2. Does it matter whether lines are in emission or absorption for the Doppler effect? Explain.

Web Projects

1. *Kids and Light.* Visit one of the many Web sites designed to teach middle and high school students about light. Read the content and try the activities. If you were a teacher, would you find the site useful for your students? Why or why not? Write a one-page summary of your conclusions.
2. *Light Bulbs.* Learn about alternatives to standard incandescent light bulbs, such as fluorescent lights, compact fluorescent lights, and halogen lights. Write a short report summarizing some of the advantages and disadvantages of each technology.
3. *Medical Imaging.* Learn about CAT scans or other technologies for medical imaging of the human body. How do they work? How are such technologies similar to those used by astronomers to learn about the universe? Write a short report summarizing your findings.

6

Telescopes

Portals of Discovery

LEARNING GOALS

6.1 Eyes and Cameras: Everyday Light Sensors
- How does your eye form an image?
- How do we record images?

6.2 Telescopes: Giant Eyes
- What are the two most important properties of a telescope?
- What are the two basic designs of telescopes?
- What do astronomers do with telescopes?

6.3 Telescopes and the Atmosphere
- How does Earth's atmosphere affect ground-based observations?
- Why do we put telescopes into space?

6.4 Telescopes and Technology
- How do we observe nonvisible light?
- How can multiple telescopes work together?

All of this has been discovered and observed these last days thanks to the telescope that I have [built], after having been enlightened by divine grace.

Galileo

We are in the midst of a great revolution in human understanding of the universe. Astonishing new discoveries about the early history of the universe, about the lives of galaxies and stars, and about planets orbiting other stars are frequent features of the daily news.

The primary fuel for this astronomical revolution has been recent and significant advances in telescope technology. A growing number of ever larger telescopes are operating around the world, and new technologies have vastly improved the quality of data that we can obtain with telescopes on the ground. Meanwhile, telescopes lofted into space are offering views of the heavens unobstructed by Earth's atmosphere while also allowing us to study wavelengths of light that do not penetrate to the ground. By studying light from across the entire spectrum, we can learn far more about the universe than we can by studying visible light alone.

Because telescopes are the portals through which we study the universe, understanding them can help us understand both the triumphs and the limitations of modern astronomy. In this chapter, we will explore the basic principles by which telescopes work, along with some of the technological advances fueling the current revolution in astronomy.

6.1 Eyes and Cameras: Everyday Light Sensors

We observe the world around us with the five basic senses: touch, taste, smell, hearing, and sight. We learn about the world by using our brains to analyze and interpret the data that our senses record. The science of astronomy progresses similarly. We collect data about the universe, and then we analyze and interpret the data. Within our solar system, we can analyze some matter directly. We can sample Earth's surface and study meteorites that fall to Earth. Occasionally, we send spacecraft to sample the surfaces or atmospheres of other worlds. Aside from these few samples collected in our own solar system, nearly all other data about the universe come to us in the form of light.

Astronomers collect light with telescopes and record light with cameras and other instruments. Thus, a good way to begin our study of telescopes is to examine the principles of eyes and cameras, our everyday light sensors.

• How does your eye form an image?

You've already seen many astronomical images in this book. Although most of these images have been made with powerful telescopes, the basic processes of image making are the same as those that occur in our eyes.

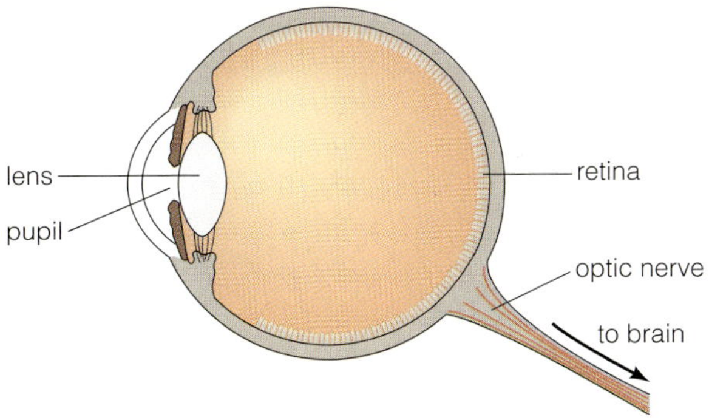

Figure 6.1 A simplified diagram of the human eye.

The eye is a remarkably complex organ, but its basic components are a *lens*, a *pupil*, and a *retina* (Figure 6.1). The retina contains light-sensitive cells (called *cones* and *rods*) that, when triggered by light, send signals to the brain via the optic nerve.

The lens of your eye creates an image on your retina because it bends light much like a simple glass lens.* To understand how that works, imagine a light wave coming toward you from a large distance. In that case, we can represent the wave's path as a simple *ray* of light traveling in a straight line toward you. The peaks and troughs of the light wave's electric and magnetic fields are perpendicular to the path of the ray. The wave slows down when it hits your eye because light generally travels more slowly through denser matter than through air. This change in speed causes the wave to bend, a phenomenon known as *refraction.* If the wave hits at an angle, the side of the wave nearest your eye slows down first, allowing the far side to catch up (Figure 6.2a). The end result is a change in the direction the light is traveling. Figure 6.2b shows an example of how the bending of light entering Earth's atmosphere from space distorts the Sun's image at sunset.

The bending of light that happens in your eye is precisely tuned to make vision possible. Light rays that enter the lens farther from the center are bent more, and rays that pass directly through the center are not bent at all. In this way, parallel rays of light, such as the light from a distant star, converge to a point called the **focus.** Figure 6.3 shows the idea for both a glass lense and an eye. The fact that parallel rays of light converge to a sharp focus explains why distant stars appear as *points* of light to our eyes or on photographs.

Light rays that are *not* parallel, such as those from a nearby object, enter a lens from different directions. These rays do not all converge at the focus, but they still follow precise rules as they are bent through the lens; some of

*Actually, both the lens and cornea of your eye are responsible for focusing an image. The cornea does about two-thirds of the work and the lens does the remaining third. For simplicity, the eye diagrams in this book show all the bending occurring at the lens.

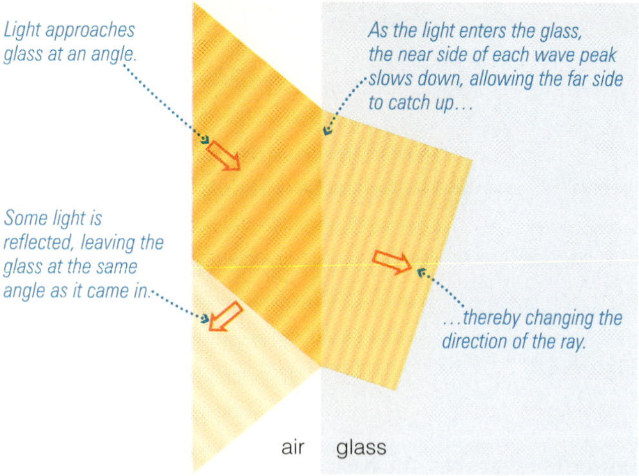

a Light that hits glass at an angle bends as it enters the glass. The wide yellow ribbons in this figure represent light waves. The darker bands on those ribbons (perpendicular to the direction of light travel) represent the positions of wave peaks.

b Earth's atmosphere also bends light slightly. The Sun looks slightly squashed at sunset because light from the lower portion of the Sun passes through more atmosphere and therefore bends slightly more than light from the upper portion.

Figure 6.2 Light bends when it moves from one substance to another, a phenomenon called *refraction*.

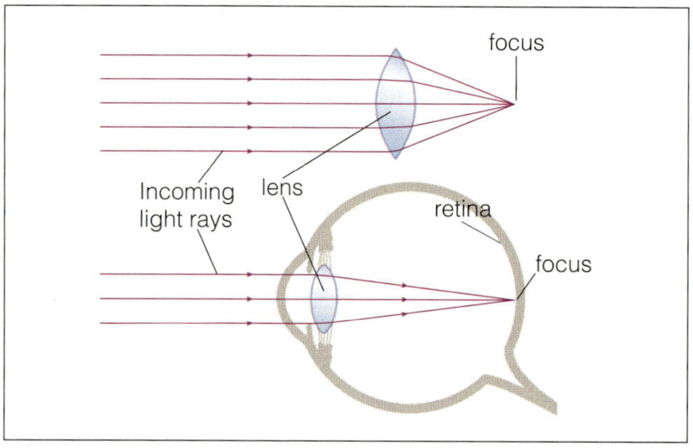

Figure 6.3 A glass lens bends parallel rays of light to a point called the *focus* of the lens. In an eye with perfect vision, rays of light are bent to a focus on the retina.

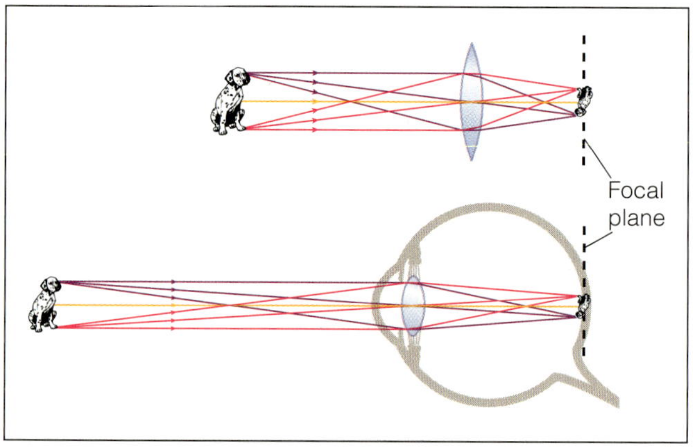

Figure 6.4 Light from different parts of an object focuses at different points to make an (upside-down) image of the object.

these rules are illustrated by the ray paths in Figure 6.4. The end result is that the rays are bent to form an **image** of the original object. The place where the image appears in focus is called the **focal plane** of the lens. In an eye with perfect vision, the focal plane is on your retina. (The retina actually is curved, rather than a flat plane, but we will ignore this detail.) The image formed by a lens is upside down but is flipped right side up by your brain, where the true miracle of vision occurs.

The pupil of the eye does not play a direct role in image formation, but it controls the amount of light that enters the eye by adjusting the size of its opening. The pupil dilates (opens wider) in low light levels, allowing your eye to gather as many photons of light as possible. It constricts in bright light so that your eye is not overloaded with light.

THINK ABOUT IT

Cover one eye and look toward a bright light. Then look immediately into a mirror and compare the openings of your two pupils. Which one is wider, and why? Why do eye doctors dilate your pupils during eye exams?

• How do we record images?

If we want to remember an image or study it in more detail, it's useful to record it with a camera. The basic operation of a camera is quite similar to that of an eye (Figure 6.5), except that the camera makes a permanent record of the image on film or by using some type of electronic detector. To keep our discussion simple, we'll use the term **detector** to mean any device that records light, including photographic

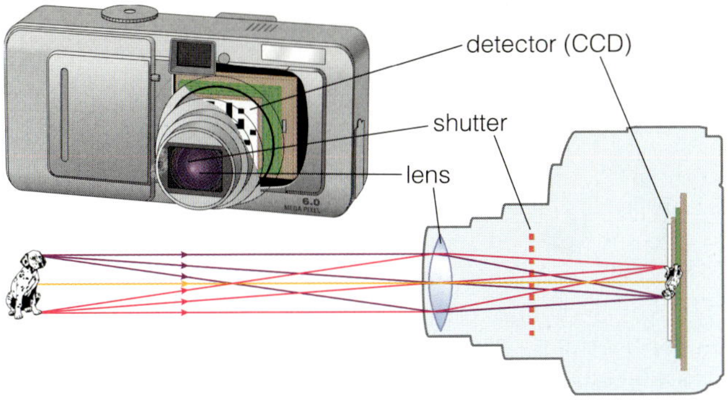

Figure 6.5 A camera works much like an eye. When the shutter is open, light passes through the lens to form an image on the detector (which may be film or an electronic device).

film and various types of digital-imaging chips. The camera lens plays the role of the lens of the eye and the detector plays the role of the retina. The camera shutter is analogous to an eyelid: Light can reach the detector only when the shutter is open. Fancier cameras even have an adjustable circular opening that controls the amount of light entering the camera just as the pupil controls the amount of light entering the eye. (The circular opening through which light enters is called the camera's *aperture*, so cameras in which the opening size is adjustable are said to have "aperture control.")

Recording an image with a camera offers at least two important advantages over simply looking at or drawing it. First, a recorded image is much more reliable and detailed than a drawing. Second, with a camera we can control the **exposure time**, the amount of time during which light collects on the detector. A longer exposure time means that more photons reach the detector, allowing the detector to record details that would be too faint to see with our eyes alone.

Thanks to the advantages of cameras over the human eye, the advent of photography in the mid-1800s spurred a leap in astronomical data collection. Until recently, these images were recorded on photographic film or glass plates coated with light-sensitive chemicals. These chemicals change in response to light, thereby recording an image.

Today, most images are recorded digitally with electronic detectors called *charge-coupled devices*, or **CCDs**. A CCD is a chip of silicon engineered to be extraordinarily sensitive to photons. The chip is physically divided into a grid of squares called *picture elements*, or **pixels** for short. When a photon strikes a pixel, it causes a bit of electric charge to accumulate. Each subsequent photon striking the same pixel adds to this accumulated electric charge. After an exposure is complete, a computer measures the total electric charge in each pixel, thus determining how many photons have struck each one. The overall image is stored on a memory chip as an array of numbers representing the results from each pixel. Many consumer camera chips now have 6 million or more pixels, and professional cameras can have significantly more.

THINK ABOUT IT

Find a digital camera (either your own or a friend's). Where is its lens? Where is its detector? Can you control its exposure time manually? How many pixels does its CCD have?

CCDs (and other types of electronic detectors) have three major advantages over photographic film. First, they are much more sensitive to light. Most photons striking film have no effect at all: Fewer than 10% of the visible photons reaching the film cause a change in the light-sensitive chemicals that coat it. In contrast, CCDs can accurately record 90% or more of the photons that strike them. Second, CCDs have a much wider *dynamic range* than photographic film, meaning that they can more easily record both dim and bright light at the same time. For example, if you try to record an image of a galaxy on film, the bright galactic center will be overexposed by the time you capture enough light to see details in the outskirts of the galaxy. A CCD may still be able to show details in both the dim and bright regions of the galaxy. Third, a digital image can be manipulated through techniques of *image processing* to bring out details that otherwise might be missed in analyzing the image.

Mastering Astronomy: Telescopes Tutorial, Lessons 1–2

6.2 Telescopes: Giant Eyes

Telescopes are essentially giant eyes that can collect far more light than our naked eyes can. They allow us to see objects that are far too faint for our eyes to see and with far more detail than our eyes can notice. In addition, we can connect scientific instruments to telescopes, allowing us to record and analyze the light they collect. For example, sophisticated cameras can make high-quality images of light collected by a telescope, and spectrographs can disperse the light into spectra that can reveal an object's chemical composition, temperature, speed, and more. In this section, we'll investigate the basic principles of telescopes.

• What are the two most important properties of a telescope?

Telescopes can differ in a number of ways, but the two most fundamental properties of any telescope are its *light-collecting area* and its *angular resolution*. Let's investigate each of these properties.

A telescope's **light-collecting area** tells us how much total light the telescope can collect at one time. Telescopes are generally round, so we usually characterize a telescope's "size" by stating the *diameter* of its light-collecting area. For example, a "10-meter telescope" has a light-collecting area that is 10 meters in diameter. Such a telescope has a light-collecting area more than a million times that of the human eye. Because area is proportional to the *square* of a telescope's diameter, a relatively small increase in diameter can

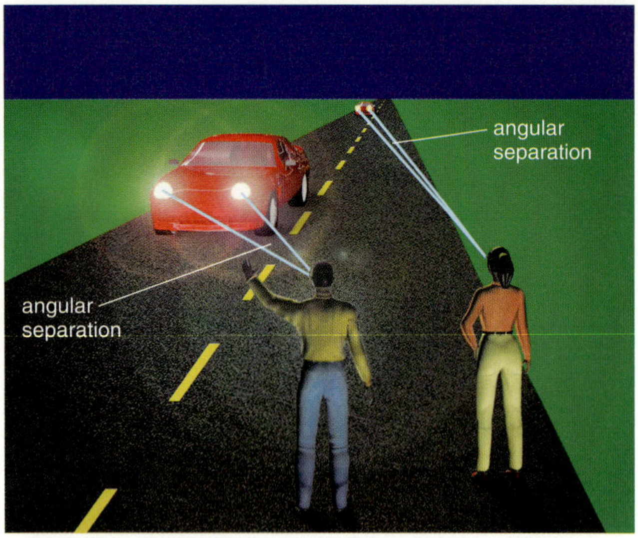

Figure 6.6 Interactive Figure This diagram shows how angular separation depends on distance. The headlights on the car have the same physical separation in both cases, but their angular separation is larger when the car is closer. Similarly, two stars separated by a particular distance will have a larger angular separation if they are nearby than if they are farther away.

COMMON MISCONCEPTIONS

Magnification and Telescopes

You are probably familiar with the way binoculars and telephoto camera lenses make objects appear larger in size—the phenomenon we call *magnification*. It's useful to know the magnification of binoculars and telephoto lenses, because it tells you what you can expect to see when you look through them. However, magnification is not a very useful way to describe a telescope's power, because magnification won't do you any good if the telescope collects too little light or has poor angular resolution. No matter how much you magnify the image, you cannot see details if the telescope does not collect enough light to show them, and you can never see details smaller than the angular resolution of the telescope. That is why magnifying an image too much just makes it look blurry, without revealing any new details, and why a telescope's light-collecting area and angular resolution are much more important than its magnification.

mean a big increase in light-collecting area. A 10-meter telescope has five times the diameter of a 2-meter telescope, so its light-collecting area is $5^2 = 25$ times as great.

THINK ABOUT IT

What is the population of your hometown? Suppose everyone in your hometown looked at the dark sky at the same time. How would the total amount of light entering everyone's eyes compare with the light collected by a 10-meter telescope? Explain.

Angular resolution gets its name because it is the smallest angle over which we can tell that two dots—or two stars—are distinct. For example, the human eye has an angular resolution of about 1 arcminute $\left(\frac{1}{60}^{\circ}\right)$, which means that two stars will appear distinct only if they have at least this much angular separation in the sky. If the stars are separated by less than 1 arcminute, our eyes will not be able to distinguish them individually and they will appear as a single star. The angular separation between two points of light depends both on their actual separation and on their distance from us [Section 2.1]; Figure 6.6 shows the idea.

Large telescopes can have amazing angular resolution. For example, the 2.4-meter Hubble Space Telescope has an angular resolution of about 0.05 arcsecond (for visible light), which would allow you to read this book from a distance of almost 1 kilometer. In principle, larger telescopes should have even better (smaller) angular resolution. However, all larger telescopes today are located on the ground, where effects of Earth's atmosphere can limit their angular resolution.

The ultimate limit to a telescope's resolving power comes from the properties of light. Because light is an electromagnetic wave [Section 5.2], beams of light can interfere with one another like overlapping sets of ripples on a pond (Figure 6.7). This *interference* causes a blurring of images that limits a telescope's angular resolution even when all

Figure 6.7 This computer-generated image shows how overlapping sets of ripples on a pond interfere with one another. The effects of the two sets of ripples add in some places, making the water rise extra high or fall extra low, and cancel in other places, making the water surface flat. Light waves also exhibit interference. (The colors in this image are for visual effect only.)

Angular Resolution

Angular resolution is the smallest angle over which we can tell that two dots are distinct. If we want to know whether two objects can be resolved, we simply need to determine their angular separation: If their angular separation is less than the angular resolution, then we will not be able to see the objects individually (they will be blurred together), and if it is more, then we will be able to tell them apart. Recall from Mathematical Insight 2.1 that we can calculate angular separation in degrees with this formula:

$$\text{angular separation} = \text{physical separation} \times \frac{360°}{2\pi \times \text{distance}}$$

With telescopes, angular resolution is usually measured in arcseconds or fractions of arcseconds. Thus, it's useful to revise the above formula to give an answer in arcseconds rather than degrees. Because there are 3,600 arcseconds in 1 degree (see Figure 2.8), we convert the units by multiplying by $\frac{3600''}{1°}$:

$$\text{angular separation} = \text{physical separation} \times \frac{360°}{2\pi \times \text{distance}} \times \frac{3600''}{1°}$$

We can simplify this formula by multiplying out the numbers; you should confirm that $360 \times 3600 \div 2\pi \approx 206{,}265$. The formula now reads:

$$\text{angular separation} = 206{,}265'' \times \frac{\text{physical separation}}{\text{distance}}$$

Remember that the physical separation and distance must both have the same units (such as kilometers or miles) in order for the angular separation to come out in arcseconds.

Example 1: Two stars in a binary star system are separated by 200 million kilometers. The system is located 20 light-years away. Can we study the stars individually with the Hubble Space Telescope's angular resolution of 0.05 arcsecond?

Solution:

Step 1 Understand: We are asked whether we can see two stars individually with the Hubble Space Telescope, which means we need to determine whether their angular separation is greater than or less than the telescope's angular resolution. The telescope can see them individually only if the angular resolution is smaller than the angular separation of the stars.

Step 2 Solve: We are given the telescope angular resolution of 0.05 arcsecond, so we need to calculate the angular separation of the two stars and see whether it is smaller or larger than 0.05 arcsecond. Note, however, that we are given the separation of the stars in kilometers and their distance in light-years. Therefore we must first put both numbers into the same units before we can use the formula we found above. Remembering that 1 light-year $\approx 10^{13}$ km (see Mathematical Insight 1.1), we find that 20 light-years $\approx 20 \times 10^{13}\text{ km} = 2 \times 10^{14}$ km. Using scientific notation, we write the physical separation of the two stars as 200 million km $= 2 \times 10^{8}$ km. We can now use our formula to find the angular separation of the two stars:

$$\begin{aligned}\text{angular separation} &= 206{,}265'' \times \frac{\text{physical separation}}{\text{distance}} \\ &= 206{,}265'' \times \frac{2 \times 10^{8}\text{ km}}{2 \times 10^{14}\text{ km}} \\ &= 0.2''\end{aligned}$$

Step 3 Explain: We have found that the angular separation of the two stars is 0.2 arcsecond, which is greater than the telescope's angular resolution of 0.05 arcsecond. We can therefore see and study the stars individually.

Example 2: If you looked at this book with a telescope that has Hubble's angular resolution of 0.05 arcsecond, how far away could you place the book and still be able to read it?

Solution:

Step 1 Understand: In this case, we can read the book if we can resolve its individual letters. So we must decide what we mean by the "angular separation" in the case of reading letters. There's a number of ways to think about this question, but here's one simple approach: You know that, in principle, we could make the letters out of tiny dots (just as they are made on a computer screen). Let's say that to read them clearly, you'd want the letters to be at least 10 dots tall (and 10 dots wide). If you measure, you'll find that the letters in this book are about 2 millimeters tall, so the dots would be separated by 0.2 millimeter. We can use this value as the physical separation of the dots. The telescope will be able to resolve the dots as long as their angular separation is at least as great as the telescope resolution of 0.05 arcsecond; we'll therefore use this value as the angular separation. We now need to calculate the distance at which the dots will have this angular separation.

Step 2 Solve: Solving the above formula for the distance, we find (you should confirm the algebra for yourself):

$$\text{distance} = 206{,}265'' \times \frac{\text{physical separation}}{\text{angular separation}}$$

We now plug in the angular separation of 0.05 arcsecond and the physical separation of the dots of 0.2 millimeter:

$$\text{distance} = 206{,}265'' \times \frac{0.2\text{ mm}}{0.05''} \approx 825{,}000\text{ mm}$$

Step 3 Explain: We've found the distance at which we could read the book, but the answer is difficult to interpret when written as nearly a million millimeters. To make the answer more meaningful, we convert it to meters:

$$825{,}000\ \cancel{\text{mm}} \times \frac{1\text{ m}}{1{,}000\ \cancel{\text{mm}}} = 825\text{ m}$$

A telescope with an angular resolution of 0.05 arcsecond should be able resolve the letters in this book as long as the book is within about 800 meters, or a little less than 1 kilometer.

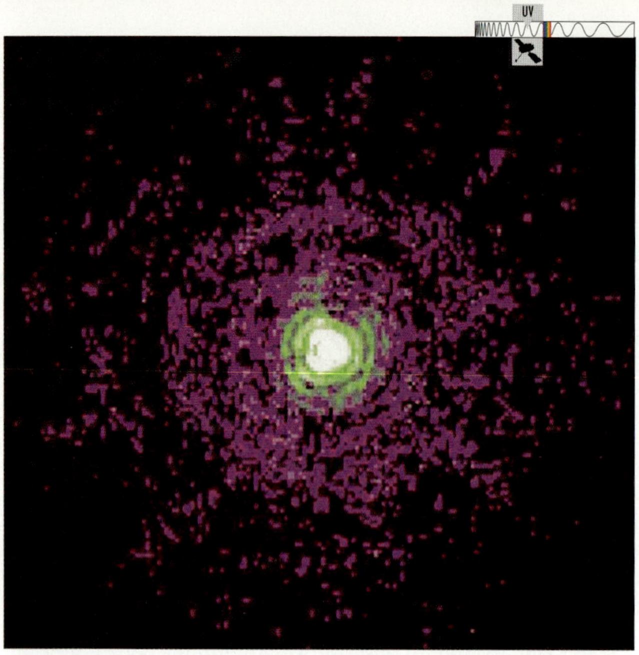

Figure 6.8 When examined in detail, a Hubble Space Telescope image of a star has rings (represented as green and purple in the figure) resulting from the wave properties of light. With higher angular resolution, the rings would be smaller.

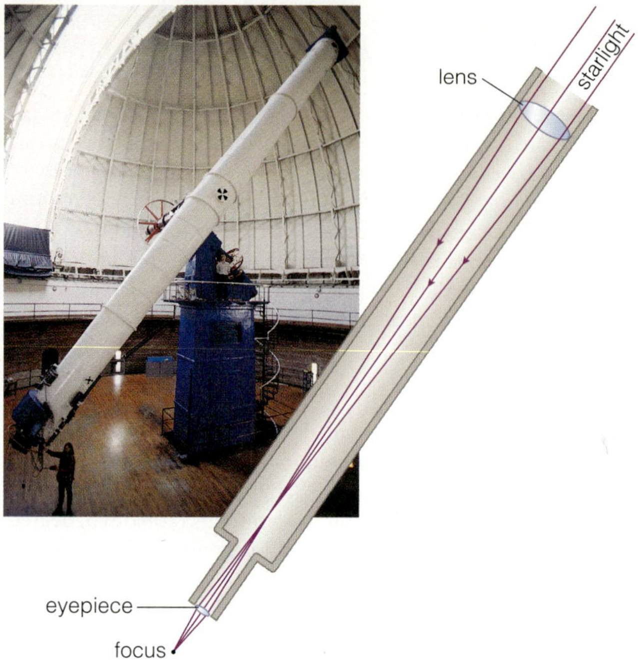

Figure 6.9 Refracting telescopes.

other conditions are perfect. That is why even a high-quality telescope in space, such as the Hubble Space Telescope, cannot have perfect angular resolution* (Figure 6.8).

The angular resolution that a telescope could achieve if it were limited only by the interference of light waves is called its **diffraction limit.** (*Diffraction* is a technical term for the specific effects of interference that limit telescope resolution.) The diffraction limit depends on both the diameter of the telescope's primary mirror and the wavelength of the light being observed (see Mathematical Insight 6.2, p. 185). For any particular wavelength of light, a larger telescope has a smaller diffraction limit, meaning it can achieve a better (smaller) angular resolution. For any particular telescope, the diffraction limit is larger (poorer angular resolution) for longer-wavelength light. That is why, for example, a radio telescope must be far larger than a visible-light telescope to achieve the same angular resolution.

• What are the two basic designs of telescopes?

Telescopes come in two basic designs: *refracting* and *reflecting.* A **refracting telescope** operates much like an eye, using transparent glass lenses to focus the light from distant objects (Figure 6.9). The earliest telescopes, including those that Galileo built, were refracting telescopes. The world's largest refracting telescope, completed in 1897, has a lens that is 1 meter (40 inches) in diameter and a telescope tube that is 19.5 meters (64 feet) long.

A **reflecting telescope** uses a precisely curved *primary mirror* to gather light (Figure 6.10). This mirror reflects the gathered light to a *secondary mirror* that lies in front of it. The secondary mirror then reflects the light to a focus at a place where the eye or instruments can observe it—sometimes through a hole in the primary mirror, and sometimes out the side of the telescope (perhaps with the aid of additional small mirrors). The fact that the secondary mirror prevents some light from reaching the primary mirror might seem like a drawback to reflecting telescopes, but in practice it is not a problem because only a small fraction of the incoming light is blocked.

Nearly all telescopes used in current astronomical research are reflectors, mainly for two practical reasons. First, because light passes *through* the lens of a refracting telescope, lenses must be made from clear, high-quality glass with precisely shaped surfaces on both sides. In contrast, only the reflecting surface of a mirror must be precisely shaped, and the quality of the underlying glass is not a factor. Second, large glass lenses are extremely heavy and can be held in place only by their edges. Since the large lens is at the top of a refracting telescope, it is difficult to stabilize refracting telescopes and to prevent large lenses from deforming. The primary mirror of a reflecting telescope is mounted at the bottom, where its weight presents a far less serious problem. (A third problematic feature of lenses, called *chromatic aberration,* occurs because a lens brings different colors of light into focus at slightly different places. This problem can be minimized by using combinations of lenses.)

For a long time, the main factor limiting the size of reflecting telescopes was the sheer weight of the glass needed

*In fact, the Hubble Space Telescope's primary mirror was made with the wrong shape, which further limited its angular resolution until corrective optics were installed in 1993. The corrective optics consist of small mirrors that correct for most of the blurring created by the primary mirror.

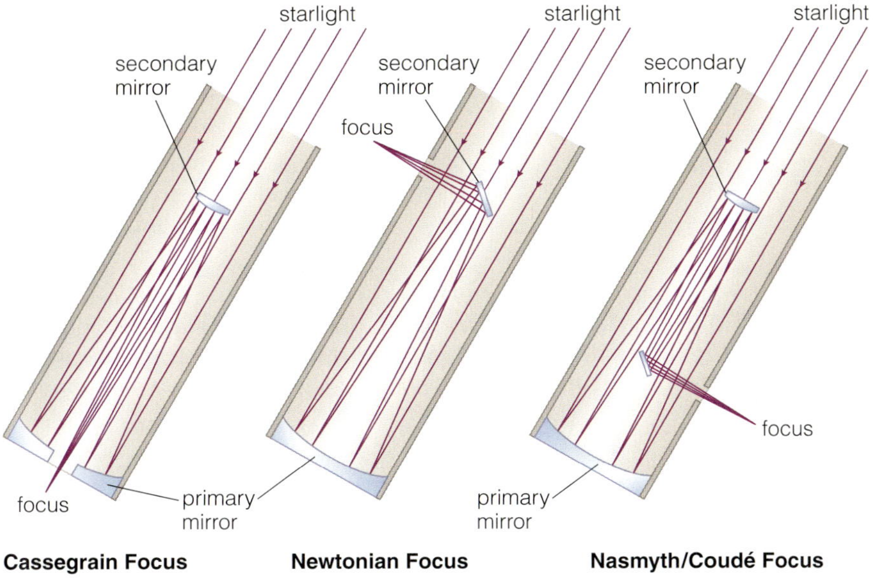

a Three variations on the basic design of a reflecting telescope. In all cases, a reflecting telescope collects light with a precisely curved primary mirror that reflects light back upward to the secondary mirror. In the *Cassegrain* design, the secondary mirror reflects the light through a hole in the primary mirror, so that the light can be observed with cameras or instruments beneath the telescope. In the *Newtonian* design, the secondary mirror reflects the light out to the side of the telescope. In the *Nasmyth* and *Coudé* designs, a third mirror is used to reflect light out the side but lower down than in the Newtonian design.

b The Gemini North telescope, located on the summit of Mauna Kea, Hawaii, is a reflecting telescope with the Cassegrain design. The primary mirror, visible at the bottom of the large lattice tube, is 8 meters in diameter. The secondary mirror, located in the smaller central lattice, reflects light back down through the hole visible in the center of the primary mirror.

Figure 6.10 Interactive Figure Reflecting telescopes.

for their primary mirrors. Recent technological innovations have made it possible to build lighter-weight mirrors, such as the one in the Gemini telescope shown in Figure 6.10b, or to make many small mirrors work together as one large one. For example, Figure 6.11 shows the primary mirror of one of the 10-meter Keck telescopes, which consists of 36 smaller mirrors that function together as one.

These new mirror-building technologies are fueling a revolution in the building of large telescopes. Before the 1990s, the 5-meter Hale telescope on Mount Palomar

Figure 6.11 This photo on the right shows the primary mirror of one of the Keck telescopes on Mauna Kea, with a man in the center for scale. The mirror is made up of 36 smaller, hexagonal mirrors, arranged in a honeycomb pattern. The photo on the left shows two Keck telescopes, with the primary mirrors visible inside the domes.

Table 6.1 Largest Optical (Visible-Light) Telescopes

Size	Name	Sponsor	Location	Opened	Special Features
10.4 m	Gran Telescopio Canarias	Spain, Mexico, U. Florida	Canary Islands	2005	Segmented primary mirror based on mirrors for Keck telescopes
10 m	Keck I and Keck II	Cal Tech, U. California, NASA	Mauna Kea, HI	1993/1996	Two identical 10-m telescopes, each with a primary mirror consisting of 36 1.8-m hexagonal segments
9.2 m	Hobby-Eberly	U. Texas, Penn State, Stanford, Germany	Mt. Locke, TX	1997	Consists of 91 1-m segments, for a total diameter of 11 m, but only 9.2 m can be used at a time; designed primarily for spectroscopy
9.2 m	South African Large Telescope	South Africa, Rutgers, UW–Madison, UNC–Chapel Hill, Dartmouth, Carnegie-Mellon, 5 others	South Africa	2004	Based on design of Hobby-Eberly telescope
2 × 8.4 m	Large Binocular Telescope	U. Arizona, Ohio State U., Italy, Germany	Mt. Graham, AZ	2006	Two 8.4-m mirrors on a common mount, giving light-collecting area of 11.8-m telescope
4 × 8.2 m	Very Large Telescope	European Southern Observatory	Cerro Paranal, Chile	2000	Four separate 8-m telescopes designed to work individually or together as the equivalent of a 16-m telescope
8.3 m	Subaru	Japan	Mauna Kea, HI	1999	Japan's first large telescope project
8 m	Gemini North and South	U.S., U.K., Canada, Chile, Brazil, Argentina	Mauna Kea, HI (North); Cerro Pachon, Chile (South)	1999	Twin telescopes, one in each hemisphere
6.5 m	Magellan I and II	Carnegie Institute, U. Arizona, Harvard, U. Michigan, MIT	Las Campanas, Chile	2000/2002	Twin 6.5-m telescopes, known respectively as the Walter Baade and Landon Clay telescopes
6.5 m	MMT	Smithsonian Institution, U. Arizona	Mt. Hopkins, AZ	2000	Replaced an older telescope in the same observatory

(outside San Diego) reigned for more than 40 years as the most powerful telescope in the world. Today, it does not even make the top-10 list for telescope size (Table 6.1). Astronomers are currently working on designs for even larger telescopes, and it's quite likely that we'll see telescopes with diameters of 30 meters or more within the next couple of decades.

• What do astronomers do with telescopes?

Cartoon images of astronomers usually show them looking through a telescope by eye (Figure 6.12). In fact, astronomers rarely have the chance to look through large telescopes by eye, because it is much more important to record observations with cameras or other equipment.

Every astronomical observation is unique, and astronomers use many different kinds of instruments and detectors to extract the information contained in the light collected by a telescope. Nevertheless, most observations fall into one of three basic categories: *imaging*, which yields photographs (images) of astronomical objects; *spectroscopy*, in which astronomers obtain and study spectra; and *timing*, which tracks how an object changes with time.* Let's look at each category in a little more detail.

Imaging At its most basic, an imaging instrument is simply a camera. Astronomers sometimes place *filters* in front of the camera that allow only particular colors or wavelengths of light to pass through. In fact, most of the richly hued

*Some astronomers include a fourth general category called *photometry*, which is the accurate measurement of light intensity from a particular object at a particular time. We do not list this as a separate category because today's detectors can generally perform photometry at the same time that they are being used for imaging, spectroscopy, or timing.

Figure 6.12 This isn't really the way professional astronomers work.

images in this and other astronomy books are made by combining images recorded through different filters (Figure 6.13).

Today, many astronomical observatories record images made from nonvisible light. For example, Figure 6.14 is an "X-ray image." What exactly does this mean, given that X rays are invisible to the eye? The easiest way to understand the idea is to think about X rays at a doctor's office. When the doctor makes an "X ray of your arm," he or she uses a machine that sends X rays through your arm. The X rays that pass through are recorded on a piece of X-ray sensitive film. We cannot see the X rays themselves, but we can see the image left behind on the film. Astronomical images work the same way. Whether a camera records visible or nonvisible light, it makes an image showing the intensity of the light.

The simplest way to display a nonvisible image is in black and white, with brighter regions of the image corresponding to brighter light coming from the object. (A doctor's X ray is actually a negative, making bones look bright because few X rays pass through them.) In astronomy, nonvisible images are often color-coded. In Figure 6.14, for example, the colors correspond to X rays of different energy. (Remember that higher energy means higher frequency and shorter wavelength.) All images recorded in nonvisible light must use some kind of color or gray-scale coding. Sometimes the colors correspond to different wavelengths (as in Figure 6.14), sometimes to different intensities of light, and sometimes to physical properties.

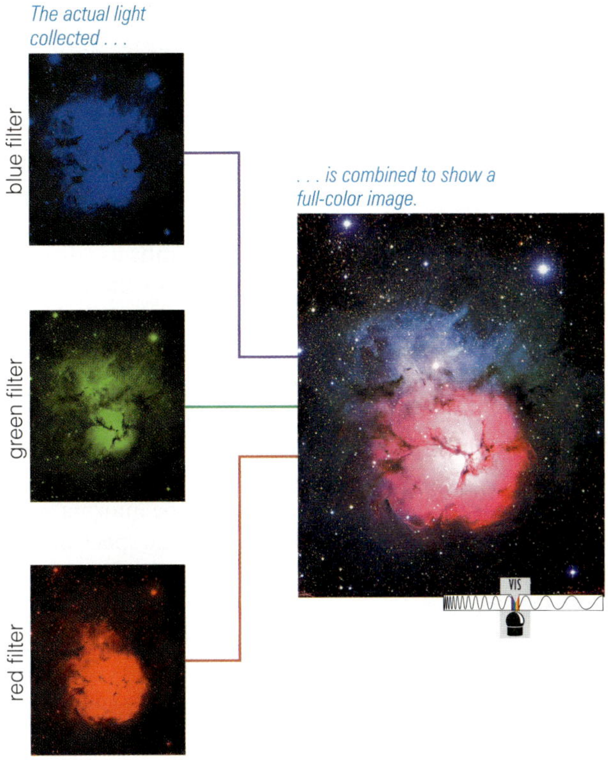

Figure 6.13 Interactive Photo In astronomy, color images are usually constructed by combining several images taken through different filters.

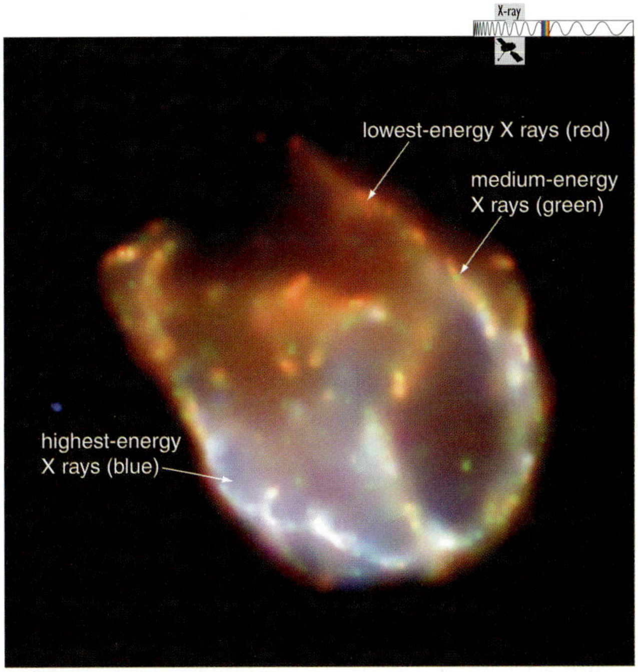

Figure 6.14 X rays are invisible, but we can color-code the information recorded by an X-ray detector to make an image of the object as it would appear in X rays. This image, from NASA's Chandra X-Ray Observatory, shows X-ray emission from the debris of a stellar explosion (a supernova remnant named N132D). The image is color-coded, with different colors representing X rays of different energy.

THINK ABOUT IT

Color-coded images are common even outside astronomy. For example, medical images from CAT scans and MRIs are usually displayed in color, even though neither type of imaging uses visible light. What do you think the colors mean in CAT scans and MRIs? How are the colors useful to doctors?

Spectroscopy Instruments called **spectrographs** use diffraction gratings (or other devices) to separate the various colors of light into spectra, which are then recorded with a detector such as a CCD (Figure 6.15).

As we discussed in Chapter 5, a spectrum can reveal a wealth of information about an object, including its chemical composition, temperature, and rotation rate. However, just as the amount of information we can glean from an image depends on the angular resolution, the information we can glean from a spectrum depends on the **spectral resolution**: The higher the spectral resolution, the more detail we can see (Figure 6.16).

In principle, astronomers would always like the highest possible spectral resolution. However, higher spectral resolution comes at a price. A telescope collects only so much light in a given amount of time, and the spectral resolution depends on how widely the spectrograph spreads out this light. If the spectrograph spreads out the light too much, the spectrum may become so dim that nothing can be recorded without a very long exposure time. Thus, for the same telescope, making a spectrum of an object requires a longer exposure time than making an image, and high-resolution spectra take longer than low-resolution spectra.

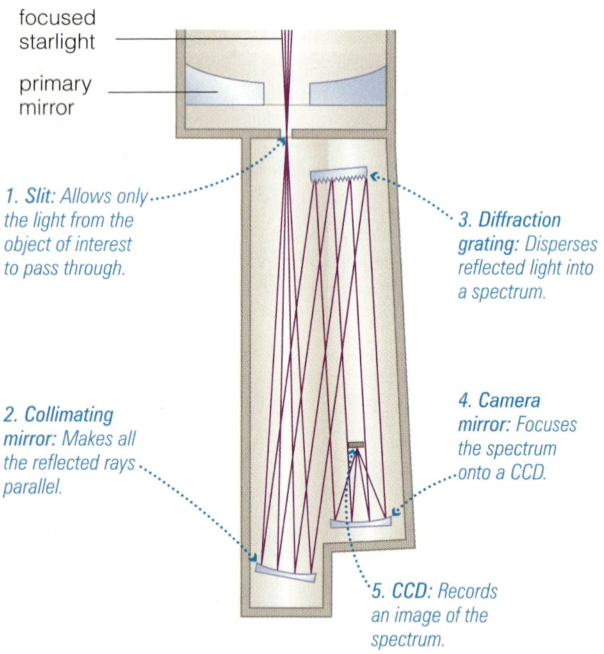

Figure 6.15 The basic design of a spectrograph. In this diagram, the spectrograph is attached to the bottom of a reflecting telescope, with light entering the spectrograph through a hole in the primary mirror. A narrow slit (or small hole) at the entrance to the spectrograph allows only light from the object of interest to pass through. This light bounces from the collimating mirror to the diffraction grating, which disperses the light into a spectrum. The dispersed light is then focused by the camera mirror and recorded by a CCD, giving an image of the spectrum.

Timing Many astronomical objects vary with time. For example, some stars undergo sudden outbursts caused by tremendous explosions. Other stars, including our own Sun, vary much less and much more slowly as sunspots or starspots cover more or less of their surfaces. Some objects vary periodically; looking for small, periodic changes in brightness is an important planet-hunting strategy [Section 13.1].

For a slowly varying object, a timing experiment may be as simple as comparing a set of images or spectra obtained on different nights. For more rapidly varying sources, specially designed instruments essentially make rapid multiple exposures, in some cases recording the arrival time of every individual photon.

The results of timing experiments are often shown as **light curves:** graphs that show how an object's intensity varies with time. For example, Figure 6.17 shows a light curve for the star Mira. Notice that the light output of Mira varies by more than a factor of 100 as it rises and falls with a period of a little less than a year.

An Astronomer's Job Using imaging, spectroscopy, and timing, astronomers can put together detailed pictures of the objects they study. However, a professional astronomer typically spends no more than a few nights a year gathering data at a large telescope. Some astronomers make no obser-

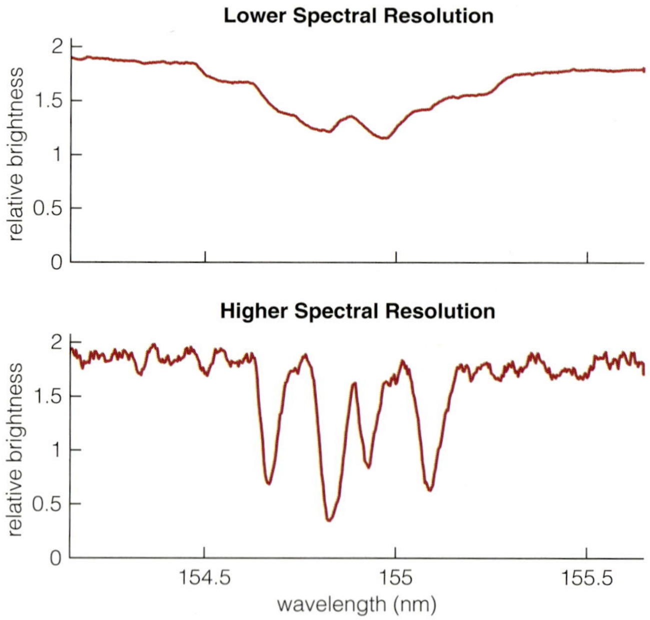

Figure 6.16 Higher spectral resolution means that we can see more details in the spectrum. The two ultraviolet spectra above both show the same object in the same wavelength band. However, we see far more detail with higher spectral resolution, including individual spectral lines that appear merged together at lower spectral resolution. (The spectrum shows absorption lines created when interstellar gas absorbs light from a more distant star.)

vations at all, instead focusing on the development of models to explain astronomical observations made by others.

The day-to-day life of an observational astronomer goes something like this: After identifying an important unanswered question, the astronomer proposes a set of observations to an organization that owns a large telescope. The astronomer must write the proposal clearly and eloquently, explaining exactly how she or he will carry out the observations and why these observations would be a good use of telescope time. Often, several astronomers with similar interests collaborate on a proposal, all planning to work together to make the observations and analyze the data. A committee of other astronomers then evaluates all the proposals that have been submitted, deciding which ones are worthy of receiving telescope time and which ones are not.

In most cases, the amount of telescope time requested in worthy proposals is much larger than the actual amount of time that is available. For example, if 100 worthy proposals each require 10 nights of observing time in the next year, there's only enough time for about one-third of them (or even fewer, since the weather isn't always perfect!). Thus, the various individuals and groups who write proposals essentially compete against one another, and only a fraction of them are awarded time on the telescope.

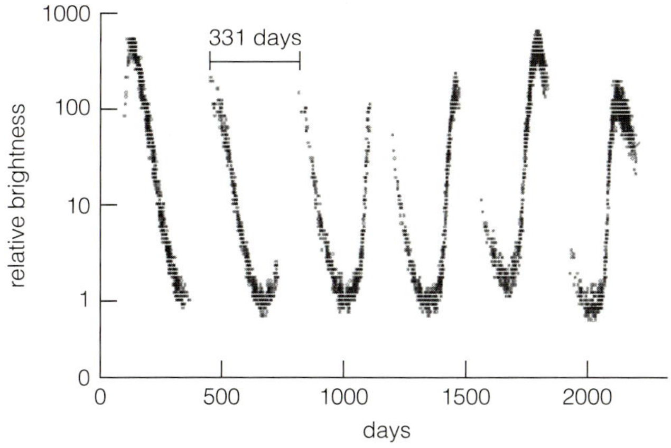

Figure 6.17 This graph shows a light curve for the variable star Mira (in the constellation Cetus), with data spanning several years. Centuries of observation show that Mira's brightness varies with an average period of 331 days.

MATHEMATICAL INSIGHT 6.2 The Diffraction Limit

A simple formula gives the diffraction limit of a telescope in arcseconds:

$$\underset{\text{(arcseconds)}}{\text{diffraction limit}} \approx 2.5 \times 10^5 \times \left(\frac{\text{wavelength of light}}{\text{diameter of telescope}}\right)$$

The wavelength of light and the diameter of the telescope must be in the same units.

Example 1: What is the diffraction limit of the 2.4-meter Hubble Space Telescope for visible light with a wavelength of 500 nanometers?

Solution:

Step 1 Understand: We are asked for the diffraction limit, which is the best angular resolution the telescope can achieve in principle. Above, we are given a formula for calculating the diffraction limit, and we already have all the information we need to use the formula.

Step 2 Solve: For light with a wavelength of 500 nanometers (500×10^{-9} meters), the diffraction limit of the Hubble Space Telescope is approximately:

$$\begin{aligned}\text{diffraction limit} &\approx 2.5 \times 10^5 \times \left(\frac{\text{wavelength}}{\text{telescope diameter}}\right)\\ &= 2.5 \times 10^5 \times \frac{500 \times 10^{-9}\ \text{m}}{2.4\ \text{m}}\\ &= 0.05\ \text{arcsecond}\end{aligned}$$

Step 3 Explain: The Hubble Space Telescope has a diffraction limited angular resolution of 0.05 arcsecond for visible light with wavelength 500 nanometers. Thus, if it actually achieves its diffraction limit, the telescope can resolve objects that are separated by more than 0.05 arcsecond. Objects separated by less will be blurred together.

Example 2: Suppose you wanted to achieve a diffraction limit of 0.001 arcsecond for visible light of wavelength 500 nanometers. How large a telescope would you need?

Solution:

Step 1 Understand: In this case we are looking for the size of the telescope with a particular diffraction limit. We have a formula for calculating the diffraction limit from the telescope size, so we need to rearrange this formula to tell us telescope size when we are given the diffraction limit. You should confirm that when we multiply both sides of the given formula by the telescope diameter and divide both sides by the diffraction limit, we get the following new formula:

$$\text{telescope diameter} \approx 2.5 \times 10^5 \times \left(\frac{\text{wavelength}}{\text{diffraction limit}}\right)$$

Step 2 Solve: Now we substitute the given values for the wavelength and the diffraction limit:

$$\begin{aligned}\text{telescope diameter} &\approx 2.5 \times 10^5 \times \frac{500 \times 10^{-9}\ \text{m}}{0.001\ \text{arcsecond}}\\ &= 125\ \text{m}\end{aligned}$$

Step 3 Explain: An optical telescope would need a mirror diameter (or separation between mirrors, for an interferometer) of 125 meters to achieve an angular resolution of 0.001 arcsecond. Note that this diameter is longer than a football field! Note also that it makes sense that we would need a telescope 50 times as large as the Hubble Space Telescope to achieve an angular resolution 50 times better than Hubble's.

Because the telescope time is so precious, it's crucial to use it efficiently. Thus, the real work begins after an observing proposal is accepted. The astronomers who wrote the proposal must prepare carefully to make sure they use the telescope time well. Then, after completing their observations, they spend even more time analyzing and interpreting their data.

6.3 Telescopes and the Atmosphere

From the time Galileo first turned his telescope to the heavens in 1609 [Section 3.3] until the dawn of the space age just a few decades ago, all astronomical telescopes were located on the ground. Even today, the vast majority of observatories are ground based, and that will probably hold true long into the future. Telescopes on the ground are much less expensive to build, operate, and maintain than telescopes in space. Nevertheless, Earth's surface is far from ideal as an observing site. In this section, we'll explore some of the problems that Earth's atmosphere poses for astronomical observations and learn why, despite the higher costs, dozens of telescopes have been lofted into Earth orbit or beyond.

• How does Earth's atmosphere affect ground-based observations?

The most obvious problems with observing from the ground are daylight and weather. Our daytime sky is bright because the atmosphere scatters sunlight, and this brightness drowns out the dim light of most astronomical objects. Thus, most astronomical observations are practical only at night. Even then, we can observe only when the sky is clear rather than cloudy. (The atmosphere does not scatter most radio waves, so radio telescopes can operate day and night and under cloudy skies.)

The constraints of daylight and weather affect the timing of observations, but by themselves do not hinder observations on clear nights. However, our atmosphere creates three other problems that inevitably affect astronomical observations: the scattering of human-made light, the blurring of images by atmospheric motion, and the fact that most forms of light cannot reach the ground at all.

Light Pollution Just as our atmosphere scatters sunlight in the daytime, it also scatters the bright lights of cities at night (Figure 6.18). That is why you cannot see as many stars when you look at the sky from a big city as you can see

---

SPECIAL TOPIC Would You Like Your Own Telescope?

Just a few years ago, a decent telescope for personal use would have set you back a few thousand dollars and taken weeks of practice to learn to use effectively. Today, you can get a good-quality telescope for just a few hundred dollars and built-in computer drives can make it very easy to use.

Before you start thinking about what telescope to buy, it's important to understand what a personal telescope can and cannot do. A telescope will allow you to look for yourself at light that has traveled vast distances through space to reach your eyes. It can be a rewarding experience, but the images in your telescope will *not* look like the beautiful photographs in this book—those are obtained with much larger telescopes and sophisticated cameras. In addition, while your telescope can in principle let you see many distant objects, including star clusters, nebulae, and galaxies, it won't allow you to find anything unless you first set it up properly. Even computer-driven telescopes, usually called "go to" telescopes, typically take 15 minutes to a half-hour to set up for each use (and longer when you are first learning).

If your goal is just to see the Moon and a few other objects with relatively little effort, you should probably go for a good pair of binoculars rather than a telescope. Indeed, it's generally a good idea for everyone to buy binoculars before buying a telescope, since binoculars will help you learn about viewing the sky and are a lot less expensive than good telescopes. Binoculars are generally described by two numbers, such as 7×35 or 12×50. The first number is the magnification; for example "7×" means that objects will look seven times closer through the binoculars than to your eye. The second number is the diameter of each lens in millimeters. As with telescopes, larger lenses mean more light and better views. However, larger lenses also tend to be heavier and more difficult to hold steady. If you buy a large pair of binoculars, you should also get a tripod to help hold it steady.

If you decide to go ahead with a telescope, the first rule to remember is that magnification is *not* the key factor to consider in telescope selection. Telescopes that are advertised by their magnification, such as "650 power," are usually junk. Forget them, and instead focus on three factors when choosing your telescope:

1. *The light-collecting area* (also called *aperture*). Most personal telescopes are reflectors, so a "6-inch" telescope means a primary mirror 6 inches in diameter.
2. *Optical quality.* A poorly made telescope won't do you much good. Your best bet if you cannot do side-by-side comparisons is to stick with a major telescope manufacturer (such as Meade, Celestron, or Orion).
3. *Portability.* A large, bulky telescope can be great if you plan to keep it on your roof, but it won't be much fun on camping trips. Depending on how you plan to use your telescope, you'll need to make trade-offs between size and portability.

Most important, remember that a telescope is an investment: It will cost real money to buy, and you will probably keep it for many years. As you should with any investment, learn all you can before you settle on a particular model. Read reviews of telescopes in magazines such as *Astronomy*, *Mercury*, and *Sky and Telescope.* Talk to knowledgeable salespeople at stores that specialize in telescopes. And find a nearby astronomy club that holds observing sessions at which you can try out some telescopes and learn from experienced telescope users. Astronomy clubs can be found almost everywhere, and you can usually find them by typing "astronomy club" and your city into a Web search engine.

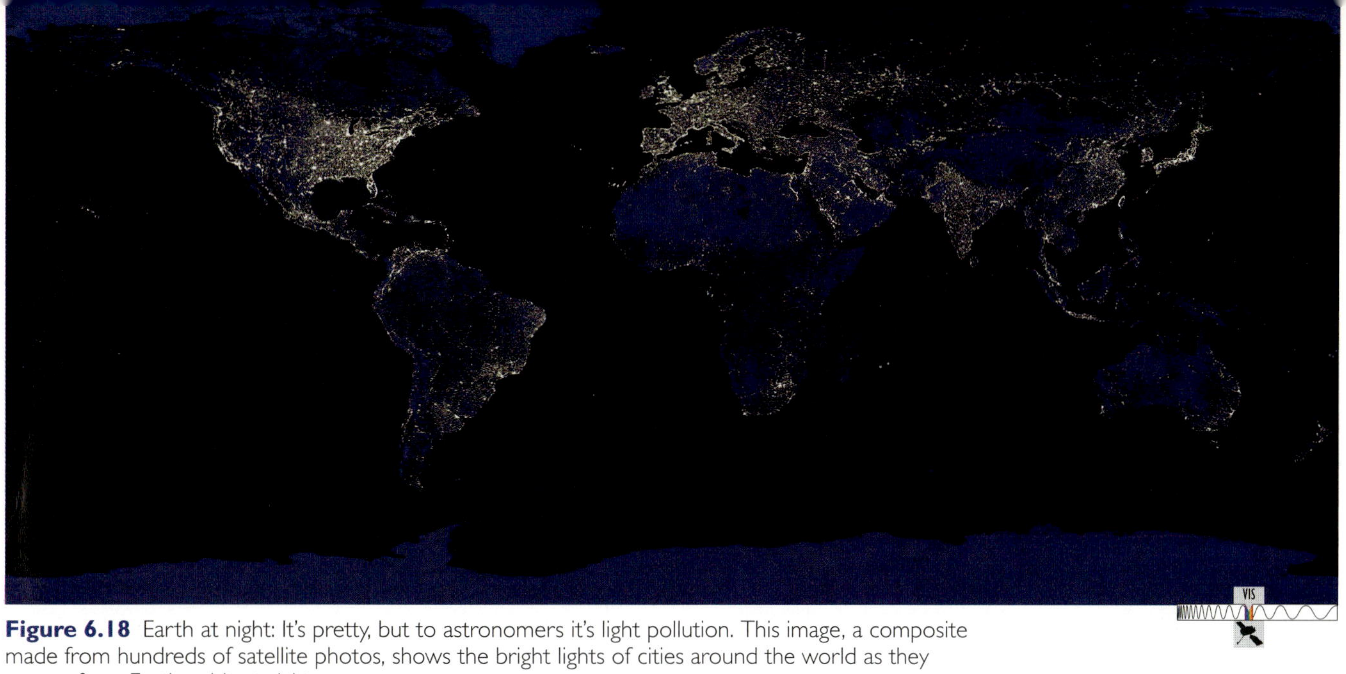

Figure 6.18 Earth at night: It's pretty, but to astronomers it's light pollution. This image, a composite made from hundreds of satellite photos, shows the bright lights of cities around the world as they appear from Earth orbit at night.

from an unpopulated area. The same problem of scattered light can seriously hinder astronomical observations. For example, the 2.5-meter telescope at Mount Wilson, the world's largest when it was built in 1917, would be much more useful today if it weren't located so close to the lights of what was once the small town of Los Angeles. The problem of scattered light at night is known as **light pollution,** because it is caused by human-made light that obscures our view of the sky.

Similar but less serious light pollution hinders many other telescopes, including the Mount Palomar telescopes near San Diego and the telescopes of the National Optical Astronomy Observatory on Kitt Peak near Tucson. Fortunately, many communities are working to reduce light pollution. Placing reflective covers on the tops of streetlights directs more light toward the ground, rather than toward the sky. Using "low-pressure" (sodium) lights also helps. Because these lights shine most brightly in just a few wavelengths of light, special filters can be used to absorb these wavelengths while transmitting most of the light from the astronomical objects under study. An extra benefit is that both reflective covers and low-pressure lights offer significant energy savings to communities that use them.

Twinkling and Atmospheric Turbulence The second major problem is somewhat less obvious but equally serious: distortion of light by the atmosphere. Winds and other air currents ensure that air in our atmosphere is constantly moving and mixing around, a phenomenon that we call **turbulence.** The ever changing motions of air due to turbulence constantly change the atmosphere's light-bending properties, so that light rays passing through the atmosphere are constantly being bent by slightly different amounts. As a result, our view of things outside Earth's atmosphere appears to jiggle around, rather like your view of things outside the water when you look at them from the bottom of a swimming pool. In the sky, we see this jiggling as the familiar twinkling of stars. Twinkling may be beautiful to the naked eye, but it causes problems for astronomers because it blurs astronomical images.

THINK ABOUT IT

Put a coin at the bottom of a cup of water. If you stir the water, the coin will appear to move around, even if it remains stationary on the bottom. What makes the coin appear to move? How is this similar to the way that our atmosphere makes stars appear to twinkle?

The blurring of images by turbulence places a severe limit on the angular resolution that ground-based telescopes can achieve. Even at the best observing sites, atmospheric turbulence typically limits the angular resolution to no better than about 0.5 arcsecond. Thus, most large telescopes cannot come anywhere near achieving the angular resolution that their size would offer in the absence of an atmosphere. For example, if you calculate its diffraction limit (see Mathematical Insight 6.2), you'll find that a 2.5-meter telescope would have angular resolution of 0.05 arcsecond in the absence of atmospheric effects—10 times better than what the atmosphere actually permits.

Astronomers can partially mitigate effects of weather, light pollution, and atmospheric distortion by choosing appropriate sites for observatories. The key criteria are that the sites be dark (limiting light pollution), dry (limiting rain and clouds), calm (limiting turbulence), and high (placing them above at least part of the atmosphere). Islands are often ideal, and the 4,300-meter (14,000-foot) summit of Mauna Kea on the Big Island of Hawaii is home to many of the world's best observatories (Figure 6.19).

In addition, a remarkable new technology called **adaptive optics** can eliminate much of the blurring caused by

Figure 6.19 Observatories on the summit of Mauna Kea in Hawaii. Mauna Kea meets all the key criteria for an observing site: It is far from big-city lights, high in altitude, and in an area where the air tends to be calm and dry.

our atmosphere. Atmospheric turbulence causes blurring because it makes the light of a star seem to dance around when viewed through a telescope. Adaptive optics essentially makes the telescope's mirrors do an opposite dance, canceling out the atmospheric distortions (Figure 6.20). The shape of the mirror (often the secondary or even a third or fourth mirror) is changed slightly many times each second to compensate for the rapidly changing atmospheric distortions. A computer calculates the necessary changes by monitoring distortions in the image of a bright star near the object under study. In some cases, if there is no bright star near the object of interest, the observatory shines a laser into the sky to create an *artificial star* (a point of light in Earth's atmosphere) that it can monitor for distortions.

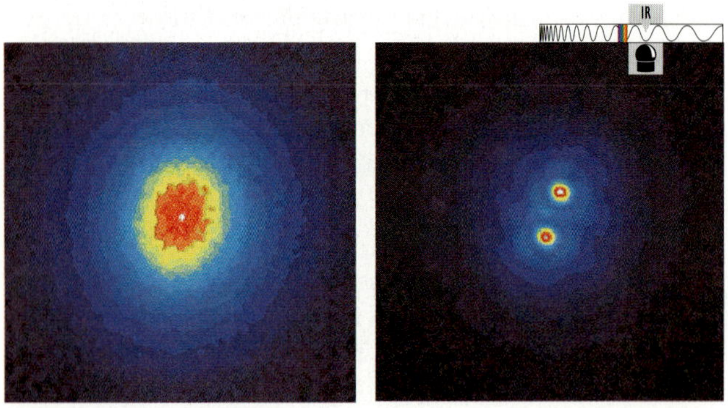

a Atmospheric distortion makes this ground-based image of a double star look like a single star.

b When the same telescope is used with adaptive optics, the two stars are clearly distinguished. The angular separation between the two stars is 0.38 arcsecond.

Figure 6.20 The technology of adaptive optics can enable a ground-based telescope to overcome most of the blurring caused by Earth's atmosphere. (Both these images were taken in near-infrared light with the Canada-France-Hawaii telescope; colors represent infrared intensity.)

• Why do we put telescopes into space?

The ultimate solution to the problems faced by ground-based observatories is to put telescopes into space, where they are above the atmosphere and thereby unaffected by daylight, weather, light pollution, and atmospheric turbulence. That is one reason why the Hubble Space Telescope (Figure 6.21) was built and why it has been so successful despite the relatively small size of its primary mirror. Notice that Hubble's 2.4-meter mirror is much smaller than the mirror sizes of the ground-based telescopes listed in Table 6.1.

However, while Hubble's visible-light images were far superior to anything possible from the ground when it was launched in 1990, the new technology of adaptive optics is helping ground-based observatories catch up rapidly. At least in some cases, ground-based observatories are now able to equal or better the image quality of the Hubble Space Telescope. So does that mean that the Hubble Space Telescope and other space observatories are becoming obsolete? No, and here's why: Earth's atmosphere poses one major problem that no Earth-bound technology can overcome—*Our atmosphere prevents most forms of light from reaching the ground at all.*

If we studied only visible light, we'd be missing much of the story that light brings to us from the cosmos. Planets are relatively cool and emit primarily infrared light. The hot upper layers of stars like the Sun emit ultraviolet and X-ray light. Some cosmic events are so violent that they can produce gamma rays that travel through space to Earth. Indeed, most astronomical objects emit light over a broad

COMMON MISCONCEPTIONS

Twinkle, Twinkle, Little Star

Twinkling, or apparent variation in the brightness and color of stars, is not intrinsic to the stars. Instead, just as water in a swimming pool bends light, Earth's atmosphere bends starlight. Air turbulence causes twinkling because it constantly changes the way the starlight is bent. Hence, stars tend to twinkle more on windy nights and at times when they are near the horizon (and therefore are viewed through a thicker layer of atmosphere).

Planets also twinkle, but not nearly as much as stars. Because planets have a measurable angular size, the effects of turbulence on any one ray of light are compensated for by the effects of turbulence on others, reducing the twinkling seen by the naked eye (but making planets shimmer noticeably in telescopes).

The blurring of images caused by twinkling is one of the primary reasons for putting telescopes into space. There, above the atmosphere, stars do not twinkle and telescopes can record sharp astronomical images.

Figure 6.21 The Hubble Space Telescope orbits Earth. Its position above the atmosphere allows it to get undistorted views of the heavens. Hubble can observe infrared and ultraviolet light as well as visible light.

COMMON MISCONCEPTIONS

Closer to the Stars?

Many people mistakenly believe that space telescopes are advantageous because their locations above Earth put them closer to the stars. You can quickly realize the error of this belief by thinking about scale. On the scale of the Voyage model solar system that we discussed in Section 1.2, the Hubble Space Telescope is so close to the surface of the millimeter-diameter Earth that you would need a microscope to resolve its altitude, while the nearest stars are thousands of kilometers away. Thus, the distances to the stars are effectively the same whether a telescope is on the ground or in space. The real advantages of space telescopes all arise from their being above Earth's atmosphere and the many observational problems it presents.

range of wavelengths. If we want to understand the universe, we must observe light all across the electromagnetic spectrum. We simply can't do this from the ground.

Figure 6.22 shows the depths to which different forms of light penetrate Earth's atmosphere. Notice that only radio waves, visible light, and small parts of the infrared spectrum can be observed from the ground. Thus, the most important reason for putting telescopes into space is to allow us to observe light that does not penetrate Earth's atmosphere. Indeed, the Hubble Space Telescope often observes in ultraviolet or infrared wavelengths that do not reach the ground, which is why it would remain a valuable observatory even if ground-based telescopes were some day to match all its visible-light capabilities.

The Hubble Space Telescope is the most famous observatory in space, but there are many others. Most of these observe parts of the electromagnetic spectrum that do not reach the ground. For example, the Chandra X-Ray Observatory is in space because an X-ray telescope would be completely useless on the ground. Table 6.2 lists some of the most important existing and planned telescopes in space.

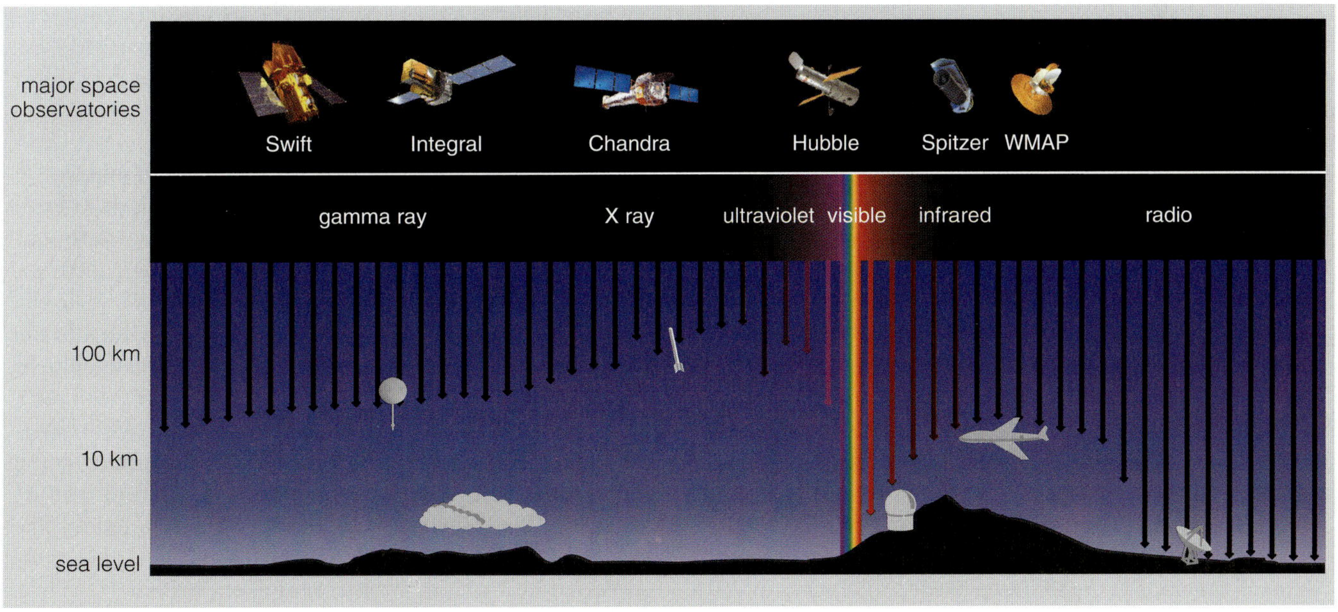

Figure 6.22 Interactive Figure This diagram shows the approximate depths to which different wavelengths of light penetrate Earth's atmosphere. Note that most of the electromagnetic spectrum—except for visible light, a small portion of the infrared, and radio waves—can be observed only from very high altitudes or from space.

Table 6.2 Selected Present and Future Major Observatories in Space

Name	*Launch Year*	*Lead Space Agency*	*Special Features*
Hubble Space Telescope	1990	NASA	Optical, infrared, and ultraviolet observations
Far Ultraviolet Spectroscopic Explorer (FUSE)	1999	NASA	Ultraviolet spectroscopy
Chandra X-Ray Observatory	1999	NASA	X-ray imaging and spectroscopy
XMM–Newton	1999	ESA*	European-led mission for X-ray spectroscopy
Wilkinson Microwave Anisotropy Probe (WMAP)	2001	NASA	Study of the cosmic microwave background
International Gamma-Ray Astrophysics Laboratory (INTEGRAL)	2002	ESA*	Gamma-ray imaging, spectroscopy, and timing
Spitzer Infrared Telescope	2003	NASA	Infrared observations of the cosmos
Swift	2004	NASA	Study of gamma-ray bursts
Kepler	2007†	ESA*	Transit search for extrasolar Earth-like planets
Planck	2008†	NASA	Study of the cosmic microwave background
Space Interferometry Mission (SIM)	2011†	NASA	First mission for optical interferometry in space
James Webb Space Telescope (JWST)	2012†	NASA	Follow-on to Hubble Space Telescope
Terrestrial Planet Finder (TPF)	2014†	NASA	Search for Earth-like planets around other stars

* European Space Agency.

† Tentative launch year.

6.4 Telescopes and Technology

Astronomers today are making new discoveries at an astonishing rate, driven largely by the availability of more and larger telescopes, including space telescopes that can observe previously inaccessible portions of the electromagnetic spectrum. However, larger telescopes are not the only fuel for the current astronomical revolution.

Some new technologies make it possible to obtain better images or spectra with existing telescopes. For example, electronic detectors are constantly improving. As a result, a relatively small telescope equipped with the latest camera technology can record images as good as those that could be captured only by much larger telescopes in the past. In addition, adaptive optics allows telescopes to overcome problems of atmospheric distortion.

Other technologies make it possible to record and analyze data more efficiently. For example, obtaining spectra of distant galaxies used to be a very time-consuming and labor-intensive task. Today, astronomers can sometimes obtain hundreds of spectra simultaneously in a single telescopic observation. Once astronomers have these spectra, they use computers to analyze the vast amount of data the spectra contain.

In this section, we'll focus our attention on two important areas of modern telescope technology that we have not discussed previously: the technology used to observe different parts of the electromagnetic spectrum and a remarkable technology that allows many small telescopes to work together to obtain higher-quality images.

• How do we observe nonvisible light?

The basic idea behind all telescopes is the same: to collect as much light as possible with as much resolution as possible. In addition, nearly all telescopes used by professional astronomers (except some telescopes for gamma-ray wavelengths) are essentially reflecting telescopes, using mirrors to bring light to a focus. Nevertheless, telescopes for most nonvisible wavelengths require some variations on the basic design used for visible-light telescopes. Let's investigate, going in order of decreasing wavelength.

Radio Telescopes Although it may take a little thought to realize it, a specialized kind of radio telescope is now the most common type of telescope in the world, so common that you'll see dozens of them on a drive through almost any neighborhood and you may already own one yourself. Why? Because every satellite dish is essentially a small radio telescope designed to collect radio waves from a satellite in Earth orbit. Just by looking at a satellite dish, you can see that it operates by the same basic principles as a reflecting telescope (Figure 6.23). The metal dish is the mirror, and it is shaped to bring the radio waves to a focus in front of the dish; that's where you see the receiver, which functions like the secondary mirror. The receiver collects the radio waves reflected by the primary mirror and sends them to the television (or some other communication device).

The only major differences between satellite dishes and astronomical radio telescopes are where they look in the

Figure 6.23 A satellite TV dish is essentially a small radio telescope.

Figure 6.24 The Arecibo radio telescope stretches across a natural valley in Puerto Rico. At 305 meters across, it is the world's largest single telescope.

sky, their sizes, and their purposes. Communication satellites have geostationary orbits in which they orbit above Earth's equator in exactly the same amount of time that Earth rotates each day (see Mathematical Insight 4.3); thus, a dish aimed at a particular satellite always points to the same spot in the local sky. Astronomical radio telescopes point toward cosmic radio sources that rise and set just like stars in the sky, so long exposures require tracking the sources across the sky. Astronomical radio telescopes are larger than satellite dishes both because cosmic radio sources are fainter than satellite signals (and thus require a larger light-collecting area) and because they are used to make images of the cosmic sources (and thus require decent angular resolution). Angular resolution is unimportant for satellite dishes, because they are not used to make images of the satellites in space; radio and television signals are encoded in the radio waves themselves, so the dish needs only to collect the radio waves and send them to a decoding device like a television.

The long wavelengths of radio waves mean that very large telescopes are necessary to achieve reasonable angular resolution. That is why radio telescopes are so much larger than visible-light telescopes. For example, the world's largest single radio telescope, the Arecibo radio dish, stretches 305 meters (1,000 feet) across a natural valley in Puerto Rico (Figure 6.24). Despite its large size, Arecibo's angular resolution is only about 1 arcminute at commonly observed radio wavelengths (for example, 21 centimeters [Section 19.2])—a few hundred times worse than the visible-light resolution of the Hubble Space Telescope. Fortunately, through an amazing technique that we'll discuss shortly (interferometry), radio telescopes can work together to achieve much better angular resolution.

If you look again at Figure 6.22, you'll see that radio waves are the only form of light besides visible light that we can observe easily from the ground. Moreover, because the atmosphere does not distort radio waves the way it distorts visible light, there's no inherent advantage to observing from space. However, "radio-wave pollution" is an even more serious impediment to radio astronomy than light pollution is to visible-light astronomy. Humans use many portions of the radio spectrum so heavily that radio signals from cosmic sources are almost completely drowned out. Astronomers therefore hope someday to put radio telescopes into deep space or on the far side of the Moon, where human signals will be blocked. In addition, because radio telescopes can be made to work together, putting them into space can in principle allow them to be spread out over a much greater distance.

Infrared Telescopes Most of the infrared portion of the spectrum is close enough in wavelength to visible light to behave quite similarly. Thus, infrared telescopes generally look much the same as visible-light telescopes. In fact, visible-light telescopes can in principle collect and focus much of the infrared spectrum; the practical limitation comes from Earth's atmosphere.

As you can see in Figure 6.22, infrared light from space generally does not reach the ground. A few portions of the infrared spectrum can be observed from high mountaintops, and many mountaintop observatories do indeed observe in the infrared. The higher you go in the atmosphere, the more infrared light becomes accessible. NASA has recently started flying an airborne infrared telescope called SOFIA (Stratospheric Observatory for Infrared Astronomy). SOFIA carries a 2.5-meter infrared telescope that looks out

Figure 6.25 This photograph shows NASA's airborne observatory, SOFIA, during a test flight prior to installation of its 2.5-meter infrared telescope. The black rectangle marks the location where the telescope has since been installed into the fuselage.

Figure 6.26 This painting shows the Spitzer Infrared Telescope, which trails millions of kilometers behind Earth in the same orbit around the Sun. The background is an artistic rendition of infrared emission from a star-forming cloud.

through a large hole cut in the fuselage of a Boeing 747 airplane (Figure 6.25). Advanced technologies make it possible for the airplane to fly smoothly despite its large hole and to keep the telescope pointed accurately at observing targets during flight.

Extreme infrared light (the longest wavelengths of infrared) poses more difficult observing challenges. Remember that all objects emit thermal radiation characteristic of their temperatures [Section 5.4]. At ordinary temperatures, the ground, an airplane, and even a telescope itself emit enough long-wavelength infrared light to interfere with any attempt to observe these wavelengths from the cosmos. The only solution to this problem is to put the telescopes into space so they get away from Earth's own heat and to cool the telescopes so they emit less and lower-energy infrared light. NASA's Spitzer Infrared Telescope (Figure 6.26), launched in 2003, is cooled with liquid helium to just a few degrees above absolute zero. Spitzer is kept far from Earth's own heat with a unique orbit: Instead of orbiting Earth like the Hubble Space Telescope, Spitzer shares Earth's orbit of the Sun but trails millions of kilometers behind our planet. A Sun shield blocks out light from the Sun.

Ultraviolet Telescopes Like infrared light, much of the ultraviolet spectrum is close enough in wavelength to visible light to behave similarly. (Very short-wavelength ultraviolet light, sometimes called *extreme ultraviolet*, behaves like X rays, which we'll discuss below.) Thus, visible-light mirrors can in principle collect and focus this ultraviolet light. However, Earth's atmosphere almost completely absorbs ultraviolet light, making most ultraviolet observations impossible from the ground.

At present, there are three major ultraviolet observatories in space: The Far Ultraviolet Spectroscopic Explorer (FUSE), designed exclusively for ultraviolet spectroscopy; the Galaxy Evolution Explorer (GALEX), a relatively small telescope that is conducting an ultraviolet survey of the entire sky; and the Hubble Space Telescope, which is capable of high-resolution imaging and spectroscopy in ultraviolet light. Hubble's continued operation depends on periodic Space Shuttle missions to repair it, update it, and boost it back up as atmospheric drag lowers its orbit. The last servicing mission was in 2002, and the observatory is unlikely to survive past about 2007 without another servicing mission. Hubble therefore faces an uncertain future, because NASA cancelled future servicing missions in the aftermath of the 2003 Columbia Space Shuttle accident. However, NASA has since agreed to reconsider this decision, and Hubble may yet be serviced in time to keep it alive for at least several more years.

X-Ray Telescopes Besides needing to be in space, X-ray observations pose another challenge: X rays have sufficient energy to penetrate many materials, including living tissue and ordinary mirrors. While this property makes X rays very useful to medical doctors, it gives astronomers headaches.

Trying to focus X rays is somewhat like trying to focus a stream of bullets. If the bullets are fired directly at a metal sheet, they will puncture or damage the sheet. However, if the metal sheet is angled so that the bullets barely graze its surface, then it will slightly deflect the bullets. Specially designed mirrors can deflect X rays in much the same way. Such mirrors are called **grazing incidence** mirrors because X rays merely graze their surfaces as they are deflected toward the focal plane. X-ray telescopes, such as NASA's Chandra X-Ray Observatory, generally consist of several nested grazing incidence mirrors (Figure 6.27).

Chandra offers the best angular resolution of any X-ray telescope yet built, but a European X-ray telescope called XMM–Newton has a larger light-collecting area. Astronomers therefore use the two observatories in the way best suited to science goals. For example, Chandra is better for

a Artist illustration of the Chandra X-ray Observatory, which orbits Earth.

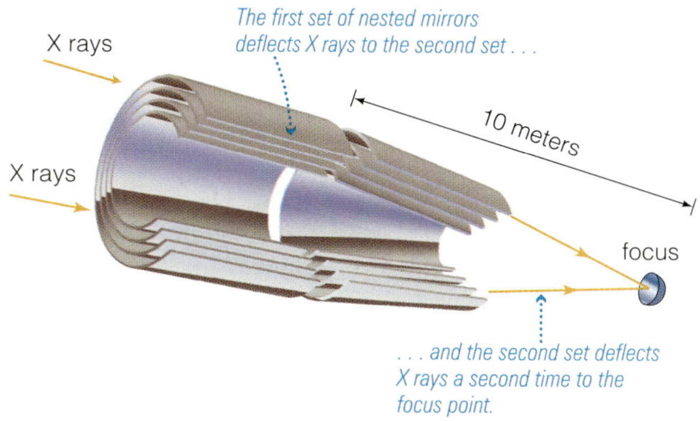

b This diagram shows the arrangement of Chandra's nested, cylindrical X ray mirrors. Each mirror is 0.8 meter long and between 0.6 and 1.2 meters in diameter.

Figure 6.27 The Chandra X-Ray Observatory focuses X rays that enter the front of the telescope by deflecting them twice so that they end up focused at the back of the telescope.

making images of X-ray sources, while XMM–Newton's larger light-collecting area allows it to obtain more detailed X-ray spectra.

THINK ABOUT IT

If you look straight down at your desktop, you probably cannot see your reflection. But if you glance along the desktop surface (or another smooth surface, such as that of a book), you should see reflections of objects in front of you. Explain how these reflections represent *grazing incidence* for visible light.

Gamma-Ray Telescopes Gamma rays can penetrate even grazing incidence mirrors and therefore cannot be focused in the traditional sense. Capturing such high-energy light at all requires detectors so massive that the photons cannot simply pass through them. The largest gamma-ray observatory to date was the 17-ton Compton Gamma Ray Observatory (Figure 6.28), which was launched in 1991 and destroyed in a controlled crash to Earth in 2000.

Gamma rays come from a number of different types of astronomical objects, but the most mysterious sources produce short bursts of gamma rays that quickly fade away [Section 18.4]. Because gamma rays are difficult to focus, Compton and other past gamma-ray telescopes were unable to pinpoint the precise locations of the sources of these gamma-ray bursts. They located the sources well enough for us to know they are in distant galaxies, but not well enough for us to identify the objects that produced them. In 2004, NASA launched a new gamma-ray observatory named Swift. In addition to its gamma-ray telescope, the Swift satellite carries X-ray and visible-light telescopes. As soon as Swift detects a burst of gamma rays, these other telescopes rapidly swing toward the source to look for corresponding bursts in X-rays and visible light. Swift thereby gets much more accurate positions for the sources of the gamma-ray bursts.

Looking Beyond Light We have learned virtually everything we know about distant objects by observing light. However, light is not the only form of information that travels through the universe, and astronomers have begun to build and use telescopes designed to observe at least three other types of "cosmic messengers." First, there's an extremely lightweight type of subatomic particle known as the *neutrino* [Sections S4.2, 14.2] that is produced by nuclear reactions,

Figure 6.28 This photograph shows the Compton Gamma Ray Observatory being deployed from the Space Shuttle in 1991. Compton revolutionized our understanding of the gamma-ray universe before its demise (in a controlled reentry into the atmosphere).

including nuclear fusion in the Sun and the reactions that accompany the explosions of distant stars. Astronomers have already had some success with "neutrino telescopes"—typically located in deep mines or under water or ice—that have provided valuable insights about the Sun and stellar explosions. Second, Earth is continually bombarded by very high-energy subatomic particles from space known as *cosmic rays* [Section 19.2]. We still know little about their origin (including how they are accelerated to such high speeds) and chemical composition, but astronomers are now using both satellites and ground-based detectors to catch and study them. Third, Einstein's general theory of relativity predicts the existence of something called *gravitational waves* [Section S3.4] that are very different in nature from light but travel at the speed of light. For decades, we've had indirect evidence that gravitational waves really exist, but until recently, direct detection of them was beyond our technological capabilities. Today, the first gravitational wave telescopes are up and running in Washington and Louisiana (as well as Italy and Germany), and astronomers hope they will eventually be able to detect gravitational waves from exotic objects like orbiting neutron stars and black holes.

• How can multiple telescopes work together?

Individual telescopes always face limits on their capabilities. Even in space, the diffraction limit places a fundamental constraint on the angular resolution of a telescope of any particular size. In addition, while astronomers would always like larger telescopes, the current state of technology and budgetary considerations create practical limits on telescope size.

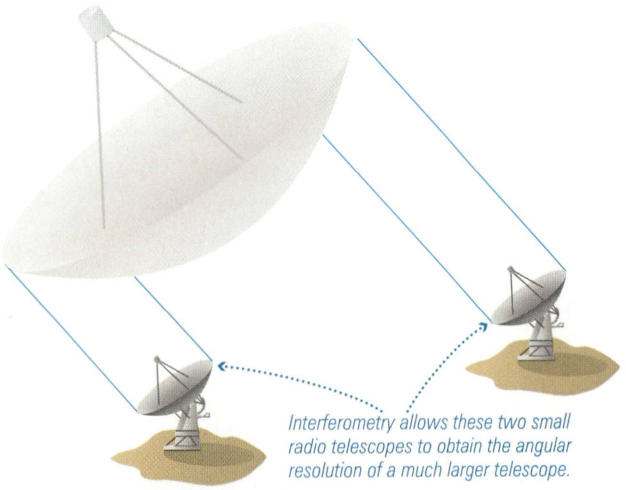

Figure 6.29 This diagram shows the basic idea behind interferometry: Smaller telescopes work together to obtain the angular resolution of a much larger telescope. Note that interferometry improves angular resolution but does *not* affect the total light-collecting area, which is simply the sum of the light-collecting areas of the individual telescopes.

Figure 6.30 The Very Large Array (VLA) in New Mexico consists of 27 telescopes that can be moved along train tracks. The telescopes work together through interferometry and can achieve an angular resolution equivalent to that of a single radio telescope almost 40 kilometers across.

These constraints ultimately limit the amount of light that we can collect with telescopes. Even if we put a bunch of telescopes together, there's no getting around the fact that their total light-collecting area is simply the sum of their individual areas. However, remember that the two key properties of a telescope are light-collecting area and angular resolution. Amazingly, astronomers have found a way to make the angular resolution of a group of telescopes far better than that of any individual telescope.

Interferometry In the 1950s, radio astronomers developed an ingenious technique for improving the angular resolution of radio telescopes: They learned to link two or more individual telescopes to achieve the angular resolution of a much larger telescope (Figure 6.29). This technique is called **interferometry** because it works by taking advantage of the wavelike properties of light that cause interference (see Figure 6.7). The procedure relies on precisely timing when radio waves reach each dish and on using computers to analyze the resulting interference patterns.

One famous example of radio interferometry, the Very Large Array (VLA) near Socorro, New Mexico, consists of 27 individual radio dishes that can be moved along railroad tracks laid down in the shape of a Y (Figure 6.30). The light-gathering capability of the VLA's 27 dishes is simply equal to their combined area, equivalent to that of a single telescope 130 meters across. But its angular resolution is equivalent to that of a much larger telescope: When the 27 dishes are spaced as widely as possible, the VLA can achieve an angular resolution that otherwise would require a single radio telescope with a diameter of almost 40 kilometers. Today, astronomers can achieve even higher

Figure 6.31 An artist's conception of a possible future lunar observatory. The Moon has no atmosphere to create distortion and has a solid surface that makes it ideal as a home for astronomical observatories.

angular resolution by linking radio telescopes around the world.

Interferometry is more difficult for shorter-wavelength (higher-frequency) light, but astronomers are rapidly learning to use the technique beyond the radio portion of the spectrum. Some success has already been achieved at infrared and visible wavelengths. Indeed, new telescopes are now often built in pairs (such as the Keck and Magellan telescope pairs) or with more than one telescope on a common mount (such as the Large Binocular Telescope) so that they can be used for infrared and visible-light interferometry. In addition, astronomers are testing technologies that may allow interferometry to be extended all the way to X rays.

The Future of Astronomy in Space It will always be cheaper and easier to do astronomy from ground-based observatories, but space is likely to play an ever increasing role in astronomy. NASA is already at work on follow-ons to the Hubble Space Telescope, such as the James Webb Space Telescope (JWST), which may launch as early as 2012. Later telescopes are likely to be interferometers, using several telescopes together to achieve the angular resolution of far larger telescopes. The potential value of such interferometers is enormous. For example, within the next two decades astronomers hope to launch a suite of small space telescopes that could work together as an interferometer to achieve the angular resolution of a single telescope hundreds of meters across [Section 13.4].

For the more distant future, many astronomers dream of an observatory on the Moon (Figure 6.31). Because the Moon has no atmosphere, it offers all the advantages of telescopes in space while also offering the ease of operating on a solid surface. Some day, astronomers may use telescopes in space or on the Moon as giant interferometers, offering views of distant objects that may be as detailed in comparison to Hubble Space Telescope images as Hubble's images are in comparison to the naked eye's.

THE BIG PICTURE

Putting Chapter 6 into Context

In this chapter, we've focused on the technological side of astronomy: the telescopes that we use to learn about the universe. Keep in mind the following "big picture" ideas as you continue to learn about astronomy:

- Technology drives astronomical discovery. Every time we build a bigger telescope, develop a more sensitive detector, or open up a new wavelength region to study, we learn more about the universe than was previously possible.
- Telescopes work much like giant eyes, enabling us to see the universe in great detail. New technologies for making larger telescopes, along with advances in adaptive optics and interferometry, are making ground-based telescopes more powerful than ever.
- For the ultimate in observing the universe, space is the place! Telescopes in space allow us to detect light from across the entire spectrum while also avoiding the distortion caused by Earth's atmosphere.

SUMMARY OF KEY CONCEPTS

6.1 Eyes and Cameras: Everyday Light Sensors

- **How does your eye form an image?** Your eye brings parallel rays of light to a **focus** on your retina. Glass lenses work similarly, so distant objects form an **image** that is in focus on the **focal plane**.

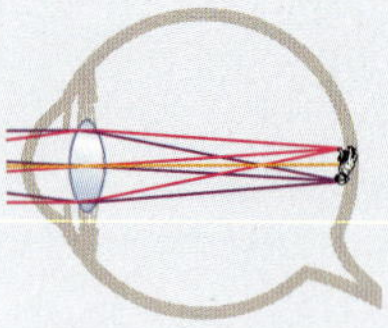

- **How do we record images?** We record images with a **detector** such as photographic film or a **CCD**. Recorded images are more reliable than just seeing by eye and can record more light if we use a longer **exposure time**.

6.2 Telescopes: Giant Eyes

- **What are the two most important properties of a telescope?** A telescope's most important properties are its **light-collecting area**, which determines how much light it gathers, and its **angular resolution**, which determines how much detail we can see in its images.

- **What are the two basic designs of telescopes?** A **refracting telescope** forms an image by bending light through a lens. A **reflecting telescope** forms an image by focusing light with mirrors.

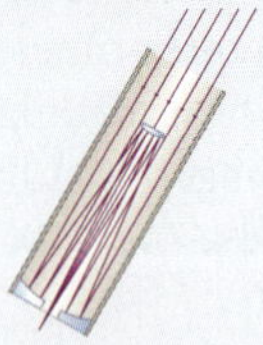

- **What do astronomers do with telescopes?** The three primary uses of telescopes are **imaging** to create pictures of distant objects, **spectroscopy** to study the spectra of distant objects, and **timing** to study how a distant object's brightness changes with time.

6.3 Telescopes and the Atmosphere

- **How does Earth's atmosphere affect ground-based observations?** Earth's atmosphere limits visible-light observations to nighttime and clear weather. **Light pollution** can lessen the quality of observations, and atmospheric **turbulence** makes stars twinkle, blurring their images. The technology of **adaptive optics** can overcome some of the blurring due to turbulence.

- **Why do we put telescopes into space?** Telescopes in space are above Earth's atmosphere and the problems it causes for observations. Most important, telescopes in space can observe all wavelengths of light, while telescopes on the ground can observe only visible light, radio waves, and small portions of the infrared.

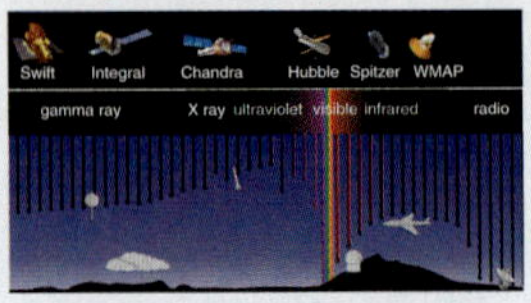

6.4 Telescopes and Technology

- **How do we observe nonvisible light?** Telescopes for nonvisible light often use variations on the basic design of a reflecting telescope. Radio telescopes use large metal dishes as their primary mirrors. Infrared telescopes are sometimes cooled to very low temperature. X-ray telescopes use **grazing incidence** reflections rather than direct reflections.

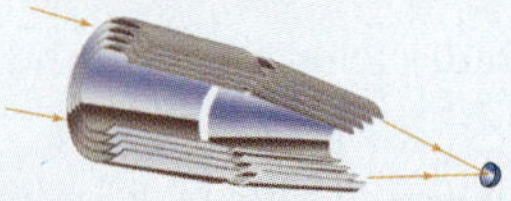

- **How can multiple telescopes work together?** The technique of **interferometry** allows multiple telescopes to be linked in a way that allows them to obtain the angular resolution of a much larger telescope.

EXERCISES AND PROBLEMS

For instructor-assigned homework go to MasteringASTRONOMY www.masteringastronomy.com

Review Questions

Short-Answer Questions Based on the Reading

1. How does your eye focus light? How is a glass lens similar? What do we mean by the *focal plane* of a lens?
2. For purposes of astronomy, what advantages does a camera have over the human eye? What advantages do *CCDs* have over photographic film?
3. What are the two key properties of a telescope, and why is each important?
4. What is the *diffraction limit*, and how does it depend on a telescope's size and the wavelength of light being observed?
5. How do *reflecting telescopes* differ from *refracting telescopes*? Which type is more commonly used by professional astronomers, and why?

6. What are the three basic categories of astronomical observation, and how is each conducted?
7. What do we mean when we speak of images made from nonvisible light, such as X-ray or infrared images? What do the colors in these images mean?
8. What do we mean by *spectral resolution*? Why is higher spectral resolution more difficult to achieve?
9. List at least three ways in which Earth's atmosphere can hinder astronomical observations, and explain why putting a telescope into space helps in each case. What problem can *adaptive optics* help with?
10. Study Figure 6.22 and describe how deeply each portion of the electromagnetic spectrum penetrates Earth's atmosphere. Based on your answers, why is space astronomy so important to our understanding of the universe?
11. How do telescopes for nonvisible wavelengths differ from those for visible light? Answer for each major wavelength band and give examples of important observatories in those bands.
12. What is interferometry, and how can it improve astronomical observations?

Test Your Understanding

Does It Make Sense?

Decide whether the statement makes sense (or is clearly true) or does not make sense (or is clearly false). Explain your reasoning. (For an example, see Chapter 1, "Does It Make Sense?")

13. The image was blurry because the photographic film was not placed at the focal plane.
14. By using a CCD, I can photograph the Andromeda Galaxy with a shorter exposure time than I would need with photographic film.
15. I have a reflecting telescope in which the secondary mirror is bigger than the primary mirror.
16. The photograph shows what appears to be just two distinct stars, but each of those stars is actually a binary star system.
17. I built my own 14-inch telescope that has a lower diffraction limit than most large professional telescopes.
18. Now that I've bought a spectrograph, I can use my home telescope for spectroscopy as well as imaging.
19. If you lived on the Moon, you'd never see stars twinkle.
20. New technologies will soon allow astronomers to use X-ray telescopes on Earth's surface.
21. Thanks to adaptive optics, the telescope on Mount Wilson can now make ultraviolet images of the cosmos.
22. Thanks to interferometry, a properly spaced set of 10-meter radio telescopes can achieve the angular resolution of a single 100-kilometer radio telescope.

Quick Quiz

Choose the best answer to each of the following. Explain your reasoning with one or more complete sentences.

23. How much greater is the light-collecting area of a 6-meter telescope than a 3-meter telescope? (a) two times (b) four times (c) six times
24. Suppose that two stars are separated in the sky by 0.1 arcsecond. If you look at them with a telescope that has an angular resolution of 0.5 arcsecond, what will you see? (a) Two distinct stars. (b) One point of light that is the blurred image of both stars. (c) Nothing at all.
25. The diffraction limit is a limit on: (a) A telescope's size. (b) A telescope's angular resolution. (c) A telescope's spectral resolution.
26. The Hubble Space Telescope obtains higher-resolution images than most ground-based telescopes because it is: (a) larger. (b) closer to the stars. (c) above Earth's atmosphere.
27. If you see the color red in an X-ray image from the Chandra X-Ray Observatory, it means: (a) the object really is red in color. (b) the red parts are hotter than the bluer parts. (c) it depends: the colors are chosen arbitrarily to represent something about the X rays recorded by the telescope.
28. The twinkling of stars is caused by: (a) variations in stellar brightness with time. (b) light pollution. (c) motion of air in our atmosphere.
29. If you wanted a radio telescope to achieve the same angular resolution as a visible-light telescope, it would need to be: (a) much larger. (b) slightly larger. (c) in space.
30. Where should you put a telescope designed for ultraviolet observations? (a) in Earth orbit. (b) on an airplane. (c) on a high mountaintop.
31. Which technology can allow a single ground-based telescope to achieve images as sharp as those from the Hubble Space Telescope? (a) adaptive optics (b) grazing incidence (c) interferometry
32. Interferometry uses two or more telescopes to achieve: (a) a light-collecting area equivalent to that of a much larger telescope. (b) an angular resolution equivalent to that of a much larger telescope. (c) both the light-collecting area and angular resolution of a much larger telescope.

Investigate Further

In-Depth Questions to Increase Your Understanding

Short-Answer/Essay Questions

33. *Image Resolution.* What happens if you take a photograph from a newspaper, magazine, or book and blow it up to a larger size? Can you see any more detail than you could see before? Explain clearly, and relate your answer to the concepts of magnification and angular resolution in astronomical observations.
34. *Type of Observation.* For each of the following, decide what type of observation (imaging, spectroscopy, timing) you would need to make. Explain clearly.
    a. Studying how a star's hot upper atmosphere changes with time.
    b. Learning the composition of a distant star.
    c. Determining how fast a distant galaxy is moving away from Earth.
35. *Telescope Location.* Where do you live? In light of the problems faced by ground-based observatories, is your location a good place for an astronomical observatory? Why or why not?
36. *Telescope Technology.* Suppose you were building a space-based observatory consisting of five individual telescopes. Which would be the best way to use these telescopes: as five individual telescopes with adaptive optics, or as five telescopes without adaptive optics linked together for interferometry? Explain your reasoning clearly.
37. *Filters.* Describe what an American flag would look like if you viewed it through a filter that transmits only red light. What would it look like through a filter that transmits only blue light?

38. *Megapixels.* Digital cameras whose detectors have more than 6 million pixels are advertised to have over 6 megapixels, and that number will surely get greater in the future. Astronomical instruments are also boasting ever greater numbers of pixels. Is it better to use those extra pixels to observe larger portions of the sky or to observe smaller details in the sky? Explain how the answer to this question depends on the type of telescope being used and recent advances in telescope technology.
39. *Lunar Observatory.* Some astronomers dream of someday having an observatory on the Moon. What advantages would a lunar observatory have over ground-based observatories on Earth? What advantages would it have over space-based observatories? What would the disadvantages be?
40. *Project: Twinkling Stars.* Using a star chart, identify 5–10 bright stars that should be visible in the early evening. On a clear night, observe each of these stars for a few minutes. Note the date and time, and for each star record the following information: approximate altitude and direction in your sky, brightness compared to other stars, color, and how much the star twinkles compared to other stars. Study your record. Can you draw any conclusions about how brightness and position in your sky affect twinkling? Explain.
41. *Project: Radio Pollution.* The rapid growth in wireless communications has vastly increased the amount of radio "noise" in the vicinity of radio telescopes. However, a few sections of the radio spectrum are protected for the sake of radio astronomy. Find out what those regions are and explain why they were chosen for special protection.
42. *Project: Personal Telescope Review.* By going to stores or looking in online or printed catalogs, find three telescopes that you could buy for under $1,000. Evaluate each one against the following criteria: light-collecting area, angular resolution, construction quality, and portability. Give each telescope a rating of 1 to 4 stars (4 is best) and state which one you would recommend for purchase to someone planning to buy a telescope in this price range.

Quantitative Problems

Be sure to show all calculations clearly and state your final answers in complete sentences.

43. *Light-Collecting Area.*
    a. How much greater is the light-collecting area of one of the 10-meter Keck telescopes than that of the 5-meter Hale telescope?
    b. Suppose astronomers built a 100-meter telescope. How much greater would its light-collecting area be than that of the 10-meter Keck telescope?
44. *Close Binary System.* Suppose that two stars in a binary star system are separated by a distance of 100 million kilometers and are located at a distance of 100 light-years from Earth. What is the angular separation of the two stars? Give your answer in both degrees and arcseconds. Can the Hubble Space Telescope resolve the two stars?
45. *Finding Planets.* Suppose you were looking at our own solar system from a distance of 10 light-years.
    a. What angular resolution would you need to see the Sun and Jupiter as distinct points of light?
    b. What angular resolution would you need to see the Sun and Earth as distinct points of light?
    c. How do the angular resolutions you found in parts (a) and (b) compare to the angular resolution of the Hubble Space Telescope? Comment on the challenge of making images of planets around other stars.
46. *Diffraction Limit of the Eye.*
    a. Calculate the diffraction limit of the human eye, assuming a wide-open pupil so that your eye acts like a lens with diameter 0.8 centimeter, for visible light of 500-nanometer wavelength. How does this compare to the diffraction limit of a 10-meter telescope?
    b. Now remember that humans have two eyes that are approximately 7 centimeters apart. Estimate the diffraction limit for human vision, assuming that your "optical interferometer" is just as good as one eyeball as large as the separation of two regular eyeballs.
47. *The Size of Radio Telescopes.* What is the diffraction limit of a 100-meter radio telescope observing radio waves with a wavelength of 21 centimeters? Compare this to the diffraction limit of the 2.4-meter Hubble Space Telescope for visible light. Use your results to explain why, to be useful, radio telescopes must be much larger than optical telescopes.
48. *Your Satellite Dish.* Suppose you have a satellite dish that is 0.5 meter in diameter and you want to use it as a radio telescope. What is the diffraction limit on the angular resolution of your dish, assuming that you want to observe radio waves with a wavelength of 21 centimeters? Would it be very useful as an astronomical radio telescope?
49. *Hubble's Field of View.* Large telescopes often have small fields of view. For example, the Hubble Space Telescope's (HST) new advanced camera has a field of view that is roughly square and about 0.06° on a side.
    a. Calculate the angular area of the HST's field of view in square degrees.
    b. The angular area of the entire sky is about 41,250 square degrees. How many pictures would the HST have to take with its camera to obtain a complete picture of the entire sky?
50. *Hubble Sky Survey?* In Problem 49, you found out how many pictures HST would require to photograph the entire sky. If you assume that it would take 1 hour for each picture, how many years would HST need to obtain photos of the entire sky? Use your answer to explain why astronomers would like to have more than one large telescope in space.
51. *Visible-Light Interferometry.* Technological advances are now making it possible to link visible-light telescopes so that they can achieve the same angular resolution as a single telescope over 300 meters in size. What is the angular resolution (diffraction limit) of such a system of telescopes for observations at a wavelength of 500 nanometers?
52. *The Size of an X-Ray Observatory.* Spacecraft designed to observe X rays from space need to be quite long because X rays are so hard to focus. Photons entering on parallel paths get deflected by the telescope's mirrors onto paths that eventually converge at a focus, but they are deflected by only about 2 degrees from their original direction. On the Chandra Observatory the largest X-ray deflecting mirrors are about 4 meters in diameter. Based on this information, estimate the length of the Chandra X-Ray Observatory.

Discussion Questions

53. *Science and Technology Funding.* Technological innovation clearly drives scientific discovery in astronomy, but the reverse is also true. For example, Newton made his discoveries in part because he wanted to explain the motions of the planets, but his discoveries have had far-reaching effects on our civilization. Congress often must make decisions between funding programs with purely scientific purposes ("basic research") and programs designed to develop new technologies. If you were a member of Congress, how would you try to allocate spending between basic research and technology? Why?
54. *A Lunar Observatory.* Do the potential benefits of building an astronomical observatory on the Moon justify its costs at the present time? If it were up to you, would you recommend that Congress begin funding such an observatory? Defend your opinions.

MEDIA EXPLORATIONS

For self-study activities go to MasteringASTRONOMY www.masteringastronomy.com

Interactive Tutorials

Tutorial Review of Key Concepts

Use the interactive **Tutorials** at **www.masteringastronomy.com** to review key concepts from this chapter.

Light and Spectroscopy Tutorial

Lesson 2 Spectroscopy
Lesson 3 Atomic Spectra—Emission and Absorption Lines
Lesson 4 Thermal Radiation

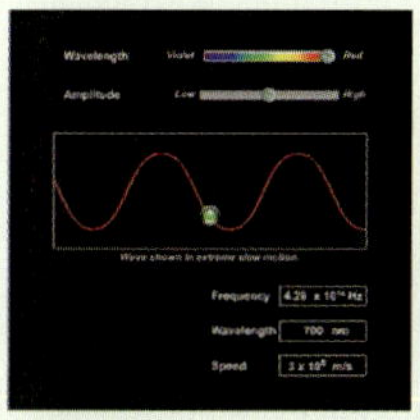
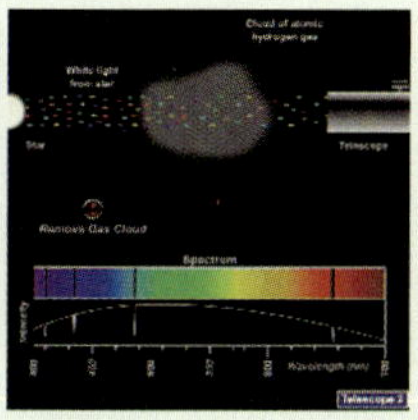
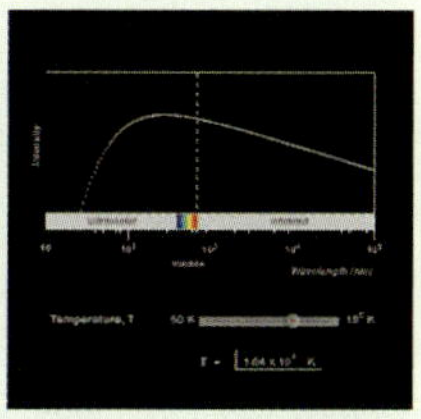

Telescopes Tutorial

Lesson 1 Optics and Light Gathering Power
Lesson 2 Angular Resolution

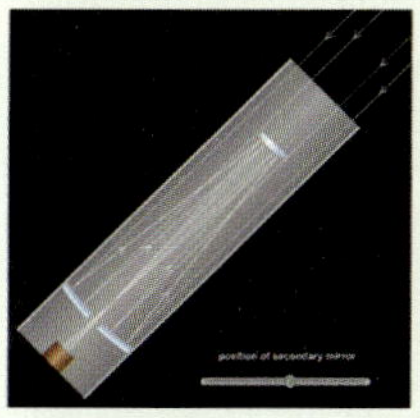
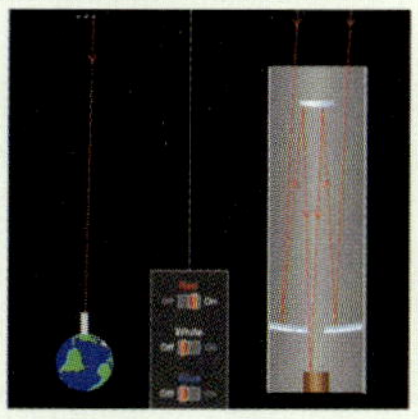
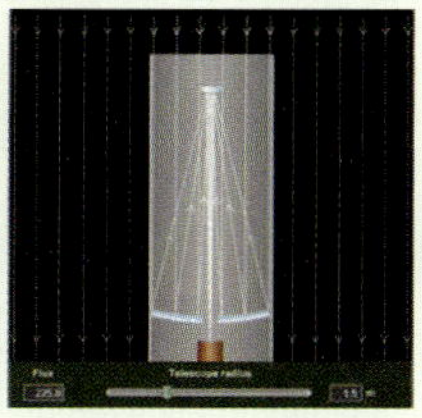

Supplementary Tutorial Exercises

Use the interactive **Tutorial Lessons** to explore the following questions.

Light and Spectroscopy Tutorial, Lessons 2–4

1. How would an object's spectrum influence the choice of telescope used to observe it?
2. Consider the spectra of hot stars as compared to those of cooler stars. Which stars would be best observed with an ultraviolet telescope? Explain.
3. Are there spectral features that can be observed only with telescopes in space? Explain.

Telescopes Tutorial, Lesson 1

1. To your eye, a star looks like a point of light in the sky, but the diagrams in this tutorial lesson show that the light from a single star enters a telescope as a series of parallel light rays. Why?
2. What is the purpose of the secondary mirror in a reflecting telescope? Why doesn't the fact that it blocks some starlight create a major problem for observations?
3. Describe how the light-collecting area of a reflecting telescope depends on its diameter.

Telescopes Tutorial, Lesson 2

1. Why do many stars that appear as single stars to the naked eye appear as pairs of stars through a telescope?
2. Describe how the angular resolution of a telescope depends on the diameter of the primary mirror.
3. Which telescope has better angular resolution: a 1-meter telescope observing ultraviolet light of wavelength 2×10^{-7} m, or a 10-meter telescope observing infrared light with wavelength 1×10^{-5} m? Explain.

Web Projects

1. *Major Ground-Based Observatories.* Take a virtual tour of one of the world's major astronomical observatories. Write a short report on why the observatory is useful to astronomy.
2. *Space Observatory.* Visit the Web site of a major space observatory, either existing or under development. Write a short report about the observatory, including its purpose, its orbit, and how it operates.
3. *Really Big Telescopes.* Several studies are under way in hopes of building telescopes far larger than any now in operation. Learn about one or more of these projects (such as OWL, the Swedish 50-m Optical Telescope, or the Thirty Meter Telescope [TMT]), and write a short report about the telescope's prospects and potential capabilities.

7

Our Planetary System

Earth, photographed from the outskirts of our solar system by the Voyager spacecraft. The "sunbeam" surrounding Earth is an artifact of light scattering in the camera.

LEARNING GOALS

7.1 Studying the Solar System
- What does the solar system look like?
- What can we learn by comparing the planets to one another?
- What are the major features of the Sun and planets?

7.2 Patterns in the Solar System
- What features of our solar system provide clues to how it formed?

7.3 Spacecraft Exploration of the Solar System
- How do robotic spacecraft work?

We succeeded in taking that picture [left], and, if you look at it, you see a dot. That's here. That's home. That's us. On it, everyone you ever heard of, every human being who ever lived, lived out their lives . . . on a mote of dust, suspended in a sunbeam.

—Carl Sagan

Our ancestors long ago recognized the motions of the planets through the sky, but it has been only a few hundred years since we learned that Earth is also a planet that orbits the Sun. Even then, we knew little about the other planets until the advent of large telescopes. More recently, the dawn of space exploration has brought us far greater understanding of other worlds. We've lived in this solar system all along, but only now are we getting to know it.

In this chapter, we'll explore our solar system like newcomers to the neighborhood. We'll begin by discussing what we hope to learn by studying the solar system, and in the process we'll take a brief tour of major features of the Sun and planets. We'll then consider some of the overall patterns we observe in the solar system—patterns that we will seek to explain in subsequent chapters. Finally, we'll discuss the use of spacecraft to explore the solar system, examining how we are coming to learn so much more about our neighbors.

7.1 Studying the Solar System

Galileo's telescopic observations began a new era in astronomy in which the Sun, Moon, and planets could be studied for the first time as *worlds*, rather than as mere lights in the sky. Since that time, people have studied these worlds in different ways. Sometimes we study them individually—for example, when we seek to map the geography of Mars or to probe the atmospheric structure of Jupiter. Other times we compare the worlds to one another, seeking to understand their similarities and differences. This latter approach is called **comparative planetology**. Astronomers use the term *planetology* broadly to include moons, asteroids, and comets as well as planets.

We will use the comparative planetology approach for most of our study of the solar system in this book. Before we can compare the planets, however, we must have a general idea of the nature of our solar system and of the characteristics of individual worlds.

Mastering Astronomy — Scale of the Universe Tutorial, Lesson 1

• What does the solar system look like?

The first step in getting to know our solar system is to visualize what it looks like as a whole. Imagine having the perspective of an alien spacecraft making its first scientific survey of our solar system. What would we see as we viewed the solar system from beyond the orbits of the planets?

Without a telescope, the answer would be "not much." Remember that the Sun and planets are all quite small compared to the distances between them [Section 1.2]—so small that if we viewed them from the outskirts of our solar system, the planets would be only pinpoints of light, and even the Sun would be just a small bright dot in the sky. But if we magnify the sizes of the planets by about a million times compared to their distances from the Sun, and show their orbital paths, we'd get the picture in Figure 7.1 (on the next page). Note that, even with this magnification, we do not see the many asteroids and comets that orbit the Sun, although the figure shows the location of the asteroid belt where many asteroids orbit. The figure does not show moons, either, although a few of the largest moons would be barely visible on the magnified scale.

As you study Figure 7.1, you'll quickly see that our solar system is *not* a random collection of worlds. For example, all the planets orbit the Sun in the same direction and in nearly the same plane, and the four inner planets are much smaller and closer together than the next four planets.

Additional patterns become clear when we look at the solar system from a viewpoint high above the ecliptic plane, as in Figure 7.2. This perspective shows us that most of the planets have nearly circular orbits, and the small white circles around all the planets except Mercury and Venus (which have no moons) show that most major moons orbit their planets in the same direction that their planets rotate. Again, the planet sizes in Figure 7.2 are magnified about a million times compared to the scale used for the orbits. The planets would be far too small to see if we tried to show them on the same scale as the orbits in this figure.

THINK ABOUT IT

Do you see any other patterns in Figures 7.1 and 7.2? For example, do you see any patterns to rotation or axis tilt? Are there exceptions to any of the general patterns you see?

The patterns revealed in Figures 7.1 and 7.2 suggest an organization to the solar system that is unlikely to have arisen by random processes. In science, we must therefore seek an explanation for the existence of these patterns. Indeed, the development of a theory that seems to explain most of the observed patterns has been one of the major triumphs of the scientific study of the solar system. We'll discuss this theory in Chapter 8, after we complete our overview of the solar system and search for additional patterns and features that a successful theory must explain.

Mastering Astronomy — Formation of the Solar System Tutorial, Lesson 1

• What can we learn by comparing the planets to one another?

We've stated that we'll take a comparative approach to our study of the worlds in our solar system, but what exactly

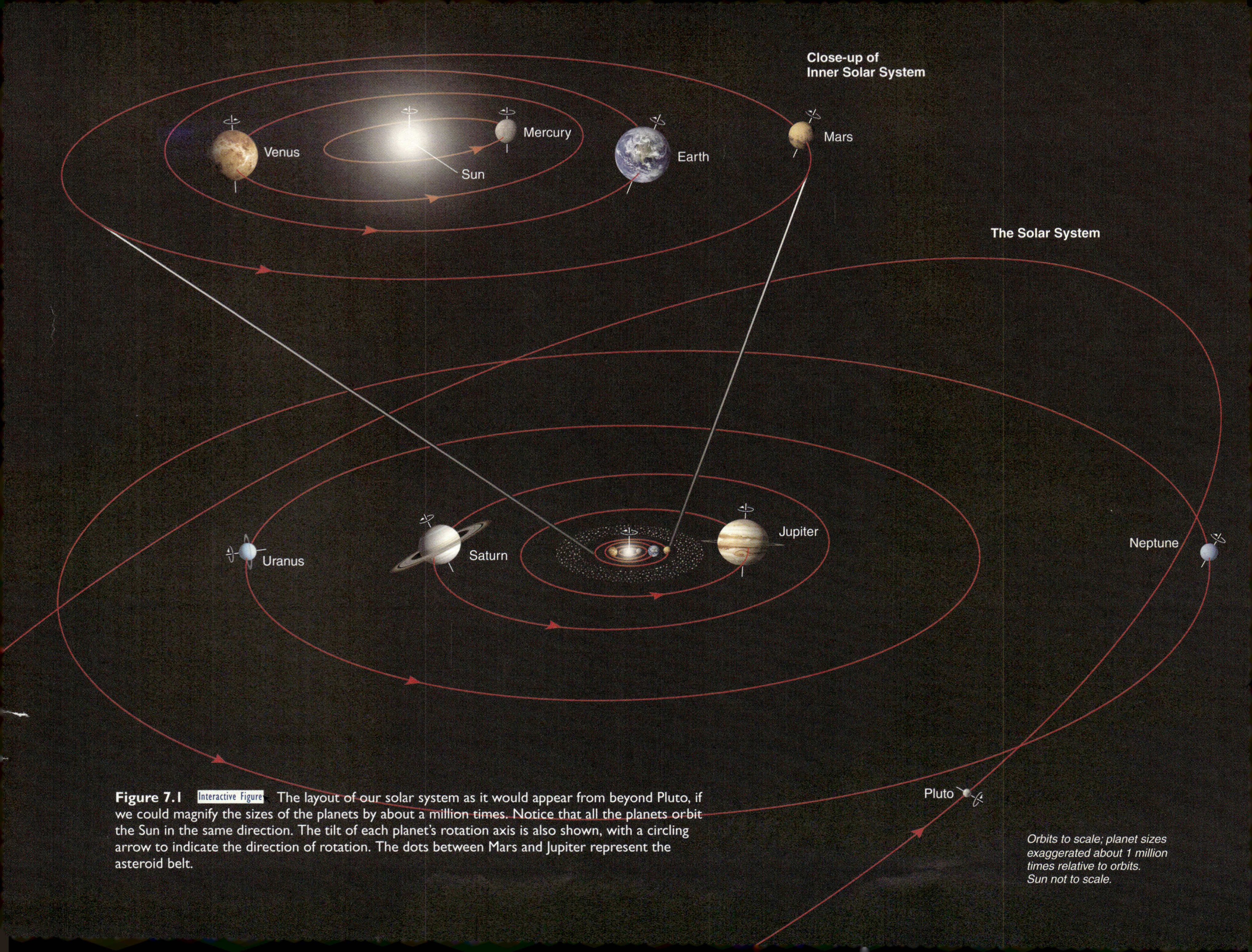

Figure 7.1 Interactive Figure The layout of our solar system as it would appear from beyond Pluto, if we could magnify the sizes of the planets by about a million times. Notice that all the planets orbit the Sun in the same direction. The tilt of each planet's rotation axis is also shown, with a circling arrow to indicate the direction of rotation. The dots between Mars and Jupiter represent the asteroid belt.

does this mean? The essence of comparative planetology lies in the idea that we can learn more about an individual world, including our own Earth, by studying it in the context of other objects in our solar system. It is much like learning more about a person by getting to know his or her family, friends, and culture.

While we still can learn much by studying planets individually, the comparative planetology approach has demonstrated its value in at least three key ways:

- Comparative study has revealed similarities and differences among the planets that have helped guide the development of our theory of solar system formation, thereby giving us a better understanding of how we came to exist here on Earth.
- Comparative study has given us deep new insights into the physical processes that have shaped Earth and other worlds—insights that can help us better understand and manage our own planet.
- Comparative study has allowed us to apply lessons from our solar system to the study of the many recently discovered planetary systems around other stars. These lessons help us understand both the general principles that govern star systems and the specific circumstances under which Earth-like planets—and possibly life—might exist elsewhere.

The comparative planetology approach should also benefit you as a student by helping you stay focused on *processes* rather than on a collection of facts. After more than four decades of planetary exploration, the "facts" can easily prove overwhelming; for example, you can find atlases with names and descriptions for hundreds of mountains, craters, and valleys on Mars alone. Your goal should not be to memorize all these features of individual worlds—though it will be useful to know names for a handful of the most significant features—but rather to think about how these features came to exist and why the features on one world are similar to or different from those on other worlds. By staying focused on this goal, you'll gain a far deeper understanding of the history of Earth and the rest of our solar system.

• What are the major features of the Sun and planets?

We cannot achieve our goal of comparing the planets unless we first know a little bit about them. The following 10 pages offer a brief tour through our solar system, beginning at the Sun and continuing to each of the planets and Pluto. The tour highlights a few of the most important features of each world we visit—just enough information so that you'll be ready for the comparative study we'll undertake in later chapters. The side of each page shows the planets to scale, using the 1-to-10-billion scale introduced in Chapter 1. The map along the bottom of each page shows the locations of the Sun and each of the planets in the Voyage scale model solar system in Washington, DC (see Figures 1.6 and 1.7), so that you can see relative distances from the Sun. Table 7.1, which follows the tour, summarizes key planetary data.

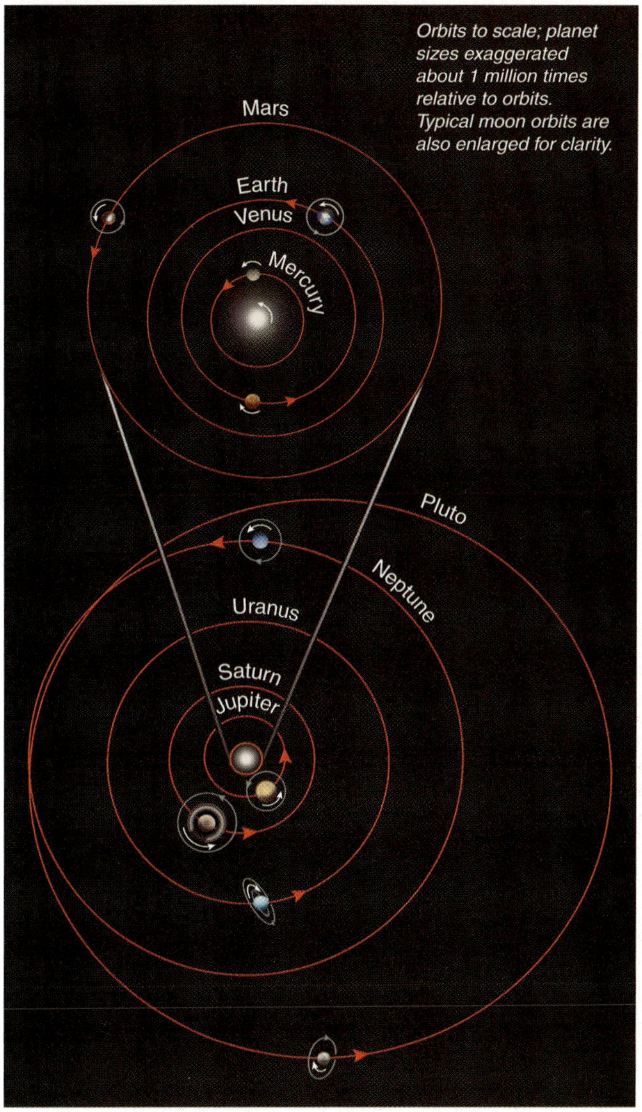

Figure 7.2 Interactive Figure View of the solar system from high above Earth's North Pole. Notice that most orbits are nearly circular. The short, curved arrow for each planet indicates the planet's direction of rotation. The small circles around the planets indicate the orbital direction of major moons. (Mercury and Venus have no moons.) Notice that most major moons orbit in the same direction that their planet rotates.

THINK ABOUT IT

As you read through the tour pages, make a brief list of the similarities and differences you notice among the planets. In addition, try to answer the following questions as you read: (1) Are all planets made of the same ingredients? (2) Are all planets "Earth-like" with solid surfaces? (3) Can you organize the planets into groups with common characteristics?

Figure 7.3 The Sun contains more than 99.9% of the total mass in our solar system.

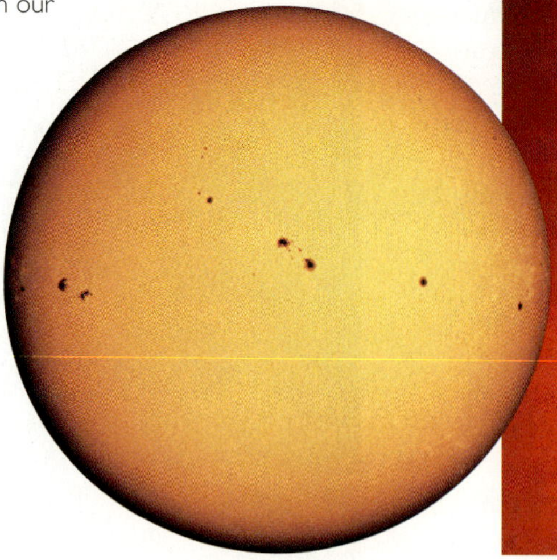

a A visible-light photograph of the Sun's surface. The dark splotches are sunspots—each large enough to swallow several Earths.

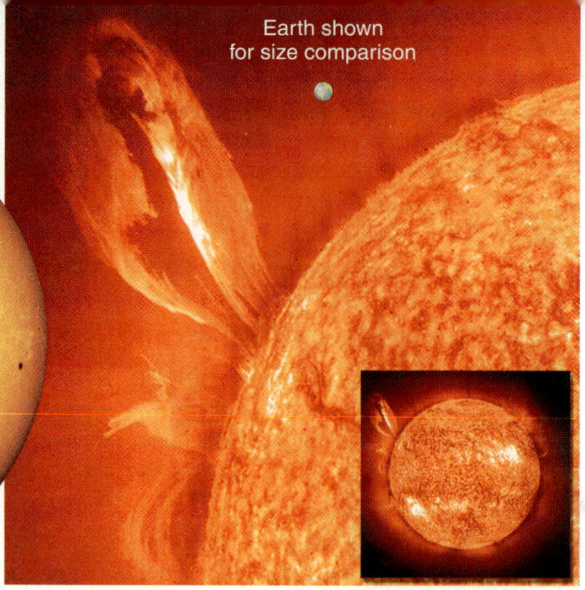

b This ultraviolet photograph, from the SOHO spacecraft, shows a huge streamer of hot gas on the Sun. The image of Earth was added for size comparison.

• The Sun

- Radius: 695,000 km $= 108\ R_{Earth}$
- Mass: 333,000 M_{Earth}
- Composition (by mass): 98% hydrogen and helium, 2% other elements

The Sun is by far the largest and brightest object in our solar system. It contains more than 99.9% of the solar system's total mass, making it more than a thousand times as massive as everything else in the solar system combined.

The Sun's surface looks solid in photographs (Figure 7.3), but it is actually a roiling sea of hot (about 5,800 K, or 6,100°C or 11,000°F) hydrogen and helium gas. The surface is speckled with sunspots that appear dark in photographs only because they are slightly cooler than their surroundings. Solar storms sometimes send streamers of hot gas soaring far above the surface.

The Sun is gaseous throughout. If you could plunge into the Sun, you'd find ever higher temperatures as you went deeper. The source of the Sun's energy lies deep in its core, where the temperatures and pressures are so high that the Sun is a nuclear fusion power plant. Each second, fusion transforms about 600 million tons of the Sun's hydrogen into 596 million tons of helium. The "missing" 4 million tons becomes energy in accord with Einstein's famous formula $E = mc^2$ [Section 4.3]. Despite losing 4 million tons of mass each second, the Sun contains so much hydrogen that it has already shone steadily for almost 5 billion years and will continue to shine for some 5 billion years more.

The Sun is certainly the most influential object in our solar system. Its gravity governs the orbits of the planets. Its heat is the primary influence on the temperatures of planetary surfaces and atmospheres. And it is the source of virtually all the visible light in our solar system—the Moon and planets shine only by virtue of the sunlight they reflect. In addition, charged particles flowing outward from the Sun (the solar wind) help shape planetary magnetic fields and can influence planetary atmospheres. Nevertheless, we can understand almost all the present characteristics of the planets without knowing much more about the Sun than what we have just discussed. We'll therefore save more detailed study of the Sun for Chapter 14, where we will study it as our prototype for understanding other stars.

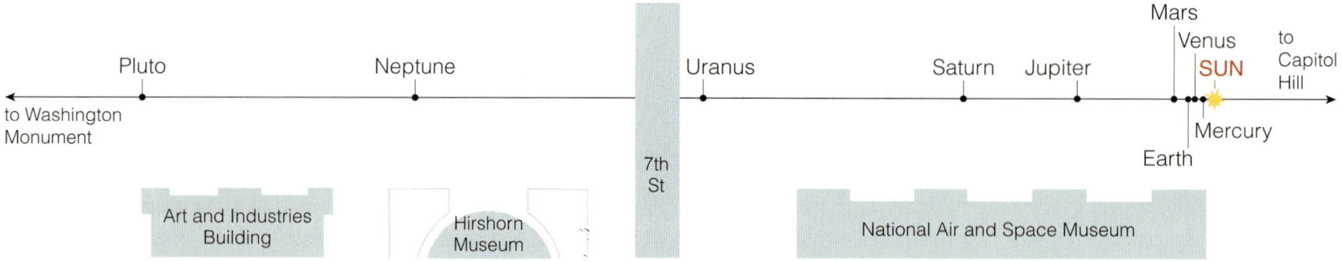

The Voyage scale model solar system represents sizes and distances in our solar system at one ten-billionth of their actual values (see Figure 1.7). The strip along the side of the page shows the sizes of the Sun and planets on this scale, and the map above shows their locations in the Voyage model on the National Mall in Washington, D.C. The Sun is about the size of a large grapefruit on this scale.

Figure 7.4 This image shows what it would look like to be orbiting a few hundred kilometers above Mercury's surface with your back toward the Sun. The view was created using imagery from NASA's *Mariner 10* spacecraft but with computer manipulation to provide color and the orbital viewpoint. The inset (right) is a composite photograph of the full disk of Mercury as seen by *Mariner 10;* the blank strip at the upper right was not photographed. (Image at left from the Voyage scale model solar system, developed by the Challenger Center for Space Science Education, the Smithsonian Institution, and NASA. Image created by ARC Science Simulations © 2001.)

• Mercury

- Average distance from the Sun: 0.39 AU
- Radius: 2,440 km = 0.38 R_{Earth}
- Mass: 0.055 M_{Earth}
- Average density: 5.43 g/cm^3
- Composition: rocks, metals
- Average surface temperature: 700 K (day), 100 K (night)
- Moons: 0

Mercury is the innermost planet of our solar system, and the smallest of the first eight planets. It is a desolate, cratered world with no active volcanoes, no wind, no rain, and no life. Because there is virtually no air to scatter sunlight or color the sky, you could see stars even in the daytime if you stood on Mercury with your back toward the Sun.

You might expect Mercury to be very hot because of its closeness to the Sun, but in fact it is a world of both hot and cold extremes. Tidal forces from the Sun have forced Mercury into an unusual rotation pattern [Section 4.5]: Its 58.6-day rotation period means it rotates exactly three times for every two of its 87.9-day orbits of the Sun. This combination of rotation and orbit gives Mercury days and nights that last about three Earth months each. Daytime temperatures reach 425°C—nearly as hot as hot coals. At night or in shadow, the temperature falls below −150°C—far colder than Antarctica in winter.

Mercury is the least studied of the inner planets. Its proximity to the Sun makes it difficult to observe through telescopes, and it has been visited by just one spacecraft: *Mariner 10,* which obtained images of one hemisphere (Figure 7.4) during three rapid flybys of Mercury in 1974–1975. The images show vast numbers of craters, ancient lava flows, and tall, steep cliffs that run hundreds of kilometers in length. As we'll discuss in Chapter 9, the cliffs may be wrinkles from an episode of "planetary shrinking" early in Mercury's history.

The influence of Mercury's gravity on *Mariner 10*'s orbit allowed scientists to determine Mercury's mass and to make inferences about its interior composition and structure. Mercury appears to be made mostly of iron, making it the most metal-rich of the planets. Scientists hope to learn much more about Mercury soon. A NASA spacecraft called *MESSENGER,* launched in 2004, will observe Mercury during three flybys in 2008 and 2009 and then begin orbiting Mercury in 2011.

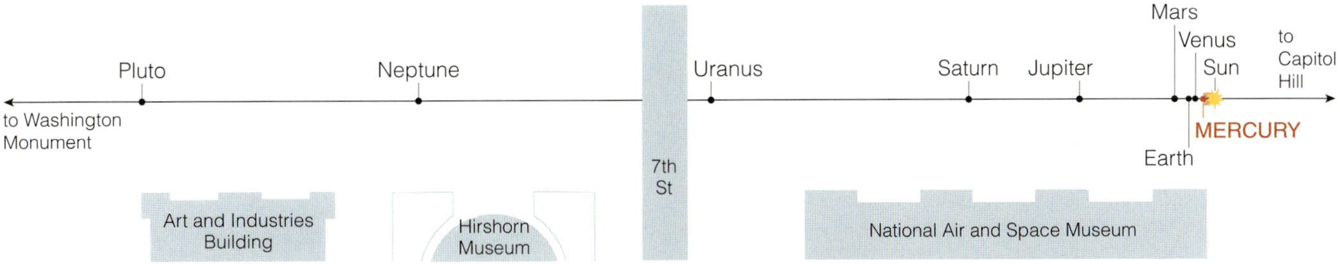

Figure 7.5 The image above shows an artistic rendition of the surface of Venus as scientists think it would appear to our eyes. The surface topography is based on data from NASA's *Magellan* spacecraft. The inset (left) shows the full disk of Venus photographed by NASA's *Pioneer Venus Orbiter* with cameras sensitive to ultraviolet light. With visible light, cloud features cannot be distinguished from the general haze. (Image above from the Voyage scale model solar system, developed by the Challenger Center for Space Science Education, the Smithsonian Institution, and NASA. Image by David P. Anderson, Southern Methodist University © 2001.)

• Venus

- Average distance from the Sun: 0.72 AU
- Radius: 6,051 km = 0.95 R_{Earth}
- Mass: 0.82 M_{Earth}
- Average density: 5.24 g/cm^3
- Composition: rocks, metals
- Average surface temperature: 740 K
- Moons: 0

Venus, the second planet from the Sun, is nearly identical in size to Earth. Prior to the era of spacecraft visits, Venus stood out largely for its strange rotation: It rotates on its axis very slowly and in the opposite direction of Earth, so that days and nights are very long and the Sun rises in the west and sets in the east instead of rising in the east and setting in the west. Its surface is completely hidden from view by dense clouds, so we knew little about it until a few decades ago, when spacecraft carrying cloud-penetrating radar finally allowed us to study Venus in detail (Figure 7.5). Because we knew so little about it, some science fiction writers used its Earth-like size, thick atmosphere, and closer distance to the Sun to speculate that it might be a lush, tropical paradise—a "sister planet" to Earth.

The reality is far different. We now know that an extreme *greenhouse effect* bakes Venus's surface to an incredible 470°C (about 880°F), trapping heat so effectively that nighttime offers no relief. Day and night, Venus is hotter than a pizza oven. All the while, the thick atmosphere bears down on the surface with a pressure equivalent to what exists nearly a kilometer (0.6 mile) beneath the ocean's surface on Earth. Even if we could somehow survive these hazards, we'd find no oxygen to breathe and no water to drink. Far from being a beautiful sister planet to Earth, Venus resembles a traditional view of hell.

The surface of Venus has mountains, valleys, and craters and shows many signs of past (and possibly present) volcanic activity. Aside from these superficial similarities, however, the geology of Venus's surface appears to be quite different from that of Earth. We will devote much of Chapters 9 and 10 to understanding how and why Venus became so unlike Earth, both in its surface geology and in its atmosphere.

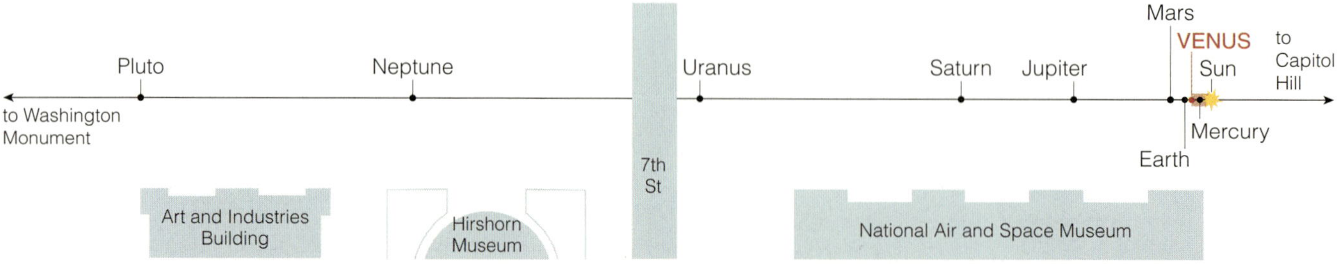

a This image (left), computer generated from satellite data, shows the striking contrast between the daylight and nighttime hemispheres of Earth. The day side reveals little evidence of human presence, but at night our presence is revealed by the lights of human activity (mostly from cities as well as agricultural, oil, and gas fires). (From the Voyage scale model solar system, developed by the Challenger Center for Space Science Education, the Smithsonian Institution, and NASA. Image created by ARC Science Stimulations © 2001.)

b Earth and the Moon, shown to scale. The Moon is about one-fourth as large as Earth in diameter, while its mass is about 1/80 Earth's mass. If you wanted to show the distance between Earth and Moon on the same scale, you'd need to hold these two photographs about 1 meter (3 feet) apart.

Figure 7.6 Earth, our home planet.

• Earth

- Average distance from the Sun: 1.00 AU
- Radius: 6,378 km = 1 R_{Earth}
- Mass: 1.00 M_{Earth}
- Average density: 5.52 g/cm^3
- Composition: rocks, metals
- Average surface temperature: 290 K
- Moons: 1

Beyond Venus, we next encounter our home planet. Although Earth is a barely visible speck when viewed on the scale of our entire solar system, it is the only known oasis of life. Even if we someday discover microscopic life on other worlds in our solar system, Earth is the only world on which humans could survive without a protective enclosure. It is the only planet with oxygen for us to breathe and ozone to shield us from deadly solar radiation. It is the only planet with abundant surface water to nurture life. Temperatures are pleasant for us because Earth's atmosphere contains just enough carbon dioxide and water vapor to maintain a moderate greenhouse effect.

Despite Earth's small size, its beauty is striking (Figure 7.6a). Blue oceans cover nearly three-fourths of the surface, broken by the continental land masses and scattered islands. The polar caps are white with snow and ice, and white clouds are scattered above the surface. At night, the glow of artificial lights clearly reveals the presence of an intelligent civilization.

Earth is the first planet on our tour with a moon. Moreover, our Moon is surprisingly large in comparison to Earth (Figure 7.6b); although it is not the largest moon in the solar system, almost all other moons are much smaller in comparison to the planets they orbit. The question of why Earth has such a large moon has long been one of the major mysteries of our solar system.

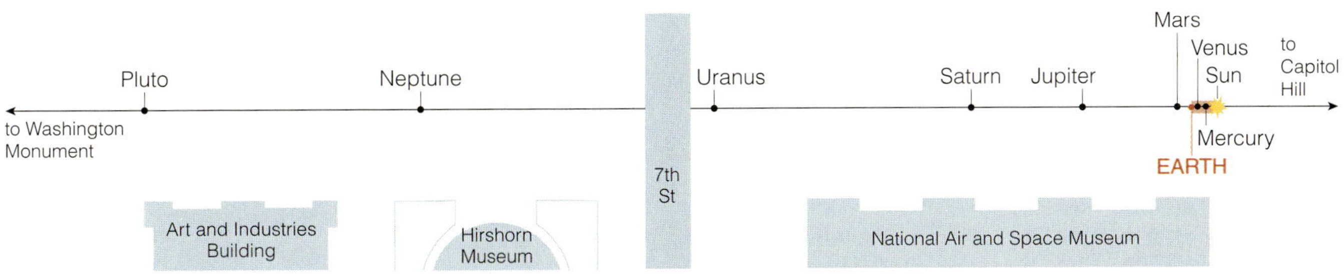

Figure 7.7 The image below shows the surface of Mars photographed by NASA's *Opportunity* rover in 2004. Notice the tracks leading away from the casing that held the rover as it descended to the surface. The globe shows the full disk of Mars; the horizontal "gash" across the center is the giant canyon Valles Marineris.

• Mars

- Average distance from the Sun: 1.52 AU
- Radius: 3,397 km = 0.53 R_{Earth}
- Mass: 0.11 M_{Earth}
- Average density: 3.93 g/cm^3
- Composition: rocks, metals
- Average surface temperature: 220 K
- Moons: 2 (very small)

The next planet on our tour is Mars, the last of the four inner planets of our solar system (Figure 7.7). Mars is larger than Mercury and the Moon but smaller than Venus and Earth. About half Earth's size in diameter, Mars has a mass about 10% that of Earth. Mars has two tiny moons, Phobos and Deimos, that look much like typical asteroids and may once have roamed freely in the asteroid belt. These moons are too small to influence Mars in any of the important ways that our Moon influences Earth (such as tides).

Mars is a world of wonders, with ancient volcanoes that dwarf the largest mountains on Earth, a great canyon that runs nearly one-fifth of the way around the planet, and polar caps made of frozen carbon dioxide ("dry ice") and water ice. Although Mars is frozen today, the presence of dried-up riverbeds, rock-strewn floodplains, and minerals that form in water offers clear evidence that Mars had at least some warm and wet periods in the distant past. Thus, Mars may once have been hospitable to life, though major flows of liquid water probably ceased at least 2–3 billion years ago.

Mars looks almost Earth-like in photographs taken by spacecraft on its surface, but you wouldn't want to visit without a space suit. The air pressure is far less than that on top of Mount Everest, the temperature is usually well below freezing, the trace amounts of oxygen would not be nearly enough to breathe, and the lack of atmospheric ozone would leave you exposed to deadly ultraviolet radiation from the Sun.

Mars is the most studied planet besides Earth. More than a dozen spacecraft have flown past, orbited, or landed on Mars, and plans are in the works for many more missions to Mars. We may even send humans to Mars within the next few decades. By overturning rocks in ancient riverbeds or chipping away at ice in the polar caps, explorers will help us learn whether Mars has ever been home to life.

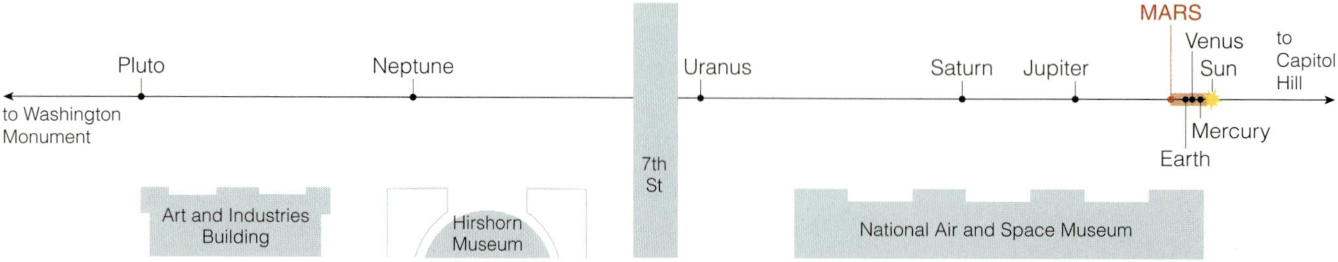

Figure 7.8 This image shows what it would look like to be orbiting near Jupiter's moon Io as Jupiter comes into view. Notice the Great Red Spot to the left of Jupiter's center. The extraordinarily dark rings discovered during the *Voyager* missions are exaggerated to make them visible. This computer visualization was created using data from both NASA's *Voyager* and *Galileo* missions. (From the Voyage scale model solar system, developed by the Challenger Center for Space Science Education, the Smithsonian Institution, and NASA. Image created by ARC Science Simulations © 2001.)

• Jupiter

- Average distance from the Sun: 5.20 AU
- Radius: 71,492 km = 11.2 R_{Earth}
- Mass: 318 M_{Earth}
- Average density: 1.33 g/cm^3
- Composition: mostly hydrogen and helium
- Cloud-top temperature: 125 K
- Moons: at least 63

To reach the orbit of Jupiter from Mars, we must traverse a distance that is more than double the total distance from the Sun to Mars, passing through the asteroid belt along the way. Upon our arrival, we find a planet much larger than any we have seen so far (Figure 7.8).

Jupiter is so different from the planets of the inner solar system that we must adopt an entirely new mental image of the term *planet*. Its mass is more than 300 times that of Earth, and its volume is more than 1,000 times that of Earth. Its most famous feature—a long-lived storm called the Great Red Spot—is itself large enough to swallow two or three Earths. Like the Sun, Jupiter is made primarily of hydrogen and helium and has no solid surface. If we plunged deep into Jupiter, the increasing gas pressure would crush us long before we ever reached its core.

Jupiter reigns over dozens of moons and a thin set of rings (too faint to be seen in most photographs). Most of the moons are very small, but four are large enough that we'd consider them planets if they orbited the Sun. These four moons—Io, Europa, Ganymede, and Callisto—are often called the *Galilean moons*, because Galileo discovered them shortly after he first turned his telescope toward the heavens [Section 3.3]. They are also planetlike in having varied and interesting geology. Io is the most volcanically active place in the solar system. Europa has an icy crust that may hide a subsurface ocean of liquid water, making it a promising place to search for life. Ganymede and Callisto may also have subsurface oceans, and their surfaces have many features that remain mysterious.

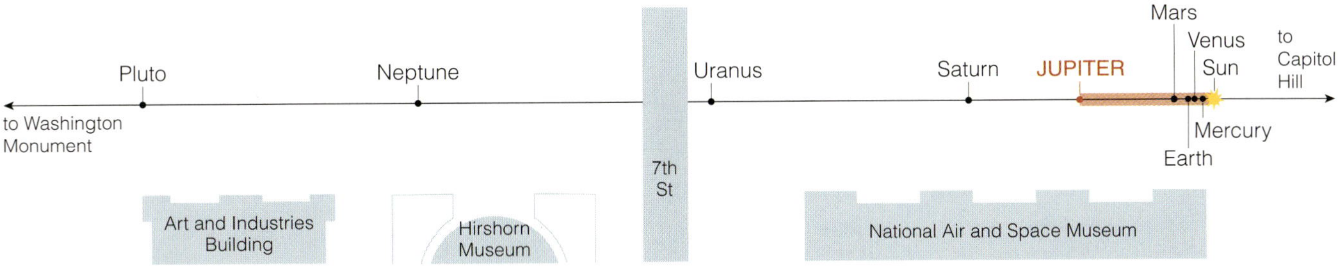

Figure 7.9 *Cassini's* view of Saturn. We see the shadow of the rings on Saturn's sunlit face, and the rings become lost in Saturn's shadow on the night side.

• Saturn

- Average distance from the Sun: 9.54 AU
- Radius: 60,268 km = 9.4 R_{Earth}
- Mass: 95.2 M_{Earth}
- Average density: 0.70 g/cm^3
- Composition: mostly hydrogen and helium
- Cloud-top temperature: 95 K
- Moons: at least 47

The journey from Jupiter to Saturn is a long one: Saturn orbits nearly twice as far from the Sun as Jupiter. Saturn, the second-largest planet in our solar system, is only slightly smaller than Jupiter in diameter. However, its lower density makes it considerably less massive than Jupiter (about one-third Jupiter's mass). Like Jupiter, Saturn is made mostly of hydrogen and helium and has no solid surface.

Saturn is famous for its spectacular rings (Figure 7.9). Although all four of the giant outer planets have rings, only Saturn's rings can be seen easily through a small telescope. The rings may look solid from a distance, but in reality they are made of countless small particles, each of which orbits Saturn like a tiny moon. If you could wander into the rings, you'd find yourself surrounded by chunks of rock and ice that range in size from dust grains to city blocks.

Like Jupiter, Saturn is orbited by many moons. Most are the size of small asteroids or comets, but a few are much larger. Perhaps the most fascinating moon is Titan, which is larger than the planet Mercury and blanketed by a thick atmosphere. On Titan's surface, you'd find an atmospheric pressure even greater than on Earth, and you could inhale air with roughly the same nitrogen content as air on Earth. However, you'd need to bring your own oxygen, and you'd certainly want a warm space suit for protection against the frigid outside temperatures. NASA's *Cassini* spacecraft, designed to explore Saturn and its rings and moons, carried a probe called *Huygens* that landed on Titan in January 2005. There, it found a bizarre landscape carved by erosion despite a temperature of −180°C.

THINK ABOUT IT

Find the current status of the *Cassini* mission (from news reports or the mission Web site). Is the spacecraft still operating? What is its next target of study? What recent discoveries has it helped us make?

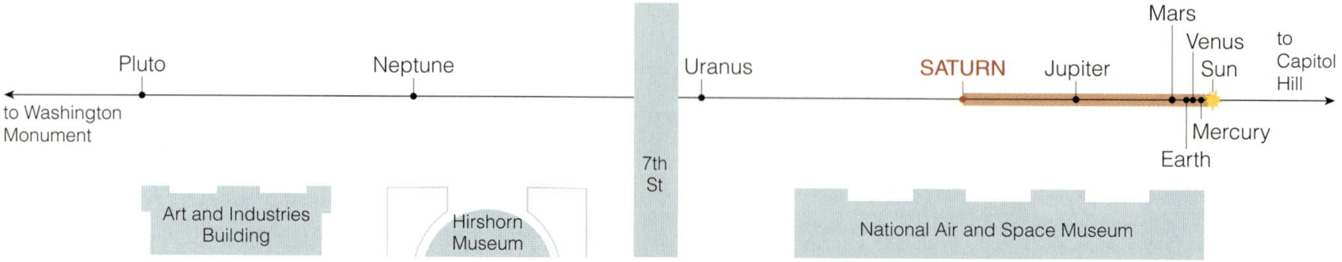

Figure 7.10 This image shows a view from a vantage point high above Uranus's moon Ariel. The ring system is shown, although it would actually be too dark to see from this vantage point. This computer simulation is based on data from NASA's *Voyager 2* mission. (From the Voyage scale model solar system, developed by the Challenger Center for Space Science Education, the Smithsonian Institution, and NASA. Image created by ARC Science Simulations © 2001.)

• Uranus

- Average distance from the Sun: 19.2 AU
- Radius: 25,559 km $= 4.0\ R_{Earth}$
- Mass: $14.5\ M_{Earth}$
- Average density: 1.32 g/cm^3
- Composition: hydrogen, helium, hydrogen compounds
- Cloud-top temperature: 60 K
- Moons: at least 27

It's another long journey to our next stop on the tour, as Uranus lies twice as far from the Sun as Saturn. Uranus (normally pronounced *YUR-uh-nus*) is much smaller than either Jupiter or Saturn but still much larger than Earth. It is made largely of hydrogen, helium, and *hydrogen compounds* such as water (H_2O), ammonia (NH_3), and methane (CH_4). Methane gas gives Uranus its pale blue-green color (Figure 7.10). Like the other giants of the outer solar system, Uranus lacks a solid surface. More than two dozen moons orbit Uranus, along with a set of rings somewhat similar to those of Saturn but much darker and more difficult to see.

The entire Uranus system—planet, rings, and moon orbits—is tipped on its side compared to the rest of the planets. This unusual orientation may be the result of a cataclysmic collision that Uranus suffered as it was forming some 4.6 billion years ago. It also makes for the most extreme pattern of seasons on any planet. If you lived on a platform floating in Uranus's atmosphere near its north pole, you'd have continuous daylight for half of each orbit, or 42 years. Then, after a very gradual sunset, you'd enter into a 42-year-long night.

Only one spacecraft has visited Uranus. *Voyager 2* flew past all four of the giant outer planets before heading out of the solar system. Much of our current understanding of Uranus comes from that mission, though powerful new telescopes are also capable of studying this planet. Scientists would love an opportunity to study Uranus and its interesting rings and moons in much greater detail, but no missions to Uranus are currently under development.

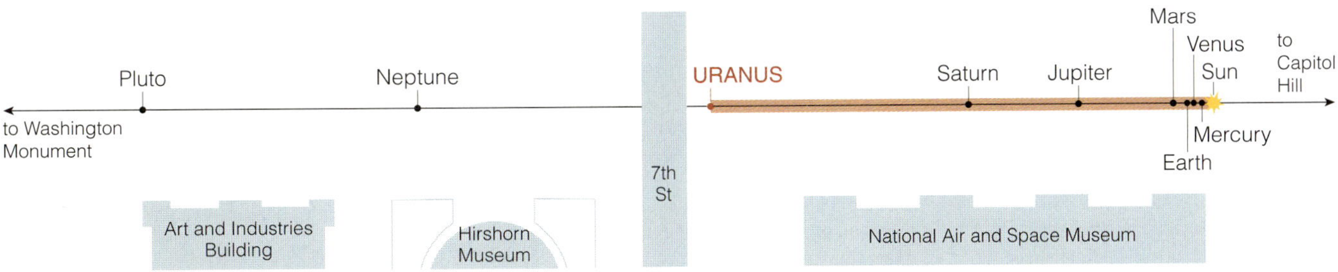

Figure 7.11 This image shows what it would look like to be orbiting Neptune's moon Triton as Neptune itself comes into view. The dark rings are exaggerated to make them visible in this computer simulation using data from NASA's *Voyager 2* mission. (From the Voyage scale model solar system, developed by the Challenger Center for Space Science Education, the Smithsonian Institution, and NASA. Image created by ARC Science Simulations © 2001.)

• Neptune

- Average distance from the Sun: 30.1 AU
- Radius: 24,764 km = 3.9 R_{Earth}
- Mass: 17.1 M_{Earth}
- Average density: 1.64 g/cm^3
- Composition: hydrogen, helium, hydrogen compounds
- Cloud-top temperature: 60 K
- Moons: at least 13

The journey from the orbit of Uranus to the orbit of Neptune is the longest yet in our tour, calling attention to the vast emptiness of the outer solar system. Nevertheless, Neptune looks nearly like a twin of Uranus, although it is more strikingly blue (Figure 7.11). It is slightly smaller than Uranus in size, but a higher density makes it slightly more massive even though the two planets share very similar compositions. Like Uranus, only the *Voyager 2* spacecraft has visited Neptune. No further missions are currently planned.

Neptune has rings and numerous moons. Its largest moon, Triton, is larger than Pluto and is one of the most fascinating moons in the solar system. Its icy surface has features that appear to be somewhat like geysers, although they spew nitrogen gas rather than water into the sky [Section 11.2]. Even more surprisingly, Triton is the only large moon in the solar system that orbits its planet "backward"—that is, in a direction opposite to the direction in which Neptune rotates. The attempt to understand how such a large moon ended up orbiting Neptune backward has given us new insights into the history of the outer solar system.

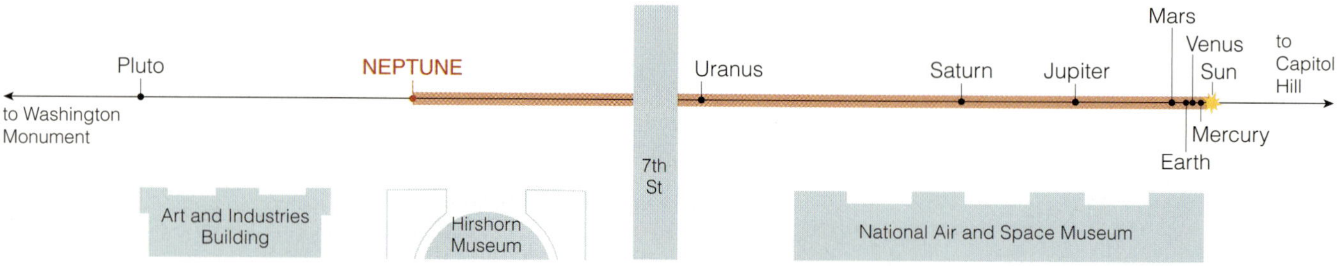

Figure 7.12 Pluto and its moons, as photographed by the Hubble Space Telescope.

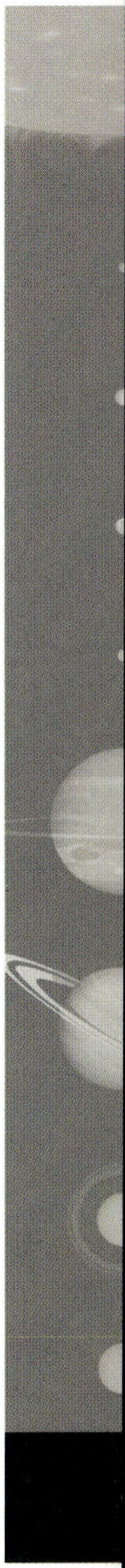

• Pluto (and Other Dwarf Planets)

- Average distance from the Sun: 39.5 AU
- Radius: 1,160 km = 0.18 R_{Earth}
- Mass: 0.0022 M_{Earth}
- Average density: 2.0 g/cm^3
- Composition: ices, rock
- Average surface temperature: 40 K
- Moons: 3

We conclude our tour at Pluto, which reigned for some 75 years as the ninth and last planet in our solar system. Pluto's average distance from the Sun lies as far beyond Neptune as Neptune lies beyond Uranus. Its great distance makes Pluto cold and dark. From Pluto, the Sun would be little more than a bright light among the stars. Pluto's largest moon, Charon, is locked together with it in synchronous rotation [Section 4.4], so that Charon would dominate the sky on one side of Pluto but never be seen from the other side.

We've known for decades that Pluto is very different in character from the first eight planets. It is much smaller and less massive than any of those planets, and its orbit is much more eccentric and inclined to the ecliptic plane. Its composition of ice and rock is also quite different from that of any of the first eight planets, although it is virtually identical to that of many known comets. Moreover, since the early 1990s, astronomers have discovered more than 1,000 objects of similar composition orbiting in Pluto's general neighborhood—the region of the solar system near and beyond the orbit of Neptune that is known as the *Kuiper belt.* Pluto is not even the largest of these Kuiper belt objects: Eris, discovered in 2005, is slightly larger in size than Pluto. Eris's 560-year orbit of the Sun is even more eccentric and inclined to the ecliptic than Pluto's.

Scientifically, these facts leave no room for doubt that both Pluto and Eris belong to a different class of objects than the first eight planets. They are just the largest known of hundreds of large iceballs—essentially large comets—located in the Kuiper belt. The only question has been one of words: Should Pluto and Eris be called "planets" or something else? In August 2006, the International Astronomical Union (IAU) voted to classify Pluto and Eris as *dwarf planets* (see Special Topic, p.9), thereby trimming the official list of planets to eight. More technically, the IAU defined a dwarf planet as any object that orbits the Sun and is large enough for its own gravity to have made it nearly round in shape, but that has not cleared its orbital neighborhood of other objects. The largest asteroid in the asteroid belt, Ceres, thereby also made the list of dwarf planets, and it is possible that dozens of other objects (mostly in the Kuiper belt) will also qualify. These are objects for which telescopic observations have so far given us size estimates but not enough resolution to confirm that they are round.

The great distances and small sizes of Pluto and other dwarf planets make them difficult to study, regardless of whether they are located in the asteroid belt or the Kuiper belt. As you can see in Figure 7.12, even the best telescopic views of Pluto and its moons reveal little detail. Better information should be coming soon. A spacecraft called *New Horizons,* launched in 2006, will fly past Pluto in mid-2015 and may then visit other objects of the Kuiper belt. Meanwhile, the *Dawn* spacecraft (scheduled for 2007 launch) should give us our first good views of large asteroids in the asteroid belt beginning in about 2011, with a pass by Ceres four years later that will closely coincide with *New Horizon's* pass by Pluto.

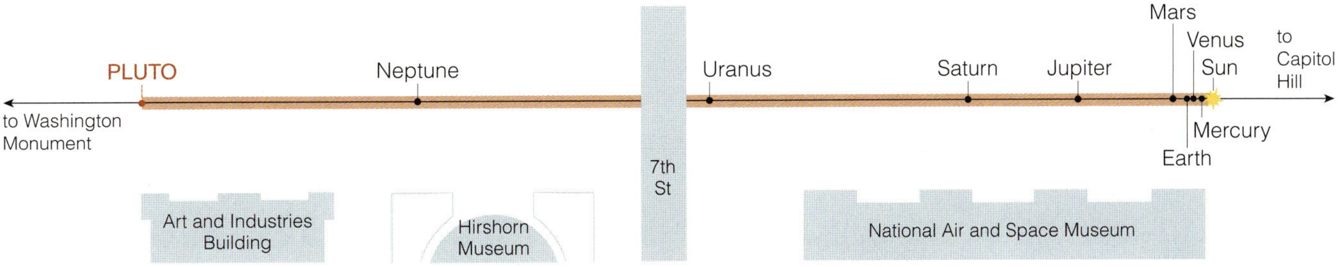

Table 7.1 Planetary Data*

Photo	Planet	Relative Size	Average Distance from Sun (AU)	Average Equatorial Radius (km)	Mass (Earth = 1)	Average Density (g/cm^3)	Orbital Period	Rotation Period	Axis Tilt	Average Surface (or Cloud Tops) Temperature†	Composition	Known Moons (2005)	Rings?
	Mercury	·	0.387	2,440	0.055	5.43	87.9 days	58.6 days	0.0°	700 K (day) 100 K (night)	Rocks, metals	0	No
	Venus	•	0.723	6,051	0.82	5.24	225 days	243 days	177.3°	740 K	Rocks, metals	0	No
	Earth	•	1.00	6,378	1.00	5.52	1.00 year	23.93 hours	23.5°	290 K	Rocks, metals	1	No
	Mars	·	1.52	3,397	0.11	3.93	1.88 years	24.6 hours	25.2°	220 K	Rocks, metals	2	No
	Jupiter	●	5.20	71,492	318	1.33	11.9 years	9.93 hours	3.1°	125 K	H, He, hydrogen compounds§	63	Yes
	Saturn	●	9.54	60,268	95.2	0.70	29.4 years	10.6 hours	26.7°	95 K	H, He, hydrogen compounds§	47	Yes
	Uranus	●	19.2	25,559	14.5	1.32	83.8 years	17.2 hours	97.9°	60 K	H, He, hydrogen compounds§	27	Yes
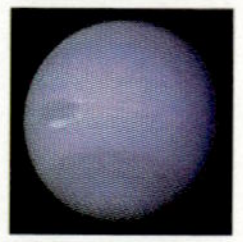	Neptune	●	30.1	24,764	17.1	1.64	165 years	16.1 hours	29.6°	60 K	H, He, hydrogen compounds§	13	Yes
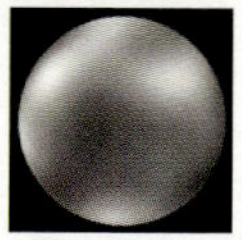	Pluto	·	39.5	1,160	0.0022	2.0	248 years	6.39 days	112.5°	40 K	Ices, rock	3	No
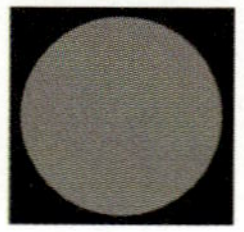	Eris	·	67.9	1,200	?	?	560 years	?	?	?	Ices, rock	1	?

*Including the dwarf planets Pluto and Eris. Appendix E gives a more complete list of planetary properties. †Surface temperatures for all objects except Jupiter, Saturn, Uranus, and Neptune, for which cloud-top temperatures are listed. §Includes water

Orbits and Kepler's Laws Tutorial, Lessons 2–4

7.2 Patterns in the Solar System

Take a little time to think about our general picture of the solar system so far. Look back at the patterns of motion you see in Figures 7.1 and 7.2. Review the tour we've taken of individual worlds and consider patterns you may have noticed about size, distance, composition, and number of moons. Identify some of these patterns in Table 7.1, along with any other patterns you might notice. Clearly, as we've said from the beginning of this chapter, our solar system is not a random collection of worlds. Rather, it is a family of worlds exhibiting many patterns and traits that would be difficult to attribute to coincidence.

• What features of our solar system provide clues to how it formed?

In science, we assume that the patterns and traits we observe among the planets are consequences of the way our solar system was born and has evolved through time. Thus, when we seek to understand the origin of our solar system, we are looking for a way to explain these patterns and traits. We can summarize them by listing four major features of our solar system:

1. **Patterns of motion among large bodies.** The Sun, planets, and large moons generally orbit and rotate in a very organized way.
2. **Two major types of planets.** The first eight planets from the Sun divide clearly into two groups: the small, rocky planets that are close together and close to the Sun, and the large, gas-rich planets that are farther apart and farther out.
3. **Asteroids and comets.** Between and beyond the planets, huge numbers of asteroids and comets orbit the Sun. The locations, orbits, and compositions of these asteroids and comets follow distinct patterns.
4. **Exceptions to the rules.** The generally orderly solar system also has some notable "exceptions to the rules." For example, Earth is unique among the inner planets in having a large moon, and Uranus has an odd, sideways tilt. We will consider these exceptions as a feature in themselves, because a successful theory of our solar system must make allowances for such exceptions even as it explains the general rules.

Because these four features will be so important to our study of the solar system in later chapters, let's investigate each of them in a little more detail.

Feature 1: Patterns of Motion Among Large Bodies Figures 7.1 and 7.2 reveal numerous patterns of motion among the large bodies of our solar system. (In this context, a "body" is simply an individual object such as the Sun, a planet, or a moon.) For example:

- All planetary orbits are nearly circular and lie nearly in the same plane.
- All planets orbit the Sun in the same direction: counterclockwise as viewed from high above Earth's North Pole.
- Most planets rotate in the same direction in which they orbit (counterclockwise as viewed from above the North Pole), with fairly small axis tilts. The Sun also rotates in this same direction.
- Most of the solar system's large moons exhibit similar properties in their orbits around their planets—for example, orbiting in their planet's equatorial plane in the same direction that the planet rotates.

We consider these orderly patterns together as the first major feature of our solar system. As we'll see in Chapter 8, our theory of solar system formation explains all these patterns as consequences of processes that occurred during the very first stages of the birth of our solar system.

Feature 2: The Existence of Two Types of Planets Our brief planetary tour showed that the four inner planets are quite different from the four large outer planets. We say that these two groups represent two distinct planetary classes: *terrestrial* and *jovian.*

The **terrestrial planets** are the four planets of the inner solar system: Mercury, Venus, Earth, and Mars. (*Terrestrial* means "Earth-like.") The four terrestrial planets are relatively small and dense, with rocky surfaces and an abundance of metals deep in their interiors. They have few moons, if any, and none have rings. We often count our Moon as a fifth terrestrial world, because its history has been shaped by the same processes that have shaped the terrestrial planets.

The **jovian planets** are the four large planets of the outer solar system: Jupiter, Saturn, Uranus, and Neptune. (*Jovian* means "Jupiter-like.") The jovian planets are much larger in size and lower in average density than the terrestrial planets. They have rings and numerous moons. Their compositions are also quite different from those of the terrestrial worlds. They are made mostly of hydrogen, helium, and **hydrogen compounds**—compounds containing hydrogen, such as water (H_2O), ammonia (NH_3), and methane (CH_4).

Because the jovian planets are made mostly of substances that are gases under earthly conditions, they are sometimes called "gas giants." They do not have solid surfaces and look like balls of gas from the outside. However, the intense pressures and temperatures of the jovian planet interiors transform these gases into forms unlike anything we ordinarily see on Earth. Table 7.2 contrasts the general traits of the terrestrial and jovian planets.

Notice that Pluto is left out in the cold, both literally and figuratively, because it is much smaller and farther from the Sun (on average) than any of the terrestrial or jovian planets. As we discussed in our tour (p. 213) and will discuss further in Chapter 12, Pluto is just one of many similar objects orbiting the Sun beyond the orbit of Neptune—which is a major reason why Pluto's official status was changed in 2006 to "dwarf planet."

Table 7.2 Comparison of Terrestrial and Jovian Planets

Terrestrial Planets	*Jovian Planets*
Smaller size and mass	Larger size and mass
Higher density	Lower density
Made mostly of rock and metal	Made mostly of hydrogen, helium, and hydrogen compounds
Solid surface	No solid surface
Few (if any) moons and no rings	Rings and many moons
Closer to the Sun (and closer together), with warmer surfaces	Farther from the Sun (and farther apart), with cool temperatures at cloud tops

Feature 3: Asteroids and Comets The third major feature of our solar system is the existence of vast numbers of small objects orbiting the Sun. These objects fall into two major groups: asteroids and comets.

Asteroids are small, rocky bodies that orbit the Sun much like planets, but they are much smaller than planets (Figure 7.13). Even the largest of the asteroids have radii of only a few hundred kilometers, which means they are dwarfed by our Moon. Most asteroids are found within the **asteroid belt** between the orbits of Mars and Jupiter (see Figure 7.1).

Comets are also small objects that orbit the Sun, but they are made largely of ices (such as water ice, ammonia ice, and methane ice) mixed with rock. You are probably familiar with the occasional appearance of comets in the inner solar system, where they may become visible to the naked eye with long, beautiful tails (Figure 7.14). These visitors, which may delight sky watchers for a few weeks or months, are actually quite rare among comets. The vast majority of comets never visit the inner solar system, instead orbiting the Sun in the extreme outer reaches of the solar system where they remain frozen and lack tails. As we'll discuss further in Chapter 8, there are two major groupings of comets in our solar system: One group orbits the Sun in the same direction and nearly the same plane as the planets, occupying a broad region beyond the orbit of Neptune that we call the *Kuiper belt* (Kuiper rhymes with *piper*); Pluto and the recently discovered Eris are just the largest known of the icy objects in the Kuiper belt. The other group of comets does not show any simple pattern to its orbits, and resides much farther from the Sun in a region called the *Oort cloud* (*Oort* rhymes with *court*).

Feature 4: Exceptions to the Rules The fourth key feature of our solar system is that the general rules have a few notable exceptions. For example, while most of the planets rotate in the same direction as they orbit, Uranus rotates nearly on its side, and Venus rotates "backward"—clockwise, rather than counterclockwise, as viewed from high above Earth's North Pole. Similarly, while most large moons orbit their planets in the same direction as their planets rotate, many small moons have much more unusual orbits.

One of the most interesting exceptions concerns our own Moon. While the other terrestrial planets have either no moons (Mercury and Venus) or very tiny moons (Mars, with two moons), Earth has one of the largest moons in the solar system.

Summary of the Four Features Figure 7.15 summarizes the four general features of the solar system that we have discussed. In Chapter 8, we will turn our attention to the task of finding a theory that can explain all these features.

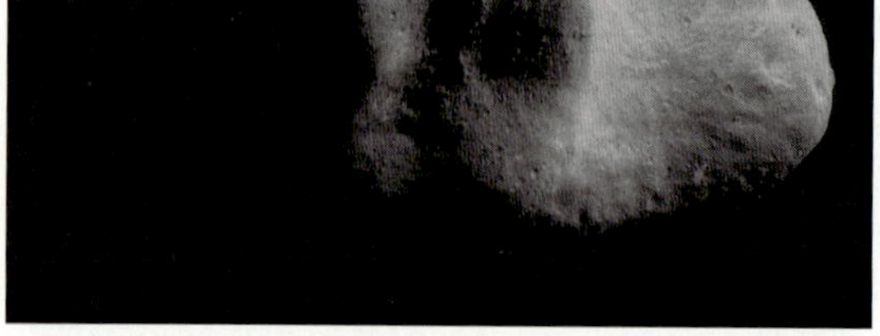

Figure 7.13 The asteroid Eros (photographed from the *NEAR* spacecraft). Its appearance is probably typical of most asteroids. Eros is about 40 kilometers in length. Like other small objects in the solar system, it is not spherical.

Figure 7.14 Comet Hale-Bopp, photographed over Boulder, Colorado, during its appearance in 1997.

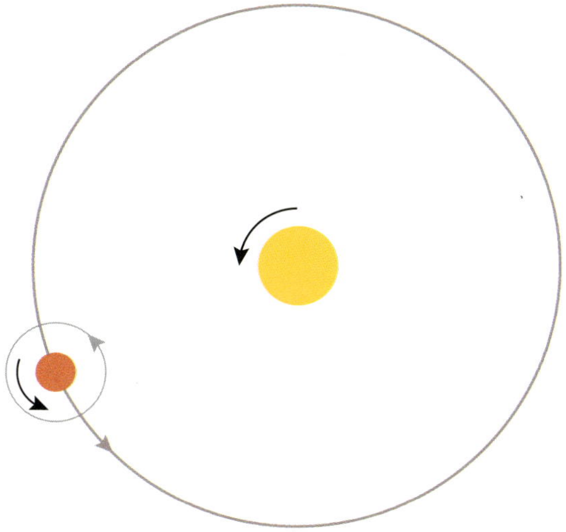

1. Large bodies in the solar system have orderly motions. All planets and most satellites have nearly circular orbits going in the same direction in nearly the same plane. The Sun and most of the planets rotate in this same direction as well.

2. The first eight planets fall into two major categories: small, rocky terrestrial planets near the Sun and large, hydrogen-rich jovian planets farther out. The jovian planets have many moons and rings made of rock and ice.

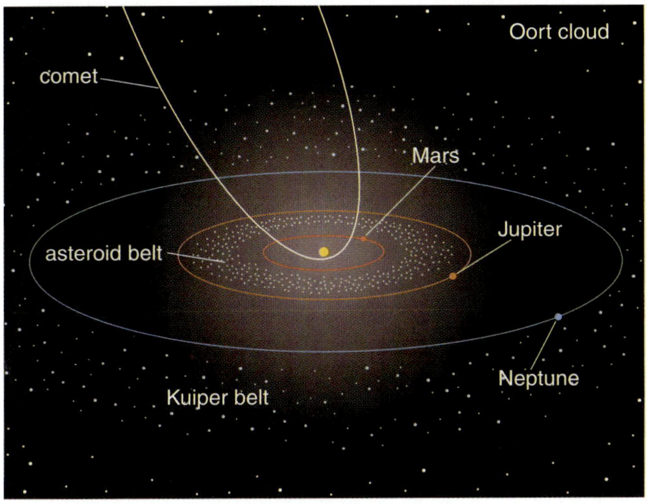

3. Swarms of asteroids and comets populate the solar system. Asteroids are concentrated in the asteroid belt, and comets populate the regions known as the Kuiper belt and the Oort cloud.

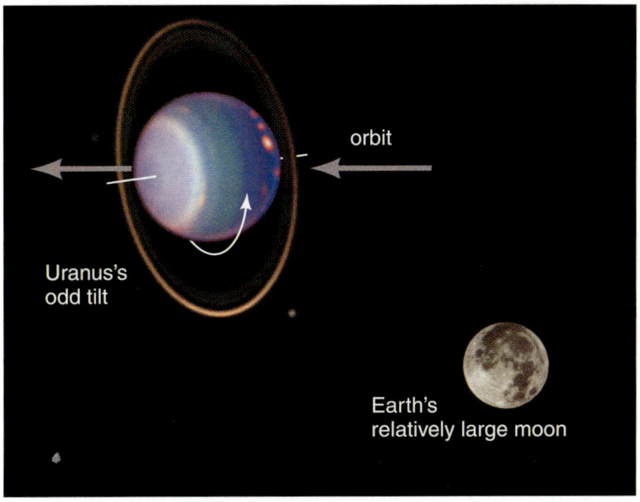

4. Several notable exceptions to these general trends stand out, such as planets with unusual axis tilts or surprisingly large moons, and moons with unusual orbits.

Figure 7.15 Four major features of the solar system.

7.3 Spacecraft Exploration of the Solar System

How have we learned so much about the solar system? Much of our knowledge comes from telescopic observations, using both ground-based telescopes and telescopes in Earth orbit such as the Hubble Space Telescope. In one case—our Moon—we have learned a lot by sending astronauts to explore the terrain and bring back rocks for laboratory study. In a few other cases, we have samples of distant worlds that have come to us as meteorites. But most of the data fueling the recent revolution in our understanding of the solar system have come from robotic spacecraft. To date, we have sent robotic spacecraft to all the terrestrial and jovian planets, as well as to many moons, asteroids, and comets. In this section, we'll briefly investigate how we use robotic spacecraft to explore the solar system.

• How do robotic spacecraft work?

The spacecraft we send to explore the planets are robots suited for long space journeys. They carry specialized equipment for scientific study. All spacecraft have computers used to control their major components, power sources such as solar cells, propulsion systems, and devices to point cameras and other instruments precisely at their targets. Robotic spacecraft operate primarily with preprogrammed instructions. They carry radios for communication, which allow them to receive additional instructions from Earth and to send home the data they collect. Most robotic spacecraft make one-way trips from Earth, never physically returning but sending their data back from space in the same way we send radio and television signals around the world.

Broadly speaking, the robotic missions we send to explore other worlds fall into four major categories:

- **Flyby:** A spacecraft on a flyby goes past a world just once and then continues on its way.
- **Orbiter:** An orbiter is a spacecraft that orbits the world it is studying, allowing longer-term study during its repeated orbits.
- **Lander or probe:** These spacecraft are designed to land on a planet's surface or probe a planet's atmosphere by flying through it. Some landers carry rovers to explore wider regions.
- **Sample return mission:** A sample return mission requires a spacecraft designed to return to Earth carrying a sample of the world it has studied.

The choice of spacecraft type depends on both scientific objectives and cost. In general, a flyby is the lowest-cost way to visit another planet, and some flybys gain more "bang for the buck" by visiting multiple planets. For example, *Voyager 2* flew past Jupiter, Saturn, Uranus, and Neptune before continuing on its way out of our solar system (Figure 7.16).

SPECIAL TOPIC How Did We Learn the Scale of the Solar System?

In this chapter, we have presented the layout of the solar system as we know it today, when we have very precise measurements of planetary sizes and distances. But how did we learn the scale of the solar system?

By the mid-1600s, Kepler's laws had given us precise values of planetary distances in astronomical units (AU)—that is, distances *relative* to the Earth-Sun distance [Section 3.3]. However, no one knew the actual value of the AU in absolute units like miles or kilometers. A number of seventeenth-century astronomers proposed ideas for measuring the Earth-Sun distance, but none were practical. Then, in 1716, Edmond Halley (best-known today for having his name on Halley's comet) hit upon the idea that would ultimately solve the problem: He realized that during a planetary *transit*, when a planet appears to pass across the face of the Sun as seen from Earth [Section S1.1], observers in different locations on Earth would see the planet trace slightly different paths across the Sun. In essence, careful observations of the transit from different places allows precise measurement of the planet's parallax, from which we can use geometry to measure its distance (see figure). Because scientists in Halley's time already knew planetary distances in AU, knowing just one planet's actual distance from parallax would allow them to calculate the actual value of the AU and hence the actual distances of all the other planets.

Only Mercury and Venus can produce transits visible from Earth. Halley realized that although we see transits of Mercury far more often than transits of Venus, the measurements would be much easier with Venus because its closer distance to Earth during a transit would give it a larger parallax. Thus, Halley advocated sending expeditions to different parts of the world to observe transits of Venus. Unfortunately, Venus transits are rare, occurring in pairs 8 years apart about every 120 years. Halley did not live to see a Venus transit, but later astronomers followed his advice, mounting expeditions to observe transits that occurred in 1761 and 1769.

The transit observations turned out to be quite difficult in practice, partly due to the inherent challenge that long expeditions posed at that time and partly because getting the geometry right required very precise timing of the beginning of Venus's transit across the Sun. Astronomers discovered that this timing was more difficult than Halley had guessed, because of optical effects that occur during a transit. Nevertheless, astronomers kept studying the data from the 1761 and 1769 transits for many decades, and by the mid-1800s these data allowed them to pin down the value of the astronomical unit to within about 5% of its modern value of 149.6 million kilometers. The next Venus transits occurred in 1874 and 1882. Photography had been invented by then, making observations more reliable, so in principle those transits could have allowed further refinement of the AU. However, by that time photography and better telescopes also made it possible to observe parallax of planets against stars, and by 1877 such observations had given us the value of the AU to within 0.2% of its modern value. More recently, a Venus transit occured in 2004 (see Figure S1.5) and another will occur in 2012.

Today, we measure the distance to Venus very precisely by bouncing radio waves off its surface with radar, a technique known as *radar ranging*. Because we know the speed of light, measuring the time it takes for the radio waves to make the round trip to Venus tells us the precise round-trip distance. We then use this distance and Venus's known distance in AU to calculate the actual value of the AU. Once we know the value of the AU, we can determine the actual distances of all the planets from the Sun, and we can determine their actual sizes from their angular sizes and distances. Indeed, we now know the layout of the solar system so well that we can launch spacecraft from Earth and send them to precise places on or around distant worlds.

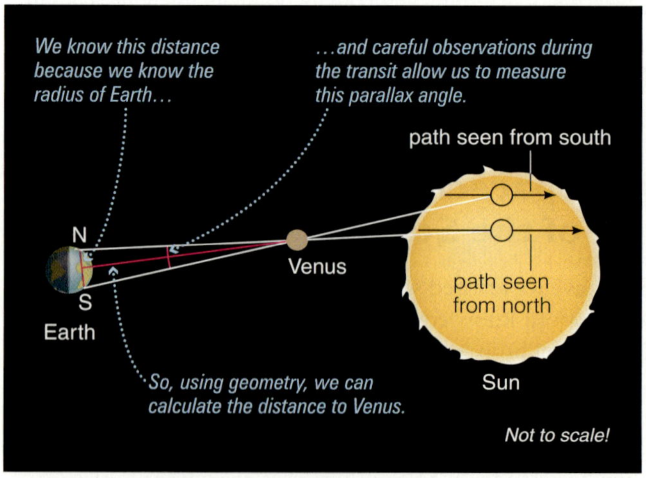

During a transit of Venus, observers at different places on Earth will see it trace slightly different paths across the Sun. The precise geometry of these events therefore allows computation of Venus's true distance, which in turn allows computation of the Sun's distance.
Adapted from *Sky and Telescope*, Feb. 2004.

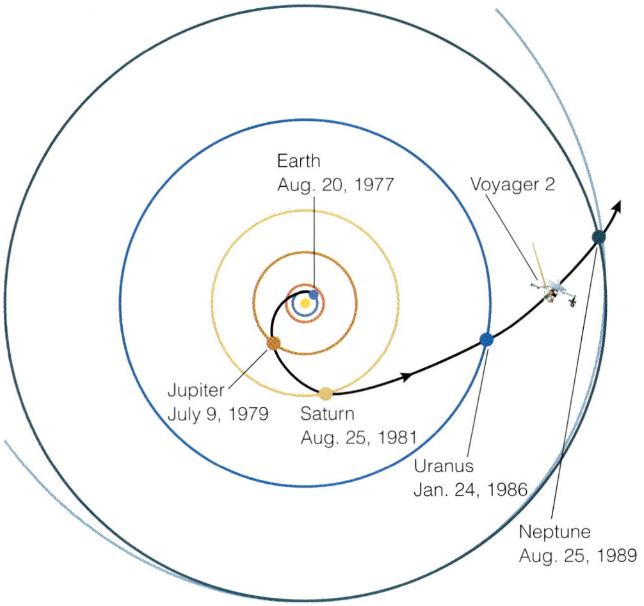

Figure 7.16 The trajectory of *Voyager 2*, which made flybys of each of the four jovian planets in our solar system.

Flybys Flybys tend to be cheaper than other missions because they are generally less expensive to launch into space. Launch costs depend largely on weight, and onboard fuel is a significant part of the weight of a spacecraft heading to another planet. Once a spacecraft is on its way, the lack of friction or air drag in space means that it can maintain its orbital trajectory through the solar system without using any fuel at all. Fuel is needed only when the spacecraft needs to change from one trajectory (orbit) to another. Moreover, with careful planning, some trajectory changes can be made by taking advantage of the gravity of other planets. If you look closely at the *Voyager 2* trajectory in Figure 7.16, you'll see that the spacecraft made significant trajectory changes as it passed by Jupiter and Saturn. In effect, it made these changes for free by using gravity to bend its path rather than by burning fuel. (This technique is known as a "gravitational slingshot," and it essentially speeds up the spacecraft significantly while slowing the planet by an unnoticeable amount.)

THINK ABOUT IT

Study the *Voyager 2* trajectory in Figure 7.16. Given that Saturn orbits the Sun every 29 years, Uranus orbits the Sun every 84 years, and Neptune orbits the Sun every 165 years, would it be possible to send another flyby mission to all four jovian planets if we launched it now? Explain.

Although a flyby offers only a relatively short period of close-up study, it can provide valuable scientific information. Flybys generally carry small telescopes, cameras, and spectrographs. Because these instruments are brought within a few tens of thousands of kilometers of other worlds (or closer), they can obtain much higher-resolution images and spectra than even the largest current telescopes viewing these worlds from Earth. In addition, flybys sometimes give us information that would be very difficult to obtain from Earth. For example, *Voyager 2* helped us discover Jupiter's rings and learn about the rings of Saturn, Uranus, and Neptune through views in which the rings were backlit by the Sun. Such views are possible only from beyond each planet's orbit.

Flybys may also carry instruments to measure local magnetic field strength or to sample interplanetary dust. The gravitational effects of the planets and their moons on the spacecraft itself provide information about object masses and densities. Like the backlit views of the rings, these types of data cannot be gathered from Earth. Indeed, most of what we know about the masses and compositions of moons comes from data gathered by spacecraft that have flown past them.

Orbiters An orbiter can study another world for a much longer period of time than a flyby. Like the spacecraft used for flybys, orbiters often carry cameras, spectrographs, and instruments for measuring the strength of magnetic fields. Missions to Venus, Mars, and Saturn's moon Titan have also used radar to make precise altitude measurements of surface features. In the cases of Venus and Titan, radar observations give us our only clear surface views, because we cannot otherwise see through the thick cloud cover on these worlds.

An orbiter is generally more expensive than a flyby for an equivalent weight of scientific instruments, primarily because it must carry added fuel to change from an interplanetary trajectory to a path that puts it into orbit around another world. Careful planning can minimize the added expense. For example, recent Mars orbiters have saved on fuel costs by carrying only enough fuel to enter highly elliptical orbits around Mars. The spacecraft then settled into the smaller, more circular orbits needed for scientific observations by skimming the Martian atmosphere at the low point of every elliptical orbit. Atmospheric drag slowed the spacecraft with each orbit and, over several months, circularized the spacecraft orbit. (This technique of using the atmosphere to slow the spacecraft and change its orbit is called *aerobraking*.) We have sent orbiters to the Moon and to the planets Venus, Mars, Jupiter, and Saturn, and to the asteroid Eros.

Landers or Probes The most "up close and personal" study of other worlds comes from spacecraft that send probes into the atmospheres or landers to the surfaces. For example, in 1995, the *Galileo* spacecraft dropped a probe into Jupiter's atmosphere [Section 11.1]. The probe collected temperature, pressure, composition, and radiation measurements for about an hour as it descended; it was then destroyed by Jupiter's high interior pressures and temperatures.

On planets with solid surfaces, a lander can offer close-up surface views, local weather monitoring, and the ability to carry out automated experiments. Some landers carry robotic rovers able to venture across the surface, such as the *Spirit* and *Opportunity* rovers that landed on Mars in

The aeroshell protects the rover from fiery temperatures as it enters the Martian atmosphere. Six minutes to landing.

With the parachute deployed three retrorockets fire their engines, suspending the lander 30–50 feet above the Martian surface.

Protected by large airbags, the lander falls away from the parachute, landing safely on Mars. The lander bounces for several minutes, traveling hundreds of meters.

After bouncing to a stop, the airbags are retracted and the lander's petals deploy, creating a ramp for the rover to descend.

The rover deploys the solar arrays, wheels, cameras, and other instruments and begins its exploration of the Martian surface.

Figure 7.17 The Mars Exploration Rovers used parachutes and air bags to land safely on Mars.

2004 and operated long beyond their planned lifetimes. Landers typically require fuel to slow their descent to a planetary surface, but clever techniques can reduce cost. For example, the recent Mars landers hit the surface of Mars at crash-landing speed but were protected by cocoons of air bags deployed on the way down. These air bags allowed the landers to bounce along the surface of Mars before finally coming to rest (Figure 7.17).

Sample Return Missions While probes and landers can carry out experiments on surface rock or atmospheric samples, the experiments must be designed in advance and must fit inside the spacecraft. These limitations make scientists long for missions that will scoop up samples from other worlds and return them to Earth for more detailed study. To date, the only sample return missions have been to the Moon. Astronauts collected samples during the *Apollo* missions, and the then–Soviet Union sent robotic spacecraft to collect rocks from the Moon in the early 1970s. Many scientists are working toward a sample return mission to Mars, and they hope to launch such a mission within the next decade or so. (A slight variation on the theme of a sample return mission is the *Stardust* mission, which collected comet dust on a flyby and will return to Earth in 2006.)

Combination Spacecraft Many missions combine more than one type of spacecraft. For example, the *Galileo* mission to Jupiter included an orbiter that studied Jupiter and its moons as well as the probe that entered Jupiter's atmosphere. The *Cassini* spacecraft, which has been orbiting Saturn since 2004, included flybys of Venus, Earth, and Jupiter during its 7-year trip to Saturn (Figure 7.18a). The spacecraft

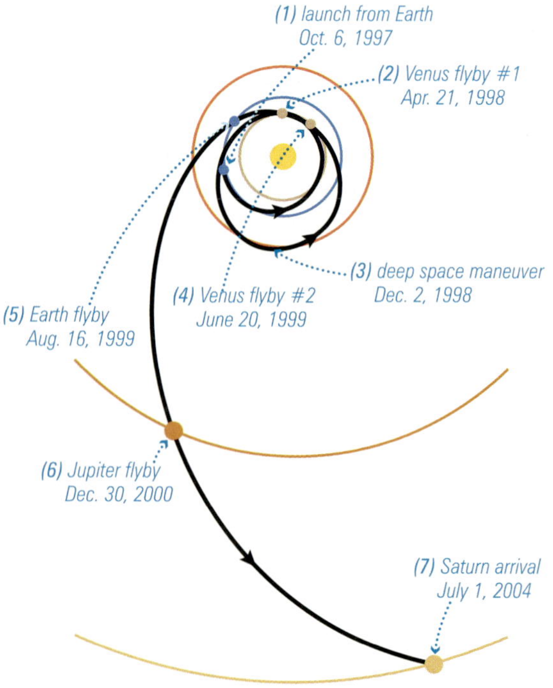

a The trajectory of *Cassini* to Saturn

b Artist's conception of the *Huygens* probe landing on Titan. From upper right to lower left: As the *Cassini* mothership passes in the distance, the probe enters the atmosphere, parachutes down, and finally lands on the surface. *Huygens* landed in January 2005, collecting data and transmitting images for several hours before its batteries ran out. (See Figure 11.29 for actual images of Titan taken by the *Huygens* probe.)

Figure 7.18 The *Cassini* mission to Saturn.

Table 7.3 Selected Robotic Missions to Other Worlds

Name	Type	Destination	Lead Space Agency	Arrival Year	Mission Highlights
New Horizons	Flyby	Pluto	NASA	2015†	First mission to Pluto, launched in 2006
Rosetta	Flyby and lander	Comet 67P/ Churyumov-Gerasimenko	ESA*	2014†	Launched in 2004, carries lander and orbiter for comet
Dawn	Flyby	Vesta, Ceres	NASA	2011,† 2015†	Study of the larger asteroids Vesta and Ceres (which qualifies as a dwarf planet)
MESSENGER	Orbiter	Mercury	NASA	2011†	Studies Mercury's surface, atmosphere, and interior
Hayabusa	Lander and sample return	Asteroid Itokawa	JAXA**	2007†	Collected asteroid sample in 2005, to return to Earth in 2007
Stardust	Flyby and sample return	Comet Wild 2	NASA	2006†	Collected dust from comet in 2004, to return to Earth in 2006
Mars Reconnaissance Orbiter	Orbiter	Mars	NASA	2006†	High-resolution mapper will identify future landing sites
Deep Impact	Flyby and "lander"	Comet Tempel 1	NASA	2005	"Lander" to hit at 10 km/sec, flyby spacecraft to study the impact
Cassini	Orbiter	Saturn	NASA	2004	Included probe (*Huygens*) that landed on Titan
Mars Exploration Rovers	Lander	Mars	NASA	2004	Rovers *Spirit* and *Opportunity* explored surface
Mars Express	Orbiter	Mars	ESA*	2004	Climate studies of Mars
Venus Express	Orbiter	Venus	ESA*	2006	Low-cost Mars Express spinoff focused on atmosphere, surface
Mars Odyssey 2001	Orbiter	Mars	NASA	2001	Detailed study of Martian surface features and composition
Near Earth Asteroid Rendezvous (NEAR)	Orbiter	Eros (asteroid)	NASA	2000	First spacecraft dedicated to in-depth study of an asteroid
Mars Global Surveyor	Orbiter	Mars	NASA	1997	Detailed imaging of surface from orbit
Mars Pathfinder	Lander	Mars	NASA	1997	Carried *Sojourner*, the first robotic rover on Mars
Galileo	Orbiter	Jupiter	NASA	1995	Dropped probe into Jupiter; close-up study of moons
Magellan	Orbiter	Venus	NASA	1990	Detailed radar mapping of surface of Venus
Voyager 1	Flyby	Jupiter, Saturn	NASA	1979, 1980	Unprecedented views of Jupiter and Saturn; continuing out of solar system
Voyager 2	Flyby	Jupiter, Saturn, Uranus, Neptune	NASA	1979, 1981, 1986, 1989	Only mission to Uranus and Neptune; continuing out of solar system
Vikings 1 and 2	Orbiter and lander	Mars	NASA	1976	First landers on Mars

*European Space Agency. **Japan Aerospace Exploration Agency. †Scheduled arrival year.

itself is an orbiter that is studying Saturn and its moons, but it also carried the *Huygens* probe that descended through the atmosphere and landed on the surface of Saturn's moon Titan (Figure 7.18b).

Exploration Past, Present, and Future Over the past several decades, studies using both telescopes on Earth and robotic spacecraft have allowed us to learn the general characteristics of all the major planets and moons in our solar system as well as the general characteristics of asteroids and comets. Telescopes will continue to play an important role in future observations, but for detailed study we will probably continue to depend on spacecraft.

Table 7.3 (p. 221) lists some significant robotic missions of the past and present. The next few years promise many new discoveries as missions arrive at their destinations. Over the longer term, all the world's major space agencies have high hopes of launching numerous and diverse missions to study our solar system. For example, nearly a dozen more missions are on the drawing board for Mars alone. If these missions come to fruition, the next couple of decades will see spacecraft that answer many specific questions about the nature of our solar system and its numerous worlds.

THE BIG PICTURE

Putting Chapter 7 into Perspective

In this chapter, we've introduced the major features of our solar system and discussed some important patterns and trends that provide clues to its formation. As you continue your study of the solar system, keep in mind the following "big picture" ideas:

- Our solar system is not a random collection of objects moving in random directions. Rather, it is highly organized, with clear patterns of motion and with most objects falling into just a few basic categories.
- Each planet has its own unique and interesting features. Becoming familiar with the planets is an important first step in understanding the root causes of their similarities and differences.
- Much of what we now know about the solar system comes from spacecraft exploration. Choosing the type of mission to send to a planet involves many considerations, from the scientific to the purely political. Many missions are currently under way, offering us hope of learning much more in the near future.

SUMMARY OF KEY CONCEPTS

7.1 Studying the Solar System

- **What does the solar system look like?** The planets are tiny compared to the distances between them. Our solar system consists of the Sun, the planets and their moons, and vast numbers of asteroids and comets. Each world has its own unique character, but there are many clear patterns among the worlds.

- **What can we learn by comparing the planets to one another?** Comparative studies reveal the similarities and differences that give clues to solar system formation, and highlight the underlying processes that give each planet its unique appearance.
- **What are the major features of the Sun and planets?** The Sun is the dominant object in our solar system. Each of the planets has unique features, but we also observe important similarities among the terrestrial worlds and among the jovian worlds, similarities that give us key clues to the processes that govern the evolution of worlds.

7.2 Patterns in the Solar System

- **What features of our solar system provide clues to how it formed?** Four major features provide clues: (1) The Sun, planets, and large moons generally rotate and orbit in a very organized way. (2) The first eight planets from the Sun divide clearly into two groups: terrestrial and jovian. (3) The solar system contains huge numbers of asteroids and comets. (4) There are some notable exceptions to these general patterns.

7.3 Spacecraft Exploration of the Solar System

- **How do robotic spacecraft work?** Spacecraft can be categorized as flyby, orbiter, lander/probe, or sample return missions. In all cases, robotic spacecraft carry their own propulsion, power, and communication systems, and can operate under preprogrammed control or with updated instructions from ground controllers.

EXERCISES AND PROBLEMS

For instructor-assigned homework go to MasteringASTRONOMY www.masteringastronomy.com

Review Questions

Short-Answer Questions Based on the Reading

1. What do we mean by *comparative planetology*? Does it apply only to planets?
2. What would the solar system look like to your naked eye if you could view it from beyond the orbit of Pluto?
3. Briefly describe the overall layout of the solar system as it is shown in Figures 7.1 and 7.2.
4. For the Sun and each of the planets in our solar system, describe at least two features that you find interesting.
5. Describe at least four orderly patterns of motion that we observe in our solar system.
6. What are the basic differences between the *terrestrial* and *jovian* planets? Which planets in our solar system fall into each group?
7. What do we mean by *hydrogen compounds*? In what kinds of planets or small bodies are they major ingredients?
8. In what ways does Pluto resemble a terrestrial planet? In what ways does it resemble a jovian planet? In what ways does it resemble neither?
9. What are *asteroids*? Where do we find most asteroids in our solar system?
10. What are *comets*? How do they differ from asteroids?
11. What is the *Kuiper belt*? What is the *Oort cloud*? How do the orbits of comets differ in the two regions?
12. Describe at least two "exceptions to the rules" that we find in our solar system.
13. Describe and distinguish between space missions that are *flybys, orbiters, landers* or *probes*, and *sample return missions*. What are the advantages and disadvantages of each mission type?
14. For a few of the most important past, present, or future robotic missions to the planets, describe their targets, types, and mission highlights.

Test Your Understanding

Does It Make Sense?

Decide whether the statement makes sense (or is clearly true) or does not make sense (or is clearly false). Explain your reasoning. (For an example, see Chapter 1, "Does It Make Sense?")

15. Pluto orbits the Sun in the opposite direction of all the other planets.
16. If Pluto were as large as the planet Mercury, we would classify it as a terrestrial planet.
17. Comets in the Kuiper belt and Oort cloud have long, beautiful tails that we can see when we look through telescopes.
18. Our Moon is about the same size as moons of the other terrestrial planets.
19. The mass of the Sun compared to the mass of all the planets combined is like the mass of an elephant compared to the mass of a cat.
20. On average, Venus is the hottest planet in the solar system—even hotter than Mercury.
21. The weather conditions on Mars today are much different than they were in the distant past.
22. Moons cannot have atmospheres, active volcanoes, or liquid water.
23. Saturn is the only planet in the solar system with rings.
24. We could probably learn more about Mars by sending a new spacecraft on a flyby than by any other method of studying the planet.

Quick Quiz

Choose the best answer to each of the following. Explain your reasoning with one or more complete sentences.

25. The largest terrestrial planet and jovian planet are, respectively: (a)Venus and Jupiter. (b) Earth and Jupiter. (c) Earth and Pluto.
26. Which terrestrial planets have had volcanic activity at some point in their histories? (a) only Earth (b) Earth and Mars (c) all of them
27. Large moons orbit their planets in the same direction the planet rotates ... (a) rarely. (b) half of the time. (c) most of the time.
28. Which of the following three kinds of objects resides closer to the Sun on average? (a) comets (b) asteroids (c) jovian planets
29. What's unusual about our Moon? (a) It's the only moon that orbits a terrestrial planet. (b) It's by far the largest moon in the solar system. (c) It's not the largest moon, but it's surprisingly large relative to the planet it orbits.
30. Planetary orbits are: (a) very eccentric (stretched-out) ellipses and in the same plane. (b) fairly circular and in the same plane. (c) fairly circular but oriented in every direction.
31. Which have more moons on average? (a) jovian planets (b) terrestrial planets (c) Terrestrial and jovian planets both have about the same number of moons.
32. The most abundant ingredient of the Sun and Jupiter is: (a) ionized metal. (b) hydrogen. (c) ammonia.
33. Are there any exceptions to the rule that planets rotate with small axis tilts and in the same direction that they orbit the Sun? (a) no (b) Venus is the only exception. (c) Venus and Uranus are exceptions.
34. The *Cassini* spacecraft: (a) flew past Pluto. (b) landed on Mars. (c) is orbiting Saturn.

Investigate Further

In-Depth Questions to Increase Your Understanding

Short-Answer/Essay Questions

35. *Planetary Tour.* Based on the brief planetary tour in this chapter, which planet besides Earth do you think is the most interesting, and why? Defend your opinion clearly in two or three paragraphs.
36. *Patterns of Motion.* In one or two paragraphs, explain why the existence of orderly patterns of motion in our solar system should suggest that the Sun and the planets all formed at one time from one cloud of gas, rather than as individual objects at different times.
37. *Two Classes of Planets.* In terms a friend or roommate would understand, write one or two paragraphs explaining why we say that the planets fall into two major categories and what those categories are. Do all planets fit in the two groups? Explain.
38. *Solar System Trends.* Study the planetary data in Table 7.1 to answer each of the following.

a. Notice the relationship between distance from the Sun and surface temperature. Describe the trend, explain why it exists, and explain any notable exceptions to the trend.
b. The text says that planets can be classified as either terrestrial or jovian, with Pluto fitting neither category. Describe in general how the columns for density, composition, and distance from the Sun support this classification.
c. Describe the trend you see in orbital periods and explain the trend in terms of Kepler's third law.

39. *Comparing Planetary Conditions.* Use the planetary data in Table 7.1 and Appendix E to answer each of the following.
a. Which column of data would you use to find out which planet has the shortest days? Do you see any notable differences in the length of a day for the different types of planets? Explain.
b. Which planets would you expect not to have seasons? Why?
c. Which column tells you how much a planet's orbit deviates from a perfect circle? Based on that column, are there any planets for which you would expect the surface temperature to vary significantly over the course of each orbit? Explain.

Quantitative Problems

Be sure to show all calculations clearly and state your final answers in complete sentences.

40. *Size Comparisons.* How many Earths could fit inside Jupiter (assuming you could fill up all the volume)? How many Jupiters could fit inside the Sun? The equation for the volume of a sphere is $V = (\frac{4}{3})\pi r^3$.
41. *Asteroid Orbit.* Ceres, the largest asteroid, has an orbital semimajor axis of 2.77 AU. Use Kepler's third law to find its orbital period. Compare your answer with Table 7.1 and name the planets that orbit just inside and outside Ceres's orbit.
42. *Density Classification.* Imagine that a new planet is discovered in our solar system with a mass of 5.97×10^{25} kilograms and a radius of 12,800 kilometers . Based just on its density, would we consider it the largest terrestrial planet or the smallest jovian planet? Explain. (*Hint:* Be careful to convert your density to units of grams per cubic centimeter in order to compare with the terrestrial and jovian planet data in this chapter.)
43. *Escape Velocity.* By studying the data in Appendix E, briefly describe how escape velocity is related to mass and radius. Is the trend what you expect based on what you learned about escape velocity in Chapter 5?
44. *Comparative Weight.* Suppose you weigh 100 pounds. How much you would weigh on each of the other planets in our solar system? Assume you can stand either on the surface or in an airplane in the planet's atmosphere. (*Hint:* Recall from Chapter 5 that weight is mass times the acceleration of gravity. The surface gravity column in Appendix E tells you how the acceleration of gravity on other planets compares to that on Earth.)
45. *Mission to Pluto.* The *New Horizons* spacecraft will take about 9 years to travel from Earth orbit to Pluto. About how fast will it be traveling on average? Assume that its trajectory is close to a straight line. Give your answer in AU/year and in km/hr.
46. *Empty Solar System.* If you could line up the Sun and all the planets end-to-end, how far would they stretch? Compare your answer to the distance between the Sun and Pluto. What does this tell you about the "emptiness" of the solar system?
47. *Planetary Parallax.* Suppose observers at Earth's North Pole and South Pole use a transit of the Sun by Venus to discover that the angular size of Earth as viewed from Venus would be 62.8 arcseconds. Earth's radius is 6378 km. Estimate the distance between Venus and Earth in km and AU. Compare your answer with information from the chapter.

Discussion Questions

48. *Where Would You Go?* Suppose you could visit any one of the planets or moons in our solar system for 1 week. Which object would you choose to visit, and why?
49. *Planetary Priorities.* Suppose you were in charge of developing and prioritizing future planetary missions for NASA. What would you choose as your first priority for a new mission, and why?

MEDIA EXPLORATIONS

For self-study activities go to MasteringASTRONOMY www.masteringastronomy.com

Interactive Tutorials

Tutorial Review of Key Concepts

Use the interactive **Tutorial** at **www.masteringastronomy.com** to review key concepts from this chapter.

Scale of the Universe Tutorial
Lesson 1 Distances of Scale: Our Solar System

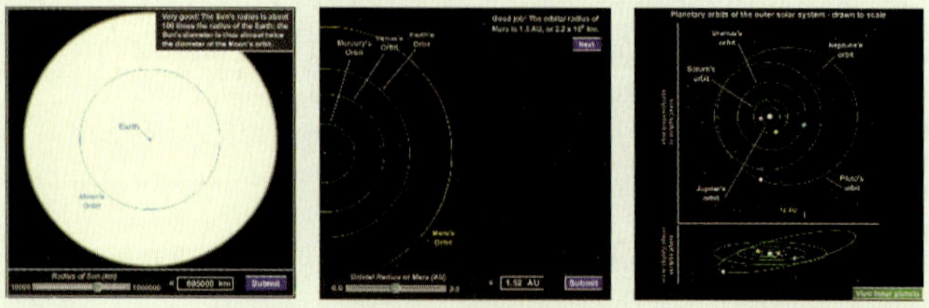

Formation of the Solar System Tutorial
Lesson 1 Comparative Planetology

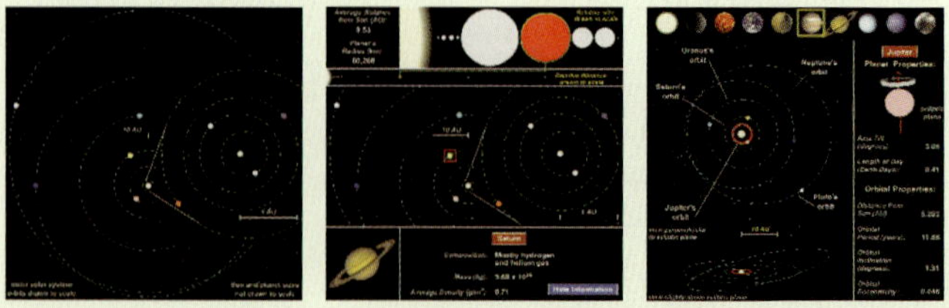

Orbits and Kepler's Laws Tutorial

Lesson 2 Kepler's First Law
Lesson 3 Kepler's Second Law
Lesson 4 Kepler's Third Law

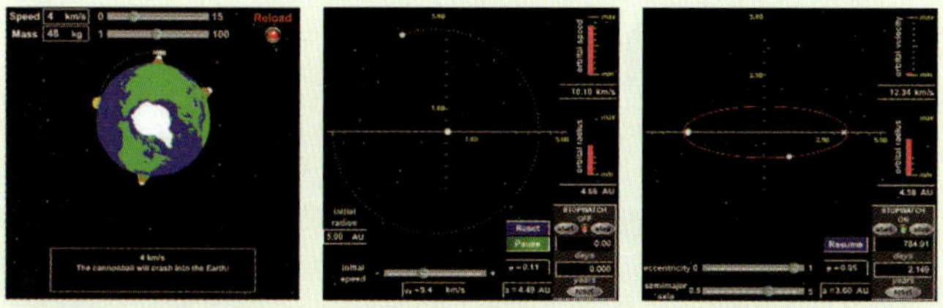

Supplementary Tutorial Exercises

Use the interactive **Tutorial Lessons** to explore the following questions.

Scale of the Universe Tutorial, Lesson 1

1. How big is an astronomical unit (AU) in comparison to more familiar units?
2. Why do we use AUs to describe distances in our solar system? Is the AU a good unit for describing distances to stars? Explain.
3. How big is the Sun relative to the Moon's orbit of Earth?
4. Explore the layout of the solar system, focusing on the scale of the various planetary orbits. Which of the key patterns discussed in this chapter are evident from the scaled view? Which require deeper study to uncover?

Formation of the Solar System Tutorial, Lesson 1

1. What features distinguish the terrestrial planets from the jovian planets?
2. What is the difference between axis tilt, orbital eccentricity, and orbital inclination? What patterns do we see in these properties among the planets?
3. Which solar system bodies constitute "exceptions to the rules," and in what ways?

Exploring the Sky and Solar System

Of the many activities available on the *Voyager: SkyGazer* CD-ROM accompanying your book, use the following files to observe key phenomena covered in this chapter.

Go to the **File: Basics** folder for the following demonstrations.

1. Saturn's Phases
2. Tracking Venus
3. Planet Panel

Go to the **File: Demo** folder for the following demonstrations.

1. Earth and Venus
2. Trailing Saturn
3. Triple Conjunction of 7 BC

Go to the **Explore** menu for the following demonstrations.

1. Solar System
2. Paths of Planets

Movies

Check out the following narrated and animated short documentaries available on **www.masteringastronomy.com** for a helpful review of key ideas covered in this chapter.

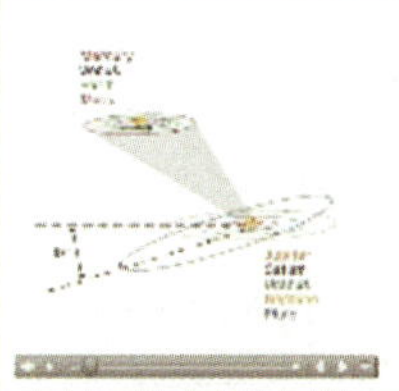

Orbits in the Solar System

History of the Solar System Movie

Web Projects

1. *Current Mission.* Visit the Web page for one of the current missions listed in Table 7.3. Write a one- to two-page summary of the mission's basic design, goals, and current status.
2. *Mars Missions.* Go to the home page for NASA's Mars Exploration Program. Write a one- to two-page summary of the plans for future exploration of Mars.

8

Formation of the Solar System

LEARNING GOALS

8.1 The Search for Origins
- What properties of our solar system must a formation theory explain?
- What theory best explains the features of our solar system?

8.2 The Birth of the Solar System
- Where did the solar system come from?
- What caused the orderly patterns of motion in our solar system?

8.3 The Formation of Planets
- Why are there two major types of planets?
- How did the terrestrial planets form?
- How did the jovian planets form?
- What ended the era of planet formation?

8.4 The Aftermath of Planet Formation
- Where did asteroids and comets come from?
- How do we explain "exceptions to the rules"?
- How do we explain the existence of our Moon?
- Was our solar system destined to be?

8.5 The Age of the Solar System
- How does radioactivity reveal an object's age?
- When did the planets form?

The evolution of the world may be compared to a display of fireworks that has just ended: some few red wisps, ashes and smoke. Standing on a cooled cinder, we see the slow fading of the suns, and we try to recall the vanished brilliance of the origin of the worlds.

G. Lemaître (1894–1966), astronomer and Catholic priest

How did Earth come to be? How old is it? Is it unique? Our ancestors could do little more than guess at the answers to these questions, and they imagined Earth to be fundamentally different from objects in the heavens. Today we know Earth as just one of many worlds in our solar system, and our solar system as just one of many planetary systems in the universe.

We seek to understand the origin of our planet from the clues we discover as we study the rest of our solar system. These clues point to a common origin for the Sun and all the planets, moons, asteroids, and comets that orbit it. By studying the clues in depth, scientists have put together a theory that explains most of the observed features of our solar system. We'll explore the theory in this chapter, using it as a case study in the way that science seeks to answer fundamental questions about our origins.

Our theory of solar system formation is important not only because it helps us understand our cosmic origins, but also because it holds the key to understanding the nature of planets. If the planets in our solar system all formed together, then their differences must be attributable to physical processes that occurred during the birth and subsequent evolution of the solar system. Thus, our study of the solar system's birth will form the basis for our comparative study of the planets in subsequent chapters. It will also help us understand and interpret recent discoveries of planets around other stars, a topic we will study in depth in Chapter 13.

8.1 The Search for Origins

The development of any scientific theory is an interplay between observations and attempts to explain those observations [Section 3.4]. Hypotheses that seem to make sense at one time might later be dismissed because they fail to explain new data. For example, ancient Greek ideas about Earth's origins probably seemed quite reasonable when people assumed that Earth was the center of the universe, but they no longer made sense after Kepler and Galileo proved that Earth is a planet orbiting the Sun.

By the end of the seventeenth century, the Copernican revolution [Section 3.3] and Newton's discovery of the universal law of gravitation [Section 4.4] had given us a basic understanding of the layout and motion of the planets and moons in our solar system. It was only natural that scientists would begin to speculate about how this system came to be. We'll spend most of this chapter discussing our modern theory of the solar system's origins. First, however, let's explore the origins of the theory and the way in which we evaluate it.

• What properties of our solar system must a formation theory explain?

Remember that a hypothesis can rise to the status of a scientific theory only if it offers a detailed, physical model that explains a broad range of observed facts. In Chapter 7, we discussed four major structural features of our solar system (see Figure 7.15). These features hold the key to testing any hypothesis that claims to explain the origin of the solar system. If the hypothesis fails to explain even one of the four features, then it cannot be correct. If it successfully explains all four, then we might reasonably assume it is on the right track. Thus, we have the following four criteria for the success of a solar system formation theory:

1. It must explain the patterns of motion we discussed in Chapter 7.
2. It must explain why planets fall into the two major categories—the small, rocky *terrestrial* planets near the Sun and the large, hydrogen-rich *jovian* planets farther out.
3. It must explain the existence of huge numbers of asteroids and comets and why these objects reside primarily in the regions we call the *asteroid belt*, the *Kuiper belt*, and the *Oort cloud*.
4. It must explain all the general patterns while at the same time making allowances for exceptions to the general rules, such as the odd axis tilt of Uranus and the existence of Earth's large Moon.

The theory can gain even further support if it makes predictions that are borne out by new observations or experiments. For example, a theory that explains the origin of *our* solar system ought to be able to make predictions about *other* solar systems. Today, we can test such predictions with the rapidly growing body of observations of young stars that appear to be in the early stages of planet formation and of actual planets discovered around other stars. As we will see in both this chapter and Chapter 13, our modern theory of solar system formation has required only minor modifications to account for these new observations, a fact that has put it on a foundation that is stronger today than ever before.

• What theory best explains the features of our solar system?

After the Copernican revolution, many scientists speculated about the origin of the solar system. However, we generally credit two eighteenth-century scientists with proposing the hypothesis that ultimately blossomed into our modern scientific theory of the origin of the solar system. Around 1755, German philosopher Immanuel Kant proposed that our solar system formed from the gravitational collapse of an interstellar cloud of gas. About 40 years later, French mathematician Pierre-Simon Laplace put forth the same idea independently (he apparently was unaware of Kant's proposal). Because an interstellar cloud is usually called a *nebula* (Latin for "cloud"), their idea became known as the *nebular hypothesis.*

The nebular hypothesis remained popular throughout the nineteenth century. By the early twentieth century, however, scientists had found a few aspects of our solar system that the nebular hypothesis did not seem to explain well—at least in its original form as described by Kant and Laplace. While some scientists sought to modify the nebular hypothesis, others looked for entirely different ideas about how the solar system might have formed.

During much of the first half of the twentieth century, the nebular hypothesis faced stiff competition from a hypothesis proposing that the planets represent debris from a near-collision between the Sun and another star. According to this *close encounter hypothesis,* the planets formed from blobs of gas that had been gravitationally pulled out of the Sun during the near-collision.

Today, the close encounter hypothesis has been discarded. Calculations showed that it could not account for either the observed orbital motions of the planets or the neat division of the planets into two major categories (terrestrial and jovian). Historically, these failures alone should have been enough to doom the hypothesis, but it lost favor for another reason as well: It required a highly improbable event. Given the vast separation between star systems in our region of the galaxy, the chance that any two stars would pass close enough to cause a substantial gravitational disruption is so small that it would be difficult to imagine it happening even in the one case needed to make our own solar system. While low probability alone could not rule out the close encounter hypothesis, it certainly did not help the case for a hypothesis that also failed on other grounds. Moreover, the low probability of forming planets in collisions leads to a testable prediction that we now know to be wrong: If planets are born only in very rare events, then planets should be very rare around other stars. The fact that we now know planets to be quite common [Section 13.1] therefore counts as an added strike against the close encounter hypothesis.

While the close encounter hypothesis was losing favor, new ideas about the physics of planet formation led to modifications of the nebular hypothesis. As scientists investigated the implications of the idea that our solar system formed from a large cloud of gas, they developed new models that led to natural explanations for all four general features of our solar system. By the latter decades of the twentieth century, so much evidence had accumulated in favor of the nebular hypothesis that it achieved the status of a scientific theory—the **nebular theory** of our solar system's birth. Moreover, as we noted earlier, the theory has gained further support by successfully predicting much of what we have discovered about other planetary systems. While recent discoveries of planets around other stars have revealed a few surprises, overall they have been consistent with what we expect if the nebular theory is correct.

Mastering ASTRONOMY Formation of the Solar System Tutorial, Lessons 1–2

8.2 The Birth of the Solar System

We are now ready to examine the nebular theory of solar system formation in more depth. In this section, we'll begin at the beginning, examining the origins of the cloud of gas that gave birth to our solar system and explaining how the gravitational collapse of this cloud led to orderly patterns of motion. Remember that the same ideas are likely to apply to the births of other star systems as well.

• Where did the solar system come from?

The nebular theory begins with the idea that our solar system was born from a cloud of gas that collapsed under its own gravity. We usually refer to this cloud as the **solar nebula**. But where did this gas come from?

Galactic Recycling According to the evidence as we understand it today, the universe was born in the Big Bang [Section 1.1]. Hydrogen and helium (and a trace of lithium) were the only chemical elements present when the universe was young. All the heavier elements have been produced since that time by stars, through nuclear fusion in their cores or the explosions that accompany their deaths. When stars die, they release much of their content back into space, including newly produced heavy elements. This material can then be recycled into new generations of stars (Figure 8.1). (We'll study the galactic recycling process in more detail in Chapter 19.)

Although this process of creating heavy elements in stars and recycling them within the galaxy has probably gone on for most of the 14-billion-year history of our universe, only a small fraction of the original hydrogen and helium has been converted into heavier elements. By the time our solar system formed about 4.6 billion years ago, about 2% (by mass) of the original hydrogen and helium in the galaxy had been converted to heavier elements; the other 98% remained hydrogen and helium. The Sun, which represents nearly all the mass in our solar system, still has this basic composition today.

Thus, the gas that made up the solar nebula was the result of billions of years of galactic recycling that occurred before our solar system was born. The Sun and planets were born from this gas, and Earth and the other terrestrial worlds ended up being made primarily from the heavier elements mixed within it. As we saw in Chapter 1, we are therefore "star stuff," because we and our planet are made of elements forged in stars that lived and died long ago.

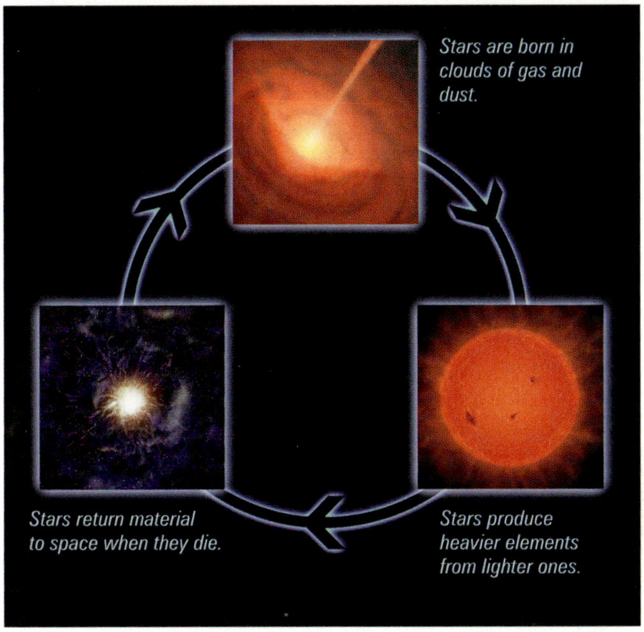

Figure 8.1 This figure, which is a portion of Figure 1.3 earlier in the book, summarizes the galactic recycling process.

THINK ABOUT IT

Could a solar system like ours have formed with the first generation of stars after the Big Bang? Explain.

Evidence from Other Gas Clouds Strong observational evidence supports the idea that our solar system was born from an interstellar cloud of gas, because stars that appear to be in the process of formation today are always found within interstellar clouds [Section 16.1].

Figure 8.2 shows the Orion Nebula, one of the most famous star-forming clouds. Some stars have already been born within this vast cloud of gas, and thousands more will be born there over the next few million years, some perhaps with planets. Moreover, each star we observe in the process of formation is surrounded by relatively dense gas that appears to be collapsing under its own gravity, suggesting that it is forming within its own "solar nebula," just as we think our own solar system formed some 4.6 billion years ago.

• What caused the orderly patterns of motion in our solar system?

The solar nebula probably began as a large, roughly spherical cloud of very cold and very low-density gas. Initially, this gas was probably so spread out—perhaps over a region a few light-years in diameter—that gravity alone may not have been able to pull it together to start its collapse. Instead, the collapse may have been triggered by a cataclysmic event, such as the impact of a shock wave from the explosion of a nearby star (a supernova).

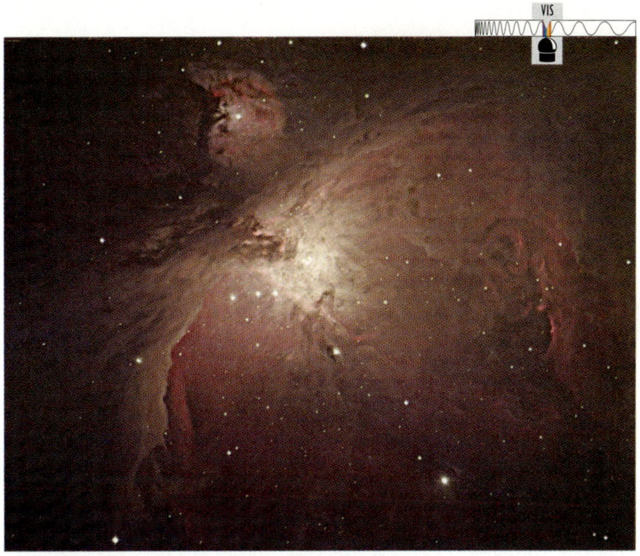

Figure 8.2 Interactive Photo This photograph shows the Orion Nebula, an interstellar cloud in which new star systems are forming. Over the next few million years, thousands of stars will be born in this gas cloud. Some of these stars may end up with their own planetary systems.

Once the collapse started, the law of gravity ensured that it would continue. Remember that the strength of gravity follows an inverse square law with distance [Section 4.4]. Because the mass of the cloud remained the same as it shrank, the strength of gravity increased as the diameter of the cloud decreased. For example, when the diameter decreased by half, the force of gravity increased by a factor of four.

Because gravity pulls inward in all directions, you might at first guess that the solar nebula would have remained spherical as it shrank. Indeed, the idea that gravity pulls in all directions explains why the Sun and the planets are spherical. However, gravity is not the only physical law that affects the collapse of a cloud of gas. When we consider everything that should have occurred in the collapsing cloud, we find that the solar nebula should have become a spinning, flattened disk surrounding a central star.

Heating, Spinning, and Flattening As the solar nebula shrank in size, three important processes altered its density, temperature, and shape, changing it from a large, diffuse (spread-out) cloud to a much smaller spinning disk (Figure 8.3):

- *Heating.* The temperature of the solar nebula increased as it collapsed. Such heating represents energy conservation in action [Section 4.3]. As the cloud shrank, its gravitational potential energy was converted to the kinetic energy of individual gas particles falling inward. These particles crashed into one another, converting the kinetic energy of their inward fall to the random motions of thermal energy (see Figure 4.12b). The Sun formed in the center, where temperatures and densities were highest.
- *Spinning.* Like an ice skater pulling in her arms as she spins, the solar nebula rotated faster and faster as it

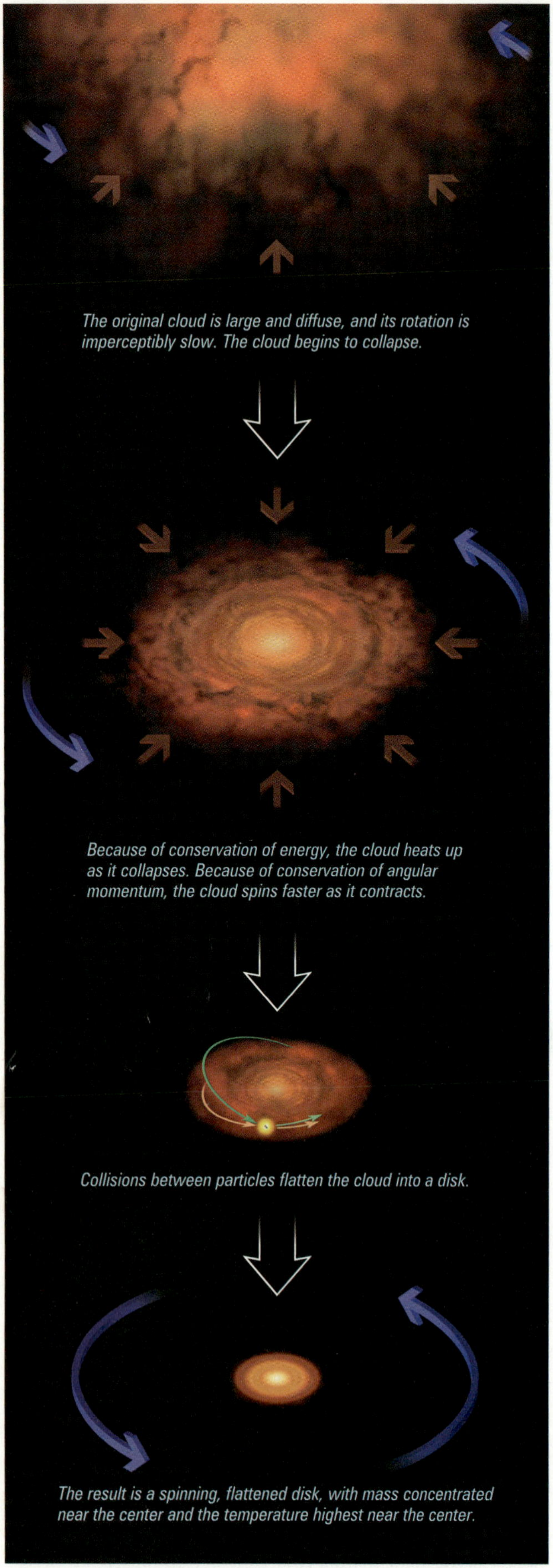

Figure 8.3 Interactive Figure This sequence of paintings shows how the gravitational collapse of a large cloud of gas causes it to become a spinning disk of matter. The hot, dense central bulge becomes a star, while planets can form in the surrounding disk.

shrank in radius. This increase in rotation rate represents conservation of angular momentum in action [Section 4.3]. The rotation of the cloud may have been imperceptibly slow before its collapse began, but the cloud's shrinkage made fast rotation inevitable. The rapid rotation helped ensure that not all the material in the solar nebula collapsed into the center: The greater the angular momentum of a rotating cloud, the more spread out it will be.

- *Flattening.* The solar nebula flattened into a disk. This flattening is a natural consequence of collisions between particles in a spinning cloud. A cloud may start with any size or shape, and different clumps of gas within the cloud may be moving in random directions at random speeds. These clumps collide and merge as the cloud collapses, and each new clump has the average velocity of the clumps that formed it. Thus, the random motions of the original cloud become more orderly as the cloud collapses, changing the cloud's original lumpy shape into a rotating, flattened disk. Similarly, collisions between clumps of material in highly elliptical orbits reduce their eccentricities, making their orbits more circular.

The formation of the spinning disk explains the orderly motions of our solar system today. The planets all orbit the Sun in nearly the same plane because they formed in the flat disk. The direction in which the disk was spinning became the direction of the Sun's rotation and the orbits of the planets. Computer models show that the planets would also have tended to rotate in this same direction as they formed—which is why most planets rotate the same way today—though the small sizes of planets compared to the entire disk allowed some exceptions to arise. The fact that collisions in the disk tended to make orbits more circular explains why most planets in our solar system have nearly circular orbits.

THINK ABOUT IT

You can create a simple analogy to the organized motions of the solar system by sprinkling pepper into a bowl of water and stirring it quickly in random directions. Because the water molecules are always colliding with one another, the motion of the pepper grains will tend to settle down into a slow rotation representing the average of the original, random velocities. Try the experiment several times, stirring the water differently each time. Do the random motions ever exactly cancel out, resulting in no rotation at all? Describe what occurs, and explain how it is similar to what took place in the solar nebula.

Testing the Model If we are correct about the processes that occurred in the solar nebula, the same processes ought to occur in other clouds of gas that collapse to form stars.

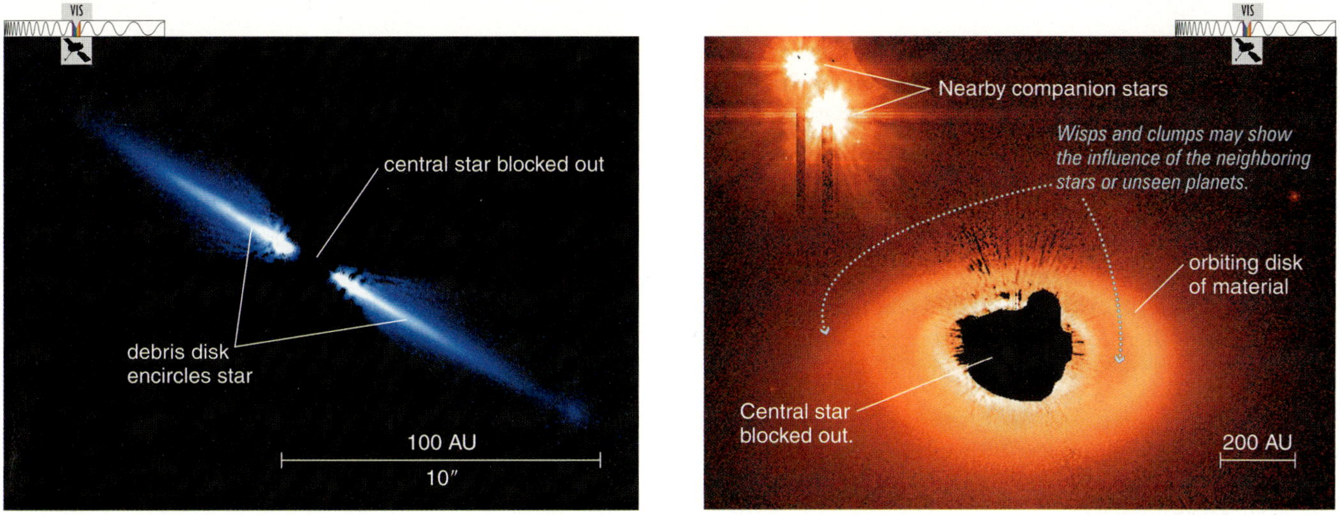

a We see this disk edge-on around the star AU Microscopii, confirming its flattened shape.

b This photo shows a disk around the star HD 141569A. The colors are not real; a black-and-white image has been tinted red to bring out faint detail.

Figure 8.4 These Hubble Space Telescope photos show flattened, spinning disks of material around other stars.

Thus, the formation of a disk in which planets can form should be a natural part of the star formation process, and we can test whether our model makes sense by searching for such disks elsewhere.

Observational evidence does indeed support the idea that our solar nebula collapsed in the way envisioned by the nebular model. The heating that occurs in a collapsing cloud of gas means that the gas should emit thermal radiation [Section 5.4], primarily in the infrared. We've detected infrared radiation from many nebulae where star systems appear to be forming today. We've even seen structures around other stars that appear to be flattened, spinning disks (Figure 8.4).

Other support for the model comes from computer simulations of the formation process. A simulation begins with a set of data representing the conditions we observe in interstellar clouds. Then, with the aid of a computer, we apply the laws of physics to predict the changes that should occur over time. These computer simulations successfully reproduce most of the general characteristics of motion in our solar system, suggesting that the nebular theory is on the right track.

Finally, evidence that our ideas about the formation of flattened disks are correct comes from many other structures in the universe. We expect flattening to occur anywhere that orbiting particles can collide, which explains why we find so many cases of flat disks, including the disks of spiral galaxies like the Milky Way, the disks of planetary rings, and the *accretion disks* surrounding many neutron stars and black holes [Section 18.3].

8.3 The Formation of Planets

The planets began to form after the solar nebula had collapsed into a flattened disk of perhaps 200 AU in diameter (about twice the present-day diameter of Pluto's orbit). In this section, we'll discuss how the planets formed within the spinning disk of gas. Along the way, we'll see how the process of planet formation led to the existence of the two basic types of planets.

• Why are there two major types of planets?

The churning and mixing of gas in the solar nebula should have ensured that the nebula had the same composition throughout: 98% hydrogen and helium plus 2% heavier elements. Yet the solar system ended up with two very different major types of planets: the small, rocky terrestrial planets and the large, gaseous jovian planets. How did the compositions of the planets end up being so different, when they were all born in the same cloud of gas? The locations of these two major classes of planets provide the key to understanding how they formed. The terrestrial planets formed in the warm, inner regions of the swirling disk and the jovian planets formed in the colder, outer regions. Let's explore how temperature differences led directly to different types of planets.

Condensation: Sowing the Seeds of Planets In the center of the collapsing solar nebula, gravity drew together enough material to form the Sun. In the surrounding disk, however, the gaseous material was too spread out for gravity alone to clump it up. Instead, material had to begin clumping in some other way and then grow in size until gravity could start pulling it together into planets. In essence, planet formation required the presence of "seeds"—solid bits of matter from which gravity could ultimately build planets.

The basic process of seed formation was probably much like the formation of snowflakes in clouds on Earth: When the temperature is low enough, some atoms or molecules in a gas may bond and solidify. (Pressures in the solar nebula were generally too low to allow the condensation of liquid droplets.) The general process in which solid (or liquid) particles form in a gas is called **condensation**—we say that the particles *condense* out of the gas. The particles that initially condense from a gas are microscopic in size, but they can grow larger with time.

Different materials condense at different temperatures. On the basis of their condensation properties, the ingredients of the solar nebula fell into four major categories, summarized in Table 8.1:

- **Hydrogen and helium gas (98% of the solar nebula).** These gases never condense under the conditions present in a nebula.
- **Hydrogen compounds (1.4% of the solar nebula).** Materials such as water (H_2O), methane (CH_4), and ammonia (NH_3) can solidify into **ices** at low temperatures (below about 150 K under the low pressure of the solar nebula).
- **Rock (0.4% of the solar nebula).** Rocky material is gaseous at very high temperatures, but condenses into solid bits of mineral at temperatures between about 500 K and 1,300 K, depending on the type of rock. (A *mineral* is a piece of rock with a particular chemical composition and structure.)
- **Metal (0.2% of the solar nebula).** Metals such as iron, nickel, and aluminum are also gaseous at very high temperatures, but condense into solid form at higher temperatures than rock—typically in the range of 1,000 K to 1,600 K.

Because hydrogen and helium gas made up 98% of the solar nebula's mass and did not condense, the vast majority of the nebula remained gaseous at all times. However, hydrogen compounds, rock, and metal could condense into solid form wherever the temperature allowed (Figure 8.5). In the innermost regions of the nebula near the forming Sun, where the temperature was above 1,600 K, it was too hot for any material to condense. Near what is now Mercury's orbit, the temperature was low enough for metals and some types of rock to condense into tiny, solid particles, but other types of rock and all the hydrogen compounds remained gaseous. More types of rock could condense, along with the metals, at the distances from the Sun where Venus, Earth, and Mars would form. In the region where the asteroid belt would eventually be located, temperatures were low enough to allow dark, carbon-rich minerals to condense, along with minerals containing small amounts of water. Hydrogen compounds could condense into ices only beyond the **frost line,** which lay between the present-day orbits of Mars and Jupiter.

The frost line marked the key transition between the warm inner regions of the solar system where terrestrial planets formed and the cool outer regions where jovian planets formed. Inside the frost line, only metal and rock could condense into solid "seeds," which is why the terrestrial planets ended up being made of metal and rock. Beyond the frost line, where it was cold enough for hydrogen compounds to condense into ices, the solid seeds were built of ice along with metal and rock. Moreover, because hydrogen compounds were nearly three times as abundant in the nebula as metal and rock combined (see Table 8.1), the total amount of solid material was far greater beyond the frost line than within it. The stage was set for the birth of two types of planets: planets born from seeds of metal and rock in the inner solar system and planets born from seeds of ice (as well as metal and rock) in the outer solar system.

Table 8.1 Materials in the Solar Nebula

A summary of the four types of materials present in the solar nebula. The squares represent the relative proportions of each type (by mass).

	Examples	*Typical Condensation Temperature*	*Relative Abundance (by mass)*
Hydrogen and Helium Gas 	hydrogen, helium	do not condense in nebula	98%
Hydrogen Compounds 	water (H_2O) methane (CH_4) ammonia (NH_3)	<150 K	1.4%
Rock 	various minerals	500–1,300 K	0.4%
Metals 	iron, nickel, aluminum	1,000–1,600 K	0.2%

THINK ABOUT IT

Consider a region of the solar nebula in which the temperature was about 1,300 K. Based on the data in Table 8.1, what fraction of the material in this region was gaseous? What were the solid particles in this region made of? After you have answered these questions, do the same for a region with a temperature of 100 K. Would the 100 K region be closer to or farther from the Sun? Explain.

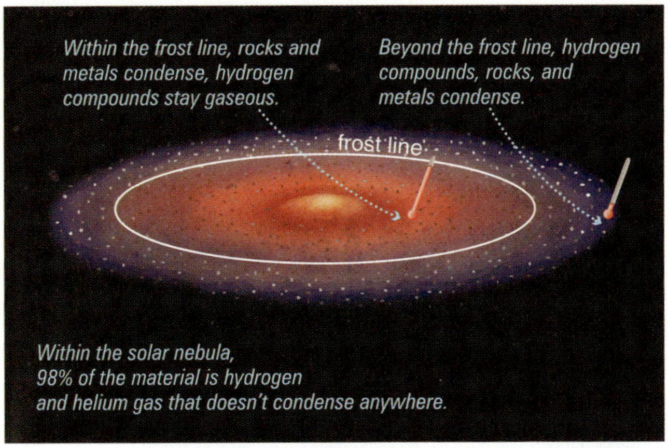

Figure 8.5 Interactive Figure Temperature differences in the solar nebula led to different kinds of condensed materials, sowing the seeds for two different kinds of planets.

• How did the terrestrial planets form?

We have seen that inner regions of the solar nebula contained tiny seeds of solid metal and rock, and we know that the inner solar system ended up with terrestrial planets made of metal and rock. Thus, the basic story of the terrestrial planets seems fairly clear: The solid seeds of metal and rock gradually grew into the full-fledged planets we see today. Because rock and metal made up such a small amount of the material in the solar nebula, the planets they built ended up relatively small in size. All that's left for us to understand are the details of how the seeds grew in size.

The process by which small "seeds" grew into planets is called **accretion** (Figure 8.6). Accretion began with the microscopic solid particles that condensed from the gas of the solar nebula. These particles orbited the forming Sun with the same orderly, circular paths as the gas from which they condensed. Individual particles therefore moved at nearly the same speed as neighboring particles, so "collisions" were more like gentle touches. Although the particles were far too small to attract each other gravitationally at this point, they were able to stick together through electrostatic forces—the same "static electricity" that makes hair stick to a comb. Small particles thereby began to combine into larger ones. As the particles grew in mass, gravity began to aid the process of their sticking together, accelerating their growth into boulders large enough to count as **planetesimals**, which means "pieces of planets."

The planetesimals grew rapidly at first. As a planetesimal grew larger, it would have both more surface area to make contact with other planetesimals and more gravity to attract them. Some planetesimals probably grew to hundreds of kilometers in size in only a few million years—a long time in human terms, but only about 1/1,000 the present age of the solar system. However, once the planetesimals reached these relatively large sizes, further growth became more difficult.

Gravitational encounters [Section 4.5] between planetesimals tended to alter their orbits, particularly those of the smaller planetesimals. With different orbits crossing each other, collisions between planetesimals tended to occur at higher speeds and hence became more destructive. Such collisions tended to produce fragmentation more often than accretion. Only the largest planetesimals avoided being shattered and thus were able to grow into full-fledged, terrestrial planets.

Theoretical evidence in support of this model comes from computer simulations of the accretion process, which support the scenario described above. Observational evidence comes from meteorites, many of which appear to be surviving fragments from the early period of condensation in our solar system [Section 9.1]. Meteorites often contain metallic grains embedded in a variety of rocky minerals (Figure 8.7), just as we expect for the planetesimals of the inner solar system. Meteorites thought to come from the

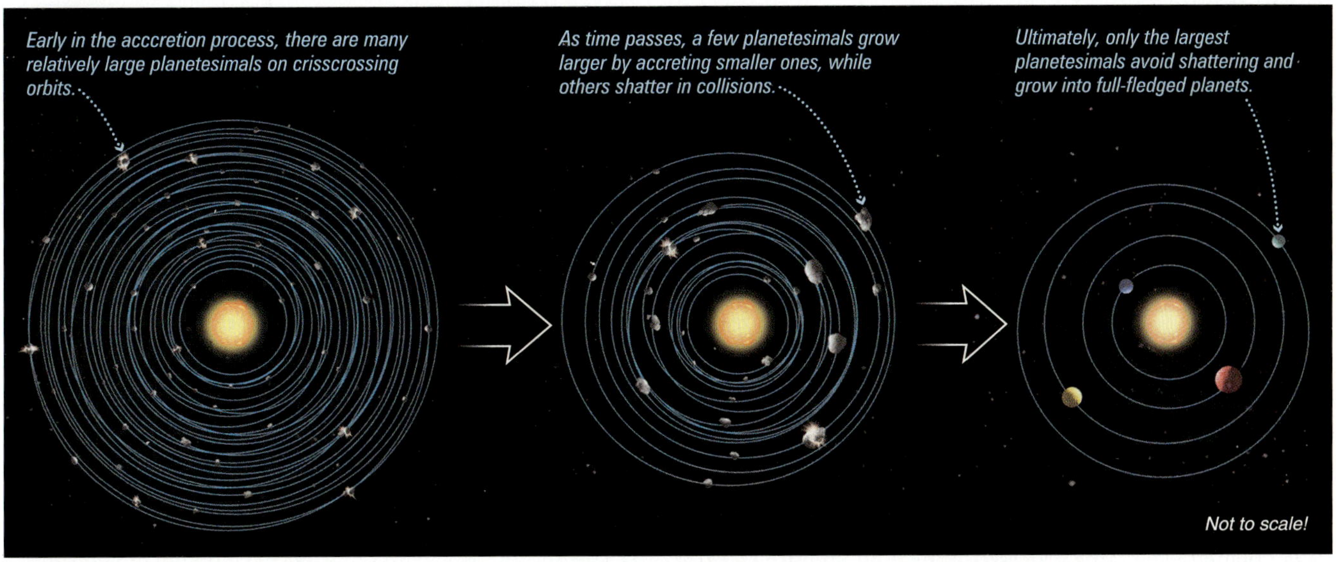

Figure 8.6 Interactive Figure These diagrams show how planetesimals gradually accrete into terrestrial planets.

Figure 8.7 Shiny flakes of metal are clearly visible in this slice through a meteorite (a few centimeters across), mixed in among the rocky material. Such metallic flakes are just what we would expect to find if condensation really occurred in the solar nebula as described by the nebular theory.

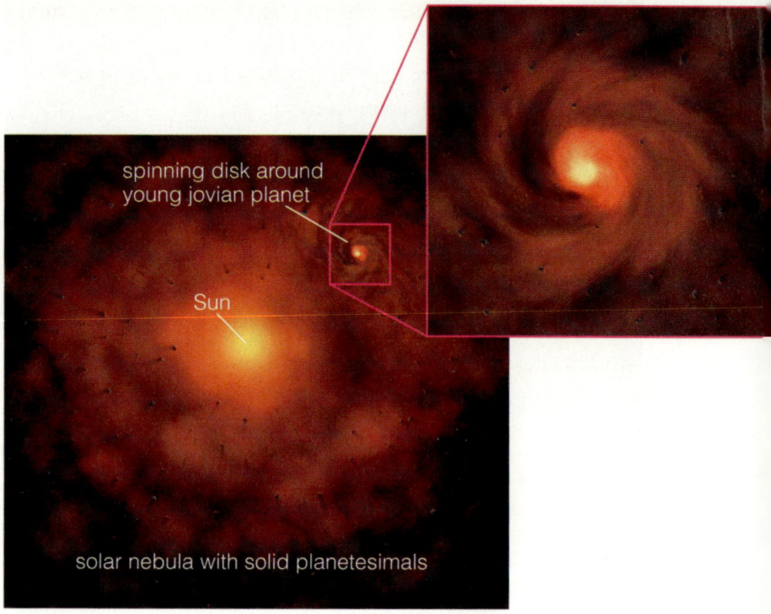

Figure 8.8 The young jovian planets were surrounded by disks of gas, much like the disk of the entire solar nebula but smaller in size. According to the leading model, the planets grew as large ice-rich planetesimals captured hydrogen and helium gas from the solar nebula. This painting shows the gas and planetesimals surrounding one jovian planet in the larger solar nebula.

outskirts of the asteroid belt contain abundant carbon-rich materials, and some contain water—again as we expect for planetesimals that formed in that region.

• How did the jovian planets form?

Accretion should have occurred similarly in the outer solar system, but condensation of ices meant that there was more solid material. Thus, the planetesimals that accreted in the outer solar system contained large amounts of ice in addition to metal and rock. The solid objects that reside in the outer solar system today, such as comets and the moons of the jovian planets, still show this ice-rich composition. However, the growth of icy planetesimals cannot be the whole story of jovian planet formation, because the jovian planets themselves are *not* made mostly of ice. Instead, they contain large amounts of hydrogen and helium gas.

The precise way in which the jovian planets formed remains a topic of scientific debate, but in the leading model they acquired their gas after beginning existence as large, ice-rich planetesimals. According to this model, some of the abundant, icy planetesimals of the outer solar system grew to masses many times that of Earth. With these large masses, their gravity became strong enough to capture and hold some of the hydrogen and helium gas that made up the vast majority of the surrounding solar nebula. As the growing planets accumulated gas, their gravity grew stronger still, allowing them to capture even more gas. Ultimately, the jovian planets grew so much that they bore little resemblance to the icy seeds from which they started, instead ending up with large abundances of hydrogen and helium gas.

This model also explains the many large moons of the jovian planets. The same processes of heating, spinning, and flattening that made the disk of the solar nebula should also have affected the gas drawn by gravity to the young jovian planets. Thus, each jovian planet probably was surrounded by its own disk of gas, spinning in the same direction that the planet rotated (Figure 8.8). Moons accreted from ice-rich planetesimals within these disks, which is why these moons tend to orbit in the same direction as their planet rotates, with nearly circular orbits lying close to the equatorial plane of their parent planet.

This model of accretion followed by gas capture explains the observed features of the jovian planets quite well. Nevertheless, some scientists have recently advanced a competing model in which disturbances (technically, instabilities) within the disk of the solar nebula could have led clumps of gas to collapse and form jovian planets without first forming an icy planetesimal. Because this model is relatively new, scientists have not yet worked out enough details to know whether it can explain all the characteristics of the jovian planets. Most planetary scientists therefore still favor the accretion and gas capture model, but good practice of science demands that we remain open to considering new ideas.

• What ended the era of planet formation?

The vast majority of the hydrogen and helium gas in the solar nebula never became part of any planet. So what became of it? Apparently, it was swept into interstellar space by the **solar wind**, which consists of charged particles (such as protons and electrons) that are continually blown off the surface of the Sun in all directions. Although the solar wind is fairly weak today, observations of other stars show that winds tend to be much stronger in young stars. Thus, the young Sun should have had a strong solar wind—strong enough to have swept huge quantities of gas out of the solar system. The radiation from the young Sun also played a role in clearing out the gas.

Sealing the Fates of Planets The clearing of the gas sealed the compositional fate of the planets. If the gas had not been cleared soon after the planets formed, it might have continued to cool until hydrogen compounds could have condensed into ices even in the inner solar system. In that case, the terrestrial planets might have accreted abundant ice, and perhaps hydrogen and helium gas as well, changing their basic nature. At the other extreme, if the gas had been blown out too early, the raw materials of the planets might have been swept away before the planets could fully form. Although these extreme scenarios did not occur in our solar system, studies of other planetary systems suggest that they may sometimes occur around other stars [Section 13.3].

Explaining the Sun's Rotation The young Sun's strong solar wind also helps explain what was once considered a surprising aspect of the Sun's rotation. According to the law of conservation of angular momentum, the spinning disk of the solar nebula should have spun fastest near its center, where most of the mass became concentrated. Thus, the young Sun should have been rotating very fast. Today the Sun rotates quite slowly, with each full rotation taking about a month. If the young Sun really did rotate fast, as our theory seems to demand, how did its rotation slow down?

Angular momentum cannot simply disappear, but it is possible to transfer angular momentum from one object to another—and then get rid of the second object. A spinning skater can slow her spin by grabbing her partner and then pushing him away. In the 1950s, scientists realized that the young Sun's rapid rotation would have generated a magnetic field far stronger than that of the Sun today. This strong magnetic field actually helped create the strong solar wind. More generally, the strong magnetic field would have made the young Sun much more active on its surface than it is today [Section 14.3]. For example, large "solar storms" and sunspots would have been much more common, causing the emission of much more ultraviolet and X-ray light. This high-energy radiation from the young Sun ionized gas in the solar nebula, creating many charged particles.

As we will discuss in more detail in Chapter 14, charged particles and magnetic fields tend to stick together. As the Sun rotated, its magnetic field dragged the charged particles along faster than the rest of the nebula, adding to their angular momentum. As the particles were gaining angular momentum, the Sun was losing it. The young Sun's strong solar wind then blew these particles into interstellar space, leaving the Sun with the greatly diminished angular momentum and much slower rotation that we see today (Figure 8.9).

Although we cannot prove that the young Sun really did lose angular momentum in this way, support for the idea comes from observations of other stars. When we look at young stars that have recently formed in interstellar clouds, we find that nearly all of them rotate rapidly and have strong magnetic fields and strong winds [Section 16.2]. Older stars, in contrast, almost invariably rotate slowly, like our Sun. This suggests that nearly all stars have their original rapid rotations slowed by transferring angular momentum to charged particles in their disks—particles that are later swept away—just as our theory suggests happened with the Sun.

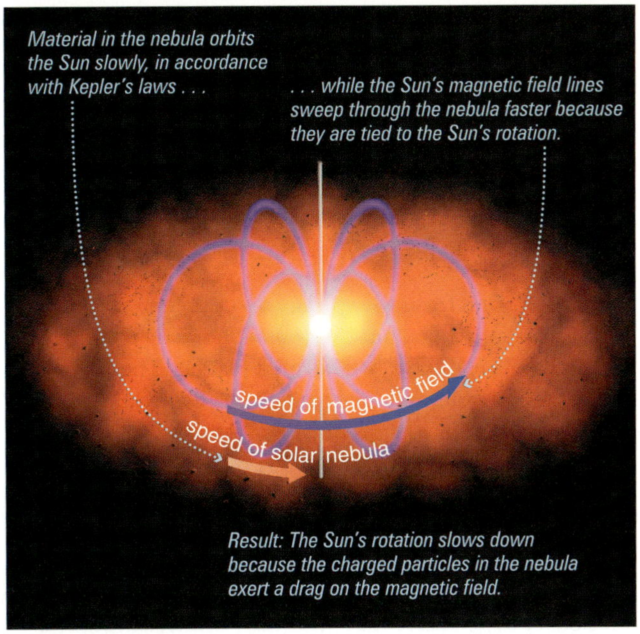

Figure 8.9 The young Sun should have rotated rapidly, but today the Sun rotates quite slowly. As shown in this diagram, the Sun probably lost angular momentum because of drag between slow-moving charged particles in the solar nebula and the Sun's rotating magnetic field (represented by purple loops).

COMMON MISCONCEPTIONS

Solar Gravity and the Density of Planets

You might think that it was the Sun's gravity that pulled the dense rocky and metallic materials to the inner part of the solar nebula, or that gases escaped from the inner nebula because gravity couldn't hold them. But this is not the case—all the ingredients were orbiting the Sun together under the influence of the Sun's gravity. The orbit of a particle or a planet does *not* depend on its size or density, so the Sun's gravity cannot be the cause of the different kinds of planets. Rather, the different temperatures in the solar nebula are the cause.

8.4 The Aftermath of Planet Formation

So far, we have seen how the nebular theory explains our solar system's patterns of motion and its two major types of planets. We still have two features of our solar system left to explain: the existence, composition, and orbital properties of the myriad asteroids and comets, and the exceptions to the general rules. These two features turn out to be closely related—both involve the planetesimals that remained among the forming planets in the early solar system.

• Where did asteroids and comets come from?

The strong wind from the young Sun cleared excess gas from the solar nebula, but many planetesimals remained scattered between the newly formed planets. These "leftovers" became asteroids and comets. We'll wait until Chapter 12 to discuss the details of why most asteroids ended up grouped in the asteroid belt while most comets ended up split between two regions, the Kuiper belt and the Oort cloud. However, we can understand the basic ideas here.

Asteroids are the rocky leftover planetesimals of the inner solar system. The four inner planets must have swept up most of the planetesimals near their orbits, leaving relatively few leftovers between them. However, no planet formed between Mars and Jupiter (a fact attributable to effects from Jupiter's strong gravity [Section 12.1]), so this region—the asteroid belt—must once have been dense with rocky planetesimals. In fact, the asteroid belt probably once contained enough planetesimals to form another terrestrial planet, but most of them ultimately crashed into the inner planets or were ejected from the solar system. Today, the total mass of the asteroids in the asteroid belt is only a tiny fraction of the mass of any terrestrial planet.

Comets are the ice-rich leftover planetesimals of the outer solar system. Comets must have initially populated the entire outer solar system, presumably out to some distance beyond Neptune. Those that orbited beyond Neptune should have been able to grow fairly large without disruption by the jovian planets. Moreover, they would have remained on their original orbits in the same plane as the orbits of the planets. These large comets thereby became the comets of the Kuiper belt, which follow the same basic orbital patterns as the planets. Pluto, which seemed a misfit when compared to the first eight planets, seems much more at home when considered as a large comet of the Kuiper belt [Section 12.3].

In contrast to the stable orbital paths for comets beyond Neptune, the comets that roamed between the jovian planets almost inevitably suffered close gravitational encounters with these planets. Ultimately, these encounters caused most of these comets to be "kicked" into orbits that carried them far from the Sun. Because the "kicks" could have sent these comets flying in random directions, their new orbits no longer displayed the original orderly motion of the disk in which the planets formed. That is why these comets now populate the Oort cloud, with randomly oriented orbits very far from the Sun.

Evidence that asteroids and comets really are leftover planetesimals comes from analysis of meteorites, spacecraft visits to comets and asteroids, and computer simulations of solar system formation. In fact, the nebular theory actually *predicts* the existence of both the Oort cloud and the Kuiper belt—a prediction first made in the 1950s. Thus, the discoveries, beginning in the 1990s, of numerous objects orbiting in the Kuiper belt represent a triumph for the nebular theory. Comets in the distant Oort cloud remain undetectable with present technology, but there is little doubt that they exist, because their presence is the only way to explain the comets that enter the inner solar system from afar.

• How do we explain "exceptions to the rules"?

We have now explained all the major features of our solar system except for the "exceptions to the rules," such as our surprisingly large Moon and the odd rotation of Uranus. Today, we think that most of these exceptions arose from collisions or close encounters with leftover planetesimals.

The Heavy Bombardment The asteroids and comets that exist today probably represent only a small fraction of the vast numbers of leftover planetesimals that roamed the young solar system. The rest are now gone. Some of these "lost" planetesimals may have been flung into deep space, but many others must have collided with the planets. The vast majority of these collisions occurred in the first few hundred million years of our solar system's history, during the period we call the **heavy bombardment.**

Every world in our solar system must have been pelted by impacts during the heavy bombardment (Figure 8.10). Some of these impacts left scars that we can still see today as impact craters. These impacts did more than just batter

Figure 8.10 Around 4 billion years ago, Earth, its Moon, and the other planets were heavily bombarded by leftover planetesimals. This painting shows the young Earth and Moon, with an impact in progress on Earth.

the planets. They also brought materials from other regions of the solar system—a fact that is critical to our existence on Earth today.

Remember that the terrestrial planets were built from planetesimals made of metal and rock. These planetesimals probably contained no water or other hydrogen compounds at all, because it was too hot for these compounds to condense in our region of the solar nebula. How, then, did Earth come to have the water that makes up our oceans? The answer is that water, along with other hydrogen compounds, must have been brought to Earth and the other terrestrial planets by the impact of planetesimals that formed farther from the Sun. We don't yet know whether these planetesimals came primarily from the outer asteroid belt, where rocky planetesimals contained small amounts of water and other hydrogen compounds, or whether they were comets containing huge amounts of ice. In either case, the water we drink and the air we breathe probably once were part of planetesimals floating beyond the orbit of Mars.

THINK ABOUT IT

Jupiter's gravity played a major role in flinging asteroids and comets into the inner solar system. Given this fact, what can you say about Jupiter's role in bringing water to Earth? How might Earth be different if Jupiter had never formed?

Captured Moons We can easily explain the orbits of most large jovian planet moons by their formation in a disk that swirled around the forming planet. However, some moons have unusual orbits—orbits in the "wrong" direction (opposite the rotation of their planet) or with large inclinations to the planet's equator. These unusual moons are probably leftover planetesimals that were captured into orbit around a planet.

It's not easy for a planet to capture a moon. An object cannot switch from an unbound orbit (for example, an asteroid whizzing by Jupiter) to a bound orbit (for example, a moon orbiting Jupiter) unless it somehow loses orbital energy [Section 4.5]. For the jovian planets, captures probably occurred when passing planetesimals lost energy to drag in the extended and relatively dense gas that surrounded these planets as they formed. The planetesimals would have been slowed by friction with the gas, just as artificial satellites are slowed by drag in encounters with Earth's atmosphere. If friction reduced a passing planetesimal's orbital energy enough, it could have become an orbiting moon. Because of the random nature of the capture process, the captured moons would not necessarily orbit in the same direction as their planet or in its equatorial plane. Computer models suggest that this capture process would have worked only on objects of a few kilometers in size. Most of the small moons of the jovian planets are a few kilometers across, supporting the idea that they were captured in this way. Mars may have captured its two small moons, Phobos and Deimos, in a similar way at a time when the planet had a much more extended atmosphere than it does today (Figure 8.11).

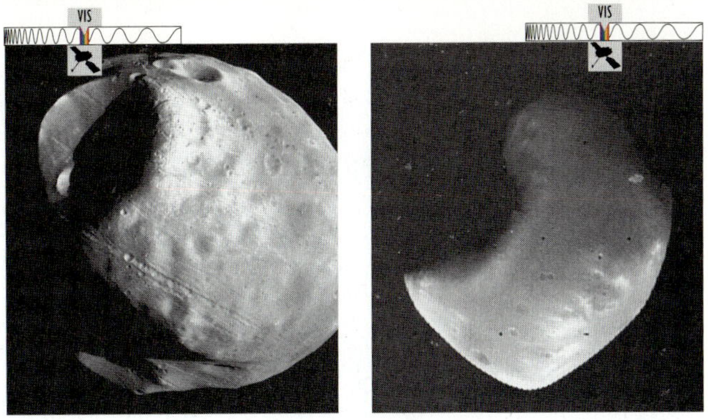

a Phobos **b** Deimos

Figure 8.11 The two moons of Mars, shown here in photos taken by the *Viking* spacecraft, are probably captured asteroids. Phobos is only about 13 kilometers across and Deimos is only about 8 kilometers across—making each of these two moons small enough to fit within the boundaries of a typical large city.

• How do we explain the existence of our Moon?

Capture processes explain the origin of many moons, but they cannot explain our own Moon. Our Moon is much too large to have been captured by a small planet like Earth. We can also rule out the possibility that our Moon formed simultaneously with Earth. If the Moon and Earth had formed together, both should have accreted from planetesimals of the same type and should therefore have approximately the same composition and density. That is not the case. The Moon's density is considerably lower than Earth's, indicating that it has a very different average composition and could not have formed in the same way or at the same time as our planet. We must look for another way to explain the origin of our surprisingly large Moon.

Giant Impacts and the Formation of Our Moon The largest leftover planetesimals may have been huge, perhaps the size of Mars. When one of these planet-size planetesimals collided with a planet, the spectacle would have been awesome. Such a **giant impact** could have significantly altered a planet's fate.

What would have happened if a Mars-size object had collided with the young Earth? Depending on exactly where and how fast the object struck Earth, the blow might have tilted Earth's axis, changed its rotation rate, or completely shattered our planet. The most interesting case arises when we consider what would have happened if the impact had blasted away rock from Earth's outer layers (mantle and crust) and sent this material into orbit around our planet. According to computer simulations, this orbiting material could have reaccreted to form our Moon (Figure 8.12).

Today, such a giant impact is the leading hypothesis for explaining the origin of our Moon. Strong support for this hypothesis comes from two features of the Moon's composition. First, the Moon's overall composition is quite similar to that of Earth's outer layers—just as we should expect if it were made from material blasted away from those layers. Second, the Moon has a much smaller proportion of easily vaporized ingredients (such as water) than Earth. This fact

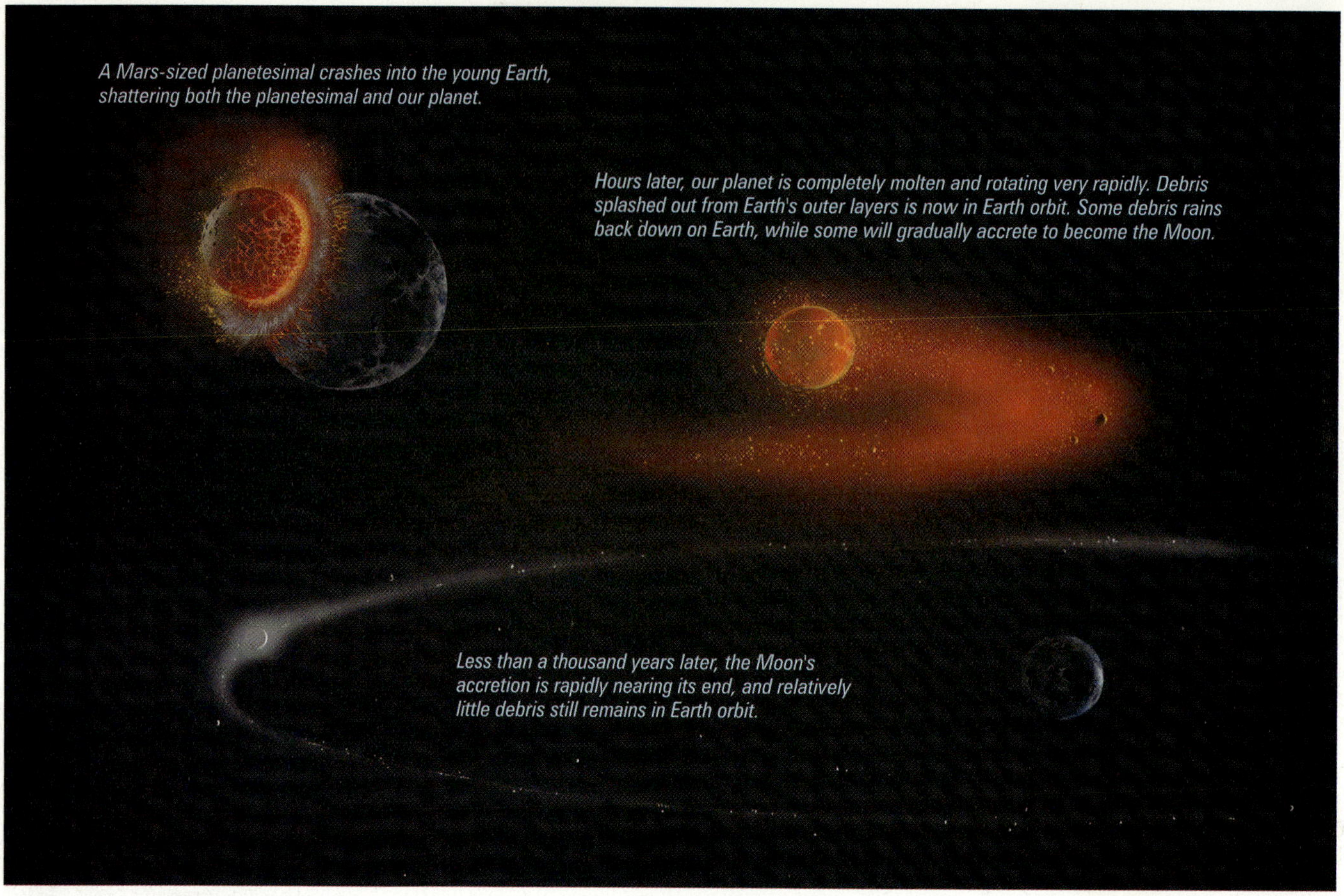

Figure 8.12 Artist's conception of the Moon's formation by a giant impact. The fact that ejected material came mostly from Earth's outer rocky layers explains why the Moon contains very little metal. The impact must have occurred more than 4.4 billion years ago, since that is the age of the oldest Moon rocks. As shown, the Moon formed quite close to a rapidly rotating Earth, but over billions of years tidal forces have slowed Earth's rotation and moved the Moon's orbit outward (see Figure 4.26).

supports the hypothesis because the heat of the impact would have vaporized these ingredients. As gases, they would not have participated in the process of reaccretion that formed the Moon.

Other Giant Impacts Giant impacts may also explain many of the other exceptions to the general trends. For example, giant impacts may have been responsible for tilting the axes of many planets (including Earth) and perhaps for tipping Uranus on its side. Pluto's moon Charon may have formed in a giant impact similar to the one that formed our Moon. Mercury's surprisingly high density may be the result of a giant impact that blasted away its rocky outer layers, leaving behind a planet made almost entirely of the metal that was once in its core.

Unfortunately, we can do little to test whether a particular giant impact really occurred billions of years ago. The difficulty in proving the giant impact hypothesis makes the idea controversial, and most planetary scientists didn't take the idea seriously when it was first proposed. But no other idea so effectively explains the formation of our Moon and other "exceptions" we've discussed. Moreover, numerous giant impacts certainly *should* have occurred, given the number of large leftover planetesimals predicted by the nebular theory. Random giant impacts are therefore considered the most likely explanation for many of the exceptions to our general planetary rules.

• Was our solar system destined to be?

We've now seen that the nebular theory accounts for all the major features of our solar system. Of course, that does not mean that it explains *everything* about the formation of our solar system. As we'll discuss in Chapter 13, discoveries of other planetary systems have already forced us to revise parts of the theory to allow for a wider range of possibilities than we observe in our own solar system. Nevertheless, the basic theory is widely accepted today because it is so strongly supported by the available evidence. Figure 8.13 summarizes our current understanding of how our solar system formed. The entire process of planet formation probably took no more than a few tens of millions of years, about 1% of the current age of the solar system.

Assuming the nebular theory is correct, was our solar nebula "destined" to form the solar system we see today? Probably not. The first stages of planet formation were orderly and inevitable according to the nebular theory.

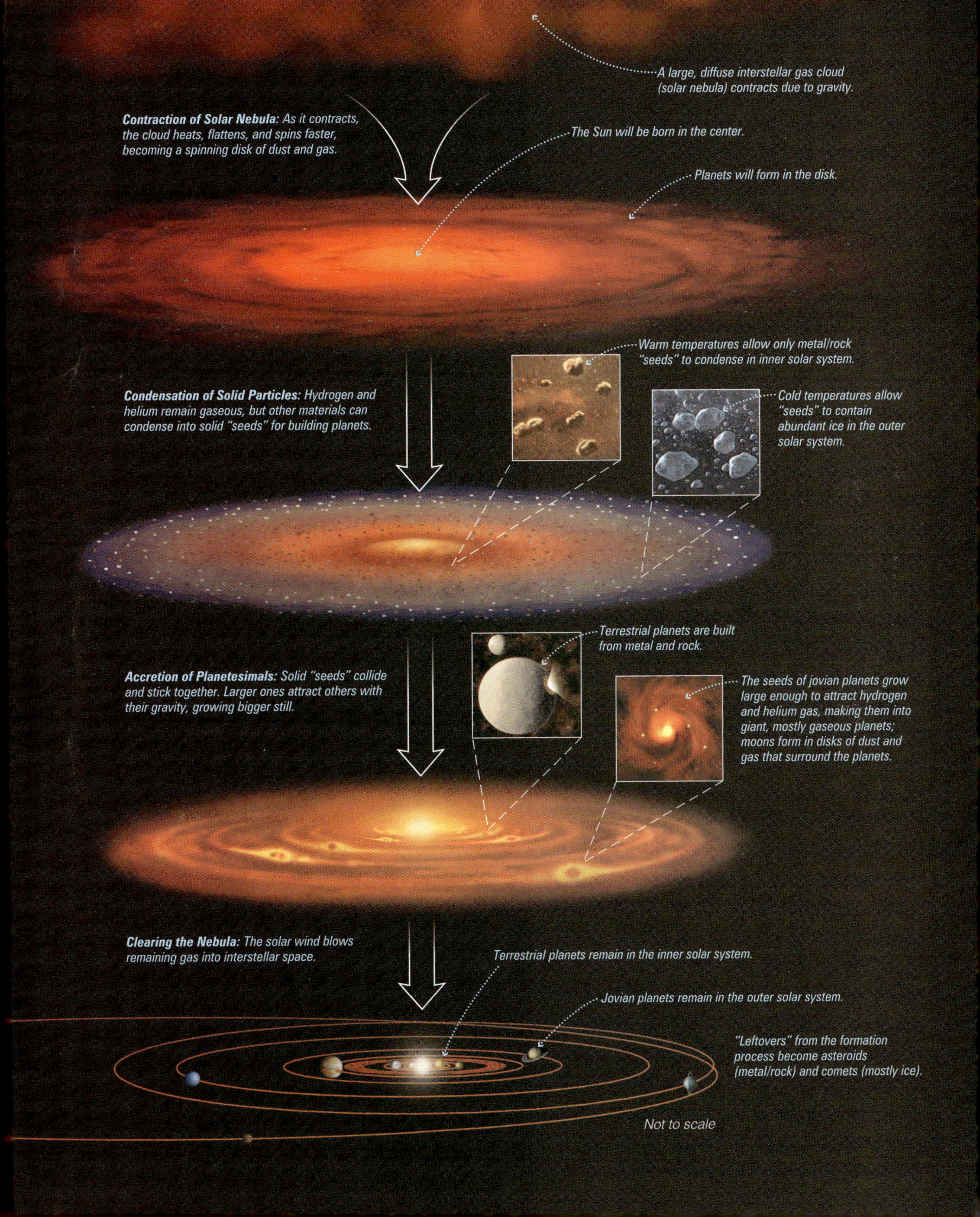

Figure 8.13 A summary of the process by which our solar system formed, according to the nebular theory.

The creation of a spinning disk, condensation within that disk, and the first stages of accretion were relatively gradual processes that probably would happen all over again if we turned back the clock. However, the final stages of accretion, and giant impacts in particular, are inherently random in nature and probably would not happen again in just the same way. A larger or smaller planet might form at Earth's location and might suffer from a larger giant impact or from no giant impact at all. However, we don't yet know whether these differences would fundamentally alter the solar system or simply change a few "minor" details—such as the possibility of life on Earth. What we do know is that, among the billions of other star systems in our galaxy, it's unlikely that any other is exactly like ours.

8.5 The Age of the Solar System

The nebular theory seems to give us a very good understanding of how our solar system was born. But *when* was it born, and how do we know? The answer is that the planets began to form through accretion about 4.6 billion years ago, a fact we learn by determining the age of the oldest rocks in the solar system.

• How does radioactivity reveal an object's age?

The first step in understanding how we learn the age of our solar system is to understand exactly what we mean by the age of a rock. A rock is a collection of a great many atoms held together in solid form. The atoms must be older than Earth, having been forged in the Big Bang or in stars that lived long ago. We cannot determine the ages of the individual atoms, because old atoms are indistinguishable from young ones. However, some atoms undergo changes with time that allow us to determine how long they have been held in place within the rock's solid structure. Thus, by the age of a rock we mean the time since its atoms became locked together in their present arrangement, which in most cases means the time *since the rock last solidified.*

The method by which we measure the age of a rock is known as **radiometric dating**. This method relies on careful measurement of the proportions of various atoms and isotopes in the rock. Remember that each chemical element is uniquely characterized by the number of protons in its nucleus. Different *isotopes* of the same element differ only in their number of neutrons [Section 5.3].

A **radioactive isotope** has a nucleus that can undergo spontaneous change, or **radioactive decay**, such as breaking apart or having one of its protons turn into a neutron. Decay can change one element into an entirely different one, with different chemical properties. For example, potassium-40 is a radioactive isotope with nuclei that decay when a proton turns into a neutron, changing the potassium-40 into argon-40. We say that potassium-40 is the *parent isotope*, because it is the original isotope before the decay, and argon-40 is the *daughter isotope* left behind by the decay process.

For any single nucleus, radioactive decay is an instantaneous event. However, in a large collection of atoms, individual nuclei will decay at different times. Thus, if we start with a collection of atoms of a particular parent isotope, we will find that it gradually transforms itself into a collection of atoms of the daughter isotope. The rate at which this transformation occurs is characterized by the parent isotope's **half-life**, the time it would take for half of the parent nuclei in the collection to decay. For example, laboratory measurements show the half-life of potassium-40 to be 1.25 billion years, which means that if we started with any particular amount of potassium-40, it would take 1.25 billion years for half of it to decay into argon-40. Every radioactive isotope has its own unique half-life, which may be anywhere from a fraction of a second to many billions of years. (We can measure half-lives even when they are billions of years because we only need to observe the decay of a tiny fraction of the nuclei to calculate how long it would take half of them to decay.)

Because we can measure decay rates so precisely, in principle it is easy to determine the age of a rock. Consider a small piece of rock that contained 1 microgram of potassium-40 and no argon-40 when it formed (solidified) long ago. Because potassium-40 has a half-life of 1.25 billion years, half the original potassium-40 would have decayed into argon-40 by the time the rock was 1.25 billion years old. Thus, after 1.25 billion years, the rock contained $\frac{1}{2}$ microgram of potassium-40 and $\frac{1}{2}$ microgram of argon-40. Half again would have decayed by the end of the next 1.25 billion years, so after 2.5 billion years the rock contained $\frac{1}{4}$ microgram of potassium-40 and $\frac{3}{4}$ microgram of argon-40. After three half-lives, or 3.75 billion years, only $\frac{1}{8}$ microgram of potassium-40 remained, while $\frac{7}{8}$ microgram had become argon-40. Figure 8.14 summarizes the gradual decrease in the amount of potassium-40 and the corresponding rise in the amount of argon-40.

We can now see the essence of radiometric dating. Suppose you find a rock that contains equal numbers of atoms of potassium-40 and argon-40. If you *assume* that all the argon came from potassium decay, then it must have taken precisely one half-life for the rock to end up with equal amounts of the two isotopes. You could therefore conclude that the rock is 1.25 billion years old (see Figure 8.14). The only question is whether you are right in assuming that the rock lacked argon-40 when it formed. In this case, knowing a bit of "rock chemistry" helps. Potassium-40 is a natural ingredient of many minerals in rocks, but argon-40 is a gas that never combines with other elements and did not condense in the solar nebula. Therefore, if you find argon-40 gas trapped inside minerals, you can be sure that it came from radioactive decay of potassium-40. (Another assumption is that no argon-40 gas has escaped from the rock. Rock chemistry tells us that as long as the rock has not been sig-

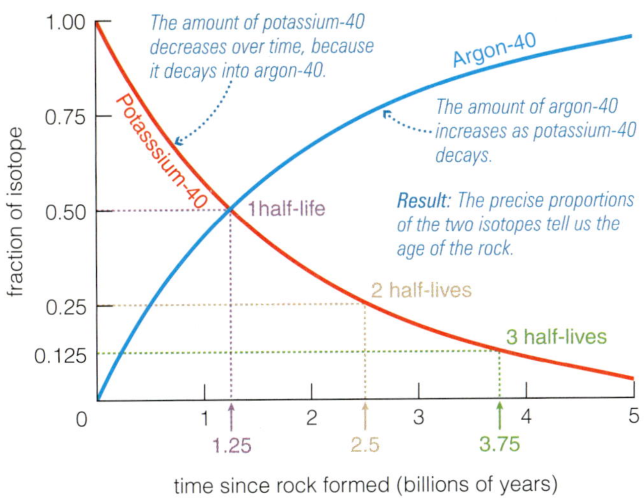

Figure 8.14 Potassium-40 is radioactive, decaying into argon-40 with a half-life of 1.25 billion years. The red line shows the decreasing amount of potassium-40, and the blue line shows the increasing amount of argon-40. The remaining amount of potassium-40 drops by half with each successive half-life.

nificantly heated since the time it formed, atoms of gases such as argon-40 will remain trapped within it.)

Rocks from the Moon offer another example of radiometric dating. Rocks that the *Apollo* astronauts brought back from the ancient lunar highlands [Section 9.3] contain minerals with a very small amount of uranium-238, which decays (in several steps) into lead-206 with a half-life of about 4.5 billion years. Lead and uranium have very different chemical behaviors, and some minerals start with virtually no lead. Laboratory analysis of such minerals in lunar rocks shows that they now contain almost equal proportions of atoms of uranium-238 and lead-206. We conclude that half the original uranium-238 has decayed, turning into the same number of lead-206 atoms. The lunar rock therefore must be about one half-life old, or about 4.5 billion years old. More precise work shows these lunar rocks to be about 4.4 billion years old.

THINK ABOUT IT

If future scientists examine the same lunar rocks 4.5 billion years from now, what proportions of uranium-238 and lead-206 will they find? Explain.

Radiometric dating is possible with many other radioactive isotopes as well. In many cases, we can date a rock that contains more than one radioactive isotope, so agreement between the ages calculated from the different isotopes gives us confidence that we have dated the rock correctly. Moreover, we can often check the results we obtain from radio-

MATHEMATICAL INSIGHT 8.1 Radiometric Dating

The amount of a radioactive substance decays by half with each half-life, so we can express the decay process with a simple formula relating the current amount of a radioactive substance in a rock to the original amount:

$$\frac{\text{current amount}}{\text{original amount}} = \left(\frac{1}{2}\right)^{t/t_{\text{half}}}$$

where t is the time since the rock formed and t_{half} is the half-life of the radioactive material. We can solve this equation for the age t by taking the logarithm of both sides and rearranging the terms. The resulting general equation for the age is:

$$t = t_{\text{half}} \times \frac{\log_{10}\left(\frac{\text{current amount}}{\text{original amount}}\right)}{\log_{10}\left(\frac{1}{2}\right)}$$

Example: You heat and chemically analyze a small sample of a meteorite. Potassium-40 and argon-40 are present in a ratio of approximately 0.85 unit of potassium-40 atoms to 9.15 units of gaseous argon-40 atoms. (The units are unimportant, because only the relative amounts of the parent and daughter materials matter.) How old is the meteorite?

Solution:

Step 1 Understand: The formula given above allows us to find the age of a rock from the current and original amounts of a parent isotope. In this case, we are told that the current amount of the parent isotope potassium-40 is 0.85 unit. We are not told the original amount, but we can figure it out from the fact that the sample now has 9.15 units of the daughter isotope argon-40. As discussed in the text, the fact that argon-40 is gaseous makes it reasonable to assume that *all* the trapped argon-40 in the meteorite is a decay product of potassium-40. Thus, the 9.15 units of argon must originally have been potassium-40, which means the original amount of potassium-40 was 0.85 + 9.15 = 10 units.

Step 2 Solve: We use the formula given above, with the current amount = 0.85 unit and the original amount = 10 units:

$$\begin{aligned} t &= 1.25 \text{ billion yr} \times \frac{\log_{10}\left(\frac{0.85}{10}\right)}{\log_{10}\left(\frac{1}{2}\right)} \\ &= 1.25 \text{ billion yr} \times \left(\frac{-1.07}{-0.301}\right) \\ &= 4.45 \text{ billion yr} \end{aligned}$$

Step 3 Explain: We have found an age of 4.45 billion years. In other words, radiometric dating of this meteorite tells us that it solidified about 4.45 billion years ago. (*Note:* This example captures the essence of radiometric dating, but real cases generally require more detailed analysis.)

SPECIAL TOPIC What Started the Collapse of the Solar Nebula?

Recall that the solar nebula probably could not have started collapsing on its own, because gravity was too weak when the gas was still spread out over a large region of space. We therefore suspect that some cataclysmic event triggered the collapse. But what type of event? Although we may never know for sure, radioactive elements in meteorites give us some clues. Radioactive elements are made only deep inside stars or in violent stellar explosions (supernovae). Thus, the presence of these elements in meteorites and on Earth underscores the fact that our solar system is made from the remnants of past generations of stars.

We have learned even more from the discovery that some meteorites contain the rare isotope xenon-129. Xenon is gaseous even at extremely low temperatures, so it could not have condensed and become trapped in the meteorites as they condensed in the solar nebula. Instead, the xenon in meteorites must be a product of radioactive decay; some other isotope must have condensed in the solar nebula, then decayed into the gaseous xenon after it was already trapped in the meteorite. Laboratory studies show that xenon-129 is a decay product of iodine-129, so we conclude that it was iodine-129 rather than xenon-129 that was present in the solar nebula. Moreover, iodine-129 has a relatively short half-life of 17 million years. Thus, no more than a few tens of millions of years could have passed between the time the iodine was produced in a stellar explosion and the time that it condensed in the solar nebula. Otherwise, too much of the iodine-129 would have decayed for significant amounts to have condensed and later decayed into xenon.

Summing up, we conclude that iodine (and presumably other materials) produced by a nearby stellar explosion ended up as part of the gas in the solar nebula, and that this explosion occurred no more than a few tens of millions of years before the collapse of the solar nebula began. This timing suggests that the shock wave from the exploding star may have been the trigger for the collapse. Once this shock wave got the collapse started, gravity took over and made the rest of the collapse inevitable.

metric dating against other methods of measuring ages, giving us even more confidence that the technique works. For example, with many fairly recent archaeological artifacts, we can confirm the ages obtained from radiometric dating by measuring ages from tree ring data. Similarly, astronomers have discovered completely independent methods for estimating the Sun's age. Although none of these methods give an age as precise as the age we obtain from radiometric dating, they all confirm that the Sun's age is in the range of 4 to 5 billion years. Overall, the technique of radiometric dating has been checked in so many ways and relies on such basic scientific principles that there is no longer any serious scientific debate about its validity.

• When did the planets form?

Radiometric dating tells us how long it has been since a rock solidified, which is not the same as the age of a planet as a whole. For example, we find rocks of many different ages on Earth. Some rocks are quite young because they formed recently from molten lava; others are much older. The oldest Earth rocks are about 4 billion years old, but even these are not as old as Earth itself, because Earth's entire surface has been reshaped through time.

Moon rocks brought back by the *Apollo* astronauts date to as far back as 4.4 billion years ago. Although they are older than Earth rocks, these Moon rocks must still be younger than the Moon itself. The ages of these rocks also tell us that the giant impact thought to have created the Moon must have occurred more than 4.4 billion years ago.

To go all the way back to the origin of the solar system, we must find rocks that have not melted or vaporized since they first condensed in the solar nebula. Meteorites that have fallen to Earth are our source of such rocks. Many meteorites appear to have remained unchanged since they condensed and accreted in the early solar system. Careful analysis of radioactive isotopes in meteorites shows that the oldest ones formed about 4.55 billion years ago, so this time must mark the beginning of accretion in the solar nebula. Because the planets apparently accreted within a few tens of millions of years, Earth and the other planets formed about 4.5 billion years ago. Thus, the age of our solar system is only about a third of the 14-billion-year age of our universe.

THE BIG PICTURE

Putting Chapter 8 into Perspective

In this chapter, we've described the current scientific theory of our solar system's formation and we've seen how this theory explains the major features we observe. As you continue your study of the solar system, keep in mind the following "big picture" ideas:

- The nebular theory of solar system formation gained wide acceptance because of its success in explaining the major characteristics of our solar system.
- Most of the general features of the solar system were determined by processes that occurred very early in the solar system's history.
- Chance events may have played a large role in determining how individual planets turned out. No one knows how different our solar system might be if it started over.
- We have learned the age of our solar system—about 4.6 billion years—from radiometric dating of the oldest meteorites. This age agrees with the ages we can estimate through a variety of other techniques, making it quite clear that we ourselves are recent arrivals on a very old planet.

SUMMARY OF KEY CONCEPTS

8.1 The Search for Origins

- **What properties of our solar system must a formation theory explain?** A successful theory must explain patterns of motion, the differences between terrestrial and jovian planets, the presence of asteroids and comets, and exceptions to the rules.

- **What theory best explains the features of our solar system?** The **nebular theory**, which holds that the solar system formed from the gravitational collapse of a great cloud of gas and dust, successfully explains all the major features of our solar system.

8.2 The Birth of the Solar System

- **Where did the solar system come from?** The cloud that gave birth to our solar system was the product of recycling of gas through many generations of stars within our galaxy. This material consisted of 98% hydrogen and helium and 2% everything else combined.

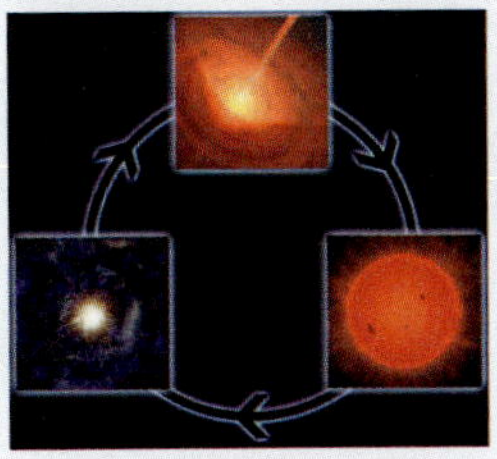

- **What caused the orderly patterns of motion in our solar system?** A collapsing gas cloud naturally tends to heat up, spin faster, and flatten out as it shrinks in size. Thus, our solar system began as a spinning disk of gas and dust. The orderly motions we observe today all came from the orderly motion of this spinning disk.

8.3 The Formation of Planets

- **Why are there two major types of planets?** The two types of planets formed from two different types of solid "seeds" that condensed from gas in the solar nebula. Within the **frost line**, temperatures were so high that only metal and rock could condense; beyond the frost line, cooler temperatures also allowed more abundant **hydrogen compounds** to condense into ice.

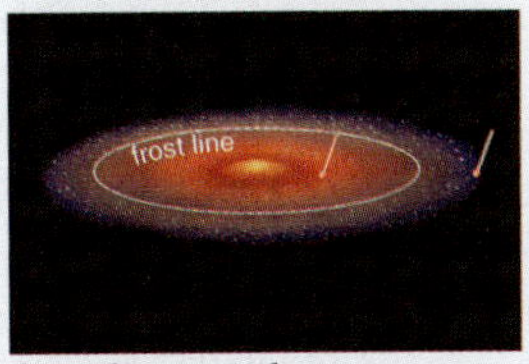

- **How did the terrestrial planets form?** Terrestrial planets formed inside the frost line, where **accretion** allowed tiny, solid grains of metal and rock to grow into **planetesimals** that ultimately merged to make the planets we see today.

- **How did the jovian planets form?** Accretion built ice-rich planetesimals in the outer solar system, and some of these icy planetesimals grew large enough for their gravity to draw in hydrogen and helium gas, building the jovian planets.

- **What ended the era of planet formation?** The young Sun had a strong solar wind, which ultimately swept away material not yet accreted onto the planets. Once the solar nebula was cleared away, planet formation ceased because there was little material left to accrete.

8.4 The Aftermath of Planet Formation

- **Where did asteroids and comets come from?** Asteroids are the rocky leftover planetesimals of the inner solar system, and comets are the ice-rich leftover planetesimals of the outer solar system.

- **How do we explain "exceptions to the rules"?** Most of the exceptions probably arose from collisions or close encounters with leftover planetesimals, especially during the **heavy bombardment** that occurred early in the solar system's history.

- **How do we explain the existence of our Moon?** Our Moon is probably the result of a **giant impact** between a Mars-size planetesimal and the young Earth. The impact blasted material from Earth's outer layers into orbit, where it reaccreted to form the Moon.

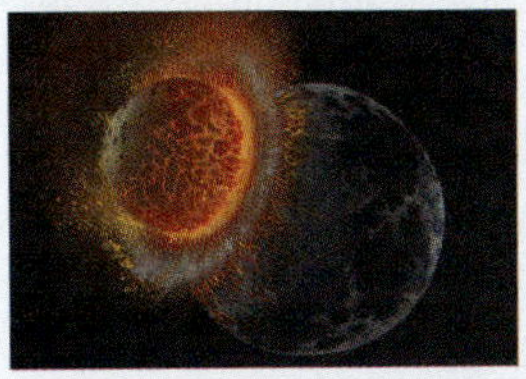

- **Was our solar system destined to be?** A solar system generally like ours was probably destined to form from the collapse of the solar nebula, but the nature of individual planets in our solar system results from significant random events that could have happened differently or not at all.

8.5 The Age of the Solar System

- **How does radioactivity reveal an object's age?** **Radiometric dating** is based on carefully measuring the proportions of radioactive isotopes and their decay products within rocks. The ratio of the two changes with time and provides a reliable measure of the rock's age.

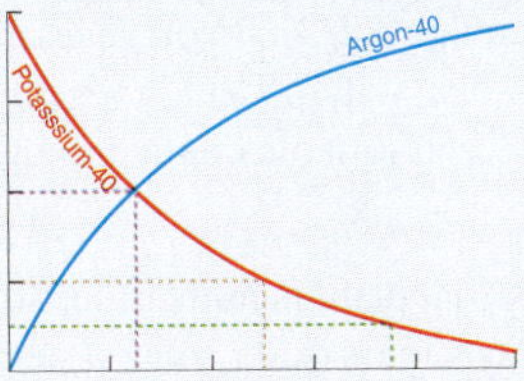

- **When did the planets form?** The planets began to accrete in the solar nebula about 4.6 billion years ago, a fact we determine from radiometric dating of the oldest meteorites.

EXERCISES AND PROBLEMS

For instructor-assigned homework go to MasteringASTRONOMY www.masteringastronomy.com

Review Questions

Short-Answer Questions Based on the Reading

1. What are the four major features of our solar system that provide clues to how it formed? Describe each one briefly.
2. What is the *nebular theory,* and why is it widely accepted by scientists today?
3. What do we mean by the *solar nebula?* What was it made of, and where did it come from?
4. Describe each of the three key processes that led the solar nebula to take the form of a spinning disk. What observational evidence supports this scenario?
5. List the four categories of materials in the solar nebula by their condensation properties and abundance. Which ingredients are present in terrestrial planets? In jovian planets? In comets and asteroids? Explain why.
6. What was the *frost line* in the solar nebula? Explain how temperature differences led to the formation of two distinct types of planets.
7. Briefly describe the process by which terrestrial planets are thought to have formed.
8. How was the formation of jovian planets similar to that of the terrestrial planets? How was it different? Why did the jovian planets end up with so many moons?
9. What is the *solar wind,* and what roles did it play in the early solar system?
10. In the context of planet formation, what are asteroids and comets? Briefly explain why we find asteroids in the asteroid belt and comets in the Kuiper belt and Oort cloud.
11. What was the *heavy bombardment,* and when did it occur?
12. How do we think the Moon formed, and what evidence supports this hypothesis? Should we be surprised that a *giant impact* could have affected our planet?
13. Describe the technique of *radiometric dating.* What is a half-life?
14. How old is the solar system, and how do we know?

Test Your Understanding

Surprising Discoveries?

Suppose we found a solar system with the property described. (These are *not* real discoveries.) In light of what you've learned about the formation of our own solar system, decide whether the discovery should be considered reasonable or surprising. Explain.

15. A solar system has five terrestrial planets in its inner solar system and three jovian planets in its outer solar system.
16. A solar system has four large jovian planets in its inner solar system and seven small planets made of rock and metal in its outer solar system.
17. A solar system has 10 planets that all orbit the star in approximately the same plane. However, five planets orbit in one direction (e.g., counterclockwise), while the other five orbit in the opposite direction (e.g., clockwise).
18. A solar system has 12 planets that all orbit the star in the same direction and in nearly the same plane. The 15 largest moons in this solar system orbit their planets in nearly the same direction and plane as well. However, several smaller moons have highly inclined orbits around their planets.
19. A solar system has six terrestrial planets and four jovian planets. Each of the six terrestrial planets has at least five moons, while the jovian planets have no moons at all.
20. A solar system has four Earth-size terrestrial planets. Each of the four planets has a single moon that is nearly identical in size to Earth's Moon.
21. A solar system has many rocky asteroids and many icy comets. However, most of the comets orbit in the inner solar system, while the asteroids orbit in far-flung regions much like the Kuiper belt and Oort cloud of our solar system.
22. A solar system has several planets similar in composition to the jovian planets of our solar system but similar in mass to the terrestrial planets of our solar system.
23. A solar system has several terrestrial planets and several larger planets made mostly of ice. (*Hint:* What would happen if the solar wind started earlier or later than in our solar system?)
24. Radiometric dating of meteorites from another solar system shows that they are a billion years younger than rocks from the terrestrial planets of the same system.

Quick Quiz

Choose the best answer to each of the following. Explain your reasoning with one or more complete sentences.

25. How many of the planets orbit the Sun in the same direction that Earth does? (a) a few (b) most (c) all
26. What fraction of the moons of the planets orbit in the same direction that their planets rotate? (a) some (b) most (c) all
27. The solar nebula was 98% . . . ? (a) rock and metal. (b) hydrogen compounds. (c) hydrogen and helium.
28. Which of the following did *not* occur during the collapse of the solar nebula? (a) spinning faster (b) heating up (c) concentrating denser materials nearer the Sun
29. What is Jupiter's main ingredient? (a) rock and metal (b) hydrogen compounds (c) hydrogen and helium
30. Which lists the major steps of solar system formation in the correct order? (a) collapse, accretion, condensation (b) collapse, condensation, accretion (c) accretion, condensation, collapse
31. Leftover ice-rich planetesimals are called (a) comets. (b) asteroids. (c) neither of the above: there were no leftover ice-rich planetesimals.
32. Why didn't a terrestrial planet form at the location of the asteroid belt? (a) There was never enough material in that part of the solar nebula. (b) The solar wind cleared away nebular material there. (c) Jupiter's gravity kept planetesimals from accreting.
33. What's the leading theory for the origin of the Moon? (a) It formed in place along with the Earth. (b) It formed from the material ejected in a giant impact. (c) It split out of a rapidly rotating Earth.
34. About how old is the solar system? (a) 4.5 million years (b) 4.5 billion years (c) 4.5 trillion years

Investigate Further

In-Depth Questions to Increase Your Understanding

Short-Answer/Essay Questions

35. *Patterns of Motion.* In one or two paragraphs, explain why the existence of orderly patterns of motion in our solar system should suggest that the Sun and the planets all formed

at one time from one cloud of gas, rather than as individual objects at different times.

36. *A Cold Solar Nebula.* Suppose the entire solar nebula had cooled to 50 K before the solar wind cleared it away. How would the composition and sizes of the planets of the inner solar system be different from what we see today? Explain your answer in a few sentences.
37. *An Early Solar Wind.* Suppose the solar wind had cleared away the solar nebula before the seeds of the jovian planets could gravitationally draw in hydrogen and helium gas. How would the planets of the outer solar system be different? Would they still have many moons? Explain your answer in a few sentences.
38. *Angular Momentum.* Suppose our solar nebula had begun with much more angular momentum than it did. Do you think planets could still have formed? Why or why not? What if the solar nebula had started with zero angular momentum? Explain your answers in one or two paragraphs.
39. *Two Kinds of Planets.* The jovian planets differ from the terrestrial planets in a variety of ways. Using phrases or sentences that members of your family would understand, explain why the jovian planets differ from the terrestrial planets in each of the following: composition, size, density, distance from the Sun, and number of satellites.
40. *Two Kinds of Planetesimals.* Why are there two kinds of "leftovers" from solar system formation? List as many differences between comets and asteroids as you can.
41. *History of the Elements.* Our bodies (and most living things) are made mostly of water: H_2O. Summarize the "history" of a typical hydrogen atom from its creation to the formation of the Earth. Do the same for a typical oxygen atom. (*Hint:* Which elements were created in the Big Bang, and where were the others created?)
42. *Our Messy Solar System.* How would Earth and other planets be different if all the "wandering planetesimals" had stayed in their location of the solar system instead of colliding or passing near other planets? Describe at least three major ways in which our solar system would differ.
43. *Pluto.* How does the nebular theory explain the origin of objects like Pluto? How was its formation similar to jovian and terrestrial planets, and how was it different?
44. *Understanding Radiometric Dating.* Imagine you had the good fortune to find a rocky meteorite in your backyard. Qualitatively, how would you expect its ratio of potassium-40 and argon-40 to be different from other rocks in your yard? Explain why, in a few sentences.

Quantitative Problems

Be sure to show all calculations clearly and state your final answers in complete sentences.

45. *Radiometric Dating.* You are analyzing rocks that contain small amounts of potassium-40 and argon-40. The half-life of potassium-40 is 1.25 billion years.
    a. You find a rock that contains equal amounts of potassium-40 and argon-40. How old is it? Explain.
    b. You find a rock that contains three times as much argon-40 as potassium-40. How old is it? Explain.
46. *Dating Lunar Rocks.* You are analyzing Moon rocks that contain small amounts of uranium-238, which decays into lead with a half-life of about 4.5 billion years.
    a. In a rock from the lunar highlands, you determine that 55% of the original uranium-238 remains, while the other 45% has decayed into lead. How old is the rock?
    b. In a rock from the lunar maria, you find that 63% of the original uranium-238 remains, while the other 37% has decayed into lead. Is this rock older or younger than the highlands rock? By how much?
47. *Carbon-14 Dating.* The half-life of carbon-14 is about 5,700 years.
    a. You find a piece of cloth painted with organic dye. By analyzing the dye, you find that only 77% of the carbon-14 originally in the dye remains. When was the cloth painted?
    b. A well-preserved piece of wood found at an archaeological site has 6.2% of the carbon-14 it must have had when it was living. Estimate when the wood was cut.
    c. Is carbon-14 useful for establishing Earth's age? Why or why not?
48. *Martian Meteorite.* Some unusual meteorites thought to be chips from Mars contain small amounts of radioactive thorium-232 and its decay product, lead-208. The half-life for this decay process is 14 billion years. Analysis of one such meteorite shows that 94% of the original thorium remains. How old is this meteorite?
49. *Collapsing Cloud.* The time it takes for a cloud 100,000 AU in diameter to collapse to form a new star turns out to be about *half* the time it would take an object to orbit the star on an elliptical orbit with a semimajor axis of 50,000 AU (half the 100,000 AU diameter). Use Kepler's third law to find the collapse time, assuming the star has the same mass as the Sun.
50. *Monster Iceballs.* The ice-rich planetesimals that formed the cores of the jovian planets were about 10 times more massive than Earth. Assuming such an iceball has a density of about 2 g/cm^3, what would its radius be? How does this compare to Earth's radius?
51. *Icy Earth.* How massive would Earth have been if it had accreted hydrogen compounds in addition to rock and metal? Assume the same proportion of the ingredients as listed in Table 8.1. What would Earth's mass be, if it had been possible to capture hydrogen and helium gas in the proportion listed in the table?
52. *Big Moons.* Use the data in Appendix E to compare Earth's Moon, Jupiter's moon Ganymede, Saturn's moon Titan, and Pluto's moon Charon. Rank them in order of their actual size (radius, in kilometers). Then rank them in order of the ratio of their size to the size of the planet they orbit. What can you conclude from the differences in the two rankings?
53. *What Are the Odds?* The fact that all the planets orbit the Sun in the same direction is cited as support for the nebular hypothesis. Imagine that there's a different hypothesis in which planets can be created orbiting the Sun in either direction. Under this hypothesis, what is the probability that nine planets would end up traveling the same direction? (*Hint:* It's the same probability as that of flipping a coin nine times and getting nine heads or nine tails.)
54. *Spinning Up the Solar Nebula.* The orbital speed of the solar nebula at Pluto's average distance from the Sun is about 5 km/sec. When the solar nebula was about 40,000 AU in radius, how fast would the same material have orbited? (Remember that when the cloud was larger, motions were more random, so we can calculate only the average.) Use the law of conservation of angular momentum (see Section 4.3).

Discussion Questions

55. *Theory and Observation.* Discuss the interplay between theory and observation that has led to our modern theory of solar system formation. What role does technology play in allowing us to test this theory?

56. *Lucky to Be Here?* Considering the overall process of solar system formation, do you think it was likely for a planet like Earth to have formed? Could random events in the early history of the solar system have prevented our being here today? What implications do your answers have for the possibility of Earth-like planets around other stars? Defend your opinions.
57. *Rocks From Other Solar Systems.* The chapter states that some of the "leftovers" from planetary formation were likely ejected from our solar system, and the same has presumably happened in other star systems. Given that fact, should we expect to find rocks in our solar system that actually formed in other star systems? How rare or common would you expect such rocks to be? As you consider your answer, be sure to consider the distance between stars. Suppose that we *did* find a meteorite identified as a leftover from another stellar system. What do you think we would learn from it?

MEDIA EXPLORATIONS

For self-study activities go to MasteringASTRONOMY www.masteringastronomy.com

Interactive Tutorials

Tutorial Review of Key Concepts

Use the interactive **Tutorial** at **www.masteringastronomy.com** to review key concepts from this chapter.

Formation of the Solar System Tutorial

Lesson 1 Comparative Planetology
Lesson 2 Formation of the Protoplanetary Disk
Lesson 3 Formation of Planets

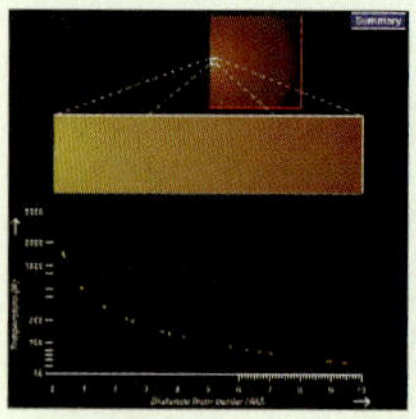
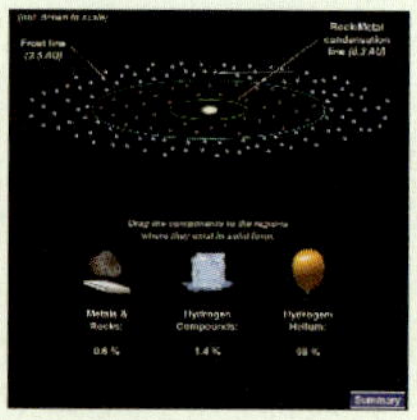
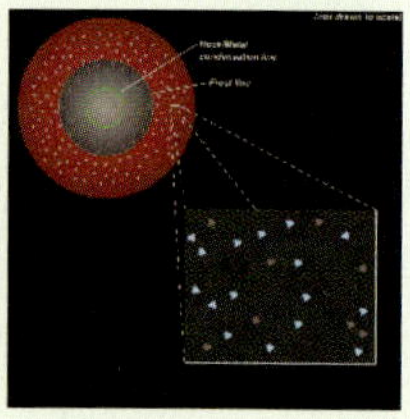

Supplementary Tutorial Exercises

Use the interactive **Tutorial Lessons** to explore the following questions.

Formation of the Solar System Tutorial, Lesson 2

1. Study the animations of the formation of the solar nebula. Why does the nebula flatten as it collapses?
2. Use the tool in the tutorial to make a graph of temperatures in the solar nebula. How does the temperature vary with the distance from the center?
3. Why didn't any planets form within 0.3 AU of the Sun?

Formation of the Solar System Tutorial, Lesson 3

1. What is the composition of planetesimals that form outside the frost line?
2. Why didn't the terrestrial planets capture hydrogen and helium gas?
3. What is Pluto made of, and why?

Exploring the Sky and Solar System

Of the many activities available on the ***Voyager: SkyGazer*** **CD-ROM** accompanying your book, use the following files to observe key phenomena covered in this chapter.

Go to the **File: Basics** folder for the following demonstrations.

1. Saturn's Phases
2. Tracking Venus
3. Planet Panel

Go to the **File: Demo** folder for the following demonstrations.

1. Earth and Venus
2. Trailing Saturn
3. Triple Conjunction of 7 BC

Go to the **Explore** menu for the following demonstrations.

1. Solar System
2. Paths of Planets

Movies

Check out the following narrated and animated short documentaries available on **www.masteringastronomy.com** for a helpful review of key ideas covered in this chapter.

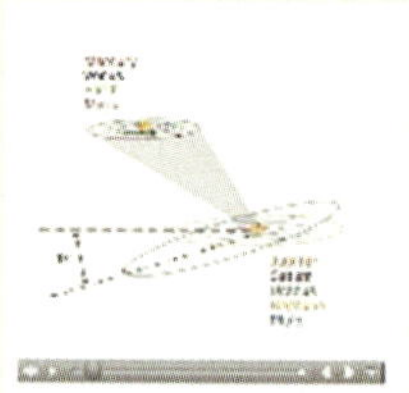

Orbits in the Solar System

History of the Solar System

Web Projects

1. *Spitzer Space Telescope.* NASA's latest "Great Observatory" operates at the infrared wavelengths especially useful for studying star and planet formation. Visit the Spitzer Web site to see if recent discoveries are confirming the nebular theory for planet formation or are requiring us to broaden our understanding of the process. Summarize your findings in a one- to two-page report.
2. *Other Solar Nebulas.* New technology enables better and better imaging of possible planet-forming regions around other stars. Search the Internet for recent images from the Hubble Space Telescope and major ground-based observatories listed in Table 6.1. Write a one- to two-page report, including pictures, describing how the new images support or expand our ideas about our own solar nebula.

9

Planetary Geology

Earth and the Other Terrestrial Worlds

LEARNING GOALS

9.1 Connecting Planetary Interiors and Surfaces
- What are terrestrial planets like on the inside?
- What causes geological activity?
- Why do some planetary interiors create magnetic fields?

9.2 Shaping Planetary Surfaces
- What processes shape planetary surfaces?
- Why do the terrestrial planets have different geological histories?
- How does a planet's surface reveal its geological age?

9.3 Geology of the Moon and Mercury
- What geological processes shaped our Moon?
- What geological processes shaped Mercury?

9.4 Geology of Mars
- How did Martians invade popular culture?
- What are the major geological features of Mars?
- What geological evidence tells us that water once flowed on Mars?

9.5 Geology of Venus
- What are the major geological features of Venus?
- Does Venus have plate tectonics?

9.6 The Unique Geology of Earth
- How do we know that Earth's surface is in motion?
- How is Earth's surface shaped by plate tectonics?
- Was Earth's geology destined from birth?

Nothing is rich but the inexhaustible wealth of nature. She shows us only surfaces, but she is a million fathoms deep.

Ralph Waldo Emerson

It's easy to take for granted the qualities that make Earth so suitable for human life: a temperature neither boiling nor freezing, abundant water, a protective atmosphere, and a relatively stable environment. But we need look only as far as our neighboring terrestrial worlds—the three other inner planets and the Moon—to see how fortunate we are. As we learned in Chapter 7, the Moon, likely made of debris from Earth, is airless and barren. Mercury is much the same. Venus is a searing hothouse, while Mars has an atmosphere so thin and cold that liquid water cannot last on its surface today.

How did the terrestrial worlds come to be so different, when all were made similarly from metal and rock that had condensed in the solar nebula? Why did Earth alone develop the conditions that permit abundant life? We can find the answers to these questions only with careful, comparative study of the planets. We'll begin the quest in this chapter by focusing on the geology of the terrestrial worlds, and then explore their atmospheres in Chapter 10. In each case, we'll start by exploring the general processes that shape the planets and then examine how these processes have played out on the individual worlds. By the end, you'll see that the histories of the terrestrial worlds were largely determined by the properties with which they were endowed at birth.

Formation of the Solar System Tutorial, Lesson 1

9.1 Connecting Planetary Interiors and Surfaces

Earth's surface seems solid and steady, but every so often it offers us a reminder that nothing about it is permanent. If you live in Alaska or California, you've probably felt the ground shift beneath you in an earthquake. In Washington State, you may have seen the recent rumblings of Mount St. Helens. In Hawaii, a visit to the still-active Kilauea volcano will remind you that you are standing on mountains of volcanic rock protruding from the ocean floor.

Volcanoes and earthquakes are not the only processes acting to reshape Earth's surface. They are not even the most dramatic: Far greater change can occur on the rare occasions when an asteroid or a comet slams into Earth. Much more gradual processes can also have spectacular effects. The Colorado River causes only small changes in the landscape from year to year, but its unrelenting flow over the past few million years carved the Grand Canyon. The Rocky Mountains were once twice as tall as they are today and have been cut down in size through tens of millions of years

Mercury is heavily cratered, but also has long, steep cliffs—one is visible here as the long curve that passes through the center of the image.

The central structure is a tall, twin-peaked volcano on Venus. Both images are based on radar data from the Magellan space craft, because Venus's thick clouds prevent us from seeing the surface in visible light.

Earth has a variety of geological features visible in this photo from orbit.

of erosion by wind, rain, and ice. Entire continents even move slowly about, completely rearranging the map of Earth every few hundred million years.

Earth is not alone in having undergone tremendous change since its birth. The surfaces of all five terrestrial worlds—Mercury, Venus, Earth, the Moon, and Mars—must have looked quite similar when they were young. All five were made of rocky material that condensed in the solar nebula, and all five were subjected early on to the impacts of the heavy bombardment [Section 8.4]. The great differences in their present-day appearances must therefore be the result of changes that have occurred through time. Ultimately, these changes must be traceable to fundamental properties of the planets.

Figure 9.1 shows global views of the terrestrial surfaces to scale, along with sample close-up views from orbit. Profound differences between these worlds are immediately obvious. Mercury and the Moon show the scars of their battering during the heavy bombardment: They are densely covered by craters except in areas that appear to be volcanic plains. Bizarre bulges and odd volcanoes dot the surface of Venus. Mars, despite its middling size, has the solar system's largest volcanoes and a huge canyon cutting across its surface, along with numerous features that appear to have been shaped by running water. Earth has surface features similar to all those on the other terrestrial worlds, and more—including a unique layer of living organisms that covers almost the entire surface of the planet.

Our purpose in this chapter is to understand how these profound differences among the terrestrial planets came to be. We will be discussing what is often called **planetary geology**; because geology is the study of Earth (*geo* means "Earth"), planetary geology is the extension of this science to other worlds. Note that we use the term *planetary geology* for the study of any solid world, even if it is a moon rather than a planet. Although we can see geological features only on planetary surfaces, most of these features are shaped by processes that take place deep beneath the surface. Thus, the first step in understanding planetary geology is to understand what planets are like on the inside.

• What are terrestrial planets like on the inside?

We cannot see inside Earth or any other planet, but we still know quite a bit about what planetary interiors are like. Our most detailed information about Earth's interior comes from **seismic waves**, vibrations that travel both through Earth's interior and along its surface after an earthquake (*seismic* comes from the Greek word for "shake"). In much the same way that shaking a gift offers clues about what's inside, the vibrations that follow earthquakes can tell us about the inside of Earth. By combining our understanding of Earth's interior with a variety of other clues, we have developed a fairly clear picture of the interiors of each of the terrestrial worlds (see box, p. 250).

Layering by Density All the terrestrial worlds have layered interiors. You are probably familiar with the idea of dividing the interior into three basic layers by density:

- **Core**: The highest-density material, consisting primarily of metals such as nickel and iron, resides in the central core.
- **Mantle**: Rocky material of moderate density—mostly minerals that contain silicon, oxygen, and other elements—forms the thick mantle that surrounds the core.

Earth's Moon

The Moon's surface is heavily cratered in most places.

Mars

Mars has impact craters like the one near the upper right, but it also has features that look much like dried up riverbeds.

Figure 9.1 Global views to scale, along with sample close-ups viewed from orbit, of each of the five terrestrial worlds. All the photos were taken with visible light, except the Venus photos, which are based on radar data.

- **Crust**: The lowest-density rock, such as granite and basalt (a common form of volcanic rock), forms the thin crust, essentially representing the world's outer skin.

Figure 9.2 shows these layers for the five terrestrial worlds. Although it is not shown in the figure, seismic studies indicate that Earth's metallic core consists of two distinct regions: a solid *inner core* and a molten (liquid) *outer core.* Because temperature increases with depth in a planet, it may seem surprising that Earth's inner core is solid while the outer core is molten. The explanation is that the inner core is kept solid by the higher pressure at this depth, even though the temperature is also higher. Venus may have a similar core structure, but without seismic data we cannot be sure.

The distinct layering of the terrestrial worlds occurs for the same reason that liquids separate by density. For example, in a mixture of oil and water, gravity pulls the denser water to the bottom, driving the less dense oil to the top. The process by which gravity separates materials by density is called **differentiation** (because it results in layers made of *different* materials). The layered interiors of the terrestrial worlds tell us that all the worlds underwent differentiation at some time in the past, which means all these worlds

SPECIAL TOPIC How Do We Know What's Inside a Planet?

Our deepest drills have barely pricked Earth's surface, penetrating much less than 1% of the way into the interior. On other worlds, we've done nothing more than examine surface rocks. How, then, can we claim to know what Earth or any other planet is like on the inside?

On Earth, much of our information about the interior comes from *seismic waves,* vibrations created by earthquakes. Seismic waves come in two basic types that are analogous to the two ways that you can generate waves in a Slinky (Figure 1). Pushing and pulling on one end of a Slinky (while someone holds the other end still) generates a wave in which the Slinky is bunched up in some places and stretched out in others. Waves like this in rock are called P waves. The *P* stands for *primary,* because these waves travel fastest and are the first to arrive after an earthquake, but it is easier to think of it as meaning *pressure* or *pushing.* P waves can travel through almost any material—whether solid, liquid, or gas—because molecules can always push on their neighbors no matter how weakly they are bound together. (Sound travels as a pressure wave and thus is quite similar to P waves.)

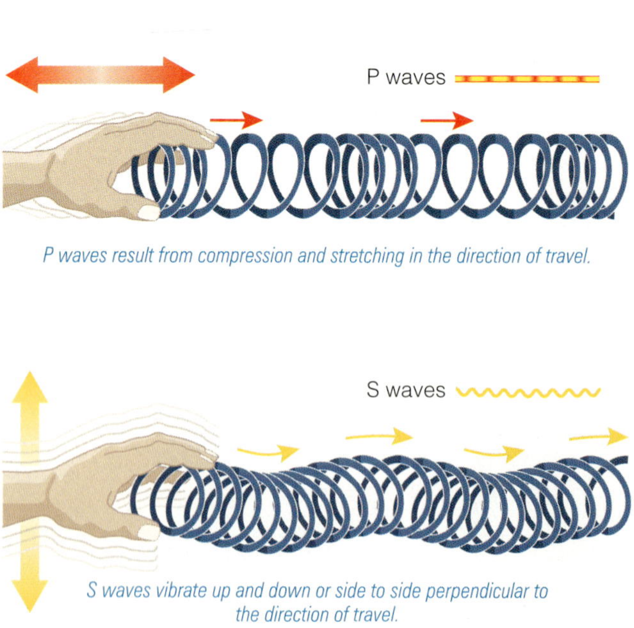

Figure 1 Slinky examples demonstrating P and S waves.

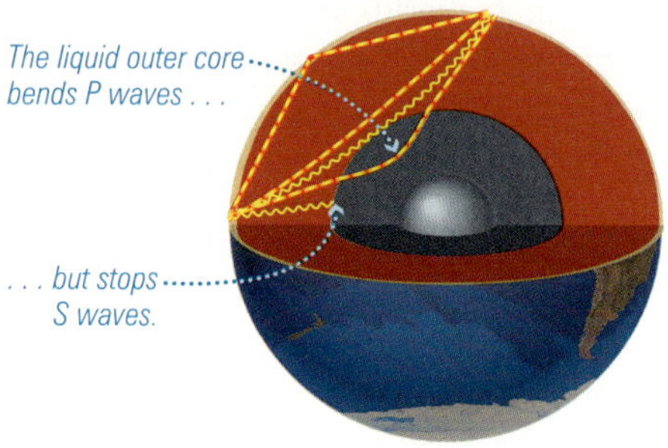

Figure 2 Because S waves do not reach the side of Earth opposite an earthquake, we infer that part of Earth's core is liquid.

Shaking a Slinky slightly up and down generates an up-and-down motion all along its length. Such up-and-down (or side-to-side) waves in rock are called S waves. The *S* stands for *secondary* but is easier to remember as meaning *shear* or *side-to-side.* S waves travel only through solids, because the bonds between neighboring molecules in a liquid or gas are too weak to transmit up-and-down or sideways forces.

The precise speeds and directions in which seismic waves travel through Earth depend on the composition, density, pressure, temperature, and phase (solid or liquid) of the material they pass through. For example, P waves reach the side of the world opposite an earthquake, but S waves do not. This tells us that a liquid layer has stopped the S waves, which is how we know that Earth has a liquid layer in its outer core (Figure 2). More careful analysis of seismic waves has allowed geologists to develop a detailed picture of what Earth looks like on the inside.

We have also used seismic waves to study the Moon's interior, thanks to monitoring stations left on the Moon by the *Apollo* astronauts. To learn about the interiors of other worlds, we use less direct clues. One key piece of information is a planet's average density, which we can calculate simply by dividing its mass by its volume. We can find a planet's mass by applying Newton's version of Kepler's third law to the orbital properties of its moons or of spacecraft that have visited it [Section 4.4] and its size through careful study of telescopic or spacecraft images. If we find that the density of the planet's surface rock is much less than its overall average density, then we know it must contain much more dense rock or metal inside. We can also learn about a planet's interior from precise measurements of its gravity, which tell us how mass is distributed within the planet; from studies of its magnetic field, which is generated deep inside the planet; and from observations of surface rocks that appear to have emerged from deep within the interior.

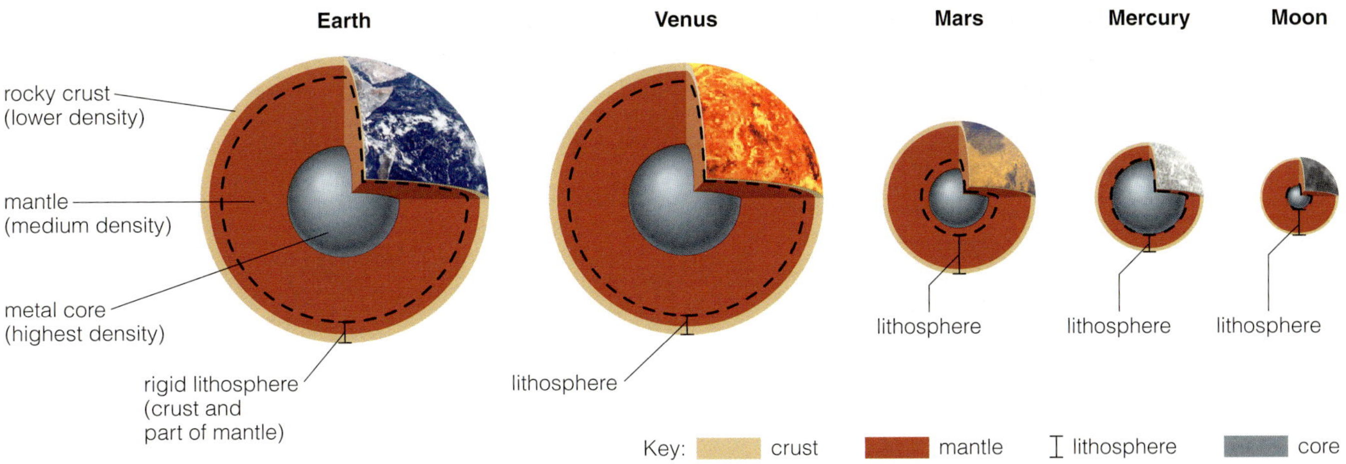

Figure 9.2 Interior structures of the terrestrial worlds, shown to scale and in order of decreasing size. The color-coded regions show the core-mantle-crust layering, which is based on density. Dashed lines indicate the lithosphere, which encompasses the crust and part of the mantle on each world and is defined by the strength of the rock rather than the density. (The thicknesses of the crust and lithosphere on Venus and Earth are exaggerated to make them visible in this figure.)

must once have been hot enough inside for their interior rock and metal to melt and separate by density. Dense metals like iron sank toward the center, driving less dense rocky material toward the surface.

Layering by Strength: The Lithosphere Although the activity of volcanoes may give the impression that Earth is molten inside, in most cases the lava actually rises upward from a fairly narrow layer of rock in the upper mantle that is only partially molten. The vast majority of Earth's interior is solid rock, and only the outer core is fully molten. However, the interior rock is not equally strong throughout, and differences in rock strength are often more important than differences in density for understanding many geological processes.

The idea that rock can vary in strength may sound strange, since we often think of rock as the very definition of strength. However, remember that rocks feel solid only because of the electrical bonds between their molecules [Section 5.3]. Although these bonds are strong, they can still break and re-form when subjected to sustained stress. Thus, over millions and billions of years, rocky material can slowly deform and flow. The long-term behavior of rock is much like that of the popular toy Silly Putty, which breaks like a brittle solid when you pull it sharply but deforms and stretches when you pull it slowly (Figure 9.3). Also like Silly Putty, rock becomes softer and easier to deform when it is warmer.

The fact that rock can deform and flow explains why small moons and asteroids are "potato-shaped" while larger worlds are spherical. The weak gravity of a small object is unable to overcome the rigidity of its rocky material, so the object retains the shape it had when it was born. For a larger world, gravity can overcome the strength of solid rock, slowly deforming and molding it into a spherical shape. Gravity will make any rocky object more than about 500 kilometers in diameter into a sphere within about 1 billion years. Larger worlds become spherical more quickly, especially if they are molten (or gaseous) at some point in their history.

In terms of rock strength, a planet's outer layer consists of relatively cool and rigid rock, called the **lithosphere** (*lithos* is Greek for "stone"), that "floats" on warmer, softer rock beneath. The lithosphere generally encompasses the

Figure 9.3 Silly Putty stretches when pulled slowly but breaks cleanly when pulled rapidly. Rock behaves just the same, but on a longer timescale.

COMMON MISCONCEPTIONS

Earth Is Not Full of Molten Lava

Many people guess that Earth is full of molten lava (more technically known as *magma*). This misconception may arise partly because we see molten lava emerging from inside Earth when a volcano erupts. However, Earth's mantle and crust are almost entirely solid. The lava that erupts from volcanoes comes only from a narrow region of partially molten material beneath the lithosphere. Indeed, the only part of Earth's interior that is fully molten is the outer core, which lies so deep within the planet that core material never erupts directly to the surface.

crust and part of the mantle, as shown by the dashed circles in Figure 9.2. Beneath the lithosphere, in the lower part of the mantle, the higher temperatures allow rock to deform and to flow much more easily. The thickness of the lithosphere is very important to geology. A thin lithosphere is brittle and can crack easily. A thick lithosphere is much stronger and can prevent volcanic eruptions and the creation of mountain ranges.

Comparing Planetary Interiors If you study the interior structures in Figure 9.2, you'll see that the basic interior structure is the same for each of the terrestrial worlds: a very thin crust, a much thicker mantle, and a central core. However, there are also some important differences, particularly in core size and in the extent to which the lithospheres reach into the mantles. With a little thought, you'll realize that these differences are linked closely with planetary size.

Notice that smaller planets generally have smaller cores, just as we would expect for worlds that were all made from metal and rock that condensed in the inner regions of the solar nebula. However, two worlds seem to deviate from the expected pattern. Mercury has a surprisingly large core, even when we account for the fact that it formed in the region of the solar nebula where the planetesimals should have been most metal-rich. That is the primary reason why scientists suspect that Mercury once suffered a giant impact [Section 8.4] that blasted away much of its original mantle and crust. The Moon's core deviates from the pattern in the opposite way, as it seems surprisingly small in comparison to the Moon's overall size. Again, the explanation may lie with a giant impact; in this case, we expect a small core if the Moon really did form from debris blasted out of Earth's outer layers (after Earth had undergone differentiation), since that debris would have contained very little high-density metal.

Lithospheric thickness is also closely related to planetary size: smaller planets tend to have thicker lithospheres. The two largest terrestrial planets, Earth and Venus, have very thin lithospheres that extend only a short way into their upper mantles. The smaller worlds—Mars, Mercury, and the Moon—have thick lithospheres that extend nearly to their cores.

• What causes geological activity?

The most profound differences between the geologies of the terrestrial worlds occur in their relative amounts of **geological activity**, a term used to describe how much ongoing change occurs on their surfaces. For example, we say that Earth is geologically active, because volcanic eruptions, earthquakes, erosion, and other geological processes continually reshape its surface. In contrast, the Moon and Mercury have virtually no geological activity, which is why their surfaces today look essentially the same as they did a billion or more years ago.

Interior heat is the primary driver of geological activity. Volcanoes can erupt on Earth because Earth is quite hot inside—hot enough that, in at least some cases, rock melts into molten lava. In contrast, the Moon has no active volcanoes, because its interior is too cool to melt rock and push it to the surface. But what makes some planetary interiors hotter than others? To find the answer, we must investigate how interiors heat up and cool off. As we'll see, we can ultimately trace a planet's internal heat, and hence its geological activity, back to its size.

How Interiors Get Hot A hot interior contains a lot of thermal energy, and the law of conservation of energy tells us that this energy must have come from somewhere [Section 4.3]. Although you might first guess that the Sun would be the heat source, this is not the case: Sunlight is the primary heat source for the *surfaces* of the terrestrial planets, but virtually none of this solar energy penetrates more than a few meters into the ground. Internal heat is a product of the planets themselves, not of the Sun. In fact, three sources of energy explain nearly all the interior heat of the terrestrial worlds (Figure 9.4):

- **Heat of accretion**: Accretion deposits energy brought in from afar by colliding planetesimals. An incoming planetesimal has a lot of gravitational potential energy when it is far away. As it approaches a forming planet, its gravitational potential energy is converted to kinetic energy, causing it to accelerate. Upon impact, much of the kinetic energy is converted to heat, adding to the thermal energy of the planet.
- **Heat from differentiation**: Differentiation also converts gravitational potential energy into thermal energy. Remember that differentiation occurs as denser materials sink and less dense materials rise in a molten interior. This process adds mass to the planet's core and reduces the mass of the outer layers. Thus, much of the planet's mass effectively moves inward, losing gravitational potential energy. The friction generated as the mass separates by density converts the lost gravitational potential energy to thermal energy, heating the interior. The same thing happens when you drop a brick into a pool: As the brick sinks to the bottom, friction with the surround-

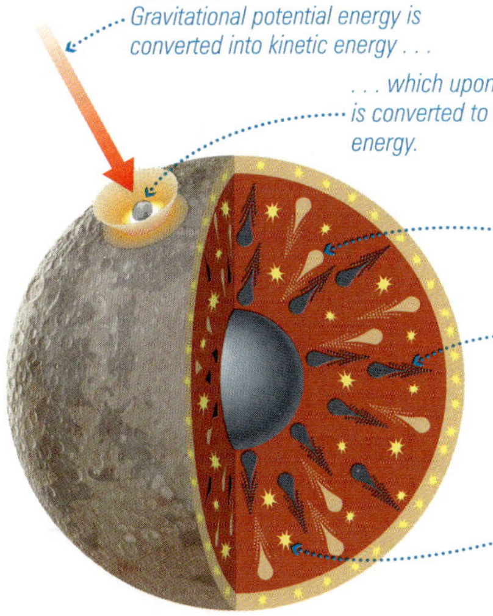

Figure 9.4 The three main heat sources for terrestrial planet interiors are accretion, differentiation, and radioactive decay. Only the latter is still a major heat source today.

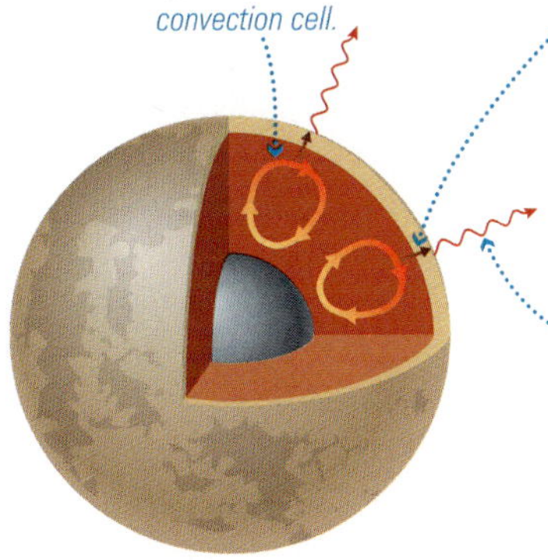

Figure 9.5 The three main cooling processes in planets. Convection can occur only in a planet that is still hot inside.

ing water heats the pool—though the amount of heat from a single brick is too small to notice.

- **Heat from radioactive decay**: Radioactive decay affects the terrestrial worlds because the rock and metal planetesimals that built them contained radioactive isotopes of elements such as uranium, potassium, and thorium. When radioactive nuclei decay, subatomic particles fly off at high speeds, colliding with neighboring atoms and heating them. In essence, this transfers some of the mass-energy ($E = mc^2$) of the radioactive nucleus to the thermal energy of the planetary interior.

Note that while accretion and differentiation deposited heat into planetary interiors only when the planets were very young, radioactive decay continues to heat the terrestrial planets to this day. Indeed, over billions of years, the total amount of heat deposited by radioactive decay has been several times as great as the amount that was deposited initially by accretion and differentiation. Nevertheless, the rate of radioactive decay declines with time, so it was a more significant heat source when the planets were young.

The combination of the three heat sources explains how the terrestrial interiors ended up with their core-mantle-crust structures. The many violent impacts that occurred during the latter stages of accretion deposited so much energy that the outer layers of the young planets began to melt. This started the process of differentiation, and the process of differentiation itself then began to release additional heat. Along with the substantial heat from early radioactive decay, the interiors became hot enough to melt and differentiate throughout.

How Interiors Cool Off Cooling a planetary interior requires transporting heat outward. Just as there are three basic heating processes for planetary interiors, there are also three basic cooling processes (Figure 9.5):

- **Convection**: Convection transports heat upward when hot material expands and rises while cooler material contracts and falls. It can occur whenever there is strong heating from below. For example, you can see convection in a pot of soup on a hot burner. (You may also be familiar with convection in weather—warm air near the ground tends to rise while cool air above tends to fall.)
- **Conduction**: Conduction is the process in which hot material transfers heat to cooler material through contact, just as a hot potato transfers its heat to your cooler hand when you pick it up. Conduction occurs through the microscopic collisions of many individual atoms or molecules. Molecules of materials in close contact are constantly colliding with one another, so the faster-moving molecules in hot material tend to transfer some of their energy to the slower-moving molecules of cooler material.
- **Radiation**: An object can also lose heat through radiation; that is, by radiating energy (light) into space. Remember that all objects emit *thermal radiation* characteristic of their temperatures [Section 5.4] and therefore are always losing some energy to space. Note that while convection and conduction can transport interior heat toward a planet's surface, the planet ultimately loses this heat to space only through radiation.

Let's start with Earth as an example of how heat moves from the deep interior to space. Convection is the most important process in Earth's deep interior. Hot rock from deep in the mantle gradually rises, slowly cooling as it makes its way upward. By the time it reaches the top of the mantle, the rock has transferred all its excess heat to its surroundings, so it is now cool and it begins to fall. The ongoing process of convection creates individual **convection cells** within the mantle. Figure 9.5 shows a few mantle convection cells as small circles, with arrows indicating the direction of flow. Keep in mind that mantle convection primarily involves solid rock, not molten rock. Because solid rock flows quite slowly, mantle convection is a very slow process. At the typical rate of mantle convection on Earth—about 1 centimeter per year—it would take about

COMMON MISCONCEPTIONS

Pressure and Temperature

Some people think that Earth's interior should be hot just because the internal pressure is so high. After all, if we compress a gas from low pressure to high pressure, it heats up. The same is not necessarily true of rock. High pressure compresses rock only slightly, so the compression causes little increase in temperature. Thus, while high pressures and temperatures sometimes go together in planets, they don't have to. In fact, after all the radioactive elements decay (billions of years from now), Earth's deep interior will become quite cool even though the pressure will be the same as it is today. The temperatures inside Earth and the other planets can remain high only if there is an internal source of heat, such as accretion, differentiation, or radioactive decay.

100 million years for a piece of rock to be carried from the base of the mantle to the top.

Mantle convection stops at the base of the lithosphere, where the rock is too strong to flow as readily as it does lower down. Thus, from the base of the lithosphere to the surface, heat continues upward primarily through conduction. (Some heat also reaches the surface through volcanic eruptions that directly carry hot rock upward.) When the heat finally reaches Earth's surface, it is radiated away into space.

The Role of Planetary Size Size is the single most important factor in planetary cooling: Just as a hot potato remains hot inside much longer than a hot pea, a large planet can stay hot inside much longer than a small one. You can see why size is the critical factor by picturing a large planet as a smaller planet wrapped in extra layers of rock. The extra rock acts as insulation, so it takes much longer for interior heat to reach the surface.

Thus, we can trace interior heat, heat transport, and geological activity all back to planetary size. The relatively small sizes of the Moon and Mercury probably allowed their interiors to cool within a billion years or so after they formed. As they cooled, their lithospheres thickened and mantle convection was confined to deeper and deeper layers. Ultimately, the mantle convection probably stopped altogether. With insufficient internal heat to drive any further movement of interior rock, the Moon and Mercury are now geologically "dead," meaning they have no more heat-driven geological activity.

In contrast, the much larger size of Earth has allowed our planet to stay quite hot inside. Mantle convection keeps interior rock in motion and the heat keeps the lithosphere thin, which is why geological activity can continually reshape the surface. Venus probably remains nearly as active as Earth, thanks to its very similar size. Mars, with a size in between that of the other terrestrial worlds, probably represents an intermediate case: It has certainly cooled significantly during its history, but it probably still retains enough internal heat for at least some geological activity. No one knows whether Mars still has any convection in its lower mantle.

THINK ABOUT IT

In general, what kind of planet would you expect to have the *thickest* lithosphere: the largest planet, the smallest planet, the planet closest to the Sun, or the planet farthest from the Sun? Why? (Assume the planet has the same composition in all four cases.)

• Why do some planetary interiors create magnetic fields?

A planet's interior structure also determines another important planetary property: the presence or absence of a **magnetic field** that can affect charged particles or magnetized objects in its vicinity. If you've ever used a compass, you know that Earth has a magnetic field that determines the direction in which the compass needle points. The global extent of Earth's magnetic field gives us a strong clue that the field is generated *inside* our planet. By examining how a planetary interior can generate a magnetic field, we'll see why Earth has a relatively strong magnetic field and the rest of the terrestrial planets do not.

You are probably familiar with the general pattern of the magnetic field created by an iron bar (Figure 9.6a). Earth's magnetic field is generated by a process more similar to that of an *electromagnet,* in which the magnetic field arises as a battery forces charged particles (electrons) to move along a coiled wire (Figure 9.6b). Earth does not contain a battery, but charged particles move with the molten metals in its liquid outer core (Figure 9.6c). Internal heat causes the liquid metals to rise and fall (convection), while Earth's rotation twists and distorts the convection pattern of these molten metals. The result is that electrons in the molten metals move within Earth's outer core in much the same way they move in an electromagnet, generating Earth's magnetic field.

We can generalize these ideas to other worlds. There are three basic requirements for a global magnetic field:

1. An interior region of electrically conducting fluid (liquid or gas), such as molten metal.
2. Convection in that layer of fluid.
3. At least moderately rapid rotation.

Earth is unique among the terrestrial worlds in meeting all three requirements, which is why it is the only terrestrial world with a strong magnetic field. The Moon has no magnetic field, either because it lacks a metallic core altogether or because its core has long since solidified and ceased convecting. Mars likewise has virtually no magnetic field today, probably because of core solidification. Venus probably has a molten core layer much like that of Earth, but either its convection or its 243-day rotation period is too slow to

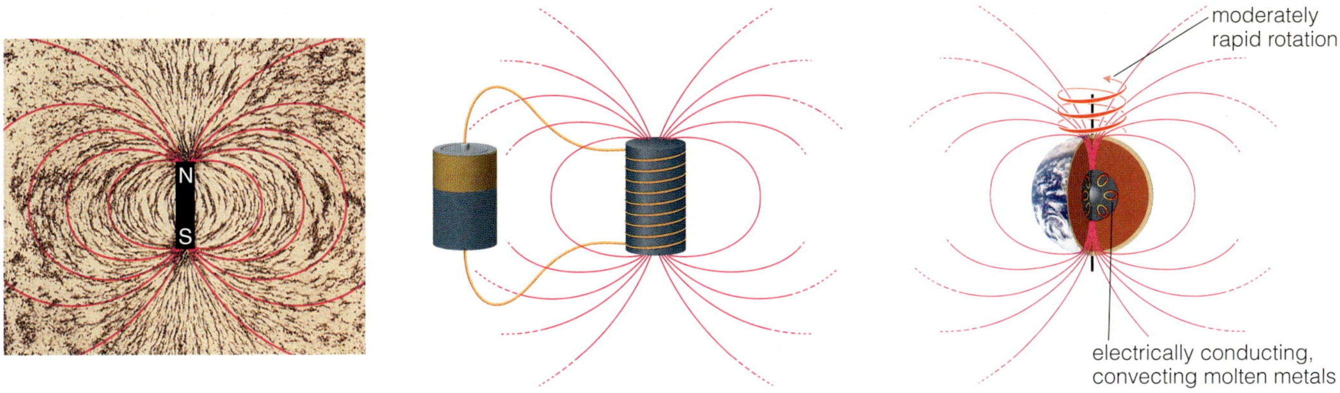

a This photo shows how a bar magnet influences iron filings (small black specks) around it. The *magnetic field lines* (red) represent this influence graphically.

b A similar magnetic field is created by an electromagnet, which is essentially a wire wrapped around a bar and attached to a battery. The field is created by the battery-forced motion of charged particles (electrons) along the wire.

c Earth's magnetic field also arises from the motion of charged particles. The charged particles move within Earth's liquid outer core, which is made of electrically conducting, convecting molten metals.

Figure 9.6 Sources of magnetic fields.

MATHEMATICAL INSIGHT 9.1 The Surface Area-to-Volume Ratio

Mathematically, we can see why large planets cool more slowly than smaller ones by thinking about their relative surface areas and volumes. The total amount of heat contained in a planet depends on the planet's total *volume.* However, this heat can escape directly to space only from the planet's surface, so the key factor in the planet's heat-loss rate is its *surface area.* As heat escapes from the surface, more heat flows upward from the interior to replace it. This process will continue until the interior is no hotter than the surface. Thus, the total time it takes for a planet to lose its internal heat must be closely related to the ratio of the surface area through which it loses heat to the volume that contains heat, which is what we call the *surface area-to-volume ratio.*

$$\text{surface area-to-volume ratio} = \frac{\text{surface area}}{\text{volume}}$$

To see exactly how the surface area-to-volume ratio affects heat loss, consider a spherical object like a planet. The surface area of a sphere is given by the formula $4\pi r^2$ and the volume is given by the formula $\frac{4}{3}\pi r^3$, where r is the radius of the sphere. Thus, for a sphere:

$$\underset{\text{(for a sphere)}}{\text{surface area-to-volume ratio}} = \frac{4\pi r^2}{\frac{4}{3}\pi r^3} = \frac{3}{r}$$

Because the radius r appears in the denominator, we conclude that *larger objects have smaller surface area-to-volume ratios.* This idea holds true not just for spheres, but for objects of any shape. Thus, if two objects start with the same internal temperature, the larger one will take longer to cool off. This principle explains many everyday phenomena. For example, crushing an ice cube does not change the total volume of the ice, but the many smaller pieces have a greater combined surface area than the original ice cube. Thus, crushing ice increases its surface area-to-volume ratio, putting more area in contact with the surrounding liquid or air. That is why crushed ice will cool a drink more quickly than an equal volume larger ice cubes.

Example 1: Compare the surface area-to-volume ratios of the Moon and Earth. How does your answer help explain why Earth's interior is much hotter than the Moon's?

Solution:

Step 1 Understand: Comparing the two surface area-to-volume ratios will tell us the relative rates at which the two worlds would lose heat if they both started with the same internal temperature. "Comparing" in this context means seeing how much larger one ratio is than the other. We know that the smaller Moon should have the larger ratio, so we'll find our answer by dividing the Moon's surface area-to-volume ratio by the Earth's surface area-to-volume ratio.

Step 2 Solve: We've already found that the surface area-to-volume ratio for a spherical world is $3/r$. Dividing the surface area-to-volume ratios for the Moon and Earth, we find:

$$\frac{\text{surface area-to-volume ratio (Moon)}}{\text{surface area-to-volume ratio (Earth)}} = \frac{3/r_{\text{Moon}}}{3/r_{\text{Earth}}} = \frac{r_{\text{Earth}}}{r_{\text{Moon}}}$$

From Appendix E, the radii of the Moon and Earth are $r_{\text{Moon}} = 1{,}738$ km and $r_{\text{Earth}} = 6{,}378$ km. Substituting these values into our equation we find:

$$\frac{\text{surface area-to-volume ratio (Moon)}}{\text{surface area-to-volume ratio (Earth)}} = \frac{6{,}378 \text{ km}}{1{,}738 \text{ km}} = 3.7$$

Step 3 Explain: We've found that the Moon's surface area-to-volume ratio is nearly four times as large as Earth's, which means the Moon would cool four times faster if both worlds started with the same temperature. However, Earth's larger size also gave it much more internal heat to begin with, which amplifies the difference in heat loss found by the surface area-to-volume ratio alone. Overall, we conclude that the Moon should have cooled many times faster than Earth, which explains why the Moon's interior is so much cooler than Earth's interior today.

generate a magnetic field. Mercury remains an enigma: It possesses a measurable magnetic field despite its small size and slow, 59-day rotation. The reason may be Mercury's huge metal core, which may still be partly molten and convecting.

The same three requirements for a magnetic field also apply to jovian planets and stars. For example, Jupiter's strong magnetic field comes from its rapid rotation and its layer of convecting, metallic hydrogen that conducts electricity [Section 11.1]. The Sun's magnetic field is generated by the combination of convection of ionized gas (plasma) in its interior and rotation.

THINK ABOUT IT

Recall that the Sun had a strong solar wind when it was young (which cleared out the remaining gas of the solar nebula [Section 8.3]) because its magnetic field was much stronger than it is today. What single factor explains why the young Sun had a much stronger magnetic field than the Sun today?

Magnetic fields can be very useful. On Earth, the magnetic field helps us navigate. Elsewhere, magnetic fields provide us with important information about planetary interiors. However, magnetic fields have virtually no effect on the structure of a planet itself. From the standpoint of planetary science, their most important role is that they can form a protective *magnetosphere* surrounding a planet, with potentially profound effects on a planet's atmosphere and any inhabitants [Section 10.1].

MasteringASTRONOMY Shaping Planetary Surfaces Tutorial, Lessons 1–3

9.2 Shaping Planetary Surfaces

Now that we have discussed how Earth and other terrestrial worlds work on the inside, we are ready to turn to their surfaces. The surfaces of the planets are remarkably flat. For example, the tallest mountains on Earth rise only about 10 kilometers above sea level, less than $\frac{1}{600}$ of Earth's radius, and are small enough that they could be represented by grains of sand on a typical globe. Nevertheless, surface features tell us a great deal about the histories of the planets.

THINK ABOUT IT

Find a globe of Earth that shows mountains in raised relief (that is, a globe that has bumps to represent mountains). Are the heights of the mountains correctly scaled relative to Earth's size? Explain.

• What processes shape planetary surfaces?

Earth offers a huge variety of geological surface features, and the variety only increases when we survey other worlds. But on closer examination, geologists have found that almost all surface features can be explained by just four major geological processes:

- **Impact cratering**: the blasting of bowl-shaped *impact craters* by asteroids or comets striking a planet's surface.
- **Volcanism**: the eruption of molten rock, or *lava,* from a planet's interior onto its surface.
- **Tectonics**: the disruption of a planet's surface by internal stresses.
- **Erosion**: the wearing down or building up of geological features by wind, water, ice, and other phenomena of planetary weather.

Let's examine each of these processes in a little more detail.

Impact Cratering The scarred faces of the Moon and Mercury (see Figure 9.1) attest to the battering that the terres-

Figure 9.7 Interactive Figure Artist's conception of the impact process.

a Meteor Crater in Arizona was created about 50,000 years ago by the impact of an asteroid about 50 meters across. The crater is more than a kilometer across and almost 200 meters deep.

b This photo shows a crater, named Tycho, on the Moon. Note the classic shape and central peak.

Figure 9.8 Impact craters.

trial worlds have taken from leftover planetesimals, such as comets and asteroids. They also immediately reveal an important feature of impact cratering: Small craters far outnumber large ones, confirming that many more small asteroids and comets orbit the Sun than large ones. While the Moon and Mercury bear the most obvious scars, all the terrestrial worlds have suffered similarly.

An impact crater forms when an asteroid or comet slams into a world with a solid surface (Figure 9.7). Impacting objects typically hit the surface at a speed between about 40,000 and 250,000 km/hr (10–70 km/s). At such a tremendous speed, the impact releases enough energy to vaporize solid rock and excavate a crater (the Greek word for "cup"). Debris from the blast shoots high above the surface and then rains down over a large area. If the impact is large enough, some of the ejected material can completely escape from the planet.

Craters generally end up being circular because an impact blasts out material in all directions, no matter the direction of the incoming object. Laboratory experiments show that craters are generally about 10 times as wide as the objects that create them and about 10–20% as deep as they are wide. For example, an asteroid 1 kilometer in diameter will blast out a crater about 10 kilometers wide and 1–2 kilometers deep. A large crater may have a central peak, which forms when the center rebounds after impact in much the same way that water rebounds after you drop a pebble into it. Figure 9.8 shows two impact craters with classic shapes, one on Earth and one on the Moon.

Although impact craters are almost always circular, details of their shapes can provide important information about geological conditions on a planetary surface. To illustrate how, Figure 9.9 contrasts three craters on Mars. The crater in Figure 9.9a has a fairly simple bowl shape,

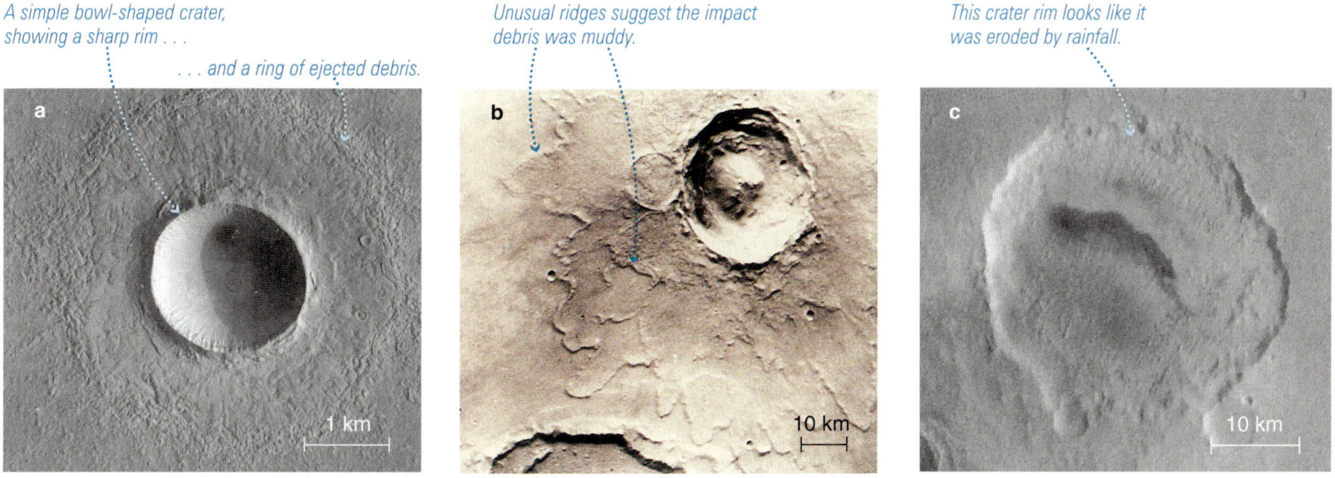

a A crater with a "standard" shape.

b This crater was probably made by an impact into icy ground.

c This crater shows evidence of erosion; see Figure 9.29b for a close-up view of this crater's floor.

Figure 9.9 Crater shapes on Mars tell us about Martian geology. (Photos from *Mars Global Surveyor* (a) and *Viking* spacecraft (b and c).)

suggesting that it formed just as we expect from the basic cratering process. The crater in Figure 9.9b also has a fairly "standard" shape, but notice the extra large bump in its center and the way it is surrounded by a pattern of what looks like mud flows. This suggests that the surface had underground water or ice that melted or vaporized on impact; the muddy debris then flowed across the surface before hardening into the pattern we see today. The crater in Figure 9.9c shows obvious signs of erosion: It lacks a sharp rim and its floor no longer has the classic bowl shape. Again, this shape tells us about Martian geological history, in this case suggesting that ancient rainfall eroded the crater and that the crater bottom was once a lake. Studies of crater shapes on other worlds provide similar clues to their surface conditions and history.

Volcanism The second major geological process is volcanism. On Earth, we generally associate volcanism with the mountains we call volcanoes. In the broader context of planetary geology, we use the term *volcanism* to refer to any eruption of molten lava onto the surface, whether it comes from a tall volcano or simply rises through a crack in a planet's lithosphere.

Volcanism occurs when underground molten rock, technically called *magma,* finds a path through the lithosphere to the surface (Figure 9.10). Molten rock tends to rise for three main reasons. First, molten rock is generally less dense than solid rock, and lower-density materials tend to rise when surrounded by higher-density materials. Second, remember that most of Earth's interior is not molten. Thus, a "chamber" of molten rock (a *magma chamber*) may be squeezed by the surrounding solid rock, which drives the molten rock upward under pressure. Third, molten rock often contains trapped gases that expand as it rises, which can make it rise much faster and lead to dramatic eruptions. The same molten rock that is called magma when it is underground is called *lava* once it erupts onto the surface.

The result of an eruption depends on how easily the molten lava flows across the surface. Lava that is "runny" can flow far before cooling and solidifying, while "thick" lava tends to collect in one place.* Broadly speaking, lava can shape three different types of volcanic features:

Figure 9.10 Interactive Figure Volcanism. This photo shows the eruption of an active volcano on the flanks of Kilauea on the Big Island of Hawaii. The inset shows the underlying process: Molten rock collects in a "magma chamber" and can erupt upward.

- The runniest lavas flow far and flatten out before solidifying, creating vast **volcanic plains** (Figure 9.11a).
- Lavas that are somewhat thicker can solidify before they can completely spread out, creating what we call **shield**

*You may hear the "thickness" of a liquid described by the technical term *viscosity.* For example, liquid water has a low viscosity that allows it to flow quickly and easily, while thick and slow-flowing honey and molasses have much higher viscosities.

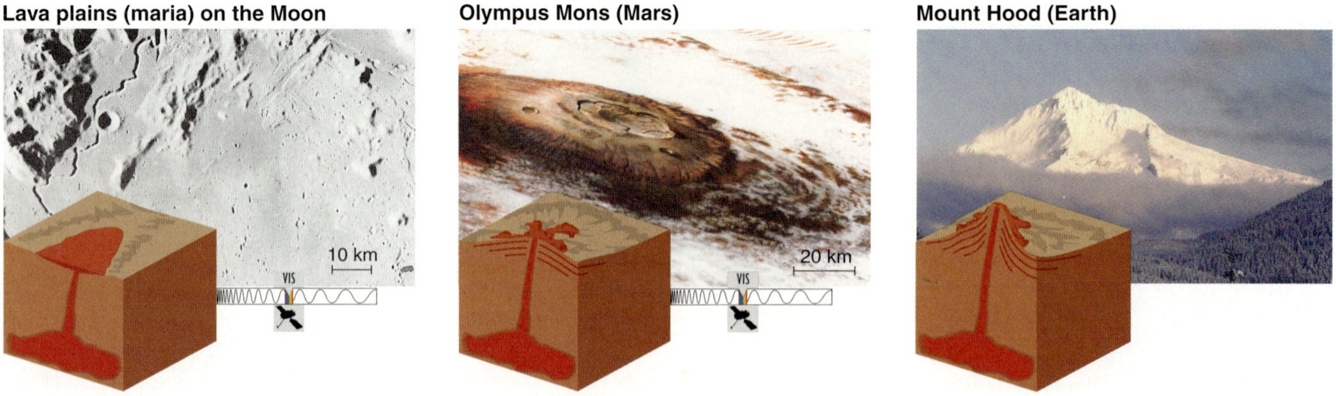

a Very runny lava makes flat lava plains like these on the moon. (The long, winding channel near the upper left was made by a river of molten lava.)

b Slightly thicker lava makes shallow-sloped shield volcanoes, such as Olympus Mons on Mars.

c The thickest lavas make steep-sloped stratovolcanoes like Oregon's Mount Hood.

Figure 9.11 The shapes of volcanic structures depend on whether the molten lava is runny or thick.

volcanoes (so named because of their shape). Shield volcanoes can be very tall but are not very steep. Examples include the mountains of the Hawaiian islands on Earth and Olympus Mons on Mars (Figure 9.11b).
- The thickest lavas can't flow very far before solidifying and therefore build up tall, steep **stratovolcanoes** (Figure 9.11c). Examples of stratovolcanoes on Earth include Mount Fuji in Japan, Mount Kilimanjaro in Tanzania, Mount Shasta in California, Mount Hood in Oregon, and Mount Rainier and Mount St. Helens in Washington State.

Both lava plains and shield volcanoes are made of **basalt**, a mixture of many different minerals that erupts from volcanoes as a high-density but fairly runny lava. All the terrestrial worlds and some jovian moons show evidence of volcanic plains and shield volcanoes, telling us that basalt is common throughout the solar system. Stratovolcanoes are made by lower-density volcanic rock that erupts as a thicker lava; this type of lava is common on Earth but apparently much rarer in the rest of the solar system.

Volcanic structures are the most obvious result of volcanism, but from our standpoint as living beings, volcanism has had a far more important effect on our planet: It explains the existence of our atmosphere and oceans. Recall that Earth and the other terrestrial planets accreted from rocky and metallic planetesimals; the water and gas in their interiors presumably came from icy planetesimals that originated in more distant reaches of the solar system and crashed into the growing terrestrial worlds [Section 8.4]. This water and gas became trapped in the interiors of the planets in much the same way the gas in a carbonated beverage is trapped in a pressurized bottle. When molten rock erupts onto the surface as lava, the release of pressure expels the trapped gases in a process we call **outgassing**. Outgassing can be either quite dramatic, as during a volcanic eruption (Figure 9.12a), or more gradual, as when gas escapes from volcanic vents (Figure 9.12b). Virtually all the gas that made the atmospheres of Venus, Earth, and Mars—and the water vapor that rained down to form Earth's oceans—originally was released from the planetary interiors by outgassing.

a This photo shows the eruption of Mount St. Helens (Washington State) on May 18, 1980. Note the tremendous outgassing that accompanies the eruption.

b This photo shows more gradual outgassing from a volcanic vent in Volcanoes National Park, Hawaii.

Figure 9.12 Examples of outgassing, which released the gases that ultimately made the atmospheres of Venus, Earth, and Mars.

Tectonics The third major geological process, tectonics, gets its name from the Greek word for "builder," *tekton*; you'll notice the same root in the word *architect*, which means "master builder." In geology, tectonics refers to "building" performed on planetary surfaces—that is, any surface reshaping that results from stretching, compression, or other forces acting on the lithosphere.

Tectonic features can arise in a variety of ways. For example, the weight of a newly formed volcano can bend or crack the lithosphere beneath it, or a plume of hot material rising from a planet's deep interior can push up on the lithosphere to create a bulge. However, most tectonic features form as a direct or indirect result of mantle convection (Figure 9.13). For example, we see compression features in places where adjacent convection cells push rock together. Such crustal compression helped create the Appalachian Mountains of the eastern United States. In other places, mantle convection may tend to pull apart the crust, making

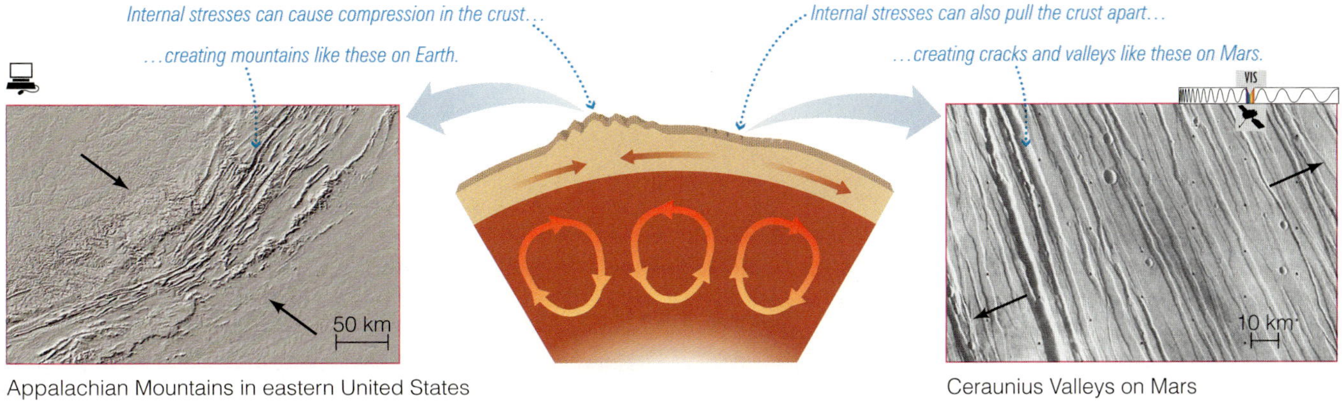

Figure 9.13 Interactive Figure Tectonic forces can produce a wide variety of features. Mountains and fractured plains are among the most common. The Mars image is a visible-light photo from orbit. The Earth image was computer generated from topographical data.

cracks and valleys. Examples of such cracks and valleys include the Guinevere Plains on Venus, the Ceraunius Valleys on Mars, and New Mexico's Rio Grande Valley. (The river named the Rio Grande came *after* the valley formed from tectonic processes.)

On Earth, the ongoing stress of mantle convection ultimately fractured Earth's lithosphere into more than a dozen pieces, or *plates*. These plates move over, under, and around each other in a process we call **plate tectonics**. The movements of the plates explain nearly all Earth's major geological features, including the arrangement of the continents, the nature of the seafloor, and the origin of earthquakes. Because plate tectonics appears to be unique to Earth among the terrestrial worlds in our solar system, we'll save further discussion of it for Section 9.6.

Erosion The last of our four major geological processes is erosion. *Erosion* is a blanket term for a variety of processes that break down or transport rock through the action of ice, liquid, or gas. The shaping of valleys by glaciers (ice), the carving of canyons by rivers (liquid), and the shifting of sand dunes by wind (gas) are all examples of erosion.

We often associate erosion with the breakdown of existing features, but erosion also builds things. Sand dunes, river deltas, and lake-bed deposits are all examples of features built by erosion. Indeed, erosion built much of the surface rock on Earth. Over long periods of time, erosion has piled sediments into layers on the floors of oceans and seas, forming what we call **sedimentary rock**. The gorgeous layered rock of the Grand Canyon is sedimentary rock that was built by erosion. Figure 9.14 shows several examples of erosion on Earth.

• Why do the terrestrial planets have different geological histories?

The terrestrial worlds show great differences in the extent to which the four geological processes have operated. The surfaces of the Moon and Mercury are dominated by impact craters, but show virtually no features of erosion. Volcanic and tectonic features are found on all the worlds, but we find tall volcanoes only on Earth, Venus, and Mars. And Earth is the only terrestrial world on which features of erosion dominate the landscape. Why do the processes operate so differently on different worlds? Ultimately, we can trace the answer back to fundamental planetary properties like size, distance from the Sun, and rate of rotation.

Figure 9.15 summarizes the key connections between a planet's fundamental properties and the geological processes of volcanism, tectonics, and erosion. Let's explore these connections in a little more detail. We'll then be ready to explore the connection between planetary properties and impact cratering, the only one of the four geological processes with an external cause.

Figure 9.14 A few examples of erosion on Earth.

a The Colorado River continues to carve the Grand Canyon after millions of years.

b Glaciers created Yosemite Valley during ice ages.

c Wind erosion wears away rocks and builds up sand dunes, such as these in a California desert.

d Erosional debris also creates geological features, as seen in this river delta.

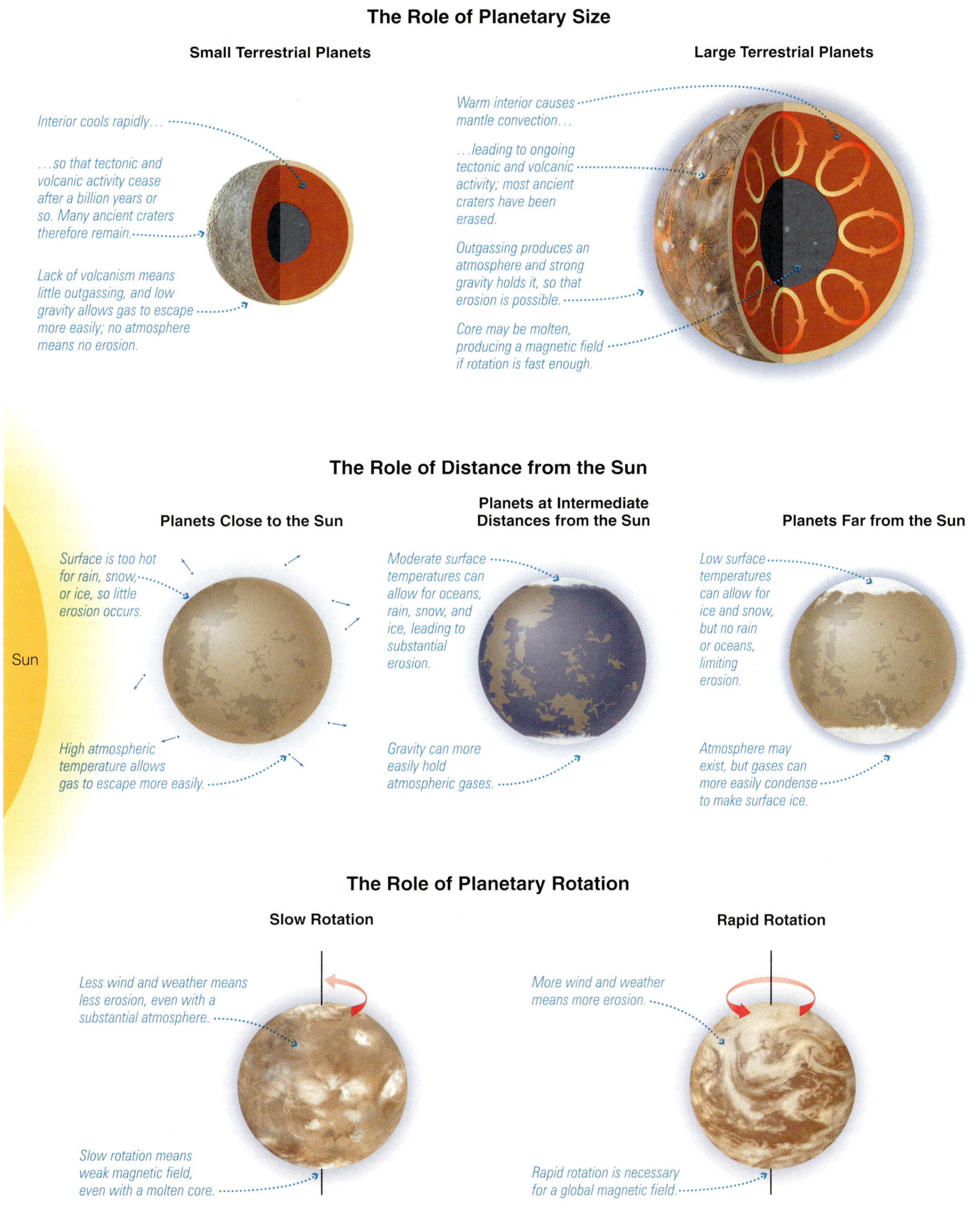

Figure 9.15 A planet's fundamental properties of size, distance from the Sun, and rotation rate are responsible for its geological history. This illustration shows the role of each key property separately. A planet's overall geological evolution depends on the combination of all the effects shown above.

Planetary Properties Controlling Volcanism and Tectonics Volcanism and tectonics both require internal heat, which is why they tend to go hand in hand: More volcanism generally also means more tectonics. The connection with internal heat also allows us to tie these processes back to planetary size. As shown in Figure 9.15, larger planets have more internal heat and hence more volcanic and tectonic activity.

All the terrestrial worlds probably had some degree of volcanism and tectonics when their interiors were young and hot. As an interior cools, however, volcanic and tectonic activity subsides. The Moon and Mercury lack ongoing volcanism and tectonics because their small size allowed their interiors to cool long ago. Earth has active volcanism and tectonics because it is large enough to still have a hot interior. Venus, nearly the same size as Earth, must still be hot inside and probably also has active volcanism and tectonics, though we lack conclusive evidence of volcanic eruptions in the past few tens of millions of years. Mars, with its smaller size, certainly has much less volcanism and tectonic activity today than it did in the distant past.

Planetary Properties Controlling Erosion The link between planetary properties and erosion is a little more complex, because erosion can occur only on a planet with a significant atmosphere. Otherwise, there can be no surface liquids, no wind, and no climate change to cause glacial movements. An atmosphere is not the only requirement for erosion, however. A stationary atmosphere that simply bears down without stirring the surface would not cause any erosion. Erosion comes from processes such as wind and rain, which we generally refer to as "weather." Thus, the basic requirements for erosion are the existence of an atmosphere and significant weather affecting the surface.

We can therefore trace a planet's level of erosion back to all three fundamental planetary properties shown in Figure 9.15. The planet's size is important because a terrestrial world can have an atmosphere only if it has had significant outgassing, and outgassing is part of volcanism. In addition, size determines the strength of gravity, and atmospheric gases are more easily lost to space on worlds with weaker gravity. The planet's distance from the Sun affects its temperature; if all else is equal (such as planetary size), the higher temperatures on a world closer to the Sun will make it much easier for atmospheric gases to escape to space, while the colder temperatures on a world farther from the Sun may cause atmospheric gases to freeze out. Distance is also important because water erosion is much more effective with liquid water than with water vapor or ice, and therefore is strongest when moderate temperature allows water vapor to condense into liquid form. Planetary rotation is important to erosion because rotation is the primary driver of winds and other weather [Section 10.2]: Faster rotation means stronger winds and storms.

The Moon and Mercury lack significant atmospheres and erosion because they lack outgassing today, and any atmospheric gases they had in the distant past have been lost to space. Mars has limited erosion; it has a very thin atmosphere because much of the water vapor and carbon dioxide outgassed in its past either escaped to space or lies frozen in its polar caps or beneath the surface. Venus and Earth probably had similar amounts of outgassing, but cooler temperatures on Earth led to condensation and the ultimate removal of most of this gas from Earth's atmosphere, though still leaving enough for plenty of erosion. Most of Venus's gas remained in its atmosphere, making its atmosphere much thicker than Earth's, but Venus has little erosion because its slow rotation gives it virtually no wind on its surface.

• How does a planet's surface reveal its geological age?

We've focused on the four geological processes and their causes individually, but they also interact with one another to shape planetary surfaces. For example, volcanism produces lava flows that can erase old impact craters, and outgassing can create atmospheres that may lead to erosion. These interactions lead to one of the most useful insights in planetary geology: We can learn about a planet's overall level of geological activity and the "geological age" of its surface simply by looking at how closely impact craters are crowded together. We consider surfaces with ongoing volcanism, tectonics, or erosion to be geologically young, because they are constantly being reshaped by these processes. In contrast, a surface that has not changed in billions of years is considered geologically old. Remember that we are speaking only of the age *of the surface as it now appears,* not the age of the planet itself: All the planets are about 4.5 billion years old [Section 8.5].

To understand how crater crowding reveals surface ages, remember that all the terrestrial worlds should have been battered by impacts during the heavy bombardment. The number of impacts on a surface of any given size should have been quite similar, regardless of a planet's size, distance from the Sun, or rotation rate. Why, then, do we see such different numbers of craters on the different worlds? Careful study of the Moon reveals the key clues. The degree of crowding among craters varies greatly from place to place on the Moon (Figure 9.16). In the regions on the Moon known as the *lunar highlands,* craters are so crowded that we see craters on top of other craters. In the regions known as the *lunar maria,* we see only a few craters on top of generally smooth volcanic plains.

The sheer number of impact craters in the lunar highlands, especially the very common small ones, tells us that no part of the Moon could have escaped heavy cratering during the early history of the solar system. Thus, the relative lack of craters in the maria tells us that some process must have *erased* these craters from the maria after the heavy bombardment ended. (We'll discuss the precise process in Section 9.3). Radiometric dating of rocks brought back from the Moon by the *Apollo* astronauts confirms this idea. Rocks from the heavily cratered lunar highlands have ages as old as 4.4 billion years. Rocks from the maria are hundreds of millions of years younger, with ages between 3.0–3.9 billion years. Because the maria contain only about 3% as many craters as the highlands even though they are just a few

a This *Apollo* photograph of the Moon shows some areas are much more heavily cratered than others. (This view of the Moon is *not* the one we see from Earth.)

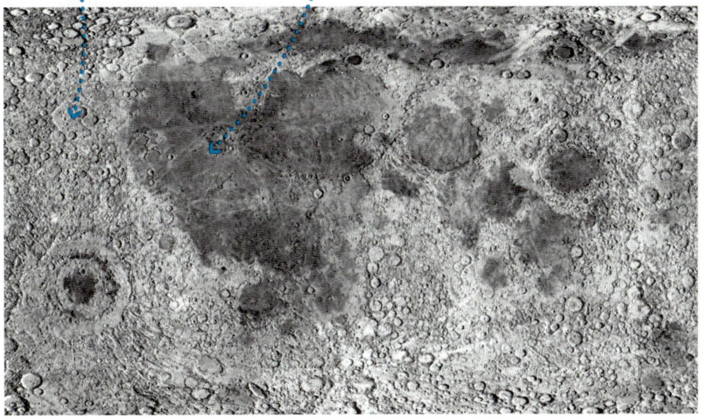

b This map shows the entire surface of the Moon in the same way a flat map of Earth represents the entire globe. Radiometric dating shows the heavily cratered lunar highlands to be a half-billion or more years older than the darkly colored lunar maria, telling us that the impact rate dropped dramatically after the end of the heavy bombardment.

Figure 9.16 These photos of the Moon show us that crater crowding is closely related to surface age.

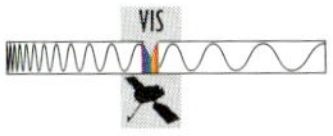

hundred million years younger, the heavy bombardment must have subsided quite early in the solar system's history. Compared to the period of heavy bombardment, relatively few impacts have occurred in the last 4 billion years.

More precise study of lunar cratering has allowed scientists to reconstruct the rate of impacts during much of the Moon's history. Because impacts are essentially random events, the same changes in impact rate over time must apply to all the terrestrial worlds. The degree of crater crowding therefore provides a good measure of the geological age of a planetary surface. A surface that is crowded with craters must have remained essentially undisturbed since the end of the heavy bombardment. A surface with few craters must have had its ancient craters erased by more recent geological processes.

Thus, the key factor determining the number of craters on a planetary surface today is its degree of past geological activity in the form of volcanism, tectonics, and erosion. All the terrestrial worlds must have been covered with impact craters in their early histories. Those that still have lots of craters today, like the Moon and Mercury, are those that have had little geological activity. Those with few craters today, like Venus and Earth, must have had extensive geological activity that has erased essentially all their ancient craters. The only craters we see on Venus and Earth are those few that have formed relatively recently (within the last few hundred million years).

The fact that more geological activity means fewer craters allows us to tie impact cratering primarily to one planetary property: size. Because the interiors of smaller planets cool more quickly, smaller worlds have had less geological activity and therefore retain more of their ancient craters. Larger worlds had at least as many impact craters when they were young as small worlds do today—in fact, larger worlds probably suffered more impacts because their stronger gravity tended to draw in and accelerate planetesimals—but their hotter interiors have supported enough geological activity to erase most of their ancient craters.

THINK ABOUT IT

Mercury is heavily cratered, much like the lunar highlands. Venus has relatively few craters. Which planet has the older surface? Is there any difference between the ages of the two planets as a whole? Explain how the surfaces of Mercury and Venus can have different ages while the planets themselves are the same age.

9.3 Geology of the Moon and Mercury

In the rest of this chapter we will investigate the geological histories of the terrestrial worlds, with the goal of explaining why each ended up as it did. Because smaller worlds have simpler geological histories, we'll start with the Moon and Mercury.

Look back at Figure 9.1, and you'll see that both these worlds are considerably smaller than Venus, Earth, or Mars. Their small size means that they long ago lost most of their internal heat, leaving them without any energy source to power ongoing geological activity. Small size also explains

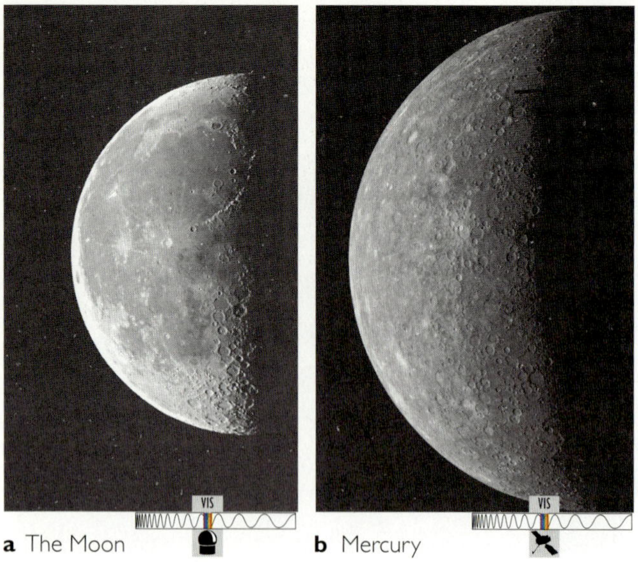

Figure 9.17 Similar views of the Moon and Mercury, shown to scale. The Mercury photo is a composite of photos from the *Mariner 10* spacecraft [Section 7.3], which photographed only parts of the surface.

their lack of significant atmospheres: Their gravity is too weak to hold gas for long periods of time, and without ongoing volcanism they lack the outgassing needed to replenish gas lost in the past.

Figure 9.17 shows the similar-looking faces of the Moon and Mercury. Their most obvious surface features are their numerous impact craters, showing that impact cratering has been by far the most important geological process on both worlds. However, closer examination shows a few features that are volcanic or tectonic in origin. Apparently, the Moon and Mercury had geological activity in the past. This should make sense: Long ago, before they had a chance to cool, these worlds were hot enough inside for at least some volcanism and tectonics.

• What geological processes shaped our Moon?

We've already discussed the fact that different parts of the Moon's surface have different geological ages. Now we are ready to examine in more detail the processes that have shaped the varying features of the Moon's surface.

Impact Cratering on the Moon We've seen that the familiar face of the full moon shows abundant cratering, but not all regions of the surface look the same. The brighter regions, called the *lunar highlands*, are heavily cratered. Other regions, known as the **lunar maria**, look smoother and darker (Figure 9.18). Indeed, the maria got their name because they look much like oceans when seen from afar; *maria* (singular, *mare*) is Latin for "seas." But the entire surface of the Moon should have been packed with craters during the heavy bombardment. So what happened to the craters that must once have been located in the regions where we now see maria?

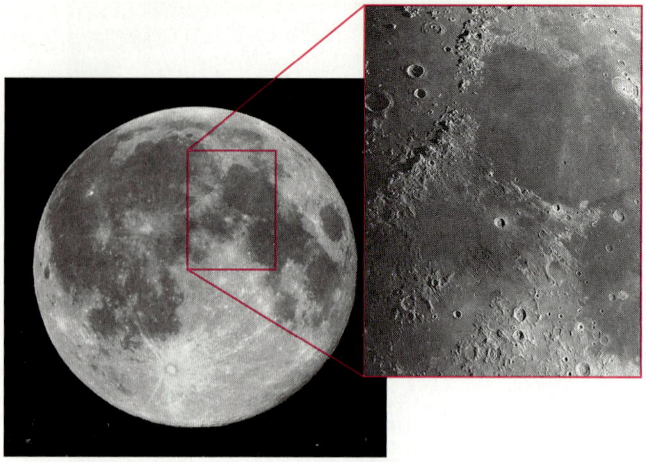

Figure 9.18 The familiar face of the Moon shows numerous dark, smooth maria.

Volcanism and Tectonics in the Lunar Maria The smooth and dark appearance of the lunar maria suggests that they were made by a flood of molten lava, and studies of moon rocks confirm this suggestion. Indeed, the maria are the Moon's most noticeable volcanic features.

Figure 9.19 shows how we think the maria formed. During the heavy bombardment, craters covered the Moon's entire surface. The largest impacts of the heavy bombardment violently fractured the Moon's lithosphere beneath the huge craters they created. However, the Moon's interior had already cooled since its formation, and there was no molten rock to flood these craters immediately. Instead, the lava floods came hundreds of millions of years later, thanks to heat released by the decay of radioactive elements in the Moon's interior. This heat gradually built up during the Moon's early history, until mantle material melted about 3 to 4 billion years ago. The molten rock then welled up through the cracks in the lithosphere, flooding the largest impact craters with lava. The maria are generally circular because they are essentially flooded craters (and craters are almost always round). Their dark color comes from the dense, iron-rich rock (basalt) that rose up from the lunar mantle as molten lava.

The flat surfaces of the maria tell us that the lunar lava spread easily and far, which means that it must have been among the runniest lava in the solar system. Its runniness can probably be traced to the Moon's formation. Recall that we think the Moon formed as the result of a giant impact of a Mars-sized body with Earth [Section 8.4]. This impact would have vaporized water and released other trapped gases, so these materials would not have become part of the Moon when it reaccreted from the giant impact debris. Water and trapped gases tend to increase the "thickness" of erupting lava in the same way that tiny bubbles increase the "thickness" of whipped cream: When you buy whipping cream in a carton, it is liquid and too thin to shape into anything, but the whipping process adds tiny air bubbles that make it thick enough so that you can build whipped cream "mountains." With no water or trapped gases, the Moon's lava would have lacked the gas bubbles

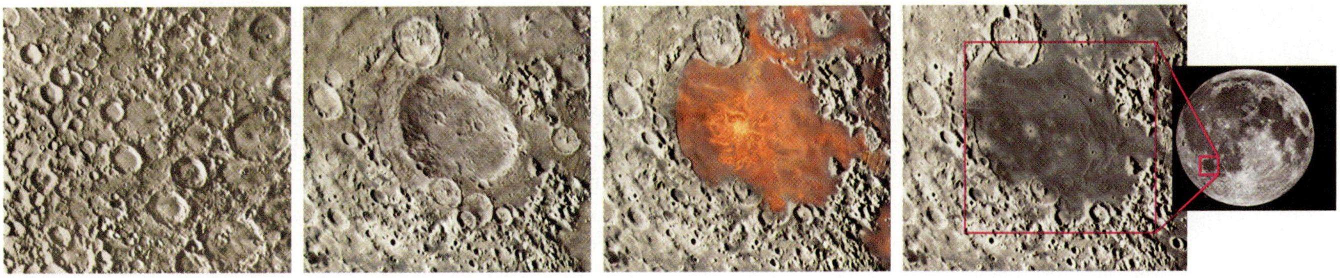

a This illustration shows the Mare Humorum region as it probably looked about 4 billion years ago, when it would have been completely covered in craters.

b Around that time, a huge impact excavated the crater that would later become Mare Humorum. The impact fractured the Moon's lithosphere and erased the many craters that had existed earlier.

c A few hundred million years later, enough heat from radioactive decay built up to melt the Moon's upper mantle. Molten lava welled up through the lithospheric cracks, flooding the impact crater.

d This photo shows Mare Humorum as it appears today, and the inset shows its location on the Moon.

Figure 9.19 The lunar maria formed between 3 and 4 billion years ago, when molten lava flooded large craters that had formed hundreds of millions of years earlier. This sequence of diagrams represents the formation of Mare Humorum.

that make lava "thick" on Earth and other worlds. The lunar lava was so runny that in places it flowed like rivers of molten rock, carving out long, winding channels (see the channel in Figure 9.11a).

The Moon's relatively few tectonic features are also found within the maria. These features, which look much like surface wrinkles, probably were created by small-scale tectonic stresses that arose when the lava that flooded the maria cooled and contracted (Figure 9.20).

Surprisingly, the Moon's far side—which we never see from Earth—has very few maria like the ones that we see on its near side (see Figure 9.16). Instead, when spacecraft first photographed the far side in the 1960s, they revealed a landscape looking almost entirely like the heavily cratered highlands of the near side. We now know that the entire far side essentially has a higher altitude than the near side. No one knows what caused this difference in altitude between the two sides, but it apparently allowed lava to well up and fill crater basins only on the near side, thus explaining why the near side has maria.

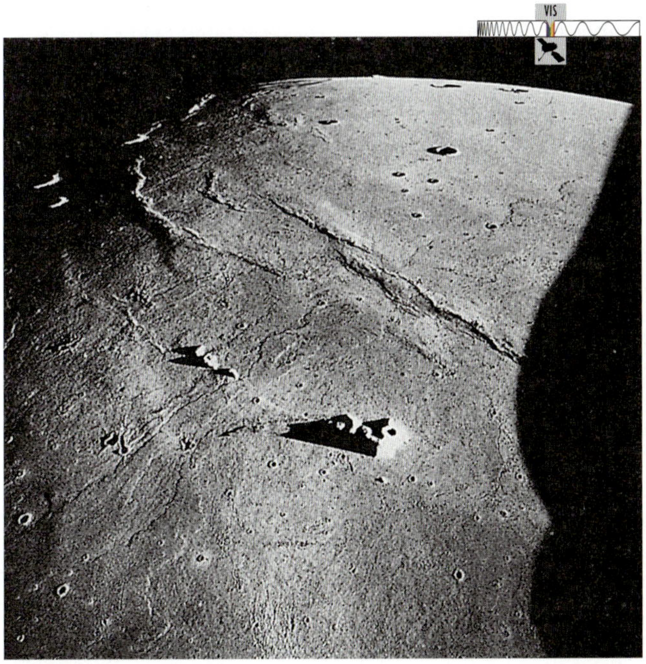

Figure 9.20 The features that look like wrinkles within maria like this one (Mare Imbrium) arose from tectonic stresses associated with the cooling and contracting of the lava that flooded the maria between 3 and 4 billion years ago.

The Moon Today: Rare Impacts and Little More The Moon's era of geological activity is long gone. Today, the Moon is a desolate and nearly unchanging place. Rare impacts may occur in the future, but we are unlikely ever to witness a major one. Little happens on the Moon, aside from the occasional visit of robotic spacecraft or astronauts from Earth (Figure 9.21a).

The only ongoing geological change on the Moon is a very slow "sandblasting" of the surface by *micrometeorites,* sand-size particles from space. These tiny particles burn up as meteors in the atmospheres of Earth, Venus, and Mars but rain directly onto the surface of the airless Moon. The micrometeorites gradually pulverize the surface rock, which explains why the lunar surface is covered by a thin layer of powdery "soil." The *Apollo* astronauts left their footprints in this powdery surface (Figure 9.21b). Pulverization by micrometeorites is a very slow process, and the astronauts' footprints will last millions of years before they are finally erased. Over billions of years, the micrometeorite impacts can gradually smooth out rough crater rims.

Aside from the micrometeorite sandblasting and rare larger impacts, the Moon has been "geologically dead" since the maria formed more than 3 billion years ago. Nevertheless, the Moon remains a prime target of future exploration. Further studies of lunar geology will help us understand both the Moon's history and the history of our solar system. The Moon may also offer an excellent location for astronomical observatories, since it has no air to obstruct our view but a stable ground on which to build large telescopes. Someday, perhaps in the not-too-distant future,

a Astronaut Gene Cernan takes the Lunar Roving Vehicle for a spin during the final *Apollo* mission to the Moon (*Apollo 17*, December 1972).

b The *Apollo* astronauts left clear footprints, like this one, in the Moon's powdery "soil." The powder is the result of gradual pulverization of surface rock by micrometeorites. Micrometeorites will eventually erase the astronauts' footprints, but not for millions of years.

Figure 9.21 The Moon today is geologically dead, but it is nevertheless a world from which we can learn a lot about the history of our solar system.

humans will build a lunar colony, with living quarters underground for protection against dangerous solar and cosmic radiation.

• What geological processes shaped Mercury?

Mercury looks so much like the Moon that it's often difficult at first to tell which world you are looking at in surface photos. The similarities extend to all the geological processes, though there are also a few differences.

Impact craters are visible almost everywhere on Mercury, indicating an ancient surface. However, Mercury's craters are less crowded together than the craters in the most ancient regions of the Moon, suggesting that molten lava later covered up some of the craters that formed on Mercury during the heavy bombardment (Figure 9.22a). As on the Moon, these lava flows probably occurred when heat from radioactive decay accumulated enough to melt part of the mantle. Although we have not found evidence of lava flows as large as those that created the lunar maria, the lesser crater crowding and the many smaller lava plains suggest that Mercury had at least as much volcanism as the Moon.

The largest single surface feature on Mercury is a huge impact crater called the *Caloris Basin* (Figure 9.22b). The Caloris Basin spans more than half of Mercury's radius, and its multiple rings bear witness to the violent impact that created it. The impact must have reverberated throughout the planet—we see evidence of violent surface shaking on the precise opposite side of Mercury from the Caloris Basin (Figure 9.22c). The Caloris Basin has few craters within it, indicating that it must have formed at a time when the heavy bombardment was already subsiding.

The most surprising features of Mercury are its many tremendous cliffs—evidence of a type of past tectonics quite different from anything we have found on any other terrestrial world (Figure 9.23). Mercury's cliffs have vertical faces up to 3 or more kilometers high and typically run for hundreds of kilometers across the surface. They probably formed when tectonic forces compressed the crust, causing the surface to crumple. Because crumpling would have shrunk the portions of the surface it affected, Mercury as a whole could not have stayed the same size unless other parts of the surface expanded. However, we find no evidence of "stretch marks" on Mercury. Can it be that the whole planet simply shrank?

Apparently so. Early in its history, Mercury's larger size and greater iron content allowed it to gain more internal heat from accretion and differentiation than did the Moon. This heat swelled the size of the large iron core. Later, as the core cooled, it contracted by perhaps as much as 20 kilometers in radius. The mantle and lithosphere must have contracted along with the core, generating the tectonic

Figure 9.22 Features of impact cratering and volcanism on Mercury. (Photos from *Mariner 10*.)

VIS

a Close-up view of Mercury's surface, showing impact craters and smooth regions where lava apparently covered up craters.

b Part of the Caloris Basin (outlined by the dashed circular ring), a large impact crater on Mercury.

c The surface on the precise opposite side of Mercury from the Caloris Basin. Seismic energy from the impact focused here and jumbled the crust.

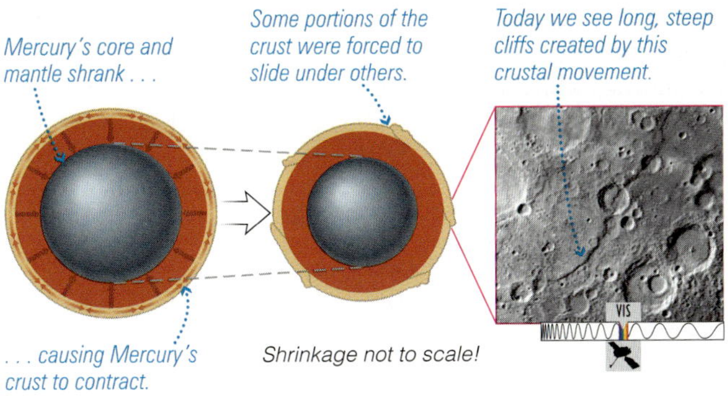

a Mercury's long cliffs probably formed when its core cooled and contracted, causing the mantle and lithosphere to shrink. This diagram shows how the cliffs probably formed as the surface crumpled.

b This cliff extends about 100 kilometers in length, and its vertical face is as much as 2 kilometers tall. (Photo from *Mariner 10*.)

Figure 9.23 Long cliffs on Mercury offer evidence that the entire planet shrank early in its history, perhaps by as much as 20 kilometers in radius.

stresses that created the great cliffs. The contraction probably also closed off any remaining volcanic vents, ending Mercury's period of volcanism.

Because Mercury is larger than the Moon, it probably retained its internal heat a bit longer. Nevertheless, it is still small enough that its interior should have cooled relatively quickly. Crater counts on Mercury suggest that all its volcanic and tectonic activity ceased within the first billion years after its formation. Like the Moon, Mercury has been geologically dead for most of its history.

9.4 Geology of Mars

Mars is the next stop on our geological tour. Because Mars is much larger than the Moon or Mercury (see Figure 9.1), we expect it to have had a more interesting and varied geological history. However, it is much smaller than Venus or Earth—its radius is about half that of Earth and its mass is only about 10% that of Earth—so we expect it to be less geologically active than our own world. Observations confirm this basic picture, though Mars's greater distance from the Sun also plays a role in its geology; Mars is about 50% farther than Earth from the Sun.

• How did Martians invade popular culture?

Have you ever noticed that people often speak of Martians but rarely of, say, Venusians or Jupiterians? It's no accident, but rather the result of misconceptions that arose from early telescopic observations of Mars.

The story begins in the late eighteenth century with the brother and sister astronomers William and Caroline Herschel. Although they are most famous for discovering the planet Uranus, they also made many observations of Mars. Their observations revealed several uncanny resemblances to Earth, including a very similar axis tilt, a day just slightly longer than 24 hours, the presence of polar caps, and seasonal variations in appearance over the course of the Martian year (about 1.9 Earth years). In 1784, William Herschel gave a talk in which he spoke with confidence about "inhabitants" of Mars.

The hypothetical Martians got a bigger break about a century later. In 1879, Italian astronomer Giovanni Schiaparelli reported seeing a network of 79 linear features on Mars through his telescope. He named these features *canali,* by which he meant the Italian word for "channels," but it was frequently translated as "canals." Inspired by what sounded like evidence of intelligent life, American astronomer Percival Lowell commissioned the building of an observatory in Flagstaff, Arizona, for the study of Mars.

The Lowell Observatory opened in 1894. Barely a year later, Lowell published detailed maps of the Martian canals and the first of three books in which he argued that the canals were the work of an advanced civilization. He suggested that Mars was falling victim to unfavorable climate changes and that the canals had been built to carry water from the poles to thirsty cities elsewhere. Lowell's work drove rampant speculation about Martians and fueled science fiction fantasies such as H. G. Wells's *The War of the Worlds* (published in 1898). The public mania drowned out the skepticism of astronomers who saw no canals through their telescopes or in their photographs (Figure 9.24).

The debate over the existence of Martian canals and cities was not entirely put to rest until 1965, when NASA's *Mariner 4* spacecraft flew past Mars and sent back photos of a barren, cratered surface. Lowell's canals were nowhere to be seen, confirming that they had been the work of a vivid imagination looking at blurred telescopic images.

The lack of ancient cities may have dampened the enthusiasm of science fiction writers, but scientists find Mars as intriguing as ever. More spacecraft have visited Mars than any other planet, and plans call for future spacecraft visits to Mars about every 26 months to coincide with optimal orbital alignments of Earth and Mars. Indeed, while we

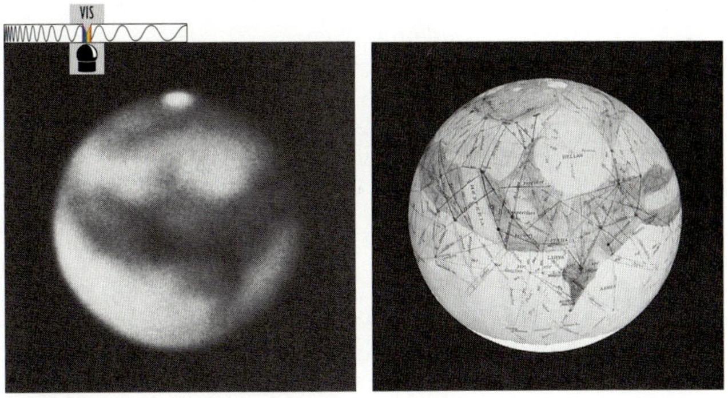

Figure 9.24 The image at left is a telescopic photo of Mars. The image at right is a drawing of Mars made by Percival Lowell. As you can see, Lowell had a vivid imagination. Nevertheless, if you blur your eyes while looking at the photo, you might see how some of the features resemble what Lowell thought he saw.

now know there have never been Martian cities, it's possible that Mars has harbored more primitive life. The search for past or present life on Mars is one of the major goals of Mars exploration.

• What are the major geological features of Mars?

Mars's total surface area is only about one-fourth that of Earth, but both planets have nearly the same amount of *land* area because Earth's surface is about three-fourths covered by water. Thus, there's plenty of territory to explore on Mars. For context, Figure 9.25 shows the full surface of Mars and the locations of features we will discuss in this book. As we'll see, close study of Mars reveals extensive evidence of all four geological processes.

Impact Cratering on Mars Aside from the polar caps, the most striking feature of Figure 9.25 is the dramatic difference in terrain around different parts of Mars. Much of the southern hemisphere has relatively high elevation and is scarred by numerous large impact craters, including the very large crater known as the Hellas Basin. In contrast, the northern plains show few impact craters and tend to be below the average Martian surface level. The differences in crater crowding clearly show that the southern highlands are a much older surface than the northern plains, which must have had their early craters erased by other geological processes.

THINK ABOUT IT

Which fundamental planetary property (size, distance from the Sun, or rotation rate) explains why Earth does not have any terrain that is as heavily cratered as the southern highlands of Mars?

Volcanism on Mars Further study suggests that volcanism was the most important of the processes responsible for erasing ancient craters on Mars, although tectonics and erosion also played a part. However, no one knows why volcanism affected the northern plains so much more than the southern highlands or why the two regions differ so much in elevation.

More dramatic evidence of volcanism on Mars comes from several towering shield volcanoes. One of these, Olympus Mons, is the tallest known volcano in the solar system (Figure 9.26; see also Figure 9.12b). Its base is some 600 kilo-

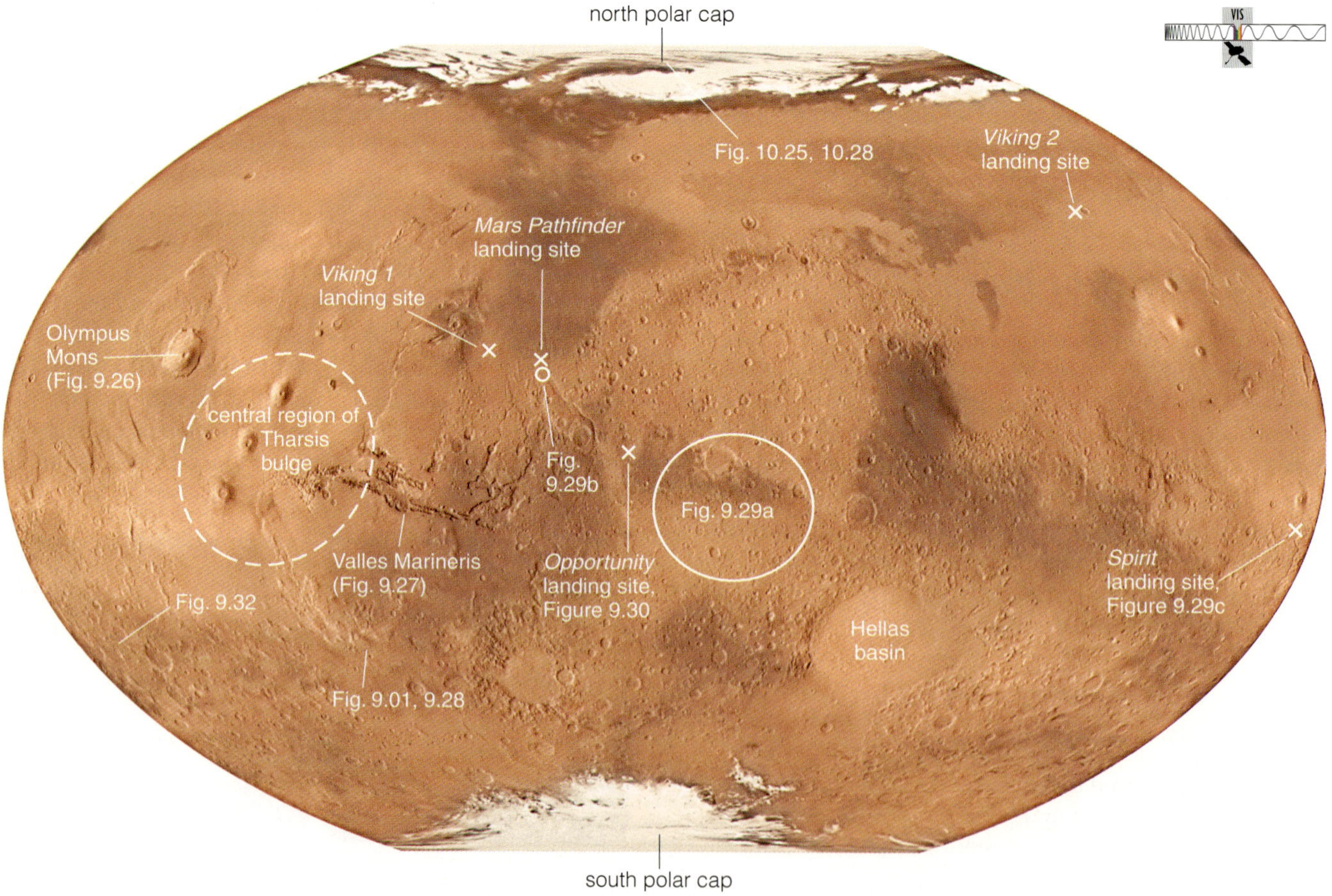

Figure 9.25 Interactive Photo This image showing the full surface of Mars is a composite made by combining more than 1,000 images with more than 200 million altitude measurements from the *Mars Global Surveyor* mission. Several key geological features are labeled, and the locations of features shown in close-up photos elsewhere in this book are marked.

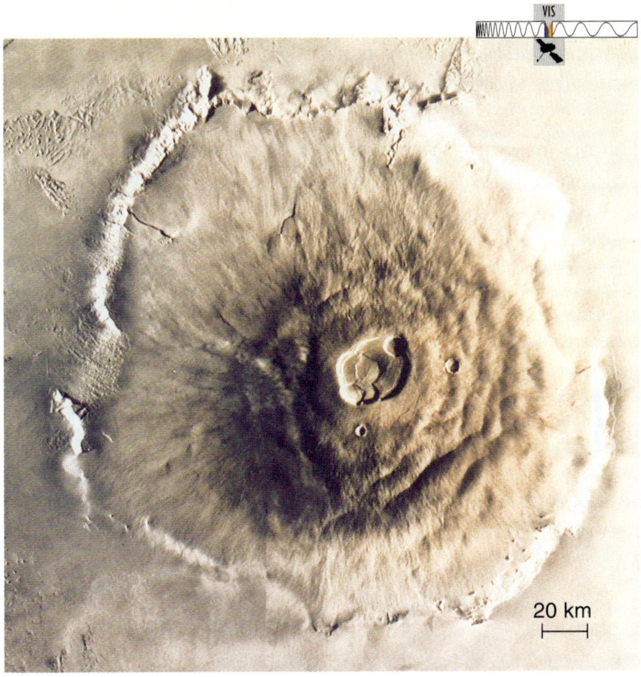

Figure 9.26 Olympus Mons, the tallest volcano in the solar system, covers an area the size of Arizona and rises higher than Mount Everest on Earth. Note the tall cliff around its rim and the central volcanic crater from which lava erupted.

meters across, large enough to cover an area the size of Arizona. Its peak stands about 26 kilometers above the average Martian surface level, or some three times as high as Mount Everest stands above sea level on Earth. Much of Olympus Mons is rimmed by a cliff that in places is 6 kilometers high.

Olympus Mons and several other large volcanoes are concentrated on or near the continent-size *Tharsis Bulge* (see Figure 9.25). Tharsis, as it is usually called, is some 4,000 kilometers across, and most of it rises several kilometers above the average Martian surface level. It was probably created by a long-lived plume of rising mantle material that bulged the surface upward and provided the molten rock for the eruptions that built the giant volcanoes.

Could volcanoes still be active on Mars? Until recently, we didn't think so. We expect Mars to be much less volcanically active than Earth, because its smaller size has allowed its interior to cool much more, and Martian volcanoes show enough impact craters on their slopes to suggest that they have been inactive for at least tens of millions of years and possibly much longer. However, geologically speaking, tens of millions of years is not that long. In addition, radiometric dating of meteorites that appear to have come from Mars (so-called Martian meteorites [Section 24.3]) shows some of them to be made of volcanic rock that solidified from molten lava as little as 180 million years ago—quite recent in the 4.5-billion-year history of the solar system. Given this evidence of geologically recent volcanic eruptions, it is likely that Martian volcanoes will erupt again someday, though not necessarily in our lifetimes. Nevertheless, the Martian interior is presumably cooling and its lithosphere thickening, so that Mars will become as geologically dead as the Moon and Mercury within a few billion years.

Tectonics and Valles Marineris Mars also has tectonic features, though none on a global scale like the plate tectonics of Earth. The most prominent tectonic feature is the long, deep system of valleys called *Valles Marineris* (Figure 9.27). Named for the *Mariner 9* spacecraft that first imaged it, Valles Marineris extends almost a fifth of the way along the planet's equator. It is as long as the United States is wide and almost four times as deep as Earth's Grand Canyon.

No one knows exactly how Valles Marineris formed. Parts of the canyon are completely enclosed by high cliffs on all sides, so neither flowing lava nor water could have been responsible. Extensive cracks on its western end run up against the Tharsis Bulge (see Figure 9.25), suggesting a connection between the two features. Valles Marineris may have formed through tectonic stresses accompanying the uplift of material that created the Tharsis Bulge, cracking the surface and leaving the tall cliff walls of the valleys.

• What geological evidence tells us that water once flowed on Mars?

Impacts, volcanism, and tectonics explain most of the major geological features of Mars, but closer examination shows ample evidence of features of erosion. For example, Figure 9.28 looks much like dry riverbeds on Earth seen from above. These channels appear to have been carved by running water, though no one knows whether the water came from runoff after rainfall, from erosion by water-rich debris flows, or from an underground source. Regardless of the specific mechanism, water was almost certainly responsible, because it is the only substance that could have been liquid under past Martian conditions and that is sufficiently abundant to have created such extensive erosion features.

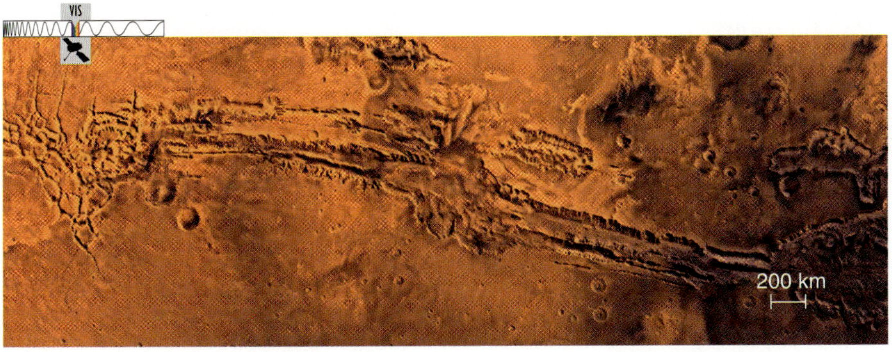

Figure 9.27 Valles Marineris is a huge system of valleys on Mars created in part by tectonic stresses.

Figure 9.28 This photo, taken by the *Viking Orbiter,* shows what appear to be dried-up riverbeds. Toward the top of the image we see many individual tributaries, which merge into the larger "river" near the lower right. Counts of craters near the channels indicate that they formed at least 2–3 billion years ago.

If you were to visit Mars, however, the idea that parts of the surface were shaped by flowing water might seem quite strange. No liquid water exists anywhere on the surface of Mars today. We know this not only because we've studied most of the surface in reasonable detail, but also because the surface conditions would not allow liquid water to be present as lakes, rivers, or even puddles. In most places and at most times, Mars is so cold that any liquid water would immediately freeze into ice. Even when the temperature rises above freezing, as it often does at midday near the equator, the air pressure is so low that liquid water would quickly evaporate [Section 5.3]. In other words, liquid water is *unstable* on Mars today: If you donned a spacesuit and took a cup of water outside your pressurized spaceship, the water would either freeze or boil away (or both) almost immediately.

Evidence for Ancient Water Flows The fact that we see water erosion features on Mars but that liquid water is absent today tells us that Mars must once have had very different surface conditions—conditions such as warmer temperatures and greater air pressure that would have allowed water to flow and rain to fall. This warmer and wetter period must have ended long ago, as you can see by looking again at Figure 9.28. Notice that a few impact craters lie on top of the channels. From counts of the craters in and near them, it appears that these channels are at least 2–3 billion years old, meaning that water has not flowed through them since that time.

Other orbital evidence also argues that Mars had rain and surface water in the distant past. Figure 9.29a shows a broad region of the ancient, heavily cratered southern highlands. Notice the indistinct rims of many large craters and the relative lack of small craters. Both facts argue for ancient rainfall, which would have eroded crater rims and erased small craters altogether. Figure 9.29b shows a close-up of a crater floor with sculpted patterns that suggest it was once the site of a lake. Ancient rains may have filled the crater, allowing sediments to build up from material that settled to the bottom. The sculpted patterns in the crater bottom may have been created as erosion by wind exposed layer upon layer of sedimentary rock, much as erosion by the water in the Colorado River exposed the layers visible in the walls of the Grand Canyon on Earth. Figure 9.29c shows a three-dimensional perspective of the surface that suggests water once flowed between two ancient crater lakes.

Figure 9.29 More orbital evidence of past water on Mars.

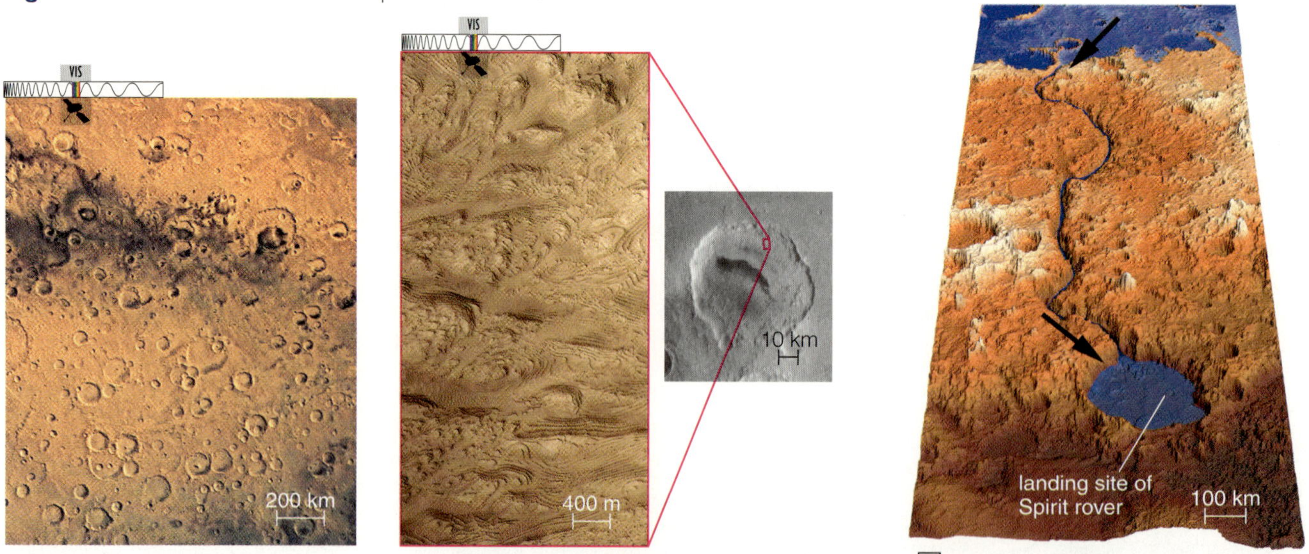

a This photo shows a broad region of the southern highlands on Mars. The eroded rims of large craters and the lack of many small craters suggest erosion by rainfall.

b This close-up view shows the floor of the crater shown earlier in Figure 9.9c. The sculpted patterns appear to be layers of sedimentary rock that were laid down at a time when the crater was filled with water.

c This computer-generated perspective view shows how a Martian valley forms a natural passage between two possible ancient lakes (shaded blue). Vertical relief is exaggerated 14 times to reveal the topography. In 2004, the *Spirit* rover landed in the crater (called Gusev) at the bottom of this picture.

Surface studies have provided additional evidence for past water on Mars. In 2004, two robotic rovers named *Spirit* and *Opportunity* landed on nearly opposite sides of Mars. *Spirit* landed in Gusev Crater, the site of the possible ancient lake shown in Figure 9.29c. *Opportunity* landed in the Meridiani Plains, where orbital spacecraft had detected spectroscopic hints of minerals that form in water. The twin rovers carried dozens of cameras, several instruments to identify rock composition, and a grinder to expose fresh rock for analysis. The rovers, designed for just 3 months of operation, were still going strong as this book was being written, more than a year and a half after their arrival.

Both rovers found solid evidence that liquid water was once plentiful on Mars. Rocks at the *Opportunity* landing site contain tiny spheres—nicknamed "blueberries" although they're neither blue nor as large as the berries we find in stores—and odd indentations suggesting that they formed in standing water, or possibly by groundwater percolating through rocks (Figure 9.30). Compositional analysis showed that the abundant "blueberries" contained the iron-rich mineral hematite, and other rocks contained the sulfur-rich mineral jarosite; both minerals form in water. Detailed chemical analysis supports the case for formation in an acidic, salty environment—possibly a sea or ocean. Moreover, a close look at the layering of the sedimentary rocks suggests a changing environment of waves and/or wind. Taken together, the orbital and surface studies provide overwhelming evidence for abundant water in Mars's past.

Martian Water Today If water once flowed over large portions of Mars, where did it all go? As we'll discuss in Chapter 10, much of the water was probably lost to space forever. However, significant amounts of water apparently still remain, frozen at the polar caps and in the top meter or so of the surface soil around much of the rest of the planet (Figure 9.31).

If water ice exists on Mars this close to the surface, even more water probably lies deeper underground. If there is still volcanic heat on Mars, this water may sometimes melt and flow. Although we have found no geological evidence to suggest that any large-scale water flows have occurred on Mars in the past billion years, orbital photographs offer tantalizing hints of smaller-scale water flows in much more recent times.

The strongest evidence for liquid water in recent times comes from photos of gullies on crater and channel walls. In Figure 9.32, for example, note the striking similarity to the gullies we see on almost any eroded slope on Earth.

a This sequence zooms in on a rock outcropping near the rover's landing site. The outcrop stands about knee-high. The close-up shows a piece of the rock about 3 centimeters across (about the width of your big toe). The layered structure, the odd indentations, and the small sphere all support the idea that the rock formed from sediments in standing water.

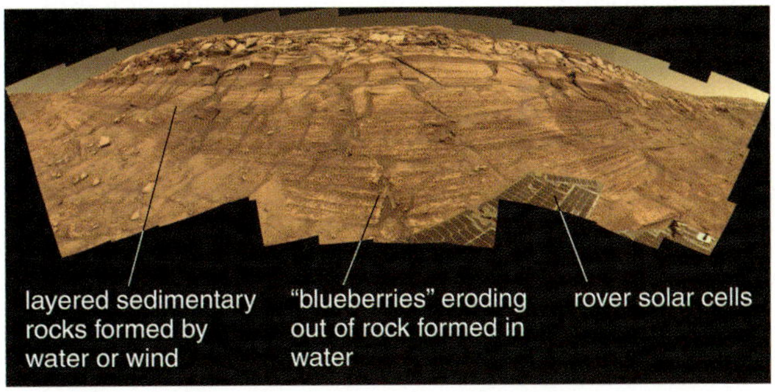

b The wall of Endurance Crater, where the impact exposed deep bedrock layers profoundly altered by water. Close-ups show the durable hematite "blueberries" eroding out of the rock and tumbling downhill in piles. Changing tilts of the rock layers hint at changing wind or waves during formation.

Figure 9.30 These photos were taken by the *Opportunity* rover, which landed on Mars in 2004.

VIS

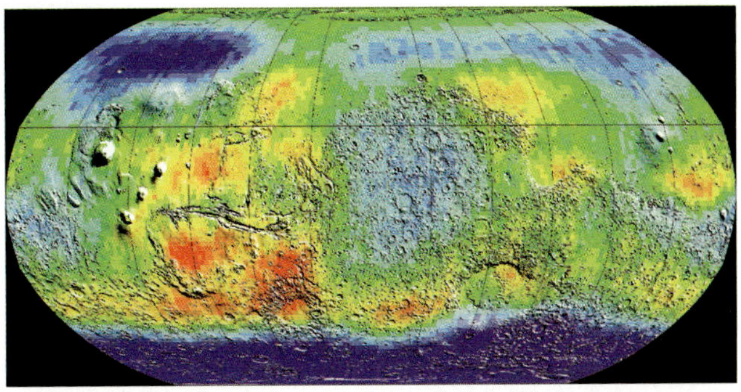

Figure 9.31 This map, made with data from *Mars Odyssey*, represents the hydrogen content of the Martian surface soil. The blue areas contain the most hydrogen, probably because they represent regions in which the top meter or so of surface soil contains water ice.

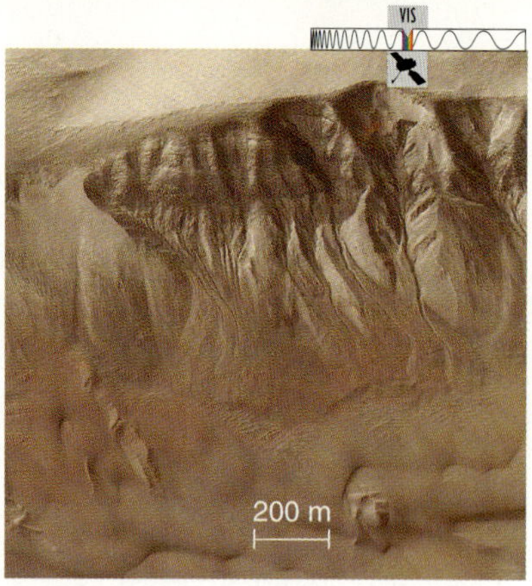

Figure 9.32 This photograph from the *Mars Global Surveyor* shows gullies on a crater wall. The gullies may have been formed by water melting under the protective cover of snowpack.

One hypothesis suggests that the gullies form when snow accumulates on the crater walls in winter and melts away in spring. Because the gullies are relatively small (note the scale bar in Figure 9.32), they should be gradually covered over by blowing sand during Martian dust storms. Gullies that are still clearly visible must be no more than a few million years old. Geologically speaking, this time is short enough to make it quite likely that water flows are still forming gullies today. Thus, while Mars's era of abundant surface water is long gone, small amounts of liquid water may still flow on occasion.

Even if water sometimes does still flow, Mars clearly was much warmer and wetter at times in the past than it is today. Ironically, Percival Lowell's supposition that Mars was drying up has turned out to be basically correct, although in a very different way than he imagined. We'll discuss the reasons for Mars's dramatic climate change in Chapter 10.

9.5 Geology of Venus

The surface of Venus is searing hot with brutal pressure [Section 7.2], making it seem quite unlike the "sister planet" to Earth it is sometimes called. However, beneath the surface, Venus must be quite similar to Earth. Venus is nearly the same size as Earth (see Figure 9.1), with a radius only about 5% smaller than Earth's. Venus and Earth are also quite close together relative to the overall size of the solar system, so both should have been built from the same kinds of planetesimals. The very similar densities of Venus and Earth (see Table 7.1) offer further evidence that both planets have the same overall composition.

With the same basic size and the same basic composition, we expect the interiors of Venus and Earth to be quite similar in structure and to retain about the same level of internal heat today. Nevertheless, we see ample evidence of "skin-deep" differences in surface geology.

• What are the major geological features of Venus?

Venus's thick cloud cover prevents us from seeing through to its surface, but we can study its geological features with radar because radio waves can pass through clouds. **Radar mapping** bounces radio waves off the surface and uses the reflections to create three-dimensional images of the surface. From 1990 to 1993, the *Magellan* spacecraft used radar to map the surface of Venus, discerning features as small as 100 meters across. Scientists have named almost all the geological features on Venus for goddesses and famous women.

Figure 9.33a shows a global map of Venus based on the *Magellan* radar observations. Three large, elevated "continents" (the three regions labeled "Terra" in Figure 9.33a) are the biggest features on the surface.

Venus has some impact craters (Figure 9.33b), but far fewer than Mercury, the Moon, or Mars. Moreover, the relatively few impact craters are distributed fairly uniformly over the entire planet, suggesting that the surface is about the same age everywhere. Precise crater counts suggest that the surface of Venus is about 750 million years old. Thus, no craters from the heavy bombardment remain, and the only craters we see on Venus are those created by impacts during the past 750 million years. Older craters must have been erased by other geological processes. Geologists have also noted an absence of small craters on Venus—the very type of crater that is most common on other worlds. Apparently, Venus lacks small craters because small impactors completely burn up as they enter Venus's thick atmosphere. Other terrestrial worlds have smaller craters (as well as large ones), because their thinner atmospheres allow smaller objects to survive the plunge to the surface.

Volcanism is clearly important on Venus. The surface shows abundant evidence of lava plains and many volcanic mountains. Some mountains are shield volcanoes (Figure 9.33c), indicating eruptions in which the lava was about as runny as the lava that formed the Hawaiian Islands on Earth. A few volcanoes have steeper sides, indicating eruptions of a higher-viscosity lava (Figure 9.33d). Volcanoes almost undoubtedly remain active on Venus, though we have not observed any eruptions. However, the composition of Venus's clouds suggests that volcanoes must still be active on geological timescales (erupting within the past 100 million years). The clouds contain sulfuric acid, which is made from sulfur dioxide (SO_2) and water. Sulfur dioxide enters the atmosphere through volcanic outgassing, but once in the atmosphere it is steadily removed by chemical reactions with surface rocks. Thus, the fact that sulfuric acid clouds still exist means that outgassing must continue to supply sulfur dioxide to the atmosphere.

The most remarkable features on Venus are tectonic in origin. Its crust is quite contorted. In some regions, the surface appears to be fractured in a regular pattern (Figure 9.33e). Other tectonic features, called *coronae* (Figure 9.33f), point directly to mantle convection on Venus. *Coronae* (Latin for "crowns") were probably formed by hot, rising plumes in the mantle. The plumes push up on the crust, forming concentric tectonic stretch marks on the surface. The plumes also force lava to the surface, explaining why numerous volcanoes are found near coronae.

The biggest difference between the geology of Venus and that of Earth is the lack of erosion on Venus. We might

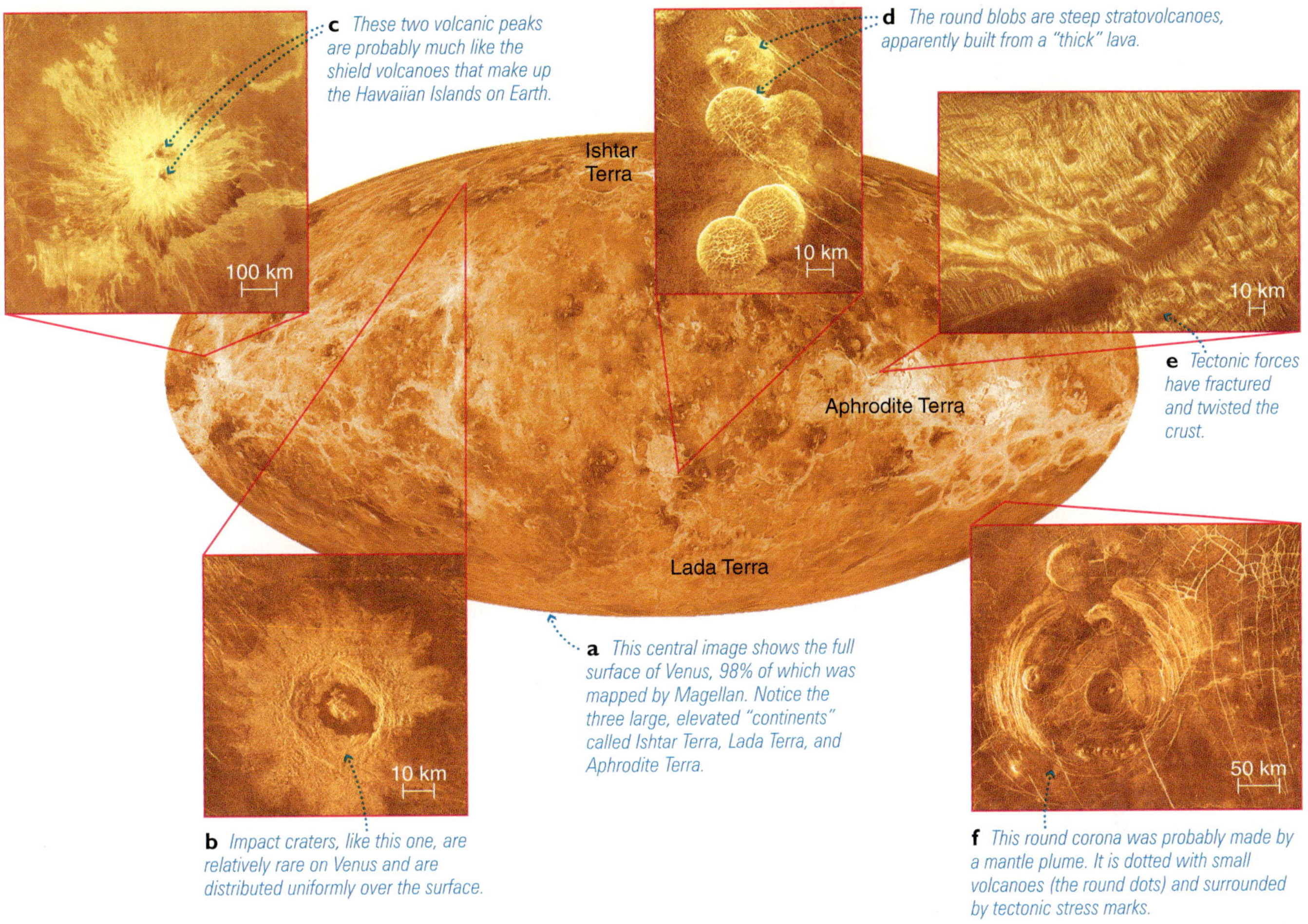

Figure 9.33 These images show the surface of Venus as revealed by radar observations from the *Magellan* spacecraft. Bright regions in the radar images represent rough areas or higher altitudes.

RADIO

naively expect Venus's thick atmosphere to produce strong erosion, but the view both from orbit and on the surface suggests that erosion is only a minor process. The then–Soviet Union landed two probes on Venus's surface in 1975. Before the intense surface heat destroyed them, the probes returned images of a bleak, volcanic landscape with little evidence of erosion (Figure 9.34). We can trace the lack of erosion on Venus to two simple facts. First, Venus is far too hot for any type of rain or snow on its surface. Second, Venus has virtually no wind or weather because of its slow rotation. Venus rotates so slowly—once every 243 days—that its atmosphere barely stirs the surface.

THINK ABOUT IT

Suppose Venus rotated as fast as Earth. How would this change its relative levels of volcanism, tectonics, and erosion? Explain.

• Does Venus have plate tectonics?

While we can easily explain the lack of erosion on Venus, another "missing feature" in Venus's geology is very surprising: Venus shows no evidence at all of Earth-like plate tectonics. Almost all the major features of Earth's geology are attributable to plate tectonics, including the ridges that extend thousands of kilometers along the seafloor, the deep ocean trenches found near continental boundaries, and long mountain ranges like the Rockies and Himalayas. Venus does not have any similar features.

VIS

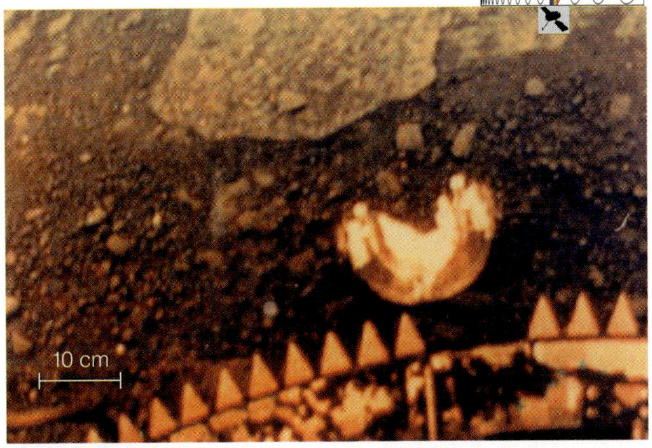

Figure 9.34 This photo from the former Soviet Union's *Venera* lander shows the surface around the base of the spacecraft (visible at lower right). This region of the planet is covered in volcanic plains formed some 750 million years ago. Many volcanic rocks are visible, hardly affected by erosion despite their age.

Moreover, Venus shows evidence of a very different type of global geological change. On Earth, plate tectonics re-sculpts the surface gradually, so that different regions have different ages. But, as we've already discussed, the uniform distribution of impact craters on Venus tells us that the entire surface is about 750 million years old. Apparently, Venus's surface was somehow "repaved" at that time.

We do not know how much of the repaving was due to tectonic processes and how much was due to volcanism, but both probably were important. It is even possible that plate tectonics played a role before and during the repaving, only to stop for some reason after the repaving episode was over. Regardless of whether Venus may have had plate tectonics in the past, its lack of plate tectonics today poses a major mystery.

Earth's lithosphere fractured into plates because of forces arising from the underlying mantle convection. The lack of plate tectonics on Venus suggests either that it has weaker mantle convection or that its lithosphere somehow resists fracturing. The first possibility seems unlikely: Because Venus is so similar in size to Earth, it should have a similar level of mantle convection. Most scientists therefore suspect that Venus's lithosphere resists fracturing into plates because it is thicker and stronger than Earth's lithosphere, though we have no direct evidence to support this hypothesis.

Even if a thicker and stronger lithosphere explains the lack of plate tectonics on Venus, we are still left with the question of why the lithospheres of Venus and Earth should differ. One possible answer is Venus's high surface temperature. Venus is so hot that any water in its crust and mantle has probably been baked out over time. Water tends to soften and lubricate rock, so its loss would have tended to thicken and strengthen Venus's lithosphere. If this idea is correct, then Venus might have had plate tectonics if it had not become so hot in the first place.

The mystery of plate tectonics is just one of many reasons why scientists would like to explore Venus in greater detail. Unfortunately, it is probably the most difficult to study of the terrestrial worlds. Its thick cloud cover always hides the surface from orbiting cameras, so only improved radar mapping will tell us more about its global geology. The harsh surface conditions make it difficult for robotic landers to survive, and sending astronauts to Venus seems almost completely out of the question. Still, we are hoping to learn more from a European mission called *Venus Express*; if all goes well, it will reach Venus in mid-2006.

THINK ABOUT IT

What is the current status of the *Venus Express* mission? If it has reached Venus, has it made any important new discoveries?

9.6 The Unique Geology of Earth

Most aspects of Earth's geology make sense in light of what we know about other terrestrial worlds. Earth's size—largest of the five terrestrial worlds—explains why our planet retains internal heat and hence remains volcanically and tectonically active. Earth's rampant erosion by water and wind is explained by the combination of our planet's size, distance from the Sun, and rotation rate: Earth's size leads to active volcanism and hence outgassing, its distance from the Sun keeps temperatures in a range in which outgassed water can be liquid, and its moderately rapid rotation drives weather that produces erosion.

The only feature of our planet's geology that stands clearly apart from the geology we see on other worlds is its unique style of tectonics. As we've seen, even similar-size Venus lacks Earth-like plate tectonics. Moreover, the importance of plate tectonics goes far beyond basic geology; as we'll discuss in Chapter 10, plate tectonics plays a crucial role in the stability of Earth's climate. Indeed, our own existence represents the culmination of a 4-billion-year chain of biological evolution that probably could not have occurred without the climate stability that we owe to plate tectonics.

• How do we know that Earth's surface is in motion?

The term *plate tectonics* refers to the scientific theory that explains much of Earth's surface geology as a result of the slow motion of *plates*—fractured pieces of the lithosphere. According to the theory, the lithosphere fractured because of stresses generated by underlying mantle convection. The plates essentially "float" over the mantle, gradually moving over, under, and around each other as convection moves Earth's interior rock. (Because the term *plate tectonics* refers to the theory, it is generally considered to be singular rather than plural, despite the *-s* on *tectonics.*)

Earth's lithosphere is broken into more than a dozen plates, with known boundaries shown in Figure 9.35. The motions of the plates are barely noticeable on human timescales—a few centimeters per year, which is about the rate at which your fingernails grow. However, geologists can now measure plate motions by comparing readings taken with the global positioning system (GPS) [Section S1.3] on either side of plate boundaries.

The GPS measurements offer the most direct evidence of plate motion. However, the overall theory of plate tectonics rests on three more significant lines of evidence: evidence of past continental motion, evidence that plates spread apart on seafloors, and a difference between the nature of Earth's crust on the seafloors and the continents. Let's investigate each of these.

Continental Motion We usually trace the origin of the theory of plate tectonics to a slightly different idea proposed in 1912 by German meteorologist and geologist Alfred Wegener: *continental drift,* the idea that continents gradually drift across the surface of the Earth. Wegener got his idea in part from the puzzlelike fit of continents such as South America and Africa (Figure 9.36). He also noted that similar types of distinctive rocks and rare fossils were found in eastern

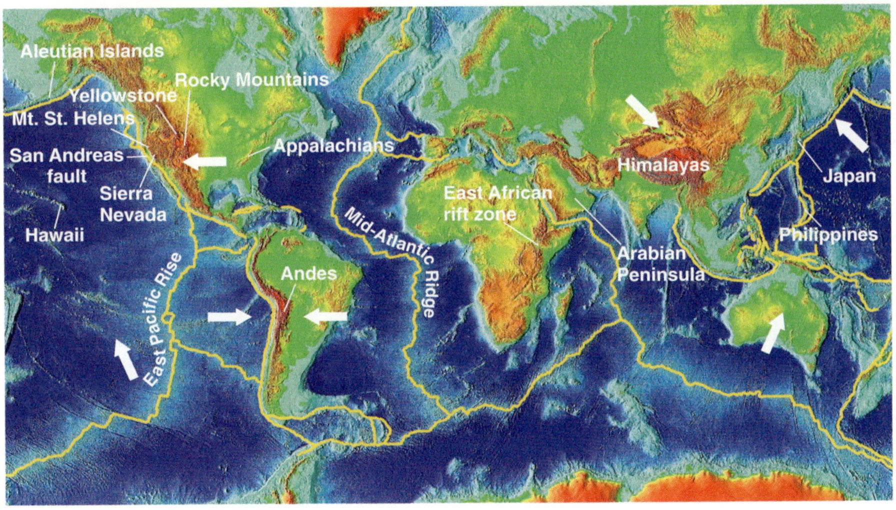

Figure 9.35 This relief map shows plate boundaries (solid yellow lines), with arrows to represent directions of plate motion. Color represents elevation, progressing from blue (lowest) to red (highest). Labels identify some of the geological features discussed in the text.

South America and western Africa, suggesting that these two regions once had been close together.

Despite these strong hints, no one at the time knew of a mechanism that could allow the continents to push their way through the solid rock beneath them. Wegener suggested that Earth's gravity and tidal forces from the Sun and Moon were responsible, but other scientists quickly showed that these forces were too weak to move continents around. As a result, Wegener's idea of continental drift was widely rejected by geologists for decades after he proposed it, even though his idea of a "continental fit" for Africa and South America ultimately gained acceptance. Today, far more extensive fossil evidence makes it very clear that the continents really were arranged differently in the past.

Seafloor Spreading In the mid-1950s, scientists began to observe geological features that suggested a mechanism for continental motion. In particular, they discovered *mid-ocean ridges* (such as the mid-Atlantic Ridge shown in Figure 9.35) along which mantle material erupts onto the ocean floor, pushing apart the existing seafloor on either side. This **seafloor spreading** helped explain how the continents could move apart with time.

Today, we understand that continents move in concert with underlying mantle convection, driven by the heat released from Earth's interior. Because this idea is quite different from Wegener's original notion of continents plowing through the solid rock beneath them, geologists no longer use the term *continental drift* and instead consider continental motion within the context of plate tectonics.

Figure 9.36 The puzzle-like fit of South America and Africa.

Seafloor Crust and Continental Crust The third line of evidence for plate tectonics comes from the fact that Earth's surface has two very different types of crust: one type found on seafloors and the other on continents (Figure 9.37). **Seafloor crust** is thinner, denser, and younger in age than **continental crust**. No such compositional difference has been found on other planets.

Seafloor crust forms a layer over the mantle that is typically only 5–10 kilometers thick. It is made primarily of basalt; recall that basalt is a mixture of many different minerals that tends to have relatively high density but is fairly runny when melted. The runniness explains why molten basalt spreads outward from volcanoes along mid-ocean ridges. Radiometric dating shows that seafloor basalt is quite young—usually less than 200 million years old—indicating that it erupted to the surface relatively recently in geological history. Further evidence of the young age of seafloors comes from studies of impact craters. Large impacts should occur more or less uniformly over Earth's surface, and the oceans won't stop a large asteroid or comet from making a huge crater. However, we find far

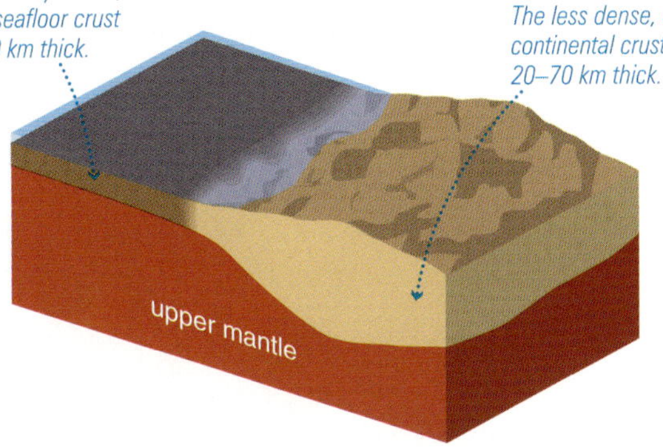

Figure 9.37 Earth has two distinct kinds of crust.

fewer large craters on the seafloor than on the continents, which means that seafloor crust must have been created more recently.

Continental crust forms a much thicker layer over the mantle than seafloor crust. It is typically between 20 and 70 kilometers thick, although it sticks up only slightly higher than seafloor crust because its sheer weight presses it down farther into the mantle below. Most of the rock found in continental crust is lower in density than the basaltic rock of the seafloor; for example, the most common type of rock on the continents, granite, is significantly less dense than basalt. Moreover, while seafloor crust is uniformly young, continental crust spans a wide range of ages. Some rocks in the continental crust date to as old as 4.0 billion years.

The two types of crust make it clear that Earth's surface must undergo continual change. New seafloor crust continually emerges at sites of seafloor spreading, while the wide age range of continental crust tells us that the continents have gradually built up over time.

• How is Earth's surface shaped by plate tectonics?

The theory of plate tectonics explains all three lines of evidence (continental motion, seafloor spreading, and two types of crust) as direct results of the way plates move about on Earth. Over millions of years, the movements involved in plate tectonics act like a giant conveyor belt for Earth's lithosphere (Figure 9.38).

Recycling of the Seafloor Mid-ocean ridges occur at places where mantle material rises upward, creating new seafloor crust and pushing plates apart. The newly formed basaltic crust cools and contracts as it spreads sideways from the central ridge, giving seafloor spreading regions their characteristic ridged shape (see Figure 9.38). Worldwide along the mid-ocean ridges, new crust covers an area of about 2 square kilometers every year, enough to replace the entire seafloor within about 200 million years—and thus explaining the less-than-200-million-year age of seafloor rocks.

Over tens of millions of years, any piece of seafloor crust gradually makes its way across the ocean bottom, then finally gets recycled into the mantle in the process we call **subduction**. Subduction occurs where a seafloor plate meets a continental plate, which is generally somewhat offshore at the edge of a sloping continental shelf. As the dense seafloor crust of one plate pushes under the less dense continental crust of another plate, it can pull the entire surface downward to form a deep *ocean trench*. At some trenches, the ocean depth is more than 8 kilometers (5 miles).

Beneath a subduction zone, the descending seafloor crust heats up and may begin to melt as it moves deeper into the mantle. If enough melting occurs, the molten rock may erupt upward. As you can see in Figure 9.38, the process of subduction tends to make the melting occur under the edges of the continents, which is why so many active volcanoes tend to be found along those edges. Moreover, the lowest-density material tends to melt first, which is why the continental crust emerging from these landlocked volcanoes is lower in density than seafloor crust. This fact explains why steep-sided stratovolcanoes, made from low-density but thick lava (see Figure 9.11c), are common on Earth's continents but rare everywhere else in the solar system: Other worlds lack anything like Earth's crustal recycling with plate tectonics and therefore generally have only volcanic plains and shield volcanoes made from basalt. (Venus may have a few stratovolcanoes, but far fewer than Earth.)

The conveyor-like process of plate tectonics undoubtedly is driven by the heat flow from mantle convection, although the precise relationship between the convection cells and the plates remains an active topic of research. For example, it is not clear whether mantle convection simply pushes plates apart at the sites of seafloor spreading or whether the plates are denser than the underlying material and pull themselves down at subduction zones.

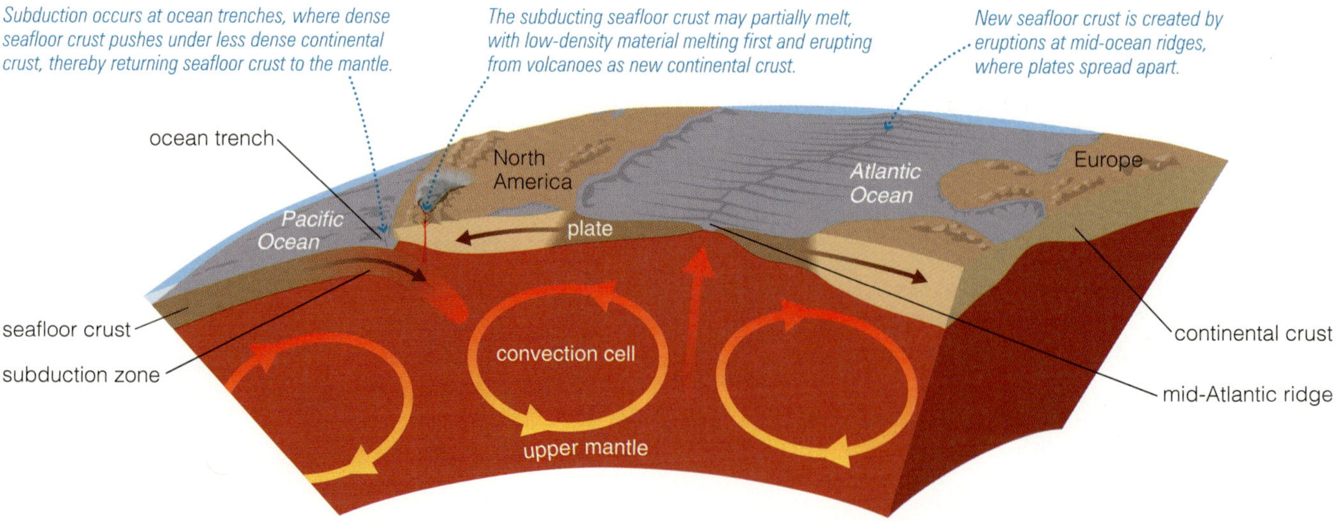

Figure 9.38 Plate tectonics acts like a giant conveyor belt for Earth's lithosphere.

Building Up Continents Note that, unlike seafloor crust, continental crust does not get recycled back into the mantle. As a result, the continents are gradually growing with time. Billions of years ago, Earth was even more of an ocean planet than it is today.

The present-day continents have been built up over billions of years, but their histories are not just a simple "piling up" of continental crust. Instead, the continents are continually reshaped by volcanism and stresses associated with plate tectonics, as well as by erosion. Figure 9.39 shows some of the complex geological history of North America, which we can use as an example of how the continents have been built.

The west coast regions of Alaska, British Columbia, Washington, Oregon, and most of California began as numerous volcanic islands in the Pacific. Long ago, plate motion carried these islands from their ocean birthplaces toward the North American mainland, where subduction continues to take the seafloor crust downward into the mantle (see Figure 9.38). Because the islands are made of lower-density continental crust, they remained on the surface and attached themselves to the edge of North America, even as the seafloor crust beneath them slid back into the mantle. Similar processes are affecting other islands today. Alaska's Aleutian Islands, a string of volcanoes located over a region where one seafloor plate is subducting under another (see the plate boundaries in Figure 9.35), are gradually growing and will someday merge together. Japan and the Philippines represent a later stage in this process; each of them once contained many small islands that merged into the fewer islands we see today. As these islands continue to grow and merge, they may eventually create a new continent or merge with an existing one.

Other portions of North America have been shaped by erosion. The Great Plains and the Midwest once were ancient seas. Erosion gradually filled these seas with sediment that hardened into sedimentary rock. The deep south formed from the buildup of sediments that were carried to these regions after eroding from other parts of the continent. Northeastern Canada features some of the oldest continental crust on Earth, worn down over time by erosion.

Tectonic processes are responsible for the mountain ranges of North America and other continents. The Sierra Nevada range in California, which lies over the region where the Pacific plate subducts under North America, formed when partially molten granite rose up from the mantle and pushed the surface rock higher. The original surface rock was sedimentary, but erosion gradually wore it away and left the granite exposed, which is why the Sierra Nevada mountains are made largely of granite.

Other mountain ranges were created by collisions between two continent-bearing plates. The Himalayas, already the tallest mountains on Earth, offer the most famous example of this process (Figure 9.40). They are still growing taller as the plate carrying the Indian subcontinent rams into the plate carrying Eurasia. Because both colliding plates are made of low-density continental crust, neither plate can subduct under the other. Instead, they push into each other and the resulting compression pushes the land upward to make mountains.

The Appalachian range in the eastern United States was also built by colliding continental plates, but in this case the collisions occurred several times. Over a period of a few hundred million years, North America apparently collided twice with South America and then with western Africa.

Figure 9.39 The major geological features of North America record the complex history of plate tectonics. Only the basic processes behind the largest features are shown here.

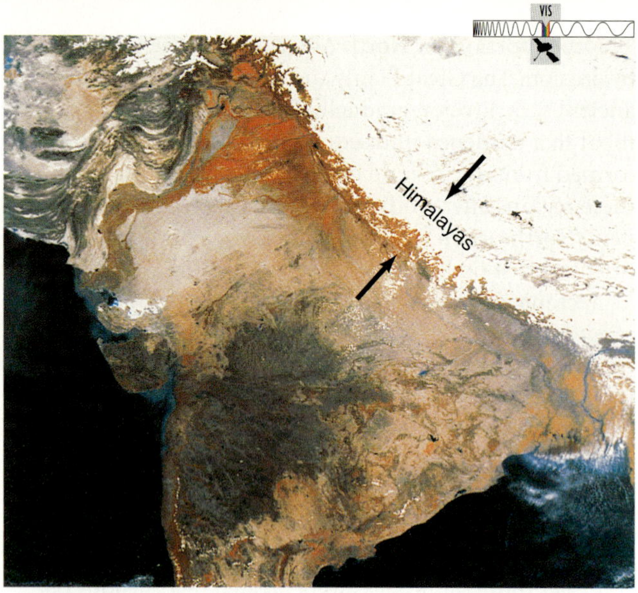

Figure 9.40 This satellite photo shows the Himalayas, which are still slowly growing as the plate carrying India pushes into the Eurasian plate. Arrows indicate the directions of plate motion.

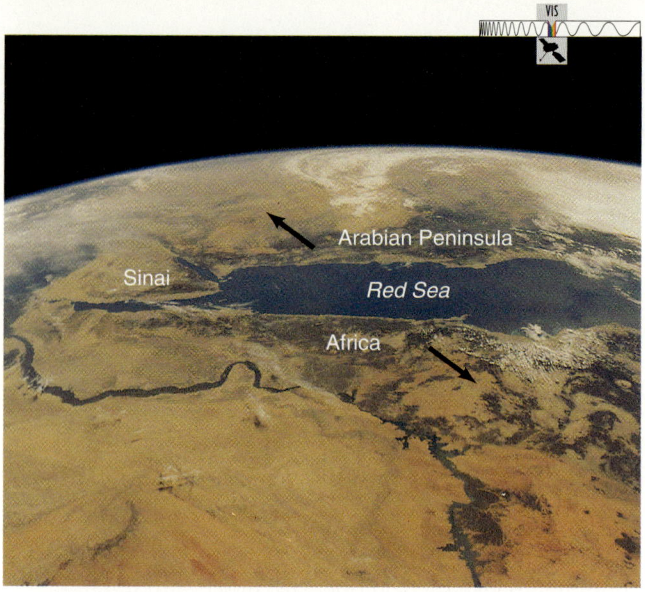

Figure 9.41 When continental plates pull apart, the crust thins and deep rift valleys form. This process tore the Arabian Peninsula from Africa, forming the Red Sea. Arrows indicate the directions of plate motion.

The Appalachians probably were once as tall as the Himalayas are now, but tens of millions of years of erosion transformed them into the fairly modest mountain range we see today. Similar processes contributed to the formation of the Rocky Mountains in the United States and Canada.

Rifts, Faults, and Earthquakes So far we have discussed features formed in places where plates are colliding, either with a seafloor plate subducting under a continental plate or with two continental plates pushing into each other. Different types of features occur in places where plates pull apart or slide sideways relative to one another.

In places where continental plates are pulling apart, the crust thins and can create a large *rift valley.* The East African rift zone is an example (see Figure 9.35). This rift is slowly growing and will eventually tear the African continent apart. At that point, rock rising upward with mantle convection will begin to erupt from the valley floor, creating a new zone of seafloor spreading. A similar process tore the Arabian Peninsula from Africa, creating the Red Sea (Figure 9.41).

Places where plates slip sideways relative to each other are marked by what we call **faults**—fractures in the lithosphere. The San Andreas Fault in California marks a line where the Pacific plate is moving northward relative to the continental plate of North America (Figure 9.42). In about 20 million years, this motion will bring Los Angeles and San Francisco together. The two plates do not slip smoothly against each other. Instead, their rough surfaces catch against each other. Stress builds up until it is so great that it forces a rapid and violent shift, causing an earthquake. In contrast to the usual motion of plates, which proceeds at a few centimeters per year, an earthquake can move plates by several *meters* in a few seconds. The movement can raise mountains, level cities, set off destructive tsunamis, and make the whole planet vibrate with seismic waves (Figure 9.43).

THINK ABOUT IT

By studying the plate boundaries in Figure 9.35, explain why the west coast states of California, Oregon, and Washington are prone to more earthquakes and volcanoes than other parts of the United States. Find the locations of recent earthquakes and volcanic eruptions worldwide. Do the locations fit the pattern you expect?

Hot Spots Not all volcanoes occur near plate boundaries. Sometimes, a plume of hot mantle material may rise in what we call a **hot spot**. The Hawaiian Islands are the result

Figure 9.42 California's San Andreas Fault marks a boundary where plates are sliding sideways as shown by the white arrows; asterisks indicate recent earthquakes. The inset photo shows a place along the San Andreas Fault where the painted lines in a road allows us to see how far the two sides of the fault moved in an earthquake.

Figure 9.43 These photos show the ancient Iranian city of Bam before and after a major earthquake in 2004. The quake killed more than 40,000 people.

of a hot spot that has been erupting basaltic lava for tens of millions of years. Plate tectonics gradually carries the Pacific plate over the hot spot, forming a chain of volcanic islands as different parts of the plate lie directly above the hot spot at different times (Figure 9.44). Today, most of the lava erupts on the Big Island of Hawaii, giving much of this island a young, rocky surface. About a million years ago, the Pacific plate lay farther to the southeast (relative to its current location), and the hot spot built the island of Maui. Before that, the hot spot created other islands, including Oahu (3 million years ago), Kauai (5 million years ago), and Midway (27 million years ago). The older islands are more heavily eroded. Midway has eroded so much that it barely rises above sea level. If plate tectonics were not moving the plate relative to the hot spot, a single huge volcano would have formed—perhaps looking somewhat like Olympus Mons on Mars (see Figure 9.26).

The movement of the plate over the hot spot continues today, building underwater volcanoes that eventually will rise above sea level to become new Hawaiian islands. The growth of a future island, named Loihi, is already well under way—prime beach real estate should be available there in about a million years or so. Hot spots can also occur beneath continental crust. For example, a hot spot is responsible for the geysers and hot springs of Yellowstone National Park. The continental plate is moving in a southwestern direction over this hot spot.

THINK ABOUT IT

Find your hometown or current home on the map in Figure 9.35 or Figure 9.39 (or both). Based on what you have learned about Earth's geology, describe the changes your home has undergone over millions (or billions) of years.

Plate Tectonics Through Time We can use the current motions of the plates to project the arrangement of continents millions of years into the past or the future. For example,

Figure 9.44 The Hawaiian Islands are just the most recent of a very long string of volcanic islands made by a mantle hot spot. The image of Loihi (lower right) was obtained by sonar, as it is still entirely underwater.

at a speed of 2 centimeters per year, a plate will travel 2,000 kilometers in 100 million years. Figure 9.45 shows several past arrangements of the continents, along with one future arrangement. Note that about 200 million years ago, the present-day continents were together in a single "supercontinent," sometimes called *Pangaea* (which means "all lands").

Mapping the sizes and locations of continents at even earlier times is more difficult. However, studies of magnetized rocks (which can record the orientation of ancient magnetic fields) and comparisons of fossils found in different places around the world have allowed geologists to map the movement of the continents much further into the past. It seems that, over the past billion years or more, the continents have slammed together, pulled apart, spun around, and changed places on the globe. Central Africa once lay at Earth's South Pole, and Antarctica once was near the equator. The continents continue to move, and their current arrangement is no more permanent than any past arrangement.

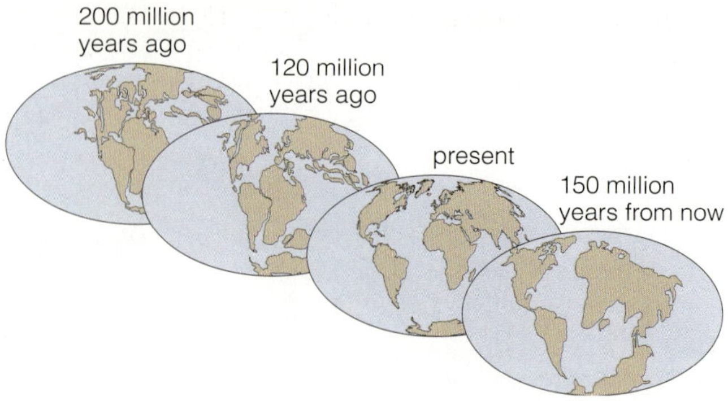

Figure 9.45 Interactive Figure Past, present, and future arrangements of Earth's continents. The present continents were all combined into a single "supercontinent" about 200 million years ago.

• Was Earth's geology destined from birth?

Now that we have completed our geological tour of the terrestrial worlds, it's interesting to ask whether fundamental planetary properties shape all geological destiny. If so, then we should be able to predict the geological features of terrestrial worlds we may someday find around other stars. If not, then we still have more to learn before we'll know whether other worlds could have Earth-like geology.

Figure 9.46 summarizes the key trends we've seen among the terrestrial worlds with respect to impact cratering, volcanism, and tectonics. All the worlds were affected similarly by impact cratering during the heavy bombardment. Volcanism and tectonics have operated to very different extents on the different worlds, but we can trace the major differences to differences in planetary size. The interiors of the smallest worlds cooled quickly, so they have not had volcanism or tectonics for billions of years. The interiors of the larger worlds have stayed hot longer, allowing volcanism and tectonics to continue for much longer time periods.

The major remaining question concerns whether fundamental properties can also explain Earth's unique plate tectonics. The lack of plate tectonics on Venus tells us that

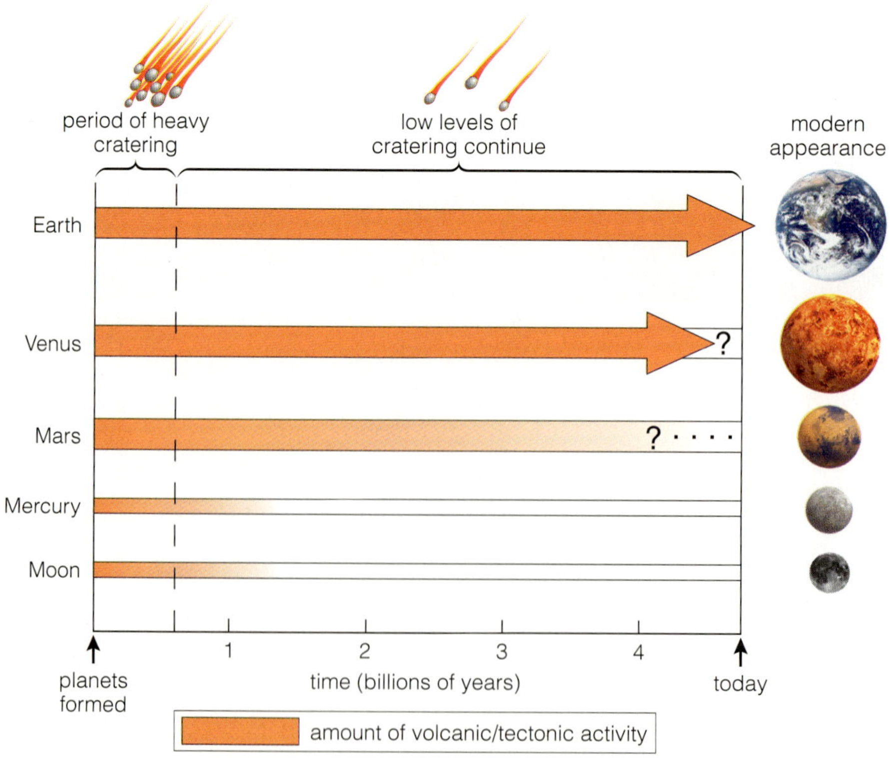

Figure 9.46 This diagram summarizes the geological histories of the terrestrial worlds. The brackets along the top indicate that impact cratering has affected all worlds similarly. The arrows represent volcanic and tectonic activity. A thicker and darker arrow means more volcanic/tectonic activity, and the arrow length tells us how long this activity persisted. Notice that the trend follows the order of planetary size: Earth remains active to this day; Venus has also been quite active, though we are uncertain whether it remains so; Mars has had an intermediate level of activity and might still have low-level volcanism; Mercury and the Moon have had very little volcanic/tectonic activity. Erosion is not shown, because it has played a significant role only on Earth (ongoing) and on Mars (where it was quite significant in the past and continues at low levels today).

we cannot attribute plate tectonics solely to size, since Venus is so similar in size to Earth. However, if we are correct in guessing that Venus lacks plate tectonics because its high surface temperature led to loss of water and lithospheric thickening, then a lower surface temperature might have led to plate tectonics on Venus as well as Earth. As we'll discuss in detail in the next chapter, we can ultimately trace Venus's high surface temperature to its nearness to the Sun. Thus, while plenty of uncertainty still remains, it seems quite likely that the broad geological histories of Earth and the other terrestrial worlds were indeed destined from birth by the fundamental properties of size, distance from the Sun, and rate of rotation.

THE BIG PICTURE

Putting Chapter 9 into Perspective

In this chapter, we have explored the geology of the terrestrial worlds. As you continue your study of the solar system, keep the following "big picture" ideas in mind:

- The terrestrial worlds all looked much the same when they were born, so their present-day differences are a result of geological processes that occurred in the ensuing 4.5 billion years.
- The extent to which different geological processes operate on the different worlds depends largely on their fundamental properties, especially their size.
- A planet's geology is largely destined from its birth—which means we should be able to predict the geology of as-yet-undiscovered planets once we know their fundamental properties.
- Earth has been affected by the same geological processes affecting the other terrestrial worlds. However, erosion is far more important on Earth than on any other terrestrial world, and Earth's unique plate tectonics may be very important to our existence.

SUMMARY OF KEY CONCEPTS

9.1 Connecting Planetary Interiors and Surfaces

- **What are terrestrial planets like on the inside?** In order of decreasing density and depth, the interior structure consists of **core, mantle,** and **crust**. The crust and part of the mantle together make up the rigid **lithosphere**. In general, a thinner lithosphere allows more geological activity.

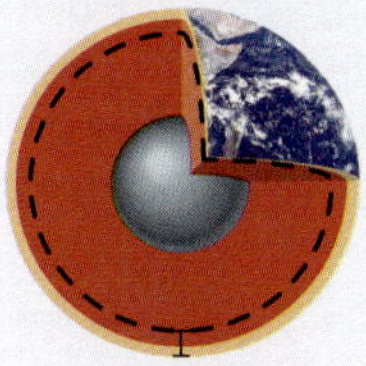

- **What causes geological activity?** Interior heat drives geological activity, by causing mantle convection, keeping the lithosphere thin, and keeping the interior partially molten. **Radioactive decay** is the main source of internal heat in terrestrial planets today. Larger planets retain internal heat longer.

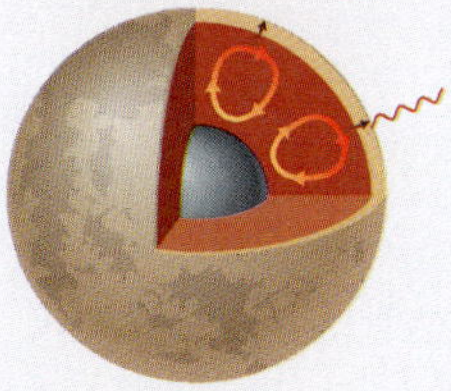

- **Why do some planetary interiors create magnetic fields?** A planetary **magnetic field** requires three things: an interior layer of electrically conducting fluid, convection of that fluid, and rapid rotation. Among the terrestrial planets, only Earth has all three characteristics.

9.2 Shaping Planetary Surfaces

- **What processes shape planetary surfaces?** The four major geological processes are **impact cratering, volcanism, tectonics,** and **erosion**.
- **Why do the terrestrial planets have different geological histories?** Fundamental planetary properties, especially size, determine a planet's geological history. Larger planets have more volcanism and tectonics, while erosion depends on a planet's size, distance from the Sun, and rotation rate.

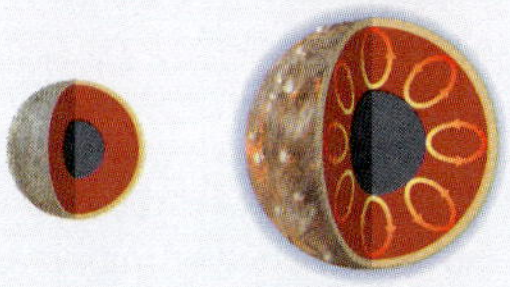

- **How does a planet's surface reveal its geological age?** Surfaces crowded with impact craters are ancient, implying that no volcanism, tectonics, or erosion has occurred recently. "Younger" surfaces with few craters have undergone some volcanic, tectonic, and/or erosional processes. Surfaces with no craters are very young, and must have ongoing geological activity.

9.3 Geology of the Moon and Mercury

- **What geological processes shaped our Moon?** The lunar surface is a combination of ancient, heavily cratered terrain and vast lava plains called the lunar **maria**. Some small tectonic features are also present. The Moon lacks erosion because of its lack of atmosphere.

- **What geological processes shaped Mercury?** Mercury resembles the Moon in being dominated by impact cratering and volcanism. It also has tremendous tectonic cliffs that probably formed when the whole planet cooled and contracted in size.

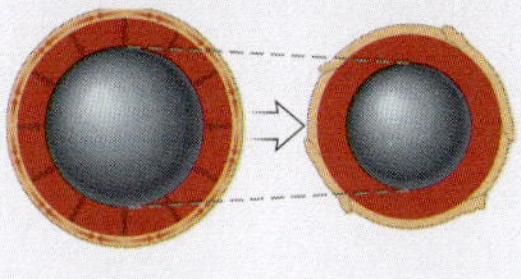

9.4 Geology of Mars

- **How did Martians invade popular culture?** Superficial similarities between the planets led to speculation about civilized Martians. Astronomer Percival Lowell thought he saw canals built by an advanced civilization, but the canals do not really exist.

- **What are the major geological features of Mars?** Mars shows evidence of all four geological processes in its past. It has the tallest volcano and the biggest canyon in the solar system, evidence of a period of great volcanic and tectonic activity. It also has abundant craters and evidence of erosion by wind and flowing water.

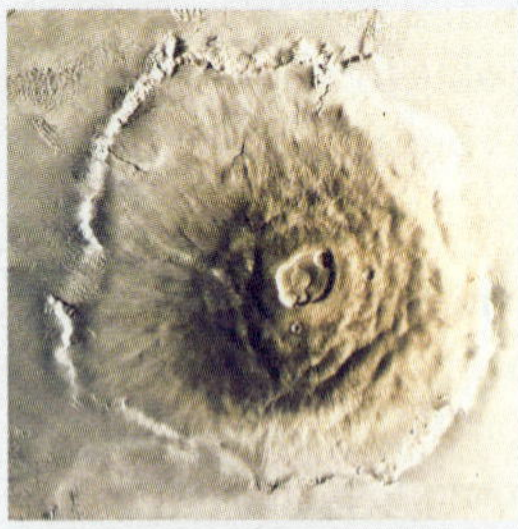

- **What geological evidence tells us that water once flowed on Mars?** Images of dry river channels and eroded craters, along with chemical analysis of Martian rocks, all show that water once flowed on Mars. Any periods of rainfall seem to have ended at least 2–3 billion years ago. Mars today still has water ice underground and in its polar caps, and could possibly have pockets of underground liquid water.

9.5 Geology of Venus

- **What are the major geological features of Venus?** Venus's surface shows evidence of major volcanic and tectonic activity in the past billion years, as expected for a planet nearly as large as Earth. This activity explains the relative lack of craters. Erosion is only a minor process despite the thick atmosphere, because of the high temperature (no rainfall) and slow rotation (little wind). Venus almost certainly remains geologically active today.

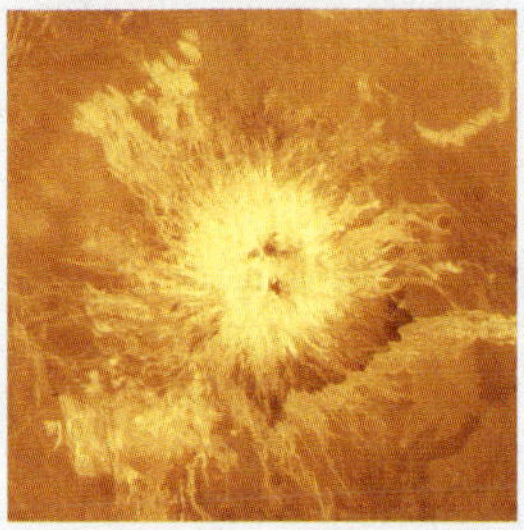

- **Does Venus have plate tectonics?** Venus may undergo occasional planetwide resurfacing, but we do not see any evidence of plate tectonics of the type we see on Earth. The lack of plate tectonics may mean that Venus has a stiffer and stronger lithosphere than Earth, perhaps because the high surface temperature has baked out any water that might have softened the lithospheric rock.

9.6 The Unique Geology of Earth

- **How do we know that Earth's surface is in motion?** The evidence for **plate tectonics** includes the way continents "fit" together, which tells us that they were close together in the past; the creation of new crust from seafloor spreading; and the compositional differences between seafloor and continental crust. Today, we can measure plate motion directly with GPS technology.

- **How is Earth's surface shaped by plate tectonics?** Plate tectonics has led to many unique features of our geology, especially **seafloor spreading** zones and the building of continents along **subduction zones**. The shifting of plates has completely changed Earth's geological appearance many times in the past billion years alone.

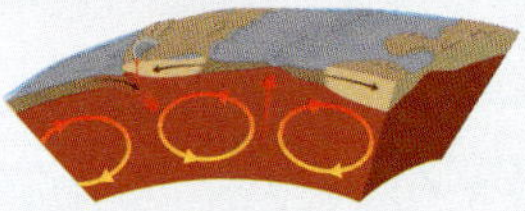

- **Was Earth's geology destined from birth?** It seems likely that the geological histories of Earth and the other terrestrial worlds were indeed destined from birth by the fundamental properties of size, distance from the Sun, and rate of rotation.

EXERCISES AND PROBLEMS

For instructor-assigned homework go to MasteringASTRONOMY www.masteringastronomy.com

Review Questions

Short-Answer Questions Based on the Reading

1. What do we mean by *planetary geology*? Briefly summarize the different geological appearances of the five terrestrial worlds.
2. What is differentiation, and how did it lead to the core-mantle-crust structures of the terrestrial worlds?
3. What do we mean by the *lithosphere*, and why isn't it listed as one of the three layers by composition? How does lithospheric thickness vary among the five terrestrial worlds?

4. Summarize the processes by which planetary interiors get hot and cool off. Why do large planets retain internal heat longer than smaller planets?
5. Why does Earth have a global *magnetic field?* Why don't the other terrestrial worlds have similarly strong magnetic fields?
6. Define each of the four major geological processes, and give examples of features shaped by each process.
7. What is *outgassing*, and why is it so important to our existence?
8. How do crater counts tell us the age of a planetary surface? Briefly explain why the Moon is so much more heavily cratered than Earth.
9. Summarize the ways in which a terrestrial world's size, distance from the Sun, and rotation rate each affect its relative level of impact cratering, volcanism, tectonics, and erosion.
10. Briefly summarize the geological history of the Moon. Be sure to explain the origin of the lunar *maria*.
11. Briefly summarize the geological history of Mercury. Be sure to explain how Mercury's great cliffs are thought to have formed.
12. Choose five features visible on the global map of Mars (Figure 9.25) and explain what we think each feature is and how it formed.
13. Explain why liquid water is not stable on Mars today, but why we nonetheless think it flowed in the distant past on Mars. Could there still be liquid water anywhere on Mars today? Explain.
14. Choose at least three major geological features of Venus and explain how we think each one formed.
15. What evidence tells us that Venus was "repaved" about 750 million years ago? What difference between the lithospheres of Venus and Earth might explain the lack of plate tectonics on Venus?
16. Describe the conveyor-like action of plate tectonics on Earth. How does this action explain the differences between seafloor crust and continental crust?
17. Briefly explain how each of the following geological features of Earth are formed: seafloors, continents, islands, mountain ranges, rift valleys, and faults.
18. To what extent do we think the geologies of the terrestrial worlds were destined from their births?

Test Your Understanding

Surprising Discoveries?

Suppose we were to make the following discoveries. (These are *not* real discoveries.) In light of your understanding of planetary geology, decide whether the discovery should be considered reasonable or surprising. Explain your answer, if possible tracing your logic back to the terrestrial world's fundamental properties of size, distance from the Sun, and rotation rate. (In some cases, both views can be defended.)

19. The next mission to Mercury photographs part of the surface never seen before and detects vast fields of sand dunes.
20. Seismographs placed on the surface of Mercury record frequent and violent earthquakes.
21. A future orbiter observes a volcanic eruption on Venus.
22. A Venus radar mapper discovers extensive regions of layered sedimentary rocks, similar to those found on Earth.
23. Radiometric dating of rocks brought back from one lunar crater shows that the crater was formed only a few tens of millions of years ago.
24. New orbital photographs of craters on Mars that have gullies also show pools of liquid water to be common on the crater bottoms.
25. Drilling into the Martian surface, a robotic spacecraft discovers liquid water deep beneath the slopes of a Martian volcano.
26. Clear-cutting in the Amazon rain forest on Earth exposes vast regions of ancient terrain that is as heavily cratered as the lunar highlands.
27. Seismic studies on Earth reveal a "lost continent" that held great human cities just a few thousand years ago but that is now buried deep underground off the western coast of Europe.
28. We find a planet in another solar system that orbits at the same distance from a Sun-like star as Earth orbits the Sun and has Earth-like plate tectonics, but it is only the size of the Moon.
29. We find a planet in another solar system that is as large as Earth but as heavily cratered as the Moon.
30. We find a planet in another solar system with Earth-like seafloor crust and continental crust but that apparently lacks plate tectonics or any other kind of crustal motion.

Quick Quiz

Choose the best answer to each of the following. Explain your reasoning with one or more complete sentences.

31. What is the most important long-lasting internal heat source responsible for geological activity? (a) accretion (b) radioactive decay (c) sunlight
32. In general, what kind of planet would you expect to have the *thickest* lithosphere? (a) the largest planet (b) the smallest planet (c) the planet farthest from the Sun
33. Which of a planet's fundamental properties has the greatest effect on its level of volcanic and tectonic activity? (a) size (b) distance from the Sun (c) rotation rate
34. What is the name of the outer rigid layer of a planet? (a) crust (b) mantle (c) lithosphere
35. Which describes our understanding of flowing water on Mars? (a) It was never important. (b) It was important once, but no longer. (c) It is a major process on the Martian surface today.
36. What do we conclude if a planet has few impact craters of any size? (a) The planet was never bombarded by asteroids or comets. (b) Its atmosphere stopped impactors of all sizes. (c) Other geological processes have wiped out craters.
37. How many of the five terrestrial worlds have surfaces being constantly reshaped by plate tectonics? (a) one (b) two (c) three or more (Be sure that you explain *why* plate tectonics occurs on this many worlds and not more.)
38. On how many of the five terrestrial worlds has erosion been an important process? (a) one (b) two (c) three or more (Be sure that you explain *why* erosion is important on this many worlds and not more.)
39. How many of the terrestrial worlds have lava plains or shield volcanoes? (a) one (b) three (c) five (Be sure to explain *why* and when these volcanic features formed.)
40. How many of the five terrestrial worlds are considered "geologically dead"? (a) none (b) two (c) four (Be sure to explain *why* these worlds became geologically dead.)

Investigate Further

In-Depth Questions to Increase Your Understanding

Short-Answer/Essay Questions

41. *Dating Planetary Surfaces.* We have discussed two basic techniques for determining the age of a planetary surface: studying the abundance of impact craters and radiometric dating of surface rocks. Which technique seems more reliable? Which technique is more practical? Explain.
42. *Comparative Erosion.* Consider erosion on Mercury, Venus, the Moon, and Mars. Which of these four worlds has the greatest erosion? Why? For each world, write a paragraph explaining its level of erosion, tracing it back to fundamental properties.
43. *Miniature Mars.* Suppose Mars had turned out to be significantly smaller than its current size—say, the size of our Moon. How would this have affected the number of geological features due to each of the four major geological processes? Do you think Mars would be a better or worse candidate for harboring extraterrestrial life? Summarize your answers in two or three paragraphs.
44. *Civilization on Mars.* Based on what we can see on the surface of Mars, does it seem possible that Mars once had a civilization with cities on the surface but that the evidence has now been erased or buried underground? Explain.
45. *Change in Fundamental Properties.* Consider Earth's size and distance from the Sun. Choose one property and suppose that it had been different (for example, smaller size or greater distance). Describe how this change might have affected Earth's subsequent geological history and the possibility of our existence today on Earth.
46. *Predictive Geology.* Suppose another star system has a rocky terrestrial planet twice as large as Earth but at the same distance from its star (which is just like our Sun) and with a similar rotation rate. In one or two paragraphs, describe the type of geology you would expect it to have.
47. *Mystery Planet.* It's the year 2098, and you are designing a robotic mission to a newly discovered planet around a nearby star that is nearly identical to our Sun. The planet is as large in radius as Venus, rotates with the same daily period as Mars, and lies 1.2 AU from its star. Your spacecraft will orbit but not land on the planet.
    a. Some of your colleagues believe that the planet has no metallic core. How could you support or refute their hypothesis?
    b. Other colleagues suspect that the planet has no atmosphere, but the instruments designed to study the planet's atmosphere fail due to a software error. However, the spacecraft can still photograph geological features. How could you use the spacecraft's photos of geological features to determine whether a significant atmosphere is (or was) present on this planet?
48. *Experiment: Geological Properties of Silly Putty.* Roll room temperature Silly Putty into a ball and measure its diameter. Place the ball on a table and gently place one end of a heavy book on it. After 5 seconds, measure the height of the squashed ball. Repeat the experiment two more times, the first time warming the Silly Putty in hot water before you start and the second time cooling it in ice water before you start. How do the different temperatures affect the rate of "squashing"? How does the experiment relate to planetary geology? Explain.
49. *Experiment: Planetary Cooling in a Freezer.* To simulate the cooling of planetary bodies of different sizes, use a freezer and two small plastic containers of similar shape but different sizes. Fill each container with cold water and put both into the freezer at the same time. Checking every hour or so, record the time and your estimate of the thickness of the "lithosphere" (the frozen layer) in the two tubs. How long does it take the water in each tub to freeze completely? Describe in a few sentences the relevance of your experiment to planetary geology. Extra credit: Plot your results on a graph with time on the x-axis and lithospheric thickness on the y-axis. What is the ratio of the two freezing times?
50. *Amateur Astronomy: Observing the Moon.* Any amateur telescope has resolution adequate to identify geological features on the Moon. The light highlands and dark maria should be evident, and shadowing is visible near the line between night and day. Try to observe the Moon near the first- or third-quarter phase. Sketch or photograph the Moon at low magnification, and then zoom in on a region of interest. Again sketch or photograph your field of view, label its features, and identify the geological process that created them. Look for craters, volcanic plains, and tectonic features. Estimate the size of each feature by comparing it to the size of the whole Moon (radius = 1,738 km).

Quantitative Problems

Be sure to show all calculations clearly and state your final answers in complete sentences.

51. *Moon and Mars.* Compare the surface area-to-volume ratios of the Moon and Mars. Based on your answer, what should you expect about interior heat on Mars compared to that on the Moon?
52. *Earth and Venus.* Compare the surface area-to-volume ratios of Venus and Earth. Based on your answer, how should you expect the amount of interior heat to compare on these two worlds?
53. *Doubling Your Size.* Just as the surface area-to-volume ratio depends on size, so can other properties. To see how, suppose that your size suddenly doubled; that is, your height, width, and depth all doubled. (For example, if you were 5 feet tall before, you now are 10 feet tall.)
    a. By what factor has your waist size increased?
    b. How much more material will be required for your clothes? (*Hint:* Clothes cover the surface *area* of your body.)
    c. By what factor has your weight increased? (*Hint:* Weight depends on the *volume* of your body.)
    d. The pressure on your weight-bearing joints depends on how much *weight* is supported by the *surface area* of each joint. How has the pressure on your weight-bearing joints changed?
54. *Lunar Footprints.* Micrometeorites will eventually erase the footprints left on the Moon by the *Apollo* astronauts. Assume that the Moon is hit by about 25 million micrometeorite impacts each day (this number comes from observations of meteors in Earth's atmosphere) and that these impacts strike randomly around the Moon's surface. Also assume that it will take about 20 such impacts to destroy a footprint. Using these assumptions, estimate how long it will take before a footprint is erased. (*Hints:* Use the Moon's surface area to determine the impact rate per square centimeter, and estimate the size of footprint.)

55. *Impact Energies.* An asteroid of 1-kilometer diameter will make a crater of about 10-kilometer diameter. How much kinetic energy does the asteroid have, if it strikes the surface at 20 km/sec? Convert your answer to megatons of TNT: 1 megaton is about 4×10^{15} joules. Compare the impactor to the largest nuclear weapons, currently less than 100 megatons of energy.
56. *Internal vs. External Heating.* In daylight, the Earth's surface absorbs about 400 watts per square meter. Earth's internal radioactivity produces a total of 3 trillion watts that leak out through our planet's entire surface. Calculate the amount of heat from radioactive decay that flows outward through each square meter of Earth's surface (your answer should have units of watts per square meter). Compare quantitatively to solar heating, and comment on why internal heating drives geological activity.
57. *Plate Tectonics.* Typical motions of one plate relative to another are 1 centimeter per year. At this rate, how long would it take for two continents 3,000 kilometers apart to collide? What are the global consequences of motions like this?
58. *More Plate Tectonics.* Consider a seafloor spreading zone creating 1 centimeter of new crust over its entire 2,000-kilometer length every year. How many square kilometers of surface will this create in 100 million years? What fraction of Earth's surface does this constitute?
59. *Planet Berth.* Imagine a planet, which we'll call *Berth*, orbiting a star identical to the Sun at a distance of 1 AU. Assume that Berth has eight times as much mass as Earth, and is twice as large as Earth in diameter.
    a. How does Berth's density compare to Earth's?
    b. How does Berth's surface area compare to Earth's?
    c. Based on your answers to (a) and (b), discuss how Berth's geological history is likely to have differed from Earth's.

Discussion Questions

60. *What Is Predictable?* We've found that much of a planet's geological history is destined from its birth. Briefly explain why, and discuss the level of detail that is predictable. For example, was Mars's general level of volcanism predictable? Could we have predicted a mountain as tall as Olympus Mons or a canyon as long as Valles Marineris? Explain.
61. *Worth the Effort?* Politicians often argue over whether planetary missions are worth the expense involved. Based on what we have learned by comparing the geologies of the terrestrial worlds, do you think the missions that have given us this knowledge have been worth their expense? Defend your opinion.
62. *Evidence of Our Civilization.* Discuss how the geological processes will affect the evidence of our current civilization in the distant future. For example, what evidence of our current civilization will survive in 100,000 years? in 100 million years? Do you think that future archaeologists or alien visitors will be able to know that we existed here on Earth?

MEDIA EXPLORATIONS

For self-study activities go to MasteringASTRONOMY www.masteringastronomy.com

Interactive Tutorials

Tutorial Review of Key Concepts

Use the following interactive **Tutorials** at **www.masteringastronomy.com** to review key concepts from this chapter.

Shaping Planetary Surfaces Tutorial

Lesson 1 The Four Geological Processes
Lesson 2 What Do Geological Processes Depend On?
Lesson 3 Planet Surface Evolution

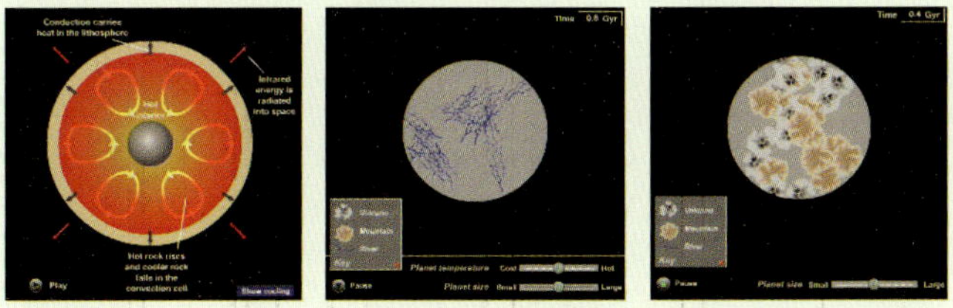

Supplementary Tutorial Exercises

Use the interactive **Tutorial Lessons** to explore the following questions.

Shaping Planetary Surfaces Tutorial, Lesson 1

1. Watch the impact cratering animation. How does the crater size depend on the size of the impactor?
2. Watch the tectonics animation. What happens in the mantle to cause compression and mountain building?
3. What happens in the mantle to cause tectonic stretching?

Shaping Planetary Surfaces Tutorial, Lesson 2

1. How has the rate of crater formation in the solar system changed over time?
2. Why do we see so many more impact craters on the Moon and Mercury than on Earth?
3. Why is erosion so much more dominant on Earth than on any of the other terrestrial worlds?

Shaping Planetary Surfaces Tutorial, Lesson 3

1. How has Earth's geology differed from that of the other terrestrial planets and why?
2. Suppose Earth had been born at half its current size. How would you expect it to have been different? Would life still be possible? Explain.
3. Suppose Earth had been born 10% larger than its current size. How would you expect it to have been different? Would life still be possible? Explain.

Exploring the Sky and Solar System

Of the many activities available on the ***Voyager: SkyGazer*** **CD-ROM** accompanying your book, use the following files to observe key phenomena covered in this chapter.

Go to the **File: Demo** folder for the following demonstrations.

1. Earth and Venus
2. Venus-Earth-Moon

Go to the **Explore** menu for the following demonstrations.

1. Solar System
2. Paths of the Planets

Movies

Check out the following narrated and animated short documentary available on **www.masteringastronomy.com** for a helpful review of key ideas covered in this chapter.

History of the Solar System

Web Projects

1. *"Coolest" Surface Photo.* Visit the Astronomy Picture of the Day Web site, and search for past images of the terrestrial worlds. Look at many of them, and choose the one you think is the "coolest." Make a printout, write a short description of what it shows, and explain what you like about it.
2. *Water on Mars.* Go to the home page for NASA's Mars Exploration Program, and look for the latest evidence concerning recent water flows on Mars. Write a few paragraphs describing the new evidence and what it tells us. How will future missions help resolve the questions?
3. *Planetary Geology.* Visit the Nine Planets Web site, which has summaries of features on each of the planets. Choose one of the terrestrial worlds, and learn about at least two geological features on it that we did not highlight in this chapter. Are the features consistent with what we expect based on the ideas of "geological destiny" discussed in this chapter? Explain.
4. *Volcanoes and Earthquakes.* Learn about one major earthquake or volcanic eruption that occurred during the past decade. Report on the geological conditions that led to the event, as well as on its geological and biological consequences.

10

Planetary Atmospheres

Earth and the Other Terrestrial Worlds

LEARNING GOALS

10.1 Atmospheric Basics
- What is an atmosphere?
- How does the greenhouse effect warm a planet?
- Why do atmospheric properties vary with altitude?

10.2 Weather and Climate
- What creates wind and weather?
- What factors can cause long-term climate change?
- How does a planet gain or lose atmospheric gases?

10.3 Atmospheres of the Moon and Mercury
- Do the Moon and Mercury have any atmosphere at all?

10.4 The Atmospheric History of Mars
- What is Mars like today?
- Why did Mars change?

10.5 The Atmospheric History of Venus
- What is Venus like today?
- How did Venus get so hot?

10.6 Earth's Unique Atmosphere
- How did Earth's atmosphere end up so different?
- Why does Earth's climate stay relatively stable?
- How might human activity change our planet?

For the first time in my life, I saw the horizon as a curved line. It was accentuated by a thin seam of dark blue light—our atmosphere. Obviously this was not the ocean of air I had been told it was so many times in my life. I was terrified by its fragile appearance.

Ulf Merbold, astronaut (Germany)

Life as we know it would be impossible on Earth without our atmosphere. This layer of gas surrounding the planet supplies the oxygen we breathe, shields us from harmful ultraviolet and X-ray radiation from the Sun, protects us from continual bombardment by micrometeorites, generates rain-giving clouds, and traps just enough heat to keep Earth habitable. We couldn't have designed a better atmosphere if we'd tried.

Despite forming under similar conditions, our neighboring planets have ended up with atmospheres utterly inhospitable to us. No other planet in our solar system has air that we could breathe or temperature and pressure conditions in which we could survive without a space suit. Mars, only slightly farther from the Sun than Earth, has an atmosphere that leaves its surface cold, dry, and barren. Venus, only slightly closer to the Sun, has an atmosphere that makes its surface conditions resemble a classical view of hell.

How did Earth end up with conditions so fortunate for life as we know it? In this chapter, we'll investigate the processes that govern planetary atmospheres and the reasons why the terrestrial worlds ended up so different. We'll also see how Earth's climate remains relatively stable, and how human activity may be threatening the very stability that we depend on.

10.1 Atmospheric Basics

All the planets of our solar system have an atmosphere to at least some degree, as do several of the solar system's larger moons. The jovian planets are essentially atmosphere throughout, because they are made largely of gaseous material. In contrast, atmospheric gases around solid bodies—including the terrestrial worlds, jovian moons with atmospheres, and Pluto—make up only a minuscule fraction of a world's total mass. Nevertheless, we can use the same principles to understand virtually all planetary atmospheres. We'll study these principles in this chapter, exploring how they explain the atmospheres of the terrestrial worlds. Later, we will see that these principles also apply to the jovian planets and other worlds with atmospheres.

The terrestrial atmospheres are even more varied than the terrestrial geologies. Figure 10.1 shows global and surface views of each of the five terrestrial worlds. Table 10.1 lists general characteristics of their atmospheres. Even a quick scan of the photos and the table reveals vast differences between the worlds.

The Moon and Mercury have so little atmosphere that it's quite reasonable to call them "airless" worlds. These worlds therefore have no wind or weather of any kind. At the other extreme, a thick atmosphere composed almost entirely of carbon dioxide completely shrouds Venus, which has surface conditions so hot and harsh that not even robotic space probes can survive there for long. Mars also has a carbon dioxide atmosphere, but its air is so thin that your body tissues would bulge painfully if you stood on the surface without wearing a full space suit.

Earth stands out in having an atmosphere that is hospitable to surface life, including human life. The composition of our atmosphere—mostly nitrogen and oxygen—

Figure 10.1 Views of the terrestrial worlds and their atmospheres from orbit and from the surface. The surface views for Mercury and Venus are artists' conceptions; the others are photos. The global views are visible-light photos taken from spacecraft. (Venus appears in gibbous phase as it was seen by the *Galileo* spacecraft during its Venus flyby en route to Jupiter.)

Table 10.1 Atmospheres of the Terrestrial Worlds

World	Composition of Atmosphere	Surface Pressure*	Average Surface Temperature	Winds, Weather Patterns	Clouds, Hazes
Mercury	helium, sodium, oxygen	10^{-14} bar	day: 425°C (797°F); night: −175°C (−283°F)	none: too little atmosphere	none
Venus	96% carbon dioxide (CO_2) 3.5% nitrogen (N_2)	90 bars	470°C (878°F)	slow winds, no violent storms, acid rain	sulfuric acid clouds
Earth	77% nitrogen (N_2) 21% oxygen (O_2) 1% argon H_2O (variable)	1 bar	15°C (59°F)	winds, hurricanes, rain, snow	H_2O clouds, pollution
Moon	helium, sodium, argon	10^{-14} bar	day: 125°C (257°F); night: −175°C (−283°F)	none: too little atmosphere	none
Mars	95% carbon dioxide (CO_2) 2.7% nitrogen (N_2) 1.6% argon	0.007 bar	−50°C (−58°F)	winds, dust storms	H_2O and CO_2 clouds, dust

* 1 bar = the pressure at sea level on Earth.

has no counterpart on any of the other terrestrial worlds. Earth's surface temperature and pressure are just right to allow for the liquid water that is so important to life. Earth's weather is far more dynamic and varied than the weather on Venus or Mars.

Indeed, Earth's atmosphere is so different from the atmospheres of the other terrestrial worlds that you might at first guess that it was made in a completely different way. However, the same basic processes are at work in all planetary atmospheres. Our primary goal in this chapter is to understand how these processes led to such profound differences among the planets. Let's begin by discussing a few basic properties of atmospheres.

• What is an atmosphere?

An *atmosphere* is a layer of gas that surrounds a world. In most cases, it is a surprisingly thin layer. On Earth, for example, about two-thirds of the air in the atmosphere lies within 10 kilometers of Earth's surface (Figure 10.2). You could represent this air on a standard globe with a layer only as thick as a dollar bill.

Figure 10.2 Earth's atmosphere, visible in this photograph from the Space Shuttle, makes a very thin layer over Earth's surface. Most of the air is in the lowest 10 kilometers of the atmosphere, visible along the edge of the planet.

The air that makes up any atmosphere is a mixture of many different gases that may consist either of individual atoms or of molecules. In general, temperatures in the terrestrial atmospheres are low enough (even on Venus) for atoms to combine into molecules. For example, the air we breathe consists of *molecular* nitrogen (N_2) and oxygen (O_2), as opposed to individual atoms (N or O). Other common molecules in terrestrial atmospheres include water (H_2O) and carbon dioxide (CO_2).

Atmospheric Pressure Collisions of individual atoms or molecules in an atmosphere create *pressure* [Section 5.3] that pushes in all directions. On Earth, for example, the nitrogen and oxygen molecules in the air fly around at average speeds of about 500 meters per second—fast enough to cross your bedroom a hundred times in 1 second. Given that a single breath of air contains more than a billion trillion molecules, you can imagine how frequently molecules collide. On average, each molecule in the air around you will suffer a million collisions in the time it takes to read this paragraph. These collisions create the pressure that pushes in all directions, and that in essence holds up the atmosphere so that it does not collapse under its own weight.

A balloon offers a good example of how pressure works in a gas. The air molecules inside a balloon exert pressure pushing outward as they constantly collide with the balloon's inside surface. At the same time, air molecules outside the balloon collide with the balloon from the other side, exerting pressure that by itself would make the balloon collapse. A balloon stays inflated when the inward pressure and the outward pressure are balanced (Figure 10.3a). (We are ignoring the tension in the rubber of the balloon walls.) Imagine that you blow more air into the balloon (Figure 10.3b). The extra molecules inside mean more collisions with the balloon wall, momentarily making the pressure inside greater than the pressure outside. The balloon therefore expands until the inward and outward pressures are again in balance.

If you heat the balloon (Figure 10.3c), the gas molecules begin moving faster and therefore collide harder and more frequently with the inside of the balloon, which also momentarily increases the inside pressure until the balloon expands. As it expands, the pressure inside it decreases and the balloon comes back into pressure balance. Conversely, cooling a balloon makes it contract, because the outside pressure momentarily exceeds the inside pressure. You can watch this contraction happen by putting a balloon in a refrigerator or freezer.

We can understand **atmospheric pressure** by applying similar principles. Gas in an atmosphere is held down by gravity. The atmosphere above any given altitude therefore has some weight that presses downward, tending to compress the atmosphere beneath it. At the same time, the fast-

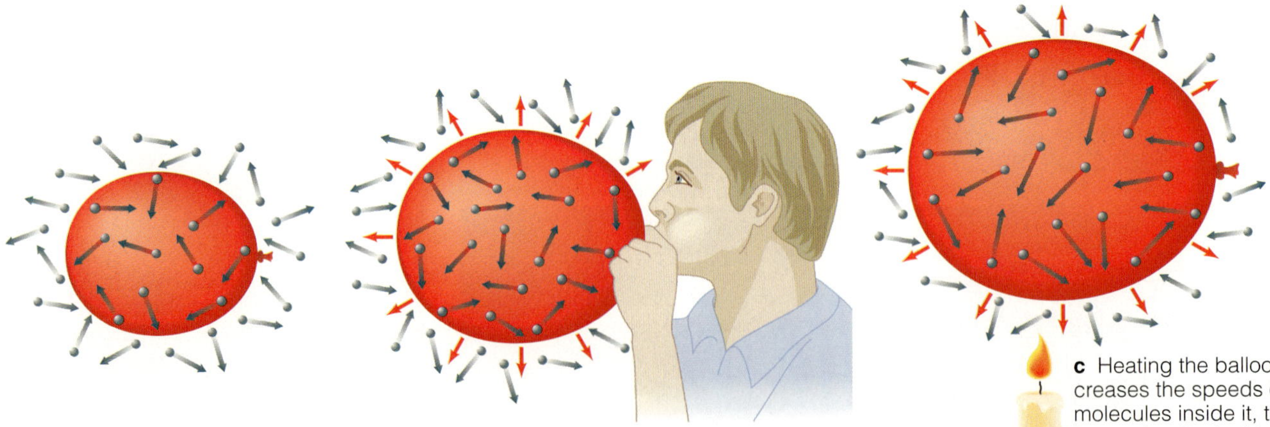

a A balloon stays inflated when the inside and outside pressures are balanced.

b Adding air molecules temporarily increases the pressure inside the balloon, so the balloon expands until pressure balance is restored.

c Heating the balloon increases the speeds of air molecules inside it, thereby increasing the inside pressure. Again, the balloon expands until the pressure balance is restored.

Figure 10.3 Gas pressure in a balloon depends on both density and temperature.

moving molecules exert pressure in all directions, including upward, which tends to make the atmosphere expand. Planetary atmospheres exist in a perpetual balance between the downward weight of their gases and the upward push of their gas pressure.

The higher you go in an atmosphere, the less the weight of the gas above you. Thus, the pressure must also become less as you go upward, which explains why the pressure decreases as you climb a mountain or ascend in an airplane. You can visualize what happens by imagining the atmosphere as a very big stack of pillows (Figure 10.4). The pillows at the bottom are highly compressed because of the weight of all the pillows above. Going upward, the pillows are less and less compressed because less weight lies on top of them.

In planetary science, we usually measure atmospheric pressure in a unit called the **bar** (as in *barometer*). One bar is roughly equal to Earth's atmospheric pressure at sea level. It is also equivalent to 1.03 kilograms per square centimeter or 14.7 pounds per square inch. In other words, if you gathered up all the air directly above any 1 square inch of Earth's surface, you would find that it weighs about 14.7 pounds.

THINK ABOUT IT

Suppose Mars had exactly the same total amount of air above each square inch of surface as does Earth. Would the atmospheric pressure be higher, lower, or the same as on Earth? Explain. (*Hint:* Remember that weight depends on the strength of gravity [Section 4.1], and that surface gravity is weaker on Mars than Earth.)

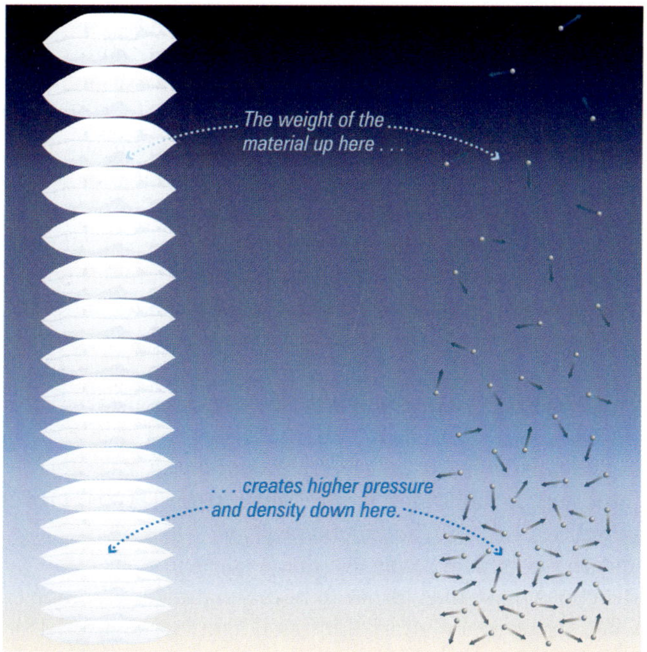

Figure 10.4 Atmospheric density and pressure decrease with altitude for the same reason that a giant stack of pillows would be more compressed at the bottom than at the top. The pressure in each layer is enough to hold up the weight of layers above, counteracting the force of gravity.

You might wonder why you don't feel Earth's atmospheric weight bearing down on you. After all, if we placed 14.7 pounds of weights on every square inch of your shoulders, you'd certainly feel the downward pressure. You don't notice the atmospheric pressure for two reasons: First, the pressure pushes in all directions, so it pushes upward and inward on you as well as downward. Second, the fluids in your body push outward with an equivalent pressure, so there is no net pressure trying to compress or expand your body. You can tell that the pressure is there, however, because you'll quickly notice any pressure changes. For example, even slight changes in pressure as you go up or down in altitude can cause your ears to "pop." More extreme changes, such as those that affect deep-sea divers when they rise too rapidly, can be deadly. You'd face even greater pressure differences if you visited other planets; as shown in Table 10.1, atmospheric pressure varies by a factor of 10^{16} among the terrestrial worlds.

Where Does an Atmosphere End? If you study Figure 10.4, you'll see that there is no clear boundary between the atmosphere and space above. Instead, the atmosphere gradually fades away as the pressure and density decrease with increasing altitude. At some point, the density becomes so low that we can't really think of the gas as "air" anymore. Collisions between atoms or molecules are rare at these altitudes, and the gas is so thin that it would look and feel as if you had already entered space. On Earth, for example, the gas is so thin above an altitude of about 60 kilometers that the sky is black even in the daytime, much like the sky on the Moon. For this reason, an altitude of about 60 kilometers is often described as "the edge of space."

However, some gas is still present even above 60 kilometers. Earth's tenuous upper atmosphere extends for several hundred kilometers above Earth's surface. The Space Shuttle, the Space Station, and many other satellites orbit Earth within these outer reaches of the atmosphere (Figure 10.5). The low-density gas may be barely noticeable under most conditions, but it still exerts drag on orbiting spacecraft. That is why satellites in low-Earth orbit slowly spiral downward, eventually burning up as they reenter the denser layers of the atmosphere [Section 4.5]. The Space Station would suffer the same fate without periodic boosts from the Space Shuttle or other rockets.

How Atmospheres Affect Planets It may be fairly obvious that Earth's atmosphere is important to our existence, but the full range of atmospheric effects is more than most people realize. To provide context for our study of planetary atmospheres, here is a brief list of the key effects of atmospheres that we will study in more detail in the rest of this chapter.

- As we've already seen, atmospheres create pressure that can in turn determine whether water can exist in liquid form on the surface.
- Atmospheres absorb and scatter light. Scattering explains why daytime skies are bright on worlds with

Figure 10.5 This photograph shows the visible glow of gas around the tail of an orbiting Space Shuttle. Some low-density gas is present even at the Shuttle's altitude of a few hundred kilometers, though most of Earth's atmospheric gas lies within 10 kilometers of the surface and altitudes above 60 kilometers are generally considered to be "space." The aurora visible in the background is additional proof of the great vertical extent of the atmosphere, since the aurora is created by interactions between charged particles and atmospheric gas at altitudes above 100 kilometers.

atmospheres and black on airless worlds like the Moon and Mercury. Absorption can keep the surface safe from dangerous radiation; for example, ozone in Earth's atmosphere prevents most solar ultraviolet radiation from reaching the ground.

- Atmospheres can create wind and weather and play a major role in long-term climate change.
- Interactions between atmospheric gases and the solar wind can create a protective *magnetosphere* around planets with strong magnetic fields.
- In some cases, atmospheres contain gases that can make planetary surfaces warmer than they would be otherwise, through what we call the **greenhouse effect**.

The greenhouse effect is arguably the most important effect that an atmosphere can have on its planet, and we'll therefore focus our attention on it now.

• How does the greenhouse effect warm a planet?

You've probably heard of the greenhouse effect, because it is an important part of the environmental problem known as *global warming* [Section 10.6]. But you may be surprised to learn that the greenhouse effect is also critical to the existence of life on Earth. Without the greenhouse effect, Earth's surface would be too cold for liquid water to flow and for life to flourish. Let's explore how the greenhouse effect works and what it does to the surface temperatures of the terrestrial worlds.

How the Greenhouse Effect Works The basic idea behind the greenhouse effect is quite simple (Figure 10.6). Some of the visible light from the Sun is absorbed by the ground (the rest is reflected). The ground returns the absorbed energy back toward space, but in the form of infrared rather than visible light. The reason planets emit infrared light is that their temperatures are too low to emit visible light. Remember that all objects emit thermal radiation by virtue of their temperatures [Section 5.4]. Planetary temperatures are in a range in which nearly all emitted energy is in the infrared.

The greenhouse effect works by "trapping" some of the infrared light that a planet emits, slowing its return to space. This trapping occurs when atmospheric gases absorb the infrared light. Gases that are particularly good at absorbing infrared light are called **greenhouse gases**. The most important greenhouse gases are water vapor (H_2O), carbon dioxide (CO_2), and methane (CH_4). These gases absorb infrared light effectively because their molecular structures make them prone to begin rotating or vibrating when struck by a photon of infrared light (see Figure 5.18).

After a greenhouse gas molecule absorbs a photon of infrared light, it reemits a photon of infrared light in some random direction. This photon is usually absorbed by another greenhouse molecule, which does the same thing. The net result is that greenhouse gases tend to slow the

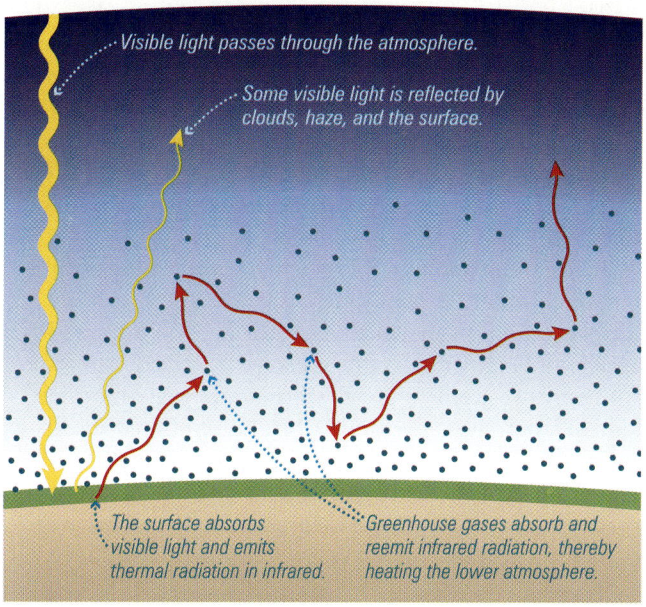

Figure 10.6 Interactive Figure The greenhouse effect. The lower atmosphere becomes warmer than it would be if it had no greenhouse gases such as water vapor, carbon dioxide, and methane.

escape of infrared radiation from the lower atmosphere, while their molecular motions heat the surrounding air. In this way, the greenhouse effect makes the surface and the lower atmosphere warmer than they would be from sunlight alone. The more greenhouse gases present, the greater the degree of surface warming.

It's important to realize that the greenhouse effect does not alter a planet's overall energy balance. The total amount of energy that a planet receives from the Sun must be precisely balanced with the amount of energy it returns to space through reflection and radiation. If it weren't balanced, the planet would rapidly heat up (if it received more energy than it returned) or cool down (if it returned more energy than it received). All the energy "trapped" by the greenhouse effect still escapes to space, just not as directly as it would otherwise. A blanket offers a good analogy: You stay warmer under a blanket not because the blanket itself provides any heat, but rather because it slows the escape of your body heat into the cold outside air. In essence, the greenhouse effect makes the total amount of energy in the lower atmosphere greater than it would be otherwise—making it warmer—while leaving the total amount of energy that escapes to space unchanged.

THINK ABOUT IT

Clouds on Earth are made of water (H_2O), which acts as a very effective greenhouse gas even when the water vapor condenses into droplets in clouds. Use this fact to explain why clear nights tend to be colder than cloudy nights.

Incidentally, the name *greenhouse effect* is a bit of a misnomer. The name comes from botanical greenhouses, usually built mostly of glass, which keep the air inside them warmer than it would be otherwise. These greenhouses trap heat simply by not letting hot air rise, which is very different from the way infrared absorption traps heat and leads to the "greenhouse effect" in planetary atmospheres.

"No Greenhouse" Planetary Temperatures We can better appreciate the importance of the greenhouse effect by considering planetary temperatures in its absence. Remember that a terrestrial planet's interior heat has very little effect on its surface temperature; sunlight is the only significant energy source for the surface [Section 9.1]. Without the greenhouse effect, a planet's average surface temperature would depend only on how much sunlight it absorbed. This absorbed sunlight would be radiated directly back to space as infrared light. Moreover, the amount of absorbed sunlight—and hence the average surface temperature—would depend on only two things:

- *The planet's distance from the Sun*, which determines the amount of energy incoming as sunlight. The closer a planet is to the Sun, the greater the intensity of the incoming sunlight.
- *The planet's overall reflectivity*, which determines the relative proportions of incoming sunlight that the planet reflects and absorbs. (You may sometimes see reflectivity described by the term *albedo.*) The higher the reflectivity, the less light absorbed and the cooler the planet.

A planet's reflectivity depends on its surface composition and color. The darker the surface, the more light it absorbs and the less it reflects. For example, clouds, snow, and ice reflect 70% or more of the light that hits them, while rocks typically reflect only about 20% of the light that hits them and absorb the other 80%.

THINK ABOUT IT

Who gets hotter in the summer sun, you wearing a white shirt or a person in black next to you? Explain.

COMMON MISCONCEPTIONS

The Greenhouse Effect Is Bad

The greenhouse effect is often in the news, usually in discussions about environmental problems, but in itself the greenhouse effect is not a bad thing. In fact, we could not exist without it, since it is responsible for keeping our planet warm enough for liquid water to flow in the oceans and on the surface. The "no greenhouse" temperature of Earth is well below freezing. Why, then, is the greenhouse effect discussed as an environmental problem? The reason is that human activity is adding more greenhouse gases to the atmosphere—which might change Earth's climate [Section 10.6]. While the greenhouse effect makes Earth livable, it is also responsible for the searing 470°C temperature of Venus—proving that it's possible to have too much of a good thing.

Both distance from the Sun and reflectivity have been measured for all the terrestrial worlds. With a little mathematics (see Mathematical Insight 10.1), it's possible to use these measurements to calculate the temperature that any planet would have without greenhouse gases. For example, such calculations show that Earth's **global average temperature**—that is, the average temperature for the entire planet—would be a chilly −16°C (+3°F) without the greenhouse effect. Because this "no greenhouse" temperature is well below the freezing point of water, we conclude that our planet would be frozen over in the absence of a greenhouse effect. Clearly, the greenhouse effect is very important to life on Earth.

Greenhouse Warming of the Terrestrial Worlds Table 10.2 lists the distance and reflectivity for each terrestrial world and compares the world's calculated "no greenhouse" temperatures to actual surface temperatures. The "no greenhouse" temperatures for Mercury and the Moon lie between their actual day and night temperatures, just as we should expect on worlds with very little atmosphere and hence no greenhouse effect. For the other planets, however, the greenhouse effect plays an important role.

The greenhouse effect is fairly weak on Mars, where the actual average surface temperature is only 6°C higher than the "no greenhouse" temperature. Earth has a moderate greenhouse effect that makes the actual global average temperature about 15°C (59°F), which is 31°C warmer than the "no greenhouse" temperature. Venus has an extreme greenhouse effect that bakes the surface to a temperature more than 500°C hotter than it would be otherwise.

If you think about it, this greenhouse warming is quite remarkable. Earth's atmosphere is made almost entirely of nitrogen and oxygen, which have no effect on infrared light and do not contribute to the greenhouse effect. (Molecules with only two atoms, especially those with two of the same kind of atom, such as N_2 and O_2, are poor infrared absorbers because they have very few ways to vibrate and rotate.) Greenhouse gases are only trace constituents of Earth's atmosphere, yet they warm our planet enough to raise the global average temperature from well below freezing to well above it.

THINK ABOUT IT

Suppose nitrogen and oxygen were greenhouse gases. Assuming our atmosphere still had the same composition and density, how would Earth be different?

MATHEMATICAL INSIGHT 10.1 "No Greenhouse" Temperatures

As discussed in the text, the "no greenhouse" temperature of a planet in our solar system depends only on the planet's distance from the Sun and its overall reflectivity. From those two factors and the Sun's brightness, we can derive the following formula for a planet's "no greenhouse" temperature:

$$T = 280\text{ K} \times \sqrt[4]{\frac{(1 - \text{reflectivity})}{d^2}}$$

where d is the distance from the Sun in AU and the reflectivity is stated as a fraction between 0 and 1. For example, if a planet reflects 20% of the incoming sunlight, its reflectivity would be stated as 0.2. The symbol $\sqrt[4]{\ }$ means the fourth root, or $\frac{1}{4}$ power.

We will not derive the formula here, but you can see how it makes sense. The term (1 − reflectivity) is the proportion of sunlight that the planet absorbs, which is the light that heats its surface. This term is divided by d^2 because the amount of energy in sunlight (per unit area) declines with the square of distance from the Sun, just as gravity declines with the square of distance. Thus, the full term $(1 - \text{reflectivity})/d^2$ represents the total amount of energy that the planet absorbs from sunlight (per unit area) each second. This energy warms the surface, and the surface returns the energy to space by thermal radiation [Section 5.4]. The energy emitted by thermal radiation depends on temperature raised to the fourth power (see Mathematical Insight 5.2). Calculating the planet's temperature therefore requires taking the fourth root of the absorbed energy, or the fourth root of $(1 - \text{reflectivity})/d^2$. The 280 K in the formula happens to be the temperature of a perfectly black planet at 1 AU from the Sun.

Example: Calculate the "no greenhouse" temperature of Mercury.

Solution:

Step 1 Understand: We can use the above formula to find a "no greenhouse" temperature by knowing a planet's distance from the Sun in AU and its reflectivity as a decimal fraction. Table 10.2 gives us both values for Mercury, except that it gives the reflectivity as a percentage. Of course, we can easily convert 12% to the decimal fraction 0.12.

Step 2 Solve: We now plug in Mercury's distance from the Sun, $d = 0.387$ AU, and its reflectivity of 0.12 to find its "no greenhouse" temperature:

$$T = 280\text{ K} \times \sqrt[4]{\frac{(1 - \text{reflectivity})}{d^2}}$$

$$= 280\text{ K} \times \sqrt[4]{\frac{(1 - 0.12)}{0.387^2}} = 436\text{ K}$$

Step 3 Explain: We have found that Mercury's "no greenhouse" temperature is 436 K. We can check this value against the value shown in Table 10.2 by converting it to Celsius. The only difference between the Celsius and Kelvin scales is their starting point: 0 Kelvin is equivalent to −273°C (see Figure 4.12). Thus, we convert the Kelvin temperature to Celsius by subtracting 273, finding that 436 K is equivalent to 436 − 273 = 163°C. Notice that this value agrees with the "no greenhouse" temperature for Mercury shown in the table. It also makes sense: Because Mercury has no greenhouse effect, its "no greenhouse" temperature should be midway between its day and night temperatures. The precise halfway point of the day and night temperatures shown in Table 10.2 is 125°C; although this is not an exact match for the 163°C that we calculated, it is close enough for us to be confident that we have the right general idea.

Table 10.2 The Greenhouse Effect on the Terrestrial Worlds

World	Average Distance from Sun (AU)	Reflectivity	"No Greenhouse" Average Surface Temperature*	Actual Average Surface Temperature	Greenhouse Warming (actual temperature minus "no greenhouse" temperature)
Mercury	0.387	12%	163°C	425°C (day), −175°C (night)	—
Venus	0.723	75%	−40°C	470°C	510°C
Earth	1.00	29%	−16°C	15°C	31°C
Moon	1.00	12%	−2°C	125°C (day), −175°C (night)	—
Mars	1.524	16%	−56°C	−50°C	6°C

*The "no greenhouse" temperature is calculated by assuming no change to the atmosphere other than lack of greenhouse warming. Thus, for example, Venus ends up with a lower "no greenhouse" temperature than Earth even though it is closer to the Sun, because the high reflectivity of its bright clouds means that it absorbs less sunlight than Earth.

• Why do atmospheric properties vary with altitude?

The greenhouse effect can warm a planet's surface and lower atmosphere, but other processes affect the temperature of the atmosphere at higher altitudes. The way in which temperature varies with altitude determines what is often called the **atmospheric structure**. Earth's atmospheric structure has four basic layers (Figure 10.7), each of which affects the planet in a distinct way:

- The **troposphere** is the lowest layer of the atmosphere. The temperature drops with altitude in the troposphere, something you've probably noticed if you've ever climbed a mountain.
- The **stratosphere** begins where the temperature stops dropping and instead begins to rise with altitude. High in the stratosphere, the temperature falls again.*
- The **thermosphere** begins at a high altitude where the temperature again starts to rise.
- The **exosphere** is the uppermost region in which the atmosphere gradually fades away into space.

The layering is shaped by the way atmospheric gas interacts with sunlight. As shown in Figure 10.7, visible light warms the surface so that the ground emits infrared light that heats the troposphere. The absorption of ultraviolet light from the Sun warms the stratosphere, while solar X rays and some solar ultraviolet light heat the thermosphere and exosphere. Let's now look at these ideas in more detail, so that we'll understand why atmospheric properties vary with altitude. Along the way, we'll also see why storms occur only in the troposphere, why ozone is so important to life on Earth, and why the sky is blue.

Interactions Between Light and Atmospheric Gases The key to understanding atmospheric structure lies in interactions between atmospheric gases and energy from the Sun. Essentially all the energy in terrestrial atmospheres comes from sunlight.

Most light coming from the Sun is visible light, but the Sun also emits significant amounts of ultraviolet light and

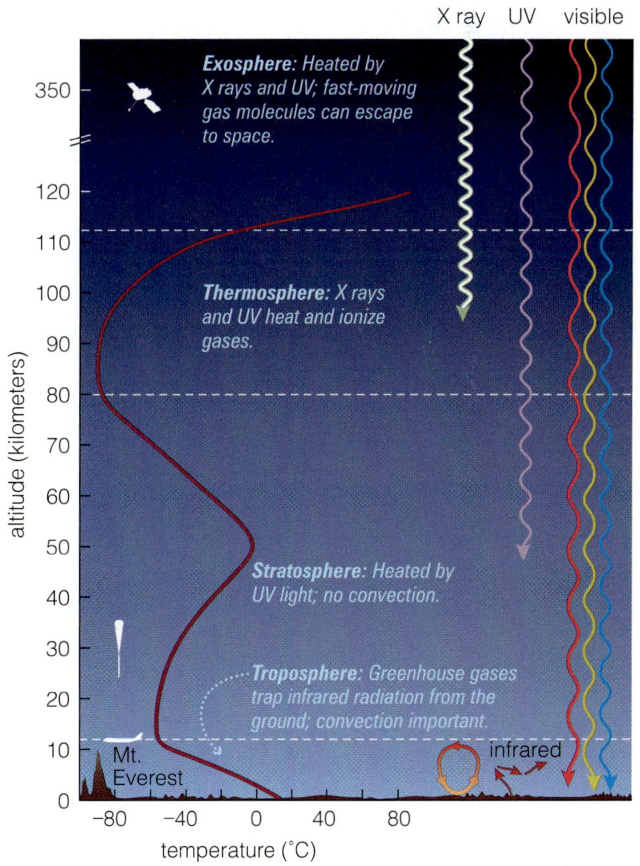

Figure 10.7 Earth's average atmospheric structure. The background color represents the air density and pressure, both of which decline rapidly with altitude. The curve shows average temperatures for each altitude. The squiggly arrows to the right represent photons and show where they are typically absorbed in the atmosphere.

*Technically, the region where the temperature falls again is called the *mesosphere,* but we will not make this distinction in this book.

X rays. The light emitted from a planet is almost entirely in the infrared. Atmospheric gases interact with each of these forms of light in different ways that are summarized in Figure 10.8:

- X rays have enough energy to knock electrons free from almost any atom or molecule; that is, they ionize [Section 5.3] the atoms or molecules they strike. Thus, virtually all atmospheric gases absorb X rays and become ionized in the process.
- Ultraviolet photons can break apart some molecules. For example, a molecule of water vapor (H_2O) can split apart if it absorbs an ultraviolet photon. Weakly bonded molecules, such as ozone (O_3), are particularly good at absorbing ultraviolet light, splitting apart in the process.
- Visible-light photons usually pass through atmospheric gases without being absorbed, but some are *scattered* so that their direction changes.
- As we discussed earlier, infrared light can be absorbed by greenhouse gases—molecules such as water vapor (H_2O), carbon dioxide (CO_2), and methane (CH_4) that easily begin rotating and vibrating.

With these ideas in mind, let's examine the reasons for Earth's atmospheric structure, working our way from the ground up.

Visible Light: Warming the Surface and Coloring the Sky The transparency of atmospheric gases to visible light explains why visible light reaches the ground and warms the surface. However, the small amount of visible-light scattering that occurs in the atmosphere has two very important effects.

First, scattering explains why the daytime sky is bright and we cannot see the stars. Without scattering, our sky would look like the daytime lunar sky, with the Sun just a very bright circle set against a black, star-studded background. Scattering also prevents shadows on Earth from being pitch black. On the Moon, shadows receive little scattered sunlight and are extremely cold and dark.

Second, the scattering of visible light explains why our sky is blue. Visible light consists of all the colors of the rainbow, but not all the colors are scattered equally. Gas molecules scatter blue light (higher energy) much more effectively than red light (lower energy). The difference in scattering is so great that, for practical purposes, we can imagine that only the blue light gets scattered. When the Sun is overhead, this scattered blue light reaches our eyes from all directions and the sky appears blue (Figure 10.9). At sunset or sunrise, the sunlight must pass through a greater amount of atmosphere on its way to us. Most of the blue light is scattered away, leaving only red light to color the sky.

Infrared Light and the Troposphere The troposphere gets its structure because of the way atmospheric gases interact with infrared light radiated upward by the ground. That is, the greenhouse effect warms the troposphere. Because the infrared light absorbed in the troposphere comes from the surface, more is absorbed closer to the ground than at higher altitudes. That is why the temperature drops with altitude in the troposphere. (The relatively small amount of infrared light coming from the Sun is mostly absorbed by atmospheric molecules before it reaches the ground and thus has relatively little effect on the atmosphere.)

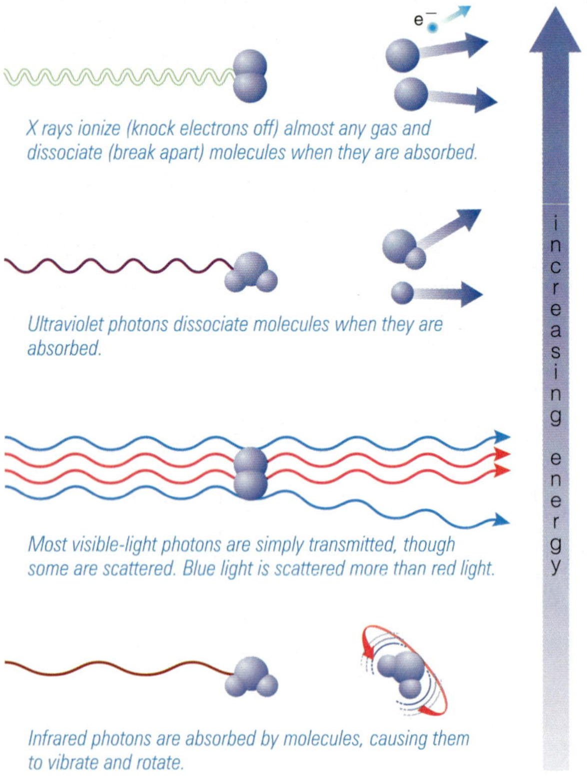

Figure 10.8 The primary effects when light of different energies strikes common atmospheric gases.

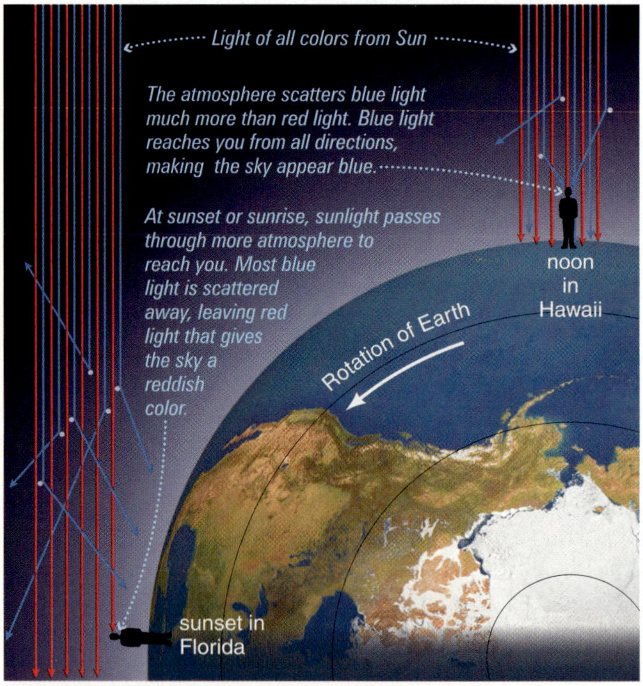

Figure 10.9 This diagram summarizes why the daytime sky is blue and sunsets (and sunrises) are red.

COMMON MISCONCEPTIONS

Why Is the Sky Blue?

If you ask around, you'll find a wide variety of misconceptions about why the sky is blue. For example, some people think the sky is blue because of light reflecting from the oceans. Anyone who lives in Nebraska will immediately know that this must be wrong—Nebraska is too far from any ocean for its sky to be influenced by such reflection. Other people claim that "air is blue." It's not obvious what someone might mean by this statement, but it's clearly wrong. If air molecules emitted blue light, then air would glow blue even in the dark. If air molecules were blue because they reflected blue light and absorbed red light, then no red light could reach us at sunset. The real explanation for the blue sky is light scattering, as shown in Figure 10.9. This explanation not only makes sense for our blue sky but also explains our red sunsets.

The drop in temperature with altitude, combined with the relatively high density of air in the troposphere, explains why the troposphere is the only layer of the atmosphere with storms. The primary cause of storms is the churning of air by convection, in which warm air rises and cool air falls. Remember that convection occurs only when there is strong heating from below [Section 9.1]. In the troposphere, the heating from the ground can drive convection. Indeed, the troposphere gets its name from this convection, since *tropos* is Greek for "turning."

Ultraviolet Light and the Stratosphere The primary source of heat in the troposphere—the absorption of infrared radiation from the ground—is not important in higher layers of the atmosphere. Once infrared radiation from the ground reaches the top of the troposphere, it travels essentially unhindered to space. Thus, we need to consider only the effects of sunlight in investigating atmospheric structure above the troposphere.

As we move upward through Earth's stratosphere, the absorption of solar ultraviolet light by ozone becomes the primary source of atmospheric heating. This heating tends to be stronger at higher altitudes, because almost all the ultraviolet light from the Sun gets absorbed before it reaches the bottom of the stratosphere. As a result, the lower stratosphere gets warmer with increasing altitude—the opposite of the situation in the troposphere.

Convection cannot occur in the lower stratosphere because heat cannot rise if the air is even hotter higher up. The lack of convection makes the air relatively stagnant and *stratified* (or layered)—with layers of warm air overlying cooler air—which is how the stratosphere gets its name. The lack of convection also means that the stratosphere essentially has no weather and no rain. Pollutants that reach the stratosphere—including the ozone-destroying chemicals known as chlorofluorocarbons (CFCs)—remain there for decades.

Note that a planet can have a stratosphere *only* if its atmosphere contains molecules that are particularly good at absorbing ultraviolet photons. Ozone (O_3) plays this role on Earth, but the lack of oxygen in the atmospheres of the other terrestrial worlds means that they also lack ozone. As a result, Earth is the only terrestrial world with a stratosphere, at least in our solar system. (The jovian planets have stratospheres due to other ultraviolet-absorbing molecules [Section 11.1].)

X Rays and the Thermosphere Nearly all gases are good X-ray absorbers, so X rays from the Sun are absorbed by the first gases they encounter as they enter the atmosphere. The density of gas in the exosphere is too low to absorb significant amounts of the incoming X rays, leaving most X rays to be absorbed in the thermosphere. Temperatures rise quite high in the thermosphere (*thermos* is Greek for "hot"), but you wouldn't feel hot there because the density and pressure are so low [Section 4.3]. Virtually no X rays penetrate beneath the thermosphere, which is why X-ray telescopes are useful only on very high-flying balloons, rockets, and spacecraft.

In addition to heating the thermosphere, X rays ionize a small but important fraction of its gas. The portion of the thermosphere that contains most of the ionized gas is called the *ionosphere.* Thus, the ionosphere is part of the thermosphere, not a separate layer. The ionosphere is very important to radio communications on Earth. The ionosphere reflects most radio broadcasts back to Earth's surface—almost as though Earth were wrapped in aluminum foil. Without reflection by the ionosphere, radio communication would work only between locations in sight of each other.

The Exosphere The exosphere is the extremely low-density gas that forms the gradual and fuzzy boundary between the atmosphere and space (*exo* means "outermost" or "outside"). The gas density in the exosphere is so low that collisions between atoms or molecules are very rare, although the high temperature means that gas particles move quite rapidly. Lightweight gas molecules sometimes reach escape velocity [Section 4.5] and fly off into space.

COMMON MISCONCEPTIONS

Higher Altitudes Are Always Colder

Many people think that the low temperatures in the mountains are just the result of lower pressures, but Figure 10.7 shows that it's not that simple. The higher temperatures near sea level on Earth are a result of the greenhouse effect, which traps more heat at lower altitudes. If Earth had no greenhouse gases, mountaintops wouldn't be so cold—or, more accurately, sea level wouldn't be so warm.

Comparative Structures of Terrestrial Atmospheres We've discussed everything we need to understand the structures of all the terrestrial atmospheres. The Moon and Mercury have so little gas that they essentially contain only an exosphere and have no structure to speak of. Figure 10.10 contrasts the structures of the lower atmospheres of Venus, Earth, and Mars. The following key ideas should make sense to you:

- All three planets have a warm troposphere at the bottom of the atmosphere, created by the greenhouse effect, and a warm thermosphere at the top, where solar X rays are absorbed.
- Only Earth has the extra "bump" of a stratosphere, because only Earth has a layer of ultraviolet-absorbing gas (ozone). Without the warming in our unique layer of stratospheric ozone, the middle altitudes of Earth's atmosphere would be almost as cold as those on Mars.
- All three planets are warmer at the surface than they would be without the greenhouse effect, although the amount of warming is far greater for Venus than for Earth or Mars.

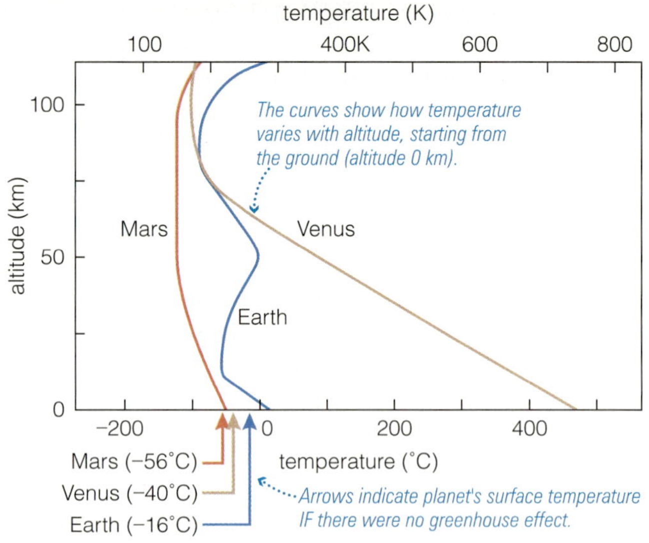

Figure 10.10 Temperature profiles for Venus, Earth, and Mars. Note that Venus and Earth are considerably warmer than they would be without the greenhouse effect. (The thermospheres and exospheres, which extend above the top of the graph, are qualitatively alike.)

THINK ABOUT IT

Would astronauts on the Moon need protection from solar X rays? What about astronauts on Mars? Explain.

Magnetospheres and the Solar Wind We have yet to consider one more type of energy coming from the Sun: the low-density breeze of subatomic charged particles that we call the *solar wind* [Section 8.3]. The solar wind does not significantly affect atmospheric structure, but it can have other important effects. The effects depend largely on whether a planet has a strong enough magnetic field [Section 9.1] to divert the charged particles of the solar wind.

On worlds that do not have strong magnetic fields, solar wind particles can directly impact the surface or atmosphere. For example, solar wind particles continually hit the upper atmospheres of Venus and Mars and the surfaces of Mercury and the Moon. In contrast, Earth's strong magnetic field creates a **magnetosphere** that acts like a protective bubble surrounding our planet (Figure 10.11a). The

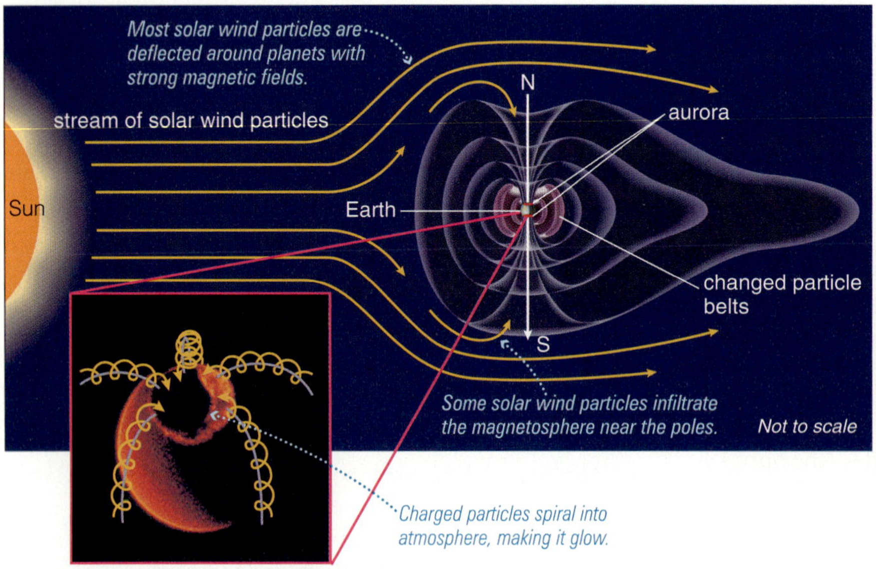

a This diagram shows how Earth's magnetosphere deflects solar wind particles. Some particles accumulate in charged particle belts encircling our planet. The inset is a photo of a ring of auroras around the North Pole; the bright crescent at its left is part of the day side of Earth.

b This photograph shows the aurora near Yellowknife, Northwest Territories, Canada. In a video, you would see these lights dancing about in the sky.

Figure 10.11 Interactive Figure Earth's magnetosphere acts like a protective bubble, shielding our planet from the charged particles of the solar wind.

magnetosphere deflects most of the charged particles from the Sun around our planet.

The magnetosphere still allows a few solar wind particles to get through, especially near the magnetic poles. Once inside the magnetosphere, these particles move along magnetic field lines, collecting in **charged particle belts** that encircle our planet. (These belts are often called the *Van Allen belts* after their discoverer.) The high energies of the particles in charged particle belts can be hazardous to spacecraft and astronauts passing through them.

Charged particles trapped in the magnetosphere also create the beautiful spectacle of light we call the **aurora** (Figure 10.11b). Variations in the solar wind can buffet the magnetosphere and give energy to particles trapped there. If a trapped particle gains enough energy, it can follow the magnetic field all the way down to Earth's atmosphere, where it collides with atmospheric atoms and molecules. These collisions cause the atoms and molecules to radiate and produce the moving lights of the aurora. Because the charged particles follow the magnetic field, auroras are most common near the magnetic poles and are best viewed at high latitudes. In the Northern Hemisphere, the aurora is often called the *aurora borealis,* or northern lights. In the Southern Hemisphere, it is called the *aurora australis,* or southern lights. The aurora can also be seen from space, where the lights can look much like surf in the upper atmosphere (see Figure 10.5).

Mastering Astronomy: Surface Temperature of Terrestrial Planets Tutorial, Lesson 1

10.2 Weather and Climate

So far, we've talked mostly about *average* conditions in planetary atmospheres, such as the average temperature on the surface and at each altitude above the surface. From our experience on Earth, however, we know that surface and atmospheric conditions constantly change. These changes are generally described as *weather* or *climate.*

Weather and climate are closely related, but they are not quite the same thing. In any particular location, some days may be hotter or cooler, clearer or cloudier, or calmer or stormier than others. The ever-varying combination of winds, clouds, temperature, and pressure is what we call **weather**.

Climate is the long-term average of weather, which means it can change only on much longer time scales. For example, Antarctic deserts have a cold, dry climate, while tropical rain forests have a hot, wet climate. Geological records show that climates can change, but most of the time the changes occur gradually over decades, centuries, or millennia.

Weather and climate can be hard to distinguish on a human time scale. For example, a drought lasting a few years may be the result of either random weather fluctuations or the beginning of a gradual change to a drier climate. The difficulty in distinguishing between random weather

SPECIAL TOPIC Weather and Chaos

Scientists today have a very good understanding of the physical laws and mathematical equations that govern the behavior and motion of atoms in the air, oceans, and land. Why, then, do we have so much trouble predicting the weather? To understand why weather is so unpredictable, we must look at the nature of scientific prediction.

Let's say you want to predict the location of a car on a road 1 minute from now. You need two basic pieces of information: where the car is now, and how it is moving. If the car is now passing Smith Road and heading north at 1 mile per minute, it will be 1 mile north of Smith Road in 1 minute.

Now suppose you want to predict the weather. Again, you need two basic types of information: (1) the current weather and (2) how weather changes from one moment to the next.

You could attempt to predict the weather by creating a "model world." For example, you could overlay a globe of Earth with graph paper and then specify the current temperature, pressure, cloud cover, and wind within each square. These are your starting points, or initial conditions. Next, you could input all the initial conditions into a computer, along with a set of equations (physical laws) that describe the processes that can change weather from one moment to the next.

Now you run your computer model to predict the weather for the next month in New York City. The model might tell you that tomorrow will be warm and sunny, with cooling during the next week and a major storm passing through a month from now. But suppose you run the model again, making one minor change in the initial conditions—say, a small change in the wind speed somewhere over Brazil. This slightly different initial condition will not change the weather prediction for tomorrow in New York City. For next week's weather, the new model may yield a slightly different prediction. For next month's weather, however, the two predictions may not agree at all!

The disagreement between the two predictions arises because the laws governing weather can cause very tiny changes in initial conditions to be greatly magnified over time. This extreme sensitivity to initial conditions is sometimes called the *butterfly effect:* If initial conditions change by as much as the flap of a butterfly's wings, the resulting prediction may be very different.

The butterfly effect is a hallmark of *chaotic systems.* Simple systems are described by linear equations in which, for example, increasing a cause produces a proportional increase in an effect. In contrast, chaotic systems are described by nonlinear equations, which allow for subtler and more intricate interactions. For example, the economy is nonlinear because a rise in interest rates does not automatically produce a corresponding change in consumer spending. Weather is nonlinear because a change in the wind speed in one location does not automatically produce a corresponding change in another location.

Despite their name, chaotic systems are not necessarily random. In fact, many chaotic systems have a kind of underlying order that explains the general features of their behavior even while details at any particular moment remain unpredictable. In a sense, many chaotic systems—like the weather—are "predictably unpredictable."

and real climate trends is an important part of the debate about human influences on the climate. We'll return to this topic in Section 10.6; for now, let's focus on understanding weather and climate generally.

• What creates wind and weather?

Wind, rain, and other weather phenomena are all driven by energy in the atmosphere, which means only planets with atmospheres can have weather. Even then, weather varies dramatically among different worlds. Earth has the most diverse weather of the terrestrial planets, so we'll use it as our example of how weather works.

The complexity of weather makes it difficult to predict, and at best the local weather can be predicted only a week or so in advance (see Special Topic box, p. 299). Nevertheless, when we look at Earth as a whole, we can identify certain general characteristics of weather.

Global Wind Patterns The wind's direction and strength can change rapidly in any particular place. On a more global scale, however, winds blow in distinctive patterns. For example, winds cause storms moving in from the Pacific to hit the west coast of the United States first and then make their way eastward across the Rocky Mountains and the Great Plains, heading to the east coast. This and other general wind patterns make up what we call Earth's **global wind patterns** (or *global circulation*).

Figure 10.12 shows Earth's global wind patterns. Notice how the wind direction changes back and forth with latitude: Equatorial winds blow from east to west, midlatitude winds blow from west to east, and high-latitude winds blow like equatorial winds from east to west. What causes this pattern of winds? Two factors explain it: atmospheric heating and planetary rotation. Let's examine each in turn.

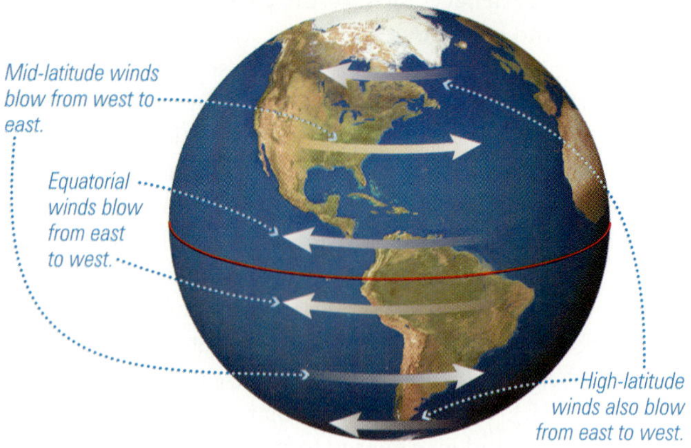

Figure 10.12 Schematic of Earth's global wind patterns. Notice that midlatitude winds blow from west to east, explaining why storms generally move across the United States from the west coast toward the east coast.

THINK ABOUT IT

Hurricanes usually form over warm, equatorial waters. Based on the wind patterns shown in Figure 10.12, which direction will a hurricane move after forming in the eastern Atlantic Ocean? Explain why these wind patterns make the southeastern United States and the Caribbean especially vulnerable to hurricanes.

Atmospheric Heating and Circulation Cells The first major factor affecting global wind patterns is atmospheric heating. Earth's equatorial regions receive more heat from the Sun than do polar regions. The excess heat makes the atmosphere expand above the equator so that it rises upward and flows toward the poles. Near the poles, cool air descends and flows toward the equator. If Earth's rotation did not influence this process, the result would be two huge **circulation cells**, one in each hemisphere (Figure 10.13). We'll see shortly that rotation causes these huge cells to split into somewhat smaller cells, but this fact doesn't change the general picture of circulation cells carrying air between the equator and the poles. (Circulation cells are often called *Hadley cells* after the man who first suggested their existence in 1735.)

The circulation cells transport heat both from lower to higher altitudes and from the equator to the poles. They therefore make Earth's polar regions much warmer than they would be in the absence of circulation. The same idea applies to different extents on Venus and Mars. On Venus, the dense atmosphere allows the circulation cells to transport so much thermal energy that temperatures are nearly the same at the equator and the poles. On Mars, the circulation cells transport very little heat because the atmosphere is so thin, so the poles remain much colder than the equator.

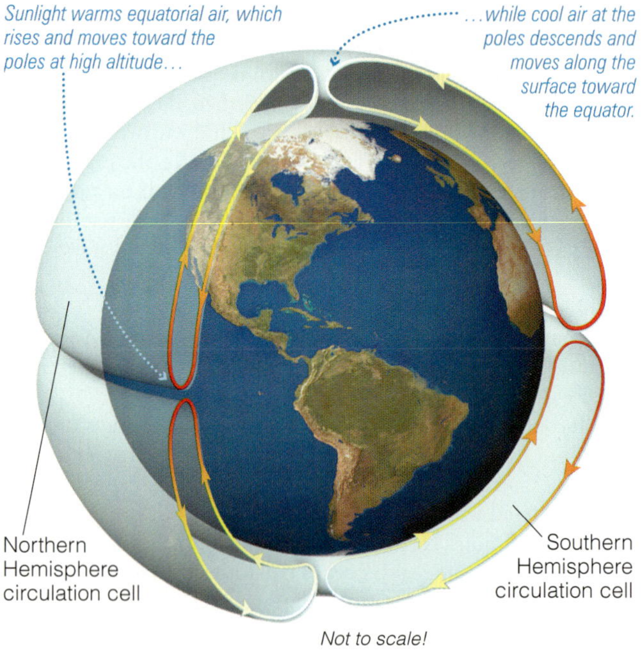

Figure 10.13 Interactive Figure Atmospheric heating creates a circulation cell in each hemisphere that carries warm equatorial air toward the poles while cool polar air moves toward the equator. This diagram shows how the circulation cells would work if our planet's rotation didn't affect them. The vertical extent of the circulation cells is greatly exaggerated; in reality, the tops of the cells are only a few kilometers above Earth's surface.

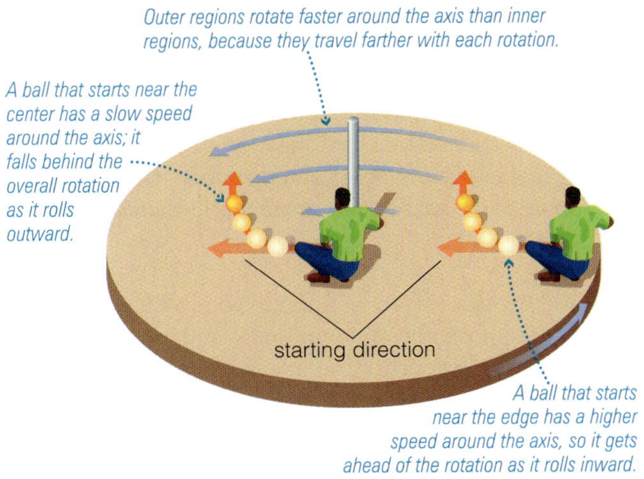

Figure 10.14 Interactive Figure The Coriolis effect on a merry-go-round rotating counterclockwise. Notice that the ball's path deviates to the *right* regardless of whether it is rolled inward or outward. If the merry-go-round were rotating clockwise, both balls would veer to the left.

Rotation and the Coriolis Effect Planetary rotation is the second major factor in global wind patterns, because rotation can alter wind patterns through what we call the **Coriolis effect** (named for the French physicist who first explained it).

It's easy to understand the Coriolis effect on a spinning merry-go-round (Figure 10.14). The outer parts of the merry-go-round move at a faster speed than the inner parts, because they have a greater distance to travel around the axis with each rotation. If you sit near the edge and roll a ball toward the center, the ball begins with your relatively high speed around the axis. As it rolls inward, the ball's high speed makes it move ahead of the slower-moving inner regions. On a merry-go-round rotating counterclockwise, the ball therefore deviates *to the right* instead of heading straight inward. The Coriolis effect also makes the ball deviate to the right if you roll it outward from a position near the center. In that case, the ball starts with your slower speed around the center and lags behind as it rolls outward to faster-moving regions. Again, it deviates to the right. (If the merry-go-round rotates clockwise rather than counterclockwise, the deviations go to the left instead of the right.)

The Coriolis effect alters the path of air on the rotating Earth in much the same way. Earth's equatorial regions travel faster than polar regions that are closer to the rotation axis (see Figure 1.12). Air moving away from the equator therefore has "extra" speed that causes it to move ahead of Earth's rotation to the east, while air moving toward the equator lags behind Earth's rotation to the west. In either case, moving air turns to the *right* in the Northern Hemisphere and to the *left* in the Southern Hemisphere, which explains why storms circulate in opposite directions in the two hemispheres (Figure 10.15). Uneven heating and cooling of the Earth's surface creates regions of slightly higher pressure ("H" on weather maps) or lower pressure ("L" on weather maps) than average. Storms generally occur around low-pressure regions, which draw air inward from surrounding regions. The Coriolis effect makes the inward-flowing air rotate counterclockwise in the Northern Hemisphere and clockwise in the Southern Hemisphere. Thus, instead of evening out the pressure quickly, the inflowing winds end up moving in a circular pattern around the low-pressure zone. That is why storms can persist for days or weeks while being carried across the globe.

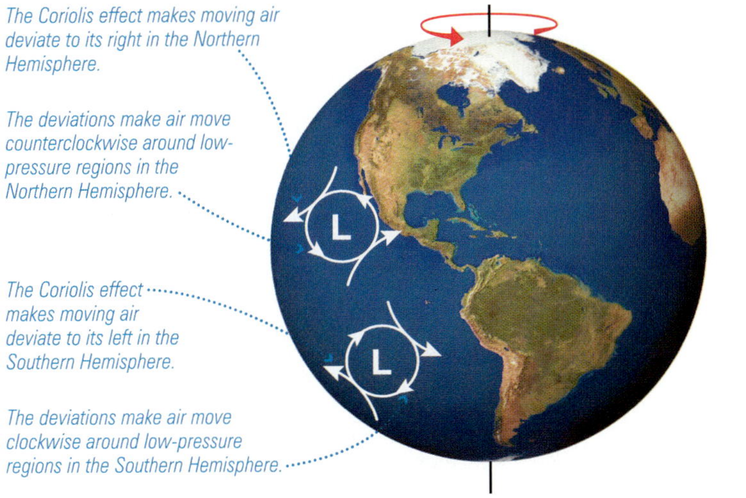

a Low-pressure regions ("L") draw in air from surrounding areas, and the Coriolis effect causes this air to circulate counterclockwise in the Northern Hemisphere and clockwise in the Southern Hemisphere.

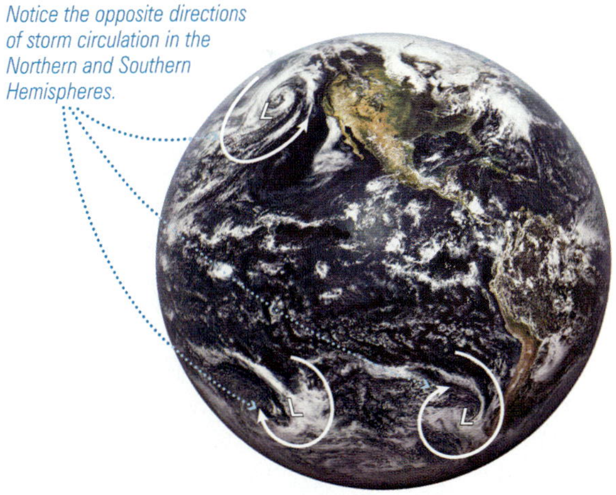

b This photograph shows the opposite directions of storm circulation in the two hemispheres. Notice the counterclockwise swirling of the Northern Hemisphere storm near the upper left and the clockwise swirls in the Southern Hemisphere.

Figure 10.15 Interactive Figure The Coriolis effect works on the rotating Earth much as it does on a merry-go-round, because regions near the equator move at faster speed around the axis than regions near the poles.

THINK ABOUT IT

The Coriolis effect also is important for long-range missiles. Suppose you are in the Northern Hemisphere on Earth and are trying to send a missile to hit a target that lies 5,000 kilometers due north of you. Should you aim directly at the target, somewhat to the left, or somewhat to the right? Explain.

The Coriolis effect plays an even more important role in shaping Earth's global wind patterns: It splits each of the two huge circulation cells shown in Figure 10.13 into three smaller circulation cells (Figure 10.16). You can understand the basic reason by considering the case of air flowing southward along the surface from the North Pole. Without rotation, this air would travel 10,000 kilometers due south to the equator. But on the rotating Earth, the Coriolis effect diverts this air to the right (westward) well before it reaches the equator. This diversion also explains the global wind directions: Surface air moves toward the equator in the cells near the equator and near the poles, so the Coriolis effect diverts this air into westward winds (see Figure 10.12). In contrast, surface air moves toward the poles in the midlatitude cells, so the Coriolis effect diverts it into winds that blow eastward. In essence, the Coriolis effect on a rotating planet tends to divert air moving north or south into east-west winds.

The Coriolis effect operates to some extent on all planets. Its strength depends on both a planet's size and rotation rate: Larger size and faster rotation both contribute to a stronger Coriolis effect. Among the terrestrial planets, Earth is the only one with a Coriolis effect strong enough to split the two large circulation cells. Venus has a very weak Coriolis effect because of its slow rotation, while Mars has a weak Coriolis effect because of its small size. The Coriolis effect is much stronger on the large and fast-rotating jovian planets, and it can therefore split their circulation cells into many more smaller cells than on Earth [Section 11.1].

Clouds and Precipitation In addition to winds, the other key components of weather are rain, snow, and hail, which together are called **precipitation** in weather reports. Precipitation requires clouds. We may think of clouds as imperfections on a sunny day, but they have profound effects on Earth and other planets. Besides being the source of precipitation, they can alter a planet's energy balance. Clouds reflect sunlight back to space, thereby reducing the amount of sunlight that warms a planet's surface, but they also tend to be made from greenhouse gases that contribute to planetary warming. Despite their significant effects, clouds generally are made from minor ingredients of the atmosphere (see Table 10.1).

On Earth, clouds are made from tiny droplets of liquid water or flakes of ice, a fact that you can feel if you walk through a cloud on a mountaintop. Thus, clouds are produced by condensation of water vapor (Figure 10.17). The water vapor enters the atmosphere through evaporation of surface water (or sublimation of ice and snow). Convection then carries the water vapor to high, cold regions of the troposphere, where it can condense to form clouds.

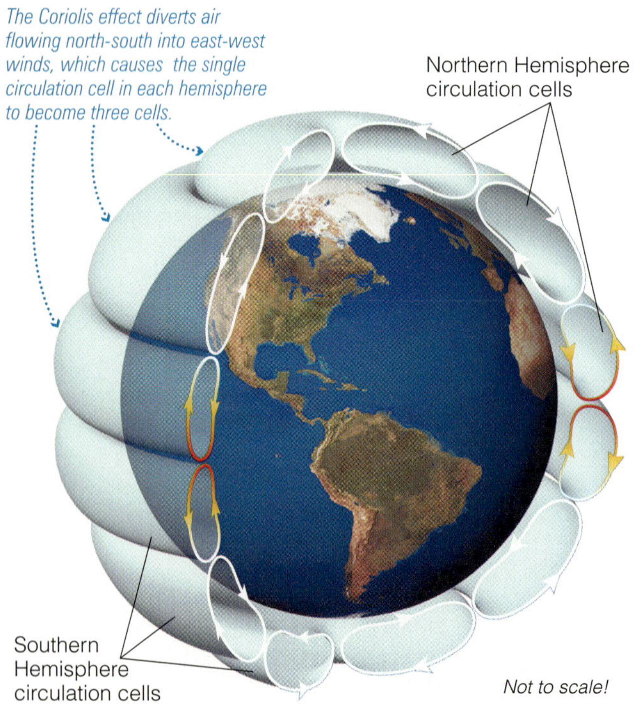

Figure 10.16 Interactive Figure On Earth, the Coriolis effect causes each of the two large circulation cells that would be present without rotation (see Figure 10.13) to split into three smaller cells.

COMMON MISCONCEPTIONS

Toilets in the Southern Hemisphere

A common myth holds that water circulates "backward" in the Southern Hemisphere—that toilets flush the opposite way, water spirals down sink drains the opposite way, and so on. If you visit an equatorial town, you may even find charlatans who, for a small fee, will demonstrate how to change the direction of water spiraling down a drain simply by stepping across the equator. This myth sounds similar to the Coriolis effect—which really does make hurricanes spiral in opposite directions in the two hemispheres—but it is completely untrue.

The Coriolis effect is noticeable only when material moves significantly closer to or farther from Earth's rotation axis, which means distances of hundreds of kilometers. It is completely negligible for objects as small as toilets or drains, or even for tornadoes or dust devils; the direction of swirl in these cases is determined by small-scale effects such as the shape of the sink or the initial direction of rotation. Thus, you can find water swirling down drains in either direction, as well as tornadoes and dust devils twisting in either direction, no matter where you are located on Earth.

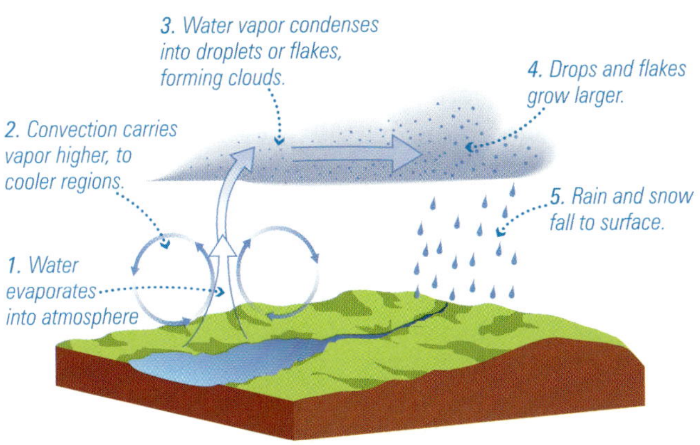

Figure 10.17 The cycle of water on Earth's surface and in the atmosphere.

Clouds can also form as winds blow over mountains, since the mountains physically push the air high enough to allow condensation. The condensed droplets or ice flakes start out very small, but gradually grow larger. If they get large enough so that the upward convection currents can no longer hold them aloft, then they begin to fall toward the surface as rain, snow, or hail.

Stronger convection means more clouds and precipitation. That is why thunderstorms are common on summer afternoons, when the sunlight-warmed surface drives strong convection. The linkage between clouds and convection also explains why Earth has lush jungles near the equator and deserts at latitudes of 20°–30° north or south. Equatorial regions experience high rainfall because they receive more sunlight, which drives strong convection. This high rainfall depletes the air of moisture as Earth's circulation cells carry it away from the equator (see Figure 10.16), leaving little moisture to fall as rain at the latitudes of the deserts.

• What factors can cause long-term climate change?

Long-term climate changes are more difficult to notice than day-to-day variations in the weather, but geological evidence proves that they occur. On Earth, we are well aware of ice ages and warm periods in our planet's past. Other planets have gone through even more dramatic changes. For example, our geological tour of Mars showed that it must have been warm and wet in the distant past but now is very cold and dry [Section 9.4].

Four major factors can lead to long-term climate change on the terrestrial worlds:

- **Solar brightening**: The Sun has grown gradually brighter with time, increasing the amount of warming sunlight that reaches the planets.
- **Changes in axis tilt**: The tilt of Earth and the other planets can change over long periods of time.
- **Changes in reflectivity**: An increase in a planet's reflectivity means a decrease in the amount of sunlight that it absorbs, and vice versa.
- **Changes in greenhouse gas abundance**: More abundant greenhouse gases result in greater warming by the atmosphere.

Figure 10.18 summarizes these four factors, which we now investigate in a little more detail. Keep in mind that more than one factor may be acting at any one time.

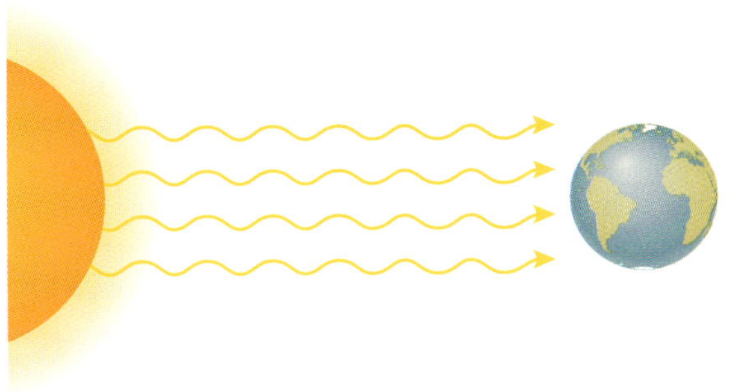

Solar brightening: As the Sun brightens with time, the increasing sunlight tends to warm the planets.

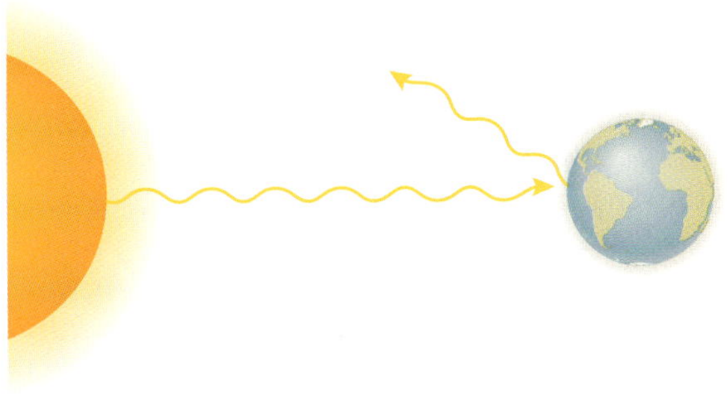

Changes in reflectivity: Higher reflectivity tends to cool a planet, while lower reflectivity leads to warming.

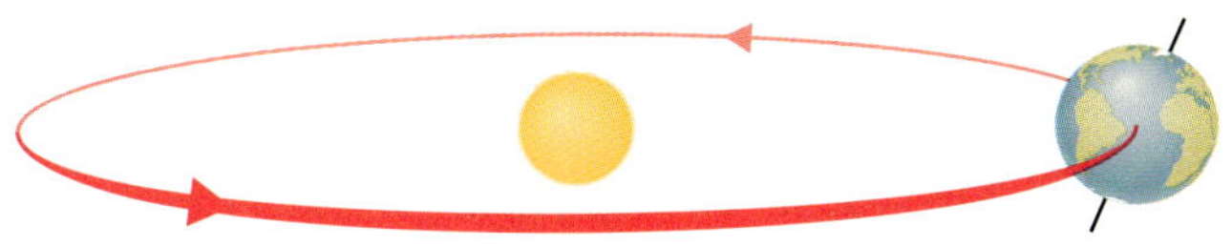

Changes in axis tilt: Greater tilt makes more extreme seasons, while smaller tilt keeps polar regions colder.

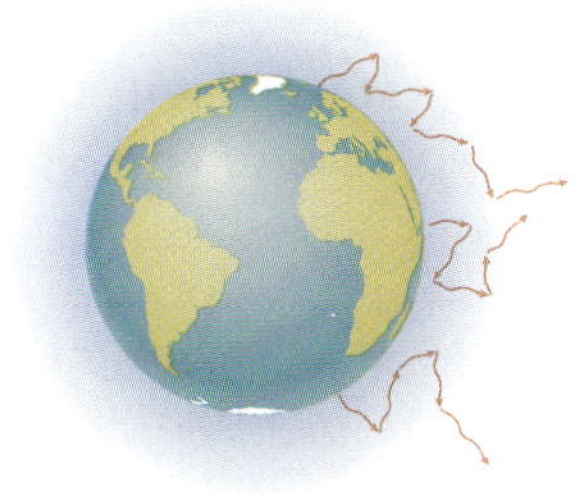

Changes in greenhouse gas abundance: An increase in greenhouse gases slows escape of infrared radiation, warming the planet, while a decrease leads to cooling.

Figure 10.18 Four major factors affecting long-term climate change.

Solar Brightening Both theoretical models of the Sun and observations of other Sun-like stars tell us that the Sun has gradually brightened with age [Section 14.2]: The Sun today is thought to be about 30% brighter than it was when our solar system was young. This brightening tends to warm climates with time, though we'd expect the effects to be noticeable only over periods of tens of millions of years or longer. Moreover, the brightening is so gradual that other climate change factors can easily overwhelm it. For example, while solar brightening alone would mean that all planets should be warmer today than they were in the past, both Mars and Earth are cooler than they were at times in their early histories.

Changes in Axis Tilt Small gravitational tugs from moons, other planets, and the Sun can change a planet's axis tilt over thousands or millions of years. For example, while Earth's current axis tilt currently is about $23\frac{1}{2}°$, the tilt varies over tens of thousands of years between about 22° and 25° (Figure 10.19). These small changes affect the climate by making seasons more or less extreme. Greater tilt means more extreme seasons, with warmer summers and colder winters. The extra summer warmth tends to prevent ice from building up, making the whole planet warmer on average. In contrast, smaller tilts mean less sunlight falls in polar regions in summer, making these regions colder.

Earth's past periods of smaller axis tilt correlate well with colder climate and ice ages, especially when considered along with other small changes in Earth's rotation and orbit. Thus, they are thought to be a primary factor in climate changes on Earth. (The cyclical changes in Earth's axis tilt and orbit are often called *Milankovitch cycles*, after the Serbian scientist who suggested their role in climate change.) As we'll discuss shortly, Mars probably experiences much more extreme changes in axis tilt and climate than Earth, and similar climate cycles may occur on some of the icy moons of the outer solar system.

Changes in Reflectivity Changes in reflectivity affect climate because they change the proportion of sunlight absorbed and reflected. If a planet reflects more sunlight, it must absorb less, which can lead to planetwide cooling. Microscopic dust particles (sometimes called *aerosols*) released by volcanic eruptions can reflect sunlight and thus cool a planet. Small but measurable planetwide cooling has been detected on Earth following a major volcanic eruption. The cooling can continue for years if the dust particles are blasted all the way into the stratosphere.

Human activity is currently altering Earth's reflectivity, although we are not sure in which direction or by precisely how much. Smog particles can act like volcanic dust, reflecting sunlight before it reaches the ground. Deforestation also increases reflectivity because it removes sunlight-absorbing plants. On the other hand, paving a road with blacktop decreases reflectivity, because black roads absorb most of the sunlight that hits them. With different human activities altering Earth's reflectivity in different directions, the overall effects are not easy to gauge.

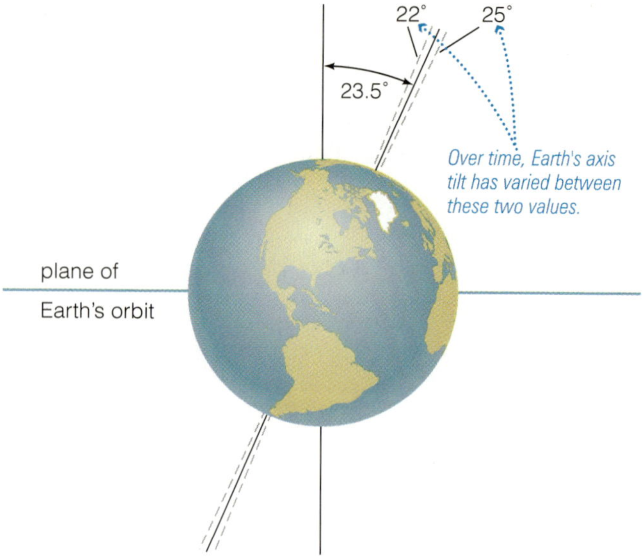

Figure 10.19 Earth's axis tilt is currently about $23\frac{1}{2}°$, but it varies over time between about 22° and 25°.

Changes in Greenhouse Gas Abundance Perhaps the most important factor in long-term climate change is a change in the abundance of greenhouse gases, which strengthens or weakens the greenhouse effect. If the abundance of greenhouse gases increases, the planet generally will warm. If the planet warms enough, increased evaporation and sublimation may add substantial amounts of gas to the planet's atmosphere, leading to an increase in atmospheric pressure. Conversely, if the abundance of greenhouse gases decreases, the planet generally will cool, and atmospheric pressure may decrease as gases freeze.

• How does a planet gain or lose atmospheric gases?

Of the factors that affect planetary climate, changes in greenhouse gas concentration appear to have had the greatest effect on the long-term climates of Venus, Earth, and Mars. But how can the greenhouse gas concentration change? In fact, such changes generally occur as part of more general changes in the abundances of atmospheric gases. Thus, before we can understand the specific reasons for long-term climate changes on the terrestrial worlds, we must investigate how atmospheres gain and lose gas.

Sources of Atmospheric Gas During their births some four and a half billion years ago, the terrestrial worlds were too small to capture significant amounts of hydrogen and helium gas before the solar wind cleared away the solar nebula [Section 8.3]. Thus, they were essentially born without atmospheres. So where did the gas in terrestrial atmospheres come from? In general, three different processes can provide gas to terrestrial atmospheres (Figure 10.20).

As we discussed in Chapter 9, volcanic *outgassing* has been the primary source of gases for the terrestrial atmospheres. Recall that the terrestrial worlds were built primarily of metal and rock, but impacts of ice-rich planetesimals

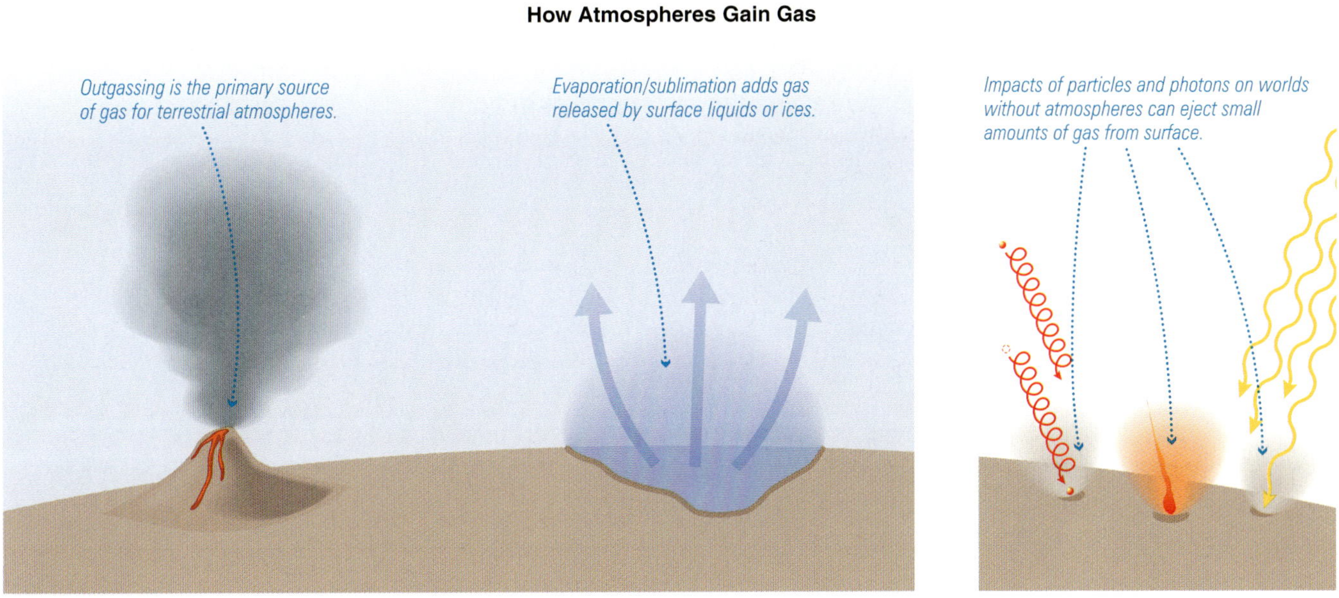

Figure 10.20 Three processes that can provide gas to terrestrial atmospheres.

(from beyond the frost line [Section 8.3]) brought in water and gas that became trapped in their interiors. Studies of volcanic eruptions show that the most common gases released by outgassing are water (H_2O), carbon dioxide (CO_2), nitrogen (N_2), and sulfur-bearing gases (H_2S and SO_2).*

Some worlds have a secondary source of atmospheric gas: evaporation of surface liquids or sublimation [Section 5.3] of surface ices into gas. After outgassing creates an atmosphere, some atmospheric gases may condense to become surface liquids or ices. For example, Earth's oceans and polar caps were made from water vapor that condensed as rain or snow, while Mars's polar caps contain both frozen carbon dioxide and water ice. Surface liquids or ices continually exchange gas with the atmosphere, and overall climate changes can alter the balance of this exchange. If a planet warms, the rates of evaporation and sublimation can increase, thereby adding gas to the atmosphere.

Outgassing and evaporation/sublimation are the only important sources of gas on planets that have significant atmospheres, such as Venus, Earth, and Mars. However, a third source can produce very small amounts of gas on worlds that are essentially unprotected by an atmosphere: Micrometeorites, solar wind particles, and high-energy solar photons can knock individual atoms or molecules free from the surface. In essence, the particles and photons affect the surface much like sandblasting, but on a much smaller scale. As a result, an "airless" world like the Moon or Mercury can end up with an extremely low-density atmosphere consisting of both the atoms and molecules released from the surface and the particles that blasted them away. This "surface ejection" process does not affect planets with more substantial atmospheres, because the atmospheres prevent small particles and high-energy solar photons from reaching the surface.

*Among these materials, only water existed in modest quantities in the solar nebula. The others were created by chemical reactions that occurred *inside* the planets after gas became trapped.

Losses of Atmospheric Gas Planets can also lose atmospheric gas through five different processes, all summarized in Figure 10.21. Two of these processes simply recycle gas from the atmosphere to the planets' surface or interior. The first, *condensation*, is the reverse of the release of gas through evaporation or sublimation; gases in the atmosphere condense and fall as rain, hail, or snow. On Mars, for example, it is cold enough for carbon dioxide to condense into solid ice, especially at the poles. That is why the Martian polar caps contain large amounts of frozen carbon dioxide (also known as "dry ice"). The second process removes gas from the atmosphere through *chemical reactions* that incorporate the gas into surface metal or rock. Rusting is a familiar example of this process: Iron rusts when it reacts with oxygen, thereby removing the oxygen from the atmosphere and incorporating it into the metal. This process is so efficient that it could remove all the oxygen in Earth's atmosphere in just a few million years; Earth retains atmospheric oxygen only because oxygen is continually resupplied by plants and other photosynthetic organisms.

The removal of gas by condensation or chemical reactions is not necessarily permanent, since it leaves open the possibility of the gas returning to the atmosphere at a later time. In contrast, the other three loss processes shown in Figure 10.21 are always permanent, because the gravity of the terrestrial planets is too weak to recapture lost gases.

The first of these permanent loss processes is the loss of gas through impacts: The same impacts that create large craters may blast significant amounts of atmospheric gas into space. All the terrestrial worlds probably lost gas in this way during the heavy bombardment; Mars probably suffered the greatest loss, because its small size compared to Venus and Earth allowed gas to escape more easily.

How Atmospheres Lose Gas

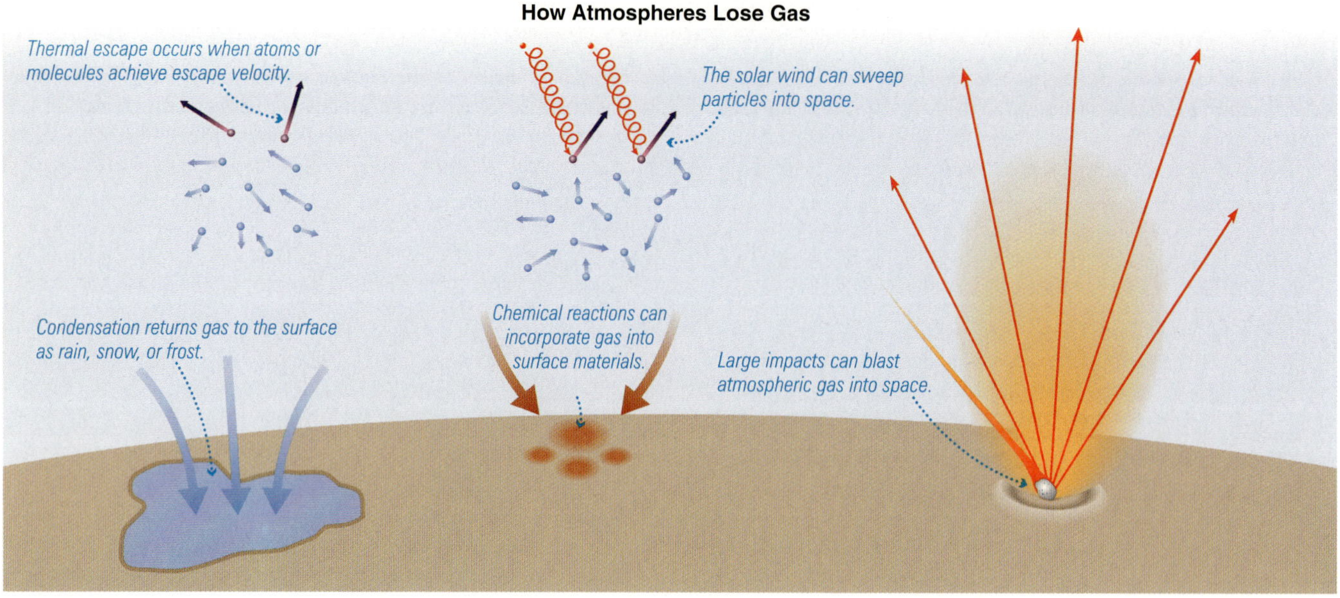

Figure 10.21 Five processes that can remove gas from terrestrial atmospheres.

The second permanent loss process primarily affects planets that lack a protective magnetosphere. Without a magnetosphere, particles from the solar wind constantly pelt the exosphere and sometimes strip ionized gas particles into space. In essence, the solar wind sweeps away some of a planet's atmospheric gas. This process has had little effect on Earth, because Earth's magnetic field creates a magnetosphere that protects our atmosphere. In contrast, as we'll discuss in Section 10.4, Mars may have lost much of its early atmosphere to stripping by the solar wind.

The third permanent loss process, called **thermal escape**, occurs when an atom or a molecule of gas in a planet's exosphere achieves escape velocity [Section 4.5] and flies off into space. In principle, thermal escape can affect any planetary atmosphere. However, its relative importance on any world depends on three distinct factors:

1. The planet's escape velocity: In general, more massive planets have higher escape velocities, making it more difficult for gases to escape.
2. The atmospheric temperature: Individual particles in a gas move at a wide range of random speeds [Section 4.3], but the higher the temperature, the faster their *average* speed and the more likely that some particles will reach escape velocity.
3. The mass of the gas particles: Because all gas particles have the same average kinetic energy at a particular temperature, lightweight gases such as hydrogen or helium move at faster average speeds than heavier ones such as oxygen or carbon dioxide (because kinetic energy depends on both mass and velocity [Section 4.3]). Lightweight gases are therefore more likely to achieve escape velocity.

All the terrestrial planets are small enough and warm enough for hydrogen and helium to escape. Thus, they were unable to hold for long any hydrogen or helium gas that they may have captured from the solar nebula when they were very young. Venus, Earth, and Mars were able to retain heavier gases in their atmospheres, which is why they have substantial atmospheres today.

10.3 Atmospheres of the Moon and Mercury

We have now covered all the basic ideas that we need to understand the atmospheric histories of the terrestrial worlds. In the rest of this chapter, we'll use the atmospheric source and loss processes to learn how and why each world ended up with its current atmosphere. We'll use the ideas we've studied about weather to understand the conditions we find on planets with atmospheres. And we'll use the four factors that affect long-term climate change to discuss how and why the climates of Venus, Earth, and Mars have followed such different paths through time. As we did in Chapter 9, we'll proceed through our tour of the terrestrial atmospheres in order of planetary size. We therefore begin with the two smallest terrestrial worlds, the Moon and Mercury.

• Do the Moon and Mercury have any atmosphere at all?

We usually don't think of the Moon and Mercury as having atmospheres, but they are not totally devoid of gas. However, their gas densities are far too low to scatter sunlight as our atmosphere does. Even in broad daylight, you would see a pitch-black sky surrounding the bright Sun. Moreover, their atmospheres are too thin to absorb much solar radiation or to have any kind of atmospheric structure such as a troposphere, stratosphere, or thermosphere. In

MATHEMATICAL INSIGHT 10.2 Thermal Escape from an Atmosphere

The amount of gas that an atmosphere loses to thermal escape depends on the speeds of the gas particles (atoms or molecules). In general, gas particles of a particular type move at a wide range of speeds. For example, Figure 1 shows the range of speeds of sodium atoms at the Moon's daytime temperature ($T = 400$ K). The peak in the figure represents the most common speed of the sodium atoms. This speed, called the *peak thermal velocity* of the sodium atoms, is given by the following formula:

$$v_{\text{thermal}} = \sqrt{\frac{2kT}{m}}$$

where m is the mass of a single atom, T is the temperature in Kelvin, and $k = 1.38 \times 10^{-23}$ joule/Kelvin is *Boltzmann's constant.* The T in the formula tells us that the peak thermal velocity increases with temperature; thus, particle speeds are generally lower for low temperatures and higher for high temperatures. The m in the formula tells us that at any particular temperature, particles of lighter gases (smaller m) move at faster speeds than particles of heavier gases.

If the peak thermal velocity of a particular type of gas is greater than the escape velocity, then most of the gas particles will quickly escape to space. However, gas can escape even if the peak thermal velocity is much less than the escape velocity, because the wide range of gas particle speeds ensures that at least some particles are moving much faster than the peak velocity. For example, Figure 1 shows that while the peak thermal velocity of sodium atoms on the Moon (about 0.5 km/s) is much less than the Moon's escape velocity (about 2.4 km/s), a small fraction of the sodium atoms are moving faster than the escape velocity. As these atoms escape, collisions between the remaining gas atoms (or between gas atoms and the Moon's surface) continually ensure that a few atoms always attain escape velocity.

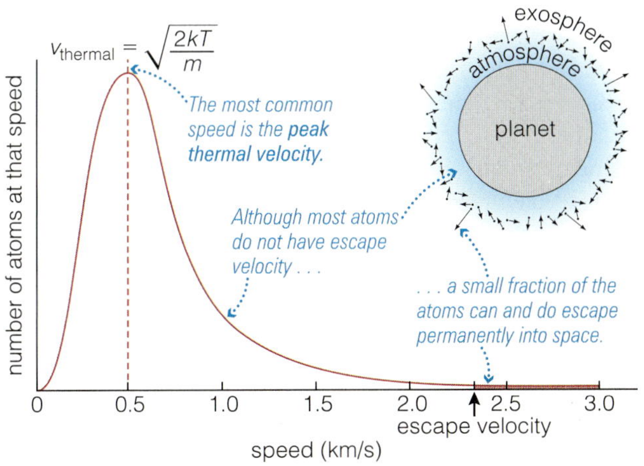

Figure 1 Interactive Figure In a gas at a given temperature, different atoms are always moving at different speeds. This plot shows the range of speeds of sodium atoms in the lunar atmosphere.

In principle, at any moment, a small fraction of the particles in any gas will be moving with a speed above the escape velocity. Some of these particles will also be moving in the right direction for them to escape (and are unlikely to suffer a collision since they are in an exosphere), which means that the gas could slowly leak away into space. However, this loss may be insignificant if the peak thermal velocity is far below the escape velocity. As a rule of thumb, we can expect thermal loss to be significant if the peak thermal velocity is more than a few percent of the escape velocity; if the peak thermal velocity of a particular type of gas particle is above about 20% of the escape velocity, then this gas will be completely lost to space in less than a few billion years (that is, in a time less than the age of our solar system).

Example: The Moon has a very thin exosphere of sodium atoms but retains virtually no hydrogen gas at all. Explain why. Useful data: The Moon's daytime temperature is $T = 400$ K; the Moon's escape velocity is about 2.4 km/s; the mass of a hydrogen atom is 1.67×10^{-27} kg; the mass of a sodium atom is 3.84×10^{-26} kg (about 23 times the mass of a hydrogen atom).

Solution:

Step 1 Understand: If we assume that the Moon loses gas through thermal escape, then the rate of escape for any particular type of gas depends on how its peak thermal velocity compares to the Moon's escape velocity. From Figure 1, we already know that the peak thermal velocity for sodium atoms is about 0.5 km/s. We don't yet know the peak thermal velocity for hydrogen, so we need to calculate it.

Step 2 Solve: From the formula and the given data, the peak thermal velocity for hydrogen atoms at the Moon's daytime temperature is:

$$v_{\text{thermal (H atoms)}} = \sqrt{\frac{2kT}{m_{H\ atom}}}$$

$$= \sqrt{\frac{2 \times \left(1.38 \times 10^{-23} \frac{\text{joule}}{\text{K}}\right) \times (400\ \text{K})}{1.67 \times 10^{-27}\ \text{kg}}}$$

$$\approx 2{,}600\ \text{m/s} = 2.6\ \text{km/s}$$

Step 3 Explain: The peak thermal velocity of the hydrogen atoms is 2.6 km/s, which is slightly greater than the Moon's escape velocity of about 2.4 km/s. Thus, hydrogen atoms quickly escape, which is why the Moon cannot retain a hydrogen atmosphere. In contrast, the peak thermal velocity for sodium atoms (about 0.5 km/s) is only about 20% of the Moon's escape velocity. Thus, while the Moon slowly loses sodium atoms to space, the rate of escape is quite slow. Moreover, micrometeorites, solar wind particles, and high-energy photons constantly blast new sodium atoms from the Moon's surface into its atmosphere, so the Moon retains a thin sodium exosphere.

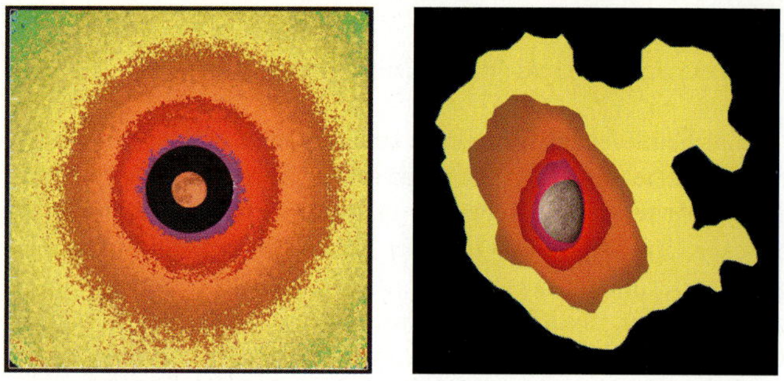

a The Moon's exosphere. **b** Mercury's exosphere.

Figure 10.22 These "images" of the atmospheres of the Moon and Mercury—which are essentially exospheres only—are based on data collected with instruments sensitive to emission lines from sodium atoms (the same emission lines that give sodium street lights their yellow color). The colors shown here are not real, but rather represent the varying gas densities from highest (purple) to lowest (black). The inset photos of the Moon and Mercury are on the same scale as the images, so you can see that their exospheres extend quite far above their surfaces.

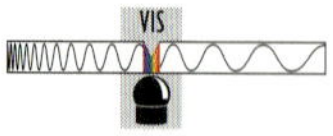

essence, the Moon and Mercury have only thin exospheres, without any other atmospheric layers.

The total amount of gas in the exospheres of the Moon and Mercury is very small. If you could condense the entire atmosphere of either the Moon or Mercury into solid form, you would have so little material that you could almost store it in a dorm room. The low density of the gas means that collisions between atoms or molecules are rare. The gas particles therefore can rise as high as their speeds allow—sometimes even escaping to space. As a result, the exospheres of the Moon and Mercury extend thousands of kilometers into space (Figure 10.22).

Source and Loss Processes on the Moon and Mercury The Moon and Mercury may once have had some gas released by volcanic outgassing, but any such gas is long gone. Recall that volcanic activity on these worlds ceased billions of years ago, so there is no ongoing source of outgassing. Some of the gas released long ago was probably lost through impacts or stripping by the solar wind, but it would have been lost to thermal escape anyway. Mercury cannot hold much of an atmosphere because its small size and high daytime temperature mean that nearly all gas particles eventually achieve escape velocity. Because it is smaller than Mercury, gases escape about as easily from the Moon even though it is also somewhat cooler than Mercury.

The only ongoing source of gas on the Moon and Mercury is the "surface ejection" that occurs when micrometeorites, solar wind particles, or high-energy solar photons knock free surface atoms and molecules. This gas never accumulates because it is lost as quickly as it is gained. Some of the relatively few particles released from the surface are blasted upward fast enough to achieve escape velocity and therefore escape directly to space. The rest bounce around like tiny rubber balls, arcing hundreds of kilometers into the sky before crashing back down to the surface. Each gas particle typically bounces a few dozen times before being absorbed back into the surface.

Ice in Polar Craters In general, we do not expect to find either liquid water or solid water ice on the Moon or Mercury. The lack of significant atmospheric pressure means that water cannot remain stable in liquid form, and daytime temperatures are high enough that any solid ice would long ago have sublimated into gas and escaped to space. However, recent observations strongly suggest that both the Moon and Mercury have some water ice at the bottoms of craters near their poles.

If you think about it, the existence of this polar ice should not be too surprising. Both the Moon and Mercury have certainly been struck many times by ice-rich comets. The ice would vaporize upon impact, and any water molecules that did not achieve escape velocity would bounce randomly around the surface until they settled down. Some molecules would have settled into craters near the poles, where they condensed into ice. Because the bottoms of these polar craters lie in nearly perpetual shadow, it is possible that water ice could remain perpetually frozen. Spacecraft observations of the Moon suggest that this is indeed the case, and radar observations of Mercury suggest that it also has ice in polar craters. Although the polar ice probably has little effect on either of these worlds, it could prove very important for future exploration. In particular, the ice in polar craters offers a potential source of water for future human colonies on the Moon.

10.4 The Atmospheric History of Mars

We began this chapter with the goal of understanding how and why the terrestrial atmospheres came to differ so profoundly. The cases of the Moon and Mercury have proved to be quite easy to understand. However, because outgassing supplied substantial early atmospheres to the three larger worlds—Mars, Venus, and Earth—their very different present atmospheres must reflect atmospheric gains and losses that have occurred through time. Again, planetary size is the most important influence on these gains and losses: Size ultimately determines the level of volcanism and outgassing on each planet and, as we'll see, size has also played a crucial role in determining which loss processes have occurred. We therefore continue our tour in size order, which brings us next to Mars.

Mars is only 40% larger in radius than Mercury, but its surface reveals a much more fascinating and complex atmospheric history. As we discussed in Chapter 9, orbital photographs and studies by robotic landers tell us that water once flowed on the Martian surface. In the distant past, Mars must have been very different from its current "freeze-dried" state.

• What is Mars like today?

The present-day surface of Mars looks much like some deserts or volcanic plains on Earth (see Figures 7.7, 9.30, and 10.1). However, its thin atmosphere makes it unlike

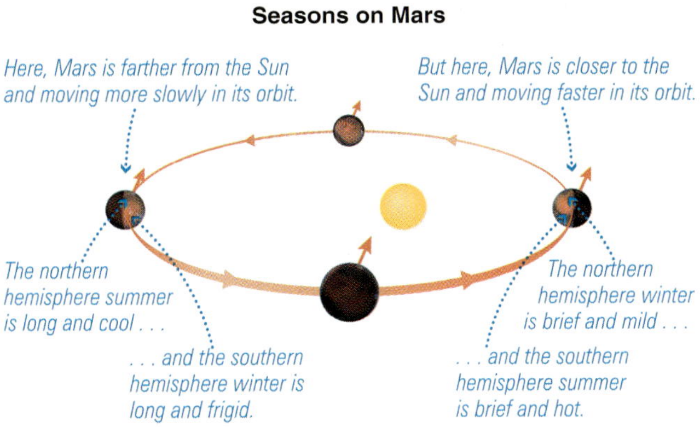

Figure 10.23 The ellipticity of Mars's orbit makes seasons more extreme (hotter summers and colder winters) in the southern hemisphere than in the northern hemisphere.

any place on Earth. The low atmospheric pressure—less than 1% of that on Earth's surface—explains why liquid water is unstable on the Martian surface [Section 9.4]: Any liquid water would rapidly disappear, with some evaporating and some freezing into ice.

The atmosphere is made mostly of carbon dioxide, but the total amount of gas is so small that it creates only a weak greenhouse effect (see Table 10.2). The temperature is usually well below freezing, with a global average of about −50°C (−58°F). The lack of oxygen means that Mars lacks an ozone layer, so much of the Sun's damaging ultraviolet radiation passes unhindered to the surface.

Martian Seasons and Winds Mars has seasons much like those on Earth because of its similar axis tilt, but they last about twice as long (because a Martian year is almost twice as long as an Earth year). On Mars, however, the shape of its orbit introduces a second important effect. Mars's more elliptical orbit puts it significantly closer to the Sun during the southern hemisphere summer and farther from the Sun during the southern hemisphere winter (Figure 10.23). Mars therefore has more extreme seasons in its southern hemisphere—that is, shorter, hotter summers and longer, colder winters—than in its northern hemisphere.

The seasonal changes create the major feature of Mars's weather: winds blowing from the summer pole to the winter pole. Polar temperatures at the winter pole drop so low (about −130°C) that carbon dioxide condenses into "dry ice" at the polar cap. At the same time, frozen carbon dioxide at the summer pole sublimates into carbon dioxide gas (Figure 10.24). During the peak of summer, nearly all the carbon dioxide may sublime from the summer pole, leaving only a residual polar cap of water ice (Figure 10.25). The atmospheric pressure therefore increases at the summer pole and decreases at the winter pole, driving strong pole-to-pole winds. Overall, as much as one-third of the total carbon dioxide of the Martian atmosphere moves seasonally between the north and south polar caps.

The direction of the pole-to-pole winds on Mars changes with the alternating seasons. Sometimes these winds initiate huge dust storms, particularly when the more extreme summer approaches in the southern hemisphere (Figure 10.26). At times, airborne dust shrouds the surface so much that no surface markings can be distinguished and large dunes form and shift with the strong winds. As the dust settles out onto the surface, it can change the color or reflectivity over vast areas, creating seasonal changes in appearance that fooled some astronomers in the late 1800s and early 1900s into thinking they saw changes in vegetation.

The Martian winds often spawn *dust devils*, swirling winds that you may have seen over desert sands or dry dirt on Earth. Dust devils look much like miniature tornadoes, but they rise up from the ground rather than coming down

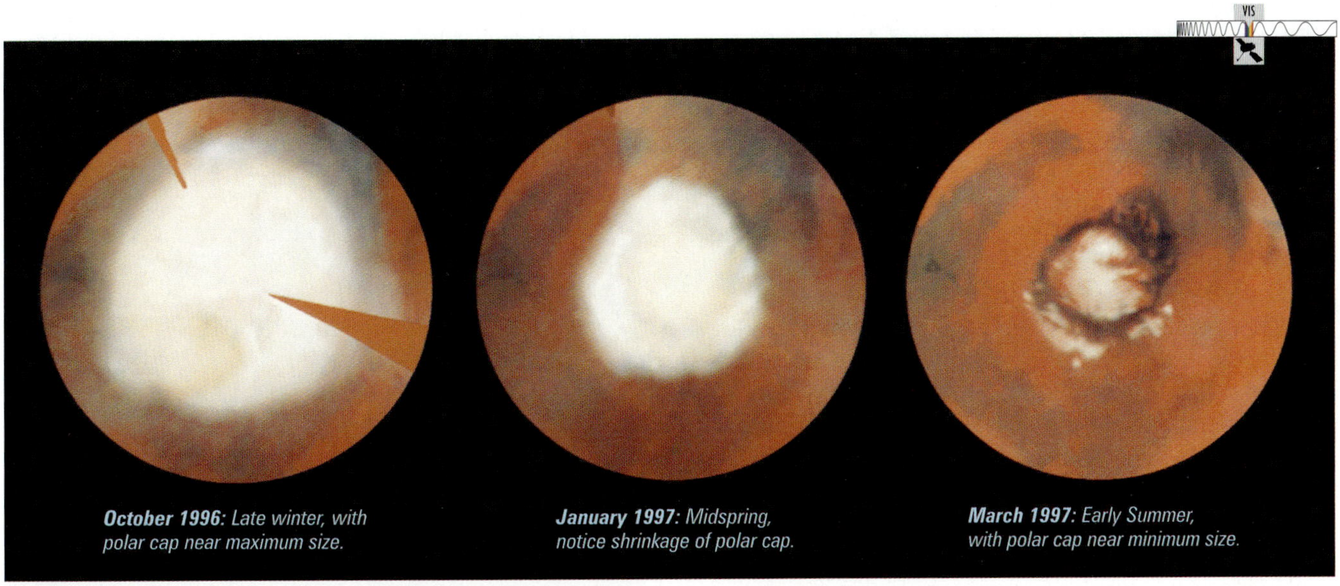

Figure 10.24 These images from the Hubble Space Telescope show the region around Mars's north pole, which is at the center of each image. Notice how the polar cap shrinks as summer approaches, due to sublimation of carbon dioxide ice. This sublimation at the summer pole occurs at the same time that carbon dioxide condenses at the winter pole. These changes drive winds from the summer pole to the winter pole.

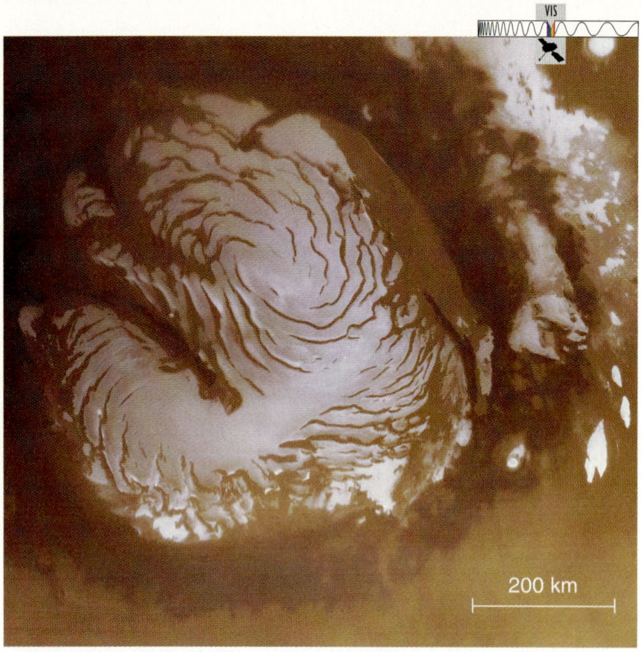

Figure 10.25 This image, from *Mars Global Surveyor,* shows the residual north polar cap during northern hemisphere summer. The white material is primarily water ice, with dust mixed in.

Figure 10.26 These two Hubble Space Telescope photos contrast the appearance of the same face of Mars in the absence (left) and presence (right) of a global dust storm.

from the sky. The air in dust devils is heated from below by the sunlight-warmed ground, and swirls because of the way it interacts with prevailing winds. Dust devils on Mars are especially common during summer in either hemisphere and can be far larger than their counterparts on Earth. The Mars rovers have even photographed dust devils (Figure 10.27), and some have apparently passed directly over the rovers without causing damage. In one case, a dust devil apparently "cleaned" dust off the solar panels of the rover Spirit, restoring power that had been lost as dust accumulated.

Color of the Martian Sky Martian winds and dust storms leave Mars with perpetually dusty air, which helps explain the colors of the Martian sky. The air on Mars scatters visible sunlight much like the air on Earth; that is, it scatters blue light more than red light. However, because there is much less air on Mars than Earth, it scatters less light. By itself this scattering would make the Martian sky a deep blue (almost black).

The dust in the Martian atmosphere has an opposite effect. Dust tends to absorb blue light, so by itself it would leave the sky with a brownish-pink color. The actual color of the Martian sky comes from the combination of the effects of air and dust. The result depends on the time of day, the season, and how much dust is in the air. Most of the time, the daytime Martian sky ends up yellow-brown. However, it sometimes ends up with other colors, such as blue or even green, particularly in the mornings and evenings.

Mars Climate and Axis Tilt The weather on Mars does not change much from one year to the next. However, changes in its axis tilt probably cause Mars to undergo longer-term cycles of climate change. Theoretical calculations suggest that Mars's axis tilt varies far more than that of Earth—from as little as 0° to 60° or more on time scales of hundreds of thousands to millions of years. This extreme variation arises for two reasons. First, Jupiter's gravity has a greater effect on the axis of Mars than on Earth, because Mars's orbit is closer to Jupiter's orbit. Second, Earth's axis is stabilized by the gravity of our relatively large Moon. Mars's two tiny moons, Phobos and Deimos (see Figure 8.11), are far too small to offer any stabilizing influence on its axis.

As we discussed earlier, changes in axis tilt affect both the severity of the seasons and the global average temperature. When Mars's axis tilt is small, the poles may stay in a perpetual deep freeze for tens of thousands of years. With more carbon dioxide frozen at the poles, the atmosphere becomes thinner, lowering the pressure and weakening the greenhouse effect. When the axis is highly tilted, the summer pole becomes quite warm, allowing substantial amounts of water ice to sublime, along with carbon dioxide, into the atmosphere. The pressure therefore increases, and Mars becomes warmer as the greenhouse effect strengthens—

Figure 10.27 This photograph shows a dust devil on Mars, photographed by the *Spirit* rover in early 2005.

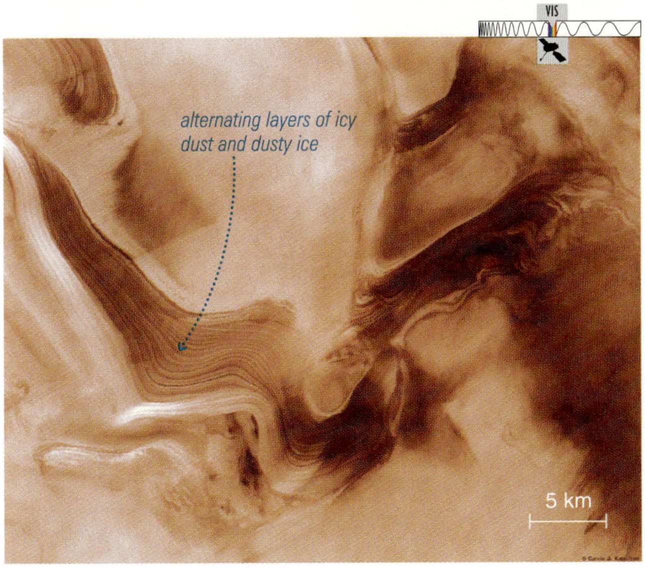

Figure 10.28 This photo from *Mars Global Surveyor* shows a region about 40 kilometers wide at the edge of a Martian polar cap. Note the alternating layers of ice and dust, which are probably the result of climate changes caused by the changing axis tilt of Mars.

although probably not by enough to allow liquid water to become stable at the surface. The Martian polar regions show layering of dust and ice that probably reflects changes in climate due to the changing axis tilt (Figure 10.28).

• Why did Mars change?

Climate change due to Mars's varying axis tilt may have had important effects, but it does not explain the most dramatic climate change in Martian history: the change from the world of flowing water that existed long ago to the cold and dry world we see today.

The geological evidence we discussed in Chapter 9 leaves little room for doubt that Mars had wetter and warmer periods, probably with rainfall, that came to an end more than 3 billion years ago. The full extent of these periods is a topic of considerable scientific debate. Some scientists think that Mars may have been continuously warm and wet for much of its first billion years of existence. Others think that Mars may have had only intermittent periods of rainfall, perhaps triggered by the heat of large impacts, and that ancient lakes, ponds, or oceans may have been completely ice-covered.

Either way, Mars must have had a much denser and warmer atmosphere to allow rainfall of any kind. Warmer temperatures would have been needed to keep water from freezing, and higher density would have been needed to make the atmospheric pressure great enough to allow for stable liquid water. Computer simulations show that Mars today could have liquid water if it had a carbon dioxide atmosphere about 400 times as dense as the current Martian atmosphere; this atmosphere would give Mars a surface pressure about three times that on Earth today. However, because the Sun was dimmer in the distant past (see Figure 10.18), even more greenhouse warming would have been needed to allow for liquid water when Mars was young. Current models are unable to account for the necessary additional warming with carbon dioxide gas alone, but additional greenhouse warming may have been provided by carbon dioxide ice clouds or atmospheric methane.

The idea that Mars once had a much thicker atmosphere and stronger greenhouse effect makes sense. Mars has plenty of ancient volcanoes, and calculations show that these volcanoes could have supplied the atmosphere with 400 times as much carbon dioxide as it has today. Moreover, if Martian volcanoes outgassed carbon dioxide and water in the same proportions as do volcanoes on Earth, Mars would have had enough water to fill oceans tens or even hundreds of meters deep.

The bigger question is not whether Mars once had a denser atmosphere but where all the atmospheric gas went. In particular, Mars must somehow have lost most of its carbon dioxide gas. This loss would have weakened the greenhouse effect until the planet essentially froze over. Some of the carbon dioxide condensed to make the polar caps and some may be chemically bound to surface rock. However, the bulk of the gas was probably lost to space.

The precise way in which Mars lost its carbon dioxide gas is not clear, but recent data suggest a close link to a change in Mars's magnetic field (Figure 10.29). Early in its history, Mars probably had molten metals in its core, much like Earth today. Circulating with convection and Mars's rotation, these metals should have produced a magnetic field and a protective magnetosphere around Mars. However, the magnetic field weakened as the small planet cooled and the core solidified, leaving the atmosphere vulnerable to solar wind particles. These solar wind particles could have stripped gas out of the Martian atmosphere and into space. Had Mars kept its magnetic field, it might have retained much of its atmosphere and might have a more moderate climate today.

Much of the water once present on Mars is also probably gone for good. Like the carbon dioxide, some water vapor may have been stripped away by the solar wind. However, Mars also lost water in another way. Because Mars lacks an ultraviolet-absorbing stratosphere, atmospheric water molecules would have been easily broken apart by ultraviolet photons. The hydrogen atoms that broke away from the water molecules would have been lost rapidly to space through thermal escape. With these hydrogen atoms gone, the water molecules could not be made whole again. Initially, oxygen from the water molecules would have remained in the atmosphere, but over time this oxygen was lost too. The solar wind probably stripped some of the oxygen away from the atmosphere, and the rest probably was drawn out of the atmosphere through chemical reactions with surface rock. This oxygen literally rusted the Martian rocks, giving the "red planet" its distinctive tint.

In summary, Mars changed primarily because of its relatively small size. It was big enough for volcanism and outgassing to release plenty of water and atmospheric gas early

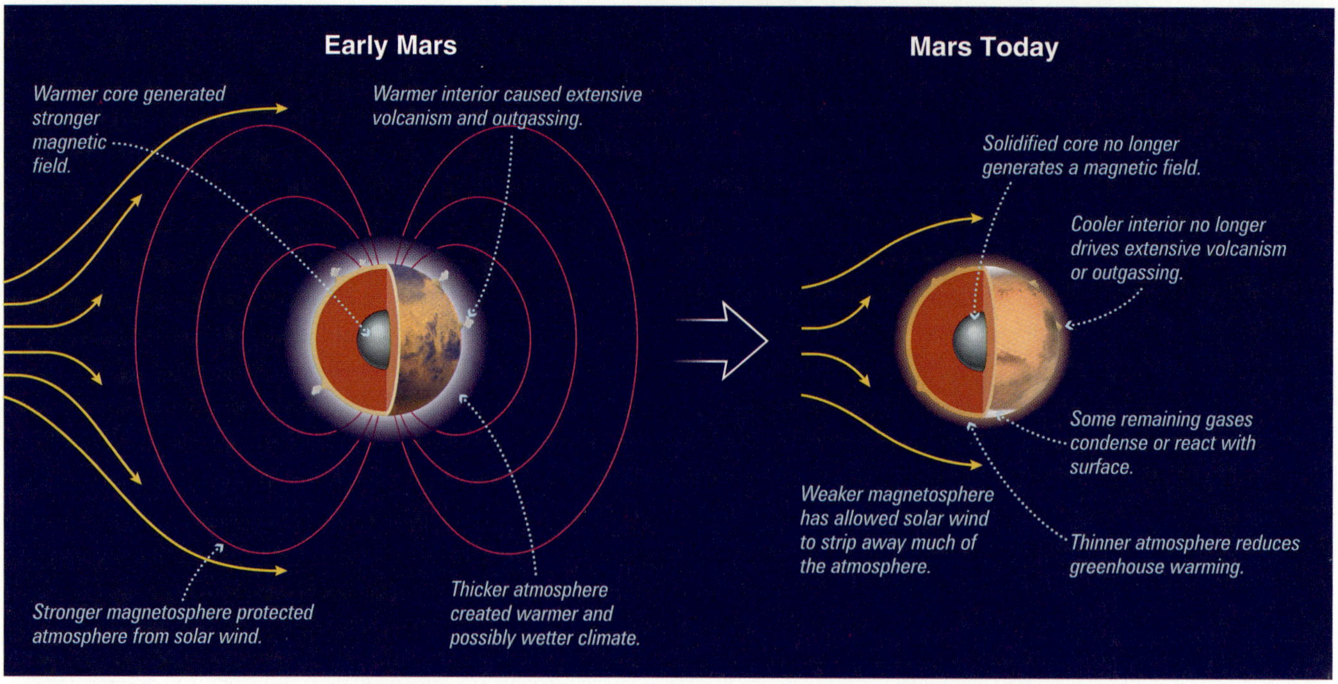

Figure 10.29 Some 3 billion years ago, Mars underwent dramatic climate change, ensuring that rain could never fall again.

in its history, but too small to maintain the internal heat needed to prevent the loss of this water and gas. As its interior cooled, its volcanoes quieted and stopped releasing gas into the atmosphere, while its relatively weak gravity and the loss of its magnetic field allowed existing gas to be stripped away to space. If Mars had been as large as Earth, so that it could still have outgassing and a global magnetic field, it might still have a moderate climate today. Mars's distance from the Sun also helped seal its fate: Even with its small size, Mars might still have some flowing water if it were significantly closer to the Sun, where the extra warmth could melt the water that remains frozen underground and at the polar caps.

The history of the Martian atmosphere holds important lessons for us on Earth. Mars apparently was once a world with pleasant temperatures and with streams, rain, glaciers, lakes, and possibly oceans. It had all the necessities for life as we know it. This once-hospitable planet turned into a frozen and barren desert at least 3 billion years ago, and it is unlikely that Mars will ever again be warm enough for its frozen water to flow. Any life that may have existed on Mars is either extinct or hidden away in a few choice locations, such as in underground water near not-quite-dormant volcanoes. As we consider the possibility of future climate change on Earth, Mars presents us with an ominous example of how drastically things can change.

THINK ABOUT IT

Some people have proposed "terraforming" Mars—that is, making it more Earth-like—by finding a way to release all the carbon dioxide frozen in its polar caps into its atmosphere. If Mars once had oceans in the distant past, could the release of gas from the polar caps allow it to have oceans again? Why or why not?

10.5 The Atmospheric History of Venus

Venus presents a stark contrast to Mars. Its thick carbon dioxide atmosphere creates an extreme greenhouse effect that bakes the surface to unbearable temperatures. It is fairly easy to understand why Venus is so different from Mars: Its larger size allowed it to retain more interior heat, leading to greater volcanism [Section 9.5]. The associated outgassing released the vast quantities of carbon dioxide that create Venus's strong greenhouse effect.

Venus becomes more mysterious when we compare it to Earth. Because Venus and Earth are so similar in size, we might naively expect both planets to have had similar atmospheric histories. Clearly, this is not the case. In this section, we'll explore how Venus's atmosphere ended up so different from Earth's, and in the process we'll learn important lessons about the habitability of our own planet.

• What is Venus like today?

If you stood on the surface of Venus, you'd feel a searing temperature higher than that of a self-cleaning oven and a tremendous pressure 90 times greater than that on Earth. A deep-sea diver would have to go nearly 1 kilometer (0.6 mile) beneath the ocean surface on Earth to feel comparable pressure. Venus's atmosphere consists almost entirely of carbon dioxide (CO_2). It has virtually no molecular oxygen (O_2), so you could not breathe the air even if you cooled it to a comfortable temperature.

Moving through the thick air near Venus's surface would feel like a cross between swimming and flying: Its density is about 10% that of water. Looking upward, you'd see a perpetually overcast sky, with only weak sunlight filtering

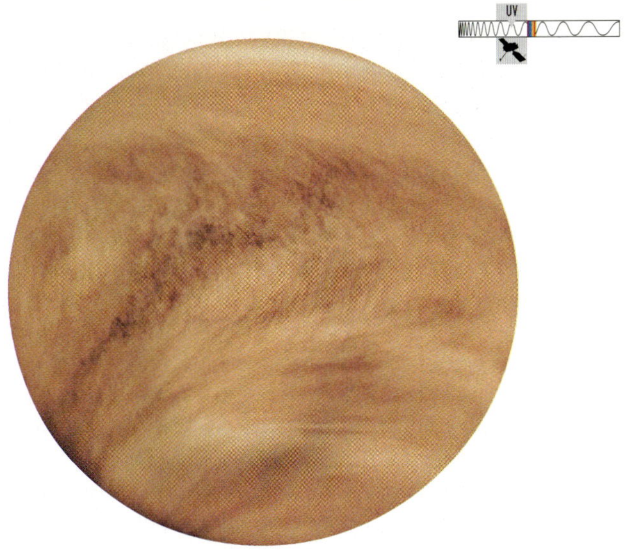

Figure 10.30 Clouds on Venus are all that can be seen in this ultraviolet image taken by the *Pioneer Venus Orbiter*.

through the thick clouds above. Because the thick atmosphere scatters nearly all the blue light away, the dimly lit sky appears reddish-orange in color.

The weather forecast for the surface of Venus today and every day is dull, dull, dull. Venus's slow rotation (243 Earth days) means a very weak Coriolis effect. As a result, Venus has little wind on its surface and never has hurricane-like storms. The top wind speeds measured by the Soviet Union's *Venera* landers were only about 6 kilometers per hour. No rain falls, because any gases that condense in the cool upper atmosphere evaporate long before they reach the ground. The weak Coriolis effect also means that Venus's atmosphere has just two large circulation cells, much like what Earth would have if rotation didn't split the cells (see Figure 10.13). The thick atmosphere makes the circulation so efficient at transporting heat from the equator to the poles that the surface temperature is virtually the same everywhere. That is, the poles are no cooler than the equator and night is just as searingly hot as day. Moreover, Venus has no seasons because it has virtually no axis tilt.* Thus, temperatures are the same year-round.

While Venus has very dull surface weather, more interesting things occur higher up in its atmosphere. Because its high surface temperature drives strong convection, Venus is covered with bright, reflective clouds (Figure 10.30). The clouds are made of droplets containing sulfuric acid (H_2SO_4), which condenses high in the troposphere where temperatures are 400°C cooler than on the surface (see Figure 10.10). The droplets sometimes fall through the upper troposphere as sulfuric acid rain, but they evaporate at least 30 kilometers above the surface because of the higher temperatures closer to the ground. The upper atmosphere of Venus also has some fast winds, though no one knows precisely why.

*In tables (such as Table 7.1), Venus's axis tilt is usually written as 177.4°. This may sound large, but notice that it is nearly 180°—the same tilt as 0° but "upside down." It is written this way because Venus rotates backward compared to its orbit, and backward rotation is equivalent to forward rotation that is upside down.

• How did Venus get so hot?

It's tempting to attribute Venus's high surface temperature solely to the fact that it is closer than Earth to the Sun. However, as we discussed earlier, Venus's cloud cover would actually make the planet quite cold without its strong greenhouse effect (see Table 10.2). Thus, the real question is why Venus has such a strong greenhouse effect.

On a simple level, the answer is Venus's thick carbon dioxide atmosphere; the total amount of carbon dioxide in Venus's atmosphere is nearly 200,000 times as great as the amount in Earth's atmosphere. However, a deeper question still remains. Given their similar sizes and compositions, we expect Venus and Earth to have had similar levels of volcanic outgassing—and the released gas ought to have had about the same composition on both worlds. Why, then, is Venus's atmosphere so different from Earth's?

The Fate of Outgassed Water and Carbon Dioxide Our understanding of planetary geology suggests that huge amounts of water and carbon dioxide should have been outgassed into the atmospheres of both Venus and Earth. Venus's atmosphere does indeed have huge amounts of carbon dioxide, but it has virtually no water. Earth's atmosphere has very little of either gas. What happened to all the outgassed water on Venus, and to the outgassed water and carbon dioxide on Earth?

We can easily account for both missing gases on Earth. The huge amounts of water vapor released into our atmosphere condensed into rain, forming our oceans. Thus, the water is still here, but mostly in liquid rather than gaseous form. The huge amount of carbon dioxide released into our atmosphere is also still here, but in solid form: Carbon dioxide dissolves in water, where it can undergo chemical reactions to make **carbonate rocks** (rocks rich in carbon and oxygen) such as limestone. Earth has about 170,000 times as much carbon dioxide locked up in rocks as in its atmosphere—which means that Earth does indeed have almost as much total carbon dioxide as Venus. Of course, the fact that Earth's carbon dioxide is mostly in rocks rather than in the atmosphere makes all the difference in the world: If this carbon dioxide were in our atmosphere, our planet would be nearly as hot as Venus and certainly uninhabitable.

Having explained what happened to the water and carbon dioxide outgassed on Earth, and knowing that Venus's carbon dioxide still remains in its atmosphere, we are left with the question of what happened to Venus's water. Venus today is incredibly dry, with a surface far too hot to have either liquid water or ice. It is even too hot for water to be chemically bound in surface rock, and any water deeper in its crust or mantle was probably baked out long ago. Measurements also show very little water in the atmosphere. Overall, the total amount of water on Venus is about 10,000 times smaller than the total amount on Earth. Moreover, the absence of water explains why Venus retains so much carbon dioxide in its atmosphere: Without oceans, carbon dioxide cannot dissolve or become locked away in carbonate rocks. Thus, if it is true that a huge amount of water was outgassed on Venus, it has somehow disappeared.

The leading hypothesis for the disappearance of Venus's water invokes some of the same processes thought to have removed water from Mars. Ultraviolet light from the Sun breaks apart water molecules in Venus's atmosphere. The hydrogen atoms then escape to space (through thermal escape), ensuring that the water molecules can never reform. The oxygen from the water molecules was lost to a combination of chemical reactions with surface rocks and stripping by the solar wind; Venus's lack of magnetic field leaves its atmosphere vulnerable to the solar wind.

Acting over billions of years, the breakdown of water molecules and the escape of hydrogen can easily explain the loss of an ocean's worth of water from Venus. Moreover, careful study of Venus's atmospheric composition offers evidence that it did indeed lose a great deal of water. Recall that most hydrogen nuclei contain just a single proton, but a tiny fraction of all hydrogen atoms (about 1 in 50,000 on Earth) contain a neutron in addition to the proton, making the isotope of hydrogen that we call *deuterium.* Water molecules that contain an atom of deuterium (called "heavy water") behave chemically just like ordinary water and can be broken apart by ultraviolet light just as easily. However, a deuterium atom is twice as heavy as an ordinary hydrogen atom and thus does not escape to space as easily when the water molecule is broken apart. If Venus lost a huge amount of hydrogen from water molecules to space, the rare deuterium atoms would have been more likely to remain behind than the ordinary hydrogen atoms. Measurements show that this is the case. The fraction of deuterium among hydrogen atoms is a hundred times higher on Venus than on Earth, suggesting that a substantial amount of water must have been broken apart and its hydrogen lost to space. We cannot determine exactly how much water Venus has lost, but it certainly seems plausible that it really did outgas as much water as Earth and then lost virtually all of it.

To summarize our discussion so far, we can successfully explain the lack of water on Venus today *if* we make a crucial assumption about water in Venus's past: The water must have all been in the atmosphere as water vapor, where ultraviolet light could break it down, rather than in liquid oceans like the water on Earth. Thus, our quest to understand Venus's high temperature leads to one more question: Why didn't Venus get oceans like Earth, which would have allowed its carbon dioxide to be trapped in carbonate rocks and prevented its water from being lost to space?

The Runaway Greenhouse Effect To understand why Venus does not have oceans, let's consider what would happen if we could somehow move Earth to the orbit of Venus.

The greater intensity of sunlight would almost immediately raise Earth's global average temperature by about 30°C, from its current 15°C to about 45°C (113°F). Although this is still well below the boiling point of water, the higher temperature would lead to increased evaporation of water from the oceans. The higher temperature would also allow the atmosphere to hold more water vapor before the vapor condensed to make rain.

The combination of more evaporation and greater atmospheric capacity for water vapor would substantially increase the total amount of water vapor in Earth's atmosphere. Because water vapor is a greenhouse gas, this added water vapor would strengthen the greenhouse effect, driving temperatures a little higher. The higher temperatures, in turn, would lead to even more ocean evaporation and more water vapor in the atmosphere—strengthening the greenhouse effect even further. In other words, we'd have a self-sustaining process in which each little bit of additional water vapor in the atmosphere would mean higher temperature and even more water vapor.

The process would rapidly spin out of control, resulting in a **runaway greenhouse effect** (Figure 10.31). Our planet would heat up until the oceans were completely evaporated and the carbonate rocks had released all their carbon dioxide back into the atmosphere. By the time the runaway process was complete, temperatures on our "moved Earth" would be even higher than they are on Venus today, thanks to the combined greenhouse effects of carbon dioxide and water vapor in the atmosphere. The water vapor would then gradually disappear, as ultraviolet light broke water molecules apart and the hydrogen escaped to space. In short, moving Earth to Venus's orbit would essentially turn our planet into another Venus.

This scenario suggests a simple explanation for why Venus is so much hotter than Earth. Even though Venus's distance from the Sun is only about 30% less than Earth's distance, the difference in distance was critical. On Earth,

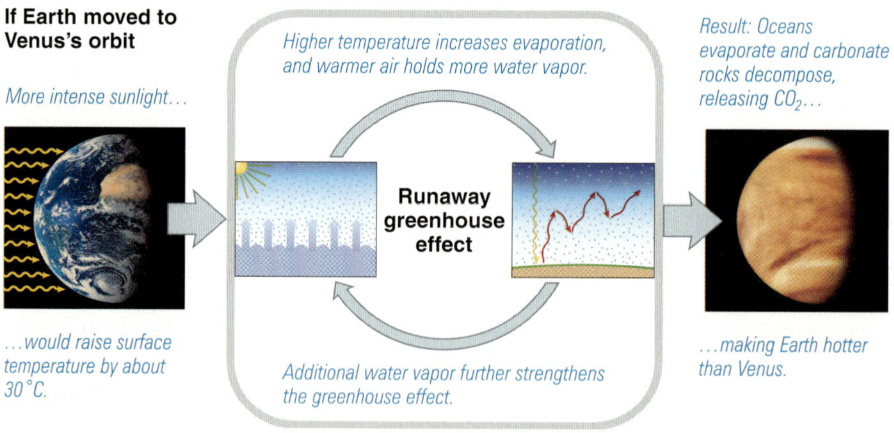

Figure 10.31 This diagram shows how, if Earth were placed at Venus's distance from the Sun, the runaway greenhouse effect would cause the oceans to evaporate completely.

it was cool enough for water to rain down to make oceans. The oceans then dissolved carbon dioxide and chemical reactions locked it away in carbonate rocks, leaving our atmosphere with only enough greenhouse gases to make our planet pleasantly warm. On Venus, the greater intensity of sunlight made it just enough warmer that oceans either never formed or soon evaporated. Without oceans to dissolve carbon dioxide and make carbonate rock, carbon dioxide accumulated in the atmosphere, leading to a runaway greenhouse effect.

The next time you see Venus shining brightly as the morning or evening "star," consider the radically different path it has taken from that taken by Earth—and thank your lucky star. If Earth had formed a bit closer to the Sun or if the Sun had been slightly hotter, our planet might have suffered the same greenhouse-baked fate.

THINK ABOUT IT

We've seen that moving Earth to Venus's orbit would cause our planet to become Venus-like. If we could somehow move Venus to Earth's orbit, would it become Earth-like? Why or why not?

A Pleasant Early Venus? Venus's closeness to the Sun may have sealed its ultimate fate, but it's possible that Venus might have been more moderate in its early history. Remember that the Sun has gradually brightened with age. Thus, sunlight was less intense on all the planets when they were young. Some 4 billion years ago, the intensity of sunlight at Venus was not much greater than it is at Earth today. Rain might well have fallen on the young planet Venus, and oceans could have formed. It's even conceivable that life could have arisen on the young Venus.

As the Sun gradually brightened, however, any liquid water or life on Venus was doomed. The runaway greenhouse effect raised the temperature so high that all the water evaporated. In the upper atmosphere, ultraviolet light broke apart the water molecules, and the hydrogen escaped to space. If Venus had oceans in its youth, the water is now gone forever.

We'll probably never know for sure whether Venus ever had oceans. The global "repaving" of Venus's surface by tectonics and volcanism [Section 9.5] would long ago have covered up any shorelines or other geological evidence of past oceans, and the high surface temperatures would have baked out any gases that might once have been incorporated into surface rock. If the climate once was pleasant, it's unlikely that any evidence survives to tell the tale.

10.6 Earth's Unique Atmosphere

Earth's atmosphere, composed mostly of nitrogen and oxygen, is the only one under which human life is possible outside an artificial environment. This atmosphere provides our planet with just enough warmth and pressure to enable water to cycle between all three phases (solid ice, liquid water, and gaseous water vapor). It protects us from harmful solar radiation and produces the weather patterns that variously bring us days of sunshine, clouds, and rain or snow. In this section, we'll discuss how and why our atmosphere is so hospitable, and we'll conclude by considering how we humans may be altering the very balances upon which we depend for survival.

• How did Earth's atmosphere end up so different?

If you think about our comparative study so far, you'll realize that most of the major physical differences in the atmospheres of the terrestrial planets are easy to understand. For example, the Moon and Mercury lack substantial atmospheres because of their small sizes, while a very strong greenhouse effect causes the high surface temperature on Venus.

Explaining differing atmospheric compositions is more challenging, especially when we consider the cases of Venus, Earth, and Mars. In all three cases, outgassing should have released the same gases. How, then, did Earth's atmosphere end up so different? We can break down this general question into four separate questions:

1. Why did Earth retain most of its outgassed water—enough to form vast oceans—while Venus and Mars lost theirs?
2. Why does Earth have so little carbon dioxide (CO_2) in its atmosphere compared to Venus, when Earth should have outgassed about as much of it as Venus?
3. Why is Earth's atmosphere composed primarily of nitrogen (N_2) and oxygen (O_2), when these gases are only trace constituents in the atmospheres of Venus and Mars?
4. Why does Earth have an ultraviolet-absorbing stratosphere, while Venus and Mars do not?

Water and Carbon Dioxide We have already answered the first two questions in our discussions of the atmospheres of Mars and Venus. On Mars, some of the outgassed water was lost after solar ultraviolet light broke water vapor molecules apart, and the rest froze and may remain in the polar caps or underground. On Venus, it was too hot for water vapor to condense, so virtually all the water molecules were ultimately broken apart, allowing the hydrogen atoms to escape to space. Earth retained its outgassed water because temperatures were low enough for water vapor to condense into rain and thereby form oceans. Evidence from tiny mineral grains (known as zircons) suggests that Earth had oceans as early as 4.3–4.4 billion years ago.

The presence of oceans also explains the low level of carbon dioxide in our atmosphere. Most of the carbon dioxide outgassed by volcanism on Earth dissolved in the oceans, where chemical reactions ultimately turned it into carbonate rocks. Even today, about 60 times as much carbon dioxide is dissolved in the oceans as is present in the atmosphere, and carbonate rocks contain some 170,000 times as much CO_2 as the atmosphere.

Nitrogen, Oxygen, and Ozone Turning our attention to the third question, it's relatively easy to explain the substantial nitrogen content (77%) of our atmosphere. Although water vapor and carbon dioxide are the two most common gases released by outgassing, the third most common gas is nitrogen. Because most of Earth's water ended up in the oceans and most of the carbon dioxide ended up in rocks, our atmosphere was left with nitrogen as its dominant ingredient. The oxygen content (21%) of our atmosphere is a little more mysterious.

Molecular oxygen (O_2) is not a product of outgassing or any other geological process. Moreover, oxygen is a highly reactive chemical that would disappear from the atmosphere in just a few million years if it were not continuously resupplied. Fire, rust, and the discoloration of freshly cut fruits and vegetables are everyday examples of chemical reactions that remove oxygen from the atmosphere (often called *oxidation reactions*). Similar reactions between oxygen and surface materials (especially iron-bearing minerals) give rise to the reddish appearance of much of Earth's rock and clay, including the beautiful reds of Arizona's Grand Canyon. Thus, we must explain not only how oxygen got into Earth's atmosphere in the first place, but also how the amount of oxygen remains relatively steady even while chemical reactions can remove it rapidly from the atmosphere.

The answer to the oxygen mystery is *life*. Plants and many microorganisms release oxygen through photosynthesis. Photosynthesis takes in CO_2 and, through a fairly complex chain of chemical reactions, releases O_2. Because chemical reactions can remove oxygen from the atmosphere, it took a long time for oxygen to accumulate in Earth's atmosphere. According to present evidence, it took at least a billion years of photosynthesis before the buildup of atmospheric oxygen began, and Earth's atmosphere probably has had enough oxygen for us to breathe only for the past few hundred million years [Section 24.1]. If you had a time machine and spun the dial to arrive randomly at any point in Earth's past, there's only about a 1 in 10 chance that you would arrive at a time when you'd be able to breathe the outside air. Today, plants and single-celled photosynthetic organisms return oxygen to the atmosphere in approximate balance with the rate at which animals and chemical reactions consume oxygen, keeping the oxygen content relatively steady.

THINK ABOUT IT

Suppose that, somehow, all photosynthetic life on Earth died out. What would happen to the oxygen in our atmosphere? Could animals, including us, still survive?

Life and oxygen also explain the presence of Earth's ultraviolet-absorbing stratosphere. In the upper atmosphere, chemical reactions involving solar ultraviolet light transform some of the O_2 into molecules of O_3, or ozone (Figure 10.32). The O_3 molecule is more weakly bound than the O_2 molecule, which allows it to absorb solar ultraviolet

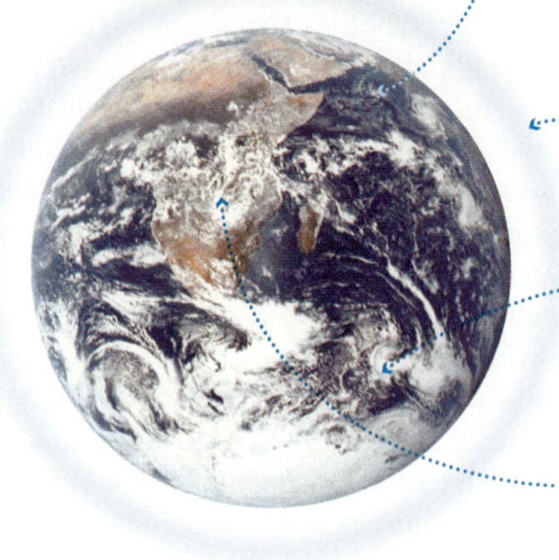

Figure 10.32 The origin of oxygen and ozone in Earth's atmosphere.

energy even better. The absorption of solar energy by ozone heats the upper atmosphere, creating the layer we call the stratosphere. This ozone layer prevents harmful ultraviolet radiation from reaching the surface. Mars and Venus lack photosynthetic life and therefore have too little O_2 and consequently too little ozone to form a stratosphere.

Maintaining Balance We have answered all four basic questions about Earth's atmosphere. However, a deeper look at what we have learned still leaves us with a major mystery. Our oceans exist because Earth has a greenhouse effect that is "just right" to keep them from either freezing or boiling away, and the presence of liquid water allowed life to arise and produce oxygen. But *why* does the amount of carbon

COMMON MISCONCEPTIONS

Ozone—Good or Bad?

Ozone often generates confusion, because in human terms it is sometimes good and sometimes bad. In the stratosphere, ozone acts as a protective shield from the Sun's ultraviolet radiation. Thus, the fact that human activity (especially the release of the chemicals known as CFCs) has been reducing stratospheric ozone is a significant environmental problem. At the same time, ozone is poisonous to most living creatures and therefore is a bad thing when it is found near Earth's surface.

In fact, ozone is one of the main ingredients in urban air pollution. It is produced as a by-product of automobiles and industry. Some people wonder whether we might be able to transport this ozone to the stratosphere and thereby alleviate the effects of ozone depletion in the stratosphere. Unfortunately, this plan won't work. Even if we could find a way to transport this ozone, all the ozone ever produced in urban pollution would barely make a dent in the amount lost from the stratosphere.

dioxide in our atmosphere stay "just right"? For example, why didn't the chemical reactions either remove *all* the carbon dioxide, or leave so much of it that the oceans would boil away? More generally, the long-term existence of Earth's oceans tells us that our planet has enjoyed remarkable climate stability—presenting a stark contrast to the dramatic climate changes that apparently occurred on Mars and Venus.

• Why does Earth's climate stay relatively stable?

Earth's long-term climate stability has clearly been important to the ongoing evolution of life—and hence to our own relatively recent arrival as a species (see Figure 1.11). Had our planet undergone a runaway greenhouse effect like Venus, life would certainly have been extinguished. If Earth had suffered loss of atmosphere and a global freezing like Mars, any surviving life would have been driven to hide in underground pockets of liquid water.

Earth's climate is not perfectly stable—our planet has endured numerous ice ages and warm periods in the past. Nevertheless, even in the deepest ice ages and warmest warm periods, Earth's temperature has remained in a range in which ocean water has remained at least partially liquid and hence could harbor life. This long-term climate stability is even more remarkable when you remember that the Sun has brightened substantially over the past 4 billion years, yet Earth's temperature has managed to stay in nearly the same range throughout this time. Apparently, the strength of the greenhouse effect self-adjusts to keep the climate stable. How does it do this?

The Carbon Dioxide Cycle The mechanism by which Earth self-regulates its temperature is called the **carbon dioxide cycle**, or the **CO_2 cycle** for short. Let's follow the cycle as illustrated in Figure 10.33, starting with the carbon dioxide in the atmosphere:

- Atmospheric carbon dioxide dissolves in rainwater, creating a mild acid.
- The mildly acidic rainfall erodes rocks on Earth's continents, and rivers carry the broken-down minerals to the oceans.
- In the oceans, calcium from the broken-down minerals combines with dissolved carbon dioxide and falls to the ocean floor, making carbonate rocks such as limestone.*
- Over millions of years, the conveyor belt of plate tectonics carries the carbonate rocks to subduction zones, and subduction carries them down into the mantle.
- As they are pushed deeper into the mantle, some of the subducted carbonate rock melts and releases its carbon dioxide, which then outgasses back into the atmosphere through volcanoes.

*During the past half billion years or so, the carbonate minerals have been made by shell-forming sea animals, falling to the bottom in the seashells left after the animals die. Without the presence of animals, chemical reactions would do the same thing—and apparently did for most of Earth's history.

The CO_2 cycle acts as a thermostat for Earth, because of the way that changes in temperature feed back into the cycle. You are probably familiar with what we generally call **feedback processes**—processes in which a change in one property amplifies (positive feedback) or counteracts (negative feedback) the behavior of the rest of the system. For example, if someone brings a microphone too close to a loudspeaker, it picks up and amplifies small sounds from the speaker. These amplified sounds are again picked up by the microphone and further amplified, causing a loud screech. This sound feedback is an example of *positive feedback*, because it automatically amplifies itself. The screech usually leads to a form of *negative feedback*: The embarrassed person holding the microphone moves away from the loudspeaker, thereby stopping the positive sound feedback.

THINK ABOUT IT

Think of at least one other everyday example each of positive feedback and negative feedback.

The CO_2 cycle has a built-in form of negative feedback that returns Earth's temperature toward "normal" whenever it warms up or cools down (Figure 10.34). The negative feedback occurs because the overall rate at which carbon dioxide is pulled from the atmosphere is very sensitive to temperature: the higher the temperature, the higher the rate at which carbon dioxide is removed.

Consider first what happens if Earth warms up a bit. The warmer temperature means more evaporation and rainfall, pulling more CO_2 out of the atmosphere. The

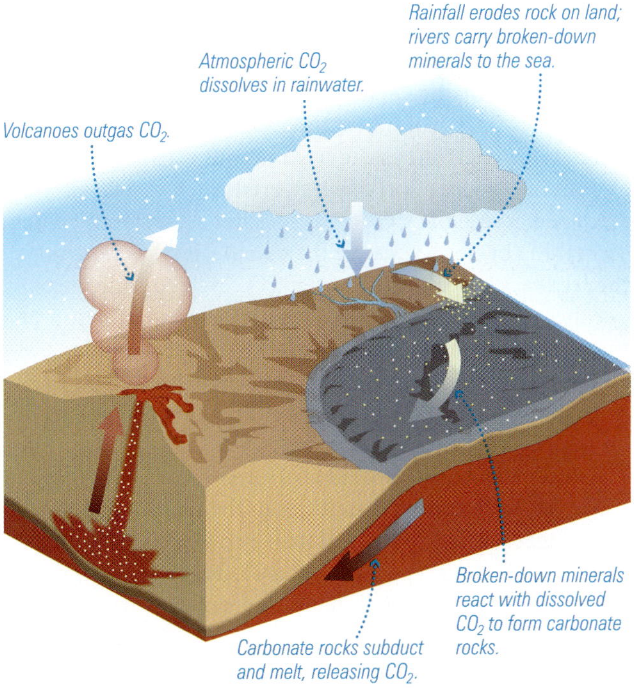

Figure 10.33 This diagram shows how the CO_2 cycle continually moves carbon dioxide from the atmosphere to the ocean to rock and back to the atmosphere. Note that plate tectonics (subduction in particular) plays a crucial role in the cycle.

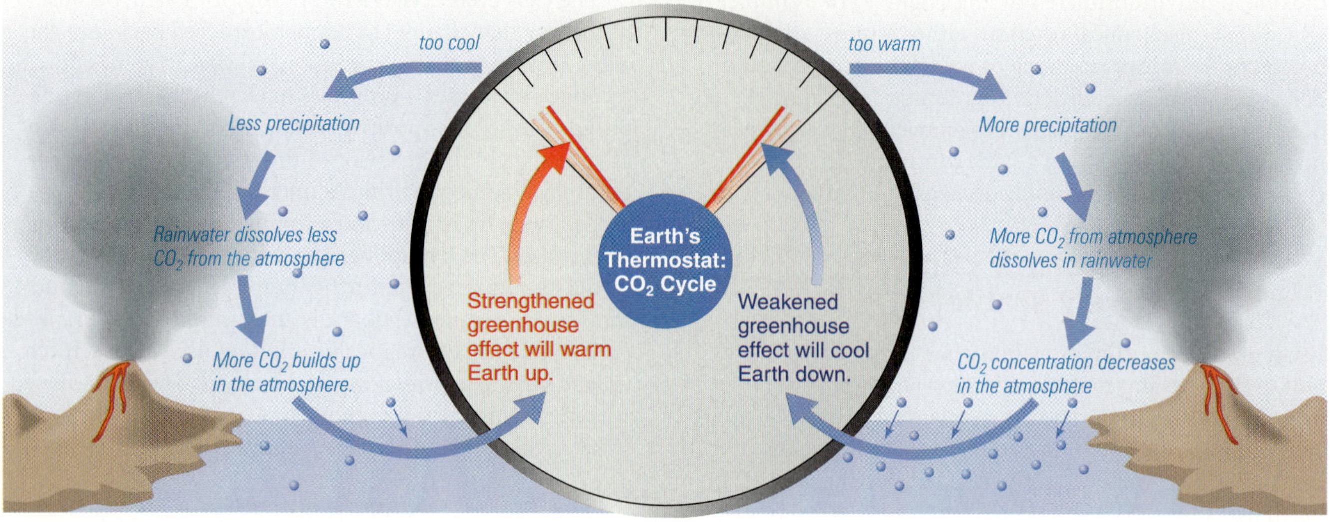

Figure 10.34 The carbon dioxide cycle acts as a thermostat for Earth through negative feedback processes. Cool temperatures cause atmospheric CO_2 to increase, and warm temperatures cause atmospheric CO_2 to decline.

reduced atmospheric CO_2 concentration leads to a weakened greenhouse effect that counteracts the initial warming and cools the planet back down. Similarly, if Earth cools a bit, precipitation decreases and less CO_2 is dissolved in rainwater, allowing the CO_2 released by volcanism to build back up in the atmosphere. The increased CO_2 concentration strengthens the greenhouse effect and warms the planet back up. Overall, the natural thermostat of the carbon dioxide cycle has allowed the greenhouse effect to strengthen or weaken just enough to keep Earth's climate fairly stable, regardless of what other changes have occurred on our planet.

Ice Ages and Other Long-Term Climate Change While Earth's climate has remained stable enough for the oceans to remain at least partly liquid throughout history, the climate has not been perfectly steady. Numerous warmer periods and numerous ice ages have occurred. Such variations are possible because the CO_2 cycle does not act instantly. When something begins to change the climate, it takes time for the feedback mechanisms of the CO_2 cycle to come into play because of their dependence on the gradual action of mineral formation in the oceans and of plate tectonics. Calculations show that the time to stabilize atmospheric CO_2 through the CO_2 cycle is about 400,000 years. That is, if the amount of CO_2 in the atmosphere were to rise due to, say, increased volcanism, it would take some 400,000 years for the CO_2 cycle to restore temperatures to their current values.*

Ice ages occur when the global average temperature drops by a few degrees. The slightly lower temperatures lead to increased snowfall that may cover continents with ice down to fairly low latitudes. For example, the northern United States was entirely covered with glaciers during the peak of the most recent ice age, which ended only about 10,000 years ago. The causes of ice ages are complex and not fully understood. Over periods of tens or hundreds of millions of years, the Sun's gradual brightening and the changing arrangement of the continents around the globe have at least in part influenced the climate. During the past few million years—a period too short for solar changes or continental motion to have a significant effect—the ice ages appear to have been strongly influenced by small changes in Earth's axis tilt and other characteristics of Earth's rotation and orbit (the *Milankovitch cycles* that we noted earlier).

Geologists have recently discovered evidence of several particularly long and deep ice ages between about 750 and 580 million years ago. During these periods, glaciers appear to have advanced all the way to the equator. If so, the global temperature must have dropped low enough for the oceans to begin to freeze worldwide. This ocean freezing would have set up a positive feedback process that would have cooled Earth even further, because ice reflects much more sunlight than water. (About 90% of incoming sunlight is reflected by ice, compared to about 5% by water.) With less sunlight being absorbed, the surface would cool further, which in turn would allow the oceans to freeze further. Geologists suspect that in this way, our planet may have entered the periods we now call **snowball Earth** (Figure 10.35). We do not know precisely how extreme the cold became. Some models suggest that at the peak of a snowball Earth period, the global average temperature may have been as low as $-50°C$ ($-58°F$) and the oceans may have been frozen to a depth of 1 kilometer or more.

How did Earth recover from a snowball phase? Even if Earth's surface got cold enough for the ocean surface to freeze completely, the interior still remained hot. As a result, volcanism would have continued to add CO_2 to the

*This time scale applies to ocean/atmosphere equilibrium only. The time scale for crust recycling is much longer, while shorter-term climate variations in atmospheric CO_2 concentration can occur through factors besides the inorganic CO_2 cycle, such as cycling of carbon dioxide by life.

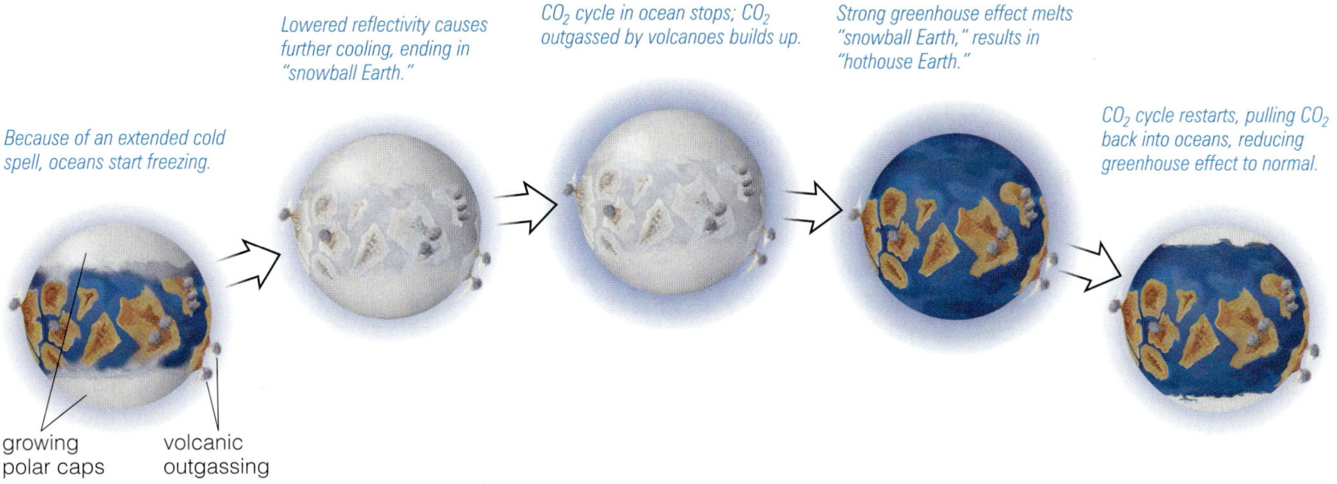

Figure 10.35 The CO_2 cycle rescues Earth from a "snowball" phase.

atmosphere. Oceans covered by ice would have been unable to absorb this CO_2 gas, and the CO_2 content of the atmosphere would have gradually built up and strengthened the greenhouse effect. As long as the ocean surfaces remained frozen, the CO_2 buildup would have continued. Scientists suspect that, over a period of 10 million years or so, the CO_2 content of our atmosphere may have increased 1,000-fold. Eventually, the strengthening greenhouse effect would have warmed Earth enough to start melting the ocean surface ice. The feedback processes that started the snowball Earth episode then moved quickly in reverse.

As the ocean surface melted, more sunlight would have been absorbed (again, because liquid water absorbs more and reflects less sunlight than ice), warming the planet further. Moreover, because the CO_2 concentration was so high, the warming would have continued well past current temperatures—perhaps taking the global average temperature to higher than 50°C (122°F). Thus, in just a few centuries, Earth would have emerged from a snowball phase into a hothouse phase. Geological evidence supports the occurrence of dramatic increases in temperature at the end of each snowball Earth episode. Earth then slowly recovered from the hothouse phase as the CO_2 cycle removed carbon dioxide from the atmosphere. This recovery would have taken the 400,000 years required for stabilization of atmospheric CO_2.

THINK ABOUT IT

Suppose Earth did not have plate tectonics. Could the planet ever recover from a "snowball" phase? Explain.

The snowball Earth episodes must have had severe consequences for any life on Earth at the time. Indeed, the end of the snowball Earth episodes roughly coincides with a dramatic increase in the diversity of life on Earth (the Cambrian explosion [Section 24.1]). Some scientists suspect that the environmental pressures caused by the snowball Earth periods may have led to a burst of evolution. If so, we might not be here today if not for the dramatic climate changes of the snowball Earth episodes.

Earth's Long-Term Future Climate Despite going through periods of ice ages, snowball Earth, and unusual warmth, Earth's climate has remained suitable for life for some 4 billion years. However, the continuing brightening of the Sun ultimately will overheat our planet.

The numerous feedback processes that affect climate make it difficult to make precise predictions about future climate. According to some climate models, the warming Sun could cause Earth to begin losing its water as soon as a billion or so years from now. If so, then life on Earth has already completed about 75% of its history on this planet. However, uncertainties in the models make it quite possible that they are wrong. The CO_2 cycle may well keep the climate steady for much more than the next billion years. If so, water and life could persist for as long as 3–4 billion years to come.

About 3–4 billion years from now, however, the Sun will have grown so warm that sunlight on Earth will be as intense as it is on Venus today. The effect will be the same as if we moved Earth to Venus's orbit today (see Figure 10.31)—a runaway greenhouse effect. The rising temperature on Earth will cause increased ocean evaporation (if the oceans still exist), and the water vapor in the atmosphere will further strengthen the greenhouse effect. The positive feedback won't stop until the oceans have evaporated away and all the carbon dioxide has been released from carbonate rocks. Our planet will become a Venus-like hothouse, with temperatures far too high for liquid water to exist.

Thus, Earth's habitability for life will come to an end between 1 and 4 billion years from now. Although this may seem depressing, remember that a billion years is a very long time—equivalent to some 10 million human lifetimes and far longer than humans have existed so far. If you want to lose sleep worrying about the future, there are far more immediate threats, including those that we will discuss next.

• How might human activity change our planet?

We humans are well adapted to the unique, present-day conditions on our planet. The amount of oxygen in our atmosphere, the average temperature of our planet, and the ultraviolet-absorbing ozone layer are just what we need to survive. We have seen that these "ideal" conditions are no accident, but instead consequences of our planet's unique geology and biology.

Nevertheless, the stories of the dramatic and permanent climate changes that occurred on Venus and Mars should teach us to take nothing for granted. Our planet may regulate its own temperature quite effectively, and life may keep the level of oxygen and ozone in the atmosphere quite steady, but it is possible for humans to interfere with these natural mechanisms.

Some of the dangers posed by human activity are already clear. For example, we learned in the 1970s that human-made chemicals known as CFCs can rise high into the atmosphere, where they destroy some of the ozone that shields our planet from dangerous ultraviolet light. In the 1980s, we found a large and ominous "hole" forming in the ozone layer over Antarctica each spring. This clear evidence of a human role in altering a crucial aspect of our planet's habitability led to international action and treaties intended to stop ozone destruction.

Another well-recognized danger comes from extinctions. According to some estimates, human activity is driving species to extinction so rapidly that up to half of today's plant and animal species may be gone by the end of this century. This rate of extinction is comparable with that in past *mass extinctions,* such as the one that wiped out the dinosaurs some 65 million years ago [Section 12.4]. Prior mass extinctions generally led to the loss of the planet's dominant species at the time—and we are the dominant species today. Although we cannot predict the consequences of current extinctions with any certainty, it is at least conceivable that we could be unwittingly clearing the way for a new set of dominant species.

The threat that causes the most debate is **global warming**—an apparent increase in Earth's global average temperature over the past century or so. The debate centers on the question of whether the warming has been caused by human input of carbon dioxide and other greenhouse gases (such as methane) into the atmosphere. The carbon dioxide cycle operates too slowly to absorb the excess greenhouse gases released by human activity, so it is reasonable to think that the warming may be occurring because these gases strengthen the greenhouse effect.

The Global Warming Debate The potential threat of global warming is a hot political issue because alleviating the threat would require finding new energy sources and making other changes that would dramatically affect the world's economy. However, a major research effort has gradually added to our understanding of the potential threat, particularly in the past decade. Although some controversy still exists about how serious the problem of global warming might be, at least three facts are now clear:

1. Measurements show that Earth has indeed warmed up over the past 50 years, by about 0.5°C. Although this may sound small, it is a significant increase in such a short time period. The warming trend has continued and gotten stronger in recent years.
2. The burning of fossil fuels and other human activity is clearly increasing the amounts of greenhouse gases in the atmosphere. The current concentration of carbon dioxide in Earth's atmosphere is significantly higher than it has been at any time during the past 400,000 years, and the concentration is rising rapidly (Figure 10.36).
3. Because we understand the basic mechanism of the greenhouse effect (see Figure 10.6), we can have no doubt that a continually rising concentration of greenhouse gases would eventually make our planet warm up.

However, while continued input of greenhouse gases would eventually cause global warming, it's difficult to prove that human activity has caused the global warming observed to date. The problem is that numerous feedback mechanisms can counter or enhance greenhouse warming on timescales of years, decades, and even centuries, and we don't yet understand all these mechanisms. For example, increased evaporation of ocean water could enhance the greenhouse effect because water vapor is also a greenhouse gas, but it could counter global warming if it caused the formation of more clouds that prevented sunlight from reaching the ground. Melting of polar ice, on the other hand, makes global warming even worse, because it changes part of Earth's surface from high reflectivity to low, leading to more absorption of solar energy.

Scientists seek to understand the problem better by creating sophisticated computer models of the climate and comparing the model predictions to actual observations. Current models suggest that global temperatures are increasing in a manner consistent with what we expect if human activity is the cause (Figure 10.37). Thus, while plenty of uncertainties still remain, most scientists are now convinced that human activity is indeed causing global warming.

Potential Consequences of Global Warming Clearly, global warming would mean a higher global average surface temperature for Earth. Beyond that simple statement, however, the issue of consequences becomes much more complicated and the uncertainties involved in prediction even greater.

Changing weather patterns would ensure that different regions of Earth experience different degrees of warming. Some regions would even become colder. Other regions might experience more rainfall or might become deserts. The greater overall warmth of the atmosphere would tend to mean more evaporation from the oceans, leading to more numerous and more intense storms.

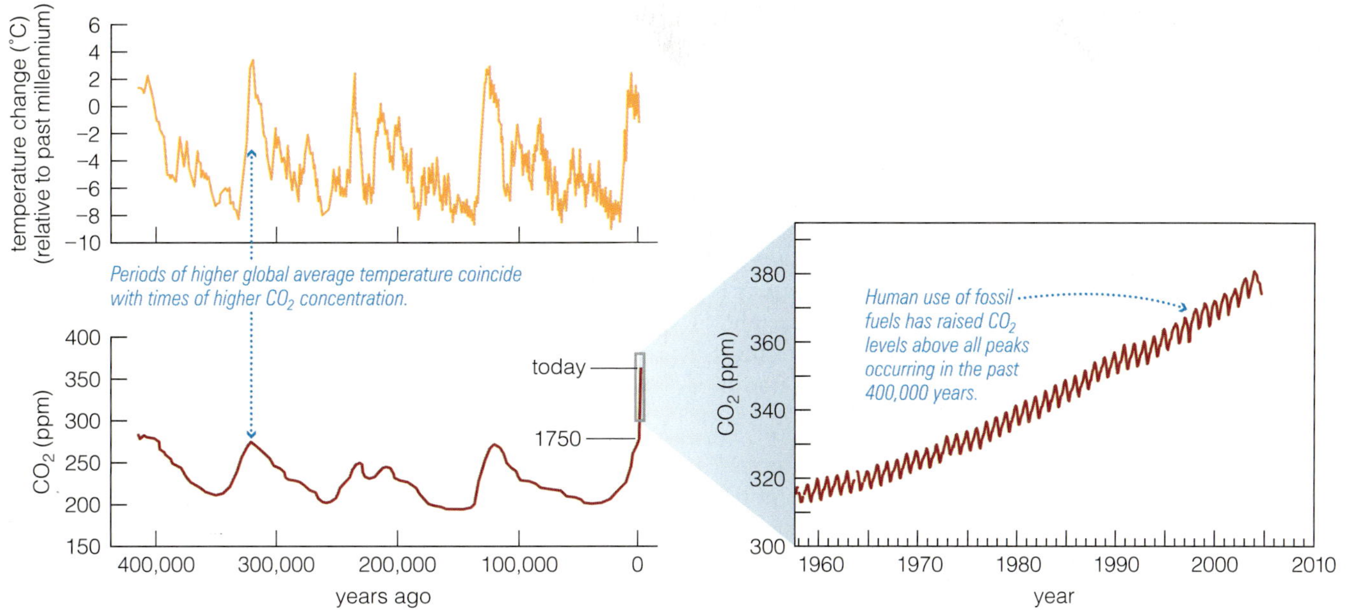

Figure 10.36 This diagram shows the atmospheric concentration of carbon dioxide and global average temperature over the past 400,000 years. Notice that both vary together: more carbon dioxide goes with higher temperature, and vice versa. The CO_2 data for the past half-century come from direct measurements (made at Mauna Loa); most of the earlier data come from studies of air bubbles trapped in Antarctic ice (ice core samples). The concentration is measured in parts per million (ppm), which is the number of CO_2 molecules among every 1 million air molecules.

Another potential threat comes from rising sea level. Water expands very slightly as it warms—so slightly that we don't notice the change in a glass of water, but enough to cause changes in sea level. Sea level has already risen some 20 centimeters in the past hundred years, and could rise another meter as the oceans warm. Such increases in sea level would threaten many coastal regions with more severe flooding. A much greater increase in sea level may occur if polar ice and glaciers melt. Such melting appears to be occurring to some extent already. For example, the famous "snows" (glaciers) of Mount Kilimanjaro are already in retreat and may be gone within the next decade or so.

Secondary effects, such as those arising from changes to ocean currents or ecological changes, pose an even more intractable problem. For example, it is difficult to know how forests and other ecosystems will respond to climate changes induced by global warming. As a result, we cannot easily predict the impact of such climate changes on food production, fresh water availability, or other issues critical to the well-being of human populations.

Given the current uncertainties, no one can predict the precise impact of global warming over the next century. However, the lesson from the solar system is clear: Dramatic and deadly change can occur unexpectedly, and we do not know how our tampering might affect the finely balanced mechanisms that control Earth's climate.

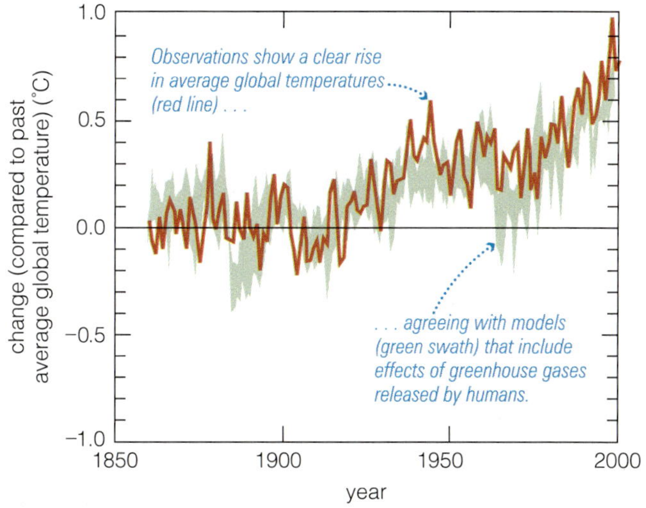

Figure 10.37 This graph compares the predictions of climate models (green swath) with observed temperature changes (red line) since about 1860. The thickness of the green swath for the models represents the range of variation between several different models. These models take into account both natural factors (such as slight changes in the Sun's brightness and the effects of volcanic eruptions) and emission of greenhouse gases by humans. The agreement is not perfect—telling us we still have much to learn—but it is good enough to give us confidence that greenhouse gases are, indeed, causing global warming.

THINK ABOUT IT

If you were a political leader, how would *you* deal with the uncertain threat of global warming?

THE BIG PICTURE

Putting Chapter 10 into Perspective

With what we have learned in this chapter and the previous chapter, we now have a complete "big picture" view of how the terrestrial worlds started out so similar yet ended up so different. As you continue your studies, keep in mind the following important ideas:

- Atmospheres affect planets in a variety of ways. For example, they absorb and scatter light, distribute heat, and create the weather that can lead to erosion. Perhaps most important, the greenhouse effect allows atmospheres to make a planet warmer than it would be otherwise.
- Atmospheric properties differ widely among the terrestrial worlds, but we can trace these differences to root causes. For example, only the larger worlds have significant atmospheres. The smaller worlds lack the internal heat needed for volcanism and the associated outgassing that releases atmospheric gases, and they also lack the gravity necessary to retain atmospheric gases.
- The climate histories of Venus and Mars suggest that major climate change should be the rule, not the exception. Mars was once warm and wet, but its small size and lack of a magnetic field caused it to lose gas and freeze over some 3 billion or more years ago. Venus may once have had oceans, but its proximity to the Sun doomed it to a runaway greenhouse effect.
- We are here to talk about these things today only because Earth somehow has managed to be the exception, a planet on which climate has remained relatively consistent. We humans are ideally adapted to Earth today, but we have no guarantee that Earth will remain as hospitable in the future, especially as we tinker with the balance of greenhouse gases that has kept our climate stable.

SUMMARY OF KEY CONCEPTS

10.1 Atmospheric Basics

- **What is an atmosphere?** An atmosphere is a layer of gas that surrounds a world and that can have profound effects on a planet's surface, including creating pressure; absorbing and scattering sunlight; creating wind and weather; interacting with the solar wind to create a magnetosphere; and creating a **greenhouse effect** that can make a planet's surface warmer than it would be otherwise.

- **How does the greenhouse effect warm a planet?** **Greenhouse gases** such as carbon dioxide, methane, and water vapor absorb infrared light emitted from a planet's surface. The absorbed photons are quickly reemitted, but in random directions. The result acts much like a blanket, warming the planet's surface.

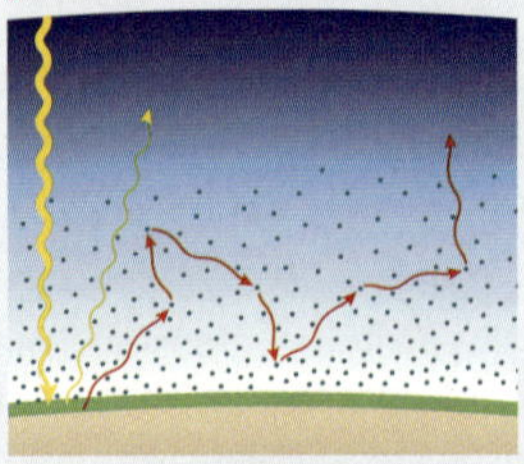

- **Why do atmospheric properties vary with altitude?** **Atmospheric structure** is determined by the way atmospheric gases interact with sunlight. On Earth, the basic structure consists of the **troposphere**, where most greenhouse warming occurs; the **stratosphere**, where **ozone** absorbs ultraviolet light from the Sun; the **thermosphere**, where solar X rays are absorbed; and the **exosphere**, the extremely low-density outer layer of the atmosphere.

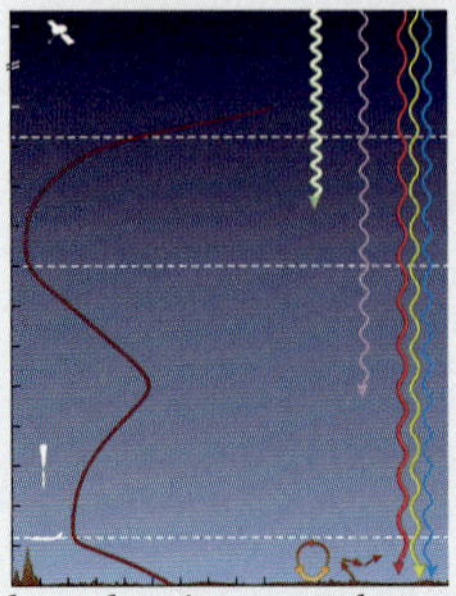

10.2 Weather and Climate

- **What creates wind and weather?** **Global wind patterns** are shaped by atmospheric heating and the **Coriolis effect** caused by a planet's rotation. Convection in the troposphere can lead to the formation of clouds and rain, hail, or snow.
- **What factors can cause long-term climate change?** The Sun has been gradually brightening since its birth more than 4 billion years ago, which tends to warm the planets. But other effects can be even more important in climate change, including changes in axis tilt, changes in reflectivity, and changes in greenhouse gas abundance.
- **How does a planet gain or lose atmospheric gases?** The three basic sources of atmospheric gas are **outgassing**, evaporation or sublimation, and the "sandblasting" of the surface on worlds with very little atmospheric gas. The five basic loss processes are condensation, chemical reactions with surface materials, the blasting away of gas by impacts, the stripping of gas by the solar wind, and **thermal escape**.

10.3 Atmospheres of the Moon and Mercury

- **Do the Moon and Mercury have any atmosphere at all?** The Moon and Mercury have only very thin exospheres consisting of gas particles released by the "sandblasting" of the surface by micrometeorites, solar wind particles, and high-energy solar photons.

10.4 The Atmospheric History of Mars

- **What is Mars like today?** Mars is cold and dry, with an atmospheric pressure so low that water is unstable. Martian weather is driven largely by seasonal changes that cause carbon dioxide alternately to condense and sublime at the poles, creating pole-to-pole winds and sometimes leading to huge dust storms.
- **Why did Mars change?** Mars's atmosphere must once have been much thicker with a much stronger greenhouse effect, so change must have occurred due to loss of atmospheric gas. Much of the lost gas probably was stripped away by the solar wind, which was able to reach the atmosphere as Mars cooled and lost its magnetic field and protective magnetosphere. Water was probably also lost because ultraviolet light could break apart water molecules in the atmosphere, and the lightweight hydrogen then escaped to space.

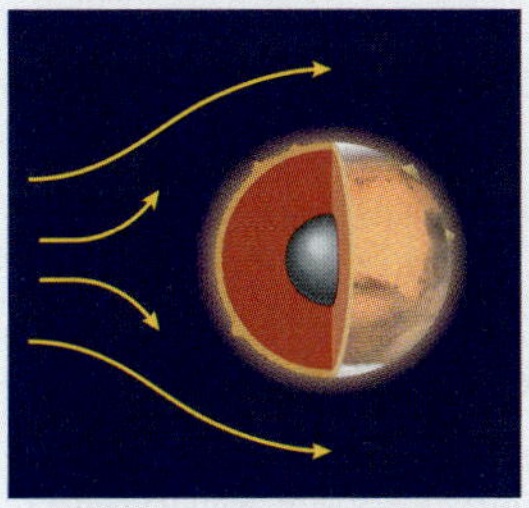

10.5 The Atmospheric History of Venus

- **What is Venus like today?** Venus has a thick atmosphere that creates its strong greenhouse effect, explaining why it is so hot. It rotates slowly and therefore has a weak Coriolis effect and weak winds, and is far too hot for rain to fall. Its atmospheric circulation keeps temperatures about the same day and night, and its lack of axis tilt means no seasonal changes.
- **How did Venus get so hot?** Venus's distance from the Sun ultimately led to a **runaway greenhouse effect**: Venus became too hot to develop liquid oceans like those on Earth. Without oceans to dissolve outgassed carbon dioxide and lock it away in carbonate rocks, all of Venus's carbon dioxide remained in its atmosphere, creating its intense greenhouse effect.

10.6 Earth's Unique Atmosphere

- **How did Earth's atmosphere end up so different?** Temperatures on Earth were just right for outgassed water vapor to condense and form oceans. The oceans dissolve carbon dioxide and ultimately lock it away in carbonate rocks, thereby keeping the greenhouse effect moderate. Nitrogen from outgassing remained in the atmosphere. Oxygen and ozone were produced by photosynthesis, which was possible because the moderate conditions allowed the origin and evolution of abundant life.
- **Why does Earth's climate stay relatively stable?** Earth's long-term climate is remarkably stable due to feedback processes that tend to counter any warming or cooling that occurs. The most important feedback process is the **carbon dioxide cycle**, which naturally regulates the strength of Earth's greenhouse effect.

- **How might human activity change our planet?** Ozone depletion can leave surface life more vulnerable to dangerous solar ultraviolet radiation, and the high rate of extinctions could have unknown consequences. The human release of greenhouse gases into the atmosphere may already be causing global warming and certainly would affect the climate if it continues.

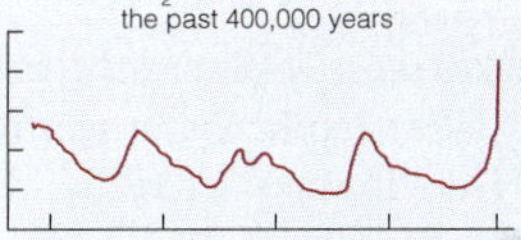

EXERCISES AND PROBLEMS

For instructor-assigned homework go to MasteringASTRONOMY www.masteringastronomy.com

Review Questions

Short-Answer Questions Based on the Reading

1. Briefly summarize the different atmospheric properties of the five terrestrial worlds. How do they differ in surface temperature and pressure? What are the different atmospheres made of?
2. Using a balloon as an analogy, explain the origin of gas pressure. What do we mean by *atmospheric pressure*? Why does pressure decrease with altitude? What is 1 *bar* of pressure?
3. Is there any atmosphere at the orbital altitude of the Space Station above Earth? Explain.
4. What does the *greenhouse effect* do to a planet? Explain the role of greenhouse gases and describe the basic mechanism of the greenhouse effect.
5. If there were no greenhouse effect, what factors would determine a planet's surface temperature? How do the "no greenhouse" temperatures of the terrestrial planets compare to their actual temperatures, and why?
6. Describe Earth's basic atmospheric structure, from the ground up. How do interactions of sunlight and gases explain the existence of each of the atmospheric layers?
7. Why is the sky blue? Why are sunrises and sunsets red?
8. Why does convection occur in the troposphere, leading to active weather, but not in the stratosphere?
9. What is ozone? How does the absence of ozone on Venus and Mars explain why these planets lack a stratosphere?
10. What is a *magnetosphere*? Briefly explain how a magnetosphere can protect a planet from the solar wind and lead to the creation of auroras.
11. What is the difference between weather and climate?
12. What are Earth's global wind patterns? Explain why our planet has circulation cells that transport warm air toward

the poles and cool air toward the equator. How does our planet's rotation affect these cells?

13. What are clouds made of? How does rain or snow form?
14. Describe each of the four factors that can lead to long-term climate change, and the effects of each factor.
15. Describe each of the processes by which a planet can gain or lose atmospheric gas. What factors determine whether gases are lost through thermal escape?
16. Why do the Moon and Mercury have so little atmospheric gas? How is it possible that they might nonetheless have water ice in polar craters?
17. Why does Mars have seasons? How do seasons on Mars differ from seasons on Earth?
18. How do we think that Mars lost atmospheric gas? What basic planetary property (such as size, distance from the Sun, or rotation rate) would have had to be different for Mars to have retained a thicker atmosphere?
19. What do we mean by a *runaway greenhouse effect?* Explain why this process occurred on Venus but not on Earth.
20. Describe several ways in which Earth's atmosphere is unique among the terrestrial worlds, and how each unique feature is important to our existence.
21. What is the *carbon dioxide cycle,* and why is it so crucial to life on Earth?
22. Briefly summarize the debate over global warming and its potential consequences.

Test Your Understanding

Does It Make Sense?

Decide whether the statement makes sense (or is clearly true) or does not make sense (or is clearly false). Explain your reasoning. (For an example, see Chapter 1, "Does It Make Sense?")

23. If Earth's atmosphere did not contain molecular nitrogen, X rays from the Sun would reach the surface.
24. If the molecular oxygen content of Earth's atmosphere increases, it will cause our planet to warm up.
25. Earth's oceans must have formed at a time when no greenhouse effect operated on Earth.
26. In the distant past, when Mars had a thicker atmosphere, it also had a stratosphere.
27. If Earth rotated faster, hurricanes would be more common and more severe.
28. Mars would still have seasons even if its orbit around the Sun were perfectly circular rather than elliptical.
29. Mars once may have been warmer than it is today, but it could never have been warmer than Earth because it is farther from the Sun than is Earth.
30. If the solar wind were much stronger, Mercury might develop a carbon dioxide atmosphere.
31. If Earth had as much carbon dioxide in its atmosphere as Venus, our planet would be too hot for liquid water to exist on the surface.
32. A planet in another solar system has an Earth-like atmosphere with plentiful oxygen but no life of any kind.

Quick Quiz

Choose the best answer to each of the following. Explain your reasoning with one or more complete sentences.

33. Which terrestrial world has the most atmosphere? (a) Venus (b) Earth (c) Mars
34. The greenhouse effect happens in which layer of an atmosphere? (a) the troposphere (b) the stratosphere (c) the lithosphere
35. What kind of light warms the stratosphere? (a) infrared (b) visible (c) ultraviolet
36. Which of the following is a strong greenhouse gas? (a) nitrogen (b) water vapor (c) oxygen
37. In which direction do hurricanes in the Southern Hemisphere rotate? (a) clockwise (b) counterclockwise (c) equally probable in either direction
38. What is the leading theory for Venus's complete lack of water in any form? (a) It formed closer to the Sun and accreted very little water. (b) It is locked away in the crust. (c) It was broken apart into H and O, which were lost from the atmosphere.
39. What kind of gas is most affected by thermal escape? (a) greenhouse gases (b) light gases (c) all gases equally
40. Which gas is the suspected cause of global warming? (a) ozone (b) water vapor (c) CO_2
41. What causes the release of oxygen gas into Earth's atmosphere? (a) outgassing (b) evaporation/sublimation (c) life
42. Where is the CO_2 that has outgassed from Earth's volcanoes? (a) in the atmosphere (b) escaped to space (c) locked up in rocks

Investigate Further

In-Depth Questions to Increase Your Understanding

Short-Answer/Essay Questions

43. *Clouds of Venus.* Table 10.2 shows that Venus's surface temperature in the absence of the greenhouse effect is lower than Earth's, even though it is closer to the Sun.
    a. Explain this unexpected result in one or two sentences.
    b. Suppose Venus had neither clouds nor greenhouse gases. What do you think would happen to the surface temperature of Venus? Why?
    c. How are clouds and volcanoes linked on Venus? What change in volcanism might result in the disappearance of clouds? Explain.
44. *Atmospheric Structure.* Study the curve showing Earth's average atmospheric structure in Figure 10.7. Sketch similar curves, and explain each curve in words, to show how the atmospheric structure would be different in each of the following cases.
    a. Suppose Earth had no greenhouse gases.
    b. Suppose the Sun emitted no ultraviolet light.
    c. Suppose the Sun had a higher output of X rays.
45. *Magic Mercury.* Suppose we could magically give Mercury the same atmosphere as Earth. Assuming this magical intervention happened only once, would Mercury be able to keep its new atmosphere? Explain.
46. *A Swiftly Rotating Venus.* Suppose Venus had rotated as rapidly as Earth throughout its history. Briefly explain how and why you would expect it to be different in terms of each of the following: geological processes, atmospheric circulation, magnetic field, and climate history. Write a few sentences about each.
47. *Coastal Winds.* During the daytime, heat from the Sun tends to make the air temperature warmer over land near the coast than over the water offshore. At night, when land cools off faster than the sea, the temperatures tend to be cooler over land than over the sea. Use these facts to predict the directions in which winds generally blow during the day and at night

in coastal regions. For example, do the winds blow out to sea or in toward land? (*Hint:* How do these conditions resemble those that cause planetary circulation cells?) Explain your reasoning in a few sentences. (Diagrams might help.)

48. *Sources and Losses.* Choose one process by which atmospheres can gain gas and one by which they can lose gas. For each process, write a few sentences that describe it and how it depends on each of the following fundamental planetary properties: size, distance from the Sun, and rotation rate.
49. *Project: Atmospheric Science in the Kitchen.*
    a. Find an empty plastic bottle, such as a water bottle with a screwtop that makes a good seal. Warm the air inside by filling the bottle partway with hot water and then shaking and emptying the bottle. Seal the bottle and place it into the refrigerator or freezer. What happens after 15 minutes? Explain why this happens, imagining that you can see the individual air molecules.
    b. You may have noticed that loose ice cubes in the freezer gradually shrink away to nothing or that frost sometimes builds up in old freezers. What technical terms used in this chapter describe these phenomena? On what terrestrial planet do these same processes play a major role in changing the atmospheric pressure?
50. *Two Paths Diverged.* By looking back to fundamental properties such as size and distance from the Sun, explain why Earth has oceans and very little atmospheric carbon dioxide, while similar-size Venus has a thick carbon dioxide atmosphere.
51. *Change in Fundamental Properties.* Consider Earth's size and distance from the Sun. Choose one property and suppose that it had been different (for example, smaller size or greater distance). Describe how this change might have affected Earth's subsequent history and the possibility of life on Earth.
52. *Feedback Processes in the Atmosphere.* As the Sun gradually brightens in the future, how can the CO_2 cycle respond to reduce the warming effect? Which parts of the cycle will be affected? Is this an example of positive or negative feedback?
53. *Earth to Mars.* Section 10.5 discusses what might happen to Earth if it were suddenly moved to the orbit of Venus. What do you think would happen to Earth if it were suddenly moved to the orbital distance of Mars? Write a few sentences explaining your answer.
54. *Global Warming.* What, if anything, should we be doing to alleviate the threat of global warming that we are not doing already? Write a one-page editorial summarizing and defending your opinion.

Quantitative Problems

Be sure to show all calculations clearly and state your final answers in complete sentences.

55. *The Mass of an Atmosphere.* What is the total mass of Earth's atmosphere? You may use the fact that 1 bar is the pressure exerted by 10,000 kilograms pushing down on a square meter in Earth's gravity. Remember that every square meter of Earth experiences this pressure from the atmosphere above it. Alternatively, you can start with the English units value for pressure of 14.7 pounds per square inch and convert to kilograms for your final answer. Remember that the surface area of a sphere of radius r is $4\pi r^2$.
56. *An Early Steam Atmosphere?* Many planetary scientists consider it likely that Earth's oceans were temporarily "boiled away" during the heavy bombardment period shortly after formation. What atmospheric pressure would result if today's oceans were converted to steam? Today's oceans cover 75% of Earth's surface to an average depth of 3.5 kilometers, and the density of water is 1,000 kilograms per cubic meter. (*Hint:* The previous problem shows how to convert a surface pressure to the total mass of an atmosphere, and this problem does the reverse.)
57. *The Role of Reflectivity.* Using Mathematical Insight 10.1, find out the maximum possible "no greenhouse" temperature for a completely black planet at 1 AU. What temperature do you find for a perfectly white planet? Explain. Finally, find out what reflectivity would be necessary to keep the average temperature exactly at the freezing point, and compare this to the Earth's actual reflectivity in Table 10.2.
58. *The Cooling Clouds of Venus.* Table 10.2 shows that Venus's temperature in the absence of the greenhouse effect is lower than Earth's, even though Venus is closer to the Sun. What would the Venus "no greenhouse" temperature be if its clouds were more transparent, giving a reflectivity the same as Earth's? What would the surface temperature be in this example if the greenhouse effect simply increased the surface temperature by the same number of degrees that it does today?
59. *Mars's Elliptical Orbit.* Mars's distance from the Sun varies from 1.38 AU to 1.66 AU. How much would this change its surface temperature? (*Hint:* Find the difference of the "no greenhouse" temperature at the two extremes.) Comment on how this affects Mars's seasons, connecting your answer to Figure 10.23.
60. *Escape from Venus.*
    a. Calculate the escape velocity from Venus's exosphere, which begins about 200 kilometers above the surface. (*Hint:* You'll need the radius of Venus and the formula in Mathematical Insight 5.3.)
    b. Calculate and compare the thermal speeds of hydrogen and deuterium atoms at the exospheric temperature of 350 K. The mass of a hydrogen atom is 1.67×10^{-27} kilograms, and the mass of a deuterium atom is about twice the mass of a hydrogen atom.
    c. In a few sentences, comment on the relevance of these calculations to the question of whether Venus has lost large quantities of water.
61. *Escape from Jupiter.* The atmospheric escape processes of this chapter also apply to the outer solar system. Jupiter's exosphere is even hotter than the Venus case of the previous problem. Will this be enough to cause even faster escape of hydrogen?
    a. Calculate the escape velocity from Jupiter's exosphere, which begins about 1,000 kilometers above the cloud tops. (*Hint:* You'll need the radius of Jupiter and the formula in Mathematical Insight 4.4.)
    b. Calculate and compare the thermal speeds of hydrogen atoms at the exospheric temperature of 800 K. The mass of a hydrogen atom is 1.67×10^{-27} kilograms.
    c. In a few sentences, state whether or not you consider the escape of hydrogen from Jupiter's hot exosphere to be significant, and why.
62. *Abundance versus Importance.* Make a table summarizing Earth's main atmospheric gases using those in Table 10.1, plus CO_2.
    a. Across the top of the table, list the five main gases in the atmosphere from most abundant to least abundant. In the next row, state the fraction of the atmosphere in percent. Find the percent values for those gases listed in

Table 10.1. For water, which can vary from place to place and time to time, use 1%. For CO_2, find the current fraction from Figure 10.35 measured in ppm. Convert ppm (part per million) to percent, remembering that 1% is "one part per hundred."

b. For each gas, decide on a "rank" in importance as a greenhouse gas. The most important gas gets a "1," etc. For those that are not greenhouse gases, put an "X." Below your table, comment on the relationship between abundance and importance as a greenhouse gas, if any.

c. Add another row of notes for each gas listing why it is important in our atmosphere, based on the content of this chapter. If its importance has not been discussed, put "X." For some gases there is more than one important aspect. Below the table, comment on whether the most abundant gases have the most important effects, in your opinion.

Discussion Questions

63. *Lucky Earth.* The climate histories of Venus and Mars make it clear that it's not "easy" to get a pleasant climate like that of Earth. How does this affect your opinion about whether Earth-like planets might exist around other stars? Explain.

64. *Terraforming Mars.* Some people have suggested that we might be able to carry out planetwide engineering of Mars that would cause its climate to warm and its atmosphere to thicken. This type of planet engineering is called *terraforming*, because its objective is to make a planet more Earth-like and thus easier for humans to live on. Discuss possible ways to terraform Mars, at least in principle. Do any of these ideas seem practical? Does it seem like a good idea? Defend your opinions.

65. *Terraforming Venus.* As a follow-up to the previous question, think of any ways in which it would be possible to terraform Venus. Again, discuss the possibilities in principle as well as their practicality.

66. *Don't Worry, Be Happy.* There are many skeptics of global warming. Research their opinions in newspapers, in magazines, and on the Internet. Do they disagree with the data, saying that Earth is not getting warmer? Do they disagree with the conclusion that humans are the primary cause? Do they disagree with the idea that action must be taken? Defend or refute their findings based on your own studies.

MEDIA EXPLORATIONS

For self-study activities go to MasteringASTRONOMY www.masteringastronomy.com

Interactive Tutorials

Surface Temperature of Terrestrial Planets Tutorial

Lesson 1 Energy Balance
Lesson 2 Role of Planet's Distance from the Sun
Lesson 3 Role of Planet's Reflectivity
Lesson 4 Role of Planet's Atmosphere

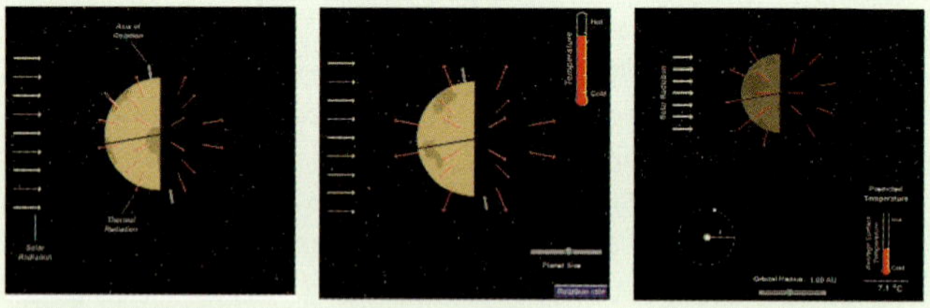

Supplementary Tutorial Exercises

Use the interactive **Tutorial Lessons** to explore the following questions.

Surface Temperature of Terrestrial Planets Tutorial, Lesson 1

1. How does the rate at which a planet emits radiation depend on the amount of sunlight it receives?
2. Does a planet's size affect its temperature? Explain.
3. How does a planet's rotation rate affect its day and night temperatures? Does rotation rate affect the average temperature? Explain.

Surface Temperature of Terrestrial Planets Tutorial, Lesson 2

1. Use the tool in Lesson 2 to predict Mercury's surface temperature. How does this compare to Mercury's actual day and night temperatures? Explain.
2. Use the tool in Lesson 2 to predict Earth's surface temperature. Why isn't this Earth's actual surface temperature? Explain.

Surface Temperature of Terrestrial Planets Tutorial, Lesson 3

1. Use the tool in Lesson 3 to predict the range of possible "no greenhouse" temperatures for Earth by moving the reflectivity slider. If Earth were black, would it be as warm as the observed temperature of +15°C?
2. Using the tool in Lesson 3, determine which causes a bigger temperature difference: changing the reflectivity from 0.0 to 0.1, or changing it from 0.9 to 1.0. Explain.

Surface Temperature of Terrestrial Planets Tutorial, Lesson 4

1. How would Earth's temperature be different if the carbon dioxide concentration doubled but all other gas concentrations remained the same?
2. A small warming could lead to increased evaporation of water. What would happen to the temperature if, along with the doubling of carbon dioxide as in question 1, the atmospheric water vapor concentration increased by 10% (to 1.1 times its current value)?
3. How might increased cloud cover offset the warming due to the greenhouse effect?
4. Increased cloud cover also means increased atmospheric water vapor. Use the tools to explore the balance between warming due to more water vapor and cooling due to more cloud cover. If we continued to add greenhouse gases to the atmosphere for centuries, what would ultimately happen to our planet's temperature? Explain.

Exploring the Sky and Solar System

Of the many activities available on the ***Voyager: SkyGazer*** **CD-ROM** accompanying your book, use the following files to observe key phenomena covered in this chapter.

Go to the **File: Basics** folder for the following demonstrations.

1. Planet Paths
2. Planet Orrery
3. Follow a Planet

Go to the **File: Demo** folder for the following demonstrations.

1. Earth and Venus
2. Hyakutake at Perihelion
3. Pluto's Orbit

Go to the **Explore** menu for the following demonstrations.

1. Solar System
2. Paths of the Planets

Movies

Check out the following narrated and animated short documentary available on **www.masteringastronomy.com** for a helpful review of key ideas covered in this chapter.

History of the Solar System

Web Projects

1. *Human Threats to Earth.* Write an in-depth research report, three to five pages in length, about current understanding and controversy regarding one of the following issues: global warming, ozone depletion, or the loss of species on Earth due to human activity. Be sure to address both the latest knowledge about the issue and proposals for alleviating any dangers associated with it. End your report by making your own recommendations about what, if anything, needs to be done to prevent damage to Earth.
2. *Spacecraft Study of Atmospheres.* Learn about a current or planned mission to study the atmosphere of one of the terrestrial worlds (possibly including Earth). Write a one- to two-page essay describing the mission and what we hope to learn from it.
3. *Martian Weather.* Find the latest weather report for Mars from spacecraft and other satellites. What season is it in the northern hemisphere? When was the most recent dust storm? What surface temperature was most recently reported from Mars's surface, and where was the lander located? Summarize your findings by writing a 1-minute script for a television news update on Martian weather.

11

Jovian Planet Systems

LEARNING GOALS

11.1 A Different Kind of Planet
- Are jovian planets all alike?
- What are jovian planets like on the inside?
- What is the weather like on jovian planets?
- Do jovian planets have magnetospheres like Earth's?

11.2 A Wealth of Worlds: Satellites of Ice and Rock
- What kinds of moons orbit the jovian planets?
- Why are Jupiter's Galilean moons so geologically active?
- What is special about Titan and other major moons of the outer solar system?
- Why are small icy moons more geologically active than small rocky planets?

11.3 Jovian Planet Rings
- What are Saturn's rings like?
- How do other jovian ring systems compare to Saturn's?
- Why do the jovian planets have rings?

Do there exist many worlds, or is there but a single world? This is one of the most noble and exalted questions in the study of Nature.

St. Albertus Magnus (1206–1280)

In Roman mythology, the namesakes of the jovian planets are rulers among gods: Jupiter is the king of the gods, Saturn is Jupiter's father, Uranus is the lord of the sky, and Neptune rules the sea. However, our ancestors could not have foreseen the true majesty of the four jovian planets. The smallest, Neptune, is large enough to contain the volume of more than 50 Earths. The largest, Jupiter, has a volume some 1,400 times that of Earth. These worlds are totally unlike the terrestrial planets. They are essentially giant balls of gas, with no solid surface on which to stand.

Why should we care about a set of worlds so different from our own? Apart from satisfying natural curiosity, studies of the jovian planets and their moons help us understand the birth and evolution of our solar system—which in turn helps us understand our own planet Earth. In addition, the jovian planets provide stepping stones to understanding the more than 150 planets so far discovered around other stars, because these planets are probably jovian in nature. In this chapter, we'll explore the jovian planet systems, first focusing on the planets themselves, then on their many moons, and finally on their beautifully complex rings.

Formation of the Solar System Tutorial, Lesson 1

11.1 A Different Kind of Planet

We toured the four jovian planets of our solar system briefly in Section 7.1. We saw that they are radically different from the terrestrial worlds in size, density, and composition. Now we are ready to explore the jovian planets in a little more depth.

In the context of human history, it's only quite recently that we recognized the vast differences between the jovian and terrestrial worlds. Jupiter and Saturn are easily visible to the naked eye, but their appearance in the night sky doesn't tell us how they differ in character from other visible planets. Uranus and Neptune were discovered only in 1781 and 1846, respectively (see box, p. 331), so they were unknown even during the Copernican revolution.

Astronomers first recognized the immense sizes of Jupiter and Saturn about 250 years ago. Recall that we need to know both angular size and distance to calculate an object's true size (see Mathematical Insight 2.1). Although Copernicus had figured out the *relative* distances to the known planets in astronomical units (AU), scientists did not establish the absolute scale of the solar system until the 1760s, when they were able to measure the true length of an astronomical unit with data from a transit of Venus (see Special Topic, p. 218). Only then could scientists calculate the true sizes of Jupiter and Saturn from their distances and angular sizes (measured through telescopes). Knowing the scale of the solar system also told scientists the distances of orbiting moons; because the orbital periods of the moons were easy to measure, the orbital distances allowed scientists to use Newton's version of Kepler's third law [Section 4.4] to calculate the masses of the jovian planets. Together, the measurements of size and mass revealed the low densities of the jovian planets, proving that these worlds are very different in nature from Earth.

The real revolution in understanding jovian planet systems began with the first spacecraft sent to visit them: *Pioneer 10* and *Pioneer 11*, which flew past Jupiter and Saturn in the early 1970s. The *Voyager* missions followed soon thereafter; *Voyager 1* flew past Jupiter and Saturn, and *Voyager 2* flew past all four jovian planets. More recently, we've learned much more about Jupiter from the *Galileo* mission, which orbited Jupiter from 1995 to 2003, and about Saturn from the *Cassini* mission, which entered Saturn orbit in 2004. The discovery of more than 150 presumably jovian planets around other stars [Section 13.2] has also heightened interest in our own solar system's jovian planets.

• Are jovian planets all alike?

Figure 11.1 shows a montage of the jovian planets compiled by the *Voyager* spacecraft, along with basic data and Earth included for scale. All four share the properties of immense size and a low-density composition of hydrogen, helium, and hydrogen compounds—just as we expect for worlds that formed beyond the frost line of the solar nebula, in the cold, outer regions of our solar system [Section 8.3]. However, a close look at the images and data in Figure 11.1 shows that the jovian planets differ substantially from one another in size, mass, density, composition, and even color. If you look closely, you'll even see subtle differences in shape, with Saturn much less like a perfect sphere than the others. As we'll see, these differences can also be traced to the birth of these planets.

Jovian Planet Composition Jupiter and Saturn are made almost entirely of hydrogen and helium, with just a few percent of their masses coming from hydrogen compounds and even smaller amounts of rock and metal. Thus, their overall compositions are much more similar to the composition of the Sun than to the compositions of the terrestrial planets. Some people have gone as far as to call Jupiter a "failed star" because it has a starlike composition but lacks the nuclear fusion needed to make it shine. Its lack of fusion is due to its mass: Although Jupiter is large for a planet, it is much less massive than any star. As a result, its gravity is too weak to compress its interior to the extreme temperatures and densities needed for nuclear fusion. Jupiter would have needed to grow to about 80 times its current

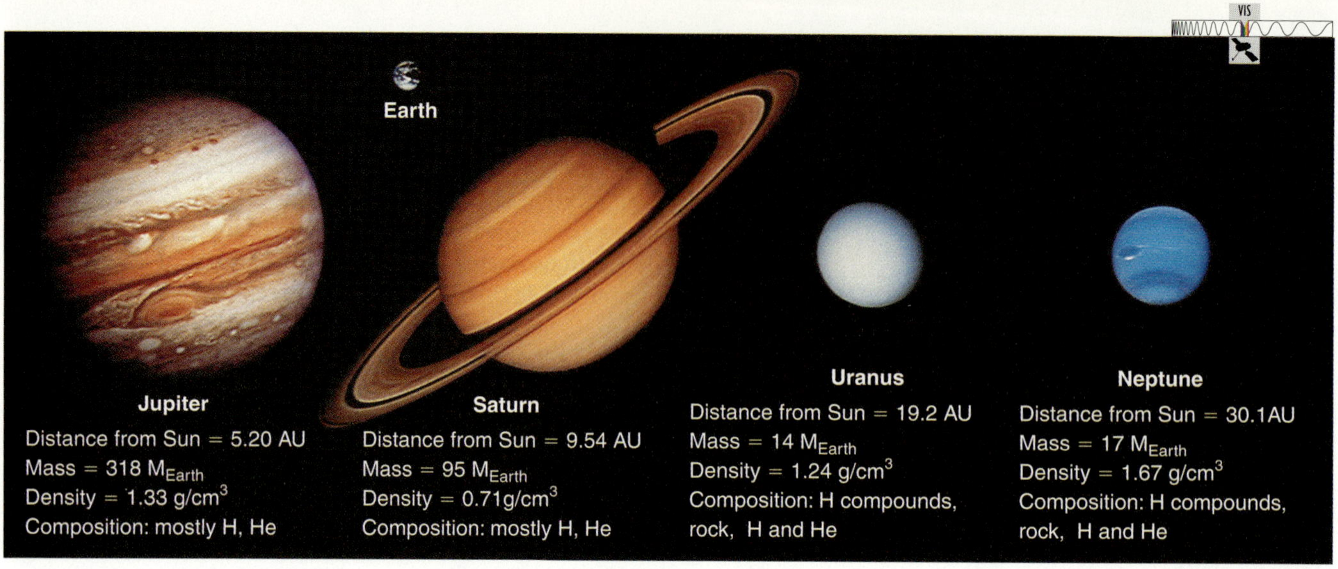

Figure 11.1 Jupiter, Saturn, Uranus, and Neptune, shown to scale with Earth for comparison.

mass to have become a star. Of course, where some people see a failed star, others see an extremely successful planet.

Uranus and Neptune are not only much smaller than Jupiter and Saturn overall, but also contain proportionally much smaller amounts of hydrogen and helium. These gases make up much less than half their total masses. Rather than being made mostly of hydrogen and helium, Uranus and Neptune are made primarily of hydrogen compounds such as water (H_2O), methane (CH_4), and ammonia (NH_3), along with smaller amounts of metal and rock.

The differences in composition among the jovian planets can probably be traced to their origins. Recall that the jovian planets are thought to have formed in a very different way from the terrestrial planets [Section 8.3]. The terrestrial planets grew only through accretion of solid planetesimals containing rock and metal. Because rock and metal were quite rare in the solar nebula, the terrestrial planets never grew massive enough for their gravity to hold much of the abundant hydrogen and helium gas that made up most of the nebula.

The jovian planets formed in the outer solar system (beyond the frost line), where it was cold enough for hydrogen compounds to condense into ices (see Figure 8.5). Because hydrogen compounds were so much more abundant than metal and rock, some of the ice-rich planetesimals of the outer solar system grew to great size. Once these planetesimals became sufficiently massive, their gravity allowed them to draw in the hydrogen and helium gas that surrounded them. All four jovian planets are thought to have grown from ice-rich planetesimals of about the same mass—roughly 10 times the mass of Earth. Thus, their differences in composition stem from the amount of hydrogen and helium gas that they captured.

Jupiter and Saturn captured so much hydrogen and helium gas that these gases now make up the vast majority of their masses. The ice-rich planetesimals from which they grew now represent only a small fraction of their overall masses—about 3% in Jupiter's case and about 10% in Saturn's case.

Uranus and Neptune pulled in much less hydrogen and helium gas. For example, notice in Figure 11.1 that Uranus's mass is about 14 times Earth's mass. Since it is thought to have grown around an ice-rich planetesimal that was about 10 times Earth's mass, the drawn-in hydrogen and helium gas must make up only about a third of Uranus's total mass. The bulk of its mass consists of material from the original ice-rich planetesimal: hydrogen compounds mixed with smaller amounts of rock and metal. The same is true for Neptune, though its higher density suggests that it may have formed around a slightly more massive ice-rich planetesimal.

We can understand why the jovian planets captured different amounts of gas by considering their distances from the Sun. As the solar system formed, the solid particles that condensed far from the Sun must have been much more widely spread out than particles that condensed nearer to the Sun. Thus, at greater distances from the Sun, it took longer for small particles to accrete into large, icy planetesimals with gravity strong enough to pull in gas from the surrounding nebula. Jupiter would have been the first jovian planet to get a planetesimal large enough for its gravity to start drawing in gas, followed by Saturn, Uranus, and Neptune. Because all the planets stopped accreting gas at the same time—when the solar wind blew all the remaining gas into interstellar space—the more distant planets had less time to capture gas and thus ended up smaller in size.

Density Differences Notice in Figure 11.1 that Saturn is considerably less dense than Uranus or Neptune. This should make sense when you compare compositions. After all, the hydrogen compounds, rock, and metal that make up Uranus and Neptune are normally much more dense than hydrogen or helium gas. By the same logic, we'd expect Jupiter to be even less dense than Saturn—but it's not. To understand Jupiter's surprisingly high density, we need to think about how massive planets are affected by their own gravity.

THINK ABOUT IT

Saturn's average density of 0.71 g/cm^3 is less than that of water. As a result, it is sometimes said that Saturn could float on a giant ocean. Suppose there really were a gigantic planet with a gigantic ocean and we put Saturn on the ocean's surface. Would it float? If not, what would happen?

Building a planet of hydrogen and helium is a bit like making one out of very fluffy pillows. Imagine assembling a planet pillow by pillow. As each new pillow is added, those on the bottom are compressed more by those above. As the lower layers are forced closer together, their mutual gravitational attraction increases, compressing them even further. At first the stack grows substantially with each pillow, but eventually the growth slows until adding pillows barely increases the height of the stack (Figure 11.2a).

This analogy explains why Jupiter is only slightly larger than Saturn in radius even though it is more than three times as massive. The extra mass of Jupiter compresses its interior to a much higher density. More precise calculations show that Jupiter's radius is almost the maximum possible radius for a jovian planet. If much more gas were added to Jupiter, its weight would actually compress the interior

SPECIAL TOPIC How Were Uranus, Neptune, and Pluto Discovered?

The planets Mercury, Venus, Mars, Jupiter, and Saturn were all known to ancient people. Each can be seen with the naked eye and each clearly wanders among the fixed stars of the constellations. In contrast, the planets (and other objects) beyond Saturn are relatively recent discoveries in human history.

Uranus was the first "discovered" planet. Although faintly visible to the naked eye, Uranus is so faint and moves so slowly in its 84-year orbit of the Sun that ancient people did not recognize it as a planet. It was even recorded as a star on some detailed early sky charts. Uranus was recognized as a planet in 1781 by English astronomers William and Caroline Herschel. William and Caroline were brother and sister, and they worked together both in building their telescopes and in much of their observing. William became the more famous of the pair by far, partly because Caroline's gender made many scientists of the time unwilling to take her seriously, but also because he was nearly 12 years older than his sister and had been the one who drew her into astronomical research. William Herschel originally suggested naming the new planet Georgium Sidus, Latin for George's star, in honor of his patron, King George III. Fortunately, the idea of "Planet George" never caught on. Instead, many eighteenth- and nineteenth-century astronomers referred to the new planet as Herschel. The modern name Uranus—after the mythological father of Saturn—was first suggested by one of Herschel's contemporaries, astronomer Johann Bode. This name was generally accepted by the mid-nineteenth century.

Neptune's discovery followed next, and it represented an important triumph for Newton's law of universal gravitation and the young science of astrophysics. By the mid-1800s, careful observations of Uranus had shown its orbit to be slightly inconsistent with that predicted by Newton's law of gravity—at least if it were being influenced only by the Sun and the other known planets. In the early 1840s in England, a student named John Adams suggested that the inconsistency could be explained if there were an unseen "eighth planet" orbiting the Sun beyond Uranus. According to the official story, he used Newton's theory to predict the location of the planet but was unable to convince British astronomers to carry out a telescopic search. However, documents from the time suggest his prediction may not have been as precise as the official history claims. Meanwhile, in the summer of 1846, French astronomer Urbain Leverrier independently made very precise calculations. He sent a letter to Johann Galle of the Berlin Observatory, suggesting a search for the eighth planet. On the night of September 23, 1846, Galle pointed his telescope to the position suggested by Leverrier. There, within 1° of its predicted position, he saw the planet Neptune. Hence, Neptune's discovery truly was made by mathematics and physics and was only confirmed with a telescope.

As a side note to this story, Leverrier had such faith in Newton's universal law of gravitation that he also suggested a second unseen planet, this one orbiting closer to the Sun than Mercury. He got this idea because other astronomers had identified slight discrepancies between Mercury's actual orbit and the orbit predicted by Newton's theory. He assumed that no one had yet seen the planet, which he called Vulcan, because it was so close to the Sun. Leverrier died in 1877, still believing that Vulcan would someday be discovered. In fact, Vulcan does not exist. Mercury's actual orbit does not match the orbit predicted by Newton's law of gravity because Newton's theory is not the whole story of gravity. About 40 years after Leverrier's death, Einstein showed that Newton's theory is only an approximation of a broader theory of gravity, known today as Einstein's *general theory of relativity.* Einstein's theory predicts an orbit for Mercury that matches its actual orbit. This match was one of the first key pieces of evidence in favor of Einstein's theory [Section S3.4].

Pluto was discovered in 1930 by American astronomer Clyde Tombaugh, culminating a search that began when astronomers analyzed the orbit of Neptune. The story of Pluto's discovery at first seemed much the same as that of Neptune's. Just as discrepancies between the predicted orbit and the actual orbit of Uranus led to the prediction that Neptune must exist, lingering discrepancies between the updated prediction for Uranus's orbit and its actual orbit suggested the existence of an even more distant "ninth planet." Tombaugh found Pluto just 6° from the position in the sky where this ninth planet had been predicted to lie, so it seemed that the search had been successful. However, while initial estimates suggested that Pluto was much larger than Earth, we now know that Pluto has a radius of only 1,160 kilometers and a mass of just 0.002 Earth mass—making it far too small to affect the orbit of Neptune. In retrospect, the supposed orbital irregularities of Neptune appear to have been errors in measurement. No "ninth planet" is needed to explain the orbit of Neptune. The discovery of Pluto was just a happy coincidence.

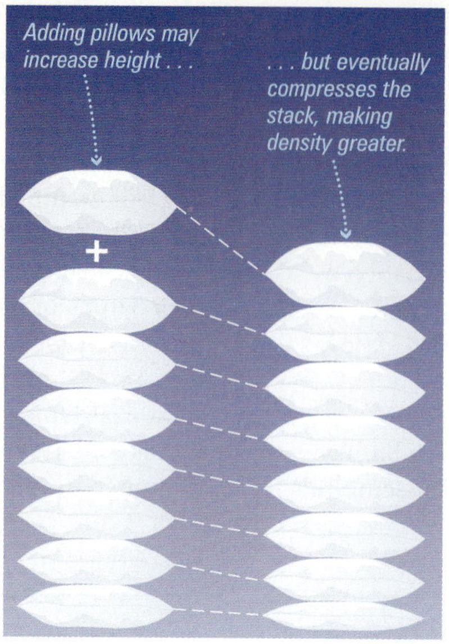

a A tall enough stack of pillows offers an analogy to massive hydrogen/helium planets: adding more pillows compresses the stack rather than increasing its height.

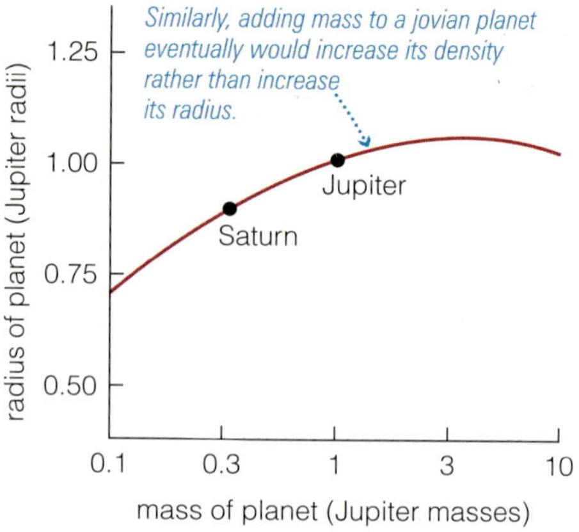

b This graph shows how radius depends on mass for a hydrogen/helium planet. Notice that Jupiter is only slightly larger in radius than Saturn, despite being three times as massive. Gravitational compression of a planet much more massive than Jupiter would actually make it smaller in size.

Figure 11.2 The relationship between mass and radius for a planet made of hydrogen and helium.

enough to make the planet *smaller* rather than larger (Figure 11.2b). Thus, some extrasolar planets that are larger in mass than Jupiter are probably smaller in size. In fact, the smallest stars are significantly smaller in radius than Jupiter, even though they are 80 times as massive.

Jovian Planet Rotation and Shape The jovian planets rotate much more rapidly than any of the terrestrial worlds. However, precisely defining their rotation rates can be difficult because they lack solid surfaces. We can measure a terrestrial rotation period simply by watching the apparent movement of a mountain or crater as a planet rotates. On jovian planets, we can observe only the movements of clouds. Cloud movements can be deceptive, because their apparent speeds are a combination of winds and planetary rotation. In addition, observations of clouds at different latitudes suggest that the jovian planets do not rotate like solid balls. Instead, their rotation rates vary with latitude: Equatorial regions complete each rotation in less time than polar regions. (The Sun also has a rotation rate that differs with latitude.)

We can measure the rotation rates of the jovian interiors by tracking emissions from charged particles trapped in their magnetospheres. This technique tells us the rotation period of the magnetosphere, which should be the same as the rotation period deep in the interior, where the magnetic field is generated. Measurements show that the jovian "day" ranges from about 10 hours on Jupiter and Saturn to 16–17 hours on Uranus and Neptune.

The rapid rotation rates of the jovian planets affect their shapes (Figure 11.3). Gravity alone would make the jovian planets into perfect spheres. Rotation makes them less like perfect spheres because it makes material bulge outward. Material near the equator, where speeds around the rotation axis are highest, is flung outward in the same way you feel yourself flung outward when you ride on a merry-go-round. The size of the equatorial bulge depends on the balance between the strength of gravity (which pulls material inward) and the rate of rotation (which pushes material outward). The balance tips most strongly toward flattening on Saturn: With its rapid 10-hour rotation period and relatively weak surface gravity, Saturn is about 10% wider at its equator than at its poles. In addition to altering a planet's appearance, the equatorial bulge exerts an extra gravitational pull that helps keep moons and rings aligned with the equator.

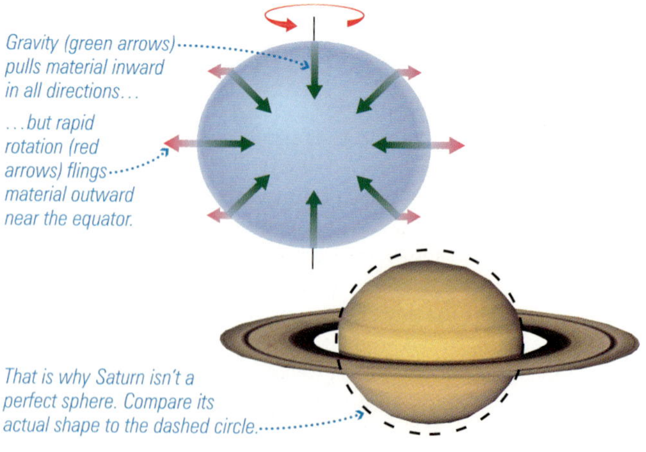

Figure 11.3 Interactive Figure Because of their rapid rotation, the jovian planets are not quite spherical. Saturn shows the biggest difference between its actual shape and a perfect sphere.

• What are jovian planets like on the inside?

The jovian planets are often called "gas giants," making it sound as if they are entirely gaseous like air on Earth. Based on their compositions, Jupiter and Saturn may seem to deserve this name; after all, they became giants primarily by capturing so much hydrogen and helium gas. The name may seem less fitting for Uranus and Neptune, since they are made mostly of materials besides pure hydrogen and helium. However, closer inspection shows that the name is a little misleading even for Jupiter and Saturn, because their strong gravity compresses most of the "gas" into forms of matter quite unlike anything we are familiar with in everyday life on Earth.

You may wonder how we can claim to know what the jovian planets are like on the inside. The answer is a combination of theoretical modeling and laboratory experiments. Just as with the terrestrial planets, detailed observations of the strength of a planet's gravity and of its magnetic field can help us put together a model of the planet's interior structure. We then use the model to predict observable phenomena, such as a planet's average density, and modify it until the predictions match the data. Today, advanced computer models successfully explain the observed sizes, densities, atmospheric compositions, and precise shapes of the jovian planets. Thus, we are confident that the interior structure found by the models is fairly close to reality. Laboratory studies also provide important data; for example, laboratory experiments have shown us how hydrogen and helium act under the tremendous temperatures and pressures that exist deep beneath the jovian cloud tops.

Inside Jupiter Let's begin discussion of the jovian interiors by using Jupiter as a prototype. Jupiter's lack of a solid surface makes it tempting to think of the planet as "all atmosphere," but you could not fly through Jupiter's interior in the way airplanes fly through air. A spacecraft plunging into Jupiter would find increasingly higher temperatures and pressures as it descended. The *Galileo* spacecraft dropped a scientific probe into Jupiter in 1995, and it collected measurements for about an hour before the ever-increasing pressures and temperatures destroyed it. *Galileo* provided valuable data about Jupiter's atmosphere, but didn't last long enough to sample the interior: It survived only to a depth of about 200 kilometers, or about 0.3% of Jupiter's radius.

In fact, while Jupiter has no solid surface, computer models tell us that it still has fairly distinct interior layers (Figure 11.4). The layers do not differ much in composition—all except the core are mostly hydrogen and helium; instead they differ in the phase (such as liquid or gas) of their hydrogen. To get a better sense of this layering, imagine plunging head-on into Jupiter in a futuristic space suit that allows you to survive the extreme interior conditions.

Near the cloud tops, you'll find the temperature to be a brisk 125 K (−148°C), the density to be a low 0.0002 g/cm^3, and the atmospheric pressure to be about 1 bar (the same as the pressure at sea level on Earth [Section 10.1]). As you plunge downward, conditions quickly become more extreme.

By a depth of 7,000 kilometers—about 10% of the way to the center—you'll find that the temperature has increased to a scorching 2,000 K and the pressure has reached 500,000 bars. The density is about 0.5 g/cm^3, or about half that of water. Under these conditions, hydrogen acts more like a liquid than a gas, which is why the layer that begins at this depth is labeled *liquid hydrogen* in Figure 11.4. Of course, like the rest of Jupiter, the layer also contains helium and hydrogen compounds.

At a depth of 14,000 kilometers, or about 20% of the way to the center, the density has become about as high as that of water (1.0 g/cm^3). The temperature is near 5,000 K, almost as hot as the surface (but not the interior) of the Sun. The pressure has reached 2 million bars. This extreme pressure forces hydrogen into a compact, metallic form. Just as is the case with everyday metals, electrons are free to move around in *metallic hydrogen*, so it conducts electricity quite effectively. This layer extends through most of the rest of Jupiter's interior and, as we'll see shortly, it is where Jupiter's magnetic field is generated.

You'll finally reach Jupiter's core at a depth of 60,000 kilometers, or about 10,000 kilometers from the center. The core temperature is some 20,000 K and the pressure is about 100 million bars. The density of the core is about 25 g/cm^3, much denser than any material you'll find on Earth's surface. The core is a mix of hydrogen compounds, rock, and metal, but these materials bear little resemblance to familiar solids or liquids because of the combination of high temperature and extreme pressure. Moreover, while the rock and metal of the terrestrial planets have separated into

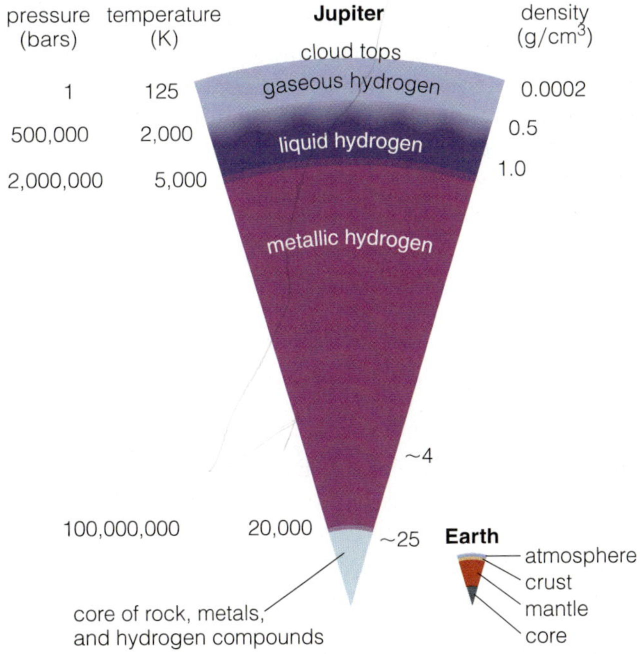

Figure 11.4 Jupiter's interior structure, labeled with the pressure, temperature, and density at various depths. Earth's interior structure is shown to scale for comparison. Note that Jupiter's core is only slightly larger than Earth but is about 10 times as massive.

layers by composition (that is, into the layers we call the core, mantle, and crust), Jupiter's core materials are probably all mixed together. The core contains about 10 times as much mass as the entire Earth, but it is only about the same size as Earth because it is compressed to such high density.

Comparing Jovian Interiors Computer models of the jovian interiors yield the somewhat surprising result that the cores of all four jovian planets are quite similar in composition and mass. Thus, the jovian planet interiors differ mainly in the hydrogen/helium layers that surround their cores. Figure 11.5 contrasts the four jovian interiors. Remember that while the outer layers are named for the phase of their hydrogen, they also contain helium and hydrogen compounds.

Saturn is the most similar to Jupiter, just as we should expect given its similar size and composition. It has the same set of four layers as Jupiter, and the layers differ from those of Jupiter only because of Saturn's lower mass and weaker gravity. The lower mass makes the weight of the overlying layers less on Saturn than on Jupiter, so you must travel deeper into Saturn to find the layer where pressure changes hydrogen from one form to another. That is why Saturn has a thicker layer of gaseous hydrogen and a much thinner and more deeply buried layer of metallic hydrogen.

Pressures within Uranus and Neptune are not high enough to form liquid or metallic hydrogen at all. Each of these two planets has only a thick layer of gaseous hydrogen surrounding its core of hydrogen compounds, rock, and metal. This core material may be liquid, making for very odd "oceans" buried deep inside Uranus and Neptune. The cores of Uranus and Neptune are larger in radius than the cores of Jupiter and Saturn, even though they have about the same mass, because they are less compressed by the lower weight of their overlying layers. The less extreme interior conditions also allowed Uranus's and Neptune's cores to differentiate, so that hydrogen compounds reside in a layer above a central layer of rock and metal.

Internal Heat When we studied the terrestrial worlds, we found that internal heat was the major driver of surface geology. The jovian planets do not have any geology, because they have no solid surfaces, but their internal heat is still important for atmospheric processes. Again, let's start by examining Jupiter's internal heat, then extend the idea to the other jovian planets.

Jupiter has a tremendous amount of internal heat, and like any hot object it gradually loses this heat to space by emitting thermal radiation [Section 5.4]. In fact, Jupiter emits almost twice as much energy as it receives from the Sun. This heat contributes significant energy to Jupiter's upper atmosphere. (For comparison, Earth's internal heat contributes only 0.005% as much energy to the surface as does sunlight.)

What keeps Jupiter so hot inside? Jupiter's large size means it loses internal heat very slowly (see Mathematical Insight 9.1), but calculations show that the remaining heat of accretion and differentiation is not enough to explain Jupiter's present-day heat loss. Radioactive decay adds further heat, but still not enough to account for all the internal heat. The most likely explanation for Jupiter's excess heat is that the planet is still slowly contracting, as if it has not quite finished forming. Contraction converts gravitational potential energy to thermal energy, so ongoing contraction represents an ongoing source of internal heat. Although we have not measured any such contraction, theoretical considerations show that it is probably occurring. Moreover, calculations tell us that contraction could easily explain Jupiter's internal heat, even if the contraction is so gradual that we have little hope of ever measuring it directly.

Regardless of the specific mechanism, Jupiter has undoubtedly lost substantial heat during the 4.5 billion years since its formation. Jupiter's interior must have been much warmer in the distant past, and this heat would have "puffed up" its atmosphere. Thus, Jupiter was larger and reflected more sunlight in the distant past, and it must have been even more prominent in Earth's sky than it is today.

Like Jupiter, Saturn emits nearly twice as much energy as it receives from the Sun, suggesting that it, too, must have some ongoing source of heat. However, Saturn's mass is too small for it to be generating this heat by contracting

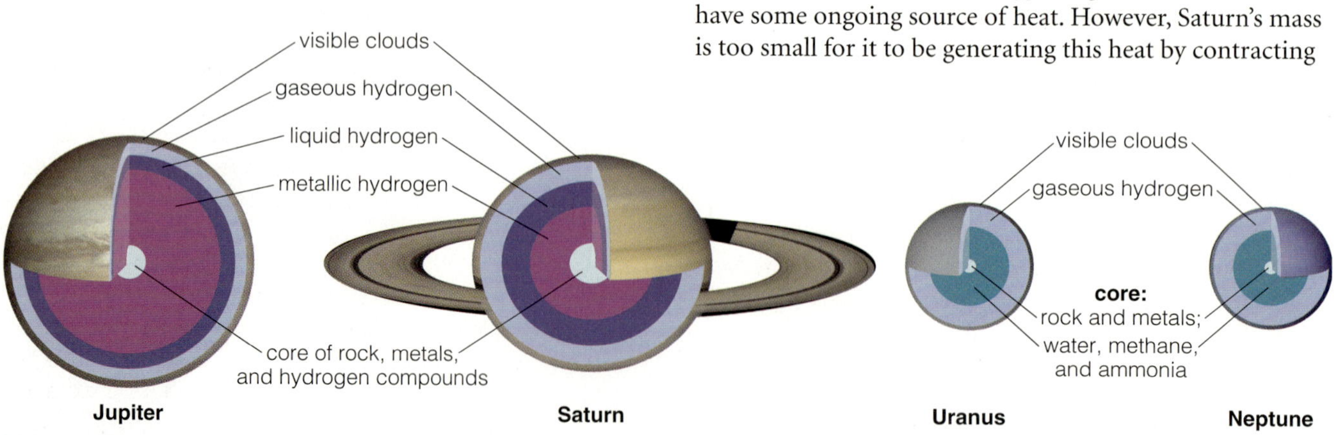

Figure 11.5 These diagrams compare the interior structures of the jovian planets, shown approximately to scale. All four planets have cores of rock, metal, and hydrogen compounds, with masses about 10 times the mass of Earth's core. They differ primarily in the depth of the hydrogen/helium layers that surround their cores. (The cores of Uranus and Neptune are differentiated into separate layers of rock/metal and hydrogen compounds.)

like Jupiter. Instead, Saturn's pressure and its lower interior temperatures may allow helium to condense into liquid form at relatively high levels within the interior. The helium droplets slowly rain down to the deeper interior. This gradual helium rain represents a sort of ongoing *differentiation* [Section 9.1], because it means that higher-density material (the helium droplets) is still sinking inside the planet. *Voyager* and *Cassini* observations have confirmed that Saturn's atmosphere is somewhat depleted of helium (compared to Jupiter's atmosphere), just as we should expect if helium has been raining down into Saturn's interior for billions of years.

Neither Uranus nor Neptune has internal conditions that would allow helium rain to form, and most of their original heat from accretion should have escaped long ago. This explains why Uranus emits virtually no excess internal energy. Neptune, however, is much more mysterious. Like Jupiter and Saturn, Neptune emits nearly twice as much energy as it receives from the Sun. The only reasonable explanation for Neptune's internal heat is that the planet is somehow still contracting, somewhat like Jupiter, thereby converting gravitational potential energy into thermal energy. However, we do not yet know why a planet of Neptune's size would still be contracting more than 4 billion years after its formation.

• What is the weather like on jovian planets?

Jovian atmospheres, by which we mean the portions of the planets that are visible to telescopes and spacecraft, have dynamic winds and weather, with colorful, cloudy skies and enormous storms. These atmospheres are made mostly of hydrogen and helium gas, mixed with small amounts of various hydrogen compounds. Because oxygen, carbon, and nitrogen are among the most common elements in the universe besides hydrogen and helium, the most common hydrogen compounds are methane (CH_4), ammonia (NH_3), and water (H_2O). Spectroscopy also reveals the presence of small amounts of more complex hydrogen compounds, such as acetylene (C_2H_2), ethane (C_2H_6), and propane (C_3H_8). Although all these hydrogen compounds together make up only a minuscule fraction of the jovian planet atmospheres, they are responsible for virtually all aspects of their appearances. Some of these compounds condense to form the clouds that are so prominent in telescope and spacecraft images. Others are responsible for the great variety of colors we see among the jovian planets. Without these compounds in their atmospheres, the jovian planets would be uniform, colorless balls of gas.

THINK ABOUT IT

Several gases in Jupiter's atmosphere—including methane, propane, and acetylene—are used as highly flammable fuels here on Earth. Jupiter also has plenty of lightning to provide sparks. Why don't these gases ignite in Jupiter's atmosphere? (*Hint:* What's missing from Jupiter's atmosphere that's necessary for ordinary fire?)

We can understand the jovian atmospheres using the same basic principles that we used with the terrestrial planet atmospheres. Of course, the different compositions, colder temperatures, and more rapid rotation rates of the jovian planets lead to some fascinating differences between terrestrial and jovian atmospheres. As we did for the jovian planet interiors, let's examine different aspects of the jovian atmospheres by starting with Jupiter as the prototype for each feature. We'll then use the general differences between the jovian planets to understand differences in their weather.

Atmospheric Structure and Clouds Telescopic observations of Jupiter, along with data returned from the *Galileo* probe during its 1995 plunge into Jupiter, tell us that the temperature structure of Jupiter's atmosphere is very similar to that of Earth (see Figure 10.7). High above the cloud tops, Jupiter's *thermosphere* consists of very-low-density gas heated to perhaps 1,000 K by solar X rays and by energetic particles from Jupiter's magnetosphere. Below the thermosphere but still above the clouds, we find Jupiter's *stratosphere.* Recall that a planet can have a stratosphere only if it has a gas that can absorb ultraviolet light from the Sun. Ozone plays this role on Earth. Jupiter lacks molecular oxygen and ozone, but has a few minor atmospheric ingredients that absorb solar ultraviolet photons. This absorption gives the stratosphere a peak temperature of around 200 K. Chemical reactions driven by the solar ultraviolet photons also create a smoglike haze that masks the color and sharpness of the clouds below. Below the stratosphere lies Jupiter's *troposphere,* where the temperature rises with depth because greenhouse gases trap both solar heat and Jupiter's own internal heat. Figure 11.6 shows how temperature varies with altitude in Jupiter's atmosphere.

In the troposphere, the warmer temperatures below drive strong convection. This convection is responsible for the thick clouds that enshroud Jupiter. Recall that clouds form when a gas condenses to make tiny liquid droplets or solid flakes. Water vapor is the only gas that can condense in Earth's atmosphere, which is why clouds on Earth are made of water droplets or ice flakes that can produce rain or snow. In contrast, Jupiter's atmosphere has several gases that can condense to form clouds. Each of these gases condenses at a different temperature, leading to distinctive cloud layers at different altitudes.

Jupiter has three primary cloud layers, which are also shown in Figure 11.6. We can understand them by imagining that we could watch gases rising through the troposphere with convection. Deep in the troposphere, the gases include three ingredients that will condense when temperatures are low enough: water (H_2O), ammonium hydrosulfide (NH_4SH), and ammonia (NH_3). The rising gas first encounters temperatures cool enough for water vapor to condense into liquid water but not cool enough for the other gases to condense. Thus, the lowest layer of clouds contains water droplets. As the remaining gas continues its rise, it next reaches an altitude at which ammonium hydrosulfide condenses to make the second cloud layer. Finally, after rising another 50 kilometers, the gas reaches an altitude

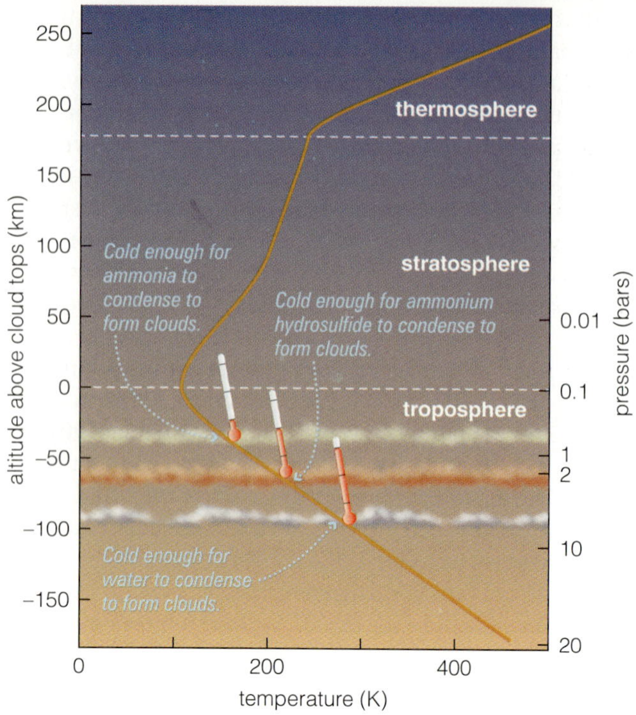

Figure 11.6 This graph shows the temperature structure of Jupiter's atmosphere. Jupiter has at least three distinct cloud layers because different atmospheric gases condense at different temperatures and hence at different altitudes.

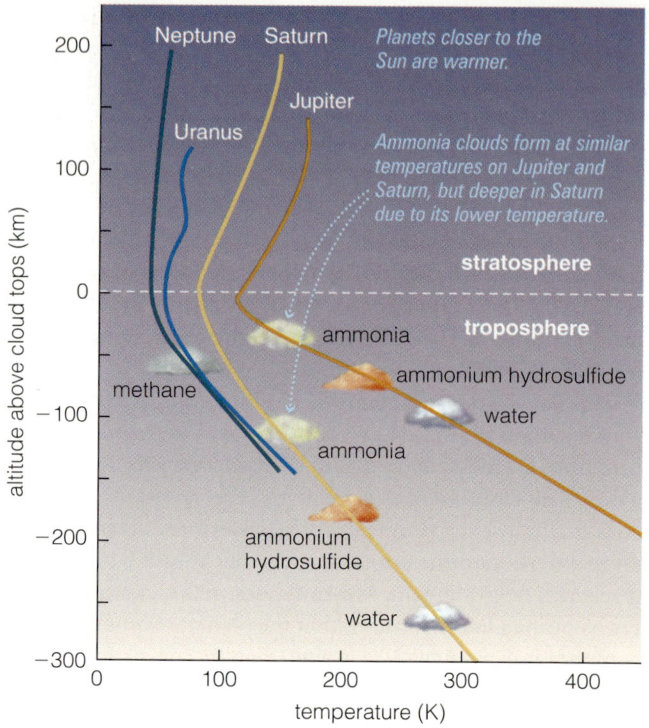

Figure 11.7 The figure contrasts the atmospheric structures and cloud layers of the four jovian planets. Note that the altitude scale is different on this figure than on Figure 11.6, allowing us to show the tropospheres and stratospheres only.

at which ammonia (NH_3) condenses to make the upper cloud layer.

Data from telescopic and spacecraft observations tell us that the atmospheric structures of the other three jovian planets are quite similar to that of Jupiter (Figure 11.7). The primary difference among them is that the atmospheres get progressively cooler with increasing distance from the Sun, just as we would expect. These temperature differences lead the planets to have their cloud layers at different altitudes.

Saturn has the same set of three cloud layers as Jupiter, but the lower overall temperatures causes these layers to lie deeper in Saturn's atmosphere. For example, to find the relatively warm temperatures at which water vapor can condense to form water clouds, we must look about 200 kilometers deeper into Saturn than into Jupiter. Saturn's cloud layers are also separated by greater vertical distances than Jupiter's, because Saturn's weaker gravity causes less atmospheric compression.

Uranus and Neptune are so cold that any cloud layers similar to those of Jupiter or Saturn would be buried too deep in their atmospheres for us to see. Thus, we do not know whether such clouds exist on these two planets. However, high in the atmospheres of Uranus and Neptune, we can see clouds made from flakes of methane snow. Methane can condense in the very cold upper tropospheres of Uranus and Neptune but not in the warmer tropospheres of Jupiter and Saturn.

Clouds and Colors The spectacular colors of the jovian planets are probably the first thing that jumps out at you when you look at the photos in Figure 11.1. Many mysteries remain about precisely why the jovian planets are so colorful, but clouds are responsible for many of the color features. Different types of clouds apparently reflect light of different colors. For example, Earth's clouds look white from space because they are made of water that reflects the white light of the Sun.

Water clouds on Jupiter and Saturn probably also reflect white light, as do the upper layer clouds made of ammonia. In contrast, the mid-layer clouds of ammonium hydrosulfide reflect brown and red light, although no one knows exactly why. Most likely, the colors in these clouds come from as-yet-unidentified compounds that are produced by chemical reactions deeper in the atmosphere and are then carried upward by convection. The fact that Saturn's clouds lie deeper in its atmosphere than Jupiter's probably explains Saturn's more subdued colors: Less light penetrates to the depths at which Saturn's clouds are found, and the light they reflect is more obscured by the atmosphere above them.

The blue colors of Uranus and Neptune come from methane gas, which is at least 20 times more abundant (by percentage) on these planets than on Jupiter or Saturn. Methane gas in the upper atmospheres of Uranus and Neptune absorbs red light, allowing only blue light to penetrate to the level at which the methane clouds exist. The methane clouds reflect this blue light upward, giving the planets their blue colors (Figure 11.8). Uranus has a lighter blue color than Neptune, probably because it has more smog-like haze to scatter sunlight before it reaches the level of the methane clouds. The extra haze is probably a result

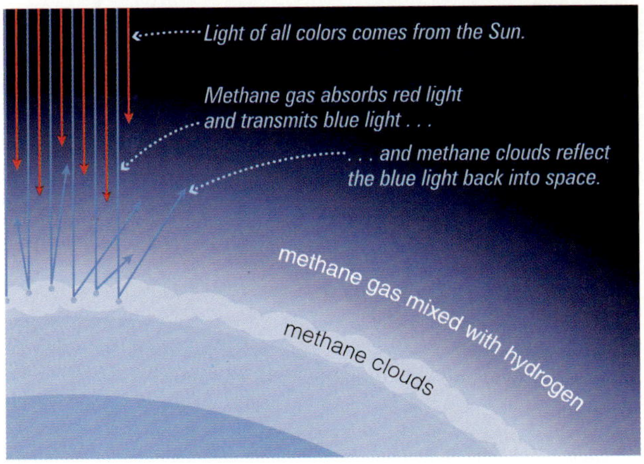

Figure 11.8 Neptune and Uranus look blue because methane gas absorbs red light but transmits blue light. Clouds of methane snowflakes reflect the transmitted blue light back to space.

of Uranus's "sideways" axis tilt, which leads to extreme seasons. With one hemisphere remaining sunlit for decades, gases have plenty of time to interact with solar ultraviolet light and to make the chemical ingredients of the haze. Continuous sunlight may also explain why Uranus has a surprisingly hot thermosphere that extends thousands of kilometers above its cloud tops. As we'll discuss later, Uranus's thermosphere has an important influence on its rings.

Global Winds and Storms on Jupiter Jupiter has dynamic weather, with strong winds and powerful storms. To understand Jupiter's weather, we must think about how its rapid rotation affects its global winds.

Jupiter has planetwide circulation cells similar to those on Earth (see Figure 10.16). As on Earth, solar heat causes equatorial air to expand and spill toward the poles, while cooler polar air flows toward the equator. Also as on Earth, the Coriolis effect splits the large equator-to-pole circulation cells into smaller cells. However, Jupiter's greater size and faster rotation make the Coriolis effect much stronger on Jupiter than on Earth. Instead of being split into just three smaller cells encircling each hemisphere, as on Earth, Jupiter's circulation cells split into many alternating bands of rising and falling air. These bands are visible as the "stripes" of alternating color in photographs of Jupiter.

The alternating colors of the stripes come from differences in the clouds that form in rising and falling air. As shown in Figure 11.9, the entire planet is blanketed with the reddish-brown, ammonium hydrosulfide clouds that make up Jupiter's middle cloud layer. In contrast, the upper layer of white ammonia clouds forms only in places where rising air carries ammonia gas to altitudes high enough and cold enough for the ammonia to condense. Thus, the white bands (sometimes called *zones*) represent regions of rising air with cold ammonia clouds, while the darker bands

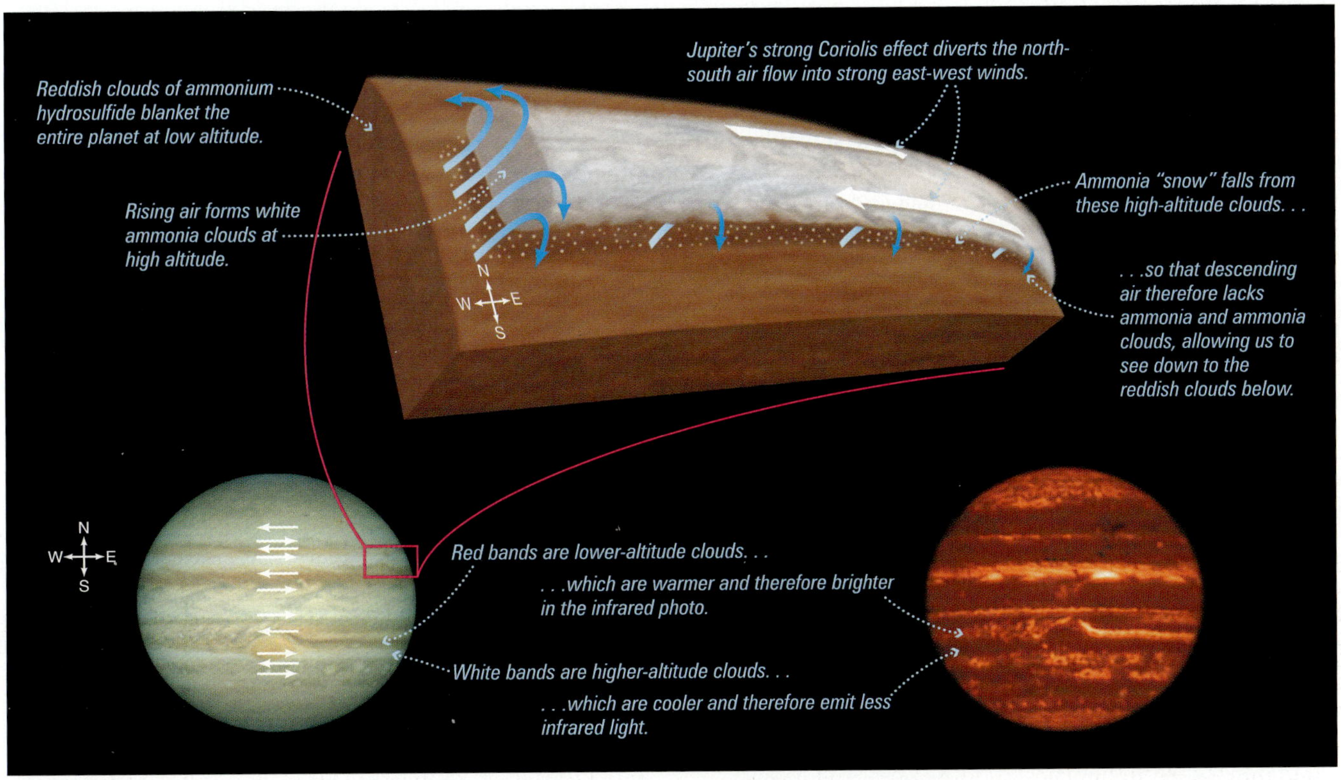

Figure 11.9 Interactive Figure Jupiter's bands of color represent alternating regions of rising and falling air: We see white ammonia clouds in regions of rising air, and see down to the underlying layer of reddish ammonium hydrosulfide clouds in regions of falling air. The white arrows on the visible-light photo indicate wind directions. The infrared photo at the right, taken nearly simultaneously with the visible-light photo, confirms that the red bands represent warmer, deeper clouds and the white bands represent cooler, higher-altitude clouds.

(sometimes called *belts*) represent regions of falling air in which we can see down to a lower, warmer cloud deck. This distinction between Jupiter's bright and dark bands is analogous to the difference on Earth between cloudy, rainy equatorial regions (regions of generally rising air) and the clear desert skies found at latitudes roughly 20°–30° north and south of the equator (regions of descending air). Notice how the infrared photo of Jupiter in Figure 11.9 confirms this idea: Warmer air emits more infrared light, so the warmer reddish bands appear brighter in the infrared photograph than the cooler white bands.

You might wonder why ammonia clouds can form only in the rising air and not in the falling air. The answer goes back to the clouds themselves. Jupiter's lower atmosphere has ammonia throughout, so there's always ammonia in the air rising upward. However, once the ammonia condenses to form clouds of ammonia ice, these ice flakes can fall back downward against the air currents. In essence, the ammonia rises up and forms clouds, then falls straight back down as ammonia "snow." Thus, the air that continues to rise above the clouds has very little ammonia left in it. This ammonia-depleted air then spills to the north and south and returns downward in the bands of falling air, explaining why these bands lack the ammonia needed to form clouds. It's analogous to the way that air descending over Earth's desert latitudes has been depleted in water by precipitation over the equator, which explains why desert regions are so dry.

The alternating bands of rising and falling air shape Jupiter's global wind patterns. As on Earth, the rising and falling air drives slow winds that are directed north or south, but Jupiter's strong Coriolis effect diverts these winds into fast east or west winds. The east-west winds have peak speeds above 400 km/hr, making hurricane winds on Earth seem mild by comparison. The winds are generally strongest at the equator and at the boundaries between bands of rising and falling air.

Jupiter's global wind patterns are sometimes interrupted by powerful storms, much as storms can interrupt Earth's global wind patterns. Of course, Jupiter's storms dwarf those that we see on Earth. Indeed, Jupiter's most famous feature—the **Great Red Spot**—is a giant storm more than twice as wide as the entire planet Earth. It is somewhat like a hurricane on Earth, except that its winds circulate around a high-pressure region rather than a low-pressure region (Figure 11.10). It is also extremely long-lived compared to storms on Earth: Astronomers have seen it throughout the three centuries during which telescopes have been powerful enough to detect it.

Other, smaller storms are constantly brewing in Jupiter's atmosphere. They are small only in comparison to the Great Red Spot. Brown ovals are low-pressure storms with their cloud tops deeper in Jupiter's atmosphere, and white ovals are high-pressure storms topped with ammonia clouds.

No one knows what drives Jupiter's storms, why Jupiter has only one Great Red Spot, or why the Great Red Spot has persisted so much longer than storms on Earth. Storms on Earth lose their strength when they pass over land, so perhaps Jupiter's biggest storms last for centuries simply because no solid surface is present to sap their energy. The bright colors of the Great Red Spot also pose a mystery: We might expect its high-altitude clouds to be white like the high-altitude ammonia clouds elsewhere, but instead, of course, they are red. The colors may be the result of chemicals formed by interactions between the storm's high-altitude gas and solar ultraviolet light, but no one really knows for sure.

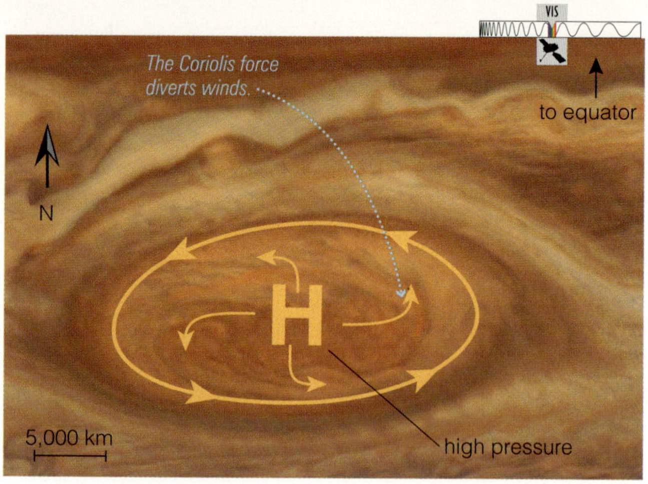

Figure 11.10 This photograph shows Jupiter's Great Red Spot, a huge high-pressure storm that is large enough to swallow two or three Earths. The overlaid diagram shows a weather map of the region.

Jupiter does not have seasons, because it has no appreciable axis tilt. In fact, Jupiter's polar temperatures are quite similar to its equatorial temperatures, presumably because heat from Jupiter's interior keeps the planet uniformly warm.

Global Winds and Storms on Saturn, Uranus, and Neptune The other jovian planets also have dramatic weather patterns (Figure 11.11). As on Jupiter, Saturn's rapid rotation creates alternating bands of rising and falling air, along with rapid east-west winds. In fact, Saturn's winds are even faster than Jupiter's—a surprise that scientists have yet to explain. We might expect seasons on Saturn, because it has an axis tilt similar to that of Earth. Some seasonal weather changes have been observed, but Saturn's internal heat keeps temperatures about the same year-round and planetwide.

Neptune's atmosphere is also banded, and we have seen a high-pressure storm, called the Great Dark Spot, similar to Jupiter's Great Red Spot. However, the Great Dark Spot did not last as long, having disappeared from view just 6 years after its discovery. Like Saturn, Neptune has an axis tilt similar to Earth's, but it has relatively little seasonal change because of its internal heat.

The greatest surprise in jovian weather comes from Uranus. When *Voyager 2* flew past Uranus in 1986, photographs revealed virtually no clouds and no banded structure like those found on the other jovian planets. Scientists attributed the lack of weather to the lower internal heat of Uranus. However, more recent observations from the Hubble Space Telescope and groundbased adaptive optics telescopes show storms raging in Uranus's atmosphere. The storms may be brewing because of the changing seasons:

a A close-up of Jupiter's atmosphere in its southern hemisphere, with the Great Red Spot visible near the top. The white oval is a low-pressure storm.

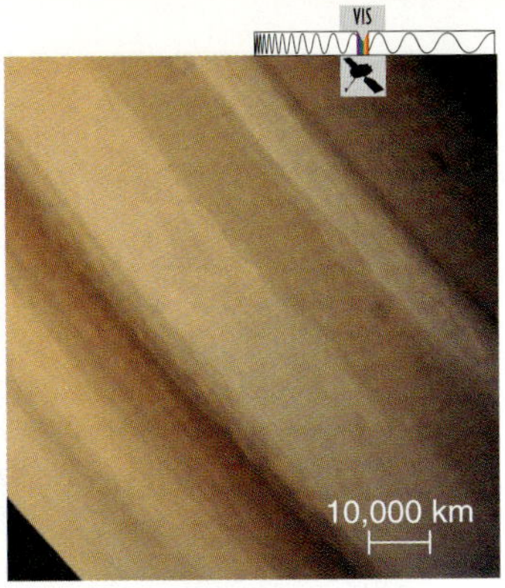

b A close-up of Saturn's atmosphere. Its banded appearance is very similar to that of Jupiter, but it has even faster winds.

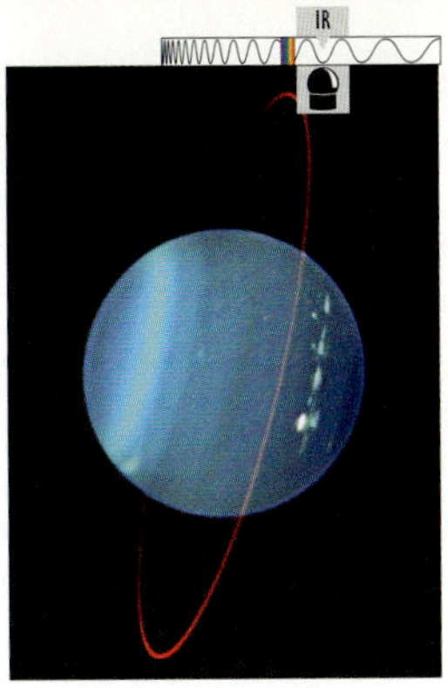

c This infrared photograph from the Keck telescope shows several storms (the bright blotches) brewing on Uranus. Uranus's ring shows up in red.

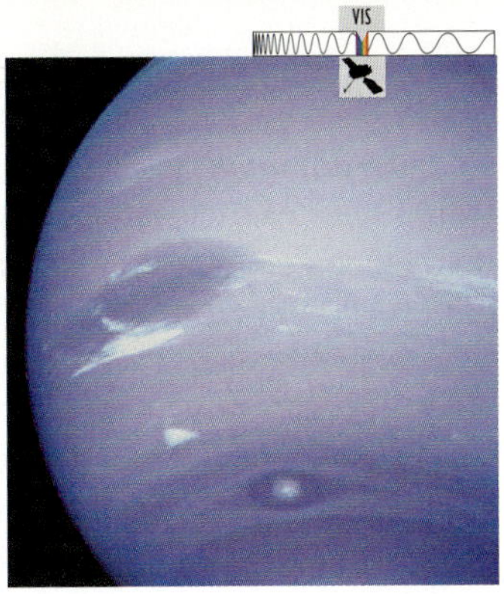

d Neptune's atmosphere has bands much like Jupiter and Saturn, and occasional strong storms. The large storm to the left of center is known as the Great Dark Spot.

Figure 11.11 Selected views of weather patterns on the four jovian planets.

Thanks to Uranus's extreme axis tilt and 84-year orbit of the Sun [Section 7.1], its northern hemisphere is just beginning to see sunlight for the first time in decades.

• Do jovian planets have magnetospheres like Earth's?

Recall that Earth has a global magnetic field generated by the movements of charged particles in our planet's metallic outer core (see Figure 9.6). The jovian planets also have global magnetic fields generated by motions of charged particles deep in their interiors. Just as with Earth, these magnetic fields create bubble-like *magnetospheres* that surround the planets and shield them from the solar wind.

Jupiter's Magnetosphere Jupiter's magnetic field is by far the strongest of the four jovian planets, and it is some 20,000 times stronger than Earth's magnetic field. As we discussed in Chapter 9, a planet can have a global magnetic field if it has: (1) an interior region of electrically conducting fluid; (2) convection in that layer of fluid; and (3) at least moderately rapid rotation. In Jupiter's case, the electrically conducting fluid region is its thick layer of metallic hydrogen (see Figure 11.4). The great extent of this region, combined with Jupiter's rapid rotation, explains Jupiter's strong magnetic field.

Jupiter's strong magnetic field also gives it an enormous magnetosphere (Figure 11.12). It begins to deflect the solar wind some 3 million kilometers (about 40 Jupiter radii)

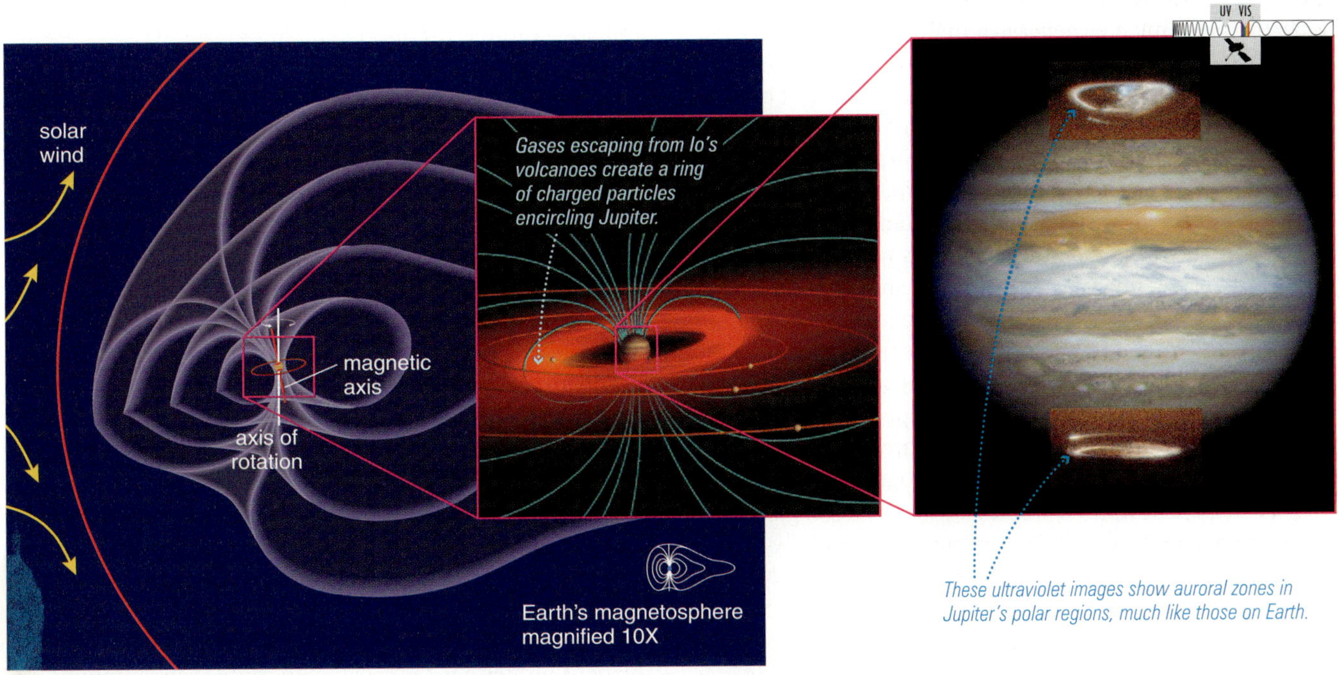

Figure 11.12 Jupiter's strong magnetic field gives it an enormous magnetosphere. Gases escaping from Io feed the donut-shaped Io torus, and particles entering Jupiter's atmosphere near its magnetic poles contribute to auroras on Jupiter. The image at the right is a composite of ultraviolet images of the polar regions overlaid on a visible image of the whole planet, all taken by the Hubble Space Telescope.

before the solar wind even reaches Jupiter. If our eyes could see this part of Jupiter's magnetosphere, it would be larger than the full moon in our sky.

Jupiter's magnetosphere traps far more charged particles than Earth's magnetosphere, largely because it has a source of particles that Earth's lacks. Nearly all the charged particles in Earth's magnetosphere come from the solar wind, but in Jupiter's case its volcanically active moon, Io, contributes many additional particles. These particles help create auroras on Jupiter. They also create belts of very intense radiation around Jupiter, which can cause damage to orbiting spacecraft.

Jupiter's magnetosphere, in turn, has important effects on Io and the other moons of Jupiter. The charged particles bombard the surfaces of Jupiter's icy moons, with each particle blasting away a few atoms or molecules. This process alters the surface materials and can even generate thin atmospheres, much like the thin atmospheres of Mercury and the Moon [Section 10.3]. On Io, bombardment by charged particles leads to the continuous escape of gases released by volcanic outgassing. The escaping gases (sulfur, oxygen, and a hint of sodium) are ionized and feed a donut-shaped charged particle belt, called the *Io torus*, that approximately traces Io's orbit (the red "donut" in Figure 11.12).

Comparing Jovian Magnetospheres The other jovian planets also have magnetic fields and magnetospheres, but theirs are much weaker than Jupiter's (although still much stronger than Earth's). The strength of each planet's magnetic field depends primarily on the size of the electrically conducting layer buried in its interior. Saturn's magnetic field is weaker than Jupiter's because it has a thinner layer of electrically conducting metallic hydrogen. Uranus and Neptune, smaller still, have no metallic hydrogen at all. Their relatively weak magnetic fields must be generated in their core "oceans" of hydrogen compounds, rock, and metal.

The size of a planet's magnetosphere depends not only on the magnetic field strength, but also on the pressure of the solar wind against it. The pressure of the solar wind is weaker at greater distances from the Sun, so the magnetospheric "bubbles" surrounding more distant planets are larger than they would be if these planets were closer to the Sun. Thus, Uranus and Neptune have moderate-size magnetospheres despite their weak magnetic fields (Figure 11.13). No other magnetosphere is as full of charged particles as Jupiter's, primarily because no other jovian planet has a satellite like Io. Because trapped particles generate auroras, Jupiter has the brightest auroras, while those of the more distant jovian planets are progressively weaker.

We still have much to learn about the jovian magnetic fields and magnetospheres. For example, we generally expect magnetic fields to be closely aligned with planetary rotation, because the magnetic fields are generated within the rotating interiors of planets. However, while this is the case for Jupiter and Saturn (whose magnetic fields are inclined to their rotation axes by 10° and 0°, respectively), it is not the case for Uranus and Neptune. *Voyager* observations showed that the magnetic field of Uranus is tipped by a whopping 60° relative to its rotation axis, and the magnetic field's center is also significantly offset from the planet's center. Neptune's magnetic field is inclined by 46° to its rotation axis. No one has yet explained these surprising observations.

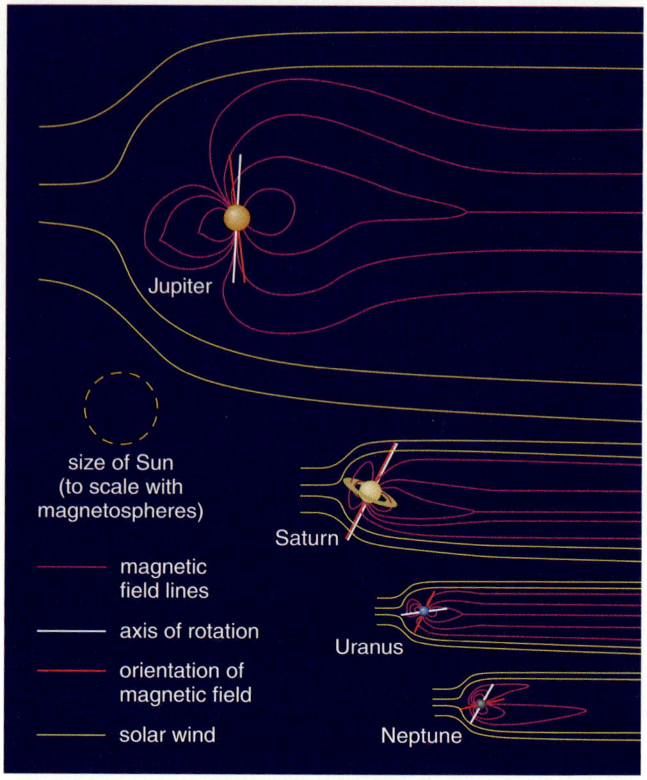

Figure 11.13 Comparison of jovian planet magnetospheres. The size of Jupiter's magnetosphere is particularly impressive in light of the greater pressure from the solar wind nearer the Sun. Note the significant tilts of the magnetic fields of Uranus and Neptune compared to their rotation axes. (Planets enlarged for clarity.)

11.2 A Wealth of Worlds: Satellites of Ice and Rock

The jovian planets are majestic and fascinating, but they are only the beginning of our exploration of jovian planet *systems*. Each of the four jovian systems includes numerous moons and a set of rings. The total mass of all the moons and rings put together is minuscule compared to any one of the jovian planets, but the remarkable diversity of these satellites makes up for their lack of size. In this section, we'll explore a few of the most interesting aspects of the jovian moons. We'll then discuss rings in the following section.

• What kinds of moons orbit the jovian planets?

We now know of more than 100 moons orbiting the jovian planets. Jupiter has the most, with more than 60 moons known to date. Because there are so many moons, it's helpful to organize them into three groups by size: small moons less than about 300 kilometers in diameter, medium-size moons ranging from about 300 to 1,500 kilometers in diameter, and large moons more than 1,500 kilometers in diameter. These categories are useful because size relates to geological activity. The large moons show evidence of active and ongoing geology, and most of the medium-size moons seem to have had geological activity in the past.

Figure 11.14 shows a montage of all the medium-size and large moons. These moons resemble the terrestrial planets in many ways. Each is spherical with a solid surface and its own unique geology. Some possess atmospheres, hot interiors, and even magnetic fields. The two largest—Jupiter's moon Ganymede and Saturn's moon Titan—are larger than the planet Mercury. Four others are larger than Pluto: Jupiter's moons Io, Europa, and Callisto and Neptune's moon Triton. However, they differ from terrestrial worlds in their compositions: Because they formed in the cold outer solar system, most of these worlds contain substantial amounts of ice in addition to metal and rock. As we will see, the fact that ice can undergo geological change at much lower temperatures than rock helps explain the surprising geological activity of many moderately sized moons.

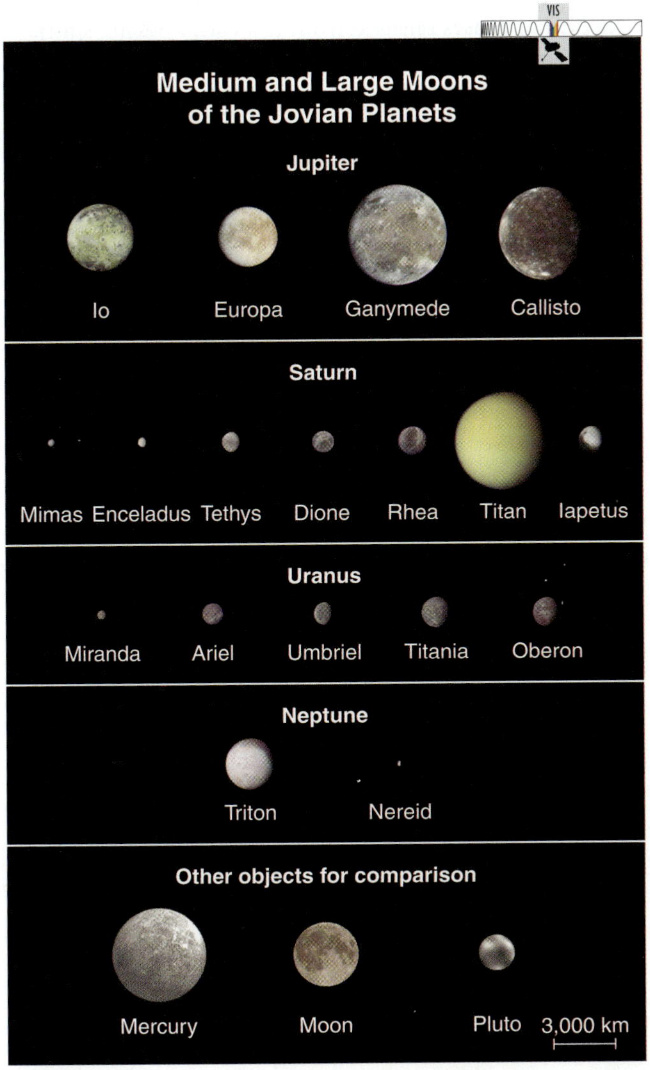

Figure 11.14 The medium-size and large moons of the jovian planets, with sizes (but not distances) shown to scale.

Most of the medium-size and large moons probably formed by accretion within the disks of gas surrounding individual jovian planets [Section 8.3]. That explains why their orbits are almost circular and lie close to the equatorial plane of their parent planet, and also why these moons orbit in the same direction in which their planet rotates. Nearly all these moons also share an uncanny trait: They always keep the same face turned toward their planet, just as our Moon always shows the same face to Earth. This synchronous rotation arose from the strong tidal forces [Section 4.5] exerted by the jovian planets, which caused each moon to end up with equal periods of rotation and orbit regardless of how fast the moon rotated when it formed.

The medium-size and large moons tend to be the most interesting, but small moons are far more numerous. Many of the small moons are probably captured asteroids or comets, and as a result they do not follow any particular orbital patterns. Dozens of the smallest moons have been discovered only within the past few years, and many more may yet be discovered. (See Table E.3 in Appendix E for a complete list of the known moons and their general properties.)

The small moons' shapes are irregular, like potatoes (Figure 11.15), because their gravities are too weak to force their rigid material into spheres. We have not studied these moons in depth, but we expect their small sizes to allow for little if any geological activity. For the most part, the small moons are just chunks of ice and rock held captive by the gravity of a massive jovian planet.

In case you are wondering, scientists originally named Jupiter's moons for the mythical lovers of the Roman god Jupiter. However, the planet Jupiter has so many moons that more recent discoveries are named for its namesake's more distant mythological relations. Saturn's moon Titan is named for the Greek gods called the Titans, who ruled before Zeus (Jupiter); other moons of Saturn are named for individual Titans. Moons of Uranus all take their names from characters in the works of William Shakespeare and Alexander Pope, while moons of Neptune are all characters related to the sea in Greek and Roman mythology.

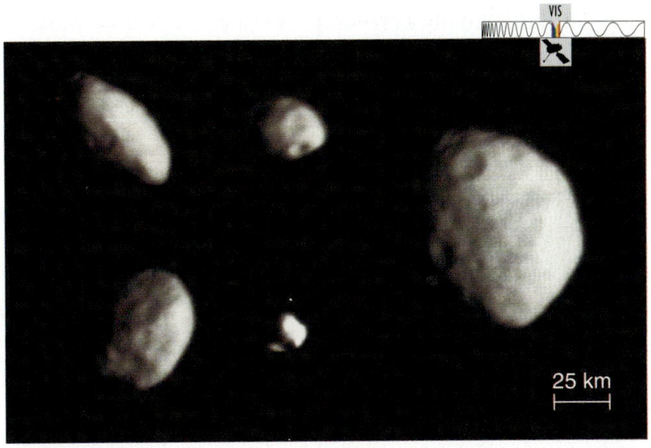

Figure 11.15 These photos from the *Voyager* spacecraft show five of Saturn's smaller moons. All are much smaller than the smallest moons shown in Figure 11.14. Their irregular shapes are due to their small size, which makes their gravities too weak to force them into spheres.

Figure 11.16 This set of photos, taken by the *Galileo* spacecraft orbiting Jupiter, shows global views of the four Galilean moons as we know them today. Sizes are shown to scale. (Io is about the size of Earth's Moon.)

• Why are Jupiter's Galilean moons so geologically active?

We are now ready to embark on a brief tour of the most interesting moons of the jovian planets. Our first stop is Jupiter. Jupiter's four largest moons, known as the *Galilean moons* because Galileo discovered them [Section 3.3], are all large enough that they would count as planets if they orbited the Sun (Figure 11.16).

Io: The Most Volcanically Active World in the Solar System For anyone who thinks of moons as barren, geologically dead places like our own Moon, Io shatters the stereotype. When the *Voyager* spacecraft first photographed Io up close more than two decades ago, we discovered a world with a surface so young that not a single impact crater has survived from past impacts. Moreover, *Voyager* cameras recorded volcanic eruptions in progress as the spacecraft passed by. We now know that Io is by far the most volcanically active world in our solar system. Large volcanoes pockmark its entire surface (Figure 11.17), and eruptions are so frequent that they have buried virtually every impact crater. Io probably also has tectonic activity, because tectonics and volcanism generally go hand in hand. However, debris from volcanic eruptions has probably buried most tectonic features.

Many of Io's frequent eruptions are surprisingly similar to those of volcanoes on Earth. Some of the taller volcanoes appear to be shield volcanoes built from flows of basalt lava, and the eruptions are accompanied by outgassing. The primary gases released from Io's volcanoes are sulfur and sulfur dioxide (SO_2). Some of this gas escapes into space, where it supplies ionized gas to the Io torus and Jupiter's magnetosphere, and some of it gives Io a very thin atmosphere. Much of the gas condenses and falls back to the surface, where the sulfur gives Io its distinctive red and orange colors and the sulfur dioxide makes a white frost. As hot lava flows across the surface (left photo in Figure 11.17), it can re-vaporize the sulfur dioxide surface ice in much the same way that lava flowing into the ocean vaporizes water on Earth. Io's low gravity and thin atmosphere allow tall plumes of this vaporized sulfur dioxide to rise upward to altitudes of hundreds of kilometers (right photo in Figure 11.17). Over a period of a few months, fallout from these tall volcanic plumes can blanket an area the size of Arizona (Figure 11.18).

Io's active volcanoes tell us that it must be quite hot inside. However, Io is only about the size of our geologically dead Moon, so it should have long ago lost any heat from its birth and it is too small for radioactivity to provide much ongoing heat. How, then, can Io be so hot inside? The only

Figure 11.17 Interactive Photo Io is the most volcanically active body in the solar system. Most of the black, brown, and red spots on the surface are recently active volcanic features. White and yellow areas are sulfur dioxide and sulfur deposits, respectively, from volcanic gases. (Photographs from the *Galileo* spacecraft; some colors slightly enhanced or altered.)

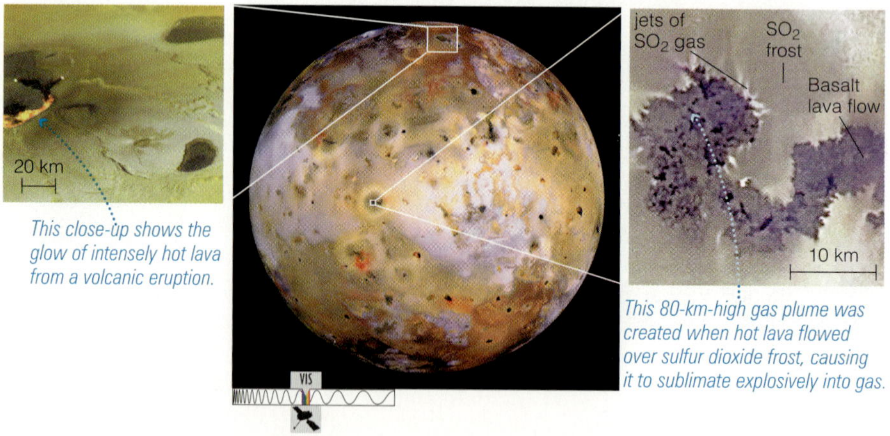

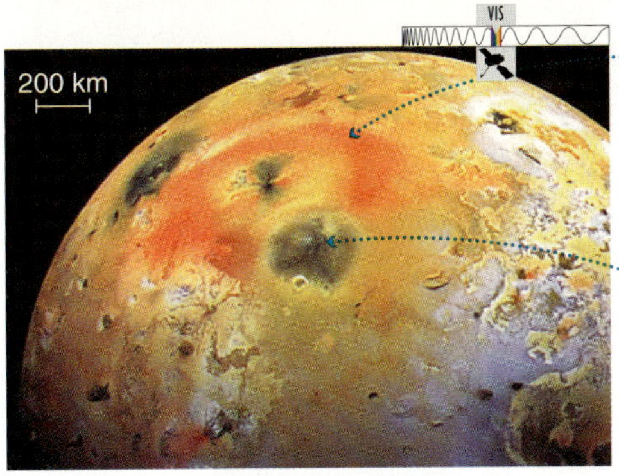

Figure 11.18 This enhanced-color photo shows how the fallout from volcanic eruptions can cover large areas of the surface in a matter of months or years.

possible answer is that some other ongoing process must be heating Io's interior. Scientists have identified this process and call it **tidal heating**, because it arises from effects of tidal forces exerted by Jupiter.

Just as Earth exerts a tidal force that causes the Moon to keep the same face toward us at all times [Section 4.5], a tidal force from Jupiter makes Io keep the same face toward Jupiter as it orbits. But Jupiter's mass make this tidal force far larger than that which Earth exerts on the Moon. Moreover, Io's orbit is slightly elliptical, so its orbital speed and distance from Jupiter vary. This variation means that the strength and direction of the tidal force change slightly as Io moves through each orbit, which in turn changes the size and orientation of Io's tidal bulges (Figure 11.19a). The result is that Io is continuously being flexed in different directions, which generates friction inside it. The flexing heats the interior in the same way that flexing warms Silly Putty. Tidal heating generates tremendous heat on Io—more than 200 times as much heat (per gram of mass) as the radioactive heat driving much of Earth's geology. This heat explains Io's volcanic activity. The energy for tidal heating ultimately comes from Jupiter's rotation, which is gradually slowing, although at a rate too small to be observed.

However, we are still left with a deeper question: Why is Io's orbit slightly elliptical, when almost all other large satellites have nearly circular orbits? The answer lies in an interesting dance executed by Io and its neighboring moons (Figure 11.19b). During the time Ganymede takes to complete one orbit of Jupiter (about 7 days), Europa completes exactly two orbits and Io completes exactly four orbits. The three moons therefore line up periodically, and the gravitational tugs they exert on one another add up over time.* Because the tugs are always in the same direction with each alignment, they tend to stretch out the orbits, making them slightly elliptical. The effect is much like that of pushing a child on a swing. If timed properly, a series of small pushes can add up to a *resonance* that causes the child to swing quite high. For the three moons, the resonance that makes their orbits elliptical comes from the small gravitational tugs that repeat at each alignment.

The phenomenon of orbital periods falling into a simple mathematical relationship is called an **orbital resonance**. As we will discuss later in this and the next two chapters, resonances are quite common in planetary systems. For example, they affect not only the Galilean moons but also planetary rings, the asteroid belt, and the Kuiper belt in our solar system, and we have even discovered resonances among the known systems of planets around other stars.

*You may wonder why periodic alignments occur. Like synchronous rotation, they are not a coincidence but rather a consequence of feedback from the tides that the moons raise on Jupiter. The periodic tugs the moons exert on one another actually work to sustain the recurring alignments.

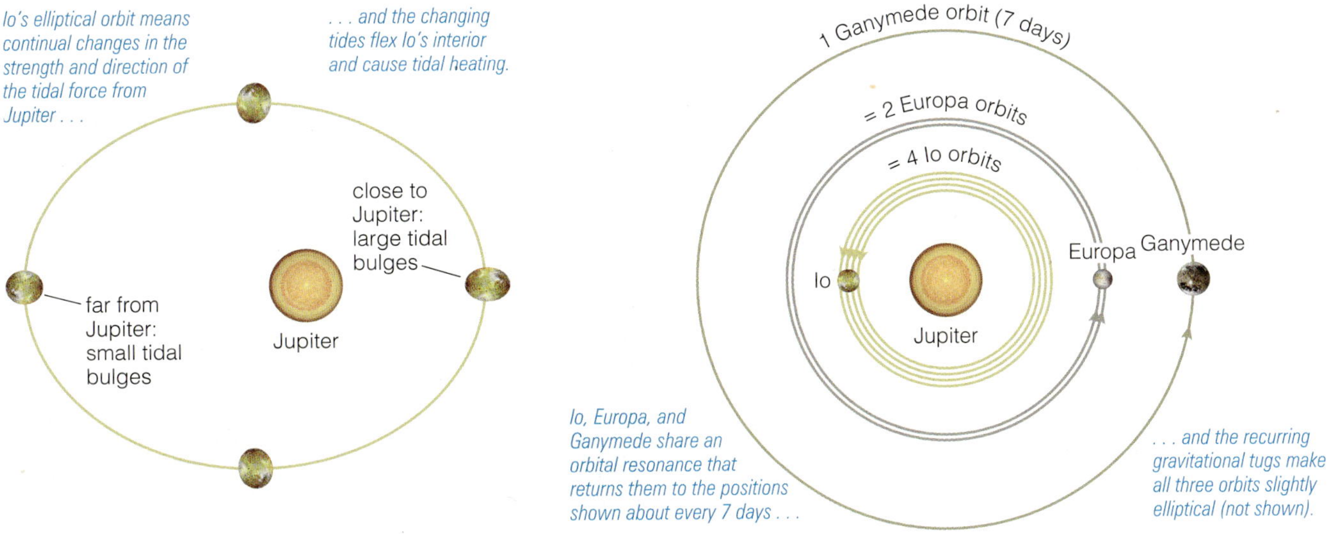

a Tidal heating arises because Io's elliptical orbit (exaggerated in this diagram) causes varying tides.

b Io's orbit is elliptical because of the orbital resonance it shares with Europa and Ganymede.

Figure 11.19 These diagrams explain the cause of tidal heating on Io. Tidal heating has a weaker effect on Europa and Ganymede, because they are farther from Jupiter and tidal forces weaken with distance.

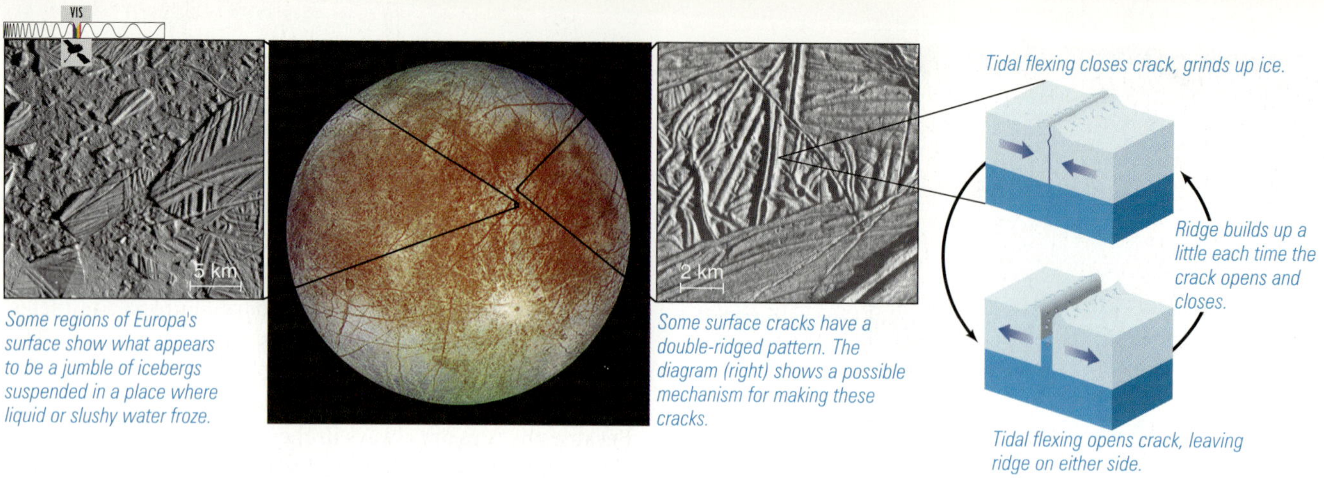

Figure 11.20 Europa's icy crust may hide a deep, liquid water ocean beneath its surface. These photos are from the *Galileo* spacecraft; colors are enhanced in the global view.

Europa: The Water World? Europa offers a stark contrast to Io. Instead of active volcanoes dotting its surface, Europa is covered by water ice (Figure 11.20). Nevertheless, its fractured, frozen surface must hide an interior made hot by the same type of tidal heating that powers Io's volcanoes—but tidal heating is weaker on Europa because it lies farther from Jupiter. Europa has only a handful of impact craters, suggesting a surface no more than a few tens of millions of years old. Clearly, active geology has covered up older craters on Europa. But what could be doing the covering up?

Scientists suspect that water (H_2O) is the agent of change on Europa's surface: either liquid water rising up from beneath the icy crust, or interior ice that is just warm enough to undergo convection that allows some of it to rise up and flow across the surface. Close-up photos of the surface support this idea. For example, some photos show what appear to be jumbled icebergs suspended in a region where liquid or slushy water has frozen (left side of Figure 11.20). Other photos show double-ridged cracks on the surface (right side of Figure 11.20) that may be created when tidal flexing allows water to well up and build up ridges.

Models of Europa's interior give us further insight into its geology. Detailed measurements of the strength of gravity over different parts of Europa from the *Galileo* spacecraft suggest that Europa has a metallic core and a rocky mantle. Above the rocky mantle, Europa has enough water to make a layer of ice about 100 kilometers thick (Figure 11.21). The upper 5 to 25 kilometers should be solidly frozen, but beneath this icy crust, tidal heating provides enough warmth to create a layer of either liquid water or relatively warm, convecting ice. If it is liquid water, then Europa's ocean may contain more than twice as much liquid water as all of Earth's oceans combined.

We do not yet have enough data to be certain that Europa's subsurface layer consists of liquid water rather than convecting ice. However, magnetic field data from the *Galileo* spacecraft strongly argue for liquid water. Europa is one of only a few moons in the solar system that have a magnetic field at all. Moreover, Europa's magnetic field changes as Jupiter rotates, suggesting that it is created (or "induced") in response to the rotation of Jupiter's strong magnetic field. In order to respond in this way to Jupiter's magnetic field, Europa would have to have a *liquid* layer of electrically conducting material. Thus, a salty ocean would fit the bill, but convecting ice would not. The data further suggest that Europa's liquid ocean must be global in extent, and that it is about as salty as Earth's oceans. Additional support for this idea comes from another of *Galileo*'s instruments, which found evidence for salty compounds on Europa's surface—possible seepage from a briny deep.

Taken together, the photographic and magnetic-field evidence makes a strong case for a deep ocean of liquid water on Europa. Perhaps as on Earth's seafloor, lava erupts from vents on Europa's seafloor, sometimes violently enough to jumble the icy crust above (Figure 11.22). And, knowing that primitive life thrives near seafloor vents on Earth, we can wonder whether Europa might also be a home to life—a possibility we will explore further in Chapter 24.

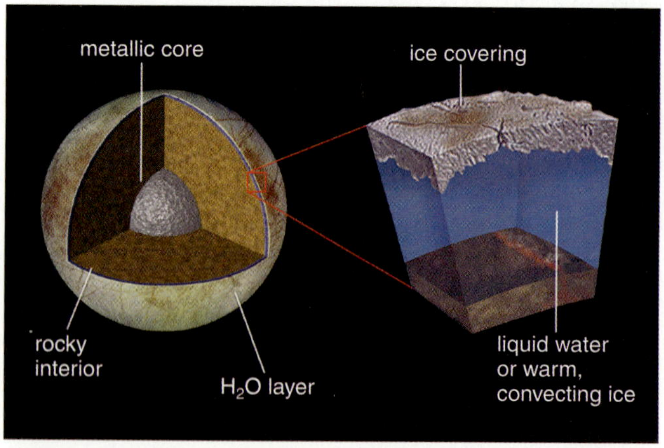

Figure 11.21 This diagram shows one model of Europa's interior structure. There is little doubt that the H_2O layer is real, but we are not yet certain that the material beneath the icy crust is actually liquid water rather than just relatively warm, convecting ice.

Figure 11.22 Tidal heating may give Europa a subsurface ocean beneath its icy crust. This artist's conception imagines a region where an undersea volcano has disrupted the crust.

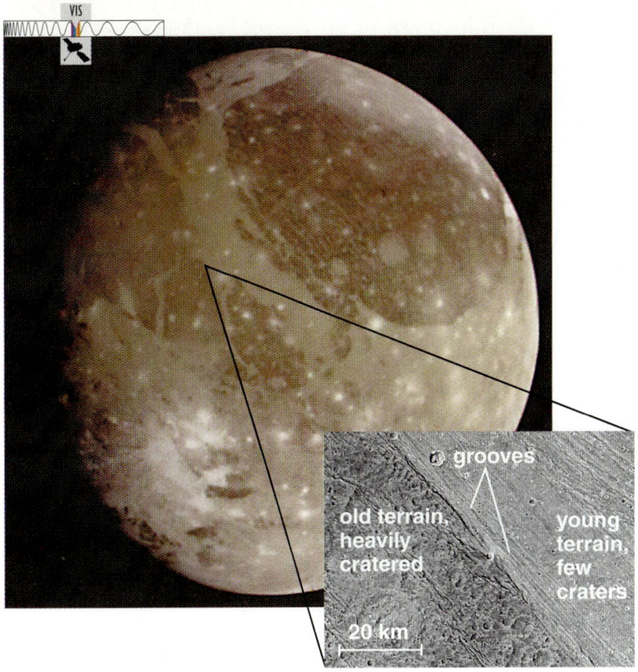

Figure 11.23 Ganymede, the largest moon in the solar system, has both old and young regions on its surface of water ice. The dark regions are heavily cratered and must be billions of years old, while the light regions are younger landscapes where eruptions of water have presumably erased ancient craters; the long grooves in the light regions were probably formed by water erupting along surface cracks. Notice that the boundary between the two types of terrain can be quite sharp.

Ganymede: King of the Moons Ganymede is the largest moon in the solar system, and its surface tells of a complex geological history. Like Europa, Ganymede has a surface of water ice. However, while Europa's surface appears relatively young everywhere, Ganymede's surface appears to have a dual personality (Figure 11.23). Some regions are dark and densely cratered, suggesting that they look much the same today as they did billions of years ago. Other regions are light-colored with very few craters, suggesting that liquid water has recently erupted and refrozen. In some cases, fairly sharp boundaries separate the two types of terrain.

The young terrain argues for occasional upwelling of liquid water or icy slush to the surface. This material would cover craters before refreezing, explaining why there are so few craters in this terrain. The long grooves are probably made by the eruption of water or slush along a crack in the surface. As the water in the crack freezes, it expands and pushes outward, creating the groove.

If liquid water occasionally wells up to the surface, could it mean that Ganymede has a subsurface ocean like that thought to exist on Europa? The case for an ocean on Ganymede is less strong than the case for an ocean on Europa but is also bolstered by magnetic-field measurements. Like Europa, Ganymede has a magnetic field that varies with Jupiter's rotation, suggesting the presence of a salty ocean beneath the surface.

One difficulty with the idea of an ocean on Ganymede is figuring out where the heat to melt subsurface ice might come from. Because Ganymede is farther from Jupiter than Europa or Io, its tidal heating is weaker and could not by itself supply enough heat to melt ice today. However, Ganymede's larger size means it should retain more heat from radioactive decay. Perhaps tidal heating and radioactive decay together provide enough heat to make a liquid layer beneath the icy surface. If so, the lesser heating on Ganymede means that its ocean probably lies at least 150 kilometers beneath the surface—much deeper than Europa's potential ocean.

Callisto: Last of the Galilean Moons The outermost Galilean moon, Callisto, looks most like what scientists originally expected for an outer solar system satellite: a heavily cratered iceball (Figure 11.24). The bright, circular patches on its surface are impact craters. They are bright presumably because large impacts blast out "clean" ice from deep underground, and this ice reflects more light than the dirtier ice that has been on the surface longer.

Craters make sense on an old surface like Callisto's, but other features are more difficult to interpret. For example, close-up photos show a dark, powdery substance concentrated in low-lying areas, leaving ridges and crests bright white (small photo in Figure 11.24). The dark powder may be debris left behind when ice sublimates into gas from Callisto's surface, much as dark material is left behind on a comet's nucleus when ice sublimates away.

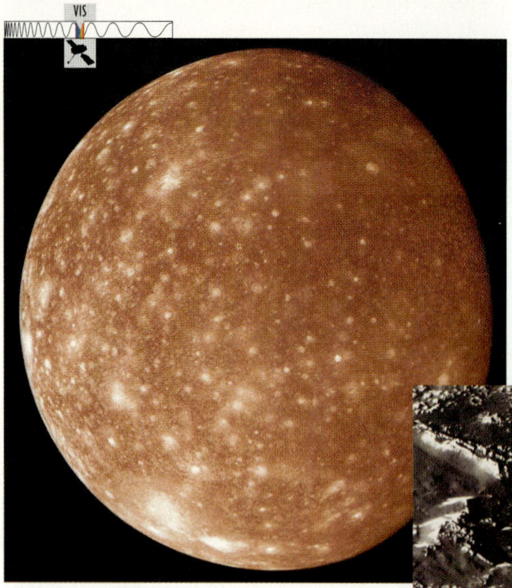

Callisto is heavily cratered, indicating an old surface that nonetheless may hide a deeply buried ocean.

Close-up photo shows a dark powder overlaying the low areas of the surface.

Figure 11.24 Callisto, the outermost of the four Galilean moons, has a heavily cratered, icy surface.

Despite its relatively large size (the third-largest moon in the solar system), Callisto lacks volcanic and tectonic features. This tells us that it lacks any significant source of internal heat. In fact, gravity measurements by the *Galileo* spacecraft showed that Callisto never underwent differentiation: Dense rock and lighter ice are thoroughly mixed throughout most of its interior, which means that its interior has never warmed significantly. The lack of heat is not surprising: Callisto has no tidal heating because it does not participate in the orbital resonances that affect the other three Galilean moons, and we do not expect its icy interior to contain enough radioactive material to supply much heat through radioactive decay.

Nevertheless, it is possible that Callisto, too, may hide a subsurface ocean. Like Europa and Ganymede, Callisto has a magnetic field that suggests the presence of a salty interior ocean. Perhaps Callisto's surface provides just enough insulation for heat from radioactive decay to melt water beneath the thick, icy crust. If so, we face the intriguing possibility that there could be three "water worlds" orbiting Jupiter—with far more total ocean than we find here on Earth.

• What is special about Titan and other major moons of the outer solar system?

Leaving Jupiter behind, we will continue our tour of moons by discussing the major moons of Saturn, Uranus, and Neptune. We'll go in planetary order, first focusing special attention on Saturn's enigmatic Titan, the second-largest moon in the solar system after Ganymede.

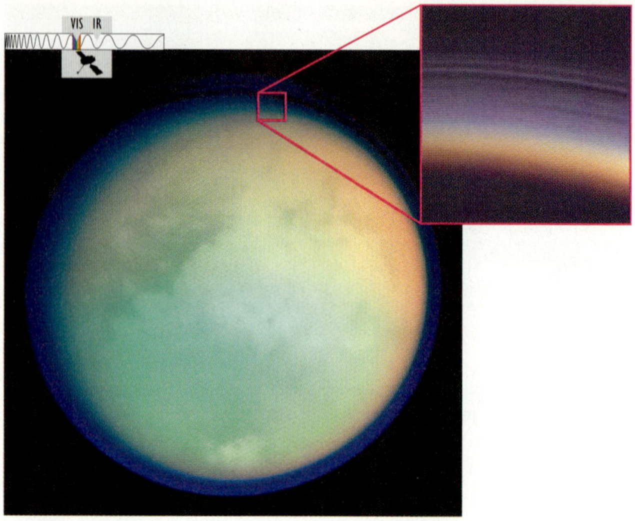

Figure 11.25 Titan, as photographed by the *Cassini* spacecraft, is enshrouded by a methane gas and tan haze. The inset shows the haze layers extending a hundred kilometers above a portion of the moon. *Cassini* was outfitted with filters designed to peer through the atmosphere at the specific near-infrared wavelengths of light that are least affected by the atmosphere or haze.

Saturn's Moon Titan: Piercing the Veil Titan is unique among the moons of our solar system in having a thick atmosphere—so thick that it hides the surface from view, except at a few specific wavelengths of light (Figure 11.25). Titan's reddish color comes from chemicals in its atmosphere, much like those that make smog over cities on Earth. The atmosphere is about 90% molecular nitrogen (N_2), not that different from the 77% nitrogen content of Earth's atmosphere. However, on Earth the rest of the atmosphere is mostly oxygen, while the rest of Titan's atmosphere consists of argon, methane (CH_4), ethane (C_2H_6), and other hydrogen compounds.

Titan's atmospheric composition is different from that of any other world in our solar system, which is not too surprising given our general understanding of atmospheric source and loss processes [Section 10.2]. Titan's icy composition supplies methane and ammonia gas through evaporation and sublimation. Solar ultraviolet light breaks down some of those molecules, releasing hydrogen atoms and leaving highly reactive compounds containing carbon and nitrogen. The hydrogen atoms can leave Titan forever by thermal escape, while the remaining molecular fragments react to make the other ingredients of Titan's atmosphere. For example, the abundant molecular nitrogen is made after ultraviolet light breaks down ammonia (NH_3) molecules, and ethane is made from methane.

The methane and ethane in Titan's atmosphere are both greenhouse gases, and therefore give Titan an appreciable greenhouse effect [Section 10.1] that makes it warmer than it would be otherwise. Still, because of its great distance from the Sun, its surface temperature is a frigid 93 K ($-180°C$). The surface pressure on Titan is about 1.5 times the sea level pressure on Earth, which would be fairly comfortable if not for the lack of oxygen and the cold temperatures.

A moon with a thick atmosphere would be intriguing enough, but we have at least two other reasons for special interest in Titan. First, its complex atmospheric chemistry probably produces numerous organic chemicals—the chemicals that are the basis of life. Few scientists think there could be actual life on Titan, due to the cold temperatures, but many hope that further study of Titan will teach us about the chemistry that may have occurred on Earth before life actually arose. Second, although it is far too cold for liquid water to exist on Titan, conditions are right for methane or ethane rain, perhaps making rivers feeding into lakes or oceans.

NASA and the European Space Agency (ESA) found Titan so intriguing that they combined forces to explore it. NASA built the *Cassini* "mother ship" that carried an ESA-built probe called *Huygens* designed to parachute to the surface for a soft landing. The probe landed successfully on January 14, 2005 (Figure 11.26). During its descent it photographed river valleys merging together, flowing down to what looks like a shoreline. On impact, instruments on the probe discovered that the landing was not so hard: The surface has a hard crust but is a bit squishy below, perhaps like sand with liquid mixed in. The view from the surface shows "ice boulders" rounded by erosion. All these results support the idea of a wet climate—but wet with liquid methane rather than liquid water.

The analysis of the data from Titan is still in its early stages, but we have already learned a great deal. The brighter regions in the photographs are icy hills, perhaps made by ice volcanoes. The dark valleys are probably created when methane rain carries down "smog particles" that concentrate on river bottoms. The vast plains into which the valleys appear to empty are indeed the low regions, but they do not appear to be liquid. They too are probably made dark from smog particles carried there by the rivers. All in all, conditions on Titan appear to be analogous to the desert southwest of the United States, where infrequent rainfall carves valleys and creates vast dry lakes called "playas" where the water evaporates or soaks into the ground.

Perhaps the most astonishing result from the *Huygens* mission is how familiar the landscape looks in this alien environment with these unfamiliar materials. Instead of liquid water, Titan has liquid methane. Instead of rock, Titan has ice. Instead of molten lava, Titan has a slush of water ice mixed with ammonia. Instead of surface dirt, Titan's surface has smog particles that rain out of the sky. Evidently, the similarities between the physical processes that occur on Titan and Earth are far more important in shaping the landscapes than the fact that the two worlds have very different compositions and temperatures. The *Cassini* mission should continue to operate in Saturn orbit until at least 2008, and scientists hope that additional radar mapping of Titan will teach us more about the surface, especially in areas that may be very different from the *Huygens* landing site.

THINK ABOUT IT

What other geological features might you expect on Titan? Remember than the *Huygens* probe saw just one location, and its view might be as limited as a single landing spot on Earth.

Saturn's Medium-Size Moons The *Cassini* mission is also teaching us more about Saturn's other moons, especially its six medium-size moons (Figure 11.27). The photographs suggest that these moons have had a complex history.

Only Mimas, the smallest of these six moons, shows little evidence of past volcanism or tectonics. It is essentially a heavily cratered iceball. One huge crater is sometimes called "Darth Crater," because of Mimas's resemblance to

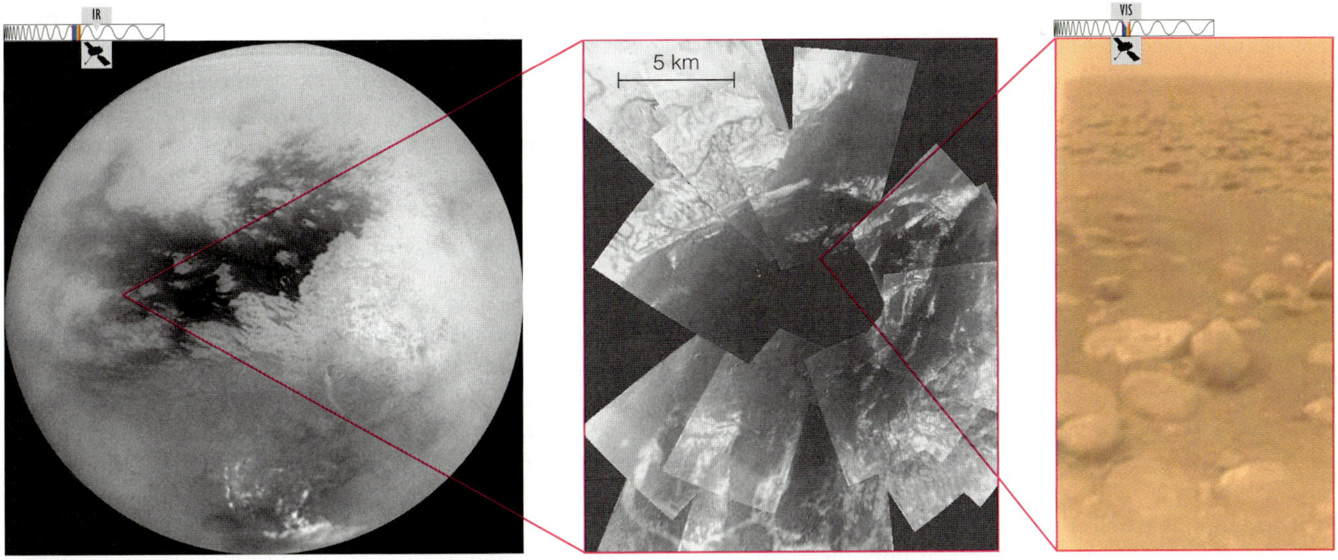

Figure 11.26 This sequence zooms in on the *Huygens* landing site on Titan. Left: a global view taken by the Cassini orbiter. Center: an aerial mosaic of images taken by the probe during descent. Right: a surface view taken by the probe after landing; the "rocks," which are 10 to 20 centimeters across, are presumably made of ices. Keep in mind that you are looking at the surface of a world more than a *billion* kilometers away.

Figure 11.27 Portraits taken by the *Cassini* spacecraft of Saturn's six medium-size moons (not to scale). All but Mimas show evidence of past volcanism and/or tectonics, and the bluish grooves in the enhanced color image of Enceladus suggests current geological activity.

the Death Star in the *Star Wars* movies. (The crater's official name is Herschel.) The impact that created this crater probably came close to breaking Mimas apart.

Saturn's other five medium-size moons also have heavily cratered surfaces, confirming that they lack global geological activity today. However, we find abundant evidence of volcanism and/or tectonics that must have occurred more recently. For example, the light regions visible in the photos may be places where icy lava once flowed. Enceladus even has strange grooves near its south pole, similar to some seen on Ganymede. These grooves are measurably warmer than the surrounding terrain, and they appear to be venting huge clouds of water vapor and ice crystals. Although tidal heating and radioactive heating are both present, neither is considered sufficient to explain this level of geological activity on Enceladus. Iapetus is just as bizarre, with some regions distinctly bright and others distinctly dark. The dark regions appear to be covered by a thin veneer of dark material, but we don't know what it is or how it got there. *Cassini* also imaged an astonishing ridge that spans more than a quarter of Iapetus' circumference, curiously aligned along the equator. No one knows its origin, but some kind of tectonic activity is the most likely explanation.

These moons are all considerably smaller than the Galilean moons of Jupiter; Enceladus, for example, is barely 500 kilometers across—small enough to fit inside the borders of Colorado. How, then, could they ever have had such active geology? Tidal heating sometimes plays a role, but the answer lies primarily with the icy compositions of these moons—a composition that makes them fundamentally different from terrestrial worlds despite the outward similarities of their solid surfaces. We'll investigate this issue in more depth after we complete our tour of the major jovian moons.

The Medium-Size Moons of Uranus Uranus does not have any moons that fall into our large category, but it has five medium-size moons: in order of orbital distance, their names are Miranda, Ariel, Umbriel, Titania, and Oberon (see Figure 11.14). Like other jovian moons, these moons are made largely of ice. Because of Uranus's great distance from the Sun, this ice includes a great deal of ammonia and methane as well as water.

Our only close look at these moons came during the *Voyager* flyby in 1986, but we learned enough to leave us with many unanswered questions. For example, Ariel and Umbriel are virtual twins in size, yet Ariel shows evidence of volcanism and tectonics, while the heavily cratered surface of Umbriel suggests a lack of geological activity. Titania and Oberon also are twins in size, but Titania appears to have had much more geological activity than Oberon. No one knows why these two pairs of similar-size moons should vary so greatly in geological activity.

Miranda, the smallest of Uranus's medium-size moons, is the most surprising (Figure 11.28). We might expect Miranda to be a cratered iceball like Saturn's similar-size moon Mimas. Instead, *Voyager* images of Miranda show tremendous tectonic features and relatively few craters.

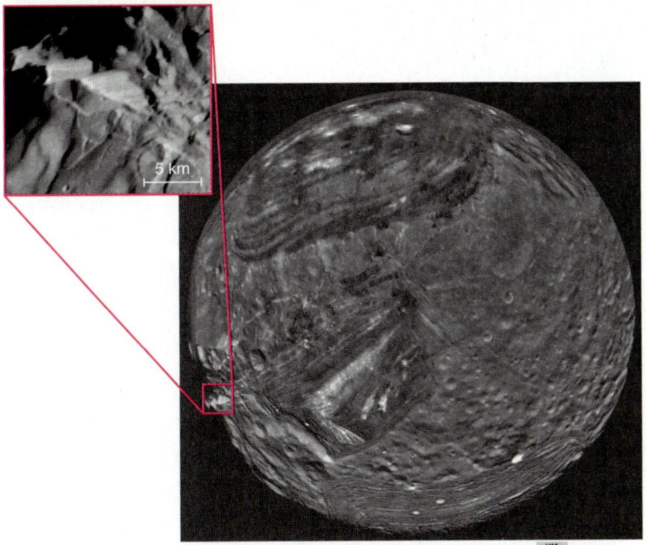

Figure 11.28 The surface of Miranda shows astonishing tectonic activity despite its small size. The cliff walls seen in the inset are higher than those of the Grand Canyon on Earth.

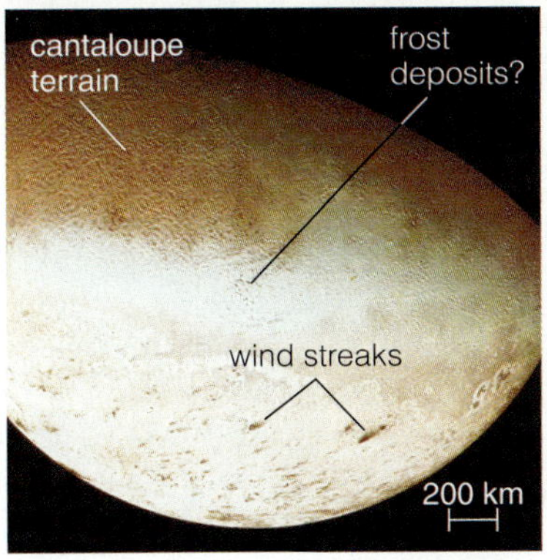

Triton's southern hemisphere as seen by *Voyager 2*.

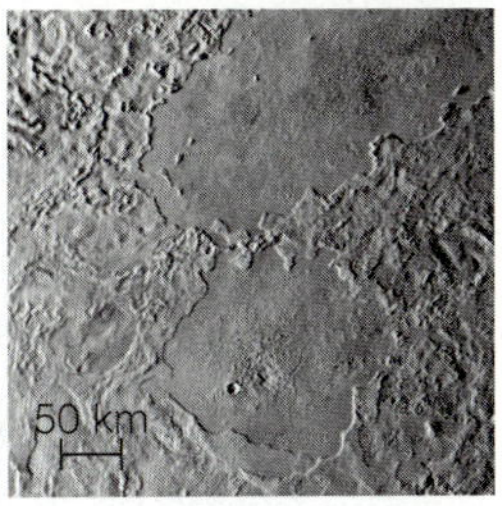

This close-up shows lava-filled impact basins similar to the lunar maria, but the lava was water or slush rather than molten rock.

Figure 11.29 Neptune's moon Triton shows evidence of a surprising level of past geological activity.

Why should Miranda be so much more geologically active than Mimas? Our best guess is that Miranda had an episode of tidal heating billions of years ago, perhaps during a temporary orbital resonance with another moon of Uranus. Mimas apparently never had such an episode of tidal heating and thus lacks similar features.

Neptune's Triton: A Captured Moon The last stop on our tour is Neptune, so distant that only two of its moons (Triton and Nereid) were known to exist before the *Voyager 2* spacecraft visited it. Nereid is Neptune's only medium-size moon, and Triton is its only large moon. Triton holds the distinction of being the coldest world in the solar system. It is even colder than Pluto because it reflects more of the weak sunlight that reaches the outskirts of the solar system.

Triton may appear to be a typical moon, but it is not. It orbits Neptune "backward" (opposite to Neptune's rotation) and at a high inclination to Neptune's equator. These are telltale signs of a moon that was captured rather than having formed in the disk of gas around its planet. However, rather than being small and potato-shaped like most captured moons, Triton is large, spherical, and icy. In fact, it is larger than Pluto. As such, Triton presents many challenges to our understanding of satellite captures. Nonetheless, Triton almost certainly once orbited the Sun rather than Neptune. Could other, similar objects still be orbiting the Sun? As we'll discuss in the next chapter, Pluto and other large objects of the Kuiper belt are strikingly similar to Triton in character, although they orbit the Sun rather than a planet.

Triton's geology is just as surprising as its origin. Although it has numerous impact craters, its surface has clearly been reshaped by geological activity (Figure 11.29). Some regions show evidence of past volcanism, perhaps an icy equivalent to the volcanism that shaped the lunar maria. Other regions show wrinkly ridges (nicknamed "cantaloupe terrain") that appear tectonic in nature; apparently, blobs of ice of different density have risen and fallen, contorting the crust. Triton even has a very thin atmosphere that has left some wind streaks on its surface.

What could have generated the heat needed to drive Triton's geological activity? The best guess is tidal heating. Triton probably had a very elliptical orbit and a more rapid rotation when Neptune first captured it. Tidal forces would have circularized its orbit, slowed it to synchronous rotation, and perhaps heated its interior enough to cause its geological activity. This idea would also explain why the geological activity has subsided with time.

• Why are small icy moons more geologically active than small rocky planets?

If you think about the role of planetary size in the geology of the terrestrial worlds, you might expect all of the jovian moons to be cold and geologically dead. After all, only two of the moons are even as large as Mercury, and Mercury has been geologically dead for more than 3 billion years. However, our tour of the moons has turned up numerous cases of spectacular past or present geological activity. In a few cases, we've been able to explain at least part of this activity as a result of heat generated by tidal heating. But we've found evidence of past activity even on moons that lack tidal heating. How is this possible?

We can trace the answer to a crucial difference between the jovian moons and the terrestrial worlds: composition. The jovian moons, by virtue of their formation in the cold outer solar system, contain large proportions of ices. Because ices can melt or deform at much lower temperatures than rock, icy worlds can experience geological activity at lower temperatures than rocky worlds. Indeed, most of the volcanism that has occurred in the outer solar system (except on Io) probably did not produce any hot lava at all. Instead, it produced icy lava that was essentially liquid water, perhaps mixed with methane and ammonia.

Figure 11.30 Jovian moons can be much more geologically active than terrestrial worlds of similar size due to their icy compositions and a heating source (tidal heating) that is not important on the terrestrial worlds.

Thus, the major lesson of our study of jovian moons is that "ice geology" is possible at much lower temperatures than "rock geology." This fact, combined with the occasional extra heating source of tidal heating, explains how the jovian moons have had such interesting geological histories despite their small sizes. In essence, the same physical properties that allowed hydrogen compounds to condense as ices in the outer solar system also allowed the worlds made from these ices to stay geologically active for long periods of time. Figure 11.30 summarizes the differences between the geology of jovian moons and that of the terrestrial worlds.

11.3 Jovian Planet Rings

We have completed our comparative study of the jovian planets and their major moons, but we have one more topic left to cover: their amazing rings. Saturn's rings have dazzled and puzzled astronomers since Galileo first saw them through his small telescope and suggested that they resembled "ears" on Saturn.

For a long time, Saturn's rings were thought to be unique in the solar system. We now know that all four jovian planets have rings. Let's begin by exploring the rings of Saturn, which we can then use as the standard for comparison when we examine the other jovian ring systems.

• What are Saturn's rings like?

You can see Saturn's rings through a backyard telescope, but learning their nature requires higher resolution (Figure 11.31). Earth-based views make the rings appear to be continuous, concentric sheets of material separated by a large gap (called the *Cassini division*). Spacecraft images reveal these "sheets" to be made of many individual rings, each separated from the next by a narrow gap. But even these appearances are somewhat deceiving. If we could wander into Saturn's rings, we'd find that they are made of countless icy particles ranging in size from dust grains to large boulders. All are far too small to be photographed even from spacecraft passing nearby.

Ring Particle Characteristics Spectroscopy reveals that Saturn's ring particles are made of relatively reflective water ice. The rings look bright where they contain enough particles to intercept sunlight and scatter it back toward us. We see gaps in places where there are few particles to reflect sunlight.

Each individual ring particle orbits Saturn independently in accord with Kepler's laws, so the rings are much like myriad tiny moons. The individual ring particles are so close together that they collide frequently. In the densest parts of the rings, each particle collides with another every few hours. However, the collisions are fairly gentle: Despite the high orbital speeds of the ring particles, nearby ring

a This Earth-based telescopic view of Saturn makes the rings look like large, concentric sheets. The dark gap within the rings is called the *Cassini division*.

b This image of Saturn's rings from the *Cassini* spacecraft reveals many individual rings separated by narrow gaps.

c Artist's conception of particles in a ring system. All the particles are moving slowly relative to one another and occasionally collide.

Figure 11.31 Interactive Photo Zooming in on Saturn's rings.

particles are moving at nearly the same speed and touch only gently when they collide.

THINK ABOUT IT

Which ring particles travel faster: those closer to Saturn or those farther away? Explain why. (*Hint:* Review Kepler's third law.)

The frequent collisions explain why Saturn's rings are perhaps the thinnest known astronomical structure. They span more than 270,000 kilometers in diameter but are only a few tens of *meters* thick. Indeed, the rings are so thin that they disappear from view when we see Saturn edge-on, as we do around the equinoxes of its 29.5-year orbit of the Sun.

To understand how collisions keep the rings thin, imagine what would happen to a ring particle on an orbit slightly inclined to the central ring plane. The particle would collide with other particles every time its orbit intersected the ring plane, and its orbital tilt would be reduced with every collision. Before long, these collisions would force the particle to conform to the orbital pattern of the other particles. Thus, any particle that moves away from the narrow ring plane is soon brought back within it. A similar idea explains why ring particles have almost perfectly circular orbits: Any ring particle with an elliptical orbit would quickly suffer enough collisions to force it into an orbit matching its neighbors.

Rings and Gaps Close-up photographs show an astonishing number of rings, gaps, ripples, and other features—the total number of features may be as high as 100,000. Scientists are still struggling to explain all the features, but some general ideas are now clear.

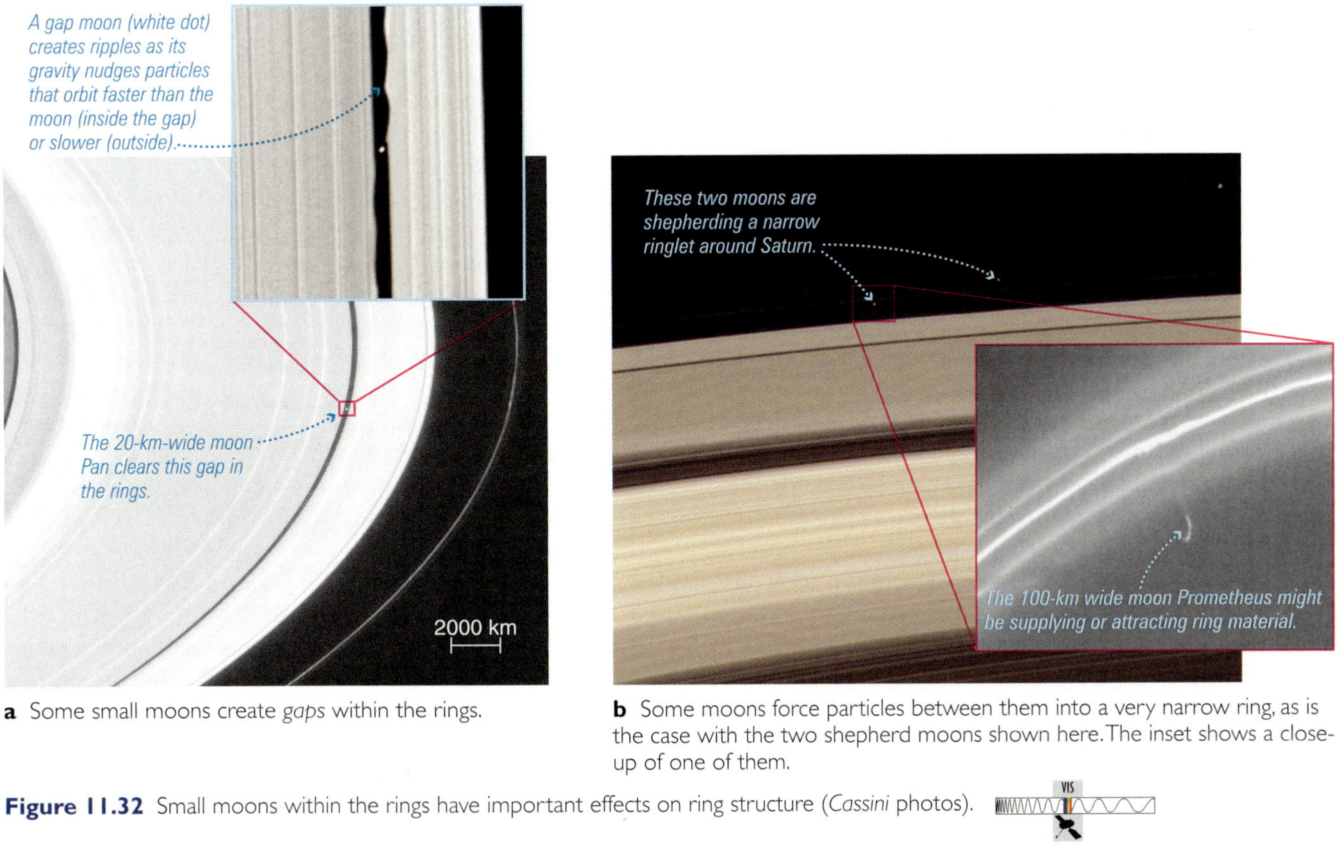

a Some small moons create *gaps* within the rings.

b Some moons force particles between them into a very narrow ring, as is the case with the two shepherd moons shown here. The inset shows a close-up of one of them.

Figure 11.32 Small moons within the rings have important effects on ring structure (*Cassini* photos).

Rings and gaps are caused by particles bunching up at some orbital distances and being forced out at others. This bunching happens when gravity nudges the orbits of ring particles in some particular way. One source of nudging comes from small moons located within the gaps in the rings themselves, which are sometimes called *gap moons*. The gravity of a gap moon can effectively keep the gap clear of smaller ring particles while creating ripples in the ring edges (Figure 11.32a). The ripples appear to move in opposite directions on the two sides of the gap, because ring particles on the inner side orbit Saturn slightly faster than the gap moon while those on the outer side orbit slightly slower than the gap moon. In some cases, two nearby gap moons can force particles between them into a very narrow ring (Figure 11.32b). The gap moons are often called *shepherd moons* in those cases, because they act like a shepherd forcing particles into line.

Ring particles also may be nudged by the gravity from larger, more distant moons. For example, a ring particle orbiting about 120,000 kilometers from Saturn's center will circle the planet in exactly half the time it takes the moon Mimas to orbit. Every time Mimas returns to a certain location, the ring particle will also be at its original location and therefore will experience the same gravitational nudge from Mimas. The periodic nudges reinforce one another and clear a gap in the rings—in this case, the large gap visible from Earth (the Cassini division). This type of reinforcement due to repeated gravitational tugs is another example of an *orbital resonance*, much like the orbital resonances that make Io's orbit elliptical (see Figure 11.19). Resonances can even create vast numbers of beautiful ripples and waves that start at the locations of resonances and spread out through the rings (Figure 11.33). Other orbital resonances, caused by moons both within the rings and farther out from Saturn, probably explain most of the intricate structures visible in ring photos.

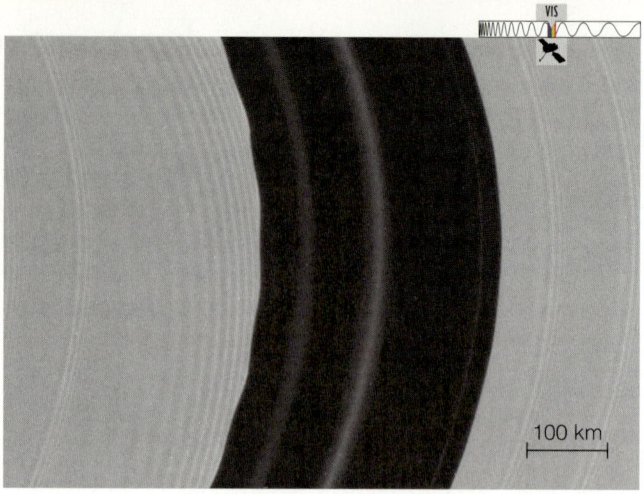

Figure 11.33 This *Cassini* image of a gap in Saturn's rings shows how the resonance with the moon Mimas creates waves that shape the rings over vast distances.

• How do other jovian ring systems compare to Saturn's?

The rings of Jupiter, Uranus, and Neptune are so much fainter than Saturn's that it took almost four centuries longer to discover them. Ring particles in these three systems are far less numerous, generally smaller, and much darker. Despite these differences, a family portrait of the jovian ring systems shows many similarities (Figure 11.34). All rings lie in their planet's equatorial plane. Particle orbits are fairly circular, with small orbital tilts relative to the equator. Individual rings and gaps are probably shaped by gap moons, shepherd moons, and orbital resonances.

Uranus's rings were first discovered in 1977 during observations of a *stellar occultation*—a star passing behind Uranus as seen from Earth. During the occultation, the star "blinked" on and off nine times before it disappeared behind Uranus and nine more times as it emerged. Scientists concluded that these nine "blinks" were caused by nine thin rings encircling Uranus. Similar observations of stars passing behind Neptune yielded more confounding results: Rings appeared to be present at some times but not at others. Could Neptune's rings be incomplete or transient?

The *Voyager* spacecraft provided some answers. *Voyager* cameras first discovered thin rings around Jupiter in 1979. After next providing incredible images of Saturn's rings, *Voyager 2* photographed the rings of Uranus as it flew past in 1986. In 1989, *Voyager 2* passed by Neptune and found that it does, in fact, have partial rings—at least when seen from Earth. The space between the ring segments is filled with dust not detectable from Earth. In addition to the dust within the individual rings of Neptune, *Voyager 2* detected vast dust sheets between the widely separated rings of both Uranus and Neptune.

The differences among the ring systems present us with unsolved mysteries. The larger size, higher reflectivity, and much greater number of particles in Saturn's rings compel us to wonder whether different processes might be at work there. The reasons for the slight tilt and eccentricity of Uranus's thin rings are not yet understood, nor do we know why Neptune's rings contain dusty regions that make them appear as partial rings when viewed from Earth. Scientists hope that we may someday learn the answers with new missions to visit Uranus and Neptune, but none are yet under development. Meanwhile, advances in ground-based observing, including the new technology of adaptive optics [Section 6.3], are providing views that in some cases rival those that we obtained from the *Voyager* spacecraft.

• Why do the jovian planets have rings?

Where do the rings come from? Did they form together with the planets out of the solar nebula, or are they a more recent phenomenon? An important clue is that the rings lie close to their planet in a region where tidal forces are very strong.

Within two to three planetary radii (of any planet), the tidal forces tugging apart an object become comparable to the gravitational forces holding it together. (This region is

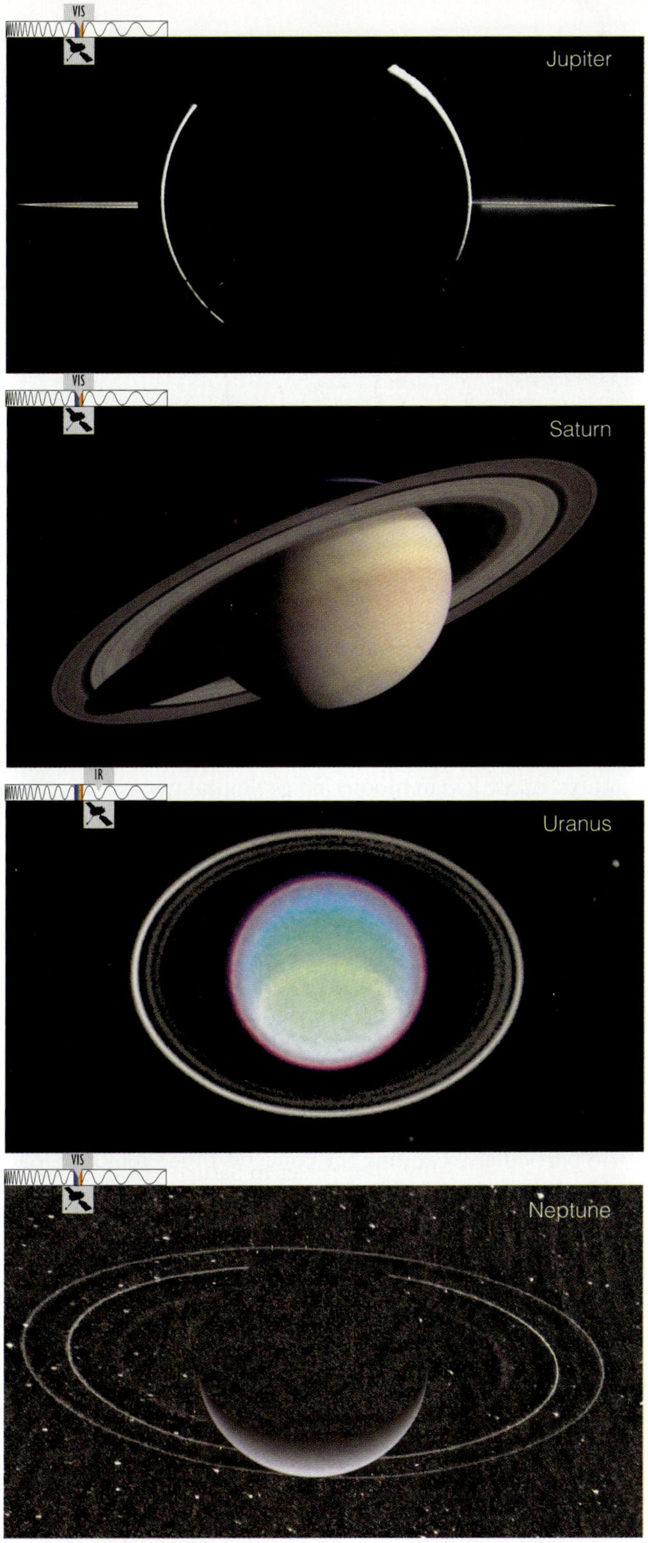

Figure 11.34 Four ring systems (not to scale). The rings differ in appearance and in the composition and sizes of the ring particles. These photos were taken with visible light except for Uranus, which was photographed in infrared light by the Hubble Space Telescope. The Saturn photo is from *Cassini* and the Jupiter and Neptune photos are from *Voyager.*

often called the *Roche tidal zone.*) Only relatively small objects held together by nongravitational forces—such as the electromagnetic forces that hold solid rock, spacecraft, and human beings together—can avoid being ripped apart by the strong tidal forces in this region. Because a large moon would be ripped apart by tidal forces in the region of the rings, scientists once suspected that Saturn's rings formed after a large moon came too close to the planet. However, moons don't simply "wander" away from their orbits, and a large moon could have been deflected toward Saturn only by a close gravitational encounter with an unusually large asteroid or comet that happened to pass nearby. Such encounters should be extremely rare. Back in the days when Saturn was the only planet known to have rings, it might have seemed possible that such a rare encounter was responsible for the rings. The discovery of rings around the other jovian planets all but ruled out this possibility, because it seems inconceivable that such an unlikely event could have occurred in all four cases.

If the rings did not originate with the destruction of a large moon, where did all the ring particles come from? Another idea that once seemed reasonable was that the ring particles might be leftover chunks of rock and ice that condensed in the disks of gas that orbited each jovian planet when it was young. This would explain why all four jovian planets have rings, because tidal forces near each planet would have prevented these chunks from accreting into a full-fledged moon. However, we now know that the ring particles cannot be leftovers from the births of the planets, because particles of the size we find in the rings today cannot survive for billions of years. Ring particles are continually being ground down in size, primarily by the impacts of the countless sand-size particles that orbit the Sun—the same types of particles that become meteors in Earth's atmosphere and cause micrometeorite impacts on the Moon [Section 9.3]. Millions of years of such tiny impacts would long ago have ground the existing ring particles to dust. The dust particles could not have survived either, because over time they are slightly slowed in their orbits by the pressure of sunlight, and this slowing eventually causes them to spiral into their planet. Ring particles may also be lost in other ways. For example, the thermosphere of Uranus extends up into the ring region, exerting atmospheric drag that also causes ring particles to spiral slowly into the planet. Between the effects of micrometeorite impacts and other processes, *none* of the abundant small particles that now occupy the jovian rings can have been there since the solar system formed more than 4 billion years ago.

We are therefore left with only one reasonable possibility for the origin of the rings: New particles are continually supplied to the rings to replace those that are destroyed. These new particles must come from a source that lies in each planet's equatorial plane. The most likely source is numerous small moons—moons the size of gap moons (see Figure 11.32)—that formed in the disks of material orbiting the young jovian planets. Like the ring particles themselves, tiny impacts are gradually grinding away these small moons, but they are large enough to still exist despite 4.5 billion years of such sandblasting.

The small moons contribute ring particles in two ways. First, each tiny impact releases particles from a small moon's surface, and these released particles become new, dust-size

Figure 11.35 This illustration summarizes the origin of rings around the jovian planets.

ring particles. Ongoing impacts ensure that some ring particles are present at all times. Second, occasional larger impacts can shatter a small moon completely, creating a supply of boulder-size ring particles. The frequent tiny impacts then slowly grind these boulders into smaller ring particles. In summary, the dust- to boulder-size particles in rings all ultimately come from the gradual dismantling of small moons that formed during the birth of the solar system (Figure 11.35).

The collisions that shatter small moons and generate a major source of ring particles must occur only occasionally and at essentially random times, which means that the number of particles in any particular ring system must change dramatically over millions and billions of years. Rings may be broad and bright when they are full of particles and almost invisible when particles are few. Thus, the brilliant spectacle of Saturn's rings may be a special treat of our epoch, one that may not have been seen a billion years ago and that may not last long on the time scale of our solar system.

THE BIG PICTURE

Putting Chapter 11 into Context

In this chapter, we saw that the jovian planets really are a different kind of planet and, indeed, a different kind of planetary system. The jovian planets dwarf the terrestrial planets. Even some of their moons are as large as terrestrial worlds. As you continue your study of the solar system, keep in mind the following "big picture" ideas:

- The jovian planets may lack solid surfaces on which geology can occur, but they are interesting and dynamic worlds with rapid winds, huge storms, strong magnetic fields, and interiors in which common materials behave in unfamiliar ways.
- Despite their relatively small sizes and frigid temperatures, many jovian moons are geologically active by virtue of their icy compositions and tidal heating. Ironically, the cold temperatures in the solar nebula led to their icy compositions and hence their geological activity.
- Ring systems probably owe their existence to small moons formed in the disks of gas that produced the jovian planets billions of years ago. The rings we see today are composed of particles liberated from those moons surprisingly recently.
- Understanding the jovian planet systems forced us to modify many of our earlier ideas about the solar system, in particular by adding the concepts of ice geology, tidal heating, and orbital resonances. Each new set of circumstances that we discover offers further opportunities to learn how our universe works.

SUMMARY OF KEY CONCEPTS

11.1 A Different Kind of Planet

- **Are jovian planets all alike?** Jupiter and Saturn are made almost entirely of hydrogen and helium, while Uranus and Neptune are made mostly of hydrogen compounds mixed with metal and rock. These differences arose because all four planets started from ice-rich planetesimals of about the same size, but captured different amounts of hydrogen and helium gas from the solar nebula.

- **What are jovian planets like on the inside?** The jovian planets have layered interiors with very high internal temperatures and pressures. All have a core about 10 times as massive as Earth, consisting of hydrogen compounds, metals, and rock. They differ mainly in their surrounding layers of hydrogen and helium, which can take on unusual forms under the extreme internal conditions of the planets.

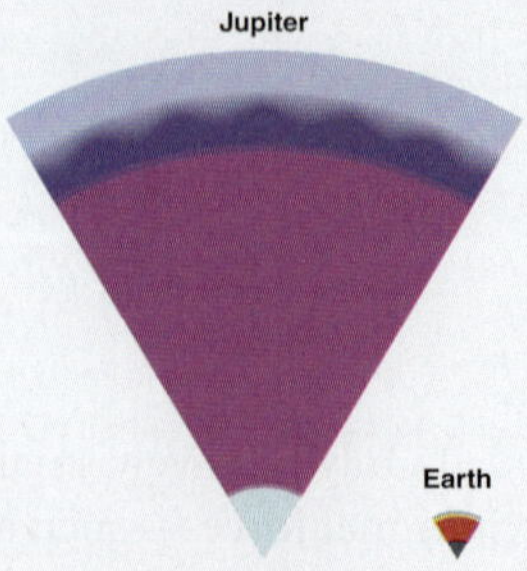

- **What is the weather like on jovian planets?** The jovian planets all have multiple cloud layers that help determine the colors of the planets, fast winds, and large storms. Some storms, such as the Great Red Spot, can apparently rage for centuries or longer.

- **Do jovian planets have magnetospheres like Earth's?** Jupiter has a magnetic field some 20,000 times stronger than Earth's, which leads to an enormous magnetosphere. Many of the particles in Jupiter's magnetosphere come from volcanic eruptions on Io. The other jovian planets also have magnetic fields and magnetospheres, but they are weaker and smaller than Jupiter's.

11.2 A Wealth of Worlds: Satellites of Ice and Rock

- **What kinds of moons orbit the jovian planets?** We can categorize the more than 100 known moons as small, medium-size, or large. Most of the medium-size and large moons probably formed with their planet in the disks of gas that surrounded the jovian planets when they were young. Smaller moons are often captured asteroids or comets.

- **Why are Jupiter's Galilean moons so geologically active?** Io is the most volcanically active object in the solar system, thanks to an interior kept hot by tidal heating—which occurs because Io's close orbit is made elliptical by orbital resonances with other moons. Europa (and possibly Ganymede) may have a deep, liquid water ocean under its icy crust, also thanks to tidal heating. Callisto is the least geologically active, since it has no orbital resonance or tidal heating.

- **What is special about Titan and other major moons of the outer solar system?** Many medium-size and large moons show a surprisingly high level of past or present volcanism or tectonics. Titan even has a thick atmosphere and ongoing erosion caused by liquid methane. Triton, which apparently was captured by Neptune, also has had relatively recent geological activity.

- **Why are small icy moons more geologically active than small rocky planets?** Ices deform and melt at much lower temperatures than rock, allowing icy volcanism and tectonics at surprisingly low temperatures. In addition, some jovian moons have a heat source—tidal heating—that is not important for the terrestrial worlds.

11.3 Jovian Planet Rings

- **What are Saturn's rings like?** Saturn's rings are made up of countless individual particles, each orbiting Saturn independently like a tiny moon. The rings lie in Saturn's equatorial plane, and they are extremely thin. Moons within and beyond the rings create many ringlets and gaps in part through orbital resonances.

- **How do other jovian ring systems compare to Saturn's?** The other jovian planets have ring systems that are much fainter in photographs. Their ring particles are generally smaller, darker, and less numerous than Saturn's ring particles.

- **Why do the jovian planets have rings?** Ring particles probably come from the dismantling of small moons formed in the disks of gas that surrounded the jovian planets billions of years ago. Small ring particles come from countless tiny impacts on the surfaces of these moons, while larger ones come from impacts that shatter the moons.

EXERCISES AND PROBLEMS

For instructor-assigned homework go to MasteringASTRONOMY www.masteringastronomy.com

Review Questions

Short-Answer Questions Based on the Reading

1. Briefly describe how the differences in composition among the jovian planets can be traced to their formation in the solar nebula.
2. Why is Jupiter so much denser than Saturn? Could a planet be smaller in size than Jupiter but greater in mass? Explain.
3. Briefly describe the interior structure of Jupiter and why it is layered in this way. How do the interiors of the other jovian planets compare to Jupiter?
4. Why does Jupiter have so much internal heat? What generates internal heat on other jovian planets?
5. Briefly describe Jupiter's atmospheric structure and cloud layers. How do the structures and clouds differ on the other jovian planets?

6. How do the cloud layers help explain Jupiter's colors? Why are Saturn's colors more subdued? Why are Uranus and Neptune blue?
7. Briefly describe Jupiter's weather patterns and contrast them with those on the other jovian planets. What is the *Great Red Spot?*
8. Why does Jupiter have such a strong magnetic field? Describe a few features of Jupiter's magnetosphere, and compare it to the magnetospheres of the other jovian planets.
9. Briefly describe how we categorize jovian moons by size. What is the origin of most of the medium-size and large moons? What is the origin of many of the small moons?
10. What are the key features of Jupiter's four Galilean moons? Explain the role of tidal heating and orbital resonances in shaping these features.
11. Describe the atmosphere of Titan. What did the *Cassini/Huygens* mission learn about Titan's surface?
12. What evidence suggests past geological activity on the medium-size moons of Saturn and Uranus? Describe a few of the mysteries that these moons pose.
13. Briefly describe Triton and explain why we think it is a captured moon.
14. Briefly explain why icy moons can have active geology at much smaller sizes than rocky worlds.
15. What are rings? What are they made of, and how do they differ among the four jovian planets? Briefly describe the effects of gap moons and orbital resonances on ring systems.
16. Explain why we think that ring particles must be replenished over time. Should the jovian planet rings always look the same?

Test Your Understanding

Surprising Discoveries?

Suppose someone claimed to make the discoveries described below. (These are not real discoveries.) Decide whether each discovery should be considered reasonable or surprising. More than one right answer may be possible, so explain your answer clearly.

17. Saturn's core is pockmarked with impact craters and dotted with volcanoes erupting basaltic lava.
18. Neptune's deep blue color is not due to methane, as previously thought, but instead is due to its surface being covered with an ocean of liquid water.
19. A jovian planet in another star system has a moon as big as Mars.
20. An extrasolar planet is made primarily of hydrogen and helium. It has approximately the same mass as Jupiter but is the same size as Neptune.
21. A new small moon orbits Jupiter outside the orbits of other known moons. It is smaller than Jupiter's other moons but has several large, active volcanoes.
22. A new moon orbits Neptune in the planet's equatorial plane and in the same direction that Neptune rotates, but it is made almost entirely of metals such as iron and nickel.
23. An icy, medium-size moon orbits a jovian planet in a star system that is only a few hundred million years old. The moon shows evidence of active tectonics.
24. A jovian planet is discovered in a star system that is much older than our solar system. The planet has no moons at all, but has a system of rings as spectacular as the rings of Saturn.
25. Radar measurements of Titan indicate that most of the moon is heavily cratered.
26. The *Cassini* mission finds 20 more moons of Saturn.

Quick Quiz

Choose the best answer to each of the following. Explain your reasoning with one or more complete sentences.

27. Which lists the jovian planets in order of increasing distance from the Sun? (a) Jupiter, Saturn, Uranus, Pluto (b) Saturn, Jupiter, Uranus, Neptune (c) Jupiter, Saturn, Uranus, Neptune
28. Why does Neptune appear blue and Jupiter red? (a) Neptune is hotter, which gives bluer thermal emission. (b) Methane in Neptune's atmosphere absorbs red light. (c) Neptune's air molecules scatter blue light, much as Earth's atmosphere does.
29. Why is Jupiter denser than Saturn? (a) It has a larger proportion of rock and metal. (b) It is more compressed by the Sun's gravity. (c) Its higher mass and gravity compress its interior.
30. Some jovian planets give off more energy than they receive because (a) fusion is taking place in their cores. (b) tidal heating is occurring. (c) they are still slowly contracting after formation.
31. The main ingredients of most satellites of the jovian planets are (a) rock and metal. (b) hydrogen compound ices. (c) hydrogen and helium.
32. Why is Io more volcanically active than our moon? (a) Io is much larger. (b) Io has a higher concentration of radioactive elements. (c) Io has a different internal heat source.
33. What is unusual about Triton? (a) It orbits its planet backward. (b) It does not keep the same face toward its planet. (c) It is the only moon with its own rings.
34. Which moon shows evidence of rainfall and erosion by some liquid substance? (a) Europa (b) Titan (c) Ganymede
35. Saturn's many moons affect the rings through (a) tidal forces. (b) orbital resonances. (c) magnetic field interactions.
36. Saturn's rings (a) have looked basically the same since they formed along with Saturn. (b) were created long ago when tidal forces tore apart a large moon. (c) are continually supplied by impacts into small moons.

Investigate Further

In-Depth Questions to Increase Your Understanding

Short-Answer/Essay Questions

37. *The Importance of Rotation.* Suppose the material that formed Jupiter came together without any rotation so that no "jovian nebula" formed and the planet today wasn't spinning. How else would the jovian system be different? Think of as many effects as you can, and explain each in a sentence.
38. *The Great Red Spot.* Based on the infrared and visible images in Figure 11.9, is Jupiter's Great Red Spot warmer or cooler than nearby clouds? Is it is higher or lower in altitude than the nearby clouds? Explain.
39. *Comparing Jovian Planets.* You can do comparative planetology armed only with telescopes and an understanding of gravity.
    a. The small moon Amalthea orbits Jupiter at about the same distance in kilometers at which Mimas orbits Saturn, yet Mimas takes almost twice as long to orbit. From this observation, what can you conclude about how Jupiter and Saturn differ? Explain.
    b. Jupiter and Saturn are not very different in radius. When you combine this information with your answer to part (a), what can you conclude? Explain.

40. *Minor Ingredients Matter.* Suppose the jovian planet atmospheres were composed only of hydrogen and helium, with no hydrogen compounds at all. How would the atmospheres be different in terms of clouds, color, and weather? Explain.
41. *Hot Jupiters.* Many of the newly discovered planets orbiting other stars are more massive than Jupiter but orbit much closer to their stars. Assuming that they would be Jupiter-like if they orbited at a greater distance from their stars, how would you expect these new planets to differ from the jovian planets of our solar system? How would you expect their moons to differ? Explain.
42. *The New View of Titan.* What planet or moon in the solar system does Titan most resemble, in your opinion? Summarize the similarities and differences in a few sentences.
43. *Galilean Moon Formation.* Look up the densities of Jupiter's four Galilean moons in Appendix E, and notice that they follow a trend with distance from Jupiter. Based on what you've learned about condensation in the solar nebula, can you suggest a reason for this trend among the Galilean moons? Next, compare the densities of the moons with the planetary densities in Table E.1. Based on the comparison, do you think it was as hot toward the center of the nebula surrounding Jupiter as it was at the center of the solar nebula? Explain.
44. *Super Jupiter.* Suppose that the solar wind had not cleared the solar nebula until much later in our solar system's history, so that Jupiter accumulated 40 times as much mass as it actually has today.
    a. Make a sketch showing how the interior layers of this "Super Jupiter" would compare to the interior layers of the real Jupiter shown in Figure 11.4.
    b. How would you expect the cloud layers and colors to differ on Super Jupiter? Explain.
    c. How would you expect the wind speeds to be different on Super Jupiter? Explain.
    d. How would you expect the magnetic field to be different on Super Jupiter? Explain.
    e. Do you think Super Jupiter would have more or less excess heat than the real Jupiter? Explain.
45. *Observing Project: Jupiter's Moons.* Using binoculars or a small telescope, view the moons of Jupiter. Make a sketch of what you see, or take a photograph. Repeat your observations several times (nightly, if possible) over a period of a couple of weeks. Can you determine which moon is which? Can you measure the moons' orbital periods? Can you determine their approximate distances from Jupiter? Explain.
46. *Observing Project: Saturn's Rings.* Using binoculars or a small telescope, view the rings of Saturn. Make a sketch of what you see, or take a photograph. What season is it in Saturn's northern hemisphere? How far do the rings extend above Saturn's atmosphere? Can you identify any gaps in the rings? Describe any other features you notice.

Quantitative Problems

Be sure to show all calculations clearly and state your final answers in complete sentences.

47. *Disappearing Moon.* Io loses about a ton (1,000 kilograms) of sulfur dioxide per second to Jupiter's magnetosphere.
    a. At this rate, what fraction of its mass would Io lose in 4.5 billion years?
    b. Suppose sulfur dioxide currently makes up 1% of Io's mass. When will Io run out of this gas at the current loss rate?
48. *Ring Particle Collisions.* Each ring particle in the densest part of Saturn's rings collides with another about every 5 hours. If a ring particle survived for the age of the solar system, how many collisions would it undergo?
49. *Prometheus and Pandora.* These two moons orbit Saturn at 139,350 and 141,700 kilometers, respectively.
    a. Using Newton's version of Kepler's third law, find their two orbital periods. Find the percent difference in their distances and in their orbital periods.
    b. Consider the two in a race around Saturn: In one Prometheus orbit, how far behind is Pandora (in units of time)? In how many Prometheus orbits will Pandora have fallen behind by one of its own orbital periods? Convert this number of periods back into units of time. This is how often the satellites pass by each other.
50. *Orbital Resonances.* Using the data in Appendix E, identify the orbital resonance relationship between Titan and Hyperion. (*Hint:* If the orbital period of one were 1.5 times the other, we would say that they are in a 3:2 resonance.) Which medium-size moon is in a 2:1 resonance with Enceladus?
51. *Titanic Titan.* What is the ratio of Titan's mass to all the other satellites of Saturn whose mass is listed in Appendix E? Calculate the strength of gravity on Titan compared to that on Mimas. Comment on how this affects the possibility of atmospheres on each.
52. *Titan's Evolving Atmosphere.* Titan's exosphere lies nearly 1,400 kilometers above its surface. What is the escape velocity from this altitude? What is the thermal speed of a hydrogen atom at the exospheric temperature of about 200 K? Use these answers (and the method of Mathematical Insight 10.2) to comment on whether thermal escape of hydrogen is likely to be important for Titan.
53. *Saturn's Thin Rings. Saturn's* ring system is over 270,000 km wide, and only a few tens of meters thick; let's assume 50 meters thick for this problem. Assuming the rings could be shrunk down so that their diameter is the width of a dollar bill (6.6 cm), how thick would the rings be? Compare your answer to the actual thickness of a dollar bill (0.01 cm).
54. *Characterizing a newly discovered planet.* In April 2005, scientists claimed to have captured light from an extrasolar planet orbiting a brown dwarf in the constellation Hydra some 230 light-years from Earth. The scientists believe the planet is about five times the mass of Jupiter.
    a. Assume the planet has the same density as Jupiter. What is the diameter of the planet?
    b. Using Mathematical Insight 2.1, calculate the angular size of the planet as viewed from Earth, in arcseconds.

Discussion Questions

55. *Jovian Planet Mission.* We can study terrestrial planets up close by landing on them, but jovian planets have no surfaces to land on. Suppose that you are in charge of planning a long-term mission to "float" in the atmosphere of a jovian planet. Describe the technology you would use and how you would ensure survival for any people assigned to this mission.
56. *Pick a Moon.* Suppose you could choose any one moon to visit in the solar system. Which one would you pick, and why? What dangers would you face in your visit to this moon? What kinds of scientific instruments would you want to bring along for studies?

MEDIA EXPLORATIONS

For self-study activities go to MasteringASTRONOMY www.masteringastronomy.com

Interactive Tutorials

Tutorial Review of Key Concepts

Use the following interactive **Tutorial** at **www.masteringastronomy.com** to review key concepts from this chapter.

Formation of the Solar System Tutorial

Lesson 1 Comparative Planetology

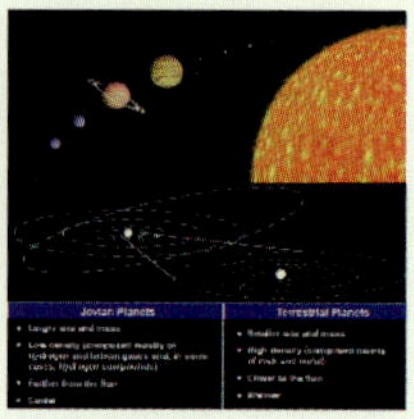

Supplementary Tutorial Exercises

Use the interactive **Tutorial Lessons** to explore the following questions.

Formation of the Solar System Tutorial, Lesson 1

1. Contrast the general features of the jovian planets with those of the terrestrial planets.
2. Compare the properties of each of the four jovian planets. What properties do they all share? In what ways are they different?

Exploring the Sky and Solar System

Of the many activities available on the ***Voyager: SkyGazer* CD-ROM** accompanying your book, use the following files to observe key phenomena covered in this chapter.

Go to the **File: Basics** folder for the following demonstrations.

1. Tracking Jupiter and Io
2. Saturn

Go to the **File: Demo** folder for the following demonstrations.

1. Backside of Jupiter
2. Locked on Dione
3. Three Moons on Jupiter

Web Projects

1. *The Galileo Mission to Jupiter.* Learn more about the *Galileo* mission and how it ended, and about ongoing scientific study of the *Galileo* data. Write a short summary of your findings.
2. *News from Cassini.* Find the latest news about the *Cassini* mission to Saturn. What is the current mission status? Write a short report about the mission's status and recent findings.
3. *Oceans of Europa.* The possibility of subsurface oceans on Europa holds great scientific interest. Investigate plans for further study of Europa, either from Earth or with future spacecraft. Write a short summary of the plans and how they might help us learn whether an ocean really exists and, if so, what it might contain.

12

Remnants of Rock and Ice

Asteroids, Comets, and the Kuiper Belt

LEARNING GOALS

12.1 Asteroids and Meteorites
- What are asteroids like?
- Why is there an asteroid belt?
- Where do meteorites come from?

12.2 Comets
- What are comets like?
- Where do comets come from?

12.3 Pluto: Lone Dog No More
- How big can a comet be?
- Are Pluto and Eris planets?

12.4 Cosmic Collisions: Small Bodies Versus the Planets
- Have we ever witnessed a major impact?
- Did an impact kill the dinosaurs?
- Is the impact threat a real danger or media hype?
- How do the jovian planets affect impact rates and life on Earth?

As we look out into the Universe and identify the many accidents of physics and astronomy that have worked to our benefit, it almost seems as if the Universe must in some sense have known that we were coming.

Freeman Dyson

Asteroids and comets might at first seem insignificant in comparison to the planets and moons we've discussed so far, but there is strength in numbers. The trillions of small bodies orbiting our Sun are far more important than their small size might suggest.

The appearance of comets has more than once altered the course of human history when our ancestors acted upon superstitions related to comet sightings. More profoundly, asteroids or comets falling to Earth have scarred our planet with impact craters and have altered the course of biological evolution. Asteroids and comets are also important scientifically: As remnants from the birth of our solar system, they have taught us a great deal about how our solar system formed.

In this chapter, we will explore asteroids, comets, and the rocks that fall to Earth as meteorites. Along the way, we'll see that Pluto, until recently considered the "ninth planet," is actually just one of many similar objects orbiting the Sun beyond Neptune—and at least one of those objects is larger than Pluto. Finally, we will explore the dramatic effects of the occasional collisions between small bodies and large planets.

12.1 Asteroids and Meteorites

The objects we've studied so far—terrestrial planets, jovian planets, and large moons—have changed dramatically since their formation. Most planets have been transformed in almost every way: Their interiors have differentiated, their surfaces have been reshaped by geological processes, and their atmospheres have changed as they have gained and lost gas. In contrast, many small bodies remain much as they were when they first formed, some 4.5 billion years ago. Comets, asteroids, and meteorites carry the history of our solar system encoded in their compositions, locations, and numbers.

Using these small bodies to understand planetary formation is a bit like picking through the trash in a carpenter's shop to see how furniture is made. By studying the scraps, sawdust, paint chips, and glue, we can develop some understanding of how the carpenter builds furniture. Asteroids, comets, and meteorites are the "scraps" left over from the formation of our solar system, so their study can give us an understanding of how the planets and larger moons came to exist.

We'll begin our study of small bodies by focusing on asteroids and meteorites in this section. The word *asteroid* means "starlike," but asteroids are starlike only in their appearance through a telescope; as we discussed in Chapter 8, asteroids are planetesimals (or fragments of planetesimals) left over from the birth of our solar system. Meteorites are pieces of rock that have fallen to the ground from space. Thus, asteroids and meteorites are closely related: Most meteorites are pieces of asteroids that orbited the Sun for billions of years before falling to Earth.

• What are asteroids like?

Asteroids are virtually undetectable by the naked eye and went unnoticed for almost two centuries after the invention of the telescope. The first asteroids were discovered only about 200 years ago, and it took 50 years to discover the first 10. Today, advanced telescopes can discover far more than that in a single night, and more than 150,000 asteroids have been cataloged. Asteroids can be recognized in telescopic images because over a short time period they move noticeably in relation to the stars (Figure 12.1).

Newly discovered asteroids first get a provisional name based on the discovery month and year; for example, the first asteroid discovered in January 2007 would be called Asteroid 2007 AA. An asteroid's orbit can be calculated from the law of gravity [Section 4.4], even after observation of only a fraction of a complete orbit, but only about 10% of the asteroids identified in images have been tracked long enough for their orbits to be calculated. In those cases, the discoverer may choose a name for the asteroid, subject to approval by the International Astronomical Union. The earliest discovered asteroids bear the names of mythological figures. More recent discoveries often carry names of scientists, cartoon heroes, pets, or rock stars.

Asteroid Sizes, Numbers, and Shapes Asteroids come in a wide range of sizes. The largest, Ceres, is just under 1,000 kilometers in diameter, which is less than $\frac{1}{3}$ the diameter of our Moon. About a dozen others are large enough that we would call them medium-size moons if they orbited a planet. Smaller asteroids are far more numerous. There are probably more than a million asteroids with diameters

Figure 12.1 Interactive Photo Because asteroids orbit the Sun, they move through our sky relative to the stars. In this long-exposure photograph, stars show up as white dots, while the motion of an asteroid relative to the stars makes it show up as a short streak.

greater than 1 kilometer, and many more even smaller in size. Despite their large numbers, asteroids don't add up to much in total mass. If we could put all the asteroids together and allow gravity to compress them into a sphere, they'd make an object less than 2,000 kilometers in diameter—much smaller than our Moon.

Asteroid shapes depend on the strength of their gravity. If an asteroid is relatively large, over time gravity can mold its solid rock into a spherical or near-spherical shape. Only Ceres is large enough for gravity to have compressed it into a roughly spherical shape—allowing Ceres to qualify as a *dwarf planet* (see Special Topic, p. 9)—while the next two largest asteroids are somewhat oblong (Pallas, 540 kilometers long, and Vesta, 510 kilometers long). For smaller asteroids, gravity is too weak to reshape the rocky material. Much like potatoes, these asteroids can have almost any shape. Indeed, some objects that appear to be single asteroids are probably two or more distinct objects held in contact by a weak gravitational attraction, while other small asteroids are little more than weakly bound piles of rubble.

The handful of asteroids that have been photographed up close by spacecraft confirm these ideas (Figure 12.2). Notice that these asteroids look more like potatoes than spheres. The numerous craters tell us that asteroids, like planets and moons, have been battered by impacts. The impact that formed Mathilde's huge, central crater (the shadowed region in Figure 12.2c) would have disintegrated a solid object so Mathilde must be a loosely bound "rubble pile" that was able to absorb the shock of the impact without completely falling apart.

Given that we've photographed so few asteroids close-up and that most asteroids are far too small for even powerful telescopes to resolve, you may wonder how we can make general statements about their sizes and shapes. It's not easy, but careful analysis of asteroid light can often teach us much about them. Let's first consider measurements of size.

Asteroids shine with reflected sunlight, so an asteroid's brightness in our sky depends on its size, distance, and reflectivity. For example, if two asteroids at the same distance have the same reflectivity, the one that appears brighter must be larger in size. We can determine an asteroid's distance from its position in its orbit. Determining reflectivity is more difficult, but we can do it by comparing the asteroid's brightness in visible light to its brightness in infrared light. The visible light is reflected sunlight, while the infrared light is thermal radiation emitted by the asteroid itself [Section 5.4]. The infrared brightness therefore depends on the asteroid's temperature, which in turn depends on how much sunlight it absorbs. Thus, the comparison between visible and infrared brightness essentially tells us the proportions of incoming sunlight that an asteroid reflects and absorbs. Once we know these proportions, the asteroid's brightness and distance tell us its size. This technique has been used to make good size estimates for more than a thousand asteroids.

THINK ABOUT IT

Suppose you discover two asteroids that are equally bright, but infrared observations tell you that Asteroid 1 is more reflective than Asteroid 2. Which one is farther away? Explain.

Determining asteroid shapes without close-up photographs is difficult but not impossible. In some cases, we can determine shapes by monitoring brightness variations as an asteroid rotates: A non-spherical asteroid with a uniformly bright surface will reflect more light when it presents its larger side toward the Sun and our telescopes. In other cases, astronomers have determined shapes by bouncing radar signals off asteroids that have passed close to Earth.

Asteroid Compositions Analysis of light can also tell us about asteroid compositions. Recall that spectra of distant objects contain spectral lines that are essentially "fingerprints" left by their chemical constituents [Section 5.4]. Thousands of asteroids have been analyzed through spectroscopy. The results are consistent with what we expect from our theory of solar system formation: Asteroids are made mostly of metal and rock, because they condensed within the frost line in the solar nebula. Those near the outskirts of the asteroid belt contain larger proportions of dark, carbon-rich material, because this material was able to condense at the

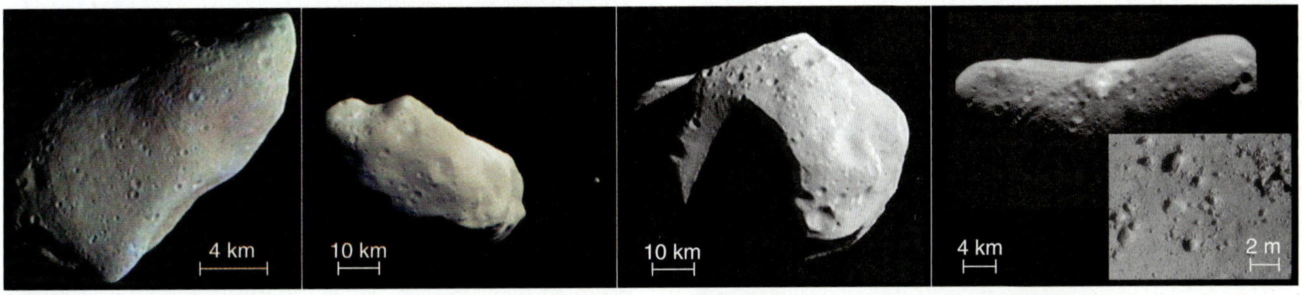

a Gaspra, photographed by the *Galileo* spacecraft.

b Ida, photographed by the *Galileo* spacecraft. The small dot to the right is Dactyl, a tiny moon orbiting Ida.

c Mathilde, photographed by the Near-Earth Asteroid Rendezvous (NEAR) spacecraft on its way to Eros.

d Eros, photographed by the NEAR spacecraft, which orbited Eros for a year before ending its mission with a soft landing on the asteroid's surface. The inset photo was taken by NEAR just before it landed.

Figure 12.2 Close-up views of asteroids studied by spacecraft.

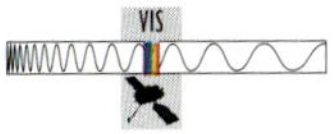

relatively cool temperatures found in this region of the solar nebula but not in the regions closer to the Sun; some even contain small amounts of water, telling us that they formed close to the frost line. A few asteroids appear to be made mostly of metals, such as iron, suggesting that they may be fragments of the iron cores of shattered worlds.

Asteroid Masses and Densities Determining an asteroid's mass and density is much more difficult than measuring its size, shape, or composition. The only direct way to measure a distant object's mass is to observe its gravitational effect on another object. We've been able to observe the gravitational effects of asteroids in only a few cases: those in which a spacecraft has passed by close enough to have its trajectory altered by the asteroid's gravity and those in which asteroids have even smaller asteroids as tiny, orbiting "moons." For asteroids with moons, such as the asteroid Ida with its tiny moon Dactyl (see Figure 12.2b) and the asteroid Eugenia in Figure 12.3, we can determine the mass of the central asteroid by applying Newton's version of Kepler's third law to the orbital characteristics of the moon [Section 4.5]. Astronomers have found a few dozen asteroid masses by this method.

Once we measure an asteroid's mass and size, we can calculate its density. The density can offer valuable insights into the asteroid's origin and makeup. For example, Mathilde—which we argued earlier must be a "rubble pile" in order to have survived the impact that made its large central crater—has a density of only 1.5 g/cm^3. This density is so low that Mathilde cannot be a solid chunk of rock, which lends further support to the idea that it is a pile of rubble held together by weak gravity. In contrast, the asteroid Eros (see Figure 12.2d) has a density of 2.4 g/cm^3, which is close to the value expected for solid rock. Eros also has a subtle pattern of surface ridges and troughs that hint at a set of parallel faults and fractures in its interior.

Asteroid Orbits The vast majority of the asteroids that have been cataloged are located in the asteroid belt between the orbits of Mars and Jupiter (Figure 12.4). All asteroids orbit the Sun in the same direction as the planets, but their orbits tend to be more elliptical and more highly inclined to the ecliptic plane (often up to 20°–30°) than those of planets.

Science fiction movies often show the asteroid belt as a crowded and hazardous place, but the asteroid belt is so large that its hundreds of thousands of asteroids are quite far apart on average. The average distance between asteroids in the asteroid belt is millions of kilometers—the equivalent of grains of sand separated by kilometers.

Not all asteroids are located in the asteroid belt. Two sets of asteroids, called *Trojan asteroids* (as in the Greek story of the Trojan War), share Jupiter's 12-year orbit around the Sun. As shown in Figure 12.4, one clump of Trojan asteroids always stays 60° ahead of Jupiter in its orbit and the other clump always stays 60° behind. The number of Trojan asteroids could be as large as the number in the asteroid belt, but the greater distance to the Trojan asteroids (and their darker surfaces) makes them more difficult to study or even count from Earth.

A relatively small number of asteroids have orbits that pass through the inner solar system—including the *near-Earth asteroids* that pass near Earth's orbit. These asteroids

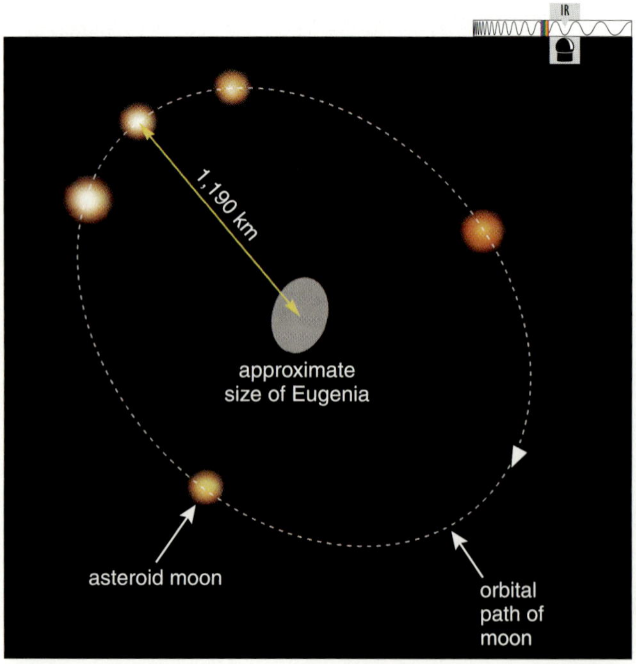

Figure 12.3 This set of images, made by a telescope using adaptive optics [Section 6.3], shows a small moon observed in five positions in its orbit around the asteroid Eugenia. The moon completes one orbit around the asteroid every 4.7 days. The asteroid itself was blocked out during the observations and is shown as the gray oval in the image.

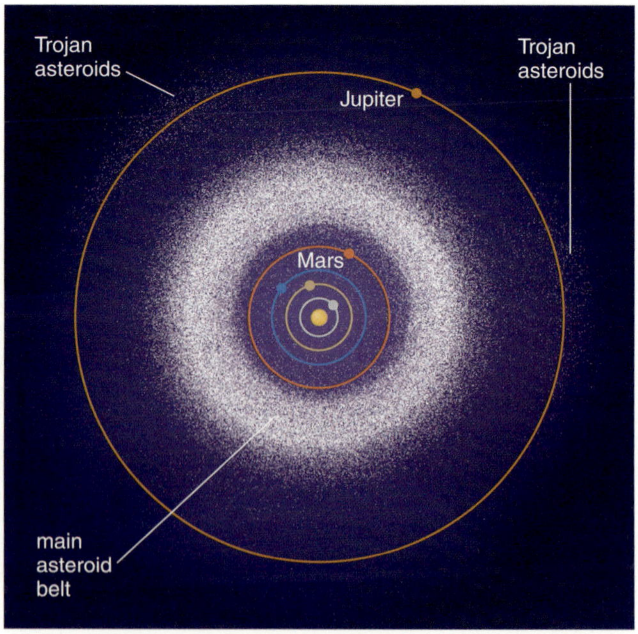

Figure 12.4 Calculated positions of 152,942 asteroids for midnight, January 1, 2004. The asteroids themselves are much smaller than the dots on this scale. The asteroids that share Jupiter's orbit, found 60° ahead of and behind Jupiter, are called *Trojan asteroids*.

COMMON MISCONCEPTIONS

Dodge Those Asteroids!

Science fiction movies often show brave spacecraft pilots navigating through crowded fields of asteroids, dodging this way and that as they heroically pass through with only a few bumps and bruises. It's great drama, but not very realistic. The asteroid belt looks crowded when we draw it on paper as in Figure 12.4, but in reality it is an enormous region of space. Despite their large numbers, asteroids are thousands to millions of kilometers apart on average—so far apart that it would take incredibly bad luck to crash into one by accident. Indeed, spacecraft must be carefully guided to fly close enough to an asteroid to take a decent photograph. Future space travelers will have plenty of dangers to worry about, but dodging asteroids is not likely to be one of them.

are probably "impacts waiting to happen." Many asteroids with similar orbits must have hit our planet in the past. In Section 12.4, we will discuss both past impacts and the future dangers these asteroids may pose.

• Why is there an asteroid belt?

The asteroid belt between Mars and Jupiter gets its name because it is where we find the majority of asteroids. But why are asteroids concentrated in this region, and why didn't a full-fledged planet form here? The answers lie with *orbital resonances* much like those that explain the elliptical orbits of Jupiter's moons Io, Europa, and Ganymede [Section 11.2] and the gaps and other structures in Saturn's rings [Section 11.3].

Remember that an orbital resonance occurs whenever two objects periodically line up with each other. Because gravity tugs at the objects in the same direction at each alignment, the effects can build up over time, just as many small pushes can make a child swing quite high on a playground swing. Objects will periodically line up—and hence have an orbital resonance—whenever one object's orbital period is a simple ratio of another object's period, such as $\frac{1}{2}$, $\frac{1}{4}$, or $\frac{2}{5}$.

In the asteroid belt, orbital resonances occur between asteroids and Jupiter, the most massive planet by far. These resonances tend to clear gaps in the asteroid belt, much as resonances clear gaps in Saturn's rings. For example, any asteroid in an orbit that takes 6 years to circle the Sun—half of Jupiter's 12-year orbital period—would receive the same gravitational nudge from Jupiter every 12 years and thus would soon be pushed out of this orbit. The same is true for asteroids with orbital periods of 4 years ($\frac{1}{3}$ of Jupiter's period) and 3 years ($\frac{1}{4}$ of Jupiter's period). Music offers an elegant analogy: When a vocalist sings into an open piano, the strings of notes "in resonance" with the voice—not just the note being sung but also notes with a half or a quarter of the note's frequency—are "nudged" and begin to vibrate.

We can see the results of these orbital resonances on a graph of the number of asteroids with various orbital periods (Figure 12.5). For example, notice the lack of asteroids with periods exactly $\frac{1}{2}$, $\frac{1}{3}$, or $\frac{1}{4}$ of Jupiter's—periods for which the gravitational tugs from Jupiter have cleared gaps in the asteroid belt. (The gaps are often called *Kirkwood gaps*, after their discoverer.)

The orbital resonances with Jupiter likely also explain why no planet ever formed between Mars and Jupiter. When the solar system was forming, this region of the solar nebula probably contained enough rocky material to form another planet as large as Earth or Mars. However, resonances with the young planet Jupiter disrupted the orbits of this region's planetesimals, preventing them from accreting into a full-fledged planet. Over the ensuing 4.5 billion years, ongoing orbital disruptions gradually kicked pieces of this "unformed planet" out of the asteroid belt altogether. Once booted from the asteroid belt, these objects either crashed into a planet or moon or were flung out of the solar system. The asteroid belt thereby lost most of its original mass, which explains why the total mass of all its asteroids is now less than that of any terrestrial planet.

THINK ABOUT IT

Why are the gaps due to orbital resonances so easy to see in Figure 12.5 but so hard to see in Figure 12.4? (*Hint:* The top axis in Figure 12.5 is the *average* orbital distance.)

The asteroid belt is still undergoing slow change today. Jupiter's gravity continues to nudge asteroid orbits, sending asteroids on collision courses with each other and occasionally the planets. A major collision occurs somewhere in the asteroid belt every 100,000 years or so. Thus, over long

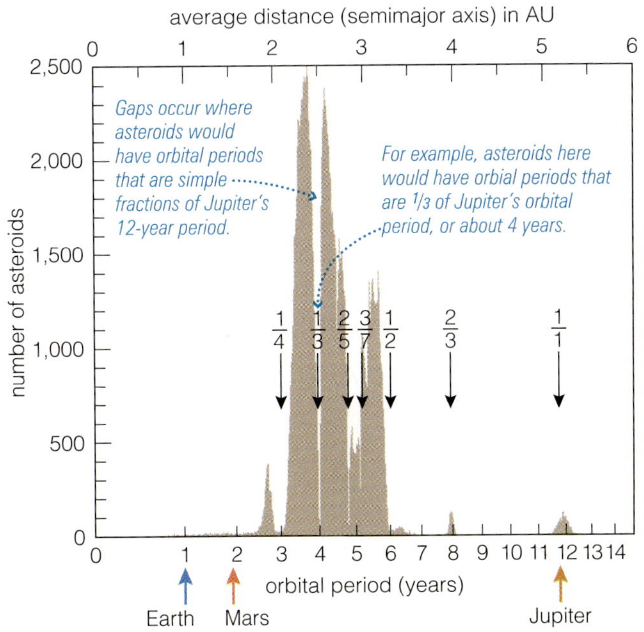

Figure 12.5 This graph shows the number of asteroids with various orbital periods, which correspond to different average distances from the Sun (labeled along the top). Notice the gaps created by orbital resonances with Jupiter.

periods of time, larger asteroids continue to be broken into smaller ones, with each collision also creating numerous dust-size particles. The asteroid belt has been grinding itself down for more than 4 billion years and will continue to do so for as long as the solar system exists.

Most orbital resonances result in gaps, but some happen to gather asteroids. Such is the case for the Trojan asteroids. Their orbits remain stable in the zones 60° ahead of and behind Jupiter, because any asteroid that wanders away from one of theses zones is nudged back into the zone by Jupiter's gravity. That explains why we see a grouping of asteroids, rather than a gap, at the $\frac{1}{1}$ resonance (where orbital periods are the same as Jupiter's) in Figure 12.5.

Figure 12.6 On March 26, 2003, a meteorite crashed through the roof of this home in Chicago. The main photo shows the hole in the ceiling; the inset shows the meteorite at rest on the floor. No one was hurt.

• Where do meteorites come from?

Because asteroids are leftovers from the birth of our solar system, studying them should teach us a lot about how Earth and the other planets formed. We can study asteroids with telescopes and spacecraft, but it would be far better if we could study an asteroid sample in a laboratory. No spacecraft has yet returned an asteroid sample, although if all goes well, the Japanese mission Hayabusa will return an asteroid sample to Earth in 2007. However, we already have pieces of asteroids nonetheless—the rocks called **meteorites** that fall from the sky.

The Difference Between Meteors and Meteorites Before we discuss what we learn from meteorites, let's be clear about what they are. In everyday language, people often use the terms *meteors* and *meteorites* interchangeably. Technically, however, they are not the same. A **meteor**, literally "a thing in the air," is only a flash of light caused by a particle entering our atmosphere at high speed, not the particle itself. Meteors are also sometimes called *shooting stars* or *falling stars*, because some people once thought they really were stars falling from the sky.

The vast majority of the particles that make meteors are no larger than peas and burn up completely before reaching the ground. Only in rare cases is a meteor caused by a chunk of rock large enough to survive the plunge through our atmosphere and leave a *meteorite* on the ground. Thus, a meteorite is an actual rock that has crashed to Earth. (The word *meteorite* means "associated with meteors.") Rocks from space that are large enough to become meteorites make unusually bright meteors, called *fireballs*, as they fall through our atmosphere. Observers find a few meteorites each year by following the trajectories of fireballs. Meteorites found in this way can be especially valuable to science, because we can trace their trajectories back to learn exactly where they came from and because they can be collected before they have much time to be contaminated by terrestrial material.

Meteorite Falls People didn't always accept that rocks fall from the sky. Stories of such events arose occasionally in human history, and sometimes even affected it. For example, stories of "fallen stars" influenced the philosophy of the ancient Greek scientist Anaxagoras (see Figure 3.13), who concluded that planets and stars are flaming rocks in the sky. Anaxagoras's assumption that meteorites fell from the heavens made him the first person in history known to believe that the heavens and Earth are made of the same materials, even though his guess about the nature of planets and stars was not quite correct. Many later scientists regarded stories of fallen stars more skeptically. Upon hearing of a meteorite fall in Connecticut, Thomas Jefferson (who took part in science as well as politics) reportedly said, "It is easier to believe that Yankee professors would lie than that stones would fall from heaven."

Today we know that rocks really do sometimes fall from the heavens (Figure 12.6). Meteorites are often blasted apart in their fiery descent through our atmosphere, scattering fragments over an area several kilometers across. A direct hit on the head by a meteorite would be fatal, but we have no reliable accounts of human deaths from meteorites. Of course, most meteorites fall into the ocean, which covers three-fourths of Earth's surface.

Unless you actually see a meteorite fall, it can be difficult to distinguish meteorites from terrestrial rocks. Fortunately, a few clues can help. Meteorites are usually covered with a dark, pitted crust resulting from their fiery passage through the atmosphere (Figure 12.7). Some have an unusually high

Figure 12.7 This large meteorite, called the Ahnighito Meteorite, is located at the American Museum of Natural History in New York. Its dark, pitted surface comes from its fiery passage through Earth's atmosphere.

Figure 12.8 There are two basic types of meteorites: primitive and processed. They are shown slightly larger than actual size. (The meteorites have flat faces because they have been sliced with rock saws.)

metal content, enough to attract a magnet hanging on a string. The ultimate judge of extraterrestrial origin is laboratory analysis: Meteorites often contain elements such as iridium that are very rare in Earth rocks, and even common elements in meteorites tend to have different ratios among their isotopes [Section 5.3] than rocks from Earth. If you suspect that you have found a meteorite, many museums will analyze a small chip free of charge.

Scientists have found and cataloged more than 20,000 meteorites, some found by accident and others by organized meteorite searches. Antarctica is one of the best places to hunt for meteorites, not because more fall there but because the icy surface makes them easier to find. Few terrestrial rocks end up on the ice, so a rock found on the ice has a good chance of being a meteorite. Moreover, the slow movement of Antarctic glaciers tends to carry meteorites along, concentrating them in small areas where the ice floes run into mountains. The majority of meteorites now in museums and laboratories were found in Antarctica.

Types of Meteorites The precise origin of meteorites was long a mystery, but in recent decades we've been able to determine where in our solar system they come from. The most direct evidence comes from the relatively few meteorites whose trajectories have been observed or filmed as they fell to the ground. In every case so far, these meteorites clearly originated in the asteroid belt.

Detailed analysis of thousands of meteorites shows that they come in two basic types, each of which can be further classified into two subtypes:

- **Primitive meteorites** (Figure 12.8a) are "primitive" in the sense of being the first (or "primary") type of meteorite to have formed. Radiometric dating [Section 8.5] shows them to be about 4.6 billion years old, making them remnants from the birth of our solar system, essentially unchanged since they first accreted in the solar nebula. Primitive meteorites come in two subtypes:
  - *Stony primitive meteorites* are composed of rocky minerals with a small but noticeable fraction of pure metallic flakes mixed in.
  - *Carbon-rich primitive meteorites* are generally similar to the stony primitive meteorites, but they also contain substantial amounts of carbon compounds. Some carbon-rich meteorites even contain a small amount of water.
- **Processed meteorites** (Figure 12.8b) apparently once were part of a larger object that "processed" the original material of the solar nebula into another form. Radiometric dating shows that processed meteorites are generally younger than the primitive meteorites by a few hundred million years. The processed meteorites can also be divided into two subtypes:
  - *Metal-rich processed meteorites* are made mostly of high-density iron and nickel mixed with smaller amounts of other metals. Thus, they resemble the terrestrial planet cores in composition.
  - *Rocky processed meteorites* have lower densities and are made of rock with compositions resembling that of terrestrial mantles and crusts. A few have compositions remarkably close to that of the basalts [Section 9.2] that erupt from terrestrial volcanoes.

The Origin of Primitive Meteorites The vast majority of all meteorites are primitive, and their structures confirm what their radiometric ages tell us: They are pieces of rock that accreted in the solar nebula and orbited the Sun for billions of years before finally falling to Earth. The individual flakes may represent the tiny particles that first condensed from the gas of the solar nebula. The small, roundish features visible in Figure 12.8a may be solidified droplets splashed out in the accretion process.*

Why are some primitive meteorites stony in composition while others are carbon-rich? The answer lies in where they formed in the solar nebula. Throughout the asteroid belt, temperatures were too warm for ices to condense, so all asteroids are made of rock and metal. However, beyond about 3 AU from the Sun, temperatures were low enough for carbon compounds to condense. Thus, the carbon-rich primitive meteorites must have accreted in what are now the outer regions of the asteroid belt, while the stony primitive meteorites formed in the inner regions. Laboratory studies of primitive meteorites confirm these origins. The

*The roundish features are called *chondrules*, so primitive meteorites are technically known as *chondrites*. Processed meteorites are called *achondrites*, because they lack the chondrules (*a*- means "not").

composition of carbon-rich meteorites matches the composition of dark, carbon-rich asteroids in the outer part of the asteroid belt. The composition of stony meteorites matches the composition of asteroids in the inner part of the asteroid belt.

Curiously, most of the meteorites collected on Earth are of the stony variety even though most asteroids are of the carbon-rich variety. Evidently, orbital resonances are more effective at pitching material our way from the inner part of the asteroid belt than from more distant regions of the belt.

The Origin of Processed Meteorites The processed meteorites tell a more complex story. Their compositions look similar to the cores, mantles, or crusts of the terrestrial worlds. Thus, they must be fragments of larger asteroids that underwent *differentiation* [Section 9.1] in which their interiors melted so that metals sank to the center and rocks rose to the surface. This idea explains why we find two types of processed meteorites.

The rocky processed meteorites are often so close in composition to volcanic rocks on Earth that they must have been made by lava flows. We conclude that these meteorites were essentially chipped off the surfaces of asteroids that once had active volcanism, probably by collisions with smaller asteroids. Perhaps as many as a dozen large asteroids were geologically active shortly after the formation of the solar system. The old ages of processed meteorites tell us that this active period was short-lived—the interiors of these asteroids must have cooled quickly.

Metal-rich processed meteorites are less common. They have core-like compositions, suggesting that they are fragments of larger asteroids that shattered in collisions. These shattered worlds not only lost a chance to grow into planets but sent many fragments onto collision courses with other asteroids and other planets, including Earth. Thus, processed meteorites essentially present us with an opportunity to study a "dissected planet." They also represent a form of direct proof that large worlds really do undergo differentiation, confirming what we infer from seismic studies of Earth.

A forthcoming NASA mission called *Dawn* should tell us more about the connection between meteorites and asteroids. If all goes well, *Dawn* will visit several asteroids for close-up study, beginning in about 2011. Both Ceres and Vesta are included in *Dawn*'s planned itinerary. Vesta may be particularly revealing, because scientists suspect that we already have pieces of it in our laboratories: Spectroscopy tells us that the rocky processed meteorite shown in Figure 12.8b matches Vesta in composition, suggesting it was chipped from Vesta's surface.

Meteorites from the Moon and Mars In a few cases, the compositions of processed meteorites do not appear to match any known asteroids. Instead they appear to match either the Moon or Mars.

Careful analysis of these meteorite compositions makes us very confident that we really do have a few meteorites that were once part of the Moon or Mars. Moderately large impacts can blast surface material from terrestrial worlds into interplanetary space. Once they are blasted into space, the rocks orbit the Sun until they come crashing down on another world. Calculations show that it is not surprising that we should have found a few meteorites chipped off the Moon and Mars in this way.

Study of these *lunar meteorites* and *Martian meteorites* is providing new insights into conditions on the Moon and Mars. In at least one case, a Martian meteorite may offer clues about whether life ever existed on Mars [Section 24.2].

12.2 Comets

Asteroids are one of the two major categories of small bodies in the solar system. The other is comets, to which we now turn our attention. Asteroids and comets have much in common. They are both leftover planetesimals from the birth of our solar system. They both come in a wide range of sizes. The primary difference between them is their composition, which reflects where they formed. Asteroids are rocky because they formed in the inner solar system where metal and rock condensed. Comets are ice-rich because they formed beyond the frost line, where abundant hydrogen compounds condensed into ice [Section 8.3]. Note that we refer to any icy leftover planetesimal orbiting the Sun as a comet, regardless of its size, whether it has a tail, or where it resides or comes from.

• What are comets like?

For most of human history, comets were familiar only from their occasional presence in the night sky. Every few years, a comet becomes visible to the naked eye, appearing as a fuzzy ball with a long tail. Indeed, the word *comet* comes from the Greek word for "hair," because comets were recognized by the long, hairlike tails they can display in our sky (Figure 12.9).

In comet photographs, the tails make it look as if the comets are racing across the sky, but they are not. If you watch a comet for minutes or hours, you'll see it staying nearly stationary relative to the stars around it in the sky. Over many days it will rise and set just like the stars, while gradually moving relative to the constellations. You may be able to see it night after night for several weeks or more, until it finally fades from view.

Today, we know that the vast majority of comets do not have tails and never venture anywhere close to Earth. Instead, they remain perpetually in the far outer reaches of our solar system, orbiting the Sun far beyond the orbit of Neptune. The comets that appear in the night sky are the rare ones that have had their orbits changed by the gravitational influences of planets, other comets, or stars passing by in the distance. Their new orbits may carry them much closer to the Sun, eventually bringing them into the inner solar system, where they may grow the tails that allow us to see them in the sky. Most of these comets will not return

a Comet Hyakutake.

b Comet Hale-Bopp, photographed at Mono Lake, CA.

Figure 12.9 Interactive Photo Brilliant comets can appear at almost any time, as demonstrated by the back-to-back appearances of comet Hyakutake in 1996 and comet Hale-Bopp in 1997.

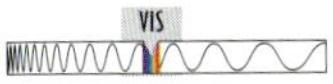

to the inner solar system for thousands of years, if ever. A few happen to pass near enough to a planet to have their orbits changed further, and some end up on elliptical orbits that periodically bring them close to the Sun.

Comets in Human History The occasional presence of a bright comet in the night sky was hard to miss before the advent of electric lights. In some cultures, these rare intrusions into the unchanging heavens foretold bad or good luck. The ancient Chinese, for example, believed that the appearance of a comet led to a time of major and tumultuous change.

Few ancient cultures made any attempt to explain comets in astronomical terms. In fact, comets were generally thought to be within Earth's atmosphere until 1577, when Tycho Brahe [Section 3.3] used observations made from different locations on Earth to prove that a comet lay far beyond the Moon. A century later, Newton correctly deduced that comets orbit the Sun. Then, in a book published in 1705, English scientist Edmund Halley (1656–1742) used Newton's law of gravitation to calculate the orbit of a comet that had been seen in 1682. He showed that it was the same comet that had been observed on numerous prior occasions, which meant that it orbited the Sun every 76 years. Halley predicted that the comet would return in 1758. Although he had died by that time, the comet returned as he predicted and it was then named in his honor. Thus, Halley did not really "discover" his comet but rather gained fame for being the first to correctly calculate a comet's orbit and predict its return. Halley's comet last passed through the inner solar system in 1986, and it will return in 2061.

Today, both professional and amateur astronomers routinely scan the skies with telescopes and binoculars in hopes of discovering a new comet. Several comets are discovered every year. The first observers (up to three) to report a comet discovery to the International Astronomical Union have the comet named after them. Thus, comets now bring a form of good luck to their discoverers. More generally, comets are good luck for astronomers seeking to understand the solar system. These leftover planetesimals teach us about the history of the outer solar system just as asteroids teach us about the history of the inner solar system.

The surprise champion of comet discovery is the *SOHO* spacecraft, an orbiting solar observatory. *SOHO* has detected more than 1,000 "Sun-grazing" comets, most on their last pass by the Sun (Figure 12.10).

Comet Composition We have learned about the composition of comets by studying their spectra. The results are just what we expect for objects that formed in the cold outer solar system: Comets are basically chunks of ice mixed with rocky dust and some more complex chemicals, and hence they are often described as "dirty snowballs."

Spectra of comets confirm a distant origin because they show the presence of compounds that could have condensed only in the cold outer regions of the solar nebula. For example, comet spectra invariably show emission features from hydrogen compounds, including water. They also show emission from carbon dioxide and carbon monoxide, gases that condensed only in the very coldest and most distant regions of the solar nebula. Comet spectra also show many more complex molecules, including some organic molecules, leading some scientists to speculate that much of the organic material that made life possible on Earth was brought here by comets.

Figure 12.10 Comet SOHO-6's final blaze of glory. The *SOHO* spacecraft observed this "Sun-grazing" comet a few hours before it passed just 50,000 kilometers above the Sun's surface. The comet did not survive its passage, due to the intense solar heating and tidal forces. In this image, the large, orange disk blocks out the Sun; the Sun's size is indicated by the white circle.

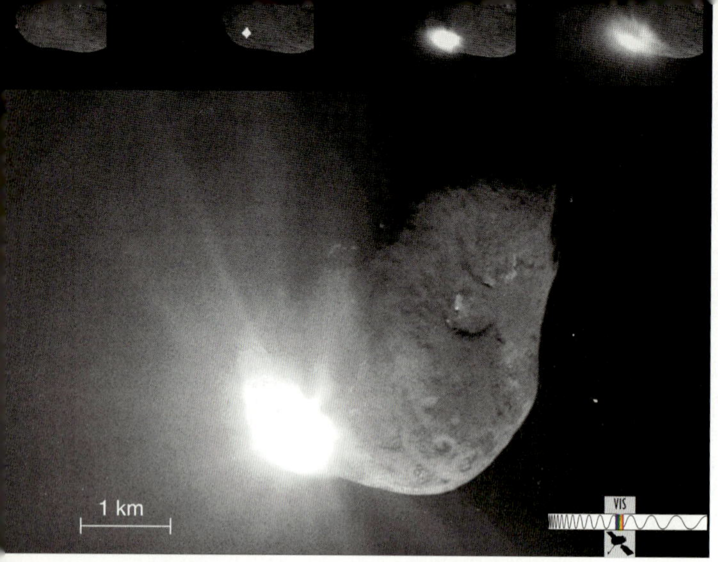

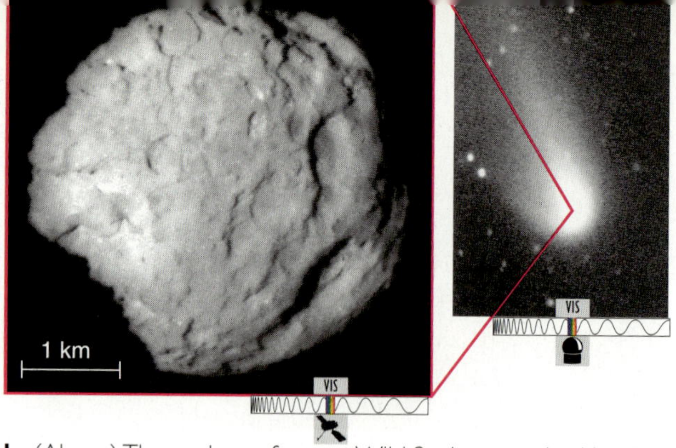

b (Above) The nucleus of comet Wild 2, photographed by the *Stardust* spacecraft along with a ground-based image of the comet. Although not visible in the *Stardust* photograph, the spacecraft found violent jets of gas and dust shooting outward from the comet's nucleus. The irregular surface probably shows effects from a combination of impacts and uneven sublimation rates in different regions.

a These photos were taken by the Deep Impact spacecraft as its 370-kg impactor crashed into Comet Tempel 1. The entire sequence, which starts just before impact at the upper left, unfolded over just 67 seconds.

Figure 12.11 Comet nuclei.

Telescopic spectroscopy tells us about the material comets shed into space, but is this material representative of their internal compositions? NASA's *Deep Impact* mission was designed to answer this and other questions about comets. The main spacecraft carried a 370-kg "impactor" that it released as it approached Comet Tempel 1. Instruments on the spacecraft then observed the results as the impactor slammed into the comet at a speed of 37,000 kilometers per hour on July 4, 2005 (Figure 12.11a). The data were only beginning to be analyzed as this book went to press, but the impactor apparently penetrated many meters into a relatively soft and porous surface before it vaporized due to the energy of the collision. The size of the ejected plume of debris exceeded predictions, giving scientists confidence that careful study will teach us a lot about what comets are made of.

We'll learn even more about comet composition when we can study comet material in the laboratory—which should be happening by the time you read this book. In early 2004, NASA's *Stardust* spacecraft passed near comet Wild 2 (pronounced "Vilt 2"), using a piece of foamed glass called "aerogel" to capture dust particles that the comet had shed into space (Figure 12.11b). *Stardust* returned safely to Earth on January 15, 2006.

THINK ABOUT IT

Do a web search on the *Stardust* mission. What have we learned from it?

The Flashy Lives of Comets Comets grow tails only as they enter the inner solar system, where they are heated by the warmth of the Sun. To see what happens, let's follow the comet path shown in Figure 12.12a, and the comet anatomy in Figure 12.12b.

Far from the Sun, the comet is completely frozen—the "dirty snowball." It begins to shed gas and dust as it approaches the Sun, but it still has its central chunk of ice, which we call the **nucleus** of the comet. Our best views of comet nuclei have come during spacecraft flybys: the first was a flyby of Halley's comet by the European Space Agency's *Giotto* spacecraft in 1986, followed more recently by the Deep Impact and Stardust missions. Despite their icy compositions, comet nuclei are darker than charcoal, reflecting less than 5% of the light that falls on them. Apparently, it doesn't take much rocky soil or carbon-rich material to darken a comet. Density measurements show that comet nuclei are not packed very solidly. For example, the density of Halley's nucleus is considerably less than the density of water (1 g/cm^3), suggesting that the nucleus is part ice and part empty space.

The scale bars in Figures 12.11 and 12.12b show that comet nuclei are typically less than about 20 km across. How can such small objects put on such spectacular shows? As a comet accelerates toward the Sun, its surface temperature increases, and ices begin to sublimate into gas that easily escapes the comet's weak gravity. Some of the escaping gas drags away dust particles from the nucleus, and the gas and dust create a huge, dusty atmosphere called a **coma**. The coma is far larger than the nucleus it surrounds. Spacecraft observations show that the release of gas from the comet nucleus can be quite violent, with jets of gas and dust shooting out at speeds of hundreds of meters per second from the interior of the nucleus into space.

The coma grows as the comet continues into the inner solar system, and tails form as extensions of the coma. Comet tails can be hundreds of millions of kilometers in length. Because they look much like exhaust trails behind rockets, many people mistakenly guess that tails extend behind comets as they travel through their orbits. In fact, comet tails generally point away from the Sun, regardless of the direction in which the comet is traveling.

Comets actually have two visible tails, one made of ionized gas, or *plasma,* and the other made of dust. The **plasma tail** consists of gas escaping from the coma. Ultraviolet light from the Sun ionizes the gas, and the solar wind then carries this gas straight outward from the Sun

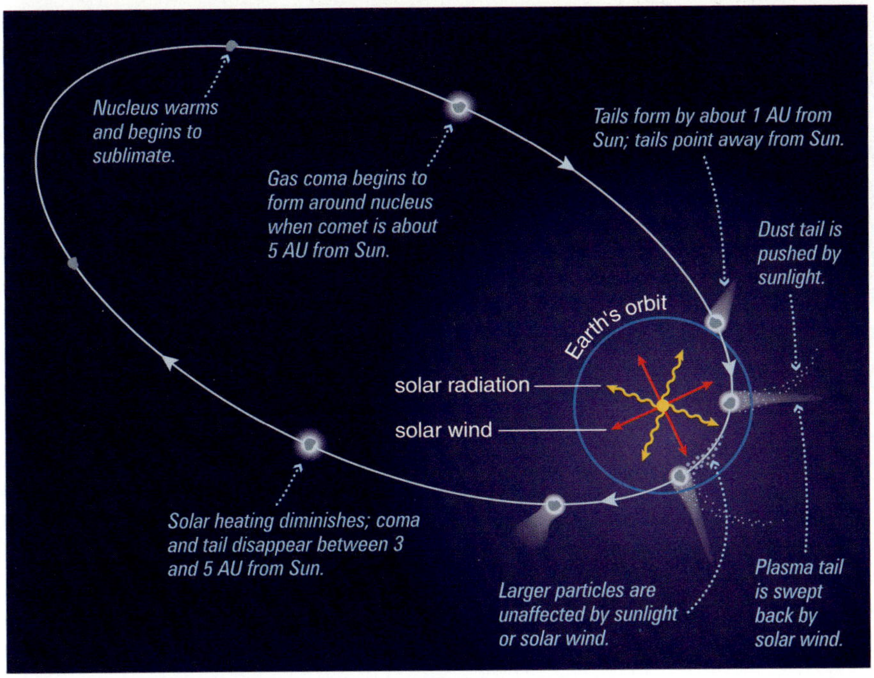

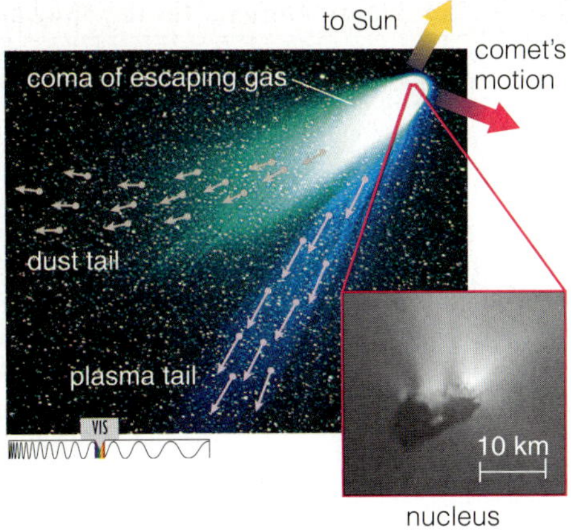

b (Above) Anatomy of a comet. The larger image shows a ground based photo of comet Hale-Bopp. The inset shows the nucleus of Halley's comet photographed by the *Giotto* spacecraft.

a (Left) This diagram shows the changes that occur when a comet's orbit takes it on a passage into the inner solar system. (Not to scale.)

Figure 12.12 A comet grows a coma and tail around its nucleus only if it happens to come close to the Sun. Most comets never do this, instead remaining perpetually frozen in the far outer solar system.

at speeds of hundreds of kilometers per second. Thus, the plasma tail extends almost directly away from the Sun at all times. The **dust tail** is made of dust-size particles escaping from the coma. These particles are not affected by the solar wind and instead are pushed away from the Sun by the much weaker pressure of sunlight itself (*radiation pressure*). Thus, while the dust tail points generally away from the Sun, it has a slight curve back in the direction the comet came from.

After the comet loops around the Sun and begins to head back outward, sublimation declines, the coma dissipates, and the tails disappear. Nothing happens until the comet again comes sunward—in a century, a millennium, a million years, or perhaps never.

Comets that repeatedly visit the inner solar system, like Halley's comet, cannot last long on the time scale of our solar system. A comet probably loses about 0.1% of its ice on every pass around the Sun, so it could not make more than a few hundred passages before losing most of its original ice. Changes in composition may end the comet's "life" even faster. In a close pass by the Sun, a comet may shed a layer of material a meter thick. Dust that is too heavy to escape accumulates on the surface. This thick, dusty deposit helps make comets dark and may eventually block the escape of interior gas, preventing the comet from growing a coma or tails on future passes by the Sun. No one is certain what happens to a comet after its ices can no longer sublime into gas and escape. Either the remaining dust disguises the dead comet as an asteroid, or the comet comes "unglued" without the binding effects of ices. In the latter case, the comet may simply disintegrate along its orbit.

Comets striking the Earth do not strew meteorites across the surface as asteroids do. The icy materials of a small comet vaporize on entry, and the rocky material in comets takes the form of dust or pebbles too small to survive the passage through Earth's atmosphere.

Comet Tails and Meteor Showers In addition to the plasma and dust in their two visible tails, the gas escaping from comets also carries away sand- to pebble-size pieces of rocky material. These particles are too big to be affected by either the solar wind or sunlight, so they drift away slowly and spread along the orbital path. These particles essentially form a third, invisible tail that follows the comet around its orbit. They are also the particles responsible for most meteors and meteor showers.

We see a meteor lighting up the sky when one of these small particles (or a similar-size particle from an asteroid) burns up in our atmosphere. The sand- to pebble-size particles are much too small to be seen themselves, but they enter the atmosphere at such high speed—up to about 250,000 km/hr (70 km/s)—that they make the surrounding air glow with heat. It is this glow that we see as the brief but brilliant flash of a meteor. The small particles are vaporized by the heat and thus never reach the ground. An estimated 25 million particles of comet dust enter the atmosphere worldwide every day, burning up as meteors and adding hundreds of tons of comet dust to Earth daily.

Even more comet dust enters the atmosphere during a **meteor shower**. Some comet dust is sprinkled throughout the inner solar system, but the "third tails" of ejected particles make the dust much more concentrated along the orbits of comets. Thus, meteor showers occur when Earth crosses through a comet's orbit. Meteor showers recur at about the same time each year because the orbiting Earth passes through a particular comet's orbit at the same time each year. For example, the meteor shower known as the *Perseids* occurs about August 12 each year—the time when

Table 12.1 Major Annual Meteor Showers

Shower Name	Approximate Date	Associated Comet
Quadrantids	January 3	?
Lyrids	April 22	Thatcher
Eta Aquarids	May 5	Halley
Delta Aquarids	July 28	?
Perseids	August 12	Swift-Tuttle
Orionids	October 22	Halley
Taurids	November 3	Encke
Leonids	November 17	Tempel-Tuttle
Geminids	December 14	Phaeton
Ursids	December 23	Tuttle

Earth passes through the orbit of a comet called Swift-Tuttle. Table 12.1 lists major annual meteor showers and their parent comet, if known.

THINK ABOUT IT

The associated "comet" for the Geminid meteor shower is an object called Phaeton, which is classified as an asteroid because we've never seen a coma or tail associated with it. How is it possible that Phaeton looks like an asteroid today but once shed the particles that create the annual Geminid meteor shower?

You can typically see a few meteors each hour on any clear night, but you may see far more during one of the annual meteor showers. During a "good" shower at a dark site, you may see dozens of meteors per hour. Remember that you are watching the effects of comet dust entering our planet's atmosphere from space. The meteors generally appear to radiate from a particular direction in the sky, for essentially the same reason that snow or heavy rain seems to come from a particular direction in front of a moving car (Figure 12.13). Because more meteors hit Earth from the front than from behind (just as more snow hits the front windshield of a moving car), meteor showers are best observed in the predawn sky, which is the time of day when part of your sky faces in the direction of Earth's motion.

• Where do comets come from?

We've already stated that comets come from the outer solar system, but we can be much more specific. By analyzing the orbits of comets that pass close to the Sun, scientists learned that there are two major "reservoirs" of comets in the outer solar system.

Most comets that visit the inner solar system follow orbits that seem almost random. They do not orbit the Sun in the same direction as the planets, and their elliptical orbits can be pointed in any direction. Moreover, their orbits show that they come from far beyond the orbits of the planets—sometimes nearly a quarter of the distance to the nearest star. These comets must come plunging sunward from a vast, spherical region of space that scientists call the **Oort cloud** (after astronomer Jan Oort; rhymes with "court"). Be sure to note that the Oort cloud is not a cloud of gas, but rather a collection of many individual comets. Based on the number of Oort cloud comets that enter the inner solar system each year, we conclude that the Oort cloud must contain about a trillion (10^{12}) comets orbiting the Sun at distances up to about 50,000 AU.

A smaller number of the comets that visit the inner solar system have a pattern to their orbits. They travel around the Sun in the same plane and direction as the planets on elliptical orbits that carry them no more than about twice

Figure 12.13 The geometry of meteor showers.

a Meteors appear to radiate from a particular point in the sky for the same reason that we see snow or heavy rain come from a single point in front of a moving car.

b This digital composite photo, taken in Australia during the 2001 Leonid meteor shower, shows meteors as streaks of light. The large rock is Uluru, also known as Ayers Rock.

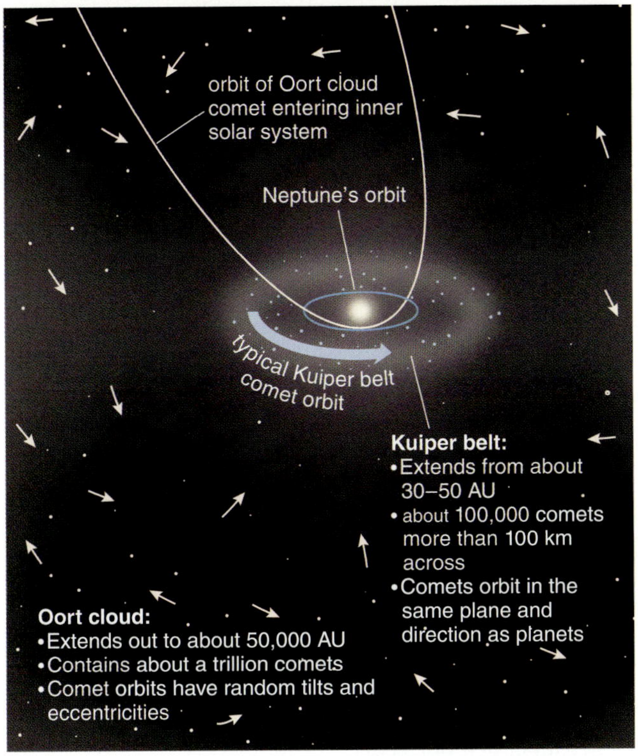

Figure 12.14 The comets we occasionally see in the inner solar system come from two major reservoirs of comets in the outer solar system: the Kuiper belt and the Oort cloud.

as far from the Sun as Neptune. These comets must come from a ring of comets that orbit the Sun beyond the orbit of Neptune. This ring is usually called the **Kuiper belt** (after astronomer Gerald Kuiper; rhymes with "piper"). Figure 12.14 contrasts the general features of the Kuiper belt and the Oort cloud.

The Origin of Comets How did comets end up in these far-flung regions of the solar system? The only answer that makes scientific sense comes from thinking about what happened to the icy, leftover planetesimals that roamed the region in which the jovian planets formed.

The leftover planetesimals that cruised the spaces between Jupiter, Saturn, Uranus, and Neptune were doomed to suffer either a collision or a close gravitational encounter with one of the young jovian planets. Recall that when a small object passes near a large planet, the planet is hardly affected but the small object may be flung off at high speed [Section 4.5]. Thus, the planetesimals that escaped being swallowed up by the jovian planets tended to be flung off in all directions. Some may have been cast away at such high speeds that they completely escaped the solar system and now drift through interstellar space. The rest ended up on orbits with very large average distances from the Sun, becoming the comets of the Oort cloud. The random directions in which they were flung explains why the Oort cloud is roughly spherical in shape. Oort cloud comets are so far from the Sun that they can be nudged by the gravity of nearby stars, preventing some of them from ever returning near the planets and sending others plummeting toward the Sun.

Gravitational encounters were much less likely for icy planetesimals that roamed in relative safety beyond the orbits of the jovian planets. These planetesimals tended to keep the orbits on which they formed in the spinning disk of the solar nebula—orbits that go in the same direction as planetary orbits and that are concentrated relatively near the ecliptic plane. Many of these planetesimals still have these orbits today, which is why they populate the ring-shaped Kuiper belt. Most remain perpetually beyond the orbit of Neptune, but gravitational resonances with the jovian planets occasionally nudge an orbit enough to send a Kuiper belt comet plunging inward toward the Sun. The Kuiper belt extends to only about twice Neptune's orbital distance, because at greater distances the density of the solar nebula was apparently too low to allow the accretion of planetesimals.

To summarize, the comets of the Kuiper belt seem to have originated farther from the Sun than the comets of the Oort cloud, even though the Oort cloud comets are now much more distant. The Oort cloud consists of leftover planetesimals that were flung outward after forming between the jovian planets, while the Kuiper belt consists of leftover planetesimals that formed and still remain in the outskirts of the planetary realm, roughly between the orbit of Neptune and twice Neptune's distance from the Sun.

Formation of the Solar System Tutorial, Lesson 3

12.3 Pluto: Lone Dog No More

You first began to learn about the solar system some time back in elementary school. You probably heard then about asteroids and comets, and you certainly would have been taught about the "nine planets." At this point in the text, we've discussed in detail the four terrestrial planets, the four jovian planets, asteroids, and comets. That leaves only one object left to discuss from your elementary school list: the "ninth planet," Pluto.

Pluto seemed a misfit among the planets almost since the day it was discovered in 1930 (see Special Topic, p. 331). We knew from the start that its 248-year orbit is more elliptical and more inclined to the ecliptic plane than that of any other planet (Figure 12.15). Note that, despite the fact that Pluto sometimes comes closer to the Sun than Neptune, there is no danger of collision: Neptune orbits the Sun precisely three times for every two Pluto orbits, and this stable orbital resonance means that Neptune is always a safe distance away whenever Pluto approaches its orbit.

As we learned more about Pluto, it proved to be a misfit in size and composition as well: It is far smaller than even the terrestrial planets, and its ice-rich composition fits neither the terrestrial nor jovian categories. These seeming oddities long made Pluto seem like a lone dog of the outer solar system. But it is not: We now know that Pluto is part of a huge pack of similar objects—essentially large comets—that orbit in its region of the solar system. In fact, Pluto is not even the largest member of this pack.

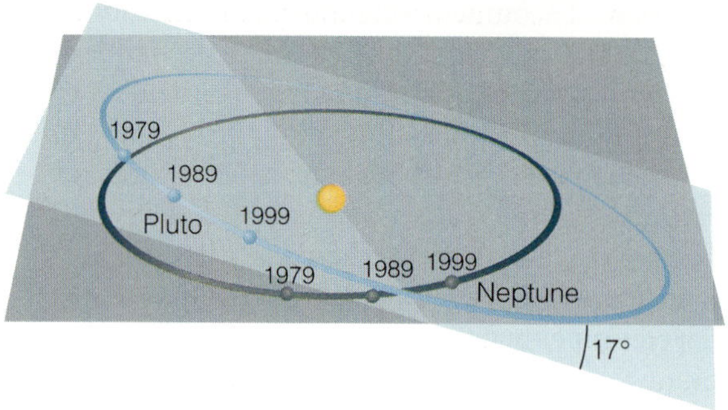

Figure 12.15 Interactive Figure Pluto's orbit is significantly elliptical and tilted relative to the ecliptic. It comes closer to the Sun than Neptune for 20 years in each 248-year orbit, as was the case between 1979 and 1999. There's no danger of a collision, however, thanks to an orbital resonance in which Neptune completes three orbits for every two of Pluto's orbits.

• How big can a comet be?

We've known since the 1950s that many of the comets that visit the inner solar system come from the reservoir that we call the Kuiper belt. We've also known that Pluto orbits the Sun near the middle of this reservoir. Over the past couple decades, scientists gradually came to realize that Pluto's location is no mere coincidence, but rather evidence of an astonishing fact: Some of the objects that populate the Kuiper belt are far larger than any of the comets that we ever see in the inner solar system. This fact forces us to reconsider just exactly what we mean by a "comet."

All the comets we've observed in the inner solar system are fairly small, with nuclei no larger than about 20 kilometers across. This shouldn't be surprising based on our understanding of comet origins, because only relatively small objects are likely to have their orbits changed enough to send them from the Kuiper belt or Oort cloud into the inner solar system. But it should make us wonder: Given that the comets we see are just rare and relatively small visitors from distant reaches of the solar system, could these outer realms hold much larger objects as well?

According to our understanding of solar system formation, the icy planetesimals of the Kuiper belt should have been able to continue their accretion as long as other icy particles were nearby. In principle, one of these planetesimals could have grown large enough to become the seed of a fifth jovian planet, beyond Neptune, but that did not occur—probably because the density of material was too low at this great distance from the Sun. Nevertheless, it's reasonable to think that many of them grew to hundreds or even thousands of kilometers in diameter.

You might think it would be easy to observe icy objects of such sizes, but it requires powerful telescope technology. Trying to see a 1000-kilometer iceball at Pluto's distance from Earth is equivalent to looking for a snowball the size of your fist that is 600 kilometers away (and in dim light, too). Aside from Pluto, the first large objects in the Kuiper belt were discovered in the early 1990s. The discoveries have come at a rapid pace ever since.

As astronomers surveyed more and more of the sky in search of large, icy objects, the record size seemed to rise with each passing year. In 2002, the record was set by an object named *Quaoar* (pronounced *kwa-o-whar*), which is more than half the diameter of Pluto. Two objects announced in 2004—2004DW and Sedna—are between about two-thirds and three-quarters of Pluto's size. It seemed only a matter of time until scientists would find an object larger than Pluto, and the time came on July 29, 2005, when Caltech astronomer Michael Brown announced the discovery of the object now named Eris (Figure 12.16). Eris is named for a Greek goddess who caused strife and arguments among humans (a commentary on the arguments its discovery caused about the defintion of "planet").

Eris currently lies about twice as far as Pluto from the Sun. Its 560-year orbit is inclined to the ecliptic plane more than twice as much as Pluto's (which is shown in Figure 12.15), and it is so eccentric that it must sometimes come closer to the Sun than Pluto. Hubble Space Telescope observations indicate that Eris is about 5% larger than Pluto. Pluto has therefore fallen to second place in the rankings of large objects in the Kuiper belt, and it may yet fall further—as of 2006, only about half the sky had been carefully searched for large Kuiper belt objects.

What should we call all these big iceballs? The terminology has been a topic of much debate, but the science behind it seems clear. By 2006, more than 1,000 icy objects had been directly observed in the Kuiper belt (Figure 12.17), allowing scientists to infer that the region contains at least 100,000 objects more than 100 kilometers across. Clearly, the objects that reside in the Kuiper belt come in a wide range of sizes. Some are so small that their orbits can be easily perturbed, sending them plunging toward the Sun so that we see them as comets, while others are as big or bigger than Pluto, which was long considered a "planet." Regardless of size, however, the members of the Kuiper belt probably all share the same basic composition of ice and rock, making them essentially comets of different sizes. That

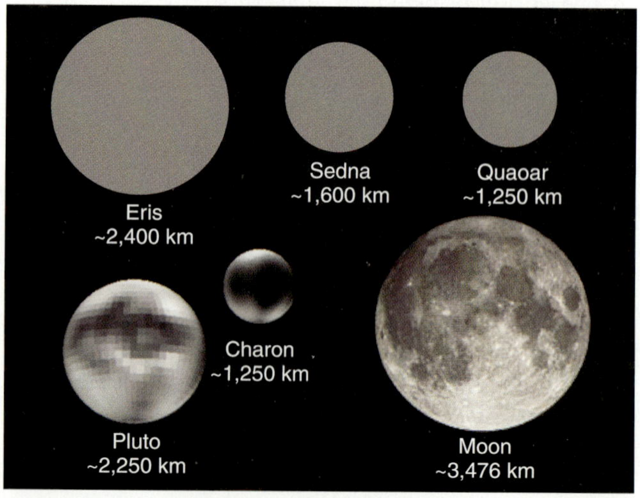

Figure 12.16 Diameters to scale of five of the largest objects in the outer solar system, along with the Moon for comparison.

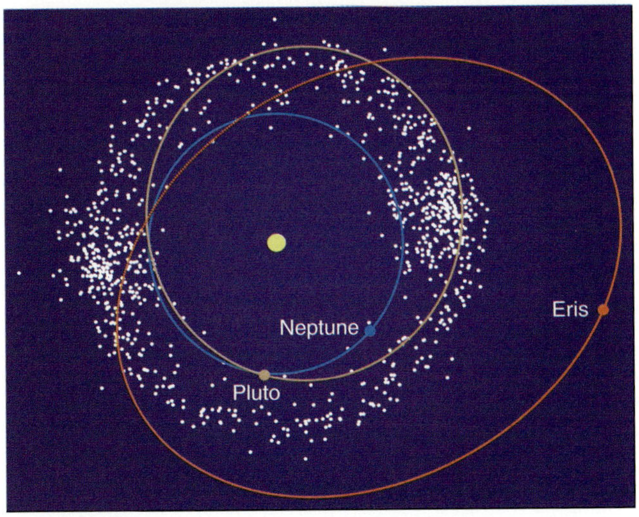

Figure 12.17 Approximate current positions of known members of the Kuiper Belt, with the orbits of Neptune, Pluto, and Eris shown for reference. Remember that many Kuiper belt comets, including Pluto and Eris, have large orbital inclinations (tilts relative to the ecliptic plane) that are not shown in this diagram.

is why we often refer to all of them as *comets* of the Kuiper belt. However, some astronomers object to calling such large objects "comets," especially since they never venture into the inner solar system to grow comas or tails. As a result, you may also hear these objects referred to as Kuiper Belt Objects (KBOs), Trans-Neptunian Objects (TNOs), or a variety of other names.

• What are the large objects of the Kuiper belt like?

Although Pluto, Eris, and other similar objects are "large" compared to most comets of the Kuiper belt, they are still quite small compared to any of the terrestrial or jovian planets. These small sizes, combined with their great distances, makes them very difficult to study. Even our best photographs show little more than fuzzy blobs; see, for example, the photo of Pluto and its moon Charon in Figure 7.12. More commonly, photographs show only a dot of light. As a result, we generally know little about these objects besides their orbits and the fact that their reflectivities and spectra suggest that they have ice-rich compositions. To understand these objects a little better, let's first consider Pluto and then turn our attention to the rest.

Pluto We've had more time to study Pluto than any other object of the Kuiper belt, but the most important reason we've been able to learn a lot about Pluto is its moon, Charon. Remember, for example, that we can determine an object's mass only if we can observe its gravitational effect on some other object. As a result, we didn't even know Pluto's mass well until Charon's discovery in 1978, after which Charon's orbital characteristics allowed us to find the combined Pluto–Charon mass with Newton's version of Kepler's third law [Section 4.4]. (Eris also has a moon, which will allow us to determine its mass in the same way.)

More precise data came from some good luck in timing: From 1985 to 1990, Pluto and Charon happened to be aligned in a way that made them eclipse each other every few days as seen from Earth—something that won't happen again for more than 100 years. Detailed analysis of brightness variations during these eclipses allowed the calculation of accurate sizes, masses, and densities for both Pluto and Charon. These measurements confirmed that Pluto and Charon are both made of ice mixed with rock—just like comets. The eclipse data even allowed astronomers to construct rough maps of Pluto's surface markings, which have not yet been explained (Figure 12.18).

Comparing Pluto and Charon leads to some interesting surprises. Charon's diameter is more than half Pluto's, its mass is about one-eighth of Pluto's, and it orbits only 20,000 kilometers from Pluto. For comparison, our Moon has a mass $\frac{1}{80}$ of Earth's and orbits 400,000 kilometers away. Thus, Charon is even larger relative to its parent planet than our Moon is to Earth. Moreover, Pluto is slightly higher in density than Charon. Together, these facts have led astronomers to guess that Charon was created by a giant impact similar to the one thought to have formed our Moon [Section 8.4]. A large comet crashing into Pluto may have blasted away its low-density outer layers, which then formed a ring around Pluto and eventually reaccreted to make Charon. The impact may also explain why Pluto rotates with its rotation axis tipped almost on its side (see Figure 7.1).

Pluto is very cold, with an average temperature of only 40 K, as we would expect at its great distance from the Sun. Nevertheless, Pluto currently has a thin atmosphere of nitrogen and other gases formed by sublimation of surface ices. The atmosphere should gradually refreeze onto the surface as Pluto's 248-year orbit carries it farther from the Sun, although recent results have puzzled astronomers by showing the opposite effect.

Despite the cold, the view from Pluto would be stunning. Charon would dominate the sky, appearing almost 10 times larger in angular size than our Moon appears from Earth. Pluto and Charon's mutual tidal pulls long ago made them rotate synchronously with each other [Section 4.5]. Thus, Charon is visible from only one side of Pluto and always shows the same face to Pluto. Moreover, the synchronous rotation means that Pluto's "day" is the same length as Charon's "month" (orbital period) of 6.4 Earth days. Viewed from Pluto's surface, Charon neither rises nor sets but instead hangs motionless as it cycles through its phases every 6.4 days. The Sun would appear more than a thousand

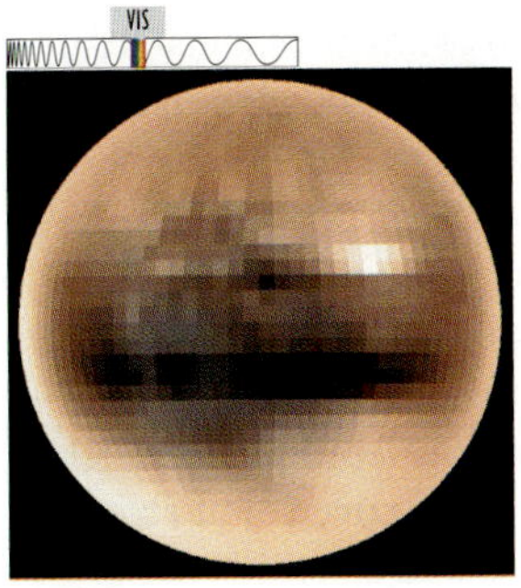

Figure 12.18 The surface of Pluto in approximate true color, as derived with computer processing from brightness measurements made during mutual eclipses between Pluto and Charon. The many squares in the image represent the smallest regions for which we could accurately estimate colors.

times fainter than it appears here on Earth and would be no larger in angular size than Jupiter appears in our skies.

Other Kuiper Belt Objects We don't know much about other large comets of the Kuiper belt, but we expect them to be similar in nature to Pluto and Charon. They all probably formed in the same region of the solar system, so we expect them to have similar compositions. Stronger evidence of similarities comes from careful study of orbits. Like Pluto, many Kuiper belt comets have stable orbital resonances with Neptune. In fact, hundreds of Kuiper belt comets have the same orbital period and average distance from the Sun as Pluto itself (and are nicknamed "Plutinos"). Several Kuiper belt comets besides Pluto are also known to have moons, and in those cases we can use Newton's version of Kepler's third law [Section 4.4] to calculate their masses and then infer their densities. The results confirm the idea that these objects have comet-like compositions of ice and rock.

We should learn much more about the objects of the Kuiper belt in about a decade, when the *New Horizons* spacecraft should flyby Pluto and perhaps a couple other Kuiper belt comets. Meanwhile, if you think back to what we learned in our tour of jovian moons in Chapter 11, you'll realize that we probably already have close-up photos of at least one *former* member of the Kuiper belt: Neptune's moon Triton (see Figure 11.29).

Recall that Triton's "backward" orbit indicates that it must be a captured object [Section 11.2]. Its location suggests that it was most likely captured from the Kuiper belt, since many Kuiper belt comets have orbits that cross Neptune's orbit. The fact that Triton appears to be quite similar in nature to Pluto lends further support to the idea that it once orbited the Sun within the Kuiper belt. Moreover, Triton is about 15% larger than Pluto (in diameter). Triton's case seems to offer additional proof that other objects bigger than Pluto were able to form in the Kuiper belt. Some may still await discovery. Others may have been destroyed in giant impacts of the type thought to have created Pluto's moon Charon. Those that did not become trapped in stable resonances may have fallen inward to be accreted by the jovian planets or been thrown outward and ejected from the solar system. All in all, it seems quite likely that the Kuiper belt, like the asteroid belt, once held more large objects than it does today.

• Are Pluto and Eris planets?

The discovery of Eris has forced astronomers to face the issue of whether it or Pluto should count as planets. In ancient times, a "planet" was an object that moved relative to the stars in the night sky, which meant the Sun and Moon were considered planets but Earth was not. After the Copernican revolution, the term "planet" was used for any object that orbited the Sun, putting Earth on the list and removing the Sun and Moon. This definition successfully accommodated the additions of Uranus and Neptune after their discoveries, but it was challenged by discoveries of asteroids. At first, some scientists considered asteroids to be planets, but as their numbers grew and their small sizes were recognized, they became "minor planets," or asteroids as we know them today.

Pluto was quickly given planetary status after its discovery in 1930, mainly because early estimates of its size were quite large. These overly generous estimates came about in part because scientists significantly underestimated its reflectivity, but they were also influenced by the circumstances of its discovery. Astronomers thought that Pluto's mass could help account for reported discrepancies in Neptune's orbit—reports that later turned out to be erroneous (see Special Topic, p. 331). Indeed, similar errors in measurement led some astronomers (most notably, Percival Lowell) to suggest the existence of yet another planet, which led to a search for a tenth large planet that became popularly known as "Planet X." Note that while Eris was sometimes called "Planet X" in the media, it bears no resemblance to the Planet X sought almost a century ago.

We now know that no jovian-size ninth or tenth planets exist in our solar system—if they did, by now we would have detected their gravitational influences. That, plus the fact that we now know Pluto to be just one of many similar objects, led many people to suggest that Pluto should be demoted in much the same way that some asteroids were demoted from an initial status as planets. Because Eris is so similar in size to Pluto, demoting Pluto would also mean that Eris is not a planet. Our solar system would have just 8 "major planets," followed by many "minor planets" that would include both large asteroids and large comets.

However, because popular culture has embraced Pluto as a planet for 75 years, other people suggested that it should stay a planet—in which case Eris would become the tenth planet. Under this proposal, Pluto's size would represent the minimum for a planet, and additional objects would make the planet list only if they prove to be larger than Pluto.

A third idea comes from people who object to what seem like arbitrary size cutoffs, instead arguing that the minimum size of a planet should be based on an object's physical characteristics. Several ideas have been proposed, but the most popular suggests that an object should be called a planet if it is large enough for gravity to make it round. By this definition, there may be dozens of Kuiper belt comets that would make the list, along with one or two of the largest asteroids.

In August 2006, astronomers debated these proposals at a meeting of the International Astronomical Union (IAU). The outcome was that the IAU officially demoted Pluto from planet to *dwarf planet*—a new category using the criterion of roundness. Eris and the asteroid Ceres also are now officially dwarf planets.

THINK ABOUT IT

What is your opinion of the IAU decision? Should Pluto or any other Kuiper belt comets be called planets? How about "planet-sized" moons, such as Triton or the Galilean moons of Jupiter? Consider several possible definitions of planet, and discuss the pros and cons of each one. With your favorite definition, how many planets would there be in our solar system?

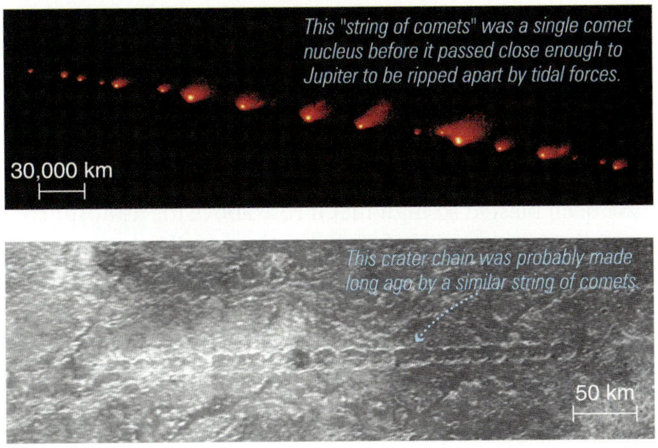

Figure 12.19 Tidal forces from Jupiter can break a single comet nucleus into a chain of smaller nuclei.

12.4 Cosmic Collisions: Small Bodies Versus the Planets

The hordes of small bodies orbiting the Sun are slowly shrinking in number through collisions with the planets and ejection from the solar system. Many more must have roamed the solar system in the days of the heavy bombardment, when most impact craters formed [Section 8.4]. Plenty of small bodies still remain, however, and cosmic collisions still occur on occasion. These collisions have had important effects on Earth and other planets.

• Have we ever witnessed a major impact?

Modern scientists have never witnessed a major impact on a solid world, but in 1994 we were privileged to witness one on Jupiter. The event was likely a rare one: Astronomers estimate that a major impact occurs on Jupiter about once every 1,000 years. The dramatic event was the impact of a comet named *Shoemaker-Levy 9*, or *SL9* for short.

Rather than having a single nucleus, SL9 consisted of a string of comet nuclei lined up in a row (Figure 12.19a). Apparently, tidal forces from Jupiter ripped apart a single comet nucleus during a previous close pass near the planet. Crater chains on Jupiter's moons are evidence that similar breakups of comets near Jupiter have occurred in the past. For example, Figure 12.19b shows a chain of craters on Callisto, presumably formed when a string of comet nuclei crashed into its icy surface.

Comet SL9 was discovered more than a year before its impact on Jupiter, and orbital calculations had told astronomers precisely when the impact would occur. Thus, when the impacts finally began, they were observed with nearly every major telescope in existence, as well as by spacecraft that were in a position to get a view. The images were astonishing (Figure 12.20). Each of the individual nuclei in comet SL9 crashed into Jupiter with an energy equivalent to that of a million hydrogen bombs. Comet nuclei barely a kilometer across left scars larger than the entire Earth. The scars lasted for months before Jupiter's strong winds finally made them fade from view.

The SL9 impacts allowed us to study both the impact process and material splashed out from deep inside Jupiter. They also provided two important sociological lessons. First, "Comet Crash Week" proved to be one of the best examples of international collaboration in history. With the aid of the Internet, scientists quickly and effectively shared data from observatories around the world. Second, extensive media coverage helped the event capture the public imagination, leading to awareness that impacts are not merely relegated to ancient geological history. If such violent impacts can happen on other planets in our lifetime, could they also happen on Earth?

• Did an impact kill the dinosaurs?

There's little doubt that major impacts have occurred on Earth in the past. Meteor Crater in Arizona (see the photo that opens this chapter, p. 359) formed about 50,000 years ago when a metallic asteroid roughly 50 meters across

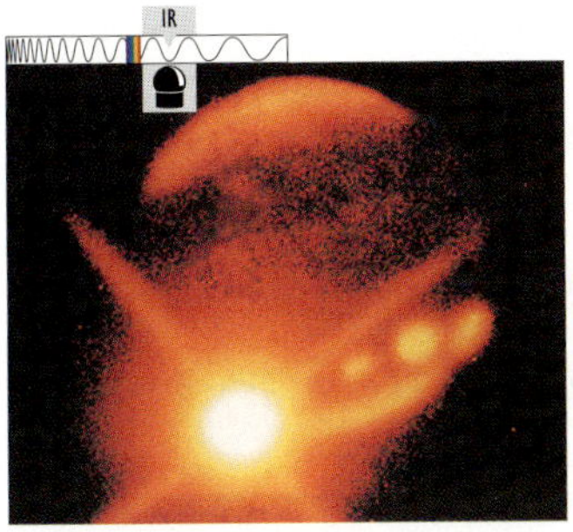

a This infrared photo shows the brilliant glow of a rising fireball from one of the impacts. Jupiter is the round disk, with the impact occuring near the lower left of the disk.

b Each of the black dots in this Hubble Space Telescope photo is a scar from the impact of one of the SL9 nuclei.

c This painting shows how the impacts might have looked from the surface of Io. The impacts occurred on Jupiter's night side.

Figure 12.20 The impacts of comet Shoemaker-Levy 9 on Jupiter allowed astronomers their first direct view of a cosmic collision.

crashed to Earth with the explosive power of a 20-megaton hydrogen bomb. Although the crater is only a bit more than 1 kilometer across, the blast and ejecta probably battered an area covering hundreds of square kilometers. Meteor Crater is relatively small and recent. Despite the fact that erosion and other geological processes have erased most of Earth's impact craters, geologists have identified more than 100 impact craters on our planet. So before we consider whether an impact might occur in our lifetimes, it's worth examining the potential consequences if it did. Clearly, an impact could cause widespread physical damage. But a growing body of evidence, accumulated over the past three decades, suggests that an impact can do much more—in some cases, large impacts may have altered the entire course of evolution.

In 1978, while analyzing geological samples collected in Italy, a scientific team led by Luis and Walter Alvarez (father and son) made a startling discovery. They found that a thin layer of dark sediments deposited about 65 million years ago—about the time the dinosaurs went extinct—was unusually rich in the element iridium. Iridium is a metal that is common in meteorites but rare on Earth's surface (because it sunk to Earth's core when our planet underwent differentiation). Subsequent studies found the same iridium-rich layer in 65-million-year-old sediments around the world (Figure 12.21).* The Alvarez team suggested a stunning hypothesis: The extinction of the dinosaurs was caused by the impact of an asteroid or comet.

In fact, the death of the dinosaurs was only a small part of the biological devastation that seems to have occurred 65 million years ago. The fossil record suggests that up to 99% of all living organisms died around that time and that up to 75% of all existing *species* were driven to extinction. This makes the event a clear example of a **mass extinction**—the rapid extinction of a large fraction of all living species. Could it really have been caused by an impact?

Evidence for the Impact There's still some scientific controversy about whether the impact was the sole cause of the mass extinction or just one of many causes, but there's little doubt that a major impact coincided with the death of the dinosaurs. Key evidence comes from further analysis of the sediment layer. Besides being unusually rich in iridium, this layer contains four other features: (1) unusually high abundances of several other metals, including osmium, gold, and platinum; (2) grains of "shocked quartz," quartz crystals with a distinctive structure that indicates they experienced the high-pressure conditions of an impact; (3) spherical rock "droplets" of a type known to form when drops of molten rock cool and solidify in the air; and (4) soot (at some sites) that appears to have been produced by widespread forest fires.

All these features point to an impact. The metal abundances look much like what we commonly find in meteorites rather than what we find elsewhere on Earth's surface. Shocked quartz is also found at other impact sites, such as Meteor Crater in Arizona. The rock "droplets" presumably were made from molten rock splashed into the air by the force and heat of the impact. The soot probably came from vast forest fires ignited by impact debris. Some debris would have been blasted so high that it rose above the atmosphere, spreading worldwide before falling back to Earth. On their downward plunge, friction would have heated the debris particles until they became a hot, glowing rain of rock.

In addition to the evidence within the sediments, scientists have identified a large impact crater that appears to match the age of the sediment layer. The crater, about 200 kilometers across, is located on the coast of Mexico's Yucatán Peninsula, about half on land and half underwater (Figure 12.22). Its size indicates that it was created by the impact of an asteroid or a comet measuring about 10 kilometers across. (It is named the *Chicxulub crater,* after a nearby fishing village.)

The Mass Extinction If the impact was indeed the cause of the mass extinction, here's how it probably happened: On that fateful day some 65 million years ago, the asteroid or comet slammed into Mexico with the force of a hundred million hydrogen bombs (Figure 12.23). It apparently hit at a slight angle, sending a shower of red-hot debris across the continent of North America. A huge tsunami sloshed more than 1,000 kilometers inland. Much of North American life may have been wiped out almost immediately. Not long after, the hot debris raining around the rest of the world ignited fires that killed many other living organisms.

The longer-term effects were even more severe. Dust and smoke remained in the atmosphere for weeks or months, blocking sunlight and causing temperatures to fall as if Earth were experiencing a global and extremely harsh winter. The reduced sunlight would have stopped photosynthesis for up to a year, killing large numbers of species throughout the food chain. This period of cold may have been followed by a period of unusual warmth. Some evidence suggests that the impact site was rich in carbonate rocks, so the impact may have released large amounts of carbon dioxide into the atmosphere. The added carbon dioxide would have strengthened the greenhouse effect, and the months of global winter may have been followed by decades or longer of global summer.

Figure 12.21 Around the world, sedimentary rock layers dating to 65 million years ago share the evidence of the impact of a comet or asteroid. Fossils of dinosaurs and many other species appear only in rocks below the iridium-rich layer.

*The layer marks what geologists call the *K-T boundary,* because it separates sediments deposited in the Cretaceous and Tertiary periods (the "K" comes from the German word for Cretaceous, *Kreide*). The mass extinction that occurred 65 million years ago therefore is often called the *K-T event.*

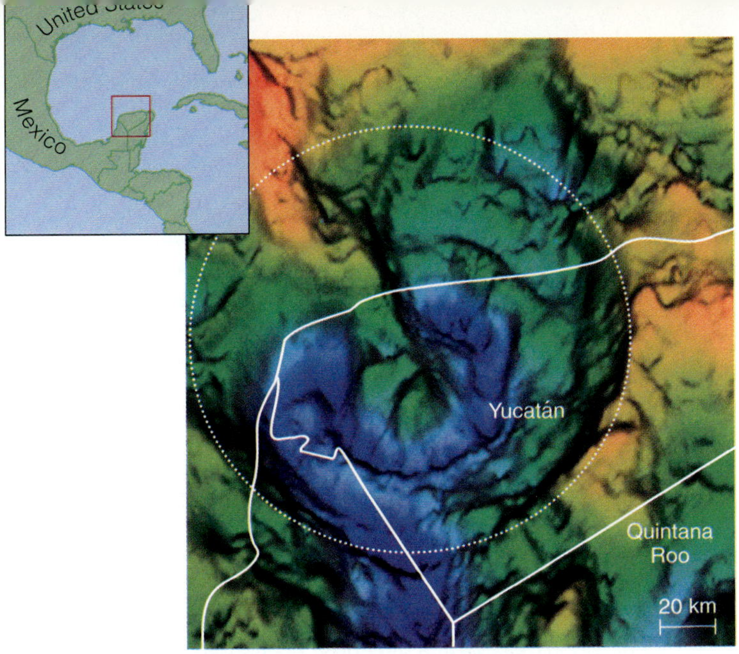

Figure 12.22 This computer-generated image, based on measurements of small local variations in the strength of gravity, shows an impact crater (dashed circle) in the northwest corner of Mexico's Yucatán Peninsula. The crater measures about 200 kilometers across, and about half of it lies underwater off the coast. (The solid white lines show the coastline and borders of Mexican states.)

Figure 12.23 This painting shows an asteroid or comet moments before its impact on Earth, some 65 million years ago. The impact probably caused the extinction of the dinosaurs, and if it hadn't occurred the dinosaurs might still rule Earth today.

The impact probably also caused chemical reactions in the atmosphere that produced large quantities of harmful compounds, such as nitrous oxides. These compounds dissolved in the oceans, where they probably were responsible for killing vast numbers of marine organisms. Acid rain may have been another by-product, killing vegetation and acidifying lakes around the world.

Perhaps the most astonishing fact is not that 75% of all species died but that 25% survived. Among the survivors were a few small mammals. These mammals may have survived in part because they lived in underground burrows and managed to store enough food to outlast the global winter that immediately followed the impact.

The evolutionary impact of the extinctions was profound. For 180 million years, dinosaurs had diversified into a great many species large and small, while mammals (which had arisen at almost the same time as the dinosaurs) had generally remained small and rodentlike. With the dinosaurs gone, mammals became the new kings of the planet. Over the next 65 million years, the small mammals rapidly evolved into an assortment of much larger mammals—ultimately including us.

Controversies and Other Mass Extinctions There seems little doubt that a major impact coincided with the mass extinction of 65 million years ago, but does it tell the entire story? The jury is still out. The fossil record can be difficult to read when we are trying to understand events that happened over just a few years rather than a few million years. Some scientists suspect that the dinosaurs were already in decline and the impact was only the last straw. Others suggest that major volcanic eruptions also may have played a role.

Was the dinosaur extinction a unique event? Measuring precise extinction rates becomes more difficult as we look to older fossils, but there appear to have been at least four other mass extinctions during the past 500 million years. None of the other mass extinctions is as closely tied to an impact as the dinosaur extinction, but impacts almost certainly played a role. Sediments from the times of other mass extinctions reveal evidence similar to that found in the iridium-rich layer tied to the death of the dinosaurs. In some cases, impact craters have been found that date to about the right times. Much more research is needed, but impacts appear to have played a major role in shaping the history of life on Earth.

• Is the impact threat a real danger or media hype?

On June 14, 2002, the 100-meter-wide asteroid 2002MN passed within 120,000 kilometers of Earth (less than a third of the Earth-Moon distance). There was no advance warning; in fact, the asteroid wasn't discovered until after it had passed by. Although an asteroid of this size would not cause global devastation, it might kill millions of people in the unlikely event that it struck a large city. How concerned should we be about the possibility of objects like this striking Earth—or, worse yet, even larger objects such as those that may have caused past mass extinctions? We can analyze the threat by examining both the frequency of past impacts and the number of objects in space that pose a potential future threat.

No large impacts have left craters in modern times, but we know of at least one smaller impact that devastated a fairly large area. In 1908, a tremendous explosion occurred over Tunguska, Siberia (Figure 12.24). Entire forests were flattened and set on fire, and air blasts knocked over people, tents, and furniture up to 200 kilometers away. Seismic disturbances were recorded at distances of up to 1,000 kilometers, and atmospheric pressure fluctuations were detected almost 4,000 kilometers away. The explosion, now estimated to have released energy equivalent to that of

Figure 12.24 This photo shows forests burned and flattened by the 1908 impact over Tunguska, Siberia.

several atomic bombs, is thought to have been caused by a small asteroid no more than about 40 meters across. Atmospheric friction caused it to explode completely before it hit the ground, so it left no impact crater. If the asteroid had exploded over a major city instead of Siberia, it would have been the worst natural disaster in human history.

Objects of similar size to that of the Tunguska event probably strike our planet every couple hundred years or so. They have gone unnoticed, presumably because they have always hit remote areas, with most striking the oceans. Nevertheless, the death toll would be enormous if such an object struck a densely populated area.

Another way to gauge the threat is to look at asteroids that might strike Earth. Astronomers have identified more than 600 asteroids larger than 1 kilometer in diameter with orbits that pass near Earth's orbit. Orbital calculations show that none of these known objects will collide with Earth in the foreseeable future. However, the vast majority of near-Earth asteroids probably have not yet been detected. Astronomers estimate that hundreds more undiscovered near-Earth asteroids are larger than 1 kilometer across and that another 50,000 are larger than 100 meters across. NASA has set a goal of detecting 90% of all large, near-Earth asteroids by 2008.

We do not have a similar way of studying the threat from comets, because they reside so far from the Sun. By the time we saw a comet plunging toward us from the outer solar system, we'd have at best a few years to prepare for the impact.

While it seems a virtual certainty that many more large impacts will batter Earth, the chance of a major impact happening in our lifetimes is quite small. Nevertheless, the chance is not zero, which is one reason why NASA is working to predict potential asteroid threats. Figure 12.25 shows how often, on average, we expect objects of different sizes to hit Earth. Objects a few meters across probably enter Earth's atmosphere every week or two, each liberating energy equivalent to that of an atomic bomb as it burns up. Larger impacts are obviously more devastating but thankfully more rare. Hundred-meter impacts that forge craters similar to Meteor Crater in Arizona probably strike only about every 10,000 years or so. Impacts large enough to cause mass extinctions come tens of millions of years apart.

If we were to find an asteroid or a comet on a collision course with Earth, could we do anything about it? Many people have proposed schemes to save Earth by using nuclear weapons or other means to demolish or divert an incoming asteroid, but no one knows whether current technology is really up to the task. We can only hope that the threat doesn't become a reality before we're ready.

THINK ABOUT IT

Study Figure 12.25. Based on the frequency of impacts large enough to cause serious damage, do you think we should be spending time and resources to counter the impact threat? Or should we focus resources on other threats first? Defend your opinion.

While near-Earth asteroids may pose real dangers, they may also offer opportunities. These asteroids bring valuable resources tantalizingly close to Earth. Iron-rich asteroids are particularly enticing, because they probably contain many precious metals that have mostly sunk to the core on Earth. In the future, it may prove technically feasible and financially profitable to mine metals from asteroids and bring these resources to Earth. It may also be possible to gather fuel and water from asteroids for use in missions to the outer solar system.

• How do the jovian planets affect impact rates and life on Earth?

Ancient people imagined that the mere movement of planets relative to the visible stars in our sky could somehow have an astrological influence on our lives. Although we no longer give credence to this ancient superstition, we now

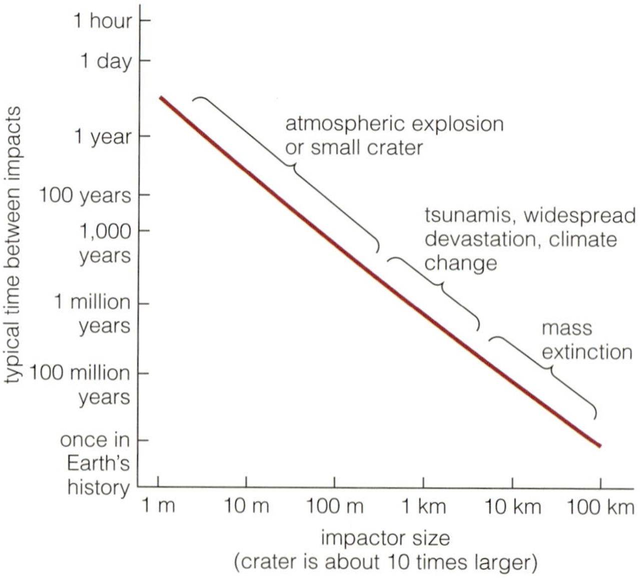

Figure 12.25 This graph shows that larger objects (asteroids or comets) hit the Earth less frequently than smaller ones. The labels describe the effects of impacts of different sizes.

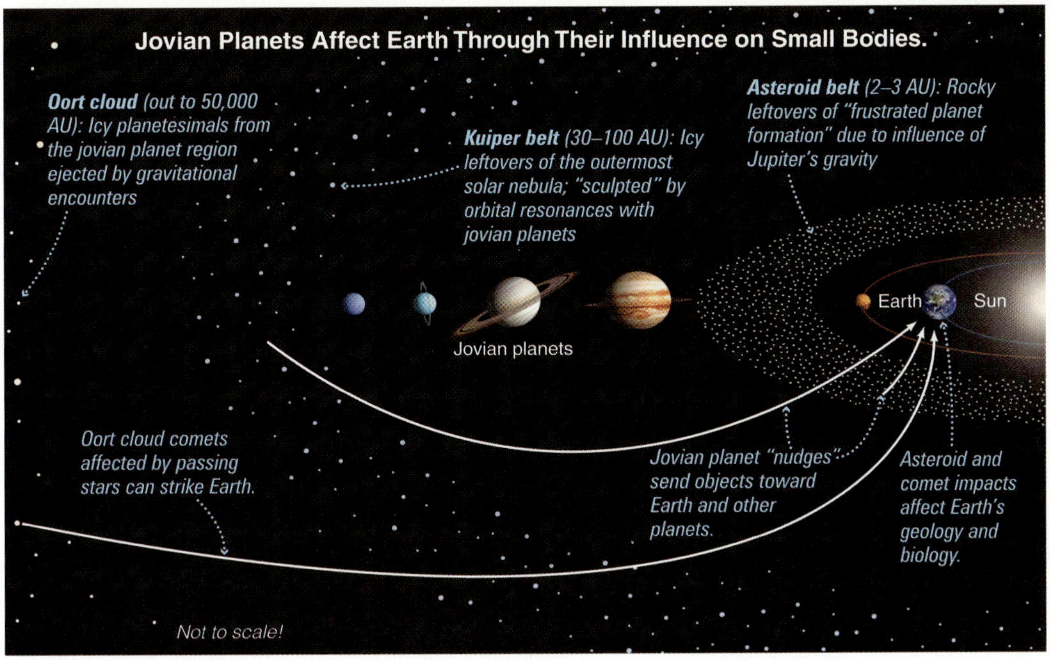

Figure 12.26 The connections between the jovian planets, small bodies, and Earth. The gravity of the jovian planets helped shape both the asteroid belt and the Kuiper belt, and the Oort cloud consists of comets ejected from the jovian planet region by gravitational encounters with these large planets. Ongoing gravitational influences sometimes send asteroids or comets heading toward Earth.

know that planets can have a real effect on life on Earth. By catapulting asteroids and comets in Earth's direction, other planets have caused cosmic collisions that helped shape Earth's destiny.

Jupiter and the other jovian planets have had the greatest effects because of their influence on the small bodies of the solar system. As we saw earlier in this chapter, Jupiter disturbed the orbits of rocky planetesimals outside Mars's orbit, preventing a planet from forming and instead creating the asteroid belt. The jovian planets ejected icy planetesimals to create the distant Oort cloud of comets, and orbital resonances with Neptune shape the orbits of many comets in the Kuiper belt. Ultimately, every asteroid or comet that has crashed to Earth since the end of the heavy bombardment was in some sense sent our way by the influence of Jupiter or one of the other jovian planets. Figure 12.26 summarizes the ways the jovian planets have controlled the motions of asteroids and comets.

Thus, because impacts of asteroids and comets have played such an important role in the history of our planet, we find a deep connection between the jovian planets and the survival of life on Earth. If Jupiter did not exist, the threat from asteroids might be much smaller, since the objects that make up the asteroid belt might instead have become part of a planet. On the other hand, the threat from comets might be much greater: Jupiter probably ejected more comets to the Oort cloud than any other jovian planet, and without Jupiter, those comets might have remained dangerously close to Earth. Of course, even if Jupiter has protected us from impacts, it's not clear whether that has been good or bad for life overall. The dinosaurs appear to have suffered from an impact, but the same impact may have paved the way for our existence. Thus, while some scientists argue that more impacts would have damaged life on Earth, others argue that more impacts might have sped up evolution.

The role of Jupiter has led some scientists to wonder whether we could exist if our solar system had been laid out differently. Could it be that civilizations can arise only in solar systems that happen to have a Jupiter-like planet in a Jupiter-like orbit? No one yet knows the answer to this question. What we do know is that Jupiter has had profound effects on life on Earth and will continue to have effects in the future.

THE BIG PICTURE

Putting Chapter 12 into Context

In this chapter we concluded the study of our own solar system by focusing on its smallest objects, and we found that these objects can have big consequences. Keep in mind the following "big picture" ideas:

- Asteroids, comets, and meteorites may be small compared to planets, but they provide much of the evidence that has helped us understand how the solar system formed.
- The small bodies are subject to the gravitational whims of the largest. The jovian planets shaped the asteroid belt, the Kuiper belt, and the Oort cloud, and they continue to nudge objects onto collision courses with the planets.
- Collisions not only bring meteorites and leave impact craters but can profoundly affect life on Earth. An impact probably was responsible for wiping out the dinosaurs, and future impacts pose a threat that we cannot ignore.

12.1 Asteroids and Meteorites

- **What are asteroids like?** Asteroids are rocky leftovers from the era of planetary formation. Most are small, and despite their enormous numbers their total mass is less than that of any terrestrial planet.

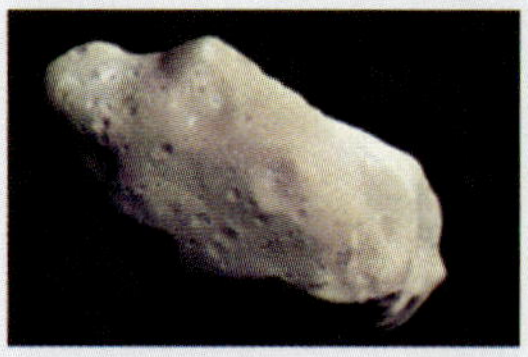

- **Why is there an asteroid belt?** Orbital resonances with Jupiter disrupted the orbits of planetesimals in the asteroid belt, preventing them from accreting into a planet. Many were ejected but some remained and make up the asteroid belt today. Most asteroids in other regions of the inner solar system accreted into one of the planets.

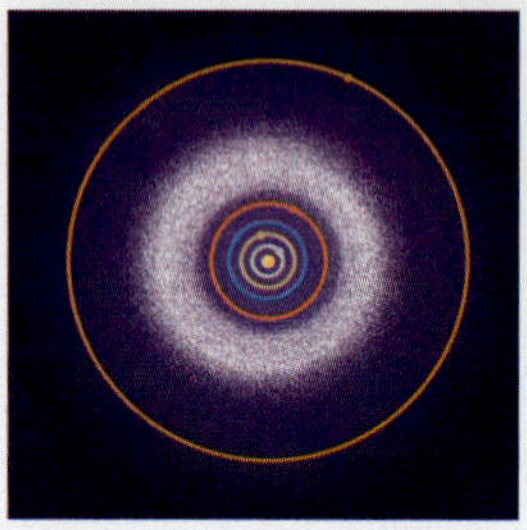

- **Where do meteorites come from?** **Primitive meteorites** are remnants of the solar nebula, containing intermixed rock and metal flakes. Some are rich in carbon compounds. **Processed meteorites** are fragments of larger asteroids and therefore can be metallic like a planet's core or rocky like its mantle or crust.

12.2 Comets

- **What are comets like?** Comets are icy leftovers from the era of planet formation, and most orbit far from the Sun. If a comet approaches the Sun, its **nucleus** heats up and its ice sublimates into gas. The escaping gases carry along some dust, forming a **coma** and two tails: a **plasma tail** of ionized gas and a **dust tail**. Larger particles can also escape, becoming the particles that cause meteors and **meteor showers** on Earth.

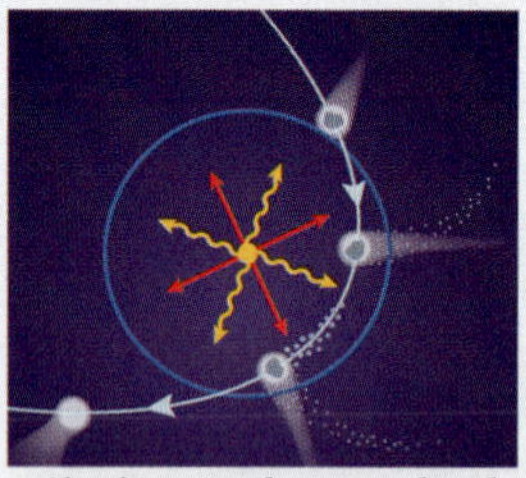

- **Where do comets come from?** Comets come from two reservoirs: the **Kuiper belt** and the **Oort cloud**. The Kuiper belt comets still reside in the region beyond Neptune in which they formed. The Oort cloud comets formed between the jovian planets, and were kicked out to great distance by gravitational encounters with the planets.

12.3 Pluto: Lone Dog No More

- **How big can a comet be?** In the Kuiper belt, icy planetesimals were able to grow to hundreds or thousands of kilometers in size. The recently discovered "Planet X" is the largest known of these objects and Pluto is the second largest.

- **What are the large objects of the Kuiper belt like?** Like smaller comets, these objects are ice-rich in composition. They orbit the Sun roughly between the orbit of Neptune and twice that distance from the Sun. Their orbits tend to be more elliptical and more inclined to the ecliptic plane than those of the terrestrial and jovian planets. Many share orbital resonances with Neptune; Pluto, for example, completes 2 orbits for every 3 Neptune orbits. A few have moons, including Pluto.

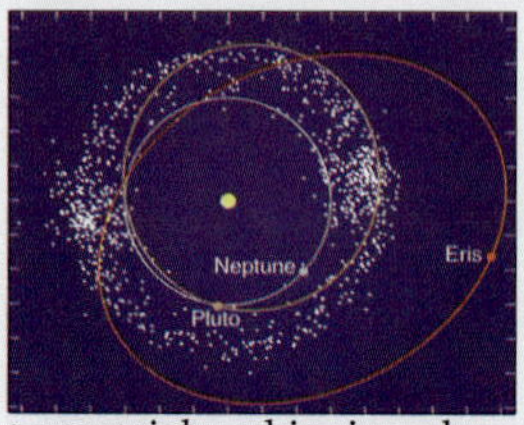

- **Are Pluto and Eris planets?** Officially, Pluto and Eris are now classified as *dwarf planets*. But no matter what you call them, their properties clearly identify them as members of the Kuiper belt.

12.4 Cosmic Collisions: Small Bodies Versus the Planets

- **Have we ever witnessed a major impact?** In 1994, we observed the impacts of comet Shoemaker-Levy 9 on Jupiter. The comet had fragmented into a string of individual nuclei, so there was a string of impacts that left Jupiter's atmosphere scarred for months.

- **Did an impact kill the dinosaurs?** It may not have been the sole cause, but a major impact clearly coincided with the **mass extinction** in which the dinosaurs died out, about 65 million years ago. Sediments from the time contain iridium and other clear evidence of an impact, and an impact crater of the right age lies near the coast of Mexico.

- **Is the impact threat a real danger or media hype?** Impacts certainly pose a threat, though the probability of a major impact in our lifetimes is fairly low. Impacts like the Tunguska event may occur every couple hundred years, and would be catastrophic if they occurred over cities.

- **How do the jovian planets affect impact rates and life on Earth?** Impacts are always linked in at least some way to the gravitational influences of Jupiter and the other jovian planets. These influences have shaped the asteroid belt, the Kuiper belt, and the Oort cloud, and continue to determine when an object is flung our way.

EXERCISES AND PROBLEMS

For instructor-assigned homework go to MasteringASTRONOMY www.masteringastronomy.com

Review Questions

Short-Answer Questions Based on the Reading

1. Briefly describe the general characteristics of asteroids—including sizes, masses, densities, and compositions—and how we measure them.
2. How does the largest asteroid compare in size to the planets? How does the total mass of all asteroids compare to the mass of a terrestrial world?
3. Where is the *asteroid belt* located, and why? Briefly explain how orbital resonances with Jupiter have affected the asteroid belt.
4. What is the difference between a *meteor* and a *meteorite?* How can we distinguish a meteorite from a terrestrial rock?
5. Distinguish between *primitive meteorites* and *processed meteorites* in terms of both composition and origin.
6. What does a comet look like when it is far from the Sun? How does its appearance change when it is near the Sun? What happens to comets that make many passes near the Sun?
7. What produces the *coma* and tails of a comet? What is the *nucleus?* Why do tails point away from the Sun?
8. Explain how meteor showers are linked to comets. Why do meteor showers recur at about the same time each year?
9. Describe the *Kuiper belt* and *Oort cloud* in terms of their locations and the orbits of comets within them. How did comets come to exist in these two regions?
10. Briefly describe Pluto and Charon. Why won't Pluto collide with Neptune? How do we think Charon formed?
11. What is Eris? How is it related to Pluto? What evidence suggests that these objects are really just large comets of the Kuiper Belt?
12. Briefly describe the impact of comet Shoemaker-Levy 9 on Jupiter.
13. Briefly describe the evidence suggesting that an impact caused the mass extinction that killed off the dinosaurs. How did the impact lead to the mass extinction?
14. How often should we expect impacts of various sizes on Earth? How serious a threat do we face from these impacts?

Test Your Understanding

Surprising Discoveries?

Suppose we found a solar system with the property described. (These are *not* real discoveries.) In light of what you've learned about the formation of our own solar system, decide whether the discovery should be considered reasonable or surprising. Explain.

15. A small asteroid that orbits within the asteroid belt has an active volcano.
16. Scientists discover a meteorite that, based on radiometric dating, is 7.9 billion years old.
17. An object that resembles a comet in size and composition is discovered orbiting in the inner solar system.
18. Studies of a large object in the Kuiper belt reveal that it is made almost entirely of rocky (as opposed to icy) material.
19. Astronomers discover a previously unknown comet that will be brightly visible in our night sky about 2 years from now.
20. A mission to Pluto finds that it has lakes of liquid water on its surface.
21. Geologists discover a crater from a 5-kilometer object that impacted Earth more than 100 million years ago.
22. Archaeologists learn that the fall of ancient Rome was caused in large part by an asteroid impact in Asia.
23. Astronomers discover three objects with the same average distance from the Sun (and the same orbital period) as Pluto.
24. Astronomers discover an asteroid with an orbit suggesting that it will collide with Earth in the year 2064.

Quick Quiz

Choose the best answer to each of the following. Explain your reasoning with one or more complete sentences.

25. The asteroid belt lies between the orbits of (a) Earth and Mars. (b) Mars and Jupiter. (c) Jupiter and Saturn.
26. Jupiter nudges the asteroids through the influence of (a) tidal forces. (b) orbital resonances. (c) magnetic fields.
27. Can an asteroid be pure metal? (a) No, all asteroids contain rock. (b) Yes, it must have formed where only metal could condense in the solar nebula. (c) Yes, it must have been the core of a shattered asteroid.
28. Did a large terrestrial planet ever form in the region of the asteroid belt? (a) No, because there was never enough mass there. (b) No, because Jupiter prevented one from accreting. (c) Yes, but it was shattered by a giant impact.
29. What does Pluto most resemble? (a) a terrestrial planet (b) a jovian planet (c) a comet
30. How big an object causes a typical shooting star? (a) a grain of sand or a small pebble (b) a boulder (c) an object the size of a car
31. Which have the most elliptical and tilted orbits? (a) asteroids (b) Kuiper belt comets (c) Oort cloud comets
32. Which are thought to have formed farthest from the Sun? (a) asteroids (b) Kuiper belt comets (c) Oort cloud comets
33. About how often does a 1-kilometer object strike the Earth? (a) every year (b) every million years (c) every billion years
34. What would happen if a 1-kilometer object struck the Earth? (a) It would break up in the atmosphere without causing widespread damage. (b) It would cause widespread devastation and climate change. (c) It would cause a mass extinction.

Investigate Further

In-Depth Questions to Increase Your Understanding

Short-Answer/Essay Questions

35. *The Role of Jupiter.* Suppose that Jupiter had never existed. Describe at least three ways in which our solar system would be different, and clearly explain why.
36. *Life Story of an Iron Atom.* Imagine that you are an iron atom in a processed meteorite made mostly of iron. Tell the story of how you got to Earth, beginning from the time you were part of the gas in the solar nebula 4.6 billion years ago. Include as much detail as possible. Your story should be scientifically accurate but also creative and interesting.
37. *Asteroid Discovery.* You have discovered two new asteroids and named them Albert and Isaac. Both lie at the same distance from Earth and have the same brightness when you look at them through your telescope, but Albert is twice as bright as Isaac at infrared wavelengths. What can you deduce

about the relative reflectivities and sizes of the two asteroids? Which would make a better target for a mission to mine metal? Which would make a better target for a mission to obtain a sample of a carbon-rich planetesimal? Explain.

38. *Asteroids vs. Comets.* Contrast the compositions and locations of comets and asteroids, and explain in your own words why they have turned out differently.
39. *Comet Tails.* Describe in your own words why comets have tails. Why do most comets have two distinct visible tails, and why do the tails go in different directions? Why is the third, invisible tail of small pebbles of interest to us on Earth?
40. *Oort Cloud vs. Kuiper Belt.* Explain in your own words how and why there are two different reservoirs of comets. Be sure to discuss where the two groups of comets formed and what kinds of orbits they travel on.
41. *Project: Tracking a Meteor Shower.* Armed with an expendable star chart and a dim (preferably red) flashlight, set yourself up comfortably to watch one of the meteor showers listed in Table 12.1. Each time you see a meteor, record its path on your star chart. Record at least a few dozen meteors, and try to determine the *radiant* of the shower—that is, the point in the sky from which the meteors appear to radiate. Does the meteor shower live up to its name?
42. *Project: Dirty Snowballs.* If there is snow where you live or study, make a dirty snowball. (The ice chunks that form behind tires work well.) How much dirt does it take to darken snow? Find out by allowing your dirty snowball to melt in a container and measuring the approximate proportions of water and dirt afterward.

Quantitative Problems

Be sure to show all calculations clearly and state your final answers in complete sentences.

43. *Adding Up Asteroids.* It's estimated that there are a million asteroids 1 kilometer across or larger. If a million asteroids 1 kilometer across were all combined into one object, how big would it be? How many 1-kilometer asteroids would it take to make an object as large as the Earth? (*Hint:* You can assume they're spherical. The equation for the volume of a sphere is $\left(\frac{4}{3}\right)\pi r^3$, where r is the radius—not the diameter.)
44. *Impact Energies.* A relatively small impact crater 20 kilometers in diameter could be made by a comet 2 kilometers in diameter traveling at 30 km/s (30,000 m/s).
    a. Assume that the comet has a total mass of 4.2×10^{12} kilograms. What is its total kinetic energy? (*Hint:* The kinetic energy is equal to $\frac{1}{2} \times m \times v^2$, where m is the comet's mass and v is its speed. If you use mass in kilograms and velocity in m/s, the answer for kinetic energy will have units of joules.)
    b. Convert your answer from part (a) to an equivalent in megatons of TNT, the unit used for nuclear bombs. Comment on the degree of devastation the impact of such a comet could cause if it struck a populated region on Earth. (*Hint:* One megaton of TNT releases 4.2×10^{15} joules of energy.)
45. *The "Near Miss" of Toutatis.* The 5-kilometer asteroid Toutatis passed a mere 1.5 million kilometers from Earth in 2004. Suppose Toutatis were destined to pass *somewhere* within 3 million kilometers of Earth. Calculate the probability that this "somewhere" would have meant that it slammed into Earth. Based on your result, do you think it is fair to call the 1992 passage a "near miss"? Explain. (*Hint:* You can calculate the probability by considering an imaginary dartboard of radius 3 million kilometers in which the bull's-eye has Earth's radius, 6,378 kilometers.)
46. *Room to Roam.* It's estimated that there are a trillion comets in the Oort cloud, which extends out to about 50,000 AU. What is the total volume of the Oort cloud, in cubic AU? How much space does each comet have in cubic AU, on average? Take the cube root of the average volume per comet, to find their typical spacing in AU. (*Hints:* For the purpose of this calculation, you can assume the Oort cloud fills the whole sphere out to 50,000 AU. The equation for the volume of a sphere is $\left(\frac{4}{3}\right)\pi r^3$, where r is the radius.)
47. *Pluto and Charon Eclipses.* From 1985 to 1990, the orbit of Pluto's moon Charon caused it to pass in front of and behind Pluto during its 6-day orbit. We can use some simple examples to see how these eclipses allowed astronomers to learn about the two objects.
    a. Assume that Pluto and Charon each reflect the same fraction of the sunlight that hits them, and that their disks reflect the light uniformly. In this case, the total amount of light reflected is proportional to the areas of their disks ($A = \pi r^2$). Using the values from Appendix E, find the ratio of light reflected by Charon to that reflected by Pluto. Draw a rough "light curve" that shows how the total light from the whole system (Pluto plus Charon) varies over the 6-day orbital period, with Charon first passing in front of Pluto for a few hours, and then behind. Label these events on the plot.
    b. Now assume that Charon is much less reflective than Pluto. On the plot for part (a), using a different color show qualitatively how the light curve would be different. (*Hint:* Which has a bigger effect: a dark object passing in front of or behind a bright object?)
48. *Comet Temperatures.* Find the "no greenhouse" temperatures for a comet at distances from the Sun of 50,000 AU (in the Oort cloud), 3 AU, and 1 AU (see Mathematical Insight 10.1). Assume that the comet reflects 3% of the incoming sunlight. At which location will the temperature be high enough for water ice to sublime (about 150 K)? How do your results explain comet anatomy? Explain.
49. *Comet Dust Accumulation.* A few hundred tons of comet dust gets added to Earth daily from the millions of meteors that enter our atmosphere. Estimate the time it would take for Earth to get 0.1% heavier at this rate. Is this mass accumulation significant for Earth as a planet? Explain.

Discussion Questions

50. *Rise of the Mammals.* Suppose the impact 65 million years ago had not occurred. How do you think our planet would be different? For example, do you think that mammals still would eventually have come to dominate Earth? Would we be here? Defend your opinions.
51. *The Status of Pluto.* Officially, Pluto is no longer a planet, but instead is a "dwarf planet." Nevertheless, some astronomers have objected to Pluto's demotion, and it remains possible that the International Astronomical Union will reconsider the official status of Pluto, Eris, and other objects. Do you think the definitions should be reconsidered? Defend your opinion.
52. *Reducing the Impact Threat.* Based on your own opinion of how the impact threat compares to other threats to our society, how much money and resources do you think should be used to alleviate it? What types of programs would you support? (Examples include programs to search for objects that could strike Earth, to develop defenses against impacts, or to build actual defenses.) Defend your opinions.

MEDIA EXPLORATIONS

For self-study activities go to MasteringASTRONOMY www.masteringastronomy.com

Interactive Tutorials

Tutorial Review of Key Concepts

Use the following interactive **Tutorial** at **www.masteringastronomy.com** to review key concepts from this chapter.

Formation of the Solar System Tutorial

Lesson 3 Formation of Planets

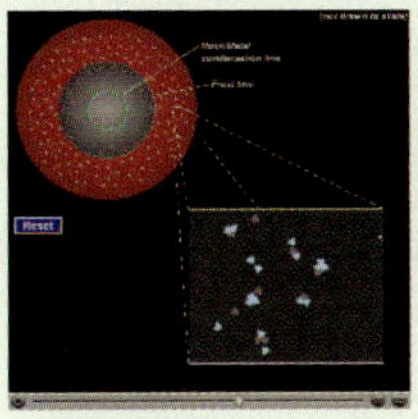

Supplementary Tutorial Exercises

Use the interactive **Tutorial Lesson** to explore the following questions.

Formation of the Solar System Tutorial, Lesson 3

1. Where were icy planetesimals able to form in the solar nebula?
2. Pluto is in the outer solar system. Why didn't it capture hydrogen and helium gas like the jovian planets?
3. Suppose the solar wind had not cleared the solar nebula until much later in the history of the solar system. Do you think Pluto or Eris would have been different? Explain.

Exploring the Sky and Solar System

Of the many activities available on the ***Voyager: SkyGazer*** **CD-ROM** accompanying your book, use the following files to observe key phenomena covered in this chapter.

Go to the **File: Basics** folder for the following demonstrations.

1. Orbit of Hale-Bopp
2. Pluto's Orbit

Go to the **File: Demo** folder for the following demonstrations:

1. Hale-Bopp Path
2. Hyakutake at Perihelion
3. Hyakutake Nears Earth
4. Pluto's Orbit

Movies

Check out the following narrated and animated short documentaries available on **www.masteringastronomy.com** for a helpful review of key ideas covered in this chapter.

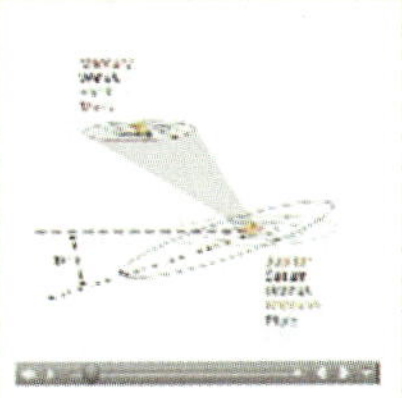

History of the Solar System Movie

Orbits in the Solar System Movie

Web Projects

1. *Recent Asteroid and Comet Missions.* Chose either the *NEAR* mission to the asteroid Eros or the *Stardust* mission to return comet material to Earth. Learn about the mission goals and status. What did it or will it accomplish? Write a one- to two-page summary of your findings.
2. *Future Asteroid and Comet Missions.* Learn about another proposed space mission to study asteroids or comets. For the mission you choose, write a short report about the mission plans, goals, and prospects for success.
3. *The New Horizons Mission to Pluto.* The *New Horizons* mission to visit Pluto was launched in 2006. Find out the current status of the mission. What are its science goals? Summarize your findings in a few paragraphs.
4. *Impact Hazards.* Many groups are searching for near-Earth asteroids that might impact the Earth. They use something called the *Torino Scale* to evaluate the possible danger posed by an asteroid based on how well we know its orbit. What is this scale? What object has reached the highest level on this scale? What were the estimated chances of impact, and when?

13

Other Planetary Systems

The New Science of Distant Worlds

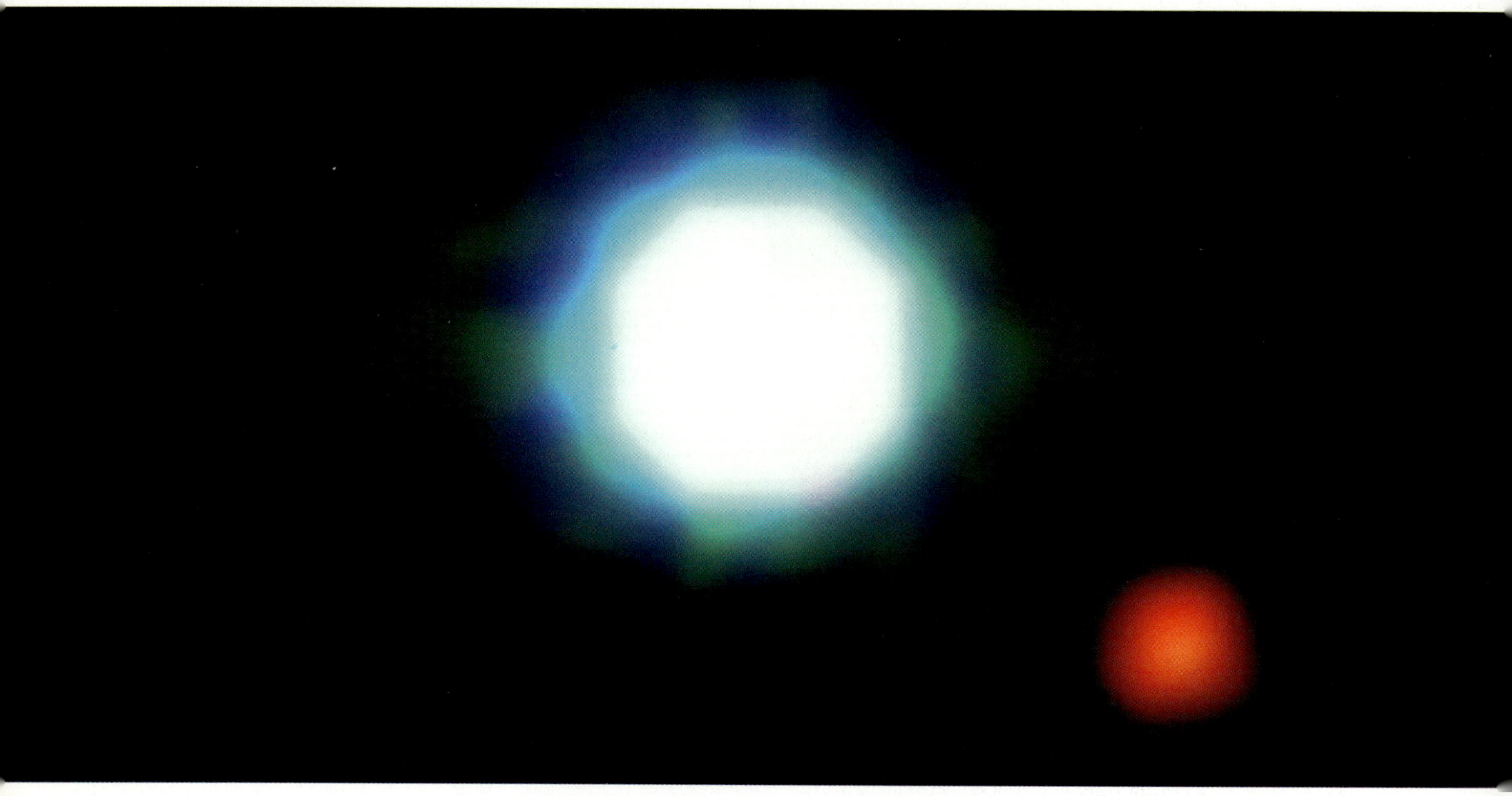

LEARNING GOALS

13.1 Detecting Extrasolar Planets
- Why is it so difficult to detect planets around other stars?
- How do we detect planets around other stars?

13.2 The Nature of Extrasolar Planets
- What have we learned about extrasolar planets?
- How do extrasolar planets compare with those in our solar system?

13.3 The Formation of Other Solar Systems
- Can we explain the surprising orbits of many extrasolar planets?
- Do we need to modify our theory of solar system formation?

13.4 Finding New Worlds
- How will we search for Earth-like planets?

How vast those Orbs must be, and how inconsiderable this Earth, the Theatre upon which all our mighty Designs, all our Navigations, and all our Wars are transacted, is when compared to them. A very fit consideration, and matter of Reflection, for those Kings and Princes who sacrifice the Lives of so many People, only to flatter their Ambition in being Masters of some pitiful corner of this small Spot."

—Christiaan Huygens, c. 1690

A little more than a decade ago, all of planetary science was based on the study of just the planets in our solar system, along with their moons and the many small bodies that orbit the Sun. Then, beginning in 1995, a dramatic change occurred. Within just 10 years, more than a hundred planets were discovered orbiting other stars. Today, the number of known planets continues to rise rapidly and we are beginning to learn about the characteristics of these distant worlds.

The discovery of planets around other stars represents a triumph of modern technology. It also has profound philosophical implications. Knowing that planets are common increases the chance that we might someday find life elsewhere, perhaps even intelligent life. Moreover, by giving us many more worlds to compare to our own, it vastly enhances our ability to learn how planets work and hence may help us better understand our home planet, Earth.

The study of other planetary systems also allows us to test in new settings the theory of our solar system's birth. Recall that the nebular theory, which we discussed in Chapter 8, successfully explains the major characteristics of our solar system. If this theory is correct, it should be able to explain the observed properties of other planetary systems as well. In this chapter, we'll focus all our attention on the exciting new science of other planetary systems. We'll begin by exploring how distant planets have been detected, then discuss the nature of these planets and how they have affected our understanding of planetary science, and finally consider future plans for detecting and studying distant planets, including missions that may soon make it possible to detect worlds similar to our own Earth.

Mastering Astronomy — Detecting Extrasolar Planets Tutorial, Lessons 1–3

13.1 Detecting Extrasolar Planets

The very idea of planets around other stars, or **extrasolar planets** for short, would have shattered the worldviews of many people throughout history. After all, cultures of the western world long regarded Earth as the center of the universe, and nearly all ancient cultures imagined the heavens to be a realm distinct from Earth.

The Copernican revolution, in which we learned that Earth is a planet orbiting the Sun, opened up the possibility that planets might also orbit other stars. Still, until quite recently, no one ever succeeded in finding a planet orbiting another star. Our goals in this section are to understand why the detection of extrasolar planets presents such an extraordinary technological challenge and how astronomers have begun to meet that challenge.

Before we begin, it's worth noting that these discoveries have further complicated the question of precisely how we define a planet [Section 12.3]. Recall that the word *planet* comes from the Greek for "wandering star" and originally referred to the seven naked-eye objects that appear to wander among the constellations in our sky: the Sun, the Moon, Mercury, Venus, Mars, Jupiter, and Saturn. At the end of the Copernican revolution, the word *planet* came to mean a world orbiting the Sun. The discoveries of Uranus in 1781 and Neptune in 1846 thereby gave us a solar system with eight known planets. Meanwhile, scientists learned that comets also orbit the Sun and began to discover other small bodies—asteroids—orbiting the Sun as well. The asteroids were often called "minor planets" to distinguish them from the eight larger planets. Then came Pluto, and more recently Eris, which beg the question of whether these large Kuiper belt objects should count as planets.

In much the same way that Pluto and Eris raise the question of a minimum size for a planet, extrasolar planets raise the question of a maximum size. As we will see shortly, many of the known extrasolar planets are considerably more massive than Jupiter. If you recall that Jupiter has a composition very similar to that of the Sun and emits more total energy than it receives from the Sun, you can see why an even larger object might begin to beg the question of when it stops being a "very large planet" and instead should be considered a "very tiny star" (more technically called a *brown dwarf* [Section 16.3]). Moreover, our new understanding of planetary systems tells us that objects the size of planets may sometimes be ejected from their birthplaces into interstellar space; should they still count as planets once they no longer orbit a star?

Astronomers have not yet agreed on the best way to answer these questions, and as a result there is no official definition of the term *planet*. For now, we'll assume that any object less than about a dozen times the size of Jupiter can count as a planet, as long as we find it orbiting a star.

• Why is it so difficult to detect planets around other stars?

We've known for centuries that other stars are distant suns (see Special Topic box, p. 386), making it seem natural to assume that they are orbited by their own planetary systems. The nebular theory of solar system formation, well established by the middle of the twentieth century, made extrasolar planets seem even more likely. As we discussed in Chapter 8, the nebular theory explains our planetary system as a natural consequence of processes that accompanied the birth of our Sun. If the theory is correct, planets should be common throughout the universe. But are they? Prior to 1995, we lacked the evidence needed to draw a conclusion. Even today, the list of known extrasolar planets leaves many important questions unanswered, including how commonly planets form and whether Earth-like planets exist elsewhere.

What makes it so difficult to detect extrasolar planets? You already know the answer, if you think back to the scale of the solar system that we discussed in Chapter 1. Recall that on a 1-to-10-billion scale, the Sun is the size of a grapefruit, Earth is a pinhead orbiting 15 meters away, and Jupiter is a marble orbiting 80 meters away. On the same scale, the distance to the nearest stars besides the Sun is equivalent to the distance across the United States. Thus, seeing an Earth-like planet orbiting the nearest star besides the Sun would be like looking from San Francisco for a pinhead orbiting just 15 meters from a grapefruit in Washington, D.C. Seeing a Jupiter-like planet would be like trying to see a marble about 80 meters from the grapefruit-size star.

The scale alone would make the task quite challenging, but it is further complicated by the fact that a Sun-like star would be a *billion times* as bright as the light reflected from any of its planets. Because even the best telescopes blur the light from stars at least a little, the glare of scattered starlight would overwhelm the small blips of planetary light.

If you think about these challenges, you might easily come to the conclusion that detection would be near impossible. Indeed, as recently as the early 1990s, many astronomers thought that we were still decades away from meeting the challenge. But human ingenuity proved greater than the pessimists had guessed. Thanks to technological advances and clever planet-hunting strategies, we have begun to discover planets orbiting other stars. Although it is too soon to know for sure, it seems ever more likely that billions of planetary systems inhabit our Milky Way Galaxy.

• How do we detect planets around other stars?

The discovery of extrasolar planets has opened a new era in planetary science. The era is generally considered to have

SPECIAL TOPIC How Did We Learn That Other Stars Are Suns?

We often take it for granted that stars are other suns—meaning objects that give off enough energy to supply light and heat to orbiting planets—but this fact is not at all obvious from looking at the night sky. After all, the feeble light of stars hardly seems comparable to the majestic light of the Sun. Most ancient observers guessed that stars were much more mundane; typical guesses suggested that they were holes in the celestial sphere or flaming rocks in the sky.

The only way to realize that stars are suns is to know that they are incredibly far away; then, a simple calculation will show that they are actually as bright as or brighter than the Sun [Section 15.1]. The first person to make reasonably accurate estimates of the distances to stars was Christiaan Huygens (1629–1695). By *assuming* that other stars are indeed suns, as many earlier astronomers had guessed, Huygens successfully estimated stellar distances. The late Carl Sagan eloquently described the technique:

> *Huygens drilled small holes in a brass plate, held the plate up to the Sun and asked himself which hole seemed as bright as he remembered the bright star Sirius to have been the night before. The hole was effectively $\frac{1}{28,000}$ the apparent size of the Sun. So Sirius, he reasoned, must be 28,000 times farther from us than the Sun, or about half a light-year away. It is hard to remember just how bright a star is many hours after you look at it, but Huygens remembered very well. If he had known that Sirius was intrinsically brighter than the Sun, he would have come up with [a much better estimate of] the right answer: Sirius is [8.6] light-years away.**

Huygens could not actually prove that stars are suns, since his method was based on the assumption that they are. However, his results explained a fact known since ancient times: Stellar parallax is undetectable to the naked eye [Section 2.4]. Recall that the lack of detectable parallax led many Greeks to conclude that Earth must be stationary at the center of the universe, but this lack also has an alternate explanation: Stars are incredibly far away. Even with his original estimate that Sirius was only half a light-year away, Huygens knew that its parallax would have been far too small to observe by naked eye or with the telescopes available at the time. Thus, Huygens "closed the loop" on the ancient mystery of the nature of stars, showing that their appearance and lack of parallax made perfect sense if they were very distant suns. This new knowledge apparently made a great impression on Huygens; as you can see from his quotation on top of p. 385, Huygens became one of the first people to express the idea that the vast size of the cosmos ought to affect our perceptions of ourselves.

Christiaan Huygens (1629–1695)

*From *Cosmos*, by Carl Sagan (Random House, 1980). Sagan demonstrates the technique in the *Cosmos* video series, Episode 7.

begun in 1995, with the discovery of a planet orbiting a star called 51 Pegasi.* The discovery was made by Swiss astronomers Michel Mayor and Didier Queloz, and soon confirmed by a team led by Geoffrey W. Marcy and R. Paul Butler of San Francisco State University. More than 100 other extrasolar planets have been discovered since that time, many of them by these same teams of astronomers.

There are two basic ways in which astronomers can identify extrasolar planets:

1. Directly: Pictures or spectra of the planets themselves constitute direct evidence of their existence.
2. Indirectly: Precise measurements of stellar properties (such as position, brightness, or spectra) may indirectly reveal the effects of orbiting planets.

Direct detection is preferable in principle, because it can tell us far more about the planet's properties. However, current telescopes are not quite up to the challenge of direct detection, at least for planets around ordinary stars. As a result, nearly all extrasolar planets discovered to date have been found by indirect techniques. Let's now explore detection techniques in a little more detail.

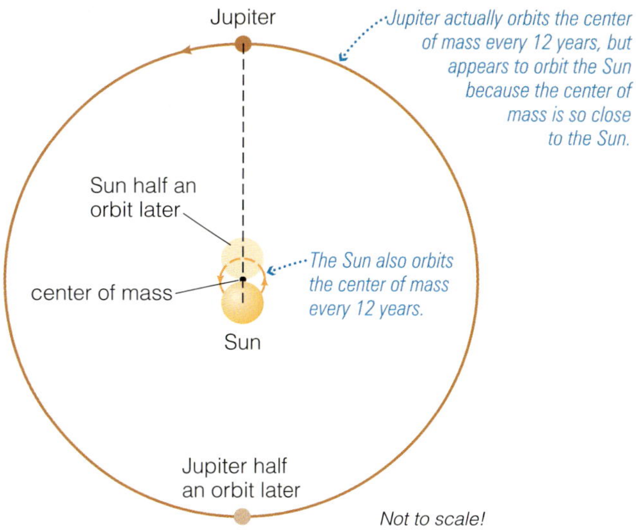

Figure 13.1 Interactive Figure This diagram shows how both the Sun and Jupiter actually orbit around their mutual center of mass, which lies very close to the Sun. The diagram is not to scale; the sizes of the Sun and its orbit are exaggerated about 100 times compared to the size shown for Jupiter's orbit.

Gravitational Tugs To date, nearly all extrasolar planets have been detected by observing the gravitational tugs they exert on the stars they orbit. This type of detection is indirect because we discover the planets by observing the stars they orbit without actually seeing the planets themselves.

Although we usually think of a star as remaining still while planets orbit around it, that is only approximately correct. In reality, all the objects in a star system, including the star itself, orbit the system's *center of mass* [Section 4.4]. To understand how this fact allows us to discover extrasolar planets, imagine the viewpoint of extraterrestrial astronomers observing our solar system from afar.

Let's start by considering only the influence of Jupiter, which exerts a much stronger gravitational tug on the Sun than the rest of the planets combined. The center of mass between the Sun and Jupiter lies just outside the Sun's visible surface (Figure 13.1). Thus, what we usually think of as Jupiter's 12-year orbit around the Sun is really a 12-year orbit around the center of mass; we generally don't notice this fact because the center of mass is so close to the Sun itself. In addition, because the Sun and Jupiter are always on opposite sides of the center of mass (otherwise it wouldn't be a "center"), the Sun must orbit this point with the same 12-year period as Jupiter. The Sun's orbit traces out only a very small circle (or ellipse) with each 12-year period, because the Sun's average orbital distance is barely larger than its own radius. Nevertheless, with sufficiently precise measurements, extraterrestrial astronomers could detect this orbital movement of the Sun. They could thereby deduce the existence of Jupiter, even without having observed Jupiter itself. They could even determine Jupiter's mass from the Sun's orbital characteristics as it goes around the center of mass. Remember that for any particular separation between the Sun and a planet, a more massive planet pulls the center of mass farther from the Sun's center, thereby giving the Sun a larger orbit and a faster orbital speed around the center of mass.

Now let's add in the effects of Saturn, which exerts the second greatest gravitational tug on the Sun. Saturn takes 29.5 years to orbit the Sun, so by itself it would cause the Sun to orbit their mutual center of mass every 29.5 years. However, because Saturn's influence is secondary to that of Jupiter, this 29.5-year period appears as an added complication on top of the Sun's 12-year orbit around its center of mass with Jupiter. In other words, every 12 years the Sun would return to *nearly* the same orbital position around its center of mass with Jupiter, but the precise point of return would make its own even tinier orbital circle (or ellipse) with Saturn's 29.5-year period. By measuring this motion carefully from afar, an extraterrestrial astronomer could deduce the existence and masses of both Jupiter and Saturn after a few decades of observing.

The other planets also exert gravitational tugs on the Sun, which create further complications to the Sun's orbital motion around the solar system's center of mass (Figure 13.2). These extra effects become increasingly difficult to measure in practice, but in principle they would allow an extraterrestrial astronomer to discover all the planets in our solar system. If we turn this idea around, you'll realize that it

*The first actual detection of extrasolar planet-size objects occurred in 1992, when precise timing measurements revealed the existence of three objects with Earth-like masses orbiting a type of "dead" star known as a pulsar [Section 18.2]. Because pulsars are created when stars die in supernova explosions, these "planets" must be either the charred remains of pre-existing planets or, more likely, objects that somehow formed from supernova debris. Either way, they are not planets in the same sense as those that form during star birth; in this chapter, we will focus only on planets around ordinary stars like our Sun.

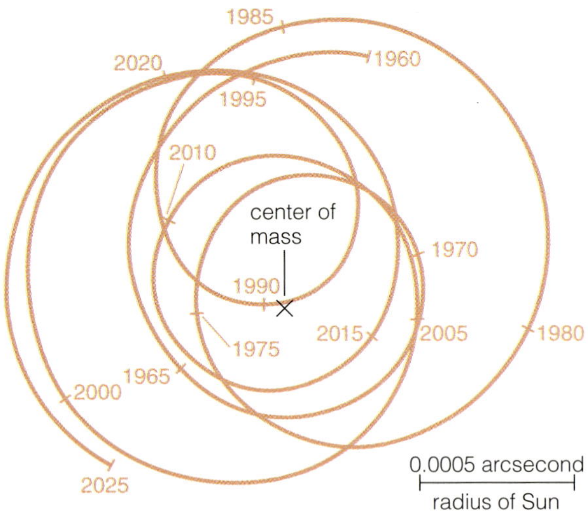

Figure 13.2 This diagram shows the orbital path of the Sun around the center of mass of our solar system as it would appear from a distance of 30 light-years away; the graph shows the motion between 1960 and 2025. Notice that the entire range of motion during this period is only about 0.0015 arcsecond, which is almost 100 times smaller than the angular resolution of the Hubble Space Telescope. Nevertheless, if alien astronomers could measure this motion, they could learn of the existence of planets in our solar system.

means we can search for planets in other star systems by carefully watching for the tiny orbital motion of a star around the center of mass of its star system.

There are two techniques for observing the small orbital motion of a star caused by the gravitational tugs of planets. The first is straightforward: We make very precise measurements of a star's position to see if it is, in fact, moving about in a way that tells us it is being tugged on by planets. This method is usually called the **astrometric technique**, because it requires very precise measurements of stellar positions in the sky (*astrometric* means "measurement of the stars"). The second scheme, known as the **Doppler technique**, detects orbital motion by searching for changing Doppler shifts in a star's spectrum. The star's orbital motion will show up with alternating blueshifts as the star moves toward us in its orbit and redshifts as it moves away. Each technique has advantages and limitations.

The Astrometric Technique The astrometric technique has been used for many decades to identify binary star systems, since two orbiting stars will move periodically around their center of mass. The technique works especially well for binary systems in which the two stars are not too close together, because the stellar motions tend to be larger in those cases. In the case of planet searches, however, the expected stellar motion is much more difficult to detect.

SPECIAL TOPIC The Names of Extrasolar Planets

The planets in our solar system have familiar names rooted in mythology. Unfortunately, there's not yet a well-accepted scheme for naming extrasolar planets. As a result, we generally refer to extrasolar planets by the star they orbit, such as "the planet orbiting the star named . . . " Worse still, the stars themselves often have confusing or even multiple names, reflecting naming schemes used in star catalogs made by different people at different times in history.

A few hundred of the brightest stars in the sky carry proper names from ancient times. Many of these names are Arabic—such as Betelgeuse, Algol, and Aldebaran—because of the work of the Arabic scholars of the Middle Ages [Section 3.2]. In the early 1600s, German astronomer Johann Bayer developed a system that gave names to many more stars. In Bayer's system, each star gets a name based on its constellation and a Greek letter indicating its ranking in brightness within that constellation. For example, the brightest star in the constellation Andromeda is called Alpha Andromedae, the second brightest is Beta Andromedae, and so on. Of course, because there are only 24 letters in the Greek alphabet, this system works for only the 24 brightest stars in each constellation. About a century later, English astronomer John Flamsteed published a more extensive star catalog in which he used numbers once the Greek letters were exhausted. For example, 51 Pegasi gets its name from Flamsteed's catalog. (Flamsteed's numbers are based on position within a constellation rather than brightness.)

As more powerful telescopes made it possible to discover more and fainter stars, astronomers developed many new star catalogs. The names we use today usually come from one of these catalogs. For example, the star HD209458 gets its name from the Henry Draper (HD) catalog and has the number 209458 in that catalog. You may also see stars with numbers preceded by other catalog names, including Gliese, Ross, and Wolf; these catalogs are also named for the astronomers who compiled them. Moreover, because the same star is often listed in several of these catalogs, a single star can have several different names. Some of the newest planets orbit stars so faint they have not been previously cataloged. They then carry the name of the observing program that discovered them, such as TrES-1 for the first discovery of Trans-Atlantic Exoplanet Survey, or OGLE-TR-132b for the 132nd object scrutinized by the Optical Gravitational Lensing Experiment.

Objects orbiting other stars usually carry the star name plus a letter denoting their order of discovery around that star. The host star itself is designated "a." Thus, for example, HD209458b is the first object discovered to be orbiting star number 209458 in the Henry Draper catalog; Upsilon Andromedae d is the third object discovered to be orbiting the twentieth brightest star (because upsilon is the twentieth letter in the Greek alphabet) in the constellation Andromeda. Unfortunately, this scheme doesn't tell you whether the orbiting object is thought to be a planet or another star. Here's hoping that astronomers soon devise a better naming system for these wonderful new worlds!

As an example, from a distance of 10 light-years, a Jupiter-size planet orbiting 5 AU from a Sun-like star would cause its star to dance slowly over a side-to-side angular distance of only about 0.003 arcsecond—approximately the width of a hair seen from a distance of 5 kilometers. Remarkably, with careful telescope calibration astronomers can now measure displacements this small, and instruments currently under development will be five to 10 times better yet. However, two other complications add to the difficulty of the astrometric technique.

First, the farther away the star, the smaller its side-to-side movement will appear. For example, while Jupiter causes the Sun to move by about 0.003 arcsecond as seen from 10 light-years away, the observed motion is only half as large when seen from 20 light-years away and one-tenth as large when seen from 100 light-years away. Thus, the astrometric technique works best for massive planets around relatively nearby stars.

The second complication arises from the time required to detect a star's motion. It is obviously easier to detect larger movements than smaller ones, and a planet with a larger orbit has a larger effect on its star. To understand why, consider what would happen if Jupiter moved farther from the Sun. Because the center of mass of the solar system is very nearly at the balance point between the Sun and Jupiter, moving Jupiter outward would also cause the center of mass to move farther from the Sun. With the center of mass located farther from the Sun, the Sun's orbit around the center of mass—and hence its side-to-side motion as seen from a distance—would be larger. Thus, the astrometric technique works best not only for massive planets, but also in particular for massive planets that are far from their stars. However, Kepler's third law tells us that a more distant planet takes longer to complete its orbit, which means the stellar motion it causes also takes longer. For example, while Jupiter causes the Sun to move around the center of mass with a 12-year period, Neptune's effects on the Sun show up with the 165-year period of Neptune's orbit. It would probably take a century or more of patient observation to be confident that stellar motion was occurring in a 165-year cycle.

As a result of these complications, the astrometric technique has been used to detect only one extrasolar planet through early 2005—and the planet had already been discovered by the Doppler technique. Nevertheless, as we'll discuss in Section 13.4, we expect the astrometric technique to be used extensively in the future.

The Doppler Technique The Doppler technique has been used for the vast majority of planet discoveries to date. An orbiting planet causes its star to move alternately slightly toward and away from us, which makes the star's spectral lines shift alternately toward the blue and toward the red (Figure 13.3a). The 1995 discovery of a planet orbiting 51 Pegasi came when this star was found to be moving with a rhythmic wobble corresponding to an orbital speed of 57 meters per second (Figure 13.3b). The four-day period of the star's motion tells us the orbital period of its planet. Thus, we know that the planet lies so close to the star that its "year" lasts only four of our days, and its surface temperature is probably over 1,000 K (Figure 13.4).

Current techniques can measure a star's velocity to within about 1 meter per second—walking speed—which corresponds to a Doppler wavelength shift of only one part in 300 million. Thus, we can find planets that exert a considerably smaller gravitational tug on their stars than the

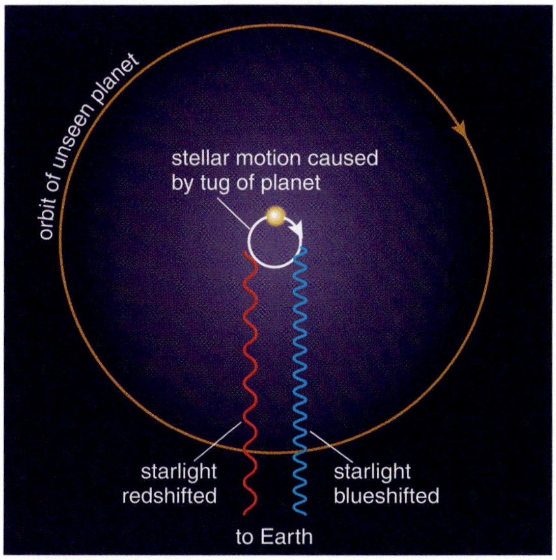

a Doppler shifts allow us to detect the slight motion of a star caused by an orbiting planet.

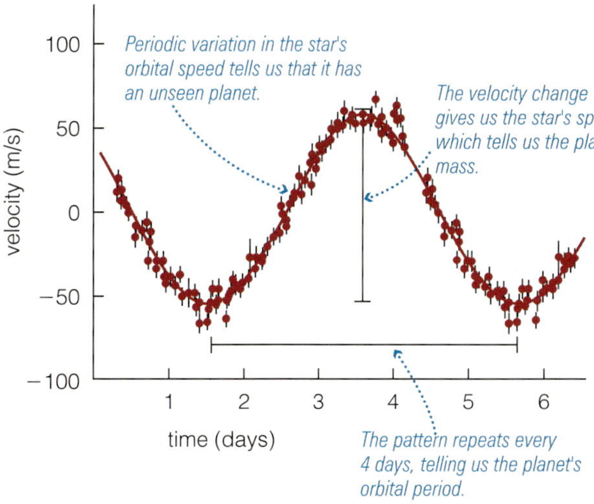

b A periodic Doppler shift in the spectrum of the star 51 Pegasi shows the presence of a large planet with an orbital period of about 4 days. Dots are actual data points; bars through dots represent measurement uncertainty.

Figure 13.3 Interactive Figure The Doppler technique for discovering extrasolar planets.

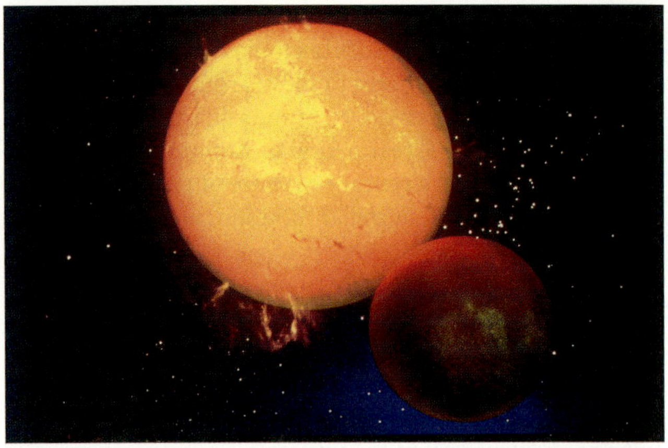

Figure 13.4 Artist's conception of the planet orbiting 51 Pegasi. As we'll discuss later in the chapter, the planet's mass is probably similar to that of Jupiter and its close orbit (about one-eighth of Mercury's orbital distance from the Sun) should give it a surface temperature above 1,000 K, making it an example of what we call a "hot Jupiter."

planet orbiting 51 Pegasi. Moreover, by carefully analyzing Doppler shift data, we can not only learn of the existence of a planet but also determine its average distance from its star, the shape of its orbit, and even its mass. After all, it is the mass of the planet that causes the star to move around the system's center of mass, so for a given orbital distance, a more massive planet will cause faster stellar motion.

We derive an extrasolar planet's orbital distance using Newton's version of Kepler's third law [Section 4.4]. Recall that for a small object like a planet orbiting a much more massive object like a star, this law expresses a relationship between the star's mass, the planet's orbital period, and the planet's average distance (semimajor axis). We generally know the masses of the stars around which we have found extrasolar planets (through methods we'll discuss in Chapter 15) and the Doppler data tell us the orbital period, so we can calculate orbital distance.

We can determine orbital shape from the shape of the Doppler data curve. A planet with a perfectly circular orbit travels at a constant speed around its star, so its data curve would be perfectly symmetric. Any asymmetry in the Doppler curve tells us that the planet is moving with varying speed and therefore must have a more eccentric ("stretched out") elliptical orbit. Figure 13.5 shows four examples of Doppler data for extrasolar planets and what we learn in each case.

THINK ABOUT IT

Study the four velocity data curves in Figure 13.5. How would each be different if the planet were: (a) closer to its star? (b) more massive? Would your answers be different if the orbit were eccentric? Do the answers depend on the orientation of the elliptical orbit?

In some cases, existing Doppler data are good enough to tell us whether the star has more than one planet. Remember that if two or more planets exert a noticeable gravitational tug on their star, the Doppler data will show the combined effect of these tugs. In 1999, such analysis was used to infer the existence of three planets around the star Upsilon Andromedae, making this the first bona fide, multiple-planet solar system known beyond our own. By early 2005, more than a dozen other multiple-planet systems had been identified.

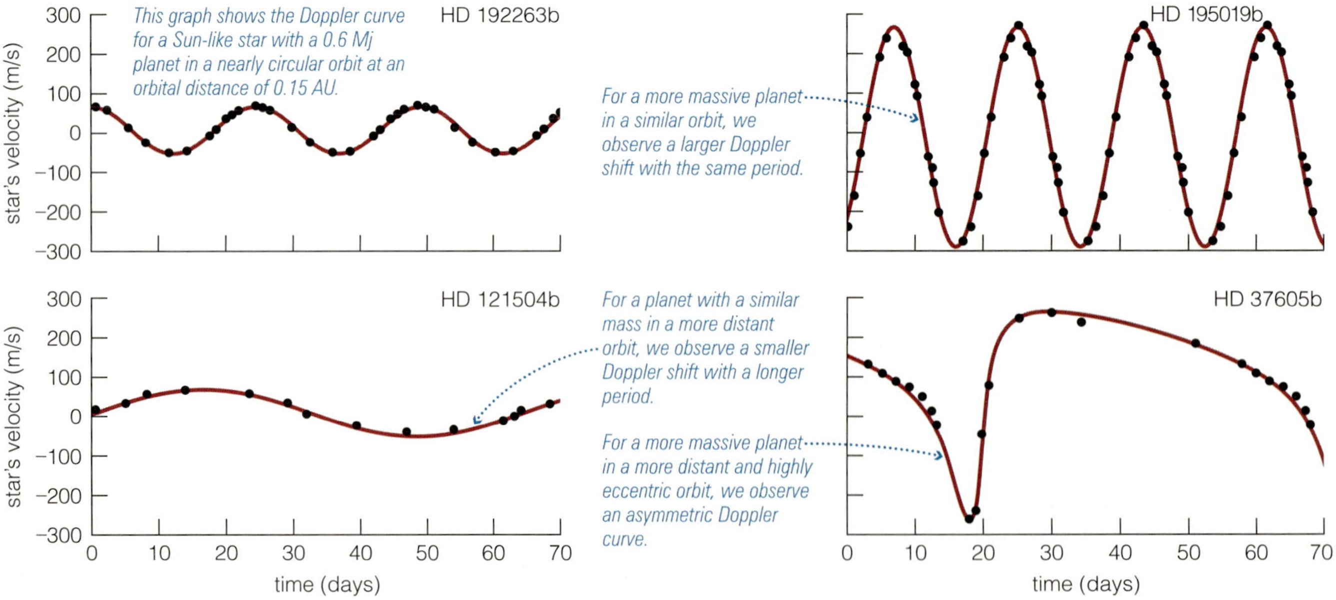

Figure 13.5 Sample data showing how measurements of Doppler shifts allow us to learn about extrasolar planets for different types of orbits. The curves are based on actual planets, whose data are listed in Appendix E.

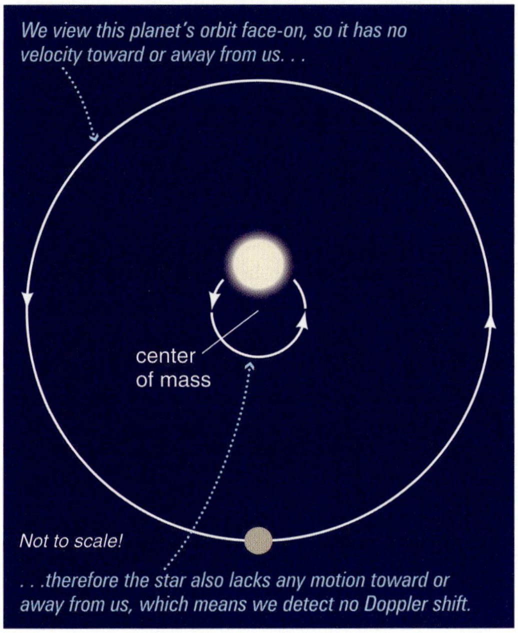

a If we view a planetary orbit face-on, we will not detect any Doppler shift at all.

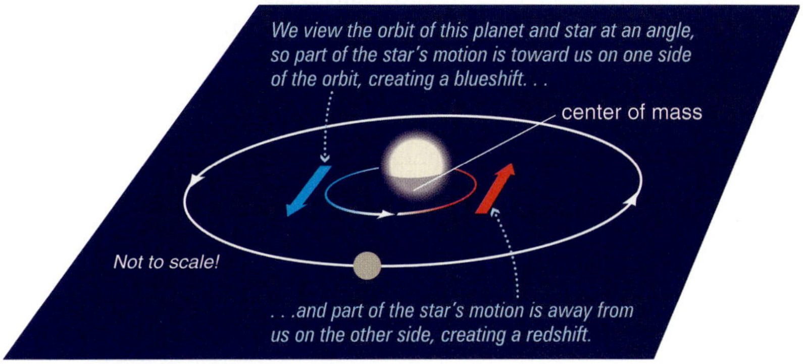

b We can detect a Doppler shift only if the planet and star have some part of their orbital velocities directed toward or away from us. The more the orbit is tilted toward edge-on, the greater the shift we'll observe.

Figure 13.6 The amount of Doppler shift we observe in a star's spectrum depends on the orientation of the planetary orbit that causes the shift.

The Doppler technique also tells us about planetary masses, though with an important caveat. Remember that Doppler shifts reveal only the part of a star's motion directed toward or away from us (see Figure 5.23). Thus, for example, a planet whose orbit we view face-on does not cause any Doppler shift in the spectrum of its star, making it impossible to detect the planet with the Doppler technique, let alone to determine its mass (Figure 13.6a). We can observe Doppler shifts in the star's spectrum only if it has a planet orbiting at some angle other than face-on (Figure 13.6b), and the Doppler shift will tell us the star's full orbital velocity only if we are viewing it precisely edge-on. As a result, a planetary mass that we infer from its star's Doppler shift will be accurate only for a planet in an edge-on orbit. In all other cases, the Doppler shift underestimates the true orbital speed of the star and therefore also causes us to underestimate the true mass of the planet. The more the orbit is tilted away from edge-on, the greater the degree of the underestimate. Thus, planetary masses inferred from the Doppler technique alone are actually *minimum* possible masses (or "lower limits") for these planets.

We can determine a precise mass only if we somehow know that we are viewing an edge-on orbit or if we have some way to measure the precise orbital angle. For example, if we also know the planet's side-to-side motion from the astrometric technique, we can combine this information with the toward and away motions from the Doppler technique to determine the orbital angle. Unfortunately, we rarely have such knowledge, which means that nearly all planetary masses found through the Doppler technique are really minimum masses rather than actual masses. Nevertheless, statistical arguments show that in two out of three cases, the planet's true mass will be no more than double this minimum mass, so the minimum mass is still a relatively good estimate.* Of course, this also means that a third of the planets discovered with the Doppler technique have masses that are more than double the minimum mass. About 6% of the time, the planet's actual mass will be more than 10 times the minimum mass, which may raise the question of whether it is truly a planet or a very small star.

The Doppler technique is very powerful, but it has limits. In particular, it is best suited to identifying massive planets that orbit relatively close to their star. This limitation arises because gravity weakens with distance, so a planet of a given size pulls harder on its star—making the star move faster—if it is closer. Moreover, because a close-in planet has a much shorter orbital period than a more distant planet, it takes a lot less time to observe the periodic Doppler shifts caused by close-in planets. For example, while it takes only a few weeks of observation to detect a planet with a 4-day period like the one orbiting 51 Pegasi,

*You can understand the statistical argument if you have studied trigonometry: The actual mass is the minimum mass divided by the *cosine* of the orbital inclination (where edge-on is an angle of 0° and face-on is an angle of 90°). Thus, the actual mass is more than double the minimum mass only for cosines smaller than $\frac{1}{2}$, which means for angles between 60° and 90°. By random chance, only about $\frac{1}{3}$ of all stars will have such large orbital angles. Similarly, the actual mass is more than 10 times the minimum mass for cosines smaller than 0.1, which means angles between about 84.3° and 90°; these angles occur randomly about 6% of the time.

it would take 12 years to observe just a single orbital cycle of a planet in an orbit like that of Jupiter around our Sun. The Doppler technique also presents a practical difficulty: Because current technology allows it to be applied to just one star at a time and requires a relatively large telescope, only a relatively small number of stars can be studied with this technique.

The limitations of the Doppler technique explain what may at first seem like surprising facts: Most of the extrasolar planets discovered to date orbit relatively close to their stars, and we have yet to discover any planets at all with Earth-like masses. Both these facts may simply be *selection effects* of the Doppler technique; that is, the technique tends to find (or "select") massive planets in close orbits much more easily than any other type of planet. Planets with masses similar to Earth would have such weak gravitational effects on their stars that we could not use the Doppler technique to find them with current technology, while planets orbiting far from their stars have such long orbital periods that it might take decades of observations to detect them. Thus, the current lack of evidence for Earth-mass planets or jovian-mass planets in distant orbits does not necessarily mean that such planets are rare.

Transits and Eclipses A third indirect way of detecting distant planets does not require the observation of gravitational tugs at all. Instead, it relies on searching for slight changes in a star's brightness caused by a planet passing in front of it or behind it.

If we were to examine a large sample of stars with planets, a small number of them—typically one in several hundred—will by chance be aligned in such a way that one or more of its planets will pass directly between us and the star once each orbit. The result is a **transit**, in which the planet appears to move across the face of the star. We occasionally witness this effect in our own solar system when Mercury

MATHEMATICAL INSIGHT 13.1 Finding Orbital Distances for Extrasolar Planets

The Doppler technique directly tells us a planet's orbital period. We can then use this period to determine the planet's orbital distance. If the planet were orbiting a star of exactly the same mass as the Sun, we could find the distance by applying Kepler's third law in its simplest form: $p^2 = a^3$. In fact, many of the planets discovered to date *do* orbit Sun-like stars, so this law gives a good first estimate of orbital distance. But for more precise work, we must use Newton's version of Kepler's third law. As discussed in Mathematical Insight 4.3, this law reads:

$$p^2 = \frac{4\pi^2}{G(M_1 + M_2)} a^3$$

In the case of a planet orbiting a star, p is the planet's orbital period, a is its average orbital distance (semimajor axis), and M_1 and M_2 are the masses of the star and planet, respectively. (G is the gravitational constant; $G = 6.67 \times 10^{-11}\ \text{m}^3/(\text{kg} \times \text{s}^2)$.) Because a star is so much more massive than a planet, the sum $M_{\text{star}} + M_{\text{planet}}$ is pretty much just M_{star}; that is, we can neglect the mass of the planet compared to the star. With this approximation, we can rearrange the equation to find the orbital distance a:

$$a \approx \sqrt[3]{\frac{GM_{\text{star}}}{4\pi^2} p_{\text{planet}}^2}$$

We will discuss how we determine stellar masses in Chapter 15; for now, we will assume the stellar masses are known so that we can calculate orbital distances of the planets.

Example 1: Doppler measurements show that the planet orbiting 51 Pegasi has an orbital period of 4.23 days, and the star's mass is 1.06 times that of our Sun. How far does the planet orbit from its star?

Solution:

Step 1 Understand: We are given both the planet's orbital period and the star's mass, so we can use Newton's version of Kepler's third law to find the planet's orbital distance. However, to make the units consistent, we need to convert the given stellar mass to kilograms and the given orbital period to seconds; we look up the fact that the Sun's mass is about 2×10^{30} kilograms.

$$M_{\text{star}} = 1.06 \times M_{\text{Sun}} = 1.06 \times (2 \times 10^{30}\ \text{kg}) = 2.12 \times 10^{30}\ \text{kg}$$

$$p = 4.23\ \text{day} \times \frac{24\ \text{hr}}{1\ \text{day}} \times \frac{3600\ \text{s}}{1\ \text{hr}} = 3.65 \times 10^5\ \text{s}$$

Step 2 Solve: We now use these values of the period and mass in the above equation to find the orbital distance a:

$$\begin{aligned} a &\approx \sqrt[3]{\frac{GM_{\text{star}}}{4\pi^2} p_{\text{planet}}^2} \\ &= \sqrt[3]{\frac{6.67 \times 10^{-11} \frac{\text{m}^3}{\text{kg} \times \text{s}^2} \times 2.12 \times 10^{30}\ \text{kg}}{4 \times \pi^2} (3.65 \times 10^5\ \text{s})^2} \\ &= 7.81 \times 10^9\ \text{m} \end{aligned}$$

Step 3 Explain: We've found that the planet orbits its star at a distance of 7.8 billion meters. It's much easier to interpret this number if we state it as 7.8 million kilometers or, better yet, convert it to astronomical units. If you recall or look up the fact that 1 AU is about 150 million kilometers or 1.50×10^{11} meters, then we can convert the orbital distance from meters to AU as follows:

$$a = 7.81 \times 10^9\ \text{m} \times \frac{1\ \text{AU}}{1.50 \times 10^{11}\ \text{m}} = 0.052\ \text{AU}$$

We now see that the planet's orbital distance is only 0.052 AU—small even compared to our closest planet, Mercury, which orbits at 0.39 AU. In fact, comparing the planet's 7.8-million-kilometer distance to the size of the star itself (presumably close to the 700,000-kilometer radius of our Sun), we estimate that the planet orbits its star at a distance only a little more than 10 times the star's radius.

or Venus crosses in front of the Sun (see Figure S1.5). Of course, other star systems are so far away that we cannot actually see a planetary dot set against the face of the star as we can for Mercury or Venus set against the face of the Sun. Nevertheless, a transiting planet will block a little of its star's light, allowing us not only to detect the planet's existence but also to calculate the planet's size in comparison to that of its star (see Mathematical Insight 13.3). In practice, we usually know stellar sizes (through methods we'll discuss in Chapter 15), which means we can determine planetary sizes by observing transits.

Detecting planets through transits requires many repeated observations, because most stars exhibit intrinsic variations in brightness. To be confident that a dip in brightness is caused by an orbiting planet rather than variability in the star itself, we need to see the same dimming repeated multiple times. If the repeated dimming occurs with a regular period, then it is very likely telling us the orbital period of a transiting planet. We can then calculate the planet's orbital distance and mass.

THINK ABOUT IT

What kind of planet is most likely to cause a transit across its star that we could observe from Earth: (a) a large planet close to its star? (b) a large planet far from its star? (c) a small planet close to its star? or (d) a small planet far from its star? Explain.

MATHEMATICAL INSIGHT 13.2 Finding Masses of Extrasolar Planets

We can find the mass of an extrasolar planet by using the law of conservation of momentum [Section 4.3]. Consider a star with a single planet, each orbiting around their common center of mass (see Figure 13.1) with the same orbital period. The system as a whole has no momentum relative to this center of mass (because it stays in a fixed place between the star and planet), so the planet's momentum must be equal to the star's momentum but in the opposite direction. Thus, if you remember that momentum is mass times velocity, then we can equate the amounts of momentum held by the star and planet:

$$M_{star}v_{star} = M_{planet}v_{planet}$$

where M stands for the mass of the star or planet, and v stands for the velocity relative to the center of mass. We can use this equation to find the planet's mass if we know the star's mass, the star's orbital velocity, and the planet's orbital velocity; we simply solve the equation for M_{planet}:

$$M_{planet} = \frac{M_{star}v_{star}}{v_{planet}}$$

The Doppler technique gives us a direct measurement of the star's velocity toward or away from us and, as discussed earlier, we generally know the star's mass. We can calculate the planet's orbital velocity from its orbital period and orbital distance; we learn the former directly from the Doppler technique and calculate the latter with the method in Mathematical Insight 13.1. Each time the planet completes an orbit, it must travel a distance of $2\pi a$, where a is the planet's average orbital distance. (Notice that we are using the formula for circumference of a circle, even though the orbits are generally elliptical.) The time it takes to complete the orbit is the orbital period p. Thus, the planet's average orbital velocity must be:

$$v_{planet} = \frac{2\pi a_{planet}}{p_{planet}}$$

You should confirm that substituting this expression for the planet's velocity into the equation above to find the planet's mass gives us the following:

$$M_{planet} = \frac{M_{star}v_{star}p_{planet}}{2\pi a_{planet}}$$

Remember that this formula with velocity data from the Doppler technique will give us the *minimum* mass of the planet unless we somehow know that we are viewing the orbital motion edge-on.

Example 1: Estimate the mass of the planet orbiting 51 Pegasi.

Solution:

Step 1 Understand: From Mathematical Insight 13.1, we already know the planet's orbital period ($p = 3.65 \times 10^5$ s), the planet's orbital distance ($a = 7.82 \times 10^9$ m), and the star's mass ($M_{star} = 2.12 \times 10^{30}$ kg). As stated earlier in the text, the star's velocity averages 57 meters per second (see Figure 13.3b). Thus, we can use the formula found above to calculate the planet's mass.

Step 2 Solve: We plug in all the values found earlier into the mass formula:

$$\begin{aligned} M_{planet} &= \frac{M_{star}v_{star}p_{planet}}{2\pi a_{planet}} \\ &= \frac{(2.12 \times 10^{30}\,\text{kg}) \times \left(57\,\frac{\text{m}}{\text{s}}\right) \times (3.65 \times 10^5\,\text{s})}{2\pi \times (7.82 \times 10^9\,\text{m})} \\ &\approx 9 \times 10^{26}\,\text{kg} \end{aligned}$$

Step 3 Explain: We have found that the minimum mass of the planet is about 9×10^{26} kilograms. This answer will be more meaningful if we convert it to Jupiter masses. From Appendix E, Jupiter's mass is 1.9×10^{27} kilograms, so the planet's minimum mass is:

$$M_{planet} = 9 \times 10^{26}\,\text{kg} \times \frac{1 M_{Jupiter}}{1.9 \times 10^{27}\,\text{kg}} = 0.47 M_{Jupiter}$$

The planet orbiting 51 Pegasi has a mass of at least 0.47 Jupiter mass, which is just under half Jupiter's mass. Remembering that most planets will have actual masses within a factor of about 2 of the minimum mass, we see that the planet probably has a mass quite similar to that of Jupiter.

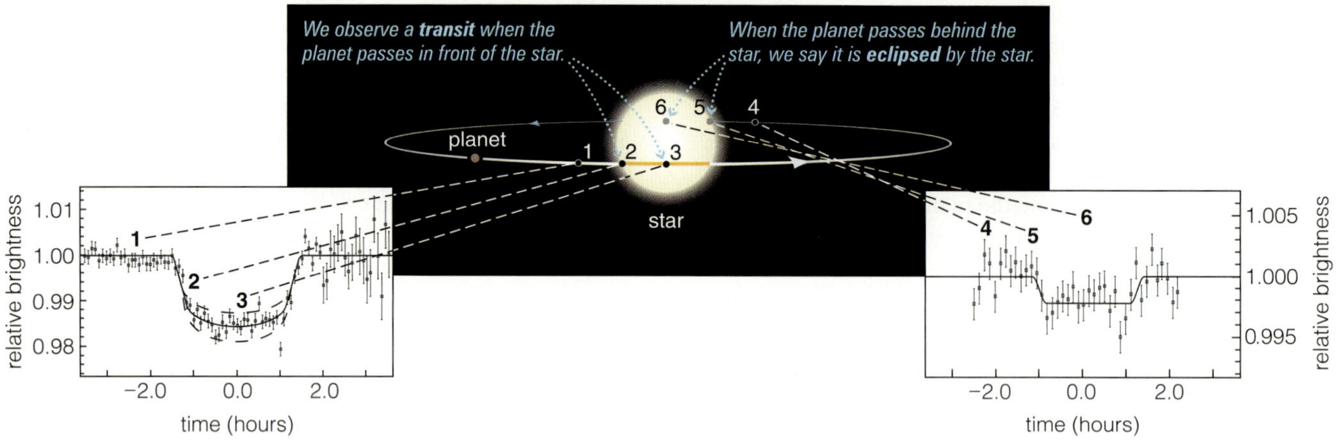

Figure 13.7 Interactive Figure The artist's conception shows the planet orbiting the star HD209458. The data shows how the star's brightness changes during transits and eclipses, which each occur once with every $3\frac{1}{2}$-day orbit. During a transit, the star's brightness drops for about 2 hours by 1.7%, which tells us how the planet's radius compares to the radius of its star. During an eclipse, the infrared signal drops by 0.25%, which tells us about the planet's thermal emission.

The first success of the transit method came during follow-up studies of a planet that had already been discovered with the Doppler technique. The planet, which orbits a star called HD209458, was already known to complete an orbit every $3\frac{1}{2}$ days. Thus, when astronomers observed the star to undergo dips in brightness every $3\frac{1}{2}$ days (and at just the times that the Doppler measurements said the planet would be moving across our line of sight), they realized they were observing repeated transits by the planet (Figure 13.7). This discovery greatly advanced our understanding of the planet itself. First, because the transits revealed that we view the planet's orbit edge-on, it told us that the mass derived from the Doppler technique was the planet's true mass and not just a minimum mass. Second, the amount of the dip in the star's brightness allowed us to determine the planet's radius and hence its volume. We were thereby able to learn the planet's density (by dividing its mass by its volume). Because the radius and density match the values expected for a jovian planet, we had acquired the first strong evidence that a planet in another star system really does resemble the jovian planets of our own solar system.

Transits can also tell us about the composition of a planet's upper atmosphere or exosphere (Figure 13.8). To see how, consider what happens if a planet like Jupiter passes in front of its star. Although Jupiter is composed of gas, from a distance it looks like a solid disk extending out as far as its cloud tops. During the transit, this "solid" disk blocks all the starlight coming from directly behind it. Now, suppose the Jupiter-like planet also has a low-density upper atmosphere extending above its cloud tops. The gas in this upper atmosphere would absorb starlight at specific wavelengths that depend on its composition. For example, if the planet's upper atmosphere contained sodium gas, the star's spectrum would show stronger sodium absorption lines

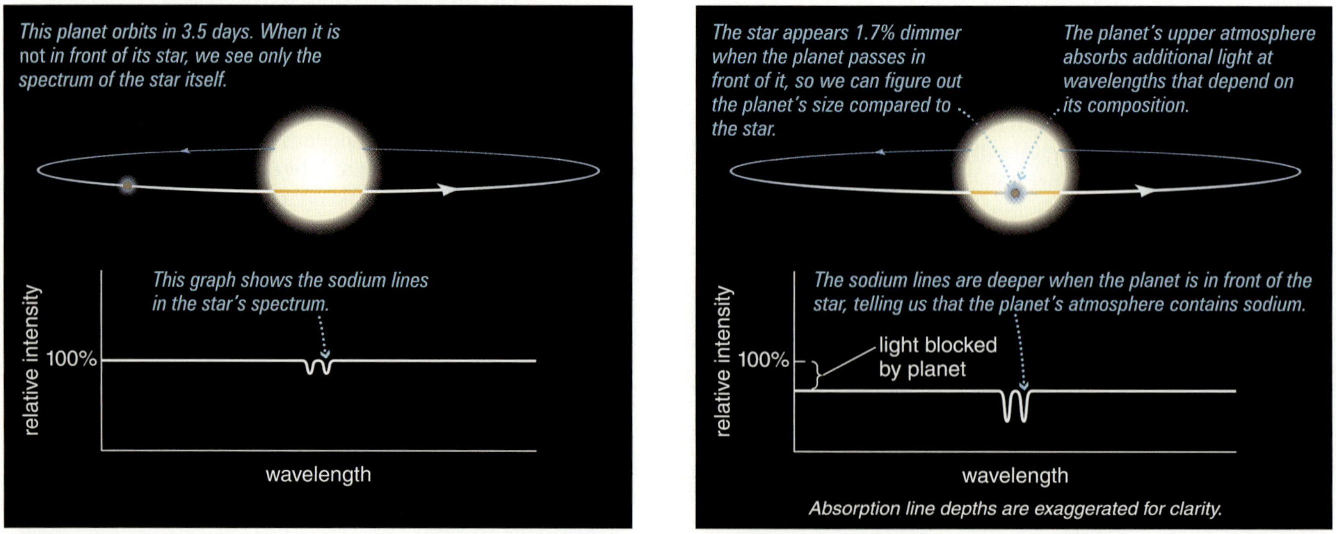

Figure 13.8 This diagram shows how transit observations can give us information about the composition of an extrasolar planet's extended upper atmosphere or exosphere.

during the transit than at other times. We'd thereby learn that the planet contains sodium in its upper atmosphere.

Planets that pass in front of their stars in transits can also pass behind their stars, in which case the star blocks the light from the planet. Such an event is called an **eclipse** (see Figure 13.7), because the star blots out the light from the planet in much the same way that the Moon can blot out the light of our Sun during a total solar eclipse [Section 2.3]. Observing an eclipse is much like observing a transit; in both cases, we measure the total light from the star and planet, searching for a small dip in brightness. Because the star is so much brighter than the planet, the dip in brightness is smaller during an eclipse than a transit. To maximize the effect of the eclipse, astronomers therefore observe at infrared wavelengths, because planets emit most of their own radiation in the form of infrared thermal emission while stars emit more at visible wavelengths. Even so, the total drop in light was only 0.25% when the Spitzer Space Telescope observed an eclipse of the planet orbiting the star HD209458. Nonetheless, astronomers were able to determine the total amount of thermal emission from the planet, confirming its expected temperature of over 1,100 K. In fact, much as for planets in our own solar system, the infrared brightness can be used to determine how much starlight the planet reflects or absorbs (see Mathematical Insight 10.1), and the crude infrared spectrum can be used to identify gases in the planet's atmosphere (see Figure 5.14). The first successful observations of this type were announced in 2005, and it is likely that this technique will be improved and applied to more planets by the time you are reading this book.

The method of observing extrasolar planets through transits and eclipses has some unusual strengths and weaknesses. The most obvious weakness is that it can work only for the small fraction of planetary systems whose orbits are oriented edge-on to the Earth. A second weakness is that the method is biased in favor of planets with short orbital periods—and hence with orbits close to their stars—both because these planets transit more frequently and because we must observe repeated transits before we can be confident of a discovery. Counterbalancing these weaknesses is a very important strength: With sufficiently precise measurements of stellar brightness, the transit method ought to be able to detect planets far smaller than is currently possible with the astrometric or Doppler techniques. Indeed, as we'll discuss in Section 13.4, NASA is currently preparing a mission (called *Kepler*) specifically intended to search for transits by Earth-size planets.

Perhaps the most interesting aspect of the transit method is that it can work with both large and small telescopes. With a large telescope, searches for transits can successfully monitor stars at much greater distances than is possible with the astrometric or Doppler techniques. At the opposite extreme, a telescope as small as 4 inches in diameter has been used to discover an extrasolar planet, and it's relatively easy to confirm for yourself some of the transits already detected. What was once considered impossible can now be assigned as homework (see Problem 50 at the end of the chapter).

MATHEMATICAL INSIGHT 13.3 Finding Sizes of Extrasolar Planets

While finding the masses of extrasolar planets was a breakthrough suggesting that we were detecting jovian planets, it wasn't until astronomers measured extrasolar planet sizes that we gained confidence in their jovian nature. Transits provide the needed size information in a simple geometric fashion.

Like other extrasolar planet measurements, the technique relies on a good understanding of the host star. In particular, we need to know the physical size of the star's disk even though we cannot resolve it with our telescopes. If the star's light can be considered uniform over its disk, then the fractional light drop that occurs during a transit will be:

$$\begin{matrix}\text{fraction of}\\ \text{light blocked}\end{matrix} \approx \frac{\text{area of planet's disk}}{\text{area of star's disk}} = \frac{\pi r_{\text{planet}}^2}{\pi r_{\text{star}}^2} = \frac{r_{\text{planet}}^2}{r_{\text{star}}^2}$$

Note that we used the formula for the area of a circle (πr^2) because from a distance the planet and star should look like circular disks. The equation should also make sense if you look at Figure 13.7a.

We can now rearrange the equation to find the planet's radius; you should confirm that solving for r_{planet} gives us the following formula:

$$r_{\text{planet}} \approx r_{\text{star}} \times \sqrt{\text{fraction of light blocked}}$$

Example 1: What is the radius of the planet orbiting the star HD209458? Needed data: The star's radius is 1.15 times that of our Sun, or about 800,000 kilometers, and the planet blocks 1.7% of the starlight during transit.

Solution:

Step 1 Understand: As shown in Figure 13.7b, transits of this planet cause an approximately 1.7% dip in its star's brightness. Thus, the fraction of the star's light that is blocked is $1.7\% = 0.017$. We now have all the information we need to use the above equation to find the planet's radius.

Step 2 Solve: We plug the numbers into the equation above to find:

$$\begin{aligned} r_{\text{planet}} &\approx r_{\text{star}} \times \sqrt{\text{fraction of light blocked}} \\ &= 800{,}000\ \text{km} \times \sqrt{0.017} \\ &\approx 100{,}000\ \text{km} \end{aligned}$$

Step 3 Explain: The planet's radius is close to 100,000 kilometers. From Appendix E, Jupiter's radius is about 71,500 kilometers. Thus, the planet's radius is about $\frac{100{,}000}{71{,}500} \approx 1.4$ times as large as Jupiter's. In other words, the planet is about 40% larger than Jupiter in radius (or diameter).

Direct Detection The indirect planet-hunting techniques we have discussed so far have started a revolution in planetary science by demonstrating that our solar system is just one of many planetary systems. However, these indirect techniques tell us relatively little about the planets themselves, aside from their orbital properties and sometimes their masses and radii. To learn more about their nature, we need to be able to observe the planets themselves so that we can look for details in images of their surfaces and atmospheres, or study their spectra to learn about their composition and properties.

Unfortunately, it is still not possible to obtain images or spectra of planets around Sun-like stars, primarily because of the incredible glare of the stars themselves. However, in at least one case as of early 2005, astronomers have observed a possible planet around a much fainter star. The "star" is actually a type of object known as a *brown dwarf* [Section 16.3]; brown dwarfs have masses between those of very large planets and those of very small stars and they emit very little visible light. The "candidate planet" has been photographed in the infrared by both the European Southern Observatory's Very Large Telescope (see the photo that opens this chapter) and the Hubble Space Telescope (Figure 13.9). Notice that the infrared emission from the brown dwarf itself was subtracted from the Hubble photo, allowing us to see the candidate planet more clearly. The infrared data tell us that the object is relatively cool in temperature, making it seem much more like a bona fide planet than a brown dwarf or cool star. The Very Large Telescope used adaptive optics to "unblur" effects of Earth's atmosphere [Section 6.3], thereby obtaining picture quality comparable to that from Hubble. Moreover, the Very Large Telescope was able to obtain an infrared spectrum of the candidate planet. The spectrum showed evidence of water molecules, which is consistent with what we would expect for a jovian planet. (Remember that water is one of the most common molecules in jovian planet atmospheres, and forms one of their cloud layers [Section 11.1].)

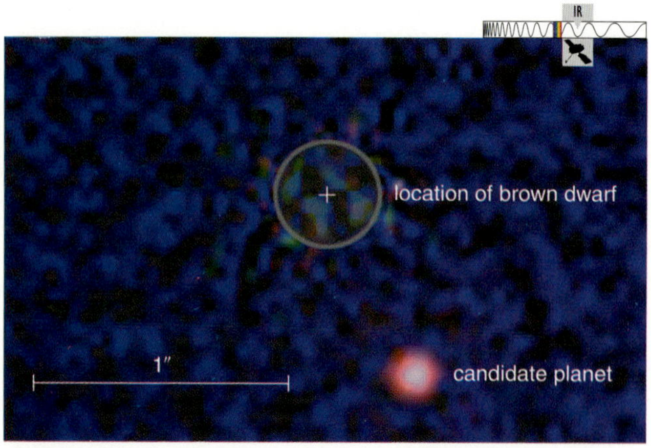

Figure 13.9 This infrared image from the Hubble Space Telescope shows a "candidate planet" in orbit around the brown dwarf known as 2M1207. The light from the brown dwarf (located within the dotted circle) was subtracted out. The red, green, and blue colors represent different wavelengths of infrared light. The candidate planet is brightest at the longest wavelengths, indicating that it is relatively cool in temperature—just as we would expect for a jovian planet.

Other Planet-Hunting Strategies The astonishing success of recent efforts to find extrasolar planets has led astronomers to think of many other possible ways of enhancing the search. One example is a project known as the Optical Gravitational Lensing Experiment (OGLE), a large survey of thousands of distant stars. Although it was not originally designed with planet detection in mind, OGLE has already detected several planets by observing transits. It has also succeeded in detecting at least one by the method in its name—*gravitational lensing*, an effect predicted by Einstein's general theory of relativity that occurs when one object's gravity bends or brightens the light of a more distant object [Sections S3.4, 22.2]. Another strategy looks not for planets themselves or their influence on stars, but rather for their effects on the vast disks of dust that surround many stars. If a planet is present within a dust disk, it can exert small gravitational tugs on the dust particles that will show up as gaps, waves, or ripples in the disk. Such is the case for the star Beta Pictoris, which is surrounded by a dust disk with ripples indicating that it also has one or more planets. Other astronomers are searching for the thermal emission from the impacts of accreting planetesimals, and even for the kinds of emission known to come from the magnetospheres of our jovian planets. As we learn more about extrasolar planets, new search methods are sure to arise.

Mastering Astronomy — Detecting Extrasolar Planets Tutorial, Lessons 1–3

13.2 The Nature of Extrasolar Planets

The mere existence of planets around other stars changes our perception of our place in the universe, because it shows that our planetary system is not unique. In addition, these planets provide us with two new ways to expand our understanding of planetary science.

First, the nature of the individual planets themselves gives us an opportunity to learn more about the range of possible planets. For example, we know that the planets of our own system come in two basic types: terrestrial and jovian. Learning about the nature of extrasolar planets may tell us whether these are the only two categories of planet or whether there are others that we do not see in our own solar system.

Second, studying the arrangements of other planetary systems can tell us whether the layout of our solar system is common or rare, and thereby shed light on whether the nebular theory really does explain the origin of our solar system as neatly as we have presumed. For example, in our solar system the terrestrial planets are all located close to the Sun and the jovian planets are much farther away—an observation that the nebular theory successfully explains.

Can the nebular theory explain the layouts of other solar systems equally well?

In this section, we'll focus our attention on the nature of the extrasolar planets themselves, discussing what we have learned about their orbits, masses, sizes, and compositions. We'll then be prepared to turn our attention in the next section to the questions of how other planetary systems compare to our own and how any differences may have arisen.

• What have we learned about extrasolar planets?

By mid-2005, we already knew of more than 160 extrasolar planets. This number is large enough that we can begin to search for patterns, trends, and groupings that might give us insight into how these planets compare to the planets of our own solar system and how they formed. Before we go into detail, let's summarize the planetary properties that we can learn with current detection techniques:

- *Orbital period:* All three indirect techniques that we discussed (astrometric, Doppler, and transits) tell us the orbital period of detected planets.
- *Orbital distance:* Once we know orbital period and the mass of the star, we can calculate orbital distance by using Newton's version of Kepler's third law.
- *Orbital shape:* We need data spanning an entire orbit to determine orbital shape—that is, whether the orbit is nearly circular or a more eccentric ellipse. Both the astrometric and Doppler techniques can provide the needed data. Transits give us data only for the small part of the orbit during which the planet passes in front of its star and hence do not tell us orbital shape.
- *Mass:* We can determine an extrasolar planet's mass from its orbital period, the mass of its star, and the speed at which it makes its star orbit their mutual center of mass. In principle, we can learn a star's full orbital speed with the astrometric technique, while the Doppler technique tells us only the part of the speed directed toward or away from us. Thus, the Doppler technique tells us a minimum mass for the planet, and gives us a precise mass only if we also know the angle of the orbit to our line of sight. We cannot learn mass from transits alone.
- *Size (radius):* We can learn a planet's size only by observing transits. The dip in the star's brightness during a transit tells us the fraction of its light blocked by the planet, which allows us to calculate the planet's radius.
- *Density:* We can calculate a planet's average density from its size and mass. Because we get size only from transits, we can determine density only for planets that produce transits and for which we also have mass data from the astrometric or Doppler techniques.
- *Composition:* We learn composition from spectra. Transits can provide limited information about the composition of a planet's upper atmosphere if the star shows absorption during the transit that is not present at other times. Eclipses can also provide limited spectral information. More detailed information about composition requires spectra from direct detections.

We are now ready to see what we've learned about the extrasolar planets discovered to date.

Orbits Much as Johannes Kepler first appreciated the true layout of our own solar system [Section 3.3], we can now step back and see the layout of many other solar systems. Figure 13.10a shows the orbits of the first 146 known extrasolar planets all superimposed on each other, as if they were all orbiting a single star; the dots indicate the minimum masses of the planets found through the Doppler technique.

Despite the crowding of the orbits when viewed this way, at least two important facts should jump out at you. First, notice that only a handful of these planets have orbits that take them beyond about 5 AU, which is Jupiter's distance from our Sun. Most of them orbit very close to their host star. Second, notice that many of the orbits are clearly elliptical, rather than nearly circular like the orbits of planets in our own solar system. These facts are even easier to see if we display the same information on a graph (Figure 13.10b). Look first at the green dots representing planets in our own solar system; notice that they are located at the distances you should expect and all but Mercury have very small eccentricity, meaning nearly circular orbits. Now look at the red dots representing extrasolar planets. Quite a few of these planets orbit their stars more closely than Mercury orbits the Sun, and none are located as far from their stars as the jovian planets of our solar system. Many also have large orbital eccentricities, telling us that their elliptical orbits have very stretched-out shapes. As we'll see shortly, both facts give us important clues about the nature of these extrasolar planets.

THINK ABOUT IT

Should we be surprised that we haven't found many planets orbiting as far from their stars as Saturn, Uranus, and Neptune orbit the Sun? Why or why not?

At least 15 stars have so far been found to contain two or more planets, and one system has four known planets (Figure 13.11). This is not surprising, since our own solar system and our understanding of planet formation suggest that any star with planets is likely to have multiple planets. We will probably find many more multiple-planet systems as observations improve. However, one fact about these other planetary systems is surprising: Already, we've found at least seven systems that seem to have planets in orbital resonances [Section 11.2] with each other. For example, three systems have one planet that orbits in exactly half the time as another planet. In our own solar system we've seen the tremendous importance of orbital resonances in sculpting planetary rings, stirring up the asteroid belt, and even affecting the orbits of Jupiter's moons. Orbital resonances may also have profound influences for extrasolar planets, as we'll discuss in Section 13.4.

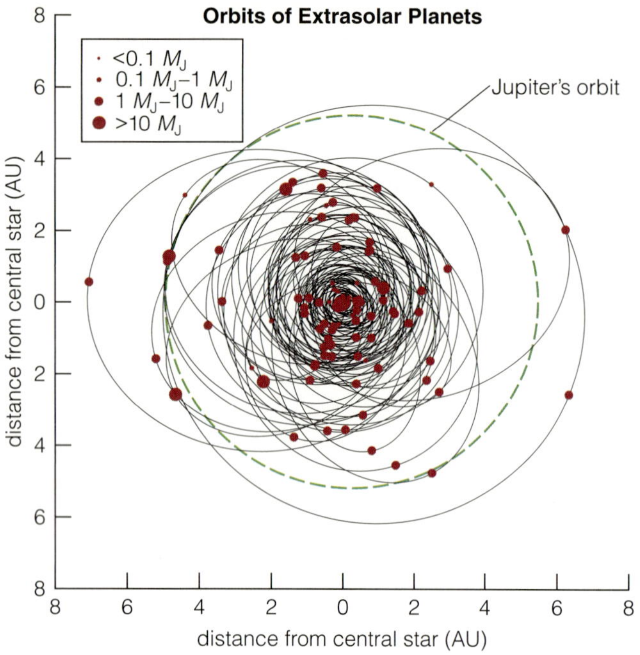

a This diagram shows all the orbits superimposed on each other, as if all the planets orbited a single star. The sizes of the dots indicate the minimum masses of the planets; the dots are located at the aphelion (farthest) point for each orbit.

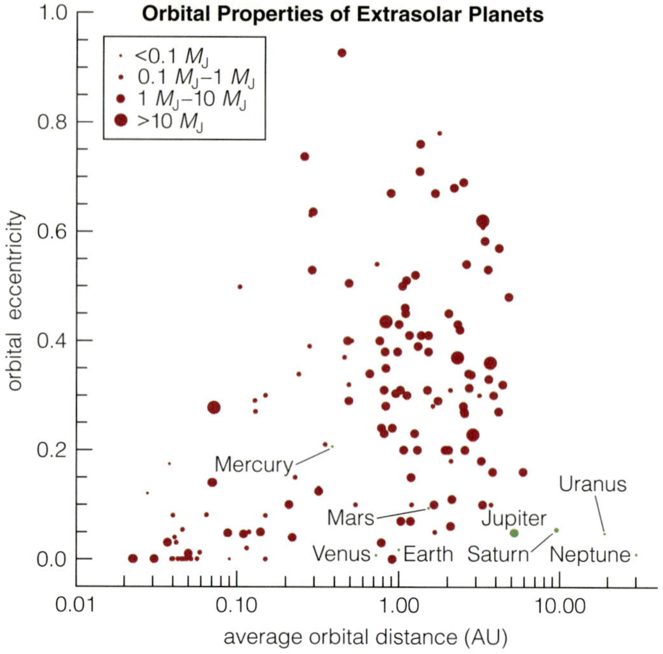

b The data from (a) are shown here as a graph. Each red dot represents an extrasolar planet, and the green dots represent the planets of our own solar system. Planets that orbit closer to their stars are farther to the left on this diagram, and the planets with more circular orbits are lower down on the diagram. Notice that many extrasolar planets orbit quite close to their stars and most have larger orbital eccentricities than the planets of our solar system.

Figure 13.10 Orbital properties of the first 146 known extrasolar planets.

Masses Look again at Figure 13.10. The sizes of the dots indicate the approximate minimum masses of these planets; they are minimums because nearly all have been found with the Doppler technique. The mass data are easy to see if we display them as a bar chart (Figure 13.12). We see clearly that the planets we've detected in other star systems are generally quite massive. Most are more massive than Jupiter, and even the least massive ones are almost as massive as Uranus and Neptune. The least massive extrasolar planet found through mid-2005 has a mass about 7.5 times that of Earth. None are anywhere near as low in mass as Earth, which has a mass of about 0.003 Jupiter mass. Thus, if we go by mass alone, it seems that all the known extrasolar planets must be jovian rather than terrestrial in nature.

Should we be surprised by the lack of planets with masses like the terrestrial worlds? Not yet. Remember that nearly all these planets have been found indirectly by looking for the gravitational tugs they exert on their stars. Massive planets exert much greater gravitational tugs and are therefore much easier to detect. Thus, as we discussed earlier, the abundance of very massive planets among these early discoveries is probably a "selection effect" arising because current planet-finding techniques detect massive planets far more easily than lower mass planets.

Sizes and Densities The masses of the known extrasolar planets suggest they are jovian in nature, but mass alone cannot rule out the possibility of "supersize" terrestrial planets—that is, very massive planets made of metal or rock. To check whether the planets are jovian in nature, we also need to know their sizes, from which we can calculate their densities. If their sizes and densities are consistent with those of the jovian planets in our solar system, then we'd have good reason to think that they really are jovian in nature.

Unfortunately, as we've already discussed, the vast majority of known extrasolar planets have been detected by the Doppler technique, which gives us reasonable mass estimates but no information about size. However, in the relatively few cases for which we also have size data from transits, the sizes and densities do indeed seem consistent with what we expect for jovian planets.

In fact, the main surprise in the size and density data to date is that most extrasolar planets actually seem to be larger and less dense than the jovian planets of our own solar system. For example, the first planet for which we obtained size data (the planet orbiting the star HD209458) is apparently lower in mass than Jupiter (0.63 Jupiter mass) but larger than Jupiter in radius (1.43 Jupiter radii). Its density is 0.3 gram per cubic centimeter, or about one-fourth of Jupiter's density. Could that mean this planet is made of different, lower-density ingredients than Jupiter? That hardly seems possible, since Jupiter is mostly made of hydrogen and helium gas—the lowest-density planetary building materials. Instead, there's probably a much simpler explanation. This

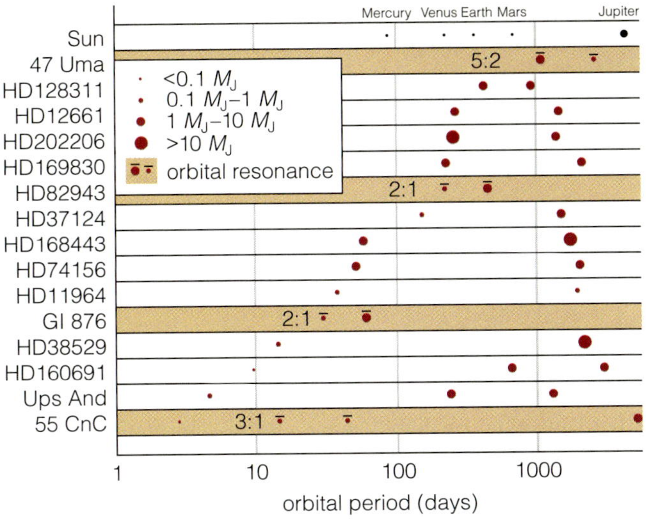

Figure 13.11 This diagram shows the orbital distances and approximate masses of the planets in the first 15 multiple-planet systems discovered. The four highlighted systems are the ones with the best data to show planets in orbital resonances. For example, the "5:2" for 47 Uma indicates that the inner of the two indicated planets completes exactly five orbits while the outer planet completes two orbits. When reading orbital periods from the graph, be sure to notice that the axis is exponential, so that each tick mark represents a period 10 times longer than the previous one.

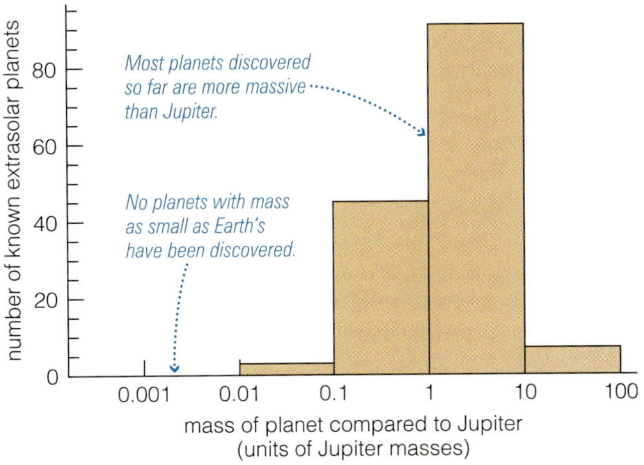

Figure 13.12 This bar chart shows the number of planets of different masses among the first 146 extrasolar planets discovered. The masses are approximate and are based on minimum masses found with the Doppler technique. Notice that the axis uses an exponential scale so that the wide range of masses can all fit on the graph.

planet orbits very close to its star: Its average orbital distance is only 0.047 AU, which is only 12% of Mercury's distance from the Sun. Thus, we expect this planet to be very hot, both because it receives so much heat from its star and because the warm environment slows the escape of the internal heat that jovian planets acquire during formation [Section 11.1]. Calculations show that its temperature is probably more than 1,000 K at its cloud tops, and it should be much warmer than Jupiter throughout its interior. This high temperature puffs up its hydrogen and helium gas, explaining why it is larger in size and lower in density than Jupiter. We now have size and density information for at least eight extrasolar planets (see Appendix E.4), confirming that most close-in worlds are larger and lower in density than Jupiter. One exception is a planet discovered in mid-2005 around the star HD149026. This planet appears to have a Saturn-like mass but a Neptune-like density of about 1.5 g/cm^3 (about twice the density of Saturn). Astronomers deduce that it must contain a proportion of dense materials—rock, metal, and hydrogen compounds—more like that of Neptune than Saturn, despite its Saturn-like mass.

Compositions We have even less direct information about the compositions of extrasolar planets. Nevertheless, the available data also support the idea that these planets are jovian in nature.

Our first data about an extrasolar planet's composition came from the same transiting planet for which we first measured size. During the transits of HD209458, the Hubble Space Telescope detected additional absorption by hydrogen and sodium that is not present at other times (see Figure 13.8). From the amount of extra absorption, astronomers concluded that the planet has an extended upper atmosphere containing abundant hydrogen—just as we should expect for a jovian planet—and a trace of sodium. The next case of composition information came from the "candidate planet" shown in Figure 13.9 and the photo that opens this chapter. As we discussed earlier, a spectrum of this planet revealed the presence of atmospheric water, consistent with what we find in the jovian planets of our own solar system.

• How do extrasolar planets compare with those in our solar system?

Thanks to our detailed understanding of our own jovian planets (Chapter 11), we can take the limited information at hand and make some educated guesses about the nature and behavior of extrasolar planets. In particular, do planets in other star systems fit the same terrestrial and jovian categories as the planets in our solar system? The answer is a tentative "yes," with the recognition that technological limitations mean we have not yet identified even a single terrestrial planet in another star system.

For the known extrasolar planets, we have no reason to doubt that they are similar in nature to the jovian planets of our solar system. Although the only evidence for this claim comes from planetary mass in the vast majority of cases, the few cases for which we also have data on size, density, or composition lend strong support. After all, these few cases are essentially a random sample of the full set of known extrasolar planets, so the fact that these cases all appear to be jovian in nature makes it likely that most or all of the known extrasolar planets are jovian.

We have found one surprising characteristic among these extrasolar planets: All the jovian planets of our solar system orbit far from the Sun on nearly circular orbits, while many of the extrasolar planets orbit quite close to their stars or have highly elliptical orbits. We'll discuss how these planets might have gotten into these surprising orbits in the next section, but it does not change the fact that they are probably jovian planets made mostly of hydrogen and helium gas. Because their close orbits make them much higher in temperature than Jupiter, these planets have been nicknamed "hot Jupiters" (see Figure 13.4).

Assuming we are correct about the nature of the hot Jupiters, we can use models to predict a lot about their outward appearance and characteristics, even though we've never actually photographed one. We simply need to ask how Jupiter would be different if it had somehow migrated into an orbit very close to the Sun. The models show that, if Jupiter orbited just 0.05 AU from the Sun (1% of its actual distance), the extra heat would make it about 50% larger in radius—consistent with what we found for the transiting planet around HD209458.

What would the hot Jupiter look like? It would probably have clouds much like the real Jupiter but of a different type. Recall that Jupiter has multiple cloud layers made from droplets or ice flakes of compounds such as ammonia, methane, and water. The temperature on the hot Jupiter would be far too high for these gases to condense. However, other ingredients may be able to form clouds at high altitudes, including some materials that we usually think of as solids. The models suggest that a hot Jupiter with a temperature above about 1,000 K would have clouds of "rock dust" containing common minerals (especially the mineral enstatite, or $MgSiO_3$).

The hot Jupiter would probably also have a striped appearance much like that of the real Jupiter. In fact, we'd probably be able to see the stripes even on the night side, because the planet's high temperatures would make it glow at visible wavelengths. The stripes should be present because the atmospheric circulation of a hot Jupiter should still be driven by the same basic principles that apply to terrestrial and jovian planets in our solar system [Sections 10.2, 11.1]. Because the hot Jupiter is so close to its star, intense starlight will warm one side of the planet. If the planet rotates fast enough to generate a strong Coriolis effect, it should create patterns of planet-circling winds much like those on Jupiter. Can we expect a hot Jupiter to rotate rapidly? In fact, we probably already know the rotation rates of all the hot Jupiters. They are so close to their stars that strong tidal forces have almost certainly caused them to rotate synchronously with their orbits, just as the Moon rotates synchronously with its orbit around Earth [Section 4.5]. That is, their rotation periods are the same as their orbital periods, and they always show the same face toward their stars. Because the orbital periods of many of these hot Jupiters are just a few days, we expect their rotation periods also to be a few days. Although these periods mean slower rotation than that of jovian planets in our solar system, they should still be fast enough to create strong Coriolis effects and planet-circling winds on such massive planets. Figure 13.13 summarizes the similarities and differences expected between the real Jupiter and "hot Jupiters."

Figure 13.13 A summary of the expected similarities and differences between the real Jupiter and extrasolar "hot Jupiters" orbiting Sun-like stars.

13.3 The Formation of Other Solar Systems

The discovery of extrasolar planets presents us with an opportunity to test our theory of solar system formation. Can our existing theory explain other planetary systems, or will we have to go back to the drawing board? The discoveries have already presented at least one significant challenge: The surprising orbits of the "hot Jupiters" and many other extrasolar planets.

Recall that, according to the nebular theory, planets were born in a spinning disk of material surrounding our Sun [Section 8.3]. The spinning disk was made mostly of hydrogen and helium, but a small fraction of the material consisted of hydrogen compounds (water, ammonia, methane) and an even smaller fraction of metal and rock. Terrestrial planets formed in the warm inner regions of the disk, where only metal and rock could condense into solid form. Jovian planets formed in the cold outer regions of the disk, where temperatures were low enough for hydrogen compounds to condense into ices. Because hydrogen compounds were several times as abundant as rock and metal, the icy planetesimals in the outer solar system grew to much larger sizes than the rocky planetesimals of the inner solar system. They grew large enough so that their gravity began to draw in hydrogen and helium gas from the surrounding nebula, which explains how they ultimately became jovian planets made mostly of hydrogen, helium, and hydrogen compounds. We would expect that planets should form more easily in nebulas with a higher proportion of rock and ice, and in fact more planets have been found around stars richer in the elements that make these ingredients. The nebular theory explains the nearly circular orbits of the planets in our solar system as a direct consequence of the orderly, circular motion of the spinning disk in which the planets formed.

Thus, the nebular theory clearly predicts that jovian planets should form only in the cold outer regions of star systems and that they should be born with nearly circular orbits. Known extrasolar planets, many with close-in or highly elliptical orbits, present a direct challenge to these ideas.

• Can we explain the surprising orbits of many extrasolar planets?

The nature of science demands that we question the validity of a theory whenever it is challenged by any observation or experiment [Section 3.4]. If the theory cannot explain the new observations, then it must be revised or discarded. The surprising orbits of many of the extrasolar planets have indeed caused scientists to reexamine the nebular theory of solar system formation.

The questioning began almost immediately upon the discovery of the first extrasolar planets. The close-in orbits of these massive planets made scientists wonder whether something might be fundamentally wrong with the nebular theory. For example, could it be possible that jovian planets *can* form close to their stars? Astronomers addressed this question by studying many possible models of planet formation and re-examining in detail the entire basis of the nebular theory. Several years of such re-examination did not turn up any good reasons to discard the basic theory. While it's still possible that a major flaw has gone undetected, it seems more likely that the basic outline of nebular theory is correct. Scientists therefore suspect that the extrasolar jovian planets were indeed born with circular orbits far from their stars, and that those that now have close-in or eccentric orbits underwent some sort of "planetary migration" or suffered gravitational interactions with other massive objects.

Planetary Migration If the hot Jupiters formed in the outer regions of their star systems and then migrated inward, how did these planetary migrations occur? You might think that drag within the solar nebula could cause planets to migrate, much as atmospheric drag can cause satellites in low-Earth orbit to lose orbital energy and eventually plunge into the atmosphere; however, calculations show this drag effect to be negligible. A more likely scenario is that waves propagating through a gaseous disk lead to migration (Figure 13.14). A planet's gravity can nudge particles in the disk and thereby create waves that propagate through the disk, causing material to bunch up as the waves pass by. The "bunched up" matter in the wave peaks, in turn, exerts an extra gravitational nudge back on the planet. This gravitational nudge can cause the planet to migrate inward.

Computer models confirm that waves in the nebula can cause young planets to spiral slowly toward their star. In our own solar system, this migration is not thought to have played a significant role because the solar wind cleared out the gas before it could have much effect. But planets may form earlier in at least some other solar systems, allowing time for jovian planets to migrate substantially inward. In

Figure 13.14 This figure shows a simulation of waves created by a planet embedded in a dusty disk of material surrounding its star.

some cases, the planets may form so early that they end up spiraling into their stars. Indeed, astronomers have observed some stars that have an unusual assortment of elements in their outer layers, suggesting that they may have swallowed planets

The migration hypothesis seems to work well in computer models, but as this book went to press it faces a new challenge: the discovery of a hot Jupiter in a triple-star system. The radiation from the adjacent stars should have cleared the gas from the disk quite early, in which case it is difficult to see how migration could have occurred. It's too soon to tell if this discovery requires rethinking the migration hypothesis overall, or if this is just an unusual case. One thing is certain: We are learning far more about solar system formation from the wealth of new worlds than we could from our solar system alone.

Encounters and Resonances A possible explanation for eccentric orbits is close gravitational encounters [Section 4.5] between young jovian planets in the outer regions of a disk. Close gravitational encounters between two massive planets can send one planet out of the star system entirely while the other is flung inward into a highly elliptical orbit. Alternatively, a jovian planet could migrate inward as a result of multiple close encounters with much smaller planetesimals. The jovian planets in our solar system probably migrated by this latter mechanism. As comets were ejected into the Oort cloud [Section 12.2], the jovian planets must have migrated inward, as a consequence of the law of conservation of energy.

Gravitational interactions can also affect orbits through resonances. Recall that Jupiter's moons Io, Europa, and Ganymede share orbital resonances that cause their orbits to be more elliptical than they would be otherwise (see Figure 11.20b). Models show that similar resonances between jovian planets could also cause their orbits to become more eccentric, possibly explaining the eccentricities found even for some extrasolar planets orbiting at Jupiter-like distances. In some cases, the resonance could even lead to planetary migration or to a planet's being ejected from its system altogether. In other cases, planetary migration might force two planets into a resonance that they did not have originally. The fact that we've already discovered orbital resonances in several multiple-planet systems (see Figure 13.11) lends support to the idea that resonances could play an important role in other solar systems.

• Do we need to modify our theory of solar system formation?

We began this section by asking whether the nebular theory of solar system formation can still hold up in light of our discoveries of planets around other stars. As we've seen, it probably can—provided that we allow for planets undergoing orbital changes after their births.

The bottom line is that discoveries of extrasolar planets have shown us that the nebular theory was incomplete. It explained the formation of planets and the simple layout of a solar system such as ours. However, it needs new features—such as planetary migration and gravitational encounters or resonances—to explain the differing layouts of other solar systems. A much wider range of solar system arrangements now seems possible than we had guessed before the discovery of extrasolar planets.

THINK ABOUT IT

Look back at the discussion of the nature of science in Chapter 3, especially the definition of a scientific theory. Does our theory of solar system formation qualify as a scientific theory, even though we have recently learned that it needs modification? Does this mean the theory was "wrong" before the modifications were made? Explain.

Given the fact that the majority of planetary systems discovered to date have planets that seem to have undergone migration or other orbital changes, and that we have not yet found any other planetary system with a layout like our own, it's natural to wonder whether our own solar system is the unusual one. Could it be that the neat layout of our solar system is the result of some sort of extreme cosmic luck, and that most or all other planetary systems undergo far greater changes after their planets are born? First impressions from the current data might seem to support this idea, but a more careful look shows that we cannot yet draw such a conclusion.

Among the thousands of Sun-like stars that astronomers have so far examined in search of extrasolar planets, only about 1 in 10 show evidence of planets around them. While this could mean that planetary systems are relatively rare, a more likely hypothesis is that planets are present but more difficult to detect in the other 9 in 10 systems. Remembering that current technology makes it easier to find massive planets in close-in orbits, this latter hypothesis would suggest that in most solar systems, jovian planets orbit far from their stars—just as in our own solar system. Thus, current data are fully consistent with the hypothesis that most solar systems are more like our own than they are like the ones discovered to date. In essence, our early discoveries of extrasolar planets may be much like a glimpse of animals in a rain forest: The jungle appears full of brightly colored parrots and frogs, but far more animals fail to catch our eye. As time goes on and technology improves, we may begin to find more systems like our own.

If solar systems like our own are indeed common, then the real challenge is to explain precisely how and under what conditions a system may end up like the "unusual" ones discovered to date. This challenge has led planetary scientists to look more closely at the question of exactly how the planets in our own solar system would have interacted with one another when the solar system was young. These studies are only in preliminary stages at present, but they are already causing some scientists to wonder whether migration and gravitational interactions were more important in our solar system than previously thought. It may be that all planetary systems experience these processes to some extent, but that in most cases they stop before jovian planets end up with close-in or eccentric orbits.

All this discussion might seem fairly detailed, but in fact it has profound implications for the way we view our place in the universe. If solar systems like ours are indeed common, then it seems reasonable to imagine that Earth-like planets—and perhaps life and civilizations—might also be common. But if our solar system is a rarity or even unique, then Earth might be the lone inhabited planet in our galaxy or even the universe. We'll discuss this important issue in more depth in Chapter 24, when we turn our attention to life in the universe. In the end, however, the only way to resolve the questions is to gather more data. Thus, we turn now to the future of planetary science, and the question of how we will gather the data needed to find out whether planets like ours are rare or common.

13.4 Finding New Worlds

We have entered a new era in planetary science, one in which our understanding of planetary processes can be based on far more planets than just those of our own solar system. Although our current knowledge of extrasolar planets and their planetary systems is still quite limited, ingenious new observing techniques, dedicated observatories, and ambitious space telescopes should broaden our understanding dramatically in the coming years and decades.

In this section we'll focus on the more dramatic improvements that space missions and new ground-based methods will provide. These techniques will not only permit the discovery of Earth-like planets (if they exist), but will also give us the ability to map and study planets around other stars in far greater detail than we can now.

• How will we search for Earth-like planets?

There's probably no bigger question in planetary science than whether Earth-like planets exist around other stars. NASA's Spitzer Space Telescope has already seen the infrared glow from dust created by the collisions of rocky planetesimals in other accreting solar systems, and we have no reason to doubt that terrestrial planets have formed around other stars. Thus, we have good reason to think that Earth-like planets should be out there, but are they? When you consider the fact that searching for an Earth-size planet around even the nearest stars is like searching for a pinhead from thousands of miles away, you might be tempted to think it would be an impossible task. Remarkably, however, missions currently in preparation should be capable of doing just that. It is likely that by the end of this decade we will have surveyed tens of thousands of star systems and will have learned the definitive answer to the question of whether Earth-size planets are rare or common. Let's examine just a few of the upcoming missions that should help us answer age-old questions about our place in the universe.

Transit Missions: Kepler *and* COROT As we discussed in Section 13.1, we should in principle be able to detect planets the size of Earth or even smaller by searching for transits—the slight dips in stellar brightness that would occur when one of these planets passes in front of its star. In practice, the search for transits by Earth-size planets poses three major technological challenges. First, the dips in brightness caused by Earth-size planets will be very small and therefore require extraordinarily precise measurement. For example, viewed from afar, a transit of our own Sun by Earth would dim the Sun's light by only about 0.008%—not quite one part in 10 thousand. Second, because stars can vary in brightness for reasons besides transits, we can be confident that we've detected a planet only if the transit repeats with a regular period. For planets with sizes and orbital periods like those of the terrestrial worlds in our solar system, this means searching for transits that last no more than a few hours and recur anywhere from every couple of months to every couple of years. Clearly, we are likely to miss the transits unless we continuously monitor stars both day and night year-round. The third challenge comes from the fact that only a tiny fraction of planetary systems will by chance be oriented in such a way that their planets pass in front of their star from our vantage point on Earth. For example, only about 1 in 200 star systems should by chance have an orientation that would allow us to see a transit by an Earth-size planet in an Earth-like orbit. We therefore must monitor thousands of stars in order to have a reasonable expectation of just a few successes.

All three challenges should be met by a NASA mission called *Kepler*, tentatively scheduled for 2008 (Figure 13.15). *Kepler* will be launched with escape velocity

Figure 13.15 Artist's conception of NASA's *Kepler* mission to find Earth-like planets around other stars.

into an orbit around the Sun, so that Earth will not get in the way of its observations. It will then be pointed so that its telescope will essentially stare continuously at the same portion of the sky (in the constellation Cygnus) for the next 4 years. Its field of view is wide enough so that it will be able to monitor about 100,000 stars, measuring their brightnesses about every 15 minutes. Its cameras are sensitive enough so that it should be able to detect transits of Earth-size planets around Sun-like stars, and transits of planets as small as Mercury around somewhat dimmer stars. Because we don't know how common Earth-size planets are, it's difficult to predict how many *Kepler* should find. However, if we assume that our solar system is typical with its two Earth-size planets (Venus and Earth), calculations show that *Kepler* should detect about 50 such planets during its 4 years of observations. Of course, it should be even more successful at detecting larger planets that block more of their star's light, and should therefore greatly add to our current collection of known extrasolar planets.

While NASA is working on *Kepler*, the European Space Agency is building its own mission capable of searching for transits. The European mission, called *COROT*, is actually scheduled for launch ahead of *Kepler*, in mid-2006. It is not sufficiently sensitive to detect planets as small as Earth, but it should be able to detect planets just a few times larger.

THINK ABOUT IT

Find the current status of the *Kepler* and *COROT* missions. Has either been launched yet? If so, has it found any planets yet? If not, when is launch expected?

Astrometric Missions: GAIA *and* SIM Recall from Section 13.1 that the astrometric technique can in principle be used to find the slight side-to-side motions of stars in our sky caused by the gravitational tugs of orbiting planets. To date, this technique has not yielded any planetary discoveries, primarily because current telescopes cannot measure stellar positions with sufficient accuracy. For example, if you look back at Figure 13.2, you'll see that from 30 light-years away, we'd need position measurements accurate to less than a milliarcsecond (a thousandth of an arcsecond) to notice the effects of Jupiter or Saturn on the Sun's position.

The European Space Agency's *GAIA* mission, slated for launch in 2010, has the ambitious goal of performing astrometric observations of a billion stars in our galaxy with an accuracy of 10 microarcseconds. (A microarcsecond is one-millionth of an arcsecond.) *GAIA* will include a telescope dedicated to obtaining spectra for Doppler measurements. By simultaneously making position and Doppler measurements, *GAIA* will allow us to precisely determine a star's motion in all three dimensions, thereby allowing us to compute very accurate masses for orbiting planets. Furthermore, *GAIA* is capable of planetary transit detections. All told, it may discover thousands of new worlds. Part of *GAIA*'s power comes from its 180 CCD detectors [Section 6.1], with a total area equivalent to a square 0.8 meter on a side.

The only way to achieve higher astrometric precision is through interferometry [Section 6.4], in which two or more telescopes work together to obtain the angular resolution of a much larger telescope. Astronomers are making rapid progress in adapting interferometry to ground-based infrared and visible-light telescopes. For example, the twin Keck telescopes on Mauna Kea may soon be capable of detecting at least some planets through the astrometric technique. A NASA mission tentatively slated for launch in 2011 should do even better: the *Space Interferometry Mission (SIM)*. As currently planned, *SIM* will be capable of measuring stellar positions to a precision of just 1 microarcsecond—10 times better than *GAIA*'s precision and good enough to detect stellar motion caused by Earth-size planets around the nearest few dozen stars and motion due to Jupiter-size planets orbiting stars as far as 3,000 light-years away.

SIM will also serve as a test bed for a new technology, called *nulling interferometry*, designed to cancel (or "null") the light from a star so that we can more easily see its orbiting planets. This technology should in principle make possible direct detection of extrasolar planets around relatively nearby stars. *SIM* itself may be able to make crude images of a few Jupiter-size planets, and follow-on missions may give us our first pictures and spectra of terrestrial worlds around other stars.

Direct Detection: TPF *and* Darwin Astronomers are optimistic that the missions described above and the ever-expanding searches conducted with ground-based telescopes will discover thousands of new extrasolar planets and answer the question of whether Earth-*size* planets are rare. But these admirable goals fall short of answering the more deep-seated question of whether other planets are Earth-*like*. To answer this question, we need actual images or spectra of distant terrestrial worlds, so that we can learn whether they are geologically dead like Mercury and the Moon, frozen like Mars or overheated like Venus, or "just right" like Earth.

To accomplish the needed direct detection, NASA is in early planning stages for two missions in the *Terrestrial Planet Finder (TPF)* project. One mission, designated *TPF-C* (with the "C" standing for coronographic), would use a 5-meter telescope equipped with specialized baffles and masks to block out starlight so that we can see surrounding planets. This mission could be launched as early as 2014. The second and more ambitious mission, designated *TPF-I* ("I" for interferometry), envisions five spacecraft flying in tightly controlled formation as an interferometer (Figure 13.16), using nulling interferometry to cancel out light from the star so that we can see planets. Scientists hope that *TPF-I* will launch by 2020. The European Space Agency is contemplating a similar mission called *Darwin* on about the same time scale, so NASA and the European Space Agency may combine efforts in this grand endeavor. Either way, the *TPF-I* or *Darwin* missions would have the ability to obtain infrared spectra of Earth-sized planets around relatively nearby stars, thereby allowing us to search for spectral signatures of life-sustaining atmospheres, and possibly, of life itself.

Figure 13.16 This painting shows one possible configuration for NASA's planned mission called *TPF-I*, with five telescopes flying in formation so that they can obtain crude images and spectra of extrasolar planets.

THINK ABOUT IT

How do you think the discovery of other Earth-like planets would change our view of our place in the universe? Would it matter whether we knew that the planets had life? Defend your opinions.

THE BIG PICTURE

Putting Chapter 13 into Perspective

In this chapter, we have explored one of the newest areas of astronomy—the study of solar systems beyond our own. As you continue your studies, keep in mind the following important ideas:

- With the discoveries of more than 100 extrasolar planets already in hand, we now know that planetary systems are common in the universe, although we do not yet know whether most are similar to or different from our own.
- The discovery of other planetary systems has inaugurated a new era in planetary science, one in which we have far more individual worlds to study and in which we can put our theory of solar system formation to the test.
- Because nearly all detections to date have been made with indirect techniques, we do not yet know much about the planets we have discovered. However, mass estimates combined with limited information about size and composition suggest that we have so far discovered planets similar in nature to the jovian planets of our solar system.
- It is too soon to know if Earth-like planets are rare or common, but missions currently being planned should help us answer this fundamental question within the next couple of decades.

SUMMARY OF KEY CONCEPTS

13.1 Detecting Extrasolar Planets

- **Why is it so difficult to detect planets around other stars?** The great distances to stars would by itself make it very difficult to detect **extrasolar planets**, and the problem is even more difficult because typical stars are a billion times brighter than the light reflected from any of their planets.
- **How do we detect planets around other stars?** Nearly all known extrasolar planets have been discovered indirectly. We can look for a planet's gravitational effect on its star through the **astrometric technique**, which looks for small shifts in stellar position, or the **Doppler technique**, which looks for the back-and-forth motion of stars revealed by Doppler shifts. We can also search for **transits** and **eclipses** in which a system becomes slightly dimmer as a planet passes in front of or behind its star. The Doppler technique has yielded the vast majority of extrasolar planet discoveries to date.

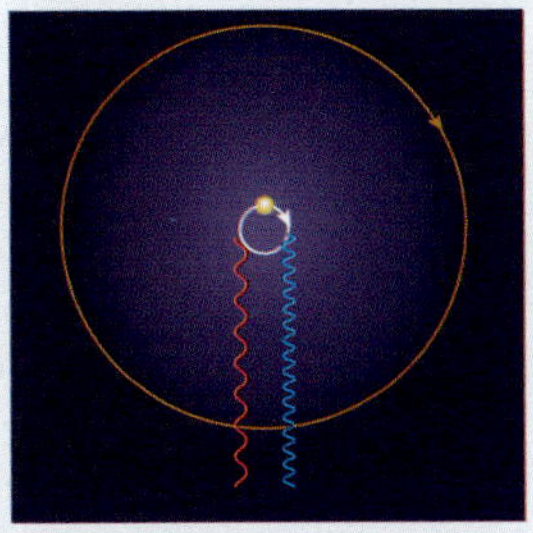

13.2 The Nature of Extrasolar Planets

- **What have we learned about extrasolar planets?** The known extrasolar planets are all much more massive than Earth. Many of them orbit surprisingly close to their stars or have large orbital eccentricities. We have limited information about sizes and compositions, but these data are consistent with the idea that the planets are jovian in nature.

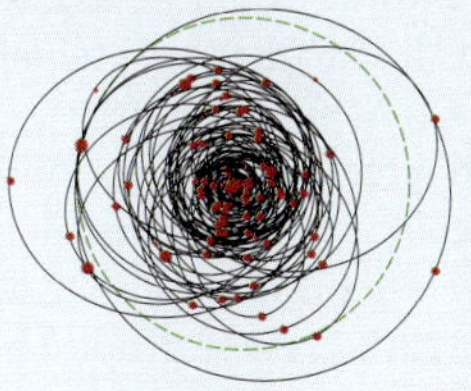

- **How do extrasolar planets compare with those in our solar system?** The known planets are probably jovian in nature. The planets that orbit close to their stars are called "hot Jupiters" because they must have very high temperatures that puff them up in size and give them a lower density than Jupiter.

13.3 The Formation of Other Solar Systems

- **Can we explain the surprising orbits of many extrasolar planets?** The planets with close-in or eccentric orbits probably were born on orbits similar to those of the jovian planets in our solar system. Their orbits then changed through some sort of planetary migration due to waves in the gaseous disk in which they formed, or through gravitational encounters or resonances with other massive planets.
- **Do we need to modify our theory of solar system formation?** Our basic theory of solar system formation seems to be sound, but we have had to modify it to allow for orbital change of the type thought to have occurred with the "hot Jupiters" or planets on very eccentric orbits.

13.4 Finding New Worlds

- **How will we search for Earth-like planets?** The first systematic attempt will be made by searching for transits with the *Kepler* mission. Later missions, such as the *Terrestrial Planet Finder*, may be able to detect such planets directly.

EXERCISES AND PROBLEMS

For instructor-assigned homework go to MasteringASTRONOMY www.masteringastronomy.com

Review Questions

Short-Answer Questions Based on the Reading

1. Why are extrasolar planets hard to detect directly?
2. What are the three major methods used to detect extrasolar planets indirectly?
3. Explain why a planet can cause its star to move slightly in the sky.
4. How does the astrometric technique work? Why hasn't it been very successful in discovering planets to date?
5. How does the Doppler technique work? Explain how it can tell us a planet's orbital period, orbital distance, and orbital eccentricity.
6. Why does the Doppler technique generally allow us to determine only minimum planetary masses rather than actual planetary masses? Should we expect these minimum masses to be close to the actual masses? Explain.
7. How does the transit technique work? Could we use this method to find planets around all stars that have them? Why or why not?
8. Have any planets been detected directly? Explain.
9. Briefly summarize all the planetary properties we can in principle measure with current techniques, and state which techniques allow us to measure each of these properties.
10. How do the orbits of known extrasolar planets differ from those of jovian planets in our solar system? Why are these orbits surprising?
11. What data suggest that the known extrasolar planets are similar in nature to the jovian planets in our solar system?
12. What do we mean by a "hot Jupiter"? How should we expect a hot Jupiter to compare to the real planet Jupiter?
13. Many extrasolar jovian planets orbit surprisingly close to their stars. How might they have ended up in these orbits?
14. Many extrasolar jovian planets have high orbital eccentricities. How might they have ended up with such eccentric orbits?
15. Based on current data, does it seem likely that our solar system has a particularly unusual layout? Explain.
16. Briefly describe how the *Kepler* mission will search for evidence of Earth-size planets.
17. How will the *SIM* and *GAIA* missions aid our search for extrasolar planets?
18. What technologies offer the hope of determining whether Earth-like planets exist around other stars? What missions may give us this capability?

Test Your Understanding

Does It Make Sense?

Decide whether the statement makes sense (or is clearly true) or does not make sense (or is clearly false). Explain your reasoning. (For an example, see Chapter 1, "Does It Make Sense?")

19. An extraterrestrial astronomer surveying our solar system with the Doppler technique could discover the existence of Jupiter with just a few days of observation.
20. The fact that we have not yet discovered an Earth-mass planet tells us that such planets must be very rare.
21. Within the next few years, astronomers expect to confirm all the planet detections made with the Doppler technique by observing transits of these same planets.
22. Although "hot Jupiters" are unlikely places to find life, they could be orbited by moons that would have pleasant, Earth-like temperatures.
23. Before the discovery of planetary migration, scientists were unable to explain how Saturn could have gotten into its current orbit.
24. It's the year 2007: Astronomers have successfully photographed an Earth-size planet, showing that it has oceans and continents.
25. It's the year 2025: Astronomers have just announced that they have obtained a spectrum showing the presence of oxygen in the atmosphere of an Earth-size planet.
26. It's the year 2040: Scientists announce that our first spacecraft to reach an extrasolar planet is now orbiting a planet around a star located near the center of the Milky Way Galaxy.

27. An extrasolar planet is discovered with a year that lasts only 3 days.
28. Later this year, scientists use the Doppler technique to identify a planet whose mass is equal to Earth's mass.
29. Astronomers announce that all the Doppler technique discoveries of extrasolar planets made to date are actually more massive brown dwarfs, and we had thought they were less massive only because we didn't realize that they have nearly face-on orbits.
30. The number of extrasolar planets increases from around 150 in 2005 to more than 1,000 by the year 2010.

Quick Quiz

Choose the best answer to each of the following. Explain your reasoning with one or more complete sentences.

31. What method has detected the most extrasolar planets so far? (a) the transit method (b) Hubble images (c) the Doppler method
32. The extrasolar planets discovered so far most resemble (a) terrestrial planets. (b) jovian planets. (c) large icy worlds.
33. How many extrasolar planets have been detected? (a) between 10 and 100 (b) between a hundred and a thousand (c) more than a thousand
34. Which one of the following can the transit method tell us about a planet? (a) its mass (b) its size (c) the eccentricity of its orbit
35. Which method could detect a planet in an orbit that is face-on to the Earth? (a) Doppler method (b) transit method (c) astrometry
36. How is the planet 51 Pegasi b different from Jupiter? (a) much closer to its star (b) much longer year (c) much more massive
37. Most known extrasolar planets are more massive than Jupiter because (a) we do not expect smaller planets to exist. (b) current detection methods are more sensitive to larger planets. (c) the Doppler method usually overestimates planet masses.
38. Which detection method can be used on a backyard telescope with a CCD system? (a) Doppler method (b) transit method (c) astrometry
39. What's the best explanation for the location of "hot Jupiters"? (a) They formed closer to their stars than Jupiter did. (b) They formed farther out than Jupiter did and then migrated inward. (c) The stars' stronger gravity pulled them closer.
40. Earth-sized planets orbiting normal stars (a) have already been discovered. (b) should be discovered in the next few years by ground-based telescopes. (c) should be discovered by 2010 by a space telescope.

Investigate Further

In-Depth Questions to Increase Your Understanding

Short-Answer/Essay Questions

41. *Why So Soon?* The detection of extrasolar planets came much sooner than astronomers expected. Was this a result of planets being different than expected, or of technology improving faster? Explain.
42. *Why Not Hubble?* Of the more than 160 extrasolar planets discovered, only one possible planet has ever been imaged by the Hubble Space Telescope. What limits Hubble's ability to image planets around other stars, and why was it possible in the case of Figure 13.9?
43. *Explaining the Doppler Method.* Explain how the Doppler method works in terms an elementary school child would understand. It may help to use an analogy to explain the difficulty of direct detection and for the general phenomenon of the Doppler shift.
44. *Comparing Methods.* What are the advantages and disadvantages of the Doppler and transit methods? What kinds of planets are easiest to detect in each case? Are there certain planets that each method cannot detect, even if the planets are very large? Explain. What advantages come if a planet can be detected by both methods?
45. *No Hot Jupiters Here.* How do we think "hot Jupiters" formed? Why didn't one form in our solar system?
46. *Resonances.* How have resonances been important in affecting extrasolar planet orbits? How are these effects similar to the effects of resonances in our solar system, and how are they different?
47. *Low-Density Planets.* Only one planet in our solar system has a density less than 1 gram per cubic centimeter, but many extrasolar planets do. Explain why in a few sentences. (Hint: consider the densities of the jovian planets in our solar system, given in Figure 11.1.)
48. *A Year on HD209458b.* Imagine you're visiting the planet that orbits the star HD209458, hovering in the upper atmosphere in a suitable spacecraft. What would it be like? What would you see, and how would it look different compared to floating in Jupiter's atmosphere? Consider factors like local conditions, clouds, how the Sun would appear, and orbital motion.
49. *Lots of Big Planets.* Many of the planets discovered so far are more massive than our most massive planet. Does this mean our solar system is unusual? If so, how or why? If not, why not?
50. *Detect an Extrasolar Planet for Yourself.* Most colleges and many amateur astronomers have the equipment necessary to detect known extrasolar planets using the transit method. All that's required is a telescope 10 or more inches in diameter, a CCD camera system, and a computer system for data analysis. The basic method is to take exposures of a few minutes' duration over a period of several hours around the times of predicted transit, and to compare the brightness of the star being transited relative to other stars in the same CCD frame (Figure 13.7). For complete instructions, see MasteringAstronomy.com.

Quantitative Problems

Be sure to show all calculations clearly and state your final answers in complete sentences.

51. *Lost in the Glare.* How hard would it be for an alien astronomer to detect the light from planets in our solar system compared to light from the Sun itself?
    a. Calculate the fraction of the total emitted sunlight that is reflected by Earth. (*Hint:* Imagine a sphere around the Sun the size of the planet's orbit (area $= 4\pi a^2$). What fraction of that area does the disk of a planet (area $= \pi r^2_{\text{planet}}$) take up? Earth's reflectivity is 29%.
    b. Would detecting Jupiter be easier or harder than detecting Earth? Comment on whether you think Jupiter's larger size or greater distance has a stronger effect on its detectability. You may neglect any difference in reflectivity between Earth and Jupiter.
52. *Transit of TrES-1.* The planet orbiting this star has been detected by both the transit and Doppler methods, so we can calculate its density and get an idea of what kind of planet it is.

a. Using the method of Mathematical Insight 13.3, calculate the radius of the transiting planet. The planetary transits block 2% of the star's light. The star TrEs-1 has a radius of about 85% of our Sun's radius.

b. The mass of the planet is approximately 0.75 times the mass of Jupiter, and Jupiter's mass is about 1.9×10^{27} kilograms. Calculate the average density of the planet. Give your answer in grams per cubic centimeter. Compare this density to the average densities of Saturn ($0.7\ g/cm^3$) and Earth ($5.5\ g/cm^3$). Is the planet terrestrial or jovian in nature? (*Hint:* To find the volume of the planet, use the formula for the volume of a sphere: $V = \left(\frac{4}{3}\right)\pi r^3$. Be careful with unit conversions.)

53. *Planet Around 51 Pegasi.* The star 51 Pegasi has about the same mass as our Sun. A planet discovered around it has an orbital period of 4.23 days. The mass of the planet is estimated to be 0.6 times the mass of Jupiter. Use Kepler's third law to find the planet's average distance (semimajor axis) from its star. (*Hint:* Because the mass of 51 Pegasi is about the same as the mass of our Sun, you can use Kepler's third law in its original form, $p^2 = a^3$ [Section 3.3]. Be sure to convert the period into years before using this equation.)

54. *One Born Every Minute?* It's possible to make a rough estimate of how often planetary systems form by making some basic assumptions. For example, if you assume that the stars we see have been born at random times over the last 10 billion years, then the rate of star formation is simply the number of stars we see divided by 10 billion years. The fraction of stars with detected extrasolar planets is at least 5%, so this factor can be multiplied in to find the approximate rate of formation of planetary systems.

a. Using these assumptions, how often does a planetary system form in our galaxy? (Our galaxy contains at least 100 billion stars.)

b. How often does a planetary system form somewhere in the observable universe, which contains at least 100 billion galaxies?

c. Write a few sentences describing your reaction to your results. Do you think the calculation is realistic? Are the rates larger or smaller than you expected?

55. *Identical Planets?* The Doppler method allows us to find a planet's mass if we know the star's maximum velocity, the planet's period, and the star's mass (Mathematical Insight 13.2).

a. Imagine two planets orbiting a 1 solar mass star with orbits edge-on to the Earth. The peak Doppler shift for each is 50 m/s, but one has a period of 3 days and the other has a period of 300 days. Calculate the two masses and say which, if either, is larger.

b. Imagine another planet in an edge-on orbit around a star with a mass half that of our Sun. How does this planet compare to those in part (a)? Show your work.

56. *Finding the Smallest Planet.* One of the goals of extrasolar planet searches is to find ever-smaller planets that might resemble those in our solar system. What is the least massive planet that can be detected at 0.1 AU from a 1-solar-mass star if the smallest velocity measurement is 1 m/s? What is the least massive planet that can be detected at 0.1 AU from a 0.5-solar-mass star? Repeat the calculations using an orbital distance of 10 AU for both stars.

57. *Finding Orbit Sizes.* The Doppler method (Mathematical Insight 13.1) allows us to find a planet's semimajor axis using just the orbital period and the star's mass.

a. Imagine that a new planet is discovered around a 2-solar-mass star that has a period of 5 days. What is its semimajor axis?

b. Another planet is discovered around a 0.5-solar-mass star with a period of 100 days. Find the semimajor axis.

58. *Search Strategies.* The kinds of planets you can detect depend on your observing strategy.

a. Imagine that your search can find planets with orbital periods between 4 and 40 days. If you are searching for planets around stars with the same mass as our Sun, over what range of semimajor axes can you detect planets?

b. Repeat the calculation above for stars half and twice the Sun's mass. What combination (massive vs. low-mass star, short- vs. long-period planet) will detect the planet with the largest semimajor axis?

59. Are *They Searching Too?* Which planet in our solar system causes the largest Doppler shift of our Sun that alien astronomers might observe? Consider the cases of Mercury, Earth, and Jupiter. How do those Doppler shifts compare to our current technological limit of about 1 m/s?

60. *Habitable Planet Around 51 Pegasi?* The star 51 Pegasi is approximately as bright as our Sun and has a planet that orbits at a distance of only 0.052 AU.

a. Suppose the planet reflects 15% of the incoming sunlight. Using Mathematical Insight 10.1, calculate its "no greenhouse" average temperature. How does this temperature compare to that of Earth?

b. Repeat part (a), but assume that the planet is covered in bright clouds that reflect 80% of the incoming sunlight.

c. Based on your answers to parts (a) and (b), do you think it is likely that the conditions on this planet are conducive to life? Explain.

61. *Tricky Transits.* The transit method requires the planet's orbit to be nearly edge-on as seen from the Earth. You can use simple geometry to find out what fraction of planetary systems are likely to have a transit. We'll make the simplifying assumptions that the stars are all the size of our Sun, and that the transit can be detected as long as the middle of the planet passes in front of any part of the star. Because there's no organization to the orbits of planets around other stars, their tilts relative to edge-on are equally likely at all angles from 0 to 90°.

a. Take the case of "hot Jupiters," which orbit at about 0.05 AU. What's the maximum tilt away from edge-on at which such a planet causes a transit? What fraction of possible tilt angles can cause a transit? If you observe 10,000 stars of which 2% have "hot Jupiters," how many planets are you likely to discover? (*Hint:* draw a diagram of the system with the orbit edge-on, with Earth to the side.)

b. Repeat the calculations above for Earth-size planets orbiting at 1 AU. Assume 10% of stars have such planets.

c. Repeat for the case of Jupiter-size planets at 5 AU, assuming 10% of stars have Jupiters. About how long would you have to wait to see three transits for each planet?

Discussion Questions

62. *So What?* What is the significance of the discovery of extrasolar planets, if any? Justify your answer in the context of this book's discussion of the history of astronomy.

63. *Is It Worth It?* The cost of the *Kepler* mission is several hundred million dollars. The cost of the *Terrestrial Planet Finder* mission is likely to be several billion dollars. Are these expenses worth it, compared to the results expected? Defend your opinion.

64. *What If?* Consider the possible outcomes of the missions described in Section 13.4. What results would change our perspective on our solar system? On the possibility of life in the universe?

MEDIA EXPLORATIONS

For self-study activities go to MasteringASTRONOMY www.masteringastronomy.com

Interactive Tutorials

Tutorial Review of Key Concepts

Use the interactive **Tutorial** at **www.masteringastronomy.com** to review key concepts from this chapter.

Detecting Extrasolar Planets Tutorial

Lesson 1 Taking a Picture of a Planet
Lesson 2 Stars' Wobbles and Properties of Planets
Lesson 3 Planetary Transits

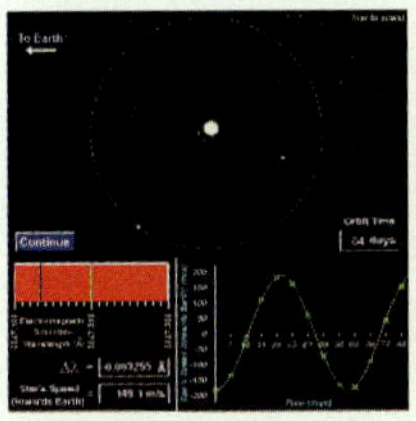
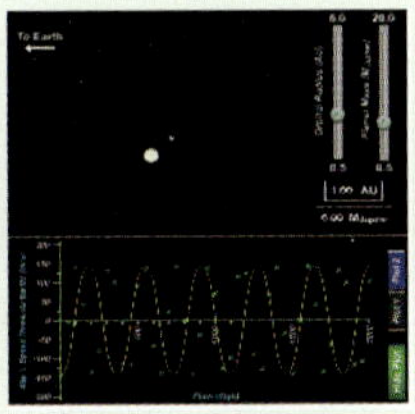
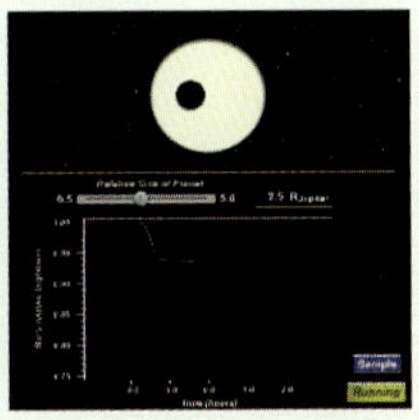

Supplementary Tutorial Exercises

Use the interactive **Tutorial Lessons** to explore the following questions.

Detecting Extrasolar Planets Tutorial, Lesson 2

1. Using the tool provided, explain how weekly measurements allow us to determine the orbital period of the extrasolar planet.
2. Use the tool to vary the mass of the planet. How does its mass affect the Doppler shifts in its star's light?
3. Use the tool to vary the orbital radius of the planet. How does the orbital radius affect the Doppler shifts in its star's light?

Detecting Extrasolar Planets Tutorial, Lesson 3

1. Under what conditions can we view a planetary transit of another star?
2. How does the change in brightness during a transit depend on the planet's properties?

Exploring the Sky and Solar System

Use the ***Voyager: SkyGazer*** **CD-ROM** accompanying your book to locate the brightest stars known to harbor planets. (These will be the stars in Appendix E.4 named with Greek letters preceding the constellation name.) Are they visible from where you live? At what time of the year and night? Print out a star chart and see if you can find the star in the sky. Bring along the information from the appendix to help you picture the planetary system in the sky.

Movies

Check out the following narrated and animated short documentary available on **www.masteringastronomy.com** for a helpful review of key ideas covered in this chapter.

History of the Solar System Movie

Web Projects

1. *New Planets.* Find the latest information on extrasolar planet discoveries. Create a personal "planet journal," complete with illustrations as needed, with a page for each of at least three recent discoveries of new planets. On each journal page, note the technique that was used to find the planet, give any information we have about the nature of the planet, and discuss how the planet does or does not fit in with our current understanding of planetary systems.
2. *Direct Detections.* In this chapter we saw only one example of a direct detection of a possible extrasolar planet. Search for new information on this and any other direct detections now known. Has the detection discussed in this chapter been confirmed (or disavowed) as a planet? Have we made any other direct detections, and if so, how? Summarize your findings with a short written report, including images of the directly detected planets.
3. *Extrasolar Planet Mission.* Visit the Web site for one of the future space missions discussed in this chapter and learn more about the mission design, capabilities, and goals. Write a short report on your findings.

14

Our Star

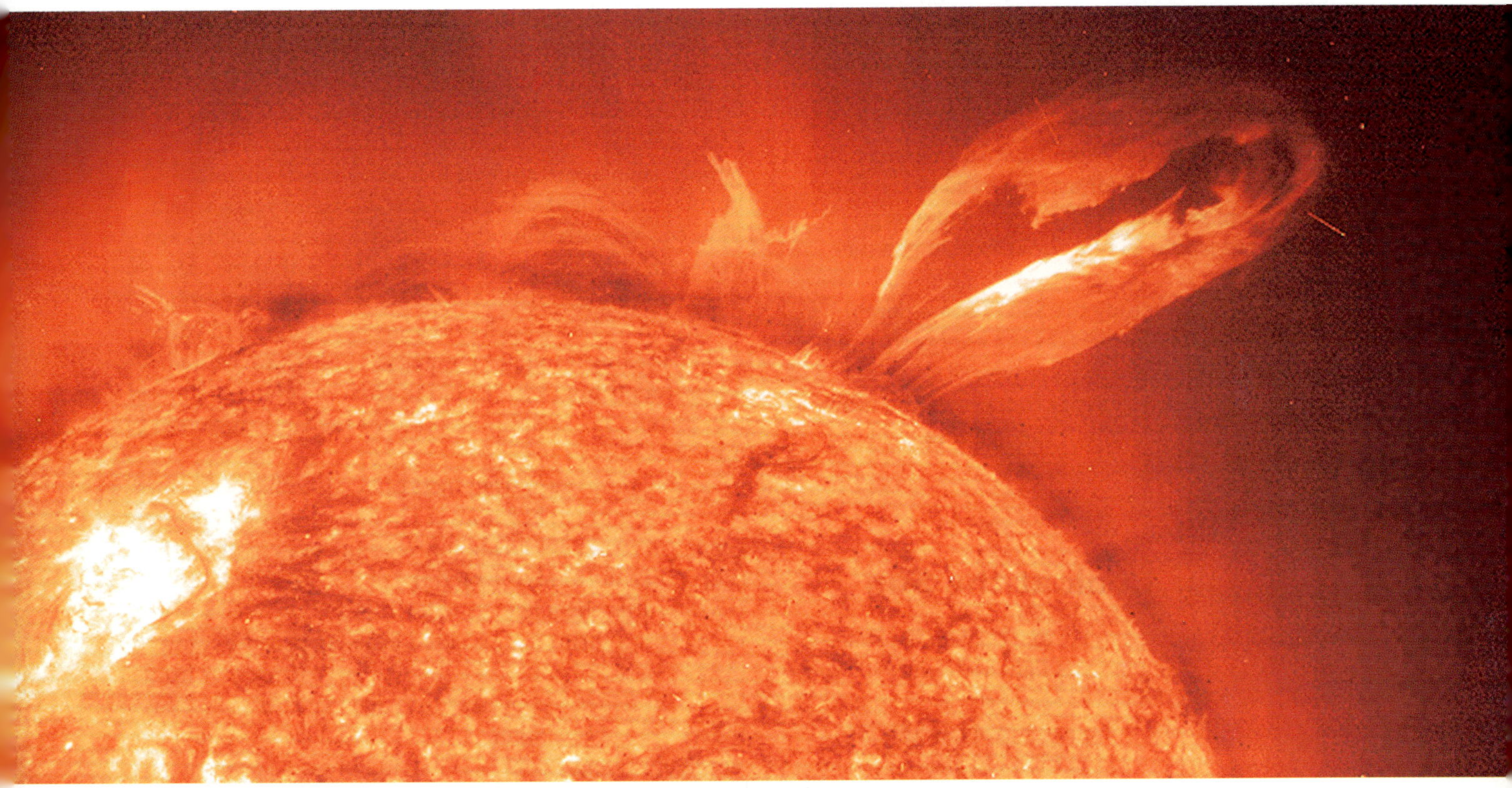

LEARNING GOALS

14.1 A Closer Look at the Sun
- Why was the Sun's energy source a major mystery?
- Why does the Sun shine?
- What is the Sun's structure?

14.2 The Cosmic Crucible
- How does nuclear fusion occur in the Sun?
- How does the energy from fusion get out of the Sun?
- How do we know what is happening inside the Sun?

14.3 The Sun-Earth Connection
- What causes solar activity?
- How does solar activity affect humans?
- How does solar activity vary with time?

I say Live, Live, because of the Sun, The dream, the excitable gift.

—Anne Sexton (1928–1974)

Astronomy today encompasses the study of the entire universe, but the root of the word *astronomy* comes from the Greek word for "star." Although we have learned a lot about the universe up to this point in the book, only now do we turn our attention to the study of the stars, the namesakes of astronomy.

When we think of stars, we usually think of the beautiful points of light visible on a clear night. But the nearest and most easily studied star is visible only in the daytime—our Sun. Of course, the Sun is important to us in many more ways than as an object for astronomical study. The Sun is the source of virtually all light, heat, and energy reaching Earth, and life on Earth's surface could not survive without it.

In this chapter, we will study the Sun in some depth. We will learn how the Sun generates the energy that supports life on Earth. Equally important, we will study our Sun as a star so that in subsequent chapters we can more easily understand stars throughout the universe.

MasteringASTRONOMY The Sun Tutorial, Lesson 1

14.1 A Closer Look at the Sun

Most of us take it for granted today that the Sun is a star—a giant ball of hot gas that generates light and shines it brightly in all directions. However, scientists discovered this fact only quite recently in human history. In this section, we'll consider the age-old question of what makes the Sun shine, and then take an imaginary plunge into the Sun that will get us better acquainted with its general features.

• Why was the Sun's energy source a major mystery?

The Sun's energy is absolutely vital to human existence. Ancient peoples certainly recognized that fact. Some worshipped the Sun as a god. Others created mythologies to explain its daily rise and set. But no one who lived before the twentieth century knew how the Sun provides us with light and heat.

Most ancient thinkers assumed the Sun to be some type of fire, perhaps a lump of burning coal or wood. The Greek philosopher Anaxagoras (c. 500–428 B.C.) imagined the Sun to be a very hot, glowing rock about the size of the Greek peninsula of Peloponnesus (comparable in size to Massachusetts). It was a reasonable suggestion for the times, since science had not yet advanced to the point where the idea could be tested, and it made him one of the first people in history to believe that objects in the heavens are composed of the same types of materials that we find on Earth.

The question of how the Sun shines became more pressing once scientists learned the Sun's true size and distance from Earth. Even the best ancient estimates of the Sun's distance had put the Sun much closer to Earth than it really is, so ancient people did not recognize the Sun's truly enormous energy output. But the Sun's size and distance were reasonably well known by the mid-1800s [see Special Topic, p. 218] making it clear that the Sun had to be generating an astonishing amount of energy to heat and illuminate the Earth from 150 million kilometers away.

What could possibly make the Sun shine so brightly? Familiar energy sources on Earth offered only two possibilities: Either the Sun was a cooling ember that had once been much hotter, or it generated energy from some type of chemical burning similar to the burning of coal or wood. Simple calculations showed that a cooling or chemically burning Sun could shine for a few thousand years, which squared well with biblically based estimates of Earth's age that were popular at the time. However, both these ideas suffered from fatal flaws. If the Sun were a cooling ember, it would have been much hotter just a few hundred years earlier, making it too hot for civilization to have existed. Chemical burning was ruled out when calculations showed that it could not possibly account for the Sun's huge energy output.

In the late 1800s, astronomers came up with what seemed at first like a much better idea: They suggested that the Sun generates energy by slowly contracting in size, a process called **gravitational contraction** (or *Kelvin-Helmholtz contraction*, after the scientists who proposed the mechanism). Recall that a shrinking gas cloud heats up because some of the gravitational potential energy of gas particles far from the cloud center is converted into thermal energy as the gas moves inward (see Figure 4.14). A gradually shrinking Sun would always have some gas moving inward, converting gravitational potential energy into thermal energy. This thermal energy would keep the inside of the Sun hot. Because of its large mass, the Sun would need to contract only very slightly each year to maintain its temperature—so slightly that the contraction would have been unnoticeable to nineteenth-century astronomers. Moreover, calculations showed that gravitational contraction could have kept the Sun shining steadily for up to about 25 million years, which many astronomers thought was plenty long enough. However, geologists disagreed: Their studies of rocks and fossils had already led them to conclude that Earth was at least hundreds of millions and possibly billions of years old, which meant that gravitational contraction could not be the mechanism by which the Sun generates its energy.

Geologists and astronomers debated Earth's age for many years, but by the beginning of the twentieth century the geological evidence had convinced most astronomers that Earth really is much older than 25 million years. Thus, astronomers were left with a major mystery: They knew the Sun had been producing energy at a great rate for at least hundreds of millions of years, but they knew of no mechanism that could explain this energy output.

Figure 14.1 An acrobat stack is in gravitational equilibrium: The lowest person supports the most weight and feels the greatest pressure, and the overlying weight and underlying pressure decrease for those higher up.

• Why does the Sun shine?

The Sun's true energy generation mechanism did not become clear until Einstein published his special theory of relativity in 1905. Einstein's theory included his famous discovery of $E = mc^2$ [Sections 4.3, S2.3]. This equation shows that mass itself contains an enormous amount of potential energy. Calculations immediately showed that the Sun's mass contained more than enough energy to account for billions of years of sunshine, if only there were some way for the Sun to convert the energy of mass into thermal energy. It took a few decades for scientists to work out the details, but by the end of the 1930s we had learned that the Sun converts mass into energy through *nuclear fusion* of hydrogen into helium [Section 1.1].

Thus, the basic answer to the question of why the Sun shines is that it generates energy through nuclear fusion. This cannot be the entire story, however, because fusion requires extremely high temperatures and densities (for reasons we will discuss in the next section). Thus, we still must answer two deeper questions. First, given that hot gases tend to expand, what keeps the Sun stable in size even while it generates so much light and heat? Second, how did the Sun get hot enough for fusion in the first place? Let's look at each of these questions, so that we can finally answer the age-old question of why the Sun shines.

The Stable Sun The high temperatures and densities that sustain nuclear fusion are found deep in the Sun's core. Thus, for the Sun to shine steadily, it must have a way of keeping the core hot and dense. It maintains these internal conditions through a natural balance between two competing forces: gravity pulling inward and pressure pushing outward. This balance is called **gravitational equilibrium** (or *hydrostatic equilibrium*).

A stack of acrobats provides a simple example of gravitational equilibrium (Figure 14.1). The bottom person supports the weight of everybody above him, so his arms must push upward with enough pressure to support all this weight. At each higher level, the overlying weight is less, so it's a little easier for each additional person to hold up the rest of the stack.

Gravitational equilibrium works much the same in the Sun, except the outward push against gravity comes from internal gas pressure rather than an acrobat's arms. The Sun's internal pressure precisely balances gravity at every point within it, thereby keeping the Sun stable in size (Figure 14.2). Because the weight of overlying layers is greater as we look deeper into the Sun, the pressure must increase with depth. Deep in the Sun's core, the pressure makes the gas hot and dense enough to sustain nuclear fusion. The energy released by fusion, in turn, heats the gas and maintains the pressure that keeps the Sun in balance against the inward pull of gravity.

THINK ABOUT IT

Earth's atmosphere is also in gravitational equilibrium, with the pressure in lower layers supporting the weight of upper layers. Use this idea to explain why the air gets thinner at higher altitudes.

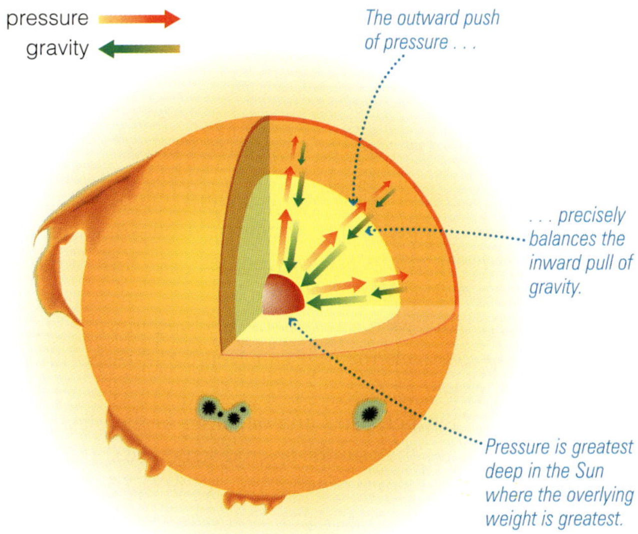

Figure 14.2 Gravitational equilibrium in the Sun: At each point inside, the pressure pushing outward balances the weight of the overlying layers.

How Fusion Started The steadiness of our Sun's gravitational equilibrium is fairly easy to understand if we think only about how it works at present: Gravity pushes inward while the energy released by fusion maintains the pressure that pushes outward. But how did the Sun become hot enough for fusion to begin in the first place?

The answer invokes the mechanism of gravitational contraction that astronomers of the late 1800s mistakenly thought might be responsible for the Sun's heat today. Recall that our Sun was born about 4.6 billion years ago from a collapsing cloud of interstellar gas [Section 8.2]. The contraction of the cloud released gravitational potential energy. Much of this energy was radiated away from the cloud's surface as thermal radiation, but the rest of it remained inside, causing the interior temperature and pressure to rise. When the central temperature and density finally grew high enough to sustain nuclear fusion, energy generation in the Sun's interior came into balance with the energy lost from the surface in the form of radiation. This energy balance stabilized the size of the Sun, bringing it into a state of gravitational equilibrium that has persisted to this day.

Calculations show that the Sun was born with enough hydrogen in its core to shine steadily and maintain its gravitational equilibrium for about 10 billion years. Thus, the Sun is only about halfway through its 10-billion-year "lifetime." About 5 billion years from now, when the Sun finally exhausts its nuclear fuel, the internal pressure will drop and gravitational contraction of the core will begin once again. As we will see in later chapters, some of the most important and spectacular processes in astronomy arise from the changes that occur as the crush of gravity begins to overcome a star's internal sources of pressure.

In summary, the answer to the question, "Why does the Sun shine?" is that about 4.6 billion years ago, *gravitational contraction* made the Sun hot enough to sustain nuclear fusion in its core. Ever since, energy liberated by fusion has maintained the Sun's *gravitational equilibrium* and kept the Sun shining steadily, supplying the light and heat that sustain life on Earth.

• What is the Sun's structure?

We've already stated that the Sun is a giant ball of hot gas. To be more precise, the Sun is a ball of *plasma*—a gas in which many of the atoms are ionized because of the high temperature [Section 5.3]. A plasma behaves much like an ordinary gas, but its many positively charged ions and freely moving electrons mean it can create and respond to magnetic fields.

The differing temperatures and densities of the plasma at different depths give the Sun the layered structure shown in Figure 14.3. To make sense of what you see in the figure, let's imagine that you have a spaceship that can somehow withstand the immense heat and pressure of the Sun, and take an imaginary journey from Earth to the center of the Sun.

The Sun from Afar As you begin your voyage from Earth, the Sun appears as a whitish ball of glowing gas. Just as astronomers have done in real life, you can use simple observations to determine basic properties of the Sun. Spectroscopy [Section 5.4] tells you that the Sun is made almost entirely of hydrogen and helium. From the Sun's angular size and distance, you can determine that its radius is nearly 700,000 kilometers, or more than 100 times the radius of Earth. Even **sunspots**, which appear as dark splotches on the Sun's surface, can be larger in size than Earth (Figure 14.4).

You can calculate the Sun's mass from its gravitational effects on the planets by using Newton's version of Kepler's third law [Section 4.4]. You'll find that the Sun's mass is 2×10^{30} kilograms—about 300,000 times the mass of Earth and more than 1,000 times the mass of all the planets in our solar system put together. You can determine the Sun's rotation rate by observing the motion of sunspots or by measuring Doppler shifts on opposite sides of the Sun.

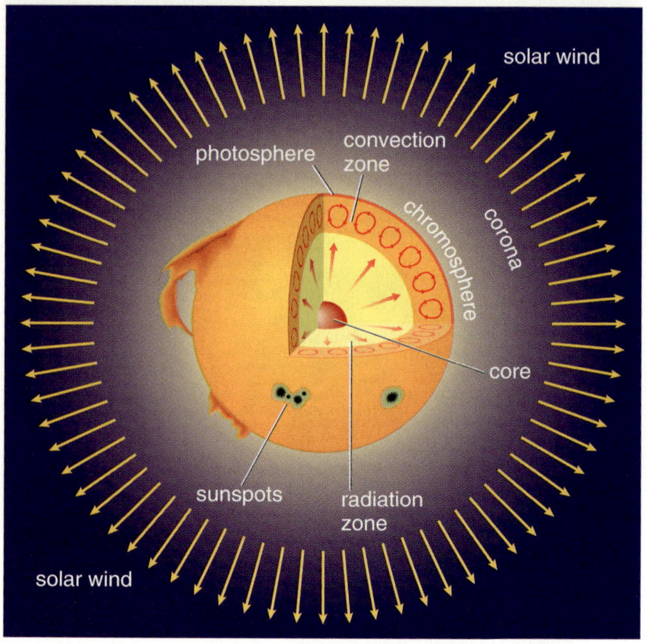

Figure 14.3 Interactive Figure The basic structure of the Sun.

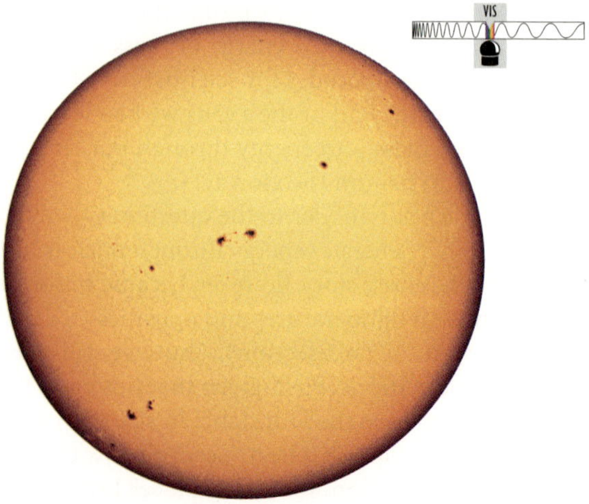

Figure 14.4 This photo of the visible surface of the Sun shows several dark sunspots, each large enough to swallow our entire planet.

You'll find that, unlike a spinning ball, the entire Sun does *not* rotate at the same rate. Instead, the solar equator completes one rotation in about 25 days, and the rotation period increases with latitude to about 30 days near the solar poles.

THINK ABOUT IT

As stated above, we can use Newton's version of Kepler's third law to determine the Sun's mass. What two properties of Earth do we need to know in order to apply this law?

The Sun releases an enormous amount of radiative energy into space, which you can measure through the window of your spacecraft. Recall that in science we measure energy in units called joules [Section 4.3] and define *power* as the *rate* at which energy is used or released [Section 5.1]. The standard unit of power is the *watt*, defined as 1 joule of energy per second; that is, 1 watt = 1 joule/s. For example, a 100-watt light bulb requires 100 joules of energy for every second it is left turned on. The Sun's total power output, or **luminosity**, is an incredible 3.8×10^{26} watts. If we could somehow capture and store just 1 second's worth of the Sun's luminosity, it would be enough to meet current human energy demands for roughly the next 500,000 years. Table 14.1 summarizes the basic properties of the Sun.

Of course, only a tiny fraction of the Sun's total energy output reaches Earth, with the rest dispersing in all directions into space. Most of this energy is radiated in the form of visible light, but after you've left the protective blanket of Earth's atmosphere, you encounter significant amounts of other types of solar radiation, including dangerous ultraviolet and X rays. Your spaceship will require substantial shielding to protect you from the serious radiation burns these high-energy forms of light cause.

Entering the Sun's Atmosphere Even at great distances from the Sun, you and your spacecraft can feel slight effects from the **solar wind**—a stream of charged particles continually blown outward in all directions from the Sun. The solar wind helps shape the magnetospheres of planets [Sections 10.1, 11.1] and blows back the material that forms the plasma tails of comets [Section 12.2]. The solar wind also cleared away the gas of the solar nebula at the end of the era of planet formation, some 4.5 billion years ago [Section 8.4].

As you approach the Sun more closely, you'll begin to encounter the low-density gas that represents what we usually think of as the Sun's atmosphere. The outermost layer of this atmosphere, called the **corona**, extends several million kilometers above the visible surface of the Sun. The temperature of the corona is astonishingly high—about 1 million K—explaining why this region emits most of the Sun's X rays. However, the corona's density is so low that your spaceship feels relatively little heat despite the million-degree temperature [Section 4.3].

Nearer the surface, the temperature suddenly drops to about 10,000 K in the **chromosphere**, the middle layer of the solar atmosphere and the region that radiates most of the Sun's ultraviolet light. Then you plunge through the lowest layer of the atmosphere, or **photosphere**, which is the visible surface of the Sun. Although the photosphere looks like a well-defined surface from Earth, it consists of gas far less dense than Earth's atmosphere. The temperature of the photosphere averages just under 6,000 K, and its surface seethes and churns like a pot of boiling water. The photosphere is also where you'll find sunspots; the wild swings of your compass needle will tell you that sunspots are actually regions with very strong magnetic fields.

Table 14.1 Basic Properties of the Sun

Radius (R_{Sun})	696,000 km (about 109 times the radius of Earth)
Mass (M_{Sun})	2×10^{30} kg (about 300,000 times the mass of Earth)
Luminosity (L_{Sun})	3.8×10^{26} watts
Composition (by percentage of mass)	70% hydrogen, 28% helium, 2% heavier elements
Rotation rate	25 days (equator) to 30 days (poles)
Surface temperature	5,800 K (average); 4,000 K (sunspots)
Core temperature	15 million K

Plunging into the Interior Up to this point in your journey, you may have seen Earth and the stars when you looked

COMMON MISCONCEPTIONS

Is the Sun on Fire?

We are accustomed to saying that the Sun is "burning," a description that conjures up images of a giant bonfire in the sky. However, the Sun does not burn in the same sense as a fire burns on Earth. Fires generate light through chemical changes that consume oxygen and produce a flame. The glow of the Sun has more in common with the glowing embers left over after the flames have burned out. Much like hot embers, the Sun's surface shines because it is hot enough to emit thermal radiation that includes visible light [Section 5.4].

Hot embers stop glowing as they cool, because cooler objects emit only infrared light and not visible light. The Sun keeps shining with visible light because energy rising from the Sun's core keeps its surface hot. The energy rising from the core is generated by nuclear fusion, so we sometimes say that it is the result of "nuclear burning"—a term that suggests nuclear changes in much the same way that "chemical burning" suggests chemical changes. Thus, while it is reasonable to say that the Sun undergoes nuclear burning in its core, the Sun's surface shines primarily with thermal radiation and not through any kind of burning fire.

back. But as you slip beneath the photosphere, blazing light engulfs you. You are inside the Sun, and incredible turbulence tosses your spacecraft about. If you can hold steady long enough to see what is going on around you, you'll notice spouts of hot gas rising upward, surrounded by cooler gas cascading down from above. You are in the **convection zone**, where energy generated in the solar core travels upward, transported by the rising of hot gas and falling of cool gas called *convection* [Section 9.1]. With some quick thinking, you may realize that the photosphere above you is the top of the convection zone and that convection is the cause of the Sun's seething, churning appearance.

About a third of the way down to the center, the turbulence of the convection zone gives way to the calmer plasma of the **radiation zone**, where energy moves outward primarily in the form of photons of light. The temperature rises to almost 10 million K, and your spacecraft is bathed in X rays trillions of times more intense than the visible light at the solar surface.

No real spacecraft could survive, but your imaginary one keeps plunging straight down to the solar **core**. There you finally find the source of the Sun's energy: nuclear fusion transforming hydrogen into helium. At the Sun's center, the temperature is about 15 million K, the density is more than 100 times that of water, and the pressure is 200 billion times that on the surface of Earth. The energy produced in the core today will take a few hundred thousand years to reach the surface.

With your journey complete, it's time to turn around and head back home. We'll continue this chapter by studying fusion in the solar core and then tracing the flow of the energy generated by fusion as it moves outward through the Sun.

The Sun Tutorial, Lessons 2–3

14.2 The Cosmic Crucible

The prospect of turning common metals like lead into gold enthralled those who pursued the medieval practice of alchemy. Sometimes they tried primitive scientific approaches, such as melting various ores together in a vessel called a *crucible.* Other times they tried magic. However, their get-rich-quick schemes never managed to work. Today we know that there is no easy way to turn other elements into gold, but it *is* possible to transmute one element or isotope into another.

If a nucleus gains or loses protons, its atomic number changes and it becomes a different element. If it gains or loses neutrons, its atomic mass changes and it becomes a different isotope [Section 5.3]. The process of splitting a nucleus into two smaller nuclei is called **nuclear fission**. The process of combining nuclei to make a nucleus with a greater number of protons or neutrons is called **nuclear fusion** (Figure 14.5). Human-built nuclear power plants rely on nuclear fission of uranium or plutonium. The nuclear power plant at the center of the Sun relies on nuclear fusion, turning hydrogen into helium.

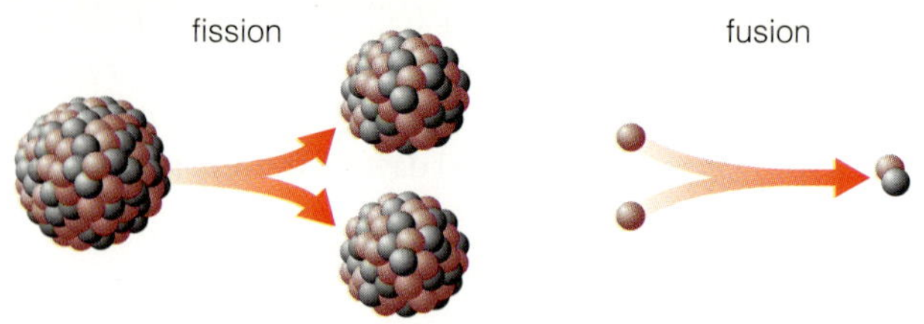

Figure 14.5 Nuclear fission splits a nucleus into smaller nuclei (not usually of equal size), while nuclear fusion combines smaller nuclei into a larger nucleus.

• How does nuclear fusion occur in the Sun?

Fusion occurs within the Sun because the 15 million K plasma in the solar core is like a "soup" of hot gas, with bare, positively charged atomic nuclei (and negatively charged electrons) whizzing about at extremely high speeds. At any one time, some of these nuclei are on high-speed collision courses with each other. In most cases, electromagnetic forces deflect the nuclei, preventing actual collisions, because positive charges repel one another. If nuclei collide with sufficient energy, however, they can stick together to form a heavier nucleus (Figure 14.6).

Sticking positively charged nuclei together is not easy. The **strong force**, which binds protons and neutrons together in atomic nuclei, is the only force in nature that can overcome the electromagnetic repulsion between two positively charged nuclei [Section S4.2]. In contrast to gravitational and electromagnetic forces, which drop off gradually as the distances between particles increase (by an inverse square law [Section 4.4]), the strong force is more like glue or Velcro: It overpowers the electromagnetic force over very small distances but is insignificant when the distances between particles exceed the typical sizes of atomic nuclei. The key to nuclear fusion, therefore, is to push the positively charged nuclei close enough together for the strong force to outmuscle electromagnetic repulsion.

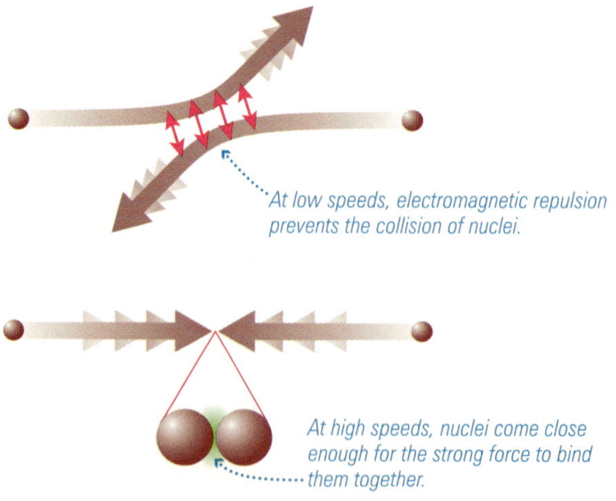

Figure 14.6 Positively charged nuclei can fuse only if a high-speed collision brings them close enough for the strong force to come into play.

The high pressures and temperatures in the solar core are just right for fusion of hydrogen nuclei into helium nuclei. The high temperature is important because the nuclei must collide at very high speeds if they are to come close enough together to fuse. (Quantum tunneling is also important to this process [Section S4.4].) The higher the

MATHEMATICAL INSIGHT 14.1 Mass-Energy Conversion in Hydrogen Fusion

The fusion of hydrogen into helium releases energy because the four hydrogen nuclei (protons) that go into the overall reaction have a slightly greater mass than the helium nucleus that comes out. Laboratory measurements show that a single proton has a mass of 1.6726×10^{-27} kg, so four protons have four times as much mass, or a total of 6.690×10^{-27} kg. Measurements show that a single helium-4 nucleus has a slightly lower mass, 6.643×10^{-27} kg. Thus, when four protons fuse to make one helium-4 nucleus, the amount of mass that "disappears" and becomes energy is:

$$6.690 \times 10^{-27}\ \text{kg} - 6.643 \times 10^{-27}\ \text{kg} = 0.047 \times 10^{-27}\ \text{kg}$$

If we divide this mass loss by the original mass of the four protons, we find the fractional loss of mass that occurs in hydrogen fusion:

$$\begin{matrix}\text{fractional mass loss} \\ \text{in hydrogen fusion}\end{matrix} = \frac{0.047 \times 10^{-27}\ \text{kg}}{6.69 \times 10^{-27}\ \text{kg}} = 0.007$$

That is, a fraction 0.007, or 0.7%, of the original hydrogen mass is converted into energy during the hydrogen fusion process. For example, when 1 kilogram of hydrogen fuses, the resulting helium weighs only 993 grams, while 7 grams of mass turns into energy. We can use these facts about mass loss during fusion with Einstein's equation $E = mc^2$ to learn about what happens in the Sun.

Example 1: How much hydrogen is converted to helium each second in the Sun?

Solution:

Step 1 Understand: The first step in understanding this question is to realize that we can figure out how much hydrogen is being converted to helium in the Sun only if we know the Sun's total energy output. Fortunately, we already know this (see Table 14.1): The Sun's luminosity is 3.8×10^{26} watts, which means it produces 3.8×10^{26} joules of energy each second. Knowing the amount of energy produced each second means we can use Einstein's equation $E = mc^2$ to calculate the amount of mass that must be turned into energy each second. Then we can figure out how much hydrogen must fuse in order to turn this amount of mass into energy, using the fact that 0.7% of the hydrogen mass becomes energy during the fusion process.

Step 2 Solve: As described above, we can use Einstein's equation to calculate the amount of mass that must be converted to energy each second in the Sun. We start by solving Einstein's equation for the mass, m:

$$E = mc^2 \quad \Rightarrow \quad m = \frac{E}{c^2}$$

Now we plug in 3.8×10^{26} joules for the energy (E) that the Sun produces each second and recall that the speed of light is $c = 3 \times 10^8$ m/s; to make sure we can work with the units properly, we use the fact that 1 joule is equivalent to 1 kg $\times$ m^2/s^2:

$$m = \frac{E}{c^2} = \frac{3.8 \times 10^{26}\,\dfrac{\text{kg} \times \text{m}^2}{\text{s}^2}}{\left(3 \times 10^8\,\dfrac{\text{m}}{\text{s}}\right)^2} = 4.2 \times 10^9\ \text{kg}$$

We've found the amount of mass converted to energy in the Sun each second. We know that this mass is only 0.7% (= 0.007) of the mass of the hydrogen that goes into the fusion process. Thus, the total mass of the hydrogen that undergoes fusion each second must be:

$$\text{mass of hydrogen fused} = \frac{4.2 \times 10^9\ \text{kg}}{0.007} = 6.0 \times 10^{11}\ \text{kg}$$

Step 3 Explain: We have found that the Sun fuses 600 billion kilograms of hydrogen each second. We also found that about 4.2 billion kilograms of this mass is converted to energy. Thus, of the 600 billion kilograms of hydrogen the Sun fuses each second, about 4 billion kilograms turns into energy and the remaining 596 billion kilograms becomes helium. The numbers may be easier to understand if we convert them to metric tons, using the fact that 1 metric ton = 1,000 kilograms. In that case, we say that the Sun fuses about 600 million metric tons of hydrogen each second, producing about 596 million metric tons of helium and turning 4 million metric tons into energy.

Example 2: How many times does the fusion reaction occur each second in the Sun?

Solution:

Step 1 Understand: There are several ways we could approach this problem, but perhaps the simplest is just to divide the total amount of mass lost each second in the Sun by the mass loss that occurs each time four hydrogen nuclei are fused into one helium nucleus. The result will tell us how many times the fusion reaction must occur each second.

Step 2 Solve: From Example 1, we know that the Sun converts a total of 4.2×10^9 kg of mass into energy each second, and from the discussion above we know that each individual fusion reaction converts 0.047×10^{-27} kg of mass into energy. We divide these masses to find the number of fusion reactions that occur each second in the Sun:

$$\begin{aligned}\begin{matrix}\text{number of} \\ \text{fusion reactions} \\ \text{(per second)}\end{matrix} &= \frac{\text{total mass lost through fusion (per second)}}{\text{mass lost in each fusion reaction}} \\ &= \frac{4.2 \times 10^9\ \text{kg}}{0.047 \times 10^{-27}\ \text{kg}} \\ &= 8.9 \times 10^{37}\end{aligned}$$

Step 3 Explain: Notice that our result, 8.9×10^{37}, is just a little less than 10^{38}. Thus, we have found that the overall hydrogen fusion reaction, which fuses four hydrogen nuclei into one helium nucleus, occurs nearly 10^{38} times per second in the Sun.

temperature, the harder the collisions, making fusion reactions more likely at higher temperatures. The high pressure of the overlying layers is necessary because without it the hot plasma of the solar core would simply explode into space, shutting off the nuclear reactions.

THINK ABOUT IT

The Sun generates energy by fusing hydrogen into helium, but as we'll see in later chapters, some stars fuse helium and even heavier elements. Do temperatures need to be higher or lower for the fusion of heavier elements than for the fusion of hydrogen? Why? (*Hint:* How does the amount of charge of a nucleus affect the difficulty of fusing it to another nucleus?)

The Proton-Proton Chain Let's investigate the fusion process in the Sun in a little more detail. Recall that hydrogen nuclei are simply individual protons, while the most common form of helium consists of two protons and two neutrons. Thus, the overall hydrogen fusion reaction transforms four individual protons into a helium nucleus containing two protons and two neutrons:

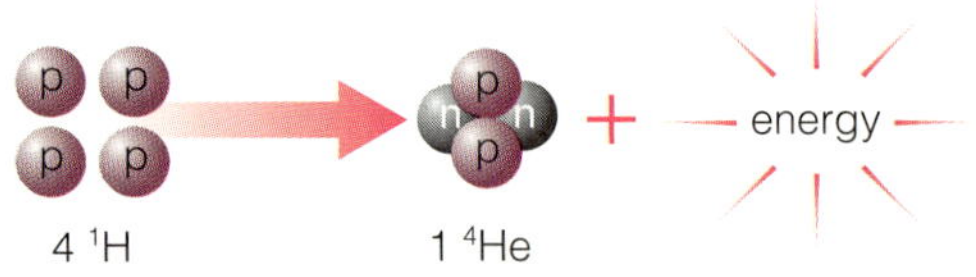

This overall reaction actually proceeds through several steps involving just two nuclei at a time. The sequence of steps that occurs in the Sun is called the **proton-proton chain**, because it begins with collisions between individual protons (hydrogen nuclei). Figure 14.7 illustrates the steps in the proton-proton chain:

1. ***Step 1.*** Two protons fuse to form a nucleus consisting of one proton and one neutron, which is the isotope of hydrogen known as *deuterium.* Note that this step converts a proton into a neutron, reducing the total nuclear charge from +2 for the two fusing protons to +1 for the resulting deuterium nucleus. The lost positive charge is carried off by a *positron* (antielectron), the antimatter version of an electron with a positive rather than a negative charge [Section S4.2]. A **neutrino**—a subatomic particle with a very tiny mass—is also produced in this step.* (The positron won't last long, because it soon meets up with an ordinary electron, resulting in the creation of two gamma-ray photons through matter-antimatter annihilation.) This step must occur twice in the overall reaction, since it requires a total of four protons.
2. ***Step 2.*** A fair number of deuterium nuclei are always present along with the protons and other nuclei in the solar core, since Step 1 occurs so frequently in the Sun (about 10^{38} times per second). Step 2 occurs when one of these deuterium nuclei collides and fuses with a proton. The result is a nucleus of helium-3, a rare form of helium with two protons and one neutron, along with the production of a gamma-ray photon. This step also occurs twice in the overall reaction.
3. ***Step 3.*** The third and final step of the proton-proton chain requires the addition of another neutron to the helium-3, thereby making normal helium-4. This final step can proceed in several different ways, but the most common is through a collision of two helium-3

*Producing a neutrino is necessary because of a law called *conservation of lepton number:* The number of leptons (e.g., electrons or neutrinos [Chapter S4]) must be the same before and after the reaction. The lepton number is zero before the reaction because there are no leptons. Among the reaction products, the positron (antielectron) has lepton number −1 because it is antimatter, and the neutrino has lepton number +1. Thus, the total lepton number remains zero.

Hydrogen Fusion by the Proton-Proton Chain

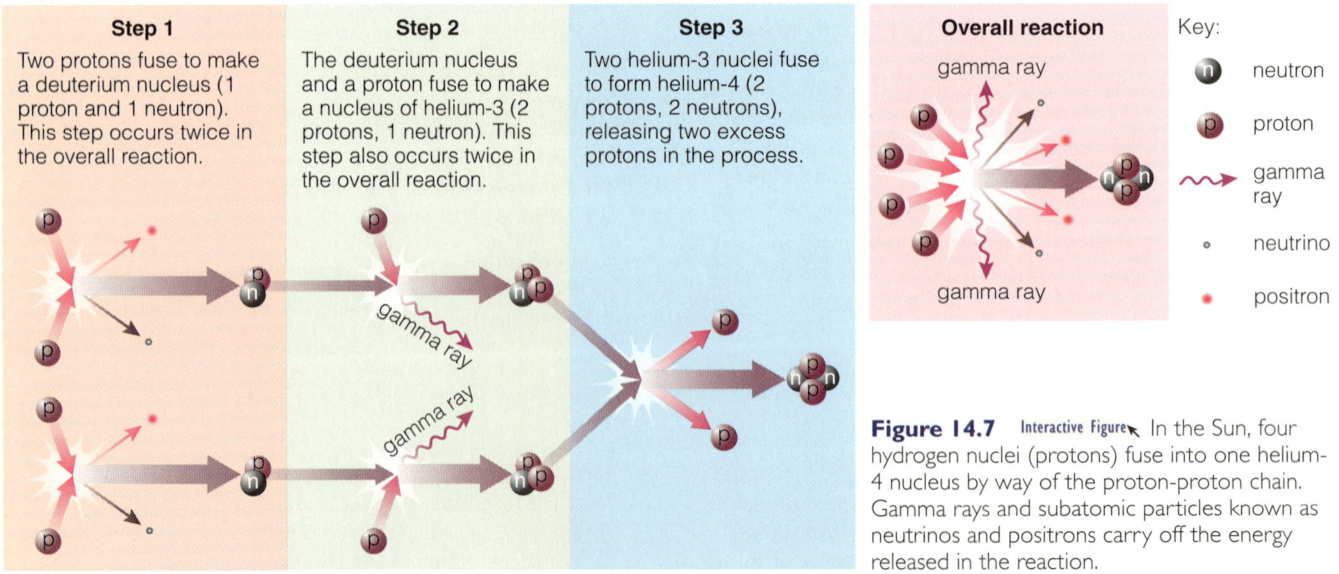

Figure 14.7 Interactive Figure In the Sun, four hydrogen nuclei (protons) fuse into one helium-4 nucleus by way of the proton-proton chain. Gamma rays and subatomic particles known as neutrinos and positrons carry off the energy released in the reaction.

nuclei. Each of these helium-3 nuclei resulted from a prior, separate occurrence of Step 2 somewhere in the solar core. The final result is a normal helium-4 nucleus and two protons.

Notice that a total of six protons enter the reaction during Steps 1 and 2, with two coming back out in Step 3. Thus, the overall reaction is just as described above, with four protons combining to make one helium nucleus. The gamma rays and subatomic particles (neutrinos and positrons) carry off the energy released in the reaction.

Fusion of hydrogen into helium generates energy because a helium nucleus has a mass slightly less (by about 0.7%) than the combined mass of four hydrogen nuclei (see Mathematical Insight 14.1). Thus, when four hydrogen nuclei fuse into a helium nucleus, a little bit of mass disappears. The disappearing mass becomes energy in accord with Einstein's formula $E = mc^2$. About 98% of the energy emerges as kinetic energy of the resulting helium nuclei and radiative energy of the gamma rays. As we will see, this energy slowly percolates to the solar surface, eventually emerging as the sunlight that bathes Earth. Neutrinos carry off the other 2% of the energy. Overall, fusion in the Sun converts about 600 million tons of hydrogen into 596 million tons of helium every second, which means that 4 million tons of matter is turned into energy each second. Although this sounds like a lot, it is a minuscule fraction of the Sun's total mass and thus does not affect the overall mass of the Sun in any measurable way.

The Solar Thermostat Nuclear fusion is the source of all the energy the Sun releases into space. If the fusion rate varied, so would the Sun's energy output, and large variations in the Sun's luminosity would almost surely be lethal to life on Earth. Fortunately, the Sun fuses hydrogen at a steady rate, thanks to a natural feedback process that acts as a thermostat for the Sun's interior. We have encountered a natural thermostat once before, when we discussed how Earth's carbon dioxide cycle keeps our planet's surface temperature relatively steady [Section 10.6]. The solar thermostat is even more efficient.

Solar energy production remains steady because the rate of nuclear fusion is very sensitive to temperature. A slight increase in the Sun's core temperature would mean a much higher fusion rate, and a slight decrease in temperature would mean a much lower fusion rate. Either kind of change in the fusion rate would alter the Sun's gravitational equilibrium—the balance between the pull of gravity and the push of internal pressure—in a way that would quickly restore the original temperature and fusion rate. To see how, let's examine what would happen if there were a small change in the core temperature (Figure 14.8).

Suppose the Sun's core temperature rose very slightly. Protons in the core would then move faster and collide more often. The rate of nuclear fusion would therefore soar, generating lots of extra energy. Because energy moves so slowly through the Sun, this extra energy would be bottled up in the core, causing an increase in the core pressure. The push of this pressure would temporarily exceed the pull of gravity, causing the core to expand and cool. This cooling, in turn, would cause the fusion rate to drop back down until the core was restored to its original size and temperature, thus returning the fusion rate back to its original value.

A slight drop in the Sun's core temperature would trigger an opposite chain of events. The reduced core tempera-

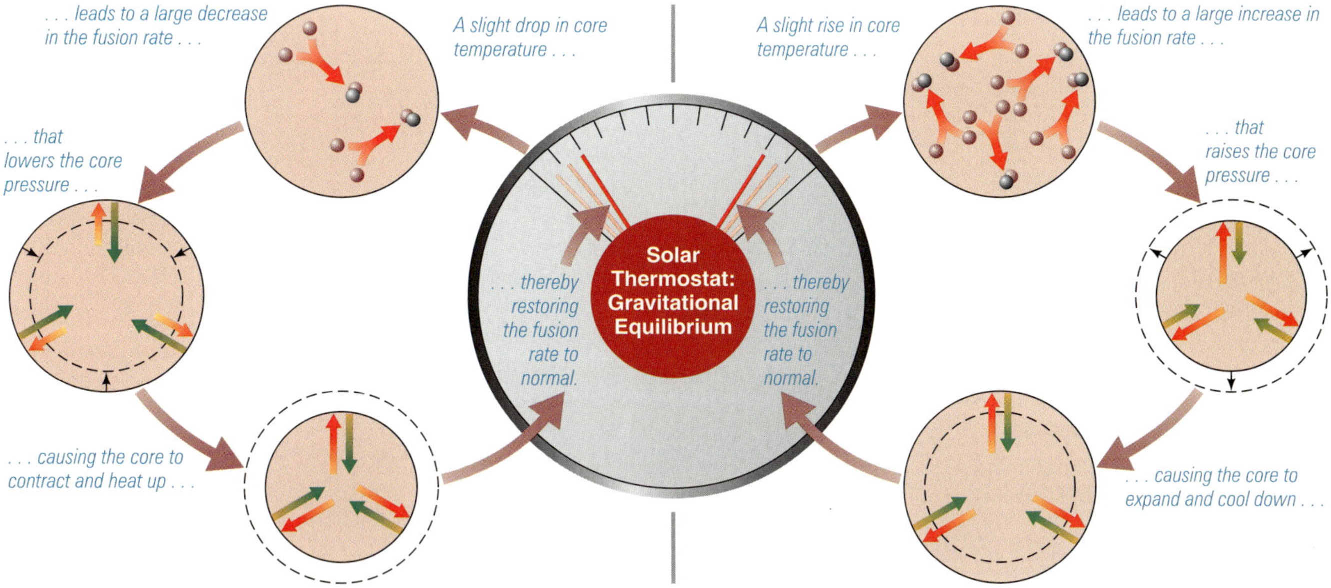

Figure 14.8 Interactive Figure The solar thermostat. Gravitational equilibrium regulates the Sun's core temperature. Everything is in balance if the amount of energy leaving the core equals the amount of energy produced by fusion. A rise in core temperature triggers a chain of events that causes the core to expand, lowering its temperature to its original value. A decrease in core temperature triggers the opposite chain of events, also restoring the original core temperature.

ture would lead to a decrease in the rate of nuclear fusion, causing a drop in pressure and contraction of the core. As the core shrank, its temperature would rise until the fusion rate returned to normal and restored the core to its original size and temperature.

Notice that the solar thermostat depends on *two* kinds of balance. The first is gravitational equilibrium—pressure must balance gravity at each point in the Sun. Second, the flow of energy through the Sun must remain balanced: The rate of energy production in the core must be the same as the rate at which energy flows outward through each layer of the Sun. If either of these forms of balance is somehow disrupted, natural feedback processes will automatically restore it, keeping the Sun's central temperature and

MATHEMATICAL INSIGHT 14.2 Pressure in the Sun: The Ideal Gas Law

The pressure that resists gravity inside the Sun comes from the thermal motions of gas particles. Recall that particles in a hot gas move more quickly and collide with more force than those in a cooler gas [Section 5.3], which means they also exert more pressure on any surface that tries to contain them. (Remember also that pressure is defined as the force per unit area exerted on any surface.) However, pressure also depends on *how many* particles are colliding with each unit area of that surface each second, which means that pressure depends on the *number density* of gas particles—the number of gas particles contained in each cubic centimeter. Mathematically, we express this relationship between temperature, density, and pressure with the **ideal gas law**, which can be written as:

$$P = nkT$$

where P represents gas pressure, n represents the number density of particles (in particles per cubic centimeter), T represents gas temperature (in Kelvin), and $k = 1.38 \times 10^{-23}$ joules/K, which is known as Boltzmann's constant. This law applies to all gases consisting of simple, freely flying particles, like those in the Sun. In the units given here, the ideal gas law gives pressure in units of joules per cubic centimeter (J/cm^3); these units of energy per unit volume are equivalent to units of force per unit area.

Example 1: The Sun's core contains about 10^{26} particles per cubic centimeter at a temperature of 15 million K. How does the gas pressure in the core of the Sun compare to the pressure of Earth's atmosphere at sea level, where there are about 2.4×10^{19} particles per cubic centimeter at a temperature of roughly 300 K?

Solution:

Step 1 Understand: In order to determine pressure, we need to know both the temperature and the number density of particles in a gas. We have that information for both the Sun's core and the Earth's atmosphere, so we have all the information we need. Because we are asked to compare the pressures, we don't need to calculate them individually; instead, we need only find their ratio by dividing the larger (the Sun's core pressure) by the smaller (Earth's atmospheric pressure).

Step 2 Solve: Dividing the Sun's core pressure by Earth's atmospheric pressure, we find that their ratio is:

$$\frac{P_{core}}{P_{atmos}} = \frac{n_{core}kT_{core}}{n_{atmos}kT_{atmos}} = \frac{10^{26}\ \text{cm}^{-3} \times 1.5 \times 10^{7}\ \text{K}}{2.4 \times 10^{19}\ \text{cm}^{-3} \times 300\ \text{K}}$$

$$= 2 \times 10^{11}$$

Step 3 Explain: We have found that the Sun's core pressure is about 200 billion (2×10^{11}) times as great as atmospheric pressure on Earth. This extreme pressure explains why the Sun can steadily fuse hydrogen into helium without blowing itself apart. The gravitational forces responsible for compressing the core to this extreme pressure are strong enough to keep that plasma in place while it releases fusion energy.

Example 2: How would the Sun's core pressure change if the Sun fused all its core hydrogen into helium without shrinking and without its core temperature changing? Use your answer to explain what should really happen as fusion occurs in the real Sun, in which the core *can* shrink and heat up. To simplify the problem, assume that the Sun's core begins with pure hydrogen that is fully ionized, so that there are two particles for every hydrogen nucleus (the proton plus one electron that must also be present for charge balance in the core), and ends as pure helium with three particles for each helium nucleus (the nucleus plus two electrons to balance the two protons in helium).

Solution:

Step 1 Understand: The core pressure will change if either the temperature or the number density of particles changes. We are told that the temperature does not change in this hypothetical example. We are also told that the core does not shrink, meaning that its volume remains constant, so a change in number density occurs only because there are fewer particles after fusion than before it. Thus, we simply need to determine how fusion changes the number of particles in the core.

Step 2 Solve: In this simplified example, the core initially contained two particles for every hydrogen nucleus and finished with three particles for every helium nucleus. However, we also know that it takes four hydrogen nuclei to make one helium nucleus. Thus, for each fusion reaction, there are eight particles before the reaction (four electrons and four protons) and three particles after it (one helium nucleus and two electrons)—meaning that there are $\frac{3}{8}$ as many particles after fusion as before it. Because the temperature and volume don't change in this example, the core pressure is also $\frac{3}{8}$ of its original value after all the hydrogen fuses into helium.

Step 3 Explain: In this simplified example, the core pressure at the end of the fusion process would have decreased to $\frac{3}{8}$ of its initial value. In reality, a decrease in core pressure would tip the balance of gravitational equilibrium in favor of gravity. The core must therefore shrink as fusion progresses in the Sun, in response to the gradually decreasing number of particles. This shrinkage decreases the volume of the core, allowing the number density of particles to stay approximately constant. The gradual shrinkage also produces a slight but gradual rise in core temperature, and therefore in the fusion rate, which is why the Sun's luminosity gradually increases with time during its hydrogen-burning life.

fusion rate steady and the Sun's overall energy output steady as well.

The Gradually Brightening Sun While the processes involved in gravitational equilibrium prevent erratic changes in the fusion rate, they also ensure that the fusion rate gradually rises over billions of years. This rise in the fusion rate explains a fact we first encountered when we discussed climate change factors in Chapter 10: the gradual brightening of the Sun with time.

Remember that each fusion reaction converts *four* hydrogen nuclei into *one* helium nucleus. Thus, the total number of *independent particles* in the solar core gradually falls with time. This gradual reduction in the number of particles causes the solar core to shrink. The slow shrinkage, in turn, gradually increases the core temperature and fusion rate, keeping the core pressure high enough to counteract the stronger compression of gravity. Theoretical models indicate that the Sun's core temperature should have increased enough to raise its fusion rate and the solar luminosity by about 30% since the Sun was born 4.6 billion years ago.

• How does the energy from fusion get out of the Sun?

The solar thermostat balances the Sun's fusion rate so that the amount of nuclear energy generated in the core equals the amount of energy radiated from the surface as sunlight. However, the journey of solar energy from the core to the photosphere takes hundreds of thousands of years.

Most of the energy released by fusion starts its journey out of the solar core in the form of photons. Although photons travel at the speed of light, the path they take through the Sun's interior zigzags so much that it takes them a very long time to make any outward progress. Deep in the solar interior, the plasma is so dense that a photon can travel only a fraction of a millimeter in any one direction before it interacts with an electron. Each time a photon "collides" with an electron, the photon gets deflected into a new and random direction. Thus, the photon bounces around the dense interior in a haphazard way (sometimes called a *random walk*) and only very gradually works its way outward from the Sun's center (Figure 14.9). The technical term for this slow, outward migration of photons is **radiative diffusion**; to *diffuse* means to "spread out" and *radiative* refers to the photons of light or radiation.

THINK ABOUT IT

Radiative diffusion is just one type of diffusion. Another is the diffusion of dye through a glass of water. If you place a concentrated spot of dye at one point in the water, each individual dye molecule begins a random walk as it bounces among the water molecules. The result is that the dye gradually spreads through the entire glass. Can you think of any other examples of diffusion in the world around you?

Energy released by fusion moves outward through the Sun's radiation zone (see Figure 14.3) primarily by way of these randomly bouncing photons. At the top of the radiation zone, where the temperature has dropped to about 2 million K, the solar plasma absorbs photons more readily (rather than just bouncing them around). This absorption creates the conditions needed for convection [Section 9.1] and hence marks the beginning of the Sun's convection zone.

Figure 14.9 A photon in the solar interior bounces randomly among electrons, slowly working its way outward.

In the convection zone, convection occurs because hot gas is less dense than cool gas. Like a hot-air balloon, a hot bubble of solar plasma rises upward through the cooler plasma above it. Meanwhile, cooler plasma from above slides around the rising bubble and sinks to lower layers, where it is heated. The rising of hot plasma and sinking of cool plasma form a cycle that transports energy outward from the base of the convection zone to the solar surface, or photosphere (Figure 14.10a). There, the density of the gas becomes so low that photons can escape to space, which is why we see the photosphere as the "surface" of the Sun.

The convecting gas gives the photosphere the mottled appearance that we see in close-up photographs (Figure 14.10b): We see bright blobs where hot gas is welling up from inside the Sun and darker borders around those blobs where the cooler gas is sinking.* If we watched a movie of the photosphere, we'd see its surface bubbling rather like a pot of boiling water. Just as bubbles in a pot of boiling water burst on the surface and are replaced by new bubbles, each hot blob lasts only a few minutes before being replaced by others bubbling upward. The average temperature of the gas in the photosphere is about 5,800 K, but the convection means that the precise temperature varies significantly from place to place.

To summarize, energy produced by fusion in the Sun's core works its way slowly through the radiation zone by random bounces of photons, then gets carried upward by convection in the Sun's convection zone. The photosphere lies at the top of the convection zone and marks the place where the density of gas has become low enough that photons can escape to space. Thus, the energy produced hundreds of thousands of years earlier in the solar core finally emerges from the Sun as thermal radiation [Section 5.4] coming from the 5,800 K gas of the photosphere. Once in space, the photons travel straight away at the speed of light, bathing the planets in sunlight.

*The blobs are formally called *granules*, and the photosphere's mottled appearance is sometimes referred to as *solar granulation*.

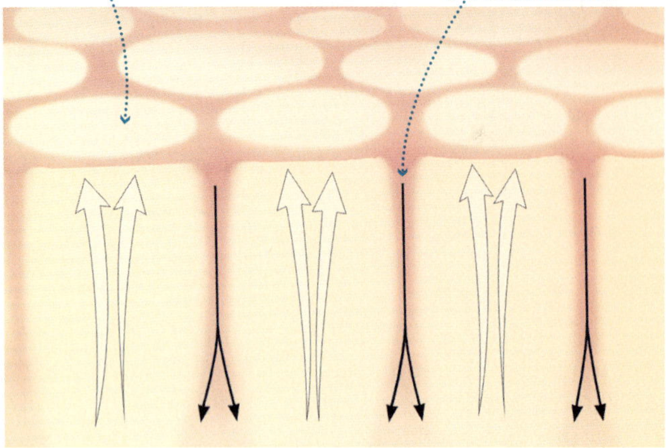

a This diagram shows convection beneath the Sun's surface: hot gas (white arrows) rises while cooler gas (orange/black arrows) descends around it.

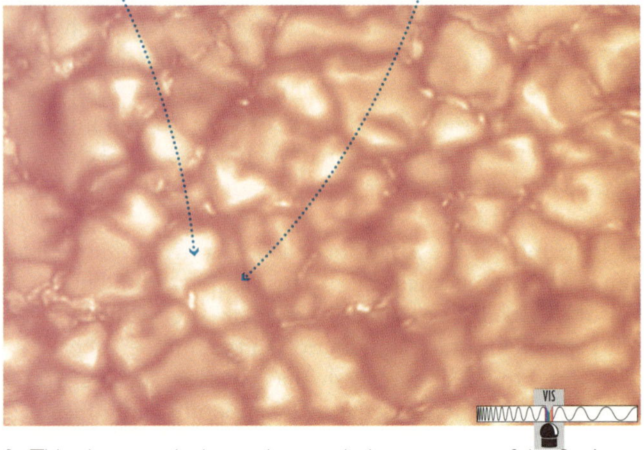

b This photograph shows the mottled appearance of the Sun's photosphere. The bright spots, each about 1,000 kilometers across, correspond to the rising plumes of hot gas in the diagram in part (a).

Figure 14.10 The Sun's photosphere churns with rising hot gas and falling cool gas as a result of underlying convection.

• How do we know what is happening inside the Sun?

We cannot see inside the Sun, so you may wonder how we can claim to know so much about what goes on underneath its surface and in its core. In fact, we can study the Sun's interior in three different ways: through mathematical models of the Sun, observations of solar vibrations, and observations of solar neutrinos.

Mathematical Models The primary way we learn about the interior of the Sun (and other stars) is by creating *mathematical models* that use the laws of physics to predict internal conditions. A basic model uses the Sun's observed composition and mass as inputs to equations that describe gravitational equilibrium, the solar thermostat, and the rate at which solar energy moves from the core to the photosphere. With the aid of a computer, we can use the model to calculate the Sun's temperature, pressure, and density at any depth. We can then predict the rate of nuclear fusion in the solar core by combining these calculations with knowledge about nuclear fusion gathered in laboratories here on Earth.

If a model is a good description of the Sun's interior, it should correctly "predict" the radius, surface temperature, luminosity, age, and many other observable properties of the Sun. Current models do indeed predict these properties quite accurately, giving us confidence that we really do understand what is going on inside the Sun.

Solar Vibrations A second way to learn about the inside of the Sun is to observe vibrations of the Sun's surface that are somewhat similar to the vibrations that earthquakes cause on Earth. These vibrations result from movement of gas within the Sun, which generates waves of pressure that travel through the Sun like sound waves moving through air. We can observe these vibrations on the Sun's surface by looking for Doppler shifts [Section 5.5]. Light from portions of the surface that are rising toward us is slightly blueshifted, while light from portions that are falling away from us is slightly redshifted. The vibrations are relatively small but measurable (Figure 14.11).

In principle, we can deduce a great deal about the solar interior by carefully analyzing these vibrations. (By analogy to seismology on Earth, this type of study of the Sun is

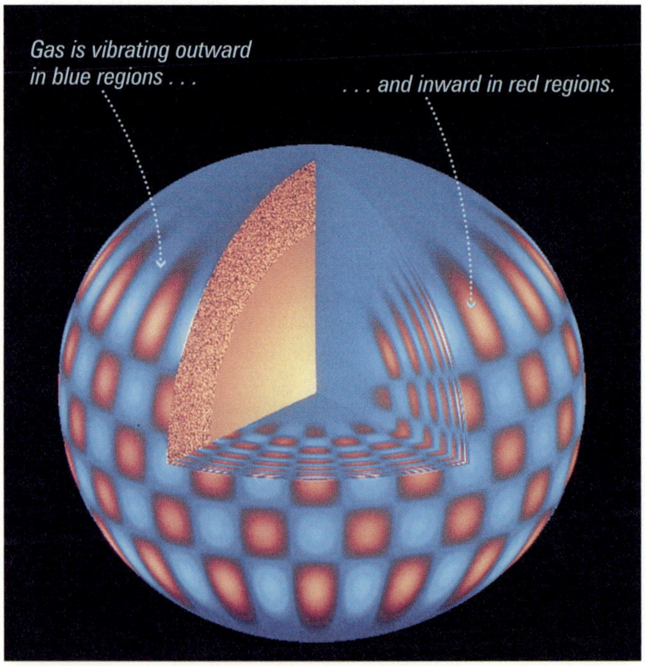

Figure 14.11 This schematic diagram shows one of the Sun's many possible vibration patterns, which we can measure from Doppler shifts in spectra of the Sun's surface.

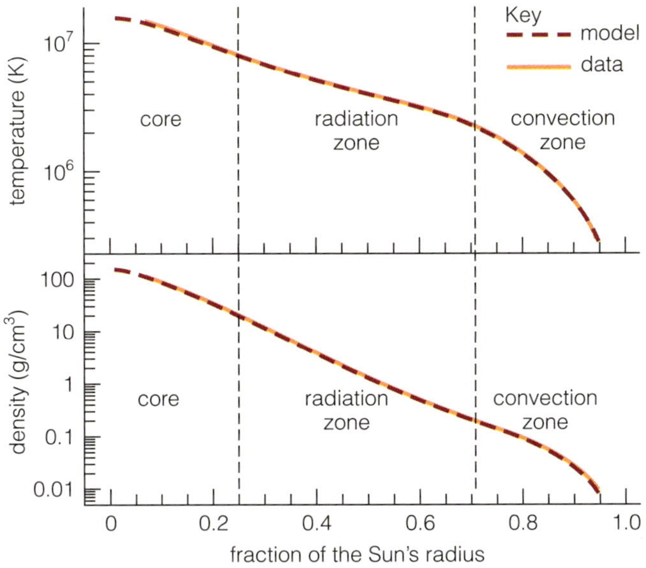

Figure 14.12 This graph shows the close agreement between predictions about the Sun's interior structure based on mathematical models and actual data obtained from observations of the Sun's surface vibrations. The agreement gives us confidence that the models are a good representation of the Sun's interior.

Figure 14.13 This tank of dry-cleaning fluid (visible underneath the catwalk), located deep within South Dakota's Homestake mine, was a solar neutrino detector. The chlorine nuclei in the cleaning fluid turned into argon nuclei when they captured neutrinos from the Sun.

called *helioseismology*—*helios* means "sun.") Results to date confirm that our mathematical models of the solar interior are on the right track (Figure 14.12). At the same time, they provide data that help to improve the models further.

Solar Neutrinos A third way to study the Sun's interior is to observe subatomic particles produced by fusion reactions in the core. Don't panic, but as you read this sentence, about a thousand trillion solar neutrinos will zip through your body. Traveling at nearly the speed of light, they reach us just minutes after being produced in fusion reactions in the Sun's core (see Figure 14.7). Fortunately, they won't do any damage, because neutrinos rarely interact with anything. Neutrinos created by fusion in the solar core fly quickly through the Sun as if passing through empty space. In fact, while an inch of lead will stop an X ray, stopping an average neutrino would require a slab of lead more than a light-year thick! Clearly, counting neutrinos is dauntingly difficult, because virtually all of them stream right through any detector built to capture them.

Nevertheless, neutrinos *do* occasionally interact with matter, and it is possible to capture a few solar neutrinos with a large enough detector. To distinguish neutrino captures from reactions caused by other particles, scientists usually place neutrino detectors deep underground in mines. The overlying rock blocks most other particles, but the neutrinos have no difficulty passing through.

The first major solar neutrino detector, built in the 1960s, was located 1,500 meters underground in the Homestake gold mine in South Dakota (Figure 14.13). The detector for this "Homestake experiment" consisted of a 400,000-liter vat of chlorine-containing dry-cleaning fluid. It turns out that, on very rare occasions, a chlorine nucleus can capture a neutrino and change into a nucleus of radioactive argon. By looking for radioactive argon in the tank of cleaning fluid, experimenters could count the number of neutrinos captured in the detector.

From the many trillions of solar neutrinos that passed through the tank of cleaning fluid each second, experimenters expected to capture an average of just one neutrino per day. This predicted capture rate was based on measured properties of chlorine nuclei and models of nuclear fusion in the Sun. However, over a period of more than two decades, neutrinos were captured only about once every three days on average. That is, the Homestake experiment detected only about one-third of the predicted number of neutrinos. This disagreement between model predictions and actual observations came to be called the **solar neutrino problem**.

For more than three decades, the solar neutrino problem was one of the great mysteries in astronomy. Other experiments found the same shortfall as the Homestake experiment. These results allowed only two possible conclusions: Either something was wrong with our understanding of fusion in the Sun, or some of the Sun's neutrinos were somehow hiding themselves from us.

Today, scientists are fairly convinced that the neutrinos were hiding. Here's how neutrinos are thought to hide: Scientists have known for decades that neutrinos come in three types, called *electron neutrinos, muon neutrinos*, and *tau neutrinos* [Section S4.2] Fusion reactions in the Sun produce only electron neutrinos, and until recently most solar neutrino detectors could detect only electron neutrinos. Thus, if some of the electron neutrinos produced by fusion change into neutrinos of the other two types on

Figure 14.14 This photograph shows the main tank of the Sudbury Neutrino Observatory in Canada, located at the bottom of a mine shaft more than 2 kilometers underground. The large sphere, 12 meters in diameter, contains 1,000 tons of ultrapure "heavy water." (Heavy water is water in which the two hydrogen atoms are replaced by deuterium, making each molecule heavier than a molecule of ordinary water.) Neutrinos of all three types can cause reactions in the heavy water, and detectors surrounding the tank can record these reactions when they occur.

their way to Earth, then our detectors would count fewer electron neutrinos than we would have expected.

Recent experiments support the idea that neutrinos can change type on their way to Earth, probably during the portion of their trip from the Sun's core to its surface. Most significantly, a new detector in Canada called the Sudbury Neutrino Observatory is capable of detecting all three neutrino types (Figure 14.14). Early results from this observatory suggest that the total number of neutrinos of all types is indeed what we expect from our models of nuclear fusion in the Sun. If these results hold up as more data are collected, it will mean not only that the solar neutrino problem has been solved, but also that neutrino counts verify that nuclear fusion occurs in the Sun in just the way our models predict.

14.3 The Sun-Earth Connection

Energy liberated by nuclear fusion in the Sun's core eventually reaches the solar surface. There, before it is ultimately released into space as sunlight, the energy stimulates a wide variety of phenomena that we can observe from Earth. Sunspots are only the most obvious of these phenomena. Because sunspots and other features of the Sun's surface change with time, they constitute what we call *solar weather* or **solar activity**. The "storms" associated with solar weather are not just of academic interest. Sometimes they are so violent that they affect our day-to-day life on Earth. In this section, we'll explore solar activity and its far-reaching effects.

• What causes solar activity?

Most of the Sun's surface churns constantly with rising and falling gas and thus looks like the close-up photo shown in Figure 14.10b. However, larger features sometimes appear, including sunspots, the huge explosions known as solar

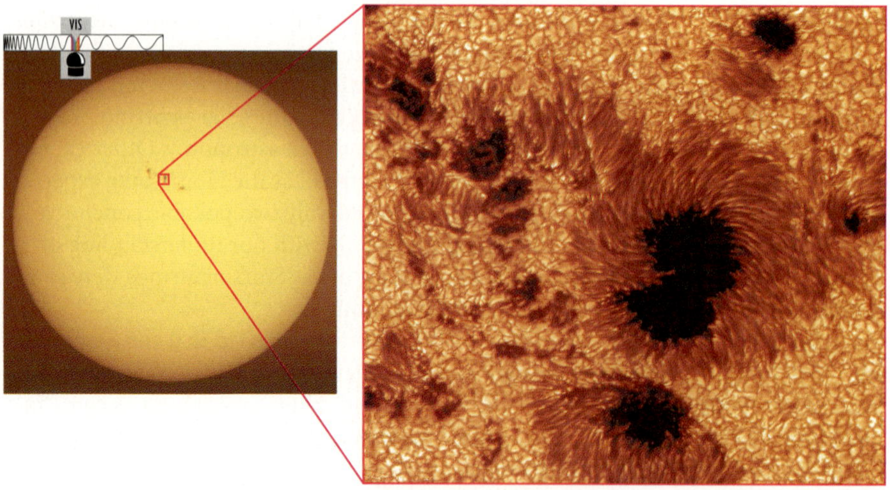

a This close-up view of the Sun's surface shows two large sunspots and several smaller ones. Each of the big sunspots is roughly as large as Earth.

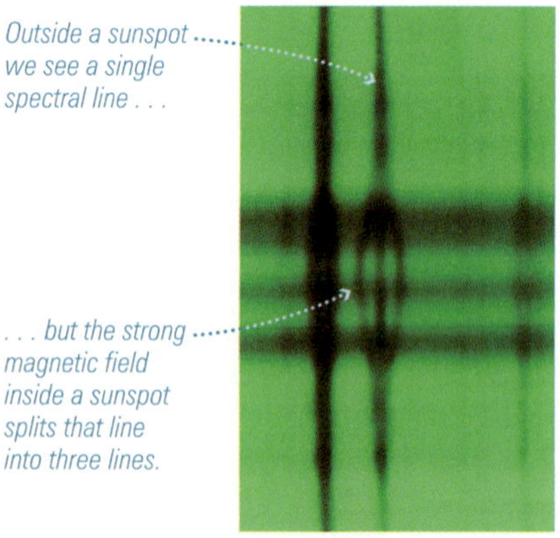

b Very strong magnetic fields split the absorption lines in spectra of sunspot regions. The dark vertical bands are absorption lines in a spectrum of the Sun. Notice that these lines split where they cross the dark horizontal bands corresponding to sunspots.

Figure 14.15 Sunspots are regions of strong magnetic fields.

flares, and gigantic loops of hot gas that extend high into the Sun's corona. All these features are created by magnetic fields, which form and change easily in the convecting plasma in the outer layers of the Sun.

Sunspots and Magnetic Fields Sunspots are the most striking features of the solar surface (Figure 14.15a). If you could look directly at a sunspot (but you should *never* stare at the Sun), you would find it blindingly bright. Sunspots appear dark in photographs only because they are less bright than the surrounding photosphere. They are less bright because they are cooler: The temperature of the plasma in sunspots is about 4,000 K, significantly cooler than the 5,800 K plasma that surrounds them.

If you think about this for a moment, you may wonder how sunspots can be so much cooler than their surroundings. Gas can usually flow easily, so you might expect the hotter gas from outside a sunspot to mix with the cooler gas within it, quickly warming the sunspot. The fact that sunspots stay relatively cool means that something must prevent hot plasma from entering them, and that "something" turns out to be magnetic fields.

Detailed observations of the Sun's spectral lines reveal sunspots to be regions with strong magnetic fields. These magnetic fields can alter the energy levels in atoms and ions and therefore can alter the spectral lines they produce. More specifically, magnetic fields cause some spectral lines to split into two or more closely spaced lines (Figure 14.15b). Wherever we see this effect (called the *Zeeman effect*), we know that magnetic fields must be present. Scientists can map magnetic fields on the Sun by looking for the splitting of spectral lines in light from different parts of the solar surface.

To understand how sunspots stay cooler than their surroundings, we must investigate the nature of magnetic fields

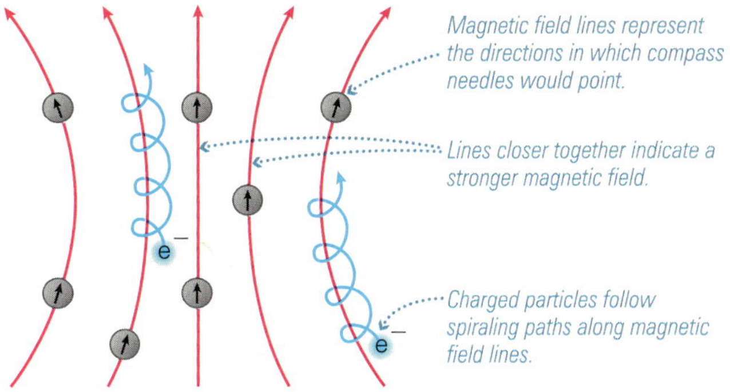

Figure 14.16 We draw magnetic field lines (red) to represent invisible magnetic fields.

in a little more depth. Magnetic fields are invisible, but we can represent them by drawing **magnetic field lines** (Figure 14.16). These lines represent the directions in which compass needles would point if we placed them within the magnetic field. The lines are closer together where the field is stronger, and farther apart where the field is weaker. Because these imaginary field lines are so much easier to visualize than the magnetic field itself, we usually discuss magnetic fields by talking about how the field lines would appear. Charged particles, such as the ions and electrons in the solar plasma, cannot easily move perpendicular to the field lines but instead follow spiraling paths along them.

Solar magnetic field lines act somewhat like elastic bands, being twisted into contortions and knots by turbulent motions in the solar atmosphere. Sunspots occur where the most taut and tightly wound magnetic fields poke nearly straight out from the solar interior (Figure 14.17a). These tight magnetic field lines suppress convection within the sunspot and prevent surrounding plasma from entering the sunspot. With hot plasma unable to enter the region, the sunspot plasma becomes cooler than that of the rest of

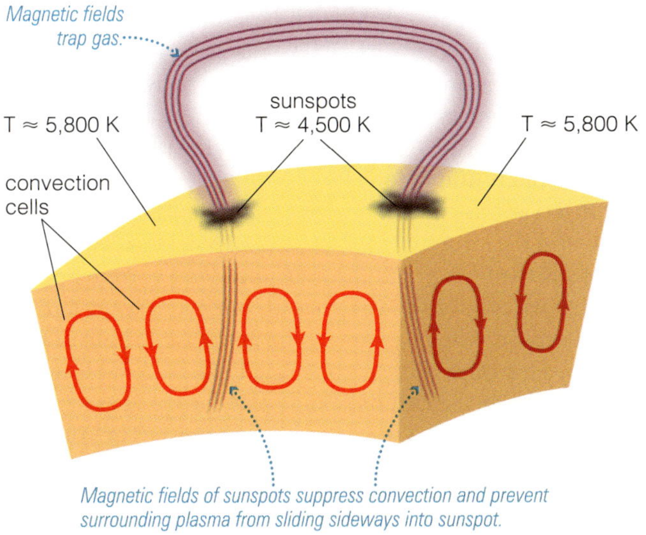

a Pairs of sunspots are connected by tightly wound magnetic field lines.

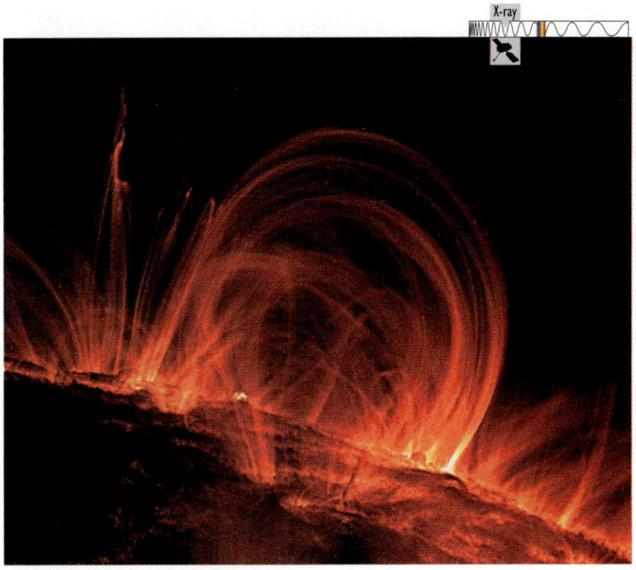

b This X-ray photo (from NASA's TRACE mission) shows hot gas trapped within looped magnetic field lines.

Figure 14.17 Strong magnetic fields keep sunspots cooler than the surrounding photosphere, while magnetic loops can arch from the sunspots to great heights above the Sun's surface.

the photosphere. Individual sunspots typically last up to a few weeks, dissolving when their magnetic fields weaken and allow hotter plasma to flow in.

Sunspots tend to occur in pairs, connected by a loop of magnetic field lines that can arc high above the Sun's surface (Figure 14.17b). Gas in the Sun's chromosphere and corona becomes trapped in these loops, making giant **solar prominences**. Some prominences rise to heights of more than 100,000 kilometers above the Sun's surface (Figure 14.18). Individual prominences can last for days or even weeks, disappearing only when the magnetic fields finally weaken and release the trapped gas.

Solar Storms The magnetic fields winding through sunspots and prominences sometimes undergo dramatic and sudden change, producing short-lived but intense storms on the Sun. The most dramatic of these storms are **solar flares**, which send bursts of X rays and fast-moving charged particles shooting into space (Figure 14.19).

Flares generally occur in the vicinity of sunspots, which is why we think they are created by changes in magnetic fields. The leading model for solar flares suggests that they occur when the magnetic field lines become so twisted and knotted that they can no longer bear the tension. They are thought to suddenly snap and reorganize themselves into a less twisted configuration. The energy released in the process heats the nearby plasma to 100 million K over the next few minutes or hours, generating X rays and accelerating some of the charged particles to nearly the speed of light.

Heating of the Chromosphere and Corona As we've seen, many of the most dramatic weather patterns and storms on the

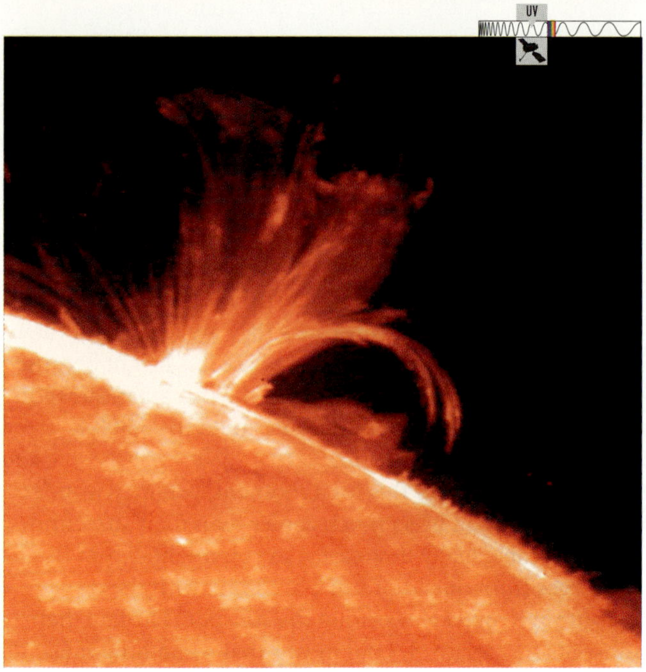

Figure 14.19 This photo (from the *TRACE* spacecraft) of ultraviolet light emitted by hydrogen atoms shows a solar flare erupting from the Sun's surface.

Sun show up when we look at the very hot gas of the Sun's chromosphere and corona. But why is that gas so hot in the first place? This question has perplexed scientists for decades.

Remember that temperatures gradually decline as we move outward from the Sun's core to the top of its photosphere. We might expect the decline to continue in the Sun's atmosphere, but instead it reverses, making the chromosphere and corona much hotter than the Sun's surface. Some aspects of this atmospheric heating remain a mystery today, but we have at least a general explanation: The Sun's strong magnetic fields carry energy upward from the churning solar surface to the chromosphere and corona.

More specifically, the rising and falling of gas in the convection zone probably shakes tightly wound magnetic field lines beneath the solar surface. The magnetic field lines carry this energy upward to the solar atmosphere, where they deposit this energy as heat. Thus, the same magnetic fields that keep sunspots cool make the overlying plasma of the chromosphere and corona hot.

Observations confirm the connection between magnetic fields and the structure of the chromosphere and corona. The density of gas in the chromosphere and corona is so low that we cannot see this gas with our eyes except during a total eclipse, when we can see the faint visible light scattered by electrons in the corona (see Figure 2.29). However, we can observe the chromosphere and corona at any time with ultraviolet and X-ray telescopes in space: The roughly 10,000 K plasma of the chromosphere emits strongly in the ultraviolet, and the million K plasma of the corona is the source of virtually all X rays coming from the Sun. Figure 14.20 shows an X-ray image of the Sun. The X-ray emission is brightest in regions where hot gas is being trapped and heated in loops of the magnetic field. In fact, the bright

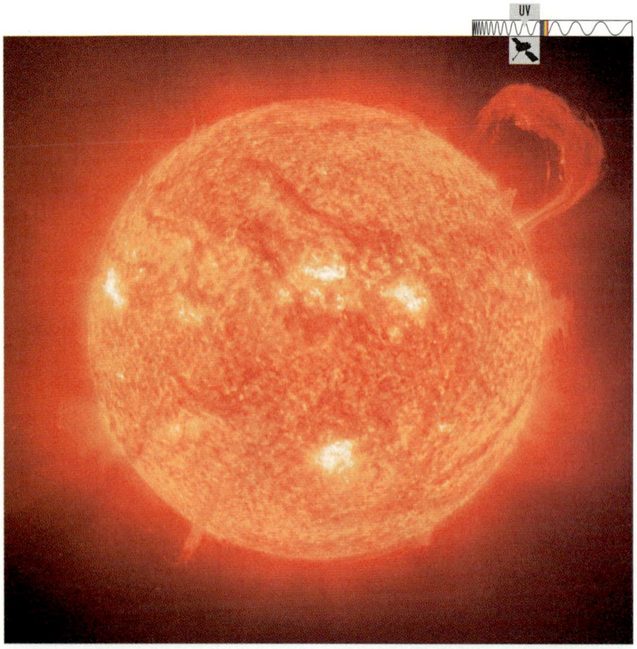

Figure 14.18 A gigantic solar prominence erupts from the solar surface at the upper right of this ultraviolet light photo (from the *SOHO* mission). The height of the prominence is over 20 times the diameter of Earth.

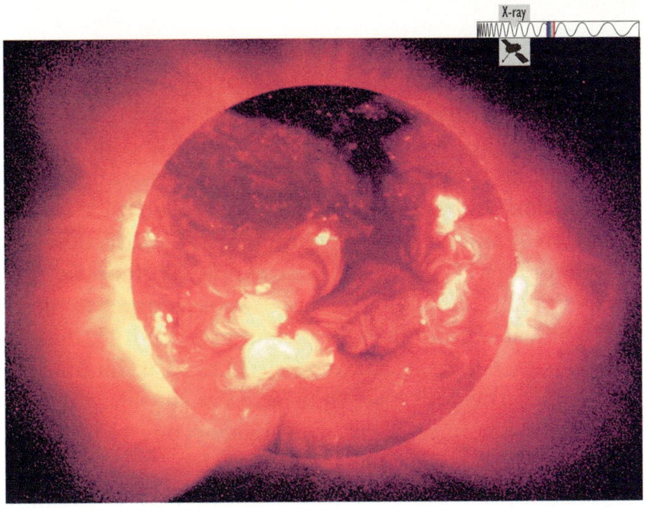

Figure 14.20 Interactive Photo An X-ray image of the Sun reveals the million-degree gas of the corona. Brighter regions of this image (yellow) correspond to regions of stronger X-ray emission. The darker regions (such as near the north pole at the top of this photo) are the coronal holes from which the solar wind escapes. (From the Yohkoh space observatory.)

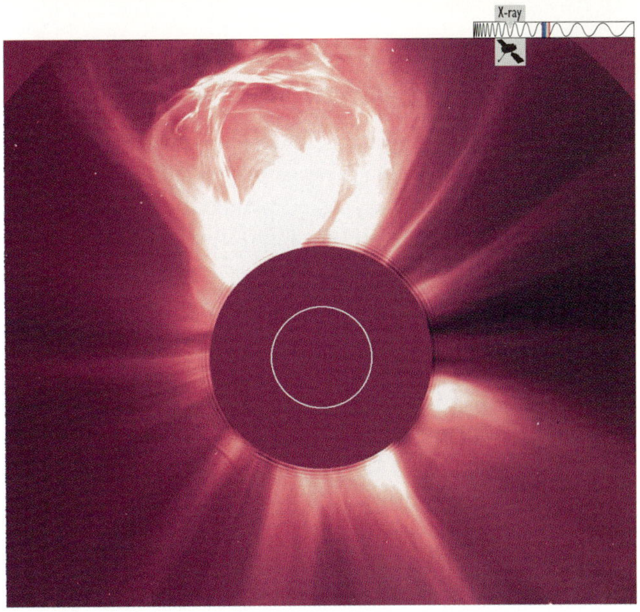

Figure 14.21 This X-ray image from the *SOHO* spacecraft shows a coronal mass ejection (the bright arc of gas headed almost straight upward as oriented here) during the solar storms of 2003. The central red disk blocks the Sun itself, and the white circle represents the size of the Sun in this picture.

spots in the corona tend to be directly above sunspots in the photosphere, confirming that they are created by the same magnetic fields.

Notice that some regions of the corona, called **coronal holes**, barely show up in X-ray images. More detailed analyses show that the magnetic field lines in coronal holes project out into space like broken rubber bands, allowing particles spiraling along them to escape the Sun altogether. These particles streaming outward from the corona are the source of the *solar wind*, which blows through the solar system at an average speed of about 500 kilometers per second and has important effects on planetary surfaces, atmospheres, and magnetospheres. Well beyond the planets, the pressure of interstellar gas must eventually halt the solar wind. The *Pioneer* and *Voyager* spacecraft that visited the outer planets in the 1970s and 1980s [Section 7.3] are still traveling outward from our solar system and may soon encounter this "boundary" (called the *heliopause*) marking the outer limit of the realm of the Sun.

The solar wind also gives us something tangible to study. In the same way that meteorites provide us with samples of asteroids we've never visited, solar wind particles captured by satellites provide us with a sample of material from the Sun. Analysis of these solar particles has reassuringly verified that the Sun is made mostly of hydrogen, just as we conclude from studying the Sun's spectrum.

• How does solar activity affect humans?

Flares and other solar storms sometimes eject large numbers of highly energetic charged particles from the Sun's corona. These particles travel outward from the Sun in huge bubbles that we call **coronal mass ejections** (Figure 14.21). The bubbles have strong magnetic fields and can reach Earth in a couple of days if they happen to be aimed in our direction. Once a coronal mass ejection reaches Earth, it can create a *geomagnetic storm* in Earth's magnetosphere. On the positive side, these storms can lead to unusually strong auroras (see Figure 10.11) that can be visible throughout much of the United States. On the negative side, they can hamper radio communications, disrupt electrical power delivery, and damage the electronic components in orbiting satellites.

During a particularly powerful magnetic storm on the Sun in March 1989, the U.S. Air Force temporarily lost track of more than 2,000 satellites, and powerful currents induced in the ground circuits of the Quebec hydroelectric system caused it to collapse for more than 8 hours. The combined cost of the loss of power in the United States and Canada exceeded $100 million. In January 1997, AT&T lost contact with a $200 million communications satellite, probably because of damage caused by particles coming from another powerful solar storm. More recently, in the fall of 2003, a series of extremely powerful solar flares once again threatened Earth's communications and electrical systems, but the systems escaped major damage, due in part to our improved preparedness for solar storms.

Satellites in low Earth orbit are particularly vulnerable during periods of strong solar activity, when the increase in solar X rays and energetic particles heats Earth's upper atmosphere, causing it to expand. The density of the gas surrounding low-flying satellites therefore rises, exerting drag that saps their energy and angular momentum. If this drag proceeds unchecked, the satellites ultimately plummet back to Earth. Thus, satellites in low orbits, including the Hubble Space Telescope and the Space Station, require occasional boosts to prevent them from falling out of the sky.

• How does solar activity vary with time?

Solar weather is just as unpredictable as weather on Earth. Individual sunspots can appear or disappear at almost any time, and we have no way to know that a solar storm is coming until we observe it through our telescopes. However, long-term observations have revealed overall patterns in solar activity that make sunspots and solar storms more common at some times than at others.

The Sunspot Cycle The most notable pattern in solar activity is the **sunspot cycle**—a cycle in which the average number of sunspots on the Sun gradually rises and falls (Figure 14.22). At the time of *solar maximum,* when sunspots are most numerous, we may see dozens of sunspots on the Sun at one time. In contrast, we may see few if any sunspots at the time of *solar minimum.* The frequency of prominences, flares, and coronal mass ejections also follows the sunspot cycle, with these events being most common at solar maximum and least common at solar minimum.

As you can see in Figure 14.22a, the sunspot cycle varies from one period to the next. Some maximums have much greater numbers of sunspots than others. The average length of time between maximums is 11 years, but we have observed it to be as short as 7 years and as long as 15 years. The locations of sunspots on the Sun also vary with the sunspot cycle (Figure 14.22b). As a cycle begins at solar minimum, sunspots form primarily at midlatitudes (30° to 40°) on the Sun. The sunspots tend to form at lower latitudes as the cycle progresses, appearing very close to the solar equator as the next solar minimum approaches. Then the sunspots of the next cycle begin to form near midlatitudes again.

A less obvious feature of the sunspot cycle is that something peculiar happens to the Sun's magnetic field at each solar minimum: The Sun's entire magnetic field flip-flops, turning magnetic north into magnetic south and vice versa. We know this because the magnetic field lines connecting pairs of sunspots (see Figure 14.17) on the same side of the solar equator all tend to point in the same direction throughout an 11-year cycle. For example, all compass needles might point from the easternmost sunspot to the westernmost sunspot in each sunspot pair north of the solar equator. However, as the cycle ends at solar minimum, the magnetic field reverses: In the subsequent solar cycle, the field lines connecting pairs of sunspots point in the opposite direction. Thus, the Sun's *complete* magnetic cycle (sometimes called the *solar cycle)* really averages 22 years, since it takes two 11-year sunspot cycles before the magnetic field is back the way it started.

Longer-Term Changes in the Sunspot Cycle Figure 14.22 shows how the sunspot cycle has varied in length and intensity during the past century, the time during which we've had the most complete records of sunspot number. However, astronomers have observed the Sun telescopically for nearly 400 years, and these longer-term observations suggest that the sunspot cycle can change even more dramatically (Figure 14.23). For example, astronomers observed virtually no sunspots between the years 1645 and 1715, a period sometimes called the *Maunder minimum* (after E. W. Maunder, who identified it in historical sunspot records).

Is it possible that the Maunder minimum is part of a longer-term cycle of solar activity, one lasting much longer than 11 or 22 years? Some scientists have hypothesized that such cycles might exist, but little evidence has been found

Figure 14.22 Sunspot cycle during the past century.

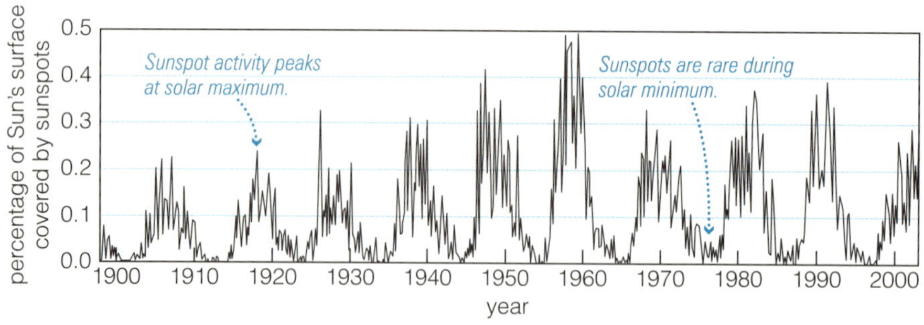

a This graph shows how the number of sunspots on the Sun changes with time. The vertical axis shows the percentage of the Sun's surface covered by sunspots. The cycle has a period of approximately 11 years.

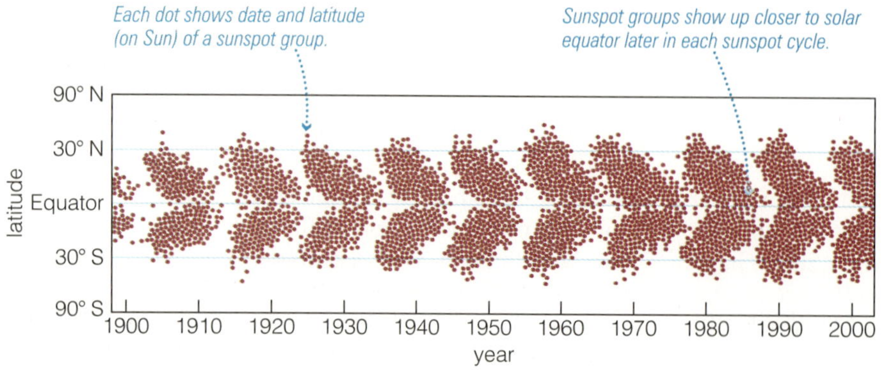

b This graph shows how the latitudes at which sunspot groups appear tend to shift during a single sunspot cycle.

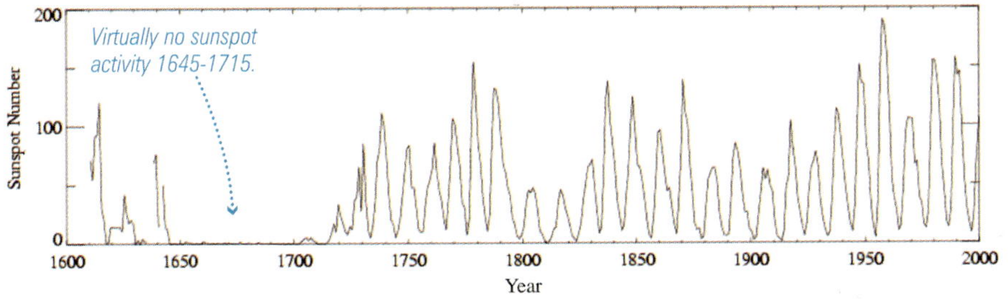

Figure 14.23 This graph reconstructs the sunspot cycle over the past 400 years, based on available data from telescopic observations.

to back up these claims. Of course, the search for long-term variations is difficult, because we have very limited observational data predating the invention of the telescope. Chinese astronomers kept records of naked-eye observations of sunspots going back almost 2,000 years, but these records are sparse and naked-eye observations may not be very reliable.

The search for long-term cycles therefore relies on less direct evidence. For example, we can make educated guesses about past solar activity from descriptions of solar eclipses recorded around the world: When the Sun is more active, the corona tends to have longer and brighter "streamers" visible to the naked eye. We can also gauge past solar activity by studying the amount of carbon-14 in tree rings. High-energy *cosmic rays* [Section 19.2] coming from beyond our own solar system produce radioactive carbon-14 in Earth's atmosphere. During periods of high solar activity, the solar wind tends to grow stronger, shielding Earth from some of these cosmic rays. Thus, production of carbon-14 drops when the Sun is more active. Because trees steadily incorporate atmospheric carbon into their rings (through their respiration of carbon dioxide), we can estimate the level of solar activity during each year of a tree's life by measuring the level of carbon-14 in the corresponding ring. These data have not yet turned up any clear evidence of longer-term cycles of solar activity, but the search goes on.

The Cause of the Sunspot Cycle The precise reasons for the sunspot cycle are not fully understood, but solar physicists believe they understand the general nature of the processes involved. The leading model ties the sunspot cycle to a combination of convection and the way in which the Sun's rotation rate varies with latitude. Convection is thought to dredge up weak magnetic fields generated in the solar interior, amplifying them as they rise. The Sun's rotation—faster at its equator than near its poles—then stretches and shapes these fields.

THINK ABOUT IT

Suppose you take a photograph of the Sun and notice two sunspots: one near the equator and one directly north of it at higher latitude. If you look again at the Sun in a few days, would you still find one sunspot directly north of the other? Why or why not? Explain how you could use this type of observation to learn how the Sun rotates.

Imagine what happens to magnetic field lines that start out running along the Sun's surface from south to north (Figure 14.24). At the equator, the lines circle the Sun every 25 days, but at higher latitudes they lag behind. As a result, the lines gradually get wound more and more tightly around the Sun. This process, operating at all times over the entire Sun, produces the contorted field lines that generate sunspots and other solar activity.

The detailed behavior of these magnetic fields is quite complex, so scientists attempt to study it with sophisticated computer models. Using these models, scientists have successfully replicated many features of the sunspot cycle, including changes in the number and latitude of sunspots and the magnetic field reversals that occur about every 11 years. However, much still remains mysterious, including why the period of the sunspot cycle varies and why solar activity is different from one cycle to the next.

Over extremely long time periods—hundreds of millions to billions of years—these theoretical models predict a gradual lessening of solar activity. Recall that, according to our theory of solar system formation, the Sun must have rotated much faster when it was young [Section 8.3], and a faster rotation rate should have meant much more activity. Observations of other stars that are similar to the Sun but rotate faster confirm that these stars are much more active. We find evidence for many more "starspots" on these stars than on the Sun, and their relatively bright ultraviolet and X-ray emissions suggest that they have brighter chromospheres and coronas—just as we would expect if they are more active than the Sun.

The Sunspot Cycle and Earth's Climate Despite the changes that occur during the sunspot cycle, the Sun's total output of energy barely changes at all—the largest measured changes have been less than 0.1% of the Sun's average luminosity. However, the ultraviolet and X-ray output of the Sun, which comes from the magnetically heated gas of the chromosphere and corona, can vary much more significantly. Could any of these changes affect the weather or climate on Earth?

Some data suggest connections between solar activity and Earth's climate. For example, the period from 1645 to 1715, when solar activity seems to have virtually ceased (see Figure 14.23), was a time of exceptionally low temperatures in Europe and North America known as the *Little Ice Age*. However, no one really knows whether the low solar activity caused these low temperatures, or whether it was just a coincidence. Similarly, some researchers have claimed that certain weather phenomena, such as drought cycles or frequencies of storms, are correlated with the 11- or 22-year

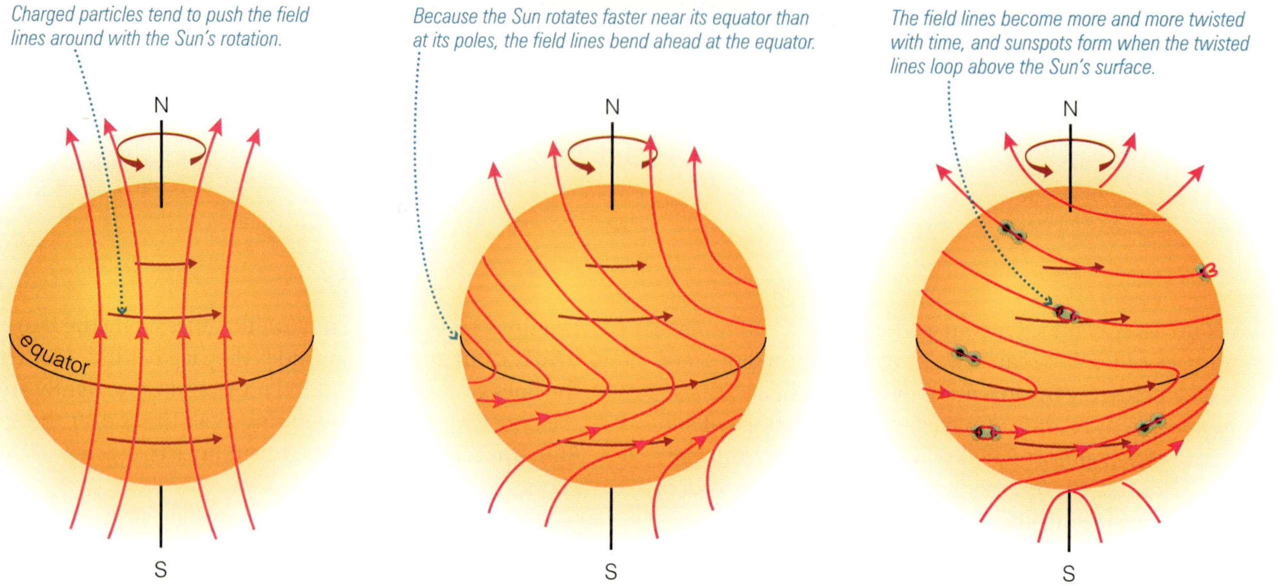

Figure 14.24 The Sun rotates more quickly at its equator than it does near its poles. Because gas circles the Sun faster at the equator, it drags the Sun's north-south magnetic field lines into a more twisted configuration. The magnetic field lines linking pairs of sunspots, depicted here as green and black blobs, trace out the directions of these stretched and distorted field lines.

cycle of solar activity. A few scientists have even claimed that changes in the Sun during the last century have had a greater impact on global warming [Section 10.6] than emissions of greenhouse gases. Again, the data supporting these claims are weak, though we cannot entirely rule them out. The bottom line is that we do not yet know the extent to which solar activity affects weather on Earth. Given the importance of the question, it is certain to remain a hot topic of scientific research.

THE BIG PICTURE

Putting Chapter 14 into Context

In this chapter, we have examined our Sun, the nearest star. When you look back at this chapter, make sure you understand these "big picture" ideas:

- The ancient riddle of why the Sun shines is now solved. The Sun shines with energy generated by fusion of hydrogen into helium in the Sun's core. After a journey through the solar interior lasting several hundred thousand years and an 8-minute journey through space, a small fraction of this energy reaches Earth and supplies sunlight and heat.
- Gravitational equilibrium, the balance between pressure and gravity, determines the Sun's interior structure and helps create a natural thermostat that keeps the fusion rate steady in the Sun. If the Sun were not so steady, life on Earth might not be possible.
- The Sun's atmosphere displays its own version of weather and climate, governed by solar magnetic fields. Some solar weather, such as coronal mass ejections, clearly affects Earth's magnetosphere. Other claimed connections between solar activity and Earth's climate may or may not be real.
- The Sun is important not only as our source of light and heat, but also because it is the only star near enough for us to study in great detail. In the coming chapters, we will use what we've learned about the Sun to help us understand other stars.

14.1 A Closer Look at the Sun

- **Why was the Sun's energy source a major mystery?** Before we learned the Sun's distance from Earth, no one realized just how much energy the Sun releases into space. Once we learned the Sun's energy output, scientists were at a loss to explain it until Einstein showed that mass itself contains a tremendous amount of energy.

- **Why does the Sun shine?** The Sun shines because **gravitational equilibrium** keeps its core hot and dense enough to release energy through nuclear fusion. The core originally became hot through the release of energy by **gravitational contraction**, as gravity made the Sun's birth cloud contract.

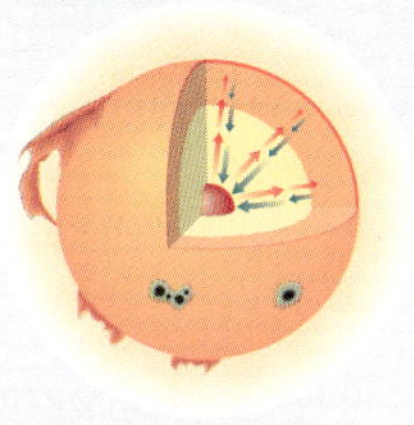

- **What is the Sun's structure?** The Sun's interior layers, from the inside out, are the **core**, the **radiation zone**, and the **convection zone**. Atop the convection zone lies the **photosphere**, the layer from which photons can freely escape into space. Above the photosphere, which is essentially the surface of the Sun, are the **chromosphere** and the **corona**.

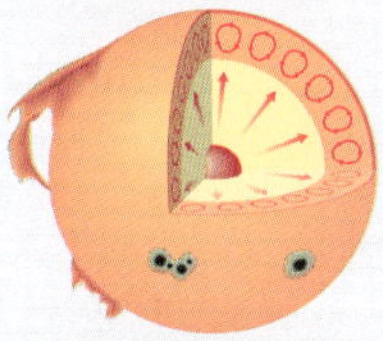

14.2 The Cosmic Crucible

- **How does nuclear fusion occur in the Sun?** The core's extreme temperature and density are just right for fusion of hydrogen into helium, which occurs via the **proton-proton chain**. Because the fusion rate is so sensitive to temperature, gravitational equilibrium acts as a thermostat that keeps the fusion rate steady.

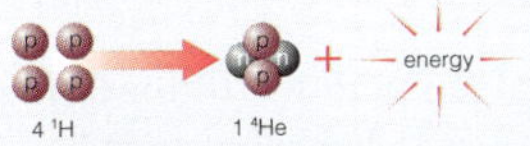

- **How does the energy from fusion get out of the Sun?** Energy moves through the deepest layers of the Sun—the core and the radiation zone—through **radiative diffusion**, in which photons bounce randomly among gas particles. After energy emerges from the radiation zone, convection carries it the rest of the way to the photosphere, where it is radiated into space as sunlight. Energy produced in the core takes hundreds of thousands of years to reach the photosphere.

- **How do we know what is happening inside the Sun?** We can construct theoretical models of the solar interior using known laws of physics and then check the models against observations of the Sun's size, surface temperature, and energy output. We also use studies of solar vibrations and solar **neutrinos**.

14.3 The Sun-Earth Connection

- **What causes solar activity?** Convection combined with the rotation pattern of the Sun—faster at the equator than at the poles—causes **solar activity** because these gas motions stretch and twist the Sun's magnetic field. These contortions of the magnetic field are responsible for phenomena such as **sunspots**, **flares**, **prominences**, and **coronal mass ejections**, and for heating the gas in the chromosphere and corona.

- **How does solar activity affect humans?** Bursts of charged particles ejected from the Sun during periods of high solar activity can hamper radio communications, disrupt electrical power generation, and damage orbiting satellites.

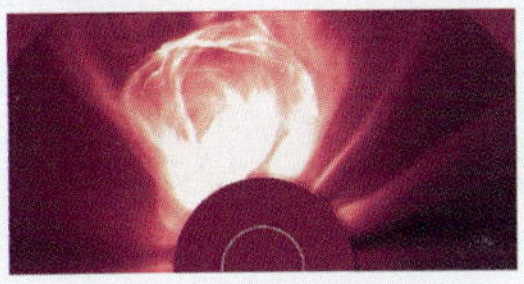

- **How does solar activity vary with time?** The **sunspot cycle**, or the variation in the number of sunspots on the Sun's surface, has an average period of 11 years. The magnetic field flip-flops every 11 years or so, resulting in a 22-year magnetic cycle. Sunspots first appear at midlatitudes at solar minimum, then become increasingly more common near the Sun's equator as the next minimum approaches. The number of sunspots can vary dramatically from one cycle to the next, and sometimes sunspots seem to be absent altogether.

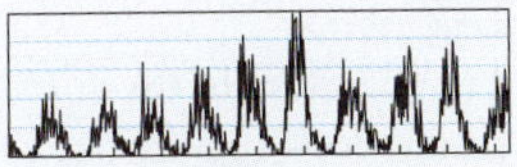

EXERCISES AND PROBLEMS

For instructor-assigned homework go to MasteringASTRONOMY www.masteringastronomy.com

Review Questions

Short-Answer Questions Based on the Reading

1. Briefly describe how *gravitational contraction* generates energy. When was it important in the Sun's history? Explain.
2. What two forces are balanced in *gravitational equilibrium?* Describe how gravitational equilibrium makes the Sun hot and dense in its core.
3. State the Sun's luminosity, mass, radius, and average surface temperature, and put the numbers into perspective that makes them meaningful.
4. Briefly describe the distinguishing features of each of the layers of the Sun shown in Figure 14.3.
5. What is the difference between nuclear *fission* and nuclear *fusion?* Which one is used in nuclear power plants? Which one does the Sun use?
6. Why does nuclear fusion require high temperatures and pressures?
7. What is the overall nuclear fusion reaction in the Sun? Briefly describe the proton-proton chain.
8. Does the Sun's fusion rate remain steady or vary wildly? Describe the feedback process that regulates the fusion rate.
9. Why has the Sun gradually brightened with time?
10. Why does the energy produced by fusion in the solar core take so long to reach the solar surface? Describe the processes by which energy generated by fusion makes its way to the Sun's surface.
11. Explain how mathematical models allow us to predict conditions inside the Sun. How can we be confident that the models are on the right track?
12. What are *neutrinos?* What was the *solar neutrino problem,* and why do we think it has now been solved?
13. What do we mean by *solar activity?* Describe some of the features of solar activity, including *sunspots, solar prominences, solar flares,* and *coronal mass ejections.*
14. Describe the appearance and temperature of the Sun's photosphere. Why does the surface look mottled? How are sunspots different from the surrounding photosphere?
15. How do magnetic fields keep sunspots cooler than the surrounding plasma? Explain.
16. Why is the chromosphere best viewed with ultraviolet telescopes? Why is the corona best viewed with X-ray telescopes? Briefly explain how we think the chromosphere and corona are heated.
17. What is the sunspot cycle? Describe how the changes that occur during the cycle, including magnetic field reversals at each solar minimum. Are there any longer-term changes that occur in solar activity?
18. Describe the leading model for explaining the sunspot cycle. Does the sunspot cycle influence Earth's climate? Explain.

Test Your Understanding

Does It Make Sense?

Decide whether the statement makes sense (or is clearly true) or does not make sense (or is clearly false). Explain your reasoning. (For an example, see Chapter 1, "Does It Make Sense?")

19. Before Einstein, gravitational contraction appeared to be a perfectly plausible mechanism for solar energy generation.
20. A sudden temperature rise in the Sun's core is nothing to worry about, because conditions in the core will soon return to normal.
21. If fusion in the solar core ceased today, worldwide panic would break out tomorrow as the Sun began to grow dimmer.
22. Astronomers have recently photographed magnetic fields churning deep beneath the solar photosphere.
23. Neutrinos probably can't harm me, but just to be safe I think I'll wear a lead vest.
24. There haven't been many sunspots this year, but there ought to be many more in about 5 years.
25. News of a major solar flare today caused concern among professionals in the fields of communications and electrical power generation.
26. By observing solar neutrinos, we can learn about nuclear fusion deep in the Sun's core.
27. If the Sun's magnetic field somehow disappeared, there would be no more sunspots on the Sun.
28. Scientists are currently building an infrared telescope designed to observe fusion reactions in the Sun's core.

Quick Quiz

Choose the best answer to each of the following. Explain your reasoning with one or more complete sentences.

29. Which of these groups of particles has the greatest mass? (a) a helium nucleus with two protons and two neutrons (b) four electrons (c) four individual protons
30. Which of these layers of the Sun is coolest? (a) photosphere (b) chromosphere (c) corona
31. Which of these layers of the Sun is coolest? (a) core (b) radiation zone (c) photosphere
32. Scientists estimate the central temperature of the Sun using (a) probes that measure changes in Earth's atmosphere. (b) mathematical models of the Sun. (c) laboratories that create miniature versions of the Sun.
33. Why do sunspots appear darker than their surroundings? (a) They are cooler than their surroundings. (b) They block some of the sunlight from the photosphere. (c) They do not emit any light.
34. At the center of the Sun, fusion converts hydrogen into (a) plasma. (b) radiation and elements like carbon and nitrogen. (c) helium, energy, and neutrinos.
35. Solar energy leaves the core of the Sun in the form of (a) photons. (b) rising hot gas. (c) sound waves.
36. How does the number of neutrinos passing through your body at night compare with the number passing through your body during the day? (a) about the same (b) much smaller (c) much larger
37. What is the most common kind of particle in the solar wind, in which the overall charge is neutral? (a) proton (b) electron (c) helium nucleus
38. Which of these things poses the greatest hazard to communications satellites? (a) photons from the Sun (b) solar magnetic fields (c) protons from the Sun

Investigate Further

In-Depth Questions to Increase Your Understanding

Short-Answer/Essay Questions

39. *The End of Fusion I.* Describe what would happen in the Sun if fusion reactions abruptly shut off.
40. *The End of Fusion II.* If fusion reactions were to suddenly shut off in the Sun, how would we be able to tell?
41. *A Really Strong Force.* How would the interior temperature of the Sun be different if the strong force that binds nuclei together were 10 times stronger?
42. *Measuring the Sun's Rotation.* Suppose you observe a sunspot from Earth, taking a picture of the Sun each day. Using these photos, you count the number of days for the sunspot to return to the same position it had in the first picture. Now suppose a friend of yours standing on Pluto does the same thing. Would you and your friend measure the same number of days for the sunspot to return to where it started? Explain your answer.
43. *Covered with Sunspots.* Describe what the Sun would look like from Earth if the entire photosphere were the same temperature as a sunspot.
44. *Inside the Sun.* Describe how scientists determine what the interior of the Sun is like. Why have we not yet sent a probe into the Sun to measure what is happening there?
45. *Solar Energy Output.* The Sun's output of visible light remains very steady, staying well within 1% of its current value for the past century. However, the Sun's maximum X-ray output can be as much as 10 times greater than its minimum output. Explain why the changes in X-ray output can be so much more pronounced than those in the output of visible light.
46. *An Angry Sun.* A *Time* magazine cover once suggested that an "angry Sun" was becoming more active as human activity changed Earth's climate through global warming. It's certainly possible for the Sun to become more active at the same time that humans are affecting Earth, but is it possible that the Sun could be responding to human activity? Can humans affect the Sun in any significant way? Explain.

Quantitative Problems

Be sure to show all calculations clearly and state your final answers in complete sentences.

47. *Chemical Burning and the Sun.* Estimate how long the Sun would last if it were merely a huge fire that was releasing chemical energy. Assume that the Sun begins with roughly 10^8 joules per kilogram, a chemical energy content typical of atomic matter.
48. *The Lifetime of the Sun.* The total mass of the Sun is about 2×10^{30} kg, of which about 75% was hydrogen when the Sun formed. However, only about 13% of this hydrogen ever becomes available for fusion in the core. The rest remains in layers of the Sun where the temperature is too low for fusion.
    a. Based on the given information, calculate the total mass of hydrogen available for fusion over the lifetime of the Sun.
    b. Combine your results from part (a) and the fact that the Sun fuses about 600 billion kilograms of hydrogen each second to calculate how long the Sun's initial supply of hydrogen can last. Give your answer in both seconds and years.
    c. Given that our solar system is now about 4.6 billion years old, when will we need to worry about the Sun running out of hydrogen for fusion?
49. *Solar Power Collectors.* This problem leads you through the calculation and discussion of how much solar power can be collected by solar cells on Earth.
    a. Imagine a giant sphere surrounding the Sun with a radius of 1 AU. What is the surface area of this sphere, in square meters? (*Hint:* The formula for the surface area of a sphere is $4\pi r^2$.)
    b. Because this imaginary giant sphere surrounds the Sun, the Sun's entire luminosity of 3.8×10^{26} watts must pass through it. Calculate the power passing through each square meter of this imaginary sphere in *watts per square meter.* Explain why this number represents the maximum power per square meter that a solar collector in Earth orbit can collect.
    c. List several reasons why the average power per square meter collected by a solar collector on the ground will always be less than what you found in part (b).
    d. Suppose you want to put a solar collector on your roof. If you want to optimize the amount of power you can collect, how should you orient the collector? (*Hint:* The optimum orientation depends on both your latitude and the time of year and day.)
50. *Solar Power for the United States.* The total annual U.S. energy consumption is about 2×10^{20} joules.
    a. What is the average power requirement for the United States, in watts? (*Hint:* 1 watt = 1 joule/s.)
    b. With current technologies and solar collectors on the ground, the best we can hope is that solar cells will generate an average (day and night) power of about 200 watts/m^2. (You might compare this to the maximum power per square meter you found in Problem 49b.) What total area would we need to cover with solar cells to supply all the power needed for the United States? Give your answer in both square meters and square kilometers.
    c. The total surface area of the United States is about 2×10^7 km^2. What fraction of the U.S. area would have to be covered by solar collectors to generate all of the U.S. power needs? In one page or less, describe potential environmental impacts of covering so much area with solar collectors. Also discuss whether you think these environmental impacts would be greater or less than the impacts of using current energy sources such as coal, oil, nuclear power, and hydroelectric power.
51. *The Color of the Sun.* The Sun's average surface temperature is about 5,800 K. Use Wien's law (see Mathematical Insight 5.2) to calculate the wavelength of peak thermal emission from the Sun. What color does this wavelength correspond to in the visible-light spectrum? Why do you think the Sun appears white or yellow to our eyes?
52. *The Color of a Sunspot.* The typical temperature of a sunspot is about 4,000 K. Use Wien's law (see Mathematical Insight 5.2) to calculate the wavelength of peak thermal emission from a sunspot. What color does this wavelength correspond to in the visible-light spectrum? How does this color compare with that of the Sun?
53. *Solar Mass Loss.* Estimate how much mass the Sun loses through fusion reactions during its 10-billion-year life. You can simplify the problem by assuming the Sun's energy

output remains constant. Compare the amount of mass lost with Earth's mass.

54. *Pressure of the Photosphere.* The gas pressure of the photosphere changes substantially from its upper levels to its lower levels. Near the top of the photosphere the temperature is about 4,500 K and there are about 1.6×10^{16} gas particles per cubic centimeter. In the middle the temperature is about 5,800 K and there are about 1.0×10^{17} gas particles per cubic centimeter. At the bottom of the photosphere the temperature is about 7,000 K and there are about 1.5×10^{17} gas particles per cubic centimeter. Compare the pressures of each of these layers and explain the reason for the trend in pressure that you find. How do these gas pressures compare with Earth's atmospheric pressure at sea level?
55. *Tire Pressure.* Air pressure at sea level produces a force of about 15 pounds per square inch. The recommended air pressure in your car tires is about 30 pounds per square inch. How does the density of gas particles inside your tires compare with the density of gas particles in the air outside your tires? What happens to gas pressure in the tire if it springs a leak and loses gas particles? How does your tire respond to this loss of gas particles? How is the tire's response like the response of the Sun's core to the slowly declining number of independent particles within it? How is the tire's response different?
56. *Your Energy Content.* The power needed to operate your body is about 100 watts. Suppose your body could run on fusion power and could convert 0.7% of its mass into energy. How much energy would be available through fusion? For how long could your body then operate on fusion power?

Discussion Questions

57. *The Role of the Sun.* Briefly discuss how the Sun affects us here on Earth. Be sure to consider not only factors such as its light and warmth, but also how the study of the Sun has led us to new understandings in science and to technological developments. Overall, how important has solar research been to our lives?
58. *The Sun and Global Warming.* One of the most pressing environmental issues on Earth is the extent to which human emissions of greenhouse gases are warming our planet. Some people claim that part or all of the observed warming over the past century may be due to changes in the Sun, rather than to anything humans have done. Discuss how a better understanding of the Sun might help us comprehend the threat posed by greenhouse gas emissions. Why is it so difficult to develop a clear understanding of how the Sun affects Earth's climate?

MEDIA EXPLORATIONS

For self-study activities go to MasteringASTRONOMY www.masteringastronomy.com

Interactive Tutorials

Tutorial Review of Key Concepts

Use the following interactive **Tutorials** at **www.masteringastronomy.com** to review key concepts from this chapter.

The Sun Tutorial

Lesson 1 Structure and Gravitational Equilibrium
Lesson 2 Fusion of Hydrogen into Helium
Lesson 3 Why Does Fusion Only Occur at High Temperatures?

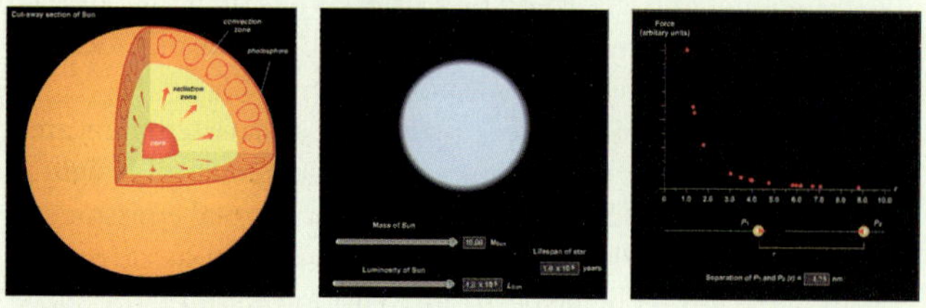

Web Projects

1. *Current Solar Weather.* Daily information about solar activity is available at NASA's Web site spaceweather.com. Where are we in the sunspot cycle right now? When is the next solar maximum or minimum expected? Have there been any major solar storms in the past few months? If so, did they have any significant effects on Earth? Summarize your findings in a one- to two-page report.
2. *Solar Observatories in Space.* Visit NASA's Web site for the Sun-Earth connection and explore some of the current and planned space missions designed to observe the Sun. Choose one mission to study in greater depth, and write a one- to two-page report on the mission status and goals and what it has taught or will teach us about the Sun.
3. *Sudbury Neutrino Observatory.* Visit the Web site for the Sudbury Neutrino Observatory (SNO) and learn how it has helped to solve the solar neutrino problem. Write a one- to two-page report describing the observatory, any recent results, and what we can expect from it in the future.
4. *Nuclear Power.* There are two basic ways to generate energy from atomic nuclei: through nuclear fission (splitting nuclei) and through nuclear fusion (combining nuclei). All current nuclear reactors are based on fission, but fusion would have many advantages if we could develop the technology. Research some of the advantages of fusion and some of the obstacles to developing fusion power. Do you think fusion power will be a reality in your lifetime? Explain.

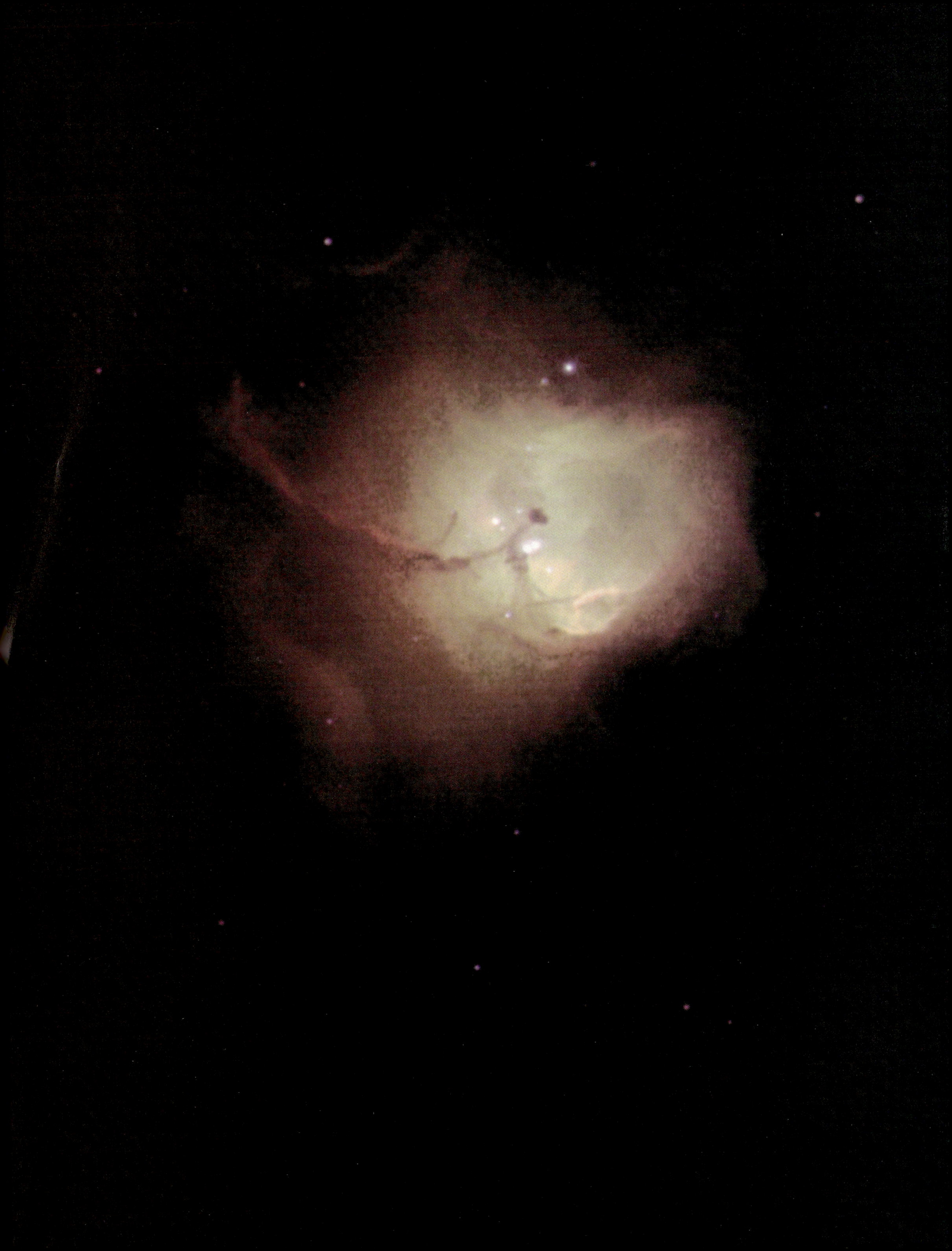

24

Life in the Universe

LEARNING GOALS

24.1 Life on Earth
- When did life arise on Earth?
- How did life arise on Earth?
- What are the necessities of life?

24.2 Life in The Solar System
- Could there be life on Mars?
- Could there be life on Europa or other jovian moons?

24.3 Life Around Other Stars
- Are habitable planets likely?
- Are Earth-like planets rare or common?

24.4 The Search for Extraterrestrial Intelligence
- How many civilizations are out there?
- How does SETI work?

24.5 Interstellar Travel and Its Implications to Civilization
- How difficult is interstellar travel?
- Where are the aliens?

We, this people, on a small and lonely planet
Travelling through casual space
Past aloof stars, across the way of indifferent suns
To a destination where all signs tell us
It is possible and imperative that we learn
A brave and startling truth.

Maya Angelou, Excerpted from
A Brave and Startling Truth

We have covered a lot of ground in this book, discussing fundamental questions about the nature and origin of our planet, our star, our galaxy, and our universe. But we have not yet discussed one of the most profound questions of all: Are we alone? The universe seems to be filled with worlds beyond imagination—more than 100 billion star systems in our galaxy alone, and some 100 billion galaxies in the observable universe—yet we do not know whether any other world has ever been home to life.

In this chapter, we will discuss the possibility of life beyond Earth. We'll begin by considering the history of life on Earth, which should help us understand the prospects for finding life elsewhere. We'll then consider the possibility of finding microbial life elsewhere in our solar system or beyond and examine efforts to search for extraterrestrial intelligence (SETI). Finally, we'll discuss the astonishing implications that the search for life may hold for the future of our own civilization.

24.1 Life on Earth

It may seem that aliens are everywhere. Popular shows and movies, such as *Star Trek* and *Star Wars,* show aliens from many different worlds traveling easily among the stars. Closer to home, supermarket tabloids routinely carry headlines about the latest alien atrocities or about alien corpses hidden by the government at "Area 51."

Despite their popularity in the media, any alien visitors have been sadly negligent in leaving scientific evidence of their trespass. Decades of scientific observation and study have not turned up a single piece of undeniable evidence that aliens have been here (see the box "Are Aliens Already Here?," p. 730). Why, then, do so many scientists believe it is worth the effort to search for life beyond Earth?

For a long time, the answer was simply that it seemed natural for other worlds to be inhabited. Belief in life on other worlds was common even among ancient Greek philosophers, and many famous scientists of the past few centuries took it as a given that intelligent beings exist on other planets. For example, Kepler suggested that the Moon was inhabited, and William Herschel (co-discoverer of Uranus with his sister Caroline) spoke of life on virtually all the planets in our solar system. Most famously, in the late nineteenth century Percival Lowell claimed to see networks of canals on Mars, which he argued were the mark of an advanced civilization [Section 9.4]. Lowell's ideas became the basis for H. G. Wells's novel *The War of the Worlds*, upon which a recent movie was based.

Hopes of finding life on our planetary neighbors were diminished by the bleak images of Mars returned by spacecraft and the discovery of the runaway greenhouse effect on Venus. For a few decades, scientific interest in extraterrestrial life waned. However, that interest has been on the upswing again during the past couple of decades. Part of the renewed interest has come from the discovery of planets around other stars [Section 13.1], which confirms that there are plenty of places to look for life beyond our own solar system. But new discoveries about life on Earth have played an even more important role. In particular, three recent developments in the study of life on Earth have made it seem much more likely that we could find life elsewhere:

- We have learned that life arose quite early in Earth's history, suggesting that life might also form quickly on other worlds with the right conditions.
- Laboratory experiments have shown that the chemical constituents thought to have been common on the young Earth combine readily into complex organic molecules. We may never know exactly how life arose on Earth, but these experiments suggest that life might arise through naturally occurring chemistry—in which case the same chemistry could have given rise to life on many other worlds.
- We have discovered microscopic living organisms that can survive in conditions similar to those found on at least some other worlds in our solar system, suggesting that the basic necessities of life may be quite common in the universe.

Because these three ideas are so important to understanding the modern science of life in the universe, let's examine each of them in greater detail. Along the way, we'll also briefly discuss the nature and history of life on Earth. We will then be prepared to consider how we might actually search for life on other worlds.

• When did life arise on Earth?

Our planet was born about 4.5 billion years ago, but we would not expect life to have had much of a chance to survive during the first several hundred million years of Earth's history. This was the time of the *heavy bombardment* [Section 8.4], when large asteroids or comets should have repeatedly struck Earth. At least a few of these impacts probably had sufficient energy to vaporize the early oceans completely, killing off any life that might already have been present on Earth. The ancestors of living organisms today could not have arisen much before the end of the heavy bombardment.

Studies of craters on the Moon suggest that the last major impacts of the heavy bombardment occurred between about 4.2 and 3.9 billion years ago. Remarkably, we now have evidence suggesting that life was thriving prior to

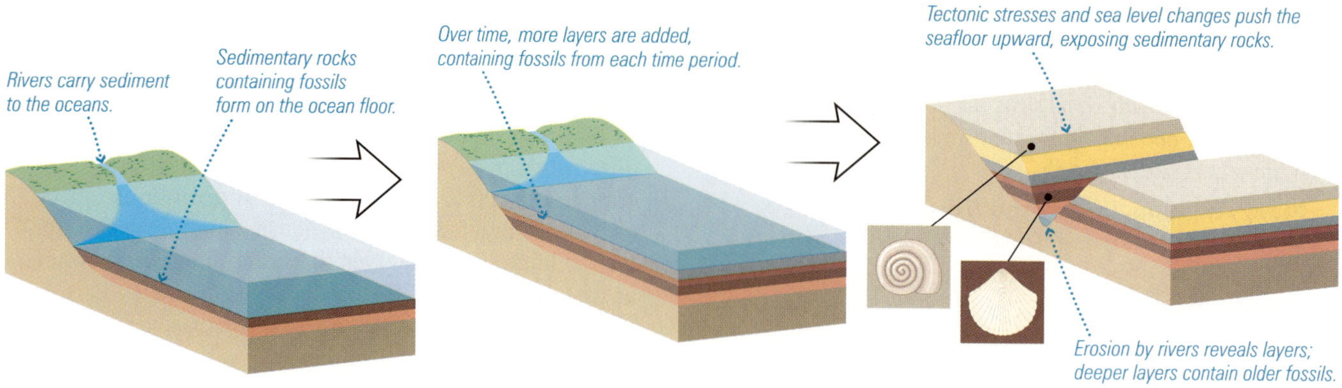

Figure 24.1 Formation of sedimentary rock. Each layer represents a particular time and place in Earth's history and is characterized by fossils of organisms that lived in that time and place.

3.85 billion years ago. If we are interpreting the evidence correctly, it means that life arose in a geological blink of an eye from the moment when conditions first allowed it. To understand this evidence and its importance, we need to discuss how scientists study the history of life on Earth.

The Geological Time Scale We learn about the history of life on Earth through the study of **fossils**, relics of organisms that lived and died long ago. Most fossils form when dead organisms fall to the bottom of a sea (or other body of water) and are gradually buried by layers of sediment. The sediments are produced by erosion on land and carried by rivers to the sea. Over millions of years, sediments pile up on the seafloor, and the weight of the upper layers compresses underlying layers into rock. Erosion or tectonic activity can later expose the fossils (Figure 24.1). In some places, such as the Grand Canyon, the sedimentary layers record billions of years of Earth's history (Figure 24.2). Some fossils are remarkably well preserved (Figure 24.3), though the vast majority of dead organisms decay completely and leave no fossils behind.

The key to reconstructing the history of life is to determine the dates at which fossil organisms lived. The *relative* ages of fossils found in different layers are easy to determine: Deeper layers formed earlier and contain more ancient fossils. Radiometric dating [Section 8.5] confirms these relative ages and gives us fairly precise absolute ages for fossils. Based on the layering of rocks and fossils, geologists divide Earth's history into a set of distinct intervals that make up what we call the **geological time scale.** Figure 24.4 shows the names of the various intervals on a timeline, along with numerous important events in Earth's history.

THINK ABOUT IT

Study Figure 24.4. How does the length of time during which animals and plants have lived on land compare to the length of time during which life has existed? How does the length of time during which humans have existed compare to the length of time since mammals and dinosaurs first arose?

Fossil Evidence for the Early Origin of Life You might wonder why the geological time scale shows so much more detail in the last few hundred million years than it does for earlier times. The answer is that fossils become increasingly difficult to find as we look deeper into Earth's history, for three major reasons: First, older rocks are much rarer than younger rocks, because most of Earth's surface is geologically young. Second, even when we find very old rocks, they often turn out to have been subject to transformations (caused by heat and pressure) that would have destroyed any fossil

Figure 24.2 The rock layers of the Grand Canyon record 2 billion years of Earth's history.

Figure 24.3 Dinosaur fossils.

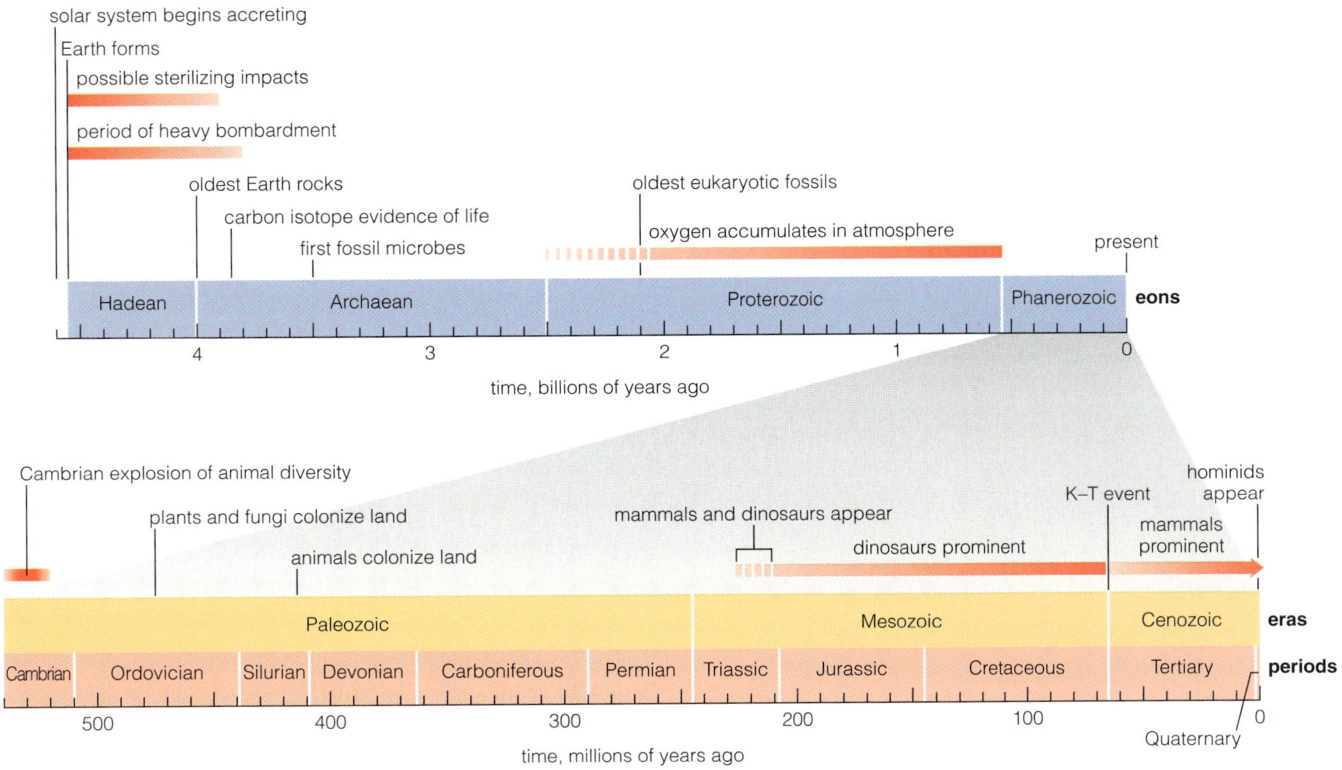

Figure 24.4 The geological time scale. Be sure to notice that the lower timeline is an expanded view of the last portion of the upper timeline. The eons, eras, and periods are defined by changes observed in the fossil record. The absolute ages come from radiometric dating. (The K–T event is the geological term for the impact linked to the mass extinction of the dinosaurs [Section 12.4].)

evidence they may have contained. Third, all life prior to a few hundred million years ago was microscopic, and microscopic fossils are much more difficult to identify than dinosaur bones.

Despite these difficulties, geologists have found a few very old rocks that suggest life was already thriving on Earth by 3.5 billion years ago, and possibly for several hundred million years before that. One strong line of evidence comes from rocks called *stromatolites*. These rocks show structure nearly identical to that found in large bacterial mats today (Figure 24.5). Radiometric dating shows that some stromatolites are 3.5 billion years old.

The fact that organisms were already advanced enough to build stromatolites by 3.5 billion years ago suggests that more primitive organisms must have lived even earlier. Some evidence supports this idea. More ancient rocks that have undergone too much change to leave fossils intact may still hold carbon that was once part of living organisms.

Carbon has two stable isotopes: carbon-12, with six protons and six neutrons in its nucleus, and carbon-13, which

a These large mats photographed at Shark Bay, Western Australia, are colonies of microbes known as "living stromatolites."

b The bands visible in this section of a modern-day mat are formed by layers of sediment adhering to different types of microbes.

c This section of a 3.5-billion-year-old stromatolite shows a structure nearly identical to that of a living mat. This structure, along with more detailed study, strongly suggests that stromatolites are the fossilized remains of microbes, including some photosynthetic ones, that lived as much as 3.5 billion years ago.

Figure 24.5 Rocks called stromatolites offer evidence of photosynthetic life as early as 3.5 billion years ago.

has one extra neutron (see Figure 5.9). Living organisms incorporate carbon-12 slightly more easily than carbon-13. As a result, the fraction of carbon-13 is always a bit lower in fossils than in rock samples that lack fossils. All life and all fossils tested to date show the same characteristic ratio of the two carbon isotopes. Dating very ancient rocks can be difficult, but some rocks that are more than 3.85 billion years old show the same ratio of carbon-12 to carbon-13, suggesting that these rocks contain remnants of life.

To summarize, fossil evidence points to life already thriving by 3.5 billion years ago, and carbon isotope evidence suggests that life was present more than 3.85 billion years ago. Thus, even under the conservative assumption that the last major impact of the heavy bombardment occurred 4.2 billion years ago, we conclude that life arose in a period of no more than about 300 million years. If the last major impact occurred 3.9 billion years ago, then life arose within a few tens of millions of years or less.

• How did life arise on Earth?

We have discussed when life arose on Earth, but how did life arise? To answer this question, we must first consider how life today differs from life at the time of its origin on Earth. Then we can ask how the first organisms may have originated.

The Theory of Evolution The fossil record clearly shows that life has gone through great changes over time (see Figure 24.4). Thus, if we are going to understand how life arose, we must understand what causes this change, so that we can trace life back to its origin. The unifying theory through which scientists understand the history of life on Earth is the **theory of evolution**, proposed by Charles Darwin (1809–1882) in his 1859 book called *The Origin of Species.*

Evolution simply means "change with time," and strong fossil evidence for evolution had already been gathered before Darwin published his theory. Indeed, most scientists of Darwin's time had already accepted the idea that species change through time; what was missing prior to Darwin was an explanation of how and why species might undergo change. In essence, the fossil record provides strong evidence that evolution *has* occurred, while Darwin's theory of evolution explains *how* it occurs.

Darwin provided extensive evidence for the occurrence of evolution and, based on this evidence, he laid out a clear model to explain how evolution occurs. As is the case with most scientific theories, the underlying logic of his model is really very simple. As described by biologist Stephen Jay Gould (1941–2002), Darwin built his model from "two undeniable facts and an inescapable conclusion":

- *Fact 1: overproduction and struggle for survival.* Any localized population of a species has the potential to produce far more offspring than the local environment can possibly support with resources such as food and shelter. This overproduction leads to a struggle for survival among the individuals of the population.
- *Fact 2: individual variation.* Individuals in a population of any species vary in many heritable traits (that is, traits passed from parents to offspring). No two individuals are exactly alike, and some individuals possess traits that make them better able to compete for food and other vital resources.
- *The inescapable conclusion: unequal reproductive success.* In the struggle for survival, those individuals whose traits best enable them to survive and reproduce will, on average, leave the largest number of offspring that in turn survive to reproduce. Therefore, in any local environment, heritable traits that enhance survival and successful reproduction will become progressively more common in succeeding generations. It is this unequal reproductive success that Darwin called **natural selection**; that is, nature "selects" the advantageous trait to be passed on to future generations.

Of course, even the strongest logic still needs evidence to back it up if it is to gain acceptance as a scientific theory [Section 3.4], and Darwin's greatest achievement may have been in compiling a large body of extensive evidence in support of his theory. By studying the variety of species in places like the Galapagos Islands, Darwin demonstrated that natural selection had given the individuals of different species important survival advantages. Moreover, his work showed that when enough small individual variations accumulate, natural selection can even give rise to an entirely new species. Darwin's evidence was so complete and so well documented that the vast majority of biologists accepted his theory within a decade of the publication of his book.

The Mechanism of Evolution In the nearly 150 years since Darwin first published his theory, ongoing research has only given further support to Darwin's idea. Perhaps the strongest support for the theory has come from the discovery of **DNA** (short for *deoxyribonucleic acid*), the genetic material of all life on Earth. As a result of this discovery, scientists now understand the mechanism of evolution on a molecular level.

As you have probably learned in past biology classes, a molecule of DNA consists of two long strands—somewhat like the interlocking strands of a zipper—wound together in the spiral shape known as a *double helix* (Figure 24.6). The instructions for assembling a living organism are written in the precise order of four chemical bases that make up the interlocking portions of the DNA "zipper": adenine (A), thymine (T), guanine (G), and cytosine (C). These bases pair up in a way that ensures that each of the two strands of a DNA molecule contains the same genetic information. Thus, by unwinding and allowing new strands to form alongside the original ones (with the new strands made from chemicals floating around inside a cell), a single DNA molecule can give rise to two identical copies of itself.

Living organisms reproduce by copying DNA molecules and passing them on to their descendants. Evolution occurs because the passing of genetic information from one generation to the next is not always perfect. An organism's DNA

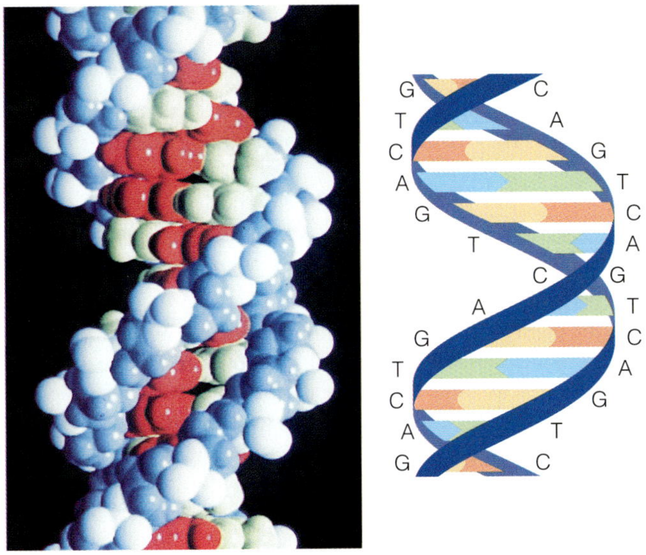

Figure 24.6 DNA is the genetic material of all life on Earth. Left: A model of a small piece of a DNA molecule. Right: This diagram shows that a DNA molecule is made from two interlocking strands wound in the shape of a double helix. The interlocking pieces (which look like ladder steps in the figure) are composed of four chemical bases, labeled by their first letters A, T, G, and C. Note that A always connects to T, and G always connects to C. Thus, each strand contains the same basic information, so when the two strands separate and "helper molecules" bring new chemical bases to attach to each one, the result is two new molecules each identical to the original one that separated. This ability of DNA to copy itself is the key to heredity.

may be altered by occasional copying errors or by external influences, such as ultraviolet light from the Sun or exposure to toxic or radioactive chemicals. Any change in the sequence of the chemical bases in an organism's DNA is called a **mutation**. Many mutations are lethal, killing the cell in which the mutation occurs. Some, however, may improve a cell's ability to survive and reproduce, allowing the cell to pass this improvement to its offspring.

Our molecular understanding of the mechanism of natural selection has put the theory of evolution on a stronger foundation than ever. While no theory can ever be proved true beyond all doubt, the theory of evolution today is as solid as any theory in science, including the theory of gravity and the theory of atoms. Indeed, biologists today routinely witness evolution occurring before their eyes among laboratory microorganisms, or over periods of just a few decades among plants and animals subjected to some kind of environmental stress. Moreover, the theory of evolution has become the underpinning of virtually all modern biology, medicine, and agriculture. For example, agricultural scientists apply the idea of natural selection to develop pest control strategies that reduce the populations of harmful insects without harming the populations of beneficial ones; medical researchers test new drugs on animals that are genetically similar to humans, because the theory of evolution tells us that genetically similar species should have at least somewhat similar physiological responses; and biologists study the relationships between organisms by comparing their DNA.

The First Living Organisms The basic chemical nature of DNA is virtually identical among all living organisms. This fact, along with other biochemical similarities shared by all living organisms (such as the fact that all organisms build proteins from the same set of amino acids and use essentially the same genetic code), tells us that all life on Earth today can trace its origins to a common ancestor. In other words, sometime before about 3.85 billion years ago, a single organism arose and gradually evolved into every other organism that has ever lived.

What did the first living organism look like? Although we may never have fossils of the earliest organisms, in principle we can learn about early life through careful studies of the DNA of living organisms. Biologists can determine the evolutionary relationships among living species by comparing the sequences of bases in their DNA. For example, two organisms whose DNA sequences differ in five places for a particular gene are probably more distantly related than two organisms whose gene sequences differ in only one place.

Many such DNA comparisons suggest that all living organisms are related in a way depicted schematically by the "tree of life" in Figure 24.7. Although details in the structure of this tree remain uncertain, it confirms that living organisms share a common ancestry. As it is currently understood, the tree of life shows that life on Earth is divided into three major groupings (known as the *domains* Bacteria, Archaea, and Eukarya). Notice that plants and animals represent only two tiny branches of the great diversity of life on Earth.

Comparison of DNA sequences also allows at least reasonable guesses as to which organisms found on Earth today most resemble the common ancestor. The answer appears to be microscopic organisms living in the deep oceans around seafloor volcanic vents called *black smokers* (after

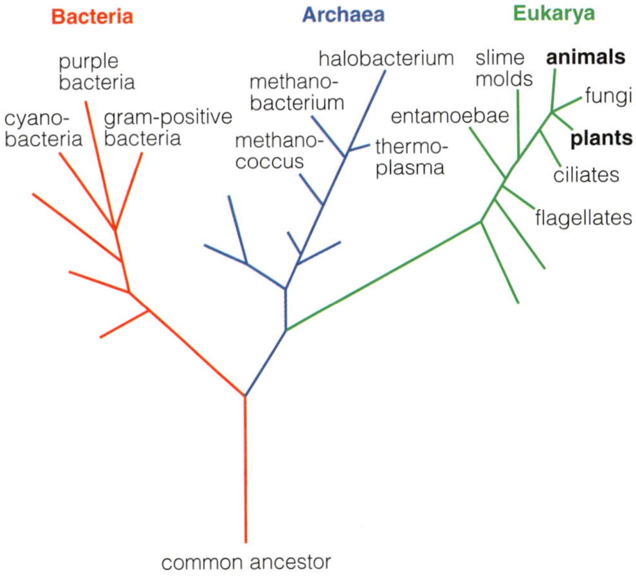

Figure 24.7 The tree of life, showing evolutionary relationships determined by comparison of DNA sequences in different organisms. Just two small branches at the upper right represent *all* plant and animal species.

the dark, mineral-rich water that flows out of them) and in hot springs in places such as Yellowstone (Figure 24.8). These organisms thrive in temperatures as high as 110°C (230°F). (The high pressures at the seafloor prevent the water from boiling despite the high temperature.) Unlike most life at Earth's surface, which depends on sunlight, these organisms get energy from chemical reactions in water heated volcanically by Earth itself.

That early organisms might have lived in such "extreme" conditions may at first seem surprising, but it makes sense

SPECIAL TOPIC Evolution and the Schools

You're probably familiar with the public debate about how evolution should be taught in schools. Much of the controversy comes from misunderstandings on both sides. On the antievolution side, there's a general lack of understanding of why scientists are near unanimous in accepting the theory of evolution. On the scientists' side, there's often a lack of respect for people's individual religious beliefs.

The heart of the debate surrounds the question of what counts as science and what does not. You can see why by considering a common question: Do you believe in creationism or evolution? This question may be philosophically and theologically interesting, but it is poorly phrased from a scientific standpoint because science is about evidence, not beliefs. A tremendous weight of evidence supports the idea that the universe is much older than the Earth, that Earth is some $4\frac{1}{2}$ billion years old, that life on Earth has existed for nearly 4 billion years and has undergone dramatic change during that time, and that this change has occurred through the mechanism of evolution by natural selection. Nevertheless, all the evidence in the world can never prove a scientific theory to be true beyond all doubt. To take a somewhat extreme case, you could accept the overwhelming evidence for evolution and still believe that God created Earth and the universe a mere 6,000 years ago by, for example, assuming that God put all the evidence for evolution in place. Because the evidence would look the same in this case as if evolution actually occurred, science cannot say anything about the validity of such a belief.

We can gain deeper insight by fitting the theory of evolution to the hallmarks of science illustrated in Figure 3.26. The first hallmark says that science seeks explanations for observed phenomena that rely on natural causes, and the theory of evolution does exactly that: It explains the fossil record and observed differences between species in terms of natural selection. Note that this hallmark does not in any way preclude the possibility of divine intervention; it just says that if divine intervention has occurred, we cannot study it scientifically. The second and third hallmarks concern the way evidence is gathered and interpreted: In science, we create and test models designed to explain our observations, and we modify or discard those models if they fail the tests. Again, the theory of evolution exhibits these hallmarks because it is continually subject to testing and verification, and it has indeed been modified as new evidence has come to light. For example, Darwin's original theory discussed natural selection, but the molecular basis for it wasn't added until the discovery of DNA. Similarly, Darwin thought that evolution would be a gradual process occurring in a large population of a given species, but more recent evidence suggests that evolution often occurs in small, isolated populations, and often in episodic bursts. The ongoing debate about the precise nature of evolution sometimes makes nonscientists think there is scientific controversy over the basic theory, but in reality it shows why the process of science has put the theory of evolution by natural selection on such a solid footing.

The fact that evolution is such a well-established and important scientific theory makes it an obvious part of the science curriculum for the schools, and few people disagree. However, many people have suggested teaching alternative ideas such as *creationism* or *intelligent design* alongside evolution in school curricula. The scientific response is very simple: Scientists take no position on the general school curriculum, but the *science curriculum* should not include these alternate ideas because they are not science. Again, we can see why by considering whether these ideas reflect the hallmarks of science.

Let's start with *creationism*, the idea that the world was divinely created. In fact, we can't even define this idea clearly enough to test it against the hallmarks of science. Many religions incorporate stories of creation, and these stories are often quite different from one another. For example, many Native American religious beliefs speak of creation in terms that bear little resemblance to the story in Genesis. Even among Christians who claim a literal belief in the Bible, there are differences of interpretation about creation. While some Biblical literalists argue that the creation must have occurred in just 6 days, as the first chapter of Genesis seems to say, others suggest that the term "day" in Genesis does not necessarily mean 24 hours and that Genesis is compatible with a much older Earth. Thus, while creationism is a perfectly valid religious belief, it does not qualify as a scientific theory.

The more recent idea of *intelligent design* holds that the complexity of life is too great to have arisen naturally, and therefore there must have been a "designer." Intelligent design can sound much like science, especially because most of its proponents accept both the idea of an old Earth and the idea that some living things have changed through time as indicated by the fossil record. However, the assumption of an intelligent designer does not yield testable predictions, because it posits that the designer is beyond our scientific comprehension; in essence, the idea of intelligent design means that there is no point in continuing to look for natural causes, because a supernatural process made the universe and life. This idea may or may not be true, but it isn't science. As a simple analogy, consider the collapse of a bridge. If an engineer declares that the collapse was an act of God, he could well be right—but that belief won't help him learn how to design a better bridge. It is the scientific quest for a natural understanding of life—embodied in the theory of evolution—that has led to the discovery of genetics, DNA, and virtually all modern medicine.

So where does this discussion leave us with regard to the question of teaching evolution in schools? Let's start with what our discussion does *not* tell us. It does not tell us that everyone needs to accept the theory of evolution—whether you choose to believe the theory is up to you. Nor does it tell us whether divine intervention has guided evolution—science does not say anything about the existence or role of God. What it *does* tell us is that the theory of evolution is a clear and crucial part of the study of science, and that without it we cannot fully understand modern science and its impact on our world.

a This photograph shows a black smoker—a volcanic vent on the ocean floor that spews out hot, mineral-rich water.

b This aerial photo shows a hot spring in Yellowstone National Park. The different colors are from different microbes that survive in water of different temperatures. For a sense of scale, note the walkway winding along the lower right.

Figure 24.8 DNA evidence suggests that the ancestor of all life on Earth resembled organisms that live today in hot water near volcanic vents or hot springs.

when we think about it more deeply. Sunlight can provide energy only to organisms capable of photosynthesis, which is a very complex molecular process. The chemical pathways for extracting energy from the heat of Earth itself are actually much simpler and therefore far more likely to have been used by the earliest organisms. Moreover, chemical reactions tend to proceed faster and with greater diversity in hotter water than cooler water, so the hot water near volcanic vents may have offered a natural laboratory in which chemistry "experiments" were constantly under way.

The Transition from Chemistry to Biology The theory of evolution tells us how a single organism can have evolved into the great diversity of life on Earth today. But where did the first organism come from? We may never know for sure, because it is unlikely that any fossil evidence could have recorded the transition from nonlife to life. However, over the past several decades, scientists have conducted many laboratory experiments designed to mimic the conditions that existed on the young Earth. These experiments suggest that life could have arisen through natural chemical reactions.

The first such experiments were performed in the 1950s (Figure 24.9), and they have been greatly refined and improved since that time. In essence, the experiments simply mix chemical ingredients thought to have been present on the early Earth and then "spark" the chemicals with electricity to simulate lightning or other energy sources. The chemical reactions that follow have produced all the major molecules of life, including all the amino acids and DNA bases. Many of these same molecules are found today in meteorites, suggesting that impacts may also have been an important source of organic molecules on the early Earth.

Laboratory experiments also show that mixing a warm dilute solution of organic molecules with naturally occurring sand or clay allows the molecules to assemble themselves into much more complex molecules. Strands of RNA (ribonucleic acid), an important molecule in all living organisms that looks much like a single strand of DNA, have been produced in the laboratory with lengths of up to nearly 100 bases. Because some RNA molecules are capable of self-replication, many biologists now presume that RNA was the original genetic material of life on Earth, with DNA coming later.

Other experiments have shown that microscopic enclosed membranes also form under conditions expected

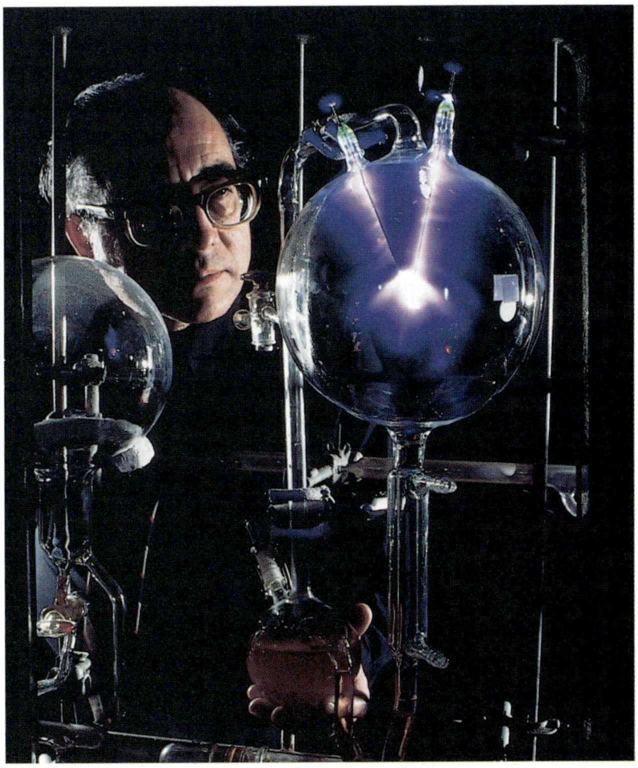

Figure 24.9 This photograph shows chemist Stanley Miller with a reproduction of the experimental setup he first used in the 1950s to study pathways to the origin of life. (He worked with Harold Urey, so the experiment is called the Miller-Urey experiment.) The flasks are filled with liquids or gases thought to have been present on the early Earth, and an electrical discharge supplies energy. The result is the production of many complex organic molecules. Scientists now think the early atmosphere was different from that used in Miller's original experiment, but similar experiments with the correct gases still yield similar results.

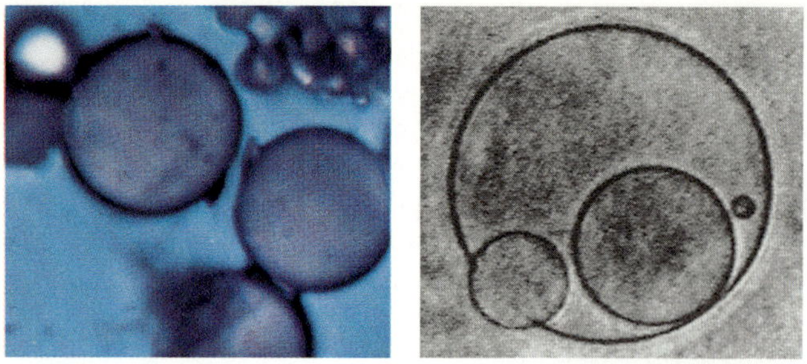

a These microscopic spheres were made by cooling a warm water solution of amino acids. They are not alive, but they exhibit many lifelike properties.

b These microscopic membranes are made from lipids that, when mixed with water, spontaneously form enclosed droplets.

Figure 24.10 These microscopic photos show enclosed membranes that self-assemble in laboratory experiments mimicking conditions on the early Earth.

on the early Earth (Figure 24.10). Self-replicating RNA molecules may have become enclosed in such membranes, making what are sometimes called "pre-cells." Those pre-cells in which RNA replicated faster and more accurately were more likely to spread, leading to a type of positive feedback that would have encouraged even faster and more accurate replication. Given millions of years of chemical reactions occurring all over the Earth, RNA might eventually have evolved into DNA, making true living organisms.

Figure 24.11 summarizes the steps by which chemistry might have become biology on Earth. We may never know whether life really arose in this way, but it certainly seems possible.

Could Life Have Migrated to Earth? Our scenario suggests that life could have arisen naturally here on Earth. However, an alternative possibility is that life arose somewhere else first—perhaps on Venus or Mars—and then migrated to Earth on meteorites. Remember that we have collected meteorites that were blasted by impacts from the surfaces of the Moon and Mars [Section 12.1]. Calculations suggest that Venus, Earth, and Mars all should have exchanged many tons of rock, especially in the early days of the solar system when impacts were more common.

The idea that life could travel through space to land on Earth, sometimes called *panspermia*, once seemed outlandish. After all, it's hard to imagine a more forbidding environment than that of space, with no air, no water, and constant bombardment by dangerous radiation from the Sun and stars. However, the presence of organic molecules in meteorites and comets tells us that the building blocks of life can indeed survive in space, and tests have shown that some microbes can survive in space for years.

In a sense, Earth, Venus, and Mars have been "sneezing" on each other for billions of years. Life could conceivably have originated on any of these three planets and been transported to the others. It's an intriguing thought, but it does not change our basic scenario for the origin of life—it simply moves it from one planet to another.

A Brief History of Life on Earth We are now ready to take a quick look at the history of life on Earth, as summarized in Figure 24.4. We will focus on events that may help us understand what we are looking for when we search for life on other worlds.

Chemical reactions probably began producing organic molecules not long after Earth's formation more than 4.5 billion years ago. Some mineral evidence suggests that Earth already had oceans by 4.3 to 4.4 billion years ago, and the early oceans would have been natural laboratories for chemistry that could have led to life. However, even if life arose at such early times, it was almost certainly extinguished by major impacts of the heavy bombardment. The common ancestor of life today arose after the impacts had subsided.

Whether through chemical reactions in the oceans or through migration from Venus or Mars, living organisms quickly took hold on Earth after the end of the heavy bombardment. Starting from the first organism, evolution rapidly diversified life on Earth.

Despite the rapid pace of evolution, nearly all life remained single-celled for at least a billion years. Land remained inhospitable even longer, because of the lack of a protective ozone layer. Continents much like those of today were surrounded by oceans teeming with life, but, despite pleasant temperatures and plentiful rainfall, the land itself was probably as barren as Mars is now. Life on land probably became possible only after atmospheric oxygen accumulated to the point where an ozone layer could form.

Nearly all the oxygen in our atmosphere was originally released through photosynthesis by single-celled organisms

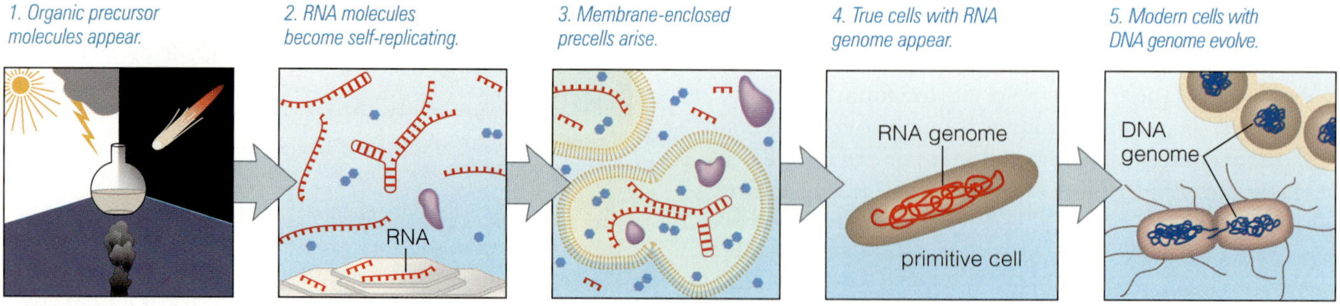

Figure 24.11 A summary of the steps by which chemistry on the early Earth may have led to the origin of life.

Figure 24.12 This microscopic photo shows a chain of modern cyanobacteria. The ancestors of these living organisms produced essentially all the oxygen in Earth's atmosphere.

known as cyanobacteria (Figure 24.12). These microbes may have been producing oxygen through photosynthesis as early as 3.5 billion years ago. However, the oxygen did not immediately begin to accumulate in the atmosphere. For more than a billion years, chemical reactions with surface rocks pulled oxygen back out of the atmosphere as fast as the cyanobacteria could produce it. But these tiny organisms were abundant and persistent, and eventually the surface rock was so saturated by oxygen that the rate of oxygen removal slowed down. At that point, some 2 billion years ago, oxygen began to accumulate in the atmosphere, though it may not have reached a level that we could have breathed until just a few hundred million years ago.

Today, we often think of oxygen as a necessity for life. However, oxygen was probably poisonous to most organisms living before about 2 billion years ago (and remains a poison to many microbes still living today). The rise of atmospheric oxygen therefore caused tremendous evolutionary pressure and may well have been a major factor in the eventual evolution of complex plants and animals.

There were undoubtedly many crucial changes as primitive microbes gradually evolved into multicellular organisms and early plants and animals, but the fossil record does not allow us to pinpoint the times at which all these changes occurred. However, we see a dramatic change in the fossil record dating to about 540 million years ago. Over the next 40 million years or so, animal life evolved from tiny and primitive organisms into all the basic body plans (phyla) that we find on Earth today. This remarkable flowering of animal diversity occurred in such a short time relative to the history of Earth that it is often called the *Cambrian explosion.* (*Cambrian* is the name geologists give to the period from about 540 million to 500 million years ago; see Figure 24.4.)

Early dinosaurs and mammals arose around the same time, some 225 to 250 million years ago, but dinosaurs at first proved more successful and dominated the landscape for well over 100 million years. Their sudden demise 65 million years ago, probably as a result of an asteroid or comet impact [Section 12.4], paved the way for the evolution of large mammals—including humans. The earliest humans appeared on the scene only a few million years ago, or after 99.9% of Earth's history to date had already gone by. Our few centuries of industry and technology have come after 99.99999% of Earth's history.

Despite our recent arrival (in geological terms), modern humans are arguably the most successful species ever to inhabit Earth. Humans survive and prosper in virtually every type of land environment. With proper equipment, we can survive underwater, and a few people have lived under the sea for extended periods of time in experimental habitats. We are even developing the technology to survive away from our planet in the inhospitable environment of space. Indeed, it is our very success that poses the greatest threat to our future survival, which now depends on our ability to limit the damage we do to the environment that makes our existence possible [Section 10.6].

• What are the necessities of life?

If we are going to look for other worlds that might harbor life, a first step is to understand the necessities of life. While it is always possible that life on other worlds could be quite different from life on Earth, it's easiest to begin the search for life by looking for conditions that would be comfortable for organisms from Earth. So what exactly does Earth life need to survive?

If we think about ourselves, the requirements for life seem fairly stringent: We need abundant oxygen in an atmosphere that is otherwise not poisonous, we need temperatures in a fairly narrow range of conditions, and we need abundant and varied food sources. However, the discovery of life in "extreme" environments—such as in the hot water near black smokers and hot springs—shows that many microbes (often called *extremophiles*) can survive under a much wider range of conditions.

The organisms living in hot water prove that at least some microbes can survive in much higher temperatures than we would have guessed. Other organisms live in other extremes. In the freezing cold but very dry valleys of Antarctica, the surface appears barren but microbes have been found living *inside* rocks, surviving on tiny droplets of liquid water and energy from sunlight. Other microscopic organisms have been found living up to several kilometers underground in water that fills pores within the subterranean rock. We have found life thriving in environments so acidic, alkaline, or salty that we humans would be poisoned almost instantly. We have even found microbes that can survive high doses of radiation, making it possible for them to live for many years in the radiation-filled environment of space.

If we compare the different forms of life on Earth, we find that life as a whole has only three basic requirements:

1. A source of nutrients (elements and molecules) from which to build living cells.
2. Energy to fuel the activities of life, either from sunlight, from chemical reactions, or from the heat of Earth itself.
3. Liquid water.

These requirements give us a basic road map for the search for life elsewhere. If we want to find life on another world, it makes sense to start by searching for worlds that offer these basic necessities.

Interestingly, only the third requirement (liquid water) seems to pose much of a constraint. Organic molecules are present almost everywhere—even on meteorites and comets. Many worlds are large enough to retain internal heat that could provide energy for life, and virtually all worlds have sunlight (or starlight) bathing their surfaces, although the inverse square law for light [Section 15.1] means that light provides less energy to worlds farther from their sun or star. Thus, nutrients and energy should be available to at least some degree on almost every planet and moon. In contrast, liquid water seems to be relatively rare, and the search for liquid water therefore drives the search for life in our universe.

24.2 Life in The Solar System

If we assume that life elsewhere would be at least a little bit like life on Earth, then the search for life begins with a search for **habitable worlds**—worlds that contain the basic necessities for life as we know it, including liquid water. The liquid water requirement rules out most of the worlds in our solar system. For example, Mercury and the Moon are barren and dry; Venus is too hot for liquid water; and Pluto, Planet X, other comets, and most of the small moons of the outer solar system are too cold. The jovian planets may have droplets of liquid water in some of their clouds, but the strong vertical winds on these planets make their clouds unlikely homes for life. That leaves two major possibilities besides Earth for habitability in our solar system: (1) Mars and (2) a few of the large moons orbiting jovian planets, most notably Europa [Section 11.2]. Let's discuss the current evidence about the potential habitability of these worlds.

SPECIAL TOPIC What Is Life?

You may have noticed that while we've been talking about life, we haven't actually defined the term. In fact, it's surprisingly difficult to draw a clear boundary between life and nonlife. Life can be so difficult to define that we may be tempted to fall back on the famous words of Supreme Court Justice Potter Stewart, who, in avoiding the difficulty of defining pornography, wrote: "I shall not today attempt further to define [it]. . . . But I know it when I see it." If living organisms on other worlds turn out to be much like those found on Earth, it may prove true that we'll know them when we see them. But if the organisms are fairly different from those on Earth, we'll need clearer guidelines to decide whether they are truly "living."

One way to seek distinguishing features of life is to study living organisms, looking for common characteristics. Given the difficulty of defining life, you probably won't be surprised to learn that there are exceptions to almost any "rule" we think of. Nevertheless, biologists have identified at least six key properties that most or all living organisms on Earth appear to share:

1. *Order:* Living organisms are not random collections of molecules but rather have molecules arranged in orderly patterns that form cell structures.
2. *Reproduction:* Organisms are capable of reproducing.
3. *Growth and development:* Living organisms grow and develop in patterns determined at least in part by heredity.
4. *Energy utilization:* Living organisms use energy to fuel their many activities.
5. *Response to the environment:* Life actively responds to changes in its surroundings. For example, simple organisms may alter their chemistry in the presence of a food source, while warm-blooded mammals may sweat, pant, or adjust blood flow to maintain a constant internal temperature.
6. *Evolutionary adaptation:* Life evolves through natural selection, as organisms pass on traits that make them better adapted to survival in their local environments.

These six properties are all important, but biologists today regard evolution as the most fundamental and unifying of them. Evolutionary adaptation is the only property that can explain the great diversity of life on Earth. Moreover, understanding how evolution works allows us to understand how all the other properties came to be. As a result, the simplest definition of life might be "something that can reproduce and evolve through natural selection."

For most practical purposes, this definition of life would probably suffice. However, some cases may still challenge this definition. For example, computer scientists can now write programs (that is, lines of computer code) that can reproduce themselves (that is, create additional sets of identical lines of code). By adding instructions that allow random changes to the programs, they can even make "artificial life" that evolves on a computer. Should this "artificial life," which consists of nothing but electronic signals processed by computer chips, be considered alive?

The fact that we have such difficulty distinguishing the living from the nonliving on Earth suggests that we should be very cautious about constraining our search for life elsewhere. No matter what definition of life we choose, there's always the possibility that we'll someday encounter something that challenges it. Nevertheless, the properties of reproduction and evolution seem very likely to be shared by most if not all life in the universe and therefore provide a useful starting point as we consider how to explore the possibility of extraterrestrial life.

THINK ABOUT IT

Is it possible for a world to be habitable but not actually have life? Is it possible for a world to have life but not be habitable? Explain.

• Could there be life on Mars?

We now have sufficiently detailed images of Mars to be quite confident that no civilizations have ever existed there. The "canals" seen by Percival Lowell were some type of mirage, presumably formed by a combination of real Martian features and his own imagination. Nevertheless, we have good reason to believe that liquid water flowed on Mars in the past [Section 9.4], making it at least seem possible that life could once have found a home there. Moreover, since Mars today contains subsurface ice, it's conceivable that life might still survive near sources of volcanic heat, where pockets of liquid water might persist underground.

Missions to Mars Mars is not only the best candidate in our solar system for life beyond Earth but is also the only place where we've begun an actual search for life. Our first attempt to search for life on Mars came with the *Viking* missions to Mars in the 1970s. Two *Viking* landers arrived on the surface of Mars in 1976. Each was equipped with a robotic arm for scooping up soil samples, which were fed into several onboard robotically controlled experiments.

Three of the *Viking* experiments were designed expressly to look for signs of life. None of these experiments could actually "see" life but rather looked for chemical changes that could be attributed to living organisms. Although all three experiments gave results that initially seemed consistent with life, further study suggested that chemical reactions could have produced the same results. Moreover, a fourth experiment, which analyzed the content of Martian soil, found no measurable level of organic molecules—the opposite of what we would expect if life were present. As a result, most scientists have concluded that no life was present at the locations where *Viking* sampled the soil.

More recent missions to Mars, including the Pathfinder lander in 1997 and the *Spirit* and *Opportunity* rovers that landed in 2004, have taken a more measured approach to the search for life (Figure 24.13). Rather than conducting direct experiments like those on *Viking*, these missions have been designed to help us better understand Martian conditions, so that we'll have a better idea of how and where to search for life with future missions.

Within a decade or so, NASA hopes to launch a mission to Mars that will bring back surface samples for study in laboratories on Earth. Later, we may send more advanced robots or humans to Mars, where they could search for fossils or living organisms in deep canyons like Valles Marineris, in ancient valley bottoms and dried-up lake beds, or in underground pockets of water near not-quite-dead volcanoes.

Figure 24.13 The *Opportunity* rover studies a rock on Mars during its mission in 2004. Such studies are helping scientists understand whether and when Mars may have been habitable.

The search will not be easy, but we should eventually learn whether life has ever existed on Mars.

Methane on Mars In 2004, scientists working with the European Space Agency's *Mars Express* mission, which entered Mars orbit in 2003, announced an unexpected and potentially important result: Instruments on the spacecraft had detected methane gas in the Martian atmosphere. This news came as a surprise to most scientists, because methane gas cannot last more than a few centuries in the Martian atmosphere before chemical reactions transform it into other gases.* Thus, if the methane detection is real, it implies that Mars has a source of ongoing methane production somewhere on or beneath its surface.

Some scientists still doubt that the data are being properly interpreted, but a few ground-based observations made to check the result seem to confirm a methane signature in spectra of Mars. Where could the methane be coming from? We know of at least three possibilities: comet impacts, volcanic activity, or life. The first possibility is highly unlikely, since impacts are such rare events and an impact would have to have occurred quite recently for methane to remain in the atmosphere. That leaves us with volcanism or life.

A few scientists have looked at correlations between the location of methane and water vapor in the Martian atmosphere and used the results to argue that we might actually be seeing the signature of microscopic life on Mars. The vast majority of scientists consider any such conclusion to be premature at best, arguing that volcanism is a more likely source for the methane. After all, as we discussed in Chapter 9, we suspect that Mars has at least some low level of ongoing volcanic activity. But even this result could have

*The methane is oxidized to form water and carbon dioxide; the oxidation occurs because there is some free oxygen in the Martian atmosphere, made by breakup of the abundant atmospheric carbon dioxide.

implications for the existence of life: If there is enough volcanic activity to produce a detectable level of methane, it may also mean there is enough volcanic heat to make pockets of liquid water in some places beneath the Martian surface, raising the possibility that life could exist in some of this liquid water.

The Debate over Martian Meteorites An entirely different approach to searching for life on Mars relies on studies of the couple of dozen known *Martian meteorites*—meteorites whose chemical and isotopic composition suggests they came from Mars [Section 12.1]. One of these meteorites, designated ALH84001, generated particular excitement when a group of scientists claimed that it contained evidence of past life on Mars (Figure 24.14).

ALH84001 was found on the Antarctic ice in 1984. Careful study of the meteorite shows that it landed in Antarctica about 13,000 years ago, following a 16-million-year journey through space after an impact blasted it from Mars. The rock itself dates to 4.5 billion years ago, indicating that it solidified shortly after Mars formed and therefore resided on Mars throughout the times when Mars may have been warmer and wetter.

Painstaking analysis of the meteorite reveals several lines of evidence that could indicate the past presence of life on Mars. For example, the rock contains layered carbonate minerals and complex organic molecules (polycyclic aromatic hydrocarbons, or PAHs) that are associated with life when found in Earth rock, as well as microscopic chains of magnetite crystals quite similar to chains made in Earth rocks by living bacteria. Most intriguingly, highly magnified images reveal rod-shaped structures that look much like recently discovered *nanobacteria* on Earth. The terrestrial nanobacteria are about $\frac{1}{100}$ the size of ordinary bacteria, and some biologists question whether they truly are living organisms. However, they appear to contain DNA, which suggests that they are indeed a form of life. If so, could the similar-looking structures in the Martian meteorite be fossil life from Mars?

It's possible, but each of the tantalizing hints of Martian life can also be explained in a nonbiological way. Subsequent studies have shown that chemical and geological processes can produce structures very similar to those found in the Martian meteorite. In addition, terrestrial bacteria have been found living inside the meteorite, indicating that it was contaminated by Earth life during the 13,000 years it resided in Antarctica. This contamination may explain the presence of the complex molecules found in the rock.

On balance, most scientists now doubt that the Martian meteorite shows true evidence of Martian life. Nevertheless, studies of ALH84001 and other Martian meteorites are continuing, and they may yet turn up surprises.

• Could there be life on Europa or other jovian moons?

After Mars, the next most likely candidates for life in our solar system are some of the moons of the jovian planets—especially Jupiter's moons Europa, Ganymede, and Callisto and Saturn's moon Titan [Section 11.2]. These moons are all large enough that they would be considered planets if they orbited the Sun independently.

The strongest of these candidates for harboring life is Jupiter's moon Europa, on which tidal heating probably creates a deep ocean beneath an icy crust (see Figure 11.21). The ice and rock from which Europa formed undoubtedly included the necessary chemical ingredients for life, and Europa's internal heating (primarily due to tidal heating) is strong enough to power volcanic vents on the sea bottom. Thus, it's easy to imagine places on Europa's ocean floor that look much like black smokers on Earth (see Figure 24.8a). If life on Earth first really arose near such undersea volcanic vents, Europa would seem to have everything needed for an origin of life.

The possibility of life on Europa is especially interesting because, unlike any potential life on Mars, it would not necessarily have to be microscopic. After all, the several kilometers of surface ice that hide Europa's ocean (if it exists) could also hide large creatures swimming within it. However, the potential energy sources for life on Europa are far more limited than the energy sources for life on Earth, mainly because sunlight could not fuel photosynthesis in the subsurface ocean. As a result, most scientists suspect

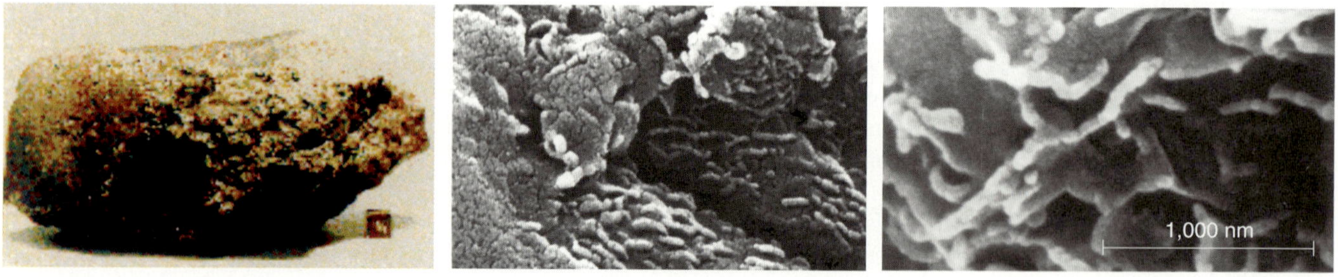

a The Martian meteorite ALH84001, before it was cut open for detailed study. The block on the right provides a sense of scale; it measures 1 cubic centimeter, about the size of a typical sugar cube.

b This photo shows rod-shaped structures found in a highly magnified slice of ALH84001. They measure about 100 nanometers in length and are as small as 10–20 nanometers in width.

c This photo shows terrestrial nanobacteria in a sample of volcanic rock from Sicily. They are close in size to the structures seen in ALH84001.

Figure 24.14 Controversial meteorite evidence for life on Mars.

that any life that might exist on Europa would probably be quite small and primitive.

As we discussed in Chapter 11, some evidence suggests that Jupiter's moons Ganymede and Callisto may also have subsurface oceans. However, these moons have much less internal heat and therefore would have even less energy for life than Europa. If they have life at all, it is almost certainly small and primitive. Nevertheless, Europa, Ganymede, and Callisto offer the astonishing possibility that Jupiter alone could be orbited by more worlds with life than we find in all the rest of the solar system.

Titan offers another enticing place to look for life. Its surface is far too cold for liquid water, but it may have lakes and rivers of liquid methane [Section 11.2]. Although many biologists think it unlikely, it is possible that these liquids could support life much as does water on Earth. Titan also appears to have many other organic molecules on its surface, and, at the very least, Titan offers an incredible natural laboratory of interesting organic chemistry.

Moreover, we can't completely rule out water-based life on Titan, even if it seems improbable. While liquid water does not exist on Titan's surface today, pockets of liquid water may have existed in the past following impacts of asteroids or comets. The heat of a large impact could have melted water ice and kept it liquid for as long as a few thousand years. If water-based life arose in such temporary pockets of water, it might have found a way to survive as the water froze, perhaps by migrating deep underground where Titan's internal heat might still keep some water liquid.

Mastering ASTRONOMY Detecting Extrasolar Planets Tutorial, Lessons 1–3

24.3 Life Around Other Stars

We already know of planets orbiting more than 150 stars besides our Sun, and it's quite likely that billions of planetary systems inhabit our galaxy. These numbers might immediately make prospects for life elsewhere seem quite good, but numbers alone don't tell the whole story. In this section, we'll consider the prospects for life on worlds orbiting other stars.

Before we begin, we must distinguish between *surface* life like that on Earth and *subsurface* life like that we envision as a possibility on Mars or Europa. While large telescopes could in principle allow us to discover surface life on distant planets, no foreseeable technology will allow us to find life that is hidden deep underground in other star systems (unless the subsurface life has a noticeable effect on the planet's atmosphere). We therefore will focus on the search for life on planets with habitable surfaces—surfaces with temperatures and pressures that could allow liquid water to exist.

• Are habitable planets likely?

We have not yet discovered any extrasolar planets that seem likely to be habitable. Presumably because of the limitations of current detection techniques [Section 13.1], nearly all the extrasolar planets found to date are much more massive than Earth, suggesting that they are more like the jovian planets of our solar system than the terrestrial planets. Such planets are unlikely candidates for life, though some of them have orbits that could conceivably allow them to have moons with life.

However, the existence of other jovian planets makes it seem quite reasonable to suppose that terrestrial planets are also common. Our technology is just not yet up to the task of finding them. Thus, if we hope to find other habitable worlds, the first task is to decide whether we should expect them to exist, and if so, to determine where to look.

Constraints on Star Systems Before we consider planets themselves, it's useful to ask how many stars have any chance of having planets with life. In other words, which stars would make good "Suns," providing sufficient heat and light to the surfaces of terrestrial planets that happen to orbit them?

The first requirement for a star to have life-bearing worlds is that it be old enough that life could have arisen. More massive stars live shorter lives, and the most massive stars live no more than a few million years [Section 17.3]. Given that life on Earth did not arise for hundreds of millions of years after our solar system was born, we can rule out any star with more than a few times the mass of our Sun. However, because less massive stars are far more common than more massive stars, the lifetime constraint rules out only about 1% of all stars.

A second requirement is that the star allow planets to have stable orbits. Surveys suggest that roughly half of all stars are in binary or multiple star systems [Section 15.1], in which stable planetary orbits are less likely than around single stars. If life is not possible in such systems, then we can rule out life around half the stars in our galaxy. Of course, the other half—still some 100 billion stars or more—remain possible homes for life. Moreover, under some circumstances, stable planetary orbits are possible in multiple star systems, so we shouldn't entirely rule out life in such systems.

A third constraint on the likelihood of finding habitable planets is the size of a star's **habitable zone**—the region in which a terrestrial planet of the right size could have a surface temperature that might allow for liquid water and life. Figure 24.15 shows the approximate sizes, to scale, of the habitable zones around our Sun, around a star with about half the mass of the Sun (spectral type K), and around a star with about $\frac{1}{10}$ the mass of our Sun (spectral type M). Although habitable planets seem possible in all three cases, the smaller size of the habitable zones around the less massive stars makes it less likely that suitable planets would have formed in these regions.

All in all, it seems that the vast majority of stars are at least potentially capable of having life-bearing planets. Moreover, even very conservative assumptions suggest enormous numbers of possibilities. For example, limiting the search for habitable planets to stars very similar to our Sun (that is, spectral type G) would still mean billions of potential other Suns in the Milky Way Galaxy.

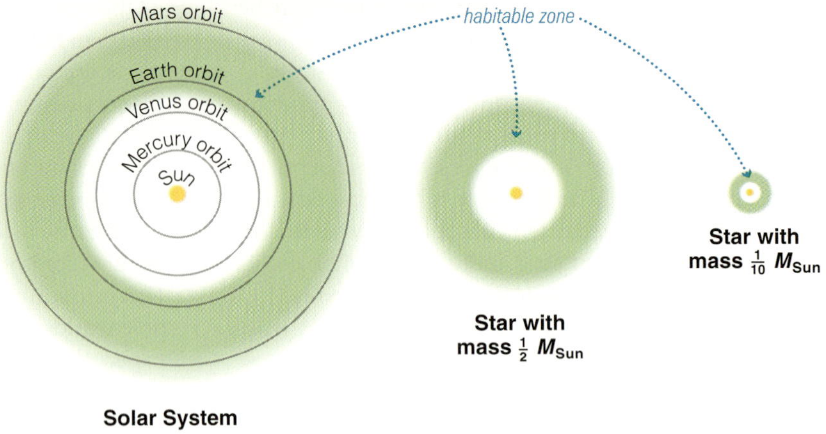

Figure 24.15 Interactive Figure The approximate habitable zones around our Sun, a star with half the mass of the Sun (spectral type K), and a star with $\frac{1}{10}$ the mass of the Sun (spectral type M), shown to scale. The habitable zone becomes increasingly smaller and closer-in for stars of lower mass and luminosity.

Finding Habitable Planets Finding Earth-size planets is a daunting technological challenge [Section 13.1]. Recall that looking for an Earth-like planet around a nearby star is like standing on the East Coast of the United States and looking for a pinhead on the West Coast. Nevertheless, advancing technology should soon put Earth-like planets within our telescopic reach. The Kepler mission, scheduled for launch in 2008, will look for transits of Earth-size planets in front of their stars (see Figure 13.15). Scientists hope that Kepler will detect hundreds of Earth-size planets, and orbital properties measured by Kepler will tell us whether these planets lie within their stars' habitable zones. The Space Interferometry Mission (SIM), which NASA hopes to launch by 2011, may also be capable of detecting Earth-size planets around some nearby stars.

We will need images or spectra to determine whether the planets really are habitable or have life. Scientists are actively working on technologies that may provide such data. If all goes well, within about a decade NASA hopes to follow SIM with the *Terrestrial Planet Finder (TPF)*, an orbiting interferometer that will be capable of obtaining low-resolution spectra and crude images (a few pixels) of Earth-like planets around nearby stars [Section 13.4].

Astronomers hope to deploy even more powerful interferometers in later decades, either in space or on the Moon. Within the lifetimes of today's college students, we could conceivably have optical interferometers with dozens of telescopes spread across hundreds of kilometers. Such telescopes will be able not only to detect Earth-like planets around nearby stars but also to obtain fairly clear images and high-resolution spectra of those planets.

Signatures of Life Images from future telescopes may tell us whether the planets have continents and oceans like Earth and perhaps even allow us to monitor seasonal changes. Spectra from future telescopes should prove even more important to the search for life. Moderate-resolution infrared spectra can reveal the presence and abundance of many atmospheric gases, including carbon dioxide, ozone, methane, and water vapor (Figure 24.16). Careful analysis of atmospheric makeup might tell us whether a planet has life.

On Earth, for example, the large abundance of oxygen (21% of our atmosphere) is a direct result of photosynthetic life. Abundant oxygen in the atmosphere of a distant world might similarly indicate the presence of life, since we know of no nonbiological way to produce an oxygen abundance as high as Earth's. Other evidence might come from the ratio of oxygen to the other detected gases. Scientists engaged in the search for life in other planetary systems are working to improve our understanding of how life influences atmospheric chemistry in hopes that we will be able to recognize particular gas combinations as unmistakable signatures of life.

• Are Earth-like planets rare or common?

We will not know for certain whether habitable planets exist or how common such planets may be until we survey many star systems with telescopes capable of detecting such

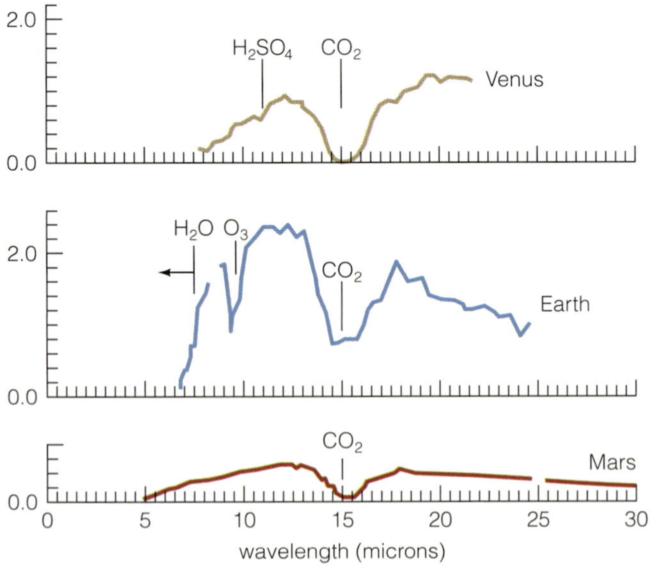

Figure 24.16 The infrared spectra of Venus, Earth, and Mars, as they might be seen from afar, showing absorption features that point to the presence of carbon dioxide (CO_2), ozone (O_3), and sulfuric acid (H_2SO_4) in their atmospheres. While carbon dioxide is present in all three spectra, only our own planet has appreciable oxygen (and hence ozone)—a product of photosynthesis. If we could make similar spectral analyses of distant planets, we might detect atmospheric gases that would indicate life.

small planets. Nevertheless, the existence of more than one Earth-size planet in our own solar system and our understanding of planetary formation make it seem reasonable to imagine that many Earth-size planets exist within the habitable zones of other stars. But should we expect these planets to have Earth-like conditions in which life could arise and evolve?

Most scientists think so, but a few have raised some very interesting questions. In essence, these scientists suggest that Earth's hospitality is the result of several rare kinds of planetary luck. According to this idea, sometimes called the "rare Earth hypothesis," the specific circumstances that have allowed life on Earth to survive and evolve into complex forms (such as oak trees and people) might be so rare that ours could be the only planet in the galaxy that harbors anything but the simplest life. Let's briefly examine some of the key issues in the rare Earth hypothesis.

Galactic Constraints Proponents of the rare Earth hypothesis suggest that Earth-like planets can form in only a relatively small region of the Milky Way Galaxy, making the number of potential homes for life far smaller than we might otherwise expect it to be. In essence, they argue that there is a fairly narrow ring at about our solar system's distance from the center of the Milky Way Galaxy that makes up a *galactic habitable zone* analogous to the habitable zone around an individual star (Figure 24.17).

According to the arguments for a galactic habitable zone, outer regions of our galaxy are unlikely to have terrestrial planets because of a low abundance of elements besides hydrogen and helium. Recall that the fraction of heavy elements varies among different stars, from less than 0.1% among the old stars in globular clusters to more than 2% among young stars in the galactic disk [Section 19.3]. Even within the galactic disk, the abundance tends to decline with distance from the center of the galaxy. Because terrestrial planets are made almost entirely of heavy elements, a lower abundance of these elements might lessen the chance that terrestrial planets could form. The inner regions of the galaxy are ruled out primarily through an argument concerning supernova rates. Supernovae are more common in the more crowded, inner regions of the galactic disk, making it more likely that a terrestrial planet would be exposed to the intense radiation from a nearby supernova. By assuming that this radiation would be detrimental to life, the proponents of a galactic habitable zone argue against finding habitable planets in the inner regions of the galaxy. Together, the constraints on finding Earth-like planets in the inner and outer regions of the galaxy leave the galactic habitable zone as a relatively narrow ring encompassing no more than about 10% of the stars in the galactic disk.

However, other scientists offer counterarguments to both sets of galactic constraints. For the heavy-element abundance, they note that Earth's mass is less than $\frac{1}{100,000}$ of the mass of the Sun. Thus, even a very small heavy-element abundance could be enough to make one or more Earth-like planets. Unless there is something about the planetary accretion process that prevents terrestrial planets from forming in systems with low heavy-element abundances, then we might find terrestrial planets around almost any star. Regarding the radiation danger from supernovae, we do not really know whether such radiation would be detrimental to life. A planet's atmosphere might protect life against the effects of the radiation. It is even possible that the radiation could be beneficial to life by increasing the rate of mutations and thereby accelerating the pace of evolution. If these counterarguments are correct, then Earth-like planets might be found throughout much or all of the galaxy.

Figure 24.17 The green ring in this diagram of the Milky Way Galaxy represents what some scientists believe is a galactic habitable zone—the only region of the galaxy in which Earth-like planets are likely to be found. However, other scientists doubt the claims that underlie the galactic habitable zone, in which case Earth-like planets could be far more widespread.

Impact Rates and Jupiter Another issue raised by rare Earth proponents is the impact rate on planets in other star systems. We have seen that Earth was probably subjected to numerous large impacts—some large enough to vaporize the oceans and sterilize the planet—during the heavy bombardment that went on during the first half-billion years or so after our planet's birth. In our solar system, the impact rate lessened dramatically after that. Might the impact rate remain high much longer in other planetary systems?

The most numerous small objects in our solar system are the trillion or so comets of the distant Oort cloud [Section 12.2]. Fortunately for us, nearly all these myriad objects are essentially out of reach, posing no threat to our planet. However, the reason they are out of reach can be traced directly to Jupiter: Recall that the Oort cloud's comets are thought to have formed among the jovian planets, later being "kicked out" to their current orbits by close encounters with these planets, especially Jupiter (see Figure 12.24). Thus, if Jupiter did not exist, many of the comets might have remained in regions of the solar system where they could pose a danger to Earth. In that case, the heavy bombardment might never have ended, and huge impacts would continue to this day. From this viewpoint, our existence on Earth has been possible only because of the "luck" of having Jupiter as a planetary neighbor.

The primary question in this case is just how "lucky" this situation might be. Our discoveries of extrasolar planets so far suggest that Jupiter-size planets are in fact quite common. However, we've also found that many large planets migrate inward [Section 13.3], perhaps disrupting terrestrial planet orbits along the way. Whether having a Jupiter in the right place for ejecting comets is lucky or typical remains an open question.

Climate Stability Another issue affecting the rarity of Earth-like planets is climate stability. Earth's climate has been stable enough for liquid water to exist throughout the past 4 billion years. This climate stability has almost certainly played a major role in allowing complex life to evolve on our planet. If our planet had frozen over like Mars or overheated like Venus, we would not be here today. Advocates of the rare Earth hypothesis point to at least two pieces of "luck" related to Earth's stable climate.

The first piece of "luck" is the existence of plate tectonics. Recall that plate tectonics plays a major role in regulating Earth's climate through the carbon dioxide cycle [Section 10.6]. Plate tectonics probably was not necessary to the origin of life, but it seems to have been very important in keeping the climate stable enough for the subsequent evolution of plants and animals. But are we really "lucky" to have plate tectonics, or should this geological process be common on similar-size planets elsewhere? We do not yet know. The apparent lack of plate tectonics on Venus [Section 9.5], which is quite similar in size to Earth, might seem to argue for plate tectonics being rare. On the other hand, some scientists suspect that the lack of plate tectonics on Venus can be traced to its runaway greenhouse effect [Section 10.5], which may have baked out subsurface water and caused Venus's lithosphere to thicken too much to allow plate movements. Venus suffered its runaway greenhouse effect because it is not quite far enough from the Sun to be within the Sun's habitable zone. In that case, it's possible that any Earth-size planet within a star's habitable zone would have plate tectonics.

The second piece of "luck" in climate stability is the existence of Earth's relatively large Moon. Models of Earth's rotation and orbit show that if the Moon did not exist, gravitational tugs from other planets would cause large swings in Earth's axis tilt over periods of tens to hundreds of thousands of years. (Such swings in axis tilt are thought to occur on Mars.) Changes in axis tilt would affect the severity of the seasons, which in turn could cause deeper ice ages and more intense periods of warmth. Given that the Moon probably formed as a result of a random giant impact [Section 8.4], we might seem to be very lucky to have the Moon and the climate stability it brings.

Again, however, there are other ways to look at the issue. Changes in axis tilt might warm or cool different parts of the planet dramatically, but the changes would probably occur slowly enough for life to adapt or migrate as the climate changed. In addition, our Moon's presumed formation in a random giant impact does not necessarily mean that large moons will be rare. At least a few giant impacts should be expected in any planetary system. Indeed, Earth may not be the only object in our own solar system that ended up with a large moon because of a giant impact. Pluto's moon (Charon) may have formed in the same way [Section 12.3]. Thus, while luck was certainly involved in Earth's having a large moon, it might not be a very rare kind of luck.

The Bottom Line The bottom line is that while the rare Earth hypothesis offers some intriguing arguments, it is too early to say whether any of them will hold up over time. For each potential argument that Earth has been lucky, we've seen counterarguments suggesting otherwise. There's no doubt that our solar system and our world have "personality"—they exhibit properties that might be found only occasionally in other star systems—but we have no clear reason to think that these special properties are essential to the existence of complex or even intelligent life. Indeed, it may be that we have missed out on some helpful phenomena that could have sped evolution on Earth. We might be less lucky than we recognize, and creatures on other worlds might regard the nature of our planet with disappointment. Until we learn much more about other planets in the universe, we cannot know whether Earth-like planets and complex life are common or rare.

24.4 The Search for Extraterrestrial Intelligence

So far, we have focused on search strategies for microbial or other nonintelligent life. However, if intelligent beings and civilizations exist elsewhere, we might be able to find them with a completely different type of search strategy. Instead of searching for hard-to-find spectroscopic signs of life, we might simply listen for signals that intelligent beings are sending into interstellar space, either in deliberate attempts to contact other civilizations or as a means of communicating among themselves. The search for signals from other civilizations is generally known as the **search for extraterrestrial intelligence**, or **SETI** for short.

• How many civilizations are out there?

SETI efforts have a chance to succeed only if other advanced civilizations are broadcasting signals that we could receive. Thus, in order to judge the chances of SETI success, we'd like to know how many civilizations are broadcasting such signals right now.

Given that we do not even know whether microbial life exists anywhere beyond Earth, it should be clear that we cannot yet know whether other civilizations exist, let alone how many there might be. Nevertheless, for the purposes of planning a search for extraterrestrial intelligence, it is useful to have an organized way of thinking about the number of civilizations that might be out there. To keep our discussion simple, let's consider only the number of potential civilizations in our own galaxy. We can always extend our estimate to the rest of the universe by simply multiplying the result we find for our galaxy by 100 billion, the approximate number of galaxies in the observable universe.

The Drake Equation The first scientific conference on the search for extraterrestrial intelligence was held in 1961 in Green Bank, West Virginia, where a pioneering search for an alien signal had recently been conducted. During the meeting, astronomer Frank Drake wrote a simple equation designed to summarize the factors that would determine the number of civilizations we might contact (Figure 24.18).

This equation is now known as the **Drake equation**, and in principle it gives us a simple way to calculate the number of civilizations capable of interstellar communication that are currently sharing the Milky Way Galaxy with us. In a form slightly modified from the original, the Drake equation looks like this:

$$\text{Number of civilizations} = N_{HP} \times f_{life} \times f_{civ} \times f_{now}$$

This equation will make sense once you understand the meaning of each factor:

- N_{HP} is the number of habitable planets in the galaxy; that is, the number of planets that are at least capable of having life.
- f_{life} is the fraction of habitable planets that actually *have* life. For example, if $f_{life} = 1$ it would mean that all habitable planets have life, and if $f_{life} = \frac{1}{1,000,000}$ it would mean that only 1 in a million habitable planets have life. Thus, the product $N_{HP} \times f_{life}$ tells us the number of life-bearing planets in the galaxy.
- f_{civ} is the fraction of the life-bearing planets upon which a civilization capable of interstellar communication *has at some time* arisen. For example, if $f_{civ} = \frac{1}{1,000}$ it would mean that such a civilization has existed on 1 out of 1,000 planets with life, while the other 999 out of 1,000 have not had a species that learns to build radio transmitters, high-powered lasers, or other devices for interstellar conversation. When we multiply this factor by the first two factors to form the product $N_{HP} \times f_{life} \times f_{civ}$, we get the total number of planets upon which intelligent beings have evolved and developed a communicating civilization at some time in the galaxy's history.
- f_{now} is the fraction of these civilization-bearing planets that happen to have a civilization *now,* as opposed to, say, millions or billions of years in the past. This factor is important because we can hope to contact only civilizations that are broadcasting signals we could receive at present.*

*For the purposes of the Drake equation, we'll assume that the term f_{now} takes into account the light-travel time for signals from other stars; for example, if a star with a civilization is 10,000 light-years away, it counts in determining f_{now} if the civilization existed 10,000 years ago, because signals broadcast at that time would just now be arriving at Earth.

Figure 24.18 Astronomer Frank Drake, with the equation he first wrote in 1961. (With Dr. Drake's approval, we use a slightly modified form of his equation in this book.)

Because the product of the first three factors tells us the total number of civilizations that have *ever* arisen in the galaxy, multiplying by f_{now} tells us how many civilizations we could potentially make contact with today. Thus, the result of the Drake equation is the number of civilizations that we might hope to contact.

Unfortunately, we don't yet know the value of any of the factors in the Drake equation, so we cannot actually calculate the number of civilizations in our galaxy. Nevertheless, the equation is a useful way of organizing our thinking, as we can see by thinking about the potential values for each of its factors.

THINK ABOUT IT

Try the following sample numbers in the Drake equation. Suppose there are 1,000 habitable planets in our galaxy, that 1 in 10 habitable planets has life, that 1 in 4 planets with life has at some point had an intelligent civilization, and that 1 in 5 civilizations that have ever existed is in existence now. How many civilizations would exist at present? Explain.

The Number of Life-Bearing Planets Let's begin with the first two factors in the Drake equation, whose product $(N_{HP} \times f_{life})$ tells us the number of life-bearing planets in our galaxy. We can make a reasonably educated guess only about the first factor, the number of habitable planets (N_{HP}). Current understanding of solar system formation (see Figure 8.13) suggests that terrestrial planets ought to form fairly easily and, as we discussed earlier, there ought to be many billions of stars in our galaxy with habitable zones large enough to have Earth-like planets. Thus, unless some of the "rare Earth" ideas prove to be correct, it seems entirely reasonable to suppose that our galaxy has billions of habitable planets.

The factor f_{life} presents more difficulty. For the moment, we have no reliable way to estimate the fraction of habitable planets upon which life actually arose. The problem is that we cannot generalize when we have only one example to study—our own Earth. Still, the fact that life arose rapidly on Earth suggests that the origin of life was fairly "easy." In that case, we might expect most or all habitable planets to also have life, making the fraction f_{life} close to 1. Of course, until we have solid evidence that life arose anywhere else, such as on Mars, it is also possible that Earth might really have been very lucky and that f_{life} might be so close to zero that life has never arisen on any other planet in our galaxy.

The Question of Intelligence Even if life-bearing planets are very common, civilizations capable of interstellar communication might not be. The fraction of life-bearing planets that at some time have such civilizations, f_{civ}, depends on at least two things: First, a planet would have to have a species evolve with sufficient intelligence to develop interstellar communication. In other words, the planet needs a species at least as smart as we are. Second, that species would have to develop a civilization with technology at least as advanced as ours.

Although we cannot really be sure, most scientists suspect that only the first requirement is difficult to meet. A fundamental assumption in nearly all science today is that we are not "special" in any particular way. We live on a fairly typical planet orbiting an ordinary star in a normal galaxy, and we assume that living creatures elsewhere—whether they prove to be rare or common—would be subjected to evolutionary pressures quite similar to those that have operated on Earth. Thus, if species with intelligence similar to ours have arisen elsewhere, we assume that they would have similar sociological drives that would eventually lead them to develop the technology necessary for interstellar communication.

If this assumption is correct, then the fraction f_{civ} depends primarily on the question of whether sufficient intelligence is rare or common among life-bearing planets. As with the question of life of any kind, the short answer to this question is we just don't know, but we can get at least some insight by considering what happened on Earth.

Look again at Figure 24.4. While life arose quite quickly on Earth, nearly all life remained microbial until just a few hundred million years ago, and it took nearly 4 billion years for us to arrive on the scene. This slow progress toward intelligence might suggest that producing a civilization is very difficult even when life is present. On the other hand, roughly half the stars in the Milky Way are older than our Sun, so if Earth's case is typical, then plenty of planets have existed long enough for intelligence to arise. Thus, the value of f_{civ} depends on whether the 4 billion years it took for humans to evolve on Earth is typical, fast, or slow. If it was unusually fast and the average time needed to develop complex species is much longer than 4 billion years, then most life-bearing planets may be covered with nothing more advanced than bacteria. If the time we took to evolve on Earth is typical or slow, then intelligent species might be quite common.

A second way to address the question of intelligence is by considering our level of intelligence in comparison to that of other animals on Earth. It's difficult to get animals to take IQ tests, but we can get a rough measure of intelligence by comparing brain mass to total body mass (a measure sometimes called the *encephalization quotient,* or EQ). Figure 24.19 shows the brain weights for a sampling of birds and mammals (including primates) plotted against their body weights. There is a clear and expected trend in that heavier animals have heavier brains. By drawing a straight line that fits these data, we can define an average value of brain mass for each body mass. Animals whose brain mass falls above the line are smarter than average, while animals whose brain mass falls below the line are less mentally agile. Keep in mind that it is the *vertical* distance above the line that tells us how much smarter a species is than the average, and that the scale goes in powers of 10 on both axes. If you look closely, you'll see that the data point for humans lies significantly farther above the line than the data point for any other species. Thus, by this measure of intelligence, we are far smarter than any other species that has ever existed on Earth.

Some people use this fact to argue that even on a planet with complex life, a species as intelligent as we are would be very rare. They say that even if there is an evolutionary drive toward intelligence in general, it takes extreme luck to reach our level of intelligence. After all, while it's evolutionarily useful to have enough intelligence to capture prey and evade other predators, it's not clear why natural selection would lead to brains big enough to build spacecraft. However, the same data can be used to reach an opposite conclusion. The scatter in the levels of intelligence among different animals tells us that some variation should be expected, and statistical analysis shows that we are not unreasonably far above the average. It might therefore be inevitable that some species would develop our level of intelligence on any planet with complex life.

Technological Lifetimes For the sake of argument, let's assume that life and intelligence are at least reasonably likely, so

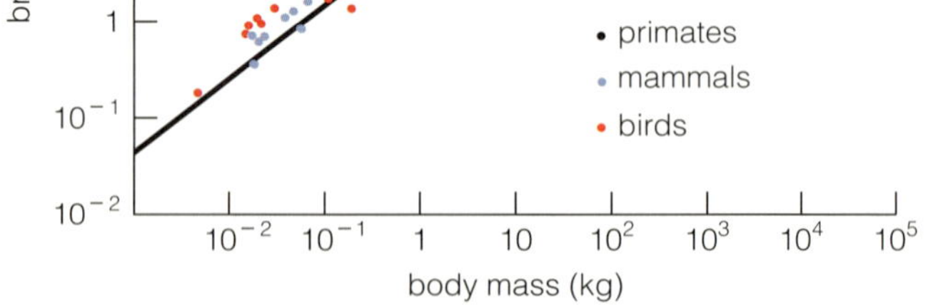

Figure 24.19 This graph shows how brain mass compares to body mass for some mammals (including primates) and birds. The straight line represents an average of the ratio of brain mass to body mass, so that animals that fall above the line are smarter than average and animals that fall below the line are less smart. Note that the scale uses powers of 10 on both axes. (Adapted from Sagan, 1977.)

that thousands or millions of planets in our galaxy have at some time given birth to a civilization. In that case, the final factor in the Drake equation, f_{now}, determines the likelihood of there being someone whom we could contact now. The value of this factor depends on the survivability of civilizations.

Consider our own example. In the roughly 12 billion years during which our galaxy has existed, we have been capable of interstellar communication via radio for only about 60 years. Thus, if we were to destroy ourselves tomorrow (saving you the unpleasantness of a final exam), then other civilizations could have received signals from us during only 60 years out of the galaxy's 12-billion-year existence, equivalent to 1 part in 200 million of the galaxy's history. If such a short technological lifetime is typical of civilizations, then f_{now} would be only $\frac{1}{200,000,000}$, and some 200 million civilization-bearing planets would need to have existed at one time or another in the Milky Way in order for us to have a good chance of finding another civilization out there now.

However, we'd expect f_{now} to be so small only if we are on the brink of self-destruction—after all, the fraction will grow larger for as long as our civilization survives. Thus, if civilizations are at all common, survivability is the key factor in whether any are out there now. If most civilizations self-destruct shortly after achieving the technology for interstellar communication, then we are almost certainly alone in the galaxy at present. But if most survive and thrive for thousands or millions of years, the Milky Way may be brimming with civilizations—most of them far more advanced than our own.

THINK ABOUT IT

Describe a few reasons why a civilization capable of interstellar communication would also be capable of self-destruction. Overall, do you believe our civilization can survive for thousands or millions of years? Defend your opinion.

• How does SETI work?

If there are indeed other civilizations out there, then in principle we ought to be able to make contact with them. Based on our current understanding of physics, it seems likely that even very advanced civilizations would communicate much as we do—by encoding signals in radio waves or other forms of light. Most SETI researchers use large radio telescopes to search for alien radio signals (Figure 24.20). A few researchers are beginning to check other parts of the electromagnetic spectrum. For example, some scientists use visible-light telescopes to search for communications encoded as laser pulses. Of course, advanced civilizations may well have invented communication technologies that we cannot even imagine. In that case, SETI efforts will not detect them.

A good way to think about our chances of picking up an alien signal is to imagine what aliens would need to do to pick up signals from us. We have been sending relatively high-power transmissions into space since about the 1950s in the form of television broadcasts. Thus, in principle, anyone within about 50 light-years of Earth could watch our old television shows (perhaps a frightening thought). However, in order to detect our broadcasts, they would need far larger and more sensitive radio telescopes than we have today. If their technology were at the same level as ours, they could receive a signal from us only if we deliberately broadcast an unusually high-powered transmission.

To date, humans have made only a few attempts to broadcast our existence in this way. The most powerful of these occasional transmissions was made in 1974 and lasted only 3 minutes. The powerful planetary radar transmitter on the Arecibo radio telescope (see Figure 6.24) was used to send a small pictorial message to the globular cluster M 13 (Figure 24.21). This target was chosen in part because it contains a few hundred thousand stars, seemingly offering a good chance that at least one has a civilization around it.

Figure 24.20 The 64-meter Parkes radio telescope in New South Wales, Australia. A SETI experiment "piggybacks" on this telescope while it is engaged in other astronomical research.

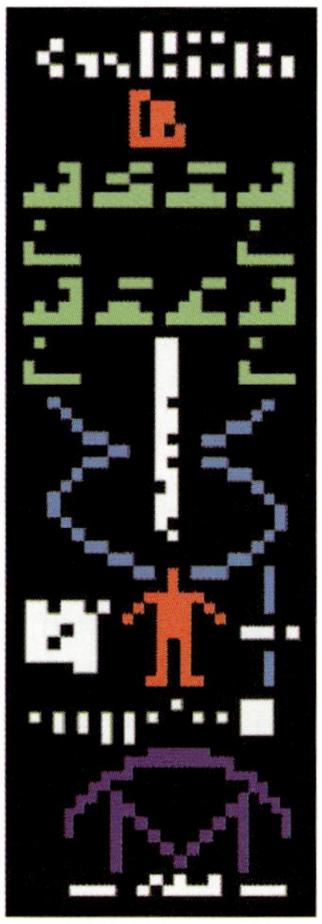

Figure 24.21 In 1974, a short message was broadcast to the globular cluster M 13 using the Arecibo radio telescope. The picture shown here was encoded by using two different radio frequencies, one for "on" and one for "off" (the colors shown here are arbitrary). To decode the message, the aliens would need to realize that the bits are meant to be arranged in a rectangular grid as shown, but that should not be difficult: The grid has 73 rows and 23 columns, and aliens would presumably know that these are both prime numbers. The picture represents the Arecibo radio dish, our solar system, a human stick figure, and a schematic of DNA and the eight simple molecules used in its construction.

However, M 13 is about 21,000 light-years from Earth, so it will take some 21,000 years for our signal to get there and another 21,000 years for any response to make its way back to Earth.

Several SETI projects under way or in development would be capable of detecting signals like the one we broadcast from Arecibo if they came from civilizations within a few hundred light-years. However, we could detect the signal only if we had the receiver tuned to the frequency of the broadcast—just as you can listen to your favorite radio station only by calling up the correct frequency on your radio dial. What radio frequency would aliens use? In the past, some astronomers made guesses about popular alien frequencies based on things such as frequencies emitted by common molecules in interstellar space. Today, SETI efforts generally seek to bypass this question by scanning millions of frequency bands simultaneously. Thus, if anyone nearby is deliberately broadcasting on an ongoing basis, we have a good chance of detecting the signals.

THINK ABOUT IT

SETI efforts are often controversial because of their cost and uncertain chances of success. However, SETI supporters say the cost is justified because contact with an extraterrestrial intelligence would be one of the most important discoveries in human history. Do you agree? Defend your opinion.

24.5 Interstellar Travel and Its Implications to Civilization

So far, we have discussed ways of detecting distant civilizations without ever leaving the comfort of our own planet. Could we ever actually visit other worlds in other star systems? A careful analysis of this question turns out to have profound implications for the future of our civilization. To see why, we first need to consider the prospects for achieving interstellar travel.

• How difficult is interstellar travel?

In many science fiction movies, our descendants travel among the stars as routinely as we jet about Earth in airplanes. They race around the galaxy in starships of all sizes and shapes, circumventing nature's prohibition on faster-than-light travel by entering hyperspace, wormholes, or warp drive. They witness firsthand incredible cosmic phenomena such as stars and planets in all stages of development, accretion disks around white dwarfs and neutron stars, and the distortion of spacetime near black holes. Along the way they encounter numerous alien species, most of which happen to look and act a lot like us.

Unfortunately, Einstein's theory of relativity tells us that real interstellar travel must be at speeds slower than the speed of light [Section S2.2], and even approaching that speed will require overcoming huge technological hurdles. Nevertheless, we have already sent out our first emissaries to the stars, and there's no reason to believe that we won't develop better technologies in the future.

The Challenge of Interstellar Travel To date, we have launched four spacecraft that will leave our solar system and eventually travel among the stars: the planetary probes *Pioneer 10, Pioneer 11, Voyager 1,* and *Voyager 2.* These spacecraft are traveling about as fast as anything ever built by humans, but their speeds are still less than $\frac{1}{10{,}000}$ the speed of light. It would take each of them some 100,000 years just to reach the next nearest star system (Alpha Centauri), but their trajectories won't take them anywhere near it. Instead, they will simply continue their journey without passing close to any nearby stars, wandering the Milky Way for millions or even billions of years to come. Each carries a greeting from Earth, just in case someone comes across one of them someday (Figure 24.22).

If we want to make interstellar journeys within human lifetimes, we will need starships that can travel at speeds close to the speed of light. We will need entirely new types of engines to reach such high speeds. The energy requirements of interstellar spacecraft may pose an even more daunting challenge. For example, the energy needed to accelerate a single ship the size of *Star Trek*'s *Enterprise* to just half the speed of light would be more than 2,000 times the total annual energy use of the world today. Clearly, interstellar travel will require vast new sources of energy. In addition, fast-moving starships will require new types of shielding to protect crew members from instant death. As a starship travels through interstellar gas at near-light speed, ordinary atoms and ions will hit it like a deadly flood of high-energy cosmic rays.

If we succeed in building starships capable of traveling at speeds close to the speed of light, the crews will face significant social challenges. According to well-tested principles of Einstein's theory of relativity, time will run much slower on a spaceship that travels at high speed to the stars than it does here on Earth [Section S2.4]. For example, in a ship traveling at an average speed of 99.9% of the speed of light, the 50-light-year round trip to the star Vega would take the travelers aboard only about 2 years—but more than 50 years would pass on Earth while they were gone. Thus, the crew will need only 2 years' worth of provisions and will age only 2 years during the voyage, but they will return to a world quite different from the one they left. Family and friends will be older or deceased, new technologies might have made their knowledge and skills obsolete, and many political and social changes may have occurred in their absence. Clearly, the crew will face a difficult adjustment when they come home to Earth.

Starship Design Despite all the difficulties, some scientists and engineers have already proposed designs that could in principle take us to nearby stars. In the 1960s a group of scientists proposed an approach, called *Project Orion,* that envisioned accelerating a spaceship with repeated detona-

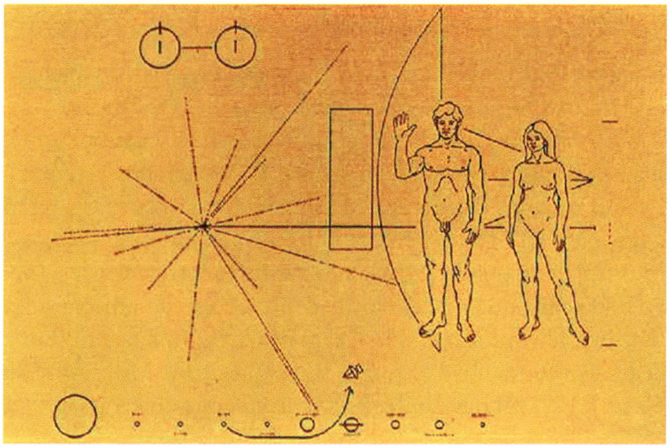

a The *Pioneer* plaque, about the size of an automobile license plate. The human figures are shown in front of a drawing of the spacecraft to give them a sense of scale. The "prickly" graph to their left shows the Sun's position relative to nearby pulsars, and Earth's location around the Sun is shown below. Binary code indicates the pulsar periods; because pulsars slow with time, the periods will allow someone reading the plaque to determine when the spacecraft was launched.

b *Voyagers* 1 and 2 carry a phonograph record—a 12-inch gold-plated copper disk containing music, greetings, and images from Earth.

Figure 24.22 Messages aboard the *Pioneer* and *Voyager* spacecraft, which are bound for the stars.

tions of relatively small hydrogen bombs. Each explosion would take place a few tens of meters behind the spaceship and would propel the ship forward as the vaporized debris hit a "pusher plate" on the back of the spacecraft (Figure 24.23). Calculations showed that a spaceship accelerated by the rapid-fire detonation of a million H-bombs could reach Alpha Centauri in just over a century. In principle, we could build an Orion spacecraft with existing technology, though it would be very expensive and would require an exception to the international treaty banning nuclear detonations in space.

A century to the nearest star isn't bad, but it still wouldn't make interstellar travel easy. Unfortunately, no available technology could go much faster. The problem is weight: Making a rocket faster requires more fuel, but adding fuel adds weight and makes it more difficult for the rocket to accelerate. Calculations show that even in the best case, rockets carrying nuclear fuel could achieve speeds no more than a few percent of the speed of light. Nevertheless, we can envision some possible future technologies that might get around this problem.

Figure 24.23 Artist's conception of the Project Orion starship, showing one of the small hydrogen-bomb detonations that would propel it. Debris from the detonation strikes the flat disk, called the pusher plate, at the back of the spaceship. The central sections (enclosed in a lattice) hold the bombs, and the front sections house the crew.

One idea suggests powering starships with engines that generate energy through matter-antimatter annihilation. Whereas nuclear fusion converts less than 1% of the mass of atomic nuclei into energy, matter-antimatter annihilation [Section 23.1] converts *all* the annihilated mass into energy. Starships with matter-antimatter engines could probably reach speeds of 90% or more of the speed of light. At these speeds, the slowing of time predicted by relativity becomes noticeable, putting many nearby stars within a few years' journey for the crew members. However, because no natural reservoirs of antimatter exist, we would have to be able to manufacture many tons of antimatter and then store it safely for the trip—capabilities that are far beyond our present means.

An even more speculative and futuristic design, known as an *interstellar ramjet,* would collect interstellar hydrogen with a gigantic scoop, using the collected gas as fuel for its nuclear engines (Figure 24.24). By collecting fuel along the way, the ship would not need to carry the weight of fuel onboard. However, because the density of interstellar gas is so low, the scoop would need to be enormous. As astronomer Carl Sagan said, we are talking about "spaceships the size of worlds."

The bottom line is that while we face enormous obstacles to achieving interstellar travel, there's no reason to think it's impossible. If we can avoid self-destruction and if we continue to explore space, it seems quite likely that our descendants will make journeys to the stars.

Figure 24.24 Artist's conception of a spaceship powered by an interstellar ramjet. The giant scoop in the front (left) collects interstellar hydrogen for use as fusion fuel.

• Where are the aliens?

Imagine that we survive and become interstellar travelers and that we begin colonizing habitable planets around nearby stars. As the colonies grow at each new location, some of the people may decide to set out for other star systems. Even if our starships traveled at relatively low speeds—say, a few percent of the speed of light—we could have dozens of outposts around nearby stars within a few centuries. In 50,000 years, our descendants would be spread among stars within a few hundred light-years of Earth. In a few million years, we could have outposts throughout the Milky Way Galaxy. We will have become a true galactic civilization.

SPECIAL TOPIC Are Aliens Already Here?

In this chapter, we have discussed contact with intelligent aliens as a possibility, not a reality. However, public opinion polls suggest that up to half the American public believes that aliens are already visiting us. What can science say about this remarkable notion?

The bulk of the claimed evidence for alien visitation consists of sightings of UFOs—unidentified flying objects. Many thousands of UFOs are reported each year, and no one doubts that unidentified objects are being seen. The question is whether they are alien spacecraft.

Aliens have long been a staple of science fiction, but modern interest in UFOs began with a widely reported sighting in 1947. While flying a private plane near Mount Rainier in Washington State, businessman Kenneth Arnold saw nine mysterious objects streaking across the sky. He told a reporter that the objects "flew erratic, like a saucer if you skip it across the water." (In fact, he may have seen meteors skipping across the atmosphere, though no one knows for sure.) He did *not* say that the objects were saucer-shaped, but the reporter nevertheless wrote up Arnold's experience as a sighting of "flying saucers." The story was front-page news throughout America, and within a decade "flying saucers" had invaded popular culture, if not our planet.

The flying saucer reports also interested the U.S. Air Force, largely out of concern that the UFOs might represent new types of aircraft developed by the then–Soviet Union. For two decades, the Air Force hired teams of academics to study UFO reports. In the overwhelming majority of cases, these experts were able to specify a plausible identification of the UFO. The explanations included bright stars and planets, aircraft and gliders, rocket launches, balloons, birds, ball lightning, meteors, atmospheric phenomena, and the occasional hoax. For a minority of the sightings, the investigators could not deduce what was seen, but their overall conclusion was that there was no reason to believe the UFOs were either highly advanced Soviet craft or visitors from other worlds. The Air Force ultimately dropped its investigations of the UFO phenomenon.

Believers discounted the Air Force denials, claiming to have other evidence of alien visitation. So far, none of this evidence has ever withstood scientific scrutiny. Photographs and film clips are nearly always too fuzzy to clearly show alien spacecraft, except in cases that are obviously faked. UFO witnesses are frequently credible (they include seasoned pilots), but generally there are several possible explanations for what they've seen besides alien spacecraft. Crop circles are easily made by pranksters. Stories of alien abductions are dramatic but cannot be verified. Pieces of metal that "UFO experts" say could not have been made by humans have turned out to be pieces of cars or refrigerators. Champions of alien visitation generally explain away the lack of clear evidence in one of two ways: government cover-ups or a failure of the mainstream scientific community to take the relevant phenomena seriously. Neither explanation is compelling.

It's certainly conceivable that a secretive government might *try* to put the lid on evidence of alien visits, though the motivation for doing so is unclear. The usual explanations are that the public couldn't handle the news or that the government is taking secret advantage of the alien materials to design new military hardware (via "reverse engineering"). Both explanations are silly. Half the population already believes in alien visitors and would hardly be shocked if newspapers announced that aliens were stacked up in government warehouses. As for reverse engineering extraterrestrial spacecraft, we should keep in mind how difficult it is to travel from star to star. Any society that could do so routinely would be technologically far beyond our own. Reverse engineering their spaceships is as unlikely as expecting Neanderthals to construct personal computers just because a laptop somehow landed in their cave. In addition, while a government might successfully hide evidence for a short time, does it really seem possible that evidence could remain secret for decades? After all, talk show fame and riches would await anyone who uncovered the conspiracy. Moreover, unless the aliens landed only in the United States, can we seriously believe that *every* government has cooperated in hiding the evidence?

Alleged disinterest on the part of the scientific community is an equally unimpressive claim. Scientists are constantly competing with one another to be the first with a great discovery, and clear evidence of alien visitors would certainly rank high on the all-time list. Countless researchers would work evenings and weekends, without pay, if they thought they could make such a discovery. The fact that few scientists are engaged in such study reflects not a lack of interest, but a lack of evidence worthy of study.

Of course, absence of evidence is not evidence of absence. Most scientists are open to the possibility that we might someday find evidence of alien visits, and many would welcome aliens with open arms. So far, however, we have no hard evidence to support the belief that aliens are already here.

Now, if we take the idea that *we* could develop a galactic civilization within a few million years and combine it with the reasonable (though unproved) idea that civilizations ought to be common, we are led to an astonishing conclusion: Someone else should already have created a galactic civilization. In fact, it should have been done a long time ago.

To see why, let's take some sample numbers. Suppose the overall odds of a civilization arising around a star are about the same as your odds of winning the lottery, or 1 in a million. Taking a low estimate of 100 billion stars in the Milky Way Galaxy, this means some 100,000 civilizations in our galaxy alone. Further, suppose we are a fairly typical civilization, so that civilizations generally arise when their stars are approaching 5 billion years old. Given that the galaxy is some 12 billion years old, the first of these 100,000 civilizations would have arisen at least 7 billion years ago. Others would have arisen, on average, about every 70,000 years. Under these assumptions, we would expect the youngest civilization besides ourselves to be some 70,000 years ahead of us technologically, and most would be millions or billions of years ahead of us.

Thus, we encounter a strange paradox: Plausible arguments suggest that a galactic civilization should already exist, yet we have so far found no evidence of such a civilization. This paradox is often called *Fermi's paradox,* after the Nobel Prize–winning physicist Enrico Fermi. During a 1950 conversation with other scientists about the possibility of extraterrestrial intelligence, Fermi responded to speculations by asking, "So where is everybody?"

This paradox has many possible solutions, but broadly speaking we can group them into three categories:

1. *We are alone.* There is no galactic civilization because civilizations are extremely rare—so rare that we are the first to have arisen on the galactic scene, perhaps even the first in the universe.
2. *Civilizations are common, but no one has colonized the galaxy.* There are at least three possible reasons why this might be the case. Perhaps interstellar travel is much harder or vastly more expensive than we have guessed, and civilizations are unable to venture far from their home worlds. Perhaps the desire to explore is unusual, and other societies either never leave their home star systems or stop exploring before they've colonized much of the galaxy. Most ominously, perhaps many civilizations have arisen, but they have all destroyed themselves before achieving the ability to colonize the stars.
3. *There IS a galactic civilization,* but it has not yet revealed its existence to us.

We do not know which, if any, of these explanations is the correct solution to the question "Where are the aliens?" However, each category of solution has astonishing implications for our own species.

Consider the first solution—that we are alone. If this is true, then our civilization is a remarkable achievement. It implies that through all of cosmic evolution, among countless star systems, we are the first piece of our galaxy or the universe ever to know that the rest of the universe exists. Through us, the universe has attained self-awareness. Some philosophers and many religions argue that the ultimate purpose of life is to become truly self-aware. If so, and if we are alone, then the destruction of our civilization and the loss of our scientific knowledge would represent an inglorious end to something that took the universe some 14 billion years to achieve. From this point of view, humanity becomes all the more precious, and the collapse of our civilization would be all the more tragic.

The second category of solutions has much more terrifying implications. If thousands of civilizations before us have all failed to achieve interstellar travel on a large scale, what hope do we have? Unless we somehow think differently than all other civilizations, this solution says that we will never go far in space. Because we have always explored when the opportunity arose, this solution almost inevitably leads to the conclusion that failure will come about because we destroy ourselves. We can only hope that this answer is wrong.

The third solution is perhaps the most intriguing. It says that we are newcomers on the scene of a galactic civilization that has existed for millions or billions of years before us. Perhaps this civilization is deliberately leaving us alone for the time being and will someday decide the time is right to invite us to join it.

No matter what the answer turns out to be, learning it is sure to mark a turning point in the brief history of our species. Moreover, this turning point is likely to be reached within the next few decades or centuries. We already have the ability to destroy our civilization. If we do so, then our fate is sealed. But if we survive long enough to develop technology that can take us to the stars, the possibilities seem almost limitless.

THE BIG PICTURE

Putting Chapter 24 into Context

Throughout our study of astronomy, we have taken the "big picture" view of trying to understand how we fit into the universe. Here, at last, we have returned to Earth and examined the role of our own generation in the big picture of human history. Tens of thousands of past human generations have walked this Earth. Ours is the first generation with the technology to study the far reaches of our universe, to search for life elsewhere, and to travel beyond our home planet. It is up to us to decide whether we will use this technology to advance our species or to destroy it.

Imagine for a moment the grand view, a gaze across the centuries and millennia from this moment forward. Picture our descendants living among the stars, having created or joined a great galactic civilization. They will have the privilege of experiencing ideas, worlds, and discoveries far beyond our wildest imagination. Perhaps, in their history lessons, they will learn of our generation—the generation that history placed at the turning point and that managed to steer its way past the dangers of self-destruction and onto the path to the stars.

24.1 Life on Earth

- **When did life arise on Earth?** Fossil evidence puts the origin of life at least 3.5 billion years ago, and carbon isotope evidence pushes this date to more than 3.85 billion years ago. Thus, life arose within a few hundred million years after the last major impact of the heavy bombardment, and possibly in a much shorter time.

- **How did life arise on Earth?** Genetic evidence suggests that all life on Earth evolved from a common ancestor, and this ancestor was probably similar to microbes that live today in hot water near undersea volcanic vents or hot springs. We do not know how this first organism arose, but laboratory experiments suggest that it may have been the result of natural chemical processes on the early Earth. Once life arose, it rapidly diversified and evolved through **natural selection**.

- **What are the necessities of life?** Life on Earth thrives in a wide range of environments, and in general seems to require only three things: a source of nutrients, a source of energy, and liquid water.

24.2 Life in The Solar System

- **Could there be life on Mars?** Mars once had conditions that may have been conducive to an origin of life. If life arose, it might still survive in pockets of liquid water underground.

- **Could there be life on Europa or other jovian moons?** Europa probably has a subsurface ocean of liquid water, and may have undersea volcanoes on its ocean floor. If so, it has conditions much like those in which life on Earth probably arose, making it a good candidate for life beyond Earth. Ganymede and Callisto might have oceans as well. Titan may have other liquids on its surface, though it is too cold for liquid water. Perhaps life can survive in these other liquids, or perhaps Titan has liquid water deep underground.

24.3 Life Around Other Stars

- **Are habitable planets likely?** Billions of stars have at least moderate-size **habitable zones** in which life-bearing planets might exist. We do not yet have the technology to search for habitable planets directly, but several planned missions should be able to begin the search soon.

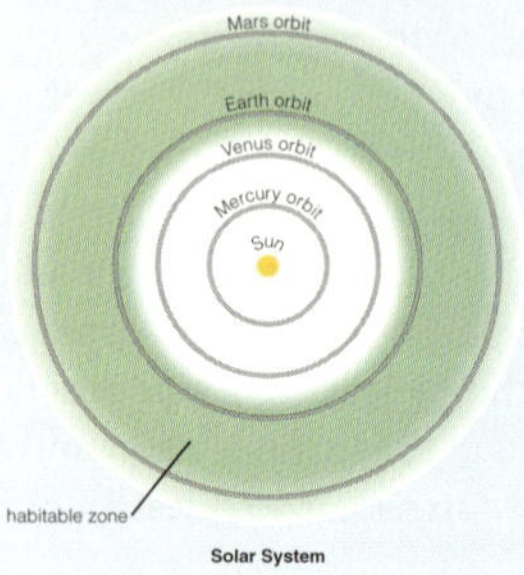

- **Are Earth-like planets rare or common?** We don't know. Arguments can be made on both sides of the question, and at present we lack the data to determine their validity.

24.4 The Search for Extraterrestrial Intelligence

- **How many civilizations are out there?** We don't know, but the **Drake equation** gives us a way to organize our thinking about the question. The equation (in a modified form) says that the number of civilizations in the Milky Way Galaxy with whom we could potentially communicate is $N_{HP} \times f_{life} \times f_{civ} \times f_{now}$, where N_{HP} is the number of habitable planets in the galaxy, f_{life} is the fraction of habitable planets that actually have life on them, f_{civ} is the fraction of life-bearing planets upon which a civilization capable of interstellar communication has at some time arisen, and f_{now} is the fraction of all these civilizations that exist now.

- **How does SETI work?** **SETI**, the search for extraterrestrial intelligence, generally refers to efforts to detect signals—such as radio or laser communications—coming from civilizations on other worlds.

24.5 Interstellar Travel and Its Implications to Civilization

- **How difficult is interstellar travel?** Convenient interstellar travel remains well beyond our technological capabilities, because of the technological requirements for engines, the enormous energy needed to accelerate spacecraft to speeds near the speed of light, and the difficulties of shielding the crew from radiation. Nevertheless, people have proposed ways around all these difficulties, and it seems reasonable to think that we will someday achieve interstellar travel if we survive long enough.

- **Where are the aliens?** It seems that we should be capable of colonizing the galaxy in a few million years or less, and the galaxy was around for at least 7 billion years before Earth was even born. Thus, it seems that someone should have colonized the galaxy long ago—yet we have no evidence of other civilizations. Every possible category of explanation for this surprising fact has astonishing implications for our species and our place in the universe.

EXERCISES AND PROBLEMS

For instructor-assigned homework go to MasteringASTRONOMY www.masteringastronomy.com

Review Questions

Short-Answer Questions Based on the Reading

1. Describe three recent developments in the study of life on Earth that make it seem much more likely that we could find life elsewhere.
2. How do we study the history of life on Earth? Describe the *geological time scale* and a few of the major events along it.
3. Summarize the evidence pointing to an early origin of life on Earth. How far back in Earth's history did life exist?
4. What is the theory of evolution, and why is it so critical to our understanding of the history of life on Earth? Briefly describe the logic that led Darwin to propose the theory of evolution.
5. What is *natural selection*, and how does it explain the evolution of life on Earth? What is *DNA*, and how do *mutations* in DNA explain the molecular basis of natural selection?
6. Give a brief overview of the history of life on Earth. What evidence points to a common ancestor for all life? How and when did oxygen accumulate in Earth's atmosphere? When did larger animals diversify on Earth?
7. Where do we think life on Earth first arose, and why?
8. Describe how laboratory experiments are helping us study the origin of life on Earth.
9. Is it possible that life migrated to Earth from elsewhere? Explain.
10. Describe the range of environments in which life thrives on Earth. What three basic requirements apply to life in all these environments?
11. What is a *habitable world?* Which worlds in our solar system seem potentially habitable, and why?
12. Briefly summarize the debate over possible fossil evidence of life in a meteorite from Mars.
13. What do we mean by a star's *habitable zone?* Do we expect many stars to be capable of having habitable planets? Explain.
14. What is the *rare Earth hypothesis?* Summarize the arguments on both sides regarding the validity of this hypothesis.
15. What is the *Drake equation?* Define each of its factors, and describe the current state of understanding about the potential values of each factor.
16. What is SETI? Describe the capabilities of current SETI efforts.
17. Summarize the factors that make interstellar travel difficult, and describe a few technologies that might someday make it possible nonetheless.
18. What is Fermi's paradox? Describe several potential solutions to the paradox, and the implications of each to the future of our civilization.

Test Your Understanding

Fantasy or Science Fiction?

Each of the following describes some futuristic scenario that, while perhaps common and entertaining, may or may not be plausible. In each case, decide whether the scenario is plausible according to our present understanding of science or whether it is unlikely to be possible. Explain your reasoning.

19. The first human explorers on Mars discover that the surface is littered with the ruins of an ancient civilization including remnants of tall buildings and temples.
20. The first human explorers on Mars drill a hole into a Martian volcano to collect a sample of soil from several meters underground. Upon analysis of the soil, they discover that it holds living microbes resembling terrestrial bacteria but with a different biochemistry.
21. In 2020, a spacecraft lands on Europa and melts its way through the ice into the Europan ocean. It finds numerous strange, living microbes, along with a few larger organisms that feed on the microbes.
22. It's the year 2075. A giant telescope on the Moon, consisting of hundreds of small telescopes linked together across a distance of 500 kilometers, has just captured a series of images of a planet around a distant star that clearly show seasonal changes in vegetation.
23. A century from now, after completing a careful study of planets around stars within 100 light-years of Earth, we've discovered that the most diverse life exists on a planet orbiting a young star that formed just 100 million years ago.
24. In 2030, a brilliant teenager working in her garage builds a coal-powered rocket that can travel at half the speed of light.
25. In the year 2750, we receive a signal from a civilization around a nearby star telling us that the *Voyager 2* spacecraft recently crash-landed on its planet.
26. Crew members of the matter-antimatter spacecraft *Star Apollo*, which left Earth in the year 2165, return to Earth in the year 2450, looking only a few years older than when they left.
27. Aliens from a distant star system invade Earth with the intent to destroy us and occupy our planet, but we successfully fight them off when their technology proves no match for ours.
28. A single great galactic civilization exists. It originated on a single planet long ago but is now made up of beings from many different planets, each of which was assimilated into the galactic culture in turn.

Quick Quiz

Choose the best answer to each of the following. Explain your reasoning with one or more complete sentences.

29. Fossil evidence suggests that life on Earth arose (a) almost immediately after Earth formed. (b) very soon after the end of the heavy bombardment. (c) about a billion years before the rise of the dinosaurs.
30. The theory of evolution is (a) a scientific theory, meaning that it is backed by extensive evidence. (b) one of several competing scientific models that all seem equally successful in explaining the nature of life on Earth. (c) essentially just a guess about how life changes through time.
31. Plants and animals are (a) the two major forms of life on Earth. (b) the only organisms that have DNA. (c) just two small branches of the diverse "tree of life" on Earth.
32. Which of the following is a reason why early living organisms on Earth could not have survived on the surface? (a) the lack of an ozone layer (b) the lack of oxygen in the atmosphere (c) the fact that these organisms were single-celled
33. According to current understanding, the key requirement for life is (a) photosynthesis. (b) liquid water. (c) an ozone layer.

34. Which of the following worlds is not considered a candidate for harboring life? (a) Europa (b) Mars (c) the Moon
35. How does the habitable zone around a star of spectral type G compare to that around a star of spectral type M? (a) It is larger. (b) It is hotter. (c) It is closer to its star.
36. In the Drake equation, suppose that the term $f_{life} = \frac{1}{2}$. What would this mean? (a) Half the stars in the Milky Way Galaxy have a planet with life. (b) Half of all life forms in the universe are intelligent. (c) Half of the habitable worlds in the galaxy actually have life, while the other half don't.
37. The amount of energy that would be needed to accelerate a large spaceship to a speed close to the speed of light is (a) about 100 times as much energy as is needed to launch the Space Shuttle. (b) more than the total amount of energy used by the entire world in a year. (c) more than the amount of energy that our Sun emits into space in a year.
38. According to current scientific understanding, the idea that the Milky Way Galaxy might be home to a civilization millions of years more advanced than ours is (a) a virtual certainty. (b) extremely unlikely. (c) one reasonable answer to Fermi's paradox.

Investigate Further

In-Depth Questions to Increase Your Understanding

Short-Answer/Essay Questions

39. *Artificial Selection.* Suppose you lived hundreds of years ago (before we knew about genetic engineering) and wanted to breed a herd of cows that provided more milk than the cows in your current herd. How would you have gone about it? How would this process of "artificial selection" be similar to natural selection? How would it be different?
40. *Statistics of One.* Much of the search for life in the universe is based on what we know from the example of life on Earth. Write a short essay discussing the pros and cons of basing our general understanding of biology on the example of Earth.
41. *Most Likely to Have Life.* Suppose you were asked to vote in a contest to name the world in our solar system (besides Earth) "most likely to have life." Which world would you cast your vote for? Explain and defend your choice in a one-page essay.
42. *Likely Suns.* Study the stellar data for nearby stars given in Appendix F, Table F.1. Which star on the list would you expect to have the largest habitable zone? Which would have the second-largest habitable zone? If we rule out multiple-star systems, which star would you expect to have the highest probability of having a habitable planet? Explain your answers.
43. *Are Earth-Like Planets Common?* Based on what you have learned in this book, form an opinion as to whether you think Earth-like planets will ultimately prove to be rare, common, or something in between. Write a one- to two-page essay explaining and defending your opinion.
44. *Solution to the Fermi Paradox.* Among the various possible solutions to the question "Where are the aliens?," which do you think is most likely? (If you have no opinion on their likelihood, which do you like best?) Write a one- to two-page essay in which you explain why you favor this solution.
45. *What's Wrong with This Picture?* Many science fiction stories have imagined the galaxy divided into a series of empires, each having arisen from a different civilization on a different world, that hold each other at bay because they are all at about the same level of military technology. Is this a realistic scenario? Explain.
46. *Aliens in the Movies.* Choose a science fiction movie (or television show) that involves an alien species. Do you think aliens like this could really exist? Write a one- to two-page critical review of the movie, focusing primarily on the question of whether the movie portrays the aliens in a scientifically reasonable way.

Quantitative Problems

Be sure to show all calculations clearly and state your final answers in complete sentences.

47. *Nearest Civilization I.* Suppose there are 10,000 civilizations existing in the Milky Way Galaxy right now. If the civilizations are randomly distributed throughout the disk of the galaxy, about how far (on average) would it be to the nearest civilization? (*Hint:* Start by finding the area of the Milky Way's disk, assuming that it is circular and 100,000 light-years in diameter. Then find the average area per civilization, and use the distance across this area to estimate the distance between civilizations.)
48. *Nearest Civilization II.* Repeat Problem 47, but this time assume that there are only 100 civilizations in the Milky Way Galaxy right now.
49. *SETI Search.* Suppose there are 10,000 civilizations broadcasting radio signals in the Milky Way Galaxy right now. On average, how many stars would we have to search before we would expect to hear a signal? Assume there are 500 billion stars in the galaxy. How does your answer change if there are only 100 civilizations instead of 10,000?
50. *SETI Signal.* Consider a civilization broadcasting a signal with a power of 10,000 watts. The Arecibo radio telescope, which is about 300 meters in diameter, could detect this signal if it is coming from as far away as 100 light-years. Suppose instead that the signal is being broadcast from the other side of the Milky Way Galaxy, about 70,000 light-years away. How large a radio telescope would we need to detect this signal? (*Hint:* You can do this problem with the inverse square law for light.)
51. *Cruise Ship Energy.* Suppose we have a spaceship about the size of a typical ocean cruise ship today, which means it has a mass of about 100 million kilograms, and we want to accelerate the ship to a speed of 10% of the speed of light.
    a. How much energy would be required? (*Hint:* You can find the answer simply by calculating the kinetic energy of the ship when it reaches its cruising speed; because 10% of the speed of light is still small compared to the speed of light, you can use the formula that tells us that kinetic energy $= \frac{1}{2} \times m \times v^2$.)
    b. How does your answer compare to total world energy use at present, which is about 5×10^{22} joules per year?
    c. The typical cost of energy today is roughly 5¢ per 1 million joules. Using this price, how much would it cost to generate the energy needed by this spaceship?
52. *Matter-Antimatter Engine.* Consider the spaceship from Problem 51. Suppose you want to generate the energy to get it to cruising speed using matter-antimatter annihilation. How much antimatter would you need to produce and take on the ship? (*Hint:* Remember that when matter and antimatter meet, they turn all their mass into energy equivalent to mc^2.)

Discussion Questions

53. *Funding the Search for Life.* Imagine that you are a member of Congress, so that your job includes deciding how

much government funding goes to research in different areas of science. How much would you allot to the search for life in the universe compared to the amount allotted to research in other areas of astronomy and planetary science? Why?

54. *Distant Dream or Near Reality -Reality?* Considering all the issues surrounding interstellar flight, when (if ever) do you think we are likely to begin traveling among the stars? Why?
55. *The Turning Point.* Discuss the idea that our generation has acquired a greater responsibility for the future than any previous generation. Do you agree with this assessment? If so, how should we deal with this responsibility? Defend your opinions.

MEDIA EXPLORATIONS

For self-study activities go to MasteringASTRONOMY www.masteringastronomy.com

Interactive Tutorials

Tutorial Review of Key Concepts

Use the following interactive **Tutorial** at **www.masteringastronomy.com** to review key concepts from this chapter.

Detecting Extrasolar Planets Tutorial

Lesson 1 Taking a Picture of a Planet
Lesson 2 Stars' Wobbles and Properties of Planets
Lesson 3 Planetary Transits

Supplementary Tutorial Exercises

Use the interactive **Tutorial Lesson** to explore the following questions.

Detecting Extrasolar Planets Tutorial, Lessons 1–3

1. Give two reasons why visual detection of planets orbiting other stars is extremely difficult.
2. As you move away from two objects, what happens to the apparent angle between them?
3. How can we detect extrasolar planets?
4. How might transits allow us to detect Earth-size planets around other stars?

Movies

Check out the following narrated and animated short documentary available on **www.masteringastronomy.com** for a helpful review of key ideas covered in this chapter.

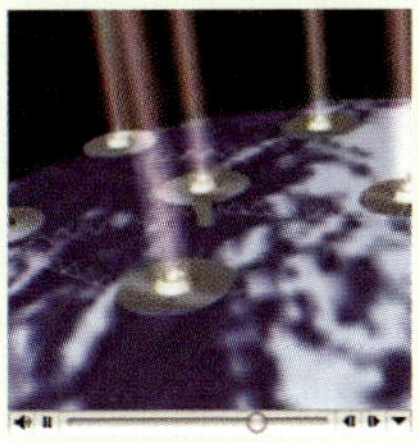

The Search for Extraterrestrial Life Movie

Web Projects

1. *Astrobiology News.* Go to NASA's astrobiology home page and read some of the recent news from the search for life in the universe. Choose one recent news article and write a one- to two-page summary of the research and how it relates to the question of life in the universe in general.
2. *Martian Meteorites.* Find information about the latest discoveries concerning Martian meteorites and whether the meteorites contain evidence of life. Choose one recent discovery that seems important, and write a short summary of how you think it alters the debate about the habitability of Mars or about life on Mars.
3. *The Search for Extraterrestrial Intelligence.* Go to the home page for the SETI Institute. Learn more about how SETI is funded and carried out. In one page or less, describe the SETI Institute and its work.
4. *Starship Design.* Find more details about a proposal for starship propulsion or design. How would the starship work? What new technologies would be needed, and what existing technologies could be applied? Summarize your findings in a one- to two-page report.
5. *Advanced Spacecraft Technologies.* NASA supports many efforts to incorporate new technologies into spaceships. Although few of them reach the level of being suitable for interstellar colonization, most are innovative and fascinating. Learn about one such NASA project, and write a short summary of your findings.

Appendixes

A Useful Numbers

Astronomical Distances

1 AU $\approx 1.496 \times 10^8$ km

1 light-year $\approx 9.46 \times 10^{12}$ km

1 parsec (pc) $\approx 3.09 \times 10^{13}$ km $\approx$ 3.26 light-years

1 kiloparsec (kpc) = 1,000 pc $\approx 3.26 \times 10^3$ light-years

1 megaparsec (Mpc) = 10^6 pc $\approx 3.26 \times 10^6$ light-years

Astronomical Times

1 solar day (average) = 24^h

1 sidereal day $\approx 23^h\,56^m\,4.09^s$

1 synodic month (average) $\approx$ 29.53 solar days

1 sidereal month (average) $\approx$ 27.32 solar days

1 tropical year $\approx$ 365.242 solar days

1 sidereal year $\approx$ 365.256 solar days

Universal Constants

Speed of light:	$c = 3 \times 10^5$ km/s $= 3 \times 10^8$ m/s
Gravitational constant:	$G = 6.67 \times 10^{-11} \dfrac{\text{m}^3}{\text{kg} \times \text{s}^2}$
Planck's constant:	$h = 6.626 \times 10^{-34}$ joule $\times$ s
Stefan–Boltzmann constant:	$\sigma = 5.7 \times 10^{-8} \dfrac{\text{watt}}{\text{m}^2 \times \text{Kelvin}^4}$
mass of a proton:	$m_p = 1.67 \times 10^{-27}$ kg
mass of an electron:	$m_e = 9.1 \times 10^{-31}$ kg

Useful Sun and Earth Reference Values

Mass of the Sun: 1 $M_{\text{Sun}} \approx 2 \times 10^{30}$ kg

Radius of the Sun: 1 $R_{\text{Sun}} \approx$ 696,000 km

Luminosity of the Sun: 1 $L_{\text{Sun}} \approx 3.8 \times 10^{26}$ watts

Mass of the Earth: 1 $M_{\text{Earth}} \approx 5.97 \times 10^{24}$ kg

Radius (equatorial) of the Earth: 1 $R_{\text{Earth}} \approx$ 6,378 km

Acceleration of gravity on Earth: $g = 9.8$ m/s^2

Escape velocity from surface of Earth: v_{escape} = 11 km/s = 11,000 m/s

Energy and Power Units

Basic unit of energy: 1 joule $= 1 \dfrac{\text{kg} \times \text{m}^2}{\text{s}^2}$

Basic unit of power: 1 watt = 1 joule/s

Electron-volt: 1 eV $= 1.60 \times 10^{-19}$ joule

B Useful Formulas

- Universal law of gravitation for the force between objects of mass M_1 and M_2, distance d between their centers:

$$F = G\frac{M_1 M_2}{d^2}$$

- Newton's version of Kepler's third law; p and a are period and semimajor axis, respectively, of either orbiting mass:

$$p^2 = \frac{4\pi^2}{G(M_1 + M_2)} a^3$$

- Escape velocity at distance R from center of object of mass M:

$$v_{escape} = \sqrt{\frac{2GM}{R}}$$

- Relationship between a photon's wavelength (λ), frequency (f), and the speed of light (c):

$$\lambda \times f = c$$

- Energy of a photon of wavelength λ or frequency f:

$$E = hf = \frac{hc}{\lambda}$$

- Stefan–Boltzmann law for thermal radiation at temperature T (in Kelvin):

$$\text{emitted power per unit area} = \sigma T^4$$

- Wien's law for the peak wavelength (λ_{max}) thermal radiation at temperature T (in Kelvin):

$$\lambda_{max} = \frac{2{,}900{,}000}{T}\text{nm}$$

- Doppler shift (radial velocity is positive if the object is moving away from us and negative if it is moving toward us):

$$\frac{\text{radial velocity}}{\text{speed of light}} = \frac{\text{shifted wavelength} - \text{rest wavelength}}{\text{rest wavelength}}$$

- Angular separation (α) of two points with an actual separation s, viewed from a distance d (assuming d is much larger than s):

$$\alpha = \frac{s}{2\pi d} \times 360°$$

- Inverse Square Law for Light:

$$\text{apparent brightness} = \frac{\text{luminosity}}{4\pi d^2}$$

(where d is the distance to the object)

- Parallax formula (distance d to a star with parallax angle p in arcseconds):

$$d\,(\text{in parsecs}) = \frac{1}{p\,(\text{in arcseconds})}$$

$$\text{or, } d\,(\text{in light-years}) = 3.26 \times \frac{1}{p\,(\text{in arcseconds})}$$

- The orbital velocity law, to find the mass M_r contained within the circular orbit of radius r for an object moving at speed v:

$$M_r = \frac{r \times v^2}{G}$$

C A Few Mathematical Skills

This appendix reviews the following mathematical skills: powers of 10, scientific notation, working with units, the metric system, and finding a ratio. You should refer to this appendix as needed while studying the textbook.

C.1 Powers of 10

Powers of 10 simply indicate how many times to multiply 10 by itself. For example:

$$10^2 = 10 \times 10 = 100$$

$$10^6 = 10 \times 10 \times 10 \times 10 \times 10 \times 10 = 1{,}000{,}000$$

Negative powers are the reciprocals of the corresponding positive powers. For example:

$$10^{-2} = \frac{1}{10^2} = \frac{1}{100} = 0.01$$

$$10^{-6} = \frac{1}{10^6} = \frac{1}{1{,}000{,}000} = 0.000001$$

Table C.1 lists powers of 10 from 10^{-12} to 10^{12}. Note that powers of 10 follow two basic rules:

1. A positive exponent tells how many zeros follow the 1. For example, 10^0 is a 1 followed by no zeros, and 10^8 is a 1 followed by eight zeros.
2. A negative exponent tells how many places are to the right of the decimal point, including the 1. For example, $10^{-1} = 0.1$ has one place to the right of the decimal point; $10^{-6} = 0.000001$ has six places to the right of the decimal point.

Multiplying and Dividing Powers of 10

Multiplying powers of 10 simply requires adding exponents, as the following examples show:

$$10^4 \times 10^7 = \underbrace{10{,}000}_{10^4} \times \underbrace{10{,}000{,}000}_{10^7} = \underbrace{100{,}000{,}000{,}000}_{10^{4+7} = 10^{11}} = 10^{11}$$

$$10^5 \times 10^{-3} = \underbrace{100{,}000}_{10^5} \times \underbrace{0.001}_{10^{-3}} = \underbrace{100}_{10^{5+(-3)} = 10^2} = 10^2$$

$$10^{-8} \times 10^{-5} = \underbrace{0.00000001}_{10^{-8}} \times \underbrace{0.00001}_{10^{-5}} = \underbrace{0.0000000000001}_{10^{-8+(-5)} = 10^{-13}} = 10^{-13}$$

Table C.1 Powers of 10

Zero and Positive Powers			*Negative Powers*		
Power	**Value**	**Name**	**Power**	**Value**	**Name**
10^0	1	One			
10^1	10	Ten	10^{-1}	0.1	Tenth
10^2	100	Hundred	10^{-2}	0.01	Hundredth
10^3	1,000	Thousand	10^{-3}	0.001	Thousandth
10^4	10,000	Ten thousand	10^{-4}	0.0001	Ten thousandth
10^5	100,000	Hundred thousand	10^{-5}	0.00001	Hundred thousandth
10^6	1,000,000	Million	10^{-6}	0.000001	Millionth
10^7	10,000,000	Ten million	10^{-7}	0.0000001	Ten millionth
10^8	100,000,000	Hundred million	10^{-8}	0.00000001	Hundred millionth
10^9	1,000,000,000	Billion	10^{-9}	0.000000001	Billionth
10^{10}	10,000,000,000	Ten billion	10^{-10}	0.0000000001	Ten billionth
10^{11}	100,000,000,000	Hundred billion	10^{-11}	0.00000000001	Hundred billionth
10^{12}	1,000,000,000,000	Trillion	10^{-12}	0.000000000001	Trillionth

Dividing powers of 10 requires subtracting exponents, as in the following examples:

$$\frac{10^5}{10^3} = \underbrace{100{,}000}_{10^5} \div \underbrace{1{,}000}_{10^3} = \underbrace{100}_{10^{5-3} = 10^2} = 10^2$$

$$\frac{10^3}{10^7} = \underbrace{1{,}000}_{10^3} \div \underbrace{10{,}000{,}000}_{10^7} = \underbrace{0.0001}_{10^{3-7} = 10^{-4}} = 10^{-4}$$

$$\frac{10^{-4}}{10^{-6}} = \underbrace{0.0001}_{10^{-4}} \div \underbrace{0.000001}_{10^{-6}} = \underbrace{100}_{10^{-4-(-6)} = 10^2} = 10^2$$

Powers of Powers of 10

We can use the multiplication and division rules to raise powers of 10 to other powers or to take roots. For example:

$$(10^4)^3 = 10^4 \times 10^4 \times 10^4 = 10^{4+4+4} = 10^{12}$$

Note that we can get the same end result by simply multiplying the two powers:

$$(10^4)^3 = 10^{4\times 3} = 10^{12}$$

Because taking a root is the same as raising to a fractional power (e.g., the square root is the same as the $\frac{1}{2}$ power, the cube root is the same as the $\frac{1}{3}$ power, etc.), we can use the same procedure for roots, as in the following example:

$$\sqrt{10^4} = (10^4)^{1/2} = 10^{4\times(1/2)} = 10^2$$

Adding and Subtracting Powers of 10

Unlike with multiplication and division, there is no shortcut for adding or subtracting powers of 10. The values must be written in longhand notation. For example:

$$10^6 + 10^2 = 1{,}000{,}000 + 100 = 1{,}000{,}100$$

$$10^8 + 10^{-3} = 100{,}000{,}000 + 0.001 = 100{,}000{,}000.001$$

$$10^7 - 10^3 = 10{,}000{,}000 - 1{,}000 = 9{,}999{,}000$$

Summary

We can summarize our findings using n and m to represent any numbers:

- To *multiply* powers of 10, *add* exponents: $10^n \times 10^m = 10^{n+m}$
- To *divide* powers of 10, *subtract* exponents: $\dfrac{10^n}{10^m} = 10^{n-m}$
- To *raise* powers of 10 to other powers, multiply exponents: $(10^n)^m = 10^{n \times m}$

C.2 Scientific Notation

When we are dealing with large or small numbers, it's generally easier to write them with powers of 10. For example, it's much easier to write the number 6,000,000,000,000 as 6×10^{12}. This format, in which a number *between* 1 and 10 is multiplied by a power of 10, is called **scientific notation**.

Converting a Number to Scientific Notation

We can convert numbers written in ordinary notation to scientific notation with a simple two-step process:

1. Move the decimal point to come after the *first* nonzero digit.
2. The number of places the decimal point moves tells you the power of 10; the power is *positive* if the decimal point moves to the left and *negative* if it moves to the right.

   Examples:

$$3{,}042 \xrightarrow[\text{3 places to left}]{\text{decimal needs to move}} 3.042 \times 10^3$$

$$0.00012 \xrightarrow[\text{4 places to right}]{\text{decimal needs to move}} 1.2 \times 10^{-4}$$

$$226 \times 10^2 \xrightarrow[\text{2 places to left}]{\text{decimal needs to move}} (2.26 \times 10^2) \times 10^2 = 2.26 \times 10^4$$

Converting a Number from Scientific Notation

We can convert numbers written in scientific notation to ordinary notation by the reverse process:

1. The power of 10 indicates how many places to move the decimal point; move it to the *right* if the power of 10 is positive and to the *left* if it is negative.
2. If moving the decimal point creates any open places, fill them with zeros.

Examples:

$$4.01 \times 10^2 \xrightarrow[\text{2 places to right}]{\text{move decimal}} 401$$

$$3.6 \times 10^6 \xrightarrow[\text{6 places to right}]{\text{move decimal}} 3{,}600{,}000$$

$$5.7 \times 10^{-3} \xrightarrow[\text{3 places to left}]{\text{move decimal}} 0.0057$$

Multiplying or Dividing Numbers in Scientific Notation

Multiplying or dividing numbers in scientific notation simply re-quires operating on the powers of 10 and the other parts of the number separately.

Examples:

$$(6 \times 10^2) \times (4 \times 10^5) = (6 \times 4) \times (10^2 \times 10^5) = 24 \times 10^7 = (2.4 \times 10^1) \times 10^7 = 2.4 \times 10^8$$

$$\frac{4.2 \times 10^{-2}}{8.4 \times 10^{-5}} = \frac{4.2}{8.4} \times \frac{10^{-2}}{10^{-5}} = 0.5 \times 10^{-2-(-5)} = 0.5 \times 10^3 = (5 \times 10^{-1}) \times 10^3 = 5 \times 10^2$$

Note that, in both these examples, we first found an answer in which the number multiplied by a power of 10 was *not* between 1 and 10. We therefore followed the procedure for converting the final answer to scientific notation.

Addition and Subtraction with Scientific Notation

In general, we must write numbers in ordinary notation before adding or subtracting.

Examples:

$$(3 \times 10^6) + (5 \times 10^2) = 3{,}000{,}000 + 500 = 3{,}000{,}500 = 3.0005 \times 10^6$$

$$(4.6 \times 10^9) - (5 \times 10^8) = 4{,}600{,}000{,}000 - 500{,}000{,}000 = 4{,}100{,}000{,}000 = 4.1 \times 10^9$$

When both numbers have the *same* power of 10, we can factor out the power of 10 first.

Examples:

$$(7 \times 10^{10}) + (4 \times 10^{10}) = (7 + 4) \times 10^{10} = 11 \times 10^{10} = 1.1 \times 10^{11}$$

$$(2.3 \times 10^{-22}) - (1.6 \times 10^{-22}) = (2.3 - 1.6) \times 10^{-22} = 0.7 \times 10^{-22} = 7.0 \times 10^{-23}$$

C.3 Working with Units

Showing the units of a problem as you solve it usually makes the work much easier and also provides a useful way of checking your work. If an answer does not come out with the units you expect, you probably did something wrong. In general, working with units is very similar to working with numbers, as the following guidelines and examples show.

Five Guidelines for Working with Units

Before you begin any problem, think ahead and identify the units you expect for the final answer. Then operate on the units along with the numbers as you solve the problem. The following five guidelines may be helpful when you are working with units:

1. Mathematically, it doesn't matter whether a unit is singular (e.g., meter) or plural (e.g., meters); we can use the same abbreviation (e.g., m) for both.

2. You cannot add or subtract numbers unless they have the *same* units. For example, 5 apples + 3 apples = 8 apples, but the expression 5 apples + 3 oranges cannot be simplified further.

3. You *can* multiply units, divide units, or raise units to powers. Look for key words that tell you what to do.

   - *Per* suggests division. For example, we write a speed of 100 kilometers per hour as:

   $$100 \frac{\text{km}}{\text{hr}} \quad \text{or} \quad 100 \frac{\text{km}}{1\ \text{hr}}$$

   - *Of* suggests multiplication. For example, if you launch a 50-kg space probe at a launch cost *of* \$10,000 per kilogram, the total cost is:

   $$50\ \cancel{\text{kg}} \times \frac{\$10{,}000}{\cancel{\text{kg}}} = \$500{,}000$$

   - *Square* suggests raising to the second power. For example, we write an area of 75 square meters as 75 m^2.

   - *Cube* suggests raising to the third power. For example, we write a volume of 12 cubic centimeters as 12 cm^3.

4. Often the number you are given is not in the units you wish to work with. For example, you may be given that the speed of light is 300,000 km/s but need it in units of m/s for a particular problem. To convert the units, simply multiply the given number by a *conversion factor:* a fraction in which the numerator (top of the fraction) and denominator (bottom of the fraction) are equal, so that the value of the fraction is 1; the number in the denominator must have the units that you wish to change. In the case of changing the speed of light from units of km/s to m/s, you need a conversion factor for kilometers to meters. Thus, the conversion factor is:

   $$\frac{1{,}000\ \text{m}}{1\ \text{km}}$$

   Note that this conversion factor is equal to 1, since 1,000 meters and 1 kilometer are equal, and that the units to be changed (km) appear in the denominator. We can now convert the speed of light from units of km/s to m/s simply by multiplying by this conversion factor:

   $$\underbrace{300{,}000 \frac{\cancel{\text{km}}}{\text{s}}}_{\text{speed of light in km/s}} \times \underbrace{\frac{1{,}000\ \text{m}}{1\ \cancel{\text{km}}}}_{\text{conversion from km to m}} = \underbrace{3 \times 10^8 \frac{\text{m}}{\text{s}}}_{\text{speed of light in m/s}}$$

   Note that the units of km cancel, leaving the answer in units of m/s.

5. It's easier to work with units if you replace division with multiplication by the reciprocal. For example, suppose you want to know how many minutes are represented by 300 seconds. We can find the answer by dividing 300 seconds by 60 seconds per minute:

   $$300\ \text{s} \div 60 \frac{\text{s}}{\text{min}}$$

However, it is easier to see the unit cancellations if we rewrite this expression by replacing the division with multiplication by the reciprocal (this process is easy to remember as "invert and multiply"):

$$300\text{ s} \div 60\,\frac{\text{s}}{\text{min}} = \underbrace{300\,\cancel{\text{s}} \times \underbrace{\frac{1\text{ min}}{60\,\cancel{\text{s}}}}_{\text{invert}}}_{\text{and multiply}} = 5\text{ min}$$

We now see that the units of seconds (s) cancel in the numerator of the first term and the denominator of the second term, leaving the answer in units of minutes.

More Examples of Working with Units

Example 1. How many seconds are there in 1 day?

Solution: We can answer the question by setting up a *chain* of unit conversions in which we start with 1 *day* and end up with *seconds*. We use the facts that there are 24 hours per day (24 hr/day), 60 minutes per hour (60 min/hr), and 60 seconds per minute (60 s/min):

$$\underbrace{1\,\cancel{\text{day}}}_{\substack{\text{starting}\\\text{value}}} \times \underbrace{\frac{24\,\cancel{\text{hr}}}{\cancel{\text{day}}}}_{\substack{\text{conversion}\\\text{from}\\\text{day to hr}}} \times \underbrace{\frac{60\,\cancel{\text{min}}}{\cancel{\text{hr}}}}_{\substack{\text{conversion}\\\text{from}\\\text{hr to min}}} \times \underbrace{\frac{60\text{ s}}{\cancel{\text{min}}}}_{\substack{\text{conversion}\\\text{from}\\\text{min to s}}} = 86{,}400\text{ s}$$

Note that all the units cancel except *seconds*, which is what we want for the answer. There are 86,400 seconds in 1 day.

Example 2. Convert a distance of 10^8 cm to km.

Solution: The easiest way to make this conversion is in two steps, since we know that there are 100 centimeters per meter (100 cm/m) and 1,000 meters per kilometer (1,000 m/km):

$$\underbrace{10^8\text{ cm}}_{\substack{\text{starting}\\\text{value}}} \times \underbrace{\frac{1\text{ m}}{100\text{ cm}}}_{\substack{\text{conversion}\\\text{from}\\\text{cm to m}}} \times \underbrace{\frac{1\text{ km}}{1{,}000\text{ m}}}_{\substack{\text{conversion}\\\text{from}\\\text{m to km}}} = 10^8\,\cancel{\text{cm}} \times \frac{1\,\cancel{\text{m}}}{10^2\,\cancel{\text{cm}}} \times \frac{1\text{ km}}{10^3\,\cancel{\text{m}}} = 10^3\text{ km}$$

Alternatively, if we recognize that the number of kilometers should be smaller than the number of centimeters (because kilometers are larger), we might decide to do this conversion by dividing as follows:

$$10^8\text{ cm} \div \frac{100\text{ cm}}{\text{m}} \div \frac{1{,}000\text{ m}}{\text{km}}$$

In this case, before carrying out the calculation, we replace each division with multiplication by the reciprocal:

$$\begin{aligned} 10^8\text{ cm} \div \frac{100\text{ cm}}{\text{m}} \div \frac{1{,}000\text{ m}}{\text{km}} &= 10^8\text{ cm} \times \frac{1\text{ m}}{100\text{ cm}} \times \frac{1\text{ km}}{1{,}000\text{ m}} \\ &= 10^8\,\cancel{\text{cm}} \times \frac{1\,\cancel{\text{m}}}{10^2\,\cancel{\text{cm}}} \times \frac{1\text{ km}}{10^3\,\cancel{\text{m}}} \\ &= 10^3\text{ km} \end{aligned}$$

Note that we again get the answer that 10^8 cm is the same as 10^3 km, or 1,000 km.

Example 3. Suppose you accelerate at 9.8 m/s^2 for 4 seconds, starting from rest. How fast will you be going?

Solution: The question asked "how fast?" so we expect to end up with a speed. Therefore, we multiply the acceleration by the amount of time you accelerated:

$$9.8\frac{\text{m}}{\text{s}^2} \times 4\text{ s} = (9.8 \times 4)\frac{\text{m} \times \cancel{\text{s}}}{\text{s}^{\cancel{2}}} = 39.2\frac{\text{m}}{\text{s}}$$

Note that the units end up as a speed, showing that you will be traveling 39.2 m/s after 4 seconds of acceleration at 9.8 m/s^2.

Example 4. A reservoir is 2 km long and 3 km wide. Calculate its area, in both square kilometers and square meters.

Solution: We find its area by multiplying its length and width:

$$2\text{ km} \times 3\text{ km} = 6\text{ km}^2$$

Next we need to convert this area of 6 km^2 to square meters, using the fact that there are 1,000 meters per kilometer (1,000 m/km). Note that we must square the term 1,000 m/km when converting from km^2 to m^2:

$$6\text{ km}^2 \times \left(1{,}000\frac{\text{m}}{\text{km}}\right)^2 = 6\text{ km}^2 \times 1{,}000^2\frac{\text{m}^2}{\text{km}^2} = 6\,\cancel{\text{km}^2} \times 1{,}000{,}000\frac{\text{m}^2}{\cancel{\text{km}^2}}$$
$$= 6{,}000{,}000\text{ m}^2$$

The reservoir area is 6 km^2, which is the same as 6 million m^2.

C.4 The Metric System (SI)

The modern version of the metric system, known as *Système Internationale d'Unites* (French for "International System of Units") or **SI**, was formally established in 1960. Today, it is the primary measurement system in nearly every country in the world with the exception of the United States. Even in the United States, it is the system of choice for science and international commerce.

The basic units of length, mass, and time in the SI are:

- The **meter** for length, abbreviated m
- The **kilogram** for mass, abbreviated kg
- The **second** for time, abbreviated s

Multiples of metric units are formed by powers of 10, using a prefix to indicate the power. For example, *kilo* means 10^3 (1,000), so a kilometer is 1,000 meters; a microgram is 0.000001 gram, because *micro* means 10^{-6}, or one millionth. Some of the more common prefixes are listed in Table C.2.

Table C.2 SI (Metric) Prefixes

Small Values			*Large Values*		
Prefix	**Abbreviation**	**Value**	**Prefix**	**Abbreviation**	**Value**
Deci	d	10^{-1}	Deca	da	10^1
Centi	c	10^{-2}	Hecto	h	10^2
Milli	m	10^{-3}	Kilo	k	10^3
Micro	μ	10^{-6}	Mega	M	10^6
Nano	n	10^{-9}	Giga	G	10^9
Pico	p	10^{-12}	Tera	T	10^{12}

Metric Conversions

Table C.3 lists conversions between metric units and units used commonly in the United States. Note that the conversions between kilograms and pounds are valid only on Earth, because they depend on the strength of gravity.

Example 1. International athletic competitions generally use metric distances. Compare the length of a 100-meter race to that of a 100-yard race.

Solution: Table C.3 shows that 1 m = 1.094 yd, so 100 m is 109.4 yd. Note that 100 meters is almost 110 yards; a good "rule of thumb" to remember is that distances in meters are about 10% longer than the corresponding number of yards.

Example 2. How many square kilometers are in 1 square mile?

Solution: We use the square of the miles-to-kilometers conversion factor:

$$(1\ \text{mi}^2) \times \left(\frac{1.6093\ \text{km}}{1\ \text{mi}}\right)^2 = (1\ \cancel{\text{mi}^2}) \times \left(1.6093^2\,\frac{\text{km}^2}{\cancel{\text{mi}^2}}\right) = 2.5898\ \text{km}^2$$

Therefore, 1 square mile is 2.5898 square kilometers.

C.5 Finding a Ratio

Suppose you want to compare two quantities, such as the average density of the Earth and the average density of Jupiter. The way we do such a comparison is by dividing, which tells us the *ratio* of the two quantities. In this case, the Earth's average density is 5.52 grams/cm^3 and Jupiter's average density is 1.33 grams/cm^3 (see Figure 8.1), so the ratio is:

$$\frac{\text{average density of Earth}}{\text{average density of Jupiter}} = \frac{5.52\ \cancel{\text{g/cm}^3}}{1.33\ \cancel{\text{g/cm}^3}} = 4.15$$

Notice how the units cancel on both the top and bottom of the fraction. We can state our result in two equivalent ways:

- The ratio of the Earth's average density to Jupiter's average density is 4.15.
- The Earth's average density is 4.15 times Jupiter's average density.

Sometimes, the quantities that you want to compare may each involve an equation. In such cases, you could, of course, find the ratio by first calculating each of the two quantities individually and then dividing. However, it is much easier if you first express the ratio as a fraction, putting the equation for one quantity on top and the other on the bottom. Some of the terms in the equation may then cancel out, making any calculations much easier.

Example 1. Compare the kinetic energy of a car traveling at 100 km/hr to that of a car traveling at 50 km/hr.

Table C.3 Metric Conversions

To Metric	*From Metric*
1 inch = 2.540 cm	1 cm = 0.3937 inch
1 foot = 0.3048 m	1 m = 3.28 feet
1 yard = 0.9144 m	1 m = 1.094 yards
1 mile = 1.6093 km	1 km = 0.6214 mile
1 pound = 0.4536 kg	1 kg = 2.205 pounds

Solution: We do the comparison by finding the ratio of the two kinetic energies, recalling that the formula for kinetic energy is $\frac{1}{2}mv^2$. Since we are not told the mass of the car, you might at first think that we don't have enough information to find the ratio. However, notice what happens when we put the equations for each kinetic energy into the ratio, calling the two speeds v_1 and v_2:

$$\frac{\text{K.E. car at } v_1}{\text{K.E. car at } v_2} = \frac{\cancel{\tfrac{1}{2}}\cancel{m_{\text{car}}}{v_1}^2}{\cancel{\tfrac{1}{2}}\cancel{m_{\text{car}}}{v_2}^2} = \frac{{v_1}^2}{{v_2}^2} = \left(\frac{v_1}{v_2}\right)^2$$

All the terms cancel except those with the two speeds, leaving us with a very simple formula for the ratio. Now we put in 100 km/hr for v_1 and 50 km/hr for v_2:

$$\frac{\text{K.E. car at 100 km/hr}}{\text{K.E. car at 50 km/hr}} = \left(\frac{100\ \cancel{\text{km/hr}}}{50\ \cancel{\text{km/hr}}}\right)^2 = 2^4 = 4$$

The ratio of the car's kinetic energies at 100 km/hr and 50 km/hr is 4. That is, the car has four times as much kinetic energy at 100 km/hr as it has at 50 km/hr.

Example 2. Compare the strength of gravity between the Earth and the Sun to the strength of gravity between the Earth and the Moon.

Solution: We do the comparison by taking the ratio of the Earth–Sun gravity to the Earth–Moon gravity. In this case, each quantity is found from the equation of Newton's law of gravity. (See Section 4.4.) Thus, the ratio is:

$$\frac{\text{Earth–Sun gravity}}{\text{Earth–Moon gravity}} = \frac{\cancel{G}\dfrac{\cancel{M_{\text{Earth}}}M_{\text{Sun}}}{(d_{\text{Earth–Sun}})^2}}{\cancel{G}\dfrac{\cancel{M_{\text{Earth}}}M_{\text{Moon}}}{(d_{\text{Earth–Moon}})^2}} = \frac{M_{\text{Sun}}}{(d_{\text{Earth–Sun}})^2} \times \frac{(d_{\text{Earth–Moon}})^2}{M_{\text{Moon}}}$$

Note how all but four of the terms cancel; the last step comes from replacing the division with multiplication by the reciprocal (the "invert and multiply" rule for division). We can simplify the work further by rearranging the terms so that we have the masses and distances together:

$$\frac{\text{Earth–Sun gravity}}{\text{Earth–Moon gravity}} = \frac{M_{\text{Sun}}}{M_{\text{Moon}}} \times \frac{(d_{\text{Earth–Moon}})^2}{(d_{\text{Earth–Sun}})^2}$$

Now it is just a matter of looking up the numbers (see Appendix E) and calculating:

$$\frac{\text{Earth–Sun gravity}}{\text{Earth–Moon gravity}} = \frac{1.99 \times 10^{30}\ \cancel{\text{kg}}}{7.35 \times 10^{22}\ \cancel{\text{kg}}} \times \frac{(384.4 \times 10^3\ \cancel{\text{km}})^2}{(149.6 \times 10^6\ \cancel{\text{km}})^2} = 179$$

In other words, the Earth–Sun gravity is 179 times stronger than the Earth–Moon gravity.

D The Periodic Table of the Elements

Key

12 — Atomic number **Mg** — Element's symbol Magnesium — Element's name 24.305 — Atomic mass*

*Atomic masses are fractions because they represent a weighted average of atomic masses of different isotopes—in proportion to the abundance of each isotope on Earth.

1 **H** Hydrogen 1.00794																	2 **He** Helium 4.003
3 **Li** Lithium 6.941	4 **Be** Beryllium 9.01218											5 **B** Boron 10.81	6 **C** Carbon 12.011	7 **N** Nitrogen 14.007	8 **O** Oxygen 15.999	9 **F** Fluorine 18.988	10 **Ne** Neon 20.179
11 **Na** Sodium 22.990	12 **Mg** Magnesium 24.305											13 **Al** Aluminum 26.98	14 **Si** Silicon 28.086	15 **P** Phosphorus 30.974	16 **S** Sulfur 32.06	17 **Cl** Chlorine 35.453	18 **Ar** Argon 39.948
19 **K** Potassium 39.098	20 **Ca** Calcium 40.08	21 **Sc** Scandium 44.956	22 **Ti** Titanium 47.88	23 **V** Vanadium 50.94	24 **Cr** Chromium 51.996	25 **Mn** Manganese 54.938	26 **Fe** Iron 55.847	27 **Co** Cobalt 58.9332	28 **Ni** Nickel 58.69	29 **Cu** Copper 63.546	30 **Zn** Zinc 65.39	31 **Ga** Gallium 69.72	32 **Ge** Germanium 72.59	33 **As** Arsenic 74.922	34 **Se** Selenium 78.96	35 **Br** Bromine 79.904	36 **Fr** Krypton 83.80
37 **Rb** Rubidium 85.468	38 **Sr** Strontium 87.62	39 **Y** Yttrium 88.9059	40 **Zr** Zirconium 91.224	41 **Nb** Niobium 92.91	42 **Mo** Molybdenum 95.94	43 **Tc** Technetium (98)	44 **Ru** Ruthenium 101.07	45 **Rh** Rhodium 102.906	46 **Pd** Palladium 106.42	47 **Ag** Silver 107.868	48 **Cd** Cadmium 112.41	49 **In** Indium 114.82	50 **Sn** Tin 118.71	51 **Sb** Antimony 121.75	52 **Te** Tellurium 127.60	53 **I** Iodine 126.905	54 **Xe** Xenon 131.29
55 **Cs** Cesium 132.91	56 **Ba** Barium 137.34		72 **Hf** Hafnium 178.49	73 **Ta** Tantalum 180.95	74 **W** Tungsten 183.85	75 **Re** Rhenium 186.207	76 **Os** Osmium 190.2	77 **Ir** Iridium 192.22	78 **Pt** Platinum 195.08	79 **Au** Gold 196.967	80 **Hg** Mercury 200.59	81 **Ti** Thallium 204.383	82 **Pb** Lead 207.2	83 **Bi** Bismuth 208.98	84 **Po** Polonium (209)	85 **At** Astatine (210)	86 **Rn** Radon (222)
87 **Fr** Francium (223)	88 **Ra** Radium 226.0254		104 **Rf** Rutherfordium (261)	105 **Db** Dubnium (262)	106 **Sg** Seaborgium (263)	107 **Bh** Bohrium (262)	108 **Hs** Hassium (265)	109 **Mt** Meitnerium (266)	110 **Uun** Ununnilium (269)	111 **Uuu** Unununium (272)	112 **Uub** Ununbium (277)						

Lanthanide Series

57 **La** Lanthanum 138.906	58 **Ce** Cerium 140.12	59 **Pr** Praseodymium 140.908	60 **Nd** Neodymium 144.24	61 **Pm** Promethium (145)	62 **Sm** Samarium 150.36	63 **Eu** Europium 151.96	64 **Gd** Gadolinium 157.25	65 **Tb** Terbium 158.925	66 **Dy** Dysprosium 162.50	67 **Ho** Holmium 164.93	68 **Er** Erbium 167.26	69 **Tm** Thulium 168.934	70 **Yb** Ytterbium 173.04	71 **Lu** Lutetium 174.967

Actinide Series

89 **Ac** Actinium 227.028	90 **Th** Thorium 232.038	91 **Pa** Protactinium 231.036	92 **U** Uranium 238.029	93 **Np** Neptunium 237.048	94 **Pu** Plutonium (244)	95 **Am** Americium (243)	96 **Cm** Curium (247)	97 **Bk** Berkelium (247)	98 **Cf** Californium (251)	99 **Es** Einsteinium (252)	100 **Fm** Fermium (257)	101 **Md** Mendelevium (258)	102 **No** Nobelium (259)	103 **Lr** Lawrencium (260)

E Planetary Data

Table E.1 Physical Properties of the Sun and Planets

Name	*Radius (Eq[a]) (km)*	*Radius (Eq) (Earth units)*	*Mass (kg)*	*Mass (Earth units)*	*Average Density (g/cm^3)*	*Surface Gravity (Earth = 1)*	*Escape Velocity (km/s)*
Sun	695,000	109	1.99×10^{30}	333,000	1.41	27.5	—
Mercury	2,440	0.382	3.30×10^{23}	0.055	5.43	0.38	4.43
Venus	6,051	0.949	4.87×10^{24}	0.815	5.25	0.91	10.4
Earth	6,378	1.00	5.97×10^{24}	1.00	5.52	1.00	11.2
Mars	3,397	0.533	6.42×10^{23}	0.107	3.93	0.38	5.03
Jupiter	71,492	11.19	1.90×10^{27}	317.9	1.33	2.36	59.5
Saturn	60,268	9.46	5.69×10^{26}	95.18	0.70	0.92	35.5
Uranus	25,559	3.98	8.66×10^{25}	14.54	1.32	0.91	21.3
Neptune	24,764	3.81	1.03×10^{26}	17.13	1.64	1.14	23.6
Pluto	1,160	0.181	1.31×10^{22}	0.0022	2.05	0.07	1.25
Planet X[b]	1,430	0.22	—	—	—	—	—

[a]Eq = equatorial.

[b] Little is known about this newly-discovered object, formally known as 2003 UB313. Even its size could be 25% larger or smaller, depending on its actual reflectivity. Its mass is therefore unknown to a factor of two or more.

Table E.2 Orbital Properties of the Sun and Planets

Name	Distance from Sun[a] (AU)	(10^6 km)	Orbital Period (years)	Orbital Inclination[b] (degrees)	Orbital Eccentricity	Sidereal Rotation Period (Earth days)[c]	Axis Tilt (degrees)
Sun	—	—	—	—	—	25.4	7.25
Mercury	0.387	57.9	0.2409	7.00	0.206	58.6	0.0
Venus	0.723	108.2	0.6152	3.39	0.007	−243.0	177.3
Earth	1.00	149.6	1.0	0.00	0.017	0.9973	23.45
Mars	1.524	227.9	1.881	1.85	0.093	1.026	25.2
Jupiter	5.203	778.3	11.86	1.31	0.048	0.41	3.08
Saturn	9.539	1,427	29.42	2.48	0.056	0.44	26.73
Uranus	19.19	2,870	84.01	0.77	0.046	−0.72	97.92
Neptune	30.06	4,497	164.8	1.77	0.010	0.67	29.6
Pluto	39.48	5,906	248.0	17.14	0.248	−6.39	112.5
Planet X[d]	67.89	10,156	560	44.0	0.438	—	—

[a] Semimajor axis of the orbit.

[b] With respect to the ecliptic.

[c] A negative sign indicates rotation is backward relative to other planets.

[d] The orbit of this newly-discovered object (formally known as 2003 UB313) is well known, but little is known about the object itself.

Table E.3 Satellites of the Solar System (as of 2005)[a]

Planet Satellite	Radius or Dimensions[b] (km)	Distance from Planet (10^3 km)	Orbital Period[c] (Earth days)	Mass[d] (kg)	Density[d] (g/cm^3)	Notes About the Satellite
Earth						**Earth**
Moon	1,738	384.4	27.322	7.349×10^{22}	3.34	*Moon:* Probably formed in giant impact.
Mars						**Mars**
Phobos	$13 \times 11 \times 9$	9.38	0.319	1.3×10^{16}	1.9	*Phobos, Deimos:* Probable captured asteroids.
Deimos	$8 \times 6 \times 5$	23.5	1.263	1.8×10^{15}	2.2	
Jupiter						**Jupiter**
Small inner moons (4 moons)	8 to 83	128–222	0.295–0.6745	—	—	*Metis, Adrastea, Amalthea, Thebe:* Small moonlets within and near Jupiter's ring system.
Io	1,821	421.6	1.769	8.933×10^{22}	3.57	*Io:* Most volcanically active object in the solar system.
Europa	1,565	670.9	3.551	4.797×10^{22}	2.97	*Europa:* Possible oceans under icy crust.
Ganymede	2,634	1,070.0	7.155	1.482×10^{23}	1.94	*Ganymede:* Largest satellite in solar system; unusual ice geology.
Callisto	2,403	1,883.0	16.689	1.076×10^{23}	1.86	*Callisto:* Cratered iceball.
Irregular group 1 (7 moons)	4–85	7,200–17,000	130–457	—	—	*Themisto, Leda, Himalia, Lysithea, Elara, and 44 others:* Probable captured moons with inclined orbits.
Irregular group 2 (48 moons)	1–30	15,900–29,500	−490 to −983	—	—	*Ananke, Carme, Pasiphae, Sinope, and 46 others:* Probable captured moons in inclined backward orbits.
Saturn						**Saturn**
Small inner moons (11)	3–89	134–212	0.574–1.1	—	—	*Pan, Atlas, Prometheus, Pandora, Epimetheus, Janus and 5 more:* Small moonlets within and near Saturn's ring system.
Mimas	199	185.52	0.942	3.70×10^{19}	1.17	*Mimas, Enceladus, Tethys:* Small and medium-size iceballs, many with interesting geology.
Enceladus	249	238.02	1.370	1.2×10^{20}	1.24	
Tethys	530	294.66	1.888	6.17×10^{20}	1.26	
Calypso & Telesto	8–12	294.66	1.888	—	—	*Calypso, Telesto:* Small moonlets sharing Tethys's orbit.
Dione	559	377.4	2.737	1.08×10^{21}	1.44	*Dione:* Medium-size iceball, with interesting geology.
Helene & Polydeuces	2–16	377.4	2.737	1.6×10^{16}	—	*Helene & Polydeuces:* Small moonlets sharing Dione's orbit.
Rhea	764	527.04	4.518	2.31×10^{21}	1.33	*Rhea:* Medium-size iceball, with interesting geology.
Titan	2,575	1,221.85	15.945	1.35×10^{23}	1.88	*Titan:* Dense atmosphere shrouds surface; ongoing geological activity.
Hyperion	$180 \times 140 \times 112$	1,481.1	21.277	2.8×10^{19}	—	*Hyperion:* Only satellite known not to rotate synchronously.

Iapetus	718	3,561.3	79.331	1.59×10^{21}	1.21	*Iapetus:* Bright and dark hemispheres show greatest contrast in the solar system.
Phoebe	110	12,952	−550.4	1×10^{19}	—	*Phoebe:* Very dark; material ejected from Phoebe may coat one side of Iapetus.
Irregular groups (25 moons)	2–16	11,400–23,400	450–930	—	—	Probable captured moons with highly inclined and/or backward orbits.
			−550 to −1,320	—	—	
Uranus						**Uranus**
Small inner moons (13 moons)	5–81	49–98	0.4–0.9	—	—	*Cordelia, Ophelia, Bianca, Cressida, Desdemona, Juliet, Portia, Rosalind, Belinda, Puck, 1986 U10, 2003 U1, 2003 U3:* Small moonlets within and near Uranus's ring system.
Miranda	236	129.8	1.413	6.6×10^{19}	1.26	*Miranda, Ariel, Umbriel, Titania, Oberon:* Small and medium-size iceballs, with some interesting geology.
Ariel	579	191.2	2.520	1.35×10^{21}	1.65	
Umbriel	584.7	266.0	4.144	1.17×10^{21}	1.44	
Titania	788.9	435.8	8.706	3.52×10^{21}	1.59	
Oberon	761.4	582.6	13.463	3.01×10^{21}	1.50	
Irregular group (9 moons)	5–95	4,280–21,000	580–2,820	—	—	*Caliban, Sycorax, Stephano, Prospero, Setebos, Trinculo, 2001 U2, 2001 U3, 2003 U3:* Probable captured moons; several in backward orbits.
Neptune						**Neptune**
Small inner moons (5 moons)	29–86	48–74	0.30–0.55	—	—	*Naiad, Thalassa, Despina, Galatea, Larissa:* Small moonlets within and near Neptune's ring system.
Proteus	$218 \times 208 \times 201$	117.6	1.121	6×10^{19}	—	
Triton	1,352.6	354.59	−5.875	2.14×10^{22}	2.0	*Triton:* Probable captured Kuiper belt object—largest captured object in solar system.
Nereid	170	5,588.6	360.125	3.1×10^{19}	—	*Nereid:* Small, icy moon; very little known.
Irregulars (5 moons)	15–27	16,600–48,600	1,870–9,412	—	—	*2002 N1, N2, N3, N4, 2003 N1:* Possible captured moons in inclined or backward orbit.
Pluto						**Pluto**
Charon	593	19.6	6.38718	1.56×10^{21}	1.6	*Charon:* Unusually large compared to its planet; may have formed in giant impact.
S/2005 P1 and S/2005 P2	32–100	—	—	—	—	Newly discovered moons outside Charon's orbit.

[a] *Note:* Authorities differ substantially on many of the values in this table.

[b] $a \times b \times c$ values for the Dimensions are the approximate lengths of the axes (center to edge) for irregular moons.

[c] Negative sign indicates backward orbit.

[d] Masses and densities are most accurate for those satellites visited by a spacecraft on a flyby. Masses for the smallest moons have not been measured but can be estimated from the radius and an assumed density.

Table E.4 Fifty Extrasolar Planets of Note (listed in order of distance from their star)

Name	Detection Methods[2]	Minimum mass (Jupiter masses)	Semimajor Axis (AU)	Period (days)	Radius (Jupiter radii)	Notes[1]
Gliese 876 d	radial velocity	0.023	0.021	1.94	—	"Hot Jupiter"; smallest known semimajor axis of 2005
OGLE-TR-56 b	radial velocity, transit	1.45	0.023	1.21	1.23	"Hot Jupiter"; first planet discovered by transit
GJ 436 b	radial velocity	0.067	0.028	2.64	—	"Hot Jupiter"; sub-Saturn mass
OGLE-TR-132 b	radial velocity, transit	1.19	0.031	1.69	1.13	"Hot Jupiter"
55 Cnc e	radial velocity	0.045	0.038	2.81	—	"Hot Jupiter"; sub-Neptune mass
TrES-1	transit, radial velocity, eclipse	0.61	0.039	3.03	1.08	"Hot Jupiter"
HD 46375 b	radial velocity	0.25	0.041	3.02	—	"Hot Jupiter"; sub-Saturn mass
OGLE-TR-10 b	transit, radial velocity	0.54	0.042	3.10	1.16	"Hot Jupiter"
HD 149026 b	radial velocity, transit	0.36	0.042	2.88	0.73	"Hot Jupiter"
HD 188753A b	radial velocity	1.14	0.045	3.35	—	"Hot Jupiter"; first planet discovered in triple-star system
HD 209458 b	radial velocity, transit, eclipse	0.69	0.045	3.52	1.32	"Hot Jupiter"; first planet and first atmosphere successfully detected by transit
HD 88133 b	radial velocity	0.22	0.047	3.41	—	"Hot Jupiter"; sub-Saturn mass
OGLE-TR-111 b	radial velocity, transit	0.53	0.047	4.02	1.00	"Hot Jupiter"
51 Peg b	radial velocity	0.47	0.052	4.23	—	"Hot Jupiter"; first exoplanet discovered around sun-like star
Ups And b	radial velocity	0.69	0.059	4.62	—	"Hot Jupiter"; in first multi-planet system discovered around sun-like star
HD 160691 d	radial velocity	0.044	0.090	9.55	—	"Hot Jupiter"; sub-Neptune mass
55 Cnc b	radial velocity	0.78	0.115	14.7	—	—
HD 190360 c	radial velocity	0.057	0.128	17.1	—	Sub-Saturn mass
Gliese 876 c	radial velocity	0.56	0.130	30.1	—	"Eccentric"[3]
HD 102117 b	radial velocity	0.14	0.149	20.7	—	Sub-Saturn mass
Gliese 876 b	radial velocity	1.94	0.208	60.9	—	—
55 Cnc c	radial velocity	0.22	0.240	43.9	—	"Eccentric"; sub-Saturn mass
HD 16141 b	radial velocity	0.23	0.350	75.6	—	Sub-Saturn mass
HD 80606 b	radial velocity	3.4	0.439	112	—	"Eccentric"
HD 82943 c	radial velocity	0.88	0.730	222	—	"Eccentric"

Ups And c	radial velocity	1.89	0.829	242	—	"Eccentric"; in first multi-planet system discovered
HR 810 b	radial velocity	1.94	0.910	311	—	—
HD 210277 b	radial velocity	1.24	1.097	436	—	"Eccentric"; planet mass inferred from surrounding disk
HD 114783 b	radial velocity	0.99	1.200	501	—	—
HD 41004A b	radial velocity	2.3	1.310	655	—	"Eccentric"; planet in a system with two stars and brown dwarf
HD 183263 b	radial velocity	3.7	1.520	634	—	"Eccentric"
HD 114386 b	radial velocity	0.99	1.620	872	—	"Eccentric"
HD 4208 b	radial velocity	0.80	1.670	812	—	—
HD 13189 b	radial velocity	14.0	1.850	472	—	"Eccentric"
Gamma Cephei b	radial velocity	1.59	2.030	902	—	First extrasolar planet discovered in close stellar binary system
47 UMa b	radial velocity	2.5	2.090	1089	—	—
HD 50554 b	radial velocity	4.9	2.380	1279	—	"Eccentric"
Ups And d	radial velocity	3.8	2.530	1284	—	"Eccentric"; in first multi-planet system discovered
HD 202206 c	radial velocity	2.4	2.550	1383	—	"Eccentric"
HD 12661 c	radial velocity	1.57	2.560	1445	—	—
HD 168443 c	radial velocity	17.1	2.870	1740	—	—
HD 37124 d	radial velocity	0.66	3.190	2295	—	—
Epsilon Eridani b	radial velocity, astrometry	0.86	3.300	2502	—	"Eccentric"; star surrounded by dust disk
HD 74156 c	radial velocity	6.2	3.400	2025	—	"Eccentric"
HD 169830 c	radial velocity	4.0	3.600	2102	—	"Eccentric"
HD 38529 c	radial velocity	12.7	3.680	2174	—	"Eccentric"
47 UMa c	radial velocity	0.76	3.730	2594	—	—
HD 50499 b	radial velocity	1.71	3.860	2583	—	—
HD 72659 b	radial velocity	3.0	4.160	3177	—	"Eccentric"
55 Cnc d	radial velocity	3.9	5.257	4517	—	"Eccentric"; largest known semimajor axis as of 2005

Notes: [1]The list includes all planets detected by two methods, most planets in multiple systems, most "Hot Jupiters," and a representative sample of other extrasolar planets. More than 100 known extrasolar planets are not listed.

[2]If two detection methods are listed, the discovery method is given first.

[3]"Eccentric" means eccentricity > 0.25

F Stellar Data

Table F.1 Stars Within 12 Light-Years

Star	Distance (ly)	Spectral Type		RA h	RA m	Dec °	Dec ′	Luminosity (L/L_{Sun})
Sun	0.000016	G2	V	—	—	—	—	1.0
Proxima Centauri	4.2	M5.5	V	14	30	−62	41	0.0006
α Centauri A	4.4	G2	V	14	40	−60	50	1.6
α Centauri B	4.4	K0	V	14	40	−60	50	0.53
Barnard's Star	6.0	M4	V	17	58	+04	42	0.005
Wolf 359	7.8	M6	V	10	56	+07	01	0.0008
Lalande 21185	8.3	M2	V	11	03	+35	58	0.03
Sirius A	8.6	A1	V	06	45	−16	42	26.0
Sirius B	8.6	DA2	—	06	45	−16	42	0.002
UV Ceti	8.7	M5.5	V	01	39	−17	57	0.0009
BL Ceti	8.7	M6	V	01	39	−17	57	0.0006
Ross 154	9.7	M3.5	V	18	50	−23	50	0.004
Ross 248	10.3	M5.5	V	23	42	+44	11	0.001
ε Eridani	10.5	K2	V	03	33	−09	28	0.37
Lacaille 9352	10.7	M1.5	V	23	06	−35	51	0.05
Ross 128	10.9	M4	V	11	48	+00	49	0.003
EZ Aquarii A	11.3	M5	V	22	39	−15	18	0.0006
EZ Aquarii B	11.3	M6	V	22	39	−15	18	0.0004
EZ Aquarii C	11.3	M6.5	V	22	39	−15	18	0.0003
61 Cygni A	11.4	K5	V	21	07	+38	42	0.15
61 Cygni B	11.4	K7	V	21	07	+38	42	0.09
Procyon A	11.4	F5	IV–V	07	39	+05	14	7.4
Procyon B	11.4	DA	—	07	39	+05	14	0.0005
Gliese 725 A	11.5	M3	V	18	43	+59	38	0.02
Gliese 725 B	11.5	M3.5	V	18	43	+59	38	0.01
GX Andromedae	11.15	M1.5	V	00	18	+44	01	0.03
GQ Andromedae	11.15	M3.5	V	00	18	+44	01	0.003
DX Cancri	11.8	M6.5	V	08	30	+26	47	0.0003
ε Indi A	11.8	K5	V	22	03	−56	45	0.26
ε Indi B	11.8	T1.0	Brown Dwarf	22	04	−56	46	—
ε Indi C	11.8	T1.0	Brown Dwarf	22	04	−56	46	—
τ Ceti	11.9	G8	V	01	44	−15	57	0.59
GJ 1061	11.9	M5.5	V	03	36	−44	31	0.0009

Note: These data were provided by the RECONS project, courtesy of Dr. Todd Henry (July, 2005). The luminosities are all total (bolometric) luminosities. The DA stellar types are white dwarfs. The coordinates are for the year 2000. The bolometric luminosity of the brown dwarfs is primarily in the infrared, and has not been measured accurately yet.

Table F.2 Twenty Brightest Stars

Star	Constellation	RA h	RA m	Dec °	Dec ′	Distance (ly)	Spectral Type		Apparent Magnitude	Luminosity (L/L_{Sun})
Sirius	Canis Major	6	45	−16	42	8.6	A1	V	−1.46	26
Canopus	Carina	6	24	−52	41	313	F0	Ib-II	−0.72	13,000
α Centauri	Centaurus	14	40	−60	50	4.4	{G2	V	−0.01	1.6
							K0}	V	1.3	0.53
Arcturus	Boötes	14	16	+19	11	37	K2	III	−0.06	170
Vega	Lyra	18	37	+38	47	25	A0	V	0.04	60
Capella	Auriga	5	17	+46	00	42	{G0	III	0.75	70
							G8}	III	0.85	77
Rigel	Orion	5	15	−08	12	772	B8	Ia	0.14	70,000
Procyon	Canis Minor	7	39	+05	14	11.4	F5	IV–V	0.37	7.4
Betelgeuse	Orion	5	55	+07	24	427	M2	Iab	0.41	38,000
Achernar	Eridanus	1	38	−57	15	144	B5	V	0.51	3,600
Hadar	Centaurus	14	04	−60	22	525	B1	III	0.63	100,000
Altair	Aquila	19	51	+08	52	17	A7	IV–V	0.77	10.5
Acrux	Crux	12	27	−63	06	321	{B1	IV	1.39	22,000
							B3}	V	1.9	7,500
Aldebaran	Taurus	4	36	+16	30	65	K5	III	0.86	350
Spica	Virgo	13	25	−11	09	260	B1	V	0.91	23,000
Antares	Scorpio	16	29	−26	26	604	M1	Ib	0.92	38,000
Pollux	Gemini	7	45	+28	01	34	K0	III	1.16	45
Fomalhaut	Piscis Austrinus	22	58	−29	37	25	A3	V	1.19	18
Deneb	Cygnus	20	41	+45	16	2,500	A2	Ia	1.26	170,000
β Crucis	Crux	12	48	−59	40	352	B0.5	IV	1.28	37,000

Note: Three of the stars on this list, Capella, α Centauri, and Acrux, are binary systems with members of comparable brightness. They are counted as single stars because that is how they appear to the naked eye. All the luminosities given are total (bolometric) luminosities. The coordinates are for the year 2000.

G Galaxy Data

Table G.1 Galaxies of the Local Group

Galaxy Name	*Distance (millions of ly)*	*Type*[a]	*RA* h	m	*Dec* °	′	*Luminosity (millions of L_{Sun})*
Milky Way	—	Sbc	—	—	—	—	15,000
WLM	3.0	Irr	00	02	−15	30	50
IC 10	2.7	dIrr	00	20	+59	18	160
Cetus	2.5	dE	00	26	−11	02	0.72
NGC 147	2.4	dE	00	33	+48	30	131
And III	2.5	dE	00	35	+36	30	1.1
NGC 185	2.0	dE	00	39	+48	20	120
NGC 205	2.7	E	00	40	+41	41	370
And VIII	2.7	dE	00	42	+40	37	240
M 32	2.6	E	00	43	+40	52	380
M 31	2.5	Sb	00	43	+41	16	21,000
And I	2.6	dE	00	46	+38	00	4.7
SMC	0.19	Irr	00	53	−72	50	230
And IX	2.9	dE	00	52	+43	12	—
Sculptor	0.26	dE	01	00	−33	42	2.2
LGS 3	2.6	dIrr	01	04	+21	53	1.3
IC 1613	2.3	Irr	01	05	+02	08	64
And V	2.9	dE	01	10	+47	38	—
And II	1.7	dE	01	16	+33	26	2.4
M 33	2.7	Sc	01	34	+30	40	2,800
Phoenix	1.5	dIrr	01	51	−44	27	0.9
Fornax	0.45	dE	02	40	−34	27	15.5
EGB0427 + 63	4.3	dIrr	04	32	+63	36	9.1
LMC	0.16	Irr	05	24	−69	45	1,300
Carina	0.33	dE	06	42	−50	58	0.4
Canis Major	0.025	dIrr	07	15	−28		—
Leo A	2.2	dIrr	09	59	+30	45	3.0
Sextans B	4.4	dIrr	10	00	+05	20	41
NGC 3109	4.1	Irr	10	03	−26	09	160
Antlia	4.0	dIrr	10	04	−27	19	1.7
Leo I	0.82	dE	10	08	+12	18	4.8
Sextans A	4.7	dIrr	10	11	−04	42	56
Sextans	0.28	dE	10	13	−01	37	0.5
Leo II	0.67	dE	11	13	+22	09	0.6
GR 8	5.2	dIrr	12	59	+14	13	3.4
Ursa Minor	0.22	dE	15	09	+67	13	0.3
Draco	2.7	dE	17	20	+57	55	0.3
Sagittarius	0.08	dE	18	55	−30	29	18
SagDIG	3.5	dIrr	19	30	−17	41	6.8
NGC 6822	1.6	Irr	19	45	−14	48	94
DDO 210	2.6	dIrr	20	47	−12	51	0.8
IC 5152	5.2	dIrr	22	03	−51	18	70
Tucana	2.9	dE	22	42	−64	25	0.5
UKS2323-326	4.3	dE	23	26	−32	23	5.2
And VII	2.6	dE	23	38	+50	35	—
Pegasus	3.1	dIrr	23	29	+14	45	12
And VI	2.8	dE	23	52	+24	36	—

[a] Types beginning with S are spiral galaxies classified according to Hubble's system (see Chapter 15). Type E galaxies are elliptical or spheroidal. Type Irr galaxies are irregular. The prefix d denotes a dwarf galaxy. This list is based on a list originally published by M. Mateo in 1998, and augmented by discoveries of Local Group galaxies made between 1998–2005.

Table G.2 Nearby Galaxies in the Messier Catalog[a,b]

Galaxy Name (M / NGC)[c]	*RA* h	m	*Dec* °	′	*RV$_{hel}$*[d]	*RV$_{gal}$*[e]	*Type*[f]	*Nickname*
M31 / NGC224	00	43	+41	16	−300 ± 4	−122	Spiral	Andromeda
M32 / NGC221	00	43	+40	52	−145 ± 2	32	Elliptical	
M33 / NGC598	01	34	+30	40	−179 ± 3	−44	Spiral	Triangulum
M49 / NGC4472	12	30	+08	00	997 ± 7	929	Elliptical/ Lenticular/Seyfert	
M51 / NGC5194	13	30	+47	12	463 ± 3	550	Spiral/Interacting	Whirlpool
M58 / NGC4579	12	38	+11	49	1,519 ± 6	1,468	Spiral/Seyfert	
M59 / NGC4621	12	42	+11	39	410 ± 6	361	Elliptical	
M60 / NGC4649	12	44	+11	33	1,117 ± 6	1,068	Elliptical	
M61 / NGC4303	12	22	+04	28	1,566 ± 2	1,483	Spiral/Seyfert	
M63 / NGC5055	13	16	+42	02	504 ± 4	570	Spiral	Sunflower
M64 / NGC4826	12	57	+21	41	408 ± 4	400	Spiral/Seyfert	Black Eye
M65 / NGC3623	11	19	+13	06	807 ± 3	723	Spiral	
M66 / NGC3627	11	20	+12	59	727 ± 3	643	Spiral/Seyfert	
M74 / NGC628	01	37	+15	47	657 ± 1	754	Spiral	
M77 / NGC1068	02	43	−00	01	1,137 ± 3	1,146	Spiral/Seyfert	
M81 / NGC3031	09	56	+69	04	−34 ± 4	73	Spiral/Seyfert	
M82 / NGC3034	09	56	+69	41	203 ± 4	312	Irregular/Starburst	
M83 / NGC5236	13	37	−29	52	516 ± 4	385	Spiral/Starburst	
M84 / NGC4374	12	25	+12	53	1,060 ± 6	1,005	Elliptical	
M85 / NGC4382	12	25	+18	11	729 ± 2	692	Spiral	
M86 / NGC4406	12	26	+12	57	−244 ± 5	−298	Elliptical/Lenticular	
M87 / NGC4486	12	30	+12	23	1,307 ± 7	1,254	Elliptical/Central Dominant/Seyfert	Virgo A
M88 / NGC4501	12	32	+14	25	2,281 ± 3	2,235	Spiral/Seyfert	
M89 / NGC4552	12	36	+12	33	340 ± 4	290	Elliptical	
M90 / NGC4569	12	37	+13	10	−235 ± 4	−282	Spiral/Seyfert	
M91 / NGC4548	12	35	+14	30	486 ± 4	442	Spiral/Seyfert	
M94 / NGC4736	12	51	+41	07	308 ± 1	360	Spiral	
M95 / NGC3351	10	44	+11	42	778 ± 4	677	Spiral/Starburst	
M96 / NGC3368	10	47	+11	49	897 ± 4	797	Spiral/Seyfert	
M98 / NGC4192	12	14	+14	54	−142 ± 4	−195	Spiral/Seyfert	
M99 / NGC4254	12	19	+14	25	2,407 ± 3	2,354	Spiral	
M100 / NGC4321	12	23	+15	49	1,571 ± 1	1,525	Spiral	
M101 / NGC5457	14	03	+54	21	241 ± 2	360	Spiral	
M104 / NGC4594	12	40	−11	37	1,024 ± 5	904	Spiral/Seyfert	Sombrero
M105 / NGC3379	10	48	+12	35	911 ± 2	814	Elliptical	
M106 / NGC4258	12	19	+47	18	448 ± 3	507	Spiral/Seyfert	
M108 / NGC3556	11	09	+55	57	695 ± 3	765	Spiral	
M109 / NGC3992	11	55	+53	39	1,048 ± 4	1,121	Spiral	
M110 / NGC205	00	38	+41	25	−241 ± 3	−61	Elliptical	

[a] Galaxies identified in the catalog published by Charles Messier in 1781; these galaxies are relatively easy to observe with small telescopes.

[b] Data obtained from NED: NASA/IPAC Extragalactic Database (http://ned.ipac.caltech.edu). The original Messier list of galaxies was obtained from SED, and the list data were updated to 2001 and M 102 was dropped.

[c] The galaxies are identified by the Messier number (M followed by a number) and by their NGC numbers, which come from the *New General Catalog* published in 1888.

[d] Radial velocity in km/s, with respect to the Sun (heliocentric). Positive values mean motion away from the Sun, and negative values are toward the Sun.

[e] Radial velocity in km/s, with respect to the Milky Way Galaxy, calculated from the RV_{hel} values with a correction for the Sun's motion around the galactic center.

[f] Galaxies are first listed by their primary type (spiral, elliptical, or irregular) and then by any other special categories that apply (see Chapter 15).

Table G.3 Nearby, X-ray Bright Clusters of Galaxies

Cluster Name	*Redshift*	*Distance*[a] *(billions of ly)*	*Temperature of Intracluster Medium (millions of K)*	*Average Orbital Velocity of Galaxies*[b] *(km/sec)*	*Cluster Mass*[c] *(10^{15} M_{Sun})*
Abell 2142	0.0907	1.26	101. ± 2	1,132 ± 110	1.6
Abell 2029	0.0766	1.07	100. ± 3	1,164 ± 98	1.5
Abell 401	0.0737	1.03	95.2 ± 5	1,152 ± 86	1.4
Coma	0.0233	0.32	95.1 ± 1	821 ± 49	1.4
Abell 754	0.0539	0.75	93.3 ± 3	662 ± 77	1.4
Abell 2256	0.0589	0.82	87.0 ± 2	1,348 ± 86	1.4
Abell 399	0.0718	1.00	81.7 ± 7	1,116 ± 89	1.1
Abell 3571	0.0395	0.55	81.1 ± 3	1,045 ± 109	1.1
Abell 478	0.0882	1.23	78.9 ± 2	904 ± 281	1.1
Abell 3667	0.0566	0.79	78.5 ± 6	971 ± 62	1.1
Abell 3266	0.0599	0.84	78.2 ± 5	1,107 ± 82	1.1
Abell 1651a	0.0846	1.18	73.1 ± 6	685 ± 129	0.96
Abell 85	0.0560	0.78	70.9 ± 2	969 ± 95	0.92
Abell 119	0.0438	0.61	65.6 ± 5	679 ± 106	0.81
Abell 3558	0.0480	0.67	65.3 ± 2	977 ± 39	0.81
Abell 1795	0.0632	0.88	62.9 ± 2	834 ± 85	0.77
Abell 2199	0.0314	0.44	52.7 ± 1	801 ± 92	0.59
Abell 2147	0.0353	0.49	51.1 ± 4	821 ± 68	0.56
Abell 3562	0.0478	0.67	45.7 ± 8	736 ± 49	0.48
Abell 496	0.0325	0.45	45.3 ± 1	687 ± 89	0.47
Centaurus	0.0103	0.14	42.2 ± 1	863 ± 34	0.42
Abell 1367	0.0213	0.30	41.3 ± 2	822 ± 69	0.41
Hydra	0.0126	0.18	38.0 ± 1	610 ± 52	0.36
C0336	0.0349	0.49	37.4 ± 1	650 ± 170	0.35
Virgo	0.0038	0.05	25.7 ± 0.5	632 ± 41	0.20

Note: This table lists the 25 brightest clusters of galaxies in the X-ray sky from a catalog by J. P. Henry (2000).

[a] Cluster distances were computed using a value for Hubble's constant of 21.5 km/sec/million light-years.

[b] The average orbital velocities given in this column are the velocity component along our line of sight. This velocity should be multiplied by the square root of 2 to get the average orbital velocity used in Mathematical Insights 22.2 and 22.3.

[c] This column gives each cluster's mass within the largest radius at which the intracluster medium can be in gravitational equilibrium. Because our estimates of that radius depend on Hubble's constant, these masses are inversely proportional to Hubble's constant, which we have assumed to be 21.5 km/sec/million light-years.

H Selected Astronomical Web Sites

The Web contains a vast amount of astronomical information. For all your astronomical Web surfing, the best starting point is the Web site for this textbook:

MasteringAstronomy™
www.masteringastronomy.com

The following are some other sites that may be of particular use. In case any of the links change, you can always find live links to these sites, and many more, on the MasteringAstronomy Web site.

Key Mission Sites

The following table lists the Web pages for major current astronomy missions.

Site	*Description*	*Web Address*
NASA's Office of Space Science Missions Page	**Direct links to all past, present, and planned NASA space science missions**	**http://spacescience.nasa.gov/missions**
Cassini/Huygens	Mission arrived at Saturn in 2004	http://saturn.jpl.nasa.gov/index.cfm
Chandra X-Ray Observatory	Latest discoveries, educational activities, and other information from the Chandra X-Ray Observatory	http://chandra.harvard.edu
Far Ultraviolet Spectroscopic Explorer (FUSE)	Ultraviolet observatory in space	http://fuse.pha.jhu.edu
Galileo	Mission orbiting Jupiter	http://www.jpl.nasa.gov/galileo
Hubble Space Telescope	Latest discoveries, educational activities, and other information from the Hubble Space Telescope	http://hubble.stsci.edu
Mars Exploration Program	Information on current and planned Mars missions	http://mars.jpl.nasa.gov
Microwave Anisotropy Probe (MAP)	Mission to study the cosmic microwave background	http://map.gsfc.nasa.gov
Spitzer Space Telescope	Infrared observatory launched in 2003	http://spitzer.caltech.edu
Stratospheric Observatory for Infrared Astronomy (SOFIA)	Airborne observatory scheduled to begin flights in 2002	http://sofia.arc.nasa.gov

Key Observatory Sites

The following table lists the Web pages leading to major ground-based observatories.

Site	*Description*	*Web Address*
World's Largest Optical Telescopes	**Direct links to most of the world's major optical observatories**	**http://www.seds.org/billa/bigeyes.html**
Arecibo Observatory (Puerto Rico)	World's largest single-dish radio telescope	http://www.naic.edu
Cerro Tololo Inter-American Observatory	Links to major observatories on site in Cerro Tololo, Chile	http://www.ctio.noao.edu
European Southern Observatory	Links to European telescope projects in Chile, including the Very Large Telescope	http://www.eso.org
Mauna Kea Observatories	Links to major observatories in Hawaii, including Keck, Gemini, Subaru, CFHT, and others	http://www.ifa.hawaii.edu/mko
Mt. Palomar Observatory	Powerful telescope near San Diego	http://www.astro.caltech.edu/palomarpublic
National Optical Astronomy Observatory	Home page for United States national observatories in Arizona, Hawaii, and Chile	http://www.noao.edu
National Radio Astronomy Observatory	Home page for United States national radio observatories, including the Very Large Array (VLA)	http://www.nrao.edu

More Astronomical Web Sites

The following Web sites are some of the authors' favorites among many other non-commercial resources for astronomy.

Site	*Description*	*Web Address*
MasteringAstronomy	**Don't forget to start here for all your astronomical Web surfing.**	**http://www.masteringastronomy.com**
American Association of Variable Star Observers (AAVSO)	One of the largest organizations of amateur astronomers in the world. Check this site if you are interested in serious amateur astronomy.	http://www.aavso.org
Astronomical Society of the Pacific	An organization for both professional astronomers and the general public, devoted largely to astronomy education.	http://www.astrosociety.org
Astronomy Picture of the Day	An archive of beautiful pictures, updated daily.	http://antwrp.gsfc.nasa.gov/apod
AstroWeb	Listing of major resources for astronomy on the Web.	http://www.stsci.edu/astroweb/astronomy.html
Canadian Space Agency	Home page for Canada's space program.	http://www.space.gc.ca
European Space Agency (ESA)	Home page for this international agency.	http://www.esa.int
The Extrasolar Planets Encyclopedia	Information about the search for and discoveries of extrasolar planets.	http://cfa-www.harvard.edu/planets
NASA Home Page	Learn almost anything you want about NASA.	http://www.nasa.gov
NASA Science News	Read the latest news from NASA; has option to subscribe to e-mail notices of news releases.	http://science.nasa.gov
The Nine Planets (University of Arizona)	A multimedia tour of the solar system.	http://www.nineplanets.org
The Planetary Society	Has more than 100,000 members who are interested in planetary exploration and the search for life in the universe.	http://planetary.org
The SETI Institute	Devoted to the search for other civilizations.	http://www.seti.org
Voyage Scale Model Solar System	Take a virtual tour of the Voyage Scale Model Solar System.	http://www.voyageonline.org

The 88 Constellations

Constellation Names (English Equivalent in Parentheses)

Andromeda (The Chained Princess)
Antlia (The Air Pump)
Apus (The Bird of Paradise)
Aquarius (The Water Bearer)
Aquila (The Eagle)
Ara (The Altar)
Aries (The Ram)
Auriga (The Charioteer)
Boötes (The Herdsman)
Caelum (The Chisel)
Camelopardalis (The Giraffe)
Cancer (The Crab)
Canes Venatici (The Hunting Dogs)
Canis Major (The Great Dog)
Canis Minor (The Little Dog)
Capricornus (The Sea Goat)
Carina (The Keel)
Cassiopeia (The Queen)
Centaurus (The Centaur)
Cepheus (The King)
Cetus (The Whale)
Chamaeleon (The Chameleon)
Circinus (The Drawing Compass)
Columba (The Dove)
Coma Berenices (Berenice's Hair)
Corona Australis (The Southern Crown)
Corona Borealis (The Northern Crown)
Corvus (The Crow)
Crater (The Cup)
Crux (The Southern Cross)

Cygnus (The Swan)
Delphinus (The Dolphin)
Dorado (The Goldfish)
Draco (The Dragon)
Equuleus (The Little Horse)
Eridanus (The River)
Fornax (The Furnace)
Gemini (The Twins)
Grus (The Crane)
Hercules
Horologium (The Clock)
Hydra (The Sea Serpent)
Hydrus (The Water Snake)
Indus (The Indian)
Lacerta (The Lizard)
Leo (The Lion)
Leo Minor (The Little Lion)
Lepus (The Hare)
Libra (The Scales)
Lupus (The Wolf)
Lynx (The Lynx)
Lyra (The Lyre)
Mensa (The Table)
Microscopium (The Microscope)
Monoceros (The Unicorn)
Musca (The Fly)
Norma (The Level)
Octans (The Octant)
Ophiuchus (The Serpent Bearer)
Orion (The Hunter)
Pavo (The Peacock)
Pegasus (The Winged Horse)
Perseus (The Hero)
Phoenix (The Phoenix)
Pictor (The Painter's Easel)
Pisces (The Fish)
Piscis Austrinus (The Southern Fish)
Puppis (The Stern)
Pyxis (The Compass)
Reticulum (The Reticle)
Sagitta (The Arrow)
Sagittarius (The Archer)
Scorpius (The Scorpion)
Sculptor (The Sculptor)
Scutum (The Shield)
Serpens (The Serpent)
Sextans (The Sextant)
Taurus (The Bull)
Telescopium (The Telescope)
Triangulum (The Triangle)
Triangulum Australe (Southern Triangle)
Tucana (The Toucan)
Ursa Major (The Great Bear)
Ursa Minor (The Little Bear)
Vela (The Sail)
Virgo (The Virgin)
Volans (The Flying Fish)
Vulpecula (The Fox)

Constellation Locations

Each of the charts on these pages shows half of the celestial sphere in projection, so you can use them to learn the approximate locations of the constellations. The grid lines are marked by right ascension and declination.

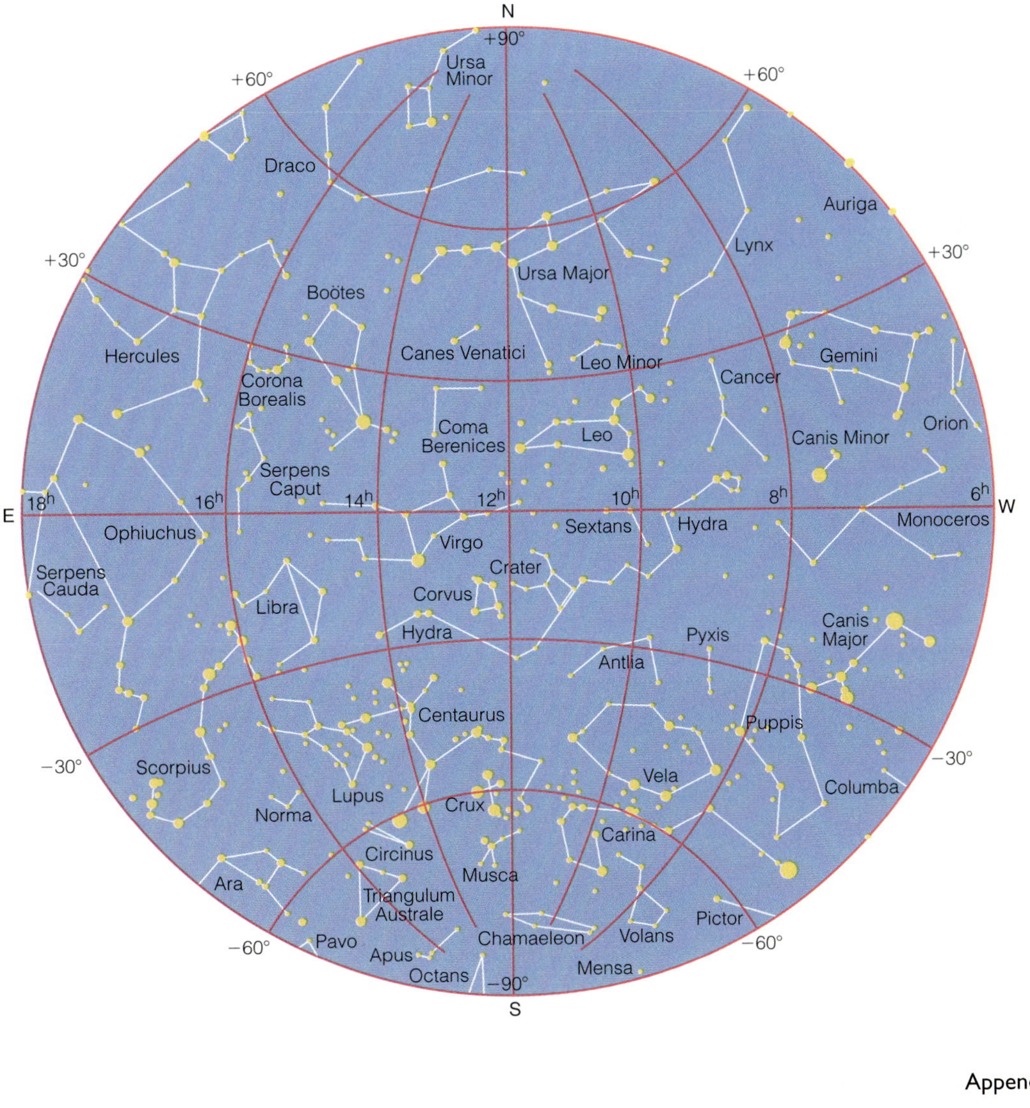

J Star Charts

How to use the star charts:

Check the times and dates under each chart to find the best one for you. Take it outdoors within an hour or so of the time listed for your date. Bring a dim flash-light to help you read it.

On each chart, the round outside edge represents the horizon all around you. Compass directions around the horizon are marked in yellow. Turn the chart around so the edge marked with the direction you're facing (for example, north, southeast) is down. The stars above this horizon now match the stars you are facing. Ignore the rest until you turn to look in a different direction.

The center of the chart represents the sky overhead, so a star plotted on the chart halfway from the edge to the center can be found in the sky halfway from the horizon to straight up.

The charts are drawn for 40°N latitude (for example, Denver, New York, Madrid). If you live far south of there, stars in the southern part of your sky will appear higher than on the chart and stars in the north will be lower. If you live far north of there, the reverse is true.

© 1999 *Sky & Telescope*

Use this chart January, February, and March.

Early January — 1 A.M.
Late January — Midnight

Early February — 11 P.M.
Late February — 10 P.M.

Early March — 9 P.M.
Late March — Dusk

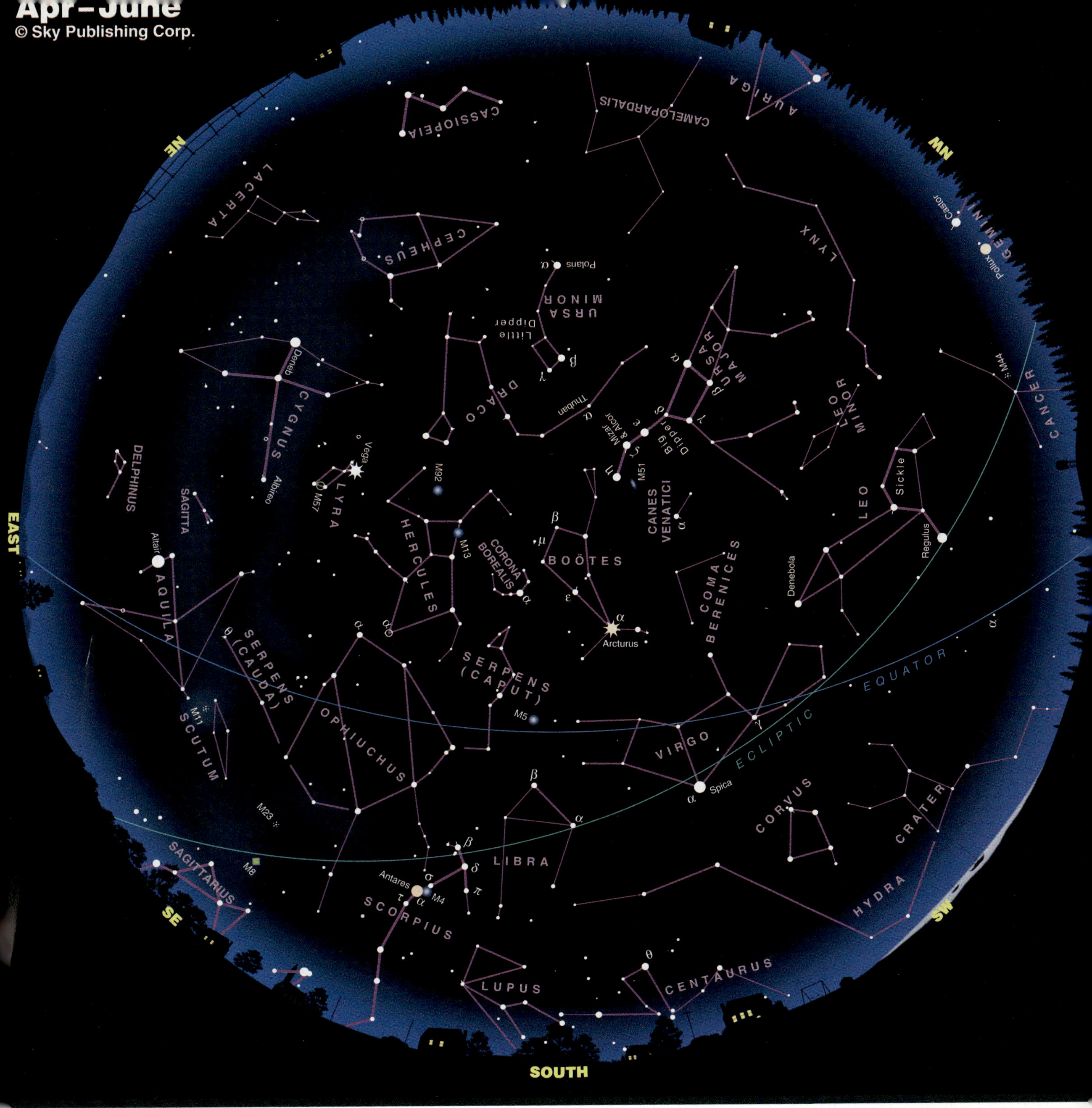

Use this chart April, May, and June.

Early April — 3 A.M.*
Late April — 2 A.M.*

Early May — 1 A.M.*
Late May — Midnight*

Early June — 11 P.M.*
Late June — Dusk

*Daylight Saving Time

Use this chart July, August, and September.

Early July — 1 A.M.*
Late July — Midnight*

Early August — 11 P.M.*
Late August — 10 P.M.*

Early September — 9 P.M.*
Late September — Dusk

*Daylight Saving Time

Use this chart October, November, and December.

Early October — 1 A.M.*
Late October — Midnight*

Early November — 10 P.M.
Late November — 9 P.M.

Early December — 8 P.M.
Late December — 7 P.M.

*Daylight Saving Time

Glossary

21-cm line A spectral line from atomic hydrogen with wavelength 21 cm (in the radio portion of the spectrum).

absolute magnitude A measure of an object's luminosity; defined to be the apparent magnitude the object would have if it were located exactly 10 parsecs away.

absolute zero The coldest possible temperature, which is 0 K.

absorption (of light) The process by which matter absorbs radiative energy.

absorption-line spectrum A spectrum that contains absorption lines.

accelerating universe The possible fate of our universe in which a repulsive force (*see* cosmological constant) causes the expansion of the universe to accelerate with time. Its galaxies will recede from one another increasingly faster, and it will become cold and dark more quickly than a coasting universe.

acceleration The rate at which an object's velocity changes. Its standard units are m/s^2.

acceleration of gravity The acceleration of a falling object. On Earth, the acceleration of gravity, designated by g, is 9.8 m/s^2.

accretion The process by which small objects gather together to make larger objects.

accretion disk A rapidly rotating disk of material that gradually falls inward as it orbits a starlike object (e.g., white dwarf, neutron star, or black hole).

active galactic nuclei The unusually luminous centers of some galaxies, thought to be powered by accretion onto supermassive black holes. Quasars are the brightest type of active galactic nuclei; radio galaxies also contain active galactic nuclei.

active galaxy A term sometimes used to describe a galaxy that contains an *active galactic nucleus.*

adaptive optics A technique in which telescope mirrors flex rapidly to compensate for the bending of starlight caused by atmospheric turbulence.

albedo Describes the fraction of sunlight reflected by a surface; albedo = 0 means no reflection at all (a perfectly black surface); albedo = 1 means all light is reflected (a perfectly white surface).

Algol Paradox A paradox concerning the binary star Algol, which contains a sub-giant star that is less massive than its main-sequence companion.

altitude (above horizon) The angular distance between the horizon and an object in the sky.

amino acids The building blocks of proteins.

analemma The figure-8 path traced by the Sun over the course of a year when viewed at the same place and the same time each day; represents the discrepancies between apparent and mean solar time.

Andromeda Galaxy (M 31; the Great Galaxy in Andromeda) The nearest large spiral galaxy to the Milky Way.

angular momentum Momentum attributable to rotation or revolution. The angular momentum of an object moving in a circle of radius r is the product $m \times v \times r$.

angular resolution (of a telescope) The smallest angular separation that two pointlike objects can have and still be seen as distinct points of light (rather than as a single point of light).

angular size (or **angular distance**) A measure of the angle formed by extending imaginary lines outward from our eyes to span an object (or between two objects).

annihilation *See* matter–antimatter annihilation

annular solar eclipse A solar eclipse during which the Moon is directly in front of the Sun but its angular size is not large enough to fully block the Sun; thus, a ring (or *annulus*) of sunlight is still visible around the Moon's disk.

Antarctic Circle The circle on the Earth with latitude 66.5°S.

antielectron *See* positron

antimatter Refers to any particle with the same mass as a particle of ordinary matter but whose other basic properties, such as electrical charge, are precisely opposite.

aphelion The point at which an object orbiting the Sun is farthest from the Sun.

apogee The point at which an object orbiting the Earth is farthest from the Earth.

apparent brightness The amount of light reaching us *per unit area* from a luminous object; often measured in units of $watts/m^2$.

apparent magnitude A measure of the apparent brightness of an object in the sky, based on the ancient system developed by Hipparchus.

apparent retrograde motion Refers to the apparent motion of a planet, as viewed from Earth, during the period of a few weeks or months when it moves westward relative to the stars in our sky.

apparent solar time Time measured by the actual position of the Sun in your local sky; defined so that noon is when the Sun is *on* the meridian.

arcminutes (or **minutes of arc**) One arcminute is 1/60 of 1°.

arcseconds (or **seconds of arc**) One arcsecond is 1/60 of an arcminute, or 1/3,600 of 1°.

Arctic Circle The circle on the Earth with latitude 66.5°N.

asteroid A relatively small and rocky object that orbits a star; asteroids are officially considered part of a category known as "small solar system bodies."

asteroid belt The region of our solar system between the orbits of Mars and Jupiter in which asteroids are heavily concentrated.

astrobiology The study of life on Earth and beyond; emphasizes research into questions of the origin of life, the conditions under which life can survive, and the search for life beyond Earth.

astrometric technique The detection of extrasolar planets through the side-to-side motion of a star caused by gravitational tugs from the planet.

astronomical unit (AU) The average distance (semimajor axis) of the Earth from the Sun, which is about 150 million km.

atmosphere A layer of gas that surrounds a planet or moon, usually very thin compared to the size of the object.

atmospheric pressure The surface pressure resulting from the overlying weight of an atmosphere.

atmospheric structure The layering of a planetary atmosphere due to variations in temperature with altitude. For example, Earth's atmospheric structure from the ground up consists of the troposphere, stratosphere, thermosphere, and exosphere.

atomic mass number The combined number of protons and neutrons in an atom.

atomic number The number of protons in an atom.

atoms Consist of a nucleus made from protons and neutrons surrounded by a cloud of electrons.

aurora Dancing lights in the sky caused by charged particles entering our atmosphere; called the *aurora borealis* in the Northern Hemisphere and the *aurora australis* in the Southern Hemisphere.

autumnal equinox *See* fall equinox

axis tilt (of a planet in our solar system) The amount by which a planet's axis is tilted with respect to a line perpendicular to the ecliptic plane.

azimuth (usually called *direction* in this book) Direction around the horizon from due north, measured clockwise in degrees. E.g., the azimuth of due north is 0°, due east is 90°, due south is 180°, and due west is 270°.

bar The standard unit of pressure, approximately equal to the Earth's atmospheric pressure at sea level.

baryonic matter Refers to ordinary matter made from atoms (because the nuclei of atoms contain protons and neutrons, which are both baryons).

baryons Particles, including protons and neutrons, that are made from three quarks.

basalt A type of dark, high-density volcanic rock that is rich in iron and magnesium-based silicate minerals; it forms a runny (easy flowing) lava when molten.

belts (on a jovian planet) Dark bands of sinking air that encircle a jovian planet at a particular set of latitudes.

Big Bang The name given to the event thought to mark the birth of the universe.

Big-Bang theory The scientific theory of the universe's earliest moments, stating that all the matter in our observable universe came into being at a single moment in time as an extremely hot, dense mixture of subatomic particles and radiation.

Big Crunch If gravity ever reverses the universal expansion, the universe will someday begin to collapse and presumably end in a Big Crunch.

binary star system A star system that contains two stars.

biosphere Refers to the "layer" of life on Earth.

BL Lac objects The name given to a class of active galactic nuclei that probably represent the centers of radio galaxies whose jets happen to be pointed directly at us.

blackbody radiation *See* thermal radiation

black hole A bottomless pit in spacetime. Nothing can escape from within a black hole, and we can never again detect or observe an object that falls into a black hole.

black smokers Structures around seafloor volcanic vents that support a wide variety of life.

blowout Ejection of the hot, gaseous contents of a superbubble when it grows so large that it bursts out of the cooler layer of gas filling the galaxy's disk.

blueshift A Doppler shift in which spectral features are shifted to shorter wavelengths, caused when an object is moving toward the observer.

bosons Particles, such as photons, to which the exclusion principle does not apply.

bound orbits Orbits on which an object travels repeatedly around another object; bound orbits are elliptical in shape.

brown dwarf An object too small to become an ordinary star because electron degeneracy pressure halts its gravitational collapse before fusion becomes self-sustaining; brown dwarfs have mass less than $0.08M_{\text{Sun}}$.

bubble (interstellar) The surface of a bubble is an expanding shell of hot, ionized gas driven by stellar winds or supernovae; inside the bubble, the gas is very hot and has very low density.

bulge (of a spiral galaxy) The central portion of a spiral galaxy that is roughly spherical (or football shaped) and bulges above and below the plane of the galactic disk.

Cambrian explosion The dramatic diversification of life on Earth that occurred between about 540 and 500 million years ago.

carbonate rock A carbon-rich rock, such as limestone, that forms underwater from chemical reactions between sediments and carbon dioxide. On Earth, most of the outgassed carbon dioxide currently resides in carbonate rocks.

carbon dioxide cycle The process that cycles carbon dioxide between the Earth's atmosphere and surface rocks.

carbon stars Stars whose atmospheres are especially carbon-rich, thought to be near the ends of their lives; carbon stars are the primary sources of carbon in the universe.

Cassini division A large, dark gap in Saturn's rings, visible through small telescopes on Earth.

CCD (charge coupled device) A type of electronic light detector that has largely replaced photographic film in astronomical research.

celestial coordinates The coordinates of right ascension and declination that fix an object's position on the celestial sphere.

celestial equator (CE) The extension of the Earth's equator onto the celestial sphere.

celestial navigation Navigation on the surface of the Earth accomplished by observations of the Sun and stars.

celestial sphere The imaginary sphere on which objects in the sky appear to reside when observed from Earth.

Celsius (temperature scale) The temperature scale commonly used in daily activity internationally. Defined so that, on Earth's surface, water freezes at 0°C and boils at 100°C.

center of mass (of orbiting objects) The point at which two or more orbiting objects would balance if they were somehow connected; it is the point around which the orbiting objects actually orbit.

central dominant galaxy A giant elliptical galaxy found at the center of a dense cluster of galaxies, apparently formed by the merger of several individual galaxies.

Cepheid *See* Cepheid variable

Cepheid variable A particularly luminous type of pulsating variable star that follows a period–luminosity relation and hence is very useful for measuring cosmic distances.

Chandrasekhar limit *See* white dwarf limit

charged particle belts Zones in which ions and electrons accumulate and encircle a planet.

chemical enrichment The process by which the abundance of heavy elements (heavier than helium) in the interstellar medium gradually increases over time as these elements are produced by stars and released into space.

chondrites Another name for primitive meteorites, named for the round chodrules within them. Achondrites, meaning "without chondrules," is another name for processed meteorites.

chromosphere The layer of the Sun's atmosphere below the corona; most of the Sun's ultraviolet light is emitted from this region, in which the temperature is about 10,000 K.

circulation cells (also called *Hadley cells*) Large-scale cells (similar to convection cells) in a planet's atmosphere that transport heat between the equator and the poles.

circumpolar star A star that always remains above the horizon for a particular latitude.

climate Describes the long-term average of weather.

close binary A binary star system in which the two stars are very close together.

closed universe The universe is closed if its average density is greater than the critical density, in which case spacetime must curve back on itself to the point where its overall shape is analogous to that of the surface of a sphere. In the absence of a repulsive force (*see* cosmological constant), a closed universe would someday stop expanding and begin to contract.

cluster of galaxies A collection of a few dozen or more galaxies bound together by gravity; smaller collections of galaxies are simply called *groups*.

cluster of stars A group of anywhere from several hundred to a million or so stars; star clusters come in two types—open clusters and globular clusters.

CNO cycle The cycle of reactions by which intermediate- and high-mass stars fuse hydrogen into helium.

coasting universe The possible fate of our universe in which the mass density of the universe is *smaller* than the critical density, so that the collective gravity of all matter cannot halt the expansion. In the absence of a repulsive force (*see* cosmological constant), such a universe would keep expanding forever with little change in its rate of expansion.

coma (of a comet) The dusty atmosphere of a comet created by sublimation of ices in the nucleus when the comet is near the Sun.

comet A relatively small, icy object that orbits a star. Like asteroids, comets are officially considered part of a category known as "small solar system bodies."

comparative planetology The study of the solar system by examining and understanding the similarities and differences among worlds.

compound (chemical) A substance made from molecules consisting of two or more atoms with different atomic numbers.

condensates Solid or liquid particles that condense from a cloud of gas.

condensation The formation of solid or liquid particles from a cloud of gas.

conduction (of energy) The process by which thermal energy is transferred by direct contact from warm material to cooler material.

conjunction (of a planet with the Sun) When a planet and the Sun line up in the sky.

conservation of angular momentum (law of) The principle that, in the absence of net torque (twisting force), the total angular momentum of a system remains constant.

conservation of energy (law of) The principle that energy (including mass-energy) can be neither created nor destroyed, but can only change from one form to another.

conservation of momentum (law of) The principle that, in the absence of net force, the total momentum of a system remains constant.

constellation A region of the sky; 88 official constellations cover the celestial sphere.

continental crust The thicker lower-density crust that makes up Earth's continents. It is made when remelting of seafloor crust allows lower-density rock to separate and erupt to the surface. Continental crust ranges in age from very young to as old as about 4 billion years (or more).

convection The energy transport process in which warm material expands and rises, while cooler material contracts and falls.

convection cell An individual small region of convecting material.

convection zone (of a star) A region in which energy is transported outward by convection.

Copernican revolution The dramatic change, initiated by Copernicus, that occurred when we learned that Earth is a planet orbiting the Sun rather than the center of the universe.

core (of a planet) The dense central region of a planet that has undergone differentiation.

core (of a star) The central region of a star, in which nuclear fusion can occur.

Coriolis effect Causes air or objects moving on a rotating planet to deviate from straight-line trajectories.

corona (solar) The tenuous uppermost layer of the Sun's atmosphere; most of the Sun's X rays are emitted from this region, in which the temperature is about 1 million K.

coronal holes Regions of the corona that barely show up in X-ray images because they are nearly devoid of hot coronal gas.

coronal mass ejections Bursts of charged particles from the Sun's corona that travel outward into space.

cosmic microwave background The remnant radiation from the Big Bang, which we detect using radio telescopes sensitive to microwaves (which are short-wavelength radio waves).

cosmic rays Particles such as electrons, protons, and atomic nuclei that zip through interstellar space at close to the speed of light.

cosmological constant The name given to a term in Einstein's equations of general relativity. If it is not zero, then it represents a repulsive force or a type of energy (sometimes called *dark energy* or *quintessence*) that might cause the expansion of the universe to accelerate with time.

cosmological horizon The boundary of our observable universe, which is where the lookback time is equal to the age of the universe. Beyond this boundary in spacetime, we cannot see anything at all.

Cosmological Principle The idea that matter is distributed uniformly throughout the universe on very large scales, meaning that the universe has neither a center nor an edge.

cosmological redshift Refers to the redshifts we see from distant galaxies, caused by the fact that expansion of the universe stretches all the photons within it to longer, redder wavelengths.

cosmology The study of the overall structure and evolution of the universe.

cosmos An alternate name for the *universe.*

critical density The precise average density for the entire universe that marks the dividing line between a recollapsing universe and one that will expand forever.

critical universe The possible fate of our universe in which the mass density of the universe *equals* the critical density. The universe will never collapse, but in the absence of a repulsive force it will expand more and more slowly as time progresses. *See* cosmological constant

crust (of a planet) The low-density surface layer of a planet that has undergone differentiation.

curvature of spacetime A change in the geometry of space produced in the vicinity of a massive object and that is responsible for the force we call gravity. The overall geometry of the universe may also be curved, depending on its overall mass-energy content.

cycles per second Units of frequency for a wave; describes the number of peaks (or troughs) of a wave that pass by a given point each second. Equivalent to *hertz.*

dark energy Name sometimes given to energy that could be causing the expansion of the universe to accelerate. *See* cosmological constant.

dark matter Matter that we infer to exist from its gravitational effects but from which we have not detected any light; dark matter apparently dominates the total mass of the universe.

daylight saving time Standard time plus 1 hour, so that the Sun appears on the meridian around 1 P.M. rather than around noon.

declination (dec) Analogous to latitude, but on the celestial sphere; it is the angular north-south distance between the celestial equator and a location on the celestial sphere.

degeneracy pressure A type of pressure unrelated to an object's temperature, which arises when electrons (electron degeneracy pressure) or neutrons (neutron degeneracy pressure) are packed so tightly that the exclusion and uncertainty principles come into play.

degenerate object An object in which degeneracy pressure is the primary pressure pushing back against gravity, such as a brown dwarf, white dwarf, or neutron star.

deuterium A form of hydrogen in which the nucleus contains a proton and a neutron, rather than only a proton (as is the case for most hydrogen nuclei).

differential rotation Describes the rotation of an object in which the equator rotates at a different rate than the poles.

differentiation The process in which gravity separates materials according to density, with high-density materials sinking and low-density materials rising.

diffraction grating A finely etched surface that can split light into a spectrum.

diffraction limit The angular resolution that a telescope could achieve if it were limited only by the interference of light waves; it is smaller (i.e., better angular resolution) for larger telescopes.

dimension (mathematical) Describes the number of independent directions in which movement is possible; e.g., the surface of the Earth is two-dimensional because only two independent directions of motion are possible (north-south and east-west).

direction (in local sky) One of the two coordinates (the other is altitude) needed to pinpoint an object in the local sky. It is the direction, such as north, south, east, or west, in which you must face to see the object. *See also* azimuth

disk component (of a galaxy) The portion of a spiral galaxy that looks like a disk and contains an interstellar medium with cool gas and dust; stars of many ages are found in the disk component.

disk population Refers to stars that orbit within the disk of a spiral galaxy. Sometimes called Population I.

DNA (deoxyribonucleic acid) The molecule that represents the genetic material of life on Earth.

Doppler effect (shift) The effect that shifts the wavelengths of spectral features in objects that are moving toward or away from the observer.

Doppler technique The detection of extrasolar planets through the motion of a star toward and away from the observer caused by gravitational tugs from the planet.

double-shell burning star A star that is fusing helium into carbon in a shell around an inert carbon core and is fusing hydrogen into helium in a shell at the top of the helium layer.

down quark One of the two quark types (the other is the up quark) found in ordinary protons and neutrons. Has a charge of $-\frac{1}{3}$.

Drake equation An equation that lays out the factors that play a role in determining the number of communicating civilizations in our galaxy.

dust (or **dust grains**) Tiny solid flecks of material; in astronomy, we often discuss interplanetary dust (found within a star system) or interstellar dust (found between the stars in a galaxy). *See also* interstellar dust grains

dust tail (of a comet) One of two tails seen when a comet passes near the Sun (the other is the plasma tail); composed of small solid particles pushed away from the Sun by the radiation pressure of sunlight.

dwarf elliptical galaxy A small elliptical galaxy with less than about a billion stars.

dwarf galaxies Relatively small galaxies, consisting of less than about ten billion stars.

dwarf planet An object that orbits the Sun and is massive enough for its gravity to have made it nearly round in shape, but that does not qualify as an official planet because it has not cleared its orbital neighborhood. The dwarf planets of our solar system include the asteroid Ceres and the Kuiper belt objects Pluto and Eris.

Earth-orbiters (spacecraft) Spacecraft designed to study the Earth or the universe from Earth orbit.

eccentricity A measure of how much an ellipse deviates from a perfect circle; defined as the center-to-focus distance divided by the length of the semimajor axis.

eclipse Occurs when one astronomical object casts a shadow on another or crosses our line of sight to the other object.

eclipse seasons Periods during which lunar and solar eclipses can occur because the nodes of the Moon's orbit are nearly aligned with the Earth and Sun.

eclipsing binary A binary star system in which the two stars happen to be orbiting in the plane of our line of sight, so that each star will periodically eclipse the other.

ecliptic The Sun's apparent annual path among the constellations.

ecliptic plane The plane of the Earth's orbit around the Sun.

ejecta (from an impact) Debris ejected by the blast of an impact.

electrical charge A fundamental property of matter that is described by its amount and as either positive or negative; more technically, it is a measure of how a particle responds to the electromagnetic force.

electromagnetic field An abstract concept used to describe how a charged particle would affect other charged particles at a distance.

electromagnetic force One of the four fundamental forces; it is the force that dominates atomic and molecular interactions.

electromagnetic radiation Another name for light of all types, from radio waves through gamma rays.

electromagnetic spectrum The complete spectrum of light, including radio waves, infrared, visible light, ultraviolet light, X rays, and gamma rays.

electromagnetic wave A synonym for light, which consists of waves of electric and magnetic fields.

electron degeneracy pressure Degeneracy pressure exerted by electrons, as in brown dwarfs and white dwarfs.

electrons Fundamental particles with negative electric charge; the distribution of electrons in an atom gives the atom its size.

electron-volt (eV) A unit of energy equivalent to 1.60×10^{-19} joule.

electroweak era The era of the universe during which only three forces operated (gravity, strong force, and electroweak force), lasting from 10^{-38} second to 10^{-10} second after the Big Bang.

electroweak force The force that exists at high energies when the electromagnetic force and the weak force exist as a single force.

element (chemical) A substance made from individual atoms of a particular atomic number.

ellipse A type of oval that happens to be the shape of bound orbits. An ellipse can be drawn by moving a pencil along a string whose ends are tied to two tacks; the locations of the tacks are the foci (singular, focus) of the ellipse.

elliptical galaxies Galaxies that appear rounded in shape, often longer in one direction, like a football. They have no disks and contain very little cool gas and dust compared to spiral galaxies, though they often contain very hot, ionized gas.

elongation (greatest) For Mercury or Venus, the point at which it appears farthest from the Sun in our sky.

emission (of light) The process by which matter emits energy in the form of light.

emission-line spectrum A spectrum that contains emission lines.

emission nebula Another name for an ionization nebula. *See also* ionization nebula

energy Broadly speaking, energy is what can make matter move. The three basic types of energy are kinetic, potential, and radiative.

equation of time Describes the discrepancies between apparent and mean solar time.

equinox See *fall equinox* and *spring equinox.*

equivalence principle The fundamental starting point for general relativity, which states that the effects of gravity are exactly equivalent to the effects of acceleration.

era of atoms The era of the universe lasting from about 500,000 years to about 1 billion years after the Big Bang, during which it was cool enough for neutral atoms to form.

era of galaxies The present era of the universe, which began with the formation of galaxies when the universe was about 1 billion years old.

era of nuclei The era of the universe lasting from about 3 minutes to about 380,000 years after the Big Bang, during which matter in the universe was fully ionized and opaque to light. The cosmic background radiation was released at the end of this era.

era of nucleosynthesis The era of the universe lasting from about 0.001 second to about 3 minutes after the Big Bang, by the end of which virtually all of the neutrons and about one-seventh of the protons in the universe had fused into helium.

erosion The wearing down or building up of geological features by wind, water, ice, and other phenomena of planetary weather.

eruption The process of releasing hot lava on the planet's surface.

escape velocity The speed necessary for an object to completely escape the gravity of a large body such as a moon, planet, or star.

evaporation The process by which atoms or molecules escape into the gas phase from a liquid.

event Any particular point along a worldline represents a particular event; all observers will agree on the reality of an event but may disagree about its time and location.

event horizon The boundary that marks the "point of no return" between a black hole and the outside universe; events that occur within the event horizon can have no influence on our observable universe.

evolution (biological) The gradual change in populations of living organisms responsible for transforming life on Earth from its primitive origins to the great diversity of life today.

exchange particle According to the standard model of physics, each of the four fundamental forces is transmitted by the transfer of particular types of exchange particles.

excited state (of an atom) Any arrangement of electrons in an atom that has more energy than the ground state.

exclusion principle The law of quantum mechanics that states that two fermions cannot occupy the same quantum state at the same time.

exosphere The hot, outer layer of an atmosphere, where the atmosphere "fades away" to space.

expansion (of universe) The idea that the space between galaxies or clusters of galaxies is growing with time.

exposure time The amount of time for which light is collected to make a single image.

extrasolar planet A planet orbiting a star other than our Sun.

Fahrenheit (temperature scale) The temperature scale commonly used in daily activity in the United States. Defined so that, on Earth's surface, water freezes at 32°F and boils at 212°F.

fall equinox (autumnal equinox) Refers both to the point in Virgo on the celestial sphere where the ecliptic crosses the celestial equator and to the moment in time when the Sun appears at that point each year (around September 21).

false-color image An image displayed in colors that are *not* the true, visible-light colors of an object.

fault (geological) A place where rocks slip sideways relative to one another.

feedback processes Processes in which a small change in some property (such as temperature) leads to changes in other properties that either amplify or diminish the original small change.

fermions Particles, such as electrons, neutrons, and protons, that obey the exclusion principle.

Fermi's paradox The question posed by Enrico Fermi about extraterrestrial intelligence—

"So where is everybody?"—which asks why we have not observed other civilizations even though simple arguments would suggest that some ought to have spread throughout the galaxy by now.

field An abstract concept used to describe how a particle would interact with a force. For example, the idea of a *gravitational field* describes how a particle would react to the local strength of gravity, and the idea of an *electromagnetic field* describes how a charged particle would respond to forces from other charged particles.

filter (for light) A material that transmits only particular wavelengths of light.

fireball A particularly bright meteor.

fission The process by which one atomic nucleus breaks into two smaller nuclei. It releases energy if the two smaller nuclei together are less massive than the original nucleus.

flare star A small, spectral type M star that displays particularly strong flares on its surface.

flat (or **Euclidean**) **geometry** Refers to any case in which the rules of geometry for a flat plane hold, such as that the shortest distance between two points is a straight line.

flat universe A universe in which the overall geometry of spacetime is flat (Euclidean), as would be the case if the density of the universe is equal to the critical density.

flybys (spacecraft) Spacecraft that fly past a target object (such as a planet), usually just once, as opposed to entering a bound orbit of the object.

focal plane The place where an image created by a lens or mirror is in focus.

foci Plural of *focus.*

focus (of a lens or mirror) The point at which rays of light that were initially parallel (such as light from a distant star) converge.

focus (of an ellipse) One of two special points within an ellipse that lie along the major axis; these are the points around which we could stretch a pencil and string to draw an ellipse. When one object orbits a second object, the second object lies at one focus of the orbit.

force Anything that can cause a change in momentum.

formation properties (of planets) In this book, for the purpose of understanding geological processes, planets are defined to be born with four formation properties: size (mass and radius), distance from the Sun, composition, and rotation rate.

fossil Any relic of an organism that lived and died long ago.

frame of reference (in relativity) Two (or more) objects share the same frame of reference if they are *not* moving relative to each other.

free-fall Refers to conditions in which an object is falling without resistance; objects are weightless when in free-fall.

free-float frame A frame of reference in which all objects are weightless and hence float freely.

frequency Describes the rate at which peaks of a wave pass by a point; measured in units of 1/s, often called *cycles per second* or *hertz.*

frost line The boundary in the solar nebula beyond which ices could condense; only metals and rocks could condense within the frost line.

fundamental forces There are four known fundamental forces in nature: gravity, the electromagnetic force, the strong force, and the weak force.

fundamental particles Subatomic particles that cannot be divided into anything smaller.

fusion The process by which two atomic nuclei fuse together to make a single, more massive nucleus. It releases energy if the final nucleus is less massive than the two nuclei that went into the reaction.

galactic cannibalism The term sometimes used to describe the process by which large galaxies merge with other galaxies in collisions. *Central dominant galaxies* are products of galactic cannibalism.

galactic disk (of a spiral galaxy) *See* disk component

galactic fountain Refers to a model for the cycling of gas in the Milky Way Galaxy in which fountains of hot, ionized gas rise from the disk into the halo and then cool and form clouds as they sink back into the disk.

galactic wind A wind of low-density but extremely hot gas flowing out from a starburst galaxy, created by the combined energy of many supernovae.

galaxy A huge collection of anywhere from a few hundred million to more than a trillion stars, all bound together by gravity.

galaxy cluster *See* cluster of galaxies

galaxy evolution The formation and development of galaxies.

Galilean moons The four moons of Jupiter that were discovered by Galileo: Io, Europa, Ganymede, and Callisto.

gamma-ray burst A sudden burst of gamma rays from deep space; such bursts apparently come from distant galaxies, but their precise mechanism is unknown.

gamma rays Light with very short wavelengths (and hence high frequencies)—shorter than those of X rays.

gap moons Tiny moons located within a gap in a planet's ring system. The gravity of a gap moon helps clear the gap.

gas phase The phase of matter in which atoms or molecules can move essentially independently of one another.

gas pressure Describes the force (per unit area) pushing on any object due to surrounding gas. *See also* pressure

general theory of relativity Einstein's generalization of his special theory of relativity so that the general theory also applies when we consider effects of gravity or acceleration.

genetic code The "language" that living cells use to read the instructions chemically encoded in DNA.

geocentric model Any of the ancient Greek models that were used to predict planetary positions under the assumption that Earth lay in the center of the universe.

geocentric universe (ancient belief in) The idea that the Earth is the center of the entire universe.

geological activity Processes that change a planet's surface long after formation, such as volcanism, tectonics, and erosion.

geological processes The four basic geological processes are impact cratering, volcanism, tectonics, and erosion.

geological time scale The time scale used by scientists to describe major eras in Earth's past.

geology The study of surface features (on a moon, planet, or asteroid) and the processes that create them.

geostationary satellite A satellite that appears to stay stationary in the sky as viewed from Earth's surface, because it orbits in the same time it takes Earth to rotate and orbits in Earth's equatorial plane.

geosynchronous satellite A satellite that orbits Earth in the same time it takes Earth to rotate, or one sidereal day.

giant impact A collision between a forming planet and a very large planetesimal, such as is thought to have formed our Moon.

giant molecular cloud A very large cloud of cold, dense interstellar gas, typically containing up to a million solar masses worth of material. *See also* molecular clouds

giants (luminosity class III) Stars that appear just below the supergiants on the H–R diagram because they are somewhat smaller in radius and lower in luminosity.

global positioning system (GPS) A system of navigation by satellites orbiting the Earth.

global warming An expected increase in Earth's global average temperature caused by human input of carbon dioxide and other greenhouse gases into the atmosphere.

global wind patterns (or **global circulation**) Wind patterns that remain fixed on a global scale, determined by the combination of surface heating and the planet's rotation.

globular cluster A spherically shaped cluster of up to a million or more stars; globular clusters are found primarily in the halos of galaxies and contain only very old stars.

gluons The exchange particles for the strong force.

grand unified theory (GUT) A theory that unifies three of the four fundamental forces—the strong force, the weak force, and the electromagnetic force (but not gravity)—in a single model.

granulation (on the Sun) The bubbling pattern visible in the photosphere, produced by the underlying convection.

gravitation (law of) *See* universal law of gravitation

gravitational constant The experimentally measured constant G that appears in the law of universal gravitation:

$$G = 6.67 \times 10^{-11} \frac{m^3}{kg \times s^2}$$

gravitational contraction The process in which gravity causes an object to contract, thereby converting gravitational potential energy into thermal energy.

gravitational encounter Occurs when two (or more) objects pass near enough so that each can feel the effects of the other's gravity and can therefore exchange energy.

gravitational equilibrium Describes a state of balance in which the force of gravity pulling inward is precisely counteracted by pressure pushing outward.

gravitational lensing The magnification or distortion (into arcs, rings, or multiple images) of an image caused by light bending through a gravitational field, as predicted by Einstein's general theory of relativity.

gravitationally bound system Any system of objects, such as a star system or a galaxy, that is held together by gravity.

gravitational potential energy Energy that an object has by virtue of its position in a gravitational field; an object has more gravitational potential energy when it has a greater distance that it can potentially fall.

gravitational redshift A redshift caused by the fact that time runs slow in gravitational fields.

gravitational time dilation The slowing of time that occurs in a gravitational field, as predicted by Einstein's general theory of relativity.

gravitational waves Predicted by Einstein's general theory of relativity, these waves travel at the speed of light and transmit distortions of space through the universe. Although not yet observed directly, we have strong indirect evidence that they exist.

gravitons The exchange particles for the force of gravity.

gravity One of the four fundamental forces; it is the force that dominates on large scales.

grazing incidence (in telescopes) Reflections in which light grazes a mirror surface and is deflected at a small angle; commonly used to focus high-energy ultraviolet light and X rays.

great circle A circle on the surface of a sphere whose center is at the center of the sphere.

greatest elongation *See* elongation (greatest).

Great Red Spot A large, high-pressure storm on Jupiter.

greenhouse effect The process by which greenhouse gases in an atmosphere make a planet's surface temperature warmer than it would be in the absence of an atmosphere.

greenhouse gases Gases, such as carbon dioxide, water vapor, and methane, that are particularly good absorbers of infrared light but are transparent to visible light.

Gregorian calendar Our modern calendar, introduced by Pope Gregory in 1582.

ground state (of an atom) The lowest possible energy state of the electrons in an atom.

group (of galaxies) A few to a few dozen galaxies bound together by gravity. *See also* cluster of galaxies

GUT era The era of the universe during which only two forces operated (gravity and the grand-unified-theory or GUT force), lasting from 10^{-43} second to 10^{-38} second after the Big Bang.

GUT force The proposed force that exists at very high energies when the strong force, the weak force, and the electromagnetic force (but not gravity) all act as one.

H II region Another name for an ionization nebula. *See* ionization nebula

habitable world A world with environmental conditions under which life could *potentially* arise or survive.

habitable zone The region around a star in which planets could potentially have surface temperatures at which liquid water could exist.

Hadley cells *See* circulation cells

half-life The time it takes for half of the nuclei in a given quantity of a radioactive substance to decay.

halo (of a galaxy) The spherical region surrounding the disk of a spiral galaxy.

Hawking radiation Radiation predicted to arise from the evaporation of black holes.

heavy bombardment The period in the first few hundred million years after the solar system formed during which the tail end of planetary accretion created most of the craters found on ancient planetary surfaces.

heavy elements In astronomy, *heavy elements* generally refers to all elements *except* hydrogen and helium.

helium-burning star A star that is currently fusing helium into carbon in its core.

helium-capture reactions Fusion reactions that fuse a helium nucleus into some other nucleus; such reactions can fuse carbon into oxygen, oxygen into neon, neon into magnesium, and so on.

helium flash The event that marks the sudden onset of helium fusion in the previously inert helium core of a low-mass star.

helium fusion The fusion of three helium nuclei into one carbon nucleus; also called the *triple-alpha reaction.*

hertz (Hz) The standard unit of frequency for light waves; equivalent to units of 1/s.

Hertzsprung–Russell (H–R) diagram A graph plotting individual stars as points, with stellar luminosity on the vertical axis and spectral type (or surface temperature) on the horizontal axis.

high-mass stars Stars born with masses above about $8M_{Sun}$; these stars will end their lives by exploding as supernovae.

horizon A boundary that divides what we can see from what we cannot see.

horizontal branch The horizontal line of stars that represents helium-burning stars on an H–R diagram for a cluster of stars.

horoscope A predictive chart made by an astrologer; in scientific studies, horoscopes have never been found to have any validity as predictive tools.

hot spot (geological) A place within a plate of the lithosphere where a localized plume of hot mantle material rises.

hour angle (HA) The angle or time (measured in hours) since an object was last on the meridian in the local sky. Defined to be 0 hours for objects that *are* on the meridian.

Hubble's constant A number that expresses the current rate of expansion of the universe; designated H_0, it is usually stated in units of km/s/Mpc. The reciprocal of Hubble's constant is the age the universe would have *if* the expansion rate had never changed.

Hubble's law Mathematically expresses the idea that more distant galaxies move away from us faster; its formula is $v = H_0 \times d$, where v is a galaxy's speed away from us, d is its distance, and H_0 is Hubble's constant.

hydrogen compounds Compounds that contain hydrogen and were common in the solar nebula, such as water (H_2O), ammonia (NH_3), and methane (CH_4).

hydrogen-shell burning Hydrogen fusion that occurs in a shell surrounding a stellar core.

hydrosphere Refers to the "layer" of water on the Earth consisting of oceans, lakes, rivers, ice caps, and other liquid water and ice.

hydrostatic equilibrium *See* gravitational equilibrium

hyperbola The precise mathematical shape of one type of unbound orbit (the other is a parabola) allowed under the force of gravity; at great distances from the attracting object, a hyperbolic path looks like a straight line.

hypernova A term sometimes used to describe a supernova (explosion) of a star so massive that it leaves a black hole behind.

hyperspace Any space with more than three dimensions.

hypothesis A tentative model proposed to explain some set of observed facts, but which has not yet been rigorously tested and confirmed.

ice ages Periods of global cooling during which the polar caps, glaciers, and snow cover extend closer to the equator.

ices (in solar system theory) Materials that are solid only at low temperatures, such as the hydrogen compounds water, ammonia, and methane.

ideal gas law The law relating the pressure, temperature, and number density of particles in an ideal gas.

image A picture of an object made by focusing light.

imaging (in astronomical research) The process of obtaining pictures of astronomical objects.

impact The collision of a small body (such as an asteroid or comet) with a larger object (such as a planet or moon).

impact basin A very large impact crater often filled by a lava flow.

impact crater A bowl-shaped depression left by the impact of an object that strikes a planetary surface (as opposed to burning up in the atmosphere).

impact cratering The excavation of bowl-shaped depressions (*impact craters*) by asteroids or comets striking a planet's surface.

impactor The object responsible for an impact.

inflation (of the universe) A sudden and dramatic expansion of the universe thought to have occurred at the end of the GUT era.

infrared light Light with wavelengths that fall in the portion of the electromagnetic spectrum between radio waves and visible light.

inner solar system Generally considered to encompass the region of our solar system out to about the orbit of Mars.

intensity (of light) A measure of the amount of energy coming from light of specific wavelength in the spectrum of an object.

interferometry A telescopic technique in which two or more telescopes are used in tandem to produce much better angular resolution than the telescopes could achieve individually.

intermediate-mass stars Stars born with masses between about 2 and 8 M_{Sun}; these stars end their lives by ejecting a planetary nebula and becoming a white dwarf.

interstellar cloud A cloud of gas and dust between the stars.

interstellar dust grains Tiny solid flecks of carbon and silicon minerals found in cool interstellar clouds; they resemble particles of smoke and form in the winds of red giant stars.

interstellar medium Refers to gas and dust that fills the space between stars in a galaxy.

interstellar ramjet A hypothesized type of spaceship that uses a giant scoop to sweep up interstellar gas for use in a nuclear fusion engine.

interstellar reddening The change in the color of starlight as it passes through dusty gas. The light appears redder because dust grains absorb and scatter blue light more effectively than red light.

intracluster medium Hot, X-ray-emitting gas found between the galaxies within a cluster of galaxies.

inverse square law Any quantity that decreases with the square of the distance between two objects is said to follow an inverse square law.

inverse square law for light The law stating that an object's apparent brightness depends on its actual luminosity and the inverse square of its distance from the observer:

$$\text{apparent brightness} = \frac{\text{luminosity}}{4\pi \times (\text{distance})^2}$$

inversion (atmospheric) A local weather condition in which air is colder near the surface than higher up in the troposphere—the opposite of the usual condition, in which the troposphere is warmer at the bottom.

ionization The process of stripping an electron from an atom.

ionization nebula A colorful, wispy cloud of gas that glows because neighboring hot stars irradiate it with ultraviolet photons that can ionize hydrogen atoms.

ionosphere A portion of the thermosphere in which ions are particularly common (due to ionization by X rays from the Sun).

ions Atoms with a positive or negative electrical charge.

Io torus A donut-shaped charged-particle belt around Jupiter that approximately traces Io's orbit.

irregular galaxies Galaxies that look neither spiral nor elliptical.

isotopes Each different isotope of an element has the *same* number of protons but a *different* number of neutrons.

jets High-speed streams of gas ejected from an object into space.

joule The international unit of energy, equivalent to about 1/4,000 of a Calorie.

jovian nebulae The clouds of gas that swirled around the jovian planets, from which the moons formed.

jovian planets Giant gaseous planets similar in overall composition to Jupiter.

Julian calendar The calendar introduced in 46 B.C. by Julius Caesar and used until it was replaced by the Gregorian calendar.

Kelvin (temperature scale) The most commonly used temperature scale in science, defined such that absolute zero is 0 K and water freezes at 273.15 K.

Kepler's first law States that the orbit of each planet about the Sun is an ellipse with the Sun at one focus.

Kepler's laws of planetary motion Three laws discovered by Kepler that describe the motion of the planets around the Sun.

Kepler's second law States that, as a planet moves around its orbit, it sweeps out equal areas in equal times. This tells us that a planet moves faster when it is closer to the Sun (near perihelion) than when it is farther from the Sun (near aphelion) in its orbit.

Kepler's third law States that the square of a planet's orbital period is proportional to the cube of its average distance from the Sun (semimajor axis), which tells us that more distant planets move more slowly in their orbits. In its original form, written $p^2 = a^3$. *See also* Newton's version of Kepler's third law

kinetic energy Energy of motion, given by the formula $\frac{1}{2}mv^2$.

Kirchhoff's laws A set of rules that summarizes the conditions under which objects produce thermal, absorption line, or emission line spectra. In brief: (1) An opaque object produces thermal radiation. (2) An absorption line spectrum occurs when thermal radiation passes through a thin gas that is cooler than the object emitting the thermal radiation. (3) An emission line spectrum occurs when we view a cloud of gas that is warmer than any background source of light.

Kirkwood gaps On a plot of asteroid semimajor axes, regions with few asteroids as a result of orbital resonances with Jupiter.

K-T event (impact) The collision of an asteroid or comet 65 million years ago that caused the mass extinction best known for wiping out the dinosaurs. K-T stands for the initials of the geological layers above and below the event.

Kuiper belt The comet-rich region of our solar system that spans distances of about 30–100 AU from the Sun; Kuiper belt comets have orbits that lie fairly close to the plane of planetary orbits and travel around the Sun in the same direction as the planets.

Kuiper belt object Any object orbiting the Sun within the region of the Kuiper belt, although the term is most often used for relatively large objects. For example, Pluto and Eris are considered large Kuiper belt objects.

Large Magellanic Cloud One of two small, irregular galaxies (the other is the Small Magellanic Cloud) located about 150,000 light-years away; it probably orbits the Milky Way Galaxy.

large-scale structure (of the universe) Generally refers to structure of the universe on size scales larger than that of clusters of galaxies.

latitude The angular north-south distance between the Earth's equator and a location on the Earth's surface.

leap year A calendar year with 366 rather than 365 days; our current calendar (the Gregorian calendar) has a leap year every 4 years (by adding February 29) except in century years that are not divisible by 400.

length contraction Refers to the effect in which you observe lengths to be shortened in reference frames moving relative to you.

lenticular galaxies Galaxies that look lens-shaped when seen edge-on, resembling spiral galaxies without arms. They tend to have less cool gas than normal spiral galaxies but more gas than elliptical galaxies.

leptons Fermions *not* made from quarks, such as electrons and neutrinos.

life track A track drawn on an H–R diagram to represent the changes in a star's surface temperature and luminosity during its life; also called an *evolutionary track.*

light-collecting area (of a telescope) The area of the primary mirror or lens that collects light in a telescope.

light curve A graph of an object's intensity against time.

light gases (in solar system theory) Refers to hydrogen and helium, which never condense under solar nebula conditions.

light pollution Human-made light that hinders astronomical observations.

light-year The distance that light can travel in 1 year, which is 9.46 trillion km.

liquid phase The phase of matter in which atoms or molecules are held together but move relatively freely.

lithosphere The relatively rigid outer layer of a planet; generally encompasses the crust and the uppermost portion of the mantle.

Local Bubble (interstellar) The bubble of hot gas in which our Sun and other nearby stars apparently reside. *See also* bubble (interstellar)

Local Group The group of about 40 galaxies to which the Milky Way Galaxy belongs.

local sidereal time (LST) Sidereal time for a particular location, defined according to the position of the spring equinox in the local sky.

More formally, the local sidereal time at any moment is defined to be the hour angle of the spring equinox.

local sky The sky as viewed from a particular location on Earth (or another solid object). Objects in the local sky are pinpointed by the coordinates of *altitude* and *direction* (or azimuth).

local solar neighborhood The portion of the Milky Way Galaxy that is located relatively close (within a few hundred to a couple thousand light-years) to our Sun.

Local Supercluster The supercluster of galaxies to which the Local Group belongs.

longitude The angular east-west distance between the prime meridian (which passes through Greenwich) and a location on the Earth's surface.

lookback time Refers to the amount of time since the light we see from a distant object was emitted. I.e., if an object has a lookback time of 400 million years, we are seeing it as it looked 400 million years ago.

low-mass stars Stars born with masses less than about $2M_{\text{Sun}}$; these stars end their lives by ejecting a planetary nebula and becoming a white dwarf.

luminosity The total power output of an object, usually measured in watts or in units of solar luminosities ($L_{\text{Sun}} = 3.8 \times 10^{26}$ watts).

luminosity class Describes the region of the H–R diagram in which a star falls. Luminosity class I represents supergiants, III represents giants, and V represents main-sequence stars; luminosity classes II and IV are intermediate to the others.

lunar eclipse Occurs when the Moon passes through the Earth's shadow, which can occur only at full moon; may be total, partial, or penumbral.

lunar maria The regions of the Moon that look smooth from Earth and actually are impact basins.

lunar month *See* synodic month

lunar phase Describes the appearance of the Moon as seen from Earth.

MACHOs Stands for *massive compact halo objects* and represents one possible form of dark matter in which the dark objects are relatively large, like planets or brown dwarfs.

magma Underground molten rock.

magnetic braking The process by which a star's rotation slows as its magnetic field transfers its angular momentum to the surrounding nebula.

magnetic field Describes the region surrounding a magnet in which it can affect other magnets or charged particles in its vicinity.

magnetic-field lines Lines that represent how the needles on a series of compasses would point if they were laid out in a magnetic field.

magnetosphere The region surrounding a planet in which charged particles are trapped by the planet's magnetic field.

magnitude system A system of describing stellar brightness by using numbers, called *magnitudes,* based on an ancient Greek way of describing the brightnesses of stars in the sky. This system uses *apparent magnitude* to describe a star's apparent brightness and *absolute magnitude* to describe a star's luminosity.

main sequence (luminosity class V) The prominent line of points running from the upper left to the lower right on an H–R diagram; main-sequence stars shine by fusing hydrogen in their cores.

main-sequence fitting A method for measuring the distance to a cluster of stars by comparing the apparent brightness of the cluster's main sequence with the standard main sequence.

main-sequence lifetime The length of time for which a star of a particular mass can shine by fusing hydrogen into helium in its core.

main-sequence stars Stars whose temperature and luminosity place them on the main sequence of the H–R diagram. Main-sequence stars are all releasing energy by fusing hydrogen into helium in their cores.

main-sequence turnoff A method for measuring the age of a cluster of stars from the point on its H–R diagram where its stars turn off from the main sequence; the age of the cluster is equal to the main-sequence lifetime of stars at the main-sequence turnoff point.

mantle (of a planet) The rocky layer that lies between a planet's core and crust.

Martian meteorite This term is used to describe meteorites found on Earth that are thought to have originated on Mars.

mass A measure of the amount of matter in an object.

mass-energy The potential energy of mass, which has an amount $E = mc^2$.

mass exchange (in close binary star systems) The process in which tidal forces cause matter to spill from one star to a companion star in a close binary system.

mass extinction An event in which a large fraction of the species living on Earth go extinct, such as the event in which the dinosaurs died out about 65 million years ago.

mass increase (in relativity) Refers to the effect in which an object moving past you seems to have a mass greater than its rest mass.

massive-star supernova A supernova that occurs when a massive star dies, initiated by the catastrophic collapse of its iron core; often called a Type II supernova.

mass-to-light ratio The mass of an object divided by its luminosity, usually stated in units of solar masses per solar luminosity. Objects with high mass-to-light ratios must contain substantial quantities of dark matter.

matter–antimatter annihilation Occurs when a particle of matter and a particle of antimatter meet and convert all of their mass-energy to photons.

mean solar time Time measured by the average position of the Sun in your local sky over the course of the year.

meridian A half-circle extending from your horizon (altitude 0°) due south, through your zenith, to your horizon due north.

metallic hydrogen Hydrogen that is so compressed that the hydrogen atoms all share electrons and thereby take on properties of metals, such as conducting electricity. Occurs only under very high-pressure conditions, such as those found deep within Jupiter.

metals (in solar system theory) Elements, such as nickel, iron, and aluminum, that condense at fairly high temperatures.

meteor A flash of light caused when a particle from space burns up in our atmosphere.

meteorite A rock from space that lands on Earth.

meteor shower A period during which many more meteors than usual can be seen.

Metonic cycle The 19-year period, discovered by the Babylonian astronomer Meton, over which the lunar phases occur on the same dates.

microwaves Light with wavelengths in the range of micrometers to millimeters. Microwaves are generally considered to be a subset of the radio wave portion of the electromagnetic spectrum.

mid-ocean ridges (on Earth) Long ridges of undersea volcanoes, along which mantle material erupts onto the ocean floor and pushes apart the existing seafloor on either side. These ridges are essentially the source of new seafloor crust, which then makes its way along the ocean bottom for millions of years before returning to the mantle at a subduction zone.

Milankovitch cycles The cyclical changes in Earth's axis tilt and orbit that can change the climate and cause ice ages.

Milky Way Used both as the name of our galaxy and to refer to the band of light we see in the sky when we look into the plane of the Milky Way Galaxy.

millisecond pulsars Pulsars with rotation periods of a few thousandths of a second.

minor planets An alternate name for *asteroids.*

model (scientific) A representation of some aspect of nature that can be used to explain and predict real phenomena without invoking myth, magic, or the supernatural.

molecular bands The tightly bunched lines in an object's spectrum that are produced by molecules.

molecular cloud fragments (or *molecular cloud cores*) The densest regions of molecular clouds, which usually go on to form stars.

molecular clouds Cool, dense interstellar clouds in which the low temperatures allow hydrogen atoms to pair up into hydrogen molecules (H_2).

molecular dissociation The process by which a molecule splits into its component atoms.

molecule Technically the smallest unit of a chemical element or compound; in this text, the term refers only to combinations of two or more atoms held together by chemical bonds.

momentum The product of an object's mass and velocity.

moon An object that orbits a planet.

mutations Errors in the copying process when a living cell replicates itself.

natural selection The process by which mutations that make an organism better able to survive get passed on to future generations.

neap tides The lower-than-average tides on Earth that occur at first- and third-quarter moon, when the tidal forces from the Sun and Moon oppose one another.

nebula A cloud of gas in space, usually one that is glowing.

nebular capture The process by which icy planetesimals capture hydrogen and helium gas to form jovian planets.

nebular theory The detailed theory that describes how our solar system formed from a cloud of interstellar gas and dust.

net force The overall force to which an object responds; the net force is equal to the rate of change in the object's momentum, or equivalently to the object's mass × acceleration.

neutrino A type of fundamental particle that has extremely low mass and responds only to the weak force; neutrinos are leptons and come in three types—electron neutrinos, mu neutrinos, and tau neutrinos.

neutron degeneracy pressure Degeneracy pressure exerted by neutrons, as in neutron stars.

neutron star The compact corpse of a high-mass star left over after a supernova; typically contains a mass comparable to the mass of the Sun in a volume just a few kilometers in radius.

neutrons Particles with no electrical charge found in atomic nuclei, built from three quarks.

newton The standard unit of force in the metric system:

$$1 \text{ newton} = 1 \frac{\text{kg} \times \text{m}}{\text{s}^2}$$

Newton's first law of motion States that, in the absence of a net force, an object moves with constant velocity.

Newton's laws of motion Three basic laws that describe how objects respond to forces.

Newton's second law of motion States how a net force affects an object's motion. Specifically: force = rate of change in momentum, or force = mass × acceleration.

Newton's third law of motion States that, for any force, there is always an equal and opposite reaction force.

Newton's universal law of gravitation *See* universal law of gravitation

Newton's version of Kepler's third law This generalization of Kepler's third law can be used to calculate the masses of orbiting objects from measurements of orbital period and distance. Usually written as:

$$p^2 = \frac{4\pi^2}{G(M_1 + M_2)} a^3$$

nodes (of Moon's orbit) The two points in the Moon's orbit where it crosses the ecliptic plane.

nonbaryonic matter Refers to exotic matter that is not part of the normal composition of atoms, such as neutrinos or the hypothetical WIMPs.

nonscience As defined in this book, nonscience is any way of searching for knowledge that makes no claim to follow the scientific method, such as seeking knowledge through intuition, tradition, or faith.

north celestial pole (NCP) The point on the celestial sphere directly above the Earth's North Pole.

nova The dramatic brightening of a star that lasts for a few weeks and then subsides; occurs when a burst of hydrogen fusion ignites in a shell on the surface of an accreting white dwarf in a binary star system.

nuclear fission The process in which a larger nucleus splits into two (or more) smaller particles.

nuclear fusion The process in which two (or more) smaller nuclei slam together and make one larger nucleus.

nucleus (of a comet) The solid portion of a comet, and the only portion that exists when the comet is far from the Sun.

nucleus (of an atom) The compact center of an atom made from protons and neutrons.

observable universe The portion of the entire universe that, at least in principle, can be seen from Earth.

Occam's razor A principle often used in science, holding that scientists should prefer the simpler of two models that agree equally well with observations. Named after the medieval scholar William of Occam (1285–1349).

Olbers' paradox Asks the question of how the night sky can be dark if the universe is infinite and full of stars.

Oort cloud A huge, spherical region centered on the Sun, extending perhaps halfway to the nearest stars, in which trillions of comets orbit the Sun with random inclinations, orbital directions, and eccentricities.

opacity A measure of how much light a material absorbs compared to how much it transmits; materials with higher opacity absorb more light.

opaque (material) Describes a material that absorbs light.

open cluster A cluster of up to several thousand stars; open clusters are found only in the disks of galaxies and often contain young stars.

open universe The universe is open if its average density is less than the critical density, in which case spacetime has an overall shape analogous to the surface of a saddle.

opposition The point at which a planet appears opposite the Sun in our sky.

optical quality Describes the ability of a lens, mirror, or telescope to obtain clear and properly focused images.

orbital energy The sum of an orbiting object's kinetic and gravitational potential energies.

orbital resonance Describes any situation in which one object's orbital period is a simple ratio of another object's period, such as 1/2, 1/4, or 5/3. In such cases, the two objects periodically line up with each other, and the extra gravitational attractions at these times can affect the objects' orbits.

orbital velocity law This law (a variation on Newton's version of Kepler's third law) allows us to use a star's orbital speed and distance from the galactic center to determine the total mass of the galaxy contained *within* the star's orbit. Mathematically, it is written:

$$M_r = \frac{r \times v^2}{G}$$

where M_r is the mass contained within the star's orbit, r is the star's distance from the galactic center, v is the star's orbital velocity, and G is the gravitational constant.

orbiters (of other worlds) Spacecraft that go into orbit of another world for long-term study.

outer solar system Generally considered to encompass the region of our solar system beginning at about the orbit of Jupiter.

outgassing The process of releasing gases from a planetary interior, usually through volcanic eruptions.

oxidation Refers to chemical reactions, often with the surface of a planet, that remove oxygen from the atmosphere.

ozone The molecule O_3, which is a particularly good absorber of ultraviolet light.

ozone depletion Refers to the declining levels of atmospheric ozone found worldwide on Earth, especially in Antarctica, in recent years.

ozone hole A place where the concentration of ozone in the stratosphere is dramatically lower than is the norm.

pair production The process in which a concentration of energy spontaneously turns into a particle and its antiparticle.

parabola The precise mathematical shape of a special type of unbound orbit allowed under the force of gravity; if an object in a parabolic orbit loses only a tiny amount of energy, it will become bound.

paradigm (in science) Refers to general patterns of thought that tend to shape scientific beliefs during a particular time period.

paradox A situation that, at least at first, seems to violate common sense or contradict itself. Resolving paradoxes often leads to deeper understanding.

parallax The apparent shifting of an object against the background, due to viewing it from different positions. *See also* stellar parallax

parallax angle Half of a star's annual back-and-forth shift due to stellar parallax; related to the star's distance according to the formula

$$\text{distance in parsecs} = \frac{1}{p}$$

where p is the parallax angle in arcseconds.

parsec (pc) Approximately equal to 3.26 light-years; it is the distance to an object with a parallax angle of 1 arcsecond.

partial lunar eclipse A lunar eclipse in which the Moon becomes only partially covered by the Earth's umbral shadow.

partial solar eclipse A solar eclipse during which the Sun becomes only partially blocked by the disk of the Moon.

particle accelerator A machine designed to accelerate subatomic particles to high speeds in order to create new particles or to test fundamental theories of physics.

particle era The era of the universe lasting from 10^{-10} second to 0.001 second after the Big Bang, during which subatomic particles were continually created and destroyed and ending when matter annihilated antimatter.

peculiar velocity (of a galaxy) The component of a galaxy's velocity relative to the Milky Way that deviates from the velocity expected by Hubble's law.

penumbra The lighter, outlying regions of a shadow.

penumbral (lunar) eclipse A lunar eclipse in which the Moon passes only within the Earth's penumbral shadow and does not fall within the umbra.

perigee The point at which an object orbiting the Earth is nearest to the Earth.

perihelion The point at which an object orbiting the Sun is closest to the Sun.

period–luminosity relation The relation that describes how the luminosity of a Cepheid variable star is related to the period between peaks in its brightness; the longer the period, the more luminous the star.

phase (of matter) Describes the way in which atoms or molecules are held together; the common phases are solid, liquid, and gas.

photon An individual particle of light, characterized by a wavelength and a frequency.

photosphere The visible surface of the Sun, where the temperature averages just under 6,000 K.

pixel An individual "picture element" on a CCD.

Planck era The era of the universe prior to the Planck time.

Planck's constant A universal constant, abbreviated *h*, with value $h = 6.626 \times 10^{-34}$ joule $\times$ s.

Planck time The time when the universe was 10^{-43} second old, before which random energy fluctuations were so large that our current theories are powerless to describe what might have been happening.

planet A moderately large object that orbits a star and shines primarily by reflecting light from its star. More precisely, according to a definition approved in 2006, a planet is an object that: (1) orbits a star (but is itself neither a star nor a moon); (2) is massive enough for its own gravity to give it a nearly round shape; and (3) has cleared the neighborhood around its orbit. Objects that meet the first two criteria but not the third, including Ceres, Pluto, and Eris, are designated *dwarf planets.*

planetary geology The extension of the study of Earth's surface and interior to apply to other solid bodies in the solar system such as terrestrial planets and jovian planet moons.

planetary nebula The glowing cloud of gas ejected from a low-mass star at the end of its life.

planetesimals The building blocks of planets, formed by accretion in the solar nebula.

plasma A gas consisting of ions and electrons.

plasma tail (of a comet) One of two tails seen when a comet passes near the Sun (the other is the dust tail); composed of ionized gas blown away from the Sun by the solar wind.

plates (on a planet) Pieces of a lithosphere that apparently float upon the denser mantle below.

plate tectonics The geological process in which plates are moved around by stresses in a planet's mantle.

polarization (of light) Light is polarized when the electric and magnetic fields of light waves are all aligned in some particular way.

Population I *See* disk population

Population II *See* spheroidal population

positron The antimatter equivalent of an electron. It is identical to an electron in virtually all respects, except it has a positive rather than a negative electrical charge.

potential energy Energy stored for later conversion into kinetic energy; includes gravitational potential energy, electrical potential energy, and chemical potential energy.

power The rate of energy usage, usually measured in watts (1 watt = 1 joule/s).

precession The gradual wobble of the axis of a rotating object around a vertical line.

precipitation Condensed atmospheric gases that fall to the surface in the form of rain, snow, or hail.

pressure Describes the force (per unit area) pushing on an object. In astronomy, we are generally interested in pressure applied by surrounding gas (or plasma). Ordinarily, such pressure is related to the temperature of the gas (*see* thermal pressure). In objects such as white dwarfs and neutron stars, pressure may arise from a quantum effect (*see* degeneracy pressure). Light can also exert pressure. (*See* radiation pressure.)

primary mirror The large, light-collecting mirror of a reflecting telescope.

prime focus (of a reflecting telescope) The first point at which light focuses after bouncing off the primary mirror; located in front of the primary mirror.

prime meridian The meridian of longitude that passes through Greenwich, England, defined to be longitude 0°.

primitive meteorites Meteorites that formed at the same time as the solar system itself, about 4.6 billion years ago. Primitive meteorites from the inner asteroid belt are usually stony, and those from the outer belt are usually carbon-rich.

processed meteorites Meteorites that apparently once were part of a larger object that "processed" the original material of the solar nebula into another form. Processed meteorites can be rocky if chipped from the surface or mantle, or metallic if blasted from the core.

proper motion The motion of an object in the plane of the sky, perpendicular to our line of sight.

protogalactic cloud A huge, collapsing cloud of intergalactic gas from which an individual galaxy formed.

proton–proton chain The chain of reactions by which low-mass stars (including the Sun) fuse hydrogen into helium.

protons Particles found in atomic nuclei with positive electrical charge, built from three quarks.

protoplanetary disk A disk of material surrounding a young star (or protostar) that may eventually form planets.

protostar A forming star that has not yet reached the point where sustained fusion can occur in its core.

protostellar disk A disk of material surrounding a protostar; essentially the same as a protoplanetary disk, but may not necessarily lead to planet formation.

protostellar wind The relatively strong wind from a protostar.

protosun The central object in the forming solar system that eventually became the Sun.

pseudoscience Something that purports to be science or may appear to be scientific but that does not adhere to the testing and verification requirements of the scientific method.

Ptolemaic model The geocentric model of the universe developed by Ptolemy in about 150 A.D.

pulsar A neutron star from which we see rapid pulses of radiation as it rotates.

pulsating variable stars Stars that alternately grow brighter and dimmer as their outer layers expand and contract in size.

quantum mechanics The branch of physics that deals with the very small, including molecules, atoms, and fundamental particles.

quantum state Refers to the complete description of the state of a subatomic particle, including its location, momentum, orbital angular momentum, and spin, to the extent allowed by the uncertainty principle.

quantum tunneling The process in which, thanks to the uncertainty principle, an electron or other subatomic particle appears on the other side of a barrier that it does not have the energy to overcome in a normal way.

quarks The building blocks of protons and neutrons, quarks are one of the two basic types of fermions (leptons are the other).

quasar The brightest type of active galactic nucleus.

radar mapping Imaging of a planet by bouncing radar waves off its surface, especially important for Venus and Titan where thick clouds mask the surface.

radar ranging A method of measuring distances within the solar system by bouncing radio waves off planets.

radial motion The component of an object's motion directed toward or away from us.

radial velocity The portion of any object's total velocity that is directed toward or away from us. This part of the velocity is the only part that we can measure with the Doppler effect.

radiation pressure Pressure exerted by photons of light.

radiation zone (of a star) A region of the interior in which energy is transported primarily by radiative diffusion.

radiative diffusion The process by which photons gradually migrate from a hot region (such as the solar core) to a cooler region (such as the solar surface).

radiative energy Energy carried by light; the energy of a photon is Planck's constant times its frequency, or $h \times f$.

radioactive decay The spontaneous change of an atom, in which its nucleus breaks apart or a proton turns into an electron, thereby changing into a different element. It also releases heat to a planet's interior.

radioactive element (or **radioactive isotope**) A substance whose nucleus tends to fall apart spontaneously.

radio galaxy A galaxy that emits unusually large quantities of radio waves; thought to contain an active galactic nucleus powered by a supermassive black hole.

radio lobes The huge regions of radio emission found on either side of radio galaxies. The lobes apparently contain plasma ejected by powerful jets from the galactic center.

radiometric dating The process of determining the age of a rock (i.e., the time since it solidified) by comparing the present amount of a radioactive substance to the amount of its decay product.

radio waves Light with very long wavelengths (and hence low frequencies)—longer than those of infrared light.

recession velocity (of a galaxy) The speed at which a distant galaxy is moving away from us due to the expansion of the universe.

recollapsing universe The possible fate of our universe in which the collective gravity of all its matter eventually halts and reverses the expansion. The galaxies will come crashing back together, and the universe will end in a fiery Big Crunch.

red giant A giant star that is red in color.

red-giant winds The relatively dense but slow winds from red giant stars.

redshift (Doppler) A Doppler shift in which spectral features are shifted to longer wavelengths, caused when an object is moving away from the observer.

reference frame (frame of reference) Two people (or objects) share the same reference frame if they are *not* moving relative to one another.

reflecting telescope A telescope that uses mirrors to focus light.

reflection (of light) The process by which matter changes the direction of light.

reflection nebula A nebula that we see as a result of starlight reflected from interstellar dust grains. Reflection nebulae tend to have blue and black tints.

refracting telescope A telescope that uses lenses to focus light.

resonance *See* orbital resonance

rest wavelength The wavelength of a spectral feature in the absence of any Doppler shift or gravitational redshift.

retrograde motion Motion that is backward compared to the norm; e.g., we see Mars in apparent retrograde motion during the periods of time when it moves westward, rather than the more common eastward, relative to the stars.

revolution The orbital motion of one object around another.

right ascension (RA) Analogous to longitude, but on the celestial sphere; it is the angular east-west distance between the vernal equinox and a location on the celestial sphere.

rings (planetary) Consist of numerous small particles orbiting a planet within its Roche zone.

Roche tidal zone The region within two to three planetary radii (of any planet) in which the tidal forces tugging an object apart become comparable to the gravitational forces holding it together; planetary rings are always found within the Roche tidal zone.

rocks (in solar system theory) Material common on the surface of the Earth, such as silicon-based minerals, that are solid at temperatures and pressures found on Earth but typically melt or vaporize at temperatures of 500–1,300 K.

rotation The spinning of an object around its axis.

rotation curve A graph that plots rotational (or orbital) velocity against distance from the center for any object or set of objects.

runaway greenhouse effect A positive feedback cycle in which heating caused by the greenhouse effect causes more greenhouse gases to enter the atmosphere, which further enhances the greenhouse effect.

saddle-shaped (or **hyperbolic**) **geometry** Refers to any case in which the rules of geometry for a saddle-shaped surface hold, such as that two lines that begin parallel eventually diverge.

Sagittarius Dwarf A small, dwarf elliptical galaxy that is currently passing through the disk of the Milky Way Galaxy.

saros cycle The period over which the basic pattern of eclipses repeats, which is about 18 years $11\frac{1}{3}$ days.

satellite Any object orbiting another object.

scattered light Light that is reflected into random directions.

Schwarzschild radius A measure of the size of the event horizon of a black hole.

science The search for knowledge that can be used to explain or predict natural phenomena in a way that can be confirmed by rigorous observations or experiments.

scientific method An organized approach to explaining observed facts through science.

scientific theory A model of some aspect of nature that has been rigorously tested and has passed all tests to date.

seafloor crust On Earth, the thin, dense crust of basalt created by seafloor spreading.

seafloor spreading On Earth, the creation of new seafloor crust at mid-ocean ridges.

search for extraterrestrial intelligence (SETI) The name given to observing projects designed to search for signs of intelligent life beyond Earth.

secondary mirror A small mirror in a reflecting telescope, used to reflect light gathered by the primary mirror toward an eyepiece or instrument.

sedimentary rock A rock that formed from sediments created and deposited by erosional processes.

seismic waves Earthquake-induced vibrations that propagate through a planet.

semimajor axis Half the distance across the long axis of an ellipse; in this text, it is usually referred to as the *average* distance of an orbiting object, abbreviated *a* in the formula for Kepler's third law.

Seyfert galaxies The name given to a class of galaxies found relatively nearby and that have nuclei much like those of quasars, except that they are less luminous.

shepherd moons Tiny moons within a planet's ring system that help force particles into a narrow ring. A variation on *gap moons.*

shield volcano A shallow-sloped volcano made from the flow of low-viscosity basaltic lava.

shock wave A wave of pressure generated by gas moving faster than the speed of sound.

sidereal day The time of 23 hours 56 minutes 4.09 seconds between successive appearances of any particular star on the meridian; essentially the true rotation period of the Earth.

sidereal month About $27\frac{1}{4}$ days, the time required for the Moon to orbit the Earth once (as measured against the stars).

sidereal period (of a planet) A planet's actual orbital period around the Sun.

sidereal time Time measured according to the position of stars in the sky rather than the position of the Sun in the sky. *See also* local sidereal time

sidereal year The time required for the Earth to complete exactly one orbit as measured against the stars; about 20 minutes longer than the tropical year on which our calendar is based.

silicate rock A silicon-rich rock.

singularity The place at the center of a black hole where, in principle, gravity crushes all matter to an infinitely tiny and dense point.

Small Magellanic Cloud One of two small, irregular galaxies (the other is the Large Magellanic Cloud) located about 150,000 light-years away; it probably orbits the Milky Way Galaxy.

snowball Earth Name given to a hypothesis suggesting that, some 600–700 million years ago, the Earth experienced a period in which it became cold enough for glaciers to exist worldwide, even in equatorial regions.

solar activity Refers to short-lived phenomena on the Sun, including the emergence and disappearance of individual sunspots, prominences, and flares; sometimes called *solar weather.*

solar circle The Sun's orbital path around the galaxy, which has a radius of about 28,000 light-years.

solar day Twenty-four hours, which is the average time between appearances of the Sun on the meridian.

solar eclipse Occurs when the Moon's shadow falls on the Earth, which can occur only at new moon; may be total, partial, or annular.

solar flares Huge and sudden releases of energy on the solar surface, probably caused when energy stored in magnetic fields is suddenly released.

solar luminosity The luminosity of the Sun, which is approximately 4×10^{26} watts.

solar maximum The time during each sunspot cycle at which the number of sunspots is the greatest.

solar minimum The time during each sunspot cycle at which the number of sunspots is the smallest.

solar nebula The piece of interstellar cloud from which our own solar system formed.

solar neutrino problem Refers to the disagreement between the predicted and observed number of neutrinos coming from the Sun.

solar prominences Vaulted loops of hot gas that rise above the Sun's surface and follow magnetic-field lines.

solar sail A large, highly reflective (and thin, to minimize mass) piece of material that can "sail" through space using pressure exerted by sunlight.

solar system (or **star system**) Consists of a star (sometimes more than one star) and all the objects that orbit it.

solar thermostat The regulation of the Sun's core temperature that comes about because pressure and gravity must balance within the Sun.

solar wind A stream of charged particles ejected from the Sun.

solid phase The phase of matter in which atoms or molecules are held rigidly in place.

solstice See *summer solstice* and *winter solstice.*

sound wave A wave of alternately rising and falling pressure.

south celestial pole (SCP) The point on the celestial sphere directly above the Earth's South Pole.

spacetime The inseparable, four-dimensional combination of space and time.

spacetime diagram A graph that plots a spatial dimension on one axis and time on another axis.

special theory of relativity Einstein's theory that describes the effects of the facts that all motion is relative and that everyone always measures the same speed of light.

spectral lines Bright or dark lines that appear in an object's spectrum, which we can see when we pass the object's light through a prismlike device that spreads out the light like a rainbow.

spectral resolution Describes the degree of detail that can be seen in a spectrum; the higher the spectral resolution, the more detail we can see.

spectral type A way of classifying a star by the lines that appear in its spectrum; it is related to surface temperature. The basic spectral types are designated by a letter (OBAFGKM, with O for the hottest stars and M for the coolest) and are subdivided with numbers from 0 through 9.

spectrograph An instrument used to record spectra.

spectroscopic binary A binary star system whose binary nature is revealed because we detect the spectral lines of one or both stars alternately becoming blueshifted and redshifted as the stars orbit each other.

spectroscopy (in astronomical research) The process of obtaining spectra from astronomical objects.

spectrum (of light) *See* electromagnetic spectrum

speed The rate at which an object moves. Its units are distance divided by time, such as m/s or km/hr.

speed of light The speed at which light travels, which is about 300,000 km/s.

spherical geometry Refers to any case in which the rules of geometry for the surface of a sphere hold, such as that lines that begin parallel eventually meet.

spheroidal component (of a galaxy) The portion of any galaxy that is spherical (or football-like) in shape and contains very little cool gas; generally contains only very old stars. Elliptical galaxies have only a spheroidal component, while spiral galaxies also have a disk component.

spheroidal galaxy Another name for an elliptical galaxy.

spheroidal population Refers to stars that orbit within the spheroidal component of a galaxy. Thus, elliptical galaxies have only a spheroidal population (they lack a disk population), while spiral galaxies have spheroidal population stars in their bulges and halos. Sometimes called Population II.

spin (quantum) *See* spin angular momentum

spin angular momentum Often simply called *spin,* it refers to the inherent angular momentum of a fundamental particle.

spiral arms The bright, prominent arms, usually in a spiral pattern, found in most spiral galaxies.

spiral density waves Gravitationally driven waves of enhanced density that move through a spiral galaxy and are responsible for maintaining its spiral arms.

spiral galaxies Galaxies that look like flat, white disks with yellowish bulges at their centers. The disks are filled with cool gas and dust, interspersed with hotter ionized gas, and usually display beautiful spiral arms.

spreading centers (geological) Places where hot mantle material rises upward between plates and then spreads sideways creating new seafloor crust.

spring equinox (vernal equinox) Refers both to the point in Pisces on the celestial sphere where the ecliptic crosses the celestial equator and to the moment in time when the Sun appears at that point each year (around March 21).

spring tides The higher-than-average tides on Earth that occur at new and full moon, when the tidal forces from the Sun and Moon both act along the same line.

standard candle An object for which we have some means of knowing its true luminosity, so that we can use its apparent brightness to determine its distance with the luminosity–distance formula.

standard model (of physics) The current theoretical model that describes the fundamental particles and forces in nature.

standard time Time measured according to the internationally recognized time zones.

star A large, glowing ball of gas that generates energy through nuclear fusion in its core. The term *star* is sometimes applied to objects that are in the process of becoming true stars (e.g., protostars) and to the remains of stars that have died (e.g., neutron stars).

starburst galaxy A galaxy in which stars are forming at an unusually high rate.

star cluster *See* cluster of stars

star-gas-star cycle The process of galactic recycling in which stars expel gas into space where it mixes with the interstellar medium and eventually forms new stars.

star system See *solar system.*

state (quantum) *See* quantum state

steady state theory A now-discredited theory that held that the universe had no beginning and looks about the same at all times.

Stefan–Boltzmann constant A constant that appears in the laws of thermal radiation, with value

$$\sigma = 5.7 \times 10^{-8} \frac{\text{watt}}{m^2 \times \text{Kelvin}^4}$$

stellar evolution The formation and development of stars.

stellar parallax The apparent shift in the position of a nearby star (relative to distant objects) that occurs as we view the star from different positions in the Earth's orbit of the Sun each year.

stellar wind A stream of charged particles ejected from the surface of a star.

stratosphere An intermediate-altitude layer of the atmosphere that is warmed by the absorption of ultraviolet light from the Sun.

stratovolcano A steep-sided volcano made from viscous lavas that can't flow very far before solidifying.

string theory New ideas, not yet well-tested, that attempt to explain all of physics in a much simpler way than current theories.

stromatolites Large bacterial "colonies."

strong force One of the four fundamental forces; it is the force that holds atomic nuclei together.

subduction (of tectonic plates) The process in which one plate slides under another.

subduction zones Places where one plate slides under another.

subgiant A star that is between being a main-sequence star and being a giant; subgiants have inert helium cores and hydrogen-burning shells.

sublimation The process by which atoms or molecules escape into the gas phase from a solid.

summer solstice Refers both to the point on the celestial sphere where the ecliptic is farthest north of the celestial equator and to the moment in time when the Sun appears at that point each year (around June 21).

sunspot cycle The period of about 11 years over which the number of sunspots on the Sun rises and falls.

sunspots Blotches on the surface of the Sun that appear darker than surrounding regions.

superbubble Essentially a giant interstellar bubble, formed when the shock waves of many individual bubbles merge to form a single, giant shock wave.

supercluster Superclusters consist of many clusters of galaxies, groups of galaxies, and individual galaxies and are the largest known structures in the universe.

supergiants (luminosity class I) The very large and very bright stars that appear at the top of an H–R diagram.

supermassive black hole Giant black hole, with a mass millions to billions of times that of our Sun, thought to reside in the centers of many galaxies and to power active galactic nuclei.

supernova The explosion of a star.

Supernova 1987A A supernova witnessed on Earth in 1987; it was the nearest supernova seen in nearly 400 years and helped astronomers refine theories of supernovae.

supernova remnant A glowing, expanding cloud of debris from a supernova explosion.

synchronous rotation Describes the rotation of an object that always shows the same face to an object that it is orbiting because its rotation period and orbital period are equal.

synchrotron radiation A type of radio emission that occurs when electrons moving at nearly the speed of light spiral around magnetic field lines.

synodic month (or **lunar month**) The time required for a complete cycle of lunar phases, which averages about $29\frac{1}{2}$ days.

synodic period (of a planet) The time between successive alignments of a planet and the Sun in our sky; measured from opposition to opposition for a planet beyond Earth's orbit, or from superior conjunction to superior conjunction for Mercury and Venus.

tangential motion The component of an object's motion directed across our line of sight.

tangential velocity The portion of any object's total velocity that is directed across (perpendicular to) our line of sight. This part of the velocity cannot be measured with the Doppler effect. It can be measured only by observing the object's gradual motion across our sky.

tectonics The disruption of a planet's surface by internal stresses.

temperature A measure of the average kinetic energy of particles in a substance.

terrestrial planets Rocky planets similar in overall composition to Earth.

theories of relativity (*special* and *general*) Einstein's theories that describe the nature of space, time, and gravity.

theory (in science) *See* scientific theory.

theory of evolution The theory, first advanced by Charles Darwin, that explains *how* evolution occurs through the process of *natural selection.*

thermal emitter An object that produces a thermal radiation spectrum; sometimes called a "blackbody."

thermal energy Represents the collective kinetic energy, as measured by temperature, of the many individual particles moving within a substance.

thermal escape The process in which atoms or molecules in a planet's exosphere move fast enough to escape into space.

thermal pressure The ordinary pressure in a gas arising from motions of particles that can be attributed to the object's temperature.

thermal pulses The predicted upward spikes in the rate of helium fusion, occurring every few thousand years, that occur near the end of a low-mass star's life.

thermal radiation The spectrum of radiation produced by an opaque object that depends only on the object's temperature; sometimes called "blackbody radiation."

thermosphere A high, hot X-ray-absorbing layer of an atmosphere, just below the exosphere.

tidal force A force that is caused when the gravity pulling on one side of an object is larger than that on the other side, causing the object to stretch.

tidal friction Friction within an object that is caused by a tidal force.

tidal heating A source of internal heating created by tidal friction. It is particularly important for satellites with eccentric orbits such as Io and Europa.

time dilation Refers to the effect in which you observe time running slower in reference frames moving relative to you.

timing (in astronomical research) The process of tracking how the light intensity from an astronomical object varies with time.

torque A twisting force that can cause a change in an object's angular momentum.

total apparent brightness *See* apparent brightness. We sometimes say "total apparent brightness" to distinguish it from wavelength-specific measures such as the apparent brightness measured in visible light.

totality (eclipse) The portion of either a total lunar eclipse during which the Moon is fully within the Earth's umbral shadow or a total solar eclipse during which the Sun's disk is fully blocked by the Moon.

total luminosity *See* luminosity. We sometimes say "total luminosity" to distinguish it from wavelength-specific measures such as the luminosity emitted in visible light or the X-ray luminosity.

total lunar eclipse A lunar eclipse in which the Moon becomes fully covered by the Earth's umbral shadow.

total solar eclipse A solar eclipse during which the Sun becomes fully blocked by the disk of the Moon.

transit A transit is an event in which a planet passes in front of a star (or the Sun) as seen from Earth. Only Mercury and Venus can be seen in transit of our Sun. The search for transits of extrasolar planets is an important planet detection strategy.

transmission (of light) The process in which light passes through matter without being absorbed.

transparent (material) Describes a material that transmits light.

triple-alpha reaction *See* helium fusion

Trojan asteroids Asteroids found within two stable zones that share Jupiter's orbit but lie 60° ahead of and behind Jupiter.

tropical year The time from one spring equinox to the next, on which our calendar is based.

tropic of Cancer The circle on the Earth with latitude 23.5°N. It is the northernmost latitude at which the Sun ever passes directly overhead (at noon on the summer solstice).

tropic of Capricorn The circle on the Earth with latitude 23.5°S. It is the southernmost latitude at which the Sun ever passes directly overhead (at noon on the winter solstice).

troposphere The lowest atmospheric layer, in which convection and weather occur.

Tully–Fisher relation A relationship among spiral galaxies showing that the faster a spiral galaxy's rotation speed, the more luminous it is; it is important because it allows us to determine the distance to a spiral galaxy once we measure its rotation rate and apply the luminosity–distance formula.

turbulence Rapid and random motion.

ultraviolet light Light with wavelengths that fall in the portion of the electromagnetic spectrum between visible light and X rays.

umbra The dark central region of a shadow.

unbound orbits Orbits on which an object comes in toward a large body only once, never to return; unbound orbits may be parabolic or hyperbolic in shape.

uncertainty principle The law of quantum mechanics that states that we can never know both a particle's position and its momentum, or both its energy and the time it has the energy, with absolute precision.

universal law of gravitation The law expressing the force of gravity (F_g) between two objects, given by the formula

$$F_g = G\frac{M_1 M_2}{d^2}$$

$$\left(G = 6.67 \times 10^{-11} \frac{m^3}{\text{kg} \times \text{s}^2}\right)$$

universal time (UT) Standard time in Greenwich (or anywhere on the prime meridian).

universe The sum total of all matter and energy.

up quark One of the two quark types (the other is the down quark) found in ordinary protons and neutrons. Has a charge of $+\frac{2}{3}$.

velocity The combination of speed and direction of motion; it can be stated as a speed in a particular direction, such as 100 km/hr due north.

vernal equinox *See* spring equinox

virtual particles Particles that "pop" in and out of existence so rapidly that, according to the uncertainty principle, they cannot be directly detected.

viscosity Describes the "thickness" of a liquid in terms of how rapidly it flows; low-viscosity liquids flow quickly (e.g., water), while high-viscosity liquids flow slowly (e.g., molasses).

visible light The light our eyes can see, ranging in wavelength from about 400 to 700 nm.

visual binary A binary star system in which we can resolve both stars through a telescope.

voids Huge volumes of space between superclusters that appear to contain very little matter.

volatiles Refers to substances, such as water, carbon dioxide, and methane, that are usually found as gases, liquids, or surface ices on the terrestrial worlds.

volcanic plains Vast, relatively smooth areas created by the eruption of very runny lava.

volcanism The eruption of molten rock, or lava, from a planet's interior onto its surface.

watt The standard unit of power in science: 1 watt = 1 joule/s.

wavelength The distance between adjacent peaks (or troughs) of a wave.

weak bosons The exchange particles for the weak force.

weak force One of the four fundamental forces; it is the force that mediates nuclear reactions; also the only force besides gravity felt by weakly interacting particles.

weakly interacting particles Particles, such as neutrinos and WIMPs, that respond only to the weak force and gravity; that is, they do not feel the strong force or the electromagnetic force.

weather Describes the ever-varying combination of winds, clouds, temperature, and pressure in a planet's troposphere.

weight The net force that an object applies to its surroundings; in the case of a stationary body on the surface of the Earth, weight = mass × acceleration of gravity.

weightless A weight of zero, as occurs during free-fall.

white dwarf limit (also called the *Chandrasekhar limit*) The maximum possible mass for a white dwarf, which is about $1.4M_{\text{Sun}}$.

white dwarfs The hot, compact corpses of low-mass stars, typically with a mass similar to the Sun compressed to a volume the size of the Earth.

white dwarf supernova A supernova that occurs when an accreting white dwarf reaches the white-dwarf limit, ignites runaway carbon fusion, and explodes like a bomb; often called a *Type Ia supernova.*

WIMPs Stands for *weakly interacting massive particles* and represents a possible form of dark matter consisting of subatomic particles that are dark because they do not respond to the electromagnetic force.

winter solstice Refers both to the point on the celestial sphere where the ecliptic is farthest south of the celestial equator and to the moment in time when the Sun appears at that point each year (around December 21).

worldline A line that represents an object on a spacetime diagram.

wormholes The name given to hypothetical tunnels through hyperspace that might connect two distant places in our universe.

X rays Light with wavelengths that fall in the portion of the electromagnetic spectrum between ultraviolet light and gamma rays.

X-ray binary A binary star system that emits substantial amounts of X rays, thought to be from an accretion disk around a neutron star or black hole.

X-ray burster An object that emits a burst of X rays every few hours to every few days; each burst lasts a few seconds and is thought to be caused by helium fusion on the surface of an accreting neutron star in a binary system.

X-ray bursts Burst of X-rays coming from sudden ignition of fusion on the surface of an accreting neutron star in an X-ray binary system.

Zeeman effect The splitting of spectral lines by a magnetic field.

zenith The point directly overhead, which has an altitude of 90°.

zodiac The constellations on the celestial sphere through which the ecliptic passes.

zones (on a jovian planet) Bright bands of rising air that encircle a jovian planet at a particular set of latitudes.

Index

Page references in italics refer to figures and tables. Page references followed by an "n" refer to footnotes.

1
A Universe of Life?

LEARNING GOALS

1.1 THE POSSIBILITY OF LIFE BEYOND EARTH
- What are we searching for?
- Is it reasonable to imagine life beyond Earth?

1.2 THE SCIENTIFIC CONTEXT OF THE SEARCH
- How does astronomy help us understand the possibilities for extraterrestrial life?
- How does planetary science help us understand the possibilities for extraterrestrial life?
- How does biology help us understand the possibilities for extraterrestrial life?

1.3 PLACES TO SEARCH
- Where should we search for life in the universe?
- Could aliens be searching for us?

1.4 THE NEW SCIENCE OF ASTROBIOLOGY
- How do we study the possibility of life beyond Earth?

Sometimes I think the universe is full of life and sometimes I don't. Either way is equally amazing.
Arthur C. Clarke

Gazing at the night sky, we can't help but be amazed at the glittering stars that silently beckon us from the depths of space. Stars are suns, much like our own Sun, each shining brightly for millions or billions of years. Many have planets, some of which may be much like the planets in our own solar system. Among these countless worlds, it may seem hard to imagine that Earth could be the only home for life. But while the possibility of life beyond Earth might seem quite reasonable, we do not yet have sufficient evidence to know whether it actually exists.

Learning whether the universe is full of life holds great significance for the way we view ourselves and our planet. If life is rare or nonexistent elsewhere, we will view our planet with added wonder. If life is common, we'll know that Earth is not quite as special as it may seem. If civilizations are common, we'll be forced to accept that we ourselves are just one of many intelligent species throughout the universe. The profound implications of finding—or not finding—extraterrestrial life make the question of life beyond Earth an exciting topic of study.

The primary purpose of this book is to give you the background needed to understand new and exciting developments in the human quest to find life beyond Earth. To do that, in coming chapters we will focus in some detail on the scientific issues that frame the search for life. First, however, let's start with a brief introduction to the subject, so that you'll understand why it has become such a hot topic of scientific research.

1.1 THE POSSIBILITY OF LIFE BEYOND EARTH

Aliens are everywhere, at least if you follow the popular media (Figure 1.1). Starships on television, such as the *Enterprise* or *Voyager*, are on constant prowl throughout the galaxy, seeking out new life and hoping it speaks English (or something close enough to English for the "universal translator"). In *Star Wars*, aliens from many planets gather at bars to share drinks and stories, and presumably to marvel at the fact that they have greater similarity in their level of technology than do different nations on Earth. Closer to home, movies like *Independence Day, Men in Black*, and *War of the Worlds* feature brave Earthlings battling evil aliens, while tabloids like the *Weekly World News* routinely carry headlines about the latest alien landings. Even serious newspapers and magazines run occasional articles about UFO sightings or about claims that the U.S. government is hiding frozen alien corpses at "Area 51." Given this intense media interest, it's probably not surprising that close to half of all Americans claim to believe that we have already been visited by beings from other worlds.

Figure 1.1

Aliens have become a part of modern culture, as illustrated in this movie poster.

Scientists are interested in aliens too, although most scientists remain deeply skeptical about reports of aliens on Earth (for reasons we'll discuss later in the book). Scientists are therefore searching for life elsewhere, looking for evidence of life on other worlds in our solar system, trying to learn whether we should expect to find life on planets orbiting other stars, and searching for signals broadcast by other civilizations. Indeed, the study of life in the universe is one of the most rapidly growing fields of active scientific research, largely because of its clear significance: The discovery of life of any kind beyond Earth would forever change our perspective on how we fit into the universe as a whole, and would undoubtedly teach us much more about life here on Earth as well.

THINK ABOUT IT Check television and movie listings to see what is currently showing that involves aliens of some sort. Do you think any of the shows portray aliens in a scientifically realistic fashion? Explain.

• What are we searching for?

When we say we are searching for *life* in the universe, just what is it that we are looking for? Is it the kind of intelligent life we see portrayed in science fiction TV shows and films? Is it something more akin to the plants and animals we see in parks and zoos? Is it tiny, bacteria-like microbes? Or could it be something else entirely?

The simple answer is "all of the above." When we search for **extraterrestrial life,** or life beyond Earth, we are looking for any sign of life, be it simple, complex, or intelligent. We don't care if it looks exactly like life we are familiar with on Earth, or if it is dramatically different. However, we can't really answer the question of what we are looking for unless we know what life *is*.

Unfortunately, defining life is no simple matter, not even here on Earth where we have bountiful examples of it. Ask yourself: What common attributes make us think that a bacterium, a beetle, a mushroom, a tumbleweed, a maple tree, and a human are all alive, while we don't think the same of a crystal, a cloud, an ocean, or a fire? If you spend just a little time thinking about this question, you'll begin to appreciate its difficulty. For example, you might say that life can move, but the same is true of clouds and oceans. You might say that life can grow, but so can crystals. Or you might say that life can reproduce and spread, but so can fire. We will explore in Chapter 5 how scientists try to answer this question and come up with a general definition of life, but for now it should be clear that this is a complicated question that affects how we search for life in the universe.

Because of this definitional difficulty, the scientific search for extraterrestrial life in the universe generally presumes a search for life that is at least somewhat Earth-like and that we could therefore recognize based on what we know from studying life on Earth. Science fiction fans will object that this search is far too limited, and they may be right—but we have to start somewhere, so we begin with what we understand.

• Is it reasonable to imagine life beyond Earth?

The scientific search for life in the universe is a relatively recent development in human history, but the idea of extraterrestrial life is not. Many ancient cultures told stories about beings living among the stars and, as we'll discuss in Chapter 2, the ancient Greeks engaged in serious philosophical debate about the possibility of life beyond Earth.

Until quite recently, however, all these ideas remained purely speculative, because there was no way to study the question of life beyond Earth scientifically. It was always possible to *imagine* extraterrestrial life, but there was no scientific reason to think that it could really exist. Indeed, the relatively small amounts of data that might have shed some light on the question of life beyond Earth were often misinterpreted. Prior to the twentieth century, for example, some scientists guessed that Venus might harbor a tropical paradise—a guess that was based on little more than the fact that Venus is covered by clouds and closer than Earth to the Sun. Mars was the subject of even more intense debate, largely because a handful of scientists thought they saw long, straight canals on the surface [Section 8.1]. The canals, which don't really exist, were cited as evidence of a martian civilization.

Today, we have enough telescopic and spacecraft photos of the planets and large moons in our solar system to be quite confident that no civilization has ever existed on any of them. The prospect of large animals or plants seems almost equally unlikely. Nevertheless, scientific interest in life beyond Earth has exploded in the past few decades. Why?

We'll spend most of the rest of the book answering this question, but we can summarize the key points briefly. First, although large, multicellular life in our solar system seems unlikely anywhere but on Earth, new discoveries in both planetary science and biology have given us some reason to think that simpler life—perhaps tiny microbes—might yet exist on other planets or moons that orbit our Sun. Second, while we've long known that the universe is full of *stars,* we've only recently gained concrete evidence telling us that it is also full of *planets,* which means there are far more places where we could potentially search for life. Third, advances both in scientific understanding and in technology now make it possible to study the question of life in the universe through established techniques of science, something that was not possible just a few decades ago. For example, we now understand enough about biology to explore the conditions that might make it possible for life to exist on other worlds, and we know enough about planets to consider which ones might be capable of harboring life. Indeed, while at present we have only limited ability to actually search for life beyond Earth, we are rapidly developing the spacecraft technology needed to search for microbes on other worlds of our solar system and the telescopic technology needed to look for signs of life among the stars.

The bottom line is that while it remains possible that life exists only on Earth, we now have plenty of scientific reasons to think that life might be widespread and that we could detect it if it is. In the rest of this chapter, we'll briefly discuss the scientific context of the search and the places where we might conduct the search, so that you will have a sense of what we will cover in more depth in the rest of the book.

1.2 THE SCIENTIFIC CONTEXT OF THE SEARCH

Almost every field of scientific research has at least some bearing on the search for life in the universe. Even seemingly unrelated fields such as mathematics and computer science play important roles. For example, we use mathematics to do the many computations that help us understand all other areas of science, and we use computers to simulate everything from the formation of stars and planets to the way in which the molecules of life interact with one another. However, three particular areas play an especially important role in framing the context of the scientific search for life: astronomy; planetary science, which includes geology and atmospheric science; and biology. In this section, we'll take a brief look at the context contributed by each of these three sciences.

• How does astronomy help us understand the possibilities for extraterrestrial life?

For most of human history, our conception of the universe was quite different from what it is today. Earth was widely assumed to be the center of the universe. Planets were lights in the sky, named for ancient gods, and no one had reason to think they could be *worlds* on which we might search for life. Stars were simply other lights in the sky, distinguished from the planets only by the fact that they remained fixed in the patterns of the constellations, and few people even considered the possibility that our Sun could be one of the stars. Moreover, with the Sun and planets presumed to be orbiting around Earth, there was certainly no reason to think that stars could have planets of their own, let alone planets on which there might be life.

When you consider the dominance that this Earth-centered, or **geocentric**, view of the universe held for thousands of years, it becomes obvious that astronomy plays a key role in framing the context of the modern search for life. We will discuss in Chapter 2 how and why the human view of the cosmos changed dramatically about 400 years ago, and we'll consider the modern astronomical context in some detail in Chapter 3. But the point should already be clear: We now know that Earth is but one tiny world orbiting one rather ordinary star in a vast cosmos, and this fact opens up countless possibilities for life on other worlds.

Astronomy provides context to the search for life in many other ways as well, but one more is important enough to mention right now: By studying distant objects, we have learned that the physical laws that operate in the rest of the universe are the same as those that operate right here on Earth. This tells us that if something happened here, it is possible that the same thing could have happened somewhere else, at least in principle. We are not the center of the universe in location, and we have no reason to think we are "central" in any other way, either.

• How does planetary science help us understand the possibilities for extraterrestrial life?

Planetary science is the name we give to the study of almost everything having to do with planets. It includes the study of planets themselves, as well as the study of moons orbiting planets, the study of how planets form, and the study of other objects that may form in association with planets (such as asteroids and comets). Planetary science helps set the context for the search for life in the universe in several different ways, but two are especially important.

First, by learning how planets form, we develop an understanding of how common we might expect planets to be. Until just about the middle of the last century, we really had no basis for assuming that many other

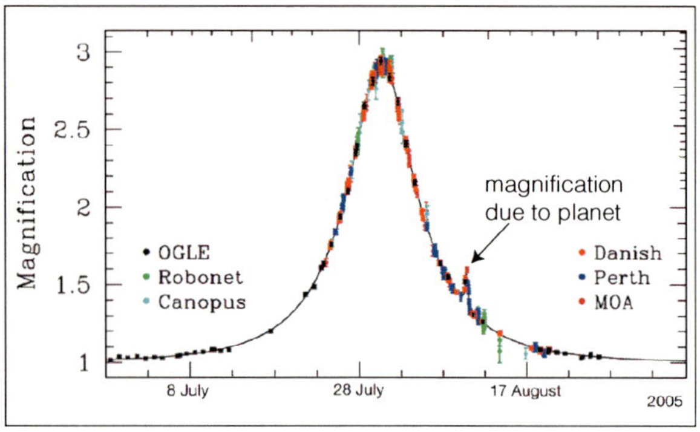

Figure 1.2

The painting (above) shows an artist's impression of what an extrasolar planet discovered in early 2006 may look like. It was the smallest extrasolar planet yet discovered, with a mass about five times that of Earth. Its odd-sounding name, OGLE-2005-BLG-390Lb, is a designation indicating that it was discovered by a technique based on Einstein's general theory of relativity (gravitational lensing, discussed in Chapter 11). The graph shows the data that led to the planet's discovery.

stars would have their own planets. Some scientists thought this likely, while others did not, and we lacked the data needed to distinguish between the two possibilities [Section 3.5]. But during the latter half of the twentieth century, a growing understanding of the processes by which our own solar system formed—much of it based on evidence obtained through human visits to the Moon and spacecraft visits to other planets—gradually made it seem more likely that other stars might similarly be born with planetary systems.

Still, as recently as 1995, no one was sure whether planets existed around other stars. That was the year in which the first strong evidence was obtained for the existence of **extrasolar planets**, or planets orbiting stars other than our Sun.* Since that time, additional discoveries of extrasolar planets have poured in at an astonishing rate, so that the known extrasolar planets now far outnumber the planets of our own solar system (Figure 1.2). Based on the statistics of these discoveries, it now seems likely that many or most stars have planets and, as we'll discuss in Chapter 11, it seems reasonable to imagine that life—and possibly even civilizations—could exist on at least some of these planets or their moons.

The second way in which planetary science shapes the context for the search for life is by helping us understand how planets work. For example, by studying planets and comparing them to one another, we have learned why some planets are rocky like Earth while others, like Jupiter, contain vast amounts of hydrogen and helium gas. We've also learned why Venus is *so* much hotter than Earth despite the fact that, in the scheme of our solar system, it is only slightly closer to the Sun. Similarly, we can now explain why the Moon is desolate and barren even though it orbits the Sun at essentially the same distance as Earth, and we have a fairly good idea of why Mars is cold and dry today, when evidence shows that it was warmer and wetter in the distant past.

This understanding of how planets work gives us deeper insight into the nature of planetary systems in general. More important to our purposes, it also helps us understand what to look for as we search for **habitable worlds**—worlds that contain the basic necessities for life. After all, given that there are far more worlds in the universe than we can ever hope to study in detail, we can improve our odds of success in finding life by constraining the search to those worlds that are the most promising. Be sure to note that when we ask whether a world is habitable, we are asking whether it offers environmental conditions under which life *could* arise or survive, not whether it actually harbors life.

Also keep in mind that when we say a world is habitable, we do not necessarily mean that plants, animals, or people could survive there. For much of Earth's history, nearly all life was microscopic, and even today, the total mass of microbes on Earth is greater than that of all plants and animals combined. The search for habitable worlds is primarily a search for places where microbes of some kind might survive, though we might find larger organisms as well.

• How does biology help us understand the possibilities for extraterrestrial life?

Astronomy, planetary science, and other science disciplines play important roles in shaping the context for the search for life in the universe,

*There was one earlier discovery, but it was of objects orbiting what we call a *neutron star*, not an "ordinary" star like our Sun.

but since we are searching for *life,* the context of biology is especially important. Just as you wouldn't look for a house to buy without knowing something about real estate, it would make no sense to search for life if we didn't know something about how life functions. The key question about the biological context of the search revolves around whether we should expect biology to be rare or common in the universe.

Wherever we have looked in the universe, we have found clear evidence that the same laws of nature are operating. We see galaxies sprinkled throughout space, and we see that the same stellar processes that occur in one place also occur in others. In situations in which we can observe orbital motions, we find that they agree with what we expect from the law of gravity. These and other observations make us confident that the basic laws of physics that we've discovered here on Earth also hold throughout the universe.

We can be similarly confident that the laws of chemistry are universal. Observations of distant stars show that they are made of the same chemical elements that we find here in our own solar system, and that interstellar gas clouds contain many of the same molecules we find on Earth. This provides conclusive evidence that atoms come in the same types and combine in the same ways throughout the universe.

The universality of physics and chemistry is what makes us confident that we will find planets and other worlds, including many that are at least potentially habitable, throughout our Milky Way Galaxy and throughout the universe. Could biology also be universal? That is, could the biological processes we find on Earth be common throughout the universe? If the answer is yes, then the search for life elsewhere should be exciting and fruitful. If the answer is no, then life may be a rarity.

Because we haven't yet observed biology anywhere else in the universe, we can't yet know whether biology is universal. However, evidence from Earth gives us reason to think that it might be. Laboratory experiments suggest that chemical constituents found on the early Earth would have combined readily into complex organic (carbon-based) molecules, including many of the building blocks of life [Section 6.2]. Indeed, scientists have found organic molecules in meteorites (chunks of rock that fall to Earth from space) and, through spectroscopy [Section 3.4], in clouds of gas between the stars. The fact that such molecules form even under the extreme conditions of space suggests that they form quite readily and may be common on many worlds.

Of course, the mere presence of organic molecules does not necessarily mean that life will arise, but the history of life on Earth gives us some reason to think that the step from chemistry to biology is not especially difficult. As we'll discuss in Chapter 6, geological evidence tells us that life on Earth arose quite early in Earth's history, at least on a geological time scale. This early origin of life on Earth suggests—but certainly does not prove—that the same process might occur on other worlds. If the transition from chemistry to biology was exceedingly improbable, we might expect that it would have required much more time. Thus, the early origin of life on Earth makes it seem reasonable to think that life would emerge just as quickly on other worlds with similar conditions.

THINK ABOUT IT Microbial life on Earth predates intelligent life like us by at least 3–4 billion years. Do you think this fact tells us anything about the likelihood of finding *intelligent* life, as opposed to finding any life, on extrasolar planets? Explain.

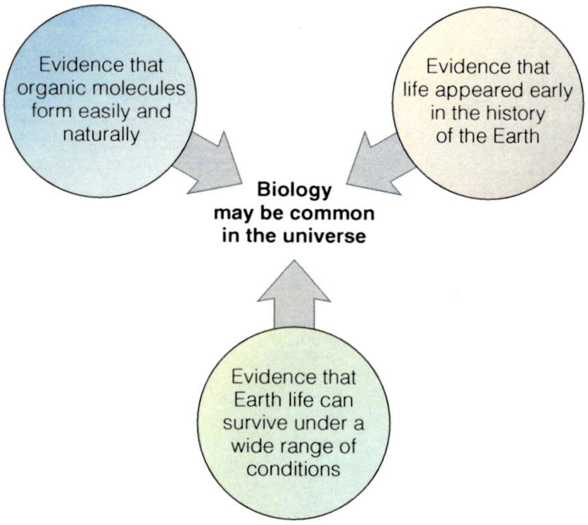

Figure 1.3

Three lines of evidence that give us at least some reason to think that biology may be common in the universe.

If life really can be expected to emerge under the right conditions, the only remaining question is the prevalence of those "right" conditions. Here, too, recent discoveries give us reason to think that biology could be common. In particular, biologists have found that microscopic life can survive and prosper under a much wider range of conditions than was believed only a few decades ago [Section 5.5]. For example, we now know that life exists in extremely hot water near deep-sea volcanic vents, in the frigid conditions of Antarctica, and inside rocks buried a kilometer or more beneath the Earth's surface. Indeed, if we were to export these strange organisms from Earth to other worlds in our solar system—perhaps to Mars or Europa—it seems possible that at least some of them would survive. This suggests that the range of "right" conditions for life may be quite broad, in which case it might be possible to find life even on planets that are significantly different in character from Earth.

In summary, we have no reason to think that life ought to be rare and several reasons to expect that it may be quite common (Figure 1.3). If life is indeed common, studying it will give us new insights into life on Earth, even if we don't find other intelligent civilizations. These enticing prospects have captured the interest of scientists from many disciplines and from around the world, giving birth to a new science devoted to the study of, and search for, life in the universe.

1.3 PLACES TO SEARCH

The study of life in the universe involves fundamental research in all the scientific areas we have already mentioned, and others as well. Indeed, as you'll see throughout this book, the study of life in the universe goes far beyond simply searching for living organisms. Still, all of this study is driven by the possibility that life exists elsewhere, so before we dive into any details, it's worth a quick overview of the places and methods we use to search for life beyond Earth.

• Where should we search for life in the universe?

The search for life in the universe takes place on several different levels. First, and in many ways foremost, it is a search for life right here on Earth. As we discussed earlier, we are still learning about the places and conditions under which life exists on Earth, and many scientists are busy searching for new species of life on our own world. After all, the more we know about life here, the better we'll be able to search for it elsewhere.

Searching Our Own Solar System Turning our attention to places besides Earth, the first place to search for life is on other worlds in our own solar system. Our solar system has a lot of worlds: It has the planets orbiting the Sun, moons orbiting planets, and huge numbers of smaller objects such as asteroids and comets.

Figure 1.4 shows a portrait of some of our best current views of the planets in our solar system. Note that it is *not* to scale, since its purpose is to show each planet as we know it today from spacecraft or through telescopes; you can turn to Figure 3.3 to see the sizes correctly scaled. Note also that we have included both Pluto and the recently discovered Eris (which is slightly larger than Pluto), though these two objects are now considered "dwarf planets" (see Special Topic 1.1).

The photos alone make clear how different Earth is from every other planet in our solar system. Ours is the only planet with oceans of liquid

Figure 1.4

A "family portrait" of the planets that orbit our Sun, shown in order of increasing average distance from the Sun; the photos are *not* shown to scale. Pluto and Eris are included, though they are now considered "dwarf planets."

water on its surface, a fact that provides an instant clue about why Earth is home to so much life: Water is crucial to all life on Earth. Indeed, as we'll discuss in Chapters 5 and 7, we have good reason to think that liquid water is always a requirement for life, though it's possible that a few other liquids might work in place of water.

Given that we are primarily looking for life that is at least somewhat Earth-like, the need for water or some other liquid places constraints on where we might find life. Among the planets, Mars is by far the most promising candidate. As we'll discuss in detail in Chapter 8, strong evidence tells us that the now-barren surface of Mars (Figure 1.5) once had flowing water, making it seem reasonable to imagine life having arisen on Mars at that time. Mars still has significant amounts of water ice, so it is even possible that life exists on Mars today, perhaps hidden away in places where volcanic heat keeps underground water liquid. Past or present life seems much less likely on any of the other planets, though we can't rule

Figure 1.5

The surface of Mars, photographed by NASA's *Spirit* rover from a perch in the Columbia Hills. The martian surface is dry and barren today, but strong evidence points to liquid water on its surface in the distant past.

Figure 1.6

This photograph shows Jupiter and two of its moons: Io is the moon in front of Jupiter's Great Red Spot, and Europa is to the right. Scientists suspect that Europa has a deep ocean beneath its surface of ice, making it a prime target in the search for life in our solar system.

it out completely; we'll discuss these dim planetary prospects for life in Chapter 7.

Aside from the planets, the most promising abodes for life in the solar system are a few of the large moons. At least six moons are potential candidates for life, with the best candidate being Jupiter's moon Europa (Figure 1.6). Current evidence strongly suggests that Europa hides a deep ocean of liquid water under its icy crust. Indeed, if we are interpreting the evidence correctly, the Europan ocean may have considerably more water than all of Earth's oceans combined [Section 9.2]. Because we suspect that life on Earth got started in the deep oceans [Section 6.1], Europa may well have all the conditions needed both for life to have arisen and for its ongoing survival. Two other moons of Jupiter—Ganymede and Callisto—also show some evidence of subsurface oceans, though the evidence is less strong and other considerations (primarily availability of energy) give them poorer prospects for harboring life. A fourth candidate for a life-bearing moon is Titan, which orbits Saturn and is the only moon in the solar system with a substantial atmosphere. Titan is far too cold for liquid water on its surface, but it apparently has liquid methane. It may also have liquid water deep underground. At the least, studies of Titan [Section 9.4] show that it has interesting organic chemistry, even if it does not have life. The fifth and sixth moons for which we have found evidence

SPECIAL TOPIC 1.1: HOW MANY PLANETS ARE THERE IN OUR SOLAR SYSTEM?

Everyone in college today grew up in an era when we were taught that there are nine planets in our solar system. Well, as you've probably heard, astronomers recently voted to demote Pluto from being a regular planet to being a "dwarf planet," leaving only eight official planets in our solar system. But don't despair if you're a Pluto fan. Counting planets is about words, not about science.

A little background: Pluto was discovered in 1930, at a time when astronomers were searching for an object thought to be causing slight perturbations to the orbits of other planets around the Sun. Neptune itself had been discovered in just this way in 1846, after scientists used perturbations in Uranus's orbit to predict the existence and location of an "eighth planet." Neptune was found as soon as astronomers pointed telescopes to the calculated position, which is why we sometimes say that Neptune was discovered with physics and mathematics, and only confirmed with a telescope.

Over the ensuing decades, a few scientists claimed to see ongoing orbital discrepancies and embarked on a search for a "ninth planet" that might be causing them. Pluto was found during this search, though about 12 full-moon-widths away from the predicted position. And though hailed as a planet upon discovery, its status gradually became suspect, as we learned that its mass is much less than 1% that of Earth, and its orbit is much more tilted and elongated than that of any of the other planets. Worse, reanalysis of past observations suggested that the claimed orbital discrepancies had simply been errors in measurement, making Pluto a solution to a nonexistent problem.

In 1951, by analyzing comet orbits, astronomer Gerard Kuiper (rhymes with "piper") predicted the existence of a "comet belt" beyond Neptune—now called the *Kuiper belt*—analogous to the asteroid belt between Mars and Jupiter. Telescope technology caught up with this prediction in the 1990s, and astronomers confirmed the existence of vast numbers of "Kuiper belt objects" orbiting the Sun in the general vicinity of Pluto. Scientifically, it became obvious that Pluto was much more like a large member of this group than like a small version of any of the other eight planets. Still, as long as Pluto was the largest known Kuiper belt object, most astronomers were content to leave its planetary status intact.

Then, in July 2005, astronomer Michael Brown announced the discovery of a Kuiper belt object—given the name Eris after a Greek goddess who bred discord among humans—that is slightly larger than Pluto. This discovery was a fatal blow to the status quo of nine planets; after all, if Pluto is big enough to count as a planet, then Eris must belong to the club too. But what of the dozens of Kuiper belt objects only a little smaller than Pluto, and of future discoveries yet to be made? How would we draw the line between "planet" and "large Kuiper belt object"?

Official decisions on astronomical names and definitions rest with the International Astronomical Union (IAU), an organization made up of professional astronomers from around the world. At a contentious meeting in August 2006, the IAU voted to define a "planet" as an object that: (1) orbits a star (but is itself neither a star nor a moon); (2) is massive enough for its own gravity to give it a nearly round shape; and (3) has cleared the neighborhood around its orbit. By this definition, Pluto gets kicked out of the planetary club, because it shares its orbital neighborhood with other Kuiper belt objects. Pluto, Eris, and other objects that are large enough to be round but have not cleared their orbital neighborhoods are designated "dwarf planets." The myriad smaller objects that orbit the Sun, including most asteroids and comets, become members of a class called "small solar system bodies."

The important thing to remember is that this new definition was established by vote, making it politics, not science. The politics may yet change again (some astronomers have already voiced objections to the new definition), but Pluto will remain the same. Moreover, many people are likely to keep thinking of Pluto as a planet regardless of what the IAU says, much as most people still think of Europe and Asia as separate continents, although technically both are part of Eurasia. The bottom line is that while it's good to know the current official definitions, it's better to understand the science behind them.

of subsurface liquids—possibly including liquid water—are Saturn's moon Enceladus and Neptune's moon Triton.

Searching Among the Stars In terms of numbers, there are many more places to look for life on planets and moons around other stars than in our own solar system. However, the incredible distances to the stars [Section 3.2] make searches of these worlds much more difficult. All stars are so far away that we will need great leaps in technology to have any hope at all of sending spacecraft to study their planets up close; for example, with current spacecraft, the journey to even the nearest stars would take close to 100,000 years.

With visits out of reach, telescopic searches represent our only hope of finding life on extrasolar worlds. Current telescope technology is barely able to detect the existence of extrasolar planets; indeed, so far, we can detect only rather large planets, not planets as small as Earth. But the technology is advancing rapidly. We may be capable of detecting Earth-size planets by 2010 or sooner, and within a couple of decades we may have telescopes that are able to obtain crude pictures and spectra of planets and moons around other stars. As a result, one important area of research lies in trying to figure out the photographic or spectral "signatures" that would tell us we are looking at a world with life.

• Could aliens be searching for us?

So far we have talked about searching for life that is not searching for us; that is, life that we could identify only by seeing it with our spacecraft or telescopes. But if life really is common in the universe, there could be other places like Earth where life has evolved to become intelligent enough to be interested in searching for life beyond its home world. In that case, it is possible that other civilizations might actually be broadcasting signals that we could detect. The **search for extraterrestrial intelligence**, or **SETI**, which we'll discuss in Chapter 12, focuses on the search for such signals from alien civilizations (Figure 1.7). Although we don't know whether the search will meet with success, we can be sure that the unambiguous receipt of an alien message would be one of the most significant discoveries in human history—not to mention the fact that it would also probably answer many of our other questions about life in the universe.

Figure 1.7

This 140-foot radio telescope in West Virginia was used in 1996 to search for signals from extraterrestrial civilizations.

1.4 THE NEW SCIENCE OF ASTROBIOLOGY

We have seen that the study of life in the universe is a multidisciplinary field of scientific research, involving scientists with training in many different specialties. Nevertheless, because it has become a prominent and important area of study, it would be good to give the science of life in the universe its own name. A number of different names are in use, including "exobiology" and "bioastronomy," but in this book we follow the lead of NASA and call it **astrobiology.** This term is meant to invoke the combination of astronomy (the study of the universe) and biology (the study of life), so *astrobiology* literally means "the study of life in the universe."

• How do we study the possibility of life beyond Earth?

Because astrobiology is a young science, scientists are still working to decide where to focus their research efforts. One major player in this effort is the NASA Astrobiology Institute, a collaboration involving scientists from NASA and more than a dozen other research institutions across the United States. Similar efforts are under way in other countries, including the United Kingdom, Sweden, France, Spain, Russia, and Australia. These collaborations are among the most interdisciplinary in any area of science, bringing together astronomers, biologists, geologists, chemists, and many others seeking to understand the prospects of finding life beyond Earth.

Although different groups focus their efforts in different areas, most astrobiology research is concentrated in the following three areas:

1. Studying the conditions conducive to the origin and ongoing existence of life.
2. Looking for such conditions on other planets in our solar system and around other stars.
3. Looking for the actual occurrence of life elsewhere.

Notice that astrobiology therefore includes much more than simply searching for extraterrestrial life or civilizations. At a fundamental level, astrobiology research seeks to reveal the connections between living organisms and the places where they reside. In this sense, finding no life (on Mars, for example) is just as significant a result as finding life, because either way we learn about the conditions that can lead to the presence of life, about how life evolves in conjunction with planets, and about whether life is likely to be rare or common throughout the universe.

In the rest of this book, we will focus on the same three areas listed above. After discussing the scientific context of the search in greater detail in Chapters 2 and 3, we'll turn our attention in Chapters 4 through 6 to the nature, origin, and evolution of life on Earth. This study of the history of life on our planet will help us understand the conditions under which we might expect to find life elsewhere. Next we'll discuss prospects for life elsewhere in our solar system in Chapters 7 through 10, and then discuss the prospects for finding life—including intelligent life—beyond our solar system in Chapters 11 through 13. Along the way, we'll learn what science can currently say about the future of life on Earth, we'll consider possible futures for our own species, and we'll discuss the philosophical implications of the search for—and potential discovery of—life beyond Earth.

THE BIG PICTURE
Putting Chapter 1 into Context

In this chapter, we have offered a brief overview of the ideas we will cover in more depth in the rest of the book, primarily so that you will have a sense of what to expect in the rest of your study of life in the universe. As we will in every chapter, we conclude with a brief "big picture" recap of how these ideas fit into the overall goals of the scientific study of life in the universe:

- Despite the abundance of aliens in popular media, we don't yet have any convincing evidence for life beyond Earth. Nevertheless, current understanding of astronomy, planetary science, and biology gives us good reason to think that it is at least reasonable to imagine that life may be widespread, and the discovery of life of any kind would have profound significance to our understanding of life in the universe.

- It's conceivable that life may exist on any of several worlds in our own solar system, but it's extremely unlikely that any of this life is intelligent. However, we find many more possibilities when we consider life on planets or moons around other stars. And, through the search for extraterrestrial intelligence (SETI), it is even possible that we could receive a signal from an advanced civilization.
- The prospect that life may be common in the universe has given rise to the new science of astrobiology, an exciting and interdisciplinary topic of research that focuses both on understanding the possibility of finding life elsewhere and on the actual search for life beyond Earth.

SUMMARY OF KEY CONCEPTS

1.1 THE POSSIBILITY OF LIFE BEYOND EARTH

• What are we searching for?

The search for **extraterrestrial life** is in principle a search for *any* kind of life. However, the difficulty of clearly defining life means that it's easier to focus the search on life that is at least somewhat similar to life here on Earth. This still opens a wide range of possibilities, from bacteria-like microbes to complex plants and animals.

• Is it reasonable to imagine life beyond Earth?

People have long considered the possibility of life beyond Earth, but only recently have we been able to examine this possibility through the lens of science. While we have no evidence at this time of actual life beyond Earth, our scientific understanding of the possibilities makes it reasonable to think that life could exist elsewhere.

1.2 THE SCIENTIFIC CONTEXT OF THE SEARCH

• How does astronomy help us understand the possibilities for extraterrestrial life?

Astronomy tells us that we live on just a tiny planet orbiting one rather ordinary star in a vast cosmos, and that the same physical laws that operate here also operate throughout the universe. Together these ideas suggest that there could be many other worlds with life.

• How does planetary science help us understand the possibilities for extraterrestrial life?

Planetary science helps us understand both how common we should expect other worlds to be and the conditions under which we might expect those worlds to be capable of harboring life. Based on current understanding of how planets form, we expect planets to be common around other stars—an idea that has been confirmed by recent discoveries of **extrasolar planets**. By learning how planets work, we learn the conditions that might make a **habitable world**, meaning a world that has the basic necessities for life, even if it does not actually have life.

• How does biology help us understand the possibilities for extraterrestrial life?

With no life currently known to exist beyond Earth, we do not know whether biology should be common on worlds with habitable conditions. However, three lines of evidence suggest that life could be common: (1) The fact that life arose quickly on Earth suggests that it might occur on any world that has the "right" conditions. (2) We know from observations of meteorites and interstellar clouds that organic molecules are common throughout the galaxy, suggesting that we'll find them on many other worlds. (3) The fact that life on Earth survives even under some seemingly extreme conditions suggests that life is hardy enough to survive in many other places as well.

1.3 PLACES TO SEARCH

• Where should we search for life in the universe?

The search begins right here on Earth, as we seek to learn more about the life on our own planet. Elsewhere in our solar system we can search many planets and moons, but current understanding suggests that the most promising candidates for life are Mars and Jupiter's moon Europa. In the future, as telescope technology improves, we should be able to conduct telescopic searches for life around other stars.

• Could aliens be searching for us?

If life is common in the universe, civilizations might also be common, in which case other civilizations might be conducting their own searches and broadcasting signals of their existence. We look for such signals from alien civilizations through the **search for extraterrestrial intelligence**, or **SETI**.

1.4 THE NEW SCIENCE OF ASTROBIOLOGY

• How do we study the possibility of life beyond Earth?

The science of life in the universe, or astrobiology, focuses on three major areas: (1) studying the conditions conducive to the origin and ongoing existence of life; (2) looking for such conditions on other planets in our solar system and around other stars; and (3) looking for the actual occurrence of life elsewhere. Together, these studies should help us understand the connections between living organisms and the places where they reside.

EXERCISES AND PROBLEMS

REVIEW QUESTIONS

Short-Answer Questions Based on the Reading

1. Why are scientists interested in the possibility of life beyond Earth?
2. People have long been interested in life beyond Earth. What is different today that makes this possibility seem scientifically reasonable?
3. What do we mean by a *geocentric universe?* In general terms, contrast a geocentric view of the universe with our modern view of the universe.
4. What are *extrasolar planets?* In what way does their discovery make it seem more reasonable to imagine finding life elsewhere?
5. What do we mean by a *habitable* world? Does a habitable world necessarily have life?
6. What do we mean by the "universality" of physics and chemistry? Although we don't know yet whether biology is similarly universal, what evidence makes it seem that it might be?
7. Besides Earth, what worlds in our solar system seem most likely to have life? Why?
8. Could we actually detect life on an extrasolar planet and moon with current technology? Explain.
9. What is the *search for extraterrestrial intelligence* (SETI)?
10. What do we mean by *astrobiology?* What other terms are sometimes used to describe it? What are the major areas of research in astrobiology?

TEST YOUR UNDERSTANDING

Quick Quiz

Choose the best answer to each of the following. Explain your reasoning with one or more complete sentences.

11. An *extrasolar planet* is: (a) a planet that is larger than our Sun; (b) a planet that orbits a star other than our Sun; (c) a planet located in another galaxy.
12. A *habitable planet* is: (a) a planet that has oceans like the Earth; (b) a planet that has life of some kind; (c) a planet that may or may not have life, but that has environmental conditions under which it seems that life could arise or survive.
13. By a *geocentric* view of the universe, we mean: (a) the ancient idea that Earth resided at the center of the universe; (b) the idea that Earth is the only planet with life in the universe; (c) a view of the universe shaped by current understanding of geological science.
14. According to current scientific understanding, life on Earth: (a) was exceedingly improbable; (b) arose quite soon after conditions allowed it; (c) may have been inevitable, but took billions of years to arrive.

15. The correct order for the eight official planets in our solar system, from closest to farthest from the Sun, is: (a) Mercury, Venus, Earth, Mars, Saturn, Jupiter, Neptune, Uranus; (b) Mercury, Venus, Earth, Mars, Jupiter, Uranus, Neptune, Saturn; (c) Mercury, Venus, Earth, Mars, Jupiter, Saturn, Uranus, Neptune.
16. Today, the research known as SETI is conducted primarily by: (a) scanning the skies for signals from alien civilizations; (b) sending spacecraft to the planets; (c) using telescopes to observe extrasolar planets.
17. If we sent one of our current spacecraft to a nearby star (besides the Sun), the trip would take about: (a) a decade; (b) a century; (c) 100,000 years.
18. Scientists today are interested in searching for life on Mars because: (a) We see clear evidence of a past civilization on Mars. (b) Mars contains frozen water ice at its polar caps. (c) Evidence suggests that Mars had liquid water on its surface in the distant past.
19. Based on current evidence, the object in our solar system most likely to have a deep, subsurface ocean of liquid water is: (a) Mars; (b) Europa; (c) Titan.
20. Based on the way scientists view the study of astrobiology, failure to find life on any other world would mean: (a) The whole subject has been a waste of time. (b) We must have done something wrong, since life has to exist beyond Earth. (c) We have learned important lessons about the conditions that made life on Earth possible.

INVESTIGATE FURTHER

In-Depth Questions to Increase Your Understanding

Discussion Questions

21. *Aliens Among Us.* Take an informal poll of your friends or classmates. How many believe we have already been visited by aliens? On what basis do they hold their beliefs? How strong are their convictions on this issue? In light of your findings and what you've learned in this chapter, discuss whether public interest in aliens visiting Earth has any bearing on the scientific study of astrobiology.
22. *Conducting the Search.* Given the large number of possible places to look for life, how would you prioritize the search? In other words, where would you look first for life on other worlds in our own solar system, and how would you come up with a search strategy for other star systems? Explain your priorities and strategies clearly.
23. *Funding for Astrobiology.* Imagine that you are a member of Congress, so it is your job to decide how much government funding goes to research in astrobiology. What factors would influence your decision? Do you think you would increase or decrease such funding from the current level? Explain.

WEB PROJECTS

24. *Astrobiology News.* Go to NASA's Astrobiology home page and read some of the recent news from astrobiology research. Choose one recent news article, and write a one- to two-page summary of the research and how it fits into astrobiology research in general.
25. *The NASA Astrobiology Institute.* Go to the home page for the NASA Astrobiology Institute (NAI) and learn more about how it is organized and the type of research it supports. Also learn whether your school or any nearby institutions participate in the NAI. In one page or less, describe the NAI and its work and discuss the particular contributions of any institutions located near you.
26. *International Astrobiology.* Search the Web for information on astrobiology efforts outside the United States. Learn about the effort in one particular country or group of countries. What areas of research are emphasized? How do the researchers involved in the effort collaborate with other international astrobiology efforts? Write a one- to two-page report on your findings.
27. *The Search for Extraterrestrial Intelligence.* Go to the home page for the SETI Institute. Learn more about how SETI is funded and how the institute does its work. Summarize your findings in about one page.

4
The Habitability of Earth

LEARNING GOALS

4.1 GEOLOGY AND LIFE
- How is geology crucial to our existence?

4.2 RECONSTRUCTING THE HISTORY OF EARTH AND LIFE
- What can we learn from rocks and fossils?
- How do we learn the age of a rock or fossil?
- What does the geological record show?

4.3 THE HADEAN EARTH AND THE DAWN OF LIFE
- How did Earth get an atmosphere and oceans?
- Could life have existed during Earth's early history?

4.4 GEOLOGY AND HABITABILITY
- What is Earth like on the inside?
- How does plate tectonics shape Earth's surface?
- Why does Earth have a protective magnetic field?

4.5 CLIMATE REGULATION AND CHANGE
- How does the greenhouse effect make Earth habitable?
- What regulates Earth's climate?
- How does Earth's climate change over long periods of time?

4.6 THE PROCESS OF SCIENCE IN ACTION: FORMATION OF THE MOON
- How did the giant impact win out over competing models?
- Does the giant impact model count as science?

A Rock, A River, A Tree
Hosts to species long since departed
Marked the mastodon,
The dinosaur, who left dry tokens
Of their sojourn here
On our planet floor,
Any broad alarm of their hastening doom
Is lost in the gloom of dust and ages.

Maya Angelou, from *On the Pulse of the Morning*

Some people think that the vast numbers of worlds in our galaxy make it inevitable that we live in a galaxy full of life. Others think that life beyond Earth will prove to be rare or nonexistent. But no matter what you think, one fact seems indisputable: So far, we have no convincing evidence for the existence of life anyplace except right here on Earth.

This simple fact means that the scientific search for life in the universe must necessarily begin with the study of life on Earth. After all, unless we can understand why life exists here, we have little hope of understanding the prospects for finding life elsewhere. Our goal in this and the next two chapters is to understand why life thrives on our planet, so that in later chapters we can apply this understanding to the search for life on other worlds.

In this chapter, we will focus on understanding the physical conditions that make our planet habitable. We'll explore the role that geology plays in Earth's habitability, and discuss how, when, and why Earth became a suitable home for life. With this understanding under our belts, we'll then be ready to turn our attention in Chapters 5 and 6 to life itself.

4.1 GEOLOGY AND LIFE

It's easy to take for granted the qualities that make Earth so suitable for human life: a moderate temperature, abundant water, a protective atmosphere, and a relatively stable environment. But we need look only as far as our neighboring worlds of the inner solar system—the Moon and the three other terrestrial planets—to see how fortunate we are (Figure 4.1). The Moon and Mercury are airless and barren, with surfaces covered by the craters of past impacts. Venus was long hidden from view by its thick, global cloud cover, but we now know that it is a searing hothouse, with surface temperatures higher than that of a pizza oven and surface pressure nearly as great as that a kilometer deep in Earth's oceans. Mars has many features that look almost Earth-like and might indicate a hospitable past, but today Mars has an atmosphere so thin that a visiting astronaut would require a full space suit at all times, and the only surface water is frozen as ice.

Why is it that Earth alone appears to have a habitable surface among the five inner worlds of our solar system? Our distance from the Sun is

Mercury's surface looks much like that of the Moon, with many impact craters.

Cloud-penetrating radar was used to obtain this view of a twin-peaked volcano on Venus.

We see land, ocean, and snow-capped peaks in this view of Earth from orbit.

The Moon's surface is heavily cratered.

Mars lacks surface liquid water today, but features that look like dry riverbeds suggest a more hospitable past.

Figure 4.1

The inner worlds of our solar system, shown to scale, along with sample close-ups from orbiting spacecraft.

clearly an important factor, but that cannot be the whole story. After all, the Moon shares our distance from the Sun, yet lacks all of the qualities that make Earth habitable. Comparing Earth and the Moon in Figure 4.1 suggests that Earth's much larger size is important, but that can't be the whole story either, since hothouse Venus is only slightly smaller than Earth.

Scientists suspect that Earth owes its habitability primarily to a combination of its size and distance from the Sun, and perhaps to a few other factors as well. We'll devote most of this chapter to exploring how these factors have led to Earth's habitability, but first let's look a little more closely at the surprising ways in which Earth's geology is crucial to our existence.

• How is geology crucial to our existence?

Geology is a word with multiple meanings. Taken literally, it is the study of Earth (because *geo* means "Earth"), but scientists commonly extend the meaning to encompass the study of any world with a solid surface. We also use the word *geology* to describe the processes and features that shape worlds; for example, when we speak of Earth's geology, we can mean anything from the composition of our planet to the volcanoes and other processes that rework the surface.

When we consider life on relatively short time scales, such as decades or centuries, we generally don't need to think about how life interacts with our planet's geology. Except for organisms that obtain their energy directly from chemical interactions with their environment—an important class of organisms that we'll discuss further in Chapter 5—we can understand most short-term biology by considering interactions between species; only rarely are these interactions affected directly by something like a volcanic eruption. However, over longer time scales, geology and life are deeply intertwined. In fact, it is Earth's geology that has made our planet habitable, ultimately allowing not only the existence of life but also the long-term evolution of life into complex forms that include us.

Geology is important to life on Earth in many ways, but three aspects of Earth's geology stand out as being especially important:

- **Volcanism**. A volcanic eruption can be a spectacular sight, but volcanoes are important to our existence on a much deeper level: Volcanic activity releases gases trapped in Earth's interior, and these gases were the original source of Earth's atmosphere and oceans. In addition, volcanism releases heat that, we suspect, helped lead to the origin of life on our planet.

- **Plate tectonics**. Earth's surface has been shaped largely by the movement and recycling of rock between the surface and the interior. This process, called *plate tectonics,* is best known for gradually rearranging the continents of Earth, but its most profound relevance to life involves Earth's climate: According to modern understanding, plate tectonics is largely responsible for the long-term climate stability that has allowed life to evolve and thrive for some 4 billion years.

- **Earth's magnetic field**. Compass needles point north because our planet has a global magnetic field generated deep in its interior. You may know that the magnetic field has at least a few biological effects—for example, some birds use the magnetic field to help guide their migrations—but its deeper significance is to our atmosphere. The magnetic field shields Earth's atmosphere from the energetic particles of the solar wind [Section 3.3], and without this shielding, it's likely that a significant portion of our planet's atmosphere would by now have been stripped away into space.

Because these factors seem so important to life on Earth, we'd like to understand the likelihood of finding them on other worlds. To do so, we must understand how they work and how they came about, a task to which we'll devote most of this chapter. First, however, it's important to understand how we learn about these and other processes that shape our planet, which ultimately comes down to methods for reconstructing Earth's history from the clues we find written in rocks and fossils.

4.2 RECONSTRUCTING THE HISTORY OF EARTH AND LIFE

Human recorded history dates back only a few thousand years on a planet that has existed for about $4\frac{1}{2}$ billion years. To put this fact in perspective, imagine making a timeline to represent Earth's history. On a timeline the length of a football field, human civilization would appear only in a tiny sliver at the end of the line—a sliver no thicker than a piece of paper. How, then, can we possibly know anything about the long history that preceded human civilization?

The answer is that this history is recorded in rocks and **fossils**—relics of organisms that lived and died long ago—that contain clues we can read to learn about the past. Reading this history is not as easy as reading words on a page, but with proper scientific tools it can be read reliably. Our task in this section is to explore a few of the key scientific ideas that have helped us put together a chronology of geology and life on our planet.

a This solidified lava is an example of igneous rock. **b** Metamorphic rock has gone through transformations that often give a contorted appearance. **c** Sedimentary rock tends to build up in layers like those visible here.

Figure 4.2

Samples of the three basic rock types.

• What can we learn from rocks and fossils?

Recall that matter is found in three basic phases: solid, liquid, and gas. The atoms and molecules of liquids or gases are in constant motion, remixing so rapidly that we can't possibly learn much about how they were arranged in the past. But solid objects offer a different story: Because atoms and molecules are essentially locked in place in a solid, they preserve information about the time at which they first became locked together—that is, at the time the object solidified.

Many solid objects preserve past records, including ancient bones and archaeological artifacts such as pieces of pottery or cloth. But when we seek to learn about Earth's history, we must look to the rocks and fossils that make up what we call the **geological record**. (Some people use the term *fossil record* synonymously, but the latter technically refers only to relics of life.) To see how we read this record, let's begin by discussing how we classify the rocks that preserve past history.

Types of Rocks Geologists classify rocks into three basic types according to how they are made (Figure 4.2):

- **Igneous rock** is made from molten rock that cools and solidifies.
- **Metamorphic rock** is rock that has been structurally or chemically transformed by high pressure or heat that was not quite high enough to melt it.
- **Sedimentary rock** is made by the gradual compression of *sediments,* such as sand and silt at the bottoms of seas and swamps.

Note that rock can change from one type to another. For example, an igneous rock may be transformed by high pressure or heat into a metamorphic rock, and both igneous and metamorphic rock may be eroded into sediments and become part of a sedimentary rock. Sedimentary rock may then be carried deep underground, where it can melt and then resolidify as igneous rock. In fact, each of the three rock types can be transformed into the others, an idea often described as the *rock cycle* (Figure 4.3).

Because rock can be recycled among the three types, a rock's type does not necessarily tell us much about its composition. In fact, individual rocks of any of the three types usually contain a mixture of different crystals in close contact. Each individual crystal represents a **mineral**, which is the word we use to describe a crystal of a particular chemical composition and

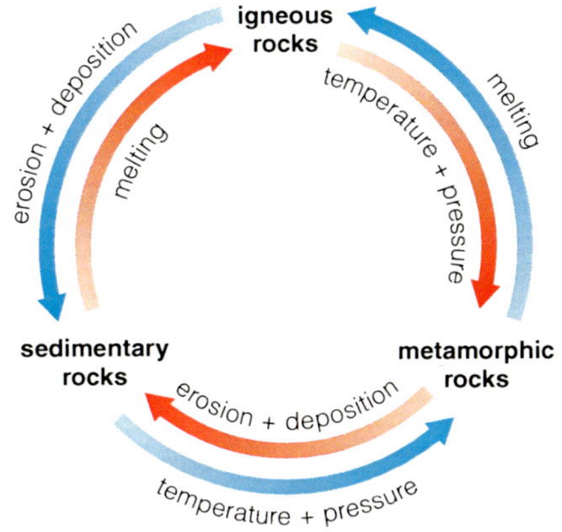

Figure 4.3

The rock cycle describes how rocks can change from one to another of the three basic types. Not shown are loops that transform rock within the individual types; for example, igneous rocks can melt and then reform as new igneous rocks.

structure. Geologists have identified more than 3,000 distinct mineral types, but we often group them by their primary constituents. For example, all minerals that contain substantial amounts of silicon and oxygen are called *silicates;* familiar silicates include quartz and feldspar. Similarly, *carbonates,* such as limestone, are minerals containing large amounts of carbon and oxygen.

Overall, a rock's type—igneous, metamorphic, or sedimentary—tells us *how* it was made, while its mineral composition tells us *what* it is made of. However, because rocks of any type may contain different mixes of minerals, geologists use many more names to subclassify rocks. For example, basalt and granite are names given to two of the most common and important subtypes of igneous rock: **Basalt** is a dark, dense igneous rock that is commonly produced by undersea volcanoes and that is rich in iron and magnesium-based silicate minerals. **Granite**, which is much lighter in color and less dense than basalt, is an igneous rock common in mountain ranges; it gets its name from its grainy appearance and it is composed largely of quartz and feldspar minerals.

Sedimentary Strata Sedimentary rock is particularly important to our study of Earth's history not only because it may contain intact fossils, but also because it forms in a way that tends to produce a record of time. The sediments that make sedimentary rock are produced primarily by erosion on land. Wind, water, and ice can all help break up solid rock into small pieces, some smaller than a millimeter across, and these small pieces (or *grains*) comprise sediments. Sediments can be carried away by rivers and deposited on floodplains or in the oceans. Over millions of years, sediments pile up on the seafloor, and the weight of the upper layers compresses underlying layers into rock. Fossils can be made when remains of living organisms are buried along with the sediments. Aquatic organisms may be buried in sediments simply because they settle to the bottom of the sea. Some land organisms form fossils when their remains are swept into bodies of water. In other cases, remains of land organisms may be buried in place by windblown silt and later compressed by sediments deposited on top of them when changing sea levels rise over them.

Sediments deposited at different times tend to look different as a result of changes in the rate of sedimentation, in the composition or grain size of sediments settling to the bottom, or in the type of organisms leaving fossils. As a result, sedimentary rock tends to be marked by distinct layers, or **strata** (singular, *stratum*). We can view these strata in sedimentary rocks that have been exposed by, for example, the gradual action of a river carving through the rock over millions of years or by a cut made through a mountain to make way for a road. Figure 4.4 summarizes the process by

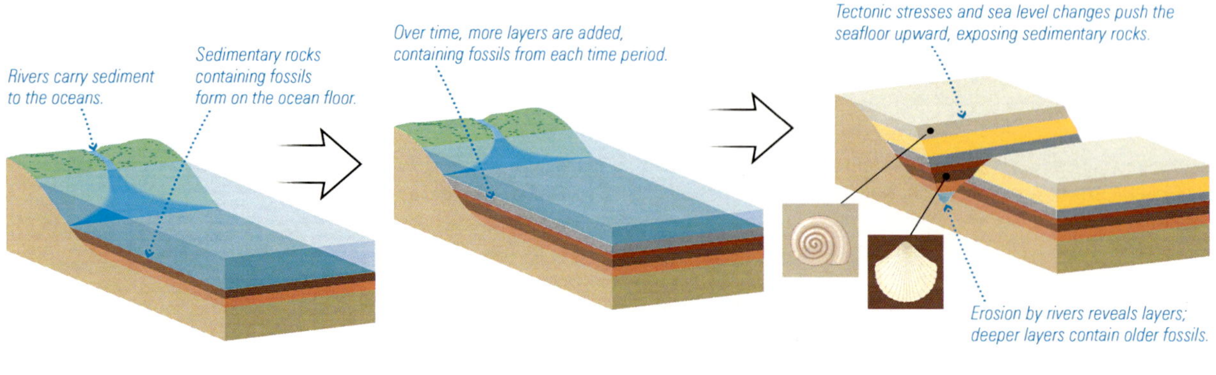

Figure 4.4

Formation of sedimentary rock. Each stratum, or layer, represents a particular time and place in Earth's history and is characterized by fossils of organisms that lived in that place and time. (Adapted from Campbell, Reece, *Biology,* Seventh Edition, © 2005.)

which sedimentary rock forms. Figure 4.5 shows part of the Grand Canyon, which was carved by the Colorado River. The rock layers of the Grand Canyon walls record more than 500 million years of Earth's history.

Because sedimentary rock builds up gradually over time, at any particular location the more deeply buried layers generally are older. This allows geologists to determine the *relative* ages of rocks and fossils buried in sediments. For example, because fossils of dinosaurs appear only in layers older than those in which we find fossils of primates, we conclude that dinosaurs lived before primates.

No single location contains strata from the entire history of Earth, but geologists have put together a fairly detailed geological record by comparing sedimentary strata from many sites around the world. Scientists correlate the strata from different sites by looking for layers with similar fossils. For example, suppose an upper stratum at one location contains fossils of the same type as those found in a lower stratum at another location. In that case, the first location must contain more ancient strata than the second location (Figure 4.6).

Figure 4.5

The walls of the Grand Canyon are exposed sedimentary rock in which the strata record more than 500 million years of Earth's history.

Rock Analysis

You can often tell a rock's type simply from its appearance. For example, a piece of recently solidified lava is obviously igneous, while a rock with an embedded seashell is probably sedimentary. Scientists, too, often start by analyzing a rock's appearance and considering where it was found, as both of these offer clues to the rock's history. But if you really want to learn about a rock, you need to analyze it in the laboratory.

Scientists can analyze rocks in a variety of ways, but three types of analysis are particularly important in reconstructing a rock's history and hence the history of our planet.

- **Mineralogical analysis** generally means identifying the minerals present in a rock.
- **Chemical analysis** generally means determining the elemental or molecular composition of a rock or mineral. For example, chemical analysis will tell you the percentages of a rock that consist of iron, silicon, carbon, or other elements.
- **Isotopic analysis** generally means determining the ratio of different *isotopes* [Section 3.4] of particular elements in a rock. For example, oxygen has three stable isotopes: oxygen-16 (eight protons and eight neutrons) is by far the most common, but it is always

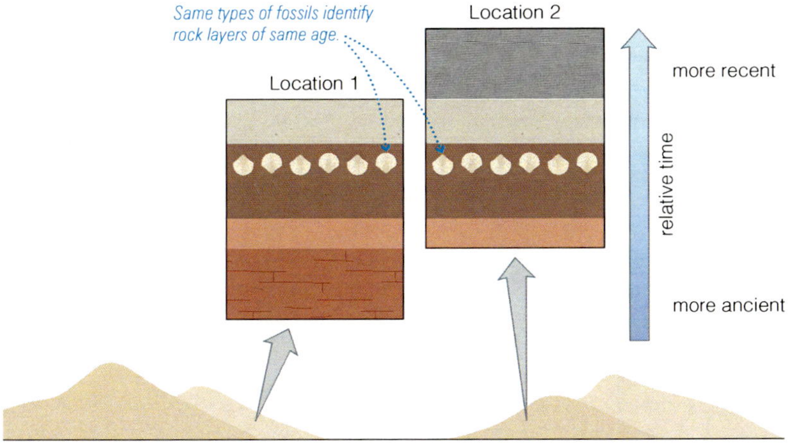

Figure 4.6

In this diagram, we imagine comparing sedimentary strata at two locations. We find that the fossils found in a particular layer near the top at Location 1 are of the same type as those found in a lower layer at Location 2. We conclude that the two sets of strata represent overlapping time periods, with the strata at Location 1 going farther back in time.

mixed with small amounts of oxygen-17 (eight protons and nine neutrons) and oxygen-18 (eight protons and ten neutrons); isotopic analysis can tell us the relative amounts of the three oxygen isotopes in a rock.

These three types of analysis are often used in tandem and each can provide clues about a rock's history; these techniques can also be applied to fossils and other solids. The importance of chemical analysis is probably fairly obvious, since it is always useful to know what an object is made of. Mineralogical analysis can tell us about the temperature and pressure conditions under which a rock formed. For example, while graphite and diamond are both minerals made of nearly pure carbon (they differ in their crystal structures), diamond forms only under much higher pressure conditions than graphite.

Isotopic analysis can be particularly illuminating, because measurements show that isotopes tend to exist in particular overall ratios in nature. For example, the overall ratio of oxygen-16 to oxygen-18 in nature is about 2,000 to 1. Thus, if we find a rock that has more than 1 in 2,000 of its oxygen atoms in the form of oxygen-18, we know that something must have happened to cause the rock to become enriched with this heavier oxygen isotope. Even more important, some isotopes turn out to be **radioactive**, meaning that their nuclei are unstable and tend to change over time in a predictable way. This peculiar property of certain isotopes means that they can serve as natural clocks that allow us to learn *when* a rock (or other solid object) formed.

• How do we learn the age of a rock or fossil?

The most reliable method for measuring the age of a rock, fossil, or other solid object is known as **radiometric dating**. This method relies on careful measurement of an object's proportions of various atoms and isotopes.

A **radioactive isotope** has a nucleus that can undergo spontaneous change, or **radioactive decay**, such as breaking apart or having one of its protons turn into a neutron. For example, carbon comes in two stable isotopes: carbon-12 (98.9% of carbon atoms), with 6 protons and 6 neutrons, and carbon-13 (1.1% of carbon atoms), with 6 protons and 7 neutrons. This stable carbon is sometimes found mixed with much smaller amounts of the radioactive isotope carbon-14, which has 6 protons and 8 neutrons. Other elements, such as uranium, are always unstable—and therefore radioactive—no matter which isotope we are dealing with. When a radioactive nucleus undergoes decay, we say that the original nucleus (before decay) is the *parent* nucleus and the changed nucleus (after decay) is the *daughter* isotope.

Radioactive decay can occur in a variety of ways. Sometimes a large atomic nucleus ejects a helium nucleus, which consists of two protons and two neutrons. (This process is often called *alpha decay.*) In that case, the remaining daughter nucleus has a lower atomic mass than its parent nucleus. For example, uranium-238 decays by ejecting a helium nucleus, leaving thorium-234 as its daughter. However, this isotope is not the final daughter of uranium-238, because thorium-234 is also radioactive. Through a chain of individual decays, eight of which involve the emission of a helium nucleus, uranium-238 ultimately decays into lead-206, which is stable and decays no further (Figure 4.7a).

In other cases, radioactive decay occurs when a nucleus spontaneously emits or absorbs an electron, causing one of its neutrons to turn

Figure 4.7

Examples of radioactive decay.

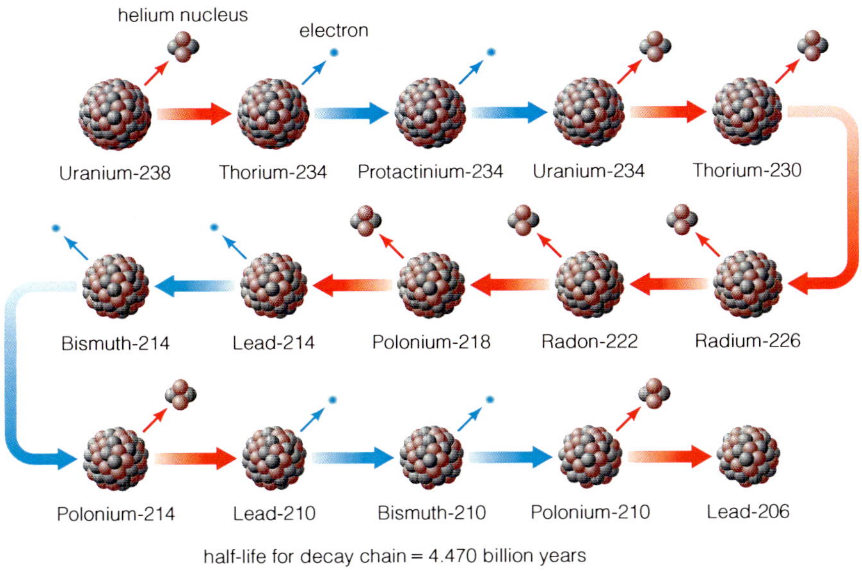

a Uranium-238 decays through a chain of individual decay processes, eight of which involve the emission of a helium nucleus, ultimately leaving lead-206 as its stable daughter isotope. The half-life for the decay chain as a whole is 4.47 billion years.

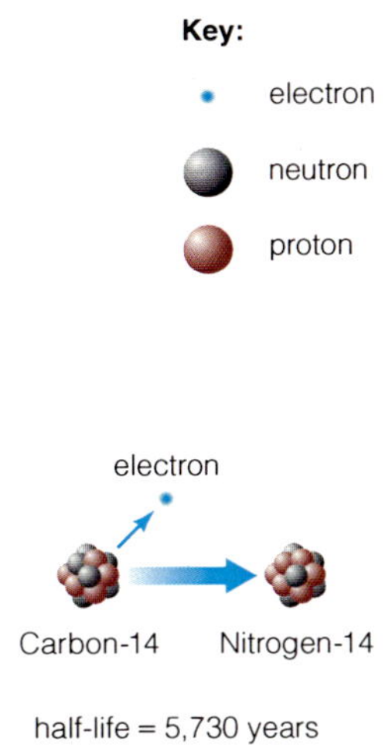

b Carbon-14 decays by emitting an electron from its nucleus, which changes one neutron into a proton to make nitrogen-14 as its stable daughter. The half-life of carbon-14 is 5,730 years.

into a proton, or a proton to turn into a neutron. (The processes are called *beta decay* and *electron capture*, respectively; an absorbed electron is one of the atom's own electrons that gets captured by the nucleus.) In these cases, the parent and daughter nuclei will both have the same atomic mass number (the same total number of protons and neutrons), but they will represent different elements because they have different numbers of protons. For example, carbon-14 decays by emitting an electron as one of its neutrons becomes a proton, so its daughter isotope has seven protons and seven neutrons and therefore is nitrogen-14 (Figure 4.7b).

Regardless of the decay process, radioactive decay always occurs at a specific and measurable rate that is different for every radioactive isotope. Thus, the basic idea behind radiometric dating is to determine the age of a rock (or other solid object) from the ratio of parent and daughter atoms within it, which depends only on the decay rate and the length of time over which the decay has been occurring. To fully understand how this technique works, we must explore the nature of radioactive decay in a bit more detail.

The Probabilistic Nature of Radioactive Decay Radioactive decay is governed by the laws of quantum physics, which means it is a probabilistic process, much like coin tossing. If you toss a single coin, you cannot determine for sure whether it will land heads or tails. All you can say is that it has a 50% chance of landing heads and a 50% chance of landing tails. However, while the probabilistic nature of coin tossing means you cannot predict the outcome for a single coin toss, it allows you to predict the outcome accurately for many coin tosses. For example, if you toss a fair coin one million times, you can be quite certain that heads will come up close to 500,000 times (but you would not expect the number to be *exactly* 500,000).

THINK ABOUT IT To convince yourself that probability can be used to make predictions involving large numbers of atoms or coins, toss a coin 100 times (or toss 100 coins all at once) and record your results. Do you get a result close to 50, as we would expect? Compare your results with those of other students, and discuss what they tell you about the nature of probability.

In the case of radioactive decay, the relevant probability describes the chance of a single nucleus decaying within a specific amount of time (such as within a year). However, because atoms are so tiny, we nearly always deal with such huge numbers of them that we can ignore the individual probabilities and focus instead on the rate at which large numbers of the radioactive nuclei decay.

For example, suppose we study a sample of 1 microgram of a radioactive substance, which despite its small weight (about one-millionth the weight of a paper clip) contains trillions of individual atoms. If we find that 1% of the atoms in the sample undergo radioactive decay within 1 year—meaning that at the end of the year we have 0.99 microgram of the parent substance remaining—we can conclude that any individual atom has a 1% probability of decaying in a 1-year period.

Note that, like coin tosses, results from the past do not affect results for the future. Just as a coin landing heads on one toss does not change the 50% probability of its landing heads on the next toss, the past history of the radioactive sample does not affect the future probability of decay. In our example, we would find that 1% of the atoms in the sample decay in any 1-year period, regardless of how we obtained the sample, how many atoms it contains, or how long we have been studying it. We can therefore reliably and reproducibly measure the decay rate for any radioactive substance by studying a sample of the substance in a laboratory.

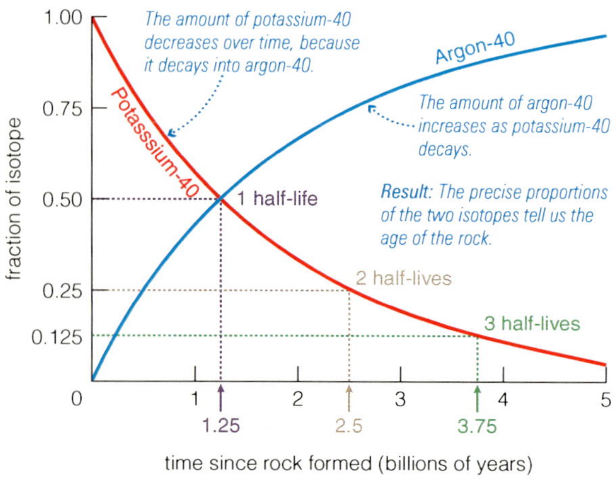

Figure 4.8

Potassium-40 is radioactive, decaying into argon-40 with a half-life of 1.25 billion years. The red line shows the decreasing amount of potassium-40, and the blue line shows the increasing amount of argon-40. The remaining amount of potassium-40 drops by half with each successive half-life.

The Concept of a Half-Life Although the probability of decay in a 1-year period is a perfectly good way to describe a substance's decay rate, we usually describe the decay rate by the substance's **half-life**—the time it would take for half the atoms in a sample of the substance to decay. That is, the amount of radioactive substance in any size sample drops by half with each half-life.

Consider the radioactive decay of potassium-40, which undergoes spontaneous change into argon-40 when its nucleus absorbs an electron to change one of its protons into a neutron. Laboratory measurements show that this decay process has a half-life of 1.25 billion years. Suppose a small piece of rock contained 1 microgram of potassium-40 and no argon-40 when it formed (solidified) long ago. The half-life of 1.25 billion years means that half the original potassium-40 would have decayed into argon-40 by the time the rock was 1.25 billion years old, so at that time the rock would have contained $\frac{1}{2}$ microgram of potassium-40 and $\frac{1}{2}$ microgram of argon-40. Half of this remaining potassium-40 would then decay by the end of the next 1.25 billion years, so after 2×1.25 billion $= 2.5$ billion years, the rock would have contained $\frac{1}{4}$ microgram of potassium-40 and $\frac{3}{4}$ microgram of argon-40. After three half-lives, or 3×1.25 billion $=$ 3.75 billion years, only $\frac{1}{8}$ microgram of potassium-40 would have remained, while $\frac{7}{8}$ microgram had become argon-40. Figure 4.8 summarizes the gradual decrease in the amount of potassium-40 and the corresponding rise in the amount of argon-40.

THINK ABOUT IT Briefly describe how it is possible for us to know that potassium-40 has a half-life of 1.25 billion years, even though we have been capable of studying potassium-40 in the laboratory for only a few decades.

The Essence of Radiometric Dating The potassium-40 example shows the essence of radiometric dating. Suppose you find a rock (or a mineral within a rock) that contains equal amounts of potassium-40 and argon-40. As long as you can be confident that the rock contained no argon-40 when it formed (which you can, as we'll discuss shortly) and has lost none during its history, you can conclude that the rock is one half-life, or 1.25 billion years, old. If instead the rock contains seven times as much argon-40 as potassium-40—that is, only $\frac{1}{8}$ of the original potassium-40 remains—you can conclude that it is three half-lives, or 3.75 billion years, old. More generally, you can get the age of any such rock from the relative amounts of potassium-40 and argon-40 and the graphs in Figure 4.8.

Radiometric dating with other isotopes works the same way. By comparing the amounts of parent and daughter isotopes in the rock (or other object), we can determine the fraction of parent isotopes that has decayed since the rock formed.* Then, based on laboratory measurements of the decay rate (half-life), we can determine the rock's age. Table 4.1 lists several of the parent–daughter isotope pairs commonly used in radiometric dating. Notice that the half-lives vary dramatically, so different pairs are useful for dating materials of different ages. For example, carbon-14 is useful only for dating objects less than about 50,000 years old, while uranium-238 is useful for dating the oldest rocks on Earth.

The major issue that sometimes arises in radiometric dating concerns knowing a rock's original composition. If we're going to determine the rock's age by comparing the amounts of a radioactive isotope and its daughter product, we must have some way of knowing how much of the daughter was originally present, if any. In the case of potassium-40, this is quite easy. The daughter product, argon-40, is a gas that does not combine with other elements and can become trapped in rock only under special circumstances. Thus, argon-40 generally cannot be part of a rock when it solidifies, so any argon-40 trapped inside a rock must have come from radioactive decay of potassium-40. Moreover, since the argon-40 atoms become trapped within the rock at the time they are produced by the decay of potassium-40, we can be similarly confident that no argon has escaped from the rock, unless the rock has undergone heating or some other transformation that could have released the trapped gas.

Radiometric dating with other substances is not always as easy, but detailed study of several isotope ratios or several different mineral grains within a single rock often allows geologists to determine a rock's original composition. Still, if a rock has undergone partial melting or a major shock during its history, or if water has removed some atoms from the rock, we may find somewhat different ages when we date the rock with different parent–daughter isotope pairs. Nevertheless, decades of experience in working with these complications have given geochemists a precise understanding of how they affect radiometric dating. As a result, it is nearly always possible to know whether an age is correct or uncertain.

Note that, while radiometric dating is an extremely powerful technique, it works only for dating solids that have not undergone significant change since they formed. Thus, for rocks, we get unambiguous ages only for igneous rocks. Metamorphic rocks are more challenging to date, because they may have undergone change that has altered their isotope ratios from what they would be due to decay alone. Still, we can often place

*The specific methodology varies in some cases. Carbon-14, for example, is continually produced in Earth's atmosphere by interactions between nitrogen-14 and high-energy particles coming from the Sun. An estimate of the original carbon-14 content of a fossil is based on the atmospheric production rate rather than on a parent–daughter isotope ratio.

TABLE 4.1 *Selected Isotopes Used for Radiometric Dating of Rocks and Fossils*

Some of the isotopes decay in several stages of parent–daughter pairs, and only the final daughter product is shown. (Source: Berkeley Laboratory Isotopes Project)

Parent Isotope	*Daughter Isotope*	*Half-Life*
Rubidium-87	Strontium-87	49.4 billion years
Lutetium-176	Hafnium-176	37.1 billion years
Thorium-232	Lead-208	14.0 billion years
Uranium-238	Lead-206	4.47 billion years
Potassium-40	Argon-40	1.25 billion years
Uranium-235	Lead-207	704 million years
Aluminum-26	Magnesium-26	717,000 years
Carbon-14	Nitrogen-14	5,730 years

fairly clear age constraints on metamorphic rocks, such as a minimum time since a particular rock formed, and learn much about the changes they have undergone. A sedimentary rock cannot be dated as a whole because it is a compressed mix of rock grains containing minerals of different ages; we can sometimes date mineral grains found within a sedimentary rock, but this tells us when the grains originally formed as opposed to when the sediments were deposited or compressed into a solid rock. To estimate ages of sedimentary rocks, we therefore generally rely on dating igneous rocks buried above them or intruding into them. Similar ideas apply when we use radiometric dating for fossils and other artifacts: A radiometric age will be reliable if we can be confident in knowing precisely how much of the original parent isotope has decayed, and any uncertainty in decay fraction will mean a corresponding uncertainty in the age.

The Reliability of Radiometric Ages Scientists today have great confidence in the reliability of ages measured through radiometric dating. The underlying processes that govern radioactive decay are well understood, and the measurements needed to determine half-lives are straightforward when done with care. Perhaps most important, in many cases scientists can use several different radioactive isotopes to measure the ages of rocks or fossils. The ages from different parent-daughter isotope pairs almost always agree for a particular rock, fossil, or mineral grain, and when they don't it is usually possible to determine a cause for the difference. This agreement shows that the technique of radiometric dating is highly reliable. Ages of different rocks and fossils from particular strata of sedimentary rock also agree, further confirming the reliability of the technique.

Additional checks come from the fact that, at least in some cases, we can estimate ages in ways that are independent of radiometric dating. For example, if radiometric ages are correct, then they should confirm the ordering of the relative ages in sedimentary strata—and they do. A more precise check can be made for relatively young archaeological artifacts: Some ancient Egyptian artifacts have dates printed on them, and the dates agree with ages found with radiometric dating; in other cases, we can confirm radiometric ages by comparing them to ages that we can obtain from tree ring data. We can even confirm the $4\frac{1}{2}$-billion-year radiometric age for the solar system as a whole by comparing it to an age based on detailed study of the Sun. Theoretical models of the Sun, along with observations of other stars, show that stars slowly expand and brighten as they age. The model ages are not nearly as precise as radiometric ages, with uncertainties of up to a billion years or so, but the current state of the Sun agrees with what we expect for a star of its mass and age.

Taken together, our theoretical understanding of radioactive decay and the various ways of confirming ages found by radiometric dating leave no room for doubt about the reliability of the technique. Today, scientists routinely use radiometric dating to determine ages of rocks and fossils, and the overall uncertainty in radiometric age dates is usually less than about 1–2%.

• What does the geological record show?

Today, we have a detailed understanding of many of the events that mark the geological record, because geologists have carefully studied and dated tens of thousands of rocks and fossils. But before we get into the specifics, it's important to remember that while radiometric dates can be quite precise even very far back in time, our understanding of Earth's prevailing

COSMIC CALCULATIONS 4.1

Radiometric Dating

The amount of a radioactive substance decays by half with each half-life, so we can express the decay process with a simple formula relating the current amount of a radioactive substance in a rock to the original amount:

$$\frac{\text{current amount}}{\text{original amount}} = \left(\frac{1}{2}\right)^{t/t_{\text{half}}}$$

where t is the time since the rock formed and t_{half} is the half-life of the radioactive material. We can solve this equation for t by taking the logarithm of both sides and rearranging the terms:

$$t = t_{\text{half}} \times \frac{\log_{10}\left(\dfrac{\text{current amount}}{\text{original amount}}\right)}{\log_{10}\left(\dfrac{1}{2}\right)}$$

Example: You chemically analyze a small sample of a meteorite. Potassium-40 and argon-40 are present in a ratio of approximately 0.85 unit of potassium-40 atoms to 9.15 units of gaseous argon-40 atoms. (The units are unimportant, because only the relative amounts of the parent and daughter materials matter.) How old is the meteorite?

Solution: Because no argon gas could have been present in the meteorite when it formed (see discussion in text), the 9.15 units of argon–40 must originally have been potassium–40 that has decayed with its half-life of 1.25 billion years. The sample must therefore have started with $0.85 + 9.15 = 10$ units of potassium-40 (the original amount), of which 0.85 unit remains (the current amount). The formula now reads:

$$\begin{aligned} t &= 1.25 \text{ billion yr} \times \frac{\log_{10}\left(\dfrac{0.85}{10}\right)}{\log_{10}\left(\dfrac{1}{2}\right)} \\ &= 1.25 \text{ billion yr} \times \left(\frac{-1.07}{-0.301}\right) \\ &= 4.45 \text{ billion yr} \end{aligned}$$

This meteorite solidified about 4.45 billion years ago.

conditions becomes less certain as we look deeper into the past. Any individual rock or fossil may give us only limited information about the conditions under which it formed, so a fuller understanding requires studying many rocks and fossils from the same time period. However, the geological record is more sparse as we look further back. Earth's surface undergoes continual change through volcanism, plate tectonics, and erosion, making older rocks comparatively rarer than younger ones. For fossils left by life, the problem is compounded by other factors, best understood if we think about how fossils form.

Fossil Formation Although we tend to think of fossils as "remains" of living organisms, most fossils contain little or no organic matter. Figure 4.9 summarizes some of the ways in which fossils are made. In general, when an organism dies and gets buried in sediments, minerals

Figure 4.9

A gallery of fossils.

a A dinosaur bone preserved in sandstone in Dinosaur National Monument, which straddles Utah and Colorado.

b A 190-million-year-old petrified (stone) tree in Arizona.

c These 375-million-year-old impressions are casts of dead organisms (called ammonites) made when minerals filled the empty space left after the organism decayed.

d This 40-million-year-old leaf still retains organic material, including DNA.

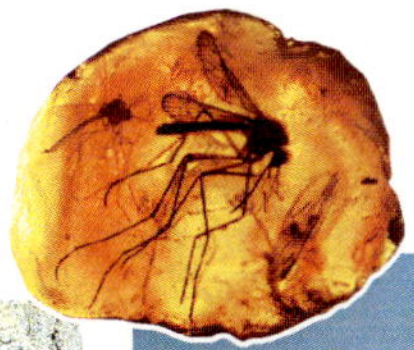

e An insect preserved in hardened tree resin (often called *amber*).

f These tusks belong to a whole 23,000-year-old mammoth discovered in Siberian ice in 1999.

g This boy is standing in a 150-million-year-old dinosaur track in Colorado.

dissolved in groundwater gradually replace organic material. Mineral-rich portions of organisms, such as bones, teeth, and shells, may be left behind, becoming fossils like those of the dinosaur bones displayed in many museums (Figure 4.9a). In some cases, the mineral replacement is complete and organisms literally turn to stone; the "stone trees" of Arizona's Petrified Forest formed in this way (Figure 4.9b). In many other cases, the organisms themselves decay, but in doing so they leave an empty mold that fills with minerals dissolved in water. The minerals may then make a cast in the shape of the dead organism (Figure 4.9c).

More rarely, some of the organic material from a dead organism may be preserved well enough to allow at least some study. Some fossil plant leaves are still green and well enough preserved for their cells to be studied with microscopes, even though they died millions of years ago (Figure 4.9d). In other rare cases, whole organisms may be preserved in tree resin (Figure 4.9e) or frozen in ice (Figure 4.9f). One of the most interesting types of fossil is left not by a dead organism but by the activity of an organism while it was alive. For example, *coprolites* are rocks that consist of petrified excrement, which can allow us to learn about an animal's diet. In other cases, scientists have found fossilized dinosaur footprints, made when mineral processes preserved impressions left by a dinosaur as it walked through soft soil or mud (Figure 4.9g). Such fossil tracks provide clues about how dinosaurs walked and can help scientists hypothesize about dinosaur behavior.

THINK ABOUT IT Molecules of DNA tend to break down rapidly and easily after an organism dies, making it difficult or impossible to identify even fragments of DNA in most fossils. What types of fossils do you think are most likely to yield intact DNA? What types are least likely to yield intact DNA? Explain.

Although fossils can be made in numerous ways, only a tiny fraction of living organisms leave behind any kind of fossil at all. The vast majority of dead organisms decay—becoming food for living organisms in the soil—long before any mineral replacement can occur. If you've ever read about criminologists exhuming the skeletons of people who have been dead for just a few years, you know that even bones and teeth usually decay quite rapidly after death.

The tiny fraction of organisms that have left fossils in relatively recent times in Earth's history still means a substantial fossil record. But fossils become rarer as we look deeper into the geological record, primarily for two reasons. First, fossils can suffer the same fates as rocks, so older fossils are more likely to have been destroyed over time by volcanism, erosion, or other geological processes. The second reason comes from the nature of past life itself. Studies of the geological record show that large plants and animals—which make the most easily discovered fossils—are relatively recent arrivals in Earth's history. For the first 90% or so of Earth's history, any living organisms were all microscopic, which means their fossilized remains must also be microscopic. Clearly, it is much more difficult to find and identify microscopic fossils than huge fossils of dinosaur bones. This fact is especially important in astrobiology, since we generally focus more on the origin and early evolution of life than on large plants and animals. Indeed, as we'll discuss in Chapter 6 , the difficulty of finding and studying microfossils (and other potential traces of life in very old rocks) has led to significant scientific controversies about exactly when life took hold on Earth.

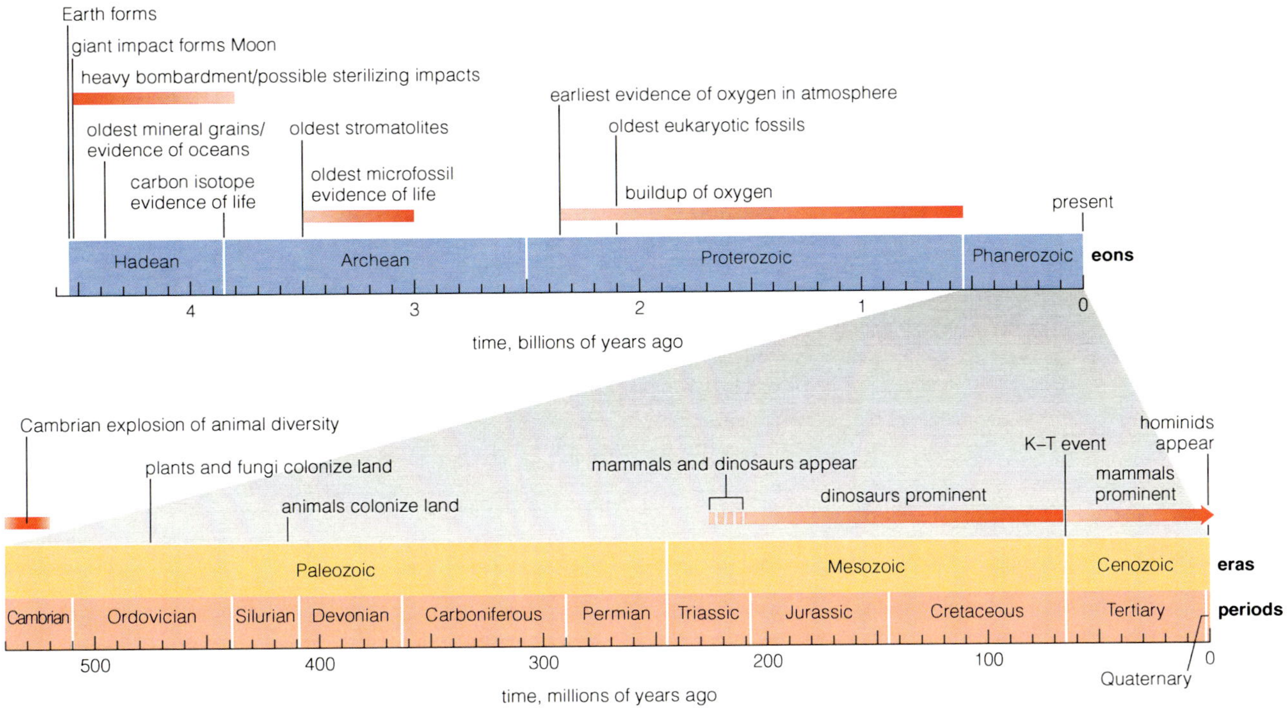

Figure 4.10

The geological time scale. Notice that the lower timeline is an expanded view of the last portion (Phanerozoic) of the upper timeline. The eons, eras, and periods are defined by changes in the rocks and fossils present in the geological record. The absolute ages come from radiometric dating. (Some of the events listed on the timeline are discussed later in this chapter, and others are discussed in Chapter 6.)

The Geological Time Scale Although the geological record gives us far more information about recent events than the distant past, geologists have nevertheless been able to piece together a fairly detailed history of our planet going nearly all the way back to the time of its formation. To help organize this history, scientists divide it into a set of distinct intervals that make up what we call the **geological time scale**. Figure 4.10 shows the names of the various intervals on a timeline, along with numerous important events that we will discuss later in this chapter and in Chapter 6. It is not necessary to memorize the names of the geological time intervals for the purposes of this book. However, the names are commonly used in books and articles about Earth or astrobiology, so it's useful to be familiar with them.

The first major division of geological time is into a set of four **eons**: the Hadean, Archean, Proterozoic, and Phanerozoic. We can understand these names by working backward in time and looking at their Greek roots. The Phanerozoic eon extends from the present back to about 540 million years ago; its name comes from the Greek for "visible life" because it is marked by the presence of fossils visible to the naked eye. The Proterozoic eon, which extends from 540 million to about 2.5 billion years ago, means the eon of "earlier life" because it shows fossils of single-celled organisms that lived before the Phanerozoic eon. The Archean eon extends from 2.5 to about 3.85 billion years ago and is named for "ancient life"; it got this name after the discovery of fossils from the first half of Earth's history. The Hadean eon got its name at a time when it was presumed that the early Earth would have had "hellish" conditions during this early time (Hades was the Greek mythological name for the underworld); however, as we'll discuss in the next section, recent evidence suggests that the Hadean may not have been quite that bad.

The fact that the geological record is much richer for more recent times is reflected in the more detailed naming system used for these times.

The most recent, or Phanerozoic, eon is subdivided into three major **eras**: the Paleozoic, Mesozoic, and Cenozoic. These names also have Greek roots and mean, respectively, "old life," "middle life," and "recent life." The three eras are further subdivided into **periods**. The periods do not follow any consistent naming scheme. For example, the Cambrian period* gets its name from the Roman name for Wales (in Great Britain), the Jurassic period gets its name from rocks found in the Jura mountains of Europe, and the Tertiary period simply means "third" period. The recent geologic periods are even further subdivided into *epochs* and *ages*, but these are not shown in Figure 4.10.

Note that the eons, eras, and periods do not have uniform lengths. For example, the Proterozoic eon is about three times the length of time of the Phanerozoic eon, and the Paleozoic era is longer than the Mesozoic and Cenozoic eras combined. The divisions in the geological time scale are determined not by duration but by specific changes in the geological record, such as where certain species disappear and other new ones appear.

The Age of Earth

Figure 4.10 starts with Earth's birth, but we cannot read the precise time at which that occurred from the geological record. The oldest intact Earth rocks date to about 4.0 billion years ago. Apparently, rocks that formed prior to this time have been remelted or reshaped so much that we cannot obtain an age through radiometric dating. So how do we know the age of our planet?

We have a variety of ways of looking back to earlier times. The most direct way is based on studies of tiny mineral grains of zirconium silicate, or *zircons* for short (Figure 4.11). Although they are found embedded in much younger sedimentary rocks, radiometric dating (based on analysis of uranium and hafnium isotopes found within them) shows that some zircons solidified as much as 4.4 billion years ago. Moreover, their isotopic content suggests that they formed at a time when continents had already begun to form, suggesting that Earth's crust had separated from interior material by about 4.5 billion years ago.

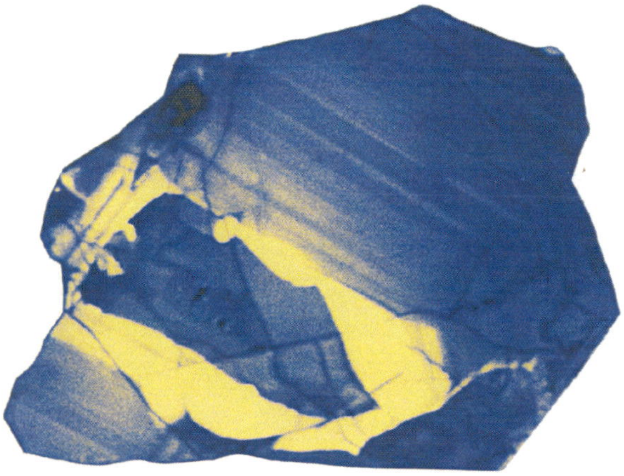

Figure 4.11

This image shows a tiny zircon crystal—about 0.2 millimeter across—that was found embedded in a rock formation in Western Australia. Radiometric dating shows it to be nearly 4.4 billion years old. The image was made with a technique called cathodoluminescence, in which an electron beam is focused on the crystal, causing it to emit visible light; the colors in the image are not real, but instead correspond to differing intensities in the emitted light.

Moon rocks also help us constrain Earth's age. Some Moon rocks brought back by the *Apollo* astronauts are considerably older than any intact Earth rocks, reflecting the fact that volcanism and other geological processes have done far less reshaping of the Moon's surface than they have of Earth's surface. Radiometric dating with isotopes of uranium and lead places the ages of the oldest Moon rocks at more than 4.4 billion years, telling us that Earth and the Moon must have formed before this time. In fact, the leading model for the Moon's formation holds that it formed as a result of a *giant impact* that blasted material out of Earth's outer layers; we'll discuss this model in detail in Section 4.6. Earth must therefore be at least slightly older than the Moon, which means older than 4.4 billion years.

While the zircons and lunar rocks set a minimum age for Earth, we can set a maximum age by dating the formation of the solar system as a whole. We do this by studying meteorites. Some meteorites are younger than the solar system as a whole, because they are fragments of asteroids that formed and later shattered in collisions. However, a large number of meteorites have a chemical structure that suggests they were among the

*The Cambrian is the earliest period in the Phanerozoic eon and was once thought to be the first period in which fossil organisms could be found. For that reason, the entire time before the Phanerozoic eon—that is, the Hadean, Archean, and Proterozoic eons—is often called the *Precambrian*.

first pieces of solid material to condense in the early history of the solar system. These meteorites also all have about the same age, offering further evidence that they represent material from the very beginning of the solar system; these date to 4.57 billion years ago (with an uncertainty of less than about 0.02 billion years).

The meteorite data allow scientists to use other techniques to further constrain the ages of both Earth and the Moon. These techniques are somewhat complex in their details, but they are based on comparisons of isotope ratios in meteorites, Earth rocks, and Moon rocks. The results show that both Earth and the Moon had formed within about 50 to 70 million

KEY GEOLOGICAL DEFINITIONS

This box summarizes the definitions of a few of the most basic geological terms used in this chapter and throughout the book.

ROCKS, MINERALS, FOSSILS:

minerals: The basic pieces of solid rock; a particular mineral is distinguished from other minerals by its chemical composition or crystal structure (or both).

rocks: Intact solids that may contain a variety of minerals. We classify rocks into three basic types by their formation process: *Igneous rock* is of volcanic origin, made when molten rock cools and solidifies. *Sedimentary rock* is made by the gradual compression of sediments, which may contain bits of other rock types as well as fossils. *Metamorphic rock* was once either igneous or sedimentary but has since been transformed (but not melted) by high heat or pressure. Rocks of any of the three types may be subclassified by the minerals they contain.

fossil: Any relic left behind by living organisms that died long ago.

TERMS RELATED TO GEOLOGICAL TIME:

geological record: The information about Earth's past that is recorded in rocks and fossils; the latter record is sometimes called the *fossil record.*

geological time scale: The time scale used to measure the history of Earth.

radiometric dating: The method of determining the age of a rock or fossil from study of radioactive isotopes contained within it.

half-life: The time it takes for half of the atoms to decay in a sample of a radioactive substance.

TERMS RELATED TO EARTH'S GEOLOGICAL HISTORY:

differentiation: A process in which materials separate by density. In Earth, differentiation led to a dense *core* made mostly of iron and nickel, a rocky *mantle* made mostly of *silicates* (minerals rich in silicon and oxygen), and a *crust* made of the lowest-density rocks.

heavy bombardment: The period of time during which the planets were heavily bombarded by leftover planetesimals, starting from the time the planets first formed and likely ending some 3.8–4.0 billion years ago.

outgassing: The process of releasing gases trapped in a planetary interior into the atmosphere.

lithosphere: The layer of cooler, more rigid rock that sits above the warmer, softer mantle rock below. It encompasses both the crust and the uppermost portion of the mantle, extending to a depth of about 100 kilometers. On Earth, the lithosphere is broken into a set of large *plates.*

seafloor crust: The relatively dense, thin, young crust found on Earth's seafloors, composed largely of the igneous rock called *basalt.*

continental crust: The thicker, lower-density crust that makes up Earth's continents. It is made when remelting of seafloor crust allows lower-density rock to separate, and typically consists of granite. Continental crust ranges in age from very young to as old as about 4.0 billion years.

plate tectonics: The geological process in which lithospheric plates move around the surface of Earth. It acts like a conveyor belt, with new seafloor crust erupting and spreading outward from mid-ocean ridges and then being recycled back into the mantle by subduction at ocean trenches. It also explains *continental drift,* because plates carry the continents with them as they move.

TERMS RELATED TO CLIMATE AND CLIMATE REGULATION:

greenhouse effect: The effect that makes a planet's surface warmer than it would be in the absence of an atmosphere. It is caused by the presence of *greenhouse gases*—such as carbon dioxide (CO_2), water vapor (H_2O), and methane (CH_4)—that can absorb infrared light emitted by the planetary surface (after the surface is heated by sunlight).

carbon dioxide cycle (CO_2 cycle): The cycle that keeps the amount of carbon dioxide in Earth's atmosphere small and nearly steady and hence keeps Earth habitable. Over time, this cycle has locked up most of Earth's carbon dioxide in *carbonate* rocks (rocks rich in carbon and oxygen) such as limestone.

ice ages: Periods of time during which Earth becomes unusually cold, so that water from the oceans freezes out as ice and covers a substantial portion of the continents.

Snowball Earth: Refers to periods of extreme ice ages that may have occurred several times before about 580 million years ago.

global warming: Usually refers to the current warming of Earth caused by human input of greenhouse gases into the atmosphere.

years (0.05 to 0.07 billion years) after the oldest meteorites formed. We therefore conclude that our planet accreted quickly in the early solar system, and the giant impact that formed the Moon happened quite early in Earth's history as well. By about 4.5 billion years ago, our planet had its Moon and must have been essentially at its current mass and size, ready for its geology to begin shaping the features that would ultimately make it our home.

4.3 THE HADEAN EARTH AND THE DAWN OF LIFE

We know comparatively little about the Hadean eon, which constitutes a little more than the first half-billion years of Earth's history. Nevertheless, it was clearly an important time in the history of our planet, and evidence that we'll discuss in Chapter 6 shows that life arose during or not long after this period of time. So if we want to understand when and how our planet became habitable and gave birth to life, we need to start back in this earliest of Earth's time periods.

• How did Earth get an atmosphere and oceans?

Before Earth could have life, it needed an atmosphere and liquid water oceans. Neither existed when Earth first formed. Our planet was too small and too warm to capture significant amounts of hydrogen and helium gas as it accreted within the solar nebula [Section 3.3], and no other gases were present in large enough quantities to make a substantial atmosphere. In fact, the presence of Earth's atmosphere and oceans once posed a mystery to solar system formation theory, because the planetesimals that condensed at Earth's distance from the Sun should have been made only of rock and metal, with no gaseous or icy content at all. How, then, did Earth obtain the water and gases that make up our oceans and atmosphere?

MOVIE MADNESS: FLINTSTONES AND DINOSAURS

The Paleolithic (early stone age) began roughly 2 million years ago. You can be sure that for any of your ancestors unlucky enough to be born in those dim and distant times, life was generally short and brutish.

That's reality. But the impression you'd get from watching Hollywood's stone age sitcom *The Flintstones* is that our predecessors enjoyed modern lifestyles and contemporary situations. The only real difference between their existence and yours was that all the Flintstones' technology was rock-based, with domesticated dinosaurs supplying the power. Cars rolled on stone wheels, smiling dinos served as cranes and derricks, and air travel was facilitated by obliging pterodactyls. Fred Flintstone and his buddy Barney were smooth-skinned and beardless (try shaving with a rock razor), and regularly took showers using elephants to provide the spray, although it's a puzzle how those bathroom pachyderms avoided being preyed on by the dinosaurs.

Needless to note, most of this doesn't make sense. *The Flintstones* is time out of joint from top to bottom, and the most serious offense is mingling our Paleolithic ancestors with dinosaurs. After all, the genus *Homo* has been around for only a few million years, and *Homo sapiens* have been strutting our stuff for only the last 100 thousand. Dinosaurs, on the other hand, disappeared about 65 million years ago.

The Flintstones may be harmless Hollywood entertainment, but some non-Hollywood folks have seriously argued that dinosaurs and humans really were contemporaries. In the 1960s, a claim was made that fossilized human footprints could be seen alongside dinosaur prints in the Paluxy riverbed near Glen Rose, Texas. If true, this would indicate that our reading of the geological record is in serious error. But careful investigation by paleontologists has shown that the supposed "man tracks" were, in fact, just more dinosaur prints in which the small, three-pronged toe marks had been filled in by mud or eroded away.

Humans never met dinosaurs, and we could never have used them to power society. Fortunately, our predecessors eventually learned to work with metals, invent engines, and learn about the world through science. It was these developments that give us a lifestyle that our stone age ancestors could never have imagined.

Models of planetary formation now show that while Earth formed primarily from "local" planetesimals of rock and metal, it should also have incorporated some planetesimals from farther out in the solar system; these planetesimals were flung inward by gravitational interactions with other planetesimals and forming planets. Some of these planetesimals came from far enough away—probably from the region currently occupied by the asteroid belt—that they contained ices or rock chemically bound with molecules of water or other common gases. As these planetesimals became part of the forming Earth, their gaseous content became trapped on or within our planet.

The young Earth therefore contained trapped gases, including water, held under pressure in the interior in much the same way that the gas in a carbonated beverage is trapped in a pressurized bottle. When molten rock erupts onto the surface as lava, the release of pressure violently expels the trapped gas in a process we call **outgassing.** Outgassing probably released most of the water vapor that condensed to form our oceans as well as most of the gas that formed our atmosphere.

Some water and gas may also have been supplied directly to the surface by impacts after Earth formed. This process may have begun to create an atmosphere even before Earth was fully formed—perhaps when it was only one-third its current radius. It is difficult to determine the relative contributions of this process and outgassing, but recent studies of comet composition suggest that outgassing was more important to the formation of the atmosphere and oceans.

Volcanism is the major source of outgassing, as you can probably guess from looking at any photo of a volcanic eruption (Figure 4.12). Some outgassing also occurs as a result of impacts—the heat of an impact can melt rock and allow gas to escape. Studies of present-day volcanoes show that the primary gases released by outgassing are water vapor (H_2O), carbon dioxide (CO_2), nitrogen (N_2), sulfur-bearing gases (H_2S or SO_2), and hydrogen (H_2). Because Earth's overall composition has not changed much since its formation, we expect that the same gases were released in early times.* The water vapor condensed as rain to fill Earth's oceans (along with water that bubbled out of the ground at volcanic vents). The gases that remained airborne made up Earth's early atmosphere.

Isotopic analysis of the oldest zircon grains (see Figure 4.11) suggests that substantial amounts of water—and probably oceans—were already present on Earth at the time these grains solidified, some 4.4 billion years ago. If so, much of Earth's atmospheric gas must have been released quite early, perhaps as a side effect of Earth's interior melting or of the giant impact thought to have formed the Moon. Taken alongside the evidence noted earlier suggesting that continents had also begun to form when the zircons solidified, we are led to the intriguing idea that Earth may have had early continents, oceans, and an atmosphere within only about 100 million years after the planet first formed.

The composition of the early atmosphere was very different from that of the atmosphere today. The early atmosphere was probably dominated by carbon dioxide† (for reasons we will discuss shortly), while today's atmosphere is dominated by nitrogen (about 78% of the atmosphere)

Figure 4.12

This photo shows the eruption of Mount St. Helens (in Washington State) on May 18, 1980. Eruptions are accompanied by a tremendous amount of outgassing. (The gas itself is generally invisible, so what you see is dust and ash expelled along with the gas.)

*Some present-day volcanoes release recycled gases, while others release "juvenile" gases (gases that have not previously been outgassed). The outgassing composition is determined from those releasing juvenile gases, since these should be indicative of the gases originally trapped in Earth's interior.

†Significant amounts of hydrogen may also have been present, a controversial idea that we'll discuss in Chapter 6.

and contains only trace amounts of carbon dioxide (less than 0.1%). More important to us, the early atmosphere contained essentially no molecular oxygen (O_2), while the present atmosphere is about 21% oxygen. Thus, we could not have breathed the atmosphere on the early Earth. Earth's present oxygen atmosphere is almost certainly a result of photosynthesis by living organisms, as we will discuss further in Chapter 6. Life therefore must have arisen in a nearly oxygen-free environment.

• Could life have existed during Earth's early history?

The Hadean eon got its name because of its presumed hellish conditions, but the recent zircon evidence tells a different story. If the evidence is being interpreted correctly, it means that Earth may have been habitable within 100 million years after its formation. Many modern-day microbes survive just fine in the absence of oxygen, and it seems likely that such organisms could have thrived almost from the moment the oceans and atmosphere first formed.

However, while the Hadean might have been reasonably balmy at most times, those calm periods must have been frequently interrupted by great violence: Volcanic eruptions were probably more frequent and larger than those of today and, even more significantly, the young Earth was bombarded by planetesimals left over from the birth of the solar system.

The Heavy Bombardment The accretion of the planets did not end suddenly. The young planets must have shared the solar system with vast numbers of "leftover" planetesimals. Some of these leftovers still survive as asteroids and comets, but many more must have crashed into the planets. The vast majority of these collisions occurred in the first few hundred million years of our solar system's history, during the period we call the **heavy bombardment.** On planets and moons with solid surfaces, such collisions leave **impact craters** as visible scars. A small telescope reveals the presence of numerous impact craters on the Moon (see Figure 4.1).

Earth must have experienced even more impacts than the Moon, both because it presents a larger target and because its stronger gravity tends to attract more objects. However, while the Moon still shows scars of craters formed throughout much of its history, craters on Earth tend to get erased with time by erosion, volcanic eruptions, and plate tectonics. Thus, if we wish to learn about the past cratering rate on Earth, we must look to the evidence recorded on the Moon.

Figure 4.13 shows a map of the Moon's entire surface. Notice that some regions are much more crowded with craters than others. The most heavily cratered regions are the **lunar highlands**, where craters are so abundant that we see overlapping crater boundaries and craters on top of other craters. Radiometric dating of Moon rocks from the lunar highlands shows them to be more ancient than rocks from other regions of the Moon; the highlands are the sources of the Moon rocks that date to more than 4.4 billion years ago, and even their youngest rocks are generally at least 4.0 billion years old. Thus, we conclude that the many craters of the lunar highlands formed more than 4 billion years ago. Clearly, the Moon was being heavily pelted from space during the first half-billion years of the solar system's history, and the same must have been true for Earth.

In contrast, we see relatively few craters in the regions known as the **lunar maria** (see Figure 3.16). The maria are huge impact craters that are smooth because they were covered over by molten lava. The large impacts fractured the lunar crust, creating cracks through which molten

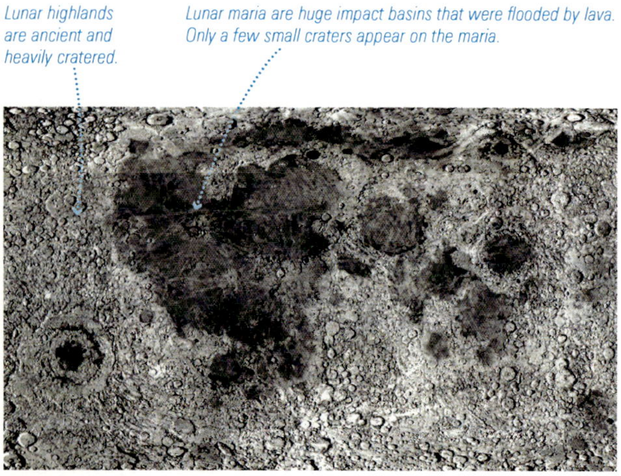

Figure 4.13

This flat map shows the entire surface of the Moon in the same way that a flat map of Earth represents the entire globe. The "lunar highlands" are heavily cratered, and radiometric dating shows that these portions of the lunar surface are quite ancient—some highland rocks date back to some 4.4 billion years ago. The lunar maria are about a half-billion or more years younger, and their relatively few craters tell us that the impact rate had dropped dramatically by the time the maria formed.

lava escaped at some later time; radiometric dating of rocks from the maria shows that the lava flows occurred between about 3.9 and 3.0 billion years ago. The lava flooded the crater basins, covering the existing craters. Thus, the craters we now see within the maria must be the result of impacts that occurred *after* the lava flows, and the relatively small number of these craters tells us that impacts have been relatively rare since the time the maria solidified.

More detailed studies of the ages of lunar craters, as well as analysis of mineral grains in meteorites and old zircons in Earth rocks, can in principle tell us more precisely when the impact rate dropped. Although we might expect that the heavy bombardment would have tapered off gradually, substantial evidence points to a **late heavy bombardment**, around 3.9 billion years ago, that may have been responsible for many of the Moon's large craters; some evidence also suggests that the time between 4.5 billion years ago and the late heavy bombardment may have been relatively quiet. We do not know why the impact rate might have spiked upward with the late heavy bombardment, but some scientists hypothesize that it may have been due to "planetary migration" in which gravitational interactions caused the orbits of the young planets to change. We suspect that similar migration may explain the surprising orbits of some extrasolar planets [Section 11.3], and planetary movement could have disrupted the orbits of asteroids and led to a period of many impacts. In any event, the lunar record shows that the heavy bombardment came to an end by about 3.8 billion years ago.

Every object in the solar system must have been similarly pelted during the heavy bombardment, so all solid surfaces should originally have been as crowded with craters as the lunar highlands. So in addition to helping us understand the early history of Earth and the Moon, these ideas allow us to estimate surface ages for worlds throughout the solar system. Those surface regions that still have abundant craters, such as much of Mercury's surface (see Figure 4.1), must therefore have undergone little change during the last 4 billion years or so. Surfaces with fewer craters must be correspondingly younger, indicating that their original craters have been erased in some way. Thus, thanks to the reconstruction of cratering rates made possible by studies of Moon rocks, planetary scientists can use simple counts of craters on other worlds to estimate the ages of the surfaces. These age estimates are not nearly as precise as radiometric dating—indeed, ages from crater counts are sometimes uncertain by as much as a billion years or so—but until we begin collecting rock samples from other worlds, they are the only age evidence we have to go by.

THINK ABOUT IT Crater abundance varies greatly in different regions of Mars. For example, the "southern highlands" are quite crowded with craters, while the "Tharsis Bulge" has volcanoes but few impact craters. Which region has the older surface? Does this imply any difference in the number of impacts that occurred in the two regions? Why or why not?

Possible Sterilizing Impacts

We have found that the Hadean Earth must have endured large impacts at a rate much higher than Earth has experienced since. What does this mean for the possibility of life during the Hadean? Remember that the Hadean lasted more than 600 million years, which means that even with "frequent" large impacts, individual events were typically separated by thousands or millions of years. There was never a time when impacts occurred so rapidly that they could have done the equivalent of hitting every living organism on the head. Given the evidence of continents and oceans all the way back to 4.4 to 4.5 billion years ago, there seems no reason why life could not have arisen during the Hadean. Thus, the question of life during the Hadean seems less a question of whether it could have *existed* than whether it could have *survived* the occasional large impacts.

The effects of impacts depend primarily on their sizes. As we'll discuss in Chapter 6, the impact of an object about 10 to 20 kilometers across is thought to have precipitated a series of global changes that caused the extinction of the dinosaurs some 65 million years ago. Fortunately for us, impacts of that size are extremely rare today. But the lunar evidence tells us that far larger impacts occurred during the heavy bombardment.

To understand the effects of very large impacts, scientists calculate the amount of energy they would release and then model how the energy would heat the planet. Such calculations tell us that the impact of an asteroid more than about 350–400 kilometers across would provide enough heat to completely vaporize Earth's oceans and raise the global surface temperature to as high as 2,000°C (3,600°F). It's unlikely that any life could survive such effects, so we conclude that an impact of that size would be a **sterilizing impact**—one that would kill off all life on the planet. Slightly smaller impacts could have vaporized the top few hundred meters of the oceans. These impacts probably would have been sufficient to kill off most life, although life near the ocean bottom (such as near deep-sea vents) or deep underground might have been able to survive.

Impacts large enough to have sterilized much or all of the planet should have been rare even in the Hadean, and we have no way to know exactly how many occurred or when they occurred. Nevertheless, studies of lunar craters and probability arguments suggest they should have happened at least a half-dozen to a dozen times during Earth's first few hundred million years of existence, with the last one probably occurring some time between about 4.2 and 3.8 billion years ago. We therefore reach an important conclusion about life on Earth: While conditions may have allowed life to arise during the Hadean, perhaps even as far back as 4.4 to 4.5 billion years ago, most or all of this life would have been wiped out by major impacts. If an impact fully sterilized the planet, life would have had to start all over again; indeed, it is possible that life arose and was subsequently extinguished multiple times on the early Earth. If an impact was just short of sterilizing, it still would have wiped out all but the forms of life adapted to live in the "safe" environments of the deep ocean or underground. Either way, only after the last major impact was life able to have free rein, and begin the long evolution that ultimately blossomed into the great variety of life on Earth today.

4.4 GEOLOGY AND HABITABILITY

Although no sterilizing impacts have occurred since the end of the heavy bombardment, smaller impacts still occasionally occur; as the case of the dinosaurs shows, these can have great effect on our planet and its life. Nevertheless, if we focus on geology, impact craters only rarely have a significant effect. Instead, virtually every known feature of Earth's surface (with the exception of the relatively few impact craters) has been shaped by one or more of the following three geological processes: volcanism, plate tectonics, and erosion. Why are these processes so important on Earth?

Remarkably, all three of these processes that continually reshape our planet's surface are directly attributable to internal heat. The connection is most obvious for volcanism, since we actually see molten rock rising up from Earth's interior, but it applies equally to plate tectonics. The connection to internal heat is a little more subtle for erosion by wind, water, and ice, but remember that these things exist only because our planet has oceans and an atmosphere, and that the oceans and atmosphere were produced by volcanic outgassing.

Even more important, as we noted in Section 4.1, volcanism and plate tectonics have played major roles in Earth's long-term habitability. Thus, if we hope to understand Earth's surface habitability, we must first understand what our planet is like on the inside and how the interior conditions drive these crucial geological processes. That is our task in this section; as we'll see, internal heat is also a major factor in creating Earth's magnetic field, which we identified in Section 4.1 as a third important ingredient in long-term habitability.

• What is Earth like on the inside?

Our deepest drills have barely pricked Earth's surface, reaching only a few kilometers down into the nearly 6,400 kilometers to Earth's center. Nevertheless, we have managed to learn a lot about our planet's interior structure. One set of clues comes from the nature of surface rocks. For example, the fact that surface rocks are considerably less dense than Earth's overall average density tells us that much denser material must reside in the interior than on the surface. Other clues come from precise measurements of Earth's gravitational field strength in different locations and studies of Earth's magnetic field.

We learn much more about Earth's internal structure from the study of **seismic waves**—waves that propagate much like sound waves both through Earth's interior and along its surface after an earthquake. The precise speed and direction of seismic waves depend on the composition, density, pressure, temperature, and phase (solid or liquid) of the material they pass through. After an earthquake, geologists record the arrival times and the strengths of seismic waves at stations distributed all around the world. At each station, the arriving seismic waves tell us something about the average state of the material through which they have passed. By comparing data from many different stations, scientists have pieced together a fairly detailed picture of Earth's internal structure.

Earth's Interior Structure

Seismic studies reveal that Earth's interior is divided into three major layers by density (Figure 4.14):

- **Core:** The highest-density material, consisting primarily of the metals nickel and iron, resides in the central core. Earth's core has two distinct regions: a solid **inner core**, surrounded by a molten (liquid) **outer core**.
- **Mantle:** Rocky material of moderate density—mostly silicate minerals rich in silicon and oxygen—forms the thick mantle that surrounds the core and makes up most of Earth's volume.
- **Crust:** The lowest-density rock, including igneous rocks such as granite and basalt, forms the thin crust that essentially represents Earth's outer skin.

The molten lava from volcanoes gives many people the incorrect impression that Earth is molten inside, but in fact the lava rises upward from a fairly narrow zone of rock in the upper mantle that is only partially molten. The vast majority of Earth's interior is solid rock, and only the

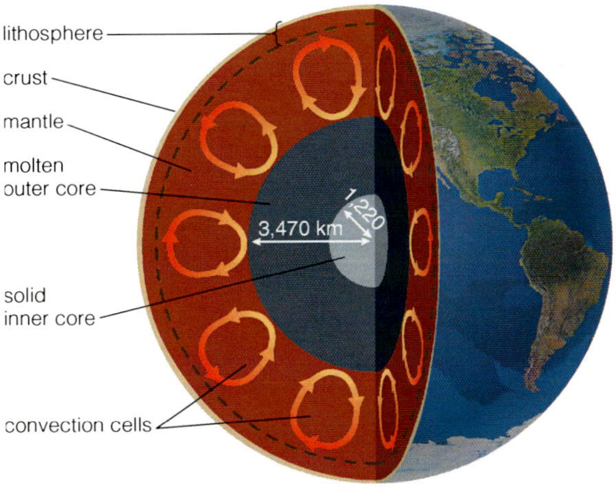

Figure 4.14

Earth's interior structure, determined from seismic studies; the layering by core-mantle-crust is based on density, while the identification of the lithosphere is based on rock strength. The circular arrows represent the general pattern of mantle convection.

outer core is fully molten. However, the interior rock is not equally strong throughout, and differences in rock strength are often more important than differences in density for understanding many geological processes.

The idea that rock can vary in strength may seem a bit surprising, since in our everyday lives we tend to think of rock as the ultimate in strength and stability (hence the phrase "solid as a rock"). However, the strength of rock depends on its temperature and the surrounding pressure, and under the conditions found in Earth's mantle, even "solid" rock can flow gradually. The popular toy Silly Putty provides a good analogy. The putty can feel pretty solid, especially when it is cold; you can even form it into a ball and bounce it. But if you put a pile of it on a table or inside its "egg" container, after a few days you'll see that it has flowed slowly outward.

In terms of rock strength, geologists define Earth's outer layer as the relatively cool and rigid rock, called the **lithosphere** (*lithos* is Greek for "stone"), that "floats" on warmer, softer rock beneath. The lithosphere encompasses the crust and the upper part of the mantle, as shown in Figure 4.14. Beneath the lithosphere, the higher temperatures allow rock to deform and to flow much more easily. In fact, the mantle rock flows with a characteristic pattern called **convection,** in which hot material expands and rises while cooler material contracts and falls. You are probably familiar with convection in other situations, as it can occur any time a substance is strongly heated from below. For example, you can see convection in a pot of soup on a hot stove, and convection is important in weather because the warm air near the ground tends to rise while the cool air above tends to fall. In the mantle, convection is driven by heat from the core. This heat causes rock near the base of the mantle to expand, giving it a tendency to rise because it becomes lighter and less dense than the rock above it. Meanwhile, rock near the top of the mantle (just below the lithosphere) can cool as its heat flows to the surface (by conduction), causing it to contract and sink. The ongoing process of convection creates *convection cells,* indicated by the circular arrows in Figure 4.14. Keep in mind that mantle convection involves the flow of solid rock, so it is quite slow: Typically, mantle rock flows at a rate of only perhaps 10 centimeters per year, slow enough that it would take about 100 million years for a particular piece of rock to be carried from the base to the top of the mantle.

Differentiation and Internal Heat Earth's interior layering tells us that it at one time underwent the process known as **differentiation,** in which materials separate according to their density. Earth must have undergone differentiation quite early in its history; as we'll discuss in Section 4.6, the giant impact that formed the Moon must have occurred after differentiation, and we already discussed the fact that this event happened quite early.* For differentiation to have occurred so rapidly, our planet must have been molten or nearly molten throughout its interior. The melting allowed material to separate by density, much as oil separates from water when you mix them and let the mixture sit.

The heat that caused rock to melt came from three main sources. First, the impacts of accretion created heat that melted the outer layers of the young Earth. Second, as denser materials sank through the molten outer layers, their gravitational potential energy was converted into ther-

---

*A second line of evidence for early differentiation comes from isotopic comparisons between meteorites and lead ore on Earth's surface, leading to the conclusion that most lead must have sunk to the core by the time our planet was about 30 million years old.

mal energy that added further heat to the interior. Third, heat is continually released by the radioactive decay of elements within Earth. This heat source is still important today, but it was even more important in early times, when there was more radioactive material to decay. Once the outer layers began to melt due to the first heat source (accretion), the second and third heat sources (sinking of dense materials and radioactive decay) ensured that our planet would completely melt and differentiate fully.

All the terrestrial worlds in our solar system underwent similar melting and differentiation, and the evidence we discussed earlier suggests that differentiation occurred when the planets were very young. Since that time, they have never again been hot enough to melt fully, and they have all been slowly cooling with time. However, different worlds have cooled at different rates.

In general, two factors determine a world's cooling rate. The first is size: Large worlds tend to retain their internal heat much longer than do smaller worlds. You can see why by picturing a large world as a smaller world wrapped in extra layers of rock. The extra rock acts as insulation, making it take longer for interior heat to escape into space. If you now add the fact that the larger world contains more heat in the first place, it's clear that a large world will take much longer to cool than a small world. The Moon's relatively small size probably allowed it to cool substantially within a billion years or so after its formation, while Earth's interior still remains hot enough to keep iron molten in the outer core.

THINK ABOUT IT Give an example from everyday life of a small object cooling faster than a larger one and of a small object warming up more quickly than a larger one. How do these examples relate to the issue of geological activity on Earth and the Moon?

The second general factor in the cooling rate is ongoing heat deposition: If a world has a source of ongoing internal heat, it will tend to cool more slowly with time. For the terrestrial worlds, the only significant source of ongoing heat is radioactive decay. Because many radioactive materials have long half-lives, they can continue to add heat to the interior for billions of years. Over time, radioactive decay has contributed several times as much heat to Earth's interior as accretion and differentiation. As we'll discuss in Chapter 9, some moons of jovian planets have other sources of ongoing heat deposition (such as *tidal heating*) that can keep them much hotter than we would otherwise expect given their relatively small sizes.

• How does plate tectonics shape Earth's surface?

Our discussion of internal heat explains most of the differences in geological activity that we see between Earth and the other terrestrial worlds. The Moon and Mercury have essentially no active, internally driven geology, because their small sizes allowed them to cool long ago. Mars is large enough to have had significant volcanism in the past, but its interior must now be much cooler than it was in its heyday. Venus is only slightly smaller than Earth, so it probably retains nearly as much internal heat; confirmation of this idea comes from the fact that Venus has few impact craters, indicating that lava flows or other processes that recycle surface material have erased craters from the more distant past. However,

Earth appears to be geologically distinct from Venus and all the other terrestrial worlds in one crucial way: Earth is the only planet with ongoing plate tectonics.

The word *tectonics* comes from the Greek word for "builder," *tekton;* you'll notice the same root in the word *architect,* which means "master builder." In geology, tectonics refers to "building" performed on planetary surfaces—that is, any surface reshaping that results from stretching, compression, or other forces acting on the lithosphere. Tectonic processes have operated to some extent on all the terrestrial worlds, but it is most important on Earth, where it operates by the distinctive mechanism of plate tectonics.

The Meaning of Plate Tectonics The term *plate tectonics* refers to the scientific theory that explains much of Earth's surface geology as a result of the slow motion of *plates*—fractured pieces of the lithosphere—driven by the underlying convection of the mantle. According to the theory, the lithosphere fractured because of stresses generated by mantle convection, and the resulting plates now essentially "float" over the mantle, gradually moving over, under, and around each other as convection moves Earth's interior rock. Because the term *plate tectonics* refers to the theory, it is generally considered to be singular rather than plural, despite the *-s* on *tectonics.*

Earth's lithosphere is broken into about a dozen plates, with known boundaries shown in Figure 4.15. Except during earthquakes, the motions of the plates are barely noticeable on human time scales—a few centimeters per year, which is about the rate at which your fingernails grow. However, geologists can now measure plate motions by comparing readings taken with the global positioning system (GPS) on either side of plate boundaries.

Evidence for Plate Tectonics The GPS measurements offer the most direct evidence of plate motion. However, the overall theory of plate tectonics rests on three additional significant lines of evidence: evidence of past continental motion, evidence that plates spread apart on seafloors, and a difference between the nature of Earth's crust on the seafloors and the continents.

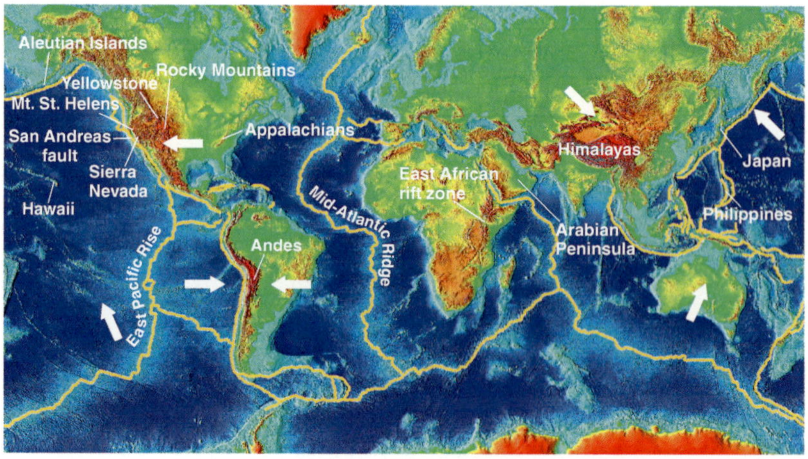

Figure 4.15

This relief map shows plate boundaries (solid yellow lines), with arrows to represent directions of plate motion. Color represents elevation, progressing from blue (lowest) to red (highest).

We usually trace the origin of the theory of plate tectonics to a slightly different idea proposed in 1912 by German meteorologist and geologist Alfred Wegener: *continental drift,* the idea that continents gradually drift across the surface of Earth. Wegener got his idea in part from the puzzlelike fit of continents such as South America and Africa (Figure 4.16). This fit had been noted by earlier mapmakers, but Wegener took the idea further. He noted that similar types of distinctive rocks and rare fossils were found in eastern South America and western Africa, suggesting that these two regions once had been close together.

Figure 4.16

The puzzle-like fit of South America and Africa.

Despite these strong hints, no one at Wegener's time knew of a mechanism that could allow the continents to push their way through the solid rock beneath and around them. Wegener suggested that Earth's gravity and tidal forces from the Sun and Moon were responsible, but other scientists quickly showed that these forces were too weak to move continents around. As a result, Wegener's idea of continental drift was rejected by all but a few geologists for decades after he proposed it, even though his evidence of a "continental fit" for Africa and South America ultimately proved correct. Today, far more extensive fossil evidence makes it clear that the continents really were arranged differently in the past, and our understanding of how rock can flow in the mantle allows us to understand the real reasons that the continents move around.

Scientists began to recognize the mechanism of continental motion through the discovery of *mid-ocean ridges* in the mid-1950s. Mantle material erupts onto the ocean floor along these ridges, such as the mid-Atlantic Ridge shown in Figure 4.15, while the existing seafloor spreads outward to either side. This **seafloor spreading** helped explain how the continents could move apart with time. Because this idea is quite different from Wegener's original notion of continents plowing through the solid rock beneath them, geologists no longer use the term *continental drift* and instead consider continental motion within the context of plate tectonics.

The third line of evidence for plate tectonics (in addition to continental motion and seafloor spreading) comes from the fact that Earth's surface has two very different types of crust (Figure 4.17). **Seafloor crust** is made primarily of the relatively high-density igneous rock called *basalt,* which commonly erupts from volcanoes like those along mid-ocean ridges and in Hawaii. Seafloor crust is typically only 5–10 kilometers thick, and radiometric dating shows that it is quite young—the average age is about 70 million years, and even the oldest seafloor crust is less than about 200 million years old. **Continental crust** is made of lower-density rock, such as granite, and its rock spans a wide range of ages from the very young all the way back to the oldest rocks found on Earth. Continental crust is much thicker than seafloor crust—typically 20–70 kilometers thick—but it sticks up only slightly higher because its weight presses it down farther onto the mantle below.

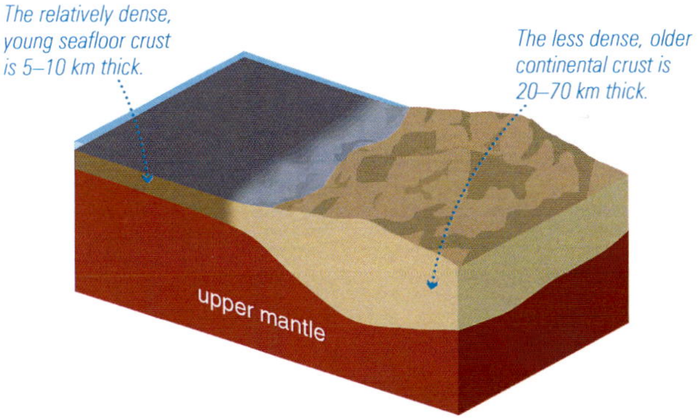

Figure 4.17

Earth today has two distinct kinds of crust.

The two types of crust make it clear that Earth's surface must undergo continual change. New seafloor crust continually emerges at sites of seafloor spreading, while the wide age range of continental crust tells us that the continents have gradually built up over time.

The Mechanism of Plate Tectonics

Today, we understand that plates move in concert with underlying mantle convection, driven by the heat released from Earth's interior. Over millions of years, the movements involved in plate tectonics act like a giant conveyor belt for Earth's lithosphere

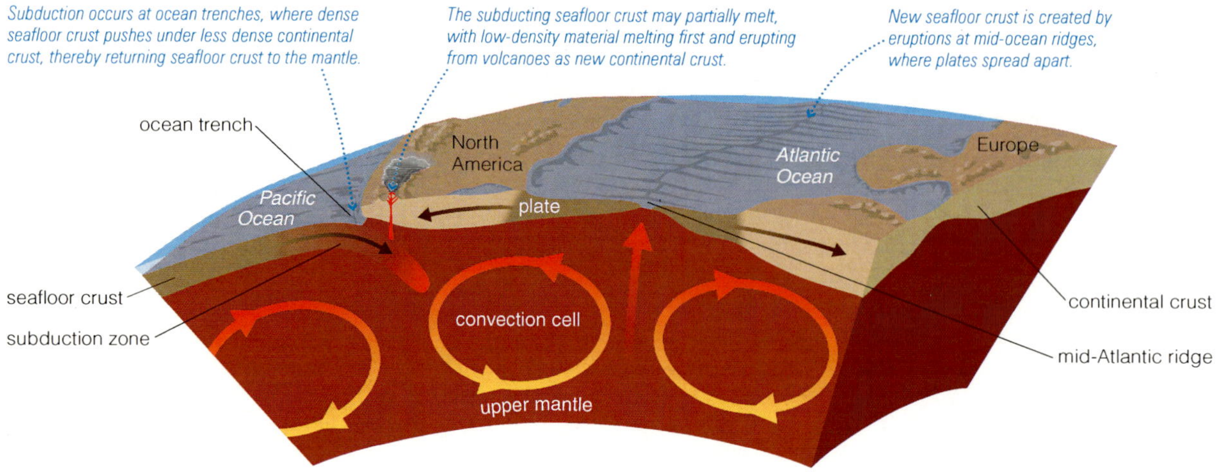

Figure 4.18 Interactive Figure

Plate tectonics acts like a giant conveyor belt for Earth's lithosphere.

(Figure 4.18). This movement explains many of the major features of Earth's geology.

Seafloor spreading occurs at mid-ocean ridges because they are places where mantle material rises upward toward the surface. As it gets close to the surface, the lower pressure allows it to partly melt; the molten material then erupts to the surface, cooling and contracting as it spreads sideways. This explains both the formation of the basaltic seafloor crust and the characteristic shapes of the ridges. Worldwide along the mid-ocean ridges, new crust covers an area of about 2 square kilometers every year, enough to replace the entire seafloor within about 200 million years—and thus explaining the geologically young age of seafloor rocks.

Over tens of millions of years, any piece of seafloor crust gradually makes its way across the ocean bottom, then finally gets recycled into the mantle in the process we call **subduction**. Subduction occurs where a seafloor plate meets a continental plate, which is generally somewhat offshore at the edge of a sloping *continental shelf*. The continental rocks are less dense than those on the seafloor; thus, as the dense seafloor crust of one plate pushes under the less dense continental crust of another plate, it can pull the entire surface downward to form a deep *ocean trench*. At some trenches, the ocean depth is more than 8 kilometers.

Beneath a subduction zone, the descending seafloor crust heats up and may begin to melt as it moves deeper into the mantle. If enough melting occurs, the molten rock may erupt upward. As you can see in Figure 4.18, the process of subduction tends to make the melting occur under the edges of the continents, which is why so many active volcanoes tend to be found along those edges. This volcanic activity explains the presence of many coastal mountain ranges (as well as many older mountain ranges that are no longer located along coasts). Moreover, the lowest-density material tends to melt first, which is why the continental crust emerging from these landlocked volcanoes is lower in density than seafloor crust.

Spreading and subduction are not the only ways in which plates interact. Two continental plates crashing into each other can push each other upward, providing a second way of creating mountain ranges. The Himalayas are still slowly growing in this way as the plate carrying India pushes into the plate carrying most of the rest of Eurasia (Figure 4.19). In places where continental plates are pulling apart, the crust thins and

Figure 4.19

This satellite photo shows the Himalayas, which are still slowly growing as the plate carrying India pushes into the Eurasian plate. Arrows indicate the directions of plate motion. (See the plate boundaries in Figure 4.15.)

can create a large *rift valley.* The East African rift zone is an example (see Figure 4.15). This rift is slowly growing and will eventually tear the African continent apart. At that point, rock rising upward with mantle convection will begin to erupt from the valley floor, creating a new zone of seafloor spreading. A similar process tore the Arabian Peninsula from Africa, creating the Red Sea (Figure 4.20).

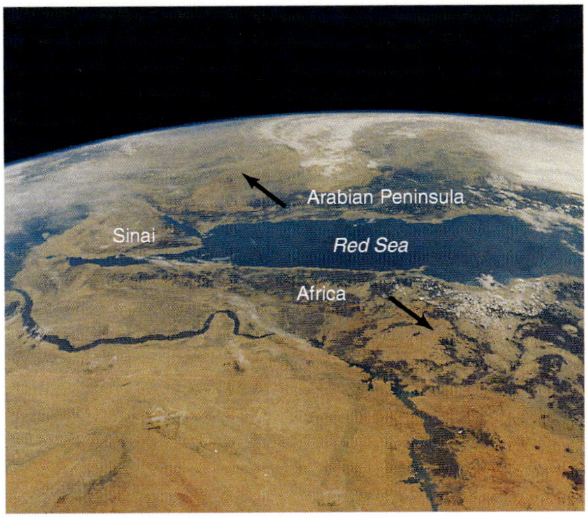

Figure 4.20

When continental plates pull apart, the crust thins and deep rift valleys form. This process tore the Arabian Peninsula from Africa, forming the Red Sea. Arrows indicate the directions of plate motion.

Places where plates slip sideways relative to each other are marked by what we call plate boundary **faults**. For example, the San Andreas Fault in California marks a line where the Pacific plate is moving northward relative to the continental plate of North America (Figure 4.21). At its current rate, this motion will bring Los Angeles (on the Pacific plate) and San Francisco Bay (on the North American plate) together in about 20 million years. The two plates do not slip smoothly against each other. Instead, their rough surfaces catch against each other. Tension builds up until it is so great that it forces a rapid and violent shift, causing an earthquake. In contrast to the usual motion of plates, which proceeds at a few centimeters per year, an earthquake can move plates by several *meters* in a few seconds. The movement can raise mountains, level cities, set off destructive tsunamis, and make the whole planet vibrate with seismic waves (much like a ringing bell).

THINK ABOUT IT By studying the plate boundaries in Figure 4.15, explain why the west coast states of California, Oregon, and Washington are prone to more earthquakes and volcanoes than most other parts of the United States. Find the locations of recent earthquakes and volcanic eruptions worldwide. Do the locations fit the pattern you expect?

Not all earthquakes and volcanoes occur near plate boundaries. Earthquakes sometimes occur along old or buried faults that are now far from plate boundaries; some of the biggest earthquakes in U.S. history occurred in 1811 and 1812 along the New Madrid fault zone, which runs through parts of Illinois, Kentucky, Missouri, Arkansas, and Tennessee. Volcanic activity may occur any place where a plume of hot mantle material rises up to make what we call a **hot spot**. The Hawaiian Islands are the result of a hot spot that has been erupting basaltic lava for tens of millions of years. Plate tectonics gradually carries the Pacific plate over the hot spot, forming a chain of volcanic islands as different parts of the plate lie directly above the hot spot at different times (Figure 4.22). Today, most of the lava erupts on or near the Big Island of Hawaii, giving much of this island a young, rocky surface. About a million years ago, the Pacific plate lay farther to the southeast (relative to its current location), and the hot spot built the island of Maui. Before that, the hot spot created other islands, including Oahu (3 million years ago), Kauai (5 million years ago), and Midway (27 million years ago). The older islands are more heavily eroded. Midway has been eroded so much that it barely rises above sea level. The movement of the plate over the hot spot continues today, building underwater volcanoes that eventually will rise above sea level to become new Hawaiian Islands. The growth of a future island, named Loihi, is already well under way—prime beach real estate should be available there in about a million years or so. Hot spots can also occur beneath continental crust. For example, a continental hot spot is responsible for the volcanism that supplies the heat for the geysers and hot springs of Yellowstone National Park.

Figure 4.21

California's San Andreas Fault marks a boundary where plates are sliding sideways as shown by the white arrows; asterisks indicate sites and years of major earthquakes. The inset photo shows a place along the San Andreas Fault where the painted lines in a road allow us to see how far the two sides of the fault moved in an earthquake.

Figure 4.22

The Hawaiian Islands are just the most recent of a long string of volcanic islands made by a mantle hot spot. The black-and-white image of Loihi (lower right) was obtained by sonar, as it is still entirely underwater. The long chain records the past 60 million years of history of the oceanic crust in the region.

Plate Tectonics Through Time We can use the current motions of the plates to project the arrangement of continents millions of years into the past or the future. For example, at a speed of 2 centimeters per year, a plate will travel 2,000 kilometers in 100 million years. Figure 4.23 shows several past arrangements of the continents, along with one future arrangement. Note that the present-day continents were once all stuck together in a single "supercontinent," sometimes called *Pangaea* (meaning "all lands"), which began to break up about 225 million years ago.

Mapping the sizes and locations of continents at even earlier times is more difficult. However, studies of magnetized rocks (which can record the orientation of ancient magnetic fields) and comparisons of fossils found in different places around the world have allowed geologists to map the movement of the continents much further into the past. It seems that, over at least the past billion years or more, the continents have slammed together, pulled apart, spun around, and changed places on the globe. Central Africa once lay at Earth's South Pole, and Antarctica once was near the equator. The continents continue to move, and their current arrangement is no more permanent than any past arrangement.

Has Earth had plate tectonics throughout its history? We are not yet certain, but we have some evidence for plate tectonics going back quite far

Figure 4.23 Interactive Figure

Selected past, present, and future arrangements of Earth's continents.

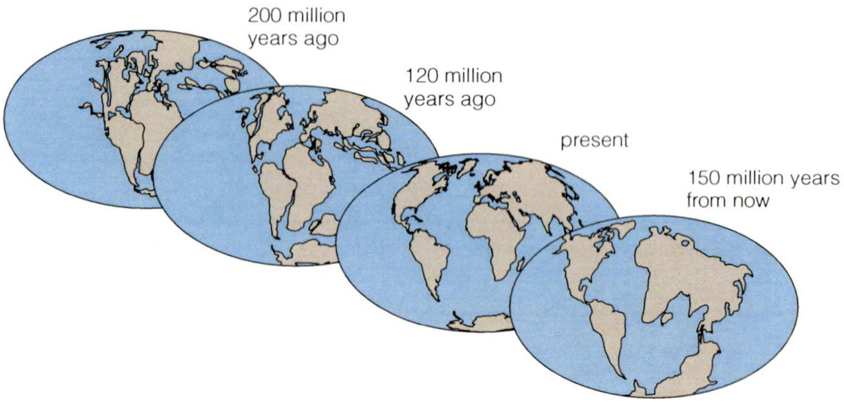

in time. The oldest fairly definitive evidence comes from seismic studies that suggest the presence of an ancient subduction zone (found in Canada) that formed some 2.7 billion years ago; subduction would be a sure sign of the conveyorlike action of plate tectonics. More controversial is the recent zircon evidence suggesting that continental crust had already begun to form nearly 4.5 billion years ago. If the evidence is being properly interpreted, it suggests that plate tectonics began just shortly after the birth of Earth, since continental crust is generally formed as a direct result of the separation of rock by density along subduction zones.

Cause and Effects of Plate Tectonics We have a fairly good understanding of how plate tectonics works, and how it is driven by Earth's internal heat and mantle convection. However, we still face at least one significant mystery: Why does plate tectonics operate only on Earth among the terrestrial worlds?

The answer is probably simple for the Moon, Mercury, and Mars: Because their small size has allowed their interiors to cool much more than Earth's interior, their lithospheres have thickened. If they have any remaining internal convection at all, it is too weak to break their thick lithospheres into plates. Venus poses a greater mystery, since it is almost the same size as Earth and therefore should have retained a similar amount of internal heat.

We still do not know why Venus lacks plate tectonics today, or whether it had plate tectonics in the past [Section 10.2]. However, we have at least one plausible hypothesis: As we'll discuss further in Chapter 10, Venus's high surface temperature has probably baked out water from its crust and upper mantle. This drying of the rock may have strengthened and thickened Venus's lithosphere so that it has resisted the fracturing that occurred on Earth; the high temperature may also make Venus's lithosphere less brittle than Earth's colder crust. If this hypothesis is correct, then we can ultimately trace the cause of plate tectonics to two factors: heat-driven mantle convection and a lithosphere thin and brittle enough to be fractured by the movement of the underlying mantle.

Whatever its cause, the effects of plate tectonics are profound. We've seen that plate tectonics is the most important geological mechanism on Earth, and it plays a key role in explaining nearly all of Earth's geological features. But its deeper significance to life lies in the fact that its recycling of rock turns out to play a crucial role in climate regulation, a topic we'll address in Section 4.5.

• Why does Earth have a protective magnetic field?

Planetary atmospheres do not necessarily last forever. All the terrestrial worlds had at least some gas released from their interiors by outgassing, but some of the gas eventually escaped to space. Atmospheric gas can be lost to space in at least three ways.

First, gas molecules move fast enough that they exceed their world's escape velocity and can simply "take off" into space. This process is often called **thermal escape**, because gas particles tend to move at higher speeds when their temperature is higher, and hence are more likely to escape when temperatures are high than low. Moreover, at any particular temperature, lightweight molecules such as hydrogen tend to move faster than heavier molecules such as oxygen; thus, light gases escape more easily. That is why none of the terrestrial worlds has any significant amounts of hydrogen gas; the large, jovian planets retain hydrogen because their large masses give them much higher escape velocities. More generally, smaller worlds have lower escape velocities, so gas escapes from them much more readily. That is why the Moon and Mercury, the two smallest of the terrestrial worlds, have become essentially airless.

Second, impacts can also blast atmospheric gas into space. Again, smaller worlds are more prone to this type of loss because of their lower escape velocities: Equivalent impacts are more likely to blast material upward with escape velocity on a smaller world than a larger world.

Third, gas can be lost through a mechanism known as **solar wind stripping**, which occurs when particles from the solar wind in effect sweep atmospheric gas particles into space. This mechanism acts slowly, but calculations suggest that without the protection of Earth's global magnetic field, over billions of years the solar wind would have swept away much of our planet's atmosphere.

Requirements for a Global Magnetic Field A **magnetic field** can affect charged particles or magnetized objects in its vicinity. For example, if you've ever used a compass, you know that Earth has a magnetic field that determines the direction in which the compass needle points. The global extent of Earth's magnetic field gives us a strong clue that the field is generated *inside* our planet.

You are probably familiar with the general pattern of the magnetic field created by an iron bar magnet (Figure 4.24a). Earth's magnetic field is generated by a process more similar to that of an *electromagnet,* in which the magnetic field arises as a battery forces charged particles (electrons) to move along a coiled wire (Figure 4.24b). Earth does not contain a battery, of course, but charged particles move with the molten metals in its liquid outer core (Figure 4.24c). Internal heat causes the liquid metals to rise and fall (convection), while Earth's rotation twists and distorts the convection pattern of these molten metals. The result is that electrons in the molten metals move within Earth's outer core in much the same way that they move in an electromagnet, generating Earth's magnetic field.

We can generalize what we know about Earth's magnetic field to other worlds. There are three basic requirements for a global magnetic field:

Figure 4.24

Sources of magnetic fields.

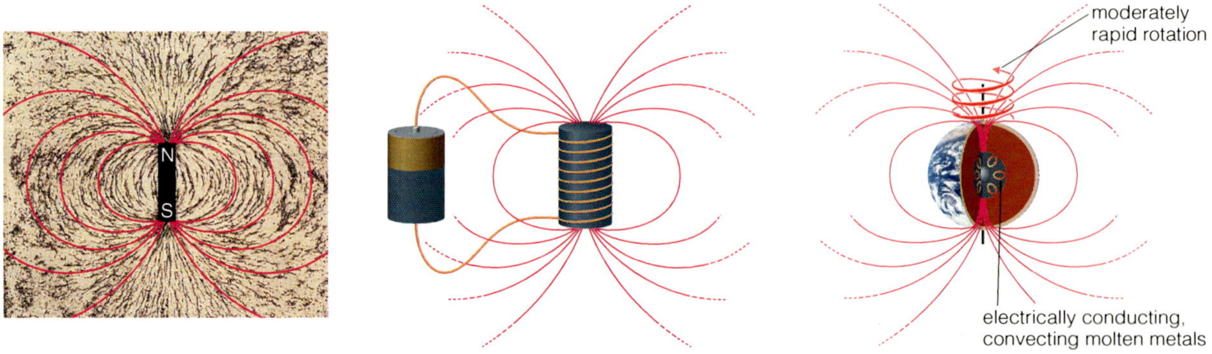

a This photo shows how a bar magnet influences iron filings (small black specks) around it. The *magnetic field lines* (red) represent this influence graphically.

b A similar magnetic field is created by an electromagnet, which is essentially a wire wrapped around a bar and attached to a battery. The field is created by the battery-forced motion of charged particles (electrons) along the wire.

c Earth's magnetic field also arises from the motion of charged particles. The charged particles move within Earth's liquid outer core, which is made of electrically conducting, convecting molten metals.

1. An interior region of electrically conducting fluid (liquid or gas), such as molten metal.
2. Convection in that layer of fluid.
3. At least moderately rapid rotation of the planet.

Earth is unique among the terrestrial worlds in meeting all three requirements, which is why it is the only terrestrial world in our solar system with a strong magnetic field. The Moon has no magnetic field, either because it lacks a metallic core altogether or because its core has long since solidified and ceased convecting. Mars likewise has virtually no magnetic field today, probably because of similar core solidification. Venus probably has a molten metal layer much like that of Earth, but either its convection or its 243-day rotation period is too slow to generate a magnetic field. Mercury poses a slight enigma: It possesses a measurable magnetic field despite its small size and slow, 59-day rotation. The reason may be tied to the fact that Mercury has a very large metal core, which may still be partly molten and convecting.

The Magnetosphere and the Solar Wind The magnetic field protects Earth's surface and atmosphere from most of the energetic particles of the solar wind because it creates a **magnetosphere** that acts like a protective bubble surrounding our planet. The magnetosphere deflects most of the solar wind particles while channeling a few toward the poles, where they can cause auroras (Figure 4.25). The magnetosphere itself is invisible, but we can map its presence with devices that work much like compass needles and we can detect particles that become trapped within it (in the charged particle belts, also known as *Van Allen belts*).

The magnetosphere generally deflects particles while they are still high above our atmosphere, and it therefore prevents the solar wind from stripping Earth's atmospheric gas away. Indeed, evidence for this protective function comes from studying our neighboring planets. As we'll discuss in Chapter 8, Mars today apparently has much less atmospheric gas than it did in the distant past, and we suspect that Mars lost much of its gas when its interior cooled to the point that it no longer generated a strong

Figure 4.25 Interactive Figure

Earth's magnetosphere acts like a protective bubble, shielding our planet from the charged particles of the solar wind.

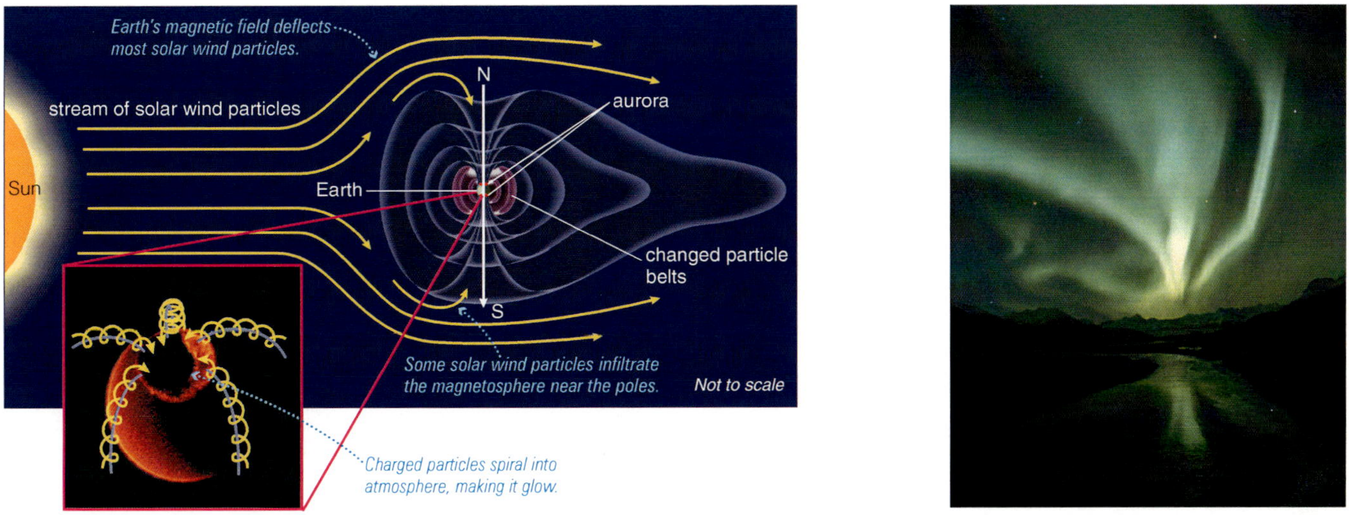

a This diagram shows how Earth's invisible magnetosphere (represented in purple) deflects solar wind particles. Some particles accumulate in charged particle belts encircling our planet. The inset is a photo of a ring of auroras around the North Pole; the bright crescent at its left is part of the day side of Earth.

b This photograph shows an aurora visible over the coast of Norway. In a video, you would see these lights dancing about in the sky.

magnetic field and protective magnetosphere. (Mars probably also lost gas due to impacts.) Careful studies of the isotopic composition of Venus's atmosphere suggest that it, too, has lost gas to solar wind stripping, as we would expect given its lack of magnetic field. However, Venus has such a thick atmosphere that its overall gas loss has been proportionally small.

4.5 CLIMATE REGULATION AND CHANGE

At the beginning of this chapter, we stated that three crucial ingredients in Earth's long-term habitability have been volcanism, plate tectonics, and the magnetic field. We have seen that volcanism's most important role has been in releasing the gases that formed our oceans and atmosphere, while the magnetic field has helped prevent atmospheric gas from being lost to space. For plate tectonics, we have seen the way it is responsible for shaping our planet's surface; in fact, plate tectonics plays an even more important role, because it helps regulate the climate.

You are probably aware that Earth's climate has not been perfectly stable through time: Even during human history our planet has experienced what we call *ice ages,* and as we look back through geological time, we find other periods of far more severe cold or warmth. Nevertheless, the climate has been sufficiently stable for life to exist continually for nearly 4 billion years. Because life on Earth needs liquid water [Section 5.3], we infer that the oceans have remained at least partially liquid throughout this long period of time. Although the temperature range in which water can be liquid may seem wide to us humans, when we compare it to temperatures found on other worlds, we realize that Earth's climate has been remarkably stable through time.

At first, Earth's stable climate might not seem surprising. After all, the primary source of heat for the atmosphere and oceans is the Sun, and Earth's orbit about the Sun should not have changed since its formation. However, theoretical models of the Sun and observations of other Sun-like stars reveal an important fact: Stars gradually brighten with age. Models suggest that the Sun today may be as much as 30% brighter than it was when Earth formed, which means the young Earth received a lot less solar warmth and light than it does today. How, then, could our planet have been warm enough for liquid water in the distant past, and why hasn't our planet overheated as the Sun brightened? To answer these and other questions about Earth's long-term climate, we must begin by investigating why Earth is warm enough for liquid water in the first place.

• How does the greenhouse effect make Earth habitable?

Most people assume that Earth is warm enough for liquid water simply because we lie at the "right" distance from the Sun, but this clearly is not the whole story. The Moon lies at the same distance, but its daytime temperatures rise to 125°C (257°F)—well above the normal boiling point of water—while its nighttime temperatures plummet to a frigid −175°C (−283°F). Moreover, the fact that the Sun has brightened with time means that even if we were at the "right" distance when Earth was young, that same distance would be too close to the now-brighter Sun.

In fact, it's fairly easy to calculate the temperature Earth would have, based solely on its distance from the Sun and the amount of incoming sunlight absorbed by its surface. Such a calculation shows that the **global average temperature**—that is, the average temperature for the entire planet—would be −1°C (+30°F), well below the freezing point of water.

But Earth is not frozen. The actual global average temperature today is about 15°C (59°F), and geological evidence shows that it has been warmer at various times in the past. Something must be making our planet much warmer than we would expect based on its distance from the Sun alone, and that something is what we call the **greenhouse effect**.

Figure 4.26 shows the basic mechanism of the greenhouse effect. Sunlight consists mostly of visible light, which passes easily through most atmospheric gases and reaches Earth's surface. Some of this visible light gets absorbed by the ground, while the rest is reflected back to space. The ground must return the energy it absorbs back to space, because if it didn't the energy would make the ground heat up rapidly. However, the fact that the ground doesn't glow in the dark tells us that the ground does not return the energy in the same visible-light form that it absorbs it. Instead, the ground returns the energy in the form of infrared light.

The greenhouse effect works by "trapping" some of the infrared light, thereby slowing its return to space. This trapping occurs because some atmospheric gases can absorb the infrared light. Gases that are particularly good at absorbing infrared light are called **greenhouse gases**. The most important greenhouse gases are water vapor (H_2O), carbon dioxide (CO_2), and methane (CH_4). These gases absorb infrared light effectively because their molecular structures make them prone to begin rotating or vibrating when struck by an infrared photon (an individual "piece" of light); diatomic molecules such as nitrogen (N_2) and oxygen (O_2) generally cannot rotate or vibrate in these ways and hence do not absorb infrared light.

After a greenhouse gas molecule absorbs the energy of an infrared photon, it quickly releases the energy by emitting a new infrared photon. However, the new photon will be emitted in some random direction that is unlikely to be the same direction from which the original photon came. The new photon is then usually absorbed by another greenhouse molecule, which does the same thing. The net result is that greenhouse gases tend to slow the escape of infrared radiation from the lower atmosphere, while their molecular motions heat the surrounding air. In this way, the greenhouse effect makes the surface and the lower atmosphere warmer than they would be from sunlight alone. A blanket offers a good analogy: You stay warmer under a blanket not because the blanket itself provides any heat, but rather because it slows the escape of your body heat into the cold outside air. The more greenhouse gases that are present, the greater the degree of surface warming. On Earth, the naturally occurring greenhouse effect is strong enough to raise the global average temperature by about 31°C from what it would be without greenhouse gases. Without this warming, our planet would be frozen over.

Incidentally, you are probably aware that the greenhouse effect is often in the news, usually portrayed in a negative light as part of an environmental problem. But as we have just seen, the greenhouse effect is not a bad thing in and of itself, since life on Earth would not exist without it. Why, then, is the greenhouse effect discussed as an environmental problem? The reason is that human activity is adding more greenhouse gases to the atmosphere, thereby strengthening the greenhouse effect and further warming our planet ("global warming"). While the precise effects of this human-induced warming are not known and are somewhat controversial [Section 10.5], we need only look to our hot neighbor Venus to see that changes to the greenhouse effect should not be taken lightly. While the greenhouse effect makes Earth livable, it is also responsible for the searing 470°C (878°F) temperature of Venus—proving that it's possible to have too much of a good thing.

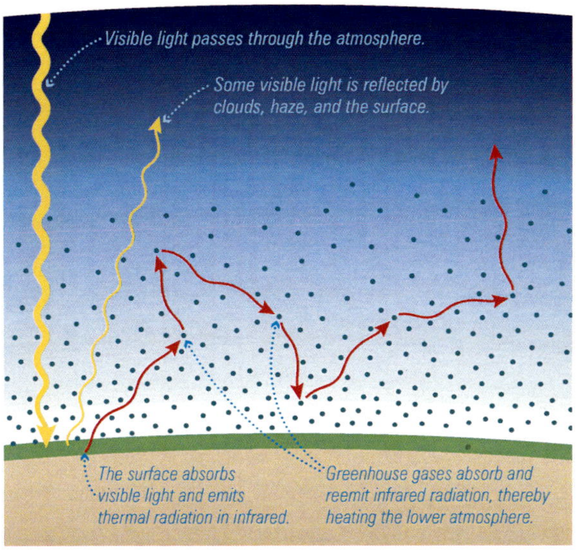

Figure 4.26 Interactive Figure

The greenhouse effect makes the surface and lower atmosphere much warmer than they would be without greenhouse gases such as water vapor, carbon dioxide, and methane.

THINK ABOUT IT Carbon dioxide makes up less than 1% of our atmosphere today, while nitrogen and oxygen together make up some 98% of our atmosphere. Why, then, do we focus on carbon dioxide when we talk about Earth's climate?

• What regulates Earth's climate?

The case of Venus leads us to a crucial question about Earth's hospitable climate. As we'll discuss in Chapter 10, Venus's extreme greenhouse effect occurs because its atmosphere contains almost 200,000 times as much carbon dioxide as Earth's atmosphere. But Venus and Earth are nearly the same size and both were made from similar materials, so volcanic outgassing should have released similar amounts of carbon dioxide on both worlds. Moreover, outgassing from modern-day volcanoes on Earth shows that they do indeed release plenty of carbon dioxide, and over time, volcanoes must have outgassed nearly as much carbon dioxide into Earth's atmosphere as we find in the atmosphere of Venus. Where, then, did all of Earth's carbon dioxide end up?

Geological studies reveal the answer: Most of Earth's carbon dioxide is locked up in **carbonate rocks**—sedimentary rocks such as limestone that are rich in carbon and oxygen. By estimating the total amount of carbonate rock on Earth, we find that these rocks contain about 170,000 times as much carbon dioxide as our atmosphere, which means that Earth does indeed have almost as much total carbon dioxide as Venus. Venus lacks carbonate rock (for reasons we'll discuss in Chapter 10), so all of its carbon dioxide remains in its atmosphere. Keep in mind that this difference between the two planets in carbon dioxide location makes all the difference in the world: If Earth's carbon dioxide were in our atmosphere rather than in carbonate rocks, our planet would be nearly as hot as Venus and certainly uninhabitable.

Of course, the fact that Earth's carbon dioxide is locked up in rocks leads to a deeper question: How did it get there? The answer lies with a mechanism closely tied to plate tectonics.

The Carbon Dioxide Cycle

The mechanism by which carbon dioxide has been removed from Earth's atmosphere and by which the current small amount of atmospheric carbon dioxide remains stable is called the inorganic **carbon dioxide cycle**, or the **CO_2 cycle** for short. Let's follow the cycle as illustrated in Figure 4.27, starting with the carbon dioxide in the atmosphere:

- Atmospheric carbon dioxide dissolves in rainwater, creating a mild acid.
- The mildly acidic rainfall erodes rocks on Earth's continents, and rivers carry the broken-down minerals to the oceans.
- In the oceans, calcium from the broken-down minerals combines with dissolved carbon dioxide and falls to the ocean floor, making carbonate minerals.* These can accumulate at the bottom of the oceans to build large quantities of rock such as limestone over time.

*During the past half-billion years or so, the carbonate minerals have been made by shell-forming sea animals, falling to the bottom in the seashells left after the animals die. Without the presence of animals, chemical reactions would do the same thing—and apparently did for most of Earth's history.

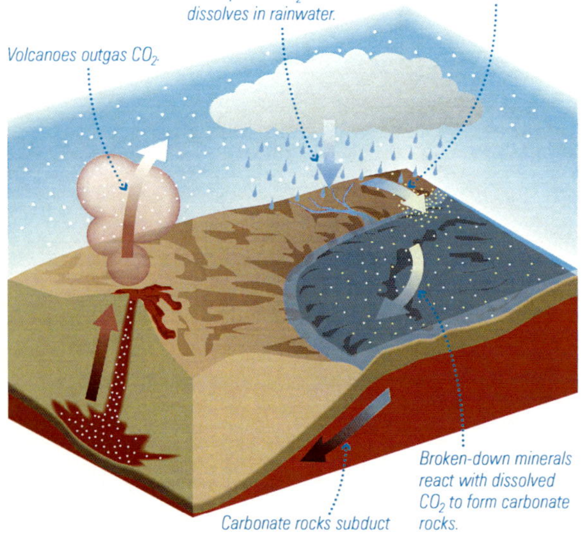

Figure 4.27

This diagram shows how the CO_2 cycle continually moves carbon dioxide from the atmosphere to the ocean to rock and back to the atmosphere. Note that plate tectonics (subduction in particular) plays a crucial role in the cycle.

- Over millions of years, the conveyor belt of plate tectonics carries the carbonate rocks to subduction zones, and subduction carries them down into the mantle.
- As they are pushed deeper into the mantle, some of the subducted carbonate rock heats up and releases its carbon dioxide, which then outgasses back into the atmosphere through volcanoes.

In summary, the reason that Earth has so little carbon dioxide in its atmosphere is that most of the carbon dioxide was dissolved in the oceans, where chemical reactions converted it to carbonate minerals. In fact, about 60 times as much carbon dioxide is dissolved in the oceans as is present in our atmosphere, though this amount still pales in comparison to the 170,000 times as much that the CO_2 cycle has locked up in rock.

The CO_2 Cycle as a Thermostat The CO_2 cycle acts as a thermostat for Earth because of the way that changes in temperature feed back into the cycle. You are probably familiar with what we generally call **feedback processes**—processes in which a change in one property amplifies (positive feedback) or counteracts (negative feedback) the behavior of the rest of the system. For example, if someone brings a microphone too close to a loudspeaker, it picks up and amplifies small sounds from the speaker. These amplified sounds are again picked up by the microphone and further amplified, causing a loud screech. This sound feedback is an example of *positive feedback,* because it automatically amplifies itself. The screech usually leads to a form of *negative feedback:* The embarrassed person holding the microphone moves away from the loudspeaker, thereby stopping the positive sound feedback.

The CO_2 cycle has a built-in form of negative feedback that returns Earth's temperature toward "normal" whenever it warms up or cools down. This negative feedback occurs because the overall rate at which carbon dioxide is pulled from the atmosphere is extremely sensitive to temperature: the higher the temperature, the higher the rate at which carbon dioxide is removed. Figure 4.28 shows how it works. Consider first what happens if Earth warms up a bit. The warmer temperature means more evaporation and rainfall, pulling more CO_2 out of the atmosphere. The reduced atmospheric CO_2 concentration leads to a weakened greenhouse effect that counteracts the initial warming and cools the planet

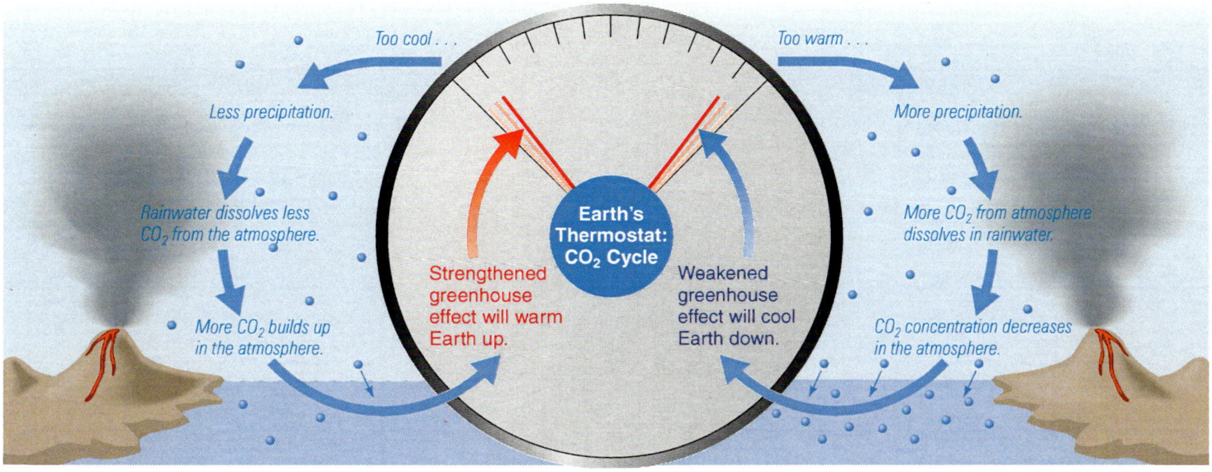

Figure 4.28

The carbon dioxide cycle acts as a thermostat for Earth through negative feedback processes. Cool temperatures cause atmospheric CO_2 to increase, and warm temperatures cause atmospheric CO_2 to decline.

back down. Similarly, if Earth cools a bit, precipitation decreases and less CO_2 is dissolved in rainwater, allowing the CO_2 released by volcanism to build back up in the atmosphere. The increased CO_2 concentration strengthens the greenhouse effect and warms the planet back up.

Overall, the natural thermostat of the CO_2 cycle has allowed the greenhouse effect to strengthen or weaken just enough to keep Earth's climate in a range that has allowed for liquid water, regardless of what other changes have occurred on our planet. And, because subduction plays a critical role in the CO_2 cycle, we now see the importance of plate tectonics to Earth's climate: Without plate tectonics, there would be no CO_2 cycle to regulate our planet's surface temperature.

• How does Earth's climate change over long periods of time?

While Earth's climate has remained stable enough for the oceans to remain at least partly liquid throughout history, the climate has not been perfectly steady. Numerous warmer periods and numerous ice ages have occurred. Such variations are possible because the CO_2 cycle does not act instantly. When something begins to change the climate, it takes time for the feedback mechanisms of the CO_2 cycle to come into play, because these mechanisms depend on the gradual action of mineral formation in the oceans and of plate tectonics. Calculations show that the time to stabilize atmospheric CO_2 through the CO_2 cycle is about 400,000 years. That is, if the amount of CO_2 in the atmosphere were to rise due to, say, increased volcanism, it would take some 400,000 years for the CO_2 cycle to restore temperatures to their current values.*

Ice Ages **Ice ages** occur when the global average temperature drops by a few degrees. The slightly lower temperatures lead to increased snowfall, which may cover continents with ice down to fairly low latitudes. For example, the northern United States was entirely covered with glaciers during the peak of the most recent ice age, which ended only about 10,000 years ago. In fact, relative to temperatures over at least the past 200 million years, we are still in an ice age. This ice age has persisted for the past 35 million years or so, with periods of deeper cold interspersed with periods of warmer temperatures, such as the present. Remarkably, recent evidence indicates that we can enter or leave a cold period very rapidly, within as short a time as a few decades.

The causes of ice ages are complex and not fully understood. Over periods of tens or hundreds of millions of years, the Sun's gradual brightening and the changing arrangement of the continents around the globe have at least in part influenced the climate. During the past few million years—a period too short for solar changes or continental motion to have a significant effect—the ice ages appear to have been strongly influenced by small, cyclical changes in Earth's rotation and orbit. These cyclical changes are often called *Milankovitch cycles,* after the Serbian scientist who suggested their role in climate change.

For example, while Earth's current axis tilt is about $23\frac{1}{2}°$, the tilt varies over time between about 22° and 25° (Figure 4.29). These small changes affect the climate by making seasons more or less extreme. Greater tilt

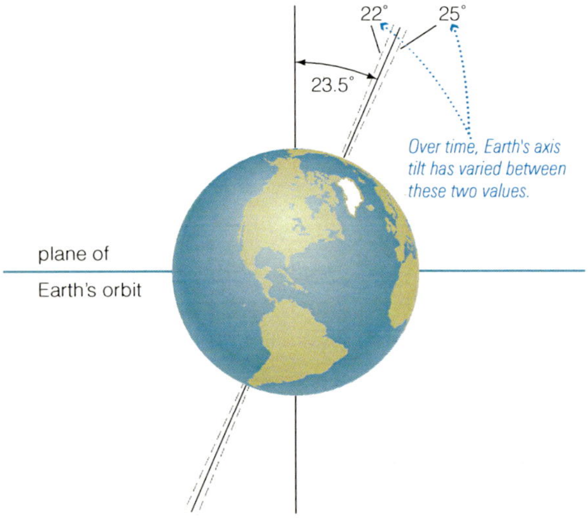

Figure 4.29

Small changes in Earth's axis tilt affect the climate: Greater tilt tends to mean a warmer climate.

*This time scale applies to ocean/atmosphere equilibrium only. The time scale for crust recycling is much longer, while shorter-term climate variations in atmospheric CO_2 concentration can occur through factors besides the inorganic CO_2 cycle, such as cycling of carbon dioxide by life or the addition of CO_2 to the atmosphere through human activity.

means more extreme seasons, with warmer summers and colder winters. The extra summer warmth tends to prevent ice from building up, making the whole planet warmer on average. In contrast, smaller tilt means less extreme seasons that tend to keep polar regions colder and darker on average, allowing ice to build up. Earth's past periods of smaller axis tilt correlate well with colder climate and ice ages, especially when considered along with other cyclical changes in Earth's rotation and orbit.

Snowball Earth Geologists have recently discovered evidence of several particularly long and deep ice ages between about 750 and 580 million years ago, and another similar set between about 2.4 and 2.2 billion years ago. During these periods, glaciers appear to have advanced all the way to the equator. If so, the global temperature must have dropped low enough for the oceans to begin to freeze worldwide. This ocean freezing would have set up a positive feedback process that would have cooled Earth even further, because ice reflects much more sunlight than water. (About 90% of incoming sunlight is reflected by ice, compared to about 5% by water.) With less sunlight being absorbed, the surface would cool, which in turn would allow the oceans to freeze further. Geologists suspect that in this way, our planet may have entered the periods we now call **Snowball Earth**. We do not know why these episodes occurred or precisely how extreme the cold became. However, some models suggest that at the peak of a Snowball Earth period, the global average temperature may have been as low as −50°C (−58°F) and the oceans may have been frozen to a depth of 1 kilometer or more.

How did Earth recover from a snowball phase? Figure 4.30 shows the current model. Even if Earth's surface got cold enough for the ocean surface to freeze completely, the interior still remained hot. As a result, volcanism would have continued to add CO_2 to the atmosphere. Oceans covered by ice would have been unable to absorb this CO_2 gas, and the CO_2 content of the atmosphere would have gradually built up and strengthened the greenhouse effect. As long as the ocean surfaces remained frozen, the CO_2 buildup would have continued. Models suggest that, over a period of 10 million years or so, the CO_2 content of our atmosphere may have increased 1,000-fold.

Eventually, the strengthening greenhouse effect would have warmed Earth enough to start melting the ocean surface ice. The feedback processes that started the Snowball Earth episode then moved quickly in reverse. As the ocean surface melted, more sunlight would have been

Figure 4.30

The CO_2 cycle rescues Earth from a snowball phase.

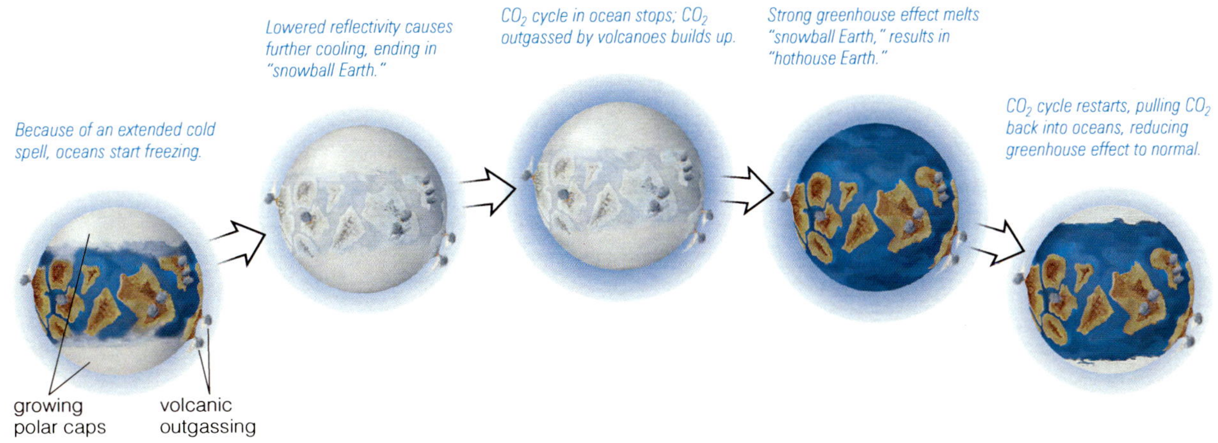

absorbed (again, because liquid water absorbs more and reflects less sunlight than ice), warming the planet further. Moreover, because the CO_2 concentration was so high, the warming would have continued well past current temperatures—perhaps taking the global average temperature to higher than 50°C (122°F). Thus, in just a few centuries, Earth would have emerged from a snowball phase into what we might call a hothouse phase. Geological evidence supports the occurrence of dramatic increases in temperature at the end of each Snowball Earth episode. Earth then slowly recovered from the hothouse phase as the CO_2 cycle removed carbon dioxide from the atmosphere.

THINK ABOUT IT Suppose Earth did not have plate tectonics. Could the planet ever recover from a snowball phase? Explain.

The Snowball Earth episodes must have had severe consequences for any life on Earth at the time. Indeed, the end of the Snowball Earth episodes roughly coincides with a dramatic increase in the diversity of life on Earth (the increase called the *Cambrian explosion* [Section 6.3]). Some scientists suspect that the environmental pressures caused by the Snowball Earth periods may have led to a burst of evolution. If so, we might not be here today if not for the dramatic climate changes of the Snowball Earth episodes.

Earth's Long-Term Habitability We have covered a lot of ground in this chapter, but it has given us a clear picture of the major factors that have kept our planet habitable for the past 4 billion or more years. Let's briefly review a few of the key points:

- Volcanic outgassing released most of the gases that made the atmosphere and the water vapor that condensed to form the oceans.
- Earth has kept its atmosphere at least in part because the magnetic field has protected atmospheric gases from being stripped away by the solar wind.
- The greenhouse effect warms our planet enough for water to be liquid, but not so much that the water would boil away.
- This moderate greenhouse effect is maintained by the self-regulating mechanism of the CO_2 cycle, which depends on plate tectonics.
- Even with the regulation provided by the CO_2 cycle, the climate still goes through changes influenced by variations in Earth's axis tilt and other properties of its rotation and orbit.
- The regulatory mechanism sometimes breaks down, leading to periods such as the Snowball Earth episodes, but the CO_2 cycle ultimately brings the climate back into balance.

These ideas should help us understand the prospects for finding other habitable worlds, especially those that might have the long-term habitability that could allow for the evolution of intelligent species and civilizations. Of course, these ideas also leave several questions unanswered. For example, should we expect to find plate tectonics and a CO_2 cycle on other worlds similar to Earth, or was some rare "luck" involved in Earth's getting these regulatory mechanisms? We might also wonder how long

Earth's climate can continue to regulate itself as the Sun brightens with time, a topic we'll discuss in Chapter 10. Finally, and perhaps of the most immediate importance, we might wonder whether we humans could alter the regulatory mechanisms enough to cause serious consequences for current life on Earth or for our own civilization. Unfortunately, the answer appears to be yes, but we'll save this discussion for Section 10.5.

4.6 THE PROCESS OF SCIENCE IN ACTION: FORMATION OF THE MOON

Earlier in this chapter, we stated without evidence the idea that the Moon formed when a "giant impact" blasted away much of the material in the young Earth's outer layers. If you think about it, this is a rather astonishing idea, since it postulates a single event at a time so far in the distant past that we have little hope of finding rocks that survived the event and could tell its tale. For this chapter's case study in the process of science in action, we'll explore how and why this remarkable idea has gained widespread acceptance in the scientific community.

• How did the giant impact model win out over competing models?

The existence of the Moon has long been puzzling. As we discussed in Chapter 3, the Moon counts as one of the "exceptions to the rules" when we consider the overall formation of the solar system, because it is far larger in comparison to its planet (Earth) than any other moon in the solar system, except for Pluto's moon Charon. So how did the Moon form?

Three Models, All Flawed During the mid-twentieth century, three competing models were advanced to explain the Moon's existence. The first held that the Moon formed along with Earth through the same process of accretion; in essence, this idea suggested that the two worlds were born together. The second model suggested that the Moon had been an independent "planet" orbiting the Sun that was somehow captured into Earth's orbit. The third model suggested that a young, molten Earth had been spinning so rapidly that it split into two pieces, casting off the piece that became the Moon.

All three models had difficulties right from the start. The joint formation model just didn't seem to work when scientists tried to calculate exactly what might have happened. If you try to build a planet and such a large moon in close proximity, gravitational interactions between them disrupt the process. Moreover, the Moon's average density is much lower than Earth's, which doesn't make sense if both worlds accreted from the same material as Earth. If the Moon had formed in this way, scientists were missing some key ingredient in their models.

The capture model seemed too improbable. It is difficult for a planet to gravitationally capture a passing object under any circumstances, because the passing object has its own orbital energy carrying it around the Sun. This energy cannot simply disappear, so an object can be captured only if it somehow loses some of its orbital energy. Captures therefore are most likely for small objects that lose orbital energy to friction with gas surrounding a planet. Mars probably captured its two small moons in this way, back at a time when it had a more extended atmosphere. Jupiter and the other jovian planets probably also captured many of their small moons this way, back when they were still surrounded by gas from the solar nebula. But Earth never had an atmosphere thick enough to have slowed an object the size of the Moon. The only way that Earth could have captured the Moon would have been if another, similarly sized object was passing by at precisely the same time as the Moon, and if the Moon and this other object had exchanged just enough energy for the Moon to end up in orbit of Earth. It's possible in principle, but highly unlikely.*

The splitting model also suffered from improbability. For example, it seemed unlikely that Earth could ever have been spinning fast enough to spin off the Moon. Still, like the other two models, it could not be ruled out completely at the time.

The *Apollo* missions to the Moon ended this debate, because study of rocks brought back by the astronauts ruled out all three models. The key finding was that the Moon rocks differed significantly in composition from that of Earth rocks. This immediately ruled out the joint formation model in which both worlds would have accreted from essentially identical material. The capture and splitting models were ruled out by chemical processing that had apparently occurred in the Moon rocks: The Moon rocks contained virtually no **volatile**, or easily vaporized, ingredients. In this context, volatile ingredients include not only things such as water but also elements such as lead and gold that vaporize at lower temperatures than other metals and rocks. Because these volatile elements should have been mixed in with other elements in any accreting object, the Moon could not have accreted first and been captured later. And because these volatiles are present in Earth, the Moon should also have had them if it split off from a spinning, molten planet.

The Giant Impact Model With all three models fatally flawed, it was back to the drawing board, which meant taking a closer look at the clues. Two key pieces of evidence soon began to stand out:

1. The Moon's average density is much smaller than Earth's, and in fact is about the same as the density of Earth's mantle. This suggests that the Moon lacks a large iron core like that of Earth and the other terrestrial planets and instead is made almost entirely from material like that of Earth's mantle.
2. The overall composition of the Moon rocks looked quite similar to the composition of Earth's mantle material, except for the lack of volatile elements. Since heating could cause volatile elements to vaporize and escape into space, the rock composition suggested that the Moon was built from mantlelike material that had been strongly heated before it collected to form the Moon.

Taken together, the evidentiary implication seemed almost obvious: The Moon was made from material that accreted in Earth orbit after first being violently blasted out of Earth's mantle. The idea that it was made from mantle material would explain the Moon's general resemblance to Earth's mantle, and the violence of being blasted out would explain the heat necessary to have allowed volatiles to vaporize and escape. But what could have blasted out a large portion of Earth's mantle?

---

*There is one case in the solar system in which a fairly large moon apparently *was* captured: The orbital characteristics of Neptune's moon Triton make it almost certain to be a captured orbit [Section 9.3]. We do not know how such a large moon (it has about one third the mass of our Moon) was captured; however, because Neptune is much more massive than Earth (about 17 times Earth's mass), a capture event would have had a higher probability.

The Moon's scarred surface suggests an answer. As we discussed in Section 4.3, the early history of the solar system was marked by the heavy bombardment, during which leftover planetesimals smashed into the young planets. Earlier in time, the remaining planetesimals were likely to be more numerous and larger. In fact, models of planetary formation suggest that the late stages of accretion must have been extraordinarily violent [Section 3.3]. Rather than the four terrestrial planets that exist in the inner solar system today, there may have been a dozen or more planet-size bodies. The current planets are the survivors of the shattering collisions that must have occurred. Thus, impacts of planet-size bodies were not only possible but even likely in the early history of the solar system. Such a "giant impact" could easily have had enough energy to blast much of Earth's mantle into space.

Today, sophisticated computer models are used to test the giant impact idea, and it seems to work. The outcome of a giant impact depends on many factors, including the mass and speed of the incoming object, and the precise place and angle at which it strikes Earth. By testing many scenarios, scientists have developed the model of the Moon's formation summarized in Figure 4.31. According to this model, a Mars-size object blasted into the young Earth. The impact must have occurred after Earth had differentiated but before the age of the oldest Moon rocks; radiometric dating and other isotopic evidence tells us that the impact occurred within about 10 to 20 million years after Earth's iron sunk to the core, which occurred more than 4.50 billion years ago. The blast shattered and melted our planet, splashing out molten debris from the mantle. Much of this material fell back to Earth, but some remained in orbit. There,

Figure 4.31

Artist's conception of the Moon's formation by a giant impact. As shown, the Moon formed quite close to a rapidly rotating Earth, but over billions of years tidal forces have slowed Earth's rotation and moved the Moon's orbit outward.

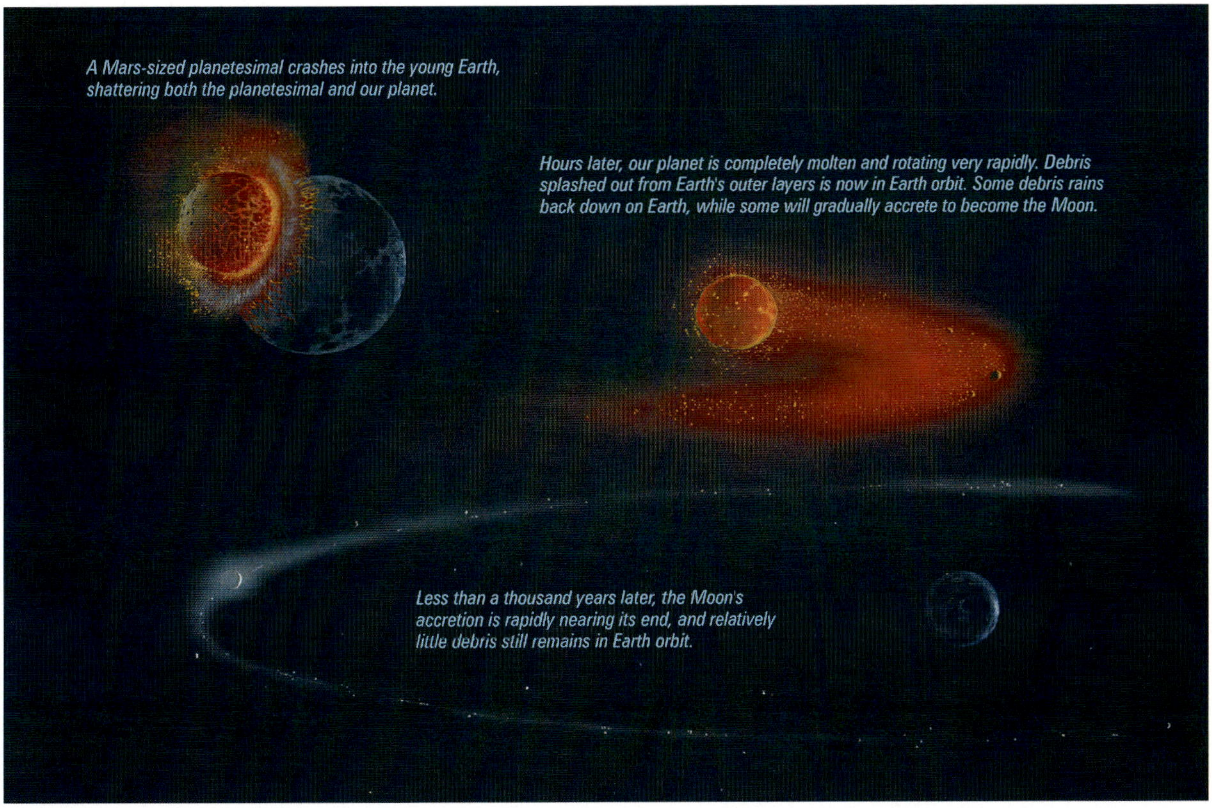

with its volatile content having vaporized and escaped, the material accreted to make the Moon.

• Does the giant impact model count as science?

The giant impact model works so well and so successfully explains the compositional differences between Earth and the Moon that it is widely accepted among planetary scientists. But is it really "scientific" to invoke a single event for which we may never have more than indirect evidence?

One way to decide whether the giant impact model should count as science is to see how it stacks up against the hallmarks of science presented in Chapter 2. Looked at this way, the giant impact model certainly qualifies as science. It invokes a natural explanation for the origin of the Moon, one that even seems likely given what we know about the collisions that must have occurred in the solar system's early history. It also makes testable predictions. The most important of these predictions are the ones about composition — the idea that the Moon formed from mantle material leads to specific predictions about the composition of Moon rocks, and these predictions match the evidence. Note that the fact that the evidence was discovered before the model was proposed is not important here: Just as Kepler came up with his model of planetary motion to explain data that Tycho had already collected, the key point is that the model has been worked out in detail and it successfully matches the observations. The model will also be subject to ongoing tests in the future. For example, the *Apollo* astronauts collected Moon rocks from only a handful of sites on the Moon. In the future, we will presumably collect rocks from many other parts of the Moon. If these rocks were to turn up compositional surprises, it would cause us to reconsider and perhaps even discard the giant impact model. Thus, the giant impact model exhibits all the hallmarks of science: It is natural and testable, and we can imagine future discoveries that would cause us to call it into question.

THINK ABOUT IT Considering the evidence for the giant impact model, do you think it qualifies as a hypothesis or a theory, or something in between? Note that even scientists disagree on this question, so be sure to defend your opinion.

The giant impact model has important consequences both to our understanding of the solar system and to the search for life in the universe. For our solar system, it may help explain other "exceptions to the rules." If a giant impact really was as likely as we have presumed, then Earth should not have been the only planet to suffer one. Indeed, we have reason to think that at least three other planets also suffered giant impacts. Mercury's surprisingly large iron core is easily explained if a giant impact also blasted away much of its mantle, but without leaving a moon behind. Pluto's moon Charon also shows characteristics we'd expect if it formed in a giant impact. A giant impact might also account for the huge axis tilt of Uranus, which rotates essentially on its side (although an alternative idea has recently been proposed). We do not yet have the direct evidence we would need to test any of these other giant impact ideas, but they seem reasonable in light of what we have learned about our own Moon.

In terms of life, the consequences of the giant impact model lie with roles the Moon has played in shaping Earth's biological history. The Moon is the primary cause of Earth's tides (the Sun contributes only about one third as much to Earth's tides as the Moon), and many living organisms have biological cycles tied to the tides. If there had been no giant impact and no Moon, these types of cycles might not have arisen. The Moon also plays a role in Earth's long-term climate stability. Recall that small changes in Earth's axis tilt can significantly change the climate, bringing on or ending ice ages. Models show that Earth's axis tilt would vary much more if we did not have the Moon; the Moon's gravity exerts a stabilizing influence on axis tilt. Indeed, as we'll see in Chapter 8, evidence suggests that Mars undergoes much more extreme changes in axis tilt, changes that are possible because it lacks a large moon. Some scientists therefore wonder if the existence of the Moon, along with the axis tilt stability that it brings, was necessary to our own evolution on Earth. If it was—and if the Moon really is the result of a random giant impact—then the possibility of finding intelligent life elsewhere may depend on the likelihood of giant impacts that result in forming large moons around terrestrial planets.

THE BIG PICTURE

Putting Chapter 4 into Context

In this chapter, we've explored the interconnections between geology and habitability, learning about the conditions that have made life on Earth possible. As you continue your studies, keep in mind the following "big picture" ideas:

- We can read Earth's geological history by studying the geological record. This history is not mere speculation. It is recorded in ways that we can verify independently in rocks and fossils from many places around the world and through the well-verified technique of radiometric dating for determining ages. While we probably won't ever know every detail of Earth's history, we already have a complete enough picture to understand the major processes and events that have shaped our planet.
- Earth's surface has been shaped by active geology driven by internal heat. This geology made life possible by causing the outgassing of the material that made our oceans and atmosphere, while also creating a magnetic field that may have helped preserve the atmosphere. Moreover, through the action of plate tectonics, geology has kept our planet's climate stable enough for water to remain liquid for the past 4 billion years or more.
- Geology plays a role in making life possible, but life, once it takes hold, can also change the conditions on a planet. The oxygen in our atmosphere is a direct consequence of life. Today life affects our planet in another way: Our advanced civilization is capable of changing the way our climate functions, with consequences that we cannot fully predict.
- We do not yet know how common habitable planets may be in the universe, but geology clearly plays a crucial role in habitability. We'll return to geological considerations many more times throughout this book.

SUMMARY OF KEY CONCEPTS

4.1 GEOLOGY AND LIFE

- **How is geology crucial to our existence?**

Geology appears to be crucial to our existence in at least three ways: **Volcanism** released most of the gas that made the atmosphere and the water vapor that condensed to form the oceans; **plate tectonics** is crucial to the climate regulation that has kept Earth habitable over the long term; and Earth's **magnetic field** has probably helped preserve the atmosphere from being stripped by the solar wind.

4.2 RECONSTRUCTING THE HISTORY OF EARTH AND LIFE

- **What can we learn from rocks and fossils?**

Rocks and fossils preserve a record of the conditions under which they formed. By studying **mineral** structure and chemical and isotopic composition, we can learn such things as when the rock or fossil formed, how it formed, and what kinds of conditions prevailed on Earth when it formed.

- **How do we learn the age of a rock or fossil?**

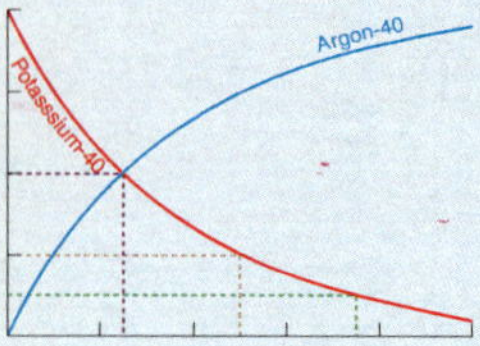

Radiometric dating is based on carefully measuring the proportions of radioactive isotopes and their decay products within rocks. The ratio of the two changes with time and provides a reliable measure of a rock's or fossil's age.

- **What does the geological record show?**

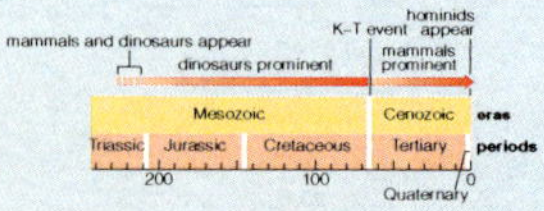

The geological record allows us to reconstruct Earth's history, which we summarize with the **geological time scale**. We divide this history into four **eons** (the Hadean, Archean, Proterozoic, and Phanerozoic), subdividing the last one into three **eras** and shorter **periods**. The time scale extends back to Earth's birth a little over 4.5 billion years ago.

4.3 THE HADEAN EARTH AND THE DAWN OF LIFE

- **How did Earth get an atmosphere and oceans?**

The water and gases that made our atmosphere were originally trapped inside our planet. They were released by volcanic outgassing. Water vapor condensed to form the oceans and the other gases made Earth's early atmosphere. Life has since transformed the atmosphere, most importantly by adding molecular oxygen.

- **Could life have existed during Earth's early history?**

Earth may have had oceans and continents early in its history, but the first half-billion years or so were the time of the **heavy bombardment**. Large impacts must have been common, and a few could have been **sterilizing impacts** that would have killed any life on Earth, or slightly smaller impacts that would have allowed only deep-sea or underground life to survive.

4.4 GEOLOGY AND HABITABILITY

- **What is Earth like on the inside?**

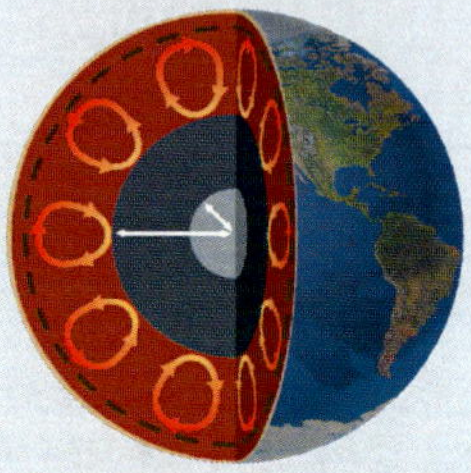

In order of decreasing density and depth, the interior structure consists of the **core**, **mantle**, and **crust**. The crust and part of the mantle together make up the rigid **lithosphere**. Internal heat allows the mantle rock to convect slowly.

- **How does plate tectonics shape Earth's surface?**

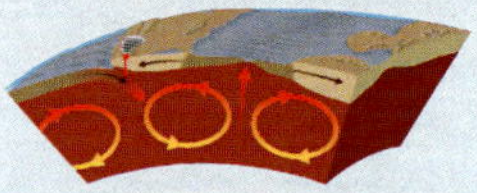

Plate tectonics has led to many unique features of our geology, especially **seafloor spreading** zones and the building of continents along **subduction zones**. The shifting of plates has completely changed Earth's geological appearance many times in the past billion years.

- **Why does Earth have a protective magnetic field?**

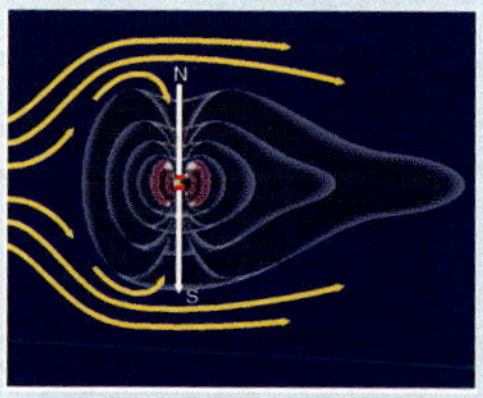

Earth's magnetic field is generated by the combination of its molten outer core, convection in that outer core, and a moderately rapid rotation rate. The magnetic field creates a **magnetosphere** that acts like a protective bubble surrounding our planet.

4.5 CLIMATE REGULATION AND CHANGE

- **How does the greenhouse effect make Earth habitable?**

Greenhouse gases such as carbon dioxide, methane, and water vapor absorb infrared light emitted from a planet's surface. The absorbed photons are quickly reemitted, but in random directions. The result acts much like a blanket, warming the planet's surface. Without the warming due to the greenhouse effect, Earth would be frozen.

- **What regulates Earth's climate?**

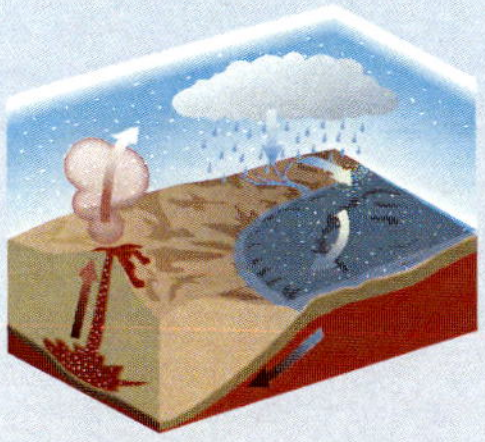

Earth's long-term climate is remarkably stable due to feedback processes that tend to counter any warming or cooling that occurs. The most important feedback process is the **carbon dioxide cycle**, which naturally regulates the strength of Earth's greenhouse effect.

- **How does Earth's climate change over long periods of time?**

Earth's temperature has remained in a range allowing liquid water at least since the end of the heavy bombardment, but there have been periods of unusual warmth or cold. Recent **ice ages** are tied to small changes in Earth's rotation and orbit. Geological evidence also shows more extreme variations in the past, including the **Snowball Earth** episodes that ended more than 500 million years ago.

4.6 THE PROCESS OF SCIENCE IN ACTION: FORMATION OF THE MOON

- **How did the giant impact model win out over competing models?**

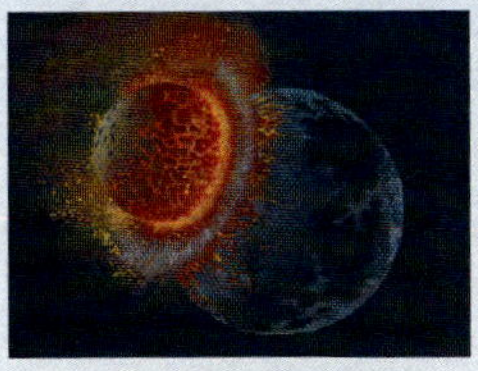

The giant impact model is the only model of the Moon's formation that successfully explains the differences in composition between Earth and the Moon. According to this model, the Moon formed from mantle material splashed out of Earth by the impact of a Mars-size object.

- **Does the giant impact model count as science?**

The giant impact model shows all the hallmarks of science: It is natural and testable, and we can imagine future discoveries that would cause us to call it into question.

EXERCISES AND PROBLEMS

REVIEW QUESTIONS

Short-Answer Questions Based on the Reading

1. Briefly describe why the study of geology is important to the study of life in the universe.
2. What do we mean by the *geological record?* Why is it important?
3. Describe the three basic types of rock. How does recycling occur among these rock types? What is a *mineral?* How do we study rocks in the laboratory?
4. How are sedimentary *strata* made, and how do they enable us to determine the relative ages of rocks and fossils?
5. Describe the technique of *radiometric dating*, and explain how we know it is reliable. Be sure to explain what we mean by a *radioactive isotope*, parent and daughter isotopes, and a *half-life*.
6. How do fossils form? Do most living organisms leave fossils? Do most fossils contain organic matter? Explain.
7. Summarize the *geological time scale*. What are *eons, eras*, and *periods?*
8. How old is Earth? How do we know?
9. Briefly describe how *outgassing* led to the origin of our oceans and atmosphere. How did Earth's early atmosphere differ from Earth's current atmosphere?
10. What was the *heavy bombardment?* When did it occur? What were *sterilizing impacts*, and when did the last one occur?
11. Briefly describe Earth's *core-mantle-crust* structure, and how it obtained this structure. What is the lithosphere? What is mantle convection?
12. Briefly describe the conveyorlike action of *plate tectonics* and the evidence for this action. How does plate tectonics account for the observed differences in the *seafloor crust* and the *continental crust?*
13. Describe how plate tectonics shapes important geological features of Earth, including mid-ocean ridges, continents, mountain ranges, rift valleys, and earthquakes. How did Hawaii form?
14. What evidence do we have for the operation of plate tectonics in Earth's distant past? Why do we think Earth has plate tectonics?
15. What are the three requirements for a planetary magnetic field, and how does Earth meet them? How does the magnetic field protect our atmosphere?
16. Briefly describe the mechanism by which the *greenhouse effect* warms a planet. What are the most common *greenhouse gases?*
17. What has happened to most of the carbon dioxide outgassed through Earth's history? Describe the *carbon dioxide cycle* and how it helps regulate Earth's climate.
18. What are *ice ages*, and what may cause them? What do we mean by *Snowball Earth* periods, and how does Earth recover from them?
19. Briefly summarize the key ways in which geology is important to Earth's long-term habitability.
20. How do we think the Moon formed, and what evidence supports this model? Why were other models ruled out? Should we be surprised that a giant impact could have affected our planet?

TEST YOUR UNDERSTANDING

Does It Make Sense?

Decide whether each statement makes sense (or is clearly true) or does not make sense (or is clearly false). Explain your reasoning clearly; because not all of these have definitive answers, your explanation is more important than your chosen answer.

21. We can expect that if there are paleontologists a few million years from now, they will find the fossil remains of almost every human who ever lived.
22. Nearly all the rocks I found in the lava fields of Hawaii are igneous.
23. The most common rock type in the strata of the Grand Canyon is sedimentary rock.
24. Although Earth contains its densest material in its core, it's quite likely that terrestrial planets in other star systems would contain their lowest-density rock in their cores and their highest-density rock in their crusts.
25. If you had a time machine that dropped you off on Earth during the Hadean eon, you'd be quickly killed by a sterilizing impact.
26. If there were no plate tectonics on Earth, our planet would be far too hot to have liquid oceans.
27. Without the greenhouse effect, there probably would be no life on Earth.
28. If nitrogen were a greenhouse gas, our planet would be far hotter than it is.
29. We can learn a lot about Earth's early history by studying the Moon.
30. Science can never determine with confidence the times or sequence of events that occurred millions or billions of years ago.

Quick Quiz

Choose the best answer to each of the following. Explain your reasoning with one or more complete sentences.

31. A rock's type (igneous, metamorphic, or sedimentary) tells us: (a) its age; (b) its chemical composition; (c) how it was made.
32. To learn a rock's age, we must: (a) determine its chemical composition; (b) identify its mineral structure; (c) measure the ratios of different isotopes within it.
33. Radiometric dating now allows us to determine Earth's age to an accuracy of about: (a) a billion years; (b) 20 million years; (c) a few thousand years.
34. The timing of the last sterilizing impact is important because it probably tells us: (a) when life could have taken long-term hold on Earth; (b) when Earth's iron sank to its core; (c) when the dinosaurs became extinct.
35. We learn about the heavy bombardment by studying: (a) craters and rocks from the Moon; (b) zircon mineral grains; (c) Earth's oldest igneous rocks.
36. Earth has retained a lot of internal heat primarily because of its: (a) distance from the Sun; (b) large iron core; (c) relatively large size.
37. Plate tectonics is best described as a process that: (a) recycles rock between Earth's surface and upper mantle; (b) brings metal from Earth's core to the surface; (c) allows continents to plow through the crust.
38. Earth has far less atmospheric carbon dioxide than Venus because: (a) Earth was born with less of this gas; (b) Earth's carbon dioxide was lost in the giant impact that formed the Moon; (c) Earth's carbon dioxide is locked up in carbonate rocks.
39. If Earth had more greenhouse gases in its atmosphere, it would: (a) heat up; (b) cool off; (c) accelerate plate tectonics.
40. *Snowball Earth* refers to: (a) one of a series of very deep ice ages that occurred more than 500 million years ago; (b) the idea that Earth would be frozen without the greenhouse effect; (c) any of the ice ages that have occurred in the past few million years.

INVESTIGATE FURTHER

In-Depth Questions to Increase Your Understanding

Short-Answer/Essay Questions

41. *Understanding Radiometric Dating.* Imagine you had the good fortune to find a meteorite in your backyard that appears to be a piece of material from the early history of the solar system. Qualitatively, how would you expect its ratio of potassium-40 and argon-40 to be different from that of other rocks in your yard? Explain why, in a few sentences.
42. *Dating Planetary Surfaces.* We have discussed two basic techniques for determining the age of a planetary surface: studying the abundance of impact craters and radiometric dating of surface rocks. Describe each technique briefly. Which technique seems more reliable? Which technique is more practical? Explain.
43. *Earth Without Differentiation.* Suppose Earth had never undergone differentiation. How would Earth be different? Write two or three paragraphs discussing likely differences. Explain your reasoning carefully.
44. *Earth Without Plate Tectonics.* Suppose plate tectonics had never begun on Earth. How would Earth be different? Write two or three paragraphs discussing likely differences. Explain your reasoning carefully.
45. *Earth Without the Moon.* Suppose the giant impact that formed the Moon had never occurred. How would you expect Earth to be different? Explain your reasoning carefully.
46. *Feedback Processes in the Atmosphere.* As the Sun gradually brightens in the future, how can the CO_2 cycle respond to reduce the warming effect? Which parts of the cycle will be affected? Is this an example of positive or negative feedback?
47. *Experiment: Geological Properties of Silly Putty.* Roll room-temperature Silly Putty into a ball and measure its diameter. Place the ball on a table and gently place one end of a heavy book on it. After 5 seconds, measure the height of the squashed ball. Repeat the experiment two more times, the first time warming the Silly Putty in hot water before you start and the second time cooling it in ice water before you start. How do the different temperatures affect the rate of "squashing"? How does the experiment relate to planetary geology? Explain.
48. *Experiment: Planetary Cooling in a Freezer.* To simulate the cooling of planetary bodies of different sizes, use a freezer and two small plastic containers of similar shape but different size. Fill each container with cold water and put both into the freezer at the same time. Checking every hour or so, record the time and your estimate of the thickness of the "lithosphere" (the frozen layer) in the two tubs. How long does it take the water in each tub to freeze completely? Describe in a few sentences the relevance

of your experiment to planetary geology. Extra credit: Plot your results on a graph with time on the x-axis and lithospheric thickness on the y-axis. What is the ratio of the two freezing times?

Quantitative Problems

Be sure to show all calculations clearly and state your final answers in complete sentences.

49. *Geological Time.* Geological time scales are often written in ways that can mask their significance. For each of the following pairs of times, state which one is larger and by how much:
    a. 25,000 years, 0.1 million years
    b. 4 million years, 0.05 billion years
    c. 0.1 billion years, 1 million years
50. *Dating Lunar Rocks.* You are analyzing Moon rocks that contain small amounts of uranium-238, which decays into lead with a half-life of about 4.5 billion years.
    a. In one rock from the lunar highlands, you determine that 55% of the original uranium-238 remains; the other 45% decayed into lead. How old is the rock?
    b. In a rock from the lunar maria, you find that 63% of the original uranium-238 remains; the other 37% decayed into lead. Is this rock older or younger than the highlands rock? By how much?
51. *Carbon-14 Dating.* The half-life of carbon-14 is about 5,700 years.
    a. You find a piece of cloth painted with organic dyes. By analyzing the dye in the cloth, you find that only 77% of the carbon-14 originally in the dye remains. When was the cloth painted?
    b. A well-preserved piece of wood found at an archaeological site has 6.2% of the carbon-14 that it must have had when it was living. Estimate when the wood was cut.
    c. Suppose a fossil is 570,000 years old, which is 100 half-lives of carbon-14. What fraction of its original carbon-14 would remain? Use your answer to explain why carbon-14 generally is not useful for dating fossils of this age.
52. *Martian Meteorite.* Some unusual meteorites thought to be chips from Mars contain small amounts of radioactive thorium-232 and its decay product, lead-208. The half-life for this decay process is 14 billion years. Analysis of one such meteorite shows that 94% of the original thorium remains. How old is this meteorite?
53. *Big Moons.* Use the data in Appendix E to compare Earth's Moon, Jupiter's moon Ganymede, Saturn's moon Titan, and Pluto's moon Charon. Rank them in order of their actual size (radius, in kilometers). Then rank them in order of the ratio of their size to the size of the planet they orbit. What can you conclude from the differences in the two rankings? Which moons seem most "unusual" compared to other moons in the solar system?
54. *Internal vs. External Heating.* In daylight, Earth's surface absorbs about 400 watts per square meter. All of Earth's internal radioactivity produces a total of 3 trillion watts, which leak out through the surface. Calculate the internal heat flow (watts per square meter) averaged over Earth's surface. Compare quantitatively to solar heating, and comment on why internal heating drives geological activity.
55. *Plate Tectonics.* Typical motions of one plate relative to another are 1 centimeter per year. At this rate, how long would it take for two continents 3,000 kilometers apart to collide? What are the global consequences of motions like this?
56. *More Plate Tectonics.* Consider a seafloor spreading zone creating 1 centimeter of new crust over its entire 2,000-kilometer length every year. How many square kilometers of surface will this create in 100 million years? What fraction of Earth's surface does this constitute?

Discussion Questions

57. *The Age of Earth.* Some people still question whether we have a reasonable knowledge of the age of Earth or the ability to date events in Earth's history. Based on what you have learned about both relative and absolute ages on the geological time scale, do you think it is reasonable for scientists to be confident of ages found by radiometric dating? Is there any scientific reason to doubt the reliability of our chronology of Earth? Explain.
58. *Plate Tectonics and Us.* Based on what you learned in this chapter, can you imagine cases in which civilizations might arise on planets without plate tectonics? Defend your opinion.
59. *Implications for Other Worlds.* Overall, do you think that Earth's geological features are likely to be rare or common on other worlds? How does your answer affect your opinion of the prospects of discovering life or civilizations on other worlds?
60. *Evidence of Our Civilization.* Discuss how the geological processes will affect the evidence of our current civilization in the distant future. For example, what evidence of our current civilization will survive in 100,000 years? in 100 million years? Do you think that future archaeologists or alien visitors will be able to know that we existed here on Earth?

WEB PROJECTS

61. *Local Geology.* Learn as much as you can about how geological features in or near your hometown were formed. Write a one- to three-page summary of your local geology.
62. *Volcanoes and Earthquakes.* Learn about one major earthquake or volcanic eruption that occurred during the past decade. Report on the geological conditions that led to the event, as well as on its geological and biological consequences.
63. *Formation of the Moon.* Scientists continue to model and study the giant impact thought to have formed the Moon. Look for and report on one recent discovery that may shed more light on how or when the giant impact occurred.

5
The Nature of Life on Earth

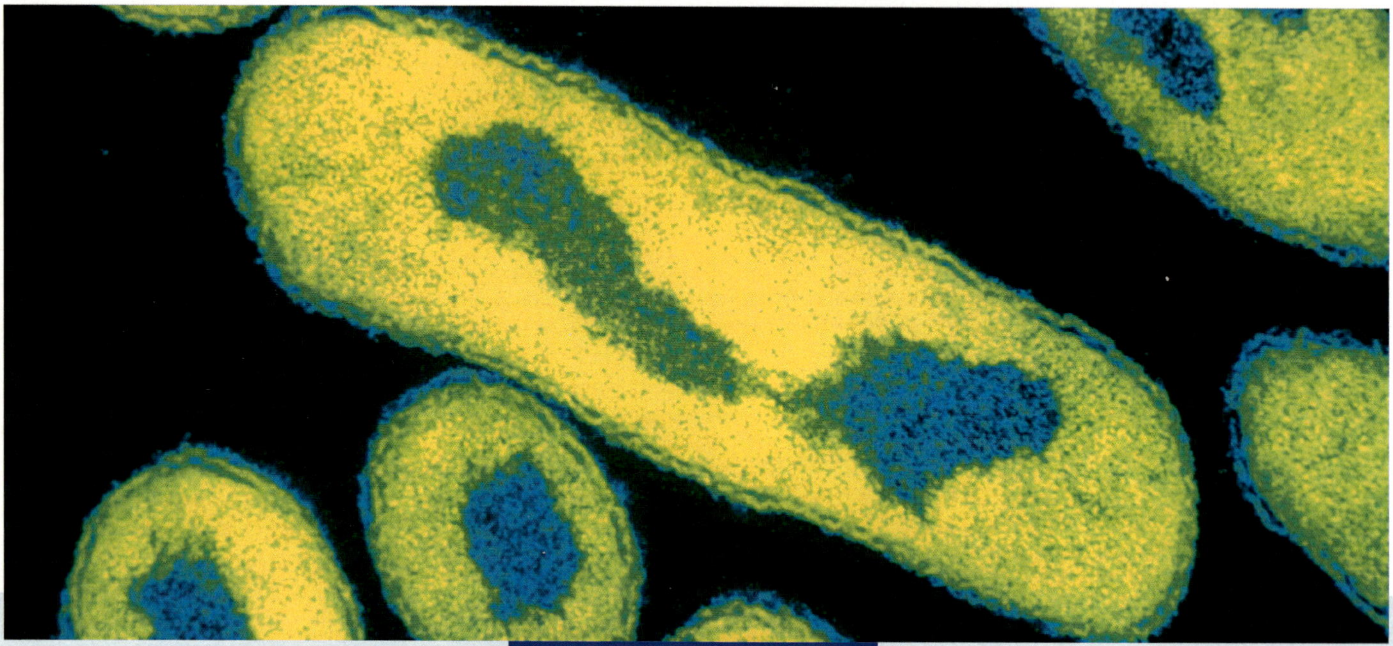

LEARNING GOALS

5.1 DEFINING LIFE
- What are the general properties of life on Earth?
- What is the role of evolution in defining life?
- What is life?

5.2 CELLS: THE BASIC UNITS OF LIFE
- What are living cells?
- What are the molecular components of cells?
- What are the major groupings of life on Earth?

5.3 METABOLISM: THE CHEMISTRY OF LIFE
- What are the basic metabolic needs of life?
- How do we classify life by its metabolic sources?

5.4 DNA AND HEREDITY
- How does the structure of DNA allow for its replication?
- How is heredity encoded in DNA?
- How does life evolve?

5.5 LIFE AT THE EXTREME
- What kinds of conditions can life survive?
- Are extremophiles really extreme?

5.6 THE PROCESS OF SCIENCE IN ACTION: EVOLUTION AS SCIENCE
- Is evolution a fact or a theory?
- Are there scientific alternatives to evolution?

There is grandeur in this view of life, with its several powers, having been originally breathed by the Creator into a few forms or into one; and that, whilst this planet has gone cycling on according to the fixed law of gravity, from so simple a beginning endless forms most beautiful and most wonderful have been, and are being, evolved.

Charles Darwin, *The Origin of Species*, 1859

To this point in the book, we've talked only generally about the search for life in the universe and about the conditions that make life possible on Earth. Now, we are ready to begin talking about life itself and how we will search for it. But what exactly are we searching for, and where should we look? To answer these questions, we need to know something about the nature of life and how life interacts with its planetary environment.

The prime difficulty we face in discussing the nature of life is that we have only one example to study: life on Earth. Life on Earth is remarkably diverse, with organisms ranging in size from the microscopic to huge plants and animals, and life of some kind is found in almost every conceivable place on and near the planet's surface. Nevertheless, all living organisms on Earth share common biochemical characteristics that tell us that all species are related to one another. Although no one knows for sure, it seems likely that other types of biochemistry should be possible, in which case life on other worlds might be quite different from life on Earth.

Still, if we want to know what life might be like on other worlds, the best place to start is to learn as much as possible about life here on this world. In this chapter, we'll explore the nature of life on Earth. Along the way, we'll see that life elsewhere would almost certainly share at least a few characteristics with life on Earth, and we'll gain a few clues about where we might most profitably focus our search for life in the universe.

5.1 DEFINING LIFE

What is life? This seemingly simple question lies at the heart of research into life in the universe. After all, if we are interested in the possibility of life elsewhere, we must know what it is that we are looking for. Unfortunately, defining life has proved to be surprisingly difficult, even when we consider only life on Earth. It is all the more difficult when we consider life elsewhere, because we cannot be sure that life on other worlds would resemble life on Earth physically or chemically. Given the difficulty of defining life, the only sensible way to proceed is by studying the one example of life that we know, hoping it will yield fundamental insights into how life operates and into the environmental conditions required to support life. In this first section, we will explore general characteristics of life on Earth and attempt to come up with at least some reasonably useful definition of life.

• What are the general properties of life on Earth?

A cat and a car have much in common. Both require energy to function—the cat gets energy from food, and the car gets energy from gasoline. Both can move at varying speeds and can turn corners. Both expel waste products. But a cat clearly is alive, while a car clearly is not. What's the difference?

In the case of a cat and a car, we can find many important differences without looking too far. For example, cats reproduce themselves, while cars must be built in factories. But as we look deeper into the nature of life, it becomes increasingly difficult to decide what characteristics separate living organisms from rocks and other nonliving materials. Indeed, the question can be so difficult to answer that we may be tempted to fall back on the famous words of U.S. Supreme Court Justice Potter Stewart, who, in avoiding the difficulty of defining pornography, wrote: "I shall not today attempt further to define [it]. . . . But I know it when I see it."* If living organisms on other worlds turn out to be much like those on Earth, it may prove true that we'll know them when we see them. But if the organisms are fairly different from those on Earth, we'll need clearer guidelines to decide whether or not they are truly "living."

One way to seek distinguishing features of life is to study living organisms, looking for common characteristics. Given the difficulty of defining life, you probably won't be surprised to learn that there are exceptions to almost any "rule" we think of. Nevertheless, biologists have identified at least six key properties that appear to be shared by most or all living organisms on Earth, all of which are summarized in Figure 5.1. Let's briefly investigate each property.

Order The materials in living organisms always exhibit some type of order. For example, the molecules in living cells are not scattered randomly about but instead are arranged in patterns that make cell structures. These structures, in turn, make possible all the other properties of life that we will discuss. Note that order alone does not make something living: A book also has order, because words are not scattered randomly on the pages, but it is not alive. The same is true for rock crystals, which have their atoms arranged in an orderly way, and even for the individual molecules of life such as proteins or DNA; these molecules clearly have order, but we consider them only to be building blocks of life, not life itself.

Nevertheless, it seems reasonable to expect that all living things will show order. In logical terms, we say that order is a *necessary condition* for life, because something cannot be alive without order. However, order is not a *sufficient condition* for life, because order alone does not make something alive.

THINK ABOUT IT The idea of necessary and sufficient conditions is important in science. To make sure you understand it, decide whether each of the following conditions is necessary or sufficient (or neither or both) for the given effect: (1) condition: breathing; effect: human survival while sleeping; (2) condition: living in New York City; effect: living in the United States; (3) condition: meeting all requirements for a college degree; effect: receiving a college degree.

---

*From Potter Stewart's concurring opinion in *Jacobellis v. Ohio,* 378 U.S. 184, 198 (1964).

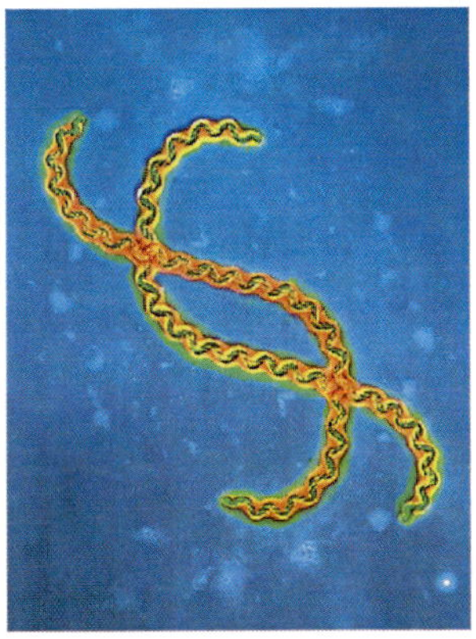

a Order: Living organisms exhibit order in their internal structure, as is apparent in this microscopic view of spiral patterns in two single-celled organisms.

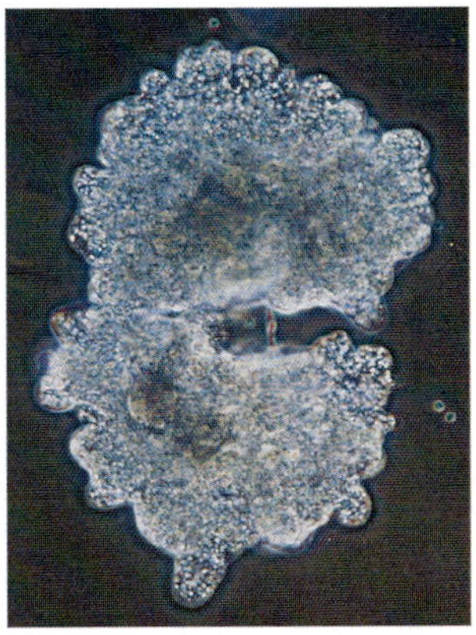

b Reproduction: Organisms reproduce their own kind. Here, a single-celled organism (an amoeba) has already copied its genetic material (DNA) and is now dividing into two cells.

c Growth and development: Living organisms grow and develop in patterns determined at least in part by heredity. Here, we see a Nile crocodile emerging from its shell.

Figure 5.1

Six key properties of life.

d Energy utilization: Living organisms use energy to fuel their many activities. These tube worms which live near a deep-sea volcanic vent, obtain energy from chemical reactions made possible in part by the heat released from the volcanic vent.

e Response to the enviornment: Life actively responds to changes in its surroundings. Here, we see a jackrabbit's ears flush with blood; the blood flow adjusts automatically to help the animal maintain a constant internal temperature by adjusting the heat loss from the ears.

f Evolutionary adaptation: Life evolves in a way that leads to organisms that are adapted to their enviornments. Here, a pygmy seahorse is camouflaged in its coral surroundings.

Reproduction Living organisms reproduce or are products of reproduction. Simple life-forms, such as bacteria, reproduce by dividing to make nearly exact copies of themselves. More complex organisms may reproduce in more sophisticated ways—including sexual reproduction, in which offspring inherit genetic material from two parents.

Reproduction seems necessary to a definition of life; after all, without reproduction, there would be no way for life as a whole to survive the deaths of individuals. Nevertheless, it's difficult to simply say that all life is capable of reproduction, because there are some exceptions. For example, a mule is sterile and cannot reproduce, but it is the product of reproduction between two closely related animals (a horse and a donkey) and is clearly alive.

Any attempt to include reproduction in the definition of life also exposes borderline cases about which even scientists disagree. For example, *viruses* are generally much smaller than bacteria and are incapable of reproducing on their own: If you place a virus by itself in a dish full of nutrients, nothing will happen. However, when a virus infects a living organism, it can reproduce by commandeering the organism's reproductive machinery for its own purposes. The fact that viruses can reproduce when they infect other organisms but not when they are on their own seems to put them somewhere between the nonliving and the living. Another borderline case concerns the infectious proteins known as *prions,* which are thought to be the agents of *mad cow disease.* Prions are still fairly mysterious, but they appear to be abnormal forms of protein molecules that somehow cause normal protein molecules to change into the abnormal prion form. In other words, they make copies of themselves by causing other molecules to change rather than by actually replicating themselves. Most biologists therefore put them on the nonliving side of the gray region between nonlife and life, though they present at least some ambiguity.

Growth and Development Living organisms grow and develop in patterns directed at least in part by **heredity**—traits passed to an organism from its parent(s). The property of growth and development appears necessary to life in that all organisms grow or develop during at least some periods in their life cycles, but it is not sufficient to constitute life. For example, fire grows and develops as it spreads through a forest, but a fire is not alive. As we will discuss later in this chapter, all life on Earth passes on its heredity through the molecules known as DNA. (Some viruses use a related molecule called RNA, but we will leave viruses and prions out as we discuss "life" in the rest of this chapter.)

KEY BIOLOGICAL DEFINITIONS

This box summarizes the definitions of a few of the most important biological terms used in this chapter and throughout the book.

TERMS RELATED TO EVOLUTION:

evolution (biological): The gradual change in populations of living organisms that is responsible for transforming life on Earth from its primitive origins to the great diversity of life today.

evolutionary adaptation: An inherited trait that enhances an organism's ability to survive and reproduce in a particular environment.

theory of evolution: The theory, first advanced by Charles Darwin, that explains how and why living organisms evolve through time.

natural selection: The primary mechanism by which evolution proceeds. More specifically, natural selection refers to the process by which, over time, advantageous genetic traits naturally win out (are "selected") over less advantageous traits because they are more likely to be passed down through succeeding generations.

species: Precise definitions vary, but for our purposes we can consider a species to be a population of organisms that is genetically distinct from other groups of organisms. Species are given two-part scientific names consisting of a genus followed by a more specific category; for example, humans belong to the genus *Homo* and are fully classified as *Homo sapiens.*

TERMS RELATED TO HEREDITY:

heredity: The characteristics of an organism passed to it by its parent(s), which it can pass on to its offspring. The term can also apply to the transmission of these characteristics from one generation to the next. Hereditary information is encoded in DNA.

gene: The basic functional unit of an organism's heredity. A single gene consists of a sequence of DNA bases (or RNA bases, in some viruses) that provide the instructions for a single cell function (such as building a protein).

genome: The complete sequence of DNA bases in an organism, encompassing all of the organism's genes along with noncoding DNA in between.

genetic code: The specific set of rules by which the sequence of bases in DNA is "read" to provide the instructions that make up genes.

DNA (deoxyribonucleic acid): The basic hereditary molecule of life on Earth. A DNA molecule consists of two strands, twisted in the shape of a double helix, along each of which lies a long sequence of *DNA bases.* The four DNA bases are adenine (A), cytosine (C), guanine (G), and thymine (T), and they can be paired across the two DNA strands only so that A pairs with T and C pairs with G.

Energy Utilization Living organisms use energy to create and maintain patterns of order within their cells, to reproduce, and to grow. Life without energy utilization is simply not possible (though some organisms can survive temporarily in dormant states). Of course, energy utilization is not sufficient to constitute life; any electrical or gas-powered appliance uses energy to function.

We can gain further insight into the importance of energy utilization by considering what is sometimes called the *thermodynamics* of life. Thermodynamics is a branch of science that deals with energy and the rules by which it operates. Recall the *law of conservation of energy* [Section 3.4], which tells us that energy can be neither created nor destroyed but only transformed from one form of energy to another; this law is sometimes referred to as the "first law of thermodynamics." The **second law of thermodynamics** states that, when left alone, the energy in a system undergoes conversions that lead to increasing disorder. Living organisms are a perfect example of this law's importance: If you place a living organism into a sealed box, it will eventually use up the available energy and therefore no longer be able to build new molecules or fuel any of the molecular processes needed for life. Its molecules will therefore become more disordered with time—for example, the molecules may decay or may lose the orderly relationships they maintain with other molecules when the organism is alive—causing the organism to die. Thus, to maintain order and survive, a living organism must have a continual source of energy that it can use to counter the tendency for disorder to take over. Living organisms get this energy from the environment, either through food or through chemical interactions with the environment. The environment, in turn, gets its energy either from an internal source such as the heat of the planet itself, or from an external source such as sunlight.

KEY BIOLOGICAL DEFINITIONS *continued*

RNA (ribonucleic acid): A molecule closely related to DNA, but with only a single strand and a slightly different backbone and set of bases, that plays critical roles in carrying out the instructions encoded in DNA.

TERMS RELATED TO THE MODERN CLASSIFICATION OF LIFE:

cell: The basic structure of all life on Earth, in which the living matter inside is separated from the outside world by a barrier called a *membrane.* Cells come in two basic types: *eukaryotes,* which contain a distinct *cell nucleus* with its own membrane, and *prokaryotes,* which lack cell nuclei.

domains of life: All known species of life fall into one of three broad domains: *Bacteria, Archaea,* and *Eukarya;* the last includes all plants and animals, as well as fungi and many microbes.

tree of life: A representation of biochemical and genetic relationships between species; the three major branches of the tree are the three domains (Bacteria, Archaea, and Eukarya).

TERMS RELATED TO CELLULAR CHEMISTRY:

organic molecule: Generally, any molecule containing carbon, but with some exceptions among molecules that are most commonly found independent of life. For our purposes in this book, the important exceptions are carbon dioxide (CO_2) and carbonate minerals, which we do not consider to be organic.

organic chemistry: The chemistry of organic molecules.

biochemistry: The chemistry of life.

amino acids: The molecules that form the building blocks of proteins. Most organisms construct proteins from a particular set of 20 amino acids, although several dozen other amino acids can be found in nature. More technically, an amino acid is a molecule containing both an *amino group* (NH or NH_2) and a *carboxyl group* (COOH).

protein: A large molecule assembled from amino acids according to instructions encoded in DNA. Proteins play many roles in cells; a special category of proteins, called **enzymes,** catalyzes nearly all the important biochemical reactions that occur within cells.

catalysis: The process of causing or accelerating a chemical reaction by involving a substance or molecule that is not permanently changed by the reaction. The unchanged substance or molecule involved in catalysis is called a **catalyst.** In living cells, the most important catalysts are the proteins known as *enzymes.*

metabolism: The many chemical reactions that occur in living organisms to provide cellular energy and nutrients.

Thus, life probably is not possible on a world that lacks a long-term source of energy input to the environment.

Response to the Environment All living organisms interact with their surroundings and actively respond in at least some ways to environmental changes. For example, some simple organisms may move to a region where the temperature is more suited to their growth, and warm-blooded mammals may sweat, pant, or adjust blood flow to maintain a constant internal temperature. Like all the other properties on our list, response to the environment is a necessary but not sufficient condition for life. Many human-made devices also respond to changes in the environment; for example, a thermostat can respond to changes in temperature by turning on heating or cooling systems.

Evolutionary Adaptation Life has changed dramatically over time as the organisms that lived billions of years ago have gradually evolved into the great variety of organisms found on Earth today. Life evolves as a result of the interactions between organisms and their environments, leading over time to evolutionary adaptations that make species better suited to their environments. When the adaptations are significant enough, organisms carrying the adaptations may be so different from their ancestors that they constitute an entirely new species.

Before we continue, it's worth noting that, like life itself, the familiar term **species** is not so easy to define in a precise way. Traditionally, a species was defined as a group of organisms that share some set of common characteristics and are capable of interbreeding with one another to produce fertile offspring. Thus, for example, horses and donkeys represent different species because the result of their interbreeding is an infertile mule. However, while this definition of species works fairly well for animals and most plants, it does not work for organisms that reproduce asexually, including microorganisms that reproduce through cell division. As a result, biologists today recognize species as groups of organisms that are genetically distinct from other groups, though the precise border between one species and another is not always clear, especially with microorganisms. Once a species is identified, it is given a scientific name that consists of two parts. The first part is the **genus**, which describes the "generic" category to which the organism belongs, while the second part distinguishes multiple species within the same genus. (You may recognize that the term *genus* is related to "generic" and that the term *species* is related to "specific.") The full name is always written in italics, with the genus capitalized. For example, humans are scientifically classified as *Homo sapiens,* meaning that we are one specific species that has been identified within the genus *Homo* (the others are all extinct). Horses and donkeys are, respectively, *Equus caballus* and *Equus asinus,* names that show that both belong to the same genus (*Equus*).

• What is the role of evolution in defining life?

All six properties of life that we have discussed are important, but biologists today regard evolutionary adaptation as the most fundamental and unifying of all these properties. It is the only property that can explain the great diversity of life on Earth, and an understanding of it allows us to understand how all the other properties of life came to be. Modern understanding of the capacity for evolutionary adaptation is described by the **theory of evolution**. Because this theory is so central to modern biology, let's briefly investigate the origin of the theory and the evidence that supports it.

An Ancient Idea The word *evolution* simply means, "change with time," and the idea that life might evolve through time goes back more than 2,500 years. The Greek scientist Anaximander (c. 610–547 B.C.) promoted the idea that life originally arose in water and gradually evolved from simpler to more complex forms. A century later, Empedocles (c. 492–432 B.C.) suggested that creatures poorly adapted to their environments would perish, foreshadowing the modern idea of evolutionary adaptation. Many of the early Greek atomists [Section 2.1] probably held similar beliefs, though the evidence is sparse. Aristotle, however, maintained that species are fixed and independent of one another and do not evolve. This Aristotelian view eventually became entrenched within the theology of Christianity, and evolution was not taken seriously again for some 2,000 years. In the mid-1700s, scientists began to suspect that many fossils represented extinct ancestors of living species. Then, in the early 1800s, French naturalist Jean Baptiste Lamarck suggested that the best explanation for the relationship between fossils and living organisms is that life-forms evolve by gradually adapting to perform successfully in their environments.

Lamarck's idea of evolution by adaptation represented the first clear attempt to explain what we now consider the "observed facts" of evolution. That is, observations of how fossils differ in different layers of the geological record and of relationships between living species make it quite clear that life has changed over time. However, Lamarck was unable to come up with a successful theory to explain *how* evolution occurs. His hypothesis concerning the mechanism of evolution, called "inheritance of acquired characteristics," suggested that organisms develop new characteristics during their lives and then pass these characteristics on to their offspring. For example, Lamarck would have imagined that weight lifting would enable a person to create an adaptation of great strength that could be genetically passed to his or her children. While this hypothesis may have seemed quite reasonable at the time, it has not stood up to scientific scrutiny and therefore has been discarded as a model of how evolution occurs. It has been replaced by a different model, proposed by the British naturalist Charles Darwin.

The Mechanism of Evolution Charles Darwin described his theory of evolution in his book *The Origin of Species,* first published in 1859. In this book, Darwin laid out the case for evolution in two fundamental ways. First, he described his observations of living organisms (made during his voyages on the HMS *Beagle*) and showed how they supported the idea that evolutionary change really does occur. Second, he put forth a new model of *how* evolution occurs, backing up his model with a wealth of evidence. In essence, the geological record and the observed relationships between species together provide strong evidence that evolution *has* occurred, while Darwin's theory of evolution explains *how* it occurs.

As is the case with most scientific theories, the underlying logic of Darwin's model is really quite simple. As described by biologist Stephen Jay Gould (1941–2002), Darwin built his model from "two undeniable facts and an inescapable conclusion":

- *Fact 1: overproduction and competition for survival.* Any localized population of a species has the potential to produce far more offspring than the local environment can support with resources such as food and shelter. This overproduction leads to a competition for survival among the individuals of the population.

- *Fact 2: individual variation.* Individuals in a population of any species vary in many heritable traits (traits passed from parents to offspring). No two individuals are exactly alike, and some individuals possess traits that make them better able to compete for food and other vital resources.
- *The inescapable conclusion: unequal reproductive success.* In the struggle for survival, those individuals whose traits best enable them to survive and reproduce will, on average, leave the largest number of offspring that in turn survive to reproduce. Therefore, in any local environment, heritable traits that enhance survival and successful reproduction will become progressively more common in succeeding generations.

It is this unequal reproductive success that Darwin called **natural selection**: Over time, advantageous genetic traits will naturally win out (be "selected") over less advantageous traits because they are more likely to be passed down through many generations. This process explains how species can change in response to their environment—by favoring traits that improve adaptation—and thus is the primary mechanism of evolution.

Evidence for Evolution by Natural Selection Darwin backed up his logical claim that evolution proceeds through natural selection by documenting cases in which related organisms are adapted to different environments or lifestyles. He found a particularly striking example among the finches of the Galápagos Islands (Figure 5.2). The 13 finch species on

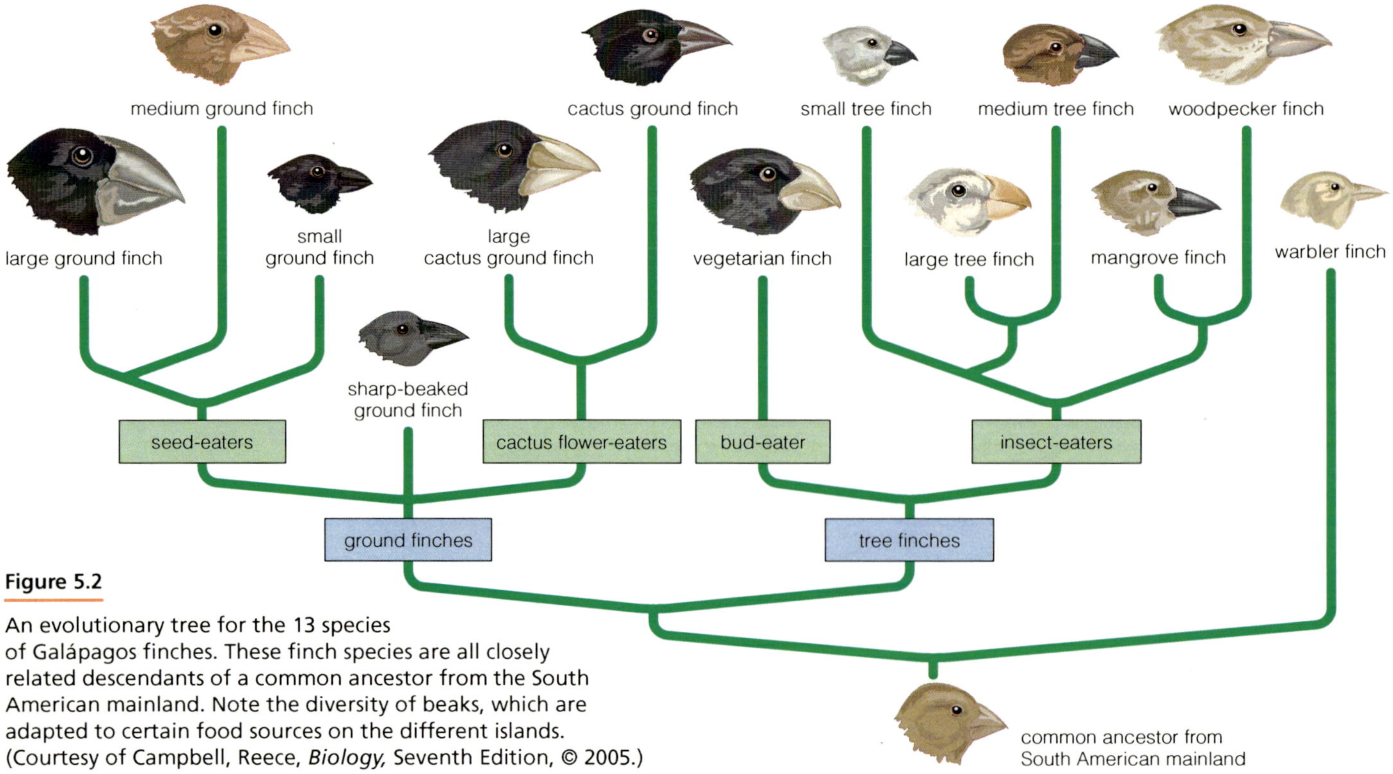

Figure 5.2

An evolutionary tree for the 13 species of Galápagos finches. These finch species are all closely related descendants of a common ancestor from the South American mainland. Note the diversity of beaks, which are adapted to certain food sources on the different islands. (Courtesy of Campbell, Reece, *Biology,* Seventh Edition, © 2005.)

these young volcanic islands closely resemble one another and mainland South American finches, but these particular species are found nowhere else in the world. Moreover, different islands have different species, and each species is adapted to particular microenvironments and lifestyles. For example, some live on the ground while some live in cacti or brush, and some have beaks adapted for cracking seeds while others have insect-catching beaks. Darwin realized that natural selection could explain this situation. He presumed that an ancestral pair of finches reached the Galápagos from the mainland (perhaps by being blown off course by winds). Over time, local populations of island finches gradually adapted to different environments and lifestyles, ultimately becoming the distinct species that he observed.

Darwin recognized similar patterns among many other species in the Galápagos and elsewhere in his round-the-world voyage. He also discovered fossils of extinct organisms that were clearly related to modern organisms, yet different in key respects. For example, in Brazil he found fossils of giant armadillos that he realized must be the ancestors of the modern armadillos found in the same region. The pieces of the puzzle gradually came together in Darwin's mind: He realized that natural selection not only explained the differences between closely related modern species like the finches, but that it also explained the fact that larger changes can occur over longer periods of time, with the result that entire species can become extinct and new ones can take their places.

Darwin also found strong support for his theory of evolution by looking at examples of *artificial selection*—the selective breeding of domesticated plants or animals by humans. Over the past few thousand years, humans have gradually bred many plants and animals into forms that bear little resemblance to their wild ancestors. For example, Figure 5.3 shows how artificial selection has created a variety of vegetables from a single common ancestor. Dogs also are examples of the power of artificial selection: Dog breeds as different as Rottweilers and Chihuahuas were bred from a common ancestor within just a few thousand years. Darwin recognized that if artificial selection could cause such profound changes in just a few

Figure 5.3

The six vegetables shown all look and taste quite different, but they were all bred by humans from the same wild ancestor (wild mustard). This is an example of artificial selection, which is much like natural selection except that humans (rather than nature) decide which individuals in each generation survive and breed offspring. (Courtesy of Campbell, Reece, *Biology,* Seventh Edition, © 2005.)

thousand years, natural selection could do far more over the many millions of years during which it has operated.

Today, we can observe natural selection occurring right before our eyes. In many places on Earth, species have changed in time spans as short as a few decades in response to human-induced environmental changes. On a microbial level, natural selection is what allows a population of bacteria to become resistant to specific antibiotics; those few bacteria that acquire a genetic trait of resistance are the only ones that survive in the presence of the antibiotic. Indeed, bacterial cases of natural selection pose a difficult problem for modern medicine, because bacteria can quickly develop resistance to almost any new drug we produce. As a result, pharmaceutical companies are constantly working to develop new antibiotics as bacteria become resistant to existing ones. Viruses can evolve even faster, which is one reason it has proved so difficult to fight viral diseases such as influenza and AIDS.

The Molecular Basis of Evolution Darwin's theory of evolution by natural selection tells us that species adapt and change by passing hereditary traits from one generation to the next. However, Darwin did not know precisely how these traits were communicated across generations, nor did he know why there is always variation among individuals or how new traits can appear in a population. Today, thanks to discoveries in molecular biology made since Darwin's time, we know the answers to all these questions. In particular, we now know that organisms are built from instructions contained in a molecule called **DNA** (short for "*d*eoxyribo*n*ucleic *a*cid"), and biologists can now trace how evolutionary adaptations are related to changes that occur through time in DNA. We'll discuss how DNA makes evolutionary adaptations possible when we discuss

SPECIAL TOPIC: CHARLES DARWIN AND THE THEORY OF EVOLUTION

Charles Robert Darwin was born into a wealthy and educated family in England on February 12, 1809. His father was a physician, and he had two famous grandfathers. His paternal grandfather, Erasmus Darwin (1731–1802), was a renowned physician and scientist who was a strong proponent of the idea that life evolved gradually. His maternal grandfather, Josiah Wedgwood, started the famous Wedgwood Pottery and China company that still bears his name. Darwin's mother died when he was just 8 years old, but his father and his extended family provided him with a generally happy childhood.

At his father's urging, Darwin enrolled in medical school at age 16. However, he was so horrified by the sight of operations, then done without anesthesia, that he left after just 2 years. He next enrolled in Christ College at Cambridge University, intending to become a minister. While there, he began to indulge a childhood love for the study of nature. Shortly after graduating in 1831, Darwin was offered the opportunity to serve as the naturalist aboard a ship of exploration—the HMS *Beagle.* Darwin was 22 years old when the *Beagle* set sail on December 27, 1831. The voyage lasted nearly 5 years, and proved to be the defining event of Darwin's life.

The *Beagle* spent much of its voyage exploring the coasts of South America and nearby islands. While the crew conducted surveys, Darwin went ashore to observe the geology and life and collected numerous specimens to take back to England. He also read extensively during the voyage, and one book that had recently been published proved particularly influential: Charles Lyell's *Principles of Geology,* which presented the case for an ancient Earth sculpted by gradual geological processes. Darwin was given the book by a friend who expected Darwin to disagree with its conclusions; instead, Darwin found that his own observations of geology gave only further credence to Lyell's theory.

Meanwhile, Darwin became intrigued by the many adaptations he observed for species in varied environments. He was particularly impressed by the animal life he observed during a 5-week stay on the Galápagos Islands, which lie approximately 1,000 kilometers (600 miles) due west of the South American coast of Ecuador. He focused special attention on the Galápagos finches, concluding that the different bird species must have evolved from a common mainland ancestor (see Figure 5.2). However, at the time he returned to England, he still did not understand how the evolutionary changes occurred.

In 1838, Darwin read Thomas Malthus's *An Essay on the Principle of Population,* in which Malthus famously argued that populations are capable of growing faster than food supplies can support. The essay helped Darwin crystallize the idea of natural selection by making clear that individuals within a population must compete for survival (the idea embodied in Fact 1 on p. 155). He then began intensive study of how humans bred domestic plants and animals, which helped him understand the variation in populations (Fact 2 on p. 156). By 1842, Darits structure and function in Section 5.4, but for now the key point is that our understanding of DNA means that we understand the specific mechanisms by which natural selection occurs.

Our detailed understanding of how evolution proceeds on a molecular level, coupled with all the other evidence for evolution collected by Darwin and others, puts the theory of evolution by natural selection on a solid foundation. That is, it is a *scientific theory* [Section 2.3] that has withstood countless tests and challenges, each of which has only strengthened its foundation. Like any scientific theory, the theory of evolution can never be proved beyond all doubt. However, as we'll discuss further in Section 5.6, no credible scientific alternative to the theory of evolution has been proposed, and the evidence in the theory's favor is overwhelming. Indeed, it is difficult to imagine any aspect of biological science that can be understood without being examined in the context of the theory of evolution. That is why evolution has become the unifying theme of all modern biology.

THINK ABOUT IT The idea that life changes through time is quite ancient, and it was already well supported by observations of fossils before Darwin was even born. Moreover, Lamarck recognized that evolution occurs as a result of adaptations about a half-century before Darwin advanced the idea of natural selection. Given these facts, explain why we credit Darwin with the theory of evolution.

• What is life?

Now that we have examined the fundamental properties of life on Earth, let's return to our original goal in this section: Can we come up with a definition of life?

SPECIAL TOPIC: CHARLES DARWIN AND THE THEORY OF EVOLUTION *continued*

win was convinced that natural selection held the key to evolution, and he began to draft the text that would eventually be published as *The Origin of Species*.

Darwin is said to have been a pleasant man. He was an ardent opponent of slavery, and while he was not a feminist by contemporary standards, he believed that women should be treated with dignity and respect. He was deeply concerned with the impact his theory would have on those who believed in the biblical story of creation. That is probably why he did not publish his theory immediately—he wanted to take time building his case, in hopes that his theory would be so strong that it would be accepted by all without anyone taking offense. Indeed, Darwin might have delayed publication indefinitely if not for a manuscript he received from another scientist, Alfred Russel Wallace, on June 18, 1858.

Wallace had been observing geology and life in Indonesia and had independently come to the same conclusion as Darwin: that life evolves through natural selection. After reading Wallace's draft paper, Darwin worried that "all my originality will be smashed." Fortunately, both Darwin and Wallace were willing to share credit. Their first papers on the theory of evolution were read back-to-back at a scientific meeting in London on July 1, 1858. A little over a year later, Darwin finally published *The Origin of Species*. All 1,250 copies in the first printing sold out on the first day. Within a decade, Darwin's theory was accepted by the vast majority of biologists, an acceptance that has grown stronger ever since.

Darwin never had a taste for arguing about evolution, leaving that to other scientists (especially Thomas Huxley, who called himself "Darwin's bulldog"). He continued his scientific work, publishing several more books on evolution and related topics. In his personal life, Darwin married a cousin, Emma. He and Emma had ten children, but two died in infancy and a third died at age 10. Darwin died on April 19, 1882, at the age of 73. A parliamentary petition won him burial in London's Westminster Abbey, where he lies next to Sir Isaac Newton.

Charles Darwin and his son William, photographed in 1842.

Based on the central role of evolution, our simplest definition might be that *life is something that can reproduce and evolve through natural selection*. This definition is probably sufficient for most practical purposes, but some cases may still challenge this definition. For example, scientists can now write computer programs (lines of computer code) that can reproduce themselves (create additional sets of identical lines of code). By adding programming instructions that allow random changes, so that the programs can compete and change through a computer analog of natural selection, scientists can even make "artificial life" that evolves on a computer.

THINK ABOUT IT Do you think computer programs that can reproduce and evolve are alive? Why or why not? Would your opinion change if these programs evolved to the point where they could write their own computer code or exchange e-mail with us? What if they wrote other programs that operated machinery to build other computers? Do you think it is possible to create true life on a computer?

Another issue with this definition concerns the origin of life. Darwin's theory of evolution does not tell us how the first life got started. For that, as we'll discuss in Chapter 6, we presume there must have been some type of molecular or *chemical evolution* (as opposed to biological evolution) that went on until the first living organism arose. The idea of chemical evolution is no more surprising than that of natural selection—certain chemical processes are energetically favored under certain circumstances, and laboratory experiments show that under the right conditions, chemicals can evolve in complexity much like life. However, it begs the question of whether we would recognize a clear distinction between, for example, the last case of chemical evolution and the first living organism capable of biological evolution. No one knows the answer; some scientists think there must have been a clear "first" living organism, while others think that the emergence of life may have been marked by a more gradual transition.

The fact that we have such difficulty distinguishing the living from the nonliving on Earth suggests that we should be cautious about constraining our search for life elsewhere. No matter what definition of life we choose, the possibility always remains that we'll someday encounter something that challenges our definition. Nevertheless, the ability to reproduce and evolve through natural selection seems likely to be shared by most, if not all, life in the universe. Thus, this idea provides a useful starting point as we consider how to explore the possibility of extraterrestrial life.

5.2 CELLS: THE BASIC UNITS OF LIFE

Now that we have discussed general properties of life, we are ready to look more specifically at the nature of life on Earth. In this section, we will explore cells, the basic units of life. We will then be prepared to consider in Section 5.3 how cells make use of energy and in Section 5.4 how cells reproduce and evolve.

• What are living cells?

All living organisms are made of **cells**—microscopic units in which the living matter inside is separated from the outside world by a barrier called

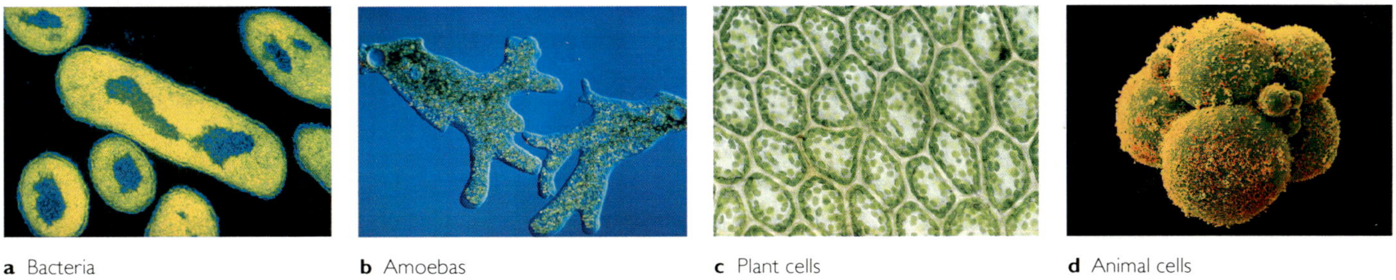

Figure 5.4

Microscopic views of four types of living cells. In each cell, a membrane separates the living matter inside the cell from the outside world.

a **membrane*** (Figure 5.4). Thus, cells are the basic structures of life on Earth. Some organisms consist only of a single cell. Other organisms, like oak trees and people, are complex structures in which trillions of cells work cooperatively, dividing various tasks among specialized cells of different types.

Given the great diversity of life on Earth, you may be surprised to learn that all living cells share a great many similarities. For example, all pass on their hereditary information in the same basic way with DNA, and many other chemical processes are nearly the same in all cells. These similarities, which we'll discuss in more detail later, are profoundly important to our understanding of the origin of life. As far as we know, there is no reason why all living cells must share these characteristics. That is, while life elsewhere might also be composed of cells, we should not expect those cells to have the same biochemistry as cells on Earth. Thus, the many similarities among all cells on our planet suggest a startling conclusion: *All life on Earth shares a common ancestor.* In other words, every living organism on Earth is related to every other one because all evolved over billions of years from the same origin of life.†

Earth Life Is Carbon-Based

Life on Earth is made from more than 20 different chemical elements. However, just four of these elements—oxygen, carbon, hydrogen, and nitrogen—make up about 96% of the mass of typical living cells. Most of the remaining mass consists of just a few other elements, notably calcium, phosphorus, potassium, and sulfur (Figure 5.5).

Given that oxygen dominates Figure 5.5, you might be tempted to say that life on Earth is "oxygen-based." However, most of the oxygen in living cells is actually part of water molecules (H_2O). The molecules that account for a cell's structure and function owe their remarkable qualities to a different element: carbon. For this reason, we say that life on Earth is **carbon-based**.

Why is carbon so important to life on Earth? The primary answer to this question lies in how carbon can combine with other elements to

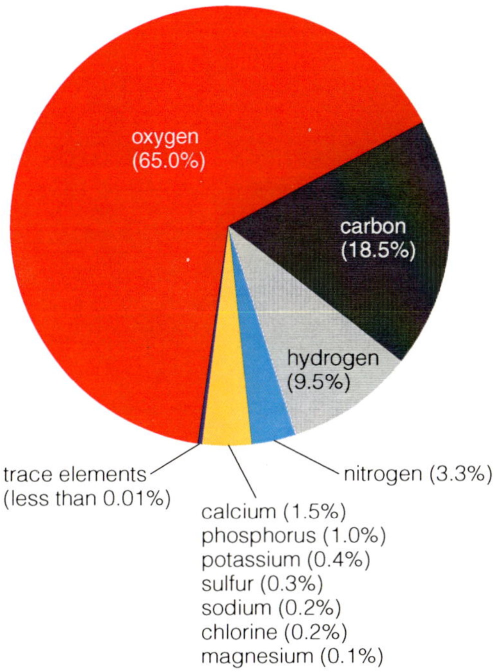

Figure 5.5

This pie chart shows the chemical composition of the human body by weight; this composition is fairly typical of all living matter on Earth. The trace elements include (in alphabetical order) boron, chromium, cobalt, copper, fluorine, iodine, iron, manganese, molybdenum, selenium, silicon, tin, vanadium, and zinc. (Adapted from Campbell, Reece, Simon, *Essential Biology,* Second Edition, © 2004.)

*Some organisms, such as some slime molds, do not perfectly fit this picture of discrete cells, because they consist of a large mass of protoplasm containing thousands of nuclei. Nevertheless, the basic idea that living tissue is contained in a package separated from the external environment still holds.

†We have not yet identified every type of living organism on Earth, so it is still conceivable that we'll someday discover organisms right here on Earth (perhaps deep underground or in other isolated ecosystems) that use a different biochemistry and hence seem to have come from a separate origin of life. If so, it would greatly expand our understanding of biology, since we'd have more than one form of life to study up close in our laboratories.

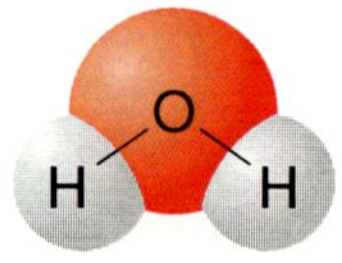

Figure 5.6

A water molecule has a single oxygen atom bonded to two hydrogen atoms. (Adapted from Campbell, Reece, Simon, *Essential Biology,* Second Edition, © 2004.)

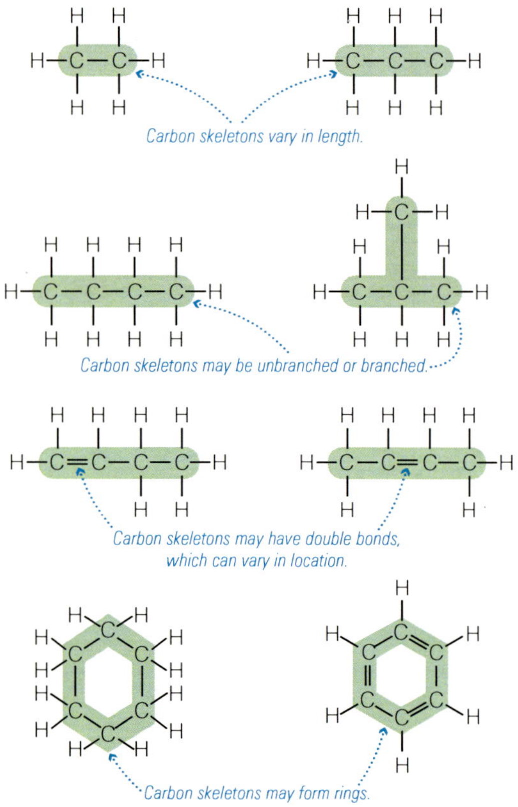

Figure 5.7

These diagrams represent several relatively simple hydrocarbons—organic molecules consisting of a carbon skeleton attached to hydrogen atoms. The carbon skeletons are highlighted in green. Each single line represents a single chemical bond; a double line represents a double bond. Note that every carbon atom has a total of four bonds (a double bond counts as two single bonds). (Adapted from Campbell, Reece, Simon, *Essential Biology,* Second Edition, © 2004.)

make complex molecules. The atoms in any molecule are linked together by what we call **chemical bonds**, which essentially involve sharing of electrons between the individual atoms of a molecule. Different elements are capable of making chemical bonds in different ways. For example, hydrogen atoms generally can bond with only one other atom at a time, while oxygen atoms generally can bond with at most two other atoms at a time. We can see these properties in a water molecule (Figure 5.6): The oxygen atom has two chemical bonds, one to each of the two hydrogen atoms, while each hydrogen atom has only a single chemical bond to the oxygen atom.

Carbon is a particularly versatile chemical element because it can bond to as many as four atoms at a time. This allows carbon atoms to link together in an endless variety of carbon "skeletons" varying in size and branching patterns (Figure 5.7). The carbon atom sometimes uses two of its bonds to link with the same atom, forming what we call a *double bond;* notice the double bonds in the lower set of molecules of Figure 5.7.

We refer to carbon molecules generically as **organic molecules**. The simplest organic molecules consist of carbon skeletons bonded only to hydrogen atoms; these simple organic molecules are often called *hydrocarbons* to reflect the fact that they contain only hydrogen and carbon. In more complex organic molecules, one or more carbon atoms are bonded to something besides hydrogen and other carbon atoms (Figure 5.8).

Silicon-Based Life (and Other Similar Ideas) When we consider the possibility of extraterrestrial life, it's natural to wonder whether it might be based on an element besides carbon. In truth, we cannot say for sure whether other elements would work. However, given the importance to life on Earth of carbon's ability to form four bonds at once, we might expect that any other elemental basis for life would have to have the same bonding capability. Among the elements common on Earth's surface—and likely to be common on other planets—silicon is the only element besides carbon that can have four bonds at once. As a result, science fiction writers have often speculated about finding silicon-based life on other worlds.

Unfortunately for science fiction, silicon has at least three strikes against it as a basis for life. First and most important, the bonds formed by silicon are significantly weaker than equivalent bonds formed by carbon. As a result, complex molecules based on silicon are more fragile than those based on carbon—probably too fragile to form the structural components of living cells. In particular, complex silicon-based molecules cannot exist long in water, which is also generally thought to be necessary to life [Section 7.1]. Second, unlike carbon, silicon does not normally form double bonds; instead, it forms only single bonds. This limits the range of chemical reactions that silicon-based molecules can engage in as well as the variety of molecular structures that can form. Third, carbon can be mobile in the environment in the form of gaseous carbon dioxide, but silicon dioxide is a solid (for example, quartz is made from silicon dioxide) that offers no similar mobility. Given the three strikes against silicon, most scientists consider it unlikely that life can be silicon-based. Moreover, observational evidence on Earth also argues against silicon: Silicon is about 1,000 times as abundant as carbon in Earth's crust, so the fact that life here is carbon-based despite the greater abundance of silicon suggests that carbon will always win out over silicon as a basis for life.

A few other elements have also been suggested as possibilities for replacing carbon on other worlds, but most scientists believe carbon's natural advantages will still win out. We have found carbon-based (organic) molecules even in space (as identified in meteorites and interstellar clouds), suggesting that carbon chemistry is so easy and so common that even if life with another basis were possible, carbon-based life probably would arise first and then reproduce so successfully that it would crowd out the possibility of any other type of life.

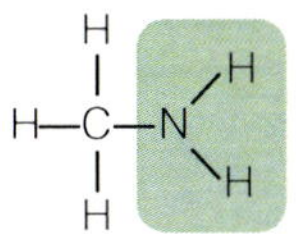

Figure 5.8

In a more complex organic molecule, at least one bond links a carbon atom to something besides hydrogen or another carbon atom. Here, one of the carbon atom's four bonds links it to an *amino group* (which consists of a nitrogen atom and two hydrogen atoms), highlighted in green. (Adapted from Campbell, Reece, Simon, *Essential Biology,* Second Edition, © 2004.)

• What are the molecular components of cells?

All the major components of cells are made from complex organic molecules. Today, biologists know the precise chemical structure of a great many of these molecules, and this knowledge has enabled them to gain a deep understanding of the biochemistry of life. If you take a course in biology, you will learn about much of this biochemistry; here we focus only on generalities about the molecules of life. The large molecular components of cells fall into four main classes: *carbohydrates, lipids, proteins,* and *nucleic acids.* Let's briefly investigate the properties of each class that are most important to life.

Carbohydrates

You're probably familiar with **carbohydrates** as a source of food energy—the sugars and starches known to athletes and dieters as "carbs." In addition to providing energy to cells, carbohydrates make important cellular structures. For example, a carbohydrate called *cellulose* forms the fibers of cotton and linen and is the main constituent of wood. Life on other worlds would presumably need molecules to play these same energy-storing and structural roles, but we do not yet know whether such molecules would have to resemble carbohydrates on Earth or if they could be very different in their chemistry. As a result, we will have little more to say about carbohydrates in this book.

Lipids

Like carbohydrates, **lipids** can store energy for cells. The types of lipids that store energy are more commonly known as *fats.* Thus, despite the bad reputation of fat, it is actually critical to living cells. Lipids also play a variety of other roles in cells on Earth, but from the standpoint of life in the universe, perhaps their most important role is as the major ingredients of cell membranes. That is, lipids form the barriers that make it possible for cells to exist. Moreover, as we'll discuss in more detail in Chapter 6, the membrane-forming role of lipids is thought to have played a critical role in the origin of life: Lipids can spontaneously form membranes in water and probably did so on the early Earth. Other organic molecules would have been trapped inside the space formed by the membranes, making what were in essence tiny chemical factories. These tiny chemical factories may have facilitated the chemical reactions that ultimately led to the origin of life.

Proteins: Key Evidence for a Common Ancestor of Life

The molecules called **proteins** are often described as the workhorses of cells, because they participate in such a vast array of functions. Some proteins serve as structural elements in cells. Others, called **enzymes**, are crucial to nearly all the important biochemical reactions that occur within cells—including the copying of genetic material (DNA)—because they serve as *catalysts* for these reactions. A **catalyst** is any substance (not necessarily a single molecule) that either facilitates a chemical reaction that would not otherwise occur or accelerates a reaction that would otherwise occur much more slowly; the catalyst itself is not changed by the process. Enzymes are catalysts because they greatly accelerate the reactions in which they are involved, even though they enter and leave the reactions essentially unchanged. Moreover, because an enzyme is left unchanged after it catalyzes a reaction, a single enzyme can catalyze a specific chemical reaction many times without needing to be rebuilt.

All proteins, whether they serve as enzymes or in other roles, are large molecules built from long chains of smaller molecules called **amino acids**. The "amino" in amino acids refers to the *amino group* that they all share—a nitrogen atom bonded to two hydrogen atoms and a carbon atom (see Figure 5.8); amino acids also always contain what is called a *carboxyl group* (COOH). Different types of amino acids are distinguished by the different sets of atoms also bonded to the central carbon.

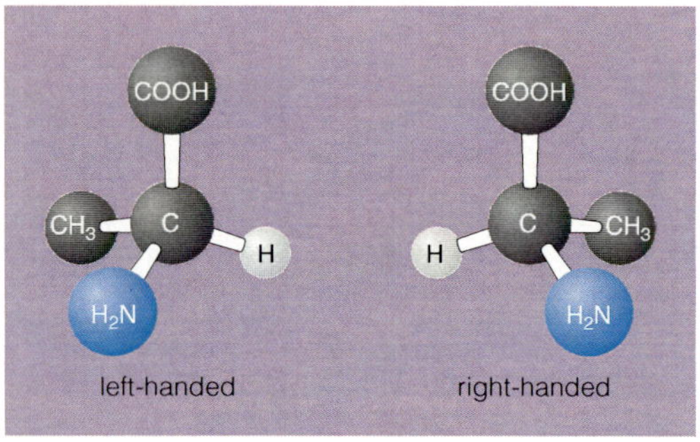

Figure 5.9

Any particular amino acid comes in two forms, distinguished by their *handedness.* These diagrams show the left- and right-handed versions of the amino acid *alanine.* Notice that the two versions are mirror images of each other.

The nature of the amino acid chains that make proteins in living organisms provides important evidence supporting the idea that all life on Earth shares a common ancestor. Biochemists have identified more than 70 different amino acids, but most life on Earth builds proteins from only 20 of them. (Two additional amino acids are known to be used in rare cases by particular microorganisms, and scientists suspect that other cases of rare amino acids may yet be discovered.) If life on Earth had more than one common ancestor, we might expect that different organisms would use different sets of amino acids, but they don't. Moreover, naturally occurring amino acids come in two slightly different forms, distinguished by their **handedness** (or *chirality*): The "left-handed" and "right-handed" versions are mirror images of each other (Figure 5.9). Amino acids found in nonbiological circumstances generally consist of a mix of the left- and right-handed versions, but living cells use only the left-handed versions of amino acids to build proteins. Again, the fact that all life on Earth makes use of the same versions of amino acids suggests a common ancestor. Carbohydrates provide some similar evidence, as life on Earth uses mainly the right-handed versions of sugars.

THINK ABOUT IT Large impacts can blast meteorites into space, allowing rocks from one world to travel to another. As we'll discuss in Chapter 6, some scientists hypothesize that microscopic life might survive such impacts and thereby might have migrated between the inner worlds of our solar system. Suppose we discover life on Mars and, while it also has proteins, we find that it builds them from a different set of amino acids than does life on Earth, and they are all the right-handed versions. Would this support or contradict the hypothesis that life migrated between Earth and Mars?

Nucleic Acids Perhaps no cellular molecule is more famous than DNA, which is the basic hereditary material of all life on Earth. A second important nucleic acid, RNA (short for "*r*ibo*n*ucleic *a*cid"), helps carry out the instructions contained in DNA. Thus, the nucleic acids DNA and RNA are responsible for allowing cells to function according to precise, heritable instructions. Changing a cell's DNA changes the inherent nature of an organism; indeed, it is changes to DNA that allow species to evolve. We do not know whether other types of molecules could replace nucleic acids in life elsewhere, but it is difficult to imagine life existing in any form without a molecule or molecules to serve the hereditary functions of DNA and RNA. These molecules are so important that we'll devote Section 5.4 to discussing them in more detail.

• What are the major groupings of life on Earth?

Up until just a few decades ago, we were generally able to classify life only by its outward appearance. If you look around, appearances suggest what most people believed for thousands of years—that life exists in the two basic forms that we call plants and animals. The first evidence of a different reality surfaced around the same time as the Copernican revolution. While Galileo turned his telescopes to the heavens, other scientists began to employ similar lens technology to study the microscopic world. The precise origin of the microscope is not known, but the first practical microscopes used for scientific study were built by the Dutch scientist Anton Van Leeuwenhoek (1632–1723; last name pronounced *"LAY-ven-hook"*). During decades of observations beginning around 1674, Leeuwenhoek discovered the world of microscopic life. He was the first to realize that drops of pond water are teeming with microorganisms—a discovery now repeated by almost every elementary school student. He also discovered bacteria and studied the microscopic structure of many plant and animal cells.

With hindsight, it may seem surprising that anyone could have thought that all microorganisms might just be tiny plants or animals, since we now know that microorganisms are far more genetically diverse than larger organisms. But that is exactly what happened. If you look at an old-enough biology textbook, you'll see that life was classified into two "kingdoms": the plants and the animals, and microbes were generally just stuck into one of those two. In the 1960s, biologists expanded the list to five kingdoms, with two (*protista* and *monera*) reserved for microorganisms; the third new addition was *fungi,* by then recognized to be different from plants. However, as our understanding of biochemistry improved during the ensuing decades, biologists began to consider whether life could be classified by its cell structure or biochemistry (including genetics), rather than by its outward appearance. Today, we know that classification by the biochemistry of cells gives us much deeper insights into the relationships between different living species than does classification by appearances alone.

Microscopic Life

Plant and animal cells look different under a microscope; for example, plant cells have what we call a *cell wall* that helps protect the cell membrane, while animal cells do not. However, if you examine many different types of cells under a microscope, you'll find that plant and animal cells overall don't seem that different from one another, at least in comparison to the variety of other types of cells that you'll see. From their microscopic structural appearances, living cells on Earth appear to come in two basic types distinguished by the presence or absence of a **cell nucleus**—an internal membrane that effectively walls off the genetic material (DNA) from the rest of the cell. Organisms that lack cell nuclei are called **prokaryotes** and organisms that have cell nuclei are called **eukaryotes** (Figure 5.10).

All prokaryotes are single-celled organisms (though some of these organisms form colonies in which they live or work together), but they nonetheless represent an incredibly wide range of different species. Examples of prokaryotes include the well-known bacteria *E. coli* and *Salmonella,* both of which are common causes of food poisoning. Eukaryotes may be either single-celled—amoebas are examples—or multicellular. All multicellular organisms are eukaryotes, which means the eukaryotes encompass all plants and animals.

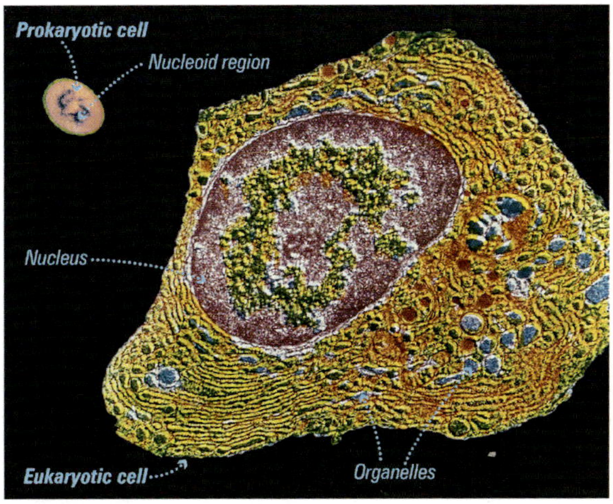

Figure 5.10

These microscopic photographs contrast a typical prokaryotic cell and a typical eukaryotic cell. The eukaryotic cell (an amoeba) is about ten times larger in diameter than the prokaryotic cell (a bacterium); this size difference is typical, though prokaryotic cells come in a fairly wide range of sizes. The eukaryotic cell is also much more complex, with numerous specialized structures and a *cell nucleus* containing its DNA. The prokaryotic cell lacks a distinct nucleus, though its DNA is still concentrated near the center.

Because we are more familiar with plants and animals than with microbes (meaning any single-celled organism), most people assume that microscopic life is a "minor" part of life on Earth. Also, because we tend to associate bacteria with disease, many people assume that microbes are generally harmful. In fact, both assumptions are wrong. Although we humans like to think of ourselves as the dominant form of life on Earth, this is not the case. Much as the Copernican revolution proved that we are not the center of the universe, recent inventories of life show that we are not even the center of Earth's "biological universe." We do not yet know the total amount of microscopic life on Earth, because we do not know how far life extends underground and we are only beginning to get accurate measurements of the amount of life in seawater and soil, but even these limited measurements make clear that tiny microbes are far more dominant in terms of mass and volume than we are (see Cosmic Calculations 5.1). Moreover, most microbes (including most bacteria) are harmless to humans, and many are crucial to our survival. For example, bacteria in our intestines provide us with important vitamins, and bacteria living in our mouths prevent harmful fungi from growing there. Other microbes play crucial roles in cycling carbon and other vital chemical elements between organic matter and the soil and atmosphere; for example, microbes are responsible for decomposing dead plants and animals. Indeed, plant and animal life would be doomed if microbes somehow disappeared from Earth. In contrast, microbes could survive just fine without plants and animals, as they did during most of the history of life on Earth [Section 6.3].

The Three Domains of Life on Earth Because of the visible distinction between cells with and without nuclei, biologists long assumed that prokaryotes and eukaryotes must represent two fundamentally different

COSMIC CALCULATIONS 5.1

The Dominant Form of Life on Earth

We can use estimation to show that microbes far outweigh human beings on Earth. We first estimate the total mass of the approximately 6 billion human beings on Earth. If an average person is 50 kilograms (110 pounds), the total human mass is about:

$$6 \times 10^9 \text{ persons} \times 50\frac{\text{kg}}{\text{person}} = 3 \times 10^{11} \text{ kg}$$

We next estimate the mass of microbes in the oceans. The density of microbes varies significantly with location and depth, but a rough average is *1 billion* (10^9) microbes per liter of water. Multiplying this value by the total volume of ocean water (a Web search reveals this to be about 1.4×10^9 km^3) gives us an estimate of the total number of microbes in the ocean:

$$\text{Total microbes} \approx \underbrace{10^9\frac{\text{microbes}}{\text{liter}}}_{\text{density of microbes in ocean water}} \times \underbrace{(1.4 \times 10^9 \text{ km}^3)}_{\text{volume of ocean water}} \times \underbrace{\left(10^3\frac{\text{m}}{\text{km}}\right)^3 \times \left(10^3\frac{\text{liter}}{\text{m}^3}\right)}_{\text{convert from km}^3 \text{ to liters}}$$

$$= 10^9\frac{\text{microbes}}{\text{liter}} \times (1.4 \times 10^9 \text{ km}^3) \times \left(10^9\frac{\text{m}^3}{\text{km}^3}\right) \times \left(10^3\frac{\text{liter}}{\text{m}^3}\right)$$

$$= 1.4 \times 10^{30} \text{ microbes}$$

To find the total mass of microbes, we next need to know the typical microbe mass. A typical bacterium measures about 1 micrometer on a side, which means it has a volume of about 1 cubic micrometer. There are 1 million (10^6) micrometers per meter, so the volume of a bacterium is:

$$1\ \mu\text{m}^3 \times \left(10^{-6}\frac{\text{m}}{\mu\text{m}}\right)^3 = 10^{-18} \text{ m}^3$$

Because life is made mostly of water, we can use the density of water (1,000 kg/m^3) as the density of a microbe. Multiplying the microbe volume by the density, we estimate that the typical microbe mass is:

$$10^{-18} \text{ m}^3 \times 10^3\frac{\text{kg}}{\text{m}} = 10^{-15} \text{ kg}$$

We combine our results to find the total mass of microbes in the oceans:

$$\text{total mass of microbes} \approx (\text{number of microbes}) \times (\text{mass per microbe})$$

$$= 1.4 \times 10^{30} \text{ microbes} \times 10^{-15}\frac{\text{kg}}{\text{microbe}}$$

$$= 1.4 \times 10^{15} \text{ kg}$$

We can compare to the mass of human beings by dividing:

$$\frac{\text{total mass of microbes}}{\text{total mass of humans}} \approx \frac{1.4 \times 10^{15} \text{ kg}}{3 \times 10^{11} \text{ kg}} \approx 5{,}000$$

The total mass of microbes in the oceans is roughly 5,000 times that of all humans combined.

lines of life. However, analysis of cellular biochemistry has shown that this is *not* the case. Instead, based on biochemical and genetic similarities and differences, biologists now classify all life into three broad **domains** known as the **Bacteria**, the **Archaea**, and the **Eukarya**. The biochemical and genetic relationships between these domains are illustrated in what biologists usually call the **tree of life**. Figure 5.11 represents the basic structure of the tree of life as it is known today.

The tree of life is proving to be an extremely valuable organizational tool for biologists, and there are at least three key ideas that emerge from it that you should remember as you continue your study of astrobiology:

1. The idea of classifying life into three domains by biochemistry and genetics gives us an organizational scheme that is fundamentally different from the older idea of classifying life into "kingdoms" based solely on structural and physiological differences among species. Notice, for example, that all plants and animals are now seen to represent just two small branches of domain Eukarya. (If it helps, you can think of the domains as "superkingdoms" that represent a higher level of structure than the old kingdoms.)
2. The tree gives us a new and fundamentally different view of relationships among microbes. Bacteria and Archaea used to be grouped together as prokaryotes because they all lack cell nuclei, but we now place them in different domains because they differ in basic features of their biochemistry. For example, Bacteria and Archaea have different types of lipid structures in their cell membranes and they make their cellular proteins in different ways. Indeed, Archaea appear to be somewhat more closely related to Eukarya than to Bacteria.
3. The true diversity of life on Earth is found almost entirely within the microscopic realm. Biochemically and genetically, we humans (and all other animals) are much more closely related to mushrooms than most microbes are related to one another. Moreover, because the vast majority of microscopic species probably remain undiscovered, we are likely to find much more diversity as we continue to study biology. We will certainly discover many more branches within the three domains, and it is even possible that entirely new domains will be discovered in the future.

In addition, the branch lengths shown in the tree of life represent the amounts of genetic difference between different species. Thus, as we trace the branches back toward the "root" shown in Figure 5.11, we are presumably looking back to species that split from common ancestors at earlier times. The closer the we get to the root, the closer we must be to finding an organism that resembles a common ancestor of all life on Earth. As we'll discuss in Chapter 6, this idea is being used to probe possible scenarios for the origin of life on Earth.

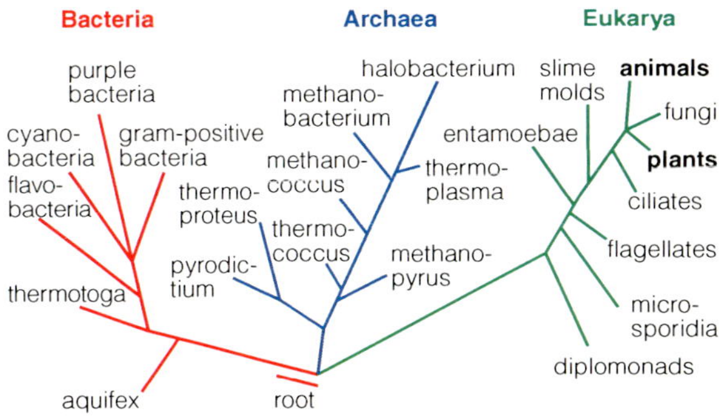

Figure 5.11

The tree of life has three major domains: Bacteria, Archaea, and Eukarya; note that all plants and animals represent just two small branches of the domain Eukarya. (Only a few of the many known branchings within each domain are shown in this diagram; it also remains possible that additional domains will be discovered.)

5.3 METABOLISM: THE CHEMISTRY OF LIFE

Why are cells so important to life on Earth? More to the point, is it possible that life elsewhere might exist without having a fundamental organizational unit like the cell? To answer these questions, we must understand the processes that take place inside living cells. These processes, which are all chemical in nature, make up what we call **metabolism**. More specifically, metabolism is a blanket term that refers to the many chemical reactions that occur in living organisms and are involved in providing energy or nutrients to cells.

• What are the basic metabolic needs of life?

Most of the important chemical reactions that occur in cells share a common characteristic: Without the help provided by the cell itself, the reactions would occur too slowly to be useful for life. In this sense, a cell's primary purpose is to serve as a tiny chemical factory in which desired chemical reactions occur much more rapidly than they could otherwise, thereby making it possible to turn simple molecules into the great variety of complex organic molecules needed by living organisms. As is also the case in many factories, cellular work sometimes involves breaking down molecules as well as building them. Like any manufacturing process, this biochemical manufacturing process requires two basic things:

1. *A source of raw materials* with which to build new products. In the case of living cells, the key raw materials are molecules that provide the cell with carbon and other basic elements of life.
2. *A source of energy* to fuel the metabolic processes that break down old molecules and manufacture new ones.

Given the large variety of molecules involved in metabolic processes, you might think that cells would need an equally large variety of sources of raw materials and energy to survive. However, cells have the ability to build incredible variety from a limited set of starting materials. Part of this ability comes from the remarkable variety of enzymes in living cells. Each enzyme is specialized to catalyze particular chemical reactions needed in cellular manufacturing. The remarkable diversity of enzymes in living organisms today is a testament to the power of evolution. The instructions for enzyme creation are encoded in DNA and hence have been evolving for billions of years.

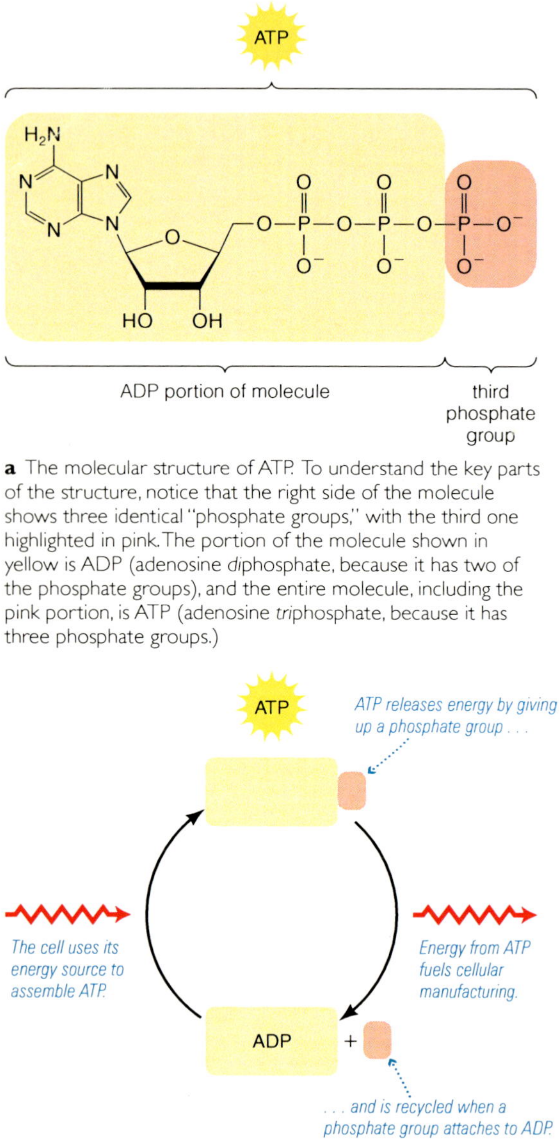

a The molecular structure of ATP. To understand the key parts of the structure, notice that the right side of the molecule shows three identical "phosphate groups," with the third one highlighted in pink. The portion of the molecule shown in yellow is ADP (adenosine *di*phosphate, because it has two of the phosphate groups), and the entire molecule, including the pink portion, is ATP (adenosine *tri*phosphate, because it has three phosphate groups.)

b Cells recycle ATP. The ATP molecule gives up energy when it splits into ADP (yellow) and a phosphate group (pink). Energy input puts the ATP molecule back together.

Figure 5.12

(Adapted from Campbell, Reece, Simon, *Essential Biology,* Second Edition, © 2004.)

The Role of ATP

Another reason why cells can produce so much variety from so little input is that, regardless of where they get their energy, all cells put the energy to work in the same basic way. Every living cell uses the same molecule, called **ATP** (short for "*a*denosine *tri*phosphate"), to store and release energy for nearly all its chemical manufacturing (Figure 5.12a). This usage of ATP vastly simplifies the manufacturing process, because it means that a cell needs an outside energy source only for the purpose of producing ATP, rather than for producing the full variety of organic molecules in cells. Once ATP is produced, it can be used to provide energy for any cellular reaction. Moreover, the nature of ATP makes it completely recyclable. Each time a cell draws energy from a molecule of ATP, it leaves a closely related by-product, called ADP (short for "*a*denosine *di*phosphate"), that can be easily turned back into ATP (Figure 5.12b).

The fact that all life on Earth uses the same molecule (ATP) for energy storage offers further evidence for a common origin of life. There's no known reason why other molecules could not fill the role of ATP. Thus, the fact that all living cells use ATP suggests that they all evolved from a common ancestor that made use of this remarkable molecule.

Carbon Sources and Energy Sources

Because carbon compounds are the primary raw materials needed for life, the needs of metabolism essentially come down to the need for a carbon source and an energy source.

We humans, like all animals, meet both of these needs with food. The food we eat gets digested and carried in molecular form by our bloodstreams to individual cells. There, the cells make use of the molecules from our food sources. Some of the molecules are used as the carbon source for cellular manufacturing, while others undergo chemical reactions that release energy the cell can use to fuel its ATP cycle. Of course, not all organisms get their carbon and their energy by eating other organisms. Plants, for example, get energy from sunlight, and some microorganisms get energy from chemical reactions that take place inside rocks or in hot springs. Let's next turn our attention to the variety of carbon and energy sources used by different forms of life on Earth.

• How do we classify life by its metabolic sources?

Living cells on Earth get their carbon and energy from a surprisingly wide variety of sources. Indeed, as we'll see, this wide variety makes life elsewhere seem more likely. For example, we already know of worlds within our own solar system, such as Mars and Europa, that may well have the necessary materials and energy source for metabolism, suggesting that life could at least in principle exist on those worlds.

Today, astrobiologists classify life into four major categories by its metabolic sources, summarized in Table 5.1. We can understand their rather long and technical names by looking first at their carbon sources and then at their energy sources.

Carbon Sources: Autotrophs and Heterotrophs Cells need a source of carbon from which to build the skeletons of their organic molecules. In the broadest sense, cells can get their carbon in either of two ways:

1. Some cells get carbon by consuming preexisting organic compounds, that is, by eating. For example, we humans acquire carbon by eating plants or other animals. Any organism that gets its carbon by eating is called a **heterotroph;** the word comes from *hetero* meaning "others" and *troph* meaning "to feed." All animals are heterotrophs, as are many microscopic organisms.
2. Some cells get carbon directly from the environment by taking in carbon dioxide from the atmosphere or carbon dioxide dissolved in water. An organism that gets its carbon directly from the environment is called an **autotroph,** meaning "self-feeding." For example, trees and most other plants are autotrophs.

If you look at the first column of Table 5.1, you'll see that the first two entries are both autotrophs and the second two are both heterotrophs. However, in each case the entry carries a prefix of either *photo* (meaning "light") or *chemo* (meaning "chemicals"). These prefixes describe the energy sources that go with the carbon sources of either eating (heterotroph) or environmental carbon dioxide (autotroph).

Energy Sources: Light or Chemicals Broadly speaking, the energy source that a living cell uses to make ATP can be one of three things (see Section 9.4 for more details about chemical energy for life):

1. Some cells get energy directly from sunlight, using the process we call *photosynthesis.* For example, plants acquire their energy from sunlight. Organisms that get energy from sunlight are given the prefix *photo.*

TABLE 5.1 *Metabolic Classifications of Living Organisms*

Metabolic Classification	*Carbon Source*	*Energy Source*	*Examples*
photoautotroph	carbon dioxide	sunlight	plants, photosynthetic bacteria
chemoautotroph	carbon dioxide	inorganic chemicals (e.g., iron, sulfur, ammonia)	some bacteria and archaea, especially in extreme environments
photoheterotroph	organic compounds	sunlight	some bacteria and archaea
chemoheterotroph	organic compounds	organic compounds	animals, many microbes

2. Some cells get energy from food; that is, they take chemical energy from organic compounds they've eaten and use it to make their own ATP. Since the energy comes from chemical reactions, these organisms get the prefix *chemo*.
3. Some cells get energy from *inorganic* chemicals—chemicals that do *not* contain carbon—in the environment. This type of energy source is different in character from organic food, and cells that get energy directly from the environment require neither sunlight nor other organisms to survive. However, because the energy still comes from chemical reactions, these organisms also get the prefix *chemo*.

The Four Metabolic Classifications We can now put the carbon and energy sources together to understand the four metabolic classifications in Table 5.1. The first row of the table shows the **photoautotrophs**, which get their energy from sunlight (*photo*) and their carbon from carbon dioxide in the environment (making them autotrophs). This category therefore includes plants, as well as microorganisms that obtain their energy through photosynthesis.

The second row shows the **chemoautotrophs**, which obtain energy from chemical reactions (*chemo*) involving inorganic chemicals and carbon from environmental carbon dioxide (making them autotrophs). These are in some ways the most amazing organisms, because they need neither organic food nor sunlight to survive. For example, the archaea known as *Sulfolobus* (a genus that includes many distinct species) live in volcanic hot springs and obtain energy from chemical reactions involving sulfur compounds. As is the case with *Sulfolobus*, chemoautotrophs are often found in environments where most other organisms could not survive. For much the same reason, they may also be the organisms most likely to be found on other worlds, since a wider range of conditions seems suitable to them than to other forms of life.

The third row shows the **photoheterotrophs**, which get energy from sunlight (*photo*) but get their carbon by consuming other organisms or the remains of such organisms (making them heterotrophs). This category is much rarer, but some prokaryotes do indeed get their carbon by eating organic compounds while making ATP with energy from sunlight. Examples include bacteria known as *Chloroflexus*, which obtain their carbon from other bacteria but their energy from photosynthesis. These organisms live in lakes, rivers, hot springs, and some aquatic environments very high in salt content.

The fourth row shows **chemoheterotrophs**, which get both their energy and their carbon from food. This category therefore includes us and all other animals, as well as many microorganisms. That is, we are chemoheterotrophs because we extract chemical energy (*chemo*) from food and carbon from eating (making us heterotrophs).

THINK ABOUT IT Classify each of the following into one of the four metabolic categories listed in Table 5.1: (a) an organism that gets its energy from chemicals near an undersea volcano and gets its carbon from carbon dioxide dissolved in water; (b) a tomato plant; (c) a fly.

Metabolism, Water, and the Search for Life The four metabolic classifications are quite general, so they ought to apply equally well to life elsewhere as to life on Earth. Moreover, any type of complex metabolism requires the existence of some kind of structure that allows carbon and energy to come together to manufacture (and break down) the molecules needed for life. Thus, unless we are failing to imagine an entirely different potential mode of operation, it seems likely that all living organisms must have a fundamental structure that functions much like cells on Earth. This crucial observation means that we can search for life in the universe by searching for cells rather than having to search for a much broader variety of possible structures.

This leaves us with one final ingredient to consider in metabolism: liquid water. On Earth, water plays three key roles in metabolism. First, metabolism requires that organic chemicals be readily available for reactions. Liquid water makes this possible by allowing organic chemicals essentially to float within the cell (because the chemicals dissolve in water). Second, metabolism requires a means of transporting chemicals to and within cells, and of transporting waste products away; water is the medium of this transport. Third, water plays a role in many of the metabolic reactions within cells; for example, water molecules are necessary for the reactions that store and release energy in ATP.

All living cells on Earth depend on liquid water to play these three roles, and this dependence limits the conditions under which we find life on our planet: We find life only in places where it is neither too cold nor too hot for liquid water to exist. Indeed, while we've seen that life on Earth can use a variety of different carbon and energy sources, liquid water is one thing that no organism on Earth can survive without. (Some organisms can become dormant and grow or metabolize in the absence of liquid water, but they cannot survive permanently in such conditions.) Does this need for liquid water also apply to life on other worlds? Certainly, some kind of liquid seems necessary, but we'll save discussion of possibilities other than water for Chapter 7.

5.4 DNA AND HEREDITY

In the previous two sections, we studied two key features of life on Earth that are likely to be crucial to life anywhere else as well: the structural units of cells and the metabolic processes that keep cells alive. A third feature that seems generally needed for all life is some means of storing information; that is, a set of "operating instructions" for the cell and a way of passing these instructions down through the generations. This information is what we generally call an organism's *heredity.*

All living things on Earth encode their hereditary information in the molecule known as DNA (although some viruses use RNA). That is, DNA

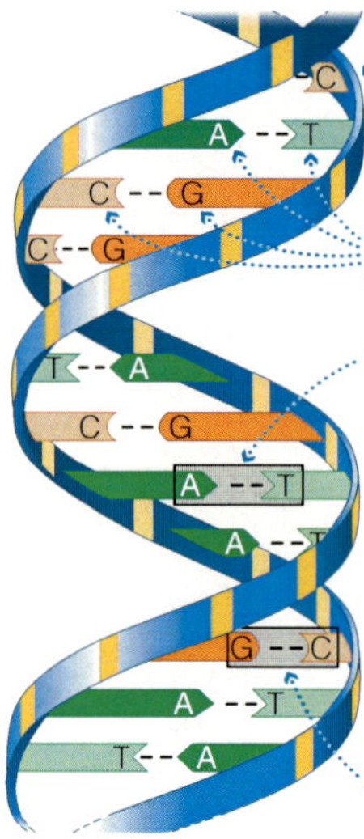

Figure 5.13

This diagram represents a DNA molecule, which looks much like a zipper twisted into a spiral. The important hereditary information is contained in the "teeth" linking the strands. These "teeth" are the DNA bases. Only four DNA bases are used, and they can link up between the two strands only in specific ways: T attaches only to A, and C attaches only to G. (The color coding is arbitrary and is used only to represent different types of chemical groups; in the backbone, blue and yellow represent sugar and phosphate groups, respectively.) (Adapted from Campbell, Reece, Simon, *Essential Biology,* Second Edition, © 2004.)

holds the "operating instructions" for living organisms on Earth. DNA also allows organisms to reproduce, because it can be accurately copied. In this section, we will explore how DNA determines the nature of an organism and allows reproduction. We will also discuss how rare errors in the copying of DNA can lead to evolutionary adaptations, thus giving us the molecular-level understanding of natural selection that we first discussed in Section 5.1.

• How does the structure of DNA allow for its replication?

The molecular structure of DNA, a *double helix,* is one of the most familiar scientific icons of our time (Figure 5.13). A helix is a three-dimensional spiral, such as you would make by extending a Slinky toy; a double helix has two intertwined strands, each in the shape of a helix. The structure looks much like a zipper twisted into a spiral. The fabric edges of the zipper represent the "backbone" of the DNA molecule, while the zipper teeth that link the two strands represent molecular components called **DNA bases**. The chemical structure of the backbone is interesting and important in its own right, but it is the DNA bases that hold the key to heredity. Life on Earth makes use of only four DNA bases: adenine (abbreviated A), guanine (G), thymine (T), and cytosine (C).

The key to DNA's ability to be duplicated by cellular machinery lies in the way the four DNA bases pair up to link the two strands: T can pair up only with A, while C can pair up only with G. Figure 5.13 shows this pairing by representing the different bases with different shapes. For example, the shape of A, which is depicted as ending with an open triangle, fits only into the notch in T. Similarly, what is shown as the curved end of G fits only into the curved notch in C. These diagrams are only schematic representations—there aren't literally notches and curves of these shapes—but the real chemical bases work much the same way: Their actual shapes and sizes determine how they pair up.

The process by which DNA is copied, called **DNA replication,** is illustrated in Figure 5.14. Step 1 begins with the complete double helix. In Step 2, the two strands separate, "unzipping" the links between the paired bases. Step 3 shows how, once the strands have been "unzipped," each strand can serve as a template for making a new strand. Because the "teeth" of each new strand must link to the existing strands according to the base pairing rules—T goes only with A, and C goes only with G—each new strand will be *complementary* to an existing one. (By saying that two strands are complementary, we mean that, while they are not identical, they contain the same information because knowing the base sequence on one strand automatically tells us the base sequence on the other strand.) The end result, shown in Step 4, is two identical copies of the original DNA molecule. When a cell divides, one copy goes to each daughter cell. Because cell division is the key to passing down genetic material from one generation to the next, DNA replication explains the basis of heredity.

Although the DNA copying process is easy in principle, the actual mechanics are fairly complex. More than a dozen special enzymes are involved in the various steps, performing tasks such as unzipping the double helix, making sure the correct bases pair up, checking for and correcting any errors in the copying process, and rezipping the new DNA molecules. This complexity is one reason why errors sometimes occur in DNA replication; as we'll discuss shortly, these errors are crucial to evolution. The com-

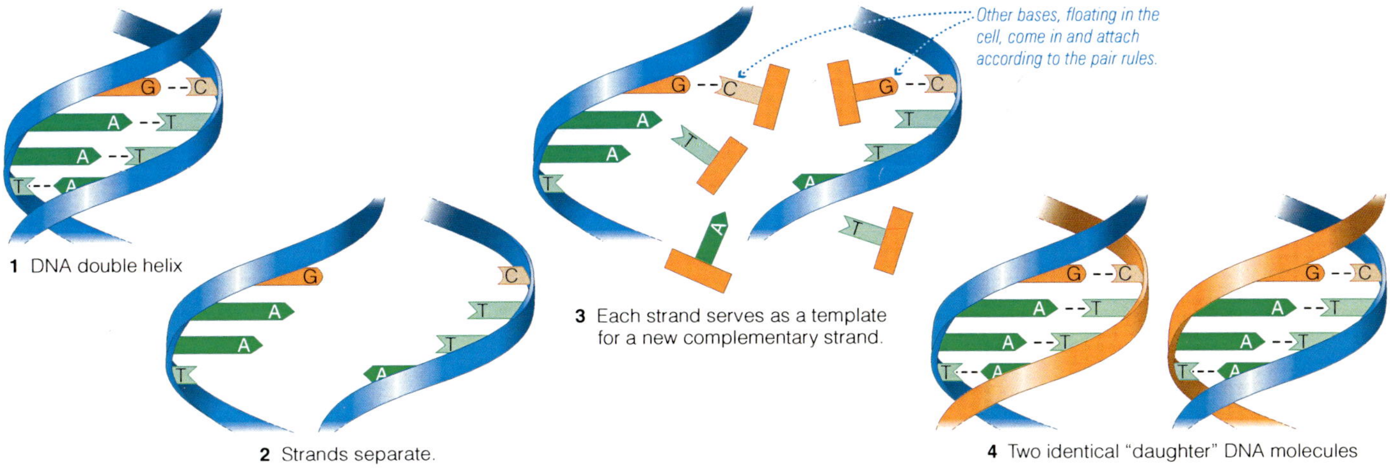

Figure 5.14

DNA replication. DNA copies itself by "unzipping" its two strands, each of which then serves as a template for making a new, complementary strand built in accord with the base pairing rules: A goes only with T, and C goes only with G. The end result is two identical copies of the original DNA molecule. (Adapted from Campbell, Reece, Simon, *Essential Biology,* Second Edition, © 2004.)

plexity of replication also makes it extremely unlikely that DNA could have been the original hereditary molecule for life on Earth. Instead, most biologists believe that DNA evolved from a simpler self-replicating molecule—probably a form of RNA—that carried hereditary information in the earliest living organisms [Section 6.2]. However DNA evolved, it has proved remarkably successful—it is now the hereditary material for every known organism on Earth. The basic copying process shown in Figure 5.14 explains how all life on our planet, from the smallest bacteria to humans, passes its genetic information from one generation to the next.

• How is heredity encoded in DNA?

Besides having the ability to be replicated, DNA also determines the structure and function of the cells within any living organism. In essence, the "operating instructions" for a living organism are contained in the precise arrangement of chemical bases (A, T, C, and G) in the organism's DNA. Today, biologists have technology that allows them to rapidly determine the sequence of bases in almost any strand of DNA. This technology has been used to determine the DNA sequences that code for many cell functions, as well as to determine the complete DNA sequences of many living organisms. For example, the Human Genome Project, completed in 2003, was a 13-year effort in which scientists ultimately determined the order of all three billion bases that make up the DNA of a human being. (In humans, this DNA is spread among the 46 *chromosomes* found in normal human cells.)

Genes and Genomes Within a large DNA molecule, isolated sequences of DNA bases represent the instructions for a variety of cell functions. For example, a particular sequence of bases may contain the instructions for building a protein, for building a piece of RNA, or for carrying out or regulating one of these building processes. The instructions representing any individual function—such as the instructions for building a single protein—make up what we call a **gene.** A gene is the basic functional unit of an organism's heredity—a single gene consists of a sequence of DNA bases (or RNA bases, in some viruses) that provides the instructions for a single cell function.

Interestingly, among plants, animals, and other Eukarya, most of the DNA is *not* part of any gene; that is, much of the DNA does not appear to carry the instructions for any particular cell function. For example, this so-called **noncoding DNA** (sometimes called "junk DNA") makes up more than 95% of the total DNA in human beings, and similarly large fractions of the DNA of many other eukaryotes. Biologists suspect that most of this noncoding DNA represents evolutionary artifacts—pieces of DNA that may once have had functions in ancestral cells but that no longer are important, much like the appendix is an organ that no longer plays an important role in our bodies. However, recent discoveries suggest that at least some of the noncoding DNA may play important roles that are not yet fully understood.

The complete sequence of DNA bases in an organism, encompassing all of the organism's genes as well as all its noncoding DNA, is called the organism's **genome.** Different organisms have genomes that vary significantly both in total length (number of bases) and in their numbers of genes. For example, some simple microbes have DNA that extends only a few hundred thousand bases and contains only a few hundred genes.* We humans have a genome that contains an estimated 20,000 to 25,000 genes among its sequence of some three billion DNA bases. Note that, genetically speaking, we are by no means the most complex organisms on Earth. Rice, for example, has significantly more genes than we do—recent sequencing of the rice genome shows that it has about 37,000 genes, though it has a shorter total DNA sequence. Other organisms have far more DNA than people. For example, the simple plant known as the "whisk fern" (*Psilotum nudum*) has more than 70 times as many bases in its genome as humans, though most of this extra DNA is probably noncoding.

Every member of a particular species has the same basic genome. However, there is always some variation among individuals. For example, while all human beings have the same set of genes, the genes of different individuals differ slightly; for example, they may vary here or there in their precise sequence of DNA bases. These differences in the genes of individuals explain why we are not all identical, and they are also the source of the individual variation that underlies the theory of evolution (see Fact 2 on p. 156). Moreover, with a few exceptions, every cell in a living organism contains the same set of genes. Different cell types, such as muscle cells or brain cells, differ only because they *express,* or actually use, different portions of their full set of genes. Thus, the DNA found in almost any cell in any organism contains the complete instructions for building an organism of that species. This fact underlies the science of *cloning,* in which a single cell from a living organism is used to grow an entirely new organism with an identical set of genes.

The Genetic Code A strand of DNA contains a long, unbroken sequence of DNA bases; for example, a particular sequence might contain the bases ACTCAGCTTCAACGG. . . . For a sequence like this to be useful as the instructions for a cell function, there must be a set of rules for how to "read" the sequence. That is, the rules must specify how to break the long sequence into individual "words," as well as where to start reading and where to stop reading the words that represent the instructions for a single gene. The set of rules for reading DNA is called the **genetic code** (Figure 5.15). More specifically, genetic "words" consist of three DNA bases in a row. For the purpose of protein building, each word represents either a particular amino acid or a "start reading" or "stop reading" instruction.

*Many viruses are far simpler, with just a few thousand bases and a handful of genes. Mitochondria within eukaryotic cells, which are thought to have had free-living ancestors, are also much simpler than the simplest bacteria sequenced to date. For example, human mitochondria have fewer than 17,000 DNA base pairs, representing fewer than 40 genes.

first base	second base: T	second base: C	second base: A	second base: G	third base
T	TTT Phenylalanine	TCT Serine	TAT Tyrosine	TGT Cysteine	T
T	TTC Phenylalanine	TCC Serine	TAC Tyrosine	TGC Cysteine	C
T	TTA Leucine	TCA Serine	TAA stop	TGA stop	A
T	TTG Leucine	TCG Serine	TAG stop	TGG Tryptophan	G
C	CTT Leucine	CCT Proline	CAT Histidine	CGT Arginine	T
C	CTC Leucine	CCC Proline	CAC Histidine	CGC Arginine	C
C	CTA Leucine	CCA Proline	CAA Glutamine	CGA Arginine	A
C	CTG Leucine	CCG Proline	CAG Glutamine	CGG Arginine	G
A	ATT Isoleucine	ACT Threonine	AAT Asparagine	AGT Serine	T
A	ATC Isoleucine	ACC Threonine	AAC Asparagine	AGC Serine	C
A	ATA Isoleucine	ACA Threonine	AAA Lysine	AGA Arginine	A
A	ATG Met or start	ACG Threonine	AAG Lysine	AGG Arginine	G
G	GTT Valine	GCT Alanine	GAT Aspartic acid	GGT Glycine	T
G	GTC Valine	GCC Alanine	GAC Aspartic acid	GGC Glycine	C
G	GTA Valine	GCA Alanine	GAA Glutamic acid	GGA Glycine	A
G	GTG Valine	GCG Alanine	GAG Glutamic acid	GGG Glycine	G

Figure 5.15

The genetic code. This table shows how three-base-pair "words" of DNA code for particular amino acids or a start or stop instruction. For example, you can find the "word" CAG by looking along the left for C as the first base, along the top for A as the second base, and along the right for G as the third base; you'll then see that CAG codes for the amino acid glutamine. Notice that in most cases, the first two letters alone determine the amino acid; for example, if the first two letters are CT, the amino acid is always leucine regardless of the third letter. This suggests that the current genetic code evolved from an earlier version that used only two-letter "words" rather than three-letter "words." (Adapted from Campbell, Reece, Simon, *Essential Biology,* Second Edition, © 2004.)

Because the genetic words consist of three DNA bases in a row and there are four DNA bases to choose from (A, C, T, G), the total number of words in the genetic code is $4^3 = 4 \times 4 \times 4 = 64$; you can see all 64 words spelled out in Figure 5.15. Notice that this is significantly more than the number of amino acids used to make proteins, which is 20 for most organisms. Thus, the genetic code contains a fair amount of redundancy. For example, the genetic words ACC and ACA both represent the same amino acid. Moreover, a close examination of the genetic code also offers a hint about the likely evolution of DNA: The codes for most amino acids really depend on just the first two bases in the three-base genetic words. For example, all four of the three-base words starting with AC—ACC, ACA, ACT, and ACG—code for the same amino acid (threonine). This suggests that the genetic code once depended only on two-base words rather than three-base words. Most biologists now believe that early life-forms used only a two-base language, which later evolved into the current three-base language of the genetic code.

THINK ABOUT IT Note that a two-base language would allow only $4 \times 4 = 16$ possible words—not enough for all the amino acids used by living organisms today. What does this imply about proteins in early life-forms? Does it seem likely that even earlier organisms might have used only a one-base language? Explain.

Another important feature of the genetic code is that it is the same in nearly all living organisms on Earth. Only a few organisms show any variations at all on this code, and the variations that do exist are minor. (Variations in the genetic code are also found in mitochondria, structures within eukaryotic cells that contain their own DNA.) Nevertheless, the fact that some variations occur tells us that not all the specifics of the genetic code were inevitable. If we think of the genetic code as a language, the fact that nearly all organisms use the same genetic code is as if everyone on Earth spoke the same language, even though other languages are possible. This common language of the genetic code is further evidence for a common ancestor of all life on Earth.

The Role of RNA While the sequence of bases in a gene holds the instructions for its function, the actual implementation of these instructions is quite complex. As with DNA replication, many enzymes are involved in carrying out genetic instructions. In addition, the molecule RNA plays a particularly important role in these functions. A molecule of RNA is quite similar in structure to a *single* strand of DNA, except that it has a slightly different backbone and one of its four bases is different from one of the DNA bases (RNA uses a base called *uracil* (U) in place of DNA's thymine (T)).

RNA participates in carrying out genetic instructions in several different ways. For example, in the process of building a protein, a molecule of RNA is first assembled along one strand of DNA, essentially transcribing the DNA instructions for use in another part of the cell. (This process is called *transcription.*) This molecule of RNA then goes to a site in the cell where amino acids are actually assembled into proteins. Other molecules of RNA collect amino acids from within the cell and bring them to this site, where yet another form of RNA helps attach the amino acids into the chains that make proteins. (This second process is called *translation,* because it effectively *translates* the genetic instructions into an actual protein.) If you take a biology course, you will learn much more about the many roles of RNA.* However, for our purposes in this book, it is not necessary to know much more about RNA except the fact that it is vital for carrying out the instructions encoded in DNA.

• How does life evolve?

One of the most remarkable aspects of our current knowledge of DNA is that it has allowed us to further confirm Darwin's theory of evolution by natural selection. In particular, while Darwin had to base his theory on the variation in populations that he could directly observe, we now know precisely how such variation occurs at the molecular level. The key to this knowledge lies in understanding how DNA molecules gradually change through time. Based on what we've already said about DNA replication and protein building, we can see how and why changes in DNA occur.

Mutations: The Molecular Basis of Evolution Despite its complexity, the process of DNA replication proceeds with remarkable speed and accuracy. Some microbes can copy their complete genomes in a matter of minutes, and copying the complete three-billion-base sequence in human DNA takes a human cell only a few hours. In terms of accuracy, the copying process generally occurs with less than one error *per billion* bases copied. Nevertheless, errors do sometimes occur. For example, the wrong base may occasionally get attached in a base pair, such as linking C to A rather than to G. In other cases, an extra base may be accidentally inserted into a gene, a base may be deleted, or an entire sequence of bases might be duplicated or eliminated. Absorption of ultraviolet light or nuclear radiation or the action of certain chemicals (carcinogens) can also cause mistakes to occur. Any change in the base sequence of an organism's DNA is called a **mutation.**

Mutations can affect proteins in a variety of ways. Some mutations have no effect at all. For example, suppose a mutation causes the genetic

---

*Readers familiar with the different types of RNA will recognize that messenger RNA (mRNA) is involved in transcription, transfer RNA (tRNA) is responsible for collecting amino acids, and both of these meet up to assemble proteins on structures called ribosomes, which are built from proteins and ribosomal RNA (rRNA).

word ACC to change to ACA in a gene that makes a protein. Because both of these words code for the same amino acid (threonine; see Figure 5.15), this mutation will not change the protein made by the gene. Other single-base mutations—such as changing ACC to CCC—will change a single amino acid in a protein. In some cases, such a change will alter a protein only slightly, hardly affecting its functionality. But in other cases, the change can be much more dramatic. For example, the cause of sickle-cell disease (Figure 5.16), which kills some 100,000 people each year worldwide, can be traced to a single mutation in the gene that makes hemoglobin in which the base A changed to the base T in just one place within the gene.* Mutations that add or delete a base within a gene tend to have the most dramatic effects on protein structure. The reason is that the genetic code has no "punctuation" or spacing between words; instead of saying something like "the fat cat ate the rat," for example, it says "thefatcatatetherat." Thus, if a letter (base) is added to or deleted from such a sequence, the result will be nonsense from that point on. For example, inserting an "a" so that the sequence becomes "the*a*fatcatatetherat" would cause it to be read as "the *a*fa tca tat eth era t."

Mutations that change proteins are often lethal, because the cell may not be able to survive without the correctly structured protein. However, if the cell survives, the mutation will be copied every time its DNA is replicated. In that case, the mutation represents a permanent change in the cell's hereditary information. If the cell happens to be one that gets passed to the organism's offspring—as is always the case for single-celled organisms and can be the case for animals if the mutation occurs in an egg or sperm cell—the offspring will have a gene that differs from that of the parent. It is this process of mutation that leads to variation among individuals in a species. Each of us differs slightly from all other humans because we each possess a unique set of genes with slightly different base sequences.

Figure 5.16

These microscopic views contrast a normal human red blood cell with a blood cell found in patients with sickle-cell disease. The sickle shape makes it easier for the blood cells to clog tiny blood vessels, which can lead to debilitating disease. Sickle-cell disease occurs in people whose gene for hemoglobin differs from the "normal" gene in just a single DNA base. (Adapted from Campbell, Reece, Simon, *Essential Biology,* Second Edition, © 2004.)

THINK ABOUT IT Ultraviolet radiation from the Sun can cause mutations in the DNA of skin cells. Based on what you've learned, explain why this is potentially dangerous (and, indeed, is the cause of skin cancer). How would sunscreen help prevent such mutations?

Mutations therefore provide the basis for evolution. Given that each individual of a species possesses slightly different genes, it is inevitable that some genes will provide advantageous adaptations to the environment. As we discussed in Section 5.1, the combination of individual variation and population pressure leads to natural selection, in which the advantageous adaptations will preferentially be passed down through the generations. Thus, what was once a random mutation in a single individual can eventually become the "normal" version of the gene for an entire species. In this way, species evolve through time. Notice that, while we often associate the word *mutation* with harm, evolution actually proceeds through the occasional beneficial mutation. Although such beneficial mutations may be relatively rare compared to other mutations, natural selection allows these beneficial mutations to propagate preferentially, so tremendous changes can accrue over time.

---

*Humans have two copies of each gene, and sickle-cell disease generally occurs only in people who have the sickle-cell mutation in both copies of the gene. From an evolutionary standpoint, this mutation remains prevalent in the population because it actually confers an advantage—malaria resistance—to people with only one copy of the mutated gene.

Evolution sometimes occurs in an even more dramatic way: In some cases, organisms can transfer entire genes to other organisms, a process called *lateral gene transfer.* This process is one of the primary ways that bacteria gain resistance to antibiotics. We humans have also learned to use this process for our benefit through what we call *genetic engineering,* in which we take a gene from one organism and insert it into another. For example, genetic engineering has allowed us to produce human insulin for diabetic patients: The human gene for insulin is inserted into bacteria, and these bacteria produce insulin that can be extracted and used as medicine. Lateral gene transfer can change a species more rapidly than individual mutations, but mutations are still the underlying basis, since they created the genes in the first place.

DNA and Life on Other Worlds It is difficult to imagine life that does not have heredity, because it seems crucial for any form of life to have some means of storing its operating instructions and passing them on to its offspring. We've seen that DNA is the carrier of heredity for all life on Earth, though as we'll discuss in Chapter 6, we have good reason to believe that very early life on Earth used RNA for this role. Should we expect DNA, or RNA, to also be the heredity molecule for life elsewhere? We do not yet know whether other, quite different molecules might be able to carry hereditary information in the same way as DNA. However, it seems a near certainty that any form of life anywhere else will have some molecule that plays the same functional role that DNA plays on Earth.

5.5 LIFE AT THE EXTREME

We've discussed all the fundamental characteristics of life on Earth: the basic structure of cells, the metabolism of cells, and the means by which cells store and pass on their heredity. We've also discussed why these characteristics seem likely to be shared, at least in a general sense, by any life we find elsewhere. In essence, our bottom-line conclusion is that life

MOVIE MADNESS: THE ANDROMEDA STRAIN

Most movie aliens are complex critters—"bugeyed monsters" that come to Earth in high-speed, hi-tech spacecraft to undertake a campaign of havoc and destruction. But a different threat is offered by *The Andromeda Strain* (1971) as the U.S. military sends a satellite into space to scoop up some simple alien life—green goo attached to small meteors—hoping that microbe-sized aliens might be useful for biological warfare.

You might argue that, as a scheme for developing new weaponry, this approach is a long shot. But not in the movies. The satellite returns to Earth, and some of the alien microbes escape into the air to promptly wipe out a small New Mexico town. (Aliens nearly always prefer landing in the American Southwest.) Unaware of the military's game plan, a team of cranky civilian scientists is brought together in a top-secret Nevada bio lab to figure out what's afoot and stop the infection from space.

Most of the movie is taken up with the step-by-step analysis performed by the scientists as they methodically try to isolate and understand the meteor-borne microbes. They find that the space critters have no nucleic acids, no amino acids, and none of the usual features of biology on Earth. In fact, these extraterrestrial entities are crystals—the entire organism is all made of the *same* molecule (like salt, except with a much more complicated molecule). This is a bit hard to swallow, but there's more: each time the alien bugs reproduce (roughly every few seconds), they also mutate. And unlike for earthly life, the mutations are not random. Each cell mutates simultaneously and identically to every other.

By the end of the film, this peculiar property makes all the weighty work of the scientists irrelevant. The alien microbes mutate into a harmless form and drift out to sea. But think about it: any species in which all individuals mutate the same way at the same time will be unable to adapt to a changing environment. Every microbe is, and remains, identical. You should feel sorry for the Andromeda strain, because it has no Darwinian evolution and, consequently, no chance of long-term survival. As a strategy for life, it strains credulity.

In real life, NASA has a program to ensure that materials brought to Earth by future spacecraft will be carefully handled to forestall any chance of contamination by alien microbes. This news should be a relief to residents of the Southwest.

elsewhere ought to share a lot of common features with life on Earth. But don't be tempted to think this means that life elsewhere will look like us—that is, like humans or any other animals. In fact, most life on Earth does not look much like "us." We've already noted that microbes are far more common than multicellular Eukarya. Perhaps even more startling, in recent decades biologists have discovered that life can survive in an astonishing variety of environments that would be lethal to humans.

• What kinds of conditions can life survive?

Deep on the ocean floor are places where volcanic activity releases hot water and rock into the surrounding ocean. Minerals and other dissolved chemicals can cause the water to turn black, which is why some of the vents and their plumes of hot water are called *black smokers* (Figure 5.17). The water coming out of such vents can be heated to temperatures above 350°C (660°F), far above the normal boiling point of 100°C (212°F). However, the ocean pressure at these depths is so great that the water remains liquid despite its high temperature.

If you took any "ordinary" organism and placed it in the extremely hot water near a volcanic vent, it would die quickly because the high temperature would cause many of its critical cell structures to fall apart. Yet in recent decades scientists have discovered life—mostly microbes of the domain Archaea*—thriving in the extremely hot water around black smokers and other vents. For example, an organism called *Pyrolobus fumarii,* which was actually discovered *in* the walls of a black smoker (its name means "fire lobe of the chimney"), can grow in water heated to as high as 113°C (235°F). And in 2003, researchers discovered another species of Archaea living near volcanic vents that can grow in even hotter water. This species does not yet have an official name but is being called "Strain 121" because it can grow in water as hot as 121°C (250°F); in addition, it can survive in the lab for up to 2 hours at temperatures of 130°C (266°F). Both *P. fumarii* and Strain 121 are chemoautotrophs that get their carbon from dissolved carbon dioxide and their energy from inorganic chemical reactions that occur in the hot water. Similar organisms thrive in hot springs on Earth's surface, such as in the springs around Yellowstone National Park (Figure 5.18).

Organisms that survive in extremely hot water are sometimes called *thermophiles,* meaning "lovers of heat" (the suffix *phile* means "lover") or, for those living at the highest temperatures, *hyperthermophiles.* More generally, organisms that survive in extreme environments of any kind are called **extremophiles**, or "lovers of the extreme." Extremophiles are quite varied, though most of them are members of the domains Bacteria or Archaea. Some can live in "normal" as well as extreme conditions, while others can survive only in the extreme conditions. For example, many hyperthermophiles die when brought to "normal" temperatures because their enzymes have evolved to function only at the high temperatures in which they live. Indeed, many extremophiles are anaerobic (meaning they live without oxygen), and they are poisoned by the oxygen on which our own lives depend.

Hot environments are not the only extreme conditions favored by some organisms. The dry valleys of Antarctica receive so little rain or

Figure 5.17

This photograph shows a black smoker—a volcanic vent on the ocean floor that spews out extremely hot, mineral-rich water. Organisms like *P. fumarii* and Strain 121 survive here in water above the normal boiling temperature.

Figure 5.18

A hot spring in Yellowstone National Park; to judge its size, notice the walkway winding along the lower right. The different colors in the water are from different bacteria that survive in water of different temperatures.

*Many known species of Archaea live in extreme environments, but extreme conditions are *not* a general feature of domain Archaea. Indeed, some of the most common organisms in "ordinary" environments on Earth are Archaea. For example, some 20–50% of the living cells in cool ocean water typically represent various species of Archaea.

Figure 5.19

The main photo shows a cold, dry valley in Antarctica. These valleys are among the driest deserts on Earth (the ice is runoff from surrounding regions), yet they are still home to life. The inset shows a slice of rock (about 1.8 centimeters across) from a dry valley. The colored zones contain microbes that live inside the rock in the airspaces between tiny mineral grains. The organisms are dormant and frozen for most of the year, but can grow during the approximately 500 hours per year when sunlight warms the rock above the freezing point.

snowfall that they are among the driest deserts on Earth, and temperatures in these valleys rarely rise above freezing (Figure 5.19). Nevertheless, there is life in the dry valleys living *inside* rocks. We often think of rocks as solid, but most rocks are composed of individual mineral grains packed together, leaving small spaces between the grains. Even in the dry valleys, these spaces within the rocks occasionally contain water from the rare rain- or snowfall. Sunlight can penetrate up to a few millimeters into the rock before being completely absorbed, so the layers just below the rock's surface can have temperatures slightly above freezing despite the freezing temperatures around them. Amazingly, there are microbes that survive in these tiny pockets of liquid water inside rocks in freezing cold valleys.

Other extremophiles live in conditions far too cold, acidic, alkaline, or salty for "ordinary" life to survive, and some may offer examples of the types of organisms we might find on other worlds. For example, there are some species that can survive at temperatures as low as $-20°C$, as long as even a thin film of liquid water is available. (These cold-loving organisms are called *psychrophiles,* essentially the opposite of *thermophiles.*) Some microbes can even survive high doses of radiation. A bacterial species known as *Deinococcus radiodurans* can survive radiation more than 1,000 times that which would be lethal to humans and other animals. These remarkable organisms actually thrive in radioactive waste dumps! They could survive the radiation exposure on a world without ozone and even in space, and they can survive extremely dry conditions as well.

One particular group of extremophiles is of special interest for its possible relevance to life on Mars. Microbes called *endoliths* (meaning "within rocks") can live several kilometers below the surface of Earth in water that fills the pores within rock. One community of endoliths, discovered in 1995, consists of bacteria living deep beneath the surface of Oregon and Washington in a rock formation known as the Columbia River Basalt. These organisms are chemoautotrophs that get their energy for metabolism from chemical reactions between the water and the surrounding rock, and they get their nutrients from chemicals within the rock itself and from carbon dioxide that has filtered down from the surface. Although these microbes clearly share the same common ancestors as the rest of terrestrial life, they tell us something significant about life's ability to survive in remarkably diverse environments. In particular, the subsurface environment in which they live almost certainly exists in similar form on other planets, such as Mars, even when the surface may be too cold for liquid water. Moreover, endoliths may be quite common. No one knows exactly how many of them exist here on Earth, but some estimates suggest that the total mass of subsurface organisms living in rock may exceed that of all the life on Earth's surface.

Another amazing adaptation to extreme conditions is found in **endospores**—special "resting" cells produced by some bacteria (Figure 5.20). Endospores allow the organisms that create them to become dormant, neither growing nor dying in extremely inhospitable conditions. (The nondormant organisms are not necessarily extremophiles; some live under more "normal" conditions.) For example, endospores of the bacterium *Bacillus anthracis,* which causes the deadly disease anthrax, can survive a complete lack of water, extreme heat or cold, and most poisons. Some endospores can survive even in the vacuum of space, which is why planetary scientists worry that our interplanetary spacecraft could potentially contaminate other worlds with life from Earth. Moreover, some endospores can remain dormant for centuries—and perhaps even longer—raising the possibility that life could survive journeys aboard meteorites that are blasted off one planet and land on another [Section 8.4].

• Are extremophiles really extreme?

From our human point of view, the environments in which extremophiles survive truly are extreme. But if an extremophile could think, it would probably claim that its environment is quite normal and that ours is the one that is extreme. So who's right, humans or the extremophiles?

In some sense, it's all just relative. Any species would naturally consider its own environment to be normal and others to be extreme. A more important question we might ask is which environment is more common. Surprisingly, if we look at the history of Earth, so-called extreme environments have been much more common than an environment suitable for humans. Earth's atmosphere may have contained oxygen at a level suitable for human life for only the past few hundred million years, or about 10% of Earth's history [Section 6.3]. Indeed, for the first couple of billion years after life first arose on Earth, extremophiles were the only organisms that could survive. Even today, it's an open question whether extremophiles are more or less common than organisms that live in conditions favorable to humans. All in all, extreme life appears to be much more the norm than is life that lives in an environment suitable to humans.

The study of extremophiles has several important implications for the search for extraterrestrial life, but two are particularly important. First, the fact that extremophiles apparently evolved earlier than other forms of life [Section 6.1] suggests that we should begin the search for life elsewhere by searching for similar extreme organisms. Second, the fact that extremophiles can survive such a broad range of conditions suggests that life may be possible in many more places than we would have guessed only a few decades ago: Any world containing an environment in which some type of extremophile might survive becomes a good candidate for the search for life.

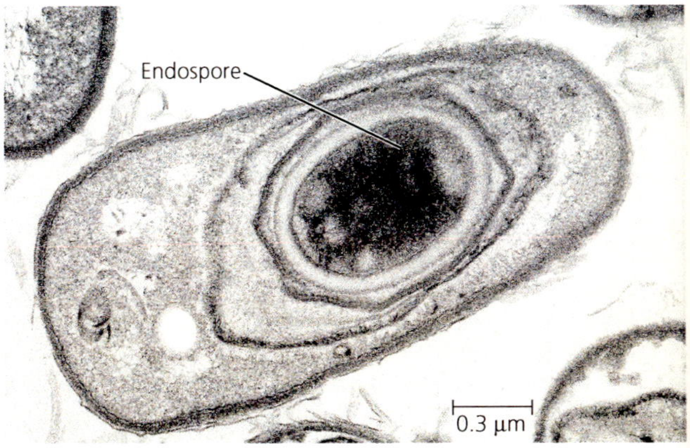

Figure 5.20

This microscopic photo shows an endospore created by the bacterium *Bacillus anthracis*. There are actually two cells here, one inside the other. The outer cell produced the specialized inner cell, which is the endospore. The endospore has a thick, protective coat. Its interior is dehydrated, and no metabolism occurs. Under harsh conditions, the outer cell may disintegrate, but the endospore can survive. When the environment becomes more hospitable, the endospore absorbs water and resumes growth.

5.6 THE PROCESS OF SCIENCE IN ACTION: EVOLUTION AS SCIENCE

In this chapter and throughout this book, we treat evolution as an established fact, a position consistent with official statements by virtually every scientific society, including the National Academy of Sciences, the American Association for the Advancement of Science, the National Association of Biology Teachers, and the American Astronomical Society. However, if you live in the United States, you've almost certainly heard about public battles in which the scientific idea of evolution is portrayed as being controversial. How can an idea so well accepted in science be considered so differently by many among the general public? The answer lies in differences between the way science works and the way that people often seek knowledge in their daily lives, and especially in the difference between the evidence-based approach of science and the faith-based approach of religion. Once this difference is understood, much of the supposed conflict between science and religion is not an issue, which explains why the vast majority of religious denominations see no inherent conflict between their faiths and the science of evolution. For this chapter's case study in the process of science, we'll explore why most scientists and most theologians agree that evolution and faith can coexist without difficulty.

• Is evolution a fact or a theory?

By this point in the book, you should already recognize that the question of whether evolution is a fact or a theory offers a false choice, much like asking whether gravity is a fact or a theory [Section 2.4]. Gravity is a fact in that objects really do fall down and planets really do orbit the Sun, but we use the *theory* of gravity to explain exactly how and why these things occur. The theory of gravity is not presumed to be perfect and indeed has at least one known flaw (its inconsistency with quantum mechanics on very small scales). Moreover, Newton's original theory of gravity is now considered only an approximation to Einstein's improved theory of gravity, which itself will presumably be found to be an approximation to a more complete theory that has not yet been discovered.

The same idea holds for evolution. Nearly all scientists consider evolution to be a fact, because both the geological record and observations of modern species make clear that living organisms really do change with time. We use the *theory* of evolution to explain how and why these changes occur. For example, we use the theory of evolution to understand the changes in species recorded in the geological record, the genetic relationships between modern species, and the way that bacteria can rapidly acquire resistance to antibiotics.

The theory of evolution clearly explains the major features of life on Earth, but scientists still debate the details of the theory. For example, there is considerable debate about the rate at which evolution proceeds: Some scientists suspect that evolution is "punctuated" with periods of rapid change followed by long periods in which species remain quite stable, while others suspect that evolution proceeds at a steadier pace. This debate can be quite heated between individual scientists, but it does not change the overall idea that life evolves and it will eventually be settled by additional evidence. Similarly, scientists often debate the precise relationships between species, especially those that are extinct. For example, we do not yet have enough evidence to put together a complete evolutionary tree for relationships among all dinosaur species, and the relationship between extinct dinosaurs and modern birds is not yet fully understood.

Indeed, we can draw an analogy between Darwin's original theory of evolution by natural selection and Newton's original theory of gravity: Just as Newton's theory captured the main features of gravity but proved to be incomplete, Darwin's theory clearly captures the main features of evolution but is not complete. Moreover, like Newton's theory, Darwin's also has been modified and improved with time. Just as Einstein's general theory of relativity allowed us to understand gravity in more realms than Newton's original theory (by refining it, not refuting it), so does our modern understanding of DNA and mutations allow us to understand biological changes and relationships beyond those that Darwin was able to understand or was even aware of. And like the theory of gravity, the theory of evolution remains a work in progress. We expect to learn more about evolution as we continue to study relationships between species, how DNA works (especially the noncoding regions), and the biochemistry of life. Perhaps someday we'll even be able to broaden the theory through the study of *comparative evolution*, in which we'll explore the similarities and differences among living organisms on multiple worlds.

Does all this mean that you need to *believe* that evolution really occurred? From the standpoint of learning science, what counts is *understanding* evolution; belief is up to you. Remember, all the evidence in the world can never prove any scientific theory true beyond all doubt [Section 2.3]. Even if we had a complete geological record, with precise dating of every species that ever lived, you could still choose to believe, for example, that the fossils had been placed there intact at some single moment in the past, rather than having been deposited as the evidence suggests. All we can say from a scientific viewpoint is that a tremendous wealth of evidence points to the idea that life on Earth has evolved through time, that these evolutionary changes have been driven by natural selection, and that natural selection occurs on a molecular level as genes are modified through mutations of DNA.

• Are there scientific alternatives to evolution?

In recent years, the public controversy over evolution has centered largely around the question of whether it should be the only idea about our origins taught in science classes or whether it should be taught alongside other, competing ideas. Other ideas certainly offer different visions of how we came to exist; for example, the idea that God created Earth and the universe a mere 6,000 years ago is obviously quite different from the idea that life has evolved gradually over the past 4 billion years. The question is whether this idea or other competing ideas qualify as science.

The best way to determine whether any alternatives to evolution qualify as science is to consider them against the hallmarks of science that we discussed in Chapter 2. Let's start by showing that the theory of evolution *does* satisfy the standards of science. The first hallmark states that science seeks explanations for observed phenomena that rely solely on natural causes. The theory of evolution clearly does this, as it explains the geological record and observed relationships between species through the mechanism of natural selection and other natural causes.

The second hallmark states that science progresses through the creation and testing of models. Our understanding of evolution has indeed progressed in this way. The idea of evolution won out over Aristotle's competing idea of species that never changed. As the fact of evolutionary change gained acceptance, the first model proposed to explain these changes came from Lamarck (see Section 5.1); his model was later discarded because Darwin's alternative model explained the observations so much more successfully. Our current, molecular model of evolution is a refinement of Darwin's original model, and we can expect further refinements to the theory in the future as continued study turns up new evidence.

The theory of evolution also satisfies the third hallmark, which states that a scientific model must make testable predictions that would lead us to revise or abandon the model if the predictions do not agree with observations. Our modern, molecular theory of evolution clearly qualifies. For example, it predicts that diseases can and will evolve in response to medicines designed to combat them, a prediction borne out in the rapid way that many diseases acquire drug resistance. It also predicts that genetically similar species should respond to medicines in similar ways, a prediction confirmed by the fact that we can test many medicines in other primates and they do indeed have effects similar to those they have in humans. The theory of evolution also provides a road map that we can use to modify organisms through genetic engineering; in this sense, every genetically engineered grain of rice or kernel of corn represents a success of the predictive abilities of the theory of evolution.

In fact, even Darwin's original theory made testable predictions. For natural selection to be possible, Darwin had to assume that living organisms have some way of passing on their heritable traits from parent to offspring. So although he did not predict the existence of DNA per se, his theory clearly predicted that some type of mechanism had to exist to carry the hereditary information. Similarly, now that we know about DNA and the genetic code, the theory of evolution predicts that closely related species should be genetically similar, a prediction that has been confirmed in just the past few years by genome sequencing. For example, in the ordering of their base sequences, the DNA of humans and chimpanzees is 98.5% identical. If we were not closely related to chimpanzees, we would not expect such similar genomes.

Now that we have established that evolution qualifies as science, we next turn to the question of whether any of the alternative models that have been suggested for inclusion in the classroom might also qualify as science. Since the time that Darwin first published his theory, the main alternatives have been religious ideas about creation. Here, we run into an immediate problem: There are so many different religious ideas about creation that we can't even define the potential alternatives clearly. For example, many Native American religious beliefs speak of creation in terms that bear little resemblance to the Judeo-Christian tradition found in Genesis. Even among people who claim a literal belief in the Bible, there are differences in interpretation about creation. Some biblical literalists argue that the creation must have occurred in just 6 days, as the first chapter of Genesis seems to say, while others suggest that the term "day" in Genesis does not necessarily mean 24 hours and therefore that the story in Genesis is compatible with a much older Earth and with evolution.

Nevertheless, a few groups have tried to claim that scientific evidence supports some alternative to evolution. In the 1980s, an idea called "creation science" emerged and its proponents tried to find scientific evidence to support the idea that Earth was created a mere 6,000 years ago. However, to support this "young Earth" view, they not only had to reject evolution, but they also had to reject the tremendous weight of evidence that supports an old Earth and an old universe—evidence based on such things as radiometric dating of rocks, astronomical measurements of distances to other stars and galaxies (since their light has obviously had time to reach us), and even tree ring data that go back more than 6,000 years. For all this evidence to be wrong, we would have to have fundamental errors in our basic understanding of the laws of nature, an idea that seems implausible, given the many successes of modern physics and chemistry.

More recently, some people have begun to advance an idea called "intelligent design," or ID; this idea holds that living organisms are too complex to be explained by natural selection, and so must have been designed by some transcendent entity or power. For example, proponents of this idea point to features of the human eye as suggesting design rather than natural processes, and some believe they see evidence of "digital code" in the arrangement of the bases in the genomes of living organisms.

For the vast majority of scientists, the primary problem with these claims of intelligent design is that they do not seem to stand up to scientific scrutiny. The features that the ID proponents cite as evidence for design are to most scientists well explained by natural selection. Nevertheless, good scientists will always allow the possibility that evidence of design might someday be found. Moreover, even if no such evidence is found, absence of evidence would not preclude a role for a Designer.

The greater problems with intelligent design from a scientific perspective show up when we test it against the hallmarks of science. In particular, ID clearly fails to meet the first hallmark—that science seeks explanations for observed phenomena that rely solely on natural causes. The very idea of a transcendent Designer implies something that natural processes cannot explain, no matter whether the Designer is or isn't explicitly named as God. As a result, some ID proponents have sought to redefine science to allow nonnatural explanations. The problem with such a redefinition is that it would render science impotent. As a simple analogy, consider the collapse of a bridge. If you choose to believe that the collapse was an act of God, you might well be right—but this belief won't help you design a better bridge. We learn to build better bridges only by assuming that collapses happen through natural causes that we can understand and learn from. In precisely the same way, it is the scientific quest for a natural understanding of life that has led to the discovery of relationships between species, genetics, DNA, and much of modern agriculture and medicine. Many of the scientists who made these discoveries, including Charles Darwin himself (see the quotation at the beginning of this chapter), believed deeply that they could see God's hand in creation. But if they had let their belief stop them from seeking natural explanations, they would have discovered nothing.

Intelligent design also fails to meet the second and third hallmarks of science, because it does not offer a predictive model that can be tested. The assumption of a Designer might or might not be correct, but it does not tell us how life would be different from what we'd see if there were no Designer. Moreover, as we've discussed, scientists continually modify the theory of evolution as new evidence requires. In the unlikely event that we found evidence that strongly contradicted the current theory—for example, fossil evidence proving that people and dinosaurs existed at the same time—scientists would willingly discard the theory and go back to the drawing board. In contrast, because most proponents of intelligent design are motivated by their religious faith, their belief in ID is unshakable.

The bottom line is that science and faith are different things, and the relative worth of one does not override the worth of the other. Whether you choose to believe the theory of evolution is up to you, and if you do believe it, you can choose whether to believe that it occurred through random chance or with the help of a guiding hand. But whatever your beliefs, the theory of evolution is a clear and crucial part of modern science, and it is integral to an understanding and appreciation of modern biology. And more important for the discussion in this book, the theory of evolution frames our ideas about how to search for life beyond Earth. No competing model offers any similar scientific benefits.

THE BIG PICTURE

Putting Chapter 5 into Context

In this chapter, we have surveyed the nature of life on Earth and explored some of the implications of this survey for the search for life elsewhere. As you continue in your studies, keep in mind the following "big picture" ideas:

- If we are going to search for life, it's useful to think about just what it is we are searching for. Defining life turns out to be surprisingly difficult, but at a minimum it seems that life must be capable of reproducing and evolving. Thus, evolution plays a central role in the definition of life as well as in our understanding of life on Earth.
- Life on Earth has at least three key features that are likely to be shared by any life we find elsewhere: (1) Life has a fundamental structural unit, which we call the cell; (2) living cells undergo metabolism, by which we mean chemical reactions that keep the cells alive; and (3) living cells have a heredity molecule, which is DNA for life on Earth, that allows them to store their operating instructions and to pass these instructions to their offspring.

- Life on Earth survives under a much wider range of conditions than we would have guessed a few decades ago, suggesting that life elsewhere might similarly be found in a fairly broad range of environments. This fact greatly increases the number of worlds on which we might hope to find life.

SUMMARY OF KEY CONCEPTS

5.1 DEFINING LIFE

• What are the general properties of life on Earth?

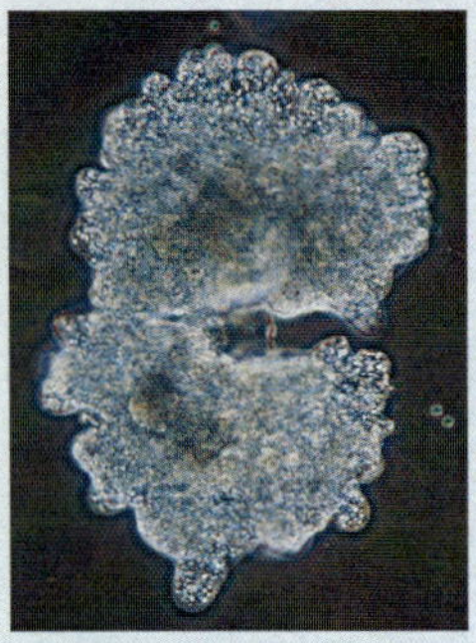

Six key properties of life on Earth are order, reproduction, growth and development, energy utilization, response to the environment, and evolutionary adaptation.

• What is the role of evolution in defining life?

The **theory of evolution**, which holds that life changes over time through the mechanism of **natural selection**, is the unifying principle of modern biology. It holds this central role because it successfully explains all the other properties of life, the observations we make in the geological record, and the observations we make of relationships between living organisms.

• What is life?

No known definition of life works in all circumstances, but for most purposes the following definition will suffice: Life is something that can reproduce and evolve through natural selection.

5.2 CELLS: THE BASIC UNITS OF LIFE

• What are living cells?

Cells are the basic units of life on Earth, as they serve to separate the living matter inside them from the outside world. The barrier that marks this separation is called the **cell membrane**.

• What are the molecular components of cells?

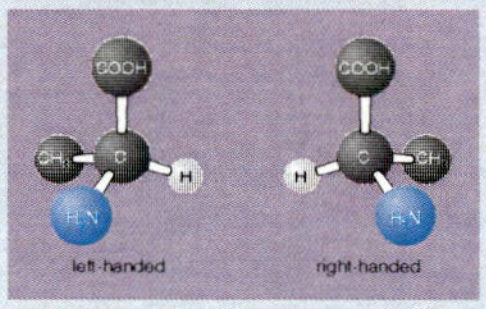

The major molecular components of cells fall into four main classes: **Carbohydrates** are sugars and starches that provide energy and build many cellular structures; **lipids**, which include fats, are the main ingredients of structures including cell membranes and also store cellular energy; **proteins** play a vast number of roles in cells, and the proteins known as **enzymes** act as **catalysts** to facilitate biochemical reactions; **nucleic acids**, which include **DNA** and **RNA**, are most important for the roles they play in heredity. Commonalities between the molecules used in different organisms, such as the fact that all life on Earth builds proteins from only left-handed versions of **amino acids**, provide strong evidence for the idea that all life evolved from a common ancestor.

• What are the major groupings of life on Earth?

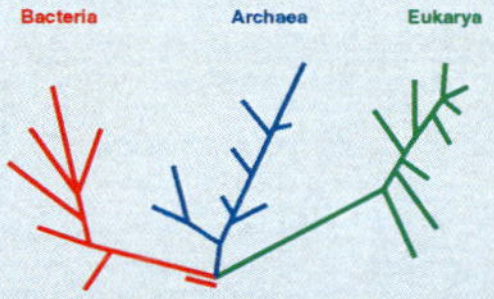

Modern biologists classify life into three **domains**: **Bacteria**, **Archaea**, and **Eukarya**. The **tree of life** shows relationships among species within the three domains; note that all plants and animals are just two small branches of the Eukarya.

5.3 METABOLISM: THE CHEMISTRY OF LIFE

• What are the basic metabolic needs of life?

(1) A source of raw materials to build cellular structures, with carbon as the most important of these materials. (2) A source of energy to fuel metabolic processes.

• How do we classify life by its metabolic sources?

We classify life by its carbon source as either a **heterotroph** that gets its carbon by eating or an **autotroph** that takes carbon directly from the environment in the form of atmospheric or dissolved carbon dioxide. We then subclassify these categories by energy source, using the prefix *photo* for life that gets energy from sunlight and the prefix *chemo* from life that gets energy either from eating or from inorganic chemical reactions.

5.4 DNA AND HEREDITY

- **How does the structure of DNA allow for its replication?**

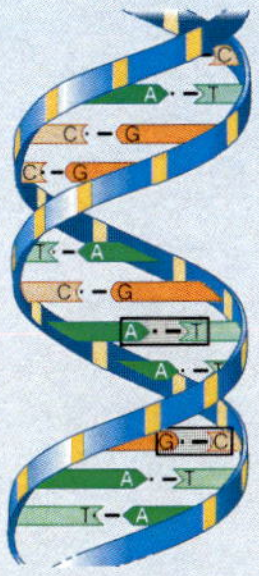

The double helix of DNA consists of two strands connected by the DNA bases. The bases connect according to precise pairing rules (T attaches only to A, and C attaches only to G), so that when the strands separate, each can serve as a template for making a new DNA molecule that is identical to the original one.

- **How is heredity encoded in DNA?**

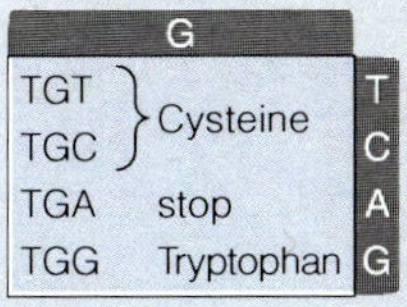

The precise sequence of the bases in a DNA molecule contains the instructions for assembling proteins and other cell functions. A segment of DNA that codes for a single cell function or protein is called a **gene**, and the complete base sequence of an organism represents its **genome**. The "language" used to translate from the base sequence into proteins is called the **genetic code**.

- **How does life evolve?**

Life evolves because the copying of DNA is not perfect, although the error rate is quite small. The occasional, random copying errors, called **mutations**, can change the instructions in DNA. Most mutations either are lethal or have no effect at all, but a few carry benefits that can then be transmitted to offspring when the DNA replicates.

5.5 LIFE AT THE EXTREME

- **What kinds of conditions can life survive?**

Many living organisms can survive in a surprisingly wide range of conditions. These **extremophiles** include microbes that can survive in temperatures above the normal boiling point of water, in the dry deserts of Antarctica, deep underground in the tiny pores of rocks, and even under exposure to high levels of radiation.

- **Are extremophiles really extreme?**

The conditions that we consider extreme are, overall, probably more typical of the conditions found on Earth during most of its history than the conditions we enjoy on the surface today. Many other worlds may have similar conditions, suggesting that extremophiles may in fact be more common in the universe than life similar to plants and animals.

5.6 THE PROCESS OF SCIENCE IN ACTION: EVOLUTION AS SCIENCE

- **Is evolution a fact or a theory?**

This question offers a false choice, because fact and theory are not considered to be opposites in science: Evolution is a well-confirmed theory based on a wide variety of observational and experimental evidence.

- **Are there scientific alternatives to evolution?**

While there are many alternative explanations for our existence, including ideas such as creation science or intelligent design, none of these ideas qualifies as a *scientific* alternative to evolution.

EXERCISES AND PROBLEMS

REVIEW QUESTIONS

Short-Answer Questions Based on the Reading

1. Briefly describe the six key properties that appear to be shared by most living organisms on Earth.
2. What is *natural selection?* Summarize the logic by which Darwin came to the "inescapable conclusion" that evolution occurs by natural selection. Describe some of the evidence that supports Darwin's *theory of evolution.*
3. Briefly describe the evidence that points to a single common ancestor for all life on Earth.
4. Why do we say that living cells are *carbon-based?* Briefly discuss whether life elsewhere could be based on something besides carbon.
5. Briefly describe each of the four main classes of cellular molecules: *carbohydrates, lipids, proteins, and nucleic acids.* What are *enzymes,* and where do they fit into this picture?
6. What are *amino acids?* What do we mean by their *handedness?* How do amino acids offer further evidence for a common ancestor for all life on Earth?
7. What are the three *domains* of life? Which domain do we belong to?
8. What do we mean by the *tree of life?* List three important ideas that we learn from the tree and that differ from older ideas about biology.
9. What is *metabolism,* and what are the two basic metabolic needs of any organism? Explain the four metabolic classifications listed in Table 5.1.
10. Why is water so important to life on Earth? List the three major roles that water plays in metabolism.
11. Describe the double helix structure of *DNA.* How does a DNA molecule replicate?
12. What is a *gene?* A *genome?* The *genetic code?*

13. What are *mutations*, and what effects can they have? Briefly explain why mutations represent the molecular mechanism of natural selection.

14. What are *extremophiles?* Give several examples of organisms that live in extreme environments. What are the implications of the existence of extremophiles for the search for extraterrestrial life?

15. Describe several ways in which the theory of evolution is analogous to the theory of gravity.

16. Explain how evolution exhibits each of the three hallmarks of science, and discuss why alternatives such as creationism and intelligent design do not show these hallmarks.

TEST YOUR UNDERSTANDING

Surprising Discoveries?

Suppose we found an organism on Earth with the characteristics described. In light of our current understanding of life on Earth, should we be surprised to find such an organism existing? Why or why not? Explain clearly; because not all of these have definitive answers, your explanation is more important than your chosen answer.

17. A single-celled organism that builds proteins using 45 different amino acids.

18. A single-celled organism that lives deep in peat bogs, where no oxygen is available.

19. A multicellular organism that reproduces without passing copies of its DNA to its offspring.

20. A single-celled organism that can survive in a dormant state even in the complete absence of any liquid water.

21. A multicellular organism that can grow and reproduce even in the absence of water.

22. A bacterium with cells that lack the molecule ATP.

23. A species of Archaea that lives in the 1,000°C molten rock of a volcano.

24. A species of Archaea that lives in the walls of a nuclear reactor.

25. Two different animal species whose genomes are more than 99% identical.

26. A species of Bacteria that has a genome 99% identical to that of humans.

Quick Quiz

Choose the best answer to each of the following. Explain your reasoning with one or more complete sentences.

27. Which of the following is *not* a key property of life? (a) the maintenance of order in living cells; (b) the ability to evolve over time; (c) the ability to violate the second law of thermodynamics.

28. *Natural selection* is the name given to: (a) the occasional mutations that occur in DNA; (b) the mechanism by which advantageous traits are preferentially passed on from parents to offspring; (c) the idea that organisms can develop new characteristics during their lives and then pass these to their offspring.

29. Which of the following is *not* considered a key piece of evidence supporting a common ancestor for all life on Earth? (a) the fact that all life on Earth is carbon-based; (b) the fact that all life on Earth uses the molecule ATP to store and release energy; (c) the fact that all life on Earth builds proteins from the same set of left-handed amino acids.

30. An organism's heredity is encoded in: (a) DNA; (b) ATP; (c) lipids.

31. An enzyme consists of a chain of: (a) carbohydrates; (b) amino acids; (c) nucleic acids.

32. Which of the following is not a source of energy for at least some forms of life on Earth? (a) inorganic chemical reactions; (b) energy release from plutonium; (c) consumption of preexisting organic compounds.

33. People belong to domain: (a) Eukarya; (b) Archaea; (c) Bacteria.

34. Which of the following mutations would you expect to have the greatest effect on a living cell? (a) a mutation that changes a single base in a region of noncoding DNA; (b) a mutation that changes the third letter of one of the three-base "words" in a particular gene; (c) a mutation that deletes one base in the middle of a gene.

35. Generally speaking, an *extremophile* is an organism that: (a) thrives in conditions that would be lethal to humans and other animals; (b) could potentially survive in space; (c) is extremely small compared to most life on Earth.

36. Based on what you have learned in this chapter, it seems reasonable to think that life could survive in each of the following habitats *except:* (a) rock beneath the martian surface; (b) a liquid ocean beneath the icy crust of Jupiter's moon Europa; (c) within ice that is perpetually frozen in a crater near the Moon's south pole.

INVESTIGATE FURTHER

In-Depth Questions to Increase Your Understanding

Short-Answer/Essay Questions

37. *Rock Life?* How do you know that a rock is not alive? In terms of the properties of life discussed in this chapter, clearly describe why a rock does not meet the criteria for being alive.

38. *The History of Evolution.* Many people assume that Charles Darwin was the first person to recognize that life evolves, but this is not true. Write a few paragraphs summarizing the history of ideas about evolution and explaining why we give Darwin credit for the theory of evolution even though he was not the first person to realize that evolution occurs.

39. *Genetic Variation.* One of the underlying facts (Fact 2 on p. 156) that explains natural selection is that individuals in a population of any species vary in many heritable traits. Based on what you have learned about the molecular basis of evolution, explain why individuals of the same species are *not* expected to be genetically identical.

40. *Artificial Selection.* Suppose you lived hundreds of years ago (before we knew about genetic engineering) and wanted to breed a herd of cows that provided more milk than cows in your current herd. How would you have gone about it? Explain, and describe how your breeding would have worked in terms of the idea of artificial selection. How does this breeding offer evidence in favor of the idea of natural selection?

41. *Ingredients of Life.* Study the ingredients of life as shown in Figure 5.5, and consider them in light of what you've learned about the overall chemical composition of the universe. Would you expect the ingredients to be rare or common on other worlds? Explain.

42. *A Separate Origin?* Suppose that we someday discover life on Mars. How might we be able to determine whether it shares a

common origin with life on Earth (perhaps suggesting that life traveled on meteorites between the two planets) or whether it is from a completely separate origin of life? Explain clearly.

43. *Dominant Life.* While most of us tend to think of ourselves as the dominant form of life on Earth, biologists generally argue that the dominant life consists of microbes of the domains Archaea and Bacteria. In two to three paragraphs, explain why microbes seem more dominant than us.

44. *The Human Power to Destroy.* We clearly have the ability to destroy ourselves today, perhaps as the result of nuclear war, or perhaps through some type of environmental catastrophe. But is there anything we could do with our current abilities that would allow us to wipe out *all* life on Earth? Explain why or why not.

45. *The Search for Life.* Based on what you have learned about life on Earth, what are we searching for when we search for life elsewhere? For example, are we searching only for worlds with surface oceans and oxygen-rich atmospheres like Earth, or for something else? Write one to three paragraphs describing the types of worlds that we can consider as potential homes for life.

46. *Evolution and God.* Does the theory of evolution preclude the existence of God? Clearly explain your answer.

Quantitative Problems

Be sure to show all calculations clearly and state your final answers in complete sentences.

47. *Atomic Numbers in Life.* A typical bacterium has a volume of about 1 cubic micrometer. A typical atom has a diameter of about 0.1 nanometer. Approximately how many atoms are in a bacterium?

48. *Oxygen Atoms in People.* Figure 5.5 shows that oxygen makes up about 65% of the mass of a human being. A single oxygen atom has a mass of 2.66×10^{-26} kg. (a) Use this fact to estimate the number of oxygen atoms in *your* body. (*Hint:* If you know your weight in pounds, you can convert to kilograms by dividing by 2.2.) (b) Compare your answer to the number of stars in the observable universe (which is roughly 10^{22}).

49. *Cellular Energy.* A typical eukaryotic cell, such as a cell in the human body, uses about 2×10^{-17} joule of energy each second. The breakdown of a single molecule of ATP (in which a phosphate separates from ATP to make ADP; see Figure 5.12) releases about 5×10^{-20} joule of energy. (a) About how many molecules of ATP must be broken down and reassembled each second to keep a eukaryotic cell alive? (b) How many times does this ATP recycling occur each *day* in a typical cell? (c) The human body has roughly 10^{14} cells. Approximately how many cycles of the ATP reaction occur each day in your body?

50. *The Genetic Code.* Suppose that, as evidence suggests, very early life on Earth used a genetic code that consisted of only two-base "words" rather than three-base "words." Could such life have made use of the same set of 20 amino acids that life uses today? Explain, using quantitative arguments.

Discussion Questions

51. *Science and Religion.* Science and religion are often claimed to be in conflict. Do you believe this conflict is real and hence irreconcilable, or is it a result of misunderstanding the differing natures of science and religion? Defend your opinion.

52. *Computer Life.* Although scientists have already developed computer programs capable of reproducing themselves and evolving, few people consider such programs to be alive. But consider future developments in computing and robotic technology. Do you think we'll ever make something out of electronics that is truly alive? Could it also be intelligent? If so, what civil rights should we give to such "artificial" intelligent life?

53. *Genetic Engineering and Future Evolution.* For billions of years, evolution has proceeded through mutations and natural selection. Today, however, we have the ability to deliberately alter DNA in what we call "genetic engineering." How do you think this ability will affect the future evolution of life? How will it affect future human evolution on Earth? Based on your answers, should we expect extraterrestrial civilizations to be naturally evolved or to be products of their own genetic engineering? Discuss and defend your opinions.

54. *Gene Transfer and GMOs.* In some cases, organisms can transfer entire genes to other organisms. This fact causes some people to worry that organisms that we have genetically engineered—commonly referred to as GMOs, for "*g*enetically *m*odified *o*rganisms"—may transfer their genes to other organisms in unexpected ways. For example, a crop engineered with a gene that gives it resistance to some pest may transfer its gene to weeds, giving them the same resistance. Discuss how GMOs might affect other organisms. Overall, what, if any, controls do you think the government should put on the use of GMOs?

WEB PROJECTS

55. *The Dover Opinion.* In December 2005, a U.S. District Court issued its opinion on a case concerning the teaching of intelligent design (ID), deciding that ID does not belong alongside evolution in science classes. The full text of the opinion, commonly called the Dover opinion, is available online. Read the opinion, and discuss its implications to the ongoing public controversy about what belongs in science classes.

56. *Darwin on Evolution.* You can find online the entire text of Charles Darwin's *The Origin of Species.* Read the final chapter, in which Darwin addresses potential criticisms of his theory. Evaluate how well he has presented his case. How much stronger does the theory seem today than at the time Darwin first described it in 1859? Summarize your conclusions in a one-page essay.

57. *Extreme Life.* Look for information about a recent discovery of a previously unknown type of extremophile. Describe the organism and the environment in which it lives, and discuss any implications of the finding for the search for life beyond Earth. Summarize your findings in a one-page report.

6
The Origin and Evolution of Life on Earth

LEARNING GOALS

6.1 SEARCHING FOR LIFE'S ORIGINS
- When did life begin?
- What did early life look like?
- Where did life begin?

6.2 THE ORIGIN OF LIFE
- How did life begin?
- Could life have migrated to Earth?

6.3 THE EVOLUTION OF LIFE
- What major events have marked evolutionary history?
- Why was the rise of oxygen so important to evolution?

6.4 IMPACTS AND EXTINCTIONS
- Did an impact kill the dinosaurs?
- Did impacts cause other mass extinctions?
- Is there a continuing impact threat?

6.5 HUMAN EVOLUTION
- How did we evolve?
- Are we still evolving?

6.6 THE PROCESS OF SCIENCE IN ACTION: ARTIFICIAL LIFE
- How might we create artificial life?
- Should we create artificial life?

Small as is our whole system compared with the infinitude of creation, Brief as is our life compared with the cycles of time, we are so tethered to all by the beautiful dependencies of law, that not only the sparrow's fall is felt to the uttermost bound but the vibrations set in motion by the words that we utter reach through all space and the tremor is felt through all time.

Maria Mitchell (1818–1889), first woman elected to the American Academy of Arts and Sciences

We explored the habitability of our planet in Chapter 4, learning how and why Earth has remained a suitable home for life during the past 4 billion or more years. Then, in Chapter 5, we saw how evolution allowed life to take advantage of this habitability, as natural selection led to a great diversity of life that now exists in almost every place on our planet that has even the tiniest amounts of liquid water. But we have yet to discuss the deepest questions of all: How did life arise in the first place, and what events shaped life's evolution into its current diversity of species?

The complete answers to these questions still elude us, but scientists have put together a fairly detailed outline of what is most likely to have occurred. We know that the early Earth was a natural laboratory for organic chemistry, and we have at least some ideas about how this chemistry might have led to life. The geological record shows us how life evolved subsequently, and it is full of important surprises. For example, plants and animals appeared only relatively recently, and we've discovered that external forces, such as the impacts of asteroids or comets, can dramatically alter the course of evolution. In this chapter, we'll survey current ideas about the origin and subsequent evolution of life on Earth, in effect studying how we ourselves came to be.

6.1 SEARCHING FOR LIFE'S ORIGINS

The geological record details much of the history of life on Earth, and the theory of evolution tells us how life has changed from the forms that existed long ago to those found on our planet today. However, neither the geological record nor ideas of biological evolution are likely to tell us precisely how life first arose on Earth. The geological record becomes increasingly incomplete as we look further back in time, and no known rocks survive from Earth's first half-billion years. The theory of evolution explains how one species can evolve into others, but does not tell us how the first living organisms came to be. Thus, any attempt to understand the origin of life must be based on careful study of limited clues about what existed in Earth's early history, along with laboratory experiments that can help us reconstruct the processes that may have occurred on the young Earth. In this section, we'll explore the clues that tell us about early life on Earth, and in the next section we'll explore how laboratory experiments shed light on a possible mechanism for the origin of the first life.

a These large mats at Shark Bay, Western Australia, are colonies of microbes known as "living stromatolites"; they stand about knee-high. Microbes near the top generate energy through photosynthesis.

b The bands visible in this section of a modern-day mat are formed by layers of sediment adhering to different types of microbes.

c This section of a 3.5-billion-year-old stromatolite shows a structure nearly identical to that of a living mat.

Figure 6.1

Rocks called stromatolites offer evidence of photosynthetic life as early as 3.5 billion years ago.

• When did life begin?

Perhaps the first thing we'd like to know about the origin of life on Earth is when it occurred. The only way to approach this question is through the study of fossils. If we find a fossil of a particular age, then we know life already existed at that time. For example, a 3.5-billion-year-old fossil tells us that life on Earth arose *before* 3.5 billion years ago. Because the geological record is incomplete and because we may not yet have discovered the oldest intact fossils, we do not know exactly how long life has existed on Earth. Nevertheless, three lines of fossil evidence all point to the idea that, geologically speaking, life arose quite early in Earth's history.

Stromatolites The first line of evidence comes from **stromatolites** (from the Greek for "rock beds"), rocks that are characterized by a distinctive, layered structure. In size, shape, and interior structure, ancient stromatolites look virtually identical to sections of mats formed today by colonies of microbes sometimes called "living stromatolites" (Figure 6.1). Living stromatolites contain layers of sediment intermixed with different types of microbes. Microbes near the top generate energy through photosynthesis, and those beneath use organic compounds left as waste products by the photosynthetic microbes. The living stromatolites grow in size as sediments are deposited over them, forcing the microbes to migrate upward in order to remain at the depths to which they are adapted. This gradual migration creates the layered structures.

The similarity of structure between the ancient stromatolites and the modern-day mats suggests a similar origin, implying that stromatolites are fossil remnants of early life. Chemical analysis offers further support for this conclusion, as stromatolites contain chemical and isotopic evidence of biological origin. Thus, while it is possible for geological processes of sedimentation to mimic the layering that we see in stromatolites, we are fairly confident that they offer evidence for the existence of microbial colonies as far back as 3.5 billion years ago. Moreover, if the microbes that made the stromatolites are like the microbes in the living mats today, then the implication is that at least some of these ancient microbes produced energy by photosynthesis. Because photosynthesis is a fairly sophisticated metabolic process, we presume that it must have taken at least a moderately long time for this process to evolve in living organisms. In other words, if we are correct in concluding that stromatolites tell us that photosynthetic life already existed some 3.5 billion years ago, then we can infer that more primitive life must have existed even earlier, and that the origin of life itself substantially predates this.

Microfossils The second line of evidence comes from individual fossilized cells. Finding ancient microscopic fossils, or microfossils, is quite challenging, both because rocks become increasingly rare with age and because the oldest rocks have been altered by geological processes in ways that tend to destroy microfossils within them. Moreover, unlike dinosaur bones that are fairly "obvious" fossils, it can be quite difficult to determine whether an interesting-looking microscopic structure is biological or mineral in origin. As a result, claims of ancient microfossils often generate significant scientific controversy, with competing hypotheses attempting to explain their origin.

Not surprisingly, the greatest controversy surrounds the oldest claimed microfossils, which come from a 3.5-billion-year-old rock formation in northwestern Australia. Microscopic photos of sections of this rock show structures that look like individual cells (Figure 6.2). This structure, along with chemical analysis showing the presence of organic carbon and an assumption that the structures resided in sedimentary rock from a shallow sea, originally led the discoverers of these structures (a team led by William Schopf of UCLA) to propose that they represented fossils of early photosynthetic organisms. However, subsequent analysis has shown that the rock is not from a shallow sea as originally assumed, but instead must have formed near a deep-sea volcanic vent similar to a black smoker (see Figure 5.17). The structures therefore cannot be fossils of photosynthetic microbes (since sunlight does not penetrate to the deep sea), and some scientists have used this fact to argue that the structures are not fossils at all, but rather were formed by some nonbiological process. There is also a possibility that, while these structures were found associated with rock known to be 3.5 billion years old, they may have come to this location more recently via flowing water from younger rock. Nevertheless, their shape and organic content still offer a reasonable argument for biological origin, so they may yet prove to represent ancient microfossils of deep-sea microbes. Only further study, and perhaps the discovery of similarly ancient structures in other places, will settle the controversy.

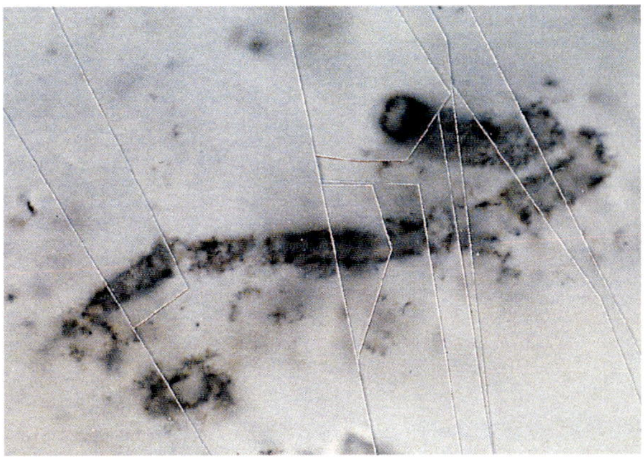

Figure 6.2

Microfossils of ancient living cells? This microscopic photograph shows structures that some researchers believe to be ancient fossil cells dating to 3.5 billion years ago. Other researchers, however, argue that the structures were formed by nonbiological processes.

THINK ABOUT IT Consider the controversy over these microfossils in light of what you have learned about the process of science in this book. Do you think the controversy indicates strength or weakness in our methods of scientific inquiry? Defend your opinion.

Other microfossils date the origin of life almost as far back with less controversy. At two locations in southern Africa, rocks that date to between 3.2 and 3.5 billion years old also show what appear to be fossilized cells. Although some researchers argue that these might also have nonbiological origin, so far they seem more likely to be true fossil cells. Moreover, these microfossils may lend further support to the hypothesis of biological origin for the microfossils in the 3.5-billion-year-old Australian rock. Despite the wide separation between Australia and Africa today, these particular African and Australian rocks appear to share similar origins, suggesting that they were geologically linked in the past. Thus, if one set of these rocks holds microfossils, we would expect the same to be true of the other set.

Microfossils in rocks dating to between 2.7 and 3.0 billion years old show more conclusive evidence for life: They contain particular molecules (such as a variety of hydrocarbons) that almost certainly indicate

Figure 6.3

This ancient rock formation on the island of Akilia (off the coast of southern West Greenland) may hold the oldest known evidence for life on Earth.

biological origin. Overall, we conclude that microfossil evidence clearly points to the existence of life before about 3.0 billion years ago and may well tell us that life existed before 3.5 billion years ago.

Isotopic Evidence The third line of evidence for an early origin of life comes from isotopic analysis of some of the most ancient rocks on Earth. Living organisms can change the ratios of isotopes from their background, nonliving values. For example, carbon has two stable isotopes: carbon-12, with six protons and six neutrons in its nucleus, and carbon-13, which has one extra neutron. Carbon-12 is far more common, but any inorganic carbon sample always contains a small proportion of carbon-13 atoms mixed in with the more numerous carbon-12 atoms. (Typically, carbon-13 accounts for about 1 out of every 89 carbon atoms.) When living organisms metabolize carbon, they incorporate carbon-12 atoms into cellular molecules slightly more easily than they do carbon-13 atoms. As a result, living organisms—and fossils of living organisms—always show a slightly lower fraction of carbon-13 atoms than that found in inorganic material.

On an island off the coast of Greenland, this lower carbon-13 ratio has been found in rocks that are more than 3.85 billion years old,* suggesting that the rocks contain remnants of life from that time (Figure 6.3). However, the claim of biological origin (made by a team led by the University of Colorado's Stephen Mojzsis) has been challenged by a few other scientists. The rocks are metamorphic, meaning they have been transformed substantially by high pressure or heat, which would explain why no intact microfossils remain within them. For these rocks to have contained life, we must presume that they were sedimentary before they underwent the metamorphic transformation; if they were volcanic (igneous) rocks, then it is much more difficult to see how they could have been home to or preserved evidence for living microbes. Scientists are still debating the nature of these rocks and hence the question of whether their carbon isotope ratio shows evidence of life. While the controversy is by no means settled, recent evidence seems to be swaying the debate in favor of the hypothesis that they really do contain evidence that life already existed by 3.85 billion years ago.

The new evidence falls into two basic categories. First, if the Greenland rocks hold evidence of life, we would expect to find similar evidence in other rocks dating to the same general time. This is indeed the case, as other rocks dating to some 3.8 billion years ago have been found to show similar carbon isotope ratios. Second, life can also alter the isotopic ratios of other elements, such as iron, nitrogen, and sulfur. Recent studies show that these isotopes also are present in characteristic ratios for life within the ancient rocks, just as we would expect if the carbon isotope data are a result of biological origin. Further study should eventually resolve the issue, one way or the other.

Implications of the Early Origin of Life We have seen that the stromatolite evidence suggests the presence of fairly advanced life by nearly 3.5 billion years ago. The microfossil evidence, while more controversial, is consistent with this dating. Carbon isotope evidence, if it stands up to further scrutiny, pushes the existence of life back to at least 3.85 billion years ago. Given the fact that the geological record is so sparse for such

*We say that the rocks are *older* than 3.85 billion years because the rocks containing the isotopic evidence are sediments that cannot be dated. However, they are cut through by igneous rocks that date to 3.85 billion years, which means the sediments must be even older.

early times, we would expect to find evidence of life in such ancient rocks only if life had already been widespread on Earth. Thus, if we are interpreting all the data correctly, it is likely that life first arose considerably earlier than 3.85 billion years ago. Geologically speaking, this would mean life arose quite early in Earth's history (see Figure 4.10).

In addition, remember from Chapter 4 that the young Earth probably experienced numerous large impacts during the heavy bombardment, some of which may have been large enough to have extinguished any life on Earth at the time, or at least to have extinguished life near the surface. We cannot know exactly when the last such impact might have occurred, but studies of cratering rates on the Moon suggest it was most likely between about 4.2 and 3.9 billion years ago.

Overall, we conclude that life arose and colonized our planet in a period of no more than a few hundred million years, and perhaps in as little as 100 million years or less. By itself, this early origin of life proves nothing about life elsewhere, since it is always possible that Earth was the lucky beneficiary of a highly improbable event. However, if we assume that what happened here would be typical of what might happen elsewhere, then the early origin of life is profoundly important: It suggests that we could expect life to arise similarly rapidly on any other world with similar conditions. Because we expect many other worlds to have conditions similar to those that prevailed on the young Earth, this idea gives us reason to think that life might be quite common in the universe.

• What did early life look like?

The earliest living organisms presumably went extinct long ago, replaced by others with evolutionary adaptations that allowed them to outcompete their ancestors. Nevertheless, just as we know that sharks and alligators are evolutionarily much older than primates (because the geological record shows them coexisting with dinosaurs while primates emerged much later), it seems reasonable to think that some modern-day microbes are more closely related than others to the earliest living organisms. Study of these more primitive species should help us understand what early life looked like, which in turn should help us investigate the question of how life arose.

Mapping Evolutionary Relationships The geological record tells us a great deal about how life evolved, but it is not the only way to study evolution. Because living species have evolved from common ancestors, the base sequence in the DNA of living organisms provides a sort of map of the genetic changes that have occurred through time. Thus, by comparing the genomes of different organisms, we should be able to reconstruct the evolutionary history of much of life on Earth.

To understand how the technique works, consider the DNA of an organism that long ago became the common ancestor of all life today. Mutations created variations on this DNA, and each new species therefore had slightly different DNA sequences than did the older species from which it evolved. Lateral gene transfer [Section 5.4] may also have been common among early living organisms, changing genomes even more rapidly. Over millions and billions of years, continuing evolution led to new species with DNA molecules increasingly different from the DNA of the common ancestor. But, always, the new molecules were built by changes to the older ones so that, in principle, the changes are traceable in the precise base sequences of living organisms.

Determining the sequence of bases in an organism's DNA is a difficult and time-consuming task, and to date only a small fraction of known species have had their complete genome sequences determined. In many more cases, biologists have compared smaller pieces of the DNA of many species.* By comparing the DNA sequences in similar genes among many different species, biologists can map the evolutionary history of the genes. For example, two species with very similar DNA sequences probably diverged relatively recently in evolutionary history, while two species with very different DNA sequences probably diverged much longer ago.

These types of DNA sequence comparisons are what has enabled biologists to map out the relationships shown in the tree of life (see Figure 5.11). That is, DNA studies tell us that life can be divided into the three domains that we discussed in Chapter 5—Bacteria, Archaea, and Eukarya—and also tell us about the branching patterns within each domain. Despite the many uncertainties that remain in the tree of life, the branching patterns still tell us a lot about evolutionary history. For example, the fact that animals and plants represent two branches that split off in about the same place from other Eukarya tells us that all animals and plants are quite similar genetically, at least in comparison to organisms on most other branches. Moreover, organisms on branches located closer to the "root" of the tree must contain DNA that is evolutionarily older, suggesting that they more closely resemble the organisms that lived early in Earth's history.

Conclusions About Early Life The genetic studies that have led to mapping the tree of life tell us that species near the root of the tree must be more similar than other species to the common ancestor of life on Earth. Although it is not shown explicitly in the version of the tree in Figure 5.11, it appears that the organisms closest to the root are extremophiles such as those living near deep-sea volcanic vents or in hot springs like those in Yellowstone [Section 5.5]. Recall that these organisms are adapted to life in hot water (making them *hyperthermophiles*) and use chemical energy rather than photosynthesis to fuel their metabolism. In other words, it seems likely that the earliest life on Earth resembled the hot-water extremophiles that still survive today.†

• Where did life begin?

We can combine what we've learned about the probable nature of early organisms with geological considerations to come up with ideas about *where* life first arose on Earth. Let's begin by considering locales that we can probably rule out.

It seems unlikely that life could have arisen on the land surface. The early atmosphere contained practically no molecular oxygen, so our planet could not have had a protective layer of ozone. Ozone (O_3) is a form of oxygen produced in the upper atmosphere by interactions between ordinary oxygen (O_2) and ultraviolet light from the Sun. Today, ozone shields Earth's surface from the Sun's dangerous ultraviolet radiation. Before the ozone layer existed, any surface life would have been exposed to high lev-

---

*A particularly common technique relies on determining the sequence of bases in molecules of ribosomal RNA (rRNA), which tells us the sequence of the DNA that coded for it; rRNA is thought to evolve more slowly than most genes, so it provides a good proxy for overall evolutionary changes.

†One important caveat: A not-quite-sterilizing impact could have extinguished all life except that living near deep-sea vents. In that case, we'd still find that present-day life evolved from a hyperthermophile ancestor, even though earlier life might have had greater variety.

els of this radiation. While we can't rule out the possibility that life might have arisen in such an environment—indeed, some organisms today (such as *D. radiodurans* [Section 5.5]) can survive high-radiation conditions—the environment would have been much more hospitable underwater (because water also absorbs ultraviolet light) or in rocks beneath the surface.

The next idea, popular a few decades ago, suggested shallow ponds as the first home for life. If the ponds were full of organic compounds (an idea we'll discuss in the next section), these compounds would have become more concentrated as the ponds evaporated, making it easier for them to combine into more complex molecules that could lead to life. However, more recent studies suggest that shallow ponds would have lacked a source of chemical energy sufficient to support the origin of life, and any life in these ponds would probably also have suffered lethal effects from ultraviolet radiation and/or high salt concentrations.

Given the DNA evidence suggesting that early organisms resembled modern-day extremophiles that live near deep-sea vents and in hot springs, it now seems reasonable to favor these locations as the first home of life. They offer plenty of chemical energy to fuel chemical reactions that might have led to life, and they are protected from the Sun's ultraviolet radiation by the water that lies above them. Of course, we cannot rule out the possibility that life originated elsewhere, such as within rocks, and later adapted to these hot-water environments.

We have at least some reason to favor deep-sea vents over surface hot springs as the earlier home for life. Smaller impacts were more common than larger ones during the heavy bombardment, so partial vaporization of the oceans should have been more common than complete vaporization. No place on Earth was completely safe, but the deep ocean was clearly safer than springs at or near the surface. We may never know for certain where life first arose on Earth, but deep-sea vents look like a good bet.

6.2 THE ORIGIN OF LIFE

Even the simplest living organisms today seem remarkably advanced. Metabolic processes involve many intricate molecules and enzymes working together. The complex chemistry of DNA and RNA is deeply intertwined with the proteins and enzymes that help in making them [Section 5.4]. Indeed, every cellular component and process depends on many other components and processes. Given the complexity and interdependency of these processes, it might at first seem difficult to conceive of ways in which they might have come to be. However, over the past few decades, laboratory experiments have given us insights into the chemical processes that likely occurred on the early Earth. While these experiments have not yet told us precisely how life first arose—and it's possible that they never will—we'll see in this section that they give us good reason to think that life may have started through natural, chemical processes.

Before we begin, it's worth noting two important caveats. First, the laboratory experiments generally try to re-create the chemical conditions that should have prevailed on the early Earth, an assumption that makes sense if life originated here. However, it is conceivable that life migrated to Earth from another world—a possibility we will also discuss in this section—in which case it might have arisen in a somewhat different chemical environment. Second, we have not said anything about the possibility that life arose through any kind of divine intervention; as we have discussed in prior chapters, that possibility falls outside the realm of science and instead is a matter of personal faith.

Figure 6.4

Stanley Miller poses with a re-creation of the original Miller–Urey experiment. The flask to the left contains liquid water that represents the sea; the gases, representing the atmosphere, consist of water vapor, methane, ammonia, and hydrogen. The central flask was supplied with energy in the form of electrical discharges. Below this flask, the gas was cooled so that it could condense and flow back into the flask with the liquid water. Chemical reactions in the experiment produced a wide variety of organic molecules.

• How did life begin?

Life today is based on the chemistry of a wide variety of organic molecules, making it logical to assume that the first life was somehow assembled from organic molecules produced by chemical reactions on the early Earth. Such reactions do not occur naturally today, because Earth's oxygen-rich atmosphere prevents complex organic molecules from forming readily outside living cells: Oxygen is such a highly reactive gas that it tends to attack chemical bonds, removing electrons and destroying organic molecules. However, the oxygen in our atmosphere is a product of life, produced by photosynthesis, which means it could not have been present before life arose. It therefore seems reasonable to think that the young Earth might have been much like a giant laboratory for organic chemistry.

The Miller–Urey Experiment As early as the 1920s, some scientists recognized that Earth's early atmosphere should have been oxygen-free, and they hypothesized that sunlight-fueled chemical reactions could have led to the spontaneous creation of organic molecules. (This idea was proposed independently by Russian biochemist A. I. Oparin and British biologist J. B. S. Haldane.) This hypothesis was put to the test in the 1950s in a famous experiment credited to Stanley Miller and Harold Urey, now known as the **Miller–Urey experiment** (Figure 6.4).

The original Miller–Urey experiment used small glass flasks to simulate chemical conditions that they thought represented those on the early Earth. One flask was partially filled with water to represent the sea and heated to produce water vapor. Gaseous methane and ammonia were added and mixed with the water vapor to represent the atmosphere. These gases flowed into a second flask, where electric sparks provided energy for chemical reactions. Below this flask, the gas was cooled so that it could condense to represent rain and then was cycled back into the water flask. The water soon began to turn a murky brown, and a chemical analysis (performed after letting the experiment run for a week) showed that it contained many amino acids and other organic molecules. Thus, the experiment appeared to support the idea that organic compounds would have formed spontaneously and in vast quantities on the early Earth, covering our planet and filling at least the top layers of the oceans with a rich "organic soup" containing all the building blocks of life.

However, other scientists soon realized that the methane and ammonia gas mixture used by Miller and Urey probably was *not* representative of Earth's early atmosphere, so the significance of their results was called into question. That the relevance of the Miller–Urey experiment is an even hotter scientific controversy today demonstrates the way that science often progresses in fits and starts.

The major issue concerns exactly what the early atmosphere was made of and, in particular, how much hydrogen it contained. When analyzed in more detail, the Miller–Urey experiment shows that hydrogen is a key ingredient needed to form large quantities of organic compounds; oxygen in any form, even when tied up in carbon dioxide, tends to suppress the formation of organic compounds. Today, we assume that Earth's early atmosphere was produced by volcanic outgassing [Section 4.2]. Outgassing may have occurred at a somewhat higher rate in the distant past than it does today, but we expect the gases to have been similar: largely water (H_2O), carbon dioxide (CO_2), nitrogen (N_2), and methane (CH_4), along with sulfur-bearing gases and hydrogen (H_2). The water condensed to form the oceans, leaving the other gases as potential ingredients for the early atmosphere. The key question is how much hydrogen remained relative to carbon dioxide. If there was a lot of hydrogen, then the Miller–Urey results might still be valid, suggesting that our young planet did indeed have a rich organic soup. If there was little hydrogen, then the oxygen in carbon dioxide would have kept the amount of organic material far lower than the Miller–Urey experiment suggests.

Prior to 2005, it was generally assumed that hydrogen would have escaped from the early atmosphere at the same rate that it escapes today, which is rapid enough that little hydrogen would have been present at any time. However, new calculations, still subject to verification, suggest that conditions in Earth's early upper atmosphere could have slowed the escape considerably, allowing hydrogen to make up as much as 30% of Earth's early atmosphere. Until this issue is settled, we will not know whether spontaneous chemical reactions would have turned Earth's surface and oceans into a rich organic laboratory or left them with only small amounts of locally produced organic compounds.

Other Sources of Organic Molecules Although the jury is still out on precisely how much organic material might have been made through processes like those in the Miller–Urey experiment, we know of at least two other potential sources of organic molecules on the early Earth.

The first of these potential "other" sources is chemical reactions near deep-sea vents. As these undersea volcanoes heat the surrounding water, a variety of chemical reactions can occur between the water and the minerals. These chemical reactions would have occurred spontaneously in the conditions thought to have prevailed in the early oceans, and they should have resulted in the production of the same types of organic molecules thought to have been necessary for the origin of life.

The other additional source of organic molecules may have been material from space. Analysis of meteorites shows that they often contain organic molecules, including complex molecules such as amino acids. Similarly, telescopic and spacecraft studies of comets suggest that they, too, contain organic molecules. Apparently, organic molecules can form under the conditions present in interplanetary space and can survive the plunge to Earth. Moreover, recent research shows that the building blocks of life should also have been produced as ultraviolet light from the young Sun caused chemical reactions on the surfaces of dust grains in the spinning disk of the solar nebula. This dust, laden with organic molecules, could have "rained down" on the young Earth. The many impacts of the heavy bombardment could have brought additional organic material from asteroids and comets. The heat and pressure generated by the impacts may have further facilitated the production of organic molecules in Earth's atmosphere and oceans. All in all, a lot of organic material could have come to Earth from space.

It's likely that all three sources of organic molecules—chemical reactions in the atmosphere, chemical reactions near deep-sea vents, and material from space—played a role in shaping the chemistry of the early Earth. More important, given three different ways of obtaining organic molecules, it seems likely that at least parts of the early Earth would have contained substantial amounts of the organic molecules needed for life. The likelihood of organic molecules forming near deep-sea vents is particularly encouraging, given the evidence suggesting that life may have originated in such locales.

The Transition from Chemistry to Biology The existence of at least three potential sources of organic molecules suggests that at least some locations on the young Earth had all the building blocks needed to make life. The next question, then, is how these building blocks might have assembled themselves to make a living cell.

Variations on the original Miller–Urey experiment have since produced all the essential building blocks of life, but, to paraphrase the late Carl Sagan, these building blocks represent only the notes of the music of life, not the music itself. Viewed in terms of simple probability, the likelihood of a set of simple building blocks ramming themselves together to form a complete living organism is at least as small as that of letting monkeys loose in a roomful of musical instruments and hearing Beethoven's Ninth Symphony. It simply wouldn't happen, even if the experiment was repeated over and over again for billions of years. There must have been at least a few intermediate steps—each involving a chemical pathway with a relatively high probability of occurring—that eased the transition from chemistry to biology.

One way to explore the transition is to work backward from organisms living today. Heredity today is shaped by DNA, which serves this function primarily because of its ability to replicate. Thus, we presume that early life must also have had a self-replicating molecule. However, this molecule probably was not DNA: Double-stranded DNA seems far too complex, and its replication far too intertwined with RNA and proteins, to have been the genetic material of the first living organisms. We are therefore looking for a molecule that is simpler than DNA but still capable of making fairly accurate copies of itself. The most obvious candidate is RNA.

RNA World RNA is much simpler than DNA because it has only one strand rather than two and its backbone structure requires fewer steps in its manufacture. But it still possesses hereditary information in the ordering of its bases, and in principle it can serve as a template for making copies of itself. However, for a while there seemed to be a problem with this idea. In modern organisms, neither DNA nor RNA can replicate itself. Both require the help of enzymes. These enzymes are proteins that are made from genetic instructions contained in DNA and carried out with the help of RNA. This fact seemed to present a "chicken and egg" dilemma: RNA cannot replicate without enzymes, and the enzymes cannot be made without RNA.

A way around this dilemma was discovered in the early 1980s by Thomas Cech and his colleagues at the University of Colorado, Boulder. They found that RNA can catalyze biochemical reactions in much the same way as enzymes (work for which Cech shared the Nobel Prize in 1989). We now know that RNA molecules play this type of catalytic role in many cellular functions, and we call such RNA catalysts *ribozymes* (by analogy to enzymes). Follow-up work has shown that some RNA molecules can at least partially catalyze their own replication. These discoveries have led biologists to envision that modern, DNA-based life may have arisen from an earlier **RNA world**, in which RNA molecules served both as genes and as chemical catalysts for copying and expressing those genes.

How might an RNA world have gotten started? The first requirement would have been the spontaneous production of self-replicating strands of RNA. Even under the most optimistic assumptions, the concentration of organic molecules on the early Earth would have been far too low to allow those building blocks to assemble spontaneously into full-fledged RNA molecules. RNA assembly almost certainly would have required some sort of catalytic reaction to facilitate it. Here, again, laboratory experiments offer evidence for such a process.

Experiments show that several types of inorganic minerals can facilitate the self-assembly of complex, organic molecules. Minerals of the type that geologists call *clay** may have been especially important to the origin of life. Clay is extremely common on Earth and in the oceans, where it forms through simple weathering of silicate minerals; indeed, the oldest zircon grains [Section 4.2] suggest the widespread abundance of clays more than 4.4 billion years ago, so we expect clay to have been common at the time of the origin of life. Moreover, clay minerals contain layers of molecules to which other molecules, including organic molecules, can adhere. When organic molecules stick to the clay in this way, the mineral surface structure can force them into such close proximity that they react with one another to form longer chains. Laboratory experiments with this process have already produced strands of RNA up to nearly 100 bases in length. Although these strands are still quite primitive compared to what might be needed to start any sort of RNA life, the experiments suggest that clays in Earth's early oceans could have served as natural chemical "factories" that could potentially have produced self-replicating (or partially self-replicating) molecules of RNA. Indeed, given the millions of years and the countless grains of clay that could have facilitated chemical reactions, it seems reasonable to *expect* that self-replicating RNA molecules would at some point have been produced.

Assuming that self-replicating RNA did indeed come to exist, it would have been rapidly modified by a molecular analog to natural selection: The RNA molecules that replicated faster and more accurately soon came to dominate the population. Copying errors introduced mutations, ensuring the production of many variations of successful molecules. This allowed the molecular evolution to continue, as those RNA molecules best suited to the particular environmental conditions were more likely to continue replicating. Molecular evolution in the RNA world would have progressed much faster if the RNA and other complex molecules could have been confined within structures similar to living cells. The advantages of confining organic molecules within such "pre-cells" (also often called *vesicles*) are twofold. First, keeping the molecules concentrated and close together should have increased the rate of reactions among them, making it more likely that cooperative relationships like those in modern cells would evolve. Second, an enclosure would have isolated its contents from the outside in a way that should have facilitated natural selection among RNA molecules. For example, suppose a particular self-replicating RNA molecule assembled amino acids into a primitive enzyme that sped up replication (Figure 6.5a). If the enzyme floated freely within the ocean water, it might just as easily have helped the replication of other RNA molecules as of the one that made it. But inside a pre-cell, the enzyme would help only the RNA that made it, giving this RNA an advantage over less capable RNA molecules in other pre-cells (Figure 6.5b).

How might such pre-cell enclosures have formed? Once again, laboratory experiments suggest that they may have appeared easily and naturally in at least two ways. First, if we cool a warm-water solution of amino acids, they can form bonds among themselves to make an enclosed, spherical structure (Figure 6.6a). Although they are not alive, these structures exhibit many lifelike properties. For example, they can grow in size by absorbing more short chains of amino acids until they reach an unstable size, at which point they split to form "daughter" spheres. They can also

*In this context, clay refers to silicate minerals with a particular physical structure; this mineralogical definition is somewhat different from what you may think of as clay in the context of pottery or sculpture.

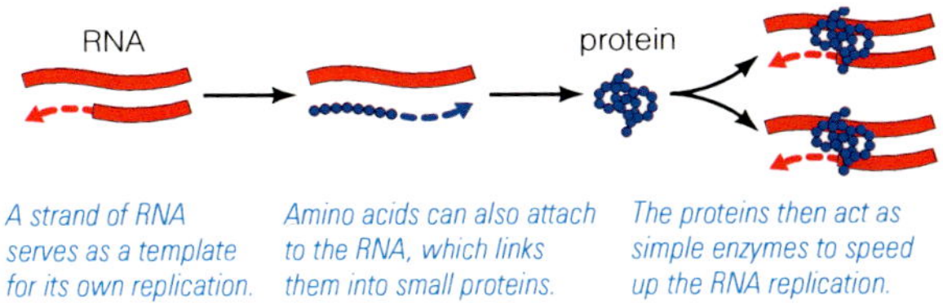

a This diagram shows a self-replicating RNA molecule that has evolved the capability to produce a primitive enzyme that helps its own replication.

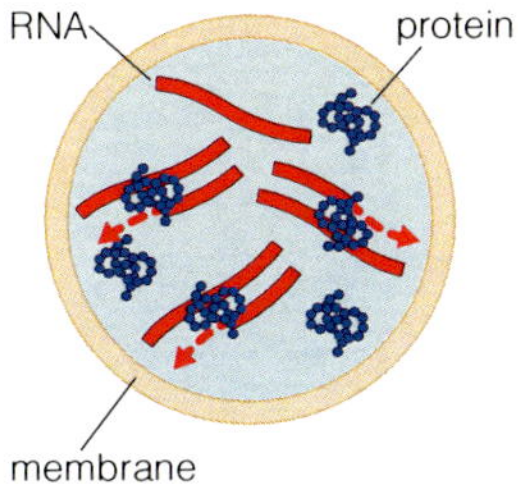

b If the RNA and the enzyme are isolated from the outside environment inside a pre-cell, then only the molecules in this particular pre-cell will benefit from the new enzyme, a fact that can speed up the molecular evolution.

Figure 6.5

Self-replicating RNA could have rapidly evolved through a molecular analog to natural selection. (Adapted from Campbell, Reece, Simon, *Essential Biology,* Second Edition, © 2004.)

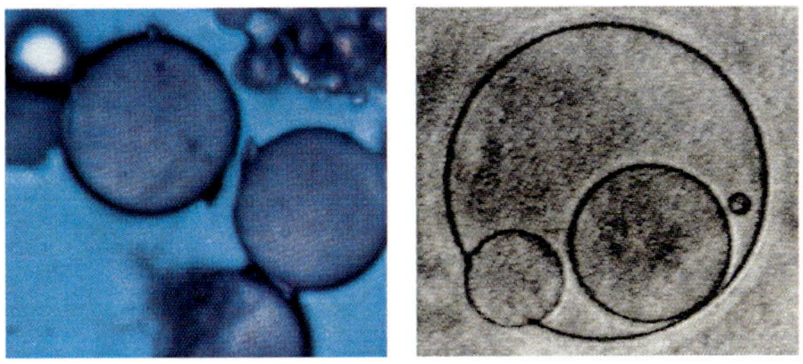

a These microscopic spheres were made by cooling a warm-water solution of amino acids. They are not alive, but they exhibit many lifelike properties.

b These microscopic membranes are made from lipids that, when mixed with water, spontaneously form enclosed droplets.

Figure 6.6

Laboratory versions of pre-cells. (Adapted from Campbell, Reece, Simon, *Essential Biology,* Second Edition, © 2004.)

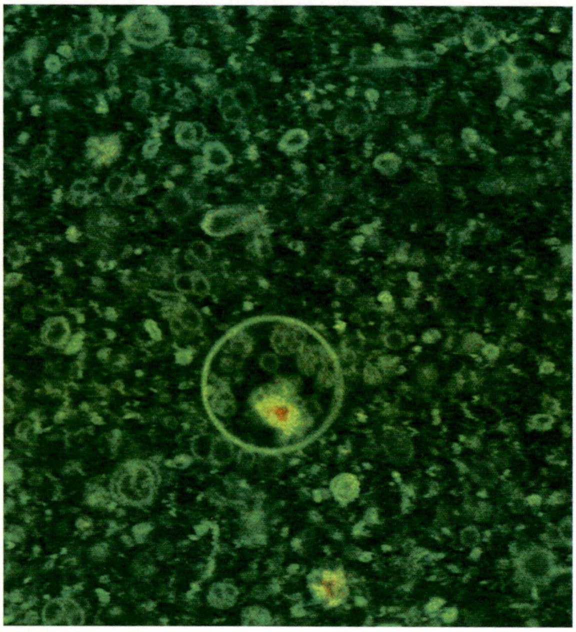

Figure 6.7

This microscopic photo (made with the aid of fluorescent dyes) shows short strands of RNA (red) contained within a lipid pre-cell (green circle), both of which formed with the aid of catalysis by clay minerals beneath them.

selectively allow some types of molecules to cross into or out of the enclosure. Some of these spheres even store energy in the form of an electrical voltage across their surfaces, which can be discharged in a way that facilitates reactions inside them. The second type of membrane forms spontaneously when we mix lipids with water (Figure 6.6b). (Recall that lipids are one of the four major types of cellular components [Section 5.2].)

Perhaps even more significantly, recent experiments show that lipid pre-cells can form on the surface of the same clay minerals that help assemble RNA molecules, sometimes with RNA inside them (Figure 6.7). This work, which we'll discuss in more detail in Section 6.6, gives further credence to the idea that an RNA world may have been born on the early Earth with the catalytic assistance of clay minerals.

If and when the RNA world got started, the molecular natural selection would have inevitably led RNA molecules to gain complexity and evolve more efficient replication pathways. At some point, the RNA pre-cells would have become sufficiently good at reproducing and evolving that we would have considered them to be "alive," although the process was probably gradual: There might never have been a particular moment when we would have been able to say that the "first living cell" had appeared on the scene. At this point, biological natural selection could take over, and it seems easy to understand why the RNA world would have eventually given way to the present DNA world: DNA is a more flexible hereditary material and is less prone to copying errors than RNA. Thus, because DNA is structurally so similar to RNA, it seems natural that the RNA molecules serving hereditary functions would have eventually evolved DNA, while other RNA molecules would have continued the cellular roles that they still play today.

Putting It All Together Let's review the sequence we've discussed here as the possible explanation for the origin of life (Figure 6.8).

1. Through some combination of atmospheric chemistry, chemistry near deep-sea vents, and molecules brought to Earth from space, the early Earth had at least localized areas with significant amounts of organic molecules that could serve as building blocks for more complex organic molecules.
2. More complex molecules, including short strands of RNA, grew from the organic building blocks, probably with the aid of reactions catalyzed by clay minerals. Some of the RNA molecules became capable of self-replication.
3. Membranes formed spontaneously, perhaps also with the aid of clay minerals, and enclosed some of these complex molecules in "pre-cells" that facilitated the development of cooperative molecular interactions.

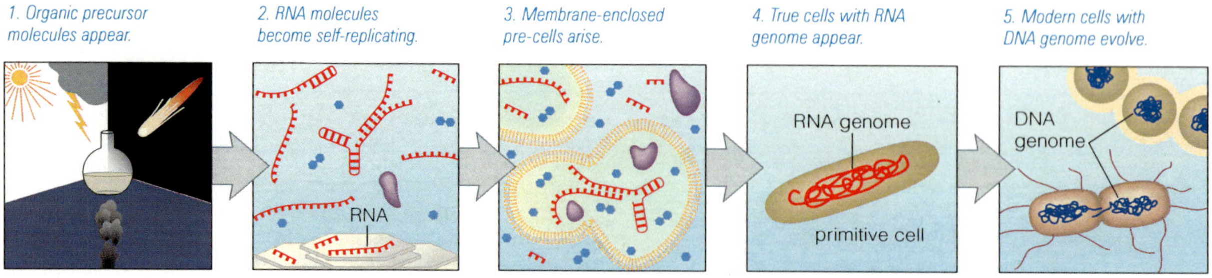

Figure 6.8

A summary of the steps by which chemistry on the early Earth may have led to the origin of life. (Adapted from Campbell, Reece, Simon, *Essential Biology,* Second Edition, © 2004.)

4. Natural selection among the RNA molecules in pre-cells gradually led to an increase in complexity, until eventually some of these structures became true living organisms.
5. Natural selection then rapidly improved and diversified life. DNA became the favored hereditary molecule, and life has continued to evolve ever since.

We may never know for certain whether life actually originated in this way, in some similar way, or in some completely different way. Nevertheless, this scenario seems quite reasonable and perhaps even "easy," given geological time scales. It seems especially reasonable given that a number of different components of the scenario have been demonstrated in laboratory experiments. Even if life did not originate in this way, it seems that it could have—which suggests that the actual path to life must have been equally easy, or else it would have followed the path we've described. In summary, we have good reason to believe that the origin of life was a *likely* consequence of conditions on the early Earth, in which case it might be equally likely that life arose on many other worlds.

THINK ABOUT IT We've noted that the probability of life's arising by randomly mixing simple organic building blocks is so small as to seem impossible. Yet, in the scenario we've described, the likelihood of getting life seems quite good. In your own words, describe why these two probabilities are so different.

• Could life have migrated to Earth?

Although our scenario suggests that life could have arisen easily and naturally here on Earth, it is also possible that life arose somewhere else first—for example, on Venus or Mars—and then migrated to Earth within meteorites.

COSMIC CALCULATIONS 6.1

Bacteria in a Bottle I: Lessons for Early Life

Once the first organisms took hold, how quickly could they have spread and evolved? A thought experiment* offers insight into this question. Suppose you place a single bacterium in a nutrient-filled bottle at noon, and that this species is capable of replicating by cell division every minute. The original bacterium grows until it divides into two bacteria at 12:01. These two bacteria divide into 4 bacteria at 12:02, which divide into 8 bacteria at 12:03, and so on. Then the number of bacteria in the bottle at any time t minutes after noon is:

$$\#\text{ bacteria at } t \text{ minutes after } 12{:}00 = 2^t$$

We'll explore general characteristics of this **exponential growth** (t is in the exponent) in Cosmic Calculations 6.2. Here, to understand how rapidly early life could have spread, let's consider the volume of a bacterial colony. A typical bacterium is 10^{-7} m (0.1 micrometer) across, which means it has a volume of about $(10^{-7}\text{ m})^3 = 10^{-21}\text{ m}^3$. So the volume of bacteria at any time t minutes after noon is

$$\text{bacterial volume} = 2^t \times 10^{-21}\text{ m}^3$$

*This thought experiment is adapted from one created by Professor of Physics Albert A. Bartlett of the University of Colorado.

Our two formulas tell us that after 60 minutes the number of bacteria is $2^{60} \approx 1 \times 10^{18}$, or a *million trillion*; and their volume is about $2^{60} \times 10^{-21}\text{ m}^3 \approx 0.001\text{ m}^3$, or 1 cubic centimeter. By this point in a small bottle, the bacteria will be close to exhausting their nutrients. But let's imagine they could somehow continue to multiply. By the end of the second hour, they would number an astonishing $2^{120} \approx 1.3 \times 10^{36}$, and their volume would be $2^{120} \times 10^{-21}\text{ m}^3 \approx 1.3 \times 10^{15}\text{ m}^3$—large enough to cover the surface of the Earth to a depth of about 2 meters (see Problem 50 at end of chapter). Continuing the calculations, you'd find that the bacteria would exceed the total volume of the world's oceans (about $1.3 \times 10^{18}\text{ m}^3$) at $t = 130$ minutes. Note that changing the doubling time from one minute to a year hardly matters; a time of $t = 130$ years rather than 130 minutes is still geologically insignificant.

Although the bacteria could not really continue this hypothetical growth, the implication should be clear: The first self-replicating organisms would have spread rapidly as far as conditions allowed, leaving the door wide open for biological evolution through natural selection.

The idea that life could travel through space to land on Earth, sometimes called *panspermia*, once seemed outlandish. After all, it's hard to imagine a more forbidding environment than that of space, where there's no air, no water, and constant bombardment by dangerous radiation from the Sun and stars. However, the presence of organic molecules in meteorites and comets tells us that the building blocks of life can survive in the space environment, and we've already discussed some Earth microbes that are capable of surviving at least moderate periods of time in space [Section 5.5]. It therefore seems possible that life could migrate from one planet to another, if it can hitch a suitable ride.

Figure 6.9

Chemical analysis of this meteorite, known as ALH84001, indicates that it came from Mars. The small block shown for scale to the lower right is 1 cubic centimeter, about the size of a typical sugar cube. (We will discuss this particular meteorite further in Section 8.5.)

The Possibility of Migration

We know that meteorites can and do travel from one world to another. Among the more than 20,000 meteorites that scientists have identified and cataloged, careful chemical analysis has so far revealed about three dozen with compositions that clearly suggest that they came from Mars (Figure 6.9); even more have been found that come from the Moon. Apparently, these meteorites were blasted from their home worlds by large impacts, then followed orbital trajectories that eventually caused them to land on Earth. Observations of these meteorites, along with theoretical calculations based on the amount of material blasted into space by impacts, suggest that over time the inner planets have exchanged many tons of rock. In a sense, Earth, Venus, and Mars have been "sneezing" on each other for billions of years, offering the possibility of microscopic life hitchhiking between worlds on one of the meteorites.

For a living microbe to arrive intact on Earth after such a journey, it would have to survive at least three potentially lethal events: the impact that blasts it off the surface of its home world, the time it spends in the harsh environment of interplanetary space, and the fiery plunge through our atmosphere. Examination of martian meteorites suggests that neither the first nor last of these events poses insurmountable obstacles. The interiors of martian meteorites (such as ALH84001) show only minimal disruption, suggesting that microbes inside these rocks could survive both the initial impact and the later fall to Earth. The larger question is whether they could survive their time in space.

The chance of surviving the trip between planets probably depends on how long the meteorite spends in space. Once a rock is launched into space, it orbits the Sun until its orbit carries it directly into the path of another planet. Most meteorites will orbit for many millions of years before reaching Earth, even if they come from a world as nearby as Venus or Mars. It seems highly unlikely that living organisms could survive in space for millions of years. However, a few meteorites are likely to be launched into orbits that cause them to crash to Earth during one of their first few trips around the Sun. For example, calculations suggest that about 1 in 10,000 meteorites may travel from Mars to Earth in a decade or less. Because experiments in Earth orbit have already shown that some terrestrial microbes can survive at least 6 years in space, it seems quite reasonable to imagine microbes from Mars arriving safely on Earth.

While migration between planets seems possible, similar considerations almost certainly rule out the possibility of migration from other star systems. Under the best of circumstances, meteorites from planets around other stars would spend millions or billions of years in space before reaching Earth; any living organisms would almost surely be killed by exposure to cosmic rays during this time. Moreover, calculations suggest that the probability of a rock from another star system hitting Earth is extremely low, which also explains why we have never yet found a meteorite from beyond our own solar system.

Reasons to Consider Migration Given the reality that the inner planets exchange rocks in the form of meteorites, the key question probably is not whether life *could* migrate through space but whether we have any reason to suppose it originated elsewhere rather than right here on Earth. Many scientists have debated this question, with the debate taking many different twists. Today, most ideas about migrating life fall into one of two broad categories.

The first broad idea suggests that life does not form as easily as we have imagined, at least under the conditions present on the early Earth. In this view, the only explanation for life on Earth (other than invoking the supernatural) would be migration from elsewhere. Although this idea in some sense only moves the problem of life's origin to another place, it at least allows for the possibility that another world had conditions that were more conducive to rapid development of life, or that life arose on a world that offered more time. The primary drawback to this idea is that other worlds in our solar system should have had no more time available than Earth, and we know of no compelling reason why any of these worlds would have offered better conditions for an origin of life. The only way to get significantly more time for an origin of life is to suppose that life migrated from another star system, but we have already explained why that possibility seems highly unlikely.

The second broad idea suggests that life forms so easily that we should expect to find life originating on any planet with suitable conditions. In that case, the origin of life in our solar system would have occurred on whichever planet got those conditions *first;* for example, if the very early Venus or very early Mars had suitable conditions for life before Earth, life from one of those worlds might have migrated to Earth and taken hold on our planet as soon as conditions allowed. In essence, this idea suggests that life never got the chance to originate indigenously on Earth because life from another planet got here first.

Implications of Migration to the Search for Life Beyond Earth

While ideas about microbes migrating *to* Earth are speculative, it seems a near-certainty that microbes *from* Earth have many times made the journey to Mercury, the Moon, Venus, and Mars. After all, there have been many large (but not sterilizing) impacts on Earth in the almost 4 billion years during which we know life existed here, offering plenty of opportunity for hitchhiking microbes. Thus, if it were possible for Earth life to survive on any of these other worlds, we should actually *expect* to find it there. As we'll discuss in Chapter 7, we can almost certainly rule out the possibility of survival on the Moon and Mercury, and probably on Venus as well. Mars, however, may well have habitats that could provide at least temporary refuge to terrestrial microbes, and Mars may have been globally habitable in the distant past.

The likelihood of such interplanetary migration raises at least two important issues in astrobiology. First, if we someday find life on Mars, we will have to wonder if it is native or if it arrived there from Earth. The only way we may ever be confident that Mars life is not transplanted Earth life will be if its biochemistry is too different from that of terrestrial life to allow for a common ancestor. Even in that case, we might wonder whether the martian life could have descended from a species that arose on Earth very early but was extinguished by a sterilizing impact before the end of the heavy bombardment. After all, if life arose and was extinguished multiple times on the early Earth, there is no reason to expect the biochemistry to have been the same with each separate origin of life.

Second, the possibility of life migrating among the planets raises the question of whether we could ever distinguish between an indigenous origin of life on Earth and an origin based on migration from elsewhere. To date, no one knows how we might distinguish between these possibilities. It's conceivable that a fossil record from Mars might suggest an earlier origin of life there, though even then we might not be certain that this life came to Earth. Venus poses a more intractable problem: As we'll discuss in Chapter 10, it is possible that Venus once had oceans and a habitable climate in which life might have arisen. However, Venus now is so hot that any fossil record would almost certainly have been destroyed by the heat and subsequent geological activity.

Despite these potential uncertainties, the major lessons of our study of life's origins still hold: One way or another, life arose on Earth quite soon after conditions first allowed it, and even if life migrated here from another world, we have good reason to think that it evolved naturally, through chemical processes that favor the creation of complex, organic molecules and the subsequent molecular evolution of self-replicating molecules.

6.3 THE EVOLUTION OF LIFE

Regardless of how or where it originated, life on Earth has been evolving throughout the 4 billion or so years during which it has existed on Earth. Careful studies of the geological record provide the key data with which we attempt to re-create the evolutionary time scale, while genome comparisons offer data that help us map relationships between species. In this section, we'll briefly retrace the history of life on Earth as it is now understood, which should in turn help us understand the possibilities for finding similarly complex life on other worlds.

• What major events have marked evolutionary history?

Reconstructing 4 billion years of history from limited clues is an obviously difficult task. Nevertheless, we have identified at least a few of the key events that have marked the evolution of early life into its current diversity, and that help us understand our own origins. As you read here, you may wish to look back at the geological time scale in Figure 4.10, which indicates the major events.

Early Microbial Evolution The earliest organisms must have been quite simple, but they undoubtedly had at least a few enzymes and a rudimentary metabolism. Their cells must have resembled modern prokaryotic cells, lacking cell nuclei and other complex structures that we now find only in eukaryotes. Moreover, because the atmosphere at that time was essentially oxygen-free, all early life must have been **anaerobic**, meaning that it did not require molecular oxygen; by contrast, we are **aerobic** organisms, because we cannot survive without molecular oxygen. Based on the hypothesis that life originated near deep-sea vents or hot springs, we expect that the first microorganisms were *chemoautotrophs* (see Table 5.1)—organisms that obtained their carbon from carbon dioxide dissolved in the oceans and their energy from chemical reactions involving inorganic chemicals. Some modern Archaea that appear to be fairly close to the root of the tree of life, such as those thriving in hot sulfur springs, obtain their energy in this way through chemical reactions involving hydrogen, sulfur, and iron compounds. Because similar compounds were abundant on the early Earth, perhaps especially so in hot springs and near deep-sea vents, it seems reasonable to assume that early life used the available inorganic chemical energy. (See Section 9.4 for discussion of possible chemical pathways.)

Natural selection probably caused rapid diversification among the early life-forms. Modern DNA replication involves a variety of enzymes that help keep the mutation rate low. Early organisms, with a much more limited set of enzymes, probably experienced many more errors in DNA copying. Because more errors mean a higher mutation rate, evolution would have been rapid among early microbes. As life diversified, many new metabolic processes evolved, making some of the new organisms biochemically quite different from their ancestors. Because of the rapid pace of early evolution, it may not have taken long to establish many of the major branches in the tree of life (see Figure 5.11).

Fossil evidence supports the idea of rapid diversification. Recall that stromatolites suggest the presence of organisms that obtained energy by photosynthesis some 3.5 billion years ago, and some of the oldest microfossils also resemble modern photosynthetic organisms. Because photosynthesis is a complex metabolic pathway, its early emergence indicates that early evolution was rapid.

Photosynthesis probably evolved through multiple steps. At first, some organisms may have developed light-absorbing pigments that absorbed excess light energy—especially ultraviolet—that was harmful to life near the ocean surface. Over time, some of these pigments evolved to enable the cell to make use of the absorbed solar energy. Modern organisms known as purple sulfur bacteria and green sulfur bacteria may be much like the early photosynthetic microbes. These organisms use hydrogen sulfide (H_2S) rather than water (H_2O) in photosynthesis and therefore do not produce any oxygen. Photosynthesis using water, which produces oxygen as a by-product, probably came later and ultimately caused the buildup of oxygen in Earth's atmosphere. The timing of the oxygen buildup is still uncertain, but it probably did not get underway much before about 2.4 billion years ago.

The rise of oxygen created a crisis for life, because oxygen attacks the bonds of organic molecules. Many species of microbes probably went extinct, and those that survived had to somehow avoid the detrimental effects of oxygen. Some avoided these effects because they lived in (or migrated to) underground locations where the oxygen did not reach them. We still find many anaerobic microbes in such locales today, living in soil or deeper underground in rocks. Others survived because the oxygen content of the atmosphere rose gradually, allowing them time to evolve new metabolic processes and protetctive mechanisms that enabled them to thrive rather than die in the presence of oxygen. Plants and animals, including us, still use the metabolic processes that evolved in response to the "oxygen crisis" faced by living organisms some 2 billion or more years ago.

The Evolution of Eukarya The evolution of Eukarya was the crucial first step in our own eventual evolution; we are, after all, members of this domain. Moreover, although the "complexity" of cellular processes is sometimes a subjective judgment, overall the structure of eukaryotic cells seems better suited to the evolution of advanced organisms than that of prokaryotic cells. Even single-celled eukaryotes exhibit much more diversity in structure than exists among all prokaryotes, and multi-celled eukaryotes enjoy diversity far beyond that. Because more variations are possible on complex structures than on simple ones, the complexity of eukaryotic cells allowed for the selection of many more adaptations than were possible in prokaryotic cells. Indeed, multicellularity appears to have evolved independently in several different branches of Eukarya, suggesting that the complex structure of eukaryotic cells opened the door for the evolution of more advanced organisms. If we ever find life elsewhere that is more like us than like simple microbes, it's likely that it will have cell structures more like those of modern Eukarya than of Bacteria or Archaea.

When did Eukarya arise? Despite the fact that modern Eukarya have more complex cellular structures than Bacteria and Archaea, genome studies do not suggest any substantial differences in the evolutionary ages of the three domains. That is, it is quite likely that members of all three domains—Bacteria, Archaea, and Eukarya—split from a common ancestor early in Earth's history. Early Eukarya could not yet have had cell nuclei and other complex intracellular structures. These must have come later, and the oldest known fossils that clearly show cell nuclei date to about 2.1 billion years ago. However, because cell nuclei do not fossilize well, it's possible that Eukarya began to have such structures much earlier but we are unable to recognize them in the fossil record.

Modern, complex Eukarya probably evolved through a combination of at least two major adaptations that arose in their simpler ancestors. First, some early species of Eukarya may have developed specialized infoldings of their membranes that compartmentalized certain cell functions, ultimately leading to the creation of a cell nucleus (Figure 6.10a). Second, some relatively large ancestral host cell absorbed smaller prokaryotes with specialized functions. The absorbed prokaryotes developed a **symbiotic relationship** with the host cell, that is, a relationship in which both the invading organism and the host organism benefit from living together (Figure 6.10b).

The key evidence favoring symbiosis (the development of a symbiotic relationship) comes from the existence of structures within eukaryotic cells that look much like tiny prokaryotes trapped inside. The most important of these "cells within cells" are **mitochondria**, the cellular organs in which oxygen helps produce energy (by making molecules of ATP), and **chloroplasts**, structures in plant cells that produce energy by photosynthesis. Mitochondria and chloroplasts even have their own DNA and reproduce themselves within their eukaryotic homes. Moreover, DNA sequences in mitochondria and chloroplasts clearly group them with domain Bacteria rather than with Eukarya. Thus, it seems a near-certainty that mitochondria and chloroplasts were originally free-living bacteria. Assuming that these bacteria had already evolved their ability to make efficient use of oxygen (in the case of mitochondria) or to carry out photosynthesis (in the case of chloroplasts), a symbiotic relationship might have developed quite easily. The host cell would benefit from the energy produced by the incorporated bacteria, while the bacteria would benefit from the protection offered by the host cell. Further support for this hypothesis comes from the many cases in which living organisms today form symbiotic relationships. Other evidence comes from the fact that some present-day bacteria show striking similarities to mitochondria and chloroplasts, suggesting that they share a common ancestry.

The Cambrian Explosion We have seen that life on Earth existed at least 3.5 billion years ago (and perhaps hundreds of millions of years be-

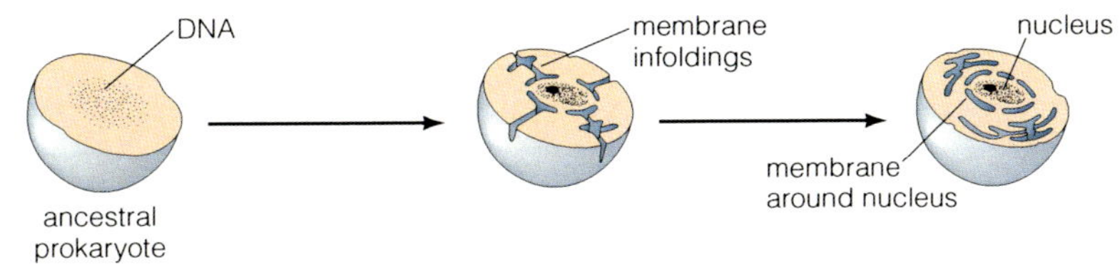

a Some large prokaryotes may have developed specialized membrane infoldings that compartmentalized certain cell functions, ultimately leading to the creation of a cell nucleus.

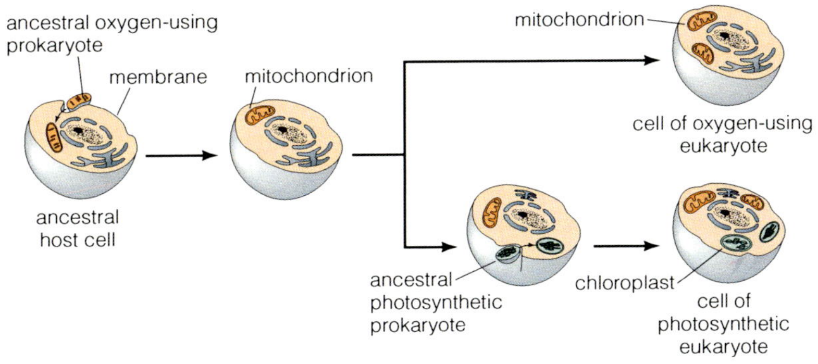

b Mitochondria and chloroplasts may have evolved as small prokaryotes invaded a larger host cell, forming a symbiotic relationship.

Figure 6.10

Hypotheses concerning the origin of eukaryotes. (Adapted from Campbell, Reece, Taylor, Simon, *Biology Concepts & Connections,* Fifth Edition, © 2006.)

fore that), and all three domains of life were well established by at least 2.1 billion years ago. However, the fossil record tells us that all this life remained microscopic (aside from microbes organized into colonies) until much later. The earliest fossil evidence for complex, multicellular organisms—all of which are Eukarya—dates to only about 1.2 billion years ago. Thus, microbes had our planet to themselves for more than the 2 billion years after the origin of life. Even today, the total biomass of microbes far exceeds that of multicellular organisms like fungi, plants, and animals [Section 5.2]. But mass isn't everything, and we have a special interest in understanding the evolution of multicellular life, even if it is comparatively rare on our world.

In particular, we have a special interest in animal evolution: Animals may represent only one small branch on the tree of life, but it's *our* branch. Moreover, we generally assume that extraterrestrial intelligence, if it exists, will belong to animal-like beings from other worlds. The fossil record suggests that animal evolution progressed slowly at first, with relatively little change seen between fossils from 1.2 billion years ago and those from a half-billion years later. But then something quite dramatic happened.

In the broadest sense, biologists classify animals according to their basic "body plans." For example, the basic body plan shared by mammals and reptiles is fundamentally different from that of insects. Animals are grouped by body plan into what biologists call **phyla** (singular, *phylum*), the next level of classification below kingdoms (such as the plant kingdom and the animal kingdom [Section 5.2]). Mammals and reptiles both belong to the phylum *Chordata,* which represents animals with internal skeletons. Insects, crabs, and spiders belong to the phylum *Arthropoda,* which represents animals with body features such as jointed legs, an external skeleton, and segmented body parts. Classifying animals into phyla is an ongoing effort by biologists, but modern animals appear to comprise about 30 different phyla, each representing a different body plan.

Remarkably, nearly all of these different body plans, plus a few others that have gone extinct, make their first known appearance in the geological record during a period spanning only about 40 million years—less than about 1% of Earth's history. This remarkable flowering of animal diversity appears to have begun about 545 million years ago, which corresponds to the start of the Cambrian period (see Figure 4.10).* Hence, it is called the **Cambrian explosion**.

THINK ABOUT IT One early trend in the evolution of multicellular organisms was a trend toward larger size. Briefly discuss how larger size might have conferred an evolutionary advantage.

The fact that the Cambrian explosion marks the only major diversification of body plans in the geological record presents us with two important and related questions: Why did the Cambrian explosion occur so suddenly, at least in geological terms, yet so long after the origin of eukaryotes, and why hasn't any similar diversification happened since?

No one knows the answers to these questions, but we can identify at least four possible contributing factors. First, the oxygen level in our atmosphere may have remained well below its present level until about the time of the Cambrian explosion. Thus, the dramatic change in animal life may have occurred at least in part because oxygen reached a critical level for the survival of larger and more energy-intensive life-forms.

A second factor, which some scientists think may have been much more important, was the evolution of genetic complexity. As eukaryotes evolved, they developed more and more genetic variation in their DNA, which opened up ever more possibilities for further variation. Perhaps the Cambrian explosion marks a point in time when organisms had become sufficiently complex that a great diversity of forms could evolve over a short time.

The third factor may have been climate change. Recall that geological evidence points to a series of snowball Earth episodes [Section 4.5] ending around the time of the Cambrian explosion. The extreme climate conditions that marked these episodes may have exerted evolutionary pressure that aided the diversification of life and then fueled the Cambrian explosion when the environmental conditions eased.

The fourth factor may have been the absence of efficient predators. Early predatory animals were probably not very sophisticated, so some adaptations that later might have been snuffed out by predation were given a chance to survive if they arose early enough. Thus, the beginning of the Cambrian period may have marked a window of opportunity for many different adaptations to gain a foothold in the environment.

This last idea may partly explain why no similar explosion of diversity has taken place since the Cambrian. Once predators were efficient and widespread, it may have been much more difficult for entirely new body forms to find an available environmental niche. In addition, the fact that certain body forms were already selected during the Cambrian explosion may have limited other options. That is, while more body plans may have been possible than actually arose, once these were in existence there may not have been clear evolutionary pathways to others. Alterna-

*Although the Cambrian explosion is among the most vivid events in the geological record, evidence indicates that the lineages we first see in the Cambrian explosion actually began evolving earlier. The Cambrian explosion didn't involve only animals; other groups, such as algae, also diversified.

tively, perhaps the various body forms that arose during the Cambrian explosion represent the full range of forms possible, at least within the constraints of the genetic variability available on Earth. In any case, we and nearly all other animals living today can trace our ancestry to species that arose during the Cambrian explosion.

The Colonization of Land Because most early microfossils are found in sediments that were originally deposited in the oceans, it's difficult to know exactly when life first migrated onto land. However, given the wide variety of environments in which microbes survive today and the fact that many different genetic lineages seem to have appeared quite early in evolutionary history, it's likely that microbial life quickly established itself wherever it could find liquid water and protection from ultraviolet radiation. Plenty of such locations are available on land—including underground and anyplace where water can pool under a shelter of overhanging rock—so it is hard to imagine reasons why microbial life would not have taken hold on land quite early. However, the situation is different for multicellular organisms. While microbes may have thrived on land, larger organisms, including all animals, remained confined to the oceans (and other bodies of water) even after the Cambrian explosion.

For larger organisms, surviving on land was more difficult than surviving in the oceans, primarily because it required evolving a means of obtaining water and mineral nutrients without simply absorbing them from surroundings. The timing of the development of the ozone layer may also have played a role in the late colonization of land. Recall that ozone is a molecule (O_3) made from oxygen atoms, so a protective ozone layer could not exist until the atmosphere contained some threshold level of oxygen. Uncertainties in both the oxygen levels through time and the level needed for a substantial ozone layer make it difficult to know when the ozone layer first appeared. But until it did, life on any exposed land surface would have been difficult or impossible.

Fossil evidence shows that plants (and perhaps fungi as well) were the first large organisms to develop the means to live on the land. The colonization of land by plants appears to have begun about 475 million years ago. DNA evidence suggests that plants evolved from a type of alga. Some ancient algae might have survived in salty shallow-water ponds or along lake edges. Because such locales occasionally dry up, natural selection would have favored adaptations, such as thick cell walls, that allowed the algae to survive during periods of dryness. Cell walls would have given the organisms structure that would have helped them survive on land. The first fully land-based organisms would have had even more advantages, because there were no land animals around to eat them. Large plants gradually developed complex bodies with some parts specialized for energy collection above ground (where sunlight is available) and other parts specialized for collecting water and nutrients from the soil.

Once plants moved onto the land, it was only a matter of time until animals followed them out of the water. Within about 75 million years, amphibians and insects were eating land plants. By the beginning of the Carboniferous period, about 360 million years ago, vast forests and abundant insects thrived around the world (Figure 6.11). These Carboniferous forests were important not only as a major step in evolution, but also because they became an important part of our modern economy. Much of the land area of the continents was flooded by shallow seas during the Carboniferous period, hindering the decay of dead plants. Thick layers of dead organic matter piled up in the stagnant waters. Over millions of years, as these layers were buried, pressure and heat gradually converted the

Figure 6.11

This painting, based on fossil evidence, shows a forest of the Carboniferous period. The large trees with straight trunks are seedless plants called lycophytes. The tree on the left with feathery branches is a horsetail. The plants near the forest floor are ferns. The dragonfly was about the size of a modern raven.

organic matter to coal. Nearly all the coal that has helped fuel our industrial age is, in fact, the remains of these forests of the Carboniferous period.

THINK ABOUT IT Based on the preceding discussion, explain why coal is called a "fossil fuel." What other fossil fuels do we use to generate energy?

• Why was the rise of oxygen so important to evolution?

We are oxygen-breathing animals, so there's no question that the rise of oxygen was critical to our eventual emergence. More generally, the rise of oxygen was important to evolution because oxygen can react so strongly with organic molecules. While these reactions can kill organisms that are not adapted to oxygen's presence, they also offer the possibility of much more efficient cellular energy production (that is, making molecules of ATP [Section 5.3]) than is possible through anaerobic processes. Thus, as aerobic organisms evolved, they were able to develop adaptations that demanded much more energy than would have been available to their anaerobic ancestors. The rise of oxygen ignited an explosion of eukaryotic diversification and, as we've discussed, may have helped fuel the Cambrian explosion.

The Origin of Atmospheric Oxygen Molecular oxygen is a highly reactive gas that would disappear from the atmosphere in just a few million years if it were not continually resupplied by life. Fire, rust, and the discoloration of freshly cut fruits and vegetables are everyday examples of **oxidation reactions**—chemical reactions that remove oxygen from the atmosphere. Many elements and molecules can participate in oxidation reactions. Today, most reactions that remove oxygen from the atmosphere occur in living organisms that use oxygen, including ourselves. Before oxygen-breathing organisms evolved, oxidation reactions primarily involved volcanic gases, dissolved iron in the oceans, and surface minerals (especially those containing iron) that could react with oxygen. Such reactions essentially "rust" the minerals, causing them to turn reddish in color. In the oceans, oxidation reactions with dissolved iron create minerals that precipitate to the bottom, forming "red beds" on the

a The blue-green color of this lake (in Cape Cod, Massachusetts) is the result of a population explosion, or "bloom" of cyanobacteria.

b This micrograph shows individual cyanobacteria from the lake.

Figure 6.12

Cyanobacteria split water and release oxygen in photosynthesis and are thought to have been responsible for the rise of oxygen in Earth's atmosphere.

ocean floor. On land, the reddish color of much of Earth's rock and clay is a direct result of oxidation reactions involving surface minerals.

The fact that free oxygen would not last long without life tells us that our atmosphere must have been essentially oxygen-free before life existed. Moreover, while today we recognize plants as a major source of oxygen, we know that plants arrived relatively recently on the evolutionary scene. Where, then, did the oxygen come from? Remarkably, it seems that we owe our oxygen atmosphere to microscopic bacteria sometimes called "blue-green algae" but more technically known as **cyanobacteria** (Figure 6.12). Fossil evidence shows that cyanobacteria were producing oxygen by at least 2.7 billion years ago, and perhaps for hundreds of millions of years before that. It took at least 2 billion years for oxygen to build up in the atmosphere to its present levels, but in the end the oxygen we breathe originally entered the air through the action of microscopic cyanobacteria.

Timing of the Oxygen Rise The precise timing of the rise of oxygen is difficult to study, because we have no direct way to sample air from hundreds of millions or billions of years ago. However, we can learn about oxygen content from a variety of other clues. For example, fossils of oxygen-breathing organisms can tell us a minimum amount of oxygen that must have been in the atmosphere for them to survive. Careful study of rock chemistry offers even more clues.

Studies of rocks that are between about 2 and 3 billion years old, especially rocks of a type called *banded iron formations* (Figure 6.13), show that the atmosphere during that time contained less than 1% of the amount of oxygen it contains today. The banded iron formations were made from iron-containing minerals dissolved in the oceans, and such iron minerals cannot dissolve if there is substantial oxygen in the atmosphere and ocean. More recent studies, based on sulfur isotope ratios in ancient rocks, constrain the timing of the rise of oxygen more tightly. Atmospheric oxygen alters the atmospheric chemistry of sulfur compounds in the atmosphere (SO_2 and H_2S from volcanic outgassing) in ways that change the ratios of sulfur isotopes that end up in surface rock. The oldest rocks showing sulfur isotopes in a ratio that indicates the presence of atmospheric oxygen are about 2.35 billion years old. If these data are

Figure 6.13

This rock is an example of a banded iron formation (BIF) formed more than 2 billion years ago. Such rocks could have formed only before the atmosphere contained significant amounts of oxygen. The pen is included for scale.

being properly interpreted, it means that the abundance of atmospheric oxygen must have been less than 20 parts per million (0.002%) up until 2.35 billion years ago. At that point, sometimes called the "great oxidation event," oxygen began to build up in the atmosphere.

The timing of the great oxidation event poses a mystery, however: Given that cyanobacteria began to produce oxygen at least 2.7 billion years ago—or at least 350 million years before the great oxidation event at 2.35 billion years ago—what took so long? We do not yet know the details, but presumably the oxygen was being removed from the atmosphere by nonbiological processes, such as oxidation of surface rock and ocean minerals, as rapidly as the cyanobacteria could make it.

An even greater mystery concerns what happened once the buildup began. Other isotopic evidence suggests that the "great oxidation event" wasn't really that great, and that oxygen levels remained far below modern levels for at least the next billion years, and perhaps all the way up to about the time of the Cambrian explosion. Although a variety of hypotheses have been proposed to explain this slow buildup of oxygen, we do not yet have sufficient data to decide which ones might work and which ones don't.

Regardless of the reason, it now seems likely that oxygen levels remained far too low for complex animals until somewhere near the time of the Cambrian explosion, which occurred when Earth was already nearly 4 billion years old. The oxygen-breathing animals that evolved at that time probably needed oxygen levels of at least 10% of the modern value. It's possible that the oxygen level was higher than that, but the first clear evidence of an oxygen level near its current value appears in the geological record only about 200 million years ago. That is when we first find charcoal in the geological record, implying that enough oxygen was present in the atmosphere for fires to burn. Thus, if you had a time machine and could randomly spin the dial to take you back to any point in Earth's history, you'd have less than about a 1 in 10 chance—and perhaps only about a 1 in 20 chance—of appearing at a time recent enough so that you could step out and breathe the air.

THINK ABOUT IT What does the late appearance of substantial atmospheric oxygen tell you about the difference between our planet's being habitable *in general* and being habitable *for us*?

Implications for Life Elsewhere The importance of oxygen to advanced life on Earth and the timing of its rise could potentially have important implications for life on other worlds. Our study of the origin of life gives us reason to think that life might be common on worlds with conditions like those of the early Earth. But the fact that it took so long for oxygen to build up in the atmosphere on Earth should make us wonder about the likelihood of getting oxygen-breathing life on other worlds. Could Earth have been "lucky" to ever get conditions that allowed the buildup of oxygen? If so, perhaps life on most other worlds would never evolve past microscopic forms; life might then be common, but advanced or intelligent life quite rare. Alternatively, maybe Earth was "unlucky" in having conditions that prevented the oxygen buildup for so long. In that case, other worlds might have complex plants and animals by the time they are just 1 to 2 billion years old, instead of having to wait until they are 4 billion years old. For the time being, we have no way to distinguish between these and other, intermediate possibilities. We will therefore continue our study with the assumption that Earth has been "typical," until and unless we learn otherwise.

6.4 IMPACTS AND EXTINCTIONS

Once animals colonized the land, the evolutionary path that led to humans becomes much clearer. Reptiles evolved from amphibians; early dinosaurs and mammals followed by about 245 million years ago. But the fossil record shows that the path was not smooth. In particular, the fossil record shows a number of striking transitions in the nature of living organisms. The most famous transition in the fossil record defines the boundary between the Cretaceous and Tertiary periods, which dates to about 65 million years ago. Dinosaur fossils exist below this boundary, but not above it. Somehow, after some 180 million years as Earth's dominant animals, the dinosaurs went extinct in a geological blink of the eye. Significant evidence now points to the idea that this extinction, and perhaps others, may have been caused by the impact of asteroids or comets crashing into Earth.

Figure 6.14

Meteor Crater in Arizona was created about 50,000 years ago by the impact of an asteroid about 50 meters across. Because the asteroid hit Earth at high speed, it left a crater more than 1 kilometer across and almost 200 meters deep (compare to the size of the parking lot and buildings at the crater's left). The K–T impact was about 200 times larger.

• Did an impact kill the dinosaurs?

There's no doubt that major impacts have occurred on Earth in the past. Meteor Crater in Arizona (Figure 6.14) formed about 50,000 years ago when a metallic asteroid roughly 50 meters across crashed to Earth with the explosive power of a 20-megaton nuclear bomb. Although the crater is only a bit more than 1 kilometer across, the blast and ejecta probably battered an area covering hundreds of square kilometers. Meteor Crater is relatively small and recent, and it is a popular tourist stop because it is so obvious. But it is not alone: Geologists have identified more than 150 impact craters on our planet, and many others have presumably been destroyed by erosion and other geological processes. Could larger impacts have altered the entire course of evolution?

The K–T Boundary Layer In 1978, while analyzing geological samples collected in Italy, a scientific team led by Luis and Walter Alvarez (father and son) made a startling discovery. They found that the thin layer of sediments that marks the Cretaceous–Tertiary boundary, called the **K–T boundary** for short (the *K* comes from the German word for "Cretaceous," *Kreide*), is unusually rich in iridium—an element that is rare on Earth's surface (because Earth's iridium sank to the core when our planet underwent differentiation [Section 4.4]) but more common in meteorites. Subsequent studies found the same iridium-rich sediment marking the K–T boundary at many other sites around the world (Figure 6.15). The Alvarez team proposed a stunning hypothesis: The extinction of the dinosaurs was caused by the impact of an asteroid or comet. They calculated that it would take an asteroid about 10–15 kilometers in diameter to produce as much iridium as is apparently distributed worldwide in the K–T boundary layer.

In fact, the death of the dinosaurs was only a small part of the biological devastation that seems to have occurred 65 million years ago. The geological record suggests that up to 99% of all living plants and animals died around that time and that up to 75% of all existing plant and animal *species* were driven to extinction. This makes the event a clear example of a **mass extinction**—the rapid extinction of a large percentage of all living species. Could it really have been caused by an impact?

Figure 6.15

Around the world, sedimentary rock layers that mark the 65-million-year-old K–T boundary share evidence of the impact of a comet or asteroid. Fossils of dinosaurs and many other species appear only in rocks below the iridium-rich layer.

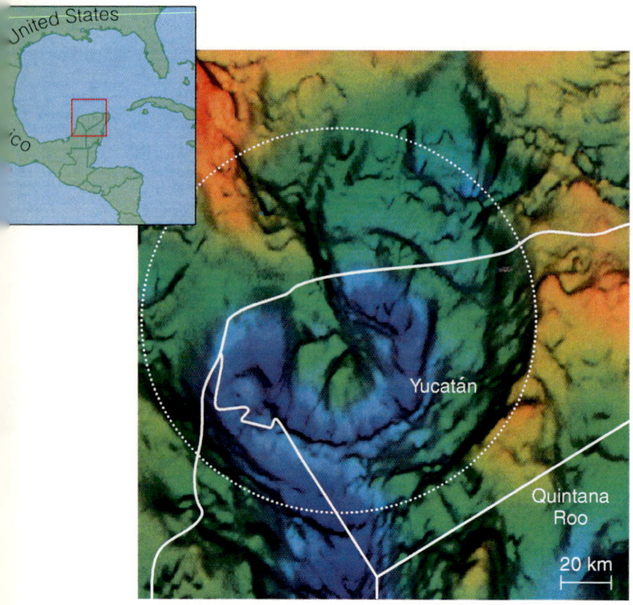

Figure 6.16

This computer-generated image, based on measurements of small local variations in the strength of gravity, reveals a buried impact crater (dashed circle) in the northwest corner of Mexico's Yucatán Peninsula. The crater measures about 200 kilometers across, and about half of it lies underwater off the coast. (The solid white lines show the coastline and borders of Mexican states.)

Figure 6.17

This painting shows an asteroid or comet moments before its impact on Earth, some 65 million years ago. The impact, known as the K–T impact, probably caused the extinction of the dinosaurs, and if it hadn't occurred, the dinosaurs might still rule Earth today.

More Evidence for the Impact There's still some scientific controversy about whether the impact was the sole cause of the mass extinction or just one of many causes, but there's little doubt that a major impact coincided with the death of the dinosaurs. Key evidence comes from further analysis of the K–T sediment layer. Besides being unusually rich in iridium, this layer contains four other unusual features: (1) high abundances of several other metals, including osmium, gold, and platinum; (2) grains of "shocked quartz," quartz crystals with a distinctive structure that indicates they experienced the high-temperature and high-pressure conditions of an impact; (3) spherical rock "droplets" of a type known to form when drops of molten rock cool and solidify in the air; and (4) soot (at some sites) that appears to have been produced by widespread forest fires.

All these features point to an impact. The metal abundances look much like what we commonly find in meteorites rather than what we find elsewhere on Earth's surface. Shocked quartz is also found at other known impact sites, such as Meteor Crater in Arizona. The rock "droplets" presumably were made from molten rock splashed into the air by the force and heat of the impact. The soot probably came from vast forest fires ignited by impact debris. Some debris would have been blasted so high that it rose above the atmosphere, spreading worldwide before falling back to Earth. On their downward plunge, friction would have heated the debris particles until they became a hot, glowing rain of rock. Like falling hot coals, the particles ignited fires as they reached the ground.

In addition to the evidence within the sediments, scientists have identified a large, buried impact crater that appears to match the age of the sediment layer. The crater, about 200 kilometers across, is located on the coast of Mexico's Yucatán Peninsula, about half on land and half underwater (Figure 6.16). Its size indicates that it was created by the impact of an asteroid or a comet measuring about 10 kilometers across, large enough to account for the iridium and other metals. (It is named the *Chicxulub crater*, after a nearby fishing village.)

The Mass Extinction If the impact was indeed the cause of the mass extinction, here's how it probably happened: On that fateful day some 65 million years ago, the asteroid or comet slammed into Mexico with the force of a hundred million hydrogen bombs (Figure 6.17). It apparently hit at an angle, sending a shower of red-hot debris across the continent of North America. A huge tsunami sloshed more than 1,000 kilometers inland. Much of North American life may have been wiped out almost immediately. Not long after, the hot debris raining around the rest of the world ignited fires that killed many other living organisms. Indeed, the entire sky may have been bright enough to roast most life on land.

Dust and smoke remained in the atmosphere for weeks or months, blocking sunlight and causing temperatures to fall as if Earth were experiencing a global and extremely harsh winter. The reduced sunlight would have stopped photosynthesis for up to a year, killing large numbers of species throughout the food chain. This period of cold may have been followed by a period of unusual warmth: Some evidence suggests that the impact site was rich in carbonate rocks, so the impact may have released large amounts of carbon dioxide into the atmosphere. The added carbon dioxide would have strengthened the greenhouse effect, so that the months of global winter immediately after the impact might have been followed by decades or longer of global summer.

The impact probably also caused chemical reactions in the atmosphere that produced large quantities of harmful compounds, such as nitrous oxides. These compounds dissolved in the oceans, where they probably were responsible for killing vast numbers of marine organisms. Acid rain may have been another by-product, killing vegetation and acidifying lakes around the world.

Perhaps the most astonishing fact is not that up to 75% of all plant and animal species died but that some 25% survived. Among the survivors were a few small mammals. These mammals may have survived in part because they lived in underground burrows and managed to store enough food to outlast the global winter that immediately followed the impact.

The evolutionary impact of the extinctions was profound. For 180 million years, dinosaurs had diversified into a great many species large and small, while most mammals (which had arisen at almost the same time as the dinosaurs) had remained small and rodentlike. With the dinosaurs gone, mammals became the new animal kings of the planet. Over the next 65 million years, the small mammals rapidly evolved into an assortment of much larger mammals—ultimately including us. So had it not been for the K–T impact, dinosaurs might still rule Earth.

Note that, while this general picture seems to make sense, the jury is still out on whether it tells the entire story. The geological record can be difficult to read when we are trying to understand events that happened over just a few years rather than a few million years. Some scientists suspect that the dinosaurs were already in decline and the impact was only the last straw. Others suggest that major volcanic eruptions also may have played a role. A few have deeper doubts about the role of an impact. Only additional data will settle the remaining questions.

• Did impacts cause other mass extinctions?

The K–T extinction seems quite clear, but it's generally difficult to measure past extinction rates precisely. The primary problem is that identifying the extinction of a species requires finding its *last* occurrence in the fossil record, which means we can be misled if we've yet to find a more recent occurrence of the species. Nevertheless, we have enough data to be sure that extinction rates vary considerably with time.

Figure 6.18 shows current data on the extinction rate for plants and animals over the past 500 million years. The data reveal at least five major

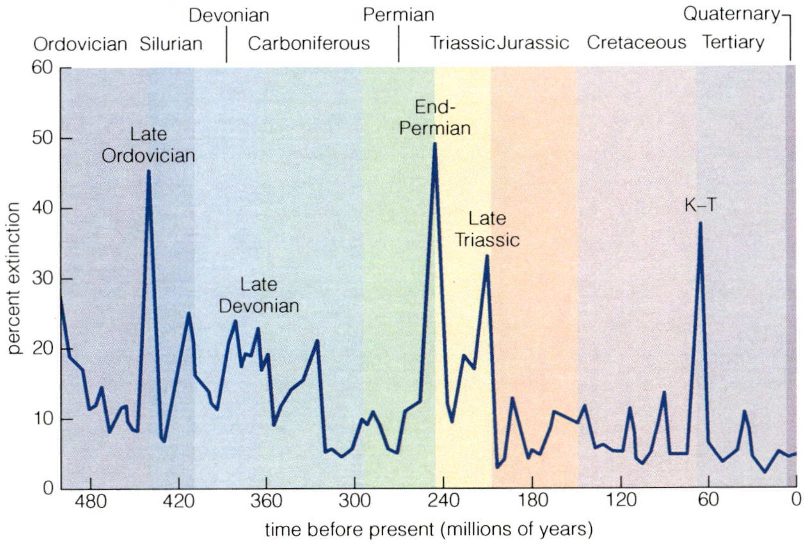

Figure 6.18

This graph shows data concerning the approximate percentage of plants and animals to go extinct with time over the past 500 million years. Peaks represent mass extinctions, with names shown for five major events. (The data actually are for families, a higher level of classification than genus and species.) (Adapted from Campbell, Reece, *Biology,* Seventh Edition, © 2005.)

mass extinctions, including the K–T extinction, and numerous smaller extinction events. Could some of these extinctions also have been caused by impacts?

Impacts and Other Hypotheses Simple probability makes impacts a reasonable hypothesis. On average, impacts the size of the K–T event should happen about every 100 million years or so. And, indeed, impacts appear to have coincided with at least some of the other mass extinctions. For example, the sedimentary strata marking the most severe mass extinction—the Permian extinction about 245 million years ago—show many of the same features as the strata marking the K–T boundary. In 2004, a research team (led by Luann Becker of UC Santa Barbara) claimed to find evidence off the northwest coast of Australia for a crater dating to about this time. Other geologists are still working to verify this claim. In any event, the K-T example suggests that we should consider impacts as possible causes for the Permian and other extinctions, even if we never identify craters left by the impacts. Remember that most impacts occur in the oceans (because oceans cover nearly $\frac{3}{4}$ of Earth's surface), and seafloor crust is almost completely recycled in about 200 million years [Section 4.4]; this recycling would destroy any evidence of a crater on the seafloor.

Besides impacts, scientists have come up with a number of other hypotheses about possible causes of mass extinctions. For example, some geologists hypothesize that episodes of unusually active volcanism may have led to climate change and high extinction rates. Other types of climate change may also have caused mass extinctions. For example, if there have been more recent periods of cooling similar to the known snowball Earth episodes (which ended more than 500 million years ago [Section 4.5]), the cold climate may have driven many species to extinction.

Another set of hypotheses envisions extinctions tied to changes in the mutation rate. While many mutations occur simply as the result of copying "errors" within cells, others are caused by external influences. For example, ultraviolet light can cause mutations, which is why sun exposure can lead to skin cancer. If the concentration of Earth's ozone layer varies with time, then the amount of ultraviolet light reaching the surface would also vary. Perhaps some of the extinctions occurred when the ozone layer thinned, allowing solar ultraviolet light to cause many more mutations.

Mutations can also be caused by high-energy particles that stream continuously from the Sun (making up the *solar wind*). Recall that Earth's magnetosphere deflects most of these particles, preventing them from reaching the surface [Section 4.4]. However, studies of magnetized rocks show that Earth's magnetic field varies significantly in strength with time, and sometimes reverses itself entirely, with the north magnetic pole becoming the south magnetic pole and vice versa. These magnetic reversals occur every few million years on average, and the magnetic field may disappear altogether for thousands of years while a reversal is in progress. The mutation rate might spike upward during this time due to the absence of the normal protection from high-energy particles. Although magnetic field reversals happen much more frequently than mass extinctions, some reversals may have occurred at times when life was more susceptible to major change and thus might have played a role in extinction events.

Supernovae and Gamma-Ray Bursts Some scientists hypothesize that more distant events could trigger mass extinctions on Earth, including *supernovae*, the explosions of massive stars [Section 3.2]. Supernovae are rare events. Out of the more than 100 billion stars in the Milky Way Galaxy, we expect only about one star per century to explode in a supernova. Most of these supernovae occur far from our solar system. Nevertheless, because the Sun orbits the center of the galaxy independently of other stars, different sets of stars make up our galactic neighborhood at different times. Simple probability calculations suggest that our planet must occasionally be located within a few tens of light-years of an exploding star. Supernovae can generate prodigious numbers of very-high-energy particles called *cosmic rays*. Thus, when a supernova occurs near Earth, we might expect a big upward spike in the number of cosmic rays reaching Earth and causing mutations.

A related idea suggests that mass extinctions could be caused by *gamma-ray bursts*—bursts of gamma rays [Section 3.4] from space that last just minutes or less—most of which are produced by unusually powerful supernovae. Atmospheric models suggest that a gamma-ray burst occurring within a few thousand light-years of Earth could generate enough gamma rays to destroy half of Earth's ozone layer, thereby leading to a massive die-off through exposure to solar ultraviolet light. Probability arguments suggest that Earth should have been exposed to such a nearby gamma-ray burst at least once in the past billion years, and a few scientists have attempted to link a gamma-ray burst to the Ordovician mass extinction some 450 million years ago.

Lessons from Mass Extinctions Whether or not we ultimately learn the causes of past mass extinctions, they clearly have had tremendous effects on the evolution of life. With each mass extinction, many of the dominant species on the planet have disappeared, creating changes in environmental conditions and predator–prey relationships. These changes allow new species to evolve over the millions of years that follow. Just as the K–T event apparently paved the way for the rise of mammals, other extinctions may have caused similarly critical junctures in the evolutionary path that made our present existence possible.

The topic of mass extinctions also holds a cautionary lesson for our species today. Human activity is driving numerous species toward extinction. The best-known cases involve relatively large and wide-ranging animals, such as the passenger pigeon (extinct since the early 1900s) and the Siberian tiger (nearing extinction). But most of the estimated 10 million or more plant and animal species on our planet live in localized habitats, and most of these species have not even been cataloged. The destruction of just a few square kilometers of forest may mean the extinction of species that live only in that area. According to some estimates, human activity is driving species to extinction so rapidly that half of today's species could be gone within a few centuries or less. On the scale of geological time, the disappearance of half the world's species in just a few hundred years would certainly qualify as another of Earth's mass extinctions, potentially changing the global environment in ways that we are unable to predict.

THINK ABOUT IT The geological record suggests that the dominant animal species are nearly always victims in a mass extinction. If we are causing a mass extinction, do you think we will be victims of it? Or will we be able to adapt to the changes so that we survive even while many other species go extinct? Defend your opinion.

Figure 6.19

Damage from the 1908 impact over Tunguska, Siberia, shown in a photo taken many years later.

• Is there a continuing impact threat?

The discovery that at least one mass extinction is tied to an impact has spurred scientific concern over whether our civilization might be vulnerable to future impacts. How serious is this threat?

Small particles hit Earth almost continuously, burning up in our atmosphere as **meteors** (sometimes called "shooting stars," a misleading name that arose long before we knew their true source). Most meteors are caused by particles no bigger than a pea. The particle itself is too small for us to see, but it enters the atmosphere at such high speed (typically between about 45,000 and 250,000 km/hr) that it burns up and heats the surrounding air, producing the meteor flash. An estimated 25 million particles enter our atmosphere and burn up as meteors each day. Interestingly, these particles add a total of about 20,000 to 40,000 tons to Earth's mass each year. This sounds like a lot in human terms, but it is negligible compared to Earth's total mass of 6 *billion trillion* (6×10^{21}) tons.

Somewhat larger objects entering our atmosphere may be heated to the point where they explode, producing an extraordinarily bright flash called a *fireball.* (Some so-called UFO sightings are actually fireballs.) If debris from the explosion hits the ground, we may find some of it as the rocks we call *meteorites.* An impacting object more than a few meters across may survive the plunge through our atmosphere intact, carrying so much energy that it excavates a visible crater, such as Meteor Crater (see Figure 6.14).

No large impacts have left craters since the dawn of human civilization, but we know of at least one smaller impact that devastated a fairly large area. In 1908, a tremendous explosion occurred over Tunguska, Siberia, flattening and setting fire to the surrounding forest (Figure 6.19). Seismic disturbances were recorded up to 1,000 kilometers away, and atmospheric pressure fluctuations were detected at distances of almost 4,000 kilometers. The explosion, now estimated to have released energy

MOVIE MADNESS: ARMAGEDDON

In 1994, a lot of people who believed the dinosaurs were the victims of a unique accident changed their minds. That was the year Comet Shoemaker–Levy 9 smacked into Jupiter, leaving entrance wounds the size of planet Earth.

Folks realized that death by rock isn't all that improbable. If it happened to the dinos, it could happen to us. An errant asteroid a dozen miles across might someday careen into our planet and raise enough dust and burn enough forests to darken the world for years. We'd all slowly starve.

Alerted to this possibility for havoc and destruction, Hollywood lost little time in showing how ingenuity and some gutsy guys (with the emphasis on the latter) could save us even if Nature hurls a large space rock our way. TV specials and theatrical films soon appeared, showing Earth under mortal threat from ballistic boulders.

In the film *Armageddon,* the incoming object is as big as Texas, and a mere few weeks away. That's a real slap in the face for astronomers. Picture this: An asteroid as big as Ceres (the largest rock in the asteroid belt) is headed our way, and the astronomers find the darn thing only when it's as close as Mars?

The end of the world is nigh, but not to worry. NASA is in high gear to divert this king-size clod and decides the best thing to do is to blow it into two large pieces with a nuclear bomb. The two pieces will diverge slightly and sail harmlessly by opposite sides of Earth. Needless to say, geeky NASA personnel aren't up to this kind of macho mission, so the space agency recruits a bunch of oil-rig roughnecks to plant and detonate the bomb. The NASA folk refer to this group of gritty misfits as "the wrong stuff."

In fact, it's a bad idea to try to blow up an incoming asteroid. The chances are that you'd only turn a single shell into buckshot. More practical schemes envision fastening some sort of rocket engine to the side of the rock and slowly nudging it out of the way. Another approach, at least for asteroids that we find when still far from Earth, is to paint the asteroid white and let the gentle pressure of sunlight do the job. Neither scheme involves roughnecks (or human pilots at all).

The threat, of course, is real. Astronomers estimate that rocks comparable to the one that did in the dinosaurs (together with countless other species) will slam into Earth roughly every 50 million years. But by carefully keeping tabs on those asteroids that cross Earth's orbit, we can see disaster coming. We'll have years to mount a defense.

The bottom line is that this is one kind of disaster we can probably avoid. After all, unlike the dinos, we've got a space program.

equivalent to that of several atomic bombs, is thought to have been caused by a small asteroid no more than about 40 meters across. Atmospheric friction caused it to explode completely before it hit the ground, so it left no impact crater. We have also witnessed one far larger impact on another world: In 1994, astronomers recorded the impact of Comet Shoemaker–Levy 9 (SL9) as it slammed into Jupiter; the comet had broken into pieces before the impact, and each piece crashed into Jupiter with energy equivalent to that of a million hydrogen bombs (Figure 6.20).

Objects of similar size to that of the Tunguska event probably strike our planet every century or so. They have gone unnoticed, presumably because they have always hit remote areas, with most striking the ocean. Nevertheless, the death toll would be enormous if such an object struck a densely populated area. Larger impacts are correspondingly rarer. Based on statistics of past impacts, Figure 6.21 shows how often, on average, we expect Earth to be hit by objects of different sizes.

Note that while the chance that our civilization is in any imminent danger of a major impact seems relatively low, it is not negligible. Moreover, the question is not *whether* a future impact will occur, but *when*. Many known asteroids have orbits around the Sun that cross Earth's orbit or pass near enough to Earth's orbit that they may someday be gravitationally perturbed into an Earth-crossing orbit. These objects have a real chance of striking Earth at some future time. Some of these objects are in the 1-kilometer size range, and a few are as big as the 10-kilometer K–T impactor. NASA has set a goal of detecting 90% of all large, near-Earth asteroids by 2008. Unfortunately, we do not yet have a similar way of studying the threat from comets, because they reside so far from the Sun. By the time we saw a comet plunging toward us from the outer solar system, we'd have at best a few years to prepare for the impact.

If we were to find an asteroid or a comet on a collision course with Earth, could we do anything about it? Many people have proposed schemes to save Earth by using nuclear weapons or other means to demolish or divert an incoming asteroid, but no one knows whether current technology is really up to the task. We can only hope that the threat doesn't become a reality before we're ready.

THINK ABOUT IT Study Figure 6.21. Based on the frequency of impacts large enough to cause serious damage, do you think we should be spending time and resources to counter the impact threat? If so, how much should we invest relative to what we invest to combat other threats? Defend your opinion.

6.5 HUMAN EVOLUTION

We've traced the course of evolution from the origin of life through the extinction of the dinosaurs. We've seen that evolution took many surprising twists and turns to that point. The subsequent evolution of mammals and humans was just as interesting. In this section, we'll briefly investigate the pathway that led to our emergence as the first species on Earth capable of learning about its own origins.

• How did we evolve?

We are primates, as are all the great apes, monkeys, and prosimians (such as lemurs). The ancestor of all of today's primates lived in the trees, and many of the traits that make us so successful evolved as adaptations to

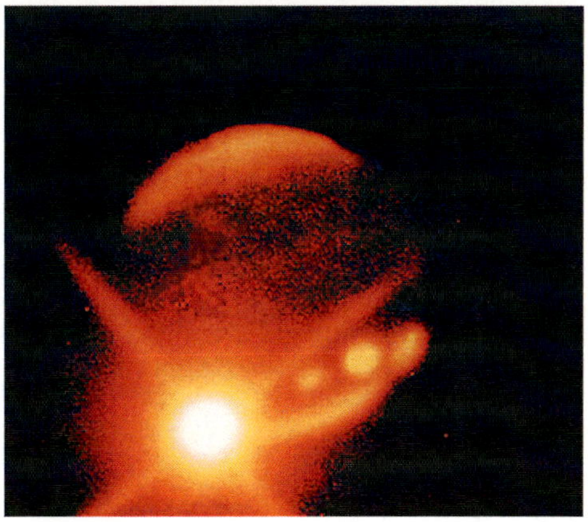

Figure 6.20

This infrared photo shows the brilliant glow of a rising fireball created when one of the pieces of Comet Shoemaker–Levy 9 (SL9) crashed into Jupiter (the round disk in the background) in 1994. Although overexposure exaggerates the size of the fireball, you can get a sense of scale by remembering that about ten Earths could fit side by side across Jupiter's diameter.

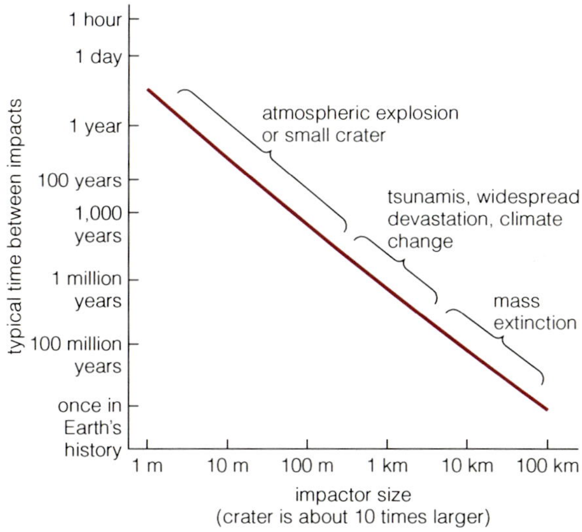

Figure 6.21

This graph shows that larger objects (asteroids or comets) hit Earth less frequently than smaller ones. The labels describe the effects of impacts of different sizes.

Figure 6.22

The evolutionary history of the major primate branches. Notice, for example, that the common ancestor of modern humans, chimpanzees, and gorillas lived between about 6 and 8 million years ago. (Adapted from Campbell, Reece, Simon, *Essential Biology*, Second Edition, © 2004.)

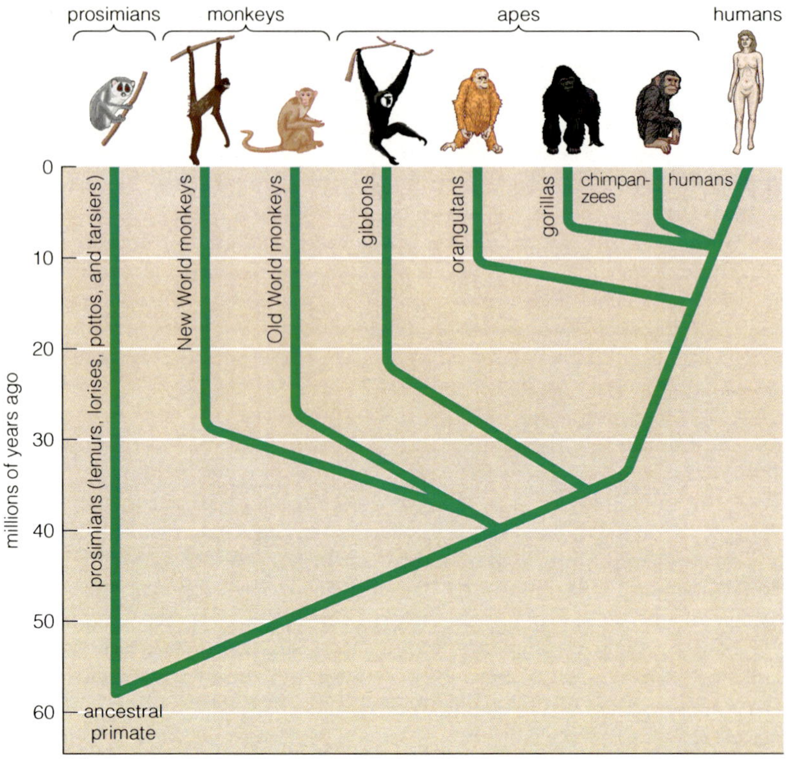

Figure 6.23

This famous type of illustration suggests that humans evolved along a simple pathway from apes. However, it is almost completely wrong.

tree life. For example, the limber arms that allow us to throw balls and work with tools evolved so that our ancestors could swing through trees, and our dexterous hands evolved to hang from branches and manipulate food. The eyes of primates are close together on the front of the face, providing overlapping fields of view that enhance depth perception—an obvious advantage when swinging from branch to branch. For the same reason, primates developed excellent eye–hand coordination.

Parental care is essential for young animals in trees, and primates evolved close parent–child bonds. These bonds, in turn, made it possible for primates to be born in a much more helpless state than the babies of most other types of animals. Although many primate species, including us, eventually moved down from the trees, most primates continue to nurture their young for a long time. This trait reaches its extreme in humans. Human babies are nearly helpless at birth and require parental care for more years than the offspring of any other species.

Contrary to a common myth, humans did not evolve *from* gorillas or other modern apes. Rather, modern apes and humans share a common ancestor that is now extinct. Figure 6.22 shows the evolutionary history of major primate branches. Our closest living relatives, chimpanzees and gorillas, shared a common ancestor with us just a few million years ago.

The Emergence of Humankind Even after hominids (human ancestors) diverged from the ancestors of chimpanzees and gorillas, human evolution followed a remarkably complex path. Indeed, one of the most pervasive but incorrect myths about human evolution is that it followed a simple pathway from stooped apes to upright humans (Figure 6.23).

The reality is that there have been numerous hominid species, some of which may be part of the lineage of modern humans and others that may have come to evolutionary dead ends. The exact timing of the first hominids is not known, but a fossil skull recently excavated in Chad and likely to be hominid dates to between about 6 and 7 million years ago. The skull of *Sahelanthropus tchadensis,* nicknamed Toumaï (Figure 6.24), shows features intermediate between those of apes and humans, along with features suggesting that Toumaï walked upright. Many researchers now suspect that Toumaï was the first true hominid, but more confirming fossil evidence will be needed to seal the case.

The earliest fossil skulls that appear to be identical to those of modern humans are about 100,000 years old. However, even then our ancestors shared the planet with at least one other hominid species, usually called *Neandertals,* that was quite similar and may have had a slightly larger brain on average than *Homo sapiens.* The Neandertals disappeared about 30,000 years ago, and no one yet knows why. In 2004, scientists unearthed evidence of yet another recent hominid species, *Homo floresiensis,* that lived on an Indonesian island as recently as about 12,000 years ago. These people apparently stood no more than about a meter tall, and some scientists question whether they were truly a separate species or whether the fossils represent aberrant *Homo sapiens.* Figure 6.25 summarizes hominid lineages from Toumaï to modern *Homo sapiens.* (*H. floresiensis* is not shown, since their status is still not well accepted.)

Deciphering the rest of the details of human ancestry is a rich field of research, and much remains subject to scientific debate. We will not discuss such details in this book, but before we leave the topic, it's worth dispelling two additional common myths. First, there is no longer a "missing link" in human evolution. While a few mysteries may always remain, we now know enough from the geological record and genome comparisons to see a clear path from the earliest microbes to ourselves. Second,

Figure 6.24

This computer reconstruction, based on the actual fossils, shows the skull of *S. tchadensis,* or Toumaï, first excavated in Chad in 2001.

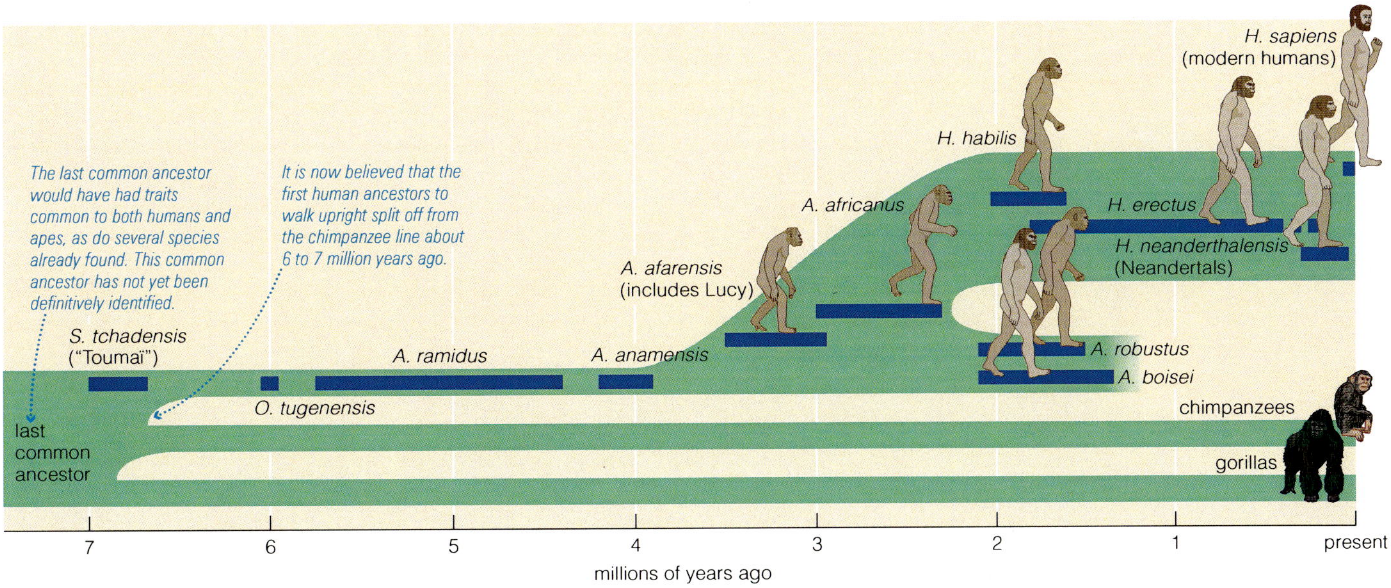

Figure 6.25

Human evolution over about the past 7 million years since our lineage separated from that of modern gorillas and chimpanzees.

despite the many species of hominids that have come and gone, all modern humans are members of the same species. That is, while people often focus on outward differences between races such as skin color or hair texture, all human genomes are virtually identical. Moreover, any small racial differences that might once have arisen have since been spread across races by the widespread interbreeding of our ancestors. Any remaining genetic differences between human races are generally much smaller than the genetic variation among the individuals in each race.

Implications for Extraterrestrial Intelligence The fact that modern gorillas, chimps, and humans all evolved from the same ancestor has at least two important implications for understanding the possibility of extraterrestrial intelligence. First, it shows that relatively small genetic differences can make a big difference in species success. More than 98% of the DNA sequences that make up the human genome are identical to the sequences that make up the chimpanzee genome. Thus, a relatively small genetic difference is all that separates our success on this planet from the current predicament of chimpanzees, which survive naturally in only a few isolated locations in Africa. Second, it suggests that the evolution of intelligence is a complex process. Gorillas and chimpanzees have been evolving from our common ancestor just as long as we have, but we are the only species building cities and radio telescopes. This fact raises the question of whether advanced intelligence is an inevitable outcome of evolution. We'll discuss this question further in Chapter 12.

• Are we still evolving?

Given the very recent arrival of *Homo sapiens* on the evolutionary time scale, it's natural to wonder whether we are still evolving. There is some scientific debate on this issue, but overall it seems unlikely that we have undergone any significant evolutionary changes in the past 40,000 years or so. That is, if we could sequence the genome of a human from that period, it would be virtually identical to the genome of anyone living today. Nevertheless, we have clearly gone through dramatic changes as a species.

Cultural and Technological Evolution These dramatic changes are not due to biological evolution, but rather to what we might call **cultural evolution**—changes that arise from the transmission of knowledge accumulated over generations. In other words, we humans can transmit our history, using both spoken and written language, which thereby allows us to learn from what has been done before. The know-how to build tractors, computers, and spaceships is stored not in our genes but in the cumulative product of hundreds of generations of human experience. Although some other species, including chimpanzees, demonstrate aspects of culture, humans are probably unique in having reached the point where cultural evolution is far more important to our changing nature than is biological evolution.

Because biological evolution is driven by random mutations, it tends to proceed at a slow and relatively steady rate. Cultural evolution, in contrast, tends to accelerate over time. The development of agriculture and written language took tens of thousands of years, while less than two centuries separate the beginning of the industrial revolution from the first walk on the Moon. More recently, we have begun to develop another new type of evolution that is accelerating even more rapidly—**technological evolution**.

You are probably familiar with the way new computers get faster each year. These speed increases come from an interesting coupling of technology and science: Increased computing power enables scientists to make new discoveries, which in turn lead to further increases in computing power, and so on. Similar couplings of technology and science affect almost every field of human knowledge. One of the most striking cases involves the way technology is helping us understand biological evolution. As we've discussed, much of our current knowledge about the origin and evolution of life comes from studies of genomes that are made possible by technological advances such as machines that can read DNA sequences.

Evolution Among Humans and Other Intelligent Civilizations

The same type of technology that allows us to read genomes also now allows us to reengineer living organisms (genetic engineering) in such a way that we may soon outpace nature in developing new species. It seems inevitable that we will also develop the capability to change human DNA, thus opening the door to attempts to "improve" our species. The moral and ethical dimensions of this power are already profound, and you will undoubtedly have to confront these issues many times during your life. For our purposes in this book, perhaps the primary lesson—whether good or bad—is that advanced civilizations can alter the course of evolution through their choosing, rather than remaining subject to the random processes of natural selection. Indeed, modern medicine has already taken us out of the realm of Darwinian evolution, because we routinely save individuals who would have died earlier in generations past. The course of our future evolution is in our own hands. If other advanced civilizations exist, they must similarly control their own destinies.

COSMIC CALCULATIONS 6.2

Bacteria in a Bottle II: Lessons for the Human Race

Recall the thought experiment of the bacteria in a bottle in Cosmic Calculations 6.1, in which we start with a single bacterium at 12:00 that replicates each minute. Suppose that bacterial growth completely fills the bottle at 1:00, exhausting all nutrients so all the bacteria die. Let's explore this issue with a series of simple questions.

Question 1: The tragedy occurs when the bottle is full at 1:00, just 60 minutes after the first bacterium started the colony at 12:00. When was the bottle *half*-full?

Answer: Most people guess 12:30, halfway through the hour of growth. But this is incorrect: The bacterial population doubled every minute, so the bottle went from half-full at 12:59 to full at 1:00.

Question 2: You are a mathematically sophisticated bacterium, and at 12:56 you recognize the impending disaster. You warn your fellow bacteria that the end is just four minutes away unless they slow their growth dramatically. Will anyone believe you?

Answer: Because the bottle was full at 1:00, it was $\frac{1}{2}$-full at 12:59, $\frac{1}{4}$-full at 12:58, $\frac{1}{8}$-full at 12:57, and $\frac{1}{16}$-full at 12:56—which means there's still 15 times as much unused bottle as used when you give your warning. Your warning may go unheeded.

Question 3: Just before the disaster strikes, a bacterial space program discovers 3 more bottles in the lab. With an immediate and massive population redistribution program among the original and 3 new bottles, how much more time will the bacteria buy?

Answer: You may be tempted to think that 3 more bottles should give the colony 3 more hours, but in fact it gives them only 2 more minutes: Because the growth occurs through doubling, the colony will fill 2 bottles at 1:01 and 4 bottles at 1:02. In fact, *nothing* could allow the growth to continue much longer, because the doublings would soon lead the bacteria to impossible volumes (see Problem 51 at the end of the chapter).

We can draw several general conclusions. First, because populations of living organisms tend to grow exponentially, numbers can rise very rapidly. This explains the inevitable population pressure that helped Darwin realize the role of natural selection (see Fact 1 on p. 155). Second, exponential growth must always be a short-term, temporary phenomenon; for living organisms, the growth typically stops due to predation or a lack of sufficient nutrients or energy. Third, these laws about growth apply to all species—our intelligence cannot make us immune to simple mathematical laws. This is a critical lesson, because human population has been growing exponentially for the past few centuries (see Figure 12.14). Of course, our intelligence gives us one option not available to bacteria. Exponential growth can stop only through some combination of an increase in the death rate or a decrease in the birth rate. Unlike bacteria, we can *choose* to stop our exponential growth with changes to our birth rate before we "fill" our planet.

6.6 THE PROCESS OF SCIENCE IN ACTION: ARTIFICIAL LIFE

In this chapter, we have discussed some of the laboratory experiments through which scientists seek to understand the origin of life. Some researchers are attempting to go even further, by trying to actually create life in the lab. Their efforts are rather different from those of the fictional Dr. Frankenstein, who sewed together dead body parts and jolted a living, humanlike creature into existence with high voltage. Today's researchers are trying to put together novel organisms that can reproduce and grow, but on a microbial scale.

Success could have a variety of implications. For example, it would help us understand how life likely got started on Earth, as well as help us decide how best to look for life in extraterrestrial environments such as subsurface aquifers on Mars or in a subterranean ocean of Europa. We might be able to evolve organisms that could provide us with a variety of benefits, medical and otherwise. Of course, it might also be possible to create dangerous organisms, forcing us to confront the ethical dilemmas of our work. For this chapter's case study in the process of science in action, we turn our attention to current work in artificial life.

• How might we create artificial life?

It's still far beyond our means to create from elementary chemicals a bacterium that's comparable to those found in nature. Nonetheless, hundreds of researchers are working to spawn A-life ("artificial life") by either rearranging bits and pieces of existing organisms or trying to build an extremely simple living cell in the lab. The former scheme is referred to as a "top-down" approach, and is similar to hot-rodding, where a car is stripped down and rebuilt with different components, producing a vehicle that is completely unlike any existing auto.

Engineering New Species from Existing Organisms Craig Venter, the man behind the private-sector effort that first sequenced the human genome, has embarked on a top-down program to make designer organisms. His goal, shared by many others, is to create microbes that do useful things, such as fighting malaria or converting atmospheric carbon dioxide into methane, a trick that could be valuable for reducing our dependence on fossil fuels. As his basic building block, he starts with *Mycoplasma genitalium,* a familiar bacterium often responsible for urinary tract infections. *M. genitalium* has a tiny genome even for a microbe (about 580,000 base pairs, compared with about three billion base pairs for a human). It has only 486 proteins, and research has suggested that it could live with only about half that number. So Venter's team is trying to remove as much of the organism's genes as they can, while still leaving it viable and able to reproduce. Once they have stripped down the *Mycoplasma's* genome to the fewest number required to survive, they will insert this minimalist DNA into the hollow shell of another *M. genitalium* cell from which the original DNA has been removed. By this method, the team hopes to produce a bare-bones microbe—the simplest thing (other than viruses) that is still alive. They can then customize this elemental organism for specialized tasks by fitting it out with some additional genes. In 2002, A-life researchers managed to create a polio virus by lashing together a short, synthetic strand of DNA. Venter's bare-bones *M. genitalium,* if he can make it work, will have a strand many dozens of times longer.

Note that this is not really creating life from scratch, but merely streamlining and modifying an existing microbe. It's a more efficient method for producing desirable organisms than simply breeding them and selecting those that have the desired properties. That scheme depends on mutation to produce a range of characteristics, and you simply cull those that are, by chance, closer to what you want. But Venter's approach deliberately introduces genes that will result in the desired behavior. Top-down engineering promises to give us entirely new species of great practical value, and it will also undoubtedly teach us much about how cells work. However, it's less likely to make clear how life got started on Earth, and thereby help us understand whether terrestrial biology is a fortunate accident, or a virtually inevitable development. The bottom-up approach to A-life more directly addresses that question.

Making Life from Raw Ingredients As we discussed in Section 6.2, it seems likely that the young Earth had plenty of organic building blocks for life, either produced by chemical reactions on Earth or brought here from space. The many possible ways in which the building blocks could have formed or arrived has led some people to believe that the creation of life in the laboratory should be just around the corner.

Few researchers think it's that easy. But several have dared to take a bottom-up approach to making A-life in the lab. One of these is Jack Szostak, a Harvard University geneticist who was working on the genes of yeast. He decided to switch gears when he read of the work of Thomas Cech, who discovered that RNA could serve as both a blueprint for reproduction and a catalyst for making proteins. Cech suggested that RNA may have been the basis of life before DNA arrived on the scene, an idea we discussed earlier as "RNA world." This vision of an RNA world encouraged Szostak to leave his yeast cells behind, and try building synthetic organisms out of strands of this older nucleotide.

There were already research hints as to how RNA life might have arisen, since prior lab experiments had shown that short strands of RNA can form when a dilute organic solution washes over the surface of clay or rock. If, by chance, one of those RNA strands could reproduce, it would have staying power. Those that could do this most quickly, and with the fewest number of copying errors, would soon dominate their environment. In this way, a robust form of RNA life could have evolved. But did it really happen that way?

Szostak seeks to gain insight into this question by attempting to produce simple RNA-based cells that can replicate. If he succeeds, he will have created a new life-form after beginning with only nonliving materials—a rather different accomplishment than the top-down A-life efforts described above. He begins with a sample of RNA fragments: thousands of trillions of short nucleotide (base) sequences. These are taken from some simple RNA strands that are known to have special talents. They can, for example, grab onto a particular type of molecule, or duplicate parts of it. He then puts his fragment collection into a test tube, and uses a technique called "in vitro selection" to fast-forward Darwinian evolution. He gives his collection the opportunity to chemically react, and then screens the samples to find, for example, those that have managed to replicate a short sequence of RNA. Keep in mind that these are merely organic molecules: They're not life. Szostak then filters out the winners in this test, and makes trillions of copies of those (some copy errors, or mutations, are allowed in this process). He runs the experiment again with the imperfectly cloned winners. Those that are even better, or at least faster, are retained for the next round—and so forth through dozens of cycles. In essence, Szostak is trying to emulate what might have happened on the early Earth by compressing time with laboratory evolution. It is somewhat analogous to producing large kernels of corn by repeatedly using seeds from only the largest ears of the crop.

So far, Szostak's lab has made strands that can replicate other RNA sequences, although they are short and the replication isn't always very accurate. But the big prize is an RNA strand that can really sift through the material of an organic solution and build a copy of itself. If Szostak succeeds, it will be possible to watch this self-replicating molecule evolve in the laboratory, because the sample will soon be dominated by the type of RNA that makes copies of itself, rather than of other RNA.

We've noted that freely floating organic molecules aren't cells. The molecules Szostak hopes to build need to be confined by a wall to keep aggressive compounds in the cruel outside world from dismantling them, while at the same time allowing the building blocks necessary for reproduction to enter their mini-habitat. Szostak therefore also seeks natural processes that could make pre-cells to enclose the RNA. He has already had some success: The clay-catalyzed lipid membranes that we discussed earlier, with RNA inside them (see Figure 6.7), were produced in his lab. When these blobs pass through porous rock-like materials, they split into smaller bubbles, with much of their RNA content still present. This is an obvious candidate for a mechanism that might have produced the first living cells, and it's remarkable that clay can both help assemble complex RNA and facilitate the development of cell membranes.

Overall, while Szostak doesn't yet have RNA that will make a complete protein or reproduce itself, he reckons that the creation of a replicating strand is well within reach. If so, we may soon know whether the origin of life could have occurred as easily as we have imagined under conditions that existed on the young Earth.

• Should we create artificial life?

There's reason to think that the top-down approach to A-life is likely to work sooner, as it starts from living organisms such as *Mycoplasma genitalium,* and modifies it with genes from other life. This is at least somewhat analogous to the Frankenstein scenario, of building something new out of existing parts, albeit at the level of single-cell life. The payoff in producing nature's own nanobots for specific tasks is obvious. We've mentioned the possibility of generating hydrocarbons for fuel, but A-life could also be used to target cancer cells or clean up toxic waste. Still, there's the danger that someone might eventually use this technology to engineer deadly organisms that have no natural enemies, a potent agent for biowarfare. Although Szostak's work is less intent on creating useful life and more focused on learning how biology began, it too poses a potential risk for misuse.

These dangers are recognized by the researchers themselves. While they're not overly worried about the threat A-life might pose (they point out that it would be extremely fragile and would have a difficult time living outside the laboratory environment), the scientists have occasionally convened panels to consider the ethical implications of their work. Do humans have the moral right to create new types of life? We don't seem to mind the development of hybrid corn or improved cattle, but the production of an entirely new species is likely to be more controversial.

There's an alternative to the A-life scenario that avoids most of these ethical dilemmas, and that is to build virtual life by modeling the functions of living cells with computer software. This might not seem particularly interesting, but in fact it would allow us to do biological experiments at the keyboard, without the difficulty (and potential danger) of using real microbes as lab rats. Indeed, it's somewhat analogous to testing aircraft with computer simulations of the air flow over their wings and fuselage, as opposed to building scale models and putting them in wind tunnels.

Software life could be useful for determining the effects of new drugs, or even for predicting the consequences of removing or adding genes (genetic engineering). *Escherichia coli* (*E. coli*) is a bacterium that resides in your gut, and is one of the best studied of all organisms. But it has more than 4,000 genes, of which approximately 1,000 have functions that are still not understood. Consequently, creating a computer model of a real *E. coli* is still beyond us. However, borrowing an idea from the top-down researchers, computer scientists are now programming a stripped-down model of the bacterium, with only about a thousand genes. This effort might bear fruit in the near future, and even though the model is only an approximation to the actual microbe, it might still be useful in understanding how life works. Building a complete computer model of *E. coli*, while not on the immediate horizon, would surely encourage scientists to modify it to more closely resemble human cells. Suppose that could be done. Then, even aside from the insight it would give into how biology functions, it would permit us to, for example, quickly evaluate drugs for fighting cancer without the necessity of testing them on laboratory animals or people.

Of course, even if the life is only programmed on a computer, once we know the necessary DNA sequences, it would be possible to put it together for real. Thus, one way or other, we are likely to be forced to confront the ethical dilemmas of creating artificial life. Indeed, with all the research that is already under way, it may be too late to put the genie back in the bottle, even if we wanted to. The future of biological science—and perhaps of our species as a result—will largely depend on the ethical choices that we make as modern biotechnology continues its technological evolution.

THE BIG PICTURE

Putting Chapter 6 into Context

In this chapter, we have completed our overview of life on Earth. We have built on our understanding of Earth's habitability (Chapter 4) and of the nature of life (Chapter 5) to develop a modern picture of the origin and evolution of life. As you continue in your studies, keep in mind the following "big picture" ideas:

- We may never know precisely how life arose on Earth. However, we have found *plausible* scenarios for the origin of life based on natural, chemical processes. These scenarios are based on solid evidence found in the geological record, in comparisons of genomes between species, and in laboratory experiments.
- Earth has supported life for most or all of the past 4 billion or more years, but life remained relatively simple for most of this period. Animal life exploded in diversity only about 545 million years ago (with the Cambrian explosion), and plants colonized the land only about

475 million years ago. If aliens had observed Earth during about the first 90% of its history, they would have found a planet that was home to nothing more than microscopic species.

- The course of evolution has been drastically changed at least several times by mass extinctions, and at least one of these mass extinctions is clearly linked to an asteroid or comet impact. Thus, as we'll discuss in more detail later in the book, the likelihood of impacts must be an important consideration in considering the habitability of other planets.
- From the time of the first living organism to today, the evolution of life on our planet has been shaped by natural selection. However, we have developed or are on the verge of developing the capability to engineer existing species, including ourselves, and perhaps to create entirely new species. This power must be available to any advanced civilization and therefore is an important consideration in the search for extraterrestrial intelligence.

SUMMARY OF KEY CONCEPTS

6.1 SEARCHING FOR LIFE'S ORIGINS

• When did life begin?

Stromatolite and sulfur isotope evidence tells us that life existed by about 3.5 billion years ago; microfossil evidence is consistent with this view. Carbon and other isotopic evidence may push the time back to 3.85 billion years ago. Life must have existed even earlier than the oldest fossil evidence of it, though we do not know exactly how much earlier.

• What did early life look like?

Although early life is long gone, genetic comparisons allow us to determine which modern organisms are evolutionarily oldest, suggesting that they are most similar to early life. These studies suggest that early life resembled modern extremophiles living in hot water near deep-sea vents or in hot springs.

• Where did life begin?

Based on the genetic studies and availability of chemical energy, deep-sea vents or hot springs are good possibilities. Because deep-sea vents are protected from impacts that might vaporize surface hot springs, they seem the most likely place for the origin of life.

6.2 THE ORIGIN OF LIFE

• How did life begin?

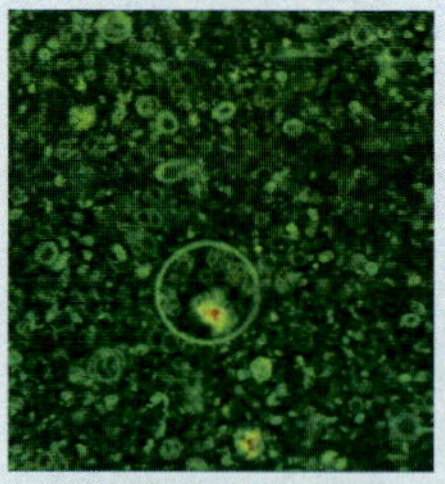

According to laboratory studies, the most likely scenario holds that organic molecules, either produced chemically or brought here from space, were found in ocean locations where clay and other minerals were common. Clay helped catalyze the building of RNA strands that became enclosed in lipid pre-cells. Some RNA molecules were able to partially or completely self-replicate, allowing natural selection among them to improve their replication until true life emerged.

• Could life have migrated to Earth?

If life originated first on Venus or Mars, it may have migrated to Earth when impacts blasted rocks from one world to another. Meteorites from Mars show that life could in principle survive the journey in some cases, and then take hold near deep-sea vents or elsewhere. Longer migrations, such as from planets around other stars, are highly unlikely.

6.3 THE EVOLUTION OF LIFE

- **What major events have marked evolutionary history?**

Life probably diversified rapidly after its origin, but remained microscopic for more than 2 billion years. Keys to the eventual transition included the origin of oxygen-producing photosynthesis, which released the oxygen now in our atmosphere, and the evolution of cell nuclei and other complex structures in eukaryotes. Multicellular animals diversified in the **Cambrian explosion**, starting about 545 million years ago. Plants and animals migrated onto land not long after.

- **Why was the rise of oxygen so important to evolution?**

Aerobic processes, using oxygen, offer the possibility of much more efficient cellular energy production than anaerobic processes, and thus can lead to much greater evolutionary diversification. The precise timing of the rise of oxygen is not well known, but began before about 2.35 billion years ago and probably did not reach levels near those of the present until the time of the Cambrian explosion or later.

6.4 IMPACTS AND EXTINCTIONS

- **Did an impact kill the dinosaurs?**

It may not have been the sole cause, but a major impact clearly coincided with the **mass extinction** that killed the dinosaurs about 65 million years ago. Sediments from the time contain iridium and other evidence of an impact, and a crater of the right age lies buried beneath the Yucatán coast of Mexico.

- **Did impacts cause other mass extinctions?**

At least five mass extinctions have occurred in the past 500 million years. Although only the K-T extinction is clearly linked to an impact, some evidence suggests that impacts have played a role in other extinctions. Other possible causes include periods of active volcanism, severe climate change, and changes to mutation rates, possibly influenced by changes in Earth's magnetic field or by distant supernovae or gamma-ray bursts.

- **Is there a continuing impact threat?**

Impacts certainly pose a threat, though the probability of a major impact in our lifetimes is small. Impacts like the Tunguska event occur more frequently (about once a century), and would be catastrophic if they occurred over cities.

6.5 HUMAN EVOLUTION

- **How did we evolve?**

Humans share a common ancestor with modern gorillas and chimpanzees, an ancestor that lived between about 6 to 8 million years ago. The earliest *Homo sapiens* emerged about 100,000 years ago.

- **Are we still evolving?**

Genetically, humans have probably changed little in at least the past 40,000 years. However, we are now changing in new ways, through **cultural evolution** and **technological evolution**.

6.6 THE PROCESS OF SCIENCE IN ACTION: ARTIFICIAL LIFE

- **How might we create artificial life?**

Scientists are seeking to make artificial life in two basic ways. A "top-down" approach starts with existing organisms and genetically strips them down, then tries to rebuild them into new species. A "bottom-up" approach starts in the laboratory with the raw ingredients of life, and seeks to reproduce life in much the same way that it presumably originated billions of years ago.

- **Should we create artificial life?**

Researchers trying to make A-life in the lab are confident they will succeed within a few years. There has already been discussion about whether this is ethically desirable, even though there could be important practical benefits. In addition to these moral concerns, some people worry that A-life might be the source of new diseases or toxins for which terrestrial life would have little resistance.

EXERCISES AND PROBLEMS

REVIEW QUESTIONS

Short-Answer Questions Based on the Reading

1. What are the three lines of fossil evidence that point to an early origin of life on Earth? Discuss each line and what it tells us about when life arose. What are the implications of an early origin to the possibility of life elsewhere?
2. How do studies of DNA sequences allow us to reconstruct the evolutionary history of life? What living organisms appear to be most closely related to the common ancestor of all present life?
3. Based on current evidence, what locations on Earth seem likely for the origin of life? What locations can we rule out?

4. What was the *Miller–Urey experiment*, and how did it work? Why is its relevance now subject to scientific debate? How else might Earth have obtained the organic building blocks of life?
5. What do we mean by an "RNA world," and why do scientists suggest that such a world preceded the current "DNA world"?
6. Briefly summarize current ideas about the sequence of events through which life may have originated on Earth. What role(s) might clay or other inorganic materials have played?
7. Briefly discuss the possibility that life migrated to Earth. Also discuss the possibility that Earth life might have migrated to other worlds, and the implications of possible migration to the search for life elsewhere.
8. Why do we think that evolution would have proceeded rapidly at first, and what fossil evidence supports this conclusion?
9. Briefly discuss the early evolution of life, from the first organisms to the development of photosynthesis and oxygen production.
10. How do we think that eukaryotes evolved? What time constraints can we place on when eukaryotes first got cell nuclei?
11. What was the *Cambrian explosion?* Briefly discuss ideas about what might have caused it and why no similar event has happened since.
12. How and when did life colonize land? Why did it take so long after the origin of life in the oceans?
13. How do we know that the early Earth could not have had an oxygen atmosphere? Where did the oxygen in our atmosphere come from? How did the introduction of oxygen affect early life?
14. Summarize the history of the oxygen buildup as it is understood today, and describe key mysteries that still remain. When did oxygen reach current levels?
15. What was the *K–T impact*, and how is it thought to have led to the demise of the dinosaurs? What evidence supports this scenario? How did this event pave the way for our existence?
16. Briefly discuss the evidence for other mass extinctions, and list a few of their possible causes.
17. Discuss the threat that future impacts may pose to us and our planet, and how we know that the threat is real.
18. Describe several adaptations that evolved so primates could live in trees and that have proved useful to us as humans.
19. When did hominids arise, and when did modern humans arise? What is the significance of the Toumaï skull, assuming it is a hominid?
20. Briefly describe and clarify a few common misconceptions about human evolution.
21. What do we mean by cultural and technological evolution? What implications do they have for extraterrestrial intelligence?
22. Briefly describe two main approaches to creating artificial life. Then describe the possibility and potential benefit of constructing computer programs that can mimic the biological functions of cells, and the ethical aspects of making artificial life.

TEST YOUR UNDERSTANDING

Would You Believe It?

Each of the following statements describes a hypothetical future discovery. In light of our current understanding of Earth and evolution, briefly discuss whether each discovery seems plausible or surprising. Explain clearly; because not all of these have definitive answers, your explanation is more important than your chosen answer.

23. We discover evidence of life, in the form of a particular ratio of carbon-12 to carbon-13, in rock that was originally formed in sediments and is 3.9 billion years old.
24. We discover an intact fossil of a eukaryotic cell, with a cell nucleus, that is 3.0 billion years old.
25. We discover a preserved, 3.5-billion-year-old microfossil that apparently had a genome genetically just like that of many modern animals.
26. We discover clear evidence that life arose on a high mountaintop, not in the oceans.
27. We discover a fossil of a large dinosaur that lived approximately 750 million years ago.
28. We discover that, contrary to present belief, oxygen was abundant in Earth's atmosphere at the time when life arose.
29. We discover a crater from the impact of a 10-kilometer asteroid that dates to about 2,500 years ago.
30. We discover an asteroid about 3 kilometers across that is on a collision course with Earth.
31. We find fossil remains of an early primate that lived about 50 million years ago and was, from all appearances, identical to a modern gorilla.
32. The first life created in the laboratory has an RNA genome, rather than a DNA genome.

Quick Quiz

Choose the best answer to each of the following. Explain your reasoning with one or more complete sentences.

33. The origin of life on Earth most likely occurred: (a) before 4.5 billion years ago; (b) between about 4.5 and 3.5 billion years ago; (c) between about 3.0 and 2.5 billion years ago.
34. The first living organisms probably were: (a) cells without nuclei that used RNA as their genetic material; (b) cells with nuclei that used RNA as their genetic material; (c) cells with nuclei that used DNA as their genetic material.
35. The importance of the Miller-Urey experiment is that: (a) It proved beyond doubt that life could have arisen naturally on the young Earth. (b) It showed that natural chemical reactions can produce building blocks of life. (c) It showed that clay can catalyze the production of RNA.
36. "RNA world" refers to: (a) the possibility that life migrated from Mars; (b) the idea that RNA was life's genetic material before DNA; (c) the idea that early life was made exclusively from RNA, needing no other organic chemicals.
37. Early life arose in an oxygen-free environment, but if any of these microbes had somehow come in contact with oxygen, the most likely effect would have been: (a) nothing at all; (b) to increase their metabolic rates; (c) to kill them.
38. The oxygen in Earth's atmosphere was originally released by: (a) outgassing from volcanoes; (b) plants; (c) cyanobacteria.
39. The *Cambrian explosion* refers to: (a) a dramatic increase in animal diversity beginning about 545 million years ago; (b) the impact that killed the dinosaurs; (c) the sudden emergence of eukaryotic life in the geological record dating to about 2.1 billion years ago.

40. Which statement about Earth's ozone layer is *not* true? (a) It protects us from dangerous solar radiation. (b) It did not exist when life first arose on Earth. (c) It first formed a few hundred million years after life colonized the land.

41. The hypothesis that an impact killed the dinosaurs seems: (a) well supported by geological evidence; (b) an idea that once made sense but now can be ruled out; (c) just one of dozens of clear examples of impacts causing mass extinctions.

42. Which of the following is a likely benefit of creating artificial life? (a) We'll gain insight into the origin of life on Earth. (b) The new life could eliminate our dependence on fossil fuels. (c) The technique could allow us to bring vanished species back to life.

INVESTIGATE FURTHER

In-Depth Questions to Increase Your Understanding

Short-Answer/Essay Questions

43. *A Brief History of Life on Earth.* Take all the ideas about the origin and evolution of life on Earth and try to condense them into a one- to three-page essay on the history of life on Earth. Or, if you prefer, try to capture the ideas in a poem.

44. *Geology and Life.* In Chapter 4, we discussed the role of plate tectonics and the CO_2 cycle in climate regulation on Earth. Suppose neither of these processes had ever operated. Could life still have arisen on Earth as discussed in this chapter? If so, how far could evolution have progressed before the lack of climate regulation would have blocked further major developments? Write a one-page essay summarizing and explaining your answers.

45. *Keys to Our Existence.* Identify and describe four crucial events in evolutionary history without which our current existence would have been highly unlikely. Explain your reasoning clearly.

46. *Extinction and Oxygen.* Suppose we somehow kill off a large fraction of the photosynthetic life on Earth. What consequences would this have for the oxygen content of our atmosphere? Explain your reasoning.

47. *Impact Movie Review.* View one of the Hollywood movies concerning the threat of an impact to our civilization, such as *Deep Impact* or *Armageddon.* Based on what you have learned in this chapter, write a one- to two-page critical review in which you include discussion of whether the impact scenario is realistic.

48. *Artificial Life Review.* Numerous science fiction stories and movies involve the creation of artificial life. Review one such story or movie, and identify at least three ideas in it that either do or do not meet the standards of being testable by science. Describe each in detail.

Quantitative Problems

Be sure to show all calculations clearly and state your final answers in complete sentences.

49. *Bacterial Evolution.* Suppose that a mutation occurs in about 1 out of every 1 million bacterial cells, and suppose that you have a bacterial colony in a bottle like that described in Cosmic Calculations 6.1 (in which the bacteria divide each minute). Given the number of bacteria in the bottle after 1 hour, approximately how many bacteria would have some type of mutation? What does this tell you about why bacteria often evolve resistance to new drugs?

50. *Deep in Bacteria.* In Cosmic Calculations 6.1, we calculated that the volume of bacteria after 120 doublings would be 1.3×10^{15} m^3. The total surface area of Earth is about 5.1×10^{14} m^2. Use these facts to calculate the average depth of the bacteria at that time, if we spread them evenly over Earth's entire surface.

51. *Bacterial Universe.* Suppose the bacteria described in Cosmic Calculations 6.1 and 6.2 could continue to multiply and spread out.
    a. Recall that the observable universe extends about 14 billion light-years in all directions from us. Calculate its volume in cubic light-years, then convert your answer to cubic meters.
    b. How long would it take for the bacteria to reach this volume? (*Hint:* You can proceed by trial and error, testing different values of t in the formula.)
    c. Even assuming that nutrients and energy were available, why couldn't the bacteria really grow this fast?

52. *Human Population Growth.* During the twentieth century, human population grew with a doubling time of about 40 years, reaching about six billion in 2000. Suppose this growth rate continued. What would human population be in 2200? in 2600? Do these populations seem possible on Earth? Explain.

53. *Impact Energy.* Consider a comet about 2 kilometers across with a mass of 4×10^{12} kg. Further assume that it crashes into Earth at a speed of 30,000 meters per second (about 67,000 miles per hour).
    a. What is the total energy of the impact, in joules? (*Hint:* The "kinetic energy" formula tells us that the impact energy in joules will be $\frac{1}{2} \times m \times v^2$, where m is the comet's mass in kilograms and v is its speed in meters per second.)
    b. A 1-megaton nuclear explosion releases about 4×10^{15} joules of energy. How many such nuclear bombs would it take to release as much energy as the comet impact?
    c. Based on your answers, comment on the degree of devastation the comet might cause.

54. *Impact Probability.* Impacts the size of the Tunguska event occur about once every century or two on average. Estimate the probability that the next impact will occur over a major city, thereby killing hundreds of thousands or millions of people. Be sure to explain all the numbers in your estimate clearly.

Discussion Questions

55. *Our Bacterial Ancestry.* Some of Darwin's early detractors complained that evolution implied we were descended from monkeys or apes. In fact, as we saw in this chapter, our evolution is built on far more primitive organisms. The oxygen we breathe was produced by bacteria and is processed in our cells by mitochondria that probably represent bacteria living symbiotically within us. Does our relationship to bacteria affect the way we should view ourselves as a species? Defend your opinion.

56. *The Missing Link.* As we discussed in this chapter, there is no longer a critical "missing link" in human evolution, and the theory of evolution has never suggested that humans evolved *from* present-day apes. Nevertheless, huge numbers of Americans profess belief in both of these myths about evolution. Why do you think these erroneous claims continue to be popular? What can or should be done to better educate the public?

57. *Evolution by Choice.* Consider the technology we are likely to have in the near future that would enable us to genetically engineer our own species, allowing us to choose the path of our future evolution. How do you think society can or should regulate the use of this awesome power? Do you think its potential benefits outweigh its risks, or vice versa? Overall, do you think

it likely that advanced civilizations, if they exist, have engineered their own evolution? Defend your opinions.

WEB PROJECTS

58. *The Origin of Life.* NASA's Astrobiology home page frequently covers new discoveries about the origin and evolution of life. Learn about one recent important discovery, and write a short essay summarizing the discovery and how it affects our understanding of how life might have evolved on Earth.
59. *Impact Programs.* The discovery that impacts could pose a threat to our civilization has led to calls for new programs to help alleviate the threat. In a few cases, legislation has even been proposed to implement such programs. Learn about one proposal for dealing with the impact threat. Write a short essay explaining the proposal and discussing your opinion of its merits.
60. *Extinction.* Learn more about how biologists estimate the rate at which human activity is driving species to extinction. What conclusions can we draw about the present rate of extinction? Based on this rate, are we in danger of causing a mass extinction comparable to past mass extinctions on Earth? Summarize your findings in a one-page essay.
61. *Artificial Life.* Find the Web sites of three research groups who are trying to produce artificial life, and write a one-page summary of each of their goals and methods.

7
Searching for Life in Our Solar System

LEARNING GOALS

7.1 ENVIRONMENTAL REQUIREMENTS FOR LIFE
- Where can we expect to find building blocks of life?
- Where can we expect to find energy for life?
- Does life need liquid water?
- What are the environmental requirements for habitability?

7.2 A BIOLOGICAL TOUR OF THE SOLAR SYSTEM: THE INNER SOLAR SYSTEM
- Does life seem plausible on the Moon or Mercury?
- Could life exist on Venus or Mars?

7.3 A BIOLOGICAL TOUR OF THE SOLAR SYSTEM: THE OUTER SOLAR SYSTEM
- What are the prospects for life on jovian planets?
- Could there be life on moons or other small bodies?

7.4 THE PROCESS OF SCIENCE IN ACTION: SPACECRAFT EXPLORATION OF THE SOLAR SYSTEM
- How do robotic spacecraft work?
- What are some key spacecraft missions for astrobiology?

... for what can more concern us than to know how this world which we inhabit is made; and whether there be any other worlds like it, which are also inhabited as this is?

Bernard Le Bovier De Fontenelle, *Conversations on the Plurality of Worlds,* 1686

Having studied Earth's habitability and life, we now turn to the search for life elsewhere in our solar system. Because there are many places to look—including other planets and their moons, and thousands of known asteroids and comets—we need a strategy to help focus our efforts on the worlds most likely to be habitable.

The first step in such a strategy is to determine where it makes sense to look, so we begin this chapter by discussing the environmental requirements that we expect to be necessary for life on any world. With those requirements in mind, we'll then take a biological tour of the solar system, seeking to determine where the requirements might be met. This will enable us to decide which worlds deserve the greatest attention, both in research and in our studies in this book.

Keep in mind that, for our solar system at least, we are looking primarily for microbes or other simple life. We have already learned enough to be confident that no other advanced civilization has ever arisen in our solar system. Still, the discovery of life of any kind would be profound, both to our understanding of biology and to philosophical considerations of our place in the universe. Even if we don't find life in our own solar system, the search itself will teach us much about the characteristics that can make a planet habitable and will thereby help us when we extend the search to other planetary systems.

7.1 ENVIRONMENTAL REQUIREMENTS FOR LIFE

There's no place like home—at least, not within our own solar system. No world besides Earth has an atmosphere that we could breathe or abundant surface water that we could drink. No other world has a combination of surface temperature and pressure under which we could survive outside without a space suit. Few worlds have atmospheres that offer any protection from dangerous ultraviolet radiation from the Sun or from high-energy particles from space. Indeed, without undertaking major engineering projects to build self-contained environments (or greatly altering the basic conditions of a planet through "terraforming" [Section 8.4]), we have no hope of long-term survival on any other world in our solar system.

However, when we discuss habitability, we generally mean an environment in which life of *some* kind might survive, not necessarily human life. This greatly broadens the possibilities. After all, we could not have survived even on our own planet for much of its history, yet life flourished just the same. Past and present life on Earth has managed to thrive in a far greater variety of environments than we ourselves can endure [Section 5.5]. If we are going to identify potentially habitable worlds in our solar system, we must specify the range of environments that we can consider acceptable for life. We've touched on some of these ideas in previous chapters. Here, we'll try to tie them all together into a clear list of environmental requirements for life.

• Where can we expect to find building blocks of life?

Perhaps the most obvious requirement for life is a set of chemical elements with which to make the components of cells. Life on Earth uses about 25 of the 92 naturally occurring chemical elements, although just four of these elements—oxygen, carbon, hydrogen, and nitrogen—make up about 96% of the mass of living organisms (see Figure 5.5). Thus, a first requirement for life might be the presence of most or all of the elements used by life.

This requirement probably can be met by almost any world. Recall that essentially all chemical elements besides hydrogen and helium were produced by stars [Section 3.2]. Although all of these "heavy elements" are quite rare compared to hydrogen and helium, they are found just about everywhere. Moreover, the elements oxygen, carbon, and nitrogen—arguably the most crucial elements for life—are the third-, fourth-, and sixth-most-abundant elements in the universe, respectively. The proportions of heavy elements varies: While they make up about 2% of the chemical content (by mass) of our solar system, they make up less than 0.1% of the mass in some very old star systems. Nevertheless, every star system we've studied has at least some amount of all the elements used by life.

The nature of solar system formation gives us additional reason to expect the elements of life to be common on other worlds. Recall that, according to the nebular theory of solar system formation, the planets were built when solid particles condensed from gas in the solar nebula, and these particles then accreted into planetesimals and ultimately into planets, moons, asteroids, and comets. The first step in this process—condensation—affects only the heavier elements or hydrogen compounds containing heavy elements, because pure hydrogen and helium always remains gaseous [Section 3.4]. As long as condensation and accretion can occur,* we expect the resulting worlds to contain the elements needed for life.

Note that this basic argument doesn't change even if we allow for life quite different from life on Earth. Life on Earth is carbon-based and, as we discussed in Chapter 5, we have good reason to think that life elsewhere would also be carbon-based. However, we can't absolutely rule out the possibility of life with another chemical basis. The set of elements (or their relative proportions) used by life based on some other element might be somewhat different from that used by carbon-based life on Earth. But the elements are still products of stars that should be found everywhere. No matter what kind of life we are looking for, we are likely to find the necessary elements on almost every planet, moon, asteroid, and comet in the universe.

---

*Observations demonstrate that condensation and accretion can occur in star systems like ours, in which the proportion of heavy elements is about 2% by mass. We do not yet know if these processes also occur in systems with much smaller abundance of heavy elements.

A somewhat stricter requirement is the presence of these elements in molecules that can be used as ready-made building blocks for life, just as the early Earth probably had at least moderate abundances of amino acids and other complex molecules [Section 6.2]. Recall that Earth's organic molecules likely came from some combination of three sources: chemical reactions in the atmosphere, chemical reactions near deep-sea vents in the oceans, and molecules brought to Earth from space. The first two sources can occur only on worlds with atmospheres or oceans, respectively. But the third source should have brought similar molecules to nearly all worlds in our solar system.

Studies of meteorites and comets suggest that organic molecules are widespread among both asteroids and comets. Because every world was pelted by asteroids and comets during its early history [Section 4.3], every world should have received at least some organic molecules; interplanetary dust may also have contained organic molecules that rained down on young worlds. However, organic molecules tend to be destroyed by solar radiation on surfaces unprotected by atmospheres. Moreover, while these molecules might stay intact beneath the surface—as they evidently do on asteroids and comets—they probably cannot react with each other unless some kind of liquid or gas is available to move them about. Thus, given that it makes sense to start our search with worlds on which organic molecules are likely to be involved in chemical reactions, we should concentrate on worlds that have either an atmosphere, or a surface or subsurface liquid medium such as water, or both.

• Where can we expect to find energy for life?

In addition to a source of molecular building blocks, life requires an energy source to fuel metabolism [Section 5.3]. Recall that life on Earth uses a wide variety of energy sources. Some organisms get energy directly from sunlight through photosynthesis. Others get energy by consuming organic molecules (for example, by eating photosynthetic organisms) or through chemical reactions with inorganic compounds of iron, sulfur, or hydrogen.

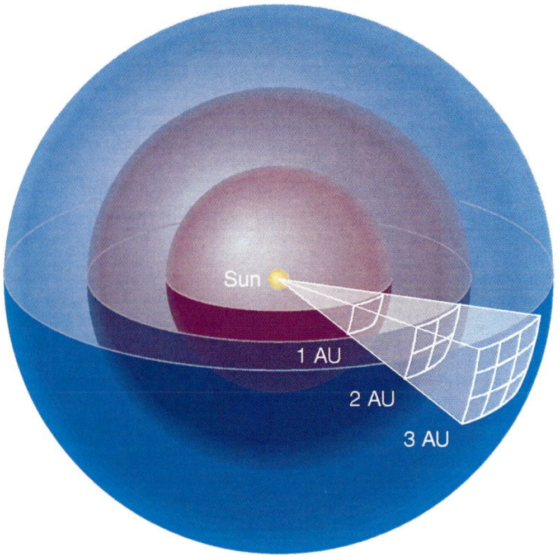

Figure 7.1

Any given amount of sunlight is spread over a larger area with increasing distance from the Sun. As shown in this diagram, the area over which the sunlight is spread increases with the square of the distance: At 2 AU the sunlight is spread over an area $2^2 = 4$ times as large as at 1 AU, and at 3 AU the sunlight is spread over an area = 9 times as large as at 1 AU. (Recall that 1 AU is the average Earth–Sun distance, or about 150 million kilometers.) Thus, the energy contained in sunlight (per unit area) *decreases* with the square of the distance from the Sun.

Sunlight is available everywhere in our solar system, though it becomes much weaker with increasing distance from the Sun. The energy available in sunlight decreases with the *square* of the distance from the Sun (Figure 7.1). For example, if we could put a leaf of a particular size on a world twice as far from the Sun as Earth, the leaf would receive only one-fourth as much energy as the same leaf on Earth (over the same period of time). At 10 times Earth's distance from the Sun—roughly the distance of Saturn—it would receive only $\frac{1}{10^2} = \frac{1}{100}$ of the energy it would receive on Earth. Photosynthetic life on such a world would have to be either much larger than life on Earth (giving it a larger surface area for collecting light), much more efficient at collecting solar energy, or much slower in its metabolism and reproduction. For the far outer solar system, sunlight almost certainly is too weak to support life.

Chemical energy sources also place constraints on life. Chemical reactions can occur under a wide variety of circumstances, but only if the potential reactants are brought into contact with each other. This means that the ongoing reactions needed to provide energy for life can occur only on worlds where materials are being continually mixed. On a practical level, this probably requires either an atmosphere to mix gases or a liquid medium to mix materials on or below a world's surface—the same requirements we found for obtaining the building blocks of life.

• Does life need liquid water?

In addition to a source of building blocks and a source of energy, there is one more ingredient that is essential to all life on Earth: liquid water. Recall that water plays at least three vital roles for life on Earth [Section 5.3]: It dissolves organic molecules, making them available for chemical reactions within cells; it allows for the transport of chemicals into and out of cells; and it is involved directly in many of the metabolic reactions that occur in cells. It is difficult to imagine life in the absence of a liquid substance to play these roles. But could these roles be fulfilled by some liquid other than water?

Potential Liquids for Life No one knows whether other liquids could support life in the absence of liquid water, but there are a number of constraints to consider. For example, a substance that might fulfill the roles of water must, like water, be fairly common. On Earth, the only liquid besides water commonly found is molten rock, which is so hot that it's difficult to imagine life surviving within it.* However, several other common substances might take liquid form on colder worlds.

Table 7.1 lists the temperature range over which water and three other potential candidates—ammonia, methane, and ethane—remain liquid. The given temperature ranges are those that apply under the atmospheric pressure on Earth. At different pressures, or with the presence of dissolved minerals, the ranges can be different. For example, salt water can remain liquid at temperatures slightly below 0°C—sea water freezes at about −2°C and water fully saturated with salt freezes at −21°C—and under sufficient pressure water can remain liquid at temperatures well above 100°C. (Near deep-sea vents, the pressure keeps water liquid at temperatures as high as about 375°C.) Despite such variation in melting and boiling temperatures, the ranges in Table 7.1 provide a useful comparison.

TABLE 7.1 *Potential Liquids for Life*

Freezing and boiling points (under 1 atmosphere of pressure) for common substances that may be found in liquid form in our solar system. The last column gives the width of the liquid range, found by subtracting the freezing point from the boiling point.

Substance	*Freezing Temperature*	*Boiling Temperature*	*Width of Liquid Range*
Water (H_2O)	0°C	100°C	100°C
Ammonia (NH_3)	−78°C	−33°C	45°C
Methane (CH_4)	−182°C	−164°C	18°C
Ethane (C_2H_6)	−183°C	−89°C	94°C

THINK ABOUT IT Oil (petroleum) is also found in liquid form on Earth. Why doesn't oil seem likely as a liquid medium for the origin of life? (*Hint:* Remember that oil is a fossil fuel.)

Advantages of Water Although we cannot rule out the other liquids in Table 7.1, water has at least three advantages that make it seem far more suitable as a liquid medium of life. First, as you can see in the table, water remains liquid over a wider and higher range of temperatures. (The liquid range for ethane is almost as wide, but at much lower temperatures.) A wider range makes it more likely that the substance can stay liquid through changes in the weather or climate. A higher temperature range facilitates chemical reactions. As a general rule, chemical reactions proceed more rapidly at higher temperatures because the molecules themselves move more rapidly. Typically, the rate of a given chemical reaction doubles with each 10°C increase in temperature. Thus, chemical reactions in water should generally proceed much more rapidly than similar reactions in liquid ammonia, methane, or ethane. Any life using these other liquids would therefore need to have a much slower metabolism

*Rock becomes molten only at temperatures above about 700°C. Given that the most extreme high-temperature extremophiles yet known can't survive in water heated above 130°C, it seems difficult to imagine life at 700°C.

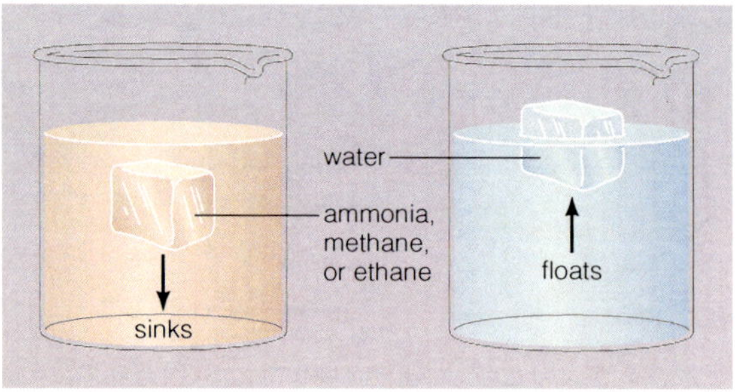

Figure 7.2

Most substances are denser as solids than as liquids, so when solid and liquid forms exist together, the solid form sinks. Water is an exception; solid water in the form of ice is less dense than liquid water, and so ice floats in liquid water.

than life on Earth. The slower rate of reactions may also make it less likely that life would arise in the first place in these other liquids, because the origin of life probably requires many complex chemical reactions.

The second advantage of liquid water involves an oddity in the way water freezes. Most substances are denser as solids than as liquids, but water is a rare exception (Figure 7.2): Ice is *less* dense than liquid water, which is why ice floats. This property helps life survive on Earth. In the winter, when surface temperatures are low enough for water to freeze, floating ice forms a layer on the tops of lakes and seas. This layer of ice insulates the water beneath, allowing it to remain liquid—which allows life to survive within it. No other liquid in Table 7.1 shares this property with water. That is, all the others sink when frozen.

The flotation of solid ice may be even more important to long-term climate stability. During periods when Earth cools a bit, such as the ice ages [Section 4.5], lower temperatures allow more ice to form on the surface. If this ice sank, the surface would still be covered with liquid water, which in turn would freeze and sink. This process would continue until no liquid water was left at the surface—that is, until lakes and oceans were completely frozen. Instead, because ice floats, a cool period thickens the insulating layer of ice on the surfaces of lakes and oceans, making it less likely that they will freeze completely.

The third advantage of liquid water over the other liquids in Table 7.1 comes from the way that electrical charge is distributed within water molecules (Figure 7.3). Within individual water molecules, the electrons tend to be distributed in a way that makes one side have a net positive charge and the other side have a net negative charge. (Molecules with such charge separation are called *polar* molecules; the term *polar* comes from the positive and negative charges being concentrated at opposite ends, or "poles," of the molecule's axis and not from anything related to temperature.) This charge separation affects the way in which water dissolves other substances. Molecules and salts that also have charge separations dissolve in water easily. Molecules that do not have any charge separation—such as molecules of oil—do not dissolve in water.

On Earth, the charge separation property of water is critical to life. Living cells have membranes that do not dissolve in water, so the membranes effectively protect the interior contents of cells. If we place living cells in liquid ethane, methane, or ammonia—molecules with less charge separation than water—their membranes may come apart. Charge separation also makes possible the formation of a special type of chemical bond, called a *hydrogen bond,* that is important to the biochemistry of life on Earth. (The formation of hydrogen bonds as water freezes also explains why ice is less dense, and hence why it floats, because these bonds force the molecules into a slightly expanded structure.) Thus, among the liquids listed in Table 7.1, Earth-like living cells can survive only in liquid water.

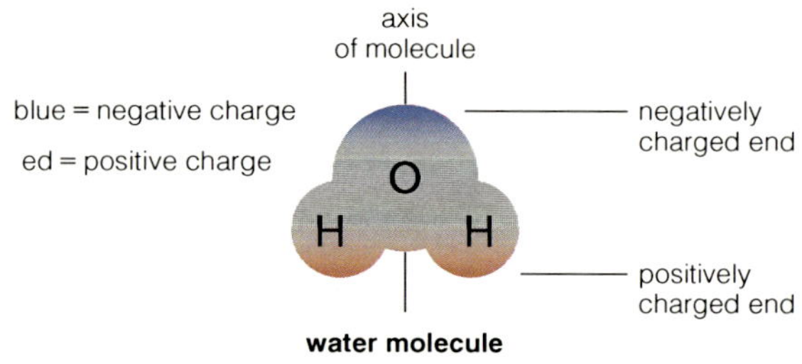

Figure 7.3

Within individual water molecules, the electrons tend to be distributed in a way that makes one side have a net positive charge and the other side have a net negative charge.

The Bottom Line on Water We've identified three advantages of water over other liquid candidates for life: (1) a wider and higher range of temperatures over which it remains liquid; (2) the fact that solid water floats; and (3) the fact that the charge separation of water molecules allows for types of chemical bonds that are not possible with the other liquids. These advantages make a strong case for the need for liquid water as a basis for life, but the case is not definitive. For example, it's possible to imagine circumstances under which the third advantage might be turned around. If life elsewhere had cell membranes made of molecules with different charge separation properties, they might dissolve in water but not in ammonia, methane, or ethane. We do not know if this is possible, but if it is, then life might exist in one of those other liquids while finding water to be poisonous or too hot.

The bottom line is that, while water clearly is an excellent medium for supporting life, we do not know if other liquids could also support life. It's difficult to tell whether the advantages we've cited for water are truly universal or are specific only to terrestrial life. Given this uncertainty, along with the fact that other liquids probably exist on (or within) some of the worlds in our solar system, we should not be too quick to rule out the potential for those worlds to support life. In general, however, water is much more abundant than other liquids both in our solar system and elsewhere (largely due to the fact that oxygen is the third-most-abundant element in the universe, after hydrogen and helium), so it remains our prime candidate for a liquid medium.

• What are the environmental requirements for habitability?

We began this section with the goal of making a list of environmental requirements for life that can help us decide which worlds to focus on as we begin the search for possible abodes of life in our solar system. In the broadest terms, we've found that the environment must satisfy three major requirements:

1. It must have a source of molecules from which to build living cells.
2. It must have a source of energy to fuel metabolism.
3. It must have a liquid medium—most likely liquid water—for transporting the molecules of life.

The first requirement is probably met by most if not all worlds. The second requirement is somewhat more limiting, but there are still plenty of worlds that should have sufficient sunlight or chemical energy for life. The third requirement—the need for a liquid—is probably the most stringent. Moreover, any world that meets this third requirement stands a good chance of meeting the first two as well. A liquid like water can facilitate chemical reactions with inorganic planetary materials, which offers at least a potential source of energy for life—and, indeed, a source tapped by many microbes on Earth. Thus, based on our current understanding of life, we can consolidate the requirements into a single "litmus test" for habitability: *A world can be habitable only if it has a liquid medium, probably meaning liquid water but possibly including one of the other liquids listed in Table 6.1.*

This requirement for a liquid certainly narrows the possibilities for life in our solar system, but not as much as you might at first guess. The wide variety of habitats in which we find both liquid water and life on Earth, including the deep ocean and in rocks buried deep underground, tells us that habitability requires only the presence of a liquid *somewhere,* not necessarily on the surface. As we'll discuss in the rest of this chapter, a large fraction of the worlds in our solar system have probably met this condition at least at some point in the past, and many still do.

There are at least two important caveats to our discussion. First, we may need to distinguish between the ability of a world to support an origin of life and its ability to support life that might have arrived from elsewhere. For example, while most scientists have concluded that life could not have arisen indigenously in Jupiter's atmosphere, some scientists have imagined organisms that could survive there if they somehow arrived from somewhere else (see Section 7.3). Second, we must remember that everything we know about life comes from the study of life on only a single world—our own. We've already noted the difficulty of knowing whether the advantages of water as a liquid medium are truly universal or specific to life on Earth. Other apparent attributes of life may similarly be less crucial elsewhere than they are on Earth. It is therefore possible that our discussions of habitability are based on too narrow a view of life, in which case our litmus test may be too narrow as well. Indeed, science fiction writers have imagined all sorts of bizarre life-forms existing under conditions far outside those we've examined here. Nature might be even more inventive. Nevertheless, a search for life must start somewhere, and it makes sense to begin by looking for the conditions that seem most likely to support life. If it turns out that our initial search is too narrow, we can always expand it in the future.

7.2 A BIOLOGICAL TOUR OF THE SOLAR SYSTEM: THE INNER SOLAR SYSTEM

Imagine living a century from now, when we have sent orbiters and landers to every major world in our solar system and when humans may even be living and working on other worlds. At that time, a "life in the universe" course might begin with a true biological tour of the solar system. One world at a time, we would discuss its habitability, the characteristics of its life or why it does not have its own indigenous life, and the prospects of survival (or the results of actual experiments) for any life from Earth that we might transport there.

We cannot undertake such a complete biological tour today. However, we have already learned enough to say which worlds seem most likely to be habitable—that is, to meet the environmental requirements for life—and which worlds seem least likely to be capable of harboring any indigenous life. This information will help guide the search for life in our solar system, because it will help us determine where to focus our efforts. And, as a practical matter, it will help us decide where to devote our attention in this book. In this section, we'll focus on our neighbors in the inner solar system: our Moon, and the terrestrial planets Mercury, Venus, and Mars.

• Does life seem plausible on the Moon or Mercury?

Although Mercury is a planet and the Moon orbits planet Earth, the two worlds share many characteristics. Both are pockmarked with craters (Figure 7.4). Both are much smaller than Earth (see Figure 4.1)—so small that by now they have lost most of their internal heat, leaving them with no ongoing volcanism and without significant tectonic activity. The lack of volcanism means no outgassing to release gases into an atmosphere, and their small sizes mean weak gravity that has long since allowed any past atmospheric gases to escape to space. That is why both worlds are essentially airless.* Mercury and the Moon also share the distinction of being among the places least likely to be habitable in the solar system, primarily because neither world is likely to have any liquids anywhere.

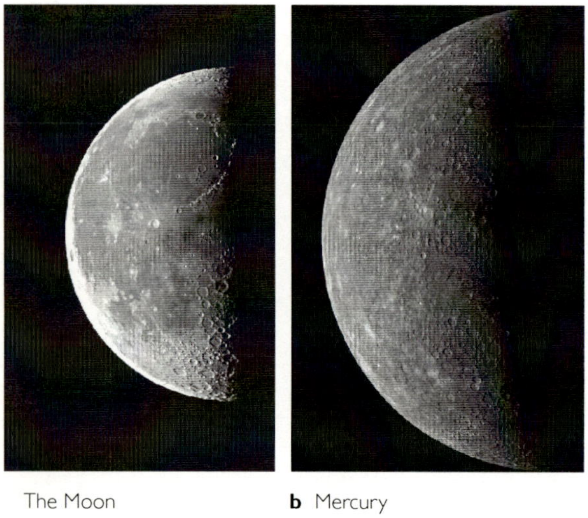

The Moon **b** Mercury

Figure 7.4

Similar views of the Moon and Mercury, shown to scale. See Figure 4.1 for a size comparison to Earth.

*Technically, the Moon and Mercury have extremely thin atmospheres created by diffusion of atoms and molecules from the interior and by release of gas molecules by impacts (mostly microscopic impacts of high-energy particles and photons), but these atmospheres are far thinner than even what we would consider to be an excellent vacuum in laboratories on Earth.

The Moon As we discussed in Section 4.6, the Moon has a far smaller proportion of water and other volatiles than Earth, probably as a result of vaporization that occurred in the giant impact thought to have formed the Moon. The only place the Moon may have any significant amount of water is in craters near its poles. The bottoms of these polar craters are perpetually shielded from sunlight, making them extremely cold and dark. As a result, ices deposited over the eons by impacts (or possibly by outgassing from the lunar interior early in the Moon's history) may lie frozen at the crater bottoms (Figure 7.5). However, even if such ices are present in lunar craters, they exist only because the crater bottoms have never warmed enough to melt them into liquid form. Thus, ice in lunar craters probably isn't relevant to the Moon's habitability, though it may be useful to *us* as a source of water for future colonies on the Moon.

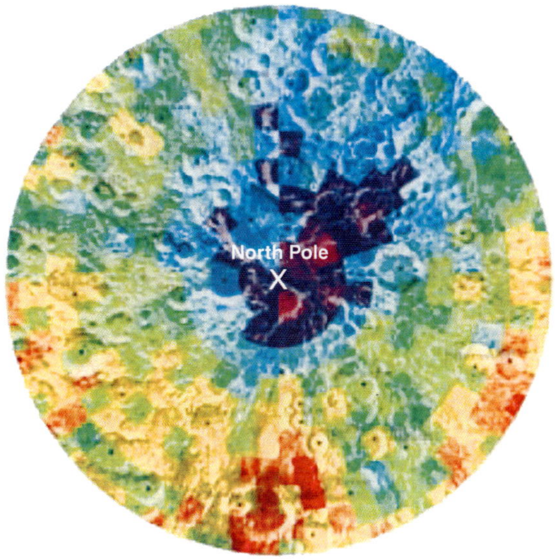

Figure 7.5

This color-coded map shows evidence for water ice in craters near the Moon's north pole. The portion of the Moon shown includes all latitudes north of 70°N, and the colors represent the concentration of hydrogen atoms measured by NASA's Lunar Prospector mission in the late 1990s. (The measurements were actually made by neutron detection techniques.) The dark blue and purple regions have the highest hydrogen concentrations, which are probably due to the presence of water ice on or just below (within a half-meter of) the surface. Note that these regions are very near the pole, where some crater bottoms are permanently shielded from the Sun. Similar results were found for the Moon's south pole.

Mercury We are less certain about the presence of water on Mercury. Because it formed so close to the Sun, Mercury must have accreted primarily from planetesimals that lacked any ices. But, as was the case with Earth, a few planetesimals from farther out in the solar system may have brought water and other ices to Mercury during its formation. If so, some of these materials may be chemically bound in subsurface rock on Mercury. However, Mercury's high average density suggests that, like Earth, it was rocked by a giant impact some time shortly after it underwent differentiation. The giant impact is thought to have ejected much of Mercury's original mantle into space, leaving behind a planet consisting predominantly of an iron-rich core. If it really occured, this impact likely would have ejected any water that accreted during the planet's formation.

Even if Mercury has water in its surface rock, it's unlikely that any of this water could ever take liquid form, because Mercury's surface temperatures are nearly always far above the boiling point or far below the freezing point. The combination of a 58.6-day rotation period and an 87.9-day orbital period gives Mercury days and nights that last about three Earth months each. As the closest planet to the Sun, the dayside temperatures get quite hot—about 425°C. But with no atmosphere to retain heat, temperatures plummet to a frigid −175°C during the long nights. Some observational evidence suggests that Mercury, like the Moon, may have ices brought by impacts in crater bottoms near its poles. (We should learn more from the MESSENGER mission to Mercury; see Section 7.4.) As is the case for the Moon, some of Mercury's polar craters have bottoms that probably have not been exposed to sunlight for billions of years, and hence where temperatures are perpetually far below freezing.

The lack of liquid water may make the Moon and Mercury lost causes when it comes to hope for finding life, but they are still important to our general understanding of planetary systems. In particular, studies of the Moon and Mercury teach us a great deal about the origin and history of our solar system and help us learn why some worlds are habitable and others are not. Thus, while the Moon and Mercury are almost certainly lifeless, they are still important targets of study for planetary scientists.

• Could life exist on Venus or Mars?

Prospects for past or present life look far better when we turn to our nearest planetary neighbors, Venus and Mars. Venus has been called our "sister planet," because it is the nearest planet to Earth in distance and is nearly identical to Earth in size. Mars is considerably smaller than Earth (see Figure 4.1), but spacecraft photos of its surface reveal an eerily Earth-like

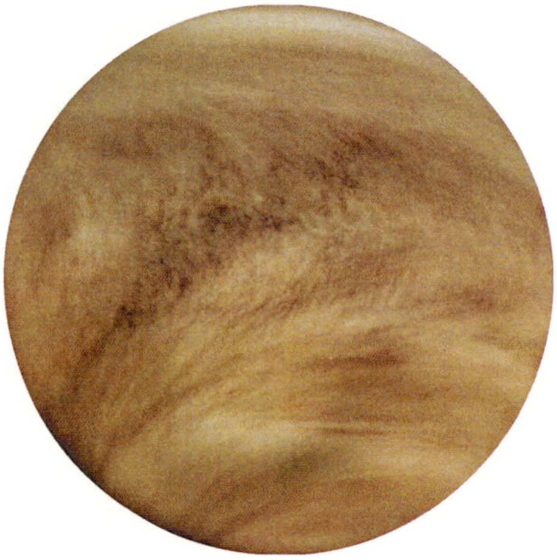

Figure 7.6

Clouds are all that can be seen in this ultraviolet image of Venus from the *Pioneer Venus Orbiter;* no surface features can be seen at all.

landscape (see Figure 1.5). Both planets have at least some potential for supporting past and present life, although Mars looks far more promising.

Venus Venus is completely enshrouded in thick clouds (Figure 7.6), and past generations of scientists could only guess about the surface conditions. Some imagined a pleasantly warm planet. Although Venus is closer to the Sun, it is not *that* much closer—about two thirds as far from the Sun as is Earth. Thus, if Venus had an Earth-like atmosphere, we might expect it to be only a little hotter than Earth. In fact, calculations show that with an Earth-like atmosphere, Venus's global average temperature would be about 35°C (95°F). This type of thinking led past generations of science fiction writers to speculate that Venus might be a lush, tropical paradise.

THINK ABOUT IT A good way to think about the expected temperature difference between Venus and Earth is to consider the 1-to-10-billion scale model of our solar system from Chapter 3. Recall that at this scale, the Sun is about the size of a grapefruit, and Venus and Earth are located about 10 and 15 meters away, respectively. Imagine that you have a grapefruit that is extremely hot. If the heat feels comfortable at a distance of 15 meters, how would it feel at 10 meters? Discuss the difference qualitatively, and relate it to the expected temperature difference between Venus and Earth.

Such dreams evaporated in 1962, when the *Mariner 2* spacecraft flew past Venus and discovered that its surface is scorching hot. (Some earlier radio observations from Earth had hinted at the high temperature, but other explanations for the data seemed possible prior to the spacecraft observations.) The surface temperature is an incredible 470°C (about 880°F)—easily hot enough to melt lead. Moreover, this temperature persists planetwide, both day and night. All the while, a thick atmosphere bears down on the surface with a pressure 90 times that on Earth's surface—equivalent to the pressure at a depth of nearly 1 kilometer (0.6 mile) in the oceans. Besides Venus's crushing pressure and searing temperature, the atmosphere of Venus contains sulfuric acid and other chemicals that are toxic to us. Indeed, landers sent to Venus by the Soviet Union provided data for only a brief period before being disabled by the high temperatures (Figure 7.7). Far from being a beautiful sister planet to Earth, Venus resembles a traditional view of hell.

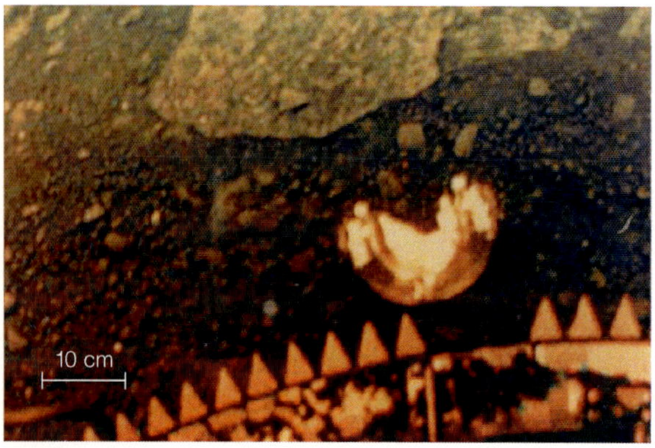

Figure 7.7

This photograph shows a small patch of the surface of enus; part of the lander that took the picture is in the foreground. The Soviet Union sent several landers to enus during the 1970s and early 1980s. Because of the high temperatures and corrosive atmosphere, none of them functioned for more than about 2 hours after landing.

What causes such extreme conditions on Venus? The answer is the *greenhouse effect*—the same effect that makes our own planet so comfortable [Section 4.5], but operating much more strongly. Recall that, in the absence of the greenhouse effect, Earth would be frozen over. Our planet is habitable because a moderate greenhouse effect traps enough heat to raise temperatures above the freezing point of water. The greenhouse effect operates in precisely the same way on Venus, but it raises temperatures much more.

The strong greenhouse effect on Venus (often called a *runaway greenhouse effect,* for reasons we will discuss in Chapter 10) comes primarily from carbon dioxide in its atmosphere. Earth has a modest greenhouse effect because carbon dioxide makes up only a fraction of 1% of our atmosphere. In contrast, carbon dioxide makes up more than 96% of Venus's far thicker atmosphere. Interestingly, we can explain this difference in atmospheric carbon dioxide by contrasting the fate of carbon dioxide gas on each planet. Both planets have outgassed similar total amounts of carbon dioxide over the course of their histories. However, while almost all of the carbon dioxide outgassed on Earth is now locked up in carbonate rocks or dissolved in the oceans (170,000 times as much as is in our atmosphere [Section 4.5]), all of Venus's outgassed carbon dioxide remains in its atmosphere. Thus, Venus's high temperature results from the extremely strong greenhouse effect produced by its atmospheric carbon dioxide, and the high pressure results from the sheer amount of this gas.*

The high surface temperature all but rules out the possibility of life on the surface of Venus today. It is far too hot for liquid water, let alone liquid ammonia, methane, or ethane. However, the surface may have been habitable in the past. Recall that carbon dioxide gets locked up in Earth's carbonate rocks through the mechanism of the carbon dioxide cycle (see Figure 4.27). This cycle depends on both plate tectonics and the presence of oceans: Carbon dioxide dissolves in the oceans, where it reacts chemically to form carbonate minerals. Venus has no similar cycle because it lacks oceans (and apparently lacks plate tectonics as well) and therefore has no mechanism for removing carbon dioxide from the atmosphere. But the situation may have been quite different in the distant past. As we'll discuss further in Chapter 10, the fact that the Sun should have been much dimmer earlier in its history may have allowed Venus to have oceans prior to about 4 billion years ago. In fact, if you consider "Earth-like" to mean conditions found on Earth today, it's conceivable that ancient Venus may have been more "Earth-like" than Earth itself was at that time.

If Venus once had oceans—a big "if"—it's possible that life arose at that time. And if so, there's one place on Venus where microbes might still survive: in the clouds. At altitudes of about 50 kilometers above the surface, the greenhouse effect is far weaker and droplets containing liquid water can and do exist. The clouds are extremely acidic, but their sulfur content could allow chemical reactions that might provide sufficient energy for extremophiles adapted to survive in this environment. The European Space Agency's *Venus Express* mission, which arrived in Venus orbit in April 2006, may be able to search for evidence of such life, if it exists.

If life does not exist in the Venusian clouds, we may never know whether Venus had life in the distant past. Crater counts suggest that Venus's entire surface is less than about a billion years old, meaning that volcanism or tectonic processes have reshaped or paved over rocks from earlier times. Thus, if life arose and survived until the runaway greenhouse effect made Venus uninhabitable, any fossil evidence would almost certainly have been destroyed long ago.

Mars While Venus overheated early on, Mars went in the other direction. There's little doubt that Mars once had flowing water with a thicker and warmer atmosphere, but some 3 billion years ago, climate change caused the planet in effect to freeze over. Any remaining surface water froze, and Mars has since lost so much atmospheric gas that the surface pressure is too low for liquid water. Nevertheless, the evidence of past water makes Mars a prime candidate for past life. Moreover, Mars retains enough internal heat that liquid water may be possible underground, in which case life might still survive there.

---

*Calculations show that carbon dioxide alone accounts for most but not all of Venus's high temperature. Other greenhouse gases—notably sulfur dioxide (SO_2), carbon monoxide (CO), and hydrochloric acid (HCl)—also make significant contributions.

Indeed, Mars seems such a good candidate for past or present life that, if it turns out never to have had it, we might have to revisit our assumptions about the likelihood of life arising under the "right" conditions. Because Mars is such a good candidate for habitability and life, we'll defer discussion of it for now, and devote all of Chapter 8 to its possibilities.

7.3 A BIOLOGICAL TOUR OF THE SOLAR SYSTEM: THE OUTER SOLAR SYSTEM

Beyond Mars, you might expect that the weakening strength of sunlight would make life increasingly less likely. But there are still plenty of worlds with internal heat that could keep some water liquid and provide energy for life. In this section, we'll continue our biological tour in the outer solar system, starting with the jovian planets, then examining a few of their large moons, and finally considering the many small bodies of our solar system, including Pluto and other Kuiper belt objects, comets, and asteroids.

• What are the prospects for life on jovian planets?

The four jovian planets—Jupiter, Saturn, Uranus, and Neptune—are very different from the terrestrial worlds. They are far more massive, lower in density, and composed largely of hydrogen, helium, and hydrogen compounds like water, methane, and ammonia. Figure 7.8 shows the bulk properties of the jovian planets, with Earth shown for scale.

The jovian planets are also different from the terrestrial planets in their interior structures, which scientists have deduced from observa-

MOVIE MADNESS: *2001—A SPACE ODYSSEY*

s the year 2001, and five clean-cut astronauts and a conniving computer named HAL are on their way to Jupiter. Yes, we know, this didn't really happen, and 2001 is now ancient history. But this Stanley Kubrick epic is a cinema classic, with music and scenes that remain a part of our popular culture, and special effects that were ahead of their time when the movie came out in 1968. If you haven't seen it, you should.

So why are astronauts and a wayward computer going to this behemoth world? Is it to study Jupiter's complex, churning atmosphere? Umm, no, that's not it. Are they on a mission to examine the giant planet's imposing magnetic field? Er, negative on that. Perhaps they're searching for signs of simple life on the moons Europa, Callisto, and Ganymede, all of which could have unseen oceans beneath their crusty exteriors? Actually, no, that's not it either.

No, these jovian rocket jockeys are headed to the king of the planets because some unknown race of aliens has been messing with our solar system. Four million years ago, some kindly extraterrestrials took a look at Earth and decided our planet's simian population had potential. They planted a dark gray monolith that, when touched by the local apes, converted them from a bunch of howling half-wits into tool-using primates.

This is an amazing tale (and one that, if it were true, would confound all paleontologists), but the aliens weren't content merely to have introduced intelligence on Earth. They also buried a second monolith on the Moon, figuring that a brainy species might eventually develop enough technology to leave its planet and find this weird artifact. The lunar monolith, in turn, directed us to Jupiter.

So that's why the interplanetary spacecraft *Discovery One* is on its way to the biggest world of the solar system. The hope, it seems, is that by journeying to this massive planet, *Homo sapiens* can finally learn whatever profound secrets these altruistic aliens wish to share.

2001 is as much a space oddity as a space odyssey. Forget the fact that the message found at Jupiter is mostly a puzzling psychedelic light show. Ignore the fact that the ship's smooth-talking, onboard computer is more scheming than Machiavelli. No, in the end the most distressing thing about *2001* (other than the dismaying truth that in 2001 the real space program wasn't even close to sending manned missions to Jupiter) is that our past and our destiny are simply some extraterrestrials' science-fair experiment. We're no more than a bunch of choreographed puppets.

Someday it's quite likely that we will travel to many other worlds of the solar system, not just in our mind's eye or in the virtual worlds of our theaters and planetaria, but in honest-to-goodness spacecraft. However, we will make these voyages of discovery based on our own curiosity and quest for knowledge—and not because some control-freak aliens pointed to the gas-giant worlds of the outer solar system with their monolith fingers and said, "Go there."

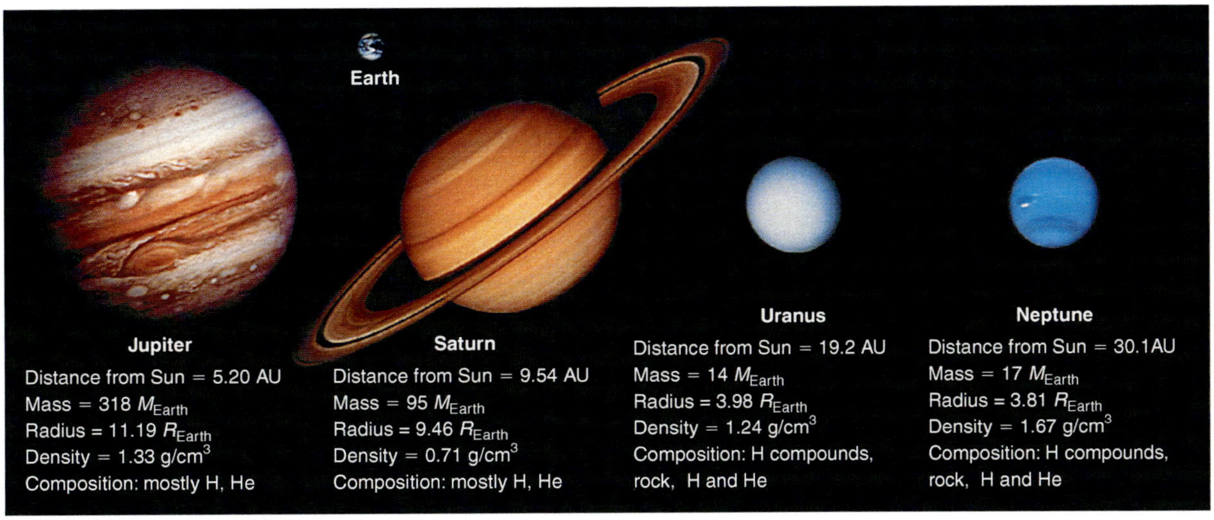

Figure 7.8

The four jovian planets of our solar system, shown to scale. Note that the tables list distance, mass, and radius in terms of Earth units; that is, M_{Earth} is Earth's mass and R_{Earth} is Earth's radius; 1 AU is Earth's distance from the Sun.

tions and theoretical models of how their materials behave, given the planetary masses and sizes. These planets lack anything resembling Earth's solid surface. Instead, as shown in Figure 7.9, their outer layers contain visible clouds surrounding extended layers of gaseous hydrogen, mixed with helium and hydrogen compounds. Deeper in their interiors, the high pressure compresses the hydrogen into liquid form, and within Jupiter and Saturn the pressure becomes so high that the hydrogen takes on a metallic form that was only recently synthesized in laboratories on Earth. Near their centers, each has a core of rock, metal, and hydrogen compounds, but the pressure is so great that these materials would be in a phase different from anything we ever see on Earth. If you plunged into any one of these worlds, you would just continue downward until you were crushed by the increasing pressure. Ultimately, your remains would sink into a hot sea of liquid hydrogen.

Because these planets are so far from the Sun, temperatures in their upper atmospheres are extremely cold. However, observations show that all must be quite hot in their deep interiors. Because these worlds lack solid surfaces like those of the terrestrial worlds, this heat cannot fuel any geological activity. However, the heat ensures that at some altitudes their atmospheres are warm enough for liquid water, and chemical reactions—some powered by frequent lightning that has been observed in

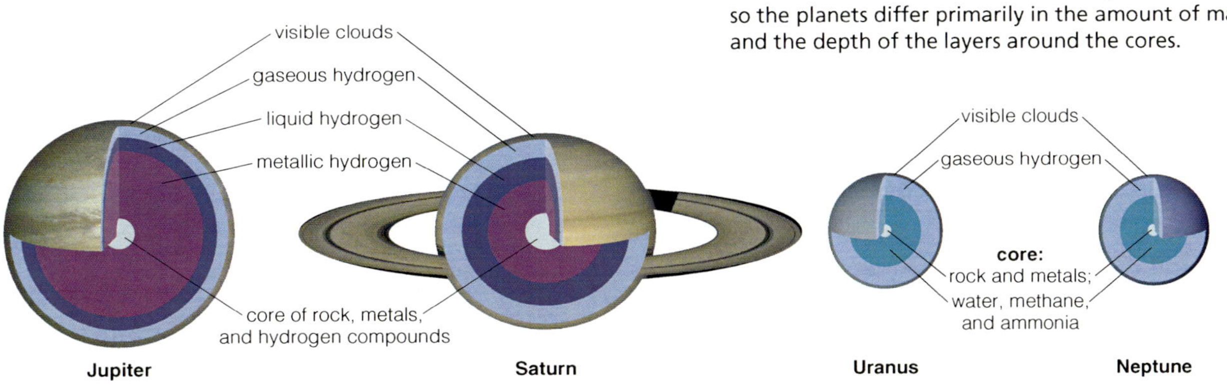

Figure 7.9

These diagrams compare the interior structures of the jovian planets, shown approximately to scale. All four have cores of rock, metal, and hydrogen compounds; the cores of Uranus and Neptune are differentiated into separate layers of rock/metal and hydrogen compounds. All four cores have about the same mass (ten Earth masses), so the planets differ primarily in the amount of material and the depth of the layers around the cores.

jovian planet atmospheres—could conceivably provide energy for life. So is it reasonable to imagine life in the jovian atmospheres?

Jupiter and Saturn Let's start by considering Jupiter, the largest and nearest of the four jovian planets. As Figure 7.9 shows, Saturn is so similar in its interior structure that the same considerations should apply to it as well.

Jupiter's temperature is far below freezing at its cloud tops, but the temperature rises rapidly with depth. Indeed, Jupiter has several cloud layers, each formed as different types of gases condense. Clouds containing droplets of liquid water can form at a depth of a little over 100 kilometers into Jupiter's atmosphere, which is just a little over 1% of the way from the highest clouds to the center. Given the comfortable temperature and the presence of liquid water, we might therefore wonder if Jupiter could be habitable and host life at this depth in its atmosphere.

Unfortunately, Jupiter's atmosphere appears to present a fatal problem when we consider the possibility of life originating there: Jupiter has strong, vertical winds with speeds that would make a hurricane seem like a gentle breeze in comparison. Any complex organic molecules that might form would quickly be carried to depths at which the heat would destroy them, making it difficult to see how any life could arise. We might imagine microbes reaching Jupiter on meteorites from elsewhere, but again the vertical winds make their survival seem impossible. Such microbes would be thrown quickly onto a nonstop elevator ride between cloud layers that are unbearably cold and others that are insufferably hot.

The only way that anyone has imagined life surviving in Jupiter's atmosphere is by supposing that it might have some sort of buoyancy that allowed it to stay at the right altitude while the vertical winds rushed by it. However, such buoyancy would require large gas-filled sacs, making the organisms themselves enormous. Given that we cannot envision a way for microbes to survive, there seems to be no way for large, buoyant organisms to evolve in the first place. They might survive if they arrived on Jupiter from elsewhere, but we know of no way that such large organisms could survive a journey through space, even if they existed elsewhere. As a result, most scientists do not consider Jupiter to be habitable. For essentially the same reasons, Saturn is also unlikely to be habitable.

Uranus and Neptune Uranus and Neptune are probably just as unlikely as Jupiter and Saturn as candidates for habitability. Their atmospheres are much colder than those of Jupiter and Saturn, due mainly to their greater distance from the Sun. If they have clouds in which liquid water droplets can form, we have not yet been able to detect them. Even if they do have such clouds deep in their atmospheres, vertical winds similar (though slower) to those on Jupiter and Saturn would also make life unlikely.

Uranus and Neptune have at least one possible zone of habitability that Jupiter and Saturn lack: their outer cores of water, methane, and ammonia. Theoretical models suggest that these materials may be in liquid form, making for very odd "oceans" in the deep interiors of Uranus and Neptune. The high pressures, strange mix of liquids, and lack of any obvious way to extract energy from these "oceans" makes life seem highly unlikely, but we cannot rule it out. However, at present we lack any technology that could allow us to search for life in the cores of these giant planets; even if life might exist there, the search for it will have to await new technological developments.

• Could there be life on moons or other small bodies?

Each of the four jovian planets reigns over a set of moons that make their systems look much like solar systems in miniature. Together, the jovian planets have at least 150 moons, and additional small moons undoubtedly still await discovery. The outer solar system also contains a myriad of other small bodies [Section 3.3], starting with the asteroids of the asteroid belt, moving out to the Kuiper belt with its "dwarf planets" like Pluto and Eris as well as many smaller comets, and on to the Oort cloud that may contain as many as a trillion icy comets. Could any of these myriad worlds be habitable?

Large Moons Perhaps surprisingly, the answer is yes, at least for a few of the largest moons, such as Jupiter's moons Europa, Ganymede, and Callisto, and Saturn's moon Titan. This answer is surprising because even these large moons are fairly small by planetary standards. None are as large as Mars, and if they were similar to the terrestrial planets in composition and in their heat sources, they would be too small to have retained internal heat that could support enough geological energy to fuel life. However, the moons of the jovian planets differ from the terrestrial planets in an important way: They tend to contain a great deal of ice, because they formed in parts of the solar system where it was cool enough for ices to condense from the solar nebula (see Figure 3.22). Because ices melt at

COSMIC CALCULATIONS 7.1

Newton's Version of Kepler's Third Law

How do we know the masses of distant objects? In many cases, we can use a modified version of Kepler's third law ($p^2 = a^3$). Recall that this law applies only to objects orbiting the Sun (see Cosmic Calculations 2.1). However, Newton found that Kepler's original law was just a specific case of a more general law, usually called *Newton's version of Kepler's third law*:

$$p^2 = \frac{4\pi^2}{G(M_1 + M_2)}a^3$$

where M_1 and M_2 are the masses of the two objects, p is their orbital period, and a is the distance between their centers. The term $4\pi^2$ is simply a number ($4\pi^2 \approx 4 \times 3.14^2 \approx 39.44$), and $G = 6.67 \times 10^{-11} \frac{\text{m}^3}{\text{kg} \times \text{s}^2}$ is the gravitational constant.

This law gives us the power to measure the masses of distant objects. Any time we measure an orbiting object's period (p) and orbital distance (a), Newton's equation allows us to calculate the sum $M_1 + M_2$ of the two objects involved in the orbit. If one object is much more massive than the other, we essentially learn the mass of the massive object, as the following example shows.

Example: Use the fact that Earth orbits the Sun in 1 year at an average distance of 150 million kilometers (1 AU) to calculate the mass of the Sun.

Solution: Newton's version of Kepler's third law becomes:

$$p_{\text{Earth}}^2 = \frac{4\pi^2}{G(M_{\text{Sun}} + M_{\text{Earth}})}a_{\text{Earth}}^3$$

Because the Sun is much more massive than Earth, the sum of their masses is nearly the mass of the Sun alone: $M_{\text{Sun}} + M_{\text{Earth}} \approx M_{\text{Sun}}$. Using this approximation, we find:

$$p_{\text{Earth}}^2 \approx \frac{4\pi^2}{GM_{\text{Sun}}}a_{\text{Earth}}^3$$

To find an expression for the mass of the Sun we multiply both sides by M_{Sun} and divide both sides by $(p_{\text{Earth}})^2$:

$$M_{\text{Sun}} \approx \frac{4\pi^2 a_{\text{Earth}}^3}{Gp_{\text{Earth}}^2}$$

Because G is given above with units of seconds and meters, we must use the same units for Earth's orbital period ($p_{\text{Earth}} = 1 \text{ year} \approx 3.15 \times 10^7$ seconds) and average orbital distance ($a_{\text{Earth}} = 1 \text{ year} \approx 1.5 \times 10^{11}$ m). We find:

$$M_{\text{Sun}} \approx \frac{4\pi^2(1.5 \times 10^{11}\ \text{m})^3}{\left(6.67 \times 10^{-11} \frac{\text{m}^3}{\text{kg} \times \text{s}^2}\right)(3.15 \times 10^7\ \text{s})^2} = 2.0 \times 10^{30}\ \text{kg}$$

Simply by knowing Earth's orbital period and distance from the Sun and the gravitational constant G, we have used Newton's version of Kepler's third law to find that the Sun's mass is about 2×10^{30} kilograms.

much lower temperatures than do metal or rock, these moons can have internally driven "ice geology"—such as internal convection involving gradual movement of icy material—with much less internal heat than is needed for the "rock geology" of the terrestrial planets.

In addition, at least some of the jovian moons—most notably Jupiter's moons Io and Europa—have an ongoing source of internal heat quite different from any heat source on the terrestrial planets. (We'll discuss this source, *tidal heating,* in Chapter 9.) The available heat can in principle melt ice, allowing the possibility of liquid water in their interiors. In at least one case, Europa, current evidence strongly suggests a deep, subsurface ocean of liquid water. Numerous other moderate- to large-size moons in the outer solar system also show evidence of past or present geological activity that could have allowed (or still allow) for some liquid medium.

The prospects for habitability are so good for some of these large moons that we'll defer their discussion, in order to give them our full attention in Chapter 9. The prospects are much poorer for the more numerous but much smaller moons, so we can consider them along with other small bodies of the solar system.

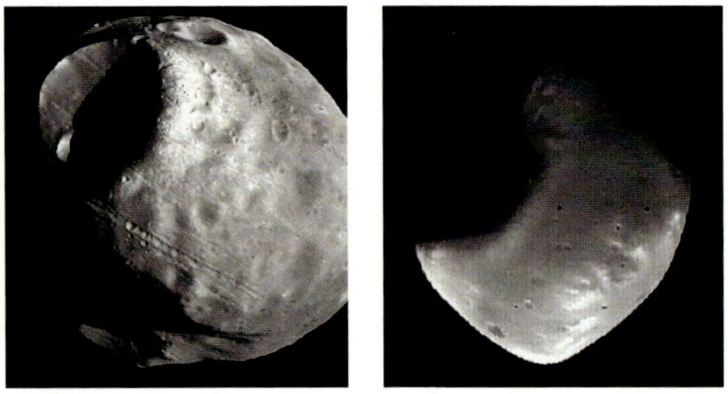

Phobos **b** Deimos

Figure 7.10

Mars has two small moons, Phobos (13 kilometers across) and Deimos (8 kilometers across), that are probably captured asteroids and much like many of the small moons of the outer solar system. Their small sizes make them unlikely to have any liquid water or life.

Small Bodies Our brief biological tour now brings us to the numerous smaller bodies in our solar system: small and medium-size moons, asteroids, comets, and the large Kuiper belt objects [Section 3.3]. We'll also consider in this group the two small moons of Mars, Phobos and Deimos (Figure 7.10), since they probably were once asteroids orbiting the Sun independently. We suspect they were captured by Mars early in its history, when it had an extended atmosphere that created the friction necessary to slow the passing asteroids enough for them to end up as moons.

The small sizes of all these objects probably rule out present life, because they almost certainly lack any liquid medium today. These bodies are too small to have any leftover internal heat that might melt any ices contained within them. Moreover, the comets and Kuiper belt objects generally orbit so far from the Sun that they are in a perpetual state of deep freeze. The lone exceptions are when impacts occur, which might melt small amounts of ice into water for short periods of time, and those rare comets that have had their original orbits perturbed enough to send them plunging into the inner solar system. As these comets come sunward, the same processes that cause their ices to vaporize into gas could conceivably melt small amounts of ice into liquid water. However, given that such periods would last only a short time on each orbit, and that such comets spend only small amounts of their multibillion-year histories on orbits that take them close to the Sun, present life does not seem likely.

The prospects for life on small bodies may have been better in the past, though it still seems unlikely. Recall that we have substantial evidence of complex organic molecules in both asteroids and comets, and studies of meteorites show that many of them must have contained liquid water during the earliest history of the solar system. The liquid water may have persisted over time periods as long as a few tens of millions of years. Could life have arisen during these time periods? We cannot rule it out, but we have found no evidence of past life in meteorites from the asteroid belt, and most scientists consider it unlikely that life could have originated on any of the countless small bodies in the solar system.

Tour Recap We have completed our brief biological tour of the solar system. Although every world, large and small, has the raw chemical elements needed for life, the possibilities for life are much more limited when we focus on the environmental requirements we have found from the study of life on Earth. In particular, the need for liquid water, or possibly some other liquid medium, means we can effectively rule out finding life on the numerous small worlds of our solar system. We have similarly ruled out life on Mercury and the Moon, and found life to be unlikely on the jovian planets. That leaves the slim chances for life in the atmosphere of Venus, and the much better chances for life on Mars and a few of the moons of the jovian planets. We will discuss these cases further in the coming chapters.

It's important to remember that just because a world is an unlikely candidate for harboring life does not mean it's uninteresting or unimportant to our understanding of life in the universe. In particular, every world has at least something to teach us about how planetary systems form and why some worlds are habitable and others are not. This in turn helps us understand the prospects for life not only in our own solar system, but among other star systems as well. We still have much to learn about these issues, which is why every world in our solar system will remain a target of scientific interest for decades or even centuries to come.

7.4 THE PROCESS OF SCIENCE IN ACTION: SPACECRAFT EXPLORATION OF THE SOLAR SYSTEM

We have discussed a lot of details about the different worlds in our solar system during our biological tour, and we'll cover even more details in the next three chapters. How have we learned so much about all these objects?

Much of our knowledge comes from telescopic observations, using both ground-based telescopes and telescopes in Earth orbit such as the Hubble Space Telescope. In one case—our Moon—we have learned a lot by sending astronauts to explore the terrain and bring back rocks for laboratory study. In a few other cases, we have samples of distant worlds that have come to us as meteorites; we also have samples of comet dust returned by the *Stardust* mission. But most of the data fueling the recent revolution in our understanding of the solar system have come from robotic spacecraft. To date, we have sent robotic spacecraft to all of the eight official planets as well as to many moons, asteroids, and comets; the spacecraft called *New Horizons* is currently en route to Pluto. For this chapter's case study in the process of science in action, we'll briefly explore how these robotic spacecraft work.

• How do robotic spacecraft work?

The spacecraft we send to explore the planets are robots suited for long space journeys and jam-packed with specialized equipment for scientific study (Figure 7.11). All spacecraft have power sources such as solar cells, propulsion systems, devices to point cameras and other instruments precisely at their targets, and computers that control their major components. Robotic spacecraft operate primarily with preprogrammed instructions. They carry radios for communication, which allow them to receive additional instructions from Earth and to send home the data they collect. Most robotic spacecraft make one-way trips from Earth, never physically returning but sending their data back from space in the same way we send radio and television signals around the world.

Figure 7.11

The *Cassini* spacecraft before launch. It is now nearly a billion miles away as it orbits Saturn. Notice major components, such as rocket thrusters at the bottom, the communications dish at the top, and various scientific instruments arrayed all around the main skeleton of the spacecraft. The *Huygens* probe (which landed on Titan) was not yet attached when this photo was taken. (For details on what you are seeing, go to the *Cassini* Web site.)

Broadly speaking, the robotic missions we send to explore other worlds fall into four major categories:

- **Flyby:** A spacecraft on a flyby goes past a world just once and then continues on its way.
- **Orbiter:** An orbiter is a spacecraft that orbits the world it is studying, allowing longer-term observation during its repeated orbits.
- **Lander** or **probe:** These spacecraft are designed to land on a planet's surface or probe a planet's atmosphere by flying through it. Some landers have carried rovers to explore wider regions.
- **Sample return mission:** A sample return mission requires a spacecraft designed to return to Earth carrying a sample of the world it has studied.

The choice of spacecraft type depends on both scientific objectives and cost. In general, a flyby is the lowest-cost way to visit another planet, and some flybys gain more "bang for the buck" by visiting multiple planets. For example, *Voyager 2* flew past Jupiter, Saturn, Uranus, and Neptune before continuing on its way out of our solar system (Figure 7.12).

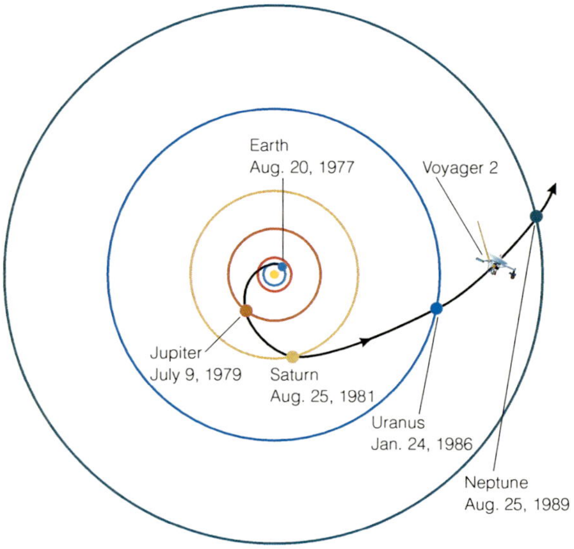

Figure 7.12

The trajectory of *Voyager 2,* which made flybys of each of the four jovian planets in our solar system.

Flybys Flybys tend to be cheaper than other missions because they are generally less expensive to launch into space. Launch costs depend largely on weight, and onboard fuel is a significant part of the weight of a spacecraft heading to another planet. Once a spacecraft is on its way, the lack of friction or air drag in space means that it can maintain its orbital trajectory through the solar system without using any fuel at all. Fuel is needed only when the spacecraft needs to change from one trajectory (orbit) to another. Moreover, with careful planning, some trajectory changes can be made by taking advantage of the gravity of the planets.

For example, if you look closely at the *Voyager 2* trajectory in Figure 7.12, you'll see that the spacecraft made significant changes in path as it passed by Jupiter and Saturn. In effect, it made these changes for free by using gravity to bend its trajectory rather than by burning fuel. (This technique is known as a "gravitational assist," and it essentially speeds up the spacecraft significantly while slowing the planet by an unnoticeable amount.) The boost in speed can be quite significant: During its February 2007 flyby of Jupiter, the speed of the *New Horizons* spacecraft should increase by about 20%, shaving more than 3 years off the time it would otherwise take to reach Pluto.

THINK ABOUT IT Study the *Voyager 2* trajectory in Figure 7.12. Given that Saturn orbits the Sun every 29 years, Uranus orbits the Sun every 84 years, and Neptune orbits the Sun every 165 years, would it be possible to send another flyby mission to all four jovian planets if we launched it now? Explain.

Although a flyby offers only a relatively short period of close-up study, it can provide valuable scientific information. Flybys generally carry small telescopes, cameras, and spectrographs. Because these instruments are brought within a few tens of thousands of kilometers (or closer) of other worlds, they can obtain much higher-resolution images and spectra than even the largest current telescopes viewing these worlds from Earth. In addition, flybys sometimes give us information that would be difficult to obtain from Earth. For example, *Voyager 2* helped us discover Jupiter's rings and learn about the rings of Saturn, Uranus, and Neptune through views in which the rings were backlit by the Sun. Such views are possible only from beyond each planet's orbit.

Flybys may also carry instruments to measure local magnetic field strength or to sample interplanetary dust. The gravitational effects of the planets and their moons on the spacecraft itself provide information about object masses and densities. Like the backlit views of the rings, these types of data cannot be gathered from Earth. Indeed, most of what we know about the masses and compositions of moons comes from data obtained by spacecraft that have flown past them.

Orbiters An orbiter can study another world for a much longer period of time than a flyby. Like the spacecraft used for flybys, orbiters often carry cameras, spectrographs, and instruments for measuring the strength of magnetic fields. Missions to Venus, Mars, and Saturn's moon Titan have also used radar to make precise altitude measurements of surface features. Radar works by sending radio waves from the spacecraft to bounce off the surface: The time it takes for the bounced signals to return to the spacecraft tells how far they traveled, allowing measurements of surface altitude. The precise way in which the signal is reflected by the surface can tell us about the terrain. In the cases of Venus and Titan, radar observations give us our only detailed surface views, because we cannot otherwise see through the thick cloud cover on these worlds.

An orbiter is generally more expensive than a flyby for an equivalent weight of scientific instruments, primarily because it must carry added fuel to change from an interplanetary trajectory to a path that puts it into orbit around another world. Careful planning can minimize the added expense. For example, recent Mars orbiters have saved on fuel costs by carrying only enough fuel to enter highly elliptical orbits around Mars. The spacecraft then settled into the smaller, more circular orbits needed for scientific observations by skimming the martian atmosphere at the low point of every elliptical orbit. Atmospheric drag slowed the spacecraft with each orbit and, over several months, circularized the spacecraft orbit. (This technique of using the atmosphere to slow the spacecraft and change its orbit is called *aerobraking*.)

To date, we have sent orbiters to the Moon and to the planets Venus, Mars, Jupiter, and Saturn, as well as to an asteroid called Eros (Figure 7.13). At present, orbiters are studying Venus, Mars, and Saturn. Another orbiter, called MESSENGER (which stands for "MErcury Surface, Space ENvironment, GEochemistry, and Ranging"), is on its way to Mercury. Launched in 2004, MESSENGER won't reach Mercury orbit until 2011. The reason for the long trip is that MESSENGER needs to slow down in order to enter an orbit around Mercury, and it would have been prohibitively expensive to launch it with enough fuel to slow itself down with rocket thrusts. Therefore, scientists planned a trajectory that allows it to use the gravity of planets to slow down, essentially doing the reverse of the type of speed boost picked up by the *New Horizons* spacecraft as it goes past Jupiter. By the time it reaches Mercury orbit, MESSENGER will have circled the Sun 15 times while passing once by Earth, twice by Venus, and three times past Mercury.

Figure 7.13

The asteroid Eros, photographed by the NEAR spacecraft. NEAR orbited Eros for a year before ending its mission with a soft landing on the surface. The inset photo was taken just before the spacecraft landed.

Landers and Probes The most "up close and personal" study of other worlds comes from spacecraft that send probes into the atmospheres or landers to the surfaces. For example, in 1995, the *Galileo* spacecraft—which orbited Jupiter for more than 5 years—dropped a probe into Jupiter's atmosphere. The probe collected temperature, pressure, composition, and radiation measurements for about an hour as it descended, teaching us a

The aeroshell protects the rover from fiery temperatures as it enters the Martian atmosphere. Six minutes to landing.

With the parachute deployed, three retrorockets fire their engines, suspending the lander 30–50 feet above the Martian surface.

Protected by large airbags, the lander falls away from the parachute, landing safely on Mars. The lander bounces for several minutes, traveling hundreds of meters.

After bouncing to a stop, the airbags are retracted and the lander's petals deploy, creating a ramp for the rover to descend.

The rover deploys the solar arrays, wheels, cameras, and other instruments and begins its exploration of the Martian surface.

Figure 7.14

This artist's conception shows how the *Spirit* and *Opportunity* rovers used parachutes and air bags to land safely on Mars.

great deal about Jupiter's winds and atmospheric conditions before it was destroyed by Jupiter's high interior pressures and temperatures. The *Cassini* spacecraft, currently in orbit of Saturn, also carried a probe, called *Huygens,* that descended to the surface of Saturn's moon Titan while studying the atmosphere on its way down. We'll discuss the findings from *Huygens* in Chapter 9, when we cover Titan in more detail.

On planets with solid surfaces, a lander can offer close-up surface views, local weather monitoring, and the ability to carry out automated experiments. Some landers carry robotic rovers able to venture across the surface, such as the *Spirit* and *Opportunity* rovers that landed on Mars in 2004 and operated long beyond their planned lifetimes. Landers typically require fuel to slow their descent to a planetary surface, but clever techniques can reduce cost. For example, the recent Mars landers hit the surface of Mars at crash-landing speed, but were protected by cocoons of air bags deployed on the way down. These air bags allowed the landers to bounce along the surface of Mars before finally coming to rest (Figure 7.14).

Sample Return Missions While probes and landers can carry out experiments on surface rock or atmospheric samples, the experiments must be designed in advance and the instrumentation must fit inside the spacecraft. These limitations make scientists long for missions that will scoop up samples from other worlds and return them to Earth for more detailed study. To date, sample return missions have been sent only to the Moon. Astronauts collected rock samples during the *Apollo* missions, and the then–Soviet Union sent robotic spacecraft to collect rocks from the Moon in the early 1970s. Many scientists are working toward a sample return mission to Mars, and they hope to launch such a mission within the next decade or so. A slight variation on the theme of a sample return mission is the *Stardust* mission, which collected comet dust on a flyby and then returned it to Earth in early 2006.

• What are some key spacecraft missions for astrobiology?

Every space mission provides at least some insight into issues important to astrobiology and, today, most missions are planned with at least some astrobiology considerations in mind. Table 7.2 lists some of the most significant robotic missions of the past and present.

A quick study of the table should make clear that we are living during an exciting time. Many missions that are already operating should provide

TABLE 7.2 *Selected Robotic Missions to Other Worlds*

Name	*Type*	*Destination*	*Lead Space Agency*	*Arrival Year*	*Mission Highlights*
New Horizons	Flyby	Pluto	NASA	2015†	First mission to Pluto, launched January 2006
Rosetta	Flyby and lander	Comet 67P/ Churyumov-Gerasimenko	ESA*	2014†	Launched in 2004, carries lander and orbiter for comet
MESSENGER	Orbiter	Mercury	NASA	2011†	Studies Mercury's surface, atmosphere, and interior
Mars Reconnaissance Orbiter	Orbiter	Mars	NASA	2006	High-resolution mapper will identify future landing sites
Venus Express	Orbiter	Venus	ESA*	2006	Studying Venus from orbit
Stardust	Flyby and sample return	Comet Wild 2	NASA	2006	Collected dust from comet in 2004, returned to Earth in 2006
Hayabusa	Lander and sample return	Asteroid Itokawa	JAXA**	2005	Collected asteroid sample in 2005, to return to Earth in 2007 (return may not be possible due to spacecraft problems)
Deep Impact	Flyby and "lander"	Comet Tempel 1	NASA	2005	"Lander" hit at 10 km/sec, flyby spacecraft studied the impact
Cassini	Orbiter	Saturn	NASA	2004	Included probe (*Huygens*) that landed on Titan
Mars Exploration Rovers	Lander	Mars	NASA	2004	Rovers *Spirit* and *Opportunity* explored surface
Mars Express	Orbiter	Mars	ESA*	2004	Climate studies of Mars
Mars Odyssey 2001	Orbiter	Mars	NASA	2001	Detailed study of Martian surface features and composition
Near Earth Asteroid Rendezvous (NEAR)	Orbiter	Eros (asteroid)	NASA	2000	First spacecraft dedicated to in-depth study of an asteroid
Mars Global Surveyor	Orbiter	Mars	NASA	1997	Detailed imaging of surface from orbit
Mars Pathfinder	Lander	Mars	NASA	1997	Carried *Sojourner*, the first robotic rover on Mars
Galileo	Orbiter	Jupiter	NASA	1995	Dropped probe into Jupiter; close-up study of moons
Magellan	Orbiter	Venus	NASA	1990	Detailed radar mapping of surface of Venus
Voyager 1	Flyby	Jupiter, Saturn	NASA	1979, 1980	Unprecedented views of Jupiter and Saturn; continuing out of solar system
Voyager 2	Flyby	Jupiter, Saturn, Uranus, Neptune	NASA	1979, 1981, 1986, 1989	Only mission to Uranus and Neptune; continuing out of solar system
Vikings 1 and 2	Orbiter and lander	Mars	NASA	1976	First landers on Mars

* European Space Agency. ** Japan Aerospace Exploration Agency. † Scheduled arrival year.

new insights into the prospects for finding life in the solar system. For example, numerous missions are currently studying Mars, which is a prime target because it seems so likely to have been habitable in the past and perhaps to be habitable underground even today. *Venus Express* may shed light on the possibility of life in Venus's atmosphere. Studies of the dust returned by the *Stardust* mission should teach us both about comets and about the origin of our solar system, while studies of the aftermath of the *Deep Impact* blast into Comet Tempel 1 should tell us about comet structure.

Over the longer term, all the world's major space agencies have high hopes of launching numerous and diverse future missions to study our solar system. Of course, these hopes are subject to the vagaries of politics and budgets for scientific exploration. If they come to fruition, however, the next couple of decades will see spacecraft that answer many fundamental questions about the habitability of worlds in our solar system, and perhaps even tell us whether any of them have supported—or still support—life.

THE BIG PICTURE
Putting Chapter 7 into Context

In this chapter, we have discussed the general requirements for habitability, and taken a biological tour of the solar system as we know it today. As you continue in your studies, keep in mind the following "big picture" ideas:

- The general environmental requirements for life are much broader than the requirements for complex beings such as humans. The search for life beyond Earth is not necessarily a search for complex animals or intelligence, but in most cases a search for simpler life-forms and microbes.
- Our solar system contains a vast number of worlds, but most of them are unlikely to have life because they lack a liquid medium of any kind. Nevertheless, a few worlds may meet the criteria for habitability in at least some regions of their surfaces, subsurfaces, or atmospheres. More worlds may have had liquid water or other liquids in the distant past.
- Much of our current knowledge about the solar system and the potential for life comes from studies conducted by robotic spacecraft. We are living during a time when many spacecraft are simultaneously exploring different worlds in our solar system, and more missions are being planned for the future.

SUMMARY OF KEY CONCEPTS

7.1 ENVIRONMENTAL REQUIREMENTS FOR LIFE

Where can we expect to find building blocks of life?
The chemical elements needed for life should be present on almost any world, but a smaller number of worlds will contain more complex organic molecules that can serve as building blocks for life. Still, the fact that these building blocks are present in asteroids and comets suggests that we'll find them in many places.

- **Where can we expect to find energy for life?**

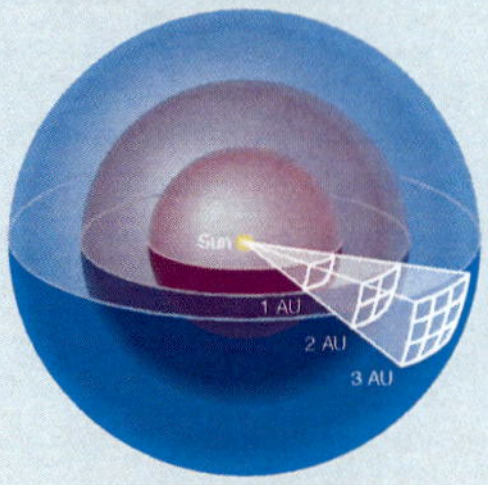

Energy for life can come from sunlight or chemical reactions. Sunlight weakens with distance from the Sun, and is unlikely to be sufficient at large distances. Chemical energy is probably available in many more places, and is likely on any world with a substantial atmosphere or a liquid medium that can mix and support chemical reactions.

• **Does life need liquid water?**

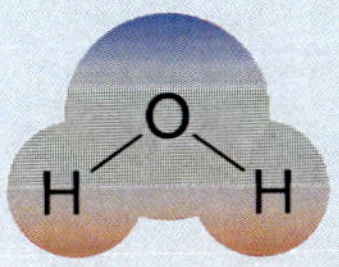

Life almost certainly requires some liquid, and water has at least three advantages over other liquids: a wider and higher range of temperatures in which it is liquid; the fact that ice floats; and the type of chemical bonding made possible by charge separation within water molecules. Nevertheless, we can't completely rule out other liquids, such as liquid ammonia, methane, or ethane.

• **What are the environmental requirements for habitability?**
Life requires a source of molecules from which to build living cells, a source of energy for metabolism, and a liquid medium for transporting chemicals. In practice, these requirements probably come down to a need for liquid water, so the possibility of liquid water is the main requirement we search for in looking for habitable worlds.

7.2 A BIOLOGICAL TOUR OF THE SOLAR SYSTEM: THE INNER SOLAR SYSTEM

• **Does life seem plausible on the Moon or Mercury?**
The Moon and Mercury are probably not habitable, since neither has liquid water nor any other liquid medium for life.

• **Could life exist on Venus or Mars?**

It is far too hot for liquid water to exist on or under the surface of Venus, making life seem unlikely. However, life might be possible high in Venus's atmosphere, where clouds contain droplets of water. Mars almost certainly had habitable conditions in the distant past, and might still have habitable regions underground.

7.3 A BIOLOGICAL TOUR OF THE SOLAR SYSTEM: THE OUTER SOLAR SYSTEM

• **What are the prospects for life on jovian planets?**

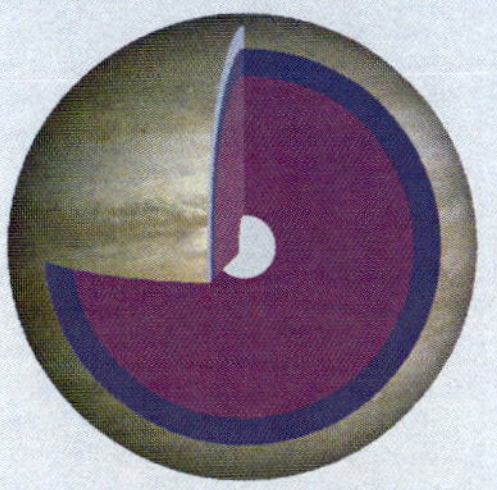

The jovian planets have depths at which liquid water can exist in their atmospheres, but strong vertical winds make life seem unlikely. Uranus and Neptune may have "oceans" of water and other liquids in their deep interiors, but at present we have no way to search such depths for life.

• **Could there be life on moons or other small bodies?**
A few large moons may contain liquid water or other liquids, and thus seem like potential candidates for life. Smaller moons and other small bodies probably do not have any liquids at present, though many may have had liquid water in the distant past.

7.4 THE PROCESS OF SCIENCE IN ACTION: SPACECRAFT EXPLORATION OF THE SOLAR SYSTEM

• **How do robotic spacecraft work?**

Spacecraft can be categorized as flyby, orbiter, lander/probe, or sample return missions. In all cases, robotic spacecraft carry their own propulsion, power, and communication systems, and can operate under preprogrammed control or with updated instructions from ground controllers.

• **What are some key spacecraft missions for astrobiology?**
All spacecraft missions have at least some connection to astrobiology. Among the most important missions at present are those studying Mars, Titan, and Venus.

EXERCISES AND PROBLEMS

REVIEW QUESTIONS

Short-Answer Questions Based on the Reading

1. Why do we expect the elements of life to be widely available on other worlds? How does the requirement of organic building blocks further constrain the prospects of habitability?
2. How does the strength of sunlight vary with distance from the Sun? Discuss the implications for photosynthetic life.
3. Under what conditions does it seem reasonable to imagine a chemical energy source for life?
4. Why is a liquid medium important for life? Why does water seem the most likely liquid medium for life? Briefly discuss a few other potential liquids for life.
5. Summarize the three major environmental requirements for life. Overall, what "litmus test" seems appropriate for constraining our search for habitable worlds, and why?
6. Why do the Moon and Mercury seem unlikely to be habitable? Does evidence for ice in lunar craters affect the answer? Explain.
7. Why is Venus so much hotter than Earth? How does this heat affect the possibility of life on Venus? Explain why Venus may

nonetheless have been habitable in the past and might still be habitable in some of its clouds.

8. Why does Mars seem such a good candidate for life?
9. Briefly discuss the possibility of life on Jupiter and Saturn.
10. In the context of habitability, how do the cases of Uranus and Neptune differ from those of Jupiter and Saturn? Explain.
11. What characteristics make some of the large moons of jovian planets seem like potential candidates for habitability?
12. What are the prospects for habitability of the many small bodies in the solar system, and why?
13. Describe and distinguish between space missions that are *flybys, orbiters, landers* or *probes*, and *sample return missions*. What are the advantages and disadvantages of each mission type?
14. For a few of the most important past, present, or future robotic missions to the solar system, describe their targets, types, and mission highlights.

TEST YOUR UNDERSTANDING

ould You Believe It?

Each of the following gives a statement that a future explorer might someday make. In each case, decide whether the claim seems plausible in light of current knowledge. Explain clearly; because not all of these have definitive answers, your explanation is more important than your chosen answer.

15. On the smallest moon of Uranus, my team discovered a vast, subsurface ocean of liquid water.
16. New spacecraft images show lakes of liquid water on both Pluto and its moon Charon.
17. By drilling about a kilometer down into the Moon's surface, we are now able to pump up liquid water for the new Moon colony.
18. I was part of the first group of people to land on Venus, where we found huge, ancient cities that had been hidden from view by cloud cover.
19. We sent a robotic airplane into the atmosphere of Jupiter, but we could not keep it at a steady altitude and it was quickly ripped apart.
20. On a moon of Neptune, we discovered photosynthetic life with a metabolism that operates nearly a hundred times faster than that of any photosynthetic organism on Earth.
21. We deposited bacteria that get energy from chemical reactions with sulfur compounds into the upper clouds of Venus, and they are surviving.
22. The drilled sample showed no signs of life on asteroid B612, but we found many complex organic molecules.
23. We cut holes in the frozen methane surface of the lake on Titan, so that we could search for swimming organisms in the liquid methane underneath it.
24. The drilled sample from Mars brought up rock that contained microscopic droplets of liquid water.

Quick Quiz

Choose the best answer to each of the following. Explain your reasoning with one or more complete sentences.

25. Oxygen and carbon are: (a) rarer than almost all other elements; (b) found only on worlds close to a star; (c) the third- and fourth-most-abundant elements in the universe.
26. On an asteroid that is twice as far as Earth from the Sun, the strength of sunlight would be: (a) twice as great as on Earth; (b) $\frac{1}{2}$ as great as on Earth; (c) $\frac{1}{4}$ as great as on Earth.
27. Compared to liquid water, liquid methane is: (a) colder; (b) hotter; (c) denser.
28. Frozen lakes often have liquid water beneath their icy surfaces primarily because: (a) Earth's internal heat keeps the water liquid; (b) ice floats and provides insulation to the water below; (c) sunlight penetrates the ice and warms the water below.
29. Temperatures on Mercury are: (a) always very hot; (b) very hot in the day and very cold at night; (c) about the same as those on Venus.
30. On Venus, liquid water: (a) does not exist anywhere; (b) exists only deep underground; (c) exists only high in the atmosphere.
31. The reason that Venus is so much hotter than Earth is: (a) it has many more volcanoes; (b) its closer distance to the Sun makes sunlight dozens of times stronger; (c) its thick, carbon dioxide atmosphere creates a far stronger greenhouse effect.
32. Life is probably not possible in Jupiter's atmosphere because: (a) it is too cold there; (b) there is no liquid water at all; (c) winds are too strong.
33. Which ingredient are you most likely to find if you randomly choose a small moon of one of the jovian planets to examine? (a) water ice; (b) organic molecules; (c) an abundance of heavy metals, such as gold.
34. The *Cassini* spacecraft: (a) flew past Pluto; (b) landed on Mars; (c) is orbiting Saturn.

INVESTIGATE FURTHER

In-Depth Questions to Increase Your Understanding

Short-Answer/Essay Questions

35. *Solar System Tour.* Based on the brief tour in this chapter, which world in our solar system (besides Earth) do you think is most likely to have life? Explain why.
36. *Galileo Spacecraft.* In 2003, scientists deliberately ended the *Galileo* mission to Jupiter by causing the spacecraft to plunge into Jupiter's atmosphere. They did this to avoid any possibility that the spacecraft might someday crash into Europa, which could potentially have "contaminated" this moon with microbes from Earth. Do you think that the scientists should also have been worried about contaminating Jupiter itself? Why or why not?
37. *Greenhouse Effect.* The text (in Chapter 4) makes the statement that the greenhouse effect on Venus proves "that it is possible to have too much of a good thing." Explain this statement in two or three paragraphs.
38. *Transplanting Life.* Suppose you could genetically engineer organisms on Earth in any way that you chose. What, if any, features could you give them that would enable them to survive on: (a) the Moon or Mercury; (b) Venus; (c) Jupiter; (d) a comet?
39. *Mission Plan.* Suppose you could send a robotic mission of any type (flyby, orbiter, probe/lander) to any one of the places dis-

cussed in this chapter as being *unlikely* to be habitable. Which place and type of mission would you choose, and why? Defend your choice clearly in a one- to two-page essay.

40. *Science Fiction Life.* Choose a science fiction book or movie that describes some form of alien life that falls outside the bounds of the type of life we have considered in this chapter. Write a one- to two-page critical review in which you discuss the plausibility of the life-form.

Quantitative Problems

Be sure to show all calculations clearly and state your final answers in complete sentences.

41. *Understanding Newton's Version of Kepler's Third Law I.* Imagine another solar system, with a star of the same mass as the Sun. Suppose there is a planet in that solar system with a mass twice that of Earth orbiting at a distance of 1 AU from the star. What is the orbital period of this planet? Explain. (*Hint:* The calculations for this problem are so simple that you will not need a calculator.)
42. *Understanding Newton's Version of Kepler's Third Law II.* Suppose a solar system has a star that is four times as massive as our Sun. If that solar system has a planet the same size as Earth orbiting at a distance of 1 AU, what is the orbital period of the planet? Explain. (*Hint:* The calculations for this problem are so simple that you will not need a calculator.)
43. *Earth Mass.* The Moon orbits Earth in an average time of 27.3 days at an average distance of 384,000 kilometers. Use these facts to determine the mass of Earth. (*Hint:* You may neglect the mass of the Moon, since its mass is only about $\frac{1}{80}$ of Earth's.)
44. *Jupiter Mass.* Jupiter's moon Io orbits Jupiter every 42.5 hours at an average distance of 422,000 kilometers from the center of Jupiter. Calculate the mass of Jupiter. (*Hint:* Io's mass is very small compared to Jupiter's.)
45. *Pluto/Charon Mass.* Pluto's moon Charon orbits Pluto every 6.4 days with a semimajor axis of 19,700 kilometers. Calculate the *combined* mass of Pluto and Charon. Compare this combined mass to the mass of Earth, which is about 6×10^{24} kg. Can you determine the individual masses of Pluto and Charon from the given data? Explain.
46. *Mission to Pluto.* The *New Horizons* spacecraft will take about 9 years to travel from Earth orbit to Pluto. About how fast will it be traveling on average? Assume that its trajectory is close to a straight line. Give your answer in AU/year and in km/hr. (*Hint:* You can find needed data in Appendix D.)

Discussion Questions

47. *Bizarre Forms of Life.* Discuss some potential forms of life that have appeared in science fiction and fall outside the general types of life that we've discussed in this chapter. Which forms seem more plausible, and why? Overall, do you think the definition of life we've used in this chapter is too constraining? Defend your opinion.
48. *Artificial Life.* Imagine that future humans decide to breed new organisms tailored to as many different environments as possible. Discuss some of the places in our solar system where we could potentially plant such artificially created species, even if life probably would not arise naturally in those places. Do you think it likely that we will someday develop life-forms for other worlds? What are the philosophical ramifications of being able to custom-tailor life for worlds that don't have any natural life?
49. *Future Astrobiology Course.* Imagine that you are living a century from now and are taking a course about life in our solar system. Based on the current rate of exploration and reasonable rates for the future, how much more do you think we will know then about life in our solar system than we know now? Speculate about some of the discoveries you think may occur in the next century.

WEB PROJECTS

50. *Project Apollo.* Learn more about NASA's *Apollo* project, the only set of missions that has ever sent humans to another world. Describe the goals and objectives of each of the *Apollo* missions. Which were successful, and which were not? What lessons does *Apollo* offer for future attempts to send humans to the Moon and beyond? Summarize your findings and your opinions about lessons for the future in a one- to two-page essay.
51. *Planetary Missions.* Visit the Web page for one of the missions listed in Table 7.2. Write a one- to two-page summary of the mission's basic design, goals, and current status.

8
Mars

LEARNING GOALS

8.1 FANTASIES OF MARTIAN CIVILIZATION
- How did Mars invade popular culture?

8.2 A MODERN PORTRAIT OF MARS
- What is Mars like today?
- What are the major geological features of Mars?
- What geological evidence tells us that water once flowed on Mars?

8.3 THE CLIMATE HISTORY OF MARS
- Why was Mars warmer and wetter in the past?
- Why did Mars change?
- Is Mars habitable?

8.4 SEARCHING FOR LIFE ON MARS
- Is there any evidence of life on Mars?
- How do we plan to search for life on Mars?
- Should we send humans to Mars?

8.5 THE PROCESS OF SCIENCE IN ACTION: MARTIAN METEORITES
- Is there evidence of life in martian meteorites?

I remember being transfixed by the first lander image to show the horizon of Mars. This was not an alien world, I thought. I knew places like it in Colorado and Arizona and Nevada. There were rocks and sand drifts and a distant eminence, as natural and unselfconscious as any landscape on Earth. Mars was a place. I would, of course, have been surprised to see a grizzled prospector emerge from behind a dune leading his mule, but at the same time the idea seemed appropriate.

Carl Sagan, in *Cosmos*

The idea of a civilization on Mars was once taken so seriously that the term *Martians* became nearly synonymous with alien life. Spacecraft sent to Mars in the 1960s shattered this fictional image of a world of cities and sophisticated beings, revealing a cold, dry, cratered landscape. But while we can rule out a martian civilization, the existence of past or present microbial life on Mars remains a subject of intense scientific investigation and debate.

Substantial evidence suggests that water once flowed on Mars, and it seems likely that Mars once had surface or subsurface environments similar to those in which life thrived on the early Earth. If life arose on Mars (or was transplanted there on meteorites from the early Earth), we may be able to find its fossil remains. It's even possible that life still survives somewhere on Mars, perhaps underground where volcanic heat can keep some water liquid.

Whether or not we someday discover martian life, by studying Mars we are sure to learn a great deal about planetary habitability. Based on current hypotheses concerning the origin of life on Earth, Mars appears to have met all of the requirements for life to have arisen there as well. If we find life, it will help us understand more about biology in general. If we don't find life, Mars will still teach us much about the range of conditions under which life can arise. In this chapter, we'll discuss our past and present understanding of the habitability of Mars and the nature of our search for life on the red planet.

8.1 FANTASIES OF MARTIAN CIVILIZATION

Shining brightly and noticeably red in the nighttime sky, Mars has long captured the human imagination. It figures prominently in many ancient myths and legends. But the idea that Mars might harbor an advanced civilization arose only a little over 200 years ago. Before we discuss what we know about Mars today, let's go back in time and look at how Mars came to be so closely associated with the possibility of extraterrestrial life. For modern-day context, look at Figure 8.1, which shows a high-resolution image of one side of Mars as revealed by recent spacecraft observations, and look back at Figure 1.5, which shows the surface of Mars from the *Spirit* rover.

Figure 8.1

The full disk of Mars, as photographed by orbiting spacecraft. The horizontal "gash" across the center is the giant canyon Valles Marineris.

• How did Mars invade popular culture?

The story begins with the noted English astronomer William Herschel (1738–1822). Though best known for discovering the planet Uranus, Herschel made numerous other astronomical discoveries, usually with help from his sister and fellow astronomer Caroline Herschel (1750–1848). The Herschels often observed Mars through their telescopes, noting its polar ice caps and discovering that the length of its day (24 hours 37 minutes) is similar to an Earth day. In a talk presented to Britain's Royal Society in 1784, William Herschel also claimed that Mars possessed an atmosphere and that consequently, "its inhabitants probably enjoy a situation in many respects similar to our own." With the mention of "inhabitants," the possibility of living beings on the red planet had been broached by a respected scientist in an academic setting, and Martians were assumed to exist. (It should be noted that Herschel was not overly particular when it came to populating the cosmos. As far as he was concerned, everything in the solar system was inhabited, including the Moon and the Sun.)

During the following century, Mars rose to the top of the astronomical charts. Herschel had spoken of permanent spots on the planet, and in the 1860s Angelo Secchi in Rome noted some peculiar, streaky features as well. But it was the work of Giovanni Schiaparelli, director of the Milan Observatory, that transformed the perception of Mars for the next hundred years. In 1877, Schiaparelli claimed to see a network of 79 linear features that he called *canali,* by which he meant the Italian word for "channels." However, it was often translated incorrectly as "canals." Coming amid the excitement that followed the 1869 opening of the Suez Canal, Schiaparelli's discovery soon inspired visions of artificial waterways built by a martian civilization. Schiaparelli himself remained skeptical of such claims, and it's not clear whether he even thought the *canali* contained water. But his work caught the imagination of a young Harvard graduate, Percival Lowell (1855–1916).

Lowell, whose degree was in mathematics, came from a wealthy and distinguished New England family. His brother Abbott became famous as a president of Harvard, and his sister Amy gained fame as a poet. After spending a few years as a businessman and as a traveler in the Far East, Percival Lowell turned to astronomy. Impassioned by his belief in the martian canals and enabled by his wealth, Lowell commissioned the building of an observatory. He was aware that not all astronomers could make out Schiaparelli's *canali,* but rather than questioning the existence of the canals, he instead guessed that they were difficult to observe because most astronomical observatories were affected by poor "seeing" (a term used to describe the clarity and stability of the air above an observatory). Seeing varies from place to place and from night to night and tends to be best in places with dry air and high altitude. After considerable research, Lowell chose Flagstaff, Arizona, for his new observatory, because he believed it had the best overall seeing conditions in the United States. The Lowell Observatory opened in 1894 and is still operating today. Over a period of two decades, Lowell meticulously mapped nearly 200 canals that he claimed to see on Mars, identifying each of them with a Latin name.

In 1895, Lowell published his first detailed maps of the canals, as well as the first of three books in which he explained not only his belief that the canals were constructed by a martian civilization but also his conclusions about the nature of that civilization. Lowell correctly deduced that Mars was generally arid, and he guessed that the polar caps were made of water ice (in reality, they also contain carbon dioxide ice). He therefore imagined that the canals were built to carry water from the poles to thirsty cities nearer the equator. From there it was a short leap to imagine the Martians as an old civilization on a dying planet. The global network of canals convinced Lowell that the Martians were citizens of a single, global nation. Such ideas provided the "scientific" basis for H. G. Wells' *The War of the Worlds*, published in 1898.

THINK ABOUT IT Think of as many popular references to a civilization of "Martians" as you can. For example, consider science fiction stories or movies, television shows, advertisements, or music that refers to intelligent beings on Mars. What do these references tell us about the influence of science on the public imagination?

Lowell was an effective advocate for the canals, but they do not really exist. Even in his own time, other scientists shot holes through most of Lowell's claims. Numerous astronomers, including some who had initially accepted the existence of canals, soon began to question Lowell's observations. These other astronomers could not see the canals when they put their eyes to the telescope—not even when using telescopes larger than Lowell's—and the canals failed to show up in photos made through telescopes as well. Moreover, while some canal-like features may have seemed to be present when seeing was poor, they were absent when seeing was best—the opposite of what we would expect if they were really there.

Other scientists questioned Lowell's basic assumptions and interpretations. Writing in 1907, Alfred Russel Wallace—the codiscoverer with Darwin of evolution by natural selection [Section 5.1]—used physical arguments to suggest that Mars must be too cold for liquid water to flow and that the polar caps must be dry ice (frozen carbon dioxide) rather than water ice. He also pointed out a major flaw in Lowell's interpretation of the canals: Lowell's canals all followed straight-line paths for hundreds or thousands of miles, but real canals would be built to follow natural contours of topography (for example, to go around mountains). In summarizing this argument, Wallace wrote that such a network of canals, "as Mr. Lowell describes, would be the work of a body of madmen rather than of intelligent beings."*

What was Lowell seeing? In only a few cases do his canals correspond to real features on Mars. For example, the canal he claimed to see most often (which he called *Agathodaemon*) coincides with the location of the huge canyon network now known as Valles Marineris (see Figure 8.1). A few other canals also roughly follow the contours of real features on Mars, but most of the canals were pure fantasy. Figure 8.2 compares a telescopic photo of Mars with one of Lowell's maps of the same regions. You can probably see how the dark and light regions match up in the photo and the drawing, but seeing any canals requires a vivid imagination.

Lowell's story illustrates both the pitfalls and the triumphs of modern science. The pitfall is that individual scientists, no matter how upstanding and dedicated, may still bring personal biases to bear on their scientific work. In Lowell's case, he was so convinced of the existence of canals and Martians that he simply ignored all evidence to the contrary. But the story's ending shows why modern science ultimately is so successful. Despite Lowell's stature, other scientists did not accept his claims on faith. Instead, they sought to confirm his observations and to test his

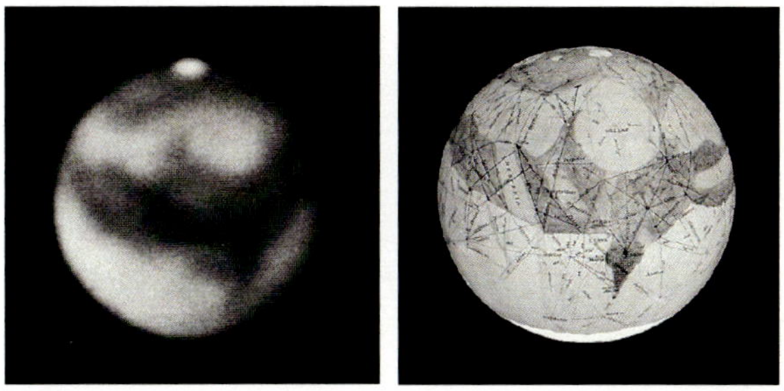

Figure 8.2

Can you see how the markings on Mars in the telescopic photo on the left might have resembled the geometrical features in the drawing by Percival Lowell on the right? Try squinting your eyes.

*Excerpted in K. Zahnle, "Decline and Fall of the Martian Empire," *Nature*, vol. 412, July 12, 2001.

underlying assumptions. They found that Lowell's claims fell short on all counts. As a result, Lowell became an increasingly isolated voice as he continued to advocate a viewpoint that was clearly wrong.

The story also illustrates the divide that often persists between scientists and the general public. Although nearly all scientists had abandoned belief in Martians within the first few decades of the twentieth century, among the public the canal myth persisted much longer. Public belief in Martians remained widespread enough to create a famous panic during Orson Welles's 1938 radio broadcast of *The War of the Worlds*—many people thought an invasion was actually under way (Figure 8.3). Books and movies about martian civilization remained popular throughout the twentieth century. Even today, a few die-hard proponents claim to see evidence of a past martian civilization, but recent high-resolution photographs make their claims increasingly implausible.

The New York Times.

Copyright, 1938, by The New York Times Company.

NEW YORK, MONDAY, OCTOBER 31, 1938.

Radio Listeners in Panic, Taking War Drama as Fact

Many Flee Homes to Escape 'Gas Raid From Mars'—Phone Calls Swamp Police at Broadcast of Wells Fantasy

A wave of mass hysteria seized thousands of radio listeners throughout the nation between 8:15 and 9:30 o'clock last night when a broadcast of a dramatization of H. G. Wells's fantasy, "The War of the Worlds," led thousands to believe that an interplanetary conflict had started with invading Martians spreading wide death and destruction in New Jersey and New York.

The broadcast, which disrupted households, interrupted religious services, created traffic jams and clogged communications systems, was made by Orson Welles, who as the radio character, "The Shadow," used to give "the creeps" to countless child listeners. This time at least a score of adults required medical treatment for shock and hysteria.

and radio stations here and in other cities of the United States and Canada seeking advice on protective measures against the raids.

The program was produced by Mr. Welles and the Mercury Theatre on the Air over station WABC and the Columbia Broadcasting System's coast-to-coast network, from 8 to 9 o'clock.

The radio play, as presented, was to simulate a regular radio program with a "break-in" for the material of the play. The radio listeners, apparently, missed or did not listen to the introduction, which was: "The Columbia Broadcasting System and its affiliated stations present Orson Welles and the Mercury Theatre on the Air in 'The War of the Worlds' by H. G. Wells."

Figure 8.3

This front-page story from the *New York Times* described the panic caused by Orson Welles's 1938 broadcast of *The War of the Worlds.*

8.2 A MODERN PORTRAIT OF MARS

The public debate about martian canals and cities was not entirely put to rest until NASA began sending spacecraft to Mars. In 1965, NASA's *Mariner 4* spacecraft flew to within 6,000 miles of the martian surface, transmitting a few dozen television-quality images of the landscape below. When the photos were given to President Lyndon Johnson, he looked at them with surprise. Mars's surface was littered with craters, not canals. The red planet looked as lifeless as the Moon. Measurements of the atmospheric pressure and temperature made from the spacecraft indicated a cold, dry planet seemingly incapable of supporting life.

Nevertheless, all was not lost when it came to the potential for life on Mars. There was no evidence of any intelligent beings, but the thin atmosphere and the polar caps left open the possibility of the existence of microbes or perhaps even some primitive plants or animals. On July 20, 1976, seven years to the day after Neil Armstrong's history-making walk on the Moon and nearly a century since Schiaparelli's description of *canali,* the thin skies above Mars were pierced by a NASA space probe. *Viking Lander 1* touched down without difficulty on the Chryse Planitia, a sprawling, rock-strewn plain about 1,300 kilometers north of the martian equator. It was followed nearly two months later by its twin, *Viking Lander 2,* which landed almost halfway around the planet. Meanwhile, two *Viking* orbiters studied the planet from above.

Figure 8.4

The surface of Mars photographed by the *Viking Lander 2* in 1979, showing a thin coating of ice on the rocks and soil. The inset shows a working model (actually, a spare spacecraft) of the *Viking* landers, identical to those that landed on Mars, on display at the National Air and Space Museum in Washington, D.C.

When the *Viking* landers' cameras opened their eyes in the frigid martian air, they found a bleak landscape with red dust and scattered rocks. No creatures stared back at the cameras, and no plants were huddled in the weak sunlight. For months the images continued to come in, but the view scarcely changed. Nothing grew other than some occasional patches of frost, and nothing moved other than windblown dust (Figure 8.4). Though neither lander could move from the spot at which it had settled, each had a robotic arm with which it collected soil for some onboard experiments designed to look for microbes (see Section 8.4).

The *Viking* orbiters and landers provided a wealth of scientific data about Mars. But they also left many questions unanswered, and the scientific community was itching for follow-up missions. Unfortunately, budgetary and political considerations, along with the failure of two Russian missions to Mars (*Phobos 1* and *2*) and one American mission (*Mars Observer*), all conspired to stop spacecraft exploration of Mars for some 20 years. The long mission drought did not end until July 4, 1997, with

Figure 8.5

The view from the *Pathfinder* lander (partially visible in the foreground); the scattered rocks were probably carried to the site by an ancient flood. The little rover, *Sojourner*, is at the upper right, studying a rock that scientists named Yogi.

the landing of *Pathfinder* and its little rover, *Sojourner* (Figure 8.5), which was named for Sojourner Truth, an African American heroine of the Civil War era. Although *Sojourner* could travel only a few tens of meters from *Pathfinder*, this was enough to check the chemical composition of many nearby rocks.

The *Pathfinder* mission marked the beginning of a new era of intensive scientific study of Mars. By early 2006, four orbiters and two roving landers were all studying Mars (see Table 7.2). By combining data from these new missions with earlier *Viking* and *Pathfinder* data and telescopic studies, we are beginning to put together a realistic portrait of the past and present habitability of Mars.

• What is Mars like today?

The present-day surface of Mars may look much like some deserts or volcanic plains on Earth, but its thin atmosphere makes it unlike any place on Earth. Table 8.1 summarizes some basic Mars data. Note that the surface temperature is usually well below freezing, while the atmospheric pressure is less than 1% that on the surface of Earth—making the air so thin that no human could survive outside for more than a few minutes without a pressurized space suit. The air contains only trace amounts of oxygen, so we could not breathe it. The lack of oxygen also means that Mars lacks an ozone layer, so much of the Sun's damaging ultraviolet radiation passes unhindered to the surface.

Nevertheless, martian conditions are much less extreme than those on the Moon (mainly because of the moderating effects of the atmosphere), and it's easy to imagine future astronauts living and working in airtight research stations while occasionally donning space suits for outdoor excursions. Martian surface gravity is about 40% that on Earth, so everyone and everything would weigh about 40% of Earth weight. Astronauts could walk around easily even while wearing space suits with heavy backpacks. It would also be easy to adapt to patterns of day and night, since the martian day is only about 40 minutes longer than an Earth day.

TABLE 8.1 *Basic Mars Data*

Average Distance from Sun	1.52 AU = 227.9 million km
perihelion distance	206 million km
aphelion distance	249 million km
Orbital Period	1.881 Earth yr
Equatorial Radius	3,397 km
Mass (Earth = 1)	0.107
Rotation Period	24 hr 37 min
Surface Gravity (Earth = 1)	0.38
Atmospheric Composition	95% CO_2, 2.7% N_2, 1.6% argon
Average Surface Temperature	−50°C
Average Surface Pressure	0.007 bar*

* 1 bar ≈ sea level pressure on Earth

The Lack of Surface Liquid Water

The low atmospheric pressure explains one of the key facts relevant to the search for life on Mars: No liquid water exists anywhere on the surface of Mars today. We know this not only because we've studied most of the surface in reasonable detail, but also because the surface conditions would not allow liquid water to be present as lakes, rivers, or even puddles. In most places and at most times, Mars is so cold that any liquid water would immediately freeze into ice. Even when the temperature rises above freezing, as it often does at midday near the equator, the air pressure is so low that liquid water would quickly evaporate. In other words, liquid water is *unstable* on Mars today: If you donned a space suit and took a cup of water outside your pressurized spaceship, the water would almost immediately either freeze or evaporate away (or some combination of both). However, as we'll discuss shortly, Mars almost certainly had liquid water in the past, and still has ample water ice and some water vapor, and perhaps even pockets of liquid water underground.

SPECIAL TOPIC: THE FACE ON MARS

Among the tens of thousands of photographs snapped as part of the *Viking* missions to Mars in the late 1970s, one achieved special fame. The *Viking 1* orbiter snapped a picture showing a feature on Mars that looked remarkably like a human face (Figure 1). The "face on Mars" soon became a cottage industry, with proponents arguing that it must be the work of a lost Martian civilization or of alien beings visiting our solar system.

To scientists, it seemed a near-certainty that the human likeness was a coincidence of geology and camera angle. However, the people promoting an artificial origin for the "face" claimed that NASA was conspiring to hide the truth—despite the fact that it was first described as a face by a NASA scientist in a NASA press release. A few proponents even claimed that a failed NASA mission to Mars (*Mars Observer* in 1992) had actually achieved orbit successfully and was imaging the face while NASA hid the results from the public.

Today, the truth is clear for all to see: Figure 2 compares the original *Viking* photo with a more recent and higher-resolution photo from *Mars Global Surveyor.* The "face" was a product of shadows and the human tendency to see patterns where none exist—just as we see faces in clouds, a man in the Moon, and canals on Mars.

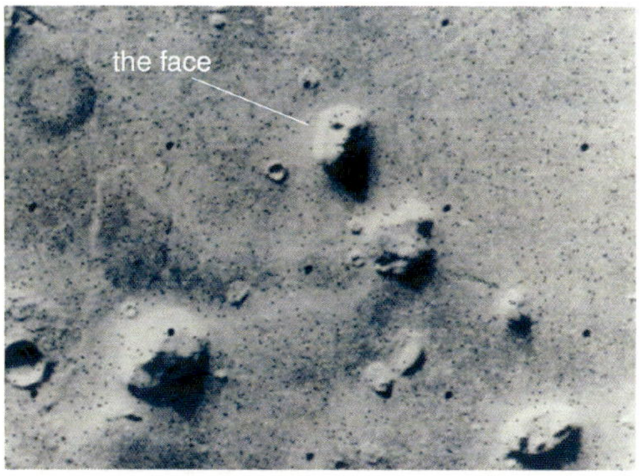

Figure 1

This *Viking 1* image shows the "face on Mars" and the surrounding area. The face is on a tall mesa about 1 kilometer across. Some of the other features were claimed to represent a "fortress" and "pyramids." The black dots are instrument artifacts.

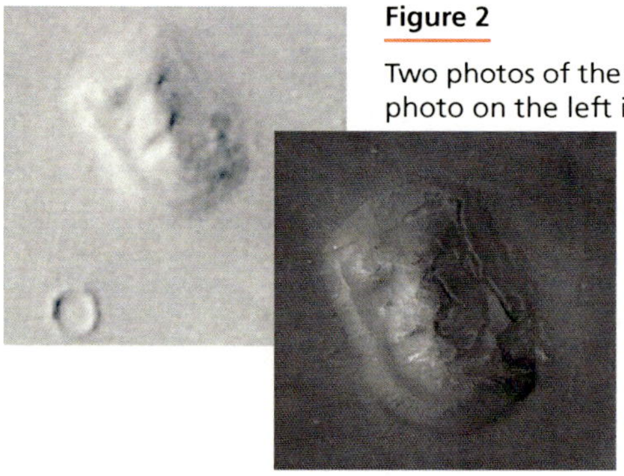

Figure 2

Two photos of the "face on Mars." The photo on the left is another image from *Viking 1* in 1976 (a blow-up view of the face in Figure 1). The much higher-resolution photo on the right is from the *Mars Global Surveyor* mission.

A few die-hard proponents still claim either that it was a face that has been heavily altered by natural erosion or that there is still a NASA conspiracy to hide the truth (though Mars is also now being studied by a European mission, so the conspiracy would have to be global). It's hard to blame them for their desire to see evidence of an extraterrestrial civilization. After all, there's a certain appeal to all the adventures that past science fiction writers imagined taking place among the inhabitants of Mars. But if you look hard enough, you'll always find coincidental patterns: For example, in searching through hundreds of other photos from Mars missions, NASA scientists came upon a crater that seems to show a "happy face" on Mars (Figure 3).

Figure 3

This *Viking* image shows a crater on Mars that appears to show a "happy face."

ticularly when the more extreme summer approaches in the southern hemisphere (Figure 8.10). At times, airborne dust shrouds the surface so much that no surface markings can be distinguished and large dunes form and shift with the strong winds.

The martian winds often spawn *dust devils,* swirling winds that you may have seen over desert sands or dry dirt on Earth. Dust devils look much like miniature tornadoes, but they rise up from the ground rather than coming down from the sky. The air in dust devils is heated from below by the sunlight-warmed ground, and dust devils swirl because of the way the rising air interacts with prevailing winds. Dust devils on Mars are especially common during summer in either hemisphere and can be far larger than their counterparts on Earth. The Mars rovers have even photographed dust devils (Figure 8.11), and some have apparently passed directly over the rovers without causing damage. In one case, a dust devil apparently "cleaned" dust off the solar panels of the rover *Spirit,* restoring power that had been lost as dust accumulated.

Winds generated by the temperature differences near the edges of the polar caps sometimes initiate local dust storms. Occasionally, local dust storms in low southern latitudes expand during summer into huge dust storms that enshroud the entire planet (see Figure 8.10). At times the martian surface becomes almost completely obscured by airborne dust. As the dust settles out, it can change the surface appearance over vast areas (for example, by covering dark regions with brighter dust); such changes fooled past astronomers into thinking they were seeing seasonal changes in vegetation.

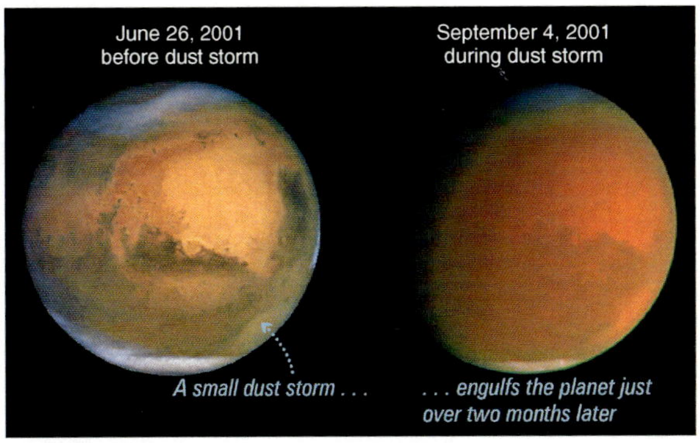

Figure 8.10

These two Hubble Space Telescope photos contrast the appearance of the same face of Mars in the absence (left) and presence (right) of a global dust storm.

Color of the Martian Sky Martian winds and dust storms also leave Mars with perpetually dusty air, which helps explain the colors of the martian sky. The air on Mars scatters visible sunlight much like the air on Earth; that is, it scatters blue light more than red light (which is why our sky is blue). However, because there is much less air on Mars than Earth, it scatters less light. Thus, by itself this scattering would make the martian sky a deep blue. The dust in the atmosphere has an almost opposite effect. Dust tends to absorb blue light, so by itself it would leave the sky with a brownish-pink color. The actual color of the martian sky comes from the combination of the effects of air and dust. The result depends on the time of day, the season, and how much dust is in the air. Most of the time, the daytime martian sky ends up yellow-brown. However, it sometimes ends up with other colors, such as blue or even green, particularly in the mornings and evenings.

• What are the major geological features of Mars?

The surface of Mars may be desolate and barren today, but it was not always so. Many surface features appear to have been shaped by liquid water, leading scientists to conclude that Mars must once have had a much more hospitable climate. But before we discuss the evidence for surface water and ideas about the climate history of Mars, it's useful to get our bearings by looking at the large-scale geographic features of the planet.

A Map of Mars Figure 8.12 shows the full surface of Mars, with the poles at the top and bottom and the equator running horizontally across the middle (in much the same way that an atlas shows the full globe of Earth). Mars is about half as large in diameter as Earth, so its surface area

Figure 8.11

This photograph shows a dust devil on Mars, photographed by the *Spirit* rover in early 2005.

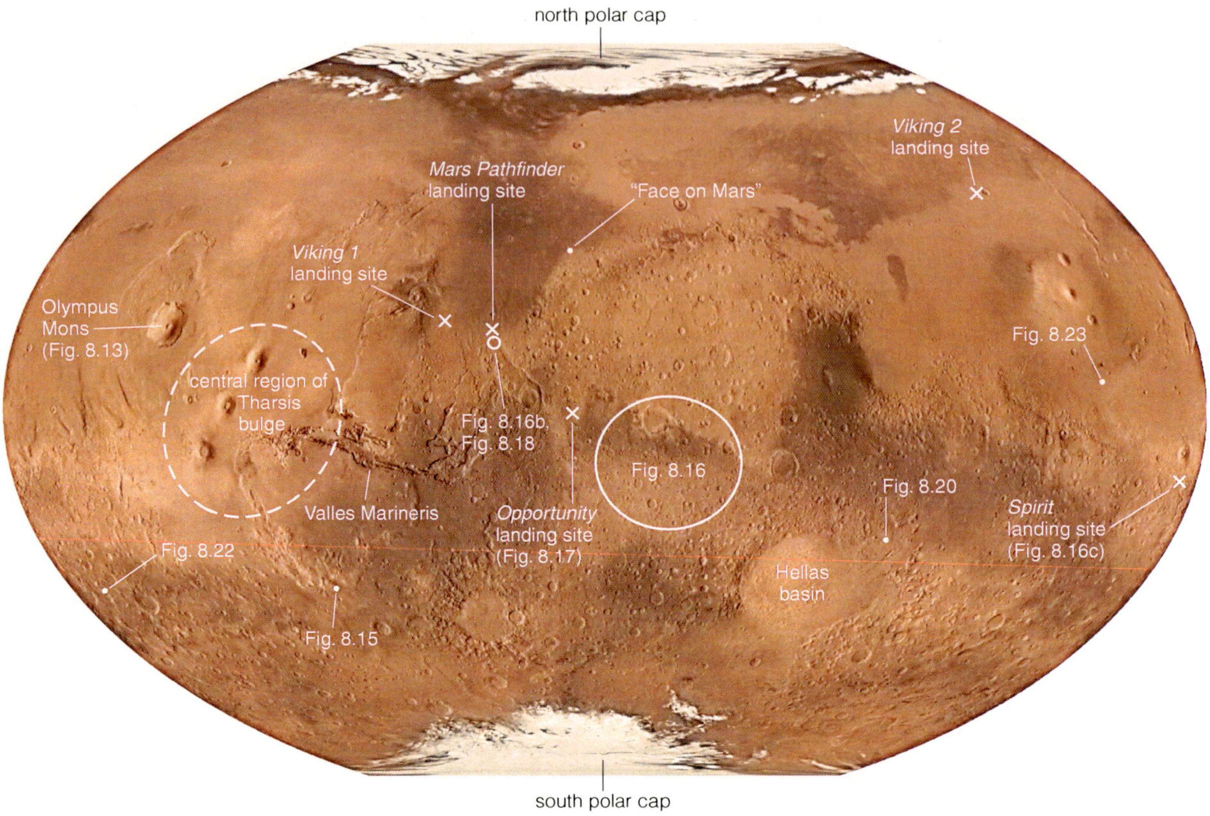

Figure 8.12 Interactive Figure

This image showing the full surface of Mars is a composite made by combining more than 1,000 images with more than 200 million altitude measurements from the *Mars Global Surveyor* mission. Several key geological features are labeled, and the locations of features shown in close-up photos elsewhere in this book are marked.

is about one fourth that of Earth (because surface area is proportional to the square of the radius). Thus, if you placed this map next to a map of Earth on the same scale, the Earth map would be four times as large in area. Moreover, because water covers about three fourths of Earth's surface, the total land area of Mars is about the same as the total land area of Earth.

THINK ABOUT IT Try to find a map of Earth that has about four times the area (twice as long and wide) as Figure 8.12. (If you cannot find a map of the right size, use any Earth map and try to photocopy it at the appropriately scaled size.) Compare the size of various Earth features, such as continents and oceans, to that of various Mars features.

Many key geographical features are clearly visible in Figure 8.12. After the polar caps, the most striking feature is probably the dramatic difference in terrain around different parts of Mars. Much of the southern hemisphere has relatively high elevation and is scarred by numerous large impact craters, including the very large crater known as the Hellas Basin. In contrast, the northern plains show few impact craters and tend to be below the average martian surface level. This difference in crater crowding tells us that the southern highlands are a much older surface than the northern plains: Recall that all worlds should have been heavily scarred with craters during the heavy bombardment, so surface regions that still show these craters must be quite old (3.8 billion years or more) while surface regions that have few craters must have had their early craters erased by subsequent geological processes [Section 4.3].

Volcanism was probably the most important of the processes responsible for erasing ancient craters on Mars, although tectonics and erosion also played a part. The northern plains show features that are characteristic of lava flows, suggesting that eruptions of an extremely fluid lava covered up the older impact craters. Interestingly, we can see faint "ghost" craters in some of these regions, suggesting that the lava flows were not thick enough to completely erase the underlying features and confirming that the entire planet was once densely cratered. Plenty of mysteries still remain, however. For example, no one knows why volcanism should have affected the northern plains so much more than the southern highlands or why the two regions differ so much in elevation.

Other evidence of volcanism on Mars comes from its towering volcanoes. You can probably recognize the volcanoes in Figure 8.12 by their dome shapes and central calderas (the "craters" in the centers of volcanoes). Many are concentrated on or near the continent-size Tharsis Bulge. Tharsis, as it is usually called, is some 4,000 kilometers across, and most of it rises several kilometers above the average martian surface level. It was probably created by a long-lived plume of rising mantle material that bulged the surface upward and provided the molten rock for the eruptions that built the giant volcanoes. The Tharsis volcanoes dwarf any found on Earth. Olympus Mons (Figure 8.13), located near the northwest edge of Tharsis, is probably the largest mountain in the solar system. Its base is some 600 kilometers across, large enough to cover an area the size of Arizona, and is rimmed by a cliff that in places is 6 kilometers high. Its peak stands about 26 kilometers above the average martian surface level, or some three times as high as Mount Everest stands above sea level on Earth.

East of Tharsis and just south of the equator is the long, deep system of valleys called *Valles Marineris*. Named for the *Mariner 9* spacecraft that first imaged it, Valles Marineris is as long as the United States is wide and almost four times as deep as the Grand Canyon. No one knows exactly how Valles Marineris formed. However, if you look closely at Valles Marineris in Figure 8.12, you'll see that its west end shows extensive cracks where it runs up against the Tharsis Bulge, suggesting a connection between the two features. Valles Marineris may have formed through tectonic stresses accompanying the uplift of material that created the Tharsis Bulge, cracking the surface and leaving the tall cliff walls of the valleys. A few features of the valleys may have been shaped by flowing water, and in some places the canyon walls show evidence of layering that may have been caused either by numerous lava flows or by deposits of sediments. Moreover, the canyon is so deep in places that some of its walls probably were once several kilometers underground, where they may have been exposed to subsurface liquid water. For this reason, Valles Marineris may be one of the best places to look for fossil evidence of past martian life.

Martian Geological History

The varying terrain and geographical features of Mars provide important clues to its geological history. In particular, detailed comparisons of crater counts on different parts of Mars allow us to divide the surface broadly into regions of three different ages (Figure 8.14). The oldest and most heavily cratered regions—which include the heavily cratered southern highlands—must still be essentially unchanged from the way they looked at the end of the heavy bombardment,

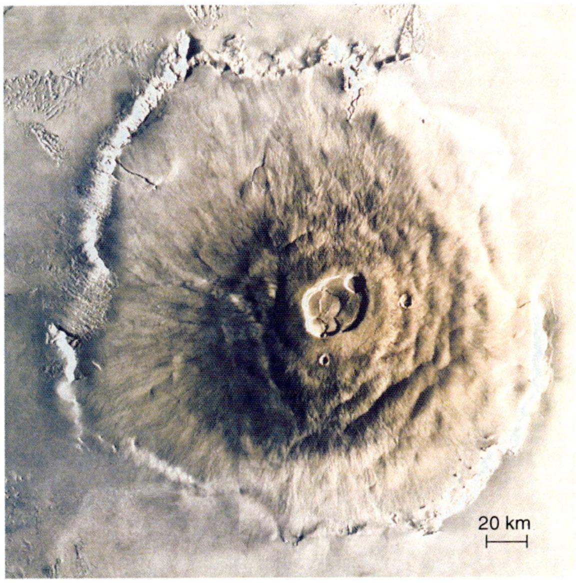

Figure 8.13

Olympus Mons, photographed from orbit. For perspective on its size compared to the entire planet, you should also look at this giant volcano toward the far left of Figure 8.12.

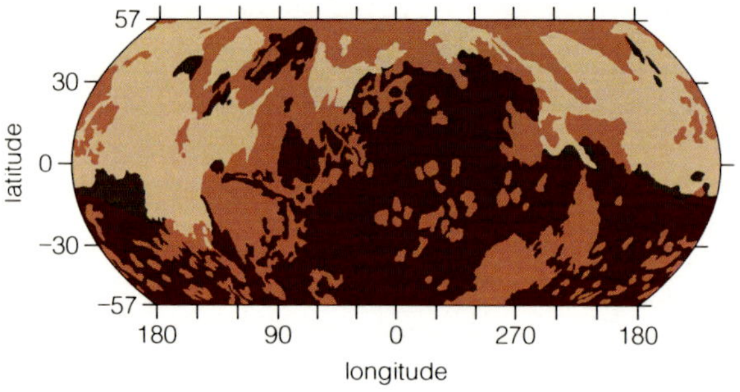

Figure 8.14

This simplified map of Mars shows how different surface regions have different ages based on crater counts; the color coding is summarized in Table 8.2 below. (Compare this simplified map to the more detailed view in Figure 8.12.)

TABLE 8.2 *Geological Eras on Mars*

Era	*Color Code in Figure 8.14*	*Time*
Early (Noachian)	(dark brown)	4.6–3.8 billion years ago
Middle (Hesperian)	(medium brown)	3.8–1.0 billion years ago
Recent (Amazonian)	(light tan)	1.0 billion years ago to the present

some 3.8 billion years ago. The youngest regions are those with the fewest craters, such as the regions around the volcanoes of Tharsis. Crater counts suggest that these young regions are less than about 1 billion years old. In between, Mars has large patches of "middle-aged" surface where crater counts suggest an age from about 1 billion to 3.8 billion years old. For simplicity, we will refer to these different age surfaces as dating to the early, middle, and recent eras of Mars, as summarized in Table 8.2.* Keep in mind that ages based on crater counts are quite rough; we will get more precise ages only when we are able to collect rocks from the different regions and determine their ages through radiometric dating [Section 4.2].

THINK ABOUT IT Does Earth's surface have regions of different ages similar to what we find on Mars? Explain.

By examining the types of geological features that appear on surfaces of different ages on Mars, we can get an idea both of what processes helped shape the surface and of when they operated. For example, we can look at features that indicate volcanic eruptions, such as lava flows or volcanoes, and deduce the history of volcanism. Such studies show that frequency of volcanic eruptions on Mars has decreased steadily since at least about 3.5 billion years ago, just as we would expect for a planet small enough to have lost much of its internal heat by now.

Nevertheless, Mars may not yet be geologically "dead," a fact that has important implications to the possibility of martian life. Although martian volcanoes show enough impact craters on their slopes to suggest that they have been inactive for at least tens of millions of years, analysis of martian meteorites (meteorites that appear to have come from Mars [Section 6.2]) offers a different perspective. Radiometric dating of these meteorites shows some of them to be made of volcanic rock that solidified from molten lava as little as 180 million years ago—quite recent in the 4.5-billion-year history of the solar system. Given this evidence of geologically recent volcanic eruptions, it is likely that martian volcanoes will erupt again someday, though not necessarily in our lifetimes. In that case, Mars must still retain at least some internal heat, and almost certainly enough to melt some underground ice into liquid water.

• What geological evidence tells us that water once flowed on Mars?

Large-scale features of Mars are important to our overall understanding of the planet, but features that might have been carved by liquid water are of particular interest for our investigation of Mars's habitability. The martian surface shows ample evidence of liquid water in its past, which is the prime reason we consider Mars a candidate for past or present life.

Evidence from Orbit The first evidence of past water came from photos taken by *Mariner 9* and the *Viking* orbiters, some of which showed features that that look much like dry riverbeds on Earth seen from above

*Planetary scientists sometimes refer to the early period of Mars as the *Noachian* (pronounced *"no-AH-ki-an"*) era, the middle period as the *Hesperian* era, and the recent period as the *Amazonian* era.

(Figure 8.15). These channels were almost certainly carved by running water, though no one knows whether the water came from runoff after rainfall, from erosion by water-rich debris flows, or from an underground source. From counts of the craters in and near the channels, it appears that they are at least 2–3 billion years old, meaning that water has not flowed through them since that time. Nevertheless, they tell us an important story about the martian past: The branching nature of the channels suggests that they were carved gradually, so liquid water must have been stable at or just below the surface at that time. Because the low temperature and low atmospheric pressure make liquid water unstable at the surface today, we conclude that Mars must have had a much warmer and thicker atmosphere during at least some times in its distant past.

Other orbital evidence also argues that Mars had rain and surface water long ago. Figure 8.16a shows a broad region of the ancient, heavily cratered southern highlands. Notice the indistinct rims of many large craters and the relative lack of small craters. Both facts argue for ancient rainfall, which would have eroded crater rims and erased small craters altogether. Figure 8.16b shows a close-up of a crater floor with sculpted patterns that suggest it was once the site of a lake. Ancient rains may have filled the crater, allowing sediments to build up from material that settled to the bottom. The sculpted patterns in the crater bottom may have been created as erosion by wind exposed layer upon layer of sedimentary rock, much as erosion by the water of the Colorado River exposed the layers visible in the walls of the Grand Canyon on Earth. Figure 8.16c shows a three-dimensional perspective of a surface region where a channel between two craters suggests that they were once lakes with water flowing between them.

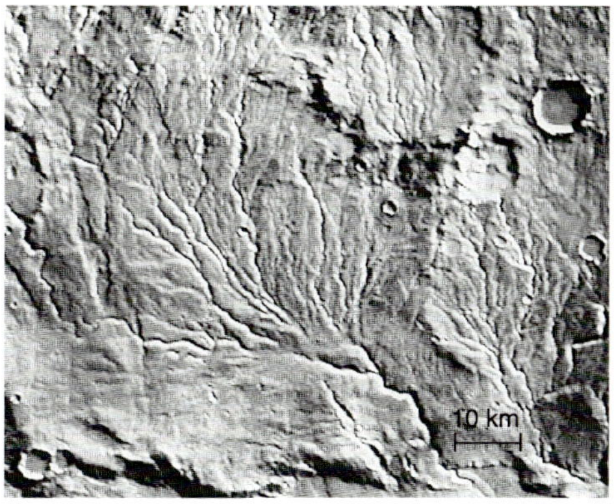

Figure 8.15

This photo, taken by a *Viking* orbiter, shows what appear to be dried-up riverbeds. Toward the top of the image we see many individual tributaries, which merge into the larger "river" near the lower right. Counts of craters near the channels indicate that they formed at least 2–3 billion years ago.

Surface Evidence Surface studies have provided additional evidence for past water on Mars. The *Pathfinder* landing site was in the midst of

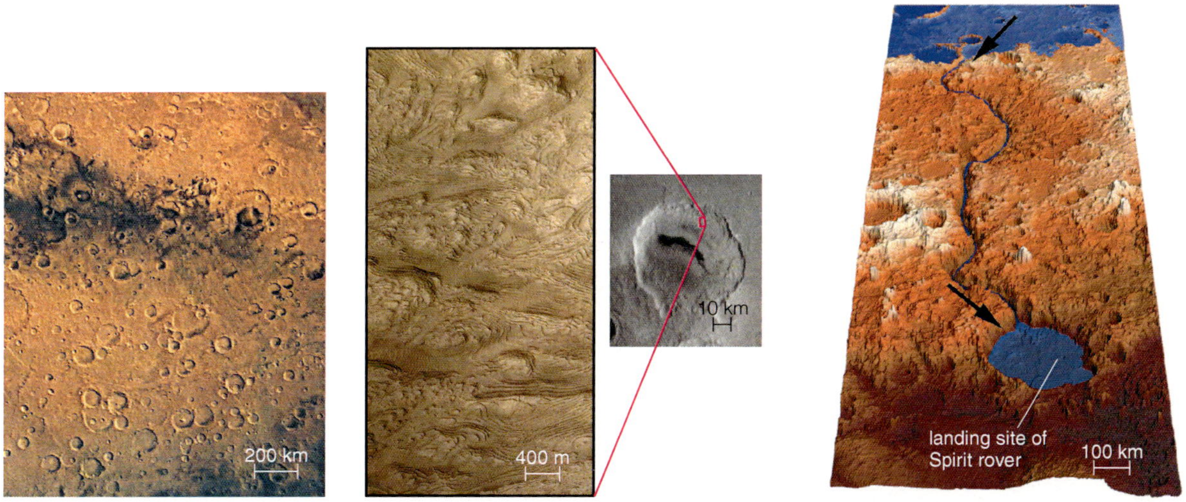

Figure 8.16

More orbital evidence of past water on Mars.

a This photo shows a broad region of the southern highlands. The eroded rims of large craters and the relative lack of small craters suggest erosion by rainfall.

b The close-up view of this crater floor shows sculpted patterns that probably represent layers of sedimentary rock, presumably laid down at a time when water filled the crater.

c This computer-generated perspective shows how a martian valley forms a natural passage between two possible ancient lakes (shaded blue); the lower crater is Gusev, where the *Spirit* rover landed. Vertical relief is exaggerated 14 times to reveal the topography.

a Zooming in on a knee-high rock outcrop near the rover's landing site. The close-up shows a piece of the rock about 3 cm across. The layered structure, odd indentations, and small sphere all support the idea that the rock formed from sediments in standing water.

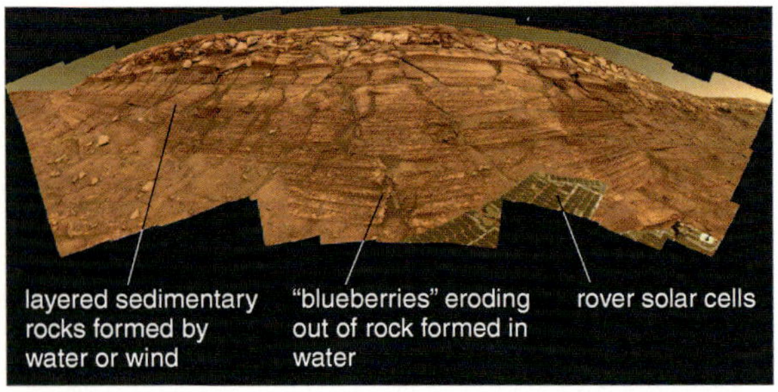

b The wall of Endurance Crater shows bedrock layers altered by water. Close-ups show hematite "blueberries" eroding out of the rock. The changing tilts of the rock layers hint at changing wind or waves during deposition.

Figure 8.17

Photos from the *Opportunity* rover on Mars.

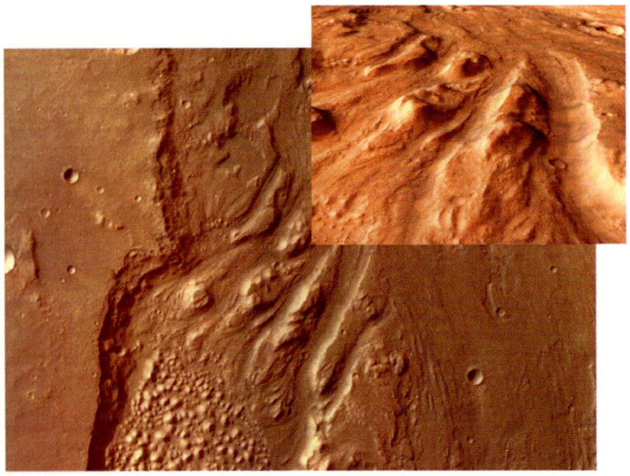

Figure 8.18

This image from Europe's *Mars Express* orbiter shows outflow channels likely carved by floodwaters. The inset is a perspective view of the central region.

what appears from orbit to be a great floodplain. The surface view from *Pathfinder* (see Figure 8.5) supports the idea that a flood passed through the region, because rocks are scattered and stacked against each other in the same way that we see them after floods on Earth.

Even stronger evidence comes from the recent rover missions to Mars. The *Spirit* rover has explored Gusev crater, suspected site of an ancient lake (see Figure 8.16c). *Opportunity* landed in the Meridiani Plains, where orbital spacecraft had detected spectroscopic hints of minerals that form in water. The twin rovers carried dozens of cameras, several instruments to identify rock composition, and even a grinder to expose fresh rock for analysis.

Both rovers found strong evidence that liquid water was once plentiful on Mars. Rocks at the *Opportunity* landing site contain tiny spheres—nicknamed "blueberries" although they're neither blue nor as large as the berries we find in stores—and odd indentations suggesting that they formed in standing water, or possibly by groundwater percolating through rocks (Figure 8.17). Compositional analysis shows that the abundant "blueberries" contain the iron-rich mineral hematite, and other rocks contain the sulfur-rich mineral jarosite. Both of these minerals form in water, and chemical analysis seems to support the case for formation in an acidic, salty environment such as a sea or ocean. In addition, the layering of the sedimentary rocks suggests a changing environment of waves and/or wind, furthering the case for a past sea or ocean in the Meridiani Plains.

The specific case for water in the Meridiani Plains is not fully closed, because some scientists have proposed volcanic processes, or even asteroid impacts, that might mimic the features more commonly attributed to water. Nevertheless, taken in combination with all the other evidence we have discussed, it seems a near-certainty that Mars had abundant flowing water in its distant past.

Possible Floods and Oceans Craters in younger regions of the martian surface show far less erosion than do the ancient craters of the southern highlands, suggesting that little or no rain has fallen during the past 2–3 billion years or more. Nevertheless, many surface features show evidence of more recent water flows. This is not as far-fetched as it may sound. Although liquid water is unstable on the surface of Mars today, a large quantity of water that suddenly erupts onto the landscape may take a short time to freeze or evaporate completely. Thus, a catastrophic release of floodwaters—from beneath the surface, for example—could survive long enough to carve channels and other surface features.

Some features on surfaces dating to middle or recent eras on Mars do indeed appear to have been shaped by catastrophic floods. As noted earlier, the *Pathfinder* landing site was in the midst of a region (called Ares Vallis) that shows many features suggesting past flooding, though it is also possible that glaciers were responsible (Figure 8.18). Assuming that the channels were carved by liquid water, we can trace the flow upstream to its source region. Because no surface reservoir is visible there, the water must have emerged from underground. Moreover, the floods appear to have occurred at least sporadically throughout the past few billion years, suggesting that liquid water has been present in the martian crust for most or all of the planet's history.

Some large volcanoes show water-carved channels on their flanks, suggesting a link between volcanic heating and some of the floods. Figure 8.19 shows a volcano with numerous downhill channels flowing outward in

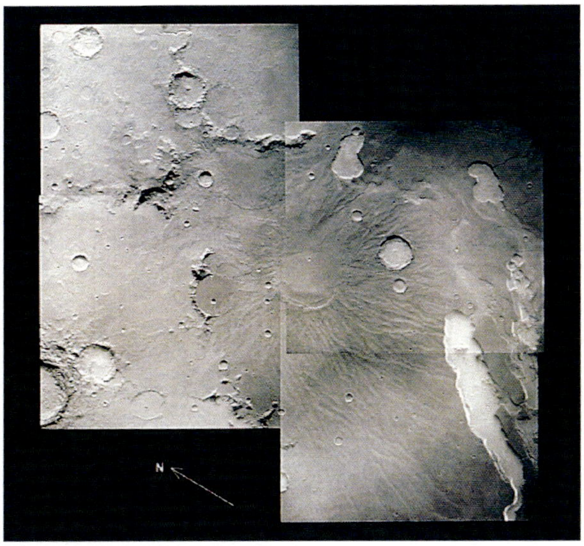

Figure 8.19

These photos show the volcano Hadriaca Patera (center), with its caldera just to the right of the border of the overlapping photos. Note the many channels flowing downhill from the caldera and the wide flood channel (called Dao Vallis) toward the lower right.

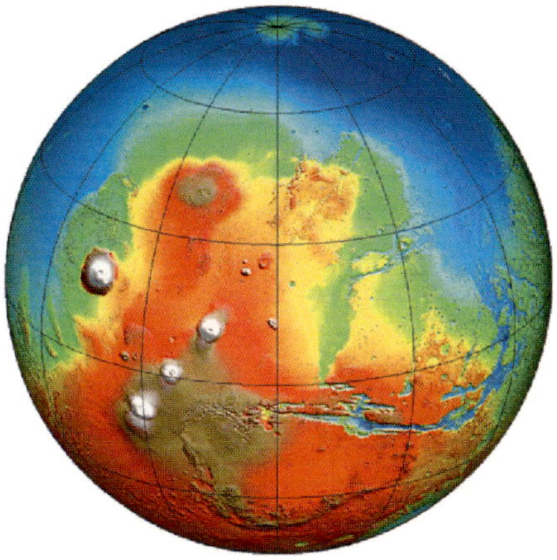

Figure 8.20

This map shows Mars color-coded by elevation. Blue areas are the farthest below the average surface level, and red and brown areas are the highest above it. Note that the entire north polar region is quite low in elevation, suggesting the possibility that this low-lying region once held an ocean. The evidence for an ocean is still weak, however.

all directions from its central caldera. Toward the lower right, we see a much wider channel that was probably carved by floodwaters released during one or more eruptions. These features suggest the past existence of underground pockets of liquid water near volcanoes—a potential habitat for life. If martian volcanoes remain active today, such pockets of water might still exist.

Most of the large flood channels appear to have drained into the northern plains, where the water may have collected before freezing, evaporating, or being absorbed into the surface. Could this mean that the northern lowlands once contained an ocean of liquid water? The topography suggests the possibility (Figure 8.20), but scientists are actively debating whether the geological features seen from orbit support the idea. Some features along the edges of the low regions look like shorelines, but we'll probably need surface studies from future landers to decide whether an ocean really existed.

Martian Water Today Mars may lack surface liquid water today, but a lot of water still exists in the form of ice. We see water ice in the polar caps, and enough ice sublimates into gas that we find water vapor and ice crystals in the martian atmosphere, which sometimes form clouds in the martian sky. Moreover, orbital studies strongly suggest that a significant amount of water ice is present in the martian subsoil around much of the planet (Figure 8.21).

Could any of this ice still melt into liquid water and flow on Mars today? Evidence for this intriguing possibility comes from orbital photos

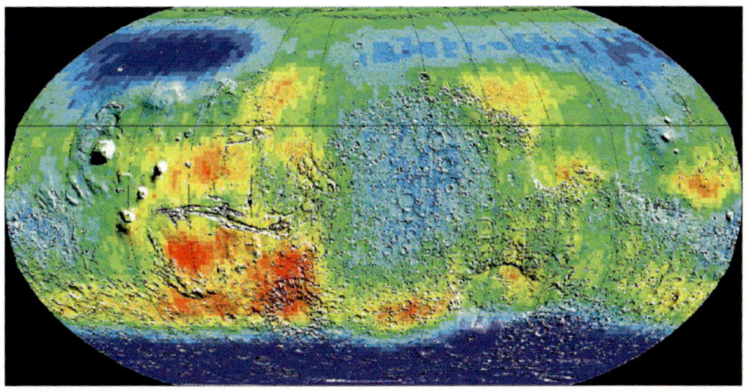

Figure 8.21

This map, made with observations from *Mars Odyssey,* represents the hydrogen content of the martian surface soil. The blue areas contain the most hydrogen, probably because they represent regions in which the top meter or so of the surface soil contains water ice.

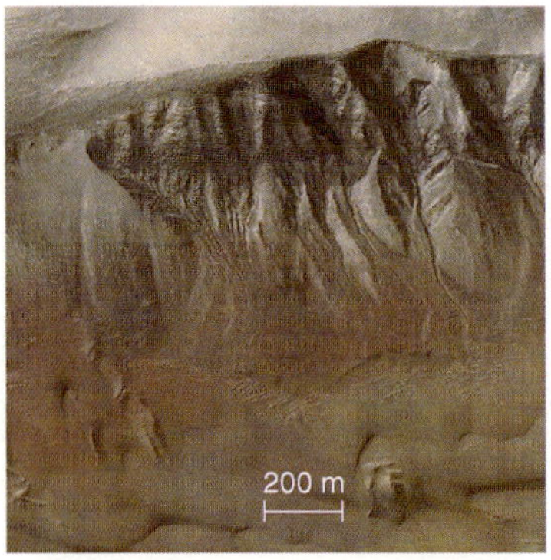

Figure 8.22

This orbital photograph from *Mars Global Surveyor* shows gullies on a crater wall. The gullies may have been formed by water melting under the protective cover of snowpack.

showing gullies on crater and channel walls that look strikingly similar to the gullies we see on almost any eroded slope on Earth (Figure 8.22). One hypothesis suggests that the gullies form when snow accumulates on the crater walls in winter and melts away in spring. Because the gullies are relatively small (note the scale bar), they should be gradually covered over by blowing sand during martian dust storms. Thus, gullies that are still clearly visible must be no more than a few million years old. Geologically speaking, this time is short enough to make it quite likely that water flows are still forming gullies today. In addition, recent images from *Mars Express* show features that some scientists interpret as "frozen seas" of ice just beneath the surface (Figure 8.23). The lack of craters on the ice plates themselves tells us that they are geologically young, opening the possibility that they might occasionally be exposed to the surface where they could potentially melt.

Even if water does still flow on occasion, Mars clearly was much warmer and wetter at times in the past than it is today. Ironically, Percival Lowell's supposition that Mars was drying up has turned out to be basically correct, although in a different way than he imagined. Thus, it is time to turn our attention to the question of why Mars changed so much, and what this change might mean for the possibility of life.

8.3 THE CLIMATE HISTORY OF MARS

From the time that *Mariner 9* and *Viking* first revealed evidence of past water on Mars, scientists have actively debated the strength of that evidence. At first, some scientists argued that channels might have been carved by lava flows or liquid carbon dioxide, but more detailed studies made this idea seem implausible. Today, scientists debate the various roles of liquid water, mud, ice, and volcanism in shaping particular features, but overall the case seems all but closed: One way or other, liquid water must have been stable or nearly stable on Mars during at least some periods of time prior to about 2–3 billion years ago. For that to have been possible, both the atmospheric pressure and temperature must have been significantly higher than they are today. Mars in the past offered a much more hospitable climate than it does today, and perhaps one in which life could have arisen and taken hold.

• Why was Mars warmer and wetter in the past?

It's easy to conclude that Mars must have been warmer and wetter in the past, but more challenging to explain why. The basic answer presumably lies with the greenhouse effect. Recall that the greenhouse effect can make a planet's surface much warmer than it would be otherwise. A moderate greenhouse effect keeps our own planet Earth from freezing over [Section 4.5], while an extremely strong greenhouse effect is responsible for the blistering temperatures on Venus [Section 7.2].

Figure 8.23

The plate-like features in this orbital photograph from *Mars Express* resemble ice patterns seen in Antarctica, leading some scientists to suggest that they represent a "frozen sea" covered by a thin coating of ash or dust.

Today, Mars has such a thin atmosphere that it has only a weak greenhouse effect, despite the fact that 95% of its atmosphere is composed of the greenhouse gas carbon dioxide (see Table 8.1). But if Mars had a thicker atmosphere in the past—as it must have, since that is the only way to explain the past stability of liquid water—the greenhouse effect would have been much stronger. In fact, computer simulations show that Mars today could have liquid water if it had a carbon dioxide atmosphere about 400 times as dense as the current martian atmosphere; this atmosphere would give Mars a surface pressure about three times that on Earth today and a temperature above freezing.

Could a much thicker carbon dioxide atmosphere be the explanation for the warmer climate of the martian past? On one level, the idea seems reasonable: Mars has plenty of ancient volcanoes, and calculations show that these volcanoes could have easily supplied the atmosphere with 400 times as much carbon dioxide as it has today. Moreover, if martian volcanoes outgassed carbon dioxide and water in the same proportions as do volcanoes on Earth, Mars would have also had enough water to fill oceans tens or even hundreds of meters deep. However, there's one problem with this explanation: While a 400-times-denser atmosphere could allow for liquid water on Mars today, remember that the Sun was dimmer in the distant past [Section 4.5]. Thus, even more greenhouse warming would have been needed to allow for liquid water when Mars was young.

Current climate models are unable to account for the necessary additional warming with carbon dioxide gas alone, so scientists are seeking other ways to explain the past climate. At least three general categories of explanation have been proposed. One set of hypotheses suggests that the greenhouse effect may have been further enhanced by clouds or by other gases, such as methane. Another hypothesis suggests that the young Sun might not have been quite as dim as is generally assumed: The brightness of the young Sun is calculated from stellar models that use the Sun's current mass. If the Sun somehow lost a small fraction of its original mass over time—which might have happened if the solar wind had been much stronger and carried more gas away—then it would not have been quite so dim at early times. A third hypothesis suggests that Mars never had an extended period of warmth, but instead had only intermittent periods of rainfall and warmth, perhaps triggered by the heat of large impacts.

We do not yet know which, if any, of these hypotheses is correct. However, none of them work unless Mars had a substantially thicker atmosphere and stronger greenhouse effect to begin with. In other words, we cannot be sure if the climate was ever quite "Earth-like," but it was certainly once much warmer and wetter than it is today.

• Why did Mars change?

Given that Mars must once have had a much denser atmosphere with a much stronger greenhouse effect, we can explain the current extremely different conditions only if Mars somehow lost a vast quantity of carbon dioxide gas. This loss would have weakened the greenhouse effect until the planet essentially froze over. Where did all this gas go? Some of the carbon dioxide condensed to make the polar caps, some may be chemically bound to surface rock, and some still makes up the martian atmosphere today. However, the bulk of the gas was probably lost to space.

Loss of Carbon Dioxide and Water The precise way in which Mars lost its carbon dioxide gas is not clear, although some gas was almost certainly lost by being blasted away by large impacts. However, recent data suggest that an even more important loss mechanism was tied to a change in Mars's magnetic field (Figure 8.24). Early in its history, Mars probably had molten metals in its core, much like Earth today [Section 4.4]. Circulating with convection and Mars's rotation, these metals should have produced a magnetic field and a protective magnetosphere around Mars. However, the magnetic field weakened as the small planet cooled and the core solidified, leaving the atmosphere vulnerable to solar wind particles.

COSMIC CALCULATIONS 8.1

The Surface Area–to–Volume Ratio

The total amount of heat contained in Mars or any other planet depends on the planet's *volume*, but this heat can escape to space only from the planet's *surface*. As heat escapes, more heat flows upward from the interior to replace it, until the interior is no hotter than the surface. Thus, the time it takes for a planet to lose its internal heat is related to the ratio of the *surface area* through which it loses heat to the *volume* that contains heat:

$$\text{surface area–to–volume ratio} = \frac{\text{surface area}}{\text{volume}}$$

A spherical planet (radius r) has surface area $4\pi r^2$ and volume $\frac{4}{3}\pi r^3$, so the ratio becomes:

$$\underset{\text{(for a sphere)}}{\text{surface area–to–volume ratio}} = \frac{4\pi r^2}{\frac{4}{3}\pi r^3} = \frac{3}{r}$$

Because r appears in the denominator, we conclude that *larger objects have smaller surface area–to–volume ratios.* (Although we've considered a sphere, this idea holds for objects of any shape.)

Example: Compare the surface area–to–volume ratios of the Moon and Earth.

Solution: Dividing the surface area–to–volume ratios for the Moon and Earth, we find:

$$\frac{\text{surface area–to–volume ratio (Moon)}}{\text{surface area–to–volume ratio (Earth)}} = \frac{3/r_{\text{Moon}}}{3/r_{\text{Earth}}} = \frac{r_{\text{Earth}}}{r_{\text{Moon}}}$$

The radii of the Moon and Earth are $r_{\text{Moon}} = 1{,}738$ km and $r_{\text{Earth}} = 6{,}378$ km:

$$\frac{\text{surface area–to–volume ratio (Moon)}}{\text{surface area–to–volume ratio (Earth)}} = \frac{6{,}378\text{ km}}{1{,}738\text{ km}} = 3.7$$

The Moon's surface area–to–volume ratio is nearly four times as large as Earth's, which means the Moon would cool four times faster if all else were equal. In fact, Earth has retained heat much longer, because its larger size gave it more heat to begin with and because Earth has a higher proportion of radioactive elements. (See Problem 51 for a similar analysis of Mars.)

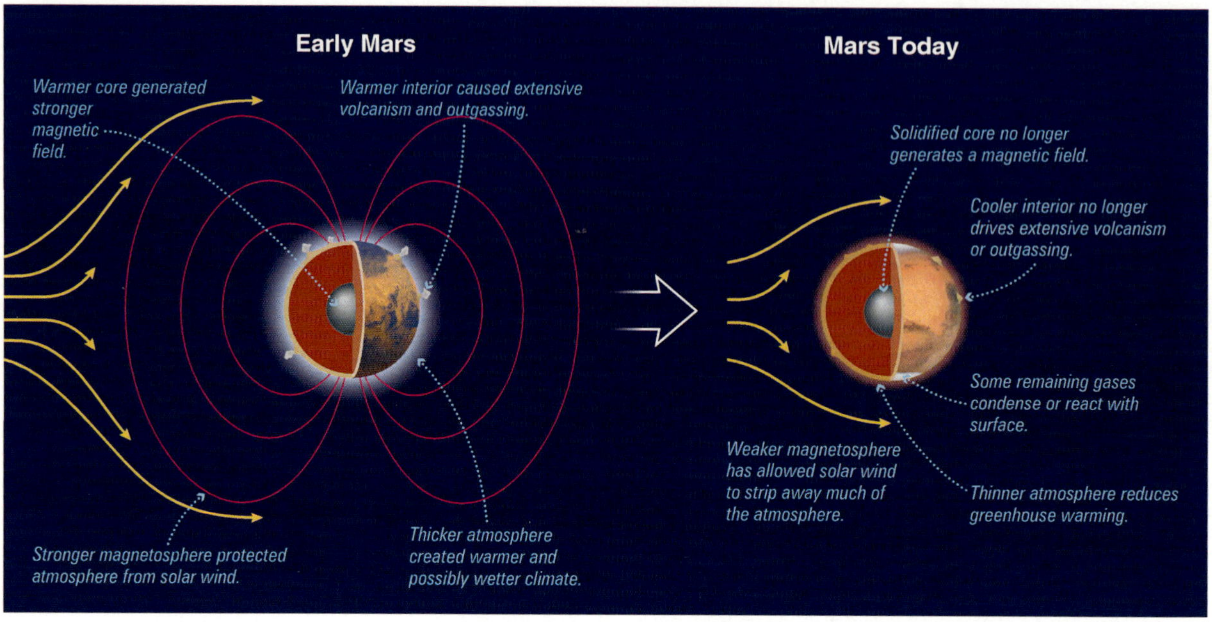

Figure 8.24

Some 3 billion years ago, Mars underwent dramatic climate change, probably because it lost its global magnetic field, leaving its atmosphere vulnerable to the solar wind.

These solar wind particles could have stripped gas out of the martian atmosphere and into space. Had Mars kept its magnetic field, it might have retained much of its atmosphere and might have a more Earth-like climate today.

Much of the water once present on Mars is also probably gone for good. Like the carbon dioxide, some water vapor may have been stripped away by the solar wind. However, Mars also lost water in another way. Because Mars lacks an ultraviolet-absorbing stratosphere, atmospheric water molecules would have been easily broken apart by ultraviolet light from the Sun. The hydrogen atoms that broke away from the water molecules would have been lost rapidly to space through thermal escape—the process in which low-mass gas atoms can reach escape velocity and escape into space [Section 4.4]. With these hydrogen atoms gone, the water molecules could not be made whole again. Initially, oxygen from the water molecules would have remained in the atmosphere, but over time this oxygen was lost too. The solar wind probably stripped some of the oxygen away from the atmosphere, and the rest probably was drawn out of the atmosphere through chemical reactions with surface rock. This oxygen literally rusted the martian rocks, giving the "red planet" its distinctive tint.

In summary, Mars changed primarily because of its relatively small size. It was big enough for volcanism and outgassing to release plenty of water and atmospheric gas early in its history, but too small to maintain the internal heat needed to prevent this loss of water and gas. As its interior cooled, its volcanoes quieted and stopped releasing gas into the atmosphere, while its relatively weak gravity and the loss of its magnetic field allowed existing gas to be stripped away to space. If Mars had been as large as Earth, so that it could still have outgassing and a global magnetic field, it might still have a pleasant climate today. Mars's distance from the Sun also helped seal its fate: Even with its small size, Mars might still have some flowing water if it were significantly closer to the Sun, where the extra warmth could melt the water that remains frozen underground and at the polar caps.

Mars Climate and Axis Tilt With the gas that once warmed the planet now gone, there is little hope that Mars will ever again have the warm, wet climate it had before about 2–3 billion years ago. However, recent studies suggest that Mars still undergoes significant climate change over geologically short periods of hundreds of thousands of years. This climate change arises from changes in the martian axis tilt, and it may have significant implications to the potential habitability of Mars.

Recall that Earth experiences long-term climate cycles, such as ice ages, due to small changes in its rotation and orbit, including small changes in axis tilt. Earth's axis tilt doesn't change much—varying only between about 22° and 25°—because our large Moon exerts a gravitational pull that stabilizes it. Mars lacks a large moon, and its two tiny moons (Phobos and Deimos) are far too small to offer any stabilizing influence on its axis. In addition, because Mars is closer than Earth to Jupiter, Jupiter's gravity more strongly perturbs Mars as it orbits the Sun. Together, calculations suggest that the lack of a stabilizing moon and the effects of Jupiter should cause Mars to experience wild swings in its axis tilt over periods of between 100,000 and 1 million years.

Models suggest that the martian axis tilt varies over time from 0° to as much as 80°, which means that the current 25° is significantly smaller than the average. These changes in tilt would have dramatic effects on the climate (Figure 8.25). When Mars's axis tilt is small, the poles may stay in a perpetual deep freeze for tens of thousands of years. With more carbon dioxide frozen at the poles, the atmosphere becomes thinner, lowering the pressure and weakening the greenhouse effect. When the axis is highly tilted, the summer pole becomes warm enough to allow substantial amounts of water ice to sublimate, along with carbon dioxide, into the atmosphere. The pressure therefore increases, and Mars becomes warmer as the greenhouse effect strengthens. The martian polar regions show layering that probably reflects changes in climate due to the changing axis tilt (Figure 8.26).

Even at the greatest tilts, the atmospheric pressure probably does not become high enough to allow liquid water to pool in surface lakes or ponds. Nevertheless, models suggest that liquid water might form just beneath the surface or at rock/ice boundaries on the surface whenever the tilt is greater than about 40°—and because a 40° value of tilt is probably about the average over time, Mars could have such zones of liquid water during most epochs.

• Is Mars habitable?

Mars clearly has the elements needed for life, and energy is available for life in the form of sunlight (on the surface) and chemical energy (underground). Thus, the question of whether Mars is habitable hinges ultimately on the availability of liquid water.

The geological evidence strongly suggests that Mars once had abundant liquid water at its surface, meaning that Mars *was* habitable some time before about 2–3 billion years ago. The only question is whether it was habitable for a long period of time—making an indigenous origin of life seem plausible—or only for shorter, intermittent periods. If water was

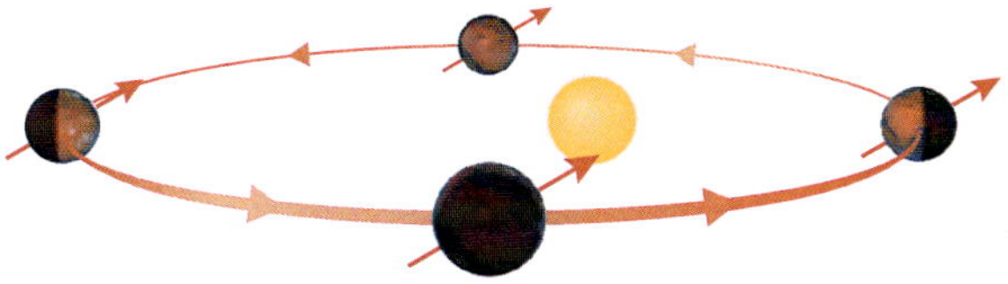

a When the axis is highly tilted, the summer pole receives fairly direct sunlight and becomes quite warm.

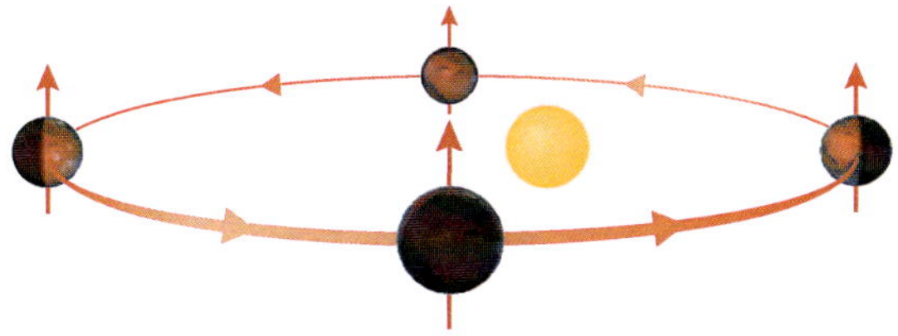

b When the axis tilt is small, the poles receive little sunlight at any time of year.

Figure 8.25

Mars's axis tilt probably varies dramatically, causing climate change because of the effect on the seasons.

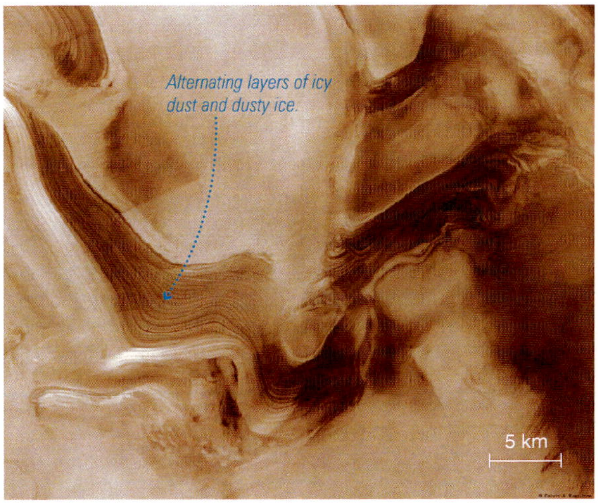

Figure 8.26

This photo from Mars orbit shows a region about 40 kilometers wide at the edge of the north polar cap. Note the alternating layers of ice and dust, which are probably the result of climate changes caused by the changing axis tilt of Mars.

present only intermittently, such as after impacts, then Mars may only intermittently have been habitable. But if the water was present for millions or tens of millions of years, then young Mars may have been quite similar to the young Earth. In that case, Mars may have been habitable during the same period of time in which life arose on Earth [Section 6.1].

The lack of liquid water means the *surface* of Mars is not habitable at the current time. However, we've seen evidence suggesting that pockets of liquid water could still exist underground, kept warm by remaining volcanic heat. In that case, Mars may currently have underground zones of habitability. Moreover, the climate changes tied to the changing martian axis tilt imply that, averaged over geological time, Mars may still have small amounts of surface water at rock/ice boundaries about half the time. On Earth, we find microbes that live in thin films of liquid water at such boundaries, which opens the intriguing possibility that Mars could have surface habitable zones during most epochs.

Unless we are drastically misinterpreting the evidence, the conclusion seems clear. The surface of Mars was habitable during some periods of its early history, and it might still sometimes be habitable when the axis tilt is greater, while the subsurface probably contains habitable zones even today. Given the apparent habitability of Mars, it is time for us to turn our attention to the search for actual life.

8.4 SEARCHING FOR LIFE ON MARS

While we have some confidence in the past and present habitability of Mars, we do not yet know whether Mars has ever actually had life. The only way to learn whether life existed in the past is to search for fossil

MOVIE MADNESS: MISSIONS TO MARS

There was a time, not so long ago, when the term *Martian* was just about synonymous with "space alien." You could frequently meet the Martians at the local cinema, where they were busy invading Earth and ruining everyone's whole day.

The classic example of this type of smooth move was in *War of the Worlds,* which has so far spawned one scary radio play and two moderately scary films. In H. G. Wells's story, sophisticated Martians abandon their turf to grab ours. Mars, you see, was drying up and dying. Earth, on the other hand, was a world with abundant water: a sanctuary for our desperate neighbors.

But space probes of the last four decades have shown that the martian surface is apparently lifeless, and there are no obvious indications of technologically sophisticated inhabitants, either dead or alive. So Hollywood, ever flexible, has switched gears. Now it is Earthlings who go to Mars—often to find hidden signs of habitation that would startle astrobiologists.

In the film *Red Planet,* our descendants try to rebuild Mars into a kinder, gentler world in order to escape environmental disaster on Earth. Robotic craft are sent to melt the polar caps and sow the planet with blue-green algae in a barely plausible bid to produce a warm, breathable atmosphere. In the course of this terraforming project, visiting humans stumble across some complicated life-forms—indigenous Martians—which look like economy-size lice. The lice eat everything from space suits to spacemen, and frankly it's a puzzle why they haven't eaten Mars itself.

As implausible as this may be, astronauts in another space opera, *Mission to Mars,* find something even less reasonable. While checking out an odd mountain in Mars's Cydonia region, they discover that it's really a massive alien "face," disguised by dust and rock. Venturing inside, the astronauts eventually learn that the ancient martian civilization that built the face was wiped out by an asteroid a half-billion years ago. Just before abandoning their planet, the Martians had launched a rocket to seed Earth with their DNA. This molecular emigration supposedly produced the Cambrian explosion that began the reign of multicellular life on Earth. One has to wonder how the Martians could be confident that the jellyfish and trilobites that resulted from their seeding would eventually evolve to humans that would look pretty much like ... the long-gone Martians!

Life on the Red Planet may, indeed, have once existed, and perhaps still does. But our current understanding of conditions on Mars strongly suggests that this life would never have resembled either voracious lice or us. It's probable that any real Martians would be visible only in a microscope.

evidence in martian rocks, and the only way to learn whether life exists today is to find it. To date, only very limited searches for life have been carried out on Mars, but much more is planned for the future.

• Is there any evidence of life on Mars?

The discovery of life on Mars would forever alter our view of life in the universe, so it should be no surprise that many scientists are working hard in hopes to be the first to find evidence for such an important discovery, if life on Mars actually exists. As we'll discuss, some scientists already claim to have found such evidence. But are they interpreting data correctly, or are they engaged in the same type of wishful thinking that led Percival Lowell astray?

The answer is the subject of heated scientific debate, but at the moment the vast majority of scientists are skeptical of claimed evidence for martian life. Nevertheless, it is worth examining the claims, both to illustrate why there is scientific controversy and because they may point us toward ways of resolving the question in the future.

The claims of evidence of life fall into three main categories: claims based on results from the *Viking* landers, recent claims based on evidence of methane in the martian atmosphere, and claims based on studies of martian meteorites found on Earth. We'll examine the first two categories here; we'll save discussion of the martian meteorites for Section 8.5.

Figure 8.27

This pair of before (left) and after (right) photos from the *Viking 2* lander show how the robotic arm pushed away a small rock on the martian surface. (You can see the entire robotic arm in the inset of Figure 8.4.)

The Viking Experiments One obvious way to search for life on Mars is to study the soil to see whether it contains living microbes. This type of search was first carried out by the two *Viking* landers in 1976.

Each of the landers was equipped with several onboard, robotically controlled experiments, along with a robotic arm for scooping up soil samples (Figure 8.27) to test in the experiments; the arm could even push aside rocks to get at shaded soil that was less likely to have been sterilized by ultraviolet light from the Sun. Three experiments were designed expressly to look for signs of life. A fourth was designed to analyze the general content of martian soil.

The first of the three biology experiments (called the *carbon assimilation experiment*) mixed a sample of martian soil with carbon dioxide (CO_2) and carbon monoxide (CO) gas brought from Earth. In some runs of the experiment, the soil was also mixed with water. The aim was to see if any of the carbon dioxide or carbon monoxide would become incorporated into the soil, as would be the case if living organisms were using either gas as a source of carbon in their metabolism. The carbon dioxide and carbon monoxide from Earth could be distinguished from the same gases in the martian atmosphere because they had been "tagged" with radioactive carbon-14 (rather than the far more common, stable isotope carbon-12). Sure enough, the carbon assimilation experiment found that the carbon-14 was incorporated into the soil, a result that seemed to suggest that metabolism was occurring. As a further test, however, the experiment was repeated with soil first heated for 3 hours to 175°C (347°F)—hot enough to break the chemical bonds between carbon and other atoms and presumably kill any carbon-based organisms that might have been present. The results showed that the tagged carbon was still incorporated into the soil, making it seem more likely that the incorporation was due to some sort of geochemical process rather than to biological activity.

The second biology experiment (the *gas exchange experiment*) mixed martian soil with a "broth" containing organic nutrients from Earth. The experiment looked for gases that might be released by the respiration of martian microbes, including hydrogen (H_2), nitrogen (N_2), oxygen (O_2), methane (CH_4), and carbon dioxide (CO_2). Again, the initial results were promising for life: As soon as the soil was exposed to the nutrients, oxygen was released into the chamber. However, further analysis suggested that the oxygen was coming from chemical rather than biological reactions. The oxygen release was somewhat characteristic of photosynthesis, but it occurred in the dark rather than in sunlight, making photosynthesis seem implausible. Moreover, the oxygen was present even when the soil was exposed only to water vapor, rather than to the nutrients, a result inconsistent with metabolism. Finally, in a result similar to that of the first biology experiment, the reactions continued even when the soil was heated to temperatures that should have easily killed any organisms present.

The third biology experiment (the *labeled release experiment*) also mixed martian soil with organic nutrients. The nutrients were tagged with radioactive carbon-14 and sulfur-35 so that, if they were consumed by martian microbes, by-products of metabolism and respiration would be released as gases and be detectable by virtue of their radioactivity. Thus, if life were present, we would expect the level of radioactivity to rise as the organisms consumed the nutrients and released the radioactive gases, and then level off as the nutrients were used up. This is precisely what happened. Moreover, unlike the cases for the first two experiments, heating of the soil in this experiment produced results consistent with life: Heating the soil to 50°C (122°F) substantially reduced the amount of radioactivity, and heating it to 160°C (320°F) eliminated any sign of the tagged isotopes in the chamber gas. This is just what we would expect if the heat were killing living organisms that caused the reactions.

It is this third experiment that leads a small fraction of scientists to think that *Viking* found evidence of life on Mars. However, most scientists are skeptical because of the results of a fourth experiment (the *gas chromatograph/mass spectrometer*) that measured the abundance of organic molecules in the martian soil. This experiment started by heating the soil to temperatures as high as 500°C (930°F)—high enough to kill any organisms, break apart any organic molecules they contained, and release the molecules as gases. The gases were passed through a *gas chromatograph,* a device that separates different gases as they pass through it, and then analyzed with a *mass spectrometer,* an instrument that measures the masses of molecules. The results showed no measurable level of organic molecules in the martian soil at all. (The only organic molecules detected were solvents left over from the cleaning of the instrument before it was launched from Earth.) Moreover, based on the sensitivity of the experiment, these results implied that organic molecules could not be present in the tested martian soil at a concentration of more than a few parts per billion—some ten million times lower than the concentration in typical Earth soil. This apparent lack of organic material seemed to rule out a biological explanation for any of the results in the other three experiments, suggesting that the seemingly positive result of the third experiment was more likely produced by some chemical process mimicking a biological result.

Methane on Mars In 2004, scientists working with the European Space Agency's *Mars Express* mission announced an unexpected and potentially important result: Instruments on the spacecraft had detected methane gas in the martian atmosphere. This news came as a surprise to most scientists, because methane gas cannot last more than a few centuries in the martian atmosphere before chemical reactions transform it into other gases.* Thus, if the methane detection is real, it implies that Mars has a source of ongoing methane production somewhere on or beneath its surface.

Some scientists still doubt that the data are being properly interpreted, but a few ground-based observations made to check the result seem to confirm a methane signature in spectra of Mars. Where could the methane be coming from? We know of at least three possibilities: comet impacts, volcanic activity, or life. The first possibility is highly unlikely, since impacts are such rare events and an impact would have to have occurred quite recently for methane to remain in the atmosphere. That leaves us with volcanism or life.

A few scientists have looked at correlations between the location of methane and water vapor in the martian atmosphere and used the results to argue that we might actually be seeing the signature of microscopic life on Mars. However, volcanism also seems a reasonable explanation, especially since other evidence suggests that Mars has at least some low level of ongoing volcanic activity. Nevertheless, even this result could have implications for the existence of life: If there is enough volcanic activity to produce a detectable level of methane, it may also mean there is enough volcanic heat to make pockets of underground liquid water, raising the possibility of subsurface life.

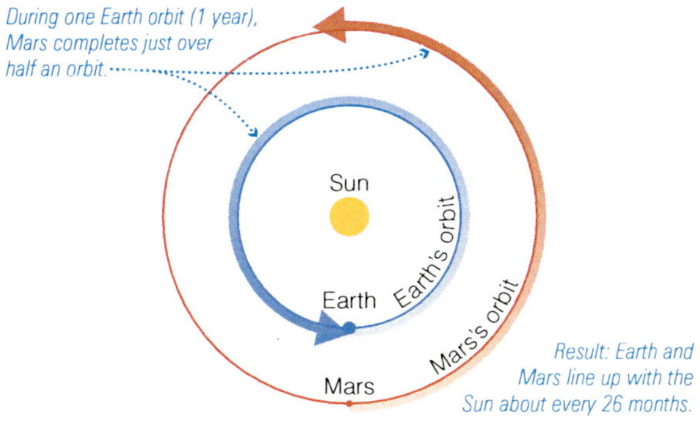

Figure 8.28

Mars takes almost 2 years to orbit the Sun compared to Earth's 1 year. Thus, if you start from an alignment as shown here, after 1 year Earth will be back in the same place, but Mars will be only halfway around. It therefore takes a second year before the two planets line up again. Using Mars's more precise orbital period of 1.88 years, we find the alignments occur every 780 days, or about 26 months.

• How do we plan to search for life on Mars?

The world scientific community has ambitious plans for continuing the search for life on Mars through new space missions. The orbits of Earth and Mars bring the two planets to closest approach about every 26 months (Figure 8.28), and scientists hope to take advantage of every upcoming alignment to send new and ever more sophisticated spacecraft to Mars.

Upcoming Mission Plans The most recent alignment, which made Mars bright in our skies during the fall of 2005, was used to time the launch of the *Mars Reconnaissance Orbiter* (*MRO*). *MRO* reached Mars orbit in March 2006 and is capable of much higher-resolution images of the surface than prior orbiters. It is also equipped with instruments to study the mineral content of the surface from orbit and to search for subsurface water ice. Its findings should help scientists choose sites for future landing missions.

The next planned lander, a NASA mission called *Phoenix,* is scheduled to be launched in August 2007, with arrival at Mars in May 2008 (Figure 8.29). *Phoenix* will not be a rover, but will have a robotic arm for scooping soil and onboard experiments for analyzing it. It will land in the north polar region, where it should be able to dig into surface ice and search for evidence of buried organic material. For the following launch window to Mars, in fall 2009, NASA is building the *Mars Science Laboratory,* a rover that will be considerably larger and more sophisticated than the *Spirit* and *Opportunity* rovers. If all goes well, the *Mars Science Laboratory* will arrive on the martian surface in May 2010. It will combine the capa-

Figure 8.29

This painting shows the *Phoenix* lander, using rocket thrusters to land on Mars. The landing is planned for May 2008.

*The methane is oxidized to form water and carbon dioxide; the oxidation occurs because there is always some amount of free oxygen in the martian atmosphere that is made by the breakup of atmospheric carbon dioxide molecules.

bilities of the earlier rovers with onboard experiments for analyzing soil and rock in much more detail.

Although any of these missions could in principle find evidence of life, they are designed primarily to make further assessments of the habitability of Mars. A more thorough robotic search for life will require more sophisticated and expensive instruments, so it seems reasonable to start by deciding where and how best to look before actually conducting the search. Nevertheless, scientists are already working on plans for robotic missions that could conduct an actual search for life in the next decade, perhaps by drilling deeper into the surface, or by exploring larger regions of the terrain. An even better way to search for life would be to bring martian rock back to laboratories on Earth, and scientists hope that a sample return mission could occur as early as a decade from now.

Preventing Contamination—In Both Directions Given the likelihood that some Earth organisms could survive in at least a few locations on Mars, it's important to make sure that our robotic missions don't accidentally contaminate Mars with life from Earth. Otherwise, microbes that hitched a ride from Earth aboard a spacecraft might fool us into thinking we'd found evidence for martian life. The possibility of contamination also poses an ethical issue: It's at least conceivable that terrestrial life could outcompete any indigenous martian life, driving the martian life to extinction. Do we have a right to do something that could endanger native life on another planet? Clearly, the best way to avoid these problems is to prevent contamination in the first place. An international treaty, signed in 1967, requires that any spacecraft sent to Mars must have a less than 1 in 1,000 chance of causing contamination. Today, scientists strive for even lower contamination probabilities by sterilizing spacecraft before they are launched.

Similarly, but in the other direction, the prospect of a sample return mission has caused some people to fear we might unleash dangerous martian microbes on Earth. Could martian microbes cause disease for which we are unprepared or outcompete terrestrial organisms on our own turf? We cannot completely rule out any danger, but it is quite unlikely, because disease-causing microbes are highly adapted to the species they infect. For example, diseases that infect plants generally do not infect animals. Indeed, "species jumping" by diseases is quite rare and generally occurs only between species that are evolutionarily close. HIV (the virus that causes AIDS), for example, is thought to have jumped from chimpanzees to humans, but on an evolutionary level this is a fairly small jump between different species of primates. Thus, even if martian microbes were accidentally released and subsequently survived on Earth, it's unlikely that they would cause disease. In addition, because martian meteorites must frequently land on Earth, life that hides in martian rocks has almost certainly reached Earth already. The fact that we do not see any harmful effects from this "natural contamination" makes it unlikely that any martian life can harm Earth life.

Nevertheless, it pays to be cautious, given the high stakes involved, and samples brought back from Mars will surely be transported in sealed containers that would not break open even if they were to crash on Earth. Once here, they will be quarantined, and subject to biological tests such as exposing terrestrial microbes to them. Biologists already know how to deal with dangerous terrestrial microbes, such as the Ebola virus, and scientists are developing protocols to ensure safe handling of any harmful martian organisms.

THINK ABOUT IT Should we allow samples from Mars to be brought to Earth, or should they be studied only in space, such as on the Space Station or at a Moon colony? Defend your opinion.

• Should we send humans to Mars?

A long-term dream of many people, and a part of NASA's vision for coming decades, is to send humans to Mars. Sending people is far more difficult than sending robots. Even with the most advanced rockets that anyone now has on the drawing board for the next couple of decades, the trip to Mars would take at least 4 months. That means a human mission would have to carry not only the weight of the astronauts and their living quarters, but also enough food, air, and water to last the trip. Shielding against dangerous radiation would also be necessary, which means having an onboard "storm cellar" in case a violent flare erupts on the Sun. Once on Mars, astronauts might conceivably get water from the subsurface ice and chemically extract oxygen from martian water or rock, but they would still need food, which would have to be either taken along with them or sent separately aboard other spacecraft. Astronauts would also need to come home, which means they'd need fuel for the return journey. That would add far more weight to the human mission, unless we can manufacture the fuel for the return mission on Mars rather than bring it from Earth (an idea that is being actively explored). Still, no matter how you look at it, the enormous amount of stuff required for a human mission ensures it would cost at least as much as dozens of robotic missions, and of course it would pose many dangers to the crew.

Scientific Pros and Cons While the cost and inherent danger of sending humans to Mars would be very high, the scientific payoff could potentially be even higher. We humans are far more capable than any robot, and a team of scientists with vehicles for traveling around the planet and equipment for drilling into the crust might well answer our questions about martian life long before they would be answered by robotic explorers. However, sending humans to Mars also has at least one significant scientific drawback: It vastly complicates the issue of avoiding contamination by terrestrial organisms. People are veritable warehouses of microbes: The number of bacteria in the average person's mouth, for example, is far greater than the number of people who have ever lived. We harbor microbes on our skin, in our breath, in our food, and in our excrement. Preventing all these microbes from escaping into the martian environment during an extended stay on the planet would be nearly impossible.

The scientific pros and cons of sending humans to Mars are fairly clear, but the history of the space program shows that human exploration has rarely been driven by science. The manned space program began for political reasons, largely as part of a "race" between the United States and the Soviet Union. For example, while the *Apollo* program provided valuable scientific data about the Moon, its primary purpose was to prove to the world that the Americans could get there before the Soviets. If we decide to send humans to Mars, the decision will also probably be based more on social and political considerations than scientific ones.

Terraforming Mars Some people dream of eventually establishing permanent colonies on Mars. For the near future, any such colonies would have to be self-contained environments, and no one would dare venture outside without a space suit. But for the more distant future, some people wonder if we might be able to alter the martian environment in ways that would make it more hospitable to us. Such changes go by the name **terraforming,** because they would tend to make the planet more Earth-like.

Proposals to terraform Mars envision changing the environment so that the atmospheric pressure and temperature become greater. Raising the temperature might be accomplished by adding a greenhouse gas to the atmosphere, while raising the pressure would require adding a large amount of gas to the atmosphere. One suggestion involves manufacturing chlorofluorocarbons (CFCs), which are strong greenhouse gases, and releasing them into the martian atmosphere. If we could strengthen the greenhouse effect enough, the warmer temperatures might begin to release frozen carbon dioxide from the martian polar caps and elsewhere beneath the surface, which would further increase the atmospheric pressure and strengthen the greenhouse effect. There still wouldn't be oxygen to breathe, but if the pressure rose enough, we might be able to walk around on Mars carrying only an oxygen tank (and some protection from ultraviolet radiation) rather than having to wear a full, pressurized space suit. Such conditions might also allow plants to survive outdoors, making it much easier to grow food, and eventually increasing the concentration of atmospheric oxygen.

The idea might just work, but putting it into practice wouldn't be easy. Because CFCs tend to be broken apart by sunlight, we would have to manufacture them continually and in great abundance in order to start the greenhouse warming. Calculations suggest that we would need a manufacturing capability about a million times greater than our recent CFC-manufacturing capability on Earth and would need to keep it up for a *few hundred thousand years* before the surface warmed enough to drive substantial quantities of carbon dioxide into the atmosphere. Thus, if it is possible at all, we have plenty of time to consider the ethical issues of terraforming, which could be quite significant if Mars turns out to have life: Do we have a right to alter a planet in a way that could harm its native life?

THINK ABOUT IT A similar ethical issue surrounds endangered species on Earth. Some people say that we have no right to drive any species to extinction—an idea that was embodied in the U.S. Endangered Species Act. Others say that potential extinctions must be weighed against the human and economic costs of preventing them. Where do you stand on this issue? Does your answer affect your opinion of whether it would be ethical to terraform Mars? Explain.

Interestingly, some of the ethical issues involved in Mars colonization were explored by science fiction writers well before the idea of terraforming ever arose. In particular, back in the days when people believed in canals and a dead or dying martian civilization, many stories dealt with the conditions under which humans might colonize Mars. So for our last word on the topic of human colonization, we turn to a science fiction story called "The Million-Year Picnic," written in 1946 by Ray Bradbury. It tells the story of a human family who escapes to Mars just as people on Earth are finishing off our civilization through hatred and war. On Mars, the family finds plenty of water and the vacant cities left by extinct Martians. The story ends with the family on the banks of a canal, where one of the children asks his father about a promise made earlier:

> *"I've always wanted to see a Martian," said Michael. "Where are they, Dad? You promised."*
>
> *"There they are," said Dad, and he shifted Michael on his shoulder and pointed straight down.*
>
> *The Martians were there. Timothy began to shiver.*
>
> *The Martians were there—in the canal—reflected in the water. Timothy and Michael and Robert and Mom and Dad.*
>
> *The Martians stared back up at them for a long, long silent time from the rippling water. . . .*

8.5 THE PROCESS OF SCIENCE IN ACTION: MARTIAN METEORITES

As we briefly noted earlier, one claim of evidence for life on Mars comes from study of rocks from Mars that have fallen to Earth—the so-called *martian meteorites* [Section 6.2]. The story begins in 1984, when a team of American scientists scooped up a 1.9-kilogram meteorite (see Figure 6.10) from the Allan Hills region of Antarctica. It was cataloged as "ALH84001": "ALH" for Allan Hills, "84" for the year in which it was found, and "001" to indicate that it was the first meteorite found on the expedition. It did not immediately draw special attention, but an analysis a decade later showed that it was one of the rare martian meteorites. It then proved itself special even among this small group of rocks, and was subject to intense study. In 1996, a team of researchers (led by David McKay at NASA) made an astonishing claim: They said that ALH84001 holds fossil evidence of past life on Mars. Because this claim would be so important if true, and because it has proved so controversial, we use it as this chapter's case study of the process of science in action.

• Is there evidence of life in martian meteorites?

To evaluate the claims about ALH84001, we must begin by understanding the rock. Scientists are fairly confident (though with some doubts) that ALH84001 really is a meteorite from Mars. It is definitely not an Earth rock, because its relative abundance of the isotopes oxygen-16, oxygen-17, and oxygen-18 are significantly different from that found in terrestrial rocks. But neither does it match what we'd expect from a piece of an asteroid or a rock from the Moon. Most important, gas trapped within ALH84001 appears very similar to that of the martian atmosphere in its chemical and isotopic composition—and distinctly different from any other known source of gas in our solar system—leading to the suspicion that it came from Mars. As discussed in Chapter 6, about three dozen such martian meteorites have been found to date, each presumably launched into space as a consequence of a large impact on Mars.

ALH84001 was singled out for more intense study than other martian meteorites for a simple reason: While other known martian meteorites are geologically young, radiometric dating showed ALH84001 to be a piece of igneous rock that solidified about 4.5 billion years ago. Thus, it formed quite shortly after Mars was born, and must have resided on Mars throughout the times when liquid water flowed on the surface. Scientists

TABLE 8.3 *The History of Meteorite ALH84001*

Time	*Event*
4.5 billion years ago	Solidifies from molten rock in the southern highlands of Mars
4.0–4.5 billion years ago	Affected by nearby impacts, but not launched into space
3.9 billion years ago	Infiltrated by water, leading to the formation of carbonate grains within the rock
16 million years ago	Blasted into space by an impact on Mars
13,000 years ago	Falls to Earth in Antarctica
December 27, 1984	Found by scientists
October 1993	Recognized as a martian meteorite
August 1996	Announcement that ALH84001 contains possible evidence of martian life

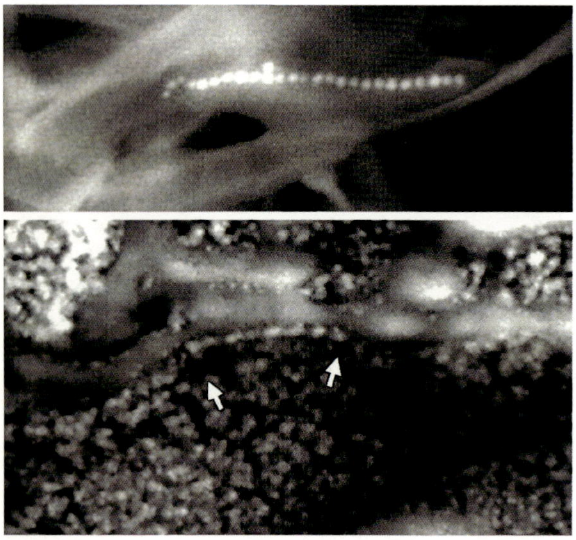

Figure 8.30

Top: Microscopic chains of magnetite crystals produced by bacteria on Earth. Bottom: Similar chains of magnetite crystals found in the carbonate grains of ALH84001. Does the similar appearance of the martian crystals suggest a similar biological origin?

therefore wondered if it might tell us something about the past habitability of Mars.

History of the Meteorite Careful study tells us quite a lot about the history of ALH84001. Radiometric dating tells us its age, while study of its structure reveals evidence of later shocks, probably due to the effects of impacts that occurred long before the one that ultimately launched it into space. The meteorite also contains carbonate grains (about 0.1 to 0.2 millimeter in diameter) that date to about 3.9 billion years ago and tell us that the rock must have been infiltrated by liquid water from which the carbonate minerals precipitated out—evidence that is at least consistent with the idea that Mars once had flowing water.

We can determine the timing of the impact that blasted ALH84001 into space by looking for effects of exposure to *cosmic rays,* high-energy particles that leave telltale chemical signatures on anything unprotected by an atmosphere. The results tell us that ALH84001 spent about 16 million years in space, which means the impact that started its journey occurred on Mars about 16 million years ago. By studying decay products from radioactive isotopes produced by the cosmic rays, we learn when cosmic rays stopped disturbing the meteorite—which must be when it fell to Earth and gained the protection of Earth's atmosphere. Such analysis shows that ALH84001 landed in Antarctica about 13,000 years ago. Table 8.3 summarizes the history of ALH84001.

Evidence of Life The claimed evidence of life in ALH84001 comes from detailed studies of its carbonate grains and the surrounding rock. In brief, four types of evidence have been cited as pointing to the existence of biology on Mars:

- The carbonate grains have a layered structure, with alternating layers of magnesium-rich, iron-rich, and calcium-rich carbonates. On Earth, this type of layering generally occurs only as a result of biological activity.
- The carbonate grains contain complex organic molecules known as *polycyclic aromatic hydrocarbons,* or *PAHs.* These molecules can be produced by both biological and nonbiological processes, and they have indeed been found in many meteorites that are not from Mars. However, their abundance in ALH84001 is much higher than that in other meteorites, and on Earth these molecules are most commonly produced by the decay of dead organisms or by reactions between such decay products and the environment (for example, in the burning of fossil fuels).
- Under a microscope, we see crystals of the mineral magnetite within the iron-rich layers of the carbonate grains. The sizes, shapes, and arrangements of these crystals appear to match those of magnetite grains that on Earth occur only when made by bacteria (Figure 8.30).
- Most intriguingly, highly magnified images of the carbonate grains reveal rod-shaped structures that look much like recently discovered *nanobacteria* on Earth (Figure 8.31). Terrestrial nanobacteria are about $\frac{1}{100}$ the size of ordinary bacteria, and some biologists question whether they truly are living organisms. However, they appear to contain DNA, which suggests that they are indeed a form of life. If so, could the similar-looking structures in the martian meteorite be fossil life from Mars?

While none of these lines of evidence alone would prove biological activity, the original investigators argued that, on the whole, biology seemed a much more likely explanation than nonbiological processes. They felt that it would be a "simpler and thus better" scientific explanation (in effect invoking Occam's razor [Section 2.3]) if only a single process—biology—could account for each observation than if a different process was required to explain each result.

Alternative Explanations The four lines of evidence for fossil life in ALH84001 might seem to make a strong case for past life on Mars. However, other scientists have proposed alternative, nonbiological mechanisms that might have produced each observed phenomenon. Let's look at these alternatives in the same order that we presented the evidence:

- There may be nonbiological ways to get layered carbonate. For example, several pulses of hot water with different dissolved elements might have passed through the rock and laid down the different mineral layers.
- Other meteorites prove that PAHs can be produced by chemical rather than biological processes, and their high abundance might also be explained by terrestrial contamination during the time the rock resided in Antarctica.
- The resemblance between the magnetite crystals in the meteorite and those made by bacteria on Earth may be coincidental, and some scientists have proposed nonbiological ways in which the crystals and chains might have been formed.
- Similarly, the rod-shaped structures don't tell us anything other than that they *look* a lot like nanobacteria. Looks alone do not prove anything, and most scientists suspect that structures with this appearance could be produced nonbiologically; they might even be artifacts of the way the meteorite slices were processed in the laboratory.

One of the most important issues in evaluating the evidence is whether some of it could be the result of terrestrial contamination. Scientists have developed control tests that check for contamination during handling and laboratory analysis of the meteorite, and no such contamination appears to have occurred. However, the rock spent 13,000 years sitting in Antarctica before scientists found it; could Earth life have contaminated it during that time?

At first, scientists were fairly confident that no contamination had occurred on the Antarctic ice, because analysis showed that ALH84001 contains a higher abundance of organic molecules deep within the rock than nearer the surface. If the organic molecules represented terrestrial contamination, we would expect just the opposite. Nevertheless, subsequent analysis found modern, living terrestrial bacteria inside the rock! This clear contamination probably explains the presence of PAHs in the rock. Though it may not affect the other lines of evidence, it certainly weakens any case to be made about martian life in ALH84001.

THINK ABOUT IT Given the fact that ALH84001 has apparently been contaminated by terrestrial bacteria, do you think we could ever be sure that a martian meteorite holds evidence of life on Mars? Defend your opinion.

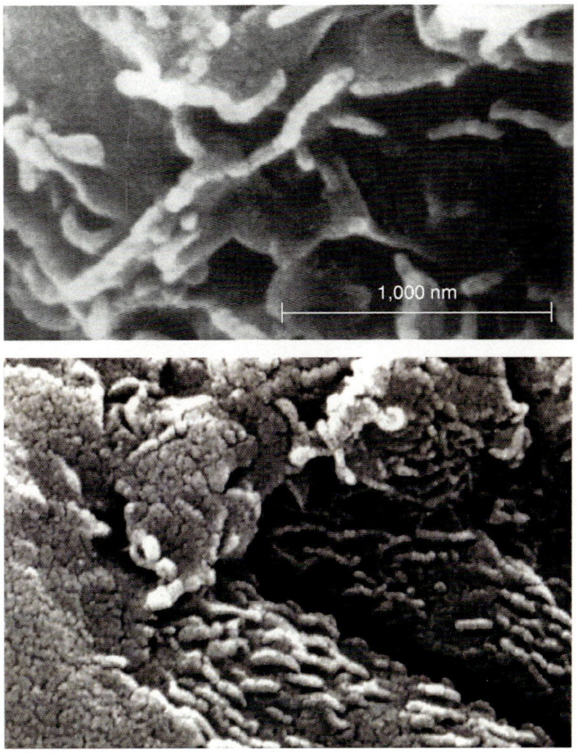

Figure 8.31

Top: Terrestrial nanobacteria in a sample of volcanic rock from Sicily. Bottom: Rod-shaped structures, each about 100 nanometers in length, found in ALH84001. Are they fossils of nanobacteria from Mars?

Summary of the Controversy The extraordinary debate over whether we already have discovered evidence of life beyond Earth will undoubtedly continue as scientists subject ALH84001 and other martian meteorites to further scrutiny. The techniques that have been applied in the analysis to date are the best available, but new technologies will allow the martian meteorites to be studied in even more detail over the coming years. Still, unless a surprising and major breakthrough occurs, it is unlikely that the meteorites by themselves can provide definitive evidence of life.

It is also possible that further study will prove that ALH84001 does not contain fossil life. But even in that case, we still won't know whether there has ever been life on Mars. Most Earth rocks do not contain fossils. Thus, whether or not life existed on Mars, it wouldn't be surprising if the first ancient Mars rock we examine lacks fossils. In all likelihood, a final verdict on the question of life on Mars will have to wait at least until future space missions scoop up and return to Earth samples of rock from a variety of locations on Mars.

THE BIG PICTURE

Putting Chapter 8 into Context

In this chapter, we have discussed past fantasies about martian civilization, our current understanding of the habitability of Mars, and the search for life on the Red Planet. As you continue your studies, keep in mind the following "big picture" ideas:

- Mars holds a special allure not only because of legitimate scientific questions, but also because past fantasies led many people to imagine a martian civilization. Mars and Martians became deeply embedded in modern popular culture, helping generate great public interest in Mars exploration.
- Different regions of the martian surface appear to be almost frozen in time, representing different eras in the planet's history. As a result, we can piece together at least a partial story of Mars from its earliest times to the present. We find a planet that has gone through dramatic change. Its surface, once warm and wet, is now dry and frozen.
- According to present understanding, Mars almost certainly was a habitable planet in the past and may still have habitable zones underground. This makes Mars a prime target in the search for life beyond Earth.

SUMMARY OF KEY CONCEPTS

8.1 FANTASIES OF MARTIAN CIVILIZATION

- **How did Mars invade popular culture?**

Superficial similarities between Mars and Earth led to speculation about martian civilization. Astronomer Percival Lowell thought he saw canals built by an advanced society, but the canals do not really exist.

8.2 A MODERN PORTRAIT OF MARS

- **What is Mars like today?**

Mars is cold and dry, with an atmospheric pressure so low that water is unstable. Martian weather is driven largely by seasonal changes that cause carbon dioxide alternately to condense and sublime at the poles, creating pole-to-pole winds and sometimes leading to huge dust storms.

- **What are the major geological features of Mars?**

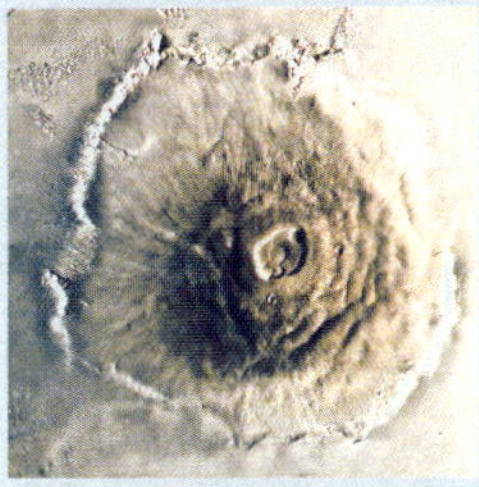

Mars has regions that are densely cratered and must be very old, and other regions with fewer craters that must be much younger. Giant volcanoes dot certain regions of Mars, and we also see evidence of past tectonics, which probably created Valles Marineris.

- **What geological evidence tells us that water once flowed on Mars?**

Orbital images of eroded craters, dry river channels, and floodplains all point to past water flows, and supporting evidence is found in chemical analysis of Martian rocks. Any periods of rainfall seem to have ended at least 2–3 billion years ago, but some flooding may have occurred later. Mars today still has water ice underground and in its polar caps, and could possibly have pockets of underground liquid water.

8.3 THE CLIMATE HISTORY OF MARS

- **Why was Mars warmer and wetter in the past?**

Mars's atmosphere must once have been much thicker with a much stronger greenhouse effect, though we do not yet know whether this made Mars warm and wet for a long, extended period of time or only intermittently.

- **Why did Mars change?**

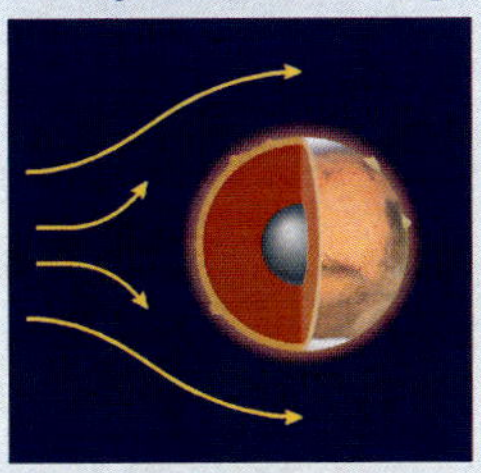

Change must have occurred due to loss of atmospheric gas, which weakened the greenhouse effect. Some gas was probably blasted away by impacts, but more probably was stripped away by the solar wind, which was able to reach the atmosphere as Mars cooled and lost its magnetic field and protective magnetosphere. Water was probably also lost because ultraviolet light could break apart water molecules in the atmosphere, and the lightweight hydrogen then escaped to space.

- **Is Mars habitable?**

Mars almost certainly had a habitable surface during its wet periods more than 2–3 billion years ago. Its surface or near-surface might still sometimes be habitable at times when its axis tilt is greater than it is now, and the subsurface may still have habitable regions today.

8.4 SEARCHING FOR LIFE ON MARS

- **Is there any evidence of life on Mars?**

The *Viking* experiments produced results that some scientists think may be evidence of life, but nonbiological explanations seem more likely. Recent observations of methane in the atmosphere may be due to volcanism. Overall, there are some possible hints of life on Mars, but no definitive evidence.

- **How do we plan to search for life on Mars?**

Space scientists plan an ongoing series of Mars missions, timed for the close approaches of Mars to Earth that occur about every 26 months.

- **Should we send humans to Mars?**

Scientifically, humans could probably answer questions about life much more quickly than robotic missions, but humans also pose a risk of contamination. Ultimately, the question will probably be decided by considerations beyond science alone.

8.5 THE PROCESS OF SCIENCE IN ACTION: MARTIAN METEORITES

- **Is there evidence of life in martian meteorites?**

Four lines of evidence have been presented as suggesting the presence of past life in a martian meteorite, but each also has a potential nonbiological explanation. As a result, we do not have a definitive answer to the question, and more study and more evidence are needed.

EXERCISES AND PROBLEMS

REVIEW QUESTIONS

Short-Answer Questions Based on the Reading

1. Briefly summarize the evidence from little over a century ago, both real and imagined, that led to widespread belief in a martian civilization.
2. What would it be like to walk on Mars today? Briefly discuss the conditions you would experience.
3. Why isn't liquid water stable at the martian surface today? What happens to water ice that melts on Mars?
4. How do martian seasons differ from Earth seasons? Briefly explain why seasonal changes on Mars are accompanied by pole-to-pole winds.
5. Give a brief overview of the geography of Mars, explaining the nature of key features such as the Tharsis Bulge, Olympus Mons, Valles Marineris, the southern highlands, and the northern lowlands.
6. How do we know that different regions of the martian surface date to different eras in the past? What have we learned about changes in martian volcanism during the past eras?
7. Summarize the evidence suggesting that Mars must have been warm and wet, possibly with rainfall, in its distant past.
8. What evidence suggests that water might still flow at or beneath the martian surface today? Why do we think that Mars might still have subsurface liquid water today?
9. How must the martian atmosphere have been different when the surface was warmer and wetter? What uncertainties remain in our understanding of the cause of the higher temperatures in the past?
10. What happened to the gas that made up the much thicker atmosphere of the distant past, and what happened to the abundant water? What role did Mars' small size play in these changes?
11. How does the tilt of the martian axis change with time, and why does it change so much? What implications do these changes have for the martian climate?
12. Based on all the geographic and geological evidence, summarize the current view about the past and present habitability of Mars.
13. Briefly summarize the *Viking* experiments and their results. Do the results constitute evidence of life? Explain.
14. What is the potential significance of atmospheric methane to the search for life on Mars? Are there other explanations for the methane?
15. Briefly summarize plans for Mars exploration over the next few years. Why do we send missions to Mars only about every 26 months?
16. Discuss the issue of biological contamination in either direction between Earth and Mars. How serious is each issue? What steps can we take to prevent contamination in each direction?
17. Summarize the scientific pros and cons of sending humans to Mars. What other considerations are likely to play a role in decisions about such missions?
18. What do we mean by *terraforming* Mars? Why might it be tempting for future human colonists? Is it something we could do within our lifetimes?
19. How do we know that ALH84001 really came from Mars, and how have we learned its history?
20. Briefly summarize the possible evidence of past life discovered in studies of ALH84001 and why this evidence generates controversy.

TEST YOUR UNDERSTANDING

Surprising Discoveries?

Suppose we were to make the following discoveries. (These are *not* real discoveries.) In light of your understanding of Mars, decide whether the discovery would be considered plausible or surprising. Explain clearly; because not all of these have definitive answers, your explanation is more important than your chosen answer.

21. The first human explorers on Mars discover that the surface is littered with the ruins of an ancient civilization, including remnants of tall buildings and temples.
22. We discover a string of active volcanoes in the heavily cratered southern highlands.
23. We find underground pools of water on the slopes of one of the Tharsis volcanoes.
24. We discover that Mars was subjected to global, heavy rainfall about 1 billion years ago.
25. Photos from future orbiters show that new gullies have formed alongside some of the ones already seen in crater walls from orbiting spacecraft.
26. We find a lake of liquid water filling a small crater close to one of the dry river channels.
27. The first fossils discovered on Mars come from the canyon walls of Valles Marineris.
28. A sample return mission finds fossil evidence not only of martian microbes, but also of larger photosynthetic plants that live on the exposed surfaces of martian rocks.
29. We discover evidence that the martian polar caps have in the past extended more than twice as far toward the equator as they do now.
30. We find rocks on Mars showing clearly that the planet once had a global magnetic field nearly as strong as Earth's magnetic field.

Quick Quiz

Choose the best answer to each of the following. Explain your reasoning with one or more complete sentences.

31. When we say that liquid water is *unstable* on Mars, we mean that: (a) a cup of water would shake uncontrollably; (b) it is impossible for liquid water to exist on the surface; (c) any liquid water on the surface would quickly either freeze or evaporate.
32. Mars's seasonal winds are driven primarily by: (a) dust; (b) sublimation of carbon dioxide ice; (c) sublimation of water ice.
33. Olympus Mons is: (a) a giant volcano; (b) a huge canyon network; (c) a continent-size plateau.

34. We can recognize the oldest surface regions of Mars by the fact that they have: (a) the most impact craters; (b) the most volcanoes; (c) the most evidence of past water flows.

35. Minerals in surface rock studied by the martian rovers seem to tell us that: (a) they probably formed in water; (b) they probably were formed by impacts; (c) they hold fossil evidence of life.

36. Rivers on Mars: (a) have never existed; (b) existed in the past but are dry today; (c) continue to have flowing water today.

37. Which must be true if Mars was warmer and wetter in the past? (a) Mars was once closer to the Sun. (b) Mars once had a much thicker atmosphere. (c) Mars must somehow have avoided the effects of the heavy bombardment.

38. Which of the following fundamental properties of Mars could explain why it once had a global magnetic field but later lost it? (a) its small size; (b) its larger distance than Earth from the Sun; (c) a rotation rate that is slightly slower than Earth's.

39. Under the leading scenario, if Mars once had much more carbon dioxide in its atmosphere, most of this carbon dioxide is now: (a) gone, because it escaped to space; (b) frozen at the polar caps; (c) locked up in the form of carbonate rocks, just like on Earth.

40. The *Viking* experiments found: (a) no evidence of life on Mars; (b) clear evidence of life on Mars; (c) some results consistent with life, but others that were inconsistent with life.

INVESTIGATE FURTHER

In-Depth Questions to Increase Your Understanding

Short-Answer/Essay Questions

41. *Hold Your Breath.* If you held your breath, would it be safe to walk outside on Mars? Why or why not?

42. *Miniature Mars.* Suppose Mars was significantly smaller than its current size—say, the size of our Moon. How would this have affected its potential habitability? Explain.

43. *Larger Mars.* Suppose Mars was significantly larger than its current size—say, the same size as Earth. How would this have affected its potential habitability? Explain.

44. *Civilization on Mars.* Based on what we can see on the surface of Mars, does it seem possible that Mars once had a civilization with cities on the surface but that the evidence has now been erased or buried underground? Explain.

45. *Martian Fossil Hunting.* On Earth, we cannot find fossil evidence of life dating to times prior to about 3.8 billion years ago. If life ever existed on Mars, is it possible that we would find older fossils than we find on Earth? Explain.

46. *Future Landing Site.* Suppose you were in charge of a mission designed to land on Mars. Assume the mission carries a rover that can venture up to about 50 kilometers from the landing site. What landing site would you choose? Write a one-page summary of why you think your site is a good target for a future mission.

47. *Human Mission Requirements.* Assume that a mission will carry humans to Mars on a journey that takes a few months in each direction and allows the explorers to spend about 2 years on the martian surface. Make a list of key provisions that would be needed for the mission, explaining the purpose of each item. In addition, briefly discuss whether you think any of these provisions could be found or manufactured on Mars rather than having to be brought from Earth.

48. *Terraforming Mars.* Make a list of the pros and cons of terraforming Mars, assuming that it is possible. Overall, do you think it would be a good idea? Write a short defense of your opinion.

49. *Mars Movie Review.* Watch one of the many science fiction movies that concern trips to Mars (such as *Total Recall* or *Red Planet*). In light of what you now know about Mars, does the movie give a realistic view of the planet? Are the plot lines that concern Mars plausible? Write a critical review of the movie, focusing on these issues.

50. *Martian Literature.* Read a book of science fiction about Mars, such as H. G. Wells's *The War of the Worlds*, Ray Bradbury's *The Martian Chronicles*, or any of the Edgar Rice Burroughs books about Martians. Write a critical review of the book, being sure to consider whether it still merits interest in light of current scientific understanding of Mars.

Quantitative Problems

Be sure to show all calculations clearly and state your final answers in complete sentences.

51. *Interior Heat.* Compare the surface area–to–volume ratios (that is, total surface area divided by total volume) of the Moon, Earth, and Mars. What does your answer tell you about how quickly each world should have cooled with time? What does your answer tell you about the implications of planetary size to habitability?

52. *Mars's Elliptical Orbit.* Mars's distance from the Sun varies from 1.38 AU to 1.66 AU. How much does this change the globally averaged strength of sunlight over the course of the martian year? Give your answer as a percentage by which sunlight at perihelion (the orbital point closest to the Sun) is stronger than that at aphelion (the farthest orbital point). Comment on how this affects the martian seasons. (*Hint:* Remember that light follows an inverse square law; see Figure 7.1.)

53. *Atmospheric Mass of Earth.* What is the total mass of Earth's atmosphere? Use the fact that, under Earth's gravity, the sea level pressure of 1 bar is equivalent to 10,000 kilograms pushing down on each square meter of the surface. Also remember that the surface area of a sphere of radius r is $4\pi r^2$.

54. *Atmospheric Mass of Mars.* The weaker gravity of Mars means that 1 bar of pressure on Mars would be that exerted by about 25,000 kilograms pushing down on each square meter of the martian surface. Based on this approximation, the atmospheric pressure on Mars (see Table 8.1), and the size of Mars, estimate the total mass of Mars's atmosphere. Compare to Earth's atmospheric mass from Problem 53.

55. *Past Gas on Mars.* Models suggest that Mars today could have liquid water on its surface if the atmosphere were about 400 times as dense as it actually is. What would the atmospheric mass be in that case? How does this compare to the present mass of Earth's atmosphere? Does it seem plausible that Mars might once have had this much gas? Explain why or why not.

Discussion Questions

56. *The Role of the Martians.* Percival Lowell may have been sadly mistaken in his beliefs about Martians, but he succeeded in generating intense public interest in Mars. If he had never made

his wild claims about canals and civilization, do you think we would be exploring Mars with the same fervor today? Defend your opinion.

57. *Lessons from Mars.* Discuss the nature of the climate change that occurred on Mars some 3 billion years ago. Do you think this climate change holds any important lessons for us as we consider potential climate changes that humans are causing on Earth? Explain.

58. *Human Exploration of Mars.* Should we send humans to Mars? If so, when? How much would you be willing to see spent on such a mission? Would you volunteer to go yourself? Discuss these questions with your classmates, and try to form a class consensus regarding the desirability and nature of a human mission to Mars.

WEB PROJECTS

59. *Martian Photo Journal.* By now, we have many thousands of photos of Mars taken both on the surface and from orbit, and virtually all of them can be found on the Web. Make your own photo journal of "Mars's Greatest Photo Hits" by choosing ten of your favorite photos. For each one, write a short descriptive caption and explain why you chose it.

60. *Current Mars Missions.* Pick one of the Mars missions that is currently operating and visit its Web site. Write a short report about the mission's history, goals, and accomplishments to date.

61. *Future Mars Missions.* Pick one of the Mars missions that is being planned or considered for the future, and visit its Web site. Write a short report about the purpose of the mission and its current status.

9
Life on Jovian Moons

The atmosphere of Titan.

LEARNING GOALS

9.1 THE MOONS OF THE OUTER SOLAR SYSTEM

- What are the general characteristics of the jovian moons?
- Why do we think that some moons could harbor life?

9.2 LIFE ON JUPITER'S GALILEAN MOONS

- Does Europa have an ocean?
- Could Europa have life?
- Could other moons of Jupiter have life?

9.3 LIFE AROUND SATURN, AND BEYOND

- Could Titan have life?
- Could other moons of Saturn have life?
- Could moons of Uranus or Neptune have life?

9.4 THE PROCESS OF SCIENCE IN ACTION: CHEMICAL ENERGY FOR LIFE

- What is the role of disequilibrium in life?
- What types of chemical reactions supply energy for life?

I was drawn by the sirens of Titan
Carried along by their call
Seeking for a way to enlighten
Searching for the sense of it all
Like a kiss on the wind I was thrown to the stars
. . .
I was drawn by the sirens of Titan (as are we all)
As are we all

Al Stewart, from his song "The Sirens of Titan," based on the Kurt Vonnegut novel

Jupiter orbits the Sun at more than five times Earth's distance, and the other jovian planets (Saturn, Uranus, and Neptune) lie much farther from the Sun. Sunlight in this distant realm is faint—too weak to provide much warmth. Nevertheless, several of the moons in these frigid outer reaches of our solar system are now considered to be possible places to find life beyond Earth. In this chapter, we will investigate these distant worlds.

We'll begin by examining the general characteristics of these moons and how and why it might be possible for some of them to have liquid water (or other liquids). We'll then turn our attention to the most promising potential abodes of life. Looking first at the moons of Jupiter, we'll discuss the evidence suggesting that Europa has a subsurface ocean that might contain more water than the oceans of Earth, making it a prime target in the search for life. We'll also investigate the likelihood that conditions similar to those on Europa might also be found on two other moons of Jupiter: Ganymede and Callisto. We'll then move further outward into the solar system, discussing the latest developments concerning the possible habitability of moons around Saturn, Uranus, and Neptune, including Saturn's remarkable moon Titan, which is blanketed by an atmosphere thicker and denser than Earth's.

Aside from addressing our general curiosity about the habitability of jovian moons, our discussion in this chapter will lead to an intriguing possibility: It's conceivable that these cold and distant moons could be the most numerous homes to life in our solar system. If any of these moons do indeed prove to harbor living things, the possibilities for finding biology elsewhere in the universe will be greatly broadened.

9.1 THE MOONS OF THE OUTER SOLAR SYSTEM

In previous chapters, we discussed life on Earth in some detail and investigated the prospects for finding life on Mars. We now turn our attention to more distant reaches of the solar system, notably the realm of the four jovian planets—Jupiter, Saturn, Uranus, and Neptune. As we discussed in Chapter 7, the jovian planets themselves seem unlikely to be habitable.

However, these planets are orbited by many moons, which we call **jovian moons** because they orbit jovian planets.

The idea of finding life in the cold outer solar system once seemed far-fetched, but several of the jovian moons now seem potentially habitable. In this section, we'll introduce the major moons and explore the mechanisms thought to make it possible for at least some of these moons to have liquid water (or other liquids) on or within them, thus creating possible habitats for life.

• What are the general characteristics of the jovian moons?

As of early 2006, the four jovian planets were known to have at least 150 moons between them, more than 20 times that of all other planets combined. While some of the larger of these moons have been known to astronomers for nearly four centuries, many of the smaller ones have been found only with the improved telescopes and space probes of the last few decades. Probes in particular have given us the chance to study these worlds from close by. While previous generations knew jovian moons only as points of light, today we have photographs of many of them that show enough detail that we can name individual craters and other features. One of the biggest surprises to emerge from our modern reconnaissance of these moons is learning that some of them might be homes to life.

Discovering the Moons Among Galileo's many discoveries [Section 2.2], one of the most notable was finding the four large moons of Jupiter. Having heard of the telescope's recent invention, Galileo built his own homemade versions beginning in 1609. At the time, the telescope was thought to be primarily a toy, or possibly useful for defense. Galileo, however, did something different with his telescopes: He turned them toward the sky. On January 7, 1610, while gazing at Jupiter, he saw what at first seemed to be three small stars in its vicinity. What intrigued Galileo was that the three were close to one another and in a line. The following night, he looked again and was surprised to note that the three stars had moved relative to Jupiter, but not in the direction or to the extent expected of stars. A few days later he noted a fourth point of light, and within a week he realized that these four "stars" always stayed close to Jupiter and were clearly in orbit around it (Figure 9.1). In March 1610, Galileo published his results in a pamphlet he called *The Starry Messenger*, claiming to have found four bodies moving around the giant planet "as Venus and Mercury around the Sun." These four bodies are what we now call the **Galilean moons** of Jupiter; proceeding outward from that planet, we know them individually as Io, Europa, Ganymede, and Callisto (Figure 9.2).

Other scientists soon discovered additional moons in the outer solar system. The accomplished Dutch scientist Christiaan Huygens (1629–1695) found the largest of Saturn's moons—Titan—in 1656. Before the close of the seventeenth century, Giovanni Domenico Cassini (1625–1712), an Italian astronomer who became director of the Paris Observatory, had discovered four more moons around the ringed planet.*

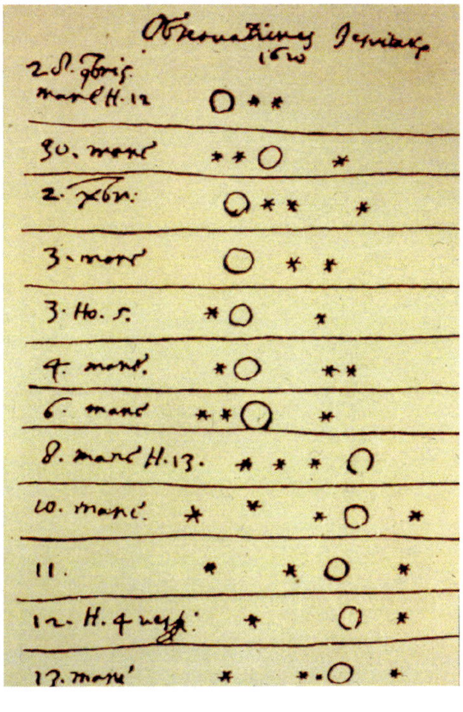

Figure 9.1

A page from Galileo's notebook written in 1610. His sketches show four "stars" near Jupiter (the circle), but in different positions at different times (and sometimes hidden from view). Galileo soon realized that the "stars" were moons orbiting Jupiter.

*Saturn's rings were first sighted by Galileo, but the resolution of his telescope was too low for him to make out what they were. Huygens was the first to realize that the rings do not touch Saturn's surface, and Cassini showed that the rings were not solid but instead were marked by a dark division, which we still call the *Cassini division*.

Figure 9.2

This set of photos, taken by the *Galileo* spacecraft, shows global views of the four Galilean moons as we know them today. Sizes are shown to scale.

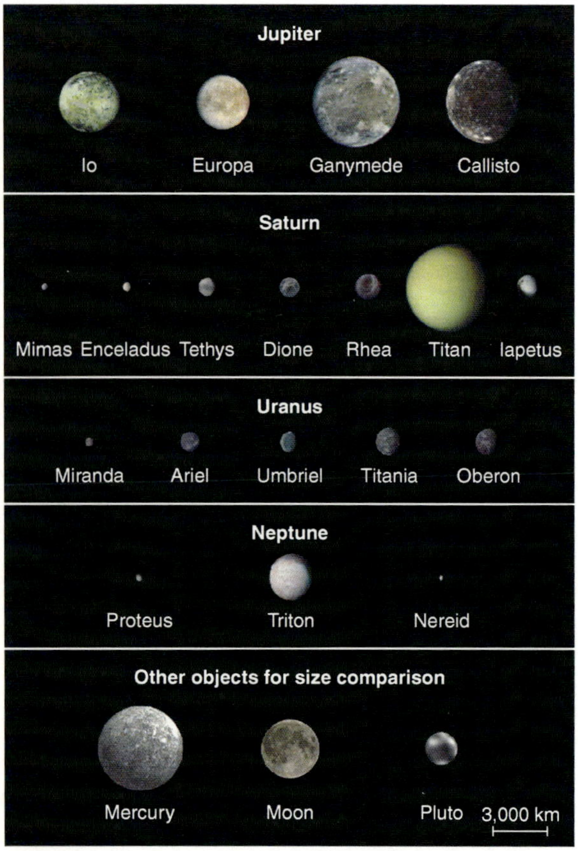

Figure 9.3

The larger moons of the jovian planets, with sizes (but not distances) shown to scale. Mercury, our own Moon, and Pluto are included for comparison.

Even today, astronomers continue to make new discoveries of moons orbiting the jovian planets, though all of the larger moons have surely been discovered by now. The new discoveries involve small moons—usually no more than a few kilometers across—detected with the aid of new telescopes or spacecraft. For example, nearly two dozen small moons were discovered in photographs taken by the *Voyager* probes in the 1970s and 1980s, and as recently as 2004 and 2005 astronomers used a telescope in Hawaii, as well as the *Cassini* spacecraft and the Hubble Space Telescope, to find 16 new moons of Saturn and two new moons of Pluto.

Sizes and Orbits of the Moons The jovian moons come in a wide range of sizes. While many small ones are not much bigger than a single mountain on Earth, others are much larger. The two largest—Jupiter's moon Ganymede and Saturn's moon Titan—are bigger than the planet Mercury. The three other Galilean moons (Io, Europa, and Callisto) and Neptune's moon Triton are bigger than Pluto. Figure 9.3 shows a montage of jovian moons larger than about 300 kilometers in diameter.

Almost all the moderate- and large-size moons orbit their planet in much the same way that planets orbit the Sun: They orbit nearly in the equatorial plane of their host world, moving in the same direction as their planet's spin. In this sense they resemble miniature solar systems, which suggests that they were formed in a smaller-scale version of the same processes that gave birth to the planets [Section 3.3]. As shown in Figure 9.4, the gravity of the jovian planets drew in gas and dust from the surrounding solar nebula. Like the solar nebula as a whole, this gas and dust formed a swirling disk-shaped cloud around each jovian planet. Condensation and accretion then built moons that shared the orbital properties of the original disks of gas and dust.

The small jovian moons differ from their larger brethren both in appearance and in the properties of their orbits. Most have an irregular shape and often resemble peanuts, potatoes, or other snack foods (Figure 9.5). This is hardly surprising: The lesser gravity of these small objects is too weak to force the rigid material of which they're composed into spheres.

Some small moons may have formed "in place" like most of the larger moons, and others come in groups that share common orbital characteristics, suggesting that they may be fragments of larger moons that broke apart. But many of the smaller moons also have orbits that are highly elliptical or inclined to the equator of their host planet; in some cases, the moons orbit backward relative to their host planet's spin. These orbital characteristics are telltale signs of moons that are captured asteroids or comets, which probably originally orbited the Sun independently. As we discussed briefly in Section 4.6, it's not easy for a planet's gravity to grab a wayward asteroid or comet: Because energy must be conserved, the passing object must somehow lose some of its orbital energy or else it would simply fly on by and continue its orbit around the Sun. The leading hypothesis holds that captures occurred when the jovian planets were young, at a time when they had large, extended atmospheres that could have served to slow down small bodies as they passed nearby, thereby removing orbital energy and allowing the objects to be captured.

Surprisingly, one large moon also has orbital characteristics suggesting it is a captured object: Neptune's moon Triton, which orbits backward relative to its planet's rotation. This orbit makes it a near-certainty that Triton once orbited the Sun as an independent object, although capturing such a large object poses a trickier problem than capturing smaller moons. Recent research suggests that Triton could have been captured if it had once been orbited by its own satellite. In that case, it could have been captured while passing close to Neptune, because the excess orbital energy that Triton had to lose could have been carried off by the satellite as it was ejected from its orbit. Regardless of the specific mechanism of the capture, for several decades scientists have assumed that Triton was once one of the *Kuiper belt objects* (KBOs) of which we now consider Pluto to be a member [Section 3.3]. Indeed, the fact that Triton is larger than Pluto is one reason why some astronomers suspected that Pluto might not be the largest such object in the outer solar system, a suspicion confirmed with the 2005 discovery of Eris—an ice-rich world that is larger than Pluto and nearly a hundred times as far from the Sun as Earth.

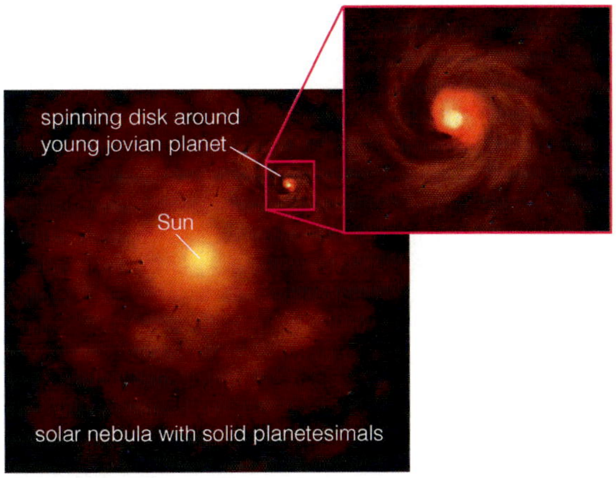

Figure 9.4

The young jovian planets are thought to have been surrounded by disks of gas and dust much like the solar nebula as a whole but smaller in size.

Composition of the Moons

As we discussed in Chapter 3, the outer solar system was cold enough to allow ices to condense along with metal and rock. We therefore would expect the jovian moons to be made of a mixture of ice and rock, and that is indeed the case for most of them. The average densities of most jovian moons are significantly lower than that of Earth, reflecting the fact that they contain substantial quantities of ice, which is low in density. Within individual moon systems, we see variations in composition reflecting the fact that the moons formed at different distances from a hotter, central planet. For example, the Galilean moons show a decrease in density with distance from Jupiter, just as we would expect if they formed from a swirling cloud of gas that was hotter in the center than in its outer regions: Io's density indicates that it has virtually no water in any phase, Europa is mostly rock with water ice (and perhaps a liquid ocean) only in its outer layers, and Ganymede and Callisto have more significant amounts of water ice relative to their amounts of rock.

In addition to the compositional variation within individual moon systems, there is also variation in the composition of moons as we move from one planet to the next; these differences came about because of temperature differences in the overall solar nebula. Water ice condensed easily at the temperatures near Jupiter, but methane and other ices condensed only at the colder temperatures found at significantly greater distances from the Sun. As a result, Jupiter's moons contain significant quantities of water ice but no other ices. Moons of the more distant planets contain higher overall proportions of ice than of rock, and contain not only water ice but also methane and other ices.

Figure 9.5

A montage of six of Saturn's smaller moons (not to scale). Because they are not spherical, the sizes in parentheses represent approximate lengths along their longest axes.

Synchronous Rotation of the Moons

Nearly all jovian moons share an uncanny trait: Like our own Moon, they always keep the same face turned toward their planet. This behavior, called **synchronous rotation**,

Figure 9.6

The fact that we always see the same face of the Moon means that the Moon must rotate once in the same amount of time that it takes to orbit Earth once, an idea you can understand by walking around a model of Earth while imagining that you are the Moon. The same idea applies to the synchronous rotation of jovian moons.

a If you do not rotate while walking around the model, you will not always face it.

b You will face the model at all times only if you rotate exactly once during each orbit.

means that each moon completes exactly one rotation around its axis while it makes one orbit around the planet. You can see how this works with a simple demonstration (Figure 9.6). Place a ball on a table to represent a planet like Earth, and walk around the ball so that your head represents an orbiting moon. If you do not rotate as you walk around the ball, you'll be facing away from it by the time you are halfway around your orbit. The only way you can face the ball at all times is by completing exactly one rotation while you complete one orbit.

Synchronous rotation is not a coincidence: Rather, it develops naturally for any moon that orbits relatively close-in to its planetary parent, and it arises as a consequence of the same gravitational effects that lead to tides on Earth. Recall that the strength of gravity declines with distance, following an inverse square law [Section 2.4]. As a result, the Moon's gravitational pull on the near side of Earth is slightly stronger than that on the far side. This *difference* in the strength of gravity is the cause of the tides (Figure 9.7). Note that there are two tidal bulges, one facing the Moon and one opposite, which is why there are two high tides per day as Earth rotates through the bulges. Simply put, the reason why there are two bulges is that the oceans on the side nearest to the Moon are being pulled out from Earth, while the oceans on the opposite side bulge because Earth is being pulled away from under them. A better way to look at tides is to recognize that the gravitational attraction toward the Moon gets progressively weaker with distance throughout Earth and that this varying attraction tends to stretch the entire Earth—land and ocean—along the Earth–Moon line. The tidal bulges are more noticeable for the ocean than for the land only because liquids flow more readily than solids. (Recall that solids *can* flow, though slowly, which is why Earth's solid mantle can undergo convection [Section 4.4].)

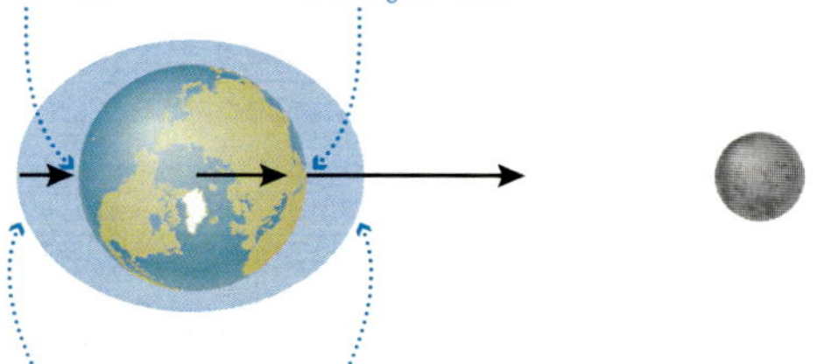

Figure 9.7

Tides on Earth are created by the varying force of attraction between different parts of Earth and the Moon. There are two daily high tides as any location on Earth rotates through the two tidal bulges. The diagram highly exaggerates the tidal bulges, which actually raise the oceans only about 2 meters and the land only about a centimeter.

THINK ABOUT IT The Sun exerts a stronger gravitational force on Earth than does the Moon—after all, Earth orbits the Sun, not the Moon. So why is the Moon rather than the Sun primarily responsible for Earth's tides? (The Sun's tidal effect on Earth is only about one third as strong as the Moon's.) Do you think other planets can have any significant effect on tides on Earth? Explain.

Because tidal forces stretch Earth itself, the process necessarily creates some friction, called **tidal friction**. As shown in Figure 9.8, this tidal friction has important consequences for Earth and the Moon, because it

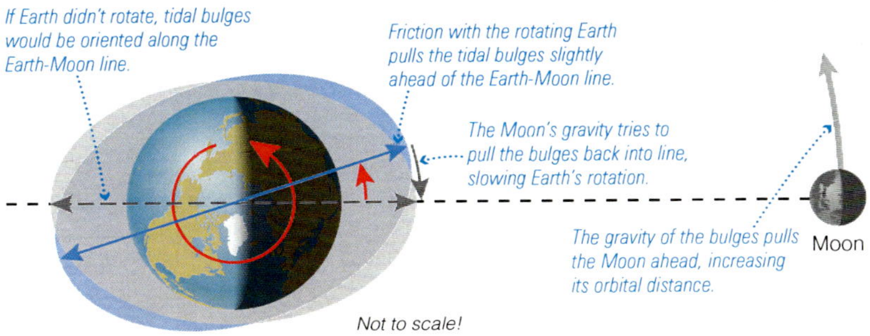

Figure 9.8

Earth's rotation pulls its tidal bulges slightly ahead of the Earth–Moon line, leading to gravitational effects that very gradually slow Earth's rotation and increase the Moon's orbital distance.

allows Earth's rotation to pull the bulges slightly ahead of the Earth–Moon line. This slight misalignment of the tidal bulges with the Earth–Moon line means the Moon's gravity is always pulling back on the bulges, causing Earth's rotation to slow gradually. At the same time, the gravity of the bulges pulls the Moon slightly ahead in its orbit, causing the Moon to move gradually farther from Earth. These effects are barely noticeable on human time scales; for example, tidal friction increases the length of Earth's day by only about 1 second every 50,000 years. (On short time scales, this effect is overwhelmed by other effects on Earth's rotation.) But the effects add up over billions of years. Early in Earth's history, a day may have been only 5 or 6 hours long and the Moon may have been one tenth or less its current distance from Earth.

We can understand the Moon's synchronous rotation by turning the situation around. Just as the Moon raises tides on Earth, Earth must raise tides on the Moon—in fact, much stronger tides, because Earth's gravity is much stronger than the Moon's. If the Moon rotated rapidly, these tides would generate substantial tidal friction that would tend to slow the Moon's rotation. The tidal friction would disappear only when the Moon's rotation slowed so much that its tidal bulges always stayed on the Earth–Moon line—which would mean that the Moon was always keeping the same face to Earth in synchronous rotation. In other words, no matter how fast the Moon may have been rotating at its birth, it was inevitable that tidal friction would slow this rotation until it was synchronous. At that point, the synchronous rotation was permanently "locked in," because there was no more tidal friction to slow the Moon's rotation further.

Tidal friction similarly explains the synchronous rotation of the jovian moons.* If at its birth a moon was rotating faster than it orbited, tidal friction slowed its rotation until the rotation and the orbit were synchronized. Similarly, if a moon was once spinning slower than it orbited, tidal friction accelerated its rotation until it was synchronized with its orbit. Given the large sizes of the jovian planets compared with their moons, synchronous rotation likely set in for the close-in moons within no more than a few million years.

• Why do we think that some moons could harbor life?

Most of the jovian moons are almost certainly lifeless. As we discussed in Chapter 7, they are too small and too far from the Sun to have any reasonable likelihood of having liquid water or other liquids. However, the

*Europa shows evidence of a very slight deviation from perfect sychronous rotation, which may be an effect of a subsurface ocean.

situation may be different for a few of the larger moons. One large moon, Titan (which we'll discuss in Section 9.3), appears to have a temperature and atmospheric pressure that at least sometimes allow liquid methane to form lakes or rivers on its surface. In addition, several moons may have just enough internal heat to keep water liquid in their interiors—making their interiors potentially habitable.

Tidal Heating Based on what we've learned about internal heat on the terrestrial planets, it might seem surprising to find much internal heat in any of the jovian moons. After all, only Ganymede and Titan are even as large as Mercury, and we know that Mercury's interior has cooled enough over time that it no longer supports any active volcanism. But when the *Voyager 1* spacecraft passed Jupiter in 1979, we got definitive proof that at least some distant moons have plenty of internal heat. Confirming a theoretical prediction made just weeks before *Voyager*'s arrival,

COSMIC CALCULATIONS 9.1

The Strength of the Tidal Force

Recall from Section 2.4 that the force of gravity acting between two objects is:

$$F_g = G\frac{M_1 M_2}{d^2}$$

The tidal force that a planet exerts on a satellite is the *difference* between the gravitational force on the near and far sides of that satellite. One way to get a sense of this difference is to consider a "test mass"—say, a 1-kg rock—on the satellite's surface. If d is the distance between the center of the planet and the center of a satellite of radius r_{sat}, then on the near side of the satellite the distance from the planet's center and the 1-kg rock is $d - r_{sat}$ and on the far side it is $d + r_{sat}$. We then calculate the gravitational force in both positions and subtract to get the tidal force acting on the satellite per kilogram of mass.

Example: Calculate the tidal force that Earth exerts on the Moon (per kilogram) and compare to the effect of the Moon's own gravity. Useful data: $M_{Earth} = 5.97 \times 10^{24}$ kilograms, $r_{Moon} = 1.74 \times 10^6$ meters, the Earth–Moon distance is $d = 3.84 \times 10^8$ meters, and $M_{Moon} = 7.35 \times 10^{22}$ kilograms. As long as you use G in standard units and the masses and distances in kilograms and meters, your answer will come out in *newtons* ($1\text{ N} \approx 0.225$ lb).

Solution: We find the tidal force F_{tidal} that Earth exerts on the rock by subtracting the gravitational force F_g when the rock is on the far side of the Moon (where the rock's gravitational attraction to Earth is weaker) from the gravitational force when it is on the near side (where the gravitational attraction is stronger):

$$F_{tidal} = F_g(\text{on near side}) - F_g(\text{on far side})$$

Let's first consider the near-side term:

$$\begin{aligned} F_g(\text{near side}) &= G\frac{M_{Earth}M_{rock}}{(d_{Earth\text{-}Moon} - r_{Moon})^2} \\ &= \left(6.67 \times 10^{-11}\frac{\text{m}^3}{\text{kg} \times \text{s}^2}\right)\frac{(5.97 \times 10^{24}\text{ kg})(1\text{ kg})}{(3.84 \times 10^8\text{ m} - 1.74 \times 10^6\text{ m})^2} \\ &= 0.00273\text{ N} \end{aligned}$$

For the far-side term, we find:

$$\begin{aligned} F_g(\text{far side}) &= G\frac{M_{Earth}M_{rock}}{(d_{Earth\text{-}Moon} + r_{Moon})^2} \\ &= \left(6.67 \times 10^{-11}\frac{\text{m}^3}{\text{kg} \times \text{s}^2}\right)\frac{(5.97 \times 10^{24}\text{ kg})(1\text{ kg})}{(3.84 \times 10^8\text{ m} + 1.74 \times 10^6\text{ m})^2} \\ &= 0.00267\text{ N} \end{aligned}$$

The difference gives us the tidal force (per kilogram):

$$\begin{aligned} F_{tidal} &= F_g(\text{on near side}) - F_g(\text{on far side}) \\ &= 0.00273\text{ N} - 0.00267\text{ N} = 0.00006\text{ N} \end{aligned}$$

For comparison, the Moon's gravitational force acting on the rock is:

$$\begin{aligned} F_g &= G\frac{M_{Moon}M_{rock}}{r_{Moon}^2} \\ &= \left(6.67 \times 10^{-11}\frac{\text{m}^3}{\text{kg} \times \text{s}^2}\right)\frac{(7.35 \times 10^{22}\text{ kg})(1\text{ kg})}{(1.74 \times 10^6\text{ m})^2} \\ &= 1.6\text{ N} \end{aligned}$$

Note that this 1.6 N gravitational force (which is the weight of the rock on the Moon and is equivalent to 1.6 N × 0.225 lb/N = 0.36 pounds) is more than 15,000 times greater than the 0.00006 N tidal force that Earth exerts on the rock. Clearly, the tidal force is quite small in comparison to the gravitational force, though over time it has been large enough to bring the Moon into synchronous rotation. (The tidal force was stronger when the Moon was closer to Earth.)

the spacecraft photographed active volcanic eruptions on Io, the innermost of the four Galilean moons. The *Galileo* orbiter then provided much more detailed views (Figure 9.9). How did scientists predict this surprising volcanism? They realized that jovian moons can have a type of internal heating different from that found on any of the terrestrial worlds, and that Io would be the extreme example of this heating.

Recall that Earth and the other terrestrial worlds were all hot inside at their births, with heat from accretion, differentiation, and radioactive decay [Section 4.4]. Over time, heat has escaped, and only radioactive decay continues to supply new heat. In small worlds, like the Moon and Mercury, the supply of new heat has not been enough to overcome the loss of escaped heat. If the same sources were all that could heat the jovian moons, they probably would be cool by now too. To explain the internal heating of Io, we must consider a different process. Scientists call this **tidal heating**, because it arises from effects of tidal forces.

Io's tidal heating arises from a combination of two factors: (1) its proximity to Jupiter means it experiences a strong tidal force from the massive planet; and (2) Io has a slightly elliptical orbit, which causes the strength and direction of the tidal force to change slightly as Io moves through each orbit. Figure 9.10 shows the result: The constantly changing orientation of Io's tidal bulges means that Io is continuously being flexed in different directions, which generates friction inside it in much the same way that flexing warms Silly Putty* Tidal heating generates tremendous heat on Io—more than 200 times as much heat (per gram of mass) as the radioactive heat driving much of Earth's geology. This heat makes Io the most volcanically active place in the solar system. Material from Io's volcanic vents reaches temperatures of more than 2,000°C (3,600°F), and eruptions often spew plumes of sulfur and other gases to heights of hundreds of kilometers.

Although the combination of the elliptical orbit and Jupiter's tidal force seem to explain tidal heating, there is still another question to ask: Why is Io's orbit slightly elliptical, when most large satellites have nearly circular orbits? The answer lies with an "orbital resonance" that occurs between Io, Europa, and Ganymede: The three moons periodically line up because, during the time Ganymede takes to complete one orbit of Jupiter (about 7 days), Europa completes two orbits and Io completes four orbits (Figure 9.11). The effect is much like that of pushing a child on a swing. If timed properly, a series of small pushes can add up to a *resonance* that causes the child to swing quite high. For the three moons, the periodic alignments mean they experience repeated gravitational tugs that act much like the repeated pushing of the child on the swing (hence the term *orbital resonance*). In this case, these gravitational tugs cause the orbits to be more elliptical than they would be otherwise.

Like synchronous rotation, we expect orbital resonances to arise quite naturally. The orbital resonances of Io, Europa, and Ganymede probably came about because of a feedback with tides that the moons raise on Jupiter. Just as our Moon raises tides on Earth that are gradually causing the Moon to move farther from Earth (and slowing Earth's rotation), the Galilean moons also raise tides in Jupiter's atmosphere that cause the moons to move farther from Jupiter. The effect is greatest on the closest-in moon, Io, so its orbit should have tended to move outward until it

*If you are looking for the ultimate source of the energy that drives tidal heating, it comes from Jupiter's rotation, which is gradually slowing much like Earth's rotation, although at a rate too small to be observed.

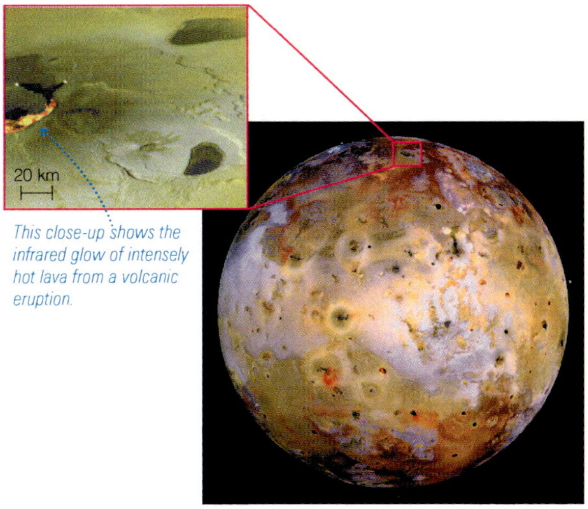

Figure 9.9 Interactive Photo

Io is the most volcanically active body in the solar system. Most of the black, brown, and red spots on the surface are related to recently active volcanoes.

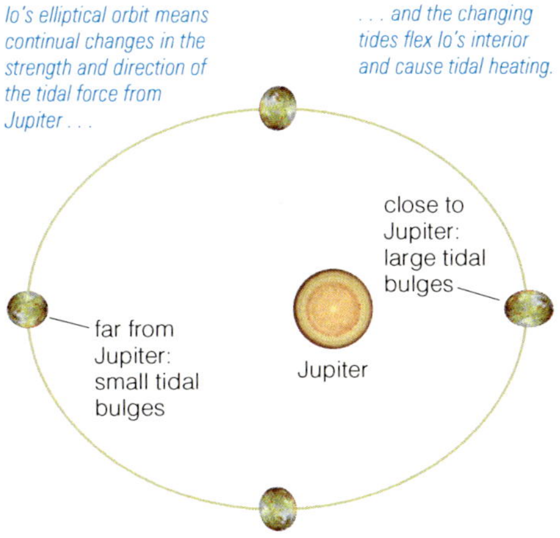

Figure 9.10

Tidal heating arises on Io from the combination of its elliptical orbit and the strong tidal force exerted on it by Jupiter. The bulges and orbital eccentricity are exaggerated.

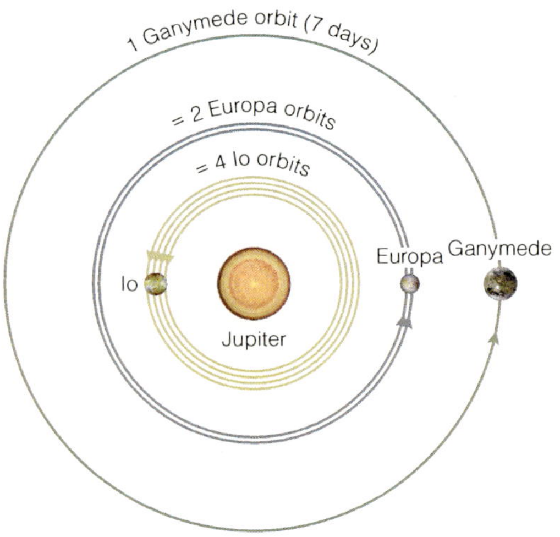

Figure 9.11

Io, Europa, and Ganymede share an orbital resonance that returns them to the positions shown every 7 days, and the recurring gravitational tugs explain why all three orbits are slightly elliptical. The effect on Io is greatest, since it is closest to Jupiter.

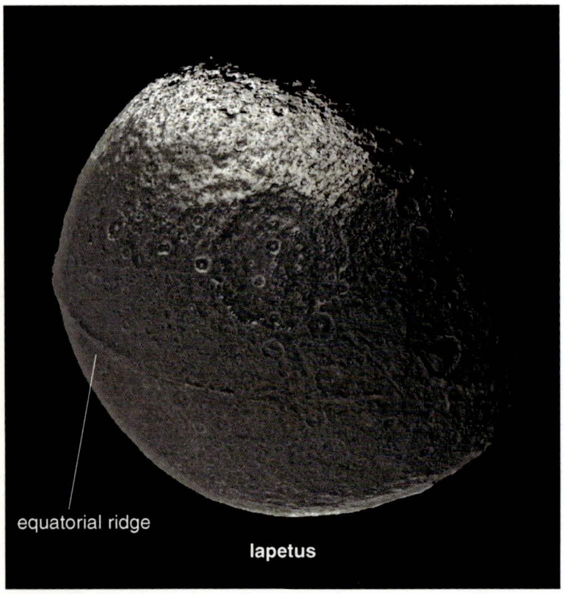

Figure 9.12

Saturn's moon Iapetus (about 1,440 kilometers in diameter) has a steep ridge running along its equator, suggesting that it must once have been surprisingly warm inside.

achieved a resonance with the next moon out, Europa. Then Io and Europa continued to move outward in lockstep until they achieved a resonance with the third moon, Ganymede. This is presumably how Io, Europa, and Ganymede came to have orbital resonances among themselves. Indeed, these moons are now moving slowly outward in lockstep toward Callisto, though it will be many billions of years before they reach a resonance with this fourth moon.

Other Heat Sources Tidal heating certainly explains Io's volcanic activity and probably explains Europa's internal heat. As we'll see in the next two sections, however, it may not be enough to explain the heating that apparently occurs or has occurred in the past on a few other jovian moons. Saturn's moon Iapetus, for example, has a striking equatorial bulge that indicates it must once have been warm enough to be fluid inside (Figure 9.12), but tidal heating does not seem to explain how it became hot enough for its interior to melt (or become sufficiently "soft" to flow). One possible explanation lies with radioactive decay, although there is great scientific debate over whether or how Iapetus and other moons could have incorporated enough radioactive material during their formation to explain the level of heating that they have apparently experienced.

9.2 LIFE ON JUPITER'S GALILEAN MOONS

Tidal heating has turned Io into a veritable hell, and its lack of water and extreme volcanic activity essentially rule it out as a home for life. But tidal heating has somewhat lesser effects on Europa and Ganymede (and no effect on Callisto, since it does not participate in the orbital resonance). In the case of Europa, in particular, it is possible that the level of tidal heating is "just right" for life. Without tidal heating, Europa would be wrapped in solid ice. However, its appearance offers hints that this moon has a subsurface ocean that makes it a possible home to life. In this section, we'll discuss the prospects of habitability and life for Europa, as well as the prospects for the outer two Galilean moons, Ganymede and Callisto.

• Does Europa have an ocean?

Even before the *Voyager* spacecraft snapped the first detailed photos of Europa in 1979, scientists already knew that Europa is covered with an ice shell. Spectroscopic observations made from Earth indicated the presence of water ice on the surface, although the amount was unknown. Theoretical studies, based on our ideas about the formation of moons, suggested that Europa was likely built of a mixture of rock and water. If heat had caused the moon to differentiate sometime in the past, then the less dense water would have migrated to the surface. The *Voyager* spacecraft confirmed at least part of this speculation: Europa's exterior is bright white and entirely ice-covered. Also, its surface is remarkably smooth, with few features rising higher than about a kilometer. Faint ridge lines—giant cracks in the ice—crisscross the surface, looking almost like a spider web of highways.

The *Galileo* orbiter revealed Europa in much more detail (Figure 9.13). Notice the small number of recognizable impact craters, which tells us that Europa's surface has been repaved in recent geological times. However, it is extremely difficult to tell whether the resurfacing has been caused by the action of liquid water or of ice that is relatively warm and therefore soft enough to flow (as glaciers flow on Earth).

By using the *Galileo* spacecraft to measure subtle variations in Europa's gravitational field, scientists were able to determine its internal layering. Crudely speaking, Europa seems to consist of a central metallic (probably iron) core, overlaid with a thick sheath of silicate rock and an 80–170-kilometer-thick outer skin of water or water ice. From the gravity measurements alone, the water layer could contain just about any combination of solid ice, liquid water, or slush (partially melted ice), because all of these have about the same density (1 g/cm^3). As a result, over the past couple of decades scientists have debated which form of water is most likely. The debate is not yet fully settled, but as we'll discuss shortly, careful analysis of data from the *Galileo* spacecraft suggests that the most likely answer is the model shown in Figure 9.14: Europa probably has a brittle, icy crust underlaid by a layer of warmer ice that can flow easily and undergo convection, and a liquid water ocean beneath that. Surface temperatures on Europa are brutally cold (typically −150°C), so its icy crust must be as stiff as granite.

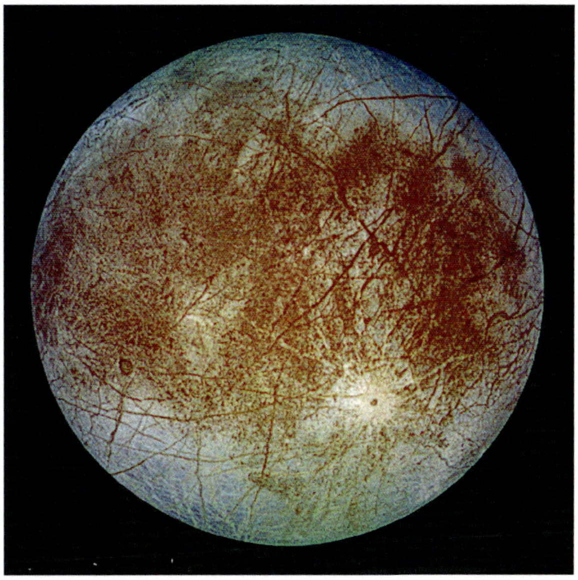

Figure 9.13

A global view of Europa, as seen from the *Galileo* spacecraft. Colors are enhanced to bring out subtle details.

Evidence for a Liquid Ocean The spacecraft pictures provide strong evidence suggesting that liquid water, not just more ice, underlies Europa's icy skin. One key piece of evidence comes from the surprising lack of impact craters. Europa has only about a few dozen large craters (10 kilometers or more across), a number that should accumulate in only a few tens of millions of years, not in the billions of years since the solar system's formation. Clearly, something has erased the craters on Europa. An obvious candidate is resurfacing by an occasional breakthrough of subsurface water or slushy ice.

Another suggestive piece of evidence for an underground ocean is the appearance of Europa's so-called *chaotic terrain*, which resembles photos of an arctic ice pack (Figure 9.15). The surface is clogged with iceberg-like blocks—some as small as football fields and others as large as a city—crisscrossed by ridge lines and suspended in what appears to have been a slushy ocean that froze. These blocks are often separated, as if they have rafted away from their original positions. Imagine putting your palms down on an assembled jigsaw puzzle and then spreading them slightly. Gaps will form between the pieces, and the picture printed on them will become disjointed. Europa's ridge lines similarly form a telltale picture on the ice, a picture that in places has fractured into small, jostled pieces. Individual ridges can be traced by the eye from one block to another, clearly showing the motion that must have occurred. Further evidence for an ocean comes from features suggesting that liquid water has gushed from below and frozen in spaces between some of the ridges (Figure 9.16).

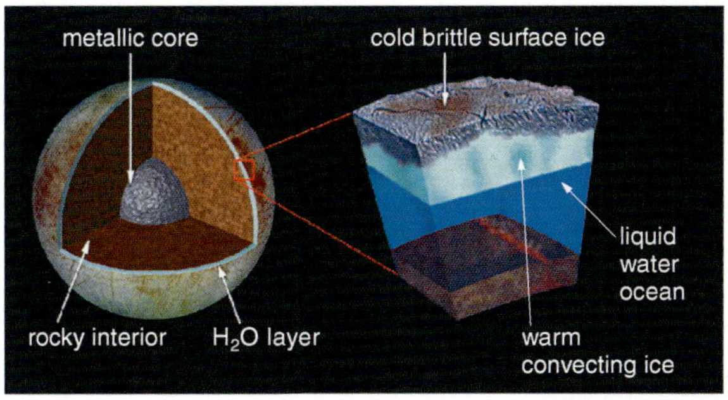

Figure 9.14

These artist's drawings show the leading model for Europa's interior layering. Notice the liquid water ocean under the outer layers of ice.

While the photographic evidence suggests that Europa's ice-pack exterior has been churned by an underlying ocean, it still leaves a shadow of a doubt. Could it be that the ocean that produced the tortured and broken surface existed millions of years ago and is now frozen, which would mean that Europa's intriguing face is merely the frozen remnant of an earlier time? Perhaps the icy crust sits atop relatively warm soft or slushy ice, and the surface features are the result of this warm ice convecting upward from below. In that case, Europa's frigid skin might extend all

Figure 9.15

Chaotic terrain on Europa. This landscape suggests that liquid water (or slushy ice) has welled up from below, breaking apart the surface and then freezing in place. The pieces can be mentally put back into place by matching up the details of the ice blocks. However, such a reassembly indicates that some of the pieces are missing; they presumably sank, melted, or crumbled during the reshuffle.

Figure 9.16

The smooth, dark terrain just to the right of the center of this photo mosaic could represent an area where liquid water broke through the surface and covered over some of the ridged landscape.

the way to the moon's rocky interior. But there's another line of evidence, arguably even more convincing than the photos, that makes the case for a liquid ocean.

In 1996, a magnetometer aboard the *Galileo* spacecraft detected a magnetic field near Europa. Because moons seldom have a magnetic field (though, as we'll discuss shortly, Ganymede does), researchers asked themselves what was causing magnetism on Europa. Jupiter has a strong magnetic field (which causes it to be a relatively powerful source of natural radio static), and Europa orbits within this field. Because Jupiter's magnetic equator is tilted with respect to Europa's orbit, its field at Europa's position is constantly changing as the giant planet spins on its axis. Just as a moving magnet produces an electric current in a coil of wire, so too could Jupiter's changing field induce currents in an electrical conductor within Europa. Of course, there's no giant coil of wire in Europa, but a salty ocean would conduct electricity in much the same way. The currents in such an electrically conducting ocean would act to set up a magnetic field that opposes Jupiter's—and thus would change as Jupiter's field changes. Additional measurements with *Galileo*'s magnetometer showed that Europa's magnetic field does indeed change as Jupiter spins. Most researchers consider this the best evidence yet that a liquid, salty ocean exists under Europa's icy surface. The magnetometer data require that Europa's subsurface water be global in extent, not limited to just a few isolated liquid pockets. They also imply that this ocean could be as salty as Earth's seas. Some of *Galileo*'s instruments also found evidence for what appear to be salts on Europa's surface—possible seepage from a briny deep.

If we tally up the evidence for an ocean on Europa, we can list the following:

- The relatively small number of craters implies that the moon's surface is young, perhaps only a few tens of millions of years old, indicating that it has been recently repaved.
- Various features on the surface (chaotic and flooded terrain) suggest that liquid water sometimes wells up from below.
- Europa has a magnetic field that is likely caused by currents produced in something that conducts electricity—like a salty ocean—as Jupiter's magnetic field changes.
- Calculations show that tidal heating can supply enough heat to keep most of Europa's ice melted beneath a solid ice crust.

While no single piece of evidence would make an overwhelmingly strong case by itself, together the evidence gives us good reason to suspect that the model in Figure 9.14 is correct, and that a liquid ocean really does exist beneath the surface of Europa. Still, the case is not definitive. For that, we'll need new space missions.

Proving the Case for an Ocean NASA scientists are hoping to send an orbiter to Europa, equipped with instruments that would include a laser altimeter and long-wavelength radar. The altimeter would be used to seek proof that an ocean lurks under the ice. The gravitational tug-of-war that Europa undergoes every 3.6 days in its elliptical path around Jupiter causes distortions in Europa's shape. If a deep, liquid ocean exists under the ice, Europa's surface would regularly bulge in and out by up to

30 meters. If there is only solid ice beneath the surface, however, then Europa won't stretch so easily and the altitude change in the bulge would be only about 1 meter. The difference can be reliably measured by the altimeter, and should settle the case regarding the existence of the ocean. The radar can also probe the surface, looking for radio reflections from a subsurface water interface. This same technique was used in Antarctica to discover a lake (Lake Vostok) that lies hidden beneath a 5-kilometer-thick deck of ice.

If there is an ocean on Europa, how big might it be? The current model (shown in Figure 9.14) suggests that the frozen crust (including the region of warmer, convecting ice) is about 20–25 kilometers thick. Given that our models of Europa's interior indicate that the total thickness of the water/ice layer is between about 80 and 170 kilometers, a 20–25-kilometer-thick crust leaves plenty of room for a deep ocean. In fact, if a global, subsurface ocean really exists, it could easily be 100 kilometers or more deep—some ten times as deep as the deepest ocean trenches on Earth. A 100-kilometer-deep europan ocean would contain roughly twice as much water as all of Earth's oceans combined.

• Could Europa have life?

The oceans on Earth teem with life, and in Chapter 6 we discussed reasons why many scientists suspect that life on Earth first arose in the oceans, and most likely near deep-sea volcanic vents. If Europa really does have a deep water ocean, tidal heating has probably kept it liquid for billions of years. Might life have arisen in the europan ocean as it arose on Earth, and could Europa be home to life today?

In Section 7.1, we identified the three key environmental requirements for life: (1) a source of elements and molecules from which to build living organisms, (2) a source of energy for metabolism and growth, and (3) a liquid medium for transporting the molecules of life. If Europa has an ocean, then it clearly meets the third requirement. As we discussed in Chapter 7, nearly all worlds probably contain at least some quantity of the 25 elements that make up life on Earth. Europa probably is no exception; its rock/water composition likely has all these elements, thereby satisfying the first requirement. Moreover, these elements would certainly be present at the rock/water boundary of the ocean floor, which is where we might expect any life to have arisen. That leaves open the question of whether Europa satisfies the second requirement: Does it have an energy source for life?

You might at first guess that the existence of a liquid water ocean would automatically mean there is energy for life. After all, if there is liquid water, there must be enough energy from tidal heating (and radioactive decay) to keep the water temperature above freezing. However, the existence of energy in the form of heated water is not by itself enough to allow the energy to be put to use. That is why fish can't live off ocean heat alone.

In general, extracting energy from a reservoir of warm water is possible only if there is also an adjacent "sink" of much colder water and a substantial difference in temperature between the two. And while we can build machines to take direct advantage of such differences in temperature, life has an additional requirement: There must be chemicals present that can react to generate energy for biological use. In practice, this additional requirement is almost certainly met when we have temperature differences at a rock/water boundary, because molecules in the rock and water can react together in a variety of useful ways. (We'll discuss some of the specific types of reactions that can supply energy to life in Section 9.4.)

Thus, to decide whether there might be life in a europan ocean, we must ask whether there is enough energy in a useful form to support biology. We can separate this question into two parts. First, is there enough energy to support an origin of life? Second, if so, is there enough energy to support a reasonable total biomass of ongoing life? Let's begin by considering the question of energy for an origin of life.

Energy for an Origin of Life From an energy standpoint, the idea that life on Earth might have begun near deep-sea vents makes sense. The volcanic vents heat the water near them to very high temperatures, creating a large temperature difference with the surrounding cooler water. This temperature difference would have facilitated chemical reactions between the water and the rock erupted from the vents, leading to the formation of complex organic molecules and perhaps to the origin of life. We might therefore wonder whether the same type of vents exist on Europa, thereby providing the energy needed for an origin of life.

We cannot see through Europa's icy crust, which is why we don't even know for certain that a liquid ocean lies beneath it. Clearly, then, we have no way at present to know whether Europa has volcanic activity on an ocean floor. Nevertheless, the possibility of volcanic vents seems reasonable. Our models of Europa's interior suggest that it should have a rocky ocean floor, and tidal heating and the decay of radioactive elements may provide enough energy to melt interior rock that could erupt into the ocean. Europa might even have large undersea volcanoes. In that case, it certainly seems plausible to imagine an independent origin of life on Europa.

Figure 9.17

On Earth, we find abundant life near deep-sea volcanic vents. However, while the vent supplies energy used by some microbes, most of the life around it—including the tube worms visible in this photo—actually gets energy from materials that filter down from above.

Energy to Support Ongoing Life If life did arise in a europan ocean, how widespread could it be today? While deep-sea vents might lead to enough energy for an origin of life, they could not by themselves support more than a small total biomass because they simply don't make enough energy available to organisms. This fact might surprise you if you think about the great communities of life that live near deep-sea vents on Earth today (Figure 9.17), but most of this life actually gets its energy from materials that filter down from above, such as dead organisms and oxygen produced by photosynthetic life near the surface. Only a small fraction of the life near Earth's deep-sea vents lives solely off energy from the vents themselves. Thus, for life to be similarly abundant or widespread on Europa, it would need some other energy source in addition to the chemical reactions near deep-sea vents.

On Earth, photosynthesis is the best-known source of energy for life, and the one that ultimately powers humans via the food chain. Photosynthesis could not directly provide energy for life in a europan ocean, because sunlight cannot penetrate through more than a few tens of meters of ice. However, if pockets of liquid water exist near the top of the ice crust, then photosynthesis might provide energy to organisms living in these pockets, and this energy might then filter downward as organisms (dead or alive) cycle between the crust and the ocean. Even in this case, however, remember that because Europa is about five times as far from the Sun as Earth, sunlight is about $5^2 = 25$ times weaker at Europa than at Earth. Overall, it seems unlikely that photosynthesis would play any role on Europa, and in the best case it could support only a much smaller abundance of life than it supports on Earth.

Before the origin of photosynthesis on Earth, biochemical reactions may have been facilitated by energy sources such as lightning, ultraviolet radiation from the Sun, and heat released by impacts of asteroids and comets. Unfortunately, none of these energy sources is likely to be useful to life in europan seas. Europa has no atmosphere for lightning, ultraviolet light does not penetrate the ice, and the time during which impacts were frequent enough to provide significant energy ended some 4 billion years ago.

There is, however, another possible scheme for producing energy on Europa's icy surface. High-energy particles accelerated and trapped in Jupiter's magnetic field, as well as ultraviolet light from the Sun, regularly slam into the surface ice. These particles and photons hit the surface with enough energy to break up molecules in the ice, leading to the production of small quantities of other molecules such as hydrogen peroxide (H_2O_2), molecular oxygen (O_2), and hydrogen (which quickly escapes); this process explains why Europa has an extremely thin atmosphere (not noticeable to the eye, but detected by instruments). These molecules can facilitate energy-producing reactions, and they should be mixed into the top meter of the europan surface by the frequent churning caused by small meteorites. If all these molecules were ultimately to end up in the ocean below, they could provide energy to support life there. Unfortunately, we don't know how much of the outer ice actually gets cycled into the water below, or how often. In addition, the total amount of energy that might be available in this way is at least ten thousand times less than the amount of energy that photosynthesis generates on Earth, so an ocean on Europa could not support anywhere near as much life as do the oceans on our planet.

One other known process might yield some energy for life on Europa. Some of the potassium contained in the rocky material that makes up this moon would dissolve in the ocean (and be frozen in the ice above). The energy from the natural decay of one of potassium's radioactive isotopes would produce both hydrogen and oxygen molecules. Rough estimates suggest that these molecules could then facilitate chemical reactions that might support a small biomass.

The Case for Possible Life on Europa Summarizing the case for life on Europa, we can say:

- There is good, indirect evidence that a liquid-water ocean exists.
- We expect the elements needed for life to be present in that ocean and on its floor.
- There are possible energy sources to support life, but the total available energy is small compared to the energy available for life on Earth.

Taken together, the evidence makes Europa seem like a good candidate for the possibility of life. That said, we should caution that it seems unlikely that sufficient sources of energy could exist in the europan ocean to support macro-fauna—the complex sea creatures of our aquariums, for example. If life exists in Europa's oceans, it is probably quite simple and small, perhaps analogous to the most primitive single-celled organisms that have existed on Earth.

THINK ABOUT IT Given the various uncertainties about a liquid ocean and available energy on Europa, do you consider it likely or unlikely that we will find life there? Why?

The only way to find out for sure whether life exists on Europa will be to go there. Although the next mission to Europa, still in its planning phases (and currently delayed due to NASA's overall budge woes), will almost certainly be an orbiter designed to determine whether the ocean really exists, subsequent missions could land on the surface. If a lander found any water that might have been brought up from below, we could analyze it for various organic molecules such as amino acids. The lander might also melt and filter a sample of surface ice—preferably a large sample since we expect the abundance of any life on Europa to be low—in search of evidence of life. Note that we'd need to be careful to sterilize the spacecraft so that we don't accidentally transport biology from Earth to Europa and then "discover" it there. Indeed, in 2003, NASA deliberately ended the *Galileo* mission by causing the spacecraft to plunge into Jupiter's atmosphere, expressly to prevent any possibility that the spacecraft might someday crash into and contaminate Europa with hitchhiking microbes from Earth. If the results from a lander are encouraging—for example, if they show the presence of organic molecules in the ice—we could then dream and scheme about probes that would manage to work their way through the ice and explore the eternally dark ocean depths where alien life might swim.

• Could other moons of Jupiter have life?

Europa is the most likely of the Galilean moons to be habitable. However, both Ganymede and Callisto are also composed of significant amounts of water ice. Could they, too, have underground oceans, and hence the possibility of life?

MOVIE MADNESS: *2010: THE YEAR WE MAKE CONTACT*

The monoliths are back, and they've brought plenty of buddies.

In the film *2010,* a sequel to the classic *2001,* humans are once again prompted by some enigmatic black slabs to head for the outer solar system. It seems that an unseen race of aliens is trying to make an important point about Jupiter and its moons. So we oblige them by sending yet another batch of confused astronauts on a billion-kilometer joy ride.

However, unlike the mission in *2001,* this crotchety crew actually returns to Earth, bringing with them a few sage words of alien advice: "All these worlds are yours except Europa. Attempt no landing there."

What's the deal? After all, this is *our* solar system. So why are some unknown, unseen entities marking Europa off-limits ? That's like Mom forbidding you to go into a basement closet. Of course that's the one place you'll find most interesting, and Arthur C. Clarke, the author of the novel *2010,* knew that there was, indeed, an interesting closet in the jovian system. The *Voyager* spacecraft had shown Europa to be an ice-covered world that could have a huge liquid ocean, and maybe even life.

In *2010,* this possibility is subtly exploited. The astronauts learn that there are chlorophyll-equipped critters somewhere below the europan ice. Alas, life in such a deep, dark habitat is a bit of a drag. So just as the humans arrive, the sophisticated aliens who built the monoliths decide to reengineer Jupiter, turning it into a mini-sun. They do this for the benefit of the primitive life-forms on Europa. The new star eventually warms their moon and converts it to something that looks like Earth during the Mesozoic era. The Europans, we presume, are destined to crawl out of their formerly ice-capped seas and find a monolith that will promote them to intelligent beings, the way our simian ancestors were improved in *2001.*

We lose Jupiter, but we gain a second, dimmer sun in our skies (no doubt a headache for astronomers and a source of confusion for migratory birds). But one wonders why these unseen extraterrestrials are so keen to meddle in the biological evolution of other worlds. Perhaps they just want our distant descendants to have some intelligent company right here in the solar system. It's a nice thought, but in the meantime someone needs to tell the space agencies that it's "hands off" Europa for the next few hundred millennia!

Ganymede Ganymede, the third from Jupiter of the four Galilean moons, is the largest moon in the solar system. Like Europa, Ganymede has a surface of hard, brittle ice. However, while Europa's impact craters have been mostly erased by other geological processes, Ganymede appears to have both young and old surface regions, sometimes separated by remarkably sharp boundaries (Figure 9.18). Some regions are dark and densely cratered, suggesting that they look much the same today as they did billions of years ago. Other regions are light-colored with few craters, suggesting that they are geologically younger and have had their ancient craters erased; the young regions also exhibit strange grooves. The most likely explanation for the younger regions of ice is that they were created by "water eruptions" that occurred when internal heat caused ice below the surface to melt, erupt as watery "lava," and then freeze. The grooves were probably made by tectonic stresses that stretched the icy crust.

The idea that water sometimes erupts onto Ganymede's surface implies that partial melting must occasionally occur in the underlying ice, but could Ganymede have a full-fledged ocean like that thought to exist on Europa? The *Galileo* spacecraft turned its magnetometer to Ganymede just as it did to Europa, and made two intriguing discoveries. First, Ganymede apparently has its own intrinsic magnetic field—one generated within the moon—which may indicate that it has a molten, convecting core somewhat like Earth's outer core. Second, the magnetometer data showed that a small part of Ganymede's magnetic field varies with Jupiter's 10-hour rotation, just like Europa's magnetic field. Again, this is interpreted to indicate a field that is produced in electrically conducting material under the surface—most likely a salty ocean. Another bit of evidence for an underground ocean is the presence of salts on Ganymede's surface, which could conceivably be brine brought up from below. Overall, a liquid water ocean seems fairly likely to exist beneath Ganymede's frozen surface, although the case for its existence is less strong than the case for an ocean within Europa.

What source of heat could be keeping water melted under Ganymede's surface? Because of its greater distance from Jupiter, Ganymede has much less tidal heating than Europa. However, Ganymede's larger size also means it should retain heat better than Europa. Perhaps the heat produced over time by the combination of tidal heating and radioactive decay has been just enough to sustain a layer of subsurface liquid water.

While the possibility of a subsurface ocean is encouraging from the standpoint of habitability, the lesser heating on Ganymede means that the ice cover would be much thicker than on Europa—probably at least 150 kilometers thick. This would make finding life in a subsurface ocean far more difficult and the transport of possible nutrients from the surface considerably less efficient. In addition, the pressure in Ganymede's interior is high enough to create high-density forms of ice that likely lie beneath any liquid water ocean. As a result, it is also unlikely that Ganymede has a rock/water boundary at its ocean bottom like that on either Earth or Europa, a fact that would further reduce the available energy for life. Life, if it exists on Ganymede at all, would probably be less abundant and less evolved than seems possible on Europa. It might also be so deep below the surface that we'd have little hope of gaining access to it.

Callisto Callisto is the farthest out of the four Galilean moons. Figure 9.19 shows that its entire surface is densely pockmarked by craters (the bright, circular patches) that must date back to the heavy bombardment. Other surface features are more difficult to interpret. For example, the close-up photo shows a dark, powdery substance that is concentrated

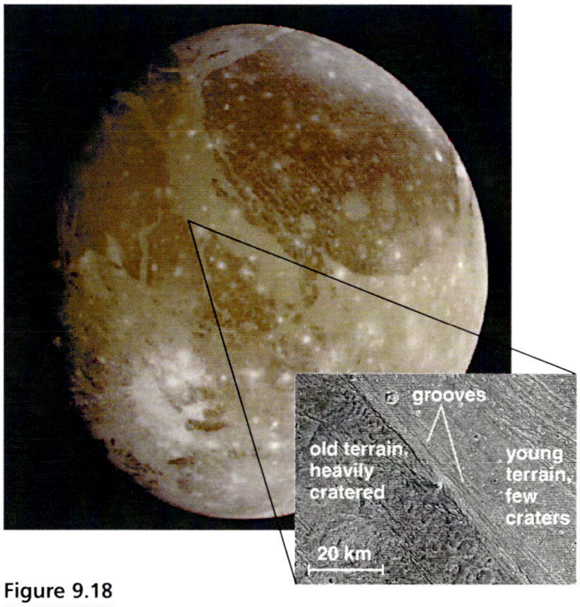

Figure 9.18

Ganymede, the largest moon in the solar system, has both old and young regions on its surface of water ice. The dark regions are heavily cratered and must be billions of years old, while the light regions are younger landscapes where tectonic faulting and eruptions of liquid water or slush have presumably erased ancient craters; the long grooves in the light regions were probably formed by tectonic stresses. Notice that the boundary between the two types of terrain can be quite sharp.

Figure 9.19

Callisto is heavily cratered, indicating an old surface that nonetheless may hide a deeply buried ocean. The inset shows how a dark powder appears to cover low-lying areas of the surface.

in low-lying areas, leaving ridges and crests bright white. The dark powder may be debris left behind when ice sublimates into gas from Callisto's surface.

Gravity measurements suggest that Callisto is mostly a ball of mixed ice and rock, overlaid by several hundred kilometers of water ice. Because its interior doesn't seem to be fully differentiated, we conclude that Callisto was never very warm inside and that neither radioactive decay nor tidal heating ever warmed this moon enough to melt it through. Surprisingly, however, the Galileo spacecraft found an induced magnetic field for Callisto, too, suggesting—as for Europa and Ganymede—the presence of a salty, subsurface ocean.

If a subsurface ocean exists on Callisto, what heat source could keep it liquid? Unlike the other three Galilean moons, Callisto doesn't participate in the orbital resonances that cause tidal heating, meaning that the warmth required to maintain a liquid ocean would almost surely have to come from radioactive decay. This meager heat source might be sufficient due to the insulating properties of a thick, icy skin. In addition, the water might contain salts and ammonia that act like antifreeze to help keep it liquid at low temperatures.

If Callisto really does have a deep, unseen ocean, we arrive at the astonishing possibility that three of Jupiter's large moons could have liquid water oceans and, perhaps, life. From the standpoint of intelligent aliens studying our solar system, they might have more examples of differing biology to explore by spending time near Jupiter than on Earth. Nevertheless, energy considerations make it unlikely that any of these moons has life in the abundance or diversity that we find for life on Earth. Callisto probably had even less energy available to life than Europa or Ganymede, and would therefore be the least likely to have life in its ocean.

THINK ABOUT IT Suppose we had the technology to send landers that could somehow reach subsurface oceans on any of Jupiter's moons. Which moon would you choose to explore first? Why?

9.3 LIFE AROUND SATURN, AND BEYOND

Prospects for habitability and life dim considerably as we go beyond Jupiter. Although a few outer solar system moons have tidal heating, we do not find more examples quite as extreme as those of, say, Io and Europa. The greater distance from the Sun also dampens the possibility of obtaining energy from sunlight, and cold temperatures would presumably slow any metabolic reactions. Nevertheless, the outer solar system offers some intriguing prospects for life—or at least for interesting organic chemistry that could be a precursor to life.

• Could Titan have life?

The best-studied candidate for habitability beyond Jupiter is Saturn's moon Titan. The second-largest moon in the solar system after Ganymede, Titan is the only solar system moon to have a substantial atmosphere. In fact, Titan's atmosphere is even thicker than Earth's. The surface pressure is about 1.5 times that on Earth, which means that if you could visit Titan, the pressure would feel fairly comfortable even without a space suit. The temperature, however, would not. Here, where sunlight is nearly 100 times as weak as on Earth, the surface temperature is a frigid −180°C (−290°F).

Moreover, while the atmosphere is 90% nitrogen (N_2)—not so different from the 77% nitrogen content of Earth's atmosphere—there is no appreciable oxygen to breathe.

While Titan's low temperatures make life there seem unlikely, some interesting chemistry is taking place on this frigid, smoggy world. As a result, it was selected as a prime target for the *Cassini* mission that reached Saturn orbit in 2004. By the time *Cassini* completes its planned 74-orbit "tour" of the Saturn system in 2008, it will have flown past Titan 44 times at a distance near enough to map surface regions in some detail. In addition, *Cassini* carried a probe, called *Huygens,* that parachuted to a soft landing on Titan's surface in January 2005. The results from *Huygens* and the ongoing observations from *Cassini* are rapidly revolutionizing our understanding of Titan.

THINK ABOUT IT Saturn's distance from Earth varies between about 1.2 and 1.6 billion kilometers (depending on whether Earth is on the same or opposite side of the Sun); in other words, the *Huygens* probe successfully landed on a world more than *a billion kilometers away.* What does this fact tell you about modern space technology? Defend your opinion.

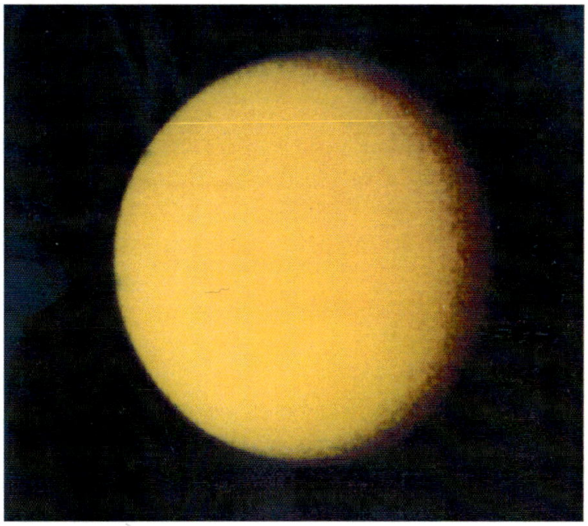

Figure 9.20

Titan, as photographed by *Voyager 2.* It is enshrouded by a reddish smog, and the mystery of what lies beneath encouraged scientists to make Titan a prime target of the *Cassini* mission.

A Moon Shrouded in Mystery Titan's atmosphere was first discovered in 1944, when spectroscopic observations from Earth showed the presence of methane. However, the amount of atmosphere was not immediately known, in part because methane is not the dominant gas (nitrogen is). Because scientists originally included the hazy atmosphere when measuring Titan's size, it was once thought to be the largest moon in the solar system. The two *Voyager* spacecraft, which passed by Saturn and its moons in 1980 and 1981, found that Titan's girth had been overestimated because it is puffed out by a 200-kilometer-thick atmosphere. The solid part of Titan has a radius 60 kilometers smaller than that of Ganymede. Cameras aboard the *Voyager* spacecraft were capable of photographing fine detail on Titan, but they saw nothing of the surface* because their vision was blocked by the opaque, reddish haze—in essence, smog (Figure 9.20). Most of the visible smog is due to chemical by-products formed when ultraviolet light from the Sun breaks apart molecules of methane.

Titan's gravitational pull on the *Voyager* spacecraft allowed researchers to accurately determine its mass, which turns out to be nearly twice that of our Moon. Knowing its mass, combined with its size, permits us to compute its average density, which is nearly the same as Callisto's (about 1.9 gm/cm^3). This suggests that Titan is made up of roughly equal volumes of rock and ice, similar to other large moons of the outer solar system.

Despite the fact that the *Voyager* cameras were frustrated by smog, these spacecraft managed to tell us much about conditions within Titan's atmosphere. *Voyager 1* was given a trajectory that allowed it to sail behind Titan, so that its radio signal would pass through the smog on its way back to Earth. This ingenious experiment provided data that were used to determine both the temperature and composition of the atmosphere. The composition measurements proved especially intriguing. Besides its 90% nitrogen content, the rest of the atmosphere is composed of (in order of abundance) methane (CH_4), argon, and ethane (C_2H_6). There are lesser

---

*Recent reprocessing of *Voyager* images shows that they did just barely detect the surface, demonstrating how improving image-processing technology can give new life to old data.

Figure 9.21

Titan unmasked. This image of Titan was made with *Cassini*'s infrared cameras, which are able to penetrate the thick smoggy atmosphere. The image uses black and white to show infrared contrast, since we cannot see infrared light. Compare the detail in this image to the featureless *Voyager* photo in Figure 9.20.

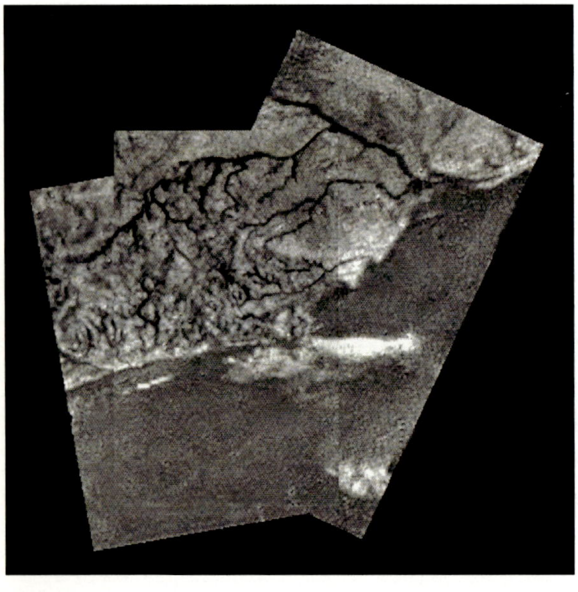

Figure 9.22

The *Huygens* probe made the photos in this mosaic when it was still several kilometers above Titan's frigid surface. The tributary-like dark channels flowing to the flat area below were probably carved by liquid methane. The icy hills reach heights of approximately 100 meters.

quantities of propane (C_3H_8), acetylene (C_2H_2), hydrogen cyanide (HCN), and carbon dioxide (CO_2). What accounts for this mixture of hydrocarbons, which reads like an oil company's product line?

To begin with, once ammonia (NH_3) from Titan's icy interior made it into the atmosphere, energetic ultraviolet light from the Sun would have broken it apart, allowing it to react to form nitrogen (N_2) and hydrogen (H_2). The nitrogen molecules were heavy enough to stay put and became the principal ingredient of Titan's air. The much lighter hydrogen escaped into space. When methane (CH_4) entered the atmosphere, it too was broken apart by ultraviolet light into hydrogen (which, again, escaped) and the simpler compounds CH and CH_2. Products of the methane breakdown then reassembled themselves into more complex hydrocarbons, especially ethane (C_2H_6). Eventually, this should have led to an atmosphere so saturated with ethane that a nonstop drizzle of ethane rain began to fall.

So far, so good. But the fact that measurements from both *Voyager* and Earth-bound telescopes showed that there's still lots of methane gas in the air was surprising; we might expect that it would all have been converted into other molecules long ago. This puzzle can be solved if we assume that a replenishment source—a reservoir of methane—slowly feeds new gas into the atmosphere. The source was hypothesized to be large pools of slowly evaporating methane on the surface, where the −180°C temperature should be just warm enough to allow methane to be in a liquid state (see Table 7.1). Some of the evaporating methane would remain in the atmosphere, some would be converted to ethane by ultraviolet light and chemical processes, and some might even rain down.

Thus, the *Voyager* studies suggested that Titan could have a drizzle of rain made up of methane or ethane droplets—in essence, liquid natural gas (which is largely made up of methane and ethane)—and perhaps even lakes or oceans of liquid methane on the surface. Clearly, this was a world that called for further study.

Origin of the Atmosphere Given that we usually think of moons as airless worlds, like our own Moon, how is it that Titan can have such a thick atmosphere? In fact, the main reason why moons generally have little or no atmosphere is their small size, which results in weak gravity that is unable to hold on to substantial amounts of gas and in little internal heat that does not lead to much outgassing. Titan can hold its atmosphere because of its relatively large size, along with extremely cold temperatures that make it less likely that gas molecules can attain escape velocity. A subtler question is why Titan has a thick atmosphere while Jupiter's moon Ganymede, which is even larger (and still quite cold), does not.

Two explanations for the difference have been suggested. First, recall that at Saturn's distance from the Sun, ices such as methane and ammonia should have been able to condense in the early solar system, but we expect only water ice at Jupiter's distance. Thus, while the outer crusts of Europa, Ganymede, and Callisto are composed largely of water ice, Titan's outer layer should have substantial amounts of methane and ammonia ice. These compounds can evaporate or sublimate into gas at lower temperatures than does water (see Table 7.1). Thus, if internal temperatures rose on Titan during differentiation, methane and ammonia ice might have turned to gas, bubbled out of the crust, and built up an atmosphere.

A second possible reason for Titan's atmosphere is that comets and asteroids hitting a moon of Saturn are traveling at lower speeds than are those that fall onto the moons of Jupiter (both because Jupiter's stronger gravity accelerates incoming objects more and because the Sun's gravity accelerates objects more at Jupiter's distance than at Saturn's). When such bodies slam into an atmosphere, they can blast away much of the atmosphere. If Ganymede once had an atmosphere, it would have been more likely than Titan's atmosphere to have been blown away by impacts.

The Cassini–Huygens Mission The intriguing prospect of a smoggy moon with frigid seas of liquid methane was a major incentive for launching the 2-ton *Cassini–Huygens* spacecraft to Saturn in 1997 (see Figure 7.11). After a circuitous, 7-year journey that took it twice past Venus, back past Earth, and then on beyond Jupiter—with each planetary pass giving it a gravitational slingshot energy boost [Section 7.4] that was necessary because of its relatively low launch speed—*Cassini* reached Saturn orbit in July 2004. With infrared cameras that can "see" through the smoggy atmosphere, *Cassini* quickly gave us much clearer views of Titan (Figure 9.21).

Our closest views of Titan came from the *Huygens* probe. Released from the *Cassini* "mother ship" on Christmas Day, 2004, the probe spent 21 days coasting toward Titan. On January 14, 2005, the probe entered Titan's smoggy atmosphere. A series of parachutes was deployed to ease *Huygens* through its $2\frac{1}{2}$-hour descent. As the descent proceeded, the probe radioed back information about the composition and temperature of Titan's atmosphere, and also snapped hundreds of photos of the landscape below.

The aerial views from *Huygens* show fantastic topographic details. Some are clearly reminiscent of coastal areas on Earth, with hills laced by branching channels that meander down to flat areas that look like rivers or lakes (Figure 9.22). However, unlike Earth, the liquid responsible for carving the channels is thought to be liquid methane, not water. Moreover, while there are indications of liquid action everywhere, *Huygens* found no sign of flowing or pooled liquids at the time of its descent. This may be because *Huygens* landed during a dry spell, or perhaps even in a desert-like area that is only very occasionally rinsed by liquids during wet seasons.

Although *Huygens* was designed to float in case it settled onto a lake of liquid hydrocarbons, it actually landed on a hard surface, reminiscent of a dry streambed, strewn with granite-hard chunks of ice (Figure 9.23), most likely water ice. It hit the surface at about the speed of a bicycle. For another $1\frac{1}{2}$ hours (until its battery expired), the parked probe continued to send back data. This was the first time that a spacecraft from Earth landed on a moon other than our own.

The *Huygens* probe did not find any liquids in its descent region, but the *Cassini* orbiter continues to look for signs of methane lakes or rivers. Some hints have already turned up; for example, the smooth feature in Figure 9.24 may well be a lake, though further observations will be needed to confirm this hypothesis. However, *Cassini* has ruled out the idea that Titan might be largely covered with liquid seas of methane or ethane, which some scientists had deemed likely prior to the spacecraft's arrival.

We will surely learn much more about Titan before the scheduled end of the *Cassini* mission in 2008, but a few new ideas are worth noting. Further analysis of Titan's atmosphere has provided new insight into its origin. In particular, aside from the argon previously identified by *Voyager*, *Huygens* found no other "noble gases" (gases of elements in the last column of the periodic table [Appendix C]) such as krypton, xenon, or neon. These nonreactive gases are present in Earth's atmosphere, presumably as a consequence of a rain of comets billions of years ago. Once the comets smashed into our world, these elements, which are relatively heavy, were trapped. The fact that they are absent on Titan (the exception, argon, arises

Figure 9.23

The view from the *Huygens* touchdown site. The small objects visible in this photo are probably water ice. They measure 10–15 centimeters in diameter, and have some indication of being "smoothed" by tumbling in liquid. It may be that this is a currently dry streambed for liquid methane. The inset shows an artist's conception of the *Huygens* probe when it came to rest.

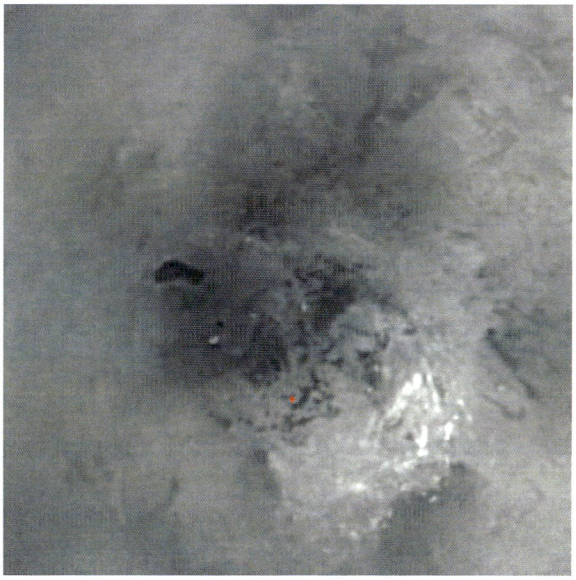

Figure 9.24

A lake on Titan? This photograph of Titan's south pole, made with an infrared camera on the *Cassini* spacecraft, shows a kidney bean-shaped dark feature reminiscent of a lake, possibly filled with liquid methane or other hydrocarbons. However, the dark area might also be a depression in the surface, covered with dark, solid material.

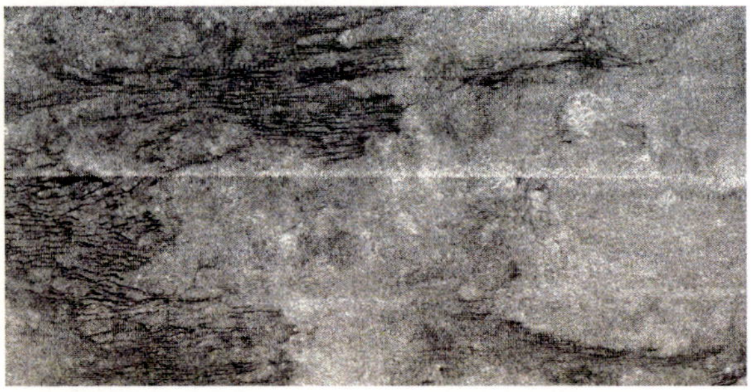

Figure 9.25

Dunes on Titan. This radar image shows a region of Titan's surface about 300 kilometers from top to bottom. The dark streaks are thought to be windblown dunes, possibly made of hydrocarbon sediments.

from the radioactive decay of potassium [Section 4.2]) suggests that no gas was supplied by comets. Instead, we infer that Titan's smoggy atmosphere was outgassed from its interior.

Outgassing suggests either some type of volcanism or sublimation of interior ices, but Titan's interior should not be warm enough to melt rock and drive volcanoes like those on Earth. Instead, Titan may have some sort of low-temperature volcanism in which the eruptions are driven by melting ices of water, methane, or ammonia; alternatively, these ices may sublimate into the atmosphere as gas without first melting. *Cassini* images should eventually settle the case. The spacecraft has already discovered a small feature in the moon's southern hemisphere that is releasing methane gas into the atmosphere, presumably from an underground supply of the gas trapped when the moon was born. Some researchers suspect that the feature is an "ice volcano," an idea that conjures up visions of "icy volcanism" (sometimes called *cryovolcanism*) looking much like volcanism on Earth aside from the far lower temperatures. Other researchers suspect that the gas is being released by sublimation rather than by anything that would look like a volcano. Either way, the release of methane from the interior may explain the mystery of Titan's atmospheric methane, without the need for pools of liquid methane like those hypothesized before the *Cassini* mission.

Perhaps the most exciting discovery to date concerns the new perspective we have gained on Titan's alien environment. Titan's landscape looks remarkably familiar to Earth's, yet it is shaped by very different materials. Instead of liquid water, Titan has liquid methane. Instead of rock, Titan has ice. Instead of molten lava, Titan has a slush of water ice mixed with ammonia. Instead of surface dirt, Titan's surface has smog particles that rain out of the sky. Titan even has wind-sculpted dunes, found in patterns similar to those found on Earth but possibly made of organic hydrocarbons rather than sand or snow (Figure 9.25). Evidently, the similarities between the physical processes that occur on Titan and Earth are far more important in shaping the landscapes than the fact that the two worlds have different compositions and temperatures.

THINK ABOUT IT Visit the *Cassini* Web site, and take a brief look at some of the latest discoveries. Can you find any that shed further light on any of the ideas discussed in the preceding paragraphs?

The Possibility of Life The temperature on Titan is so low that any surface water would be solid ice, but, as we've discussed, there is strong evidence for liquid hydrocarbons having flowed on Titan. In Section 7.1 we noted that water might not be the only liquid that could support life, though it has some clear advantages over its competitors. In particular, because methane and ethane liquids are colder than liquid water by some 200°C, any life using these liquids would probably have much slower chemical reaction rates, and hence a slower metabolism, than life using liquid water; many biologists doubt that such "slow" life is possible. (A possible exception would be if molecules of life could have weaker chemical bonding than we find for life on Earth, in which case the reaction rates might not be so limiting.) In addition, methane and ethane are far less able than water to dissolve other compounds or to facilitate the type of chemistry that might lead to life. Consequently, the outlook for biology on Titan is bleak.

It is not, however, completely hopeless. The ultraviolet light that hits Titan's atmosphere produces a wide range of organic molecules (the main contributors to the observed smog). Over billions of years, some of these compounds should have accumulated as a deep layer of organic sediment on the surface—these sediments may be the material in the dunes (see Figure 9.25). Occasional impacts by comets or asteroids would provide enough heat locally to melt any water ice and create pockets of warm water that might persist for a thousand years or so. While it is not clear that life could form in such a short time period, some interesting chemistry would certainly have occurred along the way. Moreover, if Titan really has icy volcanism, it might hint at the existence of liquid, subsurface aquifers of a water/ammonia mixture; in that case, it's conceivable that there could be volcanic "hot springs" where temperatures might rise slightly above 0°C. Perhaps life could have originated in such places and gained a foothold.

Another suggested possibility for powering the metabolism of simple life comes from chemical reactions in the upper atmosphere. Among other things, these reactions produce acetylene—an energy-rich, heavy compound that accumulates on the surface (the *Huygens* probe detected it). When acetylene reacts with oxygen, it releases a lot of energy (think of welding torches). On bitterly cold Titan, acetylene would react more slowly with hydrogen, releasing energy and producing methane as a waste product. Whether there is life that actually feeds on acetylene in this manner is, of course, only speculation, but at least a possible food source is available. One further possibility for life on Titan is deep beneath the surface. Models of Titan's interior suggest that, much like Ganymede or Callisto, it could have a liquid ocean far beneath its icy crust. However, this ocean would probably consist of a very cold ammonia/water mixture, rather than just liquid water, and the cold temperature would make life seem less likely.

Even if Titan lacks biology, studying this moon will give us valuable insights into chemical processes that might have been important to the beginnings of life on the early Earth. For example, scientists would like to know whether amino acids on Titan form equally in both right- and left-handed forms or if they exist mostly in left-handed forms like those used by life on Earth (see Figure 5.9). Moreover, we often give lip service to the fact that our ideas about the conditions that would produce life might be too conservative, but we don't often discuss alternatives. If future missions find living things on this small and hostile world—a place where the liquid environment and energy sources are quite different from those on Earth today—it will most assuredly be life as we *don't* know it.

• Could other moons of Saturn have life?

No other moon of Saturn is even close in size to Titan or to any of Jupiter's Galilean moons. Rhea, the next largest of Saturn's moons after Titan, is less than one third of Titan's size in diameter and less than one half the diameter of Europa (smallest of the Galilean moons). We expect such relatively small moons to retain far less radioactive heat than larger moons, and any tidal heating also should be at least slightly less effective. Nevertheless, several of Saturn's moons show evidence of past geological activity and one—Enceladus—has surprised scientists by showing that it is still geologically active.

Cassini photos of Enceladus show a moon with some very young surface regions. The bluish "tiger stripes" in Figure 9.26 (which are not really

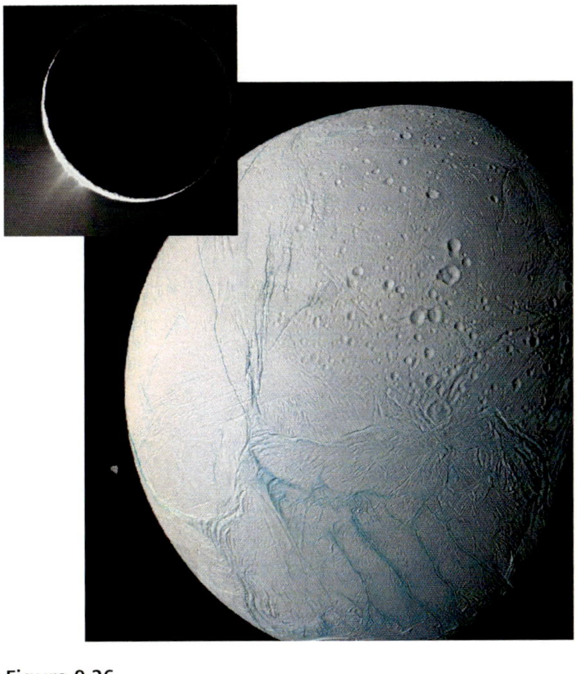

Figure 9.26

Active Enceladus. The blue "tiger stripes" in the main photo are regions of fresh ice that must have recently emerged from below. The colors are exaggerated; the image is a composite made at ultraviolet, visible, and infrared wavelengths. The inset shows Enceladus backlit by the Sun, with fountains of ice particles (and water vapor) clearly visible as they spray out to the lower left.

blue in color, since the image was made with ultraviolet and infrared as well as visible light) show regions near the moon's south pole that are measurably warmer than the surrounding terrain, and close-up examination suggests that they are cracks or grooves through which material can well up from below. These regions appear to be covered by "fresh" ice—ice that has solidified on the surface within no more than the past few thousand years, and possibly within just the past few decades or less. Moreover, images taken by *Cassini* as it looked at Enceladus backlit by the Sun show that fountains of ice particles and water vapor are spraying out from the tiger stripe regions (inset in Figure 9.26). This is clear evidence of icy volcanism on Enceladus. Spectral analysis of the tiger stripes also shows some simple organic molecules.

When you put it all together, Enceladus seems to have a subsurface liquid—probably an ammonia/water mixture—that drives its icy volcanism, at least some simple organic molecules, and enough heat to power all this activity. The astonishing conclusion: Enceladus may have subsurface habitable zones. The low temperature expected for a liquid ammonia/water mixture may make life seem unlikely, but we cannot rule it out. This unexpected finding has at least two important implications for the search for life in our solar system and the universe. First, while we know that Enceladus has some tidal heating, we are not yet certain how to explain all the geological activity that we see. Apparently, we don't yet fully understand the way that moons can be heated, and as a result conditions for liquid water (or lower-temperature ammonia/water mixtures) might be more widespread than we had imagined. Second, the fact that we have been so surprised by Enceladus should tell us that other surprises are likely to await us. Our basic ideas about where to look for life are probably still valid as starting points, but it might be wise to keep an open mind about other places as well.

• Could moons of Uranus or Neptune have life?

As we have progressed through this chapter, we have gradually moved down the scale of potential habitability. Europa seems reasonably likely to be habitable; Ganymede and Callisto seem somewhat less likely; and Titan and Enceladus, with their cold temperatures, seem to stretch to the limit the prospects for habitability. Nevertheless, our surprising findings make us ask whether there could be still other habitable places in our solar system.

After the Galilean moons and Titan, the next largest moon in our solar system is Neptune's moon Triton. As we discussed earlier, Triton orbits Neptune "backward," suggesting that it was somehow captured from what was once an independent orbit around the Sun. The question of how such a capture might have occurred is quite interesting, but here we are more concerned with the issue of potential habitability. Triton was photographed close-up only once, by *Voyager 2* in 1989. Its icy surface is an enigmatic mix of terrain that in some places is smooth and in others is crinkled into patterns resembling the skin of a cantaloupe (Figure 9.27). There are few impact craters. On the basis of the crater count on Triton today, its last resurfacing must have been no more than 10–100 million years ago. Clearly, Triton has had internal heat, possibly left over from tidal heating that would have occurred as it was being captured into its present orbit, and probably with an additional contribution from radioactive decay. Some researchers think this heat may be sufficient to occasionally cause ice volcanoes to erupt from a liquid ocean beneath the surface. The liquid would be much colder than ordinary liquid water, and

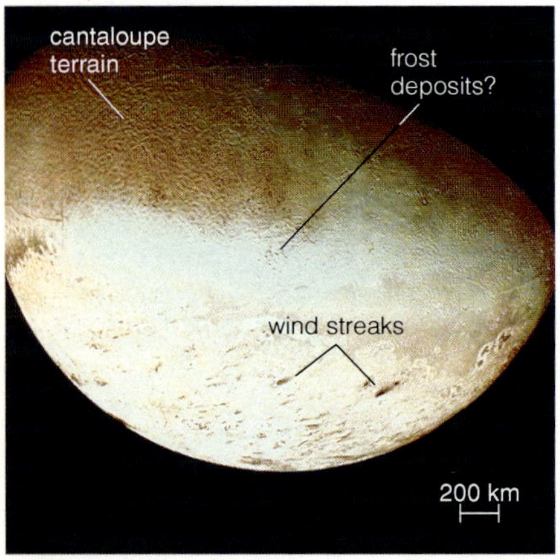

Figure 9.27

The southern hemisphere of Neptune's moon Triton, photographed by *Voyager 2* in 1989.

probably would consist of water mixed with ammonia, methane, or other melted ices.

Perhaps the most important idea emerging from our inspection of Triton is that even a moon in Neptune's distant realms, where surface temperatures are horrifically cold (colder than −230°C, or −382°F), might be geologically active and could possibly hide a liquid ocean and a refuge for life. And, given the surprise of finding the possibility of a subsurface liquid mixture on a moon as small as Enceladus, we might just find similar surprises if and when we send future spacecraft to orbit and study the satellite systems of Uranus and Neptune.

Overall, while we do not yet know if any of the moons of the jovian planets in our solar system has life, we've found at least six that seem at least potentially habitable. The lesson is clear: If similar moons are also numerous around the planets of other stars, such moons might be the most common habitats for life in the universe.

9.4 THE PROCESS OF SCIENCE IN ACTION: CHEMICAL ENERGY FOR LIFE

In our discussions of Europa, we touched on the fact that simply having a heat source is not by itself enough to make energy available for life. To support an origin of life, there must also be a chemical pathway by which complex molecules can be made, and to support ongoing life there must be chemical reactions that life can tap to fuel metabolism.

Not so long ago, the only known chemical pathways for metabolism were those used by plants and animals. However, as scientists have studied microbes living in "extreme" environments on Earth [Section 5.5], they have discovered many other metabolic pathways. These chemical pathways typically occur at interfaces between rock and water (such as in water-infiltrated rock underground), or through reactions with minerals in the hot water near deep-sea vents. If there is life on Europa or other jovian moons, it is likely to get its energy from the same types of chemical reactions. We therefore focus this chapter's case study in the process of science in action to the ways that life can extract chemical energy. As you'll see, this topic illustrates the interdisciplinary nature of astrobiology, in which biologists discover new forms of life on Earth, chemists and biochemists figure out how these organisms obtain energy, and planetary scientists then seek other worlds that might offer the same types of energy sources.

• What is the role of disequilibrium in life?

From a chemical energy standpoint, the basic requirement for life is a situation in which chemicals naturally exist in a state that is "unbalanced," which we describe as a state of **disequilibrium**. The idea is similar to that of a scale on which you place objects on both sides to see which is heavier. If the two sides weigh the same, the scale is balanced. But if you then add a little extra to one side, the scale quickly tips because it is no longer balanced. In a similar way, if there is disequilibrium among chemicals, they will start to react just like the scale starts to tip. With a scale, the movement quickly stops, because it can move only so far. But with chemicals, the reactions can continue as long as the disequilibrium remains. The idea will be clearer if we go into a bit of chemistry.

Any mixture of atoms and molecules can naturally undergo chemical reactions that may rearrange the atoms in such a way as to form or break chemical bonds. Left to themselves, however, chemical reactions will ultimately come to an **equilibrium** that represents a balance between the reacting atoms and molecules and the product atoms and molecules. For example, molecular hydrogen and oxygen can react together to make water. We can write this chemical reaction as:

$$H_2 + \tfrac{1}{2}O_2 \longleftrightarrow H_2O$$

The double arrow indicates that the reaction can proceed in both directions, and the $\frac{1}{2}$ in front of the O_2 indicates that the reaction requires only half as many oxygen molecules as hydrogen molecules.

If we begin by mixing hydrogen and oxygen, at first the reaction will proceed only toward making water molecules, since there is no water present initially. But eventually the reaction will proceed at equal rates in both directions, at which time we will have chemical equilibrium. In some cases, nearly all the hydrogen and oxygen will be converted to water, while in other cases little of it may be converted to water. The relative amounts of hydrogen, oxygen, and water at equilibrium depend on the external circumstances (such as pressure, temperature, and the presence of other chemicals). For example, the reaction between H_2 and O_2 needs a "push" to get it started. In a room filled with these two gases, this push might come from lighting a match. The energy from the match gets the reaction started; as H_2 and O_2 combine to form H_2O, additional energy is released that can then trigger more molecules to combine. This sequence can occur extremely rapidly, and the amounts of energy released can cause an explosion.

Now, imagine that the reaction between hydrogen and oxygen is occurring in a small flask that is inside a large room and that the reaction has come to equilibrium at the room's temperature. Suppose we do something suddenly that disturbs the equilibrium, such as adding excess hydrogen and oxygen molecules. Because the excess hydrogen and oxygen means that the reaction is no longer in equilibrium, we will have created a state of *disequilibrium.* This disequilibrium will cause the rate of water formation to speed up until the equilibrium is restored. Note that, once we force the chemicals into their disequilibrium, the rest of the process is completely natural: The reaction rate changes automatically in such a way as to bring the relative amounts of hydrogen, oxygen, and water back to their equilibrium values. Under the right circumstances, these reactions moving back toward equilibrium can release chemical energy that might be used by life to fuel metabolism.

No one is adding chemicals to the mixture of life, so the key to making chemical energy available for life lies in having some *natural* set of circumstances that can create and maintain a state of disequilibrium. In that case, the ongoing disequilibrium means that the reactions are always trying to move back toward equilibrium, thereby offering a continuous source of chemical energy. Natural processes that maintain chemical disequilibrium turn out to be quite common on Earth, and probably exist to some extent on any geologically active world.

For example, chemical disequilibrium inevitably exists near deep-sea volcanic vents, because mixing between the high-temperature vent water and the surrounding low-temperature ocean water creates conditions in which minerals and water will undergo chemical reactions. Because the vents continually release hot water, the disequilibrium can be maintained for long periods of time. Another place where we find ongoing disequilibrium is at interfaces between rock and water. The rock and water will naturally undergo chemical reactions, and as long as there is a supply of water that continuously circulates and comes in contact with the rock, the reactions will remain out of equilibrium. Thus, both deep-sea vents and rock/water interfaces offer places where chemical disequilibrium can provide ongoing energy that could be utilized to create complex molecules. This energy can also string complex molecules together into complicated structures—a process that may have been important to the origin of life—or to support the metabolism of living organisms.

• What types of chemical reactions supply energy for life?

Chemical disequilibrium offers the *potential* of providing chemical energy for life, but realizing this potential requires chemical reactions that life can actually use. A particular class of chemical reactions turns out to be especially important for life on Earth—reactions called **redox reactions**. Redox reactions involve an exchange or reshuffling of electric charge (which occurs through movement of electrons) between the reacting atoms or molecules.

Let's consider again what happens when hydrogen and oxygen combine to make water. Viewed on a molecular level, the reaction occurs in two steps. First, a hydrogen molecule decomposes into two protons (hydrogen nuclei) and two electrons:

$$H_2 \longrightarrow 2\,H^+ + 2\,e^-$$

(H^+ represents a single proton, which is positively charged, and e^- represents a single negatively charged electron.) Next, the two protons and two electrons combine with an oxygen atom (half of an oxygen molecule) to make a water molecule:

$$\tfrac{1}{2}O_2 + 2\,H^+ + 2\,e^- \longrightarrow H_2O$$

Viewed in this way, the production of water is a redox reaction, because electrons are effectively transferred from hydrogen to oxygen. Because hydrogen gives up the electrons, we say that it is the *electron donor* for the overall reaction. Because the oxygen takes on the electrons, we say that it acts as the *electron acceptor.*

In accepting electrons, the electrical charge of the oxygen is *reduced* (because electrons are negatively charged); hence the first three letters in *red*ox refer to this process of **reduction** of electrical charge. The charge of the hydrogen is increased, but because this increase occurs as the result of action by oxygen, we say that the hydrogen has become *oxidized.* In fact, oxygen is so efficient at grabbing electrons from other chemicals (atoms or molecules) that chemists have come to use the term **oxidation** to describe the process of losing electrons in general, even when the electrons are lost to something besides oxygen. Thus, the overall process of making water from hydrogen and oxygen, $H_2 + \tfrac{1}{2}O_2 \longrightarrow H_2O$, is called a redox reaction because the oxygen gets *reduced* while the hydrogen gets *oxidized.*

A redox reaction always involves the transfer of one or more electrons from an electron donor (which becomes oxidized) to an electron acceptor (which becomes reduced). The transfer of electrons gives off energy that can then drive other chemical reactions, including the biochemical reactions of life.

Redox Reactions on Earth Most of the key energy-generating chemical reactions used by life on Earth are redox reactions. For example, the basic process of aerobic respiration in animals involves combining a sugar acquired by eating, such as glucose ($C_6H_{12}O_6$), with oxygen acquired by breathing. The reaction makes carbon dioxide and water, and releases energy in the process:

$$C_6H_{12}O_6 + 6\,O_2 \longrightarrow 6\,CO_2 + 6\,H_2O + \text{energy}$$

This is a redox reaction because the glucose donates electrons (it is oxidized) while the oxygen accepts electrons (it is reduced).*

Many cellular energy-generating processes proceed through chains of redox reactions, sometimes called *electron transport chains* because a series of redox reactions means a series of electron transfers. In photosynthesis, for example, the chain begins when chlorophyll absorbs sunlight. The energy from the sunlight creates disequilibrium in the cell, and this disequilibrium then offers energy that the cell utilizes through a chain of redox reactions.

Possible Redox Reactions for Other Worlds Redox reactions are especially important when we consider the prospects for life in extreme environments, either on Earth or on other worlds. Many Earth organisms use fairly simple redox reactions as their primary source of energy. For example, bacteria known as *Thiobacillus ferrooxidans,* which can thrive in highly acidic conditions such as in mine tailings, obtain energy by oxidizing iron:

$$2\,Fe^{+2} + \tfrac{1}{2}O_2 + 2\,H^+ \longrightarrow 2\,Fe^{+3} + H_2O$$

(Fe^{+2} and Fe^{+3} represent iron atoms missing two and three electrons, respectively.) In this case, the iron is the electron donor (it is oxidized) and the oxygen is the electron acceptor (it is reduced). Note that neither preexisting organic molecules nor sunlight is needed for this reaction, which means reactions like this one could have been used by early life-forms on Earth—and might be used by life living underground on other worlds.

Many other redox reactions produce energy for various microbes on Earth, including reactions involving molecular hydrogen and sulfur. These are especially important when we consider possible energy sources for an origin of life. For example, both iron and sulfur are common in the disequilibrium environments of hot springs and deep-sea vents, and thus could be involved in redox reactions that might ultimately lead to life. At rock/water interfaces, which exist any place there is liquid water underground, chemical reactions between water and iron in rock can produce molecular hydrogen, which can then be used in redox reactions for biochemistry.

As a result of the understanding of this chemistry gained by studying life on Earth, we now have reason to think that life could exist on many other worlds. In particular, any geologically active world with liquid water may have places where chemical disequilibrium persists for long periods of time, such as near underwater volcanic vents or anywhere where rock and water come into contact. At these places, redox reactions can provide energy that could power biochemical reactions that might ultimately

*You can often recognize what is being oxidized or reduced in redox reactions, even without knowing precisely how electrons are rearranged. In the aerobic respiration reaction, for example, the C in glucose is being oxidized because it ends up being combined with a greater number of O atoms. Glucose has equal numbers of C and O atoms, but CO_2 has twice as many O as C atoms and so is more oxidized. The oxygen is being reduced because it ends up being combined with more hydrogen; it has no H atoms on the left side of the reaction, but has two H atoms per O atom on the right. Thus, for example, the C in CH_4 is more reduced than the C in CO_2, which is more oxidized.

lead to life, and that could support life once it arises. That is why places such as underground pockets of liquid water on Mars or the subsurface ocean of Europa seem so promising as potential abodes of life.

THE BIG PICTURE

Putting Chapter 9 into Context

In this chapter, we have considered the moons of the outer solar system and found that several of them might offer conditions suitable for life. As you continue in your studies, keep in mind the following "big picture" ideas:

- Our own Moon, a now-dead relic of an early collision, may have misled us into thinking that only planets can harbor life. In fact, moons exhibit enormous variety. Some of the moons of the outer solar system are as large as small planets, a few might have liquid oceans, and one even has a substantial atmosphere. Life might well be possible in such places.
- The solar system moon most likely to be habitable, Europa, is kept warm inside by a mechanism quite different from that which warms Earth's interior. Tidal heating, the result of orbital resonances that occur among three large moons of Jupiter, can provide a continuous source of heat for billions of years. Because orbital resonances can arise quite naturally, tidal heating may be common among moons of jovian planets throughout the universe.
- The icy moons of the outer solar system force us to rethink our basic concept of "habitability." If any of these moons are indeed homes to life, then the range of habitability is much broader than we might have guessed from studies of terrestrial worlds.
- From a chemical energy standpoint, life requires conditions in which there is a natural and ongoing source of chemical disequilibrium. Such conditions probably exist on almost any geologically active world that has liquid water, either at underwater volcanic vents or simply at the interfaces of rock and even tiny amounts of liquid water.

SUMMARY OF KEY CONCEPTS

9.1 THE MOONS OF THE OUTER SOLAR SYSTEM

• What are the general characteristics of the jovian moons?
The moons of jovian planets range greatly in size, from a few kilometers across to somewhat larger than Mercury. They tend to have ice mixed in with their rock: water ice for all the jovian moons; plus ammonia, methane, and other ices for moons of the more distant jovian planets. Nearly all are in **synchronous rotation**, keeping one side perpetually turned toward their host planet.

- **Why do we think that some moons could harbor life?**

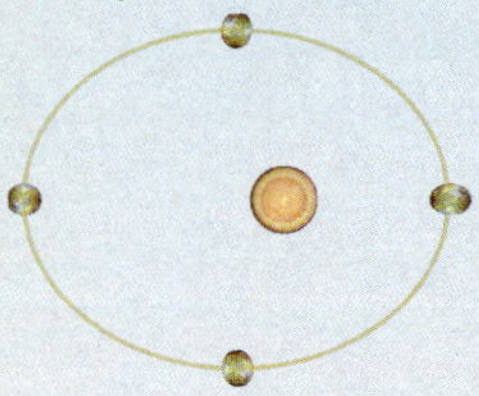

Some moons apparently retain substantial internal heat as a result of **tidal heating**, along with radioactive decay. Tidal heating explains the tremendous volcanic activity on Io and the heating thought to melt subsurface ice on Europa. A few other moons may also have liquid water or other liquids, and thus would seem to meet the minimum requirements for life.

9.2 LIFE ON JUPITER'S GALILEAN MOONS

- **Does Europa have an ocean?**

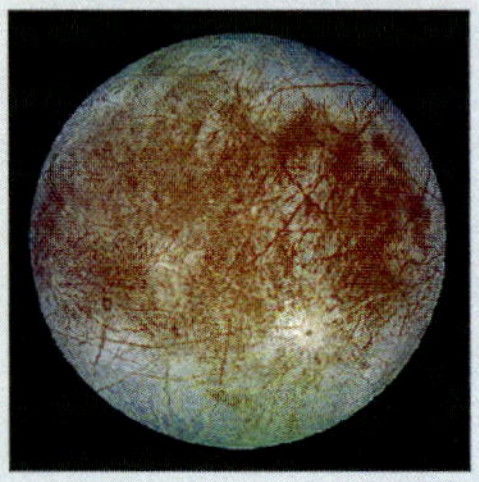

Photos of Europa's surface show numerous features, including a relative lack of impact craters, suggesting that liquid or slush from below has occasionally gushed through and repaved the surface. In addition, Europa has a magnetic field that makes sense only if we assume it has a salty ocean. These observational data, combined with the known tidal heating of Europa, make it likely that it has a subsurface ocean that may contain twice as much water as the oceans of Earth.

- **Could Europa have life?**

While it's probable that Europa has both a liquid water environment and the elements necessary for life, possible energy sources for life are much more limited than on Earth. Volcanic vents on the ocean floor could provide some energy, perhaps enough for life to have arisen but probably not enough to support life in great abundance. A few other energy sources may contribute additional energy, but overall we would expect any life on Europa to be simple and small.

- **Could other moons of Jupiter have life?**

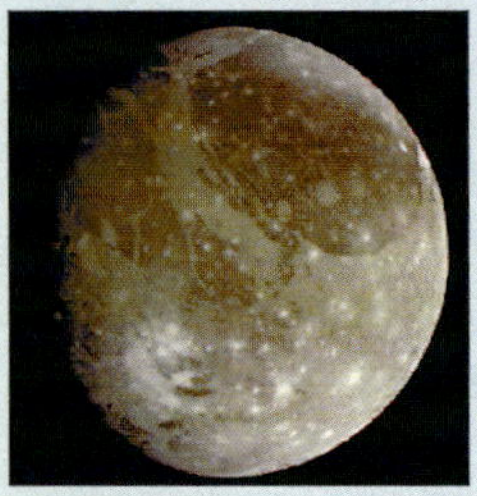

Ganymede, the solar system's largest moon, is somewhat akin to Europa, in that it has an icy crust, at least part of which seems to undergo "repaving" by liquid or slush from below. A change in this moon's magnetic field as it orbits Jupiter also suggests a subsurface, salty ocean. Callisto, the farthest out of the Galilean moons, undergoes no tidal heating and appears to have an ancient surface. Nonetheless, magnetic measurements suggest a possible hidden ocean. Thus, both Ganymede and Callisto could conceivably offer conditions for life, although energy sources are even more limited on these moons than on Europa.

9.3 LIFE AROUND SATURN, AND BEYOND

- **Could Titan have life?**

Titan has an atmosphere even thicker than Earth's, but it's largely composed of nitrogen, argon, ethane, and methane. The *Huygens* probe showed that fluid hydrocarbons have coursed over Titan's landscape, but the bitterly cold temperatures on this moon would greatly slow chemical reactions, making metabolism difficult and decreasing the chances for life. However, it is possible that Titan sometimes has surface or near-surface pockets of liquid water, and a subsurface ocean of a cold ammonia/water mixture. Some energy sources for life might also be available.

- **Could other moons of Saturn have life?**

Much to scientists' surprise, *Cassini* observations show that the relatively small moon Enceladus must have a subsurface liquid that drives the fountains of ice and water vapor observed to be emerging from the moon. Thus, it is possible that Enceladus has zones of habitability. While we have no direct evidence for similar possibilities on other moons of Saturn, the case of Enceladus tells us not to rule them out too quickly.

- **Could moons of Uranus or Neptune have life?**

Life seems less likely on such distant moons, but it is still possible that some could have habitable zones beneath their surfaces. Neptune's moon Triton shows evidence of icy volcanism, suggesting it might have liquid beneath its surface. Other moons seem like much longer shots, but the lesson of Enceladus tells us we should study them further before concluding that they lack liquids that might sustain life.

9.4 THE PROCESS OF SCIENCE IN ACTION: CHEMICAL ENERGY FOR LIFE

- **What is the role of disequilibrium in life?**

Life as we know it can exist only in places where natural conditions maintain a state of chemical **disequilibrium**. This disequilibrium can cause chemical reactions that may be used to create complex molecules, to string complex molecules together into complicated structures, or to support metabolism of living organisms.

- **What types of chemical reactions supply energy for life?**

Life on Earth gains energy from **redox reactions** in which one molecule gains electrons and another loses them. The redox reactions used by terrestrial life include those that occur near deep-sea vents and at underground rock/water interfaces, suggesting that the same types of reactions might be used by life on worlds that have similar conditions, such as Mars (underground rock/water interfaces) and Europa (deep-sea vents).

EXERCISES AND PROBLEMS

REVIEW QUESTIONS

Short-Answer Questions Based on the Reading

1. Briefly explain how the larger *jovian moons* tend to differ in general from the smaller ones. How does the formation process of the moons explain these differences?
2. Briefly describe the cause of the tides on Earth, why they lead to *tidal friction*, and how tidal friction affects Earth's rotation and the orbit of the Moon.
3. What is *synchronous rotation*, and why is it so common among the jovian moons?
4. What is *tidal heating?* Briefly explain how it can arise and persist as a result of *orbital resonances*. How does tidal heating affect Io?
5. Describe the evidence suggesting that Europa has a liquid water ocean beneath its icy crust. How might future observations confirm this idea?
6. What energy sources might be available to life on Europa? Overall, what can we say about the likelihood and abundance of life on Europa?
7. Describe the evidence for subsurface oceans on Ganymede and Callisto. What are the prospects for life on these worlds?
8. Why was Titan chosen for such intense study by the *Cassini/Huygens* mission? Why is it surprising to find methane in Titan's atmosphere?
9. Based on recent data, describe the general nature of Titan and discuss its prospects for life.
10. What makes recent discoveries about Enceladus so surprising? Could Enceladus be habitable? What lessons does Enceladus hold for our more general search for life in the universe?
11. Could Triton be habitable? Briefly discuss the possibility of finding habitable moons around Uranus or Neptune.
12. What do we mean by chemical *equilibrium* and chemical *disequilibrium?* Why is disequilibrium necessary for life?
13. What are *redox reactions?* Give a couple of examples.
14. Based on our understanding of the chemistry of life, where should we expect such chemistry to be possible? What are the implications of this idea to the search for life beyond Earth?

TEST YOUR UNDERSTANDING

Evaluate the Claims

Each of the following statements makes some claim. Evaluate the claim, writing a few sentences describing why you think it is valid or invalid (or clearly true or false). Explain clearly; because not all of these have definitive answers, your explanation is more important than your chosen answer.

15. Io is riddled with volcanoes because of its proximity to Jupiter's strong magnetic field.
16. Europa is likely to have fishlike organisms the size of whales swimming in its ocean.
17. While Europa, Ganymede, and Callisto are all candidate locations for life, we expect that the most abundant and diverse life would be found within Callisto.
18. The fact that our Moon keeps one side always facing Earth is an astonishing coincidence.
19. Titan is simply too cold to have any life.
20. Triton might have life that uses liquid ammonia, rather than liquid water, as its transport medium.
21. Io doesn't have a significant atmosphere because it lacks a source of outgassing.
22. Orbital resonances like that between Io, Europa, and Ganymede are the results of extremely rare accidents, so we would not expect tidal heating to be important in other planetary systems.
23. If there is life on Enceladus, it probably gets its energy from sunlight.
24. If our solar system is typical, then other star systems might have an average of five to ten worlds on which liquid water (or a mixture of water and some other liquid) exists in at least some places.

Quick Quiz

Choose the best answer to each of the following. Explain your reasoning with one or more complete sentences.

25. The moons of Saturn may have large amounts of ammonia and methane ice, while those of Jupiter do not because: (a) Methane and ammonia come only from comets that exist in the Oort cloud. (b) Jupiter's strong magnetic field encourages water ice to form. (c) The greater cold at Saturn's distance from the Sun means that ices of ammonia and methane could condense there but not at Jupiter.
26. Which statement about synchronous rotation is true? (a) It can develop only on moons that are born with slow rotation. (b) It occurs commonly as a result of tidal forces exerted on moons by their parent planet. (c) It can develop only on moons with liquid oceans.
27. Io is covered in volcanoes while Europa is covered in ice because: (a) Io is larger in size than Europa. (b) Io receives much more sunlight than Europa. (c) Io is subject to stronger tidal heating than Europa.
28. Which of the following is *not* an indication of liquid water beneath Europa's icy surface? (a) the moon's changing magnetic field; (b) the moon's average density; (c) surface features that look like jumbled icebergs.
29. Photosynthesis is an unlikely source of energy for life in europan seas primarily because: (a) The moon's ice cover is too thick. (b) Sunlight at the distance of Jupiter is too weak. (c) There is no soil on which plants could grow.
30. It's assumed that, even if Europa has life, the total amount of that life will be small. That's because: (a) Europa is only about as big as our own Moon, and consequently there's not much room for life. (b) The ocean will be cold, slowing down metabolism. (c) There are likely to be only limited sources of energy for life.
31. The chances for life on Titan's surface are considered slim, mainly because: (a) There's little oxygen in the atmosphere. (b) The liquid methane and ethane rain would be lethal. (c) The surface temperature is far below the freezing point of water.

32. Where might we find liquid water on Titan? (a) in lakes and rivers that appear to exist on the surface; (b) beneath the surface near sources of "icy volcanism"; (c) as droplets in high-altitude clouds.

33. Why were scientists so surprised to find active geology on Enceladus? (a) because it is so small in size; (b) because it lacks any possibility of tidal heating; (c) because it is so far from the Sun.

34. Chemical disequilibrium is likely to be present at all the following places *except:* (a) volcanic vents on ocean floors; (b) in solid ice exposed to the extreme cold of space; (c) underground aquifers where thin films of liquid water move over rock.

INVESTIGATE FURTHER

In-Depth Questions to Increase Your Understanding

Short-Answer/Essay Questions

35. *Lessons for Life.* Considering everything we've learned about the possible habitability of jovian moons, make a list of what you think are the three most important lessons that apply to the search for life in *other* solar systems. Describe the importance of each lesson clearly, and conclude by summarizing how the study of jovian moons has changed our perspective about life in the universe.

36. *Exploring Europa 1.* Although Europa is a promising place to look for life, penetrating its thick, icy crust will be difficult. Suggest a possible way of making a spacecraft that could enter the europan ocean. If it is technically feasible, do you think we should do it soon, or wait until we have further evidence of life? Defend your opinion.

37. *Exploring Europa 2.* One suggestion for determining whether Europa has life is to send an orbiter that passes close to the surface of the planet, drops a "bowling ball" that makes a crater on the surface, and then catches the ice thrown up by the impact. This sample ice would be brought back to Earth for analysis. Briefly describe what we would be looking for in such an experiment, and what it might teach us. How do you think such an experiment would affect the cost of a mission to Europa compared to, say, a mission that orbits but does not return to Earth? Explain.

38. *Europan Fish.* On Earth, fish breathe oxygen that is dissolved in the ocean. Do you expect that we will find dissolved oxygen in Europa's ocean? Why or why not? Based on your answer, if we could somehow transport fish to Europa, is it possible that they could survive in the europan ocean? What other types of life from Earth might survive on Europa?

39. *Latest Cassini Results.* Find the latest press release or discovery from the *Cassini* mission to Saturn (as posted on NASA's *Cassini* Web site). What has been discovered? Does the discovery alter any of the ideas discussed in this chapter? Explain.

40. *Life on Titan.* Several possibilities have been suggested for the support of life on Titan, including acetylene that falls onto the surface, and the possible presence of subsurface, liquid water aquifers. If you could plan one new Titan lander, how would you design it to search for life?

41. *Top-Priority Mission.* Suppose that you were chosen to design one robotic mission, and one mission only, to search for life in the outer solar system. Which world would you investigate? Defend your choice.

42. *Migrating Life.* As we discussed in Chapter 6, there is a decent likelihood that life might at some point have traveled from Earth to Mars (or vice versa) on meteorites. Discuss the likelihood that life from Earth or Mars could have made its way to, and taken root on, Europa or other jovian moons.

43. *Movie Review.* A number of science fiction movies have concerned jovian moons, including *2010, Outland,* and *Gattaca.* Choose one, watch it, and write a critical review. Be sure to comment on the accuracy of any scientific content in the movie.

44. *The Sirens of Titan.* The moon Titan plays the title role in Kurt Vonnegut's novel *The Sirens of Titan.* Although the book never intended to give a realistic portrayal of Titan, it helped popularize the moon. Read the novel, and write a short critical review, focusing on how Vonnegut's portrayal of Titan differs from reality.

Quantitative Problems

Be sure to show all calculations clearly and state your final answers in complete sentences.

45. *Orbital Resonances I.* Using the data in Appendix D, identify an orbital resonance relationship between Titan and Hyperion. (*Hint:* If the orbital period of one were 1.5 times the other, we would say that they are in a 3:2 resonance.)

46. *Orbital Resonances II.* Using the data in Appendix D, identify an orbital resonance that affects Enceladus. In light of the *Cassini* mission findings about Enceladus, how might this resonance be important?

47. *Tidal Force on the Moon.* In Cosmic Calculations 9.1, we found the tidal force that Earth exerts on the Moon today. Following a similar procedure, calculate the tidal force Earth would have exerted on the Moon shortly after the Moon formed, when it was only about $\frac{1}{10}$ its current distance from Earth. How much greater was that tidal force than it is today? What does this tell you about how a factor-of-10 change in distance affects the tidal force?

48. *Tidal Force on Io.* Using the procedure from Cosmic Calculations 9.1, calculate the tidal force exerted on Io by Jupiter. Compare to the tidal force that Earth exerts on the Moon, and comment on the implications to Io's volcanism.

49. *Tidal Force on Europa and Ganymede.* Using the procedure from Cosmic Calculations 9.1, calculate the tidal force exerted on Europa and Ganymede by Jupiter. Compare to the tidal force exerted on Io (from Problem 48), and comment on the expected strength of tidal heating on each world.

50. *Astrology Explained (or Not).* Recall that astrology claims that the positions of the planets in the sky at a person's moment of birth forever influence the person's life. In an attempt to "explain" this claim, some astrologers have proposed that the source of the influence could be tidal effects from the planets. (a) Using the method of Cosmic Calculations 9.1, calculate the tidal force exerted by Jupiter on a baby being born on Earth. For the purposes of the calculation, assume the baby is 50 centimeters long, so that the distance of the baby's "near" side to Jupiter is Jupiter's distance *minus* 25 centimeters (0.25 meter), and the distance to the baby's "far" side from Jupiter is Jupiter's distance *plus* 25 centimeters; for Jupiter's distance, use 780 million kilometers (which is about the average). (b) Jupiter, of course, is not the

This dead of midnight is the noon of thought,
And Wisdom mounts her zenith with the stars.

Anna Letitia Barbauld (1743–1825),
A Summer's Evening Meditation

Although the other worlds of our solar system may have life, there's little doubt that none will have the diversity and abundance of life found on Earth. Why is this? After all, Mars and Venus are roughly similar in size and composition to our own planet, and both have atmospheres; yet conditions on those worlds are dramatically different from those on Earth.

On one level, this question seems easy to answer: Earth has abundant liquid water on its surface, and the other planets do not. But as we probe deeper, we find that the underlying causes for *why* Earth is teeming with life, and other worlds are not, are more subtle. For example, Mars had surface liquid water in the past, but no longer does—implying that the potential for life can change with time.

In this chapter, we'll take a broad approach to the question of habitability in our solar system, seeking to understand the precise factors—such as size and distance from the Sun—that make Earth so different from its neighboring worlds. In doing so, we'll gain deep insight into the likelihood of finding Earth-like worlds around other stars. After all, we can expect Earth-like planets to be common only if they can exist over a fairly wide range of sizes and distances. Learning what makes Earth habitable will help us understand how its habitability might change in the future, either through long-term, natural changes that will ultimately occur in our solar system or through human-induced changes that could have much more immediate and serious consequences for our own civilization.

10.1 THE CONCEPT OF A HABITABLE ZONE

We all know of urban legends—for example, that small pet alligators flushed down toilets have supposedly grown to adulthood and are roaming the sewers of New York. Indeed, urban legends occur across a wide variety of subject areas, and astrobiology is no exception. So here's another one to think about: If Earth moved a mile closer to the Sun, would we all burn up? According to an urban legend that has spread widely over the past few years, the answer is yes, and the legend also holds that we'd freeze if our planet moved just a mile farther away.

But in truth, we are no more likely to burn up (or freeze) if Earth moved a mile either closer or farther from the Sun than city sanitation employees are likely to become an alligator's meal while at work. The legend is nonsense, as you can realize just by thinking about Earth's orbit around the Sun: The orbit is an ellipse, not a circle, and Earth's distance from the Sun varies from a minimum of about 147.1 million kilometers each January to a maximum of about 152.1 million kilometers each July.

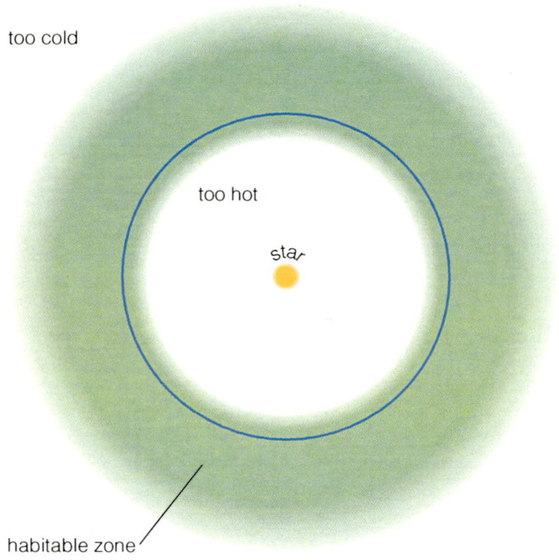

Figure 10.1

A planet with abundant liquid water on its surface can in principle exist only within some particular range of distances from a star. This range defines the star's habitable zone, and the size of the range depends on characteristics of the star.

Thus, according to the legend, our whole planet would burn up each January and freeze each July. In reality, this 5-million-kilometer variation in distance has virtually no effect on the weather at all, a fact that becomes obvious when you remember that the Northern Hemisphere has summer during the time that Earth is *farthest* from the Sun and winter when Earth is closest to the Sun. If the distance variation were important, the whole planet would have to be significantly warmer in January than in July, and it's not.

But like most urban legends, this one holds at least one deeply buried kernel of truth, because there must be some distance from the Sun that would be too hot for life to survive on Earth, and some distance at which it would be too cold. The distance isn't a mile, nor even a few million miles, but in principle we should be able to determine what it is. We'll devote much of this chapter to determining this distance.

• How does a planet's location affect its prospects for life?

If there are distances from the Sun at which our planet would become too hot or too cold for life, then there must also be some range of distances that is "just right" for a world like Earth. This range defines what we call the Sun's **habitable zone;** that is, the habitable zone is the range of distances from the Sun within which we could in principle move our planet without fundamentally changing the characteristics that make it home to abundant life (Figure 10.1). Other stars also have their own habitable zones, meaning distances at which a world similar in size to Earth, and with a similar atmosphere, would be habitable, although the sizes of habitable zones are different for different types of stars [Section 11.1].

Note that this definition captures the essence of what we mean by a habitable zone, but there are several important caveats to keep in mind:

1. The concept of a habitable zone is based on the range of distances at which *worlds similar to Earth* could exist. In other words, a habitable zone is a zone in which it is possible for a world to have abundant liquid water on its surface.
2. Simply being in a star's habitable zone is *not* sufficient to make a world habitable. The Moon presents an obvious case in point: As a companion to our planet, it is located at essentially the same distance from the Sun as Earth, but it is not habitable.
3. Habitable zones evolve with time. In particular, because stars like the Sun tend to brighten as they age, we expect the habitable zone to move outward over time.

In summary, *at any particular time, a star's habitable zone is the range of distances around it at which a planet could potentially have surface temperatures that would allow for abundant liquid water.*

We have considered a number of places besides Earth in our solar system that might either have life now or have had life in the past, including Mars, Europa, and Titan. However, none of these places seem likely to have *surface* liquid water or life at present; instead, their life would be in underground locations where liquid water might be present. This distinction may not be that important for the search for life within our own solar system, because future space missions should eventually allow us to search for life beneath the surface of worlds like Mars and Europa. However, the distinction between surface life within a habitable zone and subsurface life beyond it is important when we consider the challenges of finding life in other star systems.

THINK ABOUT IT The surface pressure and temperature on Mars are too low for liquid water to exist there today. Does this imply that Mars is beyond the outer boundary of the Sun's habitable zone? Why or why not?

We are unlikely to be able to travel to the stars anytime soon [Section 13.1], so telescopic images and spectra offer our only realistic hope of finding life on extrasolar planets or moons. Surface life may well create spectral signatures in a planet's atmosphere that would allow us to detect it [Section 11.2], but subsurface life is unlikely to cause enough atmospheric change to make an unmistakable spectral signature. Thus, while we may be able to find subsurface life (if it exists) within our own solar system, we have little hope of finding it on planets or moons around other stars. So the search for life beyond the realm of the Sun is by necessity a search for surface life on worlds within the habitable zones of their stars.

• Could life exist outside the habitable zone?

There are several ways that habitability might present itself outside the habitable zone. One is in small pockets of subsurface groundwater, such as that which may still exist on Mars today [Section 8.3]. Because this water would be kept liquid primarily by geological conditions rather than solar heat, a Mars-like planet could have subsurface habitability even if the planet itself lies beyond the outer edge of the habitable zone. It's not far-fetched to imagine that Mars-like planets could be more common than Earth-like planets.

The possibility of life in subsurface oceans, such as that thought to exist on Europa [Section 9.2], could make habitability even more common. We expect moons in the outer regions of any solar system to contain large amounts of water ice, because water is the most common of the ices that can condense in regions far from a forming star. Orbital resonances like those that contribute to the tidal heating [Section 9.1] of Io, Europa, Ganymede, and Enceladus could occur in any case where a large planet has multiple moons, and the heating could potentially melt subsurface ice. Heat from radioactive decay might also contribute to melting interior ice. Because tidal heating and radioactive decay can supply heat in the absence of sunlight, internally heated moons could exist around planets orbiting at almost any distance from a star (except very close-in, where all ice would melt and evaporate away).

Another intriguing possibility concerns *surface* habitability on Earth-size planets outside habitable zones. Theoretical calculations suggest that an Earth-size planet's own internal heat could keep its surface warm enough for liquid water *if* the surface were protected against heat loss by a thick hydrogen atmosphere. Such a hydrogen atmosphere is not possible on an Earth-size planet in our solar system, because solar heat would cause the hydrogen to escape into space fairly rapidly. But, as we will discuss further in Chapter 11, it is possible that Earth-size planets are sometimes ejected from solar systems in the process of forming, and sent into interstellar space. When an Earth-size planet is born, it might have a hydrogen atmosphere for a short time, particularly if it forms at a greater distance from its star or around a star cooler than our Sun. If such a planet is ejected into interstellar space before it loses this atmosphere, its thick hydrogen envelope might remain for many billions of years. Indeed, such "free-floating Earths" could conceivably be quite common in interstellar space, though their relatively small size and low brightness would make them extremely difficult to detect. If they exist, such planets could have surface oceans, as well as geothermal and chemical energy much like that sustaining life underground and around deep-sea vents on Earth. Although the total available energy would probably not sustain a huge amount of life, at least some might be possible.

We could also conceive of habitability outside the habitable zone if life can use a liquid medium other than water, such as liquid ethane, which as you'll remember from Table 7.1 has a much lower freezing point. Any liquid will evaporate rapidly when atmospheric pressure is low, so the presence of surface liquids of any type requires an atmosphere. In general, this means we can hope to find surface liquids only on worlds large enough to hold significant atmospheres, but not so large that they become giants like the jovian planets, where strong vertical winds probably preclude life. In our solar system, Titan is the only world that meets this criterion. Nevertheless, moons like Titan might be relatively common among other star systems. Thus, if life based on other liquids is possible, there could be many habitable worlds similar to Titan.

In summary, it's quite possible that the majority of habitable worlds in the universe are not located within stellar "habitable zones." From this standpoint, the concept of a habitable zone might seem obsolete because it doesn't account for the potential habitability of Europa-like or Titan-like moons, of Mars-like subsurface water, or of "free-floating Earths." Nevertheless, while life might be common on such worlds, it would be extremely difficult to detect. Also, it seems extremely unlikely that such worlds could give rise to complex life, let alone advanced civilizations. Thus, the concept of a habitable zone is still quite useful in searching for life beyond Earth, and it is critical when we are searching for intelligent life.

10.2 VENUS: AN EXAMPLE IN POTENTIAL HABITABILITY

One of our primary goals in this chapter is to define the boundaries of the habitable zone in our own solar system, so that we can then extend the definition to the search for planets within the habitable zones of other stars. A useful first step in defining these boundaries is to determine which planets currently lie within the Sun's habitable zone. We can clearly rule out Mercury, which is too close to the Sun, and all the planets beyond the asteroid belt, which are too far. That narrows the possibilities down to Venus, Earth, and Mars.

We have already explored the cases of Earth and Mars in some depth. Earth is obviously within the habitable zone. Mars must be at least near the borderline of the habitable zone, since it apparently had liquid water on its surface in the past. But what about Venus?

• Why is Venus so hot?

The surface of Venus is far too hot for liquid water [Section 7.2], but we can't directly attribute this heat to Venus's distance from the Sun. Venus orbits the Sun at a distance about 72% of Earth's distance; because the intensity of sunlight follows an inverse square law (see Figure 7.1), this makes the intensity of sunlight a little less than twice as great at Venus as at Earth (because $\frac{1}{0.72^2} \approx 1.9$). We'd expect this extra sunlight to make Venus warmer than Earth, but not nearly by enough to account for its searing, 470°C (878°F) surface temperature. Calculations show that if Venus had the same atmosphere as Earth, its average temperature would be only about 30°C hotter than Earth's—making Venus somewhat like the tropical planet envisioned in old science fiction.

As we discussed in Chapter 7, the immediate cause of Venus's high temperature is an extreme greenhouse effect produced primarily by atmospheric carbon dioxide. Venus has about 200,000 times as much carbon dioxide in its atmosphere as Earth, and carbon dioxide is a greenhouse gas that traps heat near a planet's surface [Section 4.5]. However, given their similar sizes and compositions, we expect Venus and Earth to have had similar levels of volcanic outgassing—and the released gas ought to have had about the same composition on both worlds. We therefore are left with a deeper question about Venus's extreme heat: Why is Venus's atmosphere so different from Earth's?

Our understanding of planetary geology suggests that huge amounts of water and carbon dioxide should have been outgassed into the atmospheres of both Venus and Earth. Venus's atmosphere does indeed have huge amounts of carbon dioxide, but it has virtually no water. Earth's atmosphere has only small amounts of either gas. Thus, we can divide the question of why the planets have such different atmospheres into two parts: (1) Why did Venus keep its atmospheric carbon dioxide and Earth did not? (2) Where did all the outgassed water go on each planet?

The Abundance of Carbon Dioxide on Venus Let's start with the question about the carbon dioxide. In fact, Venus and Earth both have similar total amounts of carbon dioxide. The difference is in where it is located. On Venus, the carbon dioxide is all in its atmosphere. On Earth, nearly all the carbon dioxide is locked up in carbonate rocks or dissolved in the oceans [Section 4.5], through the action of the carbon dioxide cycle (see Figure 4.27). This cycle is possible because carbon dioxide dissolves in water, where it can undergo chemical reactions to make carbonate minerals (minerals rich in carbon and oxygen) such as limestone. Venus lacks a similar cycle because it has no liquid water in which carbon dioxide can dissolve. We conclude that the difference in carbon dioxide is a direct consequence of a difference in liquid water: Earth's carbon dioxide became incorporated into rock because Earth has oceans and a carbon dioxide cycle; Venus's carbon dioxide is all in its atmosphere because it has no oceans and hence no carbon dioxide cycle.

The Lack of Water on Venus The differences between Venus and Earth in carbon dioxide are attributable to differences in water, so we must next ask what has happened to outgassed water on each planet. For Earth, the answer is easy: The water outgassed from volcanoes is still on Earth, but most of it is in the oceans. Venus, however, is nearly totally lacking in water in any phase. The surface is far too hot for either ice or liquid water. It is even too hot for water to be chemically bound in surface rock, and any water deeper in its crust or mantle was probably baked out long ago. The only place where Venus could conceivably have much water is in its atmosphere, but measurements show that there's little there, either. Overall, the total amount of water on Earth is about 10,000 times more than the amount on Venus.

Broadly speaking, there are two possible explanations for the lack of water on Venus. Either Venus never had much water in the first place, or Venus once had more water but somehow lost it. We cannot absolutely rule out the first explanation, but it seems unlikely. At the time the planets were forming, the temperature of the protoplanetary disk would not

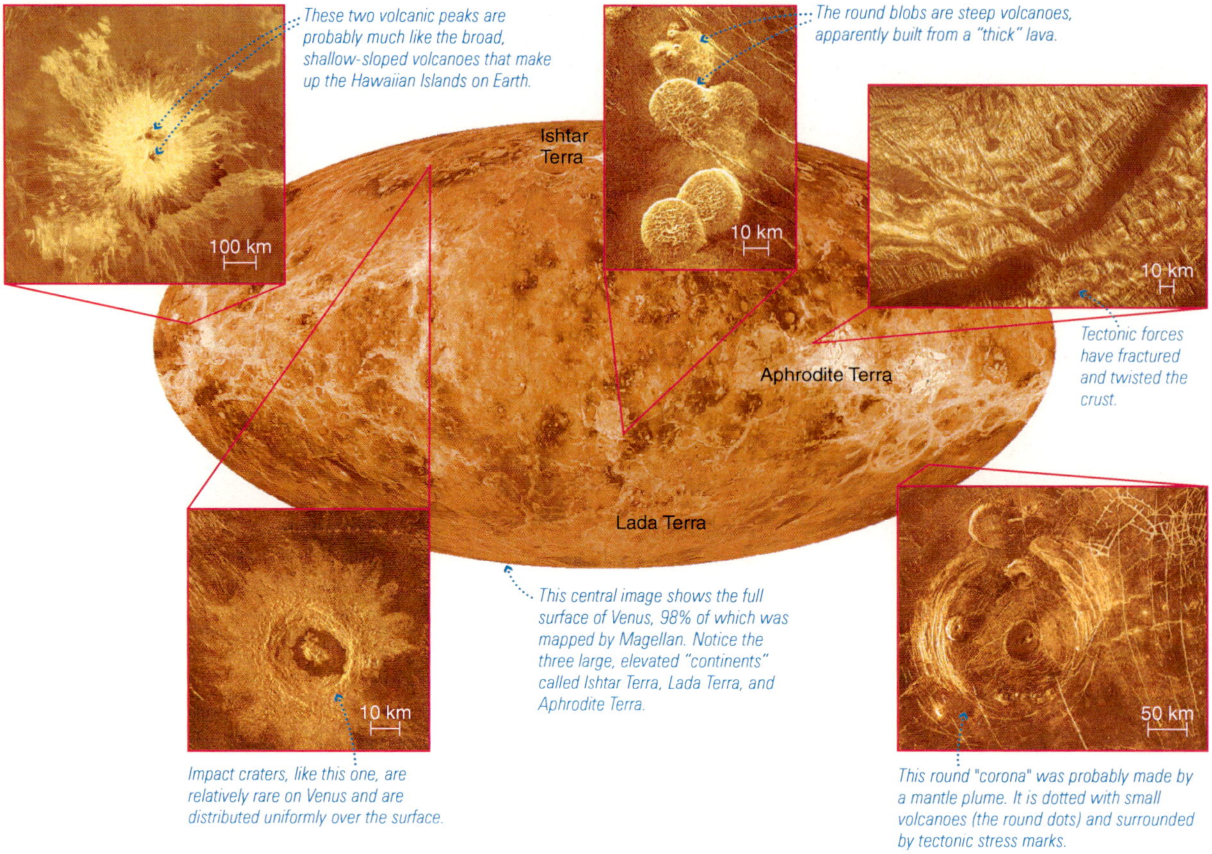

Figure 10.2

The surface of Venus, as revealed by radar observations from the *Magellan* spacecraft. Bright regions in the radar images represent rough areas or higher altitudes.

have been very different at the orbits of Venus and Earth, so both planets should have accreted from planetesimals of similar composition. As we discussed in Chapter 3, these planetesimals probably contained little or no water, because temperatures in the inner solar system were too high for water vapor to condense into solid particles of ice. The water on Earth must have come from planetesimals that originated in more distant parts of the solar system and had their orbits perturbed in such a way that they ended up crashing into our planet. Simple physical arguments suggest that such collisions should have been equally common on Venus, so Venus should have obtained water in this same way. Thus, the more likely scenario is that Venus started out with nearly as much water locked into its crust and mantle as Earth.

As on Earth, trapped water on Venus would have been released into the atmosphere through outgassing. Radar mapping of Venus's surface shows plenty of evidence of ongoing geological activity such as tectonics and volcanism (Figure 10.2). Venus has relatively few impact craters, which tells us that the craters due to large impacts must have been erased by other geological processes. Moreover, the few impact craters are distributed fairly uniformly over the planet's surface, suggesting that the surface is about the same age everywhere. Precise crater counts indicate that the surface is about 750 million years old, implying that the entire surface was somehow "repaved" at that time. We also see numerous tectonic features on Venus, as well as volcanoes. We have not witnessed any volcanic eruptions on Venus, but the composition of the clouds suggests that the volcanoes must still be active. The clouds contain sulfuric acid, which is made from sulfur dioxide (SO_2) and water. Sulfur dioxide enters the atmosphere through volcanic outgassing, but once in the atmosphere it is steadily removed by chemical reactions with surface rocks; it would all be removed within the geologically short time of 100 million years. The existence of sulfuric acid clouds therefore implies that outgassing must continue to supply sulfur dioxide to the atmosphere. All this evidence of past and present outgassing means that lots of water vapor should have entered Venus's atmosphere in the past.

The leading hypothesis for the disappearance of Venus's water invokes some of the same processes thought to have removed water from Mars [Section 8.3]: Ultraviolet light from the Sun broke apart water molecules in Venus's atmosphere. The hydrogen atoms then escaped to space (through *thermal escape*), ensuring that the water molecules could never reform. The oxygen from the water molecules was lost to some combination of chemical reactions with surface rocks and stripping by the solar wind. Venus's upper atmosphere is vulnerable to the solar wind because Venus lacks a protective magnetic field, probably due to its slow rotation. Recall that at least moderately rapid rotation is one of the requirements for a global magnetic field [Section 4.4]; Venus, which takes 243 days to complete a single rotation on its axis, does not meet this requirement. Without a magnetic field, the solar wind can strip atmospheric gas away, and probably has stripped away at least as much or more gas from Venus as it stripped away from Mars. However, because Venus has such a thick atmosphere, this gas loss would barely be noticeable.

Acting over billions of years, the breakdown of water molecules and the escape of hydrogen can easily explain the loss of an ocean's worth of water from Venus. Indeed, because Venus would have lost any surface water it once had, outgassed water could not be recycled back into the mantle the way it is recycled on Earth. Venus should therefore have lost water from its interior continuously throughout its history, explaining why its crust and mantle must by now be extremely dry—a situation quite different from that on Earth, where significant amounts of water are chemically bound to crust and mantle rock. As we discussed in Chapter 4, this dryness may have strengthened and thickened Venus's lithosphere, explaining why Venus lacks Earth-like plate tectonics.

Proving that Venus really did lose so much water is difficult, but evidence comes from the gases that didn't escape. Recall that most hydrogen nuclei contain just a single proton, but a tiny fraction of all hydrogen atoms (about 1 in 50,000 on Earth) contain a neutron in addition to the proton, making the isotope of hydrogen that we call *deuterium* (see Figure 3.24). Water molecules that contain an atom of deuterium (called "heavy water") behave chemically just like ordinary water and can be broken apart by ultraviolet light just as easily. However, a deuterium atom is twice as heavy as an ordinary hydrogen atom and thus does not escape to space as easily when the water molecule is broken apart. If Venus lost a huge amount of hydrogen from water molecules to space, the rare deuterium atoms would have been more likely to remain behind than the ordinary hydrogen atoms. Measurements show that this is the case. The fraction of deuterium among hydrogen atoms is a hundred times higher on Venus than on Earth, suggesting that a substantial amount of water must have been broken apart and its hydrogen lost to space.

The deuterium ratio does not allow us to determine exactly how much water Venus has lost, because Venus must be subjected to occasional comet impacts. The comets bring water that enters the atmosphere, where it ultimately gets broken down and its hydrogen escapes just as with Venus's original water. Because comet water contains deuterium in a ratio slightly higher, but comparable to that found on Earth (measurements indicate that comets have about twice the deuterium-to-hydrogen ratio as does Earth), the addition of comet water tends to lower the atmospheric ratio of deuterium to hydrogen on Venus. As a result, the water loss that we can infer from the current ratio of deuterium to hydrogen in the venusian atmosphere is only a *lower limit* on the actual water loss. The measured deuterium-to-hydrogen ratio implies that Venus has lost at least the equivalent of a global layer of water several meters deep. However, because this value is a lower limit, it is likely that Venus had much more water —possibly enough to have given it oceans if the water had not been lost to space.

The Runaway Greenhouse Effect We have explained how Venus might have lost its water, but why didn't Earth lose its water in the same way? The answer is the ocean itself. On Earth, most of the outgassed water vapor condensed into rain, forming the oceans, long before ultraviolet radiation could break apart much of the water vapor. The short-wavelength ultraviolet light that tends to break apart water molecules does not penetrate far into the atmosphere, let alone penetrate the ocean surface, so water in the oceans is protected. (Today, ultraviolet light is also absorbed by the ozone layer, but the ozone layer didn't exist when our planet was young.) To understand why Venus was unable to protect its water in a similar way, let's consider what would happen if we moved Earth to Venus's distance from the Sun.

Figure 10.3 summarizes what would probably occur. The greater intensity of sunlight would almost immediately raise the global average temperature by about 30°C, from its current 15°C (59°F) to about 45°C (113°F). Although this is still well below the boiling point of water, the higher temperature would lead to increased evaporation of water from the oceans. It would also allow the atmosphere to hold more water vapor before the vapor condensed to make rain (think of how much more humid hot days are than cold days). The combination of more evaporation and greater atmospheric capacity for water vapor would substantially increase the total amount of water vapor in Earth's atmosphere. Now, remember that water vapor, like carbon dioxide, is a greenhouse gas. The added water vapor would therefore strengthen the greenhouse effect, driving temperatures a little higher. The higher temperatures, in turn, would

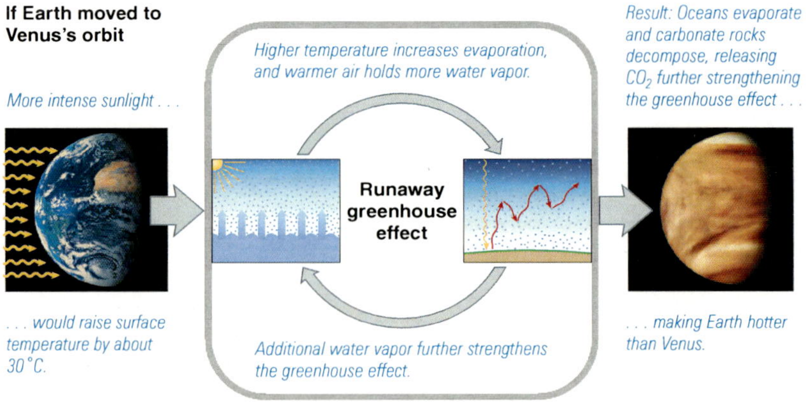

Figure 10.3

This diagram shows what scientists suspect would happen if Earth were placed at Venus's distance from the Sun: A runaway greenhouse effect would cause the oceans on Earth to evaporate completely.

lead to even more ocean evaporation and more water vapor in the atmosphere—strengthening the greenhouse effect even further. In other words, we'd have a "positive feedback loop" in which each little bit of additional water vapor in the atmosphere would mean higher temperatures and even more water vapor. The process would careen rapidly out of control, resulting in what we call a **runaway greenhouse effect.**

The runaway process would cause our planet to heat up until the oceans were completely evaporated and the carbonate rocks had released all their carbon dioxide back into the atmosphere. By the time the process was complete, temperatures on our "moved Earth" would be even higher than they are on Venus today, thanks to the combined greenhouse effects of the released carbon dioxide and water vapor in the atmosphere. The water vapor would then gradually disappear, as ultraviolet light broke water molecules apart and the hydrogen escaped to space. In short, moving Earth to Venus's orbit would essentially turn our planet into another Venus.

If it is correct, this scenario suggests a simple explanation for the difference between Earth and Venus. Venus is about 30% closer to the Sun than Earth, and this difference was enough to be critical. On Earth, it was cool enough for water to rain down to make oceans in which carbon dioxide could dissolve and undergo chemical reactions that locked it away in carbonate rocks. As a result, our atmosphere was left with only enough greenhouse gases to make our planet pleasantly warm. On Venus, the greater intensity of sunlight made it just enough warmer that oceans either never formed or soon evaporated. Without oceans to dissolve carbon dioxide and make carbonate rock, carbon dioxide accumulated in the atmosphere, leading to a runaway greenhouse effect, which resulted in the extreme temperatures of Venus today.

THINK ABOUT IT Moving Earth to Venus's orbit would probably cause our planet to become Venus-like. If we could somehow move Venus to Earth's orbit, would it become Earth-like? Why or why not?

Venus and the Boundary of the Habitable Zone The fact that moving our own planet to Venus's orbit would lead it to the same fate implies that *any* planet at this distance would be doomed to a runaway greenhouse effect (assuming it had an atmosphere from outgassing). In other words, if our scenario is correct, it seems impossible for a planet to have a habitable surface with abundant liquid water at Venus's distance from the Sun. We can therefore conclude that Venus does *not* lie within the Sun's habitable zone today.

In contrast, Earth clearly *does* lie within the habitable zone, which is why our planet has not suffered a runaway greenhouse effect. Indeed, models show that a runaway greenhouse effect could not occur on the present-day Earth, at least through natural processes (as opposed to human-induced processes), an idea supported by our planet's climate history. Whenever the atmospheric concentration of greenhouse gases has increased on Earth in the past—as it has numerous times, perhaps most extremely during "hothouse" phases that follow Snowball Earth episodes [Section 4.5]—our greater distance from the Sun has helped prevent the temperature from increasing to the point at which a runaway process would occur. Putting the lessons of Earth and Venus together, we conclude that the inner boundary of the Sun's habitable zone must currently lie beyond the orbit of Venus but within the orbit of Earth.

• Could Venus have once been habitable, and could life still exist there?

Venus's closeness to the Sun may have sealed its ultimate fate, but it's possible that Venus might have had a more moderate climate in its early history. Remember that the Sun has gradually brightened with age. Thus, sunlight was less intense on all the planets when they were young. Some 4 billion years ago, the intensity of sunlight at Venus was not much greater than it is at Earth today. Rain might well have fallen on the young planet Venus, and oceans could have formed. It's even conceivable that life could have arisen on the young Venus, or been transported there on meteorites from Earth.

As the Sun gradually brightened, any liquid water or life on Venus's surface was doomed. The runaway greenhouse effect raised the temperature so high that all the water evaporated and, as we've discussed, the water is now gone forever. Moreover, any evidence of past oceans is probably also gone. Recall that Venus's entire surface appears to have been "repaved" by tectonics and volcanism; this repaving would have covered up any shorelines or other geological evidence of past oceans. Rocks from Venus probably can't tell us anything about an oceanic past either: The high temperatures would have baked out any gases that might once have been incorporated into surface rock, thereby likely eliminating the evidence that might have told us whether the rocks formed in water or ever held life.

Nevertheless, the fact that the Sun was once dimmer makes it reasonable for us to think that Venus could have had a more Earth-like early history. If it did, and if life arose, could any life still survive? The high temperatures make it seem implausible to imagine life anywhere either on Venus's surface or underground. However, as we discussed in Chapter 7, the temperature drops significantly with altitude in the atmosphere, and Venus has high-altitude clouds containing tiny amounts of liquid water (in highly acidic form). Though the prospects of finding life in these clouds are admittedly slim, it is at least within the realm of possibility to imagine that life once arose in venusian oceans and, as the runaway greenhouse effect set in, successfully adapted to life in the sky.

10.3 SURFACE HABITABILITY FACTORS AND THE HABITABLE ZONE

The case of Venus tells us that distance from the Sun is a critical factor in surface habitability, since Venus seems to have been doomed to suffer a runaway greenhouse effect no matter how habitable it may have been in its youth. But we know that distance from the Sun is not the only habitability factor, because we have seen that Mars's small size probably explains how it lost liquid water that existed on its surface in the past. In this section, we'll summarize what we have learned about surface habitability factors, and then use these lessons to consider the present-day boundaries of the Sun's habitable zone.

• What factors influence surface habitability?

Our comparative studies of Venus, Earth, and Mars have given us deep insight into the factors that make these three worlds so different. Let's build on the discussions in both this chapter and prior chapters, so that we will gain additional insight into the prospects of finding worlds with habitable surfaces around other stars.

The Role of Distance from a Central Star As we've just discussed, the first factor that affects surface habitability is a planet's distance from its home star, in our case the Sun. The effects of subtle differences in distance can result in dramatic variations in habitability, as the case of Venus shows. As we've seen, a planet at Venus's distance with an Earth-like atmosphere would be only a little warmer than Earth; the reason that Venus is not habitable is that this "a little warmer" turns out to be warm enough to start a runaway greenhouse process that heats the planet far more. Thus, when we consider the minimum distance at which surface habitability is possible, we must account not simply for solar heating but also for resultant processes that can lead to the evaporation of surface water. Similarly, when we consider the maximum distance at which surface habitability is possible, we must consider processes that might weaken greenhouse warming, causing surface water to freeze.

There's another obvious factor that will influence the range of distances that form the habitable zone: the central star's luminosity (brightness). As we will discuss in Chapter 11, stars come with a wide range of luminosities: A few stars are much more luminous than the Sun, about 10% of all stars have luminosities similar to that of the Sun, and the vast majority of stars are considerably dimmer than the Sun. Clearly, a brighter star must have a wider and more distant habitable zone than the Sun, while a dimmer star must have a narrower and closer-in habitable zone. The wider habitable zones of brighter stars would seem to increase the odds of finding planets within these zones, but another problem may make biology rare or nonexistent in these cases: The brightest stars turn out to have extremely short lifetimes—millions of years rather than billions of years—and hence may not offer enough time to nurture biology [Section 11.1]. The dimmer stars, with their narrow habitable zones, are less likely to have planets in these zones. However, they may make up for this problem in their sheer numbers, since they are so much more common than stars like the Sun (see Cosmic Calculations 10.1).

The Role of Planetary Size Clearly, distance from the Sun is not the only factor affecting surface habitability. If it were, then the Moon would be habitable because it is the same distance from the Sun as is Earth. More to the point, if distance were all that mattered, then Mars should have become more habitable as the Sun grew brighter with time; instead, it froze over. Mars's current lack of surface habitability appears to be less the result of its distance from the Sun than of its small size. As we discussed in Chapter 8, its small size allowed its interior to cool more quickly than that of a larger planet like Earth. Its core presumably cooled to the point at which it no longer had a molten layer and therefore could no longer generate a global magnetic field. The martian atmosphere was then left vulnerable to stripping by the solar wind, and through this and other mechanisms, it lost too much gas to support a strong greenhouse effect. Without a strong greenhouse effect, Mars effectively froze. The cooler interior also means that if martian volcanoes remain active at all, they do not erupt frequently enough to resupply lost atmospheric gases.

Is there a minimum or maximum size that makes it possible for a planetary surface to remain habitable over long time periods? While it's uncertain how we should set limits on maximum size, a minimum size clearly exists. We have traced Earth's long-term habitability directly to the climate regulation provided by the carbon dioxide cycle, which in turn depends on the cycle of plate tectonics [Section 4.5]. Thus, we have at least some reason to think that plate tectonics is necessary for long-term habitability on any world, in which case the size requirement for surface habitability is a size that allows plate tectonics to exist.

We do not fully understand how planetary size is related to plate tectonics, but we are fairly confident that planets that are too small will not have tectonics, and that Mars falls below the minimum. We see no evidence of ongoing plate tectonics on Mars, and this is probably because the interior cooling attributable to its small size has allowed its lithosphere to thicken so much that it cannot break into plates. But while Mars—with a radius about half that of Earth—seems too small, we don't know how much bigger a planet would need to be so that it would have both sufficient internal warmth to power tectonics and a crust thin enough to be pushed around by the convective currents below. The case of Venus illustrates the problem. Recall that, given the closeness in size of Venus and Earth, we might expect to find plate tectonics operating on both planets, but there is no evidence of Earth-like plate tectonics on Venus. Could it be that Venus's slightly smaller size was just small enough to make the difference, much as its slightly nearer distance to the Sun led to its runaway greenhouse effect? It's possible, but a more likely hypothesis ties Venus's lack of plate tectonics to its runaway greenhouse effect. According to this hypothesis, high temperatures baked water out of the rock in Venus's crust and upper mantle, making the crust too stiff to allow subduction and thereby suppressing plate tectonics. If this explanation is correct, then Venus may well have had plate tectonics if it had not been so close to the Sun, and might have had plate tectonics in its early history when the Sun was dimmer.

At this point, all we can say with confidence is that Earth is of sufficient size for long-term surface habitability and Mars is not. Beyond that, we don't know how large a planet must be to allow for the presence of liquid water over extended periods of time. Factors of which we are still unaware may well contribute to the occurrence or absence of plate tectonics and the ability to sustain an appropriate climate.

The Role of an Atmosphere A third crucial factor in surface habitability, after size and distance, is the necessity of an atmosphere. Without sufficient atmospheric pressure, liquid water cannot be stable and abundant regardless of other factors. Moreover, an atmosphere is necessary to protect a planetary surface against harmful solar radiation. So while a world without an atmosphere might still shelter life underground, it seems implausible that any biology would creep, crawl, or fly across its landscape.

The presence or absence of a significant atmosphere sometimes depends on size. In the case of Mars, for example, we have traced its loss of surface habitability to loss of atmospheric gas, which most likely was tied to its small size and loss of magnetic field. However, there may be several other ways in which a planet either might never get or would subsequently lose an atmosphere, even if it were of the right mass and at the right distance from its star.

Remember that the early atmospheres of Venus, Earth, and Mars are all thought to have been produced through outgassing. If a planet lacked trapped gases in its interior, then no outgassing would be possible. Given that most of the gases within the terrestrial worlds are assumed to have been brought by impacts of objects from more distant reaches of the solar system, and that these objects were presumably cast onto collision courses by gravitational interactions, terrestrial planets might form without gases in a star system either that lacks gas-bearing planetesimals or in which these planetesimals for some reason never get perturbed inward. Neither situation seems likely in light of what we know about solar system formation [Section 3.3], but we can't rule them out.

For a planet that has outgassing and sufficient size to hold its gas gravitationally, we know of at least two possible ways in which it might nonetheless lose its atmosphere. First, large impacts can in principle blast significant amounts of atmospheric gas into space. This process may have played a role on Mars. If large impacts are more common in some other star systems—which may be the case in systems that lack a planet comparable in size to Jupiter at the right distance [Section 11.3]—these impacts could cause significant gas loss on Earth-size planets. Second, as we have discussed, the solar wind can strip atmospheric gas from any planet that lacks a global magnetic field, and we don't expect to find magnetic fields on slowly rotating planets. Venus has so much atmospheric gas that loss to solar wind stripping has been insignificant in comparison. Earth, however, might well have become uninhabitable (at least on its surface) if not for the protection that the magnetic field has offered against billions of years of stripping by the solar wind.

Summary of Habitability Factors We have identified three major factors for surface habitability:

1. The planet must be neither too close nor too far from its star; that is, it must be within its star's habitable zone.
2. The planet must be large enough to retain internal heat and have plate tectonics for climate regulation. We don't know the precise minimum size required, but it is certainly larger than Mars.
3. The planet must have enough of an atmosphere for liquid water to be present on its surface. This means it must have had gases trapped in its interior, so that an atmosphere could form through outgassing, and that it has not since lost too much of this atmospheric gas to impacts or solar wind stripping. Protection against the latter may require a global magnetic field, which in turn may require at least moderately rapid rotation.

The latter two requirements depend on factors intrinsic to the planet itself. The first depends on where it forms around its star, which brings us to the topic of habitable zone boundaries.

THINK ABOUT IT Using the three habitability factors, explain why the Moon is not habitable.

• Where are the boundaries of the Sun's habitable zone today?

The boundaries of the Sun's present-day habitable zone depend on the range of distances at which a planet of suitable size and with sufficient atmospheric pressure could have liquid water on its surface. The inner boundary marks the place where, if a planet were any closer to the Sun, a runaway greenhouse effect would be triggered. The outer boundary marks the place where, if a planet were any farther from the Sun, surface water would freeze.

The case of Venus has shown us that the inner boundary of the habitable zone does not lie as far inward as Venus's orbit. The case of Mars should tell us something about the outer boundary, though it is more ambiguous: Geological evidence makes it clear that Mars once had liquid water on its surface, in which case it must once have been within the Sun's habitable zone—and since the habitable zone moves outward with time, it would then still have to be in the habitable zone today. However, as we discussed in Chapter 8, models of the martian climate cannot yet fully account for the conditions that might have led Mars to have been quite so warm in its distant past, a fact that has led some scientists to suggest that Mars may have had liquid water only intermittently. In that case, Mars might never have been within the Sun's habitable zone, but only was close enough that special events (such as impacts) made its surface temporarily habitable.

The ambiguity in the case of Mars means we won't be able to put precise numbers on the boundaries of the habitable zone. Nevertheless, we can place some general constraints on the possible locations of the boundaries.

The Inner Boundary To locate the inner boundary more precisely than just saying it lies between Venus and Earth, we can consider theoretical models of what would happen to Earth if we moved it to various distances closer to the Sun. These calculations suggest that a runaway greenhouse effect would occur if we placed Earth anywhere inside of 0.84 AU from the Sun, or about halfway between the orbits of Venus and Earth. (Recall that 1 AU, or astronomical unit, is Earth's average distance from the Sun, or about 150 million kilometers.)

However, another factor might cause temperatures to spin out of control even beyond 0.84 AU from the Sun. According to some models, a moderate additional warming would allow water vapor to circulate to much higher altitudes in Earth's atmosphere (into the stratosphere). At these high altitudes, water molecules would be above much of the ozone layer and hence could be broken apart by ultraviolet light from the Sun. The hydrogen would then escape to space, causing Earth to lose this water and thereby allowing more water to rise into the upper atmosphere and be lost in turn. Over time, this **moist greenhouse effect**—so-called because the upper atmosphere would become moist with water, at least until all the water was lost—might cause Earth to lose its oceans. Note that the moist greenhouse effect leads to water loss not because the temperature is outside the range in which liquid water could exist, but rather because water that evaporates from the surface can rise into the upper atmosphere, where it can be lost to space. This is a less dramatic version of the scenario we discussed when we imagined placing Earth at Venus's orbital distance, but it could be just as lethal in the long run. An Earth-like planet might suffer this moist greenhouse water loss anywhere within 0.95 AU of the Sun, but there may be other factors that would mitigate the problem: Models that describe the onset of the moist greenhouse effect have numerous recognized uncertainties (such as the effects of clouds) that might push this distance inward.

In summary, the inner boundary of the present-day habitable zone in our solar system may be at 0.84 AU if we allow for only a simple runaway greenhouse effect, but as far out as 0.95 AU if we allow for water loss by a moist greenhouse effect.

The Outer Boundary The outer boundary of the present-day habitable zone is the distance from the Sun at which even a strong greenhouse effect could not warm a planet enough to keep liquid water from freezing. At first, we might guess that Mars is beyond the outer boundary, since the temperature is too cold for liquid water at its surface today. However, if Mars were larger and had retained a thick atmosphere, it might have enough greenhouse warming to still have a habitable surface. Thus, if it is possible in principle for a planet to keep a thick atmosphere at such distances, the outer boundary of the habitable zone could lie beyond the orbit of Mars.

Calculations suggest that this is indeed the case. If we allow for a thick atmosphere with a strong greenhouse effect, the outer boundary of the present-day habitable zone lies at about 1.7 AU, well outside Mars's average orbital distance of 1.52 AU. However, there is at least one potential problem that could bring the outer boundary in closer. If the atmosphere of a planet is too cold, the atmospheric carbon dioxide that produces greenhouse warming will condense into snowflakes and fall to the surface. This carbon dioxide snow might limit how much carbon dioxide could reside in the atmosphere, thus preventing the atmosphere from staying thick enough for a strong greenhouse effect. This scenario might occur if the atmosphere lacked dust or any other greenhouse gas to help keep the middle atmosphere warm. In that case, the outer boundary of the present habitable zone might lie at only about 1.4 AU, or just inside the orbit of Mars. Note that this case would also imply that Mars never was in the habitable zone, so it could have had surface liquid water only intermittently in the past.

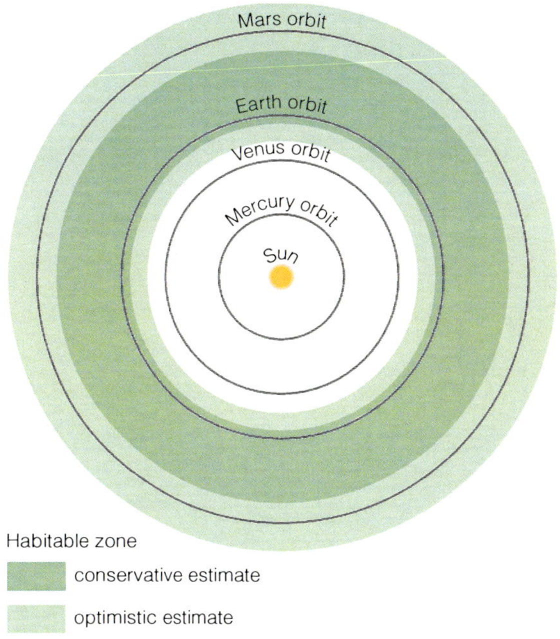

Figure 10.4

Boundaries of the Sun's habitable zone today. The narrower set of boundaries represents a model based on the more conservative assumptions, while the wider set represents the most optimistic scenarios.

The Extent of the Habitable Zone We have found two estimates each for the distances of the inner and outer boundaries of the Sun's present habitable zone. Using the more optimistic estimates, the present-day habitable zone extends from about 0.84 to 1.7 AU. Using the more conservative estimates, it extends only from about 0.95 to 1.4 AU. Both sets of boundaries are shown in Figure 10.4. Note that, even in the more conservative case, the Sun's present-day habitable zone represents a fairly wide region in the inner solar system.

Keep in mind that we don't yet know enough about how planetary atmospheres work to know which estimate is correct—or if the truth lies somewhere in between. In addition, there might be processes that can affect the atmospheric temperature that we haven't yet discovered, and therefore have not accounted for in our calculations of the habitable zone boundaries. Thus, these estimates of the boundaries of the habitable zone should be considered just that—estimates. They might be significantly refined as we learn more in the future.

THINK ABOUT IT About how much wider is the more optimistic view of the present-day habitable zone than the more conservative view? If planets form at random locations in the inner portions of other stars' solar systems, how would these two different views affect the likely number of planets within habitable zones around Sun-like stars in the Milky Way Galaxy?

10.4 THE FUTURE OF LIFE ON EARTH

Because the boundaries of the habitable zone are calculated under the assumption that we have a planet of suitable size, they depend only on the amount of heat and light put out by the Sun. Thus, when the Sun was dimmer in the past, the habitable zone must have been closer to our Sun.

COSMIC CALCULATIONS 10.1

Chances of Being in the Zone

If all other factors are equal, the likelihood of finding planets in a star's habitable zone depends on the width of the zone.

Example: The Sun's habitable zone is (optimistically) calculated to extend from about 0.84 to 1.7 AU. Consider a smaller star (of spectral type M [Section 11.1]), in which the habitable zone extends only from 0.05 to 0.1 AU. How does the probability of finding a habitable planet in this star's habitable zone compare to the probability around a Sun-like star? Given that these types of small stars outnumber Sun-like stars by approximately eight to one in our galaxy, for which class of stars should we expect more worlds in habitable zones?

Solution: The width (radius) of a habitable zone, R_{HZ}, is:

$$R_{HZ} = R_{outer} - R_{inner}$$

For a Sun-like star:

$$R_{HZ} = 1.7 - 0.84 = 0.86 \text{ AU}$$

For the smaller star:

$$R_{HZ} = 0.1 - 0.05 = 0.05 \text{ AU}$$

The probability of being in the habitable zone for the small star is only $0.05/0.86 = 0.058$ times that for a Sun-like star. However, because these stars are eight times as common, the *total number* of worlds in habitable zones around these small stars would be about $8 \times 0.058 = 0.46$, or 46%, as great as for worlds around Sun-like stars. (Of course, we have not considered factors besides the size of the habitable zone that may also be important.)

In the future, when the Sun will be brighter than it is today, the habitable zone will lie farther from the Sun. In this section, we'll discuss the way in which the Sun's habitable zone evolves with time.

• How will the Sun's habitable zone change in the future?

To determine how the habitable zone moves over time, we need to know how the Sun brightens as it ages. Fortunately, the process that causes the Sun to brighten is well understood.

The Sun shines by fusing hydrogen into helium. Each fusion reaction converts four hydrogen nuclei into one helium nucleus. Thus, the total number of *independent* particles in the solar core gradually falls with time. This gradual reduction in the number of particles causes the solar core to shrink, because there are fewer particles to generate the pressure that supports the core against the weight of overlying layers of the Sun. The slow shrinkage, in turn, gradually increases the core temperature, much as a bicycle tire pump gets warm when you compress the air in it by pushing on its piston. The gradual temperature increase causes a corresponding increase in the fusion rate, which is why the Sun slowly brightens.

Theoretical models indicate that the Sun's core temperature should have increased enough to raise its fusion rate and the solar luminosity by about 30% since the Sun was born $4\frac{1}{2}$ billion years ago. The models also allow scientists to predict the Sun's future luminosity. Observational data support the models' conclusions, as stars of particular masses do indeed vary somewhat in brightness, with older stars being brighter for any particular mass.

Figure 10.5 shows the results of calculations for the boundaries of the habitable zone from the Sun's birth until it exhausts its core hydrogen fuel some 5 billion years from now. Notice how the habitable zone gradually moves outward from the Sun, just as we would expect.

The horizontal swaths in Figure 10.5 show the distances from the Sun at which conditions have remained habitable from 4 billion years ago to the present. This zone is often called the **continuously habitable zone,** because it has been habitable at all times since the end of the heavy bombardment about 4 billion years ago. The width of the continuously habitable zone is fairly narrow if we use the more conservative assumptions, and substantially wider with the more optimistic assumptions. In fact, under the optimistic assumptions, both Earth and Mars are in the continuously habitable zone. Note also that the continuously habitable zone is defined for habitability only up to the present. If we look billions of years into the future, the habitable zone continues to move outward and the continuously habitable zone becomes narrower.

THINK ABOUT IT Was Venus ever in the habitable zone? Is it in the continuously habitable zone? Under the more conservative assumptions, about when will the continuously habitable zone move beyond Earth's orbit? Explain.

• How long can life survive on Earth?

Earth has remained habitable for some 4 billion years, allowing plenty of time for life to evolve, diversify, and ultimately make our own human existence possible. However, the continuing evolution of the habitable zone means that Earth's days of habitability must eventually come to an end.

Before we begin, it's important to note that the demise of Earth's habitability is not something worth losing sleep over. Even under the most pessimistic scenarios, our planet will remain habitable for many hundreds of millions of years to come. Under more optimistic scenarios, Earth has a couple of billion years of remaining habitability. Compared to the length of time our civilization has existed so far, these are incredibly long time scales. If our species survives for this long, we will have had plenty of time to find ways to move to other planets in our solar system or in other star systems, or to otherwise prevent our perishing with Earth.

THINK ABOUT IT Given the more immediate threats to our civilization, is it even worth thinking about what will happen to our planet millions or billions of years from now? Defend your opinion.

The End of Earth's Habitability

Thanks to the climate regulation provided by the carbon dioxide cycle, Earth has remained habitable even as our Sun has brightened by some 30% over the past 4 billion years. As the Sun continues to brighten, the carbon dioxide cycle should continue to keep the climate relatively pleasant for us for at least hundreds of millions of years to come. Eventually, however, the warming Sun will cause the cycle to break down.

If you study Figure 10.5, you'll see that under the more conservative assumptions, the inner boundary of the habitable zone will move beyond Earth's orbit in about a billion years. Thus, if these assumptions are correct, the end of habitability on Earth will come about a billion years from now. Recall that these conservative assumptions are based on the idea that a *moist greenhouse* effect will cause the oceans to evaporate away. In other words, about a billion years from now, water vapor will begin to circulate into Earth's upper atmosphere, where ultraviolet light will break apart water molecules and allow the hydrogen to escape into space, as it did with Venus. As water is lost from the upper atmosphere, more water will evaporate to take its place, until the oceans are completely gone. At that point, it seems unlikely that any life could continue to survive on Earth. Of course, no one is yet sure whether the moist greenhouse problem will really arise in about a billion years. Many feedback mechanisms in Earth's climate are not yet well understood. For example, increased cloud cover may reduce the amount of sunlight reaching Earth's surface, preventing the onset of the moist greenhouse.

Under more optimistic assumptions, Earth may remain habitable until some 3–4 billion years from now. At that time, the Sun will have grown so bright that Earth will finally suffer the fate of Venus—a runaway greenhouse effect. The rising temperature on Earth will cause increased ocean evaporation, and the buildup of water vapor in the atmosphere will increase the greenhouse effect further (see Figure 10.3). The positive feedback won't stop until the oceans have evaporated away and all the carbon dioxide has been released from carbonate rocks. Our planet will become a Venus-like hothouse, with temperatures far too high for liquid water to exist.

We know of no natural phenomena that can prevent this runaway greenhouse effect from occurring. However, if we imagine that our descendants have become sufficiently advanced in their technology, there

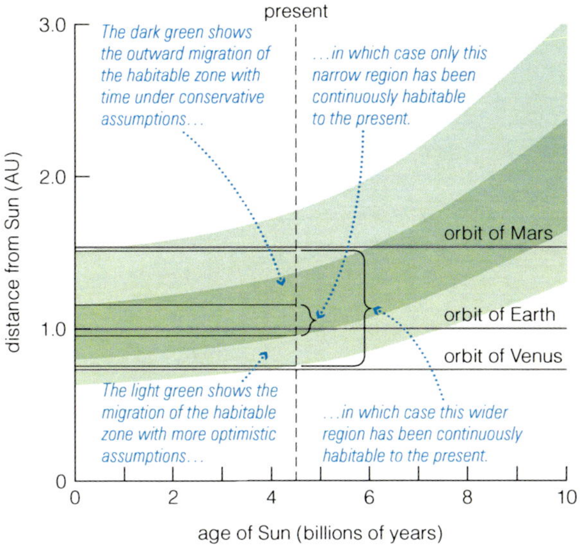

Figure 10.5

This graph shows the Sun's habitable zone through time. The narrower region represents the habitable zone based on the more conservative assumptions, and the wider region represents the habitable zone based on the more optimistic assumptions. The horizontal swaths represent the zone that has been continuously habitable from 4 billion years ago to the present, again under the conservative (narrower) and optimistic (wider) assumptions.

are several ways they could survive in our solar system. They might protect Earth itself by building a giant sunshade in space to reduce the intensity of the light reaching Earth from the brightening Sun. Or they might simply move: When the habitable zone first moves past the orbit of Earth, Mars will be well within it, so perhaps our descendants will be able to make Mars livable with some advanced terraforming [Section 8.4]. If they are truly powerful, they might even find a way to move our planet gradually outward from the Sun to keep it within the habitable zone; by slowly moving Earth to the orbit of Mars over the next 3–4 billion years, humans could stay home and our planet would stay habitable. Another possibility is to relocate the population to large, artificial habitats constructed in space. Still, at best all these solutions can be only temporary, because the Sun itself eventually will die.

Death of the Sun The Sun's gradual death has already begun, and in about 5 billion years it will completely run out of hydrogen in its central core. Then, over the following billion years or so, the Sun will undergo a dramatic transformation. During the first few hundred million years of this period, the Sun will gradually swell to about 100 times (or more) its present radius, and its surface temperature will drop (changing its color from yellow to red), becoming what we call a *red giant* star. Even though the Sun's surface temperature will decrease, its much greater size will pump greater energy into space than now. At its peak, the red giant will be about 1,000 times as luminous as the Sun is today, and Earth's surface temperature will rise to 700°C (1,292°F) or higher. Any remaining oceans will evaporate, and even underground life will be baked to death during this period as subsurface water boils away.

In its final death throes, the Sun will expel its outer layers into space, creating a *planetary nebula** (Figure 10.6). All that will remain of the Sun will be its hot central core; no more nuclear fusion will occur. This remaining core, known as a *white dwarf* star, will then gradually cool over time. The violence that accompanies the planetary nebula ejection will probably destroy Earth. Even if our planet somehow survives this event and continues to orbit the white dwarf Sun, the white dwarf's light will eventually become so feeble that Earth's charred surface will face a future of perpetual, frigid darkness.

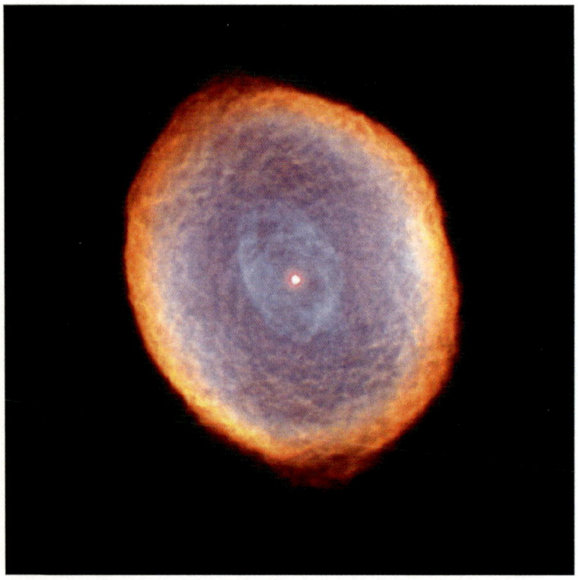

Figure 10.6

This Hubble Space Telescope photograph shows the Spirograph Nebula, an example of a planetary nebula. The gas of the nebula was expelled from a Sun-like star that had reached the end of its life. The central white dot is the white dwarf star that remains, which is essentially just the hot core of the now-dead star. Some 5 billion years from now, our Sun will suffer the same fate as the star that created this nebula.

Could We Still Survive? For those undaunted by the thought of humans or other intelligent Earth beings surviving some 5 billion years into the future, it's natural to wonder whether we could also survive the death of the Sun. The obvious solution to the Sun's death is to move to another star system. Stars that are being born today might offer great homes to us in 5 billion years. When these stars die, we can move on to others born still later. As long as there are new stars with habitable planets, we could potentially survive by migrating to new homes.

However, even this type of long-term migration has its limits. The recycling of stellar material cannot continue forever, because dying stars return to space only part of the gas from which they were made. Over time, the galaxy will contain less and less interstellar gas. About 50 billion years from now, the Milky Way Galaxy will contain so little gas that new stars will no longer be born. What then?

*Despite the name, these structures have nothing to do with planets; the term comes from the planetlike appearance of some of these nebulae when seen through a small telescope.

Ever since Edwin Hubble first discovered that our universe is expanding, we have wondered whether the expansion will continue forever or someday stop, causing the universe to collapse back in on itself. In the past few years, astronomers have been surprised to learn that the expansion appears to be accelerating, in which case it seems that the fate of the universe is to expand forever at a greater and greater clip. However, you should keep in mind that forever is an extremely long time. It remains possible that we will someday discover other surprises that will change our view of the fate of the universe.

If the universe continues to expand after all star formation has ceased, then life will be able to continue only around those long-lived stars that still have habitable planets. But even the longest-lived stars will eventually run out of hydrogen to fuse. Within a few hundred billion years, all existing stars will burn out and die, and the now-brilliant galaxies will fade into darkness. On much longer time scales, interactions among the burned-out stars will send many of them flying into the vastness of intergalactic space, while the rest will converge toward their galactic centers, merging into gigantic black holes. At this point, the story becomes even more speculative. If the so-called "grand unified theories" of physics are correct, the stellar corpses will eventually disintegrate into swarms of subatomic particles. Meanwhile, the giant black holes will slowly evaporate into energy and particles in a process first described by the noted British physicist Stephen Hawking. At some point in the far distant future, the universe will consist of nothing but a dilute sea of subatomic particles and photons of light, each separated from others by immense distances that will grow larger as the universe endlessly expands. Our current epoch of a universe filled with stars and galaxies will have been just a fleeting moment in an eternity of darkness.

This end in darkness may seem a bit depressing, but it is, after all, inconceivably far in the future. Nevertheless, it is fair to ask whether it is truly the end or instead could be followed by something else. Some serious scientists already argue that there might be ways by which an intelli-

SPECIAL TOPIC 10.1: FIVE BILLION YEARS

The Sun's demise in about 5 billion years might at first seem worrisome, but 5 billion years is an extremely long time. It is longer than Earth has yet existed, and human time scales pale by comparison. A single human lifetime, if we take it to be about 100 years, is only 2×10^{-8}, or two hundred-millionths, of 5 billion years. Because 2×10^{-8} of a human lifetime is about 1 minute, we can say that a human lifetime compared to the life expectancy of the Sun is roughly the same as 60 heartbeats in comparison to a human lifetime.

What about human creations? The Egyptian pyramids have often been described as "eternal." But they are slowly eroding due to wind, rain, air pollution, and the impact of tourists, and all traces of them will probably have vanished within a few hundred thousand years. While a hundred thousand years may seem like a long time, the Sun's remaining lifetime is some 50,000 times longer.

On a more somber note, we can gain perspective on billions of years by considering evolutionary time scales. During the past century, our species has acquired sufficient technology and power to destroy human life totally, if we so choose. However, even if we suffer that unfortunate fate, some species (including many insects) are likely to survive. Would another intelligent species ever emerge on Earth? There is no way to know, but we can look to the past for guidance. Many species of dinosaurs were biologically quite advanced, if not actually intelligent, when they were suddenly wiped out about 65 million years ago. Some small, rodentlike mammals survived, and here we are 65 million years later. We therefore might guess that another intelligent species could evolve some 65 million years after a human extinction. If these beings also destroyed themselves, another species could evolve 65 million years after that, and so on. But even at 65 million years per shot, Earth would have some 15 more chances for an intelligent species to evolve in 1 billion years—the length of time our planet will remain habitable under fairly conservative scenarios. Under more optimistic estimates for long-term habitability, there could be 60 or more periods—each as long as the period separating us from the dinosaurs—still to come before our planet dies. That is a lot of potential opportunities for the evolution of a species wise enough to avoid its self-destruction and to move on to other star systems by the time the Sun finally dies. Perhaps we ourselves will prove to be so wise.

gent civilization could survive even as the universe dies. For a lighthearted viewpoint, we turn to science fiction. Isaac Asimov, in his story "The Last Question," begins with a couple of people asking a supercomputer whether there is a way to reverse the decline of the universe and thereby avert a cold and dark end. The computer responds that the available data are insufficient to answer the question. Over billions of years, computers advance and humankind survives, making the question ever more important as the universe approaches its cold, dark future. By the end of the story, the computer exists solely in hyperspace, outside the time and space of our universe. The universe has reached a state of ultimate darkness, with nothing left alive. Meanwhile, the computer, which Asimov calls AC, whirs on in the timelessness of hyperspace until finally it learns how to reverse the decay of the universe:

> *For another timeless interval, AC thought how best to do this. Carefully, AC organized the program.*
>
> *The consciousness of AC encompassed all of what had once been a Universe and brooded over what was now Chaos. Step by step, it must be done.*
>
> *And AC said, "LET THERE BE LIGHT!"*
>
> *And there was light—*

10.5 THE PROCESS OF SCIENCE IN ACTION: GLOBAL WARMING

Earth may someday be doomed to a runaway greenhouse effect, but it won't be happening any time soon, even taking into account human actions. Nevertheless, considerable evidence suggests that Earth's global average temperature is currently on the rise, a trend referred to as **global warming**. You're undoubtedly aware that global warming is a hot political topic that has generated significant controversy. Most of the contro-

MOVIE MADNESS: THE TIME MACHINE

This film, whose resemblance to H. G. Wells's original story is slightly less than marginal, features a brilliant, young New York professor from the early years of the twentieth century who travels back and forth in time, hoping to retrieve a lost girlfriend. To do so, he uses a time machine that looks as if it were kit-built from a discarded lighthouse lens, a laboratory vacuum tank, and a surplus adding machine from the Spanish-American War.

Forget the questionable physics; this story is actually about human evolution. The bulk of the film takes place in the far future, 800,000 years hence. Thanks to an industrial accident involving some retirement home developers and the Moon, all civilization on Earth has been destroyed. Humans have evolved into two distinct races: the lovable (if somewhat simple) Eloi, and the disagreeable, ugly Morlocks. The Eloi live the life of carefree forest folk, building modest structures from bamboo and hemp. The Morlocks are slightly more sophisticated (they can smelt metal) but for some reason live underground. They venture to the surface only to hunt down the Eloi, drag them to their subterranean digs, and eat them for dinner.

Does it make sense that the most technically advanced beings on the planet will be underground heavies? In our consideration of habitable zones, we've looked at surface environments, where the raw materials of a planet are exposed to the abundant energy radiated by its sun. As we know, the energy for metabolism is tough to find below the surface, which explains why the Morlocks count on the Eloi to eat the products of photosynthesis (veggies, for instance), and then they simply eat *them.* But this is a dangerously fragile lifestyle for the Morlocks, as they are dependent on a single surface species.

Aside from such dietary considerations, it's discouraging to think that, after a length of time several times as far into the future as we've come since the first *Homo sapiens,* the best our descendants can do is erect bamboo huts. What about expansion to other planets and travel to the stars? If *The Time Machine* is an accurate vision of what happens to thinking species, then we needn't bother with our SETI experiments (Chapter 12). After all, it's difficult to build radio transmitters out of bamboo.

versy concerns whether it is occurring naturally or as a result of human activity and, if the latter, what (if anything) we should do about it. Thus, for this chapter's case study in the process of science in action, we'll investigate how researchers seek to understand global warming and its potential consequences.

• What is the evidence for global warming?

The basic science behind global warming is surprisingly simple: We know that carbon dioxide is a greenhouse gas that can enhance the greenhouse effect and therefore cause a rise in a planet's surface temperature, and we know that human activity (such as the burning of fossil fuels) is adding carbon dioxide and other greenhouse gases to Earth's atmosphere. It might therefore seem natural to conclude that our activity will cause Earth's climate to warm up. However, we also know from our study of Earth's climate regulation mechanisms that there are many feedback loops that can make the reality much less straightforward than this simple analysis would suggest. As a result, the obvious starting point for the scientific study of global warming is to find out if it is indeed occurring as we might expect and, if so, whether we can tie the warming to the carbon dioxide emissions of our civilization.

Evidence of Recent Warming You might think that determining whether our planet has been warming up during the past century or two would require nothing more than collecting temperature data from old newspapers. However, remember that global warming refers to an increase in the *average* temperature of our whole planet. We don't expect all localities to warm by the same average amount; indeed, it's possible that some places will get colder even as the planet as a whole warms up.

Today, orbiting satellites provide data that allow us to determine the global average temperature quite accurately, because they give us a view of our entire planet. We can validate these records with "ground truth" measurements recorded at more than 7,000 weather stations around the world, along with measurements of ocean temperatures generally obtained by measuring the temperature of water collected by ships' intake valves. As a result, we have reliable temperature data for the approximately three decades for which we have satellite observations of our planet.

The data become somewhat less reliable for years prior to the satellite era. Getting a good estimate of the global average temperature requires having many local measurements from around the world, and fewer such measurements are available as we look deeper into the past couple of centuries. Moreover, most past temperature records were kept in cities, which have tended to become warmer over time for reasons independent of global warming, such as through changes that occur as vegetation is paved over and as local utilities generate more heat. Scientists can often account for this "urban heat island" effect, but even then are left primarily with data for land temperatures and few records of temperatures over the oceans, which cover three fourths of Earth's surface. As a result, scientists have devoted a lot of effort in the past few years to examining past temperature data in detail. Through techniques of statistical analysis, it is possible to reconstruct a fairly reliable temperature history for most of the past two centuries, though remember that the uncertainties become larger as we look further back.

Figure 10.7 ▶

Average global temperatures from 1860 through 2005. Notice the clear global warming trend of the past few decades. The "30-year average" is the average temperature for the years 1950–1980. Data was taken from the National Center for Atmospheric Research.

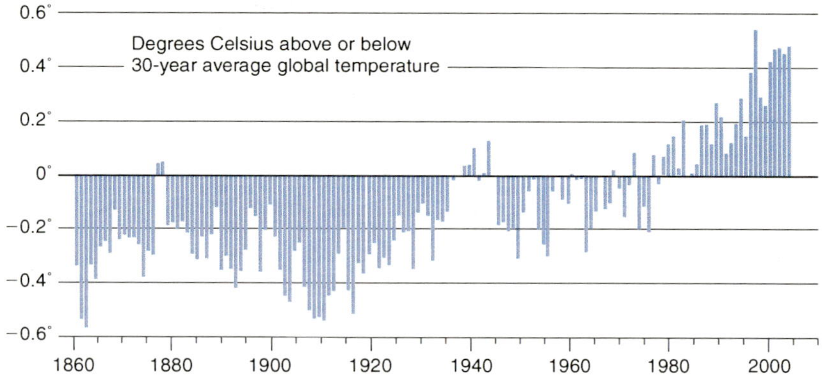

a In 2002, satellite imagery showed Antarctica's Larsen B ice shelf breaking up. The melting and fragmentation of this frozen shelf of ice, as large in area as Rhode Island and more than 200 meters thick, was the most dramatic example to date of warming in the southern continent.

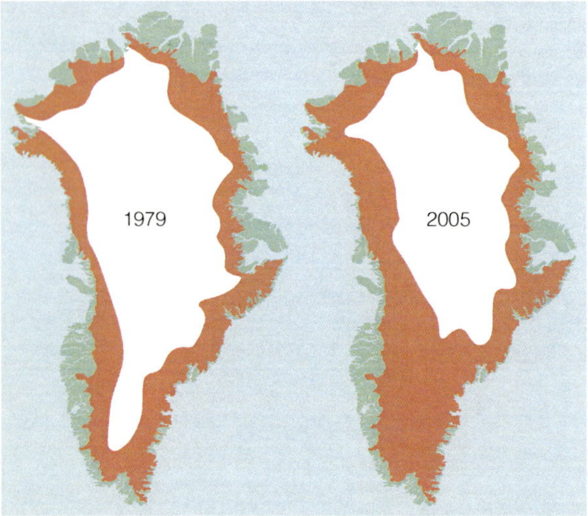

b These maps contrast the extent of the year-round Greenland ice sheet, shown in white, in 1979 and 2005. The brown area indicates the region that melts during the warm season. Notice that the melt region has expanded significantly, extending both further inland (to higher elevations) and further north.

Figure 10.8

Evidence of recent warming at the poles.

Figure 10.7 shows the reconstructed history of Earth's global average temperature since 1860. Despite the uncertainties, the overall conclusion is clear: Global average temperatures have risen by about 0.8°C (1.4°F) in the past century. Moreover, most of the warming (about 0.6°C) has occurred in just the last 30 years, the period for which the data are most reliable. The warming trend also seems to be accelerating. The year 2005 was the hottest since people began keeping records, and 9 of the previous 10 years also made the top-10 list.

THINK ABOUT IT Look up global temperature data for this year. How do the data from this year compare to those from 2005?

A temperature increase of 0.8°C may not sound like much, and it might not be so important if it were uniform everywhere. But it is not, and some regions are warming much more than others due to differences in topography, wind direction, and other factors. In general, polar regions are warming much more than equatorial regions, and the Northern Hemisphere is warming more rapidly than the Southern Hemisphere. (The greater proportion of land in the north is responsible for this, because water takes a great deal of energy to either heat or cool.) As a result, glaciers are on the retreat around the world, and polar ice appears to be melting. Orbiting satellites routinely observe massive ice sheets breaking off the continent of Antarctica (Figure 10.8a), and the area of the north polar ice cap has decreased by 20% from its extent little more than a decade ago, along with significant melting of Greenland's ice sheet (Figure 10.8b).

Correlation of Warming with Increased Carbon Dioxide Given the clear evidence of warming over recent decades, the next question is whether we can tie that warming to increased concentration of carbon dioxide (or other greenhouse gases) in the atmosphere. To establish causality in any case like this, we generally proceed in two steps. First, we look for a *correlation;* in this instance, evidence that temperatures rise and fall with carbon dioxide concentration. If we find a correlation, then we next ask whether the correlation implies causality; in this case, whether the temperature rise is *caused* by the carbon dioxide rise. Keep in mind that correlations do not automatically imply causality: Often, they may be coincidental or attributable to some underlying commonality. For example, there is a correlation between the number of churchgoers in a city and the number of beer drinkers in a city, but that obviously doesn't mean that going to church causes people to drink beer; instead, it just reflects the fact that both numbers tend to go up as population increases.

We can see evidence of a correlation by looking at past data for temperatures and carbon dioxide concentration. Direct measurements of carbon dioxide concentration have been made only since 1958, but we can get longer-term data about both temperature and CO_2 from other sources, including tree rings and ice cores drilled out of the Antarctic ice sheet. Ice cores are the frozen equivalent of sedimentary rock, made up of accumulated layers of ancient, compressed snow. The thickness of these layers—like tree rings—shows the year-to-year history of snowfall. Trapped bubbles of air in the ice can be analyzed, and oxygen isotopes within them give clues to past temperatures. When temperatures are colder, the heavier isotopes aren't transported through the air as easily as when temperatures are warm. Thus, by measuring the ratio of the heavier and lighter isotopes at each layer of the ice cores, researchers can determine relative temperatures in the distant past. Similar analysis allows reconstruction of past levels of atmospheric carbon dioxide.

As of early 2006, ice core data are available for temperatures and carbon dioxide concentration going back some 650,000 years; Figure 10.9 shows the data for the past 400,000 years, along with the direct record of carbon dioxide concentration collected over the past few decades. Notice that there is indeed a correlation between temperature and carbon dioxide concentration: Periods of higher temperature tend to also be periods of higher carbon dioxide concentration. Moreover, the data show that both temperature and carbon dioxide concentration vary substantially and naturally with time. Average temperatures have risen or fallen by as much as 10°C several times during the last 650,000 years, and carbon dioxide concentration has varied naturally between less than 200 and

Figure 10.9

This diagram shows the atmospheric concentration of carbon dioxide and global average temperature over the past 400,000 years. Notice that both vary together: More carbon dioxide goes with higher temperature, and vice versa. The CO_2 data for the past half-century come from direct measurements (made at Mauna Loa; the up and down wiggles reflect annual season cycles). Most of the earlier data come from studies of air bubbles trapped in Antarctic ice (ice core samples). The concentration is measured in parts per million (ppm), which is the number of CO_2 molecules among every one million air molecules. (The ice core data have recently been extended back to 650,000 years; these newer data are not shown but still lead to the same conclusions.)

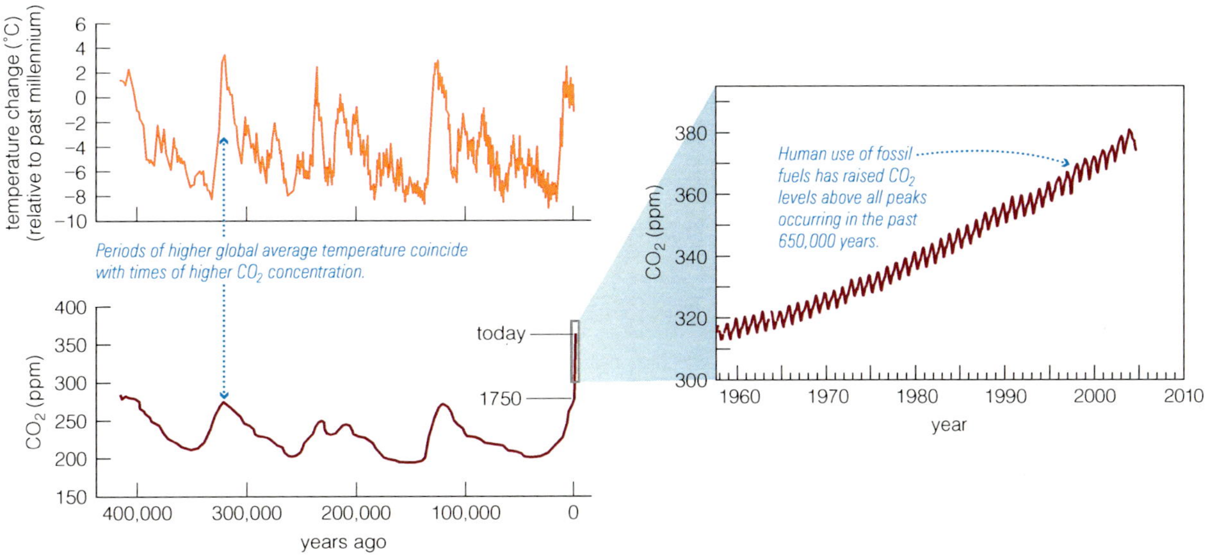

nearly 300 carbon dioxide molecules per million molecules of air (ppm). Considered against these natural changes, the temperature rise of 0.8°C over the past century may not seem all that great. But the recent rise in carbon dioxide concentration is a different story: The atmospheric CO_2 concentration is now significantly higher than it has been at any time in the past 650,000 years, and it is continuing to rise rapidly.

Global Warming Models and Uncertainties We have seen that a rising carbon dioxide concentration correlates with a rising temperature, but does that mean it *causes* the temperature rise? More to the point, do we have any reason to think that recent global warming is a result of the increased carbon dioxide put into the atmosphere by human activity, or could it just be part of the natural variation in temperature that we know has occurred in the past?

Given our understanding of the basic mechanism of the greenhouse effect, we cannot doubt that a continually rising concentration of greenhouse gases would eventually make our planet heat up. But for geologically short time scales—anything less than tens of thousands of years—there are feedback mechanisms that could potentially alter the cause and the effect. Scientists seek to understand the mechanisms and answer our questions about global warming by creating sophisticated computer models of the climate and comparing the model predictions to actual observations. If the models can correctly mimic the real climate, then they can also be used to understand how much of the temperature increase is due to human activity rather than to natural factors, and to predict how the climate will change in the future as we continue to pump greenhouse gases into the atmosphere.

Earth's climate is incredibly complex, and many uncertainties remain in attempts to model the climate on computers. Nevertheless, today's models are the result of decades of work and refinement: Each time a model of the past failed to match real data, scientists sought to understand the missing (or incorrect) ingredients in the model and then tried again with improved models. And while even today's models are not perfect, they now match real climate data quite well, giving scientists confidence that the models do indeed have predictive value. Figure 10.10 compares model data and real data, showing good agreement. Moreover, the models suggest that global temperatures are increasing in a manner consistent with what we expect if human activity is the cause: If you include the effects of the greenhouse gases put into the atmosphere by humans, the models agree with the data, but if you leave out these effects, the models fail. Thus, while plenty of uncertainties still remain, the ability to evaluate the component of warming due to our own actions continues to improve. Today, most scientists are convinced that human activity is indeed causing global warming.

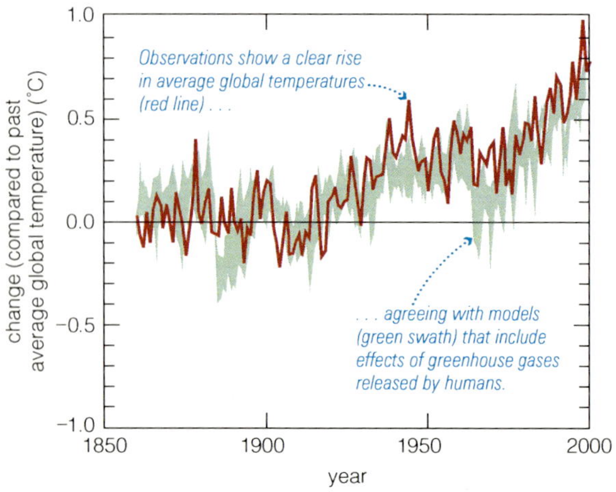

Figure 10.10

This graph compares the predictions of climate models (green swath) with observed temperature changes (red line) since about 1860. The thickness of the green swath for the models represents the range of variation between several different models. These models take into account both natural factors (such as slight changes in the Sun's brightness and the effects of volcanic eruptions) and emission of greenhouse gases by humans. The agreement is not perfect—telling us that we still have much to learn—but it is good enough to give us confidence that greenhouse gases are, indeed, causing global warming.

• What are the potential consequences of global warming?

Given that we appear to be causing global warming, the next question to ask is how it might affect us. If we can predict the future effects, then we can decide what (if anything) to do about it.

The models tell us that if current trends in the greenhouse gas concentration continue—that is, if we do nothing to slow our emissions of carbon dioxide and other greenhouse gases—the warming trend will continue to accelerate. By the end of this century, the global average temperature will be 3°C to 5°C (6°F to 10°F) higher than it is now, giving our children and grandchildren the warmest climate that any generation of *Homo sapiens* has ever experienced.

The consequences of global warming are not simply hotter weather. That alone would be important, of course, because it would shift where we could grow crops. Imagine if Northern Hemisphere wheat couldn't be grown in the American Midwest, but only in Canada, northern Europe, or Siberia. But the more serious consequences could be in changes to weather patterns brought on by the overall warming trend.

Predicting changes to local or regional weather is more difficult than predicting changes to the global average temperature, but a few consequences of global warming seem clear. Different regions of Earth will warm by different amounts; some regions will even become colder. Other regions might experience more rainfall or might become deserts. The greater overall warmth of the atmosphere would tend to mean more evaporation from the oceans, leading to more numerous and more intense storms.

Polar regions will warm the most, leading to increased melting of polar ice. This is clearly threatening to the species of these high-latitude regions (polar bears, which depend on an abundance of ice floes, are already under pressure), but it also warms the oceans everywhere, and changes their salt content as melting ice pours more fresh water into the sea. The fact that the waters of the Gulf of Mexico are at their warmest level in at least a century may be contributing to the greater strength of hurricanes that have recently blown out of the Caribbean, though it is difficult to tie specific storms to a general warming trend.

Sea level, which has already risen some 20 centimeters in the past hundred years, will continue to rise, with potentially major consequences for coastal habitats and cities. Sea level is rising primarily because water expands slightly as it warms, with an additional contribution from water added by the melting of glaciers and landlocked ice sheets. (Floating ice does not affect sea level as it melts.) By the year 2100, depending on the model used to make the projection, the oceans will have risen by between another couple dozen centimeters to a meter or more. This might be enough to flood many coastal cities, and obliterate large areas of such low-lying countries as Bangladesh. It would also make coastal storm surges far worse than they are today. Some researchers also worry that the influx of large quantities of fresh water into the (salty) seas may change the way water circulates in the oceans; changes to the flow of the Gulf Stream, for example, could have drastic climate consequences for western Europe and parts of the United States.

Other consequences could come from ecological changes, though these are even more difficult to predict. For example, it is difficult to know how forests and other ecosystems will respond to climate changes induced by global warming. As a result, we cannot easily predict the impact of such climate changes on food production, fresh water availability, or other issues critical to the well-being of human populations. Nevertheless, significant changes are well within the realm of possibility, even if we cannot be certain they will occur.

The bottom line is that we are in effect conducting a dramatic experiment on our planet in which we increase its greenhouse gas concentration far more than it would increase naturally. We do not know what the outcome of this experiment will be, or how easy or difficult it will be for us to deal with the consequences of our experiment. However, the cases of Venus and Mars show that major climate change can occur even without any human intervention. While we may not be about to do anything quite so dramatic to our own planet, we really do not know how our tampering might affect the finely balanced climate mechanisms upon which our civilization has been built.

THINK ABOUT IT If you were a political leader, how would *you* deal with the uncertain threat of global warming?

THE BIG PICTURE

Putting Chapter 10 into Context

In this chapter, we have tied together much of what we learned in other chapters to examine the concept of a habitable zone and its evolution over time. As you continue in your studies, keep in mind the following "big picture" ideas:

- The habitable zone refers to the region around a star in which a planet of suitable size (often taken to be that of Earth) could have liquid water on its surface. Despite its name, the habitable zone is not the only region around a star where habitability is possible.
- Venus, Earth, and Mars may all have been habitable in the early history of the solar system. However, only Earth has remained habitable to this day. Understanding why the climates of Venus and Mars have changed and why Earth's climate has remained comparatively stable can help us understand our climate and how we can protect it.
- The habitable zone is gradually migrating outward from the Sun. Eventually, it will lie beyond Earth's orbit, rendering our planet uninhabitable. Still, with sufficient technology, we can imagine our descendants finding a way to remain within the habitable zone until the Sun dies and then surviving by moving to other star systems.
- Given that an advanced civilization could find a way to survive as its star dies and could certainly find ways to overcome other natural threats such as asteroid impacts, it seems that nature will not impose intractable problems on us for billions of years to come. If we do not survive, it is far more likely to be the result of our own actions than the result of any natural catastrophe.

SUMMARY OF KEY CONCEPTS

10.1 THE CONCEPT OF A HABITABLE ZONE

- **How does a planet's location affect its prospects for life?**

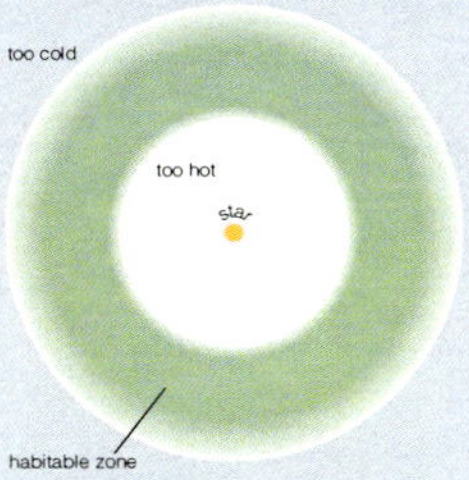

At any particular time, a star's **habitable zone** is the range of distances around it at which a planet could *potentially* have surface temperatures that would allow for abundant liquid water.

- **Could life exist outside the habitable zone?**

There could be many worlds that have underground or underwater life fueled by energy other than sunlight, in which case these worlds need not be (and are unlikely to be) in habitable zones. However, outside our solar system it would be difficult to detect life on such worlds, and it seems unlikely that complex, intelligent life could arise in such environments.

10.2 VENUS: AN EXAMPLE IN POTENTIAL HABITABILITY

- **Why is Venus so hot?**

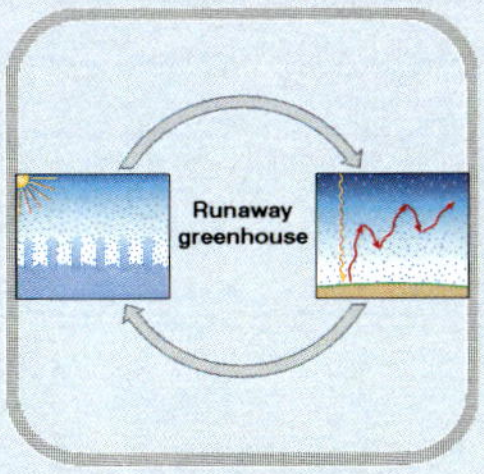

Venus's distance from the Sun ultimately led to a **runaway greenhouse effect**: Venus became too hot to develop (or keep) liquid oceans like those on Earth. Without oceans to dissolve outgassed carbon dioxide and lock it away in carbonate rocks, all of Venus's carbon dioxide remained in its atmosphere, creating its intense greenhouse effect.

- **Could Venus have once been habitable, and could life still exist there?**

Early in its history, when the Sun was some 30% dimmer than it is today, Venus may have been within the Sun's habitable zone and hence could have had rain, oceans, and perhaps life. We are unlikely ever to know for sure, because geological evidence of such a period would by now be destroyed. But if Venus once had oceans and life, it is conceivable that life could still survive among liquid droplets in high-altitude clouds.

10.3 SURFACE HABITABILITY FACTORS AND THE HABITABLE ZONE

- **What factors influence surface habitability?**

According to present understanding, a planet can have a habitable surface only if it is within its star's habitable zone, is large enough to retain internal heat and have plate tectonics, and has enough of an atmosphere for liquid water to be stable on its surface.

- **Where are the boundaries of the Sun's habitable zone today?**

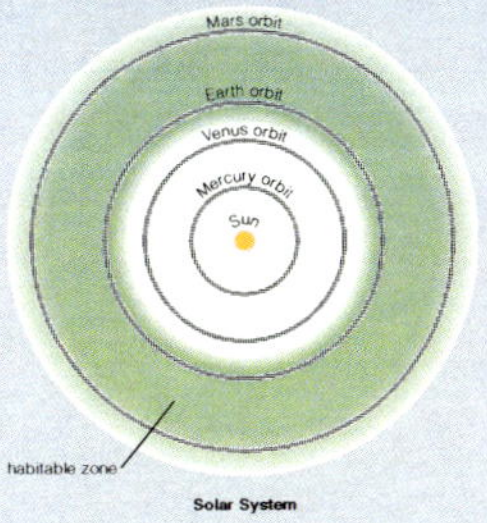

Using the most optimistic assumptions, the boundary currently extends from a distance of about 0.84 AU to 1.7 AU. Under more conservative assumptions, the boundary extends from about 0.95 AU to 1.4 AU.

10.4 THE FUTURE OF LIFE ON EARTH

- **How will the Sun's habitable zone change in the future?**

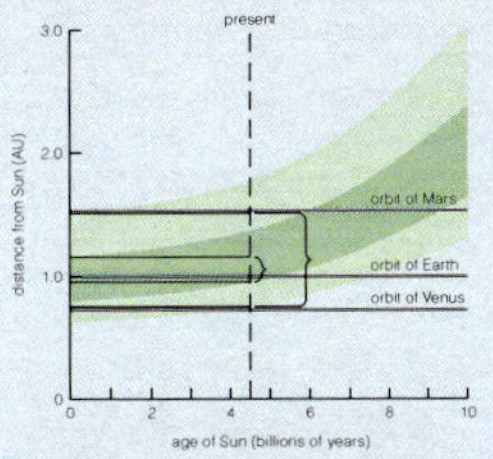

As the Sun ages, its luminosity gradually increases. As a result, the habitable zone gradually moves outward with time.

- **How long can life survive on Earth?**

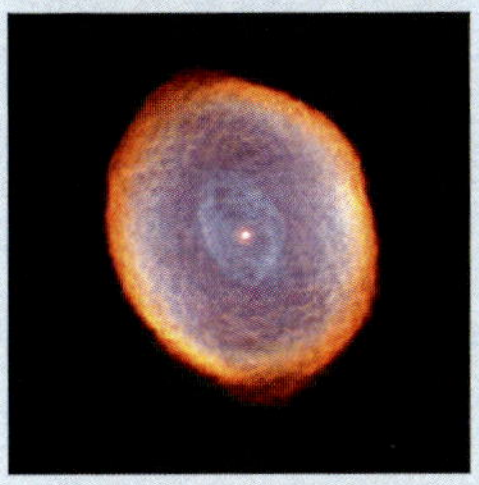

At minimum, Earth should remain habitable for another several hundred million years. By about a billion years from now, a **moist greenhouse effect** could cause Earth's oceans to evaporate away, though natural feedback processes might prevent this from occurring so soon. In 3–4 billion years, the Sun will become bright enough so that our planet will certainly be subject to a runaway greenhouse effect, ending surface habitability.

10.5 THE PROCESS OF SCIENCE IN ACTION: GLOBAL WARMING

- **What is the evidence for global warming?**

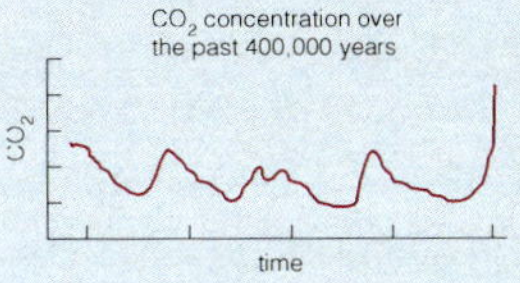

Temperature data make clear that the global average temperature is on the rise, having risen about 0.8°C over the past hundred years and 0.6°C over the past 30 years. The rise in average global temperature has been accompanied by a rise in the atmospheric CO_2 concentration, which is due to human activity. Past climate data show that the current CO_2 concentration is significantly higher than the concentration at any time in the past 650,000 years, and climate models strongly suggest that this higher concentration is indeed the cause of global warming.

• **What are the potential consequences of global warming?** Continued global warming could raise the average worldwide temperature by 3°C to 5°C during this century. Regional climate changes will be greater, and we can expect increased polar melting and a rise in sea level. Additional heat should increase ocean evaporation, which may lead to more numerous and more intense storms. Many other serious effects could also occur, but precise consequences are difficult to predict.

EXERCISES AND PROBLEMS

REVIEW QUESTIONS

Short-Answer Questions Based on the Reading

1. What is a *habitable zone*, and how is the idea useful? Is a planet in the habitable zone necessarily habitable? Explain.
2. Describe several ways in which it may be possible to have habitability outside the habitable zone. Is it possible that habitable worlds are more common outside this zone than within it? Explain.
3. Why do we think that Venus should have outgassed similar amounts of carbon dioxide and water vapor as Earth? Where is Venus's carbon dioxide today? Where is Earth's?
4. How much water is present on Venus today? How do we think that Venus lost water, and what evidence supports the idea that Venus really did lose water this way?
5. What is a *runaway greenhouse effect*, and why did it occur on Venus but not on Earth? What does this fact tell us about the inner boundary of the Sun's habitable zone?
6. Could Venus ever have had oceans and, if so, could we find geological evidence that they existed? Explain.
7. How do we expect the habitable zones of brighter stars to compare to that of the Sun?
8. Why is planetary size important to habitability? What does the case of Mars tell us about the minimum size? Can we draw any conclusions about size from the case of Venus? Explain.
9. How is an atmosphere important to surface habitability? What factors besides size and distance from the Sun might influence whether a world has an atmosphere sufficent to allow life?
10. What factors affect the location of the inner boundary of the habitable zone? Be sure to explain and consider the role of a possible *moist greenhouse effect* in such calculations.
11. What factors affect the location of the outer boundary of the habitable zone? Briefly summarize the current boundaries of our Sun's habitable zone under both the more optimistic and the more conservative scenarios.
12. Why does the Sun gradually brighten, and how does this brightening affect the location of the habitable zone over time? What do we mean by a *continuously habitable zone?*
13. How and when will Earth become uninhabitable? Why? Could humans still survive? Explain.
14. Briefly describe the eventual fates of the Sun and of the universe and what these fates might mean to our descendants (if anyone survives that long).
15. How do we determine global average temperatures over the past couple centuries? What do the data show?
16. What do ice core data tell us about the past climate and the role of greenhouse gases? How does the current concentration of atmospheric carbon dioxide compare to the concentration over past times for which we have ice core data?
17. What is the role of climate modeling in understanding global warming? How do scientists improve their models with time? Based on current models, what can we say about the cause of global warming?
18. Describe several potential future consequences of global warming, and discuss why the precise consequences are difficult to predict.

TEST YOUR UNDERSTANDING

Does It Make Sense?

Decide whether each statement makes sense or does not make sense. Explain clearly; because not all of these have definitive answers, your explanation is more important than your chosen answer.

19. If Venus were just a little bit smaller, its climate would be Earth-like.
20. Venus is not in the habitable zone now, but it may have been in the past.
21. Venus is not in the habitable zone now, but a few billion years from now it will be.
22. If we could somehow start plate tectonics on Venus, its surface would cool and it would regain the oceans it had in the past.
23. Mars will someday undergo a runaway greenhouse effect and become extremely hot.
24. We are not yet certain, but it is quite likely that Earth has suffered through a runaway greenhouse effect at least once in the past 4 billion years.
25. While the habitable zone of the Sun migrates outward over time, the habitable zones of other Sun-like stars might instead migrate inward over time.
26. If Earth someday becomes a moist greenhouse, it will have a climate that is humid but still quite comfortable.
27. The fact that additional atmospheric CO_2 must cause a greenhouse effect is all we need to know to link global warming to human activity.
28. Earth has been warmer in the past than it is today, so there is nothing to worry about with global warming.

Quick Quiz

Choose the best answer to each of the following. Explain your reasoning with one or more complete sentences.

29. The habitable zone refers to: (a) the regions of a planet where good weather allows life to exist; (b) the range of distances from a star where a planet's surface temperature is always above the freezing point of water; (c) the range of distances from a star within which water could exist in liquid form on a suitably sized planet.
30. A planet that is *not* within a habitable zone *cannot* have: (a) life; (b) subsurface oceans; (c) abundant liquid water on its surface.
31. Venus's atmosphere has much more carbon dioxide than Earth's because: (a) Venus was born in a region of the solar system where more carbon dioxide gas was present. (b) Venus lacks oceans in which carbon dioxide can be dissolved. (c) Venus has volcanoes that outgas much more carbon dioxide than those on Earth.
32. What is the likely reason for Venus's lack of water in any form? (a) The planet accreted little water during its birth. (b) The water is locked away in the crust. (c) The water was in the atmosphere, where molecules were broken apart by ultraviolet light from the Sun.
33. If Earth were to be moved to Venus's orbit, it would probably: (a) stay about the same temperature, thanks to the small amount of CO_2 in Earth's atmosphere; (b) become a tropical paradise; (c) suffer a runaway greenhouse effect and become even hotter than Venus is today.
34. The inner boundary of the Sun's habitable zone today is: (a) inside the orbit of Venus; (b) between Venus and Earth; (c) outside the orbit of Earth.
35. As the Sun ages, the habitable zone will: (a) move outward and grow wider; (b) move outward but get narrower; (c) stay about the same as it is now.
36. Which of the following could cause Earth to become uninhabitable in about 1 billion years? (a) a moist greenhouse effect; (b) a runaway greenhouse effect; (c) the death of the Sun.
37. Global warming means that: (a) Earth's average temperature is increasing; (b) every place on Earth is getting warmer; (c) Earth will soon have a greenhouse effect.
38. The current concentration of atmospheric carbon dioxide on Earth is: (a) higher than the concentration at any time during the past 650,000 years; (b) higher than in the past century, but lower than at many other times during past millennia; (c) gradually rising, but only about average for the time period during which we have ice core data.

INVESTIGATE FURTHER

In-Depth Questions to Increase Your Understanding

Short-Answer/Essay Questions

39. *Are Habitable Zone Planets Common?* Based on what you have learned so far about solar system formation and habitable zones, should we expect to find many planets in habitable zones? Explain.
40. *No Plate Tectonics.* Suppose plate tectonics magically stopped on Earth, but other geological processes (such as volcanism) continued. Would Earth's *surface* get warmer or cooler? Explain.
41. *Continuously Habitable Zone.* Is Earth in a zone that will remain continuously habitable from the Sun's birth to its death? Is any planet? Explain.
42. *Planetary Changes.* For each situation described, write two or three paragraphs explaining why the planet would or would not be habitable today.
    a. A planet the size of Mars located at the distance of Venus.
    b. A planet the size of Mars located at the distance of Earth.
    c. A planet the size of Venus located at the distance of Earth.
    d. A planet the size of Earth located at the distance of Mars.
43. *A Billion Years.* At minimum, it appears that our planet will remain habitable for at least the next billion years, give or take a couple hundred million years. How long is a billion years? Think of some ways to put this time period into perspective.
44. *Venus's History.* Many people are not surprised to learn that Venus is hotter than Earth, given that it is closer to the Sun. Explain why we cannot attribute its heat to distance from the Sun alone. How do we explain Venus's high temperature?
45. *Habitable Moons.* As we'll discuss in Chapter 11, some of the newly discovered extrasolar planets are Jupiter-like in size but are located at Earth-like distances from Sun-like stars. These planets are unlikely to be habitable themselves. Could they have moons with habitable surfaces? Explain.
46. *Greenhouse Lessons.* While it seems unlikely that human activity could cause a runaway greenhouse effect on Earth, we could still cause the climate to warm substantially. Do you think we can learn anything valuable about our potential effects on Earth's climate by studying the climate histories of Venus and Mars? If so, what? Defend your opinion.
47. *Global Warming.* Briefly summarize the evidence suggesting that global warming is occurring and is a result of human activity. Then write a short essay outlining what, if anything, we should do about it.

Quantitative Problems

Be sure to show all calculations clearly and state your final answers in complete sentences.

48. *Stellar Habitable Zone.* Consider a star slightly smaller and less luminous than the Sun, with a habitable zone that extends from 0.3 to 0.5 AU. How does the size of this star's habitable zone compare to that of the Sun, and what are the implications for the likelihood of finding planets within the zone?
49. *Massive Stellar Habitable Zone.* Consider a star that is more massive and more luminous than the Sun, with a habitable zone that extends from 2.5 to 4 AU. How does the size of this star's habitable zone compare to that of the Sun? Should we consider life to be likely around such a star? Explain.
50. *Strength of Sunlight at Earth.* The power of sunlight reaching the top of Earth's atmosphere is 1,370 watts per square meter. The amount of power flowing outward through Earth's surface caused by radioactivity within Earth is estimated to be 3 trillion watts. Which one—sunlight or radioactive heat—is providing more energy to Earth's surface, and by how much? What does your answer tell you about the relative importance of these two energy sources for life? (*Hint:* You'll need to convert the radioactive power number from a total to an amount per square meter of surface; remember that the surface area of a sphere of radius r is $4\pi r^2$.)
51. *Strength of Sunlight at Venus and Mars.* The solar energy reaching the top of Earth's atmosphere is 1,370 watts per square meter.

What is the comparable energy: (a) at the distance of Venus: (b) at the distance of Mars? (*Hint:* Remember that light follows an inverse square law [see Figure 7.1]; you'll need to look up distances in AU for Venus and Mars.)

52. *Energy Use by Cars.* There are approximately 800 million automobiles in use worldwide . If the average auto runs 1 hour a day, at 50 horsepower, what is the resultant total average power use by cars, in watts? How does this amount of power compare to the amount Earth gets from sunlight? Note that 1 horsepower is 746 watts and 1 watt = 1 joule/second. (*Hint:* You'll first need to figure out the average number of cars running at any one time, and once you find the total power for the cars, you'll need to divide by Earth's surface area in square meters so you can compare it to the solar power given in Problem 50.)

53. *Atmospheric Mass of Venus.* The atmospheric pressure on Venus is 90 bars. What is the total mass of Venus's atmosphere? You may use the fact that 1 bar is the pressure exerted by 10,000 kilograms pushing down on a square meter in Earth's gravity, and assume that Venus's gravity is essentially the same as Earth's.

54. *Melting Greenland.* Greenland is approximately 700 × 2,400 km in size, with an average ice cover that's about 1.5 kilometers thick. Suppose all the Greenland ice were to melt. By approximately how much would this raise the level of Earth's oceans? Assume that oceans cover 70% of Earth, and that water and ice are approximately the same density. (*Hint:* Dividing the volume of melted water by the surface area of the oceans will tell you how much the ocean depth will increase.)

Discussion Questions

55. *The Fate of Life in the Universe.* Consider the evidence suggesting that life is just a fleeting phase in the long-term history of the universe. Assuming this to be the case, how do you think it should influence our perspective on our own place in the universe? Why?

56. *The Politics of Global Warming.* The current scientific case for global warming seems quite strong, but the topic nonetheless generates significant political controversy. Why do you think that is the case? Do you consider global warming to be primarily a scientific issue or a political issue? Explain.

57. *Dealing with Uncertainty.* One of the difficulties in deciding what to do about global warming is the fact that the precise consequences are uncertain. In general, how do you think we as a society should deal with issues for which consequences are potentially severe but highly uncertain? How would you deal with this situation in the particular case of global warming? Explain.

WEB PROJECTS

58. *Kyoto Protocol.* What is the Kyoto protocol concerning global warming? What are its pros and cons? Do you think it is valuable and should be followed? Explain.

59. *Global Warming Scenarios.* Research data showing how the amount of global warming might differ depending on whether we decrease, keep the same, or increase our future carbon dioxide emissions. Summarize your findings.

60. *Global Warming Skeptics.* Compare the arguments from Web sites that claim that: (a) global warming is caused by fossil fuel use; and (b) there is no convincing proof that this is the case. How would you evaluate these arguments?

61. *Long-Term Survival.* Read about some exotic ideas concerning how advanced civilizations might survive as the universe grows cold and dark. Do you think any of these ideas make sense, or are they just wishful thinking? Write a short essay describing one of these ideas and your opinion of it.

11
Habitability Outside the Solar System

LEARNING GOALS

11.1 DISTANT SUNS

- How do stellar life cycles affect the possibility of habitable planets?
- How do we categorize stars?
- Which stars would make good suns?

11.2 EXTRASOLAR PLANETS: DISCOVERIES AND IMPLICATIONS

- How do we detect planets around other stars?
- What are the properties of extrasolar planets discovered so far?
- How could we detect life on extrasolar planets?

11.3 THE POSSIBILITY THAT EARTH IS RARE

- Are Earth-like planets rare or common?

11.4 THE PROCESS OF SCIENCE IN ACTION: CLASSIFYING STARS

- How and why did the Hertzsprung–Russell diagram develop?
- What can we learn from the Hertzsprung–Russell diagram?

Then felt I like some watcher of the skies, when a new planet swims into his ken.
John Keats (1795–1821)

So far we have focused on life on Earth and the prospect of life existing elsewhere in our solar system. We have found that while Earth is the only planet known to harbor life, several other worlds in our solar system are—or at some time in the past have been—potentially habitable. If we can translate this situation to the billions of other star systems in the Milky Way Galaxy and, by extension, to the billions of other galaxies that pepper the universe, then the total number of habitable worlds must be enormous. The existence of a vast number of such worlds would offer the prospect that among them we would find not only microbes, but also complex creatures and intelligent beings.

Unfortunately, current data are not sufficient for us to be sure that such an immense number of habitable worlds really exists. We have detected many extrasolar planets, but not yet planets that could be Earth-like. Is this simply because our technology is not yet up to the task, or could it be that such planets are rare?

We will devote this chapter primarily to exploring the question of whether habitable planets are common or rare and how we might someday observe—or even identify life on—such planets if they exist. This understanding will then help us frame the questions of whether other civilizations exist and how we might someday contact them if they do, topics we'll discuss in the chapters that follow.

11.1 DISTANT SUNS

Perhaps the first and most obvious prerequisite to finding a planet on which life might arise is selecting a "sun" that can provide sufficient light and heat to support habitable worlds. Not all stars are potential suns in this way. To understand what kinds of stars might be orbited by potentially habitable planets, we must first investigate the nature of stars in more depth.

• How do stellar life cycles affect the possibility of habitable planets?

In ancient times, almost any light in the sky was considered to be a star, and in some cases we still use this historical language. For example, we often refer to meteors as "shooting stars," even though they really are just bits of interplanetary dust entering our atmosphere. Asteroids got their name, which means "starlike," because that is how they appear when first seen through a telescope, even though they are actually chunks of rock in our own solar system. Our modern definition of a star is a large ball of gas that generates energy by nuclear fusion in its hot central core (see the definitions on p. 51). Thus, our Sun shines as a star because it fuses hydrogen into helium in its core.

Recall that a star goes through a "life cycle" that begins with its formation in a giant cloud of gas (see Figure 3.6). Before the center gets hot enough to ignite nuclear fusion, we refer to the unfinished star as a *protostar* (*proto* comes from the Greek for "earliest form of"). A star is "born" when nuclear fusion begins in its central core. A star "dies" when it finally ceases to produce energy by any kind of fusion.

All stars spend most of their lives (about 90% of the time from star birth to star death) fusing hydrogen into helium. Stars shine fairly steadily during this period, brightening gradually as they age (for the same reason the Sun is slowly brightening [Section 10.4]). Core hydrogen fusion may continue for millions or billions of years (depending on the star's mass, as we'll discuss shortly), but eventually the central core will be so depleted of hydrogen that the fusion cannot continue.

Perhaps surprisingly, a star that exhausts its core hydrogen begins to grow larger and brighter, becoming a *giant* or *supergiant* star. For example, when our own Sun becomes a red giant some 4–5 billion years from now, it will grow so large that it will engulf some of the inner planets. The reason an aging star grows so large and luminous is the immense pressure exerted on the core by the weight of the star's overlying gas. During the time that the central core burns hydrogen, the energy generation helps the core resist the pressure and maintain its size. Once the hydrogen in the central core runs out, the pressure causes the core to shrink and therefore rise in temperature. As this occurs, a layer surrounding the central core—a layer that still contains unburned hydrogen—ignites with nuclear fusion. In fact, this layer becomes so hot that the total rate of fusion is higher than it was while the central core fused hydrogen. The higher rate of fusion means a greater release of energy, which causes the star to increase in size and emit more light.

Stars can begin to fuse heavier elements during their giant or supergiant stages. Our own Sun will someday produce carbon by fusing the helium in its core, and more massive stars can ultimately create all the other elements (through a combination of core fusion reactions and reactions that occur in their final death throes). As we discussed in Chapter 3, this stellar manufacturing explains the existence of all the elements in the universe except the original hydrogen and helium that were spawned by the Big Bang, which is why we say we are made of "star stuff."

Eventually, the star will reach a point where it can fuse no other elements, and at that point the star dies. Relatively low-mass stars like our Sun end their lives comparatively gently, ejecting their outer layers into space as planetary nebulae and leaving behind the type of dead star that we call a *white dwarf* (see Figure 10.6). Higher-mass stars die in titanic explosions called *supernovae* (see Figure 3.8), in which their cores collapse to form stellar remnants that may be either *neutron stars* or *black holes*.

From the point of view of an astronomer, all of the various stages of stellar life are interesting. But in terms of the search for habitable planets, we are generally concerned only with stars that are in the long-lasting, hydrogen-fusing stage of their lives. Giants and supergiants may still have planets, but the rapidly changing nature of these stars (for example, bloating up in just a few million years or less) makes it unlikely that any planets could remain habitable for long. Dead stars may also have planets—indeed, the first confirmed discovery of extrasolar planets involved three small worlds orbiting a neutron star. But because a neutron star is no longer generating continuous sunshine of the type produced by ordinary stars, it would be an unlikely host for habitable planets.

• How do we categorize stars?

Stars vary significantly in their properties, such as mass, temperature, light output, and composition. The most luminous stellar beacons shine with a brilliance as great as a million Suns, while other stars are so dim that they are invisible to the naked eye even when relatively nearby. Once astronomers recognized the substantial differences among stars, their first task (like that of botanists and zoologists in organizing plants and animals) was to find a useful scheme for categorizing this diverse collection of objects.

Finding such a categorization was not trivial. Stars are so far away that even when viewed with large telescopes they appear as no more than points of light. The most obvious characteristic of a star is how bright it appears in the sky, and indeed a system for describing the brightness of stars visible to the naked eye was developed by the ancient Greeks (and is still used today). Unfortunately, ranking a star by how bright it appears in Earth's sky doesn't tell us how bright it *really* is, because the total amount of light it emits into space—which defines its **luminosity**—also depends on distance. An extremely luminous star that is far away may look dim, while a modestly luminous star that is nearby may shine quite brightly. For example, the brightest star in our sky, Sirius, is only a moderately luminous star that happens to be relatively close. (Sirius is about 8.6 light-years distant, making it the seventh-nearest star to Earth [other than the Sun].)

Refinement of the telescope eventually led to techniques for measuring stellar distances, which enabled astronomers to classify stars according to their true luminosities. But to learn more about their properties required yet another technological step: the invention of spectroscopy [Section 3.4]. In the late 1800s, astronomers at Harvard College Observatory, under the directorship of Edward Pickering (1846–1919), began a massive effort to study stellar spectra and thereby determine other characteristics of the stars.

Figure 11.1

This photograph shows how, by placing a prism in front of a telescope's lens, astronomers recorded the spectra of many stars at once. The individual spectra were somewhat crude, but were adequate for identifying the stronger hydrogen and helium lines that are the basis of the spectral type classification.

The Spectral Sequence

Making a detailed stellar spectrum is a tedious process, but at the end of the nineteenth century astronomers took an important step forward when they invented a method to record the spectra of many stars at once. To do this, astronomers mounted a glass prism in front of a telescope's objective lens and then photographed a patch of sky containing a large number of stars. On the resulting photo, the image of each star was spread out into a rainbow-like streak, and the most obvious spectral lines could be seen (Figure 11.1).

Once the technique was developed, it wasn't long before thousands of stellar spectra were in hand. The next step was to study this wealth of data and try to make sense of what it was telling us. Pickering needed help with these tasks, so he began to hire assistants whom he called "computers." The job clearly required people well trained in physics and astronomy, but, in part because the task was seen as somewhat tedious, Pickering found few takers among the men graduating from what was then the all-male Harvard College. Because at that time women with equivalent educations faced enormous obstacles to securing good positions in science, a job with Pickering represented a rare chance for career advancement for women. So Pickering recruited women who had studied physics and astronomy at colleges such as Wellesley and Radcliffe. Although the work was indeed tedious, it was also cutting-edge research, and the women astronomers of Harvard College Observatory were to make many great discoveries (Figure 11.2).

At first, the astronomers found it difficult to make sense of the spectra. Pickering suggested a scheme in which stars were classified by the visibility of hydrogen lines in their spectra (that is, spectral lines caused by the element hydrogen), using type A to designate stars with the strongest hydrogen lines, type B for those with slightly fainter lines, and so on down the alphabet to type O for stars showing the weakest lines. Following this suggestion, one of Pickering's first "computers," Williamina Fleming (1857–1911), classified more than 10,000 stellar spectra by 1890. But the work was just beginning.

In 1896, Pickering hired Annie Jump Cannon (1863–1941), who in the course of her career would personally classify the spectra of more than 400,000 stars. Within a few years of being hired, she realized that Pickering's sequence of spectral types A to O included some redundancies and, more importantly, fell into a much more natural order than he had supposed. She concluded that there were only seven major spectral types, which could be logically ordered as **OBAFGKM**, the *spectral sequence* that legions of astronomy majors have memorized using the politically incorrect mnemonic "Oh, Be A Fine Girl, Kiss Me." Cannon also subdivided each type by number; for example, stars of spectral type G could be subclassified as G0, G1, G2, and so on to G9, with G0 being most similar to the F stars and G9 being nearly the same as a K0 star. Our Sun is now classified as spectral type G2. The astronomical community adopted Cannon's system of stellar classification in 1910.

The stellar classifications clearly were telling us something important about the nature of stars, but no one yet knew just what that was. The answer finally came in 1925, in the dissertation of another woman working at Harvard College Observatory, Cecilia Payne-Gaposchkin (1900–1979). Relying on insights from what was then the newly developing science of quantum mechanics, Payne-Gaposchkin showed that the differences in the spectral types reflected differences in the surface temperatures of the stars. A later review of twentieth-century astronomy called her work "undoubtedly the most brilliant Ph.D. thesis ever written in astronomy."

Figure 11.2

Women astronomers pose with Edward Pickering at Harvard College Observatory in 1913. Annie Jump Cannon is in the back row near the right edge of the door.

The Role of Stellar Mass

This research quickly led to another interesting discovery. One might expect that stars could have many combinations of size, temperature, mass, and composition. The truth is much simpler, a fact that was first revealed when astronomers began to make graphs of the relationship between luminosity and surface temperature. These types of graphs, called *Hertzsprung–Russell diagrams,* helped astronomers unlock the secrets of stars; we will discuss these important tools in Section 11.4.

All stars are born with basically the same composition—because they form in interstellar clouds, they are born with the same compositions as the clouds, which are almost entirely (98% or more) made of hydrogen and helium. For that reason, the physics that determines a star's characteristics is relatively straightforward. During the hydrogen-burning phase of its life, a star's surface temperature (and hence its spectral type) and total luminosity are determined almost entirely by one thing: the star's mass. Table 11.1 lists typical properties for stars of each of the seven major spectral types.

TABLE 11.1 *Typical Properties for Hydrogen-Burning Stars of the Seven Major Spectral Types*

Numbers given in solar units are values in comparison to the Sun; for example, a mass of 60 solar units means 60 times the mass of the Sun. Note that the Sun is a G star. (More specifically, the Sun's spectral type is G2.)

Spectral Type	*Approximate Percentage of Stars in This Class*	*Surface Temperature (°C)*	*Luminosity (solar units)*	*Mass (solar units)*	*Lifetime (years)*
O	0.001%	50,000	1,000,000	60	500 thousand
B	0.1%	15,000	1,000	6	50 million
A	1%	8,000	20	2	1 billion
F	2%	6,500	7	1.5	2 billion
G	7%	5,500	1	1	10 billion
K	15%	4,000	0.3	0.7	20 billion
M	75%	3,000	0.003	0.2	600 billion

Stellar Lifetimes Table 11.1 also shows that the lifetimes of stars vary considerably, which we can understand by studying the mass and luminosity columns. A star's mass essentially tells us how much hydrogen fuel it has available for fusion, while its luminosity tells us how brightly the star shines and hence how rapidly it is fusing its hydrogen. Note that there is an enormous range in luminosity: The most luminous stars are nearly a billion times as luminous as the dimmest. The range in mass is much smaller. Thus, for example, an O star has some 60 times as much hydrogen fuel as a G star like the Sun (from the mass column) but burns it at a rate a million times faster than the Sun (from the luminosity column). Therefore, it will go through all of its available fuel much faster than does the Sun. The rule is a general one: *The more massive the star, the shorter its lifetime.* As shown in the last column of the table, the range of lifetimes extends from just a few hundred thousand years to hundreds of billions of years.

THINK ABOUT IT Many generations of massive O and B stars have lived and died in the history of the universe. Is it also true that many generations of G, K, and M stars have lived and died? Explain. (*Hint:* How old is the universe?)

• Which stars would make good suns?

We can use the properties listed in Table 11.1 to investigate the question of which types of stars might make suitable suns for habitable planets. We can immediately rule out stars of type O from their short lifetimes alone. Recall that, in our solar system, it took millions of years for the rocky inner planets to form by accretion [Section 3.3]. With lifetimes of less than a million years, O stars simply don't last a long enough time to allow the formation of Earth-size planets. We can probably also rule out stars of type B. Although the 50-million-year lifetime of a typical B star is probably long enough to allow planets to form, the star's death would probably occur before the process of accretion had settled down enough for life to take hold.

Spectral Types A and F Stars of types A and F, with lifetimes of 1–2 billion years, would seem to offer enough time for both the formation of planets and the beginnings of biology. After all, evidence suggests that life took hold on our own world within a few hundred million years or less of its formation [Section 6.1]. Stellar types A and F are hotter than the Sun, so their habitable zones would be wider and would lie at greater distances from their central star. By itself, the wider habitable zone could make habitable planets even more likely than they are around Sun-like stars.

One potential problem is that, because of their higher temperatures, A and F stars emit many times as much ultraviolet light as does the Sun. Biology on Earth, and perhaps biology in general, is vulnerable to high-energy ultraviolet light, which easily breaks chemical bonds in complex organic molecules. The intense radiation might well keep planetary surfaces sterile. However, there are at least two ways around this problem.

First, ultraviolet radiation does not penetrate far into the ground, oceans, or ice. If life on Earth emerged near volcanic vents in the deep oceans, as some evidence suggests, the same thing might happen on a watery planet around an A or F star. Similarly, worlds with a subsurface ocean, such as Europa may have, would offer life plenty of protection.

Second, even though our Sun emits far less ultraviolet light than A or F stars, it still emits enough to make Earth's surface sterile if not for the shielding provided by the ozone layer in the atmosphere. Planets around an A or F star might enjoy similar shielding if they had either a sufficiently thick atmosphere or an atmosphere containing sufficient oxygen. In the latter case, the additional ultraviolet light would split more atmospheric O_2 molecules, producing single oxygen atoms that would then combine to form ozone. In other words, a higher dosage of ultraviolet radiation could result in more ozone shielding.

Overall, A and F stars, despite their energetic nature, seem quite capable of hosting habitable planets. However, given the fact that complex plants and animals on Earth did not arise until our planet was some 4 billion years old, the 1- to 2-billion-year lifetime of A and F stars suggests that life on these planets would most likely be much simpler than life on Earth.

THINK ABOUT IT In light of current evidence concerning the past oxygen content of Earth's atmosphere (see Section 6.3), does it seem likely that planets around A or F stars could have ozone layers? Why or why not?

Spectral Types G, K, and M Although the possibility of habitable planets around A and F stars is intriguing, Table 11.1 suggests that statistically it's not too important: Together, these types make up only about 3% of stars. G stars, like our Sun, make up another 7%—and our own existence proves that G stars can have habitable planets. But if we want to know whether the *majority* of stars could have worlds capable of supporting life, we need to consider the smaller K and M types, which make up some 90% of all stars in the universe.

K and M stars have long lifetimes that impose no limit on their ability to have planets on which life could evolve for many billions of years. Instead, the primary issue for life around these stars is the size of their habitable zones. Because these stars are dimmer than the Sun, temperatures on orbiting planets will be lower in these systems than on planets

Figure 11.3

The approximate habitable zones around our Sun, a typical K star, and a typical M star, shown to scale. The habitable zone becomes increasingly smaller and closer-in for stars of lower luminosity. (*Note:* As discussed in Section 10.4, the habitable zone around any star moves outward with time; the zones shown here are for stars similar in age to the Sun.)

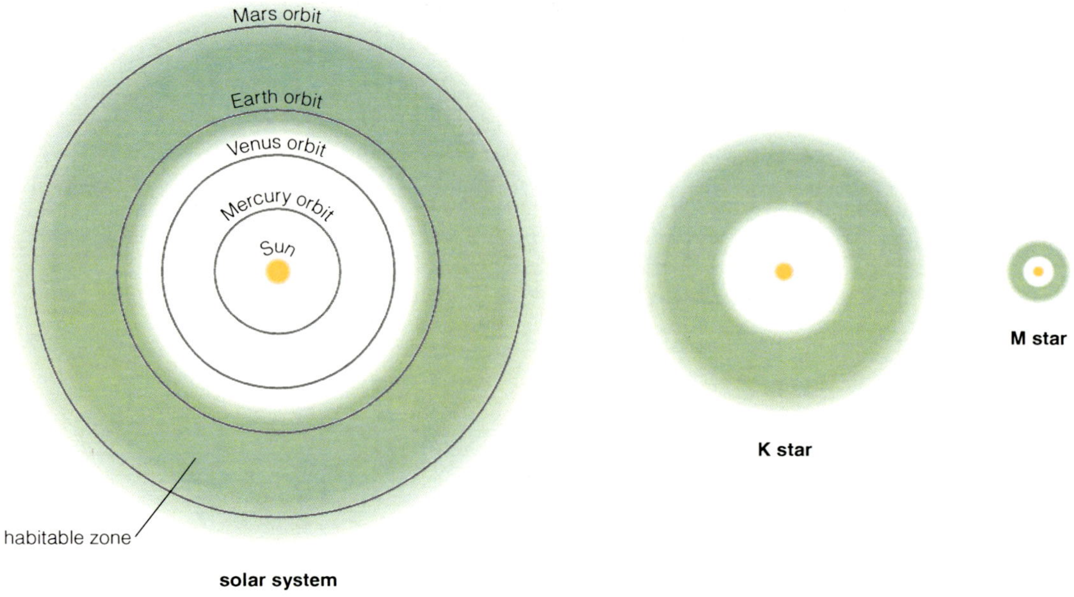

the same distances from the Sun in our solar system. More specifically, recall that the energy in starlight falls off with the square of the distance from the star (see Figure 7.1). Thus, for a planet to receive as much radiant energy as Earth from a K star having one-fourth the Sun's luminosity, the planet would have to orbit its star at one-half Earth's distance from the Sun (because $\left(\frac{1}{2}\right)^2 = \frac{1}{4}$). For a star with a luminosity 1% of the Sun's, a similar planet would need to orbit at a distance one-tenth of Earth's distance from the Sun, which means it would follow an orbit about one-fourth the size of Mercury's.

The same reasoning tells us that the boundaries of a dim star's habitable zone would also be much closer to the star than are the boundaries of the habitable zone in our solar system.* In general, the habitable zone becomes progressively smaller and closer-in for stars of lower luminosity (Figure 11.3). Around a typical M star, for example, the habitable zone would extend only over a region roughly equivalent to from one-tenth to one-fourth Mercury's distance from the Sun in our solar system. The small size of the habitable zone certainly decreases the probability that planets would be found within this region (see Cosmic Calculations 10.1).

On the other hand, the overwhelming majority of stars are M stars, due both to their long lives—every M star ever born is still shining—and to the fact that small stars are born in larger numbers than are more massive stars. Even if their individual habitable zones are modest, there are enough M stars to host a large number of inhabited worlds. For years, this optimistic thought was moderated by two objections. First, planets orbiting close to a star get locked into synchronous rotation [Section 9.1], with one side of the planet perpetually facing the star (much as the Moon always keeps one face turned to Earth). Once this happens, the side away from the star becomes perpetually dark, and the atmosphere might be

*The boundaries of the habitable zone are also affected to some extent by the star's surface temperature, which determines the wavelength at which it puts out most of its light.

expected to freeze out. Second, small stars have frequent and energetic flares*—sudden bursts of intense light and radiation—that might cook any complex life.

Recent research, however, suggests that neither of these objections is necessarily fatal. Even a modestly thick atmosphere containing carbon dioxide would circulate heat from the bright to the dark side on a synchronously locked world, keeping temperatures relatively uniform and allowing liquid water to exist over much of the planet's surface. As for flares, their dangerous ultraviolet light might actually cause the production of a protective layer of atmospheric ozone, in much the way we have described for worlds orbiting stars of types A and F. And, of course, underwater or underground life would be protected in any case. The bottom line is that M stars probably can support life. If they do, given the fact that they live so long, these stars might possibly house the galaxy's oldest biology.

Figure 11.4

This infrared image shows many brown dwarfs (circled) in the constellation Orion.

Brown Dwarfs Table 11.1 shows that an "average" (meaning middle of the range) M star has a mass about 20% that of the Sun; although it is not shown in the table, the smallest M stars have masses about 8% that of the Sun. Given that these low-mass M stars are the most plentiful of all stars, you might wonder why there aren't stars of even lower mass. The answer is that in an object with a mass less than 8% that of the Sun, gravity isn't strong enough to compress the core to high enough temperatures to sustain hydrogen fusion. As a result, such "substellar" objects, called **brown dwarfs**, are very dim and difficult to detect. Nevertheless, they have much hotter surfaces than planets, which means they emit moderate amounts of infrared radiation that can allow us to detect them if they are nearby. Thus, thanks to improved infrared telescopes, we now know that brown dwarfs are quite common (Figure 11.4). Typical brown dwarfs have masses between about 10 and 80 times that of Jupiter (the larger number is equivalent to 8% the mass of the Sun). In essence, brown dwarfs occupy a fuzzy range of masses between those of the largest planets and the smallest true stars.

Because they are so dim, brown dwarfs have no habitable zones at all. However, recent infrared observations have shown that many young brown dwarfs have disks of material surrounding them, suggesting that planets are forming in their neighborhood, and one brown dwarf appears to have a Jupiter-size planet orbiting it (Figure 11.5). Thus, while planets with habitable surfaces do not seem possible around brown dwarfs, it is possible that brown dwarfs could have planets orbited by Europa-like moons with subsurface oceans and, perhaps, life.

Figure 11.5

This infrared image from the European Southern Observatory's Very Large Telescope shows a brown dwarf called 2M1207 (blue) and what is probably a Jupiter-size planet in orbit around it (red).

Multiple Star Systems Based on our discussion so far, we have good reason to believe that the vast majority of stars are, at least in principle, capable of having habitable planets. However, there's another potential complicating factor that we have not yet considered: Many stars are not loners like the Sun, but rather are members of **multiple star systems**, in which two or more stars orbit each other closely. For example, the nearest star system to our own, Alpha Centauri, is actually a triple star

*The occurrence and size of flares, sunspots, and other types of surface "activity" on stars is related to the strength of a star's magnetic field. A star's magnetic field strength depends on its rotation rate and the depth of convection in the star's outer layers. Small stars have deeper convection, so if they rotate quickly then they can have very strong magnetic fields associated with large flares and other surface activity.

system: Its largest and brightest member ("Alpha Centauri A") is a type G star like our Sun, its second-largest ("Alpha Centauri B") is a K star that is separated from the G star by roughly the distance that separates our Sun and the planet Uranus, and its third star ("Proxima Centauri") is an M star that orbits the other two at a distance of about one-fifth of a light-year. Can multiple star systems have habitable planets? The answer to this question hinges on whether a planet in a multiple star system can have a stable orbit that keeps it within a habitable zone.

Let's consider possible planetary orbits for the most common cases, which are **binary star systems** with just two stars. (Systems with three or more stars are much less common.) Broadly speaking, there are three possible situations to consider for planetary orbits in a binary star system:

- A planet could orbit around both stars (Figure 11.6a). If the planet orbits the stellar pair at a distance considerably greater than the separation between the two stars, then gravitationally the two stars act much as one, and the planet can orbit without disruption around its distant twin masters. Computer simulations indicate that a stable orbit is generally possible for planets that orbit at a distance of more than about five times the separation of the stars. (The exact distance depends on the details of the system, such as the relative masses of the two stars and the eccentricity of their orbits.)
- A planet could orbit one star or the other (Figure 11.6b). If the two stars are themselves widely separated, then a planet near either one can orbit steadily because it will feel little disturbing effect from the second star. Computer simulations indicate that stability becomes possible when the two stars are separated by more than about five times the planet's orbital distance.
- We might imagine a planet trying to orbit between the two stars (Figure 11.6c). However, if the distance between the planet and at least one of the stars is not sufficiently different from the distance between the two stars, its orbit will not be stable. The planet will experience competing gravitational tugs from both stars that will ultimately fling it out of the system (or send it crashing into one of the stars).

As an example of the first stable case, imagine that our Sun had a companion star at Jupiter's distance (which is 5.2 times that of Earth's

Figure 11.6

Three orbital possibilities in a binary star system.

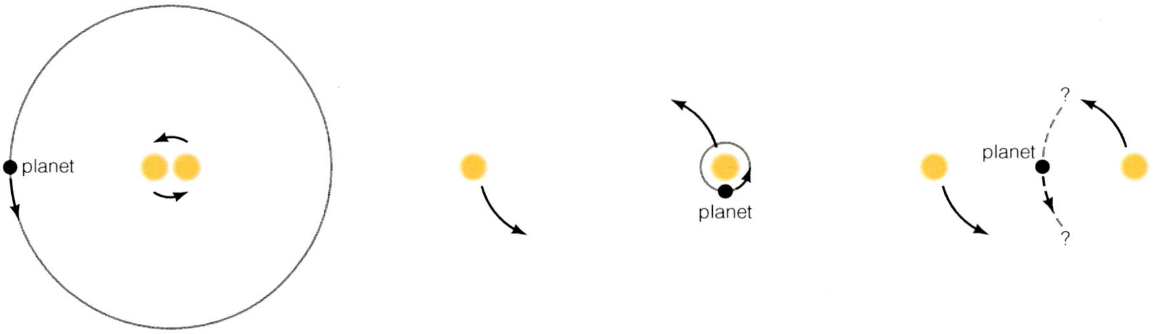

a *Potentially stable:* The planet orbits both stars; the radius of the planet's orbit is much larger than the separation between the stars.

b *Potentially stable:* The planet orbits one star; the radius of the planet's orbit is much smaller than the separation between the stars.

c *Not stable:* The orbit of the planet is neither much larger nor much smaller than the separation between the stars distance). In that case, Earth's orbit (as well as those of Mercury and Venus) would be stable. As an example of the second case, imagine that our Sun had a companion star at Earth's position. The orbits of the inner planets would not be stable, but Jupiter's orbit could be.

Given these orbital possibilities, what can we say about habitability in binary star systems? In fact, both stable cases seem to allow for habitable planets. In the first case, the habitable zone would be some region surrounding the two stars together. In the second case, the more distant star would probably have little influence on the planet's climate, so the habitable zone would be defined solely by the star that the planet orbits. Similar conclusions probably apply to other multiple star systems.

This is encouraging news in our search for life in the cosmos, because the majority (about 60%) of stars like the Sun are in multiple systems. Until recently, it was believed that the majority of *all* stars shared this penchant for partners. However, new research has shown that, among the galaxy's least massive (and dimmest) stars—the M stars—only about one-fourth are multiple. Since 75% of galactic stars are M stars, this means that, overall, about one-third of the galaxy's stars are multiple.

Summary of the Types of Stars That Might Have Habitable Planets

We began this section with the goal of determining what types of stars might potentially have habitable planets. Perhaps surprisingly, we have found relatively few limits. The only stars that we can rule out completely—the O and B stars—are rare. G stars like our Sun clearly can have habitable planets. Even though only 7% of the stars in our galaxy are G stars, 7% of a few hundred billion stars is still an impressive tally. The small size of the habitable zone around the most common types of stars—dim, low-mass K and M stars—suggests that habitable planets might be rare around these stars, but the great abundance of such stars might compensate. Even multiple star systems seem reasonably capable of having habitable planets.

Our studies of stars suggest that many or most stars could *potentially* have orbiting habitable worlds. Whether they *do* may depend on a number of other factors. To understand these factors, we must investigate the discoveries of extrasolar planets and what these discoveries have taught us about the formation and nature of other planetary systems.

11.2 EXTRASOLAR PLANETS: DISCOVERIES AND IMPLICATIONS

In Chapter 3, we discussed both theoretical and observational reasons why we expect that planets are common. On the theoretical side, current ideas about how stars and planets form suggest that the same general processes that gave birth to the planets of our solar system ought to occur around most other young stars. For example, it seems almost inevitable that a collapsing cloud of gas and dust will form a star surrounded by a spinning disk of debris (see Figure 3.19).

On the observational side, two types of evidence point to planets being common. First, we have seen flat disks of material orbiting many other newborn stars (see Figure 3.20). The disks of young stars often contain a few percent of the mass of their stars, which is more than sufficient for constructing planets. (The combined mass of all the worlds in our solar system amounts to less than 0.2% of the Sun's mass.) Based on the observations made to date, it appears that at least a quarter to half of all young stars are surrounded by such disks. Although this fact does not tell us directly whether planets are likely to form in such disks, it certainly lends support to our theory of how stars and planets form.

The second and more clear-cut form of observational evidence comes in actual planets discovered around other stars. As of mid-2006, nearly 200 extrasolar planets had been found around more than 160 different stars. For 19 of these stars, multiple planets (as many as three) had been found.

THINK ABOUT IT By the time you read this book, the number of known extrasolar planets will have grown larger still. Do a quick Web search on "extrasolar planets" to find the current number. How many more extrasolar planets have been discovered since mid-2006?

These discoveries tell us that planets must be at least moderately common, but they do not yet tell us anything directly about the possibility of habitable planets, because nearly all the extrasolar planets discovered are much closer in mass to the jovian planets in our solar system than to terrestrial worlds like Earth. This fact is probably just due to technological limitations; detecting any extrasolar planet is quite difficult, and small ones are correspondingly harder to find than larger ones. Improving technology will likely make it possible for us to detect Earth-size planets soon, and in the meantime we are learning important lessons about potential habitability from the orbital characteristics of the larger planets already discovered. In this section, we will investigate planet detection techniques and the astrobiological implications of what we have learned so far.

• How do we detect planets around other stars?

The first clear-cut discovery of a planet around another Sun-like star—a star called 51 Pegasi—came in 1995. The discovery was made by Swiss astronomers Michel Mayor and Didier Queloz and was soon confirmed by a team led by Geoffrey W. Marcy and R. Paul Butler. Many more extrasolar planets have been discovered since that time, and the hunt for worlds around other stars has become an international effort involving many teams of astronomers. Several different planet-finding strategies have been used or proposed. If we strip away the details, however, there are really only two basic ways to search for extrasolar planets:

1. *Directly:* Pictures or spectra of the planets themselves constitute direct evidence of their existence.
2. *Indirectly:* Precise measurements of stellar properties (such as position, brightness, or spectra) may indirectly reveal the effects of orbiting planets, even without seeing the planets themselves.

Until technology makes direct searches routine, we can rely only on indirect searches that look at stars, seeking effects that can be attributed to orbiting planets.

Direct detection is preferable in principle, because it can tell us far more about the planet's properties. However, current telescopes are just reaching the point where direct detection is possible, and only a handful of such detections have been made to date. Let's therefore start by considering the indirect techniques responsible for most extrasolar planet discoveries, and then come back to direct detection.

Gravitational Tugs To date, nearly all extrasolar planets have been detected by observing the gravitational tugs they exert on the stars they orbit.

This type of detection is indirect because we discover the planets by observing the stars they orbit without actually seeing the planets themselves.

Although we usually think of a star as remaining still while planets orbit around it, that is only approximately correct. In reality, all the objects in a star system, including the star itself, orbit the system's **center of mass**. To understand this concept, think of a waiter carrying a tray of drinks. To carry the tray, he places his hand under the spot at which it balances—its center of mass. If the tray has a heavy glass of water off to one side, he will place his hand a little to that side of the tray's center. Because the Sun is far more massive than all the planets combined, the center of mass of our solar system lies very close to the Sun—but not exactly at the Sun's center.

We can see how this fact allows us to discover extrasolar planets by imagining the viewpoint of extraterrestrial astronomers observing our solar system from afar. Let's start by considering only the influence of Jupiter, which exerts a much stronger gravitational tug on the Sun than the rest of the planets combined. The center of mass between the Sun and Jupiter lies just outside the Sun's visible surface (Figure 11.7). Thus, what we usually think of as Jupiter's 12-year orbit around the Sun is really a 12-year orbit around their mutual center of mass; we generally don't notice this fact because the center of mass is so close to the Sun itself. In addition, because the Sun and Jupiter are always on opposite sides of the center of mass (otherwise it wouldn't be a "center"), the Sun must orbit this point with the same 12-year period as Jupiter. The Sun's orbit traces out only a very small circle (or ellipse) with each 12-year period, because the Sun's average orbital distance is barely larger than its own radius, and with sufficiently precise measurements, extraterrestrial astronomers could detect this orbital movement of the Sun. They could thereby deduce

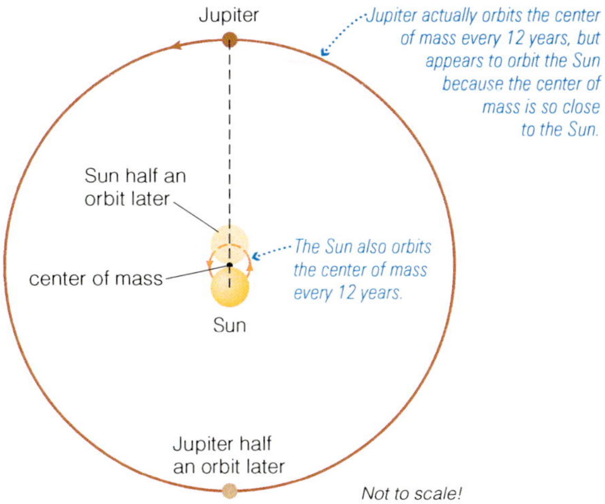

Figure 11.7

This diagram shows how both the Sun and Jupiter actually orbit around their mutual center of mass, which lies very close to the Sun. The diagram is not to scale; the sizes of the Sun and its orbit are exaggerated about 100 times compared to the size shown for Jupiter's orbit.

SPECIAL TOPIC 11.1: THE NAMES OF EXTRASOLAR PLANETS

The planets in our solar system have familiar names rooted in mythology. Unfortunately, there's not yet a well-accepted scheme for naming extrasolar planets. As a result, we generally refer to extrasolar planets by the star they orbit, such as "the planet orbiting the star named. . . ." Worse still, the stars themselves often have confusing or even multiple names, reflecting naming schemes used in star catalogs made by different people at different times in history.

A few hundred of the brightest stars in the sky carry proper names from ancient times. Many of these names are Arabic—such as Betelgeuse, Algol, and Aldebaran—because of the work of the Arabic scholars of the Middle Ages. In the early 1600s, German astronomer Johann Bayer developed a system that gave names to many more stars. In Bayer's system, each star gets a name based on its constellation and a Greek letter indicating its ranking in brightness within that constellation. For example, the brightest star in the constellation Andromeda is called Alpha Andromedae, the second brightest is Beta Andromedae, and so on. Of course, because there are only 24 letters in the Greek alphabet, this system works for only the 24 brightest stars in each constellation. About a century later, English astronomer John Flamsteed published a more extensive star catalog in which he used numbers once the Greek letters were exhausted. For example, 51 Pegasi gets its name from Flamsteed's catalog. (Flamsteed's numbers are based on position within a constellation rather than brightness, so a star may have a different designation in Bayer's system than in Flamsteed's.)

As more powerful telescopes made it possible to discover more and fainter stars, astronomers developed many new star catalogs. The names we use today usually come from one of these catalogs. For example, the star HD209458 gets its name from the Henry Draper (HD) catalog and has the number 209458 in that catalog. You may also see stars with numbers preceded by other catalog names, including Gliese, Ross, and Wolf; these catalogs are also named for the astronomers who compiled them. Moreover, because the same star is often listed in several of these catalogs, a single star can have several different names. Some of the newest planets orbit stars so faint that they have not been previously cataloged. They then carry the name of the observing program that discovered them, such as TrES-1 for the first discovery of Trans-Atlantic Exoplanet Survey, or OGLE-TR-132b for the 132nd object scrutinized by the Optical Gravitational Lensing Experiment.

Objects orbiting other stars usually carry the star name plus a letter denoting their order of discovery around that star. The host star itself is designated "a." Thus, for example, HD209458b is the first object discovered to be orbiting star number 209458 in the Henry Draper catalog; Upsilon Andromedae d is the third object discovered to be orbiting what should be the 20th brightest star (because upsilon is the 20th letter in the Greek alphabet) in the constellation Andromeda (although it isn't the 20th brightest because Bayer's measurements were not all correct). Unfortunately, this scheme doesn't tell you whether the orbiting object is thought to be a planet or another star.

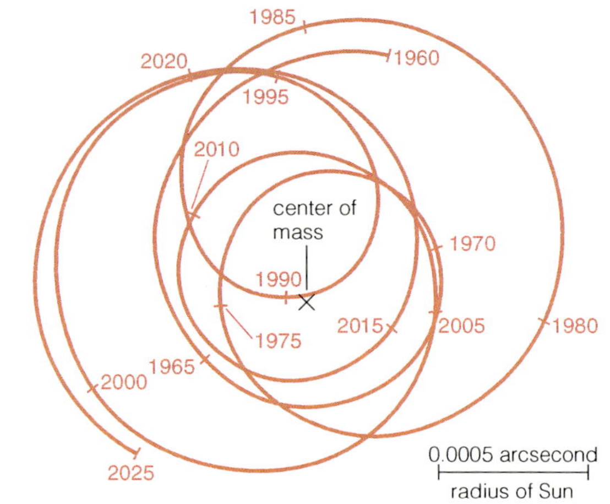

Figure 11.8

This diagram shows the orbital path of the Sun around the center of mass of our solar system as it would appear from a distance of 30 light-years away; the graph shows the motion between 1960 and 2025. Notice that the entire range of motion during this period is only about 0.0015 arcsecond, which is almost 100 times smaller than the angular resolution of the Hubble Space Telescope. Nevertheless, if alien astronomers could measure this motion, over time they could learn of the existence of planets in our solar system even without being able to detect the planets themselves.

the existence of Jupiter, even without having observed Jupiter itself. They could even determine Jupiter's mass from the Sun's orbital characteristics as it goes around the center of mass. Remember that for any particular separation between the Sun and a planet, a more massive planet pulls the center of mass farther from the Sun's center, thereby giving the Sun a larger orbit and a faster orbital speed around the center of mass.

Now let's add in the effects of Saturn, which exerts the second-greatest gravitational tug on the Sun. Saturn takes about 29 years to orbit the Sun, so by itself it would cause the Sun to orbit their mutual center of mass about every 29 years. However, because Saturn's influence is secondary to that of Jupiter, this 29-year period appears as an added complication on top of the Sun's 12-year orbit around its center of mass with Jupiter. In other words, every 12 years the Sun would return to *nearly* the same orbital position around its mutual center of mass with Jupiter, but the precise point of return would make its own even tinier orbital circle (or ellipse) with Saturn's 29-year period. By measuring this motion carefully from afar, an extraterrestrial astronomer could deduce the existence and masses of both Jupiter and Saturn after a few decades of observing.

The other planets also exert gravitational tugs on the Sun, which create further complications to the Sun's orbital motion around the solar system's center of mass (Figure 11.8). These extra effects become increasingly difficult to measure in practice, but in principle they would allow an extraterrestrial astronomer to discover all the planets in our solar system. If we turn this idea around, you'll realize that it means we can search for planets in other star systems by carefully watching for the tiny orbital motion of a star around the center of mass of its star system.

There are two techniques for observing the small orbital motion of a star caused by the gravitational tugs of planets. The first is straightforward: We make precise measurements of a star's position to see if it is, in fact, moving about in a way that tells us it is being tugged on by planets. This method is usually called the **astrometric technique**, because it requires measuring stellar positions in the sky (*astrometric* means "measurement of the stars"). The second scheme, known as the **Doppler technique**, detects orbital motion by searching for slight shifts in the positions of spectral lines in a star's spectrum. We next consider the advantages and limitations of these two techniques.

The Astrometric Technique The astrometric technique has been used for many decades to identify binary star systems, since two orbiting stars will move periodically around their center of mass. The technique works especially well for binary systems in which the two stars are not too close together, because the stellar motions tend to be larger in those cases. In the case of planet searches, however, the expected stellar motion is much more difficult to detect.

As an example, from a distance of 10 light-years, a Jupiter-size planet orbiting 5 AU from a Sun-like star would cause its star to dance slowly over a side-to-side angular distance of only about 0.003 arcsecond—approximately the width of a human hair seen from a distance of 5 kilometers. Remarkably, with careful telescope calibration, astronomers can now measure displacements this small, and instruments currently under development will be five to ten times better yet. However, two other complications add to the difficulty of the astrometric technique.

First, the farther away the star, the smaller its side-to-side movement will appear. For example, while Jupiter causes the Sun to move by about 0.003 arcsecond as seen from 10 light-years away, the observed motion is only half as large when seen from 20 light-years away and one-tenth as large when seen from 100 light-years away. Thus, the astrometric technique works best for massive planets around relatively nearby stars.

The second complication arises from the time required to detect a star's motion. It is obviously easier to detect larger movements than smaller ones, and a planet with a larger orbit has a larger effect on its star. To understand why, consider what would happen if Jupiter moved farther from the Sun. Because the center of mass of the solar system is nearly at the balance point between the Sun and Jupiter, moving Jupiter outward would also cause the center of mass to move farther from the Sun. With the center of mass located farther from the Sun, the Sun's orbit around the center of mass—and hence its side-to-side motion as seen from a distance—would be larger. Thus, the astrometric technique works best not only for massive planets, but also in particular for massive planets that are far from their stars. However, Kepler's third law [Section 2.2] tells us that a more distant planet takes longer to complete its orbit, which means that the stellar motion it causes also takes longer. For example, while Jupiter causes the Sun to move around the center of mass with a 12-year period, Neptune's effects on the Sun show up with the 165-year period of Neptune's orbit. It would probably take a century or more of patient observation to be confident that stellar motion was occurring in a 165-year cycle.

As a result of these complications, the astrometric technique has been used to detect only two of the 170 extrasolar planets known through early 2006. However, we expect the astrometric technique to prove much more valuable in the future. Astronomers are making rapid progress in developing a technology called *interferometry* that allows two or more telescopes to work together to achieve much higher angular resolution than either telescope individually. This technology has been used with radio telescopes for decades, and it is now being applied successfully to infrared and visible-light telescopes. Several major observatories, including the twin Keck observatories on Mauna Kea, may soon be capable of using interferometry to detect at least some planets through the astrometric technique. In addition, both NASA and the European Space Agency are working on space missions that will be able to search for planets using the astrometric technique. The European mission, called *GAIA,* is slated for launch in 2010 and has the goal of performing astrometric observations of a billion stars in our galaxy with an accuracy of 10 *micro*arcseconds. (A microarcsecond is one-millionth of an arcsecond.) NASA hopes to follow not long after with its *Space Interferometry Mission* (*SIM*). As currently planned, *SIM* will be capable of measuring stellar positions to a precision of just 1 microarcsecond—ten times better than *GAIA*'s precision and good enough to detect stellar motion caused by Earth-size planets around the nearest few dozen stars and motion due to Jupiter-size planets orbiting stars as far as 3,000 light-years away.

The Doppler Technique The *Doppler technique* offers another way to find the small movement of a star caused by orbiting planets, and it has proved very successful: The vast majority (more than 90%) of the extrasolar planets found to date have been found with this technique. Rather than looking for the side-to-side motion of a star against the background of the sky, as in the astrometric techniques, the Doppler technique involves studying a star's spectrum to look for telltale signs that the star is moving around a center of mass. The technique is based on searching for small shifts in the wavelengths of lines in the spectrum caused by what we call the **Doppler effect.**

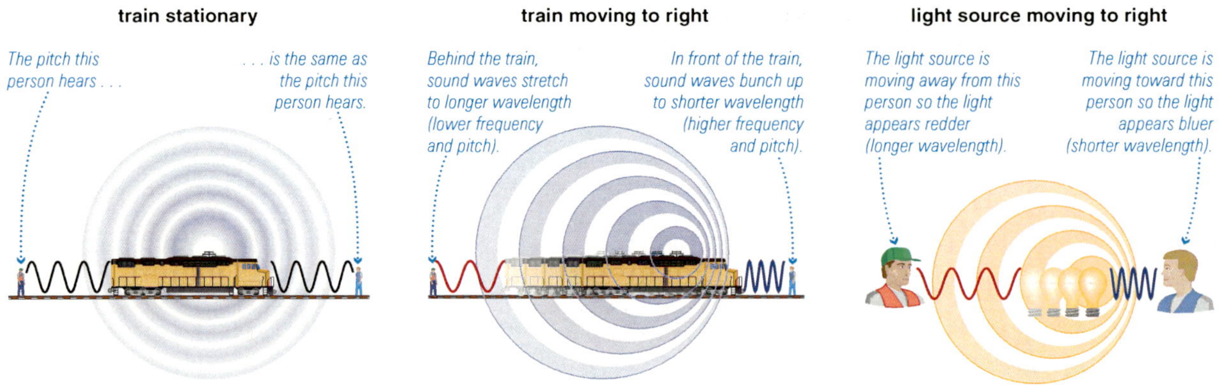

a The whistle sounds the same no matter where we stand near a stationary train.

b For a moving train, the sound you hear depends on whether the train is moving toward you or away from you.

c We get the same basic effect from a moving light source (although the shifts are usually too small to notice by eye).

Figure 11.9

The Doppler effect. Each circle represents the crests of sound (or light) waves going in all directions from the source. For example, the circles from the train might represent waves emitted 0.001 second apart.

You've probably noticed the Doppler effect on the sound of a train whistle near train tracks. (You can also notice the Doppler effect with emergency sirens or even just with the "buzz" of a fast car as it goes past you.) If the train is stationary, the pitch of its whistle sounds the same no matter where you stand (Figure 11.9a). But if the train is moving, the pitch will sound higher when the train is coming toward you and lower when it's moving away from you. Just as the train passes by, you can hear the dramatic change from high to low pitch—a sort of "weeeeeeee–ooooooooooh" sound. To understand why, we have to think about what happens to the sound waves coming from the train (Figure 11.9b). When the train is moving toward you, each pulse of a sound wave is emitted a little closer to you. The result is that waves are bunched up between you and the train, giving them a shorter wavelength and higher frequency (pitch). After the train passes you by, each pulse comes from farther away, stretching out the wavelengths and giving the sound a lower frequency.

The Doppler effect causes similar shifts in the wavelengths of light (Figure 11.9c). If an object is moving toward us, then its entire spec-

Figure 11.10

The Doppler technique for discovering extrasolar planets.

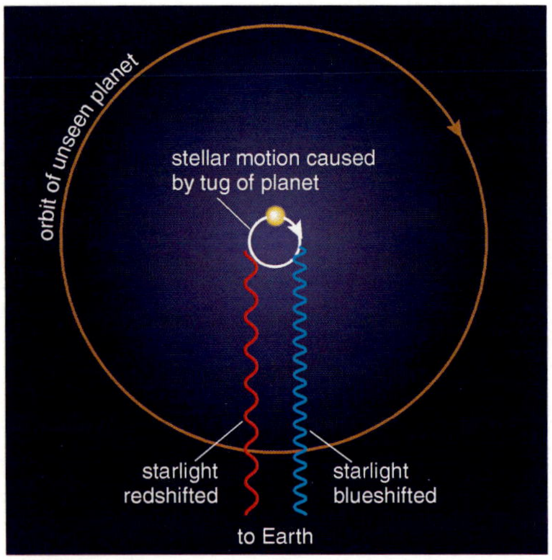

a Doppler shifts allow us to detect the slight motion of a star caused by an orbiting planet.

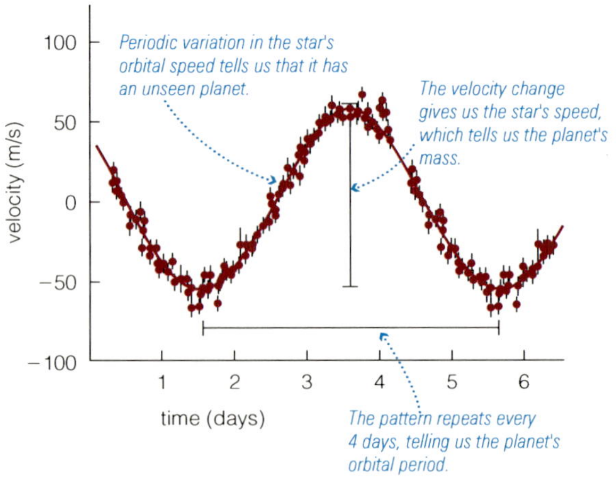

b A periodic Doppler shift in the spectrum of the star 51 Pegasi shows the presence of a large planet with an orbital period of about 4 days. Dots are actual data points; bars through dots represent measurement uncertainty.

trum—including the spectral lines within it—is shifted to shorter wavelengths (the spectral lines serve as reference lines for measuring the shift). Because shorter wavelengths of visible light are bluer, the Doppler shift of a star coming toward us is called a **blueshift.** If an object is moving away from us, its light is shifted to longer wavelengths; we call this a **redshift,** because longer wavelengths of visible light are redder. The faster the object is moving, the greater the amount of its blueshift or redshift.

The Doppler effect can be used to find planets in the following manner: If an orbiting planet causes its star to move alternately slightly toward and away from us (as part of its orbital motion), the starlight should shift alternately toward the blue and toward the red (Figure 11.10a). The 1995 discovery of a planet orbiting 51 Pegasi occurred when this star was found to be moving with a rhythmic wobble corresponding to an orbital speed of 57 meters per second (Figure 11.10b). The 4-day period of the star's motion tells us the orbital period of its planet.

Current techniques can measure a star's velocity to within about 1 meter per second—walking speed—which corresponds to a Doppler wavelength shift of only 1 part in 300 million. Thus, we can find planets that exert a considerably smaller gravitational tug on their stars than the tug of the planet orbiting 51 Pegasi. Moreover, by carefully analyzing Doppler shift data, we can learn of the existence of a planet and also determine its orbital properties, including its average distance from its star and the shape of its orbit (Figure 11.11). In some cases, the Doppler data are good enough to tell us whether a star has more than one planet.

The Doppler technique also tells us about planetary masses, but with an important caveat. Doppler shifts reveal only the part of a star's motion directed toward or away from us. Thus, for example, a planet whose orbit we view face-on does not cause any Doppler shift in the spectrum of its star, making it impossible to detect the planet with the Doppler technique, let alone to determine its mass (Figure 11.12a). We can observe Doppler shifts in the star's spectrum only if we view its planet orbiting at some

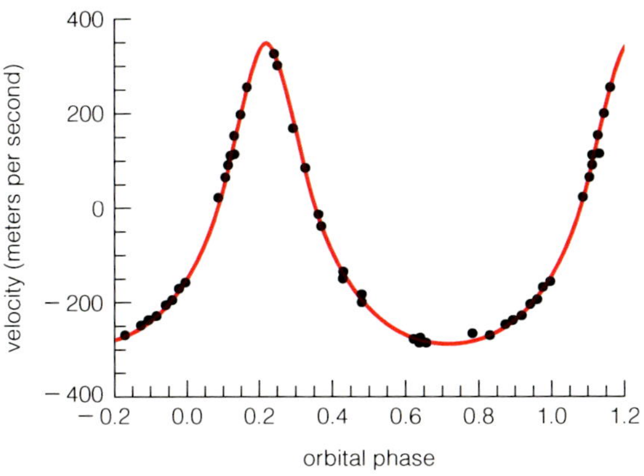

Figure 11.11

The Doppler shifts measured for the star 70 Virginis (from the work of Marcy and Butler). Note the uneven nature of the change in velocity, which tells us that the planet causing the Doppler shift is in a highly eccentric (elliptical) orbit.

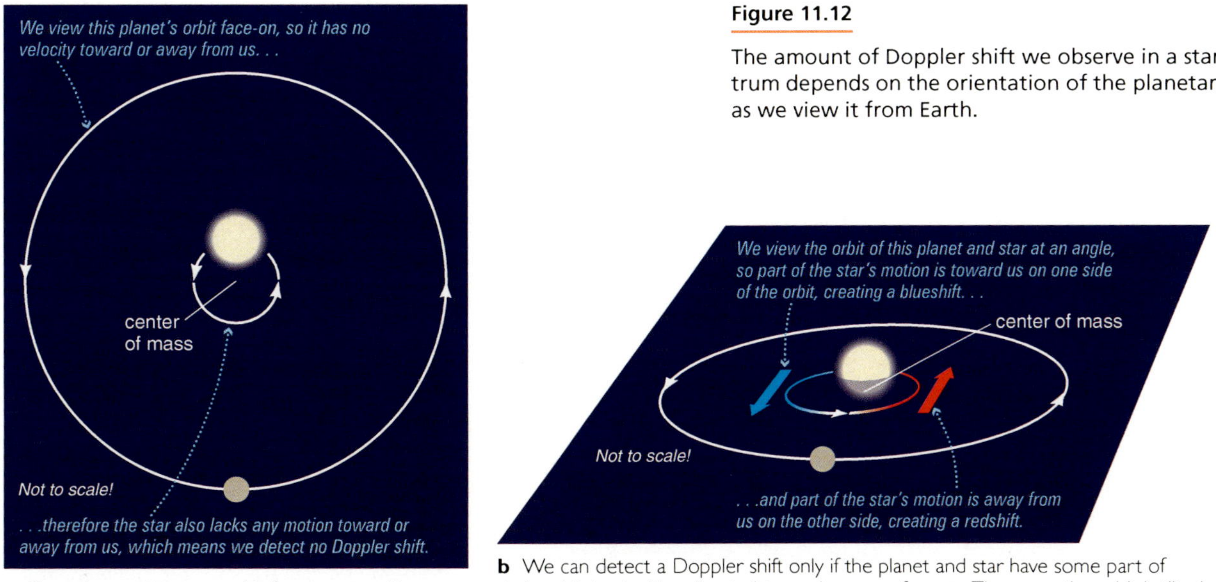

a If we view a planetary orbit face-on, we will not detect any Doppler shift at all.

b We can detect a Doppler shift only if the planet and star have some part of their orbital velocities directed toward or away from us. The more the orbit is tilted toward edge-on, the greater the shift we'll observe.

Figure 11.12

The amount of Doppler shift we observe in a star's spectrum depends on the orientation of the planetary orbit as we view it from Earth.

angle other than face-on (Figure 11.12b), and the Doppler shift will tell us the star's full orbital velocity only if we are viewing it precisely edge-on. As a result, a planetary mass that we infer from its star's Doppler shift will be accurate only for a planet that we see in an edge-on orbit. In all other cases, the Doppler shift underestimates the true orbital speed of the star and therefore also causes us to underestimate the true mass of the planet. The more the orbit is tilted away from edge-on, the greater the degree of the underestimate. Thus, planetary masses inferred from the Doppler technique alone are actually *minimum* possible masses (or "lower limits") for these planets.

We can determine a precise mass only if we somehow know that we are viewing an edge-on orbit or if we have some way to measure the precise orbital angle. For example, if we also know the planet's side-to-side motion from the astrometric technique, we can combine this information with the toward and away motions from the Doppler technique to determine the orbital angle. Unfortunately, we rarely have such knowledge, which means that nearly all planetary masses found through the Doppler technique are really minimum masses rather than actual masses. Nevertheless, statistical arguments show that in two out of three cases, the planet's true mass will be no more than double this minimum mass, so the minimum mass is still a relatively good estimate.* Of course, this also means that a third of the planets discovered with the Doppler technique have masses that are more than double the minimum mass. About 6% of the time, the planet's actual mass will be more than ten times the minimum mass, which may raise the question of whether it is truly a planet or a very small star.

The Doppler technique is powerful, but it has limits. In particular, it is best suited for identifying massive planets that orbit relatively close to their star. This limitation arises because gravity weakens with distance, so a planet of a given size pulls harder on its star—making the star move faster—if it is closer. Moreover, because a close-in planet has a much shorter orbital period than a more distant planet, it takes a lot less time to observe the periodic Doppler shifts caused by close-in planets. For example, while it takes only a few weeks of observation to detect a planet with a 4-day period like the one orbiting 51 Pegasi, it would take 12 years to observe just a single orbital cycle of a planet in an orbit like that of Jupiter around our Sun. The Doppler technique also presents a practical difficulty: Current technology allows it to be applied to just one star at a time (with a particular telescope), so only a relatively small number of stars can be studied with this technique.

The limitations of the Doppler technique explain what may at first seem like surprising facts: Most of the extrasolar planets discovered to date orbit relatively close to their stars, and we have yet to discover any planets at all with Earth-like masses. Both these facts may simply be *selection effects* of the Doppler technique; that is, the technique tends to find (or "select") massive planets in close orbits much more easily than any other type of planet. Planets with masses similar to Earth would have such weak gravitational effects on their stars that we could not use the

---

*You can understand the statistical argument if you have studied trigonometry: The actual mass is the minimum mass divided by the *cosine* of the orbital inclination (where edge-on is an angle of 0° and face-on is an angle of 90°). Thus, the actual mass is more than double the minimum mass only for cosines smaller than $\frac{1}{2}$, which means for angles between 60° and 90°. By random chance, only about one-third of all stars will have such large orbital angles. Similarly, the actual mass is more than ten times the minimum mass for cosines smaller than 0.1, which means angles between about 84.3° and 90°; these angles occur randomly about 6% of the time.

Doppler technique to find them with current technology, while planets orbiting far from their stars have such long orbital periods that it might take decades of observations to detect them. Thus, the current lack of evidence for Earth-mass planets or jovian-mass planets in distant orbits does not necessarily mean that such planets are rare.

Transits A third indirect way to detect distant planets does not require the observation of the effects of gravitational tugs at all. Instead, it relies on searching for slight changes in a star's brightness caused by a planet passing in front of it. If we were to examine a large sample of stars with planets, a small number of them—typically one in several hundred—will by chance be aligned in such a way that one or more of its planets will pass directly between us and the star once each orbit. The result is a **transit**, in which the planet appears to move across the face of the star. We occasionally witness this effect in our own solar system when Mercury or Venus crosses in front of the Sun. The result, seen through a telescope, is a small black dot transiting the Sun's bright disk (Figure 11.13). This effect is impossible for us to photograph for other stars, which appear in our telescopes only as points of light. However, as the planet crosses the star's face, it blocks a little of the star's light. The resulting tiny decrease in the star's brightness may be measurable.

Figure 11.13

On June 8, 2004, a transit of Venus was visible throughout much of Europe and Asia. In this photo, Venus appears as a black dot. It blocked out approximately 0.1% of the Sun's light for about 6 hours as it slowly transited across our home star. In principle, we should be able to detect such drops in brightness when extrasolar planets cross in front of their stars. Note that the amount of sunlight blocked out is about ten times as much as would be observed from a great distance, since—being so close to us—Venus looks relatively large. (Think of how the Moon can completely block the Sun as seen from Earth, but would be a very tiny dot against the Sun if viewed from much farther away.)

THINK ABOUT IT What kind of planet is most likely to cause a transit across its star that we could observe from Earth: (a) a large planet close to its star; (b) a large planet far from its star; (c) a small planet close to its star; or (d) a small planet far from its star? Explain.

COSMIC CALCULATIONS 11.1

Finding Masses of Extrasolar Planets

An object's momentum is defined as its mass m times its velocity v; like angular momentum [Section 3.4], momentum must be conserved. Consider a star with a single planet. Because the center of mass remains stationary between them (see Figure 11.7), the system has no momentum relative to this center of mass. The star's momentum $(m_{star} \times v_{star})$ must therefore be equal in magnitude (but opposite in direction) to the planet's momentum $(m_{planet} \times v_{planet})$:

$$m_{star}v_{star} = m_{planet}v_{planet}$$

Solving, the planet's mass is:

$$m_{planet} = \frac{m_{star}v_{star}}{v_{planet}}$$

We generally know the star's mass from its spectral type. The Doppler technique can tell us the star's velocity (v_{star}) and the planet's orbital period p; we can use the latter to find the planet's velocity (v_{planet}) *if* we already know its average orbital distance a_{planet}. Fortunately, we can generally calculate a with Newton's version of Kepler's third law (see Cosmic Calculations 7.1). Once we know a_{planet}, we find v_{planet} by realizing that each orbit of the planet represents a distance of $2\pi a$ covered in a time p (and remembering that velocity is distance divided by time):

$$v_{planet} = \frac{2\pi a_{planet}}{p_{planet}}$$

We now substitute this expression into the mass equation from above:

$$m_{planet} = \frac{m_{star}v_{star}p_{planet}}{2\pi a_{planet}}$$

If we use velocity data from the Doppler technique, this formula tells us the *minimum* mass of the planet.

Example: Estimate the mass of the planet orbiting 51 Pegasi, which has an orbital period of 4.23 days ($p = 3.65 \times 10^5$ s) and an average orbital distance $a = 7.82 \times 10^9$ m.

Solution: The data in Figure 11.10b show the star's orbital velocity to be 57 m/s. Plugging in the values to solve for the planet's minimum mass, we get:

$$\begin{aligned} m_{planet} &= \frac{m_{star}v_{star}p_{planet}}{2\pi a_{planet}} \\ &= \frac{(2.12 \times 10^{30}\text{ kg}) \times \left(57\frac{\text{m}}{\text{s}}\right) \times (3.65 \times 10^5\text{ s})}{2\pi \times (7.82 \times 10^9\text{ m})} \\ &\approx 9 \times 10^{26}\text{ kg} \end{aligned}$$

This is about half of Jupiter's mass (which is 1.9×10^{27} kg).

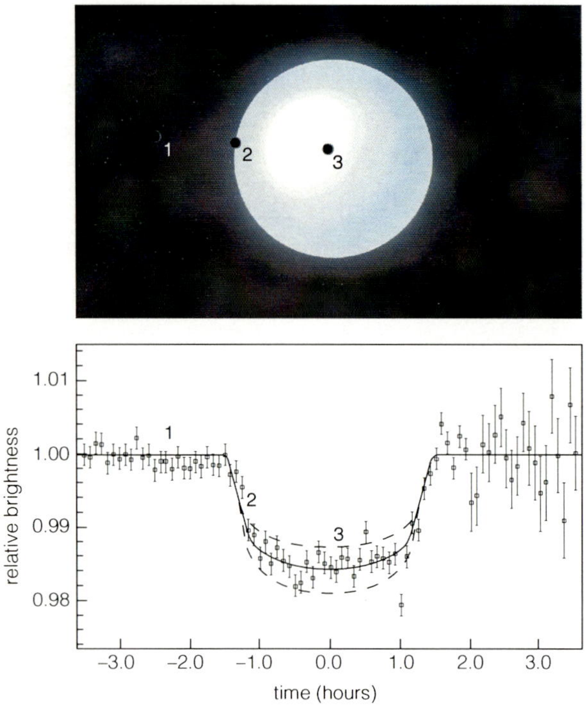

Figure 11.14

Careful measurements of the brightness of the star called HD209548 revealed that an orbiting jovian planet passes directly in front of it as seen from Earth; the numbers 1–3 on the art (top) correspond to the indicated positions on the graph.

Detecting planets through transits requires many repeated observations, because most stars exhibit intrinsic variations in luminosity. To be confident that a dip in brightness is caused by an orbiting planet rather than by variability in the star itself, we need to see the same dimming repeated multiple times. If the repeated dimming occurs with a regular period, then it is likely telling us the orbital period of a transiting planet. We can then calculate the planet's orbital distance and mass, and the amount of the drop in the star's brightness allows us to calculate the planet's radius. For example, observations show that a planet transits across the face of the star HD209548 every $3\frac{1}{2}$ days, causing a 1.7% drop in the star's brightness for about 2 hours (Figure 11.14). This tells us that the area of the planet's visible disk must be 1.7% of the area of the star's visible disk; if you remember that the area of a circle is πr^2, you'll realize that the planet's radius must be $\sqrt{0.017} = 0.13$, or 13%, of the star's radius. These data have told us that the planet has a mass, radius, and density consistent with what we expect if it really is jovian in nature. More detailed studies of how the starlight changes both during the transit and when the planet passes behind the star (called an *eclipse*) have even allowed astronomers to learn a little about the planet's atmosphere.

For an Earth-size planet, the dimming caused by a transit would be less than about 0.01% of a star's normal brightness, an amount that is difficult to measure with ground-based telescopes affected by atmospheric turbulence. Space-based telescopes, however, should be able to uncover such subtle dimming. NASA already has a mission in development, called *Kepler,* designed to look for transits caused by Earth-size (or even smaller) planets through careful monitoring of the brightness of 100,000 stars during the course of 4 years. After its launch (planned for 2008), *Kepler* should tell us whether or not solar systems in which small planets exist are plentiful. It therefore should give us our first good inventory of extrasolar planetary systems and of Earth-size planets.

Gravitational Lensing and Other Indirect Techniques All of the planet discovery techniques we've discussed so far favor the detection of planets that orbit close to their star. Ground-based searches are also most sensitive to large planets. However, a technique known as **gravitational lensing** is able to detect even Earth-size planets in large orbits, and can find them despite the fact that they're in star systems tens of thousands of light-years away.

Gravitational lensing makes use of one of the most important predictions of Einstein's general theory of relativity: Mass distorts space. The consequence is that light rays bend slightly when passing near any large mass. The light rays are traveling in straight lines, but because matter warps space, a straight line appears bent. The effect is noticeable only in the presence of strong gravitational fields, so you'd have a hard time measuring the additional bending of a flashlight beam as it, say, grazed the rooftop of your house. But the Sun has sufficient gravity to bend starlight that passes near it, an effect first measured during a solar eclipse only a few years after Einstein first published his theory. In essence, the Sun, and for that matter, any star, acts like a lens that both magnifies and distorts the light from any object that may be behind it.

Two international teams of astronomers are using this subtle effect to search for planets in the star-choked inner regions of the Milky Way. They do so by monitoring the brightness of approximately 100 million stars located in the rich star fields near the galaxy's central regions. Because they use telescopes around the globe, they can monitor these stars continuously. About twice a day, on average, one star will, by chance, slowly pass in front of another. The two stars are not likely to be close to one another in space; they simply line up by chance for a few weeks because of the motions of stars and the rotation of the galaxy. When these alignments occur, the nearer star acts as a lens to increase the brightness (usually by a factor of two or three) of the star farther away. If the nearer star also has a planet, there's a chance that it will be suitably aligned to act as a lens as well, producing a secondary brightening of the lensed background star.

As of early 2006, three extrasolar planets had been found using this method. While that's only a tiny fraction of the known extrasolar worlds, the gravitational lensing scheme broke the record for the least-massive planet detected around a normal star—finding one that is only five times the mass of Earth (see Figure 1.2).

These successes have inspired astronomers to seek other ways of searching for planets indirectly. For example, another strategy is to search not for the planets or their effect on their host stars, but for visible disturbances of the large (and therefore, somewhat easier to see) disks of dust in which they form. The disk around the star Beta Pictoris, for example, shows ripples and gaps that are the clear fingerprints of planets within the disk. Other astronomers are searching for the thermal emission from the impacts of accreting planetesimals, and even for the kinds of emission known to come from the magnetospheres of jovian planets. As we learn more about extrasolar planets, new search methods are sure to arise.

Direct Detection While upcoming missions like *Kepler* or the *Space Interferometry Mission* (*SIM*) may find a few Earth-*size* planets by using indirect techniques, they still won't tell us whether these planets are Earth-*like* in terms of habitability. For that we need to study the planets directly.

To see why direct detection of extrasolar planets is so difficult, recall the 1-to-10-billion scale model of the solar system that we discussed in Chapter 3 (see Figure 3.3). Remember that on this scale, the Sun is the size of a grapefruit, Earth is a pinhead orbiting 15 meters away, and Jupiter is a marble orbiting 80 meters away—while the distance to the nearest stars is equivalent to the distance across the United States. Thus, seeing an Earth-like planet orbiting the nearest star besides the Sun would be like looking from San Francisco for a pinhead orbiting just 15 meters from a grapefruit in Washington, D.C., and seeing a Jupiter-like planet wouldn't be much easier. The task is further complicated by the fact that a Sun-like star would be a *billion* times as bright as the light reflected from any of its planets. Because even the best telescopes blur the light from stars at least a little, the glare of scattered starlight tends to overwhelm the small blips of planetary light.

Because of these difficulties, only three planets so far have actually been photographed by direct imaging (Figure 11.5 shows one of them); that is, by finding them in photographs made with big telescopes. All these planets are large (many times Jupiter's mass), and all are at great distances from their home star (more than the distance from the Sun to Pluto), making it easier to spot them. But it's presently impossible to find planets similar to those of our own solar system with these straightforward, direct techniques.

One way to improve on this discouraging situation is to observe at infrared wavelengths. Planets are too cool to emit any visible light of their own, so they shine only by the light they reflect from their stars, but they emit reasonable amounts of infrared light. They are still far dimmer than

Figure 11.15

This artist's rendering shows one possible configuration for the second of NASA's planned *Terrestrial Planet Finder* (*TPF*) telescopes. It consists of five separate infrared telescopes flying in formation so that they can work together as an interferometer.

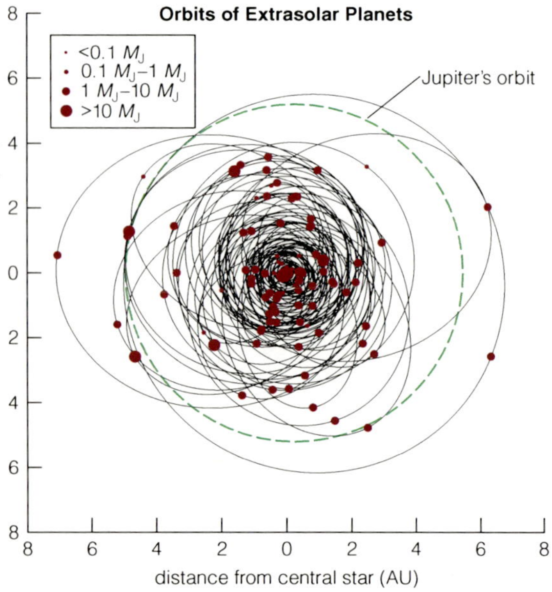

Figure 11.16

This diagram shows the orbits of the first 146 known extrasolar planets superimposed on each other, as if all the planets orbited a single star. The sizes of the dots indicate the minimum masses of the planets; the dots are located at the aphelion (farthest) point for each orbit.

their stars, but while the Sun is a billion times brighter than Jupiter in visible light, it is only about 100,000 times brighter in the (far) infrared. This substantial improvement makes direct detection much more practical with infrared light than with visible light, and the planets photographed to date have all been imaged in the infrared. In addition, scientists are developing new technologies such as *nulling interferometry*, designed to cancel (or "null") the light from a star so that we can more easily see its orbiting planets. NASA plans to test nulling technology on the *Space Interferometry Mission* (*SIM*), which may thereby be able to make crude images of a few Jupiter-size planets, though probably not of Earth-size planets.

NASA's planned follow-up mission to *SIM*, the *Terrestrial Planet Finder* (*TPF*), is tentatively envisioned as a double-barreled approach to directly imaging extrasolar planets, consisting of two, completely separate space-based telescopes. The first of these would be a more or less standard telescope, but one that is outfitted with a *coronagraph*—a small disk within the telescope that can block the light from a distant star so that we can see surrounding planets. Years after the coronagraph telescope is in orbit, *TPF*'s second mission would leave the launchpad; as currently envisioned, it would have five infrared telescopes flying in tightly controlled formation as an interferometer (Figure 11.15), using nulling interferometry to cancel out light from the star so that we can see planets. The European Space Agency (ESA) is considering a similar mission, called *Darwin*, on roughly the same time scale, and NASA and ESA are discussing possible ways of combining the two missions, perhaps with other international partners as well. In principle, either *TPF* or *Darwin* should have the power to conduct a conclusive search for Earth-size planets around 150 nearby, Sun-like stars. If the planets are there, these missions should see them. The images will be very low resolution—perhaps no more than three or four dots (pixels) for a typical planet—but the collected light should be sufficient to allow us to obtain reasonable spectra of the planets. As we'll discuss shortly, spectroscopy may allow us to determine whether a planet is habitable, and perhaps even whether it has life.

Later in this century, astronomers hope for even more powerful interferometers either in space or on the surface of the Moon. It's certainly possible that, within the lifetimes of today's college students, we'll have optical interferometers with dozens of telescopes spread across hundreds of kilometers. Such telescopes will not only be able to detect Earth-like planets around nearby stars but will also be able to obtain fairly clear images of those planets. We might see whether they have continents and oceans like Earth and perhaps be able to watch seasonal changes. Just a few centuries ago, we knew so little of the heavens that our ancestors could still imagine planets simply as "wandering stars." A mere decade ago, we had good images of the planets in our solar system but still had no conclusive evidence regarding the existence of planets elsewhere in the universe. Before the current century is out, we may have arguably better knowledge of planets around distant stars than the ancient Greeks had of our own Earth.

• What are the properties of extrasolar planets discovered so far?

Given our discussion of planet detection techniques, you probably won't be surprised by the fact that most of the planets discovered to date around hydrogen-burning stars are far larger than Earth. However, scientists have been surprised by the orbital characteristics of many of these planets.

Hot Jupiters Figure 11.16 shows the orbits of the first 146 known extrasolar planets all superimposed on each other, as if they were all orbiting a single star; the dots indicate the minimum masses of the planets found through the Doppler technique. Figure 11.17 compares the eccentricities (a technical measure of how much the orbit differs from a circle) of the planetary orbits to their distances, with the giant planets of our own solar system shown as well. If you study these figures carefully, you'll notice the following important facts:

- While most of the discovered planets have masses similar to or greater than that of Saturn or Jupiter, many orbit quite close to their central stars—some are closer to their star than Mercury is to the Sun.
- Most of the planets that are not close-in to their stars are in highly elliptical orbits, quite unlike the nearly circular paths followed by most of the planets of our own solar system. This means that the planets are much closer to their stars at some times than at others.

Recall that our theory of solar system formation suggests that massive jovian planets should form only in the cooler, more distant regions of the swirling disk around a young star (see Figure 3.22). In the warm inner regions of the disk, we expect to find rocky, terrestrial worlds. Thus, the presence of massive planets so close to their stars came as a big surprise to astronomers.

Could it be that these heavyweights are large rocky worlds? It's difficult to imagine how terrestrial planets could become so big, though we can't absolutely rule this out in all cases. However, in several cases we now have data from transits demonstrating that the planets are jovian rather than terrestrial in density, and we have every reason to suspect that the same is true for the other known massive extrasolar planets. Because the close orbits of some of the planets make them much higher in temperature than Jupiter, these planets have been nicknamed "hot Jupiters" (Figure 11.18).

Could "hot Jupiters" be habitable? It seems extremely unlikely. Their atmospheres would probably have the same strong vertical turbulence that led us to decide that Jupiter probably could not give rise to life [Section 7.3], and their proximities to their stars and elliptical orbits might only exacerbate their problems. Nevertheless, it's possible that these planets have numerous large moons, as do the jovian planets in our solar system. Some of these moons could be habitable, especially around the extrasolar planets that orbit within the habitable zones of their stars.

Implications for Solar System Formation Theory If our theory of solar system formation tells us that jovian planets can form only in the outer regions of the disks that surround newborn stars, then how can we explain the existence of "hot Jupiters" or the eccentric orbits followed by so many of these worlds? After all, the orbits in any planetary disk should tend to become circular due to an "averaging out" of the numerous collisions involved in planetary accretion.

One possibility that scientists must always consider is that something is fundamentally wrong with our model of solar system formation. That is, perhaps jovian planets *can* form in the inner solar system. Upon discovering the "hot Jupiters," scientists indeed began to consider this possibility. However, they failed to find any fundamental flaws in our basic understanding of how planets form. So they considered other ideas and now suspect that the unusual orbits followed by most of the known extrasolar

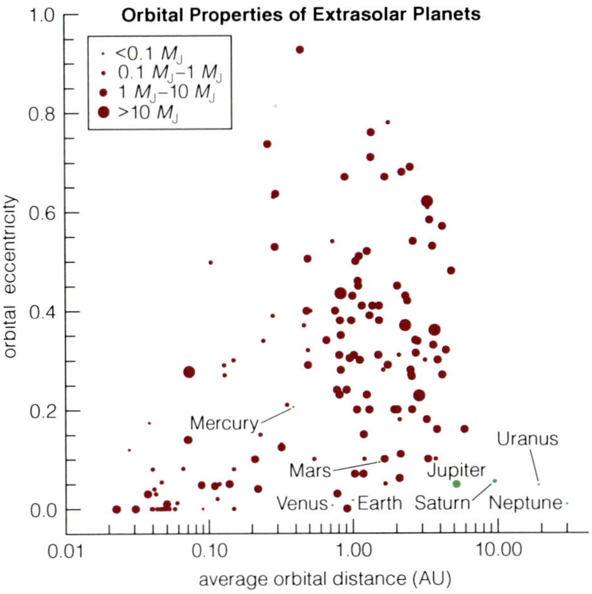

Figure 11.17

This graph shows the orbital radii (average distance from star) and eccentricities for the first 146 known extrasolar planets. Each red dot represents an extrasolar planet, and the green dots represent planets of our own solar system. Planets that orbit closer to their stars are farther to the left on this diagram, and the planets with more circular orbits are lower down. Notice that many extrasolar planets orbit quite close to their stars and most have larger orbital eccentricities than the planets of our solar system. (Eccentricity is a measure of how much an orbit differs from a circle; it is 0 for a perfect circle and approaches 1 for highly elliptical orbits.)

Figure 11.18

Artist's impression of the massive planet orbiting the star 51 Pegasi. At an orbital radius eight times closer to its star than Mercury is to the Sun, the temperatures on this planet are expected to be 1,000°C or greater. Planets like this are called "hot Jupiters."

Figure 11.19

This figure shows a simulation of waves created by a planet embedded in a dusty disk of material surrounding its star; these waves may cause the planet to migrate inward.

planets are the result of changes that occurred in these systems after the planets first formed. That is, these planets most likely did form in the outer regions of their solar systems, and only later migrated inward or were deflected into eccentric orbits.

How might planetary migration occur? You might think that drag within the solar nebula could cause planets to migrate, much as atmospheric drag can cause satellites in low-Earth orbit to lose orbital energy and eventually plunge into the atmosphere; however, calculations show this drag effect to be negligible. A more likely scenario is that waves propagating through a gaseous disk lead to migration (Figure 11.19). A planet's gravity and motion tend to disturb the otherwise smooth disk material, generating waves that propagate through the disk. The waves cause dust to bunch up into ridges as they pass by, and these ridges exert their own gravitational pull on the planet, robbing it of energy and causing it to move inward.

Computer simulations of this rather subtle interplay of an orbiting planet and a disk of dust confirm that waves in the nebula can cause young planets to spiral slowly toward their star. In our own solar system, this migration is not thought to have played a significant role, probably because the solar wind cleared out the gas before it could have much effect. But planets may form earlier in at least some other solar systems, allowing time for jovian planets to migrate substantially inward. In some cases, the planets may form so early that they end up spiraling into their stars. Indeed, astronomers have observed some stars that have an unusual assortment of elements in their outer layers, suggesting that they may have swallowed planets.

Another possible way to account for some of the observed extrasolar planets invokes close encounters between young jovian planets in the outer regions of a disk. Such an encounter might kick one planet out of the star system entirely while sending the other inward in a highly elliptical orbit. Other possibilities include the idea that a jovian planet could migrate inward as a result of multiple close encounters with much smaller planetesimals (there is evidence that the jovian planets in our solar system have migrated a little bit by this mechanism). Orbital resonances like those that make the orbits of Jupiter's Galilean moons elliptical (see Figure 9.11) might also affect planets. Several of the multi-planet systems already discovered have orbital resonances between their planets, lending support to the idea that these resonances might play an important role in the architecture of many planetary systems.

The bottom line is that our discoveries of extrasolar planets have shown us that our prior theory of solar system formation was incomplete. It explained the formation of planets well, and also the simple layout of a solar system like ours. However, it needs modification in order to explain the differing layouts of other solar systems. While we are not yet sure of the precise mechanism, it now seems clear that planets can move from their birthplaces, making possible a much wider range of solar system arrangements than we had guessed before the discovery of extrasolar planets.

THINK ABOUT IT Look back at the discussion of the nature of science in Section 2.3, especially the definition of a scientific theory. Why does our theory of solar system formation qualify as a scientific theory even though we know that it needs modification to account for migration? Does this mean that the theory was "wrong" as we understood it before? Explain.

Implications for Habitable Worlds Planetary migration may have important implications for the question of whether solar systems of the type we have recently found can harbor habitable planets. As we have discussed, it's unlikely that the large jovian planets could themselves support life. But could there also be smaller and potentially habitable terrestrial planets in these systems? Observationally we cannot say, because our current telescopes aren't able to detect Earth-size worlds. However, theoretical work suggests that even if terrestrial-size planets were born in these systems, they could face severe problems, because the migration of a large planet might disrupt the inner solar system.

If the migration occurs before terrestrial planets have finished forming, the material that would have accreted onto the terrestrial worlds might be swallowed instead by the larger world. Even if the formation process is essentially complete, gravitational encounters between big planets and small ones nearly always send the smaller ones scattering. As a large planet migrated inward, it would tend to fling less massive planets inward toward its star or outward to interstellar space. This suggests that terrestrial planets would not stay within the habitable zones of stars that also have "hot Jupiters" or other massive planets in highly elliptical orbits. Indeed, these same gravitational encounters might also disrupt and disperse any potentially habitable moons.

However, some recent research has called this discouraging scenario into question, and suggests that, even if giant planets spiral inward through a field of Earth-size worlds, not all of these smaller bodies would be scattered hither and yon. And in any case, the prospects look quite good for habitable planets around stars where planetary migration does not occur. Without migration, we expect rocky planets to form and to remain on stable orbits, and many of these orbits should be within the habitable zones of their stars.

In addition, even though the evidence so far might seem to point to the systems with substantial migration as being common, there's an important selection effect we must keep in mind. Remember that nearly all of the planets discovered have been found with the Doppler technique and that this technique works best for massive planets in close orbits. Therefore, it is possible that the reason we are finding so many systems with "hot Jupiters" is that they are the easiest ones to find with our current technology, not necessarily because they are the most common. Of the several hundred Sun-like stars that have been examined so far, approximately one in ten shows evidence of one planet or more—which means that we have not detected planets around the vast majority of Sun-like stars. In many cases, the reason may be that these stars have planetary systems more like our own than like the ones with "hot Jupiters," making them undetectable to our present technology. As the number of known extrasolar planets continues to grow, we will gain more insight about whether our solar system's layout is typical. But until we are able to make searches capable of finding small planets with upcoming missions like *Kepler*, *SIM*, and *TPF*, the question remains wide open.

• How could we detect life on extrasolar planets?

If Earth-size planets exist around other stars, we should begin to find them within the next several years and get crude images and spectra within the next several decades, thanks to improved technology and new space missions. Let's explore how these capabilities might allow us

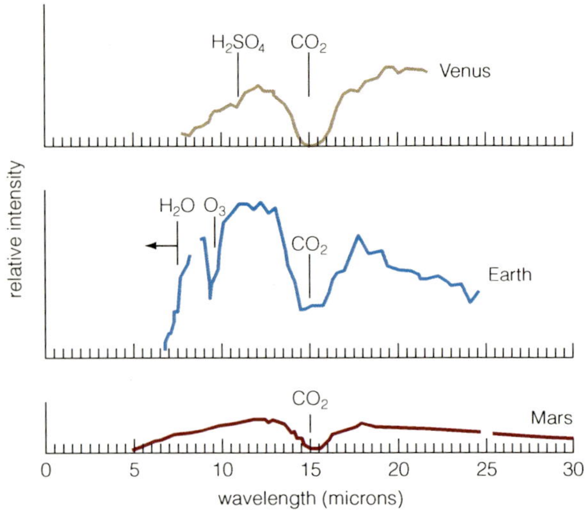

Figure 11.20

The infrared spectra of Venus, Earth, and Mars, as they might be seen from afar, showing absorption features that point to the presence of various gases in their atmospheres. While carbon dioxide is present in all three spectra, only our own planet has appreciable oxygen (and hence ozone)—a product of photosynthesis. If we could make similar spectral analyses of distant planets, we might possibly detect atmospheric gases that would indicate life.

to determine whether the planets are habitable, and perhaps whether they have life.

Suppose a telescope actually shows us a crude image—only a few pixels—of a distant world. What might we learn? To begin with, by simply watching its changing brightness, we could gain some information about the ratio of ocean to land, because seas are darker than continents. We might also find that the light from such a world changes from day to day, due to clouds, or from season to season, because of snow or ice.

A spectrum would allow us to measure many other properties. For example, even a fairly crude spectrum should allow us to gauge the surface temperature of the planet. If we can collect even more light from a distant planet, we could make a more detailed spectral analysis, one that could suggest far more convincingly that a planet is home to life. For instance, we might gain information on the types of minerals or ices on its surface. If we find that the planet has characteristics like those of Earth or Mars, we would know that it is habitable in principle though perhaps not whether it actually harbors life.

With spectra from infrared telescopes (such as *TPF* or *Darwin*), we could search for the absorption or emission features of gases in the atmosphere of a planet. Several of these gases will be easy to detect if they are present, including carbon dioxide, ozone, methane, and water vapor (Figure 11.20). While the mere presence of such gases would not necessarily point to life, their precise abundances and the combinations in which they occur could provide stronger evidence of whether life is present.

For example, Earth's atmosphere has large amounts of oxygen, the result of photosynthesis. If we found abundant oxygen in the air of another world, we would have reason to suspect the presence of life, particularly if the ratio of oxygen to the other detected gases seemed incompatible with nonbiological chemistry. To some extent, the same is true of methane, which is present in Earth's atmosphere today largely thanks to the "exhaust" gases produced by livestock and rice paddies. In early times, before photosynthesis raised the oxygen level, the atmosphere may have been altered by another metabolic process, known as *methanogenesis*, through which microbes expel methane rather than oxygen. The first billion years or so of Earth's biological history might therefore have been marked by the presence of atmospheric methane.

The bottom line is that for billions of years, life on Earth has been making its presence known to anyone with a telescope large enough to find our world and make a spectrum of its reflected light. In principle, we could identify life on other worlds in the same way. Perhaps in the next few decades we will discover abundant oxygen or methane on a distant world visible to us as no more than a dot in a telescope, providing an exciting and encouraging clue that it harbors life.

11.3 THE POSSIBILITY THAT EARTH IS RARE

We have discussed the current status of our search for extrasolar planets, along with the question of whether potentially habitable planets should be common or rare. We have found that our current knowledge is still quite incomplete. It may be that a large fraction of stars have habitable planets, or it may be that the fraction of stars with such planets is fairly low. Nevertheless, the example of our own solar system tells us that the fraction is greater than zero. Given the enormous number of stars in the galaxy, it seems highly likely that, one way or another, there are large numbers of

Earth-size planets within the habitable zones of stars. But does Earth-*size* necessarily mean Earth-*like* in terms of habitability?

• Are Earth-like planets rare or common?

Most scientists suspect that a significant percentage of Earth-size planets are likely to be Earth-like as well. However, some scientists have questioned this idea, suggesting that Earth may be the fortunate beneficiary of several kinds of planetary luck. According to this idea, sometimes called the "rare Earth hypothesis," the specific circumstances that have made it possible for evolution to progress beyond microbes to complex creatures (such as oak trees or humans) might be so rare that ours might be the only inhabited planet in a galaxy that harbors anything but the simplest life. This suggestion would have profound implications if true, particularly for the efforts to search for extraterrestrial intelligence that we will discuss in the next chapter. Let's briefly examine some of the key issues in the rare Earth hypothesis.

Figure 11.21

The green ring in this diagram of the Milky Way Galaxy highlights what some scientists hypothesize to be a galactic habitable zone—the only region of the galaxy in which Earth-like planets can form. However, other scientists doubt the claims that underlie this hypothesis in which case Earth-like planets could be far more widespread.

Galactic Constraints Proponents of the rare Earth hypothesis suggest that Earth-like planets can form in only a relatively small region of the Milky Way Galaxy, making the number of potential homes for life far smaller than we might otherwise expect it to be. In essence, they argue that there is a relatively narrow ring at about our solar system's distance from the center of the Milky Way Galaxy that makes up a *galactic habitable zone* analogous to the habitable zone around an individual star (Figure 11.21).

According to the arguments for a galactic habitable zone, outer regions of our galaxy are unlikely to have terrestrial planets because of a low abundance of elements besides hydrogen and helium. Recall that the fraction of heavy elements varies among different stars, from less than 0.1% among the old stars in globular clusters to more than 2% among young stars in the galactic disk [Section 3.2]. Even within the galactic disk, the abundance tends to decline with distance from the center of the galaxy. Because terrestrial planets are made almost entirely of heavy elements, a lower abundance of these elements might lessen the chance that terrestrial planets could form. The inner regions of the galaxy are ruled out primarily through an argument concerning supernova rates. Supernovae are more common in the more crowded, inner regions of the galactic disk, making it more likely that a terrestrial planet would be exposed to the intense radiation from a nearby supernova. By assuming that this radiation would be detrimental to life, the proponents of a galactic habitable zone argue against finding habitable planets in the inner regions of the galaxy. Together, the constraints on finding Earth-like planets in the inner and outer regions of the galaxy leave the galactic habitable zone as a relatively narrow ring encompassing no more than about 10% of the stars in the galactic disk.

However, other scientists offer counterarguments to both sets of galactic constraints. For the heavy-element abundance, they note that Earth's mass is less than $\frac{1}{100,000}$ of the mass of the Sun. Thus, even a very small heavy-element abundance could be enough to make one or more Earth-like planets. Unless there is something about the planetary accretion process that prevents terrestrial planets from forming in systems with low heavy-element abundances, then we might find terrestrial planets around almost any star. Regarding the radiation danger from supernovae, we do not really know whether such radiation would be detrimental to life. A planet's atmosphere might protect life against the effects of the radiation. It is even possible that the radiation could be beneficial to life by increasing the rate of mutations and thereby accelerating the pace of evolution. If these counterarguments are correct, then Earth-like planets might be found throughout much or all of the galaxy.

Impact Rates and Jupiter Another issue raised by rare Earth proponents concerns the rates of impacts by asteroids and comets on planets in other solar systems. Recall that Earth was probably subjected to numerous large impacts—some possibly large enough to vaporize the oceans and sterilize the planet—during the first few hundred million years after our planet was born [Section 4.3]. In our solar system, the impact rate lessened dramatically after that. Might the impact rate remain high much longer in other solar systems?

Although the scars of impact craters on numerous planets and moons bear witness to huge numbers of impacts in our solar system's history, the number of *potential* impacts is far higher. Thousands of asteroids still roam the region between Mars and Jupiter. Of even greater significance, by studying the orbits of comets that enter the inner solar system, astronomers have concluded that as many as a trillion comets must orbit the Sun at distances far beyond the orbit of Pluto, making up the *Oort cloud* [Section 3.3]. Fortunately for us, these myriad objects are essentially out of reach, posing no threat to our planet. However, we have good reason to believe that their great distances can be traced directly to gravitational effects of Jupiter.

The Oort cloud comets are thought to be the survivors among an even greater number of comets that formed originally in the region of the solar system where the jovian planets were born. During the early history of our solar system, the vast majority of these comets would have experienced close encounters with a jovian planet—most of them with Jupiter.

MOVIE MADNESS: *STAR WARS*

It's "a long time ago" in someone's far-off galaxy, and the political situation is turning ugly.

The premise of the original *Star Wars,* a cinematic space opera that, like relatives, just keeps on returning (with six installments to date), is that a galaxy-wide republic has been hijacked and converted to an autocratic, evil empire by despots in gray flannel suits. This may sound vaguely reminiscent of the story of Rome, but unlike what happened to that ancient civilization, this political shift has encouraged a serious rebellion, a war among the stars. The rebels are led by Princess Leia (you can tell she's a princess because her hair is done up like twin Danish pastries), and her strategy is to take out the empire's headquarters—an enormous, spherical spacecraft known to its friends as "the Death Star." The Death Star packs weaponry that can explode a planet in seconds (calculate, if you wish, the energy required to do that!). On the other hand, the rebels have "the force" on their side—a mystical ability to change the odds of every situation based on moral merit and self-discipline.

Most of *Star Wars* is battle of the sort that's familiar to any movie fan, except that the bad guys wear brittle, white plastic suits and fly spacecraft that look like box kites. But *Star Wars* offers some interesting peeks into life as it might be elsewhere in the cosmos. The rebels have their base of operations on a large planet's moon, a not-impossible scenario since hefty moons could be habitable. Luke Skywalker, the young hero, hails from a world circling a close double star. Not a problem—research has shown that planetary orbits around double stars could be stable.

There are peculiar anachronisms in *Star Wars,* however. The Death Star is obviously extremely high-tech, and yet the principals occasionally face off using souped-up swords. Everyone jets around in spacecraft that are somehow capable of exceeding speed-of-light travel by jumping into hyperspace, and yet we often see aliens saddled up onto giant, dinosaurlike creatures—much as in *The Flintstones.*

All of that can be forgiven. But *Star Wars* takes its biggest literary license in the fact that dozens of alien races are all living contemporaneously (although it's clear that the human types are in charge). In the movie's famous cantina scene, which takes place in the wretched port city of Mos Eisley, aliens of all shapes and colors get together to do business and get drunk. In fact, the chance that any two intelligent species (let alone dozens) appear on the galactic scene within 100,000 years of one another is quite small. If there are other societies out there, they will either be far behind us or enormously beyond our level. We won't be sharing dance music and booze with them in a seedy extraterrestrial dive.

And besides that, why does a republic have a princess, anyhow?

Recall that a close encounter between a big planet and a small object like a comet will tend to send the small object flying off in a new direction. Many comets must have ultimately crashed into the Sun or into the planets of the inner solar system. The rest were flung out to the great distances at which they now reside or out of the solar system entirely.

Thus, if Jupiter did not exist, the comets might have remained in the part of the solar system where they could pose a danger to Earth. In that case, the heavy bombardment might never have ended, and huge impacts would continue to this day. From this viewpoint, our existence on Earth has been possible only because of the "luck" of having Jupiter as a planetary neighbor.

The primary question is just how "lucky" this situation might be. Our discoveries of extrasolar planets show that Jupiter-size planets are in fact quite common, though if most of them migrate inward, then giant planets in outer orbits might be rare. Until we learn much more about the layout of other planetary systems, we have no reason to conclude that we are particularly "lucky" to have Jupiter as a planetary companion.

THINK ABOUT IT The story of life on Earth is replete with disasters that served to stress terrestrial species, resulting in the rapid evolution of new, more complex organisms. For example, the K–T impact [Section 6.4] apparently led to the demise of the dinosaurs and opened the door for the rise of mammals. More recently, ice ages are thought to have played a major role in the evolution of modern humans. Do you think it's possible that a higher rate of impacts could be good rather than bad for life on another planet? Explain.

Climate Stability Another issue affecting the rarity of Earth-like planets concerns climate stability. Recall that, in comparison to Venus and Mars, Earth has had a remarkably stable climate. This climate stability has almost certainly played a major role in allowing complex life to evolve on our planet. If our planet had frozen over like Mars or overheated like Venus, we would not be here today. Advocates of the rare Earth hypothesis point to at least two pieces of "luck" with regard to Earth's stable climate.

The first piece of luck concerns the existence of plate tectonics. As we discussed in Chapter 4, plate tectonics plays a major role in the carbon dioxide cycle and hence in regulating Earth's climate. The existence of this climate-regulating mechanism is especially important given that the Sun, like all stars, brightens as it ages. The Sun is about 30% more luminous now than when Earth formed. Yet, thanks to plate tectonics and the carbon dioxide cycle, Earth has remained habitable. Plate tectonics probably was not necessary to the origin of life, but it seems to have been quite important in keeping the climate stable long enough for the evolution of plants and animals. But are we really "lucky" to have plate tectonics, or are such processes inevitable on Earth-size planets in Earth-like orbits? As far as we are aware, there is nothing particularly unusual about Earth's size, composition, or orbit. Therefore, we have no reason to believe that planets with similar characteristics should be rare and no reason to think that geological processes should be any different in other cases. On the other hand, Venus is quite similar in size to Earth but apparently lacks plate tectonics. This may be the result of Venus's runaway greenhouse effect [Section 10.2], or it may be due to factors that we do not yet understand. Nevertheless, unless some unknown, special circumstance has encouraged long-lasting plate tectonics on Earth, it seems likely that such processes will be found on other, similar planets.

The second piece of luck regarding climate stability concerns the existence of the Moon. As we discussed in Chapter 8, Mars undergoes dramatic climate changes because the tilt of its axis varies over a significant range (see Figure 8.25). Earth's tilt varies much less, contributing to climate stability. Models of Earth's rotation and orbit show that if the Moon did not exist, gravitational tugs from other planets, especially Jupiter, would cause large swings in Earth's axis tilt over periods of tens to hundreds of thousands of years. In other words, if the Moon did not exist, Earth's spin axis would be subject to the same large swings in its tilt as occur on Mars. Given that the Moon likely formed as a result of a random, giant impact [Section 4.6], it might seem that we are "lucky" to have our Moon and the climate stability it brings.

Again, however, there are other ways to look at this issue. Changes in axis tilt might warm or cool different parts of the planet dramatically, but the changes would probably occur slowly enough for life to adapt or migrate as the climate changed. In addition, our Moon's presumed formation in a random giant impact does not necessarily mean that large moons will be rare. At least a few giant impacts should be expected in any planetary system. Indeed, Earth may not be the only object in our own solar system that ended up with a large moon because of a giant impact: Pluto's largest moon (Charon) may have formed in the same way (along with its two smaller moons). Collisions are also thought to be responsible for the many moons found around other Kuiper belt objects. Thus, while luck was certainly involved in Earth's having a large moon, it might not be a rare kind of luck.

Summarizing the Arguments The bottom line is that while the rare Earth hypothesis offers some intriguing arguments, it is too early to say whether any of them will hold up over time. For each potential argument that Earth has been lucky, we've seen counterarguments suggesting otherwise. There's no doubt that our solar system and our world have "personality"—that is, they exhibit properties that might be found only occasionally in other star systems—but we have no clear reason to think that these special properties are essential to the existence of complex or even intelligent life. Indeed, it may be that we have missed out on some helpful phenomena that could have sped evolution on Earth. We might be less lucky than we recognize, and creatures on other worlds might regard the nature of our planet with disappointment. Until we learn much more about other planets in the universe, we cannot know whether Earth-like planets and complex life are common or rare.

11.4 THE PROCESS OF SCIENCE IN ACTION: CLASSIFYING STARS

Much of this book is devoted to studying individual planets, and for obvious reasons. Planets have personality: The appearance, composition, and size of any given world depend on where and how it was formed, and what happened to it after its birth. Just because two planets are the same size hardly means that they have the same characteristics (think of Earth and Venus).

But stars are much more like clones. You might think that this is simply because all stars (other than the Sun) are quite far away. Consequently, they appear only as featureless, bright pinpoints when seen through our telescopes. But in fact, stars don't just look similar; they really are similar. Discovery of that fact has helped us puzzle out one of astronomy's fundamental mysteries: What is the life history of a star? In this section, we'll discuss a tool that was developed nearly a century ago, the **Hertzsprung–Russell (H–R) diagram**, which is used to classify the various types of stars. This diagram is as fundamental to astronomy as the periodic table is to chemistry.

• How and why did the Hertzsprung–Russell diagram develop?

Classifying objects is a familiar tactic in science. Long before biologists knew anything about DNA or even that cells exist, they were busy setting up categories for the life they could observe (this is called *taxonomy,* which comes from the Greek *tassein,* meaning "to classify," and *nomos,* meaning "a science or study"). Separating animals into reptiles, amphibians, insects, mammals, marsupials, and so on, requires no more equipment than a good eye and a pencil, but doing so eventually gives insight into the evolutionary processes that produced animal life. Even if you don't yet understand the underlying reasons for an observed diversity of objects, classifying them is a good first step.

In the late nineteenth and early twentieth centuries, astronomers were engaged in their own taxonomy projects. Even a casual observer could notice that not all stars are the same brightness. Until methods were developed for measuring colors, apparent brightness was the only known parameter that differed from star to star (other than position in the sky). A star's brightness depends on: (1) its distance, (2) its intrinsic luminosity, and (3) the dimming effects of any interstellar dust that might lie between us and the star. Even ignoring this last effect, which for nearer stars is generally not important, astronomers still have to sort out the distance of a star if they want to derive its intrinsic luminosity from its apparent brightness. They are keen to do this because, while the latter is mere happenstance, the former is a physical property of the star. This problem is akin to determining the wattage of a light bulb seen at a distance. We can easily measure its apparent brightness, but we need to know how far away it is if we want to know its intrinsic luminosity, or wattage.

In addition to luminosity, there's a second property of stars that, by the end of the nineteenth century, could be quickly determined: color. Because the cones of our retinas do not work well in low light, looking at stars with the naked eye is a poor way to judge their hue. But any color photograph of the stars of the night sky will show a range from blue to red. As discussed in Section 11.1, the spectral properties of stars simply reflect the temperatures of their outer gas layers. So knowing a star's spectral type will tell you its surface temperature (for example, the yellow Sun, a type G star, has a surface temperature around 5,500°C, as listed in Table 11.1).

In the first years of the twentieth century, Danish astronomer Ejnar Hertzsprung, who had collected data on the distances of nearby stars, noticed something peculiar: If he arranged the stars by color, from blue-white to red (spectral types B through M), the arrangement was also a sequence in intrinsic luminosity. That is, the hottest (most blue) stars were more luminous than the cooler stars. However, if he included stars that were farther away, he found that the distant red stars were, on average, intrinsically much more luminous than the nearer ones. Since he realized that the more distant sample of red stars undoubtedly included intrinsically faint objects he couldn't see, this suggested to Hertzsprung that there might be two categories of red stars: small ones (the intrinsically fainter stars) and giants (the intrinsically more luminous stars).

Hertzsprung's results were surprising because we might expect that stars could have any combination of luminosity and color. Why were the red stars separating into distinct categories? Beginning in 1911, Hertzsprung began to expand this work by plotting the colors and luminosities of stars in two clusters, the Pleiades and the Hyades. Using clusters of stars saved him from having to measure their individual distances, because all the stars in a cluster are more or less at the same distance (just as the people in Chicago are all at approximately the same distance from people in New York). He soon found that the stars were, indeed, not scattered at random across his plot, but fell into two broad groups: (1) most were in a swath running diagonally through the diagram, which was later named the **main sequence**; and (2) nearly all of the remainder were in a part of the diagram he called the *giants.*

The American astronomer Henry Norris Russell, working independently, was making a similar analysis. However, rather than examining stars in clusters, he used stars whose distances were known from trigonometric measures. Of course, this meant that his diagram was based on relatively close stars, those for which geometric calculation could be used to determine distances. But the results were quite similar to Hertzsprung's. By the 1920s, Hertzsprung's and Russell's results were codified in the H–R diagram, a graphic found in every astronomy textbook ever since (Figure 11.22).

• What can we learn from the Hertzsprung–Russell diagram?

While one might have expected to find stars all over the H–R diagram, it turned out that they were confined to only a few regions. You might compare this to characterizing humans on the basis of two easily measured parameters: height and weight. If you actually take these measures for a large number of people, and plot them on a graph with height on one axis and weight on the other, you'll find that there's a pretty narrow range within which most of the data points fall. Very short, very heavy people are rare, as are very tall, very light folk. The clustering of the data along a diagonal swath relating height and weight is a consequence of the fact that, in a general way, all people are the same. The Hertzsprung–Russell diagram suggests that stars, too, might be generally similar.

First, let's describe the parts of a modern H–R diagram where large numbers of stars are found. Note that today's diagrams are based on data of a quantity and quality unavailable a century ago. The major occupied regions are the following:

- The main sequence, which is the S-shaped line that curves from top left to bottom right. Stable stars burning hydrogen in their cores—the overwhelming majority of all stars—fall on this line. Our Sun lies about midway along the main sequence, near the center of the diagram, which is the reason the Sun is often referred to as an "average star," even though it is actually more massive than the majority of stars (the K and M stars on the lower part of the main sequence).
- At the top right are the giants and their somewhat bigger brethren, the supergiants. These stars are mostly cool and red, but very luminous. In fact, giants and supergiants are stars that used to be on the main sequence but are now at the end stages of their lives, having already exhausted their core hydrogen (see the discussion of giants and supergiants in Section 11.1).

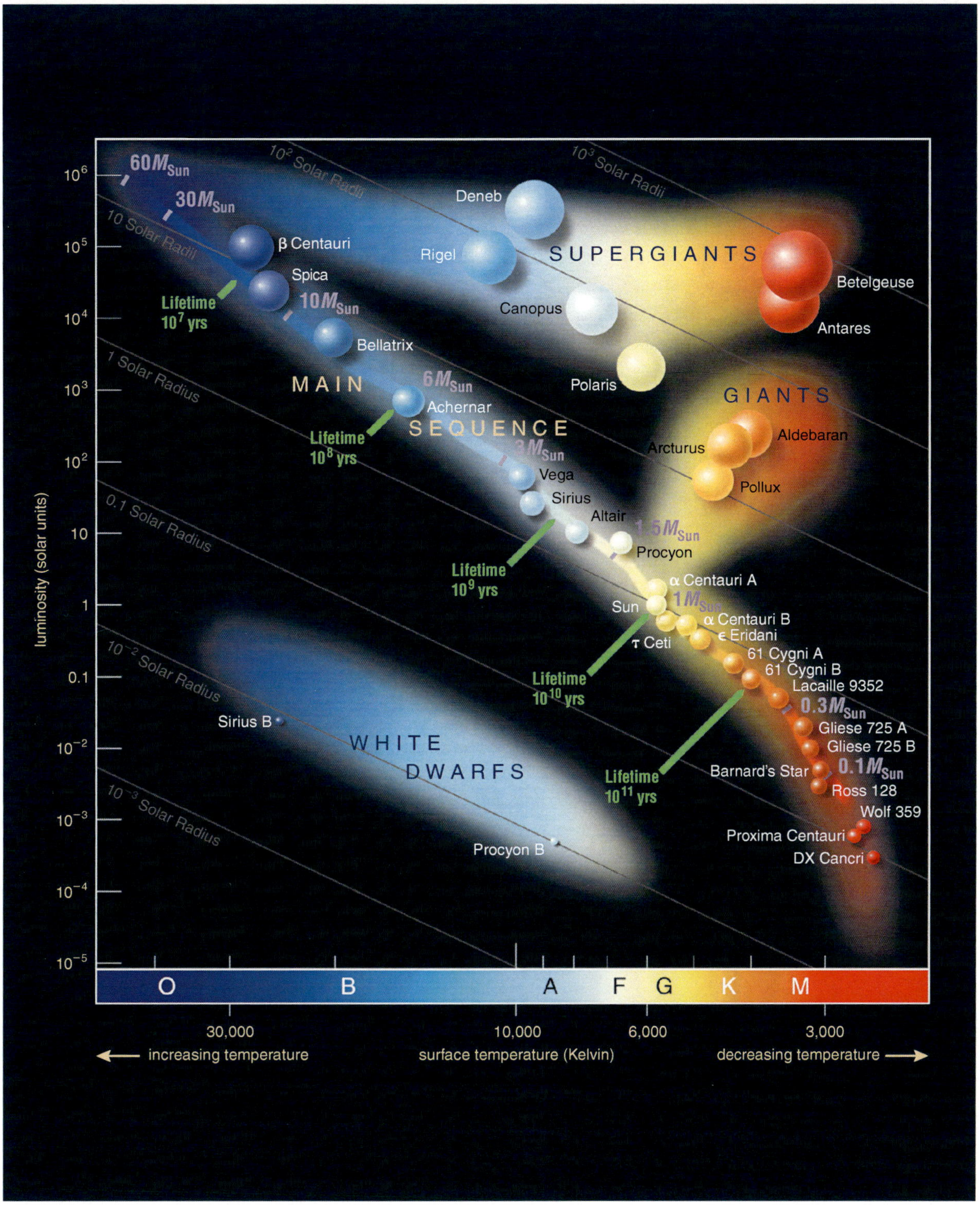

Figure 11.22

An H–R diagram, one of astronomy's most important tools, shows a star's surface temperature or spectral type along the horizontal axis and its luminosity along the vertical axis; the fact that surface temperature increases to the left (rather than to the right) is a consequence of the historical development of the spectral sequence OBAFGKM. Several of the brightest stars in the sky are plotted here, along with a few of those closest to Earth. They are not drawn to scale—the diagonal lines, labeled in solar radii, indicate how large they are compared to the Sun. The lifetime and mass labels apply only to main-sequence stars. (Star positions on this diagram are based on data from the Hipparcos satellite.)

- At the bottom left are the hot but dim white dwarfs. White dwarfs are the collapsed remains of stars like the Sun, stars that have exhausted their fuel and are slowly turning into ashen, stellar corpses.

The fact that more than nine out of ten stars are on the main sequence tells us that stars spend about 90% of their lives burning hydrogen in their cores, and their arrangement on this line tells us something about the character of stars during this long phase of their lives: The hotter they are, the more luminous they are (that is, the more energy they pump into space). This may seem obvious, and you may wonder why page space is being expended on this diagram. But it might have been true that the universe was crowded with lots of hot stars of relatively low luminosity (say, the luminosity of the Sun). A blue acetylene torch is hot, but is of low luminosity compared to a red bonfire. However, the existence of the main sequence suggests that all the stars populating it are basically built the same way, and Russell concluded that there was one variable, one parameter, that determined where on the main sequence a star would lie. That parameter turns out to be a star's mass. As we move from bottom right to top left on the main sequence, the stars become more massive (notice the mass labels in Figure 11.22). The main sequence is actually a *mass* sequence.

Think of how stars come about. Stars are formed by collapsing clouds of gas, and the composition of this gas is mostly hydrogen and helium. In this book we make a great fuss about the elements heavier than helium, because these are necessary for planets, plants, and people. But from the standpoint of making a star, which is destined to fuse hydrogen into helium in the main phase of its existence, these impurities are of small consequence. Stars form when clouds of gas collapse under their own weight, squeezing their inner cores to higher temperatures and pressures until nuclear fusion begins. This produces the heat that stops further collapse, and the resulting size and temperature are simply determined by the mass of the star itself. The greater the mass, the greater the pressure within the star. The increased pressure raises the density of matter at the core, which increases the rate at which the particles there bump into one another and fuse. So the rate at which fuel is burned—and therefore the star's luminosity—goes up as the overall mass does. In 1926, Russell and German astronomer Heinrich Vogt stated this explicitly, noting that "the evolution of stars is fixed at their birth by two inherent fundamental parameters, mass and chemical composition." This Russell–Vogt theorem is simply a reiteration of a fact that is evident from the existence of the main sequence: All hydrogen-burning stars are made from the same blueprint. They are all cousins differing only in their mass.

What about the giants and supergiants? Despite the fact that they are not on the main sequence, these luminous heavyweights do not violate the idea that mass is the most important property. They have simply changed their chemical composition by having lived long enough to burn up much of their central hydrogen fuel. They have moved on to a different "engine" for producing energy: fusing helium or even heavier elements. Paradoxically, the temporary energy boost provided by this switch in fuel causes them to swell up to the point where their surface temperatures are lowered (they become redder), even though their total luminosity has increased. From a graphical point of view, they've moved off the main sequence to take up a rather short residence in the giant or supergiant region of the H–R diagram. Eventually they exhaust these fuels and either collapse to become a white dwarf (at the bottom left of the diagram) or, if they began life as a star several times as massive as the Sun, explode in a supernova and leave behind either a neutron star or a black hole. There is no location on the H–R diagram for such pathological stellar remnants, so the heavier giants eventually, and rather suddenly, disappear from the chart.

The H–R diagram was originally just an organizational strategy, like classifying plants and animals. But we have seen that organizing leads to insights; indeed, the H–R diagram remains one of the most important tools of professional astronomers. The lesson for the process of science is clear: When you're not sure what's going on, start by organizing what you do know, and it may take you down the path to discovery.

THE BIG PICTURE

Putting Chapter 11 into Context

In this chapter, we have discussed the search for habitable planets beyond our own solar system. As you continue in your studies, keep the following "big picture" ideas in mind:

- Thanks to rapidly accumulating discoveries of planets around other stars, there's no longer any doubt that massive planets, at least, are common. However, we still do not know whether habitable planets are common or rare.
- Our discoveries of extrasolar planets have provided valuable new information about the processes involved in the birth of solar systems, forcing us to modify our existing theory to account for planetary migration, and possibly our ideas of how giant planets form.
- Although we do not yet have an answer regarding the existence of Earth-size or Earth-like planets around other stars, several planned missions will address this question. Within the next few years, we should know whether other Earth-size worlds exist. Within a few decades, we may have crude images or spectra that will tell us whether these worlds are also Earth-like.

SUMMARY OF KEY CONCEPTS

11.1 DISTANT SUNS

- **How do stellar life cycles affect the possibility of habitable planets?**

The most basic requirement for a star to be able to support life on orbiting planets is that it be stable for long enough to allow that life to originate. This limits us to stars that are in the long-lived, hydrogen-burning phase of their lives. Giants and supergiants are stars that have used up the bulk of their core hydrogen fuel, and undergo substantial change on time scales of millions of years. Consequently, they are not thought to be suitable stars for hosting planets with life.

- **How do we categorize stars?**

The **spectral sequence OBAFGKM** runs from hot to cool in terms of the surface temperatures of stars. For stars in their hydrogen-burning phases, this sequence also runs in mass order. O stars are the hottest, most massive, most luminous, and shortest lived. The Sun is a G star, and the far end of the scale has the cool, dim, low-mass, and long-lived M stars.

- **Which stars would make good suns?**

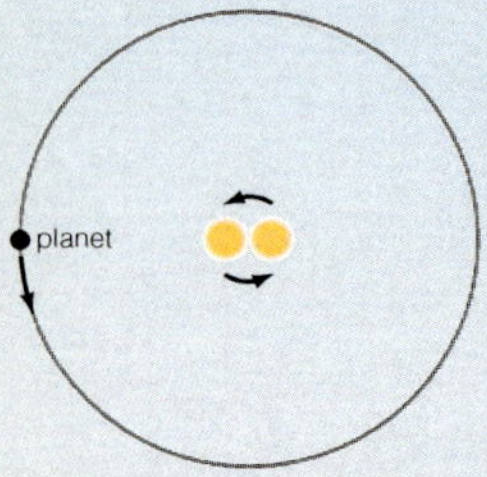

The lifetimes of O and B stars seem too short to permit life to arise on any surrounding planets. A and F stars may live long enough for life to evolve, though probably not long enough for intelligent life. G stars like the Sun clearly can have habitable planets. K and M stars have small habitable zones, but make up for that in their sheer numbers, since they are the most common types of stars. Many stars are members of multiple star systems, but they may still have habitable planets in stable orbits.

11.2 EXTRASOLAR PLANETS: DISCOVERIES AND IMPLICATIONS

- **How do we detect planets around other stars?**

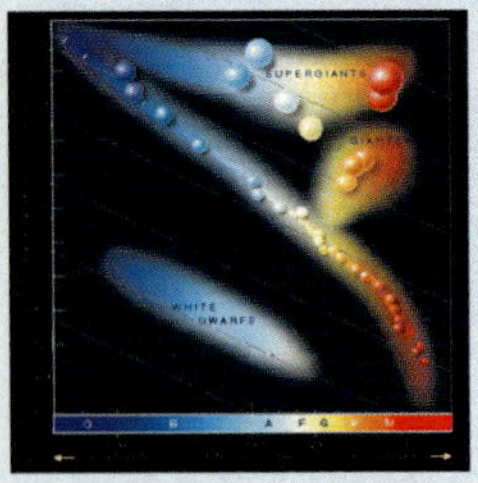

There are two fundamental methods for detecting extrasolar planets: direct observation or measuring their indirect effects. To date, almost all discoveries have been indirect. We can look for a planet's gravitational effect on its star through the **astrometric technique**, which looks for small shifts in stellar position, or the **Doppler technique**, which looks for the back-and-forth motion of stars revealed by Doppler shifts. We can also search for **transits** in which a system becomes slightly dimmer as a planet passes in front of its star. A few planets have been found by gravitational lensing, an effect predicted by Einstein's general theory of relativity.

- **What are the properties of extrasolar planets discovered so far?**

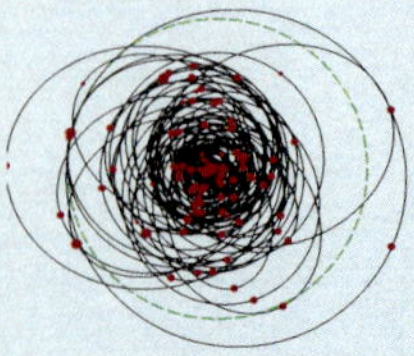

The known extrasolar planets are all at least several times as massive as Earth, and most are Jupiter-like in mass. Many orbit surprisingly close to their stars—which is why we call them "hot Jupiters"—or have large orbital eccentricities. We have only limited information about sizes and compositions, but these data are consistent with the idea that the planets are jovian in nature.

- **How could we detect life on extrasolar planets?**

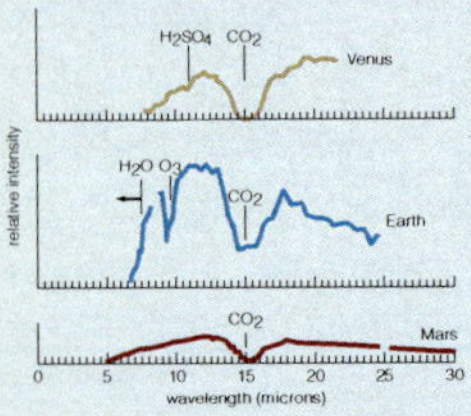

Future telescopes should allow us to obtain crude images or spectra of planets within stellar habitable zones. An image of an extrasolar planet—even if only a few pixels in size—might indicate the presence of snow or clouds, and would tell us the planet's rotation period. Spectroscopic analysis could tell us much more, and might reveal combinations of atmospheric gases, such as oxygen and methane, that would be good evidence for life.

11.3 THE POSSIBILITY THAT EARTH IS RARE

- **Are Earth-like planets rare or common?**

We don't know. Some of the key questions are whether our galaxy, like a star, has a relatively narrow habitable zone; whether the role Jupiter has played in lowering our solar system's impact rate is rare or critical to life; and whether Earth's relatively stable climate, due largely to plate tectonics and our large Moon, is rarely found on otherwise similar worlds. Arguments can be made on both sides of each question, and at present we lack the data to determine which side is correct.

11.4 THE PROCESS OF SCIENCE IN ACTION: CLASSIFYING STARS

- **How and why did the Hertzsprung–Russell diagram develop?**

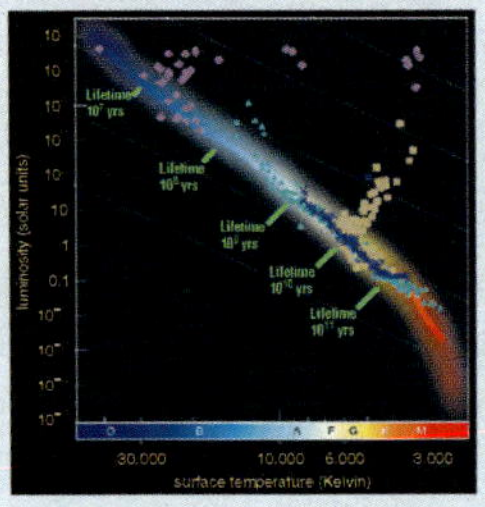

By the beginning of the twentieth century, astronomers had methods for measuring both the intrinsic luminosities and colors (which are related to surface temperatures) of stars. Ejnar Hertzsprung in Denmark and Henry Norris Russell in the United States independently realized that by classifying stars by plotting these two easily observed quantities, the stars clustered into groupings. The graph based on these quantities is now called the **Hertzsprung–Russell diagram**.

- **What can we learn from the Hertzsprung–Russell diagram?**

In the H–R diagram, most stars fall along a continuous swath, called the **main sequence**, that runs from hot, luminous stars at the upper left to cool, dim stars at the lower right. Other stars—giants and supergiants—clump in the part of the diagram where stars are luminous but have cool surface temperatures. The stellar corpses known as white dwarfs are dim but hot, so they are found in the lower left of the diagram. Study of the H–R diagram helped astronomers realize that mass is a star's most fundamental property, and the organizational power of the diagram makes it one of astronomy's most powerful tools.

EXERCISES AND PROBLEMS

REVIEW QUESTIONS

Short-Answer Questions Based on the Reading

1. Briefly summarize the life cycles of stars from birth to death. At which stage of stellar life does it seem possible that life-bearing planets could exist? Why?
2. How was the *spectral sequence* OBAFGKM discovered? Based on Table 11.1, briefly summarize how the properties of hydrogen-burning stars change as you look down the sequence from O to M.
3. For the seven major spectral types OBAFGKM, discuss whether stars of each type are likely to have habitable planets, and explain why or why not. Could *brown dwarfs* be orbited by habitable worlds? Explain.
4. What are *multiple star systems?* What are *binary star systems?* Briefly discuss the prospects of finding habitable planets in multiple star systems.
5. Briefly explain why we are confident that planets are common, even while we do not yet know whether *habitable* planets are common.
6. Explain why a planet can cause its star to move slightly in the sky. What do we mean by the *center of mass* of a system?
7. Describe the two techniques—the *astrometric technique* and the *Doppler technique*—by which we can measure gravitational effects of planets on stars and contrast their advantages and limitations. What future technologies or missions should help us extend the power of these techniques?
8. What is a *transit?* How can we use transits to find extrasolar planets? Briefly describe the significance of transits for the upcoming *Kepler* mission.
9. Why does *gravitational lensing* occur, and how can it allow us to detect a planet?
10. Why is direct detection of extrasolar planets so difficult? Why do infrared observations help? What future technologies or missions should improve our ability to detect planets directly?
11. Briefly discuss the general properties of the known extrasolar planets. What evidence suggests they are jovian rather than terrestrial in nature? What do we mean by "hot Jupiters"?
12. Why are the "hot Jupiters" surprising? What have they taught us about the formation of planetary systems? Discuss the prospects of finding habitable worlds in systems with "hot Jupiters."
13. How might future images and spectroscopy allow us to determine whether distant planets are habitable or have life?
14. What is the *rare Earth hypothesis?* Briefly summarize the arguments used to advance it and the counterarguments against each of these.
15. What is the *Hertzsprung–Russell diagram?* How does a star in the upper left section differ from one in the lower right?
16. Briefly summarize the characteristics of stars in each of the three major regions of the H–R diagram—the *main sequence,* the giants and supergiants, and the white dwarfs.

TEST YOUR UNDERSTANDING

Would You Believe It?

Suppose that, on the date indicated, you saw the following headlines. (These are not real discoveries.) In each case, decide whether the headline is believable in light of what we currently know about extrasolar planets and our technological capabilities. Explain clearly; because not all of these have definitive answers, your explanation is more important than your chosen answer.

17. Date: February 16, 2009. Headline: Astronomers Conclude That Earth-Size Planets Don't Exist.
18. Date: January 9, 2010. Headline: Astronomers Discover Earth-Like World Orbiting Massive Star of Spectral Type B.
19. Date: June 19, 2008. Headline: Spectrum Reveals Unmistakable Evidence of Life on a "Hot Jupiter."
20. Date: November 7, 2010. Headline: New Images Show Oceans on Extrasolar Planet.
21. Date: November 7, 2050. Headline: New Images Show Oceans on Extrasolar Planet.
22. Date: July 20, 2007. Headline: Giant Planet Found in Our Solar System Just Beyond Pluto.
23. Date: October 1, 2009. Headline: Astronomers Announce First Detection of an Earth-Size Planet Around Another Star.
24. Date: September 15, 2035. Headline: Sun-Like Star Has Three Planets with Life.
25. Date: March 30, 2027. Headline: More than One-Third of Stars Have Habitable Planets.
26. Date: December 13, 2010. Headline: Planet Orbiting a White Dwarf Has Surface Oceans and Oxygen Atmosphere.

Quick Quiz

Choose the best answer to each of the following. Explain your reasoning with one or more complete sentences.

27. Compared to a star of spectral type K, a star of spectral type A is generally: (a) hotter, more luminous, and more massive; (b) hotter, more luminous, and less massive; (c) cooler, dimmer, and less massive.
28. Stars of types O and B are unlikely to have planets with life because: (a) they have short stellar lives; (b) their intense ultraviolet light would sterilize any planets; (c) they don't have enough heavy elements.
29. How does the habitable zone around a star of spectral type M compare to that around a star of spectral type G? (a) It is larger and farther from its star. (b) It is hotter and much brighter. (c) It is smaller and closer to its star.
30. About how many extrasolar planets have been detected to date? (a) between 10 and 100; (b) between 100 and 1,000; (c) more than 1,000.
31. How have we detected most extrasolar planets discovered to date? (a) transits; (b) Hubble Space Telescope images; (c) the Doppler technique.

32. Which technique will the *Kepler* mission use to search for Earth-size planets around other stars? (a) transits; (b) the astrometric technique; (c) the Doppler technique.

33. Nearly all the extrasolar planets discovered to date are most likely: (a) terrestrial planets; (b) jovian planets; (c) large, icy worlds.

34. What is considered to be the most likely explanation for the close-in orbits of "hot Jupiters"? (a) They formed closer to their stars than Jupiter did. (b) They formed far from their stars, like Jupiter, but then migrated inward. (c) They are actually giant planets made of metal and rock.

35. Jupiter has had an important effect on life on Earth because: (a) Jupiter's heat has helped supply energy to life. (b) Jupiter's gravity helped clear the inner solar system of objects that could cause impacts. (c) Without Jupiter, Earth could not have a stable orbit around the Sun.

36. The main sequence on an H–R diagram represents stars that are: (a) in the final stages of their lives; (b) fusing hydrogen into helium in their cores; (c) all extremely low in mass.

INVESTIGATE FURTHER

In-Depth Questions to Increase Your Understanding

Short-Answer/Essay Questions

37. *Stars with Habitable Planets.* Based on what you've learned about stars in this chapter, make your best estimate of the fraction of all stars around which you'd expect to find planets in habitable zones. Clearly explain how you come up with your estimate. What data will we need to determine whether your estimate is accurate?

38. *Why Not Hubble?* Briefly explain why the Hubble Space Telescope has not been able to obtain images of many planets around other stars. What future technologies and missions should give us such images?

39. *Comparing Methods.* What are the advantages and disadvantages of the Doppler and transit methods? What kinds of planets are easiest to detect in each case? Are there certain planets that each method cannot detect, even if the planets are very large? Explain.

40. *Lots of Big Planets.* Many of the planets discovered so far are more massive than our most massive planet. Does this mean that our solar system is unusual? If so, how or why? If not, why not?

41. *Finding Planets.* Arrange the following planetary systems in order of decreasing detectability. Write a sentence defending each system's position in your list, and describe a scheme that might be able to detect such worlds. (a) Earth-size planets in Earth-size orbits; (b) Jupiter-size planets in Jupiter-size orbits; (c) Earth-size planets in Jupiter-size orbits; (d) Jupiter-size planets in Earth-size orbits; (e) Jupiter-size planets in interstellar space.

42. *Are Earth-Like Planets Common?* Based on what you have learned in this chapter, form an opinion as to whether you think Earth-like planets will ultimately prove to be rare, common, or something in between. Write a one- to two-page essay explaining and defending your opinion.

43. *No Hot Jupiters Here.* How do we think "hot Jupiters" formed? Why didn't one form in our solar system?

44. *Life on a Synchronously Rotating Planet.* Planets in the habitable zones of M stars are likely to rotate synchronously with their orbits, because these habitable zones are so close-in. Computer simulations suggest that on a synchronously rotating planet with a thick atmosphere, winds would carry heat from the side continually facing the star to the back, dark side. If so, there would be a ringlike zone between the light and dark halves that might be habitable. What are some of the adaptations you would expect for life in this zone? Explain.

45. *Ages of Stars on the H–R Diagram.* The giants and supergiants on the H–R diagram are stars in the last stages of their lives. Does this mean they are *older* than most main-sequence stars? Why or why not?

46. *Nightfall.* Read the short story "Nightfall" by Isaac Asimov. If such a planet really exists, do you think the scenario described is realistic? Why or why not? Summarize and defend your opinions in a one- to two-page essay.

Quantitative Problems

Be sure to show all calculations clearly and state your final answers in complete sentences.

47. *Number of Stars with Habitable Planets.* Assume that the Milky Way Galaxy has 500 billion stars (a reasonable estimate). Based on the statistics given in this chapter, approximately how many stars would be Sun-like? How many would be K stars? How many would be M stars? If you assume that an average G star has one planet in its habitable zone, while only one in five K stars and one in ten M stars has such a planet, how many total planets would you expect to find in habitable zones in the Milky Way Galaxy?

48. *Transit of HD209548.* The star HD209548, which has a transiting planet, is roughly the same size as our Sun, which has a radius of about 700,000 kilometers. The planetary transits block 1.7% of the star's light.
    a. Calculate the radius of the transiting planet. (*Hint:* The brightness drop tells us that the planet blocks 1.7% of the *area* of the star's visible disk; the formula for the area of a circle is $\pi \times r^2$.)
    b. The mass of the planet is approximately 0.6 times the mass of Jupiter, and Jupiter's mass is about 1.9×10^{27} kg. Calculate the average density of the planet. Give your answer in grams per cubic centimeter. Compare this density to the average densities of Saturn (0.7 g/cm^3) and Earth (5.5 g/cm^3). (*Hint:* To find the volume of the planet, use the formula for the volume of a sphere: $V = \frac{4}{3} \times \pi \times r^3$; be extremely careful with unit conversions.)

49. *Lost in the Glare.* How hard would it be for an alien astronomer to detect the light from planets in our solar system compared to light from the Sun itself?
    a. Calculate the fraction of the total emitted sunlight that is reflected by Earth. (*Hint:* Imagine a sphere around the Sun the size of the planet's orbit (area $= 4\pi a^2$). What fraction of that area does the disk of a planet (area $= \pi r_{\text{planet}}^2$) take up? Earth's reflectivity is 29%.
    b. Would detecting Jupiter be easier or harder than detecting Earth? Comment on whether you think Jupiter's larger size or greater distance has a stronger effect on its detectability.

You may neglect any difference in reflectivity between Earth and Jupiter.

50. *Planet Around 51 Pegasi.* The star 51 Pegasi has about the same mass as our Sun. A planet discovered around it has an orbital period of 4.23 days. The mass of the planet is estimated to be 0.6 times the mass of Jupiter. Use Kepler's third law to find the planet's average distance (semimajor axis) from its star. (*Hint:* Because the mass of 51 Pegasi is about the same as the mass of our Sun, you can use Kepler's third law in its original form, $p^2 = a^3$ [Section 2.2]. Be sure to convert the period into years before using this equation.)

51. *Finding a Planetary Mass.* Using the Doppler technique, you discover a planet that is causing its star to move at a maximum speed of 14 meters per second. The planet has an orbital period of 56 days and an average orbital distance of 55 million kilometers from its star. What is the planet's mass? (*Hint:* See Cosmic Calculations 11.1.)

52. *The Doppler Formula.* The amount of Doppler shift for light or radio waves can be calculated from this formula:

$$\frac{\text{wavelength shift}}{\text{rest wavelength}} = \frac{v}{c}$$

The rest wavelength is the wavelength of a particular spectral line in an object that is not moving (relative to us), v is the velocity of the star from which we observe a wavelength shift, and c is the speed of light (c = 300,000,000 m/sec). Suppose that, in a particular star, a spectral line with a rest wavelength of 600 nm is found to be shifted by 0.1 nm (toward the blue). How fast is that star moving toward us, in meters per second? (1 nm $= 10^{-9}$ m.)

53. *Finding a Center of Mass.* In the simple case of a two-body system—for example, a star and a single planet—the position of their center of mass can be determined from:

$$m_{\text{star}} r_{\text{star}} = m_{\text{planet}} r_{\text{planet}}$$

where m_{star} and m_{planet} are the masses of the star and the planet, respectively, and r_{star} and r_{planet} are the distances of the star and the planet from their center of mass. Consider the Sun and Jupiter, which are separated by 780 million kilometers. Using the fact that the Sun's mass is about 1,000 times the mass of Jupiter, about how far is the center of mass from the center of the Sun? How does this distance compare to the Sun's radius of 700,000 kilometers?

Discussion Questions

54. *Future Mission.* A wealthy benefactor has just given you a large grant to search for Earth-like planets around other stars. What would you do? Explain.

55. *The Copernican Principle and Rare Earth.* The Copernican revolution taught us that our planet is not the center of the universe, as had been generally believed before that time. Taking this lesson to heart, we have since assumed that our planet is not "central" or "special" in any way but rather that we are on a fairly typical planet in a fairly typical place in the universe. This principle, often called the *Copernican Principle* or the *Principle of Mediocrity,* has been borne out many times since. For example, we have learned that we are not near the center of our galaxy and that the universe has no center at all. Do you consider this principle to be in conflict with the rare Earth hypothesis? If so, does this make the rare Earth hypothesis any less scientific? Defend your opinions.

WEB PROJECTS

56. *New Planets.* Find the latest information about extrasolar planet discoveries. Create a personal "planet journal," complete with illustrations as needed, with a page for each of at least three recent discoveries of new planets. On each journal page, be sure to note the technique that was used to find the planet, give any information we have about the nature of the planet, and discuss how it does or does not fit in with our current understanding of extrasolar planets in general.

57. *The Kepler Mission.* The *Kepler* mission is the first funded mission designed expressly to look for Earth-size planets around other stars. Go to the *Kepler* Web site and learn more about the mission. Write a one- to two-page summary of the mission's goals and its current status.

58. *Planet-Finding Interferometers.* Other future missions will use interferometry to learn about extrasolar planets. Go to the Web site for one future interferometry mission under consideration, such as *SIM, TPF,* or *Darwin.* For the mission you choose, write a one- to two-page summary of the mission's goals and its current status.

12
The Search for Extraterrestrial Intelligence

LEARNING GOALS

12.1 THE DRAKE EQUATION
- What is the Drake equation?
- How well do we know the terms of the Drake equation?

12.2 THE QUESTION OF INTELLIGENCE
- Even if life is widespread, would intelligence be common?
- Will intelligence inevitably spawn technology?

12.3 SEARCHING FOR INTELLIGENCE
- How did SETI begin?
- How do we search for intelligence today?
- What happens if SETI succeeds?

12.4 THE PROCESS OF SCIENCE IN ACTION: UFOS AND ALIENS ON EARTH
- What have we learned from UFO sightings?
- Have aliens left any verifiable evidence of visitation?
- Is there a case for alien visits?

The probability of success is difficult to estimate;
but if we never search, the chance of success is zero.

P. Morrison and G. Cocconi, *Nature*, 1959

There are approximately 2,500 stars within 50 light-years of Earth. If any of these nearby stellar systems have inhabited planets, the inhabitants could already know that we exist by picking up our high-frequency radio, radar, and television transmissions. These broadcasts—unintentional evidence of our technological society—are moving into space at the speed of light and are currently washing over star systems at the rate of almost one new star system a day. By using sufficiently powerful radio telescopes, others could learn that we're here.

We are only beginning to signal our presence, but other civilizations may have been doing something similar for a long time. In the Milky Way Galaxy alone, there could be tens of billions of Earth-like worlds, and some of these could be filling the interstellar voids with their broadcasts. Using both radio and optical telescopes, scientists are attempting to find such transmissions. These experiments, called the *search for extraterrestrial intelligence* (SETI), are the primary topic of this chapter.

12.1 THE DRAKE EQUATION

The search for extraterrestrial intelligence (SETI) differs in a fundamental way from all the other searches for life that we have discussed in this book. Those searches are concerned not just with finding life itself, but with finding evidence that might point to its existence elsewhere—such as whether habitable planets are common or rare or whether our understanding of the origin of life would allow for life to arise on Mars. In contrast, SETI restricts itself to seeking clear and conclusive evidence of technologically advanced life.

Indeed, if SETI is successful in receiving and perhaps interpreting a message from a distant civilization, it might give us answers to many or most of the other questions we have discussed. To begin with, the discovery of a distant civilization would immediately prove that life is not unique to Earth. Because it is likely that any extraterrestrials we might detect with SETI experiments would be more advanced than we are, there's at least a possibility that we could learn a great deal from their transmissions, if we could understand them.

The principal goal of this chapter is to explore the methods by which scientists are now searching for evidence of other civilizations. First, however, it's worth asking whether our current understanding of life in the universe gives us any good reason to believe that the search may be successful. To some extent, the answer to this question may not matter, as stated so eloquently in the quotation that opens this chapter. That is, our innate scientific curiosity inspires us to search even if we cannot be certain that the search will ever be successful. But although we may not know the probability of success, recent advances in astrobiology allow us to say a lot

Figure 12.1

Astronomer Frank Drake, with the equation he first wrote in 1961. He is currently a Senior Scientist at the SETI Institute in California.

about the factors that might influence this probability. As in all of science, knowledge about such factors can provide important guidance to research efforts. In this section, we will identify the major factors that relate to the probability of success of SETI efforts—factors that are summarized succinctly by a simple formula known as the *Drake equation.*

• What is the Drake equation?

Early in the search for evidence of other civilizations, scientists realized that it would be useful to try to estimate the chances of finding something. In 1961, the first scientific conference on the search for extraterrestrial intelligence was held in Green Bank, West Virginia, at the radio observatory where a pioneering hunt for an alien signal had recently been conducted. (We will discuss this search, called Project Ozma, in Section 12.3.) There were only about a dozen attendees at the conference—just about the entire world's complement of people with a professional interest in the subject at the time—but their expertise ranged across the disciplines of astronomy, biology, and engineering. In setting the meeting's agenda, astronomer Frank Drake (Figure 12.1) tried to summarize the factors that would determine whether any attempt to detect intelligent extraterrestrials could succeed. He came up with a simple equation that, at least in principle, could be used to calculate the number of civilizations existing elsewhere in our galaxy (or in the universe at large) *from which we could potentially get a signal.* Note that this definition limits what we mean by "civilization" in this context. For example, the ancient Greeks had a remarkable civilization, but theirs doesn't count under this definition because they never developed radio or other technologies that could be used to effectively communicate across space.

The **Drake equation**, as it is now called, does not give us a definitive answer for the number of transmitting civilizations. Rather, it lays out the factors that are important in determining this number. The equation has played a guiding role in research bearing on life in the universe, because much of it deals with life in general and not just the smaller fraction of life that is intelligent enough to produce signals. Many of the factors in the Drake equation have been discussed earlier in this book as part of our search for biology on other worlds. Consequently, we expect that—with improvements in our knowledge of these factors—our estimate of the number of signaling civilizations will get better. However, we needn't await such improvements before we embark on a SETI search.

To keep our discussion of the Drake equation simple, we'll focus only on the number of civilizations in our own galaxy. We can always extend our estimate to the rest of the universe simply by multiplying the result we find for our galaxy by 100 billion, the approximate number of galaxies in the observable universe [Section 3.2]. We limit our focus for practical reasons, too. Any signals from other galaxies will be severely weakened by distance, making them far harder for us to detect. In addition, to stay consistent with key ideas already discussed, we'll consider a slight modification of Drake's original equation. (For the original version, see the box on p. 402.) In modified form, the Drake equation looks like this:

$$\text{Number of civilizations} = N_{HP} \times f_{life} \times f_{civ} \times f_{now}$$

Let's examine each term to see how this equation tells us the number of civilizations in the Milky Way Galaxy capable of interstellar communication:

- N_{HP} is the number of habitable planets in the galaxy. It is the first term because we assume that a prerequisite to having life or a civilization is having a habitable planet on which that life can evolve.
- f_{life} is the fraction of habitable planets that actually *have* life. For example, if $f_{life} = 1$, it would mean that all habitable planets have life; if $f_{life} = \frac{1}{1,000,000}$, it would mean that only 1 in a million habitable planets has life. Thus, the product $N_{HP} \times f_{life}$ tells us the number of life-bearing planets in the galaxy.
- f_{civ} is the fraction of the life-bearing planets on which a civilization capable of interstellar communication *has at some time* arisen. For example, if $f_{civ} = \frac{1}{1,000}$, it would mean that such a civilization has existed on 1 out of 1,000 planets with life, while the other 999 out of 1,000 have not had a species intelligent enough to build radio transmitters, high-powered lasers, or other devices for interstellar conversation. When we multiply this term by the first two terms to form the product $N_{HP} \times f_{life} \times f_{civ}$, we get the total number of planets on which intelligent beings have evolved and developed a civilization at some time in the galaxy's history.
- f_{now} is the fraction of the civilization-bearing planets that happen to have a civilization *now*, as opposed to, say, millions or billions of years in the past. This term is important because it tells us how many civilizations we could potentially talk to,* since there's no point in listening for civilizations that are long gone. Because the previous three terms told us the total number of civilizations that have *ever* arisen in the galaxy, multiplying by f_{now} tells us how many civilizations we could potentially make contact with today. For example, if the first three terms were to tell us that ten million planets in the galaxy have at some time had a communicating civilization but f_{now} turns out to be one in five million, then only two civilizations could be expected to exist today. As we will see shortly, the value of f_{now} must depend largely on how long civilizations survive once they arise.

To summarize, the Drake equation gives us a way to calculate the number of civilizations capable of interstellar communication that are currently sharing the Milky Way Galaxy with us. As such, it provides a useful way of organizing our thinking about the problem, because it tells us exactly what numbers we need to know to learn the answer. Indeed, it suffers from only one significant drawback: We don't know precise values for any of its terms!

THINK ABOUT IT Try the following sample numbers in the Drake equation. Suppose there are 1,000 habitable planets in our galaxy, that 1 in 10 habitable planets has life, that 1 in 4 planets with life has at some point had an intelligent civilization, and that 1 in 5 civilizations that have ever existed is in existence now. How many civilizations would exist at present? Explain.

*For the purposes of the Drake equation, we'll assume that the term f_{now} takes into account the light-travel time for signals from other stars; for example, if a star with a civilization is 10,000 light-years away, it counts in determining f_{now} if the civilization existed 10,000 years ago, because signals broadcast at that time would just now be arriving at Earth.

• How well do we know the terms of the Drake equation?

The only term in the Drake equation for which we can make even a reasonably educated guess is the number of habitable planets, N_{HP}. As we discussed in Chapter 11, the detections of extrasolar planets to date show that planets of some kind are likely to be quite common. While we cannot yet be sure that this also means that *habitable* planets are common, so far we have no good reason to think otherwise. Moreover, the example of our own solar system suggests that it is possible to have more than one habitable planet per system, since we now suspect that Mars was habitable at some point in its past. Given that there are several hundred billion stars in the Milky Way Galaxy, it seems entirely reasonable to suppose that there could be 100 billion or more habitable planets. Nevertheless,

SPECIAL TOPIC 12.1: FRANK DRAKE AND HIS EQUATION

If anyone could be called the Father of SETI research, Frank Drake is that person. As a young man, Drake learned electronics in the Navy. He then studied for an advanced degree in astronomy, and on graduation took a job at the new National Radio Astronomy Observatory (NRAO) in Green Bank, West Virginia. In the late 1950s, the Observatory was busy with the construction of a large radio telescope, a project that would take many years. Consequently, the Observatory opted to buy an "off-the-shelf" instrument that could be used right away. This telescope boasted a 26-meter (85-foot) reflector, which at the time was larger than most of the world's operating radio telescopes.

Once this telescope was up and running, the Observatory staff was encouraged to suggest interesting experiments for its use. Drake had already been thinking about the possibility of interstellar communication by radio, and he proposed a simple experiment to search for alien signals from two nearby star systems. This became Project Ozma, the first modern SETI search. A year later, in 1961, Drake organized a conference at Green Bank to discuss the possibility that such an experiment could actually find an extraterrestrial transmission. His "agenda" for that conference became known as the Drake equation.

Drake's original form for his equation is:

$$N = R_* \times f_{\text{planet}} \times n_e \times f_{\text{life}} \times f_{\text{intell}} \times f_{\text{civ}} \times L$$

N is the number of transmitting civilizations in our galaxy. R_* is the galactic birthrate of stars suitable for hosting life, in stars per year. For example, because there are roughly 100 billion stars in the Milky Way and the galaxy is approximately 10 billion years old, R_* is approximately 10 per year (making the rather crude assumption that all stars are suitable). The term f_{planet} is the fraction of such stars having planets; n_e is the number of planets per solar system that have an environment favorable for life; f_{life} is the fraction of such planets on which life actually evolved; f_{intell} is the fraction of inhabited worlds that develop intelligent life; f_{civ} is the fraction of planets having intelligent beings that produce a civilization capable of interstellar communication; and L is the lifetime that such civilizations are "on the air," broadcasting signals. You might want to compare these terms with the more compact factors used in the equation on page 400.

Using some admittedly optimistic estimates for the first six terms in the equation, Drake suggested that we could reasonably guess that they would multiply to approximately one per year. This is the "birthrate" of civilized societies in the Milky Way. To find out how many are broadcasting now, we need only multiply this rate by L, the number of years during which they broadcast. This is analogous to determining the number of students attending a college: It's the number per year that enter as freshmen (the entrance rate) times the number of years they spend as students (typically four). If the birthrate for civilizations is taken to be one per year, the Drake equation becomes simply $N = L$; that is, the number of transmitting civilizations is simply the average lifetime (in years) of a transmitting society.

Unfortunately, L is dependent on sociology rather than on astronomy or biology—making it far more difficult to determine its value. Attempts to estimate L usually try to guess what the one technological civilization we know—our own—is likely to do. Only a half-century after inventing radio, we also developed atomic weapons. To some people, this suggests that L might be very short—only a few centuries or less. On the other hand, we can be optimistic and assume that we will survive our own technology and exist as a society for millions of years into the future. Perhaps one of the most important things we could learn from a SETI detection is that not all technologically sophisticated societies are doomed to early self-destruction.

Frank Drake's personalized license plate.

the actual number might be far smaller, and we won't know for sure until we begin to get data about habitable planets—which should begin to happen in the course of the next 10 years, as new technology makes it possible to detect small, rocky worlds and perhaps even to analyze their atmospheres for gases that might betray the presence of life [Section 11.2].

The rest of the terms in the Drake equation present more difficulty. For the moment, we have no rational way to estimate the fraction f_{life} of habitable planets on which life actually arose. The problem is that we cannot generalize when we have only one example to study—our own Earth. Still, we are not completely without guidance. The fact that life apparently arose rapidly on Earth [Section 6.1] suggests that the origin of life was fairly "easy," in which case we might expect that most habitable planets would also have life, making the fraction f_{life} close to 1. However, until we have solid evidence that life arose anywhere else, such as on Mars, it remains possible that Earth was somehow extremely lucky and that f_{life} is so close to 0 that life has never arisen on any other planet in our galaxy.

Similarly, we have little basis on which to guess the fraction f_{civ} of life-bearing planets that eventually develop a civilization. On the one hand, the fact that life flourished on Earth for almost 4 billion years before the rise of humans might suggest that it is very difficult to produce a civilization even when there is life. On the other hand, given that the majority of all stars in the Milky Way are older than our Sun, there has been plenty of time for evolution to work on numerous planets. Any evolutionary drive toward intelligence might inevitably lead to huge numbers of civilizations, even if it takes a long time on any given world. This question of whether intelligence is a rare accident or an inevitable result of evolution is so important to the issue of the search for extraterrestrial civilizations that we will devote the next section to investigating it.

The final term in the equation, f_{now}, is particularly interesting because it is related to the survivability of civilizations. Consider our own example. In the roughly 13 billion years during which our galaxy has existed, we have been capable of interstellar communication via radio for only about 50 years, or only about 1 part in 250 million of our galaxy's history. If we were to destroy ourselves tomorrow (saving students the unpleasantness of a final exam), our technological "lifetime"—the length of time we could make ourselves known to other star systems—would be only 50 years. If this is typical of other civilizations, then our chances of finding a signal from any one of them at any random time would be only 1 in 250 million. In that case, even if there have been hundreds of millions of civilizations in the galaxy's history, no more than a few would be detectable now.

Of course, we have not yet destroyed ourselves, and we may be severely underestimating the fraction f_{now}. For example, suppose civilizations have been arising for the past 10 billion years and routinely survive for 1 billion years. Then the chance of a signal reaching us from any given civilization at a random time is 1 in 10. If $f_{\text{now}} = \frac{1}{10}$, then there may be numerous communicating civilizations out there now, even if civilizations arise rather infrequently. To take some numbers, suppose that only 1 in 10 million stars ever gets a planet with a civilization. In a galaxy of 100 billion stars, this would mean that only about 10,000 civilizations ever arise. But if $\frac{1}{10}$ of them are here *now*, then there are some 1,000 civilizations we could potentially find. Four decades ago, considerations like these led Frank Drake to conclude that the typical lifetime of civilizations must be one of the primary factors—perhaps even *the* primary factor—in the potential success of SETI efforts.

The Drake equation is mathematically simple, but its chain of terms is only as strong as its weakest link. If we know one term poorly, there is no way to improve our estimate of the number of civilizations by knowing other terms well. For example, while we might have better estimates of most of the factors in the equation as our knowledge of astronomy and biology improves, f_{now} depends on sociological factors—that is, the behavior of alien civilizations. Do they quickly self-destruct, or do they survive for long periods? The only way we can make realistic estimates of f_{now} is by actually detecting extraterrestrial societies. Thus, as long as we have this uncertainty in f_{now}, we will face a corresponding uncertainty in knowing the total number of signaling worlds even if astrobiology research someday allows us to pin down precise values for terms such as N_{HP} and f_{life}.

As a result of this "weakest link" problem, as well as the fact that all of the terms are still highly uncertain, we cannot draw any definitive conclusions from the Drake equation. Indeed, some of the factors have such a large uncertainty that the numbers we enter into the equation (choosing numbers within the range consistent with our present knowledge) can give us anything from an optimistic view that a large fraction of stars in our galaxy have intelligent, communicating beings to a pessimistic view that we would have to search a large number of galaxies to find even one other example of intelligence. The main value of the Drake equation, then, is in pointing out what factors are important and underscoring the implications of our lack of knowledge about particular factors. Therefore, it can be used to help us recognize what the issues are and where we remain ignorant, and thus steer us to areas in which more research is needed.

THINK ABOUT IT The Drake equation assumes that each transmitting civilization has sprung up independently on its own habitable planet. Is this a reasonable assumption, or might it be too limiting? Defend your opinion.

COSMIC CALCULATIONS 12.1

The Distance Between Signaling Societies

How far is it to the nearest other world with technologically advanced beings? We don't know, of course, but if we use the Drake equation to estimate the number of civilizations, we can then compute their average separation.

We start by estimating the volume V of space available for civilizations in the Milky Way Galaxy, assuming that civilizations are confined to the galaxy's disk:

$$V = \pi R^2 \times T$$

where R is the disk radius and T is the disk thickness.

Suppose the Drake equation tells us that there are N technological civilizations in our galaxy. If we assume that these civilizations are spread randomly, then the average volume of space that contains just one civilization must be the total volume of the galaxy divided by the number of civilizations, V/N. If we consider this volume per civilization to be a cube in which each side measures d light-years, then d is also the distance from the center of one cube to the center of the next, which means it is the average distance between civilizations. The volume of a cube with side length d is d^3, so $d^3 = V/N$. Solving for d, we find:

$$d = \left(\frac{V}{N}\right)^{1/3} = \left(\frac{\pi R^2 \times T}{N}\right)^{1/3}$$

Example: Suppose that $N = 10{,}000$. What is the average distance between civilizations?

Solution: The disk of the galaxy has radius $R = 50{,}000$ light-years and thickness $T = 1{,}000$ light-years, so its volume is:

$$V = \pi R^2 \times T = \pi \times (50{,}000\ \text{ly})^2 \times (1{,}000\ \text{ly}) \approx 8 \times 10^{12}\ \text{ly}^3$$

We use this volume to calculate d:

$$d = \left(\frac{V}{N}\right)^{1/3} = \left(\frac{8 \times 10^{12}\ \text{ly}^3}{10{,}000}\right)^{1/3} \approx 900\ \text{ly}$$

If there are 10,000 civilizations in the disk of our galaxy, the average distance between these civilizations is nearly 1,000 light-years.

12.2 THE QUESTION OF INTELLIGENCE

Our discussion of the Drake equation has pointed out several key factors that we must understand better if we want to know how many other civilizations exist. We have already discussed (in earlier chapters) the uncertainties surrounding the number of habitable planets and the origin of life. However, we have not yet discussed the ideas that influence our estimate of f_{civ}, which describes the probability that life will eventually give rise to intelligence and a technologically adept civilization. In this section, we turn our attention to this important topic.

• Even if life is widespread, would intelligence be common?

The probability of a SETI success—receiving a signal—depends on how many civilizations are out there and broadcasting, and intelligence is a prerequisite to a civilization. Thus, SETI is likely to be successful only if intelligent life is widespread. But is it? Broadly speaking, there are two opposing schools of thought on this question.

One school considers intelligence that is comparable to our own (in other words, one that is able to develop both science and technology) to be unlikely. From this point of view, biology might be widespread but the evolution of technological intelligence extremely rare. Life has existed on our planet for at least 3.5 billion years. But only in the last few million years has our genus *Homo* developed the capability to understand its environment, and only within the last half-millennium have we come to understand the nature of Earth and begun our exploration of the cosmos. At the very least, the late appearance of *Homo sapiens* on Earth suggests that a long period of evolution must precede the emergence of technologically intelligent creatures.

Moreover, as we discussed in Chapter 6, our existence seems to have resulted from a number of chance events. For example, the Cambrian explosion that gave rise to the "body plans" of all modern animals (including those of our phylum, Chordata) might have been the result of environmental stress introduced by Snowball Earth episodes or a massive asteroid strike, and the rise of mammals might never have occurred if the K–T impact had not wiped out the dinosaurs (Figure 12.2). The chance nature of these events and the many other forks in the evolutionary road suggest to some people that the appearance of technological intelligence on Earth was an enormously improbable event.

The second school of thought holds the opposite view. It proposes that there is evolutionary pressure for intelligence; that is, that various evolutionary mechanisms consistently encourage an increase in intelligence for a wide range of species. If this is true, then some technologically intelligent species would still have evolved on Earth even if a different sequence of past events had prevented the existence of humans. Because we are the only species on our planet that has ever developed an advanced civilization, no other known species directly supports the view that the evolution of technological intelligence is likely. Instead, those who adopt this viewpoint look more generally at the process of evolution for evidence that some of the workings of natural selection promote greater intellectual capability.

Convergent Evolution The evolutionary argument in favor of widespread intelligence is based on the phenomenon of **convergent evolution**, the tendency of organisms of different evolutionary backgrounds

Figure 12.2

The path of evolution on Earth was severely affected by chance events such as the K–T impact 65 million years ago, which wiped out the dinosaurs and most other species. This artist's impression depicts the 10-kilometer-wide asteroid as it approached Earth. If this rock had arrived at Earth's orbit 20 minutes earlier (or 20 minutes later), it would have missed our planet. Would intelligent beings still have eventually arisen?

that occupy similar ecological niches to come to resemble one another. In such cases, natural selection often produces analogous adaptations. One example of convergent evolution is the shape of large marine predators: Dolphins and sharks evolved independently from earlier mammals and fish, respectively, but both have a similar streamlined body form. The obvious reason for this is that being shaped like a torpedo makes for greater speed underwater—and speed has clear survival value for a predator. We say that the evolution of originally quite different animals has *converged* on this optimized underwater shape.

Eyesight offers another example. Vision is a useful adaptation, and most multicellular animals have some ability to see. However, the eye was not a unique evolutionary invention. Studies of evolutionary relationships show that eyes evolved independently at least eight different times. Indeed, the design of the human eye is by no means the best. Compared to the eyes of some other animals, ours have "flaws," such as the fact that the nerves in our eyes come together in a bundle before exiting through the back of our eyeballs, resulting in a small blind spot. The several independent origins of eyes suggest that evolution tends to *converge* toward developing some kind of eye to provide vision.

THINK ABOUT IT All crabs may look quite similar, but DNA studies show that various crab species evolved from very different ancestors; for example, some crab ancestors were shrimplike and others were lobsterlike. Based on this information, how is a crablike body, with its round shape and unusual manner of walking, an example of convergent evolution? Why might this body type be a natural evolutionary development?

Like speed or eyesight, intelligence—the way an animal processes information—is subject to natural selection. If there are ecological roles for which keener intelligence has survival value, then we would expect a convergent evolution to greater brain power for animals in these niches. Intelligence would be just as likely to emerge as any other generally useful adaptation. In this case, evolution would tend to raise the level of intelligence to at least some degree in a great many species (much as eyes developed in many species), which in turn would increase the chance that some of these species would evolve even higher intelligence. With more players in the game, the chance of producing human-style intelligence would be greater. Thus, if we could establish the presence of an evolutionary trend favoring intelligence, we would be encouraged to think that the emergence of technological intelligence elsewhere is not enormously improbable.

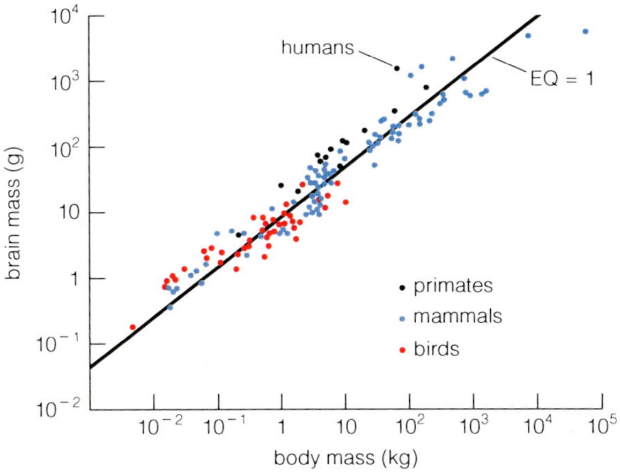

Figure 12.3

This graph shows how brain mass compares to body mass for some mammals (including primates) and birds. The straight line represents an "average" of the ratio of brain mass to body mass, which we define as an encephalization quotient of EQ = 1. Animals that fall above the line have an EQ greater than 1 and are therefore "smarter" than average. Animals that fall below the line have an EQ less than 1 and are "intellectually challenged." Note that the scale uses powers of 10 on both axes. (Adapted from Sagan, 1977.)

Measuring Intelligence

In principle, we can test whether intelligence is evolutionarily favored by measuring the brain power of a variety of animal species over time. But how can we measure the mental ability of animals? Few creatures are eager or able to take IQ tests, but we can resort to a simpler measure of raw brain power based on brain mass.

Figure 12.3 plots brain mass against body mass for a sample of birds and mammals (including primates). There is a clear and expected trend: Heavier animals have heavier brains. By drawing a straight line that fits these data, we define an average brain mass for each body mass. Those animals whose brain mass falls on this line are said to have an **encephalization quotient (EQ)** of 1, which means a typical allotment of mental ability for creatures of their size. They have enough brain mass, we can assume, for a basic set of behaviors. If their brain mass falls above this line, then it seems reasonable to suppose that they are capable of more elaborate behavior. Creatures whose brain mass falls below the line are presumably less mentally agile than average animals in this sample. The EQ, the brain mass relative to the value on the line EQ = 1, serves as an indicator of general intelligence.

Although other indicators of intelligence have been suggested (the amount of brain folding, for example, or the number of neural connections), measuring EQ has the advantage of being fast and easy. For example, notice that the EQ = 1 line shows that the average species with body mass 1 kilogram (read off the horizontal axis) has a brain mass of about 10 grams (along the vertical axis). We'd therefore say that a species with the same body mass but a brain weight of 20 grams has EQ = 2, meaning it has twice as much brain as average for its size. Similarly, a species with body mass 1 kilogram and brain mass 5 grams has EQ = $\frac{1}{2}$, meaning it has only half as much brain as average for its size. Besides this ease of computation, EQ also has the advantage of being something we can estimate for extinct species, since we can often estimate their total body masses and we can gauge their brain masses from the volume of their fossil skulls.

Admittedly, EQ is a simple measure, and it might be likened to judging the computational ability of computers by weighing their CPU chips. Nonetheless, EQ seems to correlate well with complex behavior. Carnivorous animals that need to hunt down their meals generally have higher EQs than leaf eaters, and animals that lavish care on their offspring score higher than those that ignore them.

Exploring the Evolution of Intelligence

What happens when we use EQ to investigate the premise that, on Earth, there has been a trend toward increasing brainpower over time? If we look at contemporary species, we find that although humans don't sport the most massive brains (whales, for instance, have much larger brains than we do), we do

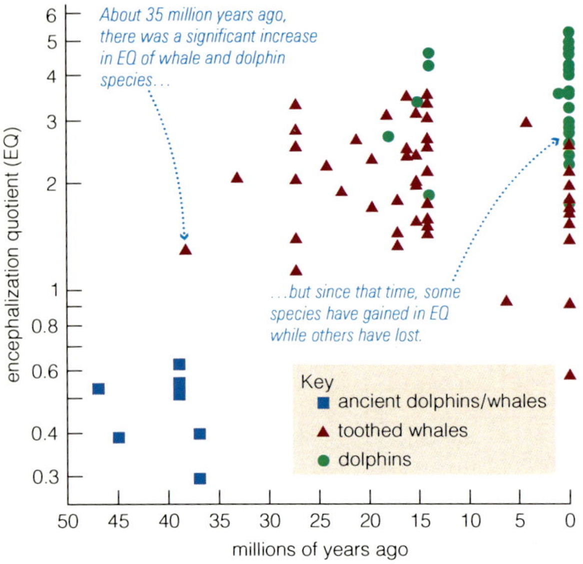

Figure 12.4

This graph shows the change in encephalization quotient (EQ) with time for toothed whales and dolphins. Note that over the course of the last 50 million years, some EQs increased substantially, while others remained relatively unchanged and some even went down. The fact that some dolphins have EQs comparable to our own suggests that high intelligence has survival value in many species, and may be a common evolutionary development. (Adapted from Marino, L., McShea, D., Uhen, M. D. 2004, *The Anatomical Record.* 281A: 1247–1255. Used with permission.)

have the biggest brains in relation to body mass. Our species' encephalization quotient is 7, meaning that our brains are 7 times more massive than would be expected for an "average" mammal of the same body weight. This EQ not only exceeds that of chimps and dolphins, whose EQs are 2.5 and 4 or 5, respectively, but it is higher than that of any other known species, alive or extinct. Thus, these numbers confirm what we already know: We're the cleverest critters on the planet.

However, a look to the past shows that we were not the only creatures evolving toward greater intelligence. Biologists recently measured the EQs for a range of toothed whales and dolphins, and noted how these values changed during the past 50 million years (Figure 12.4). They did this by using computer tomography to determine the brain volumes of long-dead animals whose fossilized skulls were sitting in museums. The animals' weight was estimated by measuring the size of some of the bones around the eye sockets, a parameter known to be strongly correlated with body mass. With data in hand, they then computed the EQs of 200 specimens, representing 62 cetacean species. What they found was that these animals experienced a major improvement in EQ 35 million years ago when they developed the navigation scheme known as echolocation (using audible "pings" as sonar), and some species underwent another big shift 15 million years ago. Not all the EQs went up, and today's cetaceans have a range of EQ from 0.2 to 5. But the smarter ones are not far behind our own EQ of 7. Humans are not closely related to dolphins in an evolutionary sense, so the fact that they have also developed large brains suggests that there is real survival value in cleverness, and that there are many ways that nature can produce it.

Another way to approach the question of the evolution of intelligence is to ask what factors might encourage its appearance and whether those factors are likely to be commonplace. To begin with, a good many "preadaptations" seem necessary. High-performance brains like those of birds and mammals need a vigorous metabolism, so intelligence is most likely to arise in warm-blooded animals. In addition, a relatively large body size is necessary to house a large brain (although recent research with insects suggests that, by miniaturizing neurons, nature might in principle be able to pack a lot of intelligence into small packages). A long period of parental care of offspring is probably also a necessary precondition for intelligence, which wouldn't be that useful without a time to learn from parents. These preconditions have existed for a wide range of species on Earth during at least the last 50 million years. Because this is the same period in which intelligence seems to have risen most dramatically, we have some hint that intelligence has indeed been evolutionarily favored.

Still, we might wonder about precisely which evolutionary mechanisms will select for intelligence. After all, having a big brain is metabolically expensive; in humans, for example, roughly one fourth of our metabolism is for the benefit of our brain, even though that brain accounts for only about 2% of body weight. Consequently, big brains will occur only in species in which the cost is rewarded with real survival advantages. One survival benefit might come from the interactions among individuals of social species. Dolphins and primates are highly social animals. Success in a social environment is enhanced by intelligence, because there is survival value in being able to judge the mood and meaning of fellow creatures. Social position in your troop or pod depends on how savvy and canny you are. Moreover, an elevated social position often allows you to have first choice in mates, so clever, high-ranking individuals will tend to produce clever, high-ranking offspring. We might reasonably expect intelligent aliens, if they exist, also to have evolved in interactive social environments.

Competition between species can also ratchet up intellect. For example, large carnivores and their prey encourage improvements in one another's intelligence. When a lioness stalks gazelles for dinner, she's more likely to catch a gazelle that's less aware of its surroundings or less cagey in devising escape maneuvers. The lion gets a meal and in the process inadvertently raises the average intelligence of the surviving gazelles. The rise in gazelle intelligence makes getting a meal tougher for less mentally agile lions, which then preferentially drop out of the gene pool. The less alert, less cunning of both species are weeded out, and the smarter members survive.

In summary, several common circumstances in the animal world naturally select for brain power, and there is evidence that more than one group of animals on our planet has been on the track to high general intelligence. This argues against the idea that the appearance of intelligence is some special, extraordinary accident of evolution on Earth and suggests that, given time and a competitive environment, many intelligent species will arise on any inhabited world. The probability of many candidates means at least some possibility that one or more of these species will develop humanlike brainpower. On the other hand, increased intelligence comes at a cost in terms of resources to support it (such as a high metabolism or carrying around a heavy head), and it is unclear that these resources would not be better utilized by, say, evolving the ability to run faster or fight more fiercely. Moreover, many species *didn't* evolve intelligence over the last 50 or 100 million years. In the end, the example of terrestrial life alone does not tell us whether there is an evolutionary imperative toward humanlike intelligence. And, even if there is such an imperative, we still must ask whether high intelligence necessarily leads to technical competence and the ability to communicate across interstellar distances.

• Will intelligence inevitably spawn technology?

It might seem natural that a species with sufficient brainpower will eventually develop science and technology. However, there are both physiological and sociological counterarguments to this idea.

On the physiological side, suppose dolphins were as intelligent as we are. Their lack of hands and their need to live in water would prevent them from building anything resembling modern technology. They might be smart and have sophisticated social structures, but without telescopes they wouldn't know much about astronomy, and without radios or lasers they wouldn't be able to talk to the dolphins of other worlds. Wolves have mouths, elephants have trunks, and ravens sport beaks and feet, but none of these animals have overall body designs that would allow them to manipulate complex tools—no matter how smart they might be.

On the sociological side, remember that *Homo sapiens* emerged in more or less its current form about 150,000 years ago. Certainly, by 20,000 or 30,000 years ago our ancestors had essentially the same level of intelligence as we do. Yet our ability to communicate through space is less than a century old, and it derives from the emergence of science—itself an endeavor that has developed only gradually over the past 2,500 years. Many different cultures developed mathematics and astronomy to varying degrees, but only the cultural line that emerged from ancient Greece ultimately led to modern science [Section 3.1]. Was this a lucky accident, or was it inevitable that some culture would develop modern technology?

Again, a single example—what happened on our own planet—cannot guarantee that science and technology will be common in the cosmos, even if high intelligence is. The only way we can answer the question of how frequent or rare technological civilizations might be is to find evidence of them on other habitable worlds. We'll now turn our attention to experiments that are trying to do just that.

12.3 SEARCHING FOR INTELLIGENCE

Receiving a message from another civilization could be one of the most important events of human history. We would know that we are not alone in the universe. We could conceivably learn a great deal about science and sociology. We could even learn how other civilizations have successfully survived periods in their history in which they had the power to destroy themselves. How might we receive such a message? In this section, we'll investigate the context in which we engage in SETI efforts. We'll begin with some historical background, then discuss the types of signals that SETI might be able to detect. We'll also explore a few possible ways of detecting extraterrestrial intelligence that don't involve communication signals.

• How did SETI begin?

Shortly after its invention, radio was recognized as a possible means of extraterrestrial communication. After all, radio travels at the speed of light and can easily bridge the airless voids of space. If extraterrestrials are using radio, then we might detect their presence without anyone having to leave the home planet.

Early Claims of Success As the twentieth century dawned, two pioneers of wireless technology became convinced (wrongly) that they had heard aliens on the airwaves. One of these pioneers was Guglielmo Marconi (1874–1937), generally celebrated as the man who made radio practical. The other was Nikola Tesla (1856–1943).

Tesla was both eccentric and brilliant. He was a prolific inventor (among other things, he made the first fluorescent lamp), but his most enduring legacy was the use of alternating current (AC) for distributing electrical power. AC—produced by having a voltage that cycles positive and negative 60 times a second—is commonplace today, but it was a radical concept when Tesla first proposed its use for a Niagara Falls generating station. The advantage of AC is that transformers can easily raise or lower the voltage, increasing it for low-loss transmission over the great distances from the power plant to your neighborhood and then lowering it for final distribution to your home. If you look at the power poles on any residential street that has aboveground electrical lines, you'll see the "step-down" transformers that provide the 115 volts (in the United States) feeding the wall sockets of the houses.

Tesla promoted AC because of these benefits, but he tried to go even further. He thought it might be possible to bring power to the people with "induction," rather than with copper cables, sending the energy through the air as low-frequency radio waves. To demonstrate his idea, he built a 60-meter-tall transmitting tower, wrapped with wire, in Colorado Springs, Colorado. His device, intended to radiate electrical energy, was an outsized example of what is now known as a Tesla coil. One night in 1901, the inventor noted that his apparatus was picking up electrical disturbances, which he ascribed to interplanetary communication. He wrote, "The feeling is constantly growing on me that I had been the first to hear the greeting of one planet to another."* We now believe that he was listening to an atmospheric phenomenon known as "whistlers," electrical noise created by distant lightning discharges.

That same year, Marconi successfully sent a radio signal across the Atlantic, proving that this invention was useful for communication over large distances. It took little imagination to guess that radio might also serve for sending messages into space. In the early 1920s, Marconi stated that he, too, had picked up signals that came from extraterrestrial sources. These experiments culminated in 1924, during one of the periods when Earth and Mars are closest to each other in their orbits. Marconi and others encouraged anyone with a radio set to listen for martian broadcasts, and even the U.S. Army joined in the search (Figure 12.5). There was considerable optimism that something would be detected, and a cryptographer was standing by in case the signals required decoding. Alas, while signals were picked up, there was no evidence that they were from Mars or any other extraterrestrial source. In retrospect, the signals that Marconi heard may also have been "whistlers" from distant lightning, or possibly garbled U.S. Navy broadcasts of which he was unaware.

Figure 12.5

An Army operator listening for martian radio signals, as pictured in *Radio Age*, October 1924.

In retrospect, these early experiments were doomed. We now know that there are no Martians sophisticated enough to assemble radio transmitters. In addition, we realize that these pioneer listening attempts were all made at the wrong spot on the radio dial. They were conducted at relatively low frequencies that have problems penetrating Earth's ionosphere—a layer of the upper atmosphere that consists of particles ionized by sunlight. The ionosphere acts as a radio "mirror," reflecting low-frequency radio emissions. Indeed, it was the ionosphere that bounced Marconi's 1901 transatlantic signal from England to Canada, thus allowing communication despite Earth's curved surface. The apparatus used by both Tesla and Marconi operated at frequencies too low to penetrate the ionosphere and therefore was insensitive to any cosmic broadcasts.

Although the radio pioneers might have mistaken thunderstorms for Martians, their enthusiasm for radio's use as a long-distance communication medium was ahead of its time. After World War II, when high-frequency radio equipment became widely available, it was possible to listen at frequencies that *could* penetrate the ionosphere. In addition, improvements in antennas and receivers made such experiments enormously more sensitive than the attempts made early in the century. The stage was set for a scientific approach to SETI.

Origins of Modern SETI The beginning of modern SETI is generally attributed to two physicists working at Cornell University. In 1959, Giuseppe Cocconi and Philip Morrison wondered how difficult it would be to send radio signals over interstellar distances. They made simple calculations showing that communication between nearby stars was possible using technology no more advanced than our own.

Cocconi and Morrison realized that the galaxy is considerably older than our solar system and consequently that there could be civilizations among the stars that have been around far longer than ours—perhaps surviving for many millions of years. For such long-lived societies, the

*Original source: Nikola Tesla, "Talking with the Planets," *Collier's Weekly* 26, no. 19, 4 (1901); taken from S. J. Dick, *The Biological Universe*, Cambridge University Press, 1996.

construction of powerful radio beacons that could be used to send "hailing signals" to other star systems should be a simple matter.

Cocconi and Morrison advocated looking for such beacons using large radio telescopes aimed at nearby stars. The exact arrangement of the planets around the stars under scrutiny was irrelevant, because the area of sensitivity of any reasonable-size radio telescope would encompass all of a star's possible planetary environs. The only other technical question concerned which frequencies to monitor with the radio receivers.

The radio portion of the electromagnetic spectrum (see Figure 3.30) extends over a wide range of frequencies. When we build a radio receiver, it is sensitive only to a particular set of frequencies within this wide range, which we refer to as its **band** of operation. For example, if you look at a typical FM radio, you'll see that it is designed to cover the band of frequencies from about 87 to 108 MHz. (Recall that the basic unit of frequency, hertz [Hz], is equivalent to waves [or cycles] per second. Thus, 1 megahertz [MHz] is a frequency of one million cycles per second.)

Of course, radio stations do not broadcast their signals over this entire band—if they did, no matter where you tuned your radio you'd hear the overlapping sounds of all the radio stations that were transmitting. Instead, radio stations limit their signals to only a narrow range of frequencies. For example, if your favorite radio station broadcasts at "97.3 on your FM dial," then you must tune your receiver so that it picks up only a small range of radio frequencies centered on 97.3 MHz. That range of transmitted frequencies is called the **bandwidth** of the signal. As a practical matter, the bandwidth is governed by how much information the broadcast contains. For example, most television signals (Figure 12.6) have a bandwidth 500 times wider than that of an AM radio station, because the TV signal contains picture (video) information in addition to sound (audio).

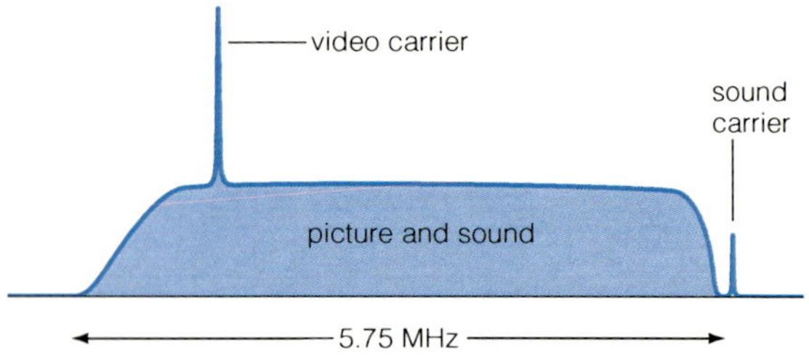

Figure 12.6

The spectrum of a standard-definition television signal, showing how the energy of transmission is spread over a bandwidth of nearly 6 MHz. The large bandwidth is necessary because the signal carries both sound and picture information. However, about one third of the total power in the TV signal is concentrated in the extremely narrow "spike" known as the *video carrier.* This carrier has a bandwidth of less than 1 Hz, or less than one six-millionth of the total bandwidth. Because the video carrier has so much of the total signal power concentrated into such a narrow bandwidth, it would be the easiest part of the signal to detect from a great distance. Hence, it is a good example of the type of narrow-band signal that SETI experiments seek. Note that a high-definition TV signal would, in principle, have even greater bandwidth, although by using digital techniques to compress the video information, this bandwidth bloat can be minimized.

Terrestrial broadcasts are made with a wide range of bandwidths depending on their purpose, but a hailing signal designed to get the attention of someone light-years away would likely include a very narrow bandwidth signal component, with much of the transmitter energy concentrated at one spot on the radio dial. This would make the signal easier to pick out against the background noise that is naturally produced by interstellar gas, distant galaxies, and the radio receiver itself. But in what part of the radio spectrum would we expect to find such a narrow-band hailing signal? In truth, we cannot say for certain, since a wide range of radio frequencies is serviceable. Because searching all these frequencies would be a hopelessly difficult task, Cocconi and Morrison proposed that SETI experimenters should tune their receivers near a spot on the dial that every scientifically literate society would know: 1,420 MHz. This is the frequency at which neutral hydrogen gas—the major constituent of the thin material that floats between the stars—produces natural radio static. Radio astronomers often use this frequency to study the distribution of interstellar gas in galaxies. Because it is such a useful spot on the dial, astronomers throughout the universe (of whatever species) would have this frequency marked on the receivers of their radio telescopes, and a transmitting beacon tuned near this frequency would be likely to attract the attention of others.

Having set out all the important principles of a radio SETI search, Cocconi and Morrison tried to interest radio astronomers in their idea. They approached those at England's Jodrell Bank Observatory but got little response. However, Frank Drake, then a young astronomer at the Green Bank radio observatory in West Virginia, had independently reached many of the same conclusions as Cocconi and Morrison. In the spring of 1960, using the 26-meter (85-foot) radio dish at Green Bank, Drake began an experiment much like the one Cocconi and Morrison had in mind (Figure 12.7). He conducted a weeks-long search for 1,420 MHz radio signals from two nearby, Sun-like stars: Epsilon Eridani and Tau Ceti. (These stars are approximately 12 light-years distant.) Drake whimsically named his search Project Ozma, after the fictional princess in the books by Frank Baum. It was the first modern SETI experiment.

Drake's search failed to detect alien signals, but it fired the imagination of many in the science community and ultimately led to a small SETI research program run by the American space agency, NASA. The NASA researchers spent many years studying the feasibility and technology of SETI. Their research led to the construction of specialized receiving systems that could be fitted to large, existing radio telescopes. In 1992, the NASA search began. But a year later, with the data collection barely under way, the program was cancelled by the U.S. Congress. Since then, scaled-down SETI programs have continued with private funding in the United States and as small university research efforts elsewhere in the world.

Figure 12.7

The 26-meter (85-foot) radio telescope used by Frank Drake in his pioneering 1960 SETI search, Project Ozma. This instrument was the first to be built at the National Radio Astronomy Observatory in Green Bank, West Virginia. Drake scrutinized two nearby star systems for signals near 1,420 MHz in frequency.

• How do we search for intelligence today?

Many of the SETI techniques in use today are elaborations on the earlier schemes proposed by Cocconi and Morrison, and first tried by Drake. However, SETI has also branched out to search for signs of intelligence in other ways as well. To understand the modern search for extraterrestrial intelligence, we first need to consider the types of signals we might be able to detect.

Categories of Signals

Drake's Project Ozma searched for a deliberately broadcast signal at a specific frequency that would be known to astronomers anywhere. It was therefore a search for an interstellar hailing signal, or beacon, intentionally sent so that others might detect it. But there are other possible signals we might hope to find. Broadly speaking, alien signals could fall into any one of the following three categories:

1. *Signals used for local communication on the world where intelligent beings live.* Our own radio and television signals fall into this category, because they are designed for our own use and not for interstellar communication. Another local use for radio signals is radar. Advanced civilizations might use radar to locate comets that pose a potential threat to their planet, for example.
2. *Signals used for communication between a civilization's home world and some other site,* such as a colony or spacecraft on another world. We have used relatively weak signals of this type to communicate with our interplanetary spacecraft. Such signals would be far stronger if they were being used, say, to communicate between colonies on planets in different star systems light-years apart.
3. *Intentional signal beacons,* such as the type searched for by Project Ozma, purposefully designed to get the attention of other societies.

In principle, SETI can search for all three types of signals, but in practice, our ability to receive signals depends on the sensitivity of our equipment. To get a rough idea of what we might be able to detect with our current technology, let's consider our own signals as an example.

The first commonplace, high-power, high-frequency transmissions from Earth were our early television broadcasts. These transmissions began in earnest during the 1950s, so they are now some 50 light-years away in space—followed by all television broadcasts since. Thus, in principle, any civilization within about 50 light-years could watch our old television shows (although it's unclear whether they would *like* them!). However, while television transmitters are fairly powerful, their antenna systems are designed to spread the signal over a wide angle, so that they reach homes in all directions from the transmitting station. This means that the strength of any signal that is, by chance, aimed at any given star is quite weak. And, like all light, these signals continue to weaken with distance (following an inverse square law, as shown in Figure 7.1). If another civilization has the same sort of receiving technology that we do, they could detect our television signals only if they were within about 1 light-year of us, and no stars are that close (the nearest are more than 4 light-years away). Turning the situation around, we see that we are not yet capable of detecting signals in the first category listed above, unless for some reason they are being broadcast with much more power than we use for our own television signals. (Some of our military radars, which employ large antennas to narrowly focus their transmissions, could be detected as far away as a few tens of light-years.) The situation is no better for signals in the second category, at least if such signals are comparable in strength to those we currently use for communicating with our own interplanetary spacecraft.

Our technology is rapidly improving, and in the future we might be able to detect alien signals in any of the three categories. But for the moment, at least, our best chance of detection is with signals in the third category. Beacon signals should be the easiest to detect because they would deliberately be made strong enough to be heard across interstellar distances. They should also be the least difficult to interpret, because they presumably would be designed for easy decoding.

Decoding a Signal How might an alien civilization design a signal so that we could decode it? In the book and movie *Contact,* author Carl Sagan supposed that the aliens might make it easy for us to recognize their message by playing back to us one of our own television transmissions. However, because our broadcasts have only had enough time to make it a relatively short distance into space (roughly 50 light-years), this strategy could work only if a civilization happened to be near enough to have already found such a signal and returned it to Earth. Fortunately, there are other ways that aliens could make it easy for us.

For example, aliens might choose to broadcast a strong, narrow-band signal. Most natural radio emissions have fairly broad bandwidths, so a signal confined to one narrow spot on the radio dial would immediately offer a hint that it might be artificial (made by intelligent beings). If the signal was also flashing on and off, or switching between two nearby frequencies, we would suspect that it was a coded message. We would undoubtedly record the pattern and try to analyze what was being "said."

The first thing to do would be to look for repetition. If aliens really wanted us to detect and decode their signal, they would repeat the entire broadcast many times, since otherwise we'd have to be listening at just the right moment to catch the signal. Knowing the total length of the message might help us greatly in figuring it out. If the total number of flashes or frequency changes was, for example, 1,679, then we might note that this is the product of two prime numbers, 23 and 73. (A prime number is one that can be divided only by 1 and itself without any remainder. The prime numbers are 1, 3, 7, 11, . . .) We could then arrange the message in a 23-by-73 grid and look for pictures or other figures.

a Arecibo has the world's largest single-dish radio antenna, shown here, with a diameter of 305 meters (1,000 feet).

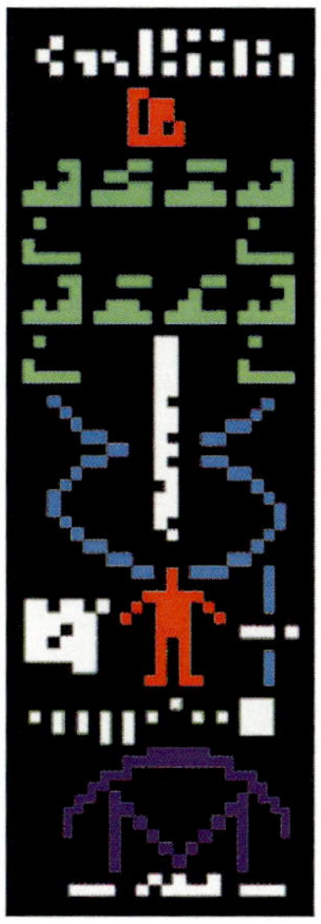

Figure 12.8

In 1974, the Arecibo radio telescope in Puerto Rico was used to send a 3-minute broadcast toward the globular cluster M13.

b The pictorial message broadcast in 1974, as it looks once you realize that the message is intended to be laid out as a rectangular grid with 73 rows and 23 columns (each of these numbers is a prime number, which should help enable any alien recipients to guess the layout). The colors are shown only to make the components clearer; the actual picture was sent in "black and white."

This simple approach is one we have taken ourselves in one of the few deliberate broadcasts that we have made to the stars. In 1974, the powerful planetary radar transmitter on the Arecibo radio telescope (Figure 12.8a) was fired up and used to send a 3-minute message to the object M13, a globular cluster containing a few hundred thousand stars. The message consisted of 1,679 bits, and each bit was represented by one of two radio frequencies. If someone picks ups the signal and recognizes that 1,679 is the prime number product 23 × 73, then they could arrange the bits in a rectangular grid with 73 rows and 23 columns. The resulting graphic, shown in Figure 12.8b, represents the Arecibo radio dish, our solar system, a human stick figure, and a schematic of DNA and the eight simple molecules used in its construction. However, because we did not repeat the signal, any aliens living around stars in M13 will have only one chance to receive it as it passes by their planet at the speed of light. That is one reason the signal was sent to a globular cluster: The cluster's several hundred thousand stars would seem to improve the odds of someone receiving the signal. If it is received, it won't be soon. M13 is about 21,000 light-years from Earth, so it will take our signal some 21,000 years to get there and another 21,000 years for any response to make its way back to Earth.

THINK ABOUT IT Some people consider the 1974 broadcast to have been a dangerous exercise that might attract the unwanted attention of hostile aliens. In general, do you think it is "safe" for us to broadcast messages to the stars? Why or why not?

Of course, alien messages could be far more sophisticated and more difficult to decode. In *Contact*, the pictures were three-dimensional, not flat, and in that case we should be looking for messages whose length is the product of three rather than two prime numbers. Alternatively, the message might be encoded in ways similar to the schemes used for sending files on the Internet. Given the enormous variety of possible ways a message could be transmitted, it might be that we would never be able to understand an alien broadcast. However, if advanced societies truly wished to get in touch, then they would undoubtedly go to some trouble to make their messages simple enough to be understandable to any civilization able to build the telescopes necessary to receive them.

Finally, it's important to remember that beacon signals will exist only if other societies make a deliberate decision to broadcast them. If no one does anything more than send occasional short signals like ours in 1974, the chances of detection will be quite small, even if many societies are out there listening. As a result, some people have asked whether we should undertake our own more deliberate effort to send a beacon signal into space. While this question continues to intrigue both researchers and the public, the consensus has been that we should focus on receiving signals first. After all, as we've seen in our consideration of the Drake equation, the chance that anyone will pick up a signal depends on how long broadcasting civilizations are "on the air." There's little point in transmitting signals for only a few weeks or even a few years. A broadcasting project would require long-term investment and a great deal of patience. Perhaps it is too soon for us to consider such a project; we developed radio technology only in the past century. There may be galactic civilizations that have had the ability to transmit signals into space for hundreds of millennia or longer. As the new kids on the block, it might make more sense to listen first.

Radio SETI Today Let's assume that someone really is broadcasting a signal that we could recognize and potentially decode. How should we go about searching for it? Today, just as in the early days of Drake's Project Ozma, most SETI searches are attempts to detect *radio* transmissions from other worlds.

A large radio antenna—almost always a radio telescope constructed for more conventional research purposes—is used to try to pick up the hoped-for cosmic signals. A low-noise amplifier at the antenna's focus

MOVIE MADNESS: *CONTACT*

Do the aliens know we're here? They might, if they're not too far away.

Contact, the movie based on Carl Sagan's novel about getting in touch with our celestial pals, starts out with a nifty sequence in which the camera backs away from Earth, slips by the Moon and planets, and eases out into the galaxy and beyond. During this high-speed countermarch, we hear the sounds of radio programs that have reached each of these cosmic outposts. With every step outward, we move back in time. (OK, there's some cinematic license here: As the camera passes Saturn, we've regressed to 1950s rock and roll. Saturn is never more than 88 light-*minutes* from Earth!)

In fact, less than 100 light-years out, Earth really does go "silent." Easy evidence for the presence of *Homo sapiens* extends only this far—about one tenth of 1% of the distance to the far edges of the galaxy.

Fortunately for the cash-starved SETI researchers in *Contact,* some friendly aliens have an outpost around the bright star Vega, a mere 25 light-years away. They've tuned in one of our early TV broadcasts and, apparently intrigued, have replied. Their response is picked up by the film's heroine, Ellie Arroway, as she uses a pair of earphones to monitor the cosmic static received by a large radio telescope. (More license here: Radio receivers for SETI sport hundreds of millions of channels. Ellie should either have donned a few hundred million headsets or left the listening to computers.)

The alien reply signal, which sounds like a pile driver hitting a pod of whales, contains an original 1936 broadcast (that way we know that the extraterrestrials are deliberately beaming to us) interweaved with construction details for ... well ... some sort of large device.

Faced with a SETI detection, the government goes nuts, and so do a lot of the citizenry. Some are ecstatic about the possibility of alien company; others see the news as heralding the apocalypse. Meanwhile, the scientists build the device—a multibillion-dollar machine that looks like the ultimate theme park attraction, but in fact is a wormhole transporter (is there any difference?).

It's nice to think that advanced aliens would want to improve our lot by sending us plans for high-tech hardware. But this seems an unlikely message from space. After all, if we could somehow contact the Neanderthals and give them plans for a personal computer, do you think they could ever, *ever* build it?

boosts the signal levels before they are further processed. The processing usually consists of digitally "slicing" the incoming, wide-band signal (typically spanning hundreds of megahertz) into many narrow-band channels. This processing is based on the assumption that some strong narrow-band component will be present in any deliberately broadcast extraterrestrial signal.

During the search, the radio telescope may be either pointed in select directions, such as toward individual stars, or swept across the heavens to study a larger section of the sky. The former technique, known as a **targeted search**, proceeds on the assumption that not all locations in space are equally probable sites for intelligent life. For example, Drake put forth the reasonable hypothesis that civilizations were most likely to exist on planets around Sun-like stars, and consequently he targeted his search at this type of star. The sweep technique, known as a **sky survey**, makes no assumptions about where intelligent aliens might be located.

Several radio SETI projects are currently under way. Table 12.1 summarizes key features of four major projects. The Inner Galactic Plane Survey is conducted on a new telescope, still under construction, called the Allen Telescope Array, which is located in northern California. SERENDIP uses the Arecibo telescope (see Figure 12.8a), while Southern SERENDIP employs the Parkes radio telescope in Australia (Figure 12.9). SETI Italia makes observations concurrently with radio astronomy projects that are being conducted with the Medicina radio telescope. For each project, the search type is indicated, along with the number of narrow-band channels being examined and the overall band of frequencies covered in each search. The last column of the table gives an indication of each experiment's sensitivity: It lists the minimum transmitter power that alien broadcasters would need to be using for us to detect their signals, assuming the aliens are located 100 light-years away and use a transmitting antenna 100 meters in diameter. The Inner Galactic Plane Survey, for example, could detect such an extraterrestrial broadcasting setup if the transmitter had a power of 70 megawatts or more. That's more than most radio and television broadcasts on Earth, but not that much more: Some of our own broadcasting stations have power greater than a megawatt, although they do not use a 100-meter antenna to narrowly focus their transmissions.

TABLE 12.1 *Current Radio SETI Projects*

Name	*Institution*	*Telescope*	*Type*	*Total Number of Channels*	*Width of Single Channel*	*Band Covered*	*Detectable Power for 100-Meter Transmitter at 100 Light-Years*
Inner Galactic Plane Survey*	SETI Institute	Allen Telescope Array	Survey	450 million	1 Hz	1,390–1,720 MHz	70 megawatts
SERENDIP	University of California, Berkeley**	Arecibo Radio Telescope (305 m)	Sky Survey	168 million	0.6 Hz	1,370–1,470 MHz	1 megawatt
Southern SERENDIP	SETI Australia Centre	Parkes Radio Telescope (64 m)	Sky Survey	58 million	0.6 Hz	1,418–1,421 MHz	1 megawatt
SETI Italia	Istituto di Radioastronomia, Bologna	Medicina radio telescope (32 m)	Sky Survey	24 million	0.6 Hz	Bands centered at 1.4, 2.8, 6.4, and 22.4 thousand MHz	Typically 30 megawatts

* Values given here are for the Allen Telescope Array capabilities at the end of 2006, when it will have 42 of its 6-meter antennas in operation. As more antennas are added (eventually reaching a total of 350), the sensitivity will increase and the observing mode will switch to a targeted search of stars within 1,000 light-years.

** About 2.5% of the data collected by the SERENDIP project is being distributed over the Internet for processing on a downloadable screen saver. This project (called SETI@home) has involved more than five million home computer users.

Figure 12.9

The 64-meter Parkes radio telescope in New South Wales, Australia. A SETI experiment "piggybacks" on this telescope while it is engaged in other astronomical research. Piggyback schemes avoid competition for telescope time and therefore provide SETI experiments with a lot of data. On the other hand, the SETI astronomers—who are only "along for the ride"—have no say in where the telescope is aimed. Since there are stars in every direction, however, this type of sky survey could still chance on a transmitting civilization.

While today's radio SETI experiments are vastly more sensitive and comprehensive than Drake's pioneering Project Ozma, they have been limited by two important constraints. First, all projects until now have used telescopes that are also in service for other astronomical research. Thus, they get only limited observing time. In some cases, such as SERENDIP, Southern SERENDIP, and SETI Italia, the SETI studies "piggyback" on other observations, which means that the choice of where the telescope points is dictated by the needs of these other research programs. Second, terrestrial radio interference, especially from radar and orbiting satellites, is becoming an increasing problem for SETI efforts.

The Allen Telescope Array (Figure 12.10) offers a solution to the first problem. Construction is now under way for this telescope in northern California. This instrument will eventually consist of 350 small (6-meter) radio dishes with a combined collecting area equivalent to that of a single 100-meter telescope. One major advantage of an array, as compared to a single-dish radio telescope, is that it can be electronically focused on several stars simultaneously, thus speeding up the search. The Array is a joint venture of the SETI Institute and the University of California, Berkeley, and it will be used full time for SETI observations. It can also be used simultaneously for conventional radio astronomy studies.

Solving the second problem, terrestrial radio interference, is more difficult. This interference greatly hinders SETI searches, because telecommunications satellites, aircraft, and Earth-bound radar produce lots of narrow-band signals—exactly the type being sought. These signals are so strong that they are picked up by radio telescopes no matter which direction they are pointed in. SETI researchers have devised various tricks

Figure 12.10

The Allen Telescope Array, now being constructed by the SETI Institute and the University of California at Berkeley in Hat Creek, California. This array of small (6-meter-diameter) dishes will eventually consist of 350 antennas (inset). The Inner Galactic Plane Survey is being conducted using a small subset of this final number.

to sort out terrestrial signals from possible extraterrestrial ones, but this problem will continue to worsen. Perhaps in the future we will be able to place radio telescopes on the far side of the Moon, where they would be shielded from Earth's noisy radio presence.

Optical SETI and Beyond The fact that most SETI searches have used radio telescopes may seem logical, since we often think of radio as something that we "listen to," but it is not the only possibility. Remember that radio is a form of light and that we "listen" to it only after the light is converted to sound by a radio receiver. In essence, the radio station uses electronics to convert the information content of the sound (for example, music) into long-wavelength light waves—wavelengths far beyond what the human eye can perceive. These light waves travel through the air to your receiver, where electronics converts the information back into sound. It just so happens that the wavelengths of light that we use for this process are in the radio part of the spectrum (see Figure 3.30), which is why we say we are "listening to the radio." Broadcast television is also encoded in radio waves, but in this case the encoding is used for pictures in addition to sound.

The common use of radio waves for encoding sound or images is a technological choice, not a requirement. In principle, any form of light would do—and often does. For example, if your computer or television is hooked up to a fiber-optic cable network, you are receiving information transmitted in the form of infrared or visible-light waves bouncing through the fiber optics, rather than in the form of radio waves. Thus, it's possible that alien civilizations might choose to communicate with visible light or other forms of light besides radio, and some SETI efforts are designed to search for such signals.

The idea of communicating between worlds via visual signals has a long and interesting history. In fact, scientists considered using light as a communication medium even before radio's invention. During the

COSMIC CALCULATIONS 12.2

Sensitivity of SETI Searches

The distance from which a SETI experiment could detect an alien transmission depends on the sensitivity of the receiver and on the strength of the signal. The inverse square law for light [Section 7.1] tells us that if aliens broadcast a signal with power P, it will have weakened by a factor d^2 by the time it reaches Earth, where d is the distance of the broadcasting civilization. Thus, the strength of the signal S we receive at Earth is this diminished power $\frac{P}{d^2}$ multiplied by the area of the receiving radio dish, A_r, and a constant of proportionality, k:

$$S = k \times A_r \times \frac{P}{d^2}$$

Example: A SETI search using the 300-m-diameter Arecibo radio telescope can pick up a 10-million-watt signal from 1,000 light-years away (if the transmitting antenna is also 300 m in diameter). If we were to detect a similar signal coming from 25,000 light-years away, how powerful would the alien transmitter have to be?

Solution: Since all else is held constant in this problem, the only change is that the second case is 25 times farther away than the first case. Thus, for a given transmitted power P, the signal strength S will be reduced by a factor of $\frac{1}{25}^2 = \frac{1}{625}$. To get the same S, the transmission power from this star would have to be 625 times stronger, or 625×10 million = 6 billion watts.

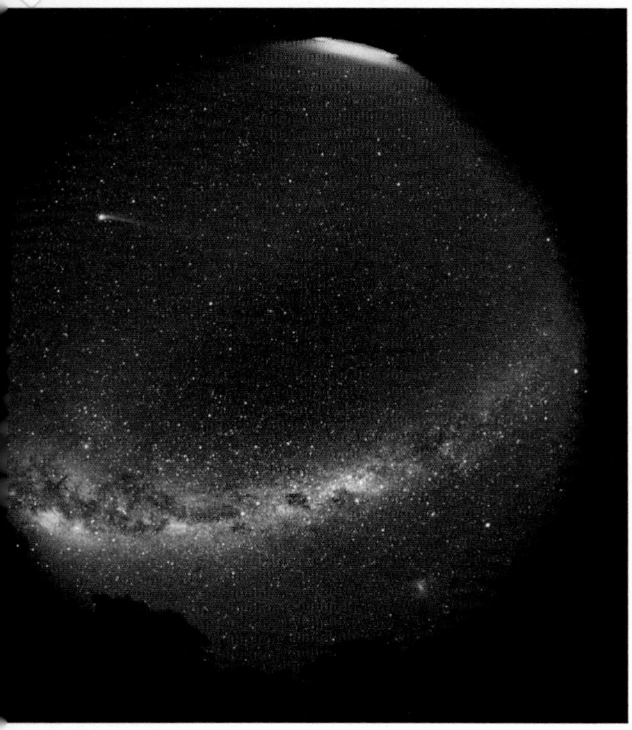

Figure 12.11

A "fish-eye" photograph of the Milky Way in the Australian sky. The dark rift running through the center of the Milky Way is dark because starlight is blocked from our view by interstellar dust. (Comet Hyakutake is visible near the upper left in this 1996 photo.)

nineteenth century, several scientists advanced proposals intended to put us in touch with sophisticated societies that were imagined to live on either the Moon or Mars. The eminent mathematician Karl Gauss is said to have suggested planting trees in the form of a right triangle in Siberia, bounded by clear-cut squares. This greenery graphic presumably would be seen by other solar system inhabitants, proving that Earthlings are at least smart enough to know the theorem of Pythagoras. While this story may be apocryphal, Gauss is definitely known to have proposed using 100 mirrors, each about 1 meter on a side, to shine the light of gas lamps into space. "This would be a discovery even greater than that of America, if we could get in touch with our neighbors on the Moon," he claimed. Another nineteenth-century suggestion was to dig trenches in the Sahara in various geometric shapes, fill them with water and oil, and ignite them at night.

Despite these old ideas, until fairly recently optical SETI did not draw much attention from researchers. Part of the reason had to do with estimates of the energy needed. In general, it requires far less energy to send an interstellar signal via radio than via visible light, so researchers guessed that radio would be the aliens' technology of choice. In addition, while radio waves travel through interstellar space with ease, visible light tends to be absorbed by tiny grains of interstellar dust that float between the stars. This dust creates the dark rift through the center of the Milky Way as we see it in our sky (Figure 12.11). When we look in different directions within the galactic plane, the dust completely blocks our view of visible light beyond a few thousand light-years—a relatively short distance compared to the 100,000-light-year diameter of our galaxy. The fact that visible light can be used to send signals for only limited distances through the galaxy seemed to argue against its use as a beacon for interstellar communication.

Researchers no longer believe that either problem is a serious strike against optical signaling. The limitations of interstellar dust are real, but signals that can travel up to a few thousand light-years could still be quite useful. After all, many millions of stars lie within 1,000 light-years of Earth. Moreover, if longer-distance communication is needed, using infrared rather than visible light would largely circumvent the problem, because infrared light penetrates the interstellar dust. The energy–cost issue argues in favor of radio only if we consider transmitters that beam their signals in all directions into space. For focused communication, such as sending a signal to a particular nearby star, visible light can easily be concentrated into a narrow beam (by using a large lens or mirror). Whereas radio is ideal for a radio station that wants its signal to be picked up by people living in all directions from the station, visible light from a flashlight (or a laser beam) is a better medium if you want to signal someone in a particular place. SETI researchers have concluded that visible light might well be useful for interstellar communication and that optical SETI thus could be a worthwhile enterprise.

How might an alien civilization communicate optically? Simply turning a continuously shining, high-powered laser on someone else's star system isn't particularly effective. A continuous beam could be lost in the glare of starlight from the transmitter's home sun unless it's extremely bright. In addition, a light that is always on doesn't convey much information (think of Morse code). Short bursts of laser light, say a billionth of a second long, would work better, because they can momentarily outshine the starlight even with relatively modest laser power. A series of such bursts—a pulse train—could contain patterns that make up a message. Such arguments have led us to imagine that advanced alien societies might be using automated, high-powered laser transmitters to send pulsed messages to thousands of nearby stars. The messages, perhaps only a few seconds long, could be repeated every several dozen hours. If a civilization in our galactic neighborhood is sending laser messages, then we might hope to find the laser "pings" by monitoring nearby stars.

Optical SETI efforts today are attempting to do just that. Like their radio counterparts, these efforts are passive—that is, they are searching for incoming signals but are not transmitting anything outward—and are systematically checking out the vicinities of local stars. The experiments look for short pulses of light that are bright enough to outshine the background shower of photons naturally produced by the alien world's host star. Photomultiplier tubes, sensitive light detectors able to see short flashes, are affixed to conventional mirror telescopes to hunt for these photon bursts. Figure 12.12 shows the setup for an optical SETI experiment at the Lick Observatory, near San Jose, California. This experiment is quite sensitive: If someone is sending out laser pulses with a transmitter no more sophisticated than we ourselves could build, the Lick experiment could pick it up from as far away as about 500 light-years. Researchers at Harvard University have constructed a telescope whose sole job is to scan the sky for short, bright flashes of light sent from other worlds.

THINK ABOUT IT It has been suggested that one benevolent use for our nuclear weapons would be to rocket them into space, line them up in geometric patterns, and then detonate them as a "broadcast signal" to distant alien societies. Do you think this would make an effective SETI signal? Why or why not?

Beyond radio and optical SETI, almost any other form of light might also be used for interstellar communication. We've already noted the potential value of using infrared signals in place of visible signals in order to penetrate interstellar dust. Similarly, distant civilizations might choose to communicate with ultraviolet light or X rays. However, these other forms of light have no obvious advantages over radio, infrared, and optical options, and they are more expensive in terms of the energy cost per "bit"

Figure 12.12

An optical SETI experiment being carried out on the 1-meter Nickel telescope at the Lick Observatory, on Mount Hamilton near San Jose, California. The photomultiplier tube detector, which is designed to react to bursts of light, is contained in the white box attached to the back of the telescope.

Figure 12.13

In the movie *2001: A Space Odyssey,* American astronauts dig up a clearly artificial monolith on the Moon. This is a signaling device left behind by an advanced society, and placed where it could be found only when humans had reached a high enough level of technical sophistication.

of information. As a result, SETI researchers feel justified in concentrating at least our current, early searches on radio and optical frequencies.

What about signals using more exotic technologies? For example, some people have suggested that advanced civilizations might send messages via the ghostlike subatomic particles called *neutrinos* or via the so-called *gravity waves* that Einstein predicted but that we have not yet detected directly.* We cannot rule out these possibilities, but we might wonder what the point would be. After all, neither of these schemes would communicate any faster than light, and they would make both transmitters and receivers far harder to build. Of course, it is also possible and perhaps even likely that sufficiently advanced civilizations have developed communication technologies that we could not detect. These technologies might involve physics that we have not yet discovered, or they might use ways of disguising signals so that our detectors cannot distinguish them from natural background signals. Thus, if other civilizations *want* to keep their communications secret from us, they probably can. SETI's hopes rest with civilizations that either are looking for contact or don't care if their communications are intercepted by others. In either of these cases, sticking with the simplest available technologies would seem to make the most sense.

Searching for Artifacts There might be other ways of detecting extraterrestrial civilizations besides picking up signals traveling through interstellar space. These possibilities apply if civilizations have advanced far beyond our own capabilities and are able either to travel between the stars or to undertake major "astro-engineering" projects that would be visible at a great distance.

If other civilizations have achieved interstellar travel, then they might have visited our solar system. In that case, they may have left artifacts behind, either accidentally or deliberately. Some people claim that we already have such artifacts from UFOs, but as we'll discuss in Section 12.4, these claims do not meet the standards of science. There are more plausible ways to imagine aliens leaving artifacts, and many science fiction writers have considered them. For example, in the classic movie *2001: A Space Odyssey,* based on the story by Arthur C. Clarke, highly advanced aliens buried a monolith on the Moon (Figure 12.13). When humans found the monolith, it notified the extraterrestrials.

It might be a while before we can search the Moon and the planets for artifacts such as buried monoliths, but a few researchers have suggested that aliens wishing to get our attention might leave calling cards in places where we could find them somewhat more easily. In particular, some people believe we should look especially hard at the so-called Lagrange points of the Earth–Moon system, named for the eighteenth-century French mathematician Joseph-Louis Lagrange (1736–1813). At these five positions in space, the effects of gravity from Earth and the Moon "cancel" in such a way that, if you floated weightlessly at one of these five points, you wouldn't be tugged toward either body. Figure 12.14 shows the five Lagrange points for the Earth–Moon system. Note that three of them are on a line passing through the centers of Earth and the Moon and that the other two are located 60° to either side of the Moon. These latter two positions, known as L4 and L5, are of particular interest. Unlike the other three, they are "stable," which means that if you started at L4 or L5 but

*However, we have strong indirect evidence for the existence of gravity waves, and a new instrument called the Laser Interferometer Gravitational Wave Observatory (LIGO) may detect them within just a few years.

then drifted slightly away, the competing effects of gravity from Earth and the Moon would bring you back to your starting point. This effect is much like keeping a marble in a shallow bowl; if the marble is moved a bit off-center, it still rolls back to the bottom of the bowl. As a result, L4 and L5 would be obvious places to leave artifacts in cold storage for possible retrieval after long periods of time. Two decades ago, a limited survey of the Lagrange points was made with small telescopes. The hope was to find bright objects that might indicate the presence of parked artifacts. Although the search didn't turn up anything clearly artificial, these locations might be worth a more thorough reconnaissance in the future.

If interstellar travelers have not left artifacts for us to find, we might still be able to detect their powerful, interstellar spacecraft. As we will discuss in Chapter 13, spacecraft capable of traveling at speeds close to the speed of light will likely have enormous engines powered by energy sources such as nuclear fission, nuclear fusion, or matter–antimatter drives. These engines would leave telltale signs of their operation—signs that might be detectable at distances of hundreds or even thousands of light-years away. Although only a limited search for distant rockets has been made, this type of phenomenon might be inadvertently discovered in the course of more conventional astronomical research.

Sufficiently advanced civilizations might also betray their presence through tremendous feats of "astro-engineering." In terms of a civilization's ability to exploit natural energy resources, the twentieth-century Russian physicist Nikolai Kardashev suggested that there might be three distinct categories of civilization:

1. **Planetary (or Type I) civilizations,** which use the resources of their home planet.
2. **Stellar (or Type II) civilizations,** which corral the resources of their home star.
3. **Galactic (or Type III) civilizations,** which employ the resources of their entire galaxy.

We are in the first category, since we exploit (often with abandon) only the meager resources of our home planet. The lights of our cities and the heat from our homes and factories, while considerable, are feeble on a cosmic scale and would be extraordinarily difficult to detect at the distances of the stars. Civilizations in the third category are so advanced that they might be difficult to imagine. As Arthur C. Clarke has stated, "Any sufficiently advanced technology is indistinguishable from magic."

However, we might be able to detect civilizations in the second category. Such a civilization, for example, might decide to capitalize on solar energy in a big way. A star like the Sun puts out a great deal of power. If we could capture just 1 second's worth of the Sun's total energy output, it would be enough to supply today's world demand for energy for approximately the next million years. But nearly all the Sun's energy escapes into space, and Earth intercepts only the tiny bit that heads our way. In principle, a technologically adept civilization could capture *all* of its star's energy by fashioning a large, thin-walled sphere (possibly built from a dismantled outer planet) around its solar system and covering the inner surface with solar cells or their equivalent. Such spheres are called **Dyson spheres**, after physicist Freeman Dyson, who proposed their possible existence (Figure 12.15).

Because a Dyson sphere would absorb all the light of its interior star, you might think it would be invisible; however, the laws of physics dictate that waste heat must escape from the sphere. This heat would be radiated as infrared light that we could detect with specialized telescopes.

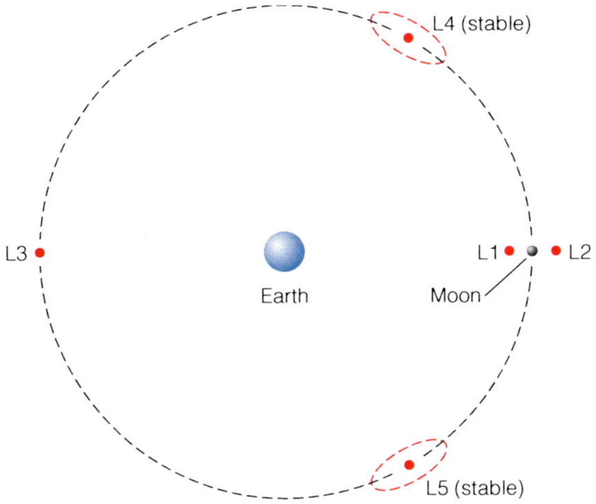

Figure 12.14

The five Lagrange points of the Earth–Moon system. The points L4 and L5, with the stable regions schematically indicated by the dashed ovals, are most attractive for the long-term parking of artifacts or probes. As the Moon orbits Earth, the Lagrange points also orbit, so that they remain in the same relative position with respect to Earth and the Moon.

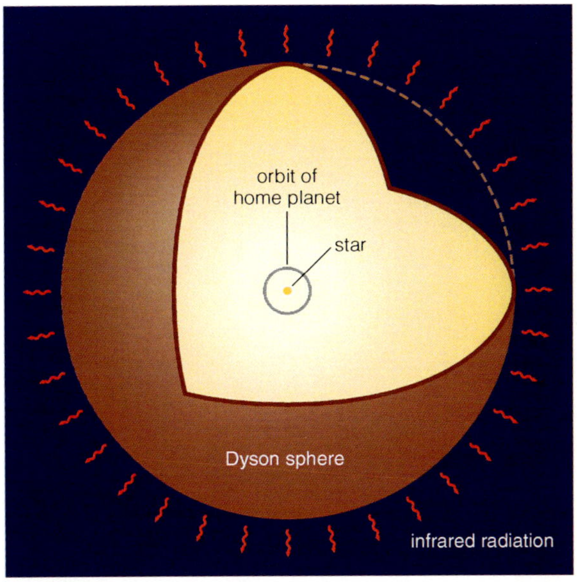

Figure 12.15

Schematic representation of a Dyson sphere, constructed exterior to an inhabited planet, with waste heat escaping from its outer surface in the form of infrared light.

Figure 12.16

A SETI experiment at the Arecibo radio telescope in Puerto Rico. The observer, Jill Tarter (the model for the fictional Ellie Arroway in the novel and movie *Contact*), monitors the progress of the observations on computer workstations. It is the computers that do the "listening," since many millions of channels are monitored simultaneously.

Thus, we could discover the presence of a stellar civilization by finding the infrared signature of a Dyson sphere. Limited searches have already been undertaken, so far to no avail.

• What happens if SETI succeeds?

Until now, no SETI experiment has turned up a confirmed extraterrestrial transmission. Several intriguing signals have been reported, however. Perhaps the best known was found in 1977 by the (now-defunct) automated SETI search at the Ohio State Radio Observatory. When an Ohio State astronomer examined the data from the night's observing, in the form of a computer printout, he found a signal so impressively strong that he wrote "Wow" on the printout margin. The "Wow" signal, made popular by its appealing nomenclature, was never seen again despite repeated attempts at the same frequency and the same sky position. Consequently, SETI researchers do not consider it a detection of extraterrestrial intelligence but instead presume it was some sort of terrestrial interference.

Despite their continued failure to find a persistent, verifiable signal, however, those engaged in SETI research remain optimistic (Figure 12.16). The motivation for this optimism is the rapid improvement in the abilities of the telescopes and detectors used in SETI. Because much of the hardware required for SETI is built with digital electronics, SETI research benefits from the rapid growth in computer capability. The density of transistors placed on silicon chips has been doubling approximately every 18 months for decades (a phenomenon sometimes called *Moore's law*, after Gordon Moore, a cofounder of Intel Corporation). This exponential improvement in electronics motivates the optimism typical of SETI researchers. As an example, Frank Drake has estimated the number of transmitting civilizations, N, as 10,000. If this is so, and if 10% of all stars are suitable for habitation, then one in a million suitable stars will have a transmitting society. The Allen Telescope Array, which is designed to take advantage of technological improvements, will be able to examine approximately a million targets by the year 2025. This suggests that, if Drake's assumptions are correct, a detection might be made within a few decades. Continued progress in astronomy—and in particular, the possibility of learning which extrasolar planets might have atmospheres that indicate the presence of life—could shorten this time even more by improving our choice of star systems to examine.

What will happen if we really do receive an artificial signal from the stars? A signal detection would need to be thoroughly verified by continued observations both at the discovery telescope and at other observatories. The nature of the signal itself will be evidence of artificiality, but scientists need to be sure it isn't due to terrestrial interference, equipment failure, or a college prank. It might take many weeks of careful work by excited astronomers before they are fully confident of a genuine detection.

Once a detection has been verified, the news would be released to the outside world. Given the potential implications of the discovery of an extraterrestrial civilization, this step must be taken with some care. For this reason, SETI research groups have agreed to follow a protocol known as the *Declaration of Principles Concerning Activities Following the Detection of Extraterrestrial Intelligence.* The protocol has no force of law, but it lays out a reasonable course of action. In particular, if a signal has been found and verified as truly extraterrestrial, astronomers around the world would be notified so that they could swing their telescopes in the direction of the signal and learn as much as possible. Governments and the public would also be informed directly. The open nature of research and the necessity to confirm any detection by using other telescopes dictate that the discovery could not be kept secret or covered up. The evidence, after all, is accessible to anyone with access to a large telescope. A recent addition to the protocol addresses the matter of sending a reply to any detected signal. The SETI community has suggested that any deliberate response should represent a consensus of the world's population, not just the wishes of whatever group made the detection.

Suppose we found more than just a signal, but also a message—bits of information. Might we be able to decipher it? In fiction, messages from aliens often have a mathematical bent, because we assume that mathematics will be developed by any technological society. The fictional aliens send us the value of pi or numbers from a well-known mathematical series. This is usually done so that we will recognize that the signal is from intelligent beings. But this sort of labeling wouldn't be needed, because we are seeking narrow-band radio signals or pulsed laser light that clearly would be artificial. In addition, using mathematics as a "language" would be a cumbersome way to express things such as political systems, religious beliefs, art, or even physical appearance. Our own messages to space have been pictorial (see Figure 12.8b and Figure 13.2), and pictures might be a better way to communicate to unfamiliar societies.

Any SETI success in itself would be an astonishing event, because it would prove that we are not alone in the universe. Depending on the nature of the signal, however, we might learn far more. If a civilization is broadcasting to us, it is likely to be hundreds or even thousands of light-years distant, and two-way communication would be tedious at best. A broadcasting society therefore might not expect a reply and might simply send a one-way message, perhaps some version of their encyclopedia or information from their internet. Such a message might take us a while to figure out but would be worth the trouble. From our discussion of the Drake equation, we learned that the chances of detecting another society increase with the average age of technological civilizations. Consequently, if we do find a signal, the chances are great that it comes from a civilization far older than our own. The knowledge we might gain from such an advanced society would be enormous.

Finally, we should also be aware of the possibility that the signal we find will not be a beacon intended for societies like ours, and might contain a message that we can never decode. Even in that case, we would at least have learned that we are neither alone nor particularly special.

12.4 THE PROCESS OF SCIENCE IN ACTION: UFOS AND ALIENS ON EARTH

A fundamental assumption of all SETI efforts is that our best hope of detecting an alien civilization lies with looking beyond Earth for signals or other evidence of their existence. But what if the aliens are already right here on our planet? Remarkably, public opinion polls since the 1960s have consistently shown that about half the American public believes that we are being visited by alien spaceships, and many are convinced that hard evidence of such visits exists. If this were true, then in principle we could learn about extraterrestrial intelligence without ever resorting to the use of expensive telescopes and complex receivers. Unfortunately, when we examine these claims of alien visits using the accepted principles of science, they prove to be far from convincing.

• What have we learned from UFO sightings?

The bulk of the claimed evidence for alien visitation consists of UFO sightings—many thousands of UFOs are reported each year. No one doubts that unidentified objects are being seen. Even many astronomers have seen objects in the sky that they cannot identify, and some of the most interesting UFO reports have come from seasoned pilots whose credibility is not in question. However, the mere fact that something is unidentified does not automatically make it an alien spacecraft. The real question is not whether UFOs are seen, but whether these sightings are actually of visitors from the stars.

We can gain some insight into this question by examining both the history of UFO sightings and the history of scientific examination of UFO claims. The first modern report of a UFO as an alien spacecraft was made by businessman Kenneth Arnold in June 1947. Arnold was flying a private plane near Mount Rainier, in the state of Washington, when he saw nine shiny objects having wings, but no tails, that appeared to be streaking across the sky at nearly 2,000 km/hr. A reporter for the United Press wrote up Arnold's experience as a sighting of "flying saucers," and the story became front-page news throughout America.

Within a decade, "flying saucers" had invaded popular culture, if not our planet. They were often seen in books and movies, and the term is still used today. This is ironic, given that Arnold didn't actually describe the objects he saw as saucer shaped, or even round. Three decades after the sighting, he explained that this impression was the result of a newspaperman's error. When, in 1947, the United Press reporter asked him how the objects moved, Arnold answered that they "flew erratic, like a saucer if you skip it across the water." He was describing their motion, not their shape. In fact, several later investigators have suggested that the objects seen by Arnold were meteors, streaks of light caused by small particles entering Earth's atmosphere.

Figure 12.17

A UFO. This photo, which shows far more detail than most pictures made of supposed visitors from other worlds, has the familiar saucer shape made popular by a reporter's error. In fact, this photo is faked, and the saucer is only a lamp shade.

Despite the reporter's misunderstanding, the idea of alien disks buzzing the countryside caught the imagination of the public. It also intrigued the U.S. Air Force, which spent two decades conducting investigations into the nature of UFOs. The military interest was prompted largely by Cold War concerns that UFOs might be new types of aircraft being developed by the Soviet Union.

In the 1950s and 1960s, teams of academics met to study the most interesting of the UFO reports. In the overwhelming majority of cases, these experts were able to plausibly identify the UFOs. They included bright stars and planets, aircraft and gliders, rocket launches, balloons, birds, ball lightning, meteors, atmospheric phenomena, and the occasional hoax. For a minority of the sightings, the investigators could not deduce what had been seen, but their overall conclusion was that there was no reason to believe that UFOs were either highly advanced Soviet craft or visitors from other worlds. The Air Force ultimately dropped its investigation of the UFO phenomenon.

However, some believers felt that the inquiries were either incomplete or part of a ruse organized by the government to put people off the scent of alien visitation. The number of UFO sightings increased (of course, the number of human-made objects in the sky was also growing during these years), and countless books, photos, and film clips were offered as evidence (Figure 12.17). Yet little of the photographic material was compelling. Some of it was ambiguous (is that a distant spacecraft or a nearby bird or bug?), and some was clearly faked. Moreover, the fact that about 90% of the well-documented UFO cases were explainable as earthly phenomena only encouraged some people to point to the 10% that weren't clearly explained. However, this is not a valid argument for the idea that the unexplained cases represent alien spaceships. After all, if a metropolitan police department solves 90% of the murder cases in a large city—all committed by humans against other humans—it doesn't suggest that the other 10% were committed by, say, aliens.

We can see the problems with UFO "evidence" even more clearly when we evaluate the sightings using the methods of science [Section 2.3]. Recall that modern science seeks natural causes to explain observed phenomena. Although aliens presumably would be "natural" if they really exist, invoking them to explain unidentified objects is no different from invoking ghosts, spirits, or Greek gods and goddesses. Any of these things might be the real explanation for UFOs, but simply saying so does not give us any insight into their true nature. We can learn something only if we can make a model to describe the phenomena of UFOs, and then test the model against additional observations. Reports of UFO sightings do not allow us to do either: Various UFO sighting reports are so different from one another that it's difficult to create any testable hypothesis about the designs of the supposed alien spacecraft, the types of engines that they use to traverse interstellar distances, or the behavior of their occupants. We are left with just individual, eyewitness reports—and as we discussed in Chapter 2, these reports offer no way for other people to evaluate their validity.

The bottom line is that UFO reports tell us nothing more than the fact that people sometimes see things in the sky that they are unable to identify or explain. But the fact that an object has not been identified does not even make it unidentifiable—it might just be that the person making the report did not recognize something that another person might have considered obvious—and it certainly does not make it an alien spacecraft. If we want to establish evidence of alien visitation on Earth, we'll need evidence that's much more solid.

THINK ABOUT IT Suppose you had a year and an unlimited budget with which to investigate the claim that UFOs represent alien spacecraft. What sorts of experiments would you conduct? What types of evidence might convince you—and the scientific community—that fast-moving lights in the sky are nonterrestrial craft?

• Have aliens left any verifiable evidence of visitation?

If aliens land in Paris, hold press conferences, shake hands with crowds, and provide artifacts from their home world, scientists would surely accept their visit as a fact. But short of that, what could convince scientists that aliens have been here? According to the hallmarks of science, we'd need something that anyone could examine, at least in principle, and that stands up to intense and continued scrutiny. While it's conceivable that photographic evidence might do—for example, if the same UFO was photographed from many different observatories—some sort of artifact or other hard evidence would be even better. Some people claim that such evidence already exists, but to date none of it has withstood serious scientific scrutiny. Let's investigate a few of the more famous claims.

Figure 12.18

Debris recovered in Roswell, New Mexico, in 1947. In this photo, made by newspaper photographer James Johnson, General Roger Ramey is showing the debris to reporters in Fort Worth, Texas.

Crashed Aliens in Roswell Perhaps the most celebrated case of supposedly alien artifacts concerns the so-called Roswell incident of 1947. The story began in early July 1947, only a few weeks after the nationwide coverage of Kenneth Arnold's "flying saucers." A rancher found some crash remnants in a pasture close to the city of Roswell, New Mexico. He reported the debris to the local sheriff, who then passed on the information to the Roswell Army Air Field, which was nearby. Several military personnel drove out to the ranch, picked up the debris, and explained to the local papers that they had recovered the remains of a "flying disk."

This dismaying story was quickly quashed. Only a day later, an Air Force officer from Fort Worth, Texas, where the debris had been flown, held a press conference in which he dismissed the flying disk notion (Figure 12.18). He stated that the debris was merely a crashed weather balloon, substantially ending interest in the incident at the time. But in 1978, UFO investigator Stanton Friedman began looking into the events at Roswell. Friedman claimed that the debris was from a spacecraft and that alien occupants had been picked up as well. He believed that the government was covering up both the fact of the mishap and its extraterrestrial victims.

The possibility of alien bodies lent this story an appeal beyond the routine sightings of lights in the sky. However, despite the claim of hard evidence, most of the story is based on testimony recorded by Friedman and others—and remember that they conducted their interviews more than *three decades* after the event. Given that long time lag, it's perhaps unsurprising that different witnesses contradicted one another, that some supposed witnesses to the crashed "saucer" had originally claimed not to have seen it, and that a few were caught in flat-out lies. Clearly, the eyewitness testimony by itself could not be considered rigorous enough to prove something as profoundly important as alien visitation.

But what of the crash remains and alien bodies? Something *did* crash at Roswell, and despite the Air Force claim, it was not a weather balloon. Instead, declassified government records tell us that the crash was of a then-secret military balloon experiment designed to detect Soviet nuclear tests. The highly classified operation, known as Project Mogul, used balloon trains to carry the detection devices aloft; the idea was that the balloons would hover at constant altitude near the borders of the Soviet Union and listen for sound waves produced by distant nuclear blasts. Project Mogul was being tested at Roswell in the late 1940s, and the principal scientist behind it has confirmed that the debris recovered by the rancher was from his experiment. Moreover, photos taken of the recovered materials by the Air Force match contemporary photos of Project Mogul balloon trains.

As for the alien bodies or more convincing spacecraft debris, believers can claim only that the government has maintained a tight cover-up, storing the evidence securely in some secret location (such as "Area 51" in Nevada) since the 1947 crash. But consider what that would entail: For some 60 years, through administrations with widely varying political agendas, hundreds of scientists and other military personnel would have had access to such a secret. Yet not a shred of actual evidence has emerged to public view, despite the fact that anyone who produced such evidence would become instantly famous on the talk-show circuit. At minimum, this conspiracy claim seems highly improbable.

In short, claims that the Roswell debris was from a downed alien craft are based on eyewitness accounts made long after the fact, and the supposed evidence of both craft and alien bodies could be real only if you believe the government capable of a highly efficient, six-decade cover-up. The nature of these claims renders them untestable and therefore unscientific, and they seem even more difficult to believe once you realize that we have an alternate explanation—the military experiment—backed by verifiable data in the form of the declassified records and photographs. It might be more interesting to believe that small people from another world just happened to make a navigational error a few weeks after flying saucer mania first swept the country, but the evidence suggests only a crashed Air Force project.

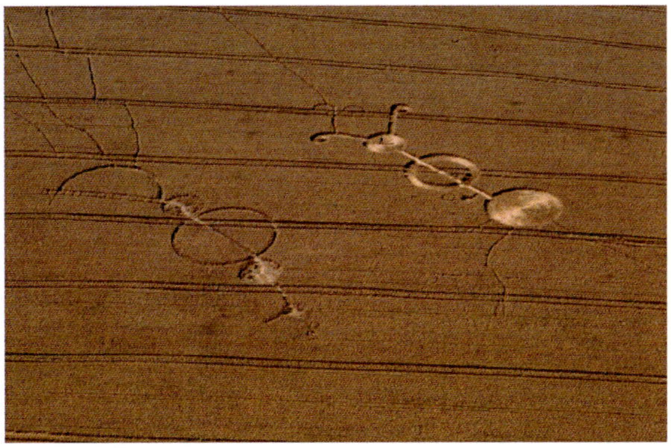

Figure 12.19

Crop circles. These relatively simple patterns have been "cropping up" in southern England for several decades and are claimed by some to be manifestations of alien activity. However, the designs are easily produced in a few hours by a small group of motivated students using nothing more than boards and ropes, which suggests a terrestrial origin. In addition, we might wonder why sophisticated extraterrestrials would travel many light-years simply to carve graffiti in our wheat.

Crop Circles, Abductions, and More UFOs and crashed aliens can nearly always be readily explained in terms that don't involve visitors from other worlds. The same is true for other highly publicized events interpreted as proof of an extraterrestrial presence in the neighborhood, including abductions, crop circles, and miscellaneous phenomena such as mutilated cattle and goat-eating chupacabras.

Crop circles, geometric patterns made in wheat fields, have generated a great deal of interest among the public (Figure 12.19). For several decades, such patterns have appeared each summer in England. According to some, they are the work of visiting extraterrestrials who are using this method to communicate to us. The proof offered for their nonhuman origin is the speed with which they appear (overnight) and the condition of the matted-down wheat, which is claimed to be inconsistent with simple trampling by humans. But there are many indications that these patterns are merely pranks, not alien messages. The continuing increase in complexity of the patterns over the years suggests that their creators are getting better at their craft—something we would not expect if aliens are using technology brought from distant stars. The fact that the crop patterns are made only at night also suggests human subterfuge. Moreover, many of the crop circles have been acknowledged as pranks, and the human pranksters have demonstrated for news reporters how the crop circles can be made quickly and easily. (Competitions in crop circle making have even been held!) In one interesting case reported in a documentary about crop circles, a self-proclaimed "expert" on the alien origins of crop circles acknowledged that some crop circles were pranks but claimed that not all of them could be. He explained to interviewers why a particular set of circles could not have been made by humans—and then watched in dismay as the interviewers showed him a video of experienced pranksters making the very ones he had just labeled "alien."

Cattle mutilations and the reputed killings of various small animals by short, bipedal creatures called chupacabras have also been attributed to alien activity, despite a centuries-long history of attacks on farm animals by both earthly predators and humans. The only evidence offered to connect these attacks to extraterrestrial beings is uncertainty over who or what is doing the killing. As always, the lack of a confirmed culprit does not mean we are seeing the work of aliens.

A more interesting phenomenon involves claims of alien abduction of humans. Although small in overall percentage of the population, a substantial number of Americans claim to have been alien abductees, often supposedly stolen away by extraterrestrials while asleep (far fewer claim to have been disturbed during waking hours). The victims state that they were either taken aboard spacecraft for observation and unwholesome experiments, or simply watched while they lay immobilized in their beds. These claims may seem difficult to explain away, but a clue to their likely origin comes from research showing that similar phenomena have been described since ancient times in a multitude of cultures. In the past, the accused culprits have been witches, ghost babies, and goblins. Today, the molester of choice is frequently a visitor from the stars.

According to many psychologists, such abduction experiences are probably attributable to **sleep paralysis,** which occurs during REM (rapid eye movement) sleep. During REM sleep, the body is naturally paralyzed, so that we don't thrash physically along with the movements we make in our dreams. Sometimes, this paralysis persists for a few minutes after the brain has started waking up, and it can give a person the alarming sensation of being awake in a paralyzed body. Visions and other sensations often occur in this state—including strange imaginings that might plausibly include witches, ghosts, goblins, or wide-eyed extraterrestrials.

Surveys suggest that sleep paralysis is experienced by approximately half of all people at some point in their lives. It therefore seems quite reasonable to imagine that a small percentage might believe they had experienced some type of alien abduction. Proponents of alien abduction dispute this idea, noting that a few victims were fully awake and alert during their experience. However, in some cases, daydreams are known to produce sensations similar to those produced by sleep paralysis. Once again, we are left to accept either a simple and earthly explanation (sleep paralysis) or an extraordinary one (abduction by aliens). If you recall our discussion of *Occam's razor* in Chapter 2, you'll understand why scientists prefer the simple explanation, and would accept the extraordinary one only if presented with very strong evidence. As Carl Sagan said, "Extraordinary claims require extraordinary evidence."

Figure 12.20

Hundreds of lines and patterns, generally obvious only from the air, are etched in the sand of the Nazca desert in Peru. This aerial photo shows the large figure of a hummingbird. Some people have claimed that these patterns must have been created or inspired by aliens, though it would not have been difficult for the local people to have made them on their own.

Ancient Visitations The vast majority of those who believe in extraterrestrial visitation claim that the aliens and their craft are among us now. However, some people have suggested that there is evidence for alien visits in the dim past of human history, or even earlier.

The idea that Earth might have been visited seems plausible, especially if civilizations turn out to be common in our galaxy. Our planet might well be of interest to alien scientists, and we'd have no way of knowing whether alien ships studied Earth in the billions of years that preceded the evolution of *Homo sapiens.* Such visits might even have continued after the rise of civilization. But while a few people have claimed that this has happened, the evidence they offer is far from compelling.

The evidence offered for past alien visitation consists of things such as ancient drawings that supposedly show alien visitors or their spacecraft, and archaeological wonders supposedly too sophisticated for our ancestors to have made on their own. These archaeological claims generally make little sense. For example, desert markings on the Nazca plains of Peru (Figure 12.20), claimed by some to be landing strips and markers for alien craft, could just as well be the work of the Nazca Indian culture that occupied the area about 2,000 years ago. No special technology would have been needed to make the markings, and patterns visible only from the air seem more likely to have been intended for gods than for aliens. That some of these drawings look somewhat like astronauts hardly improves the case (Figure 12.21), as they could as easily represent a god or a person wearing a ceremonial headdress.

Claims of alien origin for structures such as the Egyptian pyramids are nothing less than an insult to ancient people, because they seem to suggest that people of these cultures were incapable of constructing such impressive structures on their own. In fact, while the pyramids were a remarkable achievement, their construction was well within the technical expertise of the kingdom of the pharaohs. All other known ancient structures—including Stonehenge, the Mayan pyramids of Central America, and the massive stone heads of Easter Island—were also within the technical capability of the cultures that created them.

Perhaps the one thing going for the claims of ancient visitation is that they at least attempt to rely on archaeological artifacts that can in principle be studied by anyone. In that sense, they can be subjected to scientific study. The problem with these claims is that when scientists *have* studied them, the claims have failed to withstand scrutiny.

• Is there a case for alien visits?

Despite the lack of evidence of alien visits past or present, many people continue to believe they have occurred. Champions of alien visitation generally explain away the lack of compelling evidence in one of two ways: government cover-ups or a failure of the mainstream scientific community to take the relevant phenomena seriously. Let's consider both of these arguments in a little more detail.

Government conspiracy and cover-up are popular notions, particularly in the United States. It's certainly conceivable that a secretive government might try to put a lid on the best spacecraft videos and reconnaissance images from orbiting satellites, although the motivation for doing so is unclear. The usual explanations are that the public couldn't handle the news or that the government is taking secret advantage of the alien materials to design (via "reverse engineering") new military hardware. But both explanations are silly. Half the population already believes in alien visitors and would hardly be shocked if newspapers announced tomorrow that aliens were stacked up in government warehouses. As for the reverse engineering of extraterrestrial spacecraft, we should keep in mind the difficulty of traveling from star to star. Any society that could do so would be technologically far superior to our own. Our reverse engineering their spaceships is as unlikely as Neanderthals constructing personal computers just because a laptop somehow landed in their cave. In addition, while a government might successfully hide evidence for a short time, the evidence is unlikely to remain secret for decades (six decades, in the case of the Roswell claims). And, unless the aliens landed only in the United States, can we seriously believe that *every* government would cooperate in hiding the evidence?

The alleged disinterest of the scientific community is also an unimpressive claim. Scientists are constantly competing with one another to be first with a great discovery and thus are aggressive in pursuing any important new phenomenon. Clear evidence that aliens exist and are (or have been) on our planet would be hailed as one of the most important discoveries of all time. Countless researchers would work evenings and weekends, without pay, if they thought such a discovery was possible. The fact that few scientists are engaged in such study reflects not a lack of interest but a lack of evidence worthy of study.

Indeed, lack of physical evidence is the foremost reason given by scientists to justify their skepticism regarding UFOs and alien visits. Airport radars and orbiting satellites have never yielded proof of the passage of an alien spacecraft. No one has ever walked into a research lab with an alien artifact, such as a piece of material that we could not manufacture with our current technology—not even a fragment. In Earth orbit, the U.S. Air Force tracks thousands of pieces of "space junk" from our own satellites and shuttles, including discarded rockets, exploding bolts, a Hasselblad camera, paint chips, and the occasional astronaut glove. But no one has ever found a single piece of space junk that can be attributed to aliens. Most scientists are open to the possibility that we might someday find

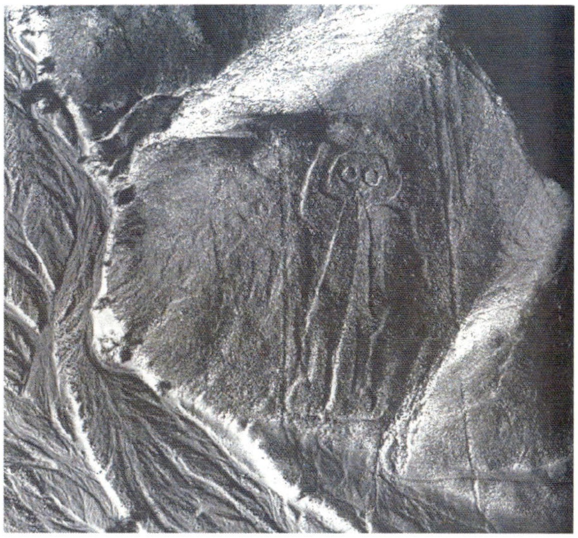

Figure 12.21

This 15-meter long etching in the Nazca plains shows an "owl man" that some people claim to be a drawing of an astronaut—presumably one of the astronauts that some people believe used the "runways" also found etched in the desert.

evidence of alien visits, and many would welcome aliens with open arms. But so far, the evidence is simply lacking. Extraterrestrial visitors to Earth remain a routine fixture of cinema and television, but in real life they are like ghosts—a pervasive and attractive idea that the vast majority of scientists treat with skepticism.

Of course, "absence of evidence is not evidence of absence," meaning that a lack of evidence for some claim doesn't make the claim untrue. In the case of alien visitation, it's conceivable that aliens really are here, and that we lack the evidence to prove it because they don't want us to know. If you wish to believe that, no one will stop you—just don't claim that your belief is based on science.

Finally, it is important to distinguish between claims that aliens are visiting us now (or visited Earth in the past) and the possibility that alien civilizations might exist. When, after World War II, space travel moved from the theoretical to the practical, it was only natural to assume that what we were trying to do—travel to other worlds—is routinely done by other civilizations. However, as we'll discuss in the next chapter, there is a great difference between journeys within the solar system and jaunts to the stars. While the former are straightforward, the latter are both enormously difficult and extremely costly in terms of energy. Nonetheless, interstellar travel doesn't violate physics, and if civilizations are common, it is certainly possible and perhaps even likely that some might have been inspired to voyage from star to star.

THE BIG PICTURE

Putting Chapter 12 into Context

In this chapter, we have explored the rationale and the methods of the search for extraterrestrial intelligence. As you continue your study, keep in mind the following "big picture" ideas:

- SETI is both a part of and distinct from other efforts in astrobiology research. Its justifications and methods depend on what we learn more generally about life in the universe. However, whereas other astrobiology research makes slow and steady progress, SETI offers the potential to give us absolute proof in one fell swoop that we are not alone—but only if we receive a clear signal from another civilization.
- We do not yet know enough about life in the universe to make a reasonable estimate either of the number of civilizations that might exist or of our odds of achieving success in SETI efforts. Nevertheless, we have no hope of success unless we try, and contact with another civilization would surely be one of the greatest discoveries in human history.
- The telescopic search for distant civilizations would be rather superfluous if aliens are already here among us. However, despite the many sightings of UFOs and other phenomena supposedly caused by aliens visiting Earth, no compelling evidence for such visits has ever been found. Scientifically, SETI represents our only current hope of detecting other civilizations.

SUMMARY OF KEY CONCEPTS

12.1 THE DRAKE EQUATION

- **What is the Drake equation?**

The **Drake equation** gives us a way to organize our thinking about the question of the number of civilizations in the Milky Way Galaxy. In its modified form, it says that the number of civilizations with whom we could potentially communicate is $N = N_{HP} \times f_{life} \times f_{civ} \times f_{now}$, where N_{HP} is the number of habitable planets in the galaxy, f_{life} is the fraction of habitable planets that actually have life on them, f_{civ} is the fraction of life-bearing planets on which a civilization capable of interstellar communication has at some time arisen, and f_{now} is the fraction of all these civilizations that exist now.

- **How well do we know the terms of the Drake equation?**

We don't know the values of any of the terms well. We have some data from extrasolar planets that can allow us to make at least an educated guess about the first term, N_{HP}; it seems likely to be quite large, perhaps 100 billion or more habitable planets in our galaxy. For the rest of the terms, we have only the example of Earth to look to, making any guesses far more uncertain.

12.2 THE QUESTION OF INTELLIGENCE

- **Even if life is widespread, would intelligence be common?**

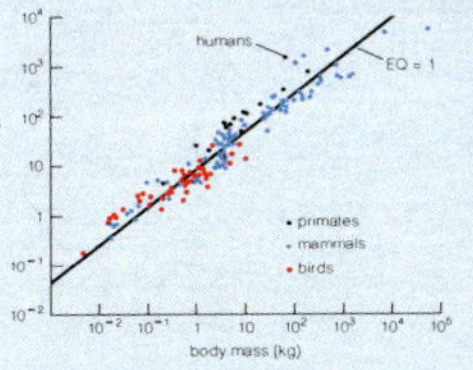

We really don't know, because we have only the example of Earth. Nevertheless, evolutionary studies indicate at least some drive toward intelligence, so it is at least plausible to imagine intelligence appearing on any planet with life, at least if given enough time.

- **Will intelligence inevitably spawn technology?**

Certainly, there are some species with physical limitations, such as lack of hands, that would seem to prevent the development of technology. But our own case is ambiguous: It took us a long time to develop technology, but we do not know if this means the development was a fortunate accident or something destined to have happened eventually.

12.3 SEARCHING FOR INTELLIGENCE

- **How did SETI begin?**

Although there were some attempts at radio contact with aliens in the early twentieth century, in retrospect we know that these efforts were doomed because they used frequencies that are blocked by Earth's ionosphere and because they focused on nearby worlds like Mars, where complex life is unlikely to exist. The origin of modern SETI is generally credited to ideas proposed by physicists Giuseppe Cocconi and Philip Morrison. Frank Drake's Project Ozma was the first organized search.

- **How do we search for intelligence today?**

SETI today is conducted primarily by searching for either radio or optical signals transmitted by distant civilizations. There may be other means of interstellar communication, but it seems reasonable to suppose that radio or optical signals will be used by at least some, if not all, other technological societies. Current signal detection efforts are probably sensitive enough to find only deliberately broadcast beacon signals.

- **What happens if SETI succeeds?**

The scientific and technological issues of SETI are important but may well pale in comparison to the societal issues if a signal is found. As a result, an important part of SETI work involves thinking about what will happen if the search ultimately proves successful.

12.4 THE PROCESS OF SCIENCE IN ACTION: UFOS AND ALIENS ON EARTH

- **What have we learned from UFO sightings?**

We've learned that people sometimes see things in the sky that they cannot identify or explain, but we have not found any evidence pointing toward an alien origin for such sightings.

- **Have aliens left any verifiable evidence of visitation?**

Although many people have made claims of hard evidence from alien visits, none of these claims has ever withstood scientific scrutiny.

- **Is there a case for alien visits?**

By the tenets of science, there is no current case for alien visits to Earth, either past or present. Keep in mind, however, that absence of evidence is not evidence of absence. If civilizations really are common, then it is conceivable that some aliens have come our way.

EXERCISES AND PROBLEMS

REVIEW QUESTIONS

Short-Answer Questions Based on the Reading

1. What is the value of the *Drake equation?* Define each of its terms, and describe the current state of understanding regarding the possible values of each term.
2. What is *convergent evolution?* How does this idea suggest that intelligence would tend to be an evolutionary imperative?
3. Briefly explain the idea of the *encephalization quotient* (*EQ*). How does it suggest that humans are indeed intelligent? What does it tell us about intelligence among other animal species?
4. Describe a few physiological and sociological factors that might influence whether an intelligent species can develop technology for interstellar communication.
5. Briefly describe early attempts at interplanetary communication by Marconi and Tesla. Why were these attempts doomed from the start?
6. Briefly discuss early SETI efforts. What do we mean by the *bandwidth* of a signal, and why does SETI concentrate on a search for narrow-bandwidth signals?
7. What are the three general categories of broadcast signals that might be detected at great distance? What are the current prospects for detecting each type of signal through SETI efforts?
8. Why do SETI researchers assume that beacon signals would be designed for easy decoding, and how might we recognize them?
9. Summarize the current techniques of radio SETI and some of the major current projects.
10. Explain why it is reasonable to imagine optical or other signals, and the method behind current optical SETI efforts.
11. Briefly discuss the possibilities of finding other civilizations via artifacts or "astroengineering."
12. What are the three distinct categories of civilization (as outlined by Kardashev)? Which one(s) can we imagine detecting through their use of resources and why?
13. Briefly discuss some of the issues that would surround an actual SETI detection.
14. Discuss several types of claims about alien visitation on Earth. Why, so far at least, do they seem not to reach the level of scientific evidence?

TEST YOUR UNDERSTANDING

Evaluate the Opinions

Each of Problems 15–24 makes a clear statement of opinion. Evaluate each statement and write a few sentences describing why you agree or disagree with it. Explain clearly; not all of these have definitive answers, so your explanation is more important than your chosen answer.

15. Humans are the "crown of creation" and an inevitable result of billions of years of evolution.
16. If, for some reason, we humans were to suddenly wipe out our species, another species—possibly the raccoons—would soon evolve greater intelligence than we possessed.
17. Sea creatures, no matter how clever they are, could never master the technology required to communicate with other worlds.
18. Most of the intelligence in the universe is not biological, but artificial ("machine intelligence").
19. Because SETI researchers are "listening" to star systems that are hundreds of light-years distant, there's a good chance that by the time we hear a signal, the civilization that sent it will have disappeared.
20. No advanced society would ever construct a beacon transmitter, because it would inevitably attract attention and might be dangerous. Similarly, we should not make deliberate transmissions to the stars.
21. We should consider including an "artifact hunt" in the space program that would search on the Moon for objects left behind by advanced extraterrestrial societies.
22. Looking for signals from star systems is a poor approach, because any truly advanced civilization will have moved beyond its home planet and populated interstellar space.
23. If 10,000 people saw the same UFO, scientists would be forced to conclude that an alien visit really occurred.
24. The absence of any scientific evidence for alien visitation on Earth implies that civilizations are rare and that SETI efforts are doomed to failure.

Quick Quiz

Choose the best answer to each of the following. Explain your reasoning with one or more complete sentences.

25. The end result of a calculation with Drake equation is intended to be an estimate of: (a) the number of worlds in the galaxy on which life has arisen; (b) the number of worlds in the galaxy on which intelligence has arisen; (c) the number of worlds in the galaxy on which civilizations are transmitting signals now.
26. Which of the following statements is true about the terms in the Drake equation? (a) Astronomical research will soon give us firm values for all of the terms. (b) Some of the terms depend on sociology, and cannot be determined by astronomers alone. (c) We already know the terms of the equation to an accuracy within a factor of two.
27. The fact that marine predators like dolphins and sharks have similar shapes despite different ancestry is an example of: (a) convergent evolution; (b) narrow bandwidth; (c) spontaneous creation.
28. Which of the following would lead an animal to a higher encephalization quotient (EQ) as it evolved? (a) growth in both body size and brain size; (b) growth in body size but not in brain size; (c) growth in brain size but not in body size.
29. The *bandwidth* of a radio signal is a measure of: (a) its frequency; (b) the range of frequencies that carry information; (c) the amount of power carried by the signal.
30. Why are we more likely to be able to detect a deliberately broadcast "beacon" signal than, say, the television broadcasts of a distant civilization? (a) because we expect beacon signals to be far more common; (b) because our current technology is probably sensitive enough to detect beacons but not much

weaker television transmissions; (c) because television is a sociological phenomenon and beacons are not, so we'd expect all civilizations to have beacons but not all to have television.

31. What is the distinguishing characteristic that those doing radio SETI experiments look for? (a) a signal containing the value of pi and other mathematical constants; (b) a signal that is an echo of an earthly broadcast; (c) a signal that extends over only a narrow band of frequencies.

32. Two-way conversation with other societies is probably unlikely, even if we make contact. This is mainly because: (a) Aliens won't speak our language. (b) It might be dangerous to get in touch. (c) The time it takes for signals to cross the distance to them could be centuries or more.

33. According to the best available evidence, the famous Roswell crash of 1947 involved: (a) an alien spacecraft; (b) an Air Force balloon experiment; (c) There is no evidence that gives us any information about the crash.

34. One reason that scientists doubt that crop circles have alien origin is: (a) They are always beautiful. (b) They can be easily made by humans. (c) Their appearance is not correlated with sightings of bright lights.

INVESTIGATE FURTHER

In-Depth Questions to Increase Your Understanding

Short-Answer/Essay Questions

35. *Drake Values.* Make your own estimates of values for each of the four terms in the modified Drake equation used in this chapter. Explain how you arrived at each estimate, and then use your estimates to calculate N.

36. F_{now}. Suppose that the number of civilizations in the galaxy has been quite large, say, one million. Does that necessarily mean that other civilizations should exist right now? Explain why or why not, and describe the factors that would influence the answer.

37. *Evolution of Intelligence.* Based on your understanding of natural selection and of the evolution of humans as discussed earlier in the book, describe at least three distinct environmental factors that have contributed to the evolution of human intelligence. Explain clearly.

38. *Intelligence on Other Worlds.* Consider again the three factors you identified in Problem 37. For each one, decide whether you think the same factor would be likely to arise and select for intelligence on another world with life, and clearly explain how you reached your conclusions.

39. *Communication.* Imagine that you have to fashion a short message that would tell extraterrestrials something about human society. What would you "say" using only three simple pictures? What would you write if you could use only a half-page of English text?

40. *Talking Back.* Suppose SETI were to find a signal coming from a star system 200 light-years away. Write a one- to two-page essay describing what, if anything, we should do to establish contact. You should think about how quickly we should respond, what the response should be, and what possible dangers might be involved.

41. *Contact.* Watch the movie *Contact,* and pay careful attention to the SETI experiment described in the first third of the film. How accurately does this experiment reflect any of the current SETI search programs? Did you spot any obvious scientific or technical errors? Write a one- to two-page essay comparing *Contact* to the reality of SETI efforts.

42. *Invasions of Movie Aliens.* Choose a science fiction movie in which aliens are presumed to be visiting Earth. Identify at least three ideas in the movie that either do or do not meet the standards of being testable by science. Describe each in detail.

Quantitative Problems

Be sure to show all calculations clearly and state your final answers in complete sentences.

43. *How Many Stars to Search?* The number of star systems that a SETI search would have to investigate before achieving success depends on how common signaling societies are in the galaxy. This is the number estimated by the Drake equation. Suppose this number is $N = 1$ million. How many star systems must be checked out by SETI in order to find one signal? What if $N = 1{,}000$? Assume that there are roughly 100 billion stars in our galaxy.

44. *Distance to E.T.* Suppose there are 100,000 signaling societies in our galaxy. What is the average distance between civilizations, assuming that civilizations are spread evenly throughout the galactic disk?

45. *Actual SETI Searches.* Project Phoenix, the largest search of individual star systems for radio signals before 2005, trained its antennas on Sun-like stars up to about 150 light-years away. How many civilizations would there have to be in order for the average distance between civilizations to be 150 light-years? If the actual value of N is 10,000, would we expect Project Phoenix to have made a detection? What if the actual value of N is ten million? Explain.

46. *Power Used by E.T.* A modern SETI search using the 300-meter-diameter Arecibo radio telescope in Puerto Rico can pick up a 10-million-watt signal from 1,000 light-years away (assuming that the broadcasting aliens have a transmitting antenna that's also 300 meters in diameter). Suppose we wish to use Arecibo to search the far side of the Milky Way Galaxy (roughly 80,000 light-years away) under the same assumptions about our setup and the transmitting antenna. What would be the required power of the alien transmitter for us to detect the signal?

47. *Transmitter Used by E.T.* A modern SETI search using the 300-meter-diameter Arecibo radio telescope in Puerto Rico can pick up a 10-million-watt signal from 1,000 light-years away (assuming that the aliens have a transmitting antenna that's also 300 meters in diameter). Suppose an alien civilization is using this same transmitter setup but is on the other side of the Milky Way Galaxy (roughly 80,000 light-years away). How large an antenna would we need to hear the signal?

Discussion Questions

48. *Measuring Intelligence.* In judging the intelligence of animals, we use the *encephalization quotient* (EQ), which depends on the ratio of brain weight to body weight. Can you think of situations for which this might be a poor way to gauge intelligence? For example, could animals have some special processing needs (such as the navigation mechanism of bats) that would make their brains larger without contributing to their intelligence? Alternatively, could animals exist (on Earth or elsewhere) whose brains

were relatively lightweight but who were still highly intelligent? Defend your opinion.

49. *Detecting Signals.* SETI scientists are sometimes criticized for using "old technology" in their search for signals. Perhaps extraterrestrials have moved beyond radio and light signaling and are using something much more sophisticated. Discuss (a) the advantages of radio and light for interstellar communication and (b) any reasonable alternatives you can think of. There is always the possibility that "new physics" will provide faster or more efficient methods for signaling. Do you think this is a reason to limit current SETI efforts?

50. *Societal Reaction.* It is frequently said that the detection of a signal by SETI would revolutionize human society. Does this statement seem reasonable? Some researchers have tried to find historical events, such as the Copernican revolution or the publishing of Darwin's theories of evolution, whose impacts might compare to that of a SETI detection. Are such examples likely to be accurate in predicting how we would react? How likely do you think it is that a SETI discovery would cause either mass panic or an outbreak of universal brotherhood?

51. *Dealing with UFO Claims.* Given the large number of people who claim to have seen a UFO, you are likely to know at least one such person, now or in the future. Perhaps *you* have seen a UFO. Suppose someone who has seen a UFO believes deeply that it was an alien spacecraft. What, if anything, would you say to that person? Why?

WEB PROJECTS

52. *Current SETI Research.* Go to the SETI Institute's Web site and use links listed there to make an inventory of current SETI projects worldwide. Organize these projects according to whether they are radio or optical, and then separate targeted searches from sky surveys. Prepare a one-page summary of this table that discusses how thorough the current searches for extraterrestrial signals are. You should consider how many star systems have been looked at carefully, how wide a band (for radio searches) has been covered, and how sensitive the searches are.

53. *SETI@home.* This is a project organized by researchers at the University of California, Berkeley, to process radio SETI data on home computers. Download the free SETI@home screen saver onto your computer, and use it to analyze data collected by Project SERENDIP. Write a one-page description of the general processing scheme used by SETI@home, as well as the types of signals it is searching for.

54. *What to Do in Case of a Signal Detection?* Download the text of the protocol *Principles Concerning Activities Following the Detection of Extraterrestrial Intelligence* from the SETI Institute's Web site. Write a short discussion of what SETI scientists expect will happen if they detect an extraterrestrial signal and whether this expectation is realistic. In particular, do you think a discovery could be "covered up," or would it leak out to the public before the scientists themselves were sure of the discovery?

55. *Alien Visits.* Learn more about a claim of alien visitation to Earth, past or present. Write a one- to two-page report explaining the claim and the evidence that supports it and discussing the plausibility of the claim.

13
Interstellar Travel and the Fermi Paradox

LEARNING GOALS

13.1 THE CHALLENGE OF INTERSTELLAR TRAVEL
- Why is interstellar travel so difficult?
- Could we travel to the stars with existing rockets?

13.2 DESIGNING SPACECRAFT FOR INTERSTELLAR TRAVEL
- How might we build interstellar spacecraft with "conventional" technology?
- How might we build spacecraft that could approach the speed of light?
- Are there ways around the light-speed limitation?

13.3 THE FERMI PARADOX
- Where is everybody?
- Would other civilizations really colonize the galaxy?
- What are possible solutions to the Fermi paradox?
- What are the implications of the Fermi paradox to human civilization?

13.4 THE PROCESS OF SCIENCE IN ACTION: EINSTEIN'S SPECIAL THEORY OF RELATIVITY
- What is "relative" about relativity?
- What evidence supports Einstein's theory?

Provide ships or sails adapted to the heavenly breezes, and there will be some who will not fear even that void ...

Johannes Kepler (in a letter to Galileo), 1593

In an age of rapid technological progress, it may seem inevitable that our rockets will soon reach the depths of interstellar space. The reality, however, is that interstellar travel is much more challenging than bridging the distances to nearby moons and planets. There are engineering and physical constraints, not least of which is the cosmic speed limit—the speed of light. Nevertheless, we can envision at least some ways by which our descendants might someday travel among the stars.

The idea that humans might someday travel throughout the galaxy should make us wonder whether other civilizations have already achieved this ability. Indeed, if civilizations are common, it seems reasonable to expect that some societies—perhaps many—began colonizing the galaxy long before the earliest humans walked the Earth, and maybe even before Earth was born. This idea leads directly to the so-called *Fermi paradox:* If someone could have colonized the galaxy by now, why don't we see any evidence of a galactic civilization?

We will begin this chapter by discussing both the challenges and possibilities of interstellar travel as we understand them today. Then, with that understanding in mind, we will confront the Fermi paradox, and see why, despite the seeming innocence of the question, its solution will undoubtedly have profound implications for the future of our own civilization.

13.1 THE CHALLENGE OF INTERSTELLAR TRAVEL

Science fiction routinely portrays our descendants hurtling through the galaxy, wending their way from one star system to another as easily as we now travel from one country to the next. We have already used rockets to explore other worlds in our solar system. Could future generations travel among the stars just by building larger versions of the rockets we use today? Perhaps surprisingly, the answer is no. The chemical rockets that have sent people to the Moon are wholly inadequate for taking people to the stars.

• Why is interstellar travel so difficult?

The fact that interstellar travel is a daunting enterprise is due to a simple circumstance: the tyranny of distance. The stars are so incredibly remote that only in the nineteenth century did astronomers develop instruments of sufficient precision even to measure the distances of the closest stars besides the Sun. When it was realized just how far away these pinpoints of light are, the French philosopher Blaise Pascal was moved to write that "the eternal silence of infinite spaces" left him terrified.

Spacecraft Bound for the Stars We've considered the vast distances to the stars in earlier chapters, using the scale model of the solar system introduced in Chapter 3. Here we consider the distances from the point of view of travel. Four of our past interplanetary probes—*Pioneers 10* and *11*, and *Voyagers 1* and *2*—are currently on their way out of the solar system; the *New Horizons* spacecraft will also head toward the stars after it passes by Pluto in 2015. How long will it take these spacecraft to reach the stars?

Let's take the first, *Pioneer 10*, as an example. This spacecraft, launched in the early 1970s, took 21 months to reach its target, the planet Jupiter, before heading out of the solar system. This might seem speedy enough in view of the fact that the giant planet is never closer than 628 million kilometers from Earth. But our nearest stellar neighbor, the Alpha Centauri star system, is 70,000 times farther away than Jupiter. If *Pioneer 10* were to cover the 4.4 light-years to Alpha Centauri at the same average speed at which it traveled to Jupiter, the journey would take 115,000 years. But *Pioneer 10* was not aimed at Alpha Centauri or at any other deliberate target—its trajectory was designed to reach Jupiter and Saturn, not any particular stars beyond. If we plot its trajectory along with the motions of nearby stars, we find that the closest *Pioneer 10* will come to any star in the next million years is 3.3 light-years. In about 2 million years, the probe will reach the general neighborhood of the bright star Aldebaran, in the constellation Taurus.

You can now see why we did not equip *Pioneer 10* or any of these other probes with instruments for studying planets in other star systems. Nevertheless, because the spacecraft themselves should survive unscathed for millions of years in the near-vacuum of interstellar space, we have included messages in case any extraterrestrial beings someday find them. The *Pioneer* probes each carry a small engraved plaque bearing a drawing of a man and a woman as well as diagrams giving the layout of the solar system and our general location in the galaxy (Figure 13.1). The *Voyager* craft, launched about 5 years after the *Pioneer* craft, carry a somewhat more sophisticated message consisting of pictures, multilingual greetings, and two dozen musical selections (ranging from Chuck Berry to Bach) on a gold-plated copper record (Figure 13.2). Although you might wonder how intelligible these earthly calling cards might be to any aliens, the chances that they will ever be found are slim. They are like messages in a bottle thrown into the ocean surf and were intended more as a statement to Earthlings than to extraterrestrials.

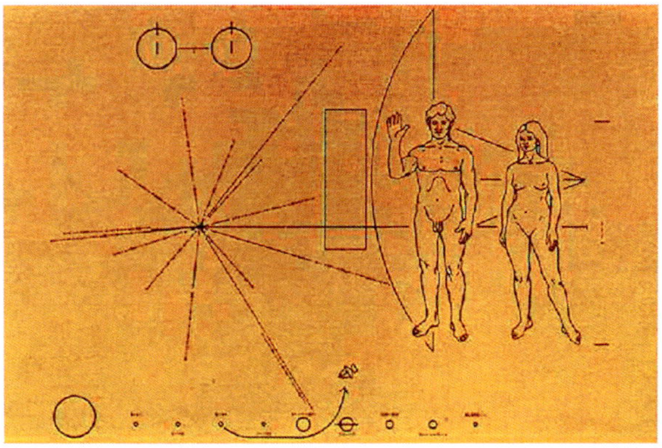

Figure 13.1

The *Pioneer* plaque, carried on both Pioneer 10 and 11 spacecraft, is about the size of an automobile license plate. The human figures are shown in front of a drawing of the spacecraft to give them a sense of scale. The "starburst" to their left shows the Sun's position relative to nearby stellar remnants known as *pulsars*, which are rapidly rotating neutron stars, and Earth's location around the Sun is shown below. Binary code indicates the pulsar periods. Because pulsars slow with time, the periods will allow someone reading the plaque to determine when the spacecraft was launched.

THINK ABOUT IT The *Pioneer* and *Voyager* "messages" are in the form of sounds, music, and pictures. But this assumes that any aliens finding these craft would have sensory organs similar to ours. Is it possible that our messages are too anthropocentric to even be recognized, or are there good reasons to think that E.T. will have eyes and ears, with characteristics similar to ours? Defend your opinion.

The Cosmic Speed Limit The *Pioneer 10* example makes the problem of interstellar travel quite clear. But it also seems to offer an obvious solution: Build spacecraft that can travel a lot faster. If it would take a little more than a hundred thousand years for *Pioneer 10* to reach the nearest stars, then a spacecraft that travels 100,000 times faster should be able to make the trip in only a little over a year.

However, this seemingly obvious solution is not allowed by the laws of physics. In particular, we know from Einstein's *special theory of relativity*

Figure 13.2

Voyagers 1 and *2* carry a phonograph record—a 12-inch gold-plated copper disk containing music, greetings, and images from Earth. One of the etchings on the disk surface gives instructions on how to play it.

that it is impossible to travel through space faster than the speed of light. (We'll discuss this theory and why it imposes a cosmic speed limit in Section 13.4.) This might not seem too limiting, given that light travels incredibly fast—about 300,000 kilometers per second (186,000 miles per second)—fast enough to reach the Moon in barely more than 1 second. But even at this remarkable speed, light takes time to travel the vast distances between the stars, which is why we measure stellar distances in light-years [Section 3.2]. As noted earlier, the nearest star system, Alpha Centauri, is about 4.4 light-years away, which means it takes light 4.4 years to reach us from this system. Because that is the fastest possible speed of travel, the best spacecraft we could hope to build would take longer than 4.4 years for the one-way trip and hence at least 8.8 years for a round trip. To make a trip across our entire galaxy—a distance of 100,000 light-years—would take any spacecraft a minimum of 100,000 years.

Could it be that Einstein's theory is wrong and that we will someday find a way to break this cosmic speed limit? Probably not. Special relativity merits the status of being a scientific theory because it is supported by an enormous body of evidence. Its predictions have been carefully tested and verified in countless experiments, so it cannot simply be "wrong." While it might someday be augmented by a more comprehensive theory, the verified results will not simply disappear; the cosmic speed limit will almost certainly remain in place.

Energy Issues Another challenge of interstellar travel is the tremendous amount of energy it would require, particularly if we wanted to send people and not just lightweight robotic probes to the stars.

Imagine that we wanted to colonize an extrasolar planet. To get a decent-size colony started, we'd need to send a fair number of people with many different sets of skills. For the sake of argument, suppose we wanted to send 5,000 people, meaning we would need a starship of similar capacity to the large starships used in the *Star Trek* television shows and movies. How much energy would such a ship require?

Interestingly, the minimum energy requirement doesn't depend on the fuel source or the ship design at all. Sending a bowling ball flying through the air takes more energy than sending a baseball flying at the same speed, regardless of whether the energy comes from your arm, from a catapult, or from some kind of gas-powered launcher. Similarly, sending either ball flying at a faster speed takes more energy. That is, the energy required to put an object in motion depends on only two things: the object's mass and the speed with which you want it to move.

We can estimate the mass of the starship by comparing it to other ships that transport large numbers of passengers. For example, the *Titanic* weighed about 18,000 kilograms per passenger (although accommodations for most passengers were hardly roomy and it carried provisions for only a couple of weeks, not many years). If we conservatively adopt the same per-person weight for our starship, we would expect its total mass to be about 100 million kilograms. Let's assume further that our starship travels at a modest 10% of light speed, which means it will take more than 40 years to reach the nearest stars. Now that we have estimated both the mass and the speed of our starship, a simple physics formula allows us to calculate the energy needed. (The required formula is the one used to compute kinetic energy, which is equal to $\frac{1}{2}mv^2$, where m is the mass of the moving object and v is its velocity.) The energy needed to get this ship to cruising speed is calculated to be 4.5×10^{22} joules, roughly equivalent to 100 times the world's current annual energy use.

In fact, we should double this value, because the amount of energy required to slow down the ship for a soft landing once we arrive at the colony is the same as that required to accelerate it to cruising speed. Thus, the total energy bill for the trip would be equivalent to at least two centuries' worth of current world energy usage. Let's put this in monetary terms: Using a typical price for home electricity in the United States (10¢ per kilowatt-hour), the energy cost of sending our craft to another star would be about $2,500,000,000,000,000,000. (To this you can add the cost of food and fresh towels for 40 years.) Clearly, unless and until we find a way to produce enormously more energy at vastly lower prices, large-scale interstellar travel will remain out of reach.

• Could we travel to the stars with existing rockets?

Today, when we think of space travel, we invariably think of rockets. But it wasn't always so. The Greeks spoke of celestial travel in the mythological flights of Phaeton and Icarus, neither of which involved rockets and both of which ended badly. As described by the Roman writer Ovid (43 B.C.–18 A.D.), Phaeton was an impetuous youth who borrowed the Sun god's chariot for a joy ride across the sky. Losing control of the horses on this ill-fated mission, Phaeton inadvertently started to set fire to Earth. Humanity's home was saved only when Zeus intervened by killing the cheeky charioteer. Icarus, the son of Daedalus, made a similar ill-advised move and became the first fictional flight victim. The feather wings Icarus used to cruise the heavens were held together with wax, and they underwent a fatal meltdown when, despite his father's warnings, he flew too close to the Sun. These stories suggest that, while they were open to the possibility of other worlds, the Greeks regarded thoughts of human flight as hubris.

The Middle Ages saw a continuation of this less-than-enthusiastic view of human flight. Medieval thinkers were not intrigued by practical travel to the heavens, reserving such jaunts for spiritual ventures. Only after the Renaissance, when the philosophy of modern scientific inquiry took hold, did humankind once again consider the possibility of breaking Earth's bonds. By 1687, Isaac Newton had produced a treatise on universal mechanics that not only described the workings of the heavens but also explained the physics required to reach them.

Newton's third law of motion states that "for every action there is an opposite and equal reaction." Envision the recoil of a gun when it is fired. The bullet moves in one direction, and the gun moves in the opposite direction. Squids, octopuses, and some other mollusks employ a similar technique in their movements. A squid takes in water that it then squirts out at higher speed behind it, thus propelling itself forward. A rocket operates slightly differently, vaporizing on-board fuel that is shot out the back. However, in both cases it is Newton's third law (or, equivalently, the law of conservation of momentum) that accounts for the forward motion.

Development of the Rocket A rocket has been described as the simplest type of engine, and even scientists as early as Newton realized it could work in empty space. Serious thought about travel to other worlds began in the nineteenth century. In the 1860s and 1870s, the French author Jules Verne wrote influential stories describing travel to the Moon. His propulsion scheme used an oversize artillery shell specially constructed for the task. While this scheme was hardly practical (the enormous acceleration of the shell when fired would turn the passengers to pancakes), Verne's writings stimulated investigation of space travel by three giants of early

COSMIC CALCULATIONS 13.1

The Rocket Equation

The rocket equation tells us how a spacecraft's final velocity, v, depends on the velocity of the exhaust gas expelled out the back, v_e, and the rocket's *mass ratio*. The mass ratio is M_i/M_f, where M_i is the mass of the rocket (including any spacecraft it is carrying) with all its fuel and M_f is the mass of the rocket after the fuel has been burned (that is, the spacecraft and any still-attached but empty fuel tanks). We can write the rocket equation in the following two equivalent forms:

$$v = v_e \ln\left(\frac{M_i}{M_f}\right) \quad \Leftrightarrow \quad \frac{M_i}{M_f} = e^{\left(\frac{v}{v_e}\right)}$$

In the equation at left, "ln" is the natural logarithm; your calculator should have a key for computing this. In the equation at right, e represents a special number with value $e \approx 2.718$; your calculator should also have a key for computing e to any power. If you are familiar with the algebra of logarithms, you can confirm that the two equations are equivalent.

Example: Suppose you want a rocket to achieve escape velocity from Earth (11 km/s) and its engines produce an exhaust velocity of 3 km/s. What mass ratio is required?

Solution: We set the rocket's final velocity to v = 11 km/s and its exhaust velocity to v_e = 3 km/s, and use the second form of the equation to find the mass ratio:

$$\frac{M_i}{M_f} = e^{\left(\frac{v}{v_e}\right)} = e^{\left(\frac{11\ \text{km/s}}{3\ \text{km/s}}\right)} = e^{\frac{11}{3}} \approx 39$$

The required mass ratio is about 39. As discussed in the text, this mass ratio cannot be achieved with a single-stage rocket, but can be reached with a multi-stage rocket.

rocketry: Konstantin Tsiolkovsky (1857–1935) in Russia, Hermann Oberth (1894–1989) in Germany, and the American Robert Goddard (1882–1945). These fledgling rocket scientists explored many of the theoretical possibilities of this type of propulsion. In particular, they worked out the so-called **rocket equation,** which describes how a vehicle's final speed depends on the propellant's velocity (see Cosmic Calculations 13.1). Both Tsiolkovsky and Goddard realized that it would be difficult for a single rocket to reach **escape velocity,** the speed necessary to overcome gravity and leave Earth behind (about 11 km/s, or 25,000 mi/hr), so they proposed the use of multistage vehicles for space flight. The three pioneers also envisioned space stations, intercontinental ballistic missiles, and ion engines.

At first, few people saw much benefit in turning these ideas into working hardware. In a 1920 technical publication, Goddard mused about the possibility that a sufficiently large rocket could reach the Moon. His speculation was immediately ridiculed by the *New York Times,* which claimed that no lunar-bound rocket could ever work since there is no air between Earth and the Moon, and thus a rocket would have nothing to "push against." However, contrary to the *Times* impression, rockets do not operate by pushing against air—or anything else. They simply employ Newton's third law, firing hot gas in one direction so that the rocket moves in the opposite direction. In fact, atmospheres hinder the performance of rockets, because they create drag that slows rockets down.

A few years later, in 1926, Goddard launched his first liquid-fueled rocket from a field in Auburn, Massachusetts (Figure 13.3a). It reached a height of 13 meters. This heroic, build-it-in-a-garage phase of rocketry was soon surpassed. The 1930s brought larger-scale efforts, particularly in Germany, where the military saw value in guided missiles. The German work culminated in the development of the V-2 rocket, used against England during World War II (Figure 13.3b). The rapid development of rocketry that followed the war was driven largely by German scientists who had been recruited by both the Russians and the Americans. The space age truly began in October 1957 with the launch of the Soviet Union's

Figure 13.3

Early rocketry.

a Robert Goddard stands by his pioneering liquid-fuel rocket in 1926. This craft reached a modest altitude: the height of a four-story building.

b The German V-2 rocket, first launched in 1942, was used against England during World War II. Note that only 16 years separate the V-2 from Goddard's first rocket.

Sputnik I, an 84-kilogram beeping metal ball—the world's first artificial satellite.

Although primarily spurred by national rivalries and military considerations, rocket development over the past five decades has allowed humans and our robot proxies to enter those tantalizing realms that had so long been beyond our grasp. What the Greeks could only dream, we can now do. Today's rockets—the direct descendants of those first envisioned nearly a century ago—are fast enough and powerful enough to allow us to explore the nearby worlds of our solar system. But could they ever be improved enough to take us to the stars?

Limitations of Chemical Rockets Even today, every rocket we use to launch spacecraft works in basically the same way as Goddard's first rocket. The engines ignite and burn a chemical fuel, such as a mixture of oxygen and kerosene. The chemical burning creates very hot gas, which is expelled through a narrow nozzle, propelling the spacecraft into orbit or to other worlds. These chemical rockets serve our current purposes fairly well (though many people dream of new technologies that would allow us to leave Earth at far lower cost). Unfortunately, they are completely inadequate for interstellar travel.

The largest chemical rocket built to date was the Saturn V, the rocket that carried the *Apollo* astronauts to the Moon (Figure 13.4). This vehicle burned liquid oxygen and kerosene, with water the major combustion product. The hot water vapor was expelled out the back at a speed of about 3 kilometers per second—roughly three times the speed of a rifle bullet. The Saturn V consisted of three separate "stages," that is, three distinct rockets perched atop one another so that each lower stage could drop away after exhausting its fuel. We can see why these multiple stages were useful—and why chemical rockets are limited—by investigating rocket mechanics in a bit more depth.

Figure 13.4

The Saturn V rocket, which was used to carry the *Apollo* astronauts to the Moon. The most powerful rocket yet built, this now 40-year-old, three-stage design weighed about 30% more than the Space Shuttle at liftoff and was capable of sending a 45,000-kilogram (50-ton) payload to the Moon. With a launchpad mass of 2.8 million kilograms for the Moon trips, its overall mass ratio was 62.

Let's start by imagining that a rocket like the Saturn V had only a single stage. In order to leave Earth behind, we need to reach Earth's escape velocity of 11 kilometers per second. Using the rocket equation developed by the pioneers of rocketry, we can calculate the **mass ratio** required to attain this speed. The mass ratio is defined as the mass of the fully fueled rocket (including any spacecraft it is carrying) divided by the rocket (and spacecraft) mass after all the fuel is burned. As shown in Cosmic Calculations 13.1, reaching escape velocity requires a mass ratio of 39, meaning that the fueled rocket on the launchpad must weigh 39 times more than the empty rocket and spacecraft alone. That is, the fuel weight would be about 38 times the weight of the spacecraft and the engines. This is clearly a discouraging requirement and one that's just about impossible to meet given the weight of tanks, fuel pumps, fins, and astronauts. Indeed, the best single-stage rockets have mass ratios of only 15 or less.

If it takes a mass ratio of 39 to leave Earth and our best rockets have mass ratios of only 15, how can we ever succeed? As the early rocket pioneers realized, the trick is to use multiple stages. By discarding each stage as its fuel is used up, the upper stages don't need to accelerate the dead weight of those below. The rocket as a whole—with all of its stages—still must weigh 39 times as much as the parts that will actually reach space. But each stage requires a much lower mass ratio, because the weight of the rocket will decrease as stages are discarded. For example, to escape Earth by means of a three-stage rocket (assuming the stages have identical mass ratios and exhaust velocities), each stage need have a mass ratio of only 3.4—well within our capabilities. (See Problem 52 at the end of the chapter.)

THINK ABOUT IT The Space Shuttle does not use stacked rockets like the Saturn V, but it still uses staging. Explain how. (*Hint:* It should be obvious if you look at a picture of the Shuttle on its launch pad.)

In principle, adding more stages can propel chemical rockets to higher speeds, but not high enough for convenient interstellar travel. For example, an oxygen-kerosene rocket like the Saturn V consisting of a stack of *100* stages (each with a mass ratio of 3.4) would reach a speed of 370 km/sec—33 times faster than the Saturn V but barely more than 0.1% of the speed of light. A trip to Alpha Centauri at this speed would still take some 4,000 years. Using more efficient chemical fuels (such as oxygen and hydrogen, rather than kerosene) can help, but by no more than about a factor of 2. Indeed, no matter what engineering refinements we consider, chemical rockets simply are not powerful enough to deliver large payloads to the stars in a reasonable length of time. Interstellar travel requires a different approach.

13.2 DESIGNING SPACECRAFT FOR INTERSTELLAR TRAVEL

Chemical rockets may be insufficient for travel to the stars, but other technologies hold greater promise. Generally speaking, we can break these technologies into two groups. The first uses "conventional" technology—that is, technology that seems within our grasp (at least if we disregard cost), even if we don't yet have it. The second group involves technologies that are theoretically possible but far beyond our present capabilities. In this section, we'll begin by investigating a few conventional technologies that would allow at least a modest degree of interstellar travel, then move on to explore more far-out ideas.

• How might we build interstellar spacecraft with "conventional" technology?

Conventional technologies for interstellar travel are based on the idea that we could adapt existing technologies to the task. None of these technologies would make interstellar travel "easy" — at best, their speeds might reduce the travel time to nearby stars to centuries or decades. Still, if we had unlimited budgets, we could in principle begin work on these technologies today.

Nuclear Rockets Chemical reactions involve shuffling the outer electrons of atoms. While these reactions can seem quite powerful (consider the drama of a Space Shuttle launch), the energy they release is insignificant compared to the amount of energy at least potentially available in the reacting materials. According to Einstein's famous formula, $E = mc^2$, any piece of matter contains an amount of energy E equivalent to its mass m multiplied by the speed of light c squared [Section 3.4], which represents an enormous amount of energy. For example, if you could turn a 1-kilogram (2.2-pound) rock completely into energy, the energy released would be equivalent to that contained in nearly 8 billion liters of gasoline—or as much gasoline as is used by all the cars in the United States in a week. However, while this energy is "there" in any piece of matter, it is very difficult to extract. Chemical reactions extract so little of it that we do not notice any change in the mass of the reacting materials.

Nuclear reactions, in contrast, can noticeably affect the mass of reacting materials. They involve changes in the dense atomic nucleus. Two basic types of nuclear reactions can be used to generate power: fission and fusion. Nuclear **fission** involves the splitting of large nuclei such as uranium or plutonium. When a uranium nucleus is split, approximately 0.07% of its mass is turned into energy. Thus, if 1,000 grams of uranium undergo fission, you'd find that the fission products (the material left over after the fission has occurred) would weigh a total of only 999.3 grams, 0.07% less than the starting weight. Although this mass loss may sound fairly small, the energy it releases dwarfs that released by chemical reactions. Nuclear fission bombs are what destroyed the Japanese cities of Hiroshima and Nagasaki at the end of World War II, and all current nuclear power plants get their energy from fission.

Nuclear *fusion*, the power source of the Sun and other stars [Section 3.2], is about ten times as efficient as fission. Fusion of hydrogen into helium converts about 0.7% of the hydrogen fuel mass into energy. The Sun, for example, fuses 600 million tons of hydrogen into helium *each second*. The resulting helium weighs 0.7% less than the original hydrogen, or about 596 million tons. The other 4 million tons of mass simply "disappears" as it becomes the energy that makes our Sun shine. We humans have managed to achieve nuclear fusion here on Earth, but only in thermonuclear bombs (or "H-bombs") and as yet not in a well-controlled, commercially useful way. This is unfortunate; not only is the fuel for fusion (hydrogen) readily available in water, but the efficiency of fusion is so great—at least compared to current energy sources such as oil, coal, and hydroelectric power—that fusion power would seem almost unlimited if we were able to tap it. For example, if we could somehow hook up a nuclear fusion plant to your kitchen sink, then by continuously fusing the hydrogen in the water flowing from the faucet we could generate more than enough power to meet all the current energy needs of the United States.* That is, with your kitchen faucet fusion plant, we could stop the drilling and importing of oil, dismantle all hydroelectric dams, shut down all coal-burning power stations, get rid of all fission power plants, and still have power to spare. And there'd be no more worries about ongoing contributions to global warming, because fusion does not release any greenhouse gases into the atmosphere.

THINK ABOUT IT Scientists have been working for decades in hopes of developing the technology for viable nuclear fusion power plants, but so far without success. As a result, governments are reluctant to spend a lot of money on further fusion research, although some is still ongoing. How much effort do you think we should put into fusion power? If we achieved it, how do you think it would change our world?

The tremendous efficiency of nuclear energy over chemical energy was bound to appeal to rocket scientists. In 1955, the U.S. Atomic Energy Commission and the U.S Air Force (and later NASA) embarked on an experiment called *Project Rover* to develop nuclear fission reactors that

---

*Actual attempts to generate fusion power use deuterium (the isotope of hydrogen with one neutron), which is present naturally in the ratio of about 1 part deuterium to 50,000 parts ordinary hydrogen. Thus, with deuterium, the needed water flow would be about 50,000 times greater than that of your kitchen faucet—but this flow (about 130,000 liters per minute) is still only about that of a small stream.

Figure 13.5

President John F. Kennedy departing the Nevada Test Site after viewing a full-scale mock-up of a nuclear-powered engine for Project Rover, December 8, 1962.

could be flown in a rocket. The idea was to use the fission reactor to generate enormous heat, which would be used to bring hydrogen gas to a temperature of millions of degrees before expelling it out the engine nozzles. (Note that the hydrogen was being used as a propellant, not for fusion.) At its peak, Project Rover employed 1,800 people and ultimately tested six fission engines. The program made substantial progress and showed that fission-powered rockets could achieve speeds at least two to three times as great as similar-size chemical rockets. By the late 1960s, NASA officials were confident that the Project Rover rockets could be used to send humans to Mars in what they hoped would be an immediate follow-on to the *Apollo* Moon landings (Figure 13.5). However, the political climate changed, and the United States abandoned its early plans for a human mission to Mars. Project Rover was terminated in 1973.

Another experimental approach, dubbed *Project Orion,* was more radical. Physicists at Nevada's Los Alamos Scientific Laboratory realized that one way to get a rocket up to much higher speed would be to toss small nuclear (fusion) bombs out the rear and let the resulting explosions push the craft forward. The bombs, released at a rate of one every few seconds or more, would drop back about 50 meters and then detonate behind a large metal "pusher plate" affixed to the tail of the rocket (Figure 13.6). This would provide an impulse to move the rocket forward. Despite suffering obvious abuse, the pusher plate wouldn't vaporize because it would be exposed to these searing explosions for only a few milliseconds at a time. The Los Alamos scientists calculated that a spaceship 1 mile long accelerated by the rapid-fire detonation of a million H-bombs could reach Alpha Centauri in just over a century. Thus, Project Orion represented the first true "starship" design to be fashioned by humans. No actual construction ever began, although in principle we could build a Project Orion–type starship with existing technology. However, this kind of starship would be very expensive and would require an exception to the international treaty banning nuclear detonations in space. Project Orion ended in 1965, due both to budget cuts and to the nuclear test ban treaty.

Another nuclear rocket design was developed in the 1970s by the British Interplanetary Society under the name *Project Daedalus* (Figure 13.7).

Figure 13.6

Artist's conception of the Project Orion starship, showing one of the small H-bomb detonations that would propel it. Debris from the detonation impacts the flat disk, called the pusher plate, at the back of the spaceship. The central sections (enclosed in a lattice) hold the bombs, and the front sections house the crew.

Figure 13.7

Artist's conception of a robotic Project Daedalus starship. The front section (upper right) holds the scientific instruments. The large spheres hold the fuel pellets for the central fusion reactor.

The idea was to shoot frozen fuel pellets of deuterium and helium-3 into a reaction chamber where they would undergo nuclear fusion. The fuel pellets, about the size of gravel, would be shot into the chamber at a rate of 250 pellets per second. There they would be encouraged to fuse by electron beams, producing a rapid-fire series of explosions that would propel the ship. Because we cannot yet build nuclear fusion reactors, this design remains beyond our current technological capabilities. Nevertheless, the proponents of Project Daedalus developed a plan for sending a robotic spacecraft to Barnard's star, a dim, type M star 6 light-years distant and the next closest star to Earth beyond the Alpha Centauri system. After 4 years of firing the engine, the craft would reach about one-tenth the speed of light and then spend the next four decades coasting to its destination. Once there, it would deploy probes and sensors to relay back photos and other data, giving us our first close-up view of another stellar system only about 50 years after its launch.

Nuclear-powered rockets are undoubtedly feasible in some form. Still, at best they would achieve speeds of about one-tenth the speed of light. Interstellar journeys would be possible, but it would take decades for them to reach even the nearest stars.

Ions, Sunlight, and Lasers The propulsion schemes described thus far involve a relatively quick acceleration of the rocket to high speed, after which the engines shut down and the craft cruises for whatever length of time it takes to reach its target. Another approach is to use a low-powered rocket whose engines keep firing continuously. The **ion engine** is an example of this approach. It works something like an old-fashioned television picture tube in that it accelerates charged particles (ions). In a television tube, electrons are fired from the back of the tube to the phosphor screen that faces the viewer. An ion rocket engine does the same, with a nozzle taking the place of the phosphor screen. Both NASA and the European Space Agency (ESA) have already used low-power ion engines successfully. These engines can be started only in space (they don't have enough thrust to lift off Earth and work best in a vacuum), but they can keep firing for long periods of time because the mass expelled per unit of time is small. Moreover, the exhaust ions are shot from the craft at tremendous speeds, and so a powerful ion rocket could in principle reach speeds approaching a percent or so of the speed of light.

Other schemes envision spacecraft that can overcome the limitations of the rocket equation by not taking along the bulky fuel. One possibility that's been considered for nearly a century is to use sunlight as power. Large, highly reflective, very thin (to minimize mass) **solar sails** could be pushed by the pressure exerted by sunlight. This pressure is so slight that we normally don't notice it, but in the vacuum of space, where friction is absent, the steady pressure of sunlight impinging on a mirrored sail could push a spacecraft to impressive speed, particularly with sails hundreds of kilometers in size. Solar sailing might well prove to be a fairly inexpensive way of navigating within the solar system, and it could even be useful for interstellar travel (Figure 13.8). Although the push from the Sun would slowly fade once such craft reached the outer solar system (at Saturn, the light intensity is less than 1% its value near Earth), a solar sailing vehicle that was started very near the Sun might achieve speeds of a few percent of light speed. It could then coast to neighboring stars in less than a century.

The fact that sunlight weakens so much with distance limits the ultimate speed of a solar sailing spacecraft. However, we could get around this problem by using a powerful laser as a power source, instead of sunlight. In principle, the laser could provide a steady and continuous "push" for the solar sail, all the way to its destination if necessary. If building large sails proves too difficult, the laser could be used to vaporize propellant on the rocket that would then move the craft forward in the usual rocket-like manner. As these craft moved light-years away, a large focusing mirror hundreds of kilometers in size would be needed at the laser base to concentrate the beam on the pinpoint target that the spacecraft had become. The primary drawback to these schemes is the power requirement. For example, accelerating a ship to half the speed of light within a few years would require a laser that uses 1,000 times as much power as all current human power consumption. Nevertheless, this approach would allow us to travel to nearby stars in a decade or two rather than many decades.

Figure 13.8

Artist's conception of a spaceship propelled by a solar sail, shown as it approaches a forming planet in a young solar system. The sail is many kilometers across. The scientific payload is at the central meeting point of the four ladder-like structures.

A laser-powered rocket would leave the passengers dependent on the efforts of those back on Earth (or wherever the laser installation was located) to keep the laser shining so they could accelerate to their desired final velocity. This might be somewhat risky given the fact that even the fastest of these transports would be en route and accelerating for decades. What if the laser crew went on strike? In addition, with no laser shining in the *opposite* direction, slowing the spacecraft to a halt at its destination (let alone returning home) would be a problem. One possibility is to use on-board propellant heated by the laser. The propellant could then be fired out the front of the craft to slow it down. Alternative braking schemes that use natural magnetic fields in space have also been suggested.

Interstellar Arks Another, less demanding approach to interstellar travel is often featured in science fiction. Forget the high-tech rocketry and accept relatively low speeds. Then deal with the resulting long travel times by putting the crew into suspended animation—hibernation, if you will—and letting them doze their way to the stars. A challenging variation on this idea is to somehow allow the travelers to live long enough to cruise the galaxy. A third suggestion is to build enormous craft that can accommodate a very large crew: in essence, an "ark." Many generations would live out their lives aboard this slow-moving vehicle before it finally reached its destination.

The difficulty with the first suggestion is that no one yet knows how to put humans to sleep for hundreds or thousands of years (and then have them wake up). However, genetic researchers have identified genes that control hibernation in animals, so it is possible that genetic engineering techniques could someday allow humans to hibernate as well.

The second suggestion, to allow the crew to live the many thousands of years necessary for interstellar travel at conventional speeds, depends on advances in medical technology. Could we somehow stop the aging process, enabling people to live such long lives that centuries or millennia of travel might seem like a walk to the corner store? We simply don't know.

The idea of building interstellar arks usually gets a skeptical reaction from sociologists. They point out that long voyages on Earth (even those that last only a few months) often end badly. Crews splinter into antagonistic factions and frequently fight for control of the ship. In addition, we might justifiably fear a deterioration in the level of expertise of the crew, with the result that the generation of folk who finally reach the target star system would neither remember why it journeyed there nor have the technical skills required to land on or colonize a world in the system.

• How might we build spacecraft that could approach the speed of light?

We have seen that the conventional approaches to interstellar travel—such as chemical, nuclear, or laser-powered rockets or a solar sail—will not bring us to the stars in anything less than decades. What we really need for interstellar travel are ships that can travel at speeds close to the speed of light. We could then reach the nearest stars in years and explore the space within a few tens of light-years of the Sun in just a few decades. Moreover, such ships would be traveling fast enough for on-board passengers to benefit from some astonishing effects of high-speed travel.

The Role of Relativity The fact that we cannot exceed the speed of light might at first make distant stars seem forever out of reach. However, the same theory that imposes the cosmic speed limit—Einstein's

TABLE 13.1 *Round-Trip Travel Time to Vega*

This table shows the time that passes on Earth and the time that passes for the crew of a spaceship on round-trip journeys at various speeds to the star Vega, a trip of 25 light-years in each direction. Speeds are given as fractions of the speed of light, c = 300,000 km/s. Note that the first-row speed of 0.00005c is equivalent to 54,000 kilometers per hour, which is roughly the speed of our fastest chemical rockets today.

Speed	*Time Measured on Earth*	*Time Measured on Ship*
$0.00005c$	1,000,000 yrs	1,000,000 yrs
$0.1c$	500 yrs	498 yrs
$0.5c$	100 yrs	86 yrs
$0.7c$	72 yrs	52 yrs
$0.9c$	56 yrs	24 yrs
$0.99c$	50 yrs	7 yrs
$0.999c$	50 yrs	2.2 yrs
$0.9999c$	50 yrs	8 months

special theory of relativity—also tells us that time is different for high-speed travelers than for people who stay at home. We'll discuss the reason this occurs in Section 13.4; here, we'll consider its implications to interstellar travel.

Imagine a trip to the star Vega, about 25 light-years away, in a spaceship traveling at constant speed of 90% of the speed of light ($0.9c$). Because light takes 25 years to travel the distance to Vega, and a ship traveling at $0.9c$ is going 90% as fast as light, the ship's travel time to Vega should be $\frac{25}{0.9} \approx 28$ years. This is indeed the time that would be measured by people staying home on Earth; that is, if the ship made the round trip at this speed, it would return home $28 \times 2 = 56$ years after it left. However, it is *not* the time that would be measured by the ship's crew.

Einstein's theory tells us that when a spaceship (or any other object) travels at close to the speed of light, its length becomes noticeably shorter in the direction of movement, its mass becomes noticeably greater, and time measured aboard proceeds noticeably more slowly than time measured by clocks at rest—a phenomenon called **time dilation**. These changes to time and space are not just speculation—they have all been carefully measured in experiments with subatomic particles that move at speeds close to the speed of light. Note that, even while the ship's time is running slow according to people back on Earth, time would feel perfectly "normal" to the crew. But less time really would pass for them. As shown in Table 13.1 (and calculated in Cosmic Calculations 13.2), only about 24 years would pass on the spaceship during the round-trip voyage to Vega at $0.9c$. In other words, this is what would really happen if a ship left on this journey in the year 2100: The ship would return in the year 2156, but the crew would have aged only 24 years. If a crew member was 20 years old when she left, she'd be 44 on her return, but her twin brother, who stayed home, would be 76.

Table 13.1 shows the benefits of relativistic travel for a hypothetical trip from Earth to Vega at various speeds. Notice that, at low speeds, there's no noticeable difference between ship time and Earth time, but the journey takes a very long time. The closer the ship gets to the speed of light, the less time that passes for the crew. Indeed, from the crew's standpoint, the trip can be made arbitrarily short simply by getting ever closer to the speed of light. But for friends left behind on Earth, the rocket can never return less than 50 years after it left.

If you study the table carefully, you might wonder if the crew of a very-high-velocity rocket would conclude that they're traveling faster than the speed of light—which would violate the cosmic speed limit of relativity. For example, at a speed of 99.99% of the speed of light ($0.9999c$), their trip takes only 8 months for a distance we said was 50 light-years. This would seem to imply a speed some 75 times the speed of light. However, special relativity also tells us that distances shrink at high speed. Once traveling at high speed, the crew would find that the distance to Vega is not 25 light-years as we measure it on Earth but instead has shrunk to a little under 0.4 light-year. Thus, they can cover this short distance in a short time, and they'll never think they are traveling faster than the speed of light.

THINK ABOUT IT Suppose you were offered the opportunity to take a trip to Vega and back at a speed of 0.9999c. How long would the round trip take, according to you? How much time would pass on Earth while you were gone? All things considered, are there any circumstances under which you would agree to take such a trip? Explain.

Incredible Journeys In fact, if we could somehow boost our spacecraft to speeds arbitrarily close to the speed of light, we could go anywhere in the universe within a human lifetime. Astronomer Carl Sagan considered a hypothetical rocket that accelerates at a steady $1g$ (or "1 gee")—an acceleration that would feel comfortably like gravity on Earth—to the halfway point of its voyage. This constant acceleration would bring the ship closer and closer to the speed of light, though it would never exceed it. (Note that the acceleration of $1g$ is constant, but the speed is not!) The ship then reverses and decelerates at $1g$ to its destination. During most of the trip, the ship would be traveling at speeds quite close to the speed of light, so time would pass quite slowly on the ship compared to time on Earth. Longer trips would mean longer periods of acceleration, bringing the ship even closer to the speed of light for most of the journey.

Calculations show that such a continuously accelerating ship could make a trip to a star 500 light-years away in only about 12 years according to those on board the ship. However, 500 years would pass on Earth. If a crew of 20-year-olds left Earth in the year 2100, they would be merely 32 years old when they reached their destination; but it would be the year 2600 on Earth (actually a bit later, since they are traveling at not quite the speed of light). If they sent a radio message back to Earth, the message would take 500 years to arrive here across the 500-light-year distance. More than 1,000 years after the crew had left, we'd get a message from people who had aged only 12 years since they'd last been seen on Earth.

Even longer trips would be possible in principle. For example, in a craft with a constant $1g$ acceleration, it would require only about 21 years of ship-board time to bridge the 28,000-light-year distance to the center of the Milky Way Galaxy, where the crew could observe firsthand the mysterious black hole thought to reside there. The 2.5-million-light-year distance to the Andromeda Galaxy could be traveled in only about 29 years of the ship's time. Thus, passengers could travel to the Andromeda Galaxy, spend 2 years studying one of its star systems and taking our first pictures of the Milky Way as it appears from afar, and return only 60 years older than when they had left. However, they would not exactly be returning "home," since 5 million years would have passed on Earth. In this sense, special relativity offers sufficiently fast travelers only a one-way "ticket to the stars." No place is out of reach—but you cannot return home to the same people and places you left behind.

In any event, while such incredible trips are allowed by the laws of physics, the energy costs would be extraordinary. Because special relativity also tells us that an object's mass increases as it approaches the speed of light, the energy cost rises just as much as time slows down. Indeed, that is one explanation for why the speed of light cannot be reached: As the ship gets closer and closer to the speed of light, its mass becomes greater and greater, so the same rocket thrust generates ever less additional speed. The mass heads toward infinity as the ship's speed approaches the speed of light—and no force in the universe can give a push to an infinite mass. That is why the ship can never reach the speed of light, no matter how powerful its engines might be, and even getting close to that speed would require amounts of energy far beyond anything we will be able to muster in the near future. Nevertheless, such practical difficulties can't stop us from speculating, and at least two potential ways of approaching the speed of light are known to exist in principle.

Matter–Antimatter Rocketry The most efficient energy source we have seen so far is nuclear fusion. But fusion converts only 0.7% of the mass of the fusing hydrogen into energy; 99.3% of the mass still remains,

COSMIC CALCULATIONS 13.2

Time Dilation

The effects of time dilation on a fast spaceship can be calculated with a simple formula if we assume that the spaceship travels at constant speed:

$$t_{\text{ship}} = t_{\text{Earth}}\sqrt{1 - \left(\frac{v}{c}\right)^2}$$

where t_{ship} is the amount of time that passes on the rocket, t_{Earth} is the amount of time that passes on Earth, v is the rocket's velocity (speed), and $c = 3 \times 10^8$ m/s is the speed of light.

Example: Consider a spaceship that travels round-trip to Vega at 90% of the speed of light. As noted in the text, the round-trip travel time measured by people on Earth is 56 years. How much time passes for passengers on the spaceship?

Solution: Because we are given that the ship travels at 90% of the speed of light, or $0.9c$, we know that $v/c = 0.9$. We plug in this value along with the time that passes on Earth, $t_{\text{Earth}} = 56$ yr:

$$\begin{aligned} t_{\text{ship}} &= t_{\text{Earth}}\sqrt{1 - \left(\frac{v}{c}\right)^2} \\ &= 56 \text{ yr} \times \sqrt{1 - 0.9^2} \\ &= 56 \text{ yr} \times \sqrt{1 - 0.81} \\ &= 56 \text{ yr} \times \sqrt{0.19} \\ &= 24.4 \text{ yr} \end{aligned}$$

This is the approximately 24-year round-trip time shown for the ship in Table 13.1.

as helium. Is there a way to turn more of the mass, or even all of it, into energy? The answer is yes, and it is called **matter–antimatter annihilation.**

Antimatter might sound like the stuff of science fiction, but it really exists. All material things are composed of "ordinary" matter, but physicists have discovered particles that are in some ways the mirror images of normal particles, differing principally in their electrical charge. The first known such particle, christened the *positron,* was discovered in 1932. It is the antimatter twin of the electron; that is, it is identical to an electron except that it has a positive rather than a negative charge. In 1955, the proton's antimatter partner was found: the antiproton. If you were to introduce an antiproton to a positron, they would form an atom of antihydrogen. This antimatter atom would behave chemically just like ordinary hydrogen, except for one thing: You wouldn't want to get near it. When matter and antimatter meet, the result is total annihilation, with 100% of the mass turning into energy. (Note that antimatter still has mass just like ordinary matter; there is no such thing as "antimass.")

The annihilation of matter and antimatter can create energy in a variety of forms. For example, annihilation of positrons with electrons produces energy as a burst of gamma rays. The annihilation of heavier particles, such as antiprotons with protons, produces a gush of particles that soon decay into neutrinos and gamma rays. These might be amenable to powering a rocket, because the flood of reaction particles could be directed out a rearward-facing nozzle. A matter–antimatter rocket could, in principle, achieve speeds of 90% of the speed of light with modest mass ratios.

While such numbers are seductive, the problem lies in rounding up and storing the required antimatter. No practical reservoirs of this material are known. Instead, we would have to manufacture the antimatter, as physicists now do with high-energy particle accelerators. However, current worldwide production of antimatter amounts to only a few billionths of a gram per year, and the energy that would be needed to manufacture larger amounts is staggering. For example, with present technology, manufacturing 1 ton of antimatter—far less than would be needed for an interstellar trip—would take more energy than humankind has used in all of history. Moreover, even if we could make the antimatter, we don't yet know of a good way to store it aboard our rockets, since it would have to be kept in some type of container in which it never touched any ordinary matter at all, since otherwise the result would be premature annihilation.

Interstellar Ramjets Another approach to achieving relativistic velocities circumvents the problems involved in carrying highly energetic fuel on board. The idea is that a starship could collect its fuel as it goes, using a giant scoop to sweep up interstellar gas. Because this gas would be mostly hydrogen, it could be funneled to a nuclear reactor, fused into helium, and then expelled out the back to propel it. Such propulsion systems are known as **interstellar ramjets** (Figure 13.9). In principle, interstellar ramjets can accelerate continuously by collecting and using fuel nonstop, getting ever closer to the speed of light.

Of course, there are practical difficulties. The typical density of the gas between the stars is only a few atoms per cubic centimeter, so the scoop would need to be hundreds of kilometers across to collect adequate supplies of fuel. As Carl Sagan said, we are talking about "spaceships the size of worlds." Another problem facing an interstellar ramjet—or any ship traveling at relativistic speeds—comes from the interstellar gas and dust itself. At 99% of the speed of light ($0.99c$), a particle the size of a sand grain packs energy equivalent to an explosion of about 100 kilo-

Figure 13.9

Artist's conception of a spaceship powered by an interstellar ramjet. The giant scoop in the front (left) collects interstellar hydrogen for use as fusion fuel.

grams of TNT. Even individual atoms encountered at this speed would be deadly, so the ship would need substantial shielding to protect both its structure and its crew.

Any type of relativistic travel, whether with matter–antimatter engines, interstellar ramjets, or some as-yet-unthought-of method, remains far beyond our current technological capabilities. But we have at least imagined ways by which an extremely advanced civilization *might* be able to travel among the stars. Whether anyone has actually done so remains unknown.

• Are there ways around the light-speed limitation?

If you are a science fiction fan, our discussion of interstellar travel may depress you. Interstellar tourism and commerce seem out of the question, even with ships that travel at speeds close to the speed of light, because of the long times involved (at least as seen from home planets and colonies). If we are ever to travel about the galaxy the way we now travel about Earth, we will need spacecraft that can somehow get us from here to there much faster than the cosmic speed limit would seem to allow. Could such spacecraft be possible?

No one really knows. However, there just might be a "loophole" in the law limiting cosmic speed. In particular, while Einstein's special theory of relativity showed that we can't travel *through* space faster than the speed of light, his general theory of relativity suggests that there might be "shortcuts" that, in effect, let us travel *outside* ordinary space in a way that greatly reduces the distances to be traveled. If so, then we might reach far-off places by taking a shortcut that lessens the distance to them.

Hyperspace In 1916, Einstein enlarged his earlier work (special relativity) to produce the general theory of relativity [Section 2.4]. In this theory, Einstein noted that we live in a four-dimensional universe, with three dimensions of space and one of time; the four dimensions together are usually called **spacetime**. The three spatial dimensions are not rigid and invariant but rather can be warped. Einstein realized that matter can

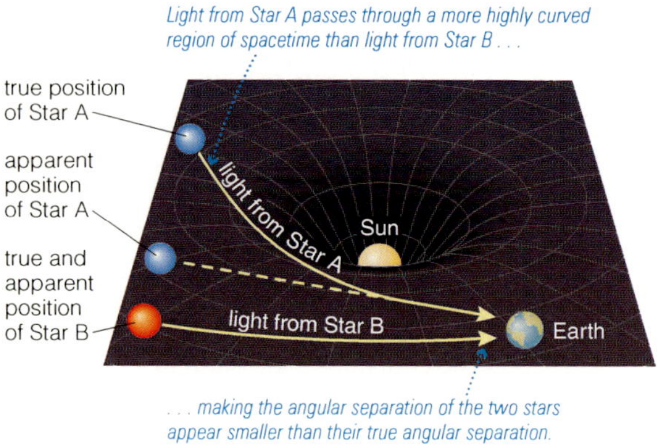

Figure 13.10

This embedding diagram shows how starlight is bent as it passes near the Sun, causing stars to appear slightly offset from their true positions in space. This effect (exaggerated in this diagram) has been measured during solar eclipses, proving that our space really is curved through hyperspace.

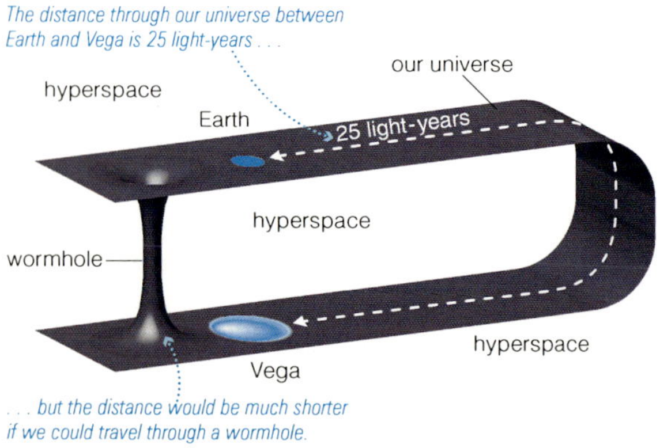

Figure 13.11

Illustration of the idea of a wormhole. Once again, we have reduced our three spatial dimensions to the flat, two-dimensional realm of a rubber sheet. Two distortions of space, one near Earth and one near Vega, have met up in hyperspace, forming a connecting tunnel. Going from Earth to Vega in ordinary space would be a 25-light-year trip. But the wormhole offers a radically shorter route—one that might be traversed in minutes without ever exceeding the speed of light.

provide the distortion. For example, any mass, such as the Sun, will produce space curvature.

How can we tell that space is "curved"? One simple test is to consider the paths of light beams, which travel through space in what we call straight lines. If space had no curvature, then two parallel light beams—for example, from two laser pointers taped side by side—would never cross. However, general relativity insists that the presence of matter can cause a warping of space that will lead these parallel beams to cross. The matter produces a distortion of the ordinary three dimensions of space into other, hypothetical dimensions that we can't see or "get into." These additional dimensions are called **hyperspace.**

To visualize this idea, physicists often resort to "embedding diagrams," such as that depicted in Figure 13.10. Space is reduced from three dimensions to two, and the resulting diagram resembles a rubber sheet; if you inhabited the world shown in the figure, you would be flat, infinitely thin, and incapable of appreciating that any dimension exists on either side of the sheet. Embedded in the sheet is a large mass, such as a star, that causes the sheet to distort into hyperspace. You cannot see the hyperspace dimensions, but you can make measurements on the rubber sheet that will tell you whether or not it is curved. In fact, during solar eclipses, we have measured changes in the apparent positions of stars whose light passes near the Sun. As Figure 13.10 shows, these changes are what we would expect if our space is curved through hyperspace.

The warping of space is usually quite small. Even the bending of starlight passing close to the Sun amounts to only a fraction of a thousandth of a degree. But if space could be distorted more dramatically, the distortion might offer us shortcuts to distant destinations.

Black Holes, Wormholes, and Warp Drive

The bizarre objects called black holes are, in fact, holes in spacetime—places where space becomes so distorted that it in effect becomes a bottomless pit. As a result, science fiction writers have sometimes imagined using black holes as shortcuts to other places. Unfortunately, this idea suffers from at least two major drawbacks. First, the only known black holes are themselves extremely far away, so getting to them in the first place would be a problem. Second, we do not know of any way we could survive a close encounter with a black hole.

However, a related phenomenon, called a **wormhole,** might be more useful. Just as a worm might shorten its trip from one side of an apple to the other by tunneling through it, so might a wormhole provide a hyperspace shortcut to a distant part of the universe. Imagine a dense mass floating somewhere near Earth, distorting the surrounding space into hyperspace. Now imagine a similar distortion occurring somewhere else in the cosmos, many light-years away. If these two distortions somehow meet up in hyperspace, they could connect two distant places in ordinary space via a short, hyperspace tunnel (Figure 13.11). Traversing this tunnel might take little time (perhaps minutes) and could short-circuit the necessity of traveling those many light-years.

While this is clearly an appealing idea, could it actually be made to work? In particular, how do we arrange for the wormhole's opportune existence? Quantum physics suggests that on the tiniest scales of the universe, in regions of space far smaller than an atomic particle, spacetime is a seething foam, constantly punctured by distortions into hyperspace. In this highly microscopic world, wormholes might be forming (and self-destructing) all the time. It's conceivable that a highly advanced society might have learned how to capture one of these natural wormholes—one that connected two places of interest—and how to quickly enlarge it to a size that would permit its use for travel. However, such wormhole construction would seem to carry an impossibly large energy cost—estimated to be a thousand times the energy released by an exploding massive star (a supernova). Moreover, even if the energy could be found, we do not yet know of a way to stabilize a wormhole against immediate collapse. All in all, we do not know enough about physics to say for sure whether wormhole travel is even possible. Given this uncertainty, we can but imagine and hope. Carl Sagan's book and movie *Contact* postulated a network of wormhole tunnels permitting fast travel throughout the universe—but even his fictional characters did not know who had built them or how.

Another possible way to travel great distances in a short time might be to exploit the warping of space by placing a dense mass in front of a spacecraft. Hanging like bait on a fishing line, this black hole on a stick would allow the shortening of spatial distances in front of the rocket. Much as you might move across a floor by scrunching a rug in front of you and straightening it out behind you, this highly unusual craft would bend space in front and leave it unaltered behind. You would continually fall into the warped space in front of the craft. This concept comes closest to what we know as the "warp drive" of science fiction. However, it would require either capturing a black hole (difficult, to be charitable) or creating one using enormous amounts of energy. In either case, the black hole would have to be carried along for the ride.

Could there be simpler ways to take advantage of hyperspace? We do not know. This leaves the door wide open for science fiction writers. In the *Star Wars* movies, a simple flip of a lever takes a ship outside our universe and into hyperspace, allowing nearly instantaneous travel to anywhere. In *Star Trek,* a command from the captain sends the ship into warp drive, apparently without the need for a black hole in front of the ship. If such schemes are at all possible, we have no inkling of how they

MOVIE MADNESS: *STAR TREK*

If you want to boldly go where no one has gone before, without spending a few hundred centuries doing it, you need warp drive.

As almost everyone knows, the various incarnations of the *U.S.S. Enterprise, Star Trek*'s famous interstellar transport, high-tail it from one part of the galaxy to another in short order thanks to a futuristic propulsion system. But what *is* warp drive, anyway?

According to the show's technical manuals, the term is merely slang for "continuum distortion propulsion"—a Latinate mouthful that describes a scheme by which powerful fields are used to distort space and allow speeds faster than light. As we discuss in the chapter, rapid travel by warping space is not an entirely nutty idea. It might be possible. And if physics were to allow it in *principle*, could our clever descendants do it in practice?

Maybe yes, maybe no. A major problem is that even if you could warp space, to do so would take enormous amounts of energy. *Star Trek* deals with this small technical detail by fueling the field-generating warp engines with antimatter (anti-hydrogen, to be precise). Combining antimatter with ordinary matter is the most efficient combustion imaginable, as the entire mass of both is converted to energy.

Of course, there's still the problem of making the antimatter, not to mention shipping it to service stations around the galaxy (being careful to keep it out of the hands of pirates—antimatter is costly). But the truly interesting thing about warp drive is the range of speeds attained. At "Warp 1" you're loping along at the speed of light. By "Warp 9"— near the top of the *Enterprise*'s speedometer—you're streaking through space at 1,000 times light speed. This means you can traverse the galaxy in a century, which is short enough to be possible, but long enough to allow you to get effectively stranded and interfere with "Starfleet Command's Prime Directive" to avoid interfering with alien cultures—which, as discussed in Section 13.3, sounds remarkably similar to one of the possible solutions of Fermi's paradox.

Mind you, you could forget warp drive entirely, and stick with the physics we know by building starships that go at 99+ percent of light speed. Special relativity would guarantee that travel times as perceived by the ship's crew would be short. They could cross the galaxy overnight, according to their own watches. But *Star Trek* has opted out of the relativistic approach for good reason, for otherwise the *Enterprise* crew would return home to find all their family and friends long dead and forgotten. Starfleet headquarters would probably be just an archaeological dig.

Better to call up Scotty in the engine room and tell him to put the pedal to the space-bending metal.

might work. But who knows what an advanced civilization might have discovered? After all, we have been studying physics in earnest for only a few centuries. Others out there may have been studying it for millions or billions of years.

Figure 13.12

Enrico Fermi (1901–1954), one of the leading physicists of the twentieth century, received the Nobel Prize in 1938 for work in understanding radioactive decay and predicting the existence of the particles known as neutrinos. By the time he was awarded his Nobel Prize, the Fascists had risen in Italy and the Nazis were in power in Germany. Abhorring these ideologies, he chose not to return to Italy after attending the Nobel Prize ceremony in Sweden. Instead, he moved to the United States, where he became a prominent figure in the Manhattan Project, which was developing the atomic bomb. Element 100 in the periodic table, fermium, was named in his honor.

13.3 THE FERMI PARADOX

We have found that practical interstellar travel is well beyond our capabilities today, but we can envision ways that more advanced civilizations might achieve it. In earlier chapters, we have found good reason to think that habitable planets could be common, some with civilizations, and that many of these civilizations could be much older than ours. These ideas lead to an idea first stated in 1950 by the Nobel Prize–winning Italian-American physicist Enrico Fermi (Figure 13.12). During a lunch at the Los Alamos National Laboratory in northern New Mexico, the conversation drifted to the possibility of extraterrestrial intelligence. The physicists present at the lunch were considering the likelihood that sophisticated cosmic societies might exist in great abundance. Fermi replied to these speculations with a disarmingly simple question: "So where is everybody?" Although serious scientific discussion of his query did not get under way for many years, its central idea is now known as the **Fermi paradox.**

• Where is everybody?

The essence of the Fermi paradox is almost as simple as Fermi's original question. It begins with the idea that neither we nor our planet should be in any way special, in which case other Earth-like planets and other advanced civilizations (meaning civilizations capable of space travel) ought to be fairly common in the galaxy. This is more or less what we conclude from the Drake equation [Section 12.1], unless the rare Earth hypothesis turns out to be correct [Section 11.3]. However, a large number of civilizations would necessarily mean many civilizations with the opportunity to develop advanced technology and interstellar travel long before we came on the scene with our rockets and radio telescopes. In that case, for reasons we'll discuss shortly, it seems that someone else should have colonized the galaxy already. But we see no evidence of such a galactic colonization effort. Thus, we have two seemingly contradictory ideas.

Statement of the Paradox Summarizing the above, we are led to the following two ideas:

1. The idea that neither we nor our planet is in any way special suggests that someone should have colonized the galaxy by now.
2. The idea of a galactic civilization implies that we should be surrounded by evidence of this civilization—but aside from unconvincing claims of extraterrestrial UFOs [Section 12.4], no such evidence exists.

By definition, the existence of two such seemingly contradictory ideas constitutes a *paradox.* But unlike some logical paradoxes (e.g., statements such as "This statement is false"), the Fermi paradox must have some solution. After all, either there is a galactic civilization out there or there isn't.

The Age of Civilizations If you look closely at the first premise of the Fermi paradox, you'll notice that it depends on the idea that, if civilizations are at all common, many should have arisen long before our own arrival.

Recall that the universe is about 14 billion years old [Section 3.2], while Earth is only $4\frac{1}{2}$ billion years old [Section 4.2]. In other words, the universe predates our planet and solar system by some $9\frac{1}{2}$ billion years. Stars began to form quite early in the history of the universe—a fact we know because the ages of the oldest stars are only a few hundred million years short of the age of the universe itself. There's some debate as to whether these early generations of stars had enough heavy elements to make Earth-like planets [Section 11.3], but little doubt that the heavy-element abundance was high enough to make Earth-like planets within a few billion years of the universe's birth. We can play with more precise estimates of these times in a variety of ways, but the bottom line is this: Unless we are misunderstanding some fundamental piece of star and planet formation, it should have been possible for Earth-like planets to have been born starting *at least 5 billion years before* our own planet was born. In other words, some other Earth-like worlds should have had a 5-billion-year head start on ours.

This 5-billion-year head start means that, if intelligent life arose on these planets in the same amount of time that it took intelligent life to arise here on Earth, the first civilizations in our galaxy should have appeared on the scene at least 5 billion years ago. In other words, if these civilizations have survived to the present day, they should be technologically ahead of us by 5 billion years.

We can take the idea a little further by making some guesses about the number of civilizations that have arisen over time. As we discussed in Chapter 12, our current understanding of the Drake equation does not allow any definitive conclusions about the number of civilizations, but it at least seems plausible to imagine that 1 in a million stars would eventually give rise to a civilization on an orbiting planet. Using a conservative estimate of 100 billion stars in the Milky Way—most far older than the Sun—this means there would have been some 100,000 civilizations in our galaxy by now. This is an astonishing idea, and one that becomes even more incredible when we put the number together with the time. If we assume that these 100,000 civilizations have arisen at random times over the past 5 billion years, then on average a civilization arises every 5 billion ÷ 100,000 = 50,000 years.

Think about what all this means. First, even with the odds of finding a civilization at only 1 in a million stars, there should still have been 100,000 civilizations that would have arisen before we came on the galactic scene. Second, it means that we are almost certainly the youngest civilization in the galaxy at present, and that on average, we'd expect the next youngest civilization to have arisen 50,000 years ago, and the third-youngest to have arisen 100,000 years ago, and so on.

Although the numbers we use for civilizations are essentially wild guesses, things don't change all that much even if we're much more conservative. If we assume that civilizations arise only around 1 in 100 million stars, rather than 1 in 1 million, we still end up with 1,000 civilizations having arisen over the past 5 billion years. And in that case, civilizations would arise about every 5 million years on average, making the next youngest civilization even more advanced relative to us.

THINK ABOUT IT Consider a couple of other variations on the above theme. How do the numbers change if the fraction of stars with civilizations is higher—say, 1 in 10,000? How do they change if the fraction of stars with civilizations is only 1 in 1 billion? Explain.

These ideas are of course speculative, but you can probably now see why Fermi asked, "So where is everybody?" And you should also be able to see why, if we ever actually meet up with another civilization, it is extremely unlikely that they will be at a technological level of development anywhere near as low as our own.

Machines Deepen the Paradox So far we have considered the idea that living beings should be out and about in our galaxy, but another possibility is that living beings could send machines outward, much as we have already begun to send robot spacecraft to explore other worlds in our solar system. Consideration of machine technology only deepens the Fermi paradox.

Beyond simply sending out individual robots such as those we can build now, we can imagine that in the future we might build much more sophisticated robots. For example, we might send robots to other worlds with programming instructions to dig up resources on arrival, use the resources to build factories, and use the factories to build spacecraft and more robots. These new robots would then go on to the next world, where they would do the same thing. Thus, these robots would be self-replicating, though in a way quite different from the self-replication of biological beings. The general idea of such self-replicating machines was first proposed by the American mathematician and computer pioneer John Von Neumann (1903–1957), so they are often called **Von Neumann machines.**

The use of Von Neumann machines would allow us to explore much farther and wider than we could by going to other worlds ourselves. Moreover, while interstellar travel poses huge barriers to us due to our limited life spans, these machines could presumably still function after journeys through space that take centuries or millennia. Once we sent the first wave of these machines to a few nearby star systems, they would gradually spread from star system to star system.

In 1981, physicist Frank Tipler used this idea of "colonization" by self-replicating Von Neumann machines to extend the Fermi paradox. In essence, Tipler argued that civilizations could effectively make their presence felt throughout the galaxy even without achieving the ability to send themselves on interstellar journeys. As soon as a civilization reached a level that allowed it to build Von Neumann machines, these machines would begin to spread through the galaxy. Because such colonization would require technology only slightly beyond our own, Tipler argued that if civilizations are common, then the galaxy would already be overrun by self-replicating machines. Because it isn't, Tipler concluded that we are alone and thus that the search for extraterrestrial intelligence (SETI) is a waste of time. Needless to say, other scientists have found plenty of fault with this conclusion, and SETI researchers are still listening hopefully for a signal from the stars. Nevertheless, it's clear that the Fermi paradox is leading us to some deep, philosophical questions about our own civilization and the possible nature of others.

• Would other civilizations really colonize the galaxy?

One obvious question built into the Fermi paradox is that of whether other civilizations really could or would colonize the galaxy. After all, if other civilizations are content to remain quietly on their home planets, or at least in their home star systems, then lots of civilizations could be out there without any having come our way. Thus, to understand the Fermi paradox at a deeper level, we must consider both the capabilities and motives that might lead other civilizations to colonize the galaxy.

Colonization Models Let's start by assuming that another civilization decided to start sending out spacecraft to colonize other habitable planets. How long would it take this civilization to colonize the entire galaxy?

The answer clearly depends on the civilization's technological capabilities. For example, if it has the technology to build spacecraft that can travel at speeds close to the speed of light, then it could add colonies throughout the galaxy fairly quickly, since trips between nearby stars would take only a few years. Perhaps surprisingly, the conclusion is not that much different if we assume much slower speeds.

Consider a civilization that has nuclear rockets such as the Project Orion or Project Daedalus rockets; as we've discussed, such rockets do not seem that far beyond our own current technological grasp. Recall that such rockets might attain speeds of about 10% of the speed of light ($0.1c$). Given that a typical distance between star systems in our region of the galaxy is about 5 light-years (the average distance is actually slightly lower than this), a nuclear spacecraft traveling at 10% of the speed of light could journey from one star system to the next in about 50 years. This trip would be possible within a human lifetime and might be practical if the colonizers have found ways to hibernate during the voyage or if they have somewhat longer life spans than we do (either naturally or through medical intervention).

After arriving at a new star system, the colonists establish themselves and begin to increase their population. Once the population has grown sufficiently, these colonists send their own pilgrims into space, adding yet more star systems to the growing civilization. Figure 13.13 shows how such colonization would gradually spread through the galaxy. The process starts at the home star system. The first few colonies are located within just a few light-years. These colonies then lead to other colonies at greater distances, as well as at unexplored locations in between. The growth tends to expand the empire around the edges of the existing empire, much like the growth of coral in the sea. For this reason, this type of colonization model is often called a *coral model.*

The overall result is a gradually expanding region in which all habitable planets are colonized. The colonization rate depends on the speed of spacecraft and the time it takes each colony to start sending their own spacecraft to other stars. For travel at 10% of the speed of light, and assuming that it takes 150 years before each colony's population grows enough to send out more colonists, calculations show that the inhabited region of the galaxy expands outward from the home world at about 1% of the speed of light (see Problem 56 at the end of the chapter). Thus, if the home star is near one edge of the galactic disk so that colonizing the entire galaxy means inhabiting star systems 100,000 light-years away, the civilization could expand through the entire galaxy in about 10 million years. The required time would be a few million years less if the home star is in a more central part of the galaxy.

For an even more conservative estimate, suppose the colonists have rockets that travel at only 1% of the speed of light and that it takes each new colony 5,000 years until it is ready to send out additional colonists. Even in this case, the region occupied by this civilization would grow at a rate of roughly $\frac{1}{1,000}$ (0.1%) the speed of light, and the entire galaxy would be colonized in 100 million years. This is still a very short time compared to the time that has been available for civilizations to arise,

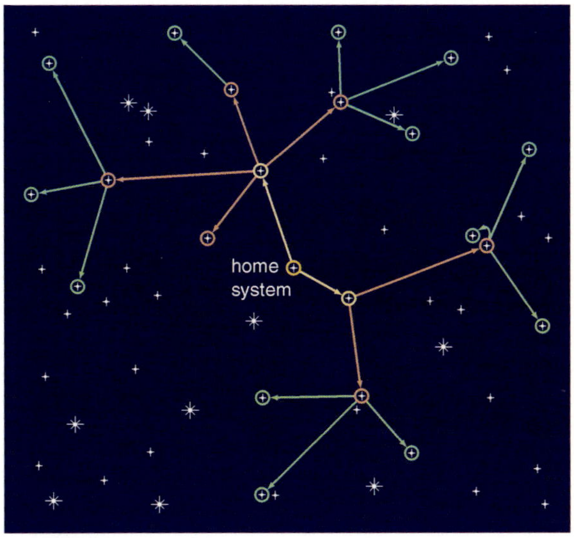

Figure 13.13

The coral model of galactic colonization. Colonization begins when the inhabitants of one star system send a few craft to nearby stars. After a time during which the new colonies grow and mature, each new colony sends a few ships with colonists to yet more distant stars, and so on. The colonization "frontier" expands at the edges, much like the growth of coral in the sea.

further deepening the mystery of why we see no evidence that anyone else has done it by now.

Motives for Colonization In developing a colonization model, we assume that other civilizations would *want* to send out colonists and colonize the galaxy. Is this a reasonable assumption?

We can address this question by considering ourselves as an example, since one of the premises of the Fermi paradox is that we are not special in any way. That is, if we would be likely to colonize the galaxy, then we should assume that others would probably act in the same way. Because we have not yet reached the technological level needed to start interstellar colonization, it's impossible to know with certainty whether we would try it if and when we achieved the capability. However, the history of the human species strongly suggests a predisposition to colonize any new territory available to us.

In many ways our entire history has been one of colonization. Modern humans arose in Africa some 100,000 years ago and almost immediately began expanding around the world. Indeed, the expansion of the human species on Earth probably looked much like the coral model we have described for galactic colonization. Early humans moved outward, gradually encompassing a larger and larger region of our planet. By about 10,000 years ago, our ancestors already lived in almost every place on Earth with a suitable environment for human life. Even after humans had effectively colonized the entire planet, attempts at colonization did not stop. For example, Europeans colonized the Americas, with devastating consequences for the people already living there.

Might our inclination to colonize subside in the future? Possibly, but recent history suggests otherwise. Already there are organizations dedicated to colonizing Mars. It doesn't matter if most people would have no interest in going, because a tiny fraction of the human population would be more than sufficient to start a new colony on another planet. Thus, it seems that if other civilizations are at all like us, they would take advantage of technological opportunities to colonize the galaxy.

Moreover, even if other civilizations don't have an inherited predisposition toward colonization, many other motives might serve to encourage it. For example, some members of an alien civilization might choose to leave their home to escape war or persecution (as did many Europeans and others coming to America, for example). Or a society might deliberately send out colonists in an attempt to make its civilization "extinction proof." For example, while our civilization could easily be wiped out in a variety of ways—from nuclear warfare to environmental catastrophe—a civilization spread among star systems would have a much more difficult time self-destructing, because a particular environmental problem affects only one planet, and the long travel times between stars would make it almost impossible to wage war on multiple planets at once. Finally, if a civilization survived long enough for its star to reach an age at which its home planet was no longer in the habitable zone of its star and life would soon be extinguished [Section 10.4], its members might have no choice but to move on in search of a new home.

THINK ABOUT IT Consider these and other possible reasons why a civilization might choose to start colonizing other star systems. In general, do you think it is reasonable to assume that other civilizations would attempt galactic colonization? Defend your opinion.

A Hard Look at Motives for Colonization Before we leave this topic, it's worth looking at a few ideas that are sometimes suggested as motives for colonization but that break down on closer examination. The most notable example of such ideas concerns the alleviation of population pressure.

After growing quite slowly for thousands of years, human population began a dramatic upward swing a few hundred years ago. Figure 13.14 shows human population over the past 12,000 years. If the current trend were to continue, today's human population of about six billion (a threshold reached in 1999) would double to 12 billion by about 2050, double again to 24 billion by 2100, and double again to 48 billion by 2150. Indeed, within just a few more centuries of such growth, we would not fit on Earth even if we all stood elbow to elbow (and of course we would have long since exhausted our ability to provide enough food for ourselves by this point). Clearly, our rapid population growth on Earth must stop soon, or we will face an unparalleled catastrophe (see Cosmic Calculations 6.2). So could colonization be the answer to our population problem?

Not a chance. Let's suppose we wanted to stabilize the population at its current size, but through colonization rather than changes in the population growth rate. Currently, we add about 100 million people to Earth each year. Thus, to keep our population stable, we'd need to move 100 million people per year off the planet. A typical Space Shuttle launch costs about $100 million and takes only six or seven people into space (and the Shuttle can't reach other planets, let alone other stars). Even if we somehow found the resources to build enough Space Shuttles or similar craft—and the fuel to launch them—the cost of sending 100 million people into space each year would be more than a quadrillion dollars, some 1,000 times the gross national product of the United States. There are not too many things that we can say are outright impossible, but solving population problems through colonization is one of them.

Another less-than-viable colonization motive is conquest. As we've seen, even if we do find other inhabited planets with intelligent beings, we are likely to be at least either many thousands of years ahead of them or many thousands of years behind them. If we are thousands of years ahead, conquering them would be like the United States conquering cave dwellers—there hardly seems much to gain. If we are thousands of years behind them, we would not be likely to prevail.

Some people suggest that we would colonize not so much through a motivation for direct conquest but rather as the consequence of our species' general tendency toward aggression. This is harder to rule out as a viable motive, but many science fiction writers (including Gene Roddenberry, creator of the *Star Trek* series) have pointed out a potential flaw in this idea: If we continue to be as aggressive and warlike as we have been in the past, we are unlikely to survive long into the future because our capacity for destruction has risen along with our level of technology. Thus, these writers argue that our surviving long enough to achieve the technology for interstellar travel will necessarily mean that we have found ways to overcome our aggressive and warlike tendencies. In that case, colonization will occur because of curiosity and a desire to explore, not because of any desire for empire building.

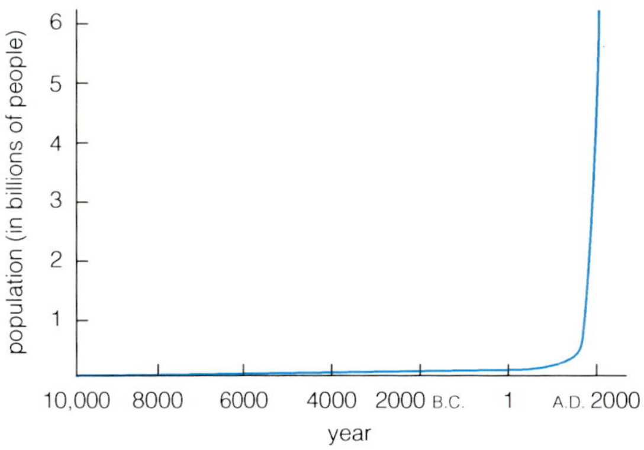

Figure 13.14

This graph shows human population over the past 12,000 years. Note the tremendous population growth that has occurred in just the past few centuries.

• What are possible solutions to the Fermi paradox?

We have now seen why the Fermi paradox is real; that is, it really does seem that if civilizations are at all common, then the galaxy should have been colonized long ago. So why don't we see any evidence of a galactic civilization? There are many possible explanations, but broadly speaking, we can group them into three major categories:

1. *We are alone.* There is no galactic civilization because civilizations are extremely rare—so rare that we are the first to have arisen on the galactic scene.
2. *Civilizations are common, but no one has colonized the galaxy.* There are at least three possible reasons why this might be the case:
   i. Technological difficulties. Interstellar travel is much harder or vastly more expensive than we have guessed, so civilizations are unable to venture far from their home worlds.
   ii. Sociological considerations. Our desire to explore is unusual, and other societies choose not to leave their home star systems. Also, for one reason or another, colonizers might run out of steam before they've conquered large tracts of galactic real estate.
   iii. Self-destruction. Many civilizations have arisen, but they have all destroyed themselves before achieving the ability to colonize the stars.
3. *There is a galactic civilization,* but it has deliberately avoided revealing its existence to us.

Let's examine each of these categories in more depth.

We Are Alone The idea that we are alone is certainly the simplest solution to the Fermi paradox. However, many people object to this solution on philosophical grounds, because it would suggest that our circumstances are very special compared to those that have arisen around any other of the more than 100 billion star systems in the galaxy. If this is true, it would go against almost everything else we have learned since the time of Copernicus. That is, while our ancestors might have imagined our planet to be the center of the universe, more recent astronomical discoveries all seem to suggest that we are not particularly special. Earth is merely a planet orbiting the Sun, our Sun is a rather ordinary star in the Milky Way Galaxy, and our galaxy is much like many other galaxies in the universe.

Of course, while the idea that we are alone might be philosophically unappealing, we cannot rule it out on scientific grounds. Indeed, proponents of the rare Earth hypothesis [Section 11.3] would not be surprised to learn that we are alone. Recall that, according to this hypothesis, the combination of circumstances that allowed for intelligent life to arise on Earth is so rare that we are likely to be the only civilization in the galaxy. The ideas that underlie the rare Earth hypothesis are controversial, but even if they prove incorrect there may be other reasons why we are alone. For example, perhaps some undiscovered law of nature has rendered civilizations impossible until quite recently. In that case, it would not be so strange to imagine that we are the first civilization, even if many others may follow.

Civilizations but No Colonization—Technological Difficulties

The second category of explanations offers the possibility that civilizations are common but colonization is not. Three possible reasons why this might be the case were listed: technological difficulties, sociological considerations, and self-destruction. Let's consider each of these, starting with technological difficulties.

We've seen that interstellar travel seems difficult but not impossible. But could we be underestimating the challenge, and could it actually be so difficult as to be essentially impossible? The energy cost of interstellar travel is sometimes suggested as an impasse. Recall that a large interstellar starship traveling at only about 10% of the speed of light would require energy comparable to what the world currently uses in a hundred years. Clearly, this requirement is prohibitive for us today, and it might be so high compared to the costs of building habitats in our own solar system that migration to other stars might always seem untenable. However, it's certainly possible that an advanced civilization could overcome this problem. The ability to produce power through nuclear fusion, for example, might allow us to generate the needed energy with relative ease. In an extreme case, a civilization capable of building a Dyson sphere [Section 12.3] would have access to all the energy produced by its star. It seems unlikely that the energy requirement alone could preclude all interstellar colonization.

A related possibility is that some other, unknown biological or physical barrier to interstellar travel exists. For example, we have assumed that in the future we'll be able to find a way to keep crews alive for the decades required to go from one star system to the next, but perhaps this is actually much more difficult than we have imagined. Possibly, some type of unknown danger lurking in space prevents intelligent beings from traveling among the stars. Science fiction writers have certainly considered such possibilities—for example, a mysterious effect that causes interstellar travelers to go insane—but they seem far-fetched in light of what we presently know about interstellar space.

It's worth noting that neither energy considerations nor lurking dangers would be enough to stop a civilization from sending out self-replicating Von Neumann machines. But there might be other reasons why no such machines are out there. For example, such machines would tend to grow in number at a rapid (exponential) rate and could in principle use up all the resources in the galaxy in just a few million years. Carl Sagan (in a paper co-written with William Newman) addressed this idea by suggesting that any civilization smart enough to build such machines would also be smart enough to recognize their dangers and therefore would not construct them in the first place.

Civilizations but No Colonization— Sociological Considerations

Let's next turn to sociological considerations. As we've discussed, it seems quite likely that, given the technological and economic opportunity, we will choose to engage in interstellar travel and galactic colonization. But could it be that we are somehow exceptional in having this desire and that other civilizations are perfectly content to stay at home? Like the "we are alone" idea, this idea suggests that we are somehow special rather than typical of intelligent beings and therefore goes against our usual assumption that there's nothing special about our circumstances. After all, we are products of the competitive forces that drive evolution by natural selection, and these forces ought to be similar on any world with life. Moreover, our colonization models show that it would take only *one* other civilization to colonize the entire galaxy in a few million years, so the lack of interest in space travel would have to apply to *every other* civilization that has ever arisen. If civilizations are common—say, if there are the 100,000 civilizations expected if 1 in a million stars has one—it's difficult to believe that not one other civilization has had similar interests to ours.

On the other hand, even with fast rockets, colonization of the entire galaxy would still take millions of years. Perhaps no civilization can maintain enthusiasm for an effort that lasts this long. The individual colonies, separated by many light-years, might evolve along different lines (either biological or cultural), shattering the unity of the empire and bringing further expansion to a halt. While these possibilities are not unreasonable, it still remains true that only one civilization needs to persevere with its colonization efforts in order to bring the entire galaxy under its wing.

One other sociological consideration suggests that advanced societies might start out like us but then "engineer" themselves in such a way so as to shut down their drive to colonize. On our planet, and presumably on others, the development of rocketry occurred at roughly the same time as the invention of nuclear weapons, chemical weapons, and other methods of mass destruction. Societies that remain aggressive are in constant danger of self-destruction. Therefore, civilizations might be motivated to find ways to reduce or channel their aggressive tendencies, perhaps through some type of genetic engineering. The oldest alien cultures, according to this line of reasoning, would have managed to rid themselves of dangerous aggression. It's conceivable that in the process they would have also chosen to focus on improving life on their home planet rather than on moving out into space.

Civilizations but No Colonization—Self-Destruction A much more sobering possibility for why the galaxy might remain uncolonized even if many civilizations have arisen—one that assumes we are completely typical of intelligent beings—is that societies inevitably self-destruct before attaining the capability for interstellar travel. While this idea is horribly tragic, it is not far-fetched. Nuclear weapons provide clear proof that the technology needed for interstellar travel can also be used to destructive ends. Similarly, any society that learns to tap energy resources would almost certainly use the most accessible energy first, which on any Earth-like planet is likely to be fossil fuels. Thus, like us, other civilizations must face the dangers posed by global warming and other environmental problems. Population growth probably also poses similar problems for all civilizations. Rapid population growth is a natural consequence of biological reproduction for a species that is no longer subject to the whims of predators or childhood diseases, which tend to hold population growth in check. From this perspective, a society can survive long enough to achieve interstellar travel only if it successfully navigates what amounts to a very difficult obstacle course—one in which each obstacle would mean the end of its civilization. Could it be that it simply can't be done?

THINK ABOUT IT What odds would you give for humanity's surviving long enough to achieve interstellar travel? Defend your choice.

There *Is* a Galactic Civilization The third category of explanations for the Fermi paradox in essence suggests that there is no paradox at all: The galactic civilization is out there, but we do not yet recognize it. Indeed, UFO buffs might claim that scientists are blind to the obvious proof of this suggestion. While we can't rule out the possibility, no evidence of alien visitation has yet withstood scientific scrutiny. Nevertheless, there are many ways by which a galactic civilization could avoid our detection.

One idea simply assumes that a galactic civilization would have no particular interest in us. After all, to a society that is millions or billions of years ahead of us, we might seem too simple to warrant attention. Of course, if civilizations are communicating or traveling among the stars, we might be able to discover them. Signaling is precisely what SETI experiments look for. The fact that we have not yet received a clearly extraterrestrial broadcast may simply be a consequence of not yet having looked at a sufficient number of stars.

Another possibility is that civilizations are aware of our presence but have deliberately chosen to keep us in the dark. This idea is sometimes called the **zoo hypothesis**, although it might better be called the "wildlife refuge" hypothesis. Just as we set aside nature reserves that are supposed to be left alone as places where wildlife can thrive without our intervention, a galactic civilization might declare planets like ours off-limits to exploration. (*Star Trek* fans might think of this as the "Prime Directive" [to avoid interfering with alien cultures] solution to the Fermi paradox.) One objection to the zoo hypothesis is that even if civilizations wanted to hide from us, we would still be able to intercept their communications among themselves. On the other hand, as we've already noted, SETI searches for signals have so far carefully investigated only a small amount of cosmic real estate and could easily have missed such communications. Alternatively, the communications might involve a technology that we have not yet developed and that therefore is undetectable to us at present.

A closely related idea suggests that a sophisticated galactic civilization might reveal itself to new societies only after they reach a certain level of technology. Perhaps the extraterrestrials place monitoring devices near star systems that show promise of emerging intelligence and patiently wait until these devices record the presence of civilization. This idea is sometimes called the **sentinel hypothesis**, after a science fiction story by Arthur C. Clarke titled "The Sentinel." The story became the basis of the book and movie *2001: A Space Odyssey*, in which a monolith buried on the Moon signals our presence when we finally dig it up (see Movie Madness in Chapter 7). Carl Sagan used a similar idea for the book and movie *Contact* (see Movie Madness in Chapter 12), in which a signaling station around the star Vega amplifies and beams back our own television broadcasts to us, leading us to our first glimpse of a galactic civilization.

• What are the implications of the Fermi paradox to human civilization?

The Fermi paradox may have its origins in a simple question—Where is everybody?—but we have seen that finding an answer is much more complex than asking the question. In fact, if we consider our possible answers in more depth, we find that each leads to astonishing implications for our own species.

Consider the first solution—that we are alone. If this is true, then our civilization is a remarkable achievement. It implies that through all of cosmic evolution, among countless star systems, we are the first piece of the universe ever to know that the rest of the universe exists. Through us, the universe has attained self-awareness. Some philosophers and many religions argue that the ultimate purpose of life is to become truly self-aware. If so, and if we are alone, then the destruction of our civilization and the loss of our scientific knowledge would represent an inglorious end to something that took the universe some 14 billion years to achieve. From this point of view, humanity becomes all the more precious, and the collapse of our civilization would be all the more tragic. Knowing this to be the case might help us learn to put petty bickering and wars behind us so that we might preserve all that is great about our species.

The second category of solutions has much more terrifying implications. If thousands of civilizations before us have all failed to achieve interstellar travel on a large scale, what hope do we have? Unless we somehow think differently than all previous civilizations, we will never go far in space. Given that we have always explored when the opportunity arose, this solution almost inevitably leads to the conclusion that failure will come about because we destroy ourselves. We can only hope that this answer is wrong.

The third solution is perhaps the most intriguing. It says that we are newcomers on the scene of a galactic civilization that has existed for millions or billions of years before us. Perhaps this civilization is deliberately leaving us alone for the time being and will someday decide the time is right to invite us to join it. If so, our entire species might be on the verge of beginning a journey every bit as incredible as that of a baby emerging from the womb and coming into the world.

You can probably now see why the Fermi paradox involves far more than a simple question. No matter what the answer turns out to be, learning it is sure to mark a turning point in the brief history of our species. Moreover, this turning point is likely to be reached within the next few decades or centuries. We already have the ability to destroy our own civilization. If we do so, then our fate is sealed. But if we survive long enough to develop technology that can take us to the stars, the possibilities seem almost limitless.

Imagine for a moment the grand view, a gaze across the centuries and millennia from this moment forward. Picture our descendants living among the stars, having created or joined a great galactic civilization. They will have the privilege of experiencing ideas, worlds, and discoveries far beyond our wildest imagination. Perhaps, in their history lessons, they will learn of our generation—the generation that history placed at the turning point and that managed to steer its way past the dangers of self-destruction and onto the path to the stars.

13.4 THE PROCESS OF SCIENCE IN ACTION: EINSTEIN'S SPECIAL THEORY OF RELATIVITY

In this chapter, we have seen that Einstein's special theory of relativity has important implications to possibilities of interstellar travel. But that is not why Einstein came up with the theory. Rather, as with the development of other scientific theories, Einstein developed the theory of relativity to explain one of the outstanding scientific mysteries of the time, which is why we will take it as this chapter's case study in the process of science in action.

The mystery concerned the speed and nature of light. In 1873, Scottish mathematician and physicist James Clerk Maxwell (1831–1879) published a paper in which he showed that light is an electromagnetic wave [Section 3.4]. His paper included a set of equations—now known as *Maxwell's equations*—that describe the nature of electromagnetic waves. These equations form the basis of our modern understanding of electricity and magnetism, and in essence you are confirming their validity every time you flip a light switch or listen to a radio or television broadcast. However, the idea that light is a wave left a fundamental question: All other types of waves—for example, sound waves, water waves, or waves on a violin string—are carried by some type of medium; what medium carries light waves through "empty" space? Maxwell and other physicists presumed that space must be filled with some medium of unknown composition that could carry the light waves—an idea that had actually been around for some time already—and this medium was known as the *ether.*

If the ether really existed, then Earth's motion around the Sun would make us move in different directions *relative* to the ether at different times of year. In 1887, A. A. Michelson (1852–1931) and E. W. Morley (1838–1923) conducted an elegant experiment—now famous as the *Michelson–Morley experiment*—designed to measure Earth's motion through the ether. The experiment relied on the idea that the speed of light would be slightly faster when it was moving in the same direction as the ether (relative to Earth), slightly slower when moving in the opposite direction, and at speeds in between for other directions. However, their experiment failed to measure any directional difference in the speed of light. In hindsight, the obvious implication was that the ether does not exist. At the time, however, this idea seemed so preposterous that Michelson and Morley went to great lengths to explain how nature might "hide" the ether's existence from human experimenters.

Einstein came up with his theory because he took the results of the Michelson–Morley experiment at face value. Instead of assuming that nature was hiding something from us, he assumed that the speed of light showed no directional variation because none exists; that is, Einstein assumed that the speed of light (through space) is a physical constant, one that will always be measured the same no matter what the motion of the light source or the observer. In doing so, he not only explained the perplexing results of the Michelson–Morley experiment, but he also cleared up some mysteries that had been associated with mathematical implications of Maxwell's equations. In other words, Einstein's theory did not come about by magic or just because he was a smart guy—it came about because there were known problems that needed to be solved, and Einstein was the first to come up with their solution.

Einstein's special theory of relativity is often portrayed as being difficult, but its basic ideas are actually quite easy to understand—though admittedly mind-boggling in their consequences. Because they are so important to our modern understanding of the nature of the universe, let's take a look at the basic ideas and their astonishing implications.

• What is "relative" about relativity?

Imagine a supersonic airplane that flies at a speed of 1,670 km/hr from Nairobi, Kenya, to Quito, Ecuador. How fast is the plane going? At first, this question sounds trivial—we have just said that the plane is going 1,670 km/hr.

But wait. Nairobi and Quito are both nearly on Earth's equator, and the equatorial speed of Earth's rotation is the same 1,670 km/hr speed at which the plane is flying. Moreover, the east-to-west flight from Nairobi to Quito is opposite the direction of Earth's rotation (Figure 13.15). Thus, if you lived on the Moon, the plane would appear to stay put *while Earth rotated beneath it.* When the flight began, you would see the plane lift off the ground in Nairobi. The plane would then remain stationary while Earth's rotation carried Nairobi away from it and Quito toward it. When Quito finally reached the plane's position, the plane would drop back down to the ground.

We have two alternative viewpoints about the plane's flight. People on Earth say that the plane is traveling westward across the surface of Earth. Observers in space say that the plane is stationary while Earth rotates

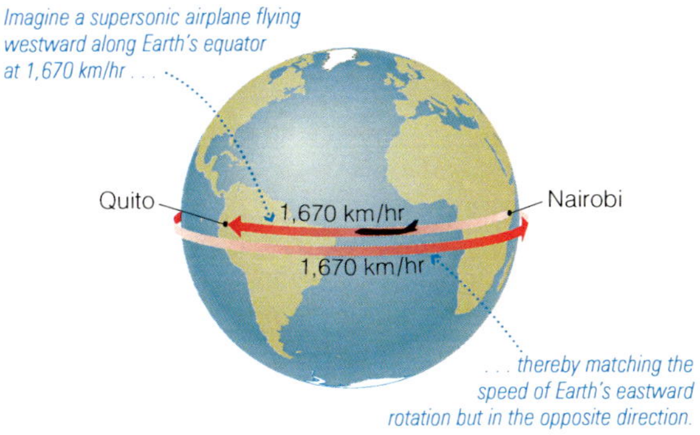

Figure 13.15

A plane flying at 1,670 km/hr from Nairobi to Quito (westward) travels precisely opposite Earth's eastward rotation. Thus, viewed from afar, the plane remains stationary while Earth rotates underneath it.

eastward beneath it. Both viewpoints are equally valid. In fact, there are many other equally valid viewpoints about the plane's flight. Observers looking at the solar system as a whole would see the plane moving at a speed of more than 100,000 km/hr—Earth's speed in its orbit around the Sun. Observers living in another galaxy would see the plane moving at about 800,000 km/hr with the rotation of the Milky Way Galaxy. The only thing all these observers would agree on is that the plane is traveling at 1,670 km/hr *relative to* the surface of Earth.

This example shows that questions like, "Who is really moving?" and "How fast are you going?" have no absolute answers. Einstein's special theory of relativity gets its name because it tells us that measurements of time and space, as well as measurements of motion, make sense only when we describe whom or what they are being measured relative to.

THINK ABOUT IT Suppose you are running on a treadmill and the readout says you are going 8 miles per hour. What is the 8 miles per hour measured relative to? How fast are you going relative to the ground? How fast would an observer on the moon see you going? Describe a few other possible viewpoints on your speed.

The Absolutes of Relativity The theory of relativity tells us that motion is always relative, but it does *not* say that *everything* is relative. In fact, the theory claims that two things in the universe are absolute:

1. The laws of nature are the same for everyone.
2. The speed of light is the same for everyone.

The first absolute, that the laws of nature are the same for everyone, is a more general version of the idea that all viewpoints on motion are equally valid. If they weren't, different observers would disagree about the laws of physics. The second absolute, that the speed of light is the same for everyone, is much more surprising. Ordinarily, we expect speeds to add and subtract. If you watch someone throw a ball forward from a moving car, you'd see the ball traveling at the speed it is thrown plus the speed of the car. But if a person shines a light beam from a moving car, you'd see it moving at precisely the speed of light (about 300,000 kilometers per second), no matter how fast the car is going. This strange fact explains why the Michelson–Morley experiment found no differences in the speed of light, and this has been experimentally verified countless times.

The Speed of Light The cosmic speed limit follows directly from this fact about the speed of light. To see why, imagine that you have just built the most incredible rocket possible, and you are taking it on a test ride. You push the acceleration button and just keep going faster and faster and faster. With enough fuel, you might expect that you'd eventually be moving faster than the speed of light. But can you?

Before we answer this question, the fact that all motion is relative forces us to answer another question: What is your speed being measured relative to? Let's begin with *your* point of view. Imagine that you turn on your rocket's headlights. Because the speed of light is the same for everyone, you must see the headlight beams traveling at the speed of light—which means they are racing away from your rocket at a speed you'll measure to be 300,000 km/s. The fact that you'll see your headlight beams racing away is true no matter how long you have been firing your rocket engines. This shouldn't be too surprising, as it's just another way of saying that you can't catch up with your own light.

Now, however, remember that everyone always measures the same speed of light. This means that people back on Earth—or anyplace else—will also say that your headlight beams are moving through space at 300,000 kilometers per second. In other words, according to anyone watching you from any place in the universe, your headlight beams are traveling at c = 300,000 km/s, and they're moving out ahead of you (Figure 13.16). Clearly, this implies that you must be traveling *slower* than the headlight beams, which means slower than the speed of light.

In case you are still not convinced, let's turn the situation around. Imagine that, as you race by some planet, a person on the planet turns on a light beam. Because the speed of light is absolute, you will see the light beam race past you at c = 300,000 km/s. The person on the planet will also see the light traveling at c = 300,000 km/s and will see the light outrace you. Again, everyone will agree that you are traveling slower than the speed of light.

The same argument applies to any moving object, and it is true with or without headlights. All light travels at the speed of light, including the light that reflects off an object (allowing us to see it) and the infrared light that even cool objects like people emit. As long as the speed of light is absolute, no material object can ever keep up with the light it emits or reflects, which means no material object can reach or exceed the speed of light. Building a spaceship to travel faster than the speed of light is not a mere technological challenge—it simply cannot be done.

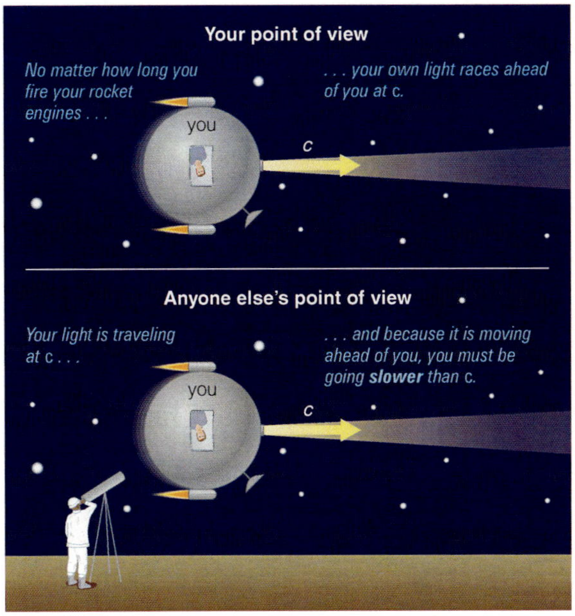

Figure 13.16

The fact that everyone always measures the same speed of light means that you cannot keep up with your own light (top), and therefore no matter what you do, other people must always conclude that you are traveling slower than light. In other words, there is no way for you ever to reach or exceed the speed of light.

The Relativity of Time and Space Like the cosmic speed limit, the other strange consequences of relativity also follow from the absoluteness of the speed of light. We will not go into details here, but you can understand the general ideas by thinking about the fact that a speed is always equal to a distance divided by a time. For example, a speed of 60 miles per hour means that you travel a distance of 60 miles in a time of 1 hour, and light's speed of 300,000 km/s means that light travels a distance of 300,000 kilometers in a time of 1 second. In our ordinary slow-moving lives, we think of times and distances as absolutes, while speed is relative—as is the case with the speed of the airplane in our Nairobi–Quito example. But Einstein's theory tells us that when it comes to light, speed is absolute—everyone always measures the same speed of light. The consequence is that time and distance must become relative.

We have already seen how time and distance are affected. Time runs slower for high-speed travelers (the phenomenon of *time dilation* that we discussed earlier in the chapter), which is why relativistic starships could allow passengers to make long trips if they traveled fast enough. Distances are also shrunk as measured by the high-speed travelers, which is why high-speed travelers to Vega would see a shorter distance than the distance we measure from Earth. Mass is also affected, so we would measure high-speed objects to have higher mass than they do when they are stationary relative to us.* Another direct consequence of these ideas—

*You can see why mass must be affected if you think about Newton's second law of motion, which states that force = mass × acceleration (see Figure 2.12). The units of acceleration are a distance divided by a time squared (such as m/s^2), so the relativity of time and distance means that even if the force (e.g., from the rocket engines) stays constant, a rocket's acceleration must decrease with increasing speed. The only way to account for this decrease when the force stays the same is to realize that the mass of the rocket must be increasing.

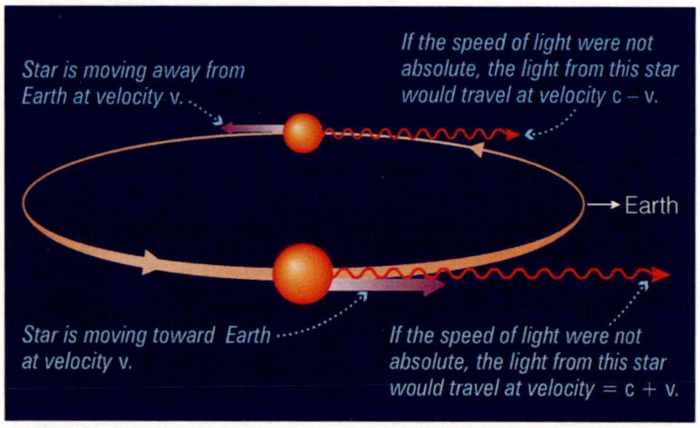

Figure 13.17

If the speed of light were *not* absolute, the speed at which light from a star in a binary system comes toward Earth would depend on its position and velocity toward us in the binary orbit. Thus, we would not see the star as a distinct point of light.

and one that can be derived with nothing more than high school algebra (though we won't do it here)—is Einstein's famous formula, $E = mc^2$.

• What evidence supports Einstein's theory?

Although we haven't gone into details here, we have stated that all the amazing consequences of relativity, including time dilation and $E = mc^2$, follow logically from the absoluteness of the speed of light. However, remember that logic alone is not good enough in science; conclusions must always remain tentative until they pass observational or experimental tests. Does relativity meet the test?

The Absoluteness of the Speed of Light The first thing we might wish to test is the surprising premise of relativity: the absoluteness of the speed of light. In principle, we can test this premise by measuring the speed of light coming from many different objects and going in many different directions and verifying that the speed is always the same. As we've already discussed, the Michelson–Morley experiment was in essence one such test, and it showed that the speed of light is indeed absolute.

In fact, we have verified this fact many other times. For example, some distant galaxies are moving away from us at speeds close to the speed of light, yet their light still arrives here traveling at the speed of light. Perhaps even more convincingly, if the speed of light were not always the same, we could not see distinct stars in binary star systems. Consider the star moving toward us in Figure 13.17. If the speed of light depended on the star's motion, its light would be coming toward us somewhat faster than the "normal" speed of light. Half an orbit later, when it was moving away from us, its light would travel to us at less than the "normal" speed of light. Thus, the light from the "fast" side of the orbit would tend to catch up with the light emitted earlier from the "slow" side, reaching us at the same time and smearing the star's image so that we'd see it in different positions all at once, instead of seeing it as a distinct star. Thus, the simple fact that we see distinct stars in binary star systems proves that the speed of light does not depend on the stars' motions; that is, light always travels at the same speed. In addition, the speed of light from the two orbiting stars in binary systems, as well as from opposite sides of the rotating Sun, has also been measured, again confirming the same result: The speed of light is always the same.

Experimental Tests of Relativity Although *we* cannot yet travel at speeds at which the effects of relativity should be obvious, tiny subatomic particles can reach such speeds, thereby allowing us to test the precise predictions of the formulas of special relativity. Let's first consider one way of testing time dilation. In machines called *particle accelerators*, physicists accelerate subatomic particles to speeds near the speed of light and study what happens when the particles collide. The colliding particles have a great deal of kinetic energy, and the collisions convert some of this kinetic energy into mass-energy that emerges as a shower of newly produced particles. Many of these particles have very short lifetimes, at the end of which they decay (change) into other particles. For example, a particle called the π^+ ("pi plus") meson has a lifetime of about 18 nanoseconds (billionths of a second) when produced at rest. But π^+ mesons produced at speeds close to the speed of light in particle accelerators last much longer than 18 nanoseconds—and by precisely the amount predicted by the time dilation formula.

The same experiments also confirm the mass increase predicted by relativity. The amount of energy released when high-speed particles collide depends on the particle masses and speeds. Just as relativity predicts, these masses are greater at high speed than they are at low speed—again by just the amount predicted by Einstein's formulas.

Particle accelerators even offer experimental evidence that nothing can reach the speed of light. It is relatively easy to get particles traveling at 99% of the speed of light in particle accelerators. However, no matter how much more energy is put into the accelerators, the particle speeds get only fractionally closer to the speed of light. Some particles have been accelerated to speeds within 0.00001% of the speed of light, but none have ever reached the speed of light.

Although the effects of relativity are obvious only at very high speeds, modern techniques of measuring time are so precise that effects can be measured even at ordinary speeds. For example, a 1975 experiment compared the amount of time that passed on an airplane flying in circles to the time that passed on the ground. Over 15 hours, the airborne clocks lost a bit under 6 nanoseconds to the ground clocks, matching the result expected from relativity. More recent experiments using the Space Shuttle have confirmed time dilation as well.

Nuclear energy also provides a test of relativity. Remember that $E = mc^2$ is a direct consequence of Einstein's theory. Every time you see film of an atomic bomb, or use electrical power from a nuclear power plant, or feel the energy of sunlight that the Sun generated through nuclear fusion, you are really experiencing direct experimental evidence of relativity.

Beyond Special Relativity All in all, special relativity is one of the best-tested theories in physics, and it has passed every experimental test to date with flying colors. That is why it merits status as a true, scientific *theory*. Of course, like any scientific theory, it is always open to further testing and refinement. Indeed, Einstein himself realized that special relativity was missing an important ingredient: It dealt successfully with motion through empty space, but not with motion affected by gravity. This missing ingredient provided much of the motivation that led Einstein to press on and publish his general theory of relativity about a decade later.

The general theory expanded on the special theory by including gravity, and in the process Einstein found that it also turned out to be an improvement on Newton's theory of gravity [Section 2.4]. It is likely that the general theory of relativity will someday need its own refinements to make it compatible with the theory of quantum mechanics. If so, it is possible that some of the ideas of special relativity may also change. But the evidence that supports the theory as we know it today is real and cannot be made to disappear. If anyone ever comes up with a better theory, the new theory will still have to explain the many experimental results that support relativity so well.

THE BIG PICTURE
Putting Chapter 13 into Context

In this chapter, we have explored the possibilities for and challenges of interstellar travel, and the surprising Fermi paradox that arises when we think about what other civilizations might already have achieved. As you continue in your studies, keep in mind the following "big picture" ideas:

- Interstellar travel may be a staple of science fiction, but it remains well beyond our current capabilities. Nevertheless, it is possible that the challenges can be surmounted and that other civilizations might already have overcome them.
- The idea that other civilizations might already have achieved interstellar travel leads to the Fermi paradox—the question of why we see no evidence of galactic civilization. This simple question is not easily dismissed, and no matter what its solution turns out to be, it has profound implications to human civilization.
- One key lesson of the Fermi paradox is that we live at a unique moment in the history of the human species. We have the ability to destroy our civilization and perhaps even to drive our species to extinction. But if we survive, our descendants might have a boundless future. From this perspective, no generation has ever borne such great responsibility.

SUMMARY OF KEY CONCEPTS

13.1 THE CHALLENGE OF INTERSTELLAR TRAVEL

• Why is interstellar travel so difficult?

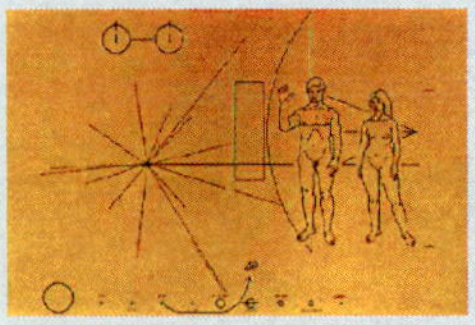

Convenient interstellar travel remains well beyond our technological capabilities. Current spacecraft would take more than 100,000 years just to traverse the distance to the nearest stars. The energy requirements for sending people on interstellar trips are enormous, far greater than all current world energy usage.

• Could we travel to the stars with existing rockets?

Nearly all rockets built to date are powered by chemical rocket engines. The **rocket equation** shows that these types of engines could not possibly get us to speeds of even 1% of the speed of light, making them impractical for interstellar travel.

13.2 DESIGNING SPACECRAFT FOR INTERSTELLAR TRAVEL

• How might we build interstellar spacecraft with "conventional" technology?

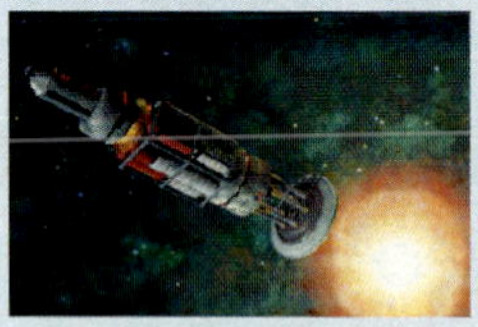

Technologies such as nuclear rocket engines, solar sails, and ion engines could in principle allow us to build starships that could travel at speeds up to about 10% of the speed of light—fast enough to reach nearby stars in less than a century.

• How might we build spacecraft that could approach the speed of light?

Several technologies that are well beyond us at present but are allowed by the laws of physics could allow starships to reach speeds close to the speed of light. These include **matter–antimatter** engines and **interstellar ramjets** that scoop up fuel as they go.

- **Are there ways around the light-speed limitation?**

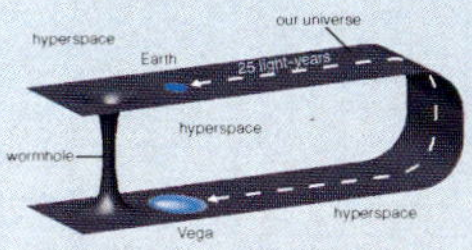

No known physical laws prevent hyperspace, **wormholes**, or warp drive from offering "loopholes" that could allow us to get from one place to another in less time than we could by traveling through ordinary space. However, we do not yet know if any of these are really possible.

13.3 THE FERMI PARADOX

- **Where is everybody?**

This seemingly simple question, known as the **Fermi paradox**, comes about because our general assumption that Earth is not unique leads us to expect that many other civilizations should by now have arisen and had the opportunity to colonize the Milky Way Galaxy. Yet we see no evidence of a galactic civilization.

- **Would other civilizations really colonize the galaxy?**

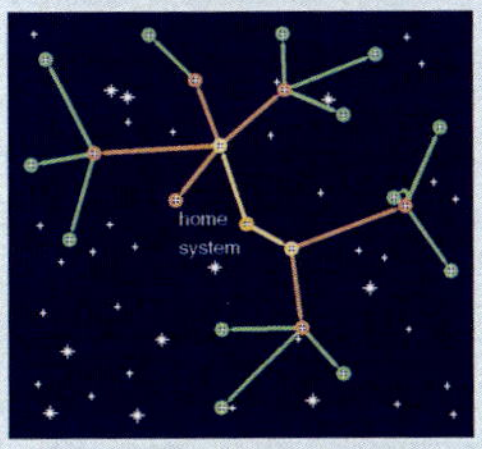

Based on the idea that other beings would have evolved in response to evolutionary pressures similar to those that led to human evolution on Earth, we should expect that other civilizations would have the same inherent drive to colonize that we seem to possess. If so, a civilization should be able to colonize the galaxy in a time that is short compared to the age of the universe, even with technology not much beyond our own.

- **What are possible solutions to the Fermi paradox?**

There are three general categories of solution to the Fermi paradox: (1) We are alone. (2) Civilizations are common, but no one has colonized the galaxy. (3) There is a galactic civilization, but it has deliberately avoided revealing its existence to us.

- **What are the implications of the Fermi paradox to human civilization?**

The first solution implies that we are the only piece of the universe that has ever attained self-awareness. The second suggests that civilizations may either change or destroy themselves before attaining the ability to travel to the stars. The third implies that we might someday meet up with a galactic civilization that predates us by millions or billions of years.

13.4 THE PROCESS OF SCIENCE IN ACTION: EINSTEIN'S SPECIAL THEORY OF RELATIVITY

- **What is "relative" about relativity?**

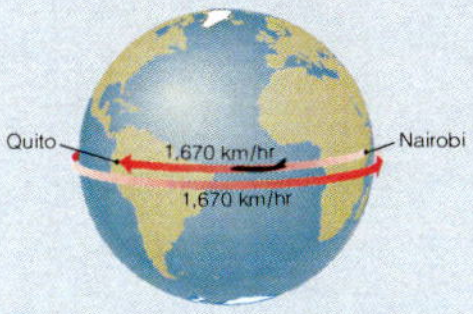

Special relativity tells us motion is relative, but everyone always agrees on the speed of light. From this it follows that different observers can measure time, distance, and mass differently, and that no material object can reach or exceed the speed of light.

- **What evidence supports Einstein's theory?**

Experiments with light confirm that its speed is always the same. Experiments with subatomic particles in particle accelerators confirm the predictions of **time dilation** and mass increase at speeds close to the speed of light, and time dilation has been verified at relatively low speeds in aircraft and spacecraft. Nuclear power plants and nuclear bombs release energy in accordance with the formula $E = mc^2$, which is also a prediction of special relativity.

EXERCISES AND PROBLEMS

REVIEW QUESTIONS

Short-Answer Questions Based on the Reading

1. Briefly describe the journeys of *Pioneers 10* and *11* and *Voyagers 1* and *2*. How do these spacecraft illustrate the challenge of interstellar travel?
2. How does the speed of light affect the possibility of interstellar travel?
3. About how much energy would it take to send enough people on a trip to start a colony around another star? Explain how you arrive at the answer.
4. How do rockets work? Briefly describe the history of rocketry.
5. What is the *rocket equation* used for? Based on the rocket equation and the *mass ratio*, briefly explain why chemical rockets are inadequate for sending people or large robotic probes to the stars.
6. Describe the proposed fusion-powered starships of Project Orion and Project Daedalus. How quickly could such ships reach the stars?
7. Discuss a few ways of reaching the stars (other than nuclear rockets) that are, at least in principle, within our current technological reach.
8. How would *time dilation* affect space travel at speeds close to the speed of light? Discuss possible ways of achieving such speeds, including matter–antimatter engines and *interstellar ramjets*.
9. Briefly discuss how Einstein's general theory of relativity might allow "shortcuts" by which we could reach distant stars in shorter times than we would expect from their measured distances. Do we know whether these shortcuts are really possible?
10. What is the *Fermi paradox?* What two seemingly contradictory ideas underlie the paradox?

11. Why does it seem that other civilizations, if they exist, should be significantly older than ours? Explain clearly.
12. What are *Von Neumann machines?* How do they affect the Fermi paradox?
13. Describe the *coral model* of galactic colonization. Why do we conclude that civilizations could have colonized the galaxy by now even with technology not much more advanced than ours?
14. Briefly discuss possible motives for galactic colonization, as well as several "motives" that don't hold up.
15. Summarize the three general categories of possible solutions to the Fermi paradox, and discuss each category in some detail.
16. Briefly discuss the profound implications of the Fermi paradox and how the answer to the paradox affects our civilization.
17. What known problems were solved when Einstein discovered the special theory of relativity?
18. Explain how the idea of an absolute speed of light leads automatically to the conclusion that no one can travel faster than light.
19. Besides the idea that you cannot reach the speed of light, what other consequences follow from the absoluteness of the speed of light?
20. Describe at least three tests that have confirmed the validity of the special theory of relativity.

TEST YOUR UNDERSTANDING

Science or Nonscience?

Each of the following describes some futuristic scenario that, while perhaps entertaining, may or may not be plausible. In each case, decide whether the scenario is plausible according to our present understanding of science or whether it is unlikely to be possible. Explain clearly; because not all of these have definitive answers, your explanation is more important than your chosen answer.

21. A brilliant teenager working in her garage discovers a way to build a rocket that burns coal as its fuel and can travel at half the speed of light.
22. Using beamed energy propulsion from a laser powered by energy produced at a windmill farm in the California desert, NASA engineers are able to send a solar sailing ship on a journey to Alpha Centauri that will take only 50 years.
23. Human colonization of the moons of Saturn occurs using spaceships powered by dropping nuclear bombs out the back of the ships.
24. In the year 2750, we receive a signal from a civilization around a nearby star telling us that the *Voyager 2* spacecraft recently crash-landed on its planet.
25. The General Rocket Corporation (a future incarnation of General Motors) unveils a new personal interstellar spacecraft that works as an interstellar ramjet with a scoop about 10 meters across.
26. Members of the first crew of the matter–antimatter spacecraft *Star Apollo,* which left Earth in the year 2165, return to Earth in the year 2450 looking only a few years older than when they left.
27. In the year 2011, we finally uncover definitive evidence of alien visits to Earth when a flying saucer crashes in the Rocky Mountains and its oxygen-kerosene fuel ignites a forest fire.
28. Aliens from a distant star system invade Earth with the intent to destroy us and occupy our planet, but we successfully fight them off when their technology proves no match for ours.
29. Aliens arrive on Earth but virtually ignore our presence, finding the diversity of earthly bacteria to be much more scientifically interesting.
30. A single great galactic civilization exists. It originated on a single planet long ago but is now made up of beings from many different planets, each of which was assimilated into the galactic culture in turn.

Quick Quiz

Choose the best answer to each of the following. Explain your reasoning with one or more complete sentences.

31. The *New Horizons* spacecraft is currently on its way to Pluto and will eventually continue out of our solar system. About how long will it take to travel the distance to the nearest stars? (a) 100 years; (b) 1,000 years; (c) 100,000 years.
32. The amount of energy that would be needed to accelerate a large spaceship to a speed close to the speed of light is (a) about 100 times as much energy as is needed to launch the Space Shuttle; (b) more than the total amount of energy used by the entire world in a year; (c) more than the amount of energy that our Sun emits into space in a year.
33. The rocket engines of our current spacecraft are powered by: (a) chemical energy; (b) nuclear energy; (c) matter–antimatter annihilation.
34. Suppose that a spaceship was launched in the year 2120 on a round-trip journey of 100 light-years, traveling at 99.99% of the speed of light. If one of the crew members was 30 years old when she left, about how old would you expect her to be on her return? (a) 31; (b) 130; (c) 29.
35. Suppose that a spaceship was launched in the year 2120 on a round-trip journey of 100 light-years, traveling at 99.99% of the speed of light. In approximately what year would the ship return to Earth? (a) 2121; (b) 2170; (c) 2220.
36. Which of the following best describes our current understanding of the possibility of fast interstellar travel through hyperspace? (a) Hyperspace travel is the method of choice for all advanced civilizations. (b) We do not know enough to say whether such travel is really possible. (c) The idea of hyperspace is pure fantasy and has no basis in reality.
37. Which of the following questions best represents the Fermi paradox? (a) Why can't we travel faster than the speed of light? (b) Why haven't we found any evidence of a galactic civilization? (c) Why haven't aliens invaded Earth and stolen our resources?
38. According to current scientific understanding, the idea that the Milky Way Galaxy might be home to a civilization millions of years more advanced than ours is: (a) a virtual certainty; (b) extremely unlikely; (c) one reasonable solution to Fermi's paradox.
39. Which of the following is *not* relative in the special theory of relativity? (a) motion; (b) time; (c) the speed of light.

40. What does the famous formula $E = mc^2$ have to do with special relativity? (a) Nothing; it comes from a different theory. (b) It is one of the two starting assumptions of special relativity. (c) It is a direct consequence of the theory, and hence a way of testing the theory's validity.

INVESTIGATE FURTHER

In-Depth Questions to Increase Your Understanding

Short-Answer/Essay Questions

41. *Distant Dream or Near-Reality?* Considering all the issues surrounding interstellar flight, when, if ever, do you think we are likely to begin traveling among the stars? Write a few paragraphs defending your opinion.

42. *What's Wrong with This Picture?* Many science fiction stories have imagined the galaxy divided into a series of empires, each having arisen from a different civilization on a different world, that hold each other at bay because they are all at about the same level of military technology. Is this a realistic scenario? Explain.

43. *Large Rockets.* Suppose we built a rocket that works much like the Space Shuttle, but was 1,000 times as large. Could this rocket get us to speeds close to the speed of light? Explain.

44. *Ticket to the Stars.* In this chapter, we've stated that relativity offers only a one-way "ticket to the stars." Explain why.

45. *Solution to the Fermi Paradox.* Among the various possible solutions we have discussed for the Fermi paradox, which do you think is most likely? (Or, if you have no opinion as to their likelihood, which do you like best?) Write a one- to two-page essay in which you explain why you favor this solution.

46. *Interstellar Travel in the Movies.* Choose a science fiction movie in which aliens (or future humans) are engaged in some type of interstellar travel. In a one- to two-page essay, briefly describe how they supposedly accomplish the travel and evaluate in depth whether the scheme seems plausible.

47. *The "Relative" in Relativity.* Many people have claimed that Einstein showed that "everything is relative." Is this true? Explain.

48. *Relativity as a Theory.* Look back at the hallmarks of science in Chapter 2. Evaluate the special theory of relativity in terms of each of the three hallmarks, showing why it meets the test of being science. Then explain why special relativity qualifies as a scientific *theory*, rather than just a hypothesis.

Quantitative Problems

Be sure to show all calculations clearly and state your final answers in complete sentences.

49. *Cruise Ship Energy.* Suppose we have a spaceship about the size of a typical ocean cruise ship today, which means it has a mass of about 100 million kilograms, and we want to accelerate the ship to a speed of 10% of the speed of light.
    a. How much energy would be required? (*Hint:* You can find the answer simply by calculating the kinetic energy of the ship when it reaches its cruising speed; because 10% of the speed of light is still small compared to the speed of light, you can use the formula that tells us that kinetic energy $= \frac{1}{2} \times m \times v^2$.)
    b. How does your answer compare to total world energy use at present, which is about 5×10^{22} joules per year?
    c. Suppose the cost of energy is 5¢ per 1 million joules. Using this price, how much would it cost to generate the energy needed by this spaceship?

50. *The Rocket Equation I.* Suppose a rocket with mass ratio $M_i/M_f = 15$ has engines that produce an exhaust velocity of 3 km/s. What is its final velocity? Is it sufficient to escape Earth? (*Hint:* See Cosmic Calculations 13.1.)

51. *The Rocket Equation II.* Suppose you want a rocket to achieve escape velocity from Earth (11 km/s) and its engines produce an exhaust velocity of 3 km/s. What mass ratio is required? Briefly explain the meaning of this mass ratio.

52. *The Multistage Rocket Equation.* The rocket equation takes a slightly different form for a multistage rocket:

$$v = nv_e \ln\left(\frac{M_i}{M_f}\right)$$

where n is the number of stages.
    a. Suppose a rocket has three stages with mass ratio $M_i/M_f = 3.4$ and engines that produce an exhaust velocity of 3 km/s. What is its final velocity? Is it sufficient to escape Earth?
    b. Suppose a rocket has 100 stages with mass ratio $M_i/M_f = 3.4$ and engines that produce an exhaust velocity of 3 km/s. What is its final velocity? Compare it to the speed of light.

53. *Relativistic Time Dilation.* Use the time dilation equation from Cosmic Calculations 13.2 to answer each of the following.
    a. Suppose a rocket travels at the escape velocity from Earth (11 km/s). Will time on the rocket differ noticeably from time on Earth? Explain.
    b. Suppose a rocket travels at a speed of $0.9c$. If the rocket is gone from Earth for 10 years as measured on Earth, how much time passes on the rocket?
    c. Suppose a rocket travels at a speed of $0.9999c$. If the rocket is gone from Earth for 10 years as measured on Earth, how much time passes on the rocket?

54. *Testing Relativity.* A π^+ meson produced at rest has a lifetime of 18 nanoseconds (1.8×10^{-8} s). Thus, in its own reference frame, a π^+ meson will always "think" it is at rest and therefore will decay after 18 nanoseconds. Suppose a π^+ meson is produced in a particle accelerator at a speed of $0.998c$. How long will scientists see the particle last before it decays? Briefly explain how an experiment like this helps verify the special theory of relativity. (*Hint:* Use the time dilation equation from Cosmic Calculations 13.2.)

55. *Long Trips at Constant Acceleration.* Consider a spaceship on a long trip with a constant acceleration of $1g$. Although the derivation is beyond the scope of this book, it is possible to show that, as long as the ship is gone from Earth for many years, the amount of time that passes on the spaceship during the trip is approximately:

$$t_{ship} = \frac{2c}{g} \ln\left(\frac{g \times D}{c^2}\right)$$

where D is the distance to the destination and ln is the natural logarithm. If D is in meters, $g = 9.8$ m/s^2, and $c = 3 \times 10^8$ m/s, the answer will be in units of seconds. Use this formula as needed to answer the following questions. Be sure to convert the distances from light-years to meters and final answers from seconds to years; useful conversions: 1 light-year $\approx 9.5 \times 10^{15}$ m; 1 yr $\approx 3.15 \times 10^7$.
    a. Suppose the ship travels to a star that is 500 light-years away. How much time will pass on the ship? Approximately how much time will pass on Earth? Explain.

b. Suppose the ship travels to the center of the Milky Way Galaxy, about 28,000 light-years away. How much time will pass on the ship? Compare this to the amount of time that passes on Earth.

c. The Andromeda Galaxy is about 2.2 million light-years away. Suppose you had a spaceship that could constantly accelerate at $1g$. Could you go to the Andromeda Galaxy and back within your lifetime? Explain. If you could make the journey, what would you find when you returned to Earth?

56. *The Coral Model of Colonization.* We can estimate the time it would take for a civilization to colonize the galaxy. Imagine that a civilization sends colonists to stars that are an average distance D away and sends them in spacecraft that travel at speed v. The time required for travel, t_{travel}, is then $t_{travel} = D/v$. Suppose that the colonists build up their colony for a time t_{col}, at which point they send out their own set of colonists to other star systems (with the same average distance and same spacecraft speed). Then the speed at which the civilization expands outward from the home star, v_{col} (for the speed of colonization), is $v_{col} = D/(t_{travel} + t_{col})$. However, this is true only if the colonization is always directed straight outward from the home star. In reality, the colonists will sometimes go to uncolonized star systems in other directions, so we will introduce a constant k that accounts for this zigzag motion. Our equation for the speed at which the civilization expands outward from the home star is:

$$v = k\frac{D}{(t_{travel} + t_{col})} = k\frac{D}{\left(\frac{D}{v} + t_{col}\right)}$$

For the purposes of this problem, assume that $k = \frac{1}{2}$ and that the average distance between star systems is $D = 5$ light-years.

a. How fast (as a fraction of the speed of light) does the civilization expand if its spacecraft travel at $0.1c$ and each colony builds itself up for 150 years before sending out the next wave of colonists? How long would it take the colonists to expand a distance of 100,000 light-years from their home star at this rate?

b. Repeat part (a), but assume that the spacecraft travel at $0.01c$ and that each colony builds itself up for 1,000 years before sending out more colonists.

c. Repeat part (a), but assume that the spacecraft travel at $0.25c$ and that each colony builds itself up for 50 years before sending out more colonists.

Discussion Questions

57. *Seeding the Galaxy.* If interstellar travel is forever impractical, are there other ways an advanced civilization might spread its culture? Clearly, communication is possible, although the speed of light makes conversations between star systems maddeningly tedious. Could a society send the information required to assemble members of its species (its "DNA," for instance) and therefore spread through the galaxy at the speed of light? Can you imagine other ways of spreading a culture without starships? Explain.

58. *Sociology of Interstellar Travel.* Suppose we somehow built a spaceship capable of relativistic travel and volunteers were being recruited for a journey to a star 15 light-years away. Would you volunteer to go? Do you think others would volunteer? In light of the effects of time dilation, discuss the benefits and drawbacks of such a trip.

59. *The Turning Point.* Discuss the idea that our generation has acquired a greater responsibility to future humans than any previous generation. Do you agree with this assessment? If so, how should we deal with this responsibility? Defend your opinions.

WEB PROJECTS

60. *Starship Design.* Find more details about a proposal for starship propulsion or design. How would the starship work? What new technologies would be needed, and what existing technologies could be applied? Summarize your findings in a one- to two-page report.

61. *Advanced Spacecraft Technologies.* NASA supports many efforts to incorporate new technologies into spaceships. Although few of them reach the level of being suitable for interstellar colonization, most are innovative and fascinating. Learn about one such NASA project, and write a short summary of your findings.

62. *Solutions to the Fermi Paradox.* Learn more about someone's pet solution to the Fermi paradox. Write a short summary of that solution, and discuss how it fits with the ideas we have discussed in this chapter.

Epilogue
Contact—Implications of the Search and Discovery

The known is finite, the unknown is infinite; intellectually we stand on an islet in the midst of an illimitable ocean of inexplicability. Our business in every generation is to reclaim a little more land.

—Thomas H. Huxley

We've covered a lot of ground in this book. We've studied life on Earth and learned about the prospects for life elsewhere in our solar system and on planets around other stars. Our scientific discussion mirrors a broader cultural phenomenon—an intense interest in ideas relating to extraterrestrial life. This interest shows up in many aspects of contemporary culture, including TV shows and movies, supermarket tabloids, and a notable public interest in the exploration of space and the search for life beyond Earth.

In this Epilogue, we'll briefly consider a few possible reasons why humans are so deeply interested in the question of whether life is present elsewhere. We'll begin with a short review of why the idea of life beyond Earth seems scientifically reasonable and then explore how the search for it helps us revisit age-old questions about the nature of humanity. We'll also discuss the philosophical and cultural consequences of finding life elsewhere—life of any kind, whether microbial or intelligent or somewhere in between.

IS THERE LIFE ELSEWHERE?

We know of only one example of life existing in the universe—life here on Earth. In this book we have discussed many reasons why it seems reasonable to think that life might exist elsewhere, but we still don't know for sure that it does. Because of this uncertainty, all we can do today is discuss the issues that could determine whether extraterrestrial life is likely or unlikely to exist and how we might search for evidence of it.

Despite these limitations, we are at a unique point in the long debate over the possibility of extraterrestrial life. We have the technological capability to explore Mars and much of the rest of our solar system, and we are rapidly developing technology that should in principle allow us to seek evidence for life on planets around other stars. After millennia of speculation about life beyond Earth, we are finally reaching the point at which we could really discover it, if it exists. This remarkable prospect calls us to discuss the philosophical and cultural consequences of finding life elsewhere. But first let's summarize the key issues in our discussion about the search for life in the universe.

• Why life seems likely

Although we don't yet know whether life exists any place besides Earth, we've seen that current science offers reasons to think it should. We can examine the nature of life on Earth—its building blocks and how it originated—and understand the environmental conditions in which life can exist. We can look at the other planets and moons in our solar system and determine whether the conditions conducive to biology exist there. And we can look for planets around other stars, learn how abundant they are and how they form, and determine whether some of them might be Earth-like planets that could be capable of supporting living things.

When we do these things, we find three key pieces of evidence that point to the idea that life should be common in the universe. We'll list them and then discuss each briefly in turn:

1. The chemical elements that make up life on Earth are common throughout the universe, and complex, carbon-bearing molecules important to life on Earth appear to form easily and naturally under conditions that should be common on many worlds.
2. Life on Earth thrives under a wide range of environmental conditions that we once considered too extreme to be capable of supporting life, and many of these types of environments are likely to be found on other planets in our own solar system and beyond.
3. It seems that life appeared on Earth quite early in its history, making it seem plausible that life is "easy."

Let's consider the first item in our list. The elements from which life is constructed (carbon, hydrogen, oxygen, nitrogen, and almost two dozen other elements) are found nearly everywhere in the universe. Hydrogen is the most abundant element by far, but the other elements used for life exist in at least modest quantities, because they have been created by nuclear fusion in earlier generations of stars and during supernova explosions of those stars. The manufactured elements were ejected by the supernova explosions into interstellar space, where they were incorporated into the gas and dust clouds out of which later stars and planets formed. Moreover, experiments in laboratories on Earth and spectroscopic observations of distant objects show that the elements used by life combine readily into the molecular building blocks of life. We have found molecules as complex as amino acids in meteorites, and we have detected numerous organic molecules even in interstellar space. We therefore expect that organic molecules should form abundantly and naturally under a range of conditions that includes those that were present on the early Earth and are likely to be present on other geologically active worlds. A wide availability of organic molecules would make it likely that the starting points for an origin of life exist on many worlds.

If the starting points for life are commonplace, the next question concerns whether available environments can allow life to arise and thrive. This is the issue addressed by the second point on our list. We have learned that life on Earth is incredibly diverse and that some organisms can thrive in a wide variety of environments that seem "extreme" to humans. For example, we have found life on Earth in the hot water near deep-sea vents, in the dry and frigid deserts of Antarctica, and inside rocks deep underground. The diversity of environments in which we find terrestrial life suggests that life could survive on any world that meets relatively simple environmental requirements—the presence of liquid water (or possibly another liquid), access to the requisite elements and molecules, and an energy source to drive metabolism. These conditions are likely to be met on any geologically active world with a rocky interior and liquid water, whether a planet or a moon. Such worlds would have the necessary elements and the potential for energy to be available via water/rock chemical reactions.

The first two points on our list tell us that the starting points for life are widely available and that life, once started, can survive in a wide range of environments. The third point tells us that getting life to start may not be difficult. Even with the present uncertainties in interpreting the geological record, life must have originated on a time scale that is very short compared to the age of Earth and to the expected ages of other habitable planets. Thus, unless Earth was somehow atypical, it seems that life might be the natural, straightforward consequence of the types of chemical reactions that can occur in planetary environments. When we put all three key pieces of evidence together, it seems reasonable to imagine life existing in at least a few other places in our own solar system and on many similar worlds throughout the universe.

• Prospects for finding life in our solar system

If life is indeed as common and wide-ranging as the evidence suggests it could be, then the first place to look for life is within our own solar system. Here in the Sun's neighborhood, Mars and Europa seem the most likely places besides Earth to harbor life, but we've also found a few other candidates.

Mars shows evidence for liquid water having been present at its surface early in its history and within its crust throughout its history. The martian atmosphere contains several of the key elements of life—notably carbon (in the form of gaseous carbon dioxide), hydrogen and oxygen (in the form of water), and nitrogen. The other necessary elements are found in surface and near-surface rocks. Energy to drive metabolism could come from chemical reactions between the water and the rocks, for example, allowing the possibility of organisms much like those that live within rocks on Earth.

Jupiter's moon Europa may have large amounts of liquid water. Although we are not yet absolutely certain, it seems likely that Europa is even more of a "water world" than Earth, with a global, 100-kilometer-thick ocean lying beneath an ice surface. We also expect that Europa has heavier elements in its rocky interior, so all the elements needed for life should be abundant. Energy for life could be supplied by chemical reactions between the water and the underlying rock, or by chemical compounds created by the impact of high-energy particles onto surface ice (driven by Jupiter's magnetic field). It seems plausible that at least microbial life could exist on Europa if its postulated ocean proves to be real.

Other places in our solar system could also potentially support life. Two other moons of Jupiter, Ganymede and Callisto, show evidence of subsurface liquid oceans similar to Europa's but at greater depth beneath their surfaces. Saturn's moon Titan apparently has liquid methane at least some of the time, and may have liquid water underground in the form of a cold, ammonia/water mixture. Saturn's moon Enceladus also seems likely to have similar liquids in its interior, and Neptune's moon Triton might have the same. These cold worlds are less likely candidates for life because chemical reactions ought to run much more slowly at cold temperatures, but we cannot rule them out.

• Prospects for finding life among the stars

Efforts that will allow us to search for life beyond our own solar system are already under way. We have already discovered almost 200 planets orbiting other stars, and while most of these are closer in size to Jupiter than to our relatively small Earth, we are rapidly developing the technology to detect smaller worlds. Although the characteristics of many of the planets discovered to date have been somewhat surprising in light of our earlier ideas about solar system formation, they do not appear to change the basic theory of planet formation and they validate the idea that planetary systems are common.

We should get more definitive answers about the nature of other planetary systems with new, space-borne telescopes planned for the next couple of decades. The *Kepler* mission should tell us whether planets the size of Earth are common. Later missions such as the *Terrestrial Planet Finder* should help us determine whether Earth-*size* planets are Earth-*like.* With these data, we should be able to answer the question of whether Earth-like planets are rare or common.

If Earth-like planets are indeed common, microbial life could be widespread throughout the galaxy. The possible widespread existence of at least microbial life begs the question of the potential for more complex or intelligent life. Our one example of intelligence here on Earth does not allow us to extrapolate or even hazard a guess as to whether intelligence should be widespread or rare; the only way to find out whether other civilizations exist is to search for them. Searches for radio or light signals from possible extraterrestrial civilizations have been going on for almost five decades. Although these SETI efforts have not yet met with success, only a small sample of the possible homes for civilization has been searched so far. As with the question of extraterrestrial life in general, the question of whether intelligent extraterrestrial life exists remains open.

• Extraterrestrial life and the human condition

We have reviewed the key scientific issues pertaining to life elsewhere. Now we turn our attention to issues that we have generally neglected to this point, including philosophical and societal issues that touch on why the search for life beyond Earth is of such broad intellectual interest to scientists and the public alike.

Why are so many people so interested in the question of whether life exists on other worlds, and what would it mean to find evidence — or a lack of evidence— for such life? We'll discuss a few possible answers to these questions, but you should recognize that the ideas we discuss may not resonate the same way with everyone. Think about what issues might be driving your own interests, which led you to take this course, and especially about how and why these personal issues might differ from the ideas we discuss.

• Our changing perspective on the world

Many people are fascinated by the question of extraterrestrial life because knowing if such life exists would likely have a profound influence on our perspective about our place in the universe. Much of what we have learned about the world around us over the past several thousand years has changed the way we interact with it. For example, a mere 10,000 years ago—a blink of an eye in the history of our planet—humans were primarily a hunter-gatherer society, living out their lives with little knowledge about what was beyond the next mountain or valley. Today, people in nearly every corner of the world recognize our species as part of an enormous cosmos. We have learned that what we do in one place on our planet affects the environment, people, and societies in all other places around the world. In addition, we now see Earth as just one of many worlds in our solar system; our solar system and the Sun as one of more than 100 billion star systems in our galaxy; and our galaxy as just one of billions of similar galaxies in the observable universe. This expansion of our world (or contraction, depending on how you perceive it!) has brought with it a need to rethink our views both of ourselves and of our relationship to the rest of the planet and universe.

Perhaps the first major transition toward a modern scientific view of ourselves (and maybe the most significant one) began nearly 500 years ago with the Copernican revolution. The work of Copernicus, Galileo, Kepler, and Newton allowed us to recognize that Earth is not at the center of the universe but instead orbits the Sun. This displacement of Earth from the center of the universe had profound philosophical and psychological effects. Earth, and by extension humanity, could no longer be viewed as the center of everything, and the world could no longer be viewed as obeying only the laws of providence rather than the laws of physics. This shift in thinking had little impact on people's day-to-day lives—we would be hard-pressed even today to think of a way in which whether Earth goes around the Sun or vice versa matters to our daily activities. However, it made a fundamental difference to our worldview. As a result, the idea initially met widespread resistance in Western society and was not fully accepted for at least a couple of hundred years following Copernicus.

A second major shift occurred in the mid-1800s with the recognition by Charles Darwin and Alfred Russel Wallace of the processes that drive the evolution of species. The basic idea of evolution was not new. By the time of Darwin, scientists already recognized that Earth was old, that fossils represented organisms that had lived in prior times, that older fossils were different from younger ones, and that both were different from current living organisms. These discoveries had already convinced many scientists that the nature of living organisms had changed over time. What Darwin and Wallace discovered was a way to explain why and how species change: through competition for survival and the process we refer to as natural selection. We now recognize that all the species present on Earth represent the product of almost 4 billion years of evolution, traceable back to the earliest history of life on Earth.

Like the Copernican revolution, recognizing the nature of evolution sparked a change in our view of ourselves and of our world. Copernicus and others showed that we are not located at the physical center of the universe, and Darwin and Wallace showed that we are not at the biological center either. Rather, we represent just one of millions of distinct species on Earth, and while we may be "dominant" over many species found in the plant and animal kingdoms, these kingdoms represent just a small part of the great diversity of life on our planet. This shift in perspective has been extremely profound, and may well explain why many people still do not accept the idea of evolution by natural selection. The intensity of the public controversy only underscores the tremendous impact that the theory of evolution has had—and continues to have—on Western society.

A third shift in perspective is taking place today with the discovery of other solar systems and the modern understanding of the potential for life to be widespread in the universe. Although many astronomers had long suspected that planets ought to be common around other stars, the first definitive evidence of extrasolar planets is slightly more than a decade old. If we ultimately confirm our guess that other worlds like ours exist or are common, we will no longer be able to regard Earth as special in any essential way. Throughout the history of scientific thought, new discoveries have increasingly displaced us from the center of the universe, both physically and metaphorically. Finding that life is common throughout the universe would lead to yet another major shift in perspective. In the rest of this Epilogue, we'll briefly explore why.

• The impact of extraterrestrial life on human perspective

People have long speculated about life beyond Earth, and many people—including many scientists—have at times been convinced that life exists on the Moon, on Mars, or on other worlds. Why, then, would the actual discovery of extraterrestrial life have a major impact on human perspective? The answer lies in the difference between guessing and knowing. As long as there is uncertainty about the existence of life on other worlds, people are free to hold a wide range of opinions. People living before the time of Copernicus could continue to believe in an Earth-centered world, but it was quite hard to continue to do so after Galileo, Kepler, and Newton offered convincing proof to the contrary. An actual discovery of life beyond Earth would force us, both as individuals and as a society, to reconsider the place of our planet and our species in the cosmos.

In contemplating the significance of finding life elsewhere, let's begin by considering what would happen if we found microbial life on Mars. The first question we would probably ask is whether the life was genetically related to terrestrial life (suggesting that it had migrated between planets on meteorites) or instead represented an independent origin of life on Mars. We could answer this question by determining the structure of the molecules that make up the martian life. Does it use DNA and RNA molecules similar to those used by terrestrial life? Does it use the same amino acids or the same proteins to carry out enzymatic reactions? Do the molecules that participate in life have the same "handedness" to them? It seems unlikely that there would be only one solution to the problems of containing and passing on the genetic information required for life to reproduce, of catalyzing the chemical reactions that make up life, and of storing and using energy in metabolism. We would therefore expect life that had an origin independent from life on Earth to have a different chemical structure.

While a discovery of microbial life on Mars that was genetically related to terrestrial life would be exciting, it probably would not have as great an impact as a discovery of life that showed evidence of an independent origin. A discovery of life from an independent origin would provide clear proof that the origin of life was not a unique event. With proof that life had originated twice in just our own solar system, we would have every reason to think it has originated many times among the many worlds around other stars and thus that life is widespread in the universe. A discovery of alien microbial life would also help us better understand life in general. We would learn more about the conditions under which life can arise and persist, as well as the conditions under which it can evolve into more complex forms. And we could begin to engage in comparative biology, in which the study of alien life would help us better understand the life on our own world.

While many people who have studied the issue of life in the universe expect that we will find microbial life to be widespread, we cannot truly envision the consequences of an actual discovery unless and until it occurs. Moreover, the history of science tells us to be prepared for surprises. For example, the real discovery of extrasolar planets proved surprising, despite their predicted existence: We learned that other solar systems can be very different from our own. If and when we do find life elsewhere, we should expect to be equally surprised about its nature.

• Extraterrestrial intelligence and the nature of humanity

The discovery of extraterrestrial life of any kind would have a profound effect on our perception of our place in the universe, but it is the potential discovery of extraterrestrial intelligence that generates the greatest public interest. What would finding intelligent life elsewhere mean to us? We discussed some of the philosophical implications in Section 13.3; here we focus on how it might affect our lives more directly.

It is difficult to predict how we would respond to such a discovery, and it is likely that people would exhibit a wide range of responses. At one

Figure E.1

A 1906 drawing used to illustrate H. G. Wells's book *The War of the Worlds,* in which Earth is invaded by hostile martians.

extreme, some people might look to extraterrestrial intelligence as a source of help for solving our problems. This view has been portrayed in many books and movies, including *Contact* and *2001: A Space Odyssey.* Interestingly, many of these stories suggest, first, that we are not able to solve our own problems without outside help and, second, that extraterrestrials will want to help us! At the other extreme, some people imagine that extraterrestrials would come to Earth and destroy our civilization, either deliberately or by accident. The martians of H. G. Wells's *War of the Worlds* came here bent on our destruction (Figure E.1). In Douglas Adams's *Hitchhiker's Guide to the Galaxy,* aliens destroy Earth as an accidental consequence of their need to build a new "interstellar bypass."

Neither extreme seems especially likely. As we've discussed, intelligent aliens capable of visiting our planet would almost certainly be far more technologically advanced than we are. It's not clear that they would have any interest in us at all (making the first extreme unlikely), and they'd seemingly have little to gain by destroying us (making the second extreme unlikely). If and when intelligent life is discovered, the reality may lie somewhere in between these two extremes—or it might be completely different.

• Significance of the search itself

A discovery of even microbial extraterrestrial life would undoubtedly bring important practical benefits. For example, studying it will help us understand what characteristics of terrestrial life are unique to Earth and what characteristics apply generally to life anywhere. If we find intelligent life elsewhere, we would learn much more about the nature of intelligence and might be exposed to cultures and societies extremely different from those of humans. If we could communicate with more advanced beings, we could possibly learn the secrets of the nature of the universe, the nature of consciousness and the mind, and technological marvels that could dramatically change our life here on Earth.

MOVIE MADNESS: *E.T.*

When cinema aliens come to Earth, it's usually wise policy to head for the storm cellar. But when wrinkly little E.T. is accidentally left behind by his planetary pals, it turns out to be good news—at least for a few suburban kids.

E.T. is the quintessential alien film—a movie that long held the record for being the most successful picture of all time. There's good reason. Appealing little E.T., who has the stature and gait of a penguin, is every kid's dream. After all, he's a friend (and really useful because with his super powers he can help you outsmart adults), and he's exclusively *your* friend. Other kids will stop kicking verbal sand in your face when you show up at the playground with a guy from another galaxy, even if he has a face like a polished redwood burl.

Aside from this childlike wish fulfillment, *E.T.* encapsulates everything we hope or think is true about intelligent extraterrestrials. To begin with, E.T. is benign and nonthreatening. Unlike evil aliens, who look like reptiles or insects, E.T. resembles a baby, with his short nose, big eyes, and wrinkled skin. He's only 2 feet high, weighs 35 pounds, and has a 25-watt fingertip. He clearly comes from a kinder, gentler planet, since his only interest in Earth's biota is its plants. (No insects had to die for this film.) Scientists who have given any thought to the true nature of advanced aliens (and those that can come to Earth are clearly advanced) have often equated technological prowess with peaceful behavior. The aliens will be friendly. This doesn't square all that well with our experience on Earth, but one can hope.

In addition, E.T. not only looks a lot like us; he acts like us, functions like us (he can get blotto on beer), and is interested in our personal lives. None of this is likely to be true, of course. He's also well adapted to terrestrial conditions, waddling around without a space suit and, indeed, without any clothes at all. In addition, nefarious federal agents wearing jackets, ties, and drab personalities are busily trying to keep the cuddly creature's visit under wraps, something that many people believe is happening in real life.

All of this may be in keeping with the public's perception of what aliens would be like. But if the search for extraterrestrial intelligence succeeds and we eventually learn the true nature of extraterrestrials, it is far more likely that they will be of a construction and temperament that are far beyond our most fevered imaginings.

However, for many people excited by the scientific search for life in the universe, the search itself is much more than a means to an end. From this viewpoint, the search is just one more critical component in our exploration of the world around us. Other components in astronomy and space science include exploring the planets and moons in our solar system as a way to understand how planets work and exploring stars and galaxies as a way to determine the nature of our universe. Other components in biology include exploring the origin and evolution of life on Earth so that we can understand how we ourselves came to exist.

In all these cases, our exploration does not seem to be driven solely by the desire to find specific answers to the scientific questions we are asking, because in each case we end up asking more questions. Instead, we seem to be driven by our inherent curiosity, our desire to understand the world around us (Figure E.2). Sometimes our curiosity leads to discoveries with practical applications, while at other times it simply helps us understand how or why the world is as it is. Understanding the world around us means learning about the broader-scale environment in which humans exist. Understanding the occurrence of planets orbiting other stars helps us understand the significance of the occurrence of planets orbiting the Sun, including Earth. Understanding the occurrence of life elsewhere allows us to understand the significance of the occurrence of life on Earth. And understanding the potential for intelligent life beyond Earth brings with it an understanding of the meaning of the existence of intelligent life here on Earth. In essence, by learning about the universe around us, we are learning about ourselves and about what it means to be human.

Exploring the universe is but one activity among many that help us learn about ourselves, including searching for new forms of life in extreme environments on Earth, exploring Earth itself, exploring the nature of human consciousness and the human mind, and exploring the human condition through literature and the arts. Each of these activities helps us learn about what it means to be alive and to be human. Thus, the exploration of the universe and the search for extraterrestrial life can be seen as nothing less than an attempt to determine the nature of humanity and our place in the universe.

Figure E.2

Earth as viewed in space from the *Apollo 17* spacecraft. Seen from this perspective, we recognize the strong connections between the terrestrial ecosystem and the planet itself, and we recognize that life, indeed, is a planetary phenomenon. As we explore the universe, we will learn whether the formations of planets around stars and the occurrence of life on planets are rare or commonplace. We may then finally answer the question of whether we are alone.

EXERCISES AND PROBLEMS

DISCUSSION QUESTIONS

1. *Is There Life Elsewhere?* After considering all the evidence to date about the possibility of extraterrestrial life, do you believe it is likely that we'll find microbial life elsewhere? Do you believe it is likely that we'll find intelligent life elsewhere? Defend your opinions, using arguments based on the full range of scientific issues discussed in this book.
2. *Microbial or Intelligent?* Do you think the implications of discovering microbial life elsewhere would be any more or less profound than those of discovering extraterrestrial intelligence? Explain your reasoning.
3. *Extraterrestrial Life and Your Religion.* Would the discovery of extraterrestrial life have any important implications for your own personal religious beliefs? Would it affect the current "official" beliefs (if any) of your religion? Explain.
4. *Extraterrestrial Life and the Debate on Evolution.* Do you think the discovery of extraterrestrial life would affect the current status of the debate over science and religion in the United States? For example, would it alter the controversy surrounding the teaching of evolution in public schools? Why or why not?
5. *Aliens and Everyday Life.* While the discovery of extraterrestrial life would surely be profound, do you think it would alter any aspect of our everyday lives? If so, how? If not, why not?
6. *The Search Itself.* Suppose we spend a fair amount of money and effort searching for life over the next few decades and ultimately find no evidence for life beyond Earth. Will the search have been a waste, a success, or something in between? Defend your opinion.

Glossary

absolute zero The coldest possible temperature, which is $0\ K = -273.15°C$.

absorption (of light) The process by which matter absorbs radiative energy.

absorption-line spectrum A spectrum that contains absorption lines.

accelerating universe The possible fate of our universe in which a repulsive force (see *cosmological constant*) causes the expansion of the universe to accelerate with time. Its galaxies will recede from one another increasingly faster, and it will become cold and dark more quickly than a coasting universe.

acceleration The rate at which an object's velocity changes. Its standard units are m/s^2.

acceleration of gravity The acceleration of a falling object. On Earth, the acceleration of gravity, designated by g, is $9.8\ m/s^2$.

accretion The process by which small objects gather together to make larger objects.

adaptive optics A technique in which telescope mirrors flex rapidly to compensate for the bending of starlight caused by atmospheric turbulence.

aerobic organisms Organisms that require molecular oxygen to survive.

albedo Describes the fraction of sunlight reflected by a surface; albedo = 0 means no reflection at all (a perfectly black surface); albedo = 1 means all light is reflected (a perfectly white surface).

Amazonian era The present era on Mars, which began about 1.0 billion years ago.

amino acids The building blocks of proteins. (More technically, an amino acid is a molecule containing both an *amino group* [NH or NH_2] and a *carboxyl group* [COOH].)

anaerobic organisms Organisms that do not require (and may even be poisoned by) molecular oxygen.

Andromeda Galaxy (M13; the Great Galaxy in Andromeda) The nearest large spiral galaxy to the Milky Way.

angular momentum Momentum attributable to rotation or revolution. The angular momentum of an object moving in a circle of radius r is the product $m \times v \times r$.

angular resolution (of a telescope) The smallest angular separation that two pointlike objects can have and still be seen as distinct points of light (rather than as a single point of light).

angular size (or **angular distance**) A measure of the angle formed by extending imaginary lines outward from our eyes to span an object (or between two objects).

annihilation See *matter–antimatter annihilation.*

Antarctic Circle The circle on Earth with latitude 66.5°S.

anthropic principle An idea that comes in a variety of forms centering around the fact that our existence is possible only because of a great number of aspects of our universe that are "fine-tuned" for life.

antielectron See *positron.*

antimatter Refers to any particle with the same mass as a particle of ordinary matter but whose other basic properties, such as electrical charge, are precisely opposite.

aphelion The point at which an object orbiting the Sun is farthest from the Sun.

apogee The point at which an object orbiting Earth is farthest from Earth.

apparent brightness The amount of light reaching us *per unit area* from a luminous object; often measured in units of $watts/m^2$.

apparent magnitude A measure of the apparent brightness of an object in the sky, based on the ancient system developed by Hipparchus.

apparent retrograde motion Refers to the apparent motion of a planet, as viewed from Earth, during the period of a few weeks or months when it moves westward relative to the stars in our sky.

Archaea One of the three domains of life; the others are Eukarya and Bacteria.

arcminutes (or **minutes of arc**) One arcminute is $\frac{1}{60}$ of 1°.

arcseconds (or **seconds of arc**) One arcsecond is $\frac{1}{60}$ of an arcminute, or $\frac{1}{3{,}600}$ of 1°.

Arctic Circle The circle on Earth with latitude 66.5°N.

Aristotelians Ancient Greek followers of Aristotle, who held that there could be only one Earth and that the heavens were a realm distinct from Earth.

asteroid A relatively small and rocky object that orbits a star; asteroids are sometimes called *minor planets* because they are similar to planets but smaller.

asteroid belt The region of our solar system between the orbits of Mars and Jupiter in which asteroids are heavily concentrated.

astrobiology The study of life on Earth and beyond; emphasizes research into questions of the origin of life, the conditions under which life can survive, and the search for life beyond Earth.

astrometric technique The detection of extrasolar planets through the side-to-side motion of a star caused by gravitational tugs from the planet.

astronomical unit (AU) The average distance (semimajor axis) of Earth from the Sun, which is about 150 million kilometers.

atmosphere A layer of gas that surrounds a planet or moon, usually very thin compared to the size of the object.

atmospheric pressure The surface pressure resulting from the overlying weight of an atmosphere.

atomic mass number The combined number of protons and neutrons in an atom.

atomic number The number of protons in an atom.

atomists Ancient Greek scholars who held that the universe is made from an infinite number of indivisible atoms.

atoms Consist of a nucleus made from protons and neutrons surrounded by a cloud of electrons.

ATP (adenosine triphosphate) The molecule that stores and releases energy for nearly all cellular processes among life on Earth.

aurora Dancing lights in the sky caused by charged particles entering our atmosphere; called the *aurora borealis* in the Northern Hemisphere and the *aurora australis* in the Southern Hemisphere.

autotroph An organism that gets its carbon directly from the atmosphere in the form of carbon dioxide.

autumnal equinox See *fall equinox.*

axis tilt (of a planet in our solar system) The amount by which a planet's axis is tilted with respect to a line perpendicular to the ecliptic plane.

Bacteria One of the three domains of life; the others are Eukarya and Archaea.

band (of sensitivity) The set of frequencies that a particular radio receiver can pick up.

bandwidth (of a transmitted signal) The range of frequencies over which a communication signal is transmitted.

The standard unit of pressure, approximately equal to Earth's atmospheric pressure at sea level.

basalt A type of dark, high-density volcanic rock that is rich in iron and magnesium-based silicate minerals; it forms a runny (easily flowing) lava when molten.

Big Bang The name given to the event thought to mark the birth of the universe.

Big Bang theory The scientific theory of the universe's earliest moments, stating that all the matter in our observable universe came into being at a single moment in time as an extremely hot, dense mixture of subatomic particles and radiation.

Big Crunch If gravity ever reverses universal expansion, the universe will someday begin to collapse and presumably end in a Big Crunch.

binary star system A star system that contains two stars.

biochemistry The chemistry of life.

biosphere Refers to the "layer" of life on Earth.

blackbody radiation See *thermal radiation.*

black smokers Structures around seafloor volcanic vents that support a wide variety of life.

blueshift A Doppler shift in which spectral features are shifted to shorter wavelengths, caused when an object is moving toward the observer.

brown dwarf An object that forms much like a star but is too low in mass to sustain nuclear fusion in its core; brown dwarfs have masses much greater than Jupiter but always less than M_{Sun}.

Cambrian explosion The dramatic diversification of life on Earth that occurred between about 540 and 500 million years ago.

carbohydrates Molecules such as sugars and starches that provide energy to cells and make important cellular structures.

carbonate rock A carbon-rich rock, such as limestone, that forms underwater from chemical reactions between sediments and carbon dioxide. On Earth, most of the outgassed carbon dioxide currently resides in carbonate rocks.

carbon-based life Life that uses molecules containing carbon for its most critical functions. All life on Earth is carbon-based.

carbon dioxide (CO_2) cycle The process that cycles carbon dioxide between Earth's atmosphere and surface rocks.

catalysis The process of causing or accelerating a chemical reaction by involving a substance or molecule that is not permanently changed by the reaction.

catalyst The unchanged substance or molecule involved in catalysis.

celestial sphere The imaginary sphere on which objects in the sky appear to reside when observed from Earth.

cell The basic structure of all life on Earth, in which the living matter inside is separated from the outside world.

Celsius (temperature scale) The temperature scale commonly used in daily activity internationally; defined so that, on Earth's surface, water freezes at 0°C and boils at 100°C.

center of mass (of orbiting objects) The point at which two or more orbiting objects would balance if they were somehow connected; it is the point around which the orbiting objects actually orbit.

charged particle belts Zones in which ions and electrons accumulate and encircle a planet.

chemical bond The linkage between atoms in a molecule.

chemical element See *element.*

chemical enrichment The process by which the abundance of heavy elements (heavier than helium) in the interstellar medium gradually increases over time as these elements are produced by stars and released into space.

chemoautotroph An organism that gets its carbon directly from the atmosphere and its energy from chemical reactions involving inorganic molecules.

chemoheterotroph An organism that gets both its energy and its carbon by consuming preexisting organic molecules; all animals are chemoheterotrophs.

chloroplasts Structures in plant cells that produce energy by photosynthesis.

civilization types A way of categorizing civilizations by whether they use resources of their planet, their star, or their galaxy. See also *galactic civilization, planetary civilization,* and *stellar civilization.*

clay Any of a variety of common silicate minerals with particular physical structures.

climate Describes the long-term average of weather.

cluster of galaxies A collection of a few dozen or more galaxies bound together by gravity; smaller collections of galaxies are simply called *groups.*

comet A relatively small, icy object that orbits a star.

compound (chemical) A substance made from molecules consisting of two or more atoms with different atomic numbers.

condensates Solid or liquid particles that condense from a cloud of gas.

condensation The formation of solid or liquid particles from a cloud of gas.

conduction (of energy) The process by which thermal energy is transferred by direct contact from warm material to cooler material.

conservation of angular momentum (law of) The principle that, in the absence of net torque (twisting force), the total angular momentum of a system remains constant.

conservation of energy (law of) The principle that energy (including mass-energy) can be neither created nor destroyed, but can change only from one form to another.

conservation of momentum (law of) The principle that, in the absence of net force, the total momentum of a system remains constant.

continental crust The thicker, lower-density crust that makes up Earth's continents. It is made when remelting of seafloor crust allows lower-density rock to separate and erupt to the surface. Continental crust ranges in age from extremely young to as old as about 4 billion years (or more).

continental drift The idea that the continents slowly move around on Earth, now known to be a result of plate tectonics.

continuously habitable zone The region around a star in which conditions could allow for surface habitability throughout the history of the star system.

convection The energy transport process in which warm material expands and rises, while cooler material contracts and falls.

convection cell A small individual region of convecting material.

convergent evolution The tendency of organisms of different evolutionary backgrounds to come to resemble one another because they occupy similar ecological niches.

Copernican revolution The dramatic change, initiated by Copernicus, that occurred when we learned that Earth is a planet orbiting the Sun rather than the center of the universe.

coral model (of colonization) A model of how a civilization might colonize the galaxy, based on growth much like that of coral in the sea.

core (of a planet) The dense central region of a planet that has undergone differentiation.

core (of a star) The central region of a star, in which nuclear fusion can occur.

cosmic microwave background The remnant radiation from the Big Bang, which we detect using radio telescopes sensitive to microwaves (which are short-wavelength radio waves).

cosmic rays Particles such as electrons, protons, and atomic nuclei that zip through interstellar space at close to the speed of light.

cosmos An alternate name for the universe.

crust (of a planet) The low-density surface layer of a planet that has undergone differentiation.

crystal A substance made from atoms arranged in precise geometrical patterns, such as in a mineral.

cultural evolution Changes that arise from the transmission of knowledge accumulated over generations.

cyanobacteria Photosynthetic bacteria thought to have been responsible for making most of the oxygen that gradually built up in Earth's atmosphere.

cycles per second Units of frequency for a wave; describes the number of peaks (or troughs) of a wave that pass by a given point each second. Equivalent to *hertz*.

dark energy Name sometimes given to energy that could be causing the expansion of the universe to accelerate.

dark matter Matter that we infer to exist from its gravitational effects but from which we have not detected any light; dark matter apparently dominates the total mass of the universe.

deuterium A form of hydrogen in which the nucleus contains a proton and a neutron, rather than only a proton (as is the case for most hydrogen nuclei).

differentiation The process in which gravity separates materials according to density, with high-density materials sinking and low-density materials rising.

disequilibrium (chemical) A state in which a mixture undergoing chemical reactions is not in equilibrium.

DNA (deoxyribonucleic acid) The molecule that represents the genetic material of life on Earth.

DNA bases The four DNA bases are adenine (A), cytosine (C), guanine (G), and thymine (T); they can be paired across the two DNA strands only so that A goes with T and C goes with G.

DNA replication The process of copying DNA molecules.

domain (of life) The highest level by which we currently classify life; the three domains are *Eukarya, Bacteria,* and *Archaea*.

Doppler effect (or **shift**) The effect that shifts the wavelengths of spectral features in objects that are moving toward or away from the observer.

Doppler technique The detection of extrasolar planets through the motion of a star toward and away from the observer caused by gravitational tugs from the planet.

Drake equation An equation that lays out the factors that play a role in determining the number of communicating civilizations in our galaxy.

dust (or **dust grains**) Tiny, solid flecks of material; in astronomy, we often discuss *interplanetary dust* (found within a star system) or *interstellar dust* (found among the stars in a galaxy).

Dyson sphere A hypothesized type of large, thin-walled sphere built to surround a star so that an advanced civilization could capture all the energy flowing out from the star; named after physicist Freeman Dyson, who proposed their possible existence.

Earth orbiters (spacecraft) Spacecraft designed to study Earth or the universe from Earth orbit.

eccentricity A measure of how much an ellipse deviates from a perfect circle; defined as the center-to-focus distance divided by the length of the semimajor axis.

ecliptic The Sun's apparent annual path among the constellations.

ecliptic plane The plane of Earth's orbit around the Sun.

ejecta (from an impact) Debris ejected by the blast of an impact.

electrical charge A fundamental property of matter that is described by its amount and as either positive or negative; more technically, it is a measure of how a particle responds to the electromagnetic force.

electromagnetic radiation Another name for light of all types, from radio waves through gamma rays.

electromagnetic spectrum The complete spectrum of light, including radio waves, infrared, visible light, ultraviolet light, X rays, and gamma rays.

electromagnetic wave A synonym for *light*, which consists of waves of electric and magnetic fields.

electron acceptor (in a redox reaction) The chemical (atom or molecule) that gains electrons in an overall chemical reaction.

electron donor (in a redox reaction) The chemical (atom or molecule) that gives up electrons in an overall chemical reaction.

electrons Fundamental particles with negative electric charge; the distribution of electrons in an atom gives the atom its size.

element (chemical) A substance made from individual atoms of a particular atomic number.

ellipse A type of oval that happens to be the shape of bound orbits. An ellipse can be drawn by moving a pencil along a string whose ends are tied to two tacks; the locations of the tacks are the foci (singular, *focus*) of the ellipse.

elliptical galaxies Galaxies that appear round in shape, often longer in one direction, like a football. They have no disks and contain little cool gas and dust compared to spiral galaxies, though they often contain extremely hot, ionized gas.

emission (of light) The process by which matter emits energy in the form of light.

emission line spectrum A spectrum that contains emission lines.

encephalization quotient (EQ) A rough measure of animal intelligence based on the ratio of an animal's brain size to its body mass.

endolith An organism that lives inside of rock.

endospore A special "resting" cell that allows some organisms to remain dormant for long periods of time.

energy Broadly speaking, energy is what makes matter move. The three basic types of energy are *kinetic, potential,* and *radiative*.

enzyme A protein that serves as a catalyst.

eons (geological) The largest divisions of time in Earth's geological history. The four eons are the Hadean, Archaean, Proterozoic, and Phanerozoic.

equilibrium (chemical) A state of balance between the reacting atoms and molecules and the product atoms and molecules in a mixture undergoing chemical reactions.

equinox See *fall equinox* and *spring equinox*.

eras (geological) The second-largest divisions of time in Earth's geological history, after eons. The Phanerozoic eon is subdivided into three eras: the *Paleozoic, Mesozoic,* and *Cenozoic*.

erosion The wearing down or building up of geological features by wind, water, ice, and other phenomena of planetary weather.

eruption The process of releasing hot lava onto the planet's surface.

escape velocity The speed necessary for an object to completely escape the gravity of a large body such as a moon, planet, or star.

Eukarya One of the three domains of life, and the one in which all plants and animals are found; the other domains are Bacteria and Archaea.

eukaryote A living organism that is a member of the domain Eukarya, and therefore is made from one or more eukaryotic cells.

eukaryotic cell A cell that contains a distinct nucleus that is separated from the rest of the cell by its own membrane.

evaporation The process by which atoms or molecules escape into the gas phase from the liquid phase.

evolution (biological) The gradual change in populations of living organisms that is responsible for transforming life on Earth from its primitive origins to the great diversity of life today.

evolutionary adaptation An inherited trait that enhances an organism's ability to survive and reproduce in a particular environment.

expansion (of universe) The idea that the space between galaxies or clusters of galaxies is growing with time.

extrasolar planet A planet orbiting a star other than our Sun.

extraterrestrial life Life that does *not* live on Earth.

emophile An organism that thrives under conditions that are extreme by human standards.

Fahrenheit (temperature scale) The temperature scale commonly used in daily activity in the United States; defined so that, on Earth's surface, water freezes at 32°F and boils at 212°F.

fall equinox (autumnal equinox) Refers both to the point in Virgo on the celestial sphere where the ecliptic crosses the celestial equator and to the moment in time when the Sun appears at that point each year (around September 21).

false color image An image displayed in colors that are *not* the true, visible-light colors of an object.

fault (geological) A place where rocks slip sideways relative to one another.

feedback processes Processes in which a small change in some property (such as temperature) leads to changes in other properties, which then either amplify or diminish the original small change.

fermions Particles, such as electrons, neutrons, and protons, that obey the exclusion principle.

Fermi's paradox The question posed by Enrico Fermi about extraterrestrial intelligence—"So where is everybody?"—that asks why we have not observed other civilizations even though simple arguments would suggest that some ought to have spread throughout the galaxy by now.

field An abstract concept used to describe how a particle would interact with a force. For example, the idea of a *gravitational field* describes how a particle would react to the local strength of gravity, and the idea of an *electromagnetic field* describes how a charged particle would respond to forces from other charged particles.

fission See *nuclear fission.*

flybys (spacecraft) Spacecraft that fly past a target object (such as a planet), usually just once, as opposed to entering a bound orbit of the object.

force Anything that can cause a change in momentum.

fossil Any relic of an organism that lived and died long ago.

fossil record (or **geological record**) The information about Earth's past that is recorded in fossils (*fossil record*) and rocks (*geological record*). Note that the terms are often used synonymously.

frequency Describes the rate at which peaks of a wave pass by a point; measured in units of 1/s, which are often called *cycles per second* or *hertz.*

frost line The boundary in the solar nebula beyond which ices could condense; only metals and rocks could condense within the frost line.

fundamental forces There are four known fundamental forces in nature: *gravity,* the *electromagnetic force,* the *strong force,* and the *weak force.*

fusion See *nuclear fusion.*

galactic civilization A civilization that employs the resources of its entire galaxy.

galaxy A huge collection of stars, anywhere from a few hundred million to more than a trillion, all bound together by gravity.

galaxy cluster See *cluster of galaxies.*

Galilean moons The four moons of Jupiter that were discovered by Galileo: Io, Europa, Ganymede, and Callisto.

gamma-ray burst A sudden burst of gamma rays from deep space; such bursts apparently come from distant galaxies, but their precise mechanism is unknown.

gamma rays Light with very short wavelengths (and hence high frequencies)—shorter than those of X rays.

gas phase The phase of matter in which atoms or molecules can move essentially independently of one another.

gas pressure Describes the force (per unit area) pushing on any object due to surrounding gas. See also *pressure.*

gene The basic functional unit of an organism's heredity: a single gene consists of a sequence of DNA bases (or RNA bases, in some viruses) that provide the instructions for a single cell function (such as for building a protein).

general theory of relativity Einstein's generalization of his special theory of relativity that makes the general theory also apply when we consider effects of gravity or acceleration.

genetic analysis The analysis of an organism's genes or genome.

genetic code The "language" that living cells use to read the instructions chemically encoded in DNA.

genetic engineering Making deliberate changes to an organism's genome.

genome The complete sequence of DNA bases in an organism, encompassing all of the organism's genes.

genus The next most precise level of classification after *species;* it is a "generic" category to which multiple species may belong.

geocentric model Any of the ancient Greek models of the universe that had Earth at the center of a celestial sphere.

geocentric universe (the ancient belief in) The idea that Earth is the center of the entire universe.

geological activity Processes that change a planet's surface long after its formation, such as volcanism, tectonics, and erosion.

geological processes The four basic geological processes are *impact cratering, volcanism, tectonics,* and *erosion.*

geological record The information about Earth's past that is recorded in both fossils and rocks. See also *fossil record.*

geological time scale The time scale used by scientists to describe major eras in Earth's past. It is divided into four *eons* (the Hadean, Archaean, Proterozoic, and Phanerozoic). The latter (Phanerozoic) eon is subdivided into three *eras* (the Paleozoic, Mesozoic, and Cenozoic), which in turn are subdivided into several *periods.* (The periods are further subdivided into *epochs* and *ages.*)

geology The study of surface features (on a moon, planet, or asteroid) and the processes that create them.

giant impact A collision between a forming planet and a very large planetesimal, such as is thought to have formed our Moon.

giants (among stars) Stars that are near the ends of their lives and that have expanded in radius to extremely large sizes.

global average temperature The average surface temperature of a planet.

global warming An expected increase in Earth's global average temperature caused by human input of carbon dioxide and other greenhouse gases into the atmosphere.

globular cluster A spherically shaped cluster of up to a million or more stars; globular clusters are found primarily in the halos of galaxies and contain only extremely old stars.

granite A light-colored and low-density igneous rock common in mountain ranges on Earth; it gets its name from its grainy appearance and it is composed largely of quartz and feldspar minerals.

gravitation (law of) See *universal law of gravitation.*

gravitational constant The experimentally measured constant G that appears in the law of universal gravitation;

$$G = 6.67 \times 10^{-11} \frac{m^3}{kg \times s^2}$$

gravitational encounter Occurs when two (or more) objects pass near enough that each can feel the effects of the other's gravity and can therefore exchange energy.

gravitational lensing The magnification or distortion (into arcs, rings, or multiple images) of an image caused by light bending through a gravitational field, as predicted by Einstein's general theory of relativity.

gravity One of the four fundamental forces; it is the force that dominates on large scales.

Great Red Spot A large, high-pressure storm on Jupiter.

greenhouse effect The process by which greenhouse gases in an atmosphere make a planet's surface temperature warmer than it would be in the absence of an atmosphere.

greenhouse gases Gases, such as carbon dioxide, water vapor, and methane, that are particularly good absorbers of infrared light but are transparent to visible light.

habitable world A world with environmental conditions under which life could *potentially* arise or survive.

habitable zone The region around a star in which planets could potentially have surface temperatures at which liquid water could exist.

Hadean eon The earliest eon in Earth's history, corresponding to times before about 4.0 billion years ago.

half-life The time it takes for half of the nuclei in a given quantity of a radioactive substance to decay.

halo (of a galaxy) The spherical region surrounding the disk of a spiral galaxy.

handedness The property of some molecules, such as amino acids, that allows them to come in two distinct forms that are mirror images of one another.

heavy bombardment The period in the first few hundred million years after the solar system formed during which the tail end of planetary accretion created most of the craters found on ancient planetary surfaces.

heavy elements In astronomy, *heavy elements* generally refers to all elements *except* hydrogen and helium.

heredity The characteristics of an organism passed to it by its parent(s), which it can pass on to its offspring. The term can also apply to the transmission of these characteristics from one generation to the next.

hertz (Hz) The standard unit of frequency for light waves; equivalent to units of 1/s.

Hertzsprung–Russell (H–R) diagram A graph plotting individual stars as points, with stellar luminosity on the vertical axis and spectral type (or surface temperature) on the horizontal axis.

Hesperian era The middle history of Mars, dating from about 3.8 to 1.0 billion years ago.

heterotroph An organism that gets its carbon by consuming preexisting organic molecules.

hot spot (geological) A place within a plate of the lithosphere where a localized plume of hot mantle material rises.

Hubble's law A mathematical expression of the idea that more distant galaxies move away from us faster.

hydrogen compounds Compounds that contain hydrogen and were common in the solar nebula, such as water (H_2O), ammonia (NH_3), and methane (CH_4).

hydrosphere Refers to the "layer" of water on Earth consisting of oceans, lakes, rivers, ice caps, and other liquid water and ice.

hyperspace Any space with more than three dimensions.

hyperthermophile An organism that thrives under conditions of extremely high temperature compared to what most organisms can tolerate.

hypothesis A tentative model proposed to explain some set of observed facts that has not yet been rigorously tested and confirmed.

ice ages Periods of global cooling during which polar caps, glaciers, and snow cover extend closer to the equator.

ices (in solar system theory) Materials that are solid only at low temperatures, such as the hydrogen compounds water, ammonia, and methane.

igneous rock Rock made when molten rock cools and solidifies.

image A picture of an object made by focusing light.

impact The collision of a small body (such as an asteroid or comet) with a larger object (such as a planet or moon).

impact crater A bowl-shaped depression left by the impact of an object that strikes a planetary surface (as opposed to burning up in the atmosphere).

impactor The object responsible for an impact.

impact sterilization The process of sterilizing a planet as a result of a large impact.

infrared light Light with wavelengths that fall in the portion of the electromagnetic spectrum between radio waves and visible light.

inner solar system Generally considered to encompass the region of our solar system out to about the orbit of Mars.

inorganic Not pertaining to life or the chemistry of carbon molecules.

intensity (of light) A measure of the amount of energy coming from light of a specific wavelength in the spectrum of an object.

interferometry A telescopic technique in which two or more telescopes are used in tandem to produce much better angular resolution than the telescopes could achieve individually.

interstellar cloud A cloud of gas and dust among the stars.

interstellar ramjet A hypothesized type of spaceship that uses a giant scoop to sweep up interstellar gas for use in a nuclear fusion engine.

inverse square law Any quantity that decreases with the square of the distance between two objects is said to follow an inverse square law.

ion engine (rocket) A rocket engine that works by accelerating charged particles and expelling them out its back.

ionization The process of stripping an electron from an atom.

ionization nebula A colorful, wispy cloud of gas that glows because neighboring hot stars irradiate it with ultraviolet photons that can ionize hydrogen atoms.

ionosphere A portion of the thermosphere in which ions are particularly common (due to ionization by X rays from the Sun).

ions Atoms with a positive or negative electrical charge.

isotopes Each different isotope of an element has the *same* number of protons but a *different* number of neutrons.

joule The international unit of energy, equivalent to about $\frac{1}{4,000}$ of a Calorie.

jovian moons The moons of jovian planets.

jovian planets Giant gaseous planets similar in overall composition to Jupiter.

Julian calendar The calendar introduced in 46 B.C. by Julius Caesar and used until it was replaced by the Gregorian calendar.

Kelvin (temperature scale) The most commonly used temperature scale in science, defined such that absolute zero is 0 K and water freezes at 273.15 K.

Kepler's first law States that the orbit of each planet about the Sun is an ellipse with the Sun at one focus.

Kepler's laws of planetary motion Three laws discovered by Kepler that describe the motion of the planets around the Sun.

Kepler's second law States that, as a planet moves around its orbit, it sweeps out equal areas in equal times. This tells us that a planet moves faster when it is closer to the Sun (near perihelion) than when it is farther from the Sun (near aphelion) in its orbit.

Kepler's third law States that the square of a planet's orbital period is proportional to the cube of its average distance from the Sun (semimajor axis), which tells us that more distant planets move more slowly in their orbits. In its original form, it is written $p^2 = a^3$. See also *Newton's version of Kepler's third law.*

kinetic energy Energy of motion, given by the formula $\frac{1}{2}mv^2$.

kingdoms (biological) Except for the three domains, kingdoms (such as the plant kingdom and the animal kingdom) are the highest classification grouping of living organisms.

K–T boundary The thin layer of dark sediments that marks the division between the Cretaceous and Tertiary periods in the fossil record (the "K" comes from the German word for "Cretaceous," *Kreide*).

K–T event (impact) The collision of an asteroid or comet 65 million years ago that caused the mass extinction best known for wiping out the dinosaurs. "K–T" stands for the initials of the geological layers above and below the event.

Kuiper belt The comet-rich region of our solar system that spans distances of about 30–100 AU from the Sun; Kuiper belt comets have orbits that lie fairly close to the plane of planetary orbits and travel around the Sun in the same direction as the planets.

Kuiper belt objects The cometlike objects located in the Kuiper belt.

Lagrange points (of the Earth–Moon system) The five positions in space where the effects of gravity from Earth and the Moon "cancel" in such a way that, if you floated weightlessly at one of these five points, you wouldn't be tugged toward either body.

late heavy bombardment An apparent increase in the impact rate near the end of the heavy bombardment, about 3.9 billion years ago.

latitude The angular north-south distance between Earth's equator and a location on Earth's surface.

light-year The distance that light can travel in 1 year, which is 9.46 trillion kilometers.

lipids Complex molecules in cells, also known *fats,* that play a variety of functions including being key components of membranes.

liquid phase The phase of matter in which atoms or molecules are held together but move relatively freely.

lithophile See *endolith.* An organism that lives inside of rock.

lithosphere The relatively rigid outer layer of a planet; generally encompasses the crust and the uppermost portion of the mantle.

Local Group The group of about 40 galaxies to which the Milky Way Galaxy belongs.

Local Supercluster The supercluster of galaxies to which the Local Group belongs.

longitude The angular east-west distance between the prime meridian (which passes through Greenwich) and a location on Earth's surface.

luminosity The total power output of an object, usually measured in watts or in units of solar luminosities ($L_{Sun} = 3.8 \times 10^{26}$ watts).

lunar maria The regions of the Moon that look smooth from Earth but actually are impact basins.

magma Underground molten rock.

magnetic field Describes the region surrounding a magnet in which it can affect other magnets or charged particles in its vicinity.

magnetic-field lines Lines that represent how the needles on a series of compasses would point if they were laid out in a magnetic field.

magnetosphere The region surrounding a planet in which charged particles are trapped by the planet's magnetic field.

main sequence The prominent line of points running from the upper left to the lower right on an H–R diagram; main-sequence stars shine by fusing hydrogen in their cores.

main-sequence stars Stars whose temperature and luminosity place them on the main sequence of the H–R diagram. Main-sequence stars release energy by fusing hydrogen into helium in their cores.

mantle (of a planet) The rocky layer that lies between a planet's core and crust.

mantle convection The flow pattern in which hot mantle material expands and rises while cooler material contracts and falls.

martian meteorite This term is used to describe meteorites found on Earth that are thought to have originated on Mars.

mass A measure of the amount of matter in an object.

mass-energy The potential energy of mass, which has an amount $E = mc^2$.

mass extinction An event in which a large fraction of the species living on Earth go extinct, such as the event in which the dinosaurs died out about 65 million years ago.

mass increase (in relativity) Refers to the effect in which an object moving past you seems to have a mass greater than its rest mass.

mass ratio (of a rocket) The ratio of the initial (launch) mass of the rocket M_i, including its fuel and any spacecraft it is carrying, to its mass M after all the fuel is burned.

matter–antimatter annihilation Occurs when a particle of matter and a particle of antimatter meet and convert all of their mass-energy to photons.

membrane (cell) A barrier that separates the inside of a cell (or cell nucleus) from the outside. Cells come in two basic types.

metabolism The many chemical reactions that occur in living organisms.

metals (in solar system theory) Elements, such as nickel, iron, and aluminum, that condense at fairly high temperatures.

metamorphic rock Rock made from igneous or sedimentary rock that gets transformed (but not melted) by high heat or pressure.

meteor A flash of light caused when a particle from space burns up in our atmosphere.

meteorite A rock from space that lands on Earth.

microwaves Light with wavelengths in the range from micrometers to millimeters. Microwaves are generally considered to be a subset of the radio-wave portion of the electromagnetic spectrum.

mid-ocean ridges (on Earth) Long ridges of undersea volcanoes, along which mantle material erupts onto the ocean floor and pushes apart the existing seafloor on either side. These ridges are essentially the source of new seafloor crust, which then makes its way along the ocean bottom for millions of years before returning to the mantle at a subduction zone.

Milankovitch cycles The cyclical changes in Earth's axis tilt and orbit that can change the climate and cause ice ages.

Milky Way Used both as the name of our galaxy and to refer to the band of light we see in the sky when we look into the plane of the Milky Way Galaxy.

Miller–Urey experiment An experiment first performed in the 1950s that was designed to learn how organic molecules might have formed naturally on the early Earth.

mineral A rocky substance with a particular chemical composition and crystal structure.

mitochondria The cellular organs in eukaryotic cells in which oxygen helps produce energy (by making molecules of ATP).

model (scientific) A representation of some aspect of nature that can be used to explain and predict real phenomena without invoking myth, magic, or the supernatural.

moist greenhouse effect A process by which a planet could lose water when the atmospheric circulation allows water vapor to rise high enough to be broken apart by ultraviolet light from the Sun.

molecule Technically the smallest unit of a chemical element or compound; in this text, the term refers only to combinations of two or more atoms held together by chemical bonds.

momentum The product of an object's mass and velocity.

moon An object that orbits a planet.

multiple star system A star system that contains two or more stars.

mutations Errors in the copying process when a living cell replicates itself.

natural selection The process by which mutations that make an organism better able to survive get passed on to future generations.

nebula A cloud of gas in space, usually one that is glowing.

nebular theory The detailed theory that describes how our solar system formed from a cloud of interstellar gas and dust.

neutrons Particles with no electrical charge that are found in atomic nuclei.

newton The standard unit of force in the metric system:

$$1 \text{ newton} = 1 \frac{\text{kg} \times \text{m}}{\text{s}^2}$$

Newton's first law of motion States that, in the absence of a net force, an object moves with constant velocity.

Newton's laws of motion Three basic laws that describe how objects respond to forces.

Newton's second law of motion States how a net force affects an object's motion. Specifically: force = rate of change in momentum, or force = mass × acceleration.

Newton's third law of motion States that, for any force, there is always an equal and opposite reaction force.

Newton's universal law of gravitation See *universal law of gravitation.*

Newton's version of Kepler's third law This generalization of Kepler's third law can be used to calculate the masses of orbiting objects from measurements of orbital period and distance; it is usually written as:

$$p^2 = \frac{4\pi^2}{G(M_1 + M_2)} a^3$$

Noachian era The era on Mars before 3.8 billion years ago.

nonscience As defined in this book, nonscience is any way of searching for knowledge that makes no claim to follow the scientific method, such as seeking knowledge through intuition, tradition, or faith.

nuclear fission The process in which a larger nucleus splits into two (or more) smaller particles.

nuclear fusion The process in which two (or more) smaller nuclei slam together and make one larger nucleus.

nucleus (of an atom) The compact center of an atom made from protons and neutrons.

nucleus (of a cell) The membrane-enclosed region of a eukaryotic cell that contains the cell's DNA.

nucleus (of a comet) The solid portion of a comet, and the only portion that exists when the comet is far from the Sun.

observable universe The portion of the entire universe that, at least in principle, can be seen from Earth.

Occam's razor A principle often used in science, holding that scientists should prefer the simpler of two models that agree equally well with observations. Named after the medieval scholar William of Occam (1285–1349).

Oort cloud A huge, spherical region centered on the Sun, extending perhaps halfway to the nearest stars, in which trillions of comets orbit the Sun with random inclinations, orbital directions, and eccentricities.

orbital energy The sum of an orbiting object's kinetic and gravitational potential energies.

orbital resonance Describes any situation in which one object's orbital period is a simple ratio of another object's period, such as $\frac{1}{2}$, $\frac{1}{4}$, or $\frac{5}{3}$. In such cases, the two objects periodically line up with each other, and the extra gravitational attractions at these times can affect the objects' orbits.

orbiters (of other worlds) Spacecraft that go into orbit of another world for long-term study.

organic chemistry The chemistry of organic molecules (whether or not the molecules are involved in life).

organic molecule This term generally refers to any molecule containing carbon, but with some exceptions among molecules that are most commonly found independent of life. For our purposes in this book, the important exceptions are carbon dioxide (CO_2) and carbonate minerals, which we do not consider to be organic.

outer solar system Generally considered to encompass the region of our solar system beginning at about the orbit of Jupiter.

outgassing The process of releasing gases from a planetary interior, usually through volcanic eruptions.

oxidation Refers to chemical reactions, often with the surface of a planet, that remove oxygen from the atmosphere.

ozone The molecule O_3, which is a particularly good absorber of ultraviolet light.

ozone depletion Refers to the declining levels of atmospheric ozone found worldwide on Earth, especially in Antarctica, in recent years.

ozone hole A place where the concentration of ozone in the stratosphere is dramatically lower than is the norm.

Pangaea A "supercontinent" that existed prior to 225 million years ago, in which all Earth's current continents were linked together.

panspermia The idea that life migrated to Earth from some extraterrestrial location.

paradigm (in science) Refers to general patterns of thought that tend to shape scientific beliefs during a particular time period.

paradox A situation that, at least at first, seems to violate common sense or contradict itself. Resolving paradoxes often leads to deeper understanding.

parallax The apparent shifting of an object against the background, due to viewing it from different positions. See also *stellar parallax.*

perigee The point at which an object orbiting Earth is nearest to Earth.

perihelion The point at which an object orbiting the Sun is closest to the Sun.

periodic table of the elements A table that lists properties of all the known elements in an organized way.

period–luminosity relation The relation that describes how the luminosity of a Cepheid variable star is related to the period between peaks in its brightness; the longer the period, the more luminous the star.

periods (geological) The third-largest divisions of time in Earth's geological history, after eons and eras.

phase (of matter) Describes the way in which atoms or molecules are held together; the common phases are *solid, liquid,* and *gas.*

photoautotroph An organism that gets its carbon directly from the atmosphere and gets its energy from sunlight through photosynthesis; plants are photoautotrophs.

photoheterotroph An organism that gets its carbon by consuming preexisting organic molecules and gets its energy from sunlight through photosynthesis.

photon An individual particle of light, characterized by a wavelength and a frequency.

phyla (singular, *phylum*) The next level of biological classification below kingdoms.

pixel An individual "picture element" in a digital picture.

planet A moderately large object that orbits a star and shines primarily by reflecting light from its star. More precisely, according to a definition approved in 2006, an object can be considered a planet only if it: (1) orbits a star; (2) is large enough for its own gravity to make it round; and (3) has cleared most other objects from its orbital path. An object that is round but has *not* cleared its orbital path, like Pluto, is designated a *dwarf planet.*

planetary civilization A civilization that uses the resources of its home planet; we are a planetary civilization by this definition.

planetary nebula The glowing cloud of gas ejected from a low-mass star at the end of its life.

planetesimals The building blocks of planets, formed by accretion in the solar nebula.

plasma A gas consisting of ions and electrons.

plates (on a planet) Pieces of a lithosphere that apparently float on the denser mantle below.

plate tectonics The geological process in which plates are moved around by stresses in a planet's mantle.

positron The antimatter equivalent of an electron. It is identical to an electron in all respects except that it has a positive rather than a negative electrical charge.

potential energy Energy stored for later conversion into kinetic energy; includes *gravitational potential energy, electrical potential energy, chemical potential energy.*

power The rate of energy usage, usually measured in watts (1 watt = 1 joule/s).

precession The gradual wobble of the axis of a rotating object around a vertical line.

precipitation Condensed atmospheric gases that fall to the surface in the form of rain, snow, or hail.

pressure Describes the force (per unit area) pushing on an object. In astronomy, we are generally interested in pressure applied by surrounding gas (or plasma).

prokaryote A living organism made from cells in which DNA is *not* confined to a distinct, membrane-enclosed nucleus. Most prokaryotes are single-celled. Prokaryotes include all the organisms in two of the three domains of life: Bacteria and Archaea.

prokaryotic cell A cell that lacks a distinct nucleus.

protein A large molecule assembled from amino acids according to instructions encoded in DNA. Proteins play many roles in cells; a special category of proteins, called *enzymes,* catalyzes nearly all of the important biochemical reactions that occur within cells.

protons Particles with positive electrical charge found in atomic nuclei; they are built from three quarks.

protoplanetary disk A disk of material surrounding a young star (or protostar) that may eventually form planets.

protoplanets A term used to describe planetesimals that have grown quite large, to planet-size.

pseudoscience Something that purports to be science or may appear to be scientific but that does not adhere to the testing and verification requirements of the scientific method.

Ptolemaic model The geocentric model of the universe developed by Ptolemy in about A.D. 150.

radar mapping Imaging of a planet by bouncing radar waves off its surface, especially important for Venus and Titan, where thick clouds mask the surface.

radial motion The component of an object's motion directed toward or away from us.

radial velocity The portion of any object's total velocity that is directed toward or away from us. This part of the velocity is the only part that we can measure with the Doppler effect.

radiative energy Energy carried by light; the energy of a photon is Planck's constant times its frequency, or $h \times f$.

radioactive decay The spontaneous change of an atom in which its nucleus breaks apart or a proton turns into an electron, thereby changing into a different element. It also releases heat to a planet's interior.

radioactive element (or **radioactive isotope**) A substance whose nucleus tends to fall apart spontaneously.

radiometric dating The process of determining the age of a rock (i.e., the time since it solidified) by comparing the present amount of a radioactive substance to the amount of its decay product.

radio waves Light with extremely long wavelengths (and hence low frequencies)—longer than those of infrared light.

rare Earth hypothesis A hypothesis holding that the specific circumstances that have made it possible for complex creatures (such as birds or humans) to have evolved on Earth might be so rare that ours may be the only inhabited planet in the galaxy that has anything but the simplest life.

red giant A giant star that is red in color; red giants are a late stage in the life of a star, occurring after the star has exhausted its central supply of core hydrogen.

redox reactions Chemical reactions that involve an exchange or reshuffling of electric charge between the reacting atoms or molecules. A redox reaction always involves the transfer of one or more electrons from an *electron donor* (which becomes oxidized) to an *electron acceptor* (which becomes reduced).

redshift (Doppler) A Doppler shift in which spectral features are shifted to longer wavelengths, caused when an object is moving away from the observer.

reduction (chemical) The process of gaining electrons—which reduces the electrical charge (because electrons carry negative charge)—in a chemical reaction.

reference frame (frame of reference) In the theory of relativity, two people (or objects) share the same reference frame if they are *not* moving relative to one another.

resonance See *orbital resonance.*

RNA (ribonucleic acid) A molecule closely related to DNA—but with only a single strand and a slightly different backbone and set of bases—that plays critical roles in carrying out the instructions encoded in DNA.

RNA world The hypothesized period during which life on Earth first evolved and used RNA, rather than DNA, as its genetic material.

rocks (in solar system theory) Material common on the surface of Earth, such as silicon-based minerals, that is solid at temperatures and pressures found on Earth but typically melts or vaporizes at temperatures of 500–1,300 K.

rock cycle The idea that rocks can be transformed between the three basic types igneous, metamorphic, and sedimentary.

rotation The spinning of an object around its axis.

runaway greenhouse effect A positive feedback cycle in which heating caused by the greenhouse effect causes more greenhouse gases to enter the atmosphere, which further enhances the greenhouse effect.

rybozymes RNA molecules that function as catalysts.

sample return mission A space mission designed to return to Earth a sample of another world.

satellite Any object orbiting another object.

science The search for knowledge that can be used to explain or predict natural phenomena in a way that can be confirmed by rigorous observations or experiments.

scientific method An organized approach to explaining observed facts through science.

scientific theory A model of some aspect of nature that has been rigorously tested and has passed all tests to date.

seafloor crust On Earth, the thin, dense crust of basalt created by seafloor spreading.

seafloor spreading On Earth, the creation of new seafloor crust at mid-ocean ridges.

search for extraterrestrial intelligence (SETI) The name given to observing projects designed to search for signs of intelligent life beyond Earth.

second law of thermodynamics The law stating that, when left alone, the energy in a system undergoes conversions that lead to increasing disorder.

sedimentary rock A rock that formed from sediments created and deposited by erosional processes. The sediments tend to build up in distinct layers, or *strata*.

seismic waves Earthquake-induced vibrations that propagate through a planet.

semimajor axis Half the distance across the long axis of an ellipse; in this text, it is usually referred to as the *average* distance of an orbiting object, abbreviated *a* in the formula for Kepler's third law.

sentinel hypothesis A possible solution to the Fermi paradox that suggests that extraterrestrials place monitoring devices near star systems that show promise of emerging intelligence, and patiently wait until these devices record the presence of civilization.

silicate rock A silicon-rich rock.

sleep paralysis The natural paralysis of the body that occurs during REM sleep; it may occasionally persist for a few minutes after the brain has started waking up, giving a person the alarming sensation of being awake in a paralyzed body. Visions and other sensations often occur in this state.

snowball Earth Name given to a hypothesis suggesting that, some 600–700 million years ago, Earth experienced a period in which it became cold enough for glaciers to exist worldwide, even in equatorial regions.

solar luminosity The luminosity of the Sun, which is approximately 4×10^{26} watts.

solar nebula The piece of interstellar cloud from which our own solar system formed.

solar sail A large, highly reflective (and thin, to minimize mass) piece of material that can "sail" through space using pressure exerted by sunlight.

solar system (or **star system**) Consists of a star (sometimes more than one star) and all the objects that orbit it.

solar wind A stream of charged particles ejected from the Sun.

solar wind stripping The stripping away of a planet's atmospheric gas by the solar wind; generally affects only planetary atmospheres that are unprotected by a global magnetic field.

solid phase The phase of matter in which atoms or molecules are held rigidly in place.

solstice See *summer solstice* and *winter solstice.*

special theory of relativity Einstein's theory that describes the effects of the facts that all motion is relative and that everyone always measures the same speed of light.

species A group of organisms that is genetically distinct from other groups; species is the most precise level of biological classification among organisms.

spectral lines Bright or dark lines that appear in an object's spectrum, which we can see when we pass the object's light through a prismlike device that spreads out the light like a rainbow.

spectral resolution Describes the degree of detail that can be seen in a spectrum; the higher the spectral resolution, the more detail we can see.

spectral type A way of classifying a star by the lines that appear in its spectrum; it is related to surface temperature. The basic spectral types are designated by a letter (OBAFGKM, with O for the hottest stars and M for the coolest) and are subdivided with numbers from 0 through 9.

spectroscopy (in astronomical research) The process of obtaining spectra from astronomical objects.

spectrum (of light) See *electromagnetic spectrum.*

speed The rate at which an object moves. Its units are distance divided by time, such as m/s or km/hr.

speed of light The speed at which light travels, which is about 300,000 km/s.

spiral galaxies Galaxies that look like flat, white disks with yellowish bulges at their centers. The disks are filled with cool gas and dust, interspersed with hotter ionized gas, and usually display beautiful spiral arms.

spring equinox (vernal equinox) Refers both to the point in Pisces on the celestial sphere where the ecliptic crosses the celestial equator and to the moment in time when the Sun appears at that point each year (around March 21).

star A large, glowing ball of gas that generates energy through nuclear fusion in its core. The term *star* is sometimes applied to objects that are in the process of becoming true stars (e.g., protostars) and to the remains of stars that have died (e.g., neutron stars).

star system See *solar system.*

stellar civilization A civilization that employs the resources of its home star (that is, not only the resources available on its home planet).

stellar evolution The formation and development of stars.

stellar parallax The apparent shift in the position of a nearby star (relative to distant objects) that occurs as we view the star from different positions in Earth's orbit of the Sun each year.

sterilizing impact An impact large enough that it would have fully vaporized Earth's oceans and killed off any life existing on Earth.

strata (rock) Layers in sedimentary rock.

stromatolites Large bacterial "colonies."

subduction (of tectonic plates) The process in which one plate slides under another.

subduction zones Places where one plate slides under another.

sublimation The process by which atoms or molecules escape into the gas phase from the solid phase.

summer solstice Refers both to the point on the celestial sphere where the ecliptic is farthest north of the celestial equator and to the moment in time when the Sun appears at that point each year (around June 21).

superclusters Superclusters consist of many clusters of galaxies, groups of galaxies, and individual galaxies and are the largest known structures in the universe.

supergiants The largest and brightest of all stars.

supernova The explosion of a star.

symbiotic relationship A relationship in which both an invading organism and a host organism benefit from living together.

synchronous rotation Describes the rotation of an object that always shows the same face to the object that it is orbiting because its rotation period and orbital period are equal.

technological evolution Change driven by the rapid development of technology.

tectonics The disruption of a planet's surface by internal stresses.

temperature A measure of the average kinetic energy of particles in a substance.

terraforming Changing a planet in such a way as to make it more Earth-like.

terrestrial planets Rocky planets similar in overall composition to Earth.

theories of relativity (**special** and **general**) Einstein's theories that describe the nature of space, time, and gravity.

theory (in science) See *scientific theory.*

theory of evolution The theory, first advanced by Charles Darwin, that explains *how* evolution occurs through the process of *natural selection.*

thermal energy Represents the collective kinetic energy, as measured by temperature, of the many individual particles moving within a substance.

thermal escape The process in which atoms or molecules in a planet's exosphere move fast enough to escape into space.

thermal radiation The spectrum of radiation produced by an opaque object that depends only on the object's temperature; sometimes called *blackbody radiation.*

thermophile An organism that thrives under conditions of high temperature compared to what most organisms can tolerate.

tidal force A force that is caused when the gravity pulling on one side of an object is larger than that on the other side, causing the object to stretch.

tidal friction Friction within an object that is caused by a tidal force.

tidal heating A source of internal heating created by tidal friction. It is particularly important for satellites with eccentric orbits such as Io and Europa.

time dilation (in relativity) Refers to the effect in which you observe time running slower in reference frames moving relative to you.

transit A transit is an event in which a planet passes in front of a star (or the Sun) as seen from Earth. Only Mercury and Venus can be seen in transit of our Sun. The search for transits of extrasolar planets is an important planet detection strategy.

tree of life An illustration of the biochemical and genetic relationships between different organisms; its three main branches are the three *domains* Bacteria, Archaea, and Eukarya.

troposphere The lowest atmospheric layer, in which convection and weather occur.

turbulence Rapid and random motion.

ultraviolet light Light with wavelengths that fall in the portion of the electromagnetic spectrum between visible light and X rays.

universal law of gravitation The law expressing the force of gravity (F_g) between two objects, given by the formula

$$F_g = G\frac{M_1 M_2}{d^2}$$

$$\left(G = 6.67 \times 10^{-11} \frac{m^3}{\text{kg} \times \text{s}^2}\right)$$

universe The sum total of all matter and energy.

velocity The combination of speed and direction of motion; it can be stated as a speed in a particular direction, such as 100 km/hr due north.

vernal equinox See *spring equinox*.

viscosity Describes the "thickness" of a liquid in terms of how rapidly it flows; low-viscosity liquids flow quickly (e.g., water), while high-viscosity liquids flow slowly (e.g., molasses).

visible light The light our eyes can see, ranging in wavelength from about 400 to 700 nm.

volatiles Refers to substances such as water, carbon dioxide, and methane that are usually found as gases, liquids, or surface ices on the terrestrial worlds.

volcanism The eruption of molten rock, or lava, from a planet's interior onto its surface.

Von Neumann machines Self-replicating machines first proposed by the American mathematician and computer pioneer John Von Neumann (1903–1957).

watt The standard unit of power in science: 1 watt = 1 joule/s.

wavelength The distance between adjacent peaks (or troughs) of a wave.

weather Describes the ever-varying combination of winds, clouds, temperature, and pressure in a planet's troposphere.

white dwarf The hot, compact corpse of a low-mass star, typically with a mass similar to that of the Sun compressed to a volume the size of the Earth.

winter solstice Refers both to the point on the celestial sphere where the ecliptic is farthest south of the celestial equator and to the moment in time when the Sun appears at that point each year (around December 21).

worldline A line that represents an object on a spacetime diagram.

wormholes The name given to hypothetical tunnels through hyperspace that might connect two distant places in our universe.

X rays Light with wavelengths that fall in the portion of the electromagnetic spectrum between ultraviolet light and gamma rays.

zircons Tiny mineral grains of zirconium silicate, usually found embedded in sedimentary rock.

zoo hypothesis A possible explanation for the Fermi paradox holding that alien civilizations are aware of our presence but have chosen to deliberately avoid contact with us.

References and Bibliography

*(*indicates articles or books that should be particularly accessible to students not majoring in science)*

General (books covering a wide range of topics relevant to many chapters in this text)

*Bennett, J., Donahue, M., Schneider, N., and Voit, M., *The Cosmic Perspective*, Addison Wesley, 2007.

*Darling, D., *The Extraterrestrial Encyclopedia*, Three Rivers Press, 2000.

*Dick, S. J., *Life on Other Worlds*, Cambridge, 1998.

*Dick, S. J., and Strick, J. E., *The Living Universe*, Rutgers, 2005.

*Grinspoon, D., *Lonely Planets: The Natural Philosophy of Alien Life*, Ecco, 2003.

*Jakosky, B., *The Search for Life on Other Planets*, Cambridge, 1998.

*Sagan, C., *Cosmos*, Random House, 1980.

*Shostak, S., *Sharing the Universe*, Berkeley Hills Books, 1998.

Chyba, C. F., and Hand, K. P., Astrobiology: The study of the living universe, *Annual Review of Astronomy and Astrophysics*, **43,** 31–74, 2005.

Lunine, J. I., *Astrobiology*, Addison Wesley, 2005.

Lunine, J. I., *Earth: Evolution of a Habitable World*, Cambridge, 1999.

Chapter 1

Approaches to Astrobiology

*Morrison, D., and Schmidt, G. K., eds., *Astrobiology Roadmap*, NASA, 1999.

Lederberg, J., Exobiology: Approaches to life beyond the Earth, *Science*, **132,** 393–400, 1960.

Chapter 2

History of Science and Astrobiology

*Asimov, I., *Asimov's Biographical Encyclopedia of Science and Technology*, Doubleday, 1982.

*Dick, S. J., *The Biological Universe*, Cambridge, 1996.

Crowe, M. J., *The Extraterrestrial Life Debate, 1750–1900*, Dover, 1999.

Crowe, M. J., *Theories of the World: From Antiquity to the Copernican Revolution*, Dover, 1990.

The Nature of Science

*Boorstin, D. J., *The Discoverers*, Vintage, 1985.

*Sagan, C., *The Demon-Haunted World: Science as a Candle in the Dark*, Ballantine, 1996.

Cleland, C. E., Historical science, experimental science, and the scientific method, *Geology*, **29,** 987–990, 2001.

Kuhn, T. S., *The Nature of Scientific Revolutions*, University of Chicago Press, 1996.

Popper, K., and Miller, D., eds., *Popper Selections*, Princeton University Press, 1985.

Chapter 3

The Big Bang, Origin of the Elements, Formation of the Solar System

*Bennett, J., *On the Cosmic Horizon*, Addison Wesley, 2001.

*Kirshner, R. P., The Earth's elements, *Scientific American*, **271,** 58–65, 1994.

*Singh, S., *Big Bang*, Harper Perennial, 2005.

*Wood, J. A., Forging the planets, *Sky & Telescope*, January 1999.

Beckwith, S. V. W., and Sargent, A. I., Circumstellar disks and the search for neighboring planetary systems, *Nature*, **383,** 139–144, 1996.

Lissauer, J. J., Planet formation, *Annual Review of Astronomy and Astrophysics*, **31,** 129–174, 1993.

Rees, M. J., Piecing together the biggest puzzle of all, *Science*, **290,** 1919–1925, 2000.

Chapter 4

Radiometric Dating and the Age of the Earth

Dalrymple, G. B., *The Age of the Earth*, Stanford, 1991.

Impact History of the Earth, Hadean Earth

Chyba, C. F., Owen, T. C., and Ip, W. H., Impact delivery of volatiles and organic molecules to the Earth, in *Hazards Due to Comets and Asteroids*, T. Gehrels, ed., University of Arizona Press, 9–58, 1994.

Harrison, T. M., et al., Heterogeneous Hadean hafnium: Evidence of continental crust at 4.4 to 4.5 Ga, *Science*, **310,** 1947–1950, 2005.

Hartmann, W. K., Ryder, G., Dones, L., and Grinspoon, D., The time-dependent intense bombardment of the primordial Earth/Moon system, in *Origin of the Earth and Moon*, R. M. Canup and K. Righter, eds., University of Arizona Press, 493–512, 2000.

Sleep, N. H., Zahnle, K. J., Kasting, J. F., and Morowitz, H. J., Annihilation of ecosystems by large asteroid impacts on the early Earth, *Nature*, **342,** 139–142, 1989.

Geology and Plate Tectonics

Anderson, D. L., Planet Earth, in *The New Solar System*, J. K. Beatty, C. C. Peterson, and A. Chaikin, eds., Sky Publishing Corporation, 111–124, 1999.

Long-Term Climate Change

*Hoffman, P. F., and Schrag, D. P., Snowball Earth, *Scientific American*, **282,** 68–75, 2000.

Kasting, J. F., and Grinspoon, D. H., The faint young sun problem, in *The Sun in Time*, C. P. Sonett, M. S. Giampapa, and M. S. Matthews, eds., University of Arizona Press, 447–462, 1991.

Formation of the Moon

Canup, R., Dynamics of lunar formation, *Annual Review of Astronomy and Astrophysics*, **42,** 441–475, 2004.

Jacobsen, S. B., The Hf-W isotopic system and the origin of the Earth and Moon, *Annual Review of Earth and Planetary Sciences*, **33,** 531–570, 2005.

Chapter 5

Basic Biology and Evolution

*Campbell, N., and Reece, J., *Biology*, Addison Wesley, 2005.

*Zimmer, C., *Evolution: The Triumph of an Idea*, Harper Collins, 2001.

Gould, S. J., Showdown on the Burgess Shale: The reply, *Natural History*, Dec. 1998–Jan. 1999.

Keenan, W. T., The chemistry of life, *Biological Science*, Norton, 47–82, 1980.

Koshland, D. E., Jr., The seven pillars of life, *Science*, **295,** 2215–2216, 2002.
Morris, S. C., Showdown on the Burgess Shale: The challenge, *Natural History*, Dec. 1998–Jan. 1999.
Sagan, C., Life, *Encyclopaedia Britannica*, 1970.

The Three Domains and the Tree of Life

*Doolittle, W. F., Uprooting the tree of life, *Scientific American*, **282,** 90–95, 2000.
Brown, J. F., and Doolittle, W. F., Root of the universal tree of life based on ancient aminoacyl-tRNA synthetase gene duplications, *Proceedings of the National Academy of Sciences*, **92,** 2441–2445, 1995.
Pace, N. R., The early branches in the tree of life, preprint for *Proceedings of a Symposium on Assembling the Tree of Life*, American Museum of Natural History, 2002.
Pace, N. R., A molecular view of microbial diversity and the biosphere, *Science*, **276,** 734–740, 1997.
Woese, C., Interpreting the universal phylogenetic tree, *Proceedings of the National Academy of Sciences*, **97,** 8392–8396, 2000.
Woese, C., The universal ancestor, *Proceedings of the National Academy of Sciences*, **95,** 6854–6859, 1998.

Life in Extreme Environments

*Madigan, M. T., and Marrs, B. L., Extremophiles, *Scientific American*, **276,** 82–87, 1997.
Nealson, K. H., The limits of life on Earth and searching for life on Mars, *Journal of Geophysical Research*, **102,** 23675–23686, 1997.
Sharma, A., et al., Microbial activity at gigapascal pressures, *Science*, **295,** 1514–1516, 2002.
Stevens, T. O., and McKinley, J. P., Lithoautotrophic microbial ecosystems in deep basalt aquifers, *Science*, **270,** 450–454, 1995.
Van Dover, C. L., et al., Biogeography and ecological setting of Indian Ocean hydrothermal vents, *Science*, **294,** 818–823, 2001.

Evolution and Creationism

*National Academy of Sciences, *Science and Creationism: A View from the National Academy of Sciences*, National Academies Press, 1999.
*Pennock, R., *Intelligent Design Creationism and Its Critics: Philosophical, Theological & Scientific Perspectives*, MIT Press, 2001.
*Pennock, R., *Tower of Babel*, MIT Press, 1999.
*Scott, E., *Evolution vs. Creationism*, University of California Press, 2005.

Chapter 6

Early Evidence for Life

Brasier, M., et al., Questioning the evidence for Earth's oldest fossils, *Nature* **416,** 76–81, 2002.
Mojzsis, S. J., Arrhenius, G., McKeegan, K. D., Harrison, T. M., Nutman, A. P., and Friend, C. R. L., Evidence for life on Earth before 3,800 million years ago, *Nature*, **384,** 55–59, 1996.
Mojzsis, S. J., Harrison, T. M., and Pidgeon, R. T., Oxygen-isotope evidence from ancient zircons for liquid water at the Earth's surface 4,300 Myr ago, *Nature*, **409,** 178–181, 2001.
Papineau, D., et al., Composition and structure of microbial communities from stromatolites of Hamelin Pool in Shark Bay, *Applied and Environmental Microbiology*, **71,** 4822–4832, 2005.
Papineau, D., et al., Nitrogen isotopic composition of ammoniated pyllosilicates: Case studies from Precambrian metamorphosed sedimentary rocks, *Chemical Geology*, **216,** 37–58, 2005.
Schopf, J. W., The oldest fossils and what they mean, in *Major Events in the History of Life*, J. W. Schopf, ed., Jones and Bartlett Publishers, 29–63, 1992.
Schopf, J. W., et al., Laser-Raman imagery of Earth's earliest fossils, *Nature*, **416,** 73–76, 2002.

Origin of Life on Earth

*Davies, P., *The Origin of Life*, Penguin Books, 2003.
Cody, G. D., et al., Primordial carbonylated iron-sulfur compounds and the synthesis of pyruvate, *Science*, **298,** 1337–1340, 2000.
Huber, C., and Wachtershauser, G., Activated acetic acid by carbon fixation on (Fe,Ni)S under primordial conditions, *Science*, **276,** 245–247, 1997.
Pace, N. R., Origin of life—Facing up to the physical setting, *Cell*, **65,** 531–533, 1991.
Shock, E. L., Amend, J. P., and Zolotov, M. Y., The early Earth vs. the origin of life, in *Origin of the Earth and Moon*, R. M. Canup and K. Righter, eds., University of Arizona Press, 527–543, 2000.

Miller–Urey Experiment

Miller, S. L., The prebiotic synthesis of organic compounds as a step toward the origin of life, in *Major Events in the History of Life*, J. W. Schopf, ed., Jones and Bartlett Publishers, 1–28, 1992.
Urey, H. C., On the early chemical history of the Earth and the origin of life, *Proceedings of the National Academy of Sciences*, **38,** 351–363, 1952.

Where Did Life Originate?

*Warmflash, D., and Weiss, B., Did life come from another world? *Scientific American*, November 2005.
Baross, J. A., and Hoffman, S. E., Submarine hydrothermal vents and associated gradient environments as sites for the origin and evolution of life, *Origins of Life*, **15,** 327–345, 1985.
D'Hondt, S., Rutherford, S., and Spivak, A. J., Metabolic activity of subsurface life in deep-sea sediments, *Science*, **295,** 2067–2070, 2002.
Ferris, J. P., Huang, C. H., and Hagan, W. J., Jr., Montmorrilonite: A multifunctional mineral catalyst for the prebiological formation of phosphate esters, *Origins of Life and Evolution of the Biosphere*, **18,** 121–133, 1988.

RNA World and Early Life

Cech, T. R., A model for the RNA-catalyzed replication of RNA, *Proceedings of the National Academy of Sciences*, **83,** 4360–4363, 1986.
Doudna, J. A., and Szostak, J. W., RNA-catalyzed synthesis of complementary-strand RNA, *Nature*, **339,** 519–522, 1989.
Joyce, G. F., The rise and fall of the RNA world, *New Biologist*, **3,** 399–407, 1991.
Kerr, R., The story of O_2, *Science*, **308,** 1730–1732, 2005.
Lazcano, A., and Miller, S. L., On the origin of metabolic pathways, *Journal of Molecular Evolution*, **49,** 424–431, 1999.
Morris, S. C., The early evolution of life, in *Understanding the Earth*, G. Brown, C. Hawkesworth, and C. Wilson, eds., Cambridge University Press, 436–457, 1992.

Impacts and Extinction Events

Alvarez, L. W., Alvarez, W., Asaro, F., and Michel, H. V., Extraterrestrial cause for the Cretaceous-Tertiary extinction, *Science*, **208,** 1095–1108, 1980.
Rampino, M. R., and Haggerty, B. M., Extraterrestrial impacts and mass extinctions of life, in *Hazards Due to Comets and Asteroids*, T. Gehrels, ed., University of Arizona Press, 827–857, 1994.

Artificial Life

*Holmes, B., The race to create life from scratch, *New Scientist*, February 12, 2005.
*Stroh, M., Life built to order, *Popular Science*, February 1, 2005.

Chapter 7

Environmental Requirements for Life

Chyba, C. F., and Hand, K. P., Life without photosynthesis, *Science*, **292,** 2026–2027, 2001.

Nealson, K. H., Post-Viking microbiology: New approaches, new data, new insights, *Origins of Life and Evolution of the Biosphere,* **29,** 73–93, 1999.

Pace, N. R., The universal nature of biochemistry, *Proceedings of the National Academy of Sciences,* **98,** 805–808, 2001.

Wald, G., The origins of life, *Proceedings of the National Academy of Sciences,* **52,** 595–611, 1964.

Life in the Solar System

Chang, S., Planetary environments and the conditions of life, *Philosophical Transactions of the Royal Society of London,* **325,** 601–610, 1988.

Soffen, G. A., Life in the solar system, in *The New Solar System,* J. K. Beatty, C. C. Peterson, and A. Chaikin, eds., Sky Publishing Corporation, 365–376, 1999.

Chapter 8

Introduction to Mars

*Squyres, S., *Roving Mars: Spirit, Opportunity, and the Exploration of the Red Planet,* Hyperion, 2005.

Carr, M. H., Mars, in *The New Solar System,* J. K. Beatty, C. C. Peterson, and A. Chaikin, eds., Sky Publishing Corporation, 141–156, 1999.

Zahne, K., Decline and fall of the martian empire, *Nature,* **412,** 209–213, 2001.

Geological and Geochemical Potential for Life on Mars

*Christensen, P. R., The many faces of Mars, *Scientific American,* July 2005.

Baker, V. R., Water and the martian landscape, *Nature,* **412,** 228–236, 2001.

Gladman, B. J., et al., The exchange of impact ejecta between terrestrial planets, *Science,* **271,** 1387–1392, 1996.

Jakosky, B. M., and Phillips, R. J., Mars volatile and climate history, *Nature,* **412,** 237–244, 2001.

Jakosky, B. M., et al., The changing picture of volatiles and climate on Mars, *Science,* **310,** 1439–1440, 2005.

Jakosky, B. M., and Shock, E. L., The biological potential of Mars, the early Earth, and Europa, *Journal of Geophysical Research,* **103,** 19359–19364, 1998.

Weiss, B. P., Yung, Y. L., and Nealson, K. H., Atmospheric energy for subsurface life on Mars? *Proceedings of the National Academy of Sciences,* **97,** 1395–1399, 2000.

Viking Search for Life on Mars

Klein, H. P., Did Viking discover life on Mars? *Origins of Life and Evolution of the Biosphere,* **29,** 625–631, 1999.

Klein, H. P., Horowitz, N. H., and Biemann, K., The search for extant life on Mars, in *Mars,* H. H. Kieffer, B. M. Jakosky, C. W. Snyder, and M. S. Matthews, eds., University of Arizona Press, 1221–1233, 1992.

ALH84001 and Implications for Life on Mars

Barber, D. J., and Scott, E. R. D., Origin of supposedly biogenic magnetite in the martian meteorite Allan Hills 84001, *Proceedings of the National Academy of Sciences,* **99,** 6556–6561, 2002.

Friedmann, E. I., Wierzchos, J., Ascaso, C., and Winklhofer, M., Chains of magnetite crystals in the meteorite ALH84001: Evidence of biological origin, *Proceedings of the National Academy of Sciences,* **98,** 2176–2181, 2001.

Golden, D. C., et al., A simple inorganic process for formation of carbonates, magnetite, and sulfides in martian meteorite ALH84001, *American Mineralogist,* **86,** 370–375, 2001.

McKay, D. S., Gibson, E. K., Jr., Thomas-Keprta, K. L., Vali, H., Romanek, C. S., Clemett, S. J., Chillier, X. D. F., Maechling, C. R., and Zare, R. N., Search for past life on Mars: Possible relic biogenic activity in martian meteorite ALH84001, *Science,* **273,** 924–930, 1996.

Treiman, A. H., A short, critical evaluation of proposed signs of ancient martian life in Antarctic meteorite ALH84001, in *Bioastronomy '99: A New Era in Bioastronomy,* G. A. Lemarchand and K. J. Meech, eds., *Astronomical Society of the Pacific,* 303–314, 2000.

Treiman, A. H., Gleason, J. D., and Bogard, D. D., The SNC meteorites are from Mars, *Planetary and Space Science,* **48,** 1213–1230, 2000.

Searching for Life on Mars

*Zorpette, G., Why go to Mars? *Scientific American,* March 2000.

Carr, M. H., and Garvin, J., Mars exploration, *Nature,* **412,** 250–253, 2001.

DeVincenzi, D. L., Race, M. S., and Klein, H. P., Planetary protection, sample return missions, and Mars exploration: History, status, and future needs, *Journal of Geophysical Research,* **103,** 28577–28585, 1998.

Mojzsis, S. J., and Arrhenius, G., Phosphates and carbon on Mars: Exobiological implications and sample return considerations, *Journal of Geophysical Research,* **103,** 28495–28511, 1998.

Westall, F., et al., An ESA study for the search for life on Mars, *Planetary and Space Science,* **48,** 181–202, 2000.

Chapter 9

Life on Moons

*LePage, A., Habitable moons, *Sky and Telescope,* 51, December 1998.

*Rothery, D. A., *Satellites of the Outer Planets,* Clarendon Press, 1992.

Irwin, L. N., and Schulze-Makuch, D., Assessing the plausibility of life on other worlds, *Astrobiology,* **1,** 143, 2001.

Europa, Ganymede, and Callisto

Chyba, C. F., and Phillips, C. B., Possible ecosystems and the search for life on Europa, *Proceedings of the National Academy of Sciences,* **98,** 801, 2001.

Kivelson, M. G., et al., Galileo magnetometer measurements: A stronger case for a subsurface ocean at Europa, *Science,* **298,** 1340–1343, 2000.

Pappalardo, R. T., Head, J. W., and Greeley, R., The hidden ocean of Europa, *Scientific American,* **281,** 54–63, 1999.

Stevenson, D. J., Jupiter and its moons, *Science,* **294,** 71–72, 2001.

Titan

Lunine, J. I., Does Titan have an ocean? A review of current understanding of Titan's surface, *Reviews of Geophysics,* **31,** 133–149, 1993.

Sagan, C., Thompson, W. R., and Khare, B. N., Titan: A laboratory for prebiological organic chemistry, *Accounts of Chemical Research,* **25,** 286–292, 1992.

Chapter 10

Venus

*Grinspoon, D., *Venus Revealed,* Addison Wesley, 1997.

Habitable Zones and Beyond

Doyle, L., ed., *Circumstellar Habitable Zones,* Travis House, 1996.

Kasting, J. F., and Catling, D., Evolution of a habitable planet, *Annual Review of Astronomy and Astrophysics,* **41,** 429–463, 2003.

Kasting, J. F., Whitmire, D. P., and Reynolds, R. T., Habitable zones around main sequence stars, *Icarus,* **101,** 108–128, 1993.

Stevenson, D. J., Life-sustaining planets in interstellar space? *Nature,* **400,** 32, 1999.

Wetherill, G. W., The formation and habitability of extra-solar planets, *Icarus,* **119,** 219–238, 1996.

Williams, D. M., Kasting, J. F., and Wade, R. A., Habitable moons around extrasolar giant planets, *Nature,* **385,** 234–236, 1997.

Global Warming

*Eilperin, J., World temperatures keep rising with a hot 2005, *Washington Post,* October 13, 2005.
*Kluger, J., Be worried, be very worried, *Time Magazine,* April 3, 2006.
Kerr, R. A., Climate change: A worrying trend of less ice, higher seas, *Science,* **311,** 1698–1701, 2006.

Chapter 11

Discovery of Extrasolar Planets, Implications for Solar System Evolution

*Croswell, K., *Planet Quest,* The Free Press, 1997.
*Doyle, L. R., Deeg, H.-J., and Brown, T. M., Searching for shadows of other Earths, *Scientific American,* **283,** 38, 2000.
*Goldsmith, D., *Worlds Unnumbered,* University Science Books, 1997.
*Malhotra, R., Migrating planets, *Scientific American,* **281,** 56–63, 1999.
Marcy, G. W., and Butler, R. P., A planetary companion to 70 Virginis, *Astrophysical Journal,* **464,** L147–L151, 1996.
Marcy, G. W., and Butler, R. P., Planets orbiting other suns, *Publications of the Astronomical Society of the Pacific,* **112,** 137, 2000.
Marcy, G. W., Cochran, W. D., and Mayor, M., Extrasolar planets around main-sequence stars, in *Protostars and Protoplanets IV,* V. Mannings, A. P. Boss, and S. S. Russell, eds., University of Arizona Press, 1285–1311, 2000.
Santos, N. C., Benz, W., and Mayor, M., Extrasolar planets: Constraints for planet formation models, *Science,* **310,** 251–255, 2005.

How to Search for Evidence of Life on Extrasolar Planets

*Angel, J. R. P., and Woolf, N. J., Searching for life on other planets, *Scientific American,* **274,** 60–66, 1966.
Ford, E. B., Seager, S., and Turner, E. I., Characterization of extrasolar terrestrial planets from diurnal photometric variability, *Nature,* **412,** 885, 2001.
Sagan, C., et al., A search for life on Earth from the Galileo spacecraft, *Nature,* **365,** 715–721, 1993.
Schindler, T. L., and Kasting, J. F., Synthetic spectra of simulated terrestrial atmospheres containing possible biomarker gases, *Icarus,* **145,** 262–271, 2000.

The Rare Earth Hypothesis

*Darling, D., Rare Earths and hidden agendas, in *Life Everywhere,* Basic Books, 2001.
*Gonzalez, G., Brownlee, D., and Ward, P. D., Refuges for life in a hostile universe, *Scientific American,* **285,** 60–67, 2001.
*Ward, P. D., and Brownlee, D., *Rare Earth: Why Complex Life Is Uncommon in the Universe,* Copernicus, 2000.

Stars

*Johnson, G., *Miss Leavitt's Stars: The Untold Story of the Woman Who Discovered How to Measure the Universe,* Norton, 2005.
*Kaler, J. B., Meet the neighbors, *Astronomy,* 46, April 2006.

Chapter 12

Nature of Intelligence and Potential for Its Occurrence

*Calvin, W. H., The emergence of intelligence, *Scientific American,* **271,** 100–107, 1994.
Emery, N. J., and Clayton, N. S., The mentality of crows: Convergent evolution of intelligence in corvids and apes, *Science,* **306,** 1903–1907, 2004.
Jerison, H. J., *Evolution of the Brain and Intelligence,* Academic Press, 1973.
Marino, L., Brain-behavior relations in primates and cetaceans: Implications for the ubiquity of factors leading to the evolution of complex intelligence, in *Astronomical and Biochemical Origins and the Search for Life in the Universe,* C. B. Cosmovici, S. Bowyer, and D. Werthimer, eds., Editrice Compositori, 553, 1997.
Marino, L., McShea, D. W., and Uhen, M. D., The origin and evolution of large brains in toothed whales, *The Anatomical Record,* **281A,** 1247–1255, 2004.
Mayr, E., The probability of extraterrestrial intelligent life, in *Extraterrestrials: Science and Alien Intelligence,* E. Regis, Jr., ed., Cambridge University Press, 23, 1985.

Artifacts

Dyson, F. J., Search for artificial stellar sources of infrared radiation, *Science,* **131,** 1667, 1960.
Freitas, R. A., Jr., and Valdes, F., The search for extraterrestrial artifacts (SETA), *Acta Astronautica,* **12,** 1027–1034, 1985.
Zubrin, R., Detection of extraterrestrial civilizations via the spectral signature of advanced interstellar spacecraft, in *Progress in the Search for Extraterrestrial Life,* ASP Conference Series, vol 74., G. S. Shostak, ed., Astronomical Society of the Pacific, 487, 1995.

SETI

*Chown, M., ET First Contact "Within 20 Years," *New Scientist,* July 23, 2004.
*Davies, P., *Are We Alone?* Basic Books, 1996.
*Drake, F. D., and Sobel, D., *Is Anyone Out There?* Delacorte Press, 1992.
*Harrison, A., *After Contact,* Plenum Press, 1997.
*Mayr, E., The search for extraterrestrial intelligence: Scientific quest or hopeful folly? The improbability of success, *The Planetary Report,* **16,** 4–7, 1996.
*McConnell, B., *Beyond Contact,* O'Reilly and Associates, 2001.
*Sagan, C., The search for extraterrestrial intelligence: Scientific quest or hopeful folly? The abundance of life-bearing planets, *The Planetary Report,* **16,** 8–13, 1996.
*Shostak, S., The Future of SETI, *Sky and Telescope,* 42, April 2001.
*Shostak, S., and Barnett, A., *Cosmic Company,* Cambridge University Press, 2003.
Cocconi, G., and Morrison, P., Searching for interstellar communications, *Nature,* **184,** 844–846, 1959.
Drake, F., Project Ozma, *Physics Today,* **14,** 40, 1961.

Scientific Study of UFOs and Alien Visitation Claims

*Alschuler, W. R., *The Science of UFOs,* St. Martin's Press, 2001.
*Clancy, S., *Abducted: How People Come to Believe They Were Kidnapped by Aliens,* Harvard University Press, 2005.
*Korff, K., *The Roswell UFO Crash: What They Don't Want You to Know,* Prometheus Books, 1997.
*Kristof, N. D., Alien abduction? Science calls it sleep paralysis, *New York Times,* July 6, 1999.
*Sturrock, P. A., *The UFO Enigma,* Warner Books, 1999.

Chapter 13

The Voyager Record

*Sagan, C., Drake, F. D., Dryan, A., Ferris, T., Lomberg, J., Sagan, L. S., *Murmurs of Earth: The Voyager Interstellar Record,* Random House, 1978.

Interstellar Travel

*Mallove, E. F., and Matloff, G. L., *The Starflight Handbook: A Pioneer's Guide to Interstellar Travel,* Wiley, 1989.
Mauldin, J., *Prospects for Interstellar Travel,* Science and Technology Series: A Supplement to Advances in the Astronautical Sciences, Vol. 80, Univelt, 1992.

Relativity, Black Holes and Worm Holes

*Thorne, K. S., *Black Holes and Time Warps: Einstein's Outrageous Legacy,* Norton, 1994.

Will We Try to Colonize Space?

*Bracewell, R., *The Galactic Club,* W. H. Freeman, 1975.

*Dyson, F., *Disturbing the Universe,* Basic Books, 1979.
*Dyson, F., *Infinite in All Directions,* Harper and Row, 1988.
*Heppenheimer, T. A., *Toward Distant Suns,* Stackpole Books, 1979.
*O'Neill, G. K., *The High Frontier: Human Colonies in Space,* Wm. Morrow, 1977.
*Sagan, C., *Pale Blue Dot: A Vision of the Human Future in Space,* Random House, 1994.

Fermi Paradox

*Finney, B., and Jones, E. M., eds., *Interstellar Migration and the Human Experience,* University of California Press, 1985.
*Webb, S., *If the Universe Is Teeming with Aliens . . . Where Is Everybody? Fifty Solutions to Fermi's Paradox and the Problem of Extraterrestrial Life,* Springer, 2002.
Ball, J. A., The zoo hypothesis, *Icarus,* **19,** 347, 1973.
Hart, M., An explanation for the absence of extraterrestrials on Earth, *Quarterly Journal of the Royal Astronomical Society,* **16,** 128, 1975.
Sagan, C., and Newman, W. I., The solipsist approach to extraterrestrial intelligence, *Quarterly Journal of the Royal Astronomical Society,* **24,** 113, 1983.
Tipler, F. J., Extraterrestrial intelligent beings do not exist, *Quarterly Journal of the Royal Astronomical Society,* **21,** 267, 1980.

Epilogue

Philosophical and Societal Issues Related to the Search for Life Elsewhere

Chyba, C. F., and Hand, K. P., Astrobiology: The study of the living universe, *Annual Review of Astronomy and Astrophysics,* 43, 31–74, 2005.
Jakosky, B. M., Philosophical aspects of astrobiology, in *Bioastronomy '99: A New Era in Bioastronomy,* G. A. Lemarchand and K. J. Meech, eds., Astronomical Society of the Pacific, 661–666, 2000.
Jakosky, B. M., and Golombek, M. P., Planetary science, astrobiology, and the role of science and exploration in society, *EOS, Transcripts of the American Geophysical Union,* **81,** 58, February 6, 2000.

Credits

CHAPTER 1 Chapter Opener "NASA, Apollo 8"

1.1 LUCASFILM/20TH CENTURY FOX / THE KOBAL COLLECTION **1.2** European Southern Observatory (ESO) **1.4** (Venus-surface) From the Voyage scale model solar system, developed by Challenger Center for Space Science Education, the Smithsonian Institution, and NASA. ©2001 David P. Anderson, Southern Methodist University; NASA; NASA/Courtesy NSSDC; NASA/JPL **1.5** NASA, JPL, Cornell **1.6** NASA, Voyager 2, NSSDC **1.7** Seth Shostak, NASA, NASA, Voyager 2, NSSDC, Seth Shostak

CHAPTER 2 Chapter Opener NASA/Johnson Space Center

2.1 David Nunuk, Anthony Ayiomamitis **2.3** NASA Earth Observing System **2.12** NASA/Goddard Institute for Space Studies, Duomo/CORBIS, "NAS, Goddard Institute for Space Studies" **2.16** NASA Earth Observing System

CHAPTER 3 Chapter Opener NASA

3.1 NASA/Goddard Institute for Space Studies, "Blakeley Kim, Benjamin Cummings Publishers, Pearson Ed.", Corel Corporation, NASA Earth Observing System, Lester Lefkowitz/CORBIS **3.2** Jeffrey Bennett, "NASA, M. Clampin (STScI)" **3.3** Jerry Lodriguss/Astropix LLC **3.4** Joe Bergeron, NASA, ESA, M. Robberto (Space Telescope Science Institute/ESA) and the Hubble Space Telescope Orion Treasury Project Team **3.5** NASA Earth Observing System **3.6** NASA **3.8** European Southern Observatory **3.11** Illustration by Matthew Perry, Jerry Lodriguss/Astropix LLC **3.13** NASA Earth Observing System **3.14** Jeffrey Bennett **3.16** "Frank Barrett, celestialwonders .com" **3.17** Niescja Turner and Carter Emmart **3.19** "NASA, JPL" **3.21** "NASA, JPL" **3.27** Getty Images, Alvis Upitis/Image Bank/Getty Images, Fred Dana/ CORBIS **3.31** "Runk/Schoenberger/Grant Heilman Photography, Inc." **3.33** NASA/Jet Propulsion Laboratory, "NASA, JPL"

CHAPTER 4

4.1 NASA, JPL, NASA, JPL, Akira Fujii (surface) NASA (globe), NASA, NASA, JPL **4.2** Dennis Flaherty/Photo Researchers, Inc, David Muench/CORBIS, Thomas Cooper/UNEP/Peter Arnold, Inc., NASA **4.5** Jeff Greenberg/Visuals Unlimited **4.9** George Gerster/Photo Researchers, Inc., Yva Momatiuk/John Eastcott/Minden Pictures, Chip Clark, Manfred Cage/Peter Arnold, Inc., Jeff Daly/Visuals Unlimited, F. Latreille/Cerpolex/Cercles Polaires Expeditions, Dr. Martin Lockley, University of Colorado **4.11** "J.W. Valley, University of Wisconsin-Madison" **4.12** "U.S. Geological Survey, Denver" **4.13** NASA/USGS, NASA **4.19** NASA **4.21** "NASA, JPL" **4.22** HURL/Chave and Malahoff, Philip Rosenberg, Philip Rosenberg, University of Hawaii **4.24** "Jules Bucher/Photo Researchers, Inc." **4.25** Cary Anderson/Aurora **4.32** NASA/JPL

CHAPTER 5 Chapter Opener "Biophoto Associates/Photo Researchers, Inc."

5.1 CNRI/SPL/Photo Researchers, Inc., Biophoto Associates/Photo Researchers, Inc., Frans Lanting/Minden Pictures, Woods Hole Oceanographic Institution, Joe McDonald/CORBIS, Fred Bavendam/Minden Pictures, "CNRI/SPL/ Photo Researchers, Inc." **5.2** "H.S. Pankratz, T.C. Beaman/Biological Photo Service" **5.3** Jack Wilburn/ Animals Animals/Earth Scenes **5.4** Biophoto Associates/Photo Researchers, Inc., Michael Walker/Photo Researchers, Inc., John Durham/Science Photo Library/Photo Researchers, Inc., Dr. Yorgos Nikas/Photo Researchers, Inc. **5.16** Stanley Flegler/Visuals Unlimited, Stanley Flegler/ Visuals Unlimited **5.17** Woods Hole Oceanographic Institution **5.18** "Douglas Faulkner/Photo Researchers, Inc." **5.19** Arne Bomblies, E. Imre Friedmann **5.B1** Richard Milner

CHAPTER 6 Chapter Opener Georgette Douwma / Nature Picture Library

6.1 Mitsuaki Iwago/ Minden Pictures, S. M. Awramik/Biological Photo Service, S. M. Awramik/Biological Photo Service **6.2** Biological Photo Services, "MSSO/SPL/Photo Researchers, Inc." **6.3** Stephen J. Mojzsis **6.4** Roger Ressmeyer/CORBIS **6.6** F.M. Menger and Kurt Gabrielson **6.7** Martin Hanczyc **6.9** NASA, Johnson Space Center **6.11** " John Weinstein, The Field Museum of National History" **6.12** T. E. Adams/Visuals Unlimited, "Dr Jeremy Burgess/ Photo Researchers, Inc." **6.14** Theodore J. Bornhorst, Courtesy of Brad Snowder **6.15** "Kirk Johnson, Denver Museum of Nature and Science" **6.16** "Virgil L. Sharpton, University of Alaska-Fairbanks", Virgil L. Sharpton **6.17** Benjamin Cummings/Pearson Education **6.19** Sovfoto/Eastfoto **6.23** "David Gifford/ SPL/Photo Researchers, Inc." **6.24** MPFT **6.25** MPFT

CHAPTER 7 Chapter Opener Jerry Lodriguss/Astropix LLC

7.1 NASA/Jet Propulsion Laboratory, NASA/Jet Propulsion Laboratory, "NASA, JPL" **7.4** Moon: Akira Fujii, Mercury: NASA, USGS, Akira Fujii, NASA/USGS **7.5** NASA, Los Alamos National Laboratory, NSSDC **7.6** NASA, NSSDC **7.7** NASA, NSSDC, NASA/JPL **7.8** NASA/JPL **7.11** NASA/JPL **7.13** "NASA, Goddard Institute for Space Studies " **7.14** "NASA, JPL", NASA/JPL-Caltech, NASA/JPL-Caltech, NASA/JPL-Caltech, NASA/JPL-Caltech, NASA/JPL-Caltech

CHAPTER 8 Chapter Opener NASA/JPL/Cornell

8.1 NASA Earth Observing System, "J. Bell (Cornell), M. Wolff (Space Science Inst.), Hubble Heritage Team (STScI/AURA), and NASA" **8.2** Lowell Observatory Photographs**,** "NASA, Dr. David E. Smith, MOLA Science Team" **8.3** Copyright 1938 The New York Times Co. Reprinted by Permission., NASA/JPL **8.4** NASA, Viking 2 Lander, NSSDC, NASA/Viking Lander, NASA/ Viking 2 Lander **8.5** NASA **8.8** NASA, NSSDC, P. James (Univ. Of Toledo), T. Clancy (Space Science Inst.), S. Lee (Univ. Colorado), NASA, P. James (Univ. Toledo), T. Clancy (Space Science Inst.), S. Lee (Univ. Colorado), NSSDC **8.9** NASA, JPL and Malin Space Science Systems **8.11** "NASA, JPL, Malin Space Science Systems " **8.12** "NASA, NSSDC, Mars Gloval Surveyor", NASA/Mars Global Surveyor/NSSDC **8.13** "NASA, USGS " **8.15** "NASA, JPL " **8.16** USGS, NASA/JPL Maline Space Science Systems, "R. P. Irwin III and G. A. Franz, National Air and Space Museum, Smithsonian Institution; a: USGS, b: NASA, JPL, c: R. P. Irwin III and G. A. Franz, National Air and Space Museum, Smithsonian Institution ", National Air and Space Museum/Smithsonian Institution **8.17** NASA/JPL, NASA/JPL, "NASA, JPL", NASA, NASA /JPL **8.18** ESA/DLR/FU Berlin (G. Neukum) **8.19** NASA/Michael Mellon **8.21** "NASA, JPL" **8.22** "NASA, the Magellan project. Additional processing by John Weiss" **8.23** ESA/DLR/FU Berlin (G. Neukum) **8.26** "NASA,JPL" **8.27** NASA/Viking Lander/NSSDC **8.29** "Corby Waste, Jet Propulsion Laboratory. NASA" **8.31** "bottom: NASA, JPL, top: R.L. Fok and F.L. Lynch", NASA/ Jet Propulsion Laboratory, "NASA, JPL" **Unnumbered Figures** 1,"NASA, JPL", NASA/JPL; 2,"NASA, JPL", NASA/JPL; 3,"NASA, JPL"

CHAPTER 9 Chapter Opener NASA/JPL/Space Science Institute

9.1 NASA/JPL/Galileo/Artist's Conception by Pam Engebretson. Drawings rendered by Pam Engebretson, from a design by Eric M. DeJong and Zareh Gorjian of JPL **9.2** NASA/JPL **9.3** NASA/JPL/Tim Parker/Bob Pappalardo **9.5** NASA/JPL **9.9** PIRL/Lunar & Planetary Laboratory (University of Arizona) and NASA, PIRL/Lunar & Planetary Laboratory (University of Arizona) and NASA **9.11** University of Arizona and NASA **9.12** NASA/JPL/Space Science Institute **9.13** NASA/JPL **9.15** NASA/JPL/Arizona State University **9.16** NASA/ JPL **9.17** Woods Hole Oceanographic Institution **9.18** NASA/JPL **9.19** NASA/

JPL, NASA/JPL/Arizona State University **9.21** NASA/JPL **9.22** NASA/JPL **9.23** NASA/JPL/Space Science Institute, NASA/JPL **9.24** NASA/JPL **9.25** NASA/JPL **9.26** NASA/JPL/Space Science Institute, NASA/JPL **9.27** NASA/JPL

CHAPTER 10 Chapter Opener NASA Earth Observing System

10.2 NASA/JPL, NASA **10.3** NASA/JPL, NASA **10.6** NASA and the Hubble Heritage Team (STScI/AURA) **10.7** "University Corporation for Atmospheric Research Adapted and updated 2006 from Karl and Trenberth (2003), courtesy K. Trenberth, and based on the HADCRUTv3 dataset from the CRU and UKMO" **10.8** NASA

CHAPTER 11 Chapter Opener Lynette R. Cook

11.1 "Dept. of Astronomy, University of Michigan at Cerro Tololo"; 3, Seth Shostak; 4,Data adapted from STARE project: (www.hao.ucar.edu/public/research/stare/hd29458.html; 5, NASA/JPL; 9,"NASA, JPL, Geoffrey Bryden" **11.2** Harvard College Observatory; 1,"Yeshe Fenner, Space Telescope Institute" **11.4** European Southern Observatory **11.5** European Southern University

CHAPTER 12 Chapter Opener Seth Shostak

12.1 Seth Shostak; 0,Seth Shostak; 1,Gordon Garradd; 2,Seth Shostak; 3,MGM / THE KOBAL COLLECTION; 6,Seth Shostak; 7,Seth Shostak; 8,Bettmann/CORBIS; 9,Christopher Cormack/CORBIS **12.2** Seth Shostak; 1,Tom Till/Photographer's Choice/Getty Images **12.5** "Radio Age, October 1924" **12.8** Seth Shostak, "Arecibo Observatory/National Astronomy and Ionosphere Center, operated by Cornell University under a cooperative agreement with the National Science Foundation" **12.9** Seth Shostak **Unnumbered Figure** Seth Shostak

CHAPTER 13 Chapter Opener Joe Bergeron

13.1 NASA/JPL; 2, CORBIS; 7, Joe Bergeron **13.2** NASA/JPL **13.3** NASA **13.4** NASA **13.5** Department of Energy **13.6** NASA/JPL **13.7** Joe Bergeron **13.8** "Joe Bergeron, Addison Wesley" **13.9** NASA/Jet Propulsion Laboratory

EPILOGUE Chapter Opener Graham Johnson/www.fiVth.com

E.1 Bettman/CORBIS **E.2** NASA

FRONT MATTER Graham Johnson/www.fiVth.com

Index

Page references in *italics* refer to figures. Page references preceded by "t" refer to tables. Page references followed by "n" refer to footnotes.